SEVENTH EDITION

Microbiology
A Systems Approach

Marjorie Kelly Cowan
Heidi Smith

Viaframe/Corbis/Getty Images

MICROBIOLOGY: A SYSTEMS APPROACH, SEVENTH EDITION

This book is printed on acid-free paper.

1 2 3 4 5 6 7 8 9 LWI 28 27 26 25 24 23

ISBN 978-1-265-46111-9 (bound edition)
MHID 1-265-46111-2 (bound edition)
ISBN 978-1-265-07867-6 (loose-leaf edition)
MHID 1-265-07867-X (loose-leaf edition)

Managing Director: *Michelle Vogler*
Portfolio Manager: *Lauren Vondra*
Product Developer: *Darlene M. Schueller*
Marketing Manager: *Tami Hodge*
Content Project Managers: *Paula Patel, Ron Nelms*
Manufacturing Project Manager: *Laura Fuller*
Content Licensing Specialist: *Lori Hancock*
Cover Image: *Viaframe/Corbis/Getty Images*
Compositor: *MPS Limited*

All credits appearing on page or at the end of the book are considered to be an extension of the copyright page.

Library of Congress Cataloging-in-Publication Data

Names: Cowan, M. Kelly, author. | Smith, Heidi (College teacher)
Title: Microbiology : a systems approach / Marjorie Kelly Cowan, Heidi Smith.
Description: Seventh edition. | New York, NY : McGraw Hill, [2024] |
Includes index.
Identifiers: LCCN 2022010085 | ISBN 9781265461119 (hardcover; alk. paper) |
ISBN 1265461112 (MHID hardcover; alk. paper) | ISBN 9781265078676
(spiral bound; alk. paper) | ISBN 126507867X (MHID spiral bound; alk. paper)
Subjects: MESH: Microbiological Phenomena | Communicable
Diseases—microbiology | Immunity
Classification: LCC QR41.2 | NLM QW 4 | DDC 579—dc23/eng/20220825
LC record available at https://lccn.loc.gov/2022010085

mheducation.com/highered

Viaframe/Corbis/Getty Images

Brief Contents

About the Authors

Viaframe/Corbis/Getty Images

Kelly Cowan has taught microbiology to pre-nursing and allied health students for over 20 years. She received her PhD from the University of Louisville and held postdoctoral positions at the University of Maryland and the University of Groningen in the Netherlands. Her campus, Miami University Middletown, is an open admissions regional campus of Miami University in Ohio. She has also authored over 25 basic research papers with her undergraduate and graduate students. For the past several years, she has turned her focus to studying pedagogical techniques that narrow the gap between under-resourced students and well-resourced students. She is past chair of the American Society for Microbiology's Undergraduate Education committee and past chair of ASM's education division, Division W.

Having a proven educator as an integrated digital author makes a *proven* learning system even better.

We are pleased to have Heidi Smith on the team. Heidi works hand-in-hand with the textbook author, creating online tools that truly complement and enhance the book's content. Because of Heidi, we offer you a robust digital learning program, tied to Learning Outcomes, to enhance your lecture and lab, whether you run a traditional, hybrid, or fully online course.

Heidi Smith leads the microbiology department at Front Range Community College in Fort Collins, Colorado. Collaboration with other faculty across the nation, the development and implementation of new digital learning tools, and her focus on student learning outcomes have revolutionized Heidi's face-to-face and online teaching approaches and student performance in her classes. The use of digital technology has given Heidi the ability to teach courses driven by real-time student data and with a focus on active learning and critical thinking activities.

Heidi is an active member of the American Society for Microbiology and participated as a task force member for the development of their Curriculum Guidelines for Undergraduate Microbiology Education. At FRCC, Heidi directs a federal grant program designed to increase student success in transfer and completion of STEM degrees at the local university as well as facilitate undergraduate research opportunities for under-represented students.

Off campus, Heidi spends as much time as she can enjoying the beautiful Colorado outdoors with her husband and four children.

Silver Oaks/McGraw Hill

Heidi Smith / Kelly Cowan

Viaframe/Corbis/Getty Images

Preface

Students:

Welcome to the microbial world! I think you will find it fascinating to understand how microbes interact with us and with our environment. The interesting thing is that each of you has already had a lot of experience with microbiology. For one thing, you are thoroughly populated with microbes right now, and much of your own genetic material actually came from viruses and other microbes. And while you have probably had some bad experiences with quite a few microbes in the form of diseases, you have certainly been greatly benefited by them as well.

This book is suited for all kinds of students and doesn't require any prerequisite knowledge of biology or chemistry. If you are interested in entering the health care profession in some way, this book will give you a strong background in the biology of microorganisms without overwhelming you with unnecessary details. Don't worry if you're not in the health professions. As the COVID-19 pandemic has shown us, a grasp of this topic is important for everyone—and it can be attained with this book.

—Kelly Cowan

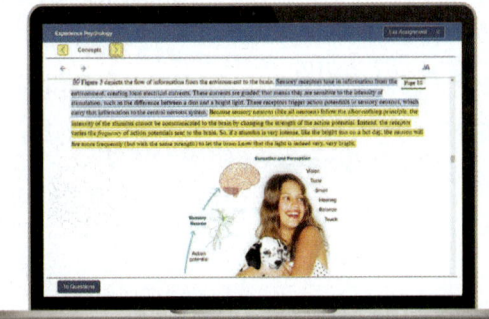

SUPPORT AT every step

Students

Get Learning that Fits You

Effective tools for efficient studying

Connect is designed to help you be more productive with simple, flexible, intuitive tools that maximize your study time and meet your individual learning needs. Get learning that works for you with Connect.

Study anytime, anywhere

Download the free ReadAnywhere® app and access your online eBook, SmartBook® 2.0, or Adaptive Learning Assignments when it's convenient, even if you're offline. And since the app automatically syncs with your Connect account, all of your work is available every time you open it. Find out more at **mheducation.com/readanywhere**

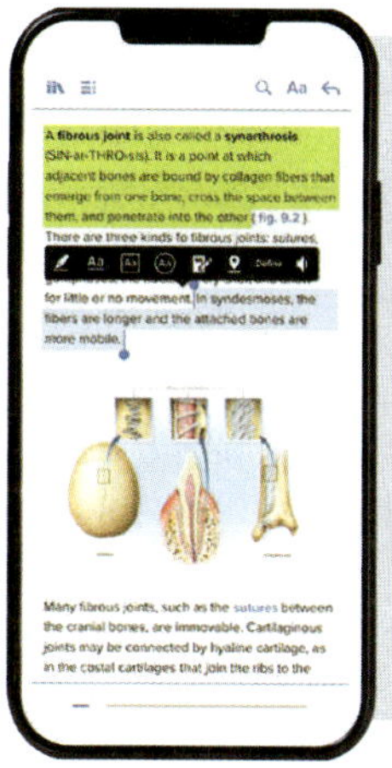

"I really liked this app—it made it easy to study when you don't have your text-book in front of you."

- Jordan Cunningham,
Eastern Washington University

iPhone: Getty Images

Everything you need in one place

Your Connect course has everything you need—whether reading your digital eBook or completing assignments for class—Connect makes it easy to get your work done.

Learning for everyone

McGraw Hill works directly with Accessibility Services Departments and faculty to meet the learning needs of all students. Please contact your Accessibility Services Office and ask them to email accessibility@mheducation.com, or visit **mheducation.com/about/accessibility** for more information.

Digital and Lab Resources for Your Success

Viaframe/Corbis/Getty Images

Need a lab manual for your microbiology course? Customize any of these manuals—add your text material—and *Create* your perfect solution!

McGraw Hill offers several lab manuals for the microbiology course. Contact your McGraw Hill learning technology representative for packaging options with any of our lab manuals.

Smith/Brown: *Benson's Microbiological Applications: Laboratory Manual in General Microbiology,* 15th edition (978-1-260-25898-1)

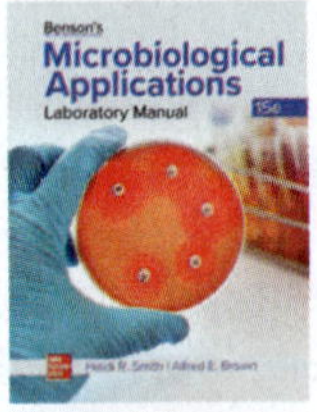
Microbiological Applications
Laboratory Manual

McGraw Hill

Morello: *Laboratory Manual and Workbook in Microbiology: Applications to Patient Care,* 12th edition (978-1-260-00218-8)

MICROBIOLOGY

McGraw Hill

Chess: *Laboratory Applications in Microbiology: A Case Study Approach,* 4th edition (978-1-259-70522-9)

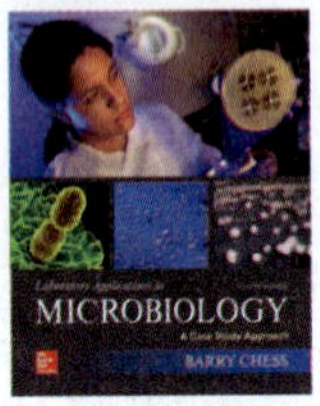
MICROBIOLOGY

McGraw Hill

Obenauf/Finazzo: *Microbiology Fundamentals: A Clinical Approach Laboratory Manual,* 4th edition (978-1-260-78609-5)

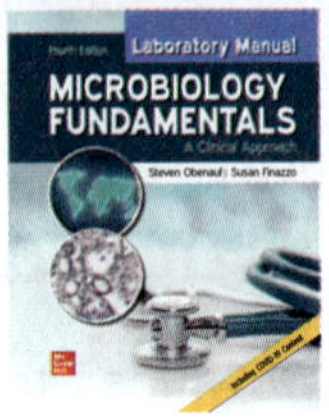
Laboratory Manual
MICROBIOLOGY FUNDAMENTALS

McGraw Hill

Explore our microbiology articles and podcasts dedicated to sharing ideas, best practices, teaching tips and more!
https://www.mheducation.com/highered/microbiology.html

NewsFlash brings the real world into your classroom! These activities in Connect tie current news stories to key concepts. After interacting with a contemporary news story, students are assessed on their ability to make the connections between real-life events and course content.

Virtual Labs Virtual Labs and Lab Simulations

While the biological sciences are hands-on disciplines, instructors are often asked to deliver some of their lab components online: as full online replacements, supplements to prepare for in-person labs, or make-up labs.

These simulations help each student learn the practical and conceptual skills needed, then check for understanding and provide feedback. With adaptive pre-lab and post-lab assessment available, instructors can customize each assignment.

From the instructor's perspective, these simulations may be used in the lecture environment to help students visualize complex scientific processes, such as DNA technology or Gram staining, while at the same time providing a valuable connection between the lecture and lab environments.

Relevancy Modules for Microbiology

With the help of our Relevancy Modules within McGraw Hill Connect, students can see how microbiology actually relates to their everyday lives. For this book, students and instructors can access the Relevancy Modules eBook at no additional cost. Auto-graded assessment questions that correlate to the modules are also available within Connect. Each module consists of videos, an overview of basic scientific concepts, and then a closer look at the application of these concepts to the relevant topic. Some topics include microbiome, immunology, microbes and cancer, fermentation, vaccines, biotechnology, global health, SARS-CoV-2, antibiotic resistance, and several others.

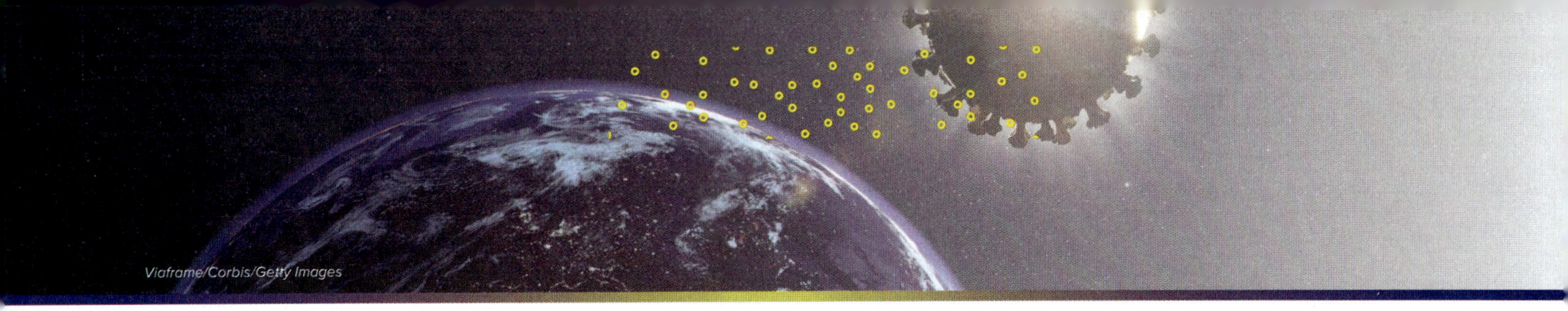

Viaframe/Corbis/Getty Images

Prep for Microbiology is designed to get students ready for a forthcoming course by quickly and effectively addressing prerequisite knowledge gaps that may cause problems down the road. This question bank highlights a series of questions, including Fundamentals of Science, Fundamentals of Math and Statistics, Fundamental Skills for the Scientific Laboratory, and Student Success, to give students a refresher on the skills needed to enter and be successful in their course! Prep maintains a continuously adapting learning path individualized for each student, and tailors content to focus on what the student needs to master in order to have a successful start in the new class.

Writing Assignment

Available within Connect and Connect Master, the Writing Assignment tool delivers a learning experience to help students improve their written communication skills and conceptual understanding. As an instructor you can assign, monitor, grade, and provide feedback on writing more efficiently and effectively.

Remote Proctoring & Browser-Locking Capabilities

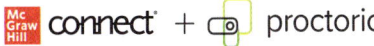

connect + proctorio

Remote proctoring and browser-locking capabilities, hosted by Proctorio within Connect, provide control of the assessment environment by enabling security options and verifying the identity of the student.

Seamlessly integrated within Connect, these services allow instructors to control the assessment experience by verifying identification, restricting browser activity, and monitoring student actions.

Instant and detailed reporting gives instructors an at-a-glance view of potential academic integrity concerns, thereby avoiding personal bias and supporting evidence-based claims.

ReadAnywhere®

Read or study when it's convenient for you with McGraw Hill's free ReadAnywhere app. Available for iOS or Android smartphones or tablets, ReadAnywhere gives users access to McGraw Hill tools including the eBook and SmartBook® 2.0 or Adaptive Learning Assignments in Connect. Take notes, highlight, and complete assignments offline—all of your work will sync when you open the app with WiFi access. Log in with your McGraw Hill Connect username and password to start learning—anytime, anywhere!

OLC-Aligned Courses

Implementing High-Quality Online Instruction and Assessment through Preconfigured Courseware

In consultation with the Online Learning Consortium (OLC) and our certified Faculty Consultants, McGraw Hill has created preconfigured courseware using OLC's quality scorecard to align with best practices in online course delivery. This turnkey courseware contains a combination of formative assessments, summative assessments, homework, and application activities, and can easily be customized to meet an individual's needs and course outcomes. For more information, visit https://www.mheducation.com/highered/olc.

Create

Your Book, Your Way

McGraw Hill's Content Collections Powered by Create® is a self-service website that enables instructors to create custom course materials—print and eBooks—by drawing upon McGraw Hill's comprehensive, cross-disciplinary content. Choose what you want from our high-quality textbooks, articles, and cases. Combine it with your own content quickly and easily, and tap into other rights-secured, third-party content such as readings, cases, and articles. Content can be arranged in a way that makes the most sense for your course, and you can include the course name and information as well. Choose the best format for your course: color print, black-and-white print, or eBook. The eBook can be included in your Connect course and is available on the free ReadAnywhere app for smartphone or tablet access as well. When you are finished customizing, you will receive a free digital copy to review in just minutes! Visit McGraw Hill Create—www.mcgrawhillcreate.com—today and begin building!

Test Builder in Connect

Available within Connect, Test Builder is a cloud-based tool that enables instructors to format tests that can be printed,

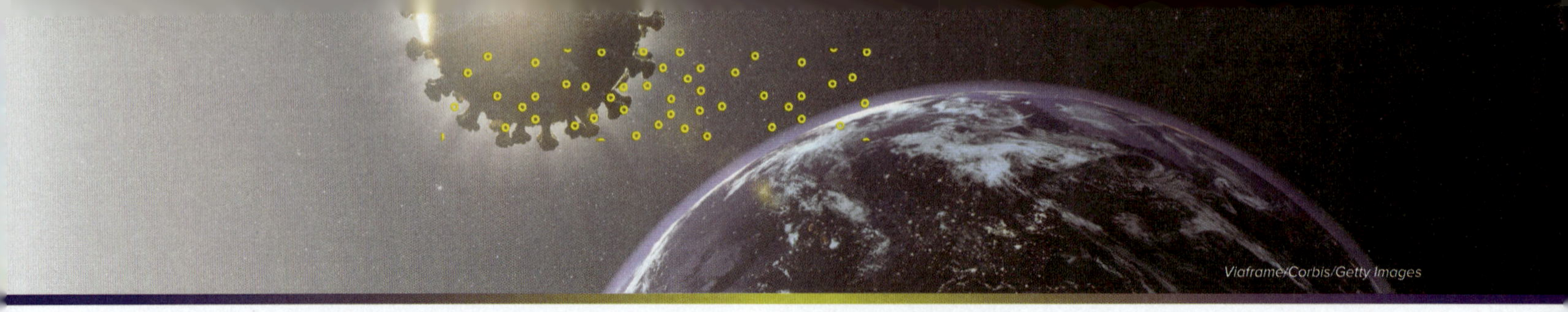

Viaframe/Corbis/Getty Images

administered within a Learning Management System, or exported as a Word document of the test bank. Test Builder offers a modern, streamlined interface for easy content configuration that matches course needs, without requiring a download.

Test Builder allows you to:

- access all test bank content from a particular title.
- easily pinpoint the most relevant content through robust filtering options.
- manipulate the order of questions or scramble questions and/or answers.
- pin questions to a specific location within a test.
- determine your preferred treatment of algorithmic questions.
- choose the layout and spacing.
- add instructions and configure default settings.

Test Builder provides a secure interface for better protection of content and allows for just-in-time updates to flow directly into assessments.

Tegrity: Lectures 24/7

Tegrity in Connect is a tool that makes class time available 24/7 by automatically capturing every lecture. With a simple one-click start-and-stop process, you capture all computer screens and corresponding audio in a format that is easy to search, frame by frame. Students can replay any part of any class with easy-to-use, browser-based viewing on a PC, Mac, or other mobile device.

Educators know that the more students can see, hear, and experience class resources, the better they learn. In fact, studies prove it. Tegrity's unique search feature helps students efficiently find what they need, when they need it, across an entire semester of class recordings. Help turn your students' study time into learning moments immediately supported by your lecture. With Tegrity, you also increase intent listening and class participation by easing students' concerns about note-taking. Using Tegrity in Connect will make it more likely you will see students' faces, not the tops of their heads.

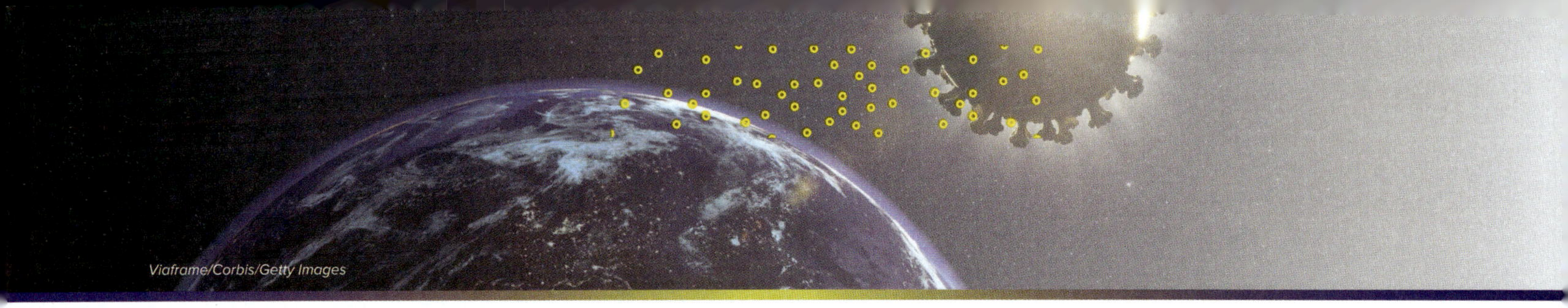

Viaframe/Corbis/Getty Images

Unique Interactive Question Types in Connect Tagged to ASM's Curriculum Guidelines for Undergraduate Microbiology and to Bloom's Taxonomy

- **Case Study:** Case studies come to life in a learning activity that is interactive, self-grading, and assessable. The integration of the cases with videos and animations adds depth to the content, and the use of integrated questions forces students to stop, think, and evaluate their understanding.
- **Media Under The Microscope:** The opening cases in the textbook help students read science articles in the popular media with a critical eye. Questions in Connect are designed to extend these cases in a manner that promotes active student learning, either at home or in the classroom.
- **Concept Maps:** Concept maps allow students to manipulate terms in a hands-on manner in order to assess their understanding of chapter-wide topics. Students become actively engaged and are given immediate feedback, enhancing their understanding of important concepts within each chapter.
- **SmartGrid Questions:** SmartGrid questions replace the traditional end-of-chapter questions, and all of these questions are available for assignment in Connect. These questions were carefully constructed to assess chapter material as it relates to all six concepts outlined in the American Society of Microbiology curriculum guidelines plus the competency of "Scientific Thinking." The questions are cross-referenced with Bloom's taxonomy of learning level. Seven concepts/competencies × three increasing Bloom's levels = a 21-question robust assessment tool.
- **What's the Diagnosis:** Specifically designed for the disease chapters of the text, this is an integrated learning experience designed to assess the student's ability to utilize information learned in the preceding chapters to successfully culture, identify, and treat a disease-causing microbe in a simulated patient scenario. This question type is true experiential learning and allows the students to think critically through a real-life clinical situation.
- **Animations:** Animation quizzes pair our high-quality animations with questions designed to probe student understanding of the illustrated concepts.
- **Animation Learning Modules:** Making use of McGraw Hill's collection of videos and animations, this question type presents an interactive, self-grading, and assessable activity. These modules take a stand-alone, static animation and turn it into an interactive learning experience for your students with real-time remediation.
- **Labeling:** Using the high-quality art from the textbook and other reliable sources, check your students' visual understanding as they practice interpreting figures and learning relationships. Easily edit or remove any label you wish!
- **Classification:** Ask students to organize concepts or structures into categories by placing them in the correct "bucket."
- **Sequencing:** Challenge students to place the steps of a complex process in the correct order.
- **Composition:** Fill in the blanks to practice vocabulary and show a critical understanding of the connections between several different concepts. These exercises may qualify as "writing across the curriculum" activities.

All McGraw Hill Connect content is tagged to Learning Outcomes for each chapter as well as topic, textbook section, Bloom's Level, and ASM Curriculum Guidelines to assist you in customizing assignments and in reporting on your students' performance against these points. This will enhance your ability to assess student learning in your courses by allowing you to align your learning activities to peer-reviewed standards from an international organization.

NCLEX® **Prep Questions:** Sample questions are available in Connect to assign to students.

Note from the Authors

Viaframe/Corbis/Getty Images

This Text's Most Important Distinguishing Features:

These are the features we feel most strongly about. They represent proven methods for enabling our students to learn and we have seen them work in the classroom. The Cowan books have always been built around logical and clear organization, a factor that is critical when nonmajors are attempting to learn a science full of new vocabulary and concepts.

- Comprehensive COVID-19 CONTENT. COVID-19 is not just covered in its disease chapter. Most chapters, even the basic science chapters, have COVID-19 content woven in as a dynamic touchstone for basic concepts.
- SYSTEMATIC ORGANIZATION of the disease chapters that groups microbes by the conditions they cause, which is the way our students will encounter diseases in the clinic.
- A new stand-alone chapter on EPIDEMIOLOGY, the importance of which has been made clear by the COVID-19 pandemic.
- OPENING CASES that teach students how to read science articles in the popular media with a critical eye.
- MICROBIOME findings in all 25 chapters—in form of Microbiome Insight boxes as well as in the text. This reinforces how game-changing the microbiome findings are.
- STUDY SMARTER: BETTER TOGETHER in each chapter that provides guidance for students' group study, either in person or online. No instructor intervention required. Research shows that well-structured group study particularly benefits under-resourced learners and students with lower levels of reading ability.
- SMARTGRIDS in each chapter. The usual end-of-chapter questions are dramatically reformatted into a 21-question grid that cross-references questions by their Bloom's level and the six core concepts of microbiology (plus the competency of scientific literacy) as identified by the American Society for Microbiology.
- CLEAN, uncluttered, and predictable sequence of chapter content. Again, this is extremely helpful to our less-prepared students.
- CONNECT UPDATES
  - CRITICAL THINKING applied through higher Bloom's level questions added to the Connect Question Bank.
  - SMARTBOOK LEARNING RESOURCES added based on heat map results showing areas where students struggle the most. Help when they need it and where they need it.
  - Ready-to-implement ACTIVE LEARNING opportunities available in every chapter in the form of case studies, scientific article reviews, and other interactive question types.

—Kelly Cowan
—Heidi Smith

Capturing Students' Attention and Learning

Viaframe/Corbis/Getty Images

Chapter Opening Case Files That Teach Students How to Judge Popular Media Articles About Science!

Each chapter opens with a revolutionary kind of case study. Titled "Media Under The Microscope," these are summaries of actual news items about microbiology topics. Students are walked through the steps of judging the relative accuracy of the popular media stories. Chapter by chapter, they learn how to critically assess the journalistic accounts. They encounter the principles of causation vs. correlation, biological plausibility, and the importance of not overstating experimental results. It is a critical need among the public today, and this textbook addresses it.

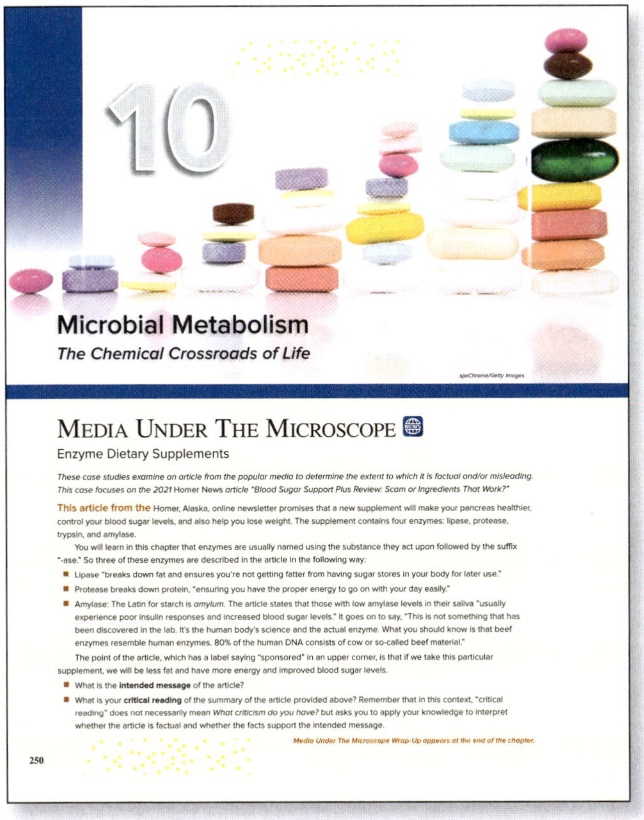

10

Microbial Metabolism

The Chemical Crossroads of Life

spxChrome/Getty Images

MEDIA UNDER THE MICROSCOPE

Enzyme Dietary Supplements

These case studies examine an article from the popular media to determine the extent to which it is factual and/or misleading. This case focuses on the 2021 Homer News *article "Blood Sugar Support Plus Review: Scam or Ingredients That Work?"*

This article from the Homer, Alaska, online newsletter promises that a new supplement will make your pancreas healthier, control your blood sugar levels, and also help you lose weight. The supplement contains four enzymes: lipase, protease, trypsin, and amylase.

You will learn in this chapter that enzymes are usually named using the substance they act upon followed by the suffix "-ase." So three of these enzymes are described in the article in the following way:

- Lipase "breaks down fat and ensures you're not getting fatter from having sugar stores in your body for later use."
- Protease breaks down protein, "ensuring you have the proper energy to go on with your day easily."
- Amylase: The Latin for starch is *amylum*. The article states that those with low amylase levels in their saliva "usually experience poor insulin responses and increased blood sugar levels." It goes on to say, "This is not something that has been discovered in the lab. It's the human body's science and the actual enzyme. What you should know is that beef enzymes resemble human enzymes. 80% of the human DNA consists of cow or so-called beef material."

The point of the article, which has a label saying "sponsored" in an upper corner, is that if we take this particular supplement, we will be less fat and have more energy and improved blood sugar levels.

- What is the **intended message** of the article?
- What is your **critical reading** of the summary of the article provided above? Remember that in this context, "critical reading" does not necessarily mean *What criticism do you have?* but asks you to apply your knowledge to interpret whether the article is factual and whether the facts support the intended message.

Media Under The Microscope Wrap-Up appears at the end of the chapter.

250

Active learning activities are assignable in Connect to extend these case files in or outside of the classroom.

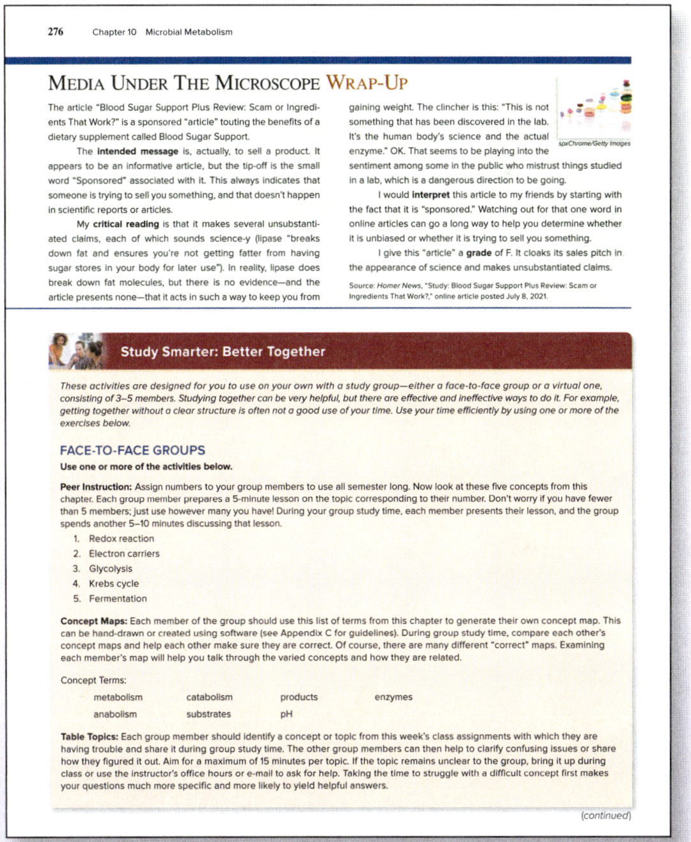

276 Chapter 10 Microbial Metabolism

MEDIA UNDER THE MICROSCOPE WRAP-UP

The article "Blood Sugar Support Plus Review: Scam or Ingredients That Work?" is a sponsored "article" touting the benefits of a dietary supplement called Blood Sugar Support.

The **intended message** is, actually, to sell a product. It appears to be an informative article, but the tip-off is the small word "Sponsored" associated with it. This always indicates that someone is trying to sell you something, and that doesn't happen in scientific reports or articles.

My **critical reading** is that it makes several unsubstantiated claims, each of which sounds science-y (lipase "breaks down fat and ensures you're not getting fatter from having sugar stores in your body for later use"). In reality, lipase does break down fat molecules, but there is no evidence—and the article presents none—that it acts in such a way to keep you from gaining weight. The clincher is this: "This is not something that has been discovered in the lab. It's the human body's science and the actual enzyme." OK. That seems to be playing into the sentiment among some in the public who mistrust things studied in a lab, which is a dangerous direction to be going.

spxChrome/Getty Images

I would **interpret** this article to my friends by starting with the fact that it is "sponsored." Watching out for that one word in online articles can go a long way to help you determine whether it is unbiased or whether it is trying to sell you something.

I give this "article" a **grade** of F. It cloaks its sales pitch in the appearance of science and makes unsubstantiated claims.

Source: *Homer News*, "Study: Blood Sugar Support Plus Review: Scam or Ingredients That Work?," online article posted July 8, 2021.

Study Smarter: Better Together

These activities are designed for you to use on your own with a study group—either a face-to-face group or a virtual one, consisting of 3–5 members. Studying together can be very helpful, but there are effective and ineffective ways to do it. For example, getting together without a clear structure is often not a good use of your time. Use your time efficiently by using one or more of the exercises below.

FACE-TO-FACE GROUPS

Use one or more of the activities below.

Peer Instruction: Assign numbers to your group members to use all semester long. Now look at these five concepts from this chapter. Each group member prepares a 5-minute lesson on the topic corresponding to their number. Don't worry if you have fewer than 5 members; just use however many you have! During your group study time, each member presents their lesson, and the group spends another 5–10 minutes discussing that lesson.

1. Redox reaction
2. Electron carriers
3. Glycolysis
4. Krebs cycle
5. Fermentation

Concept Maps: Each member of the group should use this list of terms from this chapter to generate their own concept map. This can be hand-drawn or created using software (see Appendix C for guidelines). During group study time, compare each other's concept maps and help each other make sure they are correct. Of course, there are many different "correct" maps. Examining each member's map will help you talk through the varied concepts and how they are related.

Concept Terms:

metabolism	catabolism	products	enzymes
anabolism	substrates	pH	

Table Topics: Each group member should identify a concept or topic from this week's class assignments with which they are having trouble and share it during group study time. The other group members can then help to clarify confusing issues or share how they figured it out. Aim for a maximum of 15 minutes per topic. If the topic remains unclear to the group, bring it up during class or use the instructor's office hours or e-mail to ask for help. Taking the time to struggle with a difficult concept first makes your questions much more specific and more likely to yield helpful answers.

(continued)

Photo: Caia Image/Image Source

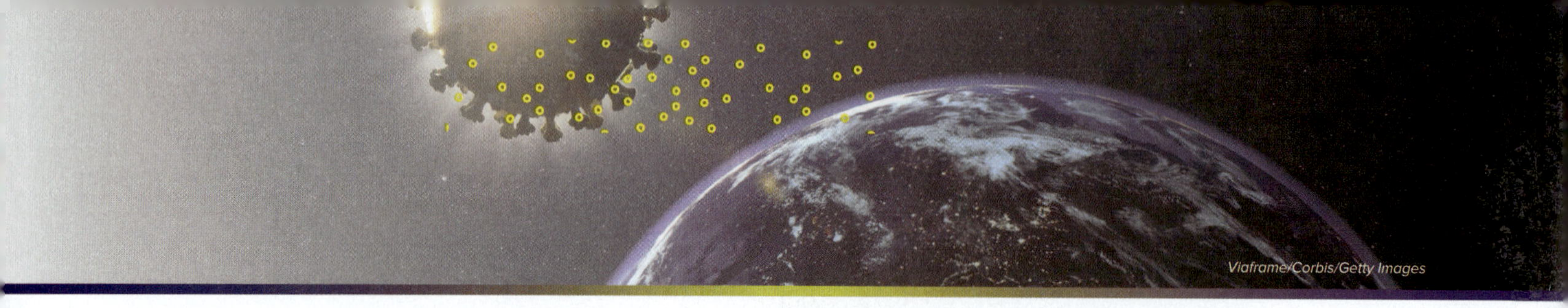

Viaframe/Corbis/Getty Images

Infographics for the Way Students Consume Data

Students have always had difficulty reading figures and graphs. Savvy data designers on the Internet have come up with a format that attracts their eyes and makes complex concepts and data more palatable to them. And more palatable means that they spend time absorbing the information.

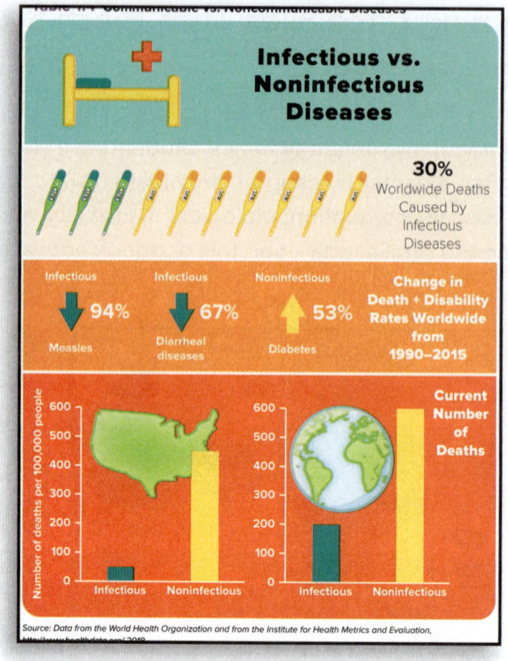

Infectious vs. Noninfectious Diseases
30% Worldwide Deaths Caused by Infectious Diseases
Infectious 94% Measles
Infectious 67% Diarrheal diseases
Noninfectious 53% Diabetes
Change in Death + Disability Rates Worldwide from 1990–2015
Number of deaths per 100,000 people
600
500
400
300
200
100
0
Infectious
Noninfectious
Current Number of Deaths
Source: Data from the World Health Organization and from the Institute for Health Metrics and Evaluation,

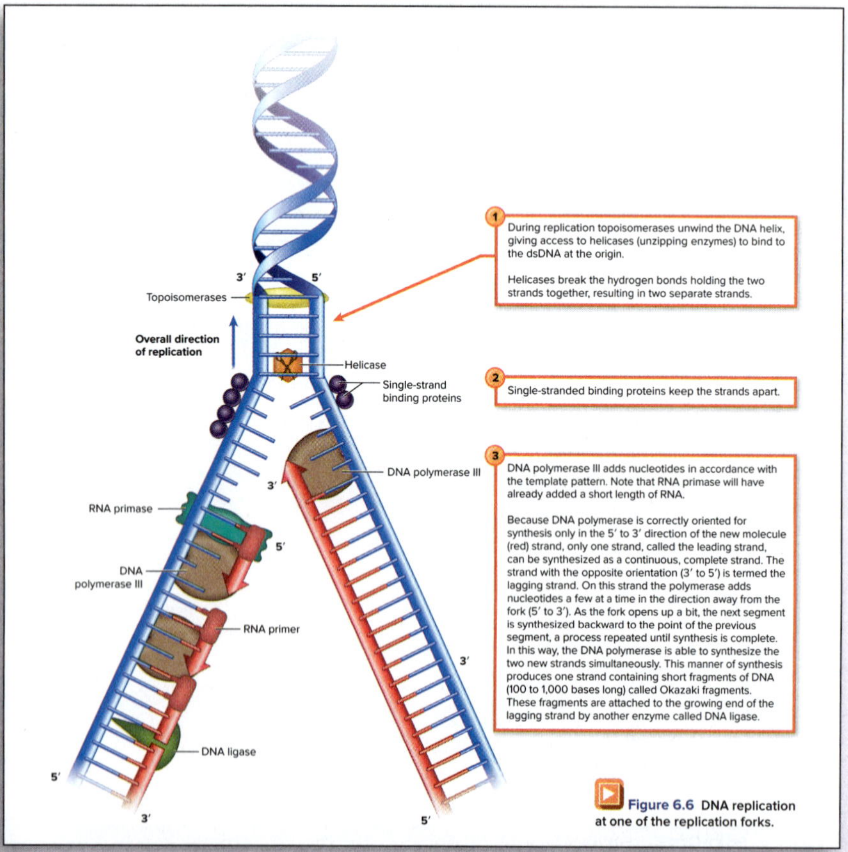

3′
5′
Topoisomerases
Overall direction of replication
Helicase
Single-strand binding proteins
DNA polymerase III
RNA primase
DNA polymerase III
RNA primer
DNA ligase
1
During replication topoisomerases unwind the DNA helix, giving access to helicases (unzipping enzymes) to bind to the dsDNA at the origin.
Helicases break the hydrogen bonds holding the two strands together, resulting in two separate strands.
2
Single-stranded binding proteins keep the strands apart.
3
DNA polymerase III adds nucleotides in accordance with the template pattern. Note that RNA primase will have already added a short length of RNA.
Because DNA polymerase is correctly oriented for synthesis only in the 5′ to 3′ direction of the new molecule (red) strand, only one strand, called the leading strand, can be synthesized as a continuous, complete strand. The strand with the opposite orientation (3′ to 5′) is termed the lagging strand. On this strand the polymerase adds nucleotides a few at a time in the direction away from the fork (5′ to 3′). As the fork opens up a bit, the next segment is synthesized backward to the point of the previous segment, a process repeated until synthesis is complete. In this way, the DNA polymerase is able to synthesize the two new strands simultaneously. This manner of synthesis produces one strand containing short fragments of DNA (100 to 1,000 bases long) called Okazaki fragments. These fragments are attached to the growing end of the lagging strand by another enzyme called DNA ligase.

Figure 6.6 DNA replication at one of the replication forks.

Figures

Many difficult microbiological concepts are best portrayed by breaking them down into stages. These figures show each step clearly marked with an orange, numbered circle and the explanation of the step, right next to it.

Viaframe/Corbis/Getty Images

Brand New Comprehensive Epidemiology Chapter

One thing the COVID-19 pandemic showed us is that all of us need a basic understanding of epidemiological principles. Kelly Cowan has taught undergraduate epidemiology for 20+ years, and explains topics such as correlation, causation, screening tests, types of studies, reproductive rates, and the natural history of disease in easy-to-understand terms.

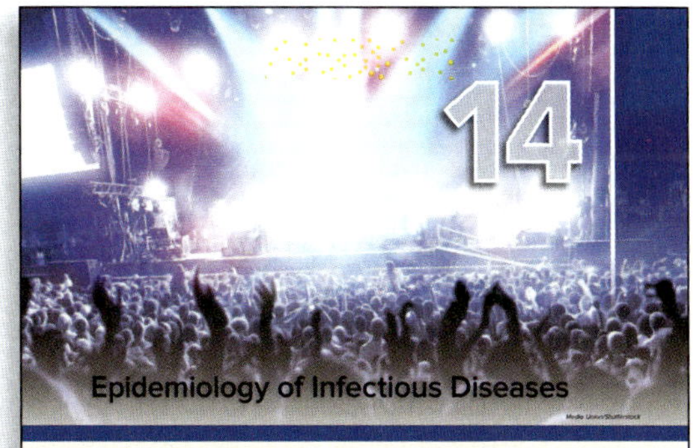
14
Epidemiology of Infectious Diseases

Media Under The Microscope

Herd Immunity

These case studies examine an article from the popular media to determine the extent to which it is factual and/or misleading. This case focuses on the 2021 American Medical Association article "What Doctors Wish Patients Knew About COVID-19 Herd Immunity."

"Herd immunity occurs when a significant portion of population becomes immune to an infectious disease, limiting further disease spread. For those who are not immune, they are indirectly protected because the ongoing disease spread is small." This is a quote from the article published on the AMA website on August 27, 2021. In this chapter you will learn about herd immunity, but you have heard about it a lot during the COVID pandemic.

The article explains that originally health officials estimated that 70% of the population would have to be immunized, or recovered from the infection, before herd immunity could be achieved.

The article continues, "Even though a person may not have immunity for whatever reason—maybe they don't have a strong immune system, they're unable to get the vaccine or children are the obvious example right now—what we do is we surround them," he said. "We insulate them from the virus by having everybody around them have immunity."

Systematic Presentation of Disease-Causing Organisms

Microbiology: A Systems Approach takes a unique approach to diseases by organizing microbial agents under the heading of the disease condition they cause. After all of them are covered, the agents are summarized in a comparative table. Every condition gets a table, whether there is one possible cause or a dozen. Through this approach, students study how diseases affect patients—the way future health care professionals will encounter them in their jobs. Other texts separate the viral causes from the bacterial causes from the fungal causes of meningitis in different parts of the chapter, as an example of why this matters.

A Note About the Chapter Organization

In a clinical setting, patients present themselves to health care practitioners with a set of symptoms, and the health care team makes an "anatomical" diagnosis—such as a *generalized vesicular rash*. The anatomical diagnosis allows practitioners to narrow down the list of possible causes to microorganisms that are known to be capable of creating such a condition (the differential diagnosis). Then the proper tests can be performed to arrive at an etiologic diagnosis (determining the exact microbial cause). The order of events is

1. anatomical diagnosis,
2. differential diagnosis, and
3. etiologic diagnosis.

In this book, we organize diseases according to anatomical diagnosis (which appears as a boxed heading). Then the agents in the differential diagnosis are each addressed. When we finish addressing each agent that could cause the condition, we sum them up in a Disease Table, whether there is only 1 possible cause or whether there are 9 or 10. Each organism is followed by a B, F, P, V, or H, standing for bacterium, fungus, protozoan, virus, or helminth. If it is a bacterium, its Gram status is also included (G+ or G−).

In the Disease Tables, you will also find a row featuring recommended treatment. Here we will identify the microbes that are on the CDC's "Threat" list for their antibiotic resistance (presented in a table in chapter 12).

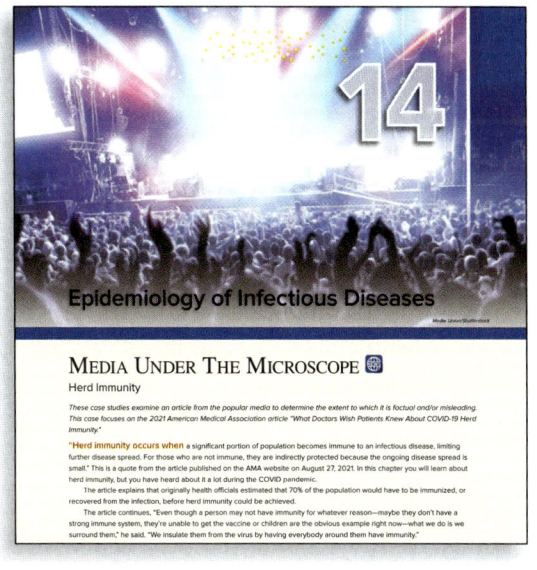

Every disease table contains national and/or worldwide epidemiological information for each causative agent.

This approach is logical, systematic, and intuitive, as it encourages clinical and critical thinking in students—the type of thinking they will be using if their eventual careers are in health care. Students learn to examine multiple possibilities for a given condition and grow accustomed to looking for commonalities and differences among the various organisms that cause a given condition.

Disease Table 19.6 Vesicular/Pustular Rash Diseases

Disease	Chickenpox	Smallpox	Hand, Foot, and Mouth Disease
Causative Organism(s)	Human herpesvirus 3 (varicella-zoster virus) V	Variola virus V	Enteroviruses, usually Coxsackie V
Most Common Modes of Transmission	Droplet contact, inhalation of aerosolized lesion fluid	Droplet contact, indirect contact	Direct and droplet contact
Virulence Factors	Ability to fuse cells, ability to remain latent in ganglia	Ability to dampen, avoid immune response	–
Culture/Diagnosis	Based largely on clinical appearance; PCR is available	Based largely on clinical appearance; if suspected, refer to CDC	Usually based on clinical presentation and history
Prevention	Live attenuated vaccine; there is also vaccine to prevent reactivation of latent virus (shingles)	Live virus vaccine (vaccinia virus)	Hand hygiene
Treatment	None in uncomplicated cases; acyclovir for high risk	Cidofovir; vaccine within 7 days of exposure	None
Distinguishing Features	No fever prodrome; lesions are superficial; in centripetal distribution (more in center of body)	Fever precedes rash; lesions are deep and in centrifugal distribution (more on extremities)	Fever prodrome; lesions in mouth first
Epidemiological Features	Chickenpox: vaccine decreased hospital visits by 88%, ambulatory visits by 59%; shingles: 1 in 3 American adults will have it at least once	Last natural case worldwide was in 1977 Category A Bioterrorism Agent	Sporadic in most of world; possibly on the rise in United States
Appearance of Lesions	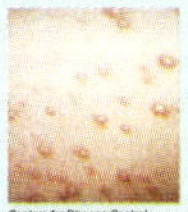 Centers for Disease Control	 Dr. Charles Farmer, Jr./CDC	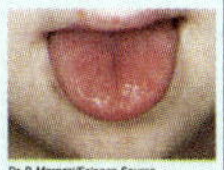 Dr. P. Marazzi/Science Source

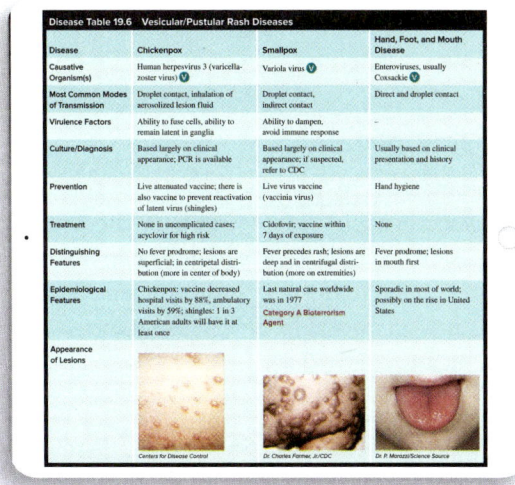

"Diagnosing Infections" Chapter

Chapter 18 brings together in one place the current methods used to diagnose infectious diseases. The chapter starts with collecting samples from the patient and details the biochemical, serological, and molecular methods used to identify causative microbes.

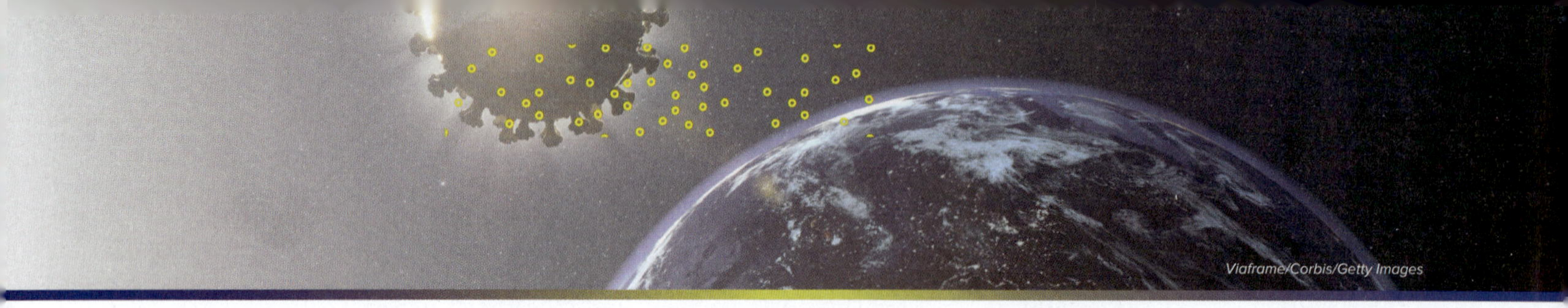
Viaframe/Corbis/Getty Images

Student-Centered Pedagogy Created to Promote Active Learning

Learning Outcomes and Assess Your Progress Questions

Every chapter in the book opens with an outline—which is a list of Learning Outcomes. Each major section of the text concludes with Assess Your Progress with the learning outcomes covered in that section. The Learning Outcomes are tightly correlated to digital material. Instructors can easily measure student learning in relation to the specific Learning Outcomes used in their course.

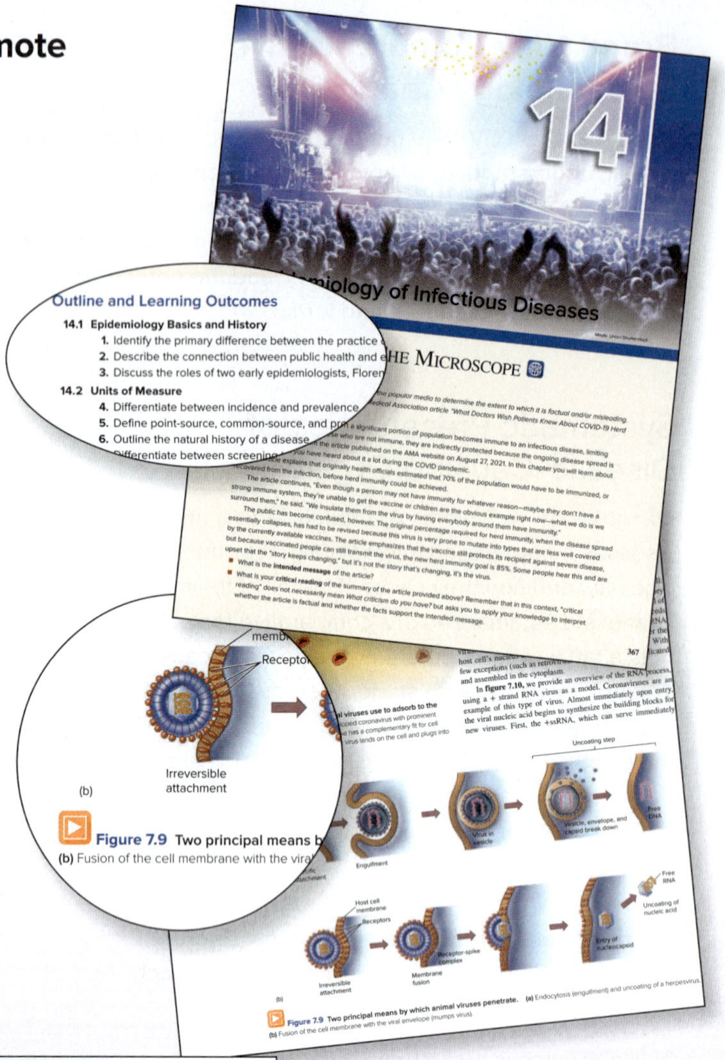
14
Outline and Learning Outcomes
14.1 Epidemiology Basics and History
1. Identify the primary difference between the practice
2. Describe the connection between public health and
3. Discuss the roles of two early epidemiologists, Floren
14.2 Units of Measure
4. Differentiate between incidence and prevalence
5. Define point-source, common-source, and pr
6. Outline the natural history of a disease
Receptor
Irreversible attachment
(b)
Figure 7.9 Two principal means b
(b) Fusion of the cell membrane with the vira
Figure 7.9 Two principal means by which animal viruses penetrate.

Animated Learning Modules

Certain topics need help to come to life off the page. Animations, video, audio, and text all combine to help students understand complex processes. Key topics have an Animated Learning Module assignable through Connect. An icon in the text indicates when these learning modules are available.

Disease Connection

Sometimes it is difficult for students to see the relevance of basic concepts to their chosen professions. So the basic science chapters contain Disease Connections, very short boxes that relate basic science topics to clinical situations.

> **Disease Connection**
>
> *Streptococcus pneumoniae* is a versatile human pathogen. It causes pneumonia, ear infections, meningitis, and other conditions. Its capsule is vital for its disease-causing capacity.

Microbiome Readings

Each chapter includes a Microbiome Insight box. These boxes are a way to emphasize the important and revolutionary ways the microbiome influences almost everything we know about human health.

INSIGHT 11.1 MICROBIOME: Hand Hygiene

How many times have you heard that one of the best ways to prevent infections in hospitals is for health care workers to wash their hands? OK, but now that we know about how important our resident microbiota is, we should ask what frequent hand washing does to that. A recent study followed 34 health care workers in a surgical intensive care unit. Twenty-four of them were registered nurses, six were respiratory therapists, and four were nurse technologists.

During a typical 12-hour shift,

- 53% of them reported washing their hands with soap and water between 6 and 20 times,
- 41% used alcohol rubs more than 20 times, and
- 62% donned more than 40 pairs of gloves.

The researchers then tested for common healthcare-associated pathogens and found that approximately 45% of the health care workers' dominant hands were positive for *Staphylococcus aureus*. They detected methicillin-resistant *Staphylococcus aureus* (MRSA) on 3.9% of the dominant hands. About 4% of hands were positive for the fungus *Candida albicans*.

These rates look very similar to the rates found in the general population—those of us who are not washing our hands 20 times a day. This might be an illustration of the difference between *colonization*—when microbes become part of your "normal" microbiota—and *contamination*—when you pick up microbes from your environment and they stay only transiently. Contaminants are usually easily washed off, while colonizers are embedded deeper in your skin oils and the top layers of epithelial cells.

One question raised, but not answered, by this study is whether altering the normal microbiome of the hand by frequent hand hygiene practices might make it (a) easier or (b) more difficult for pathogens to contaminate and/or colonize caregivers' hands.

O. Dimier/PhotoAlto

Source: Rosenthal, et al. "Healthcare Workers' Hand Microbiome May Mediate Carriage of Hospital Pathogens," *Pathogens*. 2014 Mar; 3(1): 1–13.

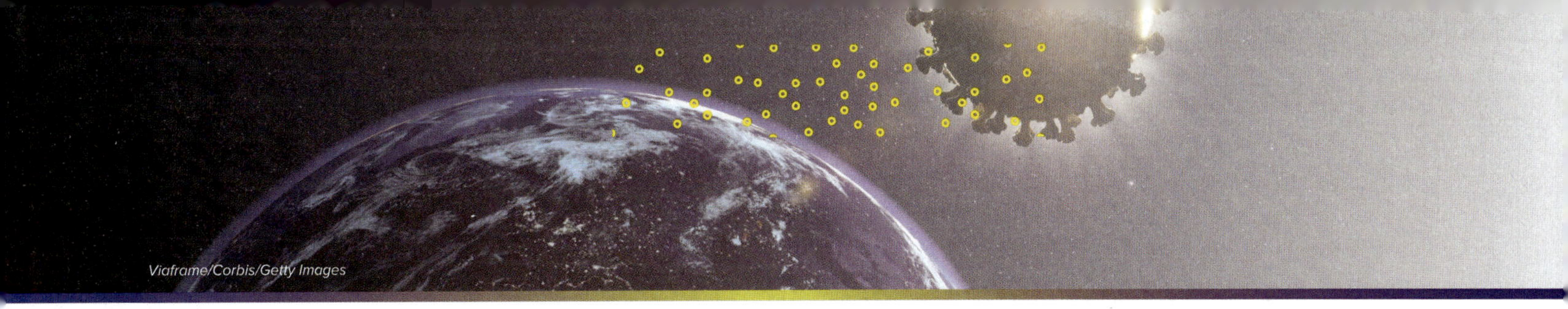

Viaframe/Corbis/Getty Images

System Summary Figures

"Glass body" figures at the end of each disease chapter highlight the affected organs and list the diseases that were presented in the chapter. In addition, the microbes are color coded by taxonomic type.

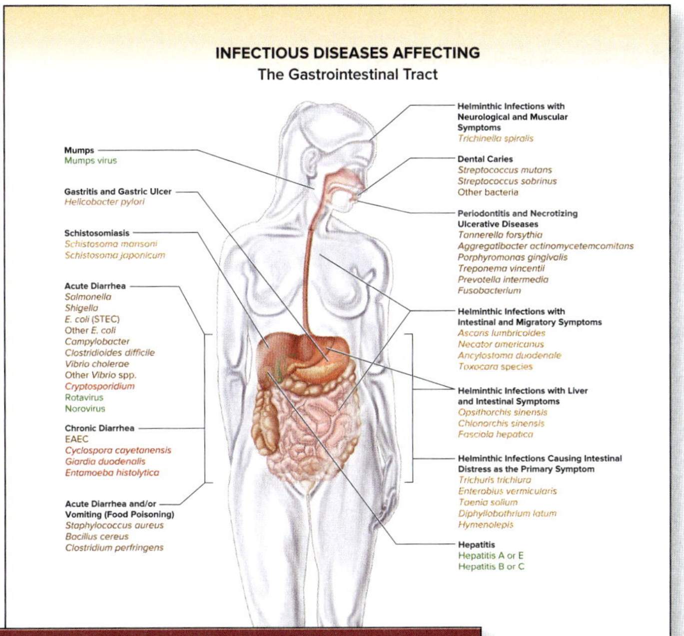

INFECTIOUS DISEASES AFFECTING
The Gastrointestinal Tract
Mumps
Mumps virus
Gastritis and Gastric Ulcer
Helicobacter pylori
Schistosomiasis
Schistosoma mansoni
Schistosoma japonicum
Acute Diarrhea
Salmonella
Shigella
E. coli (STEC)
Other E. coli
Campylobacter
Clostridioides difficile
Vibrio cholerae
Other Vibrio spp.
Cryptosporidium
Rotavirus
Norovirus
Chronic Diarrhea
EAEC
Cyclospora cayetanensis
Giardia duodenalis
Entamoeba histolytica
Acute Diarrhea and/or Vomiting (Food Poisoning)
Staphylococcus aureus
Bacillus cereus
Clostridium perfringens
Helminthic Infections with Neurological and Muscular Symptoms
Trichinella spiralis
Dental Caries
Streptococcus mutans
Streptococcus sobrinus
Other bacteria
Periodontitis and Necrotizing Ulcerative Diseases
Tannerella forsythia
Aggregatibacter actinomycetemcomitans
Porphyromonas gingivalis
Treponema vincentii
Prevotella intermedia
Fusobacterium
Helminthic Infections with Intestinal and Migratory Symptoms
Ascaris lumbricoides
Necator americanus
Ancylostoma duodenale
Toxocara species
Helminthic Infections with Liver and Intestinal Symptoms
Opsithorchis sinensis
Chlonorchis sinensis
Fasciola hepatica
Helminthic Infections Causing Intestinal Distress as the Primary Symptom
Trichuris trichiura
Enterobius vermicularis
Taenia solium
Diphyllobothrium latum
Hymenolepis
Hepatitis
Hepatitis A or E
Hepatitis B or C

Communicability vs. Deadliness Feature

Each microbe can be characterized using two important descriptors: its relative communicability and its relative deadliness. These are important epidemiologically and clinically—and usually receive only sporadic mention in textbooks—so we have created a visual feature that appears in each disease chapter, and in the epidemiology chapter.

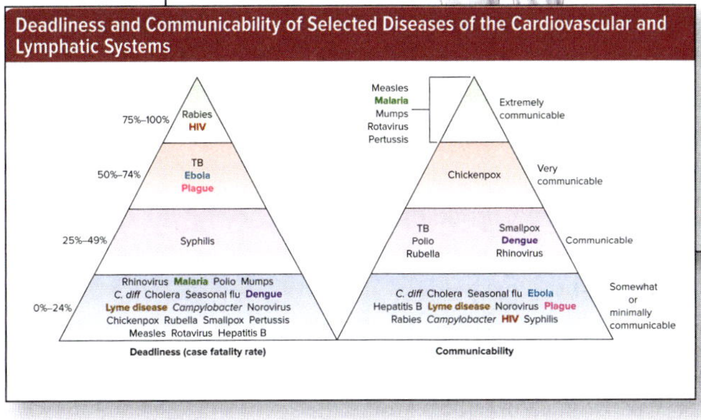

Deadliness and Communicability of Selected Diseases of the Cardiovascular and Lymphatic Systems
75%–100%
Rabies
HIV
50%–74%
TB
Ebola
Plague
25%–49%
Syphilis
0%–24%
Rhinovirus Malaria Polio Mumps
C. diff Cholera Seasonal flu Dengue
Lyme disease Campylobacter Norovirus
Chickenpox Rubella Smallpox Pertussis
Measles Rotavirus Hepatitis B
Deadliness (case fatality rate)
Measles
Malaria
Mumps
Rotavirus
Pertussis
Extremely communicable
Chickenpox
Very communicable
TB
Polio
Rubella
Smallpox
Dengue
Rhinovirus
Communicable
C. diff Cholera Seasonal flu Ebola
Hepatitis B Lyme disease Norovirus Plague
Rabies Campylobacter HIV Syphilis
Somewhat or minimally communicable
Communicability

Taxonomic List of Organisms

A taxonomic list of organisms is presented at the end of each disease chapter so students can see the taxonomic position of microbes causing diseases in that body system.

▶ Summing Up

Taxonomic Organization Microorganisms Causing Diseases of the Skin and Eyes

Microorganism	Disease	Disease Table
Gram-positive bacteria		
Staphylococcus aureus	MRSA impetigo, cellulitis, scalded skin syndrome, folliculitis, abscesses (furuncles and carbuncles), necrotizing fasciitis, bacterial conjunctivitis	MRSA skin and soft tissue infections, 19.1; Impetigo, 19.2; Cellulitis, 19.3; Scalded skin syndrome, 19.4
Streptococcus pyogenes	Impetigo, cellulitis, erysipelas, necrotizing fasciitis, scarlet fever	Impetigo, 19.2; Cellulitis, 19.3; Maculopapular rash diseases, 19.7
Clostridium perfringens	Gas gangrene	Gas gangrene, 19.5
Bacillus anthracis	Cutaneous anthrax	Large pustular skin lesions, 19.9
Streptococcus pneumoniae	Conjunctivitis	Conjunctivitis, 19.11
Gram-negative bacteria		
Chlamydia trachomatis	Neonatal conjunctivitis, bacterial conjunctivitis	Conjunctivitis, 19.11
Neisseria gonorrhoeae	Neonatal conjunctivitis, bacterial conjunctivitis, trachoma	Conjunctivitis, 19.11; Trachoma, 19.12
Wolbachia (in combination with *Onchocerca*)	River blindness	River blindness, 19.14
Haemophilus influenzae and *Moraxella*	Conjunctivitis	Conjunctivitis, 19.11
DNA viruses		
Human herpesvirus 3 (varicella) virus	Chickenpox	Vesicular or pustular rash diseases, 19.6
Variola virus	Smallpox	Vesicular or pustular rash diseases, 19.6
Parvovirus B19	Fifth disease	Maculopapular rash diseases, 19.7
Human herpesvirus 6	Roseola	Maculopapular rash diseases, 19.7
Human papillomavirus	Warts	Warts and wartlike eruptions, 19.8
Molluscum contagiosum virus	Molluscum contagiosum	Warts and wartlike eruptions, 19.8
Herpes simplex virus	Keratitis	Keratitis, 19.13
Adenovirus	Conjunctivitis	Conjunctivitis, 19.11

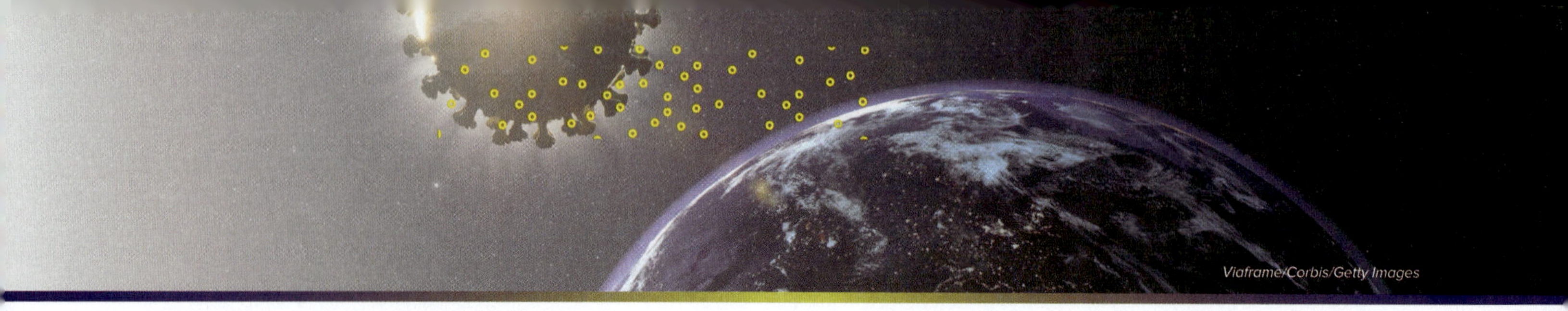

Viaframe/Corbis/Getty Images

Developing Critical Thinkers

The end-of-chapter material is linked to Bloom's Taxonomy. It has been carefully planned to promote active learning and provide review for different learning styles and levels of difficulty.

SmartGrid

This innovative learning tool distributes chapter material among the American Society for Microbiology's six main curricular concepts, plus the competency of *scientific thinking*. Each of the seven areas is probed at three different Bloom's levels. The resulting 21-question grid can be assigned by column (all multiple-choice questions about each core concept, for example) or by row (all questions related to evolution, but at increasing Bloom's level). The highest Bloom's level questions can easily be assigned as a group project or presentation topic.

These are written around the American Society for Microbiology's (ASM) Undergraduate Curriculum guidelines. More recently, ASM has created the "Microbiology in Nursing and Allied Health (MINAH) Undergraduate Curriculum Guidelines." The emphases in these two sets of guidelines are different, but the original guidelines utilized here are weighted to the emphasis in this book's content, which is appropriate for all types of microbiology students.

SmartGrid: From Knowledge to Critical Thinking

This *21 Question Grid* takes the topics from this chapter and arranges them with respect to the American Society for Microbiology's Undergraduate Curriculum guidelines—all six of the important "Concepts" as well as the important "Competency" of scientific literacy. Three questions are supplied, which cover chapter content referring to the Concept or Competency in increasing levels of Bloom's taxonomy for learning.

ASM Concept/ Competency	A. Bloom's Level 1, 2—Remember and Understand (Choose one.)	B. Bloom's Level 3, 4—Apply and Analyze	C. Bloom's Level 5, 6—Evaluate and Create
Evolution	1. The identities of microorganisms on our planet a. are mostly known. b. have nearly all been identified via microscopy. c. have nearly all been identified via culturing techniques. d. are still mostly unknown.	2. Before the availability of molecular techniques, archaea and bacteria were considered to be very closely related. Explain why that is understandable, based on what you learned about microscopy.	3. Often bacteria that are freshly isolated from a patient or the environment have thick sugary outer layers (capsules). After they are transferred many times on laboratory medium, they tend to lose their capsule. Speculate about why that might happen.
Cell Structure and Function	4. Which of these types of organisms is least likely to be identified to the genus level with light microscopy? a. bacteria b. protozoa c. fungus d. helminth	5. Obviously, *structures* of microorganisms can be revealed by various types of microscopy. Can you think of any *functions* that might also be revealed? Which ones, and by which microscopic technique?	6. Some bacteria can produce a structure called an endospore, which is used as a survival structure. It is surrounded by a very tough and impervious outer layer that keeps it protected in adverse environmental conditions. What kinds of things would you need to know about this structure if you wanted to devise a way to stain it and see it under a microscope?

High Impact Study Feature

Students benefit most from varied study and assessment methods. We've created a short set of "Terms" and "Concepts" that help students identify the most important 10 to 15 items in a chapter. If they understand these, they are well on their way to mastery.

High Impact Study

These terms and concepts are most critical for your understanding of this chapter—and may be the most difficult. Have you mastered them? In these disease chapters, the terms and concepts help you identify what is important in a different way than the comprehensive details in the Disease Tables. Your instructor will help you understand what is important for your class.

Concepts

- ☐ Defenses of nervous system
- ☐ Normal microbiota of nervous system
- ☐ Four bacterial causes of meningitis
- ☐ Other causes of meningitis
- ☐ Foodborne cause of meningitis
- ☐ Meningitis vaccines
- ☐ Gram-negative diplococci vs. gram-positive diplococci
- ☐ Difference between CJD and vCJD
- ☐ Global polio eradication
- ☐ Three types of botulism
- ☐ Differences and similarities between tetanus and botulism
- ☐ Organisms in this chapter for which there are vaccines available
- ☐ Organisms in this chapter that display significant antibiotic resistance

Terms

- ☐ Meninges
- ☐ Cerebrospinal fluid
- ☐ Blood-brain barrier
- ☐ Arbovirus
- ☐ Dead-end host
- ☐ Prion
- ☐ Progressive multifocal leukoencephalopathy
- ☐ Postinfection encephalitis
- ☐ Subacute sclerosing parencephalitis

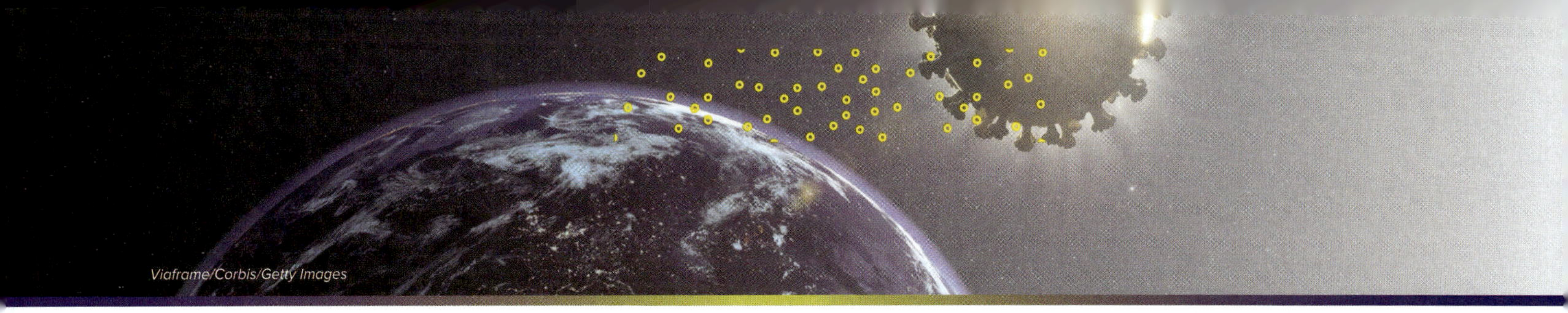

Viaframe/Corbis/Getty Images

Group Study Guide

The feature "Study Smarter: Better Together" gives students a format for their self-guided group study. We know that group study can be immensely useful for learning—but only if it is well-structured. This feature, in every chapter, helps students make the best use of their study time with their classmates, either in person or virtually, with no effort on the part of the instructor!

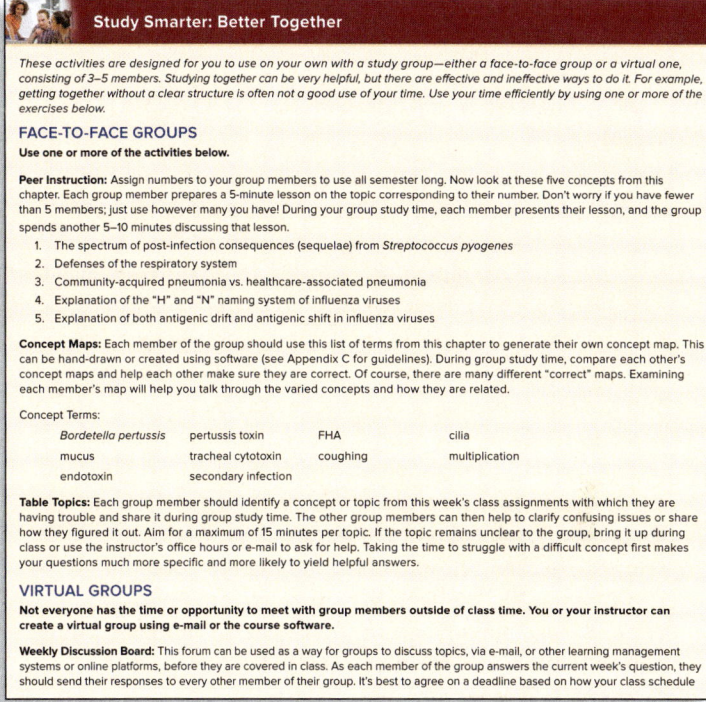

Study Smarter: Better Together

These activities are designed for you to use on your own with a study group—either a face-to-face group or a virtual one, consisting of 3–5 members. Studying together can be very helpful, but there are effective and ineffective ways to do it. For example, getting together without a clear structure is often not a good use of your time. Use your time efficiently by using one or more of the exercises below.

FACE-TO-FACE GROUPS

Use one or more of the activities below.

Peer Instruction: Assign numbers to your group members to use all semester long. Now look at these five concepts from this chapter. Each group member prepares a 5-minute lesson on the topic corresponding to their number. Don't worry if you have fewer than 5 members; just use however many you have! During your group study time, each member presents their lesson, and the group spends another 5–10 minutes discussing that lesson.

1. The spectrum of post-infection consequences (sequelae) from *Streptococcus pyogenes*
2. Defenses of the respiratory system
3. Community-acquired pneumonia vs. healthcare-associated pneumonia
4. Explanation of the "H" and "N" naming system of influenza viruses
5. Explanation of both antigenic drift and antigenic shift in influenza viruses

Concept Maps: Each member of the group should use this list of terms from this chapter to generate their own concept map. This can be hand-drawn or created using software (see Appendix C for guidelines). During group study time, compare each other's concept maps and help each other make sure they are correct. Of course, there are many different "correct" maps. Examining each member's map will help you talk through the varied concepts and how they are related.

Concept Terms:

Bordetella pertussis	pertussis toxin	FHA	cilia
mucus	tracheal cytotoxin	coughing	multiplication
endotoxin	secondary infection		

Table Topics: Each group member should identify a concept or topic from this week's class assignments with which they are having trouble and share it during group study time. The other group members can then help to clarify confusing issues or share how they figured it out. Aim for a maximum of 15 minutes per topic. If the topic remains unclear to the group, bring it up during class or use the instructor's office hours or e-mail to ask for help. Taking the time to struggle with a difficult concept first makes your questions much more specific and more likely to yield helpful answers.

VIRTUAL GROUPS

Not everyone has the time or opportunity to meet with group members outside of class time. You or your instructor can create a virtual group using e-mail or the course software.

Weekly Discussion Board: This forum can be used as a way for groups to discuss topics, via e-mail, or other learning management systems or online platforms, before they are covered in class. As each member of the group answers the current week's question, they should send their responses to every other member of their group. It's best to agree on a deadline based on how your class schedule

Visual Connections

Visual Connections questions take images and concepts learned in previous chapters or in the current one, and ask students to look at that concept in a different way, that is connected with other content. This helps students evaluate information in new contexts and enhances learning.

Visual Connections

This question uses visual images to connect content within and between chapters.

1. **From this chapter, figure 22.10.** Find the latest data for worldwide deaths due to COVID-19 and draw a bar on this graph to represent that.

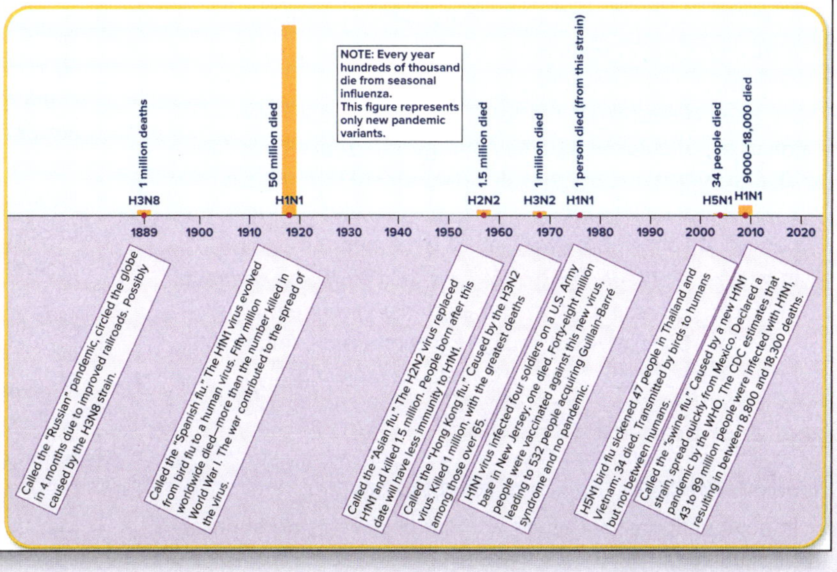

NOTE: Every year hundreds of thousand die from seasonal influenza. This figure represents only new pandemic variants.
1 million deaths
H3N8
50 million died
H1N1
1.5 million died
H2N2
1 million died
H3N2
1 person died (from this strain)
H1N1
34 people died
H5N1
9000–18,000 died
H1N1
1889
1900
1910
1920
1930
1940
1950
1960
1970
1980
1990
2000
2010
2020
Called the "Russian" pandemic, circled the globe in 4 months due to improved railroads. Possibly caused by the H3N8 strain.
Called the "Spanish flu." The H1N1 virus evolved from bird flu to a human virus. Fifty million worldwide died—more than the number killed in World War I. The war contributed to the spread of the virus.
Called the "Asian flu." The H2N2 virus replaced H1N1 and killed 1.5 million. People born after this date will have less immunity to H1N1.
Called the "Hong Kong flu." Caused by the H3N2 virus. Killed 1 million, with the greatest deaths among those over 65.
H1N1 virus infected four soldiers on a U.S. Army base in New Jersey; one died. Forty-eight million people were vaccinated against this new virus, leading to 532 people acquiring Guillain-Barré syndrome and no pandemic.
H5N1 bird flu sickened 47 people in Thailand and Vietnam; 34 died. Transmitted by birds to humans but not between humans.
Called the "swine flu." Caused by a new H1N1 strain, spread quickly from Mexico. Declared a pandemic by the WHO. The CDC estimates that 43 to 89 million people were infected with H1N1, resulting in between 8,800 and 18,300 deaths.

Viaframe/Corbis/Getty Images

Changes to the Seventh Edition

New to *Microbiology: A Systems Approach*

GLOBAL CHANGES THROUGHOUT THE SEVENTH EDITION

- With every successive edition, I (Kelly) improve readability using my studies of student literacy. This does not involve removing technical terminology. It focuses on shortening sentences and rewriting the context in which the technical terms appear using more conversational terminology, and not "professor" terminology or complex sentence structures.
- The chapter order has been changed for two reasons: (1) The genetics chapter now precedes the virology chapter because a basic understanding of genetics is critical to understanding viral genomes and replication. (2) There is a brand new chapter entitled "Epidemiology of Infectious Diseases," a subject more important than ever.
- Former Chapter 24, "Microbes and the Environment," and Chapter 25, "Applied Microbiology and Food and Water Safety," have been condensed and combined into a new Chapter 25: "Microbes in Our Environment and the Concept of One Health."
- Chapter Summaries have been redesigned, in an infographic format using less text.
- More than half of the opening case studies, "Media Under the Microscope," are brand new, covering news reports of such topics as herd immunity, COVID variants, dietary supplement claims, and vaccine misinformation.
- Throughout the book more non-male and diverse scientists are featured, with photos when possible.
- Although Cowan texts have always aimed for equity in our visual representations and language, in this edition we have redoubled our efforts. We worked hard to be inclusive of cultural diversity and gender representations. The contributions of a greater diversity of scientists is also included. This work will never stop, and we welcome comments about how we can do even better.
- In every chapter we have improved the colors and contrasts of labels, colors, and other details to assist readers who have variations in visual acuity and color vision, and to make details more visible in classroom projection.

Major chapter updates or new material

Chapter 1: The Main Themes of Microbiology

- Improved figures for types of microorganisms and burden of diseases
- Addition of CRISPR to "Recent Advances in Microbiology"
- Updated again the science on the origin of cells on the evolutionary timeline
- Added the story of a female scientist and her young daughter discovering and naming a new species

Chapter 2: The Chemistry of Biology

- Improved seven figures for better understanding
- Mentioned the existence of 22 amino acids, with 20 being the most common

Chapter 3: Tools of the Laboratory: Methods for the Culturing and Microscopy of Microorganisms

- New microbiome box about the microbiome of Leonardo da Vinci drawings
- Added super-resolution light microscopy
- Added calcofluor white KOH staining

Chapter 4: Bacteria and Archaea

- Added information about *Achromatium,* a bacterium with numerous non-identical chromosomes; used this example to discuss exceptions in all of biology

Chapter 5: Eukaryotic Cells and Microorganisms

- Included story about fungal infections in India in COVID patients
- Mention of association of *Malassezia* with pancreatic cancer

Chapter 6: Microbial Genetics

- Major overhaul; four new infographic style figures
- New information about a new DNA configuration used for regulation
- New credit given to Esther Lederburg for her work on bacteriophage
- New figure about how the COVID-19 mRNA vaccine is made
- Information on epigenetic regulation of DNA
- Increased emphasis on microRNAs

Chapter 7: Viruses and Prions

- Updated virus taxonomy from the ICTV
- Information about some viruses being found that can replicate independently
- Baltimore Classification system introduced
- COVID-19 information incorporated
- Mention made of viral cancers and oncolytic viruses used as anti-cancer agents
- Several figures enhanced

Viaframe/Corbis/Getty Images

Chapter 8: Genetic Analysis and Recombinant DNA Technology

- Reorganized for better flow and understanding
- Shifted from the phrase "genetic engineering" to "recombinant DNA technology" due to increasingly negative connotations with the first
- Added discussion of personalized medicine
- Expanded CRISPR section and included photo of the two female Nobel Prize winners
- Added information about SNP analysis being used by companies such as ancestry.com

Chapter 9: Microbial Nutrition and Growth

- New information about antagonism and synergism in the gut microbiome
- Much more emphasis on the role of microbe-microbe interactions
- Added actual photographs in place of drawings (of test tubes, plates)
- Clarified language in many places

Chapter 10: Microbial Metabolism

- New content pointing out how microbes and their hosts and neighboring microbes share their metabolisms and metabolic products
- Updated several figures to be more understandable

Chapter 11: Physical and Chemical Control of Microbes

- Information about disinfecting COVID-contaminated surfaces
- Information about why HEPA filters trap SARS-CoV-2
- Information about essential oils, and also about their unregulated nature

Chapter 12: Antimicrobial Treatment

- Updated accurate history and overview of antibiotic development, and why it has slowed down (beyond the usual profit margin excuse)
- Gram-negative bacteria as the target of important new antibiotic strategies; new figure of gram-negative envelope structure and why it is so difficult to penetrate with antibiotics
- Failed promise of high-throughput synthesis to find new antibiotics
- New approaches: nano materials; antisense RNAs; CRISPR; phages; antibiotics targeting outer membrane proteins
- Many new and heavily revised figures
- New figure on streptomycin/tuberculosis history
- New figure on the influence of antibiotics on the microbiome over the life span

Chapter 13: Microbe–Human Interactions: Health and Disease

- New information about phase 2 of the Human Microbiome Project (HMP)
- Microbiome's influence on anti-cancer drugs
- Frequent weaving-in of COVID-19; its zoonotic origin, its long-haul symptoms, etc.

Chapter 14: Epidemiology of Infectious Diseases

- Whole new chapter on epidemiology
- Ten brand new figures
- Difference between clinical practice and epidemiology
- History of epidemiology and the CDC
- Units of measure: incidence/prevalence, morbidity/mortality
- Natural history of disease
- Screening
- Causality and Hill's criteria
- Surveillance and notifiable diseases
- Healthcare-associated infections
- Basic reproduction rate and case-fatality rate
- Vaccinology—how vaccines are approved and continuously monitored and by whom (biology of vaccine remains in Adaptive Immunity chapter)

Chapter 15: Host Defenses I: Overview and Innate Immunity

- Fully converted to terms "adaptive" and "innate" replacing "specific" and "nonspecific"
- Mention of inflammation as part of long-COVID
- Discussion of new view of normal body temperature
- New research about RBCs activating innate immunity

Chapter 16: Host Defenses II: Adaptive Immunity and Immunization

- Cytokine storm defined, and discussed in the context of COVID
- New vaccine overview figure, regrouping the subtypes, because of the introduction of COVID vaccines
- Discussion of the rationale behind booster shots
- Ethical question raised about boosters for fully vaccinated in wealthy countries vs. poor countries having single-digit vaccination rates
- Information about legacy vaccines that are live attenuated being useful in a low vaccine supply situation
- Monoclonal antibody treatment (for COVID) in the category of passive immunity
- New up-and-coming vaccine technologies: including CRISPR and nanoparticles

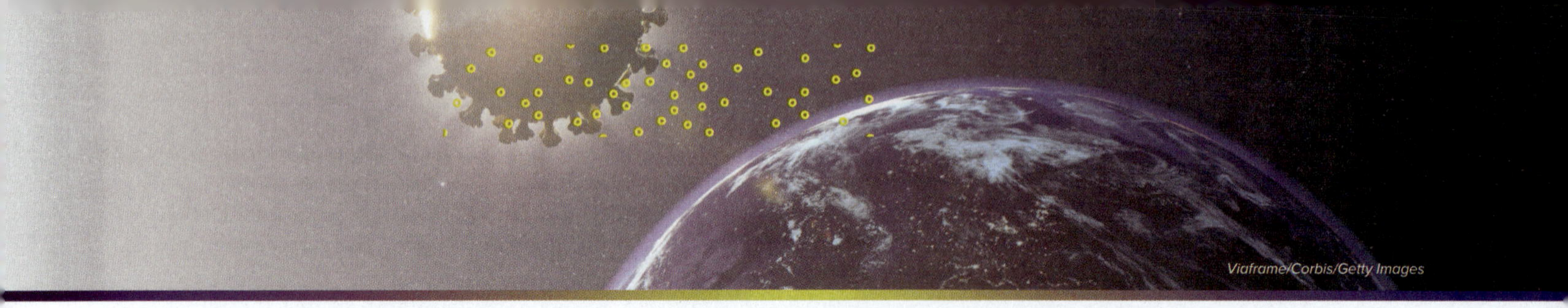
Viaframe/Corbis/Getty Images

- Discussion of a large peer-reviewed study about vaccine hesitancy among health care workers

Chapter 17: Disorders in Immunity

- Shortened and simplified discussion of types II, III, and IV hypersensitivity
- Brief paragraph about possibility that increased autoimmunity can follow COVID-19 infection
- Several figures improved or completely redone
- Brief mention about the effect of wildfires on asthma

Chapter 18: Diagnosing Infections

- Point-of-care tests defined and pointed out throughout chapter
- Bloodstream infections highlighted throughout chapter
- Added complement fixation illustration
- PCR methods grouped under the commonly used clinically heading “NAAT” (nucleic acid amplification techniques)
- PulseNet technology updated to whole genome sequencing (replacing PFGE)
- CRISPR-based diagnosis added

Chapter 19: Infectious Diseases Manifesting on the Skin and Eyes

- For this and all disease chapters, changed title wording to indicate that organization is based on where manifestations are seen
- For this and all disease chapters, changed “Culture and Diagnosis” to “Culture and/or Diagnosis” to indicate that sometimes culture is no longer the go-to or even the gold standard
- Information about new antiviral for poxviruses
- Discussion of erasure of immune memory with measles infection
- Comeback of measles and the causes
- Effect of pandemic on uptake of other childhood vaccines
- *Malassezia* implicated in pancreatic cancer

Chapter 20: Infectious Diseases Manifesting in the Nervous System

- In the Media Under The Microscopic Wrap Up, I mention that I had to be extra careful to check my own biases while critically analyzing the article because it contained a few red flags for me. I wrote that in that case, I had to redouble my efforts to consider the assertions in the article on their own merits, in case they were, in fact, reliable (modeling critical thinking)
- Added neurological symptoms during and after COVID-19
- *Coccidioides* spreading to Washington state and Utah; climate change implicated

Chapter 21: Infectious Diseases Manifesting in the Cardiovascular and Lymphatic Systems

- COVID-19 based in this chapter—with explanation why
- Link between hypertension and gut microbiome
- *C. auris* and sepsis and its pan-resistant forms
- Updates on Post-treatment Lyme Disease Syndrome
- New dengue vaccine
- New malaria vaccine for children

Chapter 22: Infectious Diseases Manifesting in the Respiratory System

- Targeting a human cell that cold viruses need as a treatment for colds
- EV-D68 as a recent cause of severe colds that may lead to acute flaccid myelitis
- Why RSV came in the summer in 2021 instead of fall and winter as usual (COVID)
- Emergence of *C. auris*
- Graph of usual influenza seasonality and death toll, including COVID-19 deaths for comparison

Chapter 23: Infectious Diseases Manifesting in the Gastrointestinal Tract

- Drastic increases in hepatitis A since 2016 and change in its epidemiology
- Vignette and photo of Rita Colwell and her groundbreaking research using saris to filter out copepod-vibrio complexes

Chapter 24: Infectious Diseases Manifesting in the Genitourinary System

- Added a demographic figure commonly used in epidemiology to drive home the role of epidemiology (and discuss *Chlamydia* rates)
- New organisms associated with urethritis (*M. genitalium* and *N. meningitidis*)

Chapter 25: Microbes in Our Environment and the Concept of One Health

- Now contains elements of former chapter 24 and chapter 25
- New climate change section outlining hantavirus, HIV, arbovirus infections, and SARS-CoV-2 and the influence of climate and behavior on them
- One Health emphasis
- Removed all cycles but C and N cycles
- Removed foodborne diseases as it was redundant with text in gastrointestinal chapter

Acknowledgments

Viaframe/Corbis/Getty Images

We are most grateful to our students who continually teach us how to communicate this subject. All the professors who reviewed manuscript or sent e-mails with feedback were our close allies as well, especially when they were free with their criticism. Jennifer Lusk contributed invaluable content to the text. Our minders at McGraw Hill are paragons of patience and professionalism: Darlene Schueller is the best editor in the business, which makes it all the more surprising that she continues to work with me on book after book. Other members of our McGraw Hill team upon whom we lean heavily are Lauren Vondra, Tami Hodge, David Hash, Paula Patel, Ron Nelms, Lori Hancock, and Betsy Blumenthal.

—Kelly Cowan
—Heidi Smith

Table of Contents

Viaframe/Corbis/Getty Images

Rich Carey/Shutterstock

Ingram Publishing/Superstock

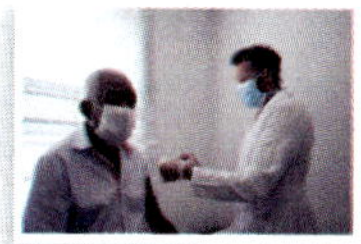

FG Trade/Getty Images

CHAPTER 7

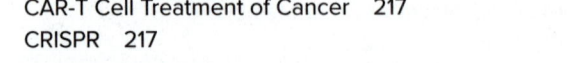

Chia-Chi Charlie Chang/NIH

CHAPTER 8

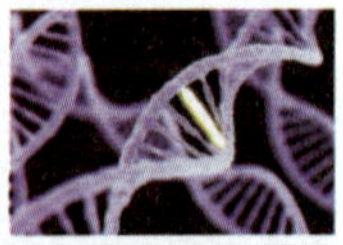
Konstantin Faraktinov/Shutterstock

CHAPTER 9

NASA

CHAPTER 10

spxChrome/Getty Images

CHAPTER 11

Kevin Griffin/123RF

CHAPTER 12

milos luzanin/Alamy Stock photo; Gary He/McGraw Hill

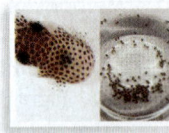

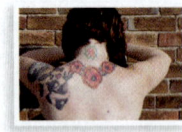

PHIL/CDC

Foodcollection/Martin Skultety/Image Source

Lipik/Shutterstock

LightField Studios/ Shutterstock

Gerhard Zwerger-Schoner/imageBROKER/Alamy Stock Photo

CHAPTER 23

Taras Vyshnya/Shutterstock

CHAPTER 24

Getty Images/Moment Open

CHAPTER 25

CHAPTER **23**

Gerhard Zwerger-Schoner/imageBROKER/Alamy Stock Photo

CHAPTER **24**

Taras Vyshnya/Shutterstock

CHAPTER **25**

Getty Images/Moment Open

NUTMEG
GROUND

1 The Main Themes of Microbiology

McGraw Hill

Media Under The Microscope

Dirty Dish Towels

These case studies examine an article from the popular media to determine the extent to which it is factual and/or misleading. This case focuses on a U.S. News and World Report *article "Study: Harmful Bacteria Is Growing on Kitchen Towels."*

A headline in *U.S. News and World Report* reads "Harmful Bacteria Is Growing on Kitchen Towels." The article summarizes a study conducted at the University of Mauritius and states that after one month of use, 49% of kitchen towels contained bacterial growth.

The researchers are cited in this article as saying that *Escherichia coli* was present in 18 of the 100 towels they studied. The study pointed out that *E. coli* is found in human feces and that this indicates fecal contamination and poor hygiene. The researchers concluded that nonhygienic practices in the kitchen could lead to food poisoning.

- What is the **intended message** of the article?
- What is your **critical reading** of the summary of the article provided above? Remember that in this context, "critical reading" does not necessarily mean *What criticism do you have?* but asks you to apply your knowledge to interpret whether the article is factual and whether the facts support the intended message.
- How would you **interpret** the news item for your nonmicrobiologist friends?
- What is your **overall grade** for the news item—taking into account its accuracy and the accuracy of its intended effect?

Media Under The Microscope Wrap-Up appears at the end of the chapter.

Outline and Learning Outcomes

1.1 The Scope of Microbiology

1. List the seven types of microorganisms studied in the field of microbiology.
2. Identify multiple professions using microbiology.

1.2 The Impact of Microbes on Earth: Small Organisms with a Giant Effect

3. Describe the role and impact of microbes on the earth.
4. Explain the theory of evolution and why it is called a theory.

1.3 Human Use of Microorganisms

5. Explain one old way and one new way that humans manipulate organisms for their own uses.

1.4 Infectious Diseases and the Human Condition

6. Summarize the relative burden of human disease caused by microbes, emphasizing the differences between developed countries and developing countries.

1.5 The General Characteristics of Microorganisms

7. Differentiate among bacteria, archaea, and eukaryotic microorganisms.
8. Identify two types of acellular microorganisms.
9. Compare and contrast the relative sizes of the different microbes.

1.6 The Historical Foundations of Microbiology

10. Make a time line of the development of microbiology from the 1600s to today.
11. List some recent microbiological discoveries of great impact.
12. Explain what is important about the scientific method.

1.7 Naming, Classifying, and Identifying Microorganisms

13. Differentiate among the terms *nomenclature, taxonomy,* and *classification.*
14. Create a mnemonic device for remembering the taxonomic categories.
15. Correctly write the binomial name for a microorganism.
16. Draw a diagram of the three major domains.
17. Explain the difference between traditional and molecular approaches to taxonomy.

1.1 The Scope of Microbiology

Microbiology is a specialized area of biology that deals with living things ordinarily too small to be seen without magnification. Such **microscopic** organisms are collectively referred to as **microorganisms** or **microbes.** In the context of infection and disease, some people call them germs, viruses, or agents; others even call them "bugs"; but none of these terms are clear. In addition, some of these terms tend to emphasize the disagreeable reputation of microorganisms. But, as we will learn throughout the course of this book, only a small minority of microorganisms actually cause harm to other living beings.

Despite that statement, this book does focus on the microorganisms that cause human disease. They can be either cellular or noncellular. The cellular microorganisms we will study are **bacteria, archaea, fungi,** and **protozoa.** Another cellular organism that causes human infections is not technically a microorganism. **Helminths** are multicellular animals whose mature form is visible to the naked eye. Acellular microorganisms causing human disease are the **viruses** and **prions. Table 1.1** gives you a first glimpse at some of these microorganisms. You will see in table 1.1 that the name "bacterium" is paired with something called "archaeon." We will say more later, but for now just know that archaea are single-celled microorganisms as well. So far, no clear evidence links them to human disease, so they will receive less emphasis in this book. There is a Note in Section 1.2 that says more about them.

The nature of microorganisms makes them both very easy and very difficult to study—easy because they reproduce so rapidly and we can quickly grow large populations in the laboratory and difficult because we usually cannot see them directly. We rely on a variety of indirect means of analyzing them in addition to using microscopes.

Microbiologists study every aspect of microbes—their cell structure and function, their growth and physiology, their genetics, their taxonomy and evolutionary history, and their interactions with the living and nonliving environment. The last aspect includes their uses in industry and agriculture and the way they interact with mammalian hosts, in particular, their properties that may cause disease or lead to benefits.

Studies in microbiology have led to greater understanding of many general biological principles. For example, the study of microorganisms established universal concepts concerning the chemistry of life, systems of inheritance, and the global cycles of nutrients, minerals, and gases. Some descriptions of different branches of study appear in **table 1.2.**

1.1 Learning Outcomes—Assess Your Progress

1. List the seven types of microorganisms studied in the field of microbiology.
2. Identify multiple professions using microbiology.

Table 1.1 Relative Sizes of Microorganisms We Will Study The size of the colored circles, not the size of the drawn organisms, are of the correct scale.

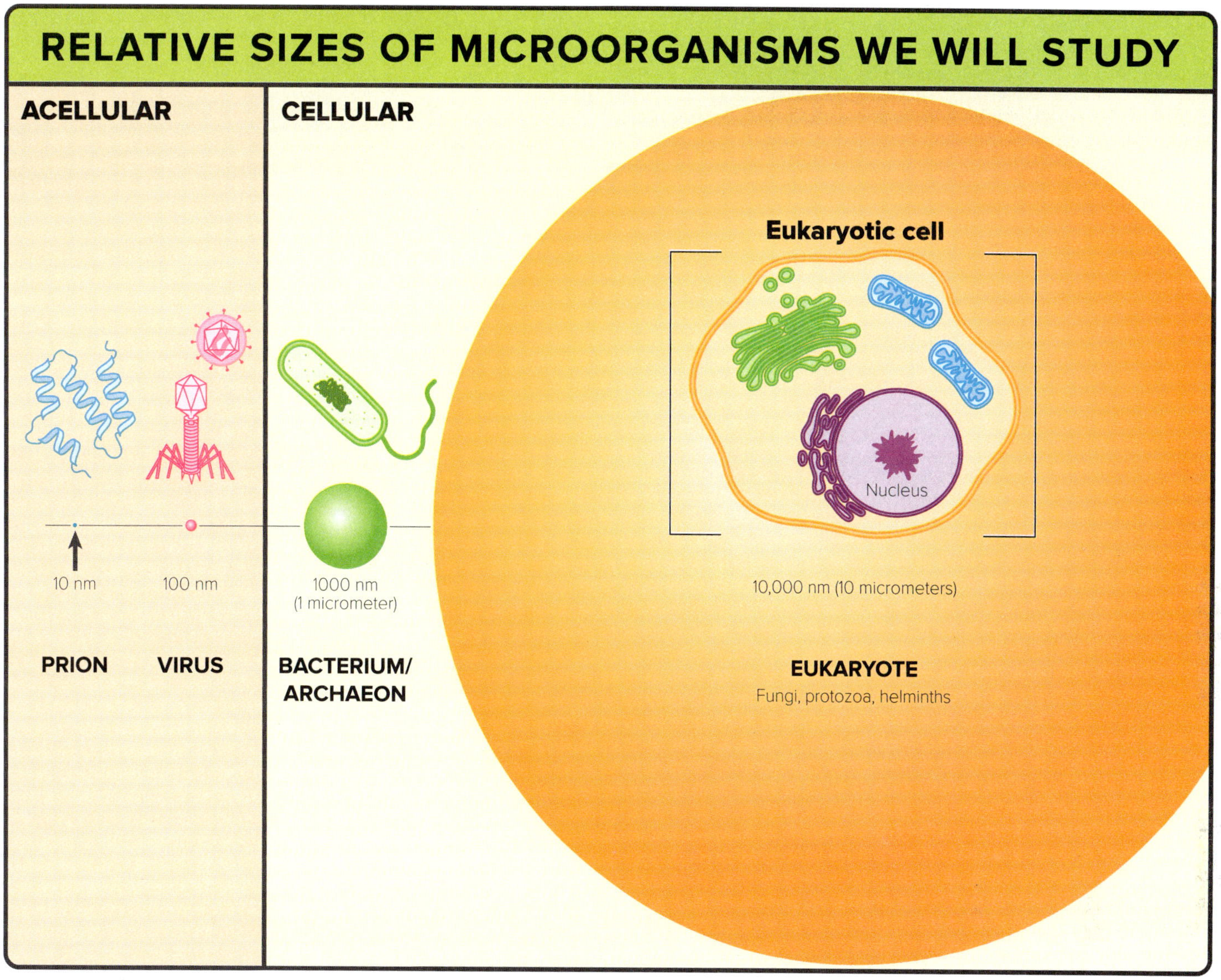
RELATIVE SIZES OF MICROORGANISMS WE WILL STUDY
ACELLULAR
CELLULAR
Eukaryotic cell
Nucleus
10 nm
100 nm
1000 nm
(1 micrometer)
10,000 nm (10 micrometers)
PRION
VIRUS
BACTERIUM/
ARCHAEON
EUKARYOTE
Fungi, protozoa, helminths

Table 1.2 Microbiology—A Sampler

A. Medical Microbiology This branch of microbiology deals with microbes that cause diseases in humans and animals. Researchers examine factors that make the microbes cause disease and mechanisms for inhibiting them.	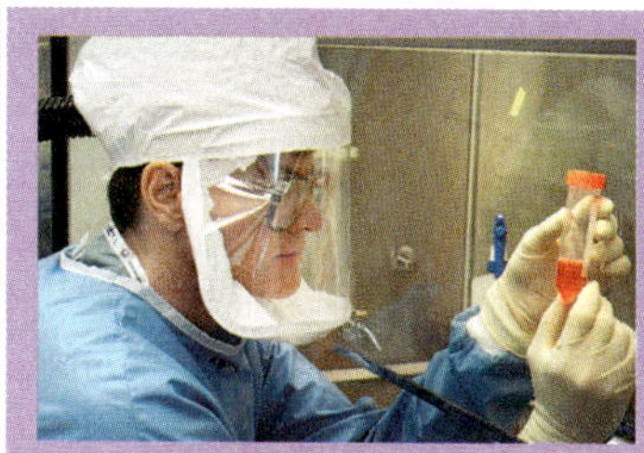	**Figure A.** A staff microbiologist at the Centers for Disease Control and Prevention (CDC) examines a culture of influenza virus identical to one that circulated in 1918. The lab is researching why this form of the virus was so deadly and how to develop vaccines and other treatments. Handling such deadly pathogens requires a high level of protection with special personal protection equipment and high-level labs. *James Gathany/CDC*
B. Public Health Microbiology and Epidemiology These branches monitor and control health and the spread of diseases in communities. Institutions involved in this work are the U.S. Public Health Service (USPHS) with its main agency; the Centers for Disease Control and Prevention (CDC) located in Atlanta, Georgia; and the World Health Organization (WHO), the medical limb of the United Nations.		**Figure B.** Two epidemiologists conducting interviews as part of the effort to curb the cholera epidemic in Haiti. Photograph taken in 2013. *Preetha Iyengar, M.D./CDC*

(continued)

Table 1.2 Microbiology—A Sampler (*continued*)

C. Immunology This branch studies the complex web of protective substances and cells produced in response to infection or cancer. It includes such diverse areas as vaccination, blood testing, and allergy. Immunologists also investigate the role of the immune system in autoimmune diseases.		**Figure C.** An immunologist and students prepare samples. *Ariel Skelley/Blend Images LLC*
D. Industrial Microbiology Industrial microbiology safeguards our food and water, and also includes biotechnology, the use of microbial metabolism to arrive at a desired product, ranging from bread making to gene therapy. Microbes can be used to create large quantities of substances such as amino acids, beer, drugs, enzymes, and vitamins.		**Figure D.** Scientists use a multispectral imaging system to inspect chickens. *Stephen R Ausmus/U.S. Department of Agriculture-ARS*
E. Agricultural Microbiology This branch is concerned with the relationships between microbes and farm animals and crops. Plant specialists focus on plant diseases, soil fertility, and nutritional interactions. Animal specialists work with infectious diseases and other associations animals have with microorganisms.		**Figure E.** Plant microbiologists examine images of alfalfa sprouts to see how microbial growth affects plant roots. *Scott Bauer/USDA*
F. Environmental Microbiology These microbiologists study the effect of microbes on the earth's diverse habitats. Whether the microbes are in freshwater or saltwater, topsoil, or the earth's crust, they have profound effects on our planet. Subdisciplines of environmental microbiology are Aquatic microbiology—the study of microbes in the earth's surface water; Soil microbiology—the study of microbes in terrestrial parts of the planet; Geomicrobiology—the study of microbes in the earth's crust; and Astrobiology (also known as exobiology)—the search for/study of microbial and other life in places off of our planet.		**Figure F.** A researcher collects samples and data in Lake Erie. *Photodiem/Shutterstock*

1.2 The Impact of Microbes on Earth: Small Organisms with a Giant Effect

For billions of years, microbes have extensively shaped the development of the earth's habitats and the evolution of other life forms. It is understandable that scientists searching for life on other planets first look for signs of microorganisms.

Scientists are constantly discovering new clues to how life formed on our planet. One view is illustrated in **figure 1.1.** But the actual events are not yet fully understood, so this is just the most current model. It is believed that soon after the earth was formed, the first ancient cells formed. From these cells, two types of single-celled organisms that are still with us today developed, the **bacteria** and the **archaea.** Those were the only types of cells on the planet for more than a billion years, at which time a much more complex cell appeared, called **eukaryotes.** "Eukary-" means *true nucleus* because these were the only cells containing a nucleus. Bacteria and archaea have no true nucleus. For that reason, some scientists have started calling them **akaryotes,** meaning *no nucleus.*

On the scale pictured in figure 1.1, humans seem to have just appeared compared to the existence of all life. Bacteria arose before even the earliest animals by billions of years. This is a good indication that humans are not likely to—nor should we try to—eliminate bacteria from our environment. They've survived and adapted to many catastrophic changes over the course of their geologic history.

Another indication of the huge influence bacteria exert is how **ubiquitous** they are. That word, ubiquitous, means that microbes

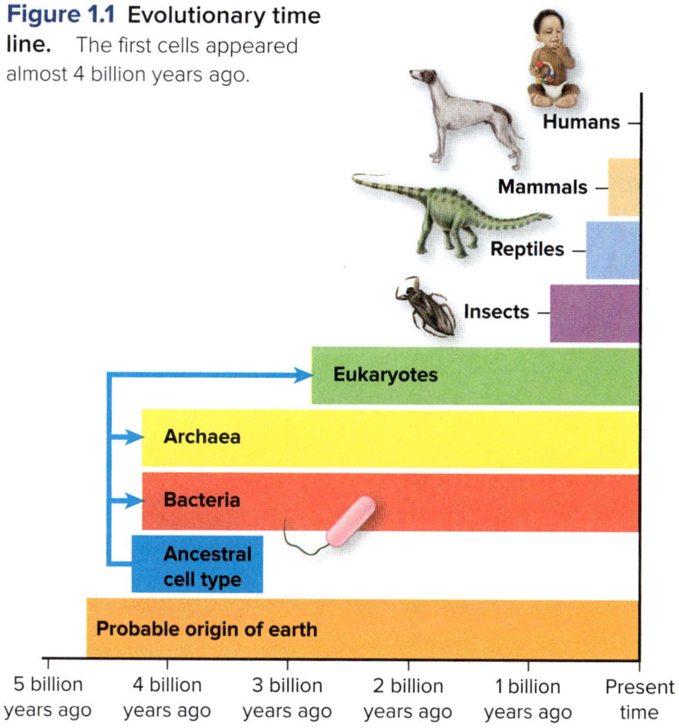
Humans
Mammals
Reptiles
Insects
Eukaryotes
Archaea
Bacteria
Ancestral cell type
Probable origin of earth
5 billion years ago
4 billion years ago
3 billion years ago
2 billion years ago
1 billion years ago
Present time

Figure 1.1 Evolutionary time line. The first cells appeared almost 4 billion years ago.

A Note About Bacteria and Archaea

We've just learned that there are three cell types: eukaryotes, bacteria, and archaea. In this book, we are going to focus on bacteria and the eukaryotes because as far as we know these groups are responsible for the majority of human diseases. We will address archaea in various sections of the book where the distinction is useful, but mainly we will refer to bacteria.

can be found nearly everywhere, from deep in the earth's crust to the polar ice caps and from oceans to inside the bodies of plants and animals. Being mostly invisible, the actions of microorganisms are usually not as obvious or familiar as those of larger plants and animals. They make up for their small size by being present in large numbers and living in places where many other organisms cannot survive. Above all, they play central roles that are absolutely essential to life.

When we point out that single-celled organisms have adapted to a wide range of conditions over the billions of years of their presence on this planet, we are talking about **evolution.** Life in its present form would not be possible if the earliest life forms had not changed constantly, adapting to their environment and circumstances. The timeframe from the far left in figure 1.1 to the far right where humans appeared involved billions and billions of tiny changes, starting with the first cell that appeared soon after the planet itself was formed.

You have no doubt heard this concept described as the "theory of evolution." Let's clarify some terms. Evolution is the accumulation of changes that occur in organisms as they adapt to their environments. It is documented every day in all corners of the planet, an observable phenomenon that is testable by science. Referring to it as the **theory of evolution** has led to great confusion among the general public. As we will explain in section 1.6, scientists use the term "theory" in a different way than the general public does. By the time a principle has been labeled a theory in science, it has undergone years and years of testing and not been disproven. This is much different than the common usage, as in "My theory is that he overslept and that's why he was late." The theory of evolution is a label for a well-studied and well-established natural phenomenon.

Microbial Involvement in Shaping Our Planet

Microbes are deeply involved in the flow of energy and food through the earth's ecosystems. (Ecosystems are communities of living organisms and their surrounding environment.) Most people are aware that plants carry out **photosynthesis,** which is the light-fueled conversion of carbon dioxide to organic material, accompanied by the formation of oxygen (called oxygenic photosynthesis). However, bacteria invented photosynthesis long before the first plants appeared, first as a process that did not produce oxygen (*anoxygenic photosynthesis*). This anoxygenic photosynthesis later evolved into oxygenic photosynthesis, which not only produced oxygen but also was much more efficient in extracting energy from sunlight. Hence, bacteria were responsible for changing the atmosphere of the earth from one without oxygen to one with oxygen. The production of oxygen also led to the use of oxygen for aerobic respiration and the formation of ozone, both of which set off an explosion in species diversification. Today, photosynthetic microorganisms (bacteria and algae) account for perhaps 70% of the earth's photosynthesis, contributing the majority of the oxygen to the atmosphere **(figure 1.2*a*).**

Another process that helps keep the earth in balance is the process of biological **decomposition** and nutrient recycling. Decomposition involves the breakdown of dead matter and wastes into simple compounds that can be directed back into the natural cycles of living things **(figure 1.2*b*).** When death occurs, the body immediately begins to decompose. Bacteria play a major role in decomposition of the body. The action of bacteria causes the conversion of soft tissues within the body to liquids and gases. The chemicals released as a result of decomposition, including hydrogen sulfide, are responsible for the pungent smell of death. If it were not for multitudes of bacteria and fungi, many chemical elements would become locked up and unavailable to organisms, and we humans would drown in our own industrial and personal wastes. In the long-term scheme of things, microorganisms are the main forces that drive the structure and content of the soil, water, and atmosphere. For example:

- The very temperature of the earth is regulated by gases, such as carbon dioxide, nitrous oxide, and methane, which create an insulation layer in the atmosphere and help retain heat. Many of these gases are produced by microbes living in the environment and in the digestive tracts of animals.
- Recent studies have found that large numbers of organisms exist within and beneath the earth's crust in sediments, rocks, and even volcanoes.

(a)

(b)

Figure 1.2 Examples of microbial habitats. **(a)** Summer pond with a thick mat of algae—a rich photosynthetic community. **(b)** Microbes play a large role in decomposing dead animal and plant matter.

(a) Jerome Wexler/Science Source; (b) Michel/Christine Denis Huot/Science Source

- Bacteria and fungi live in complex associations with plants and assist the plants in obtaining nutrients and water and may protect them against disease. Microbes form similar interrelationships with animals. For example, a rich assortment of bacteria in the stomach of cattle digests the complex carbohydrates of the animals' diets (and causes the release of methane into the atmosphere).

1.2 Learning Outcomes—Assess Your Progress

3. Describe the role and impact of microbes on the earth.
4. Explain the theory of evolution and why it is called a theory.

1.3 Human Use of Microorganisms

The diversity and versatility of microorganisms make them excellent candidates for solving human problems. By accident or choice, humans have been using microorganisms for thousands of years to improve life and even to shape civilizations. Baker's and brewer's yeasts, types of single-celled fungi, cause bread to rise and ferment sugar into alcohol to make wine and beers. Other fungi are used to make special cheeses such as blue cheese or Camembert. These and other "home" uses of microbes have been in use for thousands of years. For example, historical records show that households in ancient Egypt kept moldy loaves of bread to apply directly to wounds and lesions. This was long before penicillin was discovered in a mold called *Penicillium*. When humans figure something out through experience, rather than through research or being taught it, it is called an **empirical** finding. Empirical discoveries have shaped the development of humans through our whole history. When humans purposely manipulate microorganisms to make products in an industrial setting, it is called biotechnology. For example, some bacteria have unique capacities to mine precious metals or to create energy **(figure 1.3).**

Recombinant DNA technology is an area of biotechnology that manipulates the genetics of microbes, plants, and animals for the purpose of creating new products and genetically modified organisms (GMOs). This technology makes it possible to transfer genetic material from one organism to another and to deliberately alter DNA. Bacteria and yeasts were some of the first organisms to be genetically engineered.

Even though many citizens are very uncomfortable with GMO processes, it is also true that many people benefit from their medical, industrial, and agricultural uses. Microbes can be engineered to synthesize many critical products such as drugs and hormones. Among the genetically unique organisms that have been designed by bioengineers are bacteria that mass produce antibiotic-like substances, yeasts that produce human insulin, pigs that produce human hemoglobin, and plants that contain natural pesticides or fruits that do not ripen too rapidly.

Another way of tapping into the unlimited potential of microorganisms is the science of **bioremediation** (by′-oh-ree-mee-dee-ay″-shun). This process involves the introduction of microbes into the environment to restore stability or to clean up toxic pollutants. Microbes have a surprising capacity to break down chemicals that would be harmful to other organisms. This includes even human-made chemicals that scientists have developed and for which there are no natural counterparts.

(a)

(b)

(c)

Figure 1.3 Microbes at work. **(a)** Test tubes of yellow and green algae being grown as a possible energy source. **(b)** Microbes fermenting wine in tanks. **(c)** Workers spray nutrients on the shore of Prince William Sound in Alaska after the *Exxon Valdez* oil tanker spill (1989) in an attempt to enrich oil-degrading microbes.

(a) Dennis Schroeder/NREL/US Department of Energy; (b) Bloomberg/Getty Images; (c) Arlis/Alamy Stock Photo

Agencies and companies have developed microbes to handle oil spills and detoxify sites contaminated with heavy metals, pesticides, and other chemical wastes **(figure 1.3*c*).** One form of bioremediation that has been in use for some time is the treatment of water and sewage. Because clean freshwater supplies are dwindling worldwide, it will become even more important to find ways to reclaim polluted water.

1.3 Learning Outcome—Assess Your Progress

5. Explain one old way and one new way that humans manipulate organisms for their own uses.

1.4 Infectious Diseases and the Human Condition

One of the most fascinating aspects of the microorganisms with which we share the earth is that, despite all of the benefits they provide, they also contribute significantly to human misery as **pathogens** (path′-oh-jenz). Please understand: The vast majority of microorganisms that associate with humans cause no harm. In fact, they provide many benefits to their human hosts. It is also important to note that a diverse microbial biota living in and on humans is an important part of human well-being.

It is estimated that there are more than 2,000 different microbes that can cause various types of diseases. Infectious diseases still devastate human populations worldwide, despite significant strides in understanding and treating them. The World Health Organization (WHO) estimates there are a total of 10 billion new infections across the world every year. **Figure 1.4** depicts the 10 top causes of death per year (by all causes, infectious and noninfectious) in the United States and also worldwide. You will see in the graph where the COVID-19 death toll lies. Keep in mind, COVID-19 caused zero deaths in all years previous to 2019 (indeed, no disease, either). All of the deaths represented by the yellow bars are excess deaths by a cause that had not existed before.

We just used the terms “infectious” and “noninfectious.” We’ll explore these terms in more detail later. Generally, they refer to diseases caused by microbes (infectious) and diseases not caused by microbes (noninfectious). Note that while all microbial diseases are infectious diseases, they are not all communicable (transmitted from person to person) **(table 1.3).**

In **table 1.4** you see that diseases *not* caused by microbes are much more frequent in both the United States and the world. You will also note that the United States experiences relatively few—*relatively* few—infectious diseases compared to the number of noninfectious diseases. This graph does not account for

Disease Connection

Some of the most serious lower respiratory tract infections are influenza and pneumonia, and now COVID. Influenza and COVID infections put you at risk for developing pneumonia, caused either by the virus itself or by secondary viruses or bacteria. Of course, you can also develop pneumonia without first being infected by SARS-CoV-2 or the influenza virus.

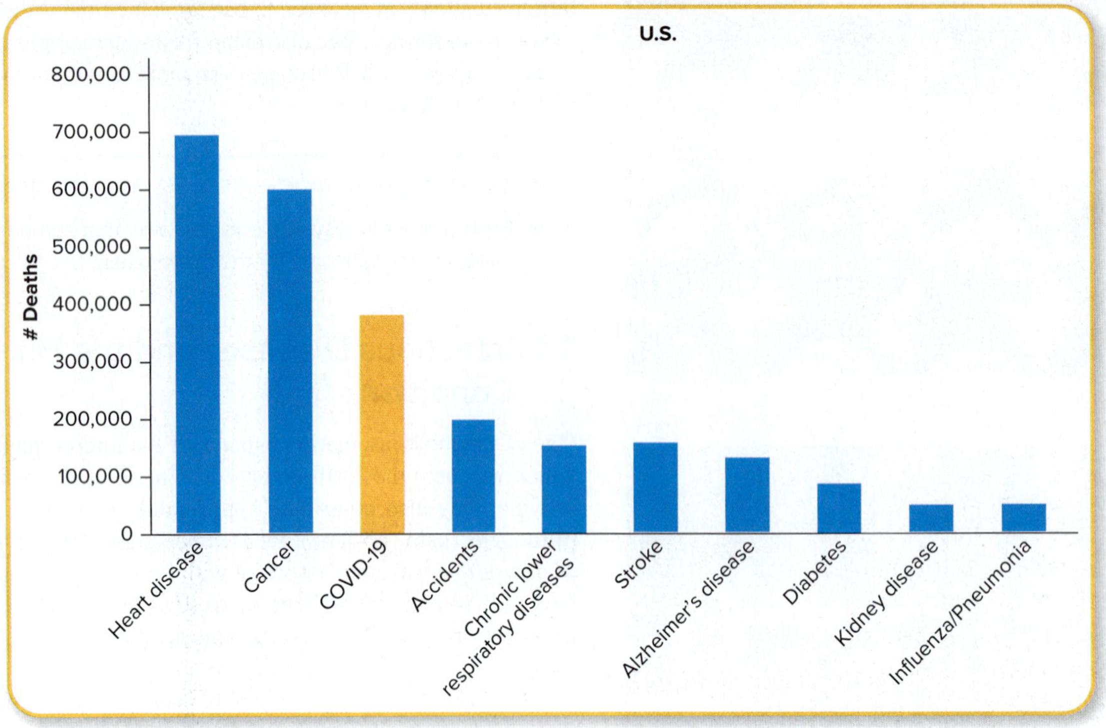
U.S.
Deaths
800,000
700,000
600,000
500,000
400,000
300,000
200,000
100,000
0
Heart disease
Cancer
COVID-19
Accidents
Chronic lower respiratory diseases
Stroke
Alzheimer's disease
Diabetes
Kidney disease
Influenza/Pneumonia

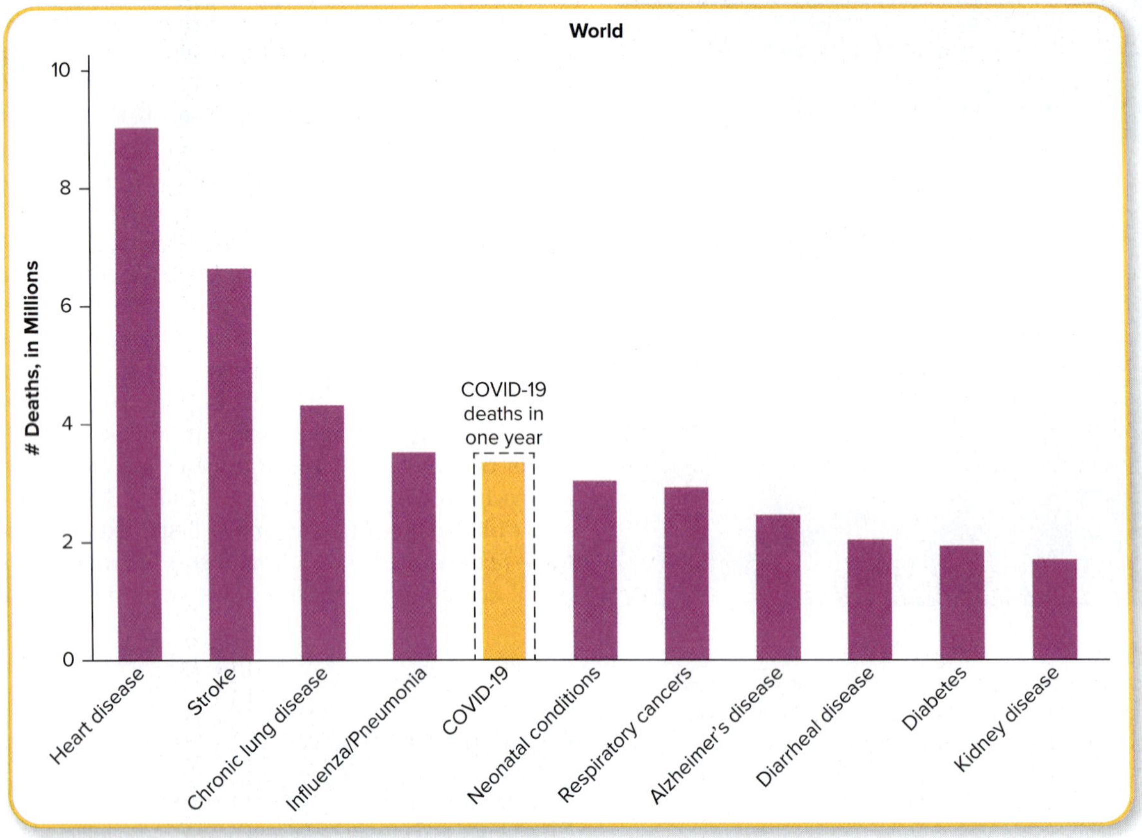
World
Deaths, in Millions
10
8
6
4
2
0
COVID-19 deaths in one year
Heart disease
Stroke
Chronic lung disease
Influenza/Pneumonia
COVID-19
Neonatal conditions
Respiratory cancers
Alzheimer's disease
Diarrheal disease
Diabetes
Kidney disease

Source: Centers for Disease Control and Prevention, Johns Hopkins University

Figure 1.4 Major causes of death in the U.S. and the world, and where COVID-19 deaths fit in. The top graph, for U.S. deaths, is data for 2020. The bottom graph, for global deaths, is data for 2019 (the latest year for which data are available.) Because COVID did not appear until December 2019, we include the 2020 level of COVID-19 deaths on this graph in yellow so you can see its impact.

Table 1.3 Terminology

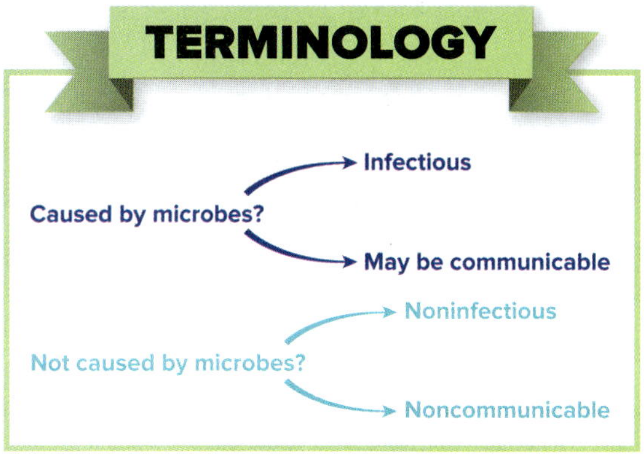
TERMINOLOGY
Caused by microbes?
Infectious
May be communicable
Not caused by microbes?
Noninfectious
Noncommunicable

Table 1.4 Infectious vs. Noninfectious Diseases

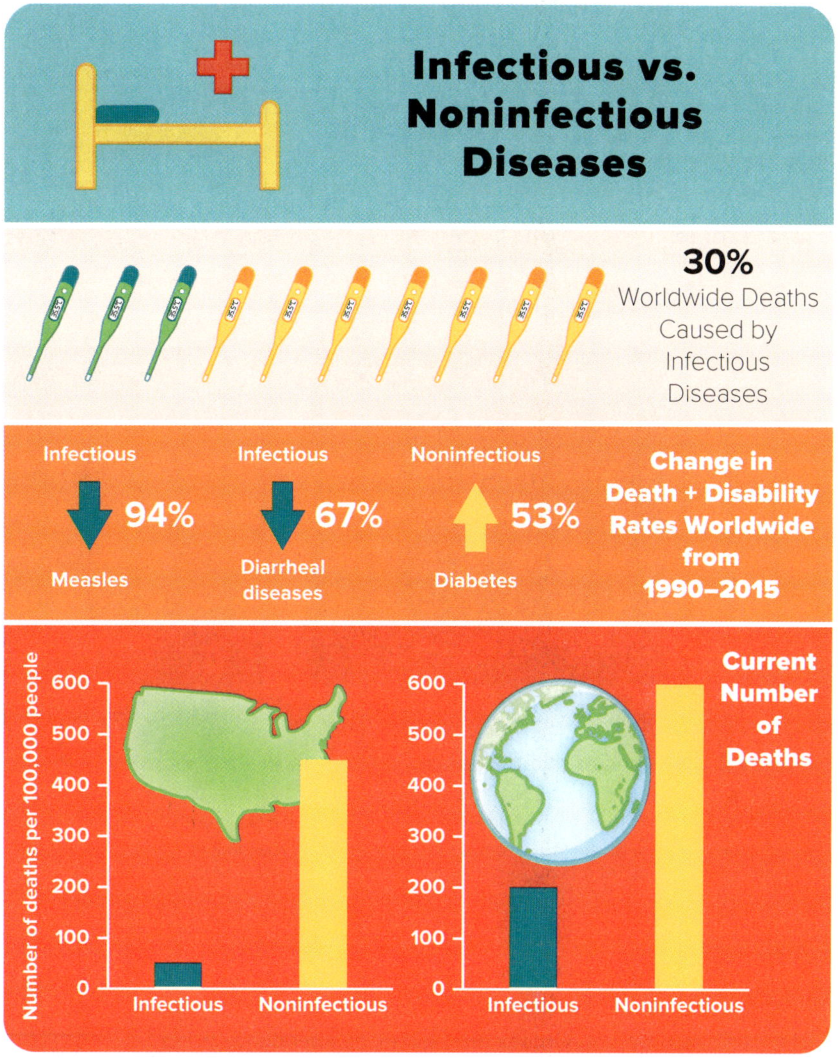
Infectious vs. Noninfectious Diseases
30% Worldwide Deaths Caused by Infectious Diseases
Infectious
94%
Measles
Infectious
67%
Diarrheal diseases
Noninfectious
53%
Diabetes
Change in Death + Disability Rates Worldwide from 1990–2015
Number of deaths per 100,000 people
600
500
400
300
200
100
0
Infectious
Noninfectious
Current Number of Deaths

Source: Data from the World Health Organization and from the Institute for Health Metrics and Evaluation, http://www.healthdata.org/ 2018.

COVID-19 deaths. You saw in figure 1.4 the effect of this one infection on death rates.

Malaria, which kills about 450,000 people every year worldwide, is caused by a microorganism transmitted by mosquitoes. Currently, the most effective way for citizens of developing countries to avoid infection with the causal agent of malaria is to sleep under a bed net because the mosquitoes are most active in the evening. Yet even this inexpensive solution is beyond the reach of many. Mothers in some parts of the world have to make nightly decisions about which of their children will sleep under the single family bed net because a second one, priced at about $10, is too expensive for them.

We are also witnessing an increase in the number of new (emerging) and older (reemerging) diseases. SARS-CoV-2, Ebola, AIDS, hepatitis C, and viral encephalitis are examples of diseases that cause severe mortality and morbidity. To somewhat balance this trend, there have also been some advances in eradication of diseases such as polio and leprosy and diseases caused by certain parasitic worms.

One of the most eye-opening discoveries in recent years is that many diseases that were not previously thought to be caused by microorganisms probably do involve microbial infection. The most famous of these is gastric ulcers, now known to be caused by a bacterium called *Helicobacter*. But there are more. An association has been established between certain cancers and various bacteria and viruses, between diabetes and the coxsackievirus, and between schizophrenia and the coxsackievirus. Diseases as different as multiple sclerosis, obsessive compulsive disorder, coronary artery disease, and even obesity have been linked to chronic infections with microbes. We're now discovering the subtler side of microorganisms. Their roles in quiet but slowly destructive diseases are now well known. These include female infertility, often caused by silent *Chlamydia* infections, and malignancies such as liver cancer (hepatitis viruses) and cervical cancer (human papillomavirus). Researchers are currently researching whether Alzheimer's disease is related to microbes found in the brains of people with the disease.

Another important development in infectious disease trends is the increasing number of patients with weakened defenses that are kept alive for extended periods. They are subject to infections by common microbes that are not pathogenic to healthy people. There is also an increase in microbes that are resistant to the drugs used to treat them.

1.4 Learning Outcome—Assess Your Progress

6. Summarize the relative burden of human disease caused by microbes, emphasizing the differences between developed countries and developing countries.

1.5 The General Characteristics of Microorganisms

Cellular Organization

As discussed in section 1.1, three basic cell lines appeared during evolutionary history. These lines—**Archaea, Eukarya,** and **Bacteria**—differ not only in how complex they are but also in their contents and function.

Helminth: Head (scolex) of *Taenia solium*

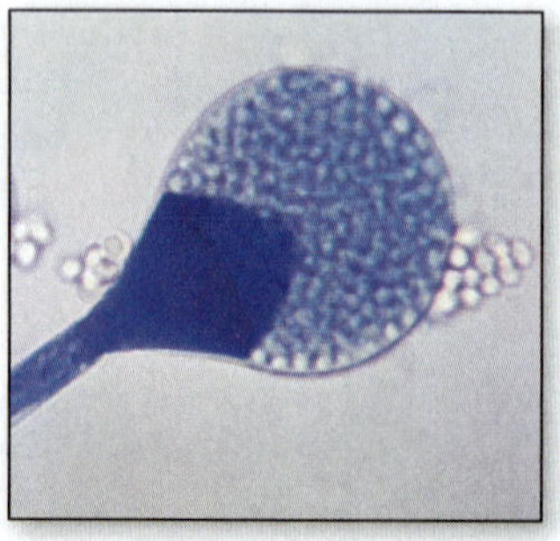

Fungus: *Mucor*

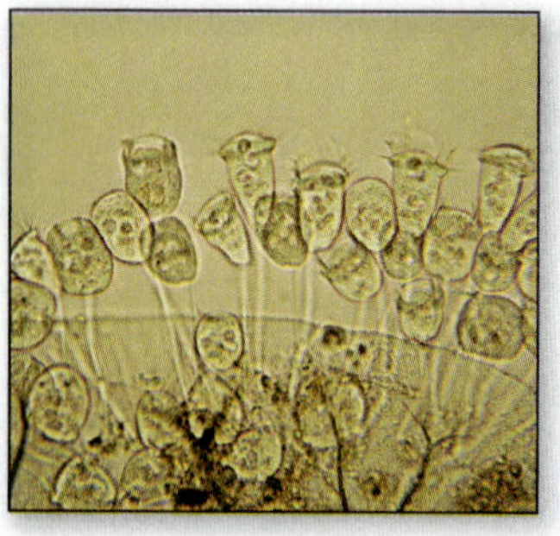

Protozoan: *Vorticella*

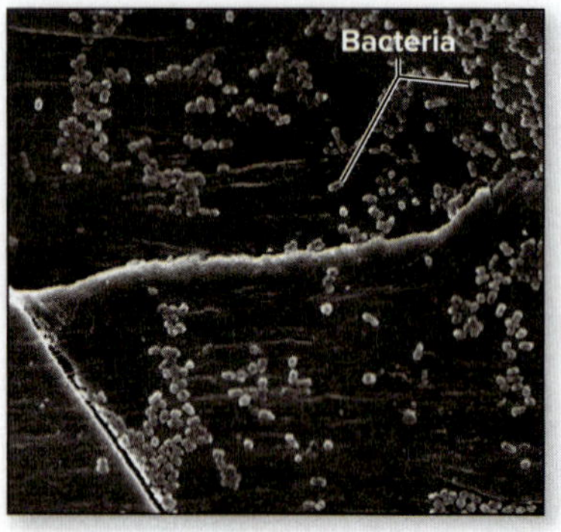
Bacteria

Bacterium: *E. coli*

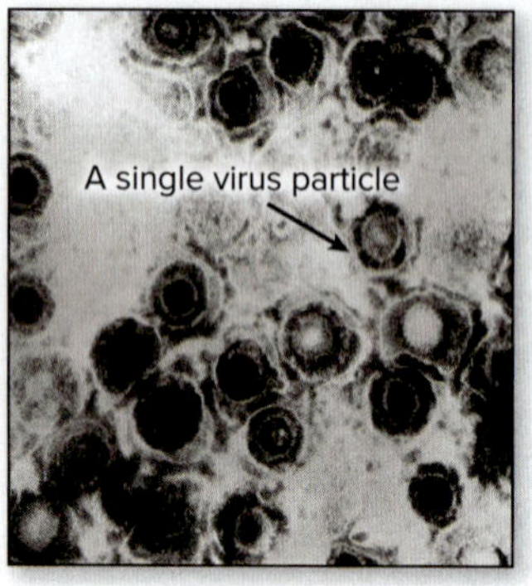
A single virus particle

Virus: Herpes simplex

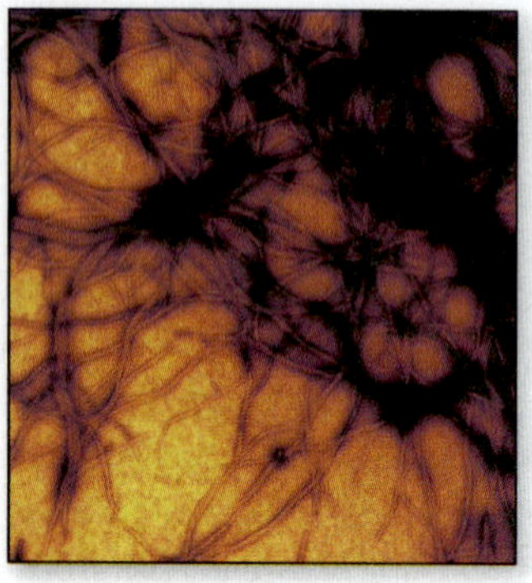

Prion

Figure 1.5 Six types of microorganisms. Archaea are not pictured here. The photographs are taken at different magnifications to help you see detailed structures, but remember that there are vast size differences among the organisms. The helminths can be visible to the naked eye, while prions, tens of thousands of times smaller, can only be seen with highly specialized microscopes.

(top left) CDC; (top middle) Dr. Lucille K. Georg/CDC; (top right) Nancy Nehring/E+/Getty Images; (bottom left) Janice Haney Carr/CDC; (bottom center) Dr. Erskine Palmer/CDC; (bottom right) Cultura/Shutterstock

A Note About Viruses and Prions

As mentioned before, **viruses** are neither independently living, nor cellular organisms. Instead, they are small particles that exist at the level of complexity somewhere between large molecules and cells. Viruses are much simpler than cells. Outside their host, they are composed essentially of a small amount of hereditary material (either DNA or RNA but never both) wrapped up in a protein covering that is sometimes enveloped by a protein-containing lipid membrane. In this extracellular state, they are individually referred to as a **virus particle** or **virion.**

Prions are highly unusual "organisms." They contain no DNA or RNA, so they have no genetic program. They are small proteins folded in intricate ways. Sometimes these prions behave like microorganisms and are transmitted from one human to another.

To make a broad generalization, bacterial and archaeal cells are about 10 times smaller than eukaryotic cells. They generally do not have many of the eukaryotic cell structures such as **organelles.** Organelles are small, double-membrane-bound structures in the eukaryotic cell that perform specific functions. The nucleus, mitochondria, and chloroplasts are organelles. All bacteria and archaea are microorganisms, but only some eukaryotes are microorganisms. Humans, after all, are eukaryotic organisms. The majority of microorganisms are single-celled (all bacteria and archaea and some eukaryotes), but some are multicellular (helminths and some fungi) **(figure 1.5).**

1.5 Learning Outcomes—Assess Your Progress

7. Differentiate among bacteria, archaea, and eukaryotic microorganisms.
8. Identify two types of acellular microorganisms.
9. Compare and contrast the relative sizes of the different microbes.

1.6 The Historical Foundations of Microbiology

If not for the extensive interest, curiosity, and devotion of thousands of microbiologists over the last 350 years, we would know little about the microscopic realm that surrounds us. Many of the discoveries in this science have resulted from the prior work of women and men who toiled long hours in dimly lit laboratories with the crudest of tools. Each additional insight, whether large or small, has added to our current knowledge of living things and processes. This section summarizes the prominent discoveries made in the past 350 years. Keep in mind that because most accounts of these discoveries were written by men, the many significant discoveries made by women have been underreported and underappreciated.

The Development of the Microscope: "Seeing Is Believing"

From very earliest history, humans noticed that when certain foods spoiled, they became inedible or caused illness, and yet other "spoiled" foods (fruit juices, cabbage) did no harm and even had enhanced flavor. Indeed, several centuries ago, there was already a sense that diseases such as the black plague and smallpox were caused by some sort of transmissible matter. But the causes of such phenomena were vague and obscure because the technology to study them was lacking. So scientists were left to speculate. One great example of a misguided understanding is a phenomenon called spontaneous generation. This incorrect concept was finally proven wrong in dramatic fashion by a very clever experiment by Louis Pasteur in the mid-1800s.

Disproving Spontaneous Generation

It is hard for us to imagine today, but for thousands of years, people thought that diseases were a curse from God or were caused by damp fogs. One very widely held belief was that plants, animals, and even people came from an invisible life-giving force. This was formally known as **abiogenesis** (a = without, bio = life, genesis = beginning), literally, "beginning in the absence of life." It was also called **spontaneous generation,** or being generated from thin air.

In 1745, a scientist named John Needham came close to disproving spontaneous generation, but his experimental design had one fatal flaw. He believed that boiling liquid broth would kill all organisms in it, so he boiled it, then immediately sealed the flasks. But the liquid became cloudy, indicating microbial growth. What nobody knew at the time was that some bacteria form endospores, extremely hardy structures that are not killed by boiling. The boiled liquid held viable endospores in it, though the experiment was interpreted to mean that life could indeed come from "the air." Another scientist, thinking that Needham's flasks must have been contaminated between boiling and sealing, boiled the liquid inside a sealed flask. It was true that no growth occurred, but abiogenesis proponents countered that the "life force" could not get in a sealed flask and that this experiment actually supported their view.

In 1859, French scientist Louis Pasteur, with a combination of skill and luck, seems to have put the belief in spontaneous generation to rest. He had to keep his flasks open to the air so that if there was a "life force" in it, it could access the boiled liquid. He designed a long, bent "swan neck" for his flasks, which meant that any microbes that fell (by gravity) into the flask would settle into the crook at the bottom of the flask neck and not make it into the broth **(figure 1.6).** In this way, he could allow the flask to remain open to the air. The results: As long as the neck of the flask was intact, no microbes grew. If the neck of the flask was broken off, so that only the straight vertical part was intact, microbes from the air fell in and the broth became cloudy. Pasteur had the immense luck that there were no endospore-forming bacteria in his broth. If there had been, even his swan-necked flask could have become cloudy. Because it did not, the idea of spontaneous generation was finally abandoned. Pasteur was lucky, indeed.

True awareness of the widespread distribution of microorganisms and some of their characteristics was finally made possible by the development of the first microscopes. These devices revealed microbes as discrete entities sharing many of the cellular characteristics of larger, visible plants and animals. Several early scientists fashioned magnifying lenses, but their microscopes lacked the optical clarity needed for examining bacteria and other small, single-celled organisms. The likely earliest record of microbes is in the works of Englishman Robert Hooke. In the 1660s, Hooke studied various everyday things like household objects, plants, protozoa, and trees. He described for the first time cellular structures in tree bark and drew sketches of "little structures" that seemed to be alive. Using a single-lens microscope he made himself, Hooke described spots of mold he found on the sheepskin cover of a book:

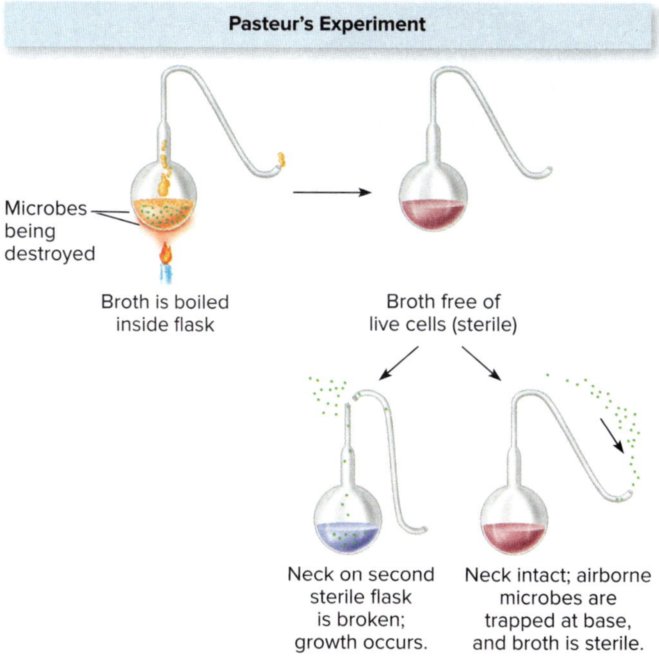
Pasteur's Experiment
Microbes being destroyed
Broth is boiled inside flask
Broth free of live cells (sterile)
Neck on second sterile flask is broken; growth occurs.
Neck intact; airborne microbes are trapped at base, and broth is sterile.

Figure 1.6 Pasteur's experiment disproving spontaneous generation.

> These spots appear'd, through a good Microscope, to be a very pretty shap'd vagetative body, which, from almost the same part of the Leather, shot out multitudes of small long cylindrical and transparent stalks, not exactly straight, but a little bended with the weight of a round and white knob that grew on the top of each of them. . . .

Figure 1.7*a* is a reproduction of the drawing he made to accompany his written observations. Hooke paved the way for even more exacting observations of microbes by Antonie van Leeuwenhoek (pronounced "Lay′-oow-un-hook"), a Dutch linen merchant and self-made microbiologist.

Imagine a dusty linen shop in Holland in the late 1600s. Women in traditional Dutch garb came in and out, choosing among the bolts of linens for their draperies and upholstery. Between customers, Leeuwenhoek retired to the workbench in the back of his shop, grinding glass lenses to ever-finer specifications so he could see with increasing clarity the threads in his fabrics. Eventually, he became interested in things other than thread counts. He took rainwater from a clay pot, smeared it on his specimen holder, and peered at it through his finest lens. He found "animals appearing to me ten thousand times less than those which may be perceived in the water with the naked eye."

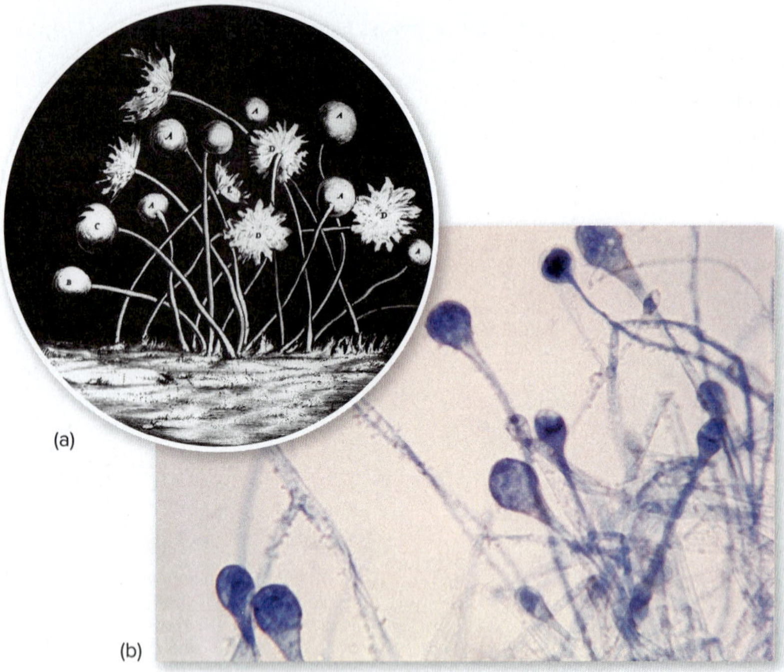
(a)
(b)

Figure 1.7 The first drawing of microorganisms.
(a) Drawing of "hairy mould" colony made by Robert Hooke in 1665.
(b) Photomicrograph of the fungus probably depicted by Hooke. It is a species of *Mucor*, a common indoor mold.
(a) Biophoto Associates/Science Source; (b) BSIP/UIG/Getty Images

He did not stop there. He scraped the plaque from his teeth, and from the teeth of some volunteers who had never brushed their teeth in their lives, and took a good, close look at that. He recorded: "In the said matter there were many very little living animalcules, very prettily a-moving. . . . Moreover, the other animalcules were in such enormous numbers, that all the water . . . seemed to be alive." Leeuwenhoek started sending his observations to the Royal Society of London, and eventually he was recognized as a scientist of great merit.

Leeuwenhoek constructed dozens of small, powerful microscopes that could magnify up to 300 times **(figure 1.8).** Considering that he had no formal training in science, his descriptions of bacteria and protozoa (which he called "animalcules") were astute and precise.

From the time of Hooke and Leeuwenhoek, microscopes became more complex and improved with the addition of refined lenses, a condenser, finer focusing devices, and built-in light sources. The prototype of the modern compound microscope, in use from about the mid-1800s, was capable of magnifications of 1,000 times or more. This was during the Victorian era, and people were so enthralled with the ability to see previously invisible worlds that microscopes became popular items in private homes. Our modern student microscopes are not greatly different in basic structure and function from those microscopes.

These events marked the beginning of our understanding of microbes and the diseases they can cause. Discoveries continue at a breakneck pace, however. In fact, this first century of the 2000s is being widely called the Century of Biology, fueled by our new abilities to study genomes and harness biological processes. Microbes have led the way in these discoveries and continue to play a large role in the new research.

To give you a feel for what has happened most recently, **table 1.5** contains some recent discoveries that have had huge impacts on our understanding of microbiology.

These examples highlight a feature of biology—and all of science—that is perhaps underappreciated. Because we have thick textbooks containing all kinds of assertions and "facts," many people think science is an iron-clad collection of facts. Wrong! Science is an ever-evolving collection of new information, gleaned from observable phenomena and synthesized with old information to come up with the current understandings of nature. Some of these observations have been confirmed so many times over such a long period of time that they are, if not "fact," very close to fact. Many other observations will be altered over and over again as new findings emerge. And that is the beauty of science.

Disease Connection

The teeth are a perfect surface for accumulating a lot of bacteria. As soon as you leave the dental hygienist's chair, your squeaky clean tooth surfaces immediately begin to accumulate proteins from the saliva. This coated surface is then colonized by streptococcal bacteria, which are then colonized by other species of bacteria, which are then colonized by more bacteria, and so on. This creates a thick community of bacteria that eventually becomes visible as plaque—especially if you never brush your teeth, as with Leeuwenhoek's subjects. This plaque can lead to cavities (known as *caries*) or gum disease.

Focus screw
Specimen holder
Lens

Figure 1.8 Leeuwenhoek's microscope. A brass replica of a Leeuwenhoek microscope. It is picked up by the bottom stem and the lens is held in front of one eye with the specimen holder facing outward.
Tetra Images/Alamy Stock Photo

Table 1.5 Recent Advances in Microbiology

The invention of the PCR technique—1980s. The polymerase chain reaction (PCR) was a breakthrough in our ability to detect tiny amounts of DNA and then amplify them into quantities sufficient for studying. It has provided a new and powerful method for discovering new organisms, diagnosing infectious diseases, and doing forensic work such as crime scene investigation. Its inventor is Kary Mullis, a scientist working at a company in California at the time. He won the Nobel Prize for this invention in 1993.

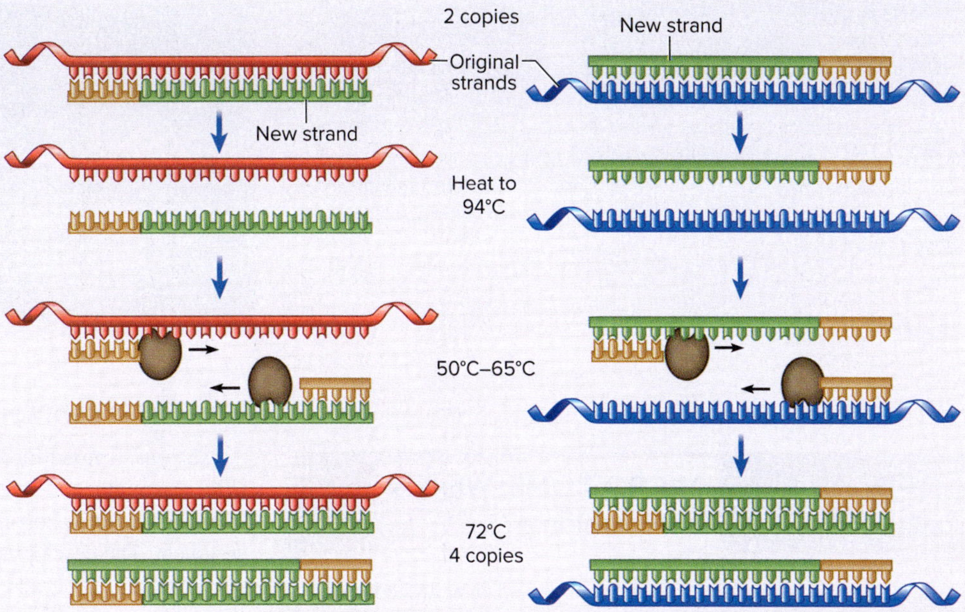
2 copies
New strand
Original strands
New strand
Heat to 94°C
50°C–65°C
72°C
4 copies

The importance of small RNAs—2000s. Once we were able to sequence entire genomes (another big move forward), scientists discovered something that turned a concept we literally used to call "dogma" on its head. The previously held "Central Dogma of Biology" was that DNA makes RNA, which leads to the creation of proteins. Genome sequencing has revealed that perhaps only 2% of DNA actually codes for a protein. Much RNA does not end up with a protein counterpart. These pieces of RNA are usually small. It now appears that they have critical roles in regulating what happens in the cell. It has led to new approaches to how diseases are treated. For example, if the small RNAs are important in bacteria that infect humans, they can be new targets for antimicrobial therapy.

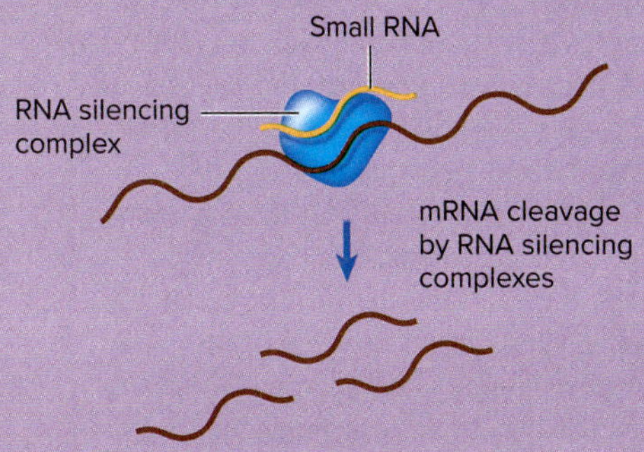
Small RNA
RNA silencing complex
mRNA cleavage by RNA silencing complexes

Genetic identification of the human microbiome—2010s and beyond. The first detailed information produced by the Human Microbiome Project (HMP) was astounding: Even though the exact types of microbes found in and on different people are highly diverse, the overall set of metabolic capabilities the bacterial communities possess is remarkably similar among people. This and other groundbreaking discoveries have set the stage for new knowledge of our microbial guests and their role in our overall health and disease. See Insight 1.1.

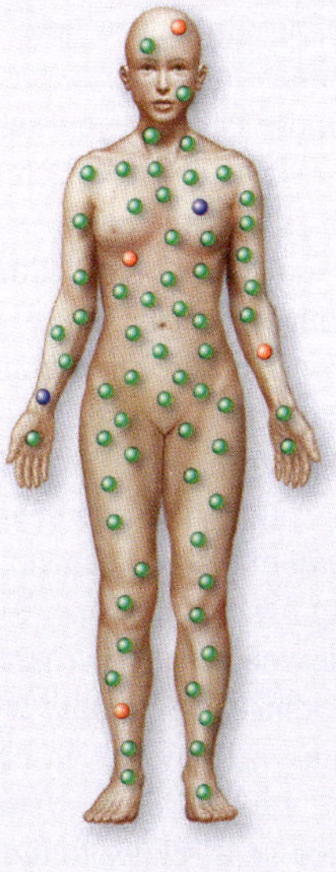

(continued)

Table 1.5 Recent Advances in Microbiology (*continued*)

CRISPR technology—2013 CRISPR (pronounced "crisper") refers to a system found naturally in bacteria and archaea that they use to repel second attacks by viruses and other invaders. It was first observed in bacteria and archaea in the 1990s. In 2013, scientists found a way to use the system to manipulate genes in any organism, in a process called **gene editing.** CRISPR technology can be used to disable a gene of choice and then insert DNA of one's choosing. This can lead to revolutionary treatments of hereditary diseases and other previously untreatable conditions.

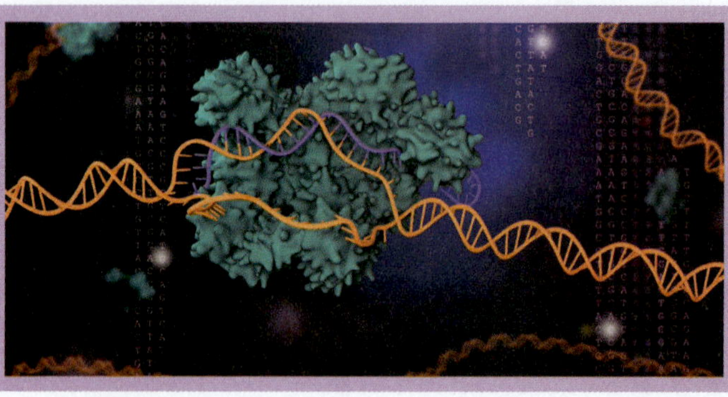

Meletios Verras/Shutterstock

INSIGHT 1.1 MICROBIOME: What Is a Microbiome?

In the past decade, a new word has popped up on newsfeeds and sites: microbiome. It refers to the sum total of all the microbes in a certain environment.

The Human Microbiome Project (HMP) began producing significant results in 2012. It used techniques to identify body microbes that did not require growing the microbes separately in the lab (a process scientists have relied on since the mid-1800s), but instead identified them on the basis of their genetic material. The HMP produced a staggering array of results, and they keep coming at breakneck pace.

We have learned that the microbiome differs based on whether you were delivered via cesarian or vaginal birth. We have learned that the gut microbiome—the microbes living in your intestinal tract—influences not just your intestinal health but also your likelihood of experiencing autoimmune disease, your weight, and even your mood! We know how the composition of the microbiome of different body systems (your skin, your eyes, your lungs) differs in health and in disease. We have learned that the microbiome *in utero* influences your embryonic development.

But there is a big note of caution here: The explosion of information about the human microbiome has led to some careless science, and even more careless journalism. One leading microbiome researcher even calls it "Microbiomania." Many journalists gloss over the difference between correlation (these two things were observed at the same time) and causation (this thing *led to* this other thing). So we get reporting that suggests that elite cyclists should be "poop-doping" (getting transplanted fecal material that contains a donor's gut microbiome) to improve their performance. There is no evidence for that, except that a study somewhere reported that an excellent cyclist's microbiome was different than normal mortals. No causation, but it made for a good headline.

In short, we have learned that the characteristics of your microbiome determine your own, human, biology—and what types of experiences you will have as an organism. In every chapter of this book, we will tell a short story in this Insight box about the microbiome as it relates to the subject matter in the chapter.

The Establishment of the Scientific Method

A serious impediment to the development of true scientific reasoning and testing was the tendency of early scientists to explain natural phenomena by a mixture of belief, superstition, and argument. The development of an experimental system that answered questions objectively and was not based on prejudice marked the beginning of true scientific thinking. These ideas gradually crept into the consciousness of the Western scientific community during the 1600s.

Here we use the term "Western" to mean European (and, later, American) science. During the Middle Ages in Europe, intellectual activity slowed down considerably for a variety of reasons, but in that same time, the Early Islamic Empire in the Middle East was experiencing a rich period of scientific, mathematical, and philosophical thinking. One particular scholar in the 10th and 11th centuries, Ibn al-Haytham (**figure 1.9**), is credited with many scientific advances, including laying the groundwork for what is known as the **scientific method.** The scientific method is a general approach taken by researchers to explain certain natural phenomena, and was refined into its present form by Western scientists in the 1600s. A

Figure 1.9 Ibn al-Haytham, a scholar of the Islamic Golden Age, lived from 965 to 1040.

World History Archive/Alamy Stock Photo

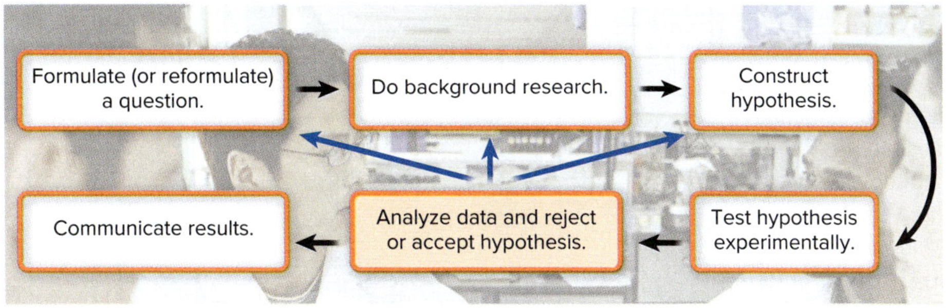

Formulate (or reformulate) a question.
Do background research.
Construct hypothesis.
Test hypothesis experimentally.
Analyze data and reject or accept hypothesis.
Communicate results.

Figure 1.10 An overview of the scientific method.
Ryan McVay/Photodisc/Getty Images

primary aim of this method is to formulate a **hypothesis,** a tentative explanation to account for what has been observed or measured. A good hypothesis should be in the form of a statement. It must be capable of being either supported or discredited by careful observation or experimentation. For example, the statement that "microorganisms cause diseases" can be experimentally determined by the tools of science, but the statement "diseases are caused by evil spirits" cannot. The scientific method is illustrated in **figure 1.10.**

Deductive and Inductive Reasoning

Science is a process of investigation using observation, experimentation, and reasoning. In some investigations, you make individual decisions by using broadly accepted general principles as a guide. This is called deductive reasoning. Deductive reasoning is the reasoning of mathematics, philosophy, politics, and ethics. Deductive reasoning is also the way a computer works. All of us rely on deductive reasoning as a way to make everyday decisions—like whether you should open attachments in e-mails from unknown senders **(figure 1.11).** We use general principles as the basis for examining and evaluating these decisions.

Inductive Reasoning

Where do general principles come from? They often come from observation of the physical world around us. If you drop an apple, it will fall whether or not you want it to and despite any laws you may pass that forbid it to do so. Science is devoted to discovering the general principles that govern the operation of the physical world.

How do scientists discover such general principles? Scientists are, above all, observers: They look at the world to understand how it works. It is from observations that scientists determine the principles that govern our physical world.

The process of discovering general principles by careful examination of specific cases is termed *inductive reasoning.* In the West, this way of thought first became popular about 400 years ago, when Isaac Newton, Francis Bacon, and others began to conduct experiments. They used the results of these experiments to infer general principles about how the world operates. Their experiments were sometimes quite simple. Newton's consisted simply of releasing an apple from his hand and watching it fall to the ground. From a host of particular observations, each no more complicated than the falling of an apple, Newton inferred a general principle—that all objects fall toward the center of the earth. This principle was a possible explanation, or hypothesis, about how the world works. You also make observations and formulate general principles based on your observations, like forming a general principle about the reliability of unknown e-mail attachments in figure 1.11. Like Newton, scientists work by forming and testing hypotheses, and observations are the materials on which they build them.

As you can see, the deductive process is used when a general principle has already been established. Inductive reasoning involves a discovery process and leads to the creation of a general principle.

A lengthy process of experimentation, analysis, and testing eventually leads to conclusions that either support or fail to support the hypothesis. If experiments do not uphold the hypothesis—that is, if it is found to be flawed—the hypothesis or some part of it is rejected. It is either discarded or modified to fit the results of the experiment. If the hypothesis is supported by the results from the experiment, it is not (or should not be) immediately accepted as fact. It then must be tested and retested. Indeed, this is an important guideline in the acceptance of a hypothesis. The results of the experiment should be published so that they can be repeated by other investigators.

In time, as each hypothesis is supported by a growing body of data and survives rigorous scrutiny, it moves to the next level of acceptance—the **theory.** A theory is a collection of statements, propositions, or concepts that explain or account for a natural event. A theory is not the result of a single experiment repeated over and over again but is an entire body of ideas that express or explain many aspects of a phenomenon. It is not just an idea or weak speculation (which is the way the word is used in everyday conversation) but a viable declaration that has stood the test of time and has yet to be disproved by serious scientific endeavors. Often, theories develop and progress through decades of research and are added to and modified by new findings. At some point, evidence of the accuracy and predictability of a theory is so compelling that the next level of confidence is reached and the theory becomes a law, or principle. For example, although we still refer to the germ *theory* of disease, so little question remains that microbes can cause disease that it has clearly passed into the realm of law. The theory of evolution falls in this category as well.

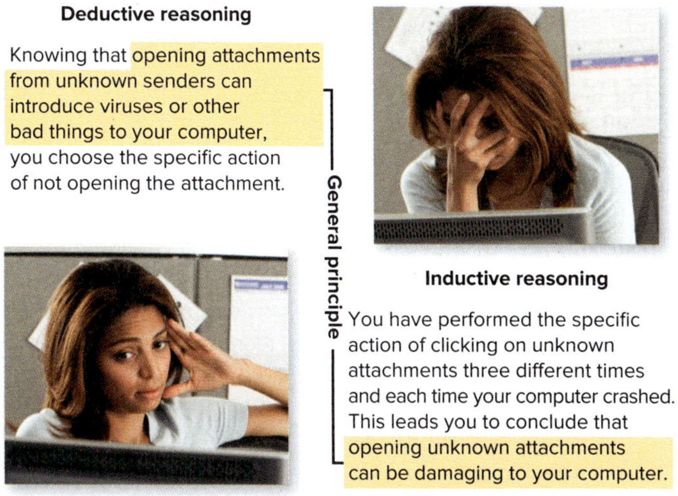

Deductive reasoning
Knowing that opening attachments from unknown senders can introduce viruses or other bad things to your computer, you choose the specific action of not opening the attachment.
General principle
Inductive reasoning
You have performed the specific action of clicking on unknown attachments three different times and each time your computer crashed. This leads you to conclude that opening unknown attachments can be damaging to your computer.

Figure 1.11 Deductive and inductive reasoning.
Tom Grill/Corbis

Science and its hypotheses and theories must progress along with technology. As advances in instrumentation allow new, more detailed views of living phenomena, old theories may be reexamined and altered and new ones proposed. But scientists do not take the stance that theories or even "laws" are ever absolutely proved. Because the COVID-19 pandemic unfolded so publicly, we were able to witness science in progress. As we studied the virus and its effects, new revelations (and new recommendations) emerged. Because those changes came so quickly compared to most scientific processes, some in the public grew frustrated with the updates and changes. But this is exactly how science must work.

The characteristics that make scientists most effective in their work are curiosity, open-mindedness, skepticism, creativity, cooperation, and readiness to revise their views as new discoveries are made.

The Development of Medical Microbiology

Early experiments on the sources of microorganisms led to the profound realization that microbes are everywhere: Not only are air and dust full of them, but the entire surface of the earth, its water, and all objects are inhabited by them. This discovery led to immediate applications in medicine. So the seeds of medical microbiology were sown in the mid to latter half of the 19th century (the 1800s) with the introduction of the germ theory of disease and the resulting use of sterile, aseptic, and pure culture techniques.

The Discovery of Spores and Sterilization

Following Pasteur's inventive work with swan-neck flasks while disproving spontaneous generation, it was not long before English physicist John Tyndall provided the initial evidence that some of the microbes in dust and air have very high heat resistance and that something stronger than boiling is required to destroy them. Later, the discovery and detailed description of heat-resistant bacterial endospores clarified the reason that heat would sometimes fail to completely eliminate all microorganisms. The modern sense of the word **sterile,** meaning completely free of all life forms (including spores) and virus particles, was established from that point on. The capacity to sterilize objects and materials is an absolutely essential part of microbiology, medicine, dentistry, and many industries.

The Development of Aseptic Techniques

From earliest history, humans experienced a vague sense that "unseen forces" or "poisonous vapors" emanating from decomposing matter could cause disease. As the study of microbiology became more scientific and the invisible was made visible, the fear of such mysterious vapors was replaced by the knowledge and sometimes even the fear of "germs." About 125 years ago, the first studies by Robert Koch clearly linked a microscopic organism with a specific disease. Since that time, microbiologists have conducted a continuous search for disease-causing agents.

At the same time that abiogenesis was being hotly debated, a few physicians began to suspect that microorganisms could cause not only spoilage and decay but also infectious diseases. It occurred to these pioneers that even the human body itself was a source of infection. Dr. Oliver Wendell Holmes, an American physician, observed that mothers who gave birth at home experienced fewer infections than did mothers who gave birth in the hospital. The Hungarian Dr. Ignaz Semmelweis showed quite clearly that women became infected in the maternity ward after examinations by physicians coming directly from the autopsy room without washing their hands.

50
CORREOS 1993
CUBA
JOSEPH
LISTER
1827 - 1912
CELEBRIDADES
DE LA CIENCIA

Figure 1.12 This is a commemorative postage stamp produced in Cuba, in the series "Celebrities of Science" (Celebridades de la Ciencia) in the 1990s, depicting the English surgeon Joseph Lister.
Torontonian/Alamy Stock Photo

The English surgeon Joseph Lister **(figure 1.12)** took notice of these observations and was the first to introduce **aseptic** (ay-sep′-tik) **techniques.** These are aimed at reducing microbes in a medical setting and preventing wound infections. Lister's concept of asepsis was much more limited than our modern precautions. It mainly involved disinfecting the hands and the air with strong antiseptic chemicals, such as phenol, prior to surgery. It is hard for us to believe, but as recently as the late 1800s, surgeons wore street clothes in the operating room and had little idea that hand washing was important. Lister's techniques and the application of heat for sterilization became the foundations for microbial control by physical and chemical methods, which are still in use today.

Disease Connection

The benefits of hand washing are widely known. Agencies such as the Centers for Disease Control and Prevention (CDC) and World Health Organization (WHO) have long promoted the importance of hand hygiene and how to perform it well. Evidence shows that proper hand washing is the single most effective method of preventing infectious disease transmission inside and outside health care settings. However, not all hand washing is created equal. Techniques for each step in the process of hand washing are supported by evidence, and when these guidelines are followed, disease transmission can be greatly reduced.

The Discovery of Pathogens and the Germ Theory of Disease

Louis Pasteur of France introduced techniques that are still used today. Pasteur made enormous contributions to our understanding of the microbial role in wine and beer formation. He invented pasteurization and completed some of the first studies showing that human diseases could arise from infection. These studies, which were based in part on groundwork laid by Islamic physicians in the Middle Ages, led to the **germ theory of disease.** Pasteur's contemporary, Robert Koch, established *Koch's postulates,* a series of proofs that verified the germ theory and could establish whether an organism was pathogenic and which disease it caused. Around 1875, Koch used this experimental system to show that anthrax was caused by a bacterium called *Bacillus anthracis.* Koch's postulates were so effective that within 25 years the causative agents of 20 other diseases were discovered. They have had to be modernized over the years but still form the basis for identifying new pathogenic microbes.

Numerous exciting technologies emerged from Koch's laboratory work. During this golden age of the 1880s, he realized that study of the microbial world would require separating microbes from each other and growing them in culture. It is not an overstatement to say that he and his colleagues invented most of the techniques that are described in chapter 3: inoculation, isolation, media, maintenance of pure cultures, and preparation of specimens for microscopic examination. Other highlights in this era of discovery are presented in later chapters on microbial control and vaccination.

1.6 Learning Outcomes—Assess Your Progress

10. Make a time line of the development of microbiology from the 1600s to today.
11. List some recent microbiological discoveries of great impact.
12. Explain what is important about the scientific method.

1.7 Naming, Classifying, and Identifying Microorganisms

Students just beginning their microbiology studies are often dismayed by the seemingly endless array of new, unusual, and sometimes confusing names for microorganisms. Learning microbial **nomenclature** is very much like learning a new language, and occasionally it may feel a bit overwhelming. But paying attention to proper microbial names is just like following a baseball game or a theater production: You cannot tell the players apart without a program! Your understanding and appreciation of microorganisms will be greatly improved by learning a few general rules about how they are named.

The science of classifying living beings into categories is **taxonomy.** Its modern form originated more than 250 years ago when Carl von Linné (also known as Linnaeus), a Swedish botanist, laid down the basic rules for *classification* and established taxonomic categories, or **taxa** (singular, *taxon*).

Von Linné realized early on that a system for recognizing and defining the properties of living beings would prevent chaos in scientific studies by providing each organism with a unique name and an exact "slot" in which to catalog it. The von Linné system has served well in categorizing the millions of different kinds of organisms that have been discovered since that time, including organisms that have gone extinct.

The primary concerns of modern taxonomy are still naming, classifying, and identifying. These three areas are interrelated and play a vital role in keeping a dynamic inventory of the extensive array of living and extinct beings. In general,

Nomenclature is the assignment of scientific names to the various taxonomic categories and individual organisms. You can remember this by recalling that "nom" means *name*.

Classification tries to arrange organisms into a hierarchy of taxa (categories).

Identification is the process of discovering and recording the traits of organisms so that they may be recognized or named and placed in an overall taxonomic scheme. Identification will be thoroughly discussed in chapter 3, so the rest of this chapter will focus on nomenclature and classification.

If you have studied biology before, you may recall that species are often defined as organisms that can successfully produce offspring together. Species also can be defined as having shared evolutionary history and ancestry. Categorizing bacteria, viruses, and other microorganisms is especially difficult because of the fact that a large amount of genetic exchange takes place *horizontally*—that is, among different organisms living together at the same time. This is as opposed to the type of genetic transfer that occurs from one generation to another. (That is called *vertical gene transfer.*) When even the cells of the same species of bacteria have a different genetic makeup because some of them have picked up genes from other species, the very idea of a species is difficult to pin down. Nevertheless, microbiologists continue to assign relatedness and categories to microorganisms.

Nomenclature: Assigning Specific Names

Many **macroorganisms** are known by a common name suggested by certain dominant features. For example, a bird species might be called a red-headed blackbird or a flowering plant species a black-eyed Susan. Some species of microorganisms are also called by informal names, including human pathogens such as "gonococcus" (*Neisseria gonorrhoeae*) or fermenters such as "brewer's yeast" (*Saccharomyces cerevisiae*), or "Iraqibacter" (*Acinetobacter baumannii*), but this is not the usual practice. If we were to adopt common names such as the "little yellow coccus," the terminology would become even more cumbersome and challenging than scientific names. Even worse, common names are notorious for varying from region to region, even within the same country. A big advantage of standardized nomenclature is that it provides a universal language, thereby enabling scientists from all countries to accurately exchange information.

The method of assigning a scientific or specific name is called the **binomial** (two-name) **system** of nomenclature. The scientific name is always a combination of the genus name followed by the species name. The genus part of the scientific name is capitalized, and the species part begins with a lowercase letter. Both should be italicized (or underlined if using handwriting), as follows:

Staphylococcus aureus

The two-part name of an organism is sometimes abbreviated to save space, as in *S. aureus,* but only if the genus name has already been stated. The source for nomenclature is usually Latin or Greek. If other languages such as English or French are used, the endings of these words are revised to have Latin endings. An international group oversees the naming of every new organism discovered, making sure that standard procedures have been followed and that there is not already an earlier name for the organism or another organism with that same name. The inspiration for names is extremely varied and often rather imaginative. A biology professor from Murray

State University in Kentucky was planting flowers in her garden with her two-year-old daughter Sylvie when Sylvie discovered something. It was a species of treehopper insect that her mother was not familiar with. It turns out that it was an as-yet-undiscoverd species, and Sylvie's mom, Dr. Laura Sullivan-Beckers, named it in honor of its discoverer, her daughter. Its binomial name is *Hebetica sylviae* **(figure 1.13).** Some species have been named in honor of a microbiologist who originally discovered the microbe or who has made outstanding contributions to the field. Other names may designate a characteristic of the microbe (shape, color), a location where it was found, or a disease it causes. Some examples of specific names, their pronunciations, and their origins are

- *Staphylococcus aureus* (staf′-i-lo-kok′-us ah′-ree-us) Gr. *staphule,* bunch of grapes; *kokkus,* berry; and *aureus*, golden. Under a microscope, this bacterium looks like a bunch of grapes, and often has a yellow color when grown on agar.
- *Campylobacter jejuni* (cam′-peh-loh-bak-ter jee-joo′-neye) Gr. *kampylos,* curved; *bakterion,* little rod; and *jejunum,* a section of intestine. This curved rod causes intestinal infection.
- *Lactobacillus sanfranciscensis* (lak′-toh-bass-ill′-us san-fran-siss-ken′-siss) L. *lacto,* milk, and *bacillus,* little rod. A bacterial species used to make sourdough bread.
- *Giardia lamblia* (jee-ar′-dee-uh lam′-blee-uh) for Alfred Giard, a French microbiologist, and Vilem Lambl, a Bohemian physician, both of whom worked on the organism. A protozoan that causes a severe intestinal infection.

Here is a helpful hint: These names may seem difficult to pronounce and the temptation is to simply "slur over them." But when you encounter the names of microorganisms in the chapters ahead, it will be extremely useful to take the time to sound them out and repeat them until they seem familiar. There are also a variety of websites that will sound out the words for you in audio. Just do an Internet search for "microbial names audio." You are much more likely to remember them that way—and they are less likely to end up in a tangled heap with all of the new language you will be learning.

Figure 1.13 A new species is discovered.
(a) The treehopper and its discoverer, Sylvie. **(b)** Sylvie with her mother, biologist Dr. Laura Sullivan-Beckers.
(a) Dr. Laura Sullivan-Beckers; (b) Robyn Pizzo

Classification: Constructing Taxonomy

The main units of a classification scheme are organized into several descending ranks. They begin with the most general all-inclusive taxonomic category as a common denominator for organisms to exclude all others. They end with the smallest and most specific category. This means that all members of the highest category share only one or a few general characteristics, whereas members of the lowest category are essentially the same kind of organism—that is, they share the majority of their characteristics. The taxonomic categories from top to bottom are **domain, kingdom, phylum** or **division,**[1] **class, order, family, genus,** and **species.** Thus, each kingdom can be subdivided into a series of phyla or divisions, each phylum is made up of several classes, each class contains several orders, and so on. Because taxonomic schemes are to some extent artificial, certain groups of organisms may not exactly fit into the main categories. In such a case, additional taxonomic levels can be created above (super) or below (sub) a taxon, giving us such categories as "superphylum" and "subclass."

In **figure 1.14,** we compare the taxonomic breakdowns of a human and a protozoan (proh′-tuh-zoh′-uhn) to illustrate the fine points of this system. Humans and protozoa are both organisms with nucleated cells (eukaryotes). So they are in the same domain, but they are in different kingdoms. Humans are multicellular animals (Kingdom Animalia), whereas protozoa are single-celled organisms that, together with algae, belong to the Kingdom

1. The term *phylum* is used for bacteria, protozoa, and animals; the term *division* is used for algae, plants, and fungi.

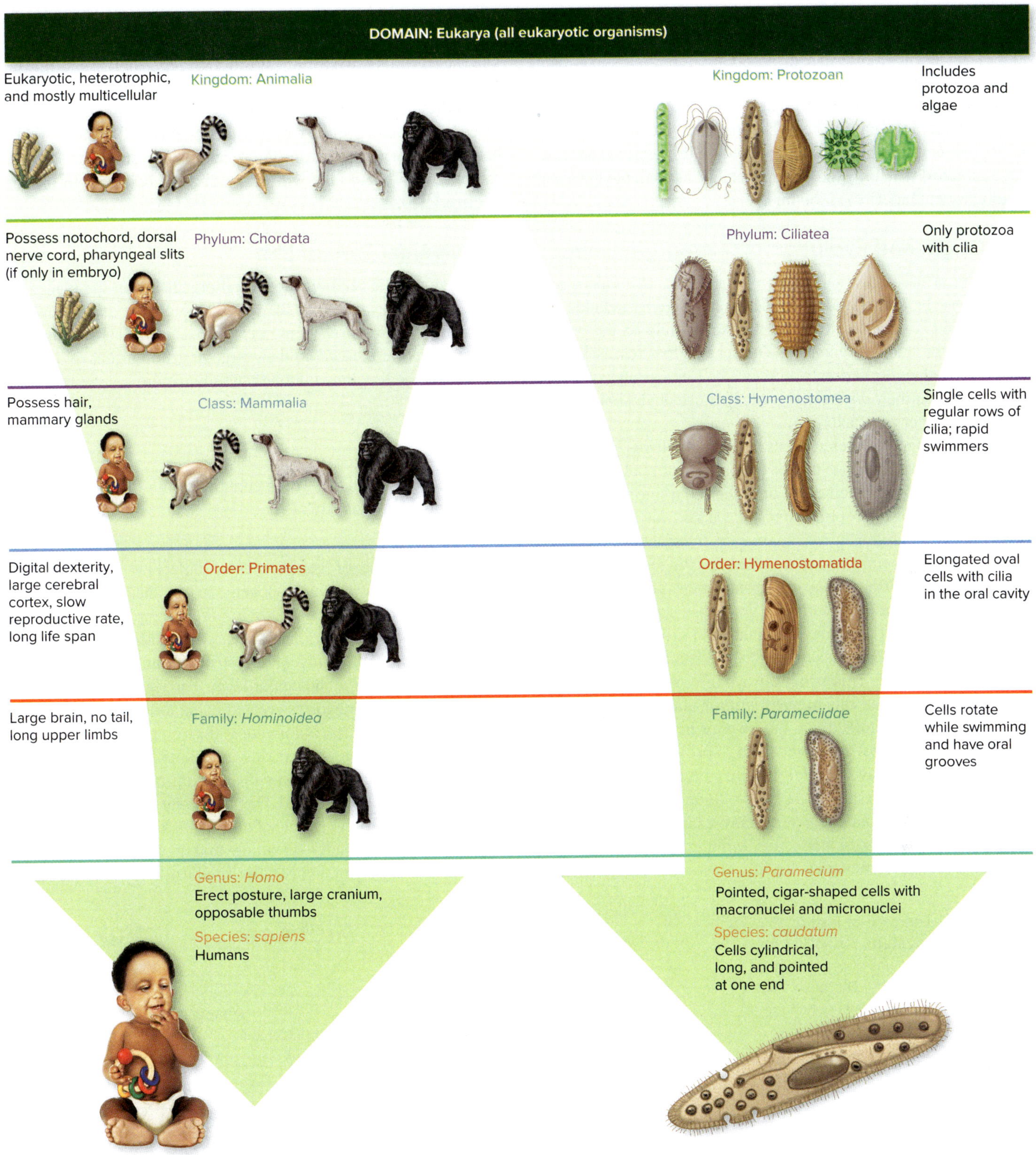
DOMAIN: Eukarya (all eukaryotic organisms)
Eukaryotic, heterotrophic, and mostly multicellular
Kingdom: Animalia
Kingdom: Protozoan
Includes protozoa and algae
Possess notochord, dorsal nerve cord, pharyngeal slits (if only in embryo)
Phylum: Chordata
Phylum: Ciliatea
Only protozoa with cilia
Possess hair, mammary glands
Class: Mammalia
Class: Hymenostomea
Single cells with regular rows of cilia; rapid swimmers
Digital dexterity, large cerebral cortex, slow reproductive rate, long life span
Order: Primates
Order: Hymenostomatida
Elongated oval cells with cilia in the oral cavity
Large brain, no tail, long upper limbs
Family: *Hominoidea*
Family: *Parameciidae*
Cells rotate while swimming and have oral grooves
Genus: *Homo*
Erect posture, large cranium, opposable thumbs
Species: *sapiens*
Humans
Genus: *Paramecium*
Pointed, cigar-shaped cells with macronuclei and micronuclei
Species: *caudatum*
Cells cylindrical, long, and pointed at one end

Figure 1.14 Sample taxonomy. Two organisms belonging to the Eukarya domain, traced through their taxonomic series; on the left, a modern human, *Homo sapiens;* on the right, a common protozoan, *Paramecium caudatum.*

Protozoa. To emphasize just how broad the category "kingdom" is, think about the fact that we humans belong to the same kingdom as jellyfish. Of the several phyla within this kingdom, humans belong to the Phylum Chordata, but even a phylum is rather broad, considering that humans share it with other vertebrates as well as with creatures called sea squirts. The next level, Class Mammalia, narrows the field considerably by grouping only those vertebrates that have hair and suckle their young. Humans belong to the Order

Primates, a group that also includes apes, monkeys, and lemurs. Next comes the Family Hominoidea, containing only humans and apes. The final levels are our genus, *Homo* (all races of modern and ancient humans), and our species, *sapiens* (meaning *wise*). Notice that for the human as well as the protozoan, the taxonomic categories in descending order become less inclusive and the individual members more closely related. In this text, we are usually concerned with only the most general (kingdom, phylum) and specific (genus, species) taxonomic levels.

The Origin and Evolution of Microorganisms

Taxonomy is used to organize all of the forms of modern and extinct life. In biology today, there are different methods for deciding on taxonomic categories, but they all rely on the degree of relatedness among organisms. The scheme that represents the natural relatedness (relation by descent) between groups of living beings is called their *phylogeny.* Biologists use phylogenetic relationships to determine taxonomy.

To understand the relatedness among organisms, we should understand some fundamentals of the process of evolution. Evolution is an important theme that underlies all of biology, including the biology of microorganisms. As we said earlier, evolution states that the hereditary information in living beings changes gradually through time and that these changes result in various structural and functional changes through many generations. The process of evolution is selective in that those changes that most favor the survival of a particular organism or group of organisms tend to be retained, whereas those that are less beneficial to survival tend to be lost. This is not always the case, but it often is. Charles Darwin called this process *natural selection.*

Evolution is founded on the two principles that (1) all new species originate from preexisting species and (2) closely related organisms have similar features because they evolved from a common ancestor. Usually, evolution progresses toward greater complexity, but there are many examples of evolution toward lesser complexity (reductive evolution). This is because individual organisms never evolve in isolation but as populations of organisms in their specific environments. The phylogeny, or relatedness by descent, of organisms is often represented by a diagram of a tree. The trunk of the tree represents the origin of ancestral lines, and the branches show offshoots into specialized groups (called clades) of organisms. This sort of arrangement places taxonomic groups with less divergence (less change in the heritable information) from the common ancestor closer to the root of the tree and taxa with more divergence closer to the top **(figure 1.15).**

A Universal Tree of Life

The phylogenetic relationships of all organisms on the planet are often depicted as "trees of life." Over the course of many years, five kingdoms of organisms were recognized. These kingdom designations were based on their structural similarities and differences, and the way they obtained their nutrition.

- Plants
- Animals
- Protista (protozoa)
- Monera (which at the time contained both Bacteria and what we now know as Archaea)
- Fungi

With the rise of genetics as a molecular science, newer methods for determining phylogeny led to the development of a differently shaped tree—with important implications for our understanding of evolutionary relatedness. Molecular genetics allowed an in-depth study of the structure and function of the genetic material at the molecular level. The findings were starkly different than the categories created using structural similarities. Two of the four macromolecules that contribute to cellular structure and function, the proteins and nucleic acids, are very well suited to study how organisms differ from one another because they can be directly compared. One particular macromolecule, the ribonucleic acid in the small subunit of the ribosome (ssuRNA), was highly "conserved"—meaning that it was nearly identical in all of the organisms within the smallest taxonomic category,

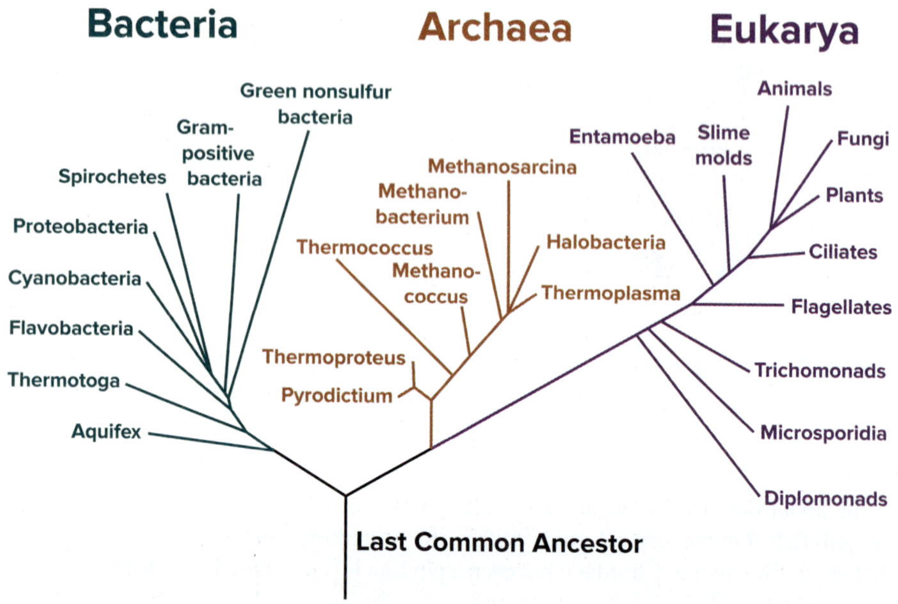
Bacteria
Archaea
Eukarya
Green nonsulfur bacteria
Gram-positive bacteria
Spirochetes
Proteobacteria
Cyanobacteria
Flavobacteria
Thermotoga
Aquifex
Methanosarcina
Methanobacterium
Thermococcus
Methanococcus
Halobacteria
Thermoplasma
Thermoproteus
Pyrodictium
Entamoeba
Slime molds
Animals
Fungi
Plants
Ciliates
Flagellates
Trichomonads
Microsporidia
Diplomonads
Last Common Ancestor

Figure 1.15 The tree of life: A phylogenetic system. A system for representing the origins of cell lines and major taxonomic groups. There are three distinct cell lines placed in superkingdoms called domains.

the species. Because of that, ssuRNA provides a "biological chronometer," or a "living record," of the evolutionary history of a given organism. Extended analysis of this molecule in akaryotic and eukaryotic cells indicated that all members of one kind of non-eukaryotic cell type had ssuRNA with a sequence that was significantly different from the ssuRNA found in the bacteria. This discovery led scientists to propose a separate taxonomic unit for this group, which they named Archaea. Under the microscope, they resembled the structure of bacteria, but molecular biology has revealed that the archaea, though seeming to be akaryotic in nature, were actually more closely related to eukaryotic cells than to bacterial cells (see table 4.1). To reflect these relationships, a new system was born. It assigns all known organisms to one of the three major taxonomic units, the **domains,** each being a different type of cell (figure 1.15).

The domains are now the highest level in hierarchy and each one can contain more than one kingdom. Take particular care to notice that Bacteria and Archaea now constitute separate domains and are not in the same kingdom, as they had been thought to be in the past. The domains are

- Bacteria
- Archaea
- Eukarya

Analysis of the ssuRNAs from all organisms in these three domains suggests that all modern and extinct organisms on earth arose from a common ancestor.

The traditional shape of the "tree of life" has given way to something else now, given our discovery of rampant horizontal gene transfer—in microbes and in larger organisms. For example, it is estimated that 40% to 50% of human DNA has been carried to humans from other species (by parasitic eukaryotes or viruses). Another example: The genome of the cow contains a great deal of snake DNA. For these reasons, most scientists like to think of a *web* as the proper representation of life these days. Nevertheless, this new scheme does not greatly affect our presentation of most microbes because we will discuss them at the genus or species level. But be aware that biological taxonomy and, more important, our view of how organisms evolved on earth are in a permanent state of transition. Keep in mind that our methods of classification or evolutionary schemes reflect our current understanding and will change as new information is uncovered.

Please note that neither viruses nor prions are included in any of the classification or evolutionary schemes because they are not cells or organisms and their position in a "tree of life" has not been definitively determined.

1.7 Learning Outcomes—Assess Your Progress

13. Differentiate among the terms *nomenclature, taxonomy,* and *classification.*
14. Create a mnemonic device for remembering the taxonomic categories.
15. Correctly write the binomial name for a microorganism.
16. Draw a diagram of the three major domains.
17. Explain the difference between traditional and molecular approaches to taxonomy.

Media Under The Microscope Wrap-Up

McGraw Hill

The news article "Study: Harmful Bacteria Is Growing on Kitchen Towels" describes a research study published in the scientific literature. The **intended message** of the news article seems to be a warning about how dangerous your dish towels are because they grow bacteria on them.

My **critical reading** takes two forms: First of all, the original research, as seen in this news article, seems to have some flaws, or flawed thinking. First of all, they found that 49% of the towels displayed bacterial growth. As you will find in this course, bacteria are everywhere, and if the towels have been in use for a month, there are certainly food sources and damp conditions that allow all of them to support bacterial growth. So I find the 49% figure suspect, and it leaves me wondering what methods they used to determine "growth."

The second part of my critical reading addresses the news article itself. First problem: The headline reads, "Bacteria Is Growing." *Bacteria* is a plural form (*bacterium* being the singular), so it is grammatically incorrect, probably stemming from a poor understanding of scientific naming—which is not unusual, but if it appears in an article about science, it does not inspire confidence. Second problem: The article implies that human feces are contaminating the towels, by using the phrasing "*E. coli* is commonly found in human feces and indicates fecal contamination and poor hygiene." In fact, the article itself states that the researchers blamed it on the preparation of meats in the kitchen. *E. coli* are found in the intestines of animals as well, and is more likely to have come from the meat, or even vegetables exposed to the excretions of farm animals, than from human intestines.

How would I **interpret** this article for my friends? I would point out the facts (described here) and tell them that sometimes news outlets try to find ways to make things sound more alarming than they actually are. I might also mention that scientific results are not written in stone after a single experiment. The way science works is that studies are repeated and the methods verified by other researchers before results are considered reliable. News outlets often grab a single study and use it to write an article to drive "clicks."

Having said all of this, my **overall grade** for the news article is a C−; questionable research reported in a questionable manner.

Source: US News and World Report, "Study: Harmful Bacteria Is Growing on Kitchen Towels," online article posted June 11, 2018.

Study Smarter: Better Together

These activities are designed for you to use on your own with a study group—either a face-to-face group or a virtual one, consisting of 3–5 members. Studying together can be very helpful, but there are effective and ineffective ways to do it. For example, getting together without a clear structure is often not a good use of your time. Use your time efficiently by using one or more of the exercises below.

FACE-TO-FACE GROUPS

Use one or more of the activities below.

Peer Instruction: Assign numbers to your group members to use all semester long. Now look at these five concepts from this chapter. Each group member prepares a 5-minute lesson on the topic corresponding to their number. Don't worry if you have fewer than five members; just use however many you have! During your group study time, each member presents their lesson, and the group spends another 5–10 minutes discussing that lesson.

1. The three biological cell types in evolutionary history
2. The role of deduction in the scientific method
3. Spontaneous generation
4. Phylogeny and taxonomy
5. The web of life

Concept Maps: Each member of the group should use this list of terms from this chapter to generate their own concept map. This can be hand-drawn or created using software (see Appendix C for guidelines). During group study time, compare each other's concept maps and help each other make sure they are correct. Of course, there are many different "correct" maps. Examining each member's map will help you talk through the varied concepts and how they are related.

Concept Terms:

microorganisms	helminths	protozoa	fungi
bacteria	viruses	akaryote	eukaryote
pathogen	organelles		

Table Topics: Each group member should identify a concept or topic from this week's class assignments with which they are having trouble and share it during group study time. The other group members can then help to clarify confusing issues or share how they figured it out. Aim for a maximum of 15 minutes per topic. If the topic remains unclear to the group, bring it up during class or use the instructor's office hours or e-mail to ask for help. Taking the time to struggle with a difficult concept first makes your questions much more specific and more likely to yield helpful answers.

VIRTUAL GROUPS

Not everyone has the time or opportunity to meet with group members outside of class time. You or your instructor can create a virtual group using e-mail or the course software.

Weekly Discussion Board: This forum can be used as a way for groups to discuss topics, via e-mail or other learning management systems or online platforms, before they are covered in class. As each member of the group answers the current week's question, they should send their responses to every other member of their group. It's best to agree on a deadline based on how your class schedule works (Saturday for the next week's topics, for example). Then, after the topic is discussed in class, each member should send a response that all group members will see with a follow-up post on the same topic. If you cover more than one chapter in a week, someone can be designated to choose which chapter Discussion Board question you will use. Or simply decide up front that you will always use the first-chapter-of-the week's question, to keep the schedule simple.

Discussion Question
What are the three domains in the tree of life, and why do scientists prefer the term "web of life" rather than "tree of life"?

Chapter Summary

THE MAIN THEMES OF MICROBIOLOGY

1.1 THE SCOPE OF MICROBIOLOGY

- There are seven types of microorganisms: bacteria, archaea, fungi, protozoa, helminths, viruses, and prions.
- Microorganisms live in every known habitat and influence many biological and physical activities on earth.
- Microbes are critical to the health of humans, and occasionally cause harm.

1.2 THE IMPACT OF MICROBES ON EARTH: SMALL ORGANISMS WITH A GIANT EFFECT

- Groups of organisms are constantly evolving to produce new forms of life.
- Microbes are crucial to the cycling of nutrients and exchange on earth.

1.3 HUMAN USE OF MICROORGANISMS

- Humans have learned how to manipulate microbes to do important work for them in industry, medicine, and caring for the environment.

1.4 INFECTIOUS DISEASES AND THE HUMAN CONDITION

- Microbiologists have identified the causative agent for many infectious diseases.
- New microbes continue to emerge and cause disease. Most often low-income countries suffer a higher toll than higher-income countries during epidemics and pandemics.

1.5 THE GENERAL CHARACTERISTICS OF MICROORGANISMS

- There are eukaryotic and akaryotic (bacterial and archaeal) microbes.
- There are also two types that are not cells at all: viruses and prions.

1.6 THE HISTORICAL FOUNDATIONS OF MICROBIOLOGY

- The invention of the microscope was crucial to the explosion of microbiological knowledge.
- Disproving the idea of spontaneous generation paved the way for true scientific progress in microbiology.

1.7 NAMING, CLASSIFYING, AND IDENTIFYING MICROORGANISMS

- Naming, classifying, and identifying organisms is called taxonomy.
- Evolutionary patterns show a weblike branching that represents the divergence of all life forms from a common ancestor.

SmartGrid: From Knowledge to Critical Thinking

This *21 Question Grid* takes the topics from this chapter and arranges them with respect to the American Society for Microbiology's Undergraduate Curriculum guidelines—all six of the important "Concepts" as well as the important "Competency" of scientific literacy. Three questions are supplied, which cover chapter content referring to the Concept or Competency in increasing levels of Bloom's taxonomy for learning.

ASM Concept/ Competency	A. Bloom's Level 1, 2—Remember and Understand (Choose one)	B. Bloom's Level 3, 4—Apply and Analyze	C. Bloom's Level 5, 6—Evaluate and Create
Evolution	**1.** Which of the following is an acellular microorganism lacking a nucleus? **a.** bacterium **b.** helminth **c.** protozoan **d.** virus	**2.** Name seven types of microorganisms that we are studying in this book and use each one in a sentence.	**3.** Defend the argument that a web of life is a more accurate representation of evolutionary relatedness than a tree of life.
Cell Structure and Function	**4.** Which of the following is a microorganism that contains organelles? **a.** prion **b.** bacterium **c.** fungus **d.** virus	**5.** Often when there is a local water main break, a town will post an advisory to boil water before ingesting it. Identify the biological basis behind the effectiveness of this procedure in minimizing illness.	**6.** Imagine a way you might design a drug that will destroy microbes without harming human cells.
Metabolic Pathways	**7.** Identify the process or environment in this list that is not affected by microorganisms. **a.** oxygen cycles **b.** recycling of dead organisms **c.** human health **d.** all of the above have microbial involvement	**8.** Discuss the current view of the evolution of eukaryotic cells.	**9.** Describe how microorganisms can accomplish bioremediation.
Information Flow and Genetics	**10.** Which of these organisms do not contain DNA? **a.** helminths **b.** fungi **c.** bacteria **d.** prions	**11.** Summarize some important facets of the human microbiome.	**12.** Suggest an argument for why eukaryotic cells have developed an enclosed nucleus unlike bacteria and archaea.
Microbial Systems	**13.** Microbes are found in which habitat? **a.** human body **b.** earth's crust **c.** oceans **d.** all of the above	**14.** Argue for or against this statement: *Microbes intend to cause human disease.*	**15.** Coevolution is a term describing the influence that two organisms occupying the same niche have on each other. Sketch a scenario for coevolution between a bacterium and a human living in the same environment.
Impact of Microorganisms	**16.** Which of the following processes can be the result of human manipulation of microbial genes? **a.** the central dogma **b.** natural selection **c.** bioremediation **d.** abiogenesis	**17.** Speculate about why scientists believe there are more microbial species that we do not yet know about than those that we do know about.	**18.** Imagine you are a guest speaker in a middle school science class. Explain to the students why human life would not be possible without microorganisms.

ASM Concept/ Competency	A. Bloom's Level 1, 2—Remember and Understand (Choose one)	B. Bloom's Level 3, 4—Apply and Analyze	C. Bloom's Level 5, 6—Evaluate and Create
Scientific Thinking	**19.** When a hypothesis has been thoroughly supported by long-term study and data, it is considered **a.** a law **b.** a speculation **c.** a theory **d.** proven	**20.** Defend the use of complicated-sounding names for identifying microorganisms.	**21.** Identify the most important component of the scientific method and defend your answer.

Answers to the multiple-choice questions appear in Appendix A.

Visual Connections

These questions use visual images or previous content to connect content within and between chapters.

1. **Figure 1.1.** Think about the origination of cells and then higher organisms. Where do you think viruses fit on this graph?

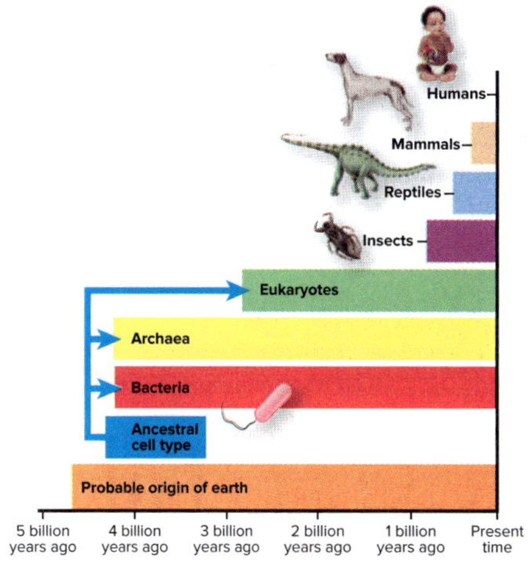
Humans
Mammals
Reptiles
Insects
Eukaryotes
Archaea
Bacteria
Ancestral cell type
Probable origin of earth
5 billion years ago
4 billion years ago
3 billion years ago
2 billion years ago
1 billion years ago
Present time

High Impact Study

These terms and concepts are most critical for your understanding of this chapter—and may be the most difficult. Have you mastered them?

Concepts

- ☐ The types of microorganisms we will study
- ☐ The three biological cell types appearing in evolutionary history
- ☐ The theory of evolution
- ☐ The role of deduction in the scientific method
- ☐ Spontaneous generation
- ☐ Relative size of the different types of microorganisms
- ☐ Phylogeny and taxonomy
- ☐ The tree of life; the web of life

Terms

- ☐ Bacteria
- ☐ Archaea
- ☐ Eukaryote
- ☐ Recombinant DNA technology
- ☐ Bioremediation
- ☐ Pathogen
- ☐ Taxa
- ☐ Nomenclature
- ☐ Phylogeny
- ☐ ssuRNA

Design Element: (College students): Caia Image/Image Source

2 The Chemistry of Biology

O. Dimier/PhotoAlto

Media Under The Microscope

Saline Drip

This opening case examines an article from the popular media to determine the extent to which it is factual and/or misleading. This case focuses on the 2018 NPR "Shots" blogpost "Why Did Sterile Salt Water Become the IV Fluid of Choice?"

You're probably aware that the go-to hydration fluid in hospitals is called "normal saline." This solution of 0.9% sodium chloride is by far the most commonly used IV drip. The article traces the history of normal saline's rise to the top. The article reveals that it was actually due to a mistake made by a Dutch scientist in the 1800s, who calculated that the concentration of salts in human blood was 0.9%. The problem is, that is incorrect. The actual concentration of salts in blood is 0.6%.

According to the article, another solution was used well before normal saline was adopted. This one, called Ringer's solution (named after its inventor), contains sodium, chloride, and potassium in concentrations that do in fact match those in human blood. It never caught on to the same degree that normal saline did. Maybe normal saline was just simpler to make.

The article describes a series of studies showing that the higher chloride content in normal saline can actually be harmful to patients, and even provides evidence that poor outcomes associated with IV fluids are 1% more common with saline than they are with Ringer's solution. Even though that difference seems tiny, the article states: "Given two solutions that are equally available, equal in cost and that millions of adults receive every year, 1 percent is pretty good. We'd be lucky to improve mortality by 1 percent with an expensive drug, much less a fluid that costs $2."

- What is the **intended message** of the article?
- What is your **critical reading** of the summary of the article provided above? Remember that in this context, "critical reading" does not necessarily mean *What criticism do you have?* but asks you to apply your knowledge to interpret whether the article is factual and whether the facts support the intended message.
- How would you **interpret** the news item for your nonmicrobiologist friends?
- What is your **overall grade** for the news item—taking into account its accuracy and the accuracy of its intended effect?

Media Under The Microscope Wrap-Up appears at the end of the chapter.

Outline and Learning Outcomes

2.1 Atoms, Bonds, and Molecules: Fundamental Building Blocks

1. Explain the relationship between atoms and elements.
2. List and define four types of chemical bonds.
3. Differentiate between a solute and a solvent.
4. Provide a brief definition of pH.

2.2 Macromolecules: Superstructures of Life

5. Name the four main families of biochemicals.
6. Provide examples of cell components made from each of the families of biochemicals.
7. Differentiate among primary, secondary, tertiary, and quaternary levels of protein structure.
8. List the three components of nucleotides.
9. Name the nitrogen bases of DNA and RNA.
10. List the three components of ATP.

2.3 Cells: Where Chemicals Come to Life

11. Recall three characteristics common to all cells.

2.1 Atoms, Bonds, and Molecules: Fundamental Building Blocks

The universe is composed of an infinite variety of substances existing in the gaseous, liquid, and solid states. All of these materials that occupy space and have mass are called **matter.** The organization of matter—whether air, rocks, animals, or bacteria—begins with individual building blocks called atoms. An **atom** is defined as a tiny particle that cannot be subdivided into smaller substances without losing its properties. Even in a science dealing with very small things, an atom's tiny size is striking. For example, an oxygen atom is only 0.0000000013 mm (0.0013 nm) in diameter, and 1 million of them in a cluster would barely be visible to the naked eye.

An atom is made of subatomic particles called protons, neutrons, and electrons. **Protons** (p^+) are positively charged; **neutrons** (n^0) have no charge (are neutral); and **electrons** (e^-) are negatively charged. The relatively larger protons and neutrons make up a central core, or *nucleus,*[1] that is surrounded by one or more electrons **(figure 2.1).** The nucleus makes up the larger mass (weight) of the atom, whereas the electron region accounts for the greater volume. To get a perspective on proportions, consider this: If an atom were the size of a baseball stadium, the nucleus would be about the size of a marble! The atom is held together by (1) the mutual attraction of the protons and electrons (opposite charges attract each other) and (2) the exact balance of proton number and electron number, which causes the opposing charges to cancel each other out. At least in theory, then, isolated intact atoms do not carry a charge.

Different Types of Atoms: Elements and Their Properties

All atoms share the same fundamental structure. All protons are identical, all neutrons are identical, and all electrons are identical. But when these subatomic particles come together in different combinations, unique types of structures called **elements** result. Each element has a characteristic atomic structure and predictable chemical behavior. To date, 118 elements, both naturally occurring and artificially produced by physicists, have been described. By convention, an element is assigned its own name with an abbreviated shorthand symbol. The elements are often depicted in a periodic table. **Table 2.1** lists some of the elements common to biological organisms, their atomic characteristics, and some of the natural and applied roles they play.

1. Be careful not to confuse the nucleus of an atom with the nucleus of a cell.

The Major Elements of Life and Their Primary Characteristics

The unique properties of each element result from the numbers of protons, neutrons, and electrons it contains, and each element can be identified by certain physical measurements.

Isotopes are variant forms of the same element that differ in the number of neutrons. These multiple forms occur naturally in certain proportions. Carbon, for example, exists primarily as carbon 12 with 6 neutrons; but a small amount (about 1%) occurs

Disease Connection

Isotopes are very useful in diagnosing certain diseases, such as cancer. Patients are injected with isotopes (forms of the element that are radioactive) called *tracers* that emit radiation from within the body. Alternatively, tracers can be inhaled or swallowed. Technicians then use special detectors to evaluate the position and concentration of the radioactive isotopes within the body. From these images, organ disease can be detected as cold spots (areas where the isotopes are only partially taken up) or hot spots (areas where the isotopes are taken up excessively). A series of images is taken over a certain period of time, and any unusual rate and/or pattern of isotope activity can indicate organ disease or malfunction.

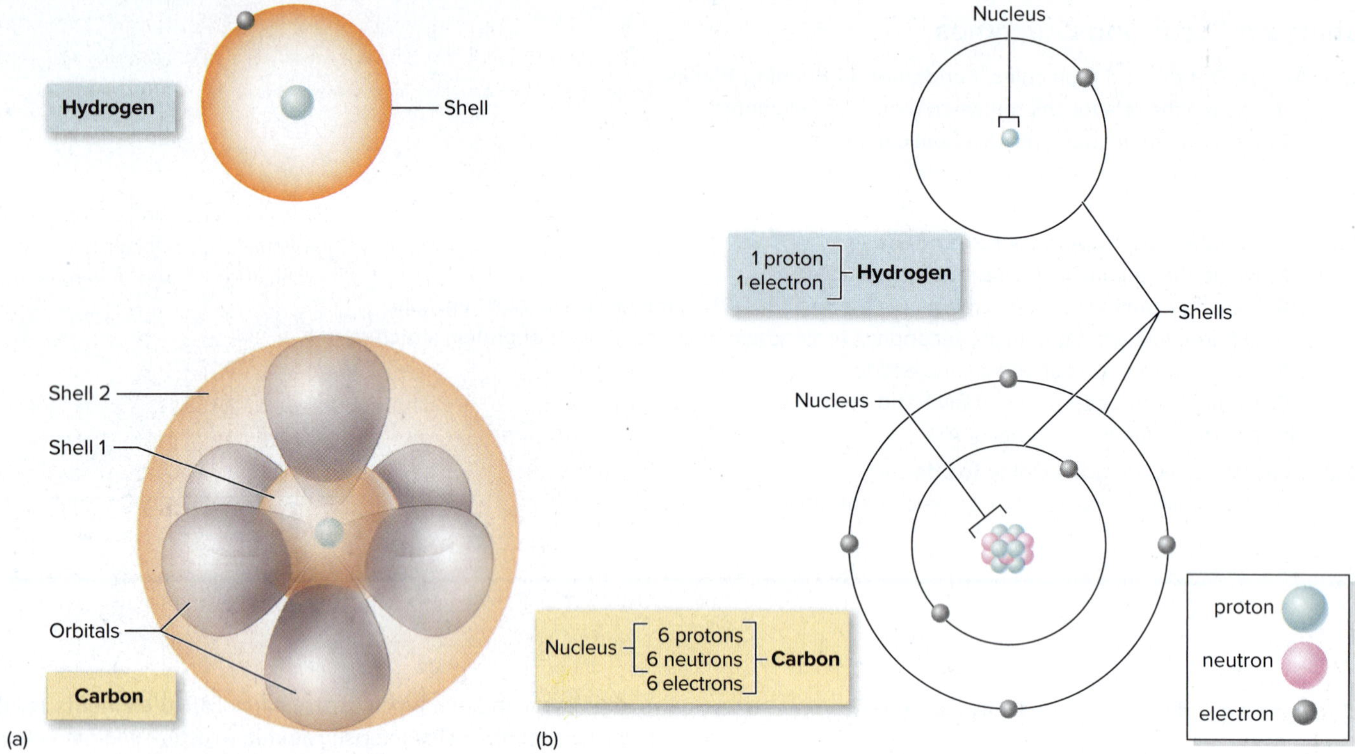
Hydrogen
Shell
Shell 2
Shell 1
Orbitals
Carbon
(a)
Nucleus
1 proton
1 electron
Hydrogen
Shells
Nucleus
Nucleus
6 protons
6 neutrons
6 electrons
Carbon
proton
neutron
electron
(b)

Figure 2.1 Models of atomic structure. **(a)** Three-dimensional models of hydrogen and carbon structures. The nucleus is surrounded by electrons in orbitals that occur in levels called shells. Hydrogen has just one shell and one orbital. Carbon has two shells and four orbitals. The shape of the outermost orbitals is a paired lobe rather than a circle sphere. **(b)** Simple models of the same atoms make it easier to show the numbers and arrangements of shells and electrons and the numbers of protons and neutrons in the nucleus. (Not to accurate scale.)

Table 2.1 The Major Elements of Life and Their Primary Characteristics

Element	Atomic Symbol‡	Atomic Mass**	Examples of Ionized Forms	Significance in Microbiology
Calcium	Ca	40.1	Ca^{2+}	Part of outer covering of certain shelled amoebas; stored within bacterial spores
Carbon	C	12.0	CO_3^{2-}	Principal structural component of biological molecules
Carbon*	C-14	14.0		Radioactive isotope used in dating fossils
Chlorine	Cl	35.5	Cl^-	Component of disinfectants; used in water purification
Cobalt	Co	58.9	Co^{2+}, Co^{3+}	Trace element needed by some bacteria to synthesize vitamins
Cobalt*	Co-60	60		An emitter of gamma rays; used in food sterilization; used to treat cancer
Copper	Cu	63.5	Cu^+, Cu^{2+}	Necessary to the function of some enzymes; Cu salts are used to treat fungal and worm infections
Hydrogen	H	1	H^+	Necessary component of water and many organic molecules; H_2 gas released by bacterial metabolism
Hydrogen*	H3	3		Has 2 neutrons; radioactive; used in clinical laboratory procedures
Iodine	I	126.9	I^-	A component of antiseptics and disinfectants; used in the Gram stain
Iodine*	I-131, I-125	131, 125		Radioactive isotopes for diagnosis and treatment of cancers
Iron	Fe	55.8	Fe^{2+}, Fe^{3+}	Necessary component of respiratory enzymes; required by some microbes to produce toxin
Magnesium	Mg	24.3	Mg^{2+}	A trace element needed for some enzymes; component of chlorophyll pigment
Manganese	Mn	54.9	Mn^{2+}, Mn^{3+}	Trace element for certain respiratory enzymes
Nitrogen	N	14.0	NO^{3-}	Component of all proteins and nucleic acids; the major atmospheric gas
Oxygen	O	16.0		An essential component of many organic molecules; molecule used in metabolism by many organisms
Phosphorus	P	31	PO_4^{3-}	A component of ATP, nucleic acids, cell membranes; stored in granules in cells
Phosphorus*	P-32	32		Radioactive isotope used as a diagnostic and therapeutic agent
Potassium	K	39.1	K^+	Required for normal ribosome function and protein synthesis; essential for cell membrane permeability
Sodium	Na	23.0	Na^+	Necessary for transport; maintains osmotic pressure; used in food preservation
Sulfur	S	32.1	SO_4^{2-}	Important component of proteins; makes disulfide bonds; storage element in many bacteria
Zinc	Zn	65.4	Zn^{2+}	An enzyme cofactor; required for protein synthesis and cell division; important in regulating DNA

Notes: ‡Based on the Latin name of the element. The first letter is always capitalized. If there is a second letter, it is always lowercase.

*Denotes the radioactive isotope of the element.

**The atomic mass or weight is equal to the average mass number for the isotopes of that element.

naturally as carbon 13 with 7 neutrons or carbon 14 with 8 neutrons. Although isotopes have virtually the same chemical properties, some of them have unstable nuclei that spontaneously release energy in the form of radiation. Such *radioactive isotopes* play a role in a number of research and medical applications. Because they emit detectable signs, they can be used to trace the position of key atoms or molecules in chemical reactions, they are tools in diagnosis and treatment, and they are even used in sterilization procedures (see the discussion of ionizing radiation in section 11.2).

Electron Orbitals and Shells

The structure of an atom can be envisioned as a central nucleus surrounded by a "cloud" of electrons that constantly rotate about the nucleus in pathways (see figure 2.1). The pathways, called **orbitals,** are not actual objects or exact locations but represent areas of space in which an electron is likely to be found. Electrons occupy energy shells, proceeding from the lower-level energy electrons nearest to the nucleus to the higher-level energy electrons in the farthest orbitals.

Electrons fill the orbitals and shells in *pairs,* starting with the shell nearest the nucleus. The first shell contains one orbital and a maximum of 2 electrons. The second shell has four orbitals and up to 8 electrons. The third shell with nine orbitals can hold up to 18 electrons, and the fourth shell with 16 orbitals contains up to 32 electrons. The number of orbitals and shells and how completely they are filled depend on the numbers of electrons so that each element will have a unique pattern. For example, helium has only a (filled) first shell of 2 electrons; oxygen has a filled first shell and a partially filled second shell of 6 electrons; and magnesium has a filled first shell, a filled second one, and a third shell that fills only one orbital, so it is nearly empty. As we will see, the chemical properties of an element are controlled mainly by the distribution of electrons in the outermost shell. Figure 2.1 and **figure 2.2** present various simplified models of atomic structure and electron maps. **Figure 2.3** presents all the elements in the periodic table.

Bonds and Molecules

Most elements do not exist naturally in pure, uncombined form but are bound together as molecules and compounds. A **molecule** is a chemical substance that results from the combination of two or more atoms. Some molecules such as oxygen (O_2) and nitrogen gas (N_2) consist of multiple atoms of the same element. Molecules that are combinations of two or more *different* elements are termed **compounds.** Compounds such as water (H_2O) and biological molecules (proteins, sugars, fats) are the predominant substances in living systems. When atoms bind together in molecules, they lose the properties of the atom and take on the properties of the combined substance.

The **chemical bonds** of molecules and compounds result when two or more atoms share, donate (lose), or accept (gain) electrons **(figure 2.4).** The number of electrons in the outermost shell of an element is known as its **valence.** The valence determines the degree of reactivity and the types of bonds an element can make. Elements with a filled (completed) outer orbital are relatively stable because they have no extra electrons to share with or donate to other atoms. For example, helium has one filled shell, with no tendency either to give up electrons or to take them from other elements, making it a stable, inert (nonreactive) gas. Elements with partially filled outer orbitals are less stable and are more likely to form some sort of bond. Many chemical reactions are based on the tendency of atoms with unfilled outer shells to gain greater stability by trying to fill up their outer shell. For example, an atom such as oxygen that can accept 2 additional electrons will bond readily with atoms (such as hydrogen) that can share or donate electrons. We explore some additional examples of the basic types of bonding in the following section.

The number of electrons in the outer shell also dictates the number of chemical bonds an atom can make. For instance, hydrogen can bind with one other atom, oxygen can bind with up to two other atoms, and carbon can bind with four.

Covalent Bonds and Polarity: Molecules with Shared Electrons

Covalent (cooperative valence) **bonds** form between atoms that share electrons rather than donating or receiving them. A simple example is hydrogen gas (H_2), which consists of two hydrogen atoms. A hydrogen atom has only a single electron, but when two of them combine, each will bring its electron to orbit about both nuclei, which creates a filled orbital (2 electrons) for both atoms, resulting in single covalent bond **(figure 2.5*a*).** Covalent bonding also occurs in oxygen gas (O_2) but with a difference. Because each atom has 2 electrons to share in this molecule, the combination creates two pairs of shared electrons, also known as a double covalent bond **(figure 2.5*b*).** The majority of the molecules associated with living things are composed of single and double covalent bonds between the most common biological elements (carbon, hydrogen, oxygen, nitrogen, sulfur, and phosphorus). **Insight 2.1** shows us that the bonding between phosphorus and oxygen to form phosphate had a lot to do with the fact that our planet has vast amounts of oxygen. Double bonds in molecules and

A Note About Mass, Weight, and Related Terms

Mass refers to the quantity of matter that an atomic particle contains. The proton and neutron have almost exactly the same mass, which is about 1.66×10^{-24} grams, a unit of mass known as a Dalton (Da) or unified atomic mass unit (U). All elements can be measured in these units. The terms *mass* and *weight* are often used interchangeably in biology, even though they apply to two different but related aspects of matter. Weight is a measurement of the gravitational pull on the mass of a particle, atom, or object. Consequently, it is possible for something with the same mass to have different weights. For example, an astronaut on the earth (normal gravity) would weigh more than the same astronaut on the moon (weak gravity). Atomic weight has been the traditional usage for biologists because most chemical reactions and biological activities occur within the normal gravitational conditions on earth. This permits use of the atomic weight as a standard of comparison. You will also see the terms *formula weight* and *molecular weight* used interchangeably, and they are in fact synonyms. They both mean the sum of atomic weights of all atoms in a molecule.

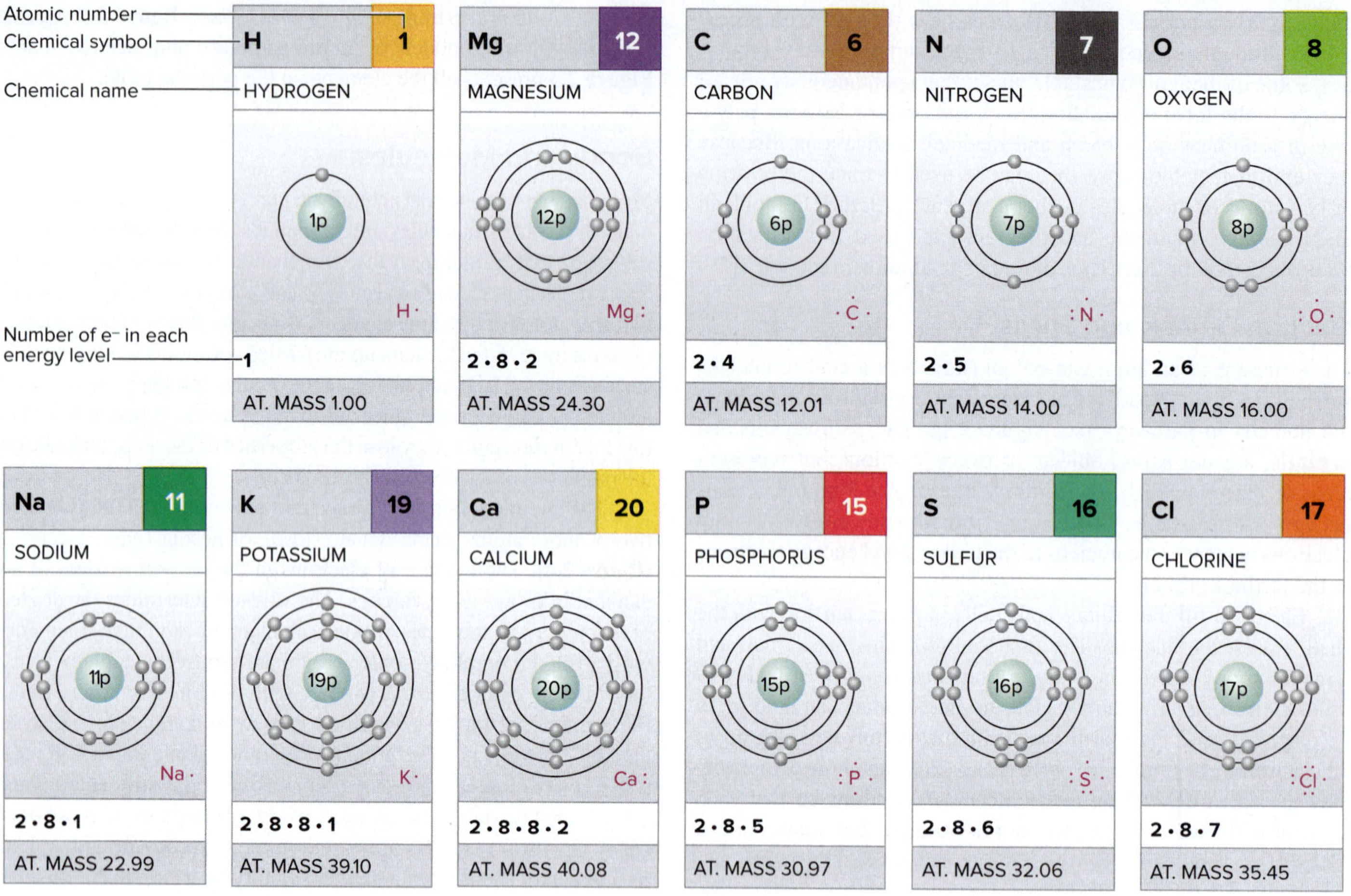
Atomic number
Chemical symbol
Chemical name
Number of e⁻ in each energy level
H 1
HYDROGEN
1p
H·
1
AT. MASS 1.00
Mg 12
MAGNESIUM
12p
Mg :
2·8·2
AT. MASS 24.30
C 6
CARBON
6p
2·4
AT. MASS 12.01
N 7
NITROGEN
7p
2·5
AT. MASS 14.00
O 8
OXYGEN
8p
2·6
AT. MASS 16.00
Na 11
SODIUM
11p
Na·
2·8·1
AT. MASS 22.99
K 19
POTASSIUM
19p
K·
2·8·8·1
AT. MASS 39.10
Ca 20
CALCIUM
20p
Ca :
2·8·8·2
AT. MASS 40.08
P 15
PHOSPHORUS
15p
2·8·5
AT. MASS 30.97
S 16
SULFUR
16p
2·8·6
AT. MASS 32.06
Cl 17
CHLORINE
17p
2·8·7
AT. MASS 35.45

Figure 2.2 Examples of biologically important atoms. Simple models show how the shells are filled by electrons as the atomic numbers increase. Notice that the outer shells of these elements are incompletely filled because they have fewer than 8 electrons.

compounds introduce more stiffness than single bonds. A slightly more complex pattern of covalent bonding is shown for methane gas (CH_4) in **figure 2.5*c*.**

Electronegativity is a term that indicates the ability to attract electrons. When atoms of different electronegativity form covalent bonds, the electrons are not shared equally and may be pulled more toward one atom than another. This pull causes one end of a molecule to assume a partial negative charge and the other end to assume a partial positive charge. A molecule with such an asymmetrical distribution of charges is termed **polar**—it has positive and negative poles. Observe the water molecule shown in **figure 2.6** and note that, because the oxygen atom is larger and has more protons than the hydrogen atoms, it will tend to draw the shared electrons with greater force toward its nucleus. This unequal force causes the oxygen part of the molecule to express a negative charge (due to the electrons being attracted there) and the hydrogens to express a positive charge (due to the protons). The fact that water is polar has a great influence on its role in biological reactions. Polarity is a significant property of many large molecules in living systems and greatly influences both their reactivity and their structure.

When covalent bonds are formed between atoms that have the same or similar electronegativity, the electrons are shared equally between the two atoms. Because of this balanced distribution, no part of the molecule has a greater attraction for the electrons. This sort of electrically neutral molecule is termed **nonpolar.**

A Note About Diatomic Elements

You will notice that hydrogen, oxygen, nitrogen, chlorine, and iodine are often shown in notation with a 2 subscript—H_2 or O_2. These elements are diatomic (two atoms), meaning that in their pure elemental state, they exist in pairs, rather than as single atoms. The reason for this phenomenon has to do with their valences. The electrons in the outer shell are configured so that they complete a full outer shell for both atoms when they bind to each other. You can see this for yourself in figures 2.3 and 2.5. Most of the diatomic elements are gases.

Ionic Bonds: Electron Transfer Among Atoms

In reactions that form **ionic bonds,** electrons are transferred completely from one atom to another and are not shared. These reactions always occur between atoms with valences that complement each other, meaning that one atom has an unfilled shell that will readily accept electrons and the other atom has an unfilled shell that will readily lose electrons. A striking example is the reaction that occurs between sodium (Na) and chlorine (Cl). Elemental

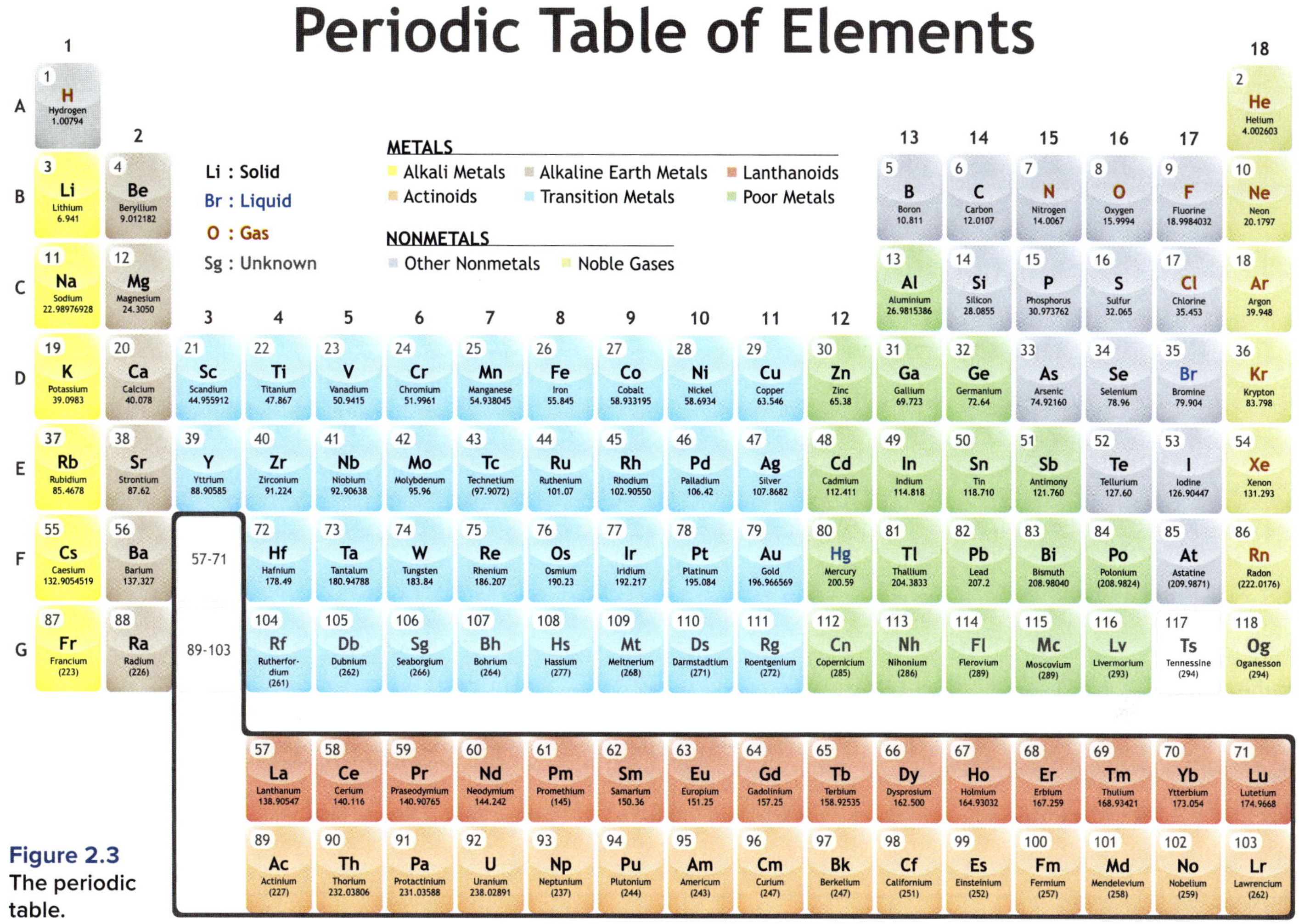
Periodic Table of Elements
1
2
3
4
5
6
7
8
9
10
11
12
13
14
15
16
17
18
A
B
C
D
E
F
G
Li : Solid
Br : Liquid
O : Gas
Sg : Unknown
METALS
Alkali Metals
Alkaline Earth Metals
Lanthanoids
Actinoids
Transition Metals
Poor Metals
NONMETALS
Other Nonmetals
Noble Gases
1 H Hydrogen 1.00794
2 He Helium 4.002603
3 Li Lithium 6.941
4 Be Beryllium 9.012182
5 B Boron 10.811
6 C Carbon 12.0107
7 N Nitrogen 14.0067
8 O Oxygen 15.9994
9 F Fluorine 18.9984032
10 Ne Neon 20.1797
11 Na Sodium 22.98976928
12 Mg Magnesium 24.3050
13 Al Aluminium 26.9815386
14 Si Silicon 28.0855
15 P Phosphorus 30.973762
16 S Sulfur 32.065
17 Cl Chlorine 35.453
18 Ar Argon 39.948
19 K Potassium 39.0983
20 Ca Calcium 40.078
21 Sc Scandium 44.955912
22 Ti Titanium 47.867
23 V Vanadium 50.9415
24 Cr Chromium 51.9961
25 Mn Manganese 54.938045
26 Fe Iron 55.845
27 Co Cobalt 58.933195
28 Ni Nickel 58.6934
29 Cu Copper 63.546
30 Zn Zinc 65.38
31 Ga Gallium 69.723
32 Ge Germanium 72.64
33 As Arsenic 74.92160
34 Se Selenium 78.96
35 Br Bromine 79.904
36 Kr Krypton 83.798
37 Rb Rubidium 85.4678
38 Sr Strontium 87.62
39 Y Yttrium 88.90585
40 Zr Zirconium 91.224
41 Nb Niobium 92.90638
42 Mo Molybdenum 95.96
43 Tc Technetium (97.9072)
44 Ru Ruthenium 101.07
45 Rh Rhodium 102.90550
46 Pd Palladium 106.42
47 Ag Silver 107.8682
48 Cd Cadmium 112.411
49 In Indium 114.818
50 Sn Tin 118.710
51 Sb Antimony 121.760
52 Te Tellurium 127.60
53 I Iodine 126.90447
54 Xe Xenon 131.293
55 Cs Caesium 132.9054519
56 Ba Barium 137.327
57-71
72 Hf Hafnium 178.49
73 Ta Tantalum 180.94788
74 W Tungsten 183.84
75 Re Rhenium 186.207
76 Os Osmium 190.23
77 Ir Iridium 192.217
78 Pt Platinum 195.084
79 Au Gold 196.966569
80 Hg Mercury 200.59
81 Tl Thallium 204.3833
82 Pb Lead 207.2
83 Bi Bismuth 208.98040
84 Po Polonium (208.9824)
85 At Astatine (209.9871)
86 Rn Radon (222.0176)
87 Fr Francium (223)
88 Ra Radium (226)
89-103
104 Rf Rutherfor-dium (261)
105 Db Dubnium (262)
106 Sg Seaborgium (266)
107 Bh Bohrium (264)
108 Hs Hassium (277)
109 Mt Meitnerium (268)
110 Ds Darmstadtium (271)
111 Rg Roentgenium (272)
112 Cn Copernicium (285)
113 Nh Nihonium (286)
114 Fl Flerovium (289)
115 Mc Moscovium (289)
116 Lv Livermorium (293)
117 Ts Tennessine (294)
118 Og Oganesson (294)
57 La Lanthanum 138.90547
58 Ce Cerium 140.116
59 Pr Praseodymium 140.90765
60 Nd Neodymium 144.242
61 Pm Promethium (145)
62 Sm Samarium 150.36
63 Eu Europium 151.25
64 Gd Gadolinium 157.25
65 Tb Terbium 158.92535
66 Dy Dysprosium 162.500
67 Ho Holmium 164.93032
68 Er Erbium 167.259
69 Tm Thulium 168.93421
70 Yb Ytterbium 173.054
71 Lu Lutetium 174.9668
89 Ac Actinium (227)
90 Th Thorium 232.03806
91 Pa Protactinium 231.03588
92 U Uranium 238.02891
93 Np Neptunium (237)
94 Pu Plutonium (244)
95 Am Americum (243)
96 Cm Curium (247)
97 Bk Berkelium (247)
98 Cf Californium (251)
99 Es Einsteinium (252)
100 Fm Fermium (257)
101 Md Mendelevium (258)
102 No Nobelium (259)
103 Lr Lawrencium (262)

Figure 2.3 The periodic table.

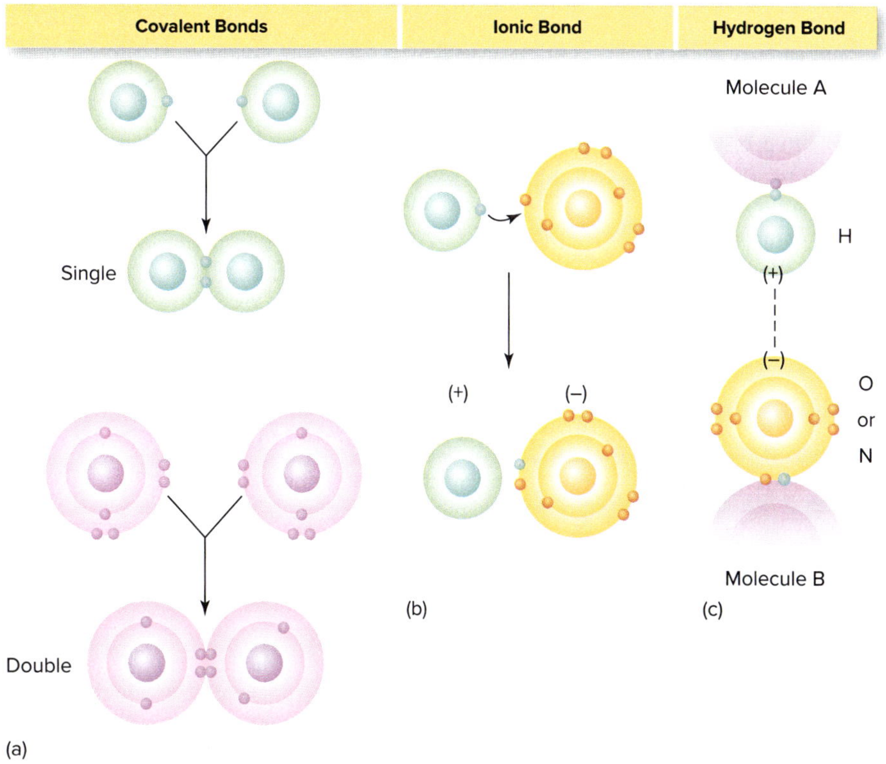
Covalent Bonds
Ionic Bond
Hydrogen Bond
Single
Double
(a)
(+)
(–)
(b)
Molecule A
H
(+)
(–)
O
or
N
Molecule B
(c)

Figure 2.4 General representation of three types of bonding. **(a)** Covalent bonds, both single and double. **(b)** Ionic bond. **(c)** Hydrogen bond. Note that hydrogen bonds are visually represented in models and formulas by dotted lines, as shown in (c).

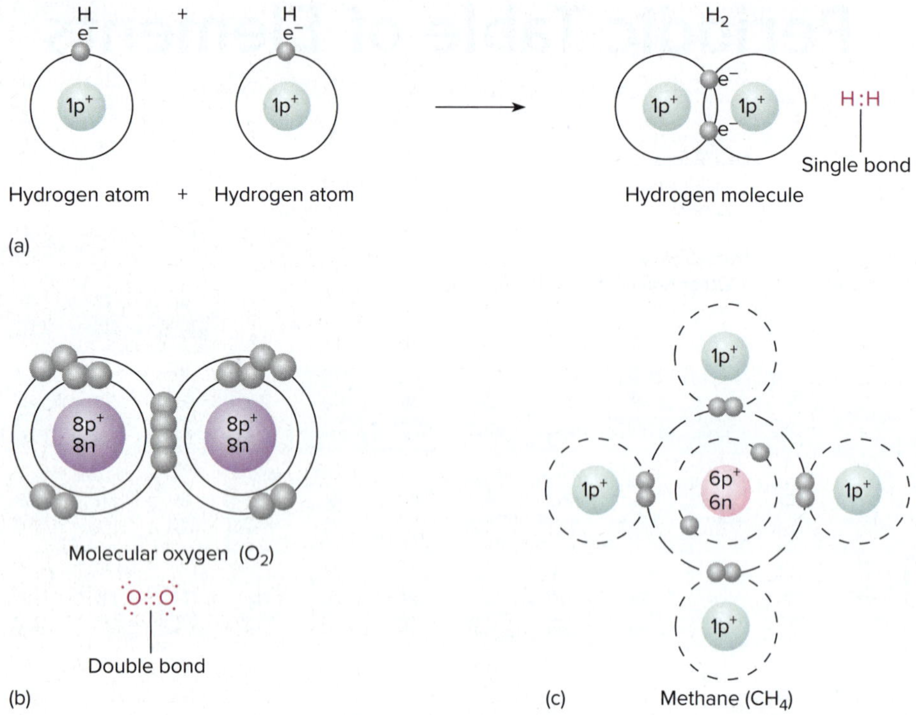
H
e⁻
+
H
e⁻
1p⁺
1p⁺
H₂
e⁻
1p⁺
1p⁺
e⁻
H:H
Single bond
Hydrogen atom + Hydrogen atom
Hydrogen molecule
(a)
8p⁺
8n
8p⁺
8n
Molecular oxygen (O₂)
O::O
Double bond
(b)
1p⁺
1p⁺
6p⁺
6n
1p⁺
1p⁺
(c)
Methane (CH₄)

Figure 2.5 Examples of molecules with covalent bonding. **(a)** A hydrogen molecule is formed when two hydrogen atoms share their electrons and form a single bond. **(b)** In a double bond, the outer orbitals of two oxygen atoms overlap and permit the sharing of 4 electrons (one pair from each) and the saturation of the outer orbital for both. **(c)** Methane. Note that carbon has 4 electrons to share and each hydrogen has one, thereby completing the shells for all atoms in the compound, and creating 4 single bonds.

INSIGHT 2.1 MICROBIOME: Thanks to the Sponge, and Its Microbiome, for Letting Us Breathe

Humans are not the only organisms that have microbiomes. In fact, scientists are cataloging the microbiomes of other organisms and also of environments. All of the earth's habitats have been greatly influenced by the microbes that live there—in other words, their microbiome.

Recently scientists made a discovery that explained at least partly the immense influx of oxygen onto the planet about 750 million years ago that enabled the Cambrian explosion. The Cambrian explosion is when there was a rapid appearance of many types of organisms and animals that we see currently on our planet, with credit going to the sudden appearance of large amounts of oxygen, mainly in the oceans covering the earth.

But where did the oxygen come from? This research suggested that it came from sponges—those sea creatures that seem to have great capacities for pumping and filtering water. But how did that lead to the introduction of all that oxygen? Because sponges live on very low levels of oxygen, and do all that pumping and filtering, they seem to have been producing oxygen in their habitats in the deep sea. But from what?

The oxygen was being produced by photosynthetic microorganisms in the vicinity of the sponges. The active pumping of the oxygen by the sponges created large deposits of phosphate (because phosphorus—abundant in the ocean—combines with oxygen to form the phosphate deposits). Scientists deduced that the large amounts of phosphate that started appearing in the ocean depths around this time are evidence for oxygen-producing bacteria—in the form of the microbiome associated with sponges. So it is quite possible that sea sponges, and their microbes, created the conditions that made oxygenic life on earth possible.

Carson Ganci/Design Pics

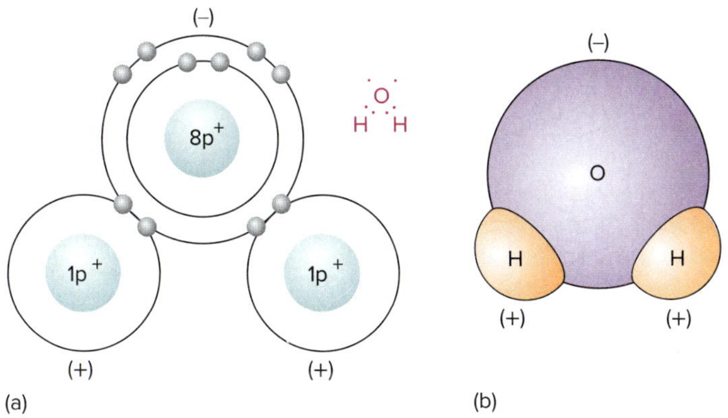
(–)
8p+
O
H H
1p+
1p+
(+)
(+)
(a)
(–)
O
H
H
(+)
(+)
(b)

Figure 2.6 Polar molecule. **(a)** A simple model and **(b)** a three-dimensional model of a water molecule indicate the polarity, or unequal distribution, of electrical charge, which is caused by the pull of the shared electrons toward the oxygen side of the molecule.

sodium is a soft, lustrous metal so reactive that it can burn flesh, and molecular chlorine is a very poisonous yellow gas. But when the two are combined, they form sodium chloride[2] (NaCl)—the familiar nontoxic table salt—a compound with properties quite different from either parent element **(figure 2.7).**

How does this change take place? Sodium has 11 electrons (2 in shell one, 8 in shell two, and only 1 in shell three), so it is 7 short of having a complete outer shell. Chlorine has 17 electrons (2 in shell one, 8 in shell two, and 7 in shell three), making it 1 short of a complete outer shell. These two atoms are very reactive with one another because a sodium atom will readily donate its single electron and a chlorine atom will naturally receive it. The outcome of this reaction is not many single, isolated molecules of NaCl but rather a solid crystal complex that interlinks millions of sodium and chloride ions **(figure 2.7*b,c*).**

Ionization: Formation of Charged Particles Molecules with ionic bonds are electrically neutral, but they can produce charged particles when dissolved in a liquid called a solvent. This phenomenon, called **ionization,** occurs when the ionic bond is broken and the atoms dissociate (separate) into unattached, charged particles called **ions (figure 2.8).** To illustrate what gives a charge to ions, let us look again at the reaction between sodium and chlorine. When a sodium atom reacts with chlorine and loses 1 electron, the sodium is left with one more proton than electrons. This imbalance produces a positively charged sodium ion (Na^+). Chlorine, on the other hand, has gained 1 electron and now has one more electron than protons, producing a negatively charged ion (Cl^-). Positively charged ions are called **cations,** and negatively charged ions are called **anions.** (A good mnemonic device is to think of the "t" in cation as a plus [+] sign and the first "n" in anion as a negative [–] sign.) Substances such as salts, acids, and bases that release ions when dissolved in water are termed **electrolytes** (because their charges enable them to conduct an electrical current). Because it is a general rule that particles of like charge repel each other and those of opposite charge attract each other, we can expect ions to interact with other ions and polar molecules. Such interactions are important in many cellular chemical reactions, in the formation of solutions, and in the reactions microorganisms have with dyes. The transfer of electrons from one molecule to another represents a significant mechanism by which biological systems store and release energy, but it can also result in the formation of products that are damaging to cells.

Hydrogen Bonding Some types of bonding do not involve sharing, losing, or gaining electrons but instead are due simply to attraction between nearby molecules or atoms. One such bond is a

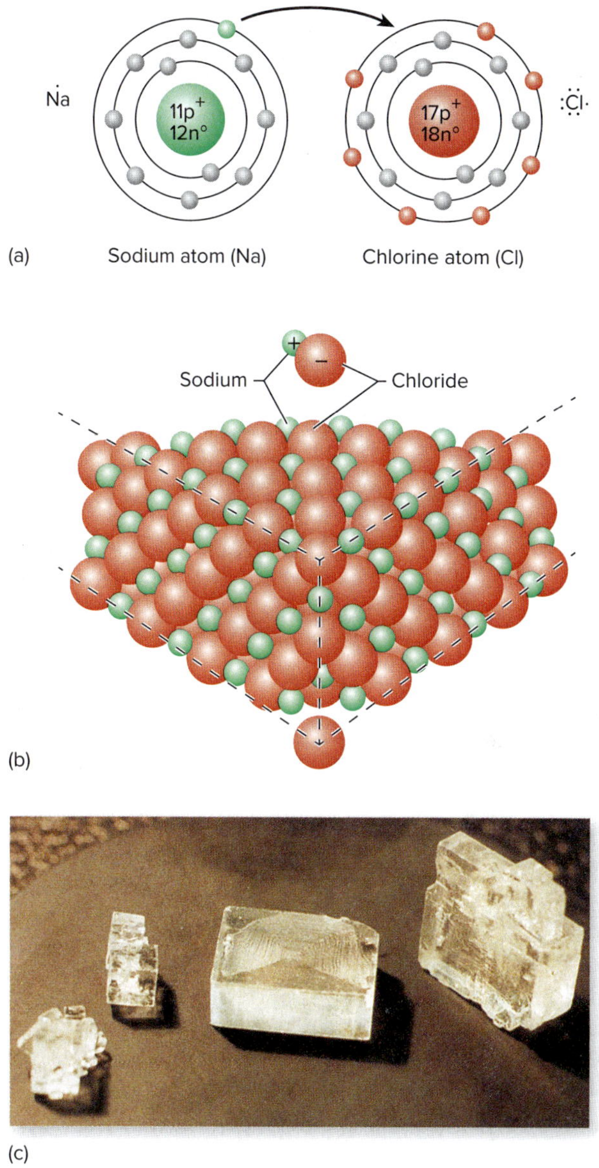
Na
11p+
12n°
17p+
18n°
:Cl·
(a)
Sodium atom (Na)
Chlorine atom (Cl)
Sodium
Chloride
(b)
(c)

Figure 2.7 Ionic bonding between sodium and chlorine.
(a) When the two elements are placed together, sodium loses its single outer orbital electron to chlorine, thereby filling chlorine's outer shell.
(b) Sodium and chloride ions form large molecules, or crystals, in which the two atoms alternate in a definite, regular, geometric pattern.
(c) Note the cubic nature of NaCl crystals at the macroscopic level.
Kathy Park Talaro

2. In general, when a salt is formed, the ending of the name of the negatively charged ion is changed to ***-ide***.

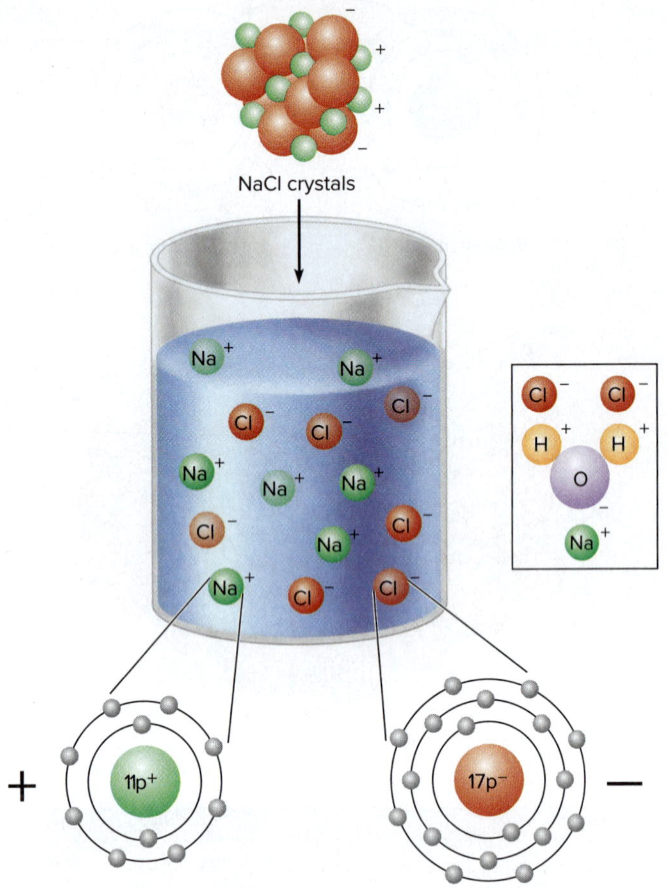
NaCl crystals
Na
Cl
H
O
11p+
17p−
+
−
Sodium ion (Na+)
(cation)
Chloride ion (Cl−)
(anion)

Figure 2.8 Ionization. When NaCl in the crystalline form is added to water, the ions are released from the crystal as separate charged particles (cations and anions) into solution. (See also figure 2.12.) In this solution, Cl^- ions are attracted to the hydrogen component of water, and Na^+ ions are attracted to the oxygen (box).

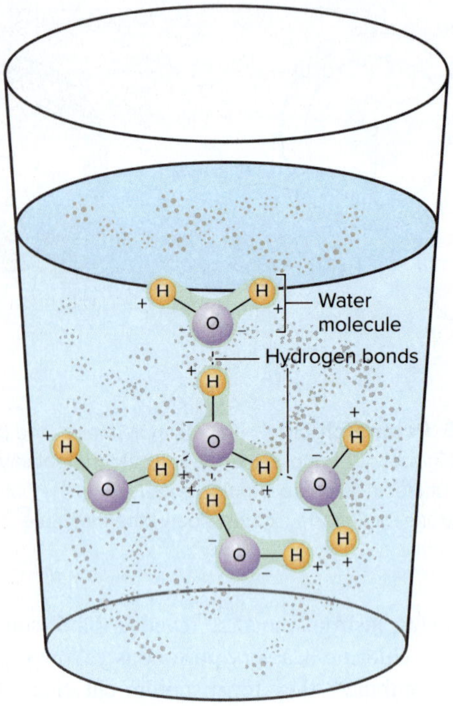
Water
molecule
Hydrogen bonds
H
O

Figure 2.9 Hydrogen bonding in water. Because of the polarity of water molecules, the mostly negatively charged oxygen end of one water molecule is weakly attracted to the mostly positively charged hydrogen end of an adjacent water molecule.

hydrogen bond, a weak type of bond that forms between a hydrogen covalently bonded to one molecule and an oxygen or nitrogen atom on the same molecule or on a different molecule. Because hydrogen in a covalent bond tends to be more positively charged, it will attract a nearby negatively charged atom and form an easily disrupted bridge with it. This type of bonding is usually represented in drawn molecular models with a dotted line. A simple example of hydrogen bonding occurs between water molecules **(figure 2.9).** More extensive hydrogen bonding is partly responsible for the structure and stability of proteins and nucleic acids, as you will see in section 2.2.

Other similar noncovalent associations between molecules are the **van der Waals forces.** These weak attractions occur between molecules that demonstrate low levels of polarity. Neighboring groups with only slight attractions will approach each other and remain associated. These forces are an essential factor in maintaining the cohesiveness of large molecules with many packed atoms.

Even though these two types of bonds, hydrogen bonds and van der Waals forces, are relatively weak on their own, they provide great stability to molecules because there are often many of them in one area. The weakness of each individual bond also provides flexibility, allowing molecules to change their shapes and also to bind and unbind to other objects relatively easily. The fundamental processes of life involve bonding and unbonding (for example, the DNA helix has to "unbond" or unwind in order for replication to occur, and hydrogen bonds and van der Waals forces are custom made for doing just that).

Chemical Shorthand: Formulas, Models, and Equations

The atomic content of molecules can be represented by a few convenient formulas. We have already been using the molecular formula, which concisely gives the atomic symbols and the number of the elements involved in subscript (CO_2, H_2O). More complex molecules such as glucose ($C_6H_{12}O_6$) can also be symbolized this way, but this way of writing it is not unique because fructose and galactose also share it. Molecular formulas are useful, but they only summarize the atoms in a compound; they do not show the position of bonds between atoms. For this purpose, chemists use structural formulas illustrating the relationships of the atoms and the number and types of bonds **(figure 2.10).** Other structural models present the three-dimensional appearance of a molecule, illustrating the orientation of atoms (differentiated by color) and the molecule's overall shape **(figure 2.11).** These are often called space-filling models, as you can get an idea of how the molecule actually occupies its space. The spheres surrounding each atom indicate how far the atom's influence can be felt, let us say. Sometimes it is also referred to as the atom's volume.

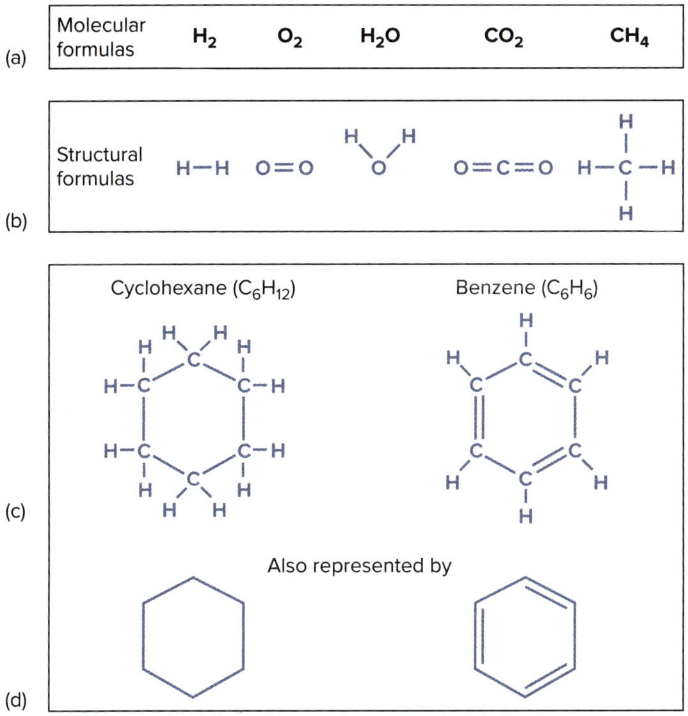
(a)
Molecular formulas
H2 O2 H2O CO2 CH4
(b)
Structural formulas
H—H O=O O=C=O
(c)
Cyclohexane (C6H12)
Benzene (C6H6)
Also represented by
(d)

Figure 2.10 Comparison of molecular and structural formulas. **(a)** Molecular formulas provide a brief summary of the elements in a compound. **(b)** Structural formulas clarify the exact relationships of the atoms in the molecule, depicting single bonds by a single line and double bonds by two lines. **(c)** In structural formulas of organic compounds, cyclic or ringed compounds may be completely labeled, or **(d)** they may be presented in a shorthand form in which carbons are assumed to be at the angles and attached to hydrogens. See figure 2.15 for structural formulas of three sugars with the same molecular formula, $C_6H_{12}O_6$.

O
H H
(a)
O C O
(b)
(c)

Figure 2.11 Three-dimensional, or space-filling, models of (a) water, (b) carbon dioxide, and (c) glucose. The red atoms are oxygen, the white ones hydrogen, and the black ones carbon.
John W. Hole

As you know, molecules are capable of changing through chemical reactions. To describe these changes, chemists use shorthand equations containing symbols, numbers, and arrows to simplify or summarize the major characteristics of a reaction. Molecules entering or starting a reaction are called **reactants,** and substances left after a reaction are called **products.** In most instances, written chemical reactions do not give the details of the exchange, in order to keep the expression simple and to save space.

In a *synthesis reaction,* the reactants bond together in a manner that produces an entirely new molecule (reactant A plus reactant B yields product AB). An example is the production of sulfur dioxide, a by-product of burning sulfur fuels and an important component of smog:

$$S + O_2 \rightarrow SO_2$$

Some synthesis reactions are not such simple combinations. When water is synthesized, for example, the reaction does not really involve one oxygen atom combining with two hydrogen atoms because elemental oxygen exists as O_2 and elemental hydrogen exists as H_2. A more accurate equation for this reaction is

$$2H_2 + O_2 \rightarrow 2H_2O$$

The equation for reactions has to be balanced—that is, the number of atoms on one side of the arrow must equal the number on the other side to reflect all of the participants in the reaction. To arrive at the total number of atoms in the reaction, multiply the prefix number by the subscript number. If no number is given, it is assumed to be 1.

In *decomposition reactions,* the bonds on a single reactant molecule are permanently broken to release two or more product molecules. A simple example can be shown for the common chemical hydrogen peroxide:

$$2H_2O_2 \rightarrow 2H_2O + O_2$$

During *exchange reactions,* the reactants trade portions with each other and release products that are combinations of the two. This type of reaction occurs between acids and bases when they form water and a salt:

$$AB + XY \rightleftharpoons AX + BY$$

The reactions in biological systems can be reversible, meaning that reactants and products can be converted back and forth. These reversible reactions are symbolized with a double arrow, each pointing in opposite directions, as in the preceding exchange reaction. Whether a reaction is reversible depends on the proportions of these compounds, the difference in energy state of the reactants and products, and the presence of **catalysts** (substances that increase the rate of a reaction). Additional reactants coming from another reaction can also be indicated by arrows that enter or leave at the main arrow:

$$\overset{CD \quad C}{X + XY \longrightarrow XYD}$$

Solutions: Homogeneous Mixtures of Molecules

A **solution** is a mixture of one or more substances called **solutes** uniformly dispersed in a dissolving medium called a **solvent.** An important characteristic of a solution is that the solute cannot be separated by ordinary settling (we call this "homogeneous"). The solute can be gaseous, liquid, or solid, and the solvent is usually a liquid. Examples of solutions are salt or sugar dissolved in water, and iodine dissolved in alcohol. In general, a solvent will dissolve a solute only if it has similar electrical characteristics as indicated by the rule of solubility, expressed simply as "like dissolves like." For example, water is a polar molecule and will readily dissolve an ionic solute such as NaCl, yet a nonpolar solvent such as benzene will not dissolve NaCl.

Water is the most common solvent in natural systems, having several characteristics that suit it to this role. The polarity of the water molecule causes it to form hydrogen bonds with other water molecules, but it can also interact readily with charged or polar molecules. When an ionic solute such as NaCl crystals is added to water, it is dissolved, thereby releasing Na^+ and Cl^- into solution. Dissolution occurs because Na^+ is attracted to the negative pole of the water molecule and Cl^- is attracted to the positive pole. For this reason, they are drawn away from the crystal separately into solution. As it leaves, each ion becomes **hydrated,** which means that it is surrounded by a sphere of water molecules **(figure 2.12).** Molecules such as salt or sugar that attract water to their surface are termed **hydrophilic.** Nonpolar molecules, such as benzene, that repel water are considered **hydrophobic.** A third class of molecules, which includes the phospholipids in cell membranes, is considered **amphipathic** because these molecules have both hydrophilic and hydrophobic properties.

The **concentration** of a solution expresses the amount of solute dissolved in a certain amount of solvent. It can be calculated by weight, volume, or percentage. A common way to calculate percentage of concentration is to use the weight of the solute, measured in grams (g), dissolved in a specified volume of solvent, measured in milliliters (mL). For example, dissolving 3 g of NaCl in 100 mL of water produces a 3% solution; dissolving 30 g in 100 mL produces a 30% solution; and dissolving 3 g in 1,000 mL (1 liter) produces a 0.3% solution.

A common way to express concentration of biological solutions is by its molar concentration, or *molarity* (M). A standard molar solution is obtained by dissolving one *mole,* defined as the molecular weight of the compound in grams, in 1 liter (1,000 mL) of solution. To make a 1 M solution of sodium chloride, we would dissolve 58 g of NaCl in 1 liter of solvent. A 0.1 M solution would require dissolving 5.8 g of NaCl in 1 liter of solvent.

Acidity, Alkalinity, and the pH Scale

Another factor with far-reaching impact on living things is their environment's relative degree of acidity or basicity (also called alkalinity). To understand how solutions become acidic or basic, let's look again at the behavior of water molecules. Hydrogens and oxygens tend to remain bonded by covalent bonds, but in certain instances, a single hydrogen can break away in its ionic form (H^+), leaving the remainder of the molecule in the form of an OH^- ion. The H^+ ion is positively charged because it is essentially a hydrogen ion that has lost its electron. The OH^- is negatively charged because it remains in possession of that electron. Ionization of water is constantly occurring, but in pure water containing no other ions, H^+ and OH^- are produced in equal amounts, and the solution remains neutral. By one definition, a solution is considered **acidic** when a component dissolved in water (acid) releases excess hydrogen ions[3] (H^+). A solution is **basic** when a component releases excess hydroxyl ions (OH^-) so that there is no longer a balance between the two ions.

To measure the acid and base concentrations of solutions, scientists use the **pH** scale, a graduated numerical scale that ranges from 0 (the most acidic) to 14 (the most basic). This scale is a useful standard for rating relative acidity and basicity. Use **figure 2.13** to familiarize yourself with the pH readings of some common substances. Because the pH scale is a logarithmic scale, each increment (from pH 2.0 to pH 3.0) represents a 10-fold change in concentration of ions. (Take a moment to glance at Appendix B to review logarithms and exponents.)

3. Actually, it forms a hydronium ion (H_3O^+), but for simplicity's sake, we will use the notation of H^+.

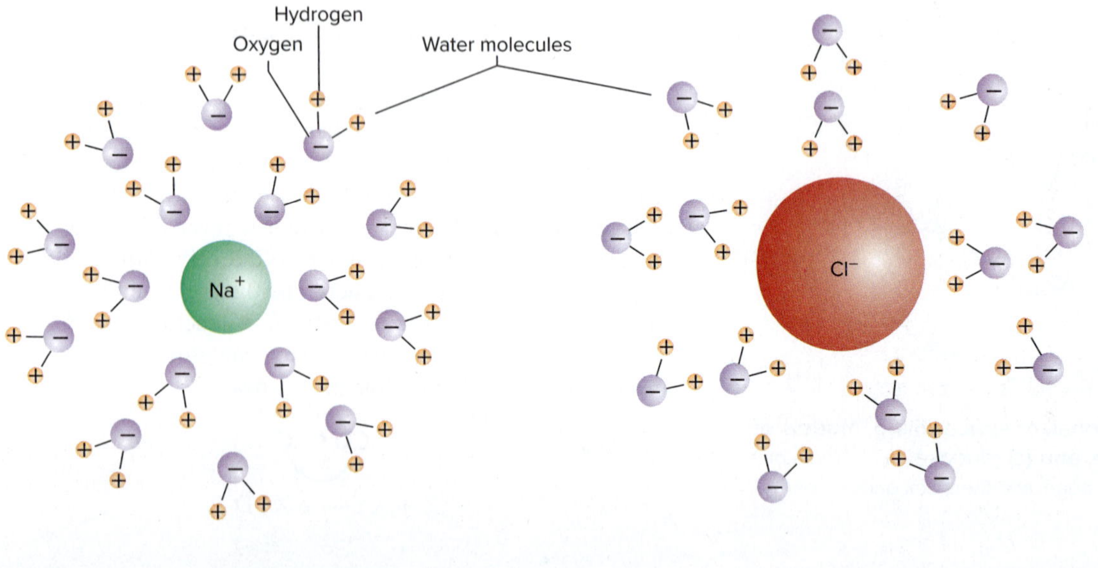
Hydrogen
Oxygen
Water molecules
Na+
Cl−

Figure 2.12 Hydration spheres formed around ions in solution. In this example, a sodium cation attracts the negatively charged region of water molecules, and a chloride anion attracts the positively charged region of water molecules. In both cases, the ions become covered with spherical layers of specific numbers and arrangements of water molecules.

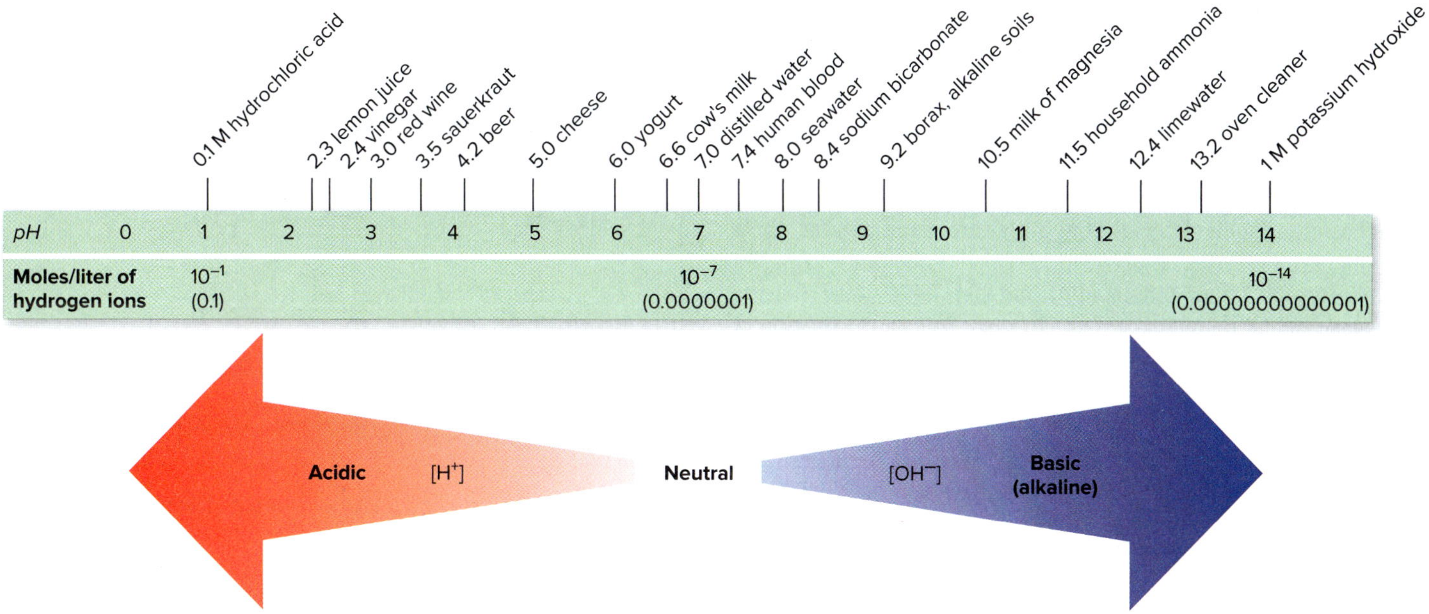
0.1 M hydrochloric acid
2.3 lemon juice
2.4 vinegar
3.0 red wine
3.5 sauerkraut
4.2 beer
5.0 cheese
6.0 yogurt
6.6 cow's milk
7.0 distilled water
7.4 human blood
8.0 seawater
8.4 sodium bicarbonate
9.2 borax, alkaline soils
10.5 milk of magnesia
11.5 household ammonia
12.4 limewater
13.2 oven cleaner
1 M potassium hydroxide
pH 0 1 2 3 4 5 6 7 8 9 10 11 12 13 14
Moles/liter of hydrogen ions
10⁻¹ (0.1)
10⁻⁷ (0.0000001)
10⁻¹⁴ (0.00000000000001)
Acidic
[H⁺]
Neutral
[OH⁻]
Basic (alkaline)

Figure 2.13 The pH scale. Shown are the relative degrees of acidity and basicity and the approximate pH readings for various substances.

Disease Connection

Although many bacteria living in or on humans require a neutral pH (around 7.0) to thrive, the bacterium responsible for gastric ulcers, *Helicobacter pylori,* is a notable exception. It lives in the stomach, which has an acidic pH of around 2. It counters this hostile environment by secreting an enzyme called urease, which cleaves the common chemical urea into carbon dioxide and ammonia. The ammonia neutralizes stomach acid, allowing the bacterium to grow and thrive.

More precisely, the pH is based on the negative logarithm of the concentration of H^+ ions (symbolized as $[H^+]$) in a solution, represented as

$$pH = -\log[H^+]$$

The quantity is expressed in moles per liter. Recall that a mole is simply a standard unit of measurement and refers to the amount of substance containing 6×10^{23} atoms.

Acidic solutions have a greater concentration of H^+ than OH^-, starting with pH 0, which contains 1.0 mole H^+/liter. Each of the subsequent whole-number readings in the scale changes in $[H^+]$ by a 10-fold reduction, so that pH 1 contains [0.1 mole H^+/liter], pH 2 contains [0.01 mole H^+/liter], and so on, continuing in the same manner up to pH 14, which contains [0.00000000000001 mole H^+/liter]. These same concentrations can be represented more manageably by exponents: pH 2 has an $[H^+]$ of 10^{-2} mole, and pH 14 has an $[H^+]$ of 10^{-14} mole **(table 2.2).** The pH units are derived from the exponent itself. Even though the basis for the pH scale is $[H^+]$, it is important to note that, as the $[H^+]$ in a solution decreases, the $[OH^-]$ increases in direct proportion. At

Table 2.2 Hydrogen Ion and Hydroxide Ion Concentrations at a Given pH

Moles per Liter of Hydrogen Ions	Logarithm	pH	Moles per Liter of OH^-
1.0	10^{-0}	0	10^{-14}
0.1	10^{-1}	1	10^{-13}
0.01	10^{-2}	2	10^{-12}
0.001	10^{-3}	3	10^{-11}
0.0001	10^{-4}	4	10^{-10}
0.00001	10^{-5}	5	10^{-9}
0.000001	10^{-6}	6	10^{-8}
0.0000001	10^{-7}	7	10^{-7}
0.00000001	10^{-8}	8	10^{-6}
0.000000001	10^{-9}	9	10^{-5}
0.0000000001	10^{-10}	10	10^{-4}
0.00000000001	10^{-11}	11	10^{-3}
0.000000000001	10^{-12}	12	10^{-2}
0.0000000000001	10^{-13}	13	10^{-1}
0.00000000000001	10^{-14}	14	10^{-0}

midpoint—pH 7, or neutrality—the concentrations are exactly equal and neither predominates, this being the pH of pure water previously mentioned.

In summary, the pH scale can be used to rate or determine the degree of acidity or basicity of a solution. On this scale, a pH below 7 is acidic, and the lower the pH, the greater the acidity. A pH above 7 is basic, and the higher the pH, the greater the basicity. Incidentally, although pHs are given here in whole numbers, more often, a pH reading exists in decimal form, for example, pH 4.5 or 6.8 (acidic) and pH 7.4 or 10.2 (basic). Because of the damaging effects of very concentrated acids or bases, most cells operate best under neutral, weakly acidic, or weakly basic conditions.

Aqueous solutions containing both acids and bases may be involved in **neutralization** reactions, which give rise to water and other neutral by-products. For example, when equal molar solutions of hydrochloric acid (HCl) and sodium hydroxide (NaOH, a base) are mixed, the reaction proceeds as follows:

$$HCl + NaOH \rightarrow H_2O + NaCl$$

Here the acid and base ionize to H^+ and OH^- ions, which form water, and other ions, Na^+ and Cl^-, which form sodium chloride. Any product other than water that arises when acids and bases react is called a salt. Many of the organic acids that function in **metabolism** are available as the acid and the salt form, depending on the conditions in the cell.

The Chemistry of Carbon and Organic Compounds

So far, our main focus has been on the characteristics of atoms, ions, and small, simple substances in living things. These substances are often lumped together in a category called **inorganic chemicals.** A chemical is called inorganic if it does not contain both carbon and hydrogen. Examples of inorganic chemicals include NaCl (sodium chloride), $Mg_3(PO_4)_2$ (magnesium phosphate), $CaCO_3$ (calcium carbonate), and CO_2 (carbon dioxide). In reality, however, most of the chemical reactions and structures of living things involve more complex molecules, termed **organic chemicals.** These are carbon compounds with a basic framework of the element carbon bonded to other atoms. Organic molecules vary in complexity. The simplest, methane (CH_4; see figure 2.5*c*), has a molecular weight of 16, and certain antibody molecules (part of our immune systems) have a molecular weight of nearly 1,000,000 and are among the most complex molecules on earth.

The role of carbon as the fundamental element of life can best be understood if we look at its chemistry and bonding patterns. The valence of carbon makes it an ideal atomic building block to form the backbone of organic molecules. It has 4 electrons in its outer orbital to be shared with other atoms (including other carbons) through covalent bonding. As a result, it can form stable chains containing thousands of carbon atoms and still has bonding sites available for forming covalent bonds with other atoms. The bonds that carbon forms are linear, branched, or ringed; and it can form four single bonds, two double bonds, or one triple bond **(figure 2.14).** The atoms with which carbon is most often associated in organic compounds are hydrogen, oxygen, nitrogen, sulfur, and phosphorus.

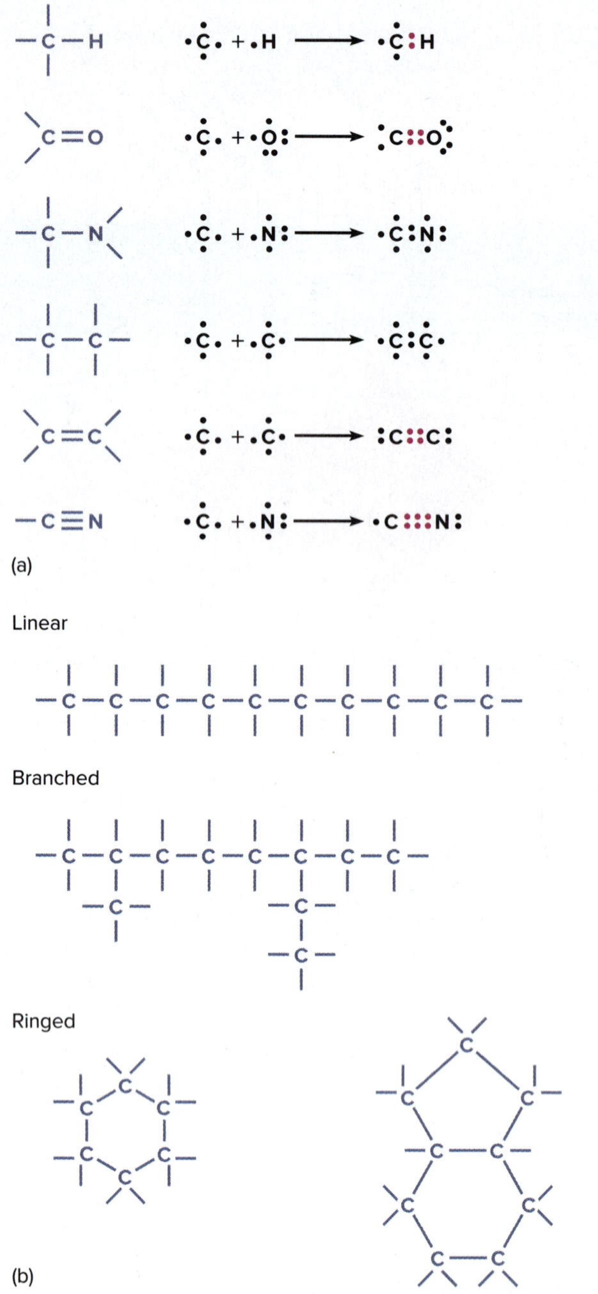
(a)
Linear
Branched
Ringed
(b)

Figure 2.14 The versatility of bonding in carbon. In most compounds, each carbon makes a total of four bonds. **(a)** Both single and double bonds can be made with other carbons, oxygen, and nitrogen; single bonds are made with hydrogen. Simple electron models show how the electrons are shared in these bonds. **(b)** Multiple bonding of carbons can give rise to long chains, branched compounds, and ringed compounds, many of which are extraordinarily large and complex.

Functional Groups of Organic Compounds

One important advantage of carbon's serving as the molecular skeleton for living things is that it is free to bind with an unending

Table 2.3 Representative Functional Groups and Classes of Organic Compounds

Formula of Functional Group	Name	Class of Compounds
	Hydroxyl	Alcohols, carbohydrates
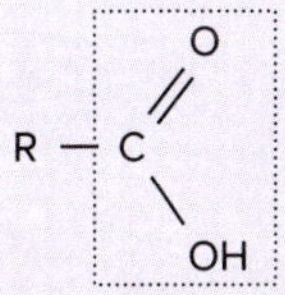	Carboxyl	Fatty acids, proteins, organic acids
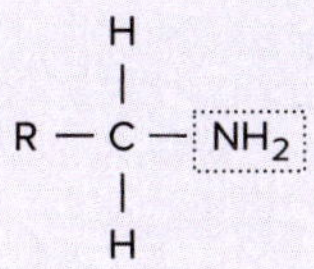	Amino	Proteins, nucleic acids
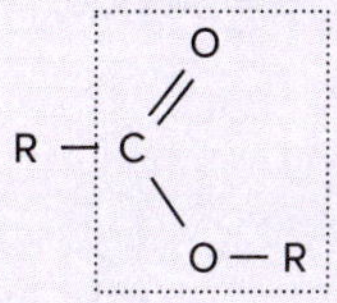	Ester	Lipids
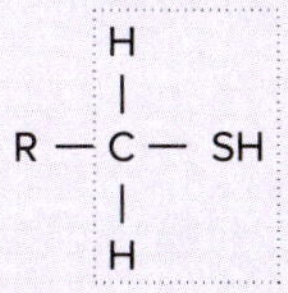	Sulfhydryl	Cysteine (amino acid), proteins
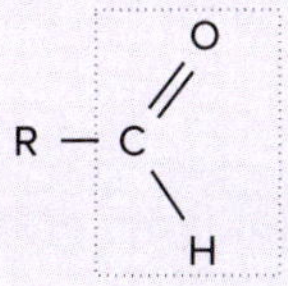	Carbonyl, terminal end	Aldehydes, polysaccharides
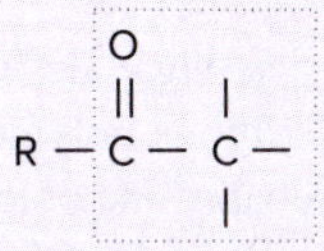	Carbonyl, internal	Ketones, polysaccharides
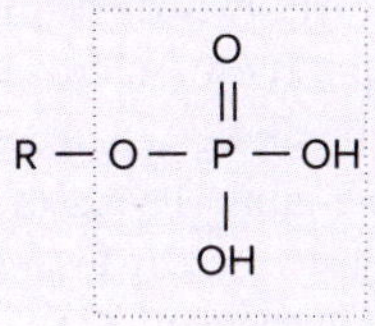	Phosphate	DNA, RNA, ATP

Note: The R designation on a molecule is shorthand for *residue,* and it indicates that what is attached at that site varies from one compound to another.

array of other molecules. These special molecular groups or accessory molecules that bind to organic compounds are called **functional groups.** Functional groups help define the chemical class of organic compounds and confer unique reactive properties on the whole molecule **(table 2.3).** Because each type of functional group behaves in a distinctive manner, reactions of an organic compound can be predicted by knowing the kind of functional group or groups it carries. Many reactions rely upon functional groups such as R—OH or R—NH_2. The R— designation on a molecule is shorthand for *residue,* and its placement in a formula indicates that the residue (functional group) varies from one compound to another.

2.1 Learning Outcomes—Assess Your Progress

1. Explain the relationship between atoms and elements.
2. List and define four types of chemical bonds.
3. Differentiate between a solute and a solvent.
4. Provide a brief definition of pH.

2.2 Macromolecules: Superstructures of Life

The chemical compounds of life are studied in the field of **biochemistry.** Biochemicals are organic compounds produced by (or components of) living things, and they include four main families: carbohydrates, lipids, proteins, and nucleic acids **(table 2.4).** The compounds in these groups are assembled from smaller molecular subunits, or building blocks, and because they are often very large compounds, they are termed **macromolecules.** All macromolecules except lipids are formed by polymerization, a process in which repeating subunits termed **monomers** are bound into chains of various lengths termed **polymers.** For example, proteins (polymers) are composed of a chain of amino acids (monomers). The large size and complex, three-dimensional shape of macromolecules enable them to function as structural components, molecular messengers, energy sources, enzymes (biochemical catalysts), nutrient stores, and sources of genetic information. In section 2.3 and in later chapters, we consider numerous concepts relating to the roles of macromolecules in cells. Table 2.4 will also be a useful reference when you study metabolism in chapter 9.

Carbohydrates: Sugars and Polysaccharides

Carbohydrates are combinations of carbon (carbo-) and water (-hydrate). Carbohydrates can be generally represented by the formula $(CH_2O)_n$, in which n indicates the number of units of this combination of atoms **(figure 2.15*a*).** Some carbohydrates contain additional atoms of sulfur or nitrogen.

There are a wide variety of different kinds of carbohydrates. The common term *sugar* **(saccharide)** refers to a simple carbohydrate such as a monosaccharide or a disaccharide. A **monosaccharide** is a simple sugar, the most basic form of carbohydrate. A monosaccharide can contain anywhere from 3 to 7 carbons. A **disaccharide** is a combination of two monosaccharides, and a **polysaccharide** is a polymer of five or more monosaccharides bound in linear or branched-chain patterns **(figure 2.15*b*).** For example, **hexoses** are composed of 6 carbons, and **pentoses** contain 5 carbons. **Glucose** (Gr. *glyko,* sweet) is the most common and universally important hexose; **fructose** is named for fruit (one place where it is found); and xylose, a pentose, derives its name from the Greek word for wood. Disaccharides are named similarly: **Lactose** (L. *lacteus,* milk) is an important component of milk; **maltose** means malt sugar; and **sucrose** (Fr. *sucre,* sugar) is common table sugar.

The Nature of Carbohydrate Bonds

The subunits of disaccharides and polysaccharides are linked by means of **glycosidic bonds.** In these bonds, carbons on adjacent sugar units are bonded to the same oxygen atom like links in

Table 2.4 Macromolecules and Their Functions

Macromolecule	Description/Basic Structure	Examples	Notes About the Examples
Carbohydrates			
Monosaccharides	3- to 7-carbon sugars	Glucose, fructose	Sugars involved in metabolic reactions; building block of disaccharides and polysaccharides
Disaccharides	2 monosaccharides	Maltose (malt sugar)	Composed of two glucoses; an important breakdown product of starch
		Lactose (milk sugar)	Composed of glucose and galactose
		Sucrose (table sugar)	Composed of glucose and fructose
Polysaccharides	Chains of monosaccharides	Starch, cellulose, glycogen	Cell wall, food storage
Lipids			
Triglycerides	Fatty acids + glycerol	Fats, oils	
Phospholipids	Fatty acids + glycerol + phosphate	Membrane components	Major component of cell membranes; storage
Waxes	Fatty acids, alcohols	Mycolic acid	Cell wall of mycobacteria
Steroids	Ringed structure	Cholesterol, ergosterol	In membranes of eukaryotes and some bacteria
Proteins			
	Amino acids	Enzymes, part of cell membrane, cell wall, ribosomes, antibodies	Serve as structural components and perform metabolic reactions
Nucleic Acids			
	Nucleotides Purines: adenine, guanine Pyrimidines: cytosine, thymine, uracil		
Deoxyribonucleic acid (DNA)	Contains deoxyribose sugar and thymine, not uracil	Chromosomes; genetic material of viruses	Determine inheritance
Ribonucleic acid (RNA)	Contains ribose sugar and uracil, not thymine	Ribosomes; mRNA, tRNA, regulatory RNAs	Facilitate expression of genetic traits

a chain **(figure 2.16).** (It's important to remember here that carbons in a molecule are assigned numbers.) As an example of how glycosidic bonds work, maltose is formed when the number 1 carbon on a glucose bonds to the oxygen on the number 4 carbon on a second glucose. Sucrose is formed when glucose and fructose bind oxygen between their number 1 and number 2 carbons; and lactose is formed when glucose and galactose connect by their number 1 and number 4 carbons. In order to form this bond, 1 carbon gives up its OH group and the other (the one contributing the oxygen to the bond) loses the H from its OH group. Because a water molecule is the product of the lost compounds, this reaction is known as **dehydration synthesis**—"de-hydration" referring to the loss of a water molecule. Three polysaccharides (starch, cellulose, and glycogen) are structurally and biochemically distinct, even though all are polymers of the same monosaccharide—glucose. The basis for their differences lies primarily in the exact way the glucoses are bound together, which greatly affects the characteristics of the end product **(figure 2.17).** The synthesis and breakage of each type of bond require a specialized catalyst called an enzyme (chapter 10).

The Functions of Polysaccharides

In cells and tissues, polysaccharides typically provide structural support and protection and also serve as nutrient and energy stores. The cell walls in plants and many microscopic algae derive their strength and rigidity from **cellulose,** a long, fibrous polymer **(figure 2.17*a*).** Because of this role, cellulose is probably one of the most common organic substances on the earth, yet it is digestible only by certain bacteria, fungi, and protozoa. These microbes, called *decomposers,* play an essential role in breaking down and recycling plant materials. Some bacteria secrete slime layers of a glucose polymer called *dextran.* This substance causes a sticky layer to develop on teeth, which leads to plaque, described in chapter 23.

Other structural polysaccharides can be conjugated (chemically bonded) to amino acids, nitrogen bases, lipids, or proteins. **Agar,** an indispensable polysaccharide in preparing solid culture media, is a natural component of certain seaweeds. It is a complex polymer of galactose and sulfur-containing carbohydrates. The exoskeletons of certain fungi contain **chitin** (ky′-tun), a polymer of glucosamine (a sugar with an amine functional group). **Peptidoglycan** (pep-tih-doh-gly′-kan) is one special class of compounds in which poly-saccharides (glycans) are linked to peptide fragments (a short chain of amino acids). This molecule provides the main source of structural support to bacterial cell walls. The cell wall of some bacteria also contains **lipopolysaccharide,** a complex of lipid and polysaccharide responsible for symptoms such as fever and shock (see section 13.2).

The outer surface of many cells has a "sugar coating" composed of polysaccharides bound in various ways to proteins (the combination is a glycoprotein). This structure, called the

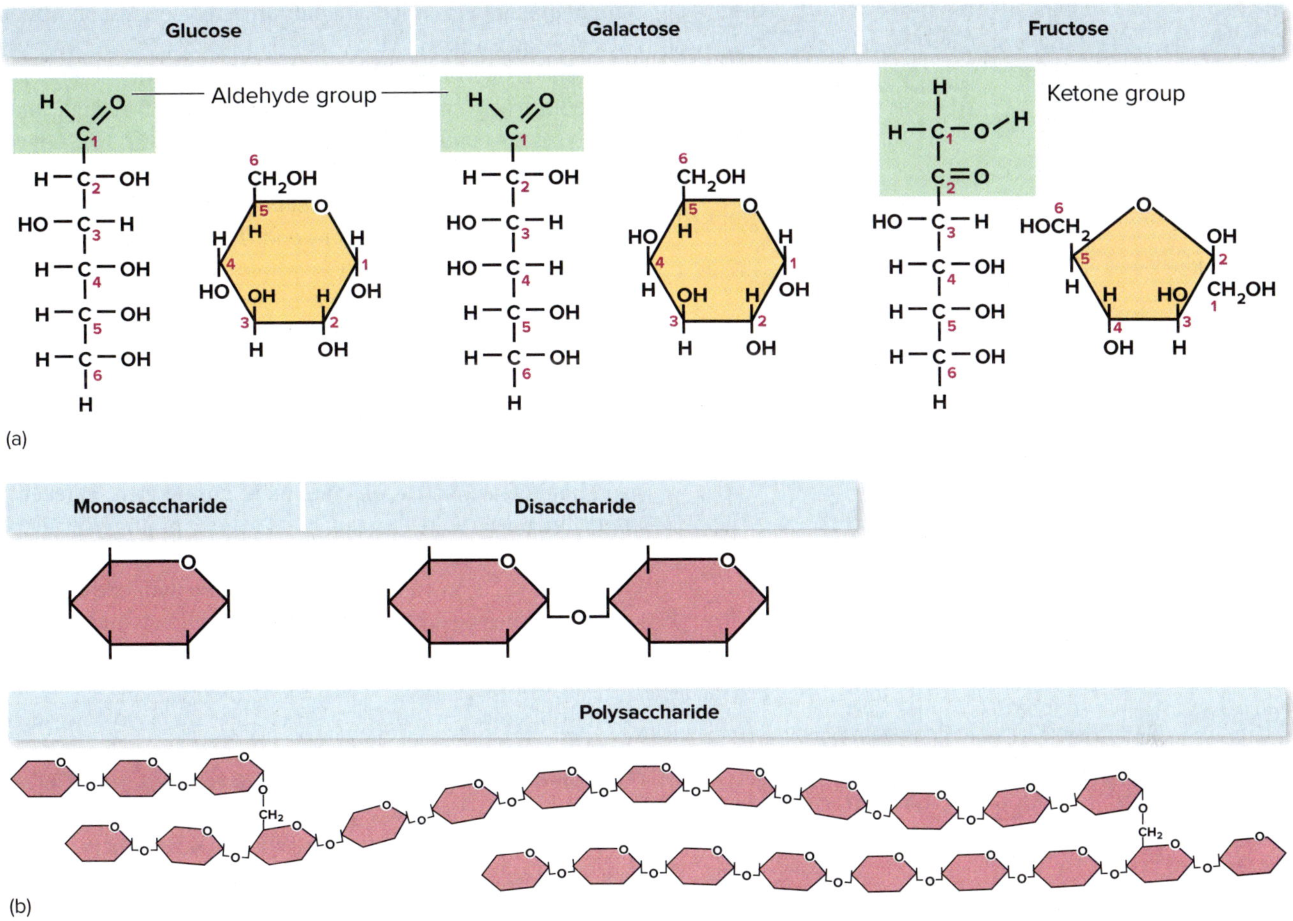
Glucose
Galactose
Fructose
Aldehyde group
Ketone group
(a)
Monosaccharide
Disaccharide
Polysaccharide
(b)

Figure 2.15 Common classes of carbohydrates. **(a)** Three hexoses with the same molecular formula and different structural formulas. Both linear and ring models are given. The linear form emphasizes aldehyde and ketone groups, although in solution the sugars exist in the ring form. Note that the carbons are numbered so as to keep track of reactions within and between monosaccharides. **(b)** Major saccharide groups, named for the number of sugar units each contains.

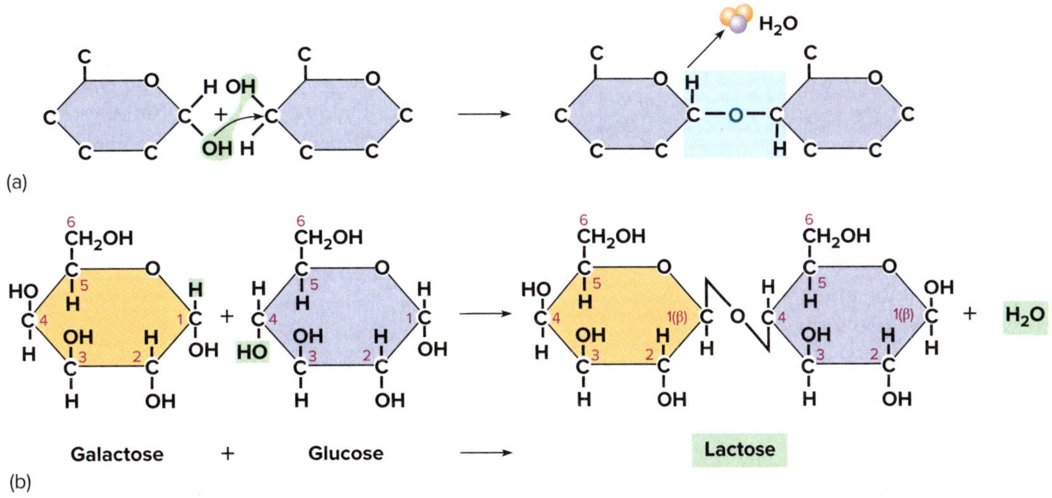
H₂O
(a)
Galactose
+
Glucose
Lactose
(b)

Figure 2.16 Glycosidic bond in a common disaccharide. **(a)** General scheme in the formation of a glycosidic bond by dehydration synthesis. **(b)** A 1,4 bond between a galactose and glucose produces lactose.

glycocalyx, functions in attachment to other cells or as a site for *receptors*—surface molecules that receive external stimuli or act as binding sites. Small glycoproteins account for the differences in human blood types, and *antibodies* are large protein and carbohydrate compounds. Viruses also have glycoproteins on their surface with which they bind to and invade their host cells.

Polysaccharides are often stored by cells in the form of glucose polymers such as **starch (figure 2.17*b*)** or **glycogen,**

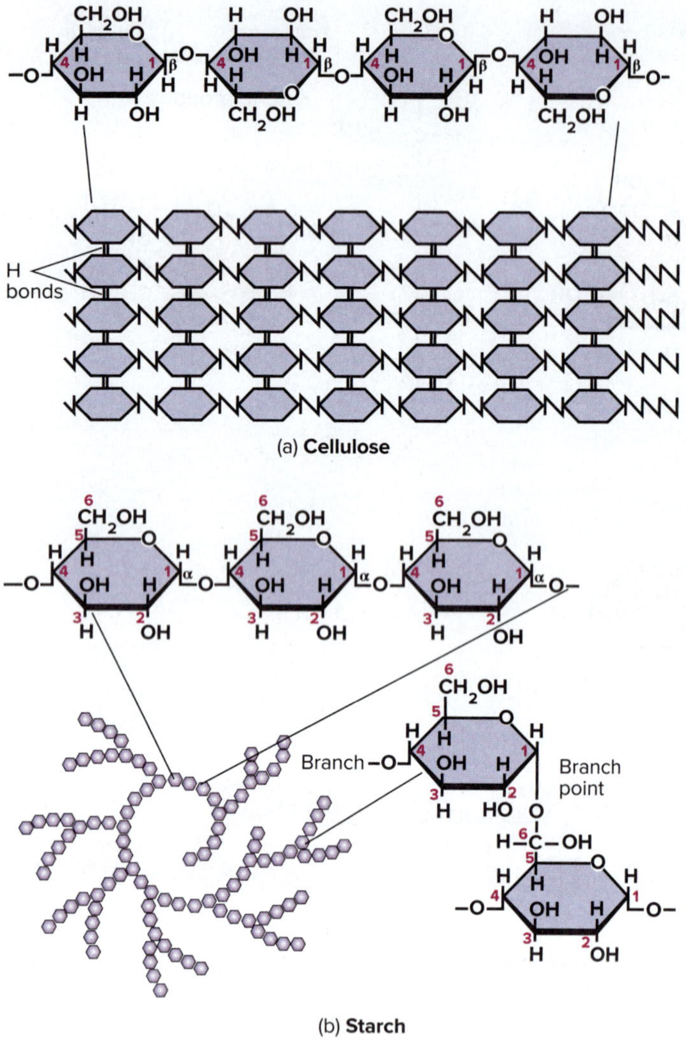
H bonds
(a) Cellulose
Branch
Branch point
(b) Starch

Figure 2.17 Polysaccharides. **(a)** Cellulose is composed of β glucose bonded in 1,4 bonds that produce lengthy, linear chains of polysaccharides that are H-bonded along their length. This is the typical structure of wood and cotton fibers. **(b)** Starch is also composed of glucose polymers, in this case α glucose. The main structure is amylose bonded in a 1,4 pattern, with side branches of amylopectin bonded by 1,6 bonds. The entire molecule is compact and granular.

but only organisms with the appropriate digestive enzymes can break them down and use them as a nutrient source. Because a water molecule is required for breaking the bond between two glucose molecules, digestion is also termed **hydrolysis.** Starch is the primary storage food of green plants, microscopic algae, and some fungi. Glycogen is a stored carbohydrate for animals and certain groups of bacteria and protozoa.

Lipids: Fats, Phospholipids, and Waxes

The term **lipid,** derived from the Greek word *lipos,* meaning fat, is an operational term for a variety of substances that are not soluble in polar solvents such as water (recall that oil and water do not mix). Lipids will dissolve in nonpolar solvents such as benzene and chloroform. The reason lipids will not dissolve in polar liquids is that they contain relatively long or complex C—H (hydrocarbon) chains that are nonpolar and thus hydrophobic. The main groups of compounds classified as lipids are triglycerides, phospholipids, steroids, and waxes.

The **triglycerides** are important storage lipids. Fats and oils are both triglycerides. They are composed of a single molecule of glycerol bound to three fatty acids **(figure 2.18). Glycerol** is a 3-carbon alcohol[4] with 3 OH groups that serve as binding sites, and fatty acids are long-chain hydrocarbon molecules with a carboxyl group (COOH) at one end. The hydrocarbon portion of a fatty acid can vary in length from 4 to 28 carbons; and, depending on the fat, it may be saturated or unsaturated. That means, if all carbons in the chain are single-bonded to 2 other carbons and 2 hydrogens, the fat is saturated. If there is at least one C═C double bond in the chain, it is unsaturated. The structure of fatty acids is what gives fats and oils (liquid fats) their greasy, insoluble nature. In general, solid fats (such as butter) are more saturated, and liquid fats (such as oils) are more unsaturated. In recent years there has been a realization that a type of triglyceride, called popularly *trans fat,* is harmful to the health of those who consume it. A trans fat is an unsaturated triglyceride with one or more of its fatty acids in a position (trans) that is not often found in nature, but it is a common occurrence in processed foods.

In most cells, triglycerides are stored in long-term concentrated form as droplets or globules. When they are acted on by digestive enzymes called lipases, the fatty acids and glycerol are freed to be used in metabolism. Fatty acids are a superior source of energy, yielding twice as much per gram as other storage molecules (starch). Soaps are potassium and sodium salts of fatty acids whose qualities make them excellent grease removers and cleaners.

Membrane Lipids

A class of lipids that serves as a major structural component of cell membranes is the **phospholipids.** Although phospholipids also contain glycerol and fatty acids, they have some significant differences from triglycerides. Phospholipids contain only two fatty acids attached to the glycerol, and the third glycerol binding site holds a phosphate group. The phosphate is in turn bonded to an alcohol, which varies from one phospholipid to another **(figure 2.19*a*).** These lipids have a hydrophilic region from the charge on the phosphoric acid–alcohol "head" of the molecule and a hydrophobic region that corresponds to the long, uncharged "tail" (formed by the fatty acids). When exposed to an aqueous solution, the charged heads are attracted to the water phase, and the nonpolar tails are repelled from the water phase **(figure 2.19*b*).** This property causes lipids to naturally assume single and double layers (bilayers), a fact that is critical to their biological significance in membranes. When two single layers of polar lipids come together to form a double layer, the outer hydrophilic face of each single layer will orient itself toward the solution, and the hydrophobic portions will become immersed in the core of the bilayer. The structure of lipid bilayers confers characteristics on membranes such as selective permeability and fluid nature.

Steroids and Waxes

Steroids are complex lipid compounds commonly found in cell membranes and animal hormones. The best known of these is the

4. Alcohols are carbon compounds containing OH groups.

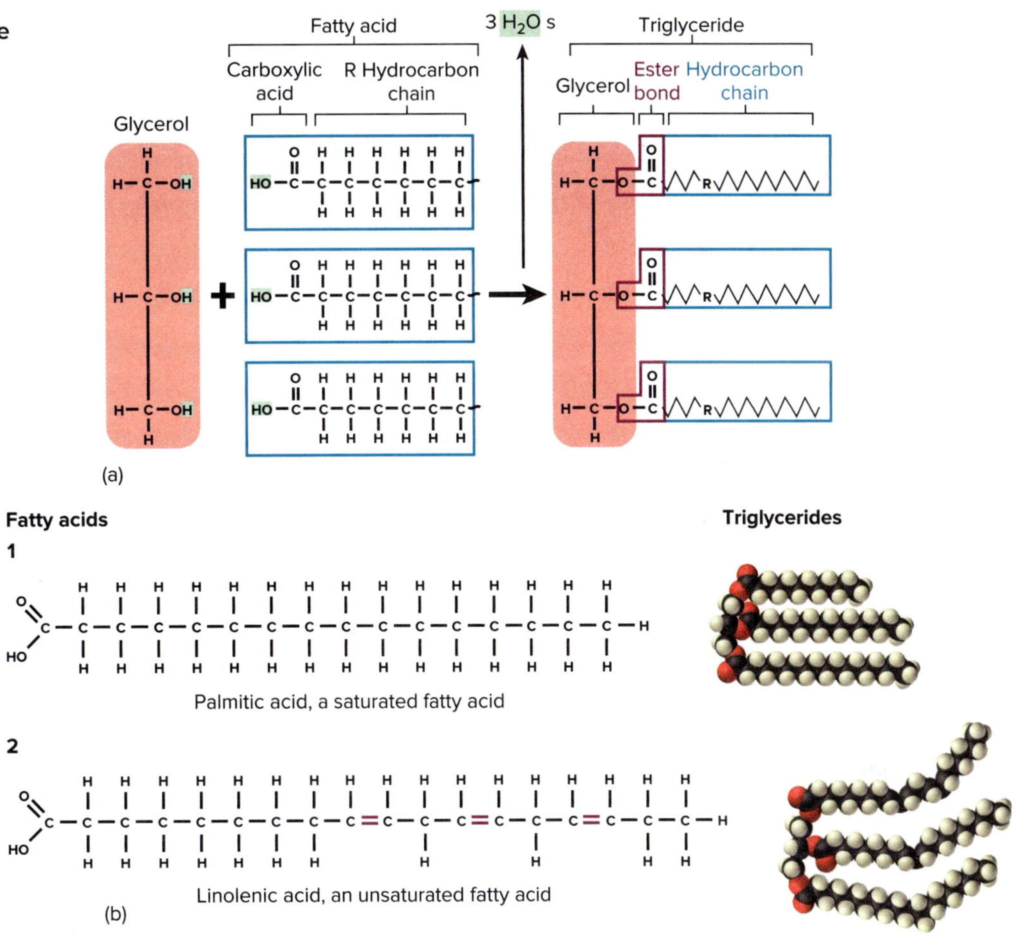
Fatty acid
3 H_2O s
Triglyceride
Carboxylic acid
R Hydrocarbon chain
Glycerol
Ester bond
Hydrocarbon chain
Glycerol
(a)
Fatty acids
Triglycerides
1
Palmitic acid, a saturated fatty acid
2
Linolenic acid, an unsaturated fatty acid
(b)

Figure 2.18 Synthesis and structure of a triglyceride. **(a)** Because a water molecule is released at each ester bond, this is another example of dehydration synthesis. The jagged lines and R symbol represent the hydrocarbon chains of the fatty acids, which are commonly very long. **(b)** Structural and three-dimensional models of fatty acids and triglycerides. (1) A saturated fatty acid has long, straight chains that readily pack together and form solid fats. (2) An unsaturated fatty acid—here a polyunsaturated one with 3 double bonds—has bends in the chain that prevent packing and produce oils (right).

sterol (meaning a steroid with an OH group) called **cholesterol (figure 2.20).** Cholesterol reinforces the structure of the cell membrane in animal cells and in an unusual group of cell-wall-deficient bacteria called the mycoplasmas. The cell membranes of fungi contain a sterol called ergosterol.

Chemically, a *wax* is formed between a long-chain alcohol and a saturated fatty compound acid. The resulting material is typically pliable and soft when warmed but hard and water resistant when cold. Think of candle wax, for example. Among living things, fur, feathers, fruits, leaves, human skin, and insect exoskeletons are naturally waterproofed with a coating of wax. Bacteria that cause tuberculosis and Hansen's disease (leprosy) produce a wax that repels ordinary laboratory stains and contributes to their damaging effects on the body.

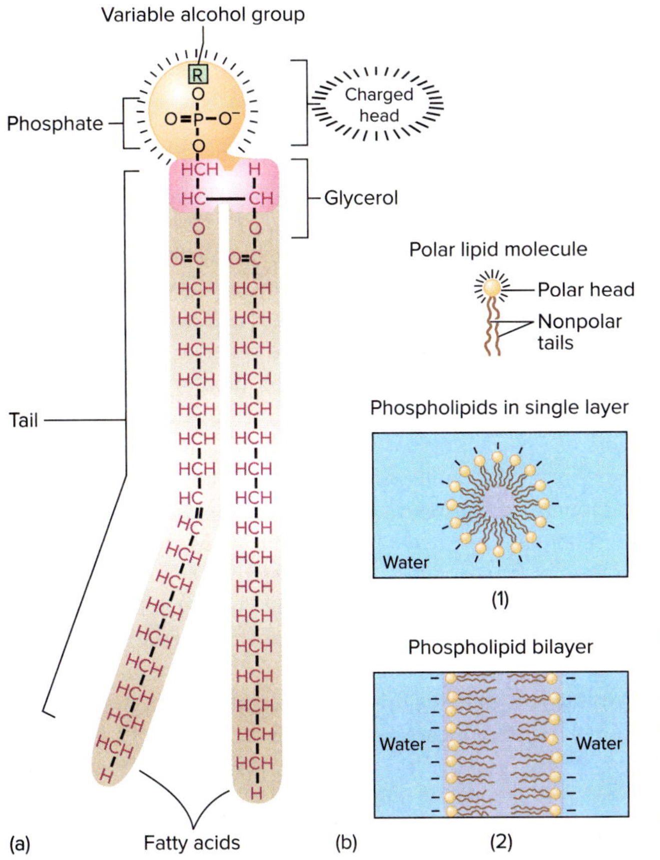
Variable alcohol group
Phosphate
Charged head
Glycerol
Tail
(a)
Fatty acids
Polar lipid molecule
Polar head
Nonpolar tails
Phospholipids in single layer
Water
(1)
Phospholipid bilayer
Water
Water
(2)
(b)

Figure 2.19 Phospholipids—membrane molecules. **(a)** A model of a single molecule of a phospholipid. The phosphate-alcohol head lends a charge to one end of the molecule. Its long, trailing hydrocarbon chain is uncharged. **(b)** The behavior of phospholipids in water-based solutions causes them to become arranged (1) in single layers that form circles called micelles, with the charged head oriented toward the water phase and the hydrophobic nonpolar tail buried away from the water phase, or (2) in double-layered phospholipid systems with the hydrophobic tails sandwiched between two hydrophilic layers.

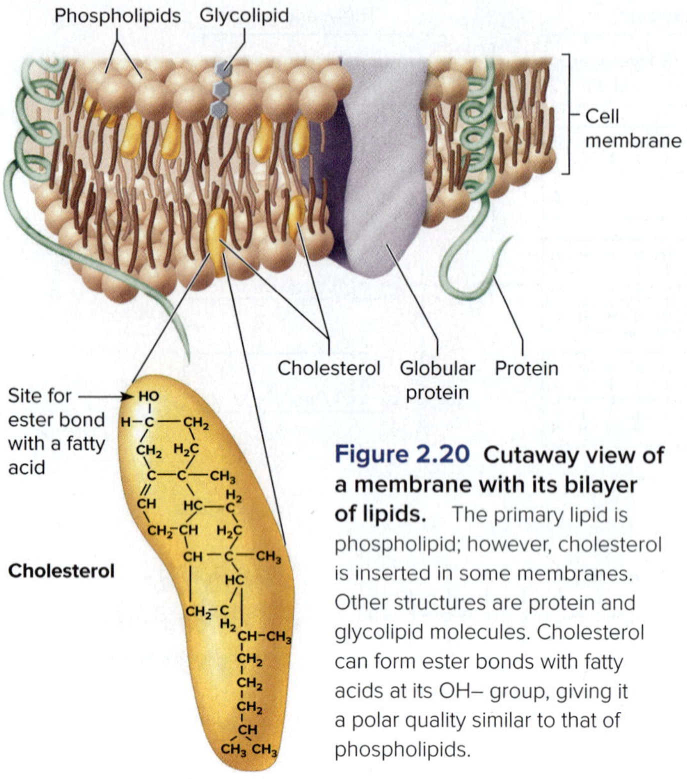
Phospholipids
Glycolipid
Cell membrane
Cholesterol
Globular protein
Protein
Site for ester bond with a fatty acid
Cholesterol

Figure 2.20 Cutaway view of a membrane with its bilayer of lipids. The primary lipid is phospholipid; however, cholesterol is inserted in some membranes. Other structures are protein and glycolipid molecules. Cholesterol can form ester bonds with fatty acids at its OH– group, giving it a polar quality similar to that of phospholipids.

Proteins: Shapers of Life

The predominant organic macromolecules in cells are **proteins.** To a large extent, the structure, behavior, and unique qualities of each living thing are a consequence of the proteins they contain. Understanding their structure will help explain a lot of things about cells. The building blocks of proteins are **amino acids,** which exist in 22 different naturally occurring forms, although 20 of these are the most common **(table 2.5).** Various combinations of these amino acids account for the nearly infinite variety of proteins. Amino acids have a basic skeleton consisting of a carbon (called the alpha, or α, carbon) linked to an amino group (NH_2), a carboxyl group (COOH), a hydrogen atom (H), and a variable R group. The variations among the amino acids are seen at the R group, which is different in each amino acid, and give the molecule its unique characteristics **(figure 2.21).** A covalent bond called a **peptide bond** forms between the amino group on one amino acid and the carboxyl group on another amino acid. As a result of peptide bond formation, it is possible to produce molecules varying in length from two amino acids to chains containing thousands of them.

Various terms are used in reference to proteins. **Peptide** usually refers to a molecule composed of short chains of amino acids, such as a dipeptide (two amino acids), a tripeptide (three), and a tetrapeptide (four). A **polypeptide** contains an unspecified number of amino acids but usually has more than 20 and is often a smaller subunit of a protein. A protein is the largest of this class of compounds and usually contains a minimum of 50 amino acids. It is common for the term *protein* to be used to describe all of these molecules; we used it in its general sense in the first sentence of this paragraph. But not all polypeptides are large enough to be considered proteins. In chapter 6 we see that protein synthesis is not just a random connection of amino acids; it is directed by information provided in DNA.

Table 2.5 Twenty Amino Acids and Their Abbreviations

Acid	Abbreviation	Type of R Group
Alanine	Ala	Nonpolar
Arginine	Arg	+
Asparagine	Asn	Polar
Aspartic acid	Asp	–
Cysteine	Cys	Polar
Glutamic acid	Glu	–
Glutamine	Gln	Polar
Glycine	Gly	Nonpolar
Histidine	His	+
Isoleucine	Ile	Nonpolar
Leucine	Leu	Nonpolar
Lysine	Lys	+
Methionine	Met	Nonpolar
Phenylalanine	Phe	Nonpolar
Proline	Pro	Nonpolar
Serine	Ser	Polar
Threonine	Thr	Polar
Tryptophan	Trp	Nonpolar
Tyrosine	Tyr	Polar
Valine	Val	Nonpolar

Note: + = positively charged; – = negatively charged.

Protein Structure and Diversity

The reason that proteins are so varied and specific is that they do not function in the form of a simple straight chain of amino acids (called the *primary structure*). A protein has a natural tendency to assume more complex levels of organization, called the secondary, tertiary, and quaternary structures **(figure 2.22).** The **primary (1°) structure** is the type, number, and order of amino acids in the chain, which varies extensively from protein to protein. The **secondary (2°) structure** arises when various functional groups exposed on the outer surface of the molecule interact by forming hydrogen bonds. This interaction causes the amino acid chain to twist into a coiled configuration called the α *helix* or to fold into an accordion pattern called a β-*pleated sheet*. Many proteins contain both types of secondary configurations. Once proteins have achieved their secondary structure, they create additional bonds between functional groups. This becomes their **tertiary structure (figure 2.22*c*).** In proteins containing the sulfur-containing amino acid **cysteine,** covalent disulfide bonds often form between sulfur atoms on two different parts of the molecule as part of the tertiary structure. These disulfide bonds are very stabilizing to the protein. Some proteins assume a **quaternary (4°) structure,** in which more than one polypeptide forms a large, multiunit protein. This is typical of antibodies and some enzymes that act in cell synthesis.

Disease Connection

Tiny changes in the primary structures of one or two proteins on the surface of the influenza virus contribute to its ability to cause disease in people year after year. These surface proteins accumulate changes in the amino acid sequence on a continual basis. The immune system, which may have been primed by previous infection or immunization to recognize a protein with a particular primary sequence, will not respond as strongly to a protein with a changed primary sequence. This is one reason you can get the flu more than once—even if you have been vaccinated in the past.

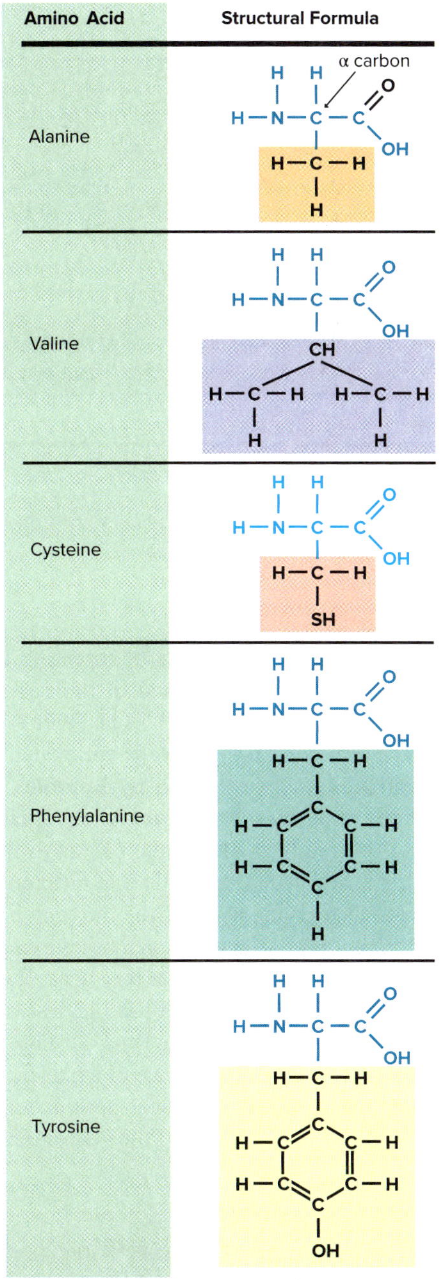
Amino Acid
Structural Formula
α carbon
Alanine
Valine
Cysteine
Phenylalanine
Tyrosine

Figure 2.21 Structural formulas of selected amino acids. The basic structure common to all amino acids is shown in blue type, and the variable group, or R group, is in the colored boxes.

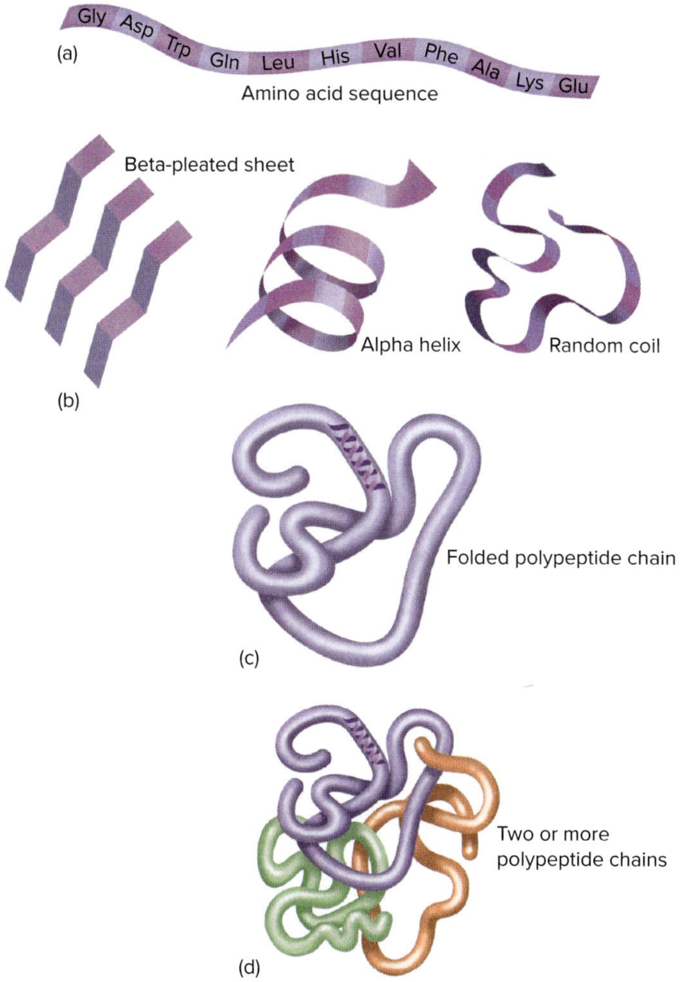
(a)
Gly Asp Trp Gln Leu His Val Phe Ala Lys Glu
Amino acid sequence
Beta-pleated sheet
Alpha helix
Random coil
(b)
Folded polypeptide chain
(c)
Two or more polypeptide chains
(d)

Figure 2.22 Stages in the formation of a functioning protein. **(a)** Its primary structure is a series of amino acids bound in a chain. **(b)** Its secondary structure develops when the chain forms hydrogen bonds that fold it into one of several configurations such as an alpha helix or beta-pleated sheet. Proteins often have several configurations in the same molecule. **(c)** A protein's tertiary structure is due to further folding of the molecule into a three-dimensional mass that is stabilized by hydrogen, ionic, and disulfide bonds between functional groups. **(d)** The quaternary structure exists only in proteins that consist of more than one polypeptide chain. Each of the chains in this protein is shown in a different color.

The most important outcome of the various forms of bonding and folding is that each different type of protein develops a unique shape, and its surface displays a distinctive pattern of pockets and bulges. As a result, a protein can react only with molecules that complement or fit its particular surface features like a lock and key. Such a degree of specificity gives rise to the functional diversity required for many thousands of different cellular activities. **Enzymes** serve as the catalysts for all chemical reactions in cells, and nearly every reaction requires a different enzyme. This specificity comes from the architecture of the binding site, which determines which molecules will fit. The same is true of antibodies. **Antibodies** are complex glycoproteins with specific regions of attachment for bacteria, viruses, and other

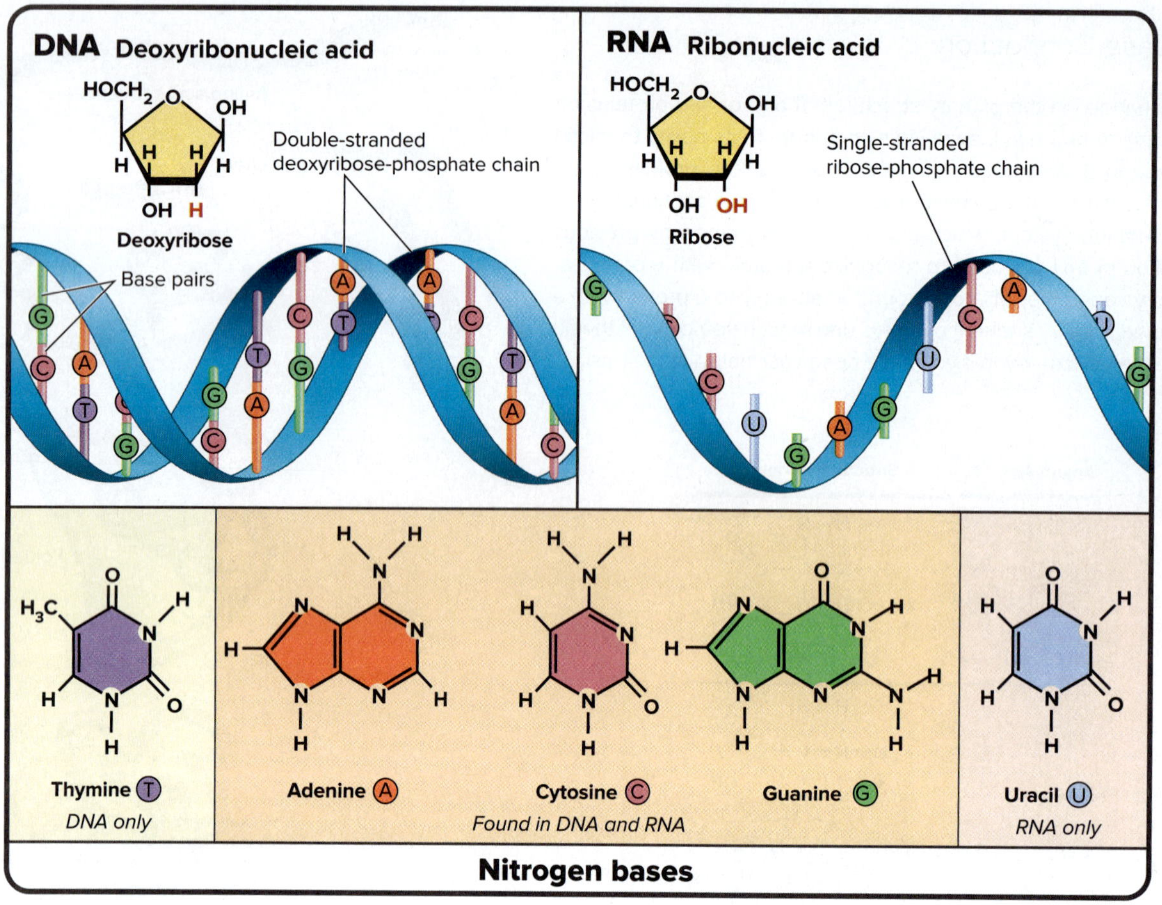
DNA Deoxyribonucleic acid
$HOCH_2$
OH
H H
H H
OH H
Deoxyribose
Double-stranded
deoxyribose-phosphate chain
Base pairs
RNA Ribonucleic acid
$HOCH_2$
OH
OH OH
Ribose
Single-stranded
ribose-phosphate chain
Thymine T
DNA only
Adenine A
Cytosine C
Guanine G
Found in DNA and RNA
Uracil U
RNA only
Nitrogen bases

Figure 2.23 The general structure of nucleic acids.

microorganisms. Certain bacterial toxins (poisonous products) react with only one specific organ or tissue; and proteins embedded in the cell membrane have reactive sites restricted to a certain nutrient. The functional three-dimensional form of a protein is termed the *native state*. If it is disrupted by some means, the protein is said to be *denatured*. Agents such as heat, acid, alcohol, and some disinfectants disrupt (denature) the stabilizing bonds and cause the molecule to become nonfunctional.

The Nucleic Acids: A Cell Computer and Its Programs

The nucleic acids are **deoxyribonucleic acid (DNA)** and **ribonucleic acid (RNA).** The universal occurrence of nucleic acids in all known cells and viruses emphasizes their important roles as informational molecules. DNA, the master computer of cells, contains a special coded genetic program with detailed and specific instructions for each organism's heredity. It transfers the details of its program to RNA, "helper" molecules responsible for carrying out DNA's instructions and translating the DNA program into proteins that can perform life functions. For now, let us briefly consider the structure and some functions of DNA, RNA, and a close relative, adenosine triphosphate (ATP).

Both DNA and RNA are polymers of repeating units called **nucleotides,** each of which is composed of three smaller units: a **nitrogen base,** a pentose (5-carbon) sugar, and a **phosphate (figure 2.23).**[5] The nitrogen base is a cyclic compound that comes in two forms: **purines** (two rings) and **pyrimidines** (one ring). There are two types of purines—**adenine (A)** and **guanine (G)**—and three types of pyrimidines—**thymine (T), cytosine (C),** and **uracil (U) (figure 2.24).** A characteristic that differentiates DNA from RNA is that DNA contains all of the nitrogen bases except uracil, and RNA contains all of the nitrogen bases except thymine. The nitrogen base is covalently bonded to the sugar **ribose** in RNA and **deoxyribose** (called that because it has one less oxygen than ribose) in DNA. Phosphate provides the final covalent bridge that connects sugars in series. So the backbone of a nucleic acid strand is a chain of alternating phosphate-sugar-phosphate-sugar molecules, and the nitrogen bases branch off the side of this backbone.

The Double Helix of DNA

DNA is a huge molecule formed by two very long polynucleotide strands linked along their length by hydrogen bonds between

5. The nitrogen base plus the pentose is called a *nucleoside*.

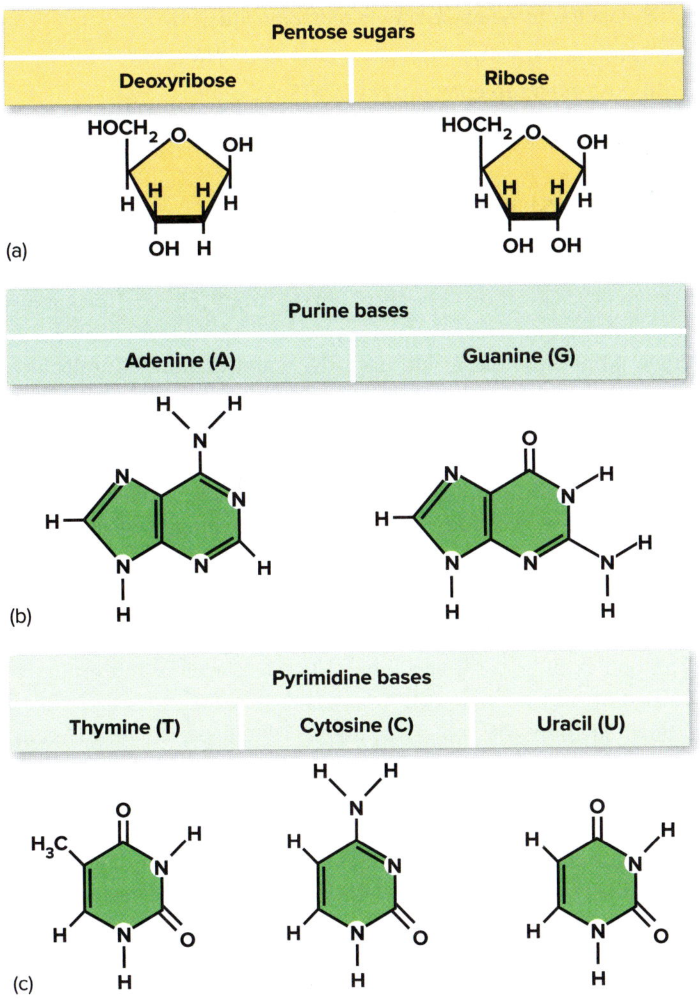
Pentose sugars
Deoxyribose
Ribose
(a)
Purine bases
Adenine (A)
Guanine (G)
(b)
Pyrimidine bases
Thymine (T)
Cytosine (C)
Uracil (U)
(c)

Figure 2.24 The sugars and nitrogen bases that make up DNA and RNA. **(a)** DNA contains deoxyribose and RNA contains ribose. **(b)** A and G purine bases are found in both DNA and RNA. **(c)** Pyrimidine bases are found in both DNA and RNA, but T is found only in DNA, and U is found only in RNA.

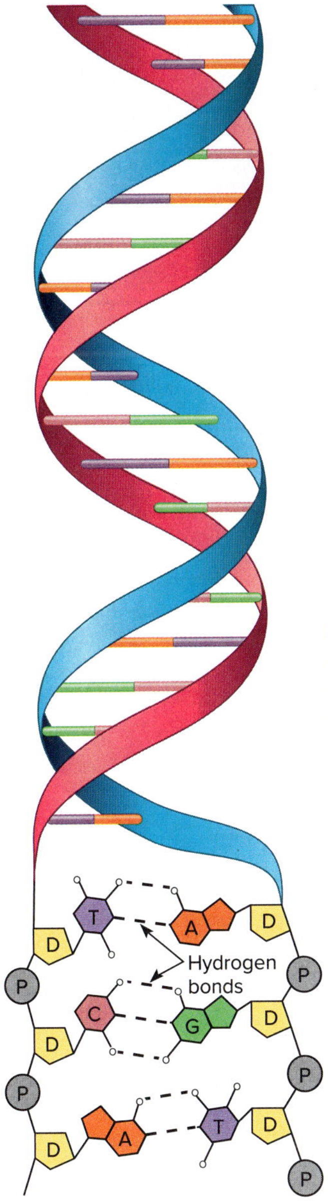
Hydrogen bonds

Figure 2.25 A structural representation of the double helix of DNA. The details of hydrogen bonds between the nitrogen bases of the two strands are shown.

complementary pairs of nitrogen bases. The pairing of the nitrogen bases occurs according to a predictable pattern: Adenine always pairs with thymine and cytosine with guanine. (Note: Here is an easy way to remember this! Adenine and thymine are *Always Together,* and guanine and cytosine make a *Good Couple.*) The bases are attracted in this way because each pair shares oxygen, nitrogen, and hydrogen atoms exactly positioned to align perfectly for hydrogen bonds **(figure 2.25).**

For ease in understanding the structure of DNA, it is sometimes compared to a ladder, with the sugar-phosphate backbone representing the side rails and the paired nitrogen bases representing the steps. Because of the way the nucleotides pair and stack up, the actual configuration of DNA is a *double helix* that looks somewhat like a spiral staircase. As is true of protein, the structure of DNA is intimately related to its function. DNA molecules are usually extremely long. The hydrogen bonds between pairs break apart when DNA is being copied, and the fixed complementary base-pairing is essential to maintain the genetic code.

RNA: Protein Synthesis and Regulation

Like DNA, RNA consists of a long chain of nucleotides. However, RNA is usually a single strand, except in some viruses. It contains ribose sugar instead of deoxyribose and uracil instead of thymine (see figure 2.23). Several functional types of RNA are formed using the DNA template through a replication-like process. Three major types of RNA are important for protein synthesis. Messenger RNA (mRNA) is a copy of a gene (a single functional part of the DNA) that provides the order and type of amino acids in a protein; transfer RNA (tRNA) is a carrier that delivers the correct amino acids for protein assembly; and ribosomal RNA (rRNA) is a major component of ribosomes. A fourth type of RNA is involved in regulating the expression of genes.

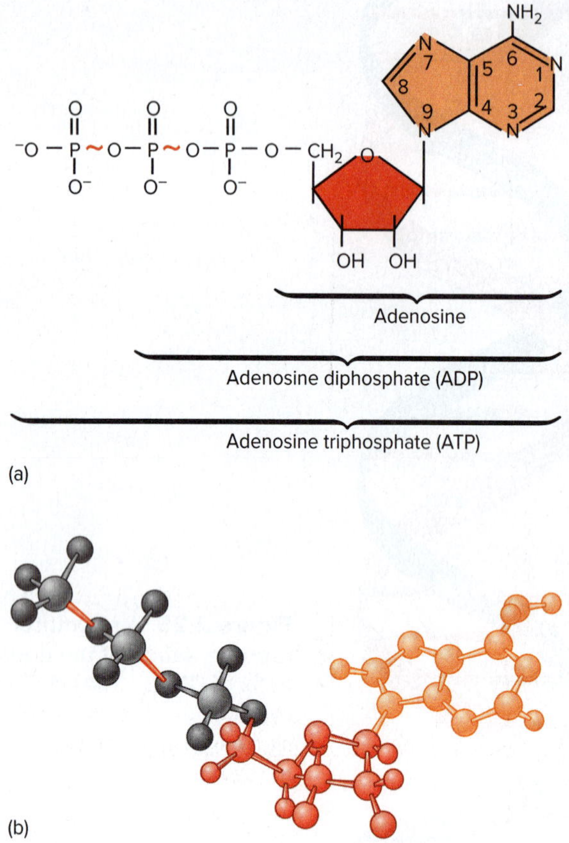
NH2
N
7
5
6
1
N
8
9
4
3
2
N
N
O
O
O
−O − P ~ O − P ~ O − P − O − CH2
O
O−
O−
O−
OH OH
Adenosine
Adenosine diphosphate (ADP)
Adenosine triphosphate (ATP)
(a)
(b)

Figure 2.26 An ATP molecule. **(a)** The structural formula. Wavy red lines connecting the phosphates represent bonds that release large amounts of energy. **(b)** A ball-and-stick model.

ATP: The Energy Molecule of Cells

A relative of RNA involved in an entirely different cell activity is **adenosine triphosphate (ATP).** ATP is a nucleotide containing adenine, ribose, and three phosphates rather than just one **(figure 2.26).** It belongs to a category of high-energy compounds (also including guanosine triphosphate [GTP]) that give off energy when the bond is broken between the second and third (outermost) phosphates. The presence of these high-energy bonds makes it possible for ATP to release and store energy for cellular chemical reactions. Breakage of the bond of the terminal phosphate releases energy to do cellular work and also generates adenosine diphosphate (ADP). ADP can be converted back to ATP when the third phosphate is restored, thereby serving as an energy depot. Carriers for oxidation-reduction activities (nicotinamide adenine dinucleotide [NAD], for instance) are also derivatives of nucleotides.

2.2 Learning Outcomes—Assess Your Progress

5. Name the four main families of biochemicals.
6. Provide examples of cell components made from each of the families of biochemicals.
7. Differentiate among primary, secondary, tertiary, and quaternary levels of protein structure.
8. List the three components of nucleotides.
9. Name the nitrogen bases of DNA and RNA.
10. List the three components of ATP.

2.3 Cells: Where Chemicals Come to Life

As we proceed in this chemical survey from the level of simple molecules to increasingly complex levels of macromolecules, at some point we cross a line from the realm of lifeless molecules and arrive at the fundamental unit of life called a **cell.** A cell is a huge aggregate of carbon, hydrogen, oxygen, nitrogen, and many other atoms. It follows the basic laws of chemistry and physics, but it is much more. The combination of these atoms produces characteristics, reactions, and products that can only be described as *living.*

Fundamental Characteristics of Cells

The bodies of some living things such as bacteria and protozoa consist of only a single cell, whereas those of animals and plants contain trillions of cells. Regardless of the organism, all cells have a few common characteristics. They tend to be spherical, polygonal, cubical, or cylindrical; and their protoplasm (internal cell contents) is encased in a cell membrane, also called a cytoplasmic membrane. They have chromosomes containing DNA and ribosomes for protein synthesis, and they are exceedingly complex in function. As you have already learned, there are three fundamentally different cell types: bacterial, archaeal, and eukaryotic.

Animals, plants, fungi, and protozoans are composed of eukaryotic cells. These cells contain a number of complex internal parts called organelles that perform useful functions for the cell involving growth, nutrition, or metabolism. Organelles are defined as cell components that perform specific functions and are enclosed by membranes. Organelles also partition the eukaryotic cell into smaller compartments. The most visible organelle is the nucleus, which contains the DNA of the cell. It is roughly ball-shaped and is surrounded by a double membrane. Other organelles include the Golgi apparatus, endoplasmic reticulum, vacuoles, and mitochondria.

Bacterial and archaeal cells may seem to be the cellular "have nots" because, for the sake of comparison, they are described by what they lack. They have no nucleus and generally no other organelles. This apparent simplicity is misleading, however, because the fine structure of these cells is actually complex. Overall, bacterial and archaeal cells can engage in nearly every activity that eukaryotic cells can, and many can function in ways that eukaryotes cannot. Chapters 4 and 5 delve deeply into the properties of bacterial and eukaryotic cells.

2.3 Learning Outcome—Assess Your Progress

11. Recall three characteristics common to all cells.

Media Under The Microscope Wrap-Up

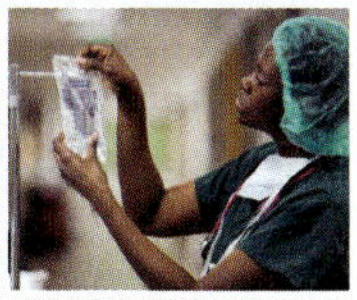

O. Dimier/PhotoAlto

The **intended message** of the article is to suggest that hospitals may have been using the wrong IV fluids for decades.

A **critical reading** of the article would include the fact that data are cited, and that multiple studies have shown similar findings. This is always a good sign in research. When different study designs yield a similar outcome, you can have more faith in the results. Also, the data are explained by the people who produced them. For example, the article quotes one of the scientists as saying, "There are 5 million patients admitted to an ICU in the United States every year. For every 100 patients treated with balanced fluids instead of saline, 1 fewer patient would experience death, new dialysis, or persistent renal problems."

Interpreting the article for your friends will require you to explain how such a seemingly small effect (1% improvement) can really be significant.

My **overall grade** for this article is a solid A. It has a single point to make, and it makes it convincingly and clearly.

Source: Clayton Dalton, "Why Did Sterile Salt Water Become the IV Fluid of Choice?," NPR, online article posted March 31, 2018.

Study Smarter: Better Together

These activities are designed for you to use on your own with a study group—either a face-to-face group or a virtual one, consisting of 3–5 members. Studying together can be very helpful, but there are effective and ineffective ways to do it. For example, getting together without a clear structure is often not a good use of your time. Use your time efficiently by using one or more of the exercises below.

FACE-TO-FACE GROUPS

Use one or more of the activities below.

Peer Instruction: Assign numbers to your group members to use all semester long. Now look at these five concepts from this chapter. Each group member prepares a 5-minute lesson on the topic corresponding to their number. Don't worry if you have fewer than 5 members; just use however many you have! During your group study time, each member presents their lesson, and the group spends another 5–10 minutes discussing that lesson.

1. Covalent and noncovalent bonds
2. Hydrogen bonding
3. Solutes and solvents
4. The pH scale
5. The four levels of structure in proteins

Concept Maps: Each member of the group should use this list of terms from this chapter to generate their own concept map. This can be hand-drawn or created using software (see Appendix C for guidelines). During group study time, compare each other's concept maps and help each other make sure they are correct. Of course, there are many different "correct" maps. Examining each member's map will help you talk through the varied concepts and how they are related.

Concept Terms:

elements	ions	atoms	molecules
protons	organic chemicals	inorganic chemicals	compounds
macromolecules	valence	neutrons	electrons

Table Topics: Each group member should identify a concept or topic from this week's class assignments with which they are having trouble and share it during group study time. The other group members can then help to clarify confusing issues or share how they figured it out. Aim for a maximum of 15 minutes per topic. If the topic remains unclear to the group, bring it up during class or use the instructor's office hours or e-mail to ask for help. Taking the time to struggle with a difficult concept first makes your questions much more specific and more likely to yield helpful answers.

(*continued on next page*)

VIRTUAL GROUPS

Not everyone has the time or opportunity to meet with group members outside of class time. You or your instructor can create a virtual group using e-mail or the course software.

Weekly Discussion Board: This forum can be used as a way for groups to discuss topics, via e-mail or other learning management systems or online platforms, before they are covered in class. As each member of the group answers the current week's question, they should send their responses to every other member of their group. It's best to agree on a deadline based on how your class schedule works (Saturday for the next week's topics, for example). Then, after the topic is discussed in class, each member should send a response that all group members will see with a follow-up post on the same topic. If you cover more than one chapter in a week, someone can be designated to choose which chapter Discussion Board question you will use. Or simply decide up front that you will always use the first-chapter-of-the week's question, to keep the schedule simple.

Discussion Question

Name three characteristics of DNA that you find to be most important or interesting, and explain why for each one.

Chapter Summary

THE CHEMISTRY OF BIOLOGY

2.1 ATOMS, BONDS, AND MOLECULES: FUNDAMENTAL BUILDING BLOCKS

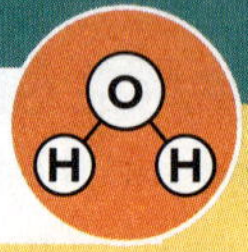
O
H
H

- **Atoms are made of protons, neutrons, and electrons.**
- **Elements are each made of a unique atom. Different elements are made of atoms with differing numbers of protons, neutrons, and electrons.**
- **Three important bonds in biology are covalent, ionic, and hydrogen bonds.**
- **Biologists define organic molecules as those containing both carbon and hydrogen.**
- **Carbon is the backbone of biological compounds because it can form single, double, and triple covalent bonds.**

2.2 MACROMOLECULES: SUPERSTRUCTURES OF LIFE

- **Macromolecules are large organic molecules built up by polymerization of smaller subunits.**
- **Carbohydrates are large molecules linked together by glycosidic bonds. They serve as protection, support, and nutrient and energy stores.**
- **Lipids are fats that are insoluble in water. They serve as cell components and nutrient and energy stores.**
- **Proteins are chains (polymers) of amino acids linked together by peptide bonds.**
- **Nucleic acids are polymers of nucleotides linked by phosphate-pentose sugar covalent bonds. Double-stranded nucleic acids are held together by hydrogen bonds. Nucleic acids are information molecules that direct cell metabolism and reproduction.**

2.3 CELLS: WHERE CHEMICALS COME TO LIFE

- **Just as the atom is the fundamental unit of matter, the cell is the fundamental element of life.**

SmartGrid: From Knowledge to Critical Thinking

This *21 Question Grid* takes the topics from this chapter and arranges them with respect to the American Society for Microbiology's Undergraduate Curriculum guidelines—all six of the important "Concepts" as well as the important "Competency" of scientific literacy. Three questions are supplied, which cover chapter content referring to the Concept or Competency in increasing levels of Bloom's taxonomy for learning.

ASM Concept/ Competency	A. Bloom's Level 1,2—Remember and Understand (Choose one)	B. Bloom's Level 3,4—Apply and Analyze	C. Bloom's Level 5,6—Evaluate and Create
Evolution	**1.** The fundamental element of life is considered to be ________, as a result of its superior bonding capabilities. **a.** nitrogen **b.** oxygen **c.** carbon **d.** magnesium phosphate	**2.** To what do many scientists attribute the explosion of many new types of organisms 750 million years ago?	**3.** Some scientists suspect that the earliest cells on the planet might have used RNA as their basic genetic code. Speculate about why DNA replaced RNA in most organisms as the primary genetic material.
Cell Structure and Function	**4.** Which of the following is a macromolecule that assembles into bilayers? **a.** protein **b.** phospholipid **c.** nucleic acid **d.** carbohydrate	**5.** Differentiate among the terms *peptide, polypeptide, amino acid,* and *protein.*	**6.** Plant cells are composed of cellulose, a complex carbohydrate exhibiting a unique bond between its glucose subunits. Provide an explanation for the fact that humans cannot digest fruits and vegetables at an efficient level.
Metabolic Pathways	**7.** Which of the following is used to store energy in cells? **a.** ATP **b.** RNA **c.** DNA **d.** each protein	**8.** Explain why breaking bonds between sugar monomers is called *hydrolysis.*	**9.** Why is it useful for a cell to be able to use acids and bases in neutralization reactions?
Information Flow and Genetics	**10.** DNA leads to RNA, which can lead to the creation of **a.** proteins. **b.** lipids. **c.** cells. **d.** oxygen.	**11.** Compare and contrast the RNA molecule with the DNA molecule.	**12.** Describe all of the types of bonds used in the DNA double helix.
Microbial Systems	**13.** RNA plays an important role in what biological process? **a.** replication **b.** protein synthesis **c.** lipid metabolism **d.** water transport	**14.** Name two types of polysaccharides that are only found in bacteria.	**15.** Conduct research to identify one bacterium that lives at low pH and another that lives at high pH. Describe the habitats both live in and what they do to survive these extreme pH levels.
Impact of Microorganisms	**16.** The first organisms on earth were **a.** oxygenic. **b.** anoxygenic. **c.** sponges. **d.** multicellular.	**17.** Some microbes are *decomposers.* Why are they crucial to life on earth?	**18.** Many archaea are *extremophiles*—organisms that thrive in extreme conditions. Evidence points to extremophiles as being the first type of cell on earth. Why is that likely?

(continued)

ASM Concept/ Competency	A. Bloom's Level 1,2—Remember and Understand (Choose one)	B. Bloom's Level 3,4—Apply and Analyze	C. Bloom's Level 5,6—Evaluate and Create
Scientific Thinking	**19.** Nonpolar molecules like benzene cannot be dissolved in **a.** anything. **b.** polar fluids, like water. **c.** nonpolar fluids, like alcohol. **d.** the lab.	**20.** Explain what *molarity* refers to.	**21.** Use table 2.2 to explain why a pH of 7 is considered neutral.

Answers to the multiple-choice questions appear in Appendix A.

Visual Connections

This question uses visual images to connect content within and between chapters.

1. **Figure 2.20.** Speculate on why sterols like cholesterol can add "stiffness" to membranes that contain them.

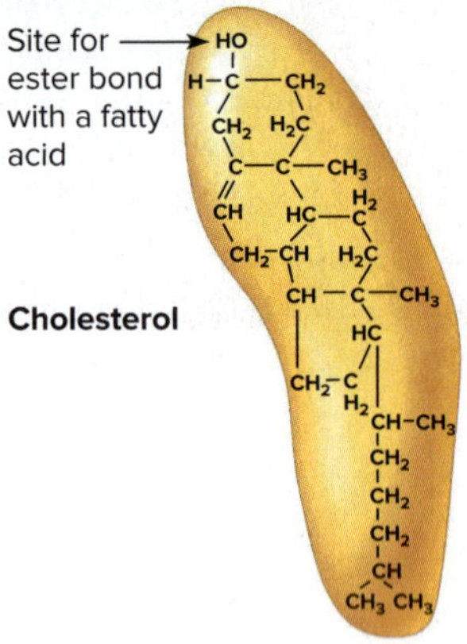

Site for ester bond with a fatty acid
Cholesterol

High Impact Study

These terms and concepts are most critical for your understanding of this chapter—and may be the most difficult. Have you mastered them?

Concepts

- ☐ The organization of the atom
- ☐ Covalent and noncovalent bonds
- ☐ Hydrogen bonding
- ☐ Solutions: solutes and solvents
- ☐ The pH scale
- ☐ Four types of macromolecules
- ☐ Four levels of structure in proteins

Terms

- ☐ Atom
- ☐ Inorganic vs. organic compounds
- ☐ Carbohydrate
- ☐ Lipid
- ☐ Nucleic acid

Design Element: (College students): Caia Image/Image Source

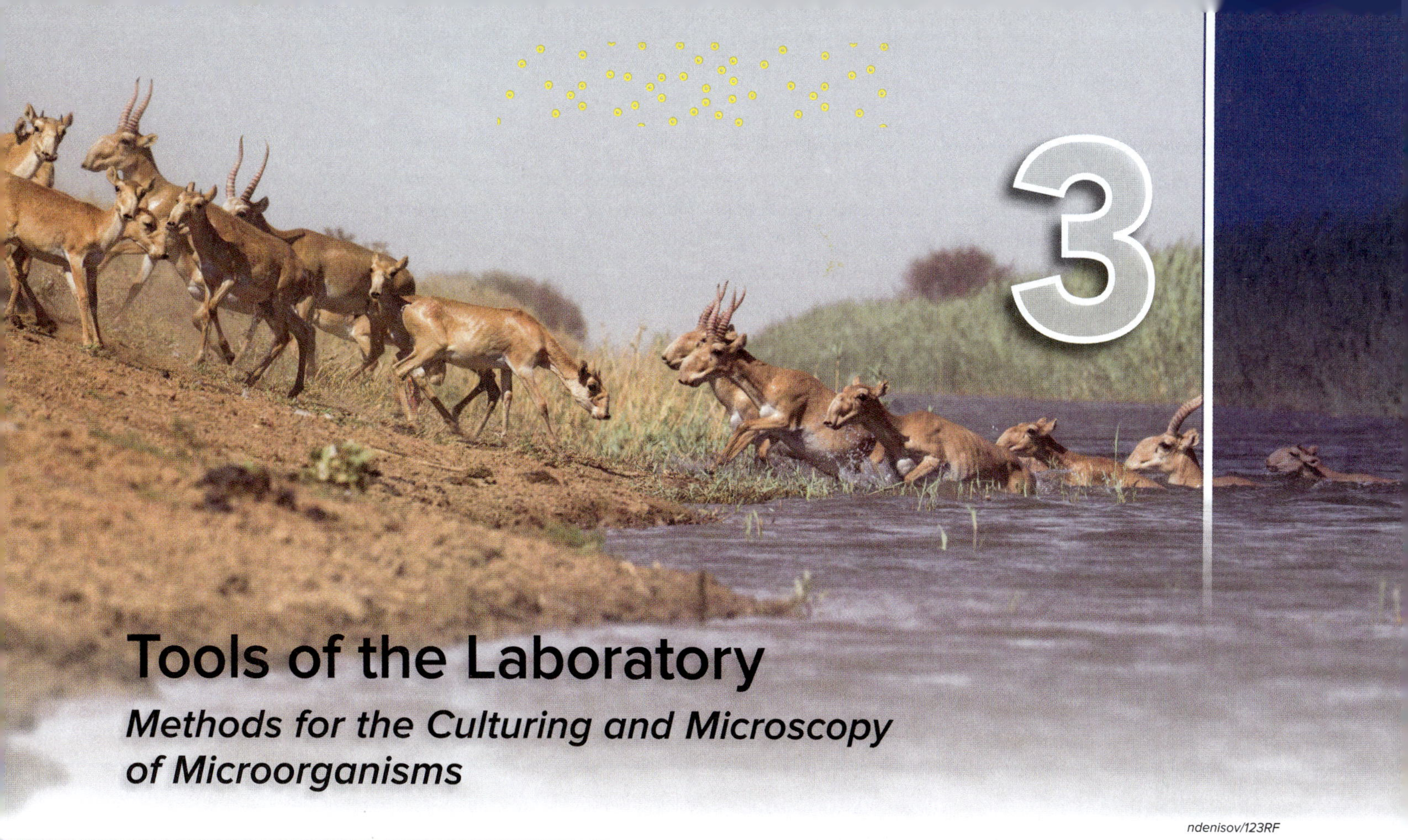
3

Tools of the Laboratory

Methods for the Culturing and Microscopy of Microorganisms

ndenisov/123RF

Media Under The Microscope

Weather Killed the Antelopes?

This opening case examines an article from the popular media to determine the extent to which it is factual and/or misleading. This case focuses on a recent New York Times *article "A Wet and Warm Spring, Then 200,000 Dead Saigas."*

This article introduces us to a type of antelope called the saiga. This endangered species is found in Kazakhstan, Mongolia, and Russia. Within the space of three weeks, every single saiga in a herd of more than 200,000 in central Kazakhstan dropped dead, across an area the size of Great Britain. This represented more than half the individuals of the entire species on the planet.

Researchers knew at the time that the cause of death was a bacterium called *Pasteurella multocida.* But they spent three years figuring out why it suddenly caused this mass mortality event. One startling feature of the die-off was that 100% of the animals died. As one of the researchers said, "I've worked with many nasty things. You always have survivors."

According to the *New York Times* article, the researchers reported in 2018 that the bacterium was a normal inhabitant of the antelopes' tonsils, usually causing no harm. They concluded that an unusual period of warm and wet weather in the 10 days preceding the die-off triggered something in the bacteria to cause them to invade the animals' intestines, leading to bloodstream infections and death. The article discussed the possible contribution of climate change. It noted that there had been a series of these die-offs among the saigas in recent years, but none before the 1980s, when these peculiar warm and wet periods began occurring.

The *New York Times* article included interviews with the researchers themselves, and they also spoke to other respected scientists in the field to get their take on the research.

Media Under The Microscope Wrap-Up appears at the end of the chapter.

- What is the **intended message** of the article?
- What is your **critical reading** of the summary of the article provided above? Remember that in this context, "critical reading" does not necessarily mean *What criticism do you have?* but asks you to apply your knowledge to interpret whether the article is factual and whether the facts support the intended message.
- How would you **interpret** the news item for your nonmicrobiologist friends?
- What is your **overall grade** for the news item—taking into account its accuracy and the accuracy of its intended effect?

Outline and Learning Outcomes

3.1 Methods of Culturing Microorganisms: The Five I's

1. Explain what the Five I's mean and what each step entails.
2. Discuss three physical states of media and when each is useful.
3. Compare and contrast selective and differential media, and give an example of each.
4. Provide brief definitions for *defined* and *complex media*.

3.2 The Microscope: Window on an Invisible Realm

5. Convert among different lengths within the metric system.
6. Describe the earliest microscopes.
7. List and describe the three elements of good microscopy.
8. Differentiate between the principles of light and electron microscopy.
9. Compare and contrast the three main categories of stains, and provide examples of each.

3.1 Methods of Culturing Microorganisms: The Five I's

Microbiologists are confronted by some unique problems. First, in the natural setting, whether in a living host or in the environment, different species of microbes are mixed together in complex associations. This can make it difficult to study the characteristics of just one of them. Second, because unseen microbes are everywhere—in the air, on every surface—it is very easy to introduce unwanted microbes when working with your intended microbe.

The "Five I's" represent five basic techniques to manipulate, grow, examine, and characterize microorganisms in the laboratory. They are inoculation, incubation, isolation, inspection, and identification (the Five I's; **figure 3.1**). These procedures are time-tested procedures to handle and maintain microorganisms.

Disease Connection

When it comes to specimen collection, timing is everything. In order to accurately identify the microorganism, and to use the most appropriate antibiotics, proper specimen collection is crucial. Ideally, cultures should be collected prior to treating a patient with antibiotics. Antibiotic treatment may eliminate microbes in the sample, affecting the accuracy of the patient's diagnosis. As treatment continues, a repeated fever or continuing symptoms may indicate the need to change the antibiotic, and additional cultures may need to be collected before administering new antimicrobial drugs.

It should be noted that in the last decade, advances in molecular techniques mean that the Five I's are no longer necessary in all cases. For example, it has become possible to identify microorganisms without growing them. But there are many things that clinical microbiologists and researchers do with microbes that require the techniques described here.

Inoculation: Producing a Culture

To grow, or **culture** (verb), microorganisms, one introduces a tiny sample (the inoculum) into a container of nutrient **medium,** which provides an environment in which they multiply. This process is called **inoculation.** To avoid introducing unwanted microorganisms to the medium, any instrument used for picking up and transferring the sample must be **sterile.** The inoculated medium is then **incubated** under appropriate conditions (next step), and the resulting growth is called a culture (noun). The source of the sample being cultured depends on the objectives of the analysis. Clinical specimens for determining the cause of an infectious disease are obtained from body fluids (blood, cerebrospinal fluid), discharges (sputum, urine, feces), anatomical sites (throat, nose, ear, eye, genital tract), or diseased tissue such as an abscess or wound. Other samples subject to microbiological analysis are soil, water, sewage, foods, air, and inanimate objects. Procedures for proper specimen collection are fully discussed in chapter 18.

Incubation

Once a container of medium has been inoculated with a specimen, it is incubated, which means it is placed in a temperature-controlled chamber (incubator) to encourage growth. Although there are microbes that grow optimally at temperatures ranging from freezing to boiling, the usual temperatures used in laboratory growth fall between

20°C and 45°C. Incubators can also control the content of atmospheric gases such as oxygen and carbon dioxide that may be required for the growth of certain microbes. During the incubation period (ranging from a day to several weeks), the microbe multiplies and produces growth that is visible to the eye **(table 3.1).** Microbial growth in a liquid medium can be seen as cloudiness (also known as turbidity), sediment, scum, or color. The most common manifestation of growth on solid media is the appearance of colonies, especially with bacteria and fungi. Colonies are actually large masses of piled-up cells.

Before we continue to cover information on the Five I's, we will take a side trip to look at media in more detail.

Media: Providing Nutrients in the Laboratory

A major stimulus to the rise of microbiology in the late 1800s was the development of techniques for growing microbes out of their natural habitats and in pure form in the laboratory. This milestone enabled the close examination of a microbe and its morphology, physiology, and genetics. The trick was figuring out which nutrients were essential for each different type of microorganism to grow in the laboratory. In fact, although we have well-developed methods of growing many different microbes, we don't know how to grow most microbes because we don't know what they need in an artificial setting.

Some microbes require only a very few simple inorganic compounds for growth. Others need a complex list of specific inorganic and organic compounds. This tremendous diversity is evident in the types of media that can be prepared. More than 500 different types of media are used in culturing and identifying microorganisms. Culture media are contained in test tubes, flasks, or Petri dishes. They are inoculated by such tools as loops, needles, pipettes, and swabs. Media are extremely varied in nutrient content and consistency and can be specially formulated for a particular purpose.

This chapter focuses on the types of nonliving media used to grow bacteria and fungi. (We also refer to these as artificial media.) Protozoa and helminths are often more complex to grow in the laboratory. Viruses will not grow on nonliving media because they absolutely require their host cells in order to reproduce themselves. For that reason, viruses are grown in cultures of live cells. Scientists have developed a method to amplify the numbers of prions in the laboratory, but it is not considered cultivation in the true sense.

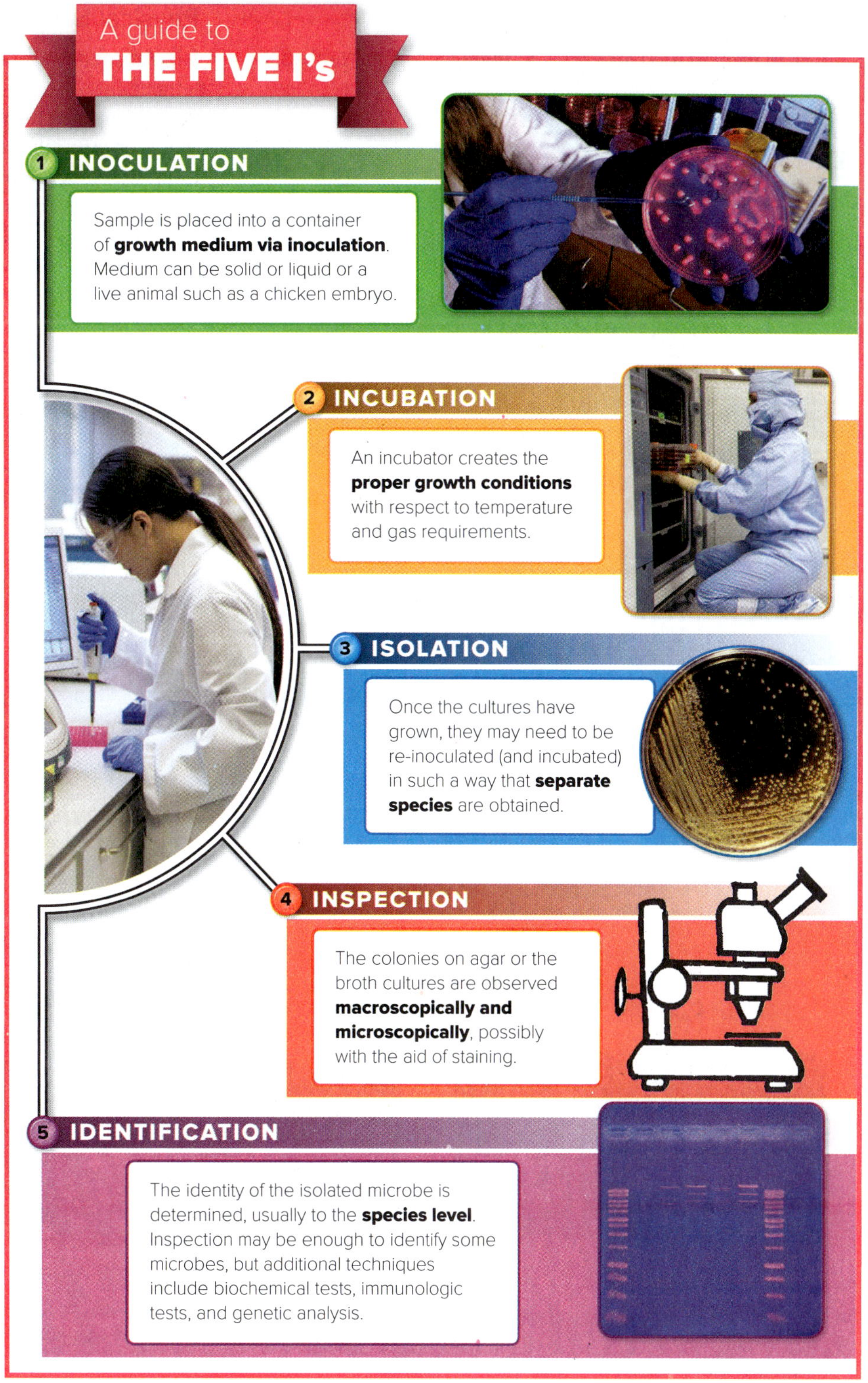
A guide to
THE FIVE I's
1 INOCULATION
Sample is placed into a container of **growth medium via inoculation**. Medium can be solid or liquid or a live animal such as a chicken embryo.
2 INCUBATION
An incubator creates the **proper growth conditions** with respect to temperature and gas requirements.
3 ISOLATION
Once the cultures have grown, they may need to be re-inoculated (and incubated) in such a way that **separate species** are obtained.
4 INSPECTION
The colonies on agar or the broth cultures are observed **macroscopically and microscopically**, possibly with the aid of staining.
5 IDENTIFICATION
The identity of the isolated microbe is determined, usually to the **species level**. Inspection may be enough to identify some microbes, but additional techniques include biochemical tests, immunologic tests, and genetic analysis.

Figure 3.1 A summary of the general laboratory techniques carried out by microbiologists. It is not necessary to perform all the steps shown or to perform them exactly in this order, but all microbiologists participate in at least some of these activities. In some cases, one may proceed right from collecting the sample to inspection, and in others, only inoculation and incubation on special media are required.

James Gathany/CDC; Monty Rakusen/Cultura/Getty Images; Javier Larrea/Pixtal/age fotostock; Centers for Disease Control and Prevention; Lisa Burgess/McGraw Hill

Types of Artificial Media

Media can be classified according to three properties **(table 3.2)**:

1. physical state,
2. chemical composition, and
3. functional type (purpose).

Table 3.1 Incubation Conditions of Various Bacterial Pathogens

Organism	Optimum Growth Temperature (°C)	Culturing Time (Days)
Listeria monocytogenes	25–30	1–2
Pseudomonas fluorescens	25–30	1–2
Streptococcus pyogenes	37	1–2
Mycobacterium tuberculosis	37	28

Physical States of Media

Liquid media are water-based solutions that do not solidify at temperatures above freezing and that tend to flow freely when the container is tilted **(figure 3.2*a*)**. These media, termed *broths, milks,* or *infusions,* are made by dissolving various solutes in distilled water. A common laboratory medium, *nutrient broth,* contains beef extract and peptone dissolved in water. Methylene blue milk and litmus milk are opaque liquids containing whole milk and dyes. Fluid thioglycollate is a slightly viscous broth used for determining the oxygen requirement of different microbes.

At ordinary room temperature, **semisolid media** exhibit a clotlike consistency **(figure 3.2*b*)** because they contain an amount of solidifying agent (agar or gelatin) that thickens them but does not produce a firm surface. Semisolid media are used to determine the motility of bacteria and to localize a reaction at a specific site.

Solid media provide a firm surface on which cells can form discrete colonies **(figure 3.2*c*)** and are useful for isolating and

Table 3.2 Three Categories of Media Classification

Physical State	Chemical Composition	Functional Type*	
1. Liquid	1. Chemically defined (synthetic)	1. General purpose	5. Reducing
2. Semisolid	2. Complex; not chemically defined	2. Enriched	6. Specimen transport
3. Solid (can be liquified)		3. Selective	7. Assay
4. Solid (cannot be liquefied)		4. Differential	8. Enumeration

*Some media can serve more than one function. For example, a medium such as brain-heart infusion is general purpose and enriched; mannitol salt agar is both selective and differential; and blood agar is both enriched and differential.

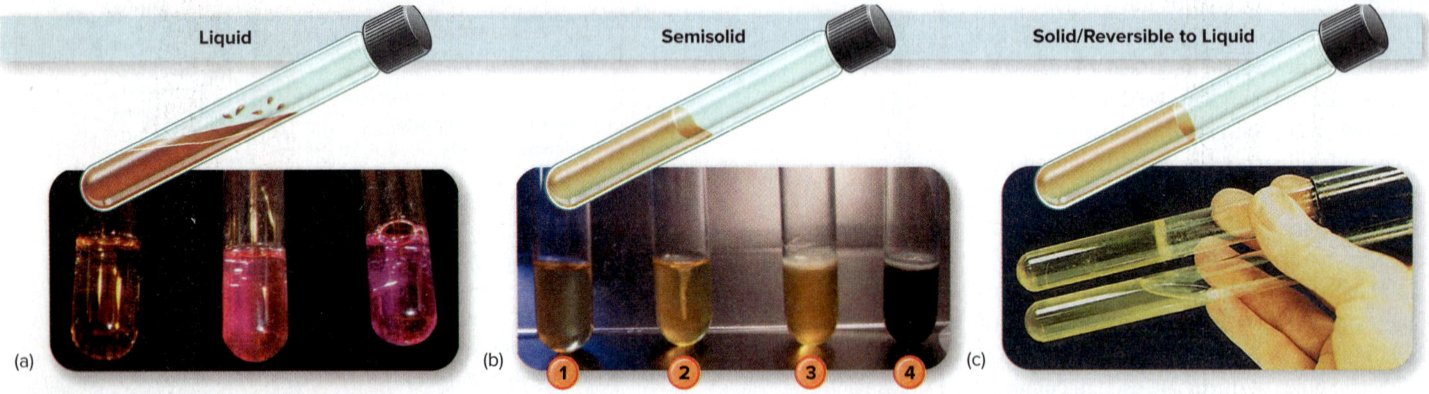
Liquid
Semisolid
Solid/Reversible to Liquid
(a)
(b)
(c)
1
2
3
4

Figure 3.2 Media in different physical forms. **(a) Liquid media** are water-based solutions that do not solidify at temperatures above freezing and that tend to flow freely when the container is tilted. Growth occurs throughout the container and will appear cloudy, or can appear as flakes or settle to the bottom of the vessel. The tubes here hold urea broth. Urea is added to the broth in order to see if the microbe being cultured contains an enzyme that digests urea and releases ammonium. If this happens, the pH of the broth is raised and the dye becomes increasingly pink. Left: uninoculated broth, pH 7; middle: weak positive, pH 7.5; right: strong positive, pH 8.0.

(b) Semisolid media are firmer than liquid media but not as firm as solid media. They do not flow freely and have a soft, clotlike consistency at room temperature. Semisolid media are used to examine the motility of bacteria and to provide a backdrop for visible reactions to occur. Here, sulfur indole motility (SIM) medium is pictured. (1) The medium is stabbed with an inoculum and incubated. Location of growth indicates nonmotility (2) or motility (3). Additionally, if H_2S gas is released, a black precipitate forms (4).

(c) Media containing 1% to 5% agar are solid enough to remain in place when containers are tilted or inverted. They are reversibly solid and can be liquefied with heat, poured into a different container, and resolidified. Solid media provide a firm surface on which cells can form discrete colonies. Nutrient gelatin contains enough gelatin (12%) to take on a solid consistency. The top tube shows it as a solid. The bottom tube indicates what happens when it is warmed or when microbial enzymes digest the gelatin and liquefy it.

All photos: Kathy Park Talaro

Table 3.3A Chemically Defined Medium for Growth and Maintenance of Pathogenic *Staphylococcus aureus*

0.25 Gram Each of These Amino Acids	0.5 Gram Each of These Amino Acids	0.12 Gram Each of These Amino Acids
Cystine	Arginine	Aspartic acid
Histidine	Glycine	Glutamic acid
Leucine	Isoleucine	
Phenylalanine	Lysine	
Proline	Methionine	
Tryptophan	Serine	
Tyrosine	Threonine	
	Valine	

Additional ingredients

0.005	mole nicotinamide	Vitamins
0.005	mole thiamine	Vitamins
0.005	mole pyridoxine	Vitamins
0.5	microgram biotin	Vitamins
1.25	grams magnesium sulfate	Salts
1.25	grams dipotassium hydrogen phosphate	Salts
1.25	grams sodium chloride	Salts
0.125	gram iron chloride	Salts

Ingredients dissolved in 1,000 milliliters of distilled water and buffered to a final pH of 7.0.

culturing bacteria and fungi. Liquefiable solid media, sometimes called reversible solid media, contain a solidifying agent that changes their physical properties in response to temperature. By far the most widely used and effective of these agents is agar, a complex polysaccharide isolated from the red alga *Gelidium.* Agar is solid at room temperature, and it melts (liquefies) at the boiling temperature of water (100°C). Once liquefied, agar does not resolidify until it cools to 42°C.

Chemical Content of Media

Media whose exact chemical compositions are known are termed *defined* (also known as *synthetic*). Such media contain pure organic and inorganic compounds that vary little from one source to another and have a molecular content specified by means of an exact formula. Defined media may contain nothing more than a few essential compounds such as salts and amino acids dissolved in water, or may be composed of a variety of defined organic and inorganic chemicals **(table 3.3).** Such standardized and reproducible media are most useful in research when the exact nutritional needs of the test organisms are known.

If even one component of a given medium is not chemically definable, the medium belongs in the *complex* category. Complex media contain extracts of animals, plants, or yeasts, including such materials as ground-up cells, tissues, and secretions. Examples are blood, serum, and meat extracts or infusions. Other nonsynthetic ingredients are milk, yeast extract, soybean digests, and peptone. Nutrient broth, blood agar, and MacConkey agar,

Table 3.3B Brain-Heart Infusion Broth: A Complex, Nonsynthetic Medium for Growth and Maintenance of Pathogenic *Staphylococcus aureus*

27.5	grams brain, heart extract, peptone extract
2	grams glucose
5	grams sodium chloride
2.5	grams disodium hydrogen phosphate

Ingredients dissolved in 1,000 milliliters of distilled water and buffered to a final pH of 7.0.

though different in function and appearance, are all complex nonsynthetic media.

Table 3.3 provides an illustration of chemically defined and complex media for the growth of *Staphylococcus* species.

Media for Different Purposes

General-purpose media are designed to grow as many different types of microbes as possible. As a rule, they are of the complex variety and contain a mixture of nutrients that could support the growth of a variety of microbial life. Examples include nutrient agar and broth, brain-heart infusion, and trypticase soy agar (TSA). An **enriched medium** contains complex organic substances such as blood, serum, hemoglobin, or special **growth factors** (specific vitamins, amino acids) that certain species require in order to grow. Bacteria that require growth factors and complex nutrients are termed **fastidious.** Blood agar, which is made by adding sterile sheep, horse, or rabbit blood to a sterile agar base **(figure 3.3*a*),** is often used to grow fastidious streptococci and other pathogens.

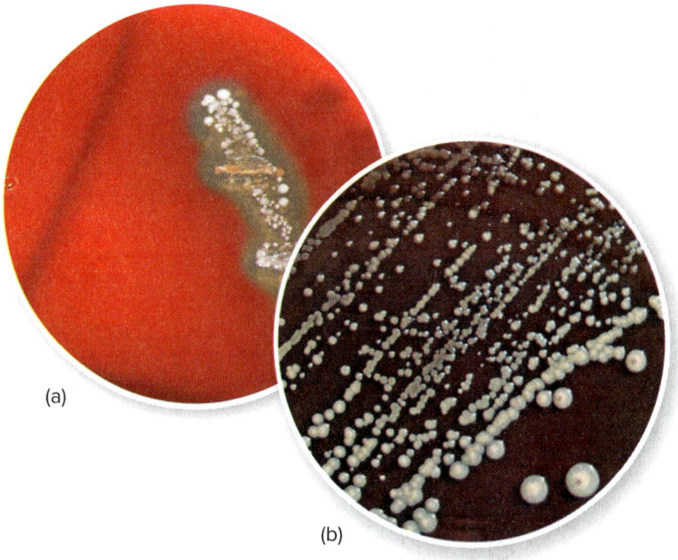

(a) (b)

Figure 3.3 Examples of enriched media. **(a)** Blood agar plate growing bacteria from the human throat. Enzymes from the bacterial colonies break down the red blood cells in the agar, leaving a clear "halo." This is called hemolysis. **(b)** Culture of *Neisseria* sp. on chocolate agar. Chocolate agar gets its brownish color from cooked blood (not chocolate) and does not produce hemolysis.

(a) Lisa Burgess/McGraw Hill; (b) Kathy Park Talaro

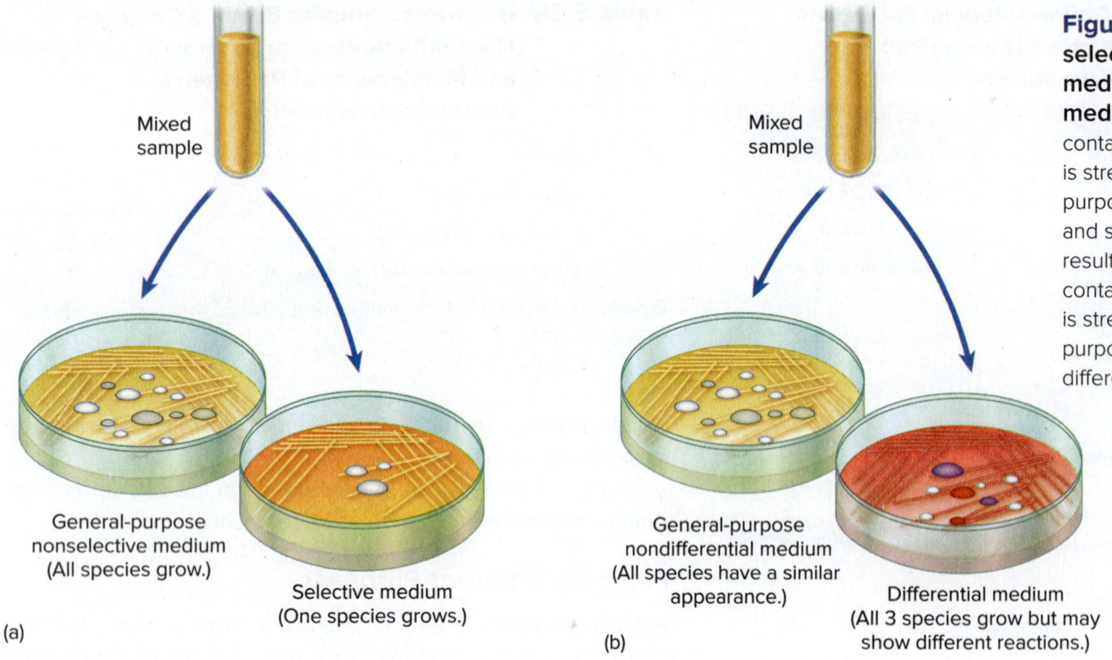
Mixed
sample
General-purpose
nonselective medium
(All species grow.)
Selective medium
(One species grows.)
(a)
Mixed
sample
General-purpose
nondifferential medium
(All species have a similar
appearance.)
Differential medium
(All 3 species grow but may
show different reactions.)
(b)

Figure 3.4 Comparison of selective and differential media with general-purpose media. **(a)** A mixed sample containing three different species is streaked onto plates of general-purpose nonselective medium and selective medium. Note the results. **(b)** Another mixed sample containing three different species is streaked onto plates of general-purpose nondifferential medium and differential medium.

Pathogenic *Neisseria* (one species causes gonorrhea) are grown on either Thayer-Martin medium or "chocolate" agar, a blood agar with added components **(figure 3.3*b*).**

Selective and Differential Media These media are designed for special microbial groups. In a single step, they can lead to the preliminary identification of a genus or even a species.

A **selective medium (figure 3.4*a*)** contains one or more agents that inhibit the growth of a certain microbe or microbes (call them A, B, and C) but not others (D). This *selects* microbe D, and only microbe D, to grow. Selective media are very important in initial isolation of a specific type of microorganism from samples containing dozens of different species—for example, feces, saliva, skin, water, and soil. They speed up isolation by suppressing the unwanted organisms and favoring growth of the desired ones.

Mannitol salt agar (MSA) **(figure 3.5*a*)** contains a high concentration of NaCl (7.5%) that inhibits most human pathogens. But the genus *Staphylococcus* grows well on this medium, and plating a mixed sample on MSA is a quick way to find those bacteria. Later we will learn about a classification system for bacteria, which divides them into two categories based on their

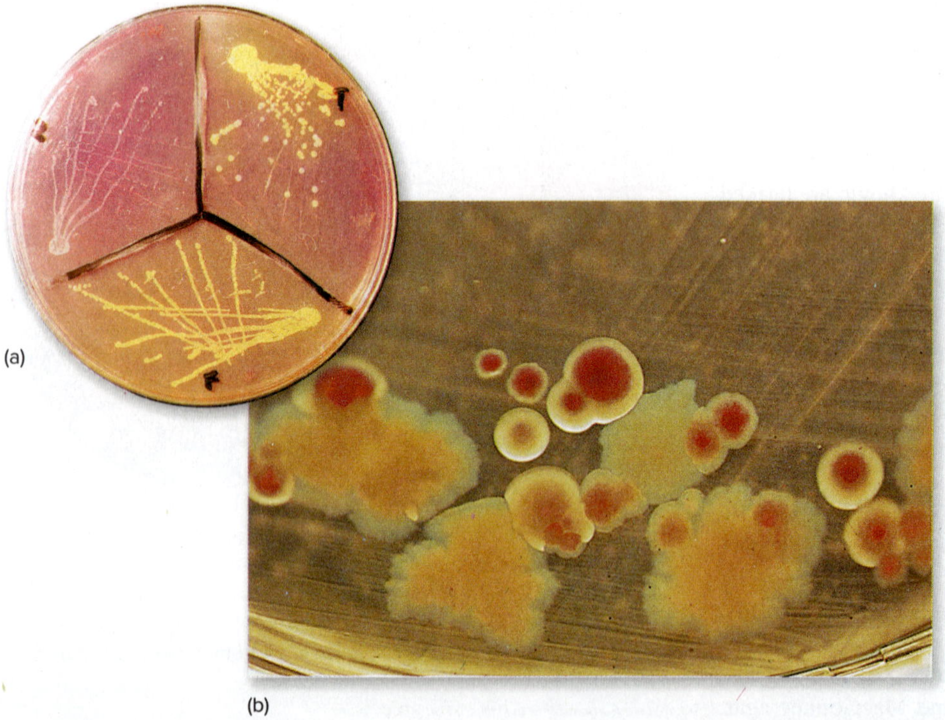
(a)
(b)

Figure 3.5 Examples of media that are both selective and differential. **(a)** Mannitol salt agar is used to isolate members of the genus *Staphylococcus*. It is selective because *Staphylococcus* can grow in the presence of 7.5% sodium chloride, whereas many other species are inhibited by this high concentration. It is also differential because it contains a dye that also identifies those species of *Staphylococcus* that produce acid (from the fermentation of mannitol) and turn the phenol red dye to a bright yellow. **(b)** MacConkey agar selects against gram-positive bacteria. It also differentiates between lactose-fermenting bacteria (indicated by a pink-red reaction in the center of the colony) and lactose-negative bacteria (indicated by an off-white colony with no dye reaction).
Kathy Park Talaro

reaction to a particular staining technique. Most bacteria are either gram-positive or gram-negative, based on this classification. If you know that you are looking for a bacterium that is gram-negative, it can be very useful to use a medium that does not allow gram-positive bacteria to grow. Media for isolating gram-negative intestinal pathogens (MacConkey agar, Hektoen enteric [HE] agar) contain bile salts, a component of feces, as a selective agent to inhibit most gram-positive bacteria **(figure 3.5*b*).** Other agents that have selective properties are dyes, such as methylene blue and crystal violet, acids, and antimicrobial drugs.

Differential media allow multiple types of microorganisms to grow but are designed to display visible differences among their colonies. Differentiation shows up as variations in colony size or color, in media color changes, or in the formation of gas bubbles or precipitates **(figure 3.4*b*).** These variations often come from the type of chemicals these media contain and the ways that microbes react to them. For example, when microbe X metabolizes a certain substance in the medium that cannot be used by organism Y, X will cause a visible change in the color of the colony or the medium and organism Y will not. Some media are sufficiently complex to show three or four different reactions **(figure 3.6).**

Dyes are frequently used as differential agents because many of them are pH indicators that change color in response to the production of an acid or a base. For example, mannitol salt agar contains phenol red, a dye that turns yellow when microbes acidify the medium by fermenting mannitol. MacConkey agar contains a dye that turns pink or red when microbes metabolize lactose in the medium.

Although blood agar is a type of enriched medium used for the growth of fastidious microbes, the presence of intact red blood cells allows it to function as a differential medium as well. **Hemolysins** are enzymes that function to lyse (break down) red blood cells for the purpose of releasing iron-rich hemoglobin for growth. When grown on blood agar, some hemolysin-producing bacteria completely lyse all red blood cells in the adjacent media, resulting in *beta-hemolysis,* a complete clearing around the bacterial colony **(figure 3.7).** Other species only partially lyse the red blood cells, producing *alpha-hemolysis,* which appears as a greening of the agar around the colony. Bacteria having no hemolysins result in no reaction in the agar, which is termed *gamma-hemolysis*.

A single medium can be both selective and differential because of its various ingredients. MacConkey agar and mannitol salt agar, for example, are both selective and differential, meaning they can suppress the growth of some organisms (i.e., be selective) while producing a visual distinction among the ones that do grow (making them differential as well).

Miscellaneous Media A **reducing medium** contains a substance (thioglycolic acid or cystine) that absorbs oxygen or slows the penetration of oxygen in a medium, which reduces its availability. Reducing media are important for growing bacteria that don't require oxygen (termed *anaerobic*) or for determining the oxygen requirements of isolates. **Carbohydrate fermentation media** contain sugars that can be fermented (converted to acids) and a pH indicator to show this reaction **(figure 3.8).**

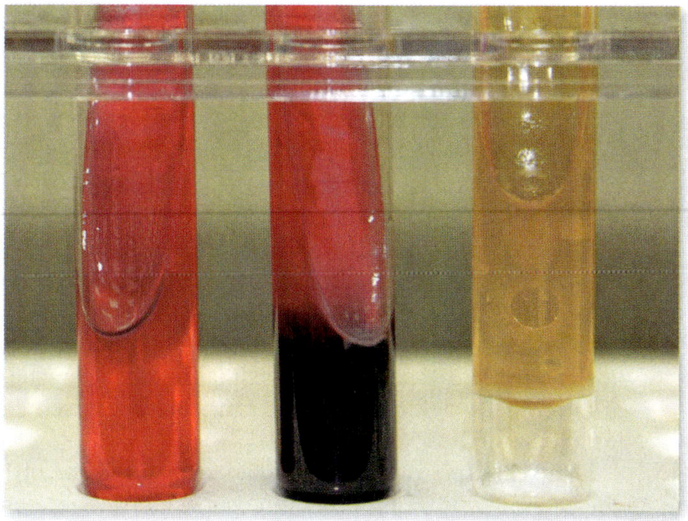

(a)

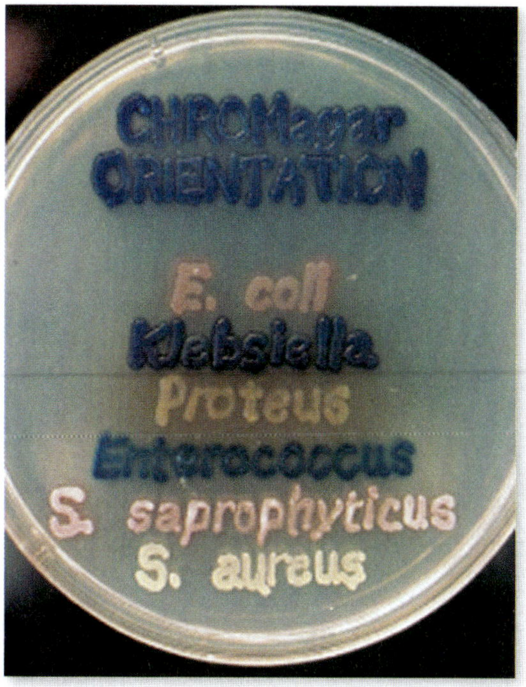
CHROMagar
ORIENTATION
E. coli
Klebsiella
Proteus
Enterococcus
S. saprophyticus
S. aureus

(b)

Figure 3.6 Media that differentiate characteristics. **(a)** Triple-sugar iron agar (TSI) in slant tubes. This medium contains three fermentable carbohydrates, along with phenol red to indicate pH changes brought about by acid production during fermentation, and a chemical (iron) that indicates H_2S gas production. Reactions (from left to right) are growth with no acid production; no acid production with H_2S gas formation (black); and acid production throughout the medium (yellow) plus gas production. **(b)** A medium developed for culturing and identifying the most common urinary pathogens. CHROMagar Orientation uses color-forming reactions to distinguish at least seven species and permits rapid identification and treatment. In the example, the bacteria were streaked so as to spell their own names.

(a) Lisa Burgess/McGraw Hill; (b) Kathy Park Talaro

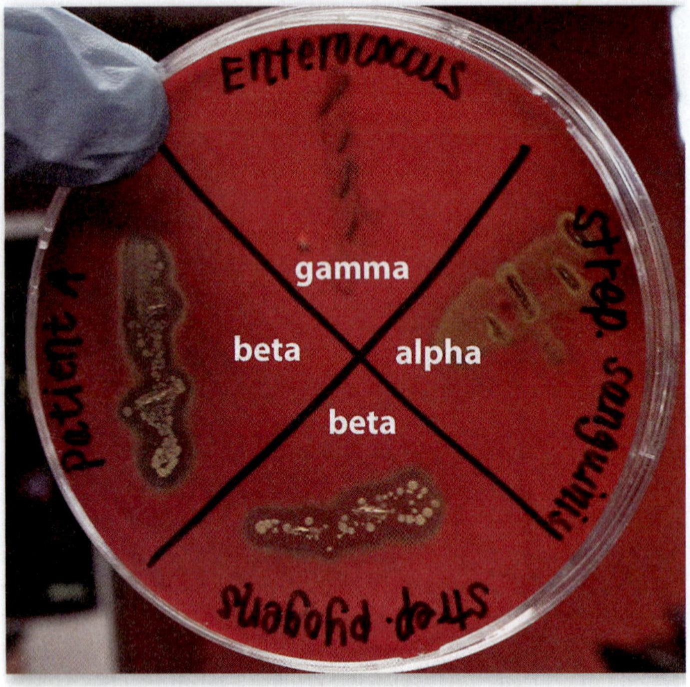
Enterococcus
gamma
beta
alpha
beta
Strep. sanguinis
Strep. pyogenes
Patient A

Figure 3.7 Types of hemolysis on blood agar. Microbes exhibiting *gamma-hemolysis,* or no change in the blood agar due to the lack of hemolysin production; *beta-hemolysis,* or a clearing of the blood agar due to complete red blood cell lysis; and *alpha-hemolysis,* or greening of the blood agar due to incomplete red blood cell lysis.
Lisa Burgess/McGraw Hill

Gas bubble
Outline of inner test tube
Control
Results

Figure 3.8 Carbohydrate fermentation in broth. This medium is designed to show fermentation (acid production) and gas formation by means of a small, upside-down test tube for collecting gas bubbles. The tube on the far left is an uninoculated negative control; the center tube is positive for acid (yellow) and gas (open space); the tube on the right shows growth but neither acid nor gas.
Dennis Strete/Fundamental Photographs, NYC

Transport media are used to maintain and preserve specimens that have to be held for a period of time before clinical analysis or to sustain delicate species that die rapidly if not held under stable conditions. **Assay media** are used by technologists to test the effectiveness of antimicrobial drugs and by drug manufacturers to assess the effect of disinfectants, antiseptics, cosmetics, and preservatives on the growth of microorganisms. (These tests are called assays.) **Enumeration media** are used by industrial and environmental microbiologists to count (enumerate) the numbers of organisms in milk, water, food, soil, and other samples.

Isolation: Separating One Species from Another

Now we are back to the five I's. The third I, **isolation,** is based on the concept that if an individual bacterial cell is separated from other cells and provided adequate space on a nutrient surface, it will grow into a discrete mound of cells called a **colony (figure 3.9).** If formed from a single cell, a colony consists of millions of offspring of that one cell and no other. Proper isolation requires that a small number of cells be inoculated into a relatively large volume or over an expansive area of medium selected to encourage the growth of the desired microbe. It generally requires the following materials: a medium that has a relatively firm surface (like agar), a Petri dish (a clear, flat dish with a cover), and inoculating tools.

There are three main ways to accomplish isolation:

- The streak plate technique
- The pour plate technique
- The spread plate technique

In the streak plate method, a small droplet of culture or sample is spread over the surface of the medium with an **inoculating loop** in a pattern that gradually thins out the sample and separates the cells spatially over several sections of the plate **(figure 3.10*a*).**

In the pour plate technique, the sample is inoculated in a step-wise fashion into a series of cooled but still liquid agar tubes so as to dilute the number of cells in each successive tube in the series **(figure 3.10*b*).** Inoculated tubes are then plated out (poured) into sterile Petri dishes and are allowed to solidify. The end result (usually in the second or third plate) is that the number of cells per volume is so decreased that cells have ample space to grow into separate colonies. One difference between this and the streak plate method is that in this technique some of the colonies will develop deep in the medium itself and not just on the surface.

With the spread plate technique, a small volume of liquid, diluted sample is dropped onto the surface of the medium and spread around evenly by a sterile spreading tool (sometimes called a "hockey stick"). Like the streak plate, cells are pushed onto separate areas on the surface so that they can form individual colonies **(figure 3.10*c*).**

In some ways, culturing microbes is analogous to gardening. Cultures are formed by "seeding" tiny plots (media) with microbial cells. Extreme care is taken to keep out weeds (contaminants). Once microbes have grown after incubation, the clinician must

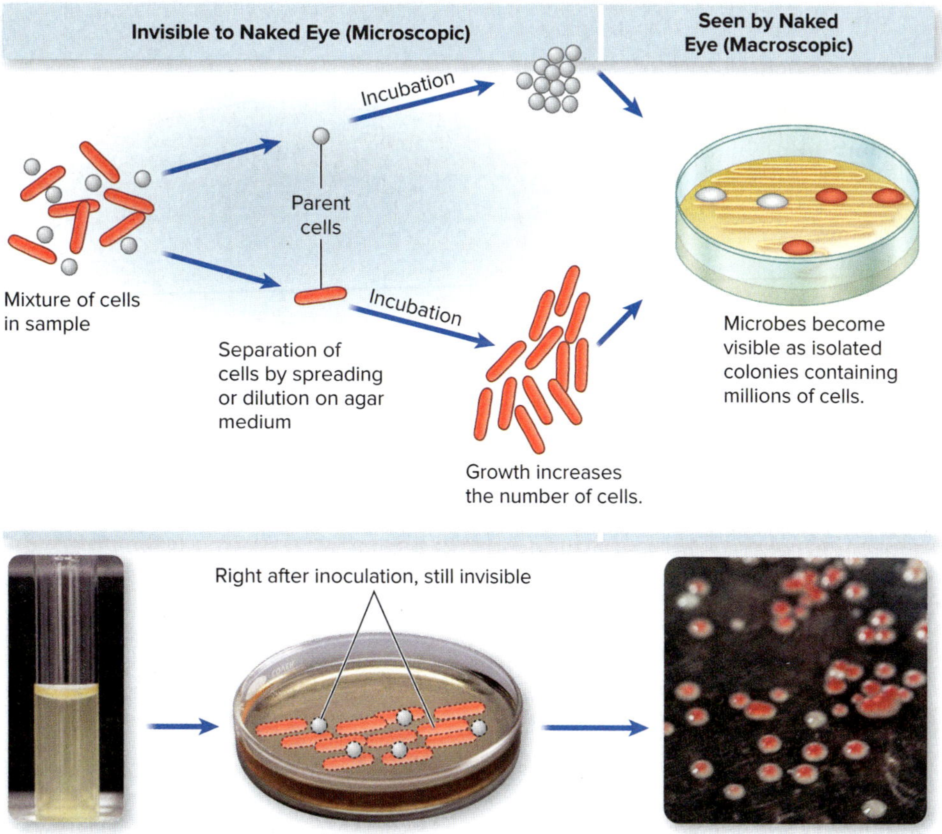
Invisible to Naked Eye (Microscopic)
Seen by Naked Eye (Macroscopic)
Incubation
Parent cells
Mixture of cells in sample
Incubation
Separation of cells by spreading or dilution on agar medium
Growth increases the number of cells.
Microbes become visible as isolated colonies containing millions of cells.
Right after inoculation, still invisible

Figure 3.9 Isolation technique. Stages in the formation of an isolated colony, showing the microscopic events and the macroscopic result. Separation techniques such as streaking can be used to isolate single cells. After numerous cell divisions during incubation, a macroscopic mound of cells, or a colony, will be formed. This is a relatively simple yet successful way to separate different types of bacteria in a mixed sample.
(a) Lisa Burgess/McGraw Hill; (b) Richard Hutchings/McGraw Hill; (c) Lisa Burgess/McGraw Hill

inspect the culture. A **pure culture** is a container of medium growing only a single known species or type of microorganism **(figure 3.11*a*).** This type of culture is most frequently used for laboratory study because it allows the systematic examination and control of one microorganism by itself. Instead of the term *pure culture,* some microbiologists prefer the term **axenic,** meaning that the culture is free of other living things except for the one being studied. A standard method for preparing a pure culture is to **subculture,** or make a second-level culture from a well-isolated colony. A tiny bit of cells is transferred into a separate container of media and incubated (see figure 3.1). Sometimes growing microbes in pure culture can tell you very little about how they act in a mixed-species environment. Being able to isolate and study them in this manner can be valuable, though, as long as you keep in mind that it is an unnatural state for them.

A **mixed culture (figure 3.11*b*)** is a container that holds two or more *identified,* easily differentiated species of microorganisms, not unlike a garden plot containing both carrots and onions. A **contaminated culture (figure 3.11*c*)** was once pure or even intentionally mixed but has since had **contaminants** (unwanted microbes of uncertain identity) introduced into it, like weeds into a garden. Contaminants may get into cultures when the lids of tubes or Petri dishes are left off for too long, allowing airborne microbes to settle into the medium. They can also enter on an incompletely sterilized inoculating loop or on an instrument that you have inadvertently reused or touched to the table or your skin.

Rounding Out the Five I's: Inspection and Identification

How does one determine what sorts of microorganisms have been isolated in cultures? Inspecting them provides a starting point. Inspection includes such things as looking at colony morphology (shape) and color, and examining the growth inside broth tubes. Some bacterial species grow evenly throughout the fluid, while others tend to grow at the top of the fluid where it meets the air or in clumps throughout the broth. Inspection also includes using the microscope to examine single cells. Generally, their microscopic appearance is of limited value because many bacteria have similar shapes. The microscope can be useful in differentiating the smaller, simpler bacterial cells from the larger, more complex eukaryotic cells. Appearance can be useful in identifying eukaryotic microorganisms to the level of genus or species because of their distinctive morphological features. We use other techniques to identify bacteria, such as *biochemical tests,* that can determine fundamental chemical characteristics including nutrient requirements, products given off during growth, presence of enzymes, and mechanisms for deriving energy.

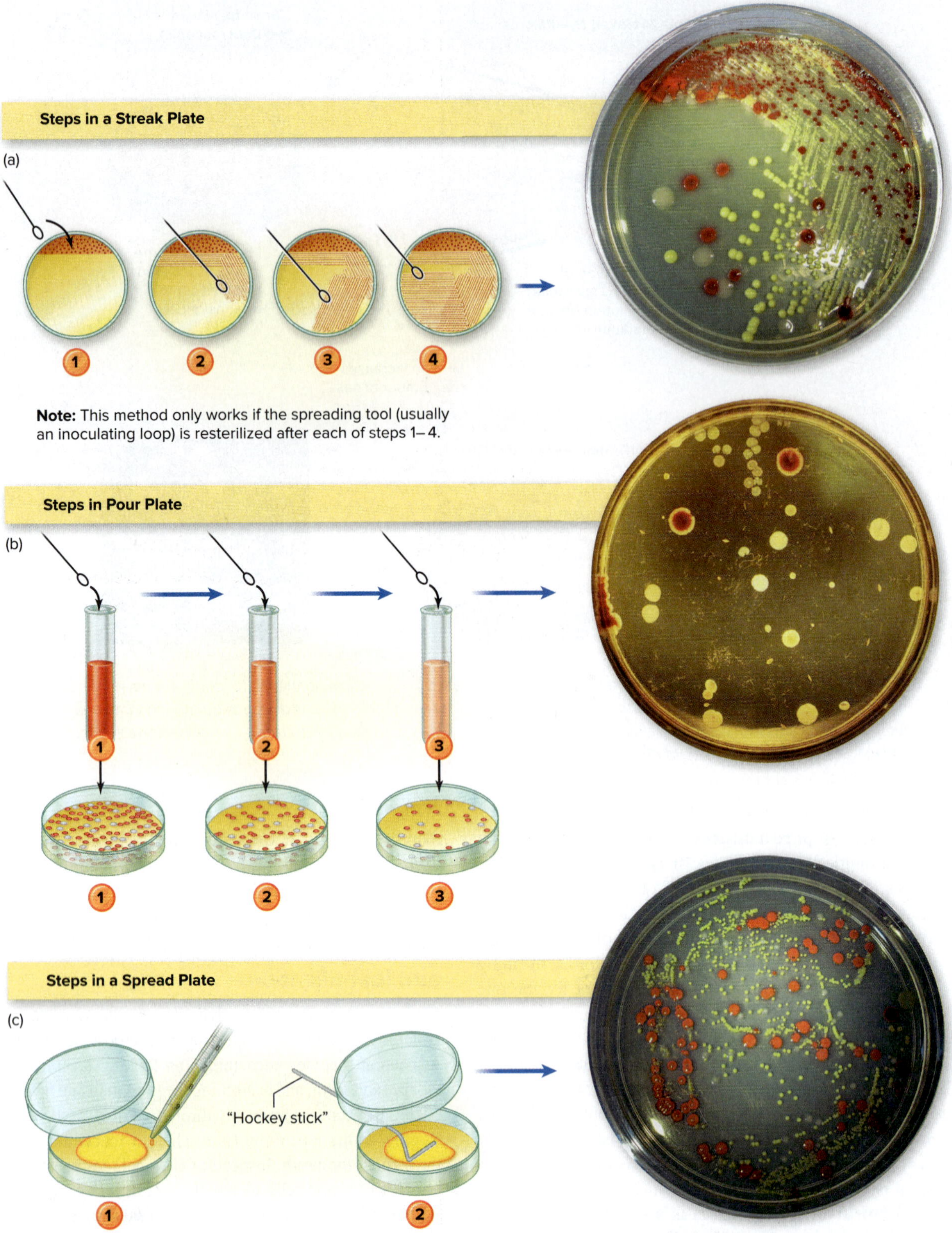
Steps in a Streak Plate
(a)
1
2
3
4
Note: This method only works if the spreading tool (usually an inoculating loop) is resterilized after each of steps 1–4.
Steps in Pour Plate
(b)
1
2
3
1
2
3
Steps in a Spread Plate
(c)
"Hockey stick"
1
2

Figure 3.10 Methods for isolating bacteria. **(a)** Steps in a quadrant streak plate and resulting isolated colonies of bacteria. Microbiologists use from 3 to 5 steps in this technique. **(b)** Steps in the pour plate method and the appearance of plate 3. **(c)** Spread plate and its result.
Kathy Park Talaro

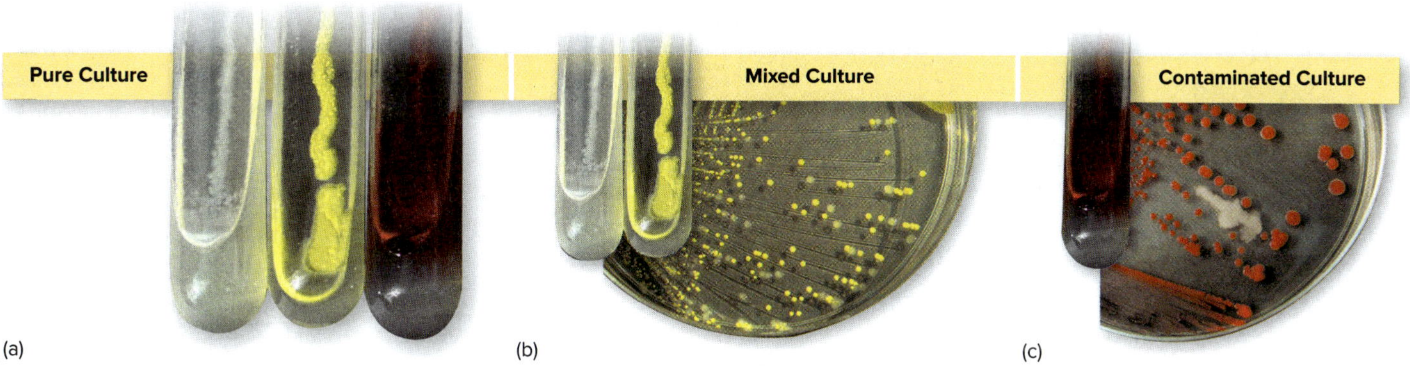
Pure Culture
Mixed Culture
Contaminated Culture
(a)
(b)
(c)

Figure 3.11 Various conditions of cultures. **(a)** Three tubes containing pure cultures of *Escherichia coli* (white), *Micrococcus luteus* (yellow), and *Serratia marcescens* (red). **(b)** A mixed culture of *M. luteus* (bright yellow colonies) and *E. coli* (faint white colonies). **(c)** This plate of *S. marcescens* was overexposed to room air, and it has developed a large, white colony. Because this intruder is not wanted, the culture is now contaminated.

Kathy Park Talaro

In addition, genetic characteristics can identify microbes based on their DNA or RNA (genotypic testing). Identification can also be accomplished by testing the isolate against known antibodies (immunologic testing). In chapter 18, we present more detailed examples of these identification methods.

3.1 Learning Outcomes—Assess Your Progress

1. Explain what the Five I's mean and what each step entails.
2. Discuss three physical states of media and when each is useful.
3. Compare and contrast selective and differential media, and give an example of each.
4. Provide brief definitions for *defined* and *complex media*.

3.2 The Microscope: Window on an Invisible Realm

Imagine Leeuwenhoek's excitement and wonder when he first viewed a drop of rainwater and glimpsed an amazing microscopic world teeming with unearthly creatures. Before learning about microscopes, we should consider how small microbes actually are.

Microbial Dimensions: How Small Is Small?

The concept of thinking small is best visualized by comparing microbes with the larger organisms of the macroscopic world and with the atoms and molecules of the molecular world **(figure 3.12).** You might also refer to table 1.1 to look at comparative sizes of microorganisms. Whereas the dimensions of macroscopic organisms are usually given in centimeters (cm) and meters (m), those of microorganisms fall within the range of millimeters (mm) to micrometers (μm) to nanometers (nm). Spend some time with **table 3.4** to help your understanding of these terms. Also, **Appendix B** offers help with exponents. The very smallest of the microbes are the prions. Then there are the viruses. They mostly range from 20 nm to about 400 nm, although there are a few types that can be as big as 800 nm or 1,500 nm. (Those viruses are as big as cells.) The smallest bacteria are around 200 nm, and the largest are as large as 1 cm. Yeasts (a single-celled form of fungus) are generally 3 to 4 μm, though some can be much larger. Protozoa are generally between 100 and 300 microns (micrometers). It is much easier to get a feel for the relative sizes by looking at a visual (figure 3.12).

Now, on to **microscopy.**

Magnification and Microscope Design

Early microscopists discovered that a clear, glass sphere could act as a lens to magnify small objects. Magnification in most microscopes results from the interaction between visible light waves and the curvature of the lens. When a beam or ray of light transmitted through air strikes and passes through the convex surface of glass, it experiences **refraction,** defined as

Table 3.4 Conversions within the Metric System

	Symbol	Factor ("Meter" multiplied by ___)	Numerically	
Giga	Gm	10^9	1 000 000 000	Billion
Mega	Mm	10^6	1 000 000	Million
Kilo	km	10^3	1 000	Thousand
Meter	m		1	
Centi	cm	10^{-2}	0.01	Hundredth
Milli	mm	10^{-3}	0.001	Thousandth
Micro	μm	10^{-6}	0.000 001	Millionth
Nano	nm	10^{-9}	0.000 000 001	Billionth

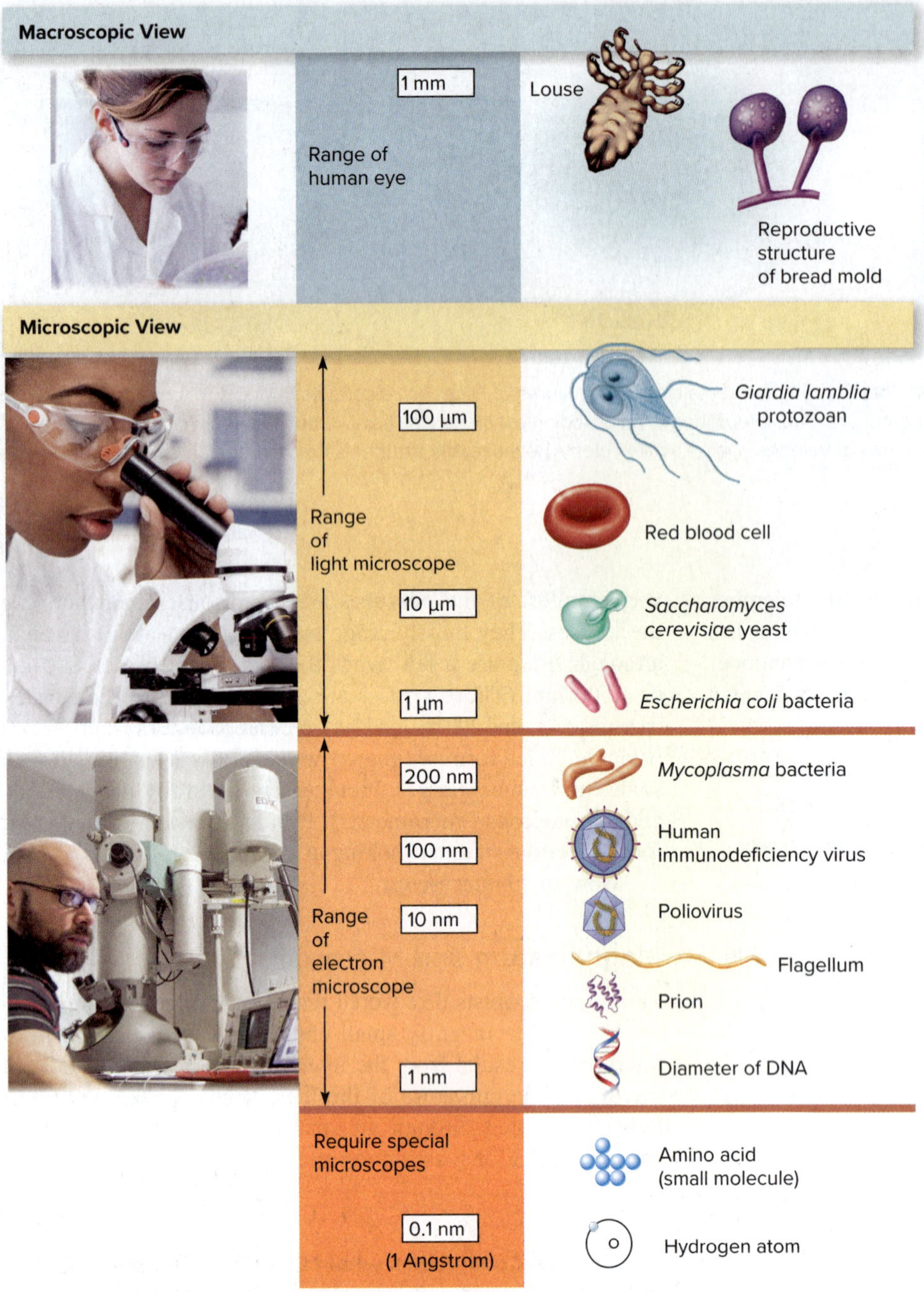
Macroscopic View
1 mm
Range of human eye
Louse
Reproductive structure of bread mold
Microscopic View
100 µm
Giardia lamblia protozoan
Range of light microscope
Red blood cell
10 µm
Saccharomyces cerevisiae yeast
1 µm
Escherichia coli bacteria
200 nm
Mycoplasma bacteria
100 nm
Human immunodeficiency virus
Range of electron microscope
10 nm
Poliovirus
Flagellum
Prion
1 nm
Diameter of DNA
Require special microscopes
Amino acid (small molecule)
0.1 nm
(1 Angstrom)
Hydrogen atom

Figure 3.12 The size of things. Common measurements encountered in microbiology and a scale of comparison from the macroscopic to the microscopic, molecular, and atomic. Most microbes encountered in our studies will fall between 100 µm and 10 nm in overall dimensions. The microbes shown are more or less to scale within size zone but not *between* size zones.

(top) Sigrid Gombert/Cultura Creative/Alamy Stock Photo; (middle) LightFieldStudios/Getty Images; (bottom) Holly Curry/McGraw Hill

the bending or change in the angle of the light ray as it passes through a medium such as a lens. The greater the difference in the composition of the two substances the light passes between (such as air and the lens), the more pronounced the refraction is. When an object is placed a certain distance from the spherical lens and illuminated with light, an image of it is formed by the refracted light. Depending upon the size and curvature of the lens, the image appears enlarged to a particular degree, which is called its power of magnification. This power is identified with a number combined with × (which is said aloud as "times"). This behavior of light is evident if one looks through an everyday object such as a glass globe or a magnifying glass **(figure 3.13).** It is basic to the function of all optical microscopes, though many of them have additional features that define, refine, and increase the size of the image. Optical microscopes are also called light microscopes.

The first microscopes were simple, meaning they contained just a single magnifying lens. Examples of this type of microscope are a magnifying glass, a hand lens, and Leeuwenhoek's basic tool shown earlier in figure 1.8. Among the refinements that led to the development of today's compound microscope were the addition of a second magnifying lens system, a lamp in the base to give off visible light to illuminate the specimen, and a special lens called the condenser that converges or focuses the rays of light to a single point on the object. The fundamental parts of a modern compound light microscope are illustrated in **figure 3.14.**

Principles of Light Microscopy

A microscope provides three properties: magnification, resolution, and contrast. Magnification of the specimen by a compound microscope occurs in two phases. The first lens in this system (the one closest to the specimen) is the objective lens, and the second (the one closest to the eye) is the ocular lens, or eyepiece **(figure 3.15).** The objective forms the initial image of the specimen, called the **real image.** When this image is projected up through the microscope body to the plane of the eyepiece, the ocular lens forms a second image, the **virtual image.** The virtual image is the one that will be received by the eye and converted to a retinal (visual) image. The magnifying power of the objectives in our everyday microscopes usually ranges from 4× to 200×, and the power of the ocular ranges from 10× to 20×. The total power of magnification of the final image formed by the combined lenses is found by multiplying the separate powers of the two lenses.

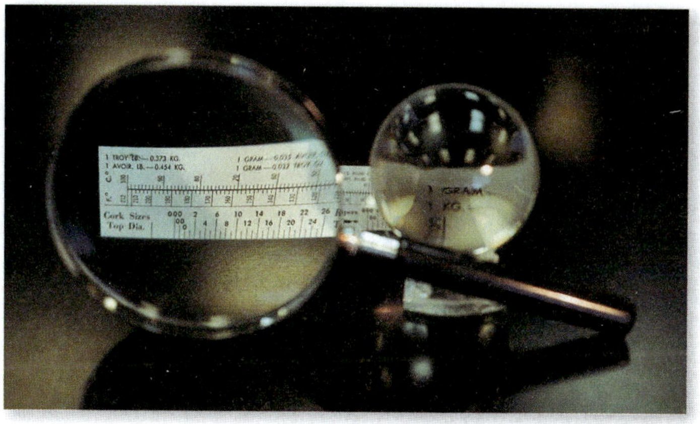

Figure 3.13 Effects of magnification. Demonstration of the magnification and image-forming capacity of clear glass "lenses." Given a proper source of light, this magnifying glass and crystal ball magnify a ruler two to three times.
Kathy Park Talaro

Power of Objective	×	Usual Power of Ocular	=	Total Magnification
10× low power objective	×	10×	=	100×
40× high dry objective	×	10×	=	400×
100× oil immersion objective	×	10×	=	1,000×

Microscopes are equipped with a nosepiece holding three or more objectives that can be rotated into position as needed. The power of the ocular usually remains constant for a given microscope. Depending on the power of the ocular, the total magnification of standard light microscopes can vary from 40× with a 4× objective (called the scanning objective) to 1,000× with the highest power objective (the oil immersion objective).

Resolution: Distinguishing Magnified Objects Clearly

As important as magnification is for visualizing tiny objects or cells, an additional optical property, resolution, is essential for seeing clearly. Resolution is the capacity of an optical system to distinguish or separate two adjacent objects or points from one another. For example, at a certain fixed distance, the lens in the human eye can resolve two small objects as separate points only as long as the two objects are no closer than 0.2 millimeter apart. The eye examination given by optometrists is in fact a test of the resolution of the human eye for various-size letters read at a particular distance. Because microorganisms are extremely small and usually very close together, they will not be seen with clarity or any degree of detail unless the microscope's lenses can resolve them. The **resolving power** of a microscope is the minimum distance two objects can be apart and still be distinguished separately.

Resolving power is determined by a combination of the objective lens characteristics and the wavelength of the light being used on the sample. The light source for optical microscopes consists of colored wavelengths in the visible spectrum. The shortest visible wavelengths are in the violet-blue portion of the

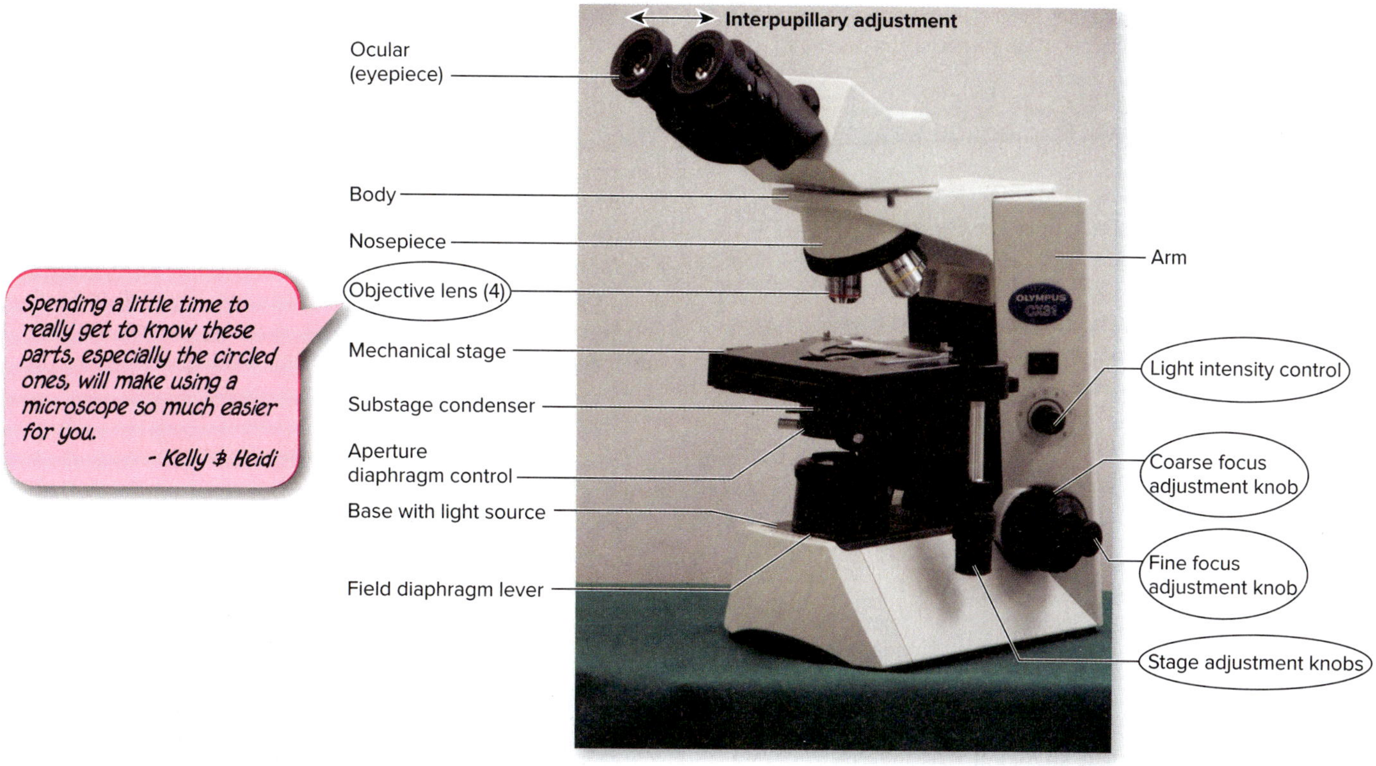
Interpupillary adjustment
Ocular (eyepiece)
Body
Nosepiece
Objective lens (4)
Mechanical stage
Substage condenser
Aperture diaphragm control
Base with light source
Field diaphragm lever
Arm
Light intensity control
Coarse focus adjustment knob
Fine focus adjustment knob
Stage adjustment knobs
Spending a little time to really get to know these parts, especially the circled ones, will make using a microscope so much easier for you.
- Kelly $ Heidi

Figure 3.14 The parts of a student laboratory microscope. This microscope is a compound light microscope with two oculars (called binocular).
James Redfearn/McGraw Hill

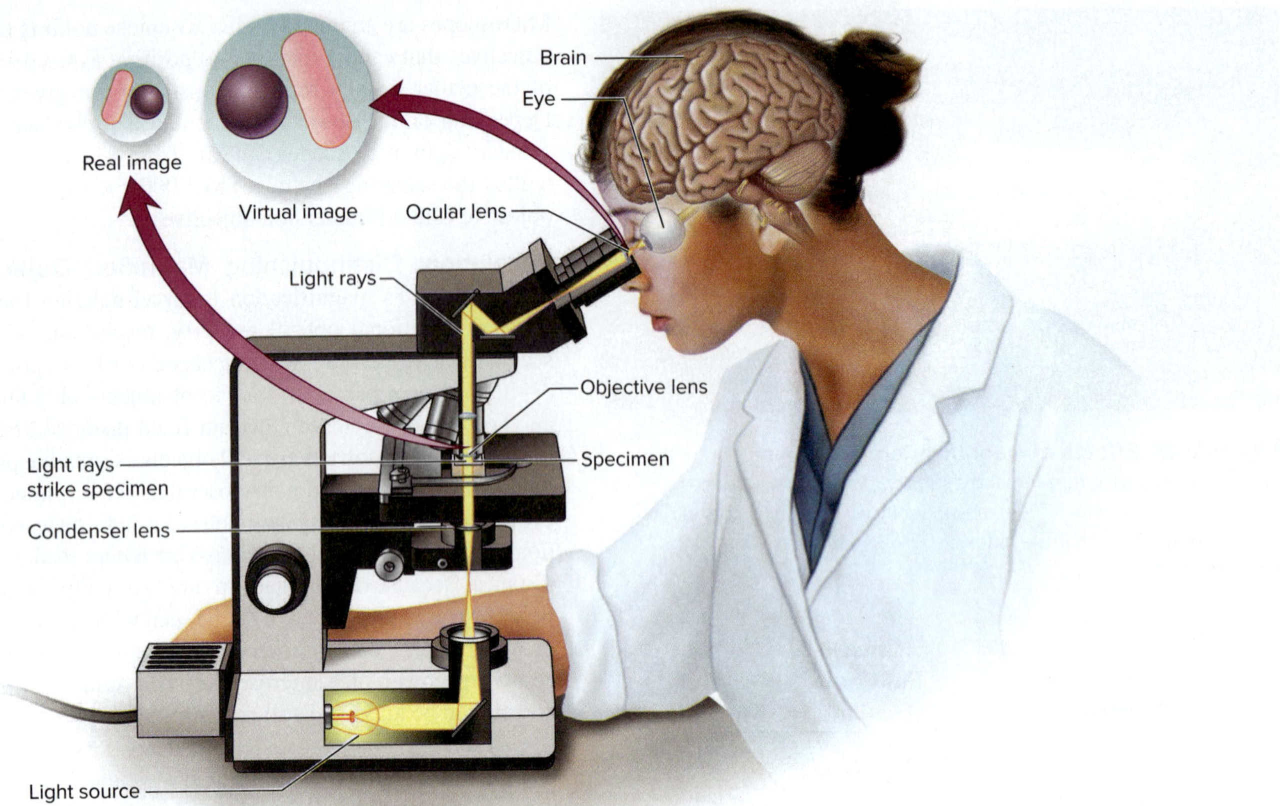
Brain
Eye
Real image
Virtual image
Ocular lens
Light rays
Objective lens
Specimen
Light rays strike specimen
Condenser lens
Light source

Figure 3.15 The pathway of light and the two stages in magnification of a compound microscope. As light passes through the condenser, it forms a solid beam that is focused on the specimen. Light leaving the specimen that enters the objective lens is refracted so that an enlarged primary image, the real image, is formed. One does not see this image, but its degree of magnification is represented by the circle on the left. The real image is projected through the ocular, and a second image, the virtual image, is formed by a similar process. The virtual image is the final magnified image that is received by the retina and perceived by the brain. Notice that the lens systems cause the image to be reversed.

A Note About Oil Immersion Lenses

In order for the oil immersion lens to provide maximum resolution, a drop of oil must be inserted between the tip of the lens and the specimen on the glass slide. Because oil has the same optical qualities as glass, the light rays will not change direction (or refract) when they come through the slide and into the oil. This prevents the scattering of the outermost light rays and creates a more focused beam of light. This can greatly enhance resolution **(figure 3.16).**

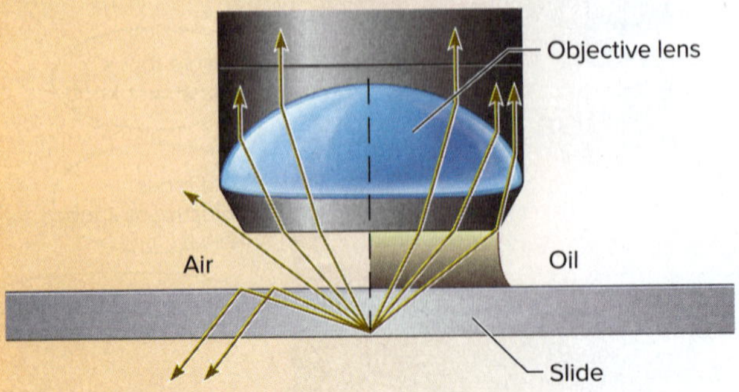
Objective lens
Air
Oil
Slide

Figure 3.16 Workings of an oil immersion lens. Without oil, some of the peripheral light that passes through the specimen is scattered into the air or onto the glass slide. This scattering decreases resolution.

spectrum (400 nanometers), and the longest are in the red portion (750 nanometers) **(figure 3.17).** Because the wavelength must pass *between* the objects that are being resolved, shorter wavelengths (in the 400–500 nanometer range) will provide better resolution **(figure 3.18).** Some microscopes have a special blue filter in place to limit the longer wavelengths of light from entering the specimen.

In practical terms, the oil immersion lens can resolve any cell or cell part as long as it is at least 0.2 micron in diameter, and it can resolve two adjacent objects as long as they are at least

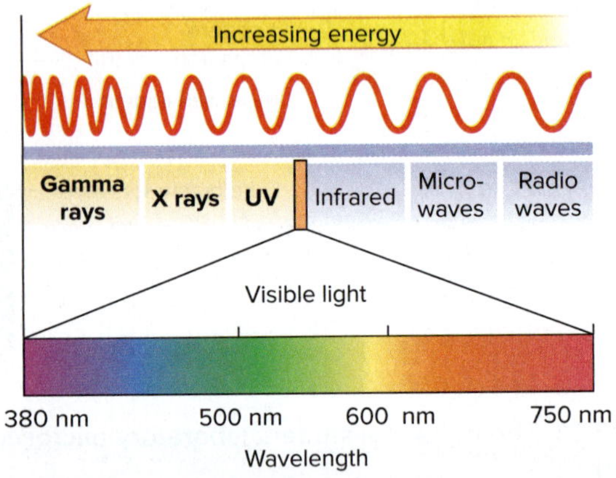
Increasing energy
Gamma rays
X rays
UV
Infrared
Micro-waves
Radio waves
Visible light
380 nm
500 nm
600 nm
750 nm
Wavelength

Figure 3.17 The electromagnetic spectrum.

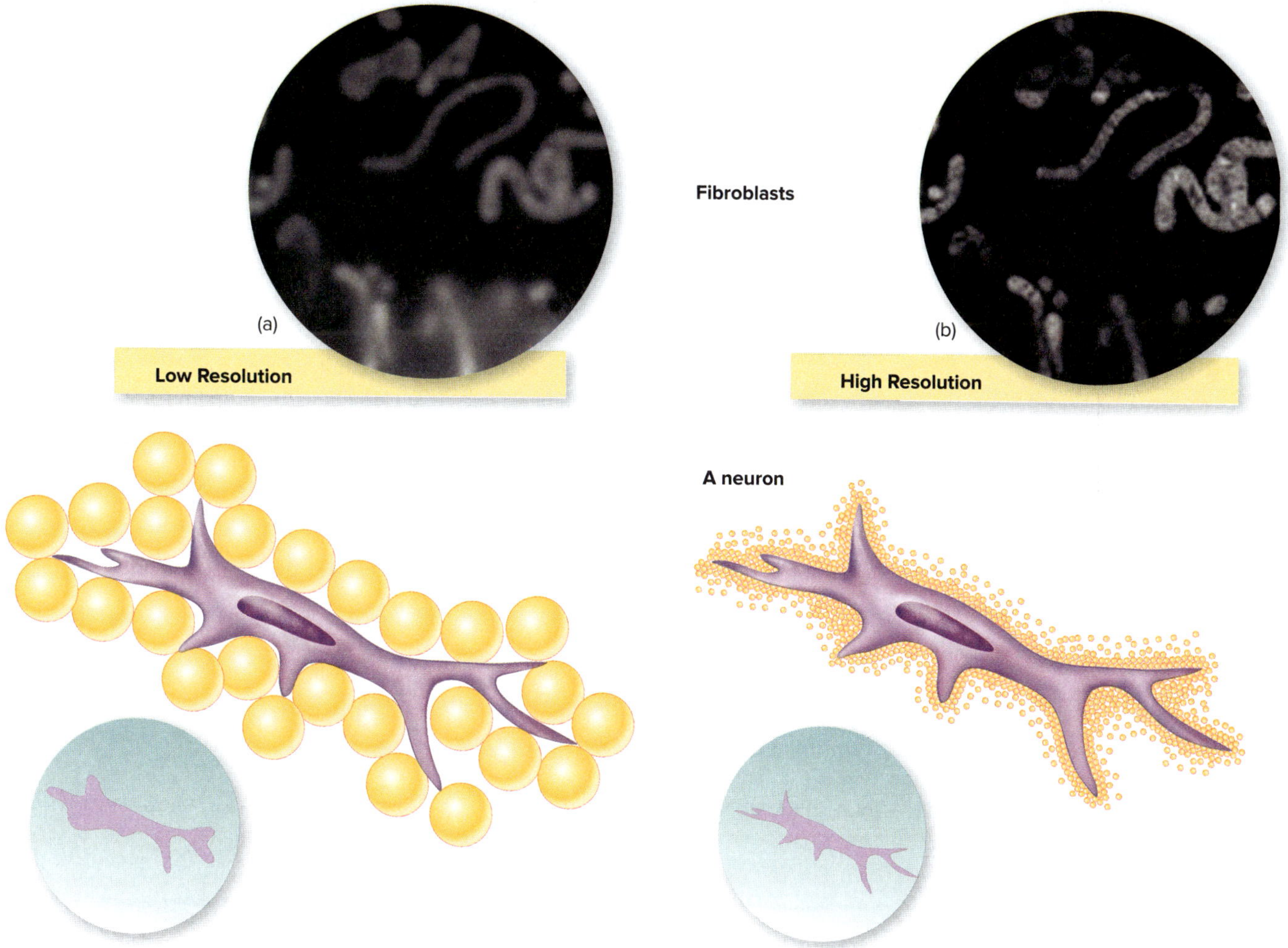
Fibroblasts
(a)
Low Resolution
(b)
High Resolution
A neuron

Figure 3.18 Effect of wavelength on resolution. A simple model demonstrates how the wavelength of light influences the resolving power of a microscope. The size of the balls illustrates the relative size of the wave. Here, a human cell (fibroblast) is illuminated with long-wavelength light **(a)** and short-wavelength light **(b)**. In **(a)**, the waves are too large to penetrate the tighter spaces and they produce a fuzzy, undetailed image.
Nikon Instruments Inc.

0.2 micron apart **(figure 3.19).** In general, organisms that are 0.5 micron or more in diameter are easily seen. This includes fungi and protozoa, some of their internal structures, and most bacteria. However, a few bacteria and most viruses are far too small to be resolved by the optical microscope and require electron microscopy. In summary, the factor that most limits the clarity of a microscope's image is its resolving power. Even if a light microscope were designed to magnify several thousand times, its resolving power could not be increased and the image it produced would simply be enlarged and fuzzy.

Contrast The third quality of a well-magnified image is its degree of contrast from its surroundings. The contrast is measured by a quality called the **refractive index.** Refractive index refers to the degree of bending that light undergoes as it passes from one medium (such as water or glass) to another medium, such as bacterial cells. The higher the difference in refractive indexes (the more bending of light), the sharper the contrast that is registered by the microscope and the eye. Because too much light can reduce contrast and burn out the image, an adjustable iris diaphragm on most microscopes controls the amount of light entering the condenser. The lack of contrast in cell components can be overcome by using special lenses or by adding stains.

Variations on the Light Microscope

There are four types of visible-light microscopes: bright-field, dark-field, phase-contrast, and interference. A fifth type of optical microscope, the fluorescence microscope, uses ultraviolet radiation as the illuminating source. (**Insight 3.1** has a story about using fluorescent microscopy to examine hundreds-of-years-old drawings by Leonardo da Vinci.) The confocal microscope, uses a laser beam. Another instrument, the super-resolution light microscope, uses two lasers. Each of these microscopes is adapted for viewing specimens in a particular way, as described in **table 3.5.**

Preparing Specimens for Optical Microscopes

A specimen for optical microscopy is generally prepared by mounting a sample on a suitable glass slide that sits on the stage between the condenser and the objective lens. How a slide specimen, or mount, is prepared depends upon (1) whether the specimen is in a living or preserved state; (2) the aims of the

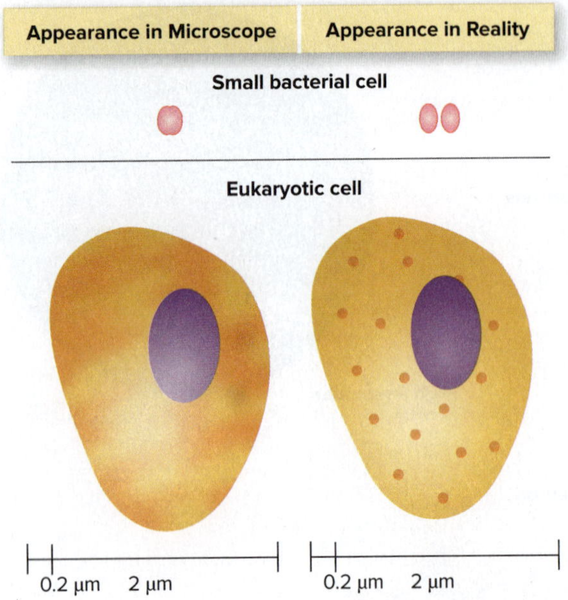

Appearance in Microscope
Appearance in Reality
Small bacterial cell
Eukaryotic cell
0.2 µm
2 µm
0.2 µm
2 µm

Figure 3.19 The importance of resolution. If a microscope has a resolving power of 0.2 µm, then the bacterial cells will not be resolvable as two separate cells. Likewise, the small specks inside the eukaryotic cell will not be visible.

user, whether to observe overall structure, identify the microorganisms, or see movement; and (3) the type of microscopy available, whether it is bright-field, dark-field, phase-contrast, or fluorescence.

Fresh, Living Preparations

Live samples of microorganisms are placed in wet mounts or in hanging drop mounts so that they can be observed as near to their natural state as possible. The cells are suspended in a suitable fluid (water, broth, saline) that temporarily maintains viability and provides space and a medium for locomotion (movement). A wet mount consists of a drop or two of the culture placed on a slide and overlaid with a coverslip. Although this type of mount is quick and easy to prepare, it has certain disadvantages. The coverslip can damage larger cells, and the slide is very susceptible to drying. Because the coverslip is not sealed around the edges, the fluid, and the microbes, can contaminate the handler's fingers. A more satisfactory alternative is the hanging drop preparation made with a special concave (depression) slide, a coverslip from which a tiny drop of sample is suspended, and petroleum jelly, like Vaseline or some other sealant, to put around the edges of the coverslip **(figure 3.20).** These types of short-term mounts provide a true assessment of the size, shape, arrangement, color, and motility of cells.

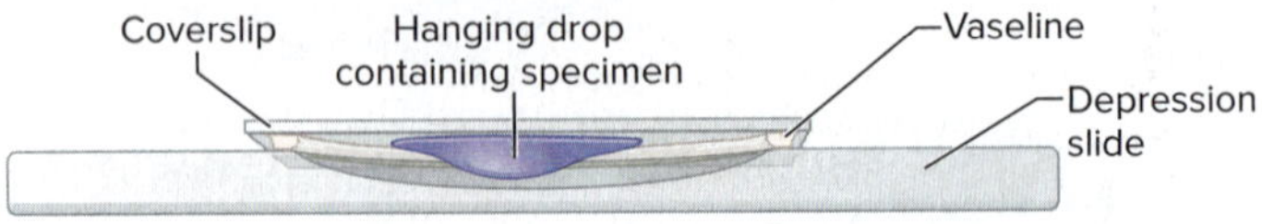

Coverslip
Hanging drop containing specimen
Vaseline
Depression slide

Figure 3.20 Hanging drop technique. Cross-section view of slide and coverslip. (Vaseline actually surrounds entire well of slide.)

INSIGHT 3.1 MICROBIOME: The Microbiome of Leonardo da Vinci's Drawings

In 2020 scientists in Italy and Austria joined forces to analyze the microbial community on the surfaces of seven of Leonardo da Vinci's most famous drawings. Curators and restorers of art have long studied the surfaces of paintings and drawings, both to recreate how they were made and with what materials, and to better understand how they deteriorate over time. Of course, these researchers had to be extremely careful with the drawings from the late 1400s. They started with a gentle micro-suction apparatus that collected dust particles and microbial cells from the surfaces, with minimal impact on the drawings. The suction machine contained membrane filters that trapped these tiny microparticles, and the membranes could then be subjected to microscopy and sophisticated DNA analysis. They used both optical microscopes and electron microscopes.

The team found that there were many species of bacteria as well as fungi on the drawing surfaces. Previous studies of deteriorating artwork had found an excess of fungi (think mold) in the microbiome of the art objects. The works in this study were not deteriorating, however, which is one reason fungi were not in the majority.

The researchers called their efforts a "bio-archive"—a method for cataloguing the conditions during the creation of the pieces, as well as how they were stored and displayed over time, and even where they were geographically located over time. It will guide preservation and restoration.

Self-portrait of Leonardo da Vinci drawn in red chalk, in the 1500s.

Pixtal/age fotostock

Table 3.5 Comparison of Types of Microscopy

Microscope	Maximum Practical Magnification	Resolution	
Visible light as source of illumination			
Bright-field 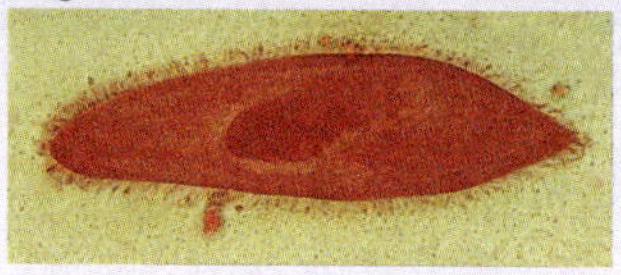 *Paramecium* (230×) *Michael Abbey/Science Source*	2,000×	0.2 µm (200 nm)	The bright-field microscope is the most widely used type of light microscope. Although we ordinarily view objects like the words on this page with light reflected off the surface, a bright-field microscope forms its image when light is transmitted through the specimen. The specimen, being denser and more opaque than its surroundings, absorbs some of this light, and the rest of the light is transmitted directly up through the ocular into the field. As a result, the specimen will produce an image that is darker than the surrounding brightly illuminated field. The bright-field microscope is a multipurpose instrument that can be used for both live, unstained material and preserved, stained material.
Dark-field 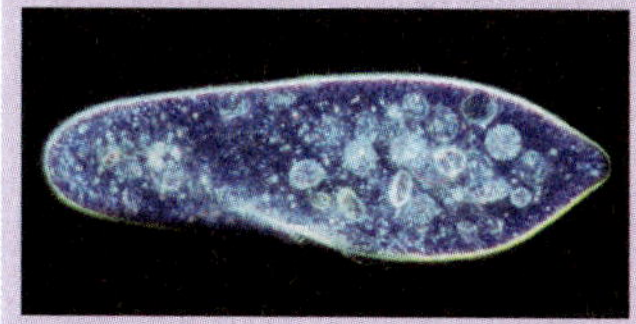 *Paramecium* (230×) *Laguna Design/Science Source*	2,000×	0.2 µm	A bright-field microscope can be adapted into a dark-field microscope by adding a special disc called a *stop* to the condenser. The stop blocks all light from entering the objective lens—except peripheral light that is reflected off the sides of the specimen itself. The resulting image is a particularly striking one: brightly illuminated specimens surrounded by a dark (black) field. The most effective use of dark-field microscopy is to visualize living cells that would be distorted by drying or heat or that cannot be stained with the usual methods.
Phase-contrast 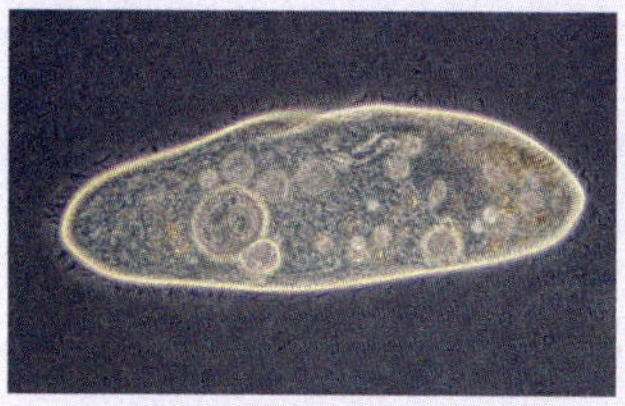 *Paramecium* (230×) *Michael Abbey/Science Source*	2,000×	0.2 µm	If similar objects made of clear glass, ice, cellophane, or plastic are immersed in the same container of water, an observer would have difficulty telling them apart because they have similar optical properties. In the same way, internal components of a live, unstained cell also lack contrast and can be difficult to distinguish. But cell structures do differ slightly in density, enough that they can alter the light that passes through them in subtle ways. The phase-contrast microscope has been constructed to take advantage of this characteristic. This microscope contains devices that transform the subtle changes in light waves passing through the specimen into differences in light intensity. For example, denser cell parts such as organelles alter the pathway of light more than less dense regions (the cytoplasm). Light patterns coming from these regions will vary in contrast. The amount of internal detail visible by this method is greater than by either bright-field or dark-field methods. The phase-contrast microscope is most useful for observing intracellular structures such as bacterial endospores, granules, and organelles, as well as the locomotor structures of eukaryotic cells such as cilia. (This image has been colorized; the actual microscopic image is black and white.)
Differential interference contrast (DIC)  *Paramecium* (230×) *Gerd Guenther/Science Source*	2,000×	0.2 µm	Like the phase-contrast microscope, the differential interference contrast microscope provides a detailed view of unstained, live specimens by manipulating the light. But this microscope has additional refinements, including two prisms that add contrasting colors to the image and two beams of light rather than a single one. DIC microscopes produce extremely well-defined images that are vividly colored and appear three-dimensional.
Ultraviolet rays as source of illumination			
Simple fluorescence 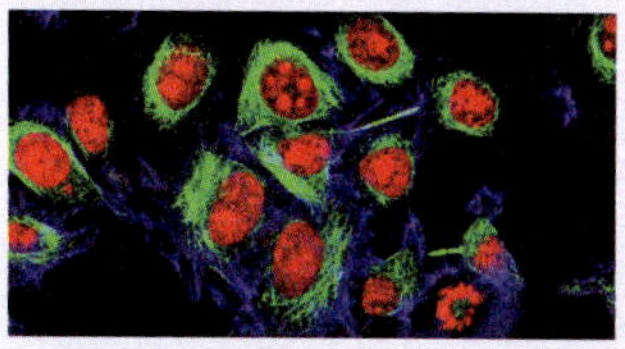 Epithelial cancer cells (400×) *Vshivkova/Shutterstock*	2,000×	0.2 µm	The fluorescence microscope is a specially modified compound microscope furnished with an ultraviolet (UV) radiation source and a filter that protects the viewer's eye from injury by these dangerous rays. The name of this type of microscopy originates from the use of certain dyes (acridine, fluorescein) and minerals that are **fluorescent.** The dyes emit visible light when bombarded by short ultraviolet rays. For an image to be formed, the specimen must first be coated with a source of fluorescence. Shining ultraviolet radiation on the specimen causes the specimen to give off light that will form its own image, usually an intense color such as red against a black field.

(*continued*)

Table 3.5 Comparison of Types of Microscopy (*continued*)

Microscope	Maximum Practical Magnification	Resolution	
Confocal 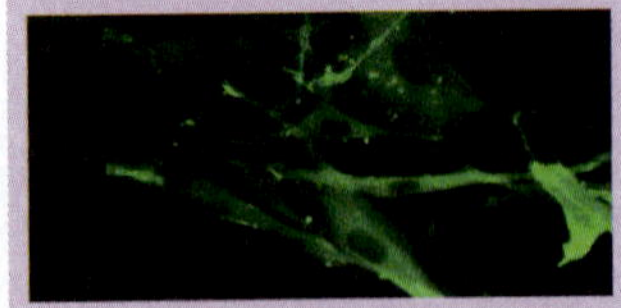 Myofibroblasts, cells involved in tissue repair (400×) *Dr. Jeremy Allen/University of Salford, Biosciences Research Institute*	2,000×	0.2 μm	The scanning confocal microscope overcomes the problem of cells or structures being too thick, a problem resulting in other microscopes being unable to focus on all their levels. This microscope uses a laser beam of light to scan various depths in the specimen and deliver a sharp image focusing on just a single plane. This enables it to capture a highly focused view at any level, ranging from the surface to the middle of the cell. It is most often used on fluorescently stained specimens, but it can also be used to visualize live unstained cells and tissues.
Super-resolution light microscopy 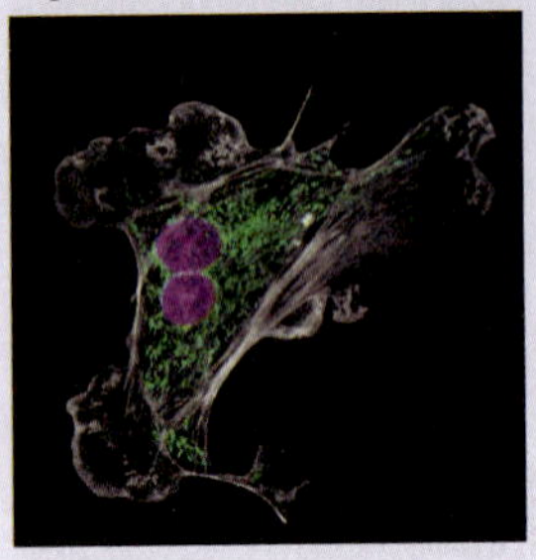 An endothelial cell. Mitochondria are green and DNA of this binucleated cell is magenta (833×) *Talley Lambert/Science Source*	2,000×	0.1 μm	This is a recently developed method of microscopy that uses two laser beams to illuminate fluorescently labeled cell components. One of the laser beams causes the desired 1 nm area to light up; the other cancels out all other fluorescence, allowing for much higher resolution.
Electron beam forms image of specimen			
Transmission electron microscope (TEM) 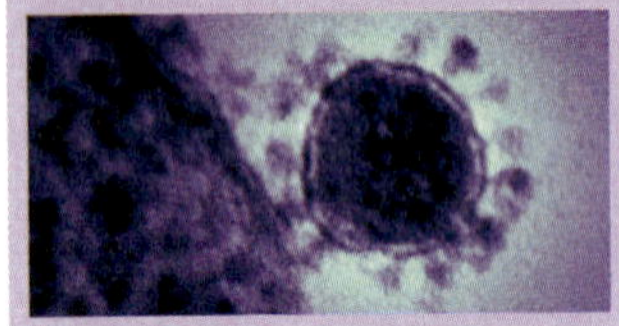 Coronavirus, causative agent of many respiratory infections (100,000×) *National Institute of Allergy and Infectious Diseases (NIAID)/NIH/USHHS)*	100,000,000×	0.5 nm	Transmission electron microscopes are the instruments of choice for viewing the detailed structure of cells and their organelles and viruses. This microscope produces its image by transmitting electrons through the specimen. Because electrons cannot readily penetrate thick preparations, the specimen must be sectioned into extremely thin slices (20–100 nm thick) and stained or coated with metals that will increase image contrast. The darkest areas of TEM micrographs represent the thicker (denser) parts, and the lighter areas indicate the more transparent and less dense parts.
Scanning electron microscope (SEM)  2 microns An algal cell showing a cell wall made of calcium disks (10,000×) *Steve Gschmeissner/Science Photo Library/Getty Images*	100,000,000×	10 nm	The scanning electron microscope provides some of the most dramatic and realistic images. This instrument is designed to create an extremely detailed three-dimensional view of all kinds of objects—from plaque on teeth to tapeworm heads. To produce its images, the SEM does not transmit electrons; it bombards the surface of a whole metal-coated specimen with electrons while scanning back and forth over it. A shower of electrons deflected from the surface is picked up with great fidelity by a sophisticated detector, and the electron pattern is displayed as an image on a television screen. The contours of the specimen resolved with scanning electron microscopy are very revealing. Areas that look smooth and flat with the light microscope display intriguing surface features with the SEM. (This image has been colorized; the actual microscopic image is black and white.)

Fixed, Stained Smears

A more permanent slide for long-term study can be created by preparing fixed, stained specimens. This involves the smear technique, that consists of spreading a thin film made from a liquid suspension of cells on a slide and air-drying it. Next, the air-dried smear is usually heated gently by a process called heat fixation that simultaneously kills the specimen and secures it to the slide. Another important action of fixation is to preserve various cellular components in a natural state with minimal distortion. Sometimes fixation of microbial cells is performed with chemicals such as alcohol and formalin.

Like images on undeveloped photographic film, the unstained cells of a fixed smear are quite indistinct, no matter how great the magnification or how fine the resolving power of the microscope. The process of "developing" a smear to create contrast and make inconspicuous features stand out requires staining techniques.

Table 3.6 Comparison of Positive and Negative Stains

Medium	Positive Staining	Negative Staining
Appearance of cell	Colored by dye 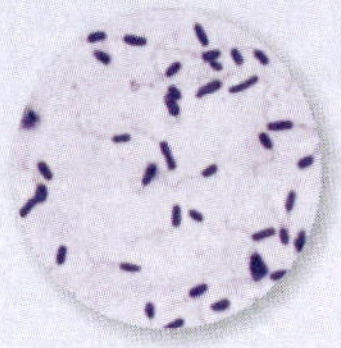	Clear and colorless 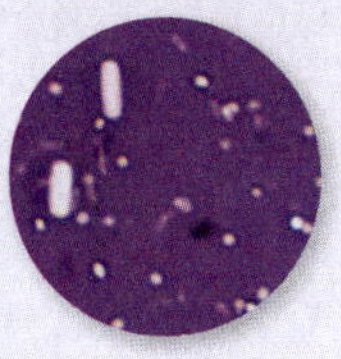
Background	Not stained (generally white)	Stained (dark gray or black)
Dyes employed	Basic dyes: Crystal violet Methylene blue Safranin Malachite green	Acidic dyes: Nigrosin India ink
Subtypes of stains	Several types: Simple stain Differential stains Gram stain Acid-fast stain Spore stain Special stains Capsule Flagella Spore Granules Nucleic acid	Few types: Capsule Spore

Photos: Lisa Burgess/McGraw Hill

Staining is any procedure that applies dyes to specimens. Dyes give color to cells or cell parts by becoming affixed to them through a chemical reaction. Dyes can be classified as basic (cationic), which have a positive charge, or acidic (anionic) dyes, which have a negative charge. Because chemicals of opposite charge are attracted to each other, cell parts that are negatively charged will attract basic dyes and those that are positively charged will attract acidic dyes **(table 3.6).** Many cells, especially those of bacteria, have a lot of negatively charged acidic substances on their surfaces and thus stain readily with basic dyes. Acidic dyes, on the other hand, tend to be repelled by cells, so they are good for negative staining (discussed next).

Negative versus Positive Staining Two basic types of staining technique are used, depending upon how a dye reacts with the specimen (summarized in table 3.6). Many procedures involve a **positive stain,** in which the dye actually sticks to the specimen and gives it color. A **negative stain,** on the other hand, is just the reverse (like a photographic negative). The dye does not stick to the specimen but settles around its outer boundary, forming a silhouette. In a sense, negative staining "stains" the fluid to produce a dark background around the cells. Nigrosin (blue-black) and India ink (a black suspension of carbon particles) are the dyes most commonly used for negative staining. The cells themselves do not stain because these dyes are negatively charged and are repelled by the negatively charged surface of the cells. The value of negative staining is its simplicity and the reduced shrinkage or distortion of cells because the smear is not heat fixed. This enables a quick assessment of cellular size, shape, and arrangement. Negative staining is also used to accentuate the capsule that surrounds certain bacteria and yeasts **(figure 3.21*c*).**

Disease Connection

Sometimes, a capsule can make all the difference. *Streptococcus pneumoniae* is a very common and dangerous pathogen with a large capsule. It colonizes the upper respiratory tract (URT) and from there can cause pneumonia, meningitis, and bloodstream infections. Scientists have shown that if the capsule is absent, the bacterium cannot attach to the URT and, therefore, causes no disease.

Simple versus Differential Staining Staining methods are classified as simple, differential, or special **(figure 3.21). Simple stains** require only a single dye and an uncomplicated procedure. **Differential stains** use two differently colored dyes, called the primary dye and the counterstain, to distinguish between cell types or parts. These staining techniques tend to be more complex and sometimes require additional chemical reagents to produce the desired reaction. Special stains are those that were developed for a single, special, purpose.

Most simple staining techniques **(figure 3.21*a*)** take advantage of the easy binding of dyes like malachite green, crystal violet, basic fuchsin, and safranin to cells. Simple stains cause all cells in a smear to appear more or less the same color, regardless of type, but they can still reveal characteristics such as shape, size, and arrangement. Another commonly used simple stain is the calcofluor-KOH stain, which is used to visualize fungi in skin, nail, or body fluid samples using a fluorescent microscope.

Types of Differential Stains A differential stain uses differently colored dyes to clearly contrast two cell types or cell parts. Common combinations are red and purple, red and green, or pink and blue. Differential stains can also point out other characteristics, such as the size, shape, and arrangement of cells. Typical examples include Gram, acid-fast, and endospore stains. Some differential staining techniques (spore stain and capsule stain) are also in the "special" category as they pinpoint a particular characteristic, such as the presence of an endospore **(figure 3.21*b*).**

Gram staining, a century-old method named for its developer, Hans Christian Gram, remains the most universal diagnostic staining technique for bacteria. It allows for quick differentiation of major categories based upon the color reaction of the cells: **gram-positive,** which stain purple, and **gram-negative,** which stain pink or red. The Gram stain is the basis of several important bacteriologic traits, including bacterial taxonomy, cell wall structure, and identification and diagnosis of infection. It can even guide the selection of the correct drug for an infection. Gram staining is discussed in greater detail in chapter 4. Please note, that the Gram staining procedure is a positive stain, in the classification of positive vs. negative staining.

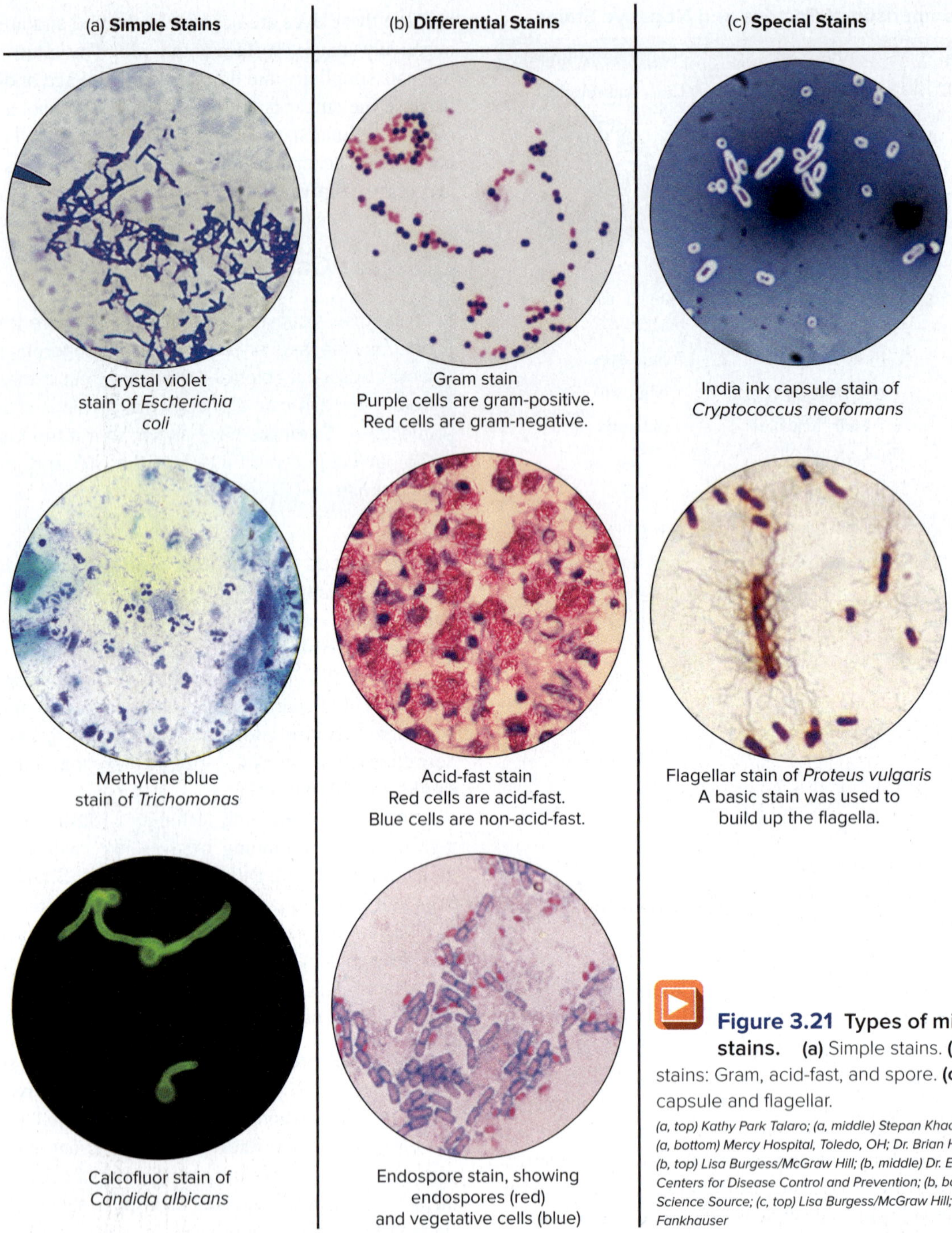
(a) Simple Stains
(b) Differential Stains
(c) Special Stains
Crystal violet stain of *Escherichia coli*
Gram stain
Purple cells are gram-positive.
Red cells are gram-negative.
India ink capsule stain of *Cryptococcus neoformans*
Methylene blue stain of *Trichomonas*
Acid-fast stain
Red cells are acid-fast.
Blue cells are non-acid-fast.
Flagellar stain of *Proteus vulgaris*
A basic stain was used to build up the flagella.
Calcofluor stain of *Candida albicans*
Endospore stain, showing endospores (red) and vegetative cells (blue)

Figure 3.21 Types of microbiological stains. **(a)** Simple stains. **(b)** Differential stains: Gram, acid-fast, and spore. **(c)** Special stains: capsule and flagellar.

(a, top) Kathy Park Talaro; (a, middle) Stepan Khadzhi/Shutterstock; (a, bottom) Mercy Hospital, Toledo, OH; Dr. Brian Harrington/CDC; (b, top) Lisa Burgess/McGraw Hill; (b, middle) Dr. Edwin P. Ewing, Jr./Centers for Disease Control and Prevention; (b, bottom) Manfred Kage/Science Source; (c, top) Lisa Burgess/McGraw Hill; (c, bottom) David Fankhauser

The **acid-fast stain,** like the Gram stain, is an important diagnostic stain that differentiates acid-fast bacteria (pink or red) from non-acid-fast bacteria (blue). This stain originated as a specific method to detect *Mycobacterium tuberculosis* in specimens. These bacterial cells have a particularly impenetrable outer wall that holds "fast" (as in *tightly* or *tenaciously*) to the dye (carbol fuchsin) even when washed with a solution containing acid or acid alcohol. This stain is used for other medically important mycobacteria such as the Hansen's disease (leprosy) bacterium and for *Nocardia,* an agent of lung and skin infections.

In the endospore stain, a dye is forced by heat into impermeable bodies called endospores. This stain is designed to distinguish between endospores and the cells that they come from (called **vegetative** cells). Gram-positive, endospore-forming members of the genus *Bacillus* (the cause of anthrax) and *Clostridium* (the cause of botulism and tetanus) are dramatic diseases that we consider in chapters 20 and 21.

Special stains **(figure 3.21*c*)** are used to emphasize certain cell parts that may not be revealed by conventional staining methods. **Capsule staining** is a method of observing the microbial capsule, an unstructured protective layer surrounding the cells of some bacteria and fungi. Because the capsule does not react with most stains, it is often negatively stained with India ink, or it may be visualized by special positive stains. The fact that not all microbes exhibit capsules is a useful feature for identifying pathogens. One example is *Cryptococcus,* which causes a serious form of fungal meningitis in patients with AIDS (see chapter 21).

Flagellar staining is a method for revealing flagella, the tiny, slender filaments used by bacteria for movement. Because the width (actually, the narrowness) of bacterial flagella lies beyond the resolving power of the light microscope, in order to be seen, they must be enlarged by depositing a coating on the outside of the filament and then staining it. Their presence, number, and arrangement on a cell are useful for identification of the bacteria.

3.2 Learning Outcomes—Assess Your Progress

5. Convert among different lengths within the metric system.
6. Describe the earliest microscopes.
7. List and describe the three elements of good microscopy.
8. Differentiate between the principles of light and electron microscopy.
9. Compare and contrast the three main categories of stains, and provide examples of each.

Media Under The Microscope Wrap-Up

ndenisov/123RF

The **intended message** of the article seems to be a fairly straightforward reporting of a mysterious, and catastrophic, epidemic event among the antelopes.

A **critical reading** of the article leads me to believe that it is indeed factual. Even though it contains sensational information (every individual in a herd dies in one three-week period), it is not *sensationalistic.* Sensational*ism* refers to the over-hyping of events or data in such a way as to scare or shock an audience. The shocking information in the article is 100% true and simply reported. Also, the fact that the reporters interviewed respected professionals who were not involved in the research is a very good sign that they were being thorough. Lastly, the *New York Times* is well respected for its science reporting. You won't necessarily know a source's reputation for its science coverage, but when you do, it can help guide your judgment about the article.

To **interpret** the article for your friends, explain that different growth conditions can cause microorganisms to express different characteristics, and in this case, the warm, moist conditions may have triggered some property to make the bacteria more aggressive.

My **overall grade** for this article is an A−. It is good reporting. The only thing I would have liked to have seen is the researchers discussing possibilities for *how* the bacteria changed.

Source: New York Times (January 17, 2018), www.nytimes.com/2018/01/17/science/saiga-deaths-bacteria.html.

Study Smarter: Better Together

These activities are designed for you to use on your own with a study group—either a face-to-face group or a virtual one, consisting of 3–5 members. Studying together can be very helpful, but there are effective and ineffective ways to do it. For example, getting together without a clear structure is often not a good use of your time. Use your time efficiently by using one or more of the exercises below.

FACE-TO-FACE GROUPS

Use one or more of the activities below.

Peer Instruction: Assign numbers to your group members to use all semester long. Now look at these five concepts from this chapter. Each group member prepares a 5-minute lesson on the topic corresponding to their number. Don't worry if you have fewer than 5 members; just use however many you have! During your group study time, each member presents their lesson, and the group spends another 5–10 minutes discussing that lesson.

1. Chemically defined versus complex media
2. Differential versus selective media
3. Streaking for isolation
4. Positive versus negative staining
5. Simple versus differential staining

Concept Maps: Each member of the group should use this list of terms from this chapter to generate their own concept map. This can be hand-drawn or created using software (see Appendix C for guidelines). During group study time, compare each other's concept maps and help each other make sure they are correct. Of course, there are many different "correct" maps. Examining each member's map will help you talk through the varied concepts and how they are related.

(continued)

Concept Terms:

inoculation	inspection	identification	isolation
incubation	medium	staining	biochemical tests
subculture			

Table Topics: Each group member should identify a concept or topic from this week's class assignments with which they are having trouble and share it during group study time. The other group members can then help to clarify confusing issues or share how they figured it out. Aim for a maximum of 15 minutes per topic. If the topic remains unclear to the group, bring it up during class or use the instructor's office hours or e-mail to ask for help. Taking the time to struggle with a difficult concept first makes your questions much more specific and more likely to yield helpful answers.

VIRTUAL GROUPS

Not everyone has the time or opportunity to meet with group members outside of class time. You or your instructor can create a virtual group using e-mail or the course software.

Weekly Discussion Board: This forum can be used as a way for groups to discuss topics, via e-mail, or other learning management systems or online platforms, before they are covered in class. As each member of the group answers the current week's question, they should send their responses to every other member of their group. It's best to agree on a deadline based on how your class schedule works (Saturday for the next week's topics, for example). Then, after the topic is discussed in class, each member should send a response that all group members will see with a follow-up post on the same topic. If you cover more than one chapter in a week, someone can be designated to choose which chapter Discussion Board question you will use. Or simply decide up front that you will always use the first-chapter-of-the week's question, to keep the schedule simple.

Discussion Question

Find a video on the Internet demonstrating the streak plate technique. (1) In your own words, provide a numbered list of instructions for performing the technique. (2) Finish this sentence: The main purpose of performing the streak plate technique is ______________________.

Chapter Summary

TOOLS OF THE LABORATORY

3.1 METHODS OF CULTURING MICROORGANISMS: THE FIVE I'S

- The Five I's summarize the kinds of laboratory methods used in microbiology. They are inoculation, incubation, isolation, inspection, and identification.
- Some microorganisms can be cultured on artificial media. Others require a living host in the laboratory. Many others are not yet culturable at all.
- Isolated colonies originate from single cells, which reproduce over and over again until there are enough of them to form a visible colony.
- Microbes are identified through their microscopic morphology, their biochemical reactions, their immunological reactivity, and their genetic characteristics.

3.2 THE MICROSCOPE: WINDOW ON AN INVISIBLE REALM

- Magnification, resolving power, and contrast are the most critical properties of a microscope.
- There are several types of optical (light) microscopes: visible light, bright-field, dark-field, phase-contrast, differential interference, fluorescence, confocal, and super-resolution light microscopes.
- Other microscopes use electrons, not light waves, as an energy source. These have both very high magnification and high resolution. The two types are scanning electron microscopes (SEM) and transmission electron microscopes (TEM).
- Stains are used to increase the contrast of specimens for microscopy. They can be categorized as simple, differential, and special.

SmartGrid: From Knowledge to Critical Thinking

This *21 Question Grid* takes the topics from this chapter and arranges them with respect to the American Society for Microbiology's Undergraduate Curriculum guidelines—all six of the important "Concepts" as well as the important "Competency" of scientific literacy. Three questions are supplied, which cover chapter content referring to the Concept or Competency in increasing levels of Bloom's taxonomy for learning.

ASM Concept/ Competency	A. Bloom's Level 1, 2—Remember and Understand (Choose one)	B. Bloom's Level 3, 4—Apply and Analyze	C. Bloom's Level 5, 6—Evaluate and Create
Evolution	**1.** The identities of microorganisms on our planet **a.** are mostly known. **b.** have nearly all been identified via microscopy. **c.** have nearly all been identified via culturing techniques. **d.** are still mostly unknown.	**2.** Before the availability of molecular techniques, archaea and bacteria were considered to be very closely related. Explain why that is understandable, based on what you learned about microscopy.	**3.** Often bacteria that are freshly isolated from a patient or the environment have thick sugary outer layers (capsules). After they are transferred many times on laboratory medium, they tend to lose their capsule. Speculate about why that might happen.
Cell Structure and Function	**4.** Which of these types of organisms is least likely to be identified to the genus level with light microscopy? **a.** bacteria **b.** protozoa **c.** fungus **d.** helminth	**5.** Obviously, *structures* of microorganisms can be revealed by various types of microscopy. Can you think of any *functions* that might also be revealed? Which ones, and by which microscopic technique?	**6.** Some bacteria can produce a structure called an endospore, which is used as a survival structure. It is surrounded by a very tough and impervious outer layer that keeps it protected in adverse environmental conditions. What kinds of things would you need to know about this structure if you wanted to devise a way to stain it and see it under a microscope?
Metabolic Pathways	**7.** A fastidious organism must be grown on what type of medium? **a.** general-purpose medium **b.** differential medium **c.** defined medium **d.** enriched medium	**8.** Write a short paragraph to differentiate among the following: *selective medium, differential medium, enriched medium.*	**9.** Alexander Fleming discovered penicillin in 1928. The first part of the story is that he noticed that agar plates upon which he was growing a bacterium were contaminated with obvious fungal colonies. Instead of throwing the plates out because they were contaminated, he examined them and realized that around the areas where the fungi were growing, there were no bacteria growing. Finish this story, using only your imagination.
Information Flow and Genetics	**10.** Viruses are commonly grown in/on **a.** animal cells or tissues. **b.** agar plates. **c.** broth cultures. **d.** all of the above.	**11.** Can you devise a growth medium with ingredients that would detect whether a culture of MRSA (methicillin-resistant *Staphylococcus aureus*) that you have been storing in your lab is still antibiotic resistant?	**12.** There is a type of differential medium that can reveal whether bacteria growing on it produce and secrete an enzyme that breaks down DNA. Why would a bacterium secrete an enzyme that destroys DNA?

(continued)

ASM Concept/ Competency	A. Bloom's Level 1, 2—Remember and Understand (Choose one)	B. Bloom's Level 3, 4—Apply and Analyze	C. Bloom's Level 5, 6—Evaluate and Create
Microbial Systems	**13.** Most of the time, microbes in natural circumstances exist **a.** as single cells. **b.** in relationship with other species. **c.** as single species. **d.** as colonies on agar.	**14.** Several bacteria live naturally in a material on your teeth called *plaque* that contains many different species, which interact with each other in significant ways. Identify some of the problems of studying one of these bacterial species after isolating it through a streak plate procedure and examining its behavior.	**15.** Archaea often grow naturally in extreme environments. Those called *extreme thermophiles* thrive in hot springs and in high-temperature hydrothermal vents in the ocean. Invent a culturing technique you could use that would be selective for these organisms.
Impact of Microorganisms	**16.** Diagnosis of infections in a hospitalized person usually requires **a.** microscopy. **b.** culture of samples from the patient. **c.** palpation of the infected area. **d.** two of the above.	**17.** After performing the streak plate procedure on a bacterial specimen, the culture was incubated. When you viewed the plate, there was heavy growth with no isolated colonies in the first quadrant. There was no other growth on the plate (in the second quadrant and beyond). Discuss possible errors that could have led to this result.	**18.** You are a scientist studying a marsh area contaminated with PCBs, toxic chemical compounds found in industrial waste. You look at a water sample using light microscopy and discover motile single cells that are about 1 micron in circumference. You hypothesize that these microbes might be useful in cleaning up the toxic waste. Because there is no instruction book for how to culture them, how might you go about creating a growth medium for them?
Scientific Thinking	**19.** If a Gram stain result is unclear, which of the following should be considered? **a.** getting a new sample **b.** restaining the same sample **c.** identity of the microbe may actually be gram-variable **d.** all of the above	**20.** Explain the difference between multiplying a meter by 10^9 and multiplying it by 10^{-9} to get a gigameter and a nanometer, respectively.	**21.** You perform the special stain for bacterial flagella on a specimen. You don't know whether or not the bacteria you are staining possess flagella. When you examine the specimen, you see no flagella on the bacteria. What are all the possibilities for why you didn't see them?

Answers to the multiple-choice questions appear in Appendix A.

Visual Connections

These questions use visual images to connect content within and between chapters.

1. **Figure 3.10.** If you were using the streak plate method to plate a very dilute broth culture (with many fewer bacteria than the broth used here), would you expect to see single, isolated colonies in area 4 or area 3? Explain your answer.

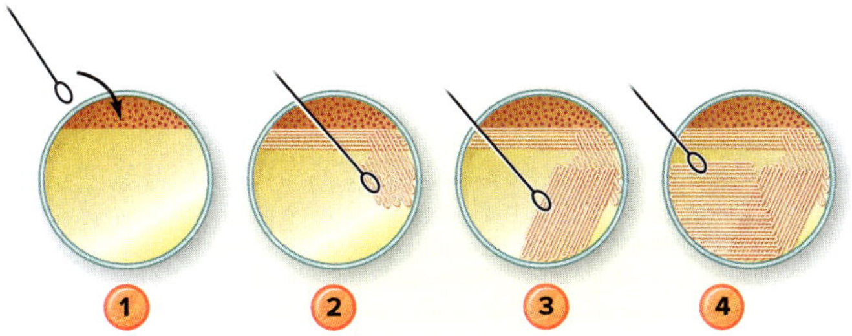
1
2
3
4

Note: This method only works if the spreading tool (usually an inoculating loop) is resterilized after each of steps 1–4.

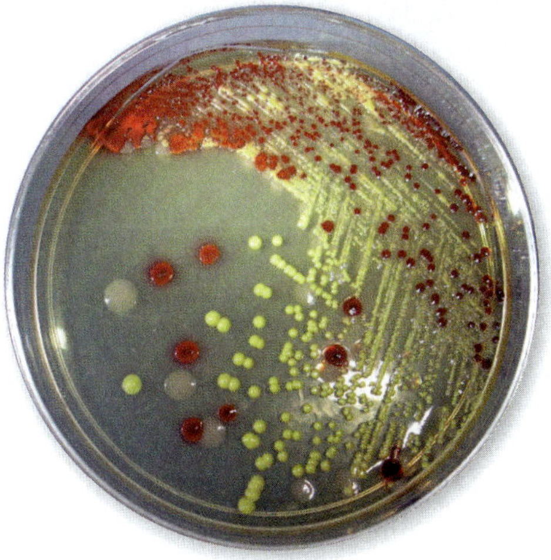

Kathy Park Talaro

High Impact Study

These terms and concepts are most critical for your understanding of this chapter—and may be the most difficult. Have you mastered them?

Concepts

- ☐ Chemically defined versus complex media
- ☐ Differential versus selective media
- ☐ Streaking for isolation
- ☐ Oil immersion lenses
- ☐ Positive versus negative staining
- ☐ Simple versus differential staining

Terms

- ☐ Sterile
- ☐ Hemolysis
- ☐ Resolution
- ☐ Contrast

Design Element: (College students): Caia Image/Image Source

4 Bacteria and Archaea

Rich Carey/Shutterstock

MEDIA UNDER THE MICROSCOPE

Plastic Eaters

This opening case examines an article from the popular media to determine the extent to which it is factual and/or misleading. This case focuses on the March 2021 article from Forbes, "The Race to Develop Plastic-Eating Bacteria."

Plastics are a major part of our everyday lives. Did you know they are man-made materials? Consider that fact and look back to the timeline in figure 1.1, and you will probably understand that there are very few microbes that have evolved the ability to digest them because they have appeared so recently on earth. For that reason, plastics linger in the environment for exceedingly long times, hundreds of years in some cases. Search for the terms "plastic" and "oceans" to see how much of a problem this is.

This article describes the discovery of a bacterium that can "eat" plastic. It is called *Ideonella sakaiensis,* and it was discovered inside a bottle recycling plant in Japan. It contains an enzyme that can digest a specific type of plastic, called PET (it stands for polyethylene terephthalate). Single-use drink bottles are made from it. Scientists have tweaked the enzyme so that it acts even better and faster. The article provides clear definitions of "enzymes" and how they work, and what the researchers did to make it better. Also, they provide a warning: If bacteria continue to evolve to be able to degrade plastic (and if scientists continue to tweak them), then structures and objects built to last a lifetime may gradually come under attack and be degraded. But having a way to deal with our plastic pollution is a major step forward.

- What is the **intended message** of the article?
- What is your **critical reading** of the summary of the article provided above? Remember that in this context, "critical reading" does not necessarily mean *What criticism do you have?* but asks you to apply your knowledge to interpret whether the article is factual and whether the facts support the intended message.
- How would you **interpret** the news item for your nonmicrobiologist friends?
- What is your **overall grade** for the news item—taking into account its accuracy and the accuracy of its intended effect?

Media Under The Microscope Wrap-Up appears at the end of the chapter.

Outline and Learning Outcomes

4.1 The Bacteria

1. List the structures all bacteria possess.
2. Identify at least four structures that some, but not all, bacteria possess.
3. Describe the three major shapes of bacteria.
4. Describe other more unusual shapes of bacteria.
5. Provide at least four terms to describe bacterial arrangements.

4.2 External Structures

6. Describe the structure and function of six different types of bacterial external structures.
7. Explain how a flagellum works in the presence of an attractant.

4.3 The Cell Envelope: The Boundary Layer of Bacteria

8. Differentiate between the two main types of bacterial envelope structure.
9. Discuss why gram-positive cell walls are stronger than gram-negative cell walls.
10. Name a substance in the envelope structure of some bacteria that can cause severe symptoms in humans.

4.4 Bacterial Internal Structure

11. Identify five structures that may be contained in bacterial cytoplasm.
12. Detail the causes and mechanisms of sporulation and germination.

4.5 The Archaea

13. List some differences between archaea and bacteria.

4.6 Classification Systems for Bacteria and Archaea

14. Differentiate between *Bergey's Manual of Systematic Bacteriology* and *Bergey's Manual of Determinative Bacteriology*.
15. Name four divisions ending in *-cutes* and describe their characteristics.
16. Define a *species* in terms of bacteria.

In chapter 1, we described bacteria and archaea as being cells with no true nucleus. Let's look at bacteria and archaea and how they differ from eukaryotes. There are two major distinctions between eukaryotic cells and non-eukaryotic cells:

- ***The way their DNA is packaged:*** Bacteria and archaea have nuclear material that is free inside the cytoplasm (i.e., they do not have a nucleus). Eukaryotes have a membrane around their DNA (making up a nucleus). Eukaryotes wind their DNA around proteins called histones, and archaea use similar proteins to do the same thing. Bacteria do not wind their DNA around proteins.
- ***Their internal structures:*** Bacteria and archaea do not have complex, membrane-bounded organelles in their cytoplasm (eukaryotes do). A few bacteria and archaea have internal membranes and microcompartments, but they are not true organelles.

Both non-eukaryotic and eukaryotic microbes are ubiquitous in the world today. Although both can cause infectious disease, the way we treat infections with bacterial and eukaryotic pathogens will be influenced by their unique cellular characteristics.

4.1 The Bacteria

We have already seen that non-eukaryotic cells were the first to appear on earth, billions of years ago. The fact that these organisms have endured for so long in such a variety of habitats is a testament to how versatile and adaptable their structure and function are.

All bacterial cells invariably have a cell membrane, cytoplasm, ribosomes, and one (or a few) chromosome(s); the majority

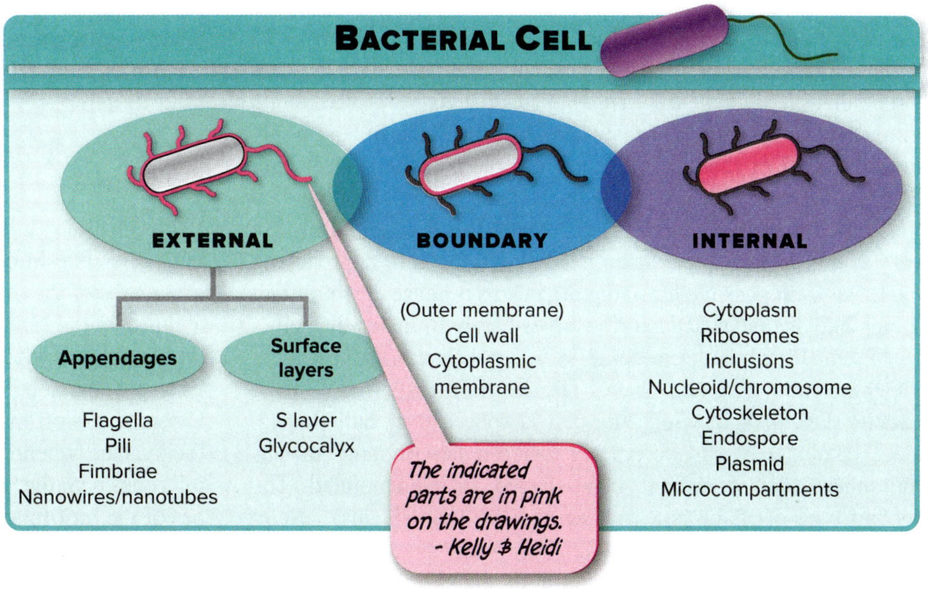
BACTERIAL CELL
EXTERNAL
BOUNDARY
INTERNAL
Appendages
Surface layers
Flagella
Pili
Fimbriae
Nanowires/nanotubes
S layer
Glycocalyx
(Outer membrane)
Cell wall
Cytoplasmic membrane
Cytoplasm
Ribosomes
Inclusions
Nucleoid/chromosome
Cytoskeleton
Endospore
Plasmid
Microcompartments
The indicated parts are in pink on the drawings.
- Kelly & Heidi

have a cell wall and some form of surface coating or glycocalyx. Specific structures that are found in some, but not all, bacteria are flagella, pili, fimbriae, an S layer, a cytoskeleton, inclusions, microcompartments, endospores, nanowires, and intracellular membranes.

The Structure of a Generalized Bacterial Cell

Bacterial cells look very simple and two-dimensional when viewed with an ordinary microscope. But the electron microscope, as well as 150 years of laboratory research, has shown us how intricate and functionally complex they actually are. **Figure 4.1** presents a three-dimensional anatomical view of a generalized, rod-shaped, bacterial cell.

Bacterial Arrangements and Sizes

In most bacterial species, each individual bacterial cell is fully capable of carrying out all necessary life activities, such as reproduction, metabolism, and nutrient processing, unlike the more specialized cells of a multicellular organism. On the other hand, sometimes bacteria *can* act as a group. When bacteria are close to one another in colonies or in biofilms, they communicate with each other through chemicals that cause them to behave differently than if they were living singly. More surprisingly, some bacteria have structures called *nanowires,* which are appendages that can be many micrometers long and are used for transferring electrons or other substances outside the cell onto metals. The wires also intertwine with the wires of neighboring bacteria and can be used for exchanging nutrients. This is not the same as being a multicellular organism, but it seems that bacteria can have it both ways: going it alone or cooperating with other microorganisms, depending on their circumstances.

Bacteria come in many different shapes, sizes, and arrangements. In terms of size, bacterial cells have an average size of about 1 micron (μm). As with everything in nature, though, there is a great deal of variation in microbial size. Until the summer of 2022, the largest non-eukaryote (akaryote) yet discovered was *Thiomargarita namibiensis,* which can be as large as 750μm **(figure 4.2).** In 2022 another *Thiomargarita* was discovered that was more than 10 times larger than *T. namibiensis*. *Thiomargarita magnifica* was found in underwater mangrove forests and is nearly 1 cm in length. On the other end of the size spectrum are the so-called ultra-small bacteria. They are between 200 and 500 nanometers. One hundred fifty of them could fit in an (average-sized) *E. coli* cell. They are very common in nature. Scientists believe they are approaching the lower limit possible for living organisms because the important things like nucleic acids and ribosomes barely fit inside the cells.

Even though bacteria come in many different shapes, there are three most common shapes **(figure 4.3).** First, there is the **coccus** (kok′-us), in which the cell is spherical or ball-shaped. Cocci (koks′-eye) can be perfect spheres, but in some species the spheres are more oval, bean-shaped, or even pointed. The second type of shape is a **rod,** or *bacillus* (bah-sil′-lus). (Note: There is also a genus named *Bacillus.*) Rods can also show different variations in shape. Depending on the species, they can be blocky, spindle-shaped, round-ended, long and threadlike (called filamentous), or even club-shaped or drumstick-shaped. When a rod is short and plump, it is called a *coccobacillus,* which, you can tell, is a combination of the words *coccus* and *bacillus.* The third general shape for bacterial cells is **curved.** If it is gently curved, it is a *vibrio* (vib′-ree-oh). A bacterium having a curled or spiral-shaped cylinder is called a *spirillum* (spy-ril′-em), a rigid helix, twisted twice or more along its axis (like a corkscrew). Another spiral cell is the *spirochete,* a more flexible form that resembles a spring.

It is somewhat common for cells of a single species to vary in shape and size. This phenomenon, called *pleomorphism,* is due to individual variations in cell wall structure caused by nutritional or slight genetic differences. For example, although the cells of *Corynebacterium diphtheriae* are generally considered rod-shaped, in culture they display variations such as club-shaped, swollen, curved, filamentous, and coccoid. The mycoplasmas are bacteria that entirely lack cell walls, and for that reason they display extreme variations in shape **(figure 4.4).**

Bacterial cells can also be categorized according to how they are arranged (see figure 4.3). The main factors influencing the arrangement of a particular cell type are its pattern of division and how the cells remain attached afterward. The greatest variety in arrangement occurs in cocci, which can be single, in pairs (diplococci), in **tetrads** (groups of four), in irregular clusters (as in staphylococci and micrococci), or in chains of a few to hundreds of cells (as in streptococci). An even more complex grouping is a cubical packet of 8, 16, or more cells called a **sarcina** (sar′-sih-nah). These different coccal groupings depend on whether the cells divide along a single plane, in two perpendicular planes, or in several intersecting planes. Because bacteria usually reproduce by splitting in two (over and over again), if the divided cells remain attached, they lead to these diverse arrangements.

Rods are less varied in arrangement because they divide only in the transverse plane (perpendicular to the axis). They occur either as single cells, as a pair of cells with their ends attached (diplobacilli), or as a chain of several cells (streptobacilli). A palisades arrangement is formed when cells of a chain remain partially attached at the ends. This hinge area can fold back, creating a side-by-side row of cells. Spirilla are occasionally found in short chains, but spirochetes rarely remain attached after division.

Biofilms

Bacteria often do not exist in a **planktonic** or single-cell form but rather live in cooperative associations that can include other organisms of the same species as well as other species of bacteria, archaea, fungi, and algae. These **biofilms** are microbial habitats with access to food, water, atmosphere, and other environmental factors that are beneficial to each type of organism living there. Often, the biofilm is stratified, with the aerobic microbes near the surface where the oxygen levels are high and the anaerobic microbes near the bottom where oxygen levels are low. Each member of the biofilm community finds its niche.

Biofilms can form on numerous inert substances, usually when the surface is moist and has developed a thin layer of organic

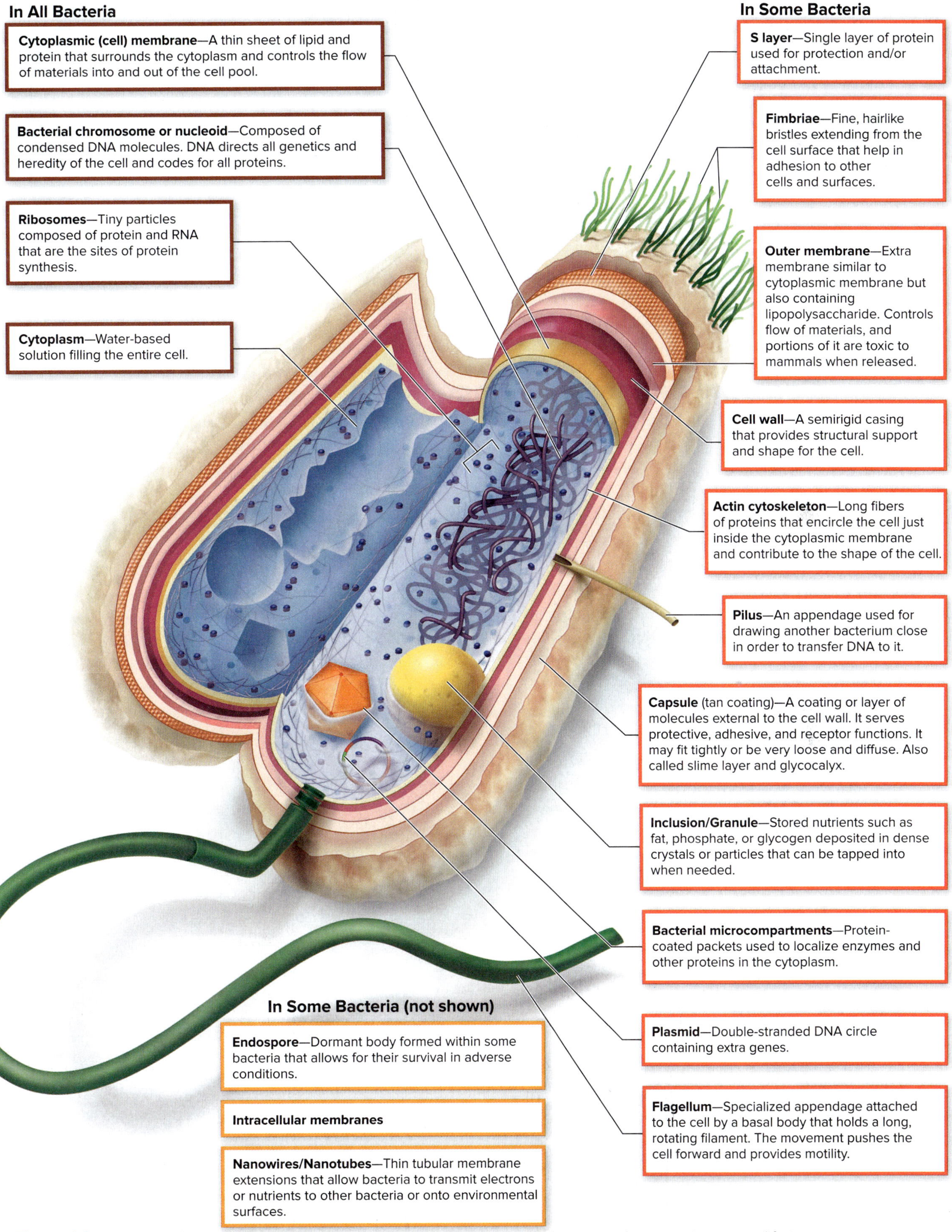
In All Bacteria
Cytoplasmic (cell) membrane—A thin sheet of lipid and protein that surrounds the cytoplasm and controls the flow of materials into and out of the cell pool.
Bacterial chromosome or nucleoid—Composed of condensed DNA molecules. DNA directs all genetics and heredity of the cell and codes for all proteins.
Ribosomes—Tiny particles composed of protein and RNA that are the sites of protein synthesis.
Cytoplasm—Water-based solution filling the entire cell.
In Some Bacteria
S layer—Single layer of protein used for protection and/or attachment.
Fimbriae—Fine, hairlike bristles extending from the cell surface that help in adhesion to other cells and surfaces.
Outer membrane—Extra membrane similar to cytoplasmic membrane but also containing lipopolysaccharide. Controls flow of materials, and portions of it are toxic to mammals when released.
Cell wall—A semirigid casing that provides structural support and shape for the cell.
Actin cytoskeleton—Long fibers of proteins that encircle the cell just inside the cytoplasmic membrane and contribute to the shape of the cell.
Pilus—An appendage used for drawing another bacterium close in order to transfer DNA to it.
Capsule (tan coating)—A coating or layer of molecules external to the cell wall. It serves protective, adhesive, and receptor functions. It may fit tightly or be very loose and diffuse. Also called slime layer and glycocalyx.
Inclusion/Granule—Stored nutrients such as fat, phosphate, or glycogen deposited in dense crystals or particles that can be tapped into when needed.
Bacterial microcompartments—Protein-coated packets used to localize enzymes and other proteins in the cytoplasm.
Plasmid—Double-stranded DNA circle containing extra genes.
Flagellum—Specialized appendage attached to the cell by a basal body that holds a long, rotating filament. The movement pushes the cell forward and provides motility.
In Some Bacteria (not shown)
Endospore—Dormant body formed within some bacteria that allows for their survival in adverse conditions.
Intracellular membranes
Nanowires/Nanotubes—Thin tubular membrane extensions that allow bacteria to transmit electrons or nutrients to other bacteria or onto environmental surfaces.

Figure 4.1 Structure of a bacterial cell. Cutaway view of a typical rod-shaped bacterium, showing major structural features.

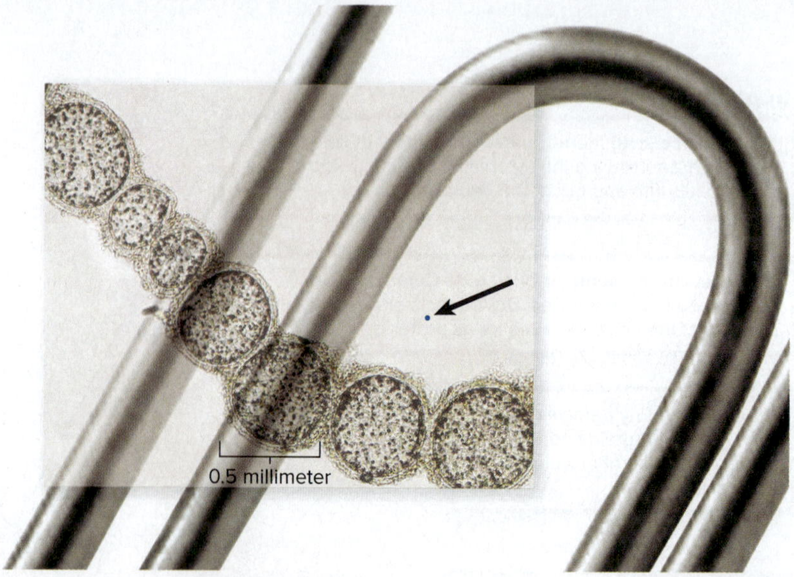
0.5 millimeter

Figure 4.2. ***Thiomargarita namibiensis.*** For comparison, these bacteria are approximately the same width as the wire of a paper clip. Also shown, as a tiny blue dot at the point of the arrow, is the approximate size of a single cell of *Staphylococcus*. It is about 0.001 millimeter (or 500x smaller) in circumference.

Courtesy Max-Planck-Institut für Marine Mikrobiologie: Holly Curry/McGraw Hill

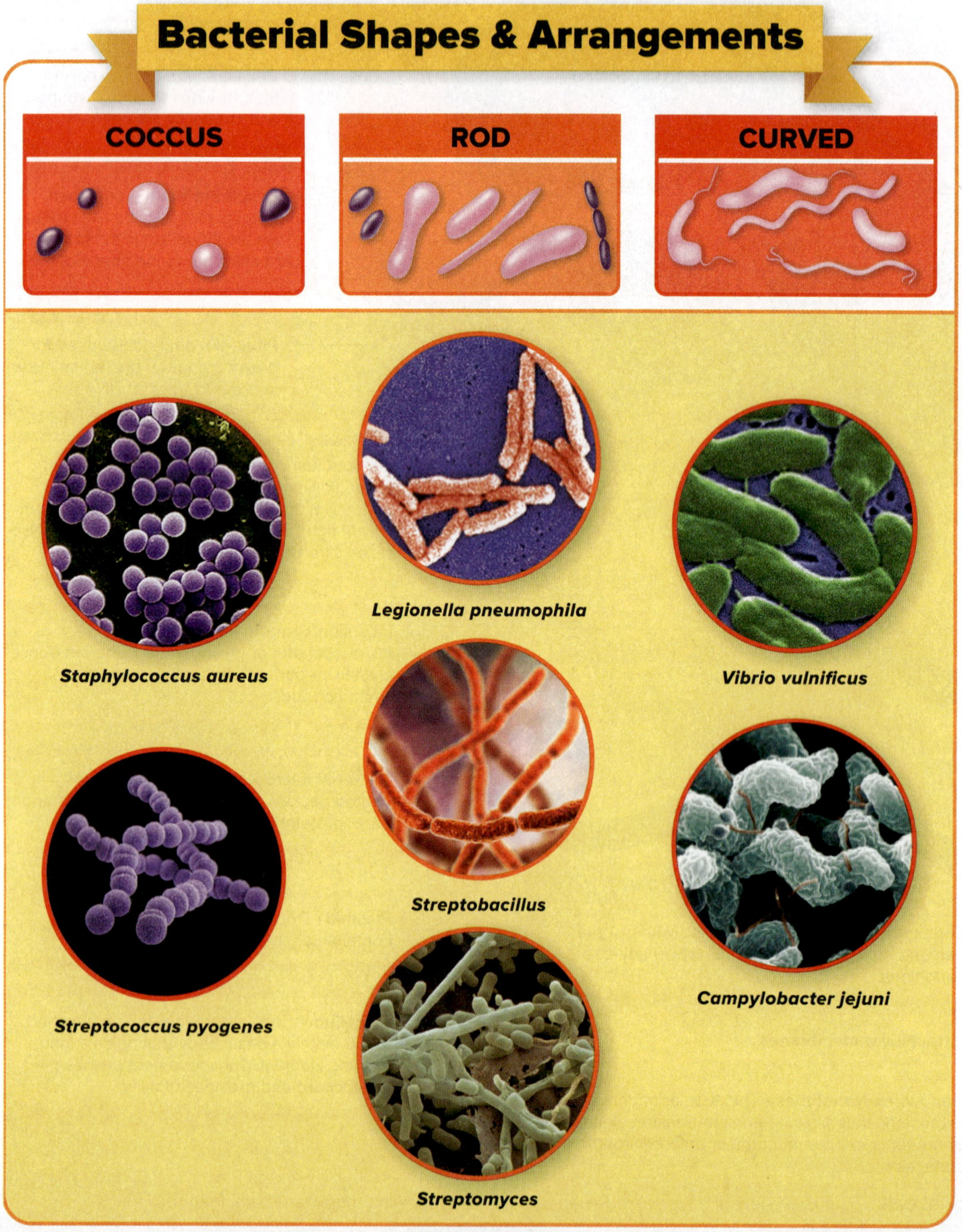
Bacterial Shapes & Arrangements
COCCUS
ROD
CURVED
Staphylococcus aureus
Legionella pneumophila
Vibrio vulnificus
Streptobacillus
Streptococcus pyogenes
Campylobacter jejuni
Streptomyces

Figure 4.3 **Bacterial shapes and arrangements.** Drawings show examples of shape variations for cocci, rods, and curved bacteria. The photos in the circles are all electron micrographs.

(1) Janice Haney Carr/CDC; (2) Melissa Brower/CDC; (3) Janice Haney Carr/CDC; (4) Kateryna Kon/Shutterstock; (5) Eye of Science/Science Source; (6) Janice Carr/CDC; (7) De Wood, Chris Pooley/USDA

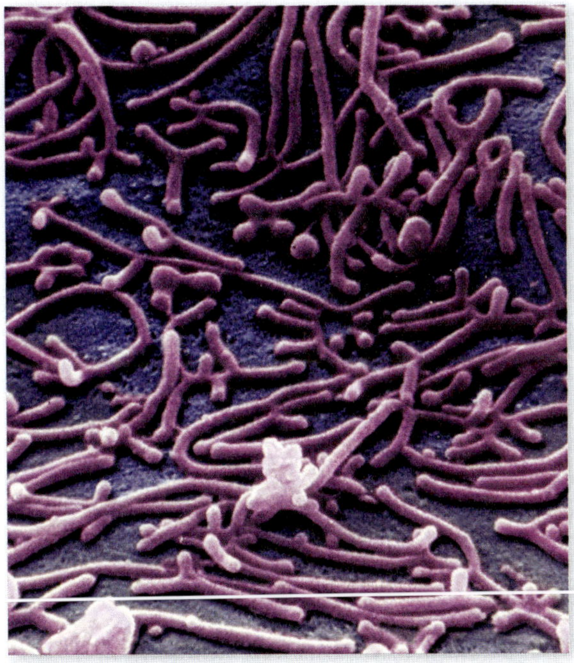

Figure 4.4 **The pleomorphic appearance of *Mycoplasma pneumoniae*, a bacterium without a cell wall.**
Don Fawcett/Science Source/Getty images

material such as polysaccharides or glycoproteins. This slightly sticky texture attracts the first single-celled "colonists" that attach and begin to multiply on the surface. As the first colonizing organisms grow, they secrete substances such as cell signal receptors, fimbriae, slime layers, capsules, and even DNA molecules that attract other microbes to the surface as well **(figure 4.5)**. This cell-to-cell communication, including a process called quorum sensing, allows for microbes of various species to grow together and secrete more extracellular matrix (shown in green in **(figure 4.5*b*)**). The biofilm can vary in thickness, depending on where it begins growing and how long it has been growing there (or how long it has been since you brushed and flossed your teeth).

Biofilms also have serious medical implications. Often, microbes will accumulate on damaged tissue such as heart valves or hard surfaces such as teeth. Bacteria also have an affinity for implanted medical devices such as IUDs, catheters, shunts, gastrostomy tubes, and urinary catheters, and readily form biofilms on these surfaces. Treating these types of infections is extremely difficult, and it has always been assumed that this was due to antibiotics being unable to penetrate the thick glycocalyx of the biofilm. Recent studies have shown that in biofilm form, microbes turn on different genes, allowing them to be impervious to antibiotic treatment. Finding novel ways to treat biofilm infections is an ongoing battle, and it is estimated that treating biofilm infections costs more than one billion dollars a year in the United States alone.

4.1 Learning Outcomes—Assess Your Progress

1. List the structures all bacteria possess.
2. Identify at least four structures that some, but not all, bacteria possess.
3. Describe the three major shapes of bacteria.
4. Describe other more unusual shapes of bacteria.
5. Provide at least four terms to describe bacterial arrangements.

4.2 External Structures

Appendages: Cell Extensions

Several different types of accessory structures sprout from the surface of bacteria. These long **appendages** are common but are not present on all species. Appendages can be divided into two major groups: those that provide motility (flagella and axial filaments) and those that provide attachment points or channels (fimbriae, pili, and nanowires).

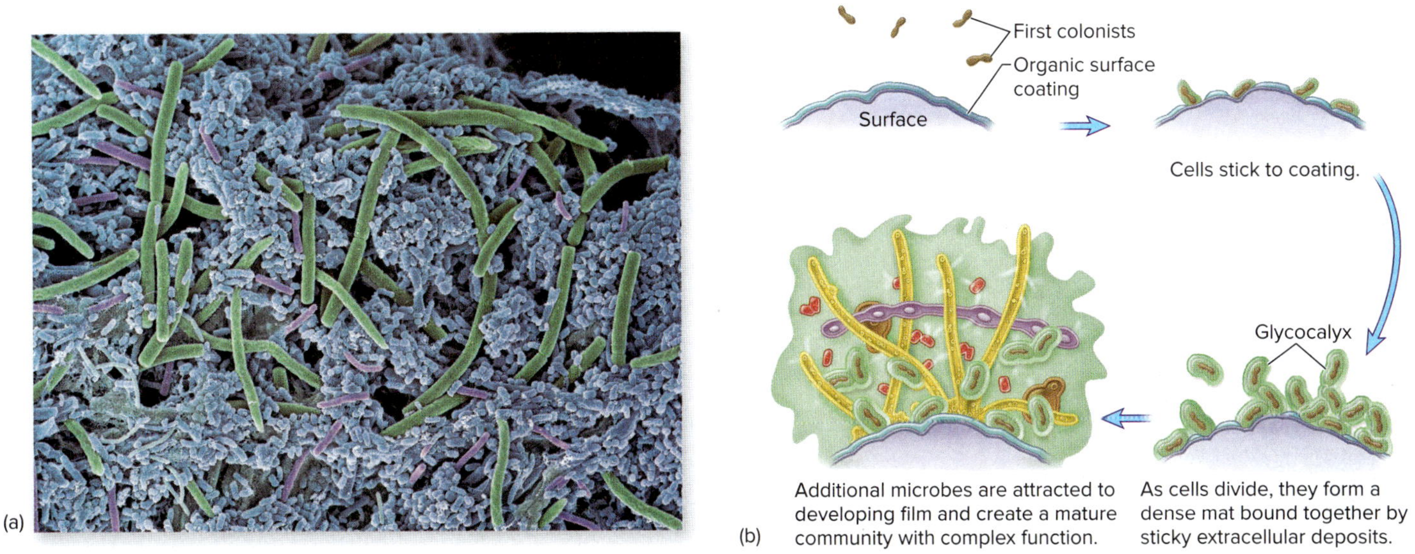
(a)
(b)
First colonists
Organic surface coating
Surface
Cells stick to coating.
Glycocalyx
Additional microbes are attracted to developing film and create a mature community with complex function.
As cells divide, they form a dense mat bound together by sticky extracellular deposits.

Figure 4.5. **Biofilms.** **(a)** Photograph of a bacterial biofilm. The blue bacteria appeared to have secreted a slime layer that aggregates them and the other species in the film. **(b)** Steps in a biofilm formation.
(a) Steve Gschmeissner/SPL/Getty Images

Figure 4.6 Details of the basal body of a flagellum in a gram-negative cell. **(a)** The hook, rings, and rod function together as a tiny device that rotates the filament 360°. **(b)** *Pseudomonas aeruginosa* seen through the light microscope after a flagellar staining procedure.

(b) Lisa Burgess/McGraw Hill

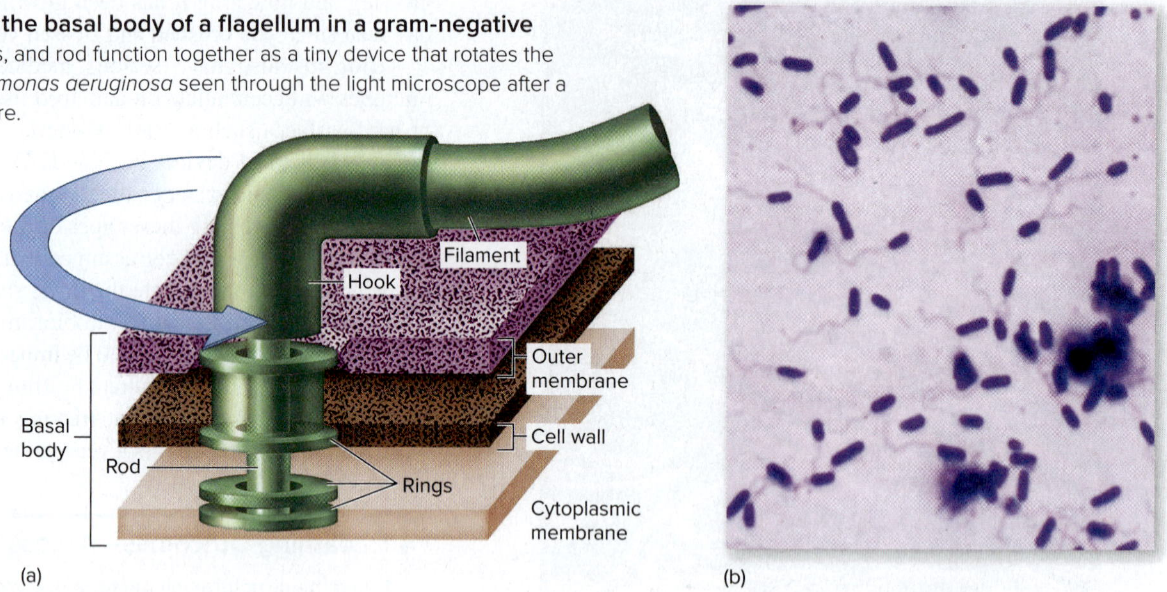
Filament
Hook
Outer membrane
Basal body
Cell wall
Rod
Rings
Cytoplasmic membrane
(a)
(b)

Disease Connection

Food poisoning, while usually short-lived, is an unpleasant experience for the suffering patient. One of the most common causative organisms is *Campylobacter jejuni,* a gram-negative monotrichous or amphitrichous polar bacterium. The flagellum contributes to the severity of disease symptoms, allowing the bacterium to burrow its way into the gastrointestinal mucus of the small intestine, where it then multiplies. The bacterium can then release toxins that cause gastrointestinal distress, including diarrhea, abdominal pain, and fever.

Flagella—Little Propellers

The bacterial **flagellum** (flah-jel′-em), an appendage of truly amazing construction, is certainly unique in the biological world. The primary function of flagella is to provide **motility,** or self-propulsion—that is, the capacity of a cell to swim freely through an aqueous habitat. There are three distinct parts: the filament, the hook, and the basal body **(figure 4.6).** The **filament,** a helical structure composed of proteins, is approximately 20 nanometers in diameter and varies from 1 to 70 microns in length. It is inserted into a curved, tubular hook. The hook is anchored to the cell by the basal body, a stack of rings firmly anchored through the cell wall to the cytoplasmic membrane and the outer membrane. This arrangement permits the hook with its filament to rotate 360°.

In general, all spirilla, about half of the rods, and a small number of cocci have flagella. Flagella vary in both number and arrangement according to two general patterns:

1. In a **polar** arrangement, the flagella are attached at one or both ends of the cell. (The ends of rod-shaped bacteria are called poles.) Three subtypes of this pattern are **monotrichous** (mah″-noh-trik′-us), with a single flagellum **(figure 4.7*a*); lophotrichous** (lo″-foh-), with small bunches or tufts of flagella emerging from the same site **(figure 4.7*c*);** and **amphitrichous** (am″-fee-), with flagella at both poles of the cell **(figure 4.7*d*).**

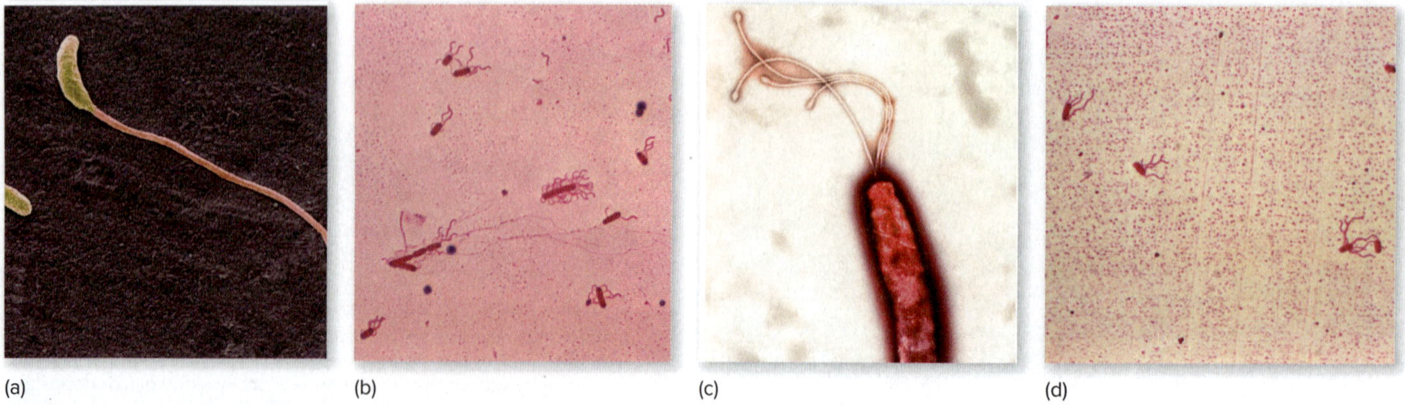
(a)
(b)
(c)
(d)

Figure 4.7 Four types of flagellar arrangements. **(a)** Monotrichous flagellum on the bacterium *Clostridium phytofermentans.* **(b)** Peritrichous flagella on *a Bacillus* species.) **(c)** Lophotrichous flagella on *Helicobacter.* **(d)** Amphitrichous flagella on *Alcaligenes.*

(a) Steve Gschmeissner/Science Photo Library/Alamy Stock Photo; (b) Dr. W.A. Clark/CDC; (c) Heather Davies/Science Photo Library/Science Source; (d) Smith Collection/Gado/Getty Images

2. In a **peritrichous** (per″-ee-) arrangement, flagella are dispersed randomly over the surface of the cell **(figure 4.7*b*).**

The presence of motility is one piece of information used in the laboratory identification or diagnosis of pathogens. Flagella are hard to visualize in the laboratory, but often it is sufficient to know simply whether a bacterial species is motile, or whether they can move. Motility can be assessed using semisolid media or through the hanging drop technique (see chapter 3).

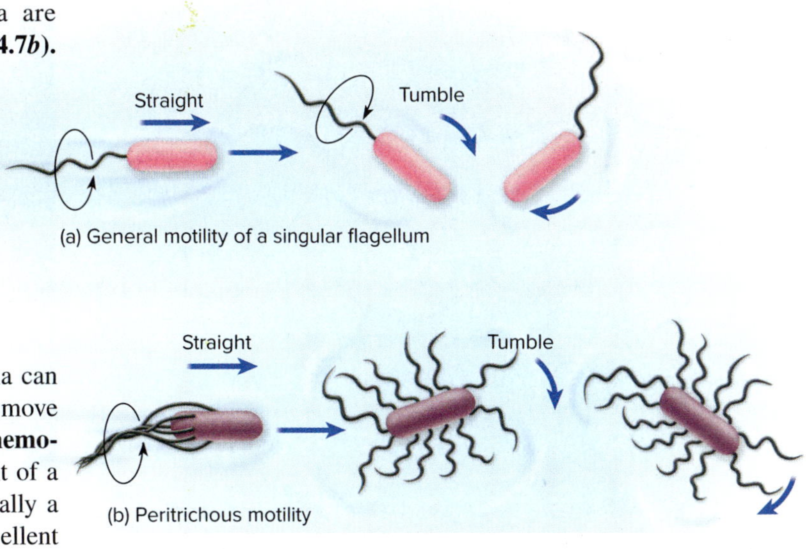
Straight
Tumble
(a) General motility of a singular flagellum
Straight
Tumble
(b) Peritrichous motility

Figure 4.8 The operation of flagella and the mode of locomotion in bacteria with polar and peritrichous flagella. **(a)** In general, when a polar flagellum rotates in a counterclockwise direction, the cell swims forward. When the flagellum reverses direction and rotates clockwise, the cell stops and tumbles. **(b)** In peritrichous forms, all flagella sweep toward one end of the cell and rotate as a single group. During tumbles, the flagella become uncoordinated.

Fine Points of Flagellar Function Flagellated bacteria can perform some rather sophisticated feats. They can detect and move in response to chemical signals—a type of behavior called **chemotaxis** (kee″-moh-tak′-sis). Positive chemotaxis is movement of a cell in the direction of a favorable chemical stimulus (usually a nutrient). Negative chemotaxis is movement away from a repellent (potentially harmful) compound.

The flagellum is effective in guiding bacteria through the environment primarily because its system for detecting chemicals is linked to the mechanisms that drive the flagellum. There are clusters of receptors in the cytoplasmic membrane that bind specific molecules in the immediate environment. When enough of these molecules attach, they transmit signals to the flagellum and set it into rotary motion. The actual "fuel" for the flagellum to turn is a gradient of protons (hydrogen ions). These are generated by the metabolism of the bacterium and bind to and detach from parts of the flagellar motor within the cytoplasmic membrane, causing the filament to rotate. If several flagella are present, they become aligned and rotate as a group **(figure 4.8).** As a flagellum rotates counterclockwise, the cell itself swims in a smooth linear direction toward the stimulus; this action is called a **run.** Runs are interrupted at various intervals by **tumbles,** during which the flagellum reverses direction and causes the cell to stop and change its course.

Alternation between runs and tumbles generates what is called a "random walk" form of motility in these bacteria. However, in response to a concentration gradient of an attractant molecule, the bacterium will begin to inhibit tumbles, permitting longer runs and overall progress toward the stimulus (figure 4.8). The movement now becomes a biased random walk in which movement is favored (biased) in the direction of the attractant. But what happens when a flagellated bacterium wants to run away from a toxic environment? In that case, the random walk then favors movement away from the concentration of repellent molecules. By delaying tumbles, the bacterium increases the length of its runs, allowing it to redirect itself away from the negative stimulus **(figure 4.9).** Again, this is called chemotaxis. Some photosynthetic bacteria exhibit **phototaxis,** movement in response to light rather than chemicals.

Periplasmic Flagella

Corkscrew-shaped bacteria called *spirochetes* (spy′-roh-keets) show an unusual, wriggly mode of locomotion caused by two or more long, coiled threads, the periplasmic flagella, also called **axial filaments.** A periplasmic flagellum is a type of internal flagellum that is enclosed in the space between the cell wall and the cytoplasmic membrane **(figure 4.10).** The filaments curl closely around the spirochete coils yet are free to contract and impart a twisting or flexing motion to the cell.

Although many archaea possess flagella, recent studies have shown that the structure is quite different from the bacterial flagellum. It is called an *archaellum* by some scientists.

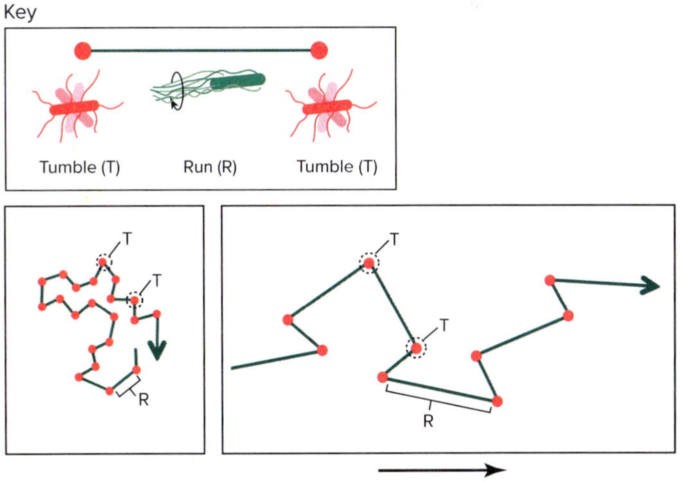
Key
Tumble (T)
Run (R)
Tumble (T)
T
T
R
T
T
R

(a) No attractant or repellent results in a "random walk" form of motion.

(b) In presence of attractant, the walk becomes "biased" toward more runs and fewer tumbles, directing the cell toward the attractant.

Figure 4.9 Chemotaxis in bacteria. **(a)** A bacterium moves via a random series of short runs and tumbles when there is no attractant or repellent. **(b)** The cell spends more time on runs as it gets closer to the attractant.

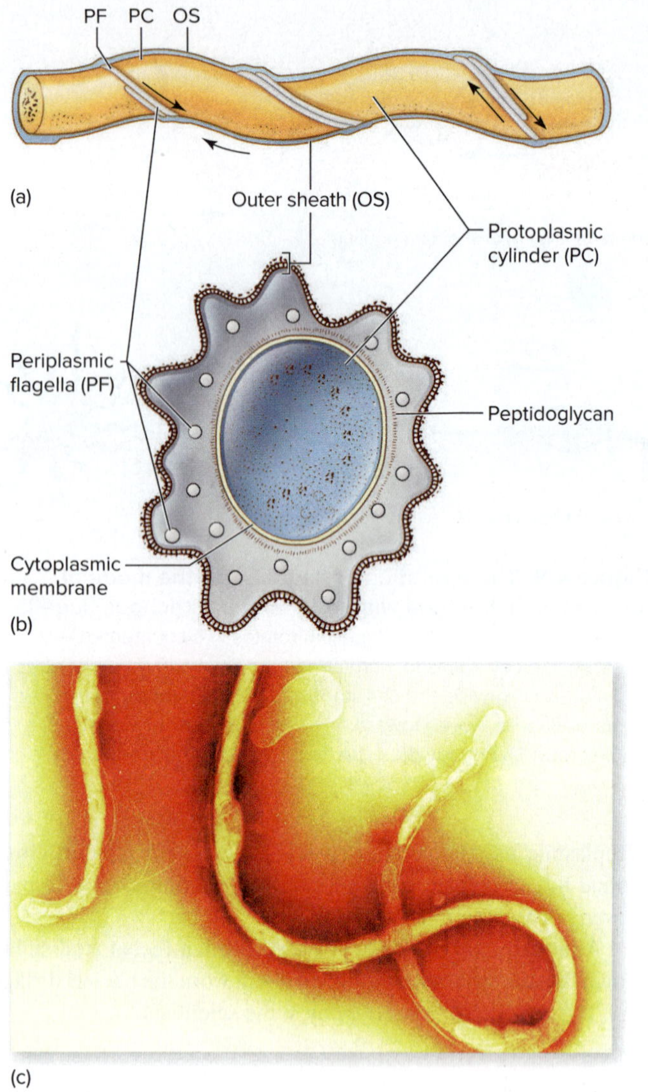
PF PC OS
(a)
Outer sheath (OS)
Protoplasmic cylinder (PC)
Periplasmic flagella (PF)
Peptidoglycan
Cytoplasmic membrane
(b)
(c)

Figure 4.10 The position of periplasmic flagella on the spirochete cell. **(a)** Longitudinal section (sideways view). **(b)** Cross section (end-on view). Contraction of the filaments results in a spinning and undulating pattern of locomotion. **(c)** Electron micrograph captures the periplasmic flagella of *Borrelia burgdorferi,* the causative agent of Lyme disease. On the left you can see that the flagella have escaped the outer membrane during preparation for microscopy. On the right they are in their correct position.

(c) NIH/Tina Carvalho, University of Hawaii at Manoa

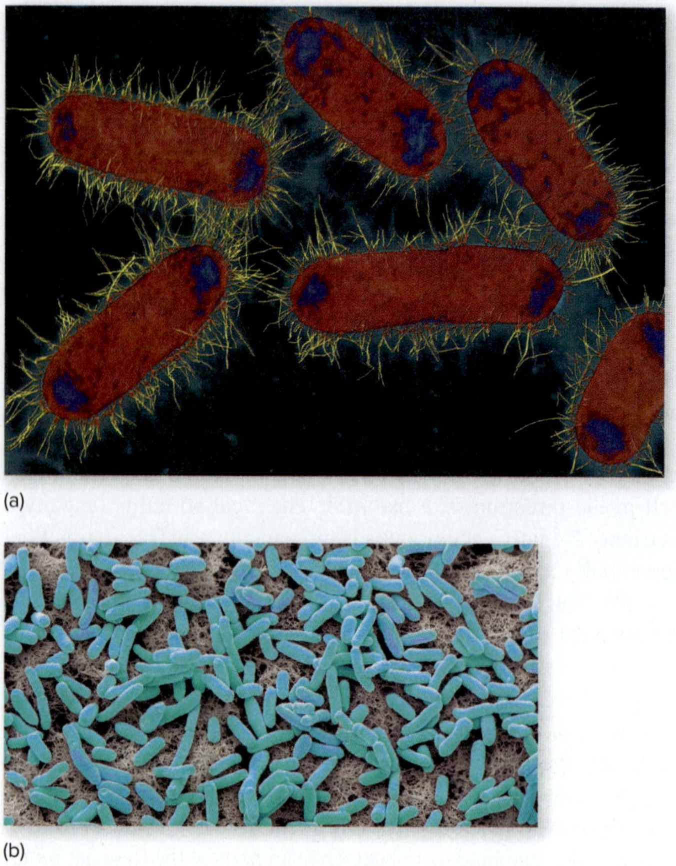
(a)
(b)

Figure 4.11 Form and function of bacterial fimbriae. **(a)** Several cells of pathogenic *Escherichia coli* covered with numerous stiff fibers called fimbriae (30,000×). Note also the dark-blue granules, which are the chromosomes. **(b)** *E. coli* cells attached to an epithelial surface using their fimbriae (fimbriae are not visible in this picture).

(a) Eye of Science/Science Source; (b) Steve Gschmeissner/Science Photo Library/Getty Images

Appendages for Attachment or Channel Formation

Although their main function is motility, bacterial flagella can be used for attachment to surfaces in some species. Two other structures, the **pilus** (pil-us; plural, *pili*) and the **fimbria** (fim′-bree-ah), are bacterial surface appendages that assist with attachment, but not motility.

Fimbriae are small, bristlelike fibers sprouting off the surface of many bacterial cells **(figure 4.11).** Their exact composition varies, but most of them are made of protein. Fimbriae have an inherent tendency to stick to each other and to surfaces. They can be responsible for the mutual clinging of cells that leads to biofilms and other thick aggregates of cells on the surface of liquids and for the microbial colonization of inanimate solids such as rocks and glass. Some pathogens, such as *Neisseria gonorrhoeae* and *Escherichia coli,* can colonize and infect host tissues because of a tight attachment between their fimbriae and epithelial cells **(figure 4.11*b*).** Mutant forms of these pathogens that lack a fimbriae, however, are unable to cause infections.

A pilus is a long, rigid, tubular structure made of a special protein, **pilin.** Its function is to connect two different cells in a process called **conjugation,** which involves partial transfer of DNA from one cell to another **(figure 4.12).** A pilus from the donor cell unites with a recipient cell, bringing it close enough for DNA transfer. Production of pili is controlled genetically, and conjugation takes place only between compatible gram-negative cells. Conjugation in gram-positive bacteria does occur but involves aggregation proteins rather than pili. There is a special type of structure in some bacteria called a Type IV pilus. Like the pili described here, it can transfer genetic material. In addition, it can act like fimbriae and assist in attachment, and it can act like flagella and make a bacterium motile.

Nanowires (also called nanotubes) are very thin, long, tubular extensions of the cytoplasmic membrane that bacteria use as

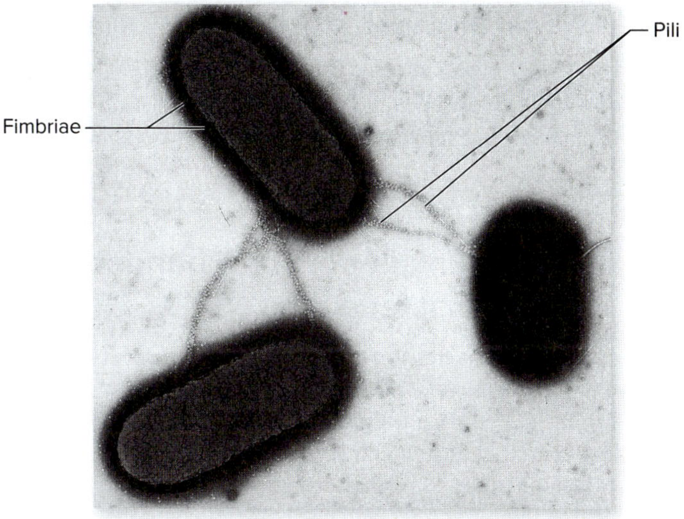
Pili
Fimbriae

Figure 4.12 Three bacteria in the process of conjugating. The conjugative pili are clearly evident, forming mutual conjugation bridges between a donor and two recipients. Fimbriae can also be seen on the two left-hand cells.

L. Caro/SPL/Science Source

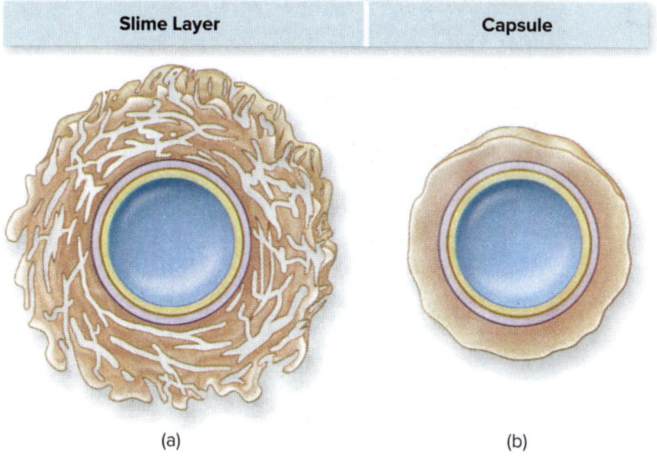
Slime Layer
Capsule
(a)
(b)

Figure 4.13 Drawings of sectioned bacterial cells to show the types of glycocalyces. **(a)** The slime layer is a loose structure that may be easily washed off. **(b)** The capsule is a thick, structured layer that is not readily removed.

channels. They are used either to transfer amino acids (i.e., food) among one another or to harvest energy by shuttling electrons from an electron-rich surface in the environment. In this second scenario, bacteria generate energy in the absence of oxygen. When we study metabolism, we will see that cells generate the necessary energy to grow by consuming nutrients and transferring electrons from the nutrients to oxygen. But many bacteria can generate their energy in the absence of oxygen by sharing electrons with substances in the environment containing iron, such as iron-rich rocks. They use long nanowires to do this, transferring electrons up and down the tubular extensions of the membrane. Scientists call this activity "breathing rock instead of oxygen."

Surface Coatings: The S Layer and the Glycocalyx

The bacterial cell surface is frequently exposed to severe environmental conditions. Bacterial cells protect themselves with either an **S layer** or a **glycocalyx,** or both. S layers are single layers of thousands of copies of a single protein linked together like a tiny chain-link fence. They are often called "the armor" of a bacterial cell. It took scientists a long time to discover them because bacteria tend to only produce them when they are in a hostile environment. The nonthreatening conditions of growing in a lab in a nutritious broth with no competitors around meant that bacteria did not produce the layer. We now know that many different species have the ability to produce an S layer, including pathogens such as *Clostridioides difficile* and *Bacillus anthracis.* Some bacteria use S layers to aid in attachment, as well.

The glycocalyx develops as a coating of repeating polysaccharide units that may or may not include protein. This protects the cell and, in some cases, helps it adhere to surfaces in its environment. The types of glycocalyces that bacteria have differ in thickness, organization, and chemical composition. Some bacteria are covered with a loose shield called a **slime layer** that evidently protects them from loss of water and nutrients **(figure 4.13*a*).** (This is different than the S layer, already described.) A glycocalyx is called a **capsule** when it is denser, thicker, and bound more tightly to the cell than a slime layer **(figure 4.13*b*).** Capsules can be viewed after a special staining technique **(figure 4.14*a*).** They are also often detectable on agar because they give their colonies a sticky (mucoid) appearance **(figure 4.14*b*).**

Specialized Functions of the Glycocalyx

Capsules are formed by many pathogenic bacteria, such as *Streptococcus pneumoniae* (a cause of pneumonia), *Haemophilus influenzae* (one cause of meningitis), and *Bacillus anthracis* (the cause of anthrax). Encapsulated bacterial cells generally have greater pathogenicity because capsules protect the bacteria against white blood cells called phagocytes. Phagocytes are a natural body defense that can engulf and destroy foreign cells, thus preventing infection. A capsular coating blocks the mechanisms that phagocytes use to attach to and engulf bacteria. By escaping phagocytosis, the bacteria are free to multiply and infect body tissues. Encapsulated bacteria that mutate to nonencapsulated forms usually lose their ability to cause disease.

Other types of glycocalyces can be important in the formation of biofilms. The thick, white plaque that forms on teeth comes in part from the surface slimes produced by certain streptococci in the oral cavity. This slime protects them from being knocked off of the teeth and provides attachment sites for other oral bacteria that, in time, can lead to dental disease. The glycocalyx of bacteria is instrumental in cementing biofilms to nonliving tissue as well. Biofilms are a major reason why foreign bodies in patients, such as plastic catheters, intrauterine devices, and implantable defibrillators, fail or cause complications. Biofilms also affect all kinds of water-delivery systems, from shower heads to dental office water systems to building HVAC systems **(figure 4.15).**

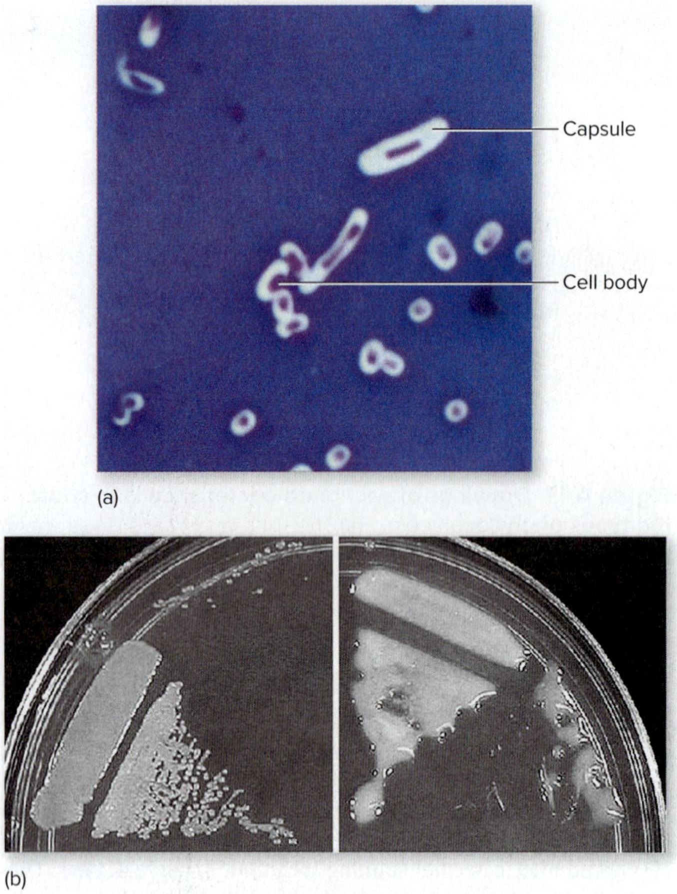
Capsule
Cell body
(a)
(b)

Figure 4.14 Encapsulated bacteria. **(a)** Negative staining reveals the microscopic appearance of large, well-developed capsules. **(b)** Colony appearance of a nonencapsulated (left) and encapsulated (right) version of a soil bacterium called *Sinorhizobium*.

(a) Lisa Burgess/BIOIMAGE INC/McGraw Hill; (b) Graham C. Walker

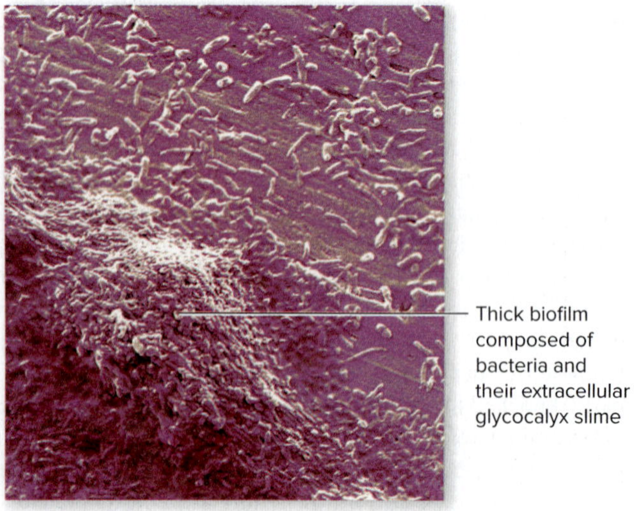
Thick biofilm composed of bacteria and their extracellular glycocalyx slime

Figure 4.15 Biofilm formation. Scanning electron micrograph of *Legionella pneumophila* typical of biofilms found inside water systems.

BSIP/UIG/Getty Images

4.2 Learning Outcomes—Assess Your Progress

6. Describe the structure and function of six different types of bacterial external structures.
7. Explain how a flagellum works in the presence of an attractant.

4.3 The Cell Envelope: The Boundary Layer of Bacteria

The boundary of a bacterium is marked by a multilayer structure that we will call the *cell envelope*. It is composed of two or three basic layers: the cell wall; the cytoplasmic membrane; and, in some bacteria, the outer membrane. The layers of the envelope are stacked one upon another and are often tightly bonded together like the outer husk and casings of a coconut. Although each envelope layer performs a distinct function, together they act as a single protective unit.

Differences in Cell Envelope Structure

More than a hundred years ago, long before the detailed anatomy of bacteria was even remotely known, a Danish physician named Hans Christian Gram developed a staining technique, the **Gram stain,** that identifies two generally different groups of bacteria. We will describe the structural differences, and later will show you how the Gram stain works. The two major groups shown by this technique are the **gram-positive** bacteria and the **gram-negative** bacteria.

Differences in the anatomy of the cell envelope are responsible for the designations *gram-positive* and *gram-negative* **(figure 4.16).** In gram-positive cells, the envelope consists of two layers: the thick cell wall, composed primarily of peptidoglycan (defined later in this section), and the cytoplasmic membrane. A gram-negative cell envelope consists of three layers: an outer membrane, a thin cell wall, and the cytoplasmic membrane.

Moving from outside to in, the outer membrane (if present) lies just under the glycocalyx. Next comes the cell wall. Finally, the innermost layer is always the cytoplasmic membrane. Because only some bacteria have an outer membrane, we discuss the cell wall first.

Structure of the Cell Wall

The **cell wall** accounts for a number of important bacterial characteristics. In general, it helps determine the shape of a bacterium, and it also provides the kind of strong structural support necessary to keep a bacterium from bursting or collapsing because of changes in osmotic pressure. In this way, the cell wall functions like a bicycle tire that maintains the necessary shape and prevents the more delicate inner tube (the cytoplasmic membrane) from bursting when it is expanded.

The cell walls of most bacteria are made from a unique macromolecule called **peptidoglycan** (PG). This compound is composed of a repeating framework of long **glycan** (sugar) chains cross-linked by short peptide (protein) fragments to provide a strong but

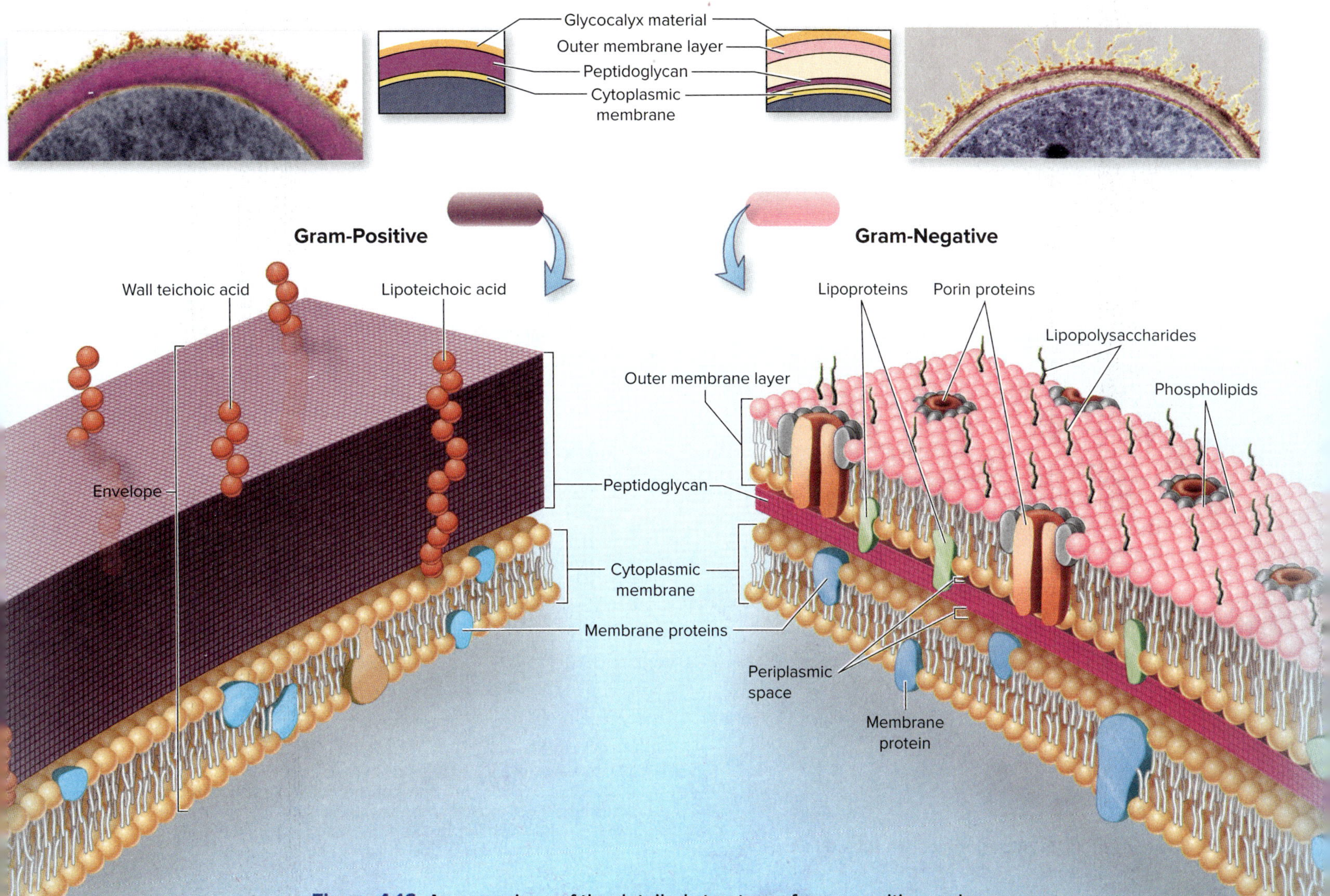
Glycocalyx material
Outer membrane layer
Peptidoglycan
Cytoplasmic membrane
Gram-Positive
Gram-Negative
Wall teichoic acid
Lipoteichoic acid
Envelope
Peptidoglycan
Cytoplasmic membrane
Membrane proteins
Lipoproteins
Porin proteins
Lipopolysaccharides
Outer membrane layer
Phospholipids
Periplasmic space
Membrane protein

Figure 4.16 A comparison of the detailed structure of gram-positive and gram-negative cell envelopes. The images at the top are electron micrographs of actual gram-positive and gram-negative cells.
Dr. Kari Lounatmaa/Science Source; Dennis Kunkel Microscopy, Inc./Science Source

flexible support framework **(figure 4.17).** The amount and exact composition of peptidoglycan vary among the major bacterial groups. The only place peptidoglycan is found on earth is here in bacterial cell walls.

Because many bacteria live in aqueous habitats with a low concentration of dissolved substances, they are constantly absorbing excess water by osmosis. Were it not for the strength and relative rigidity of the peptidoglycan in the cell wall, they would rupture from internal pressure. This function of the cell wall—and its totally unique biochemistry—has been tremendously useful in developing antimicrobial drugs. Several types of drugs used to treat infection (penicillin, cephalosporins) are effective because they target the peptide cross-links in the peptidoglycan, causing it to disintegrate. With their cell walls incomplete or missing, such cells have very little protection from **lysis** (ly′-sis), which is the disintegration or rupture of the cell. Lysozyme, an enzyme contained in tears and saliva, provides a natural defense against certain bacteria by hydrolyzing the bonds in the glycan chains and causing the wall to break down.

The Gram-Positive Cell Wall

The gram-positive cell wall is made of multiple layers of peptidoglycan, ranging from 20 to 80 nm in thickness. It also contains tightly bound acidic polysaccharides, including **teichoic acid** and **lipoteichoic acid** (see figure 4.16). Teichoic acid is a polymer of ribitol or glycerol (alcohols) and phosphate that is embedded in the peptidoglycan sheath. Lipoteichoic acid is similar in structure but is attached to the lipids in the cytoplasmic membrane. These molecules probably function in cell wall maintenance and enlargement during cell division, and they also contribute to the acidic charge on the cell surface.

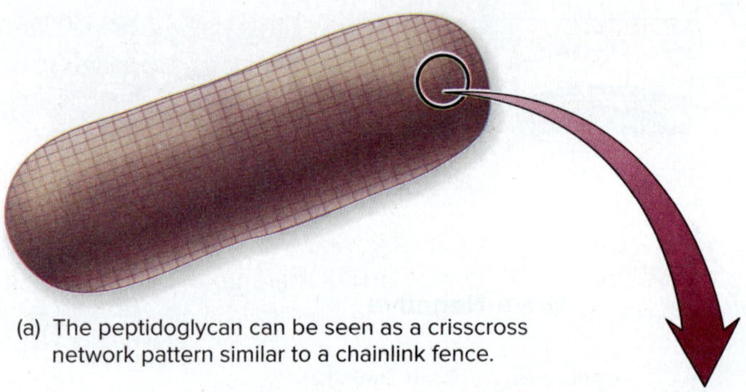

(a) The peptidoglycan can be seen as a crisscross network pattern similar to a chainlink fence.

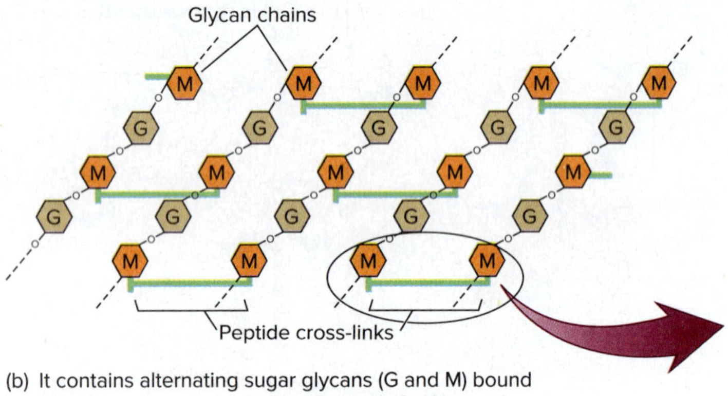
Glycan chains
M
G
Peptide cross-links

(b) It contains alternating sugar glycans (G and M) bound together in long strands. The G stands for *N*-acetyl glucosamine, and the M stands for *N*-acetyl muramic acid.

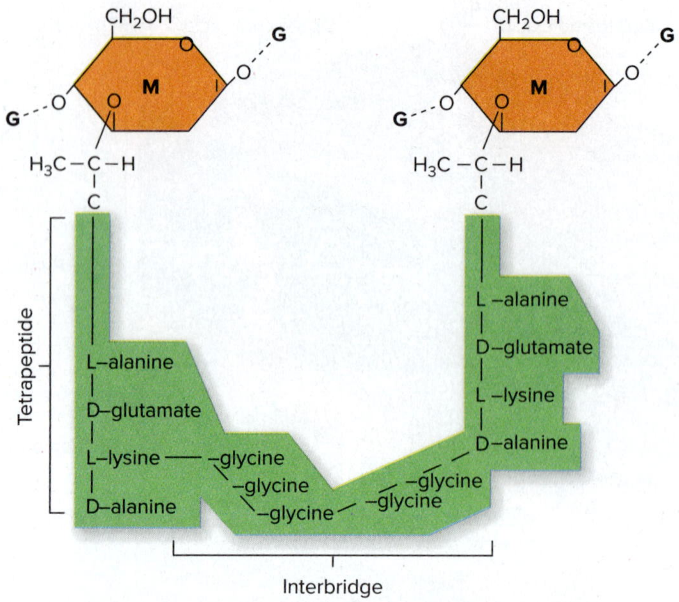
CH2OH
M
G
O
H3C–C–H
C
Tetrapeptide
L-alanine
D-glutamate
L-lysine
D-alanine
-glycine
-glycine
-glycine
-glycine
-glycine
L-alanine
D-glutamate
L-lysine
D-alanine
Interbridge

(c) A detailed view of the links between the muramic acids. Tetrapeptide chains branching off the muramic acids connect by interbridges also composed of amino acids. It is this linkage that provides rigid yet flexible support to the cell and that may be targeted by drugs like penicillin.

Figure 4.17 Structure of peptidoglycan in the cell wall.

The Gram-Negative Cell Wall

The gram-negative wall is a single, thin (1–3 nm) sheet of peptidoglycan. Although it acts as a somewhat rigid protective structure as previously described, its thinness means that gram-negative bacteria are relatively more flexible and sensitive to lysis.

Nontypical Cell Walls

Several bacterial groups lack the cell wall structure of gram-positive or gram-negative bacteria, and some bacteria have no cell wall at all. Although these unusual forms can stain positive or negative in the Gram stain, examination of their fine structure and chemistry shows that they do not really fit the descriptions for typical gram-negative or gram-positive cells. For example, the cells of *Mycobacterium* and *Nocardia* contain peptidoglycan and stain gram-positive, but the bulk of their cell wall is composed of unique types of lipids. One of these is a very-long-chain fatty acid called **mycolic acid,** known as cord factor, that contributes to the pathogenicity of this group. The thick, waxy nature imparted to the cell wall by these lipids is also responsible for a high degree of resistance to certain chemicals and dyes. Such resistance is the basis for the **acid-fast stain** used to diagnose tuberculosis and Hansen's disease (leprosy). In this stain, hot carbol fuchsin dye becomes tenaciously attached (is held fast) to these cells so that an acid-alcohol solution will not remove the dye.

Mycoplasmas and Other Cell-Wall-Deficient Bacteria

Mycoplasmas are bacteria that naturally lack a cell wall. Although other bacteria require an intact cell wall to prevent the bursting of the cell, the mycoplasma cytoplasmic membrane is stabilized by sterols and is resistant to lysis. (Recall from chapter 2 that sterols are complex lipid structures. All eukaryotic cells contain sterols, but only a few bacterial species do.) The extremely tiny **pleomorphic** mycoplasma cells are very small bacteria, ranging from 0.1 to 0.5 μm in size. They range in shape from filamentous to coccus or are doughnut-shaped. They are *not* obligate parasites and can be grown on artificial media. Mycoplasmas are found in many habitats, including plants, soil, and animals. The most important medical species is *Mycoplasma pneumoniae* **(figure 4.18),** which attaches to

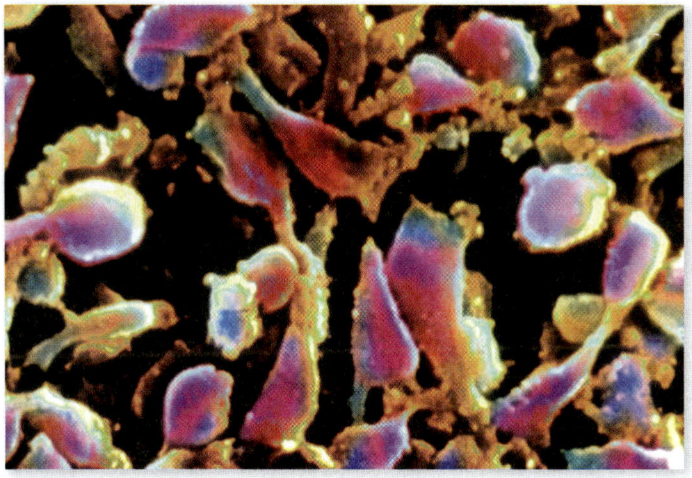

Figure 4.18 *Mycoplasma pneumoniae.*
VEM/ BSIP SA/Alamy Stock Photo

the epithelial cells in the lung and causes an atypical form of pneumonia in humans, often called walking pneumonia.

Cytoplasmic Membrane Structure

Appearing just beneath the cell wall is the **cell membrane,** which is often called the **cytoplasmic membrane.** We use that name in this book because it makes it immediately clear that we are referring to the membrane in contact with the cytoplasm. It is a very thin (5–10 nm), flexible sheet molded completely around the cytoplasm. Its general composition was described in section 2.2 as a lipid bilayer with proteins embedded to varying degrees. Bacterial cell membranes have this typical structure, containing primarily phospholipids (making up about 30% to 40% of the membrane mass) and proteins (contributing 60% to 70%). Major exceptions to this description are the membranes of mycoplasmas, which contain high amounts of sterols—rigid lipids that stabilize and reinforce the membrane, and the membranes of archaea, which contain unique branched hydrocarbons rather than fatty acids.

Some environmental bacteria, including photosynthesizers and ammonia oxidizers, contain dense stacks of internal membranes that are studded with enzymes or photosynthetic pigments. The inner membranes allow a higher concentration of these enzymes and pigments and accomplish a compartmentalization that allows for higher energy production.

Functions of the Cytoplasmic Membrane

Because bacteria have none of the eukaryotic organelles, the cytoplasmic membrane provides a site for functions such as energy reactions, nutrient processing, and synthesis. A major action of the cytoplasmic membrane is to regulate **transport,** the passage of nutrients into the cell and the discharge of wastes out of the cell. Although water and small uncharged molecules can diffuse across the membrane unaided, the membrane is a **selectively permeable** structure with special carrier mechanisms that allow many different molecules in and out.

In eukaryotic cells, important functions such as respiration and ATP synthesis take place on organelles called mitochondria. Because bacteria do not have mitochondria, they carry out these and other important activities using proteins embedded in their membranes. Enzymes in the cytoplasmic membrane also help synthesize the structural macromolecules needed to build the cell envelope and appendages. Other products (enzymes and toxins) are secreted by the membrane into the extracellular environment.

The Gram-Negative Outer Membrane

The **outer membrane** (OM) is somewhat similar in structure to the cytoplasmic membrane, except that it contains specialized types of polysaccharides and proteins. The uppermost layer of the OM "membrane sandwich" contains **lipopolysaccharide** (LPS). The polysaccharide chains extending off the surface function as cell markers and receptors. The lipid portion of LPS has been referred to as **endotoxin** because it stimulates fever and medical shock reactions in gram-negative infections such as meningitis and typhoid fever.

The innermost layer of the OM is a phospholipid layer anchored by means of lipoproteins to the peptidoglycan layer below. The outer membrane serves as a partial chemical sieve by allowing only relatively small molecules to penetrate. Access is provided by special membrane channels formed by **porin** *proteins* that completely span the outer membrane. The size of these porins can be adjusted by the cell so as to block the entrance of harmful chemicals, making them one defense of gram-negative bacteria against certain antibiotics. Review figure 4.16 to visualize the outer membrane. **Figure 4.19** gives another view of lipopolysaccharide.

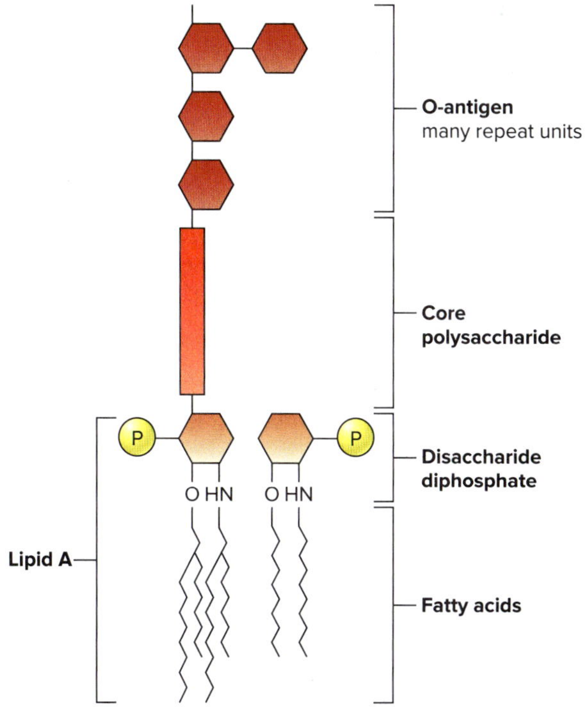
O-antigen
many repeat units
Core
polysaccharide
P
P
O HN
O HN
Disaccharide
diphosphate
Lipid A
Fatty acids

Figure 4.19 The lipopolysaccharide molecule.

The Gram Stain

The Gram stain is a differential staining procedure, the end result of which is that the bacterial cells are either purple or pink (red). We call the purple bacteria gram-positive and the pink bacteria gram-negative. That is because the primary stain is crystal violet, which, obviously, is a purple color. Cells that retain the primary stain are purple and get the label gram-positive. Thinking about that will be helpful as you try to remember which is which. Purple = positive.

Take a look at the steps described in **figure 4.20.** After a smear of bacteria on a glass slide is prepared and fixed, the first step is to apply crystal violet to the smear (primary stain), followed by the mordant (fixative), which is Gram's iodine. Alcohol is next used as a decolorizer. This is a critical step because the alcohol will dissolve the outer membrane in gram-negative cells. With their much thinner peptidoglycan layer, the crystal violet is rinsed out. Gram-positive cells have the crystal violet anchored in their thick walls due to the mordant. At this point, if you were to look at the smear with a microscope, some of the cells would be purple and some would be colorless. You might or might not be able to see faint outlines of the colorless ones without the last step. The last step is the counter stain, which adds a different color (pink or red) to the cells that had lost their color in the decolorizing step. The end result is that some species of bacteria are seen as purple, and other species are pink.

Practical Considerations of Differences in Cell Envelope Structure

Gram-positive and gram-negative bacteria differ in some significant ways. The outer membrane contributes an extra barrier in gram-negative bacteria that makes them resistant to some antimicrobial chemicals such as dyes and disinfectants, so they are generally more difficult to inhibit or kill than are gram-positive bacteria. On the other hand, alcohol-based compounds can dissolve the lipids in the outer membrane and break it down. Treating infections caused by gram-negative bacteria often requires different drugs than gram-positive infections, especially drugs that can cross the outer membrane.

The cell envelope or its parts (such as LPS) can interact with human tissues and contribute to disease. Proteins attached to the outer portion of the cell wall of several gram-positive species, including *Corynebacterium diphtheriae* (the agent of diphtheria) and *Streptococcus pyogenes* (the cause of strep throat), also have toxic properties. The lipids in the cell walls of certain *Mycobacterium* species are harmful to human cells as well. Because most macromolecules in the cell walls are foreign to humans, they stimulate antibody production by the immune system.

The good news is that the differences between gram-negative and gram-positive cells, and the differences between bacterial and human cells, provide opportunities to develop drugs that will harm one or another bacterium without harming the human host.

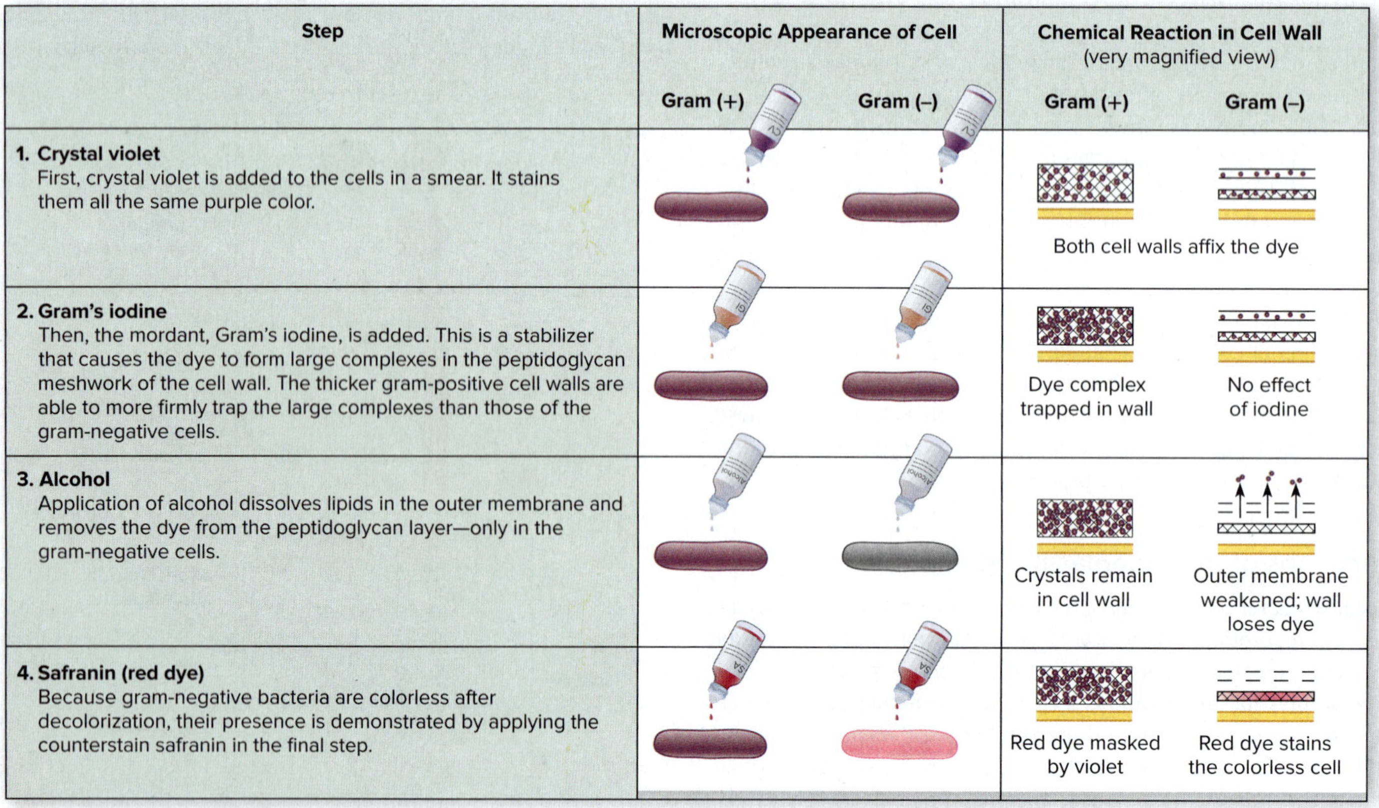
Step
Microscopic Appearance of Cell
Gram (+)
Gram (–)
Chemical Reaction in Cell Wall
(very magnified view)
Gram (+)
Gram (–)
1. Crystal violet
First, crystal violet is added to the cells in a smear. It stains them all the same purple color.
Both cell walls affix the dye
2. Gram's iodine
Then, the mordant, Gram's iodine, is added. This is a stabilizer that causes the dye to form large complexes in the peptidoglycan meshwork of the cell wall. The thicker gram-positive cell walls are able to more firmly trap the large complexes than those of the gram-negative cells.
Dye complex trapped in wall
No effect of iodine
3. Alcohol
Application of alcohol dissolves lipids in the outer membrane and removes the dye from the peptidoglycan layer—only in the gram-negative cells.
Crystals remain in cell wall
Outer membrane weakened; wall loses dye
4. Safranin (red dye)
Because gram-negative bacteria are colorless after decolorization, their presence is demonstrated by applying the counterstain safranin in the final step.
Red dye masked by violet
Red dye stains the colorless cell

Figure 4.20 Steps in the Gram Stain.

4.3 Learning Outcomes—Assess Your Progress

8. Differentiate between the two main types of bacterial envelope structure.
9. Discuss why gram-positive cell walls are stronger than gram-negative cell walls.
10. Name a substance in the envelope structure of some bacteria that can cause severe symptoms in humans.

4.4 Bacterial Internal Structure

Contents of the Cell Cytoplasm

Cytoplasm is a gelatinous solution encased by the cytoplasmic membrane. It is another important site for many of the cell's biochemical and synthetic activities. Its major component is water (70% to 80%), which serves as a solvent for the supply of molecules used for nutrients and building blocks for the cell. This mixture includes sugars, amino acids, and salts. The cytoplasm also contains larger, discrete cell masses such as the chromatin body, ribosomes, granules, and fibers resembling actin and tubulin strands that act as a cytoskeleton in bacteria that have them.

Bacterial Chromosomes and Plasmids: The Sources of Genetic Information

The hereditary material of most bacteria exists in the form of one or more circular strands of DNA designated as the **bacterial chromosome.** By definition, bacteria do not have a nucleus. Their DNA is not enclosed by a nuclear membrane but instead is aggregated in a dense area of the cell called the **nucleoid (figure 4.21).** You have already learned that DNA exists as a double helix. The bacterial chromosome, then, is a long double helix formed into a circle. There are no loose ends. Inside the cell, it is tightly coiled around special basic protein molecules so it does not look circular.

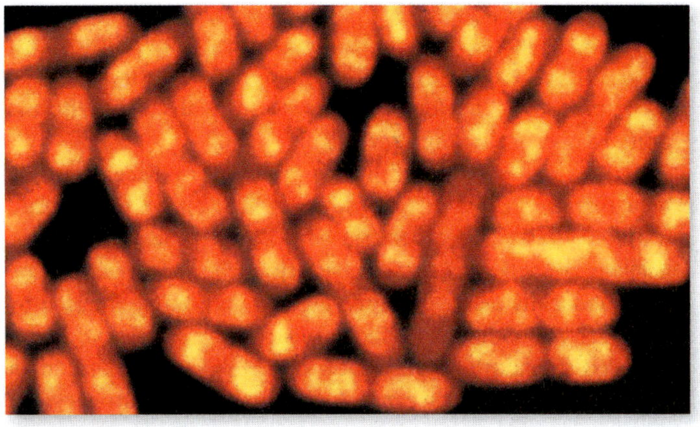

Figure 4.21 Chromosome structure. Fluorescent staining highlights the chromosomes of the bacterial pathogen *Salmonella enteritidis*. The cytoplasm is orange, and the chromosome(s) fluoresce(s) bright yellow.
Prof. E.S. Anderson/Science Source

The DNA in the chromosome is divided into sections called genes. Genes carry information required for bacterial maintenance and growth.

Let's talk about exceptions for a bit: Biology is an area that almost always displays exceptions to the norm. For example, we know that a very few species of bacteria have, indeed, been found to have a nucleus-like structure. We also know that there is one genus of bacterium that contains about 300 different chromosomes, and that these chromosomes are not identical to each other, the way bacterial chromosomes are supposed to be. This bacterium, *Achromatium,* is found in many different environments around the world, so it is very common, though it is but one rare exception to the rule.

Although the chromosome is the minimal genetic requirement for bacterial survival, many bacteria contain other nonessential pieces of DNA called **plasmids.** Plasmids exist as separate double-stranded circles of DNA, although at times they can become integrated into the chromosome. During conjugation, they may be duplicated and passed on to related nearby bacteria. During bacterial reproduction, they are duplicated and passed on to offspring. They are generally not essential to bacterial growth and metabolism, but instead they offer protective traits such as the ability to resist drugs and to produce toxins and enzymes. Because they can be readily manipulated in the laboratory and transferred from one bacterial cell to another, plasmids are an important agent in DNA technology techniques.

Ribosomes: Sites of Protein Synthesis

All cells contain thousands of tiny **ribosomes,** which are made of RNA and protein. When viewed even by very high magnification, ribosomes show up as small, spherical specks dispersed throughout the cytoplasm. They also often occur in chains called polysomes. Many are also attached to the cytoplasmic membrane. Chemically, a ribosome is a combination of a special type of RNA called ribosomal RNA, or rRNA (about 60%), and protein (40%). One method of characterizing ribosomes is by S, or Svedberg,[1] units, which describe the molecular sizes of various cell parts that have been spun down through liquids in a centrifuge and separated on the basis of molecular weight and shape. Heavier, more compact structures sediment faster and are assigned a higher S rating. Combining this method of analysis with high-resolution electron microscopy reveals that the ribosome in bacteria, which has an overall rating of 70S, is actually composed of two smaller subunits **(figure 4.22).** They fit together to form a miniature platform upon which protein synthesis is performed. Note that eukaryotic ribosomes are similar but different. Because of this, we can design drugs to target bacterial ribosomes that do not harm our own. Eukaryotic ribosomes are designated 80S. Although archaea possess 70S ribosomes, they are more similar in structure to that of 80S eukaryotic ribosomes. Please note that the values of Svedburg units are not additive. For example, the large subunit of a bacterial ribosome is 50S, and the small subunit is 30S, but the overall size is 70S.

1. Named in honor of T. Svedberg, the Swedish chemist who developed the ultracentrifuge.

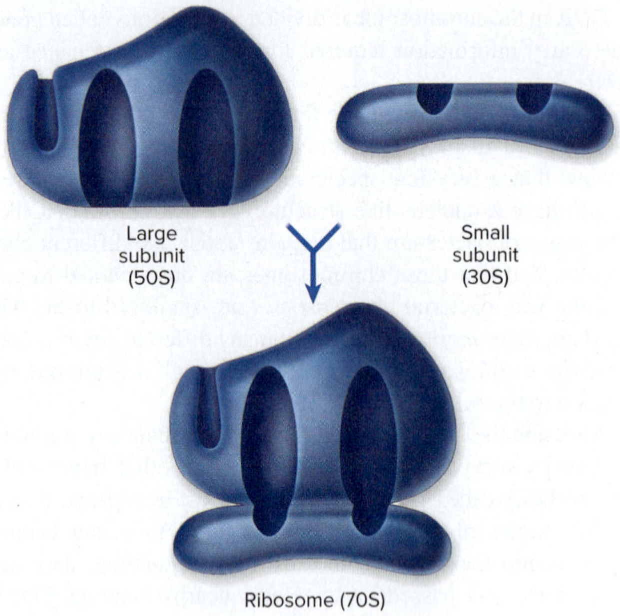
Large
subunit
(50S)
Small
subunit
(30S)
Ribosome (70S)

Figure 4.22 A model of a bacterial ribosome, showing the small (30S) and large (50S) subunits, both separate and joined.

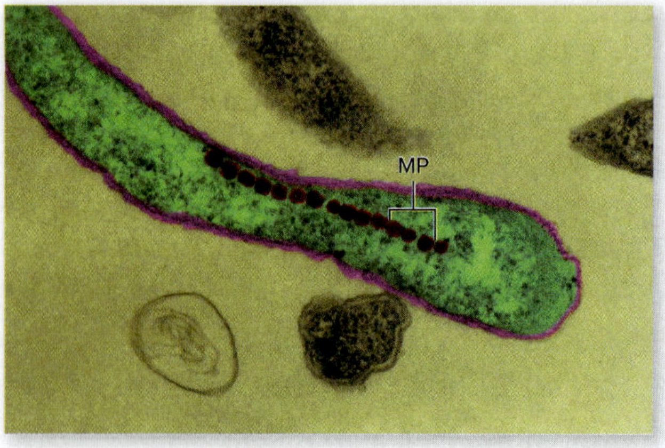
MP

Figure 4.23 Bacterial inclusion bodies. A section through *Aquaspirillum* reveals a chain of tiny iron magnets, or magnetosomes (MP in the photo). These unusual bacteria use these inclusions to orient themselves within their habitat (123,000×).

Dennis Kunkel Microscopy/Science Source

Inclusion Bodies and Microcompartments

Most bacteria are exposed to severe shifts in the availability of food. During periods of nutrient abundance, some can compensate by storing away nutrients intracellularly in **inclusion bodies,** or **inclusions,** of varying size, number, and content. As the environmental source of these nutrients becomes depleted, the bacterial cell can use its own storehouse as required. Some inclusion bodies carry condensed, energy-rich organic substances, such as glycogen and polyhydroxybutyrate (PHB), within special single-layered membranes. A unique type of inclusion found in some aquatic bacteria is gas vesicles that provide buoyancy and flotation. Other inclusions, also called granules, are crystals of inorganic compounds and are not enclosed by membranes at all. Sulfur granules of photosynthetic bacteria and polyphosphate granules of *Corynebacterium* and *Mycobacterium* are of this type. The latter represent an important source of building blocks for nucleic acid and ATP synthesis. They have been termed **metachromatic granules** because they stain a contrasting color (red, purple) in the presence of methylene blue dye.

Perhaps the most unique cell inclusion structure is involved not in cell nutrition but rather in cell orientation. Magnetotactic bacteria contain crystalline particles of iron oxide (magnetosomes) that have magnetic properties **(figure 4.23).** The bacteria use these granules to be pulled by the polar and gravitational fields into deeper habitats with a lower oxygen content.

In the early 2000s, new compartments inside bacterial cells were discovered. These were named bacterial microcompartments, or BMCs. Their outer shells are made of protein, arranged geometrically, and are packed full of enzymes that are designed to work together in pathways, thereby ensuring that they are in close proximity to one another.

The Cytoskeleton

Scientists used to think that the shape of all bacteria was completely determined by the peptidoglycan layer (cell wall). Although this is true of many bacteria, particularly the cocci, other bacteria produce long polymers of proteins that are very similar to eukaryotic **actin.** The word *cytoskeleton* derives from *cyto* (cell) and *skeleton* (supporting structure). In bacteria, these are arranged in helical ribbons around the cell just under the cytoplasmic membrane **(figure 4.24).** The fibers contribute to cell shape, perhaps by influencing the way peptidoglycan is manufactured, and function in cell division. The fibers have been found in rod-shaped and spiral bacteria. They are composed in part of proteins unique to bacterial cells, making them a potentially powerful target for future antibiotic development. Cytoskeletal proteins have also been identified in archaea.

Bacterial Endospores: An Extremely Resistant Stage

We know that the anatomy of bacteria helps them adjust rather well to difficult habitats. But of all microbial structures, nothing can compare to the bacterial **endospore** for withstanding hostile conditions and facilitating survival.

Endospores are dormant bodies produced by bacteria of the genera *Bacillus, Clostridium, Clostridioides, Sporolactobacillus,* and *Sporosarcina.* These bacteria have a two-phase life cycle—a vegetative cell and an endospore **(figure 4.25).** The vegetative cell is a metabolically active and growing cell that can be induced by environmental conditions to undergo endospore formation, a process called **sporulation.** Once formed, the endospore exists in an inert, resting condition that can be seen prominently in an endospore or Gram stain (figure 4.25). Features of endospores in the vegetative cell are somewhat useful in identifying some species because they have unique shapes and sizes. Both gram-positive and gram-negative bacteria can form endospores, but the medically relevant ones are all gram-positive. Most bacteria form only one endospore at a time; therefore, this is not a reproductive function for them.

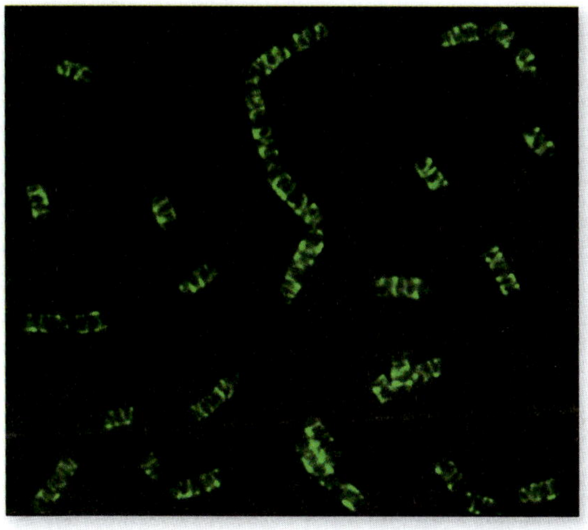

Figure 4.24 Bacterial cytoskeleton. The fibers in these rod-shaped bacteria are fluorescently stained.
Rut Carballido-Lopez/I.N.R.A. Jouy-en-Josas, Laboratoire de Génétique Microbienne

Bacterial endospores are the hardiest of all life forms, capable of withstanding extremes in heat, drying, freezing, radiation, and chemicals that would readily kill vegetative cells. Their survival under such harsh conditions is due to several factors. The heat resistance of endospores has been linked to their high content of calcium and **dipicolinic acid.** We know, for instance, that heat destroys cells by inactivating proteins and DNA and that this process depends on there being a certain amount of water in the protoplasm. Because the deposition of calcium dipicolinate in the endospore removes water and leaves the endospore very dehydrated, it is less vulnerable to the effects of heat. The endospore is metabolically inactive and highly resistant to damage from further drying. The thick, impervious cortex and endospore coats also protect against radiation and chemicals. Endospores are, for all intents and purposes, immortal, meaning they can live indefinitely. One report describes the isolation of viable endospores from a fossilized bee that was 25 million years old, infected with an endospore-forming bacterium. Microbiologists have also unearthed a viable endospore from a 250-million-year-old salt crystal.

Endospore Formation and Resistance

What causes a vegetative cell to make the switch to begin endospore formation? It is usually the fact that nutrients have started to become scarce. Carbon depletion and nitrogen depletion are common triggers. Once this stimulus has been sensed by the vegetative cell, it converts to become a sporulating cell called a **sporangium.** Complete transformation of a vegetative cell into a sporangium and then into an endospore requires 6 to 8 hours in most endospore-forming species. **Figure 4.26** illustrates some major physical and chemical events in this process.

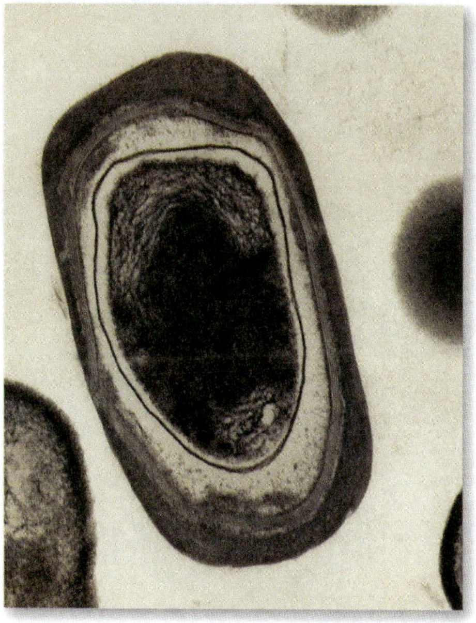

Figure 4.25 Endospore in *Bacillus subtilis*. The endospore has formed inside the vegetative cell and has not yet been released (step 7 in figure 4.26).
Dr. Tony Brain/Science Source

The Germination of Endospores

Once they are formed, the endospores can stay put (stay dormant) for thousands (or millions) of years, as we have said. The breaking of dormancy, which is called *germination,* happens in the presence of water and a specific chemical or environmental stimulus (germination agent) that has become present once again in the environment. Once it starts, it happens quite rapidly (1½ hours). Although the specific germination agent varies among species, it is generally a small organic molecule such as an amino acid or an

Disease Connection

Hospitalized patients who are diagnosed with *Clostridioides difficile,* a form of severe diarrhea caused by an endospore-forming bacterium, must be kept in isolation to prevent spreading the infection to other patients, visitors, and staff. Staff and visitors must wear gowns and gloves when entering the room and must discard soiled gowns and gloves prior to exiting the room. Endospores are difficult to eradicate using regular cleaning regimen. For that reason, rooms occupied by patients diagnosed with *C. difficile* must be terminally cleaned, meaning that every surface, including detachable ones, must be thoroughly cleaned using agents capable of killing spores before another patient can be admitted to the room.

A Note on Terminology

The word *spore* can have more than one usage in microbiology. Fungi have spores that serve as reproductive structures. The bacterial type discussed here is most accurately called an **endospore** because it is produced inside a cell (endo-). It functions in *survival,* not in reproduction, because no increase in cell number is involved in its formation. In contrast, the fungi produce many different types of spores for both survival and reproduction.

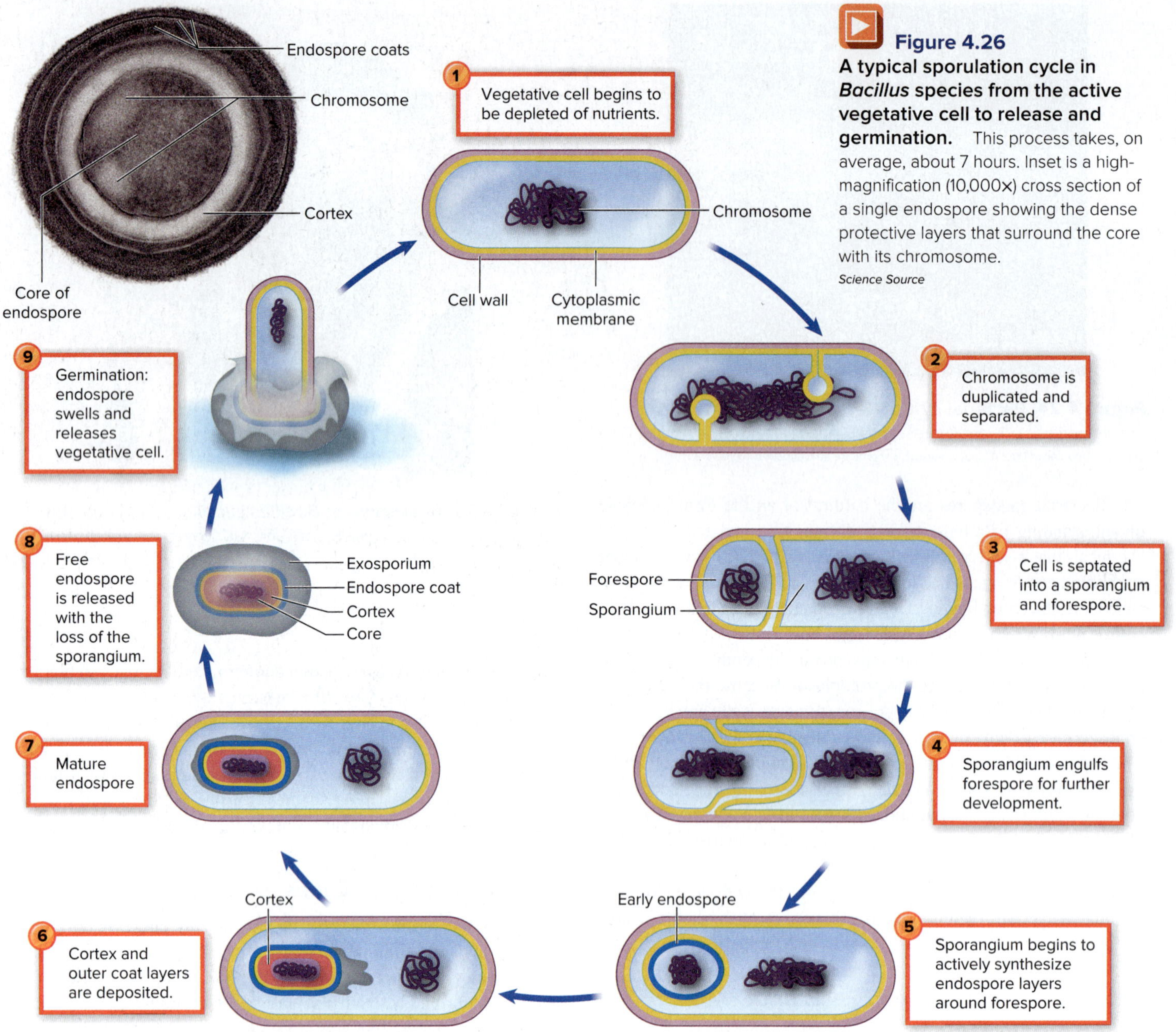

Endospore coats
Chromosome
Cortex
Core of endospore
1
Vegetative cell begins to be depleted of nutrients.
Chromosome
Cell wall
Cytoplasmic membrane
2
Chromosome is duplicated and separated.
3
Cell is septated into a sporangium and forespore.
Forespore
Sporangium
4
Sporangium engulfs forespore for further development.
5
Sporangium begins to actively synthesize endospore layers around forespore.
Early endospore
6
Cortex and outer coat layers are deposited.
Cortex
7
Mature endospore
8
Free endospore is released with the loss of the sporangium.
Exosporium
Endospore coat
Cortex
Core
9
Germination: endospore swells and releases vegetative cell.

Figure 4.26
A typical sporulation cycle in *Bacillus* species from the active vegetative cell to release and germination. This process takes, on average, about 7 hours. Inset is a high-magnification (10,000×) cross section of a single endospore showing the dense protective layers that surround the core with its chromosome.
Science Source

inorganic salt. This agent stimulates the formation of hydrolytic (digestive) enzymes by the endospore membranes. These enzymes digest the cortex of the endospore and expose the core to water. As the core rehydrates and takes up nutrients, it begins to grow out of the endospore coats. In time, it becomes a fully active vegetative cell, resuming the vegetative cycle.

Medical Significance of Bacterial Endospores

Although the majority of endospore-forming bacteria are relatively harmless, several bacterial pathogens are endospore-formers. *Bacillus anthracis* is the agent of anthrax. Its persistence in the environment in endospore form makes it an ideal candidate for bioterrorism. The genus *Clostridium* includes even more pathogens, such as *C. tetani,* the cause of tetanus (lockjaw), and *C. perfringens,* the cause of gas gangrene. When the endospores of these species are embedded in a wound that contains dead tissue, they can germinate, grow, and release potent toxins. Another toxin-forming species, *C. botulinum,* is the agent of botulism, a deadly form of food poisoning.

Because they inhabit the soil and dust, endospores are constant intruders where sterility and cleanliness are important. They resist ordinary cleaning methods that use boiling water, soaps, and disinfectants; and they frequently contaminate cultures and media. Hospitals and clinics must take precautions to guard against the potential harmful effects of endospores in wounds, especially those of *Clostridioides difficile.* Several endospore-forming species cause food spoilage or poisoning. Endospore destruction is also a particular concern of the food-canning industry. Ordinary boiling (100°C) will usually not destroy such endospores, so canning is carried out in pressurized steam at 120°C for 20 to 30 minutes. Such rigorous conditions ensure that the food is sterile and free from viable bacteria.

4.4 Learning Outcomes—Assess Your Progress

11. Identify five structures that may be contained in bacterial cytoplasm.
12. Detail the causes and mechanisms of sporulation and germination.

4.5 The Archaea

As you know, the **archaea** are considered a third cell type in a separate superkingdom (the Domain Archaea). We include them in this chapter because they share many bacterial characteristics. But they are actually more closely related to Domain Eukarya than to Bacteria. For example, archaea and eukaryotes share a number of ribosomal RNA sequences that are not found in bacteria, and their ribosomal subunit structures and process of protein synthesis are similar. **Table 4.1** outlines selected points of comparison of the three domains.

Among the ways that the archaea differ significantly from other cell types are that certain genetic sequences are found only in their rRNA and that they exhibit a unique method of DNA compaction. They also have unique membrane lipids, cell wall composition, and pilin proteins.

The archaea have unusual and chemically distinct cell walls. In some, the walls are composed almost entirely of polysaccharides, and in others, the walls are pure protein. As a group, they all lack the true peptidoglycan structure found in bacteria. Because a few archaea lack a cell wall entirely, their cytoplasmic membrane must serve the dual functions of support and transport.

The early earth is thought to have contained a hot, anaerobic "soup" with sulfuric gases and salts in abundance. Many modern archaea still live in the remaining habitats on the earth that have these same ancient conditions—the most extreme habitats in nature. These particular archaea are called extremophiles, meaning that they "love" extreme conditions in the environment.

Metabolically, the archaea exhibit incredible adaptations to what would be deadly conditions for other organisms. These hardy microbes have adapted to multiple combinations of heat, salt, acid, pH, pressure, and atmosphere. Included in this group are methane producers, hyperthermophiles, extreme halophiles, and sulfur reducers.

Members of the group called **methanogens** can convert CO_2 and H_2 into methane gas (CH_4) through unusual and complex pathways. These archaea are common inhabitants of anaerobic swamp mud, the bottom sediments of lakes and oceans, and even the digestive systems of animals and humans. The gas they produce collects in swamps and may become a source of fuel. Methane may also contribute to the "greenhouse effect," which maintains the earth's temperature and can contribute to global warming.

Other types of archaea—the extreme halophiles—require salt to grow, and some have such a high salt tolerance that they can multiply in sodium chloride solutions (36% NaCl) that would destroy most cells. They exist in the saltiest places on the earth—inland seas, salt lakes, salt mines, and salted fish. They are not particularly common in the ocean because the salt content is not high enough. Many of the "halobacteria" use a red pigment to synthesize ATP in the presence of light. These pigments are responsible for the color of the Red Sea and the red color of salt ponds **(figure 4.27).**

Table 4.1 Comparison of Three Cellular Domains

Characteristic	Bacteria	Archaea	Eukarya
Cell type	Bacterial	Archaeal	Eukaryotic
Chromosomes	Single, or few, circular	Single, circular	Several, linear
Types of ribosomes	70S	70S but structure is similar to 80S	80S
Flagella	Hook, rings, and hollow filament	Solid fimbrial-like structure called an archaellum	"9 + 2" microtubule arrangement
Contains unique ribosomal RNA signature sequences	+	+	+
Presence of peptidoglycan in cell wall	+	–	–
Cell membrane lipids	Fatty acids with ester linkages	Long-chain, branched hydrocarbons with ether linkages	Fatty acids with ester linkages
Sterols in membrane	– (Some exceptions)	–	+

Figure 4.27 Halophiles around the world. Lake Natron in the Great Rift Valley on the border of Tanzania and Kenya. Halophilic algae give the lake its color.
Paul & Paveena Mckenziee/Oxford Scientific/Getty Images

Archaea that are adapted to growth at very low temperatures are called **psychrophilic** (loving cold temperatures); those growing at very high temperatures are **hyperthermophilic** (loving high temperatures). Hyperthermophiles flourish at temperatures between 80°C and 113°C and cannot grow at 50°C. They live in volcanic waters and soils and submarine vents and are also often salt- and acid-tolerant as well. One member, *Thermoplasma,* lives in hot, acidic habitats in the waste piles around coal mines that regularly sustain a pH of 1 and a temperature of nearly 60°C. Because many archaea are unculturable, the technology of rRNA sequencing has been invaluable in the identification of these microbes. Analysis of these unique sequences has advanced not only the process of identification but also our knowledge of transcription, translation, and cellular evolution.

Archaea are not just environmental microbes. They have been isolated from human tissues such as the colon, the mouth, and the vagina **(Insight 4.1).** Recently, an association was found between the degree of severity of periodontal disease and the presence of archaean RNA sequences in the gingiva, suggesting—but not proving—that archaea may be capable of causing human disease.

4.5 Learning Outcome—Assess Your Progress

13. List some differences between archaea and bacteria.

4.6 Classification Systems for Bacteria and Archaea

Classification systems serve both practical and academic purposes. They aid in differentiating and identifying unknown species in medical and applied microbiology. They are also useful in organizing microbes and as a means of studying their relationships and origins. Since classification was started around 200 years ago, several thousand species of bacteria and archaea have been identified, named, and cataloged.

For years, scientists have had intense interest in tracing the origins of and evolutionary relationships among bacteria and archaea, but doing so has not been an easy task. One of the questions that has plagued taxonomists is, What characteristics are the most indicative of closeness in ancestry? Early bacteriologists found it convenient to classify bacteria according to shape, variations in arrangement, growth characteristics, and habitat. However, as more species were discovered and as techniques for studying their biochemistry were developed, it soon became clear that similarities in cell shape, arrangement, and staining reactions do not automatically indicate relatedness. Even though the gram-negative rods look alike, there are hundreds of different species, with highly significant differences in biochemistry and genetics. If we attempted to classify them on the basis of Gram stain and shape alone, we could not assign them to a more specific level than class. Classification schemes are increasingly using genetic and molecular traits that cannot be visualized under a microscope or in culture.

One of the most important indicators of evolutionary relatedness is comparison of the sequence of nitrogen bases in ribosomal RNA, a major component of ribosomes. More specifically, the rRNA of the small subunit of the ribosomal unit is used. This is abbreviated as SSU rRNA. Ribosomes have the same function (protein synthesis) in all cells, and they tend to remain more or less stable in their nucleic acid content over long periods. For this reason, any major differences in the sequence, or "signature," of the SSU rRNA is likely to indicate some distance in ancestry. This technique is powerful at two levels: It is effective for differentiating general group differences, allowing for the creation of branching tree diagrams showing evolutionary relatedness among microbes (see figure 1.15). It can be fine-tuned for bacterial identification at the species level (for example, in *Mycobacterium* and *Legionella*). Elements of these and other identification methods are presented in more detail in chapter 18.

The definitive published source for bacterial and archaea classification, called *Bergey's Manual,* has been in print continuously since 1923. The basis for the early classification in *Bergey's* was the **phenotypic** (not genetic) traits of bacteria, such as their shape, cultural behavior, and biochemical reactions. These traits are still used extensively by clinical microbiologists or researchers who need to quickly identify unknown bacteria. As methods for RNA and DNA analysis became available, this information was used to supplement the phenotypic information. The current version of the publication, called *Bergey's Manual of Systematic Bacteriology,* presents a comprehensive view of bacterial and archaea relatedness, combining phenotypic information with SSU rRNA sequencing information to classify them. It is now available online. (We need to remember that all classification systems are in a state of constant flux; no system is ever finished.)

INSIGHT 4.1 MICROBIOME: Archaea in the Human Microbiome

We tend to discuss archaea as "extremophiles." That's probably because the characteristics that make them able to live in extreme environments are, in fact, fascinating. So they live in hot springs, deep sea vents, and the permafrost, but do they also inhabit human bodies?

The answer is yes! A few research papers have trickled out in recent years, reporting the presence of archaea in humans. This research has provided convincing data that archaea are part of the microbiome on human skin, in the gastrointestinal tract, and in the nose and lungs. The researchers found anaerobic archaea in parts of the intestinal tract that are oxygen-free. They found ammonia-oxidizing archaea on the surface of the skin.

We will learn about how researchers study microbes without culturing them. That development led to the ability to identify the somewhat strange, but apparently vital, archaea in the human microbiome. Because no archaea have yet been directly related to disease in humans, there has been somewhat less attention paid to them by scientists. This new research shows that they may be intimately involved in maintaining our health.

Source: 2017. *mBio.* Vol. 8, no. 6. DOI: 10.11228/mBio.00824-17.

Steven P. Lynch

kali9/Getty Images

With the explosion of information about evolutionary relatedness among bacteria, the need for a *Bergey's Manual* that was more practical for identifying unknown bacteria became apparent. There is a separate book, called *Bergey's Manual of Determinative Bacteriology,* based entirely on phenotypic characteristics. It is utilitarian in focus, categorizing bacteria by traits that are easy to test in clinical, teaching, and research labs. It is widely used by microbiologists who need to identify bacteria but need not know their evolutionary backgrounds. This phenotypic classification is more useful for students of medical microbiology, as well.

Taxonomic Scheme

Bergey's Manual of Determinative Bacteriology organizes the bacteria and archaea into four major divisions. These somewhat natural divisions are based on the nature of the cell wall. The **Gracilicutes** (gras″-ih-lik′-yoo-teez) have gram-negative cell walls and are thin-skinned; the **Firmicutes** have gram-positive cell walls that are thick and strong; the **Tenericutes** (ten″-er-ik′-yoo-teez) lack a cell wall and thus are soft; and the **Mendosicutes** (men-doh-sik′-yoo-teez) are the archaea. Gracilicutes and Firmicutes contain the greatest number of species. The system used in *Bergey's Manual* organizes bacteria and archaea into subcategories such as classes, orders, and families, but these are not available for all groups.

Diagnostic Scheme

As mentioned earlier in this section, many medical microbiologists prefer an informal working system that outlines the major families

Table 4.2 A Selection of Medically Important Families and Genera of Bacteria, with Notes on Some Diseases*

I. Bacteria with Gram-Positive Cell Wall Structure	
Cocci in clusters or packets Family *Micrococcaceae: Staphylococcus* (members cause boils, skin infections)	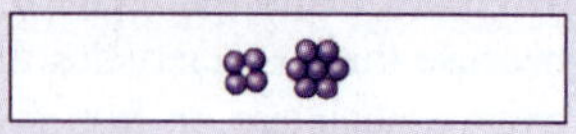
Cocci in pairs and chains Family *Streptococcaceae: Streptococcus* (species cause strep throat, dental caries)	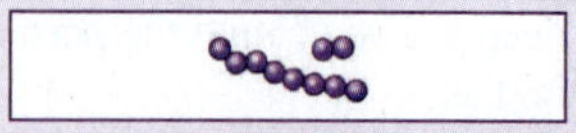
Anaerobic cocci in pairs, tetrads, irregular clusters Family *Peptococcaceae: Peptococcus, Peptostreptococcus* (involved in wound infections)	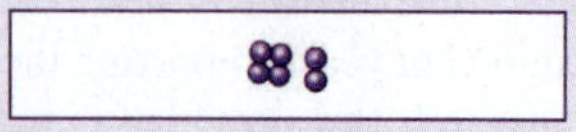
Endospore-forming rods Family *Bacillaceae: Bacillus* (anthrax), *Clostridium* (tetanus, gas gangrene, botulism)	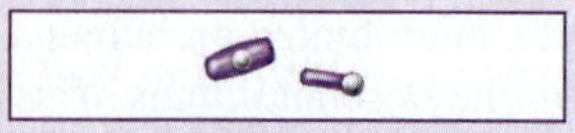
Non-endospore-forming rods Family *Lactobacillaceae: Lactobacillus, Listeria, Erysipelothrix* (erysipeloid) Family *Propionibacteriaceae: Propionibacterium* (involved in acne)	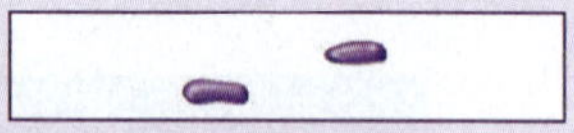
Family *Corynebacteriaceae: Corynebacterium* (diphtheria)	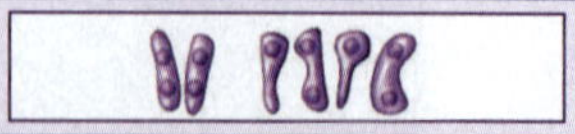
Family *Mycobacteriaceae: Mycobacterium* (tuberculosis, leprosy)	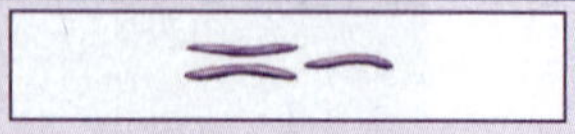
II. Bacteria with Gram-Negative Cell Wall Structure	
Aerobic cocci *Neisseria* (gonorrhea, meningitis), *Branhamella*	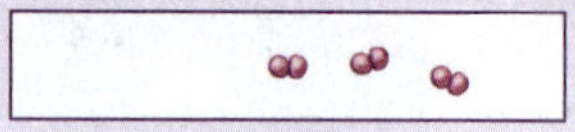
Aerobic coccobacilli *Moraxella, Acinetobacter*	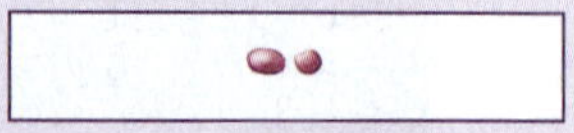
Anaerobic cocci Family *Veillonellaceae: Veillonella* (dental disease)	
Aerobic rods Family *Pseudomonadaceae: Pseudomonas* (pneumonia, burn infections) Miscellaneous: *Legionella* (Legionnaires' disease)	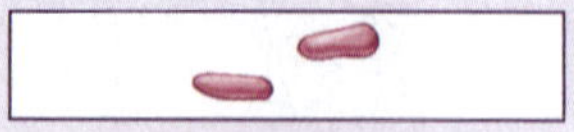
Facultative or anaerobic rods and vibrios Family *Enterobacteriaceae: Escherichia, Edwardsiella, Citrobacter, Salmonella* (typhoid fever), *Shigella* (dysentery), *Klebsiella, Enterobacter, Serratia, Proteus, Yersinia* (one species causes plague) Family *Vibrionaceae: Vibrio* (cholera, food infection), *Campylobacter, Aeromonas*	 
Anaerobic rods Family *Bacteroidaceae: Bacteroides, Fusobacterium* (anaerobic wound and dental infections)	
Helical and curved bacteria Family *Spirochaetaceae: Treponema* (syphilis), *Borrelia* (Lyme disease), *Leptospira* (kidney infection)	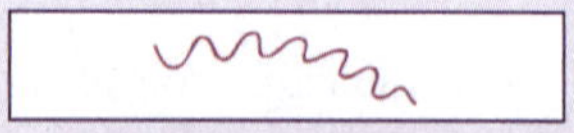
III. Bacteria with No Cell Walls	
Family *Mycoplasmataceae: Mycoplasma* (pneumonia), *Ureaplasma* (urinary infection)	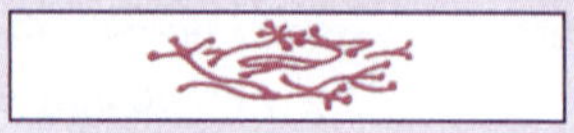

*Details of pathogens and diseases appear in chapters 19 through 24.

and genera. **Table 4.2** is an example of the phenotypic method of classification that may be used in clinical microbiology (based on phenotype, not genetics). It divides the bacteria into gram-positive, gram-negative, and those without cell walls and then subgroups them according to cell shape, arrangement, and certain physiological traits such as oxygen usage. **Aerobic** bacteria use oxygen in metabolism; **anaerobic** bacteria do not use oxygen in metabolism; and **facultative** bacteria may or may not use oxygen.

Species and Subspecies in Bacteria and Archaea

Among most organisms, the species level is a distinct, readily defined, and natural taxonomic category. In animals, for instance, a species is a distinct type of organism that can produce viable offspring only when it mates with others of its own kind. This definition does not work for bacteria and archaea primarily because they do not exhibit a typical mode of sexual reproduction. They can accept genetic information from unrelated forms, and they can alter their genetic makeup by a variety of mechanisms. Thus, it is necessary to hedge a bit when we define a bacterial species. Theoretically, it is a collection of bacterial cells, all of which share an overall similar pattern of traits, in contrast to other groups whose patterns differ significantly. Although the boundaries that separate two closely related species in a genus are in some cases arbitrary, recent sophisticated "big data" analyses of bacterial genomes confirms the differences between what we have been calling different species.

Individual members of a given species can show variations as well. Therefore, more categories within species exist, but they are not well defined. Microbiologists use terms like **subspecies, strain,** or **type** to designate bacteria of the same species that have differing characteristics. **Serotype** refers to representatives of a species that stimulate a distinct pattern of antibody (serum) responses in their hosts because of distinct surface molecules.

4.6 Learning Outcomes—Assess Your Progress

14. Differentiate between *Bergey's Manual of Systematic Bacteriology* and *Bergey's Manual of Determinative Bacteriology.*
15. Name four divisions ending in *-cutes* and describe their characteristics.
16. Define a *species* in terms of bacteria.

MEDIA UNDER THE MICROSCOPE WRAP-UP

Rich Carey/Shutterstock

The **intended message** of the article is to describe our plastics pollution problem and how bacteria may help us. It also sounded an alarm bell, that we should be vigilant to prevent unintended consequences.

A **critical reading** of the article would note these things: (1) A problem (the plastic pollution problem) is laid out; (2) there were basic definitions provided for unfamiliar terms; and (3) there were alternate scenarios presented that society should consider. All of these are hallmarks of good reporting.

Interpreting the article for your friends: This article is so well-written for the general public there is very little interpreting required.

My **overall grade** for this article is an A+. It is an excellent example of science communication for the public.

Source: Emily Flashman, *"The Race to Develop Plastic-Eating Bacteria," Forbes,* March 2021.

Study Smarter: Better Together

These activities are designed for you to use on your own with a study group—either a face-to-face group or a virtual one, consisting of 3–5 members. Studying together can be very helpful, but there are effective and ineffective ways to do it. For example, getting together without a clear structure is often not a good use of your time. Use your time efficiently by using one or more of the exercises below.

FACE-TO-FACE GROUPS

Use one or more of the activities below.

Peer Instruction: Assign numbers to your group members to use all semester long. Now look at these five concepts from this chapter. Each group member prepares a 5-minute lesson on the topic corresponding to their number. Don't worry if you have fewer than 5 members; just use however many you have! During your group study time, each member presents their lesson, and the group spends another 5–10 minutes discussing that lesson.

1. Differences between bacteria and eukaryotic cells
2. Differences between bacteria and archaea
3. Gram-positive and gram-negative envelope structure
4. *Bergey's Manual*
5. Biofilms

Concept Maps: Each member of the group should use this list of terms from this chapter to generate their own concept map. This can be hand-drawn or created using software (see Appendix C for guidelines). During group study time, compare each other's concept maps and help each other make sure they are correct. Of course, there are many different "correct" maps. Examining each member's map will help you talk through the varied concepts and how they are related.

(continued)

Concept Terms:

genus	species	serotype	domain
Borrelia burgdorferi	spirochete	gram-negative	

Table Topics: Each group member should identify a concept or topic from this week's class assignments with which they are having trouble and share it during group study time. The other group members can then help to clarify confusing issues or share how they figured it out. Aim for a maximum of 15 minutes per topic. If the topic remains unclear to the group, bring it up during class or use the instructor's office hours or e-mail to ask for help. Taking the time to struggle with a difficult concept first makes your questions much more specific and more likely to yield helpful answers.

VIRTUAL GROUPS

Not everyone has the time or opportunity to meet with group members outside of class time. You or your instructor can create a virtual group using e-mail or the course software.

Weekly Discussion Board: This forum can be used as a way for groups to discuss topics, via e-mail or other learning management systems or online platforms, before they are covered in class. As each member of the group answers the current week's question, they should send their responses to every other member of their group. It's best to agree on a deadline based on how your class schedule works (Saturday for the next week's topics, for example). Then, after the topic is discussed in class, each member should send a response that all group members will see, with a follow-up post on the same topic. If you cover more than one chapter in a week, someone can be designated to choose which chapter Discussion Board question you will use. Or simply decide up front that you will always use the first-chapter-of-the week's question, to keep the schedule simple.

Discussion Question

During the Cold War between the Soviet Union and the United States, both countries were interested in "weaponizing" *Bacillus anthracis* (anthrax) endospores to deliver via missiles and other weapons. Do you believe there is any ethical difference between using infectious agents as weapons compared to traditional weapons? Explain.

Chapter Summary

BACTERIA AND ARCHAEA

4.1 THE BACTERIA

- Bacteria and archaea are ancient forms of life, found in every niche on the planet.
- Most bacteria have one of three shapes: coccus, rod, and curved.
- The shape and arrangement of cells are key means of describing both bacteria and archaea.

4.2 EXTERNAL STRUCTURES

- The outside of bacteria can have appendages (flagella, fimbriae, pili, and nanowires) and surface coatings (S layers and glycocalyces).

4.3 THE CELL ENVELOPE: THE BOUNDARY LAYER OF BACTERIA

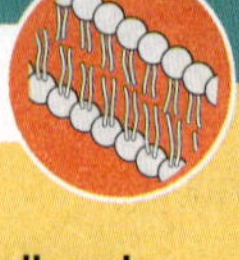

- The cell envelope is the entire boundary structure surrounding a bacterial cell.
- In gram-negative bacteria, the envelope consists of an outer membrane, the cell wall, and the cytoplasmic membrane. Gram-positive bacteria have only the cell wall and cytoplasmic membrane.
- In a Gram stain, gram-positive bacteria retain the crystal violet and stain purple. Gram-negative bacteria lose the crystal violet and stain red from the safranin counterstain.

4.4 BACTERIAL INTERNAL STRUCTURE

- All bacteria have cytoplasm, DNA, ribosomes, and structural filaments on the inside of their cell.
- Some bacteria also have inclusion bodies, microcompartments, and/or endospores.
- Endospores provide a mechanism for nearly endless survival for bacteria that manufacture them.

4.5 THE ARCHAEA

- Archaea constitute the third domain of life. They exhibit unusual properties that make them more closely related to eukaryotes than to bacteria.
- A subgroup of archaea are extremophiles, surviving in habitats that are hostile to most life.

4.6 CLASSIFICATION SYSTEMS FOR BACTERIA AND ARCHAEA

- Bacteria and archaea are formally classified by phylogenetic relationships, genetic relatedness, and phenotypic characteristics.
- A bacterial species was traditionally defined as a collection of bacterial cells that share an overall similar pattern of traits different from other groups of bacteria. Advanced genetic analyses are bearing out these species designations.

SmartGrid: From Knowledge to Critical Thinking

This *21 Question Grid* takes the topics from this chapter and arranges them with respect to the American Society for Microbiology's Undergraduate Curriculum guidelines—all six of the important "Concepts" as well as the important "Competency" of scientific literacy. Three questions are supplied, which cover chapter content referring to the Concept or Competency in increasing levels of Bloom's taxonomy for learning.

ASM Concept/ Competency	A. Bloom's Level 1, 2—Remember and Understand (Choose one.)	B. Bloom's Level 3, 4—Apply and Analyze	C. Bloom's Level 5, 6—Evaluate and Create
Evolution	**1.** Archaea **a.** are most genetically related to bacteria. **b.** contain a nucleus. **c.** cannot cause disease in humans. **d.** lack peptidoglycan in their cell walls.	**2.** Recall from chapter 1 that both bacteria and archaea appear to have evolved from the last common ancestor (LCA). Based on what you know about their habitats and structures, speculate on *why* there are differences between the two.	**3.** Suppose an argument in your city has erupted about whether bacteria can be called species. Write a letter to the editor for your local newspaper arguing one side or the other.
Cell Structure and Function	**4.** Which of the following is present in both gram-positive and gram-negative cell walls? **a.** an outer membrane **b.** peptidoglycan **c.** teichoic acid **d.** lipopolysaccharides	**5.** As a supervisor in the infection control unit, you hire a local microbiologist to analyze samples from your hospital's hot-water tank for microbial contamination. Although she was unable to culture any microbes, she reports that basic microscopic analysis revealed the presence of cells 0.8 μm in diameter that lacked a nucleus. Transmission electron microscopy showed that the cells lacked membrane-bound organelles but did contain ribosomes. Which domain of life do you hypothesize the cells represent? What further analysis would you perform?	**6.** Clearly bacteria are very different from eukaryotic cells. Yet they have similar processes and structures. Their DNA is very similar. Their ribosomes are very similar. The way they make proteins is similar. Speculate as to what led to this similarity in such radically different organisms.

(continued)

ASM Concept/ Competency	A. Bloom's Level 1, 2—Remember and Understand (Choose one.)	B. Bloom's Level 3, 4—Apply and Analyze	C. Bloom's Level 5, 6—Evaluate and Create
Metabolic Pathways	**7.** Bacterial endospores usually function in **a.** reproduction. **b.** metabolism of nutrients. **c.** survival. **d.** storage.	**8.** Creating extremely long tubes (nanotubes) out of existing cell membrane material seems like a lot of hard work. Why would this be justified for an organism in an oxygen-poor environment?	**9.** Bacteria and archaea have a much greater diversity in their metabolic capabilities than do eukaryotes. They can live at 113°C (235°F) and at incredibly high atmospheric pressures. Construct an argument for why these simple cells have developed the ability to do this.
Information Flow and Genetics	**10.** Which structure plays a direct role in the exchange of genetic material between bacterial cells? **a.** flagellum **b.** pilus **c.** capsule **d.** fimbria	**11.** Bacteria have been found to change the structures they produce when they are placed in a situation of high flow rates, especially in a narrow space, like a tube. Often these new structures allow them to attach themselves or trap other bacteria in a biofilm. Identify some advantages to this strategy from the bacterium's perspective.	**12.** Bacterial and archaeal chromosomes are not enclosed in a membrane-bound nucleus. You will learn about some advantages of this arrangement in another chapter, but can you think of some disadvantages?
Microbial Systems	**13.** Nanotubes are extensions of the ________________ that can function in ________________. **a.** membrane, genetic exchange **b.** pilus, genetic exchange **c.** flagellum, motility **d.** membrane, nutrient transfer	**14.** The results of your patient's wound culture just arrived, and Gram staining revealed the presence of pink, rod-shaped bacterial cells organized in pairs. You quickly realize that this patient could be at risk for developing fever and shock. Explain how the culture results indicated this potential risk.	**15.** We know that bacteria/archaea and their genetics are clearly influenced by their environment because they have developed membranes and structures that allow them to live in a wide variety of conditions. Do you think the environment in which they live has likewise experienced change due to *their* presence? Can you cite any examples?
Impact of Microorganisms	**16.** Find the true statement about biofilms. **a.** They are found only in outdoor environments. **b.** They are found only on artificial medical implants. **c.** They consist of many representatives of a single bacterial species. **d.** They complicate the treatment of some infections.	**17.** Suggest more than one reason why bacteria may express glycocalyces in natural circumstances but not in the laboratory.	**18.** Construct arguments agreeing with and refuting this statement: *Human infections may have originated as accidental encounters between humans and microbes that were actually meant to interact with other parts of the natural environment.*
Scientific Thinking	**19.** Which of the following would be used to identify an unknown bacterial culture that came from a patient in the intensive care unit? **a.** *Gray's Anatomy* **b.** *Bergey's Manual of Determinative Bacteriology* **c.** *Bergey's Manual of Systematic Bacteriology* **d.** *The Physician's Desk Reference*	**20.** During the Cold War between the Soviet Union and the United States, both countries were interested in "weaponizing" *Bacillus anthracis* (anthrax) endospores to deliver via missiles and other weapons. Why do you think they chose this particular organism?	**21.** During the Cold War between the Soviet Union and the United States, both countries were interested in "weaponizing" *Bacillus anthracis* (anthrax) endospores to deliver via missiles and other weapons. Do you believe there is any ethical difference between using infectious agents as weapons compared to traditional weapons? Explain.

Answers to the multiple-choice questions appear in Appendix A.

Visual Connections

These questions use visual images or previous content to connect content within and between chapters.

1. **From chapter 3, figure 3.6*b*.** Do you believe that the bacteria spelling *"Klebsiella"* or the bacteria spelling *"S. aureus"* possess the larger capsule? Defend your answer.

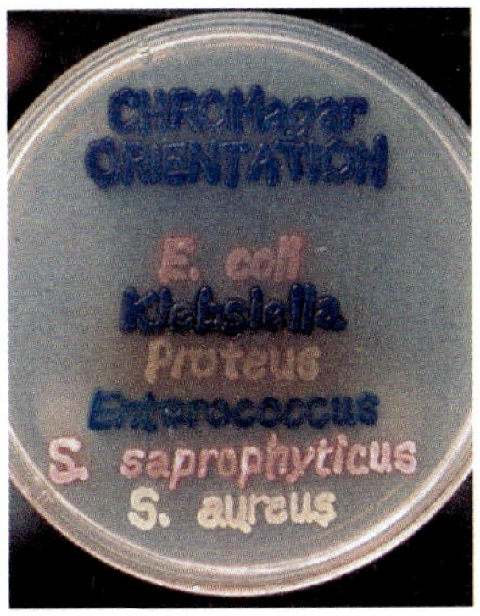
CHROMagar
ORIENTATION
E. coli
Klebsiella
Proteus
Enterococcus
S. saprophyticus
S. aureus

Kathy Park Talaro

2. **From chapter 1, figure 1.15.** Study this figure. How would it be drawn differently if the archaea were more closely related to bacteria than to eukaryotes?

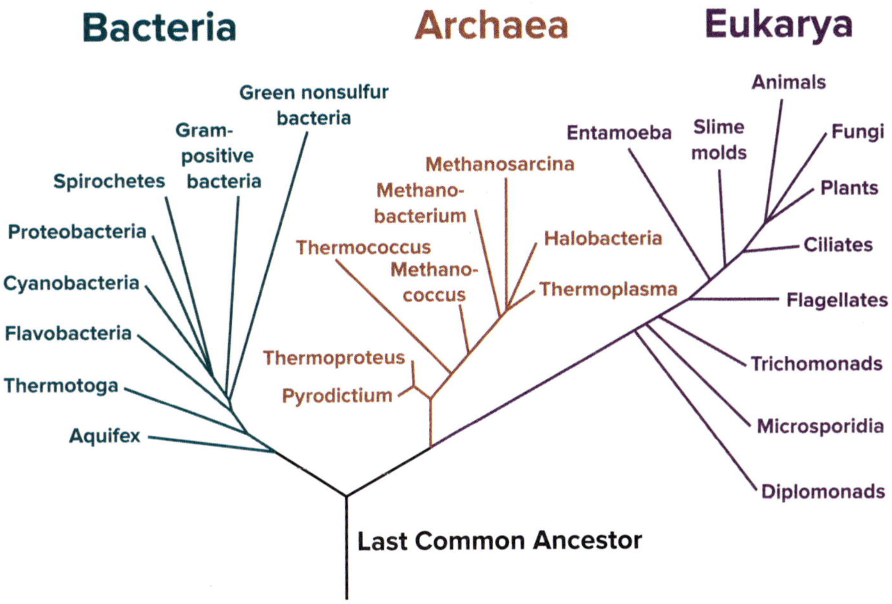
Bacteria
Archaea
Eukarya
Green nonsulfur bacteria
Gram-positive bacteria
Spirochetes
Proteobacteria
Cyanobacteria
Flavobacteria
Thermotoga
Aquifex
Methanosarcina
Methano-bacterium
Thermococcus
Methano-coccus
Halobacteria
Thermoplasma
Thermoproteus
Pyrodictium
Entamoeba
Slime molds
Animals
Fungi
Plants
Ciliates
Flagellates
Trichomonads
Microsporidia
Diplomonads
Last Common Ancestor

High Impact Study

These terms and concepts are most critical for your understanding of this chapter—and may be the most difficult. Have you mastered them?

Concepts

- ☐ Differences between bacterial and eukaryotic cells
- ☐ Differences between bacteria and archaea
- ☐ Gram-positive and gram-negative envelope structure
- ☐ *Bergey's Manuals*

Terms

- ☐ Rods/bacilli
- ☐ Cocci
- ☐ Biofilm
- ☐ Peptidoglycan
- ☐ Endospore

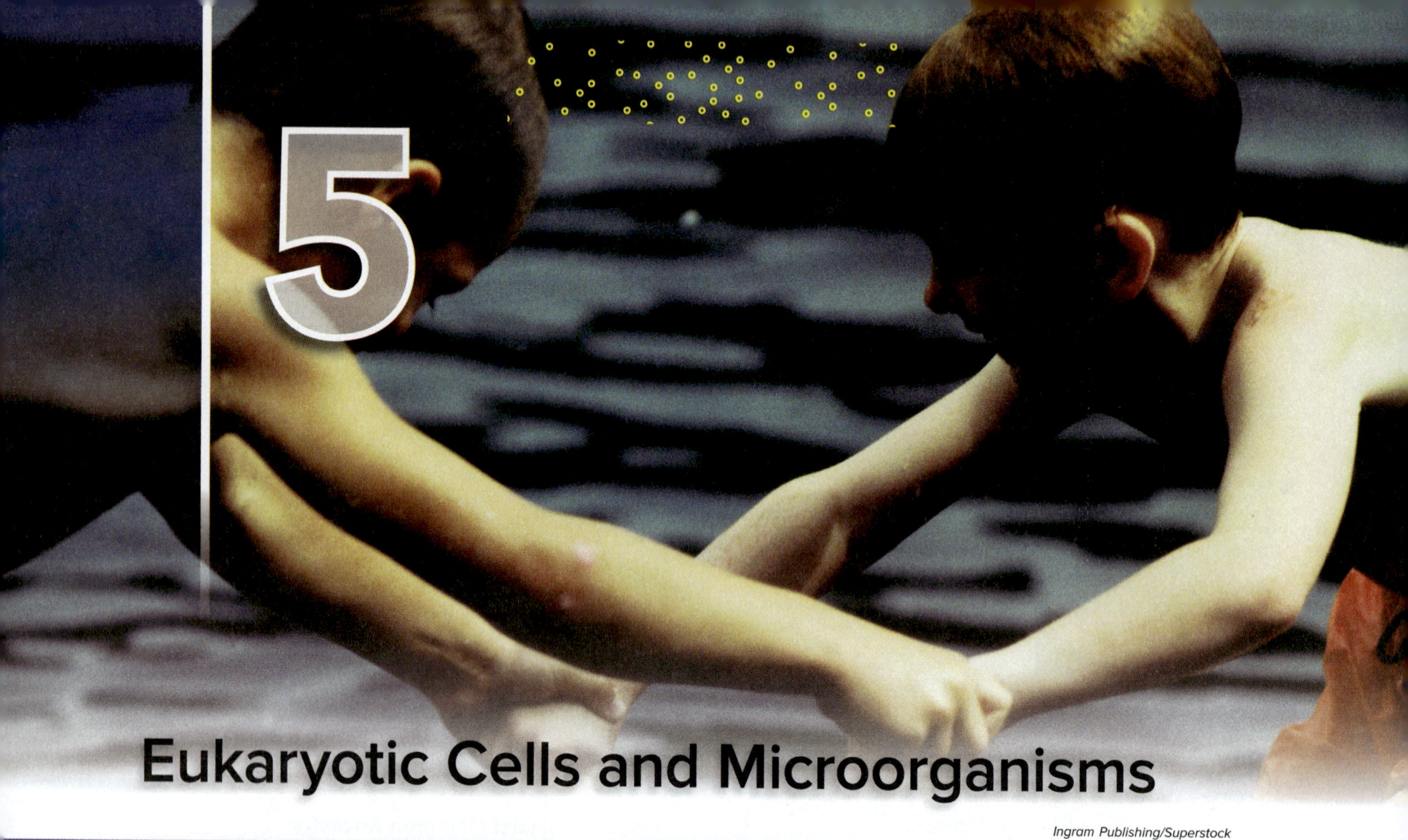

5 Eukaryotic Cells and Microorganisms

Ingram Publishing/Superstock

MEDIA UNDER THE MICROSCOPE

Amoebas in the Water

This opening case examines an article from the popular media to determine the extent to which it is factual and/or misleading. This case focuses on the September 2020 article from the online media outlet Mic, *"Brain-Eating Amoebas are Horrifying But Rare. How Scared Should I Be?"*

The amoeba *Naegleria fowleri* lives in warm freshwater, and if the freshwater is forced up a person's nose, the amoeba can invade the brain. The resulting infection is nearly 100% fatal. The article reminds us that the only danger of swimming in freshwater is if the water enters the nose with some force, when someone cannonballs into a lake, for example. The article also reminds us that the infection is extremely rare, with only 10 people contracting it in the last 10 years.

The amoeba can be present in municipal water supplies, as well. But it is harmless if it is ingested by drinking and is killed by boiling. Cases of this infection have been tied to the use of neti pots, used by some people to pass water through their nose to clear up congestion. If tap water is used, the amoeba can access the brain through the nose and cause the disease. The article cites doctors' advice that the water you use in a neti pot should be preboiled, to eliminate this protozoan.

This article discusses the fact that often news stories about deaths caused by this amoeba can be deliberately sensational. Several popular "women's" magazines ran stories with alarmist headlines implying that merely swimming in lakes could cause this disease. Others used headlines such as "Brain-Eating Amoebas in Your Neti Pot."

- What is the **intended message** of the article?
- What is your **critical reading** of the summary of the article provided above? Remember that in this context, "critical reading" does not necessarily mean *What criticism do you have?* but asks you to apply your knowledge to interpret whether the article is factual and whether the facts support the intended message.

Media Under The Microscope Wrap-Up appears at the end of the chapter.

- How would you **interpret** the news item for your nonmicrobiologist friends?
- What is your **overall grade** for the news item—taking into account its accuracy and the accuracy of its intended effect?

Outline and Learning Outcomes

5.1 Overview of the Eukaryotes

1. Relate bacterial, archaeal, and eukaryotic cells to the *last common ancestor.*
2. List the types of eukaryotic microorganisms and denote which are unicellular and which are multicellular.
3. Explain how endosymbiosis contributed to the development of eukaryotic cells.

5.2 Form and Function of the Eukaryotic Cell: External and Boundary Structures

4. Differentiate among the flagellar structures of bacteria, archaea, and eukaryotes.
5. Describe the important characteristics of a glycocalyx in eukaryotes.
6. List which eukaryotic microorganisms might have a cell wall.
7. List similarities and differences between eukaryotic and bacterial cytoplasmic membranes.

5.3 Form and Function of the Eukaryotic Cell: Internal Structures

8. Describe the main structural components of a nucleus.
9. Diagram how the nucleus, endoplasmic reticulum, and Golgi apparatus act together with vesicles during the transport process.
10. Explain the function of the mitochondrion.
11. Explain the importance of ribosomes and differentiate between eukaryotic and bacterial types.
12. List and describe the three main fibers of the cytoskeleton.

5.4 The Fungi

13. List three general features of fungal anatomy.
14. Differentiate among the terms *heterotroph, saprobe,* and *parasite.*
15. Explain the relationship between fungal hyphae and the production of a mycelium.
16. Describe two ways in which fungal spores are formed.
17. List two detrimental and two beneficial activities of fungi (from the viewpoint of humans).

5.5 The Protists

18. Note the protozoan characteristics that illustrate why they are informally placed into a single group.
19. List three means of locomotion exhibited by protozoa.
20. Explain why a cyst stage may be useful in a protozoan.
21. Give an example of a human disease caused by each of the four types of protozoa.

5.6 The Helminths

22. List the two major groups of helminths and provide examples representing each body type.
23. Summarize the stages of a typical helminth life cycle.

5.1 Overview of the Eukaryotes

Evidence from paleontology indicates that the first eukaryotic cells appeared on the earth approximately 2-3 billion years ago. It now seems clear that some of the **organelles** inside eukaryotic cells originated from more primitive cells that became trapped in them. These first eukaryotic cells were so versatile that eukaryotic microorganisms soon spread out into available habitats and adopted greatly diverse styles of living.

Becoming Eukaryotic

Now is an appropriate time to introduce the concept of **endosymbiosis.** As you recall, many scientists believe that both bacterial and eukaryotic cells emerged from an earlier, now-extinct, cell type, called the last common ancestor (LCA). You remember that eukaryotes are thought to have evolved later than bacteria and archaea. It is believed that a bacterial or archaea cell parasitized another descendant cell of the LCA **(figure 5.1)** and eventually became a permanent part of that cell as its mitochondrion. Similarly, photosynthetic bacteria are thought to have become part of the precursor eukaryotic cell, eventually becoming the chloroplast in eukaryotic plant cells. That could explain why mitochondria and chloroplasts have their own (circular) DNA, 70S ribosomes, and their own two-layer membranes. Figure 5.1 depicts one view of how this might have happened.

The first primitive eukaryotes were probably single-celled and independent, but, over time, some forms began to aggregate, forming colonies. With further evolution, some of the cells within colonies became *specialized* to perform a particular function that provided an advantage to the whole colony, such as oxygen

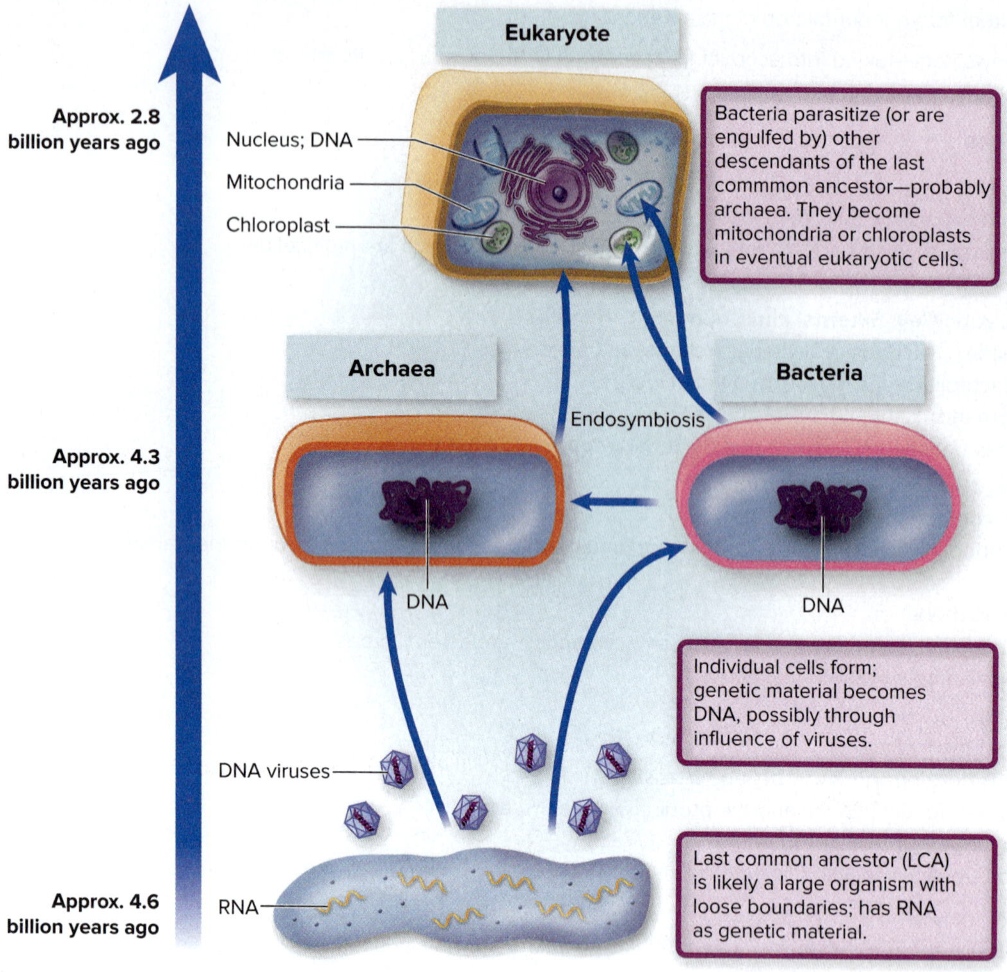
Eukaryote
Approx. 2.8 billion years ago
Nucleus; DNA
Mitochondria
Chloroplast
Bacteria parasitize (or are engulfed by) other descendants of the last commmon ancestor—probably archaea. They become mitochondria or chloroplasts in eventual eukaryotic cells.
Archaea
Bacteria
Endosymbiosis
Approx. 4.3 billion years ago
DNA
DNA
Individual cells form; genetic material becomes DNA, possibly through influence of viruses.
DNA viruses
Last common ancestor (LCA) is likely a large organism with loose boundaries; has RNA as genetic material.
Approx. 4.6 billion years ago
RNA

Figure 5.1 Endosymbiosis. Many scientists believe that bacteria and archaea developed from the last common ancestor, possibly under the influence of DNA viruses. Eukaryotes then evolved from an interaction (endosymbiosis) between an archaea-like cell (also originating from the LCA) and bacteria.

utilization, feeding, or reproduction. Complex multicellular organisms evolved as individual cells in the organism lost the ability to survive apart from the intact colony. Although a multicellular organism is composed of many cells, it is more than just a disorganized assemblage of cells like a colony. Rather, it is composed of distinct groups of cells that cannot exist independently of the rest of the body. The cell groupings of multicellular organisms that have a specific function are termed *tissues,* and groups of tissues make up *organs.*

The eukaryotes we study in medical microbiology are the fungi, the protozoa, and the helminths. All protozoa are unicellular. Many fungi are unicellular. Helminths are animals (worms) and as such are multicellular, but their egg or larval forms are unicellular **(table 5.1).** Even though we won't study algae in any depth, we will briefly look at their structure as part of this chapter. They come in unicellular and multicellular forms.

5.1 Learning Outcomes—Assess Your Progress

1. Relate bacterial, archaeal, and eukaryotic cells to the *last common ancestor.*
2. List the types of eukaryotic microorganisms and denote which are unicellular and which are multicellular.
3. Explain how endosymbiosis contributed to the development of eukaryotic cells.

Table 5.1 Eukaryotic Organisms Studied in Microbiology

Always Unicellular	May Be Unicellular or Multicellular	Always Multicellular
Protozoa	Fungi Algae	Helminths (have unicellular egg or larval forms)

5.2 Form and Function of the Eukaryotic Cell: External and Boundary Structures

The cells of eukaryotic organisms are so varied that no single member can serve as a complete, universal example—but **figure 5.2** shows a conceptual eukaryotic cell. The infographic a couple of pages ahead shows the organization of a eukaryotic cell. You can compare it to the organization for bacterial cells that you saw in chapter 4.

In general, eukaryotic microbial cells have a cytoplasmic (cell) membrane, nucleus, mitochondria, endoplasmic reticulum, Golgi apparatus, vacuoles, cytoskeleton, and glycocalyx. A cell wall, locomotor appendages, and chloroplasts are found only in

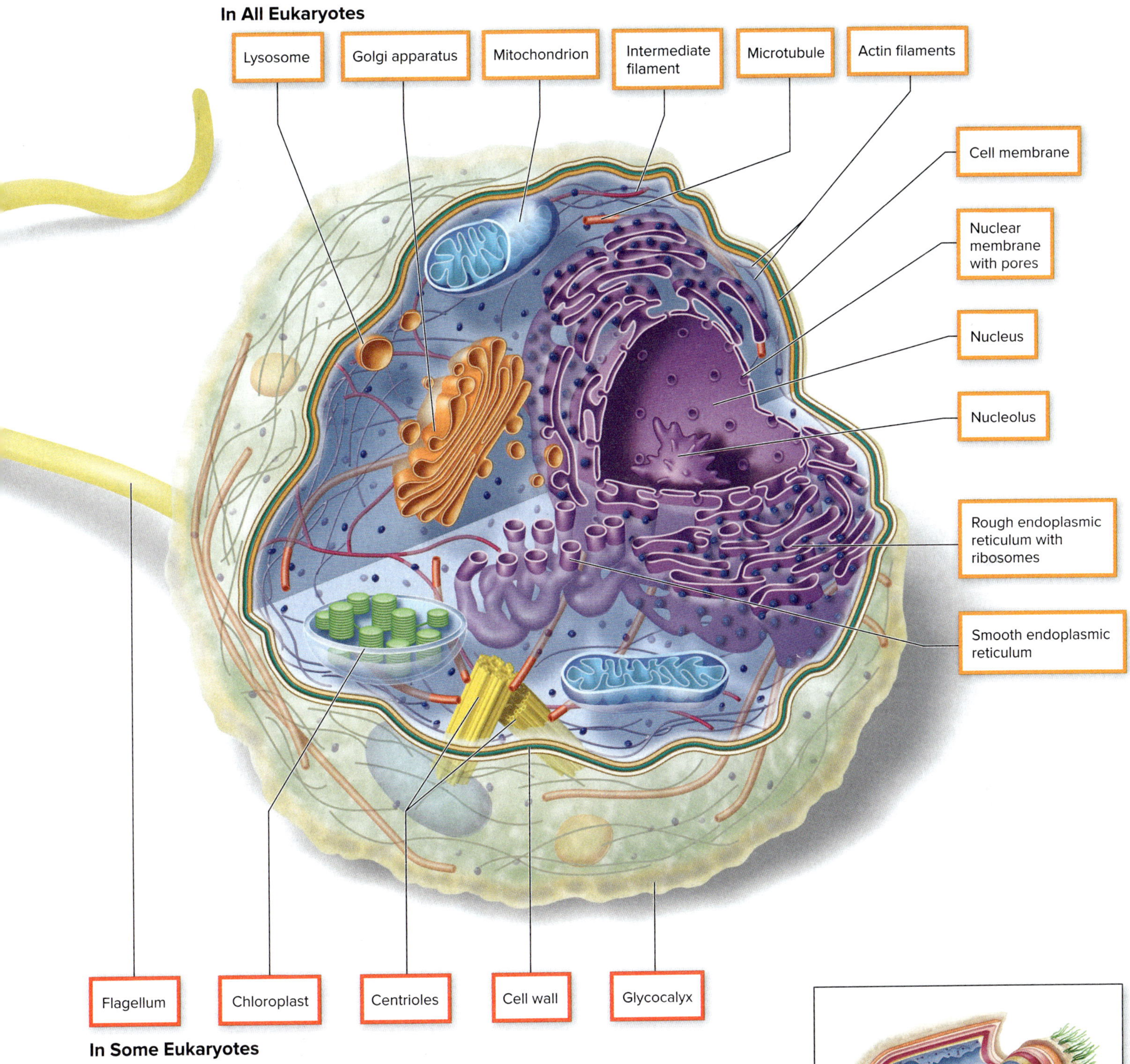
In All Eukaryotes
Lysosome
Golgi apparatus
Mitochondrion
Intermediate filament
Microtubule
Actin filaments
Cell membrane
Nuclear membrane with pores
Nucleus
Nucleolus
Rough endoplasmic reticulum with ribosomes
Smooth endoplasmic reticulum
Flagellum
Chloroplast
Centrioles
Cell wall
Glycocalyx
In Some Eukaryotes

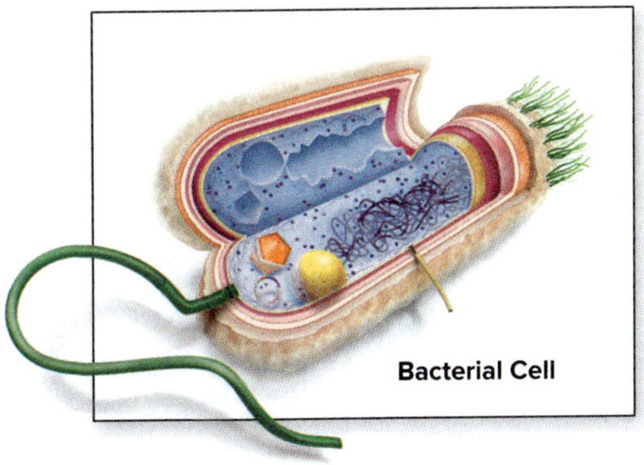
Bacterial Cell

Figure 5.2 Structure of a conceptual eukaryotic cell. The figure of a bacterial cell from chapter 4 is included here for comparison.

some groups. In the following sections, we cover the microscopic structure and functions of the eukaryotic cell. As with the bacteria, we begin on the outside and move inward through the cell.

Appendages for Movement: Cilia and Flagella

Eukaryotic flagella are much different from those of bacteria and archaea, even though they have the same name. The eukaryotic flagellum is about 10 times thicker, structurally more complex, and covered by an extension of the cell membrane. Each single

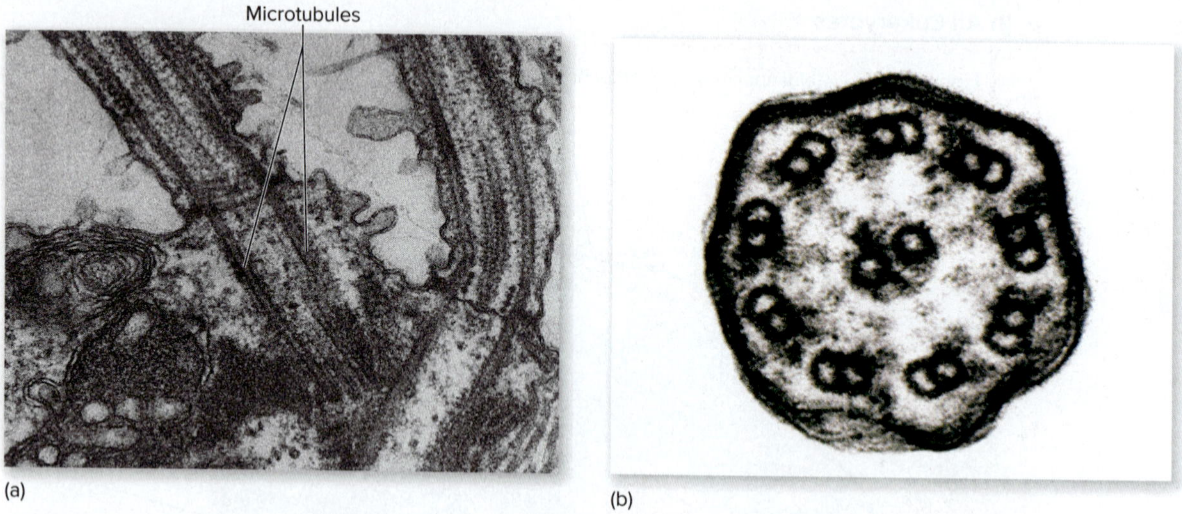
Microtubules
(a)
(b)

Figure 5.3 Microtubules in flagella. **(a)** Longitudinal section through two flagella, showing microtubules. **(b)** A cross section that reveals the typical 9 + 2 arrangement found in both flagella and cilia.

(a) Don W. Fawcett/B. Bouck/Science Source; (b) Dr. Gopal Murti/Science Source

flagellum is a long, sheathed cylinder containing regularly spaced, hollow tubules—microtubules—that extend along its entire length **(figure 5.3*a*).** A cross section reveals nine pairs of closely attached microtubules surrounding a single central pair. This scheme, called the 9 + 2 arrangement, is the pattern of all eukaryotic flagella and cilia **(figure 5.3*b*).** During locomotion, the adjacent microtubules slide past each other, whipping the flagellum back and forth. Although details of this process are too complex to discuss here, it involves expenditure of energy and a coordinating mechanism in the cell membrane. The placement and number of flagella can be useful in identifying flagellated protozoa and certain algae.

Cilia are very similar in architecture to flagella, but they are shorter and more numerous (some cells have several thousand). They are found only on a single group of protozoa and certain animal cells. In the ciliated protozoa, the cilia occur in rows over the cell surface, where they beat back and forth in regular, oarlike strokes **(figure 5.4).** Such protozoa are among the fastest of all motile cells. On some cells, cilia also function as feeding and filtering structures.

Disease Connection

As far as we know, only one ciliated protozoan causes disease in humans. *Balantidium coli* makes its home in the intestines of pigs and other mammals. If it gets transmitted to the human digestive tract, it can cause a diarrheal disease called balantidiasis.

The Glycocalyx

Most eukaryotic cells have a **glycocalyx,** an outermost boundary that comes into direct contact with the environment (see figure 5.2). This structure, which is sometimes called an

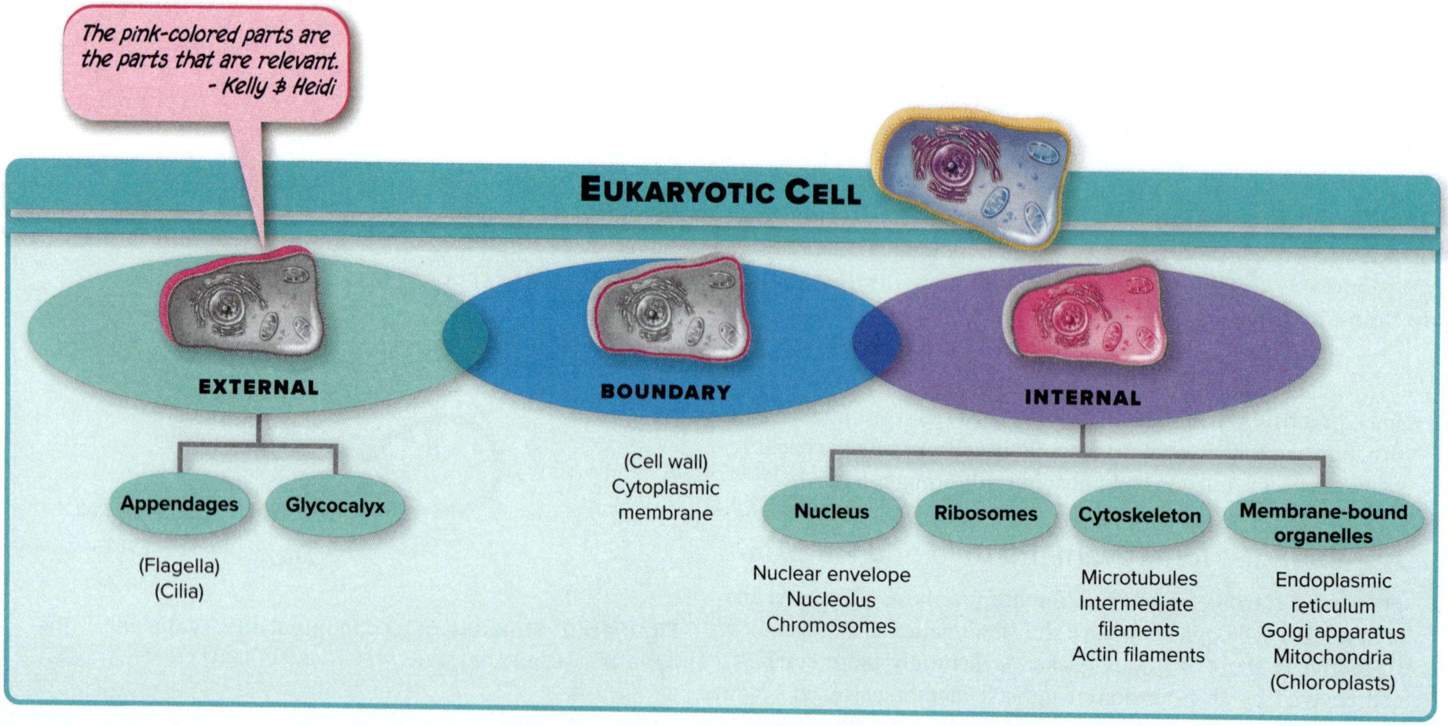
The pink-colored parts are the parts that are relevant. - Kelly & Heidi
EUKARYOTIC CELL
EXTERNAL
BOUNDARY
INTERNAL
Appendages
Glycocalyx
(Flagella)
(Cilia)
(Cell wall)
Cytoplasmic membrane
Nucleus
Ribosomes
Cytoskeleton
Membrane-bound organelles
Nuclear envelope
Nucleolus
Chromosomes
Microtubules
Intermediate filaments
Actin filaments
Endoplasmic reticulum
Golgi apparatus
Mitochondria
(Chloroplasts)

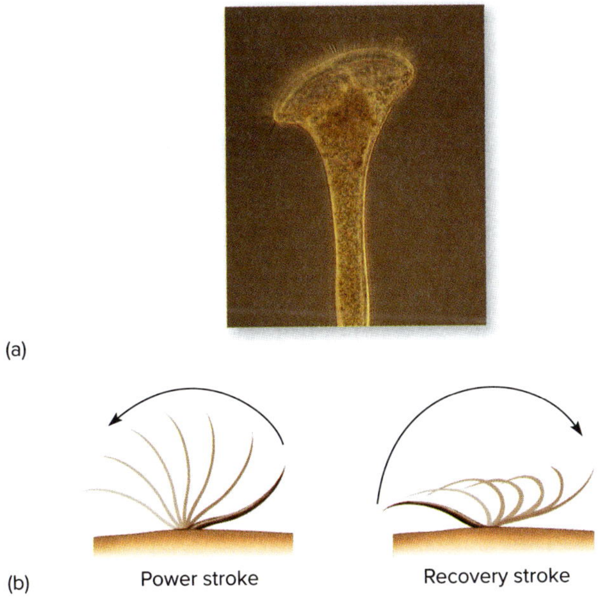
(a)
(b)
Power stroke
Recovery stroke

Figure 5.4 Structure and locomotion in ciliates. **(a)** Micrograph of the protozoan *Stentor,* with visible cilia around its top edge. **(b)** Cilia beat in coordinated waves, driving the cell forward and backward. View of a single cilium shows that it has a pattern of movement like a swimmer, with a power forward stroke and a repositioning stroke.

(a) Stephen Durr

extracellular matrix, is usually composed of polysaccharides and appears as a network of fibers, a slime layer, or a capsule much like the glycocalyx of bacteria. The glycocalyx provides protection and adherence of cells to surfaces. What lies beneath the glycocalyx varies among the several eukaryotic groups. Fungi have a thick, rigid cell wall surrounding a cell membrane, whereas protozoa and all animal cells lack a cell wall and have only a cell membrane.

Boundary Structures

The Cell Wall

Fungi and algae have cell walls. They are rigid and provide structural support and shape, but they are different in chemical composition from bacterial cell walls. Fungal cell walls have an inner layer of polysaccharide fibers composed of chitin or cellulose and an outer layer of mixed glycans **(figure 5.5).**

The Cytoplasmic Membrane

The cytoplasmic (cell) membrane of eukaryotic cells is a typical bilayer of phospholipids in which protein molecules are embedded. In addition to phospholipids, eukaryotic membranes also contain *sterols* of various kinds. Sterols are different from phospholipids in both structure and behavior, as you may recall from chapter 2. Their relative rigidity makes eukaryotic membranes more stable. This strengthening feature is extremely important in those cells that lack a cell wall. Cytoplasmic membranes of eukaryotes have the same function as those of bacteria, serving as selectively permeable barriers. Membranes have extremely sophisticated mechanisms for transporting nutrients *in* and waste and other products *out.* You'll read about these transport systems in membranes in chapter 9.

5.2 Learning Outcomes—Assess Your Progress

4. Differentiate among the flagellar structures of bacteria, archaea, and eukaryotes.
5. Describe the important characteristics of a glycocalyx in eukaryotes.
6. List which eukaryotic microorganisms might have a cell wall.
7. List similarities and differences between eukaryotic and bacterial cytoplasmic membranes.

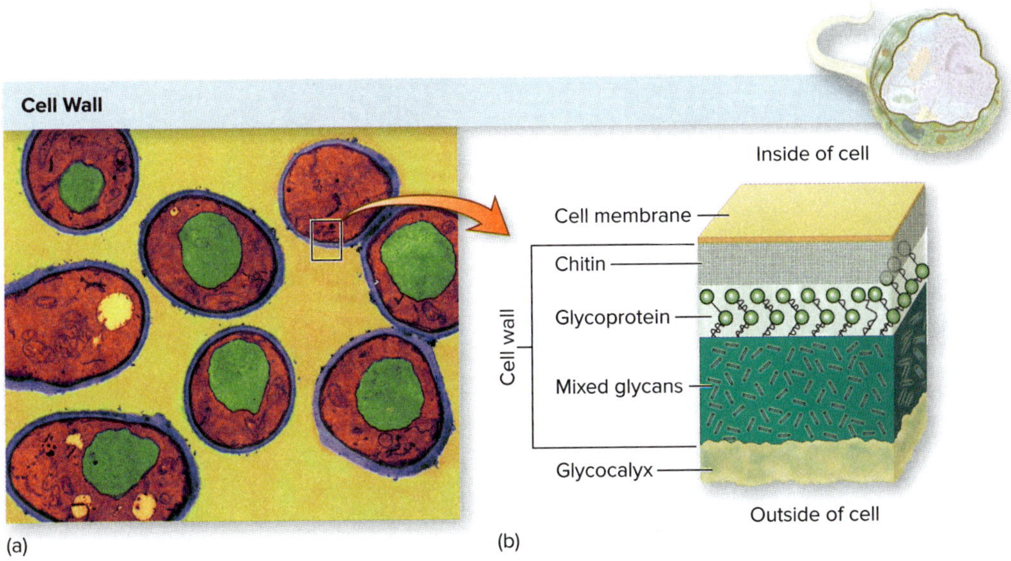
Cell Wall
Inside of cell
Cell membrane
Chitin
Glycoprotein
Mixed glycans
Glycocalyx
Cell wall
Outside of cell
(a)
(b)

Figure 5.5 Fungi and fungal cell walls. **(a)** An electron micrograph of several fungal cells. **(b)** A drawing of the section of the wall inside the square in part (a).

(a) Thomas Deerinck, NCMIR/Science Source

5.3 Form and Function of the Eukaryotic Cell: Internal Structures

Unlike bacteria and archaea, eukaryotic cells contain a number of individual membrane-bound organelles that are extensive enough to account for 60% to 80% of their volume.

The Nucleus: The Control Center

The nucleus is a compact sphere that is the most prominent organelle in eukaryotic cells. It is separated from the cell cytoplasm by an external boundary called a *nuclear envelope*. The envelope has a unique architecture. It is composed of two parallel membranes separated by a narrow space, and it is perforated with small, regularly spaced openings, called pores, formed at sites where the two membranes unite **(figure 5.6).** The nuclear pores are passageways that macromolecules use to migrate from the nucleus to the cytoplasm, and vice versa. The nucleus contains an inner substance called the *nucleoplasm* and a granular mass, the **nucleolus,** that stains more intensely because of its RNA content. The nucleolus is the site for ribosomal RNA synthesis and a collection area for ribosomal subunits. The subunits are transported through the nuclear pores into the cytoplasm for final assembly into ribosomes.

A prominent feature of the nucleoplasm in stained preparations is a network of dark fibers known as **chromatin.** Chromatin is the material of the eukaryotic **chromosomes,** large units of genetic information in the cell. The chromosomes in the nucleus of most cells are not readily visible because they are long, linear DNA molecules bound in varying degrees to **histone** proteins, and they are far too fine to be resolved as distinct structures without extremely high magnification. During **mitosis,** however, when the duplicated chromosomes are separated equally into daughter cells, the chromosomes themselves become visible as discrete bodies **(figure 5.7).** This happens when the DNA becomes highly condensed by forming coils and supercoils around the histones to prevent the chromosomes from tangling as they are separated into new cells.

The nucleus, as you've just read, contains instructions in the form of DNA. Elaborate processes have evolved for transcription and duplication of this genetic material. In addition to mitosis, some cells also undergo **meiosis,** the process by which sex cells are created. Much of the protein synthesis and other work of the cell takes place outside the nucleus in the cell's other organelles.

Endoplasmic Reticulum: A Passageway in the Cell

The **endoplasmic reticulum (ER)** is a microscopic series of tunnels used in transport and storage. Two kinds of endoplasmic reticulum are the **rough endoplasmic reticulum (RER) (figure 5.8)** and the **smooth endoplasmic reticulum (SER).** The RER comes from the outer membrane of the nuclear envelope and extends in a continuous network through the cytoplasm,

> *The cell in this corner shows the parts described in the figure colored in purple.*
> *- Kelly $ Heidi*

Figure 5.6 The nucleus. **(a)** Electron micrograph of a section through a nucleus, showing its most prominent features. **(b)** Cutaway three-dimensional view of the relationships of the nuclear envelope and pores.
(a) D Spector/Photolibrary/Getty Images

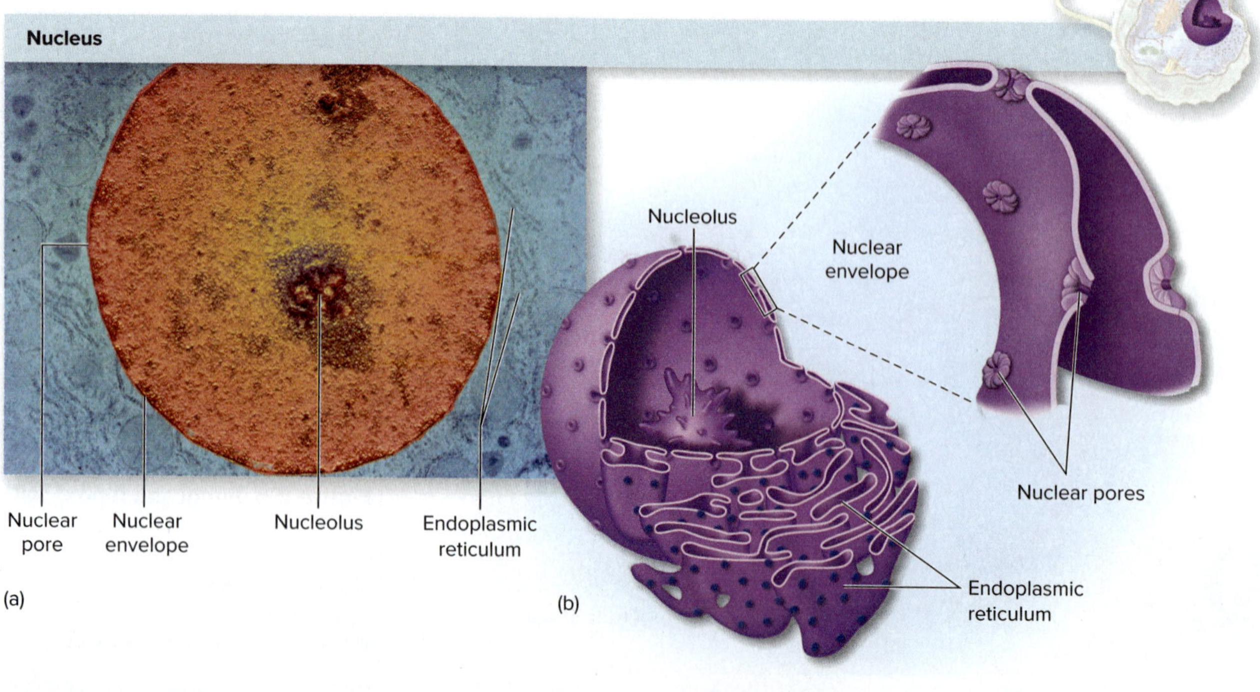
Nucleus
Nuclear pore
Nuclear envelope
Nucleolus
Endoplasmic reticulum
(a)
Nucleolus
Nuclear envelope
Nuclear pores
Endoplasmic reticulum
(b)

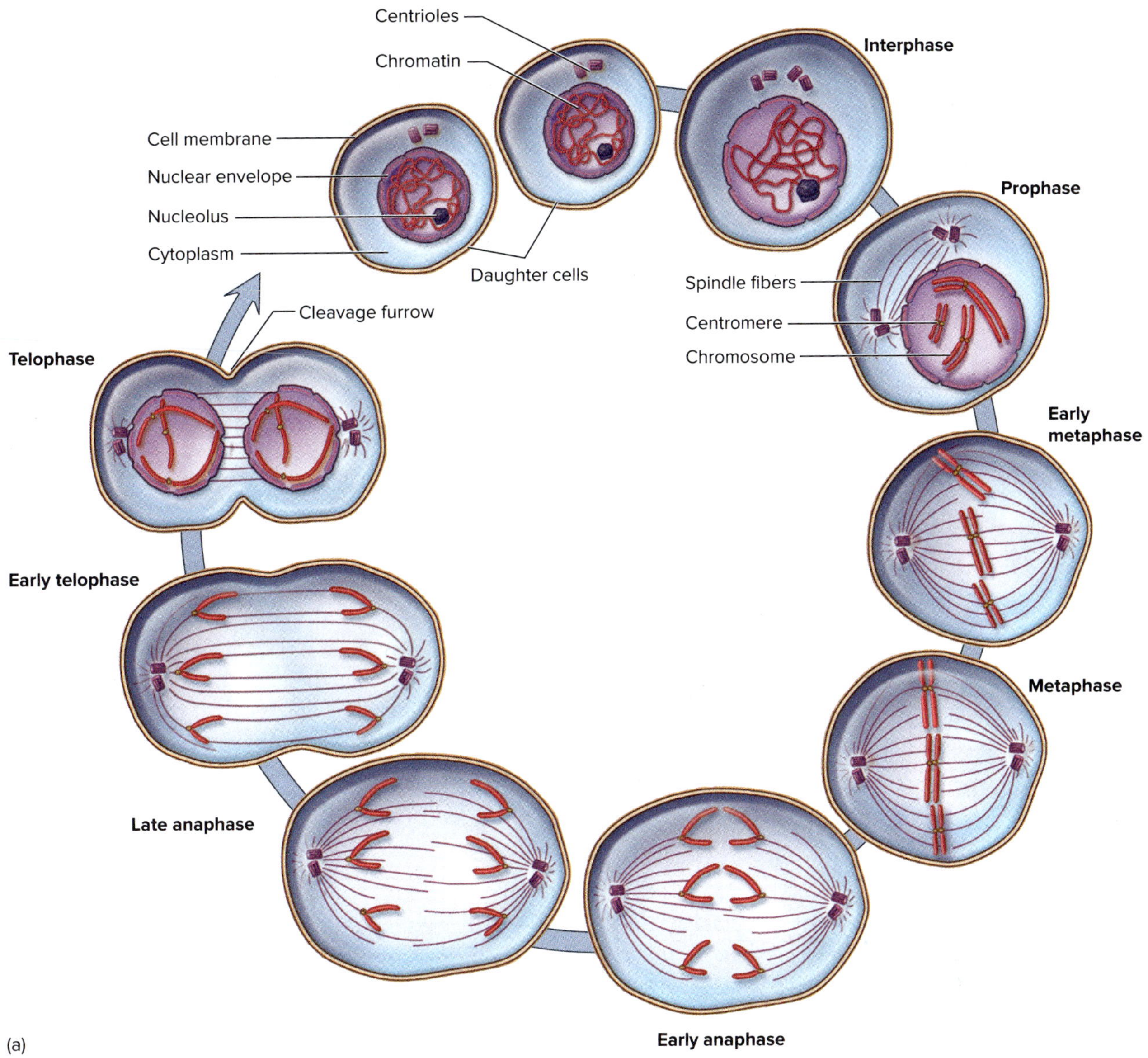

Centrioles
Chromatin
Interphase
Cell membrane
Nuclear envelope
Nucleolus
Cytoplasm
Daughter cells
Prophase
Spindle fibers
Centromere
Chromosome
Cleavage furrow
Telophase
Early metaphase
Early telophase
Metaphase
Late anaphase
Early anaphase
(a)

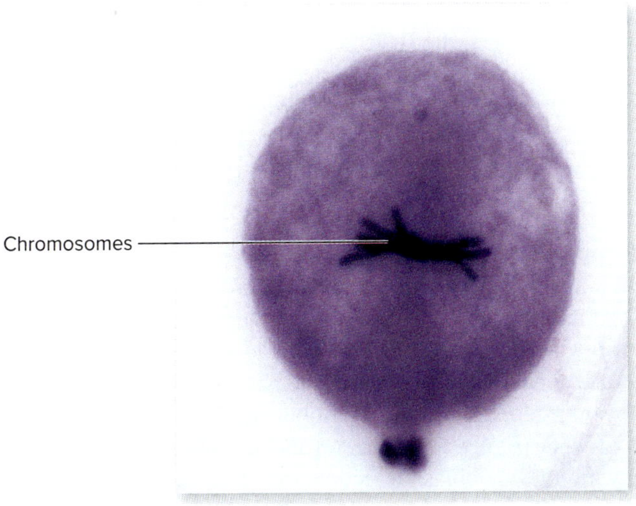

Chromosomes
(b)

Figure 5.7 Changes in the cell and nucleus that accompany mitosis in a eukaryotic cell such as a yeast. **(a)** Before mitosis (at a period called *interphase*), chromosomes are visible only as chromatin. As mitosis proceeds (known as *early prophase*), chromosomes take on a fine, threadlike appearance as they condense, and the nuclear membrane and nucleolus are temporarily disrupted. **(b)** Micrograph of a nucleus with visible chromosomes in prophase.

(b) Al Telser/McGraw Hill

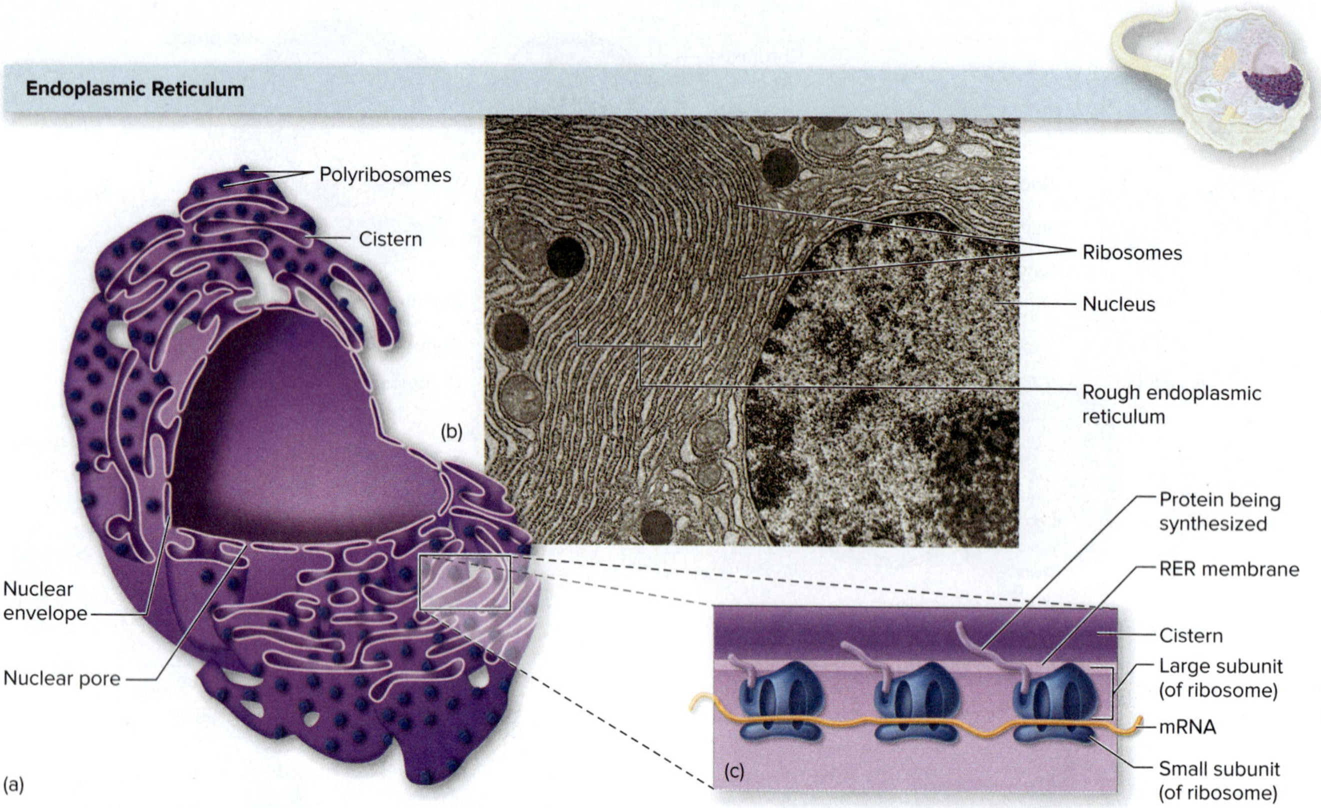
Endoplasmic Reticulum
Polyribosomes
Cistern
Ribosomes
Nucleus
Rough endoplasmic reticulum
(b)
Protein being synthesized
RER membrane
Nuclear envelope
Nuclear pore
Cistern
Large subunit (of ribosome)
mRNA
Small subunit (of ribosome)
(a)
(c)

Figure 5.8 The origin and structure of the rough endoplasmic reticulum (RER). **(a)** Schematic view of the origin of the RER from the outer membrane of the nuclear envelope. **(b)** Transmission electromicrograph of the RER. **(c)** Detail of the orientation of a ribosome on the RER membrane.
(b) Don W. Fawcett/Science Source

all the way out to the cell membrane. This architecture permits the spaces inside the RER, called *cisternae* (sis-turn′-ee), to transport materials from the nucleus to the cytoplasm and ultimately to the cell's exterior. The RER appears rough because of large numbers of ribosomes partly attached to its membrane surface. Proteins synthesized on the ribosomes are shunted into the inside space of the RER and held there for later packaging and transport. In contrast, the SER is a closed tubular network without ribosomes that functions in nutrient processing and in synthesis and storage of nonprotein macromolecules such as lipids.

Golgi Apparatus: A Packaging Machine

The **Golgi apparatus,** also called the *Golgi complex* or *Golgi body,* is the site in the cell where proteins are modified and then sent to their final destinations. It is a separate organelle consisting of a stack of several flattened, disc-shaped sacs, called cisternae. These sacs have outer membranes and cavities like those of the endoplasmic reticulum, but they do not form a continuous network **(figure 5.9).** This organelle is always closely associated with the endoplasmic reticulum. The endoplasmic reticulum buds off tiny membrane-bound packets of protein called *transitional vesicles* that are picked up by the face of the Golgi apparatus. Once inside the Golgi, the proteins are often modified by the addition of polysaccharides and lipids. The final action of this organelle is to pinch off finished *condensing vesicles* that will be conveyed to other organelles such as lysosomes or transported outside the cell as secretory vesicles **(figure 5.10).**

Nucleus, Endoplasmic Reticulum, and Golgi Apparatus: Nature's Assembly Line

As the keeper of the eukaryotic genetic code, the nucleus ultimately governs and regulates all cell activities. But because the nucleus remains fixed in a specific cellular site, it must direct these activities through a structural and chemical network (figure 5.10). This network includes ribosomes, which originate in the nucleus, and the RER, which is connected with the nuclear envelope, as well as the SER and the Golgi apparatus. Initially, a segment of the genetic code of DNA containing the instructions for producing a protein is copied into RNA, and that RNA is passed out through the nuclear pores directly to the ribosomes on the endoplasmic reticulum. Here, specific proteins are synthesized from the RNA code and deposited in the inner space of the endoplasmic reticulum. After being transported to the Golgi apparatus, the protein products are chemically modified and packaged into vesicles that can be used by the cell in a variety of ways. Some of the vesicles contain enzymes to digest food inside the cell; other vesicles are secreted to digest materials outside the

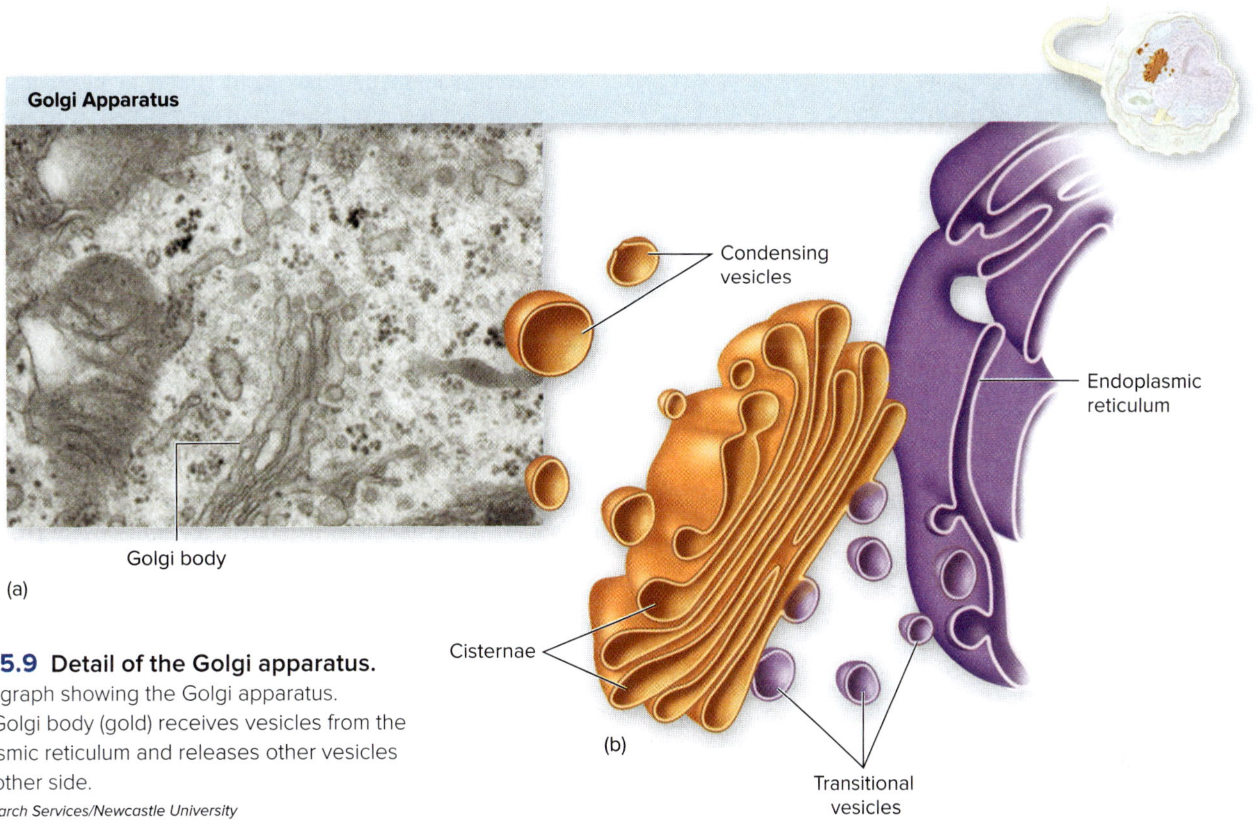
Golgi Apparatus
Golgi body
(a)
Condensing vesicles
Endoplasmic reticulum
Cisternae
(b)
Transitional vesicles

Figure 5.9 Detail of the Golgi apparatus.
(a) Micrograph showing the Golgi apparatus.
(b) The Golgi body (gold) receives vesicles from the endoplasmic reticulum and releases other vesicles from its other side.
(a) EM Research Services/Newcastle University

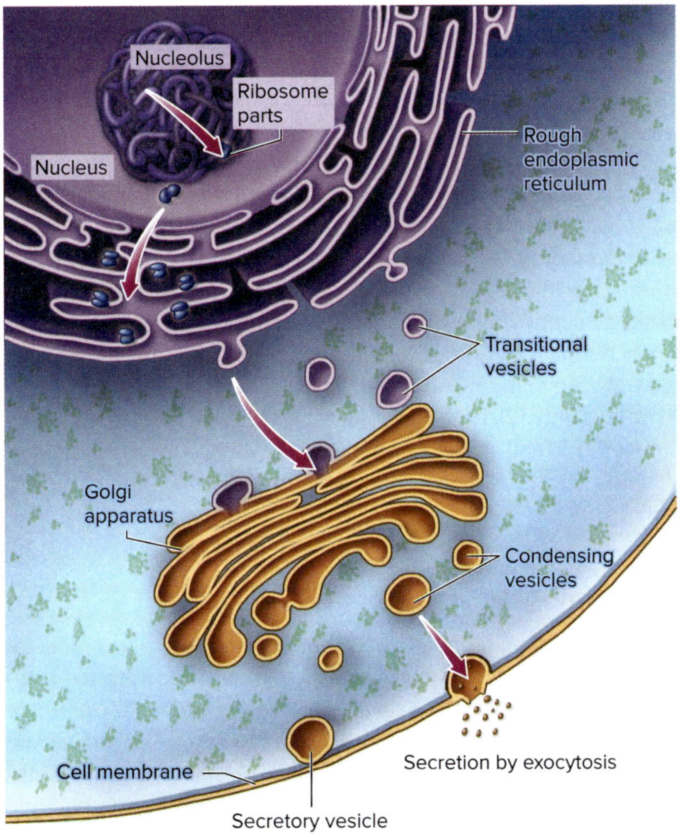
Nucleolus
Ribosome parts
Nucleus
Rough endoplasmic reticulum
Transitional vesicles
Golgi apparatus
Condensing vesicles
Cell membrane
Secretion by exocytosis
Secretory vesicle

Figure 5.10 The transport process. The cooperation of organelles in protein synthesis and transport: nucleus ⟶ RER ⟶ Golgi apparatus ⟶ vesicles ⟶ secretion.

cell, and yet others are important in the enlargement and repair of the cell wall and membrane.

A **lysosome** is one type of vesicle originating from the Golgi apparatus that contains a variety of enzymes. Lysosomes are involved in the digestion of food particles and in protection against invading microorganisms inside the cell. They also participate in digestion and removal of cell debris in damaged tissue. The formation of lysosomes involves the so-called GERL, or Golgi-endoplasmic reticulum-lysosomal, complex. Lysosomal enzymes are synthesized by the RER and are then transported through the SER. Transitional vesicles then transport the enzymes to the Golgi, where unique chemical tags on these proteins take them to the condensing vesicles that will become the primary lysosomes.

Other types of vesicles include **vacuoles** (vak′-yoo-ohlz), which are membrane-bound sacs containing fluids or solid particles to be digested, excreted, or stored. They are formed in phagocytic cells (certain white blood cells and protozoa) in response to food and other substances that have been engulfed. The contents of a food vacuole are digested through the merger of the vacuole with a lysosome. This merged structure is called a *phagolysosome* (**figure 5.11**). Other types of vacuoles are used in storing reserve food such as fats and glycogen. Protozoa living in freshwater habitats regulate osmotic pressure by means of contractile vacuoles, which expel excess water that has diffused into the cell.

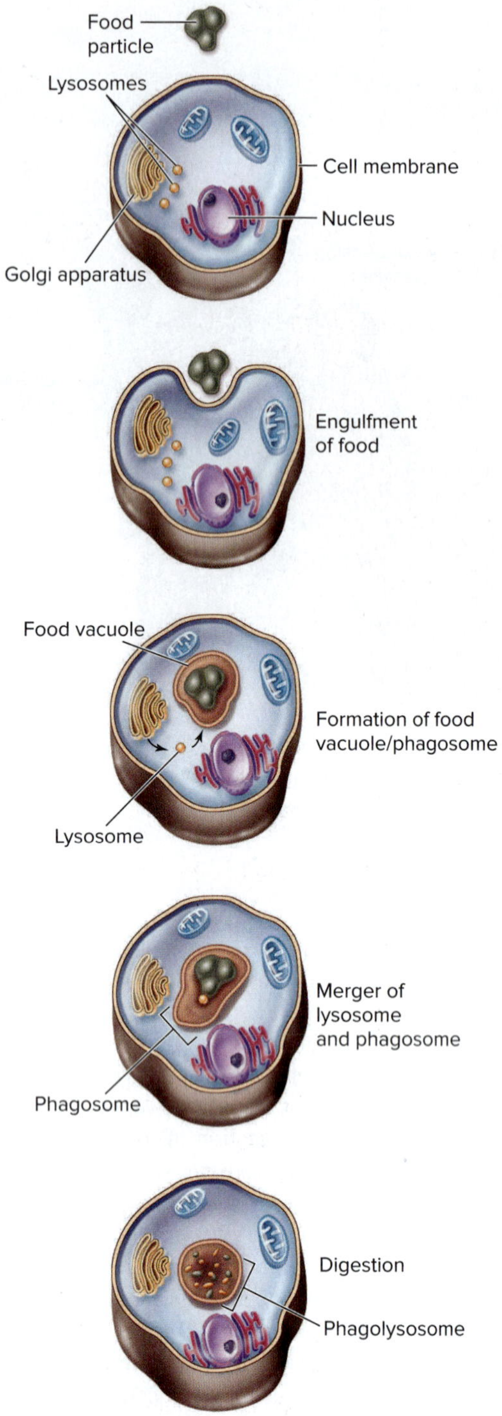
Food particle
Lysosomes
Cell membrane
Nucleus
Golgi apparatus
Engulfment of food
Food vacuole
Formation of food vacuole/phagosome
Lysosome
Merger of lysosome and phagosome
Phagosome
Digestion
Phagolysosome

Figure 5.11 The origin and action of lysosomes in phagocytosis.

Mitochondria: Energy Generators of the Cell

Although the nucleus is the cell's control center, none of the cellular activities could proceed without a constant supply of energy, the bulk of which is generated in most eukaryotes by **mitochondria** (my″-toh-kon′-dree-uh). When viewed with light microscopy, mitochondria appear as round or elongated particles scattered throughout the cytoplasm. A single mitochondrion consists of a smooth, continuous outer membrane that forms the external contour and an inner, folded membrane nestled neatly inside the outer membrane **(figure 5.12).** The folds on the inner membrane, called **cristae** (kris′-tee), may be tubular, like fingers, or folded into shelflike bands.

The cristae membranes hold the enzymes and electron carriers of aerobic respiration. This is an oxygen-using process that extracts chemical energy contained in nutrient molecules and stores it in the form of high-energy molecules, or ATP. The spaces around the cristae are filled with a complex fluid called the **matrix,** which holds ribosomes, DNA, and the pool of enzymes and other compounds involved in the metabolic cycle. Mitochondria (along with chloroplasts) are unique among organelles in that they divide independently of the cell, contain circular strands of DNA, and have bacteria-sized 70S ribosomes. These findings provide evidence that the mitochondria were bacterial cells engulfed by other cells, which were then destined to become these organelles.

While it was previously thought that all eukaryotic organisms must have mitochondria, scientists have discovered that some protozoa have pared-down versions of mitochondria (called *mitosomes*), and one species has even been found that contains no mitochondria or mitosomes.

Chloroplasts: Photosynthesis Machines

Chloroplasts are remarkable organelles found in algae and plant cells that are capable of converting the energy of sunlight into chemical energy through photosynthesis. The photosynthetic role of chloroplasts makes them the primary producers of organic nutrients upon which all other organisms (except certain bacteria) ultimately depend. Another important photosynthetic product of chloroplasts is oxygen gas. Although chloroplasts resemble mitochondria, chloroplasts are larger, contain special pigments, and are much more varied in shape.

There are differences among various algal chloroplasts, but most are generally composed of two membranes, one enclosing the other. There is a smooth, outer membrane in addition to an inner membrane. Inside the chloroplast is a third membrane folded into small, disclike sacs called **thylakoids** that are stacked upon one another into **grana.** These structures carry the green pigment chlorophyll and sometimes additional pigments as well. Surrounding the thylakoids is a substance called the **stroma (figure 5.13).** The role of the photosynthetic pigments is to absorb and transform solar energy into chemical energy, which is then used during reactions in the stroma to synthesize carbohydrates.

Ribosomes: Protein Synthesizers

In an electron micrograph of a eukaryotic cell, ribosomes are numerous tiny particles that give a dotted appearance to the cytoplasm. Ribosomes are distributed throughout the cell: Some are scattered freely in the cytoplasm and cytoskeleton; others are attached to the RER as described earlier in this section. Still others appear inside the mitochondria and in chloroplasts. Multiple ribosomes are often found arranged in short chains called *polyribosomes* (polysomes). The basic structure of eukaryotic ribosomes is similar to that of bacterial ribosomes. Both are composed of

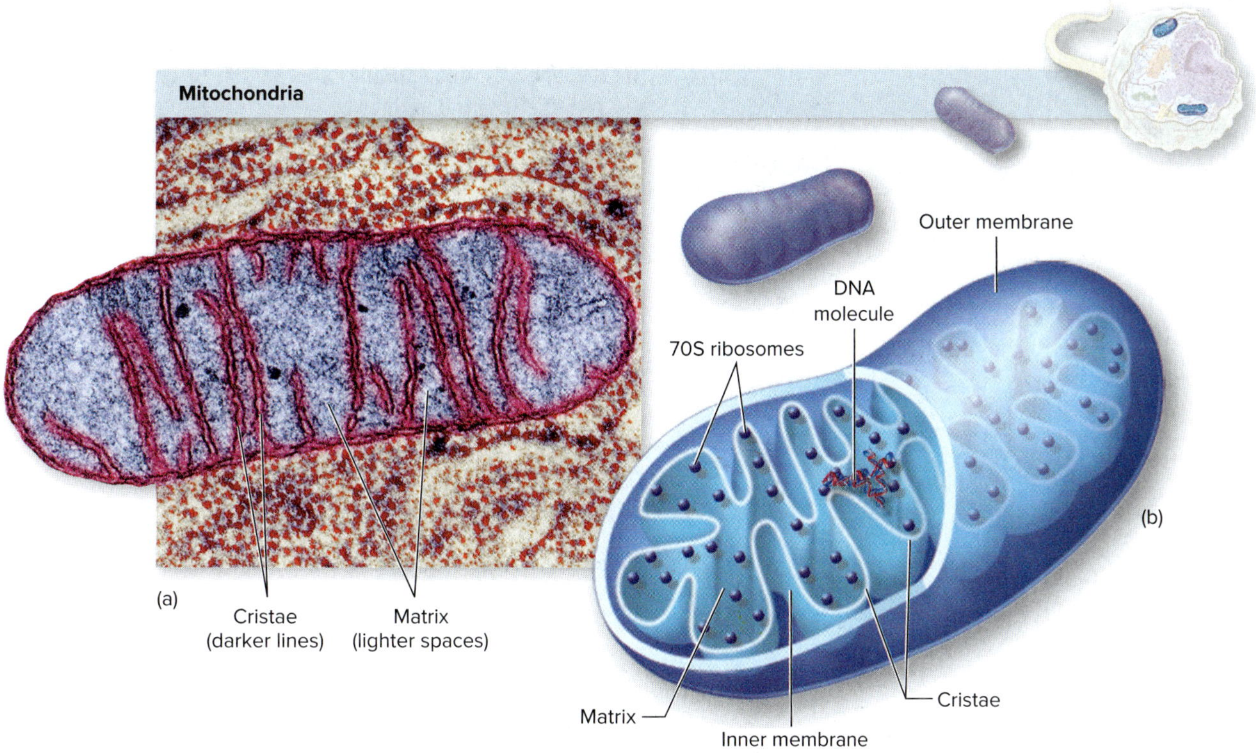
Mitochondria
(a)
Cristae
(darker lines)
Matrix
(lighter spaces)
Outer membrane
DNA
molecule
70S ribosomes
(b)
Matrix
Cristae
Inner membrane

Figure 5.12 General structure of a mitochondrion. **(a)** An electron micrograph. **(b)** A three-dimensional illustration. In most cells, mitochondria are elliptical or spherical, although in certain fungi, algae, and protozoa, they are long and filament-like.
(a) CNRI/Science Photo Library/Getty Images

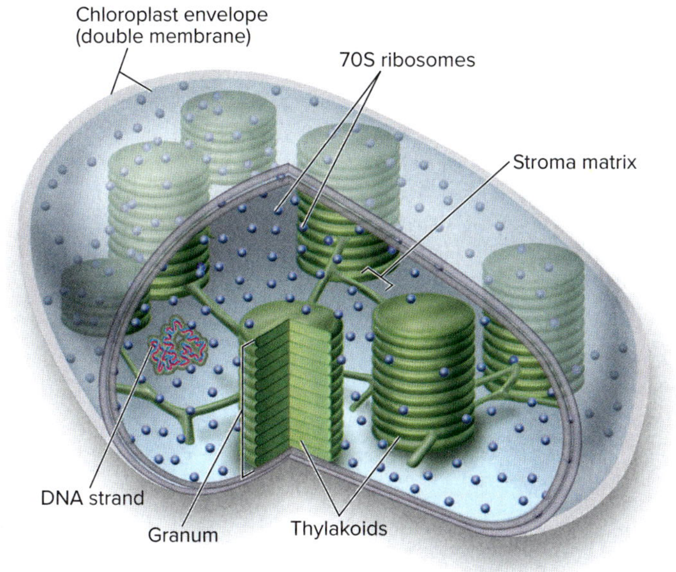
Chloroplast envelope
(double membrane)
70S ribosomes
Stroma matrix
DNA strand
Granum
Thylakoids

Figure 5.13 An algal chloroplast.

large and small subunits of ribonucleoprotein (see figure 5.8). By contrast, however, the eukaryotic ribosome (except in the mitochondrion) is the larger 80S variety that is a combination of 60S and 40S subunits. As in the bacteria, eukaryotic ribosomes are the staging areas for protein synthesis.

Disease Connection

The difference in bacterial and eukaryotic ribosome structure has important implications for our ability to fight infections with antibiotics. Because the goal of antimicrobial treatment is to harm the microbe without harming the host (made of eukaryotic cells), drugs that target parts of the ribosome that are unique to bacterial ribosomes are effective antibiotics that cause minimal harm and side effects in the host.

The Cytoskeleton: A Support Network

The cytoplasm of a eukaryotic cell is crisscrossed by a flexible framework of molecules called the cytoskeleton **(figure 5.14).** This framework appears to have several functions, such as anchoring organelles, moving RNA and vesicles, and permitting shape changes and movement in some cells. The three main types of cytoskeletal elements are *actin filaments, intermediate filaments,* and *microtubules.* **Actin filaments** are long, thin protein strands about 7 nanometers in diameter. They are found throughout the cell but are most highly concentrated just inside the cell membrane. Actin filaments are responsible for cellular movements such as contraction, crawling, pinching during cell division, and formation of cellular extensions. **Microtubules** are long, hollow tubes that maintain the shape of eukaryotic cells when they don't have cell walls and transport substances from one part of a cell to another. The spindle fibers that play an essential role in mitosis are actually microtubules that attach to chromosomes and separate them into daughter cells. As indicated earlier in this section, microtubules are also responsible for the movement of cilia and flagella. **Intermediate filaments** are ropelike structures that are about 10 nanometers in diameter. (Their name comes from their intermediate size, between actin filaments and microtubules.) Their main role is in structural reinforcement of the cell and of organelles. For example, they support the structure of the nuclear envelope.

Table 5.2 summarizes the differences between eukaryotic and bacterial cells. Viruses (discussed in chapter 7) are included as well.

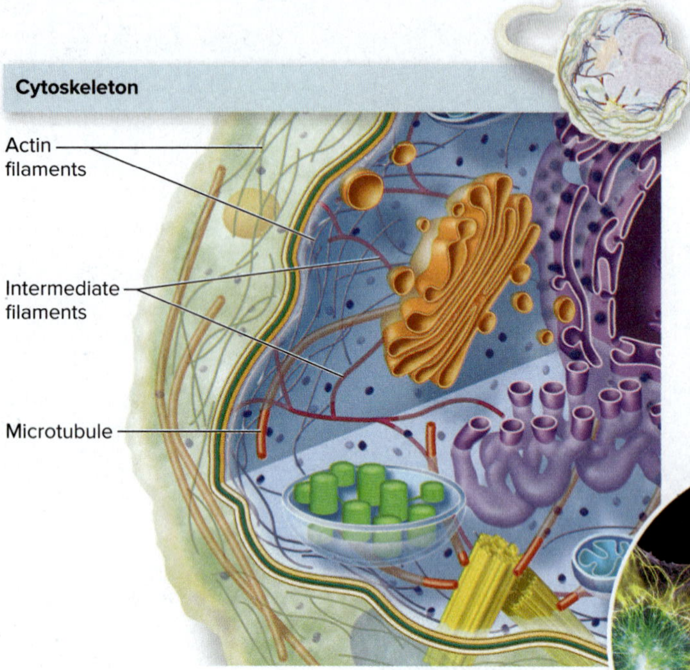
Cytoskeleton
Actin filaments
Intermediate filaments
Microtubule

(a)

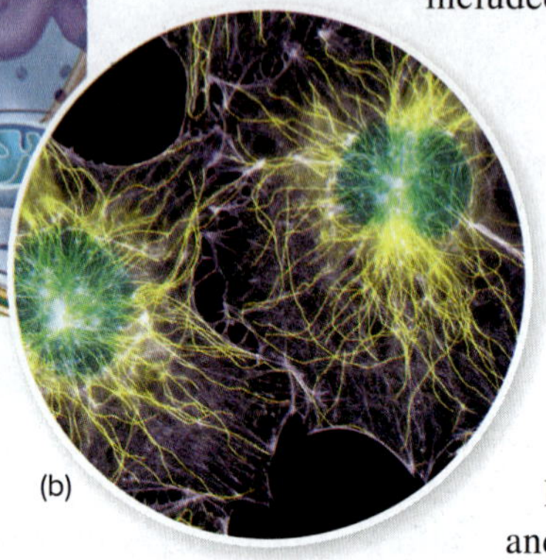

(b)

Figure 5.14 The cytoskeleton. **(a)** Drawing of microtubules, actin filaments, and intermediate filaments. **(b)** Microtubules appear white in this micrograph.

(b) Dr. Torsten Wittmann/Science Source

Survey of Eukaryotic Microorganisms

The next sections contain a general survey of the principal eukaryotic microorganisms—fungi, algae, protozoa, and parasitic worms—while also introducing elements of their structure, life history, classification, identification, and importance.

Table 5.2 The Major Elements of Life in Each Organism Type

Function or Structure	Characteristic*	Bacterial/ Archaeal Cells	Eukaryotic Cells	Viruses**
Genetics	Nucleic acids	+	+	+
	Chromosomes	+	+	–
	True nucleus	–	+	–
	Nuclear envelope	–	+	–
Reproduction	Mitosis	–	+	–
	Production of sex cells	+/–	+	–
	Binary fission	+	+	–
Biosynthesis	Independent	+	+	–
	Golgi apparatus	–	+	–
	Endoplasmic reticulum	–	+	–
	Ribosomes	+***	+	–
Respiration	Mitochondria	–	+	–
Photosynthesis	Pigments	+/–	+/–	–
	Chloroplasts	–	+/–	–
Motility/locomotor structures	Flagella	+/–***	+/–	–
	Cilia	–	+/–	–
Shape/protection	Membrane	+	+	+/–
	Cell wall	+***	+/–	– (have capsids instead)
	Capsule	+/–	+/–	–
Complexity of function		+	+	+/–
Size (in general)		0.5–3 μm****	2–100 μm	< 0.2 μm

*+ means most members of the group exhibit this characteristic; – means most lack it; +/– means some members have it and some do not.

**Viruses cannot participate in metabolic or genetic activity outside their host cells.

***The bacterial/archaeal type is functionally similar to the eukaryotic type, but it is structurally unique.

****Much smaller and much larger bacteria do exist.

5.3 Learning Outcomes—Assess Your Progress

8. Describe the main structural components of a nucleus.
9. Diagram how the nucleus, endoplasmic reticulum, and Golgi apparatus act together with vesicles during the transport process.
10. Explain the function of the mitochondrion.
11. Explain the importance of ribosomes and differentiate between eukaryotic and bacterial types.
12. List and describe the three main fibers of the cytoskeleton.

5.4 The Fungi

The Kingdom Fungi, or Myceteae, is large and filled with forms of great variety and complexity. For practical purposes, the approximately 3–4 million species of fungi can be divided into two groups: the *macroscopic fungi* (mushrooms, puffballs, gill fungi) and the *microscopic fungi* (molds, yeasts). Although the majority of fungi are unicellular, a few complex forms such as mushrooms and puffballs are considered multicellular. Cells of the microscopic fungi exist in two basic morphological types: yeasts and hyphae. A yeast cell has a round to oval shape and uses asexual reproduction. It grows swellings on its surface called *buds,* which then become separate cells. **Hyphae** (hy′-fee) are long, threadlike cells found in the bodies of filamentous fungi, or molds **(figure 5.15).** Some species form a **pseudohypha,** a chain of yeasts formed when buds remain attached in a row **(figure 5.16).** Because of its manner of formation, it is not a true hypha like that of molds. Although some fungal cells exist only in a yeast form and others occur primarily as hyphae, a few, called **dimorphic,** can take either form, depending on growth conditions such as changing temperature. This variability in growth form is particularly characteristic of some pathogenic fungi.

Fungal Nutrition

All fungi are **heterotrophic.** This means that they acquire nutrients from a wide variety of organic materials called **substrates (figure 5.17).** Most of them obtain these substrates from the remnants of dead plants and animals in soil or aquatic habitats, a trait that makes them **saprobes.** Fungi can also be **parasites,** meaning they live on the bodies of living animals or plants, although very few fungi absolutely require a living host. In general, the fungus penetrates the substrate and secretes enzymes that reduce it to small molecules that can be absorbed by the cells. Fungi have enzymes for digesting an incredible array of substances, including feathers, hair, cellulose, petroleum products, wood, and rubber. It has been said that every naturally occurring organic material on the earth can be digested by some type of fungus. Fungi are often found in nutritionally poor or adverse environments. Various fungi thrive in substrates with high salt or sugar content, at relatively high temperatures, and even in snow and glaciers. Their medical and agricultural impact is extensive. Many fungi make their home on the human body, as part of the normal human microbiome. Yet nearly 300 species of fungi can also cause human disease. Animal

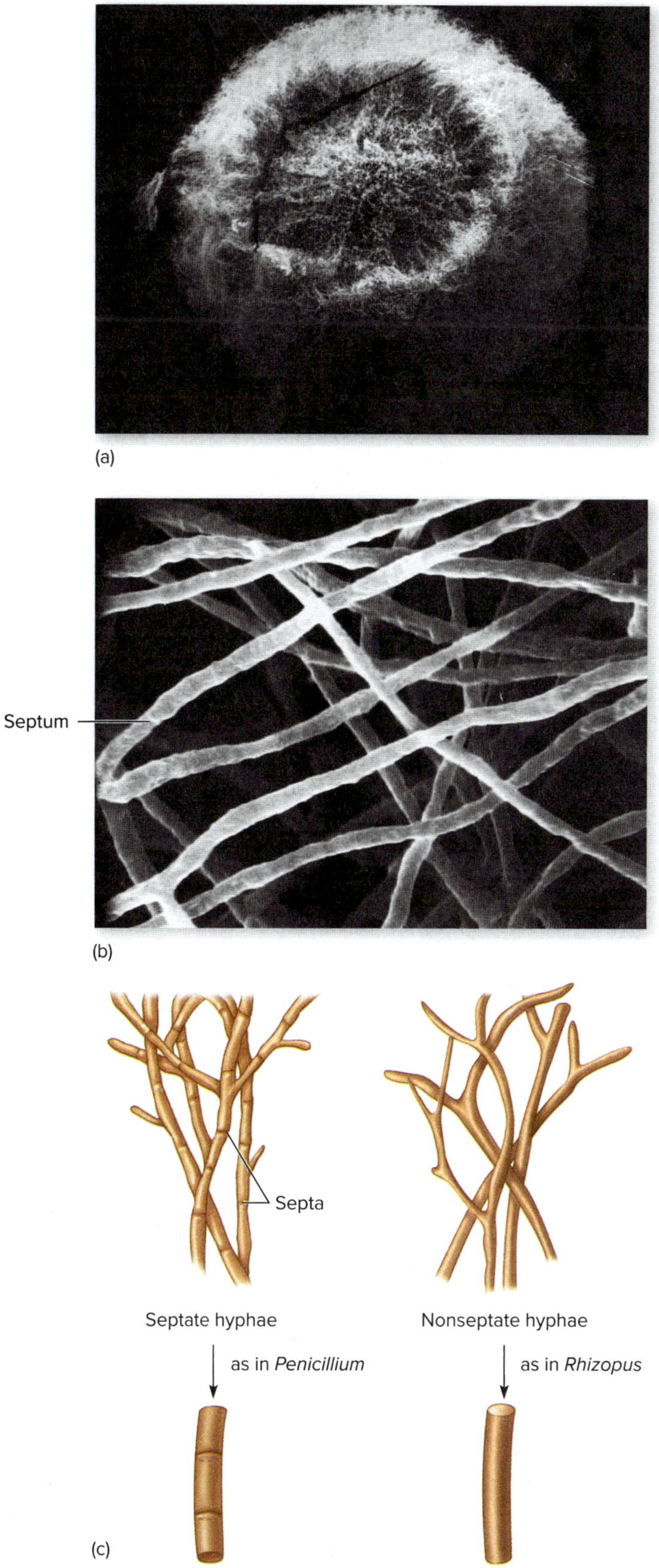
(a)
Septum
(b)
Septa
Septate hyphae
as in *Penicillium*
Nonseptate hyphae
as in *Rhizopus*
(c)

Figure 5.15 ***Diplodia maydis,* a pathogenic fungus of corn plants.** **(a)** Scanning electron micrograph of a single colony showing its filamentous texture (24×). **(b)** Close-up of hyphal structure (1,200×). **(c)** Basic structural types of hyphae.

(a–b) Dr. Judy A. Murphy

Figure 5.16 Microscopic morphology of yeasts. **(a)** General structure of a yeast cell, representing major organelles. Note the presence of a cell wall and lack of locomotor organelles. **(b)** Scanning electron micrograph of the brewer's, or baker's, yeast *Saccharomyces cerevisiae* (21,000×). **(c)** Formation and release of yeast buds and pseudohypha (a chain of budding yeast cells).

(b) Steve Gschmeissner/Science Photo Library/Getty Images

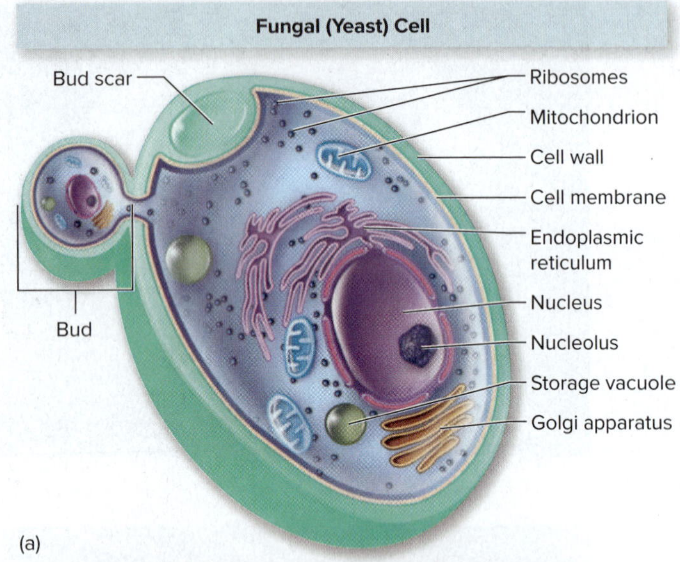
Fungal (Yeast) Cell
Bud scar
Bud
Ribosomes
Mitochondrion
Cell wall
Cell membrane
Endoplasmic reticulum
Nucleus
Nucleolus
Storage vacuole
Golgi apparatus

(a)

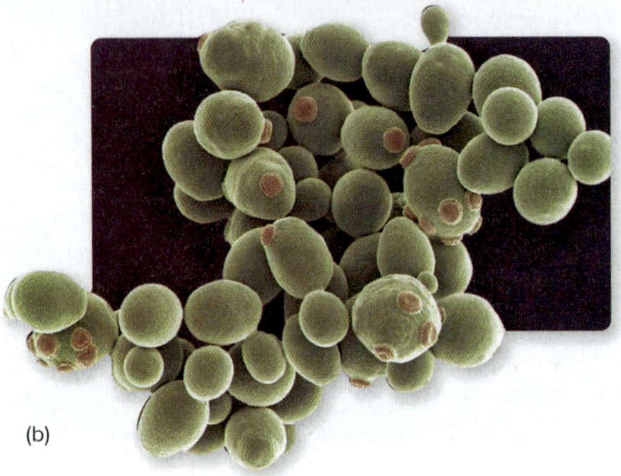

(b)

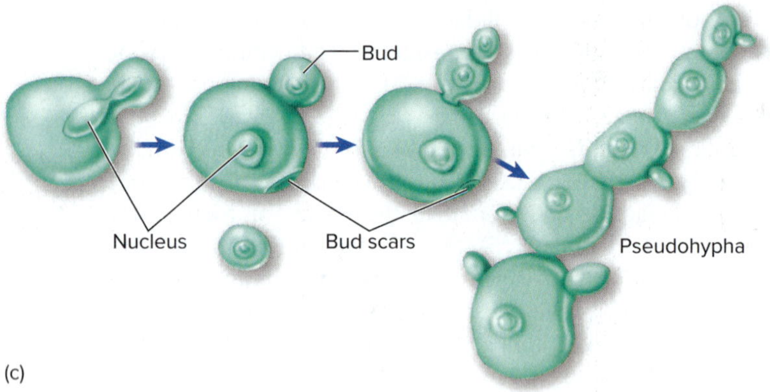
Bud
Nucleus
Bud scars
Pseudohypha

(c)

Figure 5.17 Nutritional sources (substrates) for fungi. **(a)** A fungal mycelium growing on raspberries. The fine hyphal filaments and black sporangia are typical of *Rhizopus*. **(b)** The skin of the foot infected by the fungus *Trichophyton rubrum*.

(a) Kathy Park Talaro; (b) Centers for Disease Control and Prevention

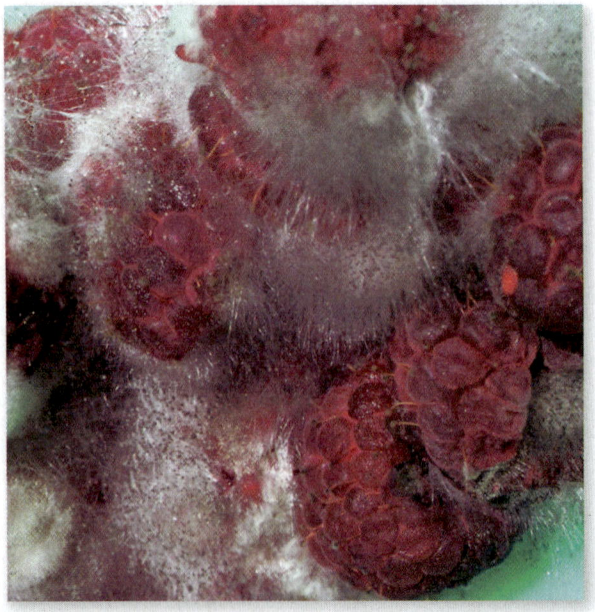

(a)

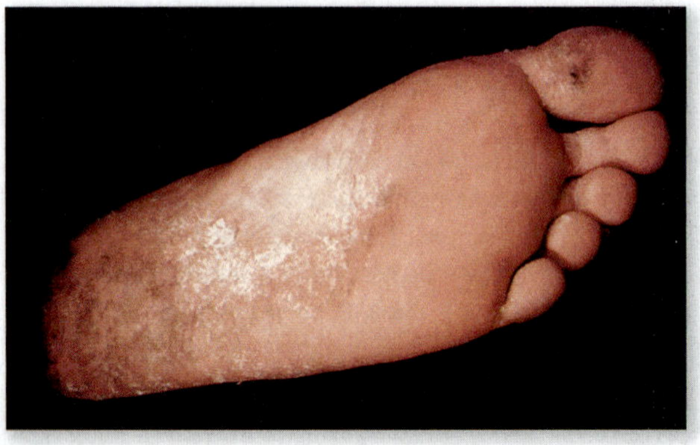

(b)

(including human) diseases caused by fungi are called **mycoses.** The toxins from fungi may cause disease in humans, and airborne fungi are a frequent cause of allergies and other medical conditions. Also, thousands of species of fungi are important plant pathogens.

Organization of Microscopic Fungi

The cells of most microscopic fungi grow in loose associations or colonies. The colonies of yeasts are much like those of bacteria in that they have a soft, uniform texture and appearance. The colonies of filamentous fungi are noted for striking cottony, hairy, or velvety textures. The woven, intertwining mass of hyphae that makes up the body or colony of a mold is called a **mycelium.**

Although the cells of hyphae contain the usual eukaryotic organelles, they also have some unique organizational features. In most fungi, the hyphae are divided into segments by cross walls, or **septa.** These fungi are described as *septate* (see figure 5.15*c*). The structure of the septa varies from solid partitions with no communication between the compartments to partial walls with small pores that allow the flow of organelles and nutrients between adjacent compartments. Nonseptate hyphae consist of one long, continuous cell *not* divided into individual compartments by cross walls. With this construction, the cytoplasm and organelles move freely from one region to another, and each hyphal element can have several nuclei.

Hyphae can also be classified according to their particular function. Vegetative hyphae (mycelia) are responsible for the visible mass of growth that appears on the surface of a substrate and penetrates it to digest and absorb nutrients. During the development of a fungal colony, the vegetative hyphae give rise to structures called reproductive, or fertile, hyphae, which branch off a vegetative mycelium. These hyphae are responsible for the production of fungal reproductive bodies called **spores.** Hyphae are illustrated in **figure 5.18.**

Reproductive Strategies and Spore Formation

Fungi have many complex and successful reproductive strategies. Most can propagate by the simple outward growth of existing hyphae or by fragmentation, in which a separated piece of mycelium can generate a whole new colony. But the primary reproductive mode of fungi involves the production of various types of spores. (Do not confuse fungal spores with the more resistant, nonreproductive bacterial endospores.) Fungal spores are responsible not only for multiplication, but also for survival, producing genetic variation, and dissemination. Because of their compactness and light weight, spores are dispersed widely through the environment by air, water, and living things. Upon encountering a favorable substrate, a spore will germinate and produce a new fungus colony in a very short time.

The fungi contain such a wide diversity of spores that they are largely classified and identified by their spores and spore-forming

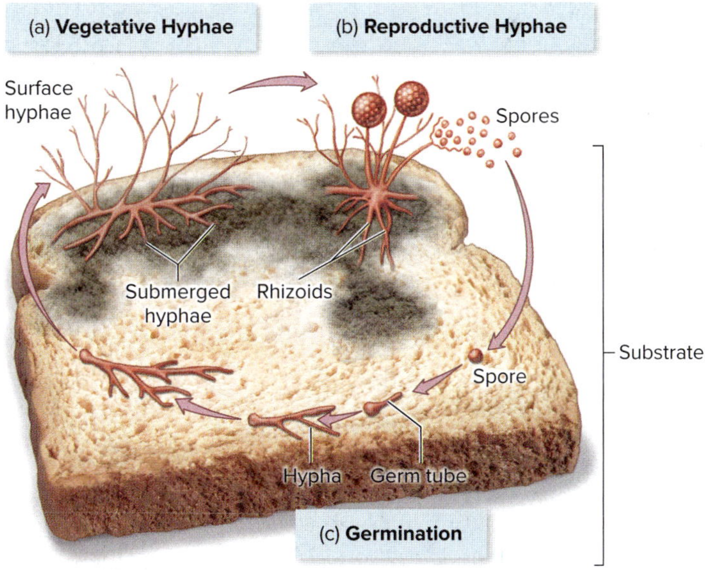

(a) Vegetative Hyphae
(b) Reproductive Hyphae
Surface hyphae
Spores
Submerged hyphae
Rhizoids
Substrate
Spore
Hypha
Germ tube
(c) Germination

Sporangia are +/− 20 µm in diameter
(d)

Figure 5.18 Functional types of hyphae using the mold *Rhizopus* as an example. **(a)** Vegetative hyphae are those surface and submerged filaments that digest, absorb, and distribute nutrients from the substrate. This species also has special anchoring structures called *rhizoids*. **(b)** Later, as the mold matures, it sprouts reproductive hyphae that produce asexual spores. **(c)** During the asexual life cycle, the free mold spores settle on a substrate and send out germ tubes that elongate into hyphae. Through continued growth and branching, an extensive mycelium is produced. So prolific are the fungi that a single colony of mold can easily contain 5,000 spore-bearing structures. Most spores do not germinate, but enough are successful to keep the numbers of fungi and their spores very high in most habitats. **(d)** Two views of *Rhizopus,* commonly known as bread mold.

(b) Rattiya Thongdumhyu/Shutterstock; (d) Richard Hutchings/McGraw Hill

structures. There are elaborate systems for naming and classifying spores, but we won't cover them. The most general subdivision is based on the way the spores arise. Asexual spores are the products of mitotic division of a single parent cell, and sexual spores are formed through a process involving the fusing of two parental nuclei followed by meiosis.

Asexual Spore Formation

There are two subtypes of asexual spores, **sporangiospores** and **conidiospores,** also called *conidia* **(figure 5.19):**

1. Sporangiospores **(figure 5.19*a*)** are formed by successive cleavages within a saclike head called a **sporangium,** which is attached to a stalk, the sporangiophore. These spores are initially enclosed but are released when the sporangium ruptures.
2. Conidiospores, or **conidia,** are free spores not enclosed by a spore-bearing sac. They develop either by the pinching off of the tip of a special fertile hypha or by the segmentation of a preexisting vegetative hypha. There are many different forms of conidia, with their own names, illustrated in **figure 5.19*b*.**

Sexual Spore Formation

Fungi can propagate themselves successfully with their millions of asexual spores. That being the case, what is the function of their sexual spores? The answer lies in important variations that occur when fungi of different genetic makeup combine their genetic material. Just as in plants and animals, this linking of genes from two parents creates offspring with combinations of genes different from that of either parent. The offspring from such a union can have slight variations in form and function that are potentially advantageous in the adaptation and survival of their species.

The majority of fungi produce sexual spores at some point. The nature of this process varies from the simple fusion of fertile hyphae of two different strains to a complex union of male and female structures and the development of special fruiting structures. The fleshy part of a mushroom is actually a fruiting body designed to protect and help disseminate its sexual spores.

Fungal Identification and Cultivation

Fungi are identified in medical specimens by first being isolated on special types of media and then being observed macroscopically and microscopically. The asexual spore-forming structures and spores are usually used to identify organisms to the level of genus and species. Other characteristics that contribute to identification are hyphal type, colony texture and pigmentation, physiological characteristics, and genetic makeup. Even as bacterial and viral identification rely increasingly on molecular techniques, fungi are some of the most strikingly beautiful life forms, and their appearance under the microscope is still heavily relied on to identify them **(figure 5.20*a,b*).**

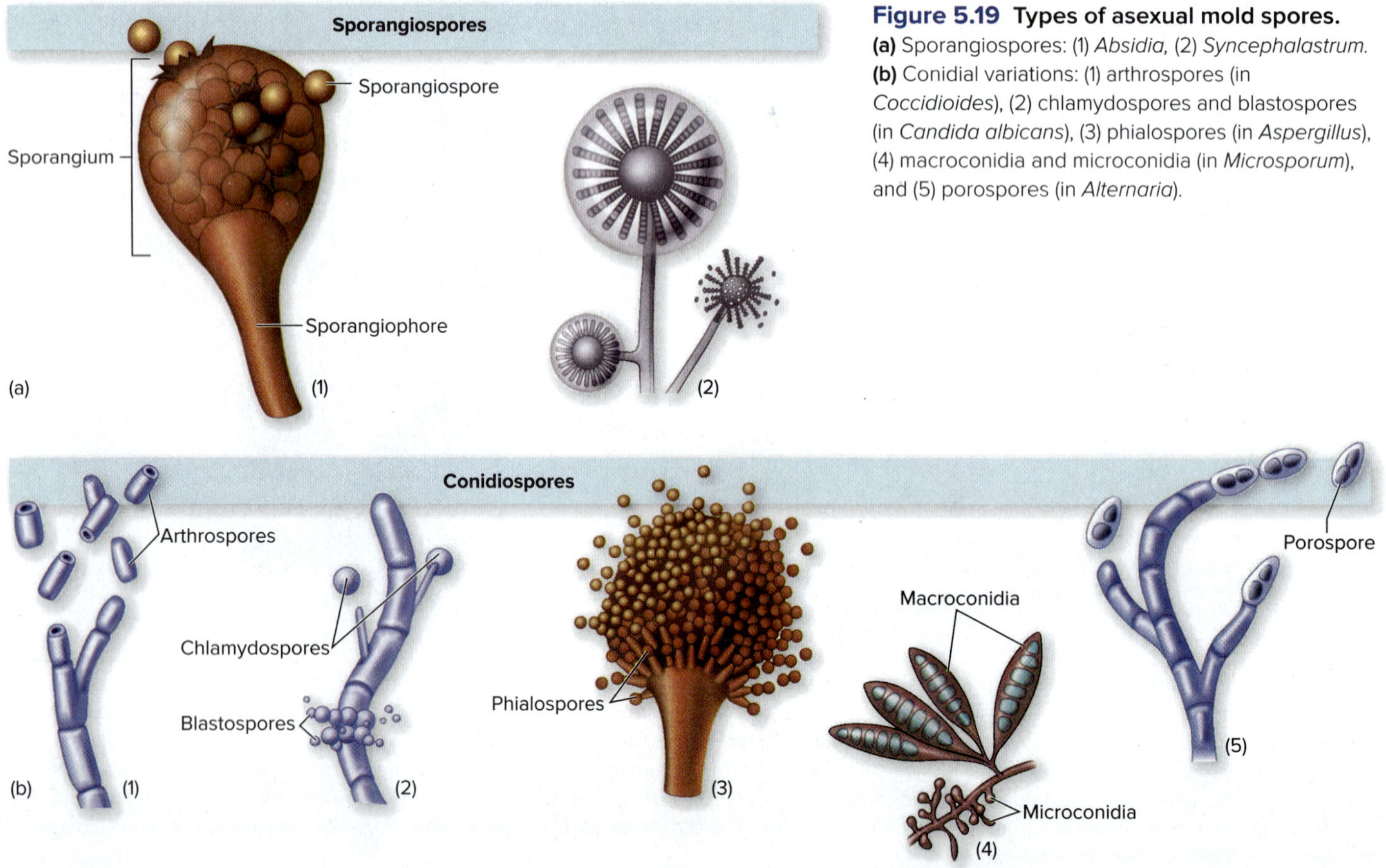
Sporangiospores
Sporangiospore
Sporangium
Sporangiophore
(a)
(1)
(2)
Conidiospores
Arthrospores
Chlamydospores
Blastospores
Phialospores
Macroconidia
Microconidia
Porospore
(b)
(1)
(2)
(3)
(4)
(5)

Figure 5.19 Types of asexual mold spores. **(a)** Sporangiospores: (1) *Absidia,* (2) *Syncephalastrum.* **(b)** Conidial variations: (1) arthrospores (in *Coccidioides*), (2) chlamydospores and blastospores (in *Candida albicans*), (3) phialospores (in *Aspergillus*), (4) macroconidia and microconidia (in *Microsporum*), and (5) porospores (in *Alternaria*).

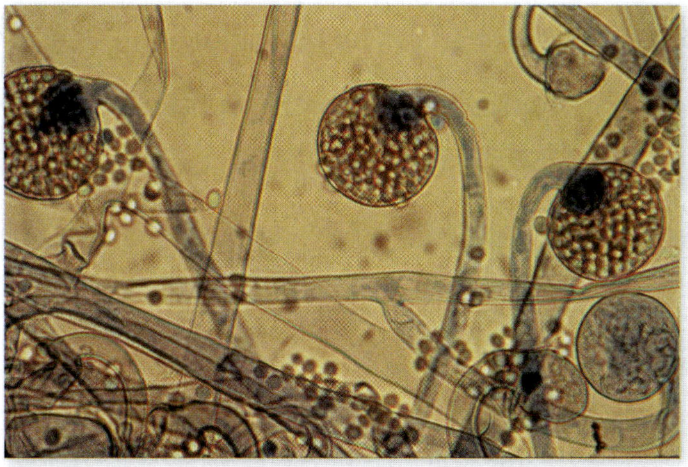

(a)

(b)

Figure 5.20 Representative fungi. **(a)** *Circinella,* a fungus associated with soil and decaying nuts. **(b)** *Aspergillus,* a very common environmental fungus that can be associated with human disease.

(a) Kathy Park Talaro; (b) Eye of Science/Science Source

The Effects of Fungi on Humans and the Environment

Nearly all fungi are free-living and do not require a host for their life cycles. They play an essential role in decomposing organic matter and returning essential minerals to the soil. The largest living thing on earth is a fungus called the honey fungus, which has underground filamentous networks that stretch for miles. On the other hand, fungi can pose problems for the agricultural industry. A number of species are pathogenic to field plants such as corn and grain, and fungi can also rot fresh produce during shipping and storage. It has been estimated that as much as 40% of the yearly fruit crop is consumed not by humans but by fungi.

Even among those fungi that are pathogenic, most human infection occurs through accidental contact with an environmental source such as soil, water, or dust. Many fungi make their home on the human body, as part of the normal human microbiome **(Insight 5.1).** Yet nearly 300 species of fungi can also cause human disease. The Centers for Disease Control and Prevention currently monitors three types of fungal disease in humans: community-acquired infections caused by environmental pathogens in the general population; hospital-associated infections caused by fungal pathogens in clinical settings; and opportunistic infections caused by pathogens infecting already weakened individuals **(table 5.3).**

This last category, opportunistic infections, has become very troubling in recent years. An important review was recently published that revealed that deaths from fungal infections have surpassed deaths due to either malaria or breast cancer worldwide. Transplant patients, cancer patients, and HIV-positive patients are particularly susceptible to opportunistic fungi. The progress of these infections in immunosuppressed people was vividly described in the report. "They'll just rot you down quick as a flash," said one of the authors. During the COVID pandemic, India struggled with massive numbers of infections in mid-2021. A fungal infection known as mucormycosis (commonly known as the black fungus) killed more than 300,000 people with COVID or recovering from it.

Disease Connection

PCP (*Pneumocystis* pneumonia) is a very common fungal infection afflicting patients with HIV/AIDS. Caused by the fungus *Pneumocystis jiroveci,* PCP results in severe inflammation and fluid accumulation in the lungs. *P. jiroveci* is likely transmitted through the air. People with normal immune systems are rarely affected with PCP—in fact, most of us have been exposed to *P. jiroveci* by the time we are 3 to 4 years of age, and our bodies have easily defeated the organism. This deadly fungal infection is considered to be an opportunistic infection because it typically attacks individuals with compromised immunity.

Fungi are involved in other medical conditions besides infections. Fungal cell walls give off chemical substances that can cause allergies. The toxins produced by poisonous mushrooms can induce neurological disturbances and even death. The mold *Aspergillus flavus* synthesizes a potentially lethal poison called aflatoxin, which is the cause of a disease in domestic animals that have eaten grain contaminated with the mold. It can also be a cause of liver cancer in humans.

On the beneficial side, free-living fungi play an essential role in decomposing organic matter and returning essential minerals to the soil. They form stable associations with plant roots that increase the ability of the roots to absorb water and nutrients. Industries have tapped the biochemical potential of fungi to produce large quantities of antibiotics, alcohol, organic acids, and vitamins. Other fungi are eaten or used as flavorings for food. The yeast *Saccharomyces* produces the alcohol in beer and wine and the gas that causes bread to rise. Blue cheese, soy sauce, and cured meats derive their unique flavors from the actions of fungi.

INSIGHT 5.1 MICROBIOME: Are Eukaryotic Microorganisms Part of Our Microbiome?

Yes! Most definitely. In this Insight, we will focus on fungi. Fungi are well known to be normal inhabitants—and sometimes pathogens—for humans. One example is the fungus that causes yeast infections of the gut, mouth, and vagina, called *Candida albicans*.

Fungal infections of the skin (and hair and nails) are quite common as well. Twenty-nine million people in the United States experience this every year. The Human Microbiome Project has finally made it possible to look at the fungal species that live normally on our skin and other surfaces.

In recent years, a research team from the National Institutes of Health (NIH) documented the fungi they found on human surfaces. They sampled 14 body sites from each of 10 healthy adults. Their DNA analysis identified more than 80 fungal genera ("genera" is the plural of genus). Previously, when these studies were conducted using culture techniques, only 18 different genera were found.

They discovered that one species, *Malassezia,* is commonly found as a normal inhabitant of skin on the head and on most of the body (the trunk). *Malassezia* has been recognized before as a cause of very superficial skin infections (see chapter 19), and it has been associated with the condition of dandruff. This study tells us that most of the time it is a normal inhabitant of our surfaces. Interestingly, in 2019 *Malassezia* was associated with pancreatic cancer. This highlights the importance of cataloguing the microbiome, to find out how prevalent certain microbes are in the "normal" biota of humans.

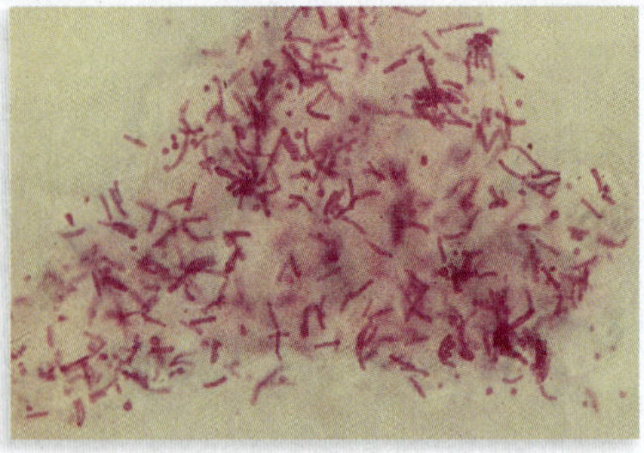

Malassezia furfur at 1,000× magnification.
Dr. Lucille K. Georg/CDC

For some reason, the body part displaying the most diverse fungal population was the heel, containing about 80 fungal genera. Toenails had about 60 genera, and the webs of the toes had about 40. Skin surfaces like the head and trunk displayed just 2 to 10 genera each.

Table 5.3 Three Categories of Fungal Infections of Humans

Source of Fungus	Who Is Affected?	Examples	Symptoms
Community-Acquired Infections			
Environmental species	General population	Histoplasmosis, caused by a fungus that thrives in bird and bat droppings	Affects lungs and respiratory tract
		Coccidioidomycosis (also called valley fever), caused by a fungus that lives in dust and soil, particularly in the SW United States	Flulike symptoms in respiratory tract and rest of body; includes headaches, night sweats, and muscle and joint pain
Hospital-Associated Infections			
Pathogens in clinical settings	People in hospitals or long-term care	Various fungi that contaminate health care facilities	———
Opportunistic Infections			
Environmental species or normal fungal biota	People who have compromised immune systems, or disrupted microbiota	Mucormycosis, caused by a fungus found in soil, leaves, compost, and rotting wood	Affects lungs and sinuses most often; skin can also be the target
		Candidiasis (commonly referred to as a yeast infection), caused by a yeast that is normal biota on human mucosal surfaces	Found at the mucosal site where the microbe normally lives in small numbers; when the normal balance of microbiota is disrupted, the yeast proliferates and causes inflammation

5.4 Learning Outcomes—Assess Your Progress

13. List three general features of fungal anatomy.
14. Differentiate among the terms *heterotroph, saprobe,* and *parasite.*
15. Explain the relationship between fungal hyphae and the production of a mycelium.
16. Describe two ways in which fungal spores are formed.
17. List two detrimental and two beneficial activities of fungi (from the viewpoint of humans).

5.5 The Protists

Algae and protozoa have traditionally been combined into the Kingdom Protista. The two major taxonomic categories of this kingdom are Subkingdom Algae and Subkingdom Protozoa. Although these general types of microbes are now known to occupy several kingdoms, it is still useful to retain the concept of a *protist* as any eukaryotic unicellular or colonial organism that lacks true tissues. We will only briefly mention algae, as they do not cause human infections for the most part.

A Note About the Taxonomy of Protists

Exploring the origins of eukaryotic cells with molecular techniques has clarified our understanding of relationships among the organisms in Domain Eukarya. The characteristics traditionally used for placing plants, animals, and fungi into separate kingdoms are their general cell type, level of organization (body plan), and nutritional type. Although these criteria often do reflect accurate differences among these organisms and give rise to the same classifications as molecular techniques, in many cases the molecular data point to new and different classifications.

Because our understanding of phylogenetic relationships is still in development, there is not yet a single official system of taxonomy for presenting all of the eukaryotes. This is especially true of the protists (which contain algae and protozoa). Genetic analysis has determined that this group, generally classified at the kingdom level, is far more diverse than previously appreciated and probably should instead be divided into several different kingdoms. Some organisms we call *protists* are more related to fungi than they are to other protists, for instance. For that reason, most scientists believe that the labels "protist" and "protozoa" are meaningless, taxonomically.

For the purposes of this book and your class, we will use the term *protozoa* to refer to eukaryotic organisms that are not animals, plants, or fungi. But be aware that the science is still developing.

The Algae: Photosynthetic Protists

The **algae** are a group of photosynthetic organisms usually recognized by their larger members, such as seaweeds and kelps. In addition to being beautifully colored and diverse in appearance, they vary in length from a few micrometers to 100 meters. Algae occur as single cells, as colonies, and in filamentous form. The larger forms can possess tissues and simple organs. **Figure 5.21** depicts various types of algae. Algal cells contain all of the eukaryotic organelles. The most noticeable of these are the chloroplasts, which contain the green pigment chlorophyll, and a number of other pigments that create the yellow, red, and brown coloration of some species.

Algae are widespread inhabitants of fresh and marine waters. They are one of the main components of the large floating community of microscopic organisms called **plankton.** In this capacity, they play an essential role in the aquatic food web and produce about 70% of the earth's oxygen. Other algal habitats include the surface of soil, rocks, and plants. Several species are even hardy enough to live in hot springs or snowbanks.

The primary medical threat from algae is due to a type of food poisoning caused by the toxins of certain marine algae. During particular seasons of the year, the overgrowth of these motile algae imparts a brilliant red color to the water, which is referred to as a *red tide.* When fish and other wildlife feed on the algae, their bodies accumulate toxins given off by the algae that can persist for several months. Paralytic shellfish poisoning is caused by eating exposed clams or other invertebrates. It is marked by severe neurological symptoms and can be fatal. Ciguatera is another serious poisioning caused by algal toxins that have accumulated in fish such as bass and mackerel. Cooking does not destroy the toxin, and there is no antidote.

Biology of the Protozoa

Although the name *protozoa* (singular, *protozoan*) comes from the Greek for "first animals," they are far from being simple, primitive organisms. The protozoa constitute a very large group

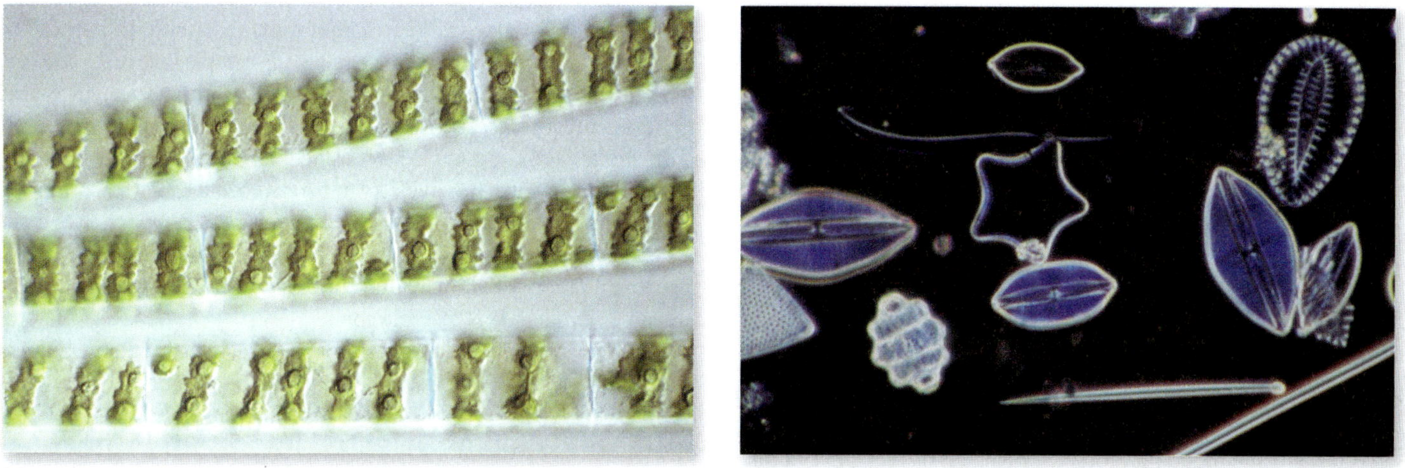

(a) (b)

Figure 5.21 Representative microscopic algae. **(a)** *Spirogyra*, a colonial filamentous form with spiral chloroplasts. **(b)** A collection of beautiful algae called diatoms shows the intricate and varied structure of their cell walls.

(a) Stephen Durr; (b) Oxford Scientific/Photodisc/Getty Images

(about 65,000 species) of creatures that, although single-celled, have startling properties when it comes to movement, feeding, and behavior. Although most members of this group are harmless, free-living inhabitants of water and soil, a few species are parasites that are collectively responsible for hundreds of millions of infections of humans each year. Before we consider a few examples of important pathogens, let us examine some general aspects of protozoan biology, remembering that the term *protozoan* is more of a convenience than an accurate taxonomic designation. As we describe them in the coming paragraphs, you will see why they are categorized together. It is because of their similar physical characteristics rather than their genetic relatedness.

Protozoan Form and Function

Most protozoan cells are single cells containing all the major eukaryotic organelles except chloroplasts. Their organelles can be highly specialized and are essentially analogous to mouths, digestive systems, reproductive tracts, and "legs"—or means of locomotion. The cytoplasm is usually divided into a clear outer layer called the **ectoplasm** and a granular inner region called the **endoplasm.** Ectoplasm is involved in locomotion, feeding, and protection. Endoplasm houses the nucleus, mitochondria, and food and contractile vacuoles. Some ciliates and flagellates even have organelles that work somewhat like a primitive nervous system to coordinate movement. Because active protozoa lack a cell wall, they have a certain amount of flexibility. Their outer boundary is a cell membrane that regulates the movement of food, wastes, and secretions. Cell shape can remain constant (as in most ciliates) or can change constantly (as in amoebas). Certain amoebas (foraminiferans) encase themselves in hard shells made of calcium carbonate. Some species have an outer layer called a **pellicle.** The size of most protozoan cells falls within the range of 3 to 300 μm. Some notable exceptions are giant amoebas and ciliates that are large enough (3 to 4 mm in length) to be seen swimming in pond water.

Nutritional and Habitat Range Protozoa are heterotrophic and usually require their food in a complex organic form. Free-living species scavenge dead plant or animal debris and even graze on live cells of bacteria and algae. Some protozoa absorb food directly through the cell membrane. Parasitic species live on the fluids of their host, such as plasma and digestive juices, or they can actively feed on tissues.

The predominant habitats for protozoa are fresh and marine water, soil, plants, and animals. Even extremes in temperature and pH are not a barrier to their existence; species are found in hot springs, ice, and habitats with low or high pH. Many protozoa can convert to a resistant, dormant stage called a *cyst.*

Styles of Locomotion Except for one group (the sporozoa), protozoa can move through fluids by means of **pseudopods** ("false feet"), **flagella,** or **cilia.** A few species have both pseudopods and flagella. Some unusual protozoa move by a gliding or twisting movement that does not appear to involve any of these locomotor structures. Pseudopods are blunt, branched, or long and pointed, depending on the particular species. The flowing action of the pseudopods results in amoeboid motion, and pseudopods also serve as feeding structures in many amoebas. (The structure and behavior of flagella and cilia were discussed earlier in this chapter.) Flagella **(figure 5.22*a*)** vary in number from one to several, and in certain species they are attached along the length of the cell by an extension of the cytoplasmic membrane called the *undulating membrane.* In most ciliates, the cilia are distributed over the entire surface of the cell in characteristic patterns. Because of the tremendous variety in ciliary arrangements and functions, ciliates are among the most diverse cells in the biological world. In certain protozoa, cilia line the oral groove and function in feeding; in others, they fuse together to form stiff props that serve as primitive rows of walking legs.

Life Cycles and Reproduction Protozoa are called **trophozoites** when they are in their motile feeding stage. This is a stage that requires ample food and moisture to remain active. A large number of species are also capable of entering into a dormant, resting stage called a **cyst** when conditions in the environment become unfavorable for growth and feeding. During *encystment,* the trophozoite cell rounds up into a sphere, and its ectoplasm secretes a tough, thick cuticle around the cell membrane **(figure 5.23).** Because cysts are more resistant than the trophozoite form to heat, drying, and chemicals, they can survive adverse periods. They can be dispersed by air currents and may even be an important factor in the spread of diseases such as amoebic dysentery. If provided with moisture and nutrients, a cyst breaks open and releases the active trophozoite. Both the cyst and trophozoite forms of protozoan pathogens can be identified through what is called O & P (ova and parasite) testing of patient stool samples. This method, combined with immunology-based tests, is currently used for disease diagnosis in cases of giardiasis and cryptosporidiosis.

The life cycles of protozoans vary from simple to complex. Several protozoan groups exist only in the trophozoite state. Many alternate between a trophozoite and a cyst stage, depending on the conditions of the habitat. The life cycle of a parasitic protozoan dictates its mode of transmission to other hosts. For example, the flagellate *Trichomonas vaginalis* causes a common sexually transmitted infection. Because it does not form cysts, it is more delicate and can only be transmitted by intimate contact between sexual partners. In contrast, intestinal pathogens such as *Cryptosporidium* and *Giardia lamblia* (figure 5.23) form cysts and are readily transmitted in contaminated water and foods.

All protozoa reproduce by relatively simple, asexual methods, usually mitotic cell division. Several parasitic species, including the agents of malaria and toxoplasmosis, reproduce asexually by multiple fission inside a host cell. Sexual reproduction also occurs during the life cycle of most protozoa. Ciliates participate in **conjugation,** a form of genetic exchange in which two cells fuse temporarily and exchange micronuclei. This process of sexual recombination yields new and different genetic combinations that can be advantageous in evolution.

Many protozoa engulf toxic bacteria and maintain them in their cytoplasm. This, in turn, can make them toxic.

Classification of Selected Important Protozoa

Taxonomists have problems classifying protozoa. We will use a functional way to categorize them, in a way that will be most useful in a clinical situation. As mentioned earlier in

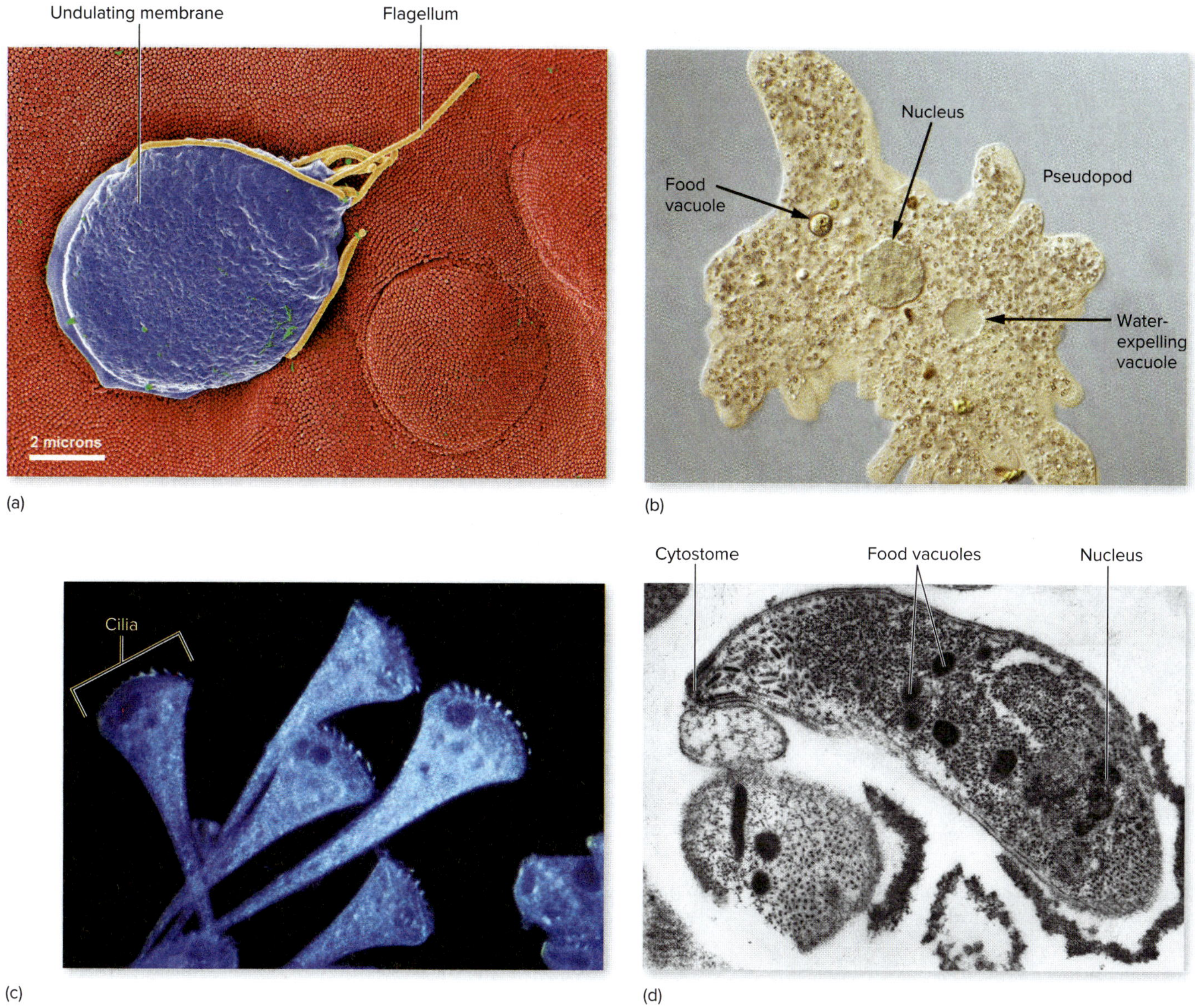

Undulating membrane
Flagellum
2 microns
(a)
Nucleus
Food vacuole
Pseudopod
Water-expelling vacuole
(b)
Cilia
(c)
Cytostome
Food vacuoles
Nucleus
(d)

Figure 5.22 Examples of the four types of locomotion in protozoa. **(a)** Using flagella: *Giardia,* displaying flagella. **(b)** Using amoeboid motion: *Amoeba,* with pseudopods. **(c)** Using cilia: *Stentor,* displaying cilia. **(d)** Sporozoan: *Cryptosporidium*. Sporozoa have no specialized locomotion organelles.

(a) Dr. Stan Erlandsen/Centers for Disease Control and Prevention; (b) Stephen Durr; (c) Oxford Scientific/Photodisc/Getty Images; (d) Michael W. Riggs

this section, protozoa can be placed in four groups based on how they move. These categories are summarized here and in figure 5.22:

Those using flagella to move. Motility is primarily by flagella alone or by both flagellar plus amoeboid motion. They have a single nucleus. Sexual reproduction, when present, is by syngamy, which is the combination of genetic material from two cells. Several parasitic forms lack mitochondria and Golgi apparatus. Most species form cysts and are free-living; the group also includes several parasites. Some species are found in loose aggregates or colonies, but most are solitary. Members include *Trypanosoma* and *Leishmania,* important blood pathogens spread by insect vectors; *Giardia* (see figure 5.22*a*), an intestinal parasite spread in water contaminated with feces; and *Trichomonas,* a parasite of the reproductive tract of humans spread by sexual contact.

Those using amoeboid motion to move. Cell form is primarily an amoeba (see figure 5.22*b*). Major locomotor organelles are pseudopods, although some species have flagellated reproductive states. They perform asexual reproduction by fission. They are mostly uninucleate. They usually can form cysts. Most amoebas are free-living and not infectious. *Entamoeba* is a pathogen or parasite of humans. Two groups have an external shell. Shelled amoebas called foraminifera and radiolarians are responsible for chalk deposits in the ocean.

Those using cilia to move. Their trophozoites are motile by cilia. Some have cilia in tufts for feeding and attachment. Most

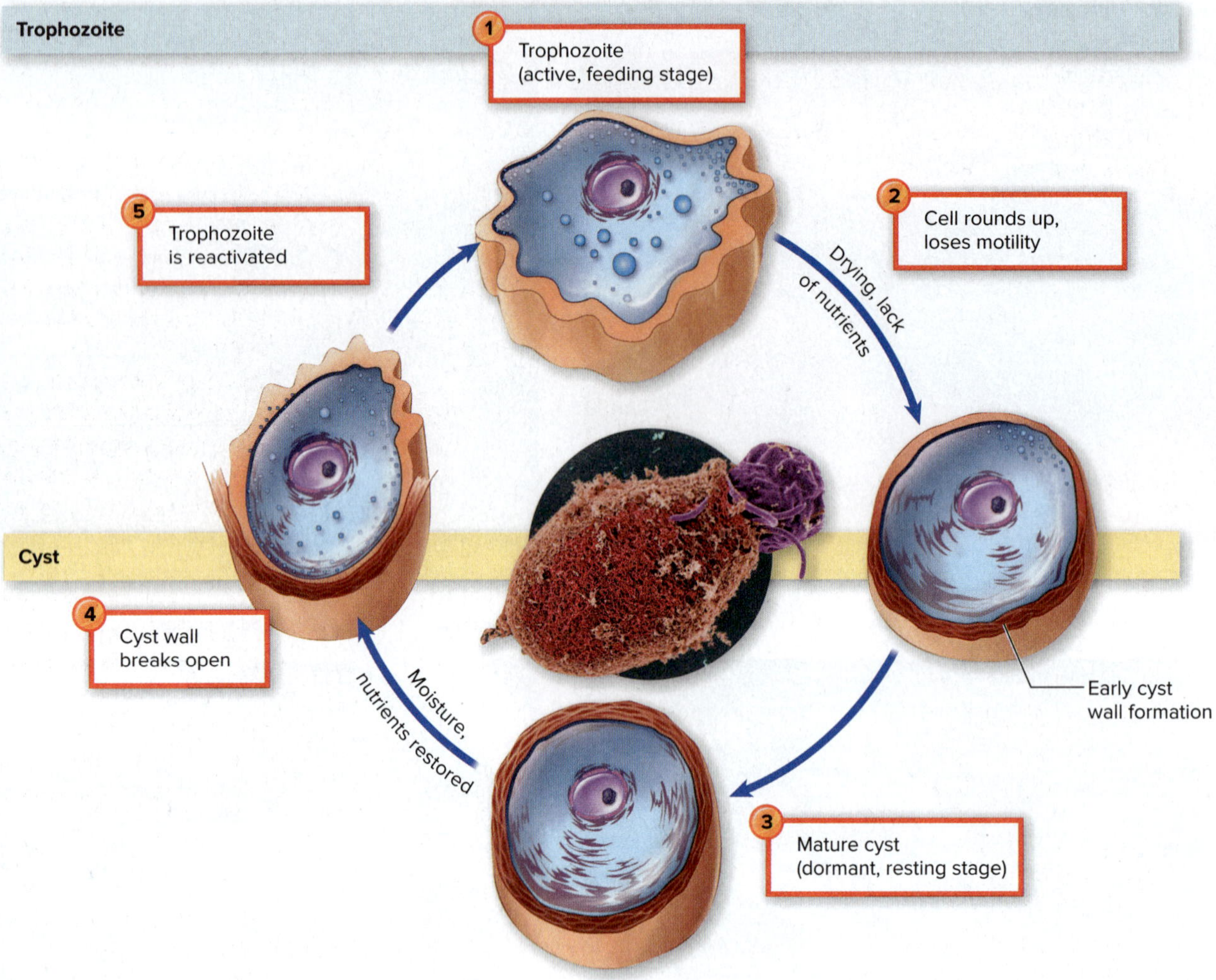
Trophozoite
1 Trophozoite (active, feeding stage)
2 Cell rounds up, loses motility
Drying, lack of nutrients
Early cyst wall formation
Cyst
3 Mature cyst (dormant, resting stage)
Moisture, nutrients restored
4 Cyst wall breaks open
5 Trophozoite is reactivated

Figure 5.23 The general life cycle exhibited by many protozoa. All protozoa have a trophozoite form, but not all produce cysts. The photo in the center shows a *Giardia* trophozoite (purple) emerging from its cyst form (orange).
Dr. Stan Erlandsen/CDC

are able to develop cysts. They have both macronuclei and micronuclei, and they divide by transverse fission. Most have a definite mouth and feeding organelle. They can show relatively advanced behavior (see figure 5.22*c*). The majority of ciliates are free-living and harmless.

Those with no motility (sporozoa). Although motility is absent in most representatives, it is exhibited by the male gametes of many members of this group. Life cycles of the apicomplexa (sporozoa) are, as the name implies, quite complex, with well-developed asexual and sexual stages. Sporozoa produce special sporelike cells called **sporozoites** (see figure 5.22*d*) following sexual reproduction, which are important in transmission of infections and exhibit a unique form of gliding motility. Most sporozoa form thick-walled zygotes called oocysts, and this entire group of organisms is parasitic. *Plasmodium,* the most prevalent protozoan pathogen, causes 100 million to 300 million cases of malaria each year worldwide. It is an intracellular parasite with a complex cycle alternating between humans and mosquitoes. *Toxoplasma gondii* causes infection (toxoplasmosis) in humans, which is acquired from cats and other animals.

Just as with bacteria and other eukaryotes, protozoa that cause disease produce symptoms in different organ systems. These diseases are covered in the disease sections of this book.

Protozoan Identification and Cultivation

The unique appearance of most protozoa makes it possible for a knowledgeable person to identify them to the level of genus—and often species—by microscopic morphology alone. Characteristics to consider in identification include the shape and size of the cell; the type, number, and distribution of locomotor structures; the presence of special organelles or cysts; and the number of nuclei. Medical specimens taken from blood, sputum, cerebrospinal fluid, feces, or the vagina are smeared directly onto a slide and observed with or without special stains. Occasionally, protozoa are cultivated on artificial media or in laboratory animals for further identification or study.

A Note About Neglected Parasitic Infections

Maybe it would come as a surprise to you that up to 25% of the world's population is infected with intestinal roundworms. As you read this chapter about fungi, protozoa, and helminths, maybe you feel confident that infections with these organisms are relatively rare in the United States. That's a mistake.

The CDC has begun a campaign against five neglected parasitic infections (NPIs) in the United States. The five infections are as follows:

Chagas disease—the trypanosome disease caused by the protozoan *Trypanosoma cruzi;*
Neurocysticercosis—caused by the tapeworm *Taenia solium;*
Toxocariasis—caused by roundworms that travel through tissues and can cause blindness;
Toxoplasmosis—caused by a protozoan with which 60 million people in the United States are infected; and
Trichomoniasis—a protozoan infection of the genital tract that leaves those infected more vulnerable to other sexually transmitted infections, including HIV. Trichomoniasis also leads to premature births by infected mothers.

Here are a few statistics:

- Neurocysticercosis is the single most common infectious cause of seizures in some areas of the United States.
- Toxocariasis is caused by dog and cat roundworms; up to 14% of the U.S. population has been exposed. About 70 people a year are blinded by this infection.
- There are up to 300,000 people infected with the protozoan causing Chagas disease.
- In the United States, 3.7 million people are infected with *Trichomonas.*

It is time to stop thinking of these infections as "other people's problems."

Important Protozoan Pathogens

Although protozoan infections are very common, they are actually caused by only a small number of species often restricted geographically to the tropics and subtropics **(table 5.4).** The study of protozoa and helminths is sometimes called *parasitology.* Although, technically, a parasite can be any organism that obtains food and other requirements at the expense of a host, the term *parasite* is most often used to denote protozoan and helminth pathogens.

Two flagellated protozoa that cause human disease are *Trypanosoma brucei* and *Trypanosoma cruzi*. *T. brucei* occurs in Africa, where it causes approximately a few thousand new cases of sleeping sickness each year. *T. cruzi,* the cause of Chagas disease, is endemic to South and Central America (although it has spread to other continents). Six to seven million people are estimated to be infected with it. Both species have long, crescent-shaped cells with a single flagellum that is sometimes attached to the cell body by an undulating membrane. Both are found in the blood during infection and are transmitted by blood-sucking vectors.

Table 5.4 Major Pathogenic Protozoa

Protozoan	Disease	Reservoir/Source
Amoeboid Protozoa		
Entamoeba histolytica	Amoebiasis (intestinal and other symptoms)	Humans, water and food
Naegleria, Acanthamoeba	Brain infection	Free-living in water
Ciliated Protozoa		
Balantidium coli	Balantidiosis (intestinal and other symptoms)	Pigs, cattle
Flagellated Protozoa		
Giardia lamblia	Giardiasis (intestinal distress)	Animals, water and food
Trichomonas vaginalis	Trichomoniasis (reproductive tract symptoms)	Human
Trypanosoma brucei, T. cruzi	Trypanosomiasis (sleeping sickness and Chagas disease)	Animals, vector-borne
Leishmania donovani, L. tropica, L. brasiliensis	Leishmaniasis (either skin lesions or widespread involvement of internal organs)	Animals, vector-borne
Nonmotile Protozoa		
Plasmodium falciparum, other *Plasmodium* species	Malaria (cardiovascular and other symptoms)	Human, vector-borne
Toxoplasma gondii	Toxoplasmosis (flulike illness)	Animals, vector-borne
Cryptosporidium	Cryptosporidiosis (intestinal and other symptoms)	Free-living, water and food
Cyclospora cayetanensis	Cyclosporiasis (intestinal and other symptoms)	Water, fresh produce

5.5 Learning Outcomes—Assess Your Progress

18. Note the protozoan characteristics that illustrate why they are informally placed into a single group.
19. List three means of locomotion exhibited by protozoa.
20. Explain why a cyst stage may be useful in a protozoan.
21. Give an example of a human disease caused by each of the four types of protozoa.

5.6 The Helminths

Tapeworms, flukes, and roundworms are collectively called *helminths,* from the Greek word meaning "worm." Adult animals are usually large enough to be seen with the naked eye, and they range from the longest tapeworms, measuring up to about 25 m in length, to roundworms less than 1 mm in length. Nevertheless, they are included among microorganisms because of their infective abilities and because a microscope is necessary to identify their eggs and larvae.

On the basis of body type, the two major groups of parasitic helminths are the flatworms (Phylum Platyhelminthes) and the roundworms (Phylum Aschelminthes, also called **nematodes**). Flatworms have a very thin, often segmented body plan **(figure 5.24*a*),** and roundworms have a long, cylindrical, unsegmented body **(figure 5.25).** The flatworm group is subdivided into the **cestodes,** or tapeworms, named for their long, ribbon-like arrangement, and the **trematodes,** or flukes, characterized by flat, oval bodies **(figure 5.24*b*).** Not all flatworms and roundworms are parasitic by nature. Many live free in soil and water. Because most disease-causing helminths spend part of their lives in the gastrointestinal tract, they are discussed in chapter 23.

General Worm Morphology

All helminths are multicellular animals equipped to some degree with organs and organ systems. In parasitic helminths, the most developed organs are those of the reproductive tract,

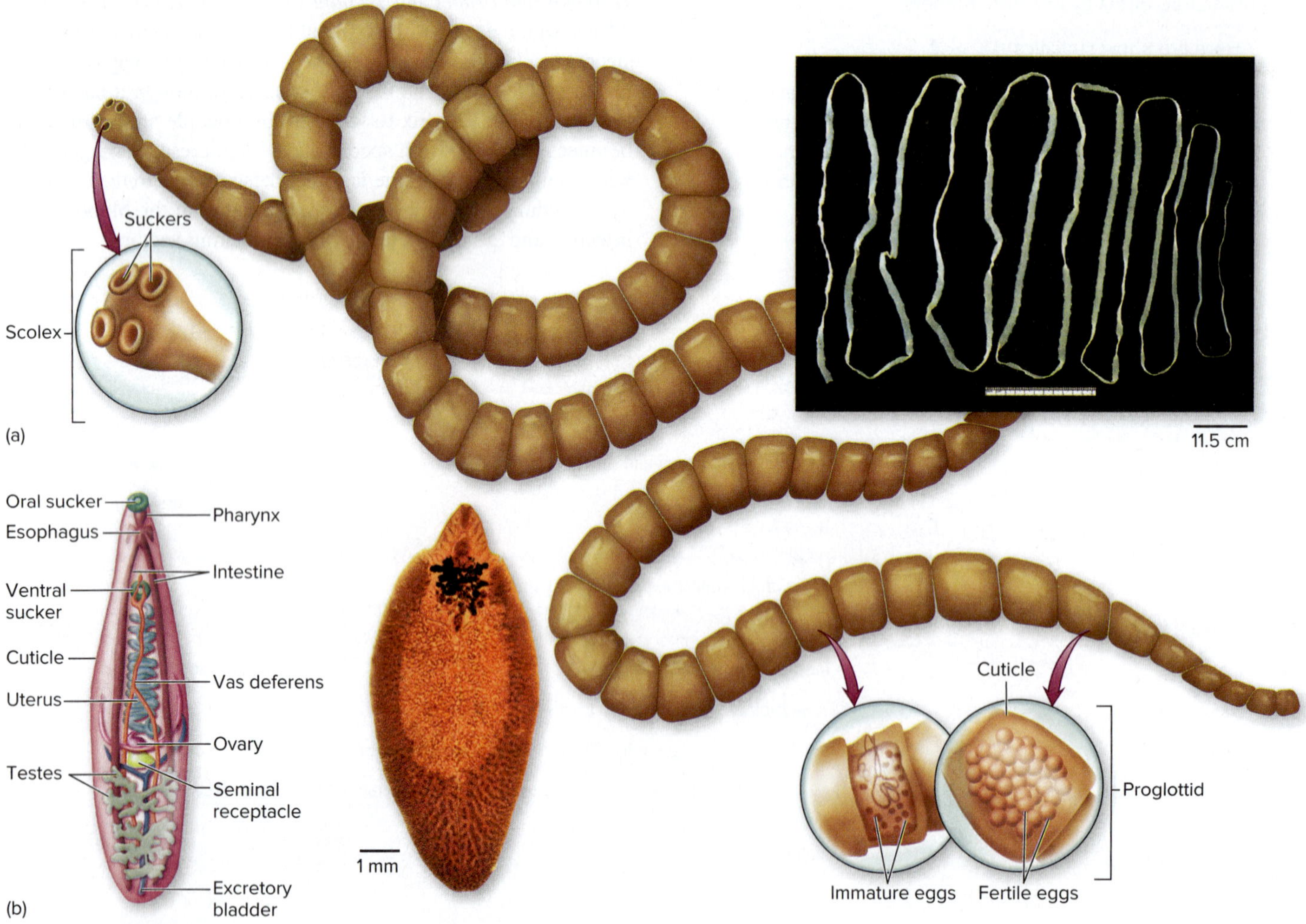
Suckers
Scolex
(a)
11.5 cm
Cuticle
Proglottid
Immature eggs
Fertile eggs
Oral sucker
Esophagus
Pharynx
Intestine
Ventral sucker
Cuticle
Vas deferens
Uterus
Ovary
Testes
Seminal receptacle
Excretory bladder
(b)
1 mm

Figure 5.24 Parasitic flatworms. **(a)** A cestode (tapeworm), showing the scolex; long, tapelike body; and magnified views of immature and mature proglottids (body segments). The photo shows an actual tapeworm. The ruler below the photo is 11.5 cm (about 4.5 inches) in length. **(b)** The structure of a trematode (liver fluke). Note the sucker that attaches to host tissue and the dominance of reproductive and digestive organs. The photo shows an actual liver fluke.

(a) Centers for Disease Control and Prevention; (b) D. Kucharski K. Kucharska/Shutterstock

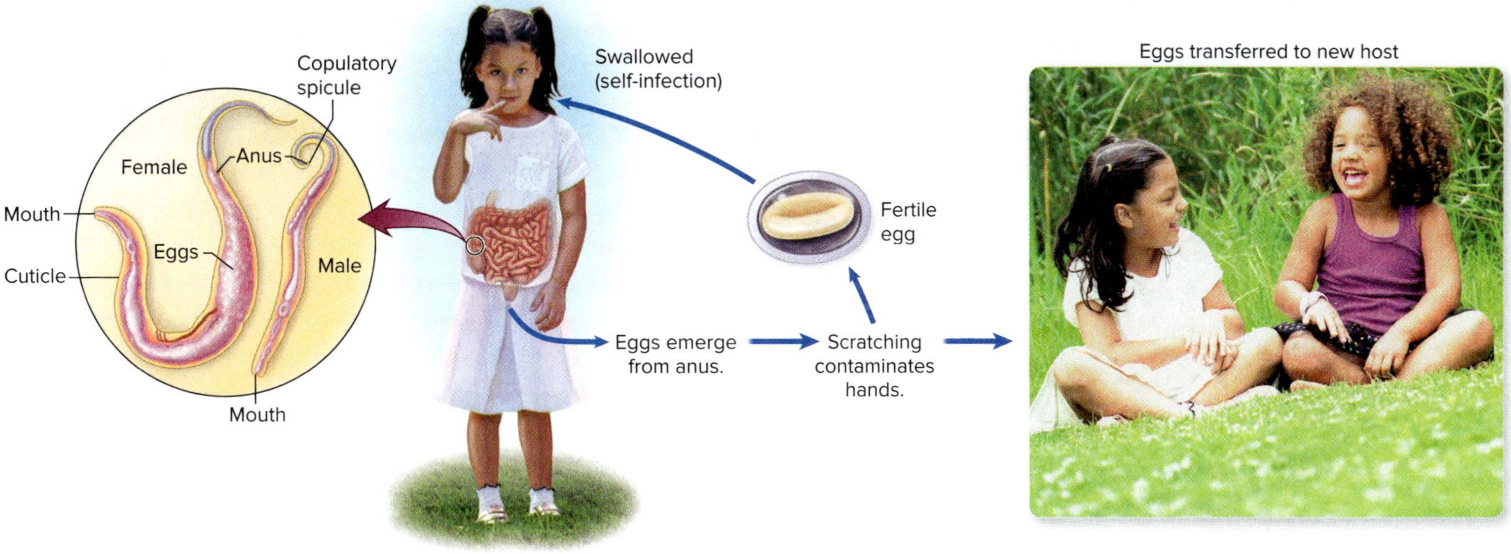
Copulatory spicule
Female
Anus
Mouth
Eggs
Cuticle
Male
Mouth
Swallowed (self-infection)
Fertile egg
Eggs emerge from anus.
Scratching contaminates hands.
Eggs transferred to new host

Figure 5.25 The life cycle of the pinworm, a roundworm. Eggs are the infective stage and are transmitted by unclean hands. Children frequently reinfect themselves and pass the parasite on to others.
Olga Enger/Shutterstock

with more primitive digestive, excretory, nervous, and muscular systems. In particular groups, such as the cestodes, reproduction is so dominant that the worms are reduced to little more than a series of flattened sacs filled with ovaries, testes, and eggs (see figure 5.24*a*). Not all worms have such extreme adaptations as cestodes, but most have a highly developed reproductive potential, thick cuticles for protection, and mouth glands for breaking down the host's tissue.

Life Cycles and Reproduction

The complete life cycle of helminths includes the fertilized egg (embryo), larval, and adult stages. In the majority of helminths, adults reproduce sexually inside a host's body. In nematodes, the sexes are separate and usually different in appearance. In trematodes, the sexes can be either separate or **hermaphroditic,** meaning that male and female sex organs are in the same worm. Cestodes are generally hermaphroditic. For a parasite's continued survival as a species, it must complete the life cycle by transmitting an infective form, usually an egg or larva, to the body of another host, either of the same or a different species. The host in which larval development occurs is the intermediate (secondary) host, and adulthood and mating occur in what is called the **definitive (final) host.** A transport host is an intermediate host that experiences no parasitic development but is an essential link in the completion of the cycle.

Humans become infected through ingestion or through penetration of tissues by the worm. The sources of the infective stage may be contaminated food, soil, or water, or other infected animals. Routes of infection are by oral intake or penetration of unbroken skin. Humans are the definitive hosts for many of the parasites listed in **table 5.5,** and in about half the diseases, they are also the sole biological reservoir. In the other cases, animals or insect vectors serve as reservoirs or are required to complete worm development. In the majority of helminth infections, the worms must leave their host to complete the entire life cycle.

Fertilized eggs are usually released to the environment and are provided with a protective shell and extra food to aid their development into larvae. Even so, most eggs and larvae are vulnerable to heat, cold, drying, and predators and are destroyed or unable to reach a new host. To counteract this formidable mortality rate, certain worms have adapted a reproductive capacity that borders on the incredible: A single female *Ascaris* worm can lay 200,000 eggs a day, and a large female can contain over 25 million eggs at varying stages of development! If only a tiny number of these eggs make it to another host, the parasite will have been successful in completing its life cycle.

A Helminth Cycle: The Pinworm

To illustrate a helminth cycle in humans, we use the example of a roundworm, *Enterobius vermicularis,* the pinworm. This worm causes a very common infestation of the large intestine. Worms range from 2 to 12 mm long and have a tapered, curved cylinder shape (see figure 5.25). The condition they cause, enterobiasis, is usually a simple, uncomplicated infection that does not spread beyond the intestine.

A cycle starts when a person swallows microscopic eggs picked up from another infected person by direct contact or by touching articles that person has touched. The eggs hatch in the intestine and then release larvae that mature into adult worms within about 1 month. Male and female worms mate, and the female migrates out to the anus to deposit eggs, which cause intense itchiness that is relieved by scratching. This contaminates the fingers, which, in turn, transfer eggs to bed sheets other inanimate objects. This person becomes a host and a source of eggs and can spread them to others in addition to reinfesting

Table 5.5 Examples of Helminths and How They Are Transmitted

Classification	Common Name of Disease or Worm	Host Requirement	Spread to Humans by:
Roundworms			
Nematodes			
Intestinal Nematodes			
Infective in egg (embryo) stage			
Ascaris lumbricoides	Ascariasis	Humans	Ingestion
Enterobius vermicularis	Pinworm	Humans	Fecal pollution of soil with eggs; ingestion of eggs
Infective in larval stage			
Trichinella spiralis	Trichina worm	Pigs, wild mammals	Consumption of meat containing larvae Burrowing of larva into tissue
Tissue Nematodes			
Onchocerca volvulus	River blindness	Humans, black flies	Fly bite
Dracunculus medinensis	Guinea worm	Humans and *Cyclops* (an aquatic invertebrate)	Ingestion of water containing *Cyclops*
Flatworms			
Trematodes			
Schistosoma japonicum	Blood fluke	Humans and snails	Ingestion of freshwater containing larval stage
Cestodes			
Taenia solium	Pork tapeworm	Humans, swine	Consumption of undercooked or raw pork
Diphyllobothrium latum	Fish tapeworm	Humans, fish	Consumption of undercooked or raw fish

himself or herself. Enterobiasis occurs most often among families and in other close living situations. Its distribution is worldwide among all socioeconomic groups, but it seems to attack children more frequently than adults.

Helminth Classification and Identification

The helminths are classified according to their shape; their size; the degree of development of various organs; the presence of hooks, suckers, or other special structures; the mode of reproduction; the kinds of hosts; and the appearance of eggs and larvae. They are identified in the laboratory by microscopic detection of the adult worm or its larvae and eggs, which often have distinctive shapes or external and internal structures. Occasionally, they are cultured in order to verify all of the life stages.

Distribution and Importance of Parasitic Worms

About 50 species of helminths parasitize humans. They are distributed in all areas of the world that support human life. Some worms are restricted to a given geographic region, and many have a higher incidence in tropical areas. Having said that, air travel, and human migration, is gradually changing the patterns of worm infections, especially of those species that do not require alternate hosts or special climatic conditions for development. The yearly estimate of worldwide cases numbers in the billions, and these are not confined to developing countries. A conservative estimate places 50 million helminth infections in North America alone.

It is important to realize that humans evolved on this planet in the constant presence of helminths—and indeed *all* types of microbes. Only very recently, in evolutionary terms, have some pockets of humankind been relatively free of helminth colonization. The absence of worm infections may in fact be leading to some of the "newer" conditions we encounter, such as autoimmunity and allergy. The reasoning here is that the part of the immune system that has evolved over millions of years to combat helminths is rather suddenly left with no job to do (in many human populations). Thus, it responds excessively or inappropriately to self antigens.

You have now learned about the variety of organisms that microbiologists study and classify. This chapter contained a very short description of the extremely complex variety of eukaryotic organisms. **Table 5.6** will help you differentiate among these and compare them to a familiar eukaryotic organism, the human. In chapter 7, you will continue the exploration of potential pathogens as you investigate the "not-quite-organisms," namely, viruses.

5.6 Learning Outcomes—Assess Your Progress

22. List the two major groups of helminths and provide examples representing each body type.
23. Summarize the stages of a typical helminth life cycle.

Table 5.6 Variation Among Eukaryotes

	Protozoa	Fungi	Algae	Helminths	Humans
Level of complexity	Always unicellular	Unicellular/ Multicellular	Unicellular/ Multicellular	Multicellular (adults) Unicellular (ova, larvae)	Multicellular
Cell wall	None	Chitin or cellulose	Cellulose	None	None
Cytoplasm	Divided (endoplasm/ ectoplasm)	Not divided	Not divided	Not divided	Not divided
Nutritional type	Heterotrophic/ Autotrophic	Heterotrophic	Heterotrophic/ Autotrophic	Heterotrophic	Heterotrophic
Motility	Flagella, cilia, pseudopodia, or none	Flagella (gametes)	Flagella (gametes)	Flagella (gametes)	Limbs
Important structures for identification	Cysts	Hyphae/Spores	Chloroplasts	Ova	None

Media Under The Microscope Wrap-Up

Ingram Publishing/ Superstock

The **intended message** of the article is twofold: It presents information about a small but dangerous threat associated with freshwater swimming, and it also debunks some of the misleading information other news articles have published.

A **critical reading** of the *Mic* article: The article immediately instills confidence as it specifically addresses how the topic has been sensationalized by other news outlets. The author also dispensed facts about how it is actually contracted and how rare it is.

To **interpret** the article for your friends, you should take an even-handed approach, explaining dangerous versus safe behaviors. This chapter educates you a bit about protozoa and amoebas, so that information can fill out your explanation to your friends.

My **overall grade** for this article is an A. It conveys an effective warning and also explains the possible misunderstandings that could have been caused by less responsible journalism.

Source: Mic (September 30, 2020), "Brain-Eating Amoebas Are Horrifying But Rare. How Scared Should I Be?," https://www.mic.com/p/brain-eating-amoebas-are-horrifying-but-rare-how-scared-should-i-be-35335509.

Study Smarter: Better Together

These activities are designed for you to use on your own with a study group—either a face-to-face group or a virtual one, consisting of 3–5 members. Studying together can be very helpful, but there are effective and ineffective ways to do it. For example, getting together without a clear structure is often not a good use of your time. Use your time efficiently by using one or more of the exercises below.

FACE-TO-FACE GROUPS

Use one or more of the activities below.

Peer Instruction: Assign numbers to your group members to use all semester long. Now look at these five concepts from this chapter. Each group member prepares a 5-minute lesson on the topic corresponding to their number. Don't worry if you have fewer than five members; just use however many you have! During your group study time, each member presents their lesson, and the group spends another 5–10 minutes discussing that lesson.

1. Last common ancestor
2. Theory of endosymbiosis
3. Protozoan locomotion
4. Complete description of one helminth's life cycle
5. Fungal asexual reproduction

(continued)

Concept Maps: Each member of the group should use this list of terms from this chapter to generate their own concept map. This can be hand-drawn or created using software (see Appendix C for guidelines). During group study time, compare each other's concept maps and help each other make sure they are correct. Of course, there are many different "correct" maps. Examining each member's map will help you talk through the varied concepts and how they are related.

Concept Terms:

Golgi apparatus	endoplasmic reticulum	cytoplasm	nucleolus
ribosomes	proteins	chloroplasts	mitochondria

Table Topics: Each group member should identify a concept or topic from this week's class assignments with which they are having trouble and share it during group study time. The other group members can then help to clarify confusing issues or share how they figured it out. Aim for a maximum of 15 minutes per topic. If the topic remains unclear to the group, bring it up during class or use the instructor's office hours or e-mail to ask for help. Taking the time to struggle with a difficult concept first makes your questions much more specific and more likely to yield helpful answers.

VIRTUAL GROUPS

Not everyone has the time or opportunity to meet with group members outside of class time. You or your instructor can create a virtual group using e-mail or the course software.

Weekly Discussion Board: This forum can be used as a way for groups to discuss topics, via e-mail or other learning management systems or online platforms, before they are covered in class. As each member of the group answers the current week's question, they should send their responses to every other member of their group. It's best to agree on a deadline based on how your class schedule works (Saturday for the next week's topics, for example). Then, after the topic is discussed in class, each member should send a response that all group members will see with a follow-up post on the same topic. If you cover more than one chapter in a week, someone can be designated to choose which chapter Discussion Board question you will use. Or simply decide up front that you will always use the first-chapter-of-the week's question, to keep the schedule simple.

Discussion Question
Why do you think the incidence of opportunistic fungal infections of humans has increased dramatically over the past 30 years?

Chapter Summary

EUKARYOTIC CELLS AND MICROORGANISMS

5.1 OVERVIEW OF THE EUKARYOTES

- **Eukaryotes are one of the three domains of life. Humans are eukaryotes, and amoebas are eukaryotes.**
- **Several eukaryotic microbes colonize mammals (and humans).**
- **Three of the major colonizers are fungi, protozoa, and helminths.**

5.2 FORM AND FUNCTION OF THE EUKARYOTIC CELL: EXTERNAL AND BOUNDARY STRUCTURES

- **Microscopic eukaryotes use flagella or cilia to move themselves or their food.**
- **The outermost boundary of most eukaryotic cells is the glycocalyx.**
- **Some eukaryotic organisms have a cell wall.**
- **The cytoplasmic membrane of eukaryotes has the same function as that of bacteria, but has a different composition.**

5.3 FORM AND FUNCTION OF THE EUKARYOTIC CELL: INTERNAL STRUCTURES

- The genome of eukaryotes is in the nucleus. Ribosomes, the sites for protein synthesis, are manufactured in the nucleolus, which is in the nucleus.
- The endoplasmic reticulum (ER) is a network of passageways extending through the cell.
- The Golgi body is a packaging center.
- The mitochondria generate energy in the form of ATP.
- The cytoskeleton determines the shape of cells and produces movement outside and inside of the cell.

5.4 THE FUNGI

- Fungi are nonphotosynthetic species with cell walls. They are either saprobes or parasites and may be uni- or multicellular.
- Fungi are heterotrophic.
- Fungi have multiple reproductive strategies.
- Fungi are often identified on the basis of their microscopic appearance.
- The three categories of fungal infection are community-acquired infections by pathogens, hospital-acquired infection by pathogens or opportunists, and community-acquired opportunistic infections.

5.5 THE PROTISTS

- There are two major protist types: algae and protozoa.
- Protozoa are unicellular heterotrophs, categorized by how they move.

5.6 THE HELMINTHS

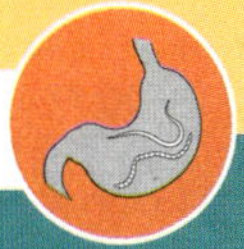

- These are animals. Their adult forms are visible to the naked eye, but their eggs and larvae are microscopic and are transmitted like other microorganisms.
- There are two major categories: the roundworms and the flatworms.

SmartGrid: From Knowledge to Critical Thinking

This *21 Question Grid* takes the topics from this chapter and arranges them with respect to the American Society for Microbiology's Undergraduate Curriculum guidelines—all six of the important "Concepts" as well as the important "Competency" of scientific literacy. Three questions are supplied, which cover chapter content referring to the Concept or Competency in increasing levels of Bloom's taxonomy for learning.

ASM Concept/ Competency	A. Bloom's Level 1, 2—Remember and Understand (Choose one.)	B. Bloom's Level 3, 4—Apply and Analyze	C. Bloom's Level 5, 6—Evaluate and Create
Evolution	**1.** Mitochondria likely originated from **a.** bacteria. **b.** invagination of the cell membrane. **c.** the LCA. **d.** chloroplasts.	**2.** Summarize the endosymbiotic theory and explain how it accounts for major structural similarities and differences between bacterial and eukaryotic cells.	**3.** The best theory of the evolution of the three domains of life (figure 5.1) envisions cellular organisms switching to a DNA vs. an RNA lifestyle very early. Create a scenario in which it would be advantageous for the last common ancestor to use RNA, while it would be advantageous for its descendants to use DNA.

(continued)

ASM Concept/ Competency	A. Bloom's Level 1, 2—Remember and Understand (Choose one.)	B. Bloom's Level 3, 4—Apply and Analyze	C. Bloom's Level 5, 6—Evaluate and Create
Cell Structure and Function	**4.** Yeasts are _____ fungi, and molds are _____ fungi. **a.** macroscopic; microscopic **b.** unicellular; filamentous **c.** motile; nonmotile **d.** water; terrestrial	**5.** Compare and contrast the structure and function of the following among bacteria and eukaryotes. **a.** ribosome **b.** flagellum **c.** glycocalyx	**6.** Using components from all three types of eukaryotic organisms in this chapter, design an ideal organism, and explain your choices.
Metabolic Pathways	**7.** The Golgi apparatus **a.** receives vesicles from the mitochondria. **b.** packages products into transitional vesicles. **c.** modifies proteins. **d.** synthesizes proteins and sterols.	**8.** Considering the role of fungi in nature, speculate on why they have such a wide array of metabolic capabilities.	**9.** Speculate on why fungi evolved to produce products such as alcohol, antibiotics, and vitamins.
Information Flow and Genetics	**10.** Fungi produce which structures for reproduction and multiplication? **a.** endospores **b.** cysts **c.** spores **d.** eggs	**11.** Write a paragraph illustrating the life of a protein, from its DNA origins to mature polypeptide, and the course of its travels within a cell throughout its synthesis.	**12.** Investigate whether there are other organisms, besides fungi, that have both sexual and asexual forms of reproduction and devise a hypothesis about why that might be advantageous.
Microbial Systems	**13.** Which of these organisms has the best potential to survive in extreme environments? **a.** fungus **b.** trophozoite form of protozoa **c.** helminth **d.** yeast	**14.** Why has the incidence of opportunistic fungal diseases in humans increased dramatically in the last 50 years?	**15.** Do you suppose any eukaryotic microbes populate biofilms in nature? Defend your answer.
Impact of Microorganisms	**16.** Which of these groups now causes more deaths on an annual basis globally than breast cancer? **a.** helminths **b.** protozoa **c.** fungi **d.** algae	**17.** Provide at least three examples illustrating both beneficial and detrimental aspects of fungi in the modern world today.	**18.** Do you suspect that the fact that humans use microbes such as fungi to manufacture massive amounts of metabolic products for our use will change the evolution of these organisms? Defend your answer.
Scientific Thinking	**19.** Which of the following would be most useful to determine whether a clinical isolate is a bacterium, fungus, or protozoan? **a.** its size and shape under a light microscope **b.** whether it has a cell wall **c.** whether it can form protective structures under stress **d.** all of the above are reliable	**20.** Why were protozoa originally considered a single group, and why is that no longer the case?	**21.** Write a paragraph that would explain the difference between heterotroph, saprobe, and parasite to a middle-school class.

Answers to the multiple-choice questions appear in Appendix A.

Visual Connections

These questions use visual images or previous content to connect content within and between chapters.

1. **From chapter 1, figure 1.15.** Which of the groups of organisms from this figure contain a nucleus?

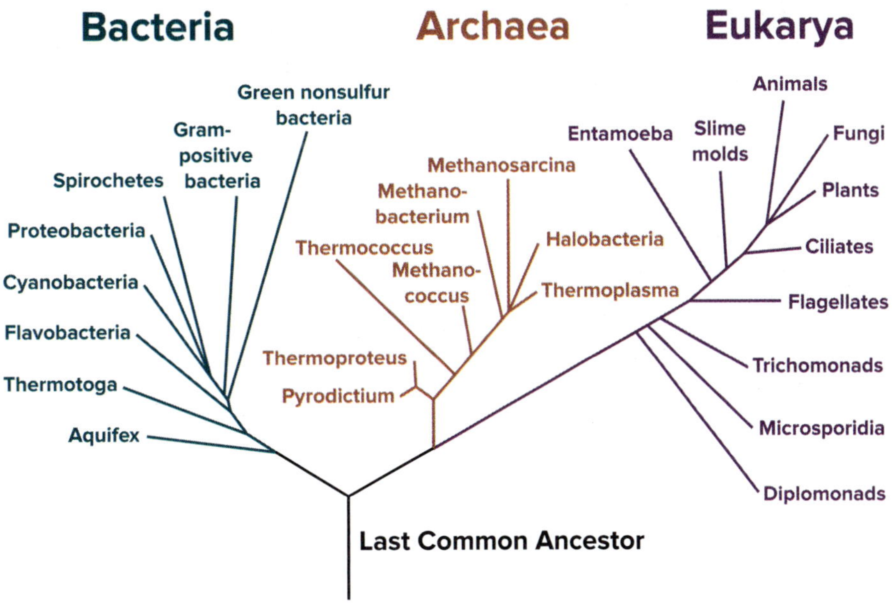
Bacteria
Archaea
Eukarya
Green nonsulfur bacteria
Gram-positive bacteria
Spirochetes
Proteobacteria
Cyanobacteria
Flavobacteria
Thermotoga
Aquifex
Methanosarcina
Methano-bacterium
Thermococcus
Methano-coccus
Halobacteria
Thermoplasma
Thermoproteus
Pyrodictium
Entamoeba
Slime molds
Animals
Fungi
Plants
Ciliates
Flagellates
Trichomonads
Microsporidia
Diplomonads
Last Common Ancestor

High Impact Study

These terms and concepts are most critical for your understanding of this chapter—and may be the most difficult. Have you mastered them?

Concepts

- ☐ Last common ancestor
- ☐ Theory of endosymbiosis
- ☐ The nucleus-RER-Golgi pathway
- ☐ Fungal reproduction
- ☐ Locomotion of protozoa
- ☐ Classification of helminths

Terms

- ☐ Nucleus
- ☐ Golgi apparatus
- ☐ Endoplasmic reticulum
- ☐ Mitochondrion
- ☐ Cytoskeleton
- ☐ Hyphae

Design Element: (College students): Caia Image/Image Source

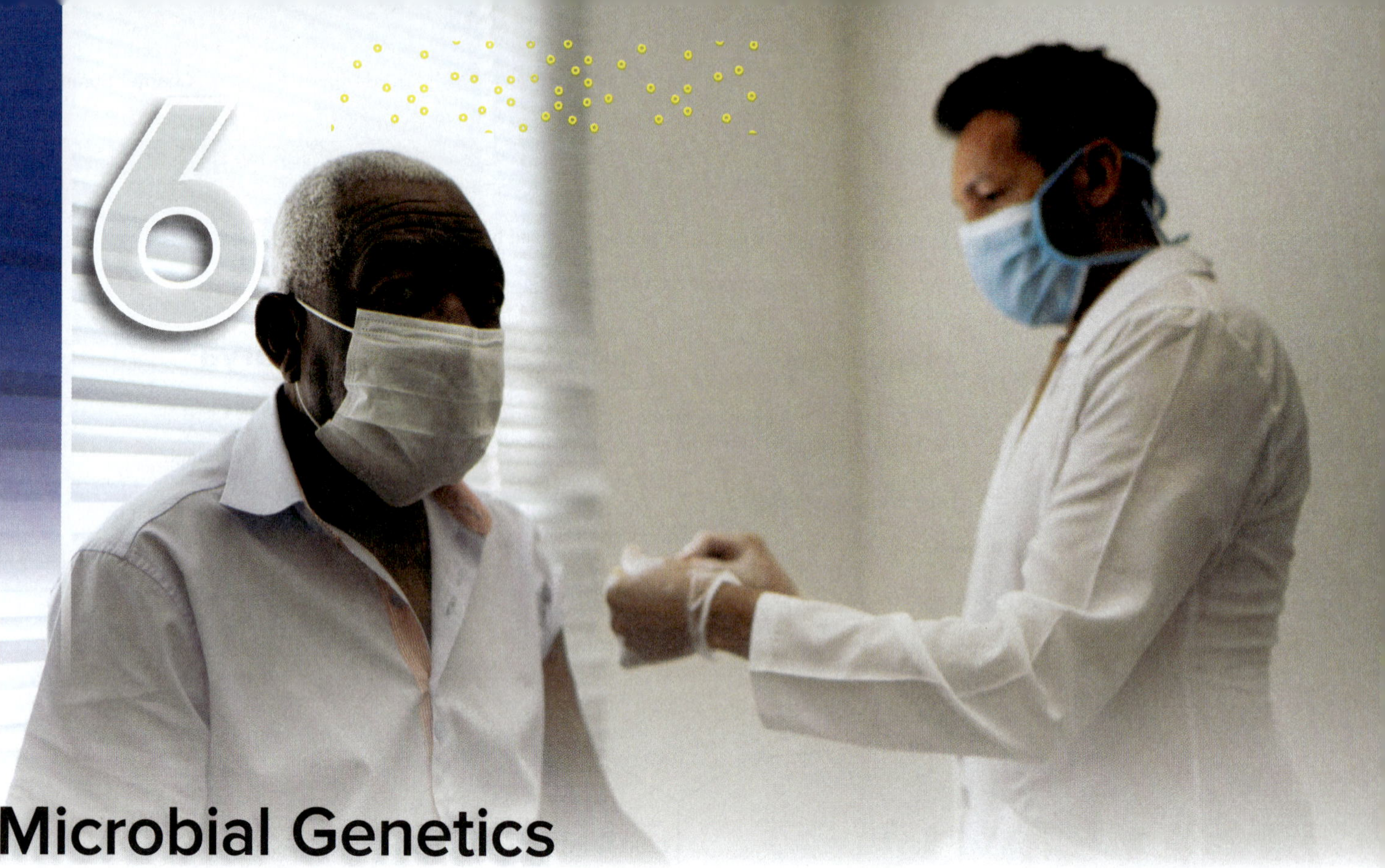

6 Microbial Genetics

FG Trade/Getty Images

Media Under The Microscope

COVID-19 Misinformation

This opening case examines an article from the popular media to determine the extent to which it is factual and/or misleading. This case focuses on the May 2020 article from Reuters, *"Fact Check: False Claim: A COVID-19 Vaccine Will Genetically Modify Humans."*

This article appearing in the online news magazine run by Reuters discussed a video that was circulating widely in mid-2020. The video claimed that the vaccine for COVID-19 changes your DNA permanently. According to the article, the video featured a self-proclaimed natural healing consultant named Andrew Kaufman who insists that the not-yet-released COVID-19 vaccines will "inject genes" and "genetically modify" humans. At the time, Mr. Kaufman was expecting the release of a DNA-based vaccine. The eventual vaccines ended up being made of RNA. Both RNA and DNA are nucleic acids that we will study in this chapter.

A scientist who watched the video commented, "That's just a myth, one often spread intentionally by anti-vaccination activists to deliberately generate confusion and mistrust." He continued,"Genetic modification would involve the deliberate insertion of foreign DNA into the nucleus of a human cell, and vaccines simply don't do that." A physician echoed this sentiment. The article also included a number of hyperlinks to fact-based websites that explained technical terms.

The article concluded with this statement: "Verdict: False. A future COVID-19 vaccine will not genetically modify humans. DNA vaccines do not integrate the DNA of the virus into the cell nucleus of its recipient."

As it turned out, the first vaccines to market were RNA based. Either way, there was no change made to host DNA.

- What is the **intended message** of the article?
- What is your **critical reading** of the summary of the article provided above? Remember that in this context, "critical reading" does not necessarily mean *What criticism do you have?* but asks you to apply your knowledge to interpret whether the article is factual and whether the facts support the intended message.
- How would you **interpret** the news item for your nonmicrobiologist friends?
- What is your **overall grade** for the news item—taking into account its accuracy and the accuracy of its intended effect?

Media Under The Microscope Wrap-Up appears at the end of the chapter.

Outline and Learning Outcomes

6.1 Introduction to Genetics and Genes: Unlocking the Secrets of Heredity

1. Define the terms *genome* and *gene*.
2. Differentiate between genotype and phenotype.
3. Diagram a segment of DNA, labeling all important chemical groups within the molecule.
4. Summarize the steps of bacterial DNA replication and the enzymes used in this process.
5. Compare and contrast the synthesis of leading and lagging strands during DNA replication.

6.2 Transcription and Translation

6. Explain how the classical view of the "central dogma" of biology has been changed by recent science.
7. Identify important structural and functional differences between RNA and DNA.
8. Illustrate the steps of transcription, noting the key elements and the outcome of mRNA synthesis.
9. List the three types of RNA directly involved in translation.
10. Define the terms *codon* and *anticodon,* and list the four known start and stop codons.
11. Identify the locations of the promoter, the start codon, and the A and P sites during translation.
12. Indicate how eukaryotic transcription and translation differ from these processes in bacteria and archaea.
13. Explain the relationship between genomics and proteomics.

6.3 Genetic Regulation of Protein Synthesis

14. Define the term *operon* and explain one advantage it provides to a bacterial cell.
15. Differentiate between repressible and inducible operons and provide an example of each.
16. List two antibiotic drugs and their targets within the transcription and translation machinery.

6.4 DNA Recombination Events

17. Explain the defining characteristics of a recombinant organism.
18. Describe three forms of horizontal gene transfer used in bacteria.

6.5 Mutations: Changes in the Genetic Code

19. Define the term *mutation* and discuss one positive and one negative example of it in microorganisms.
20. Differentiate among frameshift, nonsense, silent, and missense mutations.

6.1 Introduction to Genetics and Genes: Unlocking the Secrets of Heredity

The study of **genetics** takes place on several levels **(figure 6.1).** *Organismal genetics* observes the **heredity** of the whole organism; *cellular genetics* looks at single cells; *chromosomal genetics* examines the characteristics and actions of chromosomes; and *molecular genetics* deals with the biochemistry of the genes. All of these levels are useful areas of exploration, but we are going to examine the operation of genes at the cellular and molecular levels. We will be focusing on the genetics of microbes, because they are the subject of this book.

The Nature of the Genetic Material

For a species to survive, it must be able to make more of itself. In single-celled microorganisms, reproduction usually involves the division of the cell by means of binary fission or budding. This requires accurate duplication and separation of genetic material into each daughter cell. Before we look at how this genetic material—DNA—is copied, we look at how it is organized, starting with the general and then moving to the specific.

The Levels of Structure and Function of the Genome

The **genome** is the sum total of genetic material of an organism. **Genomics** is the study of an organism's entire genome. Although most of the genome exists in the form of chromosomes, genetic material can appear in other forms as well **(figure 6.2).** For example, bacteria and some fungi contain tiny extra pieces of DNA (plasmids), and certain organelles of eukaryotes (the mitochondria and chloroplasts) are equipped with their own DNA. The genomes of all cells are composed exclusively of DNA, but viruses contain either DNA or RNA as the principal genetic material.

In general, a **chromosome** is a distinct cellular structure composed of a neatly packaged DNA molecule. The chromosomes of eukaryotes and bacterial cells differ in several respects. The structure of eukaryotic chromosomes consists of DNA tightly wound around histone proteins, whereas a bacterial chromosome is condensed and jammed into a condensed form by means of histone*like* proteins. Eukaryotic chromosomes are located in the nucleus. They vary in number from a few to hundreds. They can occur in pairs (diploid) or singles (haploid) and they have a linear appearance. In contrast, most bacteria have a single, circular (double-stranded) chromosome, although many bacteria have multiple, circular chromosomes and some have linear chromosomes.

The chromosomes of all cells are subdivided into basic informational packets called genes. A **gene** is a length of DNA on the chromosome that provides information for a particular cell function. To be specific, a gene is a segment of DNA that contains the necessary code to make either a **protein** or an RNA molecule.

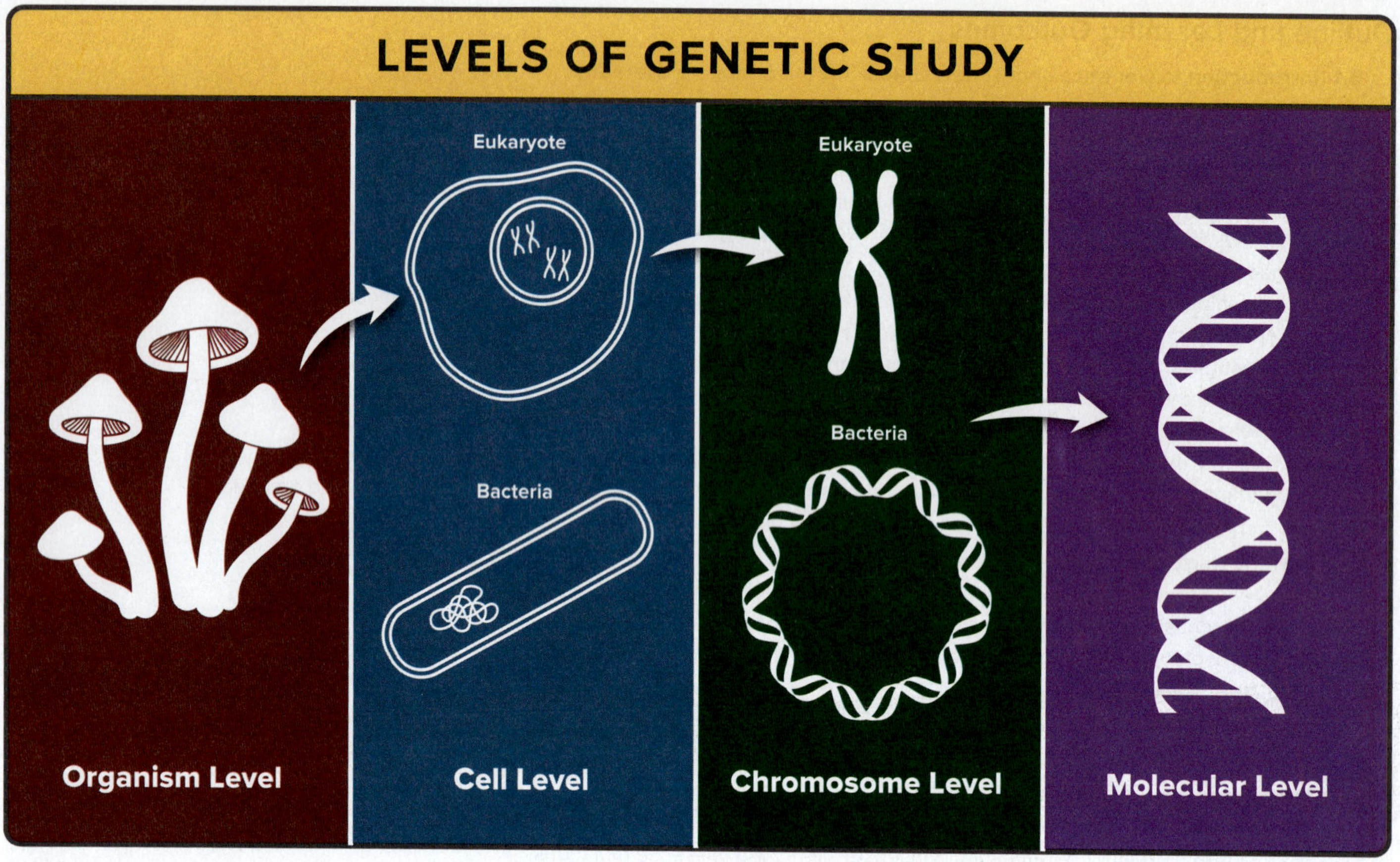
LEVELS OF GENETIC STUDY
Eukaryote
Bacteria
Eukaryote
Bacteria
Organism Level
Cell Level
Chromosome Level
Molecular Level

Figure 6.1 Levels of genetic study. The operations of genetics can be observed at the levels of organism, cell, chromosome, and DNA sequence (molecular level).

Insight 6.1 describes one of the many applications of being able to map and "use" human genes.

Genes fall into three basic categories: (1) structural genes that code for proteins, (2) genes that code for the RNA machinery used in protein production, and (3) regulatory genes that control gene expression. The sum of all of these types of genes constitutes an organism's distinctive genetic makeup, or **genotype** (jee″-noh-type). At any given time, only a portion of the genes are expressed, creating traits (certain structures or functions) referred to as the **phenotype** (fee″-noh-type). All organisms contain more genes in their genotypes than are manifested as a phenotype at any given time. In other words, the phenotype can change depending on which genes are "turned on" (expressed).

The Size and Packaging of Genomes

Genomes vary greatly in size. The smallest viruses have four or five genes; the bacterium *Escherichia coli* has a single chromosome containing about 4,000 genes; and a human cell has about 23,000 genes on 46 chromosomes. The chromosome of *E. coli* would measure about 1 mm if unwound and stretched out linearly, and yet this fits within a cell that measures just over 1 micron across, making the stretched-out DNA 1,000 times longer than the cell. Still, the bacterial chromosome takes up only about one-third to one-half of the cell's volume. How can such large molecules fit into the minuscule volume of a cell and, in the case of eukaryotes, into an even smaller compartment, the nucleus? The answer lies in the intricate coiling of DNA.

The DNA Code: A Simple Yet Profound Message

In 1953, James Watson and Francis Crick used an X-ray image produced by Rosalind Franklin to finally "picture" DNA. They discovered that DNA is a very long molecule, a type of nucleic acid, with two strands combined into a double helix **(figure 6.3).** The basic unit of DNA structure is a **nucleotide,** and a chromosome in a typical bacterium consists of several million nucleotides linked end to end. Each nucleotide is composed of **phosphate, deoxyribose** sugar, and a **nitrogenous base.** The nucleotides covalently bond to each other using a sugar–phosphate linkage that becomes the backbone of each strand. Each sugar attaches to two phosphates. One of the bonds is to the number 5′ (read "five prime") carbon on deoxyribose, and the other is to the 3′ carbon, which confers a certain order and direction on each strand **(figure 6.4).** (In cyclical carbon molecules, such as sugars, the carbons are numbered so we can keep track of them. Deoxyribose has 5 carbons numbered 1 to 5.)

The nitrogenous bases, **purines** and **pyrimidines,** attach by covalent bonds to the 1′ position of the sugar **(figure 6.4*a*).** They cross the center of the molecule and pair with appropriate complementary bases from the other strand. The paired bases are joined

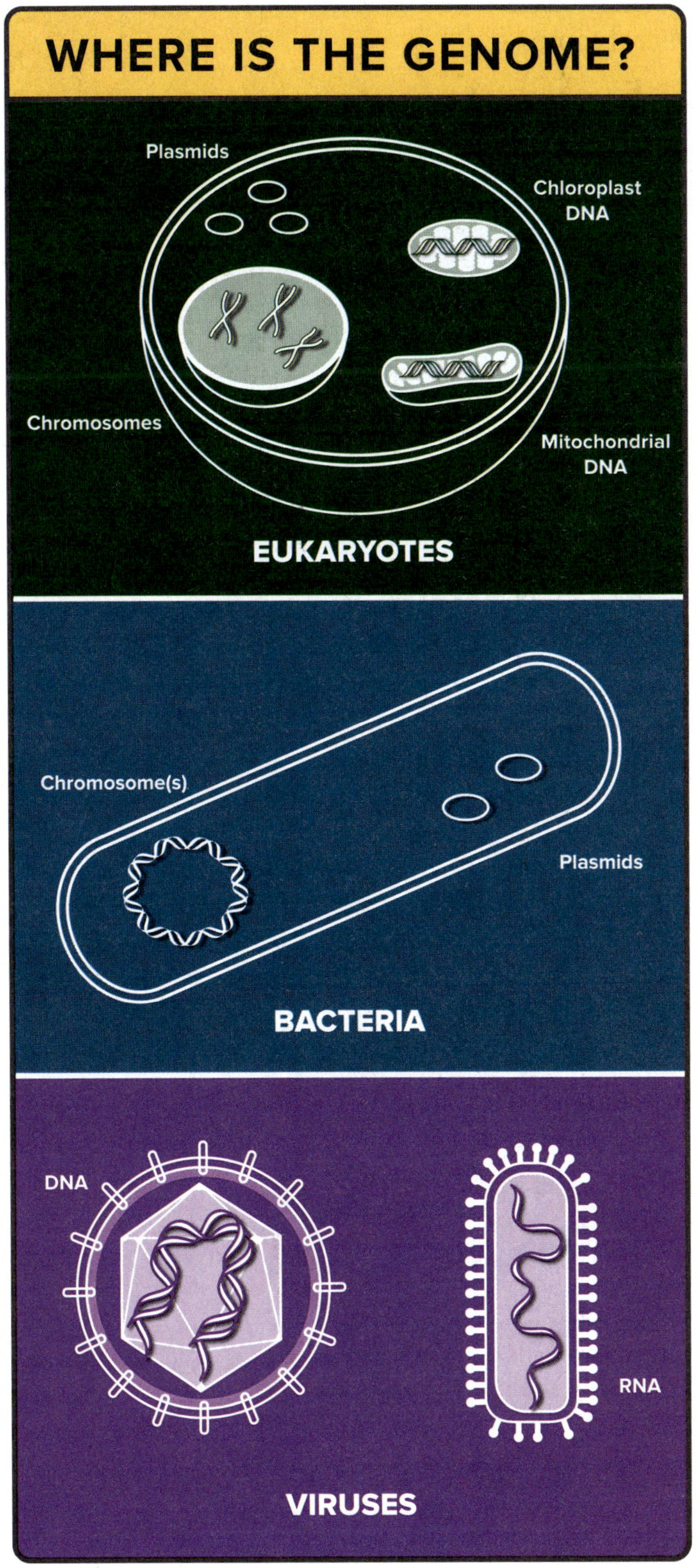
WHERE IS THE GENOME?
Plasmids
Chloroplast DNA
Chromosomes
Mitochondrial DNA
EUKARYOTES
Chromosome(s)
Plasmids
BACTERIA
DNA
RNA
VIRUSES

Figure 6.2 The location of the genome in cells and viruses (not to scale).

by hydrogen bonds. These are weak bonds that are easily broken and then reformed, allowing the molecule to be "unzipped" into its two complementary strands. This feature is very important to the processes that DNA engages in. The pairing of purines and pyrimidines is not random; it is dictated by the matching up of the same pairs of bases over and over again. In DNA, the purine **adenine (A)** always pairs with the pyrimidine **thymine (T),** and the purine **guanine (G)** always pairs with the pyrimidine **cytosine (C).** The bases are attracted to each other in this pattern because each has a complementary three-dimensional shape that matches up with its partner. Although the base-pairing partners generally do not vary, the sequence of base pairs along each DNA molecule take on an almost infinite variety of orders. (And, of course, the other strand will be complementary to the first.)

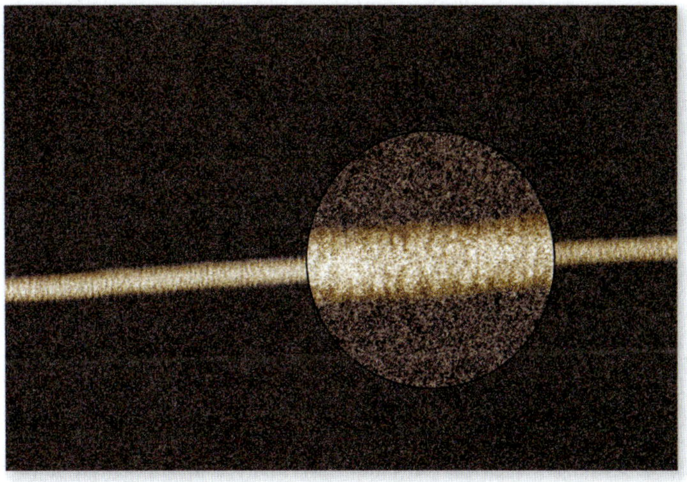

Figure 6.3 Transmission electron micrograph of DNA. The helical structure can be seen in the magnified area.
Professor Enzo Di Fabrizio, IIT/Science Source

Another important characteristic of DNA structure is the nature of the double helix itself. The two strands are not oriented in the same direction. One strand of the helix runs in the opposite direction from the other, in what is called an **antiparallel** arrangement **(figure 6.4*b*).** The position of the bond between the carbon on deoxyribose and the phosphates is used to keep track of the direction of the two sides of the helix. This means that one strand runs from the 5′ to 3′ direction, and the other runs from the 3′ to 5′ direction. This characteristic is a significant factor in DNA synthesis and protein production.

The Significance of DNA Structure

The way the nitrogenous bases line up, and base-pair, in DNA results in:

1. **Maintaining the code during reproduction.** The reliability of base pairing guarantees that the code will be retained during cell growth and division. When the two strands are separated, each one provides a template (pattern or model) for the replication (exact copying) of a new molecule **(figure 6.5).** Because the sequence of one strand automatically gives the sequence of its partner, the code can be duplicated with accuracy.
2. **Providing variety.** The order of bases along the length of the DNA strand constitutes the genetic program, or the language, of the DNA code. The message present in a gene is determined by the precise sequence of these bases, and the genome is the collection of all DNA bases that, in an ordered combination, are responsible for the unique qualities of each organism.

INSIGHT 6.1 MICROBIOME: Customizing the Microbiome

By now you know that the gut microbiome plays a major part in human health. A company in the United Kingdom called Microbiotica is banking on the fact that it can more fully characterize what individual gut microbiomes consist of, as well as discovering the type of microbiomes that lead to the most effective drug treatment for significant diseases. This type of work falls under the category of *personalized medicine,* which optimizes medical treatment through the use of an individual patient's genetic makeup, as well as the kinds of bacteria found in their gut microbiome. The company characterizes the microbiome by state-of-the-art culture techniques and gene sequencing. It is combining this information to create a massive new database that can help guide individual treatment decisions for conditions such as inflammatory bowel disease, ulcerative colitis, and immunotherapies for cancer.

The company is a spin-off of the Wellcome Trust Sanger Insititute, a British charitable institution that "uses genome sequences to advance understanding of the biology of humans and pathogens in order to improve human health." Microbiotica is using the decades of Sanger Institute innovations and gene data to zoom in on the human microbiome to develop better treatments for specific conditions.

WELLCOME
GENOME
CAMPUS

Robert Evans/Alamy Stock Photo

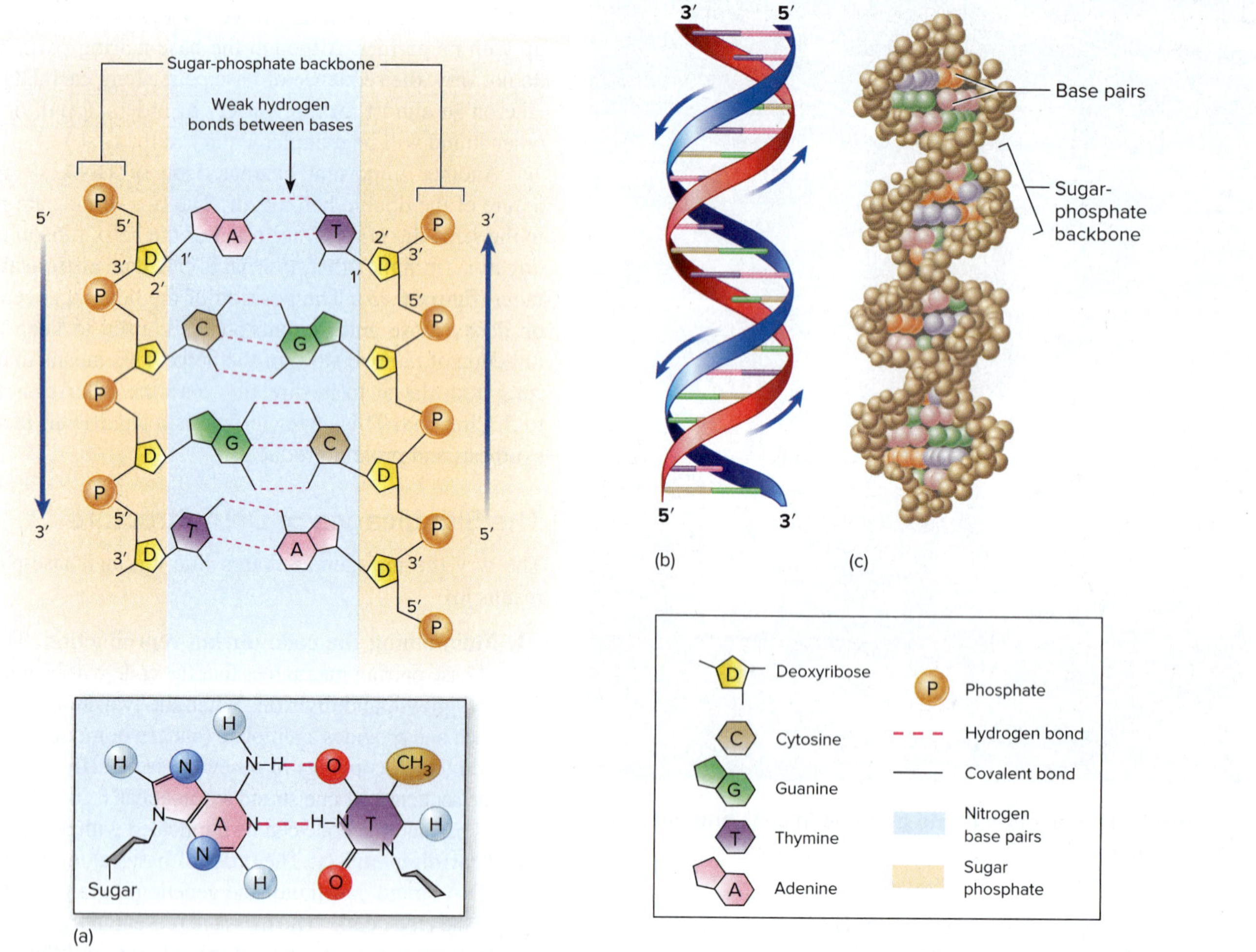
Sugar-phosphate backbone
Weak hydrogen bonds between bases
5′
3′
P
D
A
T
C
G
1′
2′
H
N
N-H
O
CH3
A
N
H-N
T
H
Sugar
O
(a)
3′
5′
5′
3′
(b)
Base pairs
Sugar-phosphate backbone
(c)
Deoxyribose
Phosphate
Cytosine
Hydrogen bond
Covalent bond
Guanine
Nitrogen base pairs
Thymine
Sugar phosphate
Adenine

Figure 6.4 Three views of DNA structure. (a) A schematic drawing to show the arrangement of the two strands the helix is made of. The 5′ and 3′ carbon positions are highlighted in green and blue, to illustrate the directionality of the strands. The inset shows details of a nitrogen base pair. **(b)** Simplified model that highlights the antiparallel arrangement. **(c)** Space-filling model that more accurately depicts the three-dimensional structure of DNA.

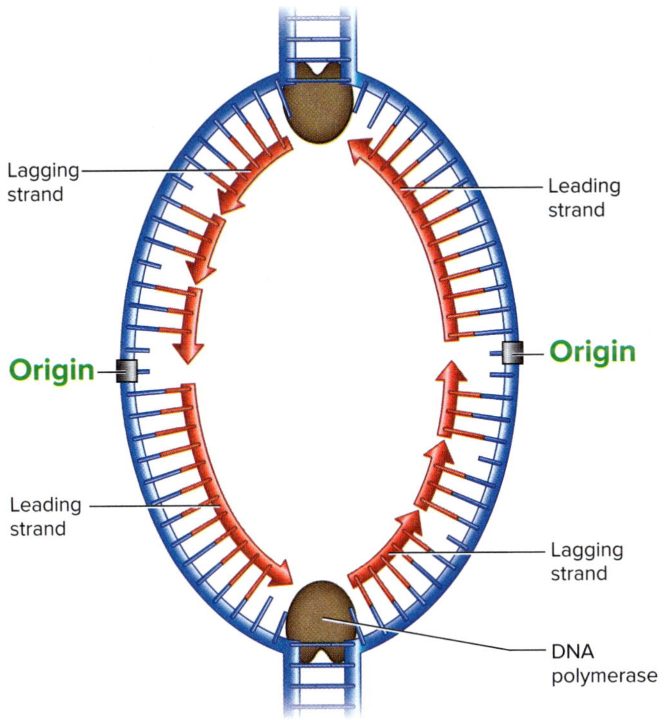
Lagging strand
Leading strand
Origin
Origin
Leading strand
Lagging strand
DNA polymerase

Figure 6.5 Simplified steps to show the semiconservative replication of DNA. Circular DNA has a special origin site where replication begins. When strands are separated, two replication forks form, and a DNA polymerase III complex enters at each fork. The DNA polymerases proceed in both directions along the DNA molecule, attaching the correct nucleotides according to the pattern of the template. An A on the template will pair with a T on the new molecule, and a C will pair with a G. The resultant new DNA molecules contain one strand of the newly synthesized DNA (shown in red) and the original template strand (shown in blue). The integrity of the code is kept intact because the linear arrangement of the bases is maintained during this process. Note that more details of the process are presented in figure 6.6.

It is tempting to ask how such a seemingly simple code can account for the extreme differences among forms as diverse as a virus, *E. coli,* and a human. The English language, based on 26 letters, can create an infinite variety of words, but how can an apparently complex genetic language such as DNA be based on just four nitrogen base "letters"? A mathematical example can explain the possibilities. For a segment of DNA that is 1,000 nucleotides long, there are $4^{1,000}$ different sequences possible. Carried out, this number would be close to 1.5×10^{602}, a number so huge that it provides nearly endless degrees of variation.

DNA Replication: Preserving the Code and Passing It On

The sequence of bases along the length of a gene is the language of DNA. For this language to be preserved for hundreds of generations, it will be necessary for the genetic program to be duplicated and passed on to each offspring. This process of duplication is called DNA replication. In the following example, we will show replication in bacteria; but, with some exceptions, it also applies to the process as it works in eukaryotes and some viruses. Early in binary fission, the metabolic machinery of a bacterium initiates the duplication of the chromosome. This DNA replication must be completed during a single generation time (around 20 minutes in *E. coli*).

The Overall Replication Process

What features allow the DNA molecule to be exactly duplicated, and how is its integrity maintained? DNA replication requires a careful orchestration of the actions of 30 different enzymes (partial list in **table 6.1**), which separate the strands of the existing DNA molecule, copy each strand, and produce two complete daughter molecules. A simplified version of replication is shown in **figure 6.6** and includes the following:

- uncoiling the parent DNA molecule;
- unzipping the hydrogen bonds between the base pairs, separating the two strands and exposing the nucleotide sequence of each strand (which is normally buried in the center of the helix) to serve as templates; and
- synthesizing two new strands by attaching the correct complementary nucleotides to each single-stranded template.

A critical feature of DNA replication is that each daughter molecule will be identical to the parent in composition, but neither one is completely new. The strand that serves as a template becomes one of the strands in the new helix. Because one parent strand is retained in every new double-stranded molecule, the process is called **semiconservative replication.** (In this instance, "semi-" means "half," as in *semicircle.*) This feature of DNA replication is central to the preservation of each organism's genetic code generation after generation.

Refinements and Details of Replication

The process of synthesizing a new daughter strand of DNA using the parental strand as a template is carried out by an enzyme called DNA polymerase III. (We will learn more about how enzymes work in chapter 10.) The entire process of replication depends on several enzymes and can be most easily understood by keeping in

Table 6.1 Some Enzymes Involved in DNA Replication and Their Functions

Enzyme	Function
Helicase	Unzipping the DNA helix
Primase	Synthesizing an RNA primer
DNA polymerase III	Adding bases to the new DNA chain; proofreading the chain for mistakes
DNA polymerase I	Removing primer, closing gaps, repairing mismatches
Ligase	Final binding of nicks in DNA during synthesis and repair
Topoisomerase I	Making single-stranded DNA breaks to relieve supercoiling at origin
Topoisomerase II (DNA gyrase) and IV	Making double-stranded DNA breaks to remove supercoiling ahead of origin and separate replicated daughter DNA molecules

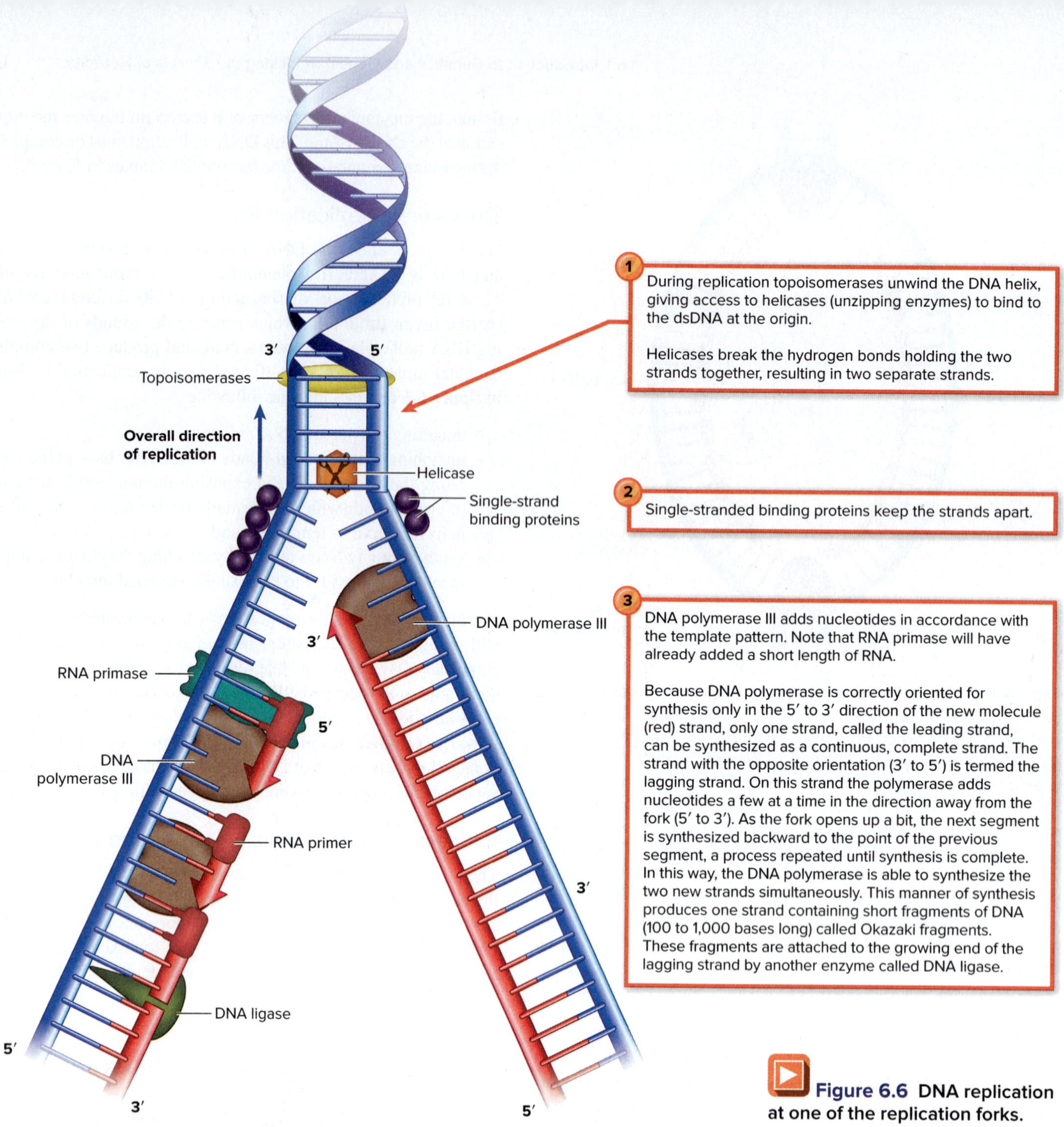
3′
5′
Topoisomerases
Overall direction of replication
Helicase
Single-strand binding proteins
DNA polymerase III
3′
RNA primase
5′
DNA polymerase III
RNA primer
DNA ligase
3′
5′
3′
5′
1 During replication topoisomerases unwind the DNA helix, giving access to helicases (unzipping enzymes) to bind to the dsDNA at the origin.
Helicases break the hydrogen bonds holding the two strands together, resulting in two separate strands.
2 Single-stranded binding proteins keep the strands apart.
3 DNA polymerase III adds nucleotides in accordance with the template pattern. Note that RNA primase will have already added a short length of RNA.
Because DNA polymerase is correctly oriented for synthesis only in the 5′ to 3′ direction of the new molecule (red) strand, only one strand, called the leading strand, can be synthesized as a continuous, complete strand. The strand with the opposite orientation (3′ to 5′) is termed the lagging strand. On this strand the polymerase adds nucleotides a few at a time in the direction away from the fork (5′ to 3′). As the fork opens up a bit, the next segment is synthesized backward to the point of the previous segment, a process repeated until synthesis is complete. In this way, the DNA polymerase is able to synthesize the two new strands simultaneously. This manner of synthesis produces one strand containing short fragments of DNA (100 to 1,000 bases long) called Okazaki fragments. These fragments are attached to the growing end of the lagging strand by another enzyme called DNA ligase.

Figure 6.6 **DNA replication at one of the replication forks.**

mind a few points concerning both the structure of the DNA molecule and the limitations of DNA polymerase III.

1. The nucleotides that need to be copied by DNA polymerase III are buried deep within the double helix. In order for the enzyme to access the DNA molecule, it must first be unwound, and then the two strands of the helix must be separated from one another.
2. DNA polymerase III is unable to *begin* synthesizing a chain of nucleotides but can only *continue* to add nucleotides to an already-existing chain.
3. DNA polymerase III can only add nucleotides in one direction, so any new strand is always synthesized in a 5′ to 3′ direction.

Figure 6.6 illustrates the details of replication. In addition to the enzymes listed in table 6.1, there are other important terms that will help you understand replication:

replication fork The place in the helix where the strands are unwound and replication is taking place. Each circular DNA molecule will have two replication forks (only one is shown in figure 6.6).

primer A length of RNA that is inserted initially during replication before it is replaced by DNA. RNA is similar to DNA but usually exists as a single strand, not a double strand.

leading strand The strand of new DNA that is synthesized in a continuous manner in the 5′ to 3′ direction.

lagging strand The strand of new DNA that must be synthesized in short segments (in a 5′ to 3′ direction) and later sealed together to form a strand in the 3′ to 5′ direction.

Okazaki fragments The short segments of DNA synthesized in a 5′ to 3′ direction, which are then sealed together to form a 3′ to 5′ strand.

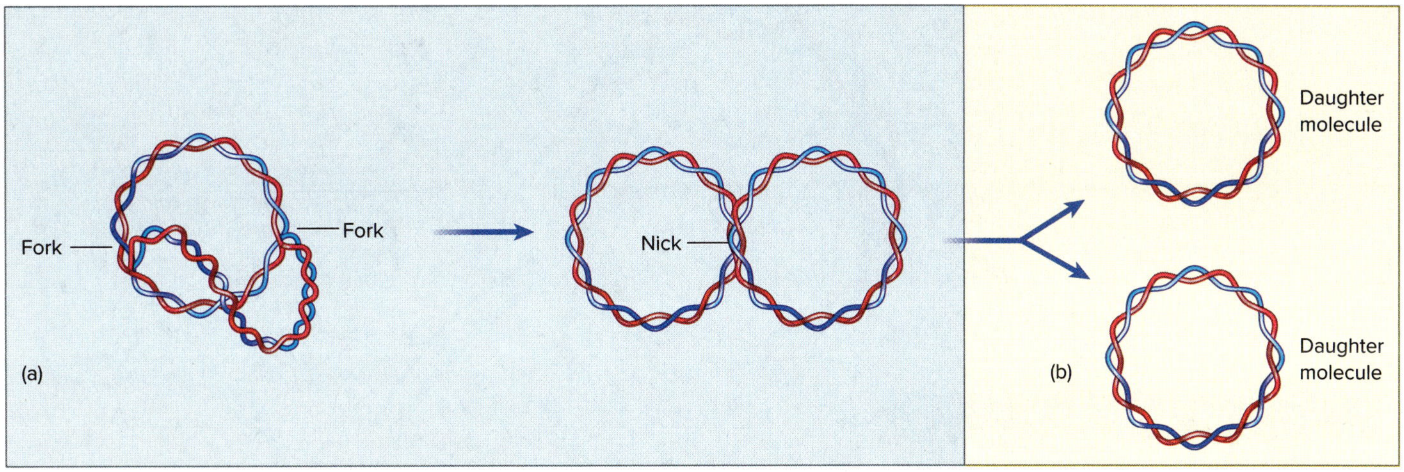
Fork
Fork
Nick
(a)
Daughter molecule
Daughter molecule
(b)

Figure 6.7 Completion of chromosome replication in bacteria. **(a)** As replication proceeds, one double strand loops away. **(b)** Final separation is achieved through action of topoisomerase IV and the final release of two completed molecules. Each of the daughter cells receives one of these during binary fission.

Elongation and Termination of the Daughter Molecules The addition of nucleotides proceeds at an astonishing pace, estimated in some bacteria to be 750 bases per second at each fork. As replication proceeds, the newly produced double strand loops away **(figure 6.7*a*).** The enzyme DNA polymerase I removes the RNA primers used to initiate DNA synthesis and replaces them with DNA. When the forks come full circle and meet, ligases move along the lagging strand to begin the initial linking of the fragments. Topoisomerase IV then causes a double-stranded DNA break that allows for the completion of synthesis and the separation of the intertwined circles into two fully replicated daughter molecules **(figure 6.7*b*).**

As in any "writing," DNA is occasionally "misspelled" when an incorrect base is added to the growing chain. Studies have shown that in bacteria such mistakes are made once in approximately 10^8 to 10^9 bases, but most of these are corrected. If not corrected, they are referred to as *mutations* (covered later in this chapter). Think of it this way: An overnight test tube culture of *E. coli* can be expected to have around 10^9 to 10^{10} bacteria per ml in it. That means that a random mutation in every single base in the *E. coli* genome will be represented in at least one cell in a millilieter of the culture. Because continued genetic integrity is very dependent on accurate replication, cells have evolved their own proofreading function for DNA. DNA polymerase III, the enzyme that elongates the molecule, can detect incorrect, unmatching bases, excise them, and replace them with the correct base. DNA polymerase I can also proofread the molecule and repair damaged DNA.

Replication of Linear DNA The replication of eukaryotic DNA is similar to that of bacteria and archaea, even though it exists in a linear form. This process also uses a variety of DNA polymerases, and replication proceeds in both directions but from multiple origins along the linear DNA molecule. Topoisomerases are used in replication to relieve the tension on the DNA as it is copied but also to recompact the DNA when the molecule is completely replicated. The synthesis of new DNA from a linear template presents a variety of challenges, however, when compared to the copying of a circular molecule of DNA. One of the most important of these is known as the "end replication problem." Due to the structure of eukaryotic DNA and the unidirectional action of DNA polymerase, the 3′ end of DNA molecules cannot be completely copied. These areas, called **telomeres,** begin to erode with each cell division. Once they shorten to a certain length, they will trigger cell death (called apoptosis).

6.1 Learning Outcomes—Assess Your Progress

1. Define the terms *genome* and *gene*.
2. Differentiate between genotype and phenotype.
3. Diagram a segment of DNA, labeling all important chemical groups within the molecule.
4. Summarize the steps of bacterial DNA replication and the enzymes used in this process.
5. Compare and contrast the synthesis of leading and lagging strands during DNA replication.

6.2 Transcription and Translation

We have explored how the genetic message in the DNA molecule is preserved through replication. Now we will move on to the precise role of DNA in the cell. Given that the sequence of bases in DNA is a genetic code, just what is the nature of this code and how is it used by the cell? Although the DNA is full of critical information, it does not perform cell processes directly. Its stored information is conveyed to RNA molecules, which carry out the instructions. The concept that genetic information flows from DNA to RNA to protein is a central theme of molecular biology **(figure 6.8*a*).** To state it clearly, the master code of DNA is first used to synthesize an RNA molecule via a process called **transcription,** and the information contained in many of these RNA molecules is then

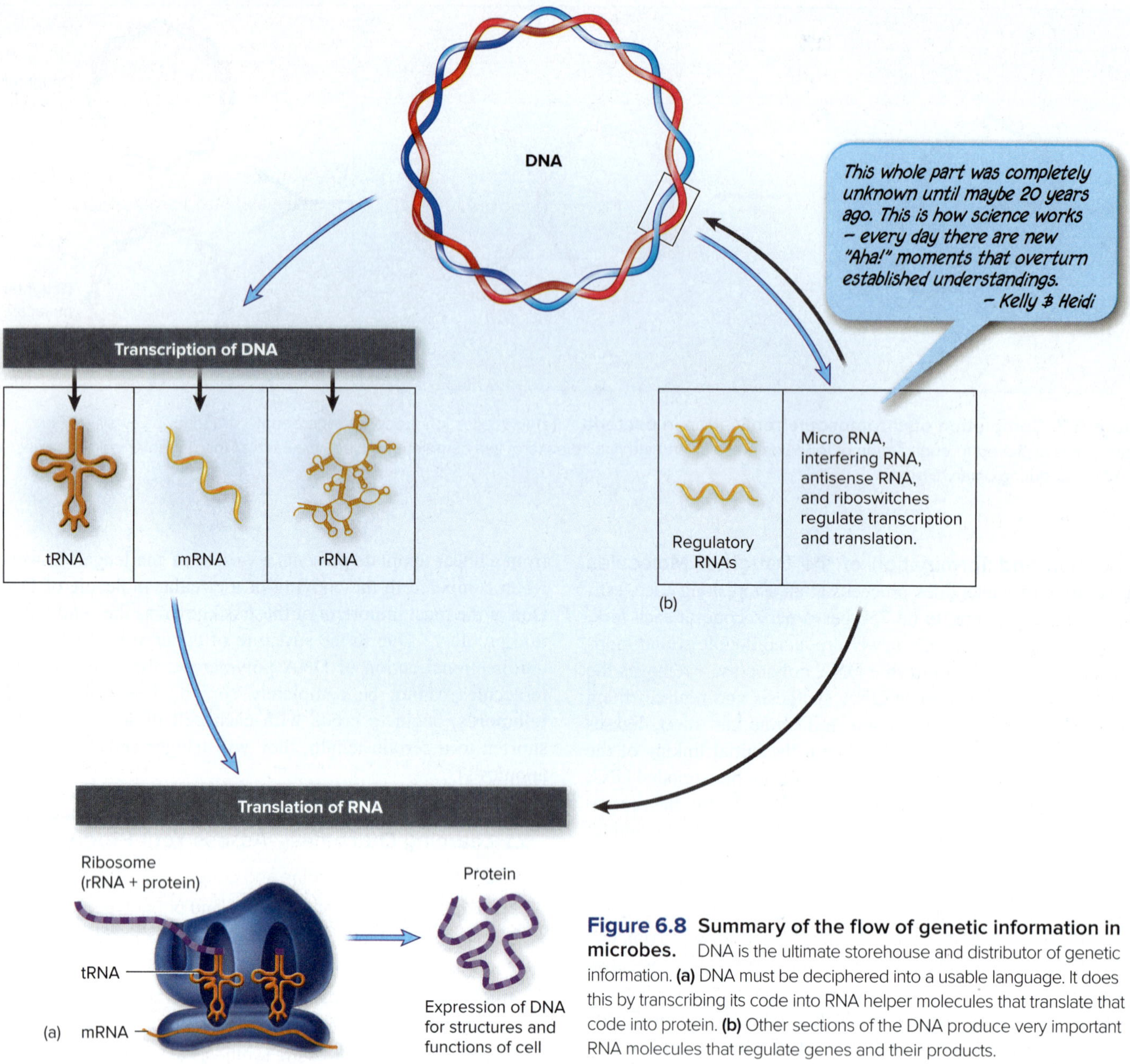
DNA
This whole part was completely unknown until maybe 20 years ago. This is how science works – every day there are new "Aha!" moments that overturn established understandings.
– Kelly $ Heidi
Transcription of DNA
tRNA
mRNA
rRNA
Regulatory RNAs
Micro RNA, interfering RNA, antisense RNA, and riboswitches regulate transcription and translation.
(b)
Translation of RNA
Ribosome (rRNA + protein)
tRNA
(a)
mRNA
Protein
Expression of DNA for structures and functions of cell

Figure 6.8 Summary of the flow of genetic information in microbes. DNA is the ultimate storehouse and distributor of genetic information. **(a)** DNA must be deciphered into a usable language. It does this by transcribing its code into RNA helper molecules that translate that code into protein. **(b)** Other sections of the DNA produce very important RNA molecules that regulate genes and their products.

used to produce proteins in a process known as **translation.** The only exceptions to this pattern are found in RNA viruses, which convert RNA to other RNA, and in retroviruses, which convert RNA to DNA.

Disease Connection

The most well-known retrovirus is HIV, which causes AIDS.

This is sometimes called the "central dogma" of biology. It represents the primary understanding of genetics during the first half century of our genetic understanding (beginning in the 1950s), but it has recently been shown to be incomplete. While it is true that proteins are made in accordance with this central dogma, there is more to the story **(figure 6.8*b*).** In addition to the RNA that is used to produce proteins, a wide variety of RNAs are used to regulate gene function.

In 1993, the discovery of short sequences of RNA that were 21 to 35 nucleotides long, termed ***micro RNA*** or *miRNA,* turned the central dogma on its ear. These tiny snippets of RNA completely changed the way scientists view DNA regulation. Sometimes these short strands of RNA can bind with the RNA being used to make proteins, and then stop the process. Micro RNAs can also bind to DNA and have regulating effects. Sometimes, the action of these small RNAs is what is called **epigenetic.** This means that they are "on top of the DNA code" and do not result in permanent DNA changes (although sometimes epigenetic changes can be passed down to subsequent generations).

Various names have been given to different types of micro RNAs, including *small interfering RNA* (*siRNA*) *riboswitches, antisense RNA,* and *piwi-interacting RNA* (*piRNA*). The system is found in many, if not most, eukaryotic organisms and viruses that infect them. Similar processes exist in bacteria as well. Many of the genetic malfunctions that cause human disease are in fact found in these regulatory RNA segments—and not in genes for proteins as was once thought. The DNA that codes for these very crucial RNA molecules was called "junk" DNA until very recently, only because scientists didn't know that they had any function.

The Gene–Protein Connection

Let's consider the relationship between genes and cell function. For instance, how does gene structure lead to the expression of traits in the individual, and what features of gene expression cause one organism to be so distinctly different from another? We know that each structural gene is a linear sequence of nucleotides that codes for a protein. Because each protein is different, each gene must also differ somehow in its composition. In fact, the language of DNA exists in the order of groups of three consecutive bases called *triplets* on one DNA strand **(figure 6.9).** Every gene differs from another in its order of triplets. An equally important part of this concept is that each triplet represents a code for a particular amino acid. When the triplet code is transcribed and translated, it dictates the type and order of amino acids in a polypeptide (protein) chain.

The final key points that connect DNA and an organism's traits follow:

1. A protein's primary structure—the order and type of amino acids in the chain—determines its characteristic shape and function (see figures 2.21 and 2.22).

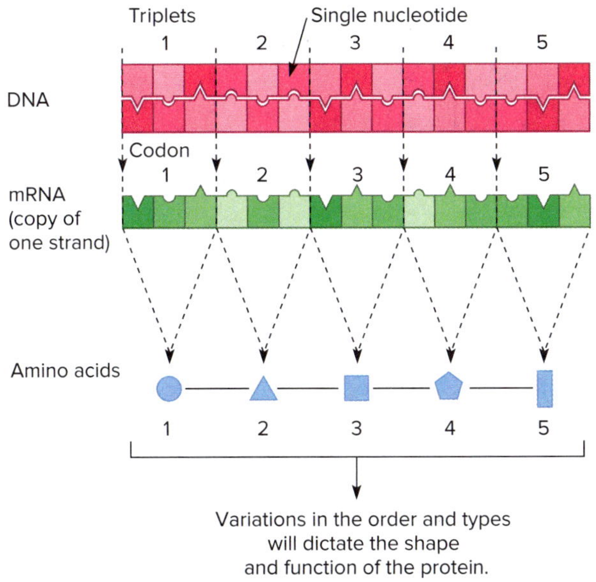
Triplets
Single nucleotide
1
2
3
4
5
DNA
Codon
1
2
3
4
5
mRNA
(copy of
one strand)
Amino acids
1
2
3
4
5
Variations in the order and types
will dictate the shape
and function of the protein.

Figure 6.9 Simplified view of the DNA–protein relationship. The DNA molecule is a continuous chain of base pairs, but the sequence must be interpreted in groups of three base pairs (a triplet). Each triplet as copied into mRNA codons will translate into one amino acid; consequently, the ratio of base pairs to amino acids is 3:1.

2. Proteins ultimately determine phenotype, the expression of all aspects of cell function and structure. Put more simply, living things are what their proteins make them. Regulatory RNAs help determine which proteins are made. **Proteomics** is the study of an organism's complete set of expressed proteins.
3. DNA is mainly a blueprint that tells the cell which kinds of proteins and RNAs to make and how to make them.

The Major Participants in Transcription and Translation

Transcription is the formation of RNA using DNA as a template. *Translation* is the creation of proteins using RNA as a template. A number of cell components are involved: most prominently, messenger RNA, transfer RNA, regulatory RNAs, ribosomes, several types of enzymes, and a storehouse of raw materials. After first examining each of these components, we will see how they come together in the assembly line of the cell.

RNAs: Tools in the Cell's Assembly Line

Ribonucleic acid is similar to DNA, but its general structure (see figure 2.23) is different in several ways:

1. It is a single-stranded molecule that exists in helical form. This single strand can assume secondary and tertiary levels of structure due to bonds within the molecule, leading to specialized forms of RNA (tRNA and rRNA—see figure 6.8*a*).
2. RNA contains **uracil (U),** instead of thymine, as the complementary base-pairing mate for adenine. This does not change the inherent DNA code in any way because the uracil still follows the pairing rules.
3. Although RNA, like DNA, contains a backbone that consists of alternating sugar and phosphate molecules, the sugar in RNA is **ribose** rather than deoxyribose.

The many functional types of RNA range from small regulatory pieces to large structural ones **(table 6.2).** All types of RNA are formed through transcription of a DNA gene, but only mRNA is further translated into another type of molecule (protein).

Messenger RNA: Carrying DNA's Message

Messenger RNA (mRNA) is a transcript (a copy) of a structural gene or genes in the DNA. It is created by a process similar to synthesis of the leading strand during DNA replication, and the complementary base-pairing rules make sure that the code will be faithfully copied in the mRNA transcript. This transcribed strand is later read as a series of triplets called **codons (figure 6.10).** The length of the mRNA molecule varies from about 100 nucleotides to several thousand.

Transfer RNA: The Adaptor Molecule

Transfer RNA (tRNA) is also a copy of a specific region of DNA; however, it differs from mRNA. It is always the same length, 75 to 95 nucleotides long, and it contains sequences of

Table 6.2 Types of Ribonucleic Acid

RNA Type	Description	Function in Cell	Translated?
Messenger (mRNA)	Dictates the sequence of amino acids in a protein	Transports the DNA master code to the ribosome	Yes
Transfer (tRNA)	A cloverleaf tRNA to carry amino acids	Brings amino acids to ribosome during translation	No
Ribosomal (rRNA)	Several large structural rRNA molecules	Forms the major part of a ribosome and participates in protein synthesis	No
Micro (miRNA), antisense, riboswitch, and small interfering (siRNA)	Regulatory RNAs	Regulation of gene expression and coiling of chromatin	No
Primer	An RNA that can begin DNA replication	Primes DNA	No
Ribozymes and spliceosomes (snRNA)	RNA enzymes, parts of splicer enzymes	Remove introns from other RNAs in eukaryotes	No

Figure 6.10 Characteristics of transfer and messenger RNA.

(a) **Transfer RNA (tRNA)**
Left: The tRNA strand loops back on itself to form intrachain hydrogen bonds. The result is a cloverleaf structure, shown here in simplified form. At its bottom is an anticodon that specifies the attachment of a particular amino acid at the 3′ end. Right: A three-dimensional view of tRNA structure.

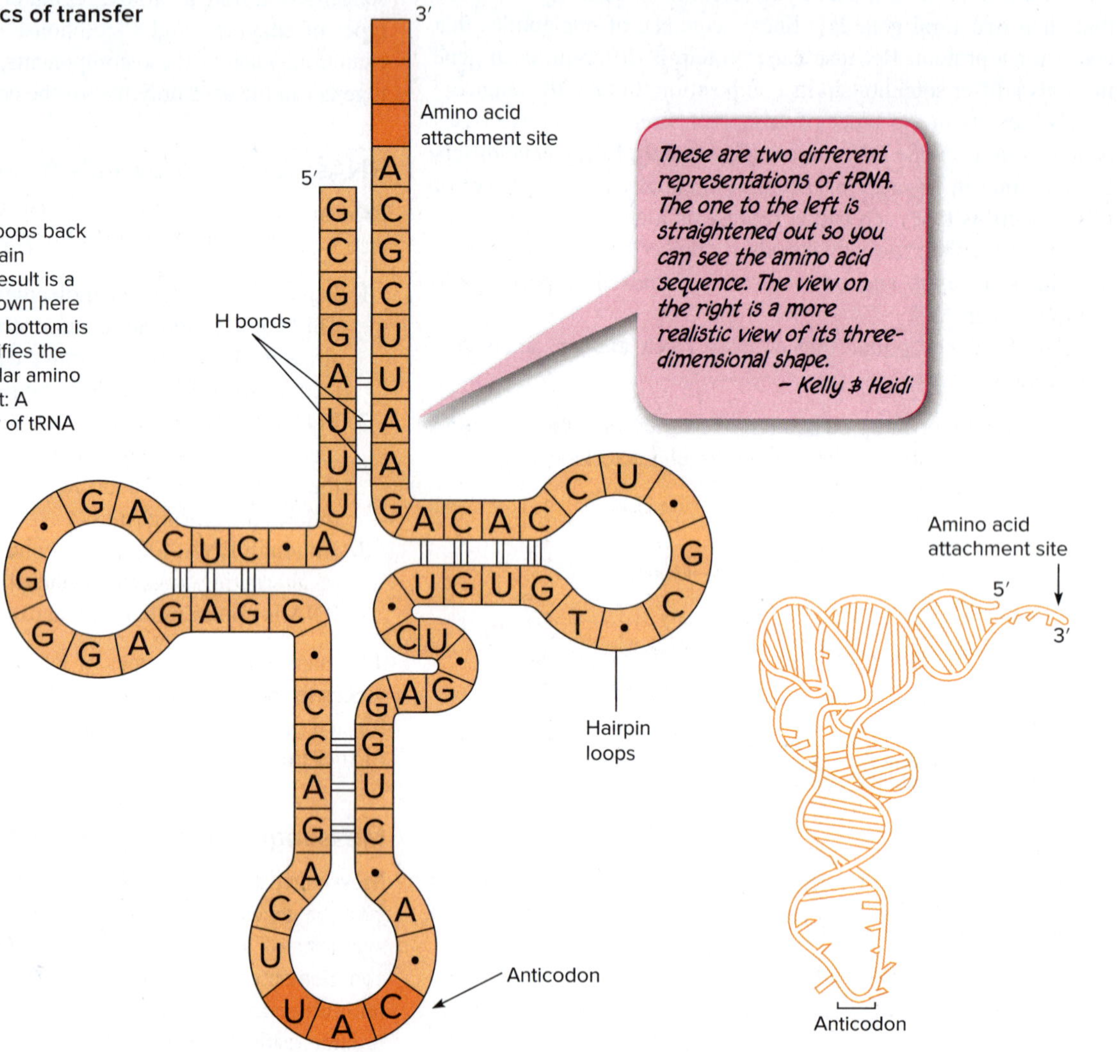
3′
Amino acid attachment site
5′
H bonds
These are two different representations of tRNA. The one to the left is straightened out so you can see the amino acid sequence. The view on the right is a more realistic view of its three-dimensional shape.
- Kelly $ Heidi
Hairpin loops
Anticodon
Amino acid attachment site
5′
3′
Anticodon

(b) **Messenger RNA (mRNA)**
A short piece of messenger RNA (mRNA) illustrates the general structure of RNA: single strandedness, repeating phosphate-ribose sugar backbone attached to single nitrogen bases, use of uracil instead of thymine.

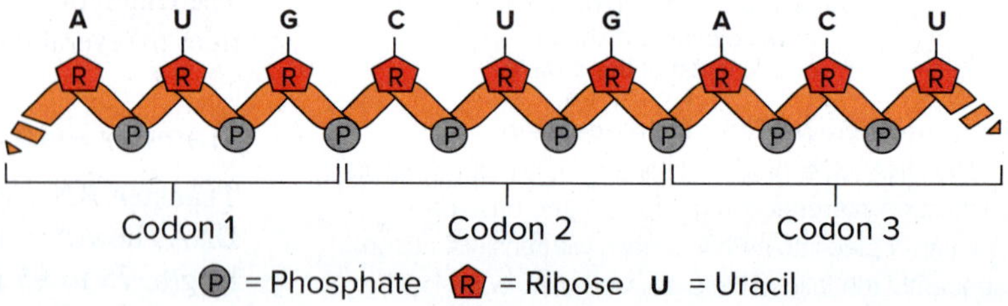
A U G C U G A C U
Codon 1
Codon 2
Codon 3
P = Phosphate
R = Ribose
U = Uracil

bases that form hydrogen bonds within itself. At these points, the molecule bends back upon itself into several *hairpin loops,* giving the molecule a secondary *cloverleaf* structure that folds even further into a complex, three-dimensional helix (as shown in figure 6.10). This compact molecule is an adaptor that converts RNA language into protein language. At the bottom loop of the cloverleaf, there is an exposed set of three bases called the **anticodon.** At the opposite end of the folded molecule is a binding site for an amino acid. For each of the 20 amino acids, there is at least one specialized type of tRNA to carry it. The anticodon matches up with its complementary codon on the mRNA molecule. This is how it acts like an adaptor. The anticodon that a tRNA molecule displays determines which amino acid it carries. So, the type of codon in mRNA determines which tRNA will bind to it, and therefore which amino acid will be brought to it. Binding of an amino acid to its specific tRNA, a process known as "charging" the tRNA, takes place in two enzyme-driven steps: First, an ATP activates the amino acid. Then the amino acid binds to the acceptor end of the tRNA. Because tRNA is the molecule that will convert the master code on mRNA into a protein, the accuracy of this step is crucial.

The Ribosome: Where It All Comes Together

The bacterial (70S) ribosome is a particle composed of tightly packaged **ribosomal RNA (rRNA)** and protein. The rRNA component of the ribosome is a long RNA molecule. It forms complex, three-dimensional shapes that contribute to the structure and function of ribosomes. The interactions of proteins and rRNA create the two subunits of the ribosome that engage in final translation of the genetic code **(figure 6.11).** A metabolically active bacterial cell can contain up to 20,000 of these tiny factories—all actively engaged in reading the genetic program, taking in raw materials, and producing proteins at an impressive rate.

Transcription: The First Stage of Gene Expression

During **transcription**, the DNA code is converted to RNA. This happens in several stages, directed by a huge and complex enzyme system called **RNA polymerase. Figure 6.12** supplies the details you will need to know about transcription. Only one strand of the DNA—the **template strand**—contains meaningful instructions for synthesis of a functioning polypeptide. The strand of DNA that serves as a template varies from one gene to another.

During elongation, which proceeds in the 5′ to 3′ direction (with regard to the growing RNA molecule), the mRNA is assembled by the addition of nucleotides that are complementary to the DNA template. Remember that uracil (U) is placed as adenine's complement. As elongation continues, the part of DNA already transcribed is rewound into its original helical form. At termination, the polymerases recognize a code in the DNA that signals the separation and release of the mRNA strand, also called the **transcript.** How long is the mRNA?

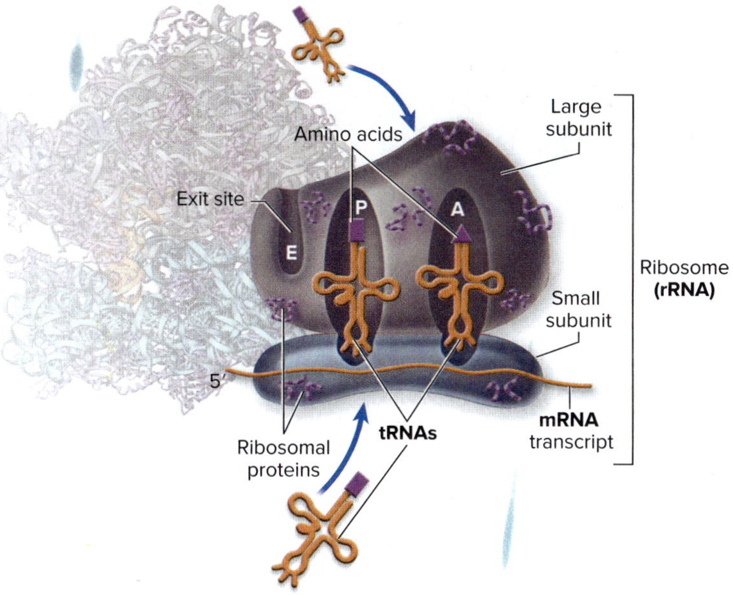
Amino acids
Large subunit
Exit site
P
A
E
Ribosome (rRNA)
Small subunit
5′
mRNA transcript
tRNAs
Ribosomal proteins

Figure 6.11 The "players" in translation. A ribosome serves as the stage for protein synthesis. Assembly of the small and large subunits results in specific sites for holding the mRNA and two tRNAs with their amino acids. This depiction of the ribosome matches the shaded depiction of the molecular view (seen in background).
(Background) Center for Molecular Biology of RNA, UC-Santa Cruz

The smallest mRNA may consist of 100 bases; an average-size mRNA may consist of 1,200 bases; and a large one may consist of several thousand.

Translation: The Second Stage of Gene Expression

In translation, RNA is read to create a string of amino acids, which becomes a protein.

The Master Genetic Code: The Message in Messenger RNA

Translation relies on a central principle: The mRNA nucleotides are read in groups of three. Three nucleotides are called a **codon,** and it is the codon that dictates which amino acid is added to the growing peptide chain. In **figure 6.13,** the mRNA codons and their corresponding amino acid specificities are given. Except in a very few cases, this code is universal, whether for bacteria, archaea, eukaryotes, or viruses.

Because there are 64 different codons[1] and only 20 different amino acids, it is not surprising that some amino acids are represented by several codons. For example, leucine can be represented by any of six different codons, whereas tryptophan is represented by a single codon. This property—that an amino acid can be represented by several codons—is called **redundancy**

1. $64 = 4^3$ (the four different codons in all possible combinations of three).

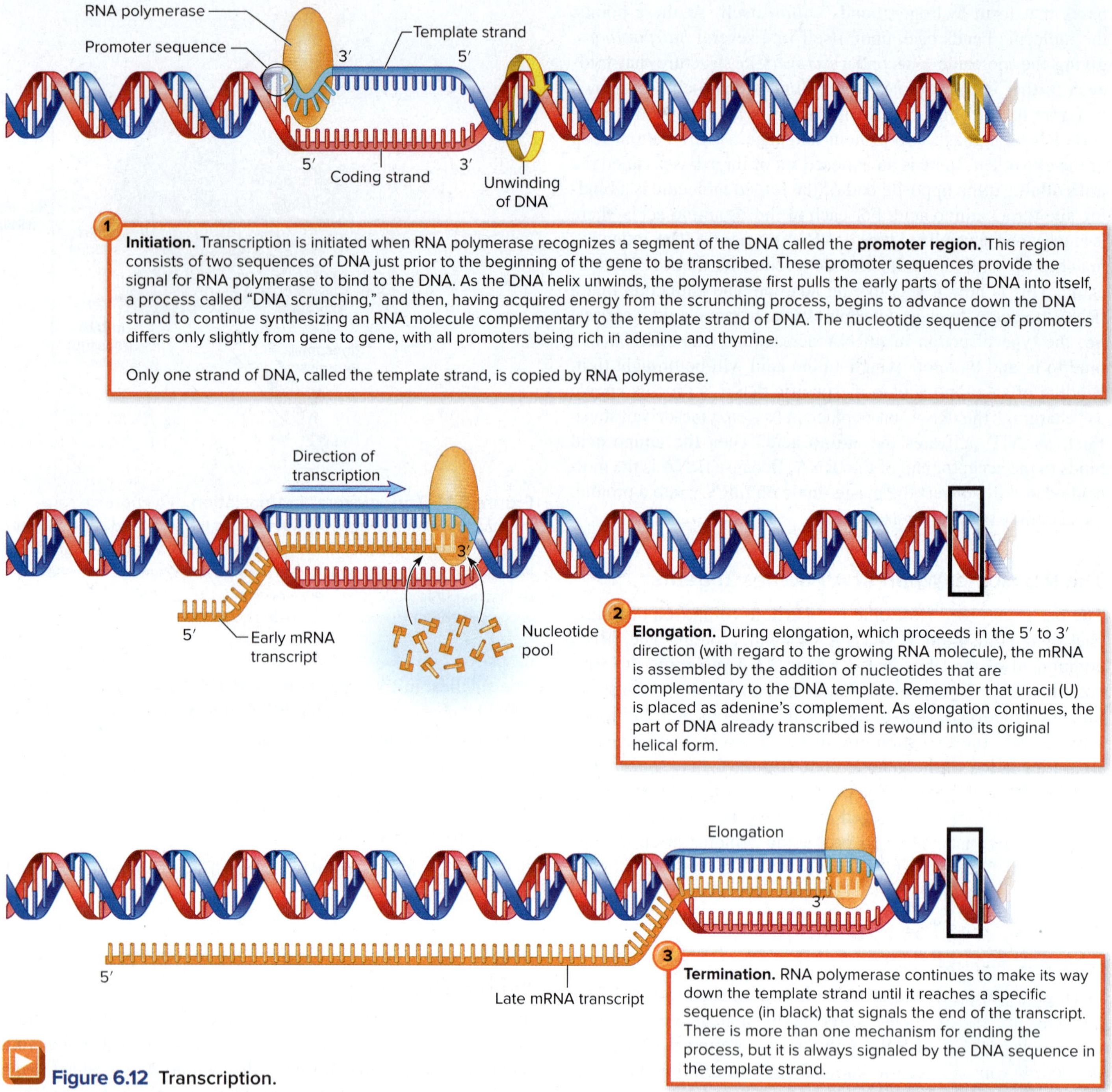
RNA polymerase
Promoter sequence
Template strand
3′
5′
5′
Coding strand
3′
Unwinding of DNA
1
Initiation. Transcription is initiated when RNA polymerase recognizes a segment of the DNA called the promoter region. This region consists of two sequences of DNA just prior to the beginning of the gene to be transcribed. These promoter sequences provide the signal for RNA polymerase to bind to the DNA. As the DNA helix unwinds, the polymerase first pulls the early parts of the DNA into itself, a process called "DNA scrunching," and then, having acquired energy from the scrunching process, begins to advance down the DNA strand to continue synthesizing an RNA molecule complementary to the template strand of DNA. The nucleotide sequence of promoters differs only slightly from gene to gene, with all promoters being rich in adenine and thymine.
Only one strand of DNA, called the template strand, is copied by RNA polymerase.
Direction of transcription
3′
5′
Early mRNA transcript
Nucleotide pool
2
Elongation. During elongation, which proceeds in the 5′ to 3′ direction (with regard to the growing RNA molecule), the mRNA is assembled by the addition of nucleotides that are complementary to the DNA template. Remember that uracil (U) is placed as adenine's complement. As elongation continues, the part of DNA already transcribed is rewound into its original helical form.
Elongation
3′
5′
Late mRNA transcript
3
Termination. RNA polymerase continues to make its way down the template strand until it reaches a specific sequence (in black) that signals the end of the transcript. There is more than one mechanism for ending the process, but it is always signaled by the DNA sequence in the template strand.

Figure 6.12 Transcription.

and allows for the insertion of correct amino acids (sometimes) even when mistakes occur in the DNA sequence, as they do on a regular basis. Also, in codons such as leucine, only the first two nucleotides are required to encode the correct amino acid, and the third nucleotide does not change its outcome. This property, called **wobble,** is thought to permit some variation or mutation without altering the message. **Figure 6.14** shows the relationship between DNA sequence, mRNA codons, tRNA anticodons, and amino acids.

All of the elements needed to synthesize a protein, mRNA, tRNAs, and amino acids, are brought together on the ribosomes **(figure 6.15).** The process has three main stages: initiation, elongation, and termination.

Initiation of Translation

The mRNA molecule leaves the DNA transcription site and is transported to ribosomes in the cytoplasm. Ribosomal subunits function by coming together and forming sites to hold the mRNA and tRNAs. The ribosome recognizes these molecules and stabilizes reactions between them. The small subunit binds to the 5′ end of the mRNA, and the large subunit carries enzymes for making peptide bonds on the protein.

First Base Position	Second Base Position: U		C		A		G		Third Base Position
U	UUU	Phenylalanine	UCU	Serine	UAU	Tyrosine	UGU	Cysteine	U
	UUC		UCC		UAC		UGC		C
	UUA	Leucine	UCA		UAA	STOP**	UGA	STOP**	A
	UUG		UCG		UAG		UGG	Tryptophan	G
C	CUU	Leucine	CCU	Proline	CAU	Histidine	CGU	Arginine	U
	CUC		CCC		CAC		CGC		C
	CUA		CCA		CAA	Glutamine	CGA		A
	CUG		CCG		CAG		CGG		G
A	AUU	Isoleucine	ACU	Threonine	AAU	Asparagine	AGU	Serine	U
	AUC		ACC		AAC		AGC		C
	AUA		ACA		AAA	Lysine	AGA	Arginine	A
	AUG	START f-Methionine*	ACG		AAG		AGG		G
G	GUU	Valine	GCU	Alanine	GAU	Aspartic acid	GGU	Glycine	U
	GUC		GCC		GAC		GGC		C
	GUA		GCA		GAA	Glutamic acid	GGA		A
	GUG		GCG		GAG		GGG		G

*This codon initiates translation.

**For these codons, which give the orders to stop translation, there are no corresponding tRNAs and no amino acids.

Figure 6.13 The genetic code: codons of mRNA that specify a given amino acid. The master code for translation is found in the mRNA codons.

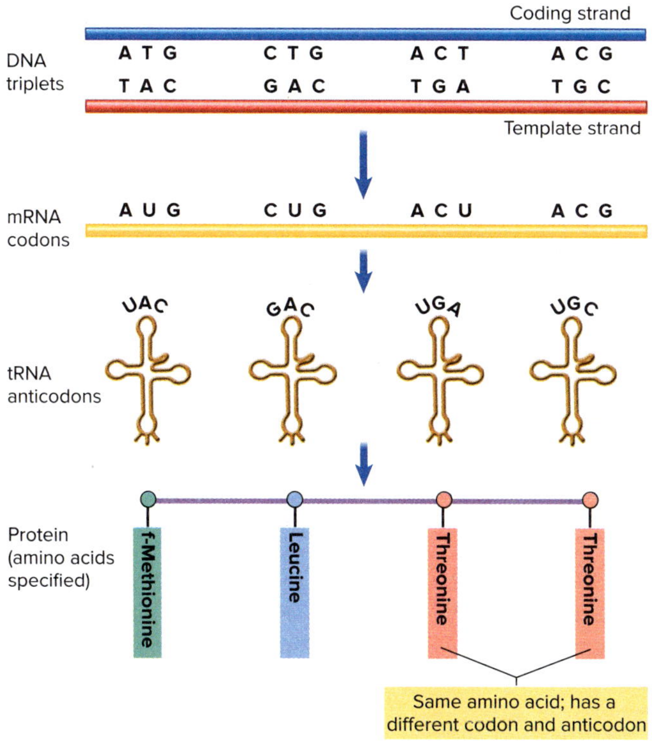

Coding strand
DNA triplets
A T G
C T G
A C T
A C G
T A C
G A C
T G A
T G C
Template strand
mRNA codons
A U G
C U G
A C U
A C G
tRNA anticodons
UAC
GAC
UGA
UGC
Protein (amino acids specified)
f-Methionine
Leucine
Threonine
Threonine
Same amino acid; has a different codon and anticodon

Figure 6.14 Interpreting the DNA code. If the DNA sequence is known, the mRNA codon can be surmised. If a codon is known, the anticodon and, finally, the amino acid sequence can be determined. Remember, it is the mRNA codons that determine the amino acids that are inserted. The reverse is not as straightforward (determining the exact codon or anticodon from amino acid sequence) due to the redundancy of the code.

The Terms of Protein Synthesis

With mRNA serving as the guide, the stage is finally set for actual protein assembly. Figure 6.15 lays out the details of translation. Several terms are important for understanding the process.

start codon The first three RNA nucleotides that signal the beginning of the message. The start codon is always AUG.

stop codon One of three codons—UAA, UAG, or UGA—that has no corresponding tRNA and therefore causes translation to be terminated; also called **nonsense codon.**

translocation The process of shifting the ribosome down the mRNA strand to read new codons.

Completion of Protein Synthesis

Before newly made proteins can carry out their structural or enzymatic roles, they often require finishing touches. Even before the peptide chain is released from the ribosome, it begins folding upon itself to achieve its biologically active tertiary conformation. Other alterations, called **posttranslational** modifications, may be necessary. Some proteins must have the starting amino acid (formyl methionine) clipped off; proteins destined to become complex enzymes have cofactors added; and some join with other completed proteins to form quaternary levels of structure.

The operation of transcription and translation is machine-like in its precision. Protein synthesis in bacteria is both efficient and rapid, as the translation of mRNA starts while transcription (i.e., the creation of the mRNA) is still occurring **(figure 6.16).** This process is called *cotranscriptional translation.* A single mRNA is long enough to be fed through more than one ribosome

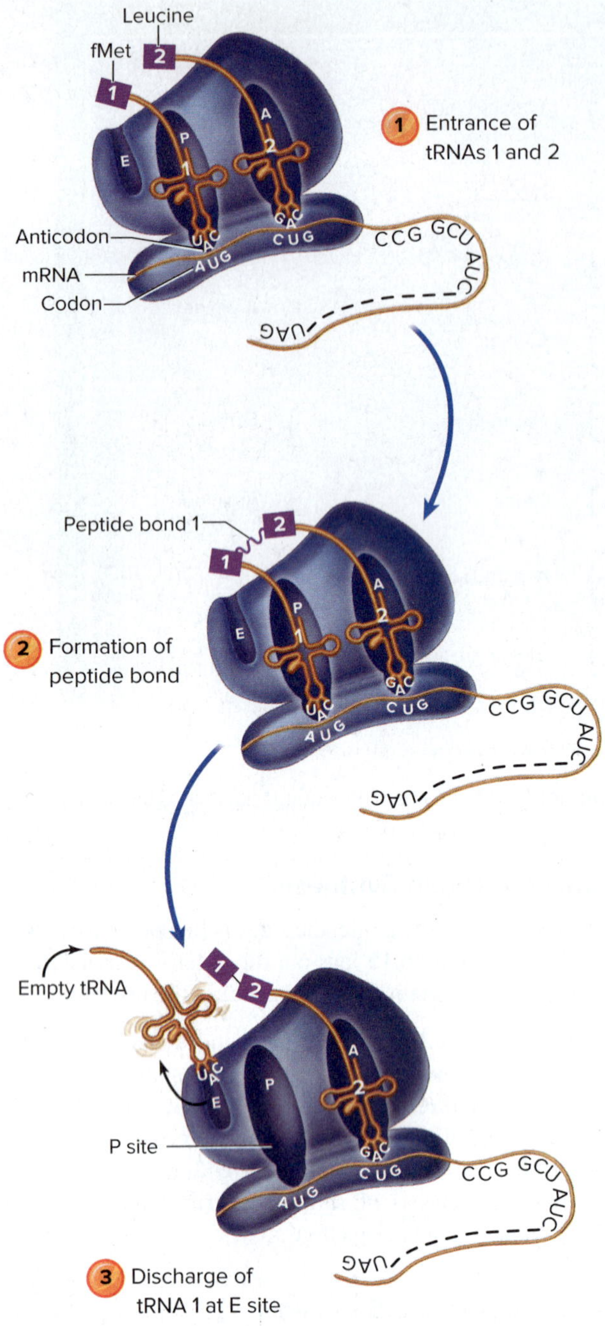
Leucine
fMet
1
2
E
P
A
Anticodon
mRNA
Codon
UAC
AUG
GAC
CUG
CCG GCU AUC
UAG
1 Entrance of tRNAs 1 and 2
Peptide bond 1
2 Formation of peptide bond
Empty tRNA
P site
3 Discharge of tRNA 1 at E site

1 The mRNA molecule leaves the DNA transcription site and is transported to ribosomes in the cytoplasm. Ribosomal subunits come together and form sites to hold the mRNA and tRNAs. The ribosome begins to scan the mRNA by moving in the 5′ to 3′ direction along the mRNA. The first codon it encounters is called the START codon, which is almost always AUG (and, rarely, GUG). With the mRNA message in place on the assembled ribosome, the next step in translation involves entrance of tRNAs with their amino acids. The pool of cytoplasm contains a complete array of tRNAs, previously charged by having the correct amino acid attached. The step in which the complementary tRNA meets with the mRNA code is guided by the two sites on the large subunit of the ribosome called the P site (left) and the A site (right). The ribosome also has an exit or E site where used tRNAs are released. (P stands for peptide site; A stands for aminoacyl [amino acid] site; E stands for exit site.)

2 Rules of pairing dictate that the anticodon of this tRNA must be complementary to the mRNA codon AUG; the tRNA with anticodon UAC will first occupy site P. It happens that the amino acid carried by the initiator tRNA in bacteria is **formyl methionine.** The formyl group provides a special signal that this amino acid is not part of the translated protein because usually fMet does not remain a permanent part of the finished protein but instead is cleaved from the finished peptide. The ribosome shifts its "reading frame" to the right along the mRNA from one codon to the next. This brings the next codon into place on the ribosome and makes a space for the next tRNA to enter the A position. A peptide bond is formed between the amino acids on the adjacent tRNAs, and the polypeptide grows in length.

Elongation begins with the filling of the A site by a second tRNA. The identity of this tRNA and its amino acid is dictated by the second mRNA codon.

3 The entry of tRNA 2 into the A site brings the two adjacent tRNAs in favorable proximity for a peptide bond to form between the amino acids (aa) they carry. The fMet is transferred from the first tRNA to aa 2, resulting in two coupled amino acids called a dipeptide.

For the next step to proceed, some room must be made on the ribosome, and the next codon in sequence must be brought into position for reading. This process is accomplished by translocation, the enzyme-directed shifting of the ribosome to the right along the mRNA strand, which causes the blank tRNA 1 to be discharged from the ribosome at the E site.

Figure 6.15 Translation.

simultaneously. This permits the synthesis of hundreds of protein molecules from the same mRNA transcript arrayed along a chain of ribosomes. This **polyribosomal complex** is indeed an assembly line for mass production of proteins. Cotranscriptional translation only occurs in bacteria and archaea because there is no nucleus and transcription and translation both occur in the cytoplasm. (In eukaryotes, transcription occurs in the nucleus.)

COVID-19 mRNA Vaccines

Now we're at the point where we can discuss the first two widely used COVID-19 vaccines in the United States in 2020 and beyond. These were both mRNA vaccines, the first of their type approved for use in humans. We will talk about vaccines in depth in chapter 16, but for now know that what is wanted in a vaccine is a piece of the virus—not able to cause infection—that can be shown to the human host so that their immune response can arm itself against it. When the actual virus appears, the host is primed and ready to release an attack against it immediately.

There are several methods to "show" a piece of the virus to the host. In these COVID-19 vaccines, the mRNA that codes for a surface protein of the virus was constructed, and then the mRNA (enclosed in a lipid bubble), was injected into human arms. The muscle cells use their translation machinery to create the virus

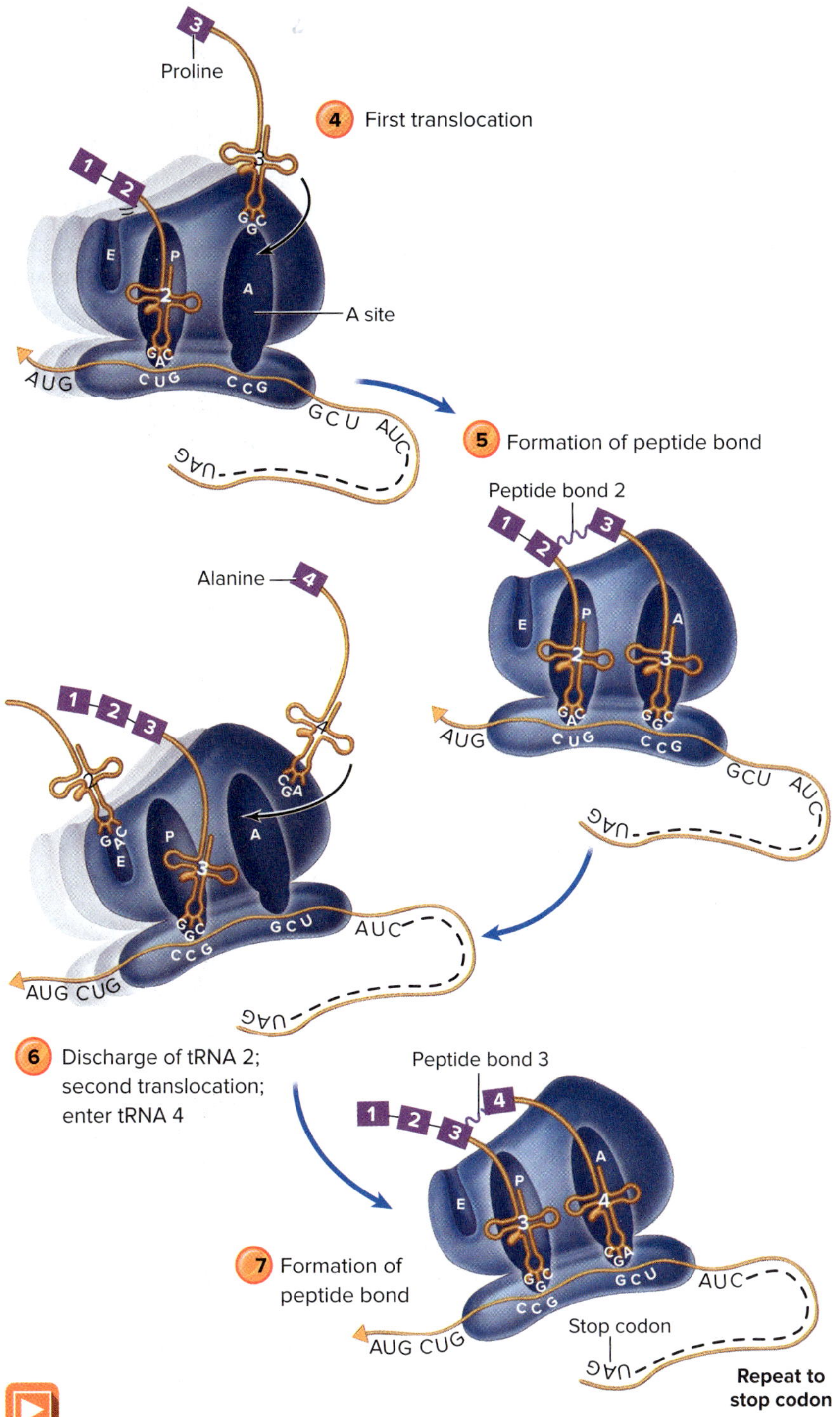
Proline
4 First translocation
A site
AUG
CUG
CCG
GCU AUC
UAG
5 Formation of peptide bond
Peptide bond 2
Alanine
6 Discharge of tRNA 2; second translocation; enter tRNA 4
Peptide bond 3
7 Formation of peptide bond
Stop codon
Repeat to stop codon

4 This also shifts the tRNA holding the dipeptide into P position. Site A is temporarily left empty. The tRNA that has been released is now free to drift off into the cytoplasm and become recharged with an amino acid for later additions to this or another protein.

The stage is now set for the insertion of tRNA 3 at site A as directed by the third mRNA codon. This insertion is followed once again by peptide bond formation between the dipeptide and amino acid 3 (making a tripeptide), splitting of the peptide from tRNA 2, and translocation.

5 This releases tRNA 2, shifts mRNA to the next position, moves tRNA 3 to position P, and opens position A for the next tRNA (which will be called tRNA 4).

6 From this point on, peptide elongation proceeds repetitively by this same series of actions out to the end of the mRNA.

7 The termination of protein synthesis is not simply a matter of reaching the last codon on mRNA. It is brought about by the presence of at least one special codon occurring just after the codon for the last amino acid. Termination codons—UAA, UAG, and UGA—are codons for which there is no corresponding tRNA. Although they are often called nonsense codons, they carry a necessary and useful message: Stop here. When this codon is reached, a special enzyme breaks the bond between the final tRNA and the finished polypeptide chain, releasing it from the ribosome.

Figure 6.15 Translation. *(Continued).*

protein, using the processes we just studied. The cells place the protein they just created on their own surfaces. The normal screening processes used by our immune systems detect it, see it as foreign, and sound the alarm, building up specific immune defenses against that protein. Later, if exposed to the actual virus, those defenses pounce on it and prevent it from causing disease. **Figure 6.17** details this process. Note that translation occurs in the cytoplasm, so the foreign mRNA never enters the nucleus, and never approaches our own DNA (see the opening Media Under The Microscope article about this).

Eukaryotic Transcription and Translation: Similar Yet Different

There are important differences in protein synthesis between eukaryotes and the noneukaryotes. Only bacteria and archaea can perform cotranscriptional translation. Although the start codon in eukaryotes is also AUG, it codes for a different form of methionine. Another difference is that eukaryotic mRNAs code for just one protein, unlike bacterial mRNAs, which often contain information from several DNA genes in series.

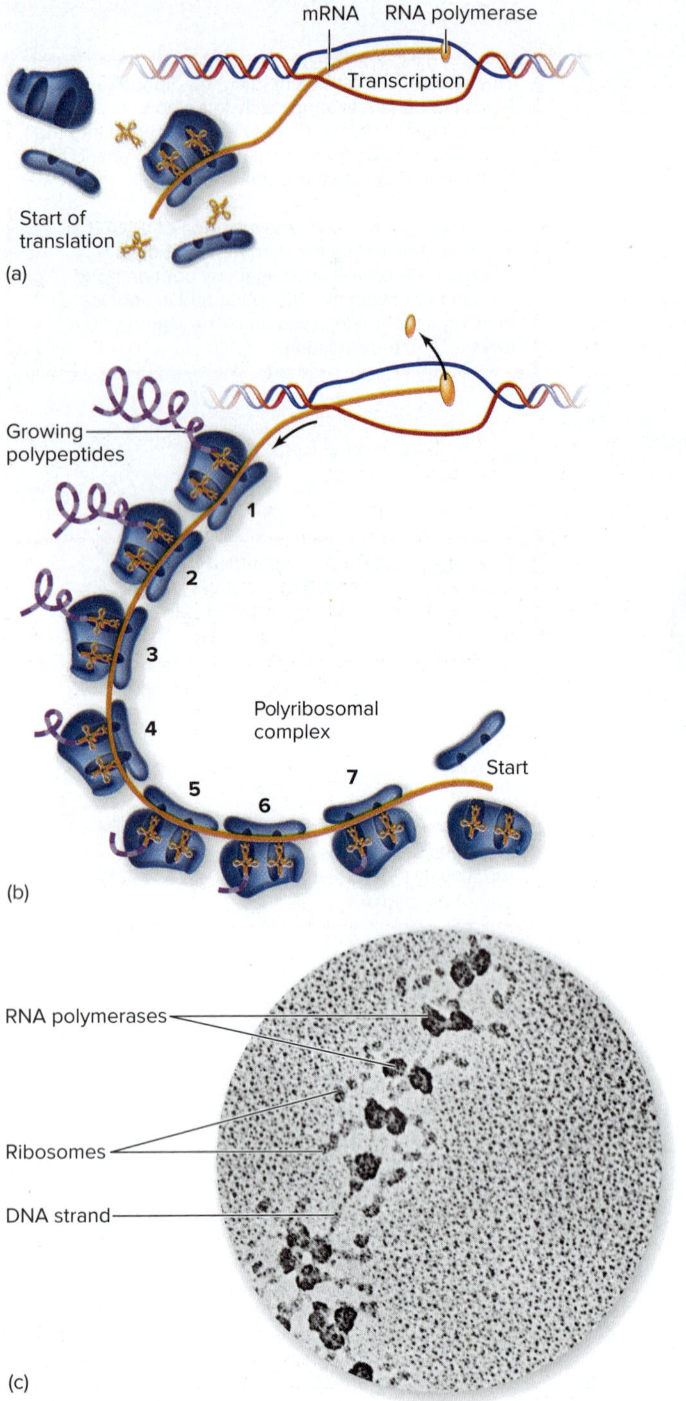
mRNA
RNA polymerase
Transcription
Start of
translation
(a)
Growing
polypeptides
1
2
3
4
5
6
7
Polyribosomal
complex
Start
(b)
RNA polymerases
Ribosomes
DNA strand
(c)

Figure 6.16 Speeding up the protein assembly line in bacteria. **(a)** The mRNA transcript encounters ribosomal parts immediately as it leaves the DNA. **(b)** The ribosomal factories assemble along the mRNA in a chain, each ribosome reading the message and translating it into protein. Many products will be well along the synthetic pathway before transcription has even terminated. **(c)** Photomicrograph of a polyribosomal complex in action.

(c) Steven McKnight and Oscar L Miller/Department of Biology/University of Virginia

Finally, we have given the simplified definition of *gene,* which works well for bacteria, but most eukaryotic genes do *not* exist as an uninterrupted series of triplets coding for a protein. A structural eukaryotic gene contains the code for a protein, but located along the gene are one to several intervening sequences of bases, called **introns,** that do not code for protein. Introns are interspersed between coding regions, called **exons,** that will be translated into protein **(figure 6.18).** We can use words as examples. A short section of colinear bacterial gene might read (using words not codons) TOM SAW OUR DOG DIG OUT. A eukaryotic gene that codes for the same portion might read TOM SAW XZKP FPL OUR DOG QZWVP DIG OUT. The recognizable words are the exons, and the nonsense letters represent the introns.

This genetic architecture requires further processing before translation. Transcription of the entire gene—both exons and introns—occurs first, producing a pre-mRNA. A series of adenosines is added to the mRNA molecule. This protects the molecule and eventually directs it out of the nucleus for translation (something bacterial cells do not have to do). Next, a structure called a **spliceosome,** containing a bit of RNA and protein, recognizes the junctions between the exons and the introns and enzymatically cuts through them. The action of this splicer enzyme loops the introns into lariat-shaped pieces, cuts them out, and joins the exons

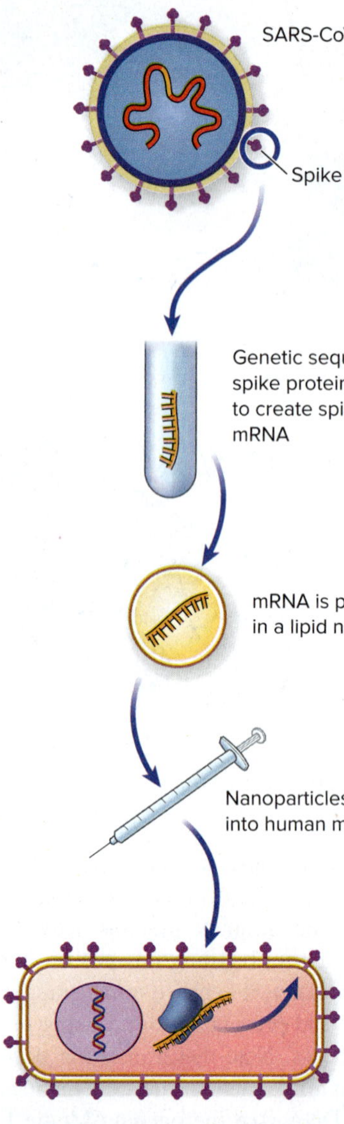
SARS-CoV-2
Spike protein
Genetic sequence of spike protein is used to create spike protein mRNA
mRNA is packaged in a lipid nanoparticle
Nanoparticles injected into human muscle
Muscle cells translate the mRNA and place the spike protein on their surfaces, alerting the immune system, which will then react to the spike proteins on the suface of SARS-CoV-2

Figure 6.17 COVID-19 mRNA vaccines.

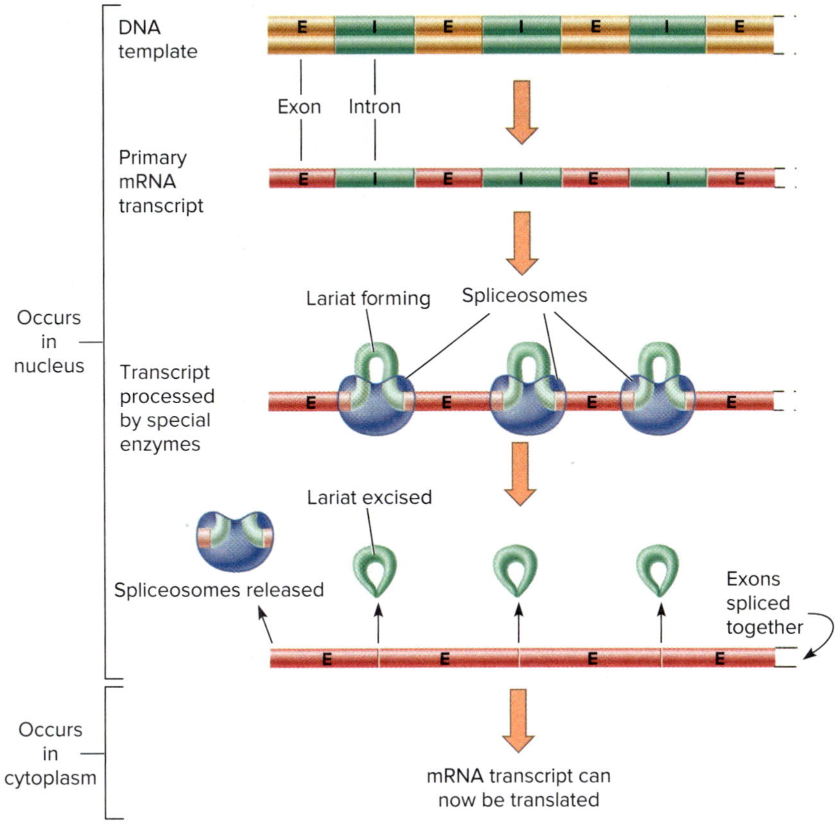

DNA template
E
I
Exon
Intron
Primary mRNA transcript
Occurs in nucleus
Lariat forming
Spliceosomes
Transcript processed by special enzymes
Lariat excised
Spliceosomes released
Exons spliced together
Occurs in cytoplasm
mRNA transcript can now be translated

Figure 6.18 The split gene of eukaryotes. Eukaryotic genes have an additional complicating factor in their translation. Their coding sequences, or exons (E), are interrupted at intervals by segments called introns (I) that are not part of that protein's code. Introns are transcribed but not translated, which means they must be removed by RNA splicing enzymes before translation.

end to end. By this means, a strand of mRNA with no intron material is produced. This completed mRNA strand can then proceed to the cytoplasm to be translated. In some eukaryotes, however, the mRNA may be alternatively spliced into multiple different mRNAs, or its sequence may be edited to produce a variety of proteins from one single gene.

6.2 Learning Outcomes—Assess Your Progress

6. Explain how the classical view of the "central dogma" of biology has been changed by recent science.
7. Identify important structural and functional differences between RNA and DNA.
8. Illustrate the steps of transcription, noting the key elements and the outcome of mRNA synthesis.
9. List the three types of RNA directly involved in translation.
10. Define the terms *codon* and *anticodon,* and list the four known start and stop codons.
11. Identify the locations of the promoter, the start codon, and the A and P sites during translation.
12. Indicate how eukaryotic transcription and translation differ from these processes in bacteria and archaea.
13. Explain the relationship between genomics and proteomics.

6.3 Genetic Regulation of Protein Synthesis

In chapter 10, we will survey the metabolic reactions in cells and the enzymes involved in those reactions. We will discuss how the expression of important genes can be increased, decreased, or stopped when appropriate. This regulation ensures that genes are active only when their products are required. This prevents the waste of energy and materials in dead-end protein synthesis. Antisense RNAs, micro RNAs, and riboswitches provide regulation in many kinds of cells. Bacteria can also place methyl groups on the adenine base in DNA, which can either activate or repress the transcription process. This methylation is epigenetic, meaning it is a modification "on top of" the DNA, and not in the DNA. But bacteria and archaea have an additional strategy: They organize collections of genes into **operons.** Operons consist of a coordinated set of genes, all of which are regulated as a single unit. Operons are categorized as either inducible or repressible. Many operons contain genes that make proteins that help metabolize nutrients. These are inducible operons meaning that the whole operon is switched on (induced) by the nutrient in question. In this way, the enzymes needed to metabolize a nutrient (lactose, for example) are only produced when that nutrient is present in the environment.

Repressible operons are those operons whose normal state is "on." They often contain genes coding for enzymes that synthesize amino acids. In the case of these operons, all of the genes in the operon are turned off (repressed) when there is enough product already synthesized by the enzyme.

The Lactose Operon: A Model for Inducible Gene Regulation in Bacteria

The best understood cell system for explaining inducible operons is the **lactose (*lac*) operon.** This system regulates lactose metabolism in *Escherichia coli.* Many other operons like this exist, and together they show us that the environment of a cell can have great impact on gene expression.

The lactose operon has three important components **(figure 6.19):**

1. the **regulator,** a gene that codes for a protein (a **repressor**) capable of repressing the operon;
2. the *control locus,* composed of two areas, the **promoter** (which is recognized by RNA polymerase) and the **operator,** a sequence that acts as an on/off switch for transcription; and
3. the *structural locus,* made up of three genes, each coding for a different enzyme needed to metabolize the lactose.

One of the enzymes, β-galactosidase, hydrolyzes the lactose into its monosaccharides; another, permease, brings lactose across the cytoplasmic membrane. The third enzyme is an acetyltransferase that helps in the metabolism of lactose.

An operon provides an efficient strategy that permits genes for a particular cell purpose to be induced or repressed in unison by a single regulatory element. The promoter, operator, and structural components usually lie adjacent to one another, but the regulator can be at a distant site.

In inducible systems like the *lac* operon, the operon is normally in an *off mode* and does not initiate transcription when the appropriate substrate is absent (figure 6.19). How is the operon maintained in this mode? The key is in the repressor protein that is coded by the regulatory gene. This relatively large molecule is **allosteric,** meaning it has two binding sites, one for the operator sequence on the DNA and another for lactose. In the absence of lactose, this repressor binds to the operator locus, thereby blocking the transcription of the structural genes lying downstream. Think of the repressor as a lock on the operator, and if the operator is locked, the structural genes cannot be transcribed. Importantly, the regulator gene lies upstream (to the left) of the operator region and is transcribed constitutively because it is not controlled as part of the operon. That means that the repressor protein is always present.

If lactose is added to the cell's environment, it triggers several events that turn the operon *on.* The binding of lactose to the repressor protein causes a conformational change in the repressor that bumps it off of the operator segment of the DNA (figure 6.19). With the operator freed up, RNA polymerase can now bind to the promoter and proceed. The structural genes are transcribed in a single unbroken transcript coding for all three gene products. (During translation, however, each protein is synthesized separately.) Because lactose is ultimately responsible for stimulating the protein synthesis, it is called the **inducer.**

As lactose is depleted, further gene expression from this operon is not necessary, so the order of events reverses. At this point, there is no longer sufficient lactose to inhibit the repressor. Now the repressor is again free to attach to the operator. The operator is locked, and transcription of the structural genes related to lactose both stop.

A small but important point about the *lac* operon is that it functions only in the absence of glucose or if the cell's energy needs are not being met by the available glucose. Glucose is the preferred carbon source because it can be used immediately in growth and does not require induction of an operon. When glucose is present, a second regulatory system ensures that the *lac* operon is inactive, regardless of lactose levels in the environment.

A Repressible Operon

Bacterial operons designed for the synthesis of amino acids, purines and pyrimidines, and many other essential cell components work on a slightly different principle—repression. Similar factors such as repressor proteins, operators, and a series of structural genes exist for this operon but with some important differences. Unlike the *lac* operon, many of these operons are normally in the *on* mode and will be turned *off* only when the cell component is no longer required. The excess cell component serves as a **corepressor** needed to block the action of the operon.

A growing cell that needs the amino acid **arginine** (*arg*) can illustsrate the operation of a repressible operon. Under these conditions, the *arg* operon is set to *on* and arginine is being actively synthesized through the action of the operon's gene products **(figure 6.20*a*).** In an active cell, the arginine will be used immediately, and the repressor will remain inactive (unable to bind the operator) because there is too little free arginine to activate it. As the cell's metabolism begins to slow down, however, the synthesized arginine will no longer be used up and will accumulate in the cytoplasm. The free arginine is then available to act as a corepressor by attaching to the repressor. This reaction changes the shape of the repressor, making it capable of binding to the operator. Transcription stops and arginine is no longer synthesized **(figure 6.20*b*).**

In eukaryotic cells, gene function can be altered by intrinsic regulatory segments similar to operons. Some molecules, called transcription factors, insert on the grooves of the DNA molecule and enhance transcription of specific genes. These transcription factors can regulate gene expression in response to environmental stimuli such as nutrients, toxin levels, or even temperature. Eukaryotic genes are also tightly regulated during tissue growth and development, leading to the hundreds of different tissue types found in higher multicellular organisms.

Recently, a new kind of DNA structure was found. It is not just a helix; there is a "knot" in the helix that involves cytosines binding to other cytosines, instead of normal base-pairing behavior. It seems to be present in eukaryotic cells to block the promoter region of genes in order to stop transcription when needed.

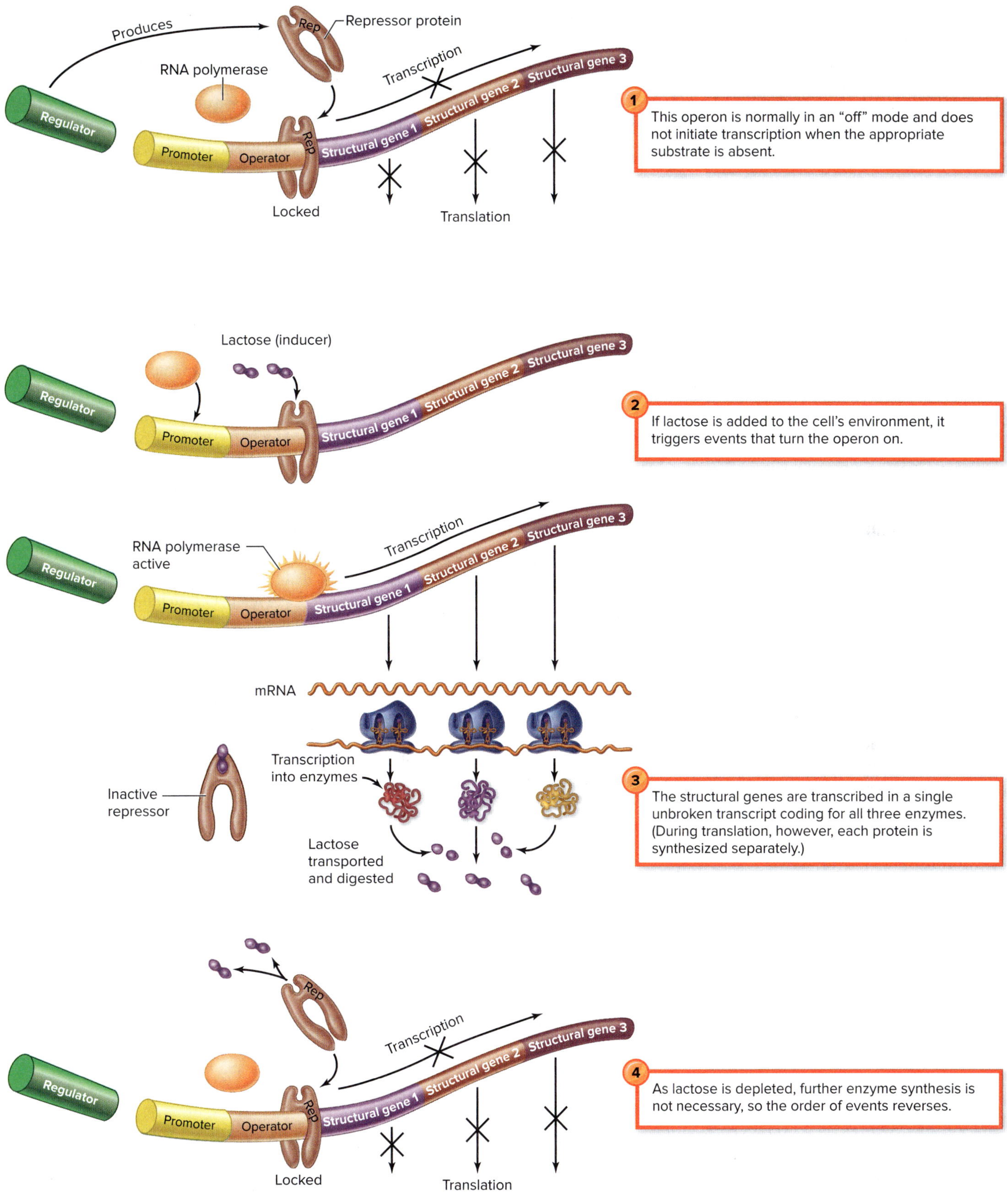
Produces
Repressor protein
Rep
RNA polymerase
Regulator
Promoter
Operator
Structural gene 1
Structural gene 2
Structural gene 3
Transcription
Locked
Translation
1
This operon is normally in an "off" mode and does not initiate transcription when the appropriate substrate is absent.
Lactose (inducer)
2
If lactose is added to the cell's environment, it triggers events that turn the operon on.
RNA polymerase active
mRNA
Transcription into enzymes
Inactive repressor
Lactose transported and digested
3
The structural genes are transcribed in a single unbroken transcript coding for all three enzymes. (During translation, however, each protein is synthesized separately.)
4
As lactose is depleted, further enzyme synthesis is not necessary, so the order of events reverses.

Figure 6.19 **The lactose (*lac*) operon in bacteria: how inducible genes are controlled by substrate.**

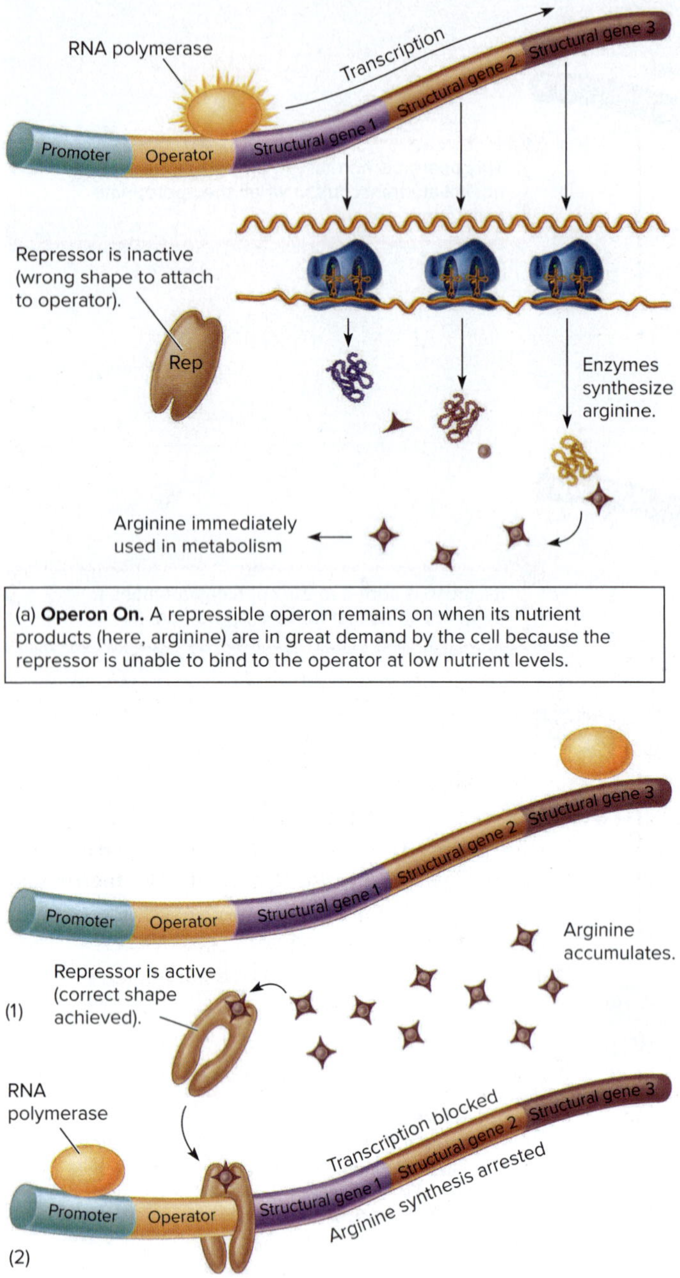
RNA polymerase
Transcription
Promoter
Operator
Structural gene 1
Structural gene 2
Structural gene 3
Repressor is inactive (wrong shape to attach to operator).
Rep
Enzymes synthesize arginine.
Arginine immediately used in metabolism
Arginine accumulates.
Repressor is active (correct shape achieved).
(1)
RNA polymerase
Transcription blocked
Arginine synthesis arrested
(2)

(a) **Operon On.** A repressible operon remains on when its nutrient products (here, arginine) are in great demand by the cell because the repressor is unable to bind to the operator at low nutrient levels.

(b) **Operon Off.** The operon is repressed when (1) arginine builds up and, serving as a corepressor, activates the repressor. (2) The repressor complex affixes to the operator and blocks the RNA polymerase and further transcription of genes for arginine synthesis.

Figure 6.20 Repressible operon: control of a gene through excess nutrient.

Phase Variation

When bacteria turn on or turn off a set of genes that leads to obvious phenotypic changes, it is sometimes called **phase variation.** Phase variation is a type of phenotypic variation that is heritable—meaning it is passed down to subsequent generations. This process involves the turning on of genes mediated by regulatory proteins, similar to operons. The term *phase variation* is most often applied to traits affecting the bacterial cell surface and was originally coined to describe the ability of bacteria to change components of their surface that marked them for targeting by the host's immune system. Because these surface molecules also influenced the bacterium's ability to attach to surfaces, the ability to undergo phase variation allowed the microbes to adapt to—and attach to—different environments. Examples of phase variation include the ability of *Neisseria gonorrhoeae* to produce attachment fimbriae and the ability of *Streptococcus pneumoniae* to produce a capsule when needed.

Antibiotics That Affect Transcription and Translation

Naturally occurring cell components are not the only agents capable of modifying gene expression. Sometimes the drugs we use to treat microbial infections are designed to react with DNA, RNA, or ribosomes and thereby alter the microbe's genetic expression. Treatment with these drugs is based on an important premise: that growth of the bacterium will be inhibited by blocking its protein-synthesizing machinery selectively, without disrupting the cell growth of the human host.

Drugs that inhibit protein synthesis may act on transcription or translation. For example, the rifamycins used in tuberculosis treatment bind to RNA polymerase, blocking the initiation step of transcription. Rifamycins are selectively more active against bacterial RNA polymerase than they are against eukaryotic RNA polymerase. Actinomycin D binds to bacterial DNA and halts mRNA chain elongation, but it also binds to human DNA. For this reason, it is very toxic to humans and never used to treat bacterial infections, though it can be used in tumor treatment.

Ribosomes are frequent targets of antibiotics. They inhibit ribosomal function and ultimately protein synthesis. The value and safety of these antibiotics again depend upon the different susceptibility of bacterial and eukaryotic ribosomes. One problem with drugs that selectively disrupt bacterial ribosomes is that the mitochondria of humans contain a bacterial type of ribosome, and these drugs may inhibit the function of the host's mitochondria. One group of antibiotics (including erythromycin and spectinomycin) prevents translation by interfering with the attachment of mRNA to ribosomes.

6.3 Learning Outcomes—Assess Your Progress

14. Define the term *operon* and explain one advantage it provides to a bacterial cell.
15. Differentiate between repressible and inducible operons and provide an example of each.
16. List two antibiotic drugs and their targets within the transcription and translation machinery.

6.4 DNA Recombination Events

Genetic recombination through sexual reproduction is an important means of genetic variation in eukaryotes. Although bacteria have no exact equivalent to sexual reproduction, they exhibit a primitive means for sharing or recombining parts of their genome. An event in which one bacterium donates DNA to another bacterium is termed **recombination,** in the end creating a new strain

Table 6.3 Types of Horizontal Gene Transfer in Bacteria

Examples of Mode	Factors Involved	Direct or Indirect*	Examples of Products of Transferred Genes
Conjugation	Donor cell with pilus Fertility plasmid in donor Both donor and recipient alive Bridge forms between cells to transfer DNA	Direct	Drug resistance; resistance to metals; toxin production; enzymes; adherence molecules; degradation of toxic substances; uptake of iron
Transformation	Free donor DNA (fragment) Live; competent recipient cell	Indirect	Polysaccharide capsule; unlimited with cloning techniques
Transduction	Donor is lysed bacterial cell; Defective bacteriophage is carrier of donor DNA; Live recipient cell of same species as donor	Indirect	Toxins; enzymes for sugar fermentation; drug resistance

**Direct* means the donor and recipient cells are in contact during exchange; *indirect* means they are not.

different from both the donor and the original recipient strain. Recombination in bacteria depends in part on the fact that bacteria contain extrachromosomal DNA (plasmids) that are capable of moving between cells. Genetic exchange has a tremendous effect on the genetic diversity of bacteria. It provides additional genes for resistance to drugs, new nutritional and metabolic capabilities, and increased virulence and adaptation to the environment.

In general, any organism that contains genes that originated in another organism is called a **recombinant.**

Horizontal Gene Transfer in Bacteria

Any transfer of DNA that results in organisms acquiring new genes that did not come from their parent organisms is called **horizontal gene transfer.** (Acquiring genes from parent organisms during reproduction is vertical gene transfer.) For decades, it has been known that bacteria engage in horizontal gene transfer. It is now becoming clear that eukaryotic organisms—including humans—also engage in horizontal gene transfer, often aided and abetted by microbes such as viruses. This revelation has upended traditional views about eukaryotic evolution, taxonomy, and even "human-ness." Remember in chapter 1 the assertion that 40% to 50% of DNA in humans comes from nonhuman species, transferred to us via viruses? Here, we will study the mechanisms used by bacteria to acquire genes horizontally.

DNA transfer between bacterial cells typically involves small pieces of DNA in the form of plasmids or chromosomal fragments. Plasmids are small, circular pieces of DNA that contain their own origin of replication and therefore can replicate independently of the bacterial chromosome. Plasmids are found in many bacteria (as well as some fungi) and typically contain, at most, only a few dozen genes. Although plasmids are not usually necessary for bacterial survival, they often carry useful traits, such as antibiotic resistance. Chromosomal fragments that have escaped from a broken bacterial cell are also commonly involved in the transfer of genetic information between cells. An important difference between plasmids and fragments is that while a plasmid has its own origin of replication and is stably replicated and inherited, chromosomal fragments must integrate themselves into the bacterial chromosome in order to be replicated and eventually passed to offspring cells.

Depending on the mode of transmission, the means of genetic recombination in bacteria is called conjugation, transformation, or transduction. **Conjugation** requires the attachment of two related species and the formation of a cellular bridge that can transport DNA. **Transformation** entails the transfer of naked DNA and requires no special vehicle. **Transduction** is DNA transfer mediated through the action of a bacterial virus **(table 6.3).**

Conjugation: Bacterial "Sex"

Conjugation is a type of genetic exchange in which a plasmid or other genetic material is transferred by a donor cell to a recipient cell via a direct connection **(figure 6.21).** Both gram-negative and gram-positive cells can conjugate. In gram-negative cells, the donor has a plasmid known as the **fertility (F) factor** that guides the synthesis of a conjugative **pilus.** The recipient cell has a recognition site on its surface. An F^+ cell has the F plasmid (donor) and an F^- cell lacks it (recipient). Contact is made when a pilus grows out from the F^+ cell, attaches to the surface of the F^- cell, contracts, and draws the two cells together (as shown in figure 6.21; see also figure 4.12). In both gram-positive and gram-negative cells, an opening is created between the connected cells, and the replicated DNA passes across from one cell to the other. The DNA probably does not pass through the pilus—that structure is used to bring the cells in contact. The actual transfer takes place through membrane secretion systems. Conjugation is a "conservative" process, in that the donor bacterium generally retains (conserves) a copy of the genetic material being transferred.

There are many different kinds of conjugative plasmids. One of the best understood plasmids is the F factor in *E. coli,* which exhibits these patterns of transfer:

1. The donor (F^+) cell makes a copy of its F factor and transmits this to a recipient (F^-) cell. The F^- cell is changed into an F^+ cell capable of producing a pilus and conjugating with other cells. No additional donor genes are transferred at this time.
2. In a variation on that process, called high-frequency recombination (Hfr), the plasmid becomes integrated into the donor chromosome before instigating transfer to the recipient cell.

The term *high-frequency recombination* was adopted to denote a cell with an integrated F factor that transmits its chromosomal genes. These genes become integrated into recipient chromosomes at a very high frequency.

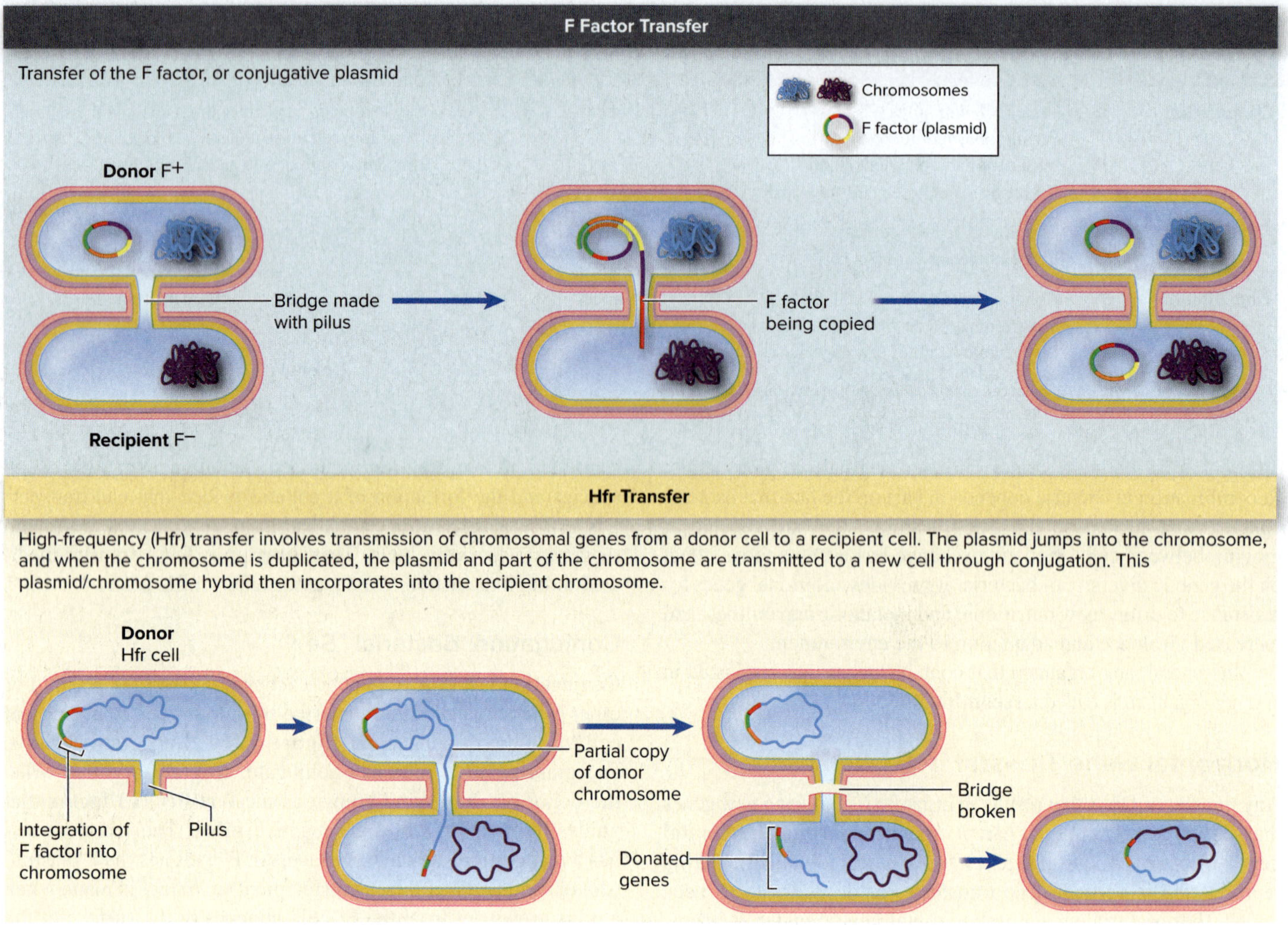
F Factor Transfer
Transfer of the F factor, or conjugative plasmid
Chromosomes
F factor (plasmid)
Donor F+
Bridge made with pilus
F factor being copied
Recipient F−
Hfr Transfer
High-frequency (Hfr) transfer involves transmission of chromosomal genes from a donor cell to a recipient cell. The plasmid jumps into the chromosome, and when the chromosome is duplicated, the plasmid and part of the chromosome are transmitted to a new cell through conjugation. This plasmid/chromosome hybrid then incorporates into the recipient chromosome.
Donor Hfr cell
Integration of F factor into chromosome
Pilus
Partial copy of donor chromosome
Bridge broken
Donated genes

Figure 6.21 Conjugation: genetic transmission through direct contact between two cells.

In Hfr transfer, the F factor can direct a more comprehensive transfer of part of the donor chromosome to a recipient cell. This transfer occurs through duplication of the DNA, after which one strand of DNA is retained by the donor, and the other strand is transported across to the recipient cell. The F factor may or may not be transferred during this process. The transfer of an entire chromosome takes about 100 minutes, but the connection between cells is ordinarily broken before this time, and rarely is the entire genome of the donor cell transferred.

Conjugation has great importance in medicine. Special **resistance (R) plasmids,** or **factors,** that carry genes for resisting antibiotics and other drugs are commonly shared among bacteria through conjugation. Transfer of R factors can give new bacteria resistance to multiple antibiotics such as tetracycline, chloramphenicol, streptomycin, sulfonamides, and penicillin. Other types of R factors carry genes for resistance to heavy metals (nickel and mercury) or for synthesizing virulence factors (toxins, enzymes, and adhesion molecules) that increase the pathogenicity of the bacterial strain. One important example of how a plasmid can add to the diease-causing capability of a bacterium is the Shiga toxin, which is a toxin produced by some *E. coli* strains. It turns a mild infection into a potentially deadly one. The Shiga toxin gene was transferred from *Shigella* species to *E. coli* species through conjugation.

Transformation: Capturing DNA from Solution

Transformation is the acceptance by a bacterial cell of small fragments of DNA from the surrounding environment **(figure 6.22).** Cells that are capable of accepting genetic material in this way are termed **competent.** The new DNA passes through the outer membrane (in gram-negatives) using special protein channels, and then moves through the cell wall via DNA-binding proteins. The DNA is then processed by the cytoplasmic membrane and transported into the cytoplasm, where some of it is inserted into the bacterial chromosome. Transformation is a natural event found in several groups of gram-positive and gram-negative bacterial species.

This phenomenon was discovered in a famous experiment that elegantly illustrates both how transformation works and how the exchange of DNA between bacteria in a single host can have real effects on the host. The experiment was conducted in the late

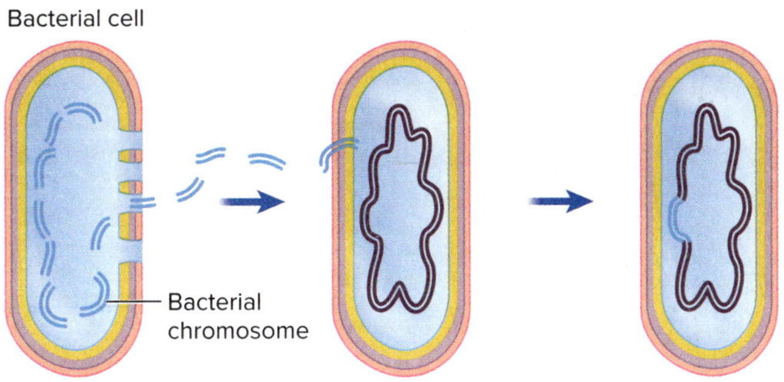
Bacterial cell
Bacterial chromosome

Figure 6.22 Transformation. DNA from dead cells is released into the environment and taken up into living cells. Some portion of the DNA may recombine into the genome of the recipient cell.

1920s by the English biochemist Frederick Griffith working with *Streptococcus pneumoniae* and laboratory mice. This bacterium can exist in two different forms: (1) those that have a capsule have a smooth (S), glassy colony appearance and are capable of causing severe disease; and (2) those that do not have a capsule have a rough (R) colony appearance and are nonpathogenic. (Recall that the capsule protects a bacterium from the phagocytic host defenses.) To set the groundwork, Griffith showed that when mice were injected with a live, virulent (S) strain, they soon died **(figure 6.23).** Mice injected with a live, nonvirulent (R) strain remained alive and healthy. Next, he tried a variation on this theme. First, he heat-killed an S strain and injected it into mice, which remained healthy. Then came the ultimate test: Griffith injected both dead S cells and live R cells into mice, with the result that the mice died from pneumococcal blood infection. If killed

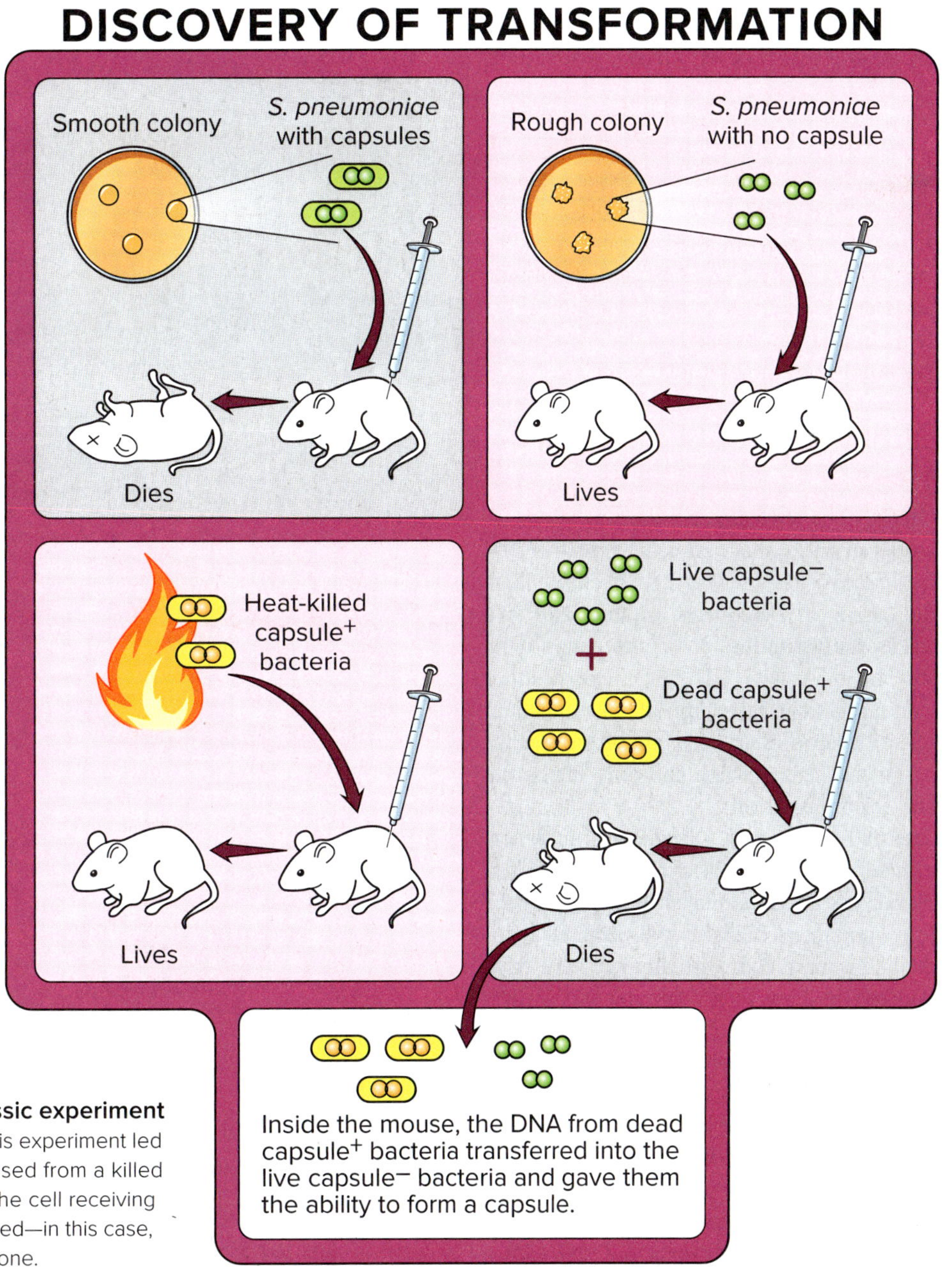
DISCOVERY OF TRANSFORMATION
Smooth colony
S. pneumoniae with capsules
Dies
Rough colony
S. pneumoniae with no capsule
Lives
Heat-killed capsule+ bacteria
Lives
Live capsule− bacteria
+
Dead capsule+ bacteria
Dies
Inside the mouse, the DNA from dead capsule+ bacteria transferred into the live capsule− bacteria and gave them the ability to form a capsule.

Figure 6.23 Griffith's classic experiment in transformation. In essence, this experiment led to the understanding that DNA released from a killed cell can be acquired by a live cell. The cell receiving this new DNA is genetically transformed—in this case, from a nonvirulent strain to a virulent one.

bacterial cells do not come back to life and the nonvirulent live strain was harmless, why did the mice die? Although he did not know it at the time, Griffith had demonstrated that dead S cells, while passing through the body of the mouse, broke open and released some of their DNA (by chance, that part containing the genes for making a capsule). A few of the live R cells subsequently picked up this loose DNA and were transformed by it into virulent, capsule-forming strains. Later studies supported the concept that a chromosome released by a lysed cell breaks into fragments small enough to be accepted by a recipient cell and that DNA, even from a dead cell, retains its genetic code.

Disease Connection

Streptococcus pneumoniae is a versatile human pathogen. It causes pneumonia, ear infections, meningitis, and other conditions. Its capsule is vital for its disease-causing capacity.

Because transformation requires no special appendages and the donor and recipient cells do not have to be in direct contact, the process is useful for certain types of recombinant DNA technology. With this technique, foreign genes from a completely unrelated organism are inserted into a plasmid, which is then introduced into a competent bacterial cell through transformation.

Transduction: The Case of the Piggyback DNA

Bacterial viruses, called bacteriophages, serve as genetic vectors (an entity that can bring foreign DNA into a cell). The process by which a bacteriophage serves as the carrier of DNA from a donor cell to a recipient cell is **transduction.** It occurs naturally in a broad spectrum of bacteria. There will be a donor bacterium and a recipient bacterium in each transduction event. The bacteria must be of the same species because any given bacteriophage can only infect one particular species of bacterium.

There are two versions of transduction. In *generalized transduction* **(figure 6.24),** random fragments of disintegrating host DNA are taken up by the phage while it is assembling itself inside the bacterial cytoplasm inside the bacterial cell. Virtually any gene from the bacterium can be transmitted through this means. In 1951 a microbiologist named Esther Lederburg discovered another kind of transduction, called *specialized transduction* **(figure 6.25).** In this process, a highly specific part of the host genome is incorporated into the virus. This specificity is explained by the prior existence of a bacteriophage DNA inserted in a fixed site on the bacterial chromosome. When activated, that DNA separates from the bacterial chromosome, carrying a

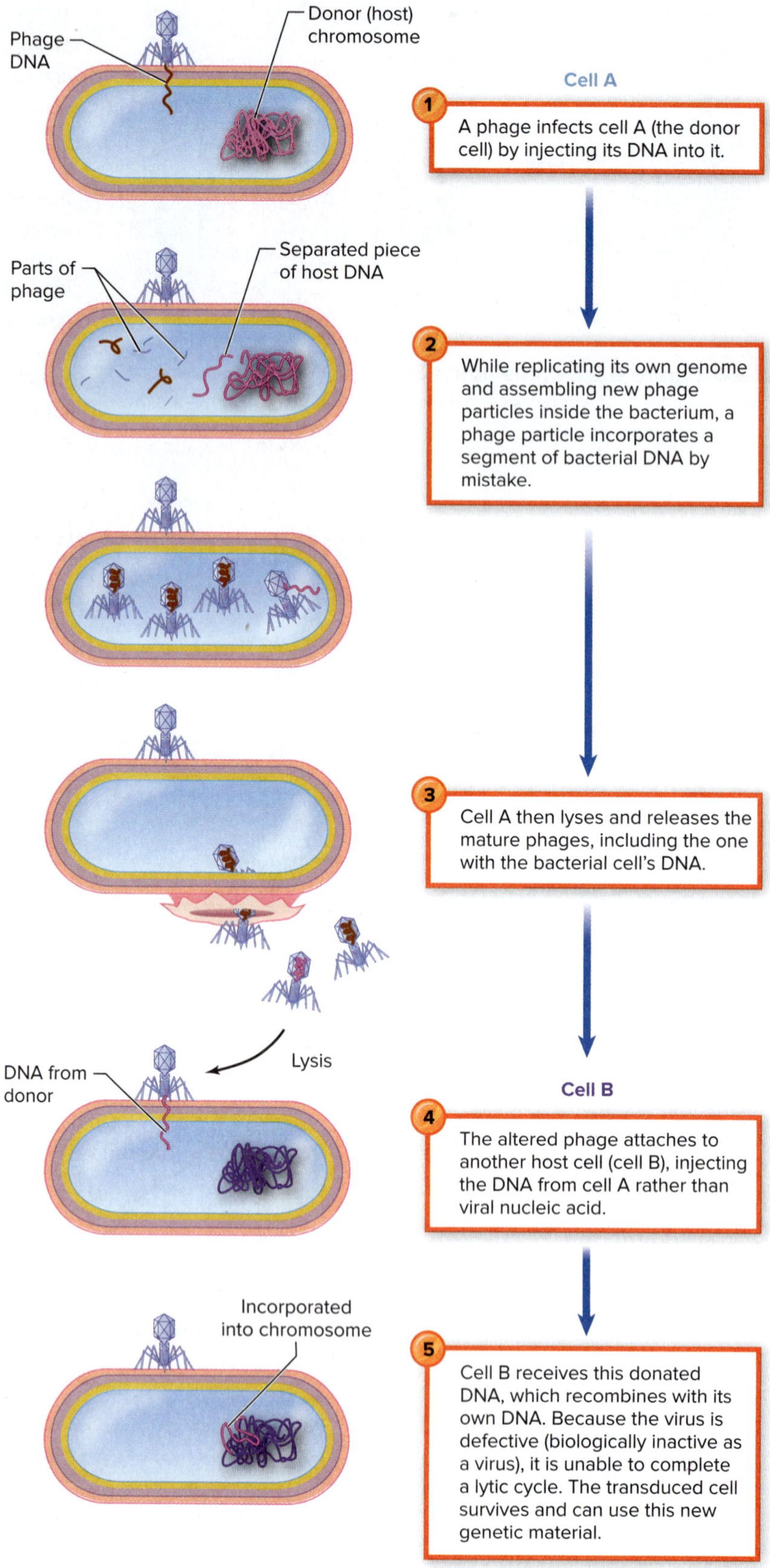
Phage DNA
Donor (host) chromosome
Cell A
1 A phage infects cell A (the donor cell) by injecting its DNA into it.
Parts of phage
Separated piece of host DNA
2 While replicating its own genome and assembling new phage particles inside the bacterium, a phage particle incorporates a segment of bacterial DNA by mistake.
3 Cell A then lyses and releases the mature phages, including the one with the bacterial cell's DNA.
Lysis
DNA from donor
Cell B
4 The altered phage attaches to another host cell (cell B), injecting the DNA from cell A rather than viral nucleic acid.
Incorporated into chromosome
5 Cell B receives this donated DNA, which recombines with its own DNA. Because the virus is defective (biologically inactive as a virus), it is unable to complete a lytic cycle. The transduced cell survives and can use this new genetic material.

Figure 6.24 Generalized transduction: genetic transfer by means of a virus carrier.

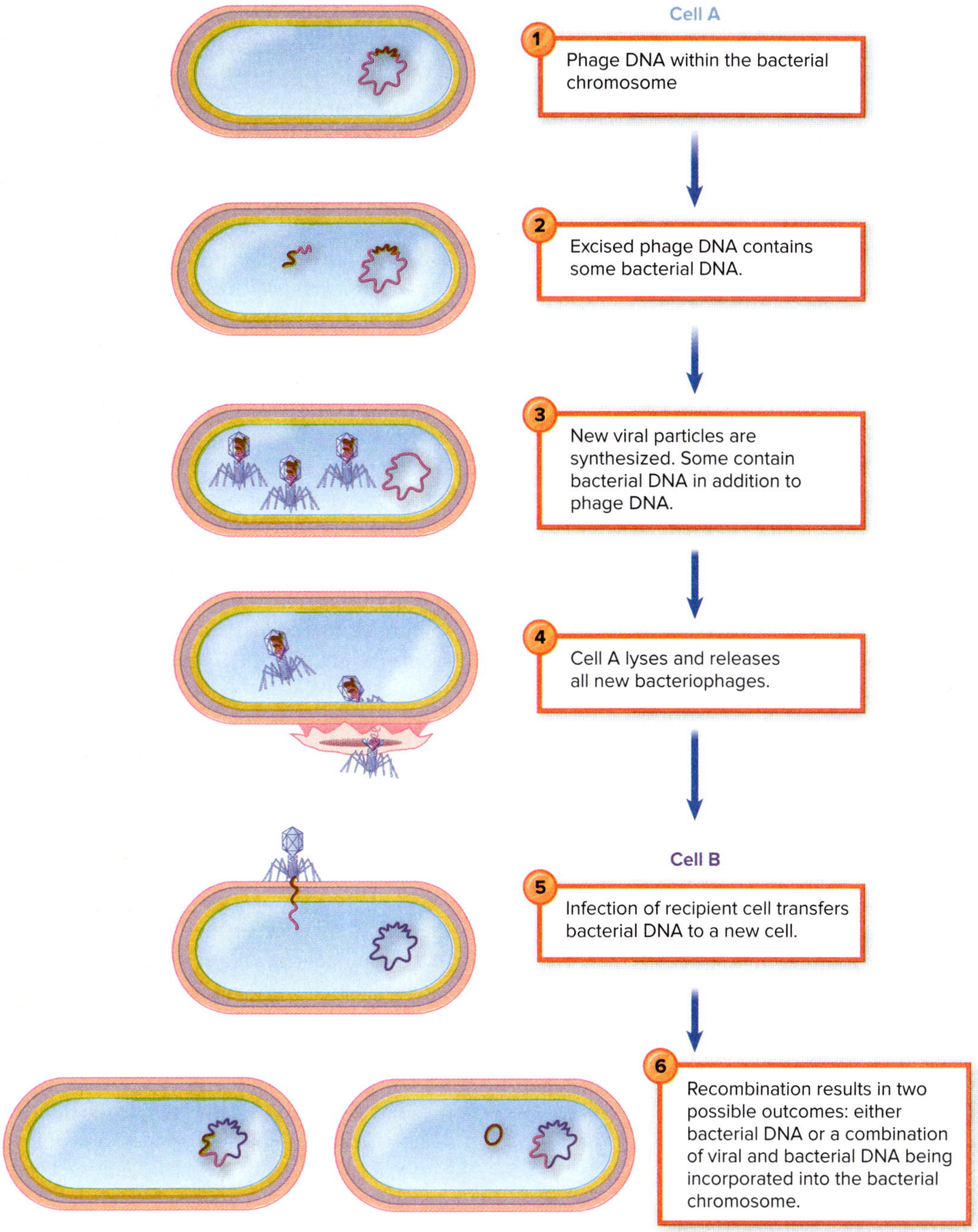
Cell A
1 Phage DNA within the bacterial chromosome
2 Excised phage DNA contains some bacterial DNA.
3 New viral particles are synthesized. Some contain bacterial DNA in addition to phage DNA.
4 Cell A lyses and releases all new bacteriophages.
Cell B
5 Infection of recipient cell transfers bacterial DNA to a new cell.
6 Recombination results in two possible outcomes: either bacterial DNA or a combination of viral and bacterial DNA being incorporated into the bacterial chromosome.

Figure 6.25 **Specialized transduction: transfer of specific genetic material by means of a virus carrier.**

small segment of host genes with it. Then these specific viral-host gene combinations are incorporated into the viral particles and carried to another bacterial cell. Lederburg's discovery of the type of bacteriophage that could do this revolutionized the study of bacterial genetics.

Several cases of specialized transduction have medical importance. The virulent strains of bacteria such as *Corynebacterium diphtheriae, Clostridium* spp., and *Streptococcus pyogenes* all produce toxins with profound physiological effects, whereas nonvirulent strains do not produce these toxins. It turns out that the toxins are produced by bacteriophage genes that have been introduced by transduction. Only those bacteria infected with a phage are toxin formers.

Transposable Elements

Another type of genetic transfer of great interest involves **transposons,** or transposable elements (TEs). TEs have the ability to shift from one part of the genome to another and so are termed "jumping genes." When the idea of their existence in corn plants was first proposed by geneticist Barbara McClintock (also in 1951), it was greeted with great skepticism because it had long been believed that the location of a given gene was set and that genes did not or could not move around. Now it is evident that jumping genes are widespread among cells and viruses.

All TEs share the general characteristic of traveling from one location to another on the genome—from one chromosomal site to another, from a chromosome to a plasmid, or from a plasmid to a chromosome. Because TEs can also be found on plasmids, they can also be transmitted from one cell to another in bacteria and a few eukaryotes. Some TEs replicate themselves before jumping to the next location, and others simply move without replicating first.

TEs contain DNA that codes for the enzymes needed to cut out and reintegrate the TE at another site in the genome. On either side of the coding region of the DNA are sequences called tandem repeats, which mark the point at which the TE is removed or reinserted into the genome. The smallest TEs consist of only these two genetic sequences and are often referred to as **insertion elements.** A type of TE called a **retrotransposon** can transcribe DNA into RNA and then back into DNA for insertion in a new location. Other TEs contain additional genes that provide traits such as antibiotic resistance or toxin production.

The overall effect of TEs—to scramble the genetic language—can be beneficial or harmful to the cell, depending upon where insertion occurs in a chromosome, what kinds of genes are relocated, and the type of cell involved. In bacteria, TEs are known to be involved in

1. changes in traits such as colony morphology, pigmentation, and antigenic characteristics;
2. replacement of damaged DNA; and
3. the transfer of drug resistance between bacteria.

Pathogenicity Islands: Special "Gifts" of Horizontal Gene Transfer?

Some of the horizontally transferred genes in bacteria have the ability to make their new hosts pathogenic, or able to cause disease. These are termed **pathogenicity islands (PAIs).** These islands contain multiple genes that are coordinated to create a new trait in the bacterium, such as the ability to scavenge iron (important for the bacterium causing the plague, *Yersinia pestis*) or the ability to produce exotoxins (seen in *Staphylococcus aureus*). The islands are usually flanked by sequences that look like genes for TE enzymes. We now know that organisms "share" their genes, sometimes in great chunks, with one another, essentially leap-frogging the evolution process by shuffling genes this way.

6.4 Learning Outcomes—Assess Your Progress

17. Explain the defining characteristics of a recombinant organism.
18. Describe three forms of horizontal gene transfer used in bacteria.

6.5 Mutations: Changes in the Genetic Code

As precise and predictable as the rules of genetic expression seem, permanent changes do occur in the genetic code. This genetic change is the driving force of evolution. For example, a pigmented bacterium can lose its ability to form pigment, or a strain of the malarial parasite can develop resistance to a drug. Any change to the nucleotide sequence in the genome is called a **mutation.** Mutations are most noticeable when the genotypic change leads to a change in phenotype. Mutations can involve the loss of base pairs, the addition of base pairs, or a rearrangement in the order of base pairs. This is different from genetic recombination, in which microbes transfer whole segments of genetic information among themselves.

A microorganism that exhibits a natural, nonmutated characteristic is known as a **wild type,** or wild strain with respect to that trait. You may ask, in a constantly changing population of microbes, what is the natural, nonmutated state? For that reason, most scientists prefer to define *wild type* as the trait present in the highest numbers in a population. If a microorganism bears a mutation, it is called a **mutant strain.** Mutant strains can show variance in morphology, nutritional characteristics, genetic control mechanisms, resistance to chemicals, temperature preference, or nearly any type of enzymatic function. Mutant strains are very useful for tracking genetic events, unraveling genetic organization, and pinpointing genetic markers.

Causes of Mutations

Mutations can be spontaneous or induced. A **spontaneous mutation** is a random change in the DNA arising from errors in replication that occur randomly. Mutation rates vary tremendously, from one mutation in 10^5 replications (a high rate) to one mutation in 10^{10} replications (a low rate). The rapid rate of bacterial reproduction allows these mutations to be observed more easily in bacteria than in most other organisms.

Induced mutations result from exposure to known **mutagens,** which are physical or chemical agents that damage DNA **(table 6.4).** Chemical mutagenic agents act in a variety of ways to change the DNA. Some agents insert completely across the DNA helices between adjacent bases to produce mutations that distort the helix. Analogs of the nitrogen bases (5-bromodeoxyuridine and 2-aminopurine, for example) are chemical mimics of natural bases that are incorporated into DNA during replication. Insertion of these abnormal bases leads to mistakes in base-pairing. Many chemical mutagens also act as carcinogens, or cancer-causing agents, when vertebrates are exposed to them.

Physical agents, especially radiation, can also alter DNA. High-energy gamma rays and X rays introduce major physical

Table 6.4 Selected Mutagenic Agents and Their Effects

Agent	Effect
Chemical	
Nitrous acid, bisulfite	Remove an amino group from some bases
Ethidium bromide	Inserts between the paired bases
Acridine dyes	Cause frameshifts due to insertion between base pairs
Nitrogen base analogs	Compete with natural bases for sites on replicating DNA
Radiation	
Ionizing (gamma rays, X rays)	Forms free radicals that cause single or double breaks in DNA
Ultraviolet	Causes cross-links between adjacent pyrimidines

changes into DNA, accumulating breaks that may not be repairable. Ultraviolet (UV) radiation induces abnormal bonds between adjacent pyrimidines that prevent normal replication. Exposure to large doses of radiation can be fatal, which is why radiation is so effective in microbial control. It can also be carcinogenic in animals. (The intentional use of UV to control microorganisms is described in another chapter.)

Categories of Mutations

Mutations range from large mutations, in which large genetic sequences are gained or lost, to small ones that affect only a single base on a gene. These latter mutations, which involve addition, deletion, or substitution of single bases, are called **point mutations.**

To understand how a change in DNA influences the cell, remember that the DNA code appears in a particular order of triplets (three bases) that is transcribed into mRNA codons, each of which specifies an amino acid. A permanent alteration in the DNA that is copied into mRNA and translated can change the structure of the protein. A change in a protein can likewise change the structure and function of a cell. Some mutations have a harmful effect on the cell, leading to cell dysfunction or death. These are called *lethal mutations. Neutral mutations* produce neither adverse nor helpful changes. Of course, mutations can also be beneficial if they provide the cell with a useful change in structure or physiology.

Any change in the code that leads to placement of a different amino acid is called a **missense mutation** (**table 6.5*b*** shows how missense mutations look). A missense mutation can

1. create a faulty, nonfunctional (or less functional) protein;
2. produce a protein that functions in a different manner; or
3. cause no significant alteration in protein function.

A **nonsense mutation,** on the other hand, changes a normal codon into a stop codon that does not code for an amino acid and stops the production of the protein wherever it occurs. A nonsense mutation almost always results in a nonfunctional protein. (**Table 6.5*d*** shows a nonsense mutation resulting from a frameshift, which is described in the next paragraph.) A **silent mutation (table 6.5*c*)** alters a base but does not change the amino acid and therefore has no effect. For example, because of the redundancy of the code, ACU, ACC, ACG, and ACA all code for threonine, so a mutation that changes only the last base will not alter the sense of the message in any way. A **back-mutation** occurs when a gene that has undergone mutation reverses (mutates back) to its original base composition.

Mutations also occur when one or more bases are inserted into or deleted from a newly synthesized DNA strand. This type of mutation, known as a **frameshift (table 6.5*d,e*),** is so named because the reading frame of the mRNA has been changed. Frameshift mutations nearly always result in a nonfunctional protein because every amino acid after the mutation is different from what was coded for in the original DNA. Also note that insertion or deletion of bases in multiples of three (3, 6, 9, etc.) results in the addition or deletion of amino acids but does not disturb the reading frame. The effects of all of these types of mutations can be seen in the table.

Repair of Mutations

Earlier, we mentioned that DNA has a proofreading mechanism to repair mistakes in replication that may otherwise become permanent. Because mutations are potentially lethal, the cell has additional systems for finding and repairing DNA that has been damaged by various mutagenic agents and processes. Most ordinary DNA damage is resolved by enzymatic systems designed to find and fix such defects.

DNA that has been damaged by ultraviolet radiation can be restored by photoactivation, or light repair. This repair mechanism requires visible light and a light-sensitive enzyme, DNA photolyase, which can detect and attach to the damaged areas. Ultraviolet repair mechanisms are successful only for a relatively small number of UV mutations. Cells cannot repair severe, widespread damage and will die. In humans, the genetic disease *xeroderma pigmentosa* is due to nonfunctioning genes for enzymes responsible for cutting out pyrimidine dimers caused by UV light. Persons suffering from this rare disorder develop severe skin cancers.

Mutations can be excised by a series of enzymes that remove the incorrect bases and add the correct ones. This process is known as

Table 6.5 Categories of Point Mutations and Their Effects

(a)	DNA	TAC	TGG	CTG	CTC	TAC	TTT...	Normal gene
	RNA	AUG	ACC	GAC	GAG	AUG	AAA...	
	Protein	Met	Thr	Asp	Glu	Met	Lys...	
(b)	DNA	TAC	TGG	CTT	CTC	TAC	TTT...	Missense mutation: leading to amino acid switch (may or may not function well)
	RNA	AUG	ACC	GAA	GAG	AUG	AAA...	
	Protein	Met	Thr	Glu	Glu	Met	Lys...	
(c)	DNA	TAC	TGG	CTA	CTC	TAC	TTT...	Base substitution: silent (no change in function)
	RNA	AUG	ACC	GAU	GAG	AUG	AAA...	
	Protein	Met	Thr	Asp	Glu	Met	Lys...	
			→ G					**Frameshift mutation**
(d)	DNA	TAC	TGC	TGC	TCT	ACT	TT...	Deletion mutation (d) ↓
	RNA	AUG	ACG	ACG	AGA	UGA	AA...	Both lead to frameshifts and can lead to premature stop codons and/or poorly functioning protein
	Protein	Met	Thr	Thr	Arg	STOP		
			Frameshift and premature stop					
(e)	DNA	TAC	TGG	GCT	GCT	CTA	CTT...	↑ Insertion mutation (e)
	RNA	AUG	ACC	CGA	CGA	GAU	GAA...	
	Protein	Met	Thr	Arg	Arg	Asp	Glu...	
			Frameshift					

excision repair. First, enzymes break the bonds between the bases and the sugar phosphate strand at the site of the error. A different enzyme subsequently removes the defective bases one at a time, leaving a gap that will be filled in by DNA polymerase I and ligase **(figure 6.26).** A repair system can also locate mismatched bases that were missed during proofreading, for example, C mistakenly paired with A, or G with T. The base must be replaced soon after the mismatch is made, or it will not be recognized by the repair enzymes.

The Ames Test

New agricultural, industrial, and medicinal chemicals are constantly being added to the environment. The discovery that many such compounds are mutagenic and that many of these mutagens are linked to cancer is significant. Although animal testing has been a standard method of detecting chemicals with carcinogenic potential, a more rapid screening system called the **Ames test** is also commonly used. In this test, the experimental subjects are bacteria whose gene expression and mutation rate can be readily observed and monitored. The premise is that any chemical capable of mutating bacterial DNA could similarly mutate mammalian (including human) DNA and is therefore potentially hazardous.

One organism commonly used in the Ames test is a mutant strain of *Salmonella typhimurium* that has lost the ability to synthesize the amino acid histidine. This defect is highly susceptible to back-mutation because the strain also lacks DNA repair mechanisms. Mutations that cause reversion to the wild strain, which is capable of synthesizing histidine, occur spontaneously at a low rate. A test agent is considered a mutagen if it enhances the rate of back-mutation beyond levels that would occur spontaneously. The procedure is outlined in **figure 6.27.** The Ames test has proved

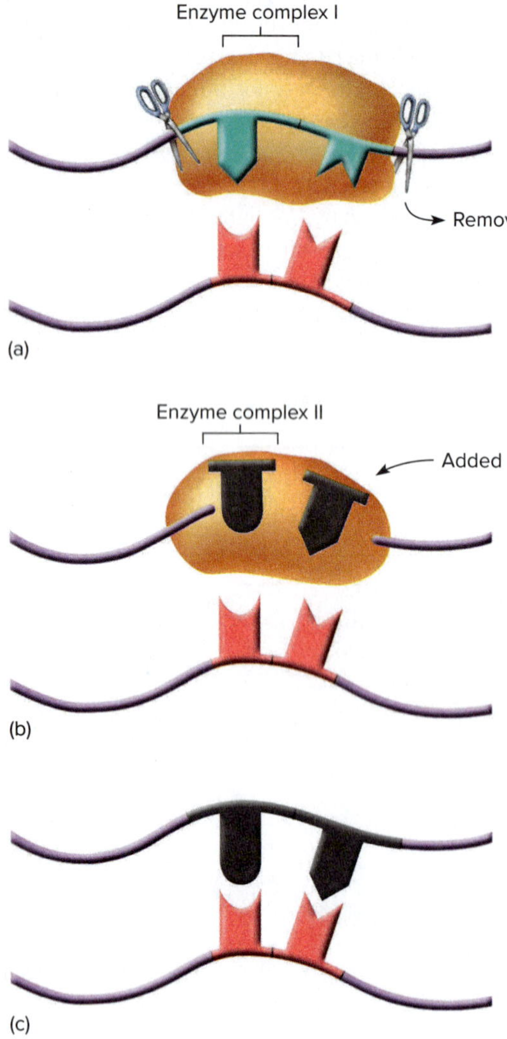
Enzyme complex I
Removed
(a)
Enzyme complex II
Added
(b)
(c)

Figure 6.26 Excision repair of mutation by enzymes. **(a)** The first enzyme complex recognizes one or several incorrect bases and removes them. **(b)** The second complex (DNA polymerase I and ligase) places correct bases and seals the gaps. **(c)** Repaired DNA.

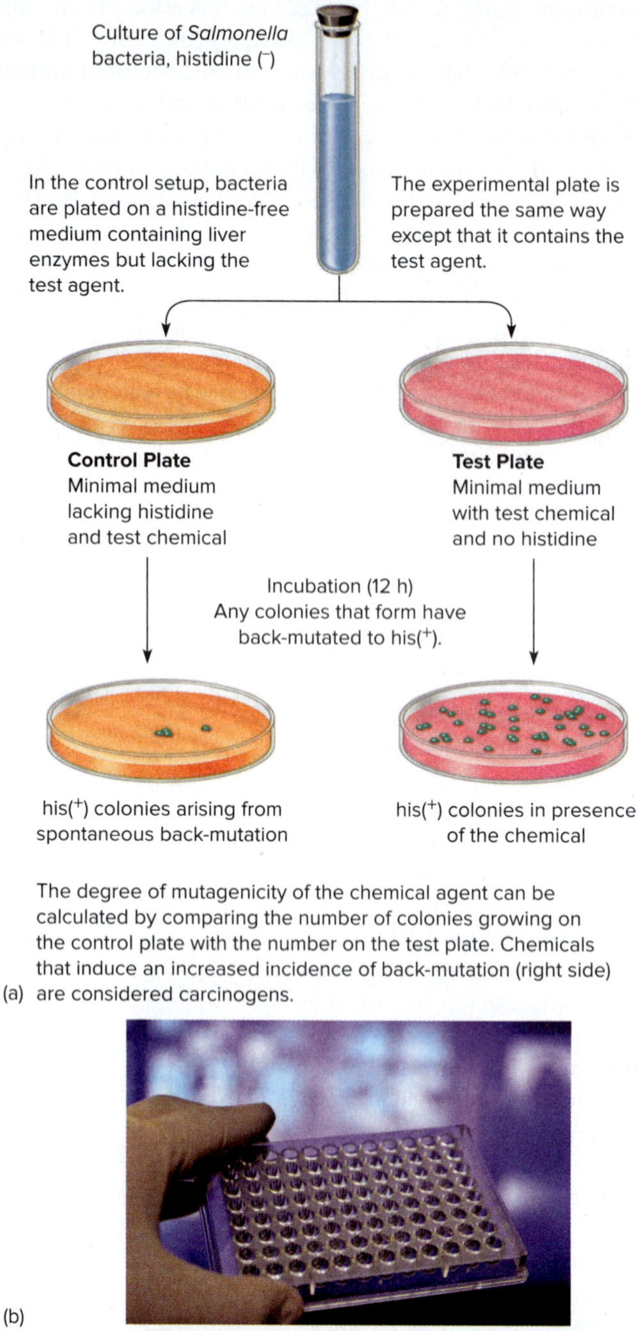
Culture of *Salmonella* bacteria, histidine (⁻)
In the control setup, bacteria are plated on a histidine-free medium containing liver enzymes but lacking the test agent.
The experimental plate is prepared the same way except that it contains the test agent.
Control Plate
Minimal medium lacking histidine and test chemical
Test Plate
Minimal medium with test chemical and no histidine
Incubation (12 h)
Any colonies that form have back-mutated to his(+).
his(+) colonies arising from spontaneous back-mutation
his(+) colonies in presence of the chemical
The degree of mutagenicity of the chemical agent can be calculated by comparing the number of colonies growing on the control plate with the number on the test plate. Chemicals that induce an increased incidence of back-mutation (right side) are considered carcinogens.
(a)
(b)

Figure 6.27 The Ames test. A typical test. Very often the test is performed in dishes called microtiter plates (b) instead of in large petri dishes.

invaluable for screening an assortment of environmental and dietary chemicals for mutagenicity and carcinogenicity without resorting to animal studies. But because many mutagens affect bacteria differently than humans, the Ames test is considered a first step that must be followed by more rigorous testing.

Positive and Negative Effects of Mutations

Many mutations are not repaired. How the cell copes with them depends on the nature of the mutation and the strategies available to that organism. Mutations are permanent and heritable and will be passed on to the offspring of organisms and new viruses. They become a long-term part of the gene pool.

If a mutation leading to a nonfunctional protein occurs in a gene for which there is only a single copy, as in haploid or simple organisms, the cell will probably die. This happens when certain mutant strains of *E. coli* acquire mutations in the genes needed to repair damage by UV radiation. Mutations of the human genome affecting the action of a single protein (mostly enzymes) are responsible for thousands of diseases.

Although most spontaneous mutations are not beneficial, a small number contribute to the success of the individual and the population by creating variant strains with alternate ways of expressing a trait. Microbes are not "aware" of this advantage and do not direct these changes; they simply respond to the environment they encounter. Those organisms with beneficial mutations can more readily adapt, survive, and reproduce. In the long-range view, mutations and the variations they produce are the raw materials for change in the population and, thus, for evolution.

Mutations that create variants occur frequently enough that any population contains mutant strains for a number of characteristics, but as long as the environment is stable, these mutants will usually never make up more than a tiny percentage of the population. When the environment changes, however, it can become hostile for the survival of certain individuals, and only those microbes bearing protective mutations will be equipped to survive in the new environment. In this way, the environment naturally selects certain mutant strains that will reproduce, give rise to subsequent generations, and, in time, be the dominant strain in the population. Through these means, a change that confers an advantage during selection pressure will likely be retained by the population. One of the clearest models for this sort of selection and adaptation is acquired drug resistance in bacteria.

6.5 Learning Outcomes—Assess Your Progress

19. Define the term *mutation* and discuss one positive and one negative example of it in microorganisms.
20. Differentiate among frameshift, nonsense, silent, and missense mutations.

Media Under The Microscope Wrap-Up

The **intended message** of the Reuters article was to debunk a popular video's message spreading false information about COVID-19 vaccines in development.

A critical reading of the article, which was essentially a critical review of the video, involves asking what the agenda of the article's authors might have been. Because the article cited both a scientist and a physician's analysis of the video, it was clearly trying to determine the facts of the matter. The article also included hyperlinks to fact-based websites that explained technical terms. All of this reliable evidence allowed us to compare the facts against the claims in the video.

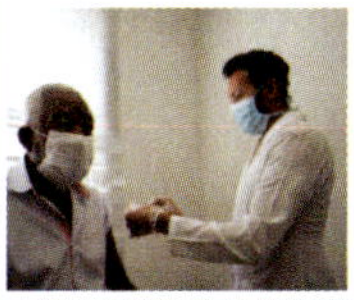

FG Trade/Getty Images

To **interpret** this article to your friends, you might need to explain to them a little bit about how DNA and RNA in a cell work. This would help them see that the facts in the article line up with what we know about cells and organisms.

My **grade** for this article is an A.

Source: *Reuters, "Fact Check: False Claim: A COVID-19 Vaccine Will Genetically Modify Humans," May 18, 2020.*

Study Smarter: Better Together

These activities are designed for you to use on your own with a study group—either a face-to-face group or a virtual one, consisting of 3–5 members. Studying together can be very helpful, but there are effective and ineffective ways to do it. For example, getting together without a clear structure is often not a good use of your time. Use your time efficiently by using one or more of the exercises below.

FACE-TO-FACE GROUPS

Use one or more of the activities below.

Peer Instruction: Assign numbers to your group members to use all semester long. Now look at these five concepts from this chapter. Each group member prepares a 5-minute lesson on the topic corresponding to their number. Don't worry if you have fewer than 5 members; just use however many you have! During your group study time, each member presents their lesson, and the group spends another 5–10 minutes discussing that lesson.

1. DNA replication
2. Transcription
3. Translation
4. Horizontal gene transfer
5. Types of mutations

Concept Maps: Each member of the group should use this list of terms from this chapter to generate their own concept map. This can be hand-drawn or created using software (see Appendix C for guidelines). During group study time, compare each other's concept maps and help each other make sure they are correct. Of course, there are many different "correct" maps. Examining each member's map will help you talk through the varied concepts and how they are related.

Concept Terms:

mRNA	tRNA	primer	rRNA
transcription	translation	DNA	DNA polymerase
RNA polymerase	Okazaki fragments	leading strand	lagging strand

Table Topics: Each group member should identify a concept or topic from this week's class assignments with which they are having trouble and share it during group study time. The other group members can then help to clarify confusing issues or share how they figured it out. Aim for a maximum of 15 minutes per topic. If the topic remains unclear to the group, bring it up during class or use the instructor's office hours or e-mail to ask for help. Taking the time to struggle with a difficult concept first makes your questions much more specific and more likely to yield helpful answers.

VIRTUAL GROUPS

Not everyone has the time or opportunity to meet with group members outside of class time. You or your instructor can create a virtual group using e-mail or the course software.

Weekly Discussion Board: This forum can be used as a way for groups to discuss topics, via e-mail, or other learning management systems or online platforms, before they are covered in class. As each member of the group answers the current week's question, they should send their responses to every other member of their group. It's best to agree on a deadline based on how your class schedule works (Saturday for the next week's topics, for example). Then, after the topic is discussed in class, each member should send a response that all group members will see with a follow-up post on the same topic. If you cover more than one chapter in a week, someone can be designated to choose which chapter Discussion Board question you will use. Or simply decide up front that you will always use the first-chapter-of-the week's question, to keep the schedule simple.

Discussion Question
Construct an argument for why tRNA has a lot of secondary structure, but mRNA does not.

Chapter Summary

MICROBIAL GENETICS

6.1 INTRODUCTION TO GENETICS AND GENES

- Nucleic acids are macromolecules that contain the blueprints of life in the form of genes.
- DNA is the genome for all cells. Some viruses can use RNA as their genomes.
- The entirety of an organism's genetic blueprint is called its genotype. Its phenotype is formed by the genes that are active at any given point in time.
- DNA copies itself by a process called semiconservative replication.

6.2 TRANSCRIPTION AND TRANSLATION

- Information in DNA is converted to proteins by the processes of transcription and translation.
- The DNA code occurs in groups of three bases and this code is copied into RNA as codons.
- Transcription uses RNA polymerase to convert the DNA codes into single-stranded RNA.
- Translation uses tRNAs and rRNA in the form of a ribosome to convert the mRNA message into proteins.
- DNA also contains a great number of non-protein-coding sequences. These sequences often serve as small RNAs with a regulatory function.
- Bacteria transcribe and translate simultaneously in the cytoplasm. Eukaryotes transcribe in the nucleus and then translate in the cytoplasm. Eukaryotes also modify their mRNA transcripts before translation by removing intron sequences.

6.3 GENETIC REGULATION OF PROTEIN SYNTHESIS

- Genes can be turned on and off by specific molecules, which can expose or hide nucleotide codes for transcribing proteins. DNA can also form folded structures that prevent transcription.
- Operons are collections of genes in bacteria that code for products with a coordinated function.
- Operons can be either in the off position, but inducible, or in the on position, and repressible.

6.4 DNA RECOMBINATION EVENTS

- In eukaryotes, genetic recombination occurs through sexual reproduction and via horizontal gene transfer.
- In bacteria, recombination occurs only through horizontal gene transfer.
- The three main types of horizontal gene transfer in bacteria are transformation, conjugation, and transduction.

6.5 MUTATIONS: CHANGES IN THE GENETIC CODE

- Changes in the genetic code can occur through recombination or mutation. Mutation means a change in the DNA nucleotide sequence.
- Mutations can be either spontaneous or induced by exposure to a mutagenic agent.
- All cells have enzymes that repair damaged DNA. When the degree of damage exceeds the ability of the enzymes to make repairs, permanent mutations occur.
- Some mutations provide benefits to their organisms, and allow them to function in environments that were previously hostile for them.

SmartGrid: From Knowledge to Critical Thinking

This *21 Question Grid* takes the topics from this chapter and arranges them with respect to the American Society for Microbiology's Undergraduate Curriculum guidelines—all six of the important "Concepts" as well as the important "Competency" of scientific literacy. Three questions are supplied, which cover chapter content referring to the Concept or Competency in increasing levels of Bloom's taxonomy for learning.

ASM Concept/ Competency	A. Bloom's Level 1, 2—Remember and Understand (Choose one.)	B. Bloom's Level 3, 4—Apply and Analyze	C. Bloom's Level 5, 6—Evaluate and Create
Evolution	**1.** Which of the following processes does not directly involve DNA? **a.** replication **b.** transcription **c.** translation **d.** conjugation	**2.** Using your knowledge of DNA from this chapter, imagine two different ways antibiotic resistance may develop in a bacterium.	**3.** Why would RNA be a more likely original nucleic acid in the first cells on the planet than DNA?
Cell Structure and Function	**4.** Which of the following is a characteristic of RNA? **a.** RNA is double-stranded. **b.** RNA contains thymine, which pairs with adenine. **c.** RNA contains deoxyribose. **d.** RNA molecules are necessary for translation and gene regulation.	**5.** List some advantages and disadvantages to a cell having DNA in the cytoplasm as opposed to in a nucleus.	**6.** Construct an argument for why tRNA contains a lot of secondary structure but mRNA does not.
Metabolic Pathways	**7.** The *lac* operon is usually in the ______ position and is activated by a/an ______ molecule. **a.** on; repressor **b.** off; inducer **c.** on; inducer **d.** off; repressor	**8.** Why would amino acid synthesis pathways be on in the default mode while lactose utilization pathways are off in the default mode?	**9.** Defend this statement: *All of biology is dependent on binding reactions.*
Information Flow and Genetics	**10.** DNA is semiconservative because the ______ strand will become half of the ______ molecule. **a.** RNA; DNA **b.** template; finished **c.** sense; mRNA **d.** codon; anticodon	**11.** Examine the DNA triplets here and determine the amino acid sequence they code for. Then provide a different DNA sequence that will produce the same protein. TAC CAG ATA CAC TCC CCT	**12.** Using the same piece of DNA in question 11 (TAC CAG ATA CAC TCC CCT) create the following: (i) a frameshift mutation; (ii) a silent mutation; and (iii) a nonsense mutation.
Microbial Systems	**13.** Which of the following is not a mechanism of horizontal gene transfer? **a.** spontaneous mutations **b.** transformation **c.** transduction **d.** conjugation	**14.** Can you imagine any scenario in which infection by a bacteriophage is of benefit to the bacterium? Explain.	**15.** If you think in terms of multiple generations of bacteria, can silent mutations in fact have an eventual effect on protein production? How?

ASM Concept/ Competency	A. Bloom's Level 1, 2—Remember and Understand (Choose one.)	B. Bloom's Level 3, 4—Apply and Analyze	C. Bloom's Level 5, 6—Evaluate and Create
Impact of Microorganisms	**16.** Bacteriophages serve as a mechanism for genetic transfer in the process called **a.** transduction **b.** conjugation **c.** transfection **d.** replication	**17.** Antibiotics that target the transcription and translation process of bacteria actually came from other microorganisms initially. Speculate on why.	**18.** Speculate on how recombinant bacteria can be used as a tool to alter organisms' traits.
Scientific Thinking	**19.** The fact that eukaryotic cells have a ________ and bacteria and archaea do not makes their processes of protein translation distinct. **a.** membrane **b.** ribosome **c.** wall **d.** nucleus	**20.** Construct an analogy using your clothes closet to describe the difference between genotype and phenotype.	**21.** What experiment could you devise to determine if a particular phage engaged in general or specialized transduction?

Answers to the multiple-choice questions appear in Appendix A.

Visual Connections

These questions use visual images or previous content to connect concepts within and between chapters.

1. **Figure 6.15, step (1).** Label each of the parts of the illustration.

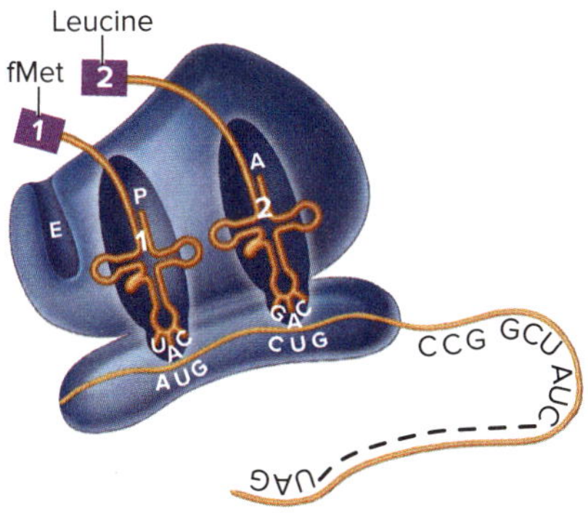
Leucine
fMet
1
2
E
P
A
AUG
CUG
CCG
GCU
AUC
UAG

2. **From chapter 4, figure 4.11*a*.** Speculate on why these cells contain two chromosomes (shown in blue).

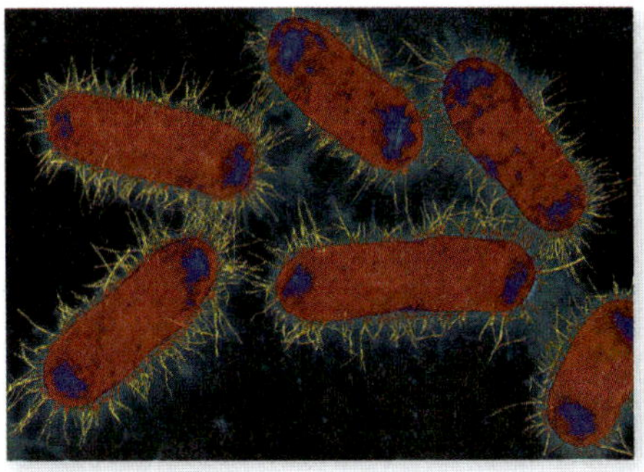

Eye of Science/Science Source

High Impact Study

These terms and concepts are most critical for your understanding of this chapter—and may be the most difficult. Have you mastered them?

Concepts

- ☐ Replication
- ☐ Transcription
- ☐ Translation
- ☐ DNA vs. RNA
- ☐ Redundancy of genetic code
- ☐ Horizontal gene transfer
- ☐ Causes of and types of mutations
- ☐ Operons

Terms

- ☐ Genome
- ☐ Chromosome
- ☐ Gene
- ☐ Antiparallel
- ☐ Semiconservative
- ☐ Introns and exons
- ☐ Conjugation
- ☐ Transformation
- ☐ Transduction

7 Viruses and Prions

Chia-Chi Charlie Chang/NIH

Media Under The Microscope

Racing to Save Her Own Life

These case studies examine an article from the popular media to determine the extent to which it is factual and/or misleading. This case focuses on the January 2020 NIH Record *article "Couple Turns Hand of Fate into Hand of Hope."*

Prions are tiny pieces of folded protein that either can be acquired in the same manner as infectious agents or can spontaneously arise through mutations in a normal brain protein. They infect the brains of mammals (including humans); remain dormant, sometimes for decades; and then cause a slow, agonizing death. There is no cure.

This article tells the story of a woman named Sonia Vallabh, formerly a Harvard-trained lawyer. Her mother was diagnosed with the mutation that causes the prion disease after displaying some symptoms and died soon after. Sonia got herself tested and found that she had the mutation as well. She would have a 90% chance of developing the disease.

She and her husband both started going to night school, studying science. Eventually they both attained PhDs and became full-time scientists at an MIT and Harvard–based institute researching prion diseases.

The article reports that they have discovered and validated a possible treatment for prion diseases. A string of synthetic RNA nucleotides (called an oligonucleotide, *oligo-* meaning "few") seems to bind to the misshapen proteins and put them out of commission. They have investment from a major pharmaceutical company, and the FDA so far has been encouraging about approval for the treatment.

- What is the **intended message** of the article?
- What is your **critical reading** of the summary of the article? Remember that in this context, "critical reading" does not necessarily mean *What criticism do you have?* but asks you to apply your knowledge to interpret whether the article is factual and whether the facts support the intended message.
- How would you **interpret** the news item for your nonmicrobiologist friends?
- What is your **overall grade** for the news item—taking into account its accuracy and the accuracy of its intended effect?

Media Under The Microscope Wrap-Up appears at the end of the chapter.

Outline and Learning Outcomes

7.1 Introduction to Viruses

1. Identify better terms for viruses than *alive* or *dead*.
2. List characteristics of viruses that distinguish them from cellular life.

7.2 The General Structure of Viruses

3. Discuss the size of viruses relative to other microorganisms.
4. Describe the function and structure(s) of viral capsids.
5. Distinguish between enveloped and naked viruses.
6. Explain the importance of viral surface proteins, or spikes.
7. Compare and contrast the composition of a viral genome to that of a cellular organism's genome.
8. Diagram the possible nucleic acid configurations exhibited by viruses.

7.3 How Viruses Are Classified and Named

9. Demonstrate how family, genus, and common names in viruses are written.

7.4 How Viruses Multiply

10. Diagram the six-step multiplication cycle of animal viruses.
11. Define the term *cytopathic effect* and provide one example.
12. Provide examples of persistent and transforming infections, describing their effects on the host cell.
13. Provide a thorough description of lysogenic and lytic bacteriophage infections.

7.5 Techniques in Cultivating and Identifying Animal Viruses

14. List the three principal purposes for cultivating viruses.
15. Describe three ways in which viruses are cultivated.

7.6 Viruses and Human Health

16. Analyze the relative importance of viruses in human infection and disease.
17. Discuss the primary reason that antiviral drugs are more difficult to design than antibacterial drugs.

7.7 Prions and Other Noncellular Infectious Agents

18. List three noncellular infectious agents besides viruses.
19. Define prions and the organ of the body that they infect.

7.1 Introduction to Viruses

Viruses are a unique group of biological entities that are known to infect every type of cell, including bacteria, algae, fungi, protozoa, plants, and animals. They are extremely abundant on our planet. Ocean waters have been found to contain 10 million viruses in a single milliliter (less than a thimbleful) of water. Lake water contains many more—as many as 250 million viruses per milliliter. We are just beginning to understand the impact of these huge numbers of viruses in our environment. The exceptional and curious nature of viruses leads to numerous questions, including the following:

1. Are they organisms, that is, are they alive?
2. What role did viruses play in the evolution of life?
3. What are their distinctive biological characteristics?
4. How can particles so small, simple, and seemingly insignificant be capable of causing disease and death?
5. What is the connection between viruses and cancer?

In this chapter, we address these questions and others.

The unusual structure and behavior of viruses have led to debates about their connection to the rest of the microbial world. One viewpoint is that because most viruses are unable to multiply independently from the host cell, they are not living. Another viewpoint proposes that even though viruses do not exhibit most of the life processes of cells, they can direct them in cells and thus are certainly more than inert and lifeless molecules. In keeping with their special position in the biological spectrum, it is best to describe viruses as either *active* or *inactive* (rather than alive or dead).

Viruses are not just agents of disease. They have many positive uses. By infecting other cells, and sometimes influencing their genetic makeup, they have shaped the way cells, tissues, bacteria, plants, and animals have evolved to their present forms. **Insight 7.1** explains how viruses are part of our normal microbiome. In addition, viruses that are no longer part of us have left some of their genetic information behind. Between 40% and 80% of the human genome may be remnants of ancient viral infections of our ancestors.

Viruses are different from their host cells in structure, behavior, and physiology. Most viruses are *obligate intracellular parasites* that in most cases cannot multiply unless they invade a specific host cell and use the cell's genetic and metabolic machinery to make and release quantities of new viruses. Other important properties of viruses are summarized in **table 7.1.**

7.1 Learning Outcomes—Assess Your Progress

1. Identify better terms for viruses than *alive* or *dead*.
2. List characteristics of viruses that distinguish them from cellular life.

INSIGHT 7.1 MICROBIOME: Are Viruses Part of the Microbiome?

Yes. The sum total of the viruses associated with your body is called the **virome.** As you are learning, viruses have to use cells as their homes. So all of the cells of your body are capable of hosting viruses. If they are part of the microbiome, and therefore part of your normal biota, they are likely to be in your cells in a dormant state, in which they are "quiet" inside your cells and not multiplying. And then there are all the viruses, called *bacteriophages,* that are infecting the bacteria that are part of your microbiome. This all adds up to perhaps 10^{15} viruses as part of your microbiome. (Written out, that's 1,000,000,000,000,000.) Scientists estimate that there are between 10^{13} and 10^{14} cells of the human body and slightly more bacterial cells in and on the average human. So the viruses win in terms of numbers.

Most people are not aware of this teeming landscape in and on our bodies. Until relatively recently, medical research was concerned only with the viruses that cause diseases—such as rabies, polio, and influenza. But sophisticated molecular techniques have been able to identify the "quiet" viruses—both those that are seemingly always quiet and those that are intermittently quiet but do become pathogenic from time to time.

For example, we know that almost every adult is infected with the Epstein-Barr virus (EBV). EBV is a herpesvirus, and virtually all people are infected with one or more of the herpesviruses. It is estimated that more than 50% of people are infected with the virus causing genital herpes, though the majority—yes, the majority—don't even know it. Most adults are infected with cytomegalovirus (CMV), which does not seem to affect adults, except to perhaps worsen illnesses with other microbes, but it can cause serious birth defects.

Having said all this, the vast majority of viruses in the human body are not pathogenic. We know that the development of the mammalian placenta was influenced by viruses. Recent research even suggests that the presence of bacteriophages in our intestinal mucosa may play a role in protecting us against bacterial infection. Because of their relatively easy movement between hosts, they have probably contributed genes from other organisms in the biosphere that have affected the development of our own physiology in significant ways.

Table 7.1 Properties of Viruses

- Are not cells
- Are obligate intracellular parasites of bacteria, protozoa, fungi, algae, plants, and animals (some exceptions have been found)
- Are inactive macromolecules outside the host cell and active only inside host cells
- Have basic structure of protein shell (capsid) surrounding nucleic acid core
- Are everywhere in nature and have had major impact on the development of biological life
- Are often ultramicroscopic in size, ranging from 20 nm to 1,500 nm
- Can have either DNA or RNA but not both
- Can have double-stranded DNA, single-stranded DNA, single-stranded RNA, or double-stranded RNA
- Carry molecules on their surface that determine specificity for attachment to host cell
- Multiply by taking control of host cell's genetic material and regulating the synthesis and assembly of new viruses
- Usually lack enzymes for most metabolic processes
- Usually lack machinery for making proteins

7.2 The General Structure of Viruses

Size Range

Many viruses are much smaller than the average bacterium. More than 2,000 bacterial viruses could fit into an average bacterial cell, and more than 50 million polioviruses can actually be found inside an infected human cell. Animal viruses (including viruses of humans) range in size from the small parvoviruses (around 0.02 μm in diameter) to viruses that are larger than small bacteria (0.4–1 μm in length). **Figure 7.1** compares the sizes of several viruses with bacteria and eukaryotic cells and molecules. The pandoravirus is as big as the *Streptococcus* bacterium (purple in figure 7.1). A virus discovered in 2014 is 50% larger than the pandoravirus, even. Some cylindrical viruses are relatively long (0.8 μm in length) but so narrow in diameter (0.015 μm) that their visibility is still limited without the high magnification and resolution of an electron microscope.

Viral architecture is most readily observed through special stains in combination with electron microscopy **(figure 7.2).**

Viral Components: Capsids, Envelopes, and Nucleic Acids

It is important to realize that viruses bear no real resemblance to cells and that they generally lack the protein-synthesizing machinery found in even the simplest cells. Their molecular structure is composed of regular, repeating subunits that give rise to their crystalline appearance. In fact, many purified viruses can form large aggregates or crystals if subjected to special treatments **(figure 7.3).** At their simplest, viruses need only a piece of genetic material and a protein coat. Viruses contain only those parts needed to invade and control a host cell: an external coating and a core containing one or more nucleic acid strands of either DNA or RNA and, sometimes, one or two enzymes. Their structure can be represented with this graphic:

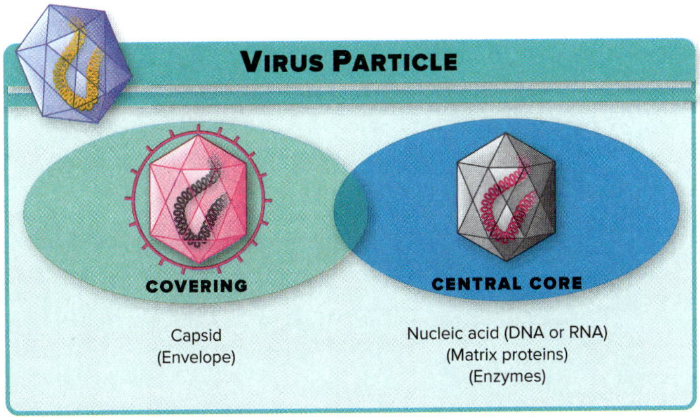
VIRUS PARTICLE
COVERING
CENTRAL CORE
Capsid
(Envelope)
Nucleic acid (DNA or RNA)
(Matrix proteins)
(Enzymes)

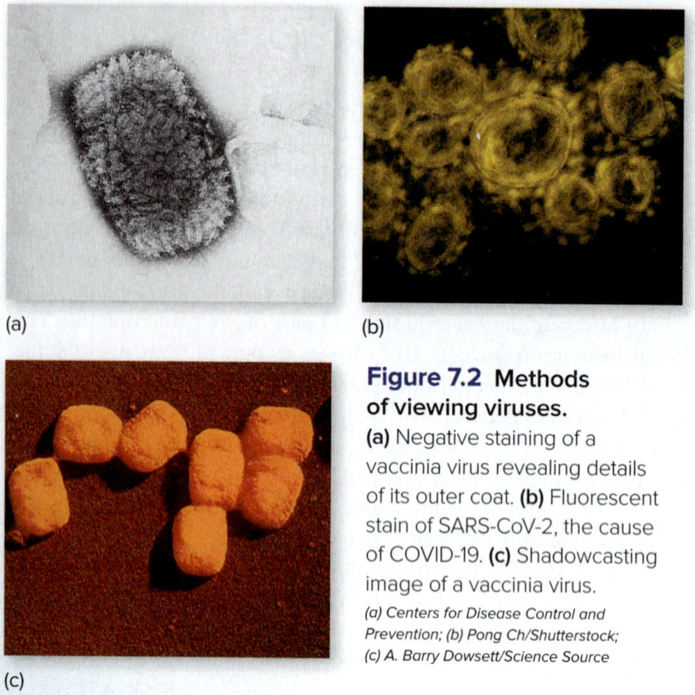
(a)
(b)
(c)

Figure 7.2 Methods of viewing viruses. **(a)** Negative staining of a vaccinia virus revealing details of its outer coat. **(b)** Fluorescent stain of SARS-CoV-2, the cause of COVID-19. **(c)** Shadowcasting image of a vaccinia virus.

(a) Centers for Disease Control and Prevention; (b) Pong Ch/Shutterstock; (c) A. Barry Dowsett/Science Source

Figure 7.1 Size comparison of viruses with a eukaryotic cell (yeast) and bacteria. A molecule of protein is included to indicate relative size of macromolecules.

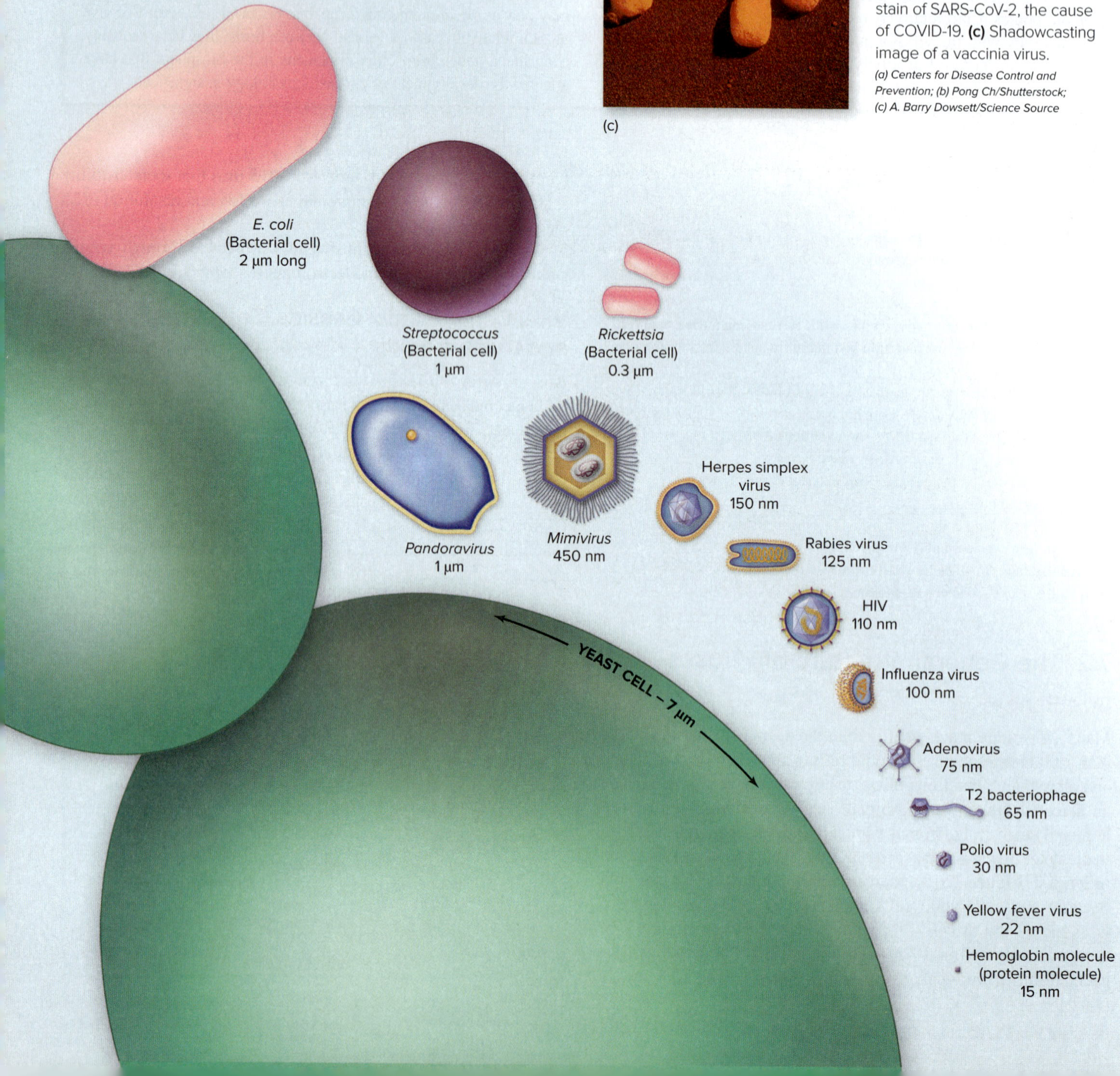
E. coli
(Bacterial cell)
2 µm long
Streptococcus
(Bacterial cell)
1 µm
Rickettsia
(Bacterial cell)
0.3 µm
Pandoravirus
1 µm
Mimivirus
450 nm
Herpes simplex virus
150 nm
Rabies virus
125 nm
HIV
110 nm
Influenza virus
100 nm
Adenovirus
75 nm
T2 bacteriophage
65 nm
Polio virus
30 nm
Yellow fever virus
22 nm
Hemoglobin molecule
(protein molecule)
15 nm
YEAST CELL – 7 µm

Figure 7.3 The crystalline nature of viruses. Highly magnified (150,000×) electron micrograph of purified poliovirus crystals, showing hundreds of individual viruses.
Omnikron/Science Source

All viruses have a protein **capsid,** or shell, that surrounds the nucleic acid in the central core. Together, the capsid and the nucleic acid are referred to as the **nucleocapsid (figure 7.4).** Members of many families of animal viruses possess an additional covering external to the capsid called an envelope, which is usually a modified piece of the host's cell membrane **(figure 7.4*b*).** Viruses that consist of only a nucleocapsid are called *naked viruses* **(figure 7.4*a*).** Both naked and enveloped viruses possess proteins (called spikes) on their outer surfaces that project from either the nucleocapsid or the envelope. They are the molecules that allow viruses to dock with their host cells. As we will see in this chapter, the enveloped viruses differ from the naked viruses in the way that they enter and leave a host cell. A fully formed virus that is able to establish an infection in a host cell is often called a **virion.**

Disease Connection

Hand sanitizer typically contains ethyl alcohol or isopropyl alcohol. Enveloped viruses are inactivated by alcohols. They are also susceptible to UV light, desiccation, soap, and other chemical formulations, which disrupt the lipid envelope. Nonenveloped (naked) viruses are generally not affected by alcohol. Examples of enveloped viruses are the HIV virus, hepatitis B virus, and the influenza virus. Nonenveloped viruses include hepatitis A and many of the enteroviruses, which cause gastrointestinal illnesses.

The Viral Capsid: The Protective Outer Shell

When a virus particle is magnified several hundred thousand times, the capsid appears as a prominent geometric feature. In general, each capsid is constructed from identical subunits called **capsomeres** that are made up of protein molecules. The capsomeres spontaneously self-assemble into the finished capsid. Depending on how the capsomeres are shaped and arranged, this assembly results in two different types for animal viruses: helical and icosahedral.

Helical capsids are summarized in **table 7.2.** The simpler **helical** capsids have rod-shaped capsomeres that bond together to form a series of hollow cylinders resembling a bracelet. During the formation of the nucleocapsid, these cylinders link with other cylinders to form a continuous helix into which the nucleic acid strand is coiled **(figure 7.5).** In electron micrographs, the appearance of a helical capsid varies with the type of virus. The nucleocapsids of naked helical viruses are very rigid and tightly wound into a cylinder-shaped package. An example is the tobacco mosaic virus, which attacks tobacco leaves (pictured in the top row of table 7.2). Enveloped helical nucleocapsids are more flexible and tend to be arranged as a looser helix within the envelope. This type of morphology is found in several enveloped human viruses, including influenza, measles, and rabies.

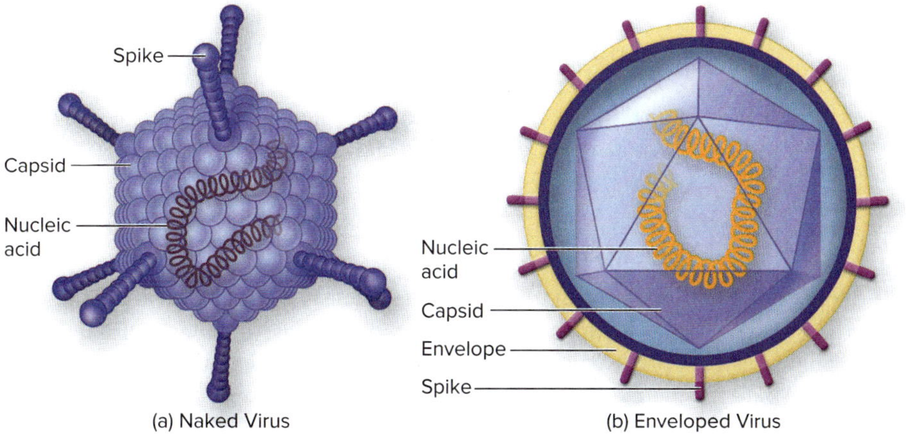
Spike
Capsid
Nucleic acid
(a) Naked Virus
Nucleic acid
Capsid
Envelope
Spike
(b) Enveloped Virus

Figure 7.4 Generalized structure of viruses. **(a)** The simplest virus is a naked virus (nucleocapsid) consisting of a geometric capsid assembled around a nucleic acid strand or strands. **(b)** An enveloped virus is composed of a nucleocapsid surrounded by a flexible membrane called an envelope.

Table 7.2 Helical Capsids

	The simpler **helical capsids** have rod-shaped capsomeres that bond together to form a series of hollow discs resembling a bracelet. During the formation of the nucleocapsid, these discs link with other discs to form a continuous helix into which the nucleic acid strand is coiled.
Naked	The nucleocapsids of naked helical viruses are very rigid and tightly wound into a cylinder-shaped package. An example is the tobacco mosaic virus, which attacks tobacco leaves (right). 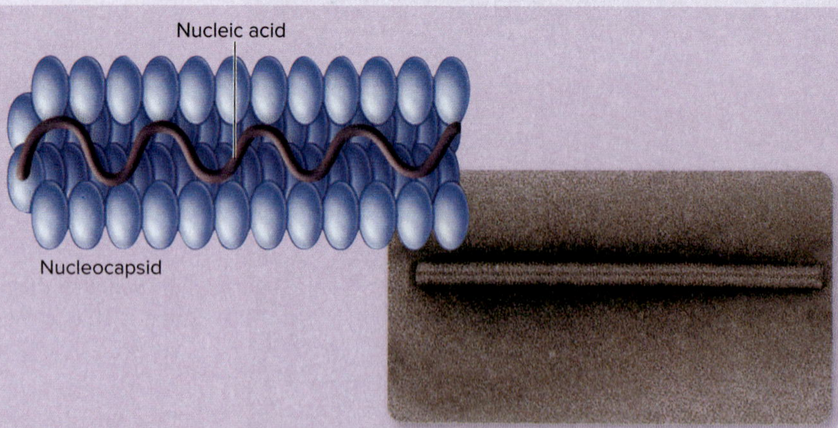 Nucleic acid Nucleocapsid
Enveloped	Enveloped helical nucleocapsids are more flexible and tend to be arranged as a looser helix within the envelope. This type of morphology is found in several enveloped human viruses, including influenza, measles, and rabies. 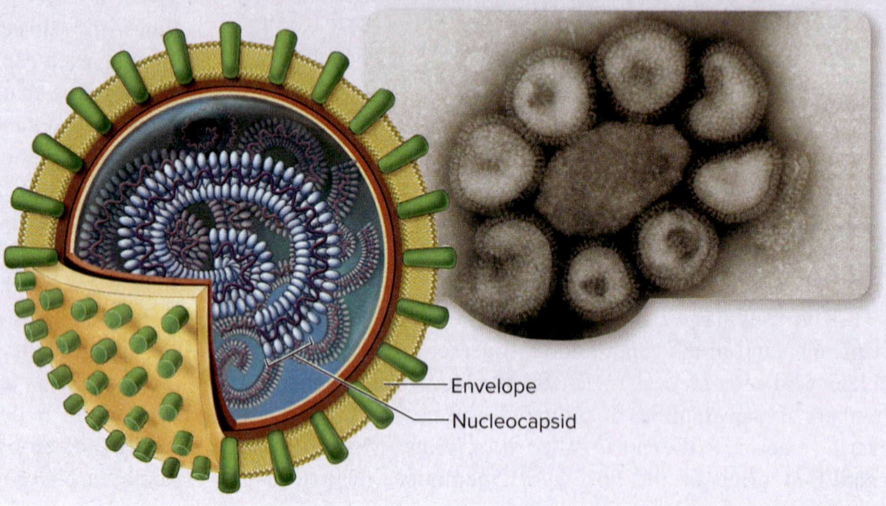 Envelope Nucleocapsid

(Top) *Omikron/Science Source;* (Bottom) *Dr. F. A. Murphy/CDC*

Disease Connection

You will see in chapter 21 that the dreaded rabies virus, which has a near 100% fatality rate in humans, has a very distinctive shape. Its characteristic bullet shape is a product of a matrix protein that lies between the helical capsid and the envelope.

The capsids of a number of virus families are arranged as an **icosahedron** (eye-koh-suh-hee′-drun)—a three-dimensional, 20-sided figure with 12 evenly spaced corners. Icosahedral capsids are summarized in **table 7.3.** The arrangements of the capsomeres vary from one virus to another. Some viruses construct the capsid from a single type of capsomere, whereas others may contain several types of capsomeres **(figure 7.6).** Although the capsids of all icosahedral viruses have this sort of symmetry, they can have major variations in the number of capsomeres. For example, a poliovirus has 32, and an adenovirus has 252 capsomeres. Individual capsomeres can look either ring- or dome-shaped, and the capsid itself can appear spherical or cubical (table 7.3). During assembly of the virus, the nucleic acid is packed into the center of this icosahedron, forming a nucleocapsid. Another factor that alters the appearance of icosahedral viruses is whether or not they have an outer envelope. Contrast the naked nucleocapsid of a papillomavirus (causes warts) with the enveloped nucleocapsid of herpes simplex (causes oral and genital herpes) **(figure 7.7).** Although most viruses have capsids that are either icosahedral or helical, there is another category of capsid that is simply called *complex*. Complex capsids, found in the viruses that infect bacteria, may have multiple types of proteins and take shapes that are not symmetrical. An example of a complex virus is shown in **table 7.4.**

The Viral Envelope

When enveloped viruses are released from the host cell, they take with them a bit of its membrane to use as an envelope. Some viruses bud off the cell membrane; and others leave via the nuclear envelope or the endoplasmic reticulum. Whichever avenue of escape, the viral envelope differs significantly from

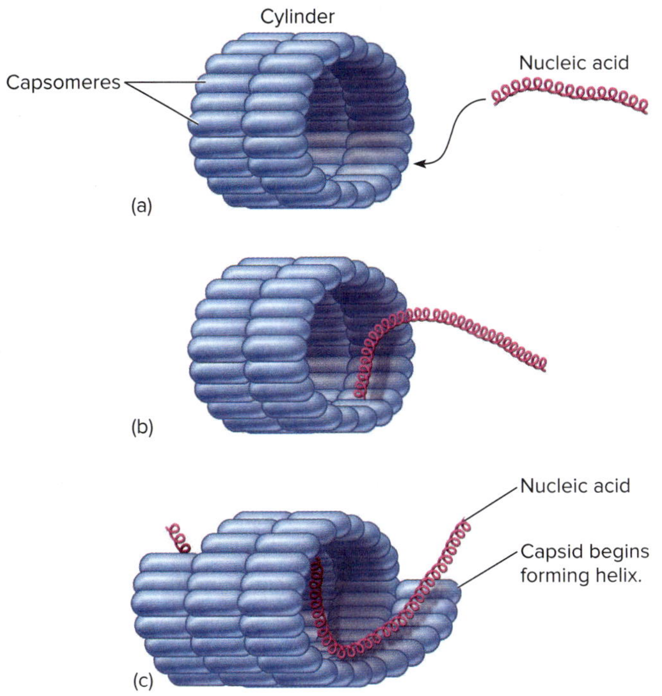
Cylinder
Capsomeres
Nucleic acid
(a)
(b)
Nucleic acid
Capsid begins forming helix.
(c)

Figure 7.5 Assembly of helical nucleocapsids. **(a)** Capsomeres assemble into hollow cylinders. **(b)** The nucleic acid is inserted into the center of the cylinder. **(c)** Elongation of the nucleocapsid progresses from one or both ends, as the nucleic acid is wound "within" the lengthening helix.

the host's membranes. Viruses place their own proteins in the membrane, which they then proceed to use as their envelope. Some of the envelope proteins attach to the capsid of the virus, and glycoproteins (proteins bound to a carbohydrate) remain exposed on the outside of the envelope. These protruding molecules, called *spikes,* are essential for the attachment of viruses to the next host cell. Because the envelope is more supple than the capsid, the surface appearance of enveloped viruses is pleomorphic, and these viruses range from spherical to filamentous in shape.

Nucleic Acids: At the Core of a Virus

The sum total of the genetic information carried by any organism is known as its **genome.** We know that the genetic information of living cells is carried by nucleic acids (DNA, RNA). Viruses, although not technically alive and definitely not cells, are no exception to this rule, but there is a significant difference. Unlike cells, which contain both DNA and RNA, viruses contain either DNA or RNA *but not both.* Because viruses must pack into a tiny space all of the genes necessary to instruct the host cell to make new viruses, the number of viral genes is often—but not always—quite small compared with that of a cell. It varies from four genes in hepatitis B virus to hundreds of genes in herpesviruses to over 2,500 in pandoraviruses. By comparison, the bacterium *Escherichia coli* has approximately 4,000 genes, and a human cell has approximately 23,000 genes. These additional genes allow cells to carry out the complex metabolic activity necessary for independent life.

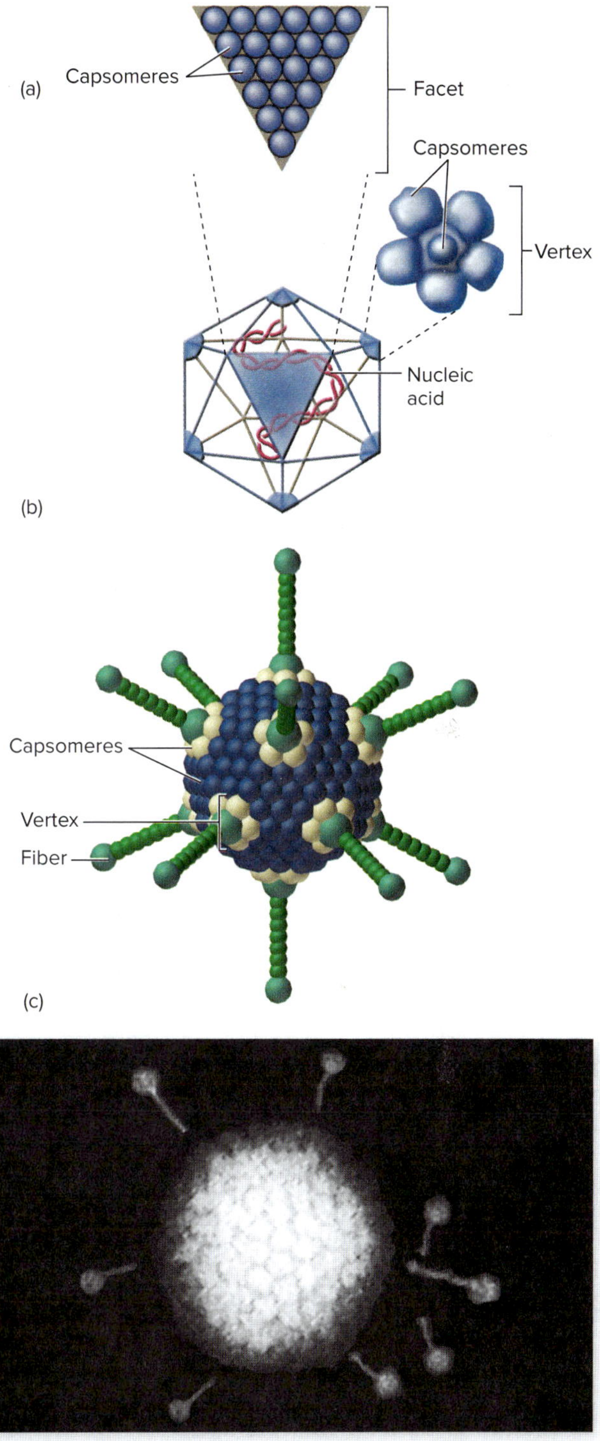
(a)
Capsomeres
Facet
Capsomeres
Vertex
Nucleic acid
(b)
Capsomeres
Vertex
Fiber
(c)
(d)

Figure 7.6 How icosahedral capsids are formed. Adenovirus is the model. **(a)** A facet or "face" of the capsid is composed of 21 identical capsomeres arranged in a triangular shape. A vertex or "point" consists of a different type of capsomere with a single penton in the center. Other viruses can vary in the number, types, and arrangement of capsomeres. **(b)** An assembled virus shows how the facets and vertices come together to form a shell around the nucleic acid. **(c)** A three-dimensional model of this virus shows fibers (spikes) attached to the pentons. **(d)** A negative stain of this virus (640,000×) highlights its texture and also fibers that have fallen off.

Dr. Linda M. Stannard, University of Cape Town/SPL/Science Source

Table 7.3 Icosahedral Capsids

	These capsids form an **icosahedron**—a three-dimensional, 20-sided figure with 12 evenly spaced corners. The arrangements of the capsomeres vary from one virus to another. Some viruses construct the capsid from a single type of capsomere, while others may contain several types of capsomeres. There are major variations in the number of capsomeres; for example, a poliovirus has 32, and an adenovirus has 252 capsomeres.
Naked	Adenovirus is an example of a naked icosahedral virus. In the photo you can clearly see the spikes, some of which have broken off. 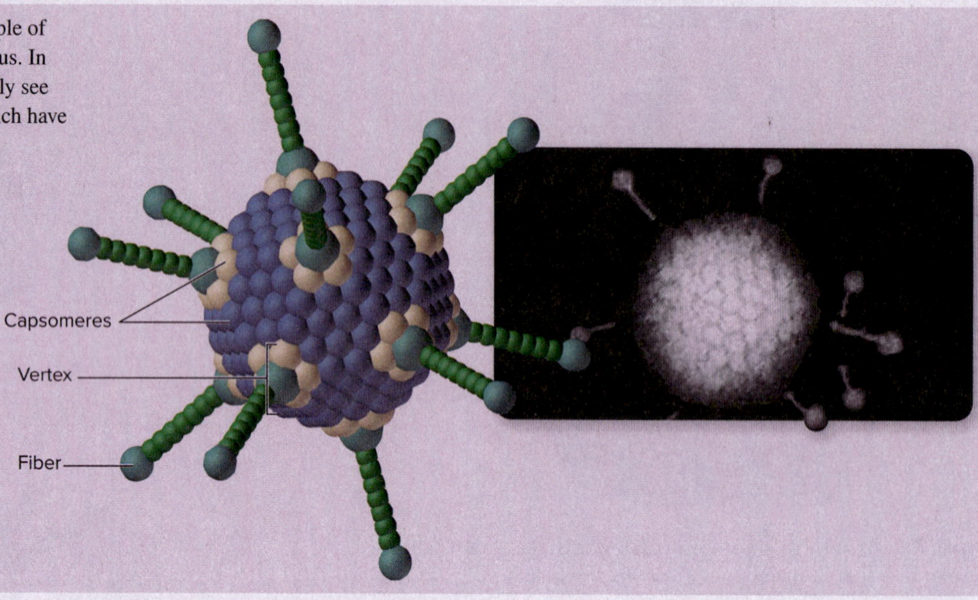 Capsomeres Vertex Fiber
Enveloped	Two very common viruses, hepatitis B virus (left) and the herpes simplex virus (right), possess enveloped icosahedrons. 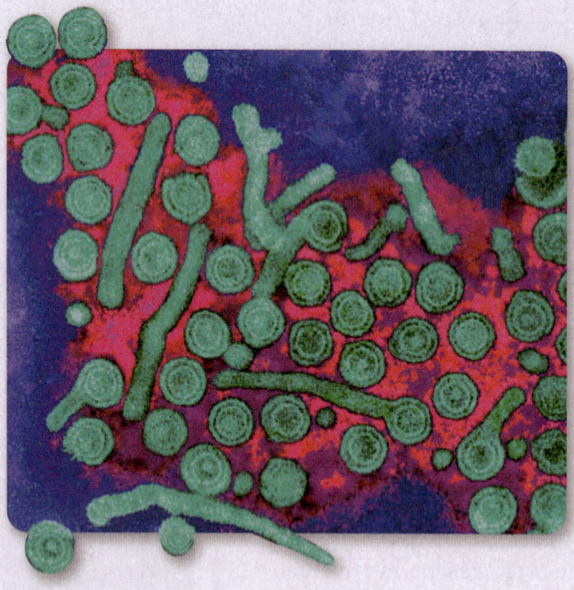 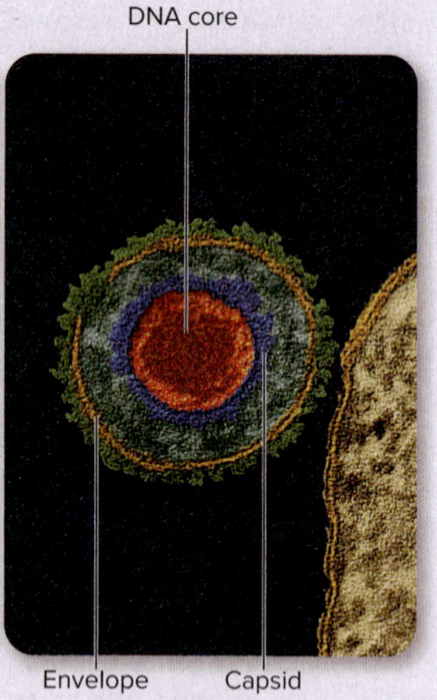 DNA core Envelope Capsid

(Adenovirus, Hepatitis B virus) *Dr. Linda M. Stannard, University of Cape Town/Science Source;* (Herpes simplex virus) Eye of Science/Science Source

In chapter 2, you learned that DNA usually exists as a double-stranded molecule and that RNA is single-stranded. Although most viruses follow this same pattern, there are important exceptions. For example, the parvoviruses contain single-stranded DNA. Reoviruses (a cause of respiratory and intestinal tract infections) contain double-stranded RNA. Furthermore, viruses exhibit wide variation in how their RNA or DNA is configured. DNA viruses can have single-stranded (ss) or double-stranded (ds) DNA; the dsDNA can be arranged linearly or in ds circles. RNA viruses can be double-stranded but are more often single-stranded. Every virus must create mRNA in order to make proteins to create copies of itself. Each different type of virus has different strategies for that, and they are classifed in a system called the Baltimore System. This classification is depicted in **table 7.5.** Single-stranded RNA genomes that are ready for immediate translation into proteins are called **positive-sense RNA.** (SARS-CoV-2—and all coronaviruses—is this type of virus.) Other RNA genomes have to be converted into the proper form to be made into proteins, and these are called **negative-sense RNA.** RNA genomes may also be *segmented,* meaning that the individual genes exist on separate pieces of RNA. The influenza

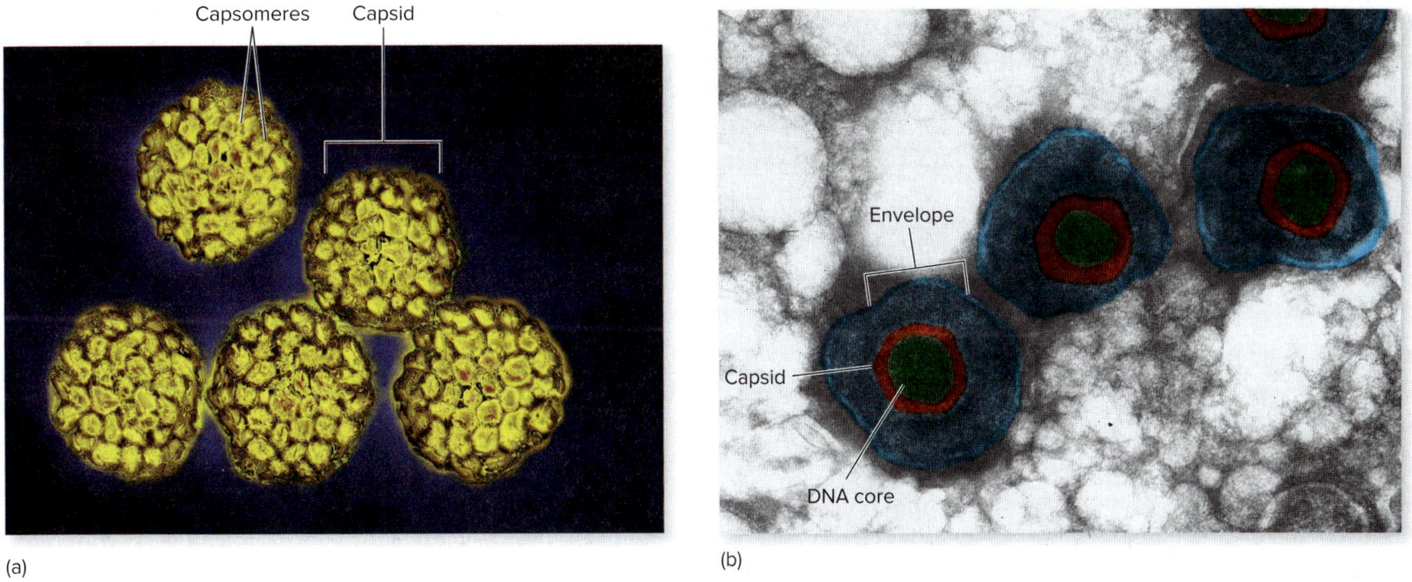
Capsomeres
Capsid
(a)
Envelope
Capsid
DNA core
(b)

Figure 7.7 Nonenveloped and enveloped viruses. **(a)** Micrograph of nonenveloped papillomaviruses with unusual, ring-shaped capsomeres. **(b)** Herpesvirus, an enveloped icosahedron (300,000×). Both micrographs have been colorized.

(a) BSIP/Universal Images Group/Getty Images; (b) Kathy Park Talaro

virus is an example of this. A special type of RNA virus is called a *retrovirus.* We'll discuss it later in this chapter.

In all cases, these tiny strands of genetic material carry the blueprint for viral structure and functions. In a very real sense, viruses are genetic parasites because they cannot multiply until their nucleic acid has reached the internal habitat of the host cell. At the minimum, they must carry their own genes for synthesizing the viral capsid and genetic material, for regulating the actions of the host, and for packaging the mature virus.

Other Substances in the Virus Particle

In addition to the protein of the capsid, the proteins and lipids of envelopes, and the nucleic acid of the core, viruses can contain enzymes for specific operations within their host cell. They may

Table 7.4 Complex Capsids

Complex capsids, only found in the viruses that infect bacteria, may have multiple types of proteins and take shapes that are not symmetrical. They are never enveloped. The one pictured on the right is a T4 bacteriophage.

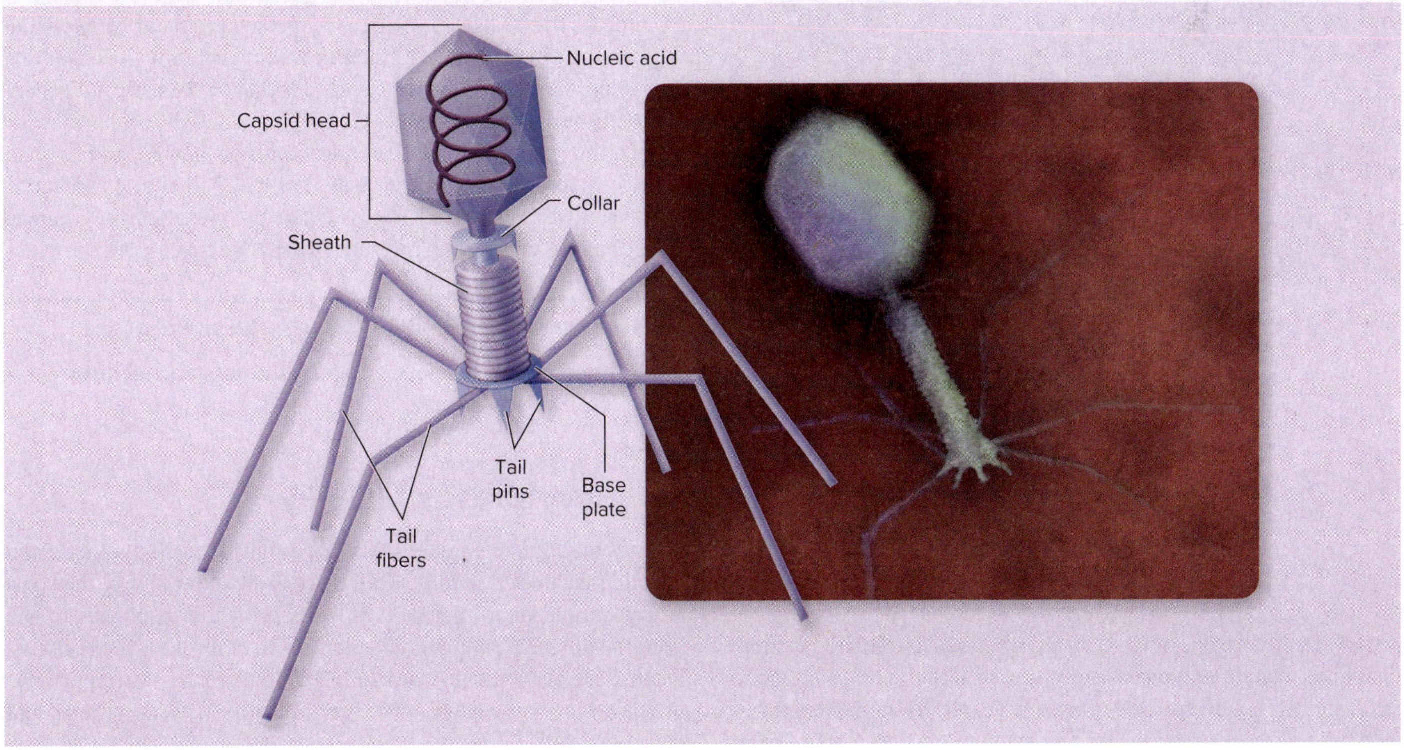
Nucleic acid
Capsid head
Collar
Sheath
Tail pins
Base plate
Tail fibers

Ami Images/Science Source

Table 7.5 Baltimore Classification System: How Each Type of Virus Makes mRNA

Types of Viral Genomes	Strategies	Examples
Class I dsDNA	→ **mRNA**	 Herpes simplex; causes herpes
Class II ssDNA	dsDNA → **mRNA**	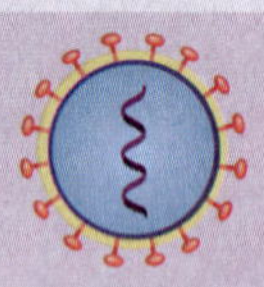 Parvovirus B19; skin rash
Class III dsRNA	→ **mRNA**	 Rotavirus; gastroenteritis
Class IV (+)ssRNA	→(−)ssRNA →**mRNA**	 Poliovirus; polio
Class V (−)ssRNA	→ **mRNA**	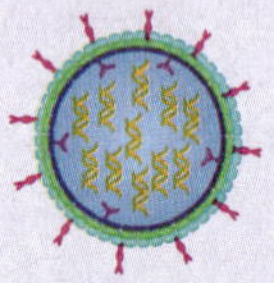 Influenza virus; influenza
Class VI ssRNA-RT	→ DNA/RNA→ dsDNA→**mRNA**	 HIV; AIDS
Class VII dsDNA-RT	→**mRNA**	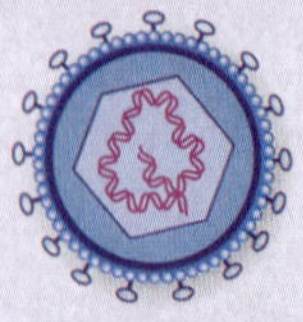 Hepatitis B virus; hepatitis B

come with preformed enzymes that are required for viral replication. Examples include **polymerases** (pol-im′-ur-ace-uz) that synthesize DNA and RNA and replicases that copy RNA. HIV comes equipped with **reverse transcriptase (RT)** for synthesizing DNA from RNA.

Some viruses can actually carry away substances from their host cell. For instance, arenaviruses pack along host ribosomes, and retroviruses "borrow" the host's tRNA molecules.

7.2 Learning Outcomes—Assess Your Progress

3. Discuss the size of viruses relative to other microorganisms.
4. Describe the function and structure(s) of viral capsids.
5. Distinguish between enveloped and naked viruses.
6. Explain the importance of viral surface proteins, or spikes.
7. Compare and contrast the composition of a viral genome to that of a cellular organism's genome.
8. Diagram the possible nucleic acid configurations exhibited by viruses.

7.3 How Viruses Are Classified and Named

It is very difficult, even for professional taxonomists, to classify viruses, and their classifications are constantly changing. In this book, we will focus on obvious characteristics of the viruses and the host cells they infect. In an informal way, we have already begun classifying viruses—as animal, plant, or bacterial viruses; enveloped or naked viruses; DNA or RNA viruses; and helical or icosahedral viruses. These introductory categories are certainly useful in organization and description, but the study of specific viruses requires a more standardized method of nomenclature. The main criteria presently used to group viruses are structure, chemical composition, and similarities in genetic makeup.

In 2020, the International Committee on the Taxonomy of Viruses issued its latest report on the classification of viruses. The committee listed 59 orders and 189 families of viruses. Examples of three of the orders of viruses are presented in **table 7.6.** Note the naming conventions: Virus orders are written with *-virales* as an ending, families are written with *-viridae* on the end, and genera end with *-virus*.

In this text, we will be using common names, rather than precise species names, for viruses. **Table 7.7** illustrates the naming system for important viruses and the diseases they cause. Note that the common names of viruses are not italicized.

7.3 Learning Outcome—Assess Your Progress

9. Demonstrate how family, genus, and common names in viruses are written.

7.4 How Viruses Multiply

Viruses are minute parasites that seize control of the synthetic and genetic machinery of cells. This process dictates the way the virus is transmitted and what it does to its host, the responses of the immune defenses, and human attempts to control viral infections. From these perspectives, we cannot overemphasize the importance of a working knowledge of the relationship between viruses and their host cells.

Table 7.6 Examples from the 59 Orders of Viruses

Order	Family	Genus	Species	Host
Caudovirales	*Salasmaviridae*	*SPO1-like virus*	Bacillus phage Nf	Bacterium
Herpesvirales	*Herpesviridae*	*Simplexvirus*	Human alphaherpesvirus 2	Human (animal)
Mononegavirales	*Rhabdoviridae*	*Lyssavirus*	Rabies virus	Animal

Table 7.7 Important Human Virus Families, Genera, Common Names, and Types of Diseases

Families	Genus of Virus	Common Name of Genus Members	Name of Disease
DNA Viruses			
Herpesviridae	*Simplexvirus*	Herpes simplex type 1 virus (HSV-1)	Fever blister, cold sores
		Herpes simplex type 2 virus (HSV-2)	Genital herpes
	Varicellovirus	Varicella zoster virus (VZV)	Chickenpox, shingles
Papillomaviridae	*Papillomavirus*	Human papillomavirus (HPV)	Several types of warts
Hepadnaviridae	*Orthohepadnavirus*	Hepatitis B virus (HBV or Dane particle)	Serum hepatitis
RNA Viruses			
Picornaviridae	*Enterovirus*	Poliovirus	Poliomyelitis
Matonaviridae	*Rubivirus*	Rubella virus	Rubella (German measles)
Flaviviridae	*Flavivirus*	West Nile virus	West Nile fever
Coronaviridae	*Coronavirus*	SARS-CoV-2	COVID-19
Filoviridae	*Ebolavirus*	Ebola virus	Ebola fever
Orthomyxoviridae	*Influenza A virus*	Influenza virus, type A (Asian, Hong Kong, and swine influenza viruses)	Influenza or "flu"
Retroviridae	*Lentivirus*	HIV (human immunodeficiency viruses 1 and 2)	Acquired immunodeficiency syndrome (AIDS)

Multiplication Cycles in Animal Viruses

The general phases in the multiplication cycle of animal viruses are

- **adsorption,**
- **penetration,**
- **uncoating,**
- **synthesis,**
- **assembly,** and
- **release** from the host cell.

The length of the entire multiplication cycle varies from 8 hours in polioviruses to 72 hours in some herpesviruses.

Adsorption and Host Range

The cycle begins when the virus encounters a susceptible host cell and attaches specifically to receptor sites on the cell membrane. (Biologists often use the word *adsorb* to mean *attach*.) The membrane receptors that viruses attach to are usually glycoproteins the cell needs for its normal function **(figure 7.8).** For example, the rabies virus binds to the acetylcholine receptor of nerve cells, and the human immunodeficiency virus (HIV) attaches to the CD4 protein on certain white blood cells. Because a virus can invade its host cell only through making an exact fit with a specific host molecule, the range of hosts it can infect in a natural setting is limited. This limitation, known as the **host range,** may be very restricted, as in the case of hepatitis B, which infects only liver cells of humans. Poliovirus, which infects both intestinal cells and nerve cells of primates (humans, apes, and monkeys) is considered moderately restrictive. The rabies virus is considered to be minimally restricted because it can infect various cells of all mammals. Cells that lack compatible virus receptors are resistant to adsorption and invasion by that virus. This explains why, for example, human liver cells are not infected by the canine hepatitis virus and dog liver cells cannot host the human hepatitis A virus. It also explains why viruses usually have tissue specificities called *tropisms* (troh′-pizmz) for certain cells in the body. The hepatitis B virus targets the liver, and the mumps virus targets salivary glands.

Penetration/Uncoating of Animal Viruses

Animal viruses exhibit some impressive mechanisms for entering a host cell. The flexible cell membrane of the host is penetrated either by the whole virus or just by its nucleic acid **(figure 7.9).**

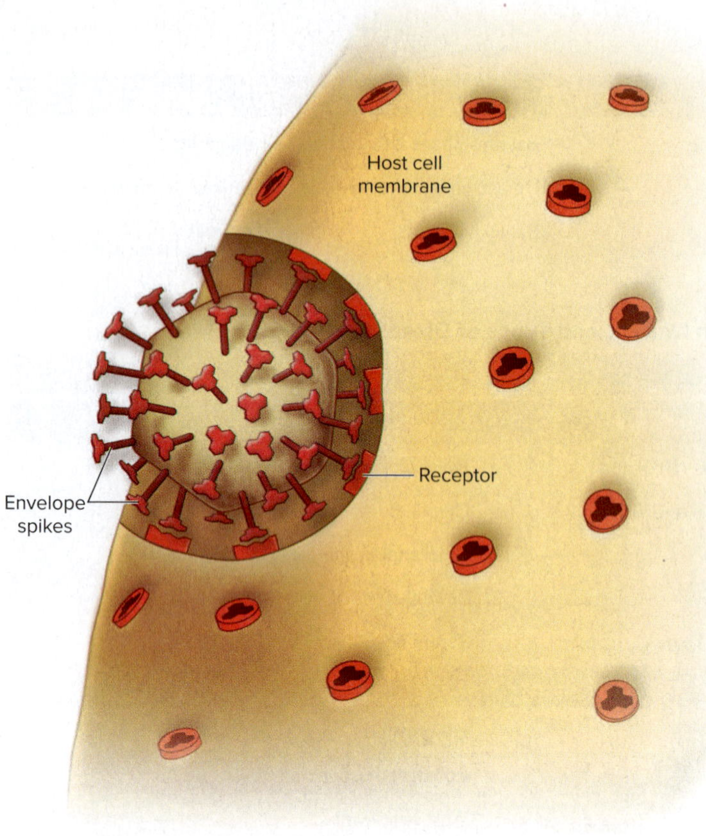
Host cell
membrane
Receptor
Envelope
spikes

Figure 7.8 The method animal viruses use to adsorb to the host cell membrane. An enveloped coronavirus with prominent spikes. The configuration of the spike has a complementary fit for cell receptors. The process in which the virus lands on the cell and plugs into receptors is termed *docking*.

In penetration by **endocytosis (figure 7.9*a*),** the entire virus is engulfed by the cell and enclosed in a vacuole or vesicle. When enzymes in the vacuole dissolve the envelope and capsid, the virus is said to be uncoated, a process that releases the viral nucleic acid into the cytoplasm. The exact manner of uncoating varies, but in many cases, the virus fuses with the wall of the vesicle. Another means of entry involves direct fusion of the viral envelope with the host cell membrane (as in influenza and mumps viruses) **(figure 7.9*b*).** In this form of penetration, the envelope merges directly with the cell membrane. By doing so, the nucleocapsid is released into the cell's interior. Uncoating will then occur later in the process.

Synthesis: Genome Replication and Protein Production

The virus really has two main jobs upon entering its host cell: make new genomic material and new proteins. In this way, they create copies of themselves. The viral nucleic acid takes control of the host's synthetic and metabolic machinery. How this proceeds will vary, depending on whether the virus is a DNA or an RNA virus. In general, the DNA viruses (except poxviruses) enter the host cell's nucleus and are replicated and assembled there. With few exceptions (such as retroviruses), RNA viruses are replicated and assembled in the cytoplasm.

In **figure 7.10,** we provide an overview of the RNA process, using a + strand RNA virus as a model. Coronaviruses are an example of this type of virus. Almost immediately upon entry, the viral nucleic acid begins to synthesize the building blocks for new viruses. First, the +ssRNA, which can serve immediately

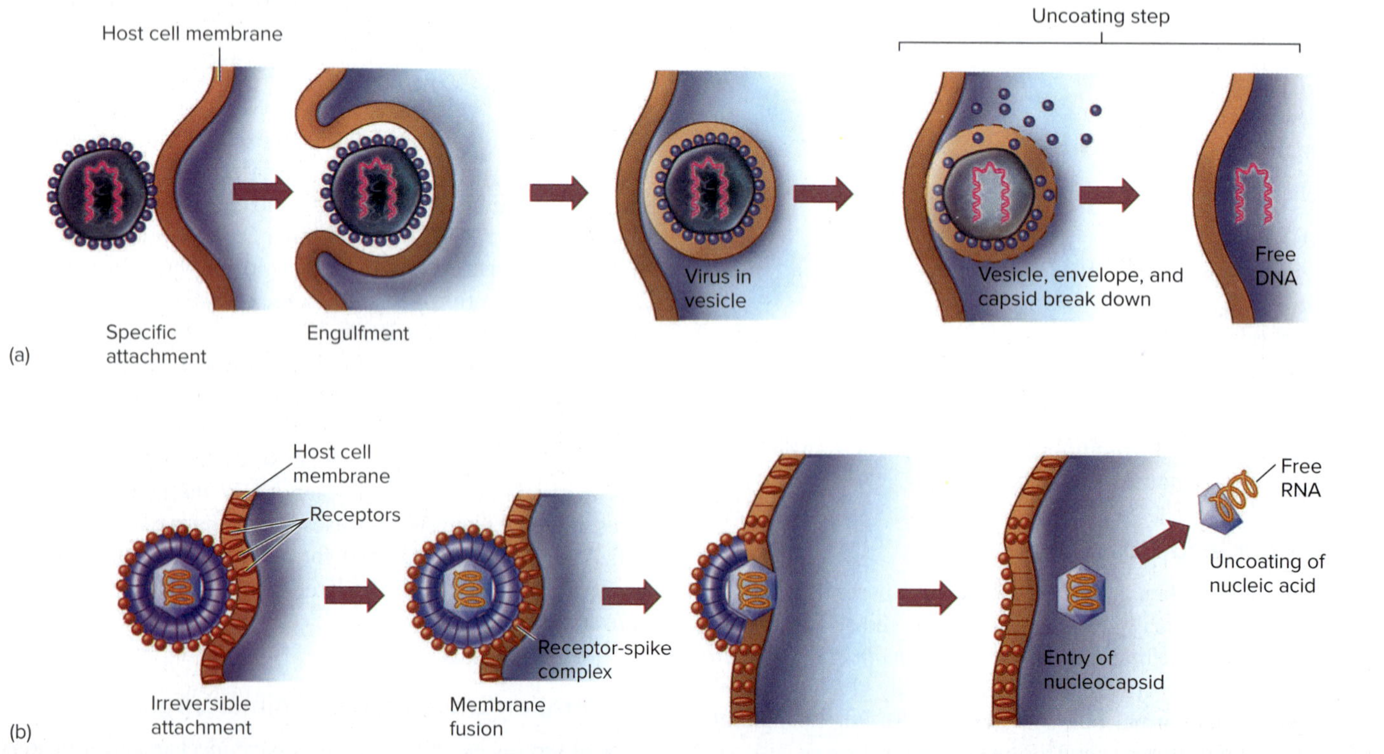
Host cell membrane
Uncoating step
Virus in
vesicle
Vesicle, envelope, and
capsid break down
Free
DNA
(a)
Specific
attachment
Engulfment
Host cell
membrane
Receptors
Free
RNA
Uncoating of
nucleic acid
Receptor-spike
complex
Entry of
nucleocapsid
(b)
Irreversible
attachment
Membrane
fusion

Figure 7.9 Two principal means by which animal viruses penetrate. **(a)** Endocytosis (engulfment) and uncoating of a herpesvirus. **(b)** Fusion of the cell membrane with the viral envelope (mumps virus).

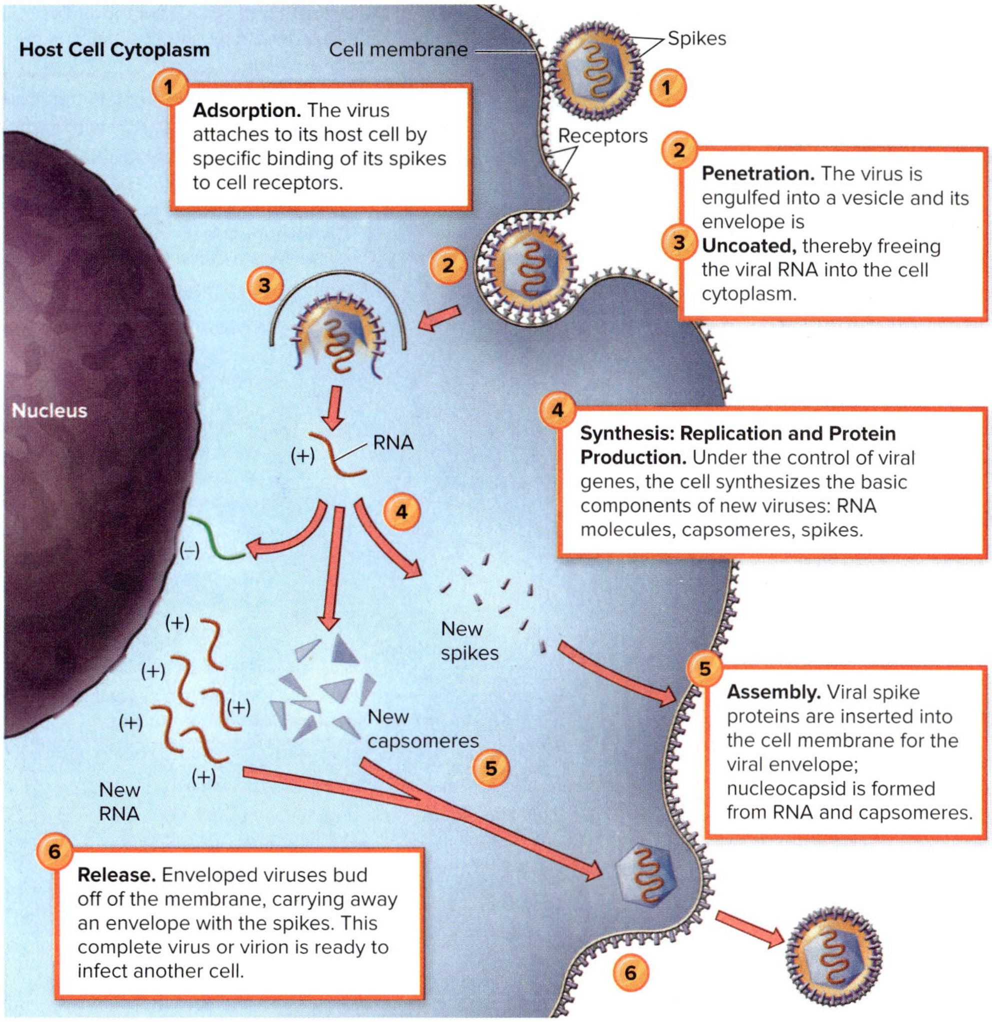

Host Cell Cytoplasm
Cell membrane
Spikes
1
Adsorption. The virus attaches to its host cell by specific binding of its spikes to cell receptors.
Receptors
2
Penetration. The virus is engulfed into a vesicle and its envelope is
3
Uncoated, thereby freeing the viral RNA into the cell cytoplasm.
Nucleus
(+) RNA
4
Synthesis: Replication and Protein Production. Under the control of viral genes, the cell synthesizes the basic components of new viruses: RNA molecules, capsomeres, spikes.
(–)
(+)
New spikes
New capsomeres
New RNA
5
Assembly. Viral spike proteins are inserted into the cell membrane for the viral envelope; nucleocapsid is formed from RNA and capsomeres.
6
Release. Enveloped viruses bud off of the membrane, carrying away an envelope with the spikes. This complete virus or virion is ready to infect another cell.

Figure 7.10 General features in the multiplication cycle of RNA animal viruses. The major events in the multiplication cycle of an enveloped + strand RNA virus.

upon entry as mRNA, starts being translated into viral proteins, especially those needed for further viral replication. The + strand is then replicated by host machinery into –ssRNA. This RNA becomes the template for the creation of many new +ssRNAs, which are used as the viral genomes for new viruses. Additional +ssRNAs are synthesized and used for late-stage mRNAs. Some viruses come equipped with the necessary enzymes for synthesis of viral components and others utilize enzymes from the host. Proteins for the capsid, spikes, and viral enzymes are synthesized on the host's ribosomes using host amino acids. Another type of RNA virus, the retroviruses, turn their RNA genomes into DNA. This step is accomplished by a viral enzyme called reverse transcriptase and has important implications in infections with these viruses, one of which is HIV.

DNA viruses generally follow the same steps of adsorption, penetration, uncoating, synthesis, assembly, and release. The steps of the synthesis process vary. **Figure 7.11** illustrates this. Replication of dsDNA viruses is divided into phases. During the early phase, viral DNA enters the nucleus, where several genes are transcribed into a messenger RNA. The newly synthesized RNA transcript then moves into the cytoplasm to be translated into viral proteins (enzymes) needed to replicate the viral DNA; this replication occurs in the nucleus. The host cell's own DNA polymerase is often involved, though some viruses (herpesvirus, for example) have their own polymerase. During the late phase, other parts of the viral genome are transcribed and translated into proteins required to form the capsid and other structures. The new viral genomes and capsids are assembled, and the mature viruses are released by budding or by host cell disintegration. Double-stranded DNA viruses interact directly with the DNA of their host cell. In some viruses, the viral DNA becomes silently integrated into the host's genome by insertion at a particular site on the host genome. This integration may later lead to the transformation of the host cell into a cancer cell and the production of a tumor.

Assembly of Animal Viruses: Host Cell as Factory

Toward the end of the cycle, mature virus particles are constructed from the growing pool of parts. In most instances, the capsid is

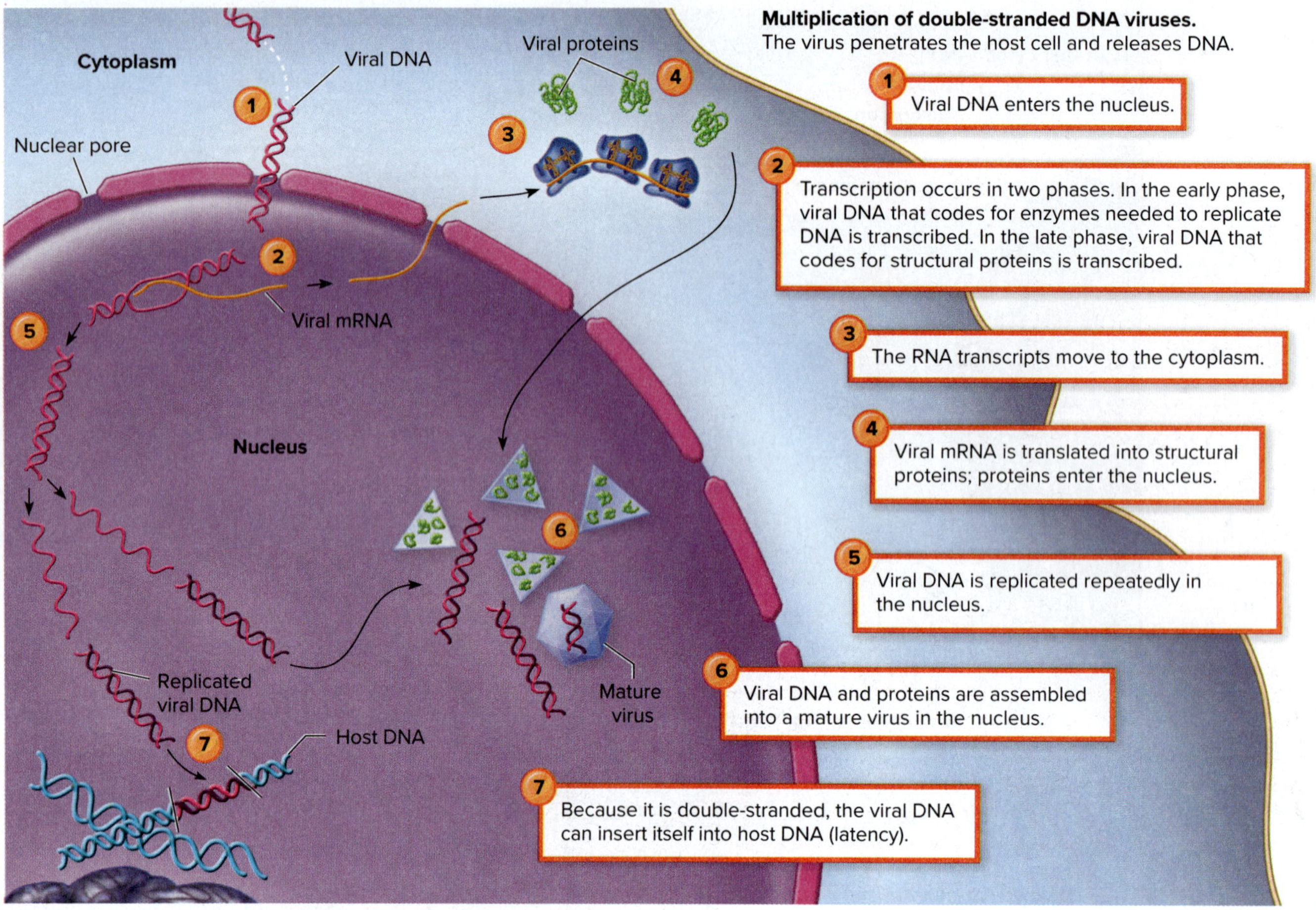
Cytoplasm
Viral DNA
Viral proteins
Nuclear pore
Viral mRNA
Nucleus
Replicated viral DNA
Host DNA
Mature virus
Multiplication of double-stranded DNA viruses.
The virus penetrates the host cell and releases DNA.
1 Viral DNA enters the nucleus.
2 Transcription occurs in two phases. In the early phase, viral DNA that codes for enzymes needed to replicate DNA is transcribed. In the late phase, viral DNA that codes for structural proteins is transcribed.
3 The RNA transcripts move to the cytoplasm.
4 Viral mRNA is translated into structural proteins; proteins enter the nucleus.
5 Viral DNA is replicated repeatedly in the nucleus.
6 Viral DNA and proteins are assembled into a mature virus in the nucleus.
7 Because it is double-stranded, the viral DNA can insert itself into host DNA (latency).

Figure 7.11 General features in the multiplication cycle of DNA animal viruses. Synthesis in a dsDNA virus.

first constructed as an empty shell that will serve as a receptacle for the nucleic acid strand. Electron micrographs taken during this time show cells with masses of viruses, often in crystalline packets **(figure 7.12).** One important event leading to the release of enveloped viruses is the insertion of viral spikes into the host's cell membrane so they can be picked up as the virus buds off with its envelope, as discussed earlier.

Release of Mature Viruses

To complete the cycle, assembled viruses leave their host in one of two ways. Nonenveloped and complex viruses that reach maturation in the cell nucleus or cytoplasm are released when the cell lyses or ruptures. Enveloped viruses are liberated by **budding,** or **exocytosis,** from the membranes of the cytoplasm, nucleus, endoplasmic reticulum, or vesicles. During this process, the nucleocapsid binds to the membrane, which curves completely around it and forms a small pouch. Pinching off the pouch releases the virus with its envelope **(figure 7.13).** Budding of enveloped viruses causes them to be shed gradually, without the sudden destruction of the cell. But regardless of how the virus leaves, most active viral infections are ultimately lethal to the cell because of accumulated damage. The number of viruses released by infected cells is variable, controlled by factors such as the size of the virus and the health of the host cell. About 3,000 to 4,000 virions are released from a single cell infected with poxviruses, whereas a poliovirus-infected cell can release over 100,000 virions. If even a small number of these virions happen to meet another susceptible cell and infect it, the potential for rapid viral proliferation is immense.

Damage to the Host Cell

The short- and long-term effects of viral infections on animal cells are well documented. Virus-induced damage to the cell that alters its microscopic appearance is termed **cytopathic effects (CPEs).** Individual cells can lose their positional orientation, undergo major changes in shape or size, or develop intracellular changes. It is common to find **inclusion bodies,** or compacted masses of viruses or damaged cell organelles, in the nucleus and cytoplasm. Because the CPEs caused by different viruses can be very distinctive, examining infected cells under a microscope can tell us which virus is infecting the cell. **Table 7.8** summarizes some prominent cytopathic effects associated with specific viruses. One very common CPE is the fusion of multiple host cells into single large cells containing multiple nuclei. These **syncytia** (singular, *syncytium*) are a result of some viruses' ability to fuse membranes. The respiratory syncytial virus (RSV) is even named for this effect.

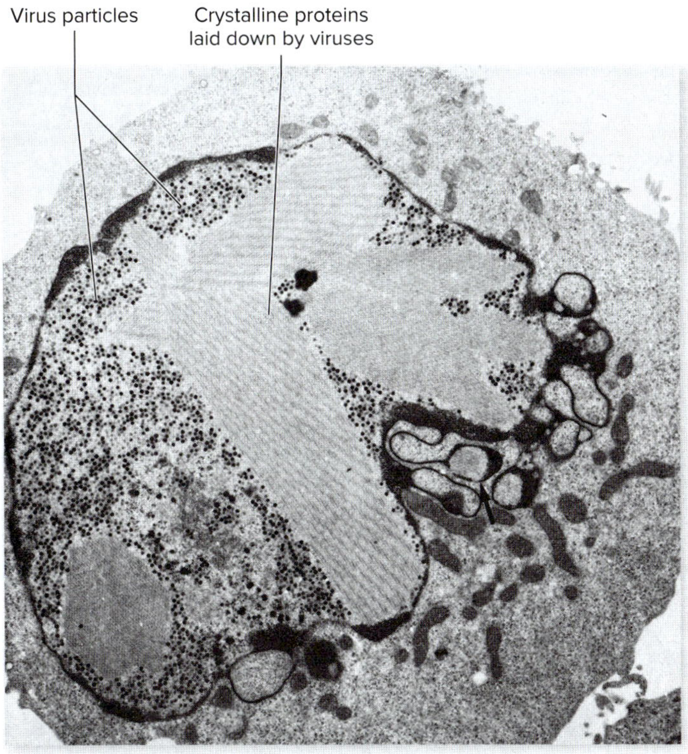
Virus particles
Crystalline proteins laid down by viruses

Figure 7.12 **Nucleus of a eukaryotic cell, containing hundreds of adenovirus virions.**
Courtesy Harold C. Smith, Department of Biochemistry and Biophysics, RNA Center & Cancer Center, University of Rochester Medical Center, Rochester, NY

Persistent Infections

Although accumulated damage from a virus infection kills most host cells, some cells maintain a carrier relationship, in which the cell harbors the virus and is not immediately lysed. These so-called *persistent infections* can last from a few weeks to the remainder of the host's life. Viruses can remain latent in the cytoplasm of a host cell, or can incorporate into the DNA of the host. When viral DNA is incorporated into the DNA of the host, it is called a **provirus.** The virus that causes roseola has been found to be passed down from parent to infant in the provirus state. One of the more serious complications occurs with the measles virus. It may remain hidden in brain cells for many years, causing progressive damage and loss of function. Several types of viruses remain in a *chronic latent state,* periodically becoming reactivated. Examples of this are herpes simplex viruses (cold sores and genital herpes) and herpes zoster virus (chickenpox and shingles). Both viruses can go into latency in nerve cells and later emerge under the influence of various stimuli to cause recurrent symptoms. NASA recently reported that herpesviruses reactivated among astronauts during spaceflight and for some time afterwards.

Viruses and Cancer

Some animal viruses enter their host cell and permanently alter its genetic material, which can lead to cancer. Experts estimate that up to 13% of human cancers are caused by viruses. (The percentage is higher in developing countries.) These viruses are

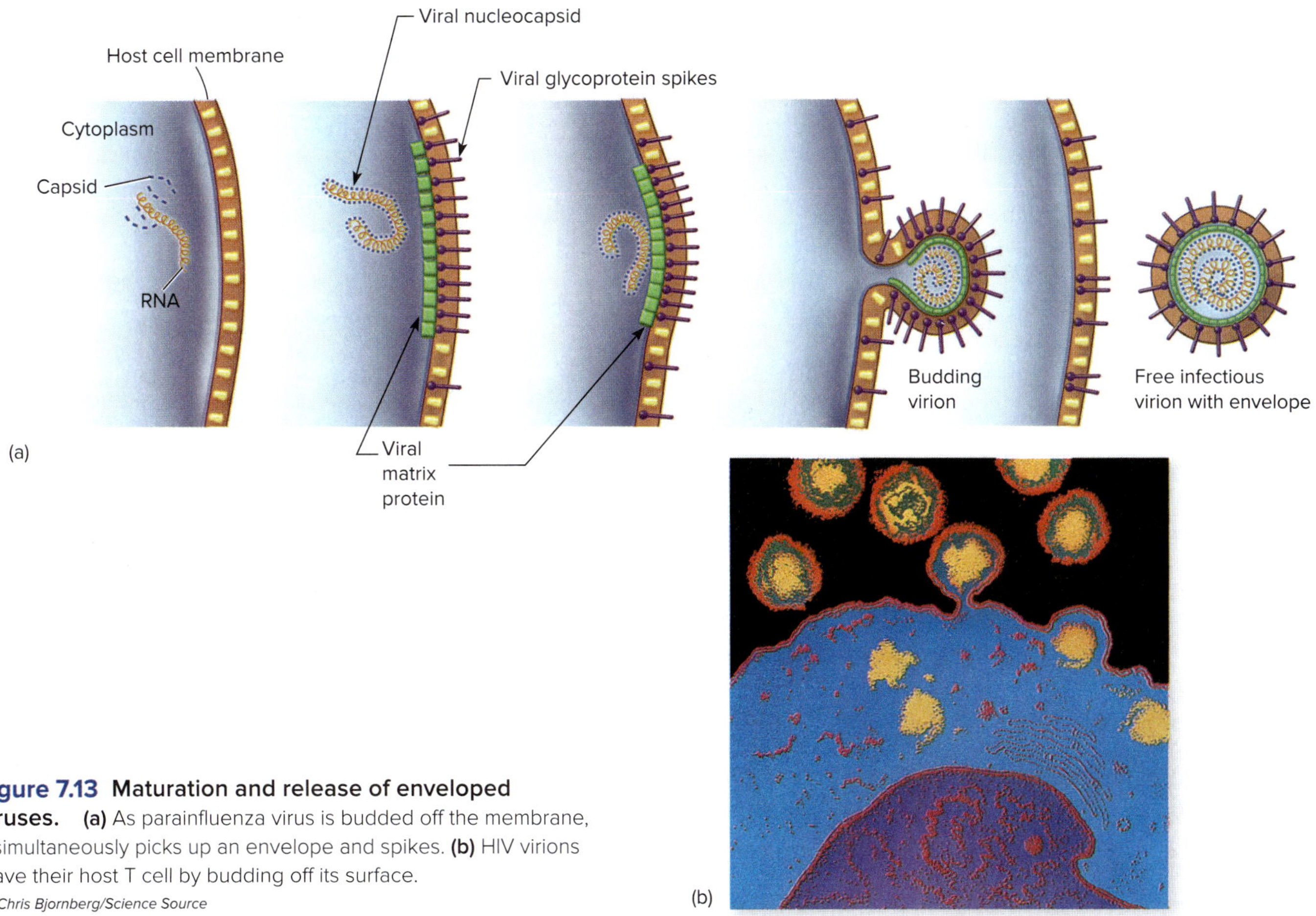
Viral nucleocapsid
Host cell membrane
Viral glycoprotein spikes
Cytoplasm
Capsid
RNA
Budding virion
Free infectious virion with envelope
(a)
Viral matrix protein
(b)

Figure 7.13 **Maturation and release of enveloped viruses.** **(a)** As parainfluenza virus is budded off the membrane, it simultaneously picks up an envelope and spikes. **(b)** HIV virions leave their host T cell by budding off its surface.
(b) Chris Bjornberg/Science Source

called *oncoviruses*, and they are said to be **oncogenic,** and their effect on the cell is called **transformation.** Viruses that cause cancer in animals act in several different ways, illustrated in **figure 7.14.** In some cases, the virus carries genes that directly cause the cancer. In other cases, the virus produces proteins that induce a loss of growth regulation in the cell, leading to cancer. Transformed cells have an increased rate of growth; alterations in chromosomes; changes in the cell's surface molecules; and the capacity to divide for an indefinite period, unlike normal animal cells. Some oncoviruses are DNA viruses such as papillomavirus (associated with cervical cancer), herpesviruses (Epstein-Barr virus causes Burkitt's lymphoma), and hepatitis B virus (liver cancer). A virus related to HIV—called HTLV-I—is involved in one type of human leukemia. These findings have spurred a great deal of speculation on the possible involvement of viruses in cancers whose causes are still unknown.

Table 7.8 Cytopathic Changes in Selected Virus-Infected Animal Cells

CYTOPATHIC EFFECTS

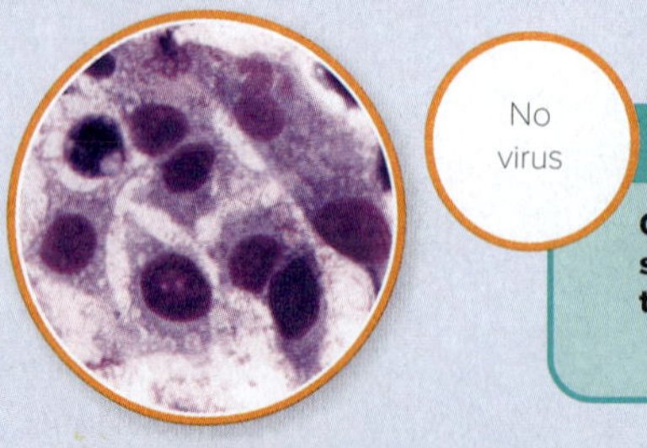 No virus	**Normal Cells**	Cells grown on a plastic surface line up in a single layer with tight connections between them.
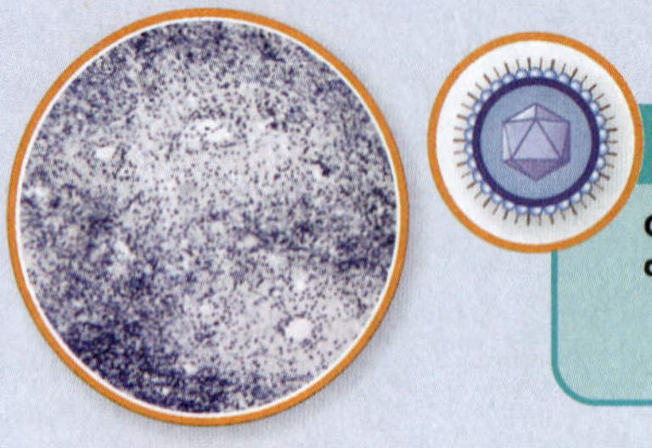	**Poliovirus**	Cells are killed completely. They shrink, detach from the surface, and lyse.
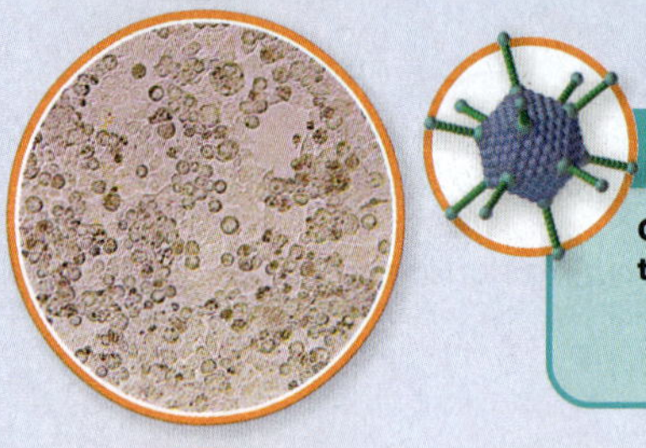	**Adenovirus**	Cells round up and partially detach from the surface. They tend to clump together.
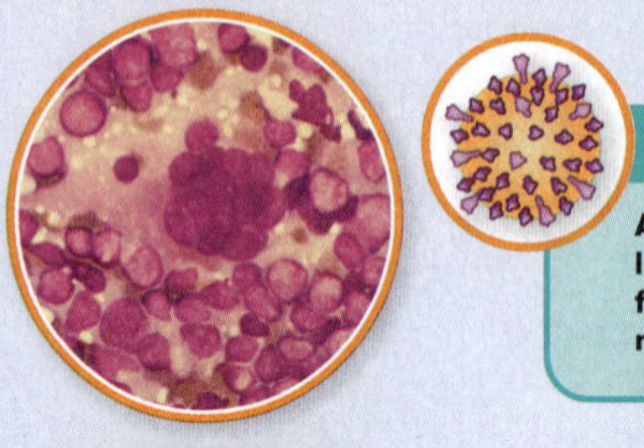	**Respiratory Syncytial Virus**	Adjacent cells fuse together, forming a large cell with many nuclei (originating from the former single cells). This multinucleate cell is called a syncytium.
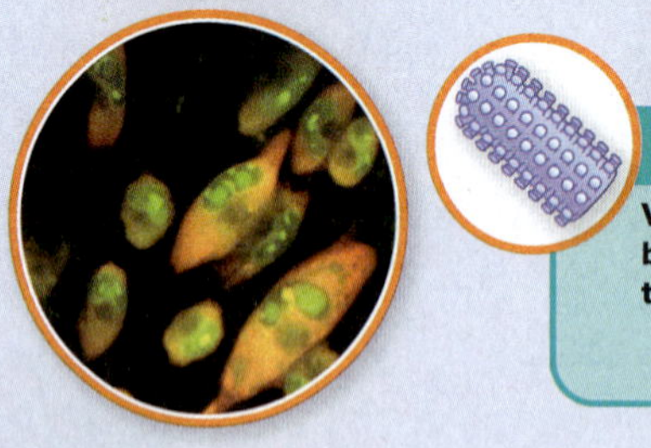	**Cytomegalovirus**	Viruses can cause formation of inclusion bodies in either the cytoplasm or nucleus of the host cell. This is cytomegalovirus.

Dr. A.J. Sulzer/CDC; CDC/Dr. Karp, Emory University; Dr. Kirsh/CDC; Dr. Edwin P. Ewing, Jr./CDC; Massimo Battaglia

Viruses That Infect Bacteria

We now turn to the multiplication cycle of another type of virus called a bacteriophage. When bacterial viruses were first discovered in 1915, it first appeared that the bacterial host cells were being eaten by some unseen parasite; hence, the name *bacteriophage* was used (*phage* coming from the Greek word for "eating"). Most bacteriophages (often shortened to **phage**) contain double-stranded DNA, although single-stranded DNA and RNA types exist as well. As far as we know, every bacterial species is parasitized by at least one specific bacteriophage. Bacteriophages are of great interest to medical microbiologists because they often make the bacteria they infect more pathogenic for humans (more about this later). Probably the most widely studied bacteriophages are those of the intestinal bacterium *Escherichia coli*—especially the ones known as the "T-even" phages such as T2 and T4. They are of the complex capsid type. They have an icosahedral capsid head containing DNA, a central tube (surrounded by a sheath), collar, base plate, tail pins, and fibers, which in combination make an efficient package for infecting a bacterial cell. It is tempting to think of these extraordinary viruses as minute spacecrafts docking on an alien planet, ready to unload their genetic cargo.

T-even bacteriophages go through stages similar to the animal viruses **(figure 7.15).** They *adsorb* to host bacteria using specific receptors on the bacterial surface. Although the entire phage does not enter the host cell, the nucleic acid *penetrates* the host after being injected through a rigid tube that the phage inserts through the bacterial membrane and wall **(figure 7.16).** This eliminates the need for *uncoating*. Entry of the nucleic acid brings host cell DNA replication and protein synthesis to a halt. Soon the host cell machinery is used for viral *replication* and synthesis of viral proteins. As the host cell produces new phage parts, the parts spontaneously *assemble* into bacteriophages.

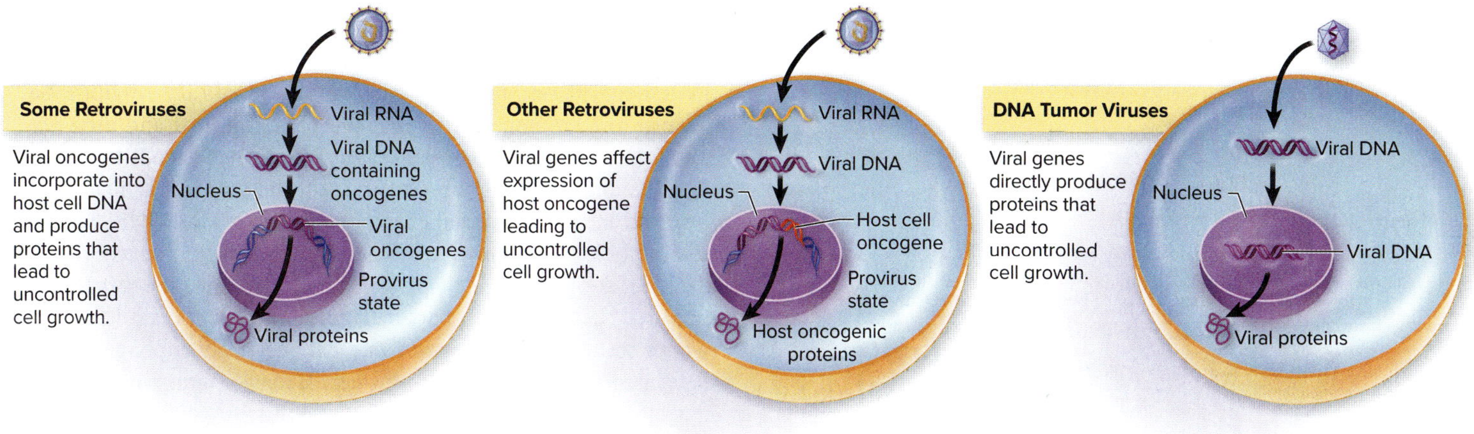
Some Retroviruses
Viral oncogenes incorporate into host cell DNA and produce proteins that lead to uncontrolled cell growth.
Viral RNA
Viral DNA containing oncogenes
Nucleus
Viral oncogenes
Provirus state
Viral proteins
Other Retroviruses
Viral genes affect expression of host oncogene leading to uncontrolled cell growth.
Viral RNA
Viral DNA
Nucleus
Host cell oncogene
Provirus state
Host oncogenic proteins
DNA Tumor Viruses
Viral genes directly produce proteins that lead to uncontrolled cell growth.
Viral DNA
Nucleus
Viral DNA
Viral proteins

Figure 7.14 **Three mechanisms for viral induction of cancer.**

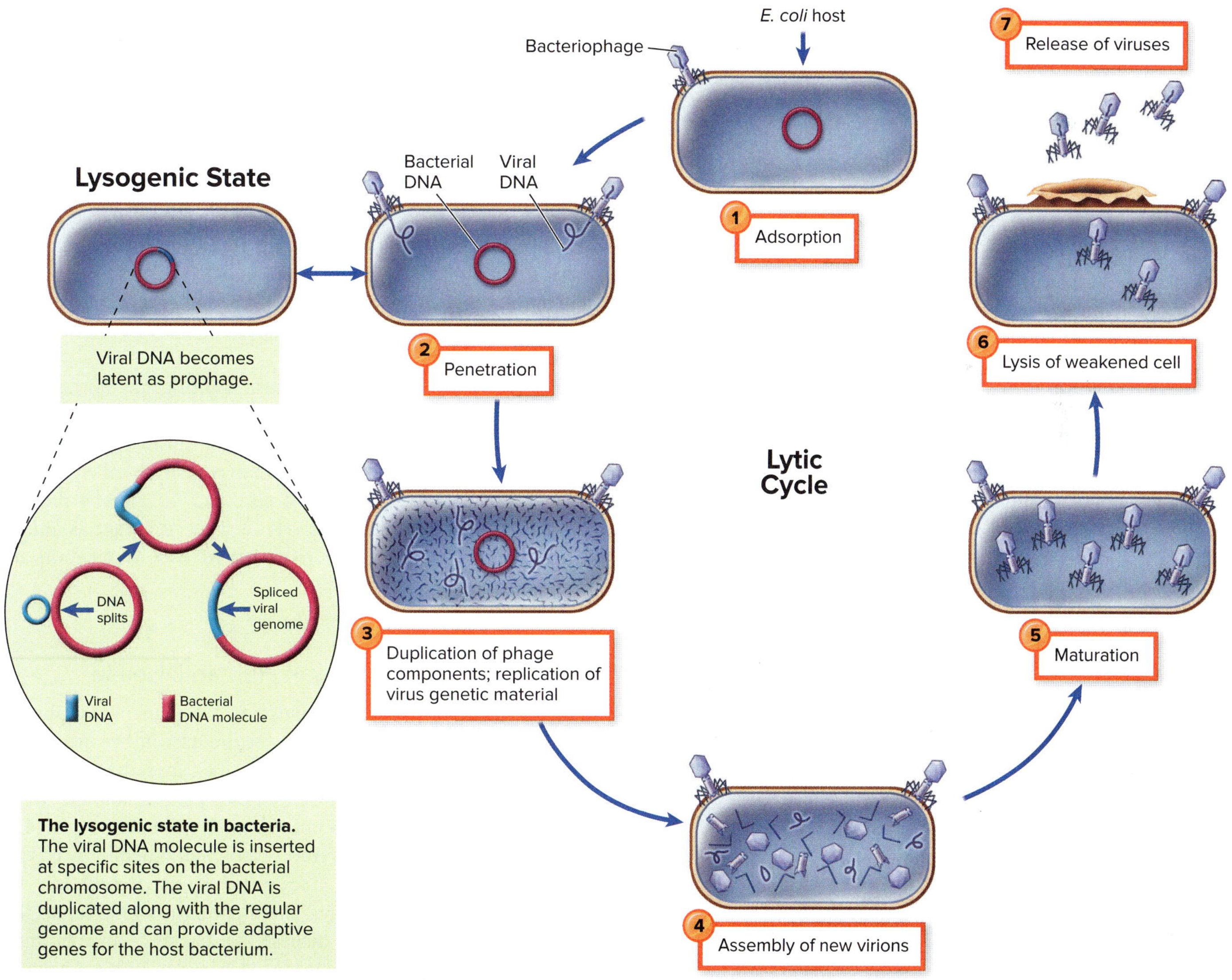
E. coli host
Bacteriophage
1 Adsorption
Bacterial DNA
Viral DNA
2 Penetration
Lysogenic State
Viral DNA becomes latent as prophage.
DNA splits
Spliced viral genome
Viral DNA
Bacterial DNA molecule
The lysogenic state in bacteria. The viral DNA molecule is inserted at specific sites on the bacterial chromosome. The viral DNA is duplicated along with the regular genome and can provide adaptive genes for the host bacterium.
3 Duplication of phage components; replication of virus genetic material
Lytic Cycle
4 Assembly of new virions
5 Maturation
6 Lysis of weakened cell
7 Release of viruses

Figure 7.15 **Events in the cycle of T-even bacteriophages.** The lytic cycle (1–7) involves full completion of viral infection through lysis and release of virions. Occasionally, the virus enters a reversible state of lysogeny (left) and its genetic material is incorporated into the host's genetic material.

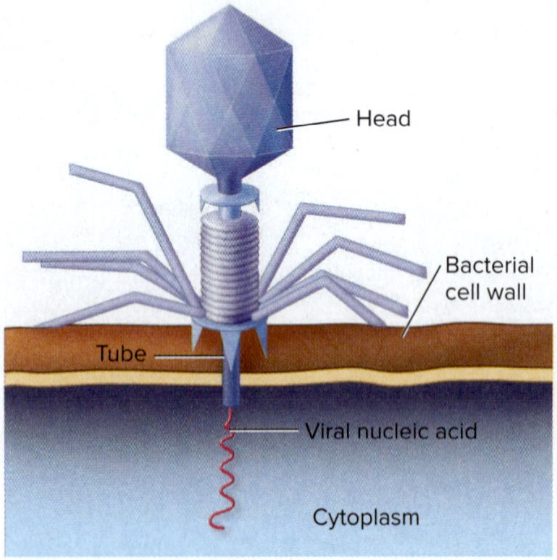
Head
Bacterial cell wall
Tube
Viral nucleic acid
Cytoplasm

Figure 7.16 Penetration of a bacterial cell by a T-even bacteriophage. After adsorption, the phage base becomes embedded in the cell wall and the sheath contracts, pushing the tube through the cell wall and releasing the nucleic acid into the interior of the cell.

An average-size *E. coli* cell can contain up to 200 new phage units at the end of this period. Eventually, the host cell becomes so packed with viruses that it **lyses**—splits open—thereby releasing the mature virions **(figure 7.17).** This process is hastened by viral enzymes produced late in the infection cycle that digest the cell envelope, thereby weakening it. When they are released, the virulent phages can spread to other susceptible bacterial cells and begin a new cycle of infection.

Bacteriophage infection may result in lysis of the cell, as just described. When this happens, the phage is said to have been in the *lytic phase* or *cycle*. Alternatively, phages can be less obviously damaging, in a cycle called the *lysogenic cycle*.

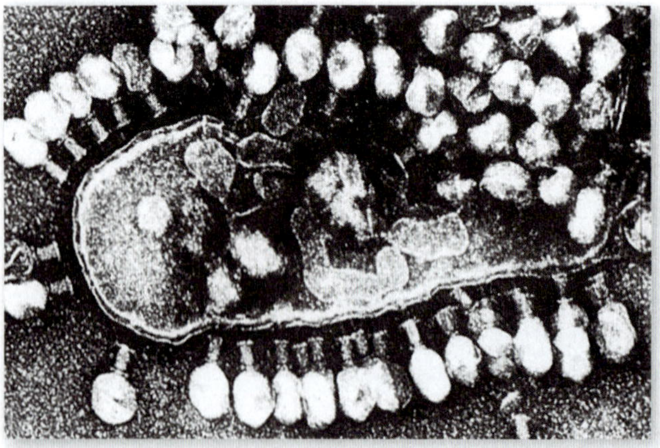

Figure 7.17 A weakened bacterial cell, crowded with viruses. The cell has ruptured and released numerous virions that can then attack nearby susceptible host cells.
Lee D. Simon/Science Source

Lysogeny: The Silent Virus Infection

Special DNA phages, called **temperate phages,** can participate in a lytic phase, but they also have the ability to undergo adsorption and penetration into the bacterial host and then *not* undergo replication or release immediately. Instead, the viral DNA enters an inactive **prophage** state, during which it is inserted into the bacterial chromosome. This viral DNA will be retained by the bacterial cell and copied during its normal cell division so that the cell's progeny will also have the temperate phage DNA (see green part of figure 7.15). This condition, in which the bacterial chromosome carries bacteriophage DNA, is termed **lysogeny** (ly-soj′-uhn-ee). Because viral particles are not produced, the bacterial cells carrying temperate phages do not lyse, and they appear entirely normal. (This will remind you of the provirus state of animal viruses.) On occasion, in a process called **induction,** the prophage in a lysogenic cell will be activated and progress directly into viral replication and the lytic cycle. Lysogeny is a less deadly form of parasitism than the full lytic cycle and is thought to be an evolutionary advancement that allows the virus to spread without killing the host.

Bacteriophages are just now receiving their due as important shapers of biological life. Scientists believe that there are more bacteriophages than all other forms of life combined in the biosphere. Viral genes linger in human, animal, plant, and bacterial genomes in huge numbers. It is estimated that 10% to 20% of DNA in bacteria is actually prophage DNA. As such, viruses can contribute what are essentially permanent traits to the bacteria, so much so that it could be said that all bacteria—indeed, all organisms—are really hybrids of themselves and the viruses that infect them.

Bacteriophages may have use in treating human bacterial infections. We will explore this possibility in chapter 12.

In 2008, a subcellular particle now termed a *virophage* revealed the existence of viruses that parasitize other viruses. When a virophage enters a host cell such as an amoeba that is already infected with a larger virus, the virophage is able to utilize genes from the other virus for its own replication and production.

The Danger of Lysogeny in Human Disease

Many bacteria that infect humans are lysogenized by phages. Sometimes, that is very bad news for the human because phage genes in the bacterial chromosome may cause the production of toxins or enzymes that cause pathology in the human. When a bacterium acquires a new trait from its temperate phage, it is called **lysogenic conversion.** The phenomenon was first discovered in the 1950s in *Corynebacterium diphtheriae,* the bacteria that cause diphtheria. The diphtheria toxin responsible for the deadly nature of the disease is a bacteriophage product. *C. diphtheriae* without the phage are harmless. Other bacteria that are made virulent by their prophages are *Vibrio cholerae,* the agent of cholera, and *Clostridium botulinum,* the cause of botulism.

Table 7.9 Comparison of Bacteriophage and Animal Virus Multiplication

	Bacteriophage	Animal Virus
Adsorption	Precise attachment of special tail fibers to cell wall	Attachment of capsid or envelope to cell surface receptors
Penetration	Injection of nucleic acid through cell wall; no uncoating of nucleic acid	Whole virus is engulfed and uncoated, or virus surface fuses with cell membrane; nucleic acid is released through uncoating
Synthesis and assembly	Occurs in cytoplasm Cessation of host synthesis Viral DNA or RNA is replicated and begins to function Viral components synthesized	Occurs in cytoplasm and nucleus Cessation of host synthesis Viral DNA or RNA is replicated and begins to function Viral components synthesized
Viral persistence	Lysogeny	Latency, chronic infection, cancer
Release from host cell	Cell lyses when viral enzymes weaken it	Some cells lyse; enveloped viruses bud off host cell membrane
Cell destruction	Immediate or delayed	Immediate or delayed

Disease Connection

Streptococcus pyogenes hardly needs any help to cause disease. This pathogen causes "strep throat," a flesh-eating disease, and serious bloodstream infections. But when it carries genes from its bacteriophage, it can cause even greater damage. Lysogenized *S. pyogenes* carries genes coding for something called erythrogenic toxin. These strains lead to postinfection problems such as scarlet fever, which results in a skin rash and a high fever, all courtesy of the bacteriophage.

The multiplication cycles of animal and bacterial viruses illustrate general features of viral multiplication in a very concrete way. The animal and bacterial cycles are compared in **table 7.9.** Because of the intimate association between the genetic material of the virus and that of the host, phages occasionally serve as transporters of bacterial genes from one bacterium to another and consequently can play a profound role in bacterial genetics. This phenomenon, *transduction,* was introduced in chapter 6.

7.4 Learning Outcomes—Assess Your Progress

10. Diagram the six-step multiplication cycle of animal viruses.
11. Define the term *cytopathic effect* and provide one example.
12. Provide examples of persistent and transforming infections, describing their effects on the host cell.
13. Provide a thorough description of lysogenic and lytic bacteriophage infections.

7.5 Techniques in Cultivating and Identifying Animal Viruses

One problem hampering earlier animal virologists was their inability to grow specific viruses routinely in pure culture—and in sufficient quantities for their studies. Early on, viruses could only be grown in an organism that was the usual host for the virus. But this method had its limitations. How could researchers have ever traced the stages of viral multiplication if they had been restricted to the natural host, especially in the case of human viruses? Fortunately, systems of cultivation with broader applications were developed, called *in vivo* (in vee′-voh) and *in vitro* (in vee′-troh) methods. Methods that use living embryos or animals are called *in vivo* methods. Another strategy is to use cells or tissues that are cultivated in the lab. These are called *in vitro* methods.

The primary purposes of viral cultivation are

1. to isolate and identify viruses in clinical specimens;
2. to prepare viruses for vaccines; and
3. to do detailed research on viral structure, multiplication cycles, genetics, and effects on host cells.

Using Live Animal Inoculation

Specially bred strains of white mice, rats, hamsters, guinea pigs, and rabbits are the usual choices for animal cultivation of viruses. Invertebrates (insects) or nonhuman primates are occasionally used as well. Because viruses can exhibit some host specificity, certain animals can propagate a given virus more readily than others. The animal is exposed to the virus by injection of a viral preparation or specimen into the brain, blood, muscle, body cavity, skin, or footpads.

Using Bird Embryos

An embryo is an early developmental stage of animals marked by rapid differentiation of cells. Birds undergo their embryonic period within the closed protective case of an egg, which makes an incubating bird egg a nearly perfect system for viral propagation. It is an intact and self-supporting unit, complete with its own sterile environment and nourishment. Furthermore, it contains several embryonic tissues that easily support viral multiplication.

Chicken, duck, and turkey eggs are the most common choices for inoculation. The virus solution must be injected through the shell, usually by drilling a hole or making a small window. Rigorous sterile techniques must be used to prevent contamination by

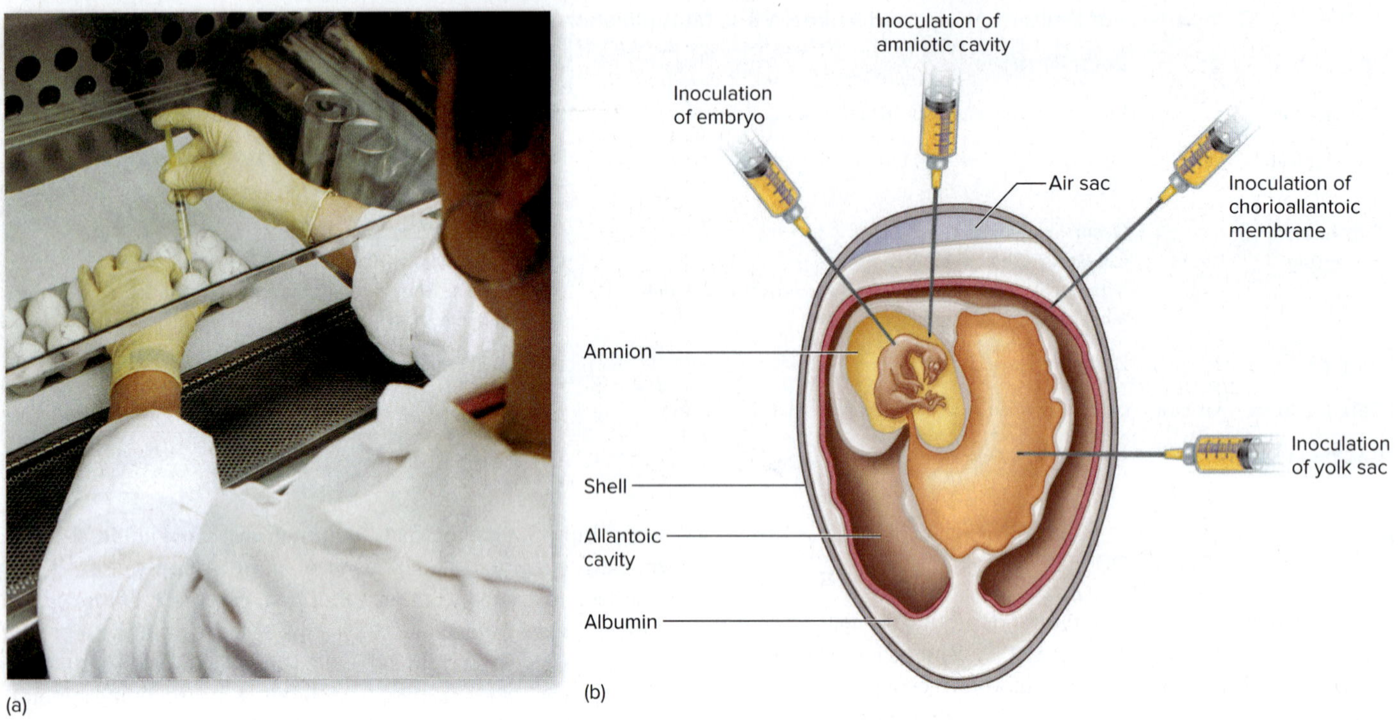
(a)
Inoculation of amniotic cavity
Inoculation of embryo
Air sac
Inoculation of chorioallantoic membrane
Amnion
Inoculation of yolk sac
Shell
Allantoic cavity
Albumin
(b)

Figure 7.18 Cultivating animal viruses in a developing bird embryo. **(a)** A technician inoculates fertilized chicken eggs with viruses in the first stage of preparing vaccines. Some influenza vaccines are prepared this way. **(b)** The shell is perforated using sterile techniques, and a virus preparation is injected into a site selected to grow the viruses. Targets include the allantoic cavity, a fluid-filled sac that functions in embryonic waste removal; the amniotic cavity, a sac that cushions and protects the embryo itself; the chorioallantoic membrane, which functions in embryonic gas exchange; the yolk sac, a membrane that mobilizes yolk for the nourishment of the embryo; and the embryo itself.

(a) Courtesy Ted Heald, State Hygienic Laboratory at the University of Iowa

bacteria and fungi from the air and the outer surface of the shell. The exact part of the embryo that is inoculated is guided by the type of virus being cultivated and the goals of the experiment **(figure 7.18).**

Using Cell (Tissue) Culture Techniques

The most important early discovery that led to easier cultivation of viruses in the laboratory was the development of a simple and effective way to grow populations of isolated animal cells in culture dishes. These types of *in vitro* cultivation systems are termed *cell culture* or *tissue culture.* (Although these terms are used interchangeably, *cell culture* is probably a more accurate description.) Animal cell cultures are grown in sterile dishes or bottles with special media that contain the correct nutrients required by animal cells to survive. The cultured cells grow in the form of a *monolayer,* a single, confluent sheet of cells that supports viral multiplication and allows visual inspection of the culture for signs of infection **(figure 7.19).**

Cultures of animal cells usually exist in the primary or continuous form. *Primary cell cultures* are prepared by placing freshly isolated animal tissue in a growth medium. The cells undergo a series of mitotic divisions to produce a monolayer on the surface of the dish. Embryonic, fetal, adult, and even cancerous tissues have served as sources of primary cultures. Primary cultures retain several characteristics of the original tissues from which they were derived, but they generally have a limited existence. Eventually, they will either die out or mutate into a line of cells that can grow continuously. Continuous cell lines tend to have altered chromosome numbers, grow rapidly, and show changes in morphology. They can be continuously subcultured, provided they are routinely transferred to fresh nutrient medium.

One way to detect the growth of a virus in culture is to observe degeneration and lysis of infected cells in the monolayer of cells. The areas where virus-infected cells have been destroyed show up as clear, well-defined patches in the cell sheet called **plaques** (figure 7.19). Plaques are essentially the visible manifestation of cytopathic effects (CPEs). This same technique is used to detect and count bacteriophages because they also produce plaques when grown in soft agar cultures of their host cells (bacteria). A plaque develops when the viruses released by an infected host cell radiate out to adjacent host cells. As new cells become infected, they die and release more viruses, and so on. As this process continues, the infection spreads gradually and symmetrically from the original point of infection, causing the macroscopic appearance of round, clear spaces that correspond to areas of dead cells.

(a)

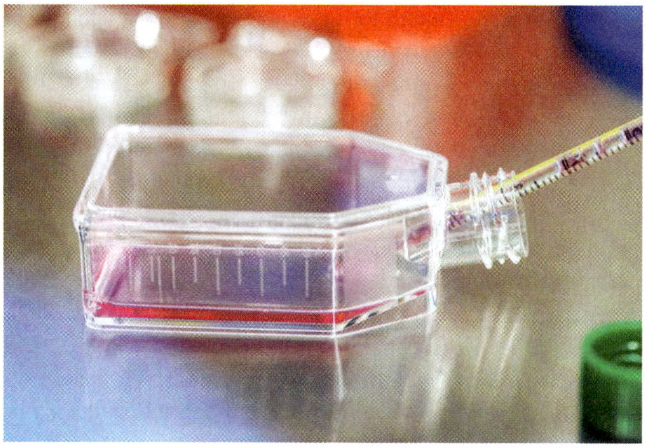

(b)

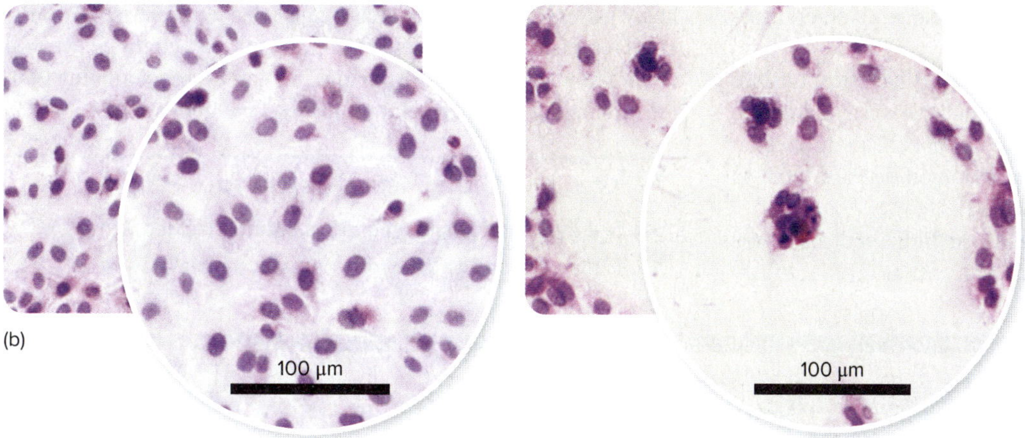
100 µm
100 µm

Figure 7.19 Appearance of normal and infected cell cultures. **(a)** A cell culture dish containing nutrient media. **(b)** When the dish is examined under a microscope, the cells are visible. On the left is an uninfected layer of animal cells. On the right, plaques in the animal cell layer are seen. These are open spaces where cells have been disrupted by viral infection.

(a) Anamaria Mejia/Shutterstock; (b) Bakonyi T, Lussy H, Weissenböck H, Hornyák A, Nowotny N./CDC

7.5 Learning Outcomes—Assess Your Progress

14. List the three principal purposes for cultivating viruses.
15. Describe three ways in which viruses are cultivated.

7.6 Viruses and Human Health

The number of viral infections that occur on a worldwide basis is nearly impossible to measure accurately. Consider widespread diseases such as COVID-19, colds, chickenpox, influenza, herpes, and warts. Although most viral infections do not result in death, some, such as rabies or Ebola, have very high mortality rates, and others can lead to long-term debility (polio, neonatal rubella). So far we know of 268 different viruses that can infect humans, and that is probably only a fraction of the real number.

The nature of viruses makes it difficult to design effective therapies against them. Because viruses are not bacteria, antibiotics aimed at disrupting bacterial cells do not work on them. Out of necessity, many antiviral drugs block virus replication by targeting the function of host cells, and for that reason can cause severe side effects. Almost all currently used antiviral drugs are designed to target one of the steps in the viral multiplication cycle you learned about earlier in this chapter. **Interferon (IFN),** a naturally occurring human cell product, can also be used with some success in treating and preventing viral infections.

Because antiviral drugs are so much trickier to design than antibacterial drugs, scientists in the 20th century turned to vaccine development to prevent the viral diseases. This is why the majority of the vaccines available today are targeted at viral diseases. The speed with which vaccines against SARS-CoV-2 were created in 2020 is a testament to decades of experience creating vaccines.

Viruses also cause some human cancers, such as liver cancer and cervical cancer. On the other hand, clinical trials of the use of viruses to target human cancers are underway.

7.6 Learning Outcomes—Assess Your Progress

16. Analyze the relative importance of viruses in human infection and disease.
17. Discuss the primary reason that antiviral drugs are more difficult to design than antibacterial drugs.

7.7 Prions and Other Noncellular Infectious Agents

Not all noncellular infectious agents are viruses. One group of unusual forms, even smaller and simpler than viruses, is called prions. Prions are implicated in chronic, persistent diseases in humans and animals. These diseases are called spongiform encephalopathies because the brain tissue removed from affected animals resembles a sponge. The infection has a long period of latency (usually several years) before the first clinical signs appear. (The opening case study is about prion disease.) Signs range from mental derangement to loss of muscle control. The diseases are progressive and universally fatal.

A common feature of these conditions is the deposition of distinct protein fibrils in the brain tissue. Researchers have determined that these fibrils are the agents of the disease and have named them **prions** (pree′-onz).

Creutzfeldt-Jakob disease (CJD) afflicts the central nervous system of humans and causes gradual degeneration and death. It is transmissible—but by an unknown mechanism. Several animals (sheep, mink, elk) are victims of similar transmissible diseases. Bovine spongiform encephalopathy (BSE), or "mad cow disease," was the subject of fears and a crisis in Europe in the late 1980s and 1990s when researchers found evidence that the disease could be acquired by humans who consumed contaminated beef. This was the first incidence of prion disease transmission from animals to humans. Several hundred Europeans developed symptoms of this variant form of Creutzfeldt-Jakob disease, leading to strict governmental controls on exporting cattle and beef products.

The fact that prions are composed primarily of protein (no nucleic acid) has certainly revolutionized our ideas of what can constitute an infectious agent. One of the most compelling questions is just how a prion could be replicated because all other infectious agents require some nucleic acid.

In 2015, researchers announced that prions were responsible for a rare but deadly brain disease called multiple-system atrophy, or MSA. It has led scientists to speculate that these misshapen proteins might be responsible for more common diseases such as Alzheimer's disease and Parkinson disease.

Other fascinating viruslike agents in human disease are defective forms called *satellite viruses* that are actually dependent on other viruses for replication. One remarkable example is the adeno-associated virus (AAV), so named because it was originally thought that it could only replicate in cells infected with adenovirus; but it can also infect cells that are infected with other viruses or that have had their DNA disrupted through other means. Another interesting satellite virus, called the delta agent, is a naked circle of RNA that is expressed only in the presence of the hepatitis B virus and can worsen the severity of liver damage.

Plants are also parasitized by viruslike agents called **viroids** that differ from ordinary viruses by being very small (about one-tenth the size of an average virus) and being composed of only naked strands of RNA, lacking a capsid or any other type of coating. Viroids are significant pathogens in several economically important plants, including tomatoes, potatoes, cucumbers, citrus trees, and chrysanthemums.

We have completed our survey of bacteria, archaea, eukaryotes, viruses, and prions and have described characteristics of different representatives of these five groups. As we continue, we will explore how microorganisms maintain themselves, beginning with nutrition and then looking into microbial metabolism.

7.7 Learning Outcomes—Assess Your Progress

18. List three noncellular infectious agents besides viruses.

19. Define prions and the organ of the body that they infect.

Media Under The Microscope Wrap-Up

Chia-Chi Charlie Chang/NIH

The news article "Couple Turns Hand of Fate into Hand of Hope" tells the story of Dr. Sonia Vallabh and her husband. The **intended message** of the article is twofold: to relate the status of the potential treatment for prion diseases they developed, and also to spotlight their personal journey to positions that allow them to study these diseases.

A **critical reading** of the article would entail looking for the scientific article in which the treatment is described. The publication, the *NIH Record*, comes from the National Institutes of Health and can be trusted because it is the national agency responsible for funding health research, and this is simply a report about one of their agency's fundees.

To **interpret** this article to your friends, you would have to explain some basic facts about prion biology. Of particular importance is the fact that prions lie dormant for years before being activated and killing you. This gave Dr. Vallabh time to study for her PhD and then make effective progress on a treatment.

My **overall grade** for the article is an A+. Not only is the science absolutely trustworthy, it is a fantastic story of human resilience, and you can't give that anything less than an A+.

Source: *NIH Record*, "Couple Turns Hand of Fate into Hand of Hope," online article posted January 24, 2020.

Study Smarter: Better Together

These activities are designed for you to use on your own with a study group—either a face-to-face group or a virtual one, consisting of 3–5 members. Studying together can be very helpful, but there are effective and ineffective ways to do it. For example, getting together without a clear structure is often not a good use of your time. Use your time efficiently by using one or more of the exercises below.

FACE-TO-FACE GROUPS

Use one or more of the activities below.

Peer Instruction: Assign numbers to your group members to use all semester long. Now look at these five concepts from this chapter. Each group member prepares a 5-minute lesson on the topic corresponding to their number. Don't worry if you have fewer than 5 members; just use however many you have! During your group study time, each member presents their lesson, and the group spends another 5–10 minutes discussing that lesson.

1. Anatomy of a virus
2. Capsid types
3. Animal virus replication cycle
4. Bacteriophage replication cycle
5. Cytopathic effects in animals and on medium

Concept Maps: Each member of the group should use this list of terms from this chapter to generate their own concept map. This can be hand-drawn or created using software (see Appendix C for guidelines). During group study time, compare each other's concept maps and help each other make sure they are correct. Of course, there are many different "correct" maps. Examining each member's map will help you talk through the varied concepts and how they are related.

Concept Terms:

adsorption	tropism	release	exocytosis
endocytosis	nucleic acids	integration	penetration
uncoating			

Table Topics: Each group member should identify a concept or topic from this week's class assignments with which they are having trouble and share it during group study time. The other group members can then help to clarify confusing issues or share how they figured it out. Aim for a maximum of 15 minutes per topic. If the topic remains unclear to the group, bring it up during class or use the instructor's office hours or e-mail to ask for help. Taking the time to struggle with a difficult concept first makes your questions much more specific and more likely to yield helpful answers.

VIRTUAL GROUPS

Not everyone has the time or opportunity to meet with group members outside of class time. You or your instructor can create a virtual group using e-mail or the course software.

Weekly Discussion Board: This forum can be used as a way for groups to discuss topics, via e-mail or other learning management systems or online platforms, before they are covered in class. As each member of the group answers the current week's question, they should send their responses to every other member of their group. It's best to agree on a deadline based on how your class schedule works (Saturday for the next week's topics, for example). Then, after the topic is discussed in class, each member should send a response that all group members will see with a follow-up post on the same topic. If you cover more than one chapter in a week, someone can be designated to choose which chapter Discussion Board question you will use. Or simply decide up front that you will always use the first-chapter-of-the week's question, to keep the schedule simple.

Discussion Question

Do you think viruses are part of the human microbiome? Why or why not?

Chapter Summary

VIRUSES AND PRIONS

7.1 INTRODUCTION TO VIRUSES

- **Viruses are not independent life forms. But they are extremely adept at manipulating their host cells for their own purposes.**
- **They are capable of infecting every type of cell on earth. Some infect other viruses also.**

7.2 THE GENERAL STRUCTURE OF VIRUSES

- Viruses have an enormous influence on our planet and on human health.
- Viruses range in size from 20 nm to 1,500 nm.
- Viruses are composed of an outer protein capsid containing either DNA or RNA. Some also have a membrane envelope around the capsid.
- Spikes on the surface of the virus capsid or envelope are critical for their attachment to host cells and determine which host cells they can infect, a phenomenon called tropism.

7.3 HOW VIRUSES ARE CLASSIFIED AND NAMED

- Viruses are grouped in various ways. The Baltimore Classification system categorizes viruses based on their genomes and how they create mRNA.
- The International Committee on the Taxonomy of Viruses lists 59 orders and 189 families of viruses. Viral classification changes rapidly.

7.4 HOW VIRUSES MULTIPLY

- Viral multiplication in a host cell involves adsorption, penetration (sometimes followed by uncoating), viral synthesis, and assembly, followed by viral release by lysis or budding.
- These events turn the host cell into a factory solely for the production of new viruses. This eventually results in the destruction of the host cell.
- Animal viruses can cause acute infections or can persist in host tissues as latent infections that can reactivate. Some latent viruses are oncogenic.
- Bacteriophages differ from animal viruses in their methods of adsorption, penetration, site of replication, and method of exit from host cells.
- Bacteriophages can enter an inactive phase called lysogeny, in which their DNA is incorporated into the host's DNA.

7.5 TECHNIQUES IN CULTIVATING AND IDENTIFYING ANIMAL VIRUSES

- Animal viruses must be studied in some type of living host cell. These can be cell cultures in dishes, bird embryos, or laboratory animals.
- Cell cultures are made of host cells grown in special sterile vessels containing growth factors.
- The host cells can either be eukaryotes (animals or plants) or bacteria.
- Virus growth in cell culture can be seen as cytopathic effects. Bacteriophage growth can be seen as clear places called plaques on a lawn of bacteria growing on agar.

7.6 VIRUSES AND HUMAN HEALTH

- Viruses are responsible for several billion infections each year. It is probable that many chronic diseases of unknown cause will eventually be connected to viral agents.
- Viral DNA is part of the human genome and is responsible for many human attributes.
- Viruses are responsible for several human cancers.
- Other viruses may be used in the future to treat human cancers.

7.7 PRIONS AND OTHER NONCELLULAR INFECTIOUS AGENTS

- Other noncellular agents of disease are the prions, which are not viruses but protein fibers; viroids, extremely small lengths of naked nucleic acid; and satellite viruses, which require the presence of larger viruses to cause disease.

SmartGrid: From Knowledge to Critical Thinking

This *21 Question Grid* takes the topics from this chapter and arranges them with respect to the American Society for Microbiology's Undergraduate Curriculum guidelines—all six of the important "Concepts" as well as the important "Competency" of scientific literacy. Three questions are supplied, which cover chapter content referring to the Concept or Competency in increasing levels of Bloom's taxonomy for learning.

ASM Concept/ Competency	A. Bloom's Level 1, 2—Remember and Understand (Choose one.)	B. Bloom's Level 3, 4—Apply and Analyze	C. Bloom's Level 5, 6—Evaluate and Create
Evolution	**1.** ___% of human DNA is thought to consist of viral DNA sequences. **a.** 0 **b.** 100 **c.** 10–20 **d.** 40–80	**2.** Discuss the influence that viruses have on the idea of "species" in bacteria.	**3.** Construct a scenario in which viral latency and lysogeny provide an evolutionary advantage to viruses.
Cell Structure and Function	**4.** The host cells that viruses can infect are determined by the **a.** receptors on the host cells. **b.** DNA in host cells. **c.** proximity of host cells. **d.** concentration of host cells in vicinity.	**5.** If viruses that normally form envelopes were prevented from budding, would they still be infectious? Why or why not?	**6.** Viruses use the host cell cytoplasmic space as their "factories" and their enzymes and other macromolecules as their "tools." Does that make them more sophisticated or less sophisticated than cells? Justify your answer.
Metabolic Pathways	**7.** The general steps in a viral multiplication cycle are **a.** adsorption, penetration, synthesis, assembly, and release. **b.** endocytosis, uncoating, replication, assembly, and budding. **c.** adsorption, uncoating, duplication, assembly, and penetration. **d.** endocytosis, penetration, replication, maturation, and exocytosis.	**8.** Compare and contrast the processes of latency and lysogeny in viruses.	**9.** Pathogenic bacteria lysogenized by phages can cause more serious disease than their counterparts that are not lysogenized. Speculate on whether it is to the bacterium's advantage to cause more serious disease or to cause less serious disease.
Information Flow and Genetics	**10.** When phage nucleic acid is incorporated into the nucleic acid of its host cell and is replicated when the host DNA is replicated, this is considered part of which cycle? **a.** lytic cycle **b.** virulence cycle **c.** lysogenic cycle **d.** cell cycle **e.** multiplication cycle	**11.** Describe the cell locations in which RNA and DNA viruses multiply.	**12.** RNA viruses tend to mutate—that is, to accumulate changes in their nucleic acid sequences—more frequently than any other organism. Speculate on why that might be and how that can be a positive trait if the virus is a human pathogen.
Microbial Systems	**13.** A virus that undergoes lysogeny is a/an **a.** temperate phage. **b.** intemperate phage. **c.** T-even phage. **d.** animal virus. **e.** DNA virus.	**14.** In figure 5.1, viruses are suggested to have contributed to cellular nucleic acid becoming DNA instead of RNA. What stages in the viral multiplication cycle could have led to this phenomenon?	**15.** Imagine and describe a scenario in which viruses in the human microbiome could exert a positive influence on human health.

ASM Concept/ Competency	A. Bloom's Level 1, 2—Remember and Understand (Choose one.)	B. Bloom's Level 3, 4—Apply and Analyze	C. Bloom's Level 5, 6—Evaluate and Create
Impact of Microorganisms	**16.** Clear patches in cell cultures that indicate sites of virus infection are called **a.** plaques. **b.** pocks. **c.** colonies. **d.** prions.	**17.** Prions are very difficult to destroy, even on surgical instruments. The diseases they cause often take years to cause symptoms. Explain how these two facts present a danger to patients undergoing brain surgery.	**18.** Construct an argument for whether humans or viruses have had more of an impact on the planet earth as it exists today.
Scientific Thinking	**19.** The number of viruses known to cause diseases in humans is in the **a.** dozens. **b.** thousands. **c.** hundreds. **d.** millions.	**20.** Write a paragraph targeted at middle-school students about why antibiotics do not work on viruses and why it is difficult to design antiviral agents.	**21.** The earliest drugs developed to treat HIV disease were "anti-reverse-transcriptase" agents. Explain why these would be effective and relatively safe.

Answers to the multiple-choice questions appear in Appendix A.

Visual Connections

This question uses visual images or previous content to connect content within and between chapters.

1. **From chapter 2, figure 2.23:** A virus containing which of these molecules would you expect to be less stable, more likely to mutate at a high rate? Why?

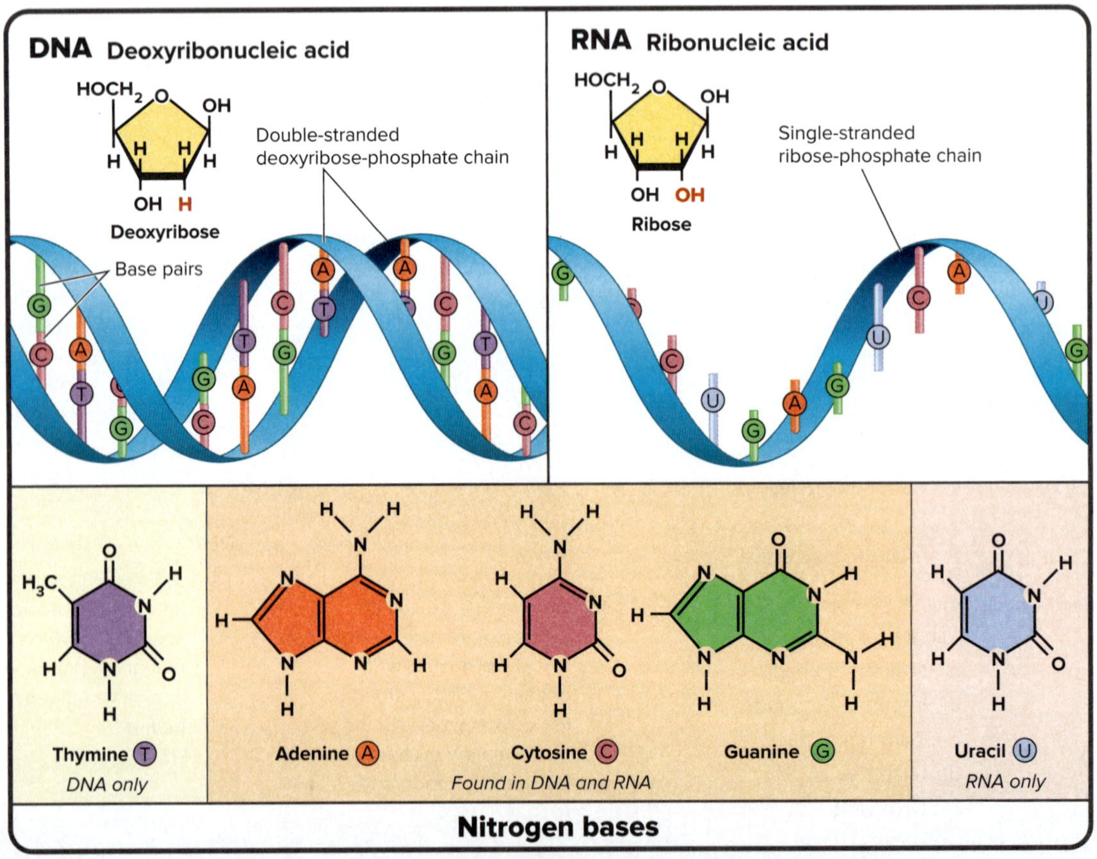

DNA Deoxyribonucleic acid
HOCH2
OH
H
OH H
Deoxyribose
Double-stranded deoxyribose-phosphate chain
Base pairs
RNA Ribonucleic acid
OH OH
Ribose
Single-stranded ribose-phosphate chain
Thymine T
DNA only
Adenine A
Cytosine C
Found in DNA and RNA
Guanine G
Uracil U
RNA only
Nitrogen bases

High Impact Study

These terms and concepts are most critical for your understanding of this chapter—and may be the most difficult. Have you mastered them?

Concepts

- ☐ Simplest structure of a virus
- ☐ Nucleic acid composition of viruses
- ☐ Capsid types
- ☐ Enveloped and naked viruses
- ☐ Animal virus multiplication cycle
- ☐ Lytic vs. lysogenic bacteriophage cycles

Terms

- ☐ Virus
- ☐ Bacteriophage
- ☐ Cytopathic effects
- ☐ Persistence
- ☐ Provirus
- ☐ Prophage
- ☐ Oncogenic
- ☐ Prions

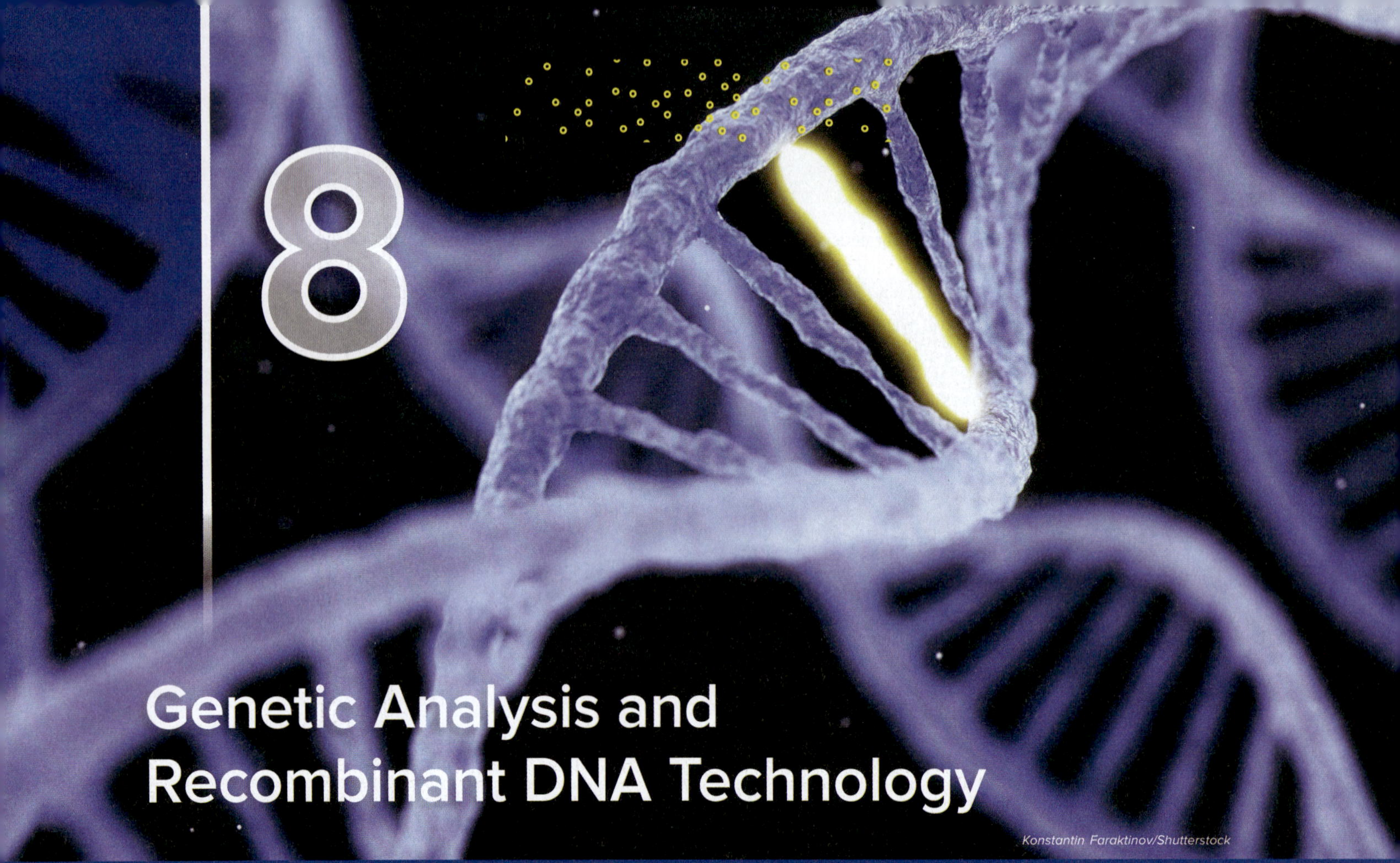

Konstantin Faraktinov/Shutterstock

8 Genetic Analysis and Recombinant DNA Technology

Media Under The Microscope

Change in One RNA Base in SARS-CoV-2 Linked to Higher Death Rate

These case studies examine an article from the popular media to determine the extent to which it is factual and/or misleading. This case focuses on the 2021 SciTechDaily *article "Genetics Research May Help Identify More Dangerous Strains of the Virus That Causes COVID-19."*

This news brief reports on research involving SARS-CoV-2, the virus that causes COVID-19. The article states that scientists analyzed over 7,000 SARS-CoV-2 isolates found in COVID-19 patients worldwide. They compared the genomic sequences of the virus in patients who died and those who survived to see if there were any differences in the sequences. The article tells us that 29,891 "locations in the genome" were assessed.

The findings reported in this article are that one base was found to be different (mutated) in one particular position in the genome among viruses that had a much higher mortality rate in their victims. It coded for a spike protein in the virus and appeared to make the virus more contagious (better at attaching) and more resistant to antibodies (that bind to the spike protein to block the attachment). This mutant strain was named the P.1 strain of the Gamma lineage. It came well before the Delta and Omicron variants, but was responsible for a deadly surge of infections in Brazil in 2021. The point here is that as new variants arise, these differences can be identified.

- What is the **intended message** of the article?
- What is your **critical reading** of the summary of the article? Remember that in this context, "critical reading" does not necessarily mean *What criticism do you have?* but asks you to apply your knowledge to interpret whether the article is factual and whether the facts support the intended message.
- How would you **interpret** the news item for your nonmicrobiologist friends?
- What is your **overall grade** for the news item—taking into account its accuracy and the accuracy of its intended effect?

Media Under The Microscope Wrap-Up appears at the end of the chapter.

Outline and Learning Outcomes

8.1 Tools and Techniques of Recombinant DNA Technology

1. Provide examples of practical applications of modern genetic technologies.
2. Explain the role of restriction endonucleases in the process of recombinant DNA technologies.
3. List the steps in the polymerase chain reaction, and discuss one disadvantage to this technique.
4. Describe how recombinant DNA is created, and discuss its role in gene cloning.

8.2 DNA Analysis

5. Describe how gel electrophoresis is used to analyze DNA.
6. Outline the general steps in DNA profiling.
7. Outline in general terms the process of DNA sequencing.
8. Discuss the significance of single nucleotide polymorphisms (SNPs) in DNA analysis.
9. Describe the uses of microarray technology.

8.3 Genetic Approaches to Healing Disease

10. Compose your own definition of personalized medicine.
11. Provide several examples of recombinant products that have contributed to human health.
12. List examples of recombinant bacteria and plants, and a purpose for each.
13. Differentiate between somatic and germline gene therapy.
14. Describe miRNAs and ways in which their discovery can impact human disease.
15. Explain how the CRISPR-Cas9 system can be used to cure genetic diseases.

8.1 Tools and Techniques of Recombinant DNA Technology

In chapter 6, we looked at the ways that microorganisms duplicate, exchange, and use their genetic information. In scientific vocabulary, this is called *basic science* because no product or application is directly derived from it. Humans, however, are quick to turn basic science into useful applications.

Seeing how microbes manipulated their own DNA allowed scientists to use these processes to accomplish goals important to human beings. Examples of human goals that have been assisted through the use of modern and not-so-modern genetic technologies can be seen in each of these scenarios:

1. Courts have, for thousands of years, relied on a description of a person's phenotype (eye color, hair color, and so on) as a means of identification. By remembering that a phenotype is the product of a particular sequence of DNA, you can quickly see how looking at someone's DNA (perhaps from a drop of blood) gives a better clue as to his or her identification.
2. We have understood for a long time that many diseases are the result of a missing or dysfunctional protein, and we have generally treated the diseases by replacing the protein as best we can, usually resulting in only temporary relief and limited success. Examples include insulin-dependent diabetes, adenosine deaminase deficiency, and blood-clotting disorders. DNA recombinant technology (sometimes called *genetic engineering*) offers the promise that fixing the underlying mutation responsible for the lack of a particular protein can treat these diseases far more successfully than we've been able to do in the past.
3. A newer technology called CRISPR offers the possibility of repairing the mutation responsible for many genetic diseases, in such a way that the ongoing generations never experience the mutation.

Information on recombinant DNA technology and its biotechnological applications is growing at such a rate that some new discovery or product is in the news on an almost daily basis. To keep this subject somewhat manageable, we present essential concepts and applications, organized under the following three topics:

- Tools and Techniques of Recombinant DNA Technology
- DNA Analysis
- Genetic Approaches to Healing Disease

DNA: The Raw Material

All of the intrinsic properties of DNA hold true whether the DNA is inside a bacterium or a test tube. But in the laboratory, we can take advantage of our knowledge of DNA chemistry to make some processes easier. It turns out that when DNA is heated to just below boiling (90°C to 95°C), the two strands separate, exposing the information contained in their bases. With the nucleotides exposed, DNA can be more easily identified, replicated, or transcribed. If heat-denatured DNA is then slowly cooled, complementary nucleotides will hydrogen bond with one another and the strands will renature, or regain their familiar double-stranded form **(figure 8.1).** As we shall see, this process is a necessary feature of the polymerase chain reaction and in the use of nucleic acid probes.

Systems for Cutting, Splicing, Amplifying, and Moving DNA

The polynucleotide strands of DNA can be clipped crosswise at selected positions by means of enzymes called **restriction endonucleases.**[1] These enzymes come from bacterial and

1. The meaning of *restriction* is that the enzymes do not act upon the bacterium's own DNA; an *endo*nuclease nicks DNA internally, not at the ends.

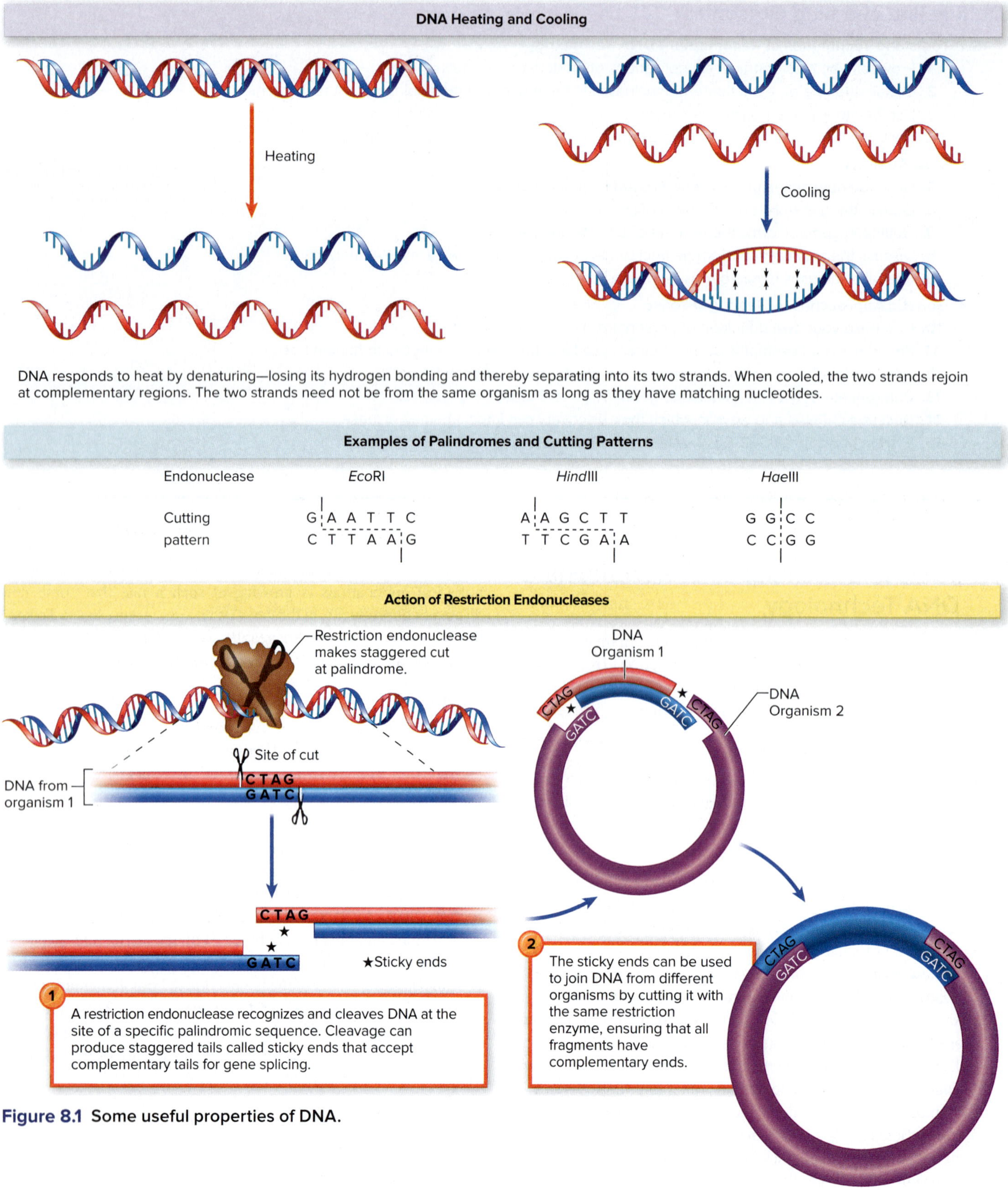
DNA Heating and Cooling
Heating
Cooling
DNA responds to heat by denaturing—losing its hydrogen bonding and thereby separating into its two strands. When cooled, the two strands rejoin at complementary regions. The two strands need not be from the same organism as long as they have matching nucleotides.
Examples of Palindromes and Cutting Patterns
Endonuclease
*Eco*RI
*Hind*III
*Hae*III
Cutting pattern
G A A T T C
C T T A A G
A A G C T T
T T C G A A
G G C C
C C G G
Action of Restriction Endonucleases
Restriction endonuclease makes staggered cut at palindrome.
Site of cut
DNA from organism 1
CTAG
GATC
★Sticky ends
1
A restriction endonuclease recognizes and cleaves DNA at the site of a specific palindromic sequence. Cleavage can produce staggered tails called sticky ends that accept complementary tails for gene splicing.
DNA Organism 1
DNA Organism 2
2
The sticky ends can be used to join DNA from different organisms by cutting it with the same restriction enzyme, ensuring that all fragments have complementary ends.

Figure 8.1 Some useful properties of DNA.

archaeal cells, and their discovery in 1971 started a genetic revolution. The enzymes recognize foreign DNA and are capable of breaking the phosphodiester bonds between adjacent nucleotides on both strands of DNA, leading to a break in the DNA strand. In bacteria and archaea in nature, this protects against the incompatible DNA of bacteriophages or plasmids. In the lab, the enzymes can be used to cleave DNA at desired sites.

Thousands of restriction endonucleases have been discovered in bacteria and archaea. Each type has a known sequence of 4 to 10 base pairs as its target, so sites of cutting can be finely controlled. Many of these enzymes have the unique property of recognizing and clipping at base sequences called **palindromes** (see figure 8.1). Palindromes are sequences of DNA that are identical when read from the 5′ to 3′ direction on one strand and the 5′ to 3′ direction on the other strand.

Endonucleases are usually named by combining the first letter of the bacterial genus, the first two letters of the species, and the endonuclease number. Thus, *Eco*RI is the first endonuclease found in *Escherichia coli* (in the R strain), and *Hind*III is the third endonuclease discovered in *Haemophilus influenzae* type d (see figure 8.1).

Endonucleases are used in the laboratory to cut DNA into smaller pieces for further study as well as to remove and insert sequences during recombinant DNA techniques, described in a later section. Most often, the enzymes make staggered, symmetrical cuts that leave short tails called "sticky ends." The enzymes cut four to five bases on the 3′ strand, and four to five bases on the 5′ strand, leaving overhangs on each end. This generates adhesive tails that will base-pair with complementary tails on other DNA fragments or plasmids (see figure 8.1). This effect makes it possible to splice genes into specific sites.

The pieces of DNA produced by restriction endonucleases are termed **restriction fragments.** Because DNA sequences vary, even among members of the same species, differences in the cutting pattern of specific restriction endonucleases give rise to restriction fragments of differing lengths, known as **restriction fragment length polymorphisms (RFLPs).** RFLPs allow the direct comparison of the DNA of two different organisms at a specific site, which, as we will see, has many uses.

Another enzyme, called a **ligase,** is necessary to seal the sticky ends together by rejoining the phosphate-sugar bonds cut by endonucleases. Its main application is in final splicing of genes into plasmids and chromosomes.

An enzyme called **reverse transcriptase** is best known for its role in the replication of HIV and other retroviruses. It also provides geneticists with a valuable tool for converting RNA into DNA. Molecules called **complementary DNA (cDNA)** can be made from messenger, transfer, ribosomal, and other forms of RNA. The technique provides a valuable means of synthesizing eukaryotic genes from mRNA transcripts **(figure 8.2).** The advantage is that the synthesized gene will not have the intervening sequences (introns) that can complicate the management of eukaryotic and archaeal genes in recombinant technology.

There is another system found in bacteria and archaea that can be exploited by scientists to alter genomes. The system is called **CRISPR,** which stands for *clustered regularly interspaced short palindromic repeats.* In the bacteria and archaea, these are short lengths of DNA with repeating nucleotides. After the repeats, short segments of spacer DNA are found. These turn out to be the vestiges of DNA left behind by "invading" bacteriophages or plasmids. The CRISPR system includes an enzyme—called Cas9—that is capable of recognizing and cutting this foreign DNA, keeping the bacterium or archaea from being invaded. It is thought to be an adaptive immune system used by bacteria. In other words, the bacteria "learn" the identity of an attacking phage by placing bits of its DNA in its own genome, and in the future can cut it up before it causes trouble. The cuts made by this system are so precise that they are compared to lasers, with restriction endonucleases being compared to scissors.

The system turns out to be highly adaptable for laboratory use, and scientists have started using CRISPR in many genetic techniques, which we will discuss later in this chapter.

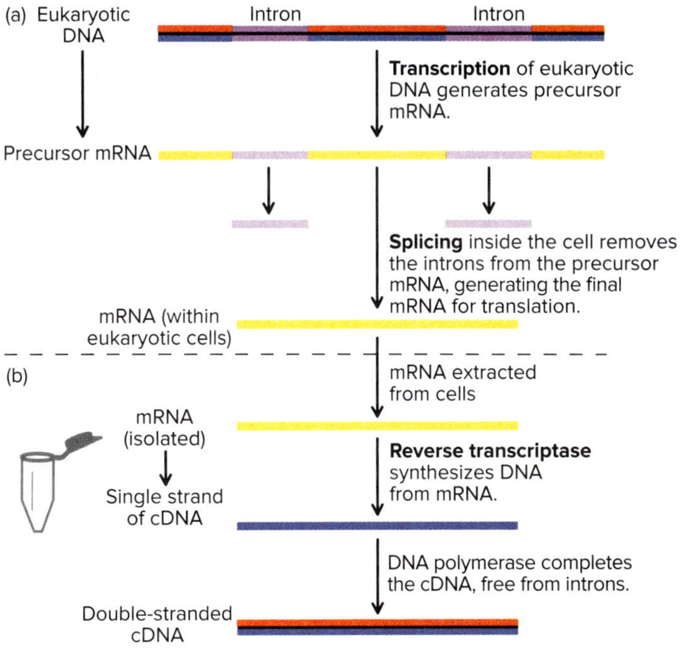
(a) Eukaryotic DNA
Intron
Intron
Transcription of eukaryotic DNA generates precursor mRNA.
Precursor mRNA
Splicing inside the cell removes the introns from the precursor mRNA, generating the final mRNA for translation.
mRNA (within eukaryotic cells)
(b)
mRNA extracted from cells
mRNA (isolated)
Reverse transcriptase synthesizes DNA from mRNA.
Single strand of cDNA
DNA polymerase completes the cDNA, free from introns.
Double-stranded cDNA

Figure 8.2 Making cDNA from eukaryotic mRNA. In order for eukaryotic genes to be expressed by a bacterial cell, a copy of DNA without introns must be cloned. The cDNA encodes the same protein as the original DNA but lacks introns.

Polymerase Chain Reaction: A Molecular Xerox Machine for DNA

Some of the techniques used to analyze DNA and RNA are limited by the small amounts of test nucleic acid available. This problem was largely solved by the invention of a simple, versatile way to amplify DNA called the **polymerase chain reaction (PCR).** This technique rapidly increases the amount of DNA in a sample without the need for growing cultures or carrying out complex purification techniques. It is so sensitive that it can detect cancer from a single cell or diagnose an infection from a single gene copy. It is comparable to being able to pluck a single DNA "needle" out of a "haystack" of other molecules and make unlimited copies of the DNA. The rapid rate of PCR makes it possible to replicate a piece of target DNA from a few copies to billions of copies in a few hours. The development of the PCR technique in 1983 is considered to be one of the great advances in molecular genetics.

To understand the idea behind PCR, it will be helpful to review figure 6.6, which describes synthesis of DNA as it occurs naturally in cells. The PCR method uses essentially the same events, with the opening up of the double strand, use of the exposed strands as templates, the addition of primers, and the action of a DNA polymerase.

Initiating the reaction requires a few specialized ingredients **(figure 8.3).** As we saw earlier, **primers** are synthetic oligonucleotides (short DNA strands) of a known sequence of 15 to 30 bases that serve as landmarks to indicate where DNA amplification will begin. To keep the DNA strands separated, processing must be

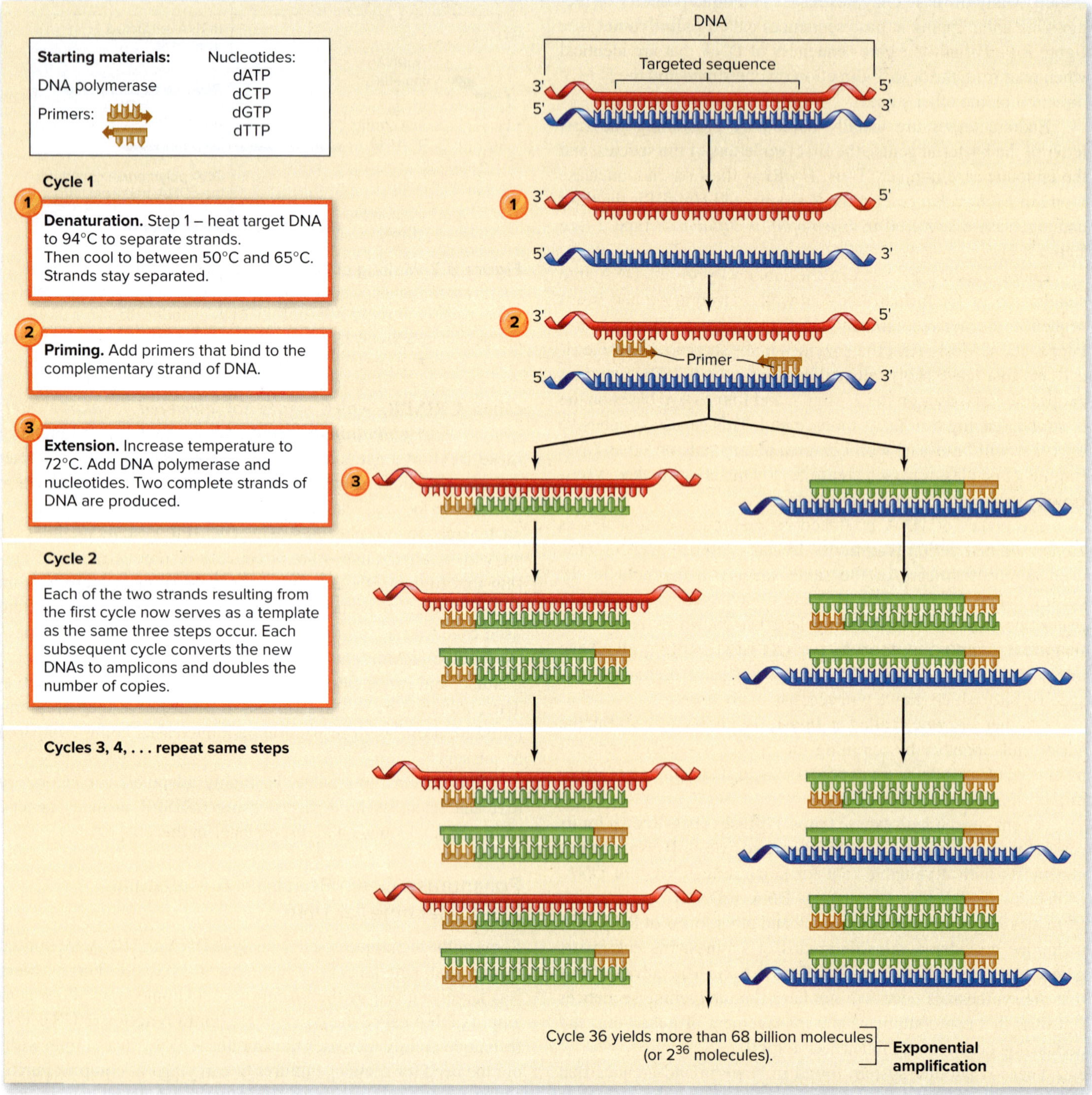
Starting materials:
DNA polymerase
Primers:
Nucleotides:
dATP
dCTP
dGTP
dTTP
DNA
Targeted sequence
3'
5'
5'
3'
Cycle 1
1
Denaturation. Step 1 – heat target DNA to 94°C to separate strands. Then cool to between 50°C and 65°C. Strands stay separated.
2
Priming. Add primers that bind to the complementary strand of DNA.
Primer
3
Extension. Increase temperature to 72°C. Add DNA polymerase and nucleotides. Two complete strands of DNA are produced.
Cycle 2
Each of the two strands resulting from the first cycle now serves as a template as the same three steps occur. Each subsequent cycle converts the new DNAs to amplicons and doubles the number of copies.
Cycles 3, 4, . . . repeat same steps
Cycle 36 yields more than 68 billion molecules (or 2^36 molecules).
Exponential amplification

Figure 8.3 The polymerase chain reaction.

carried out at a relatively high temperature. This requires the use of special **DNA polymerases** isolated from thermophilic bacteria. Examples of these unique enzymes are Taq polymerase obtained from *Thermus aquaticus* and Vent polymerase from *Thermococcus litoralis*. Enzymes isolated from these heat-loving organisms remain active at the elevated temperatures used in PCR. Another useful component of PCR is a machine called a thermal cycler that automatically performs the cyclic temperature changes.

The PCR technique operates by repetitive cycling of three basic steps: denaturation, priming, and extension. The process is fully described in figure 8.3.

It is through cyclic repetition of these steps that DNA becomes amplified. When the DNAs formed in the first cycle are denatured, they become amplicons to be primed and extended in the second cycle. Each subsequent cycle converts the new DNAs to amplicons and doubles the number of copies. The number of cycles required to produce a million molecules is 20, but the process is usually carried out to 30 or 40 cycles. One significant advantage of this technique has been its natural adaptability to automation. A PCR machine can perform 20 cycles in 2 or 3 hours.

Once the PCR is complete, the amplified DNA can be analyzed by any of the techniques discussed in this chapter. In addition a technique called real-time PCR can detect products during the reaction instead of at the end. PCR can also be adapted to analyze RNA by initially converting an RNA sample to DNA with reverse transcriptase. This cDNA can then be amplified by PCR in the usual manner.

Very recently a technique called LAMP (loop-mediated isothermal amplification) has been developed. It uses PCR principles but does not have to go through the temperature variations. It is used in one of the rapid COVID-19 diagnostic tests.

Recombinant DNA Technology and Gene Cloning

The primary intent of **recombinant DNA technology** is to deliberately remove genetic material from one organism and combine it with that of a different organism. An important objective of this technique is to form genetic **clones.** Cloning involves the removal of a selected gene from an animal, a plant, or a microorganism (the genetic donor) followed by propagating it in a different host organism. Cloning requires that the desired donor gene first be selected, cut out by restriction endonucleases, and isolated. The gene is next inserted into a **vector** (usually a plasmid or a virus) that will insert the DNA into a **cloning host.** The cloning host is usually a bacterium or a yeast that can replicate the gene and translate it into the protein product for which it codes. In the next section, we examine the elements of gene isolation, vectors, and cloning hosts and show how they participate in a complete recombinant DNA procedure.

A Note About Clones

Like so many words in biology, the word *clone* has two different, although related, meanings. In this chapter, we will discuss genetic clones created within microorganisms. What we are cloning is *genes.* We use microorganisms to allow us to manipulate and replicate genes outside of the original host of that gene. You are much more likely to be familiar with the other type of cloning—which we will call *whole-organism cloning.* It is also known as *reproductive cloning.* This is the process of creating an identical organism using the DNA from an original. Dolly the sheep was the first cloned whole organism, and many others followed in her wake. Those processes are beyond the scope of this book.

Technical Aspects

The first steps in cloning a target gene are to locate its exact site on the donor and then to isolate it. The most common strategies for doing this are as follows:

1. The DNA is removed from cells and separated into fragments by endonucleases. The correct fragment is then identified through a complicated screening process; or
2. A gene can be synthesized from isolated mRNA transcripts using reverse transcriptase (cDNA); or
3. A gene can be amplified using PCR in many cases.

Although gene cloning and isolation can be very laborious, a fortunate outcome is that, once isolated, genes can be maintained in a cloning host and vector just like a microbial pure culture. **Genomic libraries** are collections of cDNA clones that represent the entire genome of numerous organisms.

Cloning Vectors Isolated genes are not easily manipulated on their own. They are typically spliced into a cloning vector, using restriction enzymes. Plasmids are excellent vectors because they are small, well characterized, and easy to manipulate; and they can be transferred into appropriate host cells through transformation. Bacteriophages are also excellent vectors because they have the natural ability to inject DNA into bacterial hosts through transduction. Today, thousands of unique cloning vectors are available commercially. Although every vector has characteristics that make it ideal for a specific project, all vectors can be thought of as having three important attributes to consider **(figure 8.4):**

1. An origin of replication (ORI) is needed somewhere on the vector so that it will be replicated by the DNA polymerase of the cloning host.

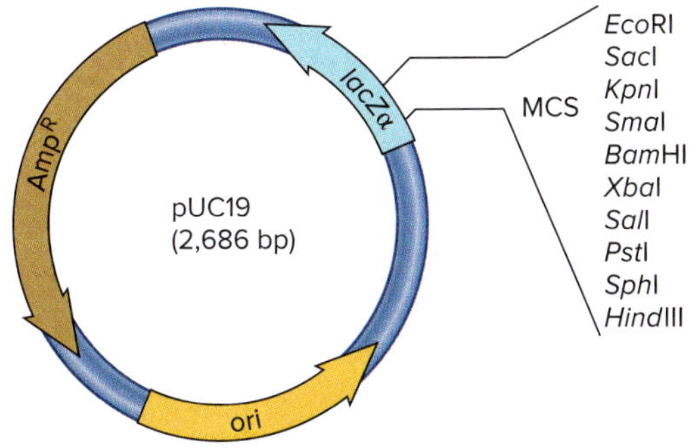
EcoRI
SacI
KpnI
SmaI
BamHI
XbaI
SalI
PstI
SphI
HindIII
MCS
lacZα
AmpR
pUC19
(2,686 bp)
ori

Figure 8.4 The cloning vector pUC19. The origin of replication is in yellow, and the ampicillin-resistance gene is in tan. The sites that can be cleaved by various restriction enzymes are indicated on the right.

2. The vector must accept DNA of the desired size. Early plasmids were limited to an insert size of less than 10 kb of DNA, far too small for most eukaryotic genes with their sizable introns. Vectors called *cosmids* can hold 45 kb, whereas complex bacterial artificial chromosomes (BACs) and yeast artificial chromosomes (YACs) can hold as much as 300 kb and 1,000 kb, respectively.
3. Vectors typically contain a gene that confers drug resistance to their cloning host. In this way, cells can be grown on drug-containing media, and only those cells that harbor a plasmid will be selected for growth.

Many vectors also have a site called a *multicloning site (MCS)*, a region of DNA that can be recognized by a wide variety of restriction enzymes.

Cloning Hosts The best cloning hosts possess several key characteristics **(table 8.1).** The traditional cloning host is *Escherichia coli.* Because this bacterium was the original recombinant host, the protocols using it are well established, relatively easy, and reliable. Hundreds of specialized cloning vectors have been developed for it. The main disadvantage with this species is that the splicing of mRNA and the modification of proteins that would normally occur in the eukaryotic endoplasmic reticulum and Golgi apparatus are unavailable in this bacterial cloning host. One alternative host for certain industrial processes and research is the yeast *Saccharomyces cerevisiae,* which, being eukaryotic, already possesses mechanisms for processing and modifying eukaryotic gene products. Certain techniques may also employ different bacteria, animal cell cultures, and even live animals and plants to serve as cloning hosts. In our coverage, we present the recombinant process as it is performed in bacteria and yeasts.

Construction of a Recombinant, Insertion into a Cloning Host, and Genetic Expression

This section illustrates one example of recombinant DNA technology, in this case, to produce a drug called alpha-2a interferon (Roferon-A). This form of interferon treats cancers such as hairy cell leukemia and Kaposi's sarcoma in AIDS patients. The human alpha interferon gene is a DNA molecule of approximately 500 bp that codes for a polypeptide of 166 amino acids. It was originally isolated and identified from human blood cells and prepared from processed mRNA transcripts that are free of introns. This step is necessary because the bacterial cloning host has none of the machinery needed to excise these nontranslated parts of a gene. The rest of the process is outlined in **figure 8.5.**

The bacterial cells are able to express the eukaryotic gene because the plasmid has been engineered to have the necessary transcription and translation recognition sequences. As the *E. coli* culture grows, it transcribes and translates the interferon gene, synthesizes the peptide, and secretes it into the growth medium. At the end of the process, the cloning cells and other chemical and microbial impurities are removed from the medium. Final processing to excise a terminal amino acid from the peptide yields the interferon product in a relatively pure form. The scale of this procedure can range from test tube size to gigantic industrial vats that can manufacture thousands of gallons of product.

Although the process we have presented here produces interferon, some variation of it can be used to mass produce a variety of hormones, enzymes, and medicinal products.

Table 8.1 Desirable Features in a Microbial Cloning Host

Rapid turnover; fast growth rate
Can be grown in large quantities using ordinary culture methods
Nonpathogenic
Genome that is well mapped
Capable of accepting plasmid or bacteriophage vectors
Maintains foreign genes through multiple generations
Will secrete a high yield of proteins from expressed foreign genes

Synthetic Biology

In recent years, researchers have staked out entirely new territory in genetic manipulation: They are creating new biological molecules and organisms from scratch. This field is called *synthetic biology.* In 2010, a scientist successfully created a self-replicating bacterial cell from four bottles of chemicals: the four nucleotides of DNA. This was a breakthrough of major proportions, as it was the first time a living, replicating cell had been synthesized from chemicals. More recently, researchers have created two new nucleotides, called X and Y. Synthetic biology uses engineering-type methods to assemble molecules and cells. Medical science is poised to be revolutionized when scientists can create precise chemicals to replace those missing in disease, assemble customized immune components, or construct biological molecules that can precisely target cancerous cells or pathogenic microbes. Synthetic biology also holds promise for alternative energy production and for offering new and different manufacturing processes. Of course, the ability of scientists to "create" life, in a sense, makes many people nervous. The scientific community, and those who monitor it, are engaged in intense conversations about the ethics—as well as security issues—of synthetic biology.

8.1 Learning Outcomes—Assess Your Progress

1. Provide examples of practical applications of modern genetic technologies.
2. Explain the role of restriction endonucleases in the process of recombinant DNA technologies.
3. List the steps in the polymerase chain reaction, and discuss one disadvantage to this technique.
4. Describe how recombinant DNA is created, and discuss its role in gene cloning.

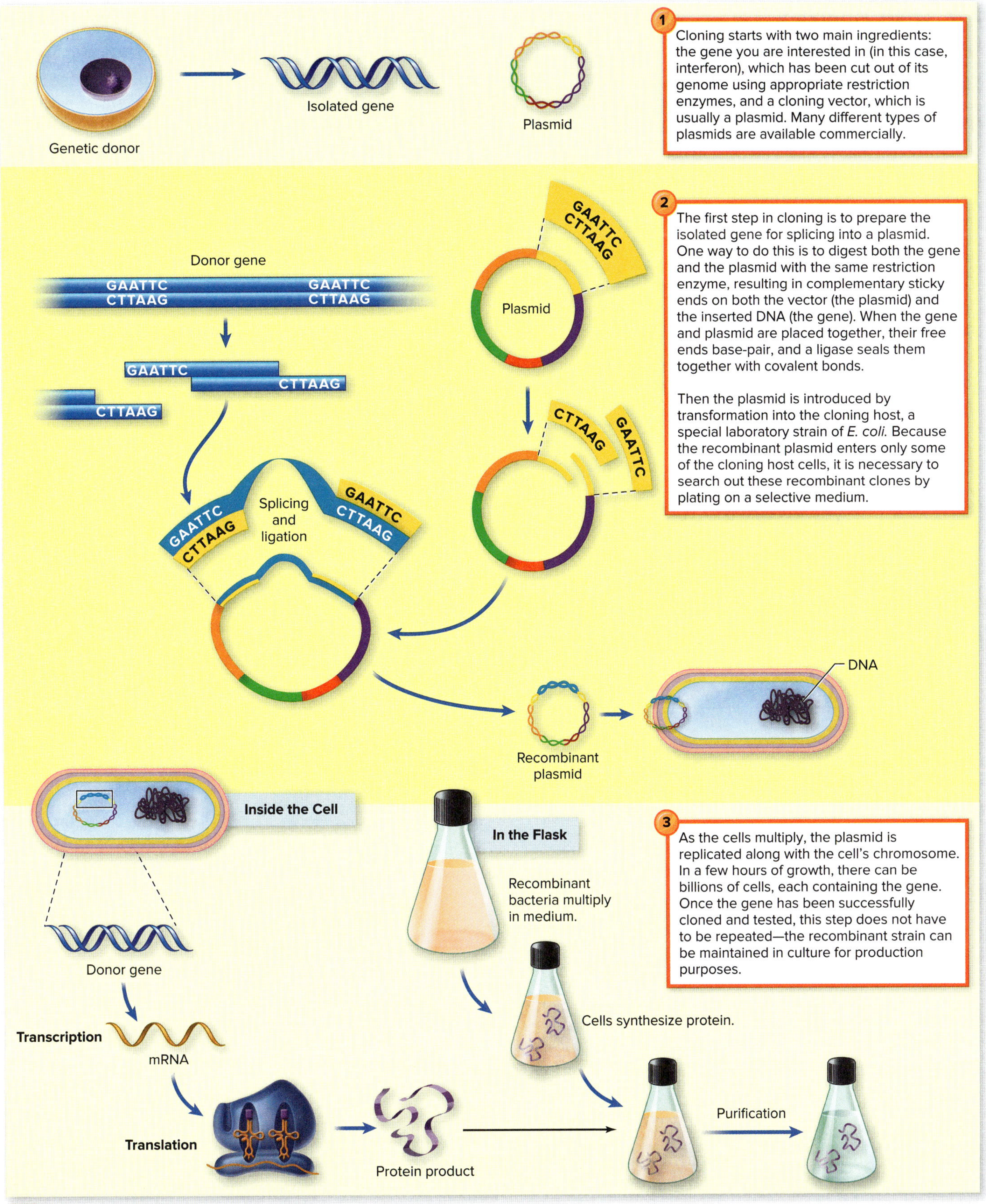
Genetic donor
Isolated gene
Plasmid
1
Cloning starts with two main ingredients: the gene you are interested in (in this case, interferon), which has been cut out of its genome using appropriate restriction enzymes, and a cloning vector, which is usually a plasmid. Many different types of plasmids are available commercially.
Donor gene
GAATTC
CTTAAG
GAATTC
CTTAAG
GAATTC
CTTAAG
CTTAAG
Plasmid
GAATTC
CTTAAG
CTTAAG
GAATTC
GAATTC
CTTAAG
Splicing
and
ligation
GAATTC
CTTAAG
2
The first step in cloning is to prepare the isolated gene for splicing into a plasmid. One way to do this is to digest both the gene and the plasmid with the same restriction enzyme, resulting in complementary sticky ends on both the vector (the plasmid) and the inserted DNA (the gene). When the gene and plasmid are placed together, their free ends base-pair, and a ligase seals them together with covalent bonds.
Then the plasmid is introduced by transformation into the cloning host, a special laboratory strain of *E. coli*. Because the recombinant plasmid enters only some of the cloning host cells, it is necessary to search out these recombinant clones by plating on a selective medium.
DNA
Recombinant
plasmid
Inside the Cell
In the Flask
Recombinant
bacteria multiply
in medium.
3
As the cells multiply, the plasmid is replicated along with the cell's chromosome. In a few hours of growth, there can be billions of cells, each containing the gene. Once the gene has been successfully cloned and tested, this step does not have to be repeated—the recombinant strain can be maintained in culture for production purposes.
Donor gene
Transcription
mRNA
Cells synthesize protein.
Translation
Protein product
Purification

Figure 8.5 Using recombinant DNA for gene cloning.

8.2 DNA Analysis

Visualizing DNA

There are many reasons scientists would want to know the characteristics of an organism's DNA. Knowing the entire genome sequence of an organism (such as a human) can reveal genetic abnormalities, ancestry, and so on. Sometimes the point is to determine if one sample of DNA is the same as another sample. If you know the DNA patterns of a pathogen causing an outbreak, for example, you can look at the DNA patterns of a microbe isolated from a patient to see if they were made ill by that same pathogen. Also, you can look at the pattern of DNA left at a crime scene and match it against a suspect's DNA.

Gel Electrophoresis

One way to produce a readable pattern of DNA fragments is through **gel electrophoresis.** In this technique, samples are placed in compartments (wells) in a soft agar gel and subjected to an electrical current **(figure 8.6*a*).** The phosphate groups in DNA give the entire molecule an overall negative charge, which causes the DNA to move toward the positive pole in the gel. The rate of movement is based primarily on the size of the fragments. The longer fragments move more slowly and remain nearer the top of the gel, whereas the shorter fragments migrate faster and are positioned farther from the wells. The DNA fragments become visible using stains, which are either incorporated into the gel material or applied when electrophoresis is finished **(figure 8.6*b*).** Electrophoresis patterns can be quite distinctive and are very useful in characterizing DNA fragments and comparing the degree of genetic similarities among samples as in a genetic fingerprint (discussed later).

Nucleic Acid Hybridization and Probes

Two different nucleic acids can **hybridize** by uniting at their complementary regions. All different combinations are possible: Single-stranded DNA can unite with other single-stranded DNA

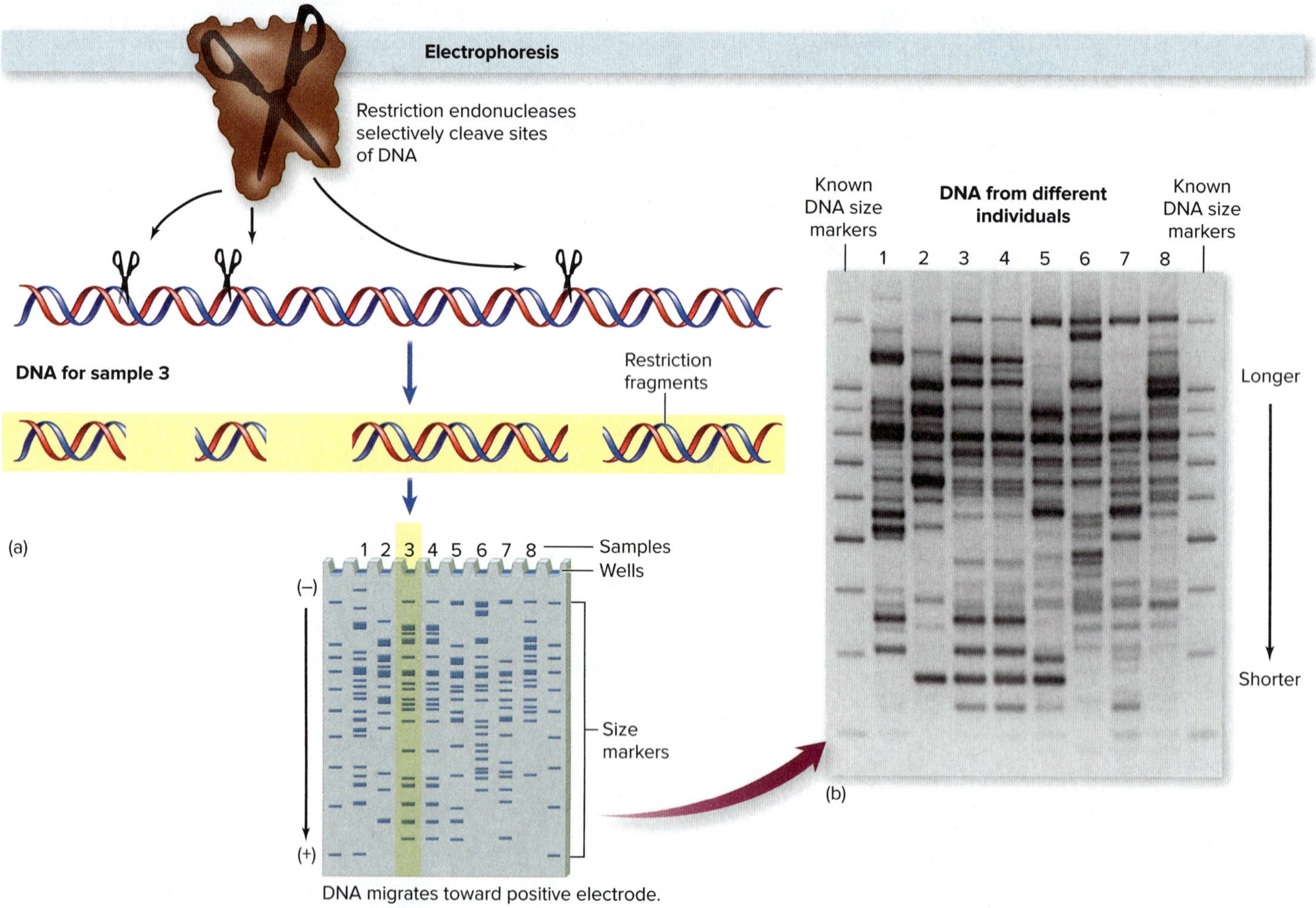
Electrophoresis
Restriction endonucleases selectively cleave sites of DNA
DNA for sample 3
Restriction fragments
(a)
1 2 3 4 5 6 7 8
Samples
Wells
(–)
(+)
Size markers
DNA migrates toward positive electrode.
Known DNA size markers
DNA from different individuals
Known DNA size markers
1 2 3 4 5 6 7 8
Longer
Shorter
(b)

Figure 8.6 Revealing the patterns of DNA with electrophoresis. **(a)** After cleavage into fragments, DNA is loaded into wells on one end of an agarose gel. When an electrical current is passed through the gel (from the negative pole to the positive pole), the DNA, being negatively charged, migrates toward the positive pole. The larger (longer) fragments, measured in numbers of base pairs, migrate more slowly and remain nearer the wells than the smaller (shorter) fragments. **(b)** An actual stained gel reveals a separation pattern of the fragments of DNA. The size of a given DNA band can be determined by comparing the distance it traveled to the distance traveled by a set of DNA fragments of known size (which are in both outer lanes).

(b) Zmeel Photography/E+/Getty Images

Disease Connection

One very useful application of gel electrophoresis is in the investigation of food-borne disease outbreaks. The CDC and public health departments use a variation of electrophoresis called pulse-field gel electrophoresis to determine if microbes isolated from food samples or ill patients are exactly the same strain and, thus, part of a common outbreak.

or RNA, and RNA can hybridize with other RNA. This property has allowed for the development of specially formulated tracers called **gene probes.** These probes consist of a short stretch of DNA of a known sequence that will base-pair with a stretch of DNA with a complementary sequence, if one exists in the test sample. So that areas of hybridization can be visualized, the probes can carry reporter molecules such as fluorescent dyes, which are visible under UV light, or luminescent labels, which give off visible light. Some probes become visible because they have an enzyme bound to them. These enzymes act on some type of nonpigmented substrate in the reaction, releasing a product that has color.

Probes are commonly used for diagnosing the cause of an infection from a patient's specimen and for identifying a culture of an unknown bacterium or virus. A simple and rapid method called a *hybridization test* does not require electrophoresis. DNA from a test sample is isolated, denatured, and placed on an absorbent filter, and then a solution containing a microbe-specific probe is added **(figure 8.7).** The blot is then developed and observed for areas of hybridization. Commercially available diagnostic kits are used to identify intestinal pathogens such as *Salmonella, Campylobacter, Shigella, Clostridioides difficile,* rotaviruses, and adenoviruses. DNA probes have also been developed for human genetic markers and some types of cancer.

With another method, called **fluorescent *in situ* hybridization (FISH),** probes are applied to intact cells and observed microscopically for the presence and location of specific genetic marker sequences on genes. *In situ* is a phrase that means "on site"—and in biology it means that organisms are examined without being removed from their habitat, or at least, as in this case, while they are still intact. These techniques can also be used to identify unknown bacteria living in natural habitats without having to culture them, and they can be used to detect RNA in cells and tissues.

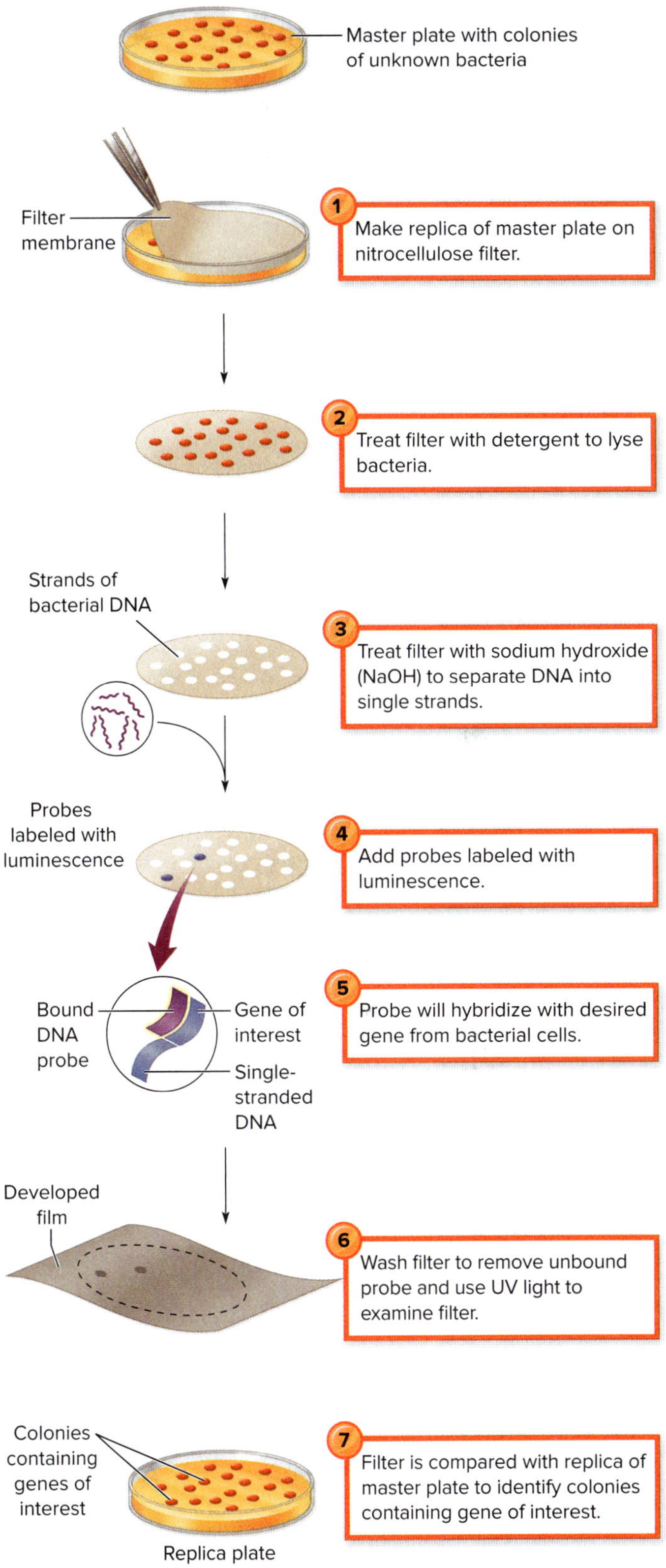

Master plate with colonies of unknown bacteria
Filter membrane
1 Make replica of master plate on nitrocellulose filter.
2 Treat filter with detergent to lyse bacteria.
Strands of bacterial DNA
3 Treat filter with sodium hydroxide (NaOH) to separate DNA into single strands.
Probes labeled with luminescence
4 Add probes labeled with luminescence.
Bound DNA probe
Gene of interest
Single-stranded DNA
5 Probe will hybridize with desired gene from bacterial cells.
Developed film
6 Wash filter to remove unbound probe and use UV light to examine filter.
Colonies containing genes of interest
Replica plate
7 Filter is compared with replica of master plate to identify colonies containing gene of interest.

Figure 8.7 **A hybridization test relies on the action of microbe-specific probes to identify an unknown bacterium or virus.**

The Size of DNA

The relative sizes of nucleic acids are usually named by the number of base pairs (bp) or nucleotides they contain. For example, the palindromic sequences recognized by endonucleases are usually 4 to 10 bp in length; an average gene in *E. coli* is approximately 1,300 bp, or 1.3 kilobases (kb); and its entire genome is approximately 4,700,000 bp, 4,700 kb, or 4.7 megabases (Mb). The DNA of the human mitochondrion contains 16 kb, and the Epstein–Barr virus (a cause of infectious mononucleosis) has 172 kb. Humans have approximately 6 billion base pairs arrayed along 46 chromosomes.

INSIGHT 8.1 MICROBIOME: Host Genetics and the Microbiome

The composition of the human microbiome shows a lot of variability from person to person. Of course, we know that humans themselves show a lot of variation, which comes from their different genetic makeup. This led scientists to wonder whether the composition of the microbiome is influenced by the host's genetics.

One good way to test this is to look at two different types of pairs of people: monozygotic (identical) twins and dizygotic (fraternal) twins. Fraternal twins have fewer genes in common than identical twins. If the microbiomes of identical twins were significantly more similar than the microbiomes of fraternal twins, it would suggest that the human genome influences what microbiome the person acquires.

To ask this question the way scientists do, you would construct a hypothesis: The degree of difference between the microbiome of fraternal twins will be no greater than the degree of difference between the microbiome of identical twins. (This is written as a null hypothesis, meaning it is a statement that there will be *no difference* between two groups.) Then you would set up an experiment to test the hypothesis, using a large number of both types of twins. This is what happened in this study. Four-hundred sixteen pairs of twins were examined.

In this study, the identical twins turned out to have more similar microbiomes than the fraternal twins. They had what they called "a hub of heritable taxa," chief among them a newly discovered bacterial group named *Christensenellaceae.*

So the hypothesis was disproven; there **was** a significant difference between the two groups. The paper's authors suggest that a person's microbiome is heritable, like having blue eyes. Only here it is a bit more indirect—a person's genotype is heritable, which determines his or her phenotype, which may determine his or her microbiome.

Halfbottle/Shutterstock

There is a saying in science, "Chance favors the prepared mind." In the case of this study, the scientists found something they were not counting on: The presence of *Christensenellaceae* was associated with low body mass index (BMI). Because this was just an association and the study could not prove causation, they did another experiment in which they deliberately exposed mice to *Christensenellaceae.* Those mice had reduced weight gain compared to mice not fed *Christensenellaceae.* So the studies continue. This is what many scientists love about their jobs: discovering surprises and finding answers to questions that practically ask themselves!

DNA Profiling

You've no doubt seen images like the one in **figure 8.8,** with its ladder-like images, on TV crime shows. These are examples of **DNA profiles** (also called fingerprints). They rely on the fact that the DNA in different individuals—even of the same species—contains small differences. This is true for humans, and it is true for bacteria of the same species. When the individual's DNA is digested with a set of restriction enzymes, the small sequence differences will cause the DNA to be cut up at different places, resulting in fragments of different lengths in the digested samples of different individuals. To actually see the array of different fragments and to compare the fragments of different individuals, gel electrophoresis is used.

One type of analysis depends on the ability of a restriction enzyme to cut DNA at a specific recognition site. If a given strand of DNA possesses the recognition site for a particular restriction enzyme, the DNA strand is cut, resulting in two smaller pieces of DNA. If the same strand from a different person does not contain the recognition site (perhaps due to a mutation many generations ago), it is not cut by the restriction enzyme and remains as a single large piece of DNA. When each of these DNA samples is digested with restriction enzymes and separated on an electrophoresis gel, the first displays two small bands, while the second displays one larger band. This is an example of a restriction fragment length polymorphism (RFLP), which was discussed earlier. All methods of DNA fingerprinting depend on some variation of this strategy to ferret out differences in DNA sequence at the same location in the genome.

A Note About the "-omics"

The ability to obtain the entire sequences of organisms has spawned new vocabulary that refers to the "total picture" of some aspect of a cell or organism.

genomics The systematic study of an organism's genes and their functions.

proteomics The study of an organism's complement of proteins (its "proteome") and functions mediated by the proteins.

metagenomics (also called "community genomics") The study of all the genomes in a particular ecological niche, as opposed to individual genomes from single species.

metabolomics The study of the complete complement of small chemicals present in a cell at any given time. Provides a snapshot of the physiological state of the cell and the end products of its metabolism.

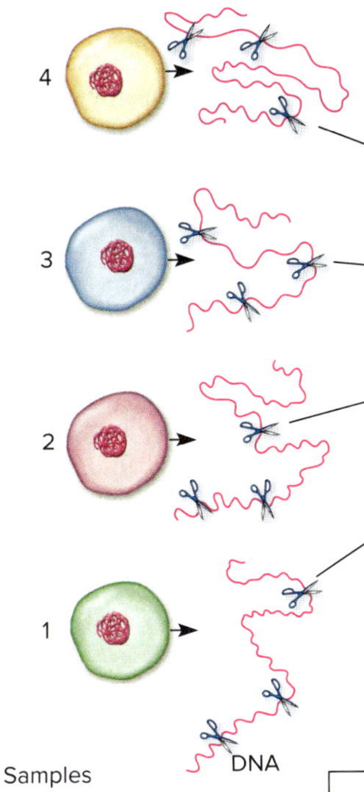
4
3
2
1
Samples
DNA

(a) Cells from different samples are processed to isolate their DNA. The DNA samples are exposed to endonucleases, which snip them at specific sites into a series of different fragments.

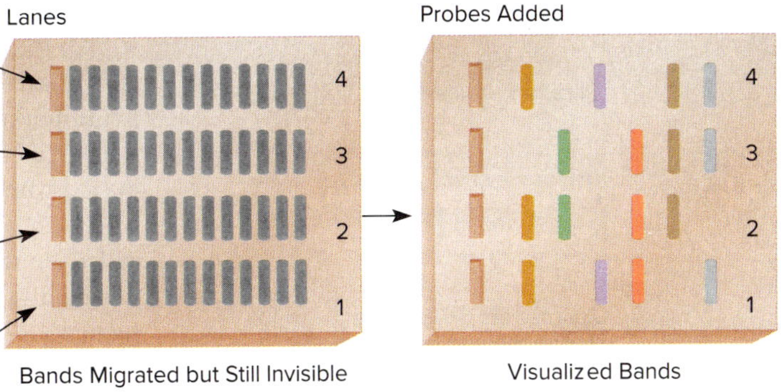
Lanes
Probes Added
4
3
2
1
Bands Migrated but Still Invisible
Visualized Bands

(b) Electrophoresis of DNA fragments separates them by size, with longer fragments nearer the wells. Although invisible to the unaided eye at this point, each well contains thousands of individual bands. After transferring the DNA fragments to a nylon membrane, specific DNA sequences are identified by hybridization of a fluorescently labeled nucleic acid probe.

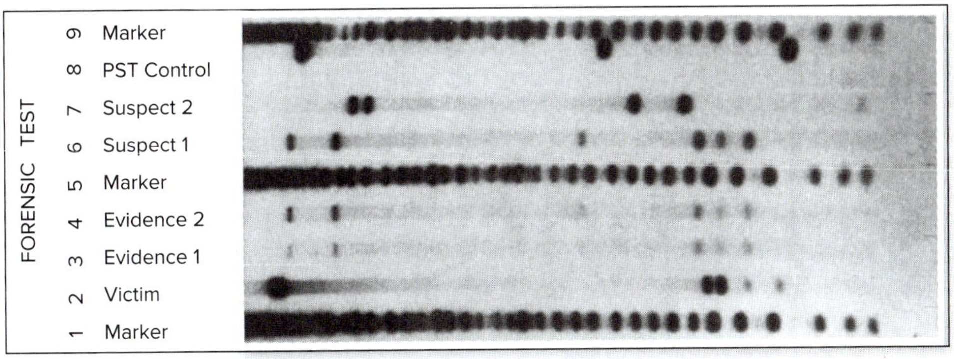
FORENSIC TEST
9 Marker
8 PST Control
7 Suspect 2
6 Suspect 1
5 Marker
4 Evidence 2
3 Evidence 1
2 Victim
1 Marker

(c) An actual human DNA profile used in a rape trial. Control lanes with known markers are in lanes 1, 5, 8, and 9. The second lane contains a sample of DNA from the victim's blood. Evidence samples 1 and 2 (lanes 3 and 4) contain semen samples taken from the victim. Suspects 1 and 2 (lanes 6 and 7) were tested. Can you tell by comparing evidence and suspect lanes which individual committed the rape?

Figure 8.8 **DNA profiles: the bar codes of life.**
Dr. Michael Baird

DNA Sequencing

By far the most detailed maps of a genome are **sequence maps,** which give an exact order of bases in a plasmid, a chromosome, or an entire genome. Genome sequencing projects have been highly successful. Genomes of thousands of organisms have been sequenced, including viruses, bacteria, and eukaryotic organisms (including humans).

Decoding the Organism's Genetic Code

One of the remarkable discoveries in this huge enterprise has been how similar the genomes of relatively unrelated organisms are. Humans share approximately 80% of their DNA codes with mice, about 60% with rice, and even 30% with the worm *C. elegans.*

The other side of the coin is that scientists have discovered that individual bacteria of the same species can contain as many as 30 additional genes not found in the originally sequenced bacterium. This has led to the understanding that we should sequence multiple representatives of the same species, to create a **pangenome**—a composite genome representing the diversity within a species.

So how is this sequencing performed? We can illustrate the process by describing an older method, called shotgun sequencing **(figure 8.9).** It is rarely used anymore, as the process has been massively automated, but it helps with understanding what happens. Shotgun sequencing can be broken down into seven steps:

1. First, the whole genome of an organism is broken down into smaller, manageable fragments.
2. The fragments are separated through gel electrophoresis.
3. Each fragment is inserted into a plasmid and is cloned into an *E. coli* cell. This produces a complete **library** of fragments. The library exists in the bacterial cultures, which can be preserved indefinitely and sampled repeatedly.

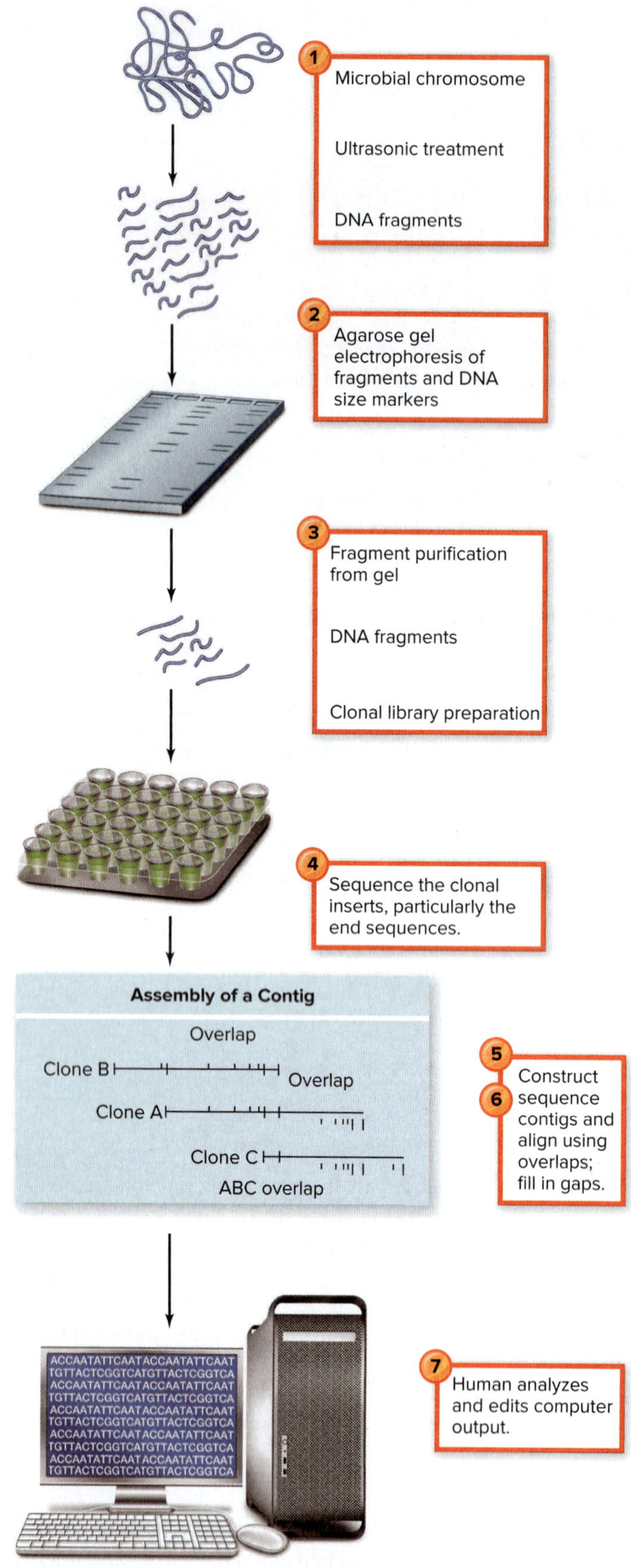
1
Microbial chromosome
Ultrasonic treatment
DNA fragments
2
Agarose gel electrophoresis of fragments and DNA size markers
3
Fragment purification from gel
DNA fragments
Clonal library preparation
4
Sequence the clonal inserts, particularly the end sequences.
Assembly of a Contig
Overlap
Clone B
Overlap
Clone A
Clone C
ABC overlap
5
6
Construct sequence contigs and align using overlaps; fill in gaps.
7
Human analyzes and edits computer output.

Figure 8.9 Whole-genome shotgun sequencing.

4. The plasmids are purified and the DNA fragments are sequenced by automated sequencers, machines that add labeled primers to each fragment. The primers usually recognize the plasmid sequences that flank the genome insert so that the machine can tell where the fragment begins and ends. Each section of the genome ends up being sequenced multiple times in an overlapping fashion.
5. A computer program takes all the sequence data and is able to find where the sequence overlaps. This automated process results in a larger, contiguous set of nucleotide sequences called **contigs.**
6. The contigs are put in the proper order to determine the entire sequence. This step is tricky. There are often gaps between the contigs, but there are a variety of methods to resolve this issue.
7. An important last step is editing. A human examines the sequence, looking for irregularities, frameshifts, and ambiguities.

In modern sequencing, similar principles are used, but the whole process is scaled up in what is called "high-throughput" genome sequencing. High-throughput sequencing (also called "deep sequencing" or "next generation sequencing") requires four steps **(figure 8.10):** (1) A library of DNA is prepared by fragmenting the DNA and fitting each fragment with some common adaptors, or short DNA sequences that are designed to match the probes in the next step. (2) The collection of sequences is placed in a flow cell, whose surface is coated with the probes that will bind with adaptor molecules on the DNA. Inside the flow cell, the conditions and reagents are provided to use one or another form of PCR to amplify each fragment many times. (3) Then all of the amplified fragments are sequenced so that each nucleotide is "read" thousands of times, each time it is present on a sequence of any length. The details of how this happens differ for each of the automated systems, but all rely on techniques that originated in the Sanger method. (4) Finally, the millions of sequence lengths—also called "reads"—are aligned using bioinformatics software. Putting all the reads together, the entire sequence can be deduced.

This whole process is mostly automated, using machines developed over the course of the last decade. Initially, sequencing the human genome took 13 years and cost about $3 billion. Now, a human genome can be sequenced in an afternoon, at a cost of not much more than $1,000 **(figure 8.11).** In January 2020, when the SARS-CoV-2 virus was first identified in China, its entire sequence was published within days.

Identifying the sequence does not necessarily tell you anything about what it does. As a result, two disciplines have grown up around managing these data: **genomics** (see **"A Note About the New -omics"**) and **bioinformatics.** The job of genomics and bioinformatics is to analyze and classify genes, determine protein sequences, and ultimately determine the function of the genes. Determining this functional information is often called **annotating** the genome. In time, well-annotated genomes will provide a complete understanding of such phenomena as normal cell function, disease, development, aging, and many other issues. In addition, they will allow us to characterize the exact genetic mechanisms behind pathogens and allow new treatments to be developed against them.

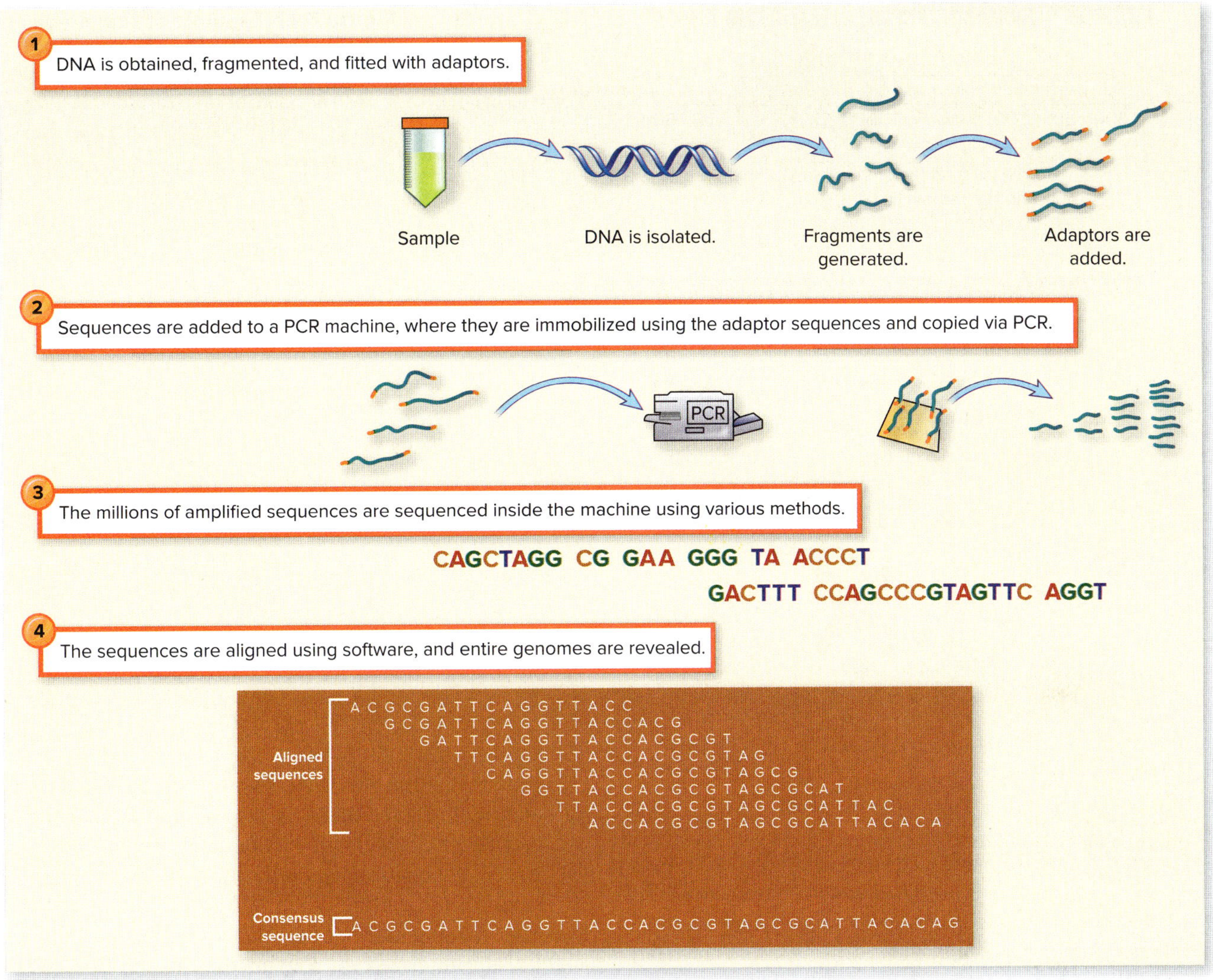
1
DNA is obtained, fragmented, and fitted with adaptors.
Sample
DNA is isolated.
Fragments are generated.
Adaptors are added.
2
Sequences are added to a PCR machine, where they are immobilized using the adaptor sequences and copied via PCR.
PCR
3
The millions of amplified sequences are sequenced inside the machine using various methods.
CAGCTAGG CG GAA GGG TA ACCCT
GACTTT CCAGCCCGTAGTTC AGGT
4
The sequences are aligned using software, and entire genomes are revealed.
Aligned sequences
A C G C G A T T C A G G T T A C C
G C G A T T C A G G T T A C C A C G
G A T T C A G G T T A C C A C G C G T
T T C A G G T T A C C A C G C G T A G
C A G G T T A C C A C G C G T A G C G
G G T T A C C A C G C G T A G C G C A T
T T A C C A C G C G T A G C G C A T T A C
A C C A C G C G T A G C G C A T T A C A C A
Consensus sequence
A C G C G A T T C A G G T T A C C A C G C G T A G C G C A T T A C A C A G

Figure 8.10 High-throughput sequencing.

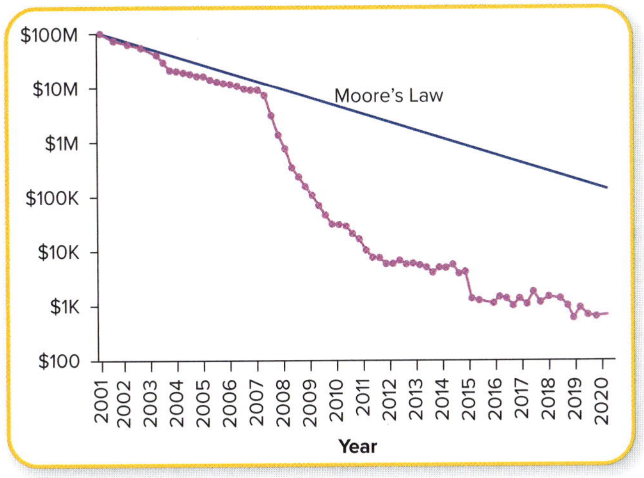
$100M
$10M
$1M
$100K
$10K
$1K
$100
Moore's Law
2001
2002
2003
2004
2005
2006
2007
2008
2009
2010
2011
2012
2013
2014
2015
2016
2017
2018
2019
2020
Year

Figure 8.11 **Cost to sequence a human genome 2001–2020.** Moore's law is a term borrowed from the computer hardware industry, predicting that the computational power of equipment doubles every year, while the costs are halved. Technologies that keep up with Moore's law are thought to be highly successful. Note how the cost of sequencing has far outperformed Moore's law.

Single Nucleotide Polymorphisms (SNPs)

One type of polymorphism recently found to be important in individual traits is called **single nucleotide polymorphism (SNP)** because only a single nucleotide is altered. This is a result of a point mutation at some point in the organism's ancestry, and it is passed on genetically. Tens of thousands of these differences at a single locus (when two different individuals are compared) are known to exist throughout the genome. The human genome contains 10 million SNPs. These variations are currently a hot area of research and commerce.

As we will see later in this chapter, the ability to identify SNPs has proven critical to the new field of personalized medicine, which is customized to a person's genetic makeup. One example is when patients' genomes are examined for SNPs that have been found to be associated with a particular disease, to determine their risk. For example, in a condition called thrombophilia (a blood-clotting disorder), a point mutation in the gene for a clotting factor (factor V) causes an arginine to become a glutamine **(figure 8.12).** This leads to increased clotting in the patient.

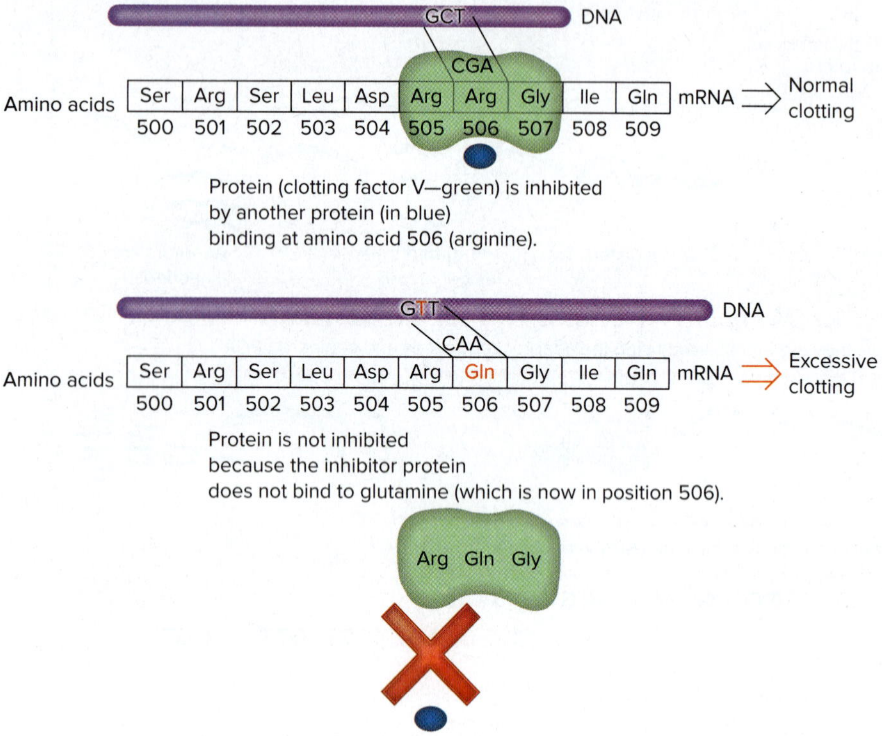
GCT
DNA
CGA
Amino acids
Ser Arg Ser Leu Asp Arg Arg Gly Ile Gln
mRNA
Normal clotting
500 501 502 503 504 505 506 507 508 509
Protein (clotting factor V—green) is inhibited
by another protein (in blue)
binding at amino acid 506 (arginine).
GTT
DNA
CAA
Amino acids
Ser Arg Ser Leu Asp Arg Gln Gly Ile Gln
mRNA
Excessive clotting
500 501 502 503 504 505 506 507 508 509
Protein is not inhibited
because the inhibitor protein
does not bind to glutamine (which is now in position 506).
Arg Gln Gly

Figure 8.12 Example of single-nucleotide polymorphism (SNP) associated with disease. When the arginine at position 506 is instead a glutamine, inhibition of the clotting factor does not occur, leading to excessive clotting.

Disease Connection

The human genome was sequenced for the first time in 2001. Even though not every gene was identified at that time, several companies sprung up to analyze consumer genomes and determine family heritage as well as carrier status with respect to hereditary diseases. Ancestry.com analyzes SNPs to determine family trees, and whole-genome sequencing to detect genetic carrier status. In 2021, the missing parts of the genome were identified and the genome is considered complete now.

Measuring Gene Expression: Microarrays

Twin advances in biology and electronics have allowed biologists to view the expression of genes in any given cell using a technique called DNA **microarray** analysis. Microarrays are able to track the expression of thousands of genes at once and are able to do so in a single efficient experiment. Microarrays consist of a "chip" made of glass, silicon, or nylon, onto which have been bound sequences from tens of thousands of different genes. A solution containing fluorescently labeled cDNA, representing all of the mRNA molecules in a cell at a given time, is added to the chip. The labeled cDNA is allowed to bind to any complementary DNA bound to the chip. Bound cDNA is then detected by exciting the fluorescent tag on the cDNA with a laser and recording the fluorescence with a detector linked to a computer. The computer can then interpret these data to determine what mRNAs are present in the cell under a variety of conditions **(figure 8.13).** In the example in this figure, you see green, red, and yellow reactions. The green fluorescent tag was added to a control population of cells, let's say a bacterium growing in fluid culture. The red fluorescent tag was added to bacteria growing in a biofilm. Genes expressed only in fluid culture appear green on the microarray; those expressed in biofilm growth are colored red; and the yellow reactions were dual-labeled, meaning those genes are expressed under both conditions. Microarray technology has also been used for the assessment of cell proteins.

Possible uses of microarrays include developing extraordinarily sensitive diagnostic tests that search for a specific pattern of gene or protein expression. As an example, being able to identify a patient's cancer as one of many subtypes (rather than just, for instance, breast cancer) allows pharmacologists and doctors to treat each cancer with the drug that will be most effective. Again, we see that genetic technology can be a very effective way to reach long-held goals.

8.2 Learning Outcomes—Assess Your Progress

5. Describe how gel electrophoresis is used to analyze DNA.
6. Outline the general steps in DNA profiling.
7. Outline in general terms the process of DNA sequencing.
8. Discuss the significance of single nucleotide polymorphisms (SNPs) in DNA analysis.
9. Describe the uses of microarray technology.

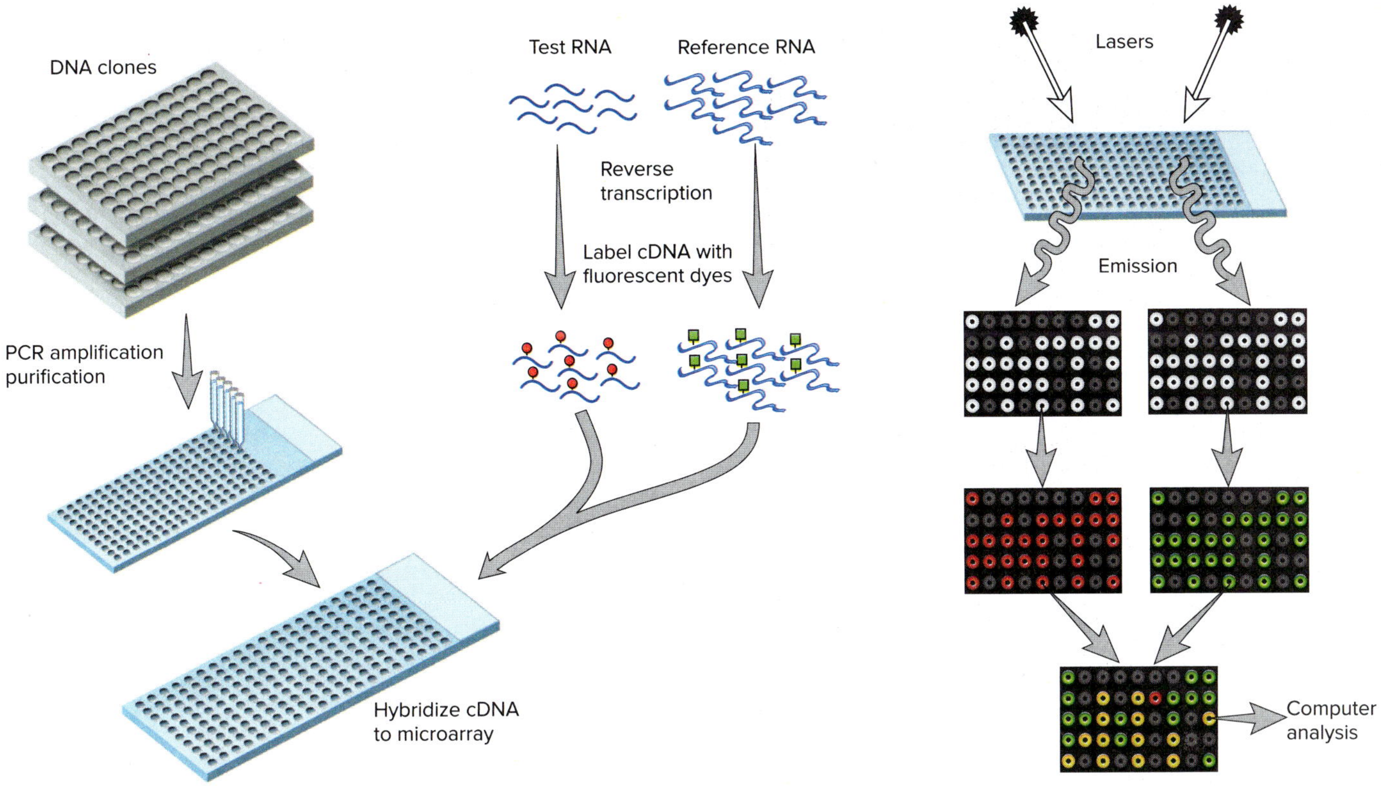

DNA clones
PCR amplification
purification
Test RNA
Reference RNA
Reverse
transcription
Label cDNA with
fluorescent dyes
Hybridize cDNA
to microarray
Lasers
Emission
Computer
analysis

Figure 8.13 Gene expression analysis using microarrays. Cloned genes from an organism are amplified by PCR; after purification, samples are applied to a chip to generate a spotted microarray. mRNA from test and reference cultures is converted to cDNA by reverse transcription and labeled with two different fluorescent dyes. The labeled mixture is hybridized to the microarray and scanned. Gray reactions represent no hybridization having occurred. Green and red reactions indicate that expression was significantly higher in either the test or reference cultures. Yellow reactions represent genes expressed equally in both reference and test RNA samples.

8.3 Genetic Approaches to Healing Disease

Using the Genome Sequence to Tailor Treatments

Because of the ready availability of whole genome sequencing of a person's DNA, scientists have started looking at how our sequence, our SNPs, and other aspects of our genome can affect the way we react to prescribed treatments. The small changes we talked about (SNPs) can make you impervious to a common drug prescribed for your condition (bad outcome), or they can make you much more affected by them (possibly good, possibly bad outcome). Now that we have the ability to map these SNPs, doctors can tailor treatments accordingly. As you can imagine, at the present time, this technology is only available to patients of financial means who can afford it. The hope with this, as with all medical advances, is that we will find a way to make it available to everyone.

Products of Recombinant DNA Technology

Recombinant DNA technology can be used to produce recombinant organisms, but it can also be used to create abundant sources of protein products or nucleotide sequences. Recombinant technology is used by pharmaceutical companies to manufacture medications that cannot be manufactured by any other means.

Recombinant DNA technology changed the outcome of diseases such as diabetes, dwarfism, and many other conditions by enabling large-scale manufacture of life-saving hormones and enzymes of human origin. Recombinant human insulin has replaced pig-derived insulin for patients with diabetes, and recombinant HGH can be administered to children with dwarfism and other conditions. HGH is also used to prevent the wasting syndrome that occurs in patients who have AIDS or cancer. In all of these applications, recombinant DNA technology has led to both a safer product and one that can be manufactured in quantities previously unimaginable. Other hormones, enzymes, and vaccines produced through recombinant DNA technology are summarized in **table 8.2.**

Genetically Modified Organisms

Recombinant organisms produced through the introduction of foreign genes are called **transgenic** or genetically modified organisms (GMOs). Foreign genes have been inserted into a variety of microbes, plants, and animals through recombinant DNA

Table 8.2 Examples of Current Medicines from Recombinant DNA Technology

Immune Treatments
Interferons—peptides used to treat some types of cancer, multiple sclerosis, and viral infections such as hepatitis and genital warts
Interleukins—types of cytokines that regulate the immune function of white blood cells; used in cancer treatment
Tumor necrosis factor (TNF)—used to treat cancer
Remicade® and Humira®—"biologics" used to treat autoimmune disorders such as rheumatoid arthritis and Crohn's disease
Hormones
Erythropoietin (EPO)—a peptide that stimulates bone marrow; used to treat some forms of anemia
Human growth hormone (HGH)—stimulates growth in children with dwarfism; prevents wasting syndrome
Enzymes
rhDNase (Pulmozyme)—a treatment that can break down the thick lung secretions of cystic fibrosis
Tissue plasminogen activating factor (tPA)—can dissolve potentially dangerous blood clots
PEG-SOD—a form of superoxide dismutase that protects against damage to brain and other tissues after surgery or severe trauma
Vaccines
Vaccines for hepatitis B, human papillomavirus, and *Haemophilus influenzae* type b meningitis
Miscellaneous
Factor VIII—needed as replacement blood-clotting factor in type A hemophilia

techniques developed especially for them. Transgenic organisms have enabled cleanup of toxic pollution in the environment, new treatments for human conditions, and improved nutrition for at-risk populations.

Recombinant Microbes: Modified Bacteria and Viruses

One of the first practical applications of recombinant DNA in agriculture was to create a genetically altered strain of the bacterium *Pseudomonas syringae.* The wild (normal) strain contains a gene that promotes ice or frost formation on moist plant surfaces. Genetic alteration of the frost gene using recombinant plasmids created a different strain that could prevent ice crystals from forming. In another case, a strain of *Pseudomonas fluorescens* has been engineered with the gene from a bacterium (*Bacillus thuringiensis*) that codes for an insecticide. These recombinant bacteria are released to colonize plant roots and help destroy invading insects. This can minimize the use of pesticide sprays and protect farm workers and people downwind of the spraying. This bacterial gene has also been added to transgenic corn and potato crops to help make them more resistant to insect pests. All releases of recombinant microbes must be approved by the Environmental Protection Agency (EPA) and are closely monitored.

The movement toward release of engineered plants into the environment has led to some controversy. Many plant geneticists and ecologists are seriously concerned that transgenic plants will share their genes for herbicide, pesticide, and virus resistance with natural plants, leading to "superweeds" that could flourish and become indestructible. Many scientists counter these fears by pointing out that non–genetically engineered plants and organisms swap genes all the time, in entirely unpredictable ways. The U.S. Department of Agriculture is carefully regulating all releases of transgenic plants. Many consumers try to avoid any foods labeled "GMO." The vast majority of scientists knowledgeable about recombinant DNA techniques consider this unnecessary. Plants and animals have been exchanging DNA from the time they appeared on the planet, and humans have been purposely cross-breeding both for centuries, selecting for new traits brought by outside genes.

Gene Therapy

We have known for decades that for certain diseases, the disease phenotype is due to the lack of a single specific protein. For example, type I diabetes is caused by a lack of insulin, leaving those with the disease unable to properly regulate their blood sugar. The initial treatment for this disease was simple: Provide diabetics with insulin isolated from a different source, in most cases the pancreas of pigs or cows. Whereas this treatment was adequate for most diabetics, it presented problems, including the large number of animals required to meet demand and the occurrence of sensitivities or allergies to animal proteins. We have already discussed the first way in which DNA technology has been used in the treatment of disease, namely, producing recombinant proteins in bacteria or yeast rather than isolating the protein from animals or humans. In fact, recombinant human insulin was the first genetically engineered drug to be approved for use in humans. The next logical step was to see if we could correct or repair a faulty gene in humans suffering from a fatal or debilitating disease. The first technology developed for this was **gene therapy.**

The inherent benefit of this therapy is to permanently cure the physiological dysfunction by repairing the genetic defect. There are various strategies for this therapy. In general, the normal gene is cloned in vectors such as retroviruses (mouse leukemia virus) or adenoviruses that are infectious but relatively harmless. In one technique, tissues can be removed from the patient and incubated with these genetically modified viruses to transfect them with the normal gene. The transfected cells are then reintroduced into the patient's body by transfusion **(figure 8.14).** Alternatively, naked DNA or a virus vector is directly introduced into the patient's tissues.

As of mid-2022, gene therapy treatments for 13 different diseases had been approved. Examples include spinal muscular

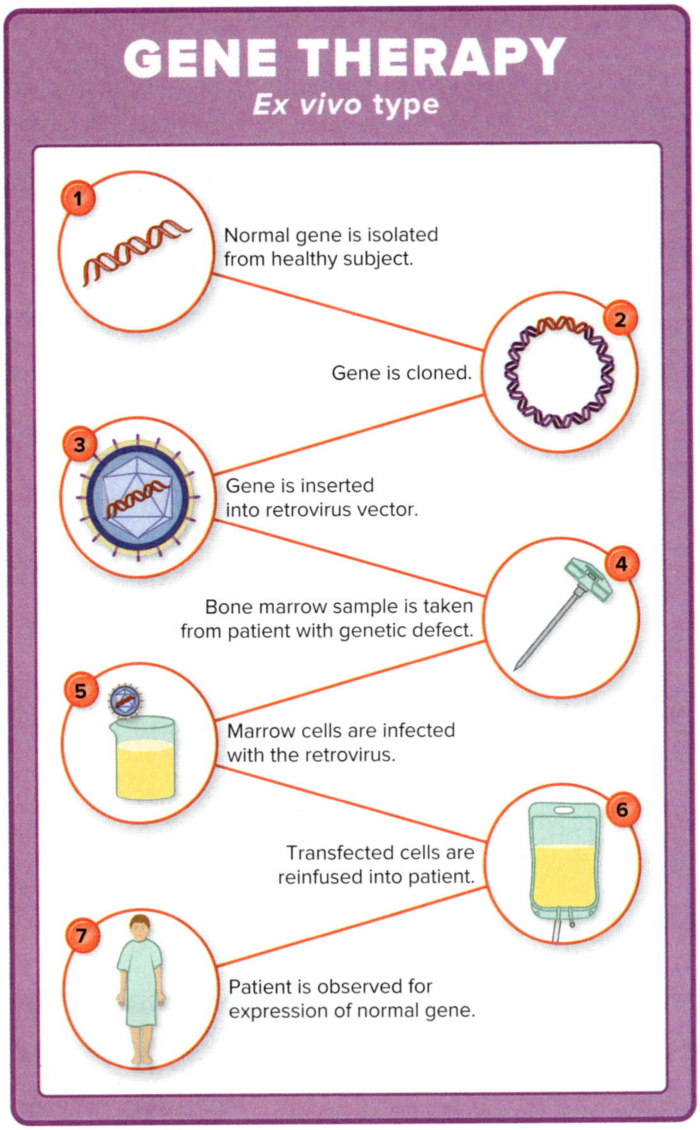
GENE THERAPY
Ex vivo type
1
Normal gene is isolated from healthy subject.
2
Gene is cloned.
3
Gene is inserted into retrovirus vector.
4
Bone marrow sample is taken from patient with genetic defect.
5
Marrow cells are infected with the retrovirus.
6
Transfected cells are reinfused into patient.
7
Patient is observed for expression of normal gene.

Figure 8.14 Protocol for the *ex vivo* type of gene therapy in humans.

atrophy, a leading cause of infant mortality, forms of inherited vision loss, and acute lymphoblastic anemia. That number is growing quickly. At last check, almost 1,100 companies were working on new gene therapies.

One form of gene therapy is called *somatic cell* gene therapy. This means that the changes are permanent in the individual who is treated, but they are not passed on to offspring. The ultimate sort of gene therapy is germline therapy, in which genes are inserted into an egg, sperm, or early embryo. In this type of therapy, the new gene will be present in all cells of the individual. The therapeutic gene is also heritable (that is, can be passed on to subsequent generations).

Small RNAs as Medicine

In the last section, we considered the use of genetic technology to replace a missing or faulty protein that is needed for normal cell function. A different problem arises when a disease results from too much of a protein. For example, Alzheimer's disease, most viral diseases, and many cancers occur when an unwanted gene is expressed. Sometimes this occurs due to a malfunctioning micro RNA.

As we have seen, cells and viruses employ a variety of micro RNAs to silence gene expression. Consequently, researchers have experimented with adding miRNA molecules to tissues or whole organisms to counter diseases. Two examples are:

- Introducing specific miRNAs into small-cell lung cancer tumors in mice blocked tumor growth.
- Micro RNAs caused regression of liver cancers in mice.

On the other hand, out-of-control miRNAs sometimes are the cause of disease. Scientists have found multiple successes in inhibiting errant miRNAs:

- Inhibiting a specific miRNA in mice stopped the growth of both breast and ovarian cancers.
- Inhibiting a micro RNA in monkeys lowered their plasma cholesterol levels.
- Multiplication of hepatitis C virus in chimpanzees was inhibited by shutting down one class of miRNAs (this strategy is now being tested in humans).

The first successful human application of interfering RNAs was conducted at the University of Tennessee in 2010. There, healthy human volunteers were given a nasal spray containing miRNA designed to silence a gene from respiratory syncytial virus (RSV). They were infected with RSV and then used the spray for 5 days. Only 44% of those who had received the interfering RNA became infected, compared with 71% of control subjects. Among those who were infected, symptoms were significantly reduced.

In summary, the discovery of miRNAs and their role in disease has opened a vast new area of research and hope for treatment, cure, and even prevention of disease.

CAR-T Cell Treatment of Cancer

In recent years, scientists have used DNA recombination techniques to alter patients' own T cells (important immune cells that battle cancer cells). They cause the patient's own T cells to express a protein on their surfaces that binds to cancer cells and cause their destruction. At this point it is an extremely expensive treatment ($475,000), so a way must be found to put it in the reach of a wide audience.

CRISPR

Earlier in this chapter we described the CRISPR-Cas9 system as it occurs naturally in bacteria. That naturally occuring system in bacteria was discovered in the early 2000s. By 2012, Jennifer Doudna, a scientist at the University of California, Berkeley, and Emmanuelle Charpentier, at the Max Planck Institute in Berlin, announced their system for using the CRISPR system to precisely edit genes in any organism **(figure 8.15).** They were awarded the Nobel Prize in Chemistry in 2020. This research will be used to precisely edit DNA and has enormous potential in medicine, agriculture, and other fields. The Nobel Committee said this in a

Figure 8.15 Jennifer Doudna (left) and Emmanuelle Charpentier (right) won a 2020 Nobel prize for developing CRISPR technology.
Dpa Picture Alliance/Alamy Stock Photo

press release: "These genetic scissors have taken the life sciences into a new epoch and, in many ways, are bringing the greatest benefit to humankind."[2] Another name for the CRISPR system is **gene drive**. There are multiple clinical trials with CRISPR using to treat a variety of diseases. One promising use is for sickle cell anemia.

In a recent experimental treatment, researchers removed red blood cells from a patient, engineered them to have a hemoglobin molecule that binds more tightly to oxygen, then returned the cells to the patient. The therapeutic potential of CRISPR is enormous, but its widespread use is still awaiting successful clinical trial data and FDA approval. In addition, many scientists are cautious, wondering whether we should attempt to "edit" germline cells in embryos using CRISPR because that could lead from the simple correction of disease genes to making designer babies to suit our preferences.

[2] Press release: The Nobel Prize in Chemistry 2020, Nobelprize.org, October 2020.

8.3 Learning Outcomes—Assess Your Progress

10. Compose your own definition of personalized medicine.
11. Provide several examples of recombinant products that have contributed to human health.
12. List examples of recombinant bacteria and plants, and a purpose for each.
13. Differentiate between somatic and germline gene therapy.
14. Describe miRNAs and ways in which their discovery can impact human disease.
15. Explain how the CRISPR-Cas9 system can be used to cure genetic diseases.

Media Under The Microscope Wrap-Up

The news article "Genetics Research May Help Identify More Dangerous Strains of the Virus That Causes COVID-19" describes a research study published in the scientific literature. The **intended message** of the article is to convey how examining the exact genetic sequence of different SARS-CoV-2 isolates can provide clues to what makes one variant more pathogenic than the other. (As an aside, the delta variant that became so important in 2021 also has been characterized with respect to which bases mutated to make it more transmissible.)

My **critical reading** of the article is that it is pretty straightforward and explains how some people get sick and die quickly, while others have mild cases, and that mutations in the virus can explain that. They mentioned that the important mutation was in the gene coding for the virus spike protein that it uses for attachment, and that the mutation also makes it less susceptible to host antibodies. All good. I have one complaint: The article said that 28,891 locations in the genome were analyzed. That left me wondering if there are 28,891 total bases in the genome or if there were more and that these were just a selection of them for some reason. (Turns out that is the total number of bases; I looked at the original research article to find out.)

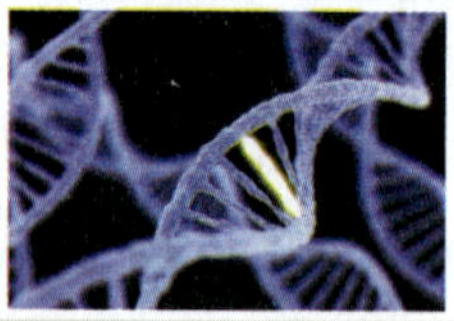

Konstantin Faraktinov/Shutterstock

To **interpret** this article to your friends, you will want to draw on your knowledge of how gene sequencing is performed (from this chapter) and tell them that it is possible to find if there are sequence difference(s) in deadly variants of the virus compared to the less deadly forms.

My **overall grade** for the article is an A−. It contains great information on a timely topic. My only quibble is not being clear on whether the total length of the genome or just selected areas were analyzed.

Source: *SciTechDaily*, "Genetics Research May Help Identify More Dangerous Strains of the Virus That Causes COVID-19," online article posted June 27, 2021.

Study Smarter: Better Together

These activities are designed for you to use on your own with a study group—either a face-to-face group or a virtual one, consisting of 3–5 members. Studying together can be very helpful, but there are effective and ineffective ways to do it. For example, getting together without a clear structure is often not a good use of your time. Use your time efficiently by using one or more of the exercises below.

FACE-TO-FACE GROUPS

Use one or more of the activities below.

Peer Instruction: Assign numbers to your group members to use all semester long. Now look at these five concepts from this chapter. Each group member prepares a 5-minute lesson on the topic corresponding to their number. Don't worry if you have fewer than 5 members; just use however many you have! During your group study time, each member presents their lesson, and the group spends another 5–10 minutes discussing that lesson.

1. Polymerase chain reaction
2. Cloning steps
3. Genome sequencing
4. Gene therapy
5. Recombinants
6. Single nucleotide polymorphisms

Concept Maps: Each member of the group should use this list of terms from this chapter to generate their own concept map. This can be hand-drawn or created using software (see Appendix C for guidelines). During group study time, compare each other's concept maps and help each other make sure they are correct. Of course, there are many different "correct" maps. Examining each member's map will help you talk through the varied concepts and how they are related.

Concept Terms:

DNA	palindrome	plasmid	vector
origin of replication	ligase	restriction endonuclease	

Table Topics: Each group member should identify a concept or topic from this week's class assignments with which they are having trouble and share it during group study time. The other group members can then help to clarify confusing issues or share how they figured it out. Aim for a maximum of 15 minutes per topic. If the topic remains unclear to the group, bring it up during class or use the instructor's office hours or e-mail to ask for help. Taking the time to struggle with a difficult concept first makes your questions much more specific and more likely to yield helpful answers.

VIRTUAL GROUPS

Not everyone has the time or opportunity to meet with group members outside of class time. You or your instructor can create a virtual group using e-mail or the course software.

Weekly Discussion Board: This forum can be used as a way for groups to discuss topics, via e-mail or other learning management systems or online platforms, before they are covered in class. As each member of the group answers the current week's question, they should send their responses to every other member of their group. It's best to agree on a deadline based on how your class schedule works (Saturday for the next week's topics, for example). Then, after the topic is discussed in class, each member should send a response that all group members will see with a follow-up post on the same topic. If you cover more than one chapter in a week, someone can be designated to choose which chapter Discussion Board question you will use. Or simply decide up front that you will always use the first-chapter-of-the week's question, to keep the schedule simple.

Discussion Question

Using your knowledge of DNA from chapters 6 and 8, imagine two different ways antibiotic resistance can develop.

Chapter Summary

GENETIC ANALYSIS AND RECOMBINANT DNA TECHNOLOGY

8.1 TOOLS AND TECHNIQUES OF RECOMBINANT DNA TECHNOLOGY

- Recombinant DNA technology uses methods that physically manipulate DNA in order to visualize, identify, and change specific DNA sequences.
- Restriction endonucleases derived from bacteria were the first and are still commonly used as "scissors" to manipulate DNA.
- The polymerase chain reaction is like a photocopying machine for DNA.
- Any organism that has another organism's DNA inserted into it is a recombinant.
- Creating recombinants in the lab is referred to as cloning, and often uses plasmids as cloning hosts.
- Synthetic biology is the ability to create organisms by putting together the right kind of "off the shelf" DNA.

8.2 DNA ANALYSIS

- Gel electrophoresis separates and displays different lengths of DNA.
- The fact that complementary bases will bind to each other is the basis for nucleic acid hybridization and probing.
- DNA can be profiled using restriction endonucleases and gel electrophoresis to compare DNA samples.
- Beginning in the 1970s, the sequence of bases in any DNA was obtainable. In 2003 the human genome was sequenced after 13 years of effort. An explosion of whole organism genome sequences has followed.
- Single nucleotide polymorphisms are a change in one base compared to a reference genome and can have major effects on the organism.

8.3 GENETIC APPROACHES TO HEALING DISEASE

- Personalized medicine is the use of a patient's DNA sequence to tailor treatments exactly to their genome.
- Genetically modifying organisms provides us with life-saving drugs and vaccines.
- Gene therapy involves inserting healthy genes into a patient whose faulty DNA is causing serious disease.
- The CRISPR-Cas9 system is revolutionizing how we treat genetic diseases.

SmartGrid: From Knowledge to Critical Thinking

This *21 Question Grid* takes the topics from this chapter and arranges them with respect to the American Society for Microbiology's Undergraduate Curriculum guidelines—all six of the important "Concepts" as well as the important "Competency" of scientific literacy. Three questions are supplied, which cover chapter content referring to the Concept or Competency in increasing levels of Bloom's taxonomy for learning.

ASM Concept/ Competency	A. Bloom's Level 1, 2—Remember and Understand (Choose one.)	B. Bloom's Level 3, 4—Apply and Analyze	C. Bloom's Level 5, 6—Evaluate and Create
Evolution	**1.** Single nucleotide polymorphisms are found in **a.** DNA. **b.** RNA. **c.** plasmids. **d.** siRNA.	**2.** Explain why restriction endonucleases are found inside bacteria.	**3.** Conduct research on CRISPR and explain in layperson's terms why there are ethical concerns around its use in recombinant DNA technology.

ASM Concept/ Competency	A. Bloom's Level 1, 2—Remember and Understand (Choose one.)	B. Bloom's Level 3, 4—Apply and Analyze	C. Bloom's Level 5, 6—Evaluate and Create
Cell Structure and Function	**4.** Which of the following is/are characteristic(s) of DNA? **a.** It uncoils when it is heated. **b.** It contains palindromes. **c.** DNA from one organism can be spliced into DNA from another organism. **d.** All of the above are characteristics of DNA.	**5.** Why did the discovery of extreme thermophilic bacteria enable the invention of the PCR technique?	**6.** The chapter did not address the use of microarrays to study proteins, except to mention their existence. Research proteome microarrays and construct a one-sentence description of them.
Metabolic Pathways	**7.** Some of the recombinant products used in medicine (table 8.2) are proteins. Recall what processes are required to produce proteins. **a.** mutations **b.** transcription **c.** translation **d.** two of the above	**8.** Discuss the intersection between the "metabolome" and phenotype.	**9.** Explain in your own words how a single nucleotide polymorphism can lead to excessive clotting.
Information Flow and Genetics	**10.** The study of the particular assortment of small molecules in a cell at any given time is called **a.** genomics. **b.** proteomics. **c.** metagenomics. **d.** metabolomics.	**11.** In the cloning procedure, why is it useful to use a vector containing an antibiotic resistance gene?	**12.** Will all single nucleotide polymorphisms result in changed phenotype? Why or why not?
Microbial Systems	**13.** The natural process of ______________ can be exploited in the lab to facilitate cloning. **a.** transformation **b.** transfection **c.** transduction **d.** all of the above	**14.** Why is it advisable to use reverse transcriptase to create DNA from an mRNA transcript when preparing eukaryotic genes for use in bacterial cells? Why not use the eukaryotic DNA directly?	**15.** Metagenomics is providing insight into the possible causes of some difficult-to-understand diseases, such as irritable bowel syndrome. Develop a possible explanation for why metagenomic approaches might be better than traditional diagnostic approaches.
Impact of Microorganisms	**16.** The creation of biological molecules and cells entirely from chemicals is called **a.** recombination. **b.** sequencing. **c.** artificial biology. **d.** synthetic biology.	**17.** You are a public health official trying to determine if 20 different people in your city were sickened by the same microbe, or if they were unrelated. Which technique from this chapter would you use, and why?	**18.** Genetically modified organisms (GMOs)—especially in foods—are considered by many in the public to be "bad." Construct a counterargument.
Scientific Thinking	**19.** How many cycles of PCR would it take to create 16 copies of a DNA molecule? **a.** two **b.** three **c.** four **d.** eight	**20.** One of the problems with early gene therapy trials was that the seemingly harmless viruses caused new malignancies in the treated person. Speculate on how that happens.	**21.** In Insight 8.1, the connection between a host's genome and its microbiome is explored. A null hypothesis is proposed. Why are many research questions posed as null hypotheses instead of "positive" hypotheses?

Answers to the multiple-choice questions appear in Appendix A.

Visual Connections

These questions use visual images or previous content to make connections to this chapter's concepts.

1. **From chapter 7, figure 7.15.** What has happened to the bacterial DNA in this illustration? What effect can this have on a bacterium? Is this temporary or permanent?

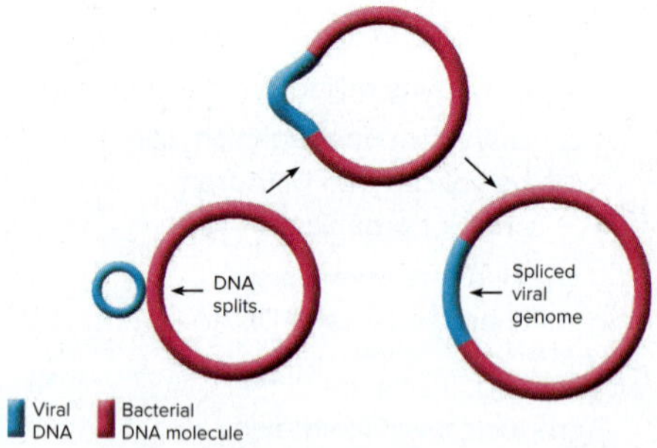
DNA splits.
Spliced viral genome
Viral DNA
Bacterial DNA molecule

High Impact Study

These terms and concepts are most critical for your understanding of this chapter—and may be the most difficult. Have you mastered them?

Concepts

- ☐ Nucleic acid hybridization
- ☐ Cloning steps
- ☐ Genome sequencing
- ☐ DNA profiles
- ☐ PCR
- ☐ Gene therapy
- ☐ CRISPR

Terms

- ☐ Restriction endonuclease
- ☐ cDNA
- ☐ Electrophoresis
- ☐ -omics
- ☐ Cloning vector
- ☐ Genetically modified organisms
- ☐ Synthetic biology
- ☐ Probes
- ☐ Single nucleotide polymorphism

Design Element: (College students): Caia Image/Image Source

9 Microbial Nutrition and Growth

NASA

Media Under The Microscope

Conan the Bacterium

These case studies examine an article from the popular media to determine the extent to which it is factual and/or misleading. This case focuses on the 2020 IndiaTimes *article "Earth Bacteria Survived in Space for Years, Proving It Can Contaminate Mars."*

Scientists from Japan placed a hardy species of bacterium called *Deinococcus* in an aluminum exposure panel outside of the International Space Station and were able to recover living bacteria after three years of the microbes sailing around in space. The scientists playfully nicknamed the microbe Conan the Bacterium. The *IndiaTimes* article describing this was published in 2020.

The article explains that the bacteria were in layers and that the outermost ones died, but those near the bottom of the stack survived. The same bacteria inside the space station did not survive because, according to the article, oxygen and moisture were lethal to them.

The bacterium's ability to withstand massive radiation was emphasized in the article. Human cells die when radiation exposure reaches 10 Grays. The bacterium reported here can withstand 10,000 Grays. This has both positive and negative implications for space travel, the article says. On the one hand, we know that using layers of bacteria, we may be able to transport microbes into space to grow them for practical purposes on other planets. On the other hand, we now know that we may be unwittingly transporting microbes to other planets and create ecological disasters there.

- What is the **intended message** of the article?
- What is your **critical reading** of the summary of the article provided above? Remember that in this context, "critical reading" does not necessarily mean *What criticism do you have?* but asks you to apply your knowledge to interpret whether the article is factual and whether the facts support the intended message.
- How would you **interpret** the news item for your nonmicrobiologist friends?
- What is your **overall grade** for the news item—taking into account its accuracy and the accuracy of its intended effect?

Media Under The Microscope Wrap-Up appears at the end of the chapter.

Outline and Learning Outcomes

9.1 Microbial Nutrition

1. List the essential nutrients of a bacterial cell.
2. Differentiate between *macronutrients* and *micronutrients*.
3. List and define four different terms that describe an organism's sources of carbon and energy.
4. Define *saprobe* and *parasite*, and explain why these terms can be an oversimplification.
5. Compare and contrast the processes of diffusion and osmosis.
6. Identify the effects of isotonic, hypotonic, and hypertonic conditions on a cell.
7. Name two types of passive transport and three types of active transport.

9.2 Environmental Factors That Influence Microbes

8. List and define five terms used to express the temperature-related growth capabilities of microbes.
9. Summarize three ways in which microorganisms function in the presence of oxygen.
10. Identify three other physical factors that microbes must contend with in the environment.
11. List and describe the five major types of microbial association.
12. Discuss characteristics of biofilms that differentiate them from free-living bacteria and their infections.

9.3 The Study of Microbial Growth

13. Summarize the steps of binary fission, and name another method of reproduction used by other bacterial species.
14. Define *doubling time,* and describe how it leads to exponential growth.
15. Compare and contrast the four phases of growth in a bacterial growth curve.
16. Identify one culture-based and one non-culture-based method used for analyzing bacterial growth.

9.1 Microbial Nutrition

Nutrition is a process by which **nutrients** are acquired from the environment and used in cellular activities such as metabolism and growth. With respect to nutrition, microbes are not really so different from humans. Bacteria digesting plastic or protozoa digesting wood in a termite's intestine seem to live radical lifestyles, but even these organisms require a constant influx of certain substances from their habitat. (There are even bacteria that live on caffeine—just as many students claim they do.) No matter how strange their diet seems, all living things require a source of elements such as carbon, hydrogen, oxygen, phosphorus, potassium, nitrogen, sulfur, calcium, iron, sodium, chlorine, magnesium, and certain other elements. But the ultimate source of a particular element, its chemical form, and how much of it the microbe needs vary among different types of organisms. Any substance, whether in elemental or molecular form, that an organism must have is called an **essential nutrient.** For microbes, the essential nutrients are carbon, hydrogen, oxygen, nitrogen, phosphorus (phosphate), and sulfur—often referred to by the acronym CHONPS.

Disease Connection

What essential nutrients do humans need to survive? Macronutrients include carbohydrates, fats, protein, and water. These are needed in large quantities to fuel the body. The micronutrients needed for humans to survive are the vitamins A, C, D, E, and K, as well as the B-complex vitamins. Minerals include calcium, potassium, chloride, phosphorus, sodium, and magnesium (also known as electrolytes), as well as copper, cobalt, zinc, chromium, iron, and fluoride (which make up the trace elements). As you can see, we actually have many essential nutrients in common with microbes!

Two categories of essential nutrients are **macronutrients** and **micronutrients.** The "macro" in macronutrients refers to the fact that these substances are required in relatively large quantities. They play principal roles in cell structure and metabolism. Examples of macronutrients are carbon, hydrogen, and oxygen. Micronutrients, or **trace elements,** such as manganese, zinc, and nickel, are used in much smaller amounts and are involved in enzyme function and maintenance of protein structure. Please remember that the terms "macro-" and "micro-" when referring to nutrients refer to the quantity needed, not to their individual sizes.

Another way to categorize nutrients is according to their carbon content. An inorganic nutrient is an atom or simple molecule that contains a combination of atoms other than carbon and hydrogen. The natural reservoirs of inorganic compounds are mineral deposits in the crust of the earth, bodies of water, and the atmosphere. Examples include metals and salts of metals (magnesium sulfate, ferric nitrate, sodium phosphate), gases (oxygen, carbon dioxide), and water **(table 9.1).** In contrast, organic nutrients contain carbon and hydrogen atoms and are usually the products of living things. They range from the simplest organic molecule, methane (CH_4), to large polymers (carbohydrates, lipids, proteins, and nucleic acids). Some microbes obtain their nutrients entirely from inorganic sources, and others require a combination of organic and inorganic sources. Parasites capable of invading and living on the human body derive all essential nutrients from host tissues, tissue fluids, secretions, and wastes.

Chemical Analysis of Microbial Cytoplasm

If we look at the chemical composition of a bacterial cell, we can deduce its nutritional requirements. **Table 9.2** lists the major contents of the intestinal bacterium *Escherichia coli.* Some of these components are absorbed in a ready-to-use form, and others must

Table 9.1 Principal Inorganic Reservoirs of Elements

Element	Inorganic Environmental Reservoir
Carbon	Air, rocks, and sediments
Oxygen	O_2 in air, certain oxides, water
Nitrogen	Air, soil, water
Hydrogen	Water, H_2 gas, mineral deposits
Phosphorus	Mineral deposits
Sulfur	Mineral deposits, volcanic sediments
Potassium	Mineral deposits, the ocean
Sodium	Mineral deposits, the ocean
Calcium	Mineral deposits, the ocean
Magnesium	Mineral deposits, geologic sediments
Chloride	The ocean
Iron	Mineral deposits, geologic sediments
Manganese, molybdenum, cobalt, nickel, zinc, copper, other micronutrients	Various geologic sediments

Table 9.2 Chemical Composition of an *Escherichia coli* Cell

	% Total Weight	% Dry Weight (minus water)		% Dry Weight
Organic Compounds			**Elements**	
Proteins	15	50	Carbon (C)	50
Nucleic acids			Oxygen (O)	20
RNA	6	20	Nitrogen (N)	14
DNA	1	3	Hydrogen (H)	8
Carbohydrates	3	10	Phosphorus (P)	3
Lipids	2	10	Sulfur (S)	1
Miscellaneous	2	4	Potassium (K)	1
			Sodium (Na)	1
Inorganic Compounds			Calcium (Ca)	0.5
Water	70		Magnesium (Mg)	0.5
All others	1	3	Chlorine (Cl)	0.5
			Iron (Fe)	0.2
			Trace metals	0.3

be synthesized by the cell from simple nutrients. The following are important features of cell composition:

- Water is the most abundant of all the components (70% of cell contents). When cell contents are considered after the removal of water, we call it "dry weight."
- Proteins are the next most prevalent chemical.
- About 97% of the dry cell weight is composed of organic compounds.
- About 96% of the dry cell weight is composed of six elements (represented by CHONPS).
- Chemical elements are needed in the overall scheme of cell growth, but most of them are available to the cell as compounds and not as pure elements (see table 9.2).
- A cell as "simple" as *E. coli* contains on the order of 5,000 different compounds, yet it needs to consume only a few types of nutrients to synthesize this great diversity.

Sources of Essential Nutrients

In their most basic form, elements that make up nutrients exist in inorganic reservoirs. These reservoirs not only serve as a permanent, long-term source of these elements but also can be replenished by the activities of organisms. In fact, the ability of microbes to keep elements cycling is essential to all life on the earth.

Carbon Sources

A **heterotroph** is an organism that must obtain its carbon in an organic form. Because organic carbon originates from the bodies of other organisms, heterotrophs are dependent on other life forms (*hetero-* is a Greek prefix meaning "other"). Among the common organic molecules that can satisfy this requirement are proteins, carbohydrates, lipids, and nucleic acids. Some organic nutrients available to heterotrophs already exist in a form that is simple enough for absorption (e.g., monosaccharides and amino acids), but larger molecules must be digested by the cell before absorption. Moreover, heterotrophs vary in their capacities to use different organic carbon sources. Some are restricted to a few substrates, whereas others (certain *Pseudomonas* species, for example) are so versatile that they can metabolize more than 100 different carbon-containing substrates.

Disease Connection

Pseudomonas aeruginosa is a gram-negative rod that is ubiquitous in the environment, due to its ability to live on so many different chemicals. It does not cause disease in healthy humans, but in immunocompromised patients, it is a major pathogen. It is involved in 90% of deaths of cystic fibrosis patients.

An **autotroph** ("self-feeder") is an organism that uses inorganic CO_2 as its carbon source. Because autotrophs have the special capacity to convert CO_2 into organic compounds, they are not nutritionally dependent on other living things—and that's where the name "self-feeder" comes from: not from feeding on itself, but by being an independent organism.

Nitrogen Sources

The main reservoir of nitrogen is nitrogen gas (N_2), which makes up 79% of the earth's atmosphere. This element is indispensable to the structure of proteins, DNA, RNA, and ATP. Such compounds are the primary nitrogen source for heterotrophs, but to be useful, they must first be degraded into their basic building blocks (proteins into amino acids or nucleic acids into nucleotides). Some bacteria and algae utilize inorganic nitrogenous nutrients (NO_3^-, NO_2^-, or NH_3). A small number of bacteria can transform N_2 into compounds usable by other organisms through the process of nitrogen fixation. Regardless of the initial form in which the inorganic nitrogen enters the cell, it must first be converted to NH_3 (ammonia), the only form that can be directly combined with carbon to synthesize amino acids and other compounds.

Oxygen Sources

Because oxygen is a major component of all organic compounds such as carbohydrates, lipids, nucleic acids, and proteins, it plays an important role in the structural and enzymatic functions of the cell. Oxygen is likewise a common component of inorganic salts such as sulfates, phosphates, nitrates, and water. Free gaseous oxygen (O_2) makes up 20% of the atmosphere. It is absolutely essential to the metabolism of many organisms.

Hydrogen Sources

Hydrogen is a major element in all organic and several inorganic compounds, including water (H_2O), salts ($Ca[OH]_2$), and certain naturally occurring gases (H_2S, CH_4, and H_2). These gases are both used and produced by microbes. Hydrogen is critical for

1. maintaining **pH,**
2. forming the vastly important **hydrogen bonds** between molecules, and
3. serving as the source of **free energy** in oxidation-reduction reactions of respiration.

Phosphorus (Phosphate) Sources

The main inorganic source of phosphorus is phosphate (PO_4^{3-}), derived from phosphoric acid (H_3PO_4) and found in rocks and oceanic mineral deposits. Phosphate is a key component of nucleic acids and is therefore essential to the genetics of cells and viruses. It is also found in ATP, an important energy molecule in cells. Other phosphate-containing compounds are phospholipids in cell membranes and coenzymes such as NAD^+ (see chapter 10). Certain environments have very little available phosphate for use by organisms and therefore limit the ability of these organisms to grow. One bacterium, *Corynebacterium,* is able to concentrate and store phosphate in metachromatic granules in the cytoplasm.

Sulfur Sources

Sulfur is widely distributed throughout the environment in mineral form. Rocks and sediments (such as gypsum) can contain sulfate (SO_4^{2-}), sulfides (FeS), hydrogen sulfide gas (H_2S), and elemental sulfur (S). Sulfur is an essential component of some vitamins (vitamin B_1) and the amino acids methionine and cysteine. Those amino acids help determine shape and structural stability of proteins by forming unique linkages called disulfide bonds.

Other Nutrients Important in Microbial Metabolism

Other important elements in microbial metabolism include mineral ions. Potassium is essential to protein synthesis and membrane function. Sodium is important for certain types of cell transport. Calcium is a stabilizer of the cell wall and endospores of bacteria. Magnesium is a component of chlorophyll and a stabilizer of membranes and ribosomes. Iron is an important component of the cytochrome proteins of cell respiration. Zinc is an essential regulatory element for eukaryotic DNA. Copper, cobalt, nickel, molybdenum, manganese, silicon, iodine, and boron are needed in small amounts by some microbes but not others. The concentration of metal ions can even influence the diseases microbes cause. For example, the bacteria that cause gonorrhea and meningitis grow more rapidly in the presence of iron ions. On the other hand, metals can also be very toxic to microbes.

Growth Factors: Essential Organic Nutrients

An organic compound such as an amino acid, nitrogenous base, or vitamin that cannot be synthesized by an organism and must be provided as a nutrient is a **growth factor.** For example, although all cells require 20 different amino acids for proper assembly of proteins, many cells cannot synthesize all of them. Those that must be obtained from food are called *essential amino acids.*

How Microbes Feed: Nutritional Types

The earth's limitless habitats and microbial adaptations are matched by an elaborate array of nutritional schemes used by microbes. (Humans do not have that flexibility; our habitat requirements are fairly narrow.). Fortunately, most organisms show consistent trends and can be described by a few general categories **(table 9.3)** and a few selected terms (see "A Note About Terminology" later in this section). The main determinants of a microbe's nutritional type are how it gets its carbon and how it gets its energy. In a previous section, microbes were defined according to their *carbon sources* as autotrophs or heterotrophs. Now we will subdivide all bacteria according to their *energy source* as **phototrophs** or **chemotrophs.** Microbes that photosynthesize (gain their energy from sunlight) are phototrophs, and those that gain energy from chemical compounds are chemotrophs. The terms for carbon and energy sources are often merged into a single word for convenience (see table 9.3). The categories described here are meant to describe only the major nutritional groups and do not include unusual exceptions.

Autotrophs and Their Energy Sources

Autotrophs derive energy from one of two possible nonliving sources: sunlight (photoautotrophs) and chemical reactions involving simple chemicals (chemoautotrophs). **Photoautotrophs** are photosynthetic—that is, they capture the energy of light rays and

Table 9.3 Nutritional Categories of Microbes by Energy and Carbon Source

Category	Energy Source	Carbon Source	Example
Autotroph			
Photoautotroph	Sunlight	CO_2	Photosynthetic organisms, such as algae, plants, most cyanobacteria
Chemoautotroph			
Chemoorganic autotrophs	Organic compounds	CO_2	Methanogens
Chemolithoautotrophs	Inorganic compounds (minerals)	CO_2	*Thiobacillus,* "rock-eating" bacteria
Heterotroph			
Photoheterotroph	Sunlight	Organic	Purple and green photosynthetic bacteria
Chemoheterotroph	Metabolic conversion of the nutrients from other organisms	Organic	Protozoa, fungi, many bacteria, animals
Saprobe	Metabolizing the organic matter of dead organisms	Organic	Fungi, bacteria (decomposers)
Parasite	Utilizing the tissues, fluids of a live host	Organic	Various parasites and pathogens; can be bacteria, fungi, protozoa, animals

transform it into chemical energy that can be used in cell metabolism **(figure 9.1).** Because photosynthetic organisms (algae, plants, some bacteria) produce organic molecules that can be used by themselves and also by heterotrophs, they form the basis for most food webs. We call them primary producers.

Chemoautotrophs are of two types: One of these is the group called **chemoorganic autotrophs.** These use organic compounds for energy and inorganic compounds as a carbon source. The second type of chemoautotroph is a group called **chemolithoautotrophs,** which require neither sunlight nor organic nutrients, relying totally on inorganic minerals. These bacteria derive energy in diverse and rather amazing ways. In very simple terms, they remove electrons from inorganic substrates such as hydrogen gas, hydrogen sulfide, sulfur, or iron and combine them with carbon dioxide and hydrogen.

Recently, scientists discovered bacteria that live solely on pure electrons, in essence *electricity* coming from rocks and metals. We'll say more about these in chapter 10. This reaction provides simple organic molecules and a modest amount of energy to drive the synthetic processes of the cell. Chemollithotrophic bacteria play an important part in recycling inorganic nutrients.

An interesting group of chemoorganic autotrophs is the **methanogens** (meth-an′-oh-genz), which produce methane (CH_4) from hydrogen gas and carbon dioxide:

$$4H_2 + CO_2 \rightarrow CH_4 + 2H_2O$$

Methane, sometimes called "swamp gas" or "natural gas," is formed in anaerobic, hydrogen-containing microenvironments of soil, swamps, mud, and even the intestines of some animals. Methanogens are archaea, some of which live in extreme habitats

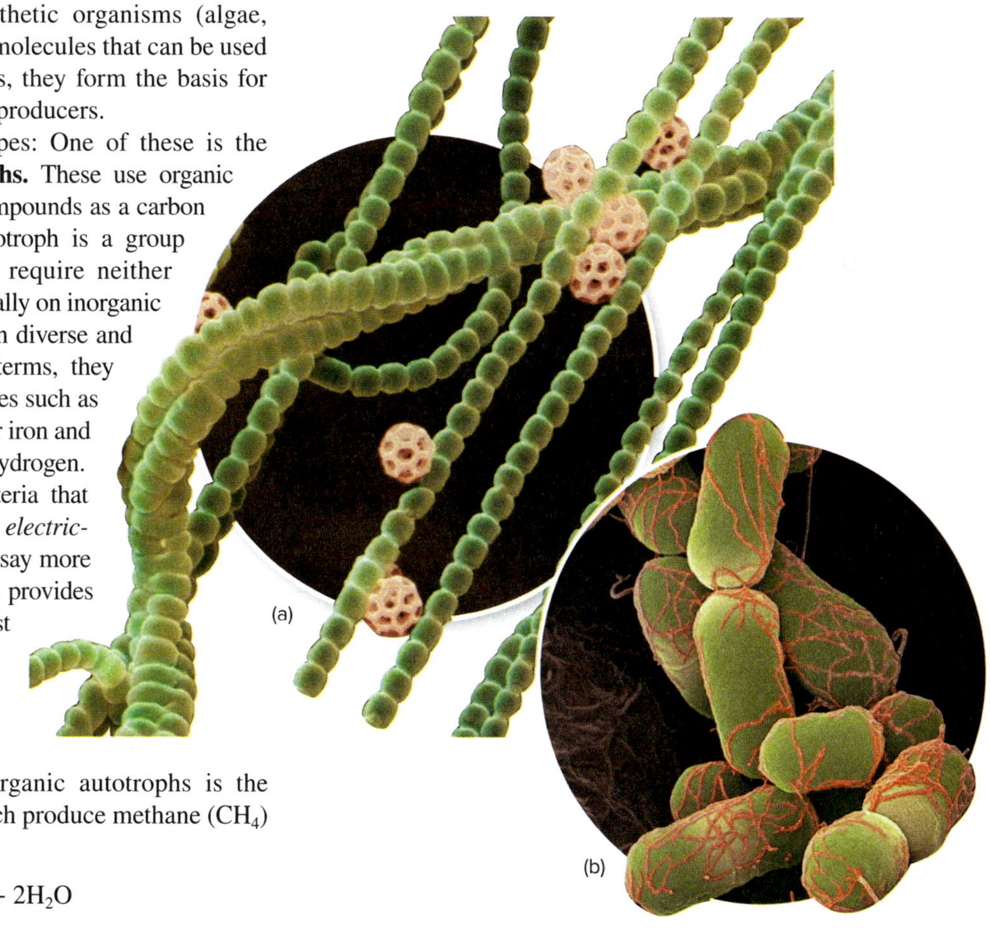
(a)
(b)

Figure 9.1 A photoautotroph and a chemoheterotroph. **(a)** *Cyanobacterium,* a photosynthetic autotroph, in green. The pink structures are spores. **(b)** *Escherichia coli,* a chemoheterotroph.
(a) Steve Gschmeissner/Science Photo Library/Getty Images; (b) Martin Oeggerli/Science Source

such as ocean vents and hot springs, where temperatures reach up to 400°C. Methane also plays a role as one of the greenhouse gases that are currently an environmental concern (see chapter 25).

Heterotrophs and Their Energy Sources

The majority of heterotrophic microorganisms are **chemoheterotrophs** that get both their carbon and energy from organic compounds. Processing these organic molecules by respiration or fermentation releases energy in the form of ATP. An example of chemoheterotrophy is **aerobic respiration,** the principal energy-yielding pathway in animals, most protozoa and fungi, and aerobic bacteria. It can be simply represented by the equation:

$$\text{Glucose } [(CH_2O)_n] + O_2 \rightarrow CO_2 + H_2O + \text{energy (ATP)}$$

This reaction is complementary to photosynthesis. It uses glucose and oxygen as reactants, and carbon dioxide is given off. In fact, the earth's balance of both energy and metabolic gases is greatly dependent on this relationship. Chemoheterotrophic microorganisms fit into one of two main categories that differ in how they obtain their organic nutrients: **Saprobes** are free-living microorganisms that feed primarily on organic detritus from dead organisms, and **parasites** ordinarily derive nutrients from the cells or tissues of a living host.

Saprobic Microorganisms Saprobes occupy a niche as decomposers of plant litter, animal matter, and dead microbes. If not for the work of decomposers, the earth would gradually fill up with organic material, and the nutrients in it would not be recycled. Most saprobes, notably bacteria and fungi, have a rigid cell wall and cannot engulf large particles of food. To compensate, they release enzymes to the extracellular environment and digest the food particles into smaller molecules that can be transported into the cell **(figure 9.2).** *Obligate saprobes* exist strictly on dead organic matter in soil and water and are unable to adapt to the body of a live host. This group includes many free-living protozoa, fungi, and bacteria. Apparently, there are fewer of these strict species than was once thought, and many supposedly nonpathogenic saprobes can infect a susceptible host. When a saprobe does infect a host, it is considered a *facultative parasite,* which means a part-time parasite. A facultative parasite will infect a host when the host is compromised. When this happens, the microbe is also described as *opportunistic.* For example, although its natural habitat is soil and water, *Pseudomonas aeruginosa* frequently causes infections in hospitalized patients. The yeast *Cryptococcus neoformans* causes a severe lung and brain infection in patients with AIDS, yet its natural habitat is the soil.

Parasitic Microorganisms Parasites live in or on the body of a host, which, by definition, they harm to some degree. Because parasites cause damage to tissues (disease) or even death, they are also called **pathogens.** Parasites range from viruses to helminths (worms). *Obligate parasites* (e.g., the leprosy bacillus and the syphilis spirochete) are unable to grow outside of a living host. Parasites that are less strict can be cultured artificially if provided with the correct nutrients and environmental conditions. Pathogenic bacteria such as *Streptococcus pyogenes* (the cause of strep throat) and *Staphylococcus aureus* can grow on artificial media.

A Note About Terminology

Much of the vocabulary for describing microbial adaptations is based on some common root words. These are combined in various ways that assist in discussing the types of nutritional or ecological adaptations, as shown in this partial list.

Root	Meaning	Example of Use
troph-	Food	Trophozoite—the feeding stage of protozoa
-phile	To love	Extremophile—an organism that has adapted to ("loves") extreme environments
-obe	To live	Microbe—to live "small"
hetero-	Other	Heterotroph—an organism that requires nutrients from other organisms
auto-	Self	Autotroph—an organism that does not need other organisms for food (obtains nutrients from a nonliving source)
photo-	Light	Phototroph—an organism that uses light as an energy source
chemo-	Chemical	Chemotroph—an organism that uses chemicals for energy, rather than light
sapro-	Rotten	Saprobe—an organism that lives on dead organic matter
halo-	Salt	Halophile—an organism that can grow in high-salt environments
thermo-	Heat	Thermophile—an organism that grows best at high temperatures
psychro-	Cold	Psychrophile—an organism that grows best at cold temperatures
aero-	Air (O_2)	Aerobe—an organism that uses oxygen in metabolism

Modifier terms are also used to specify the nature of an organism's adaptations. *Obligate* or *strict* refers to being restricted to a narrow niche or habitat, such as an obligate thermophile that requires high temperatures to grow. By contrast, *facultative* means not being so restricted but being able to adapt to a wider range of metabolic conditions and habitats. A facultative halophile can grow with or without high salt concentration.

Obligate intracellular parasitism is an extreme but relatively common mode of life. Microorganisms that spend all or part of their life cycle inside a host cell include all viruses, a few bacteria (rickettsias, chlamydias), and certain protozoa (apicomplexans). Contrary to what one may think, the cell interior is not completely without hazards, and microbes must overcome some difficult challenges. They must find a way into the cell, keep from being destroyed, not destroy the host cell too soon, multiply, and find a way to infect other cells. Intracellular parasites obtain various substances from the host cell, depending on the group. Viruses are extreme, parasitizing the host's genetic and metabolic machinery. Rickettsias steal energy from their hosts, and the malaria protozoan steals hemoglobin from its host cell.

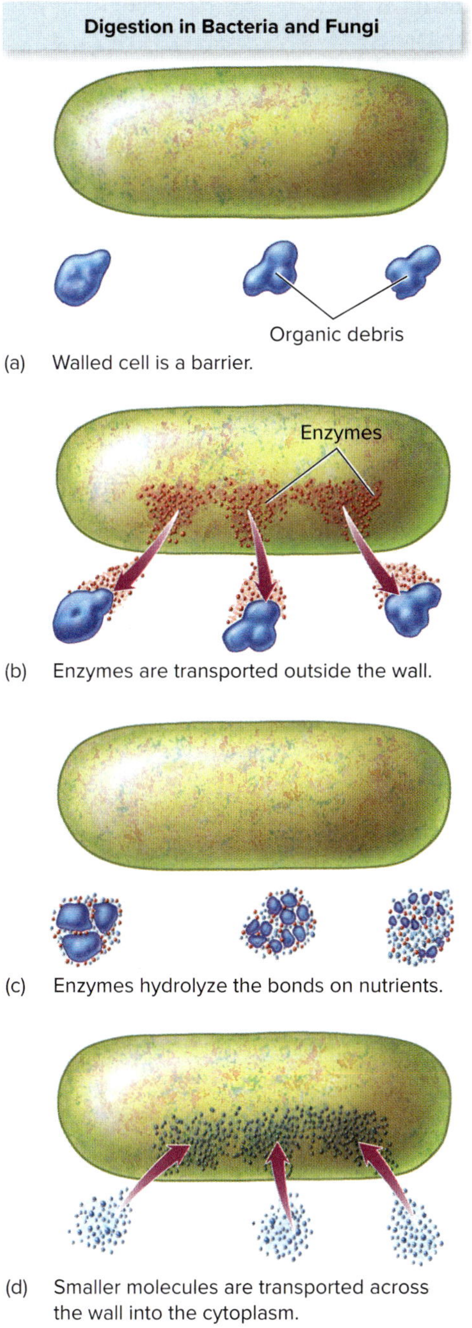
Digestion in Bacteria and Fungi
Organic debris
(a) Walled cell is a barrier.
Enzymes
(b) Enzymes are transported outside the wall.
(c) Enzymes hydrolyze the bonds on nutrients.
(d) Smaller molecules are transported across the wall into the cytoplasm.

Figure 9.2 Extracellular digestion in a saprobe with a cell wall (bacterium or fungus). **(a)** A walled cell is inflexible and cannot engulf large pieces of organic debris. **(b)** In response to a usable substrate, the cell synthesizes enzymes that are transported across the wall into the extracellular environment. **(c)** The enzymes hydrolyze the bonds in the debris molecules. **(d)** Digestion produces molecules small enough to be transported into the cytoplasm.

How Microbes Feed: Nutrient Absorption

A microorganism's habitat has to provide necessary nutrients—some abundant, others scarce—that must still be taken into the microbe. Survival also requires that cells transport waste materials out of the cell (and into the environment). Whatever the direction, transport occurs across the cell membrane, the structure specialized for this role. This is true even in organisms with cell walls (bacteria, algae, and fungi) because the cell wall is usually not a barrier to the entrance or exit of molecules. Before we talk about transport mechanisms, we'll review the basic principles of diffusion.

The Movement of Molecules: Diffusion and Transport

The driving force of transport is natural atomic and molecular movement—the tendency of atoms and molecules to be in constant random motion. The existence of this motion is evident in Brownian movement of particles suspended in liquid. It can also be demonstrated by a variety of simple observations. A drop of perfume released into one part of a room is soon smelled in another part, or a lump of sugar in a cup of tea spreads through the whole cup without stirring. This phenomenon of molecular movement, in which atoms or molecules move in a gradient from an area of higher density or concentration to an area of lower density or concentration, is **diffusion.**

Diffusion of molecules across the cell membrane is largely determined by the concentration gradient and whether the substance can permeate the membrane. But before we talk about movement of nutrients (molecules, solutes) in and out of cells, we will address the movement of water or osmosis. You may want to take a moment to review solutes and solvents in chapter 2.

The Movement of Water: Osmosis

Diffusion of water through a selectively permeable membrane is a process called **osmosis. Figure 9.3** provides a demonstration of how cells deal with various solute concentrations in aqueous solutions. In an osmotic system, the membrane is *selectively,* or *differentially, permeable,* with passageways that allow the free diffusion of water but that block certain other dissolved molecules. When this membrane is placed between solutions of differing concentrations and the solute is larger than the pores (a protein, for example), then under the laws of diffusion, water will diffuse at a faster rate from the side that has more water to the side that has less water. As long as the concentrations of the solutions differ, one side will experience a net loss of water and the other a net gain of water, until equilibrium is reached and the rate of diffusion is equalized.

Osmosis in living systems is similar to the model shown in figure 9.3. Living membranes generally block the entrance and exit of larger molecules and permit free diffusion of water. Because most cells are surrounded by some free water, the amount of water entering or leaving has a far-reaching impact on cellular activities and survival. This osmotic relationship between cells and their environment is determined by the relative concentrations of the solutions on either side of the cytoplasmic membrane **(figure 9.4).** Such systems can be compared using the terms *isotonic, hypotonic,* and *hypertonic.* (The root *-tonic* means "tension"; *iso-* means "the same," *hypo-* means "less," and *hyper-* means "over" or "more.")

Under **isotonic** conditions, the environment is equal in solute concentration to the cell's internal environment. Because diffusion of water proceeds at the same rate in both directions, there is no net

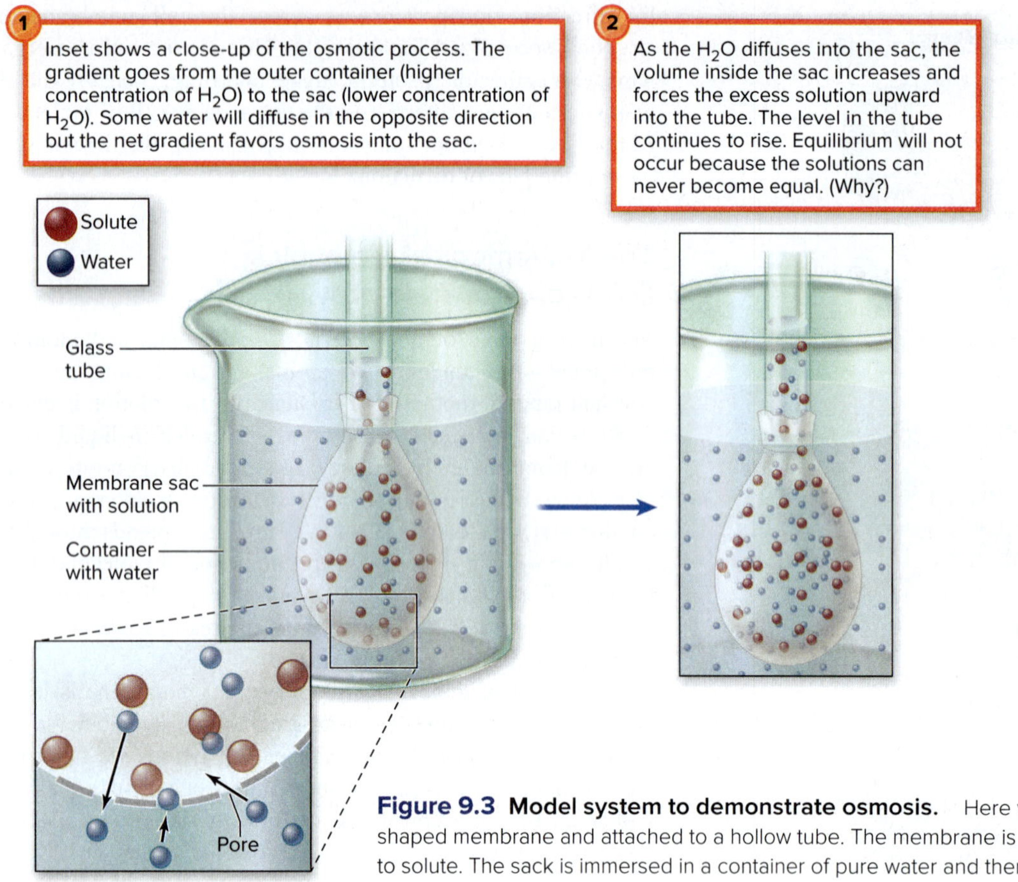
1
Inset shows a close-up of the osmotic process. The gradient goes from the outer container (higher concentration of H_2O) to the sac (lower concentration of H_2O). Some water will diffuse in the opposite direction but the net gradient favors osmosis into the sac.
2
As the H_2O diffuses into the sac, the volume inside the sac increases and forces the excess solution upward into the tube. The level in the tube continues to rise. Equilibrium will not occur because the solutions can never become equal. (Why?)
Solute
Water
Glass tube
Membrane sac with solution
Container with water
Pore

Figure 9.3 Model system to demonstrate osmosis. Here we have a solution enclosed in a sack-shaped membrane and attached to a hollow tube. The membrane is permeable to water (solvent) but not to solute. The sack is immersed in a container of pure water and then observed over time.

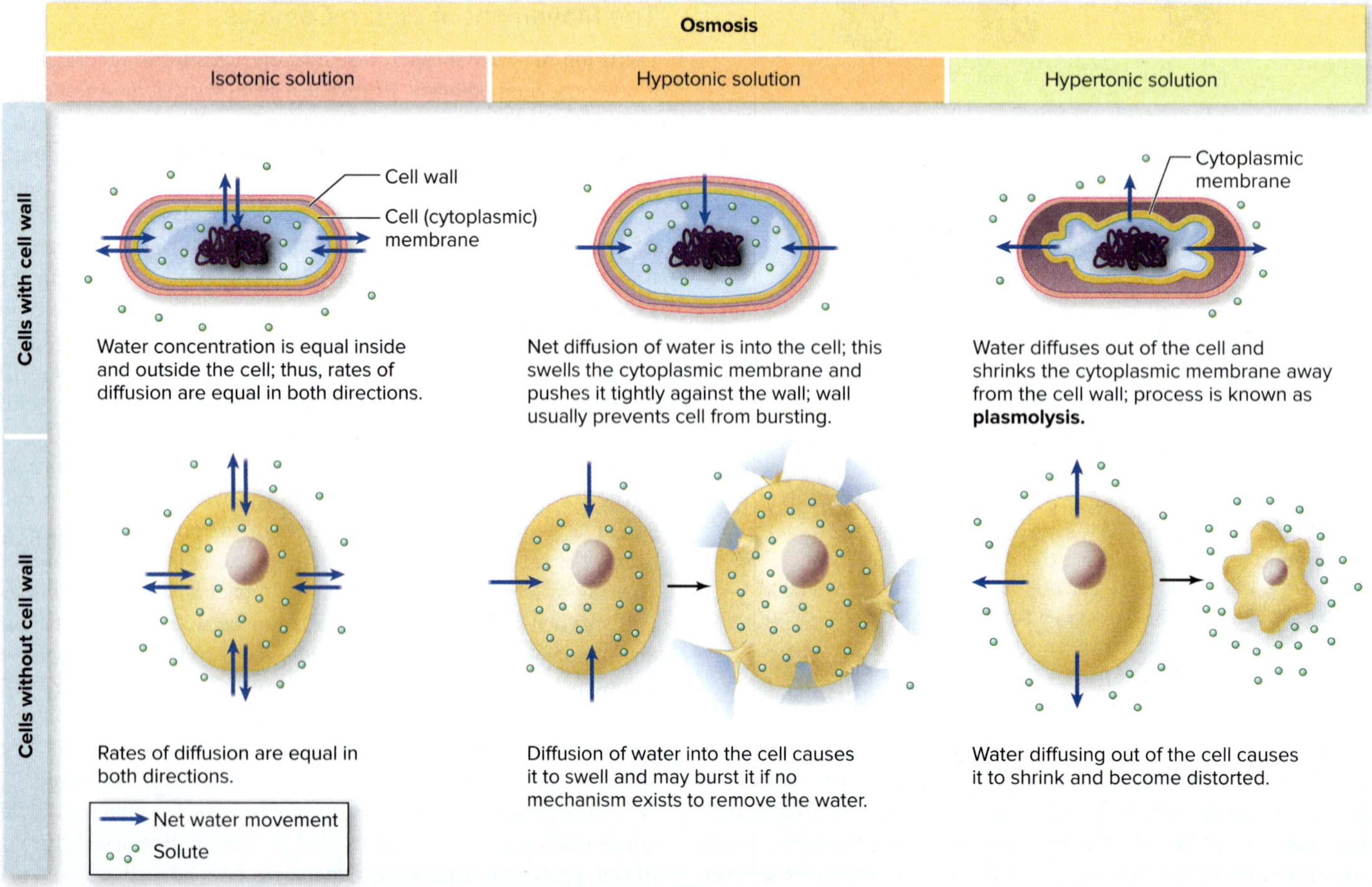
Osmosis
Isotonic solution
Hypotonic solution
Hypertonic solution
Cells with cell wall
Cell wall
Cell (cytoplasmic) membrane
Cytoplasmic membrane
Water concentration is equal inside and outside the cell; thus, rates of diffusion are equal in both directions.
Net diffusion of water is into the cell; this swells the cytoplasmic membrane and pushes it tightly against the wall; wall usually prevents cell from bursting.
Water diffuses out of the cell and shrinks the cytoplasmic membrane away from the cell wall; process is known as plasmolysis.
Cells without cell wall
Rates of diffusion are equal in both directions.
Diffusion of water into the cell causes it to swell and may burst it if no mechanism exists to remove the water.
Water diffusing out of the cell causes it to shrink and become distorted.
Net water movement
Solute

Figure 9.4 Cell responses to solutions of differing osmotic content.

change in cell volume. Isotonic solutions are generally very stable environments for cells because they are already in an osmotic steady state with the cell. Parasites living in host tissues are most likely to be living in isotonic habitats.

Under **hypotonic** conditions, the solute concentration of the external environment is lower than that of the cell's internal environment. The most hypotonic external environment for cells is pure water, because it has no solute. The net direction of osmosis is water moving from the hypotonic solution into the cell, and cells without walls swell and can burst.

A slightly hypotonic environment can be quite favorable for bacterial cells. The constant slight tendency for water to flow into the cell keeps the cytoplasmic membrane fully extended and the cytoplasm full. This is the optimum condition for the many processes occurring in and on the membrane. Slight hypotonicity is tolerated quite well by most bacteria because of their rigid cell walls.

Hypertonic conditions are also out of balance with the tonicity of the cell's cytoplasm, but in this case, the environment has a higher solute concentration than the cytoplasm. Because a hypertonic environment will force water to diffuse out of a cell, it is said to have high *osmotic pressure* or potential. The growth-limiting effect of hypertonic solutions on microbes is the principle behind using concentrated salt and sugar solutions as preservatives for food, such as with salted hams, because the microbes in them will be stopped from growing by the hypertonic state. An important note: It will help you to recall these osmotic conditions if you remember that the prefixes *iso-*, *hypo-*, and *hyper-* are all talking about the environment *outside* of the cell.

Adaptations to Osmotic Variations in the Environment

Now we'll see how specific microbes have adapted osmotically to their environments. In general, isotonic conditions pose little stress on cells, so they mainly need strategies to counteract the adverse effects of hypertonic and hypotonic environments.

A bacterium and an amoeba living in fresh pond water are examples of cells that live in constantly hypotonic conditions. The rate of water diffusing across the cell membrane into the cytoplasm is rapid and constant, and the cells would die without a way to adapt. As just mentioned, the majority of bacterial cells compensate by having a cell wall that protects them from bursting even as the cytoplasmic membrane becomes *turgid* (ter′-jid) from pressure. The amoebas (without a cell wall) adapt by deploying a water, or contractile, vacuole that moves excess water back out into the habitat like a tiny pump.

A microbe living in a high-salt environment (hypertonic) has the opposite problem and must either restrict its loss of water to the environment or increase the salinity of its internal environment. Halobacteria living in the Great Salt Lake and the Dead Sea actually take in salt to make their cells isotonic with the environment. For this reason, they have a physiological need for a high-salt concentration in their habitats.

Transporting Substances Across Membranes

It is absolutely critical that a cell be able to move polar molecules and ions across the membrane. Unfortunately, these chemicals have a greatly decreased permeability, meaning that simple diffusion will not allow this movement. The multiple ways that substances move in and out of cells are summarized in **table 9.4.** Take the time now to walk through that table.

Let's look at an additional fact about facilitated diffusion not mentioned in the table. Facilitated diffusion is also characterized by **saturation.** The rate of transport of a substance is limited by the number of binding sites on the transport proteins. As the substance's concentration increases, so does the rate of transport until the concentration of the transported substance is such that all of the transporters' binding sites are occupied. Then the rate of transport reaches a steady state and cannot move faster despite further increases in the substance's concentration. Another characteristic of these carrier proteins is that they exhibit **competition.** This is when two molecules of similar shape can bind to the same binding site on a carrier protein. The chemical with the higher binding affinity, or the chemical in the higher concentration, will be transported at a greater rate.

Neither simple diffusion nor facilitated diffusion requires energy because molecules are moving down a concentration gradient, from an area of higher concentration to an area of lower concentration.

Alternatively, some important features to note about **active transport** systems are

1. Nutrients are transported against the diffusion gradient or, if in the same direction as the natural gradient, at a rate faster than by diffusion alone;
2. Specific membrane proteins (permeases and pumps) act as carriers; and
3. Energy is required.

Examples of substances transported actively are monosaccharides, amino acids, organic acids, phosphates, and metal ions. Some freshwater algae have such efficient active transport systems that an essential nutrient can be found in intracellular concentrations 200 times that of the habitat. One important type of active transport involves specialized pumps, which can rapidly carry ions such as K^+, Na^+, and H^+ across the membrane. This activity is particularly important in the formation of ATP on the membrane and in protein synthesis. Another type of active transport, **group translocation,** couples the transport of a nutrient with its conversion to a substance that is immediately useful inside the cell.

Endocytosis: Eating and Drinking by Cells

Some eukaryotic cells move large molecules, particles, liquids, or even other cells across the cell membrane. Because the cell usually expends energy to carry out this transport, it is also a form of active transport. The substances transported do not pass physically through the membrane but are carried into the cell by **endocytosis.** First, the cell encloses the substance in its membrane, simultaneously forming a vacuole and engulfing it. Amoebas and certain white blood cells ingest whole cells or large solid matter by a type of endocytosis called **phagocytosis.** Liquids, such as oils or molecules in solution, enter the cell in a similar manner, but in this case it is called **pinocytosis.**

Table 9.4 Transport Processes in Cells

Examples	Description	Energy Requirements
Passive		
Simple diffusion	A fundamental property of atoms and molecules that exist in a state of random motion.	None. Substances move on a gradient from higher concentration to lower concentration. 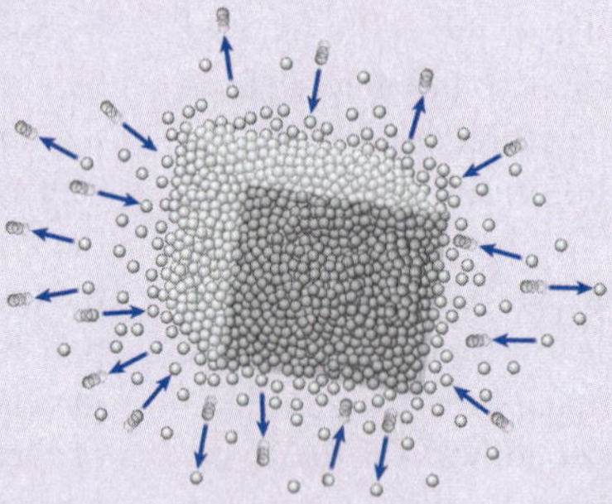
Facilitated diffusion	Molecule binds to a *specific* receptor in membrane and is carried to other side. Molecule-specific (one receptor designed for one molecule). Goes both directions. Rate of transport is limited by the number of binding sites on transport proteins.	None. Substances move on a gradient from higher concentration to lower concentration. 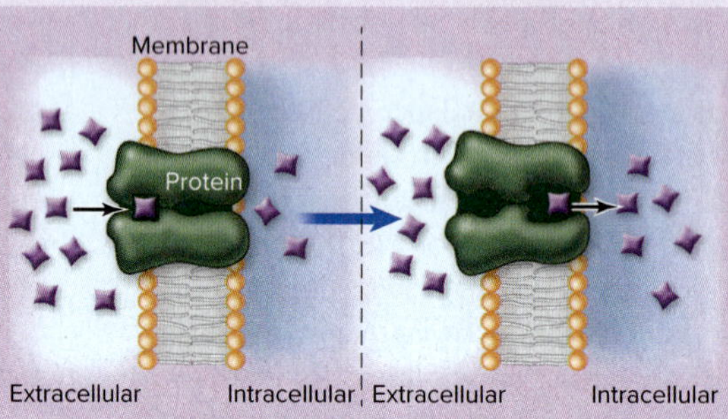Membrane Protein Extracellular Intracellular Extracellular Intracellular
Active		
Carrier-mediated active transport	Atoms or molecules are pumped into or out of the cell by specialized receptors.	Driven by ATP or the proton motive force. 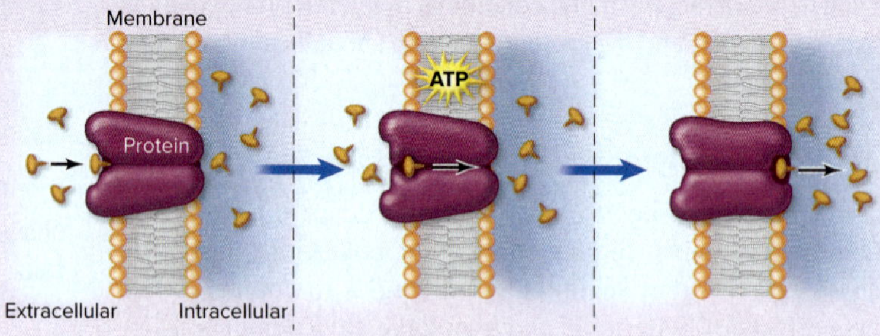Membrane Protein ATP Extracellular Intracellular
Group translocation	Molecule is moved across membrane and simultaneously converted to a metabolically useful substance.	ATP 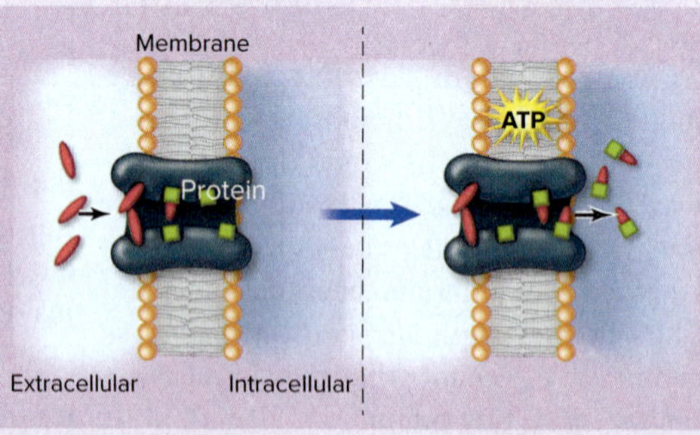Membrane Protein ATP Extracellular Intracellular
Endocytosis (bulk transport)	Mass transport of large particles, cells, and liquids by engulfment and vesicle formation. Specifically, phagocytosis moves solids into cell; pinocytosis moves liquids into cell.	ATP 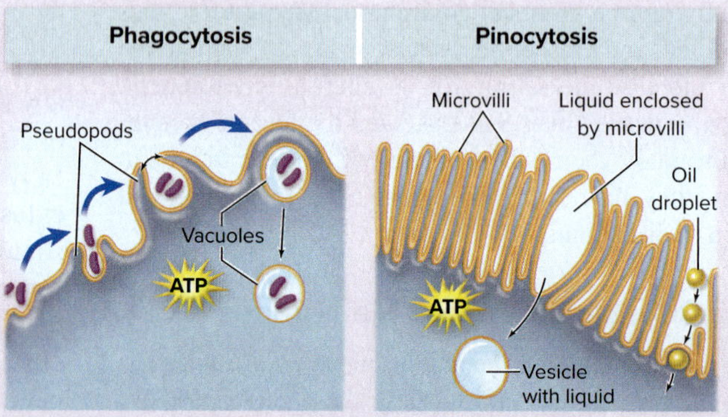Phagocytosis Pinocytosis Pseudopods Vacuoles ATP Microvilli Liquid enclosed by microvilli Oil droplet ATP Vesicle with liquid

9.1 Learning Outcomes—Assess Your Progress

1. List the essential nutrients of a bacterial cell.
2. Differentiate between *macronutrients* and *micronutrients.*
3. List and define four different terms that describe an organism's sources of carbon and energy.
4. Define *saprobe* and *parasite,* and explain why these terms can be an oversimplification.
5. Compare and contrast the processes of diffusion and osmosis.
6. Identify the effects of isotonic, hypotonic, and hypertonic conditions on a cell.
7. Name two types of passive transport and three types of active transport.

9.2 Environmental Factors That Influence Microbes

Microbes are exposed to a wide variety of environmental factors in addition to nutrients. Microbial ecology focuses on ways that microorganisms deal with or adapt to such factors as heat, cold, gases, acid, radiation, osmotic and hydrostatic pressures, and even other microbes. Adaptation involves an adjustment in biochemistry or genetics that enables long-term survival and growth. Incidentally, we should be careful to differentiate between *growth* in a given condition and *tolerance,* which implies survival without growth.

Temperature

Microbial cells are unable to control their internal temperature. Their survival is dependent on adapting to whatever temperature variations are encountered in that habitat. Microbes of different types can survive in a wide range of different temperatures. The range of temperatures for the growth of a given microbial species can be expressed as three *cardinal temperatures.* The **minimum temperature** is the lowest temperature that permits a microbe's continued growth and metabolism; below this temperature, its activities are inhibited. The **maximum temperature** is the highest temperature at which growth and metabolism can proceed. If the temperature rises slightly above maximum, growth will stop, but if it continues to rise beyond that point, the enzymes and nucleic acids will eventually become permanently inactivated (also known as *denaturation*) and the cell will die. This is why heat works so well as an agent in microbial control. The **optimum temperature** covers a small range, somewhere between the minimum and maximum, that promotes the fastest rate of growth and metabolism (rarely is the optimum a single point).

Depending on their natural habitats, some microbes have a narrow cardinal range, others a broad one. Some strict parasites will not grow if the temperature varies more than a few degrees below or above the host's body temperature. For instance, the typhus bacterium multiplies only in the range of 32°C to 38°C, and rhinoviruses (one cause of the common cold) multiply most successfully in tissues that are slightly below normal body temperature (33°C to 35°C). Other organisms are not so limited. Strains of *Staphylococcus aureus* grow within the range of 6°C to 46°C, and the intestinal bacterium *Enterococcus faecalis* grows within the range of 0°C to 44°C. At various times throughout medical history, humans have tried to induce higher temperatures in patients in attempts to cure infections.

We use particular terms to describe whether an organism grows optimally in a cold, moderate, or hot temperature range. The terms used for these ecological groups are *psychrophile, mesophile,* and *thermophile,* respectively **(figure 9.5).**

A **psychrophile** (sy′-kroh-fyl) is a microorganism that has an optimum temperature below 15°C and is capable of growth at 0°C. It is obligate with respect to cold, meaning it

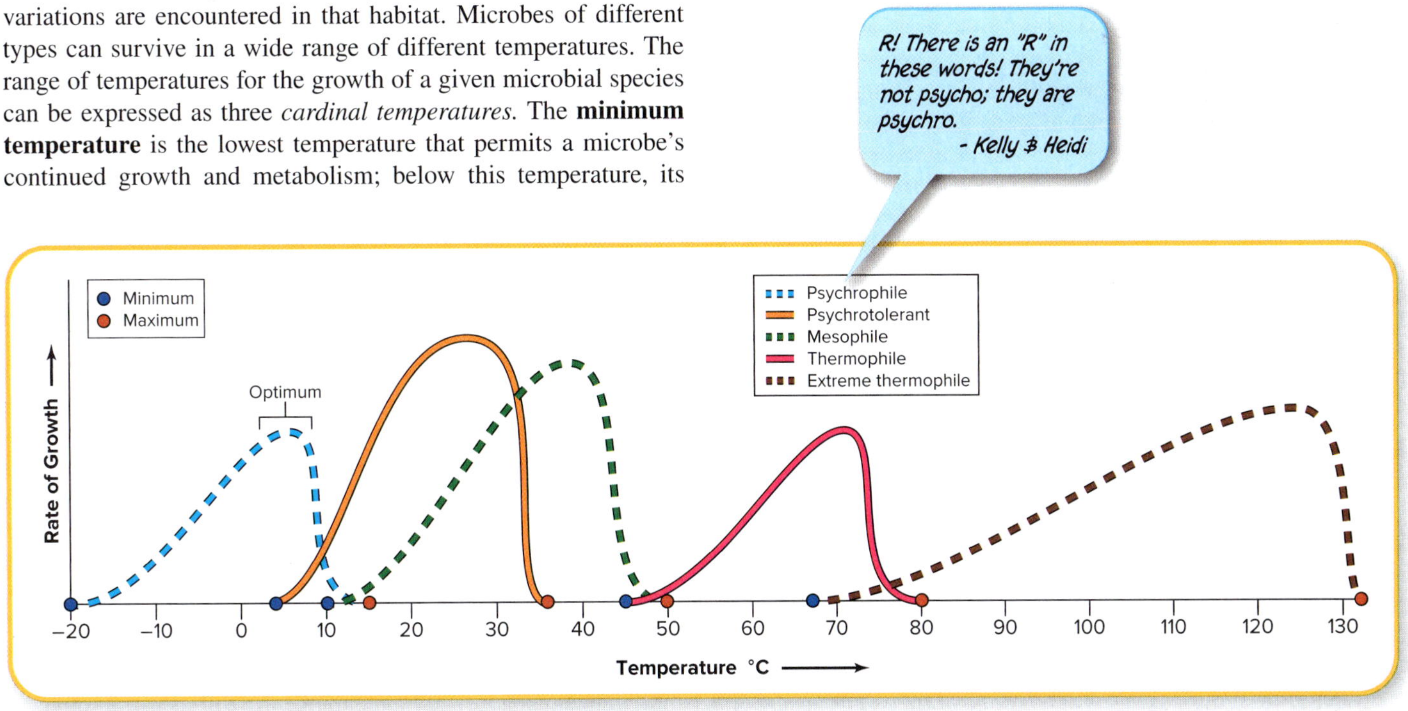
R! There is an "R" in these words! They're not psycho; they are psychro.
- Kelly $ Heidi
Minimum
Maximum
Optimum
Rate of Growth
Psychrophile
Psychrotolerant
Mesophile
Thermophile
Extreme thermophile
–20
–10
0
10
20
30
40
50
60
70
80
90
100
110
120
130
Temperature °C

Figure 9.5 Ecological groups by temperature of adaptation. Psychrophiles can grow at or near 0°C and have an optimum below 15°C. Psychrotolerant microbes have an optimum of from 15°C to 30°C. As a group, mesophiles can grow between 10°C and 50°C, but their optima usually fall between 20°C and 40°C. Generally speaking, thermophiles require temperatures above 45°C and grow optimally between this temperature and 80°C. Extreme thermophiles have optima above 80°C. Note that the ranges can overlap to an extent.

(a)
(b)

Figure 9.6 Red snow. **(a)** An early summer snowbank provides a perfect habitat for psychrophilic photosynthetic organisms like *Chlamydomonas nivalis.* **(b)** Microscopic view of this snow alga (actually classified as a "green" alga although a red pigment dominates at this stage of its life cycle).
(a) Francois Gohier/Science Source; (b) Nozomu Takeuchi

generally cannot grow above 20°C. Laboratory work with true psychrophiles can be a real challenge. Manipulations have to be done in a cold room because room temperature can be lethal to the organisms. Unlike most laboratory cultures, storage in the refrigerator incubates, rather than inhibits, them. As one may predict, the habitats of psychrophilic bacteria, fungi, and algae are lakes and rivers, snowfields **(figure 9.6),** polar ice, and the deep ocean. Rarely, if ever, are they pathogenic. A less extreme group of cold-loving bacteria, called *psychrotolerant,* grow slowly in cold but have an optimum temperature between 15°C and 30°C.

The majority of medically significant microorganisms are **mesophiles** (mez′-oh-fylz), organisms that grow at intermediate temperatures. Although an individual species can grow at the extremes of 10°C or 50°C, the optimum growth temperatures (optima) of most mesophiles fall into the range of 20°C to 40°C. Organisms in this group inhabit animals and plants as well as soil and water in temperate, subtropical, and tropical regions. Most human pathogens have optima somewhere between 30°C and 40°C (human body temperature is 37°C). Some mesophilic bacteria, such as *Staphylococcus aureus,* grow optimally at body temperature but are also psychrotolerant, meaning they can survive and multiply slowly at refrigerator temperatures, causing concern for food storage. *Listeria monocytogenes* is a human pathogen that is psychrotolerant and often grows in ice cream and refrigerated meat. *Thermoduric* microbes, which can survive short exposure to high temperatures but are normally mesophiles, are common contaminants of heated or pasteurized foods. Examples include protozoa that produce heat-resistant cysts such as *Giardia*, or sporeformers such as *Bacillus* and *Clostridium.*

A **thermophile** (thur′-moh-fyl) is a microbe that grows optimally at temperatures greater than 45°C. Such heat-loving microbes live in soil and water associated with volcanic activity, in compost piles, and in habitats directly exposed to the sun. Thermophiles vary in heat requirements, with a general range of growth of 45°C to 80°C. Most eukaryotic microbes cannot survive above 60°C, but a few thermophilic bacteria, called extreme thermophiles, grow between 80°C and 121°C (currently thought to be the temperature limit because of enzymes and cell structures). Extreme thermophiles are so heat tolerant that autoclaves can be used to isolate them in culture.

Gases

The atmospheric gases that most influence microbial growth are O_2 and CO_2. Of these, oxygen gas has the greatest impact on microbial growth. Not only is it an important respiratory gas, but it is also a powerful oxidizing agent that can exist in many toxic forms. In general, microbes fall into one of three categories:

- those that use oxygen and can detoxify it,
- those that can neither use oxygen nor detoxify it, and
- those that do not use oxygen but can detoxify it.

How Microbes Process Oxygen

As oxygen enters into cellular reactions, it is transformed into several toxic products. Singlet oxygen (written either as (O) or as $\dot{O}_2$) is an extremely reactive molecule produced by both living and nonliving processes. Notably, it is one of the substances produced by phagocytes (cells of our immune system) to kill invading bacteria. The buildup of singlet oxygen and the oxidation of membrane lipids and other molecules can damage and destroy a cell. The highly reactive superoxide ion (O_2^-), hydrogen peroxide (H_2O_2), and hydroxyl radicals (OH^-) are other destructive by-products of oxygen. To protect themselves against damage, most cells have developed enzymes whose job it is to scavenge and neutralize these chemicals. The complete conversion of superoxide ion into harmless oxygen requires a two-step process and at least two enzymes:

$$\text{Step 1.}\quad 2O_2^- + 2H^+ \xrightarrow{\text{superoxide dismutase}} H_2O_2 \text{ (hydrogen peroxide)} + O_2$$

$$\text{Step 2.}\quad 2H_2O_2 \xrightarrow{\text{catalase}} 2H_2O + O_2$$

In this series of reactions (essential for aerobic organisms), the superoxide ion is first converted to hydrogen peroxide and normal atmospheric oxygen by the action of an enzyme called superoxide dismutase. Hydrogen peroxide is also toxic to cells (which is why it is used as a disinfectant and antiseptic). It must be degraded by the enzyme catalase into water and oxygen. If a microbe is not capable of dealing with toxic oxygen by these or similar mechanisms, it is forced to live in habitats free of oxygen.

There are several different named categories of oxygen requirements. An **aerobe** (air′-ohb) (aerobic organism) can use gaseous oxygen in its metabolism and possesses the enzymes needed to process toxic oxygen products. An organism that must have oxygen is an **obligate aerobe.** Most fungi and protozoa, as well as many bacteria (and humans!), are obligate aerobes.

A **facultative anaerobe** is a microbe that does not require oxygen for its metabolism and is capable of growth in the absence of it. This type of organism metabolizes by aerobic respiration when oxygen is present. In its absence, it can toggle to an anaerobic mode of metabolism such as fermentation. Facultative anaerobes usually have both catalase and superoxide dismutase. A large number of bacterial pathogens fall into this group (e.g., gram-negative intestinal bacteria and staphylococci). A **microaerophile** (myk″-roh-air′-oh-fyl) does not grow at normal atmospheric concentrations of oxygen but requires a small amount of it in metabolism. Most organisms in this category live in a habitat (soil, water, or the human body) that provides small amounts of oxygen but is not directly exposed to the atmosphere.

An **anaerobe** (anaerobic microorganism) lacks the metabolic enzyme systems for using oxygen in respiration. Because **obligate anaerobes** (also known as strict anaerobes) lack the enzymes for processing toxic oxygen, they cannot tolerate any free oxygen in the immediate environment and will die if exposed to it. Obligate anaerobes live in habitats such as deep muds, lakes, oceans, and soil. Even though human cells use oxygen and oxygen is found in the blood and tissues, some body sites provide anaerobic pockets or microhabitats where colonization or infection can occur. One region that is an important site for anaerobic infections is the oral cavity. The pockets between the gums and the teeth can be anaerobic at the bottom of them, and this fosters the growth of microbes that cause gingivitis and periodontal disease. Another common site for anaerobic infections is the large intestine, a relatively oxygen-free habitat that harbors a rich assortment of strictly anaerobic bacteria. Growing anaerobic bacteria usually requires special media, methods of incubation, and handling chambers that exclude oxygen **(figure 9.7).**

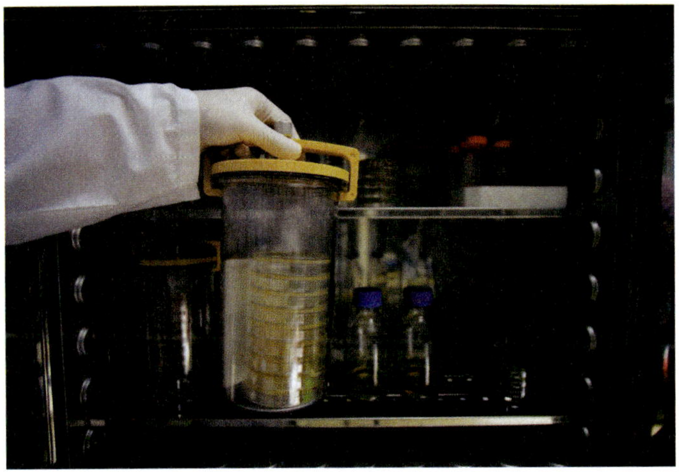

(a)

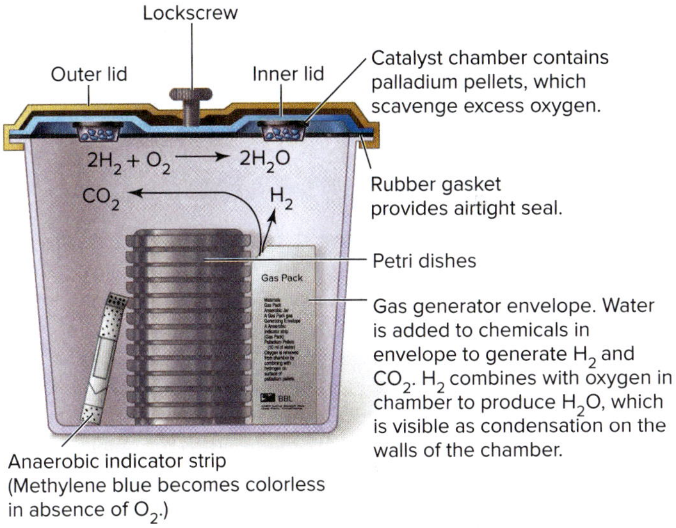
Lockscrew
Outer lid
Inner lid
Catalyst chamber contains palladium pellets, which scavenge excess oxygen.
$2H_2 + O_2 \longrightarrow 2H_2O$
CO_2
H_2
Rubber gasket provides airtight seal.
Petri dishes
Gas Pack
Gas generator envelope. Water is added to chemicals in envelope to generate H_2 and CO_2. H_2 combines with oxygen in chamber to produce H_2O, which is visible as condensation on the walls of the chamber.
Anaerobic indicator strip (Methylene blue becomes colorless in absence of O_2.)

(b)

Figure 9.7 Culturing techniques for anaerobes. **(a)** A special anaerobic environmental chamber makes it possible to handle strict anaerobes without exposing them to air. It has provisions for incubation and inspection in a completely O_2-free system. **(b)** A simpler anaerobic, or CO_2, incubator system. To create an anaerobic environment, a packet is activated to produce hydrogen gas and the chamber is sealed tightly. The gas reacts with available oxygen to produce water. Carbon dioxide can also be added to the system for growth of organisms needing high concentrations of it.

(a) Khamkhlai Thanet/Shutterstock

Disease Connection

Tetanus, also known as lockjaw, is caused by an anaerobic bacterium, *Clostridium tetani.* You are aware that you may be at risk for this infection through puncture wounds such as stepping on a nail. This reflects the oxygen requirements of the bacterium. Deep puncture wounds contain limited oxygen at the "bottom" of the wound and often are not perfused with a lot of blood (another source of oxygen). In that situation, *Clostridium* endospores can return to a growing stage and multiply.

Aerotolerant anaerobes do not utilize oxygen but can survive and grow to a limited extent in its presence. These anaerobes are not harmed by oxygen, mainly because they possess alternate mechanisms for breaking down peroxides and superoxide. Certain lactobacilli and streptococci use manganese ions or peroxidases to perform this task.

Determining the oxygen requirements of a microbe from a biochemical standpoint can be a very time-consuming process. One way to do it is to grow the microbe using reducing media (those that contain an oxygen-absorbing chemical). **Figure 9.8** depicts the growth of different microbes in tubes of fluid thioglycollate. The location of growth indicates the oxygen requirements of the microbes.

Insight 9.1 tells the tale of how microbes that were able to produce oxygen via photosynthesis were responsible for turning our planet from an anaerobic one to an aerobic one.

Now let's talk CO_2. Although all microbes require some carbon dioxide in their metabolism, *capnophiles* grow best at a

KIMAX
KIMAX

Figure 9.8 Four tubes showing three different patterns of oxygen utilization. These tubes use thioglycollate, which reduces oxygen to water, to restrict oxygen diffusion through the agar. So, whereas there is an oxygen-rich layer at the top of the agar, the oxygen concentration rapidly decreases deeper in the agar. In tube 1, the obligately aerobic *Pseudomonas aeruginosa* grows only at the very top of the agar. Tubes 2 and 3 contain two different examples of facultatively anaerobic bacteria. Many facultatives, although able to grow both aerobically and anaerobically, grow more efficiently in the aerobic mode. This is more obvious in tube 2 (*Staphylococcus aureus*) and less obvious in tube 3 (*Escherichia coli*). Tube 4 contains *Clostridium butyricum*, an obligate anaerobe.
Terese Barta

higher CO_2 tension than is normally present in the atmosphere. This becomes important in the isolation of some pathogens from clinical specimens, notably *Neisseria* (gonorrhea, meningitis), *Brucella* (undulant fever), and *Streptococcus pneumoniae.* For these bacteria, incubation is carried out in a CO_2 incubator that provides 3% to 10% CO_2 **(figure 9.7*b*).**

pH

Microbial growth and survival are also influenced by the pH of the habitat. The term *pH* was defined in chapter 2 as the degree of acidity or alkalinity (basicity) of a solution. It is expressed by the pH scale, a series of numbers ranging from 0 to 14. The pH of pure water (7.0) is neutral, neither acidic nor basic. As the pH value decreases toward 0, the acidity increases. As the pH increases toward 14, the alkalinity increases. Most organisms we are familiar with live or grow in habitats between pH 6 and 8 because strong acids and bases can be highly damaging to enzymes and other cellular substances.

Many other microorganisms live at pH extremes. Examples of obligate *acidophiles* are *Euglena mutabilis,* an alga that grows in acid pools between 0 and 1.0 pH, and *Thermoplasma,* an archaea that lacks a cell wall, lives in hot coal piles at a pH of 1 to 2, and will lyse if exposed to pH 7. Alkalinophiles, such as *Natronomonas* species, live in hot pools and soils that contain high levels of basic minerals (up to pH 12.0). Bacteria that decompose urine create alkaline conditions because ammonium (NH_4^+) can be produced when urea (a component of urine) is digested. (Ammonium raises the pH of solutions.) Metabolism of urea is one way that *Proteus* spp. can neutralize the acidity of the urine to colonize and infect the urinary system.

Osmotic Pressure

Although most microbes exist under hypotonic or isotonic conditions, a few, called **osmophiles,** live in habitats with a high solute concentration. One common type of osmophile prefers high concentrations of salt; these organisms are called **halophiles** (hay′-loh-fylz). Obligate halophiles such as *Halobacterium* and *Halococcus* inhabit salt lakes, ponds, and other hypersaline habitats. These archaea have significant modifications in their cell walls and membranes and will lyse in hypotonic habitats. *Facultative halophiles* are remarkably resistant to salt, even though they do not normally reside in high-salt environments. For example, *Staphylococcus aureus* can grow on NaCl media ranging from 0.1% up to 20%. Although it is common to use high concentrations of salt and sugar to preserve food (jellies, syrups, and brines), many bacteria and fungi actually thrive under these conditions and are common spoilage agents.

Radiation, Hydrostatic Pressure, and Moisture

Various forms of electromagnetic radiation (ultraviolet, infrared, visible light) stream constantly onto the earth from the sun. Some microbes (phototrophs) can use visible light rays as an energy source, but nonphotosynthetic microbes tend to be damaged or destroyed. Some microbial species produce yellow carotenoid pigments to protect against the damaging effects of light by absorbing and dismantling toxic oxygen. Other types of radiation that can damage microbes are ultraviolet and ionizing rays (X rays and cosmic rays). In the opening case study for this chapter, you read about bacteria that can withstand intense radiation in space. These types of radiation can be applied to control microbial growth.

The ocean depths subject organisms to various degrees of hydrostatic pressure. Deep-sea microbes called **barophiles** exist under pressures that range from a few times to over 1,000 times the pressure of our atmosphere. These bacteria are so strictly adapted to high pressures that they will rupture when exposed to normal atmospheric pressure.

All cells require water from their environment to sustain growth and metabolism. Water is the solvent for cell chemicals, and it is needed for enzyme function and digestion of macromolecules.

INSIGHT 9.1 MICROBIOME: The Great Oxidation Event and Earth's Microbiome

We have talked about how eukaryotic cells arose for the first time on earth. We also know that archaea and bacteria were the first forms of life on the planet and that they arose at a time when the planet had virtually no oxygen. How did the very first cell form, and where did the oxygen come from? It was not until what is called the *Great Oxidation Event* (GOE) occurred that life, as we know it, was possible on the earth. What is the Great Oxidation Event, and why is it significant? Until that time, atmospheric oxygen in the form of O_2 was very low—probably less than 1/10,000 of present levels. Previous to the GOE, the atmosphere contained significant amounts of methane, a greenhouse gas, which kept the earth's temperatures at high levels. The production of oxygen was a significant event in the evolution of complex life on earth. Until O_2 first accumulated in the atmosphere, the earth was too hot, too full of methane, and too acidic for life to exist.

An early archaeon probably formed by gathering a membrane around some nucleic acids and some important proteins that were part of the primordial soup, or part of the last common ancestor. This archaeon was capable of living in an environment without oxygen, and generated energy using hydrogen gas from deep thermal vents in the ocean. These types of archaea and bacteria were the only living cells present on earth until about 2.8 billion years ago (so for over 1 billion years). At that time an early form of a bacterium called *Cyanobacterium* gained the ability to produce oxygen by the process of photosynthesis.

This process eventually led to an oxygenic environment. Anaerobic organisms retreated to habitats without oxygen, and the new oxygenic conditions allowed for the multiplication of many new types of organisms, eventually leading to the rise of eukaryotic cells and the explosion of multicellular life.

Restoration Center & Damage Assessment and Restoration Program/NOAA

Geobiologists at the University of Alberta have studied oxidation of iron pyrite (FeS_2, also known as fool's gold) into iron oxide, which released acid that dissolved rocks into chromium and other metals in ancient sea beds. Their data show that chromium levels in the sea beds increased significantly about 2.48 billion years ago, about 100 million years earlier than was previously thought. These **acidophilic** microbes thrived in the extremely low pH of the ocean waters of ancient earth. The scientists found evidence of these same bacterial life forms living off of pyrite in the highly acidic wastewater of mining sites (pictured). These organisms are the most ancient oxygen producers known and may be the driving force behind the atmosphere.

A certain amount of water on the external surface of the cell is required for the diffusion of nutrients and wastes. Even in apparently dry habitats, sand or dry soil particles can retain a thin layer of water usable by microorganisms. Only dormant, dehydrated cell stages (e.g., spores and cysts) tolerate extreme drying because their enzymes go inactive.

Other Organisms

Up to now, we have considered the importance of nonliving environmental influences on the growth of microorganisms. Another profound influence comes from other organisms that share (or sometimes *are*) their habitats. In all but the rarest instances, microbes live in shared habitats, which give rise to complex and fascinating associations. Some associations are between similar or dissimilar types of microbes. Others involve multicellular organisms such as animals or plants. Interactions can have beneficial, harmful, or no particular effects on the organisms involved. They can be obligatory or nonobligatory to the members, and they often involve nutritional interactions. This outline provides an overview of the major types of microbial associations:

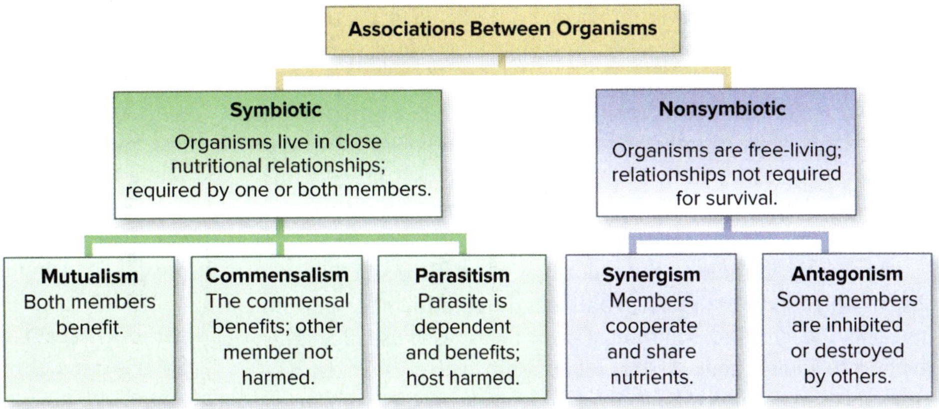
Associations Between Organisms
Symbiotic
Organisms live in close nutritional relationships; required by one or both members.
Nonsymbiotic
Organisms are free-living; relationships not required for survival.
Mutualism
Both members benefit.
Commensalism
The commensal benefits; other member not harmed.
Parasitism
Parasite is dependent and benefits; host harmed.
Synergism
Members cooperate and share nutrients.
Antagonism
Some members are inhibited or destroyed by others.

Figure 9.9 Mutualism. The small remora fish attaches itself to sharks and large fish, like the goliath grouper pictured here, and eats food particles (and probably feces) associated with larger fish. The remora receives nutrition; and the larger fish is cleaned of parasites and debris.
Rich Carey/Shutterstock

A general term used to denote a situation in which two organisms live together in a close partnership is **symbiosis,** and the members are termed *symbionts.* In everyday language, the term "symbiosis" is almost always used in a positive sense, as in, "Those roommates have a symbiotic relationship." In biology, symbiotic relationships can be either positive or negative. The term simply means a close partnership. Three main types of symbiosis occur. **Mutualism** exists when organisms live in an obligatory and mutually beneficial relationship. This association is rather common in nature because of the survival value it has for the members involved **(figure 9.9).** Many microbes that live in or on humans fall in this category. The microbes receive necessary nutrients and the host gets a variety of benefits, ranging from protection from pathogenic microbes to the healthy development of the immune system that is possible only when a robust resident microbiome is present.

In other symbiotic relationships, the relationship tends to be unequal, meaning it benefits one member and not the other. In a relationship known as **commensalism,** the member called the commensal receives benefits, while its coinhabitant is neither harmed nor benefited. As an example, some microbes can break down a substance that would be toxic or inhibitory to another microbe. Until very recently, most of the microbes living at peace with the human body were considered to be in a commensal relationship, with the microbes receiving the benefit of habitat and nutrition, while the human was indifferent to their presence. With the human microbiome project producing extensive research that demonstrates how critical our microbes are to our development, health, and welfare, we should adjust our view that most normal biota are commensals, to instead recognize that they are in a mutualistic relationship with us.

Earlier we introduced the concept of **parasitism** as a relationship in which the host organism provides the parasitic microbe with nutrients and a habitat. Multiplication of the parasite usually harms the host to some extent. As this relationship evolves, the host species may even develop tolerance for or dependence on a parasite, at which point we call the relationship commensalism or mutualism.

Antagonism is a nonsymbiotic association between free-living species that arises when members of a community compete. One example of this is when one microbe secretes chemical substances into the surrounding environment that inhibit or destroy another microbe in the same habitat. The first microbe may gain a competitive advantage by increasing the space and nutrients available to it. Interactions of this type are common in the soil, where mixed communities often compete for space and food. *Antibiosis*—the production of inhibitory compounds such as antibiotics—is actually a form of antagonism. Hundreds of naturally occurring antibiotics have been isolated from bacteria and fungi and used as drugs to control diseases.

Antagonism is also common in the human gut, where the establishment of one species or another depends on the availability of nutrients, which depends on what other microbes are already there. The exact composition of the gut microbiome is a dynamic situation with members coming, going, or becoming deeply established.

Synergism is an interrelationship between two or more free-living organisms that benefits them but is not necessary for their survival. Together, the participants cooperate to produce a result that none of them could do alone. Biofilms are the best examples of synergism.

With respect to human infections, synergistic infections can produce tissue damage that a single organism would not cause alone. Gum disease, dental caries, and gas gangrene involve mixed infections by bacteria interacting synergistically. More and more mixed infections are coming to light as we increasingly use genomic methods for identifying colonizers.

Biofilms: The Epitome of Synergy

The National Institutes of Health estimate that 80% of chronic infections are caused by biofilms **(figure 9.10).** These include chronic ear infections, prostate infections, lung infections in cystic fibrosis patients, and wound infections, as well as many others. Ordinary antibiotic treatment does not work against most biofilms (which is why the infections remain chronic).

A Note About Coevolution

Organisms that have close, ongoing relationships with each other participate in **coevolution,** the process whereby a change in one of the partners leads to a change in the other partner, which may in turn lead to another change in the first partner, and so on. This is another example of the interconnectedness of biological entities on this planet. There are many well-documented examples of the relationships between plants and insects. One of the earliest is the discovery by Charles Darwin of a plant that had a nectar tube that was 10 inches long. Knowing that the plant depended on insects for pollination, Darwin predicted the existence of an insect with a 10-inch tongue—and 41 years later one was discovered.

The plant and the insect had influenced each other's evolution over time. Mutualistic gut bacteria are considered to have coevolved with their mammalian hosts, with the hosts evolving mechanisms to prevent the disease effects of their bacterial passengers, and the bacteria evolving mechanisms to not only be less pathogenic to their hosts, but also to provide important benefits to them.

(a)

Figure 9.10 Biofilms. **(a)** SEM of a biofilm formed on a gauze bandage. The blue bacteria are methicillin-resistant *Staphylococcus aureus* (MRSA); the orange substance is the polymeric substance; and the green tubes are fibers of the gauze. **(b)** Steps in the formation of a biofilm.

(a) Ellen Swogger and Garth James/Center for Biofilm Engineering

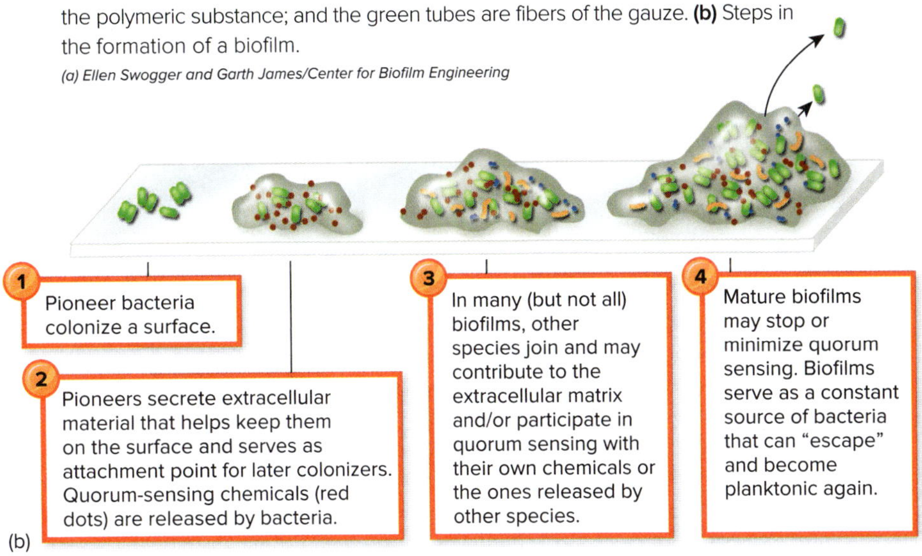
1 Pioneer bacteria colonize a surface.
2 Pioneers secrete extracellular material that helps keep them on the surface and serves as attachment point for later colonizers. Quorum-sensing chemicals (red dots) are released by bacteria.
3 In many (but not all) biofilms, other species join and may contribute to the extracellular matrix and/or participate in quorum sensing with their own chemicals or the ones released by other species.
4 Mature biofilms may stop or minimize quorum sensing. Biofilms serve as a constant source of bacteria that can "escape" and become planktonic again.

(b)

Most biofilms are mixed communities of different kinds of bacteria and other microbes. Usually there is a "pioneer" colonizer, a bacterium that initially attaches to a surface, such as a tooth or the lung tissue. Other microbes then attach either to those bacteria or to the polymeric sugar and protein substance that inevitably is secreted by microbial colonizers of surfaces. In many cases, once the cells are attached, they are stimulated to release chemicals that accumulate as the cell population grows. In this way, they can monitor the size of their own population. This is a process called **quorum sensing.** Bacteria can use quorum sensing to interact with other members of the same species as well as members of other species that are close by (in a biofilm, for example). Eventually, large complex communities are formed, in which physical and biological characteristics vary in different locations of the community. The very bottom of a biofilm may have very different pH and oxygen conditions than the surface of a biofilm, for example. It is now clearly established that microbes in a biofilm, as opposed to those in a planktonic (free-floating) state, behave and respond very differently to their environments. Different genes are even utilized in the two situations. The same chemicals that cells secrete during quorum sensing are responsible for some of this change in gene expression. At any rate, a single biofilm is usually a partnership among multiple microbial inhabitants and thus cannot be eradicated by traditional methods targeting individual infections. This kind of synergism has led to the necessity of rethinking how microbes progress from colonization to the development of true disease.

Biofilms are so prevalent that they dominate the structure of most natural environments on earth. This tendency of microbes to form biofilm communities is an ancient and effective adaptive strategy. Not only do biofilms favor microbial survival in diverse habitats, but they also offer greater access to life-sustaining conditions for the microbes.

Biofilms are known to be a rich ground for genetic transfers among neighboring cells. As our knowledge of biofilm formation and quorum sensing grows, it will likely lead to greater understanding of their involvement in infections and their contributions to disinfectant and drug resistance. It may also be the key to new drugs that successfully target these biological formations.

9.2 Learning Outcomes—Assess Your Progress

8. List and define five terms used to express the temperature-related growth capabilities of microbes.
9. Summarize three ways in which microorganisms function in the presence of oxygen.
10. Identify three other physical factors that microbes must contend with in the environment.
11. List and describe the five major types of microbial association.
12. Discuss characteristics of biofilms that differentiate them from free-living bacteria and their infections.

9.3 The Study of Microbial Growth

When microbes are provided with nutrients and the required environmental factors, they become metabolically active and grow. Growth takes place on two levels. On one level, a cell synthesizes new cell components and increases its size. On the other level, the number of cells in the population increases. This capacity for multiplication, increasing the size of the population by cell division, has tremendous importance in microbial control, infectious disease, and biotechnology. Microorganisms can reproduce in many different ways. In chapter 5, you learned how eukaryotic microbes grow. Here we focus on bacterial growth. Although bacteria can be found that grow by budding and by forming hyphae (similar to fungi), the majority of bacteria reproduce by a process called *binary fission.* But be sure to take a look at the orange "Note About Bacterial Reproduction" box.

A Note About Bacterial Reproduction—and the "Culture Bias"

By far, most of the bacteria that we know about reproduce via binary fission, as described in this chapter. But there are important exceptions. In recent years, researchers have discovered bacteria that produce multiple offspring within their cytoplasm and then split open to release multiple new bacteria (killing the mother cell). One example is *Epulopiscium*, a symbiont of surgeon fish. Most of these bacteria have never been cultured but have been studied by dissecting the animals they colonize. The long-standing belief that bacteria always multiply by binary fission is another by-product of the "culture bias"—meaning that we understand most about the bacteria that we were able to cultivate in the lab, even though there are many more bacteria that exist in the biosphere that have not yet been cultivated.

The Basis of Population Growth: Binary Fission

The division of a bacterial cell occurs mainly through **binary fission.** This means that one cell becomes two. During binary fission, the parent cell enlarges, duplicates its chromosome, and then starts to pull its cell envelope together in the center of the cell using a band of protein that is made up of proteins that resemble actin and tubulin—the protein components of microtubules in eukaryotic cells. The cell wall eventually forms a complete central septum. This process divides the cell into two daughter cells. This process is repeated at intervals by each new daughter cell in turn, and with each successive round of division, the population increases. The stages in this continuous process are shown in greater detail in **figures 9.11** and **9.12.**

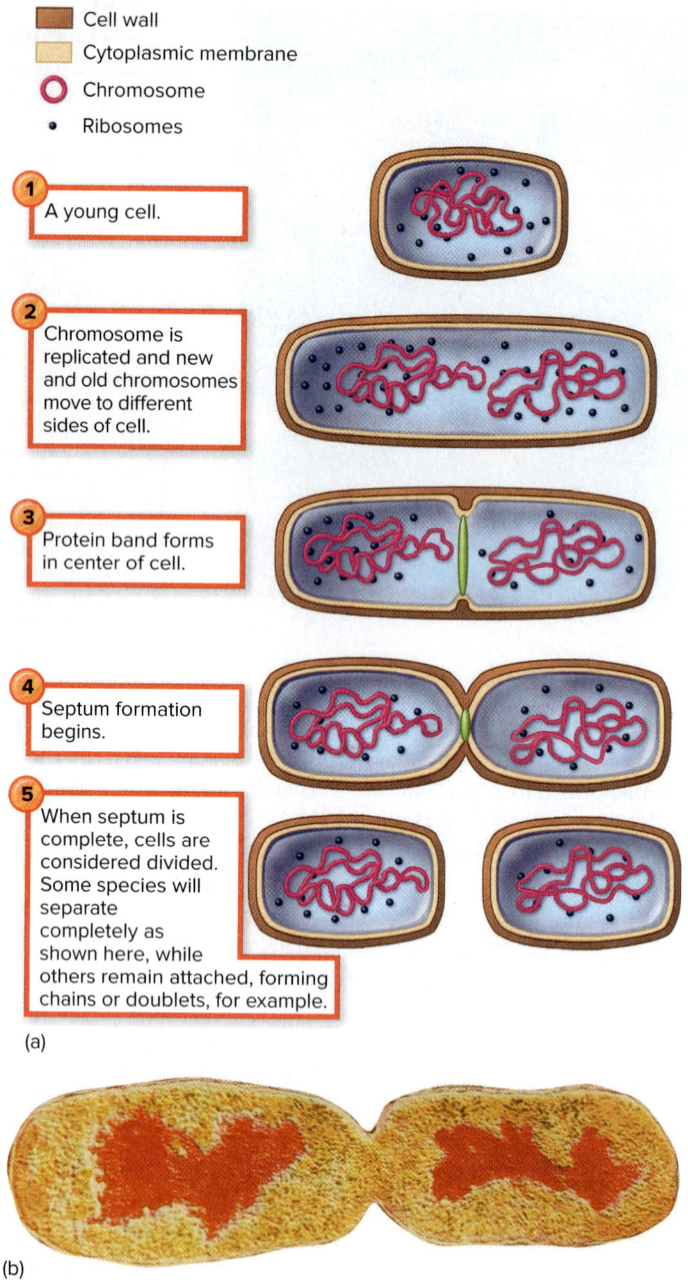
Cell wall
Cytoplasmic membrane
Chromosome
Ribosomes
1 A young cell.
2 Chromosome is replicated and new and old chromosomes move to different sides of cell.
3 Protein band forms in center of cell.
4 Septum formation begins.
5 When septum is complete, cells are considered divided. Some species will separate completely as shown here, while others remain attached, forming chains or doublets, for example.
(a)
(b)

Figure 9.11 Steps in binary fission of a rod-shaped bacterium.(a) Illustration of the process. **(b)** Micrograph of cells in the fourth stage of the process.

(b) sciencepics/Shutterstock

The Rate of Population Growth

The time required for a complete fission cycle—from parent cell to two new daughter cells—is called the **generation,** or **doubling, time.** The term *generation* has a similar meaning as it does in humans—the period between an individual's birth and the time of producing offspring. In bacteria, each new fission cycle or generation increases the population by a factor of 2, or doubles it. So, the initial parent stage consists of 1 cell, the first generation consists of 2 cells, the second 4, the third 8, then 16, 32, 64, and so on. As long as the environment remains favorable, this doubling effect can continue at a constant rate. With the passing of each generation, the population will double, over and over again.

The length of the generation time is a measure of the growth rate of an organism. Compared with the growth rates of most other living things, bacteria are extremely fast. Their average generation time is 30 to 60 minutes under optimum conditions. The shortest generation times can be 10 to 12 minutes, but longer generation times require days. For example, *Mycobacterium leprae,* the cause of Hansen's disease, has a generation time of 10 to 30 days—as long as that of some animals. Environmental bacteria commonly have generation times measured in months. Most pathogenic bacteria have relatively short doubling times. *Salmonella enteritidis* and *Staphylococcus aureus,* bacteria that cause food-borne illness, double in 20 to 30 minutes, which is why leaving food at room temperature even for a short period can cause food-borne disease. In a few hours, a population of these bacteria can easily grow from a small number of cells to several million.

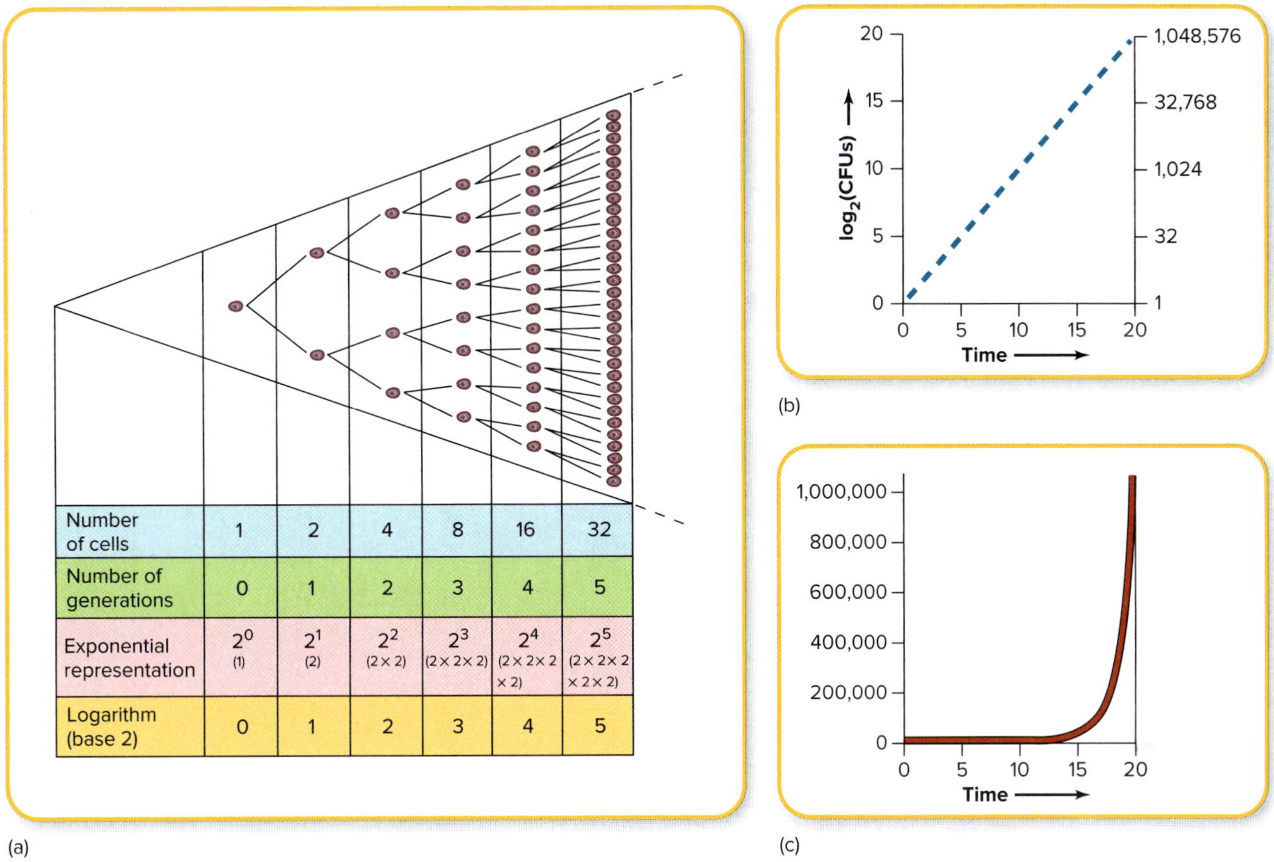
Number of cells
1
2
4
8
16
32
Number of generations
0
1
2
3
4
5
Exponential representation
2^0 (1)
2^1 (2)
2^2 (2 × 2)
2^3 (2 × 2 × 2)
2^4 (2 × 2 × 2 × 2)
2^5 (2 × 2 × 2 × 2 × 2)
Logarithm (base 2)
0
1
2
3
4
5
(a)
$\log_2$(CFUs)
20
15
10
5
0
1,048,576
32,768
1,024
32
1
0
5
10
15
20
Time
(b)
1,000,000
800,000
600,000
400,000
200,000
0
0
5
10
15
20
Time
(c)

Figure 9.12 The mathematics of population growth. **(a)** Starting with a single cell, if each product of reproduction goes on to divide by binary fission, the population doubles with each new cell division or generation. This process can be represented by logarithms (2 raised to an exponent) or by simple numbers. **(b)** Plotting the logarithm of the cells produces a straight line indicative of exponential growth, whereas **(c)** plotting the cell numbers arithmetically gives a curved slope.

Figure 9.12*a* shows several characteristics of growth: The cell population size can be represented by the number 2 with an exponent (2^1, 2^2, 2^3, 2^4); the exponent increases by one in each generation; and the number of the exponent is also the number of the generation. This growth pattern is termed **exponential.** Because these populations often contain very large numbers of cells, it is useful to express them by means of exponents or logarithms (see Appendix A). The data from a growing bacterial population are graphed by plotting the number of cells as a function of time **(figure 9.12*c*).** The cell number can be represented logarithmically or arithmetically. Plotting the logarithm number over time provides a straight line indicative of exponential growth. Plotting the data arithmetically gives a sharply curved slope. In general, logarithmic graphs are preferred because an accurate cell number is easier to read, especially during early growth phases **(figure 9.12*b*,*c*).**

Predicting the number of cells that will arise during a long growth period (yielding millions of cells) is based on a relatively simple concept. One could use the method of addition $2 + 2 = 4$, $4 + 4 = 8$, $8 + 8 = 16$, $16 + 16 = 32$, and so on; or a method of multiplication (for example, $2^5 = 2 \times 2 \times 2 \times 2 \times 2$); but it is easy to see that for 20 or 30 generations, this calculation could be very tedious. An easier way to calculate the size of a population over time is to use an equation such as

$$N_f = (N_i)2^n$$

In this equation, N_f is the total number of cells in the population at any point in the growth phase, N_i is the starting number, the exponent n denotes the generation number, and 2^n represents the number of cells in that generation. If we know any two of the values, the other values can be calculated. We will use the example of *Staphylococcus aureus* to calculate how many cells (N_f) will be present in an egg salad sandwich after it sits in a warm car for 4 hours. We will assume that N_i is 10 (number of cells deposited in the sandwich while it was being prepared). To find n, we just divide 4 hours (240 minutes) by the generation time (we will use 20 minutes). This calculation comes out to 12, so 2^n is equal to 2^{12}. Using a calculator, we find that 2^{12} is 4,096.

$$\text{Final number } (N_f) = 10 \times 4{,}096$$
$$= 40{,}960 \text{ bacterial cells in the sandwich}$$

This same equation, with modifications, is used to determine the generation time, a more complex calculation that requires knowing the number of cells at the beginning and end of a growth period. Such data are obtained through actual testing by a method discussed in the following section.

The Population Growth Curve

In reality, a population of bacteria does not maintain its potential growth rate and does not double endlessly because in most systems numerous factors prevent the cells from continuously dividing at

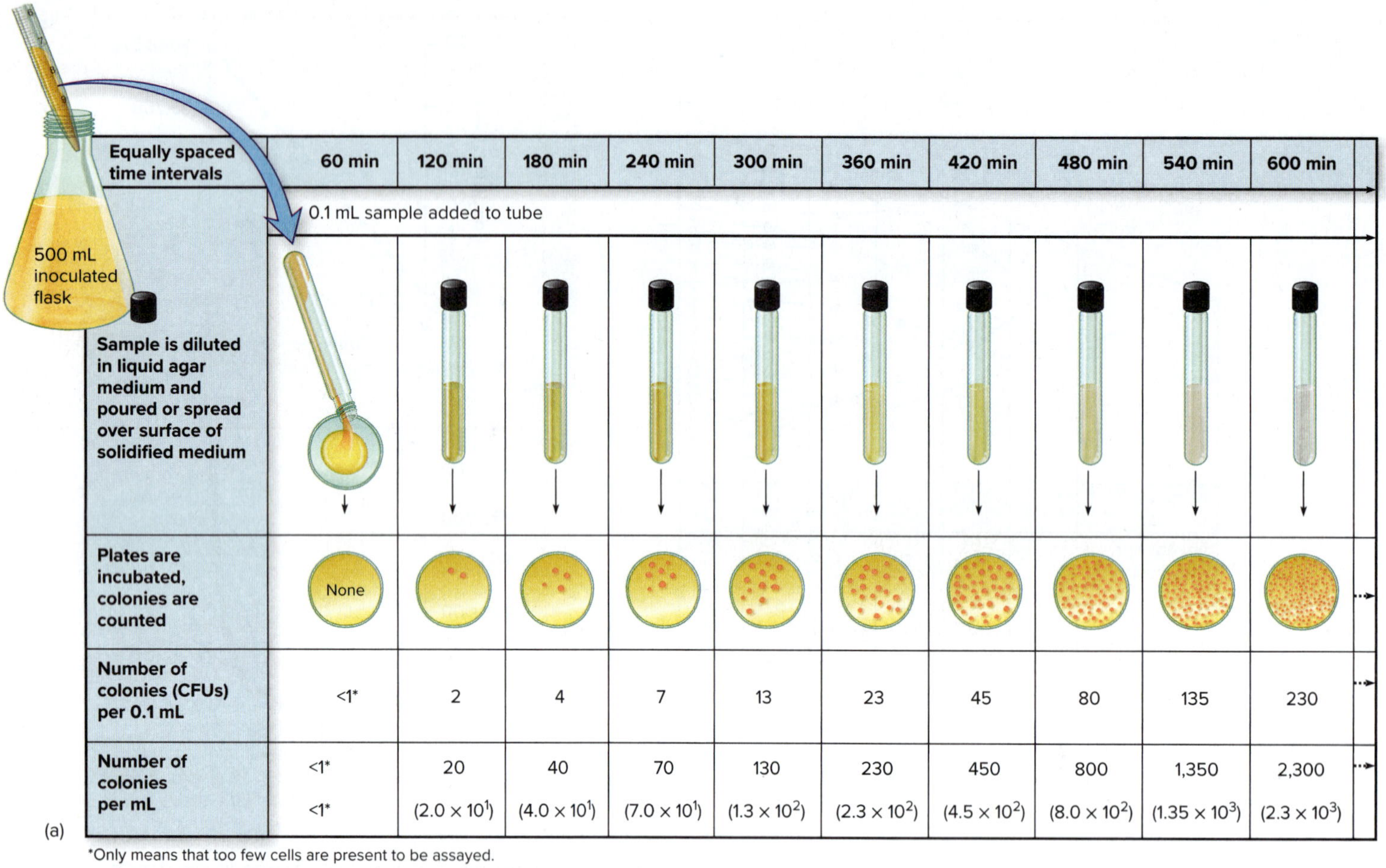
Equally spaced time intervals
60 min
120 min
180 min
240 min
300 min
360 min
420 min
480 min
540 min
600 min
500 mL inoculated flask
0.1 mL sample added to tube
Sample is diluted in liquid agar medium and poured or spread over surface of solidified medium
Plates are incubated, colonies are counted
None
Number of colonies (CFUs) per 0.1 mL
<1*
2
4
7
13
23
45
80
135
230
Number of colonies per mL
<1*
20
40
70
130
230
450
800
1,350
2,300
<1*
(2.0 x 10¹)
(4.0 x 10¹)
(7.0 x 10¹)
(1.3 x 10²)
(2.3 x 10²)
(4.5 x 10²)
(8.0 x 10²)
(1.35 x 10³)
(2.3 x 10³)
(a)
*Only means that too few cells are present to be assayed.

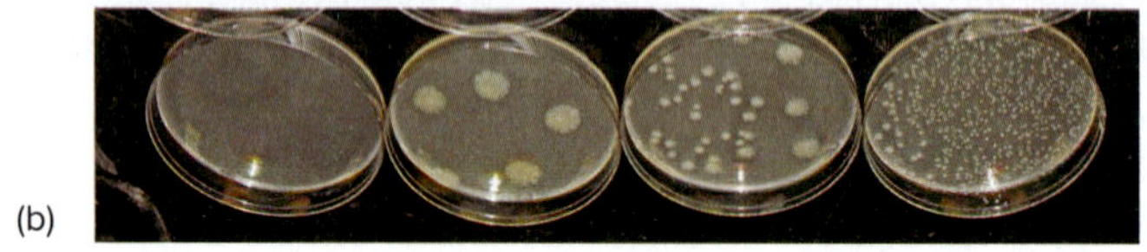
(b)

Figure 9.13 Steps in a viable plate count: batch culture method. **(a)** Illustration of the process. **(b)** Photo of actual plates with increasing numbers of CFUs.

(b) Lisa Burgess/McGraw Hill

their maximum rate. A given population typically displays a predictable pattern, or **growth curve,** over time.

The method traditionally used to observe the population growth pattern is a viable count technique, in which the total number of live cells is counted over a given time period. This is a fundamental method of laboratory microbiology. A growing population is established by inoculating a flask containing a known quantity of sterile liquid medium with a few cells of a pure culture. The flask is incubated at that bacterium's optimum temperature and timed. The population size at any point in the growth cycle is quantified by removing a tiny measured sample of the culture from the flask and plating it out on a solid medium to develop isolated colonies. This procedure is repeated at evenly spaced intervals (i.e., every hour for 24 hours).

Evaluating the samples involves a common and important principle in microbiology: One colony on the plate represents one cell or colony-forming unit (CFU) from the original sample. Because the CFU of some bacteria is actually composed of several cells (consider the clustered arrangement of *Staphylococcus,* for instance), using a colony count can underestimate the exact population size to an extent. The population size is usually expressed as "*x* CFUs per milliliter." To get a fair estimate of the total population size (number of cells) at any given point, you multiply the number of colonies arising from the fluid you removed by the dilution factor. For example, if you want to express the results as "per milliliter," but you sampled only 1/10 of a milliliter, you just multiply the number of visible CFUs × 10 **(figure 9.13).** The growth curve is determined by graphing the number for each sample in sequence for the whole incubation period.

Stages in the Normal Growth Curve

The system of culturing described in figure 9.13 is *closed,* meaning that nutrients and space are finite and there is no mechanism for the removal of waste products. Data from an entire growth period typically produce a curve with a series of phases termed the *lag phase,* the *exponential growth (log) phase,* the *stationary phase,* and the *death phase* **(figure 9.14).**

The **lag phase** is a relatively "flat" period on the graph when the population appears not to be growing or is growing at less than the exponential rate. Growth lags primarily because

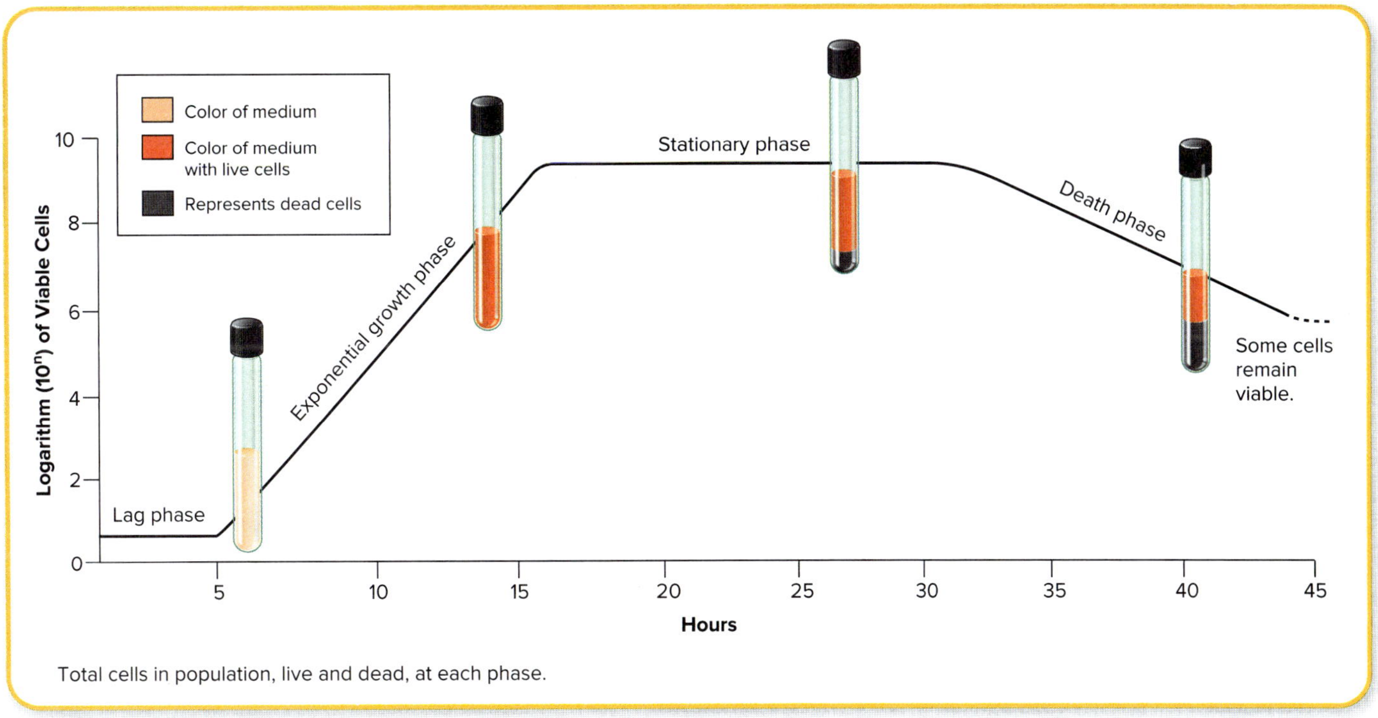
Color of medium
Color of medium with live cells
Represents dead cells
Logarithm (10^n) of Viable Cells
10
8
6
4
2
0
Lag phase
Exponential growth phase
Stationary phase
Death phase
Some cells remain viable.
5
10
15
20
25
30
35
40
45
Hours
Total cells in population, live and dead, at each phase.

Figure 9.14 The growth curve in a bacterial culture. On this graph, the number of viable cells expressed as a logarithm (log) is plotted against time. See text for discussion of the various phases. Note that with a generation time of 30 minutes, the population has risen from 10 (10^1) cells to 1,000,000,000 (10^9) cells in only 16 hours.

1. the newly inoculated cells require a period of adjustment, enlargement, and synthesis;
2. the cells are not yet multiplying at their maximum rate; and
3. the population of cells is so sparse or dilute that the sampling misses them.

The length of the lag period varies somewhat from one population to another. It is important to note that even though the population of cells is not increasing (growing), individual cells are metabolically active as they increase their contents and prepare to divide.

The cells reach the maximum rate of cell division during the **exponential growth** (logarithmic or **log**) **phase,** a period during which the curve increases dramatically. This phase will continue as long as cells have adequate nutrients and the environment is favorable.

At the **stationary growth phase,** the population enters a survival mode in which cells stop growing or grow slowly. The curve levels off because the rate of cell inhibition or death balances out the rate of multiplication. The decline in the growth rate is caused by depleted nutrients and oxygen plus accumulation of organic acids and other biochemical pollutants into the growth medium, due to the increased density of cells.

As the limiting factors intensify, cells begin to die at an exponential rate (literally perishing in their own wastes), and they are unable to multiply. The curve now dips downward as the **death phase** begins. The speed with which death occurs depends on the relative resistance of the species and how toxic the conditions are, but it is usually slower than the exponential growth phase. Viable cells often remain many weeks and months after this phase has begun. In the laboratory, refrigeration is used to slow the progress of the death phase so that cultures will remain viable as long as possible.

Practical Importance of the Growth Curve

The tendency for populations to exhibit phases of rapid growth, slow growth, and death has important implications in microbial control, infection, food microbiology, and culture technology. Antimicrobial agents such as heat and disinfectants rapidly accelerate the death phase in all populations, but microbes in the exponential growth phase are more vulnerable to these agents than are those that have entered the stationary phase. In general, actively growing cells are more vulnerable to conditions that disrupt cell metabolism and binary fission.

Understanding the stages of cell growth is crucial for work with cultures. In most cases, it is unwise to continue incubating a culture beyond the stationary phase because doing so will reduce the number of viable cells and the culture could die out completely. It is also preferable to use young cultures to do stains (an exception is the spore stain) and motility tests because the cells will show their natural size and correct Gram reaction, and motile cells will have functioning flagella.

For certain research or industrial applications, closed batch culturing with its four phases is inefficient. Instead, an automatic growth chamber called a **chemostat,** or continuous culture system, is used. This device can admit a steady stream of new nutrients and siphon off used media and old bacterial cells, thereby stabilizing the growth rate and cell number. It has the advantage of maintaining the culture in a biochemically active state and preventing it from entering the death phase.

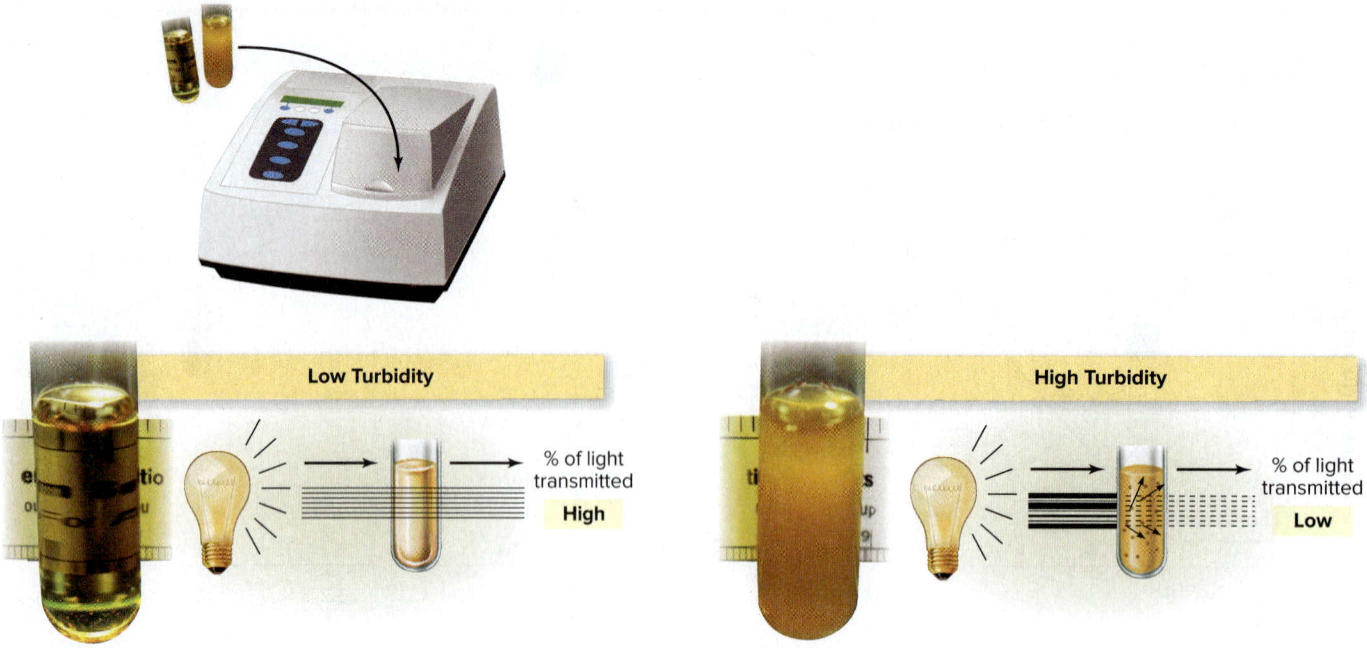
Low Turbidity
% of light
transmitted
High
High Turbidity
% of light
transmitted
Low

Figure 9.15 Turbidity measurements as indicators of growth. Holding a broth to the light is one method of checking for gross differences in cloudiness (turbidity). The broth on the left is transparent, indicating little or no growth; the broth on the right is cloudy and opaque, indicating heavy growth. The eye is not sensitive enough to pick up fine degrees in turbidity; more sensitive measurements can be made with a spectrophotometer. On the left you will see that a tube with no growth will allow light to easily pass. Therefore, more light will reach the photodetector and give a higher transmittance value. In a tube with growth, on the right, the cells scatter the light, resulting in less light reaching the photodetector and, therefore, giving a lower transmittance value.
Kathy Park Talaro

Other Methods of Analyzing Population Growth

Turbidometry

Microbiologists have developed several alternative ways of analyzing bacterial growth qualitatively and quantitatively. One of the simplest methods for estimating the size of a population is through turbidometry. This technique relies on the simple observation that a tube of clear nutrient solution becomes cloudy, or **turbid,** as microbes grow in it. In general, the greater the turbidity, the larger the population size, which can be measured by means of instruments that measure the amount of light passing through a vessel **(figure 9.15).**

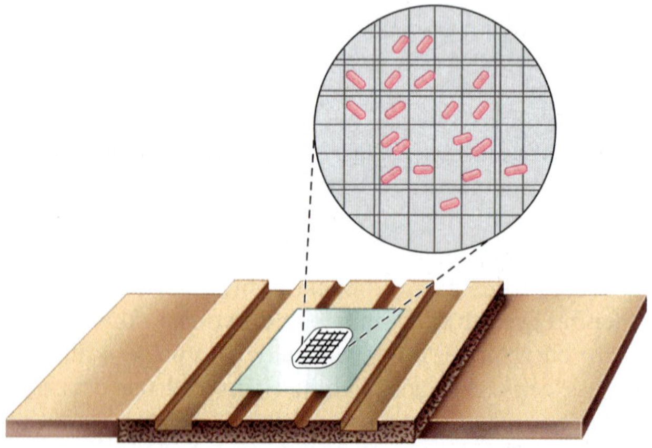

Figure 9.16 Direct microscopic count of bacteria. A small sample is placed on the grid under a cover glass. Individual cells, both living and dead, are counted. This number can be used to calculate the total count of a sample.

Counting Bacteria

Turbidity readings are useful for evaluating relative amounts of growth, but if a more quantitative evaluation is required, the viable colony count described in this chapter or some other counting procedure is necessary. The **direct (total) cell count** involves counting the number of cells in a sample microscopically **(figure 9.16).** This technique, very similar to that used in blood cell counts, employs a special microscope slide (cytometer) calibrated to accept a tiny sample that is spread over a premeasured grid. The cell count from a cytometer can be used to estimate the total number of cells in a larger sample (i.e., of milk or water). One inherent inaccuracy in this method as well as in spectrophotometry is that no distinction can be made between dead and live cells, both of which are included in the count.

Counting can be automated by sensitive devices such as the *Coulter counter,* which electronically scans a culture as it passes through a tiny pipette. As each cell flows by, it is detected and

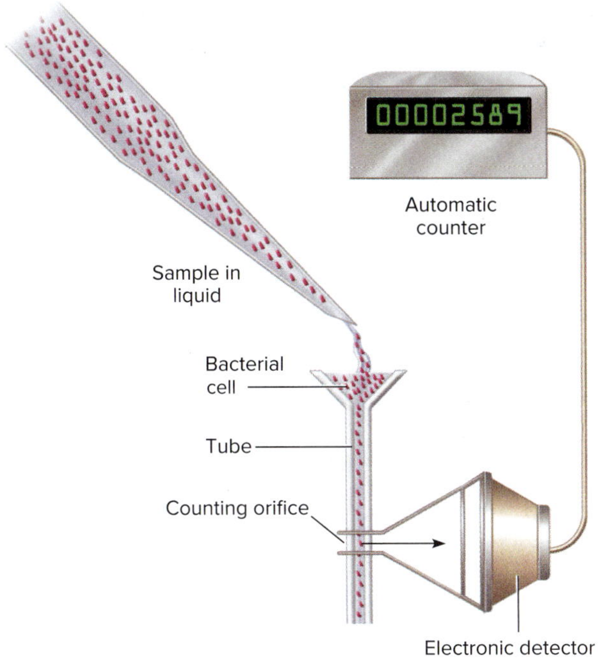
00002589
Automatic counter
Sample in liquid
Bacterial cell
Tube
Counting orifice
Electronic detector

Figure 9.17 Coulter counter. As cells pass through this device, they trigger an electronic sensor that counts them.

registered on an electronic sensor **(figure 9.17).** A *flow cytometer* works on a similar principle, but in addition to counting, it can measure cell size and even differentiate between live and dead cells. When used in conjunction with fluorescent dyes and antibodies to tag cells, it has been used to differentiate between gram-positive and gram-negative bacteria. It has been adapted for use as a rapid method to identify pathogens in patient specimens and to differentiate blood cells. More sophisticated forms of the flow cytometer can actually sort cells of different types into separate compartments of a collecting device.

New Methods

Increasingly, nonculture methods are being used for counting microbes. In chapter 6, you learned about a technique called the polymerase chain reaction (PCR), which allows scientists to quantify bacteria and other microorganisms that are present in environmental or tissue samples without isolating them and without culturing them. In addition, tests that measure ATP, the energy molecule, have been used in the food and pharmaceutical industries for some time and may hold promise for rapid quantification of microbes in other environmental samples as well.

9.3 Learning Outcomes—Assess Your Progress

13. Summarize the steps of binary fission, and name another method of reproduction used by other bacterial species.
14. Define *doubling time,* and describe how it leads to exponential growth.
15. Compare and contrast the four phases of growth in a bacterial growth curve.
16. Identify one culture-based and one non-culture-based method used for analyzing bacterial growth.

Media Under The Microscope Wrap-Up

The news article "Earth Bacteria Survived in Space for Years, Proving It Can Contaminate Mars" describes a research study published in the scientific literature.

The **intended message** of the article is to report that bacteria have survived in extreme conditions, i.e., in space—and not inside a spaceship—for years.

A **critical reading** of the article reveals to me no sensationalized claims. There is some sensational information, for sure, but it seems well explained and not oversold. I might have liked to see the authors of the article interview a scientist not involved with the project for their opinion of the science. That always strengthens confidence in the reporting.

NASA

I would **interpret** this news to my friends by explaining that radiation is one of the factors that can limit cellular life, and that these microbes can withstand 10,000 times the amount of radiation that humans can. Lastly, I would mention that we have to be careful not to contaminate pristine planets with the bacteria that hitch along on space vehicles from earth.

My **grade** for the article is a B–. Good information, well explained. A quote from an outside scientist would push it up to an A.

Source: *IndiaTimes,* "Earth Bacteria Survived in Space for Years, Proving It Can Contaminate Mars," online article posted August 27, 2020.

Study Smarter: Better Together

These activities are designed for you to use on your own with a study group—either a face-to-face group or a virtual one, consisting of 3–5 members. Studying together can be very helpful, but there are effective and ineffective ways to do it. For example, getting together without a clear structure is often not a good use of your time. Use your time efficiently by using one or more of the exercises below.

FACE-TO-FACE GROUPS

Use one or more of the activities below.

Peer Instruction: Assign numbers to your group members to use all semester long. Now look at these five concepts from this chapter. Each group member prepares a 5-minute lesson on the topic corresponding to their number. Don't worry if you have fewer than 5 members; just use however many you have! During your group study time, each member presents their lesson, and the group spends another 5–10 minutes discussing that lesson.

1. Tonicity (hypo-, iso-, hypertonic)
2. Symbiotic and nonsymbiotic relationships
3. Generation time
4. Exponential growth
5. Phases of growth

Concept Maps: Each member of the group should use this list of terms from this chapter to generate their own concept map. This can be hand-drawn or created using software (see Appendix C for guidelines). During group study time, compare each other's concept maps and help each other make sure they are correct. Of course, there are many different "correct" maps. Examining each member's map will help you talk through the varied concepts and how they are related.

Concept Terms:

symbiosis	commensalism	protection	mutualism
parasitism	disease	pathogen	normal biota

Table Topics: Each group member should identify a concept or topic from this week's class assignments with which they are having trouble and share it during group study time. The other group members can then help to clarify confusing issues or share how they figured it out. Aim for a maximum of 15 minutes per topic. If the topic remains unclear to the group, bring it up during class or use the instructor's office hours or e-mail to ask for help. Taking the time to struggle with a difficult concept first makes your questions much more specific and more likely to yield helpful answers.

VIRTUAL GROUPS

Not everyone has the time or opportunity to meet with group members outside of class time. You or your instructor can create a virtual group using e-mail or the course software.

Weekly Discussion Board: This forum can be used as a way for groups to discuss topics, via e-mail or other learning management systems or online platforms, before they are covered in class. As each member of the group answers the current week's question, they should send their responses to every other member of their group. It's best to agree on a deadline based on how your class schedule works (Saturday for the next week's topics, for example). Then, after the topic is discussed in class, each member should send a response that all group members will see with a follow-up post on the same topic. If you cover more than one chapter in a week, someone can be designated to choose which chapter Discussion Board question you will use. Or simply decide up front that you will always use the first-chapter-of-the week's question, to keep the schedule simple.

Discussion Question

When discussing bacterial growth, are we mainly talking about the growth of a single organism or the growth of the population? Explain your answer.

Chapter Summary

MICROBIAL NUTRITION AND GROWTH

9.1 MICROBIAL NUTRITION

6
C
Carbon
12.0107

- **Most organisms require six elements—carbon, hydrogen, oxygen, nitrogen, phosphorus, and sulfur—to survive, grow, and reproduce.**
- **Nutrients required in large amounts are termed macronutrients, and those required in very small amounts are micronutrients.**
- **Microbes are classified by both the chemical form of their nutrients and the energy sources they use.**

9.2 ENVIRONMENTAL FACTORS THAT INFLUENCE MICROBES

- **The environmental factors that control most microbial growth are temperature, pH, moisture, radiation, gases, and the presence of other microorganisms.**
- **Each microbe has three "cardinal" temperatures: its minimum temperature, its optimum temperature, and its maximum temperature.**
- **With regard to their temperature requirements, microorganisms are classified as psychrophiles, mesophiles, and thermophiles. Organisms that can withstand very harsh environments are called extremophiles.**
- **Most eukaryotic microbes are aerobic, while bacteria can live with or without oxygen to varying degrees.**
- **Microorganisms generally live in association with many other organisms. Their relationships range from mutualism to parasitism to antagonism, and may shift over time as resources shift.**
- **Biofilms are complex synergistic communities of microbes that behave differently than free-living microorganisms. They are probably the rule, rather than the exception, in nature.**

9.3 THE STUDY OF MICROBIAL GROWTH

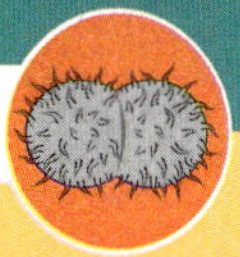

- **The splitting of a parent bacterial cell to form a pair of daughter cells is known as binary fission.**
- **Generation time is a measure of the growth rate of a microbial population. It varies in length based on the type of microbe and the environmental conditions.**
- **Microbial cultures in a batch environment exhibit four distinct stages of growth: lag, log, stationary, and death.**
- **Microbial population growth can be quantified by measuring turbidity, colony counts, and direct cell counts. Increasingly, quantitation can be accomplished with nongrowth methods.**

SmartGrid: From Knowledge to Critical Thinking

This *21 Question Grid* takes the topics from this chapter and arranges them with respect to the American Society for Microbiology's Undergraduate Curriculum guidelines—all six of the important "Concepts" as well as the important "Competency" of scientific literacy. Three questions are supplied, which cover chapter content referring to the Concept or Competency in increasing levels of Bloom's taxonomy for learning.

ASM Concept/ Competency	A. Bloom's Level 1, 2—Remember and Understand (Choose one.)	B. Bloom's Level 3, 4—Apply and Analyze	C. Bloom's Level 5, 6—Evaluate and Create
Evolution	**1.** Which descriptors are likely to have applied to the earliest microbes on the planet? **a.** chemoautotrophic **b.** thermophilic **c.** chemoheterotrophic **d.** two of the above	**2.** Can you think of any advantages gained when eukaryotes and multicellular organisms evolved to use sexual reproduction instead of binary fission? Discuss.	**3.** Write a paragraph that a middle-school student could understand about how earth's atmosphere came to contain oxygen and the role early bacteria and archaea played.

(continued)

ASM Concept/ Competency	A. Bloom's Level 1, 2—Remember and Understand (Choose one.)	B. Bloom's Level 3, 4—Apply and Analyze	C. Bloom's Level 5, 6—Evaluate and Create
Cell Structure and Function	**4.** Which of the following is true of passive transport? **a.** It requires a gradient. **b.** It uses the cell wall. **c.** It includes endocytosis. **d.** It only moves water.	**5.** Compare the effects of a hypertonic, hypotonic, and isotonic solution on an amoeba versus a bacterium.	**6.** Usually scientists looking for life on other planets try to determine first if the planet had or has evidence of water. Develop a rationale for this based on cellular chemistry.
Metabolic Pathways	**7.** An organism that can synthesize all its required organic components from CO_2 using energy from the sun is a **a.** photoautotroph. **b.** photoheterotroph. **c.** chemoautotroph. **d.** chemoheterotroph.	**8.** Provide evidence in support of or refuting this statement: Microbial life can exist in the complete absence of sunlight and organic compounds.	**9.** Develop an explanation for why biofilm bacteria use quorum sensing to regulate the size of their community.
Information Flow and Genetics	**10.** Most bacteria increase their numbers by **a.** sexual reproduction. **b.** hyphae formation. **c.** binary fission. **d.** endocytosis.	**11.** Looking at figure 9.4, explain how a cell in a hypertonic solution might have trouble dividing.	**12.** In binary fission, the parent chromosome is duplicated so that each of the two daughter cells has the same copy of the original chromosome. If this is the case, speculate on what happened to allow bacteria to become diversified into so many types?
Microbial Systems	**13.** A cell exposed to a hypertonic environment will _____ by osmosis. **a.** gain water **b.** lose water **c.** neither gain nor lose water **d.** burst	**14.** Considering figure 9.8, discuss whether tube 2 or tube 3 is more likely to grow in diverse niches or conditions. Defend your answer.	**15.** Bacteria and archaea are everywhere on the planet. Would it be to our advantage (if it were possible) to eliminate all of them, in an effort to protect the human population from current and emerging diseases? Why or why not?
Impact of Microorganisms	**16.** A pathogen would most accurately be described as a **a.** parasite. **b.** commensal. **c.** symbiont. **d.** saprobe.	**17.** How can you explain the fact that an unopened carton of milk in the refrigerator will eventually spoil even though it has never been opened?	**18.** Describe a scenario in which it would be to a pathogen's advantage to have a very long generation time.
Scientific Thinking	**19.** In a viable count, each _____ represents a _______ from the sample population. **a.** CFU, colony **b.** colony, CFU **c.** hour, generation **d.** cell, generation	**20.** If an egg salad sandwich sitting in a car on a warm day contains 40,960 bacteria after 4 hours, how many bacteria will it contain after another generation time for the bacteria?	**21.** Scientists now believe that most bacteria in nature exist in biofilms. Can you suggest some technical reasons that this was not understood for the first 150 years of microbiology?

Answers to the multiple-choice questions appear in Appendix A.

Visual Connections

These questions use visual images or previous content to connect content within and between chapters.

1. **From chapter 7, figure 7.16.** What type of symbiotic relationship is illustrated here?

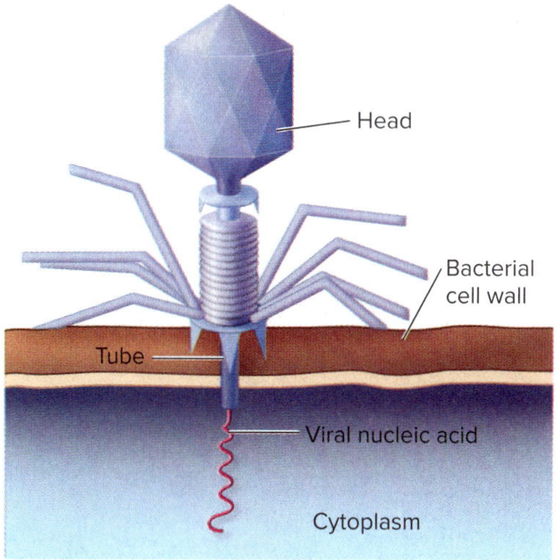

Head
Bacterial cell wall
Tube
Viral nucleic acid
Cytoplasm

High Impact Study

These terms and concepts are most critical for your understanding of this chapter—and may be the most difficult. Have you mastered them?

Concepts

- ☐ Organic versus inorganic
- ☐ Tonicity
- ☐ Oxygen requirements
- ☐ Symbiotic and nonsymbiotic relationships
- ☐ Binary fission
- ☐ Generation time
- ☐ Exponential growth
- ☐ Phases of growth

Terms

- ☐ Heterotroph
- ☐ Autotroph
- ☐ Phototroph
- ☐ Chemotroph
- ☐ Saprobe
- ☐ Parasite
- ☐ Diffusion
- ☐ Osmosis
- ☐ Psychrophile
- ☐ Mesophile
- ☐ Thermophile

Design Element: (College students): Caia Image/Image Source

10 Microbial Metabolism

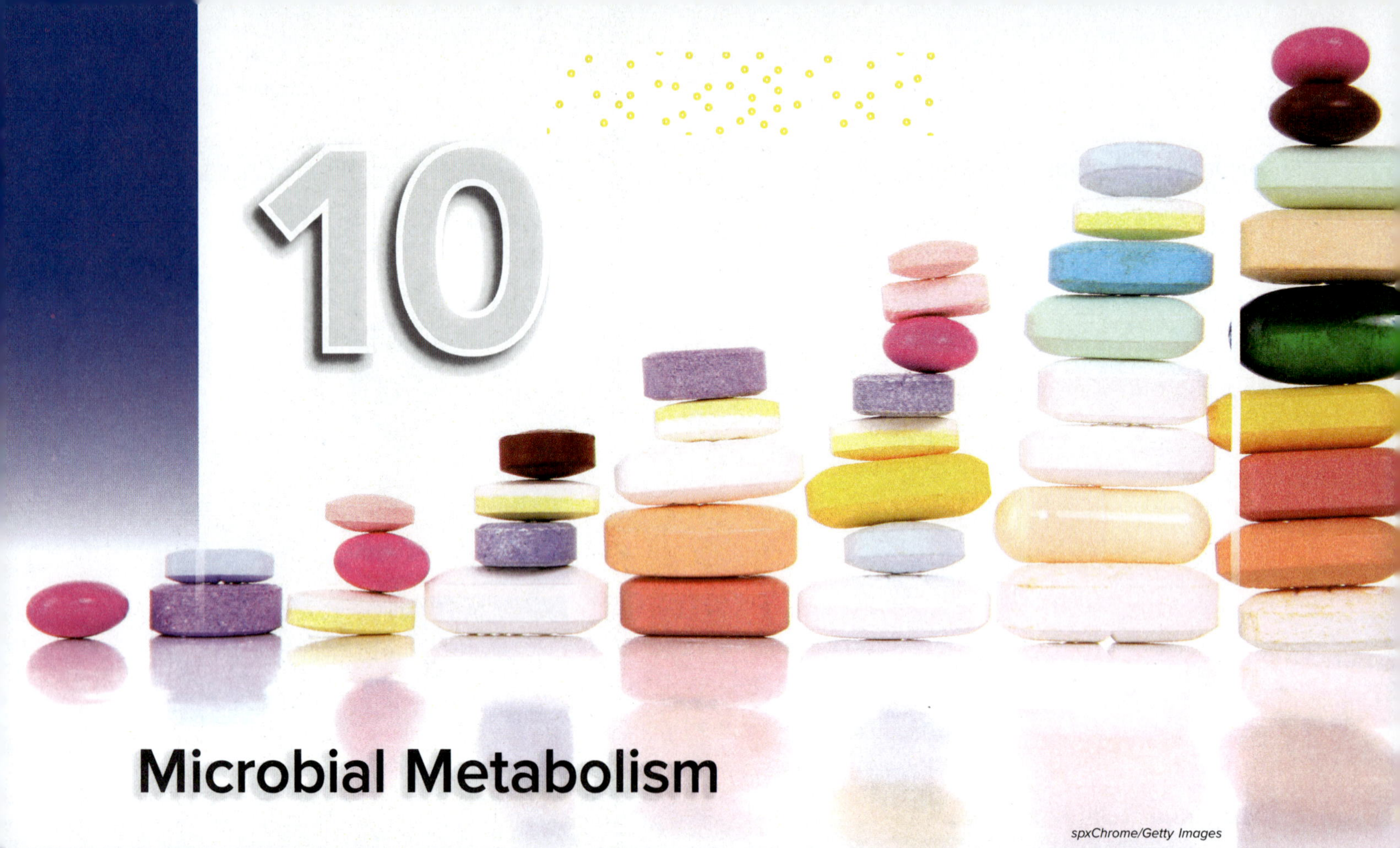

spxChrome/Getty Images

Media Under The Microscope

Enzyme Dietary Supplements

These case studies examine an article from the popular media to determine the extent to which it is factual and/or misleading. This case focuses on the 2021 Homer News *article "Blood Sugar Support Plus Review: Scam or Ingredients That Work?"*

This article from the Homer, Alaska, online newsletter, which has a label saying "sponsored" in an upper corner, says that if we take this particular supplement, we will lose weight and have more energy and improved blood sugar levels. The supplement contains four enzymes: lipase, protease, trypsin, and amylase.

You will learn in this chapter that enzymes are usually named using the substance they act upon followed by the suffix "-ase." So three of these enzymes are described in the article in the following way:

- Lipase "breaks down fat and ensures you're not getting fatter from having sugar stores in your body for later use."
- Protease breaks down protein, "ensuring you have the proper energy to go on with your day easily."
- Amylase: The Latin for starch is *amylum.* The article states that those with low amylase levels in their saliva "usually experience poor insulin responses and increased blood sugar levels." It goes on to say, "This is not something that has been discovered in the lab. It's the human body's science and the actual enzyme. What you should know is that beef enzymes resemble human enzymes. 80% of the human DNA consists of cow or so-called beef material."

- What is the **intended message** of the article?
- What is your **critical reading** of the summary of the article provided above? Remember that in this context, "critical reading" does not necessarily mean *What criticism do you have?* but asks you to apply your knowledge to interpret whether the article is factual and whether the facts support the intended message.
- How would you **interpret** the news item for your nonmicrobiologist friends?
- What is your **overall grade** for the news item—taking into account its accuracy and the accuracy of its intended effect?

Media Under The Microscope Wrap-Up appears at the end of the chapter.

Outline and Learning Outcomes

10.1 The Metabolism of Microbes

1. Describe the relationship among metabolism, catabolism, and anabolism.
2. Fully discuss the structure and function of enzymes.
3. Differentiate between an apoenzyme and a holoenzyme.
4. Differentiate between an endoenzyme and an exoenzyme, and between constitutive and regulated enzymes.
5. Diagram the four major patterns of metabolic pathways.
6. Describe how enzymes are controlled.

10.2 Finding and Making Use of Energy

7. Name the chemical in which energy is stored in cells.
8. Create a general diagram of a redox reaction.
9. Identify electron carriers used by cells.

10.3 Catabolism: Getting Materials and Energy

10. Name the three main catabolic pathways and the estimated ATP yield for each.
11. Construct a paragraph summarizing glycolysis.
12. Describe the Krebs cycle and compare the process between bacteria and eukaryotes.
13. Discuss the significance of the electron transport system.
14. State two ways in which anaerobic respiration differs from aerobic respiration.
15. Summarize the steps of microbial fermentation, and list three useful products it can create.
16. Describe how noncarbohydrate compounds are catabolized.

10.4 Biosynthesis and the Crossing Pathways of Metabolism

17. Provide an overview of the anabolic stages of metabolism.
18. Define *amphibolism*.

10.5 Photosynthesis: It Starts with Light

19. Summarize the overall process of photosynthesis in a single sentence.
20. Discuss the relationship between light-dependent and light-independent reactions.
21. Explain the role of the Calvin cycle in the process of photosynthesis.

10.1 The Metabolism of Microbes

Metabolism is a term referring to all chemical reactions and physical workings of the cell. Although metabolism entails thousands of different reactions, most of them fall into one of two general categories. The first, **anabolism,** is any process that results in synthesis of cell molecules and structures. It is a building and bond-making process that forms larger macromolecules from smaller ones, and it usually requires the input of energy. The second, **catabolism,** is the opposite of anabolism. Catabolic reactions break the bonds of larger molecules into smaller molecules and often release energy. In a cell, linking anabolism to catabolism makes sure that many thousands of processes are performed efficiently.

Another fundamental fact about metabolism is that electrons are critical to the process. In summary, a cell gathers energy by transferring electrons from an external source to internal carriers that eventually shuttle it into a series of proteins that harvest energy. Electron flow is the key. Along the way, metabolism accomplishes the following **(figure 10.1):**

1. It assembles smaller molecules into larger macromolecules needed for the cell; in this process, ATP (energy) is utilized to form bonds (anabolism).
2. It breaks down macromolecules into smaller molecules, a process that yields energy (catabolism).
3. It collects and spends energy in the form of ATP (adenosine triphosphate) or heat.

Disease Connection

Anabolism is the process of synthesizing cell molecules and structures from smaller units. Anabolic steroids are synthesized in laboratories to have the same chemical structure as the steroids found in testosterone, the male sex hormone. Anabolic steroids are often used (abused) by bodybuilders who are striving to build muscle, gain weight, and appear more masculine. They are often taken in doses as high as 100 times as the doses that are used to treat certain medical conditions, such as delayed puberty and muscle wasting caused by AIDS and other chronic conditions. Anabolic steroid use at high doses can result in damage to the heart and liver, in addition to infertility, blood clots, and psychological effects (i.e., irritability, aggression, depression).

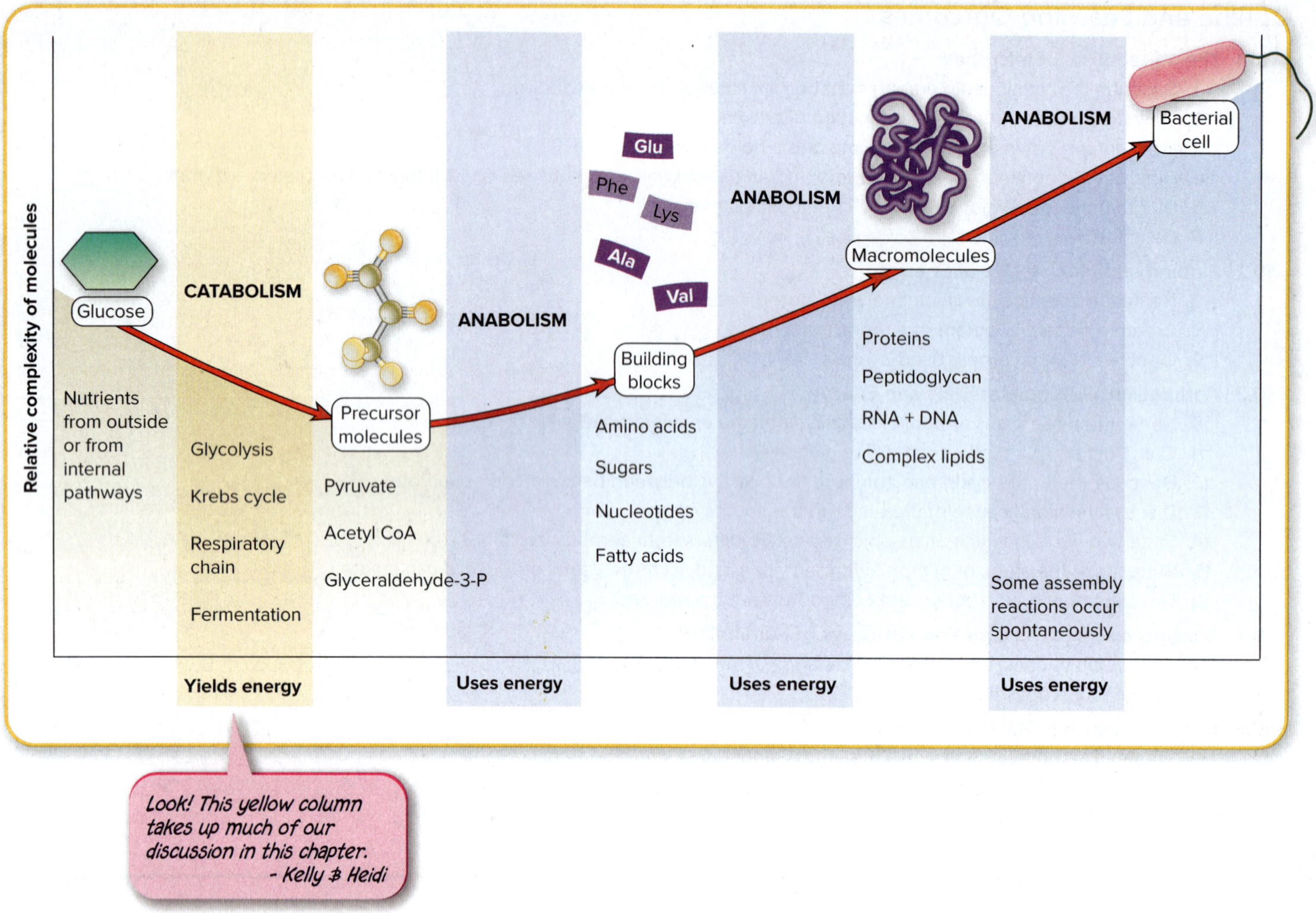
Relative complexity of molecules
Glucose
Nutrients from outside or from internal pathways
CATABOLISM
Glycolysis
Krebs cycle
Respiratory chain
Fermentation
Yields energy
Precursor molecules
Pyruvate
Acetyl CoA
Glyceraldehyde-3-P
ANABOLISM
Uses energy
Glu
Phe
Lys
Ala
Val
Building blocks
Amino acids
Sugars
Nucleotides
Fatty acids
ANABOLISM
Uses energy
Macromolecules
Proteins
Peptidoglycan
RNA + DNA
Complex lipids
ANABOLISM
Bacterial cell
Some assembly reactions occur spontaneously
Uses energy
Look! This yellow column takes up much of our discussion in this chapter. - Kelly $ Heidi

Figure 10.1 Simplified model of metabolism. Cellular reactions fall into two major categories. Catabolism (yellow) involves the breakdown of complex organic molecules to extract energy and form simpler end products. Anabolism (blue) uses the energy to synthesize necessary macromolecules and cell structures from precursors.

Enzymes: Catalyzing the Chemical Reactions of Life

A microbial cell could be viewed as a microscopic factory, complete with basic building materials, a source of energy, and a blueprint for running its extensive network of metabolic reactions. But the chemical reactions of life cannot proceed without a special class of macromolecules called **enzymes.** Enzymes are a remarkable example of **catalysts,** chemicals that increase the rate of a chemical reaction without becoming part of the products or being consumed in the reaction. It is easy to think that an enzyme *creates* a reaction, but that is not true. Chemical reactions could occur spontaneously at some point even without an enzyme—but at a very slow rate. A study of the enzyme urease shows that it increases the rate of the breakdown of urea by a factor of 100 trillion as compared to an uncatalyzed reaction. Uncatalyzed reactions do not generally occur fast enough for cellular processes. Therefore, enzymes, which speed up the rate of reactions, are indispensable to life. Other major characteristics of enzymes are summarized in **table 10.1.**

Table 10.1 Checklist of Enzyme Characteristics

- Most composed of protein; may require cofactors
- Act as organic catalysts to speed up the rate of cellular reactions
- Lower the activation energy required for a chemical reaction to proceed
- Are extremely unique from each other with respect to shape, specificity, and function
- Enable metabolic reactions to proceed at a speed compatible with life
- Have an active site for target molecules (substrates)
- Are much larger in size than their substrates
- Associate closely with substrates but do not become integrated into the reaction products
- Are not used up or permanently changed by the reaction
- Can be recycled, thus function in extremely low concentrations
- Are greatly affected by temperature and pH
- Can be regulated by feedback loops and genetic mechanisms

How Do Enzymes Work?

An enzyme speeds up the rate of a metabolic reaction, but just how does it do this? During a chemical reaction, reactants are converted to products by bonds forming or breaking. A certain amount of energy is required to initiate every such reaction, which limits its rate. This initial resistance, which must be overcome for a reaction to proceed, is measurable and is called the **activation energy** (or energy of activation). This can be overcome by

1. increasing thermal energy (heating) to increase the velocity of molecules,
2. increasing the concentration of reactants to increase the rate of molecular collisions, or
3. adding a catalyst.

In most living systems, the first two alternatives are not feasible because elevating the temperature is potentially harmful and higher concentrations of reactants are not practical. This leaves only the action of catalysts, and enzymes fill this need efficiently and potently. Enzymatic catalysts effectively lower the energy of activation, allowing a reaction to progress at a faster pace and with reduced energy input.

At the molecular level, an enzyme promotes a reaction by serving as a physical site upon which the reactant molecules, called **substrates,** can be positioned for various interactions. The enzyme is usually much larger in size than its substrate, and it presents a unique active site that accepts only that particular substrate. Although an enzyme binds to the substrate and participates directly in changes to the substrate, it does not become a part of the products, is not used up by the reaction, and can function over and over again. *Enzyme speed* is well documented. For example, the enzyme catalase converts its substrates at the rate of several million per second.

Enzyme Structure

Most enzymes are proteins, and they can be classified as simple or conjugated. Simple enzymes consist of protein alone, whereas conjugated enzymes **(figure 10.2)** contain protein and some other nonprotein molecule or molecules. The whole conjugated enzyme, sometimes referred to as a **holoenzyme,** is a combination of the protein, called the **apoenzyme** in these cases, and one or more **cofactors.** Cofactors are either organic molecules, called **coenzymes,** or inorganic elements (metal ions). For example, catalase, an enzyme that we learned in chapter 9 breaks down hydrogen peroxide, requires iron as a metallic cofactor.

There is another type of enzyme that is not made of protein, but of RNA. Named **ribozymes,** these molecules are remarkable because they are RNA molecules that catalyze reactions on other RNA. Ribozymes are thought to be remnants of the earliest molecules on earth that could have served as both catalysts and genetic material. Their discovery has lent support for what is known today as the "RNA hypothesis," which states that RNA was in fact the first genetic material within ancient cells. In natural systems, ribozymes are involved in self-splicing or cutting of RNA molecules during final processing of the genetic code.

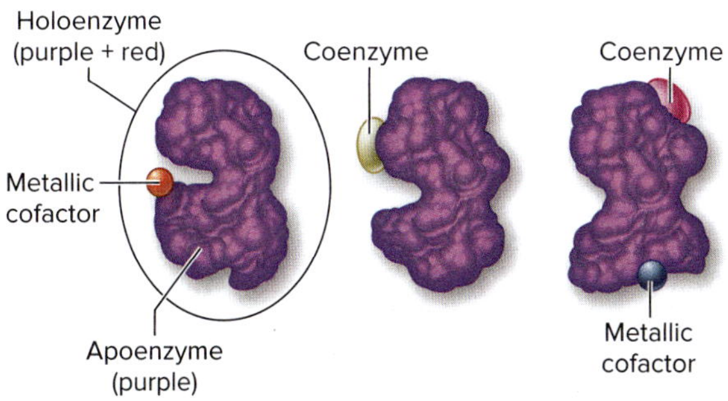
Holoenzyme
(purple + red)
Coenzyme
Coenzyme
Metallic
cofactor
Apoenzyme
(purple)
Metallic
cofactor

Figure 10.2 Conjugated enzyme structure. Conjugated enzymes have an apoenzyme (polypeptide or protein) component and one or more cofactors.

Apoenzymes: Specificity and the Active Site

Apoenzymes range in size from small polypeptides with about 100 amino acids and a molecular weight of 12,000 to large polypeptide conglomerates with thousands of amino acids and a molecular weight of over 1 million. Like all proteins, an apoenzyme exhibits levels of molecular complexity called the primary, secondary, tertiary, and—in larger enzymes—quaternary organization **(figure 10.3).** As we saw in chapter 2, the first three levels of structure arise when a single polypeptide chain undergoes an automatic folding process and achieves stability by forming disulfide and other types of bonds. The actual site where the substrate binds is a crevice or groove called the **active site,** or **catalytic site,** and there can be one or several such sites (as shown in figure 10.3). The three-dimensional shape of each site is formed by the way the amino acid chain or chains are folded. Each type of enzyme has a different primary structure (type and sequence of amino acids), variations in folding, and unique active sites.

Enzyme-Substrate Interactions

For a reaction to take place, the substrate has to nestle into the active site **(figure 10.4).** The fit is so specific that it is often described as a "lock-and-key" fit in which the substrate is inserted into the active site's pocket.

Once the enzyme-substrate complex has formed, appropriate reactions occur on the substrate, often with the aid of a cofactor, and a product is formed and released. The enzyme can then attach to another substrate molecule and repeat this action.

Cofactors: Supporting the Work of Enzymes

In chapter 9, you learned that microorganisms require specific metal ions called trace elements and certain organic growth factors. Here we see why they are needed: to assist enzymes. The metal cofactors, including iron, copper, magnesium, manganese, zinc, cobalt, selenium, and many others, help with precise functions between the enzyme and its substrate. In general, metals activate enzymes, help bring the active site and substrate close together, and participate directly in chemical reactions with the enzyme-substrate complex.

Coenzymes are one type of cofactor. The general function of a coenzyme is to remove a chemical group from one substrate molecule and add it to another substrate, thereby serving as a transient carrier of this group. Soon, we shall see that coenzymes carry and transfer hydrogen atoms, electrons, carbon dioxide, and amino groups. One of the most important kinds of coenzymes is **vitamins,** which explains why vitamins are important to nutrition and may be required as growth factors for living things. Vitamin deficiencies prevent the complete holoenzyme from forming.

Naming Enzymes

A standardized system of nomenclature and classification has been developed to keep things clear. In general, an enzyme name is composed of two parts: (1) a prefix or stem word derived from a

Figure 10.3 How the active site and specificity of the apoenzyme arise. The active site is always formed by the three-dimensional structure of the tertiary or quaternary folding, which means that amino acids that may be distant from one another in the primary structure can be adjacent in the active site.

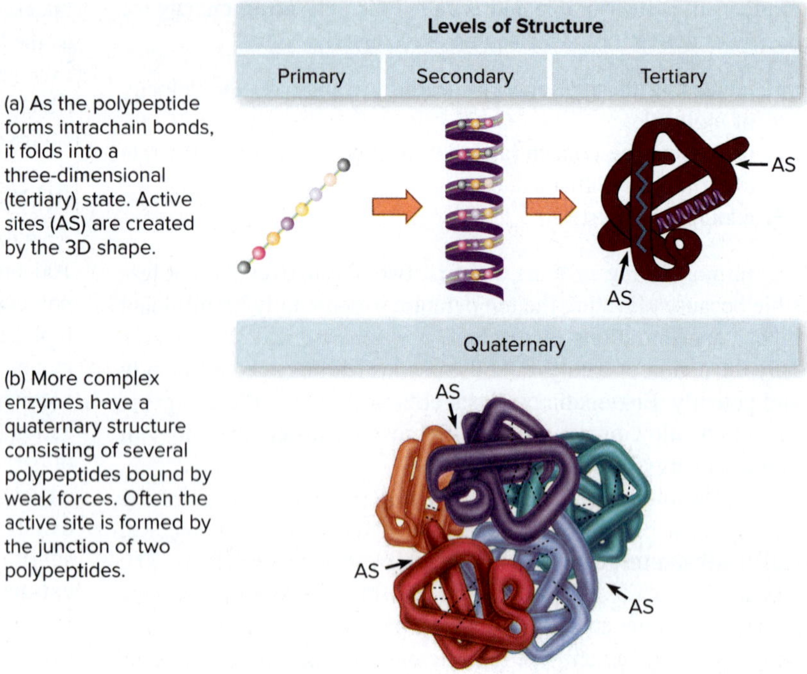
Levels of Structure
Primary
Secondary
Tertiary
(a) As the polypeptide forms intrachain bonds, it folds into a three-dimensional (tertiary) state. Active sites (AS) are created by the 3D shape.
AS
AS
Quaternary
(b) More complex enzymes have a quaternary structure consisting of several polypeptides bound by weak forces. Often the active site is formed by the junction of two polypeptides.
AS
AS
AS

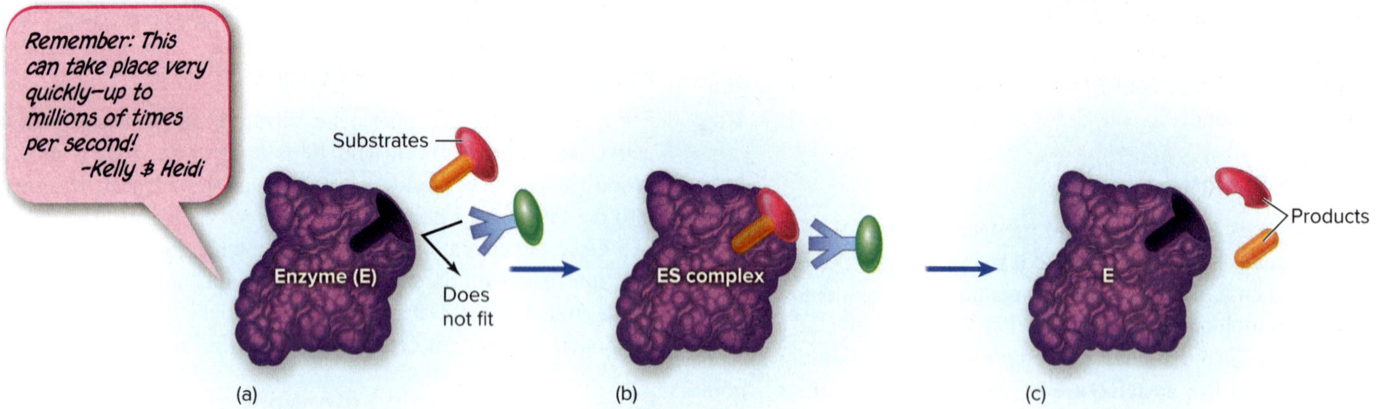
Remember: This can take place very quickly—up to millions of times per second!
-Kelly $ Heidi
Substrates
Enzyme (E)
Does not fit
ES complex
E
Products
(a)
(b)
(c)

Figure 10.4 Enzyme-substrate reactions. **(a)** When the enzyme and substrate come together, the substrate (S) must show the correct fit and position with respect to the enzyme (E). **(b)** When the ES complex is formed, it enters a transition state. During this temporary but tight interlocking union, the enzyme participates directly in breaking or making bonds. **(c)** Once the reaction is complete, the enzyme releases the products.

certain characteristic—usually the substrate acted upon or the type of reaction catalyzed, or both—followed by (2) the ending *-ase.*

This system classifies the enzyme in one of six classes, on the basis of its general biochemical action:

1. *Oxidoreductases* transfer electrons from one substrate to another, and *dehydrogenases* transfer a hydrogen from one compound to another.
2. *Transferases* transfer functional groups from one substrate to another.
3. *Hydrolases* cleave bonds on molecules with the addition of water.
4. *Lyases* add groups to or remove groups from double-bonded substrates.
5. *Isomerases* change a substrate into its isomeric[1] form.
6. *Ligases* catalyze the formation of bonds with the input of ATP and the removal of water.

1. An isomer is a compound that has the same molecular formula as another compound but differs in arrangement of the atoms.

Each enzyme is also assigned a common name that indicates the specific reaction it catalyzes. With this system, an enzyme that digests a carbohydrate substrate is a *carbohydrase;* a specific carbohydrase, *amylase,* acts on starch (amylose is a major component of starch). An enzyme that hydrolyzes peptide bonds of a protein is a *proteinase, protease,* or *peptidase,* depending on the size of the protein substrate. Some fats and other lipids are digested by *lipases.* DNA is hydrolyzed by *deoxyribonuclease,* generally shortened to *DNase.* A *synthetase* or *polymerase* bonds together many small molecules into large molecules. Other examples of enzymes are presented in **table 10.2.**

Transfer Reactions by Enzymes

Other enzyme-driven processes that involve the simple addition or removal of a functional group are important to the overall workings of the cell. Oxidation-reduction and other transfer activities are examples of these types of reactions.

Table 10.2 A Sampling of Enzymes, Their Substrates, and Their Reactions

Common Name	Official Name	Enzyme Class	Substrate	Action
Lactase	β-D-galactosidase	Hydrolase	Lactose	Breaks lactose down into glucose and galactose
Penicillinase	Beta-lactamase	Hydrolase	Penicillin	Hydrolyzes beta-lactam ring
DNA polymerase	DNA nucleotidyl-transferase	Transferase	DNA nucleosides	Synthesizes a strand of DNA using the complementary strand as a model
Lactate dehydrogenase	Same as common name	Oxidoreductase	Pyruvic acid	Catalyzes the conversion of pyruvic acid to lactic acid
Oxidase	Cytochrome oxidase	Oxidoreductase	Molecular oxygen (O_2)	Catalyzes the reduction of O_2 (addition of electrons and hydrogen)

Remember how we said that the flow of electrons is critical to metabolism? Some atoms and compounds readily give or receive electrons and participate in oxidation (the loss of electrons) or reduction (the gain of electrons). The compound that loses the electrons is **oxidized,** and the compound that receives the electrons is **reduced.** These oxidation-reduction (redox) reactions are common in the cell. Oxidoreductases remove electrons from one substrate and add them to another. Their coenzyme carriers are nicotinamide adenine dinucleotide (NAD) and flavin adenine dinucleotide (FAD). (Take note: Even if by now your eyes are glazing over at all the terms and details, this paragraph is a valuable one! If you remember the statements in this paragraph, the rest of metabolism will be a lot easier to understand.)

Location of Enzyme Action

Enzymes perform their tasks either inside or outside of the cell in which they were produced. After they are made inside the cell, **exoenzymes** are transported extracellularly, where they break down (hydrolyze) large food molecules or harmful chemicals. Examples of exoenzymes are cellulase, amylase, and penicillinase. By contrast, **endoenzymes** function inside the cell. Most enzymes of the metabolic pathways are endoenzymes **(figure 10.5).**

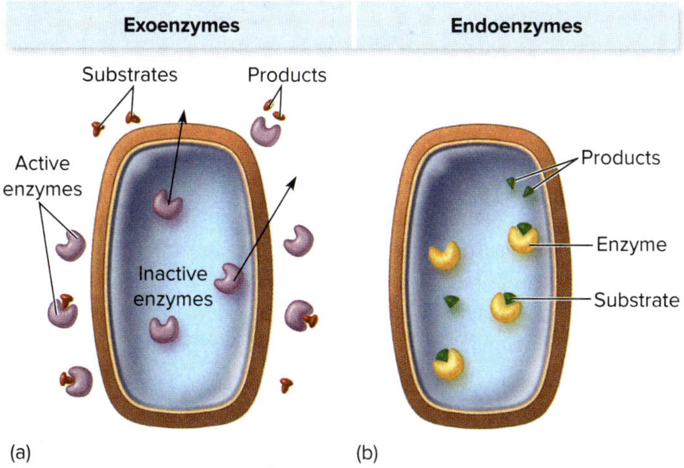
Exoenzymes
Endoenzymes
Substrates
Products
Active enzymes
Inactive enzymes
Products
Enzyme
Substrate
(a)
(b)

Figure 10.5 Types of enzymes, as described by their location of action. **(a)** Exoenzymes are released outside the cell to function. **(b)** Endoenzymes remain in the cell and function there.

Enzymes are not all produced in equal amounts or at equal rates. Some, called **constitutive enzymes (figure 10.6*a*),** are always present and in relatively constant amounts, regardless of the cellular environment. The enzymes involved in using glucose, for example, are very important in metabolism and, for that reason, are constitutive. Other enzymes are **regulated enzymes (figure 10.6*b*),** and they are either turned on (induced) or turned off (repressed) in response to changes in concentration of the substrate.

The Role of Microbial Enzymes in Disease

Many pathogens secrete unique exoenzymes that help them avoid host defenses or promote their multiplication in tissues. Because these enzymes contribute to pathogenicity, they are referred to as virulence factors. Some of them function as toxins. *Streptococcus pyogenes* (a cause of throat and skin infections) produces a streptokinase that digests blood clots and helps the bacterium invade wounds. Another exoenzyme from this bacterium is called streptolysin. In mammalian hosts, streptolysin damages blood cells and tissues. It is also responsible for lysing red blood cells used in blood agar dishes, and this trait is used for identifying the bacteria growing in culture. *Pseudomonas aeruginosa,* a respiratory and skin pathogen, produces elastase and collagenase, which digest elastin and collagen, two proteins found in connective tissue. These increase the severity of certain lung diseases and burn infections. *Clostridium perfringens,* an agent of gas gangrene, synthesizes lecithinase C, a lipase that profoundly damages cell membranes and accounts for the tissue death associated with this disease. Not all microbial enzymes digest tissues; some, such as penicillinase, inactivate penicillin and thereby protect a microbe from its effects.

Disease Connection

In competitive inhibition, enzyme activity is stopped by a molecule that resembles (mimics) the enzyme's normal substrate. There are many drugs that use the principle of competitive inhibition. Many of the drugs developed to treat HIV/AIDS prevent the enzyme HIV protease from doing its job, which is to assemble the capsids of new viruses. Certain antidepressants, diuretics, and antibiotics also act as competitive inhibitors.

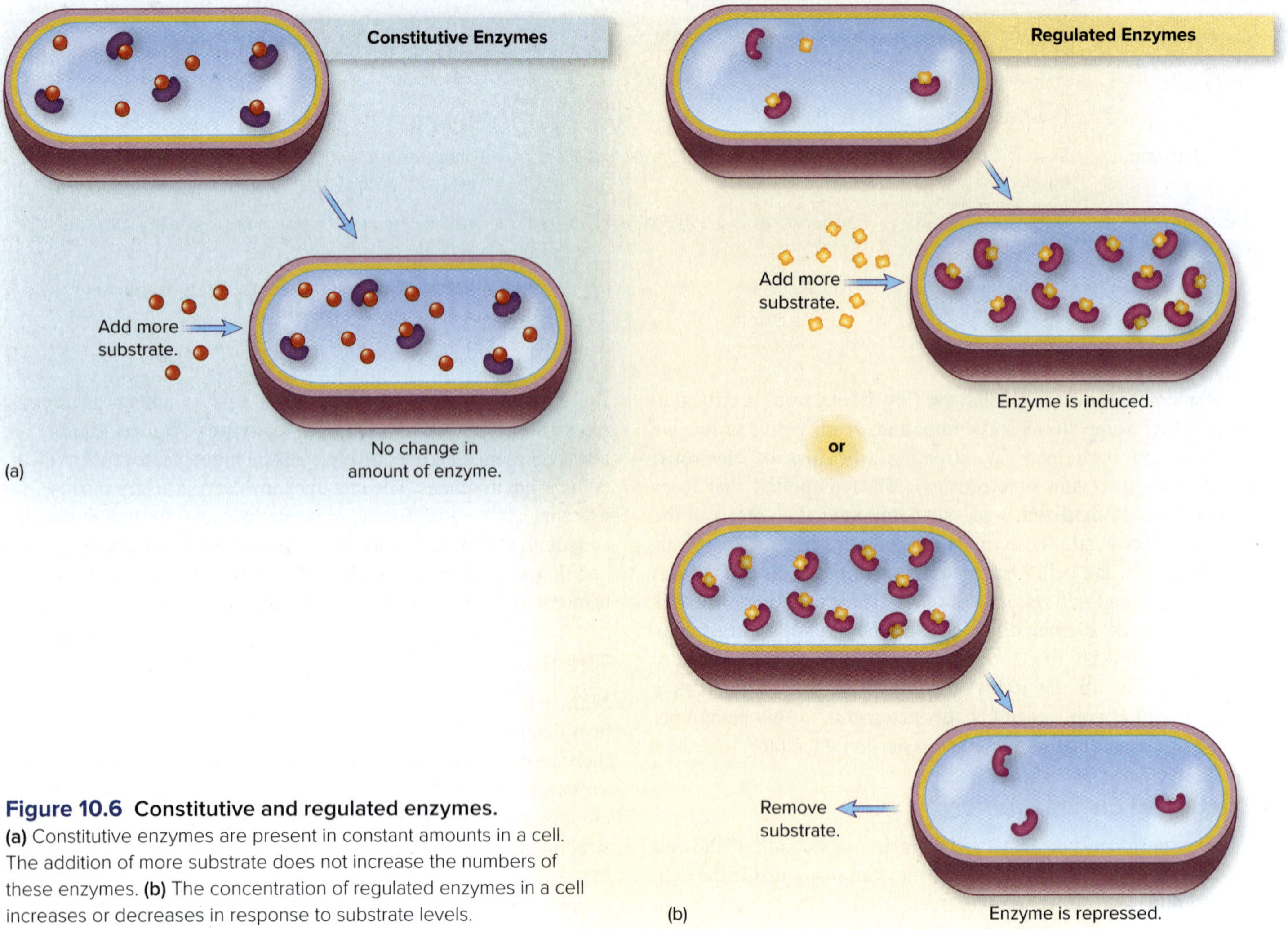
Constitutive Enzymes
Add more substrate.
No change in amount of enzyme.
(a)
Regulated Enzymes
Add more substrate.
Enzyme is induced.
or
Remove substrate.
(b)
Enzyme is repressed.

Figure 10.6 Constitutive and regulated enzymes. **(a)** Constitutive enzymes are present in constant amounts in a cell. The addition of more substrate does not increase the numbers of these enzymes. **(b)** The concentration of regulated enzymes in a cell increases or decreases in response to substrate levels.

The Sensitivity of Enzymes to Their Environment

The activity of an enzyme is highly influenced by the cell's environment. In general, enzymes operate only under the natural temperature, pH, and osmotic pressure of its organism's habitat. When enzymes are subjected to changes in these normal conditions, they tend to be chemically unstable, or **labile.** Low temperatures inhibit catalysis, and high temperatures denature the apoenzyme. **Denaturation** is a process by which the weak bonds that collectively maintain the native shape of the apoenzyme are broken. This disruption causes distortion of the enzyme's shape and prevents the substrate from attaching to the active site. Such nonfunctional enzymes block metabolic reactions and can lead to cell death. Low or high pH or certain chemicals (heavy metals, alcohol) are also denaturing agents. When enzymes are nonfunctional, metabolic reactions fail to happen and cell death can follow.

Regulation of Enzymatic Activity and Metabolic Pathways

Metabolic reactions proceed in a systematic, highly regulated manner that maximizes the use of nutrients and energy. The cell responds to environmental conditions by using those metabolic reactions that most favor growth and survival. Basically, the regulation of metabolism comes about through the regulation of enzymes via an elaborate system of checks and balances.

Metabolic Pathways

Metabolic reactions rarely consist of a single action or step. More often, they happen in a multistep series or pathway, with each step catalyzed by a different enzyme. An individual reaction can be shown in various ways, depending on the purpose at hand **(figure 10.7).** The product of one reaction is often the reactant (substrate) for the next, forming a linear chain of reactions. Many pathways have branches that provide alternate methods for nutrient processing. Others take a cyclic form, in which the starting molecule is regenerated to initiate another turn of the cycle. On top of that, pathways generally do not stand alone; they are interconnected and merge at many sites.

Direct Controls on the Action of Enzymes

The bacterial cell has many ways of directly influencing the activity of its existing enzymes. It can inhibit enzyme activity

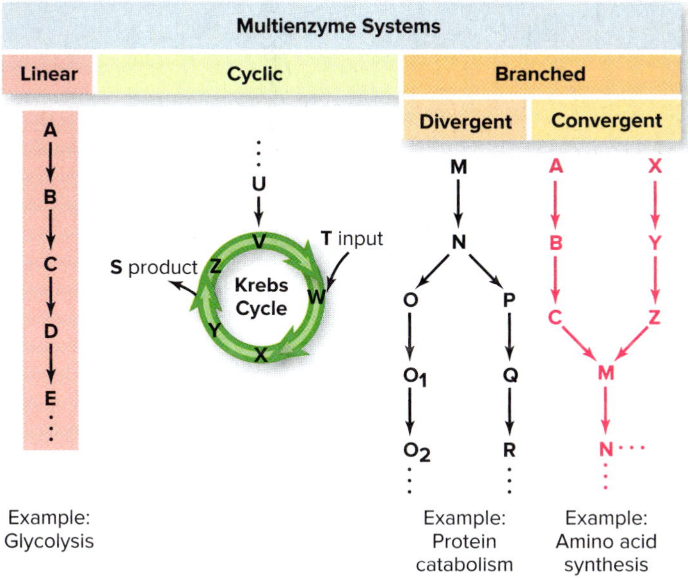
Multienzyme Systems
Linear
Cyclic
Branched
Divergent
Convergent
A B C D E
U V W X Y Z
T input
S product
Krebs Cycle
M N O P O1 Q O2 R
A B C X Y Z M N
Example: Glycolysis
Example: Protein catabolism
Example: Amino acid synthesis

Figure 10.7 Patterns of metabolism. In general, metabolic pathways consist of a linked series of individual chemical reactions that produce intermediary metabolites and lead to a final product. These pathways occur in several patterns, including linear, cyclic, and branched. Virtually every reaction in a series—represented by an arrow—involves a specific enzyme.

by supplying a molecule that resembles the enzyme's normal substrate. This "mimic" can then occupy the enzyme's active site, preventing the actual substrate from binding there. Because the mimic cannot actually be acted on by the enzyme or function in the way the product would have, the enzyme is effectively shut down. This form of inhibition is called **competitive inhibition** because the mimic is competing with the substrate for the binding site **(figure 10.8).**

Another form of inhibition can occur with special types of enzymes that have two binding sites—the active site and another area called the regulatory site (as shown in figure 10.8). These enzymes are regulated by the binding of molecules in their regulatory sites. Often, the regulatory molecule is the product of the enzymatic reaction itself. This provides a negative feedback mechanism that can slow down enzymatic activity once a certain concentration of product is produced. This is **noncompetitive inhibition** because the regulator molecule does not bind in the same site as the substrate.

Controls on Whether the Enzyme Gets Made

Controlling enzymes by controlling their synthesis is another effective mechanism because enzymes do not last indefinitely. Some wear out, some are deliberately degraded, and others are diluted with each cell division. For the reactions to continue, enzymes eventually must be replaced. This cycle works into the scheme of the cell, where replacement of enzymes can be regulated according to cell demand.

Enzyme repression is a mechanism to stop further synthesis of an enzyme somewhere along its pathway. As the level of the end product from a given enzymatic reaction has built to excess, the genes responsible for replacing these enzymes are automatically suppressed **(figure 10.9).** The response time is longer than for the direct controls described above, but its effects last longer.

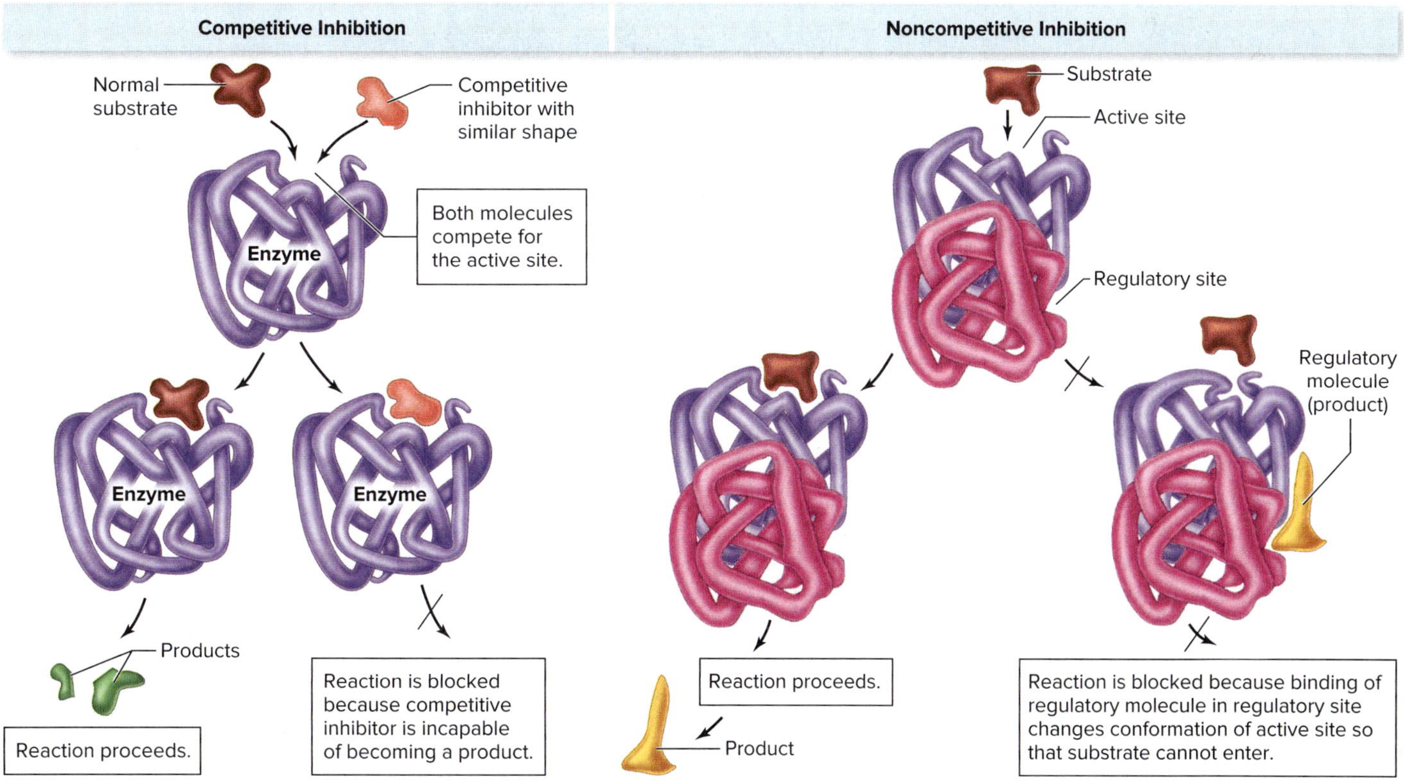
Competitive Inhibition
Normal substrate
Competitive inhibitor with similar shape
Enzyme
Both molecules compete for the active site.
Enzyme
Enzyme
Products
Reaction proceeds.
Reaction is blocked because competitive inhibitor is incapable of becoming a product.
Noncompetitive Inhibition
Substrate
Active site
Regulatory site
Regulatory molecule (product)
Reaction proceeds.
Product
Reaction is blocked because binding of regulatory molecule in regulatory site changes conformation of active site so that substrate cannot enter.

Figure 10.8 Examples of two common control mechanisms for enzyme activity.

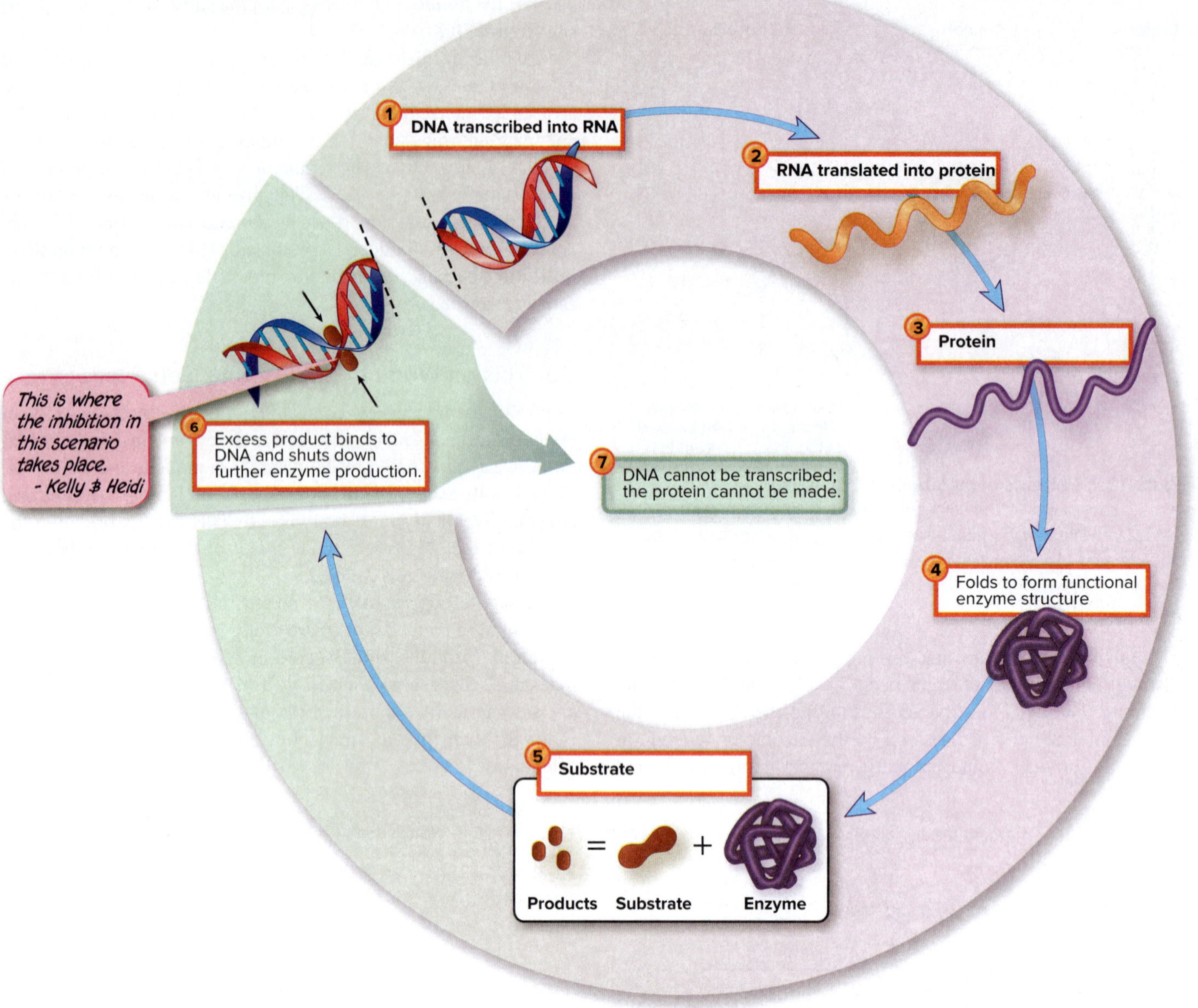
1 DNA transcribed into RNA
2 RNA translated into protein
3 Protein
4 Folds to form functional enzyme structure
5 Substrate
Products
Substrate
Enzyme
6 Excess product binds to DNA and shuts down further enzyme production.
7 DNA cannot be transcribed; the protein cannot be made.
This is where the inhibition in this scenario takes place.
- Kelly $ Heidi

Figure 10.9 One type of genetic control of enzyme synthesis: enzyme repression. 1–5: The enzyme is synthesized continuously via uninhibited transcription and translation until enough product has been made. 6, 7: Excess product reacts with a site on DNA that regulates the enzyme's synthesis, thereby inhibiting further enzyme production.

The inverse of enzyme repression is **enzyme induction.** In this process, enzymes appear (are induced) only when suitable substrates are present—that is, the synthesis of an enzyme is induced by its substrate. Both mechanisms are important genetic control systems in bacteria.

A classic model of enzyme induction is the case of the lac operon in *Escherichia coli* we examined in chapter 6. As a reminder, if *E. coli* is inoculated into a medium whose principal carbon source is lactose, it will produce the enzyme lactase to hydrolyze it into glucose and galactose. If the bacterium is subsequently inoculated into a medium containing only sucrose as a carbon source, it will cease synthesizing lactase and begin synthesizing sucrase. This response enables the organism to adapt to a variety of nutrients, and it also prevents a microbe from wasting energy by making enzymes for which no substrates are present.

10.1 Learning Outcomes—Assess Your Progress

1. Describe the relationship among metabolism, catabolism, and anabolism.
2. Fully discuss the structure and function of enzymes.
3. Differentiate between an apoenzyme and a holoenzyme.
4. Differentiate between an endoenzyme and an exoenzyme, and between constitutive and regulated enzymes.
5. Diagram the four major patterns of metabolic pathways.
6. Describe how enzymes are controlled.

10.2 Finding and Making Use of Energy

In order to carry out the work of metabolism, cells require constant input of some form of usable energy. The energy can come directly from sunlight (in photosynthesizers) or from free electrons (in electricity-harvesting bacteria). In most bacteria we examine in this book, the energy comes from organic substances, such as sugars, when their bonds are broken, releasing and transferring electrons. The energy is stored in ATP.

Energy in Cells

Not all cellular reactions are equal with respect to energy. Some release energy, and others require it. For example, a reaction that proceeds as follows:

$$X + Y \xrightarrow{\text{enzyme}} Z + \text{energy}$$

releases energy as it goes forward. This type of reaction is termed **exergonic** (ex-er-gon′-ik). Energy of this type is considered free—it is available for doing cellular work. Energy transactions such as the following:

$$\text{Energy} + A + B \xrightarrow{\text{enzyme}} C$$

are called **endergonic** (en-der-gon′-ik) because they require the addition of energy. In cells, exergonic and endergonic reactions are often coupled, so that released energy is immediately put to use.

To reiterate, cells possess specialized enzyme systems that trap the energy present in the bonds of nutrients as they are progressively broken **(figure 10.10).** During exergonic reactions, energy released by bonds is stored in certain high-energy phosphate bonds, such as in ATP. ATP stores the energy and releases it when needed for endergonic reactions. Before discussing ATP, we examine redox reactions, which provide the electrons that are critical to energy production.

A Closer Look at Oxidation and Reduction

Redox reactions always occur in pairs, with an electron donor and an electron acceptor, which constitute a *redox pair.* The reaction can be represented as follows:

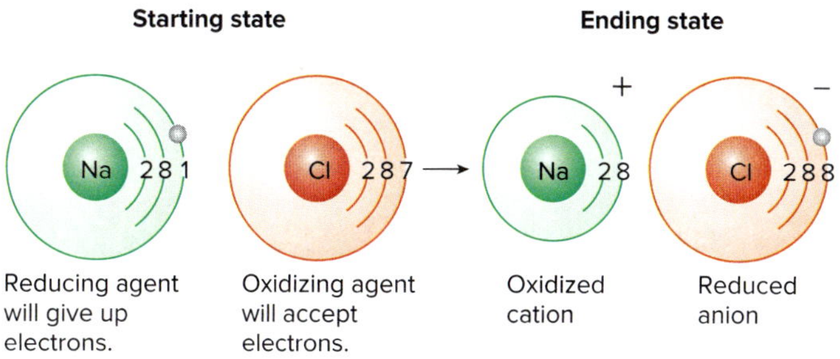
Starting state
Ending state
Na 2 8 1
Cl 2 8 7
Na 2 8
Cl 2 8 8
+
−
Reducing agent will give up electrons.
Oxidizing agent will accept electrons.
Oxidized cation
Reduced anion

Redox reactions occur in all cells and are indispensable to the required energy transformations. Important components of cellular redox reactions are enzymes called oxidoreductases. They

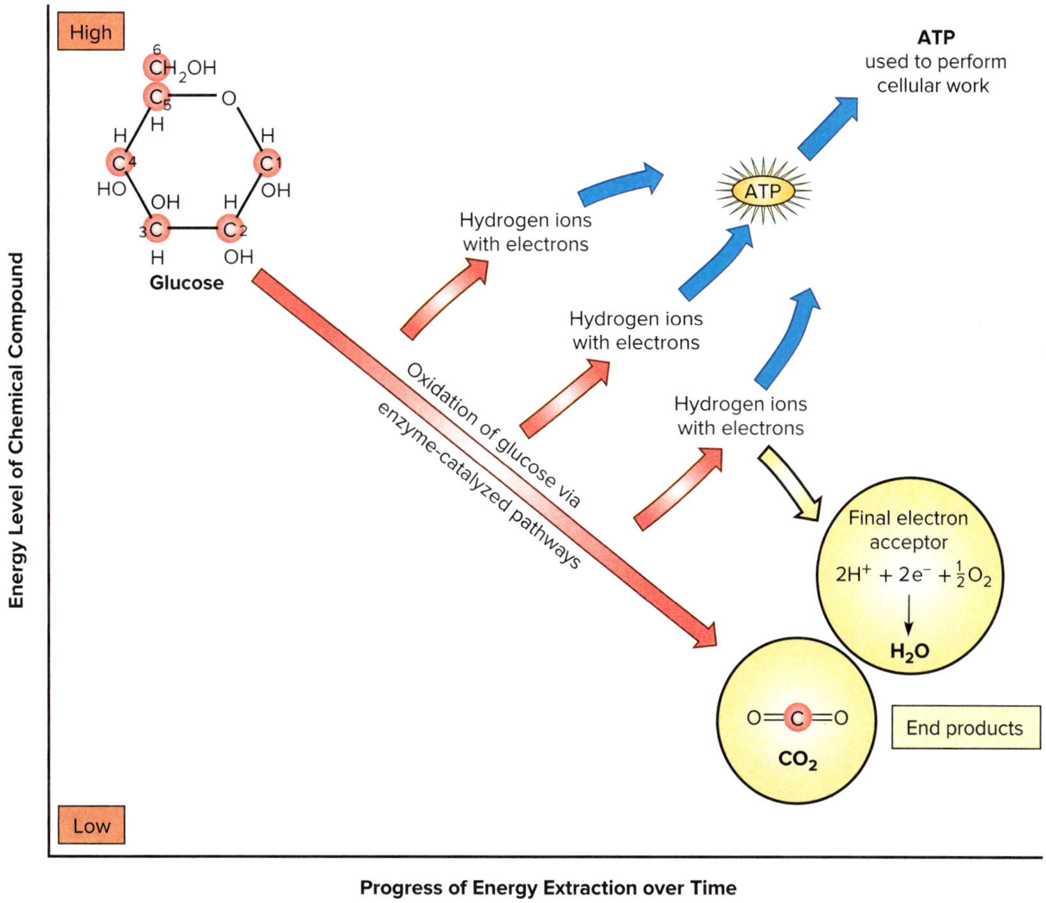
High
Low
Energy Level of Chemical Compound
Progress of Energy Extraction over Time
Glucose
Oxidation of glucose via enzyme-catalyzed pathways
Hydrogen ions with electrons
Hydrogen ions with electrons
Hydrogen ions with electrons
ATP
ATP used to perform cellular work
Final electron acceptor
$2H^+ + 2e^- + \frac{1}{2}O_2$
H_2O
CO_2
End products

Figure 10.10 A simplified model of energy production. Glucose is oxidized as it passes through sequential metabolic pathways, resulting in the removal of hydrogens and their accompanying electrons. The energy from the hydrogens and electrons is used to generate ATP. Eventually, all that is left of the carbon skeleton of glucose is the end product CO_2. Another by-product of aerobic metabolism (due to electrons and hydrogen ions combining with oxygen) is H_2O.

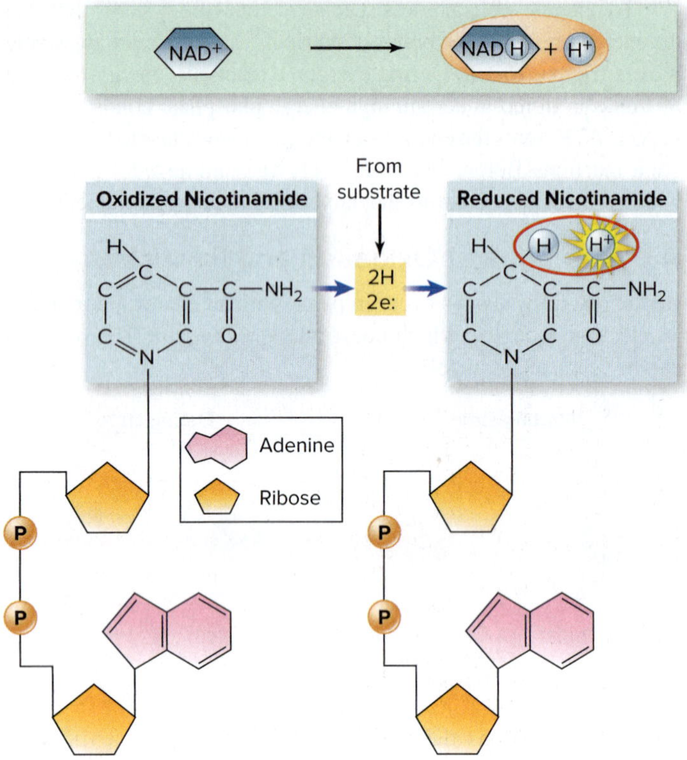
NAD+
NADH + H+
From substrate
Oxidized Nicotinamide
Reduced Nicotinamide
2H
2e:
Adenine
Ribose
P
P
P
P

Figure 10.11 Details of NAD reduction. The coenzyme NAD contains the vitamin nicotinamide (niacin) and the purine adenine attached to two ribose phosphate molecules. The principal site of action is on the nicotinamide (blue-boxed areas). Hydrogens and electrons donated by a substrate interact with a carbon on the top of the ring. One hydrogen bonds there, carrying two electrons (H:), and the other hydrogen is carried in solution as H^+ (a proton).

have coenzyme carriers called nicotinamide adenine dinucleotide (NAD) **(figure 10.11)** and flavin adenine dinucleotide (FAD). Just look one time at the whole name to appreciate that they are nucleotides, and concentrate on the blue boxes in figure 10.11. That's where the action is.

Oxidation-reduction reactions trade electrons back and forth along with the energy they contain. The newly reduced compound (the one that gains electrons) has more energy than it did in its original oxidized state. The energy now present in the electron acceptor can be captured to **phosphorylate** (add an inorganic phosphate) to ADP or to some other compound. This process stores the energy in a high-energy molecule (e.g., ATP). In many cases, the cell handles electrons not as separate entities but rather as parts of an atom such as hydrogen (which contains a proton and an electron). For simplicity's sake, we will continue to use the term *electron transfer,* but keep in mind that hydrogens are often involved in the transfer process. The removal of hydrogens from a compound during a redox reaction is called *dehydrogenation*. The job of handling these protons and electrons falls to one or more carriers, which function as short-term storehouses for the electrons until they can be transferred. As we will see, dehydrogenations are an essential supplier of electrons for the respiratory electron transport system.

Electron Carriers: Molecular Shuttles

Electron carriers resemble shuttles that are alternately loaded and unloaded, repeatedly accepting and releasing electrons and hydrogens to facilitate the transfer of redox energy. Most carriers transfer both electrons and hydrogens, but some transfer electrons only. The most common carrier is NAD, which carries hydrogens (and a pair of electrons) from dehydrogenation reactions (as shown in figure 10.11). Reduced NAD can be represented in various ways. Because 2 hydrogens are added, the actual carrier state is NADH + H^+, but this is cumbersome, so we will write it as "NADH." In catabolic pathways, electrons are extracted and carried through a series of redox reactions until the final electron acceptor at the end of a particular pathway is reached (see figure 10.10). In aerobic metabolism, this acceptor is molecular oxygen. In anaerobic metabolism, it is some other inorganic or organic compound. There are other common redox carriers: FAD, NADP (NAD phosphate), and the proteins of the respiratory chain, which are located on membranes. Double down on remembering what NAD, NADH, FAD, and FADH are, especially, and it will make the rest of this chapter - and coming chapters - much smoother for you.

Adenosine Triphosphate: Metabolic Money

Let's look more closely at the energy molecule ATP. ATP has been described as "metabolic money" because it can be earned, banked, saved, spent, and exchanged. As a temporary energy repository, ATP provides a connection between energy-yielding catabolism and the other cellular activities that require energy. Some clues to its energy-storing properties lie in its unique molecular structure.

The Molecular Structure of ATP

ATP is a three-part molecule consisting of a nitrogen base (adenine) linked to a 5-carbon sugar (ribose), with a chain of three phosphate groups bonded to the ribose **(figure 10.12).** The

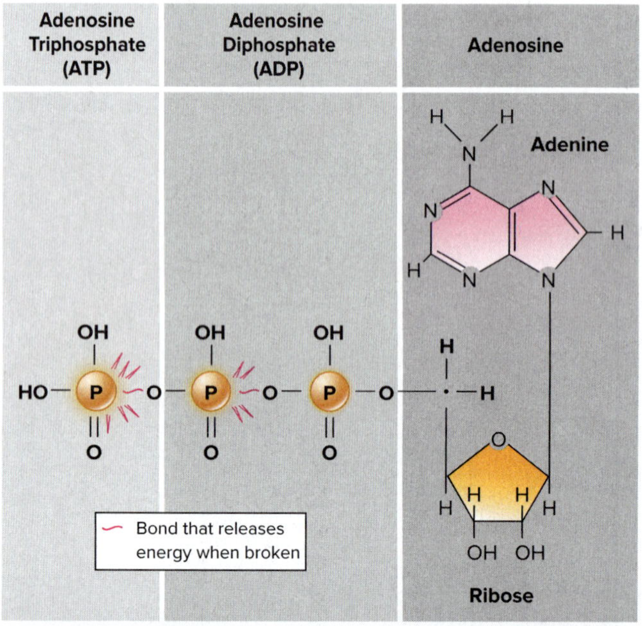
Adenosine Triphosphate (ATP)
Adenosine Diphosphate (ADP)
Adenosine
Adenine
Ribose
Bond that releases energy when broken

Figure 10.12 The structure of adenosine triphosphate (ATP). Removing the left-most phosphate group yields ADP; removing the next one yields AMP.

high energy of ATP comes from the orientation of the phosphate groups, which are bulky and carry negative charges. These negative charges are close together in the molecule, and because they are similar, they repel one another. This creates a strain, which is strongest in the bond between the last two phosphate groups. The strain on the phosphate bonds accounts for the high energy of ATP because removal of the terminal phosphates releases free energy.

Breaking the bonds between the two outermost phosphates of ATP yields adenosine diphosphate (ADP), which is then converted to adenosine monophosphate (AMP) with the breaking of the bond between the remaining two phosphates. AMP derivatives help form the backbone of RNA and are also a major component of certain coenzymes (NAD, FAD, and coenzyme A).

The Metabolic Role of ATP

ATP is the primary energy currency of the cell. When it is used in a chemical reaction, it must then be replaced. Therefore, ATP utilization and replenishment are in an ongoing cycle. Often, the energy released during ATP hydrolysis drives biosynthesis by providing an activating phosphate to an individual substrate before it is enzymatically linked to another substrate.

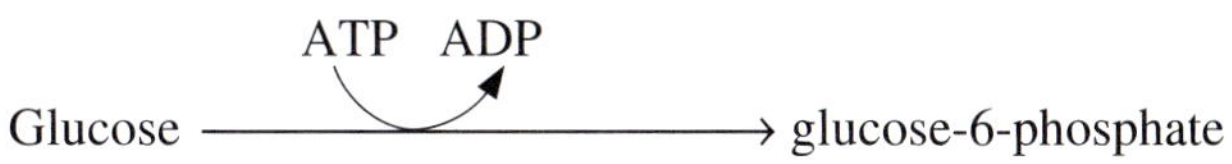
ATP
ADP
Glucose
glucose-6-phosphate

When ATP is utilized (when the terminal phosphate is removed to release energy plus ADP), ATP then needs to be regenerated. Adding the terminal phosphate back in to ADP will replenish ATP, but it requires an input of energy:

$$\text{ATP} \leftrightharpoons \text{ADP} + \text{P}_i + \text{energy}$$

In heterotrophs, the energy infusion that regenerates a high-energy phosphate comes from certain steps of catabolic pathways in which nutrients such as carbohydrates are degraded and yield energy. Some ATP molecules are formed through a process called *substrate-level phosphorylation.* In substrate-level phosphorylation, ATP is formed by transfer of a phosphate group from a phosphorylated compound (substrate) directly to ADP to yield ATP **(figure 10.13).**

Other ATPs are formed through *oxidative phosphorylation,* a series of redox reactions occurring during the final phase of the respiratory pathway. Phototrophic organisms have a system called *photophosphorylation,* in which the ATP is formed through a series of sunlight-driven reactions.

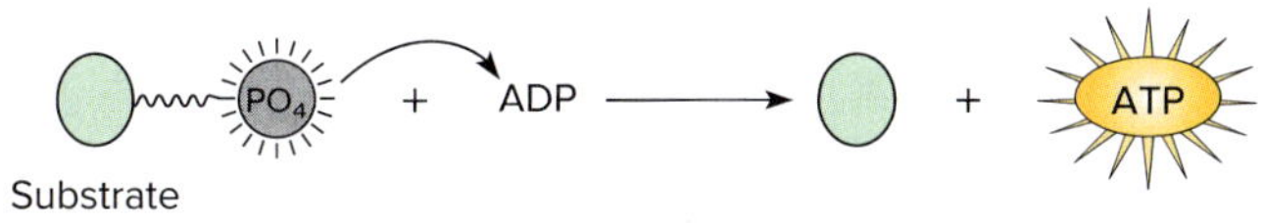
PO_4
+
ADP
+
ATP
Substrate

Figure 10.13 ATP formation by substrate-level phosphorylation. The inorganic phosphate and the substrates form a bond with high potential energy. In a reaction catalyzed enzymatically, the phosphate is transferred to ADP, thereby producing ATP.

10.2 Learning Outcomes—Assess Your Progress

7. Name the chemical in which energy is stored in cells.
8. Create a general diagram of a redox reaction.
9. Identify electron carriers used by cells.

10.3 Catabolism: Getting Materials and Energy

Now you have an understanding of all the tools a cell needs to *metabolize.* Metabolism uses *enzymes* to catalyze reactions that break down (*catabolize*) organic molecules to materials (*precursor molecules*) that cells can then use to build (*anabolize*) larger, more complex molecules that are particularly suited to them. This process is presented symbolically in figure 10.1, which is repeated as an icon next to each section in this part of the chapter so you can see which part of the overall picture we are talking about. Another very important point about metabolism is that *reducing power* (the electrons available in NADH and $FADH_2$) and *energy* (stored in the bonds of ATP) are needed in large quantities for the anabolic parts of metabolism (the blue bars in our figure). They are produced during the catabolic part of metabolism (the yellow bar).

Metabolism starts with "nutrients" from the environment, usually discarded molecules from other organisms. Cells have to get the nutrients inside. To do this, they use the transport mechanisms discussed in chapter 9. Some of these require energy, which is available from catabolism already occurring in the cell. In the next step, intracellular nutrients have to be broken down to the appropriate precursor molecules. The catabolic pathways that do this are discussed next.

Overview of Catabolism

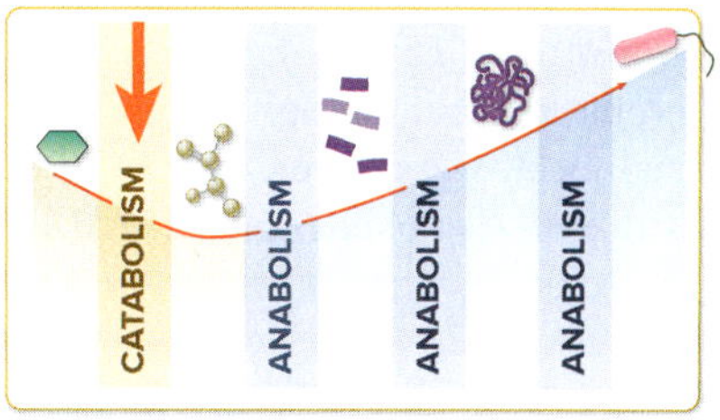
CATABOLISM
ANABOLISM
ANABOLISM
ANABOLISM

Nutrient processing is extremely varied, especially in bacteria, yet in most cases it is based on three basic catabolic pathways. Frequently, the nutrient is glucose. We will use this nutrient as an example, but there are many different nutrients that can be fed into catabolic pathways. There are several pathways that can be used to break down glucose, but the most common one is **glycolysis** (gly-kol′-ih-sis). After glycolysis, organisms use mainly three different pathways for producing the needed precursors and energy (i.e., catabolism) **(figure 10.14).**

Aerobic respiration is a series of reactions: glycolysis, the Krebs cycle, and the respiratory chain. This pathway converts glucose to CO_2 and allows the cell to recover significant amounts of energy (review figure 10.10). Aerobic respiration relies on free oxygen as the final acceptor for electrons and hydrogens and produces a large amount of ATP. Aerobic respiration is characteristic of many bacteria, fungi, protozoa, and animals. Facultative and aerotolerant anaerobes may use only the glycolysis portion to incompletely oxidize **(ferment)** glucose. In this case, oxygen

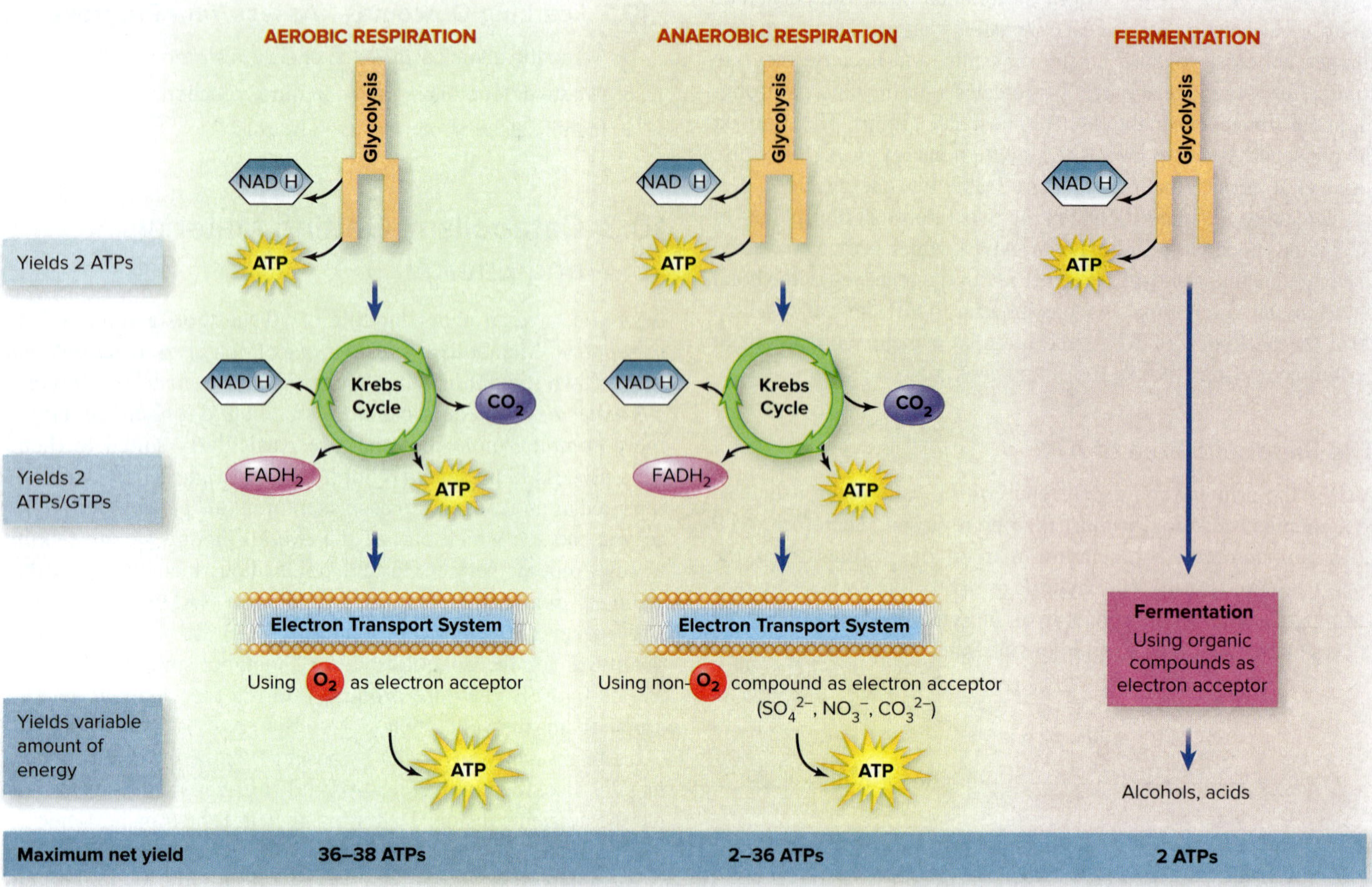
AEROBIC RESPIRATION
ANAEROBIC RESPIRATION
FERMENTATION
Glycolysis
NADH
ATP
Yields 2 ATPs
Krebs Cycle
CO2
FADH2
Yields 2 ATPs/GTPs
Electron Transport System
Using O2 as electron acceptor
Using non-O2 compound as electron acceptor (SO4 2−, NO3 −, CO3 2−)
Fermentation
Using organic compounds as electron acceptor
Yields variable amount of energy
Alcohols, acids
Maximum net yield
36–38 ATPs
2–36 ATPs
2 ATPs

Figure 10.14 Overview of the three main pathways of catabolism.

is not required, organic compounds are the final electron acceptors, and a relatively small amount of ATP is produced. Some anaerobic microorganisms metabolize by means of **anaerobic respiration.** This system involves the same three pathways as aerobic respiration, but it does not use molecular oxygen as the final electron acceptor. Instead, NO_3^-, SO_4^{2-}, CO_3^{2-}, and other oxidized compounds are utilized. Aspects of fermentation and anaerobic respiration are covered in subsequent sections of this chapter.

Aerobic Respiration

Aerobic respiration is the metabolic process in which electrons are transferred from fuel molecules such as glucose to oxygen as a final electron acceptor. This pathway is the principal energy-yielding scheme for aerobic heterotrophs, and it provides both ATP and metabolic intermediates for many other pathways in the cell, including those of protein, lipid, and carbohydrate synthesis.

Glucose: The Starting Compound

Carbohydrates such as glucose are good fuels because they are readily oxidized. That means that they are excellent hydrogen and electron donors. The enzymatic withdrawal of hydrogen from them also removes electrons that can be used in energy transfers. The end products of the conversion of these carbon compounds are energy-rich ATP and energy-poor carbon dioxide and water. Although in our discussion we use glucose as the main starting compound, other hexoses (fructose, galactose) and fatty acid subunits can enter the pathways of aerobic respiration as well.

Glycolysis: How It All Starts

Glycolysis uses several steps to convert glucose into pyruvic acid. Depending on the organism and the conditions, it may be only the first phase of aerobic respiration, or it may serve as the primary metabolic pathway (in the case of fermentation). Glycolysis provides a significant means to synthesize ATP and also to generate pyruvic acid, an essential intermediary metabolite.

Glycolysis proceeds along nine steps, starting with glucose and ending with pyruvic acid (pyruvate[2]). An overview of glycolysis will be presented here. **Figure 10.15** contains the chemical structures and a visual representation of the reactions. Each of the nine reactions is catalyzed by a specific enzyme with a specific name (not mentioned here).

2. In biochemistry, the terms used for organic acids appear as either the acid form (*pyruvic acid*) or its salt (*pyruvate*).

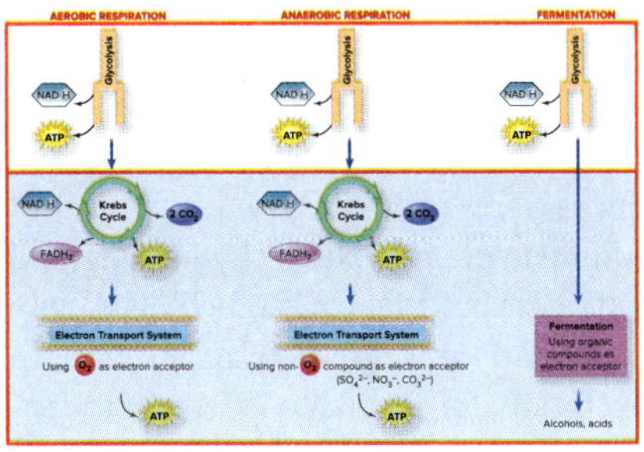
AEROBIC RESPIRATION
ANAEROBIC RESPIRATION
FERMENTATION

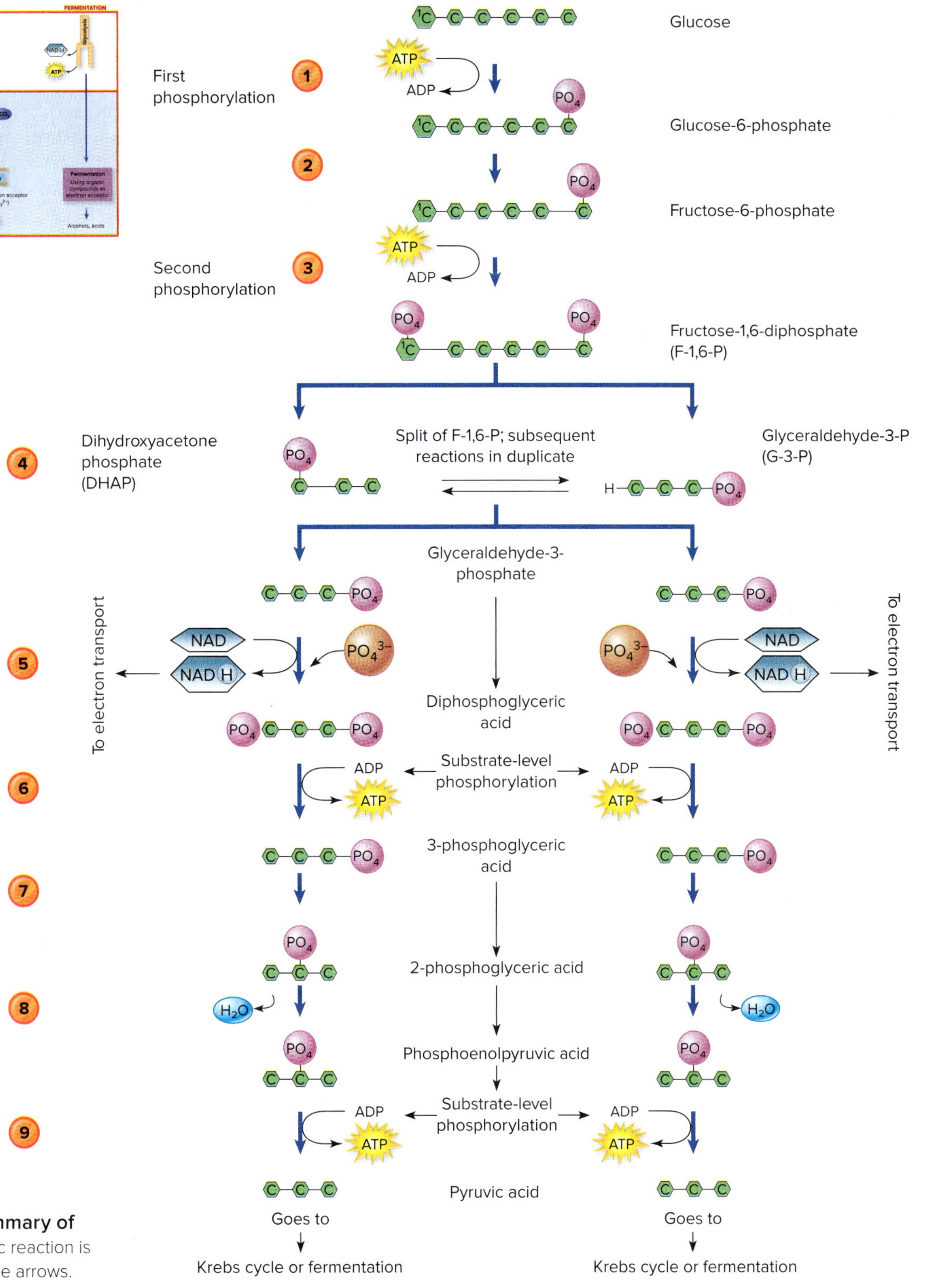
Glucose
First phosphorylation
1
ATP
ADP
PO_4
Glucose-6-phosphate
2
Fructose-6-phosphate
Second phosphorylation
3
Fructose-1,6-diphosphate (F-1,6-P)
4
Dihydroxyacetone phosphate (DHAP)
Split of F-1,6-P; subsequent reactions in duplicate
Glyceraldehyde-3-P (G-3-P)
Glyceraldehyde-3-phosphate
5
To electron transport
NAD
NAD H
PO_4^{3-}
Diphosphoglyceric acid
6
Substrate-level phosphorylation
3-phosphoglyceric acid
7
2-phosphoglyceric acid
8
H_2O
Phosphoenolpyruvic acid
9
Pyruvic acid
Goes to
Krebs cycle or fermentation

Figure 10.15 Summary of glycolysis. An enzymatic reaction is occurring at each of the blue arrows.

First, glucose is activated by adding a phosphate to it, resulting in glucose-6-phosphate. It is then converted (another reaction, another enzyme) to fructose-6-phosphate, and then another phosphate is added. The resulting molecule—fructose diphosphate—is more symmetrical and can be split into two 3-carbon molecules **(figure 10.15, 4)**. At this point, no oxidation-reduction has occurred and 2 ATPs have been used. The next step involves converting one C molecule (DHAP) into the other (glyceraldehyde-3-P; G-3-P), resulting in two G-3-Ps.

From here to the end, everything that happens in glycolysis happens twice—once to each of the 3-carbon molecules. First, the G-3-Ps each receives another phosphate. At the same time,

2 NADs in the vicinity are reduced to NADHs. These NADHs will be used in the last step of catabolism (the electron transport system) to produce ATP.

Although glycolysis is the main route to pyruvate production for most organisms, some microbes lack the enzymes for this pathway. There are alternate biochemical reactions such as the Entner-Doudoroff pathway (used by *Pseudomonas* and *Enterococcus* species) and the pentose phosphate pathway (used by some photosynthetic microbes). Our aim here is to focus on general principles, so we will restrict ourselves to glycolysis.

Disease Connection

Enterococci are bacteria that can live peacefully in the intestinal tract or the female reproductive tract, but they can occasionally cause disease in those areas or in surgical wounds. In recent years, *Enterococci* that are resistant to the antibiotic vancomycin have become problematic, particularly in hospitals. They are termed *VREs* (vancomycin-resistant enterococci). They have an extremely flexible metabolism and can use a huge variety of substrates to produce energy.

Pyruvic Acid: A Central Metabolite

Pyruvic acid occupies an important position in several pathways, and different organisms handle it in different ways **(figure 10.16).** In strictly aerobic organisms and some anaerobes, pyruvic acid enters the Krebs cycle for further processing and energy release. Facultative anaerobes can also use a fermentative metabolism, in which pyruvic acid is re-reduced into acids or other products.

The Krebs Cycle: A Carbon and Energy Wheel

In glycolysis, the oxidation of glucose yields a comparatively small amount of energy and produces pyruvic acid. Pyruvic acid is still energy-rich, containing a number of extractable hydrogens and electrons to power ATP synthesis, but this can be achieved only through the work of the second and third phases of respiration, in which pyruvic acid's hydrogens are transferred to oxygen, producing CO_2 and H_2O. In the following section, we examine the next phase of this process, the Krebs cycle (named after the scientist who first defined it). This set of reactions takes place in the cytoplasm of bacteria and is catalyzed by a group of enzymes (some of which are associated with the cytoplasmic membrane). In eukaryotic cells, this process takes place in the mitochondrial matrix.

To connect the glycolysis pathway to the Krebs cycle, for either aerobic or anaerobic respiration, the pyruvic acid is first converted to a starting compound for that cycle **(figure 10.17).** Here we have an oxidation-reduction reaction, which also releases the first carbon dioxide molecule. It involves a cluster of enzymes and coenzyme A that participate in the dehydrogenation (oxidation) of pyruvic acid, the reduction of NAD to NADH, and the decarboxylation of pyruvic acid to a 2-carbon acetyl group. The acetyl group remains attached to coenzyme A, forming acetyl coenzyme A (acetyl CoA) that feeds into the Krebs cycle.

The NADH formed during this reaction will be shuttled into electron transport and used to generate ATP via oxidative phosphorylation. *Keep in mind that all reactions described actually happen twice for each glucose because of the two pyruvates that are formed during glycolysis.*

The Krebs cycle as depicted in figure 10.17 always looks intimidating. Think of it as a series of eight reactions catalyzed by eight different enzymes.

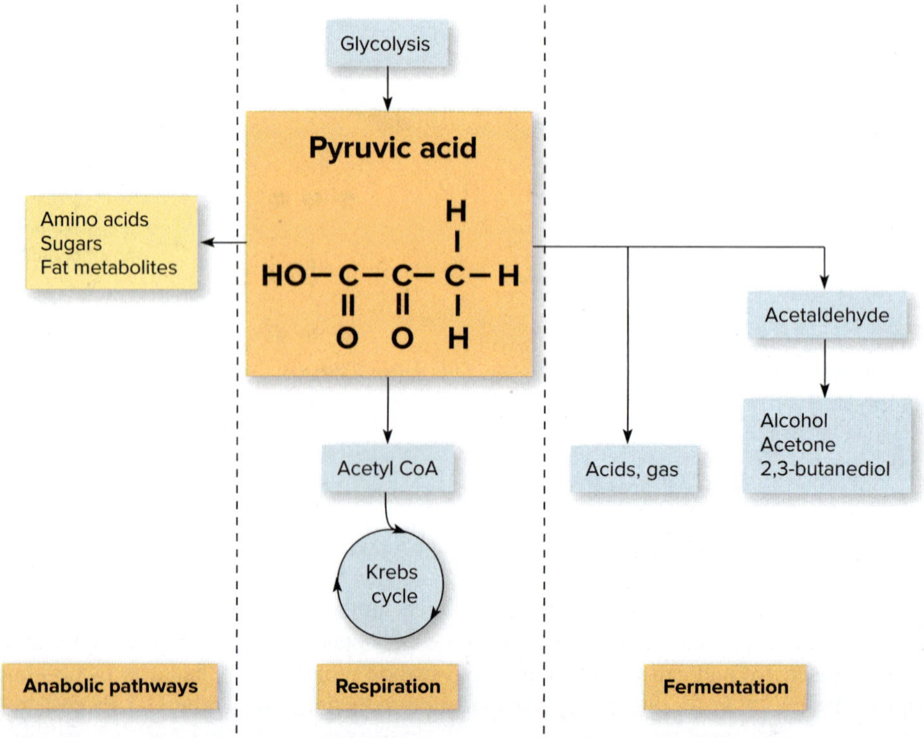
Glycolysis
Pyruvic acid
H
HO–C–C–C–H
O O H
Amino acids
Sugars
Fat metabolites
Acetaldehyde
Alcohol
Acetone
2,3-butanediol
Acetyl CoA
Acids, gas
Krebs cycle
Anabolic pathways
Respiration
Fermentation

Figure 10.16 The fates of pyruvic acid (pyruvate). This metabolite is an important hub in the processing of nutrients by microbes. It may be fermented anaerobically to several end products or oxidized completely to CO_2 and H_2O through the Krebs cycle and the electron transport system. It can also serve as a source of raw material for synthesizing amino acids and carbohydrates.

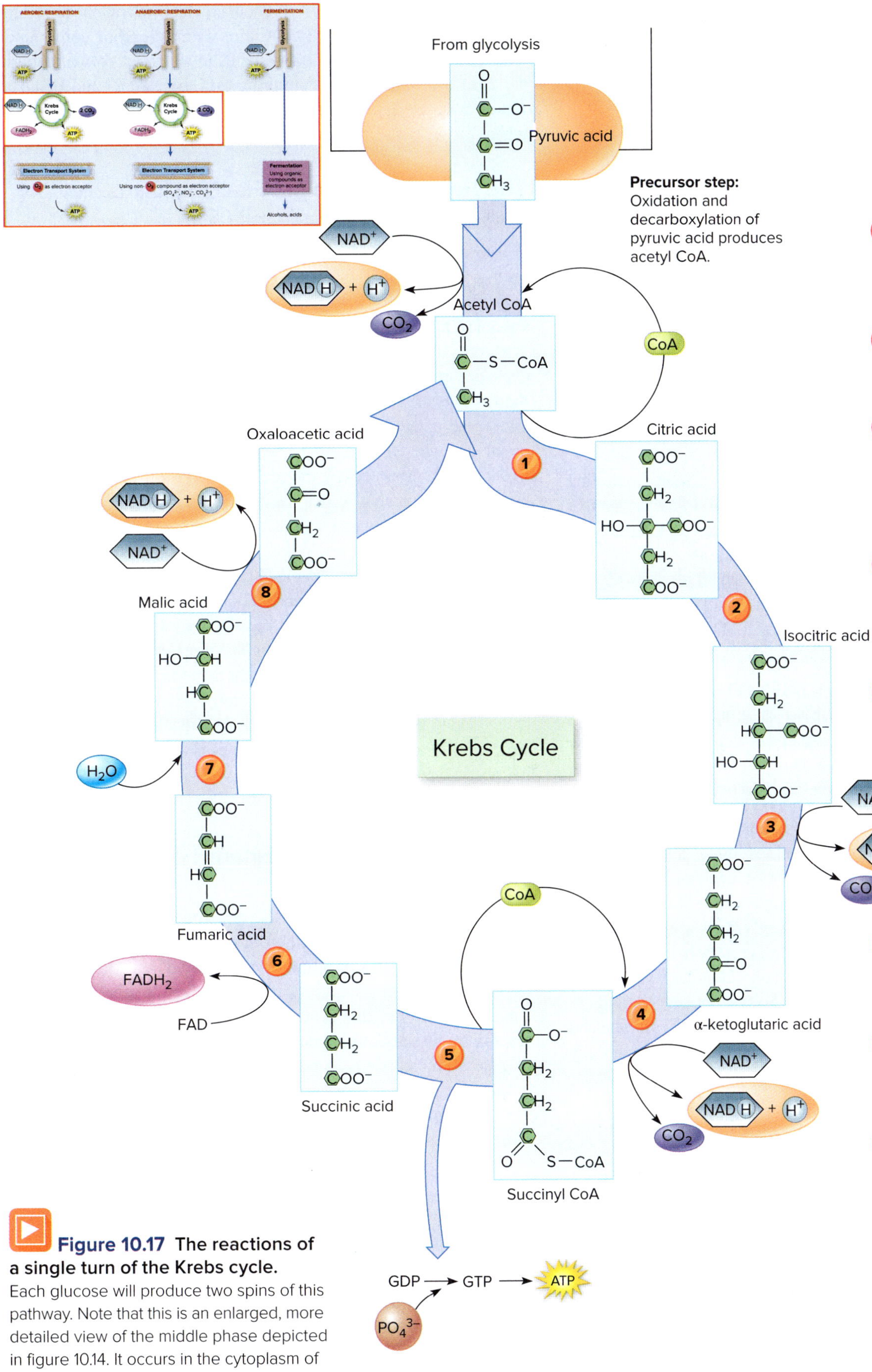
AEROBIC RESPIRATION
ANAEROBIC RESPIRATION
FERMENTATION
From glycolysis
Pyruvic acid
Precursor step:
Oxidation and decarboxylation of pyruvic acid produces acetyl CoA.
NAD+
NADH + H+
CO2
Acetyl CoA
CoA
Citric acid
Isocitric acid
α-ketoglutaric acid
Succinyl CoA
Succinic acid
Fumaric acid
Malic acid
Oxaloacetic acid
Krebs Cycle
FADH2
FAD
H2O
GDP → GTP → ATP
PO4 3−

1. The 2C acetyl CoA molecule combines with oxaloacetic acid, forming 6C citrate, and releasing CoA.
2. Citrate changes the arrangement of atoms to form isocitric acid.
3. Isocitric acid is converted to 5C α-ketoglutaric acid, which yields NADH and CO_2.
4. α-ketoglutaric acid loses the second CO_2 and generates another $NADH^+$ plus 4C succinyl CoA.
5. Succinyl CoA is converted to succinic acid and regenerates CoA. This releases energy that is captured in ATP.
6. Succinic acid loses 2 H^+ and 2 e^-, yielding fumaric acid and generating $FADH_2$.
7. Fumaric acid reacts with water to form malic acid.
8. An additional NADH is formed when malic acid is converted to oxaloacetic acid, which is the final product and can enter the cycle again, by reacting with acetyl CoA.

Figure 10.17 The reactions of a single turn of the Krebs cycle. Each glucose will produce two spins of this pathway. Note that this is an enlarged, more detailed view of the middle phase depicted in figure 10.14. It occurs in the cytoplasm of bacteria and the mitochondrial matrix of eukaryotes.

Steps in the Krebs Cycle

As you learned earlier, a cyclic pathway is one in which the starting compound is regenerated at the end. The Krebs cycle has eight steps, beginning with citric acid formation and ending with oxaloacetic acid. As we take a single spin around the Krebs cycle, it will be helpful to keep track of

- the numbers of carbons (#C) in each substrate and product,
- reactions where CO_2 is generated,
- the involvement of the electron carriers NAD and FAD, and
- where ATP is made.

The reactions in the Krebs cycle follow.

1. Oxaloacetic acid (oxaloacetate; 4C) reacts with the acetyl group (2C) on acetyl CoA, thereby forming citric acid (citrate; 6C) and releasing coenzyme A so it can join with another acetyl group.
2. Citric acid is converted to isocitric acid (isocitrate; 6C) to prepare this substrate for the decarboxylation and dehydrogenation of the next step.
3. Isocitric acid is acted upon by an enzyme complex including NAD or NADP (depending on the organism) in a reaction that generates NADH or NADPH, splits off a carbon dioxide, and leaves alpha-ketoglutaric acid (α-ketoglutarate; 5C).
4. Alpha-ketoglutaric acid serves as a substrate for the last decarboxylation reaction and yet another redox reaction involving coenzyme A and yielding NADH. The product is the high-energy compound succinyl CoA (4C).

At this point, the cycle has released 3 CO_2 molecules that balance out the original 3-carbon pyruvic acid that began the Krebs cycle. The remaining steps are needed not only to regenerate the oxaloacetic acid to start the cycle again but also to extract more energy from the intermediate compounds leading to oxaloacetic acid.

5. Succinyl CoA is the source of the one substrate-level phosphorylation in the Krebs cycle. In most microbes, it proceeds with the formation of GTP, guanosine triphosphate. GTP is readily converted to ATP. The product of this reaction is succinic acid (succinate; 4C).
6. Succinic acid next becomes dehydrogenated, but in this case, the electron and H^+ acceptor is flavin adenine dinucleotide (FAD). The enzyme that catalyzes this reaction, succinyl dehydrogenase, is found in the bacterial cytoplasmic membrane and mitochondrial cristae of eukaryotic cells. $FADH_2$ then directly enters the electron transport system. Fumaric acid (fumarate; 4C) is the product of this reaction.
7. The addition of water to fumaric acid (called hydration) results in malic acid (malate; 4C). This is one of the few reactions in respiration that directly incorporate water.
8. Malic acid is dehydrogenated (with formation of a final NADH), and oxaloacetic acid is formed. This step brings the cycle back to its original starting position, where oxaloacetic acid can react with acetyl coenzyme A.

The Krebs cycle serves to transfer the energy stored in acetyl CoA to NAD^+ and FAD by reducing them (transferring hydrogen ions to them). For that reason, the main products of the Krebs cycle are these reduced molecules (as well as 2 ATPs for each glucose molecule). The reduced coenzymes NADH and $FADH_2$ are vital to the energy production that will occur in electron transport. Along the way, the 2-carbon acetyl CoA joins with a 4-carbon compound, oxaloacetic acid, and then participates in seven additional chemical transformations while "spinning off" the NADH and $FADH_2$. That's why we called the Krebs cycle the "carbon and energy wheel."

The Respiratory Chain: Electron Transport and Oxidative Phosphorylation

We now come to the energy chain, which is the final "processing mill" for electrons and hydrogen ions and the major generator of ATP. Overall, the electron transport system (ETS) consists of a chain of special redox carriers (proteins) that receives electrons from reduced carriers (NADH, $FADH_2$) generated by glycolysis and the Krebs cycle and passes them in a sequential and orderly fashion from one redox molecule to the next. The flow of electrons down this chain allows the active transport of hydrogen ions to the outside of the membrane where the respiratory chain is located. The step that finalizes the transport process is the acceptance of electrons and hydrogen by oxygen, producing water. This process consumes oxygen. Some variability exists from one organism to another, but the principal compounds that carry out these complex reactions are NADH dehydrogenase, flavoproteins, coenzyme Q (ubiquinone), and **cytochromes.** The cytochromes contain a tightly bound metal atom at their center that is actively involved in accepting electrons and donating them to the next carrier in the series. The highly compartmentalized structure of the respiratory chain is an important factor in its function. Note in **figure 10.18** that the electron transport carriers and enzymes are embedded in the cytoplasmic membrane in bacteria. The equivalent structure for housing them in eukaryotes is the inner mitochondrial membranes pictured in **figure 10.19.** We will describe the electron transport system in both bacteria and eukaryotes.

Elements of Electron Transport: The Energy Cascade

The principal questions about the electron transport system are: How are the electrons passed from one carrier to another in the series? How does this progression result in ATP synthesis? How is oxygen (or another electron acceptor) utilized? Although the biochemical details of this process are rather complicated, the basic reactions consist of a number of redox reactions now familiar to us. In general, the carrier compounds and their enzymes are arranged in linear sequence and are reduced and then oxidized in turn.

The sequence of electron carriers in the respiratory chain of most aerobic organisms is

1. NADH dehydrogenase;
2. flavin mononucleotide (FMN);
3. coenzyme Q;
4. cytochrome *b;*
5. cytochrome c_1;
6. cytochrome *c;* and
7. cytochromes *a* and a_3, which are complexed together.

NADH from glycolysis and from the Krebs cycle enters the chain at the first carrier. This sets in motion the next six steps. With each redox exchange, the energy level of the reactants

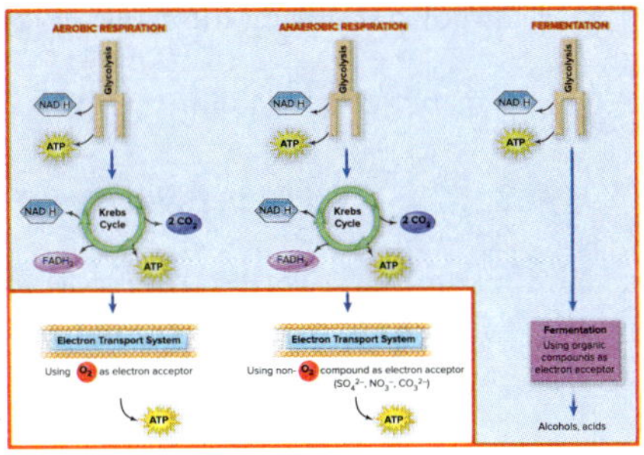

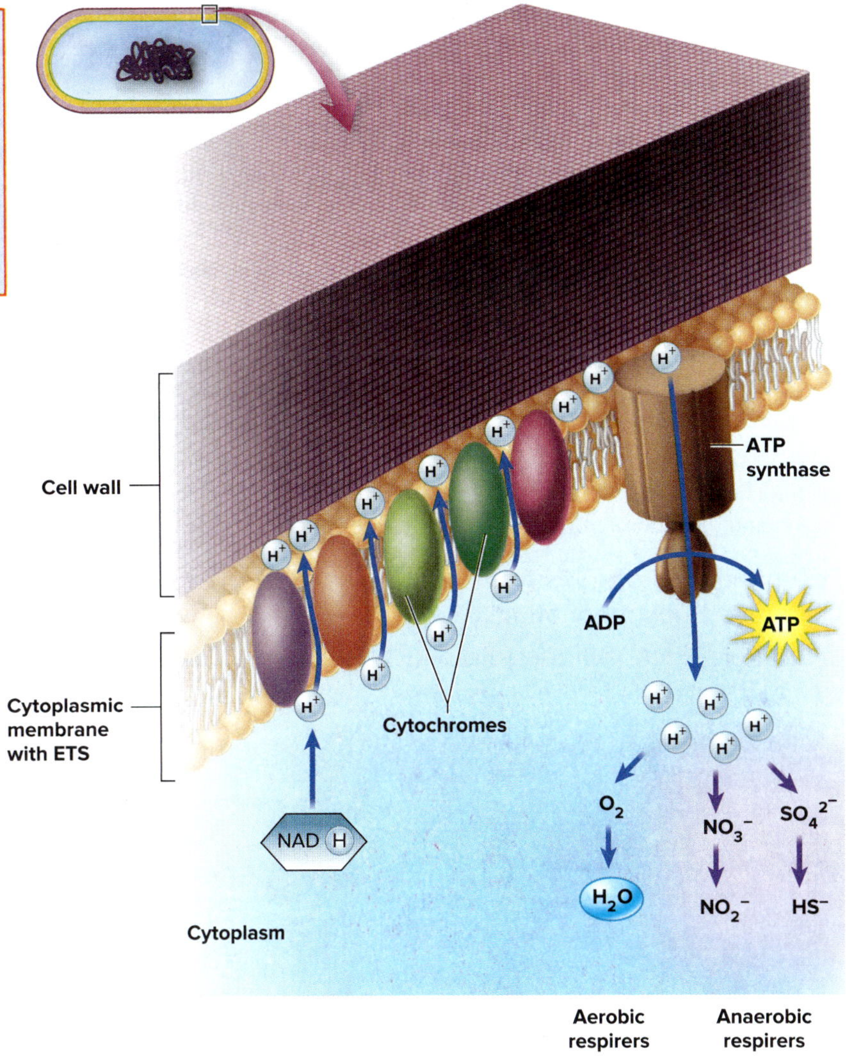
Cell wall
Cytoplasmic membrane with ETS
ATP synthase
ADP
ATP
Cytochromes
NAD H
Cytoplasm
O_2
H_2O
NO_3^-
NO_2^-
SO_4^{2-}
HS⁻
Aerobic respirers
Anaerobic respirers

Figure 10.18 The electron transport system and oxidative phosphorylation in bacterial membranes. Starting at NADH dehydrogenase, electrons brought in from the Krebs cycle by NADH are passed along the chain of electron transport carriers. Each adjacent pair of transport molecules undergoes a redox reaction. Coupled to the transport of electrons is the simultaneous active transport of H^+ into the periplasm by specific carriers. These processes set the scene for ATP synthesis and final H^+ and e^- acceptance. Note the differences in final electron acceptors in aerobic versus anaerobic respirers.

is decreased. The released energy is captured and used by the **ATP synthase** complex, stationed along the membrane at the end of the line of ETS carriers. Each NADH that enters electron transport can give rise to 3 ATPs. This coupling of ATP synthesis to electron transport is termed **oxidative phosphorylation.** Because the electrons from $FADH_2$ from the Krebs cycle enter the cycle at a later point, there is less energy to release, and only 2 ATPs are the result.

The Formation of ATP and Chemiosmosis

How exactly is electron transport linked to ATP production? We will first look at the system in bacteria, which have the components of electron transport embedded in a precise sequence on the cytoplasmic membrane. The process is called **chemiosmosis.** As the electron transport carriers shuttle electrons, they actively pump hydrogen ions (protons) into the periplasmic space, or the space between the wall and the cytoplasmic membrane, depending on whether the bacterium is gram-positive or gram-negative. This process sets up a concentration gradient of hydrogen ions called the *proton motive force* (*PMF*). The PMF consists of a difference in charge between the outside of the membrane (+) and the inside of the membrane (−) (see figure 10.18).

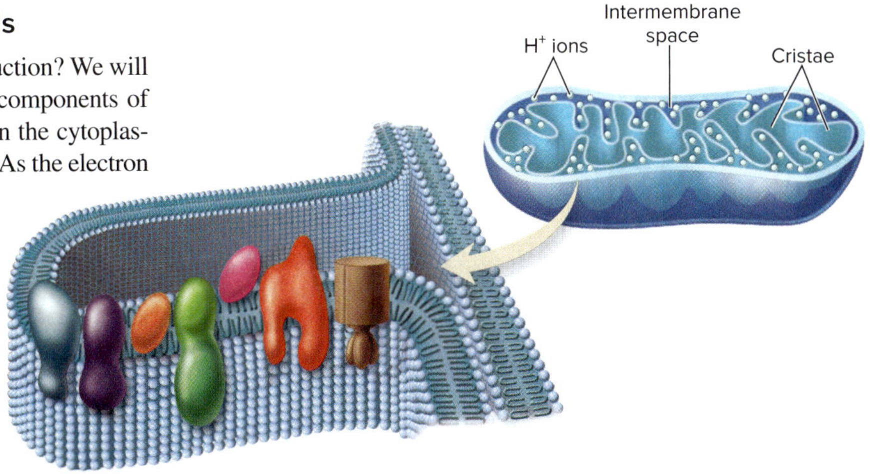
H^+ ions
Intermembrane space
Cristae

Figure 10.19 The electron transport system on the inner membrane of the mitochondrial cristae. As the carriers in the mitochondrial cristae transport electrons, they also actively pump H^+ ions (protons) to the intermembrane space, producing a chemical and charge gradient between the outer and inner mitochondrial compartments.

Separating the charge has the effect of a battery, which can temporarily store potential energy. This charge is maintained by the fact that H^+ can't cross the membrane. The only site where H^+ can diffuse into the cytoplasm is at the ATP synthase complex, which sets the stage for the final processing of H^+ leading to ATP synthesis.

ATP synthase is a complex enzyme composed of two large units (see figures 10.18 and 10.19). It is embedded in the membrane, but part of it rotates like a motor and traps chemical energy. As the H^+ ions flow through the center of the enzyme by diffusion, the other compartments pull in ADP and P_i. Rotation causes a three-dimensional change in the enzyme that bonds these two molecules, thereby releasing ATP into the cytoplasm (see figure 10.18). The enzyme is then rotated back to the start position and will continue the process.

Eukaryotic ATP synthesis occurs by means of the same overall process. However, eukaryotes have the ETS stationed in mitochondrial membranes, between the inner mitochondrial matrix and the outer intermembrane space (see figure 10.19).

Potential Yield of ATPs from Oxidative Phosphorylation

The total of five NADHs (four from the Krebs cycle and one from glycolysis) can be used to synthesize

3 ATPs per electron pair = 15 ATPs
and
15 × 2 = 30 ATPs per glucose

The single $FADH_2$ produced during the Krebs cycle results in

2 ATPs per electron pair = 2 ATPs
and
2 × 2 = 4 ATPs per glucose

Figure 10.20 summarizes the total of ATP and other products for the entire aerobic pathway. These totals are the maximum yields possible but may not be fulfilled by many organisms.

Summary of Aerobic Respiration

Originally, we presented a summary equation for respiration. We are now in a position to add up the input and output of this equation at various points in the pathways you can see and sum up the final ATP. In figure 10.20 you can see several important aspects of aerobic respiration:

1. The total possible production of ATP is 40: 4 from glycolysis, 2 from the Krebs cycle, and 34 from electron transport. However, because 2 ATPs were already spent in early glycolysis, this leaves a maximum of **38 ATPs.**

   The actual totals may be lower in certain eukaryotic cells because energy is spent in transporting the NADH produced during glycolysis across the mitochondrial membrane. Certain aerobic bacteria come closest to achieving the full total of 38

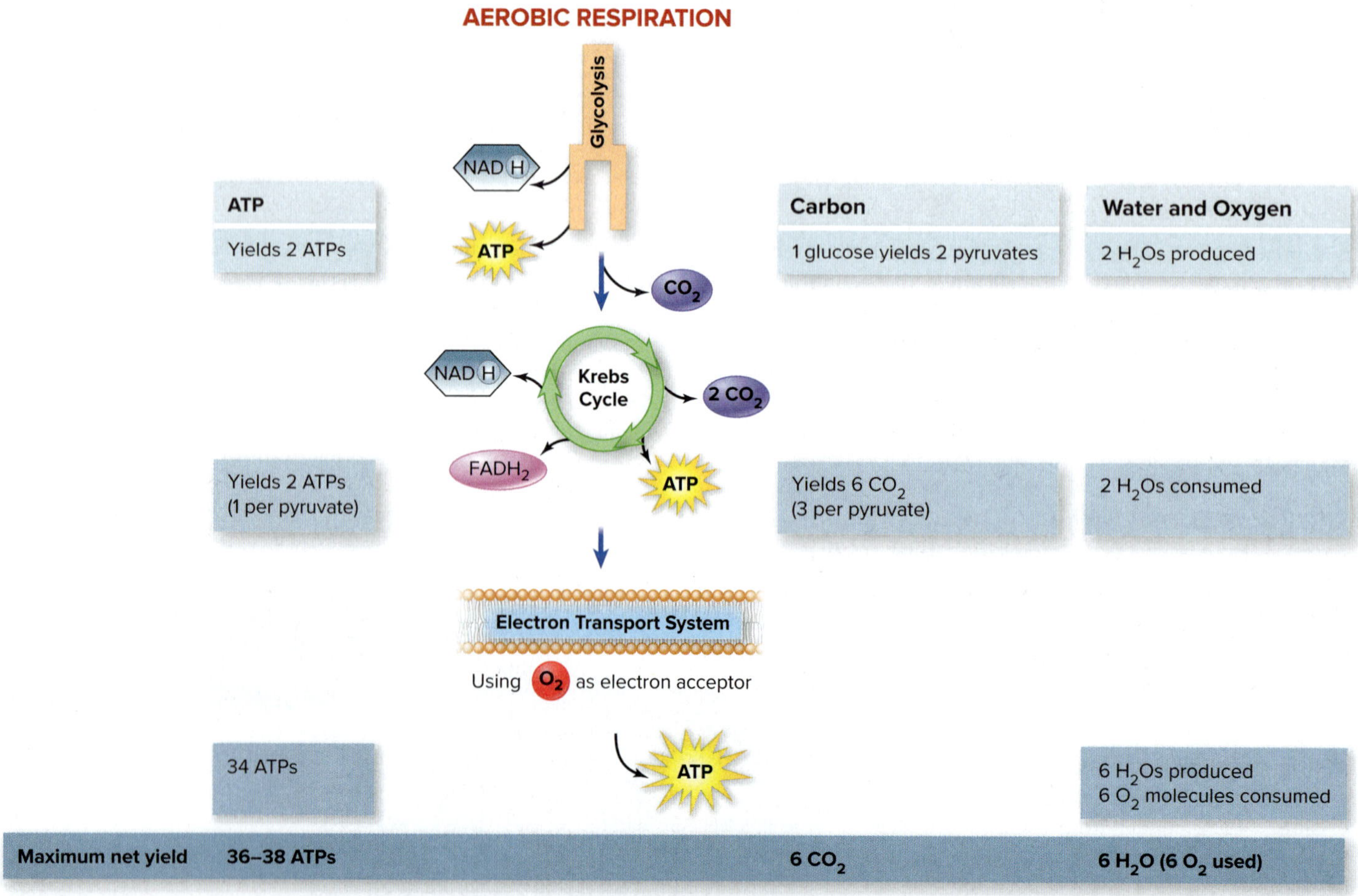
AEROBIC RESPIRATION
Glycolysis
NADH
ATP
ATP
Yields 2 ATPs
CO_2
Carbon
1 glucose yields 2 pyruvates
Water and Oxygen
2 H_2Os produced
NADH
Krebs Cycle
2 CO_2
$FADH_2$
ATP
Yields 2 ATPs (1 per pyruvate)
Yields 6 CO_2 (3 per pyruvate)
2 H_2Os consumed
Electron Transport System
Using O_2 as electron acceptor
ATP
34 ATPs
6 H_2Os produced
6 O_2 molecules consumed
Maximum net yield
36–38 ATPs
6 CO_2
6 H_2O (6 O_2 used)

Figure 10.20 Theoretic ATP yield from aerobic respiration. To attain the theoretic maximum yield of ATP, we assume a ratio of 3 for the oxidation of NADH to 2 for $FADH_2$. The actual yield is generally lower and varies between eukaryotes and bacteria and among bacterial species.

because they lack mitochondria and, for that reason, do not have to use ATP in the transport of NADH across the outer mitochondrial membrane.

2. Six carbon dioxide molecules are generated during the Krebs cycle.
3. Six oxygen molecules are consumed during electron transport.
4. Six water molecules are produced in electron transport and 2 in glycolysis, but because 2 are used in the Krebs cycle, this leaves a net number of 6.

The Terminal Step

The last step, during which oxygen accepts the electrons, is catalyzed by cytochrome aa_3, also called cytochrome oxidase. This large enzyme complex is specifically adapted to receive electrons from another cytochrome, pick up hydrogens from the solution, and react with oxygen to form a molecule of water. This reaction, though in actuality more complex, is summarized as follows:

$$2H^+ + 2e^- + \frac{1}{2}O_2 \rightarrow H_2O$$

Most eukaryotic aerobes have a fully functioning cytochrome system, but bacteria exhibit wide-ranging variations in this part of the system. Some species lack one or more of the redox steps. Others have several alternative electron transport schemes. Because many bacteria lack cytochrome oxidase, this variation can be used to differentiate among certain genera of bacteria. An oxidase detection test can be used to help identify members of the genera *Neisseria* and *Pseudomonas* and some species of *Bacillus*. Another variation in the cytochrome system is evident in certain bacteria (*Klebsiella, Enterobacter*) that can grow even in the presence of cyanide because they lack cytochrome oxidase. Cyanide will cause rapid death in humans and other eukaryotes because it blocks cytochrome oxidase, thereby completely blocking aerobic respiration, but it is harmless to these bacteria.

A potential side reaction of the respiratory chain in aerobic organisms is the incomplete reduction of oxygen to superoxide ion (O_2^-) and hydrogen peroxide (H_2O_2). These toxic oxygen products can be very damaging to cells. Aerobes have neutralizing enzymes to deal with these products, including *superoxide dismutase* and *catalase*. One exception is the genus *Streptococcus,* which can grow well in oxygen yet lacks both cytochromes and catalase. The tolerance of these organisms to oxygen can be explained by the neutralizing effects of a special peroxidase. The lack of cytochromes, catalase, and peroxidases in anaerobes as a rule limits their ability to process free oxygen and contributes to its toxic effects on them.

Anaerobic Respiration

Some bacteria have evolved an anaerobic respiratory system that functions like the aerobic cytochrome system except that it utilizes oxygen-containing ions, rather than free oxygen, as the final electron acceptor in electron transport (see figure 10.14). Of these, the nitrate (NO_3^-) and nitrite (NO_2^-) reduction systems are best known. The reaction in species such as *Escherichia coli* is represented as:

$$\underset{\text{nitrate}}{NO_3^-} + NADH \xrightarrow{\text{nitrate reductase}} \underset{\text{nitrite}}{NO_2^-} + H_2O + NAD^+$$

The enzyme nitrate reductase catalyzes the removal of oxygen from nitrate, leaving nitrite and water as products. A test for this reaction is one of the physiological tests used in identifying bacteria.

It has been found that nitrate-reducing bacteria in the mouth and gut can contribute to pulmonary and cardiovascular diseases in the host due to the overproduction of nitrate, nitrite, and other forms of nitrogen oxide, which damage blood vessels. This is a clear example of the interaction of the host microbiome on health and disease.

Fermentation

The third main catabolism pathway is **fermentation.** The definition of fermentation is *the incomplete oxidation of glucose or other carbohydrates in the absence of oxygen.* This process uses organic compounds—as opposed to O_2 or oxygen-containing ions—as the terminal electron acceptors and yields only a small amount of ATP generated in glycolysis, the beginning part of fermentation (see figure 10.14).

Over time, the term *fermentation* has acquired several looser connotations. Originally, Pasteur called the microbial action of yeast during wine production *ferments,* and to this day, biochemists use the term in reference to the production of ethyl alcohol by yeasts acting on glucose and other carbohydrates. Fermentation is also what bacteriologists call the formation of acid, gas, and other products by the action of various bacteria on pyruvic acid. The process is a common metabolic strategy among bacteria. Industrial processes that produce chemicals on a massive scale through the actions of microbes are also called fermentations. Each of these usages is acceptable for one application or another.

Without the use of an electron transport chain, it may seem that fermentation would yield only meager amounts of energy (2 ATPs maximum per glucose) and that would slow down growth. What actually happens, however, is that many bacteria can grow as fast as they would in the presence of oxygen because they speed up the rate of glycolysis. From another standpoint, fermentation permits independence from molecular oxygen and allows colonization of anaerobic environments. It also enables microorganisms with a versatile metabolism to adapt to variations in the availability of oxygen. For them, fermentation provides a means to grow even when oxygen levels are too low for aerobic respiration.

Bacteria that digest cellulose in the rumens of cattle are largely fermentative. After initially hydrolyzing cellulose to glucose, they ferment the glucose to organic acids, which are then absorbed as the bovine's principal energy source. Even human muscle cells can undergo a form of fermentation that permits short periods of activity after the oxygen supply in the muscle has been exhausted. Muscle cells convert pyruvic acid into lactic acid, which allows anaerobic production of ATP to proceed for a time. But this cannot go on indefinitely, and after a few minutes, the accumulated lactic acid causes muscle fatigue.

Disease Connection

Bacteria that cause disease in humans also can use fermentative pathways. Surprisingly, one such bacterium colonizes the lungs, where you would think oxygen would be abundant. However, in some circumstances (particularly in patients with cystic fibrosis), the bacterium *Pseudomonas aeruginosa* creates biofilms in the lungs and the biofilms become anaerobic. In these conditions, the bacterium ferments pyruvate to acetate and lactic acid to survive.

Products of Fermentation in Microorganisms

Alcoholic beverages (wine, beer, whiskey) are perhaps the most well-known fermentation products. Others are solvents (acetone, butanol), organic acids (lactic, acetic), dairy products, and many other foods. Derivatives of proteins, nucleic acids, and other organic compounds are deliberately fermented to produce vitamins, antibiotics, and even hormones such as hydrocortisone.

Fermentation products can be grouped into two general categories: alcoholic fermentation products and acidic fermentation products **(figure 10.21). Alcoholic fermentation** occurs in yeast or bacterial species that have metabolic pathways for converting pyruvic acid to ethanol. This process involves a decarboxylation of pyruvic acid to acetaldehyde, followed by a reduction of the acetaldehyde to ethanol. In oxidizing the NADH formed during glycolysis, NAD is regenerated, which allows the glycolytic pathway to continue. These processes are crucial in the production of beer and wine, though the actual techniques for arriving at the desired amount of ethanol and the prevention of unwanted side reactions are important tricks of the brewer's trade. Note that the products of alcoholic fermentation are not only ethanol but also CO_2, a gas that accounts for the bubbles in champagne and beer.

The pathways of **acidic fermentation** are extremely varied. Lactic acid bacteria ferment pyruvate in the same way that humans do—by reducing it to lactic acid. If the product of this fermentation is mainly lactic acid, as in certain species of *Streptococcus* and *Lactobacillus,* it is termed *homolactic.* The souring of milk is due largely to the production of this acid by bacteria. When glucose is fermented to a mixture of lactic acid, acetic acid, and carbon dioxide, as is the case with *Leuconostoc* and other species of *Lactobacillus,* the process is termed *heterolactic fermentation.*

Many members of the family *Enterobacteriaceae* (*Escherichia, Shigella,* and *Salmonella*) possess enzyme systems for converting pyruvic acid to several acids simultaneously. **Mixed acid fermentation** produces a combination of acetic, lactic, succinic, and formic acids, and it lowers the pH of a medium to about 4.0. *Propionibacterium* produces primarily propionic acid, which gives the characteristic flavor to Swiss cheese, while the gas (CO_2) that is released produces the holes. Some of these bacteria also

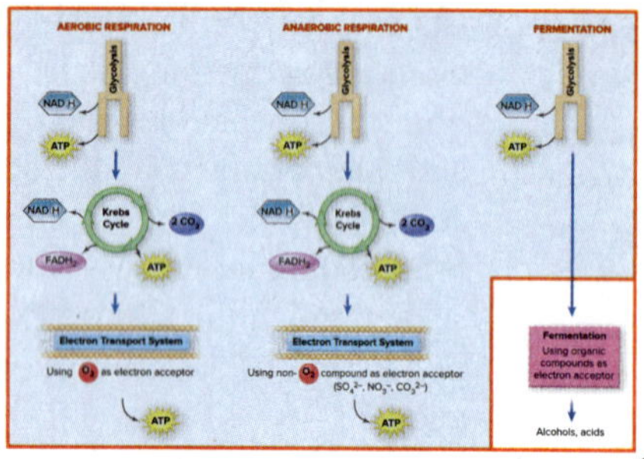

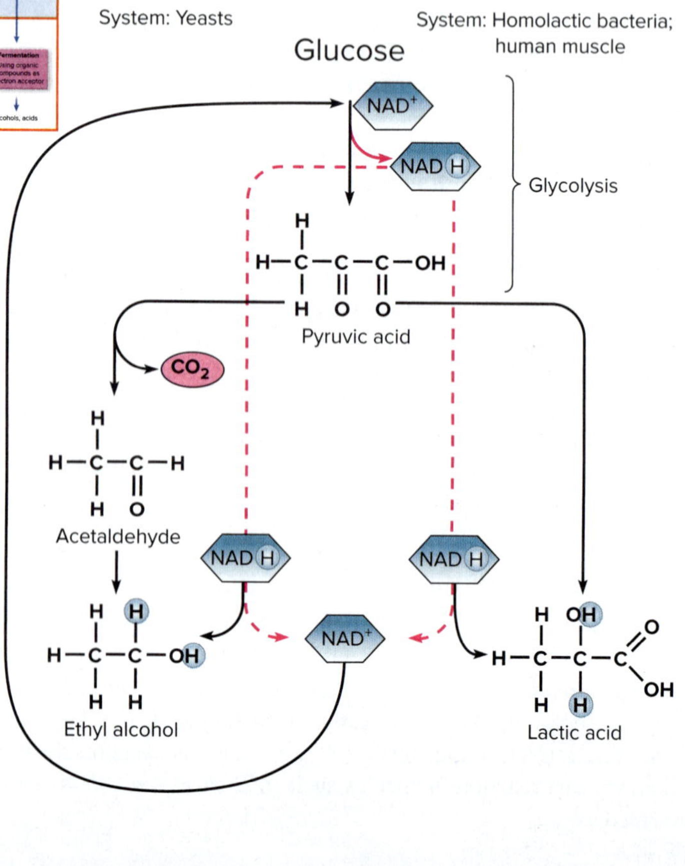
System: Yeasts
System: Homolactic bacteria; human muscle
Glucose
NAD+
NADH
Glycolysis
Pyruvic acid
CO2
Acetaldehyde
NADH
NADH
NAD+
Ethyl alcohol
Lactic acid

Figure 10.21 The chemistry of fermentation systems that produce acid and alcohol. In both cases, the final electron acceptor is an organic compound. In yeasts, pyruvic acid is decarboxylated to acetaldehyde, and the NADH given off in the glycolytic pathway reduces acetaldehyde to ethyl alcohol. In homolactic fermentative bacteria, pyruvic acid is reduced by NADH to lactic acid. Both systems regenerate NAD to feed back into glycolysis or other cycles.

further decompose formic acid completely to carbon dioxide and hydrogen gases. Because enteric bacteria commonly occupy the intestine, this fermentative activity accounts for the accumulation of some types of gas in the intestine.

Catabolism of Noncarbohydrate Compounds

We have given you one version of events for catabolism, using glucose, a carbohydrate, as our example. Other compounds serve as fuel, as well. The more complex polysaccharides are easily broken down into their component sugars, which can enter glycolysis at various points. Microbes also break down other molecules for their own use, of course. Two other major sources of energy and building blocks for microbes are lipids (fats) and proteins. Both of these must be broken down to their component parts to produce precursor metabolites and energy.

Recall from chapter 2 that fats are fatty acids joined to glycerol. Enzymes called **lipases** break these apart. The glycerol is then converted to dihydroxyacetone phosphate (DHAP), which can enter step 4 of glycolysis (see figure 10.15). The fatty acid component goes through a process called **beta oxidation.** Fatty acids have a variable number of carbons. In beta oxidation, 2-carbon units are successively transferred to coenzyme A, creating acetyl CoA, which enters the Krebs cycle. This process can yield a large amount of energy. Oxidation of a 6-carbon fatty acid yields 50 ATPs, compared with 38 for a 6-carbon sugar.

Proteins are chains of amino acids. Enzymes called **proteases** break proteins down to their amino acid components, after which the amino groups are removed by a reaction called **deamination (figure 10.22).** This leaves a carbon compound, which is easily converted to one of several Krebs cycle intermediates.

10.3 Learning Outcomes—Assess Your Progress

10. Name the three main catabolic pathways and the estimated ATP yield for each.
11. Construct a paragraph summarizing glycolysis.
12. Describe the Krebs cycle and compare the process between bacteria and eukaryotes.
13. Discuss the significance of the electron transport system.
14. State two ways in which anaerobic respiration differs from aerobic respiration.
15. Summarize the steps of microbial fermentation, and list three useful products it can create.
16. Describe how noncarbohydrate compounds are catabolized.

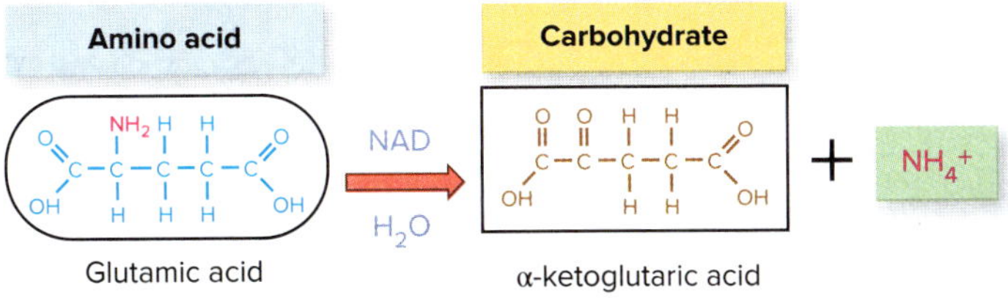
Amino acid
Carbohydrate
NAD
H_2O
NH_4^+
Glutamic acid
α-ketoglutaric acid

Figure 10.22 Deamination. Removal of an amino group converts an amino acid to an intermediate of carbohydrate metabolism. Ammonium is a by-product.

10.4 Biosynthesis and the Crossing Pathways of Metabolism

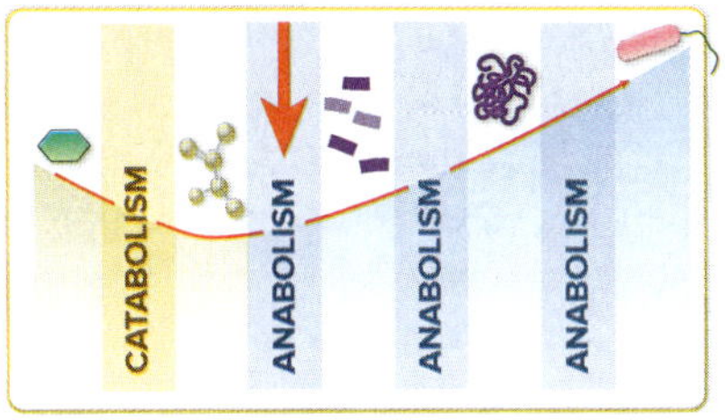
CATABOLISM
ANABOLISM
ANABOLISM
ANABOLISM

Our discussion now turns from catabolism and energy extraction to anabolic functions and biosynthesis. In this section, we present aspects of intermediary metabolism, including amphibolic pathways, the synthesis of simple molecules, and the synthesis of macromolecules.

The Efficiency of the Cell

It must be obvious by now that cells have mechanisms for careful management of carbon compounds. Rather than being dead ends, most catabolic pathways contain strategic molecular intermediates (metabolites) that can be diverted into anabolic pathways. In this way, a given molecule can serve multiple purposes, and the maximum benefit can be derived from all nutrients and metabolites of the cell pool. The property of a system to integrate catabolic and anabolic pathways to improve cell efficiency is termed **amphibolism** (am-fee-bol′-izm).

At this point in the chapter, you can appreciate a more complex view of metabolism than that presented at the beginning in figure 10.1. **Figure 10.23** includes the activities of amphibolism and so provides a more detailed picture. Cells are efficient. They can divert intermediate products from glycolysis to synthesize carbohydrates if needed. You can see that catabolism and anabolism are not always strictly separated the way figure 10.1 implies.

Amphibolic Sources of Cellular Building Blocks

Glyceraldehyde-3-phosphate can be diverted away from glycolysis and converted into precursors for amino acid, carbohydrate, and triglyceride (fat) synthesis. (A *precursor molecule* is a compound that is the source of another compound.) We have already noted the numerous directions that pyruvic acid catabolism can take. In terms of synthesis, pyruvate also plays a pivotal role in providing intermediates for amino acids. In the event of an inadequate glucose supply, pyruvate serves as the starting point in glucose synthesis from various metabolic intermediates, a process called **gluconeogenesis** (gloo′-koh-nee′-oh-gen′-uh-sis).

The acetyl group that starts the Krebs cycle is another extremely versatile metabolite that can be fed into a number of synthetic pathways. This 2-carbon fragment can be converted as a single unit into one of several amino acids, or a number of these fragments can be condensed into hydrocarbon chains that are important building blocks for fatty acid and lipid synthesis. Note that the reverse is also true—fats can be degraded to acetyl through beta oxidation and thereby enter the Krebs cycle as acetyl coenzyme A.

Pathways that synthesize the nitrogen bases (purines, pyrimidines), which are components of DNA and RNA, originate in amino acids and so can be dependent on intermediates from the Krebs cycle as well. Because the coenzymes NAD, NADP, FAD, and others contain purines and pyrimidines similar to the nucleic

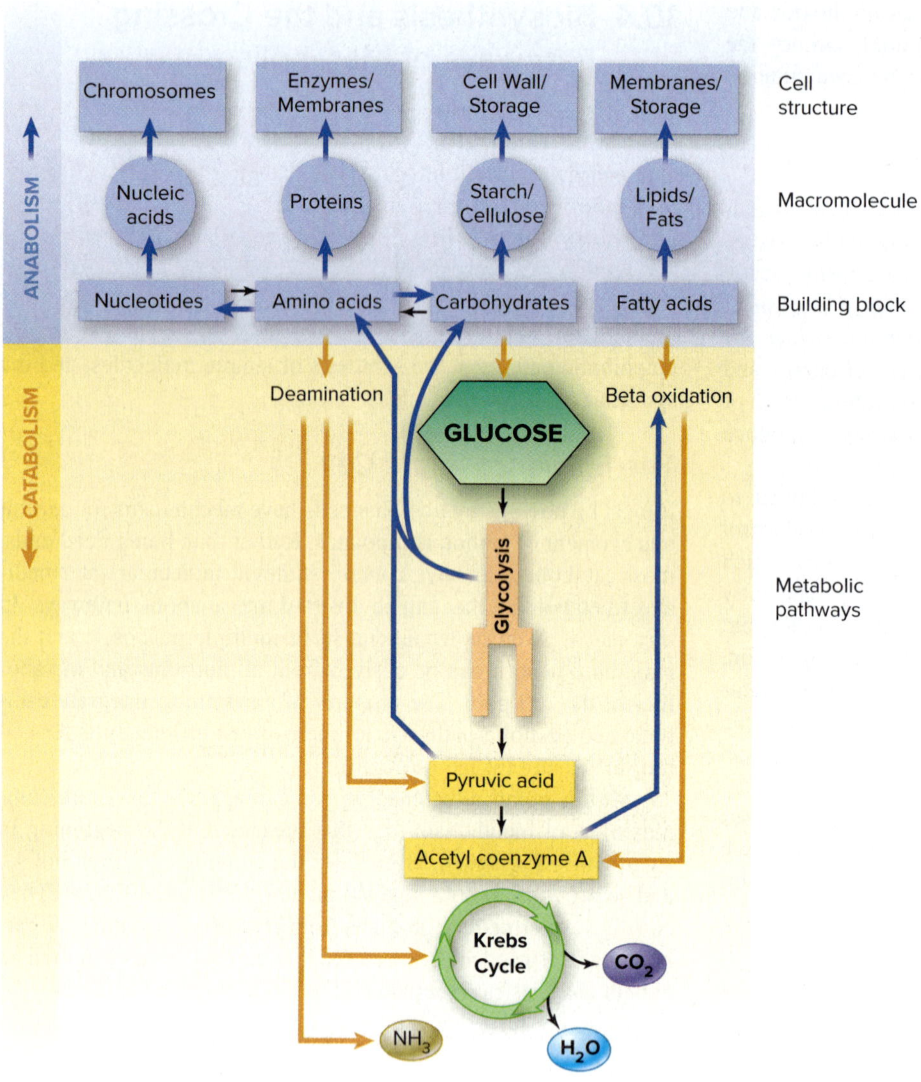
Chromosomes
Enzymes/ Membranes
Cell Wall/ Storage
Membranes/ Storage
Cell structure
Nucleic acids
Proteins
Starch/ Cellulose
Lipids/ Fats
Macromolecule
Nucleotides
Amino acids
Carbohydrates
Fatty acids
Building block
ANABOLISM
CATABOLISM
Deamination
GLUCOSE
Beta oxidation
Glycolysis
Metabolic pathways
Pyruvic acid
Acetyl coenzyme A
Krebs Cycle
CO_2
NH_3
H_2O

Figure 10.23 An amphibolic view of metabolism. A simple compound such as glucose is both broken down into smaller parts (catabolism) and modified and used as a building block in anabolism.

acids, their synthetic pathways are also dependent on amino acids. During times of carbohydrate deprivation, organisms can likewise convert amino acids to intermediates of the Krebs cycle by deamination and thereby derive energy from proteins (see figure 10.22).

We've been discussing these pathways within individual cells, but it is clear that microbes and their metabolic intermediaries and products influence other microbes and other organisms in their environment. The notions of symbiosis discussed in chapter 9 encompass the ability of adjacent organisms to use their neighbor's products.

Anabolism: Formation of Macromolecules

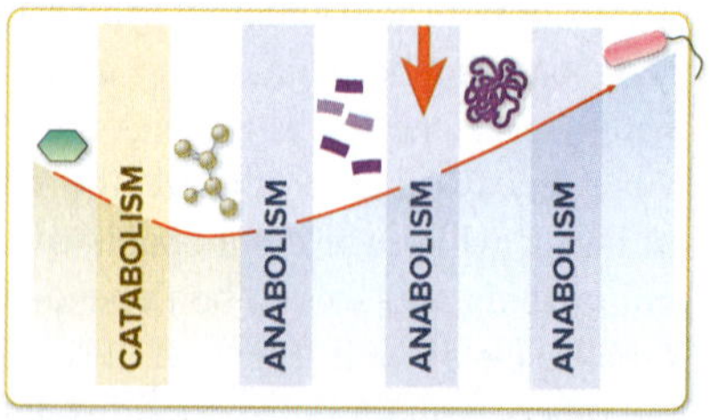
CATABOLISM
ANABOLISM
ANABOLISM
ANABOLISM

Monosaccharides, amino acids, fatty acids, nitrogen bases, and vitamins—the building blocks that make up the various macromolecules and organelles of the cell—come from two possible sources. They can enter the cell from the outside "ready to use," or they can be synthesized through various cellular pathways. The degree to which an organism can synthesize its own building blocks (simple molecules) is determined by its genetic makeup, a factor that varies tremendously from group to group. In chapter 9, you learned that autotrophs require only CO_2 as a carbon source, a few minerals to synthesize all cell substances, and no organic nutrients. Some heterotrophic organisms (*E. coli,* yeasts) are also very efficient in that they can synthesize all cellular substances from minerals and one organic carbon source such as glucose. Compare this with a strict parasite that has few synthetic abilities of its own and derives most precursor molecules from the host.

Whatever their source, once these building blocks are added to the metabolic pool, they are available for synthesis of polymers by the cell. The details of synthesis vary among the types of macromolecules, but all of them involve the formation of bonds by specialized enzymes and the expenditure of ATP.

Carbohydrate Biosynthesis

The role of glucose in metabolism and energy utilization is so crucial that its biosynthesis is ensured by several alternative pathways. Certain structures in the cell depend on an adequate supply of glucose as well. It is the major component of the cellulose cell walls of some eukaryotes and of certain storage granules (starch, glycogen). One of the intermediaries in glycolysis, glucose-6-P, is used to form glycogen. Monosaccharides other than glucose are important in the synthesis of bacterial cell walls. Peptidoglycan contains a linked polymer of muramic acid and glucosamine. Fructose-6-P from glycolysis is used to form these two sugars. Carbohydrates (deoxyribose, ribose) are also essential building blocks in nucleic acids. Polysaccharides are the predominant components of cell surface structures such as capsules and the glycocalyx, and they are commonly found in slime layers.

Amino Acids, Protein Synthesis, and Nucleic Acid Synthesis

Proteins account for a large proportion of a cell's contents. They are essential components of enzymes, the cytoplasmic membrane, the cell wall, and cell appendages. As a general rule, 20 amino acids are needed to make these proteins. Although some organisms (*E. coli,* for example) have pathways that will synthesize all 20 amino acids, others, including animals, lack some or all of the pathways for amino acid synthesis and must acquire the essential ones from their diets. Protein synthesis itself is a complex process that requires a genetic blueprint and the operation of intricate cellular machinery, as you saw in chapter 6.

Assembly of the Cell

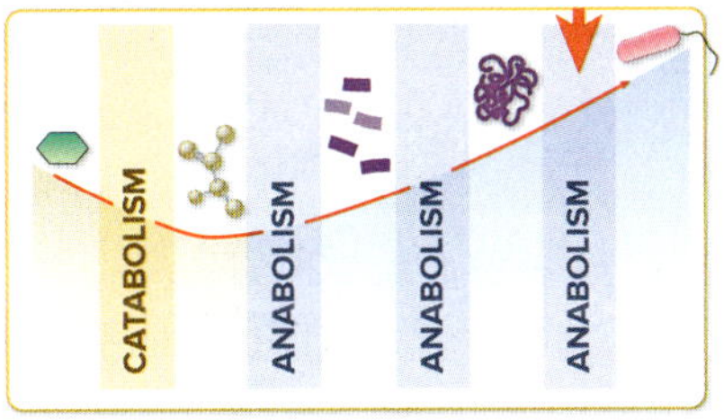
CATABOLISM
ANABOLISM
ANABOLISM
ANABOLISM

The component parts of a bacterial cell are synthesized on a continuous basis, and catabolism is also taking place as long as nutrients are present and the cell is in a nondormant state. When anabolism produces enough macromolecules to serve two cells, and when DNA replication produces duplicate copies of the cell's genetic material, the cell undergoes binary fission, which results in two cells from one parent cell. The two cells will need twice as many ribosomes, twice as many enzymes, and so on. The cell has created these during the initial anabolic phases we have described. Before cell division, the membrane(s) and the cell wall will have increased in size to create a cell that is almost twice as big as a "newborn" cell. Once synthesized, the phospholipid bilayer components of the membranes assemble themselves spontaneously with no energy input. Other assembly reactions require the input of energy. Proteins and other components must be added to the membranes. Growth of the cell wall, accomplished by the addition and coupling of sugars and peptides, requires energy input. The energy accumulated during catabolic processes provides all the energy for these complex building reactions.

10.4 Learning Outcomes—Assess Your Progress

17. Provide an overview of the anabolic stages of metabolism.
18. Define *amphibolism*.

10.5 Photosynthesis: It Starts with Light

As we mentioned earlier, the ultimate source of chemical energy in many cells comes from the sun. Most organisms depend either directly or indirectly on the sunlight's energy, which is converted into chemical energy through photosynthesis. (Some chemoautotrophs derive their energy and nutrients solely from inorganic substrates.) The other major products of photosynthesis are organic carbon compounds, which are produced from carbon dioxide through a process called carbon fixation.

With some notable exceptions, the energy that drives all life processes comes from the sun, but this source is directly available only to photosynthesizers. On land, green plants are the primary photosynthesizers. In aquatic ecosystems, algae, green and purple bacteria, and cyanobacteria fill this role. It was also recently discovered that bacteriophages that infect marine cyanobacteria provide some of the genes allowing these organisms to carry out photosynthesis.

Photosynthetic organisms use light energy to produce high-energy glucose from low-energy CO_2 and water or as you see in **Insight 10.1,** from electricity. They do this through a series of reactions involving light, pigment, CO_2, and water, which is used as a source for electrons.

Photosynthesis proceeds in two phases: the **light-dependent reactions,** which proceed only in the presence of sunlight, and the **light-independent reactions,** which proceed regardless of the lighting conditions (light or dark).

Solar energy is delivered in discrete energy packets called **photons** (also called *quanta*) that travel as waves. The wavelengths of light operating in photosynthesis occur in the visible spectrum between 400 (violet) and 700 nanometers (red). As this light strikes photosynthetic pigments, some wavelengths are absorbed, some pass through, and some are reflected. The activity that has greatest impact on photosynthesis is the absorbance of light by photosynthetic pigments. These include the **chlorophylls,** which are green; **carotenoids,** which are yellow, orange, or red; and **phycobilins,** which are red or blue-green.[3] The most important of these pigments are the bacterial chlorophylls. These molecules contain a *photocenter* that consists of a magnesium atom held in the center of a complex, ringed molecule called a *porphyrin.* As we will see, the chlorophyll molecule harvests the energy of photons and converts it to chemical energy. Other pigments such as carotenes trap

3. The color of the pigment corresponds to the wavelength of light it reflects.

INSIGHT 10.1 MICROBIOME: Electricity Eaters

This chapter describes metabolism, and states right up front that metabolism has three goals:

- break down large molecules, usually sugars, to get building blocks;
- build molecules the cell needs; and
- harvest energy to do the building work.

Put another way, electron flow—from an energy source to electron carriers to the electron transport chain—is necessary for energy to be gained.

Well, nature continues to surprise us with its many different variations on central themes. Scientists have discovered microbes that survive without using any sugars, which are the usual source of material for building blocks and energy. In fact, they metabolize and grow using only electrons from electricity. They may run the TCA cycle in reverse, creating acetyl CoA from carbon dioxide. From acetyl CoA they can create all of the molecules they need for a new cell. Then again, the very newest bacteria discovered can live (but not reproduce) with no carbon input at all—not even CO_2.

Dr. Ken Nealson, at the University of Southern California, whose lab conducted these studies, says, "This is huge. What it means is there's a whole part of the microbial world that we don't know about." In addition, there are a slew of practical applications, including waste recycling and creating biological fuel cells. Dr. Nealson is also guessing that this might be the predominant mode of life on other planets.

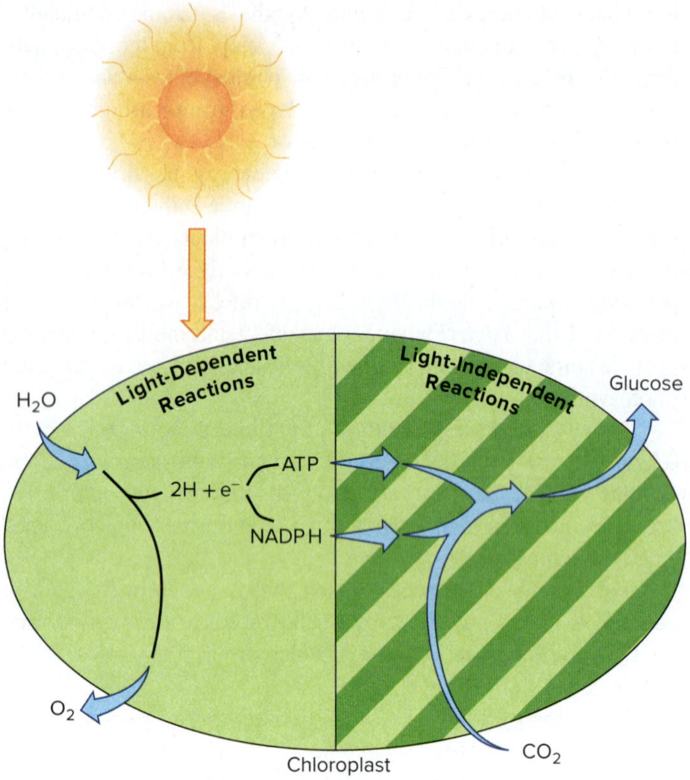

Light-Dependent Reactions
Light-Independent Reactions
H_2O
Glucose
2H + e⁻
ATP
NADPH
O_2
Chloroplast
CO_2

Figure 10.24 Overview of photosynthesis. The general reactions of photosynthesis, divided into two phases called light-dependent reactions and light-independent reactions. The dependent reactions require light to activate chlorophyll pigment and use the energy given off during activation to split an H_2O molecule into oxygen and hydrogen, producing ATP and NADPH. The independent reactions, which occur either with or without light, utilize ATP and NADPH produced during the light reactions to fix CO_2 into organic compounds such as glucose.

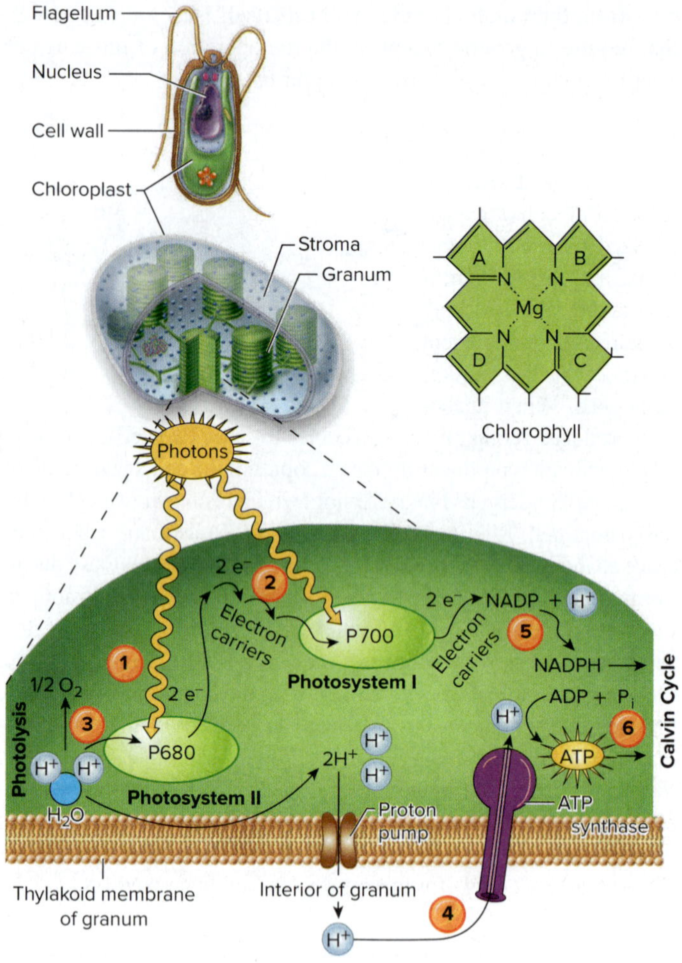

Flagellum
Nucleus
Cell wall
Chloroplast
Stroma
Granum
Chlorophyll
Mg
Photons
2 e⁻
Electron carriers
P700
Photosystem I
NADP + H⁺
NADPH
ADP + P_i
ATP
Calvin Cycle
1/2 O_2
Photolysis
P680
Photosystem II
2H⁺
H_2O
Proton pump
ATP synthase
Thylakoid membrane of granum
Interior of granum

The main events of the light reaction shown as an exploded view in one granum.

1. When light activates photosystem II, it sets up a chain reaction, in which electrons are released from chlorophyll.
2. These electrons are transported along a chain of carriers to photosystem I.
3. The empty position in photosystem II is replenished by photolysis of H_2O. Other products of photolysis are O_2 and H^+.
4. Pumping of H^+ into the interior of the granum produces conditions for ATP to be synthesized.
5. The final electron and H^+ acceptor is NADP, which receives these from photosystem I.
6. Both NADPH and ATP are fed into the stroma for the Calvin cycle.

Figure 10.25 The reactions of photosynthesis. A cell of the eukaryotic motile alga *Chlamydomonas,* with a single large chloroplast (magnified cutaway view). The chloroplast contains membranous compartments called grana where chlorophyll molecules and the photosystems for the light molecules are located.

light energy and shuttle it to chlorophyll, functioning like antennae. These light-dependent reactions are catabolic (energy-producing) reactions, which pave the way for the next set of reactions, the light-independent reactions, which need the produced energy for synthesis (anabolism). During this phase, carbon atoms from CO_2 are added to the carbon backbones of organic molecules.

The detailed biochemistry of photosynthesis is not necessary for this text, but we will provide an overview of the general process as it occurs in green plants, algae, and cyanobacteria **(figure 10.24).** Many of the basic activities (electron transport and phosphorylation) are similar to certain pathways of respiration.

Light-Dependent Reactions

The same systems that carry the photosynthetic pigments are also the sites for the light reactions. They occur in the **thylakoid** membranes of compartments called grana (singular, *granum*) in chloroplasts **(figure 10.25).** In bacteria, this occurs in specialized parts of the cytoplasmic membranes. These systems exist as two separate complexes called *photosystem I* (P700) and *photosystem II* (P680).[4] Both systems contain chlorophyll and are simultaneously activated by light, but the reactions in photosystem II help drive photosystem I. Together the systems are activated by light, then transport electrons, pump hydrogen ions, and form ATP and NADPH.

When photons enter the photocenter of the P680 system (PS II), the magnesium atom in chlorophyll becomes excited and

4. The numbers refer to the wavelength of light to which each system is most sensitive.

releases 2 electrons. The loss of electrons from the photocenter has two major effects:

1. It creates a vacancy in the chlorophyll molecule forceful enough to split an H_2O molecule into hydrogen (H^+) (electrons and hydrogen ions) and oxygen (O_2). This splitting of water, termed **photolysis,** is the ultimate source of the O_2 gas that is an important product of photosynthesis. The electrons released from the lysed water regenerate photosystem II for its next reaction with light.
2. Electrons generated by this first photoevent are immediately shunted through a series of carriers (cytochromes) to the P700 system. At this same time, hydrogen ions accumulate in the internal space of the thylakoid complex, thereby producing an electrochemical gradient.

The P700 system (PS I) has been activated by light so that it is ready to accept electrons generated by the PS II. The electrons it receives are passed along a second transport chain to a complex that uses electrons and hydrogen ions to reduce NADP to NADPH. (Recall that reduction in this sense entails the addition of electrons and hydrogens to a substrate.)

A second energy reaction involves synthesis of ATP by a chemiosmotic mechanism similar to that shown in figures 10.18 and 10.19. Channels in the thylakoids of the granum actively pump H^+ into the inner chamber, producing a charge gradient. ATP synthase located in this same thylakoid uses the energy from H^+ transport to phosphorylate ADP to ATP. Because it occurs in light, this process is termed **photophosphorylation.** Both NADPH and ATP are released into the stroma of the chloroplast, where they drive the reactions of the **Calvin cycle.**

Light-Independent Reactions

The subsequent photosynthetic reactions that do not require light occur in the chloroplast in or the cytoplasm of cyanobacteria. These reactions use energy produced by the light phase to synthesize glucose by means of the Calvin cycle **(figure 10.26).**

The cycle begins at the point where CO_2 is combined with a doubly phosphorylated 5-carbon acceptor molecule called ribulose-1,5-bisphosphate (RuBP). This process, called **carbon fixation,** generates a 6-carbon intermediate compound that immediately splits into two 3-carbon molecules of 3-phosphoglyceric acid (PGA). The subsequent steps use the ATP and NADPH generated by the photosystems to form high-energy intermediates. First, ATP adds a second phosphate to 3-PGA and produces 1,3-bisphosphoglyceric acid (BPG). Then, during the same step, NADPH contributes its hydrogen to BPG, and one high-energy phosphate is removed. These events give rise to glyceraldehyde-3-phosphate (PGAL). This molecule and its isomer dihydroxyacetone phosphate (DHAP) are key molecules in hexose synthesis leading to fructose and glucose. You may notice that this pathway is very similar to glycolysis, except that it runs in reverse (see figure 10.15). Bringing the cycle back to regenerate RuBP requires PGAL and several steps not depicted in figure 10.26.

Other Mechanisms of Photosynthesis

The **oxygenic,** or oxygen-releasing, photosynthesis that occurs in plants, algae, and cyanobacteria is the dominant type on the earth.

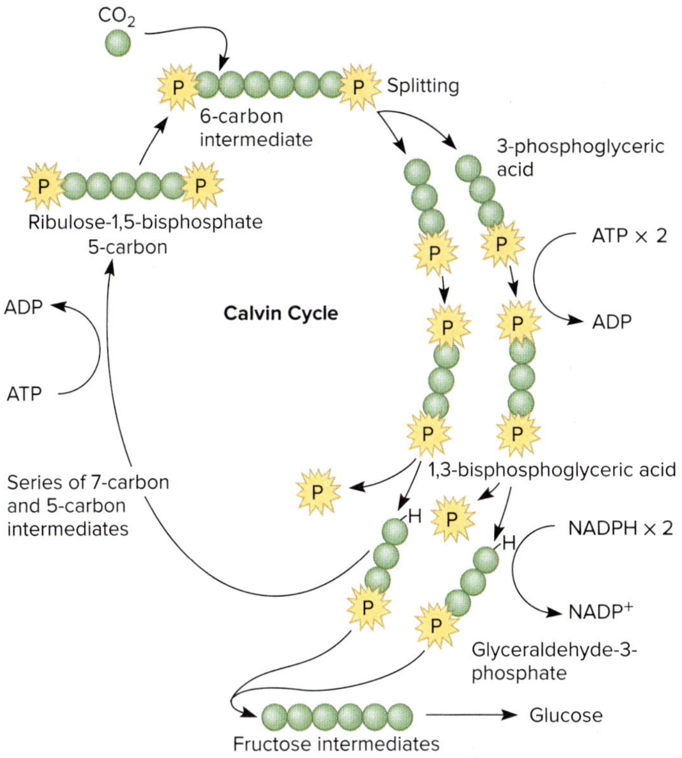
CO_2
P
P
Splitting
6-carbon intermediate
3-phosphoglyceric acid
Ribulose-1,5-bisphosphate 5-carbon
ATP × 2
ADP
ADP
ATP
Calvin Cycle
1,3-bisphosphoglyceric acid
Series of 7-carbon and 5-carbon intermediates
H
H
NADPH × 2
NADP+
Glyceraldehyde-3-phosphate
Glucose
Fructose intermediates

Figure 10.26 The Calvin cycle. The main events of the reactions in photosynthesis that do not require light. It is during this cycle that carbon is fixed into organic form using the energy (ATP and NADPH) released by the light reactions. The end product, glucose, can be stored as complex carbohydrates, or it can be used in various amphibolic pathways to produce other carbohydrate intermediates or amino acids.

Other photosynthesizers such as green and purple bacteria possess bacteriochlorophyll, which is more versatile in capturing light. They have only a cyclic photosystem I, which routes the electrons from the photocenter to the electron carriers and back to the photosystem again. This pathway generates a relatively small amount of ATP, and it may not produce NADPH. As photolithotrophs, these bacteria use H_2, H_2S, or elemental sulfur rather than H_2O as a source of electrons and reducing power. As a consequence, they are **anoxygenic** (non-oxygen-producing), and many are strict anaerobes.

While most of the mechanisms just described involve chlorophyll or bacteriochlorophyll as the light-absorbing pigment, archaea use a pigment called bacteriorhodopsin. You may recognize the root *rhodopsin,* which is a pigment present in vertebrate eyes (in the rods and cones). This type of photosynthesis does not involve electron transport but instead uses a light-driven proton pump. Through chemiosmosis, it generates ATP.

10.5 Learning Outcomes—Assess Your Progress

19. Summarize the overall process of photosynthesis in a single sentence.
20. Discuss the relationship between light-dependent and light-independent reactions.
21. Explain the role of the Calvin cycle in the process of photosynthesis.

Media Under The Microscope Wrap-Up

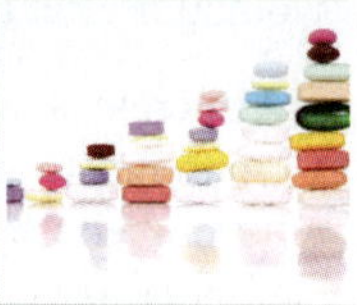
spxChrome/Getty Images

The article "Blood Sugar Support Plus Review: Scam or Ingredients That Work?" is a sponsored "article" touting the benefits of a dietary supplement called Blood Sugar Support.

The **intended message** is, actually, to sell a product. It appears to be an informative article, but the tip-off is the small word "Sponsored" associated with it. This always indicates that someone is trying to sell you something, and that doesn't happen in scientific reports or articles.

My **critical reading** is that it makes several unsubstantiated claims, each of which sounds science-y (lipase "breaks down fat and ensures you're not getting fatter from having sugar stores in your body for later use"). In reality, lipase does break down fat molecules, but there is no evidence—and the article presents none—that it acts in such a way to keep you from gaining weight. The clincher is this: "This is not something that has been discovered in the lab. It's the human body's science and the actual enzyme." OK. That seems to be playing into the sentiment among some in the public who mistrust things studied in a lab, which is a dangerous direction to be going.

I would **interpret** this article to my friends by starting with the fact that it is "sponsored." Watching out for that one word in online articles can go a long way to help you determine whether it is unbiased or whether it is trying to sell you something.

I give this "article" an **overall grade** of F. It cloaks its sales pitch in the appearance of science and makes unsubstantiated claims.

Source: *Homer News*, "Study: Blood Sugar Support Plus Review: Scam or Ingredients That Work?," online article posted July 8, 2021.

Study Smarter: Better Together

These activities are designed for you to use on your own with a study group—either a face-to-face group or a virtual one, consisting of 3–5 members. Studying together can be very helpful, but there are effective and ineffective ways to do it. For example, getting together without a clear structure is often not a good use of your time. Use your time efficiently by using one or more of the exercises below.

FACE-TO-FACE GROUPS

Use one or more of the activities below.

Peer Instruction: Assign numbers to your group members to use all semester long. Now look at these five concepts from this chapter. Each group member prepares a 5-minute lesson on the topic corresponding to their number. Don't worry if you have fewer than 5 members; just use however many you have! During your group study time, each member presents their lesson, and the group spends another 5–10 minutes discussing that lesson.

1. Redox reaction
2. Electron carriers
3. Glycolysis
4. Krebs cycle
5. Fermentation

Concept Maps: Each member of the group should use this list of terms from this chapter to generate their own concept map. This can be hand-drawn or created using software (see Appendix C for guidelines). During group study time, compare each other's concept maps and help each other make sure they are correct. Of course, there are many different "correct" maps. Examining each member's map will help you talk through the varied concepts and how they are related.

Concept Terms:

metabolism	catabolism	products	enzymes
anabolism	substrates	pH	

Table Topics: Each group member should identify a concept or topic from this week's class assignments with which they are having trouble and share it during group study time. The other group members can then help to clarify confusing issues or share how they figured it out. Aim for a maximum of 15 minutes per topic. If the topic remains unclear to the group, bring it up during class or use the instructor's office hours or e-mail to ask for help. Taking the time to struggle with a difficult concept first makes your questions much more specific and more likely to yield helpful answers.

(continued)

VIRTUAL GROUPS

Not everyone has the time or opportunity to meet with group members outside of class time. You or your instructor can create a virtual group using e-mail or the course software.

Weekly Discussion Board: This forum can be used as a way for groups to discuss topics, via e-mail, or other learning management systems or online platforms, before they are covered in class. As each member of the group answers the current week's question, they should send their responses to every other member of their group. It's best to agree on a deadline based on how your class schedule works (Saturday for the next week's topics, for example). Then, after the topic is discussed in class, each member should send a response that all group members will see with a follow-up post on the same topic. If you cover more than one chapter in a week, someone can be designated to choose which chapter Discussion Board question you will use. Or simply decide up front that you will always use the first-chapter-of-the week's question, to keep the schedule simple.

Discussion Question

Use the concept of electron carriers to explain how the electron transport chain works. Then explain why ATP synthase is so important to the life of a cell.

Chapter Summary

MICROBIAL METABOLISM

10.1 THE METABOLISM OF MICROBES

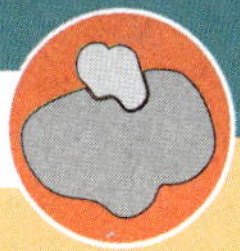

- Metabolism is the sum of cellular chemical and physical activities. It consists of anabolism, energy-requiring reactions that convert small molecules into larger molecules, and catabolism, in which large molecules are broken down and energy is produced.
- Metabolism is made possible by organic catalysts called enzymes.
- Enzymes are not consumed in the reactions they catalyze. Each enzyme acts specifically on its matching molecule or substrate.
- Enzymes are active only within narrow operating ranges of temperature, osmotic pressure, and pH.
- Enzymes can be regulated through activators or repressors binding to them, or they can be regulated during the genetic process that gives rise to enzymes.

10.2 FINDING AND MAKING USE OF ENERGY

- Energy is the capacity of a system to perform work. It is consumed in endergonic reactions and is released in exergonic reactions.
- Extracting energy requires a series of electron carriers arrayed in a redox chain between electron donors and electron acceptors.

10.3 CATABOLISM: GETTING MATERIALS AND ENERGY

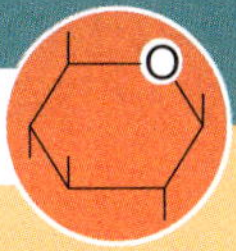

- Carbohydrates, such as glucose, are energy-rich. When catabolized, they can yield a large number of electrons per molecule.
- Glycolysis degrades glucose to pyruvic acid without requiring oxygen.
- Pyruvic acid enters into aerobic and anaerobic respiration via the Krebs cycle.
- The Krebs cycle generates ATP, CO_2, and H_2O.
- The respiratory chain completes energy extraction and yields a large amount of ATP.
- The final electron acceptor in aerobic respiration is oxygen. In anaerobic respiration, compounds such as sulfate or nitrate serve this function.
- Fermentation is an anaerobic process in which both the electron donor and final electron acceptors are organic compounds. Alcohols or acids are products of this metabolic pathway.
- Intermediates in these pathways can be made into amino acids, fatty acids, and carbohydrates, in anabolic processes.

10.4 BIOSYNTHESIS AND THE CROSSING PATHWAYS OF METABOLISM

- Cells perform amphibolically, meaning they can integrate catabolic and anabolic pathways.
- Macromolecules, such as proteins, carbohydrates, and nucleic acids, are made of building blocks from two possible sources: from outside the cell (preformed) or via synthesis in one of the anabolic pathways.

10.5 PHOTOSYNTHESIS: IT STARTS WITH LIGHT

- Photosynthesis converts the sun's energy into chemical energy and organic carbon compounds, which are produced from carbon dioxide.

SmartGrid: From Knowledge to Critical Thinking

This *21 Question Grid* takes the topics from this chapter and arranges them with respect to the American Society for Microbiology's Undergraduate Curriculum guidelines—all six of the important "Concepts" as well as the important "Competency" of scientific literacy. Three questions are supplied, which cover chapter content referring to the Concept or Competency in increasing levels of Bloom's taxonomy for learning.

ASM Concept/ Competency	A. Bloom's Level 1, 2—Remember and Understand (Choose one.)	B. Bloom's Level 3, 4—Apply and Analyze	C. Bloom's Level 5, 6—Evaluate and Create
Evolution	**1.** The electron transport system in bacteria is located on the ______________ and in eukaryotic cells on the ______________. **a.** plastid, chloroplast **b.** cytoplasmic membrane, mitochondrion **c.** cell wall, mitochondrion **d.** mitochondrion, cytoplasmic membrane	**2.** Speculate on why glycolysis, which is the beginning point of many catabolic pathways, does not require oxygen.	**3.** Provide evidence in support of or refuting the following statement: The evolution of aerobic respiration was driven by the success of photosynthetic microbes.
Cell Structure and Function	**4.** Many coenzymes are formed from **a.** metals. **b.** vitamins. **c.** proteins. **d.** substrates.	**5.** Describe the roles played by ATP and NAD in metabolism.	**6.** Explain why electron transport systems are always found in a membrane.
Metabolic Pathways	**7.** Energy is carried from catabolic to anabolic reactions in the form of **a.** ADP. **b.** high-energy ATP bonds. **c.** coenzymes. **d.** inorganic phosphate.	**8.** What is meant by the concept of the "final electron acceptor"?	**9.** Investigate the creation of lactic acid in human cells via fermentation, and write a paragraph contrasting it with microbial fermentation.
Information Flow and Genetics	**10.** Enzyme action can be blocked by competitive molecules binding in the active site, by repressors binding in a distant site, and by **a.** product binding to the DNA used to make enzymes. **b.** substrates being in high concentration. **c.** incorrect temperature conditions. **d.** two of the above.	**11.** Compare and contrast the molecular structure of ATP with the molecular structure of an RNA nucleotide.	**12.** Suggest some reasons that pyruvate is central to so many metabolic strategies.

(continued)

ASM Concept/ Competency	A. Bloom's Level 1, 2—Remember and Understand (Choose one.)	B. Bloom's Level 3, 4—Apply and Analyze	C. Bloom's Level 5, 6—Evaluate and Create
Microbial Systems	**13.** Enzymes that can be shut down or activated based on the presence of chemicals in their environment are called **a.** constitutive. **b.** repressed. **c.** holoenzymes. **d.** regulated.	**14.** Which of the three major catabolic strategies do you think the last common ancestor possessed? Defend your answer.	**15.** Write a paragraph explaining why metabolic pathways that are amphibolic are advantageous for cells.
Impact of Microorganisms	**16.** The type of microbial metabolic pathway that is most often exploited to make acids and alcohols industrially is **a.** aerobic respiration. **b.** anaerobic respiration. **c.** fermentation. **d.** none of the above.	**17.** Defend this statement: Microbes (and their metabolic strategies) are absolutely essential to plant life on earth.	**18.** Construct an argument for the possibility that scientists will continue to discover novel metabolic strategies that are not yet known.
Scientific Thinking	**19.** Which of the following is true? **a.** The suffix "-ase" indicates an enzyme. **b.** Often enzymes are named for the substrates they act upon. **c.** Enzymes are larger than their substrates. **d.** All of the above are true.	**20.** Find three enzymatic capabilities that are commonly used to identify microbial species in a clinical lab.	**21.** Polymerase chain reaction (PCR) is a lab technology that requires high temperatures to make copies of DNA. The copies are made using an enzyme called a polymerase. PCR would not have been possible before thermophilic archaea were discovered. Explain why.

Answers to the multiple-choice questions appear in Appendix A.

Visual Connections

These questions use visual images or previous content to connect content within and between chapters.

1. **From chapter 4, figure 4.16.** On the enlarged sections of both **(a)** and **(b),** draw protons in the proper compartment in such a way that it illustrates the creation of a proton motive force.

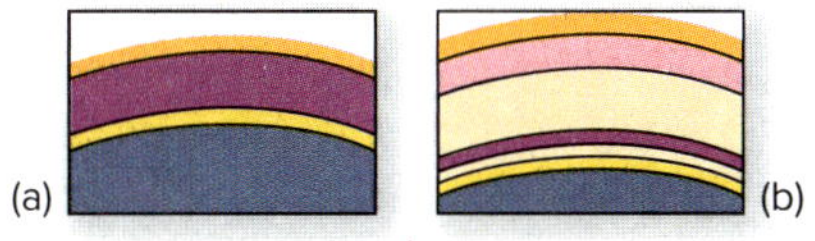
(a)
(b)

High Impact Study

These terms and concepts are most critical for your understanding of this chapter—and may be the most difficult. Have you mastered them?

Concepts

- ☐ Electron flow
- ☐ Redox reactions
- ☐ Electron carriers
- ☐ Glycolysis
- ☐ Krebs cycle
- ☐ Electron transfer system
- ☐ Aerobic and anaerobic respiration
- ☐ Amphibolism
- ☐ Photosynthesis

Terms

- ☐ Metabolism
- ☐ Catabolism
- ☐ Anabolism
- ☐ ATP
- ☐ Enzyme
- ☐ Substrate
- ☐ Fermentation
- ☐ Pyruvic acid

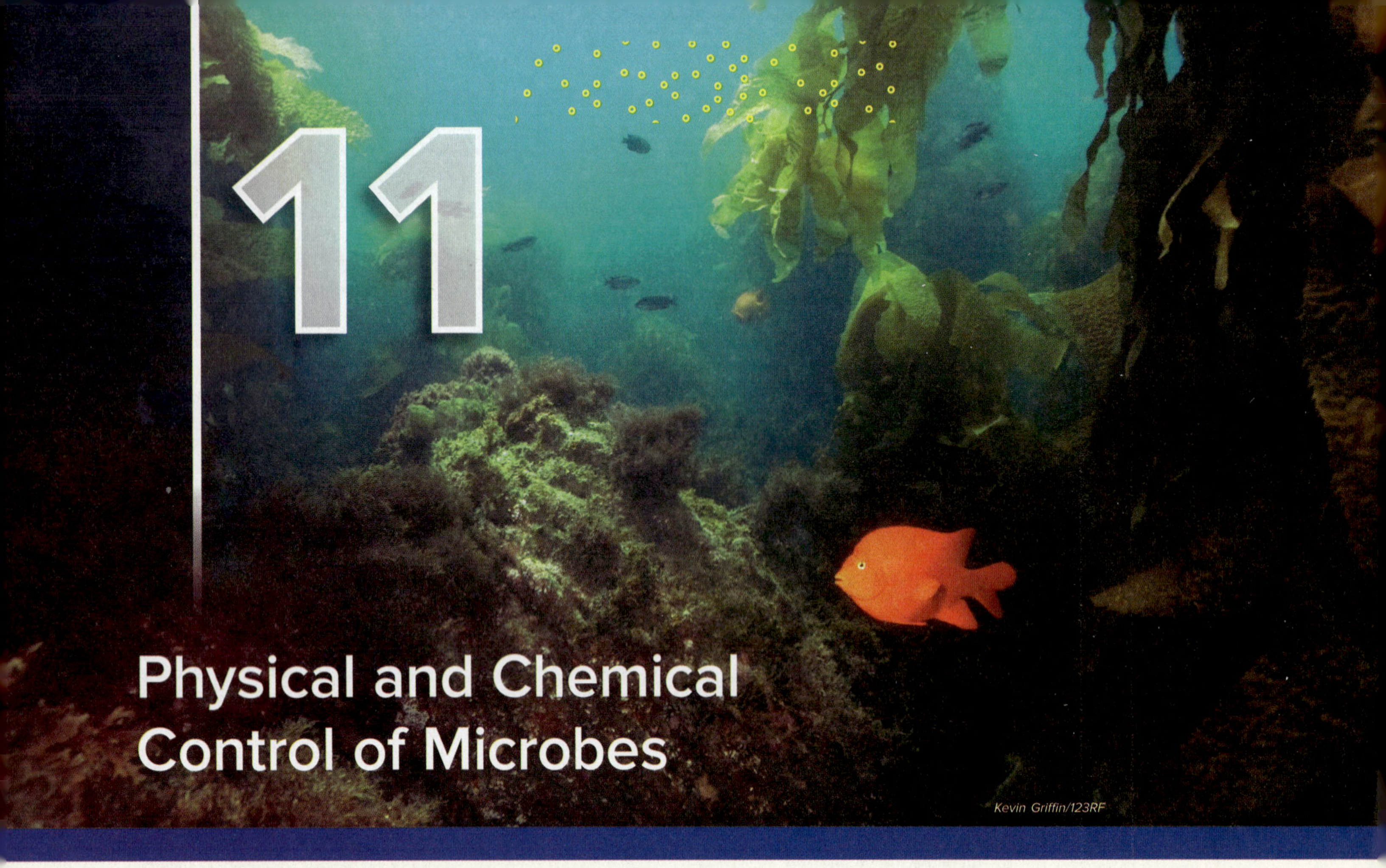

11 Physical and Chemical Control of Microbes

Kevin Griffin/123RF

Media Under The Microscope

Seaweed and Self-Cleaning Surfaces

These case studies examine an article from the popular media to determine the extent to which it is factual and/or misleading. This case focuses on the 2021 NaturalNews.com *article "Antimicrobial Compound in Seaweed Can Be Used to Develop Self-Cleaning Surfaces, New Research Finds."*

This article describes research performed by scientists from Unilever, a multinational company based in England that manufactures consumer goods. The scientists, knowing that seaweed does not become colonized by biofilms, even in very contaminated waters, had been searching for the chemical or chemicals in seaweed that prevented this from happening.

The article reveals that the chemical they found, called lactam, can deter the colonization and growth of bacteria on seaweed surfaces. The company has now created a technology that uses lactam, and hopes to make it into a product that keeps everything from kitchen countertops to clothing to ship hulls "clean." The article makes reference to the fact that biofilms are formed by bacteria communicating with one another—a phenomenon we have discussed earlier as quorum sensing. The article implies that the lactam from seaweed interrupts those signals and keeps surfaces clean.

Also, the article refers to the fact that seaweed has been found to have antimicrobial properties that are different than those of the quorum-breaking chemical lactam.

- What is the **intended message** of the article?
- What is your **critical reading** of the summary of the article provided above? Remember that in this context, "critical reading" does not necessarily mean *What criticism do you have?* but asks you to apply your knowledge to interpret whether the article is factual and whether the facts support the intended message.
- How would you **interpret** the news item for your nonmicrobiologist friends?
- What is your **overall grade** for the news item—taking into account its accuracy and the accuracy of its intended effect?

Media Under The Microscope Wrap-Up appears at the end of the chapter.

Outline and Learning Outcomes

11.1 Controlling Microorganisms

1. Distinguish among the terms *sterilization, disinfection, antisepsis,* and *decontamination*.
2. Identify the types of microorganisms that are most resistant and least resistant to control measures.
3. Compare the action of microbicidal and microbistatic agents, providing an example of each.
4. Name four categories of cellular targets for physical and chemical agents.

11.2 Methods of Physical Control

5. Name six methods of physical control of microorganisms.
6. Compare and contrast moist and dry heat methods of control, and identify multiple examples of each.
7. Define *thermal death time* and *thermal death point,* and describe their role in proper sterilization.
8. Explain three different methods of moist heat control.
9. Explain two methods of dry heat control.
10. Identify advantages and disadvantages of cold treatment and desiccation.
11. Differentiate between the two types of radiation control methods, providing an application of each.
12. Outline the process of filtration, and describe its two advantages in microbial control.
13. Identify some common uses of osmotic pressure as a control method.

11.3 Methods of Chemical Control

14. Name the desirable characteristics of chemical control agents.
15. Discuss several different halogen agents and their uses in microbial control.
16. List advantages and disadvantages to the use of phenolic compounds as control agents.
17. Explain the mode of action of alcohols and their limitations as effective antimicrobials.
18. Pinpoint the most appropriate applications of oxidizing agents.
19. Define the term *surfactant,* and explain this antimicrobial's mode of action.
20. Identify examples of some heavy metal control agents and their most common applications.
21. Discuss the advantages and disadvantages of aldehyde agents in microbial control.
22. Identify applications for ethylene oxide sterilization.

11.1 Controlling Microorganisms

Much of the time in our daily existence, we take for granted tap water that is drinkable, food that is not spoiled, shelves full of products to eradicate "germs," and drugs to treat infections. Controlling our exposure to potentially harmful microbes is a monumental concern in our lives, and it has a long and eventful history. Salting, smoking, pickling, and drying foods as well as exposing food, clothing, and bedding to sunlight were prevalent practices among early civilizations. The Greeks and Romans burned clothing and corpses during epidemics, and they stored water in copper and silver containers. During the great plague pandemic of the Middle Ages, it was common to bury corpses in mass graves, burn the clothing of plague victims, and ignite aromatic woods in the houses of the sick in the belief that fumes would combat the disease. Each of these early methods, although crude, laid the foundations for microbial control methods that are still in use today.

General Considerations in Microbial Control

The methods of microbial control used outside of the body are designed to result in four possible outcomes: sterilization, disinfection, antisepsis, or decontamination.

Sterilization is the destruction of all microbial life.

Disinfection destroys most microbial life, reducing contamination on inanimate surfaces.

Antisepsis (also called **degermation**) is the same as disinfection except a living surface is involved.

Decontamination (also called **sanitization**) is the mechanical removal of most microbes from an animate or inanimate surface.

The flowchart in **figure 11.1** summarizes the major applications and aims in microbial control.

Relative Resistance of Microbial Forms

The primary targets of microbial control are microorganisms capable of causing infection or spoilage. These are constantly present in the external environment and on the human body. This population is usually not simple or uniform. It often contains mixtures of microbes with big differences in their resistance and harmfulness. Some of the microbes that can be dangerous if not controlled include bacterial vegetative cells and endospores, fungal hyphae and spores, yeasts, protozoan trophozoites and cysts, worms, viruses, and prions. **Figure 11.2** compares the general resistance these forms have to physical and chemical methods of control.

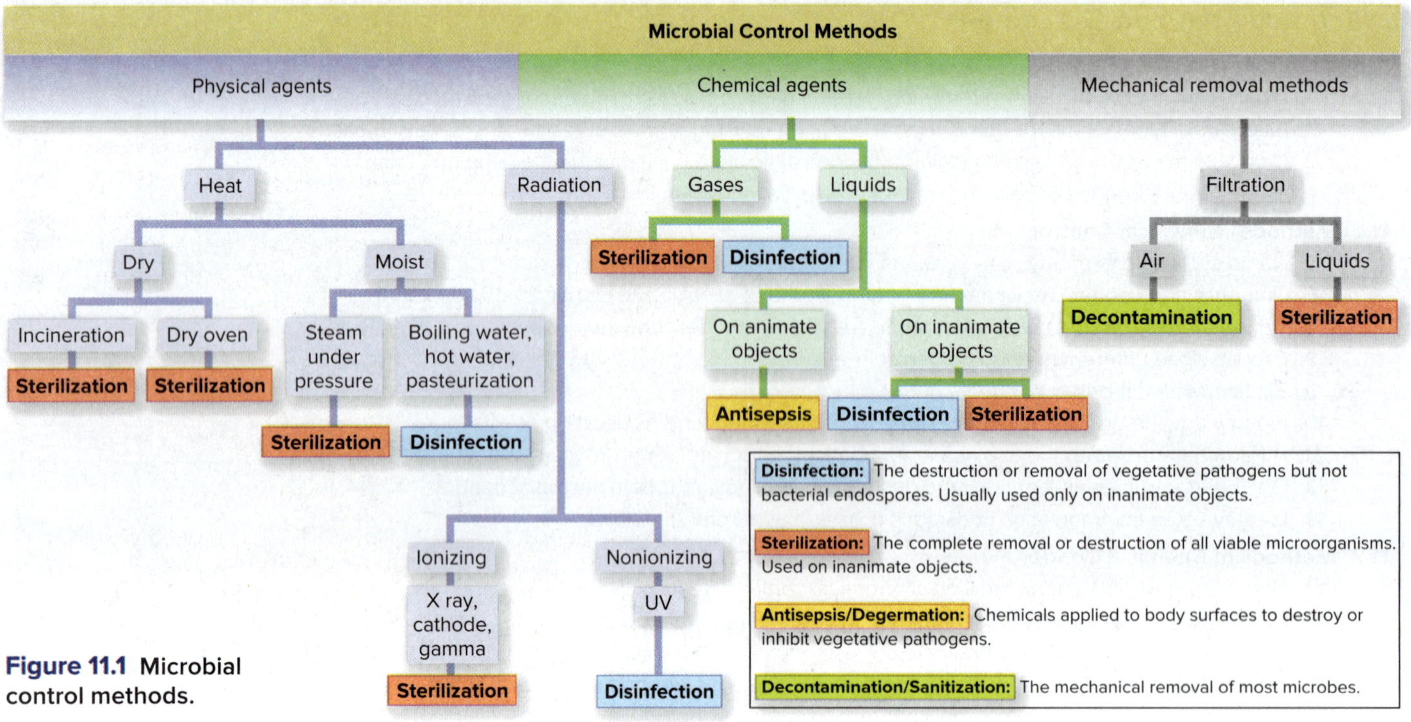
Microbial Control Methods
Physical agents
Chemical agents
Mechanical removal methods
Heat
Radiation
Gases
Liquids
Filtration
Dry
Moist
Sterilization
Disinfection
Air
Liquids
Incineration
Dry oven
Steam under pressure
Boiling water, hot water, pasteurization
On animate objects
On inanimate objects
Decontamination
Sterilization
Sterilization
Sterilization
Antisepsis
Disinfection
Sterilization
Sterilization
Disinfection
Ionizing
Nonionizing
X ray, cathode, gamma
UV
Sterilization
Disinfection
Disinfection: The destruction or removal of vegetative pathogens but not bacterial endospores. Usually used only on inanimate objects.
Sterilization: The complete removal or destruction of all viable microorganisms. Used on inanimate objects.
Antisepsis/Degermation: Chemicals applied to body surfaces to destroy or inhibit vegetative pathogens.
Decontamination/Sanitization: The mechanical removal of most microbes.

Figure 11.1 **Microbial control methods.**

Actual comparative figures on the requirements for destroying various groups of microorganisms are shown in **table 11.1.** Bacterial endospores have traditionally been considered the most resistant microbial entities, being as much as 18 times harder to destroy than their counterpart vegetative cells. Because of their resistance to microbial control, destroying them is the goal of *sterilization* because any process that kills endospores will invariably kill all less-resistant microbial forms. Other methods of control (disinfection, antisepsis) act primarily upon microbes that are less hardy than endospores.

Methods of Microbial Control

The terminology for describing and defining measures that control microbes can get a little jumbled. The official designations are those shown in figure 11.1, but in practice, people use the terms loosely. For example, someone may ask you to sterilize a patient's skin, even though this is not technically correct. It does not fit the technical definition of the term. It's best to start with a good groundwork, so let's look at concepts, definitions, and usages in antimicrobial control.

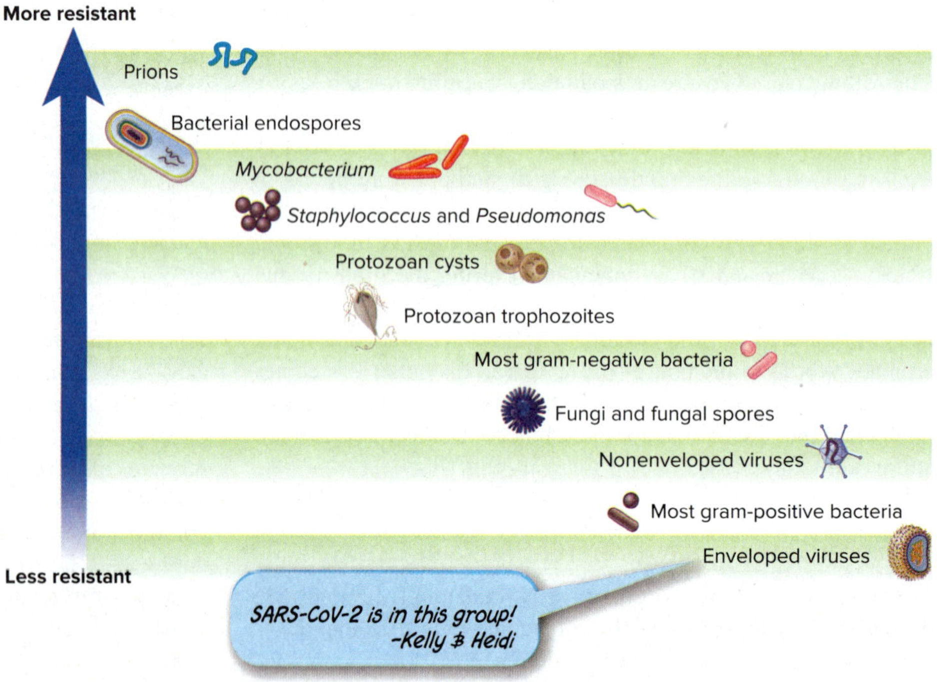
More resistant
Prions
Bacterial endospores
Mycobacterium
Staphylococcus and Pseudomonas
Protozoan cysts
Protozoan trophozoites
Most gram-negative bacteria
Fungi and fungal spores
Nonenveloped viruses
Most gram-positive bacteria
Enveloped viruses
Less resistant
SARS-CoV-2 is in this group!
-Kelly $ Heidi

Figure 11.2 **Relative resistance of different microbial types to microbial control agents.** This is a very general hierarchy; different control agents are more or less effective against the various microbes.

Table 11.1 Comparative Resistance of Bacterial Endospores and Vegetative Cells to Control Agents

Method	Required to Destroy Endospores	Required to Destroy Vegetative Forms	Endospores Are ___ More Resistant*
Heat (moist)	120°C	80°C	1.5×
Radiation (X-ray) dosage	4,000 Grays	1,000 Grays	4×
Sterilizing gas (ethylene oxide)	1,200 mg/L	700 mg/L	1.7×
Sporicidal liquid (2% glutaraldehyde)	3 h	10 min	18×

*The greater resistance of endospores versus vegetative cells given as an average figure.

Terminology

Sterilization is a process that destroys or removes all viable microorganisms, including viruses. Any material that has been subjected to this process is said to be **sterile.** These terms should be used only in the strictest sense for methods that have been proved to sterilize. An object cannot be slightly sterile or almost sterile—it is either sterile or not sterile. Control methods that sterilize are generally reserved for inanimate objects because sterilizing the human body or its parts would call for such harsh treatment that it would be dangerous and impractical.

The ability to sterilize products—surgical instruments, syringes, and some packaged foods, just to name a few—is essential to human well-being. Although most sterilization is performed with a physical agent such as heat, a few chemicals can be classified as sterilizing agents because of their ability to destroy endospores.

In many situations, sterilization is neither practical nor necessary, and only certain groups of microbes need to be controlled. Some antimicrobial agents eliminate only the susceptible vegetative states of microorganisms but do not destroy the more resistant endospore and cyst stages. Keep in mind that the destruction of endospores is not always necessary because most of the infectious diseases of humans and animals are caused by non-endospore-forming microbes.

Disinfection refers to the use of a physical process or a chemical agent (a disinfectant) to destroy vegetative pathogens but not bacterial endospores. Disinfectants are normally used only on inanimate objects because, in the concentrations required to be effective, they can be toxic to human and other animal tissue. Disinfection processes also remove the harmful products of microorganisms (toxins) from materials. Examples of disinfection include applying a solution of 5% bleach to an examining table, boiling food utensils used by a sick person, and immersing thermometers in an iodine solution between uses.

Sepsis is defined as the growth of microorganisms in the blood and other tissues. The term *asepsis* refers to any practice that prevents the entry of infectious agents into sterile tissues and thus prevents infection. Aseptic techniques commonly practiced in health care range from sterile methods to *antisepsis.* In antisepsis, chemical agents called **antiseptics** are applied directly to exposed body surfaces (skin and mucous membranes), wounds, and surgical incisions to destroy or inhibit vegetative pathogens. Examples of antisepsis (also called degermation) include preparing the skin before surgical incisions with iodine compounds, swabbing an open root canal with hydrogen peroxide, using alcohol wipes on the skin, and performing the surgical hand scrub. To recap, degermation is a synonym for antisepsis, and both are included in the term *aseptic techniques*. Sterile methods are also grouped under the term *aseptic technique*. It sounds complicated, but if you are going to be working in a clinical setting, different people will use different terms.

A Note About Prions

Prions are in a class of their own when it comes to "sterilization" procedures. This chapter defines *sterile* as the absence of all viable microbial life—but none of the procedures described in this chapter are necessarily sufficient to destroy prions. Prions are extraordinarily resistant to heat and chemicals. If instruments or other objects become contaminated with these unique agents, either they must be discarded as biohazards or, if this is not possible, enhanced sterilization procedures must be applied in accordance with CDC guidelines. The guidelines themselves are constantly evolving as new information becomes available. In the meantime, this chapter discusses sterilization using bacterial endospores as the toughest form of microbial life. When tissues, fluids, or instruments are suspected of containing prions, consultation with infection control experts and/or the CDC is recommended when determining effective sterilization conditions.

The Agents versus the Processes

The terms *sterilization, disinfection,* etc., refer to processes. There are other terms that describe the agents used in the process. Two of these are the terms *bactericidal* and *bacteristatic.* The root *-cide,* meaning "to kill," can be combined with other terms to define an antimicrobial agent aimed at destroying a certain group of microorganisms. For example, a **bactericide** is a chemical that destroys bacteria (except for those in the endospore stage). It may or may not be effective on other microbial groups. A *fungicide* is a chemical that can kill fungal spores, hyphae, and yeasts. A *virucide* is any chemical known to inactivate viruses, especially on living tissue. A *sporicide* is an agent capable of destroying bacterial endospores. A sporicidal agent can also be considered a sterilant because it can destroy the most resistant of all microbes. **Germicide** and **microbicide** are additional terms for chemical agents that kill microorganisms.

The Greek words *stasis* and *static* mean "to stand still." They can be used in combination with various prefixes to describe a condition in which microbes are prevented from multiplying but are not killed outright. Although killing or permanently inactivating microorganisms is the usual goal of microbial control, microbistasis does have meaningful applications. **Bacteristatic** agents prevent the growth of bacteria on tissues or on objects in the environment, and *fungistatic* chemicals inhibit fungal growth. The general terms for the killing or inhibition of microbes are microbicidal and microbistatic, respectively. Chemicals used to control microorganisms in the body (antiseptics and drugs) often are chosen for their microbistatic effects because the ones that are microbicidal can be highly toxic to human cells.

Decontamination (Sanitization)

Sanitization is any cleansing technique that mechanically removes microorganisms as well as other debris to reduce contamination to safe levels. A sanitizer is a compound such as soap or detergent used to perform this task.

Cooking utensils, dishes, bottles, cans, and clothing that have been washed and dried may not be completely free of microbes, but they are considered safe for normal use (sanitary). Air sanitization with ultraviolet (UV) lamps reduces airborne microbes in hospital rooms, veterinary clinics, and laboratory installations. Note that some sanitizing processes (such as dishwashing machines) may be rigorous enough to sterilize objects, but this is not true of all sanitization methods. Also note that sanitization is often preferable to sterilization. In a restaurant, for example, you could be given a sterile fork with someone else's old food on it and a sterile glass with lipstick on the rim, but it is preferable to have a sanitized glass with no remnants of the previous guest. On top of this, sterilization procedures add greatly to the cost of doing business. In other words, usefulness of a technique depends on the context.

Practical Concerns in Microbial Control

Numerous considerations should go into selecting a method of microbial control:

1. Does the item in question require sterilization, or is disinfection adequate? In other words, must endospores be destroyed, or is it necessary to destroy only vegetative pathogens?
2. Is the item to be reused or permanently discarded? If it will be discarded, then the quickest and least expensive method, which is often destructive, can be used.
3. If it will be reused, can the item withstand heat, pressure, radiation, or chemicals?
4. Is the control method suitable for a given application? (For example, UV radiation can possibly kill endospores, but it will not penetrate solid materials.) Or, in the case of a chemical, will it leave an undesirable residue?
5. Will the agent penetrate to the necessary extent?
6. Is the method cost- and labor-efficient, and is it safe?

One useful framework for determining how devices that come in contact with patients should be handled is whether they are considered *critical, semicritical,* or *noncritical.* Critical medical devices are those that are expected to come in contact with sterile tissues. Examples include a syringe needle or an artificial hip. These must be sterilized before use. Semicritical devices are those that come in contact with mucosal membranes. An endoscopy tube is an example. These must receive at least high-level disinfection and, preferably, should be sterilized. Noncritical items are those that do not touch the patient or are only expected to touch intact skin, such as blood pressure cuffs or crutches. They require only low-level disinfection unless they become contaminated with blood or body fluids.

A remarkable variety of substances can require sterilization. They range from durable solids such as rubber to sensitive liquids such as serum. An unusual case of having to sterilize an entire office building, without destroying the records and equipment in it, occurred in 2001 when the Hart Senate Office Building in Washington, D.C., was deliberately contaminated with *Bacillus anthracis* endospores. (It was eventually accomplished using sterilizing gas.) Hundreds of situations requiring sterilization confront the network of persons involved in public health care, whether technician, nurse, doctor, or manufacturer, and no method works well in every case.

Considerations such as cost, effectiveness, and method of disposal are all important. For example, disposable plastic items such as catheters and syringes that are used in invasive medical procedures have the potential for infecting tissues. These must be sterilized during manufacture by a nonheating method (gas or radiation) because heat can damage plastics. After these items have been used, it is often necessary to destroy or decontaminate them before they are discarded because of the potential risk to the handler. Steam sterilization, which is quick and sure, is a sensible choice at this point because it does not matter if the plastic is destroyed.

What Is Microbial Death?

Death is defined as the permanent termination of an organism's vital processes. Signs of life in complex organisms such as animals are often self-evident, and death is made clear by loss of nervous function, respiration, or heartbeat. In contrast, death in microscopic organisms that are composed of just one or a few cells is often hard to detect because they reveal no conspicuous vital signs to begin with. Lethal agents (such as radiation and chemicals) do not necessarily alter the overt appearance of microbial cells. Even the loss of movement in a motile microbe does not indicate death. This fact has made it necessary to develop special ways to define and delineate microbial death.

The destructive effects of chemical or physical agents occur at the level of a single cell. As the cell is continuously exposed to an agent such as intense heat or toxic chemicals, various cell structures become dysfunctional. The entire cell can sustain irreversible damage in the process. At present, the most practical way to detect this damage is to determine if a microbial cell can still reproduce when exposed to a suitable environment. If the microbe has sustained metabolic or structural damage to such an extent that it can no longer reproduce, even under ideal environmental conditions, then it is no longer viable. The permanent loss of reproductive capability, even under optimum growth conditions, has become the accepted microbiological definition of death.

Factors That Affect Death Rate

The cells of a culture can show significant variation in susceptibility to a given microbicidal agent. Death of the whole population is not instantaneous but begins when a certain threshold of microbicidal agent (some combination of time and concentration) is met. Death continues in a logarithmic manner as the time **(figure 11.3*a*)** or concentration of the agent is increased. Because many microbicidal agents target the cell's metabolic processes, active cells (younger, rapidly dividing) tend to die more quickly than those that are less metabolically active (older, inactive). Read that again—it might seem counterintuitive at first, but it is an important property to keep in mind when trying to control microbial populations inside or outside of the body. Eventually, a point is reached at which survival of any cells is highly unlikely. This point is equivalent to sterilization.

The effectiveness of a particular agent is governed by several factors besides time. These additional factors influence the action of antimicrobial agents:

1. The number of microorganisms **(figure 11.3*b*)**. A higher load of contaminants requires more time to destroy.
2. The nature of the microorganisms in the population **(figure 11.3*c*)**. In most actual circumstances of disinfection and sterilization, the target population is not a single species of microbe but a mixture of bacteria, fungi, endospores, and viruses, presenting a broad spectrum of microbial resistance.
3. The type of microbial growth. Planktonic bacterial populations are those that grow freely within fluid environments and do not attach to surfaces. In general, they are more susceptible to control agents as compared to the microbes within well-developed biofilms adhering to surfaces such as medical devices or human tissues.
4. The temperature and pH of the environment.
5. The concentration (dosage, intensity) of the agent. For example, UV radiation is most effective at 260 nm, and most disinfectants are more active at higher concentrations.
6. The mode of action of the agent **(figure 11.3*d*)**. How does it kill or inhibit the microorganism?
7. The presence of solvents, interfering organic matter, and inhibitors. Saliva, blood, and feces can inhibit the actions of disinfectants and even of heat.

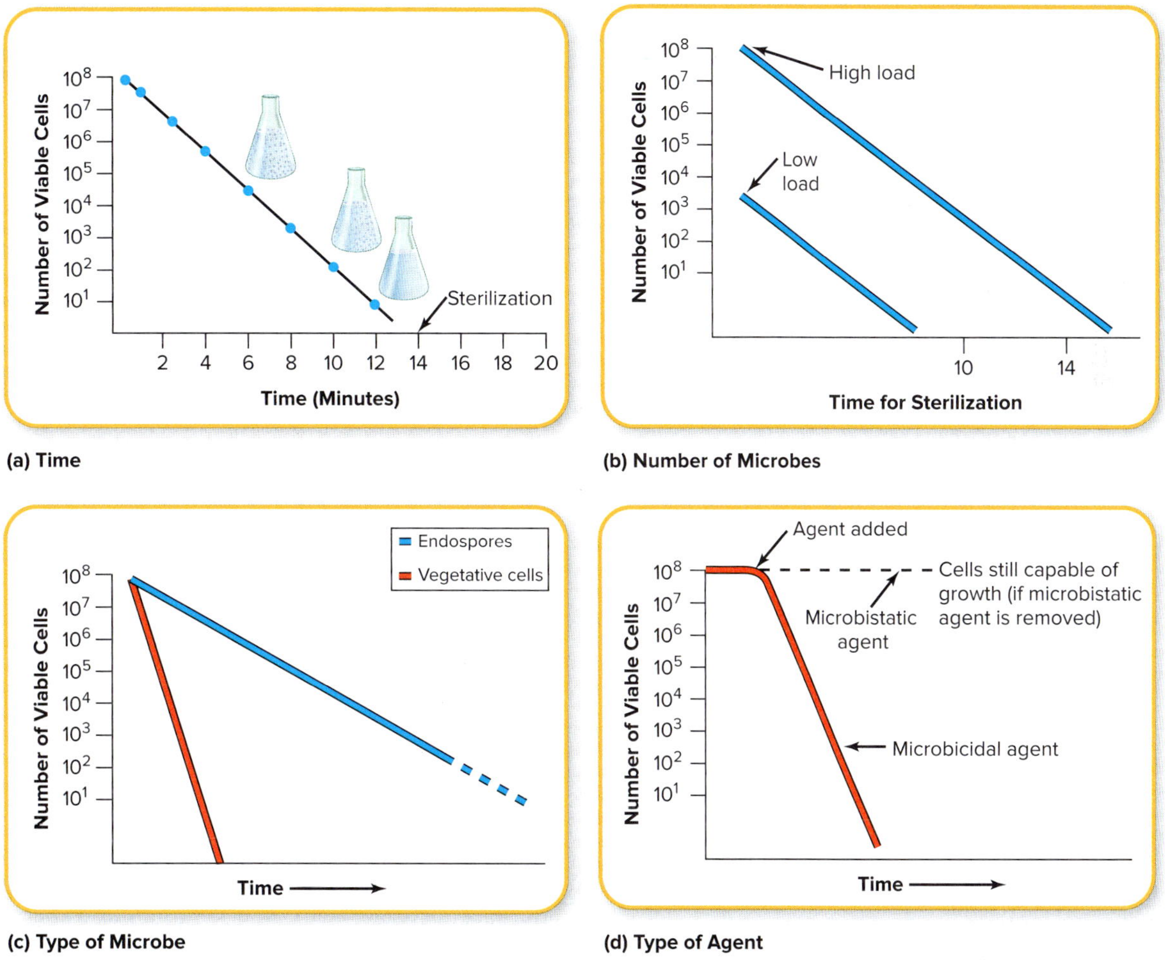
Number of Viable Cells
10^8
10^7
10^6
10^5
10^4
10^3
10^2
10^1
Sterilization
2 4 6 8 10 12 14 16 18 20
Time (Minutes)
(a) Time
High load
Low load
10 14
Time for Sterilization
(b) Number of Microbes
Endospores
Vegetative cells
Time
(c) Type of Microbe
Agent added
Cells still capable of growth (if microbistatic agent is removed)
Microbistatic agent
Microbicidal agent
Time
(d) Type of Agent

Figure 11.3 Factors that influence the rate at which microbes are killed by antimicrobial agents. **(a)** Length of time the agent is in contact with the microbes. During exposure to a chemical or physical agent, all cells of a microbial population, even a pure culture, do not die simultaneously. Over time, the number of viable organisms remaining in the population decreases logarithmically, giving a straight-line relationship on a graph. **(b)** Effect of the initial microbial load (how many microbes are present). **(c)** Relative resistance of endospores versus vegetative forms. **(d)** The difference between microbistatic and microbicidal agents.

The influence of these factors is discussed in greater detail in subsequent sections.

How Antimicrobial Agents Work: Their Modes of Action

An antimicrobial agent's adverse effect on cells is known as its *mode* (or *mechanism*) *of action*. Agents affect one or more cellular targets, inflicting damage progressively until the cell is no longer able to survive. Antimicrobial agents have a range of cellular targets, with the agents that are least selective in their targeting tending to be effective against the widest range of microbes (examples include heat and radiation). More selective agents (drugs, for example) tend to target only a single cellular component and are much more restricted as to the microbes they are effective against.

The cellular targets of physical and chemical agents fall into four general categories:

1. the cell wall,
2. the cell or cytoplasmic membrane,
3. cellular synthetic processes (DNA, RNA), and
4. proteins.

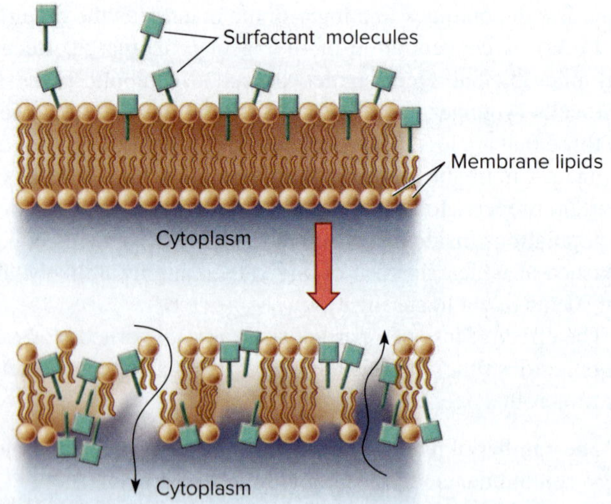

Surfactant molecules
Membrane lipids
Cytoplasm
Cytoplasm

Figure 11.4 Mode of action of surfactants on the cell membrane. Surfactants inserting in the lipid bilayer disrupt it and create abnormal channels that alter permeability and cause leakage both into and out of the cell.

The Effects of Agents on the Cell Wall

The cell wall maintains the structural integrity of bacterial and fungal cells. Several types of chemical agents damage the cell wall by blocking its synthesis, digesting it, or breaking down its surface. A cell deprived of a functioning cell wall becomes fragile and is lysed very easily. Detergents and alcohol can also disrupt cell walls, especially in gram-negative bacteria.

How Agents Affect the Cell Membrane

All microorganisms have a cell membrane composed of lipids and proteins, and many viruses have an outer (membranous) envelope. As we learned in previous chapters, a cell's membrane provides a two-way system of transport. If this membrane is disrupted, a cell loses its selective permeability and can neither prevent the loss of vital molecules nor bar the entry of damaging chemicals. Loss of those abilities leads to cell death. Detergents called **surfactants** (sir-fak′-tunts) work as microbicidal agents. Surfactants are molecules with both hydrophilic and hydrophobic regions that can physically bind to the lipid layer and penetrate the internal hydrophobic region of membranes. In effect, this process "opens up" the once tight interface, leaving leaky spots that allow damaging chemicals to seep into the cell and important ions to seep out **(figure 11.4).**

Agents That Affect Protein and Nucleic Acid Synthesis

Microbial life depends upon an orderly and continuous supply of proteins to function as enzymes and structural molecules. As we saw in chapter 6, these proteins are synthesized via the ribosomes through a process called translation. The antibiotic chloramphenicol binds to the ribosomes of bacteria in a way that stops peptide bonds from forming. In its presence, many bacterial cells are inhibited from forming proteins required in growth and metabolism and are inhibited from multiplying.

Nucleic acids are also necessary for the continued functioning of microbes. DNA must be regularly replicated and transcribed in growing cells, and any agent that affects these processes or changes the genetic code is potentially antimicrobial. Some agents bind irreversibly to DNA, preventing both transcription and translation, and others are mutagenic agents. Gamma, ultraviolet, and X radiation cause mutations that result in permanent inactivation of DNA. Chemicals such as formaldehyde and ethylene oxide also interfere with DNA and RNA function.

Agents That Alter Protein Function

A microbial cell is full of important proteins. They function properly only if they remain in a normal three-dimensional configuration called the *native state*. The antimicrobial properties of some agents come from their capacity to disrupt, or **denature,** proteins. In general, denaturation occurs when the bonds that maintain the secondary and tertiary structure of the protein are broken. Breaking these bonds will cause the protein to unfold or create random, irregular loops and coils **(figure 11.5).** One way that proteins can be denatured is through coagulation by moist heat (the same reaction seen in the irreversible solidification of the white of an egg when boiled). Chemicals such as strong organic solvents (alcohols, acids) and phenolics also coagulate proteins. Other antimicrobial agents, such as metallic ions, attach to the active site of the protein and prevent it from interacting with its correct substrate. Regardless of the exact mechanism, these losses in normal protein function can promptly stop metabolism. Most antimicrobials of this type are nonselective as to the microbes they affect.

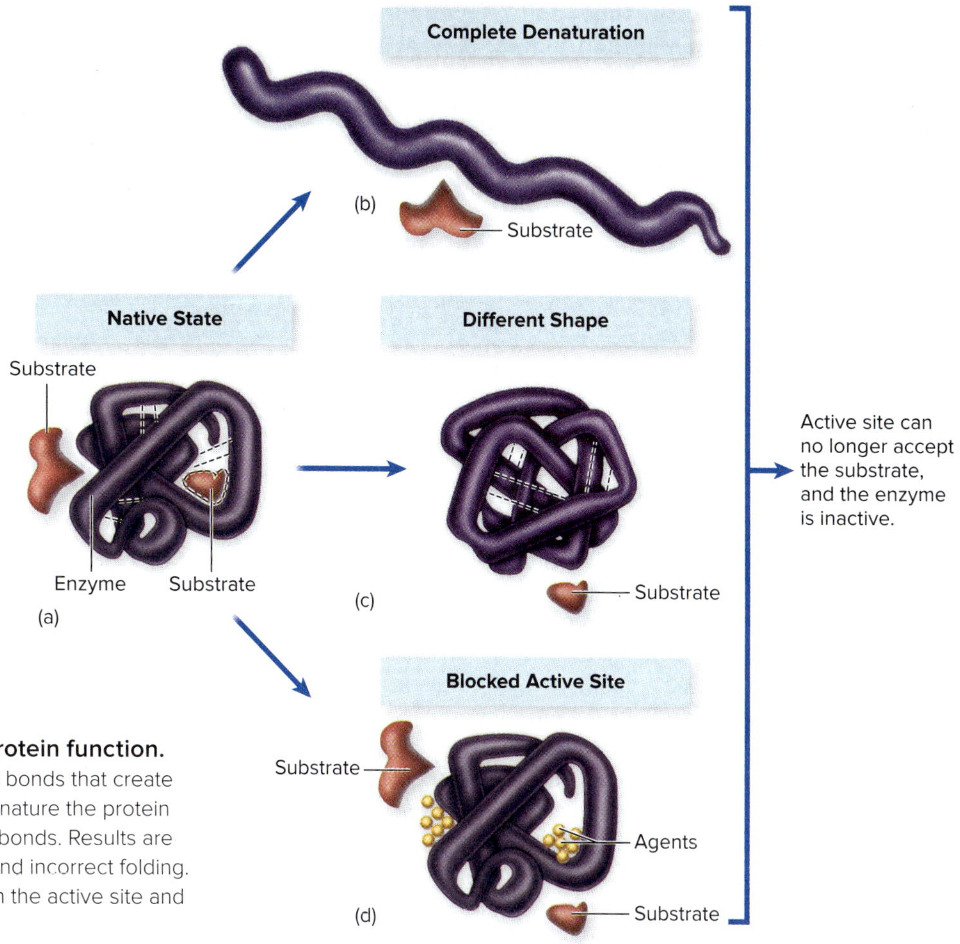
Complete Denaturation
(b)
Substrate
Native State
Substrate
Enzyme
Substrate
(a)
Different Shape
(c)
Substrate
Active site can no longer accept the substrate, and the enzyme is inactive.
Blocked Active Site
Substrate
Agents
(d)
Substrate

Figure 11.5 Modes of action affecting protein function. **(a)** The native (functional) state is maintained by bonds that create active sites to fit the substrate. Some agents denature the protein by breaking all or some secondary and tertiary bonds. Results are **(b)** complete unfolding or **(c)** random bonding and incorrect folding. **(d)** Some agents react with functional groups on the active site and interfere with bonding.

11.1 Learning Outcomes—Assess Your Progress

1. Distinguish among the terms *sterilization, disinfection, antisepsis,* and *decontamination.*
2. Identify the types of microorganisms that are most resistant and least resistant to control measures.
3. Compare the action of microbicidal and microbistatic agents, providing an example of each.
4. Name four categories of cellular targets for physical and chemical agents.

11.2 Methods of Physical Control

We can divide our methods of controlling microorganisms into two broad categories: physical and chemical. We will start with physical methods. Over the billions of years that bacteria and archaea have been on earth, they have adapted to the tremendous diversity of habitats the earth provides, including the most severe conditions of temperature, moisture, pressure, and light. Compared to them, we and other organisms are very delicate creatures indeed. For microbes that normally withstand such extreme physical conditions, our attempts at control would probably have little effect. Fortunately for us, we are most interested in controlling microbes that flourish in the same environment in which humans live, to keep them from causing human disease. The vast majority of these microbes are easily controlled by abrupt changes in their physical environment. A very common antimicrobial physical agent is heat. Other, less widely used agents include radiation, filtration, ultrasonic waves, and even cold. The following sections examine some of these methods.

Heat as an Agent of Microbial Control

A sudden departure from a microbe's optimal temperature is likely to have a detrimental effect on it. As a rule, higher temperatures (exceeding the maximum growth temperature) are microbicidal, whereas lower temperatures (below the minimum growth temperature) are microbistatic. Heat can be applied in either moist or dry forms. *Moist heat* occurs in the form of hot water, boiling water, or steam (vaporized water). In practice, the temperature of moist heat usually ranges from 60°C to 135°C. As we shall see, the temperature of steam can be regulated by adjusting its pressure in a closed container. The expression *dry heat* refers to air with a low moisture content that has been heated by a flame or an electric heating coil. In practice, the temperature of dry heat ranges from 160°C to several thousand degrees Celsius. (Remember that science usually uses the Celsius scale.)

Mode of Action and Relative Effectiveness of Heat

Moist heat and dry heat differ in their modes of action as well as in their efficiency. Moist heat operates at lower temperatures and

Table 11.2 Comparison of Times and Temperatures to Achieve Sterilization with Moist Heat and Dry Heat

	Temperature (°C)	Time to Sterilize (Min)
Moist heat	121	15
	125	10
	134	3
Dry heat	121	600
	140	180
	160	120
	170	60

shorter exposure times to achieve the same effectiveness as dry heat **(table 11.2).** Although many cellular structures are damaged by moist heat, its most microbicidal effect is the coagulation and denaturation of proteins, which quickly and permanently halt cellular metabolism.

Dry heat dehydrates the cell, removing the water necessary for metabolic reactions, and it denatures proteins. However, the lack of water actually increases the stability of some protein conformations, necessitating the use of higher temperatures when dry heat is employed as a method of microbial control. At very high temperatures, dry heat oxidizes cells, burning them to ashes. This method is the one used in the laboratory when a loop is flamed or placed in a mini-incinerator, or in industry when medical waste is incinerated.

Heat Resistance and Thermal Death: Endospores and Vegetative Cells

Bacterial endospores exhibit the greatest resistance, and vegetative states of bacteria and fungi are the least resistant to both moist and dry heat. Destruction of endospores usually requires temperatures above boiling.

Vegetative cells also vary in their sensitivity to heat. Among bacteria, the death times with moist heat range from 50°C for 3 minutes (*Neisseria gonorrhoeae*) to 60°C for 60 minutes (*Staphylococcus aureus*). It is worth noting that vegetative cells of endospore formers are just as susceptible as vegetative cells of non–endospore-formers and that pathogens are neither more nor less susceptible than nonpathogens. Other microbes, including fungi, protozoa, and worms, are somewhat similar in their sensitivity to heat. Viruses are surprisingly resistant to heat, with a tolerance range extending from 55°C for 2 to 5 minutes (adenoviruses) to 60°C for 600 minutes (hepatitis A virus). For practical purposes, all non-heat-resistant forms of bacteria, yeasts, molds, protozoa, worms, and viruses are destroyed by exposure to 80°C for 20 minutes.

Susceptibility of Microbes to Heat: Thermal Death Measurements

Adequate sterilization requires both the correct temperature and length of exposure. As we have seen, higher temperatures allow shorter exposure times, and lower temperatures require longer exposure times. A combination of these two variables constitutes the **thermal death time** (TDT), defined as the shortest length of time required to kill test microbes at a specified temperature. It is different for different species of microbes. The TDT has been experimentally determined for the microbial species that are common or important contaminants in various heat-treated materials. Another way to compare the susceptibility of microbes to heat is the **thermal death point** (TDP), defined as the lowest temperature required to kill all microbes in a sample in 10 minutes.

Many perishable substances are processed with moist heat. Some of these products are intended to remain on the shelf at room temperature for several months or even years. The chosen heat treatment must make sure the product is free of agents of spoilage or disease. At the same time, the quality of the product and the speed and cost of processing must be considered. For example, in the commercial preparation of canned green beans, one of the manufacturer's greatest concerns is to prevent growth of the bacterium that causes botulism. From several possible TDTs (that is, combinations of time and temperature) for *Clostridium botulinum* endospores, the manufacturer must choose one that kills all endospores but does not turn the beans to mush. Out of these many considerations emerges an optimal TDT for a given processing method. Commercial canneries heat low-acid foods at 121°C for 30 minutes, a treatment that sterilizes these foods. Because of such strict controls in canneries, cases of botulism due to commercially canned foods are rare.

Common Methods of Moist Heat Control

The three ways that moist heat is employed to control microbes are

1. boiling water,
2. pasteurization, and
3. steam under pressure.

These methods are described in detail in **table 11.3.**

Disease Connection

A disease known as CJD (Creutzfeldt-Jakob disease) is caused by the mysterious infectious protein known as a prion (see **A Note About Prions** earlier in this chapter). There are different forms of the disease. Some cases occur spontaneously, some are due to an inherited genetic marker, and some are transmitted from animals. In the 1980s, an outbreak of this disease occurred in cows in Great Britain and elsewhere. The condition in cattle is called *mad cow disease.* Before the existence of prions was appreciated, the disease was occasionally transmitted to patients during surgical procedures, even though the instruments being used had been sterilized after being used on the previous patient. That is because methods of sterilization used in those settings, including dry heat and autoclaving, do not destroy the prion. Current medical instrument sterilization guidelines account for the enhanced techniques required to sterilize devices that may have come in contact with prion-infected tissue. The methods include autoclaving at a higher temperature and pretreating devices with chemical sterilants before autoclaving.

A Note About Tables

In all of the tables in this chapter, agents that are capable of *sterilizing* will feature a red/pink background.

Table 11.3 Moist Heat Methods

Method	Applications
 *Charles D. Winters/McGraw Hill* **Boiling Water: Disinfection** A simple boiling water bath or chamber can quickly decontaminate items in the clinic and home. Because a single processing at 100°C will not kill all resistant cells, this method can be relied on only for disinfection and not for sterilization. Exposing materials to boiling water for 30 minutes will kill most non-endospore-forming pathogens, including resistant species such as the tuberculosis bacterium and staphylococci. Probably the greatest disadvantage with this method is that the items can be easily recontaminated when removed from the water.	Useful in the home for disinfection of water, materials for babies, food and utensils, bedding, and clothing from the sickroom
 *James King-Holmes/Science Source* **Pasteurization: Disinfection of Beverages** Fresh beverages such as milk, fruit juices, beer, and wine are easily contaminated during collection and processing. Because microbes have the potential for spoiling these foods or causing illness, heat is frequently used to reduce the microbial load or destroy pathogens. **Pasteurization** is a technique in which heat is applied to liquids to kill potential agents of infection and spoilage while retaining the liquid's flavor and food value. Ordinary pasteurization techniques require special heat exchangers that expose the liquid to 71.6°C for 15 seconds (flash method) or to 63°C to 66°C for 30 minutes (batch method). The first method is preferable because it is less likely to change flavor and nutrient content, and it is more effective against certain resistant pathogens such as *Coxiella* and *Mycobacterium*. Although these treatments inactivate most viruses and destroy the vegetative stages of 97% to 99% of bacteria and fungi, they do not kill endospores or particularly heat-resistant microbes (mostly nonpathogenic lactobacilli, micrococci, and yeasts). Milk is not sterile after regular pasteurization. In fact, it can contain 20,000 microbes per milliliter or more, which explains why even an unopened carton of milk will eventually spoil. (Newer techniques can also produce "shelf-stable" milk that has a storage life of 3 months. This milk is processed with ultrahigh temperature [UHT]—134°C—for 1 to 2 seconds.)	Milk, wine, beer, other beverages  *John A. Rizzo/Photodisc/Getty Images*
Steam Under Pressure: Autoclaving At sea level, normal atmospheric pressure is 15 pounds per square inch (psi), or 1 atmosphere. At this pressure, water will boil (change from a liquid to a gas) at 100°C, and the resultant steam will remain at exactly that temperature, which is too low to reliably kill all microbes. In order to raise the temperature of steam, the pressure at which it is generated must be increased. As the pressure is increased, the temperature at which water boils and the temperature of the steam produced both rise. For example, at a pressure of 20 psi (5 psi above normal), the temperature of steam is 109°C. As the pressure is increased to 10 psi above normal, the steam's temperature rises to 115°C, and at 15 psi above normal (a total of 2 atmospheres), it will be 121°C. It is not the pressure by itself that kills the microbes but the increased temperature it produces. Such pressure-temperature combinations can be achieved only with a special device that can subject pure steam to pressures greater than 1 atmosphere. Health and commercial industries use an **autoclave** for this purpose, **(figure 11.6)** and a comparable home appliance is the pressure cooker. The most efficient pressure-temperature combination for achieving sterilization is 15 psi, which yields 121°C. It is important to avoid overpacking or haphazardly loading the chamber, which prevents steam from circulating freely around the contents and impedes the full contact that is necessary. The length of the process is adjusted according to the bulkiness of the items in the load (thick bundles of material or large flasks of liquid) and how full the chamber is. The range of holding times varies from 10 minutes for light loads to 40 minutes for heavy or bulky ones; the average time is 20 minutes.	Heat-resistant materials such as glassware, cloth, metal instruments, liquids, paper, some media, and some heat-resistant plastics. If items are heat-sensitive (plastic Petri dishes) but will be discarded, the autoclave is still a good choice. It is ineffective for sterilizing substances that repel moisture (oils, waxes), or for those that are harmed by it (powders). **Sterilizing technique.**  *microgen/Getty Images*

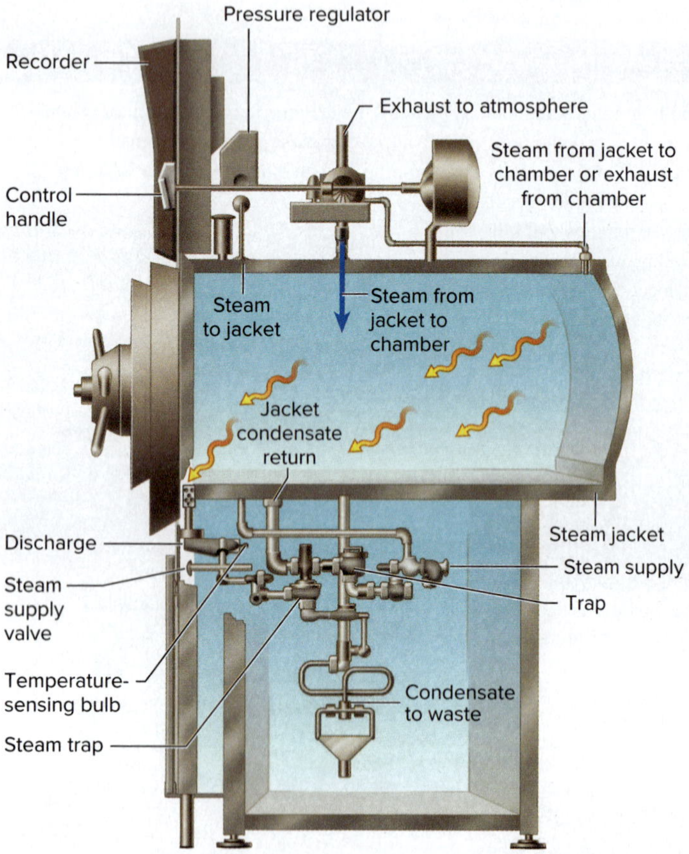
Pressure regulator
Recorder
Exhaust to atmosphere
Steam from jacket to chamber or exhaust from chamber
Control handle
Steam to jacket
Steam from jacket to chamber
Jacket condensate return
Steam jacket
Discharge
Steam supply
Steam supply valve
Trap
Temperature-sensing bulb
Condensate to waste
Steam trap

Figure 11.6 The inner workings of a standing autoclave. All autoclaves have similar features, whether they are table-top-sized or airport-hangar-sized.

Dry Heat: Hot Air and Incineration

Dry heat is not as versatile or as widely used as moist heat, but it has several important sterilization applications. The temperatures and times used in dry heat vary according to the particular method, but in general, they are higher and longer than with moist heat. **Table 11.4** describes two dry heat sterilization methods.

The Effects of Cold and Desiccation

The principal benefit of cold treatment is to slow down the growth of cultures and microbes in food during processing and storage. *It must be emphasized that cold merely slows down the activities of most microbes.* Although it is true that some microbes are killed by cold temperatures, most are not killed by gradual cooling, long-term refrigeration, or deep-freezing. In fact, freezing temperatures, ranging from −70°C to −135°C, are often used in research labs to preserve cultures of bacteria, viruses, and fungi for long periods. Some psychrophiles grow very slowly even at freezing temperatures and can continue to secrete toxic products. Ignorance of these facts is probably responsible for numerous cases of food poisoning from frozen foods that have been defrosted at room temperature and then inadequately cooked. Pathogens able to survive several months in the refrigerator are *Staphylococcus aureus; Clostridium* species (endospore formers); *Streptococcus* species; and several types of yeasts, molds, and viruses. Outbreaks of *Salmonella* food infection traced back to refrigerated foods such as ice cream, eggs, and tiramisu are testimony to the inability of cold temperatures to reliably kill pathogens.

Vegetative cells directly exposed to normal room air gradually become dehydrated, or **desiccated.** Delicate pathogens such as *Streptococcus pneumoniae,* the spirochete of syphilis, and *Neisseria gonorrhoeae* can die after a few hours of air drying, but many others are not killed, and some are even preserved. Endospores of *Bacillus* and *Clostridium* are viable for millions of years under extremely dry conditions. Staphylococci and streptococci in dried secretions and the tubercle bacillus surrounded by sputum can remain viable in air and dust for lengthy periods. Many viruses (especially nonenveloped) and fungal spores can also withstand long periods of desiccation. Desiccation can be a valuable way to preserve foods because it greatly reduces the amount of water available to support microbial growth.

It is interesting to note that a combination of freezing and drying—**lyophilization** (ly-off″-il-ih-za′-shun)—is a common method of preserving microorganisms and other cells in a viable state for many years. Pure cultures are frozen instantaneously and exposed to a vacuum that rapidly removes the water (it goes right from the frozen state into the vapor state). This method avoids the formation of ice crystals that would damage the cells. Although not all cells survive this process, enough of them do to permit future reconstitution of that culture.

As a general rule, chilling, freezing, and desiccation should not be construed as methods of disinfection or sterilization because their antimicrobial effects are erratic and uncertain, and one cannot be sure that pathogens subjected to them have been killed.

Radiation as a Microbial Control Agent

Another way that energy can serve as an antimicrobial agent is through the use of radiation. **Radiation** is defined as energy emitted from atomic activities and dispersed at high velocity through matter or space. **Figure 11.7** illustrates the different wavelengths of radiation. Although radiation exists in many states and can be described and characterized in various ways, in this discussion we consider only those types suitable for microbial control: gamma rays, X rays, and ultraviolet radiation.

The Actions of Ionizing versus Nonionizing Radiation

The actual physical effects of radiation on microbes can be understood by thinking about the process of **irradiation,** or bombardment with radiation, at the cellular level **(figure 11.8).** When a cell is bombarded by certain waves or particles, its molecules absorb some of the available energy, leading to one of two consequences: (1) If the radiation ejects orbital electrons from an atom, it causes ions to form. This type of radiation is termed **ionizing radiation.** This type of radiation causes catastrophic mutations in DNA and

Table 11.4 **Dry Heat Methods**

Method	Applications
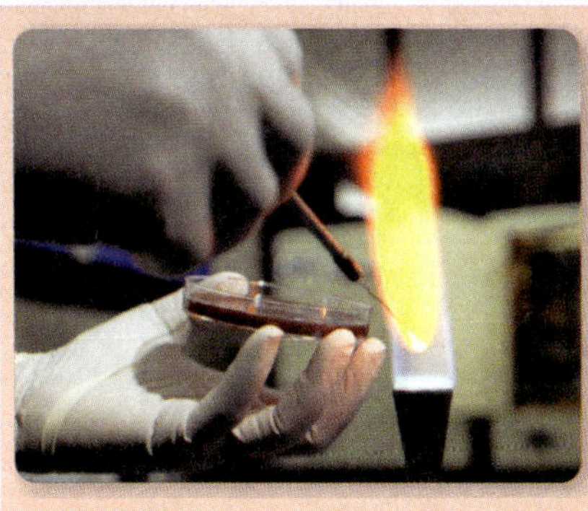 UIG/Getty Images **Incineration** in a flame is perhaps the most rigorous of all heat treatments. The flame of a Bunsen burner reaches 1,870°C at its hottest point, and furnaces/incinerators operate at temperatures of 800°C to 6,500°C. Direct exposure to such intense heat ignites and reduces microbes and other substances to ashes and gas. Incineration of microbial samples on inoculating loops and needles using a Bunsen burner is a very common practice in the microbiology laboratory. This method is fast and effective, but it is also limited to metals and heat-resistant glass materials. This method also presents hazards to the operator (an open flame) and to the environment (contaminants on needles or loops often spatter when placed in flame). Tabletop infrared incinerators have replaced Bunsen burners in many labs for these reasons. Large incinerators are regularly employed in hospitals and research labs for complete destruction of infectious materials.	Bunsen burners/small incinerators: laboratory instruments such as inoculating loops. Large incinerators: syringes, needles, culture materials, dressings, bandages, bedding, animal carcasses, and pathology samples. **Sterilizing technique.**
 RayArt Graphics/Alamy Stock Photo The **hot-air oven** provides another means of dry-heat sterilization. The so-called *dry oven* is usually electric (occasionally gas) and has coils that radiate heat within an enclosed compartment. Heated, circulated air transfers its heat to the materials in the oven. Sterilization requires exposure to 150°C to 180°C for 2 to 4 hours, which ensures thorough heating of the objects and destruction of endospores.	Glassware, instruments, powders, and oils that steam does not penetrate well. Not suitable for plastics, cotton, and paper, which may burn at the high temperatures, or for liquids, which will evaporate. **Sterilizing technique.**

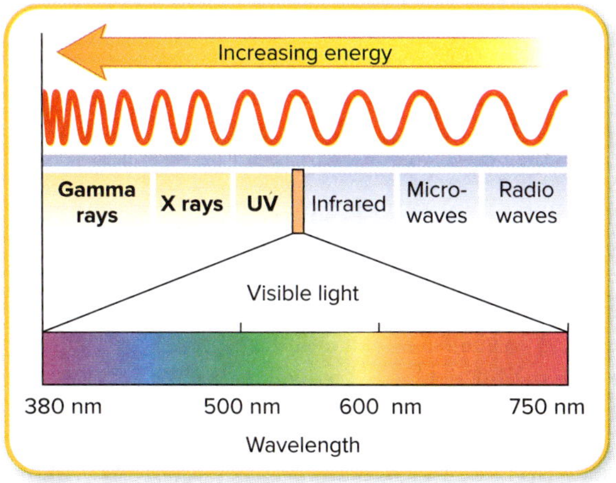
Increasing energy
Gamma rays
X rays
UV
Infrared
Micro-waves
Radio waves
Visible light
380 nm
500 nm
600 nm
750 nm
Wavelength

Figure 11.7 **The electromagnetic spectrum, showing different types of radiation.**

damages the proteins that would ordinarily repair it. Secondary lethal effects include chemical changes in organelles and the production of toxic substances. Gamma rays, X rays, and high-speed electrons are all forms of ionizing radiation. (2) **Nonionizing radiation,** best exemplified by ultraviolet (UV), excites atoms by raising them to a higher energy state, but it does not ionize them. This atomic excitation, in turn, leads to the formation of abnormal bonds within molecules such as DNA, and that is what leads to mutations **(table 11.5).**

Ionizing Radiation: Gamma Rays, X Rays, and Cathode Rays

Ionizing radiation is a highly effective alternative for sterilizing materials that are sensitive to heat or chemicals. Because it sterilizes without heating, irradiation is a type of **cold** (or low-temperature) **sterilization.**

Devices that emit ionizing rays include gamma-ray machines containing radioactive cobalt, X-ray machines more powerful than those used in medical diagnosis, and cathode-ray machines. Items are placed in these machines and irradiated for a short time with a carefully chosen dosage. The dosage of radiation is measured in *Grays* (which has replaced the older term *rads*). Depending on the application, exposure ranges from 5 to 50 kiloGrays (a kiloGray is equal to 1,000 Grays). Although all ionizing radiation can penetrate liquids and most solid materials, gamma rays are most penetrating, X rays are intermediate, and cathode rays are least penetrating.

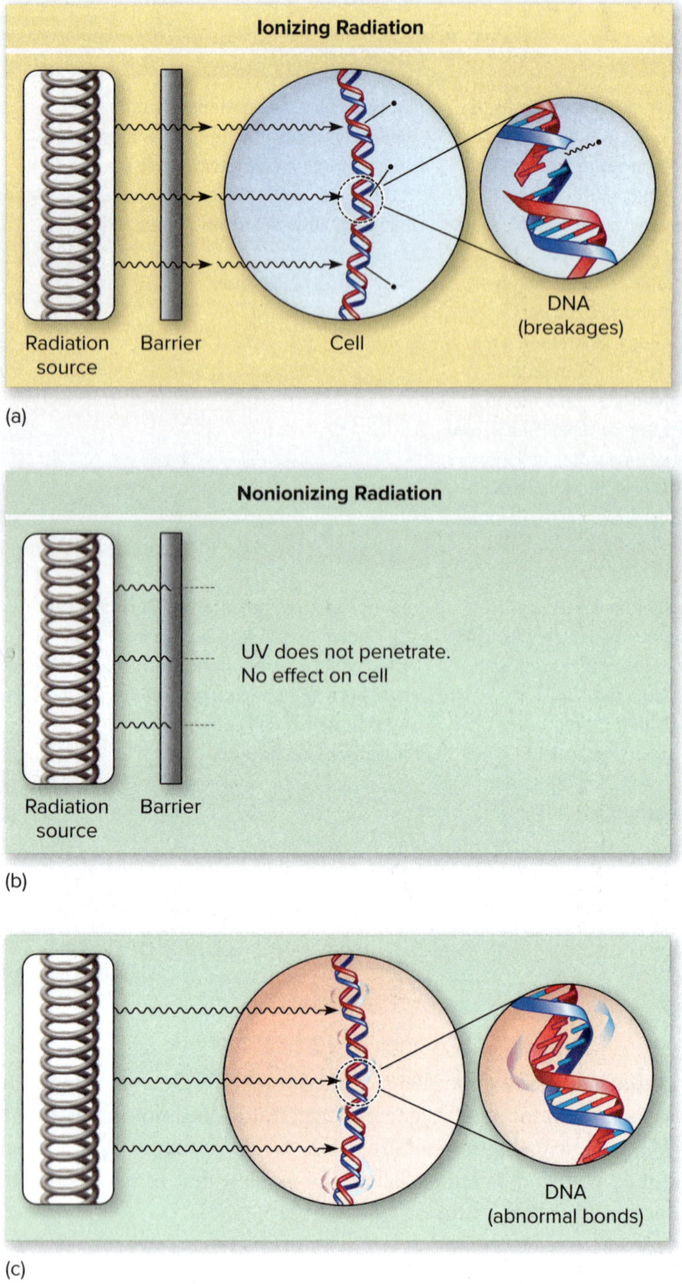
Ionizing Radiation
Radiation source
Barrier
Cell
DNA (breakages)
(a)
Nonionizing Radiation
UV does not penetrate.
No effect on cell
Radiation source
Barrier
(b)
DNA (abnormal bonds)
(c)

Figure 11.8 Cellular effects of irradiation. **(a)** Ionizing radiation can penetrate a solid barrier, bombard a cell, enter it, and dislodge electrons from molecules. Breakage of DNA creates massive mutations, and damage to proteins prevents them from repairing it. **(b)** A solid barrier cannot be penetrated by nonionizing radiation. **(c)** Nonionizing radiation enters a cell, strikes molecules, and excites them. The effect on DNA is mutation by formation of abnormal bonds.

Table 11.5 Radiation Methods

Both of these are capable of sterilizing.

Method
 *Adam Hart-Davis/Science Source* **Ionizing Radiation: Gamma Rays and X Rays** Ionizing radiation is a highly effective alternative for sterilizing materials that are sensitive to heat or chemicals. Devices that emit ionizing rays include gamma-ray machines containing radioactive cobalt.
 *Tom Pantages* **Nonionizing Radiation: Ultraviolet Rays** Ultraviolet (UV) radiation ranges in wavelength from approximately 100 to 400 nm. It is most lethal from 240 to 280 nm (with a peak at 260 nm).

Applications of Ionizing Radiation

Foods have been subject to irradiation in limited circumstances for more than 50 years. From flour to pork and ground beef to fruits and vegetables, radiation is used to kill not only bacterial pathogens but also insects and worms and even to inhibit the sprouting of white potatoes. As soon as radiation is mentioned, however, consumer concern arises that food may be made less nutritious, unpalatable, or even unsafe by its having been subjected to ionizing radiation. But irradiated food has been extensively studied, and each of these concerns has been addressed.

Irradiation may lead to a small decrease in the amount of thiamine (vitamin B_1) in food, but this change is small enough to be inconsequential. The irradiation process does produce short-lived free radical oxidants, which disappear almost immediately (this same type of chemical intermediate is produced through cooking as well). Certain foods do not irradiate well and are not good candidates for this type of antimicrobial control. The whites of eggs become milky and liquid, grapefruits get mushy, and alfalfa seeds do not germinate properly. Finally, it is important to remember that food is not made radioactive by the irradiation process. Many studies, in both animals and humans, have concluded that there are no negative effects from eating irradiated food.

Nonionizing Radiation: Ultraviolet Rays

Ultraviolet (UV) radiation ranges in wavelength from approximately 100 nm to 400 nm. It is most lethal from 240 nm to 280 nm (with a peak at 260 nm). In everyday practice, the source of UV radiation is a germicidal lamp, which generates radiation at 254 nm. Owing to its lower energy state (refer to figure 11.7), UV radiation is not as penetrating as ionizing radiation. Because UV radiation passes readily through air, slightly through liquids, and only poorly through solids, the object to be disinfected must be directly exposed to it for full effect.

As UV radiation passes through a cell, it is initially absorbed by DNA. Specific molecular damage occurs on the pyrimidine bases (thymine and cytosine), which form abnormal linkages with each other called **pyrimidine dimers (figure 11.9).** These bonds occur between adjacent bases on the same DNA strand and interfere with normal DNA replication and transcription. The results are growth inhibition and cellular death. In addition to altering DNA directly, UV radiation also disrupts cells by generating toxic photochemical products called free radicals. These highly reactive molecules interfere with essential cell processes by binding to DNA, RNA, and proteins. Ultraviolet rays are a powerful disinfecting tool for destroying fungal cells and spores, bacterial vegetative cells, protozoa, and viruses. Bacterial endospores are about 10 times more resistant to radiation than are vegetative cells, but sterilization can be achieved if the time of exposure is increased enough.

Ultraviolet treatment has proved effective in treating the surfaces of solid, nonporous materials such as walls and floors. As part of the terminal cleaning process (a process used after patients with certain infectious diseases have left a hospital room), UV machines are used **(figure 11.10)**. They reach about 10–12 feet to reach exposed surfaces and need to be moved during the process to access as many surface as possible in the room.

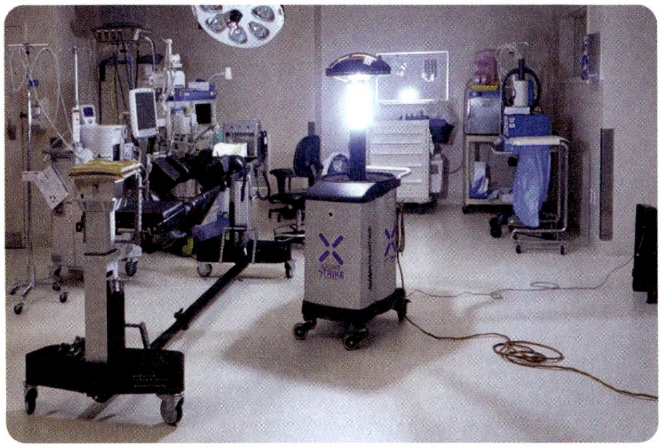

Figure 11.10 UV disinfection at a terminal cleaning. After regular cleaning has taken place, these xenon ultraviolet light machines finish the job.

Courtesy Xenex Disinfection Services

One major disadvantage of UV is its poor powers of penetration through solid materials such as glass, metal, cloth, plastic, and even paper. Another drawback to UV is the damaging effect of overexposure on human tissues, including sunburn, retinal damage, cancer, and skin wrinkling.

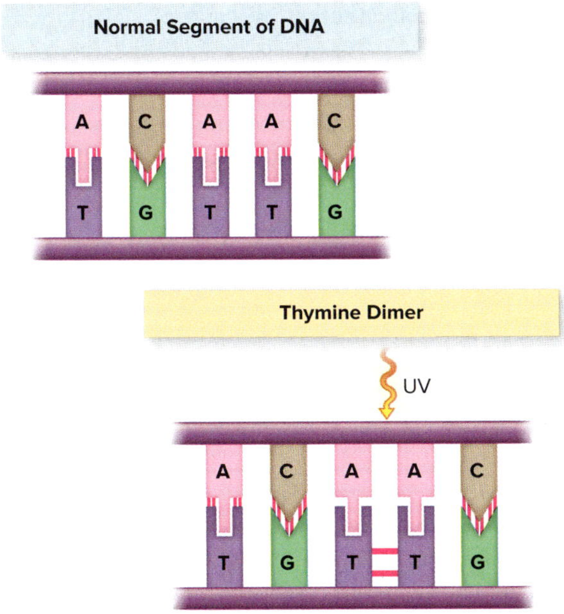
Normal Segment of DNA
A C A A C
T G T T G
Thymine Dimer
UV
A C A A C
T G T T G

Figure 11.9 Formation of pyrimidine dimers by the action of ultraviolet (UV) radiation. This shows what occurs when two adjacent thymine bases on one strand of DNA are induced by UV rays to bond laterally with each other. The result is a thymine dimer (shown on the bottom). Dimers can also occur between adjacent cytosines and thymine and cytosine bases. If they are not repaired, dimers can prevent that segment of DNA from being correctly replicated or transcribed. Extensive formation of dimers is lethal to cells.

Decontamination by Filtration: Techniques for Removing Microbes

Filtration is an effective method to remove microbes from air and liquids. In practice, a fluid is strained through a filter with openings large enough for the fluid to pass through but too small for microorganisms to pass through **(figure 11.11*a*).**

Most microbiological filters are thin membranes of cellulose acetate, polycarbonate, and a variety of plastic materials (Teflon, nylon) whose pore size can be carefully controlled and standardized. Ordinary substances such as charcoal, diatomaceous earth, or unglazed porcelain are also used in some applications. Viewed microscopically, most filters are perforated by very precise, uniform pores **(figure 11.11*b*).** The pore diameters vary from coarse (8 microns) to ultrafine (0.02 micron). Microbes and other particles that are larger than the pore diameter remain trapped on the top of the filter. Those filters with even smaller pore diameters permit true sterilization by removing viruses, and some will even remove large proteins. A sterile liquid filtrate is typically produced by suctioning the liquid through a sterile filter into a presterilized container. These filters are also used to separate mixtures of microorganisms and to count bacteria in water analysis.

Applications of Filtration

Filtration is used to prepare liquids that cannot withstand heat, including serum and other blood products, vaccines, drugs, IV fluids, and media. Filtration has been employed as an alternative method for decontaminating milk and beer without affecting their flavor. It is also an important step in water purification. It has the disadvantage of not removing smaller molecules (smaller than the pore size), such as bacterial exotoxins.

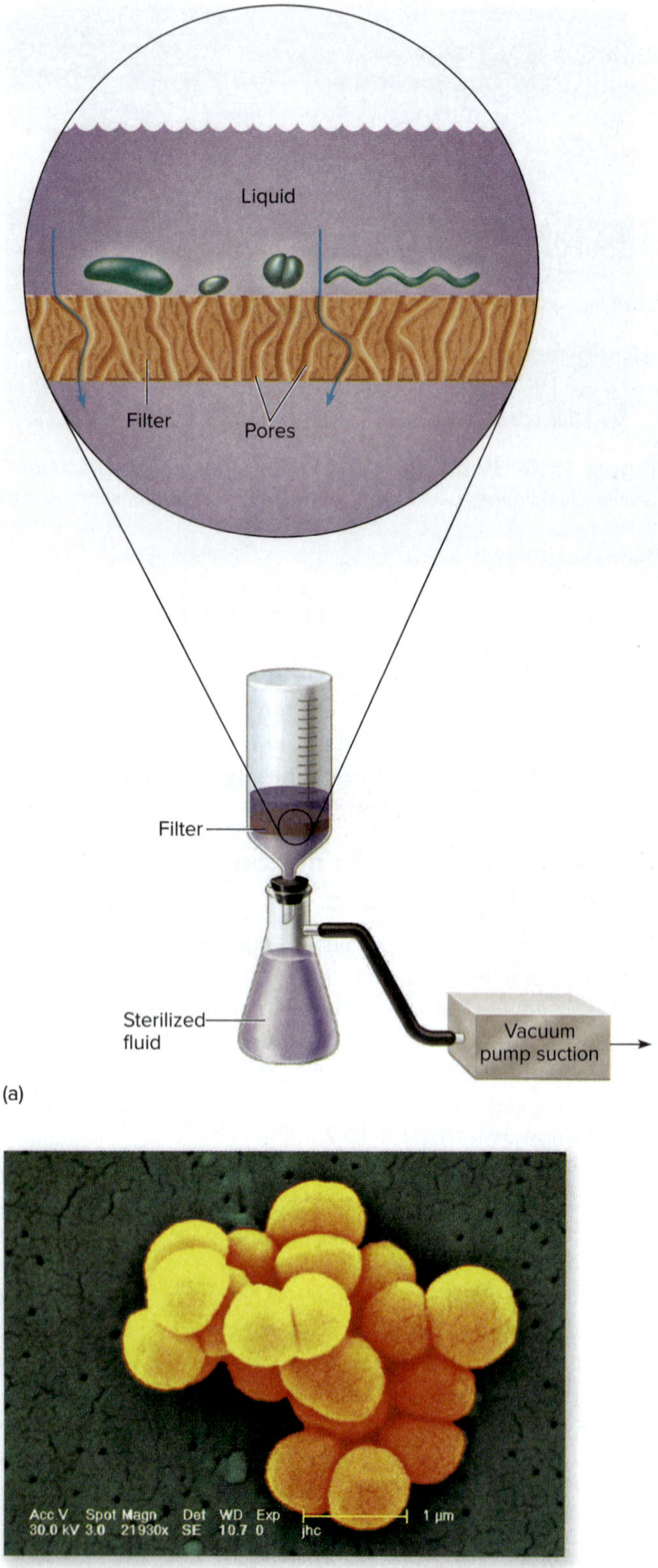
Liquid
Filter
Pores
Filter
Sterilized fluid
Vacuum pump suction
(a)
Acc V Spot Magn Det WD Exp
30.0 kV 3.0 21930x SE 10.7 0
jhc
1 µm
(b)

Figure 11.11 Membrane filtration. **(a)** Vacuum assembly for achieving filtration of liquids through suction. The circle on top shows a filter as seen in cross section, with tiny passageways (pores) too small for the microbial cells to enter but large enough for liquid to pass through. **(b)** Scanning electron micrograph of filter, showing relative size of pores and bacteria trapped on its surface.

(b) BSIP/Newscom

Filtration is also an efficient means of removing airborne contaminants that are a common source of infection and spoilage. High-efficiency particulate air (HEPA) filters are widely used to provide a flow of decontaminated air to hospital rooms and sterile rooms. HEPA masks and room air filters are an effective way to diminsh transmission of SARS-CoV-2, the virus that causes COVID. HEPA filters typically trap anything larger than 10 nanometers, and SARS-CoV-2 is approximately 125 nanometers.

Osmotic Pressure

In chapter 9, you learned about the effects of osmotic pressure on cells. This fact has long been exploited as a means of preserving food. Adding large amounts of salt or sugar to foods creates a hypertonic environment for bacteria in the foods, causing plasmolysis and making it impossible for the bacteria to multiply. People knew that these techniques worked long before the discovery of bacteria. This is why meats are "cured," or treated with high salt concentrations, so they can be kept for long periods without refrigeration. High sugar concentrations in foods like jellies have the same effect.

11.2 Learning Outcomes—Assess Your Progress

5. Name six methods of physical control of microorganisms.
6. Compare and contrast moist and dry heat methods of control, and identify multiple examples of each.
7. Define *thermal death time* and *thermal death point,* and describe their role in proper sterilization.
8. Explain three different methods of moist heat control.
9. Explain two methods of dry heat control.
10. Identify advantages and disadvantages of cold treatment and desiccation.
11. Differentiate between the two types of radiation control methods, providing an application of each.
12. Outline the process of filtration, and describe its two advantages in microbial control.
13. Identify some common uses of osmotic pressure as a control method.

11.3 Methods of Chemical Control

Chemical control of microbes probably emerged as a serious science in the 1800s, when physicians first began using lime and iodine solutions to treat wounds and to wash their hands before surgery. At the present time, more than 10,000 different antimicrobial chemical agents are manufactured. About 1,000 of them are used routinely in the health care arena and the home. A genuine need exists to avoid infection and spoilage, but the abundance of products available to "kill germs," "disinfect," "antisepticize," "clean and sanitize," "deodorize," "fight plaque," and "purify the air" indicates a preoccupation with eliminating microbes from the environment that, at times, can be excessive.

Antimicrobial chemicals can be in a liquid, gaseous, or even solid state, and they range from disinfectants and antiseptics to sterilants and preservatives (chemicals that inhibit the deterioration of substances). For the sake of convenience (and sometimes safety), many solid or gaseous antimicrobial chemicals are dissolved in water, alcohol, or a mixture of the two to produce a liquid solution. Solutions containing pure water as the solvent are termed **aqueous,** whereas those dissolved in pure alcohol or water-alcohol mixtures are termed **tinctures.**

Selecting a Microbicidal Chemical

The choice and appropriate use of antimicrobial chemical agents are of constant concern in medicine and dentistry. Here are some guidelines to be considered, and optimized to the extent possible:

1. rapid action even in low concentrations,
2. solubility in water or alcohol and long-term stability,
3. broad-spectrum microbicidal action without toxicity to human and animal tissues,
4. penetration of inanimate surfaces to sustain a cumulative or persistent action,
5. resistance to becoming inactivated by organic matter,
6. noncorrosive or nonstaining properties,
7. sanitizing and deodorizing properties, and
8. affordability and ready availability.

No chemical can completely fulfill all of those requirements, but glutaraldehyde and hydrogen peroxide approach this ideal.

Germicides are evaluated in terms of their effectiveness in destroying microbes in medical and dental settings. Echoing the language we used earlier in the chapter, referring to medical devices as critical, semicritical, or noncritical, three levels of chemical decontamination procedures exist. These are high, intermediate, and low. High-level germicides kill endospores and, if properly used, are sterilants. These germicides are used on critical items such as catheters, heart-lung equipment, and implants. These devices are not heat-sterilizable and are intended to enter the body tissues during medical procedures. Intermediate-level germicides kill fungal (but not bacterial) spores, resistant pathogens such as the tubercle bacillus, and viruses. They are used to disinfect semi-critical items (respiratory equipment, thermometers). Low levels of disinfection eliminate only vegetative bacteria, vegetative fungal cells, and some viruses. They are used to clean noncritical materials such as electrodes, straps, and pieces of furniture that touch the skin surfaces but not the mucous membranes.

Factors Affecting the Microbicidal Activity of Chemicals

Factors that affect the usefulness of a germicide include the nature of the microorganisms being treated, the nature of the material being treated, the degree of contamination, the time of exposure, and the strength and chemical action of the germicide. The strength of a germicide also plays a role in its ability to act upon a microbial population. The modes of action of most germicides are to attack the cellular targets discussed earlier: proteins, nucleic acids, the cell wall, and the cell membrane.

A chemical's strength or concentration is expressed in various ways, depending on the method of preparation. In dilutions, a small volume of the liquid chemical (solute) is diluted in a larger volume of solvent to achieve a certain ratio. For example, a common phenolic disinfectant such as Lysol is usually diluted 1:200; that is, 1 part of chemical is added to 200 parts of water by volume. Solutions such as chlorine that are effective in very diluted concentrations are expressed in parts per million (ppm). In percentage solutions, the solute is added to water to achieve a certain percentage in the solution. Alcohol, for instance, is used in percentages ranging from 50% to 95%.

Most compounds require adequate contact time to allow the chemical to penetrate and to act on the microbes present. The composition of the material being treated must also be considered, as it may greatly impact the effectiveness of these germicidal agents. Smooth, solid objects are more reliably disinfected than are those with pores or pockets that can trap soil. An item contaminated with common biological matter such as serum, blood, saliva, pus, fecal material, or urine presents a problem in disinfection. Large amounts of organic material can hinder the penetration of a disinfectant and, in some cases, can form bonds that possibly reduce its activity. Cleaning is different from disinfecting. Cleaning (physically removing organic matter) is always a necessary first step before antimicrobial chemicals can be used.

Also, common sense is an important factor in successful microbial control. For example, many hospitals use detergent wet wipes to clean surfaces. Studies have shown that when the wipes are used on a contaminated surface and then used on a second surface, the second surface becomes contaminated with whatever was on the contaminated surface. So new wipes are required for each surface. In May 2020, the World Health Organization issued guidance on the cleaning and disinfection of environmental surfaces in the context of COVID-19. The main chemicals recommended were chlorine (a halogen) and ethyl alcohol.

Germicidal Categories According to Chemical Group

Several general groups of chemical compounds are widely used for antimicrobial purposes in medicine and commerce. Prominent agents include halogens, heavy metals, alcohols, phenolic compounds, oxidizers, aldehydes, detergents, and gases. These groups are summarized in **table 11.6.** Note that each agent is described with respect to its specific forms, modes of action, indications for use, and limitations.

The following section points out some pertinent facts about each of these groups.

The Halogen Antimicrobial Chemicals

The **halogens** are fluorine, bromine, chlorine, and iodine, a group of nonmetallic elements. These elements are highly effective components of disinfectants and antiseptics because they are microbicidal and not just microbistatic, and they are sporicidal with longer exposure. For these reasons, halogens are the active ingredients in nearly one-third of all antimicrobial chemicals currently marketed.

Iodophors are complexes of iodine and alcohol. This formulation allows the slow release of free iodine and increases its degree of penetration. These compounds have largely replaced free iodine solutions in medical antisepsis because they are less prone to staining or irritating tissues. Common iodophor products marketed as Betadine, Povidone (PVP), and Isodine contain 2% to 10% of available iodine.

In 2017, the Food and Drug Administration banned many iodophors from consumer products because of unsatisfactory evidence of efficacy and issues with safety.

Table 11.6 Germicidal Categories According to Chemical Group

Agent	Target Microbes	Form(s)	Mode of Action	Indications for Use	Limitations
Halogens: chlorine  bleach *Richard Hutchings/Digital Light/McGraw Hill*	Can kill endospores (slowly); all other microbes	Liquid/gaseous chlorine (Cl_2), hypochlorites (OCl), chloramines (NH_2Cl)	In solution, these compounds combine with water and release hypochlorous acid (HOCl); denature enzymes permanently and suspend metabolic reactions	Chlorine kills bacteria, endospores, fungi, and viruses; gaseous/liquid chlorine: used to disinfect drinking water, sewage, and waste water; hypochlorites: used in health care to treat wounds, disinfect bedding and instruments, sanitize food equipment and in restaurants, pools, and spas; chloramines: alternative to pure chlorine in treating drinking water; also used to treat wounds and skin surfaces	Less effective if exposed to light, alkaline pH, and excess organic matter **Can sterilize.**
Halogens: iodine	Can kill endospores (slowly); all other microbes	Free iodine in solution (I_2) Iodophors (complexes of iodine and alcohol)	Penetrates cells of microorganisms where it interferes with a variety of metabolic functions; interferes with the hydrogen and disulfide bonding of proteins	2% iodine, 2.4% sodium iodide (aqueous iodine) used as a topical antiseptic 5% iodine, 10% potassium iodide used as a disinfectant for plastic and rubber instruments, cutting blades, etc. Iodophor products contain 2% to 10% of available iodine, which is released slowly; used to prepare skin for surgery, in surgical scrubs, to treat burns, and as a disinfectant	Can be extremely irritating to the skin and is toxic when absorbed Many iodophors banned in consumer products in 2017 **Can sterilize.**
Phenol (carbolic acid)	Some bacteria, viruses, fungi	Derived from the distillation of coal tar Phenols consist of one or more aromatic carbon rings with added functional groups	In high concentrations, they are cellular poisons, disrupting cell walls and membranes, proteins In lower concentrations, they inactivate certain critical enzyme systems	Phenol remains one standard against which other (less toxic) phenolic disinfectants are rated; the phenol coefficient quantitatively compares a chemical's antimicrobial properties to those of phenol Phenol is now used only in certain limited cases, such as in drains, cesspools, and animal quarters	Toxicity of many phenolics makes them dangerous to use as antiseptics **Many phenols banned in consumer products in 2017, including triclosan and triclocarban**
Chlorhexidine	Most bacteria, viruses, fungi	Complex organic base containing chlorine and two phenolic rings	Targets bacterial membranes, where selective permeability is lost; bacterial cell walls; and proteins, resulting in denaturation	Mildness, low toxicity, and rapid action make chlorhexidine a popular choice of agents Used in hand scrubs, prepping skin for surgery, as an obstetric antiseptic, as a mucous membrane irrigant, etc.	Effects on viruses and fungi are variable
Alcohol 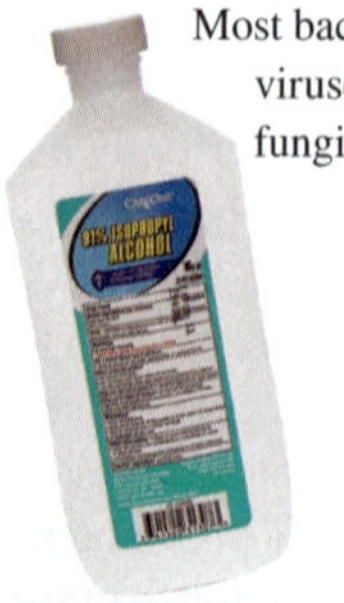 *Richard Hutchings/Digital Light/McGraw Hill*	Most bacteria, viruses, fungi	Colorless hydrocarbons with one or more —OH functional groups Ethyl and isopropyl alcohol are suitable for antimicrobial control	Concentrations of 50% and higher dissolve membrane lipids, disrupt cell surface tension, and compromise membrane integrity	Germicidal, nonirritating, and inexpensive Routinely used as skin degerming agents (70% to 95% solutions)	Rate of evaporation decreases effectiveness Inhalation of vapors can affect the nervous system

(continued)

Table 11.6 (concluded)

Agent	Target Microbes	Form(s)	Mode of Action	Indications for Use	Limitations
Oxidizing agents 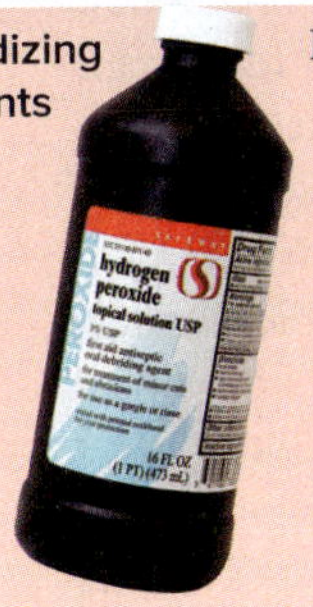 Jill Braaten/ McGraw Hill	Kill endospores and all other microbes	Hydrogen peroxide, peracetic acid	Oxygen forms free radicals (—OH), which are highly toxic and reactive to cells	As an antiseptic, 3% hydrogen peroxide used for skin and wound cleansing, mouth washing, bedsore care Used to treat infections caused by anaerobic bacteria 35% hydrogen peroxide used in low-temperature sterilizing cabinets for delicate instruments	Sporicidal only in high concentrations **Can sterilize.**
Detergents 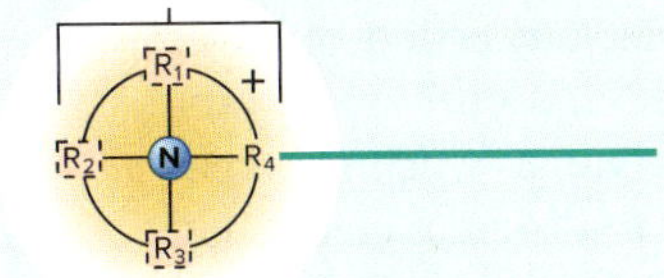 Charged Head R1 + R2 N R4 R3	Some bacteria, viruses, fungi	Polar molecules that act as surfactants Anionic detergents have limited microbial power Cationic detergents, such as quaternary ammonium compounds ("quats"), are much more effective antimicrobials	Positively charged end of the molecule binds well with the predominantly negatively charged bacterial surface proteins Long, uncharged hydrocarbon chain allows the detergent to disrupt the cytoplasmic membrane Cytoplasmic membrane loses selective permeability, causing cell death	Effective against viruses, algae, fungi, and gram-positive bacteria Rated only for low-level disinfection in the clinical setting Used to clean restaurant utensils, dairy equipment, equipment surfaces, restrooms	Ineffective against tuberculosis bacterium, hepatitis virus, *Pseudomonas*, and endospores Activity is greatly reduced in presence of organic matter Detergents function best in alkaline solutions **Some quats banned in consumer products in 2017**
Heavy metal compounds  Stephen Frisch/McGraw Hill	Some bacteria, viruses, fungi	Heavy metal germicides contain either an inorganic or an organic metallic salt; may come in tinctures, soaps, ointment, or aqueous solution	Mercury, silver, and other metals exert microbial effects by binding onto functional groups of proteins and inactivating them	Organic mercury tinctures are fairly effective antiseptics Organic mercurials serve as preservatives in cosmetics, ophthalmic solutions, and other substances Silver nitrate solutions are used for topical germicides and ointments	Microbes can develop resistance to metals Not effective against endospores Can be toxic if inhaled, ingested, or absorbed May cause allergic reactions in susceptible individuals
Aldehydes	Kill endospores and all other microbes	Organic substances bearing a —CHO functional group on the terminal carbon	Glutaraldehyde can irreversibly disrupt the activity of enzymes and other proteins within the cell Ortho-phthalaldehyde **Can sterilize.**	Glutaraldehyde kills rapidly and is broad-spectrum; used to sterilize respiratory equipment, scopes, kidney dialysis machines, dental instruments Ortho-phthalaldehyde is safer than glutaraldehyde and just as effective	Glutaraldehyde is somewhat unstable, especially with increased pH and temperature Ortho-phthalaldehyde is much more expensive than glutaraldehyde
Gaseous sterilants/ disinfectants	Ethylene oxide kills endospores; other gases less effective	Ethylene oxide is a colorless substance that exists as a gas at room temperature	Ethylene oxide reacts vigorously with functional groups of DNA and proteins, blocking both DNA replication and enzymatic actions Chlorine dioxide is a strong alkylating agent **Can sterilize.**	Ethylene oxide is used to disinfect plastic materials and delicate instruments; can also be used to sterilize syringes, surgical supplies, and medical devices that are prepackaged	Ethylene oxide is explosive—it must be combined with a high percentage of carbon dioxide or fluorocarbon It can damage lungs, eyes, and mucous membranes if contacted directly Ethylene oxide is rated as a carcinogen by the government
Acids and alkalis	Some bacteria, viruses, fungi	Organic acids in food production	pH alteration	Food manufacture, deodorants, deodorizers	Large changes in pH can be corrosive

Disease Connection

Patients confined to bed, especially patients who suffer from diabetes, sometimes develop wounds on their heel(s), caused by the pressure exerted on the heels by the mattress. When these wounds occur, they are often painted with Betadine (povidone-iodine) to prevent infection and to keep the wound dry because allowing the wound to become moist will eventually result in a large, open wound.

Phenol and Its Derivatives

Phenol (carbolic acid) is a poisonous compound derived from the distillation of coal tar. First adopted by Joseph Lister in 1867 as a surgical germicide, phenol was the major antimicrobial chemical until other phenolics with fewer toxic and irritating effects were developed. Substances chemically related to phenol are often referred to as phenolics. Solutions of phenol are now used only in certain limited cases, but phenol remains one standard against which other phenolic disinfectants are rated. The *phenol coefficient* quantitatively compares a chemical's antimicrobial properties to those of phenol. Hundreds of these chemicals are now available.

Perhaps the most widely known phenolic is *triclosan,* chemically known as dichlorophenoxyphenol. Until its ban in consumer products in 2017, it was added to dozens of products, from soaps to kitty litter. It acts as both disinfectant and antiseptic and is broad-spectrum in its effects. It is still licensed for use in health care settings, but the consumer ban came about because research established that widespread use of triclosan can lead to the development of resistance—both to triclosan itself and to unrelated antibiotics—in microbes.

Chlorhexidine

The compound chlorhexidine (Hibiclens, Hibitane, Peridex) is a complex organic base containing chlorine and two phenolic rings. Its mode of action targets both cell membranes (lowering surface tension until selective permeability is lost) and protein structure (causing denaturation). At moderate to high concentrations, it is bactericidal for both gram-positive and gram-negative bacteria but inactive against endospores. Its effects on viruses and fungi vary. It possesses distinct advantages over many other antiseptics because of its mildness, low toxicity, and rapid action. It is not absorbed into deeper tissues to any extent. It is sold in over-the-counter mouthwashes as well.

Alcohols as Antimicrobial Agents

Alcohols are colorless hydrocarbons with one or more —OH functional groups. Of several alcohols available, only ethyl and isopropyl are suitable for microbial control. Methyl alcohol is not particularly microbicidal, and more complex alcohols are either poorly soluble in water or too expensive for routine use. Alcohols are used alone in aqueous solutions or as solvents for tinctures (iodine, for example).

Alcohol's mechanism of action depends in part upon its concentration. Concentrations of 50% and higher dissolve membrane lipids, disrupt cell surface tension, and compromise membrane integrity. Alcohol that has entered the cytoplasm denatures proteins through coagulation but only in alcohol-water solutions of 50% to 95%. Alcohol is the exception to the rule that higher concentrations of an antimicrobial chemical have greater microbicidal activity. Because water is needed for proteins to coagulate, alcohol shows a greater microbicidal activity at 70% concentration (that is, 30% water) than at 100% (0% water). Absolute alcohol (100%) dehydrates cells and inhibits their growth but is generally not a protein coagulant.

Applications of Alcohols Ethyl alcohol, also called ethanol or grain alcohol, is known for being germicidal, nonirritating, and inexpensive. Solutions of 70% to 95% are routinely used as skin degerming agents because the surfactant action removes skin oil, soil, and some microbes sheltered in deeper skin layers. One limitation to its effectiveness is the rate at which it evaporates. Ethyl alcohol is occasionally used to disinfect electrodes, face masks, and thermometers, which are first cleaned and then soaked in alcohol for 15 to 20 minutes. Most alcohol-based hand sanitizers and foams contain 60% to 70% ethyl alcohol and are more resistant to evaporation, but to be microbicidal they must still be exposed to skin for 30 seconds. **Insight 11.1** addresses how often health care workers use alcohol hand rubs and raises some questions about what effect they have. Isopropyl alcohol, sold as rubbing alcohol, is even more microbicidal and less expensive than ethanol, but these benefits must be weighed against its toxicity. It must be used with caution in disinfection or skin cleansing because inhalation of its vapors can adversely affect the nervous system.

Oxidizing Agents

Hydrogen peroxide (H_2O_2) is a colorless, caustic liquid that decomposes into water and oxygen gas in the presence of light, metals, or catalase. The germicidal effects of hydrogen peroxide are due to the direct and indirect actions of oxygen. Oxygen forms hydroxyl free radicals ($^{\cdot}OH$), which, like the superoxide radical (see chapter 9), are highly toxic and reactive to cells. Although most microbial cells produce catalase to inactivate the metabolic hydrogen peroxide, it cannot neutralize the amount of hydrogen peroxide entering the cell during disinfection and antisepsis. Hydrogen peroxide is bactericidal, virucidal, fungicidal, and, in higher concentrations, sporicidal.

Applications of Hydrogen Peroxide As an antiseptic, 3% hydrogen peroxide serves a variety of needs, including skin and wound cleansing, bedsore care, and mouthwashing. It is especially useful in treating infections by anaerobic bacteria because of the lethal effects of oxygen on them. Hydrogen peroxide is also a versatile disinfectant for contact lenses, surgical implants, plastic equipment, utensils, bedding, and room interiors.

Vaporized hydrogen peroxide can also be used as a sterilant in enclosed areas. Hydrogen peroxide plasma sterilizers exist for those applications involving small industrial or medical items. For larger, enclosed spaces, such as isolators and pass-through rooms,

INSIGHT 11.1 MICROBIOME: Hand Hygiene

How many times have you heard that one of the best ways to prevent infections in hospitals is for health care workers to wash their hands? OK, but now that we know about how important our resident microbiota is, we should ask what frequent hand washing does to that. A recent study followed 34 health care workers in a surgical intensive care unit. Twenty-four of them were registered nurses, six were respiratory therapists, and four were nurse technologists.

During a typical 12-hour shift,

- 53% of them reported washing their hands with soap and water between 6 and 20 times,
- 41% used alcohol rubs more than 20 times, and
- 62% donned more than 40 pairs of gloves.

The researchers then tested for common healthcare-associated pathogens and found that approximately 45% of the health care workers' dominant hands were positive for *Staphylococcus aureus*. They detected methicillin-resistant *Staphylococcus aureus* (MRSA) on 3.9% of the dominant hands. About 4% of hands were positive for the fungus *Candida albicans*.

These rates look very similar to the rates found in the general population—those of us who are not washing our hands 20 times a day. This might be an illustration of the difference between *colonization*—when microbes become part of your "normal" microbiota—and *contamination*—when you pick up microbes from your environment and they stay only transiently. Contaminants are usually easily washed off, while colonizers are embedded deeper in your skin oils and the top layers of epithelial cells.

One question raised, but not answered, by this study is whether altering the normal microbiome of the hand by frequent hand hygiene practices might make it (a) easier or (b) more difficult for pathogens to contaminate and/or colonize caregivers' hands.

O. Dimier/PhotoAlto

Source: Rosenthal, et al. "Healthcare Workers' Hand Microbiome May Mediate Carriage of Hospital Pathogens," *Pathogens*. 2014 Mar: 3(1): 1–13.

peroxide generators can be used to fill a room with hydrogen peroxide vapors at concentrations high enough to be sporicidal.

Another compound with effects similar to those of hydrogen peroxide is ozone (O_3), used to disinfect air, water, and industrial air conditioners and cooling towers.

Chemicals with Surface Action: Detergents

Detergents are polar molecules that act as **surfactants.** Most anionic detergents have limited microbicidal power. This includes most soaps. Positively charged (cationic) detergents are much more effective, particularly the quaternary ammonium compounds (usually shortened to **quats**—their name comes from the fact that they have four R groups attached to a nitrogen atom).

The activity of cationic detergents arises from the amphipathic (two-headed) nature of the molecule. The positively charged end binds well with the predominantly negatively charged bacterial surface proteins, while the long, uncharged hydrocarbon chain allows the detergent to disrupt the cytoplasmic membrane **(figure 11.12).** Eventually, the cytoplasmic membrane loses its selectively permeable nature, leading to the death of the cell. Several other effects are seen, but the loss of integrity of the membrane is most important.

The effects of detergents are varied. When used at high enough concentrations, quaternary ammonium compounds are effective against some gram-positive bacteria, viruses, fungi, and algae. In low concentrations, they exhibit only microbistatic effects. Drawbacks to the quats include their ineffectiveness against the tuberculosis bacterium, hepatitis virus, *Pseudomonas,* and endospores at any concentration. Furthermore, their activity is greatly reduced in the presence of organic matter, and they function best in alkaline solutions. As a result of these limitations, quats are rated only for low-level disinfection in the clinical setting.

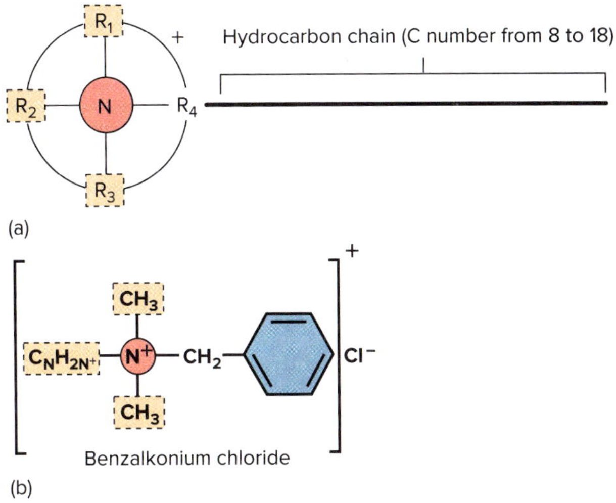

R1
R2
R3
R4
N
+
Hydrocarbon chain (C number from 8 to 18)
(a)
CH3
CH3
CNH2N+1
N+
CH2
Cl−
+
Benzalkonium chloride
(b)

Figure 11.12 The structure of detergents. **(a)** In general, detergents are polar molecules with a positively charged head and at least one long, uncharged hydrocarbon chain. The head contains a central nitrogen nucleus with various alkyl (R) groups attached. **(b)** A common quaternary ammonium detergent, benzalkonium chloride.

It was discovered in 2018 that even the high temperatures used in commercial washing machines in hospitals combined with industrial detergent did not eliminate endospores of the GI pathogen *Clostridioides difficile*. Constant vigilance and consultation with the infectious control officer in a hospital is the best way forward.

Applications of Detergents and Soaps Quats include benzalkonium chloride, Zephiran, and cetylpyridinium chloride (Ceepryn). In dilutions ranging from 1:100 to 1:1,000, quats are mixed with cleaning agents to simultaneously disinfect and sanitize floors, furniture, equipment surfaces, and restrooms. They are used to clean restaurant eating utensils, food-processing equipment, dairy equipment, and clothing. They are common preservatives for ophthalmic solutions and cosmetics. Their level of disinfection is far too low for disinfecting medical instruments.

Some quats were banned from consumer products by the FDA in 2017 because they had been found to foster resistance to antibiotics. Some of them had also been found to affect hormonal pathways in animals and humans.

Soaps are alkaline compounds made by combining the fatty acids in oils with sodium or potassium salts. In usual practice, soaps are only weak microbicides, and they destroy only highly sensitive forms such as the agents of gonorrhea, meningitis, and syphilis. The common hospital pathogen *Pseudomonas* is so resistant to soap that various species grow abundantly in soap dishes.

Soaps function primarily as cleansing agents and sanitizers in industry and the home. The superior sudsing and wetting properties of soaps help to mechanically remove large amounts of surface soil, greases, and other debris that contain microorganisms. Soaps gain greater germicidal value when mixed with agents such as chlorhexidine or iodine. They can be used for cleaning instruments before heat sterilization, degerming patients' skin, routine hand washing by medical and dental personnel, and preoperative hand scrubbing. Vigorously brushing the hands with germicidal soap over a 15-second period is an effective way to remove dirt, oil, and surface contaminants as well as some resident microbes, but it will never sterilize the skin **(figure 11.13).** However, it should be emphasized that hand washing with plain soap (not antimicrobially enhanced) and water is sufficient for the vast majority of household situations.

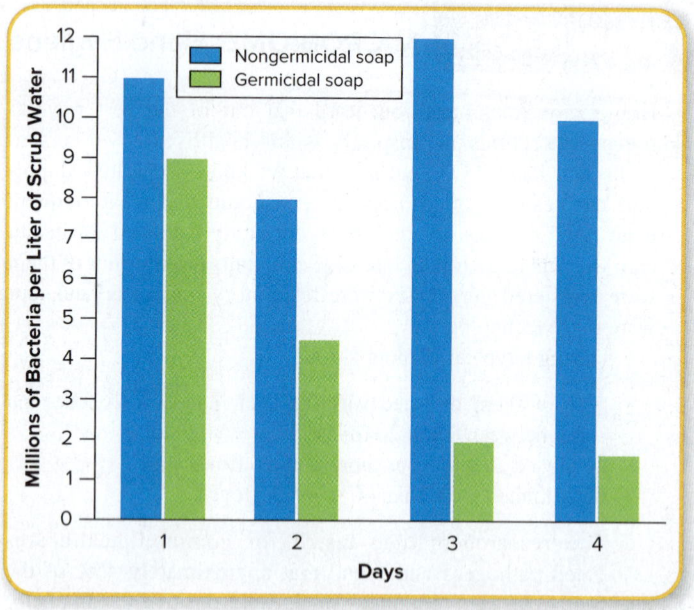
Nongermicidal soap
Germicidal soap
Millions of Bacteria per Liter of Scrub Water
0
1
2
3
4
5
6
7
8
9
10
11
12
Days
1
2
3
4

Figure 11.13 Graph showing effects of hand scrubbing. Comparison of scrubbing over several days with a nongermicidal soap versus a germicidal soap. The vertical axis shows number of live bacteria in the collected scrub water. Germicidal soap has persistent effects on skin over time, keeping the microbial count low. Without germicide, soap does not show this sustained effect.

Source: Nolte, et al. Oral Microbiology, 4/e, (c) 1982, Mosby.

Heavy Metal Compounds

Various forms of the metallic elements mercury, silver, gold, copper, arsenic, and zinc have been used in microbial control over several centuries. These are often referred to as *heavy metals* because of their relatively high atomic weight. However, from this list, only preparations containing mercury and silver still have any significance as germicides. Although some metals (zinc, iron) are actually needed by cells in small concentrations as cofactors on enzymes, the higher molecular weight metals (mercury, silver, gold) can be very toxic, even in minute quantities (parts per million). This property of having antimicrobial effects in exceedingly small amounts is called an **oligodynamic** (ol′-ih-goh-dy-nam′-ik) **action (figure 11.14).** It means "f" with a "force." Heavy metal germicides contain either one of several inorganic or an organic metallic salt, and they come in the form of aqueous solutions, tinctures, ointments, or soaps.

Mercury, silver, and most other metals exert microbicidal effects by binding onto functional groups of proteins and inactivating them, rapidly bringing metabolism to a standstill (see figure 11.5*d*). This mode of action can destroy many types of microbes, including vegetative bacteria, fungal cells and spores, algae, protozoa, and viruses (but not endospores).

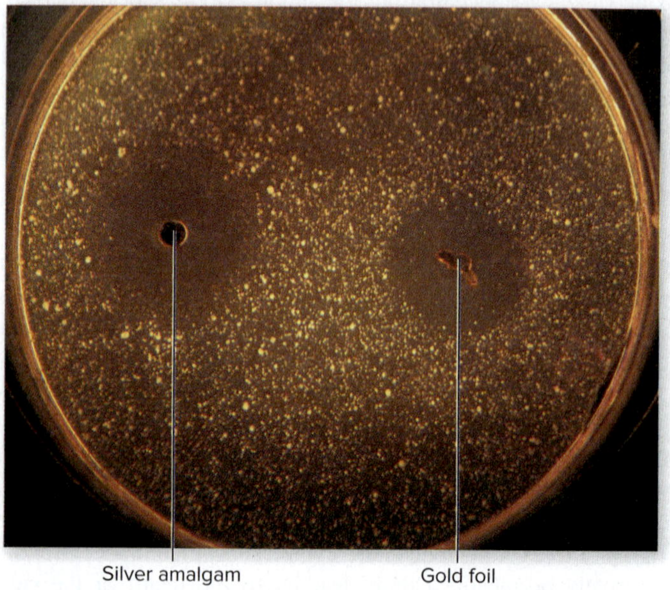
Silver amalgam
Gold foil

Figure 11.14 Demonstration of the oligodynamic action of heavy metals. A pour plate inoculated with saliva has small fragments of heavy metals pressed lightly into it. During incubation, clear zones indicating growth inhibition develop around both fragments. The slightly larger zone surrounding the amalgam on the left (sometimes used in tooth fillings) probably reflects the synergistic effect of the silver and mercury it contains.

Kathy Park Talaro

Applications of Heavy Metals A silver compound with several applications is silver nitrate ($AgNO_3$) solution. In the late 19th century, it was introduced for preventing gonococcal infections in the eyes of newborn infants who had been exposed to an infected birth canal. This preparation is not used as often now because many pathogens are resistant to it. It has been replaced by antibiotics in most instances. Solutions of silver nitrate (1% to 2%) can also be used as topical germicides on mouth ulcers and occasionally root canals. Silver sulfadiazine ointment, when added to dressings, effectively prevents infection in second- and third-degree burn patients, and pure silver is now incorporated into catheters to prevent urinary tract infections in the hospital. Colloidal silver preparations are mild germicidal ointments or rinses for the mouth, nose, eyes, and vagina. Silver ions are increasingly incorporated into many hard surfaces, such as plastics and steel, as a way to control microbial growth on items such as toilet seats, stethoscopes, and even refrigerator doors. Companies have even found ways to impregnate textiles with silver and quaternary ammonium compounds to produce antimicrobial fabrics that stay stain- and odor-free over long periods of use.

Aldehydes as Germicides

Organic substances bearing a —CHO functional group (a strong reducing group) on the terminal carbon are called aldehydes. Several common substances such as sugars and some fats are technically aldehydes. The two aldehydes used most often in microbial control are *glutaraldehyde* and *formaldehyde.*

Glutaraldehyde is a yellow liquid with a mild odor. Its mechanism of activity involves cross-linking protein molecules on the cell surface. In this process, amino acids are alkylated, meaning that a hydrogen atom on an amino acid is replaced by the glutaraldehyde molecule itself. It can also irreversibly disrupt the activity of enzymes within the cell. Glutaraldehyde is rapid and broad-spectrum and is one of the few chemicals officially accepted as a sterilant and high-level disinfectant. It destroys endospores in 3 hours and fungi and vegetative bacteria (even *Mycobacterium* and *Pseudomonas*) in a few minutes. Viruses, including the most resistant forms, appear to be inactivated after relatively short exposure times. Glutaraldehyde retains its potency even in the presence of organic matter, is noncorrosive, does not damage plastics, and is less toxic or irritating than formaldehyde. Its principal disadvantage is that it is somewhat unstable, especially with increased pH and temperature.

Formaldehyde is a sharp, irritating gas that readily dissolves in water to form an aqueous solution called **formalin.** Pure formalin is a 37% solution of formaldehyde gas dissolved in water. The chemical is microbicidal through its attachment to nucleic acids and functional groups of amino acids. Formalin is an intermediate- to high-level disinfectant, although it acts more slowly than glutaraldehyde. Formaldehyde's extreme toxicity (it is classified as a carcinogen) and irritating effects on the skin and mucous membranes greatly limit its clinical usefulness.

A third aldehyde, ortho-phthalaldehyde (OPA), is another high-level disinfectant. OPA is a pale blue liquid with a barely detectable odor and can be most directly compared to glutaraldehyde. It has a mechanism of action similar to glutaraldehyde, is stable, is nonirritating to the eyes and nasal passages, and, for most uses, is much faster acting than glutaraldehyde. It is effective against vegetative bacteria, including *Mycobacterium* and *Pseudomonas;* fungi; and viruses. Chief among its disadvantages are an inability to reliably destroy endospores and, on a more practical note, its tendency to stain proteins, including those in human skin.

Gaseous Sterilants and Disinfectants

Processing inanimate substances with chemical vapors, gases, and aerosols provides a versatile alternative to heat or liquid chemicals. Currently, those vapors and aerosols having the broadest applications are ethylene oxide (ETO), propylene oxide, and chlorine dioxide.

Ethylene oxide is a colorless substance that exists as a gas at room temperature. It is very explosive in air, a feature that can be eliminated by combining it with a high percentage of carbon dioxide or fluorocarbon. Like the aldehydes, ETO is a very strong alkylating agent, and it reacts vigorously with functional groups of DNA and proteins. Through these actions, it blocks both DNA replication and enzymatic actions. Ethylene oxide is one of a very few gases generally accepted for chemical sterilization because, when employed according to strict procedures, it is a sporicide. A specially designed ETO sterilizer called a *chemiclave,* a variation on the autoclave, is equipped with a chamber; gas ports; and it controls for temperature, pressure, and humidity. Ethylene oxide is rather penetrating but relatively slow-acting, requiring from 90 minutes to 3 hours. Some items absorb ETO residues and must be aerated with sterile air for several hours after exposure to ensure dissipation of as much residual gas as possible. For all of its effectiveness, ETO has some unfortunate features. Its explosiveness makes it dangerous to handle. Also, it can damage the lungs, eyes, and mucous membranes if contacted directly. It is rated as a carcinogen by the government.

Chlorine dioxide is another gas that has of late been used as a sterilant. Despite the name, chlorine dioxide works in a completely different way from the chlorine compounds discussed earlier in the chapter. It is a strong alkylating agent, which disrupts proteins and is effective against vegetative bacteria, fungi, viruses, and endospores. Although chlorine dioxide is used for the treatment of drinking water, wastewater, food-processing equipment, and medical waste, its most well-known use was in the decontamination of the Senate offices after the anthrax attack of 2001.

Acids and Alkalis

Conditions of very low or high pH can destroy or inhibit microbial cells, but they are of limited use due to their corrosive and hazardous nature. Aqueous solutions of ammonium hydroxide are a common component of detergents, cleansers, and deodorizers. Organic acids are widely used in food preservation because they prevent endospore germination, inhibit bacterial and fungal growth, and are generally regarded as safe to eat. For example, acetic acid (in the form of vinegar) is a pickling agent that inhibits bacterial growth; propionic acid is commonly incorporated into breads and cakes to inhibit molds; lactic acid is added to sauerkraut and olives to prevent growth of anaerobic bacteria; and benzoic and sorbic acids are added to beverages, syrups, and margarine to inhibit yeasts.

Essential Oils

In recent years, essential oils have become more prominent in western medicine. They have, of course, been used for centuries in traditional medicine. Components of essential oils (which all

Table 11.7 Active Ingredients of Various Commercial Antimicrobial Products

Product	Specific Chemical Agent	Antimicrobial Category
Lysol Sanitizing Wipes®	Hypochlorous acid	Acids
Clorox Disinfecting Wipes®	Dimethyl benzyl ammonium chloride	Detergent (quat)
Dial Antibacterial Hand Soap®	Ammonium lauryl sulfate	Surfactant/detergent
Lysol Disinfecting Spray®	Alkyl dimethyl benzyl ammonium saccharinate/ethanol	Detergent (quats)/alcohol
Wet Ones Antibacterial Moist Wipes®	Ethanol	Detergent (quat)
Noxzema Triple Clean®	Alcohol, salicylic acid	Alcohol and acid
Scope Mouthwash®	Ethanol	Alcohol
Purell Instant Hand Sanitizer®	Ethanol, propanol	Alcohols
Colgate Total Toothpaste®	Triclosan (FDA allowed this exception to the ban of triclosan)	Phenolic

come from plants) have been studied extensively, especially in countries where simple antibiotics were unavailable. But now that industrialized countries are experiencing a loss in effectiveness of antibiotics, these plant extracts are being closely considered. Some are indeed effective at stopping microbial growth, especially those that are phenolics and derivatives of phenolics. The FDA is hustling to keep up with approving those that work, and also protecting consumers from those that don't. There will certainly be a future for essential oils that are truly effective.

For a look at the antimicrobial chemicals found in some common household products, see **table 11.7.**

11.3 Learning Outcomes—Assess Your Progress

14. Name the desirable characteristics of chemical control agents.
15. Discuss several different halogen agents and their uses in microbial control.
16. List advantages and disadvantages to the use of phenolic compounds as control agents.
17. Explain the mode of action of alcohols and their limitations as effective antimicrobials.
18. Pinpoint the most appropriate applications of oxidizing agents.
19. Define the term *surfactant,* and explain this antimicrobial's mode of action.
20. Identify examples of some heavy metal control agents and their most common applications.
21. Discuss the advantages and disadvantages of aldehyde agents in microbial control.
22. Identify applications for ethylene oxide sterilization.

Media Under The Microscope Wrap-Up

Kevin Griffin/123RF

The news article "Antimicrobial Compound in Seaweed Can Be Used to Develop Self-Cleaning Surfaces, New Research Finds" describes a press release about a new antimicrobial substance.

The **intended message** of the article is to inform readers of a new possible mechanism for keeping inert surfaces free of bacterial contamination, or at least free of accumulations of bacteria known as biofilms. The chemical was discovered in seaweed and is being commercialized for consumer use.

A **critical reading** of the article raises few concerns. The mechanism seems plausible, biologically, especially because they discuss bacteria communicating with one another as one aspect of biofilm formation. They did not use the phrase "quorum sensing," but we know that is what is happening. I have two issues, though, and that is that no scientists outside of Unilever were contacted for their take on the topic, and Unilever is, of course, a for-profit company. It has a vested interest in speaking favorably of its discovery. Also, the article was based on a press release from the company and not a scientific research article, and no research publication was cited. In essence, it amounted to an advertisement, dressed up to look like an informative article.

I would **interpret it to my friends** by mentioning that biofilms (like the scum on your shower curtain) are formed due to communication among individual bacteria. Unilever claims to have found a product from seaweed that interrupts that process. But I would include the two caveats I detailed above.

My **overall grade** for the article is **C.** It seems quite plausible, but I would need more than a press release from a commercial company to make that judgment.

Source: *NaturalNews.com*, "Antimicrobial Compound in Seaweed Can Be Used to Develop Self-Cleaning Surfaces, New Research Finds," online article posted January 15, 2021.

Study Smarter: Better Together

These activities are designed for you to use on your own with a study group—either a face-to-face group or a virtual one, consisting of 3–5 members. Studying together can be very helpful, but there are effective and ineffective ways to do it. For example, getting together without a clear structure is often not a good use of your time. Use your time efficiently by using one or more of the exercises below.

FACE-TO-FACE GROUPS

Use one or more of the activities below.

Peer Instruction: Assign numbers to your group members to use all semester long. Now look at these five concepts from this chapter. Each group member prepares a 5-minute lesson on the topic corresponding to their number. Don't worry if you have fewer than 5 members; just use however many you have! During your group study time, each member presents their lesson, and the group spends another 5–10 minutes discussing that lesson.

1. -cidal vs. -static agents
2. Microbial death
3. What does "sterile" mean?
4. Which physical methods can achieve sterilization?
5. Which chemical methods can achieve sterilization?

Concept Maps: Each member of the group should use this list of terms from this chapter to generate their own concept map. This can be hand-drawn or created using software (see Appendix C for guidelines). During group study time, compare each other's concept maps and help each other make sure they are correct. Of course, there are many different "correct" maps. Examining each member's map will help you talk through the varied concepts and how they are related.

Concept Terms:

halogens	alcohol	phenolics	silver
endospores	prions	autoclaving	

Table Topics: Each group member should identify a concept or topic from this week's class assignments with which they are having trouble and share it during group study time. The other group members can then help to clarify confusing issues or share how they figured it out. Aim for a maximum of 15 minutes per topic. If the topic remains unclear to the group, bring it up during class or use the instructor's office hours or e-mail to ask for help. Taking the time to struggle with a difficult concept first makes your questions much more specific and more likely to yield helpful answers.

VIRTUAL GROUPS

Not everyone has the time or opportunity to meet with group members outside of class time. You or your instructor can create a virtual group using e-mail or the course software.

Weekly Discussion Board: This forum can be used as a way for groups to discuss topics, via e-mail, or other learning management systems or online platforms before they are covered in class. As each member of the group answers the current week's question, they should send their responses to every other member of their group. It's best to agree on a deadline based on how your class schedule works (Saturday for the next week's topics, for example). Then, after the topic is discussed in class, each member should send a response that all group members will see with a follow-up post on the same topic. If you cover more than one chapter in a week, someone can be designated to choose which chapter Discussion Board question you will use. Or simply decide up front that you will always use the first-chapter-of-the week's question, to keep the schedule simple.

Discussion Question

Triclosan and some other antimicrobial compounds have been banned from consumer products. Explain why not every chemical that can be used to kill microbes should be used.

Chapter Summary

PHYSICAL AND CHEMICAL CONTROL OF MICROBES

11.1 CONTROLLING MICROORGANISMS

- Controlling microbes in the environment and on the surface of the body calls for the use of physical and chemical agents to eliminate them or reduce their numbers.
- The final outcomes of microbial controls are named *sterilization, antisepsis, disinfection,* and *decontamination.*
- Some antimicrobial agents are -cidal (kills), and others are -static (inhibits growth).
- Microbial death is defined as the permanent loss of reproductive capacity in microorganisms.
- The four major cell targets for antimicrobial agents are the cell wall, cell membrane, biosynthesis pathways for DNA or RNA, and protein (enzyme function).

11.2 METHODS OF PHYSICAL CONTROL

- Physical methods of control include heat, cold, radiation, drying, filtration, and osmotic pressure.
- Heat is the most widely used method of physical control. It is used either in combination with water (moist heat) or as dry heat.
- The thermal death time (TDT) is the shortest length of time required to kill all microbes at a specific temperature.
- The thermal death point (TDP) is the lowest temperature at which all microbes are killed in a specified length of time (10 minutes, for example).
- The moist heat method that is capable of sterilizing is steam under pressure (autoclaving).
- The dry heat methods that are capable of sterilizing are incineration and hot air ovens.
- Both ionizing and non-ionizing radiation can sterilize under the correct conditions.
- Filtration depends on capturing microbes that are larger than the pores on the filter.

11.3 METHODS OF CHEMICAL CONTROL

- Chemical agents can be either microbicidal or microbistatic and are classified as high-, medium-, or low-level germicides.
- Objects or surfaces to be chemically disinfected in clinical settings are characterized as critical, semicritical, or noncritical based on how intimately they come in contact with patients.
- Common chemical agents are the halogens, phenols, alcohols, oxidizing agents, surfactants (detergents and soaps), aldehydes, and gaseous chemicals. Of these, the halogens, oxidizing agents, aldehydes, and gaseous chemicals are capable of sterilization under the correct conditions.

SmartGrid: From Knowledge to Critical Thinking

This *21 Question Grid* takes the topics from this chapter and arranges them with respect to the American Society for Microbiology's Undergraduate Curriculum guidelines—all six of the important "Concepts" as well as the important "Competency" of scientific literacy. Three questions are supplied, which cover chapter content referring to the Concept or Competency in increasing levels of Bloom's taxonomy for learning.

ASM Concept/ Competency	A. Bloom's Level 1, 2—Remember and Understand (Choose one)	B. Bloom's Level 3, 4—Apply and Analyze	C. Bloom's Level 5, 6—Evaluate and Create
Evolution	**1.** DNA repair mechanisms can help alleviate the effects of **a.** UV radiation. **b.** alcohol disinfection. **c.** autoclaving. **d.** dry heat.	**2.** Explain why all cells in a population do not die instantaneously when exposed to an antimicrobial agent, even when they are of a single species.	**3.** Triclosan and other antimicrobial compounds have been banned from most consumer products. Explain why we should not be using every chemical available to us to combat microbes.
Cell Structure and Function	**4.** Microbial control methods that kill ______ are able to sterilize. **a.** viruses **b.** the tuberculosis bacterium **c.** endospores **d.** cysts	**5.** Why are enveloped viruses generally more susceptible to disinfection, when they have an extra "layer"?	**6.** Construct an argument about why some antimicrobial agents are toxic for skin and human tissue.
Metabolic Pathways	**7.** Cytoplasmic enzymes are most likely to be disrupted by **a.** UV light. **b.** low temperatures. **c.** high temperatures. **d.** detergents.	**8.** Most antimicrobials that target protein function are nonselective as to the microbes they affect. Explain why that might be so.	**9.** There is a bacterium that has a metabolic pathway that enables it to survive in the presence of acid. What kind of pathway or product would you expect in this case?
Information Flow and Genetics	**10.** Transcription is targeted most directly by **a.** quats. **b.** detergents. **c.** UV radiation. **d.** alcohol.	**11.** Do you think UV radiation is a good way to disinfect living tissue? Why or why not?	**12.** Suggest a possible mechanism for how a microbe that becomes resistant to an antimicrobial chemical can transfer that to another microorganism and also transfer resistance to antibiotics in the process.
Microbial Systems	**13.** Disinfection procedures must take into account **a.** the mixture of microbes being targeted. **b.** whether the microbes in biofilms are dispersed. **c.** the amount of organic matter associated with the microbes. **d.** all of the above.	**14.** Provide a rationale for how a particular microbe could be considered a serious contaminant in one situation and completely harmless in another.	**15.** What additional considerations would you have to address if the device you were to trying to disinfect—such as a water line in a dental office—was known to contain a microbial biofilm?
Impact of Microorganisms	**16.** Sanitization is a process by which **a.** the microbial load on an object is reduced. **b.** objects are made sterile with chemicals. **c.** utensils are scrubbed. **d.** skin is debrided.	**17.** Tissue transplantation carries a risk of transferring microbes in the tissue to the recipient. What are the most appropriate ways of ensuring that does not happen?	**18.** Defend this statement: It is possible to make our environment too clean.

(continued)

ASM Concept/ Competency	A. Bloom's Level 1, 2—Remember and Understand (Choose one)	B. Bloom's Level 3, 4—Apply and Analyze	C. Bloom's Level 5, 6—Evaluate and Create
Scientific Thinking	**19.** The most versatile method for sterilizing heat-sensitive liquids is **a.** UV radiation. **b.** exposure to ozone. **c.** peracetic acid. **d.** filtration.	**20.** Precisely what is microbial death?	**21.** Devise an experiment that will differentiate between bactericidal versus bacteristatic effects.

Answers to the multiple-choice questions appear in Appendix A.

Visual Connections

This question uses a visual image to connect content within and between chapters.

1. **From chapter 4, figure 4.25.** Why would many chemical control agents be ineffective in controlling this organism?

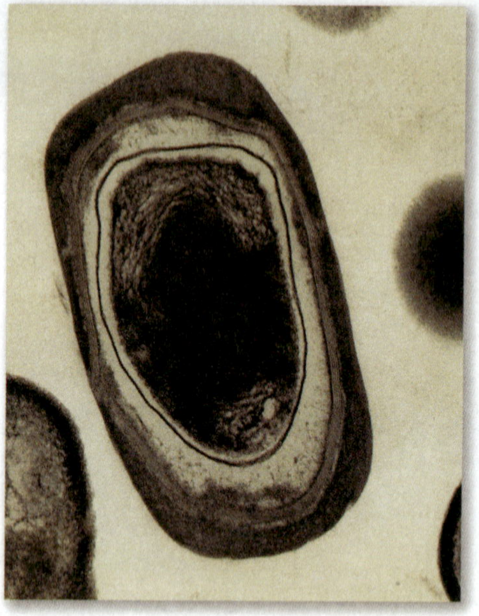

Dr. Tony Brain/Science Source

High Impact Study

These terms and concepts are most critical for your understanding of this chapter—and may be the most difficult. Have you mastered them?

Concepts

- ☐ -cidal vs. -static agents
- ☐ Microbial death
- ☐ Factors influencing microbial death
- ☐ Physical vs. chemical control
- ☐ Desirable qualities in germicidal chemicals
- ☐ Which physical methods can sterilize
- ☐ Which chemical methods can sterilize

Terms

- ☐ Sterilization
- ☐ Disinfection
- ☐ Antisepsis/degermation
- ☐ Decontamination/sanitation
- ☐ Thermal death time
- ☐ Thermal death point

Design Element: (College students): Caia Image/Image Source

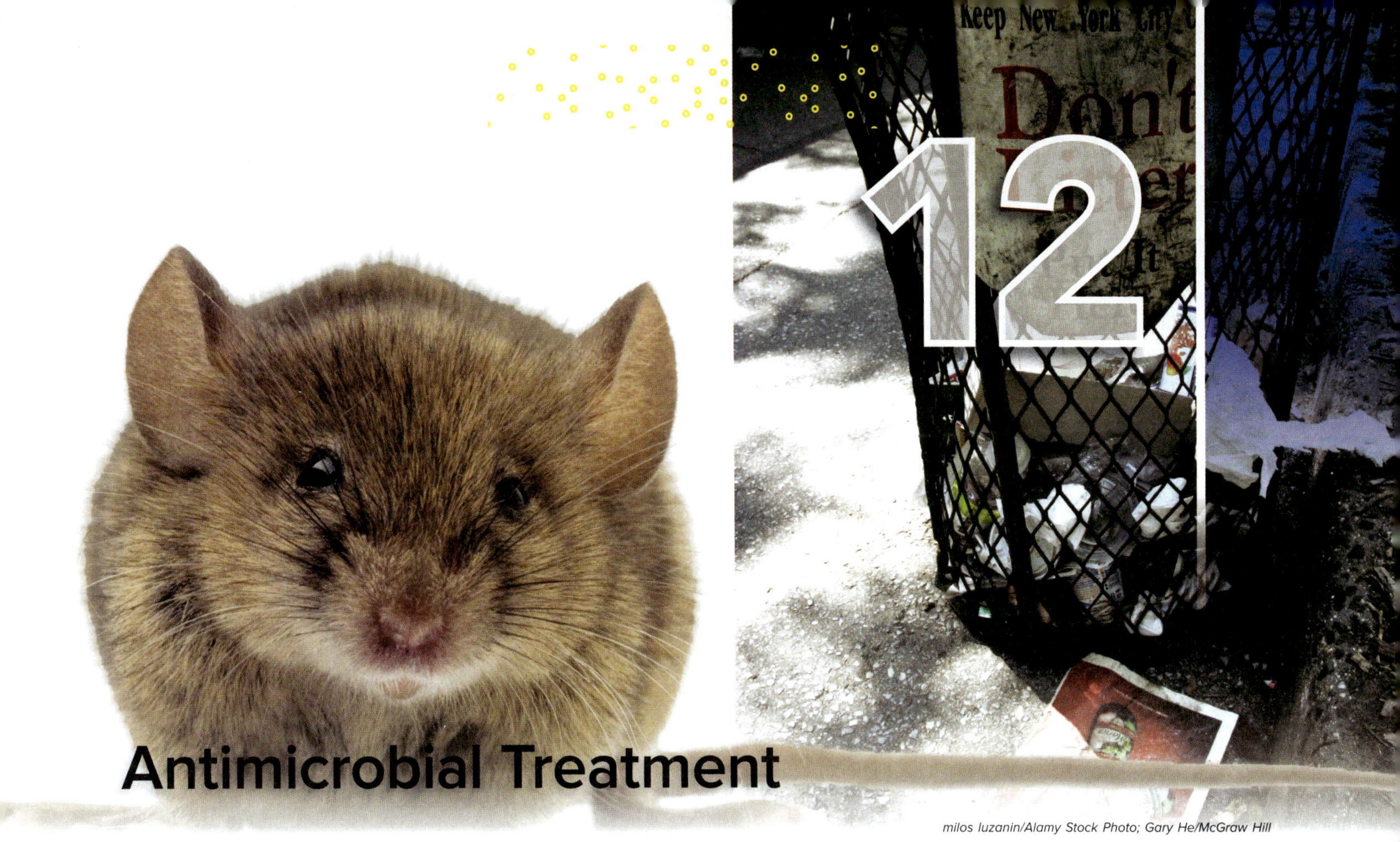

12 Antimicrobial Treatment

milos luzanin/Alamy Stock Photo; Gary He/McGraw Hill

MEDIA UNDER THE MICROSCOPE

Of Mice and Men

This opening case examines an article from the popular media to determine the extent to which it is factual and/or misleading. This case focuses on the 2018 Time.com article "NYC Mice Are Carrying Antibiotic-Resistant Germs."

This article reported the results of a study on the gut microbes of mice that were collected from residential settings in NYC. A team of researchers from Columbia University genetically analyzed the microbes found in these mice, and they discovered that almost 40% of the mice harbored pathogenic bacteria. Bacteria that can cause gastrointestinal disease, such as *Salmonella*, *E. coli*, *C. difficile*, and *Shigella,* were some examples. A deeper look showed that some of the bacteria isolated from these mice matched the exact strains that had caused recent human disease outbreaks. Many of the mice also carried antibiotic-resistant forms of these bacteria.

Dr. Ian Lipkin, the leading researcher on the project, has also conducted other studies on the microbiome of mice, and, based on his findings, he suggested that mice should certainly be considered as potential sources when disease outbreaks occur in people.

- What is the **intended message** of the article?
- What is your **critical reading** of the summary of the article? Remember that in this context, "critical reading" does not necessarily mean *What criticism do you have?* but asks you to apply your knowledge to interpret whether the article is factual and whether the facts support the intended message.
- How would you **interpret** the news item for your nonmicrobiologist friends?
- What is your **overall grade** for the news item—taking into account its accuracy and the accuracy of its intended effect?

Media Under The Microscope Wrap-Up appears at the end of the chapter.

Outline and Learning Outcomes

12.1 Principles of Antimicrobial Therapy

1. State the main goal of antimicrobial treatment.
2. Identify sources of the most commonly used antimicrobial drugs.
3. Summarize two methods for testing antimicrobial susceptibility.
4. Define *therapeutic index,* and identify whether a high or a low index is preferable in a drug.
5. Explain the concept of selective toxicity.
6. Describe the five major targets of current antimicrobial agents, and list major drugs associated with each.
7. Identify which categories of drugs are most selectively toxic, and explain why.
8. Distinguish between drug toxicity and allergic reactions to drugs.
9. Define the term *superinfection,* and summarize how it develops in a patient.

12.2 Survey of Major Antimicrobial Drug Groups

10. Distinguish between broad-spectrum and narrow-spectrum antimicrobials, and explain the significance of the distinction.
11. Trace the development of β-lactam antimicrobials, and identify which microbes they are effective against.
12. Describe the action of β-lactamases, and explain their importance in drug resistance.
13. List examples of other β-lactam antibiotics.
14. Describe common cell wall antibiotics that are not in the β-lactam class of drugs.
15. Identify the targets of several antibiotics that inhibit protein synthesis.
16. Identify the cellular target of quinolones, and provide two examples of these drugs.
17. Name two drugs that target the cellular membrane.
18. Describe the unique problems in treating biofilm infections.
19. Name the four main categories of antifungal agents, and provide one example of each.
20. List two antiprotozoal drugs and three antihelminthic drugs used today.
21. Describe three major modes of action of antiviral drugs.

12.3 Antimicrobial Resistance

22. Discuss two possible ways that microbes acquire antimicrobial resistance.
23. List five cellular or structural mechanisms that microbes use to resist antimicrobials.
24. Discuss at least three novel antimicrobial strategies that are under investigation.

12.1 Principles of Antimicrobial Therapy

A hundred years ago in the United States, one out of three children was expected to die of an infectious disease before the age of 5. Early death or severe, lifelong debilitation from scarlet fever, diphtheria, tuberculosis, meningitis, and many other bacterial diseases was a fearsome yet undeniable fact of life to most of the world's population. The introduction of modern drugs to control infections in the early to mid-1900s was a medical revolution that has added significantly to the life span and health of humans. It is no wonder that for many years, antibiotics in particular were regarded as miracle drugs. Although antimicrobial drugs have greatly reduced the incidence of certain infections, they have definitely not eradicated infectious disease and probably never will. In fact, you will read later that we are dangerously close to a post-antibiotic era, wherein the drugs we have are no longer effective.

The goal of antimicrobial treatment is simple: Administer a drug to an infected person that destroys the infective agent without harming the host's cells. In reality, this goal is difficult to achieve because many (often contradictory) factors must be taken into account. The ideal drug should be easy to administer yet be able to reach the infectious agent anywhere in the body. It should be toxic to the infectious agent (or cripple its ability to multiply) while being nontoxic to the host and must remain active in the body as long as needed yet be safely and easily broken down and excreted. Also, microbes in biofilms often require different drugs than when they are not in biofilms. In short, the perfect drug does not exist; but by balancing drug characteristics against one another, we can often achieve a satisfactory compromise **(table 12.1).**

Table 12.1 Characteristics of the Ideal Antimicrobial Drug

- Toxic to the microbe but nontoxic to the host
- Microbicidal rather than microbistatic
- Soluble in body fluids
- Remains potent long enough to act and is not broken down or excreted prematurely
- Slow or nonexistent development of antimicrobial resistance
- Complements or assists the activities of the host's defenses
- Remains active in tissues and body fluids
- Readily delivered to the site of infection
- Does not disrupt the host's health by causing allergies or predisposing the host to other infections

The Origins of Antimicrobial Drugs

Antimicrobial agents are described with regard to their origin, range of effectiveness, and whether they are naturally produced or chemically synthesized. A few of the more important terms you will encounter are found in **table 12.2.**

The first effective antibiotic was chemically synthesized in the laboratory. Its trade name was Salvarsan, and it came from chemist Paul Ehrlich's lab, who won a Nobel Prize for it. It was first marketed in 1910 for the treatment of syphilis. After that, a particular phylum of bacteria, *Actinobacteria,* was found to be a rich source of chemicals that harmed other bacteria. Members of this phylum live in soil and aquatic habitats, and produce these chemicals to inhibit the growth of other bacteria and reduce competition for resources. One genus in the phylum, *Streptomyces,* lives in the soil and is often associated with plants. Various species of *Streptomyces* produce metabolic products that are highly useful in medicine **(Figure 12.1).** These bacteria were first thought to be fungi because they produce hyphae and spores, as you can see in the righthand photo. But they are gram-positive bacteria.

After an antibiotic was isolated from *Streptomyces* in the 1940s, there was an explosion of antibiotic discoveries, largely sourced from the *Actinobacteria*. By the mid-1960s, 12 different classes of antibiotics, based on their chemical structures, had been discovered. Most of these classes contain multiple antibiotics. But the discoveries slowed and eventually stopped. An additional class, the lipopeptides, was discovered in the mid 1980s, and no new class has been discovered since. We will discuss some of the reasons for that, and our dire need for new antibiotics, as we move through this chapter.

Penicillin is often thought of as the first antibiotic, even though it became available long after Salvarsan. Salvarsan was only active against one bacterium—the one that caused syphilis, so while it was extremely useful for that disease, it was not world-changing. Penicillin comes from a fungus, *Penicillium*. Penicillin was first commercially released in 1944, during World War II, and for the first time a drug was available that stopped infections of many kinds in their tracks. It quickly became known as a miracle drug. If you take a walk in any historic cemetery, you will see gravestones of many, many children. Before the advent of penicillin, common infections were often deadly, and children were frequent victims.

My (Kelly's) youngest son was diagnosed with *Streptococcus pneumoniae* pneumonia when he was around 3 years old. He was quite ill and lethargic. He received one injection of penicillin and was back outside playing the following day. He went on to have the same infection two more times. Later, his physician and I remarked on the fact that before penicillin, he would have had only a 50% chance of surviving each time. All of us probably have similar stories. But with the passage of time, it is easy to forget how vulnerable we were, just 80 years ago.

Chemists are able to create new drugs by altering the structure of naturally occurring antibiotics. One famous example of this is the way that original, natural penicillin is made more effective against gram-negative bacteria and more resistant to host enzymes that inactivate it by adding different side chains to the molecule (more later). These are called **semisynthetic** drugs. And entirely

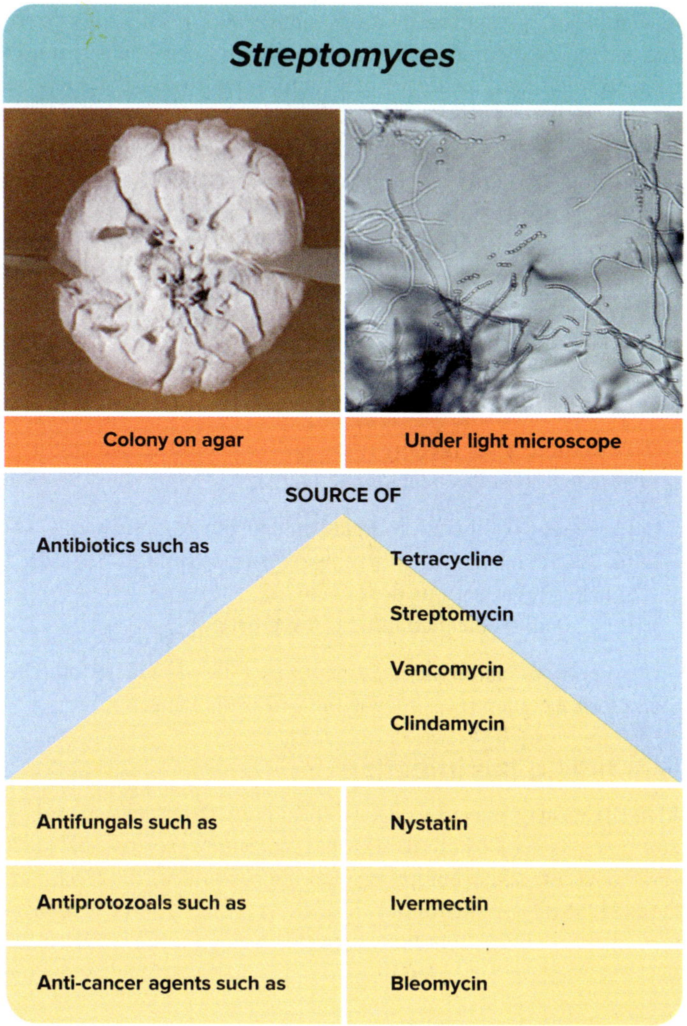
Streptomyces
Colony on agar
Under light microscope
SOURCE OF
Antibiotics such as
Tetracycline
Streptomycin
Vancomycin
Clindamycin
Antifungals such as
Nystatin
Antiprotozoals such as
Ivermectin
Anti-cancer agents such as
Bleomycin

Figure 12.1 Products of the bacterium *Streptomyces* used in medicine.

(Top left): CDC/ Dr. Ajello; (Top right) Dr. David Berd/CDC

Table 12.2 Terminology of Antimicrobials

Prophylaxis	Use of a drug to prevent imminent infection of a person at risk
Antimicrobial chemotherapy	The use of drugs to control infection
Antimicrobials	All-inclusive term for any antimicrobial drug, regardless of what type of microorganism it targets
Antibiotics	Substances produced by the natural metabolic processes of some microorganisms—or created by scientists—that can inhibit or destroy microorganisms; generally, the term is used for drugs targeting bacteria and not other types of microbes
Semisynthetic drugs	Drugs that are chemically modified in the laboratory after being isolated from natural sources
Synthetic drugs	Drugs produced entirely by chemical reactions within a laboratory setting
Narrow-spectrum (limited spectrum)	Antimicrobials effective against a limited array of microbial types—for example, a drug effective mainly on gram-positive bacteria
Broad-spectrum (extended spectrum)	Antimicrobials effective against a wide variety of microbial types—for example, a drug effective against both gram-positive and gram-negative bacteria

new molecules—**synthetic** drugs—have also been synthesized, and a few have entered clinical practice. It seems like it might be relatively easy to create new antibiotics, using the tools of whole genome sequencing, big data, and computational biology. In fact, years of work have identified a large number of potential targets in bacteria for which a "designer" drug could be manufactured. The efforts have largely not been successful because of a fundamental—and literal—barrier: bacterial membranes. Manufactured compounds have largely proved unsuccessful in permeating the single membrane of gram-positive bacteria and especially the double membrane of gram-negatives.

Starting Treatment

Before antimicrobial therapy using any type of drug can begin, it is best that at least three factors be known:

1. the identity of the microorganism causing the infection,
2. the degree of the microorganism's susceptibility (also called sensitivity) to various drugs, and
3. the overall medical condition of the patient.

Sometimes antibiotic treatment must be started before the identity of the microbe is known, but that is not ideal.

Identifying the Infection

Identification of infectious agents from body specimens should be attempted as soon as possible. It is especially important that such specimens be taken before any antimicrobial drug is given—in case the drug eliminates the infectious agent. Direct examination of body fluids, sputum, or stool is a rapid initial method for detecting and perhaps even identifying bacteria or fungi. A doctor often begins the therapy on the basis of such immediate findings or even on the basis of an informed best guess. For instance, a Gram stain of pus produced from a gonorrhea infection reveals the characteristic appearance: bean-shaped gram-negative diplococci. This appearance is unique enough (when combined with the symptoms) that the physician can assume it is *Neisseria gonorrhoeae* and begin treatment accordingly. In most cases, though, further testing is needed to identify the offending microbe. These methods may involve culturing the microbe or may instead rely on a variety of nonculture methods. Sometimes identification efforts are inconclusive. In these cases, epidemiological understanding may be required to predict the most likely agent in a given infection. For example, *Streptococcus pneumoniae* accounts for the majority of cases of meningitis in children, followed by *Neisseria meningitidis*.

Disease Connection

Salvarsan was a derivative of the poisonous chemical arsenic. It was toxic to patients, and the injections were painful. But the disease it treated, syphilis, was potentially fatal. Because Salvarsan did result in a significant cure rate, it was the first true anti-infective drug, and was hailed as a medical breakthrough and called the magic bullet. Today syphilis is often treated with penicillin G, which, when discovered, became a true magic bullet: harming the pathogen without harming the host.

Testing to See Which Drug Will Work

Determining which antimicrobial agent or agents are most effective against the microbes that are actually in the patient is essential in those groups of bacteria commonly showing resistance, such as *Staphylococcus* species, *Neisseria gonorrhoeae, Streptococcus pneumoniae, Enterococcus faecalis,* and the aerobic gram-negative enteric bacilli. However, not all infectious agents require antimicrobial sensitivity testing. Some bacteria are reliably susceptible to certain antibiotics, such as group A streptococci and penicillin. In these cases, susceptibility testing is not necessary.

Selecting the correct antimicrobial agent begins by testing the *in vitro* activity of several drugs against the infectious agent by means of standardized methods. In general, these tests involve exposing a pure culture of the bacterium to several different drugs and determining whether they inhibit the growth of the microbe.

The *Kirby-Bauer* technique is an agar diffusion test that provides useful data on antimicrobial susceptibility. In this test, the surface of a plate of special medium is spread with the test bacterium, and small discs containing a premeasured amount of antimicrobial are dispensed onto the bacterial lawn. After incubation, the zone of inhibition surrounding the discs is measured and compared with a standard for each drug **(table 12.3** and **figure 12.2).** Here are some things to note about the test:

- Each disc contains a different antibiotic.
- Here you see that the vancomycin disc ("VA") is not inhibitory at all; so the bacteria are "S" (susceptible).
- Each antibiotic has a different zone of inhibition size that determines whether a bacterium is resistant, susceptible, or of intermediate susceptibility. That means that a bacterium that exhibits a 30 mm zone of inhibition to Drug A may still be considered resistant to it, and when it exhibits a 22 mm zone of inhibition to Drug B it could be considered susceptible to it.

Table 12.3 Results of a Sample Kirby-Bauer Test

	Zone Sizes (mm) Required for:			
Drug	Susceptibility (S)	Resistance (R)	Example Results (mm) for *Staphylococcus aureus*	Evaluation*
Erythromycin	>18	<13	15	I
Penicillin G	>29	<20	10	R
Streptomycin	>15	<11	11	R
Vancomycin	>12	<9	15	S

*R = resistant, I = intermediate, S = sensitive

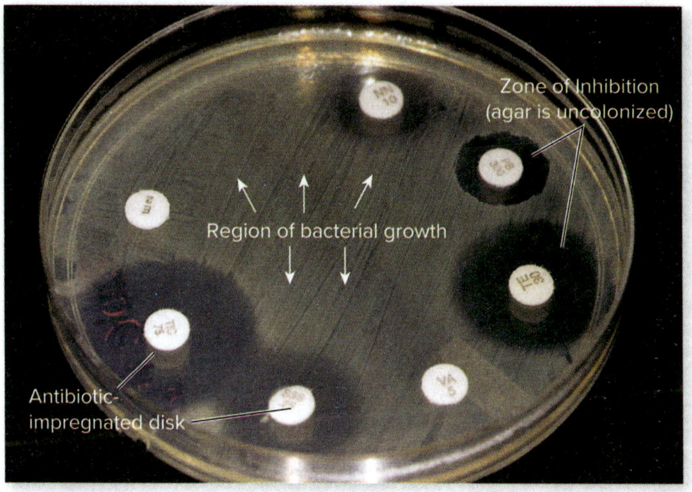
Zone of inhibition
(agar is uncolonized)
Region of bacterial growth
Antibiotic-
impregnated disk

Figure 12.2 Technique for preparation and interpretation of disc diffusion tests. Photograph of an agar dish with a bacterial isolate distributed evenly all over its surface. After inoculation, the antibiotic-containing disks are dropped onto the agar and it is incubated. The damp agar moistens the discs and antibiotic diffuses out of them in widening arcs. The concentration of the antibiotic decreases as it gets farther from the disc. If the test bacterium is sensitive to a drug, a zone of inhibition develops around its disc.
Don Rubbelke/McGraw Hill

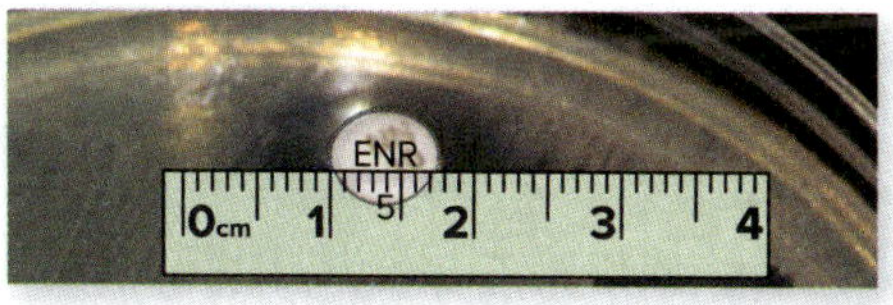
ENR
0cm
1
2
3
4

- The zone is measured from edge to edge of no growth, straight across the antibiotic disc.

This profile of antimicrobial sensitivity is called an *antibiogram*. An alternative diffusion system that provides additional information on drug effectiveness is the Etest® **(figure 12.3).**

More sensitive and quantitative results can be obtained with tube dilution tests. First, the antimicrobial is diluted serially in tubes of broth; then each tube is inoculated with a small, uniform sample of pure culture, incubated, and examined for growth (turbidity). The smallest concentration—which is the highest dilution—of drug that visibly inhibits growth is called the **minimum inhibitory concentration (MIC).** The MIC is useful in determining the smallest effective dosage of a drug and in providing a comparative index against other antimicrobials **(figure 12.4).** In many clinical laboratories, these antimicrobial testing procedures are performed in automated machines that can test dozens of drugs simultaneously.

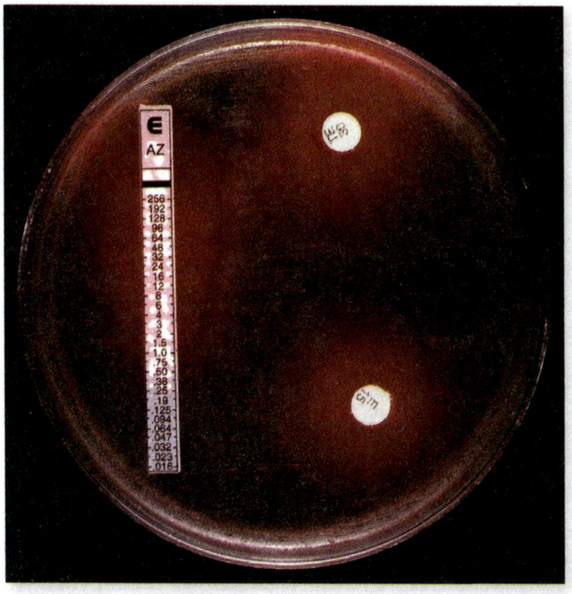
AZ

Figure 12.3 Alternative to the Kirby-Bauer procedure. Another diffusion test is the Etest®, which uses a strip to produce the zone of inhibition. The advantage of the Etest® is that the strip contains a gradient of drug (azithromycin in this photo example) calibrated in micrograms. This way, the MIC can be measured by observing the mark on the strip that corresponds to the edge of the zone of inhibition.
Dr. Richard Facklam/CDC

The MIC and Therapeutic Index

The results of antimicrobial sensitivity tests guide the physician's choice of a suitable drug. If therapy has already begun, it is used to determine if the tests bear out the use of that drug. Once therapy has begun, it is important to observe the patient's clinical response because the *in vitro* activity of the drug is not always correlated with its *in vivo* effect. In a controlled situation—such as in a hospital—when antimicrobial treatment fails, the failure is due to

1. the inability of the drug to diffuse into that body compartment (the brain, joints, skin);
2. resistant microbes in the infection that did not make it into the sample collected for testing; or
3. an infection caused by more than one pathogen (mixed), some of which are resistant to the drug.

In outpatient situations, you have to also consider the possibility that the patient did not take the antimicrobials correctly.

Disease Connection

Antibiotic failure can be caused by the inability of an antibiotic to diffuse into the target tissue. This may be due to poor circulation. For example, older patients, especially older patients with diabetes, sometimes show little to no signs of wound healing after completing a prescribed course of oral antibiotics to treat leg or foot ulcers. Why? When peripheral vascular disease (a condition that affects the blood vessels beyond the heart and brain) is present, the antibiotic may not be delivered to the lower extremities due to their poor circulation. These patients often require potent intravenous antibiotics or topical antibiotics to treat infection, rather than oral antibiotics.

Other factors influence the choice of an antimicrobial drug besides microbial sensitivity to it. The nature and spectrum of the drug, its potential adverse effects, and the condition of the patient can be critical. When more than one antimicrobial drug would be

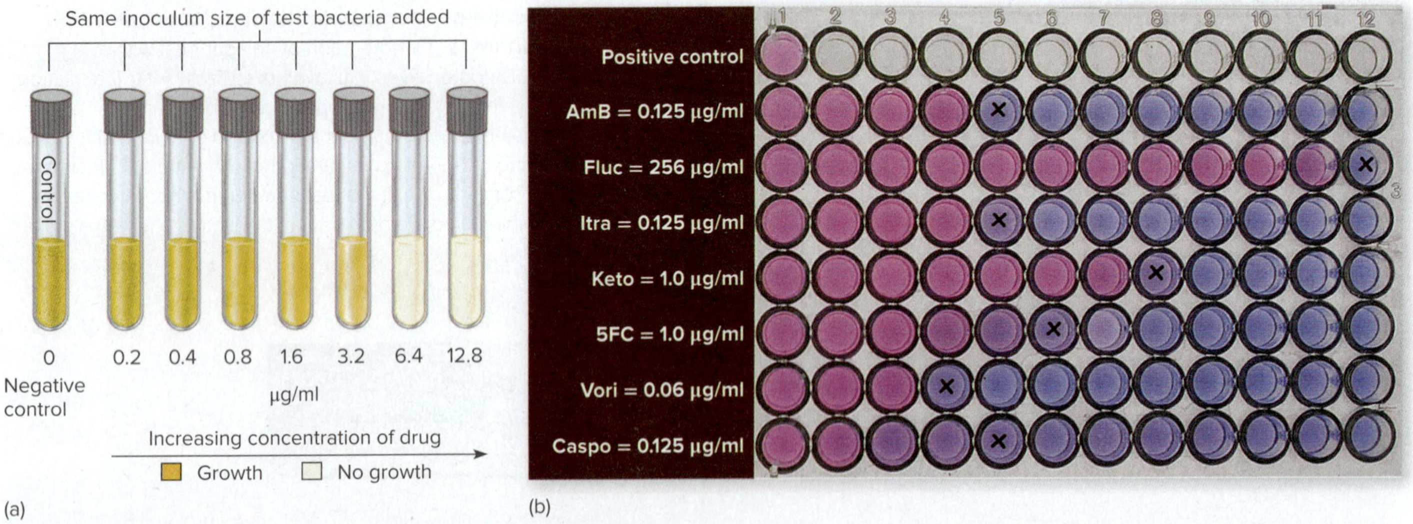

Same inoculum size of test bacteria added
Control
0
Negative control
0.2
0.4
0.8
1.6
3.2
6.4
12.8
µg/ml
Increasing concentration of drug
Growth
No growth
(a)
Positive control
AmB = 0.125 µg/ml
Fluc = 256 µg/ml
Itra = 0.125 µg/ml
Keto = 1.0 µg/ml
5FC = 1.0 µg/ml
Vori = 0.06 µg/ml
Caspo = 0.125 µg/ml
1 2 3 4 5 6 7 8 9 10 11 12
(b)

Figure 12.4 Tube dilution test for determining the minimum inhibitory concentration (MIC). **(a)** The antibiotic is diluted serially through tubes of liquid nutrient from right to left. Then all tubes are inoculated with an identical amount of a test bacterium and then incubated. The first tube on the left is a control that lacks the drug and shows maximum growth. The first tube in the series that shows no growth (no turbidity) contains the concentration of antibiotic that is the MIC. **(b)** Microbroth dilution in a multiwell plate. Here, amphotericin B, flucytosine, and several azole drugs are tested on a pathogenic yeast. Pink indicates growth, and blue indicates no growth. Numbers indicate the dilution of the MIC, and the X in each row shows the first well without growth.

(b) Dr. David Ellis

effective for treating an infection, other factors are considered. It is always best to choose the one that has the fewest effects on microbes other than the one being targeted. This decreases the potential for a variety of adverse reactions.

Because drug toxicity is of concern, it is best to choose the one with high selective toxicity for the infectious agent and low human toxicity. The **therapeutic index (TI)** is defined as the ratio of the dose of the drug that is toxic to humans to its effective dose. The closer these two figures are (the smaller the ratio), the greater the potential is for toxic drug reactions. For example, a drug that has a therapeutic index of

$$\frac{10\ \mu\text{g/ml: toxic dose}}{9\ \mu\text{g/ml (MIC)}} \quad \text{TI} = 1.1$$

is a riskier choice than one with a therapeutic index of

$$\frac{10\ \mu\text{g/ml}}{1\ \mu\text{g/ml}} \quad \text{TI} = 10$$

When a series of drugs being considered for therapy have similar MICs, the drug with the highest therapeutic index usually has the widest margin of safety. Another term, which refers to concentrations of the drug in the blood, is the **therapeutic window.** This is the range of blood level of the drug in which it is producing the desired effect without toxicity.

The physician must also take a careful history of the patient to discover any preexisting medical conditions that will influence the activity of the drug or the response of the patient. A history of allergy to a certain class of drugs precludes the use of that drug and any drugs related to it. Underlying liver or kidney disease will influence the choice of drug therapy because these organs play such an important part in metabolizing or excreting the drug. Infants, the elderly, and pregnant women require special precautions. For example, age can diminish gastrointestinal absorption and organ function, and most antimicrobial drugs cross the placenta and could affect fetal development.

Patients must be asked about other drugs they are taking because incompatibilities can result in increased toxicity or failure of one or more of the drugs. For example, the combination of aminoglycosides and cephalosporins increases toxic effects on the kidney; antacids reduce the absorption of isoniazid; and the interaction of rifampin with oral contraceptives can abolish the contraceptive's effect. Some drug combinations (penicillin with certain aminoglycosides, or amphotericin B with flucytosine) act synergistically, so reduced doses of each can be used in combined therapy. Often, patients are also taking over-the-counter medicines and supplements, many of which can affect the way antimicrobials work.

Adding to the complexity of choosing an antimicrobial is our new knowledge of the human microbiome, especially in the gut. Oral antimicrobials are chemically modified by the microbes in the gut, in different ways depending on which microbes are there. This accounts for substantial variability in how different humans respond to the drugs, and is an active area of research.

Affecting Microbes, Affecting Humans

The goal of antimicrobial drugs is either to disrupt the cell processes or structures of pathogens or to inhibit their replication. Drugs can interfere with the function of enzymes required to synthesize or assemble macromolecules, or they can destroy structures already formed in the cell. Above all, drugs should be **selectively toxic,** which means they should kill or inhibit microbial cells without simultaneously damaging host tissues. This concept of selective toxicity is central to antibiotic treatment, and the best drugs in current use are those that block the actions or synthesis of molecules in microorganisms but not in vertebrate cells. (We will see later that the drugs of the future may in fact

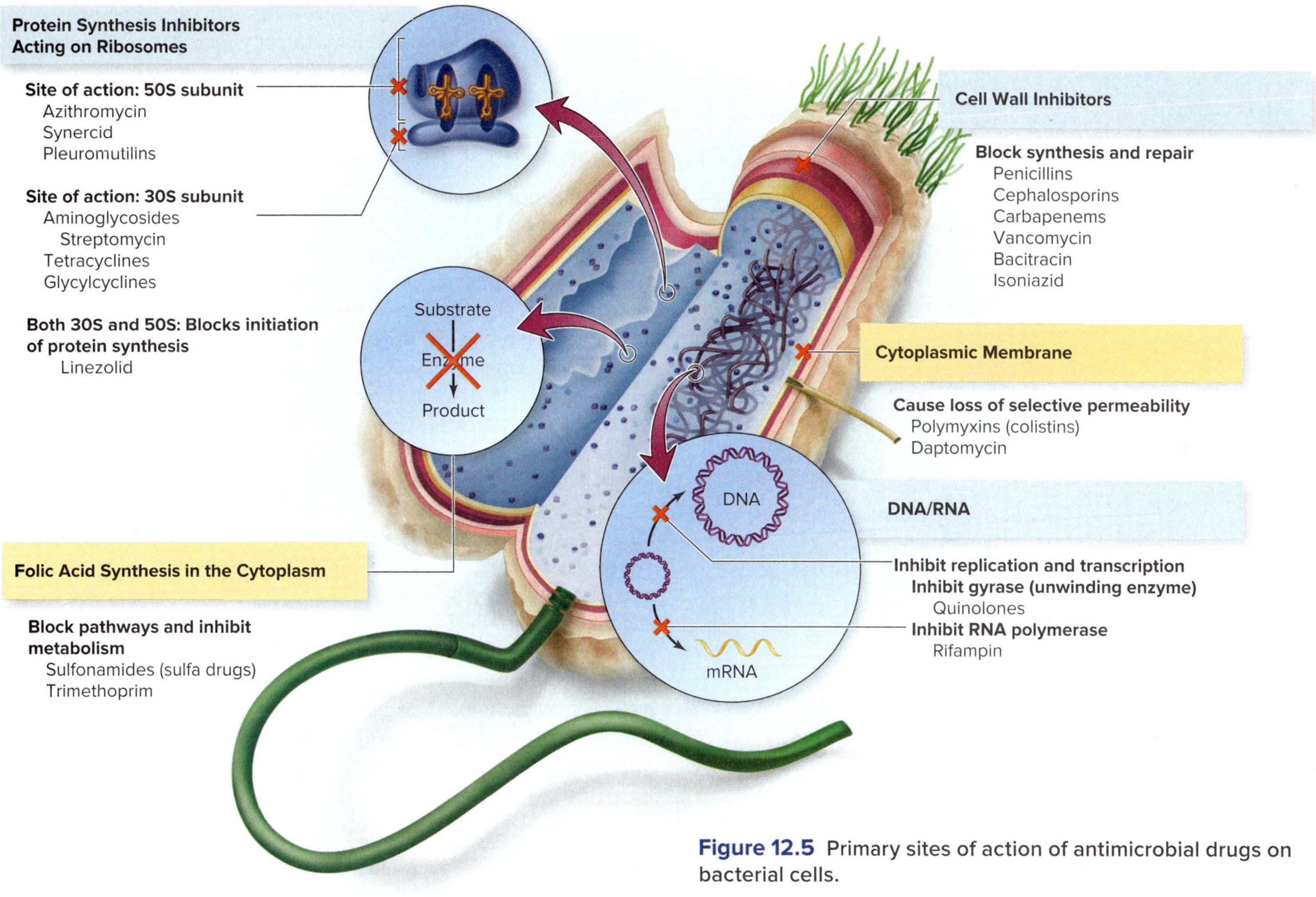
Protein Synthesis Inhibitors Acting on Ribosomes
Site of action: 50S subunit
Azithromycin
Synercid
Pleuromutilins
Site of action: 30S subunit
Aminoglycosides
Streptomycin
Tetracyclines
Glycylcyclines
Both 30S and 50S: Blocks initiation of protein synthesis
Linezolid
Substrate
Enzyme
Product
Folic Acid Synthesis in the Cytoplasm
Block pathways and inhibit metabolism
Sulfonamides (sulfa drugs)
Trimethoprim
Cell Wall Inhibitors
Block synthesis and repair
Penicillins
Cephalosporins
Carbapenems
Vancomycin
Bacitracin
Isoniazid
Cytoplasmic Membrane
Cause loss of selective permeability
Polymyxins (colistins)
Daptomycin
DNA/RNA
DNA
mRNA
Inhibit replication and transcription
Inhibit gyrase (unwinding enzyme)
Quinolones
Inhibit RNA polymerase
Rifampin

Figure 12.5 Primary sites of action of antimicrobial drugs on bacterial cells.

inhibit microbial growth by attacking some aspect of the host cell they utilize.) Examples of drugs with excellent selective toxicity are those that block the synthesis of the cell wall in bacteria (penicillins). They have low toxicity and few direct effects on human cells because human cells lack the chemical peptidoglycan and for that reason are unaffected by this action of the antibiotic. The most toxic to human cells are drugs that act upon a structure that both the infective agent and the host cell have, such as the cell membrane (an example is amphotericin B, used to treat fungal infections). As the characteristics of the infectious agent become more and more similar to those of the host cell, selective toxicity becomes more difficult to achieve and undesirable side effects are more likely to occur. We examine that subject in more detail in a later section.

Mechanisms of Drug Action on Microbes

If the goal of antimicrobial usage is to disrupt the structure or function of an organism to the point where it can no longer survive, then the first step is to identify the structural and metabolic needs of the living microbe. Once those have been determined, methods of removing, disrupting, or interfering with these requirements can be figured out. The metabolism of an actively dividing cell is marked by the production of new cell wall components (in most cells), DNA, RNA, proteins, and cell (or cytoplasmic) membrane. Consequently, current antimicrobial drugs are divided into categories based on which of these metabolic targets they affect. These categories are outlined in **figure 12.5** and include

1. inhibition of cell wall synthesis;
2. inhibition of nucleic acid (RNA and DNA) structure and function;
3. inhibition of protein synthesis, particularly involving ribosomes;
4. interference with cell membrane structure or function; and
5. inhibition of folic acid synthesis.

The Effects of Antimicrobials on Humans

Although selective toxicity is the ideal constantly being sought, antimicrobial therapy, by its very nature, involves contact with foreign chemicals that can harm human tissues. In fact, estimates indicate that at least 5% of all persons taking an antimicrobial drug experience some type of relatively serious adverse reaction to it. The major side effects of drugs fall into one of three categories: direct damage to tissues through toxicity, allergic reactions, and disruption in the balance of normal microbial biota. The damage caused by antimicrobial drugs can be short-term and reversible or permanent, and it ranges in severity from cosmetic to lethal **(table 12.4).**

Table 12.4 Major Adverse Toxic Reactions to Common Antimicrobials

Antimicrobial Drug	Primary Damage or Abnormality Produced
Antibacterials	
Penicillin G	Skin abnormalities
Cephalosporins	Inhibition of platelet function Decreased circulation of white blood cells Nephritis
Sulfonamides	Formation of crystals in kidney; blockage of urine flow Hemolysis Reduction in number of red blood cells
Polymyxins	Kidney damage Weakened muscular responses
Quinolones (ciprofloxacin, norfloxacin)	Headache, dizziness, tremors, GI distress
Antifungals	
Amphotericin B	Disruption of kidney function
Flucytosine	Decreased number of white blood cells
Antiprotozoan Drugs	
Metronidazole	Nausea, vomiting
Antihelminthics	
Pyrantel	Irritation Headache, dizziness
Antivirals	
Amantadine	Nervousness, light-headedness Nausea
AZT	Immunosuppression, anemia

Toxicity to Human Organs

Drugs most often adversely affect the following organs: the liver (hepatotoxic), kidneys (nephrotoxic), gastrointestinal tract, cardiovascular system and blood-forming tissue (hemotoxic), nervous system (neurotoxic), respiratory tract, skin, bones, and teeth. Some potential toxic effects of drugs on the body, along with the drugs that may cause them, are detailed in table 12.4.

The skin is a frequent target of drug-induced side effects. The skin response can be a symptom of drug allergy or a direct toxic effect. Some drugs interact with sunlight to cause photodermatitis, a skin inflammation. Tetracyclines are contraindicated (not advisable) for children from birth to 8 years of age because they bind to the enamel of the teeth, creating a permanent gray-brown discoloration. Pregnant women should avoid tetracyclines because they can cause liver damage. They also cross the placenta and can be deposited in the developing fetal bones and teeth. However, the most common complaint associated with oral antimicrobial therapy is diarrhea, which can progress to severe intestinal irritation or colitis. Although some drugs directly irritate the intestinal lining, the usual gastrointestinal complaints are caused by disruption of the intestinal microbiota.

Human Allergic Responses to Drugs

One of the most frequent drug reactions is **allergy.** This reaction occurs because the drug acts as an antigen (a foreign material capable of stimulating the immune system) and stimulates an allergic response. This response can be provoked by the intact drug molecule or by substances that develop from the body's metabolic alteration of the drug. In the case of penicillin, for instance, it is not the penicillin molecule itself that causes the allergic response but a product, *benzylpenicilloyl.* Allergic reactions have been reported for every major type of antimicrobial drug, but the penicillins account for the greatest number of antimicrobial allergies, followed by the sulfonamides.

People who are allergic to a drug become sensitized to it during the first contact, usually without symptoms. Once the immune system is sensitized, a second exposure to the drug can lead to a reaction such as a skin rash (hives); respiratory inflammation; and, rarely, anaphylaxis, an acute, overwhelming allergic response that develops rapidly and can be fatal. (This topic is discussed in greater detail in chapter 17.) Note: Allergy is a negative effect instigated by the human's inappropriate response to the drug. Toxicity is a direct effect of the drug on human tissue.

Suppression and Alteration of the Human Microbiota by Antimicrobials

With the explosion of knowledge about the microbiome, we have to take into consideration the effects of antibiotic usage on the human microbiome. Antibiotics may be causing a wide array of unintended effects by changing the nature of our microbiome. See **Insight 12.1** for an example of this.

Even before we had a clear picture of the absolute importance of the microbiome, it was obvious that antibiotic usage could have disruptive effects on our health. If a broad-spectrum antimicrobial is introduced into a host to treat infection, it will destroy microbes regardless of their roles as normal biota, affecting not only the targeted infectious agent but also many others in sites far removed from the original infection. When this therapy destroys beneficial resident species, other microbes that were once in small numbers can begin to overgrow and cause disease. This complication is called a **superinfection.**

Some common examples demonstrate how a disturbance in microbial biota leads to replacement biota and superinfection. A broad-spectrum cephalosporin used to treat a urinary tract infection by *Escherichia coli* will cure the infection, but it will also destroy the lactobacilli in the vagina that normally maintain a protective acidic environment there. The drug has no effect, however, on *Candida albicans,* a yeast that also resides in healthy vaginas. Released from the inhibitory environment provided by lactobacilli, the yeasts multiply and cause symptoms. *Candida* can cause similar superinfections of the oropharynx (thrush) and the large intestine.

Oral therapy with tetracyclines, clindamycin, and broad-spectrum penicillins and cephalosporins is associated with a serious

INSIGHT 12.1 MICROBIOME: Do Antibiotics Make Us Fat?

There are some who say that the rise of obesity in this country and around the world coincides in time with the discovery and rise in the use of antibiotics. Before the 1950s, Americans were all slim and dashing, right? They were if they survived childhood diseases, that is.

This is clearly just observational speculation. And here is another observation that we do not understand: For decades, farmers have been feeding their livestock antibiotics to increase their growth. Of course, that practice has come under scrutiny with the realization that antibiotic overuse contributes to the development of resistance.

Recently researchers in New York decided to put some of these speculations to the experimental test. They took five groups of mice and fed them all the same amount and kind of laboratory mouse chow. Additionally, they gave them either plain water (the control group) or water that contained antibiotics. They did this over a period of 7 weeks after the mice had been weaned (young mice, in other words). Then they analyzed the weight, fat composition, and gut microbiome of the mice.

Here is what they found: The mice that had been fed any of the antibiotic regimens had significantly higher percentages of body fat than the mice that had drunk plain water. The data are presented in the table to the right, and also in the graphs underneath it. The "Lean weight" graph shows us that there is no significant difference between any of the antibiotic groups (light green bars) and the control group (dark green bar). If there is a significant difference, there is an asterisk placed over the light green bar. In the "Body fat" graph on the right, you see that each of the antibiotic bars has an asterisk, meaning there is a significant difference between them and the control group. This was true even though their overall body weights were comparable to one another. When researchers analyzed the gut microbiome of the mice, they found that (1) the numbers of microbes in the mouse guts were comparable among groups, but (2) the types of microbes were different in the control group compared to the antibiotic groups. Specifically, the ratios of Firmicutes to Bacteroidetes (see chapter 1) were much higher in the antibiotic groups.

	Lean Weight (g)	Body Fat (%)
+ no antibiotic (control)	15.3	20.0
+ Penicillin	14.8	24.0
+ Vancomycin	15.0	21.8
+ Penicillin and Vancomycin	14.8	23.6
+ Chlortetracycline	15.2	21.5

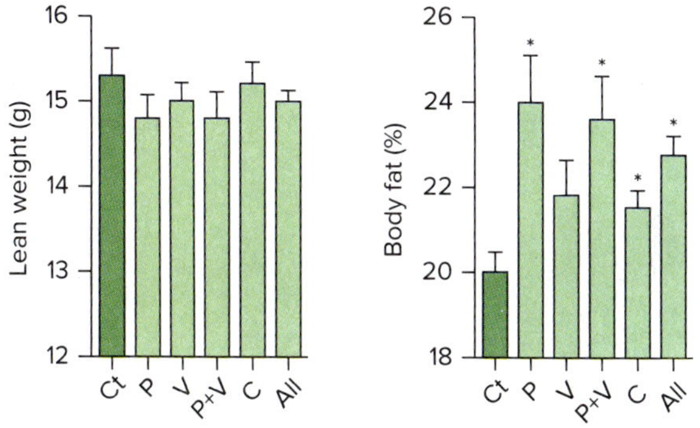

Lean weight (g)
16
15
14
13
12
Ct P V P+V C All
Body fat (%)
26
24
22
20
18
Ct P V P+V C All

Source: Cho, Ilseung et al. "Antibiotics in Early Life Alter the Murine Colonic Microbiome and Adiposity." Nature *488.7413 (2012): 621–626.*

The researchers looked at other consequences of antibiotic usage, including the metabolism of short-chain fatty acids, which have a role in carbohydrate metabolism. Those results showed differences for the antibiotic groups as well.

These are early studies, and the number of things that influence the gut microbiome and body weight are too many to count. This by no means should turn us away from antibiotics. They save lives. Physicians are now prescribing them sparingly, and we learn again, "moderation in all things."

and potentially fatal condition known as *antibiotic-associated colitis* (pseudomembranous colitis). This condition is due to the overgrowth in the bowel of *Clostridioides difficile (C. diff)*, an endospore-forming bacterium that is resistant to many antibiotics. It invades the intestinal lining and releases toxins that induce diarrhea, fever, and abdominal pain. It is very difficult to eradicate, which is why experimental treatments such as fecal transplants are sometimes considered. **Figure 12.6** gives an overview of how the microbiome can be affected by antibiotic usage throughout an individual's lifetime.

12.1 Learning Outcomes—Assess Your Progress

1. State the main goal of antimicrobial treatment.
2. Identify sources of the most commonly used antimicrobial drugs.
3. Summarize two methods for testing antimicrobial susceptibility.
4. Define *therapeutic index,* and identify whether a high or a low index is preferable in a drug.
5. Explain the concept of selective toxicity.
6. Describe the five major targets of current antimicrobial agents, and list major drugs associated with each.
7. Identify which categories of drugs are most selectively toxic, and explain why.
8. Distinguish between drug toxicity and allergic reactions to drugs.
9. Define the term *superinfection*, and summarize how it develops in a patient.

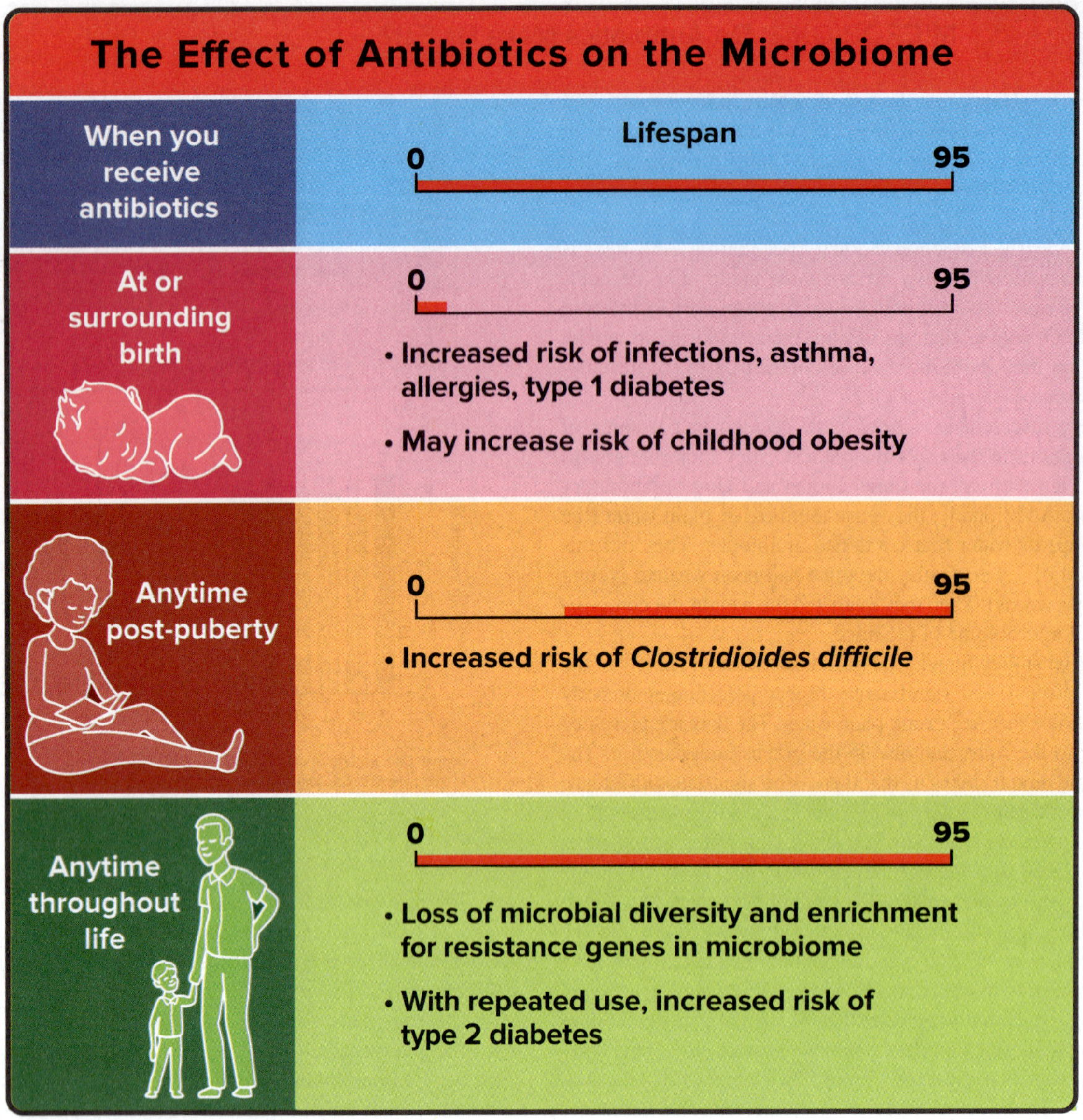
The Effect of Antibiotics on the Microbiome
When you receive antibiotics
Lifespan
0
95
At or surrounding birth
0
95
• Increased risk of infections, asthma, allergies, type 1 diabetes
• May increase risk of childhood obesity
Anytime post-puberty
0
95
• Increased risk of *Clostridioides difficile*
Anytime throughout life
0
95
• Loss of microbial diversity and enrichment for resistance genes in microbiome
• With repeated use, increased risk of type 2 diabetes

Figure 12.6 Depiction of the effects of antibiotics on the microbiome throughout life.

12.2 Survey of Major Antimicrobial Drug Groups

Earlier we said that antibiotics belonged to classes. Each class contains multiple individual drugs that are assigned to that class based on the similarity of their chemical structure. In this section we will consider antibacterial drugs in the following classes:

β-lactams	Glycopeptides
Sulfa drugs	Oxazolidinones
Aminoglycosides	Ansamycins
Tetracyclines	Quinolones
Macrolides	Streptogramins
Pleuromutilins	Lipopeptides

There are some antibiotics we will consider that aren't technically in one of these classes, or that constitute a (small) class of their own. One of the most useful ways of categorizing antimicrobials is to designate them as either **broad-spectrum** or **narrow-spectrum.** Broad-spectrum drugs are effective against more than one group of bacteria, while narrow-spectrum drugs generally target a specific group. **Table 12.5** demonstrates that tetracyclines are broad-spectrum, while polymyxin and even some members of the penicillin group are narrow-spectrum agents.

The designation of antibiotic classes has meaning to the chemists and pharmacologists who developed them. Clinicians also categorize antibiotics based on what their target in the bacterial cell is. One important thing to keep in mind is that antibiotics that are targeted to anything in the cytoplasm of the cell have a very difficult task of getting through at least one membrane.

Table 12.5 Spectrum of Activity for Antibiotics

Bacteria	Mycobacteria	Gram-Negative Bacteria	Gram-Positive Bacteria	Chlamydias	Rickettsias
Examples of diseases	Tuberculosis	Salmonellosis, plague, gonorrhea	Strep throat, staph infections*	Chlamydia, trachoma	Rocky Mountain spotted fever
Spectrum of activity of various antibiotics	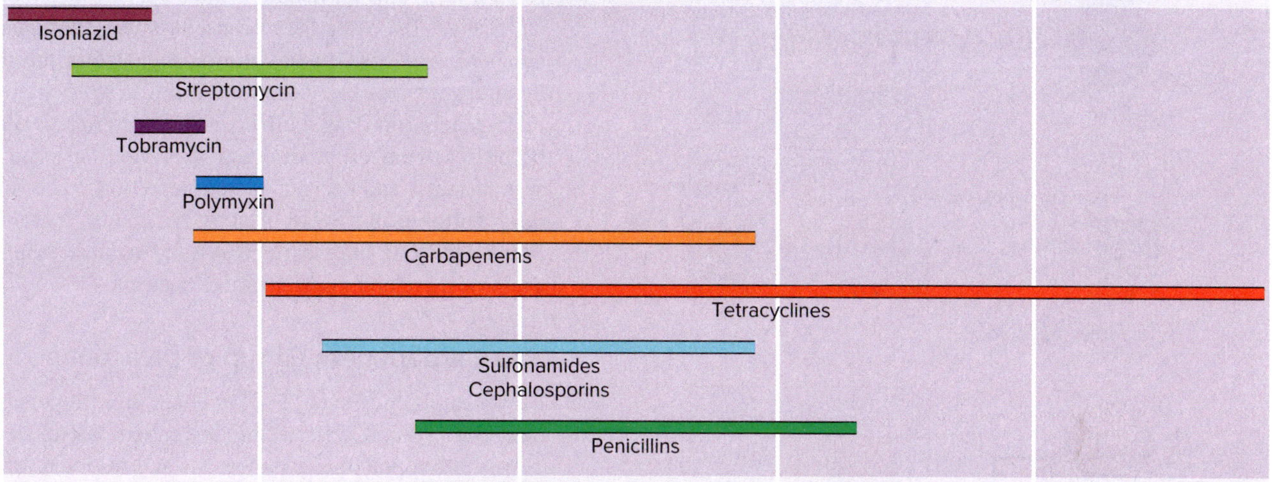Isoniazid Streptomycin Tobramycin Polymyxin Carbapenems Tetracyclines Sulfonamides Cephalosporins Penicillins				
Are there normal biota in this group?	Yes	Yes	Yes	Probably	None known

*Note that some members of a bacterial group may not be affected by the antibiotics indicated, due to acquired or natural resistance. In other words, exceptions do exist.

Because gram-negative bacteria have an extra membrane with highly negatively charged LPS on the outside, it is more difficult to find or design drugs that affect gram-negatives.

Now we will look at each target in the bacterial cell and discuss the antibiotics currently used for that target.

Antibacterial Drugs Targeting the Cell Wall

The Penicillin Group of β-Lactams

The **β-lactam** (bey′-tuh-lak′-tam) class of antibiotics is a large, diverse group of compounds, most of which end in the suffix *-cillin.* They are named for the chemical ring they all possess, called a β-lactam ring. Penicillin is the earliest representative of this class. Penicillin was discovered in 1928 when scientist Alexander Fleming noticed a contaminant on one of many agar dishes of *Staphylococcus* he had been working with. He noticed that the *Staphylococcus* colonies were not growing in the vicinity of the white fuzzy contaminant. Since the world had been searching for antibacterial compounds for some time, he wondered what the chemical from the contaminant could be, and if it could reliably inhibit the growth of bacteria. The contaminant turned out to be the fungus *Penicillium*, and the antibacterial compound that was isolated was named penicillin. Interestingly, Fleming was unable to actually isolate and produce the compound in any useful quantities. This was finally accomplished by two other scientists in Oxford in 1942. The three of them shared the 1945 Nobel Prize in Medicine for the discovery and production of penicillin. The natural product of the fungus can be used either in unmodified form or to make semisynthetic derivatives. All penicillins consist of three parts: a thiazolidine ring, a *β-lactam* ring, and a variable side chain that determines its microbicidal activity **(figure 12.7).**

Subgroups and Uses of Penicillins Penicillins G and V are the most important natural forms. Penicillin is considered the drug of choice for infections by sensitive, gram-positive cocci (some streptococci) and some gram-negative bacteria (such as the syphilis spirochete). Certain semisynthetic penicillins such as ampicillin and amoxicillin have broader spectra and can be used to treat infections by gram-negative enteric rods.

Many bacteria produce enzymes that are capable of destroying the β-lactam ring of penicillin. The enzymes are referred to as **penicillinases** or *β-lactamases,* and they make the bacteria that possess them resistant to many penicillins. Researchers have counted more than 500 different β-lactamases in bacteria. This points to how versatile bacteria can be in resisting our attacks on them. In response, scientists have created penicillinase-resistant penicillins such as methicillin, nafcillin, and cloxacillin, some of which are useful in treating infections caused by penicillinase-producing bacteria.

All of the *-cillin* drugs are relatively mild and nontoxic because of their specific mode of action on cell walls (which humans lack). The primary problems in therapy include allergy, which is

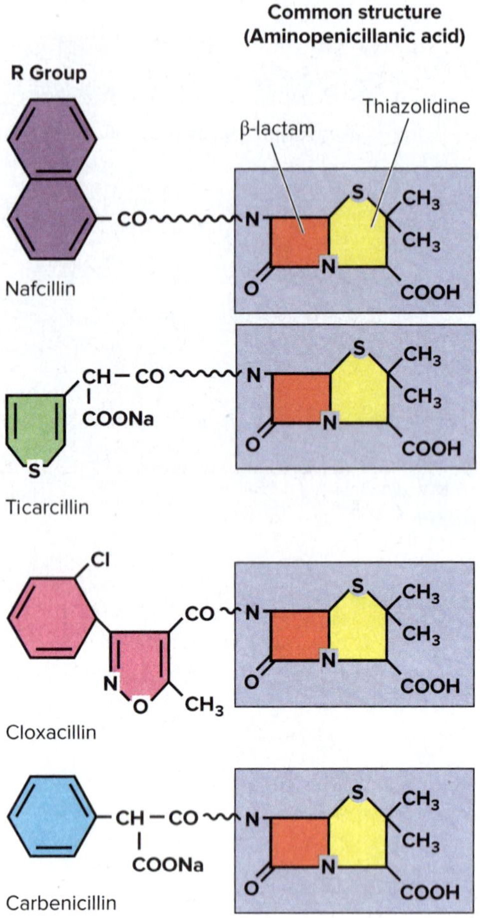
Common structure
(Aminopenicillanic acid)
R Group
Thiazolidine
β-lactam
CO
N
S
CH_3
CH_3
N
O
COOH
Nafcillin
CH–CO
COONa
S
Ticarcillin
Cl
CO
N
O
CH_3
Cloxacillin
CH–CO
COONa
Carbenicillin

Figure 12.7 Chemical structure of penicillins. All penicillins contain a thiazolidine ring (yellow) and a β-lactam ring (red), but each differs in the nature of the side chain (R group). The R group is also responsible for differences in biological activity.

altogether different from toxicity, and resistant strains of pathogens. Clavulanic acid is a chemical that inhibits β-lactamase enzymes, thereby increasing the effectiveness of β-lactam antibiotics in the presence of penicillinase-producing bacteria. For this reason, clavulanic acid is often added to semisynthetic penicillins to augment their effectiveness. For example, a combination of amoxicillin and clavulanate is marketed under the trade name Augmentin. Zosyn is a similar combination of a β-lactamase inhibitor and piperacillin that is used for a wide variety of systemic infections.

The Cephalosporin Group of β-Lactams

The cephalosporins are a group of antibiotics that were originally isolated in the late 1940s from the mold *Acremonium*, which was previously known as *Cephalosporium acremonium.* Cephalosporins are similar to penicillins. They have a β-lactam structure that can be synthetically altered and have a similar mode of action. The generic names of these compounds have the root *cef, ceph,* or *kef* in their names.

Subgroups and Uses of Cephalosporins The cephalosporins are versatile. They are relatively broad-spectrum, resistant to most penicillinases, and cause fewer allergic reactions than penicillins. Although some cephalosporins are given orally, many are poorly absorbed from the intestine and must be administered **parenterally** (par-ehn′-tur-ah-lee) by injection into a muscle or a vein.

Cephalosporins are spoken of in generations, referring to the earliest form that was isolated and utilized, and subsequent chemical changes that were then licensed for use. First-generation cephalosporins such as cephalothin and cefazolin are most effective against gram-positive cocci and a few gram-negative bacteria. A third-generation drug, ceftriaxone (Rocephin), is a semisynthetic broad-spectrum drug for treating a wide variety of respiratory, skin, urinary, and nervous system infections. The fifth-generation drug ceftobiprole exhibits activity against methicillin-resistant *Staphylococcus aureus* (MRSA) and against penicillin-resistant gram-positive and gram-negative bacteria.

The Carbapenem Group of β-Lactams

Newer antibiotics such as doripenem and imipenem belong to a new class of cell wall antibiotics called carbapenems. They are powerful but potentially dangerous and reserved for use in hospitals when other drugs are not working. Recently, the appearance of a gene in gram-negative bacteria coding for an enzyme called NDM-1 has caused great concern because it confers resistance to carbapenems and is highly transmissible from bacterium to bacterium.

Because of similarities in chemical structure among penicillins, cephalosporins, and carbapenems, allergies to penicillin often are accompanied by allergies to cephalosporins and carbapenems.

Other Drugs Targeting the Cell Wall

Bacitracin is a narrow-spectrum antibiotic produced by a strain of the bacterium *Bacillus subtilis.* Since it was first isolated, its greatest claim to fame has been as a major ingredient in a common drugstore antibiotic ointment (Neosporin) for combating superficial skin infections by streptococci and staphylococci. For this purpose, it is usually combined with neomycin (an aminoglycoside) and polymyxin.

Isoniazid (INH) is a synthetic drug that is bactericidal to *Mycobacterium tuberculosis,* but only against growing cells. It is generally used in combination with two or three additional drugs in active tuberculosis cases. **Vancomycin** is a narrow-spectrum antibiotic in the **glycopeptide** class. It is most effective in treating staphylococcal infections in cases of penicillin and methicillin resistance or in patients with an allergy to penicillins.

Antibacterial Drugs Targeting Protein Synthesis (Ribosomes)

The Aminoglycoside Drugs

Antibiotics composed of one or more amino sugars and a 6-carbon ring are referred to as **aminoglycosides (figure 12.8).** These complex compounds are the products of various species of actinomycetes in the genera *Streptomyces* and *Micromonospora.*

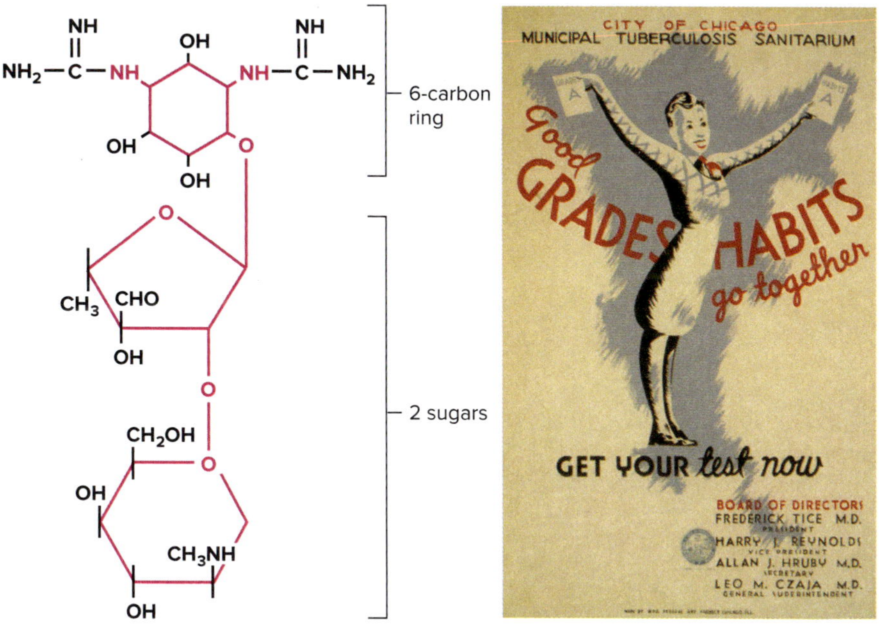
NH
NH
OH
NH_2—C—NH
NH—C—NH_2
OH
O
OH
6-carbon ring
O
CH_3
CHO
OH
O
CH_2OH
O
OH
CH_3NH
OH
2 sugars
CITY OF CHICAGO
MUNICIPAL TUBERCULOSIS SANITARIUM
Good
GRADES
HABITS
go together
GET YOUR test now
BOARD OF DIRECTORS
FREDERICK TICE M.D.
HARRY J. REYNOLDS
ALLAN J. HRUBY M.D.
LEO M. CZAJA M.D.
(a) (b)

Figure 12.8 **Streptomycin.** **(a)** Chemical structure of the antibiotic. Colored portions of the molecule show the general arrangement of an aminoglycoside. **(b)** A poster from 1939 from the City of Chicago Municipal Tuberculosis Sanitarium. The first antibiotic to treat tuberculosis (streptomycin) was not discovered until 1943.

(b) Library of Congress Prints & Photographs Division [LC-USZC2-5178]

Subgroups and Uses of Aminoglycosides The aminoglycosides are relatively broad spectrum because they inhibit the structures involved in protein synthesis **(figure 12.9).** They are especially useful in treating infections caused by aerobic gram-negative rods and certain gram-positive bacteria. Streptomycin is among the oldest of the drugs and has gradually been replaced by newer forms with less toxicity. You will notice that many aminoglycoside drugs end with the suffix *-mycin,* but this suffix is used for drugs from other families as well (such as vancomycin), so it is not a useful way of remembering which class a drug fits into.

Tetracycline Antibiotics

Tetracyclines are another class of antibiotic derived from *Streptomyces*. Their ability to bind to ribosomes and block protein synthesis accounts for the broad-spectrum effects in this class (see figure 12.9). Examples of currently popular tetracyclines are doxycycline, minocycline, glycylcyclines, and tigecycline. These last two are effective against bacteria that have become resistant to other tetracyclines.

Macrolide Antibiotics

This class of antibiotics is distinguished by a large chemical ring called a lactone **(figure 12.10).** Its structure consists of a large ring with sugars attached. These drugs are relatively broad-spectrum and of fairly low toxicity. Their mode of action is to block protein synthesis by attaching to the ribosome (see figure 12.9). Newer semisynthetic macrolides include *clarithromycin* and *azithromycin.* Both drugs are useful for middle ear, respiratory, and skin infections.

Ketolides are related to macrolides but have a different ring structure. These drugs, such as telithromycin, work on bacteria that are resistant to macrolide antibiotics.

Other Classes That Inhibit Protein Synthesis

Oxazolidinones are synthetic drugs first utilized in 2000. The first member of that class was linezolid. These drugs work by a completely novel mechanism, inhibiting the initiation of protein synthesis (see figure 12.9). Because this class of drug is not found in nature, it was hoped that resistance among bacteria would be slow to develop, but resistant strains are arising.

Drugs called **pleuromutilins** block the action of peptidyl transferase (see figure 12.9). The first representative of the pleuromutilins is retapamulin (Altabax) and is approved only for external application for skin infections such as impetigo.

Synercid is a combined antibiotic from the **streptogramin** class of drugs. It is effective against *Staphylococcus* and *Enterococcus* species that cause endocarditis and surgical infections, including resistant strains. It is one of the main choices when other drugs are ineffective due to resistance. Synercid works by binding to sites on the 50S ribosome, inhibiting translation.

Antibacterial Drugs Targeting Folic Acid Synthesis

Sulfa Drugs

These drugs are synthetic and do not originate from bacteria or fungi. Although thousands of sulfonamides have been formulated, only a few have gained any importance. Sulfisoxazole is the best agent for treating acute urinary tract infections and certain protozoan infections. Silver sulfadiazine ointment and solution are prescribed for treatment of burns **(figure 12.11)**. Another drug, trimethoprim inhibits the enzymatic step immediately following the step inhibited by sulfonamides in the synthesis of folic acid. Because of this, trimethoprim is often given in combination with sulfamethoxazole (Septra, Bactrim), to take advantage of the synergistic effect of the two drugs.

Antibacterial Drugs Targeting DNA or RNA

Even though nucleic acids in bacteria and humans are chemically similar, DNA and RNA have still proven to be fairly selective targets for antimicrobials. Fluoroquinolones, in the **quinolone** class, exhibit several ideal traits, including high potency and a broad spectrum. Even in minimal concentrations, quinolones inhibit a wide variety of gram-positive and gram-negative bacterial species. In addition, they are readily absorbed from the intestine. Just as with other drug families, there are multiple "generations" of quinolones. The first generation was typified by nalidixic acid, which is no longer available in the United States. Second-generation quinolones include ciprofloxacin and ofloxacin. Third-generation quinolones exhibit expanded activity

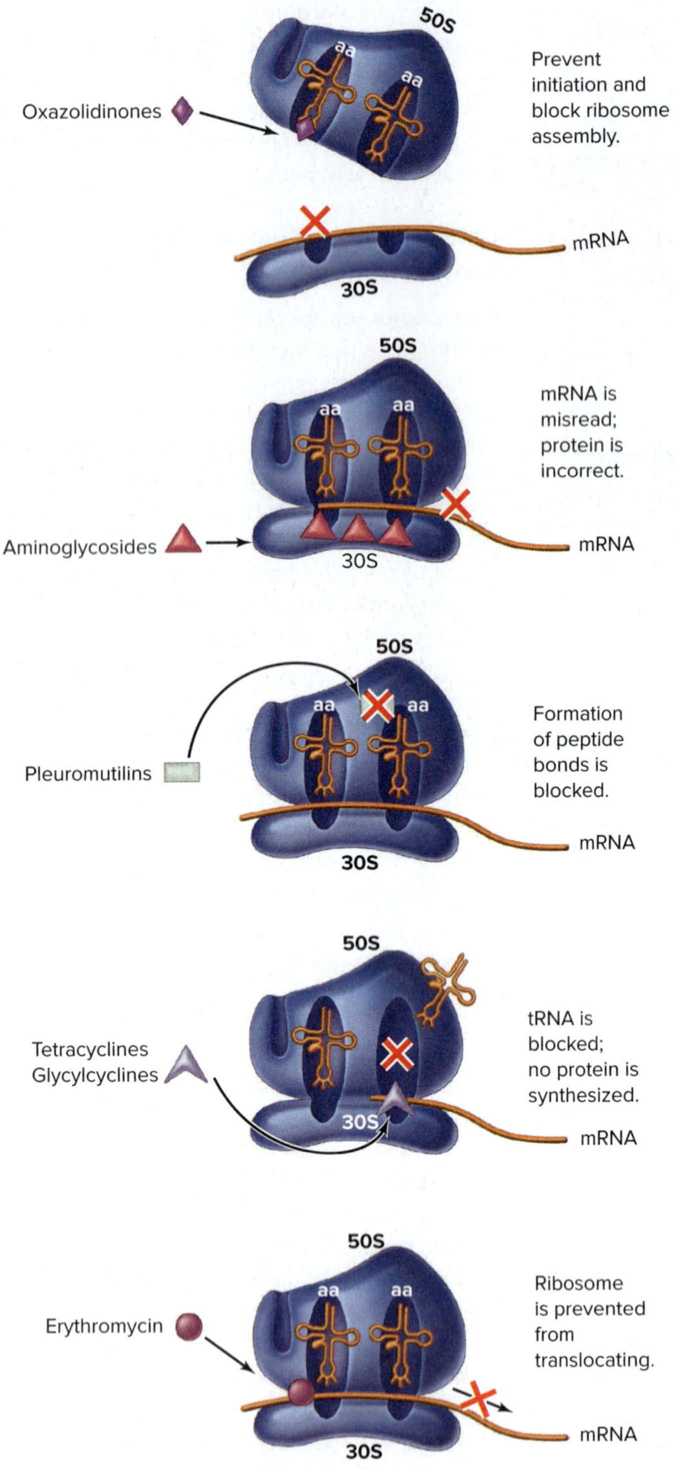
50S
aa
aa
Oxazolidinones
Prevent initiation and block ribosome assembly.
mRNA
30S
50S
aa
aa
mRNA is misread; protein is incorrect.
Aminoglycosides
30S
mRNA
50S
aa
aa
Pleuromutilins
Formation of peptide bonds is blocked.
mRNA
30S
50S
Tetracyclines
Glycylcyclines
tRNA is blocked; no protein is synthesized.
30S
mRNA
50S
aa
aa
Erythromycin
Ribosome is prevented from translocating.
mRNA
30S

Figure 12.9 Sites of inhibition on the bacterial ribosome and major antibiotics that act on these sites. All have the general effect of blocking protein synthesis. Blockage actions are indicated by a red X.

against gram-positive organisms, including some that are resistant to other drugs. The most well-known example is levofloxacin. Fourth-generation quinolones are effective against anaerobic organisms; an example is moxifloxacin. Side effects that limit the use of quinolones can include seizures and other brain disturbances.

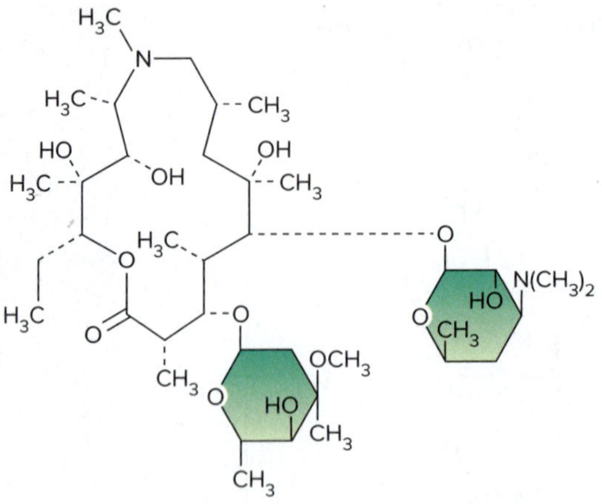

Figure 12.10 A typical macrolide antibiotic. Azithromycin: Its central feature is a large lactone ring to which two hexose sugars are attached.

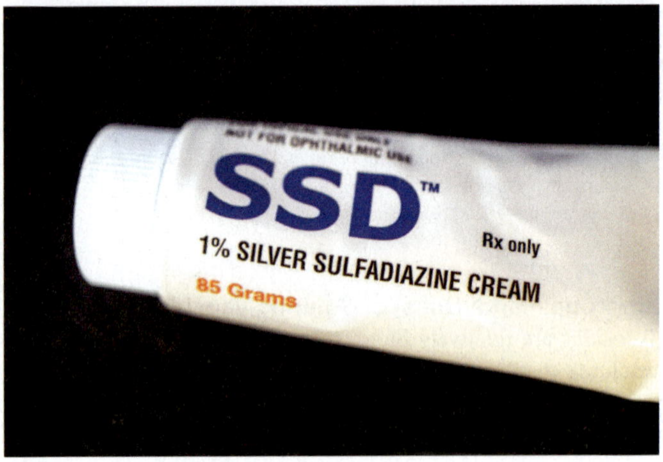
SSD
TM
Rx only
1% SILVER SULFADIAZINE CREAM
85 Grams

Figure 12.11 Sulfa drug containing silver, commonly used topically for burn wounds.

Smith Collection/Gado/Archive Photos/Getty Images

Another product of the genus *Streptomyces* is rifamycin, which is altered chemically into rifampin. It is somewhat limited in spectrum because the molecule cannot pass through the cell envelope of many gram-negative bacteria. It is mainly used to treat infections by several gram-positive rods and cocci and a few gram-negative bacteria. Rifampin figures most prominently in treating mycobacterial infections, especially tuberculosis and leprosy, but it is usually given in combination with other drugs to prevent development of resistance. It is a member of the ansamycin class of antibiotics.

Antibacterial Drugs Targeting Cell Membranes

Every cell has a membrane. As you know, gram-negative bacteria have two bilayer membranes. **Figure 12.12** illustrates a gram-negative envelope and drugs targeted to it. *Bacillus polymyxa* is the source of the **polymyxins,** narrow-spectrum peptide antibiotics with a unique fatty acid component that contributes to their

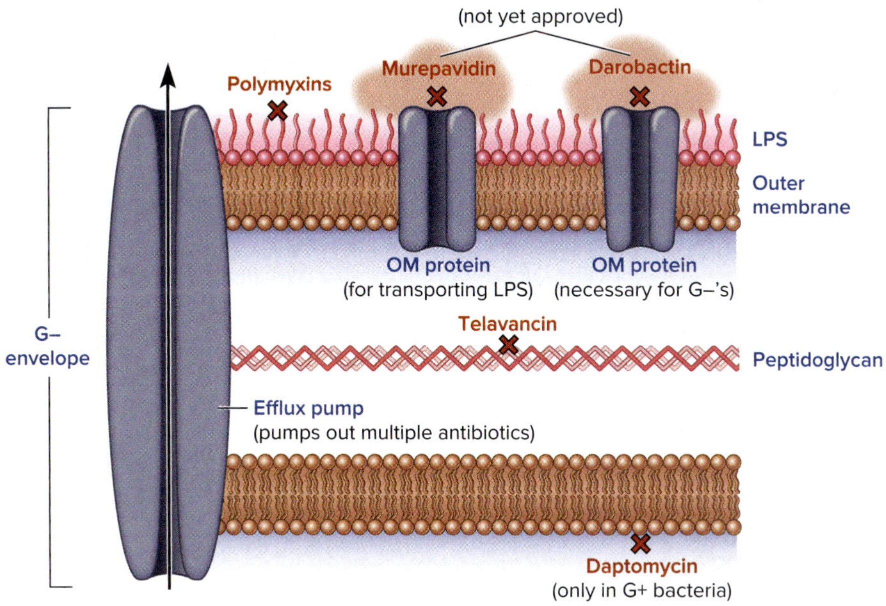

(not yet approved)
Polymyxins
Murepavidin
Darobactin
LPS
Outer membrane
OM protein
(for transporting LPS)
OM protein
(necessary for G–'s)
G– envelope
Telavancin
Peptidoglycan
Efflux pump
(pumps out multiple antibiotics)
Daptomycin
(only in G+ bacteria)

Figure 12.12 The gram-negative outer membrane and antibiotics targeted to it. Everything in shades of blue is a part of the gram-negative envelope. Items in red are drugs, both approved and not-yet approved, designed to target parts of the envelope. It has been very difficult to find or design drugs that penetrate the outer membrane and also avoid being pumped immediately out of the cell by the efflux pump. At this point (2022), polymyxins, telavancin, and daptomycin are approved and in use.

detergent activity. Only two polymyxins—B and E (also known as colistin)—have any routine applications, and even these are limited by their toxicity to the kidney. Colistin has been making a comeback, even though it is toxic. It is used as a last-resort antibiotic when bacteria are resistant to all other antibiotics.

Daptomycin (trade name Cubicin) is a **lipopeptide** made by *Streptomyces.* It is most active against gram-positive bacteria, acting to disrupt multiple aspects of membrane function.

Antibacterial Drugs and Biofilms

As you read in chapter 9, biofilm inhabitants behave differently than their free-living counterparts. One of the major ways they differ—at least from a medical perspective—is that they are as much as 1,000 times less sensitive to the same antimicrobials that work against them when they are free-living. When this was first recognized, it was assumed that it was a problem of penetration, that the (often ionically charged) antimicrobial drugs could not penetrate the sticky extracellular material surrounding biofilm organisms. While that is a factor, there is something more important contributing to biofilm resistance: the different phenotypes expressed by biofilm bacteria. When attached to surfaces, they express different genes and therefore have different antibiotic susceptibility profiles.

There are some strategies that have proven partially successful at enhancing the capability of antibiotics to combat bacteria in biofillms. One of these involves interrupting the quorum-sensing pathways that mediate communication between cells and may change phenotypic expression. Daptomycin, a lipopeptide that is effective in deep-tissue infections with resistant bacteria, has also shown some success in biofilm infection treatment. Most current research is directed at combining enzymes that can cut through the macromolecular structure of the intercellular material with the antibiotic and improve the drug's access.

Many biofilm infections can be found on biomaterials inserted into the body, such as cardiac or urinary catheters. These can be impregnated with antibiotics prior to insertion to prevent colonization. This, of course, cannot be done with biofilm infections of natural tissues, such as the prostate or middle ear.

Interestingly, it appears that treatment with some antibiotics can cause bacteria to form biofilms at a higher rate than they otherwise would. **Table 12.6** summarizes the antibacterial drugs discussed here.

Agents to Treat Fungal Infections

Because the cells of fungi are eukaryotic, they present special problems for antimicrobial treatment. The biggest issue is the similarities between fungal and human cells. This means that drugs toxic to fungal cells are also capable of harming human tissues. Therapeutic indices are generally quite low. A few agents with special antifungal properties have been developed for treating systemic and superficial fungal infections. Four main drug classes currently in use are the polyenes, the azoles, the echinocandins, and the allylamines, in addition to a few miscellaneous drugs **(table 12.7).**

Antiprotozoal and Antihelminthic Treatment

The enormous diversity among protozoan and helminth parasites and their corresponding therapies reach far beyond the scope of this textbook. Just a few of the more common drugs are surveyed here and described again for particular diseases in the organ system chapters.

Antimalarial Drugs: Quinine and Its Relatives

Quinine, extracted from the bark of the cinchona tree, was the principal treatment for malaria for hundreds of years, but it has been replaced by the synthesized quinolones, mainly chloroquine and primaquine, which have less toxicity to humans. Because there are several species of *Plasmodium* (the malaria parasite) and many stages in its life cycle, no single drug is universally effective for every species and stage, and each drug is restricted in application. A drug called artemisinin, which comes originally from a plant called sweet wormwood, used for centuries in Chinese traditional medicine, has become the staple for malaria treatment in most parts of the world. It was discovered to be effective against malaria by a woman named Tu Youyou, a Chinese scientist. She was awarded the Nobel Prize in 2015 for her breakthrough. Artemisinin combination therapy (ACT) is recommended for the treatment of certain types of malaria today and uses artemisinin with quinine derivatives or other drugs.

Table 12.6 Specific Antibacterial Drugs and Their Metabolic Targets

Drug Class/Mechanism of Action	Drug Examples	Uses/Characteristics
Drugs That Target the Cell Wall		
Penicillins 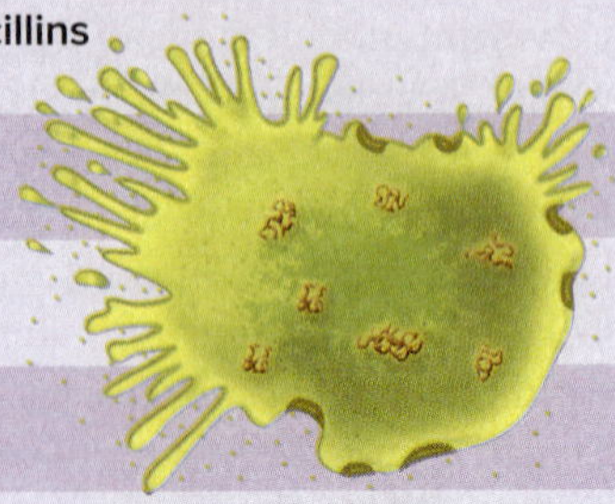 Cell lyses.	Penicillins G and V	Most important natural forms used to treat gram-positive cocci, some gram-negative bacteria
	Ampicillin, carbenicillin, amoxicillin	Have broad spectra of action, are semisynthetic; used against gram-negative enteric rods
	Methicillin, nafcillin, cloxacillin	Useful in treating infections caused by some penicillinase-producing bacteria (enzymes capable of destroying the β-lactam ring of penicillin)
	Clavulanic acid	Inhibits β-lactamase enzymes; added to penicillins to increase their effectiveness in the presence of penicillinase-producing bacteria
Cephalosporins	Cephalothin, cefazolin	First generation; most effective against gram-positive cocci, few gram-negative bacteria
	Cefaclor, cefonicid	Second generation; more effective than first generation against gram-negative bacteria such as *Enterobacter, Proteus,* and *Haemophilus*
	Cephalexin, cefotaxime	Third generation; broad-spectrum, particularly against enteric bacteria that produce β-lactamases
	Ceftriaxone	Third generation; semisynthetic, broad-spectrum drug that treats wide variety of urinary, skin, respiratory, and nervous system infections
	Cefpirome, cefepime	Fourth generation
	Ceftobiprole	Fifth generation; used against methicillin-resistant *Staphylococcus aureus* (MRSA) and against penicillin-resistant gram-positive and gram-negative bacteria
Carbapenems	Doripenem, imipenem	Powerful but potentially toxic; reserved for use when other drugs are not effective
	Aztreonam	Narrow-spectrum; used to treat gram-negative aerobic bacilli causing pneumonia, septicemia, and urinary tract infections; effective for those allergic to penicillin
Miscellaneous Drugs That Target the Cell Wall	Bacitracin	Narrow-spectrum; used to combat superficial skin infections caused by streptococci and staphylococci; main ingredient in Neosporin
	Isoniazid	Used to treat *Mycobacterium tuberculosis* but only against growing cells; used in combination with other drugs in active tuberculosis
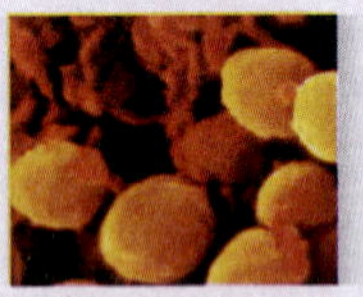 CNRI/Science Source	Vancomycin (a glycopeptide)	Narrow spectrum of action; used to treat staphylococcal infections in cases of penicillin and methicillin resistance or in patients with an allergy to penicillin
Drugs That Target Protein Synthesis (Ribosomes)		
Aminoglycosides Insert on sites on the 30S subunit and cause the misreading of the mRNA, leading to abnormal proteins	Streptomycin	Broad-spectrum; used to treat infections caused by gram-negative rods, certain gram-positive bacteria; used to treat bubonic plague, tularemia, and tuberculosis
Tetracyclines and Glycylcyclines Block the attachment of tRNA on the A acceptor site and stop further protein synthesis	Tetracycline, oxytetracycline (Terramycin)	Effective against gram-positive and gram-negative rods and cocci, aerobic and anaerobic bacteria, mycoplasmas, rickettsias, and spirochetes
Macrolides Inhibit translocation of the subunit during translation (erythromycin)	Erythromycin, clarithromycin, azithromycin	Relatively broad-spectrum, semisynthetic; used in treating ear, respiratory, and skin infections, as well as *Mycobacterium* infections in AIDS patients
Streptogramins	Quinupristin and dalfopristin (Synercid)	A combined antibiotic; effective against *Staphylococcus* and *Enterococcus* species that cause endocarditis and surgical infections, including resistant strains
Oxazolidinones	Linezolid	Synthetic drug; inhibits the initiation of protein synthesis; used to treat antibiotic-resistant organisms such as MRSA and VRE
Pleuromutilins	Altabax	Used for skin infections currently

Table 12.6 **Specific Antibacterial Drugs and Their Metabolic Targets (*continued*)**

Drug Class/Mechanism of Action	Subgroups	Uses/Characteristics
Drugs That Target Folic Acid Synthesis		
Sulfa drugs Interfere with folate metabolism by blocking enzymes required for the synthesis of tetrahydrofolate, which is needed by the cells for folic acid synthesis and eventual production of DNA, RNA, and amino acids	Sulfamethoxazole, trimethoprim	Used to treat shigellosis, acute urinary tract infections, certain protozoan infections 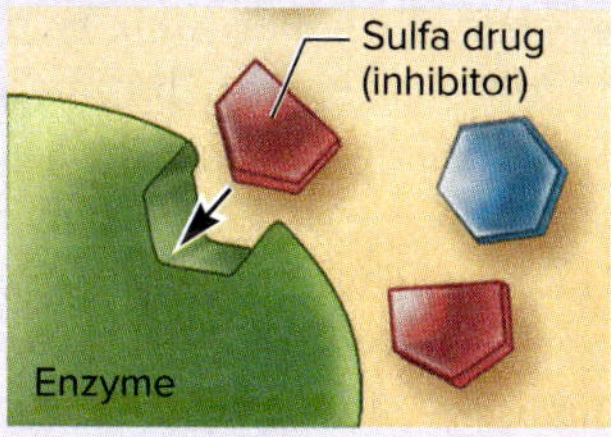 Sulfa drug (inhibitor) Enzyme
	Silver sulfadiazine	Used to treat burns, eye infections (in ointment and solution forms)
Drugs That Target DNA or RNA		
Fluoroquinolones Inhibit DNA unwinding enzymes or helicases, thereby stopping DNA transcription 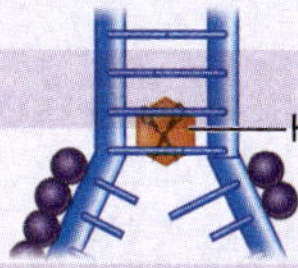 Helicase	Nalidixic acid	First generation; rarely used anymore
	Ciprofloxacin, ofloxacin	Second generation
	Levofloxacin	Third generation; used against gram-positive organisms, including some that are resistant to other drugs
Ansamycins	Rifampin	Limited in spectrum because it cannot pass through the cell envelope of many gram-negative bacilli; mainly used to treat infections caused by gram-positive rods and cocci and a few gram-negative bacteria; used to treat leprosy and tuberculosis
Drugs That Target Cell Membranes		
Polymyxins (including colistin) Interact with membrane phospholipids; distort the cell surface and cause leakage of protein and nitrogen bases, particularly in gram-negative bacteria	Polymyxin B and E	Used to treat drug-resistant *Pseudomonas aeruginosa* and severe urinary tract infections caused by gram-negative rods
	Daptomycin (a lipopeptide)	Most active against gram-positive bacteria

Table 12.7 **Agents Used to Treat Fungal Infections**

Drug Class	Drug Examples	Action
Macrolide polyenes	Amphotericin B	• Bind to fungal membranes, causing loss of selective permeability; extremely versatile • Can be used to treat skin, mucous membrane lesions caused by *Candida* species • Injectable form of the drug can be used to treat histoplasmosis and *Cryptococcus* meningitis
Azoles	Ketoconazole, fluconazole, miconazole, and clotrimazole	• Interfere with sterol synthesis in fungi • Ketoconazole—cutaneous mycoses, vaginal and oral candidiasis, systemic mycoses • Fluconazole—AIDS-related mycoses (aspergillosis, *Cryptococcus* meningitis) • Clotrimazole and miconazole—used to treat infections in the skin, mouth, and vagina
Echinocandins	Micafungin, caspofungin	• Inhibit fungal cell wall synthesis • Used against *Candida* strains and aspergillosis
Allylamines	Terbinafine, naftifine	• Inhibit enzyme critical for ergosterol synthesis • Used to treat ringworm and other cutaneous mycoses

Treatment for Other Protozoan Infections

A widely used amoebicide, metronidazole (Flagyl), is effective in treating mild and severe intestinal infections and hepatic disease caused by *Entamoeba histolytica.* Given orally, it also has applications for infections by *Giardia lamblia* and *Trichomonas vaginalis* (described in chapters 23 and 24, respectively). Other drugs with antiprotozoan activities are quinacrine (a quinine-based drug), sulfonamides, and tetracyclines.

Antihelminthic Drug Therapy

Flukes, tapeworms, and roundworms are much larger parasites than other microorganisms and, being animals, have greater similarities to human physiology. Also, the usual strategy of using drugs to block their reproduction is usually not successful in eradicating the adult worms. The most effective drugs immobilize, disintegrate, or inhibit the metabolism of all stages of the life cycle but especially nondividing helminths.

Albendazole is a broad-spectrum antiparasitic drug used in several roundworm intestinal infestations. The drug works in the intestine to inhibit the function of the microtubules of worms, eggs, and larvae. This means the parasites can no longer utilize glucose, which leads to their demise. Another compound, pyrantel, paralyzes the muscles of intestinal roundworms. Consequently, the worms are unable to maintain their grip on the intestinal wall and are expelled along with the feces by the normal peristaltic action of the bowel. Two other antihelminthic drugs are praziquantel, a treatment for various tapeworm and fluke infections, and ivermectin, a veterinary drug now used for strongyloidiasis and onchocerciasis in humans. (It has been used "off-label"—which means for infections for which it has not been approved—in low-income countries in an attempt to treat COVID-19 patients, but as of late 2021 its real efficacy has not been borne out in well-controlled studies.) Helminthic diseases are described in chapter 23 because these organisms spend at least some part of their life cycles in the digestive tract.

Antiviral Agents

Treating viral infections presents unique problems. With a virus, we are dealing with an infectious agent that relies upon the host cell for the vast majority of its metabolic functions. In currently used drugs, disrupting viral metabolism requires that we disrupt the metabolism of the host cell. Put another way, selective toxicity with regard to viral infection is difficult to achieve because a single metabolic system is responsible for the well-being of both virus and host. Although viral diseases such as measles, mumps, and hepatitis are routinely prevented by the use of effective vaccinations, relatively few viral infections have effective *treatments.*

The currently used antiviral drugs were developed to target specific points in the infectious cycle of viruses. Three major modes of action are

1. barring penetration of the virus into the host cell;
2. blocking the replication, transcription, and translation of viral molecules; and
3. preventing the maturation of viral particles.

Table 12.8 presents an overview of the most widely used antiviral drugs. The following examples provide some additional detail about the principles in table 12.8. Although antiviral drugs protect uninfected cells by keeping viruses from being synthesized and released, most are unable to destroy extracellular viruses or those in a latent state.

Fuzeon (generic name enfuvirtide), an anti-HIV drug, keeps the virus from attaching to its cellular receptor and thereby prevents the initial fusion of HIV to the host cell. Relenza and Tamiflu medications can be effective treatments for influenza A and B as well as useful prophylactics. Because one action of these drugs is to inhibit the fusion and uncoating of the virus, they must be given rather early in an infection. Also, viruses can quickly become resistant to antivirals.

Several antiviral agents mimic the structure of nucleotides and compete for sites on replicating DNA. When these "fake" nucleotides are incorporated into a growing DNA strand, all synthesis stops. Valacyclovir (Valtrex) and its relatives are synthetic purine compounds that block DNA synthesis in a small group of viruses, particularly the herpesviruses. Some derivatives of this drug are valganciclovir, famciclovir, and penciclovir.

HIV is classified as a retrovirus, meaning it carries its genetic information in the form of RNA rather than DNA (HIV and AIDS are discussed in chapter 19). Upon infection, the RNA genome is used as a template by the enzyme **reverse transcriptase (RT)** to produce a DNA copy of the virus's genetic information. Because this particular reaction is not seen outside of the retroviruses, it offers two ideal targets for chemotherapy. The first is interfering with the synthesis of the new DNA strand, which is accomplished using *nucleoside reverse transcriptase inhibitors* (nucleotide analogs), while the second involves interfering with the action of the enzyme responsible for the synthesis, which is accomplished using *nonnucleoside reverse transcriptase inhibitors.*

Azidothymidine (AZT or zidovudine), the first drug approved for HIV/AIDS in 1987, is a thymine analog that exerts its effect by incorporating itself into the growing DNA chain of HIV and terminating synthesis. It is still in use, but several newer drugs based on the same principle are in use today. Nonnucleoside reverse transcriptase inhibitors (such as nevirapine) accomplish the same goal (preventing reverse transcription of the HIV genome) by binding to the reverse transcriptase enzyme itself, inhibiting its ability to synthesize DNA.

Assembly and release of mature viral particles are also targeted in HIV through the use of protease inhibitors. These drugs (indinavir, saquinavir), usually used in combination with nucleotide analogs and reverse transcriptase inhibitors, have been shown to reduce the HIV load to undetectable levels by preventing the maturation step of virus particles in the cell. Refer to table 12.8 for a summary of HIV drug mechanisms and see chapter 21 for further coverage of this topic.

One antiviral drug that has been approved for the treatment of severe COVID-19, remdesivir, is metabolized into an analog of adenosine. When the virus is creating new nucleic acid strands, it uses the analog instead of actual adenosine, and replication is stopped.

A logical alternative to artificial drugs has been a human-based substance, **interferon (IFN).** Interferon is a glycoprotein produced primarily by fibroblasts and leukocytes in response to various immune stimuli. It has numerous biological activities, including antiviral and anticancer properties. Studies have shown that it is a versatile part of animal host defenses, having a major role in natural immunities.

12.2 Learning Outcomes—Assess Your Progress

10. Distinguish between broad-spectrum and narrow-spectrum antimicrobials, and explain the significance of the distinction.
11. Trace the development of β-lactam antimicrobials, and identify which microbes they are effective against.
12. Describe the action of β-lactamases, and explain their importance in drug resistance.

Table 12.8 Actions of Some Antiviral Drugs

Mode of Action	Examples	Effects of Drug
Inhibition of Virus Entry: Receptor/fusion/uncoating inhibitors	Enfuvirtide (Fuzeon)	Blocks HIV infection by preventing the binding of viral receptors to cell receptor (1), thereby preventing fusion of virus with cell
	Amantadine and its relatives, zanamivir (Relenza), oseltamivir (Tamiflu)	Block entry of influenza virus by interfering with fusion of virus with cell membrane (also release); stop the action of influenza neuraminidase, required for entry of virus into cell (also assembly) (2) (3)
Inhibition of Nucleic Acid Synthesis	Remdesivir	Purine analogs that terminate RNA replication in SARS-CoV-2 (4)
	Ribavirin	Purine analog, used for hepatitis C and some other diseases
	Zidovudine (AZT), lamivudine (3T3), didanosine (ddI), zalcitabine (ddC), stavudine (d4T)	Nucleotide analog reverse transcriptase (RT) inhibitors; stop the action of RT in HIV, blocking viral DNA production (5)
	Nevirapine, efavirenz, delavirdine	Nonnucleotide analog reverse transcriptase inhibitors; attach to HIV RT binding site, stopping its action (6)
Inhibition of Viral Assembly/Release	Indinavir, saquinavir	Protease inhibitors; insert into HIV protease, stopping its action and resulting in inactive noninfectious viruses (7)

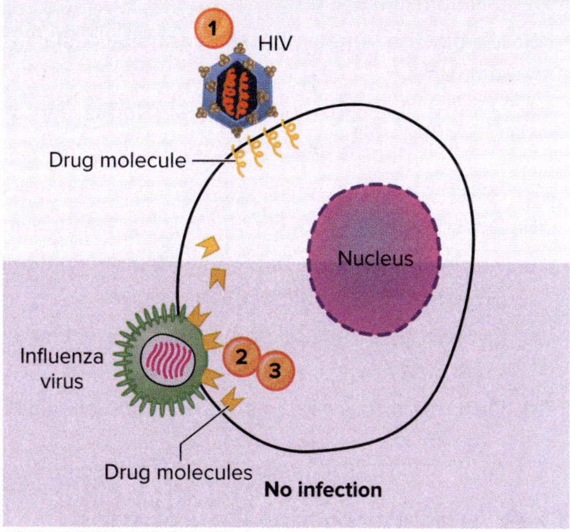
1
HIV
Drug molecule
Nucleus
Influenza virus
2
3
Drug molecules
No infection

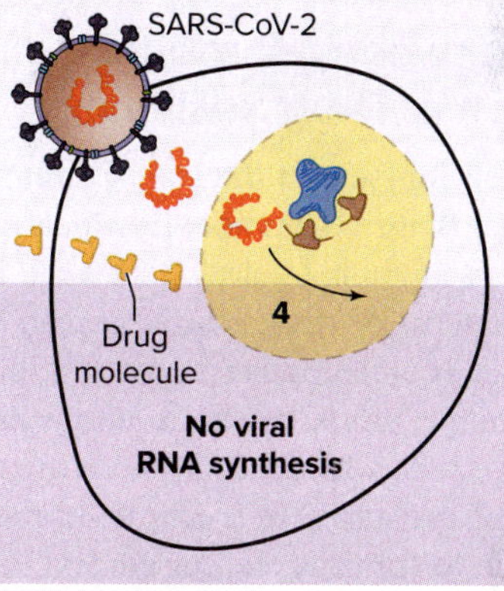
SARS-CoV-2
Drug molecule
4
No viral RNA synthesis

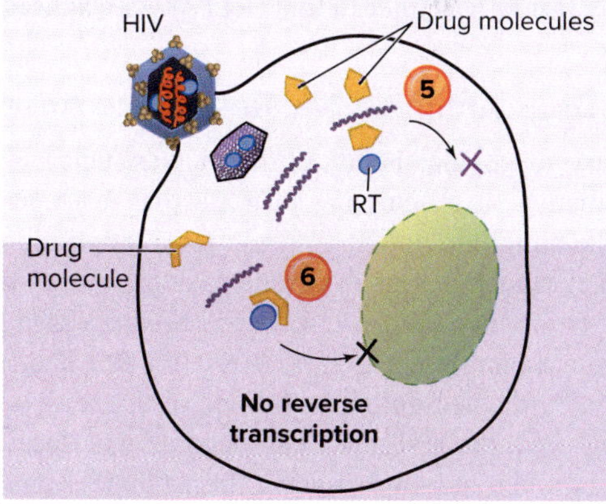
HIV
Drug molecules
5
RT
Drug molecule
6
No reverse transcription

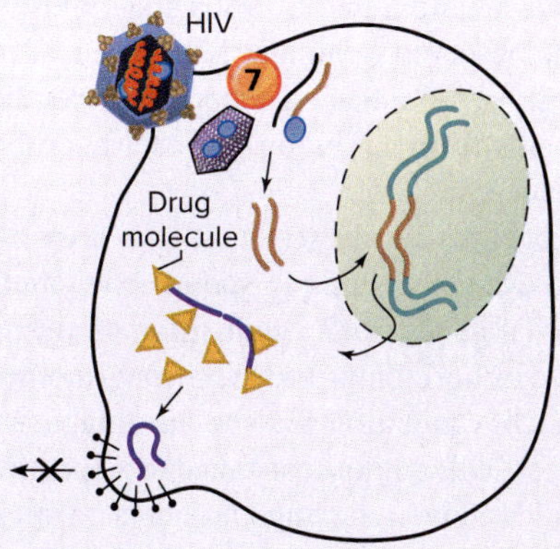
HIV
7
Drug molecule

13. List examples of other β-lactam antibiotics.
14. Describe common cell wall antibiotics that are not in the β-lactam class of drugs.
15. Identify the targets of several antibiotics that inhibit protein synthesis.
16. Identify the cellular target of quinolones, and provide two examples of these drugs.
17. Name two drugs that target the cellular membrane.
18. Describe the unique problems in treating biofilm infections.
19. Name the four main categories of antifungal agents, and provide one example of each.
20. List two antiprotozoal drugs and three antihelminthic drugs used today.
21. Describe three major modes of action of antiviral drugs.

12.3 Antimicrobial Resistance

Interactions Between Microbes and Drugs: The Acquisition of Drug Resistance

One unfortunate development in the use of antimicrobials is the growth of microbial **drug resistance,** in which microorganisms begin to tolerate an amount of drug that would ordinarily be inhibitory. The property of drug resistance can be intrinsic or acquired. Resistance is called intrinsic when it is a fixed trait of a microbe. For example, all microbes are intrinsically resistant to antibiotics they themselves produce. Of much greater importance is the acquisition of resistance to a drug by a microbe that was previously sensitive to the drug. In our context, the term *antibiotic resistance* will refer to this last type of acquired resistance.

How Does Antimicrobial Resistance Arise?

Contrary to popular belief, antibiotic resistance is not a recent phenomenon. It is tempting to think that resistance arises only after the antibiotic has been in use for a very long time. But the first bacteria that became resistant to penicillin were found in the 1940s, before the antibiotic had even been released to the public. If there is an antibiotic around, there will be a bacterium that can resist it. And when millions of doses of that antibiotic have been administered, the resistance can become widespread. The scope of the problem in terms of using the antibiotics as treatments for humans became apparent in the 1980s and 1990s, when scientists and physicians observed treatment failures on a large scale. Now it is such a large problem that an economist recently predicted that worldwide deaths from antibiotic-resistant microbes will surpass deaths from cancer by 2050.

Bacteria become newly resistant to a drug after one of the following two events occurs: (1) spontaneous mutations in critical chromosomal genes or (2) acquisition of entire new genes or sets of genes via horizontal transfer from another species. The chance that such a mutation will be advantageous is small, and the chance that it will confer resistance to a specific drug is lower still. Nevertheless, given the huge numbers of microorganisms in any population and the constant rate of mutation, such mutations do occur. The end result varies from slight changes in microbial sensitivity—which can be overcome by larger doses of the drug—to complete loss of sensitivity.

Then we have the occurrence of resistance originating through horizontal transfer from plasmids called **resistance (R) factors** that are transferred through conjugation, transformation, or transduction. Studies have shown that plasmids encoded with drug resistance are naturally present in microorganisms before they have been exposed to the drug. Such traits are "lying in wait" for an opportunity to be expressed and to confer adaptability on the species. Many bacteria also maintain transposable drug resistance sequences (transposons) that are duplicated and inserted from one plasmid to another or from a plasmid to the chromosome. Chromosomal genes and plasmids containing codes for drug resistance are faithfully replicated and inherited by all subsequent progeny. This sharing of resistance genes accounts for the rapid proliferation of drug-resistant species. As you have read in earlier chapters, gene transfers are extremely frequent in nature, with genes coming from totally unrelated bacteria, viruses, and other organisms living in the body's normal biota and the environment.

Fungi have these two options for becoming antibiotic-resistant, as well as a third option discovered in 2014. In at least some species of fungi, a small regulatory RNA known as interfering RNA—or RNAi—has been found to bind to a genetic sequence temporarily. When it is bound, the gene is silenced and the target of the antibiotic is not manufactured by the fungus. This renders it temporarily resistant to that drug. This provides the fungus more flexibility, allowing it to express the gene later when the antibiotic is no longer present. This reversible mechanism is called an **epigenetic** event, and the gene silencing is called an **epimutation.**

Specific Mechanisms of Drug Resistance

The actual changes that take place inside the cell as a result of these gene changes mostly fall into five categories. Think of the two pathways to acquring resistance as the "how" and these five mechanisms as the "what" (illustrated in **figure 12.13**):

1. New enzymes are synthesized. These inactivate the drug (only occurs through the acquisition of new genes).
2. Permeability or uptake of drug into bacterium is decreased (usually occurs via mutation).
3. Drug is immediately eliminated using a transmembrane pump (usually occurs via acquisition of new genes).
4. Binding sites for drug are decreased in number or affinity (can occur via mutation or acquisition of new genes).
5. An affected metabolic pathway is shut down or an alternative pathway is used (occurs due to mutation of gene(s) for original enzyme or enzymes).

Some bacteria can become resistant indirectly by slowing down metabolism and lapsing into dormancy. That is because most antibiotics work best on fast-growing populations. Many species of bacteria produce **persisters** under these circumstances, bacteria that are so slow-growing that they are not affected by the antibiotic. When the antibiotic goes away, the persisters can begin growing again. Persister formation is particularly common in biofilm infections. The most stark example of this is in patients

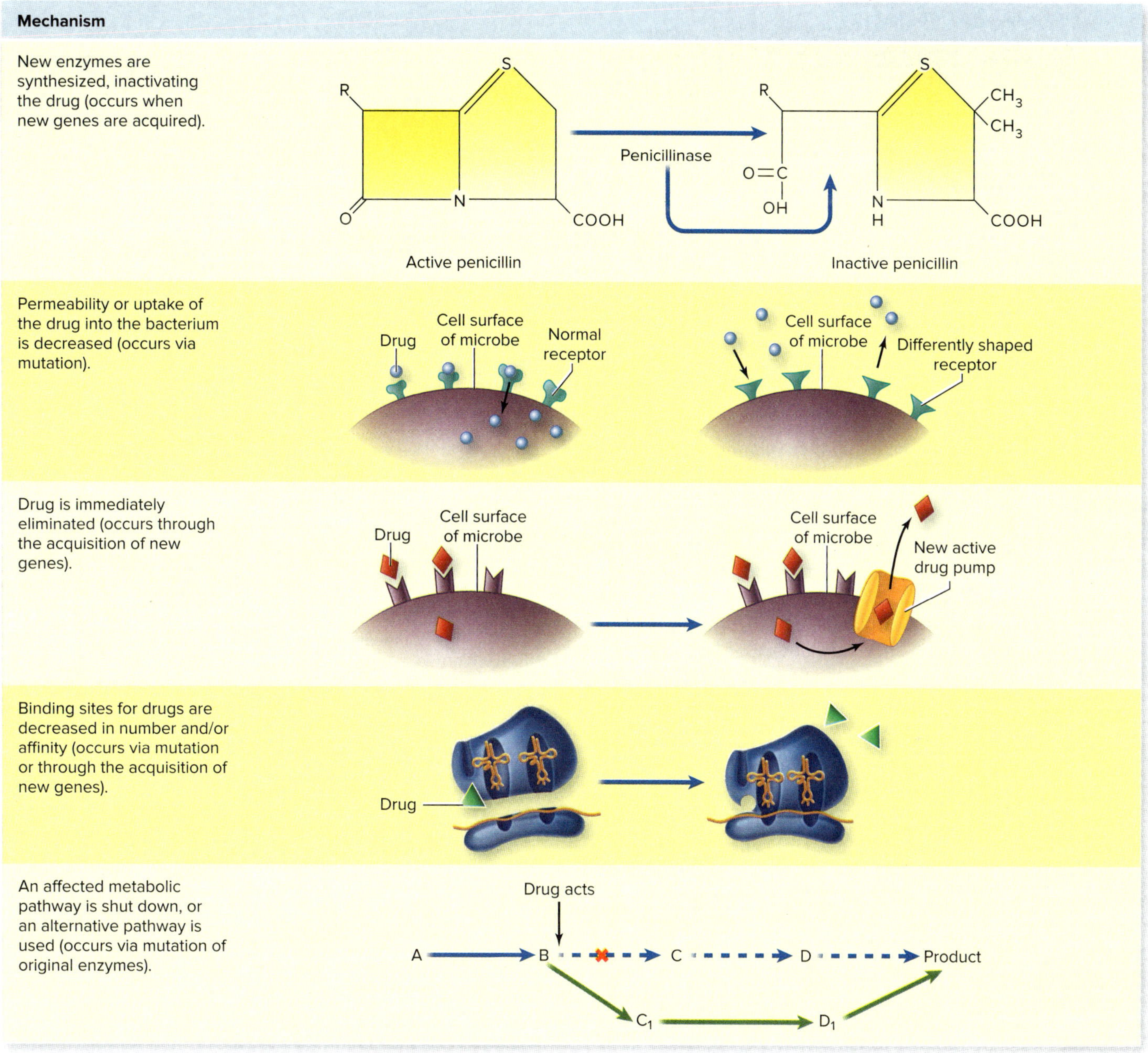
Mechanism
New enzymes are synthesized, inactivating the drug (occurs when new genes are acquired).
Penicillinase
Active penicillin
Inactive penicillin
Permeability or uptake of the drug into the bacterium is decreased (occurs via mutation).
Drug
Cell surface of microbe
Normal receptor
Differently shaped receptor
Drug is immediately eliminated (occurs through the acquisition of new genes).
New active drug pump
Binding sites for drugs are decreased in number and/or affinity (occurs via mutation or through the acquisition of new genes).
An affected metabolic pathway is shut down, or an alternative pathway is used (occurs via mutation of original enzymes).
Drug acts
Product

Figure 12.13 Mechanisms of drug resistance.

with cystic fibrosis. This condition causes buildup of mucus in the lungs, which limits access of the immune response to the *Pseudomonas aeruginosa* bacteria that colonize these patients. The bacteria are indeed in biofilms and antibiotics are not effective in eliminating the infections. This is true even though the bacteria, when examined, are still found to be susceptible to the antibiotics. The persistence of the infection is due to the biofilms producing significant numbers of persisters, which are dormant, and unaffected by the drugs.

1. Drug Inactivation Mechanisms Some microbes inactivate drugs by producing enzymes that permanently alter drug structure. One prominent example, bacterial enzymes called **β-lactamases,** hydrolyze the β-lactam ring (a critical structure) of penicillins and cephalosporins, rendering the drugs inactive. Two β-lactamases—penicillinase and cephalosporinase—disrupt the structure of certain penicillin or cephalosporin molecules, so their activity is lost. So many strains of *Staphylococcus aureus* produce penicillinase that regular penicillin is rarely a possible therapeutic choice. Different forms of β-lactamases have spread among other pathogenic bacteria as well.

2. Decreased Drug Permeability The resistance of some bacteria can be due to a mechanism that prevents the drug from entering the cell and acting on its target. Antibiotics often have to bind either to an external protein on the surface of the bacterium or to an internal structure, like a ribosomal protein. If a mutation has changed the amino acid sequence of the binding site, the antibiotic

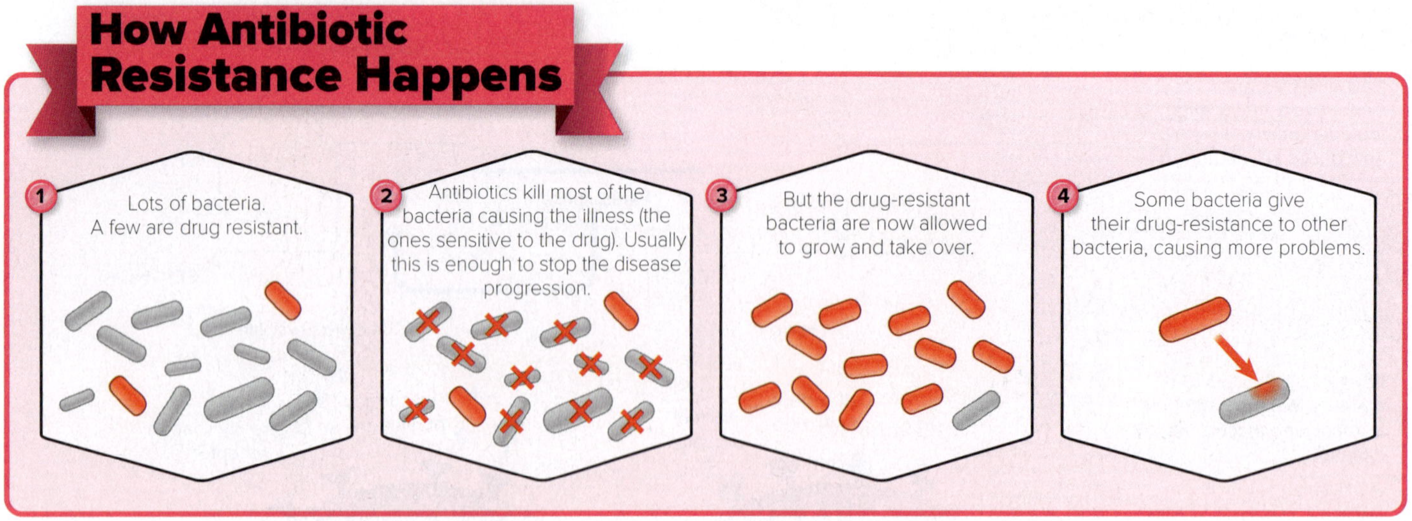
How Antibiotic Resistance Happens
1
Lots of bacteria.
A few are drug resistant.
2
Antibiotics kill most of the bacteria causing the illness (the ones sensitive to the drug). Usually this is enough to stop the disease progression.
3
But the drug-resistant bacteria are now allowed to grow and take over.
4
Some bacteria give their drug-resistance to other bacteria, causing more problems.

Figure 12.14 How antibiotic resistance happens.

will no longer be able to attach, and the bacterium will become impervious to the drug.

3. Drug Is Eliminated Many bacteria possess multidrug-resistant (MDR) pumps that actively transport drugs and other chemicals out of cells. These pumps are proteins encoded by plasmids or chromosomes. They are stationed in the cell membrane and expel molecules by a proton-motive force similar to ATP synthesis (see figure 12.13). They confer drug resistance on many gram-positive pathogens (*Staphylococcus, Streptococcus*) and gram-negative pathogens (*Pseudomonas, E. coli*). Because they lack selectivity, one type of pump can expel a broad array of antimicrobial drugs, detergents, and other toxic substances.

4. Change of Drug Receptors Bacteria can become resistant to aminoglycosides when point mutations in ribosomal proteins arise (see figure 12.13). Erythromycin and clindamycin resistance is associated with an alteration on the 50S ribosomal binding site. Penicillin resistance in *Streptococcus pneumoniae* and methicillin resistance in *Staphylococcus aureus* are related to an alteration in the binding proteins in the cell wall. Enterococci have acquired resistance to vancomycin through a similar alteration of cell wall proteins. Fungi can become resistant by decreasing their synthesis of ergosterol, the principal receptor for certain antifungal drugs.

5. Changes in Metabolic Patterns The action of drugs directed at metabolic products can be avoided by a microbe if it develops an alternative metabolic pathway or enzyme. For example, sulfonamide and trimethoprim resistance develop when microbes deviate from the usual patterns of folic acid synthesis. Fungi can acquire resistance to flucytosine by completely shutting off certain metabolic activities.

Natural Selection and Drug Resistance

Any large population of microbes is likely to contain a few individual cells that are already drug resistant because of prior mutations or transfer of plasmids. As long as the drug is not present in the habitat, the numbers of these resistant forms are likely to remain low. But if the population is subsequently exposed to this drug **(figure 12.14),** sensitive individuals are inhibited or destroyed, and resistant forms survive and proliferate. During subsequent population growth, offspring of these resistant microbes will inherit this drug resistance. In time, the replacement population will have a majority of the drug-resistant forms and can eventually become completely resistant. In ecological terms, the environmental factor (in this case, the drug) has put selection pressure on the population, allowing the more "fit" microbe (the drug-resistant one) to survive, and the population has evolved to a condition of drug resistance.

The Human Role in Antimicrobial Resistance

A recent study found that 75% of antimicrobial prescriptions are for throat, sinus, lung, and upper respiratory infections. A fairly high percentage of these are viral in origin and will have little or no benefit from antibacterial drugs. In the past, many physicians tended to use a "shotgun" antimicrobial therapy for minor infections, which involves administering a broad-spectrum drug instead of a more specific, narrow-spectrum one. This practice led to superinfections and other adverse reactions. Importantly, it also caused the development of resistance in "bystander" microbes (normal biota) that were exposed to the drug. This helped to spread antibiotic resistance to pathogens. Also, tons of excess antimicrobial drugs produced in this country are exported to developing countries, where controls are not as strict. It is common for people in these countries to self-medicate without understanding the correct medical indication. Drugs used in this way are largely ineffective, but worse yet, they are known to be responsible for the emergence of drug-resistant bacteria that subsequently cause epidemics.

The Hospital Factor

The hospital environment continually exposes pathogens to a variety of drugs. Hospitals also house susceptible patients with weakened defenses and a workforce that may not strictly adhere to universal precautions. These factors have led to penicillin resistance in nearly 100% of all *Staphylococcus aureus* strains within just 30 years.

Drugs in Animal Feeds

Nearly 70% of all antibiotics in the United States are given to livestock, both because they have been found to result in larger animals and as a general prevention of bacterial disease. The WHO issued a call to stop using medically important antibiotics in agriculture, and some countries have done well with this. Enteric bacteria such as *Salmonella, Escherichia coli,* and enterococci that live as normal intestinal biota of these animals readily share resistance plasmids and are constantly selected and amplified by exposure to drugs. These pathogens subsequently "jump" to humans and cause drug-resistant infections, often at epidemic proportions. A bill in Congress called the Preservation of Antibiotics for Medical Treatment Act has been introduced multiple times by Representative Louise Slaughter of New York, the only microbiologist in Congress. It was most recently reintroduced in 2017. As of the printing of this textbook, it was still not approved by Congress.

An Urgent Problem

Textbooks generally avoid using superlatives and exclamation points. But the danger of antibiotic resistance can hardly be overstated. The Centers for Disease Control and Prevention (CDC) issued a "Threat Report" about this issue for the first time in 2013, and it continues to monitor the situation, which it labels "potentially catastrophic."

Even though the antibiotic era began less than 80 years ago, we became so confident it would be permanent that we may have forgotten what it was like before antibiotics were available. Even routine surgeries were life-threatening without prophylactic antibiotics. Certain types of pneumonia had a 50% fatality rate. Strep throat could turn deadly overnight. Infected wounds often required amputations or led to death. Yet the effectiveness of our currently available antibiotics is declining, in some cases very rapidly. There is a real possibility that we will enter a postantibiotic era, in which some infections will be untreatable.

The CDC has categorized resistant bacteria into three groups **(Table 12.9).** This classification is used to inform the nation's plan for combating antibiotic resistance. Note that this table is not meant to contain protozoa, helminths, or viruses.

Table 12.9 Most Serious Antibiotic Resistance Threats

Most Serious Antibiotic Resistance Threats	Gram-Positive or Gram-Negative (or Other)
Urgent Threats	
Carbapenem-resistant *Acinetobacter*	G–
Candida auris	Fungus
Clostridioides difficile	G+ endospore-former
Carbapenem-resistant Enterobacterales	G–
Drug-resistant *Neisseria gonorrhoeae*	G–
Serious Threats	
Drug-resistant *Campylobacter*	G–
Drug-resistant *Candida* (other than *C. auris*)	Fungus
ESBL*-producing Enterobacterales	G–
Vancomycin-resistant *Enterococci* (VRE)	G–
Multidrug-resistant *Pseudomonas aeruginosa*	G–
Drug-resistant nontyphoidal *Salmonella*	G–
Drug-resistant *Salmonella* serotype Typhi	G–
Drug-resistant *Shigella*	G–
Methicillin-resistant *Staphylococcus aureus* (MRSA)	G+
Drug-resistant *Streptococcus pneumoniae*	G+
Drug-resistant tuberculosis	—
Concerning Threats	
Erythromycin-resistant Group A *Streptococcus*	G+
Clindamycin-resistant Group B *Streptococcus*	G+

*Extended-spectrum β-lactamase

Source: CDC, 2021.

New Approaches to Antimicrobial Therapy

The good news is that there is now an uptick in research into novel antimicrobial strategies. When antibiotic resistance first became widely problematic, the specter of "no new drugs" was very real because drug makers had drastically slowed their antibiotic development. They can perhaps be excused because there was a time from the 1950s to the 1980s when the medical world assumed that eventually all infections could be wiped out with antibiotics. That proved to be a premature assessment, as we have seen.

Until recently, most attempts to find new antibiotics were based on the old paradigm that the usual cell targets will simply need to have new drugs designed to inhibit them. Designing drugs turned out to be less useful than expected, however, because very few of them could penetrate the gram-negative envelope, and the vast majority of bacteria that are becoming untreatable are gram-negative. (Take another glimpse at table 12.9 to see that.) Many researchers are now convinced that they will need to go back to discovering natural antibiotics in bacteria because they can penetrate the membrane. It is interesting that they are now looking for bacteria that live in animals, not the soil, for these new mining adventures. Granted, the animals are very small. Nematodes harbor bacteria that produce antibiotics that antagonize competing bacteria but don't harm eukaryotic cells.

The past decade has also brought about a revolution in how we think about antibiotics. There are several new approaches, shown here:

- The use of **nanomaterials,** such as graphene oxide and particulate gold particles, causes damage to bacterial membranes without antibiotics. Other nanoparticles can generate toxic reactive oxygen and nitrogen species inside the bacterial cell. Nanomaterials can also serve as catalysts to convert bacterial components into toxic chemicals. Alternatively, nanomaterials can be used to deliver natural antibiotics more effectively to their targets. These studies are still confined to the laboratory.
- **Antisense RNAs** are a form of small RNAs that can be designed to enter a microbe, bind to its DNA or RNA, and shut down the expression of critical genes, leading to cell death in a very targeted way.

- **CRISPR technology** is being tested to target chromosomal or virulence genes in pathogens and eliminate those sequences in the microbes.
- **Bacteriophages** have been used in the past to battle infections prior to antibiotic discovery and when antibiotics were unavailable. Think about it: Phages are extremely specific to a single bacterium and harmless to human cells. "Soups" of phages for bacteria likely causing diarrhea (a potentially fatal disease in young children) were used, especially in second-world countries like the former Soviet Union, up until the 1950s, when antibiotics became widely available. Current researchers are exploring these now, as well.
- **Antibiotics specifically targeting critical gram-negative outer membrane proteins** are also in development, and probably are the closest to being commercialized (see figure 12.12).

Helping Nature Along

Other novel approaches to controlling infections include the use of **probiotics** and **prebiotics.** Probiotics are preparations of live microorganisms that are fed to animals and humans to improve their intestinal biota. This can replace microbes lost during antimicrobial therapy or simply supplement the biota that is already there. Recent years have seen a huge increase in the numbers of probiotic products sold in grocery stores **(figure 12.15).** Experts generally find these products safe, and in some cases they can be effective.

Prebiotics are nutrients that encourage the growth of beneficial microbes in the intestine. For instance, certain sugars such as fructans are thought to encourage the growth of the beneficial *Bifidobacterium* in the large intestine and to discourage the growth of potential pathogens. You can be sure that you will hear more about prebiotics and probiotics as the concepts become increasingly well studied by scientists. Clearly, the use of these agents is a different type of antimicrobial strategy than we are used to, but it may have its place in a future in which traditional antibiotics are more problematic.

A technique that is being employed in some medical circumstances is the use of fecal transplants in the treatment of recurrent *Clostridioides difficile* infections. This procedure involves the transfer of feces, containing beneficial normal biota, from healthy patients to affected patients via colonoscopy **(figure 12.16).** This is, in fact, just an adaptation of probiotics. Instead of a few beneficial bacteria being given orally, with the hope that they will establish themselves in the intestines, a rich microbiota is administered directly to the site it must colonize—the large intestine. This method has had documented success in farm animals, and studies have shown a therapeutic effect in humans as well. Also, companies are now scrambling to create "gut microbiome" pills, in attempts to eliminate the "yuck" factor. It is not clear whether these will be as effective because they have to traverse the stomach and small intestine before reaching the large intestine. At any rate, the gut microbiome seems to be so important for health, even outside the intestines, that you will probably see supplementation with healthy fecal bacteria used more broadly in the future.

DANNON
DanActive
IMMUNITY
DANNON
ACTIVIA
Yoplait
PLUS

Figure 12.15 Probiotics. On the left, popular probiotic-containing products are pictured. On the right, a common probiotic bacterium, *Lactobacillus casei,* is pictured.

(Bacterium) Steve Gschmeissner/Science Photo Library/Alamy Stock Photo; (Yogurt) Kathy Park Talaro

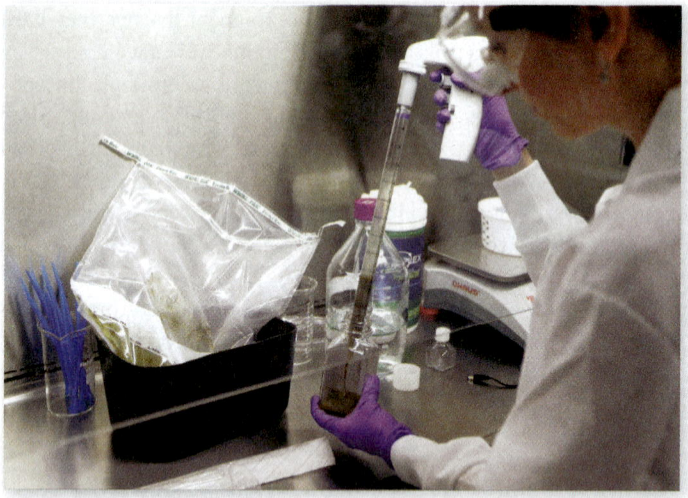

Figure 12.16 Fecal transplant. A technician prepares healthy feces for implantation into a patient.

Steven Senne/AP Images

12.3 Learning Outcomes—Assess Your Progress

22. Discuss two possible ways that microbes acquire antimicrobial resistance.
23. List five cellular or structural mechanisms that microbes use to resist antimicrobials.
24. Discuss at least three novel antimicrobial strategies that are under investigation.

Media Under The Microscope Wrap-Up

milos luzanin/Alamy Stock Photo; Gary He/McGraw Hill

The **intended message** of this article is that mice are carriers of human disease–causing bacteria and should be considered when outbreaks occur in urban populations.

Because the author of the article gives plenty of details about the reported study, my **critical reading** involves a closer look at the credibility of the researchers and a reading of the original journal article. A quick search on Dr. Ian Lipkin yields a lengthy list of his accomplishments over three decades, including involvement in early AIDS research in the 1980s, West Nile virus research during the emergence of encephalitis in 1999, and his assistance in the SARS and MERS outbreaks. He is considered a pioneer of genetic methods to identify microbes. If anyone has the credentials to study zoonotic agents and disease outbreaks, this is the guy! A look at the published study gives more clarity on the microbes isolated from the mice, including their antibiotic-resistance genes.

I would tell my **nonmicrobiologist friends** to practice good pest control in their homes when it comes to mice and other household critters to reduce potential sources of disease.

My **overall grade** for this article is a B. While it does do a good job of referencing and summarizing a credible research study, the author does create a bit of confusion regarding antibiotic resistance in the article. In one paragraph, she misleadingly refers to the mice themselves as antibiotic resistant, rather than the correct explanation of mice carrying antibiotic-resistant bacteria as indicated in her title and elsewhere in the article. Antibiotic resistance is a common topic when it comes to public health, and, unfortunately, many people misunderstand what this really means. To give this article an A, I would have expected the author to use this opportunity to better educate readers about antibiotic resistance and its impact on public health.

Source: Time.com, "NYC Mice Are Carrying Antibiotic-Resistant Germs," online article posted 4/17/2018.

Study Smarter: Better Together

These activities are designed for you to use on your own with a study group—either a face-to-face group or a virtual one, consisting of 3–5 members. Studying together can be very helpful, but there are effective and ineffective ways to do it. For example, getting together without a clear structure is often not a good use of your time. Use your time efficiently by using one or more of the exercises below.

FACE-TO-FACE GROUPS

Use one or more of the activities below.

Peer Instruction: Assign numbers to your group members to use all semester long. Now look at these five concepts from this chapter. Each group member prepares a 5-minute lesson on the topic corresponding to their number. Do not worry if you have fewer than 5 members; just use however many you have! During your group study time, each member presents their lesson, and the group spends another 5–10 minutes discussing that lesson.

1. Narrow-spectrum vs. broad-spectrum
2. Ways microbes acquire antimicrobial resistance (not *mechanisms,* see below)
3. Mechanisms of antimicrobial resistance (not *ways microbes acquire them,* see above)
4. Selective toxicity
5. Minimum inhibitory concentration

Concept Maps: Each member of the group should use this list of terms from this chapter to generate their own concept map. This can be hand-drawn or created using software (see Appendix C for guidelines). During group study time, compare each other's concept maps and help each other make sure they are correct. Of course, there are many different "correct" maps. Examining each member's map will help you talk through the varied concepts and how they are related.

Concept Terms:

selective toxicity	MIC	broad-spectrum	therapeutic index
disc diffusion	sensitivity	resistance	therapeutic window

(continued)

Table Topics: Each group member should identify a concept or topic from this week's class assignments with which they are having trouble and share it during group study time. The other group members can then help to clarify confusing issues or share how they figured it out. Aim for a maximum of 15 minutes per topic. If the topic remains unclear to the group, bring it up during class or use the instructor's office hours or e-mail to ask for help. Taking the time to struggle with a difficult concept first makes your questions much more specific and more likely to yield helpful answers.

VIRTUAL GROUPS

Not everyone has the time or opportunity to meet with group members outside of class time. You or your instructor can create a virtual group using e-mail or the course software.

Weekly Discussion Board: This forum can be used as a way for groups to discuss topics, via e-mail, or other learning management systems or online platforms, before they are covered in class. As each member of the group answers the current week's question, they should send their responses to every other member of their group. It's best to agree on a deadline based on how your class schedule works (Saturday for the next week's topics, for example). Then, after the topic is discussed in class, each member should send a response that all group members will see with a follow-up post on the same topic. If you cover more than one chapter in a week, someone can be designated to choose which chapter Discussion Board question you will use. Or simply decide up front that you will always use the first-chapter-of-the week's question, to keep the schedule simple.

Discussion Question
A friend was recently diagnosed with strep throat. One week after his treatment, his symptoms returned. Your friend tells you, "I must have become immune to the drug the doctor gave me." Discuss the validity of your friend's statement, providing support for or refuting their claim.

Chapter Summary

ANTIMICROBIAL TREATMENT

12.1 PRINCIPLES OF ANTIMICROBIAL THERAPY

- Antimicrobial drugs are produced either synthetically or from natural sources. They inhibit or destroy microbial growth in an infected host.
- Antimicrobial drugs can be either broad-spectrum or narrow-spectrum. The former are drugs that act against both gram-positive and gram-negative bacteria, and the latter act against only one of these, and possibly just one or two bacterial species.
- Bacteria and fungi are the primary sources of most currently used antibiotics. Chemical changes can improve on these.
- Important terms related to antimicrobial treatment are the minimal inhibitory concentration (MIC), therapeutic index, and therapeutic window.
- Antimicrobials are best when they are selectively toxic, meaning they are toxic to the microbe but not to the host.
- The word *antibiotic* is generally reserved for drugs that target bacteria.
- There are five main cellular targets for antibiotics in cellular microbes: cell wall synthesis, nucleic acid structure and function, protein synthesis (ribosomes), cell membranes, and folic acid synthesis.
- There are three broad categories of detrimental effects that antimicrobials can have on humans/mammals: allergic reactions, toxicity, and the alteration of the microbiome.

12.2 SURVEY OF MAJOR ANTIMICROBIAL DRUG GROUPS

- The classes of antibacterial antibiotics are β-lactams, sulfa drugs, aminoglycosides, tetracyclines, macrolides, pleuromutilins, glycopeptides, oxazolidinones, ansamycins, quinolones, streptogramins, and lipopeptides.
- Bacteria in biofilms respond differently to antibiotics than when they are free-floating. It is difficult to remove biofilm infections.
- Fungal, protozoan, and helminthic antimicrobials are not very selectively toxic because all these microbes are eukaryotic organisms.
- Antiviral drugs interfere with viral replication by blocking viral entry into cells, blocking the replication process, or preventing the assembly of viral subunits into complete virions.

12.3 ANTIMICROBIAL RESISTANCE

- Microorganisms are all naturally resistant to various drugs, but acquired antimicrobial resistance is an inevitable result of using antimicrobials.
- Bacteria naturally experience a large number of mutations every time they replicate. Some of these instantly make a bacterium resistant to a drug it was formerly sensitive to. In the presence of that antibiotic, the mutant will grow and displace the sensitive bacteria.
- Bacteria also share, via horizontal gene transfer, entire plasmids that give rise to multiprotein pumps that can expel antibiotics before they have a chance to act.
- The bacteria that are most critical in terms of antibiotic resistance are mostly gram-negatives, and new drugs are in development in an attempt to overcome the major barrier that the outer membrane presents.
- Novel strategies for antibiotics include the use of nanomaterials, both on their own and as a delivery tool for antibiotics; antisense RNAs to shut down production of various proteins; CRISPR technology; and bacteriophage therapy (a new "old" strategy).

SmartGrid: From Knowledge to Critical Thinking

This *21 Question Grid* takes the topics from this chapter and arranges them with respect to the American Society for Microbiology's Undergraduate Curriculum guidelines—all six of the important "Concepts" as well as the important "Competency" of scientific literacy. Three questions are supplied, which cover chapter content referring to the Concept or Competency in increasing levels of Bloom's taxonomy for learning.

ASM Concept/ Competency	A. Bloom's Level 1, 2—Remember and Understand (Choose one)	B. Bloom's Level 3, 4—Apply and Analyze	C. Bloom's Level 5, 6—Evaluate and Create
Evolution	**1.** Microbial resistance to drugs is acquired through **a.** conjugation. **b.** transformation. **c.** transduction. **d.** all of these.	**2.** In the "Media Under The Microscope Wrap-Up," we criticize the fact that the author of the article mentions that the mice became resistant to the antibiotic. What is wrong with that statement?	**3.** You are invited to participate in a debate about the concept of evolution. Construct an opening statement supporting evolution, using antibiotic resistance in microorganisms as an example.
Cell Structure and Function	**4.** Drugs that prevent the formation of the bacterial cell wall are **a.** quinolones. **b.** penicillins. **c.** tetracyclines. **d.** aminoglycosides.	**5.** Why does the β-lactam class of antibiotics have milder toxicity than other antibiotics?	**6.** This chapter discussed five major cellular targets for antibiotic action. Can you imagine a different one? Which one and why?

(continued)

ASM Concept/ Competency	A. Bloom's Level 1, 2—Remember and Understand (Choose one)	B. Bloom's Level 3, 4—Apply and Analyze	C. Bloom's Level 5, 6—Evaluate and Create
Metabolic Pathways	**7.** A compound synthesized by bacteria or fungi that destroys or inhibits the growth of other microbes is a/an **a.** synthetic drug. **b.** antibiotic. **c.** interferon. **d.** competitive inhibitor.	**8.** Conduct research to find out why drugs blocking folic acid synthesis are highly selective (not harmful to humans).	**9.** What kind of organisms do you suppose first possessed β-lactamases, and for what reason?
Information Flow and Genetics	**10.** R factors are ________ that contain a code for ________. **a.** genes, replication **b.** plasmids, drug resistance **c.** transposons, interferon **d.** plasmids, conjugation	**11.** You take a sample from a growth-free portion of the zone of inhibition in the Kirby-Bauer test and inoculate it onto a plate of nonselective medium. What does it mean if growth occurs on the new plate?	**12.** Develop an argument about why the CRISPR technique could potentially provide a permanent solution to antibiotic resistance, where other "fixes" have been temporary.
Microbial Systems	**13.** Treating malarial infections is theoretically difficult because **a.** the protozoal parasite is eukaryotic and therefore similar to human cells. **b.** there are several species of *Plasmodium.* **c.** no single drug can target all the life stages of *Plasmodium.* **d.** all of the above are true.	**14.** Can you think of a situation in which it would be better for a drug to be microbistatic rather than microbicidal? Discuss thoroughly.	**15.** Scientists have found evidence that exposing some bacteria to antibiotics makes them much more likely to form biofilms. Construct an argument for why this might be.
Impact of Microorganisms	**16.** Pick the true statement about antibiotics. **a.** Antibiotics were invented by chemists in the 1830s. **b.** Antibiotics are natural products of microorganisms. **c.** Antibiotics are inherently dangerous for the environment. **d.** None of the above are true.	**17.** Speculate on whether having large concentrations of antibiotics in water sources or farmland would be a positive or negative phenomenon. Defend your answer.	**18.** Imagine a postantibiotic era in which there are no drug treatments for microbial infections. What alternative strategies will make up our arsenal against morbidity and mortality due to microbes?
Scientific Thinking	**19.** An antimicrobial drug with a ________ therapeutic index is a better choice than one with a ______ therapeutic index. **a.** low; high **b.** high; low	**20.** A critically ill patient enters your emergency room, exhibiting signs and symptoms of severe septic shock. In this case, should you immediately begin treatment with a broad-spectrum drug or a narrow-spectrum drug? Explain your answer.	**21.** You are the director of the Centers for Disease Control and Prevention. What are your top three recommendations for halting the spread of antibiotic-resistant bacteria?

Answers to the multiple-choice questions appear in Appendix A.

Visual Connections

This question uses visual images to connect content within and between chapters.

1. **Figure 12.5.** Where could penicillinase affect each of these antibiotics?

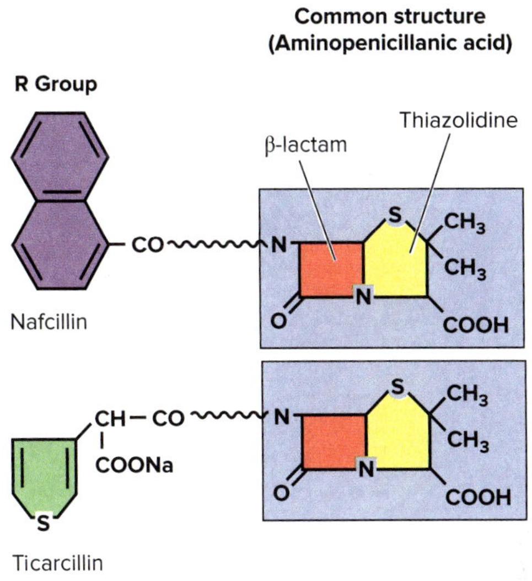
Common structure
(Aminopenicillanic acid)
R Group
Thiazolidine
β-lactam
CO
N
S
CH3
CH3
N
O
COOH
Nafcillin
CH — CO
COONa
N
S
CH3
CH3
N
O
COOH
S
Ticarcillin

High Impact Study

These terms and concepts are most critical for your understanding of this chapter—and may be the most difficult. Have you mastered them?

Concepts

- ☐ Narrow-spectrum vs. broad-spectrum
- ☐ Disc diffusion test
- ☐ Broth dilution test
- ☐ Sites of activity for antimicrobials
- ☐ Why anti-eukaryotic antimicrobials are toxic
- ☐ Ways microbes acquire antimicrobial resistance
- ☐ Mechanisms of antimicrobial resistance
- ☐ Allergy vs. toxicity

Terms

- ☐ Selective toxicity
- ☐ Susceptibility
- ☐ Resistance
- ☐ Minimum inhibitory concentration
- ☐ Therapeutic index
- ☐ Probiotics
- ☐ Prebiotics
- ☐ Superinfection

13

Microbe–Human Interactions

Health and Disease

Margaret McFall-Ngai; Jamie Foster

Media Under The Microscope

Baby Squids in Space

These case studies examine an article from the popular media to determine the extent to which it is factual and/or misleading. This case focuses on the 2021 Guardian *article "Dozens of Hawaiian Baby Squid Aboard Space Station for Study."*

Throughout this book, you have learned about the role the microbiome plays in human health. Soon you will learn that the immune system itself is regulated by the microbiome as well. And, of course, the microbiome plays a large role in regulating the entry and pathogenicity of disease-causing microbes.

The article in the *Guardian* tells us that in mid-2021, a bunch of baby squids took a ride on the International Space Station (ISS). They were there so that scientists on earth can study what happens to the squids' symbiosis with their microbiomes in microgravity. The article states that human astronauts experience dysregulation of their immune systems after some time in space. This is accompanied by a disruption in their microbiome. It is not known which causes which.

The tiny squids—which are only 3 inches long in adulthood—rely on their microbial residents to regulate their bioluminescence. Studying them may help scientists understand how human ISS passengers can be helped to maintain healthy immune systems and microbiomes.

- What is the **intended message** of the article?
- What is your **critical reading** of the summary of the article provided above? Remember that in this context, "critical reading" does not necessarily mean *What criticism do you have*? but asks you to apply your knowledge to interpret whether the article is factual and whether the facts support the intended message.
- How would you **interpret** the news item for your nonmicrobiologist friends?
- What is your **overall grade** for the news item—taking into account its accuracy and the accuracy of its intended effect?

Media Under The Microscope Wrap-Up appears at the end of the chapter.

Outline and Learning Outcomes

13.1 The Human Host and Its Microbiome

1. Differentiate among the terms *colonization, infection,* and *disease.*
2. Identify the sites where normal biota is found in humans.
3. Discuss how the Human Microbiome Project has changed our understanding of normal biota.

13.2 When Colonization Leads to Disease

4. Explain some of the variables that influence whether a microbe will cause disease in a particular host.
5. Differentiate between a microbe's pathogenicity and its virulence.
6. List the steps a microbe has to take to get to the point where it can cause disease.
7. List several portals of entry and exit.
8. Define *infectious dose,* and explain its role in establishing infection.
9. Describe three ways microbes cause tissue damage.
10. Compare and contrast major characteristics of exotoxins and endotoxin.
11. Explain what an epigenetic change is and how it can influence virulence.
12. Draw and label a curve representing the course of clinical infection.
13. Differentiate among the various types of reservoirs, providing examples of each.
14. List six different modes of horizontal transmission, providing an infectious disease spread by each.
15. List Koch's postulates, and explain alternative methods for identifying an etiologic agent.

13.1 The Human Host and Its Microbiome

It would be easy to look at the human body as a distinct entity made up of, well, human cells. And it would be easy to look at microbes as being either good or bad for the human body, based on whether they cause disease or not. In fact, that is the way infectious disease scientists viewed human biology and microbial pathogenesis until very recently. But it is much more complex than that. Each human body has trillions of microbes stably associated with it—some of them necessary, some of them beneficial, and all of them impactful, in seen and unseen ways, on the human organism. Do you remember the term *symbiotic* from chapter 9? It means a close and impactful relationship between two organisms (each of them being a *symbiont*). When scientists realized how vital resident microbes are to human biology, they came up with a new term: **holobiont.** A human plus all of its resident microbiota (its microbiome) is a holobiont. This chapter is about the holobiont, as well as microbes that in some way cause harm to the host, whether they start out as part of the holobiont or are invaders from the outside. These topics will set the scene for chapters 15 and 16, which deal with the ways the host defends itself against assault by microorganisms, and for the chapters of the book that examine diseases affecting different organ systems.

Colonization, Infection, Disease—A Continuum

Our bodies consist of more microbial cells than our own human cells. Some microbes are colonists (normal biota), some are rapidly lost (transients), and others invade the tissues. Sometimes resident biota become invaders.

Our resident microbiota, the human microbiome, **colonizes** us for the long term and does not generally cause disease. We encounter transient microbes in our environment, most of which are harmless, although occasionally they lead to **infection,** a condition in which microbes get past the host defenses, enter the tissues, and multiply. We may or may not be aware that an infection is taking place. When the cumulative effects of the infection damage or disrupt tissues and organs, the pathologic state that results is **disease.** See **figure 13.1** for an illustration of these different states. Please note: This figure looks a little bit complicated, but it truly explains all the different outcomes of contact with microbes, which is the point of this whole book.

A disease is defined as any deviation from health. There are hundreds of different human diseases, caused by such factors as infections, genetics, aging, and malfunctions of systems or organs. In this chapter, we discuss only **infectious disease**—a pathologic state caused directly by microorganisms or their products.

The Human Microbiome Project

When you consider the evolutionary time line (refer to figure 1.1) of bacteria and humans, it is quite clear that humans evolved in an environment that had long been populated by bacteria and single-celled eukaryotes. It should not be surprising, therefore, that humans do not fare well if they are separated from their microbes, either during growth and development or at any other time in their lives.

The extent to which this is true has been surprising even to the scientists studying it. While the human microbiome has been examined on a small scale for decades, the advent of non-culture methods for detecting microbes led to an explosion of research in the 2000s. The American effort launched in 2007 and is called the **Human Microbiome Project (HMP).** It proceeded in two phases. The first phase characterized the microbiome in various body sites of healthy hosts. The second phase has been termed the Integrative Human Microbiome Project and has examined the role of the microbiome in three deviations from health: pregnancy and pre-term birth, inflammatory bowel disease, and Type 2 diabetes. Previous to this international project, scientists and clinicians

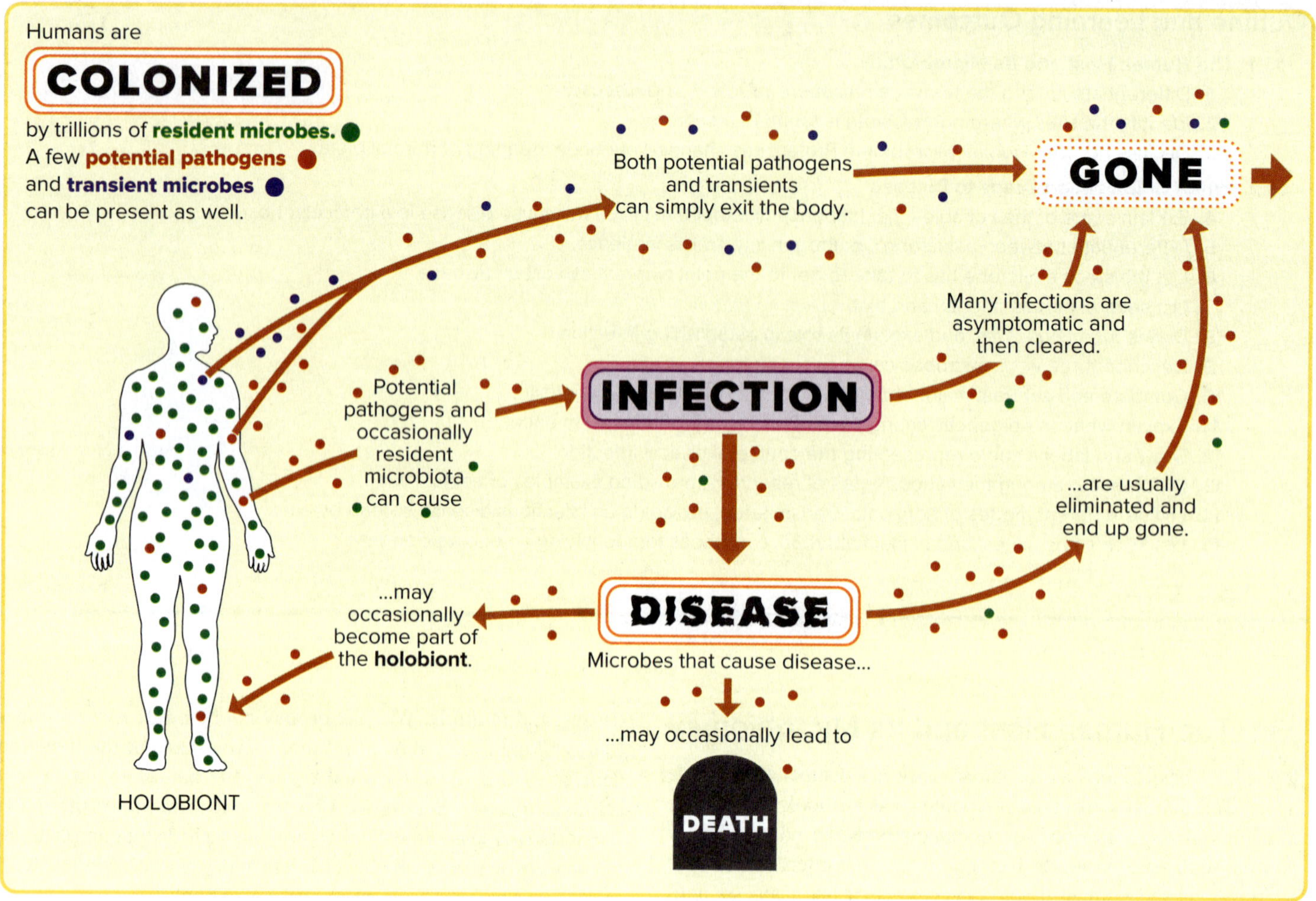
Humans are
COLONIZED
by trillions of resident microbes.
A few potential pathogens
and transient microbes
can be present as well.
Both potential pathogens and transients can simply exit the body.
GONE
Potential pathogens and occasionally resident microbiota can cause
INFECTION
Many infections are asymptomatic and then cleared.
...are usually eliminated and end up gone.
DISEASE
Microbes that cause disease...
...may occasionally become part of the holobiont.
...may occasionally lead to
DEATH
HOLOBIONT

Figure 13.1 The relationships among resident, transient, and disease-causing microbes, and the human host.

mainly relied on culture techniques to determine what the "normal biota" consisted of. That meant we only knew about bacteria and fungi that we could grow in the laboratory, which vastly undercounts the actual number and variety because many—even the majority of—microbes cannot be cultured in the laboratory, though they grow quite happily on human tissues.

The information about the human microbiome presented in this chapter reflects the new findings, which should still be considered preliminary. We will try to show you the differences between the old picture of normal biota in various organ systems and the new, emerging picture. At this point in medical history, it will be important to appreciate the transitioning view.

Several important and surprising results have already emerged. Here are just a few examples.

- Human cells contain about 21,000 protein-encoding genes, but the microbes living in/on us contain 8–20 million. Many of the proteins produced by these genes are enzymes that help us digest our food and metabolize all kinds of substances that we could not otherwise use.
- We have a lot of microbes in places we used to think were sterile. One of the most striking examples is the lungs. They were previously thought to be sterile but actually seem to contain their own sparse but diverse microbiota.
- Viruses are not traditionally discussed in the context of normal biota. However, they are certainly present in healthy humans in vast quantities, both those that infect human cells and those infecting all the cells of our resident biota. For example, researchers think that there are 100 million viruses per gram of human feces. Throughout evolutionary history, viral infections (of cells of all types) have influenced the way cells, organisms, communities, and, yes, the entire ecosystem have developed. The critical contributions of viruses are just now being rigorously studied.
- All healthy people seem to also harbor potentially dangerous pathogens in low numbers. It is a tribute to the protective properties of the normal biota that these rarely cause disease.
- Studies are showing that the relative success of viruses such as influenza and HIV in causing disease can be greatly influenced by the composition of the host microbiomes. This has implications for future treatment strategies.
- The gut microbiome can influence many aspects of human health. As an example, differences in the gut microbiome have preliminarily been associated with risk for Crohn's

Disease Connection

One reminder of the fact that even healthy people harbor small numbers of pathogenic microbes is the growing problem of *"C. diff"* infections. *Clostridioides difficile* is an endospore-forming gram-positive bacterium that lives in our guts in very small quantities, kept in check by the microbial antagonism a diverse microbiota provides. When the healthy microbiota is damaged, especially through broad-spectrum or long-term antibiotic usage, *C. difficile* can flourish, leading to severe and long-lasting intestinal disease.

disease and obesity. Though this may not seem surprising, other (preliminary) research is finding associations between the composition of the gut microbiome and heart disease, asthma, autism, diabetes, cognitive decline in the aging, and even moods.

- Abundant research now shows that the human microbiome, in the gut particularly, influences the effectiveness of anticancer treatments and other treatments directed at the host. As with all revolutionary breakthroughs, some of these early results will be overturned with further study. But one thing is absolutely clear: Our microbiome has a huge influence on our physiology.

Acquiring the Microbiota

The human body offers a seemingly endless variety of environmental niches. It displays wide variations in temperature, pH, nutrients, and oxygen tension occurring from one area to another.

Table 13.1 Current Understanding of Sites Containing Normal Microbiota

Sites Definitively Known to Harbor Normal Microbiota	
Skin and adjacent mucous membranes	External genitalia
Upper respiratory tract	Vagina
Gastrointestinal tract, including mouth	External ear canal
Outer portion of urethra, bladder, urine	External eye (lids, conjunctiva)
Breast milk	
Additional Sites Now Thought to Harbor at Least Some Normal Microbiota (or Their DNA)	
Lungs (lower respiratory tract)	
Placenta, amniotic fluid, and fetus	
Sites in Which DNA from Microbiota Has Been Detected	
Brain	
Bloodstream	

Because the body provides such a range of habitats, it should not be surprising that the body supports a wide range of microbes. **Table 13.1** outlines our current understanding of where microbes reside on our bodies and where they do not. This information is in flux, of course, as we are in the midst of a revolution in understanding what microbes are in and on our bodies, due to the new microbiome findings. See also **Insight 13.1.**

The vast majority of microbes that come in contact with the body are removed or destroyed by the host's defenses long before they are able to colonize a particular area. Of those microbes able

INSIGHT 13.1 MICROBIOME: Breast Cancer and the Breast Microbiome

As you will see in this chapter, researchers have been very clear that our "surfaces"—whether external (skin) or internal (mucous membranes lining our various tracts)—are hosts to a diverse microbiota. It is only within the last decade that we have discovered microbes in what used to be called sterile tissues. The breast is one of these tissues.

Because of its nipple and ductal tissue, the breast can acquire microbes from the environment as well as endogenously (meaning, coming from within the body). Studies have shown that women with breast cancer have higher levels of particular bacteria in the breast microbiome than healthy women do. The cancer-associated bacteria include *Staphylococcus* and *Enterobacteriaceae* species, such as *E. coli*. These microbes are known to cause DNA damage to host cells, which can lead to tumor formation. So far these studies do not prove causation, only correlation. Some scientists ask, instead, whether the bacteria are found near these tumors because the tumors provide a favorable niche for their growth. Nevertheless, new research is ongoing to look at the effects of replacing "bad" bacteria with "good" bacteria as a means to reduce breast cancer risk.

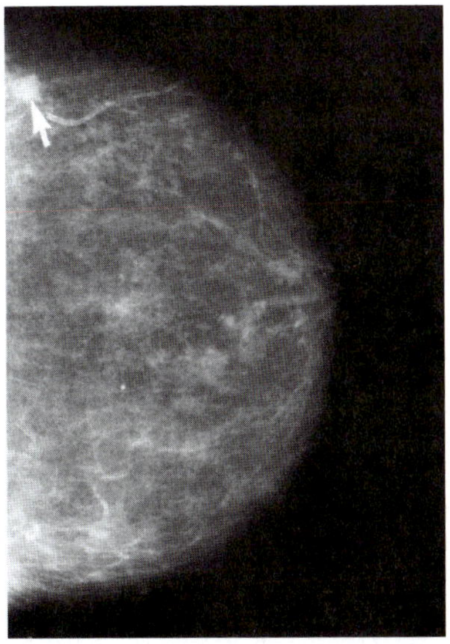

Mammogram showing cancerous tumor (arrow).
National Cancer Institute/U.S. Department of Health and Human Services (USHSS)

to establish an ongoing presence, an even smaller number are able to remain without attracting the unwanted attention of the body's defenses. This last group of organisms has evolved, along with its human hosts, to produce a complex relationship in which the effects of normal biota are generally not deleterious to the host. Recall from chapter 9 that microbes exist in different kinds of relationships with their hosts. More and more we are realizing that normal biota are in a mutualistic association with their hosts.

We now know that bacterial biota benefit the human host in many ways. The very development of our organs is influenced by the presence of resident biota. They also prevent the overgrowth of harmful microorganisms. A common example is the fermentation of glycogen by lactobacilli, which keep the pH in the vagina quite acidic and prevent the overgrowth of the yeast *Candida albicans.*

The defensive effect "good" microbes have against intruder microorganisms is called **microbial antagonism.** The microbiome exists in a steady, established relationship with the host, and its members are unlikely to be displaced by incoming microbes. This antagonism is also enabled by the chemical or physiological environment created by the resident biota, which are hostile to other microbes. In addition, members of the microbiome often can be pathogenic if they are allowed to multiply to larger numbers. Microbial antagonism is also responsible for keeping them in check.

It is also the case that hosts with compromised immune systems can very easily experience disease caused by their (previously normal) biota **(table 13.2).** This outcome is seen when AIDS patients become sick with pneumococcal pneumonia, in which the causative agent (*Streptococcus pneumoniae*) is often carried as normal biota in the nasopharynx. **Endogenous** infections (those caused by biota already present in the body) can also occur when normal biota is introduced to a site that was previously sterile.

When Does It Start?

The uterus and its contents used to be considered sterile during embryonic and fetal development. A growing number of doctors and scientists believe that fetuses are seeded with normal microbiota in utero, and that these microbes are important for healthy full-term pregnancies and the development of healthy immune systems in newborns. At any rate, we know that exposure occurs during the birth process itself, when the baby becomes colonized with the mother's vaginal biota **(figure 13.2).** (Babies born by cesarean section typically are colonized by adult skin biota.) Within 8 to 12 hours after delivery, the vaginally

Table 13.2 Factors That Weaken Host Defenses and Increase Susceptibility to Infection

- Age: the very young and the very old
- Genetic defects in immunity and acquired defects in immunity (i.e., AIDS)
- Pregnancy
- Surgery and organ transplants
- Underlying disease: cancer, liver malfunction, diabetes
- Chemotherapy/immunosuppressive drugs
- Physical and mental stress
- Other infections

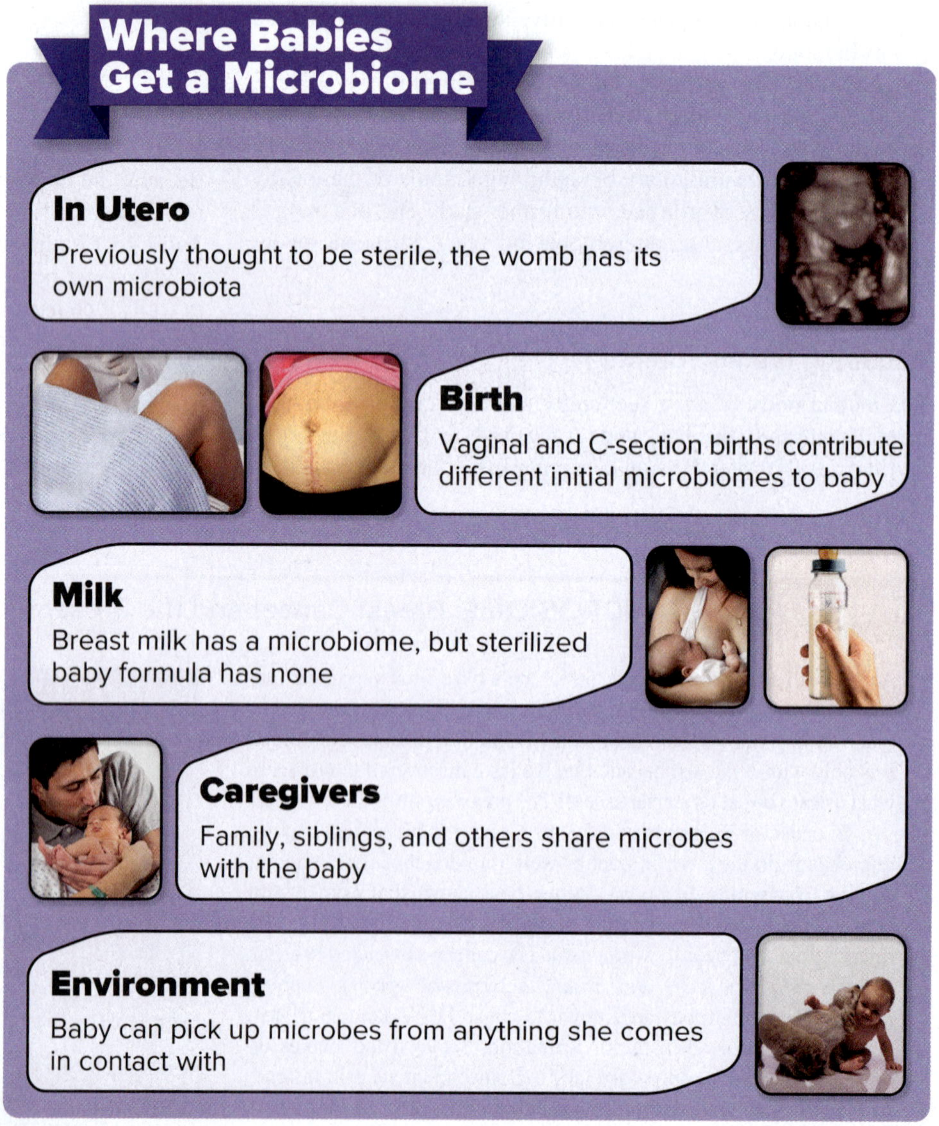
Where Babies Get a Microbiome
In Utero
Previously thought to be sterile, the womb has its own microbiota
Birth
Vaginal and C-section births contribute different initial microbiomes to baby
Milk
Breast milk has a microbiome, but sterilized baby formula has none
Caregivers
Family, siblings, and others share microbes with the baby
Environment
Baby can pick up microbes from anything she comes in contact with

Figure 13.2 The origins of microbiota in newborns.

(sonogram) Jim Connely; (legs) Adam Gault/SPL/Getty Images; (pregnant abdomen) alexmak72427/iStock/Getty Images; (infant feeding) Tetiana Mandziuk/Shutterstock; (bottle) Pixtal/SuperStock; (man kissing infant) Ariel Skelley/Marc Romanelli/Blend Images, LLC; (baby with puppies) Kwame Zikomo/Purestock/SuperStock

delivered newborn typically has been colonized by bacteria such as *Lactobacillus, Prevotella,* and *Sneathia,* acquired primarily from the birth canal. Data from the Human Microbiome Project revealed that the microbial composition of the vagina changes significantly in pregnant women. Early on, a *Lactobacillus* species that digests milk begins to populate the vagina. Immediately prior to delivery, additional bacterial species colonize the birth canal. Scientists suggest that the lactobacilli provide the newborn baby with the enzymes necessary to digest milk and that the later colonizers are better equipped to protect a newborn baby from skin disorders and other conditions. After the baby is born, the mother's vaginal microbiota returns to its former state.

The baby continues to acquire resident microbiota from the environment, notably from its diet. Throughout most of evolutionary history, of course, that has meant human breast milk. Scientists have found that human milk contains around 600 species of bacteria and a lot of sugars that babies cannot digest. The sugars *are* used by healthy gut bacteria, suggesting a role for breast milk in maintaining a healthy gut microbiome in the baby.

Indigenous Biota of Specific Regions

Microbiome research has shown that among healthy adults, the normal microbiota varies significantly. For instance, the microbiota on a person's right hand was found to be significantly different from that on the left hand. What seemed to be more important than the exact microbial profile of any given body site was the profile of proteins, especially the enzymatic capabilities. That profile remained stable across subjects, though the microbes that were supplying those enzymes could differ broadly. With that caveat, we present in **table 13.3** a summary of the types of normal, indigenous biota present in specific body sites. This table represents the results of the Human Microbiome Project with respect to bacteria, as well as information we have long had about the

Table 13.3 Life on Humans: Sites Containing Well-Established Biota and Representative Examples

Anatomical Sites	Common Genera	Remarks
Skin	**Gram-positive bacteria:** *Staphylococcus* (including *S. aureus*), *Propionibacterium, Streptococcus, Corynebacterium, Lactobacillus* **Gram-negative bacteria:** *Bacteroides, Prevotella, Haemophilus* **Fungi:** *Candida*	Skin biota varies with body location and with age; in different individuals, different genera predominate; approx. 4% of subjects carry *Staphylococcus aureus* on their skin.
Gastrointestinal Tract		
Oral cavity	**Gram-positive bacteria:** *Streptococcus* predominates; *Actinomyces, Corynebacterium* **Gram-negative bacteria:** *Haemophilus, Prevotella, Veillonella, Bacteroides, Moraxella* **Fungi:** *Candida* **Protozoa:** *Entamoeba*	More than a dozen species of *Streptococcus;* microbes colonize the epidermal layer of cheeks, gingiva, pharynx; surface of teeth; found in saliva in huge numbers.
Intestinal tract	**Gram-negative bacteria:** *Bacteroides, Prevotella* **Fewer gram-positives:** *Streptococcus, Lactobacillus* **Fungi:** *Candida*	Fecal biota consists predominantly of anaerobes; other microbes are aerotolerant or facultative. *E. coli* present in majority of subjects but in relatively low abundance.
Respiratory Tract		
Nose	**Gram-positive bacteria:** *Propionibacterium, Corynebacterium, Staphylococcus* **Gram-negative bacteria:** *Moraxella, Prevotella*	Approx. 30% of subjects carry *Staphylococcus aureus* in nose.
Throat	**Gram-positive bacteria:** *Streptococcus, Corynebacterium* **Gram-negative bacteria:** *Haemophilus, Prevotella, Veillonella, Moraxella*	Biota similar to oral cavity.
Lungs	**Gram-negative and gram-positive bacteria**	Previously thought to be sterile; asthmatic and COPD lungs colonized by different species than healthy lungs.
Vagina	**Gram-positive bacteria:** *Lactobacillus* predominates in child-bearing years; *Streptococcus* **Gram-negative bacteria:** *Prevotella* **Fungi:** *Candida*	Biota responds to hormonal changes during life, with significant changes in preparation for birth and with more variety of species after menopause.
Urinary Tract	**Gram-positive bacteria:** *Lactobacillus* (predominant) **Gram-negative bacteria:** *Prevotella, Escherichia, Enterococcus*	Recently, healthy bladder and urine were found to contain a normal microbiome.

Note: Information in this table subject to significant change as results of the Human Microbiome Project become available.

presence of fungi and other microbes in some sites. Scientists are in the process of cataloging other microorganisms via metagenomics and are just beginning to appreciate their numbers in the human microbiome. For example, we now know there are at least 100 types of fungi in the intestine and 100 million viruses per gram of feces.

13.1 Learning Outcomes—Assess Your Progress

1. Differentiate among the terms *colonization, infection,* and *disease.*
2. Identify the sites where normal biota is found in humans.
3. Discuss how the Human Microbiome Project has changed our understanding of normal biota.

13.2 When Colonization Leads to Disease

Traditionally, a microbe whose relationship with its host is parasitic and results in infection and disease has been termed a **pathogen.** There are quite a few microbes that are easy to identify as true pathogens. These are capable of causing disease in most healthy persons with normal immune systems. They are often associated with a specific, recognizable disease, which may vary in severity from mild (colds) to severe (malaria) to fatal (rabies). Examples of true pathogens include the influenza virus, plague bacterium, and malarial protozoan.

The Centers for Disease Control and Prevention categorizes these pathogens as a way of protecting people who work with them in research and clinical settings **(table 13.4).**

There are also quite a few microbes that are not thought of as true pathogens but are nonetheless known to cause disease in people with some deficit in their immunity. We characterize these people as **immunocompromised.** An example is *Pseudomonas* lung infections in people with cystic fibrosis. Also, some microbes are resident biota in one site in the body, but when transferred to another site in the body, they will cause disease there. For example, several *E. coli* strains are normal biota in the digestive tract. If they become displaced into the urinary system, they cause urinary tract infections. Sometimes these infections are called **opportunistic** because the pathogens are exploiting a new opportunity in the host.

Scientists and physicians are moving away from characterizing most microbes as a true pathogen or an opportunistic pathogen, although, as mentioned, some are very clearly one of these types. Most microbes can cause disease under the proper conditions but can coexist peaceably with their human hosts under other conditions. **Figure 13.3** illustrates this situation. There are factors on the host side to consider, as well as factors on the microbe side.

On the microbe side, the first consideration is its relative **virulence.** Although the terms *pathogenicity* and *virulence* are often used interchangeably, *virulence* is the accurate term for describing the *degree* of pathogenicity. The virulence of a microbe is indicated by its ability to (1) establish itself in the host and (2) cause damage.

A Note About Pathogens

So far, science has documented a total of 1,407 microbes that cause disease in humans. Of these, 538 are bacteria, 317 are fungi, 287 are helminths, 208 are viruses, and 57 are protozoa. Of course, we do not know how many pathogens we *do not* know about. There are plenty of conditions and diseases that have no known cause as of yet.

Table 13.4 Primary Biosafety Levels and Agents of Disease

Biosafety Level	Facilities and Practices	Risk of Infection and Class of Pathogens
1	Standard, open bench; no special facilities needed; typical of most microbiology teaching labs; access may be restricted.	Low infection hazard; microbes not generally considered pathogens and will not colonize the bodies of healthy persons; *Micrococcus luteus, Bacillus megaterium, Lactobacillus, Saccharomyces.*
2	At least level 1 facilities and practices; plus personnel must be trained in handling pathogens; lab coats and gloves required; safety cabinets may be needed; biohazard signs posted; access restricted.	Agents with moderate potential to infect; class 2 pathogens can cause disease in healthy people but can be contained with proper facilities; most pathogens belong to class 2; includes *Staphylococcus aureus, Escherichia coli, Salmonella* spp., *Corynebacterium diphtheriae;* pathogenic helminths; hepatitis A, B, and rabies viruses; *Cryptococcus* and *Blastomyces.*
3	Minimum of level 2 facilities and practices; plus all manipulation performed in safety cabinets; lab designed with special containment features; only personnel with special clothing can enter; no unsterilized materials can leave the lab; personnel warned, monitored, and vaccinated against infection dangers.	Agents can cause severe or lethal disease especially when inhaled; class 3 microbes include *Mycobacterium tuberculosis, Francisella tularensis, Yersinia pestis, Brucella* spp., *Coxiella burnetii, Coccidioides immitis,* and yellow fever, WEE, and HIV.
4	Minimum of level 3 facilities and practices; plus facilities must be isolated with very controlled access; clothing changes and showers required for all people entering and leaving; materials must be autoclaved or fumigated prior to entering and leaving lab.	Agents are highly virulent microbes that pose extreme risk for morbidity and mortality when inhaled in droplet or aerosol form; most are exotic flaviviruses; arenaviruses, including Lassa fever virus; or filoviruses, including Ebola and Marburg viruses.

Figure 13.3 Will disease result from an encounter between a (human) host and a microorganism? In most cases, all of the slider bars must be in the correct ranges and the microbe's on-off switch in the "on" position with the host's on-off switch in the "off" position in order for disease to occur. These are just a few examples and not the only options.
Dave and Les Jacobs/Kolostock/Blend Images

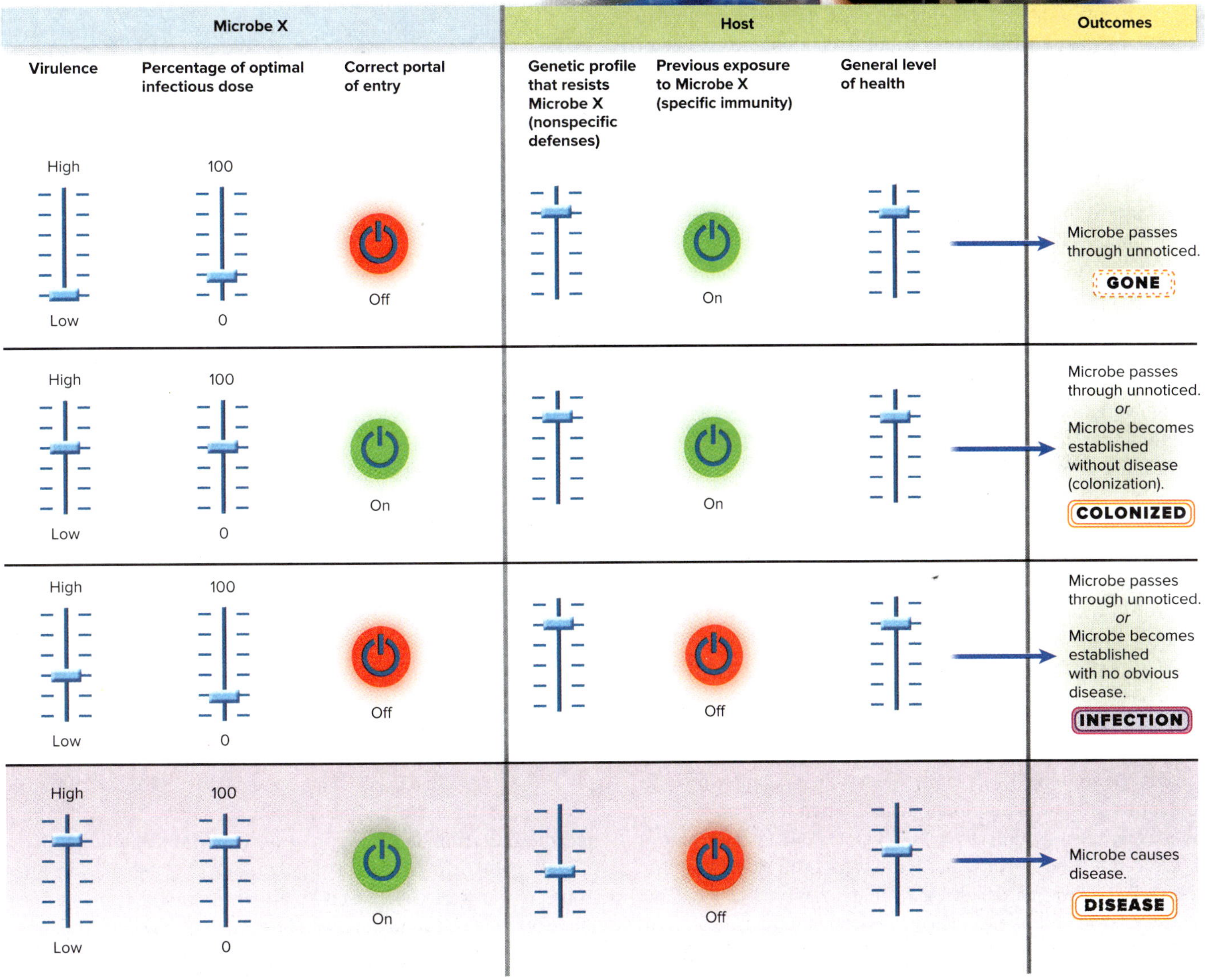
Microbe X
Host
Outcomes
Virulence
Percentage of optimal infectious dose
Correct portal of entry
Genetic profile that resists Microbe X (nonspecific defenses)
Previous exposure to Microbe X (specific immunity)
General level of health
High
Low
100
0
Off
On
Microbe passes through unnoticed.
GONE
Microbe passes through unnoticed.
or
Microbe becomes established without disease (colonization).
COLONIZED
Microbe passes through unnoticed.
or
Microbe becomes established with no obvious disease.
INFECTION
Microbe causes disease.
DISEASE

To establish themselves in a host, microbes must enter the host, negotiate the microbiome, attach firmly to host tissues, and survive the host defenses. To cause damage, microbes produce toxins or induce a host response that is actually injurious to the host. Any characteristic or structure of the microbe that contributes to the preceding activities is called a **virulence factor.** Virulence can be due to single or multiple factors. In some microbes, the causes of virulence are clearly established; in others, they are not. In the following section, we examine the effects of virulence factors while outlining the stages in the progress of an infection.

Infectious Dose and Portal of Entry

Another factor crucial to the course of an infection is the quantity of microbes in the initial dose, as you see in figure 13.3. For most agents, infection will proceed only if a minimum number, called the *infectious dose* (*ID*), is present. This number has been determined experimentally for many microbes. In general, microorganisms with smaller infectious doses have greater virulence. On the low end of the scale, the ID for *Coxiella*, the causative agent of Q fever, is a single cell. It is only about 10 cells in tuberculosis, giardiasis, and coccidioidomycosis. The ID is 1,000 bacteria for gonorrhea and 10,000 bacteria for typhoid fever, in contrast to 1,000,000,000 bacteria in cholera. Numbers below an infectious dose will generally not result in an infection. However, if the quantity is far in excess of the ID, the onset of disease can be extremely rapid.

Another factor that affects a microbe's ability to cause disease is whether it enters the host at a location that enables it to survive and thrive, its **portal of entry.** Certainly, other characteristics of microbes influence their disease-causing ability, but those pictured in figure 13.3 are three main ones.

Variability of Hosts

Now we turn to the host part of the equation. Figure 13.3 shows you three main host-related variables. (Again, there are more, many of which are still not understood.) The three variables pictured are the host's genetics, whether the host has seen the particular microbe before, and the host's overall health.

Note that different healthy individuals have widely varying responses to the same microorganism. This is determined in part by genetic variation in components of their defense systems. That is why the same infectious agent can cause severe disease in one individual and mild or no disease in another.

Why is there variation? In chapter 9, we described *coevolution* as changes in genetic composition by one species in response to changes in another. Infectious agents evolve in response to their interaction with a host (as in the case of antibiotic resistance). Hosts evolve, too. Although their pace of change is much slower than that of a microbe, changes eventually show up in human populations due to their past experiences with pathogens. One striking example is sickle-cell disease. Persons who are carriers of a mutation in their hemoglobin gene (who inherited one mutated hemoglobin gene and one normal) have few or no sickle-cell disease symptoms but are more resistant to malaria than people who have no mutations in their hemoglobin genes. When a person inherits two alleles for the mutation (from both parents), that person enjoys some protection from malaria but will suffer from sickle-cell disease. People of West African descent are much more likely to have one or two sickle-cell alleles. Malaria is endemic in West Africa. It seems the hemoglobin mutation is an adaptation of the human host to its long-standing relationship with the malaria protozoan.

In another example, researchers have found a gene that correlates with how people react to infection with the swine flu virus. The gene codes for a protein that blocks viral entry into cells. People who experienced only mild flu symptoms during the 2009 swine flu epidemic were found to have the intact gene, whereas those who became most gravely ill or who died were much more likely to have a mutated variant of this gene. During the COVID-19 pandemic, people have been identified who seem to be naturally resistant to infection. Researchers in Brazil are studying dozens of "discordant couples"—couples who live together or are married in which one of them had become infected and very ill with SARS-CoV-2, and the other remained healthy (in the absence of vaccination). They think they are narrowing down the gene or genes that differ between the two, which could lead to new prevention strategies.

In chapter 6, you learned about single nucleotide polymorphisms, variations between individuals at particular genetic locations. These SNPs can also influence your susceptibility to particular microbes.

These are examples of the variability represented by the slider bars in the column marked "genetic profile" in figure 13.3. The last human factor in figure 13.3 is general level of health of the host. If you are battling one disease or infection, your ability to respond robustly to another might be diminished. Alternatively, if you are battling Infection A, your immune system will be in an activated state, and in some instances, that can be helpful in battling Infection B. Psychological stress has also been found to negatively impact your ability to respond to infection. Again, there is still much we do not know about the interactions of the host and the infecting microbe. But figure 13.3 provides a few factors known to have an influence.

Polymicrobial Infections

From the very earliest days of infectious disease studies, in the 1800s, scientists have isolated single microorganisms to study their effects on animals or humans. Later in this chapter, you will learn about Koch's postulates, a set of rules for determining the cause of an unknown infectious condition. The postulates depend on obtaining microbes in pure culture. This procedure is extremely valuable because, as you know, most scientific experimentation requires isolating the variables and holding all but one constant.

This aspect of scientific rigor has probably kept us from understanding the roles that interacting microbes play in causing disease. Many scientists now believe that perhaps the majority of infections are **polymicrobial,** with contributions from more than one microbe. One classic set of infections is influenza (caused by a virus) and pneumonia (often caused by a bacterium). Influenza infection frequently leads to pneumonia. One of the most serious causes of pneumonia is a bacterium that is normal biota in the nose of many people. Scientists at the University at Buffalo found that influenza caused a cascade of host responses, which in turn caused the nose bacteria, which are usually resident in a biofilm there, to be set loose and travel to other sites such as the lungs and the bloodstream. That situation would definitely qualify as polymicrobial. In another example, several types of skin infections are known to be caused by either *Staphylococcus* or *Streptococcus* species. In fact, researchers have found that when these two are cultivated together with another common skin resident, *Moraxella,* both *Staphylococcus* and *Streptococcus* increase their transcription of virulence factors. Perhaps upon diagnosis, one is isolated from the skin, but it seems possible that the three of them together lead to the disease symptoms. In the next decade, many more polymicrobial causes will no doubt be discovered, which can help us look for new prevention and treatment strategies.

In the next section, we will examine the steps a microbe goes through that lead to disease.

Becoming Established: Step One—Portals of Entry

Figure 13.4 provides a schematic view of the steps a microbe takes in order to cause disease. The text sections that follow are parallel to the steps in the figure.

To initiate an infection, a microbe enters the tissues of the body by a characteristic route, the **portal of entry,** usually through the skin or a mucous membrane. The source of the infectious agent can be **exogenous,** originating from a source outside the body (the environment or another person or animal), or endogenous, already existing on or in the body (normal biota or a previously silent infection).

The majority of pathogens have adapted to a specific portal of entry, one that provides a habitat for further growth and spread. This adaptation can be so restrictive that if certain pathogens enter the "wrong" portal, they will not be successful. For instance, inoculation of the nasal mucosa with the influenza virus invariably gives rise to the flu, but if this virus contacts only the skin, no infection will result. Likewise, contact with athlete's foot fungi in small cracks in the toe webs can induce an infection, but inhaling the fungus spores will not infect a healthy individual. Occasionally, an infective agent can enter by more than one portal. For instance, *Mycobacterium tuberculosis* enters through both the respiratory and gastrointestinal tracts, and *Streptococcus* and *Staphylococcus* have adapted to invasion through several portals of entry such as the skin, urogenital tract, and respiratory tract.

Infectious Agents That Enter the Skin

The skin is a very common portal of entry. The actual sites of entry are usually nicks, abrasions, and punctures (many of which are tiny and inapparent) rather than unbroken skin. Intact skin is a very tough barrier that few microbes can penetrate. *Staphylococcus aureus* (the cause of boils), *Streptococcus pyogenes* (an agent of impetigo), the fungal dermatophytes, and agents of gangrene and tetanus gain access through damaged skin. The viral agent of cold sores (herpes simplex, usually type I) enters through the mucous membranes near the lips.

Some infectious agents create their own passageways into the skin using digestive enzymes. For example, certain helminth worms burrow through the skin directly to gain access to the tissues. Other infectious agents enter through bites. The bites of insects, ticks, and other animals can transmit a variety of viruses, rickettsias, and protozoa. An artificial means for breaching the skin barrier is the use of contaminated hypodermic needles by intravenous drug abusers. Users who inject drugs are predisposed to a disturbing list of well-known diseases: hepatitis, AIDS, tetanus, tuberculosis, osteomyelitis, and malaria. Contaminated needles often contain bacteria from the skin or environment that induce heart disease (endocarditis), lung or spine abscesses, and chronic infections at the injection site.

Although the conjunctiva, the outer protective covering of the eye, is ordinarily a relatively good barrier to infection, bacteria such as *Haemophilus aegyptius* (pinkeye), *Chlamydia trachomatis* (trachoma), and *Neisseria gonorrhoeae* have a special affinity for this tissue.

The Gastrointestinal Tract as Portal

The gastrointestinal tract is the portal of entry for pathogens contained in food, drink, and other ingested substances. These microbes are adapted to survive digestive enzymes and abrupt pH changes. The best-known enteric agents of disease are gram-negative rods in the genera *Salmonella, Shigella,* and *Vibrio,* and certain strains of *Escherichia coli.* Viruses that enter through the gut are poliovirus, hepatitis A virus, echovirus, and rotavirus. Important enteric protozoans are *Entamoeba histolytica* (amoebiasis) and *Giardia lamblia* (giardiasis).

The Respiratory Portal of Entry

The oral and nasal cavities are the gateways to the respiratory tract, the portal of entry for the greatest number of pathogens. Because there is a continuous mucous membrane surface covering the upper respiratory tract, the sinuses, and the auditory tubes, microbes can move from one site to another. The extent to which an agent is carried into the respiratory tree is based primarily on its size. In general, small cells and particles are inhaled more deeply than larger ones. Infectious agents with this portal of entry include the bacteria of streptococcal sore throat, meningitis, diphtheria, and whooping cough and the viruses of COVID-19, influenza, measles, mumps, rubella, chickenpox, and the common cold. Pathogens that are inhaled into the lower regions of the respiratory tract (bronchioles and lungs) can cause **pneumonia,** an inflammatory condition of the lung. Bacteria (*Streptococcus pneumoniae, Klebsiella, Mycoplasma*) and fungi (*Cryptococcus* and *Pneumocystis*) are a few of the agents involved in pneumonias. Other agents causing unique lung diseases are *Mycobacterium tuberculosis* and fungal pathogens such as *Histoplasma.*

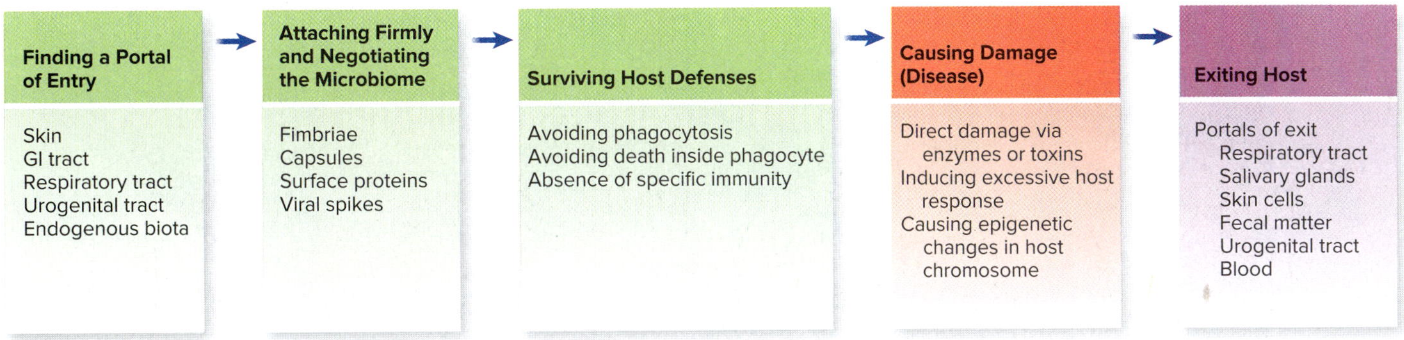
Finding a Portal of Entry
Skin
GI tract
Respiratory tract
Urogenital tract
Endogenous biota
Attaching Firmly and Negotiating the Microbiome
Fimbriae
Capsules
Surface proteins
Viral spikes
Surviving Host Defenses
Avoiding phagocytosis
Avoiding death inside phagocyte
Absence of specific immunity
Causing Damage (Disease)
Direct damage via enzymes or toxins
Inducing excessive host response
Causing epigenetic changes in host chromosome
Exiting Host
Portals of exit
Respiratory tract
Salivary glands
Skin cells
Fecal matter
Urogenital tract
Blood

Figure 13.4 The steps involved when a microbe causes disease in a host.

Urogenital Portals of Entry

The urogenital tract is the portal of entry for many pathogens that are contracted by sexual means (intercourse or intimate direct contact). **Sexually transmitted infections (STIs)** account for an estimated 4% of infections worldwide, with approximately 13 million new cases occurring in the United States each year. (These diseases are commonly called STDs, for "sexually transmitted diseases." Public health officials believe it is more accurate to call them STIs to highlight the fact that so many of the infections are silent, not causing obvious disease, even when they are causing damage to the host.)

The microbes causing STIs enter the skin or mucosa of the penis, external genitalia, vagina, cervix, and urethra. Some can penetrate an unbroken surface; others require a cut or abrasion. Some of the most common STIs are syphilis, gonorrhea, genital warts, chlamydia, and herpes. Other common sexually transmitted agents are HIV, *Trichomonas* (a protozoan), *Candida albicans* (a yeast), and hepatitis B virus. STIs are described in detail in chapter 24, with the exception of HIV (chapter 21) and hepatitis B (chapter 23).

Disease Connection

Despite its name, *Trichomonas vaginalis* infects both men and women. Previously thought to be a relatively mild infection, it is now known to enhance your risk of contracting other STIs, including HIV. A strong association has been found between *Trichomonas* infection and aggressive forms of prostate cancer. *Trichomonas* is discussed in chapter 24.

Not all urogenital infections are STIs. Some of these infections are caused by displaced organisms (as when normal biota from the gastrointestinal tract cause urinary tract infections) or by opportunistic overgrowth of normal biota ("yeast infections").

Pathogens That Infect During Pregnancy and Birth

The placenta is an exchange organ—formed by maternal and fetal tissues—that separates the blood of the developing fetus from that of the mother yet permits diffusion of dissolved nutrients and gases to the fetus. Some pathogens such as the syphilis spirochete can cross the placenta, enter the umbilical vein, and spread by the fetal circulation into the fetal tissues **(figure 13.5).**

Other infections, such as herpes simplex, can occur perinatally when the child is contaminated by the birth canal. The common infections of fetus and neonate are grouped together in a unified cluster known by the acronym **TORCH,** which medical personnel must monitor. *TORCH* stands for **t**oxoplasmosis, **o**ther diseases (syphilis, coxsackievirus, varicella-zoster virus, HIV, and chlamydia), **r**ubella, **c**ytomegalovirus, and **h**erpes simplex virus. The most serious complications of TORCH infections are spontaneous abortion, congenital abnormalities, brain damage, prematurity, and stillbirths.

Becoming Established: Step Two—Attaching to the Host and Interacting with the Microbiome

Adhesion is a process by which microbes gain a more stable foothold on host tissues. Because adhesion is dependent on binding between specific molecules on both the host and pathogen, a particular pathogen is limited to only those cells (and organisms) to which it can bind. Once attached, the pathogen is able to invade the body compartments. Bacterial, fungal, and protozoal pathogens attach most often using fimbriae (pili), surface proteins, and adhesive slimes or capsules. Viruses attach by means of specialized spikes, or glycoproteins, on their surfaces **(figure 13.6).** In addition, parasitic helminths are mechanically fastened to the portal of entry by suckers, hooks, and barbs. Adhesion methods of various microbes and the diseases they lead to are shown in **table 13.5.** Firm attachment to host tissues is almost always a prerequisite for

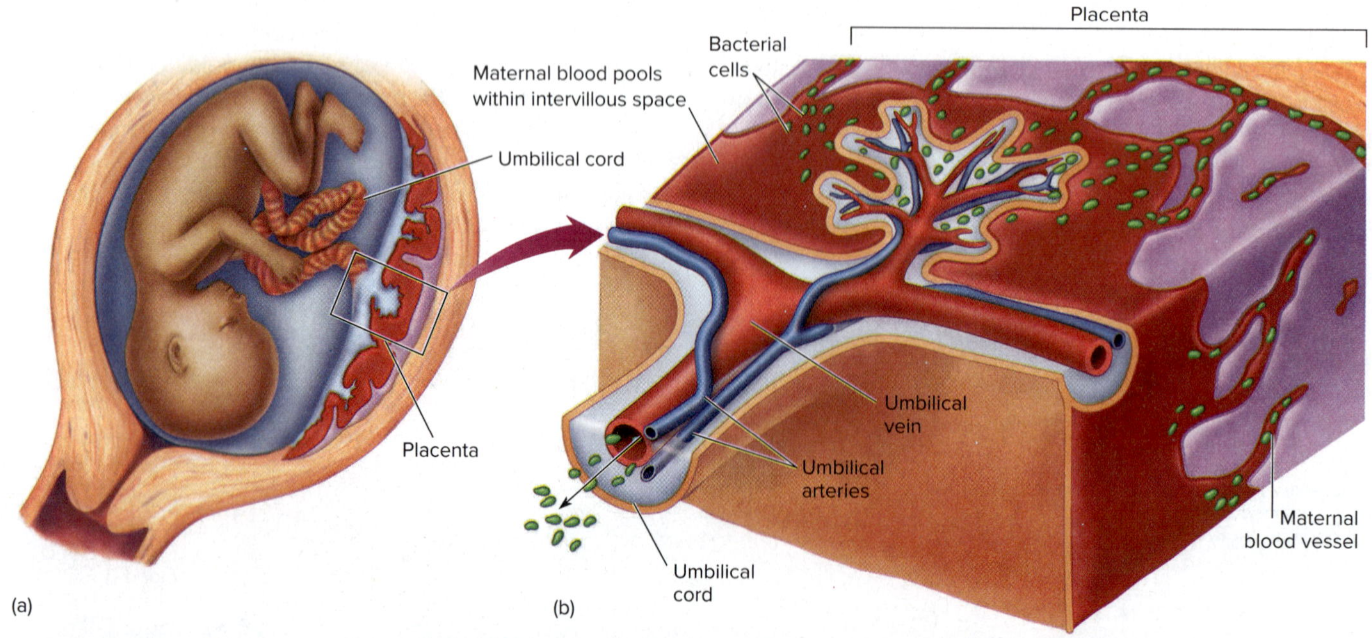

Maternal blood pools within intervillous space
Umbilical cord
Placenta
(a)
Placenta
Bacterial cells
Umbilical vein
Umbilical arteries
Umbilical cord
Maternal blood vessel
(b)

Figure 13.5 Transplacental infection of the fetus. **(a)** Fetus in the womb. **(b)** In a closer view, microbes are shown penetrating the maternal blood vessels and entering the blood pool of the placenta. They then invade the fetal circulation by way of the umbilical vein.

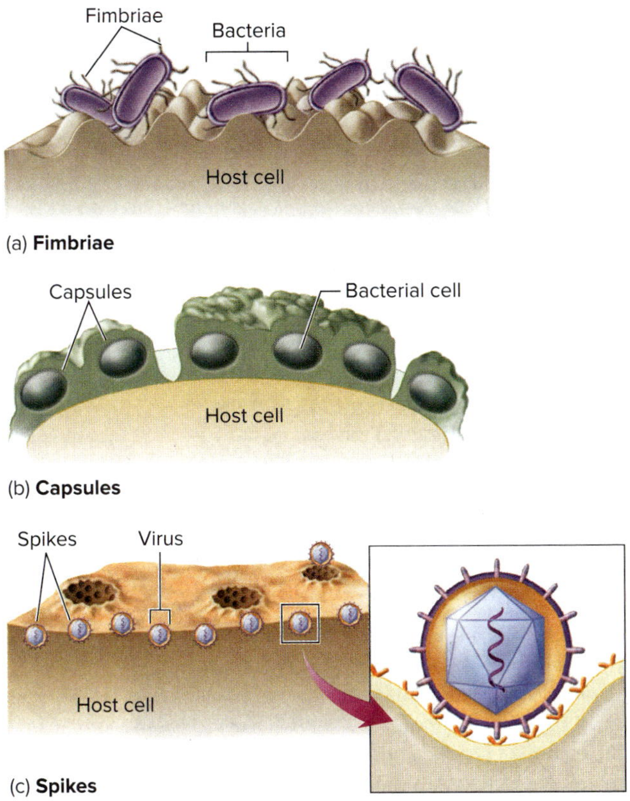
Fimbriae
Bacteria
Host cell
(a) Fimbriae
Capsules
Bacterial cell
Host cell
(b) Capsules
Spikes
Virus
Host cell
(c) Spikes

Figure 13.6 Mechanisms of adhesion by pathogens.
(a) Fimbriae—minute, bristlelike appendages. **(b)** Adherent extracellular capsules made of slime or other sticky substances. **(c)** Viral envelope spikes.

Table 13.5 Adhesive Properties of Microbes

Microbe	Disease	Adhesion Mechanism
Neisseria gonorrhoeae	Gonorrhea	Fimbriae attach to genital epithelial cells.
Escherichia coli	Diarrhea, UTIs	Fimbrial adhesion
Shigella	Dysentery	Fimbriae attach to intestinal epithelium.
Mycoplasma	Pneumonia	Specialized tip at ends of bacteria fuses tightly to lung epithelium.
Pseudomonas aeruginosa	Burn, lung infections	Fimbriae and slime layer
Streptococcus pyogenes	Pharyngitis, impetigo	Lipoteichoic acid and capsule anchor cocci to epithelium.
Streptococcus mutans, S. sobrinus	Dental caries	Dextran slime layer glues cocci to tooth surface after initial attachment.
Influenza virus	Influenza	Viral spikes attach to receptor on cell surface.
SARS-CoV-2	COVID-19	Viral spike attaches to receptors on susceptible cells.
HIV	AIDS	Viral spikes adhere to white blood cell receptors.
Giardia lamblia (protozoan)	Giardiasis	Small suction disc on underside attaches to intestinal surface.

causing disease because the body has so many mechanisms for flushing microbes and foreign materials from its tissues.

Attachment also provides proximity to other bacteria of the same and other species. If an invading bacterium can attach, it can communicate with other bacteria through quorum sensing. Research has shown that communication among microbes is critical to the establishment of infection. If quorum-sensing chemicals are blocked, the bacteria are not able to sense the presence of enough cells to mount an effective attack against the host. This forces then to sit "silently," waiting for enough members to arrive at the site of colonization, making them vulnerable to the host immune system in the meantime. Development of drugs that block this critical communication process may be on the horizon for the treatment of microbial infections.

Of course, when a new microbe enters a body site, it will also encounter the resident microbiota. The microbiome is generally very stable and "settled in" in body sites, and its presence can prevent newcomers from attaching or becoming established (microbial antagonism). In rare instances, the microbiome can actually assist disease-causing microbes. We know one thing for sure: The microbiome is part of the host environment that incoming microbes must deal with.

Becoming Established: Step Three—Surviving Host Defenses

Microbes that are not established in a normal biota relationship in a particular body site in a host are likely to encounter resistance from host defenses when first entering, especially from certain white blood cells called **phagocytes.** These cells ordinarily engulf and destroy pathogens by means of enzymes and antimicrobial chemicals.

Antiphagocytic factors are a type of virulence factor used by some pathogens to avoid phagocytes. The most aggressive strategy involves bacteria that kill phagocytes outright. Species of both *Streptococcus* and *Staphylococcus* produce **leukocidins,** substances that are toxic to white blood cells, including phagocytes. Other microorganisms secrete an extracellular surface layer (slime or capsule) that makes it physically difficult for the phagocyte to engulf them. *Streptococcus pneumoniae, Salmonella typhi, Neisseria meningitidis,* and *Cryptococcus neoformans* are notable examples. Some bacteria are adapted to survive inside phagocytes after ingestion. For instance, pathogenic species of *Legionella, Mycobacterium,* and many rickettsias are readily engulfed but are capable of avoiding further destruction. The ability to survive intracellularly in phagocytes has special significance because it provides a place for the microbes to hide, grow, and be spread throughout the body.

Pathogens also use a wide variety of **epigenetic** mechanisms to hijack host machinery or to shut it down. Epigenetic changes are changes to the DNA that affect the way it is transcribed, but they are not actual mutations in the sequence. In this case, host cell DNA is modified by adding methyl groups to it or by altering the histones around which it is wound, affecting the DNA's access to transcription enzymes. Pathogens can use these epigenetic changes as part of their crippling strategies. They can diminish the response from defensive cells, for example. The epigenetic changes can be passed down to later generations of host cells

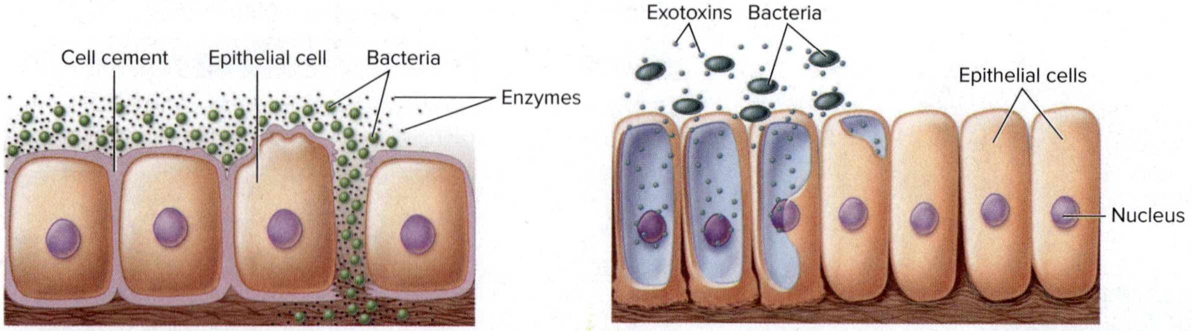
Cell cement
Epithelial cell
Bacteria
Enzymes
Exotoxins
Bacteria
Epithelial cells
Nucleus

(a) **Microbes Secrete Enzymes and Toxins**

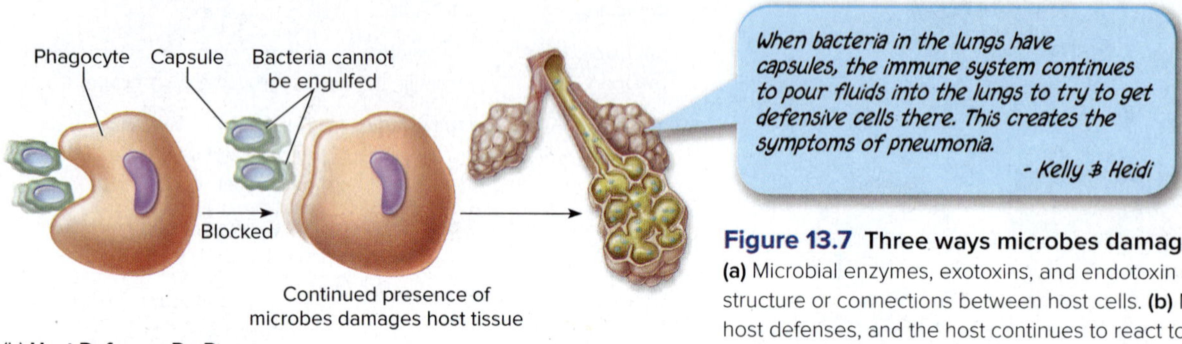
Phagocyte
Capsule
Bacteria cannot be engulfed
Blocked
Continued presence of microbes damages host tissue
When bacteria in the lungs have capsules, the immune system continues to pour fluids into the lungs to try to get defensive cells there. This creates the symptoms of pneumonia.
- Kelly $ Heidi

(b) **Host Defenses Do Damage**

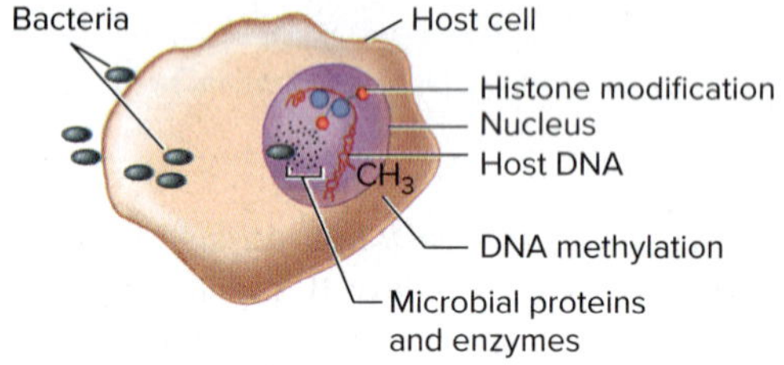
Bacteria
Host cell
Histone modification
Nucleus
Host DNA
CH3
DNA methylation
Microbial proteins and enzymes

(c) **Microbes Induce Epigenetic Changes on Host DNA**

Figure 13.7 Three ways microbes damage the host. **(a)** Microbial enzymes, exotoxins, and endotoxin disrupt host cell structure or connections between host cells. **(b)** Microbes evade initial host defenses, and the host continues to react to the presence of the microbe, causing (host) damage with its response. **(c)** Microbial products make epigenetic changes to the DNA and/or supporting structures, such as histones, altering the host genes that are expressed.

through mitosis. This may be why some infectious agents seem to have an effect on the host long after they are no longer present.

Step Four—Causing Disease

There are three main ways that microbes cause damage to hosts **(Figure 13.7):** (1) by secreting proteins (enzymes or toxins) that directly damage host cells, (2) by causing an overreaction by the body's defenses and those defenses cause host damage, and (3) by altering the host cell genome or transcription processes through epigenetic changes that temporarily or permanently disrupt normal host cell function.

A microbe's arsenal of virulence factors can be minimal or extensive. Cold viruses, for example, invade and multiply but cause relatively little damage to their host. At the other end of the spectrum, pathogens such as *Clostridium tetani* or HIV can severely damage or kill their host.

1. Direct Damage via Enzymes and Toxins

Many pathogenic bacteria, fungi, protozoa, and worms secrete **exoenzymes** that break down and inflict damage on tissues. Other enzymes dissolve the host's defense barriers and promote the spread of microbes to deeper tissues.

Examples of enzymes are:

1. mucinase, which digests the protective coating on mucous membranes and is a factor in amoebic dysentery;
2. keratinase, which digests the principal component of skin and hair and is secreted by fungi that cause ringworm; and
3. hyaluronidase, which digests hyaluronic acid, the substance that cements animal cells together. This enzyme is an important virulence factor in staphylococci, clostridia, streptococci, and pneumococci.

Some enzymes react with components of the blood. Coagulase, an enzyme produced by pathogenic staphylococci, causes clotting of blood or plasma. By contrast, the bacterial kinases (streptokinase, staphylokinase) do just the opposite, dissolving fibrin clots and expediting the invasion of damaged tissues. In fact, one form of streptokinase (Streptase) is marketed as a therapy to dissolve blood clots in patients with problems with thrombi and emboli.

A **toxin** is a specific chemical product of microbes, plants, and some animals that is poisonous to other organisms. **Toxigenicity,** the power to produce toxins, is a genetically controlled characteristic of many species and is responsible for the adverse effects of a variety of diseases generally called **toxinoses.** Toxinoses in which the toxin is spread by the blood from the site of infection are called **toxemias** (tetanus and diphtheria, for example), whereas those caused by ingestion of toxins are **intoxications** (botulism). A toxin is named according to its specific target of action: Neurotoxins act on the nervous system; enterotoxins act on the intestine; hemotoxins lyse red blood cells; and nephrotoxins damage the kidneys.

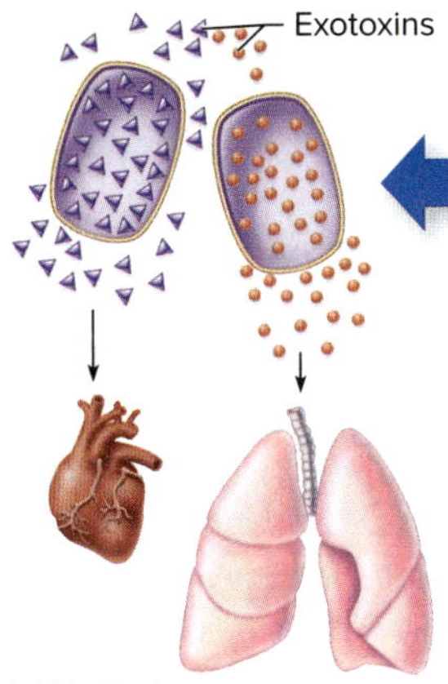
Exotoxins
(a) Exotoxins

Exotoxins	Characteristic	Endotoxin
Toxic in minute amounts	**Toxicity**	Toxic in high doses
Specific to a cell type (blood, liver, nerve); induces TNF production, resulting in fever	**Effects on the body**	Systemic: fever, inflammation
Small proteins	**Chemical composition**	Lipopolysaccharide of cell wall
Unstable	**Heat denaturation at 60°C**	Stable
Can be converted to toxoid*	**Toxoid formation**	Cannot be converted to toxoid
Stimulate antitoxins**	**Immune response**	Does not stimulate antitoxins
Usually not	**Fever stimulation**	Yes
Secreted from live cell	**Manner of release**	Released by cell via shedding or during lysis
A few gram-positive and gram-negative	**Typical sources**	All gram-negative bacteria

*A toxoid is an inactivated toxin used in vaccines.
**An antitoxin is an antibody that reacts specifically with a toxin.

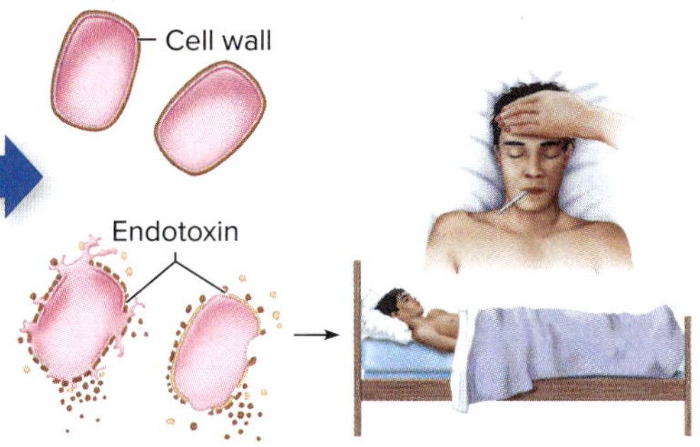
Cell wall
Endotoxin
(b) Endotoxin

Figure 13.8 The origins and effects of circulating exotoxins and endotoxin. **(a)** Exotoxins, given off by live cells, have highly specific targets and physiological effects. **(b)** Endotoxin, given off when the cell wall of gram-negative bacteria disintegrates, has more generalized physiological effects.

Another useful scheme classifies toxins according to their origins **(figure 13.8).** A toxin molecule secreted by a living bacterial cell into the infected tissues is an **exotoxin.** There are many different types of exotoxins. A toxin that is not actively secreted but is shed from the outer membrane is an **endotoxin.** There is only one form of endotoxin, which is found on all gram-negative bacteria. Other important differences between the two groups are summarized in figure 13.8.

Exotoxins are proteins with a strong specificity for a target cell and extremely powerful, sometimes deadly, effects. They generally affect cells by damaging the cell membrane and initiating lysis or by disrupting intracellular function. **Hemolysins** (hee-mahl′-uh-sinz) are a class of bacterial exotoxin that disrupts the cell membrane of red blood cells (and some other cells). This damage causes the red blood cells to **hemolyze**—to burst and release hemoglobin pigment. Hemolysins that increase pathogenicity include the streptolysins of *Streptococcus pyogenes* and the alpha (α) and beta (β) toxins of *Staphylococcus aureus* **(figure 13.9).** When colonies of bacteria growing on blood agar produce hemolysin, distinct zones appear around the colony. The type of hemolysis is often used to identify bacteria and determine their degree of pathogenicity.

The exotoxins of diphtheria, tetanus, and botulism, among others, attach to a particular target cell, become internalized, and interrupt an essential cell pathway. The consequences of cell disruption depend upon the target. One toxin of *Clostridium tetani* blocks the action of certain spinal neurons; the toxin of *Clostridium botulinum* prevents the transmission of nerve-muscle stimuli; pertussis toxin inactivates the respiratory cilia; and cholera toxin provokes profuse salt and water loss from intestinal cells. More details of the pathology of exotoxins are found in later chapters on specific diseases.

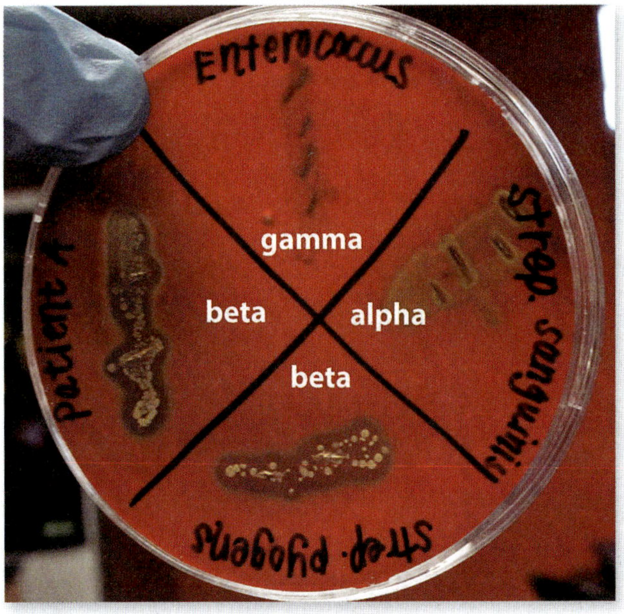
Enterococcus
gamma
beta
alpha
beta
Patient A
Strep. sanguinis
Strep. pyogenes

Figure 13.9 Different types of hemolysis by different bacteria on blood agar. Four different bacteria are streaked in four sections on the blood agar. Beta hemolysis results in complete breakdown of the red blood cells in the agar, leaving a clear halo around the colonies where the hemolysins have diffused out of the bacteria. Alpha hemolysis happens when the bacterium's hemolysins only incompletely break down the red blood cells, leaving a green tinge to the area of diffusion. *Gamma hemolysis* refers to no hemolysis (thus no hemolysins) at all.
Lisa Burgess/McGraw Hill

In contrast to the category of *exotoxins,* which contains many specific examples, *endotoxin* refers to a single substance. Endotoxin is actually a chemical called lipopolysaccharide (LPS), which is part of the outer membrane of gram-negative cell walls. Gram-negative bacteria shed these LPS molecules into tissues or into the circulation. Endotoxin differs from exotoxins because it has a variety of systemic effects on tissues and organs. Depending upon the amounts present, endotoxin can cause fever, inflammation, hemorrhage, and diarrhea. Blood infection by gram-negative bacteria such as *Salmonella, Shigella, Neisseria meningitidis,* and *Escherichia coli* are particularly dangerous because it can lead to fatal endotoxic shock.

Fungi often have toxins as well that can cause severe effects in humans. Often the toxins contaminate the environment and food, and harm humans in this way.

2. Inducing an Excessive Host Response

So far we have focused on direct virulence factors, such as enzymes and toxins. But it is probably the case that just as many microbial diseases are the result of indirect damage, or the host's excessive or inappropriate response to a microorganism. This reinforces the fact that pathogenicity is not a trait solely determined by microorganisms but is really a consequence of the interplay between microbe and host.

Long-haul COVID-19 symptoms are known to persist for months in a certain percentage of people who survived COVID -19 infection. They include fatigue, muscle pains, cardiomyopathy, and other manifestations. These continue after patients test negative for COVID-19, spurring scientists to speculate that the symptoms are caused by an overstimulated immune response.

3. Epigenetic Changes in Host Cells

This mechanism that microbes use to damage host cells is the most recently discovered. Earlier in this chapter, we discussed epigenetic methods for avoiding host cell defenses. Microbes have also been shown to shut down or activate regions of DNA in the host cell via epigenetic processes. The mechanisms include binding to host cell histones, binding to the small RNAs used for the silencing of genes, and binding to chromatin itself. These changes can harm the host cell, or change its function in some way that favors persistence of the microbe in or on it. Some pathogens have been found to secrete proteins that interact with regions of host cell DNA that are responsible for cytoskeleton structure, thereby causing disorganization of the cell. Sometimes these changes are passed on to new host cells, causing persistent symptoms. Some researchers speculate that this is one source of unexplained illnesses or symptoms where no causative microbes are found.

Patterns of Infection

Within the human body, infections show up in different patterns. In the simplest situation, a **localized infection,** the microbe enters the body and remains confined to a specific tissue **(figure 13.10*a*).** Examples of localized infections are boils, fungal skin infections, and warts.

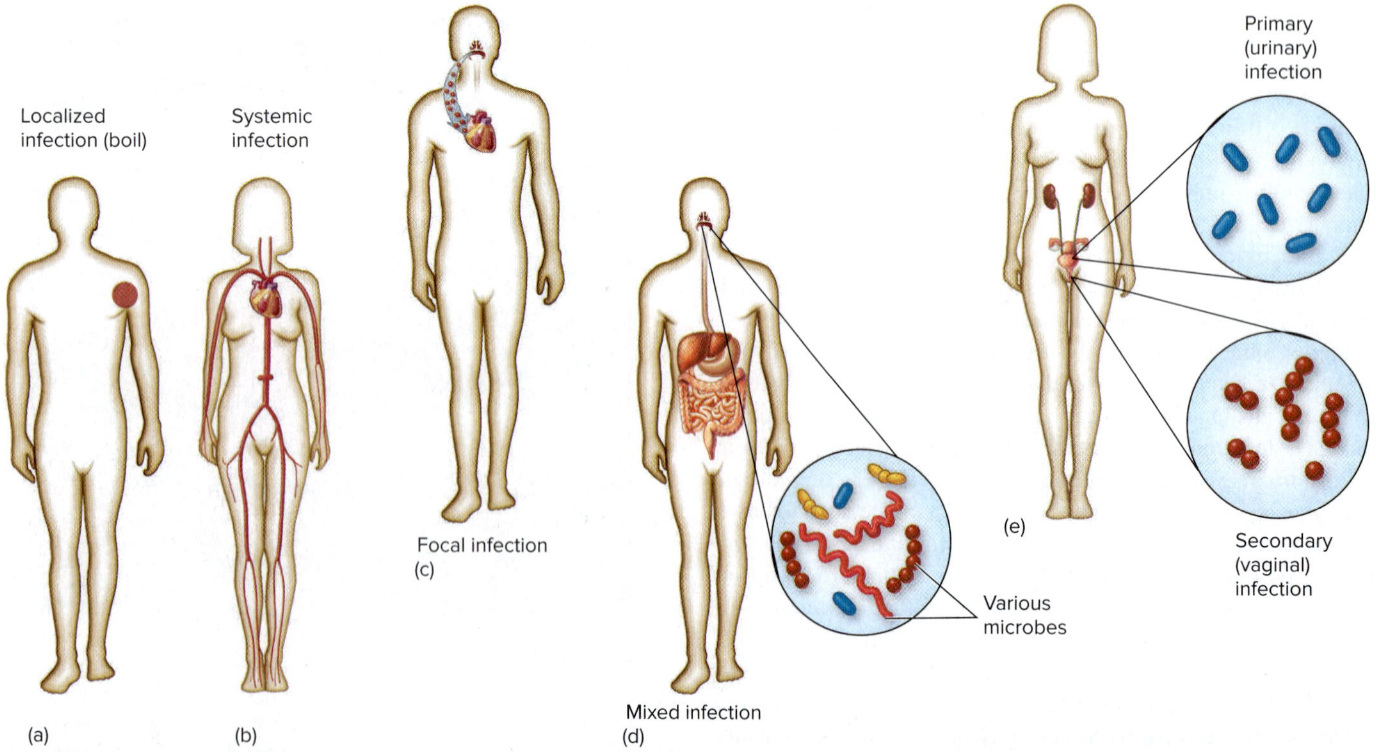
Localized infection (boil)
Systemic infection
Focal infection
(c)
Mixed infection
(d)
Various microbes
Primary (urinary) infection
Secondary (vaginal) infection
(e)
(a)
(b)

Figure 13.10 The occurrence of infections with regard to location, type of microbe, and order of infection. **(a)** A localized infection, in which the pathogen is restricted to one specific site. **(b)** Systemic infection, in which the pathogen spreads through circulation to many sites. **(c)** A focal infection occurs initially as a local infection, but circumstances cause the microbe to be carried to other sites systemically. **(d)** A mixed, or polymicrobial, infection, in which the same site is infected with several microbes at the same time. **(e)** In a primary-secondary infection, an initial infection is complicated by a second one in the same or a different location and caused by a different microbe.

Many infectious agents do not remain localized but spread from the initial site of entry to other tissues. In fact, spreading is necessary for pathogens such as rabies and hepatitis A virus, whose target tissue is some distance from the site of entry. The rabies virus travels from a bite wound along nerve tracts to its target in the brain, and the hepatitis A virus moves from the intestine to the liver via the circulatory system. When an infection spreads to several sites and tissue fluids, usually in the bloodstream, it is called a **systemic infection (figure 13.10*b*).** Examples of systemic infections are viral diseases (measles, rubella, chickenpox, and AIDS); bacterial diseases (brucellosis, anthrax, typhoid fever, and syphilis); and fungal diseases (histoplasmosis and cryptococcosis). Infectious agents can also travel to their targets by means of nerves (as in rabies) or cerebrospinal fluid (as in meningitis).

A **focal infection** is said to exist when the infectious agent breaks loose from a local infection and is carried into other tissues **(figure 13.10*c*).** This pattern is exhibited by tuberculosis or by streptococcal pharyngitis, which gives rise to scarlet fever. In the condition called toxemia,[1] the infection itself remains localized at the portal of entry, but the toxins produced by the pathogens are carried by the blood to the actual target tissue. In this way, the target of the bacterial cells can be different from the target of their toxin.

As mentioned earlier, polymicrobial diseases are more common than we think. This is called a **mixed infection (figure 13.10*d*).** Gas gangrene, wound infections, dental caries, and human bite infections tend to be mixed.

Some diseases are described according to a sequence of infection. When an initial, or **primary, infection** is complicated by another infection caused by a different microbe, the second infection is termed a **secondary infection (figure 13.10*e*).** This pattern often occurs in a child with chickenpox (primary infection) who may scratch his pox and infect them with skin inhabitant *Staphylococcus aureus* (secondary infection).

Infections that come on rapidly, with severe but short-lived effects, are called **acute infections.** Infections that progress and persist over a long period of time are **chronic infections.**

A Note About Terminology

Words in medicine have great power. A single technical term can often replace a whole phrase or sentence, thereby saving time and space in patient charting. The beginning student may feel overwhelmed by what seems like a mountain of new words. However, having a grasp of a few root words and a fair amount of anatomy can help you learn many of these words and even deduce the meaning of unfamiliar ones. Some examples of medical shorthand follow:

- The suffix *-itis* means "an inflammation" and, when affixed to the end of an anatomical term, indicates an inflammatory condition in that location. Thus, meningitis is an inflammation of the meninges surrounding the brain; encephalitis is an inflammation of the brain itself; hepatitis involves the liver; vaginitis, the vagina; gastroenteritis, the intestine; and otitis media, the middle ear. Although not all inflammatory conditions are caused by infections, many are.
- The suffix *-emia* is derived from the Greek word *haima,* meaning "blood." When added to a word, it means "associated with the blood." Thus, septicemia means sepsis (infection) of the blood; bacteremia, bacteria in the blood; viremia, viruses in the blood; and fungemia, fungi in the blood. It is also applicable to specific conditions such as toxemia, gonococcemia, and spirochetemia.
- The suffix *-osis* means "a disease or morbid process." It is frequently added to the names of pathogens to indicate the disease they cause: for example, listeriosis, histoplasmosis, toxoplasmosis, shigellosis, salmonellosis, and borreliosis. A variation of this suffix is *-iasis,* as in trichomoniasis and candidiasis.
- The suffix *-oma* comes from the Greek word *onkomas* (swelling) and means "tumor." Although it is often used to describe cancers (sarcoma, melanoma), it is also applied in some infectious diseases that cause masses or swellings (tuberculoma, leproma).

Signs and Symptoms: Warning Signals of Disease

When an infection causes pathologic changes leading to disease, it is often accompanied by a variety of signs and symptoms. A **sign** is any objective evidence of disease as noted by an observer and a **symptom** is the subjective evidence of disease as sensed by the patient. In general, signs are more precise than symptoms, though both can have the same underlying cause. For example, an infection of the brain may present with the sign of bacteria in the spinal fluid and symptom of headache; a streptococcal infection may produce a sore throat (symptom) and an inflamed pharynx (sign). A disease indicator that can be sensed and observed can qualify as either a sign or a symptom. When a disease can be identified or defined by a certain complex of signs and symptoms, it is termed a **syndrome.** Common signs and symptoms often used to assist in diagnosing infectious diseases are shown in **table 13.6.** Specific signs and symptoms for particular infectious diseases are covered in chapters 19 through 24.

Table 13.6 Common Signs and Symptoms of Infectious Diseases

Signs	Symptoms
Fever	Chills
Septicemia	Pain, ache, soreness, irritation
Microbes in tissue fluids	Malaise
Chest sounds	Fatigue
Skin eruptions	Chest tightness
Leukocytosis	Itching
Leukopenia	Headache
Swollen lymph nodes	Nausea
Abscesses	Abdominal cramps
Tachycardia (increased heart rate)	Anorexia (lack of appetite)
Antibodies in serum	Sore throat

1. This is not to be confused with toxemia of pregnancy, which is a metabolic disturbance and not an infection.

Signs and Symptoms of Inflammation The earliest symptoms of disease result from the activation of the body defense process called **inflammation.** The inflammatory response includes cells and chemicals that respond nonspecifically to disruptions in the tissue. This subject is discussed in greater detail in chapter 15, but as noted earlier, many signs and symptoms of infection are caused by the mobilization of this system. Some common symptoms of inflammation include fever, pain, soreness, and swelling. Signs of inflammation include **edema,** the accumulation of fluid in an afflicted tissue; **granulomas** and **abscesses,** walled-off collections of inflammatory cells and microbes in the tissues; and **lymphadenitis,** swollen lymph nodes.

Rashes and other skin eruptions are common signs in many diseases because they tend to mimic each other, it can be difficult to differentiate among diseases on this basis alone. The general term for the site of infection or disease is **lesion.** Skin lesions can be restricted to the epidermis and its glands and follicles, or they can extend deeper into the dermis and subcutaneous regions. The lesions of some infections undergo characteristic changes in appearance during the course of disease and, for that reason, they can fit more than one category.

Signs of Infection in the Blood Changes in the number of circulating white blood cells, as determined by special counts, are considered to be signs of possible infection. **Leukocytosis** (loo′-koh-sy-toh′-sis) is an increase in the level of white blood cells, whereas **leukopenia** (loo′-koh-pee′-nee-uh) is a decrease. Other signs of infection revolve around the occurrence of a microbe or its products in the blood. The clinical term for blood infection, **septicemia,** refers to a general state in which microorganisms are multiplying in the blood and are present in large numbers. When small numbers of bacteria are present in the blood but not necessarily multiplying, the correct term is **bacteremia. Viremia** is the term used to describe the presence of viruses in the blood, whether or not they are actively multiplying.

During infection, a normal host will show signs of an immune response in the form of antibodies in the serum or some type of sensitivity to the microbe. This fact is the basis for several serological tests used in diagnosing many infectious diseases. Such specific immune reactions indicate the body's attempt to develop specific immunities against pathogens.

Infections That Go Unnoticed In more cases than you might think, true infections go unnoticed. In other words, although infected, the host does not manifest signs or symptoms of the disease. Infections of this nature are known as **asymptomatic, subclinical,** or **inapparent** because the patient experiences no symptoms or disease and does not seek medical attention. Subclinical infections still play a major role in the transmission of infectious agents.

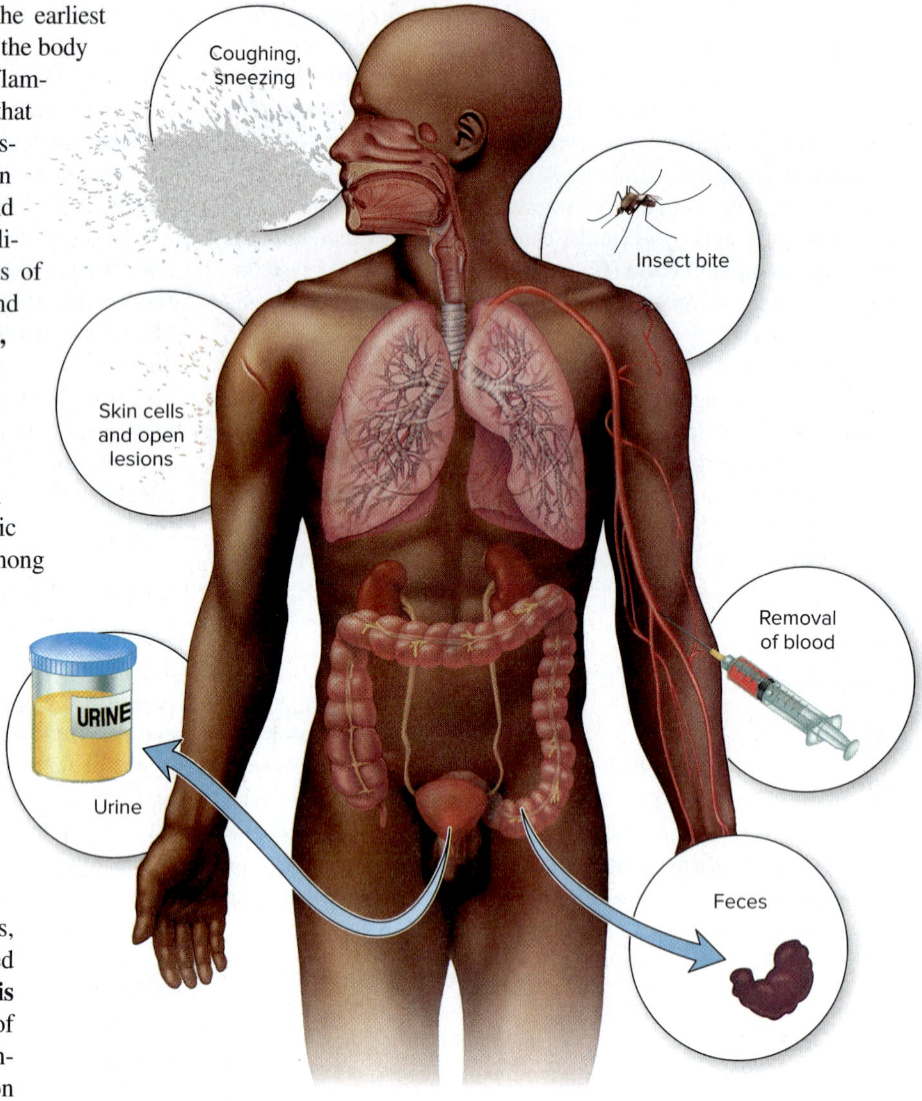
Coughing, sneezing
Insect bite
Skin cells and open lesions
Removal of blood
URINE
Urine
Feces

Figure 13.11 Major portals of exit of infectious diseases.

Step Five—Vacating the Host: Portals of Exit

Earlier, we introduced the idea that a pathogen is considered *unsuccessful* if it does not have a provision for leaving its host and moving to other susceptible hosts. With few exceptions, pathogens depart by a specific avenue called the **portal of exit (figure 13.11).** In most cases, the pathogen is shed or released from the body through secretion, excretion, discharge, or sloughed tissue. Usually there are a high number of infectious agents in these materials. This increases the likelihood that the pathogen will reach other hosts. In many cases, the portal of exit is the same as the portal of entry, but some pathogens use a different route.

Respiratory and Salivary Portals

Mucus, sputum, nasal drainage, and other moist secretions are the media of escape for the pathogens that infect the lower or upper respiratory tract. The most effective means of releasing these secretions are coughing and sneezing (see figure 13.16), although they can also be released during talking and laughing. Tiny particles of liquid released into the air form aerosols or droplets that can spread the infectious agent to other people. The agents of COVID-19, tuberculosis, influenza, measles, and chickenpox most often leave the host through airborne droplets. Droplets of saliva are the exit route for several viruses, including those of mumps, rabies, and infectious mononucleosis.

Skin Scales

The outer layer of the skin and scalp is constantly being shed into the environment. A large proportion of household dust is actually composed of skin cells. A single person can shed several billion skin cells a day. Skin lesions and their fluids can serve as portals of exit in warts, fungal infections, boils, herpes simplex, smallpox, and syphilis.

Fecal Exit

Feces are a very common portal of exit. Some intestinal pathogens grow in the intestinal mucosa and create an inflammation that increases the motility (the irritability) of the bowel. This increased motility speeds up peristalsis, resulting in diarrhea, and the fluid stool provides a rapid exit for the pathogen. A number of helminth worms release cysts and eggs through the feces. Feces containing pathogens are a public health problem when allowed to contaminate drinking water or when used to fertilize crops.

Urogenital Tract

A number of agents involved in sexually transmitted infections leave the host in vaginal discharge or semen. This is also the source of neonatal infections such as herpes simplex, *Chlamydia,* and *Candida albicans,* which infect the infant as it passes through the birth canal. Certain pathogens that infect the kidney are discharged in the urine: for instance, the agents of leptospirosis, typhoid fever, tuberculosis, and schistosomiasis.

Removal of Blood or Bleeding

Although the blood does not have a direct route to the outside, it can serve as a portal of exit when it is removed or released through a puncture made by natural or artificial means. Blood-feeding animals such as ticks and fleas are common transmitters of pathogens. The AIDS and hepatitis viruses in blood are transmitted by shared needles or through small breaks in a mucous membrane caused by sexual intercourse. Blood donation is also a means for certain microbes to leave the host, though this means of exit is now unusual because of close monitoring of the donor population and blood used for transfusions.

Long-Term Infections and Long-Term Effects

The apparent recovery of the host does not always mean that the microbe has been completely removed or destroyed by the host defenses. In certain chronic infectious diseases, the infectious agent retreats into a dormant state called **latency** after the initial symptoms clear up. Throughout this latent state, the microbe can periodically become active and produce a recurrent disease. Viral agents including herpes simplex, herpes zoster, hepatitis B, HIV, and Epstein-Barr can persist in the host for long periods. The agents of syphilis, typhoid fever, tuberculosis, and malaria also enter into latent stages. The person harboring a persistent infectious agent may or may not shed it during the latent stage. If it is shed, such persons are chronic carriers who serve as sources of infection for the rest of the population.

Some diseases leave **sequelae** (the medical term for consequences) in the form of long-term or permanent damage to tissues or organs. For example, meningitis can result in deafness, strep throat can lead to rheumatic heart disease, Lyme disease can cause arthritis, and polio can produce paralysis. The phenomenon of long-haul COVID-19, in which patients who recover suffer various systemic symptoms for weeks to months, is an example of a sequela.

The Course of an Infection

There are four distinct phases of infection and disease: the incubation period, the prodrome, the acute period, and the convalescent period **(figure 13.12).** We think of all infectious diseases as having all four of these phases, though for any given infection the lengths of the different phases can vary tremendously. The **incubation period** is the time from initial contact with the infectious agent (at the portal of entry) to the appearance of the first symptoms. During the incubation period, the agent is multiplying but has not yet caused enough damage to elicit symptoms. Although this period is relatively well defined and predictable for each microorganism, it does vary according to host resistance, degree of virulence, and distance between the target organ and the portal of entry (the farther apart, the longer the incubation period). Overall, an incubation period can range from several hours in pneumonic plague to several years in leprosy. The majority of infections, however, have incubation periods ranging between 2 and 30 days.

The earliest notable symptoms of infection usually appear as a vague feeling of discomfort, such as head and muscle aches, fatigue, upset stomach, and general malaise. This short period (1 to 2 days) is known as the **prodromal stage.** Some diseases have very specific prodromal symptoms. Other diseases have an imperceptible prodromal phase. Next, the infectious agent enters the height of the infection, often called the **acute phase,** during which it multiplies at high levels, exhibits its greatest virulence, and becomes well established in its target tissue. This period is often marked by fever and other prominent and more specific signs and symptoms, which can include cough, rashes, diarrhea,

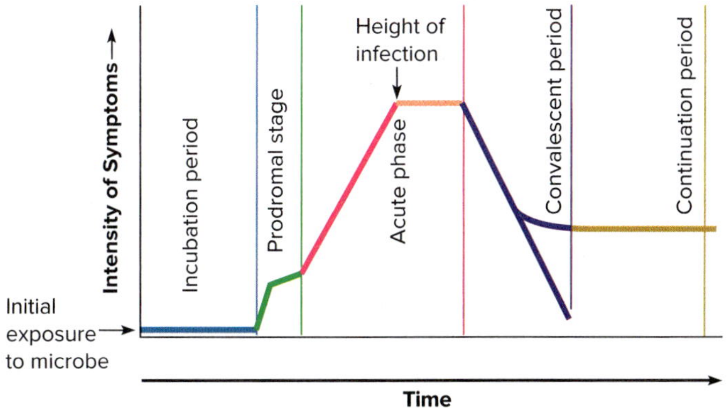
Intensity of Symptoms
Initial exposure to microbe
Incubation period
Prodromal stage
Height of infection
Acute phase
Convalescent period
Continuation period
Time

Figure 13.12 **Stages in the course of infection and disease.** The stages have different durations in different infections.

Disease Connection

Although the prodromal symptoms of many diseases are vague and nonspecific, the prodromal phase of measles is quite recognizable, if you know where to look. Two days before the onset of the red body rash, white spots called Koplik's spots appear inside the mouth next to the second molars.

loss of muscle control, swelling, jaundice, discharge of exudates, or severe pain, depending on the particular infection. The length of this period is extremely variable.

As the patient begins to respond to the infection, the symptoms decline—sometimes dramatically, other times slowly. During the recovery that follows, called the **convalescent period,** the patient's strength and health gradually return. During this period, many patients stop taking their antibiotics, even though there are still microbes in their system. This noncompliance means that bacteria with higher resistance are left behind to repopulate, putting the patient at risk for redevelopment of the infection that will not be treatable with the previously used antibiotic. Whether the infectious agent is transmissible during any given phase varies with each infection. A few agents are released mostly during incubation (measles, for example); many are released during the acute phase (*Shigella*); and others can be transmitted during all of these periods (hepatitis B).

These are the four phases all infectious diseases have. There is a fifth phase, which only some infections have: the **continuation** phase, in which *either* the organism lingers for months, years, or indefinitely after the patient is completely well *or* the organism is gone but symptoms continue. Typhoid fever is one disease in which the organism lingers after the patient has fully recovered. An example of a disease that lingers after the organism is no longer detectable is syphilis (in some cases). Chronic Lyme disease can also be put in that category.

Reservoirs: Where Pathogens Persist

In order for an infectious agent to continue to exist and be spread, it must have a permanent place to reside. The **reservoir** is the primary habitat in the natural world from which a pathogen originates. Often it is a human or animal carrier, although soil, water, and plants are also reservoirs. The reservoir can be distinguished from the infection **source,** which is the individual or object from which an infection is actually acquired. In diseases such as syphilis, the reservoir and the source are the same (the human body). In the case of hepatitis A, the reservoir (a human carrier) is usually different from the source of infection (contaminated food). **Table 13.7** describes the different types of reservoirs and how microbes travel from them to humans.

Living Reservoirs

Persons or animals with obvious symptomatic infection are obvious sources of infection, but a **carrier** is, by definition, an individual who *inconspicuously* shelters a pathogen and can spread it to others without knowing. Although human carriers are occasionally detected through routine screening (blood tests, cultures) and other epidemiological devices, they are very difficult to discover and control. As long as a pathogenic reservoir is maintained by the carrier state, the disease will continue to exist in that population and the potential for epidemics will be a constant threat. The duration of the carrier state can be short or long term, and it is important to remember that the carrier may or may not have experienced disease due to the microbe.

Several situations can produce the carrier state **(figure 13.13). Asymptomatic** (apparently healthy) **carriers** are infected but show no symptoms. A few asymptomatic infections (gonorrhea and human papillomavirus, for instance) may carry out their entire course without obvious manifestations. **Incubating carriers** spread the infectious agent during the incubation period. For example, a person infected with COVID-19 can spread the infection before symptoms appear. Recuperating patients without symptoms are considered **convalescent carriers** when they continue to shed viable microbes and transmit the infection to others. Diphtheria patients, for example, spread the microbe for up to 30 days after the disease has subsided.

An individual who shelters the infectious agent for a long period after recovery because of the latency of the infectious agent is a **chronic carrier.** Patients who have recovered from tuberculosis or hepatitis infections frequently carry the agent chronically. About 1 in 20 victims of typhoid fever continues to harbor *Salmonella typhi* in the gallbladder for several years, and sometimes for life. The most infamous of these was "Typhoid Mary," a household cook who spread the infection to hundreds of victims in the early 1900s. (*Salmonella* infection is described in chapter 23.)

The **passive carrier** state is of great concern during patient care. Medical and dental personnel who must constantly handle materials that are heavily contaminated with patient secretions and blood risk picking up pathogens mechanically and accidently transferring them to other patients. Proper hand washing, handling of contaminated materials, and aseptic techniques greatly reduce this likelihood.

Disease Connection

Health care workers can become passive carriers of methicillin-resistant *Staphylococcus aureus* (MRSA), usually in the nares (nostrils). It is estimated that up to 5% of health care workers harbor this potentially deadly bacterium. However, colonized health care workers rarely infect the patients they care for; thus, it is not recommended that asymptomatic health care workers be routinely screened or treated. But if an outbreak occurs in a hospital, the workers will likely be screened.

Animals as Reservoirs and Sources Up to now, we have lumped animals with humans in discussing living reservoirs or carriers, but animals deserve special consideration as vectors of infections. The word **vector** is used by epidemiologists to indicate a live animal that transmits an infectious agent from one host to another. The majority of vectors are arthropods such as fleas, mosquitoes, flies, and ticks, although larger animals can also spread infection—for example, mammals (rabies), birds (psittacosis), or lizards (salmonellosis).

Vectors are typically placed into one of two categories, depending on the animal's relationship with the microbe **(figure 13.14).** A **biological vector** actively participates in a pathogen's life cycle, serving as a site in which it can multiply or complete its life cycle.

Table 13.7 Reservoirs and Transmitters

Reservoirs	Transmission Examples
Living Reservoirs	
Animals (other than humans and arthropods) Mammals, birds, reptiles, etc.	Pathogens from animals • can be directly transmitted to humans; example: bats transmitting rabies to humans. • can be transmitted to humans via vectors, as with fleas passing the plague bacterium from rats to people. • can be transmitted through vehicles such as water; example: leptospirosis, which is often transmitted from animal urine to human skin via bodies of water.
Humans Actively ill	 Thinkstock/Getty Images; Ingram Micro/Media Bakery A person suffering from a cold contaminates a pen, which is then picked up by a healthy person. That is indirect transmission. Alternatively, a sick person can transmit the pathogen directly by sneezing on a healthy person.
Carriers	A person who is fully recovered from hepatitis but is still shedding hepatitis A virus in their feces may use suboptimal hand-washing technique. That person contaminates food, which a healthy person ingests (indirect transmission). Carriers can also transmit through direct means, as when an incubating carrier of HIV, who does not know they are infected, transmits the virus through sexual contact.
Arthropods Biological vectors	 CDC/James Gathany When an arthropod is the host (and reservoir) of the pathogen, it is also the mode of transmission.
Nonliving Reservoirs	
Soil **Water** **Air** **The built environment**	 Christopher Kerrigan/McGraw Hill Some pathogens, such as the TB bacterium, can survive for long periods in nonliving reservoirs. They are then directly transmitted to humans when they come in contact with the contaminated soil, water, or air.

A biological vector communicates the infectious agent to the human host by biting, aerosol formation, or touch. In the case of biting vectors, the animal can

1. inject infected saliva into the blood (the mosquito, **figure 13.14*a***),
2. defecate around the bite wound (the flea), or
3. regurgitate blood into the wound (the tsetse fly).

Mechanical vectors are not necessary to the life cycle of an infectious agent and merely transport it without being infected. The external body parts of these animals become contaminated when they come into physical contact with a source of pathogens. The agent is subsequently transferred to humans indirectly by an intermediate such as food or, occasionally, by direct contact (as in

Carrier State	Explanation	Example
Asymptomatic carriers	Infected but show no symptoms of disease	Gonorrhea, genital herpes with no lesions
Incubating carriers	Infected but show no symptoms of disease	Infectious mononucleosis
Convalescent carriers	Recuperating patients without symptoms; they continue to shed viable microbes and convey the infection to others	Hepatitis A
Chronic carriers	Individuals who shelter the infectious agent for a long period after recovery because of the latency of the infectious agent	Tuberculosis, typhoid fever
Passive carriers	Medical and dental personnel who must constantly handle patient materials that are heavily contaminated with patient secretions and blood risk picking up pathogens mechanically and accidentally transferring them to other patients	Various healthcare-associated infections

Asymptomatic
STI
Lane Oatey/Blue Jean Images/Getty Images

Microbes are multiplying.
UpperCut Images/SuperStock

Siri Stafford/Digital Vision/Getty Images

John Lund/Marc Romanelli/Blend Images LLC

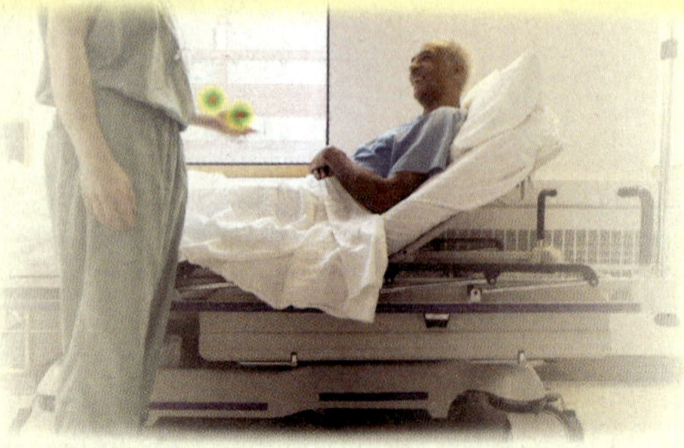
UpperCut Images/SuperStock

Figure 13.13 Carrier states.

Biological vectors are infected.
(a)

Mechanical vectors are not infected.
(b)

Figure 13.14 Two types of vectors. **(a)** Biological vectors serve as hosts during pathogen development. Examples are mosquitoes, which carry malaria; bats, which carry rabies and other viral diseases; and chickens, which can transmit their flu viruses to humans. **(b)** Mechanical vectors such as the housefly and the cockroach transport pathogens on their feet and mouthparts.

certain eye infections). Houseflies **(figure 13.14*b*)** are obnoxious mechanical vectors. They feed on decaying garbage and feces, and while they are feeding, their feet and mouthparts easily become contaminated. They also regurgitate juices onto food to soften and digest it. Flies spread more than 20 different bacterial, viral, protozoan, and helminth infections. Various flies transmit tropical ulcers, yaws, and trachoma. Cockroaches, which have similar unsavory habits, play a role in the mechanical transmission of fecal pathogens.

Many vectors and animal reservoirs spread their own infections to humans. An infection indigenous to animals but also transmissible to humans is a **zoonosis** (zoh′-uh-noh′-sis). In these types of infections, the human is essentially a dead-end host and does not contribute to the natural persistence of the microbe. Some zoonotic infections (rabies, for instance) can have multihost involvement, and others can have very complex cycles in the wild (see discussion of plague in chapter 21). Zoonotic spread of disease is promoted by close associations of humans with animals, and people in animal-oriented or outdoor professions are at greatest risk. At least 150 zoonoses exist worldwide; the most common ones are listed in **table 13.8.** Zoonoses make up a full 70% of all new emerging diseases worldwide. The zoonoses in table 13.8 are ones that continue to move from animals to humans, and not usually transmitted from human to human. When a zoonosis is caused by an organism that subsequently becomes able to be transmitted among humans, as in some influenzas and in COVID-19, potentially dangerous epidemics and pandemics can occur. It is worth noting that zoonotic infections are impossible to completely eradicate without also eradicating the animal reservoirs. Attempts have been made to eradicate mosquitoes and certain rodents, and in 2004 China slaughtered tens of thousands of civet cats that were thought (incorrectly) to be a source of the respiratory disease SARS (the one causing the first SARS pandemic, which was quickly contained).

Nonliving Reservoirs

Clearly, microorganisms have adapted to nearly every habitat in the biosphere. They thrive in soil and water and often find their way into the air. Although most of these microbes are saprobic and cause little harm and considerable benefit to humans, some are opportunists and a few are regular pathogens.

Soil harbors the vegetative forms of bacteria, protozoa, helminths, and fungi, as well as their resistant or developmental stages such as endospores, cysts, ova, and larvae. Bacterial pathogens include the anthrax bacillus and species of *Clostridium* that are responsible for gas gangrene, botulism, and tetanus. Pathogenic fungi in the genera *Coccidioides* and *Blastomyces* are spread by spores in the soil and dust. The invasive stages of the hookworm *Necator* occur in the soil. Natural bodies of water carry fewer nutrients than soil does but still support pathogenic species such as *Vibrio cholerae, Legionella, Cryptosporidium,* and *Giardia.*

Table 13.8 Common Zoonotic Infections

Disease	Primary Animal Reservoirs
Viruses	
Rabies	Mammals
Yellow fever	Wild birds, mammals, mosquitoes
Viral fevers	Wild mammals
Hantavirus	Rodents
Influenza	Chickens, birds, swine
West Nile virus	Wild birds, mosquitoes
Bacteria	
Rocky Mountain spotted fever	Dogs, ticks
Psittacosis	Birds
Leptospirosis	Domestic animals
Anthrax	Domestic animals
Brucellosis	Cattle, sheep, pigs
Plague	Rodents, fleas
Salmonellosis	Mammals, birds, reptiles, rodents
Tularemia	Rodents, birds, arthropods
Miscellaneous	
Ringworm	Domestic mammals
Toxoplasmosis	Cats, rodents, birds
Trypanosomiasis	Domestic and wild mammals
Trichinosis	Swine, bears
Tapeworm	Cattle, swine, fish

imagebroker/Alamy Stock Photo

Pixtal/age fotostock

Ingram Publishing/SuperStock

The built environment—the buildings where we live, work, and spend leisure time—can also serve as nonliving reservoirs of infection.

The Acquisition and Transmission of Infectious Agents

Infectious diseases can be categorized on the basis of how they are acquired. A disease is **communicable** when an infected host can transmit the infectious agent to another host and establish infection in that host. (Although this terminology is standard, one must realize that it is not the disease that is communicated but the microbe. Also be aware that the word *infectious* is sometimes used interchangeably with the word *communicable,* but this is not exactly accurate.) The transmission of the agent can be direct or indirect, and the ease with which the disease is transmitted varies considerably from one agent to another. If the agent is highly communicable, especially through direct contact, the disease is considered **contagious.** Influenza and measles move readily from host to host and thus are contagious, whereas Hansen's disease (leprosy) is only weakly communicable.

In contrast, a **noncommunicable** infectious disease does *not* arise through transmission of the infectious agent from host to host. The infection and disease are acquired through some other special circumstance. Noncommunicable infections occur primarily when a compromised person is invaded by his or her own microbiota (as with certain pneumonias, for example) or when an individual has accidental contact with a microbe that exists in a nonliving reservoir such as soil. Some examples are certain mycoses, acquired through inhalation of fungal spores, and tetanus, in which *Clostridium tetani* endospores from a soiled object enter a cut or wound. Infected persons do not become a source of disease to others.

Patterns of Transmission in Communicable Diseases

The routes or patterns of disease transmission are many and varied. The spread of diseases is by direct or indirect contact with animate or inanimate objects and can be horizontal or vertical. The term *horizontal* means the disease is spread through a population from one infected individual to another; *vertical* signifies transmission from parent to offspring via the ovum, sperm, placenta, or milk. Then, within the category of horizontal transmission, we divide the modes of transmission into direct contact, indirect routes, and **vector** transmission **(figure 13.15).**

Modes of Direct Transmission In order for microbes to be directly transferred, some type of contact must occur between the skin or mucous membranes of the infected person and those of the new infectee. It may help to think of this route as the portal of exit meeting the portal of entry without the involvement of an intermediate object, substance, or space. Most sexually transmitted infections are spread directly. Most obligate parasites are far too sensitive to survive for long outside the host and can be transmitted only through direct contact. The trickiest type of "contact" transmission is droplet contact, in which fine droplets are sprayed directly upon a person during sneezing or coughing (as distinguished from droplet nuclei that are transmitted over a meter or more by air). While there is some space between the infecter and the infectee, it is still considered a form of direct contact because the two people have to be in each other's presence, as opposed to indirect forms of contact.

Routes of Indirect Transmission For microbes to be indirectly transmitted, the infectious agent must pass from an infected host to an intermediate conveyor (a vehicle) and from there to another host. The original transmitter can be either openly infected or a carrier.

Indirect Spread by Fomites and Vehicles The term **vehicle** specifies any inanimate material commonly contacted by humans that can transmit infectious agents. A *common vehicle* is a single material that serves as the source of infection for many individuals. Some specific types of vehicles are food, water, various biological products (such as blood, serum, and tissue), and fomites. A **fomite** is an inanimate object that harbors and transmits pathogens. Unlike a reservoir, however, a fomite is not a continuous source of infection. The list of possible fomites is as long as your imagination allows. Probably highest on the list would be objects commonly in contact with the public, such as doorknobs, telephones, handheld remote controls, and faucet handles that are readily contaminated by touching. Shared bed linens, handkerchiefs, toilet seats, toys, eating utensils, clothing, personal articles, and syringes are other examples. Although paper money is impregnated with a disinfectant to inhibit microbes, pathogens are still isolated from bills as well as coins.

Parenteral transmission is also indirect. It refers to a puncture, in which material from the environment is deposited directly in deeper tissues. This can be intentional, in the case of contaminated needles, or unintentional, as in the case of puncture injuries.

A Note About Touching Your Face

Have you ever thought about how often you touch your face? Have you thought about why it matters? Well, your face has three major targets—portals of entry—for pathogens: your mouth, your nose, and your eyes. And your hands and fingers are often contaminated, from touching a fomite, a contaminated surface, or another person. Broadly speaking, if you could simply avoid touching your face, your chances of getting sick would fall drastically. So how often do you touch your face? Research shows it is about 15 times an hour.

Indirect Spread by Vehicles: Water, Soil, and Air as Vehicles As discussed in the section on reservoirs, soil and water harbor a variety of microbes that can sicken humans. Also, they can become temporarily contaminated with pathogens that come from humans, as in the case of water becoming contaminated during a cholera outbreak. Unlike soil and water, however, outdoor air cannot provide nutritional support for microbial growth and seldom transmits airborne pathogens. On the other hand, indoor air (especially in a closed space) can serve as an important medium for the suspension and dispersal of certain respiratory pathogens via droplet nuclei and aerosols. The distinction between indoor and outdoor air became evident during the COVID-19 pandemic, during which restaurants scrambled to create more outdoor dining spaces. Of course, even outdoors, if people are crowded together, such as at a concert, a great deal of transmission can still occur.

Droplet nuclei are dried microscopic residues created when microscopic pellets of mucus and saliva are ejected from the mouth and nose. They are generated forcefully in a sneeze or cough **(figure 13.16)** or mildly during talking or singing. The larger

Patterns of Transmission in Communicable Diseases

The rest of this table is about horizontal transmission.
- Kelly & Heidi

Mode of Transmission	Definition
Vertical	Transmission is from parent to offspring via the ovum, sperm, placenta, or milk
Horizontal	Disease is spread through a population from one infected individual to another

Direct (contact) transmission

Involves very close proximity or actual physical contact between two hosts

Types:

- Touching, kissing, sex
- Droplet contact, in which fine droplets are sprayed directly upon a person during sneezing, coughing, or speaking

Contact: kissing and sex (Epstein-Barr virus, gonorrhea)

Droplets (colds, chickenpox)

Indirect transmission

Infectious agent must pass from an infected host to an intermediate conveyor (a vehicle) and from there to another host

Infected individuals contaminate objects, food, or air through their activities

Types:

- **Fomite**—inanimate object that harbors and transmits pathogens (doorknobs, phones, faucet handles)
- **Vehicle**—a natural, nonliving material that can transmit infectious agents
  - **Air**—smaller particles evaporate and remain in the air and can be encountered by a new host; **aerosols** are suspensions of fine dust or moisture particles in the air that contain live pathogens
  - **Water**—some pathogens survive for long periods in water and can infect humans long after they were deposited in the water
  - **Soil**—microbes resistant to drying live in and can be transmitted from soil
  - **Food**—meats may contain pathogens with which the animal was infected; foods can also be contaminated by food handlers
- **Parenteral** transmission via intentional or unintentional injection into deeper tissues (needles, knives, branches, broken glass, etc.)

Special Category: oral-fecal route—using either vehicles or fomites. A fecal carrier with inadequate personal hygiene contaminates food during handling, and an unsuspecting person ingests it; alternatively, a person touches a surface that has been contaminated with fecal material and touches his or her mouth, leading to ingestion of fecal microbes

Vector transmission

Types:

- **Mechanical vector**—insect carries microbes to host on its body parts
- **Biological vector**—insect injects microbes into host; part of microbe life cycle completed in insect

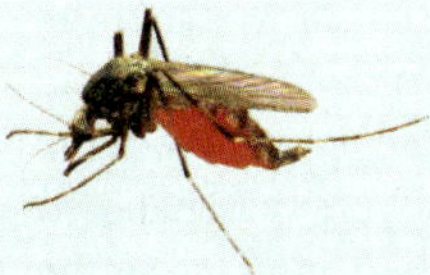

Figure 13.15 Patterns of transmission in communicable diseases. Transmission is either vertical or horizontal. Horizontal transmission is via direct, indirect, or vector transmission.

(kissing) Guas/Shutterstock; (sneezing) James Gathany/CDC; (opening door) silverlining56/E+/Getty Images; (playing soccer) Paul Bradbury/Caia Images/Glow Images; (fly) Simon Murrell/Dynamic Graphics Group/IT Stock Free/Alamy Stock Photo; (mosquito) abadonian/iStockphoto/Getty Images

Figure 13.16 The explosiveness of a sneeze. Special photography dramatically captures droplet formation in an unstifled sneeze. When such droplets dry and remain suspended in air, they become droplet nuclei.
Custom Medical Stock Photo/Alamy Stock Photo

beads of moisture settle rapidly. If these settle in or on another person, it is considered direct droplet contact, as described earlier; but the smaller particles evaporate and remain suspended for longer periods. After evaporation, microscopic pellets 1 to 4 microns in size are created. Their small size enhances their pathogenic ability to ease their passage into the lungs. They can be encountered by a new host who is geographically or chronologically distant. For that reason, they are considered transmitted by indirect contact. Droplet nuclei are implicated in the spread of hardier pathogens such as the tubercle bacillus and the influenza virus. **Aerosols** are suspensions of fine dust or moisture particles in the air that contain live pathogens. Q fever is spread by dust from animal quarters, and psittacosis is spread by aerosols from infected birds.

Outbreaks of food poisoning often result from the role of food as a common vehicle. The source of the agent can be soil, the handler, or a mechanical vector. Water that has been contaminated by feces or urine can carry *Salmonella, Vibrio* (cholera), viruses (hepatitis A, polio), and pathogenic protozoans (*Giardia, Cryptosporidium*). A type of transmission termed the *oral-fecal route* can occur in two ways. In the first, a fecal carrier with inadequate personal hygiene contaminates food during handling and an unsuspecting person ingests it. Hepatitis A, amoebic dysentery, shigellosis, and typhoid fever are often transmitted this way. Oral-fecal transmission can also involve contaminated materials such as toys and diapers. It is really a special category of indirect transmission, which specifies that the way in which the vehicle became contaminated was through contact with fecal material and that it found its way to someone's mouth.

A recent investigation of a small outbreak of diarrhea among a traveling girls' soccer team caused by norovirus showed that the first person with the infection had probably contaminated some reusable grocery bags that were stored in the hotel bathroom. Fine aerosols of her feces settled on the bags, which were then handled by other girls on the team, leading to eight of them contracting the infection. Here, the bags were the fomites, and the girls handling the bags apparently transferred the virus to their mouths.

The final type of transmission in communicable diseases is **vector** transmission (discussed earlier). Take the time to become very familiar with figure 13.15, as you will refer to its content many times during the rest of the course.

Which Agent Is the Cause? Using Koch's Postulates to Determine Etiology

An essential aim in the study of infection and disease is determining the precise **etiologic,** or causative, **agent** of a newly recognized condition. In the 1880s, Robert Koch realized that in order to prove the germ theory of disease, he would have to develop a standard for determining causation that would stand the test of scientific scrutiny. Out of his experimental observations on the transmission of anthrax in cows came a series of proofs, called **Koch's postulates,** that established these classic criteria for etiologic studies **(figure 13.17):**

1. find evidence of a particular microbe in every case of a disease,
2. isolate that microbe from an infected subject and cultivate it in pure culture in the laboratory,

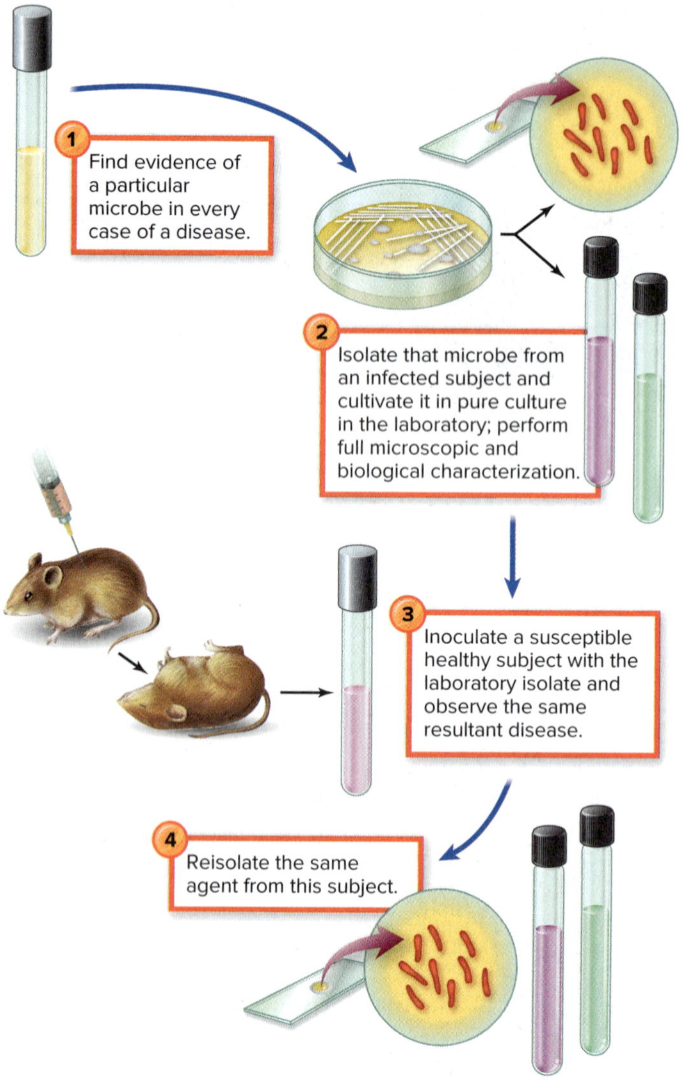
1
Find evidence of a particular microbe in every case of a disease.
2
Isolate that microbe from an infected subject and cultivate it in pure culture in the laboratory; perform full microscopic and biological characterization.
3
Inoculate a susceptible healthy subject with the laboratory isolate and observe the same resultant disease.
4
Reisolate the same agent from this subject.

Figure 13.17 Koch's postulates: Is this the etiologic agent? The microbe in the initial and second isolations (steps 2 and 4 in the figure) and the disease in the patient and experimental animal must be identical for the postulates to be satisfied.

3. inoculate a susceptible healthy subject with the laboratory isolate and observe the same disease, and
4. reisolate the laboratory isolate from this subject.

In practice, applying Koch's postulates can be complicated. Each isolated culture must be pure, observed microscopically, and identified by means of characteristic tests; the first and second isolates must be identical; and the pathologic effects, signs, and symptoms of the disease in the first and second subjects must be the same. Once established, these postulates were rapidly put to the test, and within a short time, they had helped determine the causative agents of tuberculosis, diphtheria, and plague.

Koch's postulates are reliable for many infectious diseases, but they cannot be completely fulfilled in certain situations. For example, some infectious agents, such as *M. leprae,* the cause of leprosy, are not readily isolated or grown in the laboratory. Also, there has to be a suitable animal model for the infection—a species that is also infected by the microbe and experiences a similar disease as humans. (Clearly, using humans as the test subjects is not permissible.) Mice often serve this purpose but are not suited for every disease. If there is no suitable animal model, it is very difficult to use Koch's postulates to prove the etiology. It is difficult to satisfy Koch's postulates for viral diseases because viruses usually have a very narrow host range.

Another very important reason Koch's postulates are increasingly viewed with caution is the idea, presented earlier, that perhaps the majority of human infections are polymicrobial. In those infections, Koch's postulates cannot be satisfied.

With advances in molecular biology, another alternative method for identifying an etiologic agent has been developed. There is now a set of "molecular Koch's postulates" that were formulated to establish that a gene found in a pathogen contributes to the disease-causing ability of the organism.

The general idea of Koch's postulates is that you isolate what you think the cause is, then apply it to a susceptible host population and produce the same effect. This general progression is still considered the gold standard for determining that any given factor is causing—not simply correlated with—an observed condition. Many of the media reports that present the latest "findings"—for example, that social media usage leads to shorter attention spans—are based on studies finding correlations, not causation. Koch's postulates would require measuring the attention spans of large numbers of people who had never used social media, then forcing them to use social media for some allotted time period. Then you would need to remeasure their attention spans. This would approach the gold standard of Koch's postulates, although you would not be able to completely hold all other variables constant in your subjects, unless you kept them all in your laboratory for the duration of your experiment. So you begin to see how difficult true causation is to prove.

13.2 Learning Outcomes—Assess Your Progress

4. Explain some of the variables that influence whether a microbe will cause disease in a particular host.
5. Differentiate between a microbe's pathogenicity and its virulence.
6. List the steps a microbe has to take to get to the point where it can cause disease.
7. List several portals of entry and exit.
8. Define *infectious dose,* and explain its role in establishing infection.
9. Describe three ways microbes cause tissue damage.
10. Compare and contrast major characteristics of exotoxins and endotoxin.
11. Explain what an epigenetic change is and how it can influence virulence.
12. Draw and label a curve representing the course of clinical infection.
13. Differentiate among the various types of reservoirs, providing examples of each.
14. List six different modes of horizontal transmission, providing an infectious disease spread by each.
15. List Koch's postulates, and explain alternative methods for identifying an etiologic agent.

Media Under The Microscope Wrap-Up

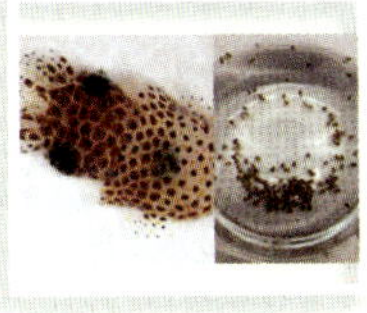

Margaret McFall-Ngai; Jamie Foster

The **intended message** of the article is that scientists are using squids as a study model to determine how human astronauts' health on board the International Space Station can be optimized. These squids have a very obvious relationship with their microbiomes (it affects their ability to glow!), so they can serve as a simple way to study how it is affected in microgravity.

My **critical reading** is that the facts as reported in the *Guardian* article seem to add up: Humans in space experience diminished immunity, and their relationship with their microbiomes also changes. The article makes a clear case for why baby squids are a good model system. I found myself asking how similar the squid physiology and immune system is to that of humans. Squid are invertebrates (mollusks).

I would explain this to my **nonmicrobiologist friends** by giving a brief explanation of the microbiome, and that it is critical to the health of humans. Apparently, squids also need a healthy microbiome to perform their functions, and so they are going into space so we can study what happens with them, in order to improve our understanding of humans spending time in space.

My **overall grade** for this article is a B. I wanted more information about why they thought a mollusk would prove useful in understanding human biology. Otherwise, it is a very thought-provoking article.

Source: *"Dozens of Hawaiian Baby Squid Aboard Space Station for Study." Guardian* online article, posted June 22, 2021.

Study Smarter: Better Together

These activities are designed for you to use on your own with a study group—either a face-to-face group or a virtual one, consisting of 3–5 members. Studying together can be very helpful, but there are effective and ineffective ways to do it. For example, getting together without a clear structure is often not a good use of your time. Use your time efficiently by using one or more of the exercises below.

FACE-TO-FACE GROUPS

Use one or more of the activities below.

Peer Instruction: Assign numbers to your group members to use all semester long. Now look at these five concepts from this chapter. Each group member prepares a 5-minute lesson on the topic corresponding to their number. Don't worry if you have fewer than 5 members; just use however many you have! During your group study time, each member presents their lesson, and the group spends another 5–10 minutes discussing that lesson.

1. Human Microbiome Project
2. Colonization vs. infection vs. disease
3. Steps a pathogen must take to cause disease
4. Three ways pathogens damage hosts
5. Koch's postulates

Concept Maps: Each member of the group should use this list of terms from this chapter to generate their own concept map. This can be hand-drawn or created using software (see Appendix C for guidelines). During group study time, compare each other's concept maps and help each other make sure they are correct. Of course, there are many different "correct" maps. Examining each member's map will help you talk through the varied concepts and how they are related.

Concept Terms:

pathogen	opportunistic infection	virulence factor	portal of entry
reservoir	normal biota	mutualism	

Table Topics: Each group member should identify a concept or topic from this week's class assignments with which they are having trouble and share it during group study time. The other group members can then help to clarify confusing issues or share how they figured it out. Aim for a maximum of 15 minutes per topic. If the topic remains unclear to the group, bring it up during class or use the instructor's office hours or e-mail to ask for help. Taking the time to struggle with a difficult concept first makes your questions much more specific and more likely to yield helpful answers.

VIRTUAL GROUPS

Not everyone has the time or opportunity to meet with group members outside of class time. You or your instructor can create a virtual group using e-mail or the course software.

Weekly Discussion Board: This forum can be used as a way for groups to discuss topics, via e-mail, or other learning management systems or online platforms before they are covered in class. As each member of the group answers the current week's question, they should send their responses to every other member of their group. It's best to agree on a deadline based on how your class schedule works (Saturday for the next week's topics, for example). Then, after the topic is discussed in class, each member should send a response that all group members will see with a follow-up post on the same topic. If you cover more than one chapter in a week, someone can be designated to choose which chapter Discussion Board question you will use. Or simply decide up front that you will always use the first-chapter-of-the week's question, to keep the schedule simple.

Discussion Question

Two people are both exposed to the same microbe in the same dose and through the same portal of entry. Provide some reasons why one gets sick and the other does not.

Chapter Summary

MICROBE–HUMAN INTERACTIONS

13.1 THE HUMAN HOST AND ITS MICROBIOME

- The development of humans and other organisms is deeply intertwined with the microbes that colonize them and the environment around them.
- Massive efforts around the world using nonculture methods to characterize microbes that colonize the human in health and disease have led to many insights about the microbiome.
- The human body and its microbiome are together called the holobiont.

13.2 WHEN COLONIZATION LEADS TO DISEASE

- The pathogenicity of a microbe refers to its ability to cause disease. Its virulence is the degree of damage it can inflict.
- There are some "true" pathogens, which cause disease in most human encounters, but the majority of microbes can be harmless, beneficial, or pathogenic depending on the circumstances.
- Every encounter between a susceptible host and a potential disease-causing organism does not result in disease. Multiple characteristics of the two partners come into play.
- Any characteristic or structure that enhances a microbe's ability to establish itself in the host and cause disease is called a virulence factor.
- To cause disease, microbes must *enter* the host, *attach* to host tissue, *avoid* host defenses, and then *cause damage*.
- There are three main ways that damage induced by a microbe occurs: using enzymes and toxins, causing the host's response to be excessive, and making epigenetic changes to host cells.
- Polymicrobial infections are very common and may well be the norm.
- There are distinct phases of infection and disease: the incubation period, the prodrome, the acute phase, the convalescent period, and, in some infections, the continuation period.
- The primary habitat of a pathogen is called its reservoir.
- A communicable disease can be transmitted from an infected host to others, but not all infectious diseases are communicable.
- The spread of infectious disease from person to person or from a reservoir to a person is called horizontal transmission. The spread from parent to offspring is called vertical transmission.
- Koch's postulates are a set of steps to determine the cause of an infectious disease. Originally created in the 1880s, they have been improved upon using molecular techniques.

SmartGrid: From Knowledge to Critical Thinking

This *21 Question Grid* takes the topics from this chapter and arranges them with respect to the American Society for Microbiology's Undergraduate Curriculum guidelines—all six of the important "Concepts" as well as the important "Competency" of scientific literacy. Three questions are supplied, which cover chapter content referring to the Concept or Competency in increasing levels of Bloom's taxonomy for learning.

ASM Concept/ Competency	A. Bloom's Level 1, 2—Remember and Understand (Choose one)	B. Bloom's Level 3, 4—Apply and Analyze	C. Bloom's Level 5, 6—Evaluate and Create
Evolution	**1.** The best descriptive term for the resident microbiota is **a.** commensal. **b.** parasitic. **c.** pathogenic. **d.** mutualistic.	**2.** In some circumstances, microbes can be quite virulent when they first infect a new species (such as in a zoonosis) but over decades of association with the new human host, cause milder and milder disease. Can you speculate about why this is evolutionarily advantageous to the pathogen?	**3.** Conduct research on germ-free mice. Use what you find to write a paragraph about the coevolution of microbes and humans.
Cell Structure and Function	**4.** Which of the following are virulence factors? **a.** toxins **b.** enzymes **c.** capsules **d.** all of the above	**5.** Why do you suppose specific adhesive structures, such as fimbriae, are critical to the disease-causing capabilities of many bacteria? Be thorough in your answer.	**6.** Discuss the role of endospores in ensuring the ongoing transmission of a bacterium in a population.
Metabolic Pathways	**7.** When the resident microbiota prevents the establishment of a pathogen, it is called **a.** disruption. **b.** a superinfection. **c.** a nonliving reservoir. **d.** microbial antagonism.	**8.** Correlate the stages in the course of an infectious disease with what you know about the growth of bacterial cultures.	**9.** *Helicobacter pylori,* a microbe that can cause disease in the stomach, produces an enzyme called urease that breaks down urea into ammonia and oxygen. What advantage does that confer on the bacterium?
Information Flow and Genetics	**10.** Most microbial exotoxins would be created using the process of **a.** DNA replication. **b.** protein synthesis. **c.** mutation. **d.** fatty acid synthesis.	**11.** Explain what you can about microbe-induced epigenetic changes to the host DNA.	**12.** Some of the most pathogenic *E. coli* possess an exotoxin gene originally found in *Shigella* bacteria. Construct a scenario for how and in what biological setting that gene was shared.
Microbial Systems	**13.** The ______ is the time between an encounter with a pathogen and the first symptoms. **a.** prodrome **b.** acute stage **c.** convalescence **d.** incubation period	**14.** A line of thinking called the "hygiene hypothesis" suggests that when children have more exposure to the environment (dirt, soil, animals), they have better health as adults. Why might this be?	**15.** At one time it was common practice to prescribe tetracycline for adolescents for months or years at a time to combat acne. Glance back at table 12.5 and predict what consequences that would have on the microbiome and the host.

ASM Concept/ Competency	A. Bloom's Level 1, 2—Remember and Understand (Choose one)	B. Bloom's Level 3, 4—Apply and Analyze	C. Bloom's Level 5, 6—Evaluate and Create
Impact of Microorganisms	**16.** Which of the following is both a reservoir and a transmitter? **a.** house fly **b.** a person with Lyme disease **c.** a mosquito **d.** none of the above	**17.** Differentiate among colonization, infection, and disease, noting a defining characteristic of each.	**18.** In question 2, we discussed pathogens that became less virulent the longer they were associated with the human host. But there is another category of pathogen that never loses any virulence over hundreds of years of association with humans. Can you deduce what circumstances make lower virulence of no advantage to a pathogen?
Scientific Thinking	**19.** T-F: All infectious diseases are also communicable. **a.** True **b.** False **c.** Need more information	**20.** You are performing Koch's postulates on sputum specimens taken from patients with a new unexplained respiratory illness. Agar cultures of all of the sputum show an abundance of a particular kind of colony, to the exclusion even of the normal variety of normal respiratory biota. No healthy subjects' cultures have this particular kind of colony. But when you administer this bacterium in aerosol form to the proper animal model, you do not get disease. What are the possible reasons for this?	**21.** Look at figure 13.3. Provide an editorial comment at the end of each row about where COVID-19 fits in the scenario.

Answers to the multiple-choice questions appear in Appendix A.

Visual Connections

This question uses visual images to connect content within and between chapters.

1. **From chapter 3, figure 3.3*a*.** What chemical is the organism in this illustration producing? How does this add to an organism's pathogenicity?

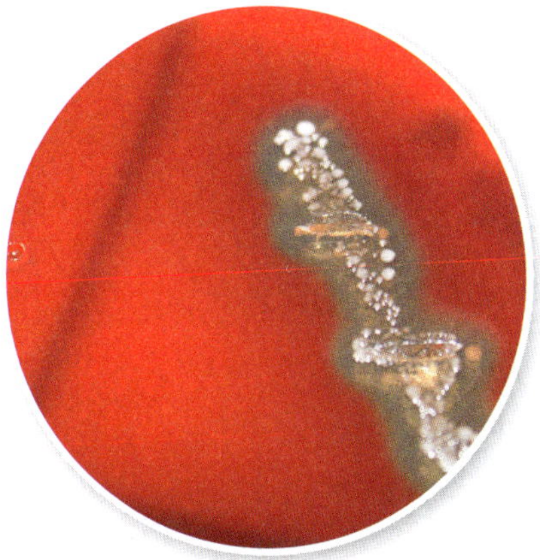

Lisa Burgess/McGraw Hill

High Impact Study

These terms and concepts are most critical for your understanding of this chapter—and may be the most difficult. Have you mastered them?

Concepts

- [ ] Human Microbiome Project
- [ ] Colonization vs. infection vs. disease
- [ ] Sites containing normal biota
- [ ] Steps a pathogen must take to cause disease
- [ ] Portals of entry and exit
- [ ] Three ways pathogens damage hosts
- [ ] Endotoxin vs. exotoxins
- [ ] Stages in the course of a disease
- [ ] Types of carriers
- [ ] Modes of transmission
- [ ] Koch's postulates

Terms

- [ ] Holobiont
- [ ] Resident microbiota
- [ ] Microbiome
- [ ] Opportunistic
- [ ] Pathogenic
- [ ] Virulence
- [ ] Virulence factors

14 Epidemiology of Infectious Diseases

Media Union/Shutterstock

MEDIA UNDER THE MICROSCOPE

Herd Immunity

These case studies examine an article from the popular media to determine the extent to which it is factual and/or misleading. This case focuses on the 2021 American Medical Association article "What Doctors Wish Patients Knew About COVID-19 Herd Immunity."

"Herd immunity occurs when a significant portion of a population becomes immune to an infectious disease, limiting further disease spread. For those who are not immune, they are indirectly protected because the ongoing disease spread is small." This is a quote from the article published on the AMA website on August 27, 2021. In this chapter you will learn about herd immunity, but you have already heard about it a lot during the COVID pandemic.

The article explains that originally health officials estimated that 70% of the population would have to be immunized, or recovered from the infection, before herd immunity could be achieved.

The article continues, "Even though a person may not have immunity for whatever reason—maybe they don't have a strong immune system, they're unable to get the vaccine or are children below a certain age—what we do is we surround them," he said. "We insulate them from the virus by having everybody around them have immunity."

The public has become confused, however. The original percentage required for herd immunity, when the disease spread essentially collapses, has had to be revised because this virus is very prone to mutate into types that are less well covered by the currently available vaccines. The article emphasizes that the vaccine still protects its recipient against severe disease, but because vaccinated people can still transmit the virus, the new herd immunity goal is 85%. Some people hear this and are upset that the "story keeps changing," but it's not the story that's changing, it's the virus.

- What is the **intended message** of the article?
- What is your **critical reading** of the summary of the article provided above? Remember that in this context, "critical reading" does not necessarily mean *What criticism do you have?* but asks you to apply your knowledge to interpret whether the article is factual and whether the facts support the intended message.

- How would you **interpret** the news item for your nonmicrobiologist friends?
- What is your **overall grade** for the news item—taking into account its accuracy and the accuracy of its intended effect?

Media Under The Microscope Wrap-Up appears at the end of the chapter.

Outline and Learning Outcomes

14.1 Epidemiology Basics and History

1. Identify the primary difference between the practice of medicine and epidemiology.
2. Describe the connection between public health and epidemiology.
3. Discuss the roles of two early epidemiologists of the modern era, Florence Nightingale and John Snow.

14.2 Units of Measure

4. Differentiate between incidence and prevalence.
5. Define point-source, common-source, and propagated epidemics, providing an indication of time frame for all three.
6. Outline the natural history of a disease.
7. Differentiate between screening tests and diagnostic tests.
8. Provide a clear distinction between correlation and causation.
9. Name and briefly define three study designs that help to contribute to understanding causation.
10. Discuss how Hill's criteria are used to bolster the case for causation.

14.3 Infectious Disease Epidemiology

11. Describe the role of surveillance in public health.
12. Name the four most important categories of healthcare-associated infection.
13. Define the basic reproduction rate and the case fatality rate with respect to infectious agents.
14. Fully explain the concept of herd immunity.
15. Provide an overview of the phases a vaccine goes through before approval by the FDA.
16. Describe the VAERS and the VICP.

14.1 Epidemiology Basics and History

Epidemiology is a field of study that has entered the public's mindset, due to the COVID-19 pandemic. The word literally means "epidemic-ology" and owes its name to the Greek terms for "upon" (epi-) and the "people" (dem). So epidemiology is the field of medicine that studies disease or adverse occurrences experienced by populations. In many ways, it is a departure from the practice of medicine with which we are all familiar. Other branches of medicine focus on addressing health and disease within an individual. For example, gastroenterologists specifically address diseases of the GI tract in individuals, oncologists manage and treat cancers in individuals, and obstetricians address all aspects of a mother's health and that of her fetus during pregnancy. You can think of epidemiology as "zooming out" quite a bit and noticing patterns of health and disease in designated populations.

Who, What, Where, When, and Why

Epidemiologists sometimes characterize their work as the 5 W's: Who, What, Where, When, and Why. Here are some situations in which they would apply these questions:

- A meat-processing plant reports that 27 of 42 staff members have experienced a similar-looking rash on their arms in the past month.
- A local luxury hotel reports that 25 of its 35 employees who work on the main floor and dining room have complained of respiratory symptoms, and 2 older personnel have been hospitalized for the symptoms.
- Hospitals around the country are experiencing admissions of patients with severe respiratory symptoms. The patients test negative for all common respiratory pathogens.
- Colon cancer has increased 52% over 10 years in one county in New Jersey.
- Adolescents decreased their use of tobacco products by 10% in a 2-year study period.

You might notice three things about this list of statements. First, they all address groups, not individuals. For example, the statement

> *The male patient had a very high blood pressure sustained over the last 90 days.*

would belong in the realm of medical practice, not epidemiology.

Second, the last two statements describe noninfectious conditions, such as cancer or even a behavior. That is because epidemiological principles are also useful to describe any type of group condition or behavior.

Third, each statement involves numbers, and rates. Epidemiology necessarily involves math, but you can get pretty far into its principles with just basic math.

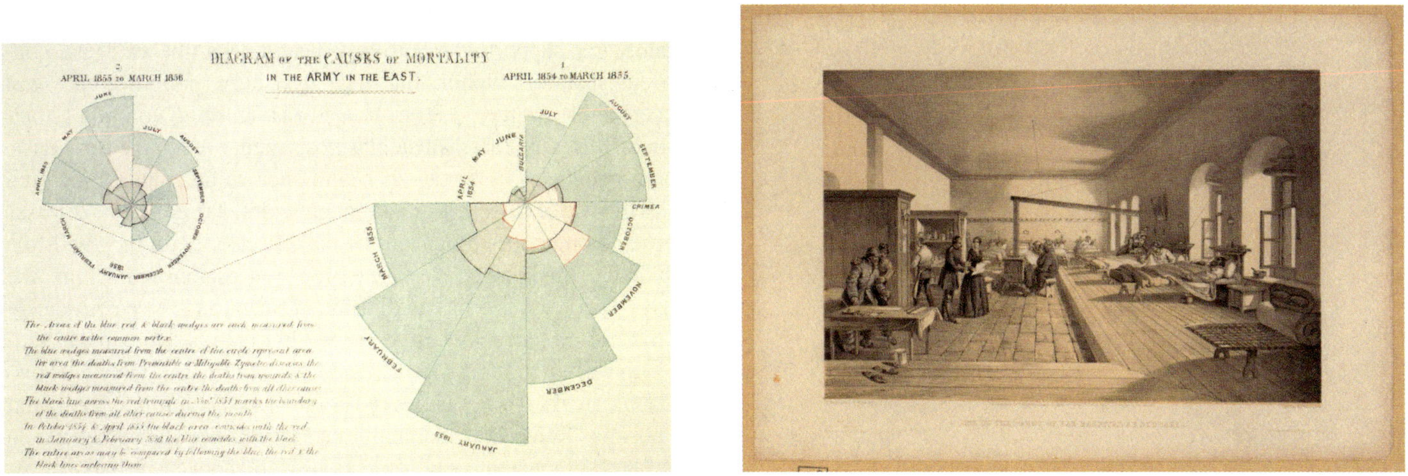
DIAGRAM OF THE CAUSES OF MORTALITY
IN THE ARMY IN THE EAST.
APRIL 1855 TO MARCH 1856.
APRIL 1854 TO MARCH 1855.

(a) (b)

Figure 14.1 The work of Florence Nightingale. **(a)** Data compiled by Florence Nightingale, an early example of epidemiologic study. **(b)** A military barracks turned field hospital in Turkey during the Crimean War. Nightingale pioneered her life-saving measures here.

(a) 914 collection/Alamy Stock Photo; (b) Library of Congress Prints & Photographs Division [LC-DIG-pga-05903]

Public Health and Epidemiology

Public health is a broad term for practices and activities designed to prevent diseases or other negative outcomes in a community, a country, or the world. Public health also encompasses activities to promote health and wellness. The information public health workers use for their work comes from epidemiological studies, largely. You might think of epidemiology as the science that informs the practice of public health. Other professions that are part of public health include social workers, policy makers, public health educators, infection control nurses in hospitals, and first responders. National public health initiatives currently include trying to limit the COVID-19 pandemic, suicide prevention, lowering the maternal mortality rate in the perinatal period, *Chlamydia* prevention, and improving opioid use detection in emergency rooms. These and many more initiatives are powered by the answers to the 5 W's, provided by epidemiological studies. The U.S. institution that is charged with protecting the public health is the Centers for Disease Control and Prevention, known as the CDC. It is part of the U.S. Public Health Service. The World Health Organization (WHO) oversees global health.

History of Epidemiology

The essence of epidemiology is noticing trends: looking at the big picture and deducing what is happening. As we discuss this science, keep in mind that the first epidemiologist was considered to be Hippocrates in 430 B.C. Then, during the height of the Islamic Golden Age, in the 1200s, Islamic scholars refined epidemiological principles, and preserved epidemiological study. Western scholars eventually built on their studies. The defining events of what we call "modern" epidemiology occurred in 1800s England. Interestingly, the year 1854 saw two critical events that were formative for epidemiology in the modern era. One of these took place in urban London, and the other on the Turkish battlefield of the Crimean War.

A Nurse and a Battlefield

If you are studying nursing, you have probably heard of Florence Nightingale. She was a nurse in England when she was sent to Turkey because British soldiers were dying at a very high rate—32 out of every 100—in a field hospital there. In addition to her nursing expertise, she had a love for mathematics, and that must have predisposed her to look at numbers and trends. She is credited with being among the first to collect and use health, disease, and death statistics for the purpose of improving public health. She noticed that the sanitary conditions were horrible in the field hospital (**figure 14.1**). The latrines were dug underneath the tents holding wounded soldiers. Most of the deaths were from dysentery, not gunshot wounds. Remember that the connection between unsanitary conditions and infections had not been firmly established by then. But her use of statistics and skill for identifying patterns led her to clean up the area and institute practices that soon reduced the percentage of soldiers dying in her field hospital to 2 of every 100, a far cry from the 32 in 100 when she arrived.

A Doctor and an Urban Epidemic

John Snow was a physician in London, England, in the 1800s. For several years, the city had been struggling with cholera cases. (Cholera is a bacterial infection leading to severe diarrhea and fluid loss. It is often deadly.) In 1854 the neighborhood of Soho in London was engulfed by a raging cholera epidemic. John Snow took it upon himself to try to put a stop to it. He had already believed for many years that cholera could be spread from water sources, but his arguments were not widely accepted. This time Snow made a map of the area and indicated the number of cases across the map (**figure 14.2**). This visual led him to a hypothesis that the water coming from a specific pump harbored the cholera bacterium.

In those days in London, a pump was the water source for a whole neighborhood. So when Snow's map showed cases clustered around a single pump—the Broad Street pump—he hypothesized that that pump was dispensing contaminated water. The town council did not believe him, but they agreed to remove the handle of that single pump. The epidemic stopped. So Dr. Snow asked two questions at the same time—Who? and Where? and his map answered those questions.

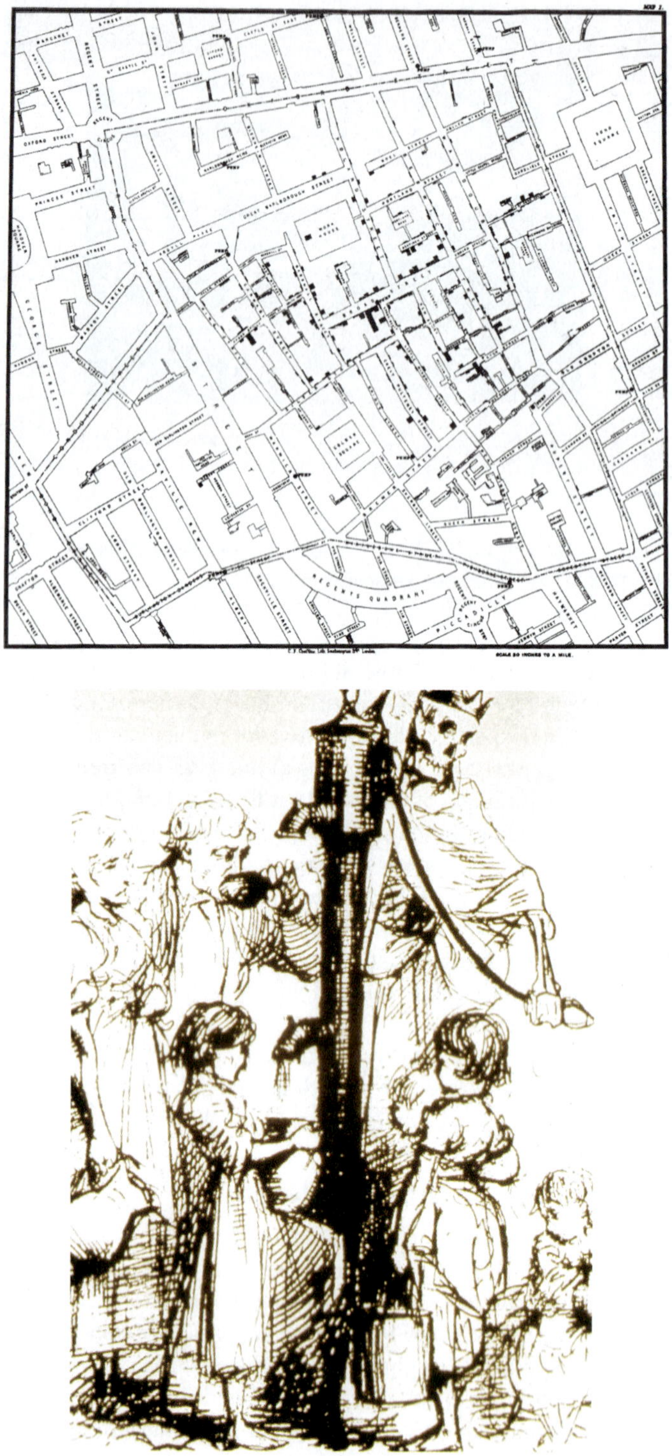

Figure 14.2 John Snow's map of cholera victims in London (1854). Each bar (top) indicated a death. The locations of water pumps are marked with a dot. Snow shut down the Broad Street pump and ended the outbreak. The bottom drawing is an artist's depiction of the "deadly" pump.

(Top) Pictorial Press Ltd/Alamy Stock Photo; (Bottom) George Pinwell/CDC

U.S. Epidemiology in the Last 75 Years

The CDC was established in 1946 with one goal: to slow down the spread of the mosquito-borne disease malaria, especially in the Deep South. Soon, its mandate expanded and it moved into its headquarters in Atlanta. It started to keep track of several infectious diseases, mainly in a supportive role to state health departments. These conditions included typhus, diarrheal dysentery, poliomyelitis, viral encephalitis, plague, Q fever, brucellosis, creeping eruption (a helminthic disease), rabies, histoplasmosis, and, even in the early days, some noninfectious threats to health such as the dangers involved with the use of insecticides and rodenticides. Nevertheless, diseases caused by microbes remained the highest priority. In the 1950s the Epidemic Intelligence Service (EIS) was established within the CDC. This corps is a special division of scientists that can be quickly activated and sent anywhere in the world to respond to outbreaks.

In early years, the EIS responded to hundreds of infectious disease outbreaks worldwide, and also an investigation into lead paint exposures. Remember that antibiotics came into widespread use in the late 1940s so that simple infections were more easily handled. At that time, the attention of the CDC turned increasingly to other types of disease, and even behaviors, most notably cigarette smoking beginning in 1964 **(figure 14.3).** That's when the push to discourage smoking started in the United States. Of course, since then, the rise of antibiotic resistance and emerging diseases have refocused epidemiologists on infectious diseases.

Recently, the CDC has been instrumental in responding to new epidemics such as the Ebola outbreaks focused in Africa beginning in 2014 and the COVID-19 pandemic beginning in 2019. A similar worldwide organization is the World Health Organization (WHO), and the CDC of the United States and other CDCs that are now active in other countries work with that body as appropriate.

SMOKING and HEALTH
REPORT OF THE ADVISORY COMMITTEE
TO THE SURGEON GENERAL
OF THE PUBLIC HEALTH SERVICE
U.S. DEPARTMENT OF HEALTH, EDUCATION, AND WELFARE
Public Health Service

Figure 14.3 Surgeon general's report on the effects of cigarette smoking. It was released on January 11, 1964.

United States. Public Health Service. Office of the Surgeon General

14.1 Learning Outcomes—Assess Your Progress

1. Identify the primary difference between the practice of medicine and epidemiology.
2. Describe the connection between public health and epidemiology.
3. Discuss the roles of two early epidemiologists of the modern era, Florence Nightingale and John Snow.

14.2 Units of Measure

The quickest and most powerful ways to describe or discover patterns is with numerical measures. Let's look at a few mathematical tools used in epidemiology. The **prevalence** of a disease is the total number of existing cases with respect to a given population. It is often thought of as a snapshot and is usually reported as the percentage of the population having a particular disease at any given time. Disease **incidence** measures the number of new cases over a certain time period (in a given population). The equations used to figure these rates are:

$$\text{Prevalence} = \frac{\text{Total number of cases in population}}{\text{Total number of persons in population}} \times 100 = \%$$

Example: The prevalence of smoking among adults in the United States is 17%.

$$\text{Incidence} = \frac{\text{Number of new cases in a designated time period}}{\text{Total number of susceptible persons}} \quad \text{(Usually reported per 100,000 persons)}$$

Example: The incidence of new tuberculosis cases in the United States in 2020 was 2.9 per 100,000.

The changes in incidence and prevalence are often followed over a seasonal, yearly, and long-term basis and are helpful in predicting trends **(figure 14.4).** Common statistics of interest to the epidemiologist are the rates of disease with regard to sex, race, or

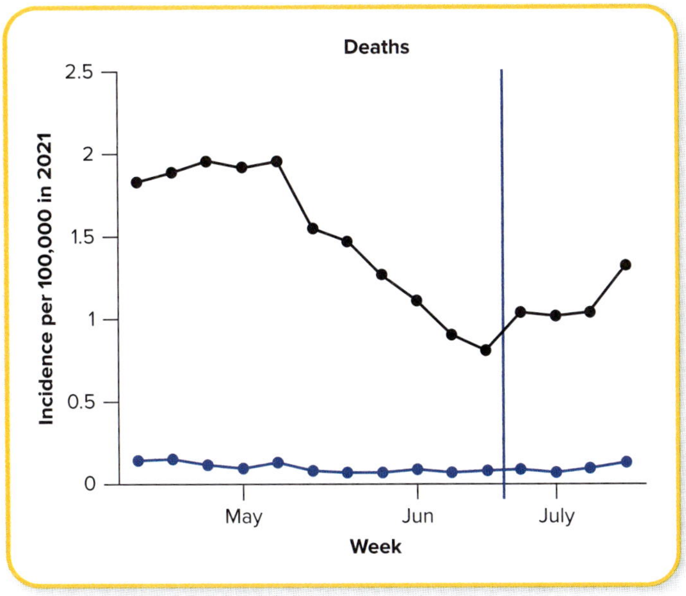
Deaths
Incidence per 100,000 in 2021
2.5
2
1.5
1
0.5
0
May
Jun
July
Week

(a) Incidence of death in the vaccinated (line on the bottom) and unvaccinated (line on the top).

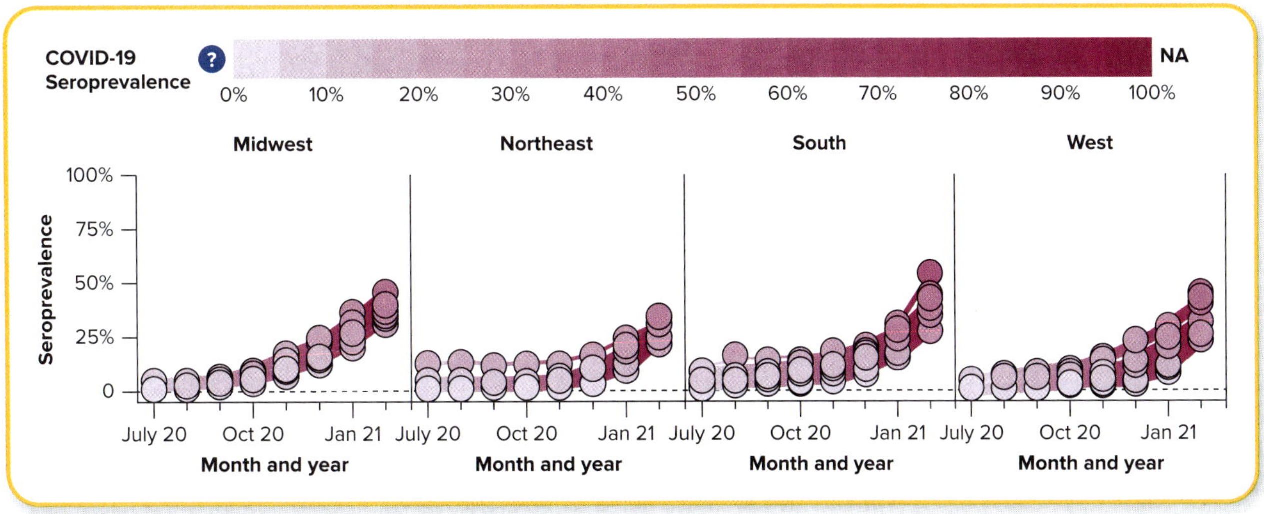
COVID-19 Seroprevalence
NA
0%
10%
20%
30%
40%
50%
60%
70%
80%
90%
100%
Midwest
Northeast
South
West
Seroprevalence
100%
75%
50%
25%
0
July 20
Oct 20
Jan 21
Month and year

(b) Prevalence (called seroprevalence since it is measured by blood tests of antibody) of COVID-19 in the four regions of the United States over a six-month period.

Figure 14.4 Incidence and prevalence graphs. **(a)** Incidence of deaths due to COVID-19. **(b)** Prevalence of positive antibody tests to COVID-19 in four regions of the United States.

CDC, Centers for Disease Control and Prevention.

geographic region. Also of importance is the **mortality rate,** which measures the number of deaths in a population due to a certain disease. Over the past century, the overall death rate from infectious diseases in the developed world has dropped, although the number of persons afflicted with infectious diseases (the **morbidity rate**) has stayed relatively high.

Disease Connection

Prevalence is affected by three factors: A new (incident) case adds to prevalence, and death and complete recovery are the only occurrences that decrease prevalence. Therefore, in the United States, HIV prevalence has *increased* even though there is much more awareness of how it is transmitted. Why? Because of our ability to treat it and keep it from killing its hosts, death removes cases from the prevalence calculation less frequently now.

Patterns

When there is an increase in disease in a particular geographic area, it can be helpful to create an epidemic curve (incidence over time) to determine if the infection is a **point-source, common-source,** or **propagated epidemic.** A *point-source epidemic,* illustrated in **figure 14.5*a*,** is one in which the infectious agent came from a single source and all of its victims were exposed to it from that source at the same time. The classic example of this is food illnesses brought on by exposure to a contaminated food item at a potluck dinner or restaurant. *Common-source epidemics* or outbreaks result from common exposure to a single source of infection that can occur over a period of time **(figure 14.5*b*).** Think of a contaminated water treatment plant that infects multiple people over the course of a week, or even of a single restaurant worker who carries hepatitis A and does not practice good hygiene. Finally, a *propagated epidemic* **(figure 14.5*c*)** results from an infectious agent that is communicable from person to person and therefore is sustained—propagated—over time in a population. Influenza is the classic example of this. The point is that each of these types of spread becomes apparent from the shape of the outbreak or the epidemic "curves." Time on the x axis is an important factor here.

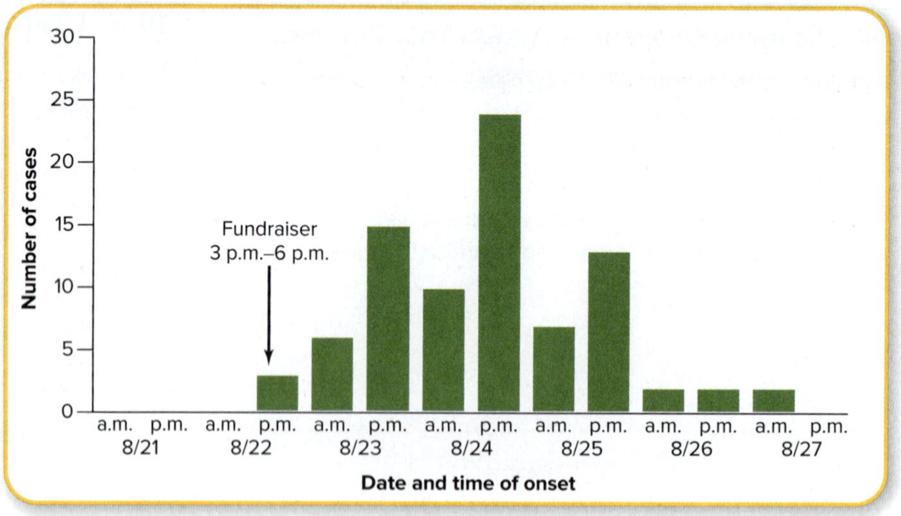
Number of cases
30
25
20
15
10
5
0
Fundraiser
3 p.m.–6 p.m.
a.m. p.m. a.m. p.m. a.m. p.m. a.m. p.m. a.m. p.m. a.m. p.m. a.m. p.m.
8/21 8/22 8/23 8/24 8/25 8/26 8/27
Date and time of onset

(a) Point-source epidemic traced to crab cakes at a fundraiser.

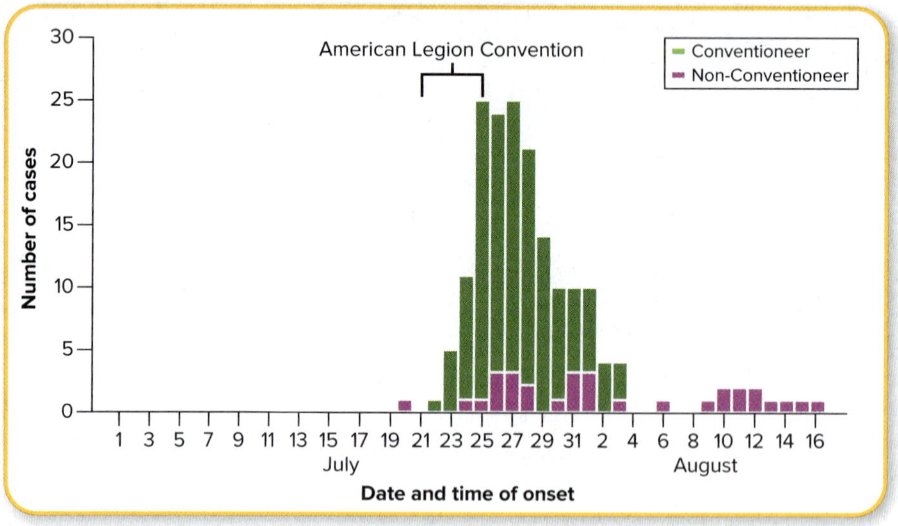
Number of cases
30
25
20
15
10
5
0
American Legion Convention
Conventioneer
Non-Conventioneer
1 3 5 7 9 11 13 15 17 19 21 23 25 27 29 31 2 4 6 8 10 12 14 16
July
August
Date and time of onset

(b) Common-source epidemic graph illustrating the first outbreak of Legionnaires' disease at the American Legion Convention in 1976 in Philadelphia.

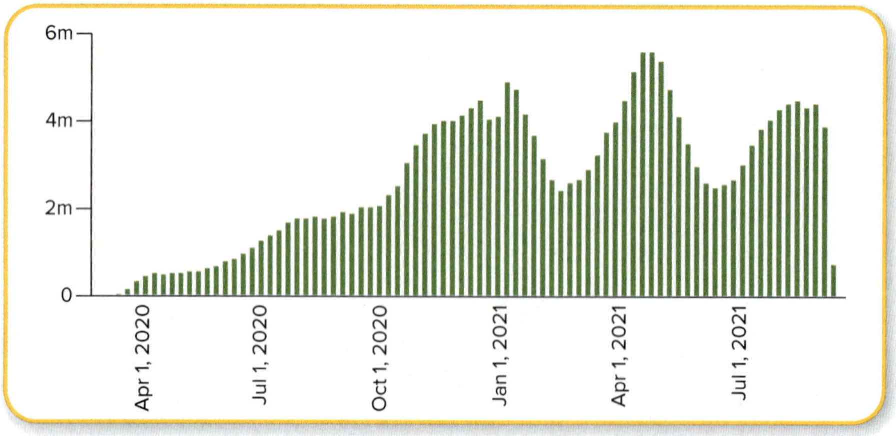
6m
4m
2m
0
Apr 1, 2020
Jul 1, 2020
Oct 1, 2020
Jan 1, 2021
Apr 1, 2021
Jul 1, 2021

(c) "Curve" representing the propagated epidemic of the SARS-CoV-19 worldwide.

Figure 14.5 Different outbreak or epidemic "curves." **(a)** Point-source epidemic; **(b)** common-source epidemic; **(c)** propagated epidemic. Notice the different spans of time represented on the x-axis.

(c) Source: World Health Organization (WHO)

An additional term, the **index case,** refers to the first patient found in an epidemiological investigation. How the cases unfurl from this case helps explain the type of epidemic it is. The index case may not turn out to be the first case—as the investigation continues, earlier cases may be found—but the index case is the case that brought the epidemic to the attention of officials. Monitoring statistics also makes it possible to define the frequency of a disease in the population. An infectious disease that exhibits a

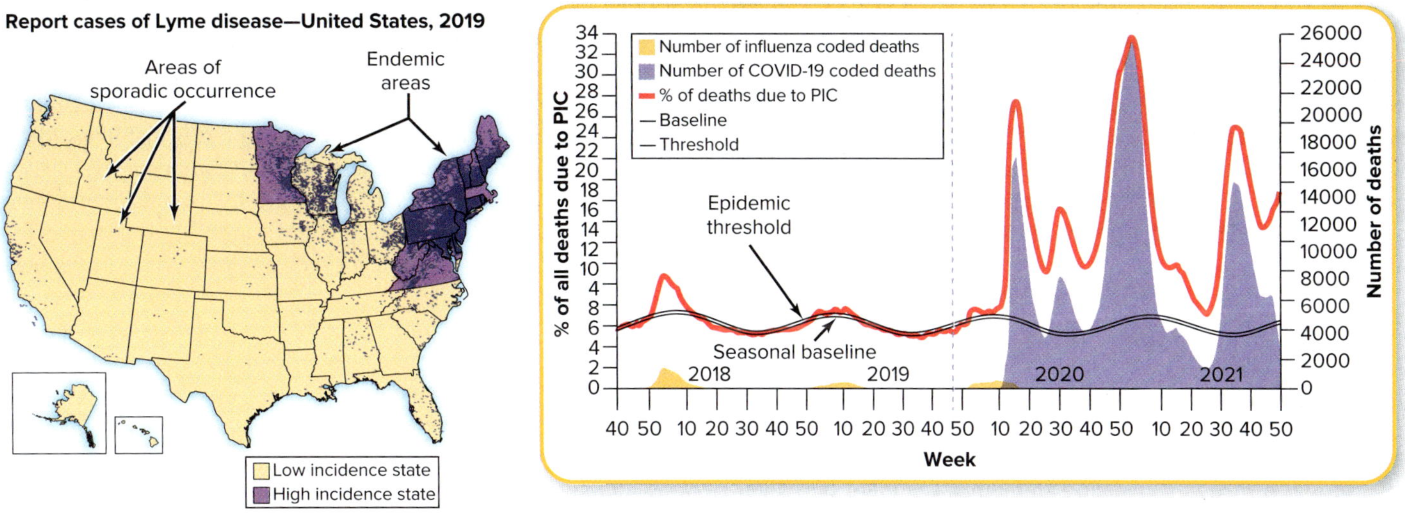
Report cases of Lyme disease—United States, 2019
Areas of sporadic occurrence
Endemic areas
Low incidence state
High incidence state
Number of influenza coded deaths
Number of COVID-19 coded deaths
% of deaths due to PIC
Baseline
Threshold
Epidemic threshold
Seasonal baseline
2018
2019
2020
2021
% of all deaths due to PIC
Number of deaths
Week

(a) Lyme disease incidence in 2019. One dot is placed randomly within county of residence for each confirmed case.

(b) To the left of the dotted line, the red line indicates the percentage of all U.S. deaths that were caused by influenza and pneumonia. To the right of the vertical dotted line, the numbers include deaths caused by COVID-19. The only epidemic levels of influenza were seen in the early winter of 2018. PIC stands for pneumonia, influenza, and COVID-19.

Figure 14.6 Patterns of infectious disease occurrence. **(a)** In an endemic occurrence, cases are concentrated in one area at a relatively stable rate. In a sporadic occurrence, a few cases occur randomly over a wide area. **(b)** An epidemic is an increased number of cases over the customary rate.
CDC, Centers for Disease Control and Prevention

relatively steady frequency over a long time period in a particular geographic locale is **endemic (figure 14.6*a*).** For example, Lyme disease is endemic to certain areas of the United States where its tick vector is found. A certain number of new cases are expected in these areas every year. Of course, in order to know whether the incidence is remaining the same or close to the same year after year, you have to plot the incidence of the disease over time. **Figure 14.6*b*** shows a common incidence graph, the incidence of influenza and pneumonia deaths (because when influenza leads to death it is usually through pneumonia). But in 2020 deaths by pneumonia started to also include deaths instigated by COVID-19, which are often caused by the sequelae, pneumonia. When a disease is **sporadic,** occasional cases are reported at irregular intervals in random locales. A single disease can be endemic in certain areas and sporadic in others. For example, in figure 14.6*a*, the occurrence of Lyme disease in New Mexico, Utah, and Idaho can be called sporadic. Some diseases, such as tetanus and diphtheria, are reported sporadically across the United States (fewer than 50 cases a year).

When statistics indicate that the prevalence of an endemic or sporadic disease is increasing beyond what is expected for that population, the pattern is described as an **epidemic.** To see this, the incidence must be visualized over time (figure 14.6*b*). The time period over which this change occurs differs for each disease. It can range from hours in food poisoning, to years in syphilis. Also, the exact percentage of increase needed before an outbreak can qualify as an epidemic is specific for each disease. Figure 14.6*b* shows the expected percentage of deaths from influenza across seasons for several years. When the actual rate significantly exceeds the "normal," or baseline, rate, indicated by the top black line, it is called an epidemic. The spread of an epidemic across continents is a **pandemic,** as exemplified by HIV, influenza, and COVID-19.

One important epidemiological phenomenon may be called the "iceberg effect," which refers to the fact that only a small portion of an iceberg is visible above the surface of the ocean, with a more massive part remaining unseen below the surface. Regardless of case reporting and public health screening, a large number of cases of infection in the community go undiagnosed and unreported. In the case of salmonellosis, approximately 40,000 cases are reported each year. Epidemiologists estimate that the actual number is more likely somewhere between 400,000 and 4,000,000. The iceberg effect can be even more lopsided for sexually transmitted infections or for infections that are not brought to the attention of reporting agencies.

The Natural History of Diseases

It is easy to think of disease as an event characterized by some unwanted symptoms. But epidemiologists think of it in a more holistic way. Each disease, whether it is infectious or noninfectious, proceeds along a timeline, as seen in **figure 14.7.** The lengths of the four major stages depicted underneath the line vary greatly, depending on the disease.

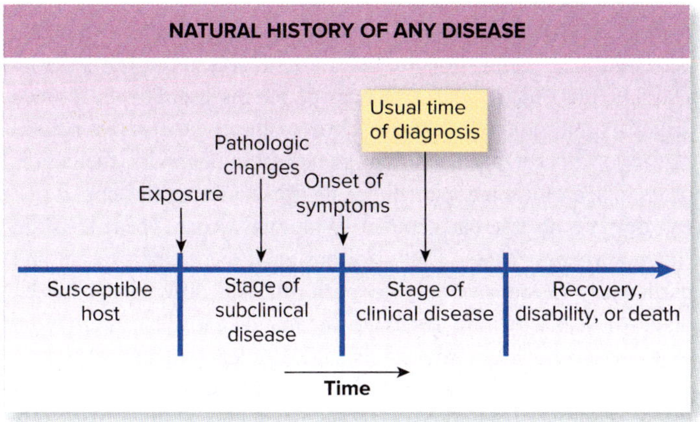
NATURAL HISTORY OF ANY DISEASE
Usual time of diagnosis
Pathologic changes
Exposure
Onset of symptoms
Susceptible host
Stage of subclinical disease
Stage of clinical disease
Recovery, disability, or death
Time

Figure 14.7 The natural history of disease.

The first stage, "susceptible host," depicts a person who has not yet encountered the agent or the circumstances that will result in an alteration in health. At some point, exposure occurs. For some period of time, the damage is not apparent. This period is termed the "subclinical disease" stage. Subclinical means that it would not be noticed by the host themselves or their health care providers. This period can be instantaneous, or of no length at all, as in the case of ingesting poison, in which the condition would appear almost immediately. More commonly, the subclinical stage can last a few hours to a few days or even weeks, and sometimes months or years. During this period, pathological changes may start. As long as they are not noticed by the host, it is still in the subclinical phase.

When symptoms become apparent, the third phase is the "clinical disease" stage, in which the most obvious effects of the disease are visible as signs and symptoms. There are only three ways for this to play out: complete recovery, remaining symptoms that we refer to as disability, and death. The course of a COVID-19 infection provides a very good illustration of these phases.

Many outcomes are possible after infection with SARS-CoV-2:

1. You remain asymptomatic, and only know you are infected if you get tested.
   In figure 14.7, this situation means that infection at the point of exposure *arrow has occurred, but the pathological consequences did not cause noticeable changes, and you advance directly to* recovery.
2. You have mild symptoms and recover fully.
   In the figure, the time you spend in the clinical disease stage *never results in advanced symptoms and you travel on to* recovery.
3. You have serious symptoms and are hospitalized.
   At this point, any of the three outcomes in the last stage can occur: You recover completely, you recover from acute disease but have some ongoing deviation from health/disability, or you die.
4. You experience mild symptoms and recover from some of them, but you continue to suffer long-term with fatigue, blood clots, myocarditis, or other symptoms.
   In the case of COVID-19, many patients experience only mild symptoms in the clinical disease stage *but then move into the last stage with lingering symptoms or even different symptoms than in the clinical stage.*

Take special note of the box "Usual Time of Diagnosis." This box occurs early in the *clinical disease stage*, soon after symptoms begin. In many cases, you are not going to seek a diagnosis before you feel any negative consequences of the disease. However, once scientists and clinicians have a clear picture of the entire natural history of a particular condition, gained from observing the course of the disease in many patients, sometimes a test is developed that can detect a disease early, before symptoms occur. These tests are called **screening tests.** These are designed to be noninvasive and particularly cheap and easy to perform. The difference between diagnosis and screening lies in the fact that diagnostic tests are often more invasive, with more hazards, and with less patient acceptance. Ideally, screening tests are deployed on healthy populations to detect a disease before it causes harm. A good example of this is getting is the mammogram. This is screening test: performed as a "first pass" on healthy people. If something comes up on the mammogram, more invasive testing is called for. Testing for prostate-specific antigen (PSA) as a screen for prostate cancer is also done via blood samples on healthy men. The essence of a screening test is that it has lower accuracy than a true diagnostic test. A positive screening test is usually followed by a more accurate diagnostic-style test. Pap smears are screening tests for cervical cancer **(figure 14.8).** Cells are collected from the cervix during a vaginal exam, and the cells are examined under a microscope to detect abnormalities. At the same time, the samples are tested for human papillomaviruses that might lead to cancer. If either of these screening procedures is positive, a biopsy of the cervix is taken at a later date.

Especially in the case of an infectious disease (as opposed to, say, cancer or heart disease), the screening can serve as the final diagnosis as well. COVID-19 screening tests are the first and usually last steps in determining if a person has COVID. The same is true with influenza tests and most sexually transmitted infection tests. But even in these cases, it is important to understand the likelihood that the tests are accurate or inaccurate.

The way to define accuracy for a screening test is to determine two different types of mistakes thath occur. One of these is how many people it calls positive who are actually negative. These are called false-positives and determine the **specificity** of a test. The opposite situation is the **sensitivity** of a test, determined by how many people who were positive were incorrectly found negative with the screening test. These results are false negatives.

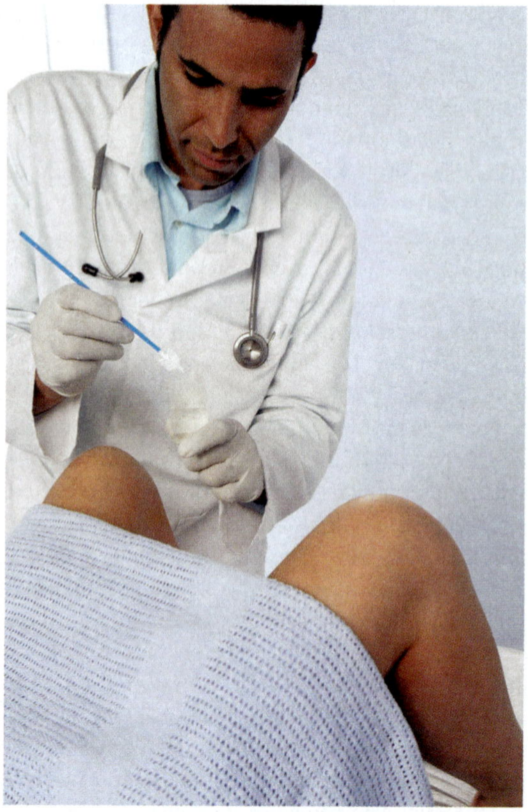

Figure 14.8 Physician places a sample from a Pap smear in liquid medium.
Adam Gault/SPL/Getty Images

Causality

In the infectious disease world, the causative agent of a particular disease is often determined by using Koch's postulates, as you saw in the last chapter. Also, molecular techniques such as PCR have identified microbes causing diseases in hosts. But finding the cause of some infectious diseases, chronic diseases, and metabolic diseases can be trickier. For one thing, many of these diseases have multiple interacting causes. Look at cardiovascular disease, for example. There can be several different causes contributing to the condition, such as diet, tobacco-smoking, and genetic predisposition.

As with Koch's postulates, it is possible to expose experimental animals to some of the possible causes and see if they develop the disease. But true proof of causation requires seeing the disease in humans. You will see many headlines and "breakthrough findings" claiming that factor *x* causes disease *y*, but the vast majority of those reports are referring to a correlation being found, not a cause.

The only sure way a factor can be found to cause an outcome is to take a healthy subject, expose them to the factor, and see that it causes the effect—the disease. In most cases of human disease, this would be unethical. For example, you cannot take a person who has never smoked cigarettes, direct them to smoke cigarettes for *x* length of time, and see if they develop COPD, or lung cancer. So, there are a variety of epidemiological strategies to approach the question ethically.

At this point, we should make a clear distinction between correlation and cause. It is quite easy to find correlations between various exposures and particular diseases. For example, you could graph the number of divorces in the state of Maine vs. the per capita consumption of margarine **(figure 14.9).** The fact that one of the variables changes in the same direction as the other variable indicates a correlation, but without more information, it does not indicate causation. Of course, you could make a graph between numbers of families with new trampolines vs. number of pediatric emergency visits, and you might also find a correlation. But, again, without further information, that would also not prove causation. You have a suspicion that it does, but it does not.

To prove causation between any given exposure and any given outcome, there are two potential paths to take: The most straightforward method is to perform an actual experiment. That means there should be two groups of people, and one of the groups should be deliberately exposed to the suspected factor, and the other group, known as the control group, will not be exposed to the suspected factor. You are probably already thinking that it doesn't sound ethical to expose healthy people to a disease-causing agent. That is 100% true. The only types of experiments like these that are performed in the U.S. are ones that test exposures that might have *positive* effects. It is ethical to test treatments using people who have diseases or conditions (of course, there are still strict criteria and controls). The intervention that is applied is one that is potentially beneficial, as opposed to one that is potentially detrimental. This is the rationale behind drug trials. The technical term for these kinds of experiments is **randomized control trials,** or RCTs.

When trying to determine whether an exposure is causative for a human disease or condition, RCTs are not appropriate. In those cases, scientists can conduct animal studies **(Insight 14.1),** and/or conduct one or more of the following types of studies:

- Cross-sectional studies: The two factors (exposure and disease) are collected at a single point in time, with no reference to which might have occurred first. An example would be if a group of adults were surveyed about their intake of alcohol and their incidence of depression. While there might be a correlation found between amount of alcohol and degree of depression, it would be impossible to determine which led to the other, and therefore which was a causative agent. But it's a start.
- Case-control studies: In these studies, a group of people are categorized as to whether they have the condition (outcome) or not, and then both groups are asked about certain exposures they have had to see if one exposure stands out in the group with the condition. An example would be enrolling a group of people in the 35–55 age group in a study and determining which ones have hearing loss and which do not. Then individuals in both of those groups will be questioned about their attendance at rock concerts during their lives.
- Cohort studies: In these studies, people without the condition in question are enrolled in the study and followed over time. They are monitored both for the appearance of a disease and for several exposure factors that are suspected of leading to that disease. For example, young healthy adult volunteers in a

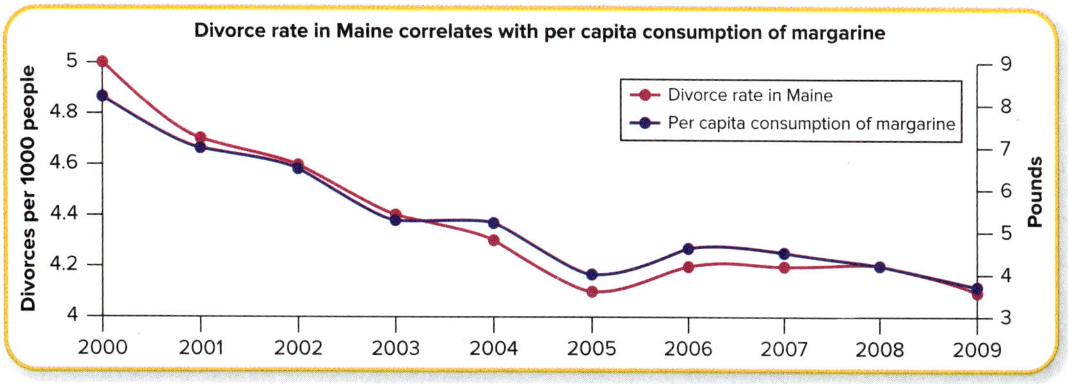
Divorce rate in Maine correlates with per capita consumption of margarine
Divorce rate in Maine
Per capita consumption of margarine
Divorces per 1000 people
Pounds
5
4.8
4.6
4.4
4.2
4
9
8
7
6
5
4
3
2000
2001
2002
2003
2004
2005
2006
2007
2008
2009

Figure 14.9 A graph correlating two completely unrelated phenomena. The correlation value for this, which measures how closely the two factors are when plotted in a certain way, is 99%, even though clearly neither one *causes* the other.

Source: Tyler J. Vigen

INSIGHT 14.1 MICROBIOME: Why Does Increased Weight Lead to Cognitive Impairment?

For some time now, scientists have documented a decline in cognitive abilities that seemed to be associated with obesity, and poor diet. Like many associations, this one has been merely correlative, and not causal, as the types of studies required to determine causation have remained difficult to conduct.

A 2020 review of these studies was conducted by a group of scientists from Australia. They noted, initially, that cognitive impairment and increased dementia risk are seen in persons with poor diet and obesity. Their study tried to look at the mechanisms for this. They analyzed 42 different research studies looking at the link. All of the research they looked at used rats as the subjects, and exposed rats to an "obesogenic diet"—one that leads to weight gain—and tested their cognitive abilities before and after doing so.

The researchers concluded that a poor diet and overeating leads to two changes: a shift in the gut microbiome composition and increased adipose tissue. Both of those outcomes lead to increased systemic inflammation, which is likely to lead to impaired cognition.

These studies, in which healthy mice are exposed to an obesogenic diet and then observed for changes in their cognition, are not ethical to perform in humans. So this is a case where experimental trials conducted in animals can help us fulfill important criteria for causation in humans using a variety of study methods.

Over Weight
Healthy Weight
Obese
Extreme Obesity

MedicalRF.com

particular city are enrolled in a study about cardiovascular disease. On a quarterly basis for the next 10 years, each volunteer gets blood work and a survey about several risk factors they may have engaged in or been exposed to. Meticulous records are kept. Some of the subjects will, in fact, develop cardiovascular disease. All of their exposures will have been documented as they occurred, and those exposures can be compared for the subjects who developed disease and the ones who did not.

Because none of these study designs are true experiments, additional criteria have to be satisfied in order to declare an exposure as causative, rather than simply correlative. These criteria are called Hill's criteria. There are nine criteria. Here are some of the main ones:

- **Strength of association.** This is dependent on statistical analysis and asks *how strongly* the exposure and the outcome are sometimes associated.
- **Consistency of association.** When multiple studies of the same and different designs provide the same result, then we can call the findings consistent.
- **Specificity of association.** This refers to the extent that only one exposure is associated with a disease, or an exposure only results in a single disease.
- **Temporality of association.** Put simply, the cause you are investigating can be shown to have occurred before the outcome. This is key, and sometimes is very difficult to establish.
- **Biologic plausibility.** Does the process that took place between exposure and outcome have a precedent in what we know in human biology? An example is the early assertion that cigarette smoking could cause lung cancer. There are many steps that go into the inception of a cancer. There are chemicals in cigarettes that could, plausibly, set them in motion. Again, it is reinforcing evidence, but not definitive evidence, on its own. But if you can't trace a biological line between the exposure and the outcome, it lacks biologic plausibility.

Fulfilling any one of these criteria certainly does not lead you to causation. And you will never fulfill all of them without an RCT. But if you can satisfy several of these criteria with your collection of non-RCT studies, you can form a hypothesis about causation.

14.2 Learning Outcomes—Assess Your Progress

4. Differentiate between incidence and prevalence.
5. Define point-source, common-source, and propagated epidemics, providing an indication of time frame for all three.
6. Outline the natural history of a disease.
7. Differentiate between screening tests and diagnostic tests.
8. Provide a clear distinction between correlation and causation.
9. Name and briefly define three study designs that help to contribute to understanding causation.
10. Discuss how Hill's criteria are used to bolster the case for causation.

14.3 Infectious Disease Epidemiology

Surveillance and Notifiable Disease

The ability to notice patterns in disease relies on **surveillance** activities. This refers to the collecting, analyzing, and reporting of data on the rates of occurrence, mortality, morbidity, and transmission of infections. Surveillance involves keeping data for

Table 14.1 Nationally Notifiable Infectious Diseases in the United States, 2021*

- *Anaplasma phagocytophilum*
- Anthrax
- Arboviral neuroinvasive and non-neuroinvasive diseases
  - California encephalitis virus disease
  - Chikungunya virus disease
  - Eastern equine encephalitis virus disease
  - Jamestown Canyon virus disease
  - Keystone virus disease
  - La Crosse disease
  - Powassan virus disease
  - St. Louis encephalitis virus disease
  - Trivittatus virus disease
  - West Nile virus disease
  - Western equine encephalitis disease
- Babesiosis
- Botulism
- Brucellosis
- Campylobacteriosis
- *Candida auris*
- Carbapenemase-producing carbapenem-resistant *Enterobacteriaceae* (CP-CRE)
- Chancroid
- *Chlamydia trachomatis*
- Cholera
- Coccidioidomycosis
- Congenital syphilis/syphilitic stillbirth
- Coronavirus disease 2019 (COVID-19)
- Crimean-Congo hemorrhagic fever
- Cryptosporidiosis
- Cyclosporiasis
- Dengue (including Dengue fever, hemorrhagic fever, and shock syndrome)
- Diphtheria
- Ehrlichiosis
- Giardiasis
- Gonorrhea
- Guanarito hemorrhagic fever
- *Haemophilus influenzae,* invasive disease
- Hansen's disease (leprosy)
- Hantavirus pulmonary syndrome
- Hemolytic uremic syndrome, post-diarrheal
- Hepatitis A, B, C
- HIV diagnosis
- Influenza-associated pediatric mortality
- Invasive pneumococcal disease (IPD)/ *Streptococcus pneumoniae*, invasive disease
- Junín hemorrhagic fever
- Lassa fever
- Legionellosis
- Leptospirosis
- Listeriosis
- Lujo virus
- Lyme disease
- Malaria
- Measles
- Meningococcal disease
- Mumps
- Novel influenza A infections
- Pertussis
- Plague
- Poliomyelitis, paralytic
- Poliovirus infection, nonparalytic
- Psittacosis
- Q fever
- Rabies, animal or human
- Rubella
- Rubella, congenital syndrome
- Salmonellosis
- Severe acute respiratory syndrome–associated coronavirus (SARS-CoV) disease
- Shiga toxin–producing *Escherichia coli* (STEC)
- Shigellosis
- Smallpox
- Snowshoe hare virus disease
- Spotted fever rickettsiosis
- Streptococcal toxic-shock syndrome
- Syphilis
- Tetanus
- Toxic shock syndrome (other than streptococcal)
- Trichinellosis
- Tuberculosis
- Tularemia
- Typhoid fever
- Vancomycin-intermediate *Staphylococcus aureus* (VISA)
- Vancomycin-resistant *Staphylococcus aureus* (VRSA)
- Varicella
- Vibriosis
- Viral hemorrhagic fevers due to:
  - Crimean-Congo hemorrhagic fever virus
  - Ebola virus
  - Marburg virus
  - New world arenaviruses (Guanarito, Machupo, Junin, and Sabia viruses)
- Yellow fever
- Zika virus infection and disease

*Reportable to the CDC; other diseases may be reportable to state departments of health.
Source: Centers for Disease Control and Prevention, 2022.

a large number of diseases seen by the medical community and reported to public health authorities. By law, certain **reportable,** or notifiable, **diseases** must be reported to health agencies; others are reported on a voluntary basis. The diseases that are reportable are either highly contagious, highly dangerous, or highly unusual, and so being aware of every occurrence is important for effective surveillance. A list of officially reportable conditions is in **table 14.1.**

A well-developed network of individuals and agencies at the local, district, state, national, and international levels keeps track of infectious diseases. Physicians and hospitals report all notifiable diseases that are brought to their attention. These reports are made either about individuals or in the aggregate, depending on the disease.

The principal government agency responsible for keeping track of infectious diseases nationwide is the CDC. The CDC publishes a weekly notice of diseases (the *Morbidity and Mortality Report*) that provides weekly and cumulative summaries of the case rates and deaths for about 50 notifiable diseases, highlights important and unusual diseases, and presents data concerning disease occurrence in the United States. It is available to anyone at www.cdc.gov/mmwr/. Ultimately, the CDC shares its statistics on disease with the WHO for worldwide tabulation and control.

Healthcare-Associated Infections: The Hospital as a Source of Disease

Infectious diseases that are acquired or develop during a hospital stay (or a stay in another health care facility, such as a rehabilitation hospital) are known as **healthcare-associated infections (HAIs).** This concept seems strange at first thought because a hospital is regarded as a place to get treatment for a disease, not a place to acquire a disease. Yet it is not uncommon for a surgical patient's incision to become infected or a burn patient to develop a case of pneumonia in the clinical setting. The CDC estimates that every day, 1 in 31 hospital patients and 1 in 43 nursing home residents has an HAI. In light of the number of admissions, this adds up to 2 to 4 million cases a year, which result in nearly 90,000 deaths. HAIs cost time and money as well as suffering. By one estimate, they amount to 8 million additional days of hospitalization a year and an increased cost of $5 to $10 billion.

So many factors unique to the hospital and nursing home environment are tied to HAIs that a certain number of infections are virtually unavoidable. After all, the hospital both attracts and creates compromised patients, and it serves as a collection point for pathogens. Some patients become infected when surgical procedures or lowered defenses permit resident biota to invade their bodies. Other patients acquire infections directly or indirectly from fomites, medical equipment, other patients, medical personnel, visitors, air, and water. The health care process itself increases the likelihood that infectious agents will be transferred from one patient to another. Indwelling devices such as catheters, prosthetic heart valves, grafts, drainage tubes, and tracheostomy tubes form ready portals of entry and habitats for infectious agents. Because such a high proportion of the hospital population receives antimicrobial drugs during their stay, drug-resistant microbes are selected for at a much greater rate than is the case outside the hospital.

The most common healthcare-associated infections involve invasive devices or surgical procedures. The top four HAIs are

- Catheter-associated urinary tract infections (CAUTI),
- Central line–associated bloodstream infections (CLABSI),
- Surgical site infections (SSI), and
- Ventilator-associated events (VA).

Two specific bacterial species are also tracked: methicillin-resistant *Staphylococcus aureus* (MRSA) and *Clostridioides difficile (C. diff)*.

The federal government has taken steps to incentivize hospitals to control HAI transmission. Starting in 2008, the Medicare and Medicaid programs stopped reimbursements to hospitals for CAUTIs, CLABSIs, and SSIs acquired during hospital care. As can be seen in **Figure 14.10,** the measures continue to help reduce the rates, although these gains are not universal and many hospitals still struggle. In the second half of of 2020, hospitals reported a 25–31% increase in ventilator usage due to the COVID-19 pandemic. By late 2020, hospitals were seeing a 45% increase in ventilator-associated HAIs.

Medical asepsis includes practices that lower the microbial load in patients, caregivers, and the hospital environment. These practices include proper hand washing, disinfection, and sanitization, as well as patient isolation. The goal of these procedures is to limit the spread of infectious agents from person to person. An even higher

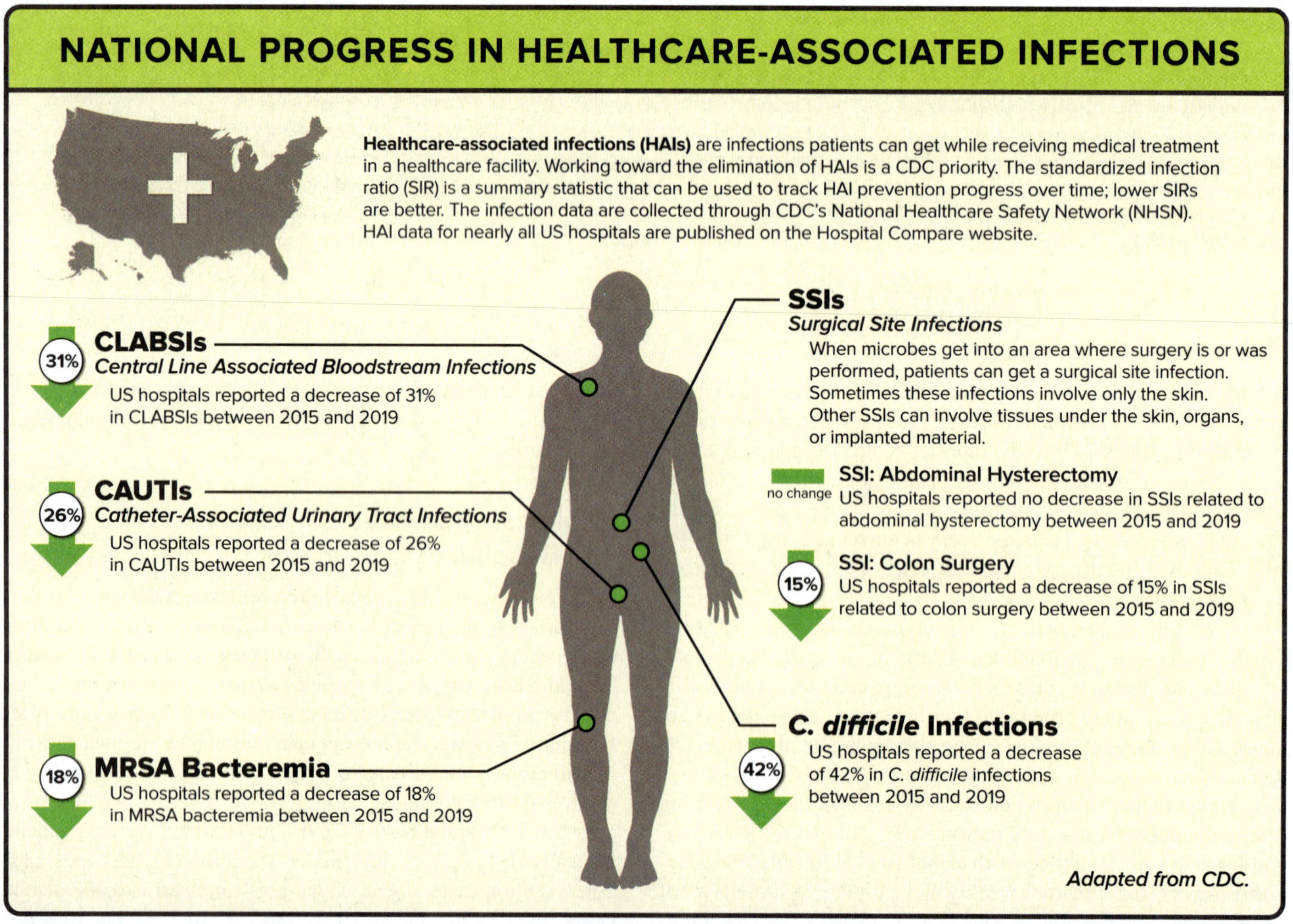

NATIONAL PROGRESS IN HEALTHCARE-ASSOCIATED INFECTIONS
Healthcare-associated infections (HAIs) are infections patients can get while receiving medical treatment in a healthcare facility. Working toward the elimination of HAIs is a CDC priority. The standardized infection ratio (SIR) is a summary statistic that can be used to track HAI prevention progress over time; lower SIRs are better. The infection data are collected through CDC's National Healthcare Safety Network (NHSN). HAI data for nearly all US hospitals are published on the Hospital Compare website.
31%
CLABSIs
Central Line Associated Bloodstream Infections
US hospitals reported a decrease of 31% in CLABSIs between 2015 and 2019
26%
CAUTIs
Catheter-Associated Urinary Tract Infections
US hospitals reported a decrease of 26% in CAUTIs between 2015 and 2019
18%
MRSA Bacteremia
US hospitals reported a decrease of 18% in MRSA bacteremia between 2015 and 2019
SSIs
Surgical Site Infections
When microbes get into an area where surgery is or was performed, patients can get a surgical site infection. Sometimes these infections involve only the skin. Other SSIs can involve tissues under the skin, organs, or implanted material.
no change
SSI: Abdominal Hysterectomy
US hospitals reported no decrease in SSIs related to abdominal hysterectomy between 2015 and 2019
15%
SSI: Colon Surgery
US hospitals reported a decrease of 15% in SSIs related to colon surgery between 2015 and 2019
C. difficile Infections
42%
US hospitals reported a decrease of 42% in C. difficile infections between 2015 and 2019
Adapted from CDC.

Figure 14.10 Trends in the most common healthcare-associated infections. The difference between 2015 and 2019 rates is indicated. (Verified data usually run about two years behind.)

Centers for Disease Control and Prevention, 2021

level of stringency is seen with *surgical asepsis,* which involves all of the strategies listed previously plus ensuring that all surgical procedures are conducted under sterile conditions. This includes sterilization of surgical instruments, dressings, sponges, etc., as well as personnel wearing sterile garments and scrupulously disinfecting the room surfaces and air. But it seems that there are always gaps.

Hospitals generally employ an *infection-control officer* who not only implements proper practices and procedures throughout the hospital but also is charged with tracking potential outbreaks, identifying breaches in asepsis, and training other health care workers in aseptic technique. Among those who benefit most from this training are nurses and other caregivers whose work, by its very nature, exposes them to needlesticks, infectious secretions, blood, and physical contact with the patient. The same practices that interrupt the routes of infection in the patient can also protect the health care worker. It is for this reason that most hospitals have adopted universal precautions that recognize that all secretions from all persons in the clinical setting are potentially infectious and that transmission can occur in either direction.

Reproducibility Rate and Case Fatality Rate

There are large differences among microbes as to their degree of communicability, which indicates how likely they are to cause epidemics. There are also large differences in how deadly they are. Epidemiologists quantify communicability by a factor called R_0 (pronounced "R-sub-zero" or "R-naught"), the basic reproduction rate. It describes how many susceptible people, on average, that one infected person will spread the infection to. The highly contagious measles virus has an R_0 of about 15, meaning that one infected person can spread the infection, on average, to 15 other individuals. The R_0 assumes those 15 people are unvaccinated for the microbe and have not experienced the infection, and therefore have no secondary immunity.

Deadliness is calculated via the case fatality rate (CFR): the numbers of persons who die of the disease ÷ the number of persons infected. This calculation is based on persons who receive no treatment. The case fatality rate for rabies is approximately 100%. The case fatality rate for cholera is about 1%.

These measures of infectious disease are approximate and can vary based on geographic location, the circumstances of an outbreak, and the variant of the microbe. But a general idea of R_0 and CFR can guide a variety of health care decisions and policies. Diseases with an extremely high R_0, for example, are the diseases for which vaccination is most needed. When we get to the disease chapters of this book, we will highlight key diseases in each chapter on a graph like **figure 14.11.** We will take both of these measures and divide them into quarters—grouping all CFRs from 0% to 25% in one quarter (at the bottom of the pyramid), and grouping R_0s into four broad categories. Practice reading the graph a little bit: Find an infection that is highly communicable but not very deadly. Now find one that is not very communicable but deadly. The good news is that none of these common infections are both highly deadly and highly communicable. And most of them are minimally communicable and minimally deadly.

It is very difficult to pin down an R_0 for a microbe causing a pandemic while the pandemic is in progress. Keep in mind that the R_0 is an estimate of spread in an unvaccinated population in which no other control measures (such as social distancing or wearing masks) have been implemented. In the case of COVID-19, early R_0 estimates ranged from 2.0 to 5.0 in different parts of the world. Epidemiologists do continue to monitor the reproductive rate as control measures are put in place, but they call it the *effective* reproductive rate. In 2021 a variant of COVID-19 called *delta* became dominant. Its R_0 seems to have been somewhere between 3.0 and 5.0. Both the original strain and the delta variant had a CFR of between 1.6 and 2.0. Soon after, a variant called *omicron* arose. Omicron was even more contagious, with estimates of an R_0 starting at 5.0. Its CFR is difficult to calculate, because so many people have been vaccinated and so many have had the disease and recovered. Frustratingly, it seems that *omicron* subvariants can cause COVID-19 disease both in people vaccinated for the orignal strain and those who have had the infection.

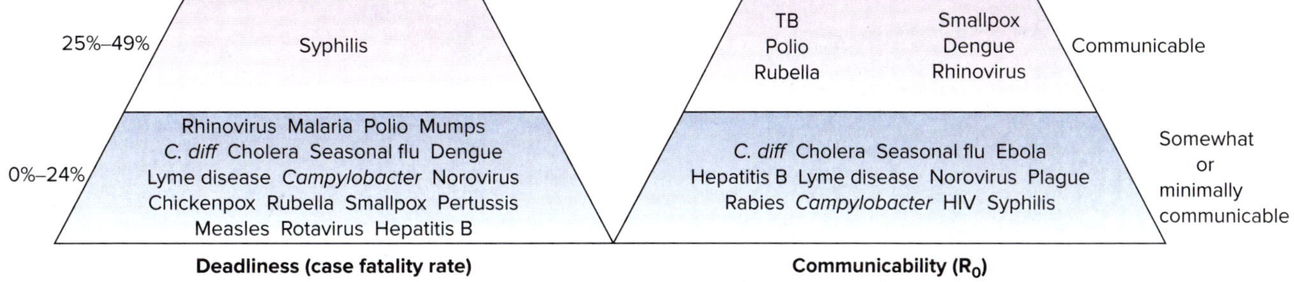
75%–100%
Rabies
HIV
50%–74%
TB
Ebola
Plague
25%–49%
Syphilis
0%–24%
Rhinovirus Malaria Polio Mumps
C. diff Cholera Seasonal flu Dengue
Lyme disease Campylobacter Norovirus
Chickenpox Rubella Smallpox Pertussis
Measles Rotavirus Hepatitis B
Deadliness (case fatality rate)
Measles
Malaria
Mumps
Rotavirus
Pertussis
Extremely communicable
Chickenpox
Very communicable
TB
Polio
Rubella
Smallpox
Dengue
Rhinovirus
Communicable
C. diff Cholera Seasonal flu Ebola
Hepatitis B Lyme disease Norovirus Plague
Rabies Campylobacter HIV Syphilis
Somewhat or minimally communicable
Communicability (R0)

Figure 14.11 The case fatality rate (CFR) and basic reproduction rate (R_0).

Vaccinology

In coming chapters we will discuss the biology of vaccines. But here we address the dynamics of vaccines in a population. The COVID-19 pandemic that began in 2020 highlighted the critical role vaccines play in controlling infectious diseases. You are probably now familiar with the phrase **herd immunity.** It was also addressed in this chapter's opening case study. In that case, herd immunity was described this way: *Herd immunity occurs when a significant portion of population becomes immune to an infectious disease, limiting further disease spread. For those who are not immune, they are indirectly protected because the ongoing disease spread is small.*

It comes down to the fact that every communicable microbe needs a certain density of susceptible subjects to maintain its transmissibility in a population **(figure 14.12).** In ideal circumstances, individuals who are immune to that particular microbe—either through having survived the disease or from being vaccinated—are a dead end. If the percentage of dead ends is high enough, the spread of the disease will simply stop. What this means is that those persons who are still susceptible to the disease because they cannot be vaccinated (such as young children) or those who have impaired immune systems are also protected by the immunity of the herd. Herd immunity that is achieved through immunization is a very important force in preventing or stopping epidemics, but it depends upon the willingness of the majority of the population to be vaccinated. Please note, that in the case of a very highly mutable virus like SARS-CoV-2, immunity in the vaccinated can become less effective over time because the virus changes. We have found in the COVID-19 pandemic that the early vaccines became less effective at preventing infection, but remained effective at preventing the worst outcomes of the disease.

Also, because SARS-CoV-2 mutates so quickly, even those who have recovered from COVID-19 can get the infection again later, when the virus is not as well recognized by the immune system. The point is that herd immunity as illustrated in figure 14.12 is the ideal, but is sometimes not the case in reality.

Despite the deviations from ideal we see with some diseases, vaccines have been amazingly effective public health tools. In fact, some have speculated that vaccination has done its job too well—at least in terms of being so effective for so long for so many dangerous childhood infections that young parents have no memory of a pre-vaccination era and do not appreciate the much greater risk of not vaccinating, compared to vaccinating. For example, in the decade before measles vaccination began, 3 to 4 million cases occurred each year in the United States. Typically, 300 to 400 children died annually from the disease and thousands died or were chronically disabled due to measles encephalitis. Put simply, childhood vaccines save the lives of 2.5 million children a year (worldwide), according to UNICEF.

Vaccines must go through many layers of trials in experimental animals and human volunteers before they are licensed

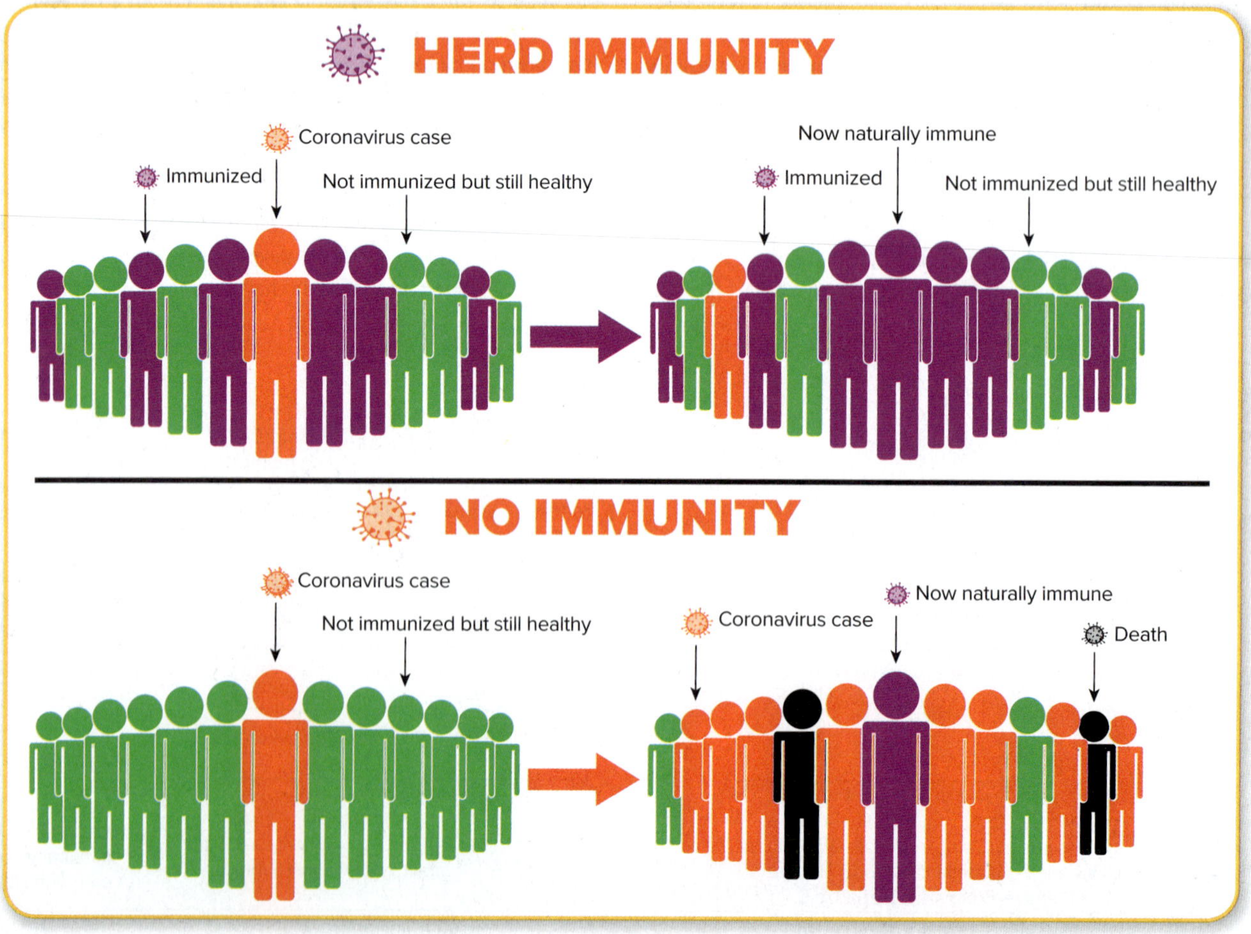
HERD IMMUNITY
Coronavirus case
Immunized
Not immunized but still healthy
Now naturally immune
Immunized
Not immunized but still healthy
NO IMMUNITY
Coronavirus case
Not immunized but still healthy
Now naturally immune
Coronavirus case
Death

Figure 14.12 An illustration of herd immunity. The orange individuals on the bottom right will also "turn purple," meaning they will become naturally immune (in an ideal scenario).

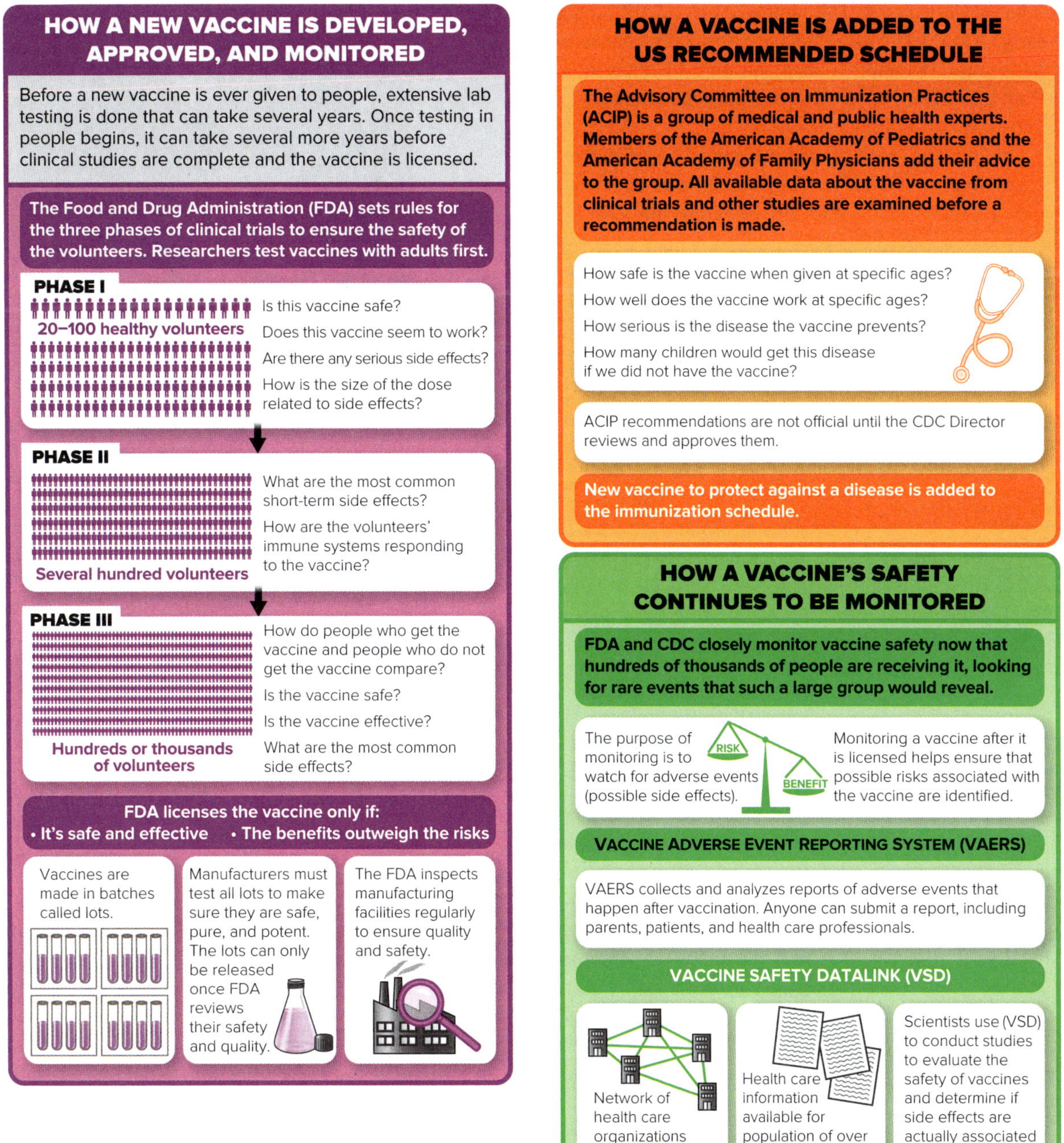

HOW A NEW VACCINE IS DEVELOPED, APPROVED, AND MONITORED
Before a new vaccine is ever given to people, extensive lab testing is done that can take several years. Once testing in people begins, it can take several more years before clinical studies are complete and the vaccine is licensed.
The Food and Drug Administration (FDA) sets rules for the three phases of clinical trials to ensure the safety of the volunteers. Researchers test vaccines with adults first.
PHASE I
20–100 healthy volunteers
Is this vaccine safe?
Does this vaccine seem to work?
Are there any serious side effects?
How is the size of the dose related to side effects?
PHASE II
Several hundred volunteers
What are the most common short-term side effects?
How are the volunteers' immune systems responding to the vaccine?
PHASE III
Hundreds or thousands of volunteers
How do people who get the vaccine and people who do not get the vaccine compare?
Is the vaccine safe?
Is the vaccine effective?
What are the most common side effects?
FDA licenses the vaccine only if:
• It's safe and effective • The benefits outweigh the risks
Vaccines are made in batches called lots.
Manufacturers must test all lots to make sure they are safe, pure, and potent. The lots can only be released once FDA reviews their safety and quality.
The FDA inspects manufacturing facilities regularly to ensure quality and safety.
HOW A VACCINE IS ADDED TO THE US RECOMMENDED SCHEDULE
The Advisory Committee on Immunization Practices (ACIP) is a group of medical and public health experts. Members of the American Academy of Pediatrics and the American Academy of Family Physicians add their advice to the group. All available data about the vaccine from clinical trials and other studies are examined before a recommendation is made.
How safe is the vaccine when given at specific ages?
How well does the vaccine work at specific ages?
How serious is the disease the vaccine prevents?
How many children would get this disease if we did not have the vaccine?
ACIP recommendations are not official until the CDC Director reviews and approves them.
New vaccine to protect against a disease is added to the immunization schedule.
HOW A VACCINE'S SAFETY CONTINUES TO BE MONITORED
FDA and CDC closely monitor vaccine safety now that hundreds of thousands of people are receiving it, looking for rare events that such a large group would reveal.
The purpose of monitoring is to watch for adverse events (possible side effects).
RISK
BENEFIT
Monitoring a vaccine after it is licensed helps ensure that possible risks associated with the vaccine are identified.
VACCINE ADVERSE EVENT REPORTING SYSTEM (VAERS)
VAERS collects and analyzes reports of adverse events that happen after vaccination. Anyone can submit a report, including parents, patients, and health care professionals.
VACCINE SAFETY DATALINK (VSD)
Network of health care organizations across the U.S.
Health care information available for population of over 9 million people.
Scientists use (VSD) to conduct studies to evaluate the safety of vaccines and determine if side effects are actually associated with vaccination.

Figure 14.13 The processes leading to approval of new vaccines.

for human use. **Figure 14.13** illustrates the process of testing vaccines, from very small studies in the beginning to post-market studies monitoring all adverse effects seen in the millions of people who receive it after it is marketed. Most vaccines can cause minor side effects, such as soreness at the injection site. More serious issues arise at a much lower rate.

The COVID-19 vaccines that were launched under Emergency Use Authorization in late 2020 and early 2021 went through these exact processes. They were sped up because of the urgency of the pandemic. Shortcuts were not taken, however. Usually the studies take years, but in the case of the COVID-19 vaccines, Phase III was started before the results of Phase II were conclusive, and everybody involved had incentives to make this a priority. The mRNA technology had been developed years earlier but not used in an approved vaccine until now. In this type of vaccine, the mRNA for a viral spike protein is injected into human muscle cells, which take up the mRNA and translate it into the viral protein and place it on their surfaces. The vaccines have since been administered to hundreds of millions of people worldwide. That provides an excellent view of the frequency of side effects. Studying the effects of vaccines after they have been widely distributed is sometimes called Phase IV of the vaccine rollout process. Experts know that nearly 100% of vaccine side effects, will be seen during the first six weeks after vaccination. With hundreds of millions doses already delivered over a period of many months, the rate of serious side effects has been exceedingly low, on the order of 1 to 2 per million doses. This means that you have a 0.0002% chance of a serious side effect from the vaccines. Compare that with the case-fatality rate of COVID-19. The numbers vary based on the variant, but early COVID-19 cases had a 1.6% to 2.0% chance of dying from the disease. Put another way, you were 11 times more likely to die from COVID-19 if you were unvaccinated than if you were vaccinated.

You will notice in the green section of figure 14.13 the acronym VAERS, standing for the Vaccine Adverse Events Reporting System. This is a centralized system for reporting any adverse event happening after a vaccination—whether it is related to the vaccine or not. Anyone can make a report—a patient, a parent, a health care provider. The CDC and the FDA use the VAERS as an early warning system to make sure they are capturing all adverse events. Most of the reported events will turn out to be unrelated to the vaccine, but no one stops any and all of them from being reported.

Similarly, there is a government program called the Vaccine Injury Compensation Program (VICP). It was established in the 1980s to provide a no-fault alternative to the traditional court systems. Anyone can file a complaint on behalf of the injured party, and it is automatically reviewed by the U.S. Department of Health and Human Services and the U.S. Department of Justice. Compensation may be awarded even if causation is not proven, and plaintiffs usually do not have to pay court or attorney fees.

Global Issues in Epidemiology

In the early 1900s, it was assumed by many that antibiotics would be the "magic bullet" that would eradicate all infectious disease from the human population. Although the mortality rates from such diseases declined dramatically after the advent of antibiotic drugs, an alarming trend was noted in the early 1980s: The incidence of infectious diseases began to increase—and it increased quite dramatically. It rose due to the appearance of a newly identified virus, HIV, but more importantly it continues to grow even today due to increasing antibiotic resistance in bacteria and *emerging* and *reemerging diseases.* Emerging diseases are caused by newly identified microbes, such as the SARS viruses and novel strains of human influenza virus. Reemerging diseases are those that have affected the human population in the past but are now becoming more prevalent due to travel, habitat invasion, or the development of drug resistance.

The CDC has a partnership with the WHO and local governments all around the world to monitor disease emergence and to spot and limit epidemics **(Figure 14.14).** This collaboration is called the Global Disease Detection (GDD) service.

When the COVID-19 pandemic struck in early 2020, the CDC coordinated with governments around the world and the WHO. We witnessed a truly global effort to work to end the pandemic. We also saw the differing effects of varying control measures in diverse countries. Studies examining the handling of the pandemic will be conducted for years to come, and will bring us rich lessons as we prepare for future events.

The CDC also monitors developing situations in health and disease and creates strategies to decrease excess morbidity and mortality. It has begun a campaign against five neglected parasitic infections (NPIs) in the United States. These are diseases that

Figure 14.14 A CDC scientist showing children in Uganda how to recognize mosquito larvae.
BK Kapella, M.D., (CDR, USPHS)/CDC

many Americans assume are common in third-world countries but rare in the United States. The five are

- Chagas disease—the trypanosome disease caused by *Trypanosoma cruzi;*
- Neurocysticercosis—caused by the tapeworm *Taenia solium;*
- Toxocariasis—caused by worms that travel through tissues and can cause blindness;
- Toxoplasmosis—a protozoan with which 60 million people in the United States are infected; and
- Trichomoniasis—a protozoal infection of the genital tract that leaves those infected more vulnerable to other sexually transmitted infections, including HIV. It also leads to premature births in infected mothers.

These diseases can affect anyone, but in the United States they are much more likely to be found in residents of poor neighborhoods and in immigrants from countries in which the diseases are prevalent. These are also the people least likely to have access to medical care. One of the consequences of that circumstance is that it is difficult to estimate infection rates, but new attempts have begun.

These are a few pertinent statistics from the CDC:

- Neurocysticercosis is the single most common infectious cause of seizures in some areas of the United States.
- Toxocariasis is caused by dog and cat roundworms; up to 14% of the US population has been exposed. About 70 people a year are blinded by this infection.
- Up to 300,000 people in the United States are currently infected with the protozoan that causes Chagas disease.
- In the United States, 3.7 million people are currently infected with *Trichomonas.*

It is time to stop thinking of these infections as "other people's problems."

Before we leave this chapter about disease surveillance and control in populations, we have to consider the threat of *bioterrorism,* the intentional or threatened use of microorganisms or toxins from living organisms to cause death or disease in humans, livestock, or plants. Although use of microbes to inflict damage on human populations or to cause political discord is not new, the stakes have become much higher with the advancement of biotechnology. The most notable bioterror event in the United States in recent times was the spread of anthrax in the United States in 2001. Beyond the targeting of humans, *agroterrorism* involves the use of microorganisms to decimate the agricultural industry. Rather than making humans ill, agroterrorists target the food supply to exert their damage. Although no documented cases have occurred, many government agencies are conducting surveillance and developing policies to prohibit this scenario from occurring in the world today.

In the future, it will take a concerted effort from the medical community, epidemiologists, and the general public to keep infectious agents in check. This will involve the development of new drugs, new education programs, increased vaccination rates worldwide, and equitable distribution of vaccines and drugs to people and countries of every socioeconomic status.

14.3 Learning Outcomes—Assess Your Progress

11. Describe the role of surveillance in public health.
12. Name the four most important categories of healthcare-associated infection.
13. Define the basic reproduction rate and the case fatality rate with respect to infectious agents.
14. Fully explain the concept of herd immunity.
15. Provide an overview of the phases a vaccine goes through before approval by the FDA.
16. Describe the VAERS and the VICP.

Media Under The Microscope Wrap-Up

Media Union/Shutterstock

The news article "What Doctors Wish Patients Knew About COVID-19 Herd Immunity" comes from the American Medical Association's website from August 27, 2021.

The **intended message** of this article is to try to explain why herd immunity is a moving target, especially when the virus involved is prone to mutations.

A **critical reading** of the article makes me think it is accurate with no propaganda or political messages. It explains the science and also provides a user-friendly definition of herd immunity. I would **interpret this article** to my friends by providing them the quote about what herd immunity is, and then explain, as you learned in this chapter, that the percentages required for herd immunity vary from microbe to microbe. Then you would want to say something about how the different variants have changed the percentage of immune people needed to approach herd immunity against COVID-19.

My **overall grade** for this article is A+. All of the science that was discussed was solid, and it was written in such a way that was nonconfrontational but extremely timely.

Source: American Medical Association, https://www.ama-assn.org/delivering-care/public-health/what-doctors-wish-patients-knew-about-covid-19-herd-immunity, online article posted August 27, 2021.

Study Smarter: Better Together

These activities are designed for you to use on your own with a study group—either a face-to-face group or a virtual one, consisting of 3–5 members. Studying together can be very helpful, but there are effective and ineffective ways to do it. For example, getting together without a clear structure is often not a good use of your time. Use your time efficiently by using one or more of the exercises below.

FACE-TO-FACE GROUPS

Use one or more of the activities below.

Peer Instruction: Assign numbers to your group members to use all semester long. Now look at these five concepts from this chapter. Each group member prepares a 5-minute lesson on the topic corresponding to their number. Don't worry if you have fewer than 5 members; just use however many you have! During your group study time, each member presents their lesson, and the group spends another 5–10 minutes discussing that lesson.

1. Difference between typical medical practice and epidemiology
2. Natural history of disease
3. Causation vs. correlation
4. Surveillance and notifiable diseases
5. How vaccines are tested before release

(continued)

Concept Maps: Each member of the group should use this list of terms from this chapter to generate their own concept map. This can be hand-drawn or created using software (see Appendix C for guidelines). During group study time, compare each other's concept maps and help each other make sure they are correct. Of course, there are many different "correct" maps. Examining each member's map will help you talk through the varied concepts and how they are related.

Concept Terms:

correlation	Hill's criteria	biological plausibility	Koch's postulates
causation	strength of association	randomized controlled trials	

Table Topics: Each group member should identify a concept or topic from this week's class assignments with which they are having trouble and share it during group study time. The other group members can then help to clarify confusing issues or share how they figured it out. Aim for a maximum of 15 minutes per topic. If the topic remains unclear to the group, bring it up during class or use the instructor's office hours or e-mail to ask for help. Taking the time to struggle with a difficult concept first makes your questions much more specific and more likely to yield helpful answers.

VIRTUAL GROUPS

Not everyone has the time or opportunity to meet with group members outside of class time. You or your instructor can create a virtual group using e-mail or the course software.

Weekly Discussion Board: This forum can be used as a way for groups to discuss topics, via e-mail or other learning management systems or online platforms, before they are covered in class. As each member of the group answers the current week's question, they should send their responses to every other member of their group. It's best to agree on a deadline based on how your class schedule works (Saturday for the next week's topics, for example). Then, after the topic is discussed in class, each member should send a response that all group members will see with a follow-up post on the same topic. If you cover more than one chapter in a week, someone can be designated to choose which chapter Discussion Board question you will use. Or simply decide up front that you will always use the first-chapter-of-the week's question, to keep the schedule simple.

Discussion Question

Using figure 14.10, discuss the probable reasons why each of the HAIs featured there is so problematic in healthcare settings.

Chapter Summary

EPIDEMIOLOGY OF INFECTIOUS DISEASES

14.1 EPIDEMIOLOGY BASICS AND HISTORY

- **Epidemiology is the science that studies health events among populations of people.**
- **Epidemiologists are trained to think of these five questions: Who? What? Where? When? and Why?**
- **The science of epidemiology informs the practice of public health.**
- **In the 1960s the public health community began to feel victorious over infectious diseases and increased its attention to noninfectious conditions, including cigarette smoking. Soon it became apparent that continued vigilance of infectious diseases was absolutely vital.**

14.2 UNITS OF MEASURE

- **Prevalence is the total number of existing cases of a condition at a defined point in time.**
- **Incidence is the number of new cases over a defined period of time.**
- **Epidemiologists speak in terms of morbidity (damage caused by a condition) and mortality (death).**
- **Outbreaks and epidemics are generally categorized as point-source, common-source, or propagated. They can also be some combination of these.**
- **Graphs and maps are important tools of epidemiology because groups of people are studied.**
- **The natural history of disease is a template for the progression from health to death, recovery, or disability caused by particular disease conditions.**
- **Screening tests are important public health tools deployed in healthy persons, generally before they are aware they have disease.**
- **Screening tests are less accurate than diagnostic tests and are characterized by their sensitivity and their specificity.**
- **One of the most critical skills for epidemiologists as well as the general public to have is the ability to distinguish between correlation and causality.**
- **The epidemiological gold standard for establishing causation is the randomized control trial (RCT).**
- **In many cases, RCTs are unethical to conduct in humans, and so several types of studies exist that, when taken together, can build a case of causation.**
- **A set of criteria, often called Hill's criteria, are applied to non-RCT studies to build the case for causation.**

14.3 INFECTIOUS DISEASE EPIDEMIOLOGY

- **Surveillance is the collecting, analyzing, and reporting of data on disease occurrence in a healthy population.**
- **Designated diseases and infections are reportable, or notifiable, meaning their incidence is registered with a state, federal, or international body.**
- **Healthcare-associated infections (HAIs) are infections acquired while in the care of the medical system, such as in a hospital or other care facility. Several categories of these are constantly monitored by hospitals and regularly reported to accrediting bodies.**
- **Each infectious agent can be described by its basic reproductive rate and its case fatality rate. These values can vary slightly in different epidemics or situations, but otherwise are characteristic of the microbe.**
- **Vaccines go through extensive human studies before they can be approved for either emergency use or conventional use. A fourth phase of evaluation continues after they are released to the general public.**

SmartGrid: From Knowledge to Critical Thinking

This *21 Question Grid* takes the topics from this chapter and arranges them with respect to the American Society for Microbiology's Undergraduate Curriculum guidelines—*In this chapter, which focuses on the impact of microbes on groups, we focus highly on "Impact of Microorganisms."* Three questions are supplied, which cover chapter content referring to the Concept or Competency in increasing levels of Bloom's taxonomy for learning.

ASM Concept/ Competency	A. Bloom's Level 1, 2—Remember and Understand (Choose one)	B. Bloom's Level 3, 4—Apply and Analyze	C. Bloom's Level 5, 6—Evaluate and Create
Evolution	**1.** Emerging infectious diseases of humans generally come from **a.** soil microbes. **b.** ocean microbes. **c.** other humans. **d.** animals.	**2.** List three possible reasons that an effective reproduction rate would change over time for a single microorganism causing a pandemic.	**3.** The CDC began in the 1940s with a mission to control the spread of malaria in the United States. Today there is very little malaria that is spread domestically in the U.S. What do you suppose changed?

ASM Concept/ Competency	A. Bloom's Level 1, 2—Remember and Understand (Choose one)	B. Bloom's Level 3, 4—Apply and Analyze	C. Bloom's Level 5, 6—Evaluate and Create
Information Flow and Genetics	**4.** You can expect that mRNA vaccines result in the production of ______________ in host cells. **a.** viral DNA **b.** viral protein **c.** host DNA **d.** host protein	**5.** In a propagated outbreak, a significant mutation in a microbial gene could cause people who have already recovered from the disease to become ill again. What effect would that have on the outbreak curve?	**6.** In what kind of an outbreak would it be most useful to have DNA analysis of the bacteria from multiple victims, in order to determine if the bacterium was identical in every case?
Impact of Microorganisms	**7.** Which of these is a designater of cumulative cases of disease in a population? **a.** incidence **b.** diagnoses **c.** propagators **d.** prevalence	**8.** Explain why healthcare-associated infections are so common and so dangerous.	**9.** Offer some key points about the difference between correlation and causation.
(This chapter is unique. We will have extra questions about the impact of microorganisms.)	**10.** Which is likely to be the most accurate? **a.** screening test **b.** diagnostic test **c.** not enough information to judge	**11.** Which phase of vaccine trials uses the fewest people, and why is that?	**12.** What is the common factor you see among the top 4 causes of healthcare-associated diseases?
	13. Which is more appropriate to calculate at the very beginning of a pandemic, the basic reproduction rate or the effective reproduction rate? **a.** basic **b.** effective	**14.** Hill's criteria for causation are a set of principles that can be used together to assess causation. Temporality is particularly important. Explain why.	**15.** Explain why performing randomized controlled trials on humans is unethical in many cases.
	16. Infectious disease surveillance involves keeping track of disease incidence in **a.** times of epidemic. **b.** when there is not an epidemic occurring. **c.** both.	**17.** Speculate as to why very few infectious diseases have a case fatality rate of more than 75%.	**18.** Would you say that the operating principles of the VAERS and the VICP "favor" the vaccine makers or patients? Provide a rationale for your answer.
Scientific Thinking	**19.** A seasonal outbreak of influenza would be what type of outbreak? **a.** point-source **b.** common-source **c.** propagated **d.** two of the above	**20.** Discuss the three possible endpoints in the natural history of disease.	**21.** You discover that a particular mutation in a surface protein of a virus makes it more likely to adhere tightly to respiratory epithelium. You have hypothesized that this mutated virus is responsible for an increase in infections by the virus. Which criterion for causality does this discovery satisfy? **a.** temporality of association **b.** strength of association **c.** biological plausibility **d.** specificity of association

Answers to the multiple-choice questions appear in Appendix A.

Visual Connections

This question uses visual images to connect content within and between chapters.

1. **From chapter 13, figure 13.11.** If the portal of exit is an insect bite vs. respiratory (coughing and sneezing), discuss what pandemic curtailment measures would be taken.

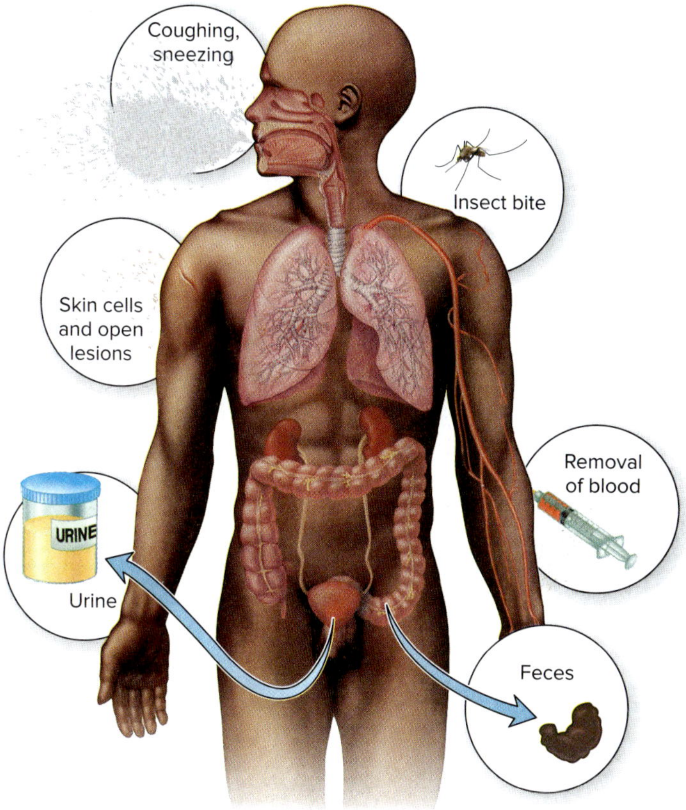
Coughing, sneezing
Insect bite
Skin cells and open lesions
Removal of blood
URINE
Urine
Feces

High Impact Study

These terms and concepts are most critical for your understanding of this chapter—and may be the most difficult. Have you mastered them?

Concepts

- ☐ Epidemiology
- ☐ Public health
- ☐ Prevalence vs. incidence
- ☐ Natural history of disease
- ☐ Hill's criteria for causation
- ☐ Herd immunity
- ☐ Phases of vaccine trials
- ☐ HAIs

Terms

- ☐ Screening
- ☐ Randomized controlled trials
- ☐ Cross-sectional studies
- ☐ Case control studies
- ☐ Cohort studies
- ☐ Basic reproductive rate
- ☐ Case fatality rate
- ☐ VAERS
- ☐ VICP
- ☐ Notifiable diseases

Design Element: (College students): Caia Image/Image Source

15 Host Defenses I

Overview and Innate Immunity

Steven Schleuning

Media Under The Microscope

Tattoos Owe Their Permanence to the Immune System

This opening case examines an article from the popular media to determine the extent to which it is factual and/or misleading. This case focuses on the recent article from Science News, *"Inked Mice Hint at How Tattoos Persist in People."*

Do you have a tattoo? You can thank one of the very important cells we will study in this chapter for it, according to this article in *Science News*. It describes a study conducted by French scientists in which the tails of mice were tattooed in a stripe pattern. The researchers then searched for where the ink particles ended up. They found that the ink is taken up by macrophages under the skin.

Macrophages are an important part of the body's defenses. They act by "gobbling up" microorganisms and foreign particles. In ideal circumstances, the macrophages break down the foreign material and discard it. Some foreign materials cannot be destroyed by the macrophages, and these stay put within the cell. The researchers in this study found the green ink in macrophages. Previous work had suggested that the permanence of tattoos was due to the fact that long-living macrophages and other cells had taken up the ink and preserved it from being washed away in the lymphatic system or with the normal sloughing of skin cells. This experiment had a twist, however. The mice that were tattooed were genetically programmed to kill their own skin macrophages. Researchers wanted to see if their tattoos would then disappear. They did not. They later found that the ink had been taken up by other, newer macrophages. This leaves open the possibility that the reason for tattoo longevity is that macrophages efficiently take up new ink or ink that is released by dying macrophages, rather than relying solely on long-living cells.

The article quotes another scientist, not involved in the study, who cautions that humans are different from mice and might well use long-living macrophages instead of passing off the ink from one generation of cells to the next. The author of the article reports that this discovery could be used to develop easier and quicker methods of tattoo removal, if the exchange of ink between old and new macrophages is interrupted.

- What is the **intended message** of the article?
- What is your **critical reading** of the summary of the article provided above? Remember that in this context, "critical reading" does not necessarily mean *What criticism do you have?* but asks you to apply your knowledge to interpret whether the article is factual and whether the facts support the intended message.
- How would you **interpret** the news item for your nonmicrobiologist friends?
- What is your **overall grade** for the news item—taking into account its accuracy and the accuracy of its intended effect?

Media Under The Microscope Wrap-Up appears at the end of the chapter.

Outline and Learning Outcomes

15.1 Defense Mechanisms of the Host: An Overview

1. Summarize the three lines of host defenses.
2. Define "marker" and discuss its importance in the second and third lines of defense.
3. Name the body systems that participate in immunity.
4. Describe the structure and function of the lymphatic system and its connection with the circulatory system.
5. Name three kinds of blood cells that function in innate immunity.
6. Connect the mononuclear phagocyte system to innate immunity.
7. Describe how and where T and B lymphocytes mature to be ready for their role in adaptive immunity.
8. Summarize the importance of cytokines, and list one pro-inflammatory and one anti-inflammatory cytokine.

15.2 The First Line of Defense

9. Identify the three components of the first line of defense.
10. Identify the four body systems that participate in the first line of defense.
11. Describe two examples of how the normal microbiota contribute to the first line of defense.

15.3 The Second Line of Defense

12. List the four major categories of innate immunity.
13. Outline the steps of phagocytosis.
14. Outline the steps of inflammation.
15. Discuss the mechanism of fever and how it helps defend the body.
16. Name four types of antimicrobial host-derived products.
17. Compose one good overview sentence about the purpose and the mode of action of the complement system and another about the purpose and mode of action of interferon.

15.1 Defense Mechanisms of the Host: An Overview

This chapter introduces the parts of the host's own biology that provide defense against infections. Topics included in this overview are the anatomical and physiological systems that detect, recognize, and destroy foreign substances and the general responses that account for an individual's long-term immunity or resistance to infection and disease.

In chapter 13, we explored the host–parasite relationship, with emphasis on the role of microorganisms in disease. In this chapter, we examine the other side of the relationship—that of the host defending itself against microorganisms. As previously stated, whether an encounter between a human and a microbe results in disease is dependent on many factors (see figure 13.3). The encounters occur constantly. In the battle against all sorts of invaders, microbial and otherwise, the body has a series of physical barriers, sends in an army of cells, and emits a flood of chemicals to protect tissues from harm.

The host defenses are a multilevel network of innate, preexisting protections and adaptive **immunities** referred to as the first, second, and third lines of defense **(figure 15.1).** The interaction and cooperation of these three levels of defense normally provide complete protection against infection. *The first line of defense* includes any barrier that blocks invasion at the portal of entry. This mostly nonspecific line of defense limits access to the internal tissues of the body. The *second line of defense* is an internalized system of protective cells and fluids and includes inflammation and phagocytosis. It acts rapidly at both the local and systemic levels once the first line of defense has been

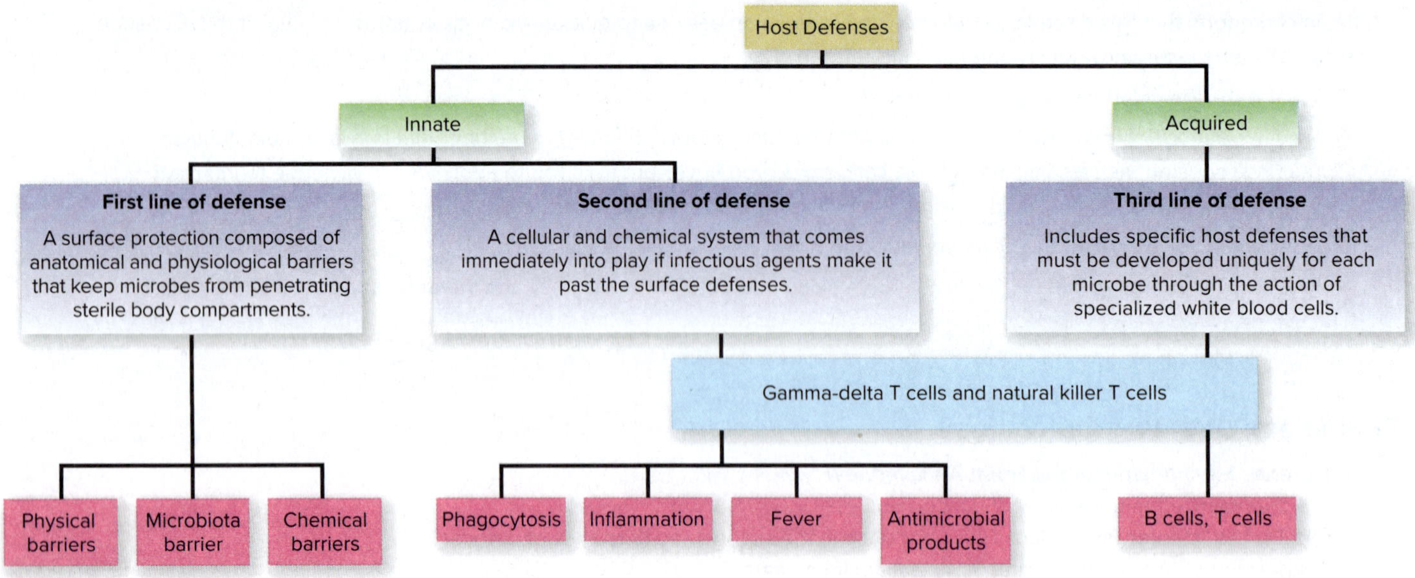
Host Defenses
Innate
Acquired
First line of defense
A surface protection composed of anatomical and physiological barriers that keep microbes from penetrating sterile body compartments.
Second line of defense
A cellular and chemical system that comes immediately into play if infectious agents make it past the surface defenses.
Third line of defense
Includes specific host defenses that must be developed uniquely for each microbe through the action of specialized white blood cells.
Gamma-delta T cells and natural killer T cells
Physical barriers
Microbiota barrier
Chemical barriers
Phagocytosis
Inflammation
Fever
Antimicrobial products
B cells, T cells

Figure 15.1 Flowchart summarizing the major components of the host defenses. Defenses are classified into one of two general categories: (1) innate or (2) acquired. These can be further subdivided into the first, second, and third lines of defense, each being characterized by a different level and type of protection. There is also a set of cells that straddle the categories of innate and acquired defenses (light blue box).

breached. The highly targeted *third line of defense* is acquired on an individual basis as each foreign substance is encountered by white blood cells called **lymphocytes.** The goal of this response is to make lymphocytes that are specifically adapted to each individual invader. The reaction with each different microbe produces unique protective substances and cells that can come into play if that microbe is encountered again. The third line of defense provides long-term immunity, which is discussed in detail in chapter 16. This chapter focuses on the first and second lines of defense.

Humans are armed with various levels of defense that do not operate separately. Most defenses overlap and are even redundant in some of their effects. This bombards microbial invaders with an entire assault force, making their survival unlikely. Because of the interwoven nature of host defenses, we will introduce basic concepts of structure and function that will prepare you for later information on adaptive reactions of the immune system.

In the body, the mandate of the immune system can be easily stated. A healthy functioning immune system is responsible for the following:

1. surveillance of the body,
2. recognition of foreign material, and
3. destruction of entities deemed to be foreign.

All cells (microbial and otherwise), as well as some particles such as pollen, display a unique mix of macromolecules on their surfaces that the immune system "senses" to determine if they are foreign or not. These chemicals are called **antigens.** Because foreign cells or particles could potentially enter through any number of portals, the cells of the immune system constantly move about the body, searching for potential pathogens. This process is carried out primarily by white blood cells, which have been trained to recognize body cells (so-called **self**) and differentiate them from any foreign material in the body, such as an invading bacterial cell **(nonself).** The ability to evaluate cells and macromolecules as either self or nonself is central to the functioning of the immune system. While foreign substances must be recognized as a potential threat and dealt with appropriately, self cells and chemicals must not come under attack by the immune defenses. Many autoimmune disorders are a result of the immune system mistakenly attacking the body's own tissues and organs. For example, in rheumatoid arthritis, the body attacks its own joints and tissues, causing pain and loss of function.

Another word for antigens is **markers (figure 15.2).** Markers, which generally consist of proteins and/or sugars, can be thought of as the cellular equivalent of facial characteristics in humans and allow the cells of the immune system to identify whether or not a newly discovered cell poses a threat. In the healthy situation, cells deemed to be self are left alone, while cells and other objects designated as foreign are marked for destruction by a number of methods, the most common of which is phagocytosis. Markers that many different kinds of microbes have in common are called **pathogen-associated molecular patterns (PAMPs).** Host cells with important roles in the innate immunity of the second line of defense use **pattern recognition receptors (PRRs)** to recognize PAMPs. There is a middle ground as well. Nonself proteins that are not harmful—such as those found in food we ingest and on commensal microorganisms—are generally recognized as such, and the immune system is signaled not to react.

Unlike many systems, the immune system does not exist in a single, well-defined site. Instead, it is a large, complex, and

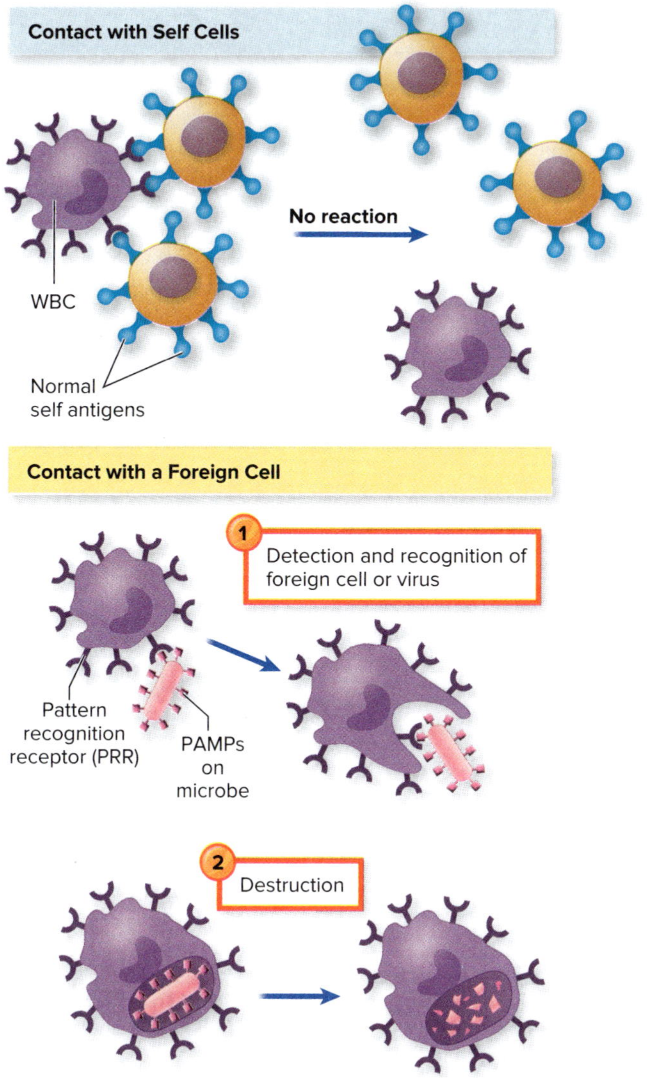
Contact with Self Cells
No reaction
WBC
Normal self antigens
Contact with a Foreign Cell
1 Detection and recognition of foreign cell or virus
Pattern recognition receptor (PRR)
PAMPs on microbe
2 Destruction

Figure 15.2 ***Search, recognize,*** **and** ***destroy*** **is the mandate of the immune system.** White blood cells are equipped with a very sensitive sense of "touch." As they sort through the tissues, they feel surface markers that help them determine what is self and what is not. When self markers are recognized, no response occurs. However, when nonself is detected, a reaction to destroy it is mounted.

diffuse network of cells and fluids that permeate every organ and tissue. It is this arrangement that promotes the surveillance and recognition processes that help screen the body for harmful substances.

Tissues, Organs, and Cells Participating in Immunity

The body is partitioned into several fluid-filled spaces called the intracellular, extracellular, lymphatic, cerebrospinal, and circulatory compartments. Although these compartments are physically separated, they have numerous connections. For effective immune responsiveness, the activities in one fluid compartment must be communicated to other compartments.

At the microscopic level, clusters of tissue cells are in direct contact with the **mononuclear phagocyte system (MPS),** which is described shortly, and the extracellular fluid (ECF). Extracellular spaces have been found to contain lymph fluid, and those spaces plus their fluid have been termed the **interstitium.** Blood and lymphatic capillaries penetrate into these tissues. This close association allows cells that originate in the MPS and ECF to diffuse or migrate into the blood and lymphatics. Any products of a lymphatic reaction can be transmitted directly into the blood through the connection between these two systems. Certain cells and chemicals originating in the blood can move through the vessel walls into the extracellular spaces and migrate into the lymphatic system.

Lymphatic System Vessels and Fluids

The **lymphatic system** is a compartmentalized network of vessels, cells, and specialized accessory organs **(figure 15.3).** It begins in the farthest reaches of the tissues as tiny capillaries that transport a special fluid (lymph) through an increasingly larger tributary system of vessels and filters (lymph nodes), and it leads to major vessels that drain back into the regular circulatory system. Some major functions of the lymphatic system are as follows:

1. to provide a route for the return of extracellular fluid to the circulatory system proper;
2. to act as a "drain-off" system for the inflammatory response; and
3. to render surveillance, recognition, and protection against foreign materials through a system of lymphocytes, phagocytes, and antibodies.

Lymphatic Fluid Lymph is a plasmalike liquid carried by the lymphatic circulation. It is formed when certain blood components move out of the blood vessels into the extracellular spaces and diffuse or migrate into the lymphatic capillaries. Like blood, it transports numerous white blood cells (especially lymphocytes) and miscellaneous materials such as fats, cellular debris, and infectious agents that have gained access to the tissue spaces.

Lymphatic Vessels The system of vessels that transport lymph is constructed along the lines of blood vessels. As the lymph is never subjected to high pressure, the lymphatic vessels appear more similar to thin-walled veins than to thicker-walled arteries. The tiniest vessels, lymphatic capillaries, accompany the blood capillaries and extend into all parts of the body except certain organs such as bone, the placenta, and the thymus. Their thin walls are easily permeated by extracellular fluid that has escaped from the circulatory system or is in the interstitium. Lymphatic vessels are found in particularly high numbers in the hands, in the feet, and around the areola of the breast.

Soon you will read about the bloodstream and blood vessels. Two overriding differences between the bloodstream and the lymphatic system should be mentioned. First, because one of the main functions of the lymphatic system is returning lymph to the circulation, the flow of lymph is in one direction only, with lymph moving from the extremities toward the heart. Eventually, lymph

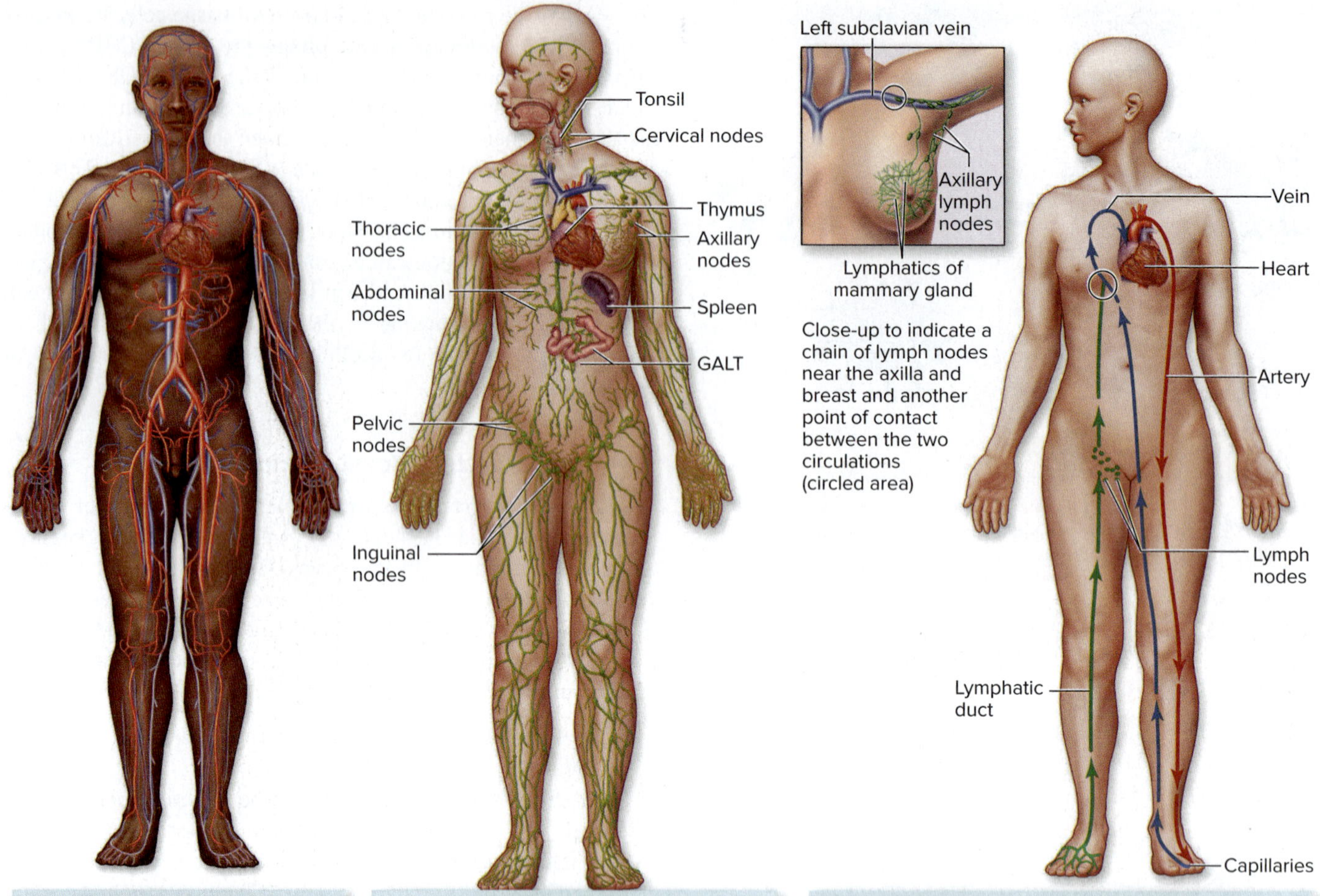

Tonsil
Cervical nodes
Thoracic nodes
Thymus
Axillary nodes
Abdominal nodes
Spleen
GALT
Pelvic nodes
Inguinal nodes
Left subclavian vein
Axillary lymph nodes
Lymphatics of mammary gland
Close-up to indicate a chain of lymph nodes near the axilla and breast and another point of contact between the two circulations (circled area)
Vein
Heart
Artery
Lymph nodes
Lymphatic duct
Capillaries

The Circulatory System

Body compartments are screened by circulating WBCs in the cardiovascular system.

The Lymphatic System

The lymphatic system consists of a branching network of vessels that extend into most body areas. Note the higher density of lymphatic vessels in the "dead-end" areas of the hands, feet, and breast, which are frequent contact points for infections. Other lymphatic organs include the lymph nodes, spleen, gut-associated lymphoid tissue (GALT), the thymus, and the tonsils.

The Lymphatic and Circulatory Systems

Comparison of the generalized circulation of the lymphatic system and the blood. Although the lymphatic vessels parallel the regular circulation, they transport in only one direction, unlike the cyclic pattern of blood. Direct connection between the two circulations occurs at points near the heart, where large lymph ducts empty their fluid into veins (circled area).

Figure 15.3 The circulatory and lymphatic systems.

will be returned to the bloodstream through the thoracic duct or the right lymphatic duct to the subclavian vein near the heart. The second difference concerns how lymph travels through the vessels of the lymphatic system. While blood is transported through the body by means of a dedicated pump (the heart), lymph is moved only through the contraction of the skeletal muscles through which the lymphatic ducts wend their way. This dependence on muscle movement helps to explain the swelling of the hands and feet that sometimes occurs during the night (when muscles are inactive) and that clears up soon after waking. It also informs the use of compression sleeves on the legs of patients who are immobilized in the hospital or long-term care.

Primary Lymphatic Organs In the lymphatic system, the sites where immune cells are born and the locations where they mature are considered **primary lymphatic organs.** Locations in the body where immune cells become activated, reside, or carry out their functions are called **secondary lymphatic organs.**

Red Bone Marrow The red bone marrow is an important intersection between the circulatory system, skeletal system, and lymphatic system. It is typically found in the internal matrix of long bones and is the site of blood cell production. All blood cells originate in the bone marrow, including B- and T-lymphocyte precursors. After B lymphocytes begin to express markers that identify them as B cells to the rest of the body, they complete their maturation process while still in the bone marrow. Once this is complete, these B cells then migrate to secondary lymphatic organs, such as the spleen or lymph nodes, where they wait to encounter foreign antigens.

The Thymus: Site of T-Cell Maturation The **thymus** originates in the embryo as two lobes in the lower neck region that fuse into a triangular structure **(figure 15.4).** Lymphocytes that originate in the bone marrow as naïve T lymphocytes migrate to the thymus to complete their maturation. Under the influence of thymic hormones, these cells develop specificity and are released into the circulation as mature T cells. The mature T cells subsequently migrate and settle in other secondary lymphoid organs, just as mature B cells do.

Secondary Lymphatic Organs One way to think of the secondary lymphatic organs is that they are where the action of immunity takes place. The lymph nodes, spleen, and various lymphoid tissues contain large numbers of T and B cells and are stationed throughout the body, ready to encounter antigens and become activated.

Lymph Nodes Lymph nodes are small, encapsulated, bean-shaped organs stationed, usually in clusters, along lymphatic channels and large blood vessels of the thoracic and abdominal cavities (see figure 15.3). Major aggregations of nodes occur in the loose connective tissue of the armpit (axillary nodes), groin (inguinal nodes), and neck (cervical nodes). Their job is to filter out materials in the lymph and to provide appropriate cells for immune reactions. In **figure 15.5,** the anatomy of a lymph node is illustrated. The outer rim of the lymph node is called the **cortex.** There are T lymphocytes in the **paracortical area** and B lymphoctyes and macrophages in the **medullary sinus.** Enlargement of the lymph nodes can provide physicians with important clues as to a patient's condition. Generalized lymph node enlargement may indicate the presence of a systemic illness, while enlargement of an individual lymph node may be evidence of a localized infection. Enlargement of lymph nodes reflects the replication of many lymphocyte clones during an adaptive immune response.

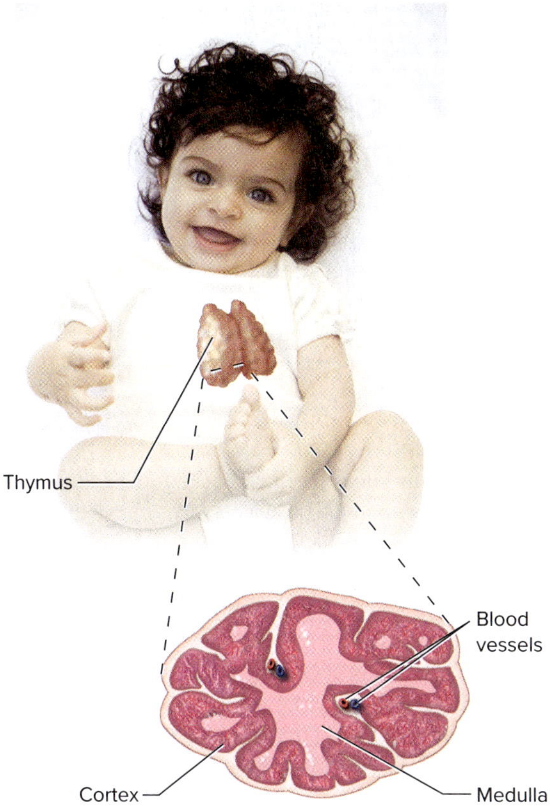
Thymus
Blood vessels
Cortex
Medulla

Figure 15.4 The thymus. Immediately after birth, the thymus is a large organ that nearly fills the region over the midline of the upper thoracic region. In the adult, however, it is proportionately smaller. The drawing shows the main anatomical regions of the thymus. Immature T cells enter through the cortex and migrate into the medulla as they mature.

Jacqueline Veissid/Photodisc/Getty Images

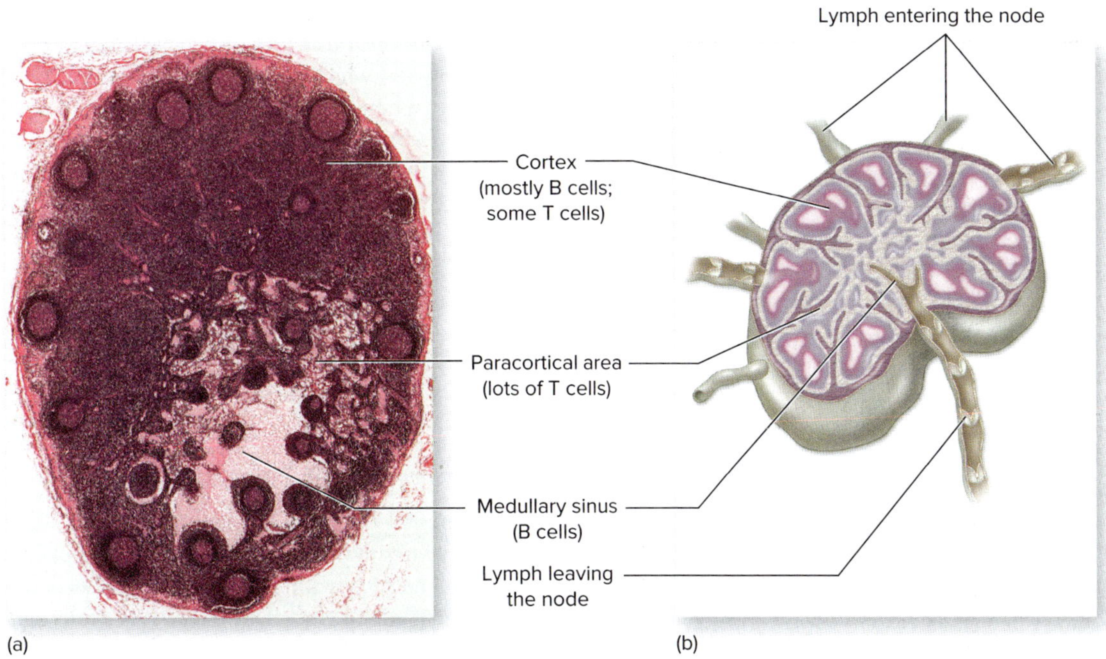
Lymph entering the node
Cortex (mostly B cells; some T cells)
Paracortical area (lots of T cells)
Medullary sinus (B cells)
Lymph leaving the node
(a)
(b)

Figure 15.5 Diagram of a lymph node. On the left **(a)** is a micrograph of a lymph node that has been cut in half. On the right **(b)** is a drawing of a lymph node.

(a) Christine Eckel/McGraw Hill

The Spleen The spleen is a lymphoid organ in the upper left portion of the abdominal cavity. It is somewhat similar to a lymph node except that it serves as a filter for blood instead of lymph. While the spleen's primary function is to remove worn-out red blood cells from circulation, its most important immunologic function is the filtering of pathogens from the blood. When they arrive in the spleen, they are engulfed by macrophages there. Although adults whose spleens have been surgically removed can live a relatively normal life, children without spleens are severely immunocompromised. The spleen also acts as a storehouse of blood that can be released in the event of hemorrhage. It can hold up to one cup of blood. For this reason, injury to the spleen can result in profuse bleeding.

Associated Lymphoid Tissues There are discrete bundles of lymphocytes at many sites on or just beneath the skin and mucosal surfaces all over the body. The tissues are often named for their location. For example, the **skin-associated lymphoid tissue** is called **SALT,** and the **mucosa-associated lymphoid tissue** is called **MALT.** The positioning of this diffuse system provides an effective first-strike potential against the constant influx of microbes and other foreign materials in food and air. In the pharynx, a ring of tissues called the **tonsils** provides an active source of lymphocytes. The breasts of pregnant and lactating women also become temporary sites of antibody-producing lymphoid tissues. The intestinal tract houses the best-developed collection of lymphoid tissue, called **gut-associated lymphoid tissue,** or **GALT.** Examples of GALT include the appendix, the lacteals (special lymphatic vessels stationed in each intestinal villus), and **Peyer's patches,** compact aggregations of lymphocytes in the ileum of the small intestine.

The Blood

The circulatory system consists of the heart, arteries, veins, and capillaries, which circulate the blood, and the lymphatic system, which includes lymphatic vessels and lymphatic organs (lymph nodes) that circulate lymph. As you will see, these two circulations parallel, interconnect with, and complement one another.

The substance that courses through the arteries, veins, and capillaries is **whole blood,** a liquid consisting of **blood cells** suspended in **plasma.** One can visualize these two components with the naked eye when a tube of unclotted blood is allowed to sit or is spun in a centrifuge. The cells' density causes them to settle into an opaque layer at the bottom of the tube, leaving the plasma, a clear, yellowish fluid, on top. Serum is essentially the same as plasma, except it is the clear fluid from clotted blood. Serum is often used in immune testing and therapy (**figure 15.6**).

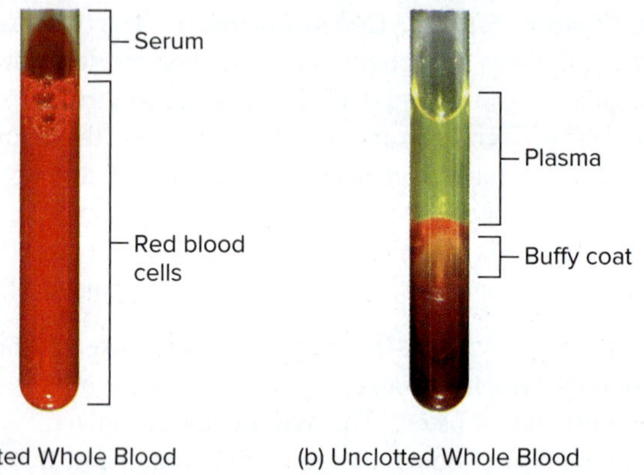
Serum
Red blood cells
Plasma
Buffy coat
(a) Clotted Whole Blood
(b) Unclotted Whole Blood

Figure 15.6 The macroscopic composition of whole blood. **(a)** When blood is allowed to clot, serum separates from the red blood cells. **(b)** When anticoagulants are present, the blood stratifies into a clear layer of plasma; a thin layer of off-white material called the buffy coat (which contains the white blood cells); and a layer of red blood cells in the bottom, thicker layer.
Martin M. Rotker/Science Source

A Survey of Blood Cells The production of blood cells is called **hematopoiesis** (hee″-mat-o-poy-ee′-sis). The primary precursor of new blood cells is a pool of undifferentiated cells called pluripotential **stem cells** in the bone marrow. "Pluripotent" means that the cells are able to become any type of blood cell that is needed. These cells are produced continually in the bone marrow and subjected to a variety of growth factors and hormones that cause them to change over time. This process is called **differentiation.** The immature stem cells change their genetic expression, which in turn causes them to express different surface markers and respond to new signals. They become more specialized over time, eventually maturing either in the bone marrow or in other sites of the body. This is the process of blood cell development.

There are two primary lines of cells that arise from stem cells and are still considered immature. These are (1) those that differentiate from a common myeloid cell, such as red blood cell precursors (erythroblasts) and platelet precursors (megakaryoblasts), and (2) those that differentiate from a common lymphoid precursor cell, such as precursors of many white blood cells (myeloblasts) and precursors of lymphocytes (lymphoblasts).

White blood cells **(leukocytes)** can generally be divided into two categories according to their staining patterns when viewed with a microscope: **granulocytes,** which have dark staining granules, and **agranulocytes,** which do not have granules and typically have large nuclei. The granules in granulocytes can be released to kill foreign cells or affect host tissues. **Figure 15.7** sorts out all these cells. They are vitally important to innate and adaptive immunity. **Insight 15.1** highlights the role of one of these cell types in a cascade of events affecting the microbiome and the risk for colon cancer.

Starting in 2018, a surprising discovery was coming to light. Until then, it was thought that red blood cells (RBCs) had little or no role in immunity, given their massive role in gas exchange in the blood. But researchers found that RBCs that encountered foreign DNA played an activating role in the innate immune system.

The Mononuclear Phagocyte System

The tissues of the body are permeated by a support network of connective tissue fibers, or a reticulum, that interconnects nearby cells and meshes with the massive connective tissue network surrounding all organs. The phagocytic cells enmeshed in this network are collectively called the mononuclear phagocyte system (MPS; **figure 15.8**). This system is intrinsic to the immune function because it provides a passageway within and between tissues and organs. The MPS is found in the thymus, where important

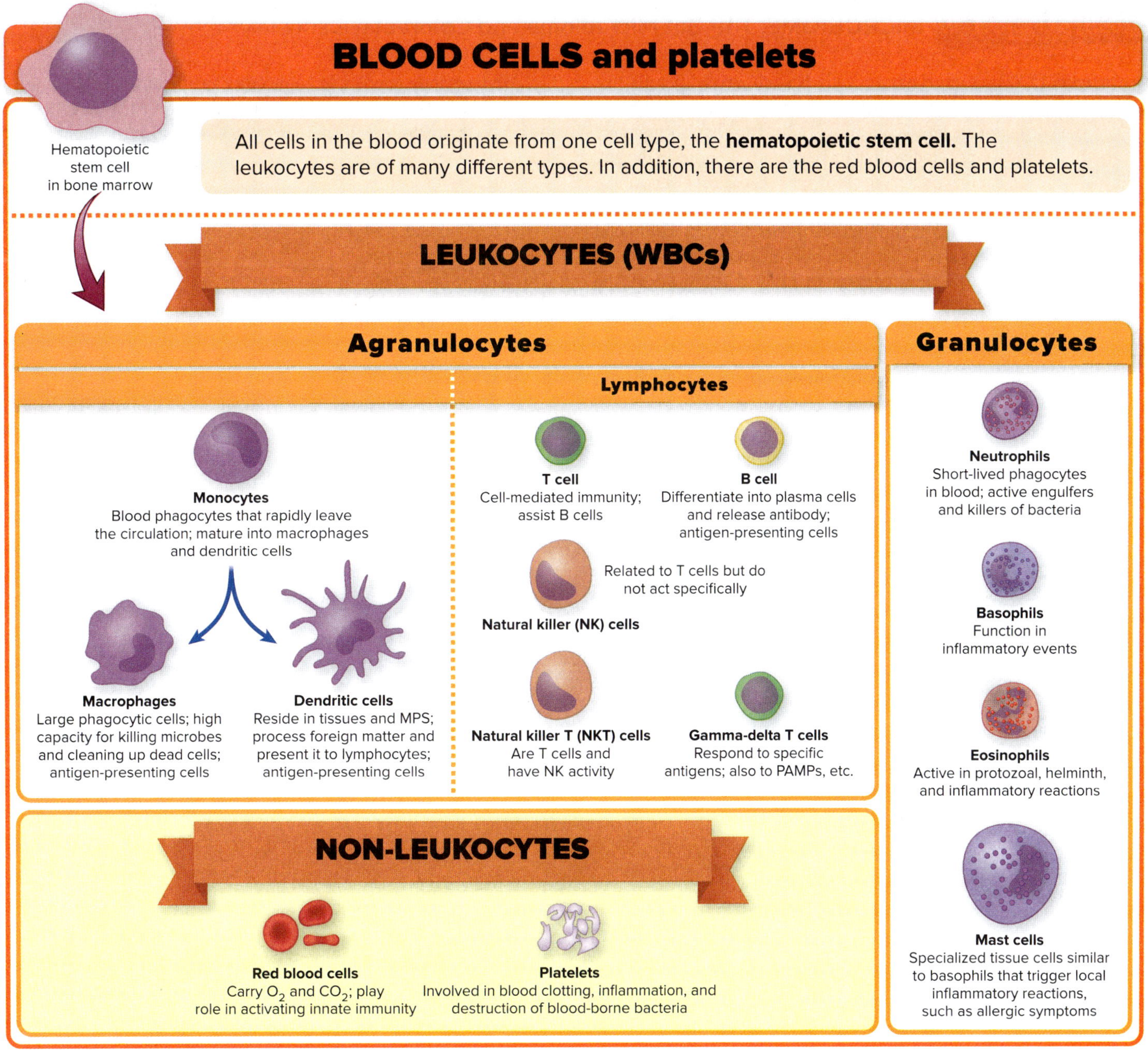
BLOOD CELLS and platelets
Hematopoietic stem cell in bone marrow
All cells in the blood originate from one cell type, the hematopoietic stem cell. The leukocytes are of many different types. In addition, there are the red blood cells and platelets.
LEUKOCYTES (WBCs)
Agranulocytes
Lymphocytes
Monocytes
Blood phagocytes that rapidly leave the circulation; mature into macrophages and dendritic cells
Macrophages
Large phagocytic cells; high capacity for killing microbes and cleaning up dead cells; antigen-presenting cells
Dendritic cells
Reside in tissues and MPS; process foreign matter and present it to lymphocytes; antigen-presenting cells
T cell
Cell-mediated immunity; assist B cells
B cell
Differentiate into plasma cells and release antibody; antigen-presenting cells
Related to T cells but do not act specifically
Natural killer (NK) cells
Natural killer T (NKT) cells
Are T cells and have NK activity
Gamma-delta T cells
Respond to specific antigens; also to PAMPs, etc.
Granulocytes
Neutrophils
Short-lived phagocytes in blood; active engulfers and killers of bacteria
Basophils
Function in inflammatory events
Eosinophils
Active in protozoal, helminth, and inflammatory reactions
Mast cells
Specialized tissue cells similar to basophils that trigger local inflammatory reactions, such as allergic symptoms
NON-LEUKOCYTES
Red blood cells
Carry O_2 and CO_2; play role in activating innate immunity
Platelets
Involved in blood clotting, inflammation, and destruction of blood-borne bacteria

Figure 15.7 Description of blood cells and platelets.

white blood cells mature, and in the lymph nodes, tonsils, spleen, and lymphoid tissue in the mucosa of the gut and respiratory tract, where most of the MPS "action" takes place. Recall that monocytes give rise to both macrophages and dendritic cells (review figure 15.7).

When differentiated macrophages and dendritic cells arrive in tissues that are near portals of entry or filtration organs, they often take up residence in these tissues permanently, waiting to attack foreign intruders **(figure 15.9).** These tissues and specialized **histiocyte** cells include the liver (Kupffer cells), lungs (alveolar macrophages), skin (Langerhans cells), brain (microglia), and others.

Disease Connection

There is a particular type of asthma induced by an excess of eosinophils in the airways. For a long time, researchers were not sure what role eosinophils played in inflammation, but now they know that eosinophils play an active role in the development of inflammation in asthma. Corticosteroids are used to treat this, like other forms of asthma. Because it is not always effective in eosinophilic asthma, other treatments are being investigated.

INSIGHT 15.1 MICROBIOME: The Gut Microbiome and Colorectal Cancer

In figure 15.7 you see two types of cells, NK cells and NKT cells, that are in the lymphocyte lineage but have no antigen receptors, so they are considered innate lymphoid cells (ILCs). They have been found to be very important in maintaining a healthy microbiome in the gut, and by doing this, they protect the host against colorectal cancer.

This class of cells has been shown to be drastically reduced as well as altered in their function in people with colorectal cancer. When these cells are disrupted in mice, the mice get aggressive forms of colon cancer. To make matters worse, the disruption in these cells leads to reduced effectiveness in cancer immunotherapies.

The researchers who uncovered this said that the ILCs communicate with T cells (from the lymphocyte lineage, and with adaptive immune properties). When communication is disrupted, generalized inflammation in the gut occurs and alters the composition of the gut microbiome. This, in turn, results in a decrease in T cells, which are critical for fighting tumors.

We have long appreciated how complex the innate and adaptive immune systems are. Only recently have we seen that their impact on the microbiome, and vice versa, is also a critical part of the health–disease continuum.

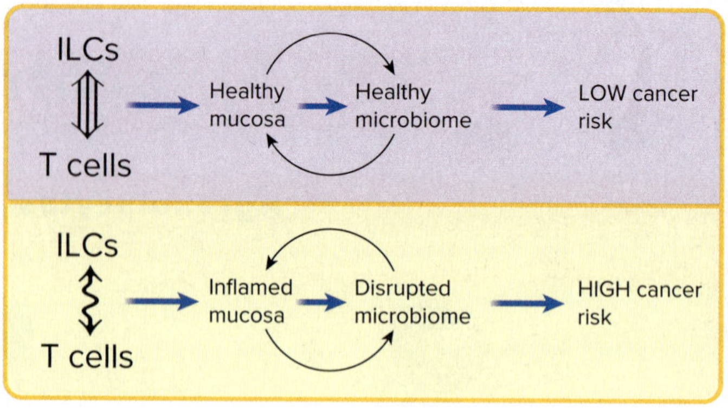
ILCs
T cells
Healthy mucosa
Healthy microbiome
LOW cancer risk
ILCs
T cells
Inflamed mucosa
Disrupted microbiome
HIGH cancer risk

Table 15.1 Types of Cytokines

	Examples	Source	Target
Pro-inflammatory cytokines, which *encourage* adaptive and innate immune responses	Interleukin-1 (IL-1)	Macrophages, B cells, dendritic cells	B cells, T cells
	Tumor necrosis factor-β (TNF-β)	T cells	Phagocytes, tumor cells
Anti-inflammatory cytokines, which *discourage* adaptive and innate immune responses	Interleukin-10 (IL-10)	T cells	B cells, macrophages
Vasodilators and vasoconstrictors, which can change the diameter of blood vessels or vessel permeability	Serotonin	Platelets and intestinal cells	Cells in peripheral and central nervous system
	Histamine	Mast cells and basophils	Blood vessels, sensory nerves, neutrophils
Growth factor cytokines, which regulate lymphocyte growth or activation	Interleukin-7 (IL-7)	Bone marrow cells, epithelial cells	Stem cells (stimulates growth of B and T cells)
	Erythropoietin	Endothelial cells	Stem cells (stimulates production of red blood cells)

Cytokines: Critical for Cell Communication

Before we specifically address details of the first or second line of defense, we need to understand the role of some potent chemicals and proteins that influence defensive responses, and that immune cells use to communicate with one another.

Hundreds of small, active molecules are constantly being secreted to regulate, stimulate, suppress, and otherwise control the many aspects of cell development, inflammation, and immunity. These substances are the products of several types of cells, including monocytes, macrophages, lymphocytes, fibroblasts, mast cells, platelets, and the endothelial cells of blood vessels.

These cell products are generally called **cytokines.** There are dozens of cytokines that have been identified. Specific cytokines will be discussed as we cover various immune processes.

Table 15.1 provides four broad categories of cytokines based on their functions. One or two examples of each functional category are provided to demonstrate the breadth of their activities.

15.1 Learning Outcomes—Assess Your Progress

1. Summarize the three lines of host defenses.
2. Define "marker" and discuss its importance in the second and third lines of defense.
3. Name the body systems that participate in immunity.
4. Describe the structure and function of the lymphatic system and its connection with the circulatory system.
5. Name three kinds of blood cells that function in innate immunity.

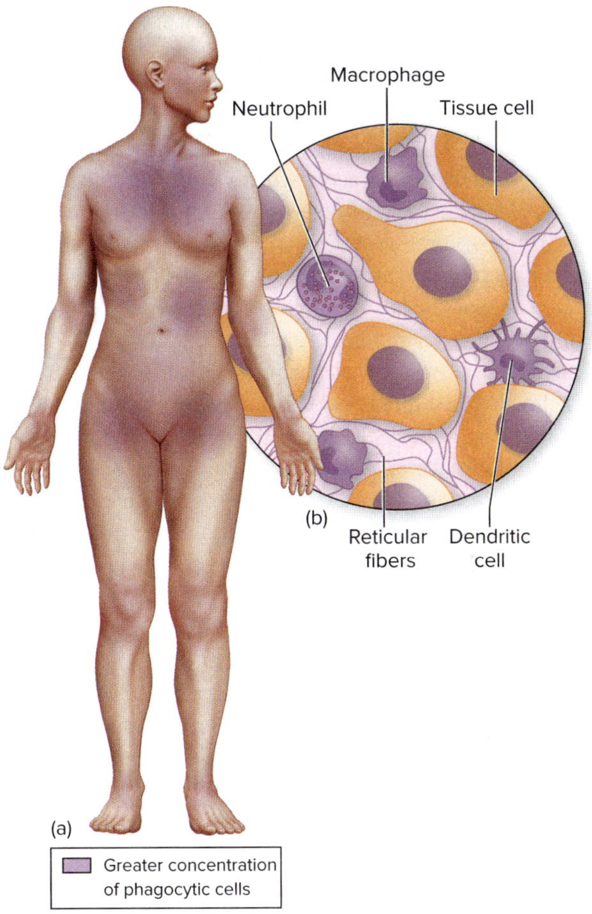
Macrophage
Neutrophil
Tissue cell
(b)
Reticular fibers
Dendritic cell
(a)
Greater concentration of phagocytic cells

Figure 15.8 The mononuclear phagocyte system is a pervasive, continuous connective tissue framework throughout the body. **(a)** The degrees of shading in the body indicate variations in phagocyte concentration (darker = greater). **(b)** This system begins at the microscopic level with a fibrous support network (reticular fibers) enmeshing each cell. This web connects one cell to another within a tissue or an organ and provides a niche for phagocytic white blood cells, which can crawl within and between tissues.

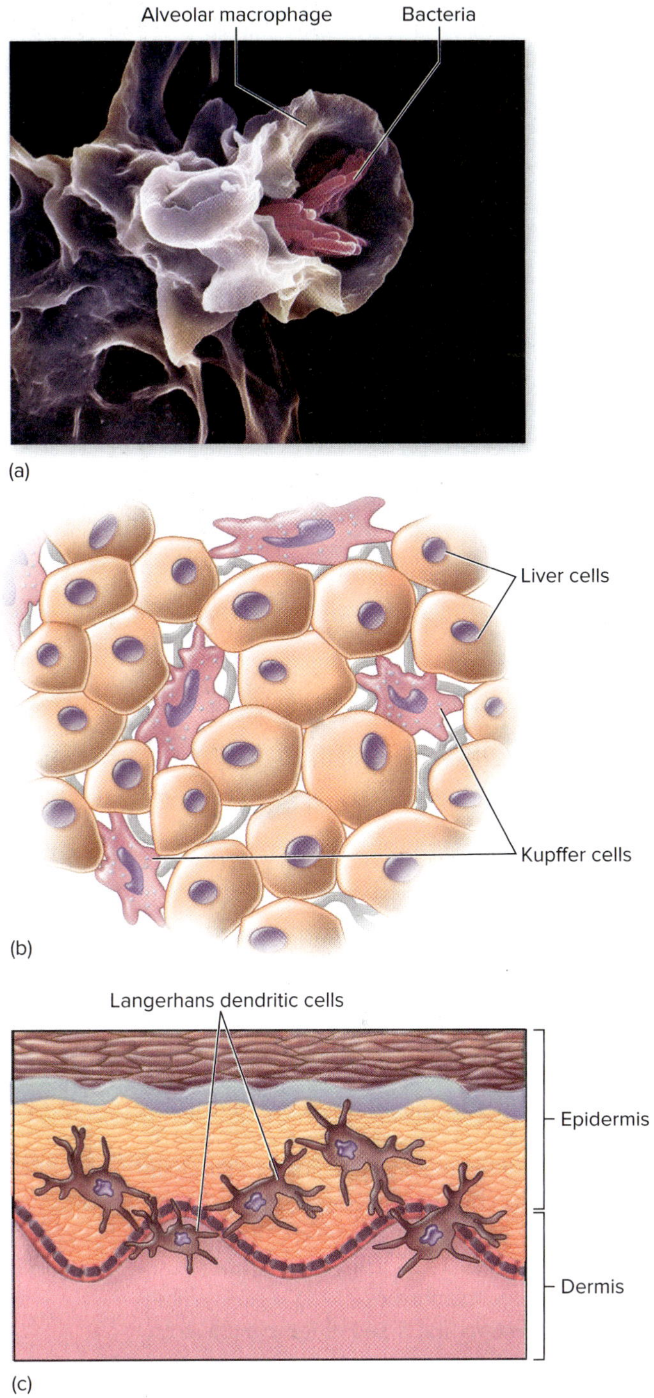
Alveolar macrophage
Bacteria
(a)
Liver cells
Kupffer cells
(b)
Langerhans dendritic cells
Epidermis
Dermis
(c)

Figure 15.9 Macrophages. **(a)** Scanning electron micrograph of an alveolar (lung) macrophage devouring bacteria. **(b)** Liver tissue with Kupffer cells. **(c)** Langerhans cells deep in the epidermis.

(a) SPL/Science Source

6. Connect the mononuclear phagocyte system to innate immunity.
7. Describe how and where T and B lymphocytes mature to be ready for their role in adaptive immunity.
8. Summarize the importance of cytokines, and list one pro-inflammatory and one anti-inflammatory cytokine.

15.2 The First Line of Defense

These are the defenses that are a normal part of the body's anatomy and physiology. These inborn defenses are either physical or chemical barriers that impede the entry of not only microbes but also any foreign agent, whether living or not **(figure 15.10).**

Physical or Anatomical Barriers at the Body's Surface

The skin and mucous membranes of the respiratory and digestive tracts have several built-in defenses. The outermost layer (stratum corneum) of the skin is composed of epithelial cells that have become compacted, cemented together, and impregnated with an insoluble protein—keratin. The result is a thick, tough layer that is highly impervious and waterproof. Few pathogens can penetrate this unbroken barrier, especially in regions such as the soles of the feet or the palms of the hands, where the stratum corneum is much thicker than on other parts of the body. It is so obvious as to be overlooked: The skin separates our inner bodies from the microbial

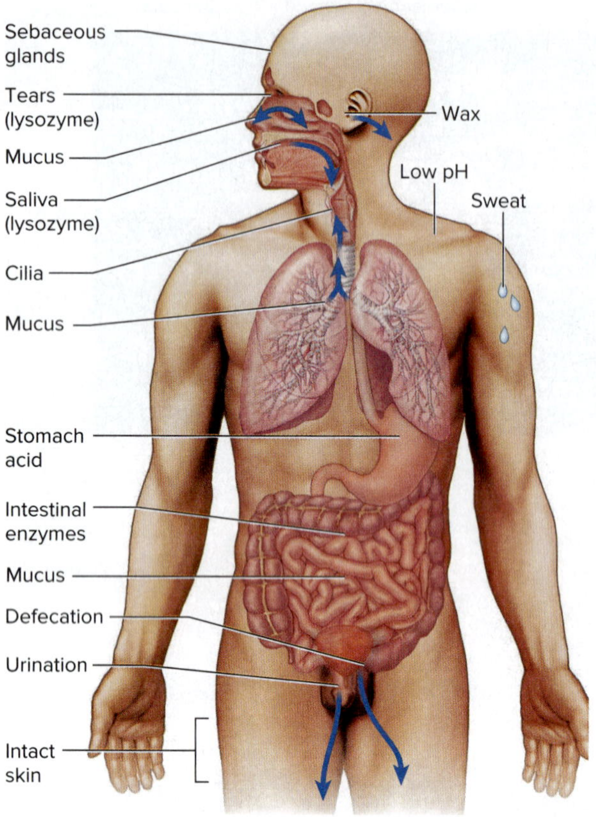
Sebaceous glands
Tears (lysozyme)
Mucus
Saliva (lysozyme)
Cilia
Mucus
Stomach acid
Intestinal enzymes
Mucus
Defecation
Urination
Intact skin
Wax
Low pH
Sweat

Figure 15.10 The primary physical and chemical defense barriers.

assaults of the environment. It is a surprisingly tough and sophisticated barrier. The keratin in the top layer of cells is a protective and waterproofing protein. In addition, outer layers of skin are constantly sloughing off, taking associated microbes with them. Other barriers associated with the skin include hair follicles and skin glands. The hair shaft is periodically shed and the follicle cells are **desquamated** (des′-kwuh-mayt-ud). The flushing effect of sweat glands also helps remove microbes.

The mucous membranes of the digestive, urinary, and respiratory tracts and of the eye are moist and permeable. They do provide barrier protection but without a keratinized layer. The mucous coat on the free surface of some membranes impedes the entry and attachment of bacteria. In addition, the mucus layer is rich with bacteriophages, and these can destroy bacteria that they encounter. Blinking and tear production (lacrimation) flush the eye's surface with tears and rid it of irritants. The constant flow of saliva helps carry microbes into the harsh conditions of the stomach. Vomiting and defecation also evacuate substances or microorganisms from the body.

The respiratory tract is constantly guarded from infection by elaborate and highly effective adaptations. Nasal hair traps larger particles. The copious flow of mucus and fluids that occurs in allergy and colds exerts a flushing action. In the respiratory tree (primarily the trachea and bronchi), a ciliated epithelium (called the ciliary escalator) moves foreign particles entrapped in mucus toward the pharynx to be removed **(figure 15.11).** Irritation of the nasal passage reflexively initiates a sneeze, which expels a large volume of air and fluids at high velocity. Similarly, the acute sensitivity of the bronchi, trachea, and larynx to foreign matter triggers coughing, which ejects irritants.

The genitourinary tract derives partial protection via the continuous trickle of urine through the ureters and from periodic bladder emptying that flushes the urethra. Vaginal secretions provide cleansing of the lower reproductive tract in females.

The vital contribution of barriers is clearly demonstrated in people who have lost them or never had them. Patients with severe skin damage due to burns are extremely susceptible to infections. Those with blockages in the salivary glands, tear ducts, intestine, and urinary tract are also at greater risk for infection. But as important as it is, the first line of defense alone is not sufficient to protect against infection. Because many pathogens find a way to circumvent the barriers by using their virulence factors, a whole set of defenses—inflammation, phagocytosis, adaptive immune responses—are brought into play.

The human microbiome forms a type of structural barrier. Because humans have evolved with—and in response to—the constant presence of our microbiome, we can consider it as an integral part of our anatomy. Even though the microbiome does not constitute an anatomical barrier, the microbial antagonism it provides can block the access of pathogens to epithelial surfaces and can

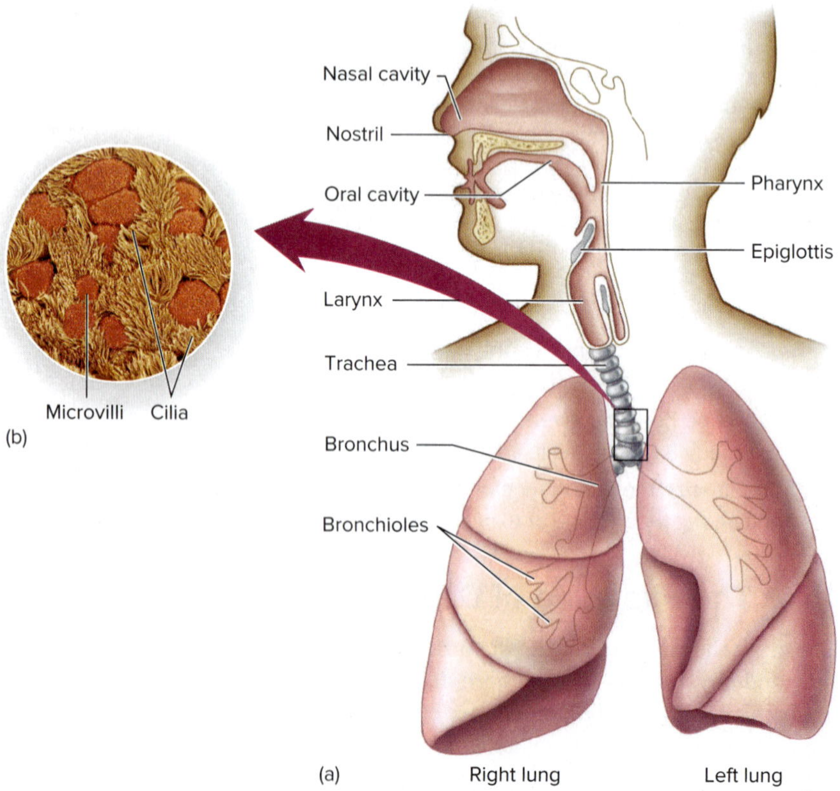
Nasal cavity
Nostril
Oral cavity
Larynx
Trachea
Bronchus
Bronchioles
Pharynx
Epiglottis
Microvilli
Cilia
(b)
(a)
Right lung
Left lung

Figure 15.11 The ciliary defense of the respiratory tree. **(a)** The epithelial lining of the airways contains a brush border of cilia to entrap and propel particles upward toward the pharynx. **(b)** Tracheal mucosa (5,000×).

(b) Susumu Nishinaga/Science Source

create an unfavorable environment for pathogens by competing for limited nutrients or by altering the local pH.

Research stemming from the Human Microbiome Project (HMP) has continued to highlight the importance of microbiota on the development of innate defenses (described in this chapter) and adaptive immunity (described in chapter 16). A robust commensal biota "trains" host defenses in such a way that commensals are kept in check and pathogens are eliminated. Evidence suggests that inflammatory bowel diseases, including Crohn's disease and ulcerative colitis, may well be results of our overzealous attempts to free our environment of microbes and to overtreat ourselves with antibiotics. The result, researchers say, is an "ill-trained" gut defense system that inappropriately responds to commensal biota. In terms of cutaneous pathogens, it has been found that the normal microbiota of the skin appears to control localized immune reactions, including T-cell activity, conferring protection against the invasion of these pathogens through the skin.

Innate Chemical Defenses

The skin and mucous membranes offer a variety of chemical defenses. Sebum from the sebacious glands exerts an antimicrobial effect, and specialized glands such as the meibomian glands of the eyelids lubricate the conjunctiva with an antimicrobial secretion. An additional defense in tears and saliva is **lysozyme,** an enzyme that hydrolyzes the peptidoglycan in the cell wall of bacteria. The high lactic acid and electrolyte concentrations of sweat and the skin's acidic pH and fatty acid content are also inhibitory to many microbes. Likewise, the hydrochloric acid in the stomach provides protection against many pathogens that are swallowed, and the intestine's digestive juices and bile can be destructive to microbes. Even semen contains an antimicrobial chemical that inhibits bacteria, and the vagina of women of childbearing age has a protective acidic pH maintained by normal biota.

15.2 Learning Outcomes—Assess Your Progress

9. Identify the three components of the first line of defense.
10. Identify the four body systems that participate in the first line of defense.
11. Describe two examples of how the normal microbiota contribute to the first line of defense.

15.3 The Second Line of Defense

Now that we have introduced the principal anatomical and physiological framework of the immune system, we address some mechanisms that play important roles in host defenses: (1) phagocytosis, (2) inflammation, (3) fever, and (4) antimicrobial products from the host. Because of the generalized nature of these defenses, they are primarily nonspecific in their effects, but they also support and interact with the adaptive immune responses described in chapter 16.

Phagocytosis: Cornerstone of Inflammation and Adaptive Immunity

By any standard, a phagocyte represents an impressive piece of living machinery, wandering through the tissues to seek, capture, and destroy a target. The general activities of phagocytes are

1. to survey the tissue compartments and discover microbes, particulate matter (dust, carbon particles, antigen-antibody complexes), and injured or dead cells;
2. to ingest and eliminate these materials; and
3. to recognize immunogenic information (markers or antigens) in foreign matter.

It is generally accepted that all cells have some capacity to engulf materials, but professional **phagocytes** do it for a living. The three main types of phagocytes are neutrophils, monocytes, and macrophages.

Neutrophils and Eosinophils

Neutrophils are general-purpose phagocytes that react early in the inflammatory response to bacteria, other foreign materials, and damaged tissue. A common sign of bacterial infection is a high neutrophil count in the blood (neutrophilia), and neutrophils are a primary component of pus. **Eosinophils** are attracted to sites of parasitic infections and antigen–antibody reactions, though they play only a minor phagocytic role.

Monocytes and Macrophages: Kings of the Phagocytes

After emigrating out of the bloodstream into the tissues due to chemical stimuli, **monocytes** are transformed by various inflammatory chemicals into macrophages. This process is marked by an increase in size and by enhanced development of lysosomes and other organelles. At one time, macrophages were classified as either *fixed* (adherent to tissue) or *wandering,* but this terminology can be misleading. All macrophages retain the capacity to move about. Whether they reside in a specific organ or wander depends upon their stage of development and the immune stimuli they receive. Specialized macrophages called histiocytes (*histio* = "tissue") migrate to a certain tissue and remain there during their life span. Examples are alveolar (lung) macrophages; the Kupffer cells in the liver; dendritic cells in the skin (see figure 15.9); and macrophages in the spleen, lymph nodes, bone marrow, kidney, bone, and brain. Other macrophages do not reside permanently in a particular tissue and drift nomadically throughout the MPS. Not only are macrophages dynamic scavengers, but they also process foreign substances and prepare them for reactions with B and T lymphocytes.

Mechanisms of Phagocytic Recognition, Engulfment, and Killing

The term *phagocyte* literally means "eating cell." But *phagocytosis* (the term for what phagocytes do) is more than just the physical process of engulfment because phagocytes also actively attack and dismantle foreign cells with a wide array of antimicrobial substances. Phagocytosis can occur as an isolated event performed by a lone phagocytic cell responding to a minor irritant in its area or as part of the orchestrated events of inflammation described in the next section. Phagocytosis occurs in steps: chemotaxis, ingestion, phagolysosome formation, destruction, and excretion **(figure 15.12).**

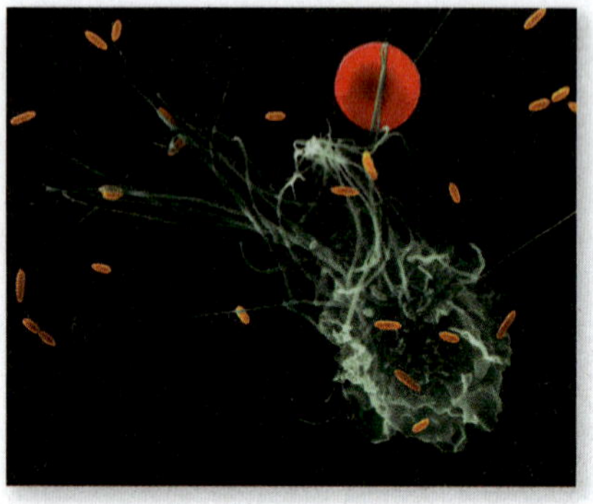

Science Photo Library/Alamy Stock Photo

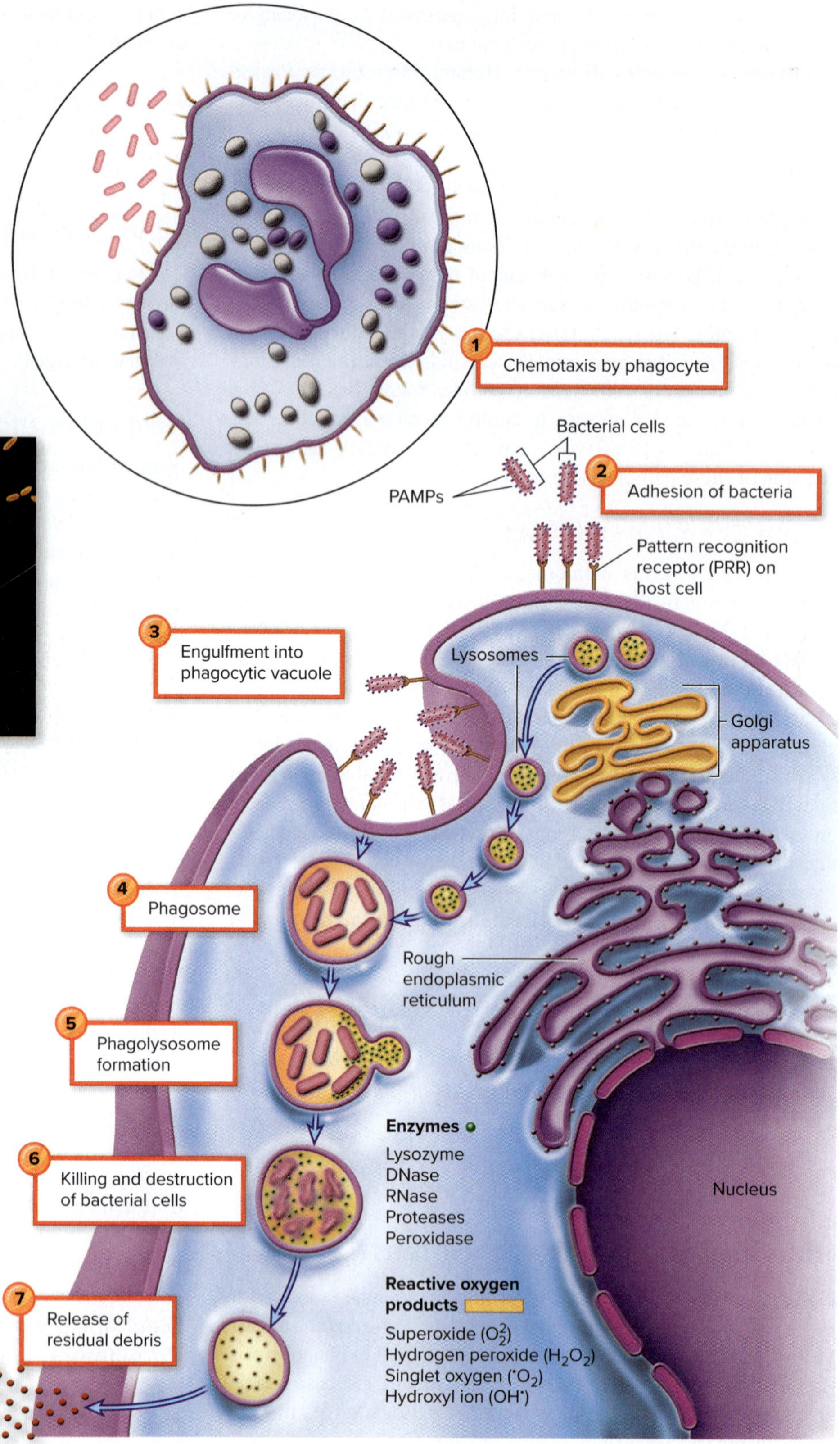
1 Chemotaxis by phagocyte
Bacterial cells
PAMPs
2 Adhesion of bacteria
Pattern recognition receptor (PRR) on host cell
3 Engulfment into phagocytic vacuole
Lysosomes
Golgi apparatus
4 Phagosome
Rough endoplasmic reticulum
5 Phagolysosome formation
6 Killing and destruction of bacterial cells
7 Release of residual debris
Enzymes
Lysozyme
DNase
RNase
Proteases
Peroxidase
Reactive oxygen products
Superoxide (O_2^-)
Hydrogen peroxide (H_2O_2)
Singlet oxygen ($^{\cdot}O_2$)
Hydroxyl ion ($OH^{\cdot}$)
Nucleus

Figure 15.12 The phases of phagocytosis. (1) Phagocyte is attracted to bacteria. (2) Close-up view of process showing bacteria adhering to phagocyte PRRs by their PAMPs. (3) Vacuole is formed around bacteria during engulfment. (4) Phagosome, a digestive vacuole, results. (5) Lysosomes fuse with phagosome, forming a phagolysosome. (6) Enzymes and toxic oxygen products kill and digest bacteria. (7) Undigested particles are released. **(Photograph)** Scanning electron micrograph of a macrophage actively engaged in devouring bacteria (10,000×).

Chemotaxis and Ingestion Phagocytes and other defensive cells are able to recognize many microorganisms as foreign because of various signal molecules that the microbes have on their surfaces. Some important examples of these are called pathogen-associated molecular patterns (PAMPs). They are molecules shared by many microorganisms—but not present in mammals—and therefore serve as "red flags" for phagocytes and other cells of innate immunity.

Bacterial PAMPs include peptidoglycan and lipopolysaccharide. Double-stranded RNA, which is found only in some viruses, is also a PAMP. On the host side, phagocytes, dendritic cells, endothelial cells, and even lymphocytes possess molecules on their surfaces—called pattern recognition receptors (PRRs)—that recognize and bind PAMPs. The cells possess these PRRs all the time, whether or not they have encountered PAMPs before. This is different from the situation with adaptive immunity. One category of PRRs is the **toll-like receptors (TLRs; figure 15.13)** (called "toll-like" because similar proteins called "toll" were originally discovered in fruit flies). The receptors not only recognize PAMPs but, upon binding, set in motion a cascade of events inside the host cell that amplifies and orchestrates a defensive response to the pathogen. This includes the formation of what is known as an inflammasome, a protein complex that promotes a fully developed inflammatory response. Alternatively, the response may be the initiation of an adaptive immune response. (There are a lot of acronyms in immunology and especially in this last paragraph. Do not let them get away from you; keep up with them. If you know what all the acronyms stand for and what they do, you are a good deal of the way there in understanding host defenses!)

There is a class of PRRs that are not part of the cell membrane of phagocytes. **Collectins** are soluble molecules that roam through blood and tissues, bind to microbial PAMPs, and mark them for phagocytic destruction.

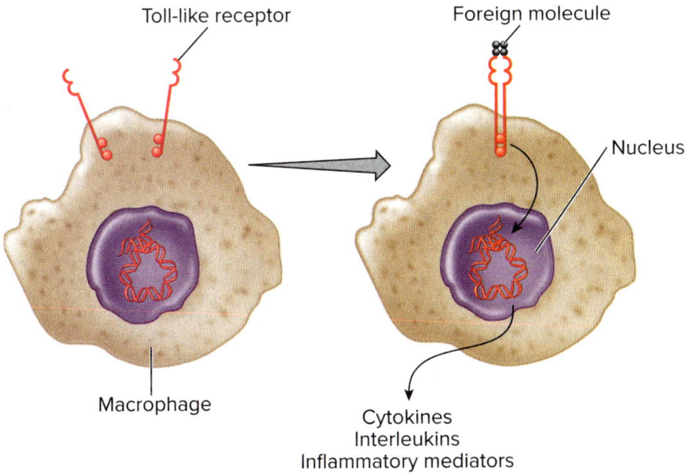
Toll-like receptor
Foreign molecule
Nucleus
Macrophage
Cytokines
Interleukins
Inflammatory mediators

Figure 15.13 Phagocyte detection and signaling with pattern recognition receptors. The example here shows the actions of a toll-like receptor that is present in the membrane of many cells of the immune system. When a molecule specific to a particular class of pathogen is recognized by a receptor, the toll-like receptors merge and bind the foreign molecule. This induces production of chemicals that stimulate an immune response.

On the scene of an inflammatory reaction, phagocytes often trap cells or debris against the fibrous network of connective tissue or the wall of blood and lymphatic vessels. Once the phagocyte has made contact with its prey, it extends pseudopods that enclose the cells or particles in a pocket and internalize them in a vacuole called a **phagosome.** It has also become apparent that neutrophils and macrophages work together to trap pathogens in a spider web–like matrix of DNA and antimicrobial proteins to enhance their destruction.

Phagolysosome Formation and Killing In a short time, **lysosomes** migrate to the scene of the phagosome and fuse with it to form a **phagolysosome.** Other granules containing antimicrobial chemicals are released into the phagolysosome, forming a potent brew designed to poison and then dismantle the ingested material. The destructiveness of phagocytosis is evident by the death of bacteria within 30 minutes after contacting this battery of antimicrobial substances.

Destruction and Elimination Systems Two separate systems of destructive chemicals await the microbes in the phagolysosome. The oxygen-dependent system (known as the respiratory burst, or oxidative burst) involves several damaging substances. Myeloperoxidase, an enzyme found in granulocytes, forms halogen ions (OCl^-) that are strong oxidizing agents. Other products of oxygen metabolism such as hydrogen peroxide, the superoxide anion (O_2^-), activated or so-called singlet oxygen ($^{\bullet}O_2$), and the hydroxyl free radical ($OH^{\bullet}$) separately and together have formidable killing power. Other mechanisms that come into play are the release of lactic acid, lysozyme, and nitric oxide (NO), a powerful mediator that kills bacteria and inhibits viral replication. Cationic proteins that injure bacterial membranes and a number of proteolytic and other hydrolytic enzymes complete the job. The small bits of undigestible debris are released from the macrophage by exocytosis.

Disease Connection

Many pathogenic bacteria have developed mechanisms to resist phagocytic digestion. The bacteria that cause tuberculosis, listeriosis, plague, and many other infections survive inside the phagocyte. This can provide them with protection from the rest of the host's defenses and allows them to be transported throughout the body.

Interestingly, recent studies suggest that in the human gut, only the epithelial cells deep in intestinal crypts express large numbers of the toll-like receptors, which function to recognize microbes. The commensal bacteria occupy the "top" of the epithelium, not the crypts. This phenomenon may help explain why commensals are tolerated, but pathogens (which are likely to colonize the crypts) are not.

Inflammation: A Complex Concert of Reactions to Injury

At its most general level, the inflammatory response is a reaction to any traumatic event in the tissues. It is so commonplace that all of us manifest inflammation in some way every day. It appears in the nasty flare of a cat scratch, the blistering of a burn, the painful lesion of an infection, and the symptoms of allergy. When close to our external surfaces, it is readily identifiable by a classic series of signs and symptoms characterized succinctly by four Latin terms: *rubor, calor, tumor,* and *dolor. Rubor* ("redness") is caused by increased circulation and vasodilation in the injured tissues; *calor* ("warmth") is the heat given off by the increased flow of blood; *tumor* ("swelling") is caused by increased fluid escaping into the tissues; and *dolor* ("pain") is caused by the stimulation of nerve endings **(figure 15.14).** A fifth symptom, loss of function, has been added to give a complete picture of the effects of inflammation. Although these manifestations can be unpleasant, they serve as an important warning that injury has taken place and set in motion responses that save the body from further injury.

It is becoming increasingly clear that some chronic diseases, such as cardiovascular disease, can be caused by chronic inflammation. While we speak of inflammation at a local site (such as a finger), inflammation can affect an entire system—such as blood vessels, lungs, skin, the joints, and so on. Some researchers believe that the very act of aging is a consequence of increasing inflammation in multiple body systems. Also, chronic inflammation is a feature of many autoimmune diseases as well as many cases of long COVID-19.

Factors that can cause inflammation include trauma from infection (the primary emphasis here), tissue injury or necrosis due to physical or chemical agents, and adaptive immune reactions. Although the details of inflammation are very complex, its chief functions are

1. to mobilize and attract immune components to the site of the injury,
2. to set in motion mechanisms to repair tissue damage and localize and clear away harmful substances, and
3. to destroy microbes and block their further invasion **(figure 15.15).**

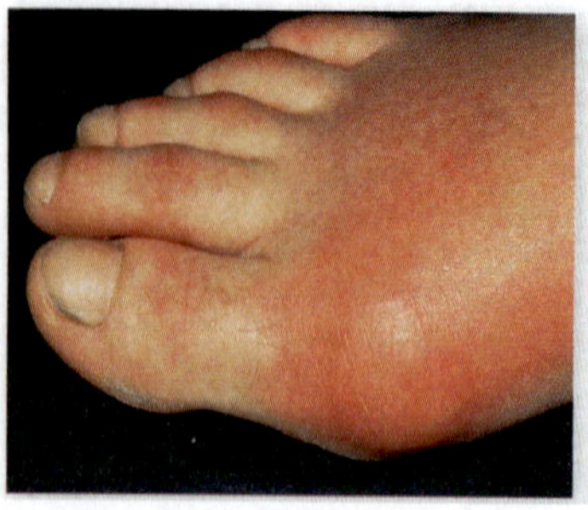
Rubor—redness
Tumor—swelling
Dolor—pain
Calor—heat

Figure 15.14 The response to injury. This classic checklist describes the reactions of the tissues to an assault. Each of the events is an indicator of one of the mechanisms of inflammation described in this chapter.
Dr. P. Marazzi/Science Source

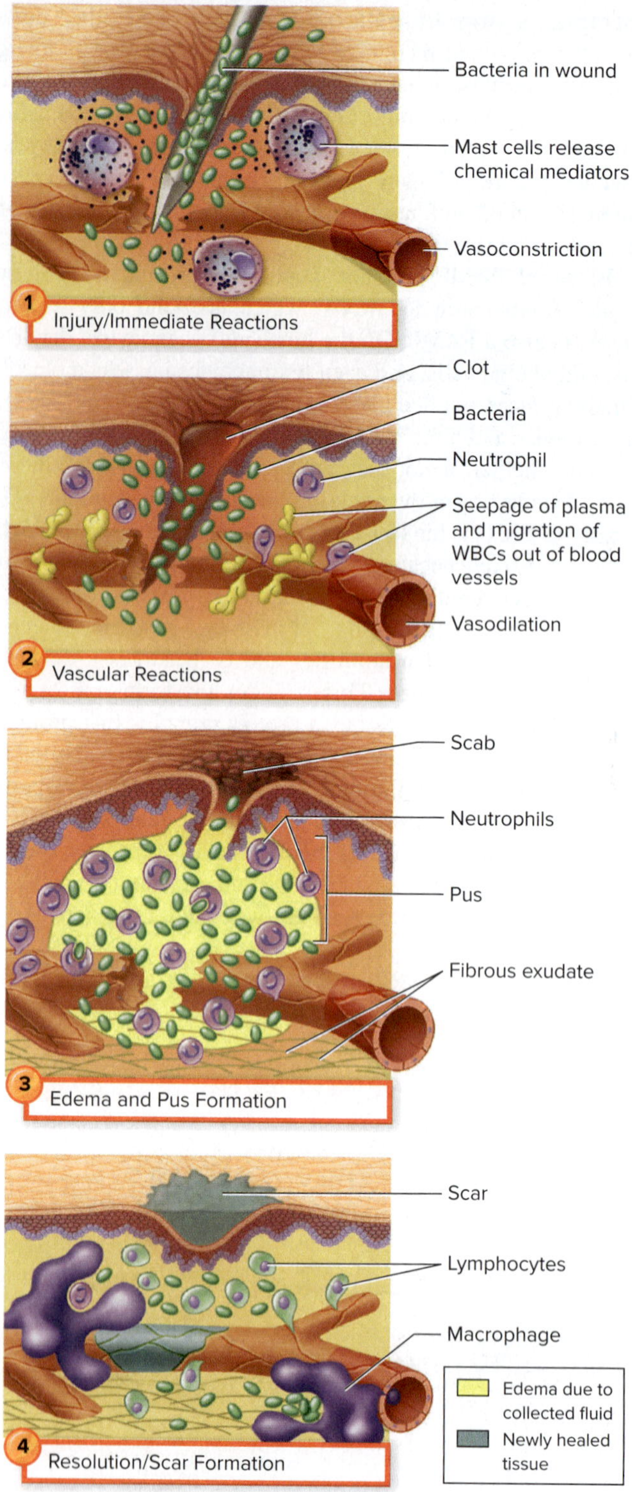
Bacteria in wound
Mast cells release chemical mediators
Vasoconstriction
1 Injury/Immediate Reactions
Clot
Bacteria
Neutrophil
Seepage of plasma and migration of WBCs out of blood vessels
Vasodilation
2 Vascular Reactions
Scab
Neutrophils
Pus
Fibrous exudate
3 Edema and Pus Formation
Scar
Lymphocytes
Macrophage
4 Resolution/Scar Formation
Edema due to collected fluid
Newly healed tissue

Figure 15.15 The major events in inflammation.
(1) Injury → Reflexive narrowing of the blood vessels (vasoconstriction) lasting for a short time → Release of chemical mediators into area.
(2) Increased diameter of blood vessels (vasodilation) → Increased blood flow → Increased vascular permeability → Leakage of fluid (plasma) from blood vessels into tissues (exudate formation).
(3) Edema → Infiltration of site by neutrophils and accumulation of pus.
(4) Macrophages and lymphocytes → Repair, either by complete resolution and return of tissue to normal state or by formation of scar tissue.

The inflammatory response is a powerful defensive reaction, a means for the body to maintain stability and restore itself after an injury. But keep in mind that inflammation also can cause a great deal of damage in the body. Inflammation may actually trap pathogens within a localized area, leading to the formation of an abscess over time. Granuloma formation is characteristic of many chronic infectious diseases, including tuberculosis and leprosy, and is due to incomplete immune activity against pathogens in localized tissue areas.

The Stages of Inflammation

The process leading to inflammation is a dynamic, predictable sequence of events that can be acute, lasting from a few minutes or hours, to chronic, lasting for days, weeks, or years. Once the initial injury has occurred, a chain reaction takes place at the site of damaged tissue, summoning beneficial cells and fluids into the injured area. As an example, we will look at an injury at the microscopic level and observe the flow of major events (as shown in figure 15.15).

Vascular Changes: Early Inflammatory Events

Following an injury, some of the earliest changes occur in the vasculature (arterioles, capillaries, venules) in the vicinity of the damaged tissue. These changes are controlled by nervous stimulation, **chemical mediators,** and cytokines released by blood cells, tissue cells, and platelets in the injured area. Some of them encourage immune reactivity. Some of them discourage or dampen that response; some act on the blood vessels in the area; and some regulate the growth or activation of white blood cells. They are absolutely critical to our defenses. You might think of them as different musical parts in a symphony, all of them necessary, and taking place in an environment of many other "instruments." Immunologists who study these chemicals may be the only ones who truly understand how they all work together. Although it may have been tempting to gloss over table 15.1, take a few minutes to go back and review what kinds of chemicals act in these processes. It will pay off in this chapter and beyond.

Although the constriction (narrowing) of arterioles is stimulated first, it lasts for only a few seconds or minutes and is followed in quick succession by the opposite reaction, vasodilation. The overall effect of vasodilation is to increase the flow of blood into the area, which allows the influx of immune components and causes redness and warmth.

Edema: Leakage of Vascular Fluid into Tissues

Some vasoactive substances cause the endothelial cells in the walls of venules to contract and form gaps, through which blood components exude into the extracellular spaces. The fluid part that escapes is called the **exudate.** Accumulation of this fluid in the tissues gives rise to local swelling and firmness, called **edema.** The exudate contains varying amounts of plasma proteins, such as globulins, albumin, the clotting protein fibrinogen, blood cells, and cellular debris. Depending on its content, the exudate may be clear, called **serous,** or it may contain red blood cells or pus. **Pus** is composed mainly of white blood cells and the debris generated by phagocytosis. In some types of edema, the fibrinogen is converted to fibrin threads that enmesh the injury site. Within an hour, multitudes of neutrophils responding chemotactically to special signaling molecules converge on the injured site (see figure 15.15, step 3).

Unique Characteristics of White Blood Cells In order for WBCs to leave the blood vessels and enter the tissues, they adhere to the inner walls of the smaller blood vessels. From this position, they are poised to migrate out of the blood into the tissue spaces by a process called **diapedesis** (dye″-ah-puh-dee′-sis).

Diapedesis, also known as *transmigration,* is aided by several related characteristics of WBCs. For example, they are actively motile and readily change shape. Diapedesis is also assisted by the nature of the endothelial cells lining the venules. They contain complex adhesive receptors that capture the WBCs and participate in their transport from the venules into the extracellular spaces **(figure 15.16).**

Another factor in the migratory habits of these WBCs is **chemotaxis,** defined as the tendency of cells to migrate in response to a specific chemical stimulus given off at a site of injury or infection (see inflammation and phagocytosis earlier in this chapter). Through this means, cells swarm from many compartments to the site of infection and remain there to perform general and specific immune functions. These basic properties are essential for the sort of intercommunication and deployment of cells required for most immune reactions (see figure 15.15).

Phagocytes migrate into a region of inflammation with a deliberate sense of direction, attracted by a gradient of stimulant products from the parasite and host tissue at the site of injury. The endpoint function of the white blood cells is the phagocytosis of microbes or other invading substances.

The Benefits of Edema and Chemotaxis Both the formation of edematous exudate and the infiltration of neutrophils are physiologically beneficial. The influx of fluid dilutes toxic substances, and the fibrin clot can effectively trap microbes and prevent their further spread. The neutrophils that aggregate in the inflamed site are immediately involved in phagocytosing and destroying bacteria, dead tissues, and particulate matter. In some types of inflammation, accumulated phagocytes contribute to pus, a whitish mass of cells, liquefied cellular debris, and bacteria.

Disease Connection

Certain bacteria (streptococci, staphylococci, gonococci, and meningococci) are especially powerful attractants for neutrophils and are thus termed **pyogenic,** or pus-forming, bacteria.

Late Reactions of Inflammation

Sometimes a mild inflammation can be resolved by edema and phagocytosis. Inflammatory reactions that are more long-lived attract a collection of monocytes, lymphocytes, and

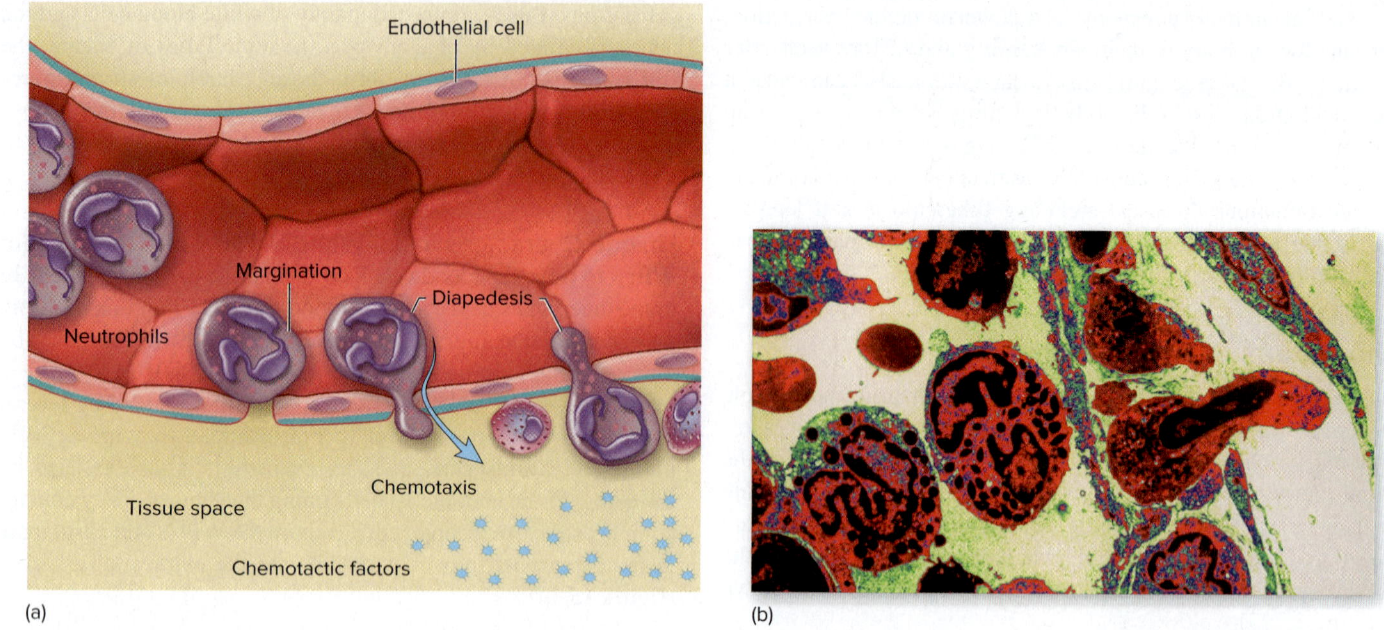
Endothelial cell
Margination
Diapedesis
Neutrophils
Chemotaxis
Tissue space
Chemotactic factors
(a)
(b)

Figure 15.16 Diapedesis and chemotaxis of leukocytes. **(a)** View of a venule depicts white blood cells squeezing themselves between spaces in the blood vessel wall via diapedesis. This process, shown in cross section, indicates how the leukocytes adhere to the endothelial wall. From this site, they are poised to migrate out of the vessel into the tissue space. **(b)** This photograph captures neutrophils in the process of diapedesis.

(b) Joubert/Science Source

macrophages to the reaction site. Clearance of pus, cellular debris, dead neutrophils, and damaged tissue is performed by macrophages, the only cells that can engulf and dispose of such large masses. At the same time, B lymphocytes react with foreign molecules and cells by producing specific antimicrobial proteins (antibodies), and T lymphocytes kill intruders directly. Late in the process, the tissue is completely repaired, if possible, or replaced by connective tissue in the form of a scar (see figure 15.15, step 4). If the inflammation cannot be relieved or resolved in this way, it can become chronic and create a long-term pathologic condition.

Fever: An Adjunct to Inflammation

An important systemic component of inflammation—and innate immunity in general—is *fever,* defined as an abnormally elevated body temperature. Although fever is a nearly universal symptom of infection, it is also associated with certain allergies, cancers, and other illnesses.

The body temperature is normally maintained by a control center in the hypothalamus region of the brain. This thermostat regulates the body's heat production and heat loss and sets the core temperature at around 36.4°C (97.5°F) with slight fluctuations (1°F) during a daily cycle. Until recently, normal body temperature was considered 98.6°F, but a great deal of recent research suggests that "normal" body temperature has fallen over the past several decades (possibly due to fewer background infections leading to less inflammation and fever in the "normal" population). Fever is initiated when substances called pyrogens (py′-roh-jenz) reset the hypothalamic thermostat to a higher setting. This change signals the musculature to increase heat production and peripheral arterioles to decrease heat loss through vasoconstriction. Fevers range in severity from low-grade (up to 38.3°C, or 100°F to 101°F) to moderate (38.8°C to 39.4°C, or 102°F to 103°F) to high (40.0°C to 41.1°C, or 104°F to 106°F). Pyrogens are either **exogenous** (coming from outside the body) or **endogenous** (originating internally). Exogenous pyrogens are products of infectious agents such as viruses, bacteria, protozoans, and fungi. One well-characterized exogenous pyrogen is endotoxin, the lipopolysaccharide found in the cell walls of gram-negative bacteria. Endogenous pyrogens are released by monocytes, neutrophils, and macrophages during the process of phagocytosis and appear to be a natural part of the immune response. Two potent pyrogens released by macrophages are interleukin-1 (IL-1) and tumor necrosis factor (TNF). Note that both of these substances are also classified as cytokines. One of their actions is to be pyrogenic. (Be sure to distinguish between the two terms *pyrogenic* and *pyogenic.*)

Benefits of Fever

The association of fever with infection strongly suggests that it serves a beneficial role. Aside from its practical and medical importance as a sign of a physiological disruption, increased body temperature has additional benefits:

- Fever inhibits multiplication of temperature-sensitive microorganisms such as the poliovirus, cold viruses, herpes zoster virus, systemic and subcutaneous fungal pathogens, *Mycobacterium* species, and the syphilis spirochete.
- Fever impedes the nutrition of bacteria by reducing the availability of iron. It has been demonstrated that during fever,

macrophages stop releasing their iron stores, which slows down several enzymatic reactions needed for bacterial growth.

- Fever increases metabolism and stimulates immune reactions and naturally protective physiological processes. It speeds up hematopoiesis, phagocytosis, and adaptive immune reactions. It increases the ability of specific lymphocytes to home in on sites of infection.

Treatment of Fever

With this revised perspective on fever, whether to suppress it or not can be a difficult decision. Some advocates feel that a slight to moderate fever in an otherwise healthy person should be allowed to run its course, in light of its potential benefits and minimal side effects. All medical experts do agree that high and prolonged fevers or fevers in patients with cardiovascular disease, head trauma, seizures, or respiratory ailments are risky and must be treated immediately with fever-reducing drugs.

Antimicrobial Products

In this section we will examine a broad group of products the host produces to battle microbial infections. We will look at four main categories: *interferons (IFN),* which are best known for targeting viruses but can also target bacteria and even tumor cells; *complement,* which targets membranes of pathogens and also of pathogen-infected host cells; *antimicrobial peptides,* which can directly kill all manner of microbes; and *restriction factors,* which inhibit the multiplication of viruses in host cells.

Interferon

Interferon (IFN) is a small protein produced naturally by certain white blood and tissue cells. It is used in therapy against certain viral infections and cancer. Although the interferon system was originally thought to be directed exclusively against viruses, it is now known to be involved also in defenses against other microbes and in immune regulation and intercommunication. Three major types are *interferons alpha* and *beta,* products of many cells, including lymphocytes, fibroblasts, and macrophages; and *interferon gamma,* a product of T cells. Interferons are also a type of cytokine (see table 15.1).

All three classes of interferon are produced in response to viruses, RNA, immune products, and various antigens. Their biological activities are extensive. In all cases, they bind to cell surfaces and induce changes in genetic expression, but the exact results vary. In addition to antiviral effects discussed in the next section, all three IFNs can inhibit the expression of cancer genes and have tumor suppressor effects. IFN alpha and beta stimulate phagocytes, and IFN gamma is a regulator of macrophages and T and B cells.

Characteristics of Antiviral Interferon When a virus binds to the receptors on a host cell, a signal is sent to the nucleus that directs the cell to synthesize interferon. After transcribing and translating the interferon gene, newly synthesized interferon molecules are rapidly secreted by the cell into the extracellular space, where they bind to other host cells. The binding of interferon to a second cell induces the production of proteins in *that* cell that inhibit viral multiplication either by degrading the viral RNA or by preventing the translation of viral proteins **(figure 15.17).** Interferon is not virus-specific, so its synthesis in response to one type of virus will also protect against other types. This is why it has been produced industrially and used as a treatment for a number of viral infections. In 2021 evidence began emerging that an interferon deficiency could cause some people to become much sicker when they are infected with SARS-CoV-2.

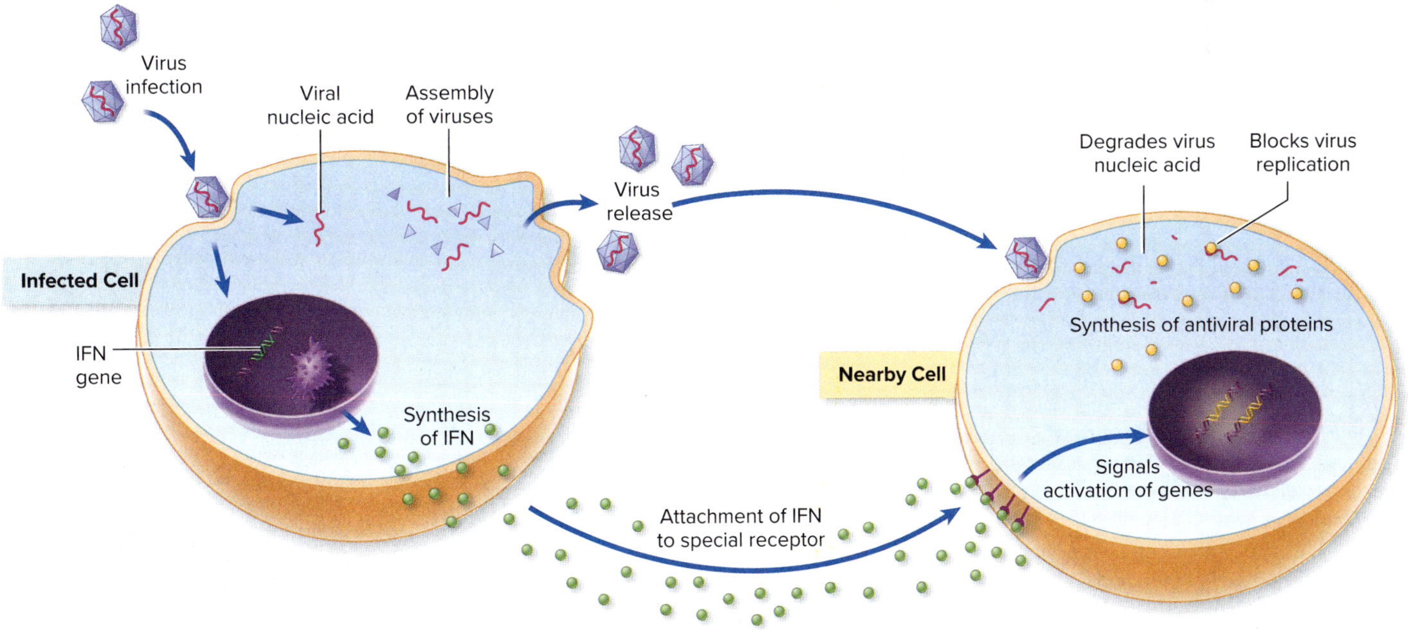
Virus infection
Viral nucleic acid
Assembly of viruses
Virus release
Degrades virus nucleic acid
Blocks virus replication
Infected Cell
IFN gene
Synthesis of IFN
Synthesis of antiviral proteins
Nearby Cell
Signals activation of genes
Attachment of IFN to special receptor

Figure 15.17 The antiviral activity of interferon. When a cell is infected by a virus, its DNA is triggered to transcribe and translate the interferon (IFN) gene. Interferon diffuses out of the cell and binds to IFN receptors on nearby uninfected cells, where it induces production of proteins that eliminate viral genes and block viral replication. Note that the original cell is not protected by IFN and that IFN does not prevent viruses from invading the protected cells.

Other Roles of Interferon Interferons are important cytokines that activate or guide the development of white blood cells. For example, interferon alpha produced by T lymphocytes activates a subset of cells called natural killer (NK) cells. In addition, one type of interferon beta plays a role in the maturation of B and T lymphocytes and in inflammation. Interferon gamma inhibits cancer cells, stimulates B lymphocytes, activates macrophages, and enhances the effectiveness of phagocytosis. It was recently discovered that interferon also reduces the amount of cholesterol in the body. Because cholesterol is used by bacteria and viruses as a nutrient, this provides another source of innate protection.

Complement

Among its many overlapping functions, the immune system has another complex and multiple-duty system called **complement.** Like inflammation and phagocytosis, complement comes into play at several levels. The complement system, named for its property of "complementing" immune reactions, consists of up to 50 blood proteins that work together to destroy bacteria and certain viruses.

The concept of a cascade reaction is helpful in understanding how complement functions. A *cascade reaction* is a sequential physiological response like that of blood clotting, in which the first substance in a chemical series activates the next substance, which activates the next, and so on, until a desired end product is reached. There are three different complement pathways, distinguished by how they become activated. The final stages of the three pathways are the same and yield a similar end result. For our discussion, we will focus on shared characteristics. **Figure 15.18** shows the three different ways that the complement system can be activated. The classical pathway is activated by antibodies bound to microbial surfaces. The lectin pathway is activated by other proteins (lectins) that bind to sugars on the

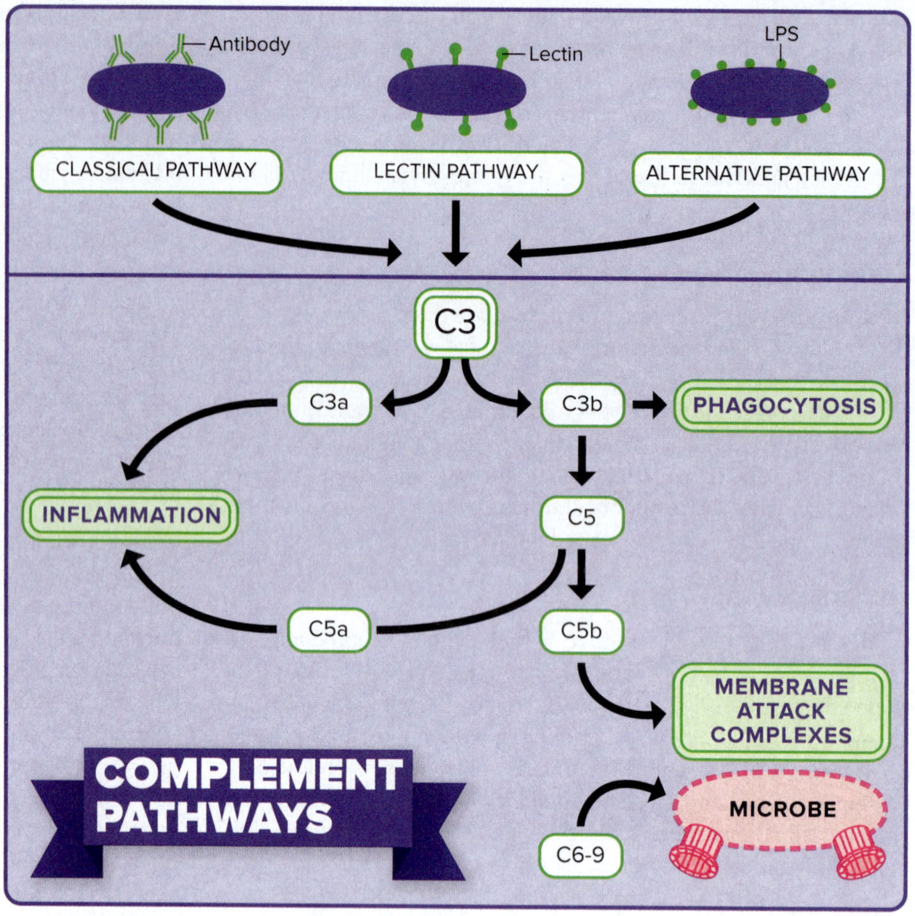

Antibody
Lectin
LPS
CLASSICAL PATHWAY
LECTIN PATHWAY
ALTERNATIVE PATHWAY
C3
C3a
C3b
PHAGOCYTOSIS
INFLAMMATION
C5
C5a
C5b
MEMBRANE ATTACK COMPLEXES
COMPLEMENT PATHWAYS
MICROBE
C6-9

(a) An overview of complement activity

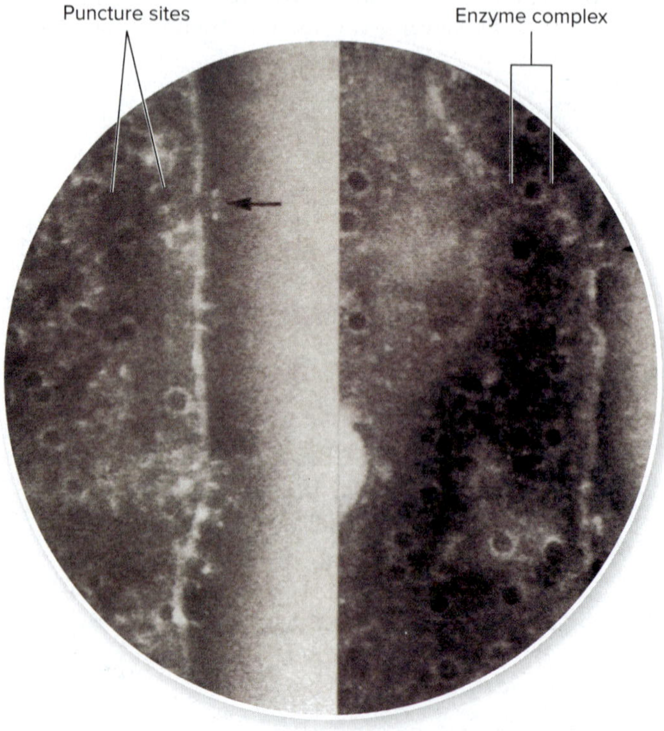

Puncture sites
Enzyme complex

(b)

An electron micrograph of a cell reveals multiple puncture sites over its surface. The lighter, ringlike structures are the actual enzyme complexes.

Figure 15.18 Complement. **(a)** Complement pathways. There are three different methods of activation. The central enzymatic reaction is when C3 is cleaved into C3a and C3b. The three possible outcomes of complement activation are the formation of membrane attack complexes, enhanced phagocytosis, and inflammation (the green boxes). The three different complement pathways are initiated in different ways, but once C3 is cleaved into C3b and C3a, and C3b becomes activated by the addition of Bb, the pathways converge and the outcome is similar. **(b)** An electron micrograph of a cell reveals multiple puncture sites on the surface. The lighter, ringlike structures are the actual enzyme complexes.

(b) Sucharit Bhakdi

surface of microbes. The alternative pathway is activated directly by repeating molecules on the microbial surface, such as LPS.

Overall Stages in the Complement Cascade In general, the complement cascade includes the four stages of *initiation, amplification and cascade, polymerization,* and *membrane attack.* At the outset, an initiator (such as microbes, or antibodies; see figure 15.18) reacts with the first complement chemical, which propels the reaction on its course. There is a recognition site on the surface of the target cell where the initial C components will bind. Through a stepwise series, each component reacts with another on or near the recognition site. Details of the pathways differ, but whether classical, lectin, or alternative, the functioning end product is a large, ring-shaped protein termed the *membrane attack complex.* This complex can digest holes in the cell membranes of bacteria, cells, and enveloped viruses, thereby destroying them.

If the target is a cell, the complement reaction causes it to disintegrate. If the target is an enveloped virus, the envelope is perforated and the virus is inactivated. The end result of complement action is multifaceted. Gram-negative bacteria and infected host cells may be lysed. Phagocytes will be attracted to the site in greater numbers. Overall inflammation will be amplified by the action of complement. In recent years, the excessive actions of complement have been implicated as aggravators of several autoimmune diseases, such as lupus, rheumatoid arthritis, and myasthenia gravis.

Antimicrobial Peptides

Antimicrobial peptides (AMPs) have only recently been appreciated. They are short proteins—of between 12 and 50 amino acids—that have the capability of inserting themselves into bacterial membranes **(figure 15.19).** Through this mechanism and others, they kill the microbes. They have names like bacteriocins, defensins, magainins, and protegrins. They are part of the innate immune system and have an effect on other actions of innate and adaptive immunity. Many researchers are looking at ways to turn these antimicrobial peptides into practical use as therapeutic drugs. Their ability to modulate immune responses would distinguish them from other antibiotics on the market and may represent a new weapon in the war against microbial drug resistance.

Host Restriction Factors

Some host cells and immune cells make molecules that can limit the ability of viruses to replicate once they are inside a host cell. These proteins and nucleic acids can bind to certain parts of the virus, prevent synthesis of new virus parts, prevent assembly of a new virus, or prevent virus release from host cells. These are

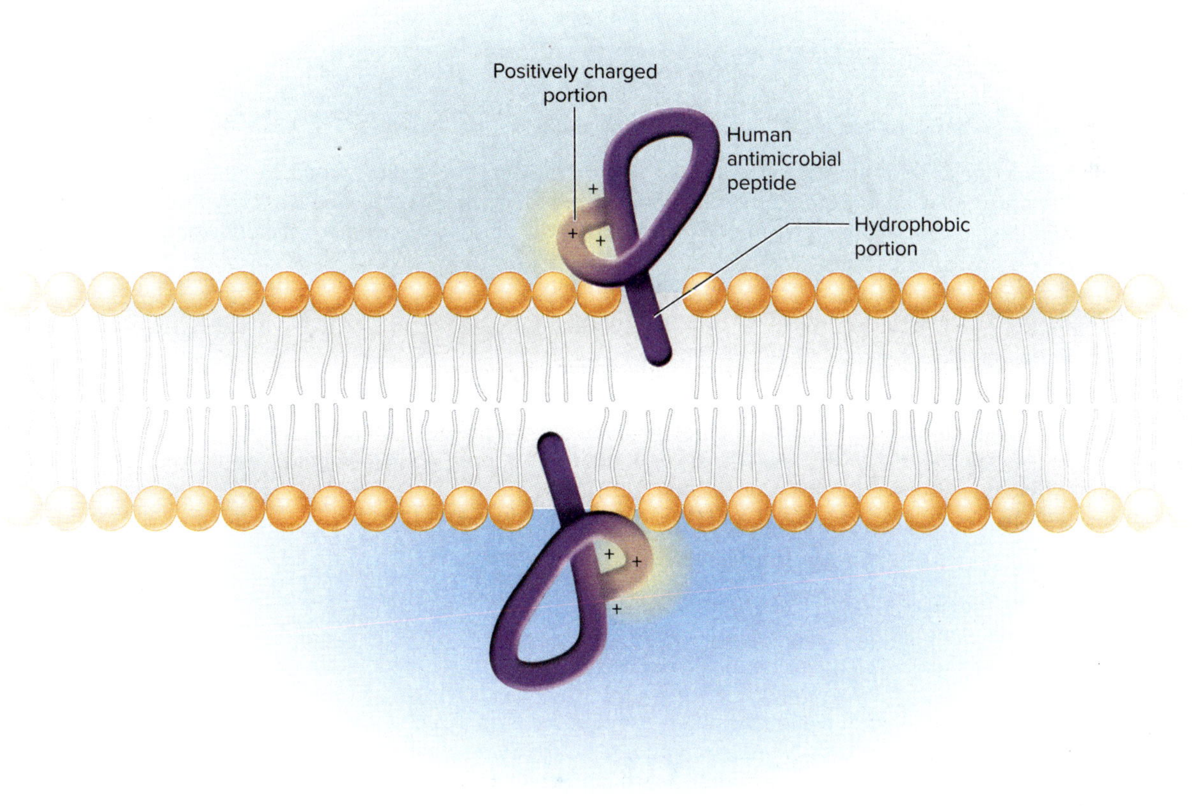
Positively charged portion
Human antimicrobial peptide
Hydrophobic portion

Figure 15.19 Antimicrobial peptides. The cytoplasmic membrane is shown (white and yellow). The purple antimicrobial peptides insert themselves in the membrane and disrupt its integrity.

collectively called **restriction factors.** One well-characterized set of restriction factors restricts the replication of retroviruses (such as HIV) by spontaneously changing cytosines to uracils and utilizing miRNAs that selectively target the viral DNA.

15.3 Learning Outcomes—Assess Your Progress

12. List the four major categories of innate immunity.
13. Outline the steps of phagocytosis.
14. Outline the steps of inflammation.
15. Discuss the mechanism of fever and how it helps defend the body.
16. Name four types of antimicrobial host-derived products.
17. Compose one good overview sentence about the purpose and the mode of action of the complement system and another about the purpose and mode of action of interferon.

Media Under The Microscope Wrap-Up

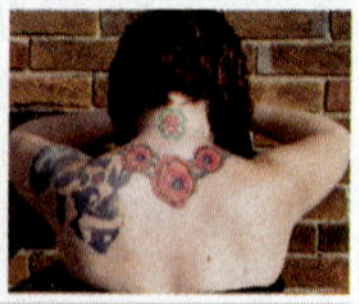
Steven Schleuning

The **intended message** of the article is to summarize a study that describes "how tattoos work" and the physiological reason they are permanent.

A **critical reading** of the article would certainly look favorably on the fact that the journalist sought out a scientist who was not involved with the study. This is always a good sign. The author of the news article is getting a second opinion, as it were. I do have a small complaint: The article states that "about 90 days later, new macrophages moved in and reabsorbed the ink." But it does not explain whether the researchers only looked after waiting 90 days or that, for some reason, it took 90 days for the new macrophages to arrive (which seems unlikely). And if that were the case, where did the ink go during that time?

Interpreting the article for your friends: You will have to describe the function of macrophages and then how their normal activities lead to the retention of ink under their skin indefinitely.

My **overall grade** for this article is a B+. I'm still wondering where the ink was during those 90 days.

Source: Dan Garisto, "Inked Mice Hint at How Tattoos Persist in People," *Science News*, March 16, 2018, www.sciencenews.org/article/inked-mice-hint-how-tattoos-persist-people.

Study Smarter: Better Together

These activities are designed for you to use on your own with a study group—either a face-to-face group or a virtual one, consisting of 3–5 members. Studying together can be very helpful, but there are effective and ineffective ways to do it. For example, getting together without a clear structure is often not a good use of your time. Use your time efficiently by using one or more of the exercises below.

FACE-TO-FACE GROUPS

Use one or more of the activities below.

Peer Instruction: Assign numbers to your group members to use all semester long. Now look at these five concepts from this chapter. Each group member prepares a 5-minute lesson on the topic corresponding to their number. Don't worry if you have fewer than 5 members; just use however many you have! During your group study time, each member presents their lesson, and the group spends another 5–10 minutes discussing that lesson.

1. Brief overview of the three lines of defense
2. The role of markers
3. How phagocytosis works
4. The steps in inflammation
5. Lymphatic system

Concept Maps: Each member of the group should use this list of terms from this chapter to generate their own concept map. This can be hand-drawn or created using software (see Appendix C for guidelines). During group study time, compare each other's concept maps and help each other make sure they are correct. Of course, there are many different "correct" maps. Examining each member's map will help you talk through the varied concepts and how they are related.

Concept Terms:

acquired immunity	adaptive immunity	innate immunity	macrophages
inflammation	fever	physical barriers	
neutrophils	leukocytes	lymphocytes	

Table Topics: Each group member should identify a concept or topic from this week's class assignments with which they are having trouble and share it during group study time. The other group members can then help to clarify confusing issues or share how they figured it out. Aim for a maximum of 15 minutes per topic. If the topic remains unclear to the group, bring it up during class or use the instructor's office hours or e-mail to ask for help. Taking the time to struggle with a difficult concept first makes your questions much more specific and more likely to yield helpful answers.

VIRTUAL GROUPS

Not everyone has the time or opportunity to meet with group members outside of class time. You or your instructor can create a virtual group using e-mail or the course software.

Weekly Discussion Board: This forum can be used as a way for groups to discuss topics, via e-mail, or other learning management systems or online platforms, before they are covered in class. As each member of the group answers the current week's question, they should send their responses to every other member of their group. It's best to agree on a deadline based on how your class schedule works (Saturday for the next week's topics, for example). Then, after the topic is discussed in class, each member should send a response that all group members will see with a follow-up post on the same topic. If you cover more than one chapter in a week, someone can be designated to choose which chapter Discussion Board question you will use. Or simply decide up front that you will always use the first-chapter-of-the week's question, to keep the schedule simple.

Discussion Question

Why do you think that the intestines have one of the body's most well-developed sets of lymphoid tissues?

Chapter Summary

HOST DEFENSES I: OVERVIEW AND INNATE IMMUNITY

15.1 DEFENSE MECHANISMS OF THE HOST: AN OVERVIEW

- The interconnecting network of host protection against microbial invasion is organized into three lines of defense: They can be summarized as the barriers, the innate defenses, and the adaptive immune response.
- The immune system is a network of vessels, organs, and cells positioned throughout the body. It is composed of primary and secondary lymphatic organs.
- Primary organs are the thymus and the red blood marrow. Secondary organs are the lymphatic vessels; lymph nodes; spleen; lymphoid tissues such as GALT, MALT, and SALT; and the tonsils.
- The blood contains both adaptive and innate defenses. Innate defenses include the granulocytes, macrophages and dendritic cells. The two major components of the adaptive immune response are the T lymphocytes, which provide adaptive cell-mediated immunity, and the B lymphocytes, which provide adaptive antibody-mediated immunity.
- The mononuclear phagocyte system consists of cells distributed throughout the body's network of connective tissue fibers.
- Cytokines are molecules produced by cells of the immune system that regulate, stimulate, suppress, and influence cell development, inflammation, and immunity.

15.2 THE FIRST LINE OF DEFENSE

- **The first line of defense is made up of physical and chemical barriers. The microbiome is part of this barrier function.**
- **Mucous membranes and the skin are physical barriers. Sweat, lysozyme, fatty acids, and pH levels are typical chemical barriers.**

15.3 THE SECOND LINE OF DEFENSE

- **Innate immune reactions are generalized responses to invasion, regardless of the type. These include phagocytosis, inflammation, fever, and an array of antimicrobial products.**
- **Phagocytosis is accomplished by macrophages along with neutrophils and a few other cell types.**
- **The four symptoms of inflammation are rubor (redness), calor (heat), tumor (swelling), and dolor (pain). Loss of function often accompanies these.**
- **Fever is another component of innate immunity. It increases the rapidity of the host immune responses and reduces the viability of many microbial invaders.**
- **There are four main types of antimicrobial products in the second line of defense: interferons, the complement cascade, antimicrobial peptides, and host restriction factors.**

SmartGrid: From Knowledge to Critical Thinking

This *21 Question Grid* takes the topics from this chapter and arranges them with respect to the American Society for Microbiology's Undergraduate Curriculum guidelines—all six of the important "Concepts" as well as the important "Competency" of scientific literacy. Three questions are supplied, which cover chapter content referring to the Concept or Competency in increasing levels of Bloom's taxonomy for learning.

ASM Concept/ Competency	A. Bloom's Level 1, 2—Remember and Understand (Choose one)	B. Bloom's Level 3, 4—Apply and Analyze	C. Bloom's Level 5, 6—Evaluate and Create
Evolution	**1.** A microorganism carries ______ markers and a B cell carries ______ markers. **a.** self, nonself **b.** nonself, self **c.** self, self **d.** nonself, nonself	**2.** Explain why the presence of lysozyme in human fluids such as tears and saliva is clear evidence of the coevolution of humans and microbes.	**3.** The pediatrician you work for has just recommended to the mother of a child with a cold and fever of 100°F *not* to treat the fever with children's Tylenol®. When the doctor leaves the room, the mother asks you to explain his recommendation. Use an evolutionary perspective to construct an answer for her.
Cell Structure and Function	**4.** Which of the following cells are lymphocytes? **a.** macrophages **b.** neutrophils **c.** red blood cells **d.** B cells	**5.** In what way is a phagocyte a tiny container of disinfectants?	**6.** Conduct research on the NK cells and NKT cells and describe some distinguishing characteristics.
Metabolic Pathways	**7.** Cytokines are secreted by which cells? **a.** macrophages **b.** B cells **c.** T cells **d.** all of these	**8.** Explain the sequence of physiological events leading to fever.	**9.** Explain the physiological events that lead to each of the four cardinal signs of inflammation.

ASM Concept/ Competency	A. Bloom's Level 1, 2—Remember and Understand (Choose one)	B. Bloom's Level 3, 4—Apply and Analyze	C. Bloom's Level 5, 6—Evaluate and Create
Information Flow and Genetics	**10.** The initial reaction to the presence of viruses in a human cell is the production of ___________ by that cell. **a.** complement **b.** interferon **c.** antiviral protein **d.** fever	**11.** If the synthesis of antiviral proteins begins after exposure of the host cell to interferons, what kind of genes are the antiviral protein genes likely to be? **a.** constitutive **b.** repressed **c.** induced **d.** all of these	**12.** If scientists were going to re-create host cell restriction factors in the lab so that they could be used as antiviral drugs, how would they go about that?
Microbial Systems	**13.** An example of an exogenous pyrogen is **a.** interleukin-1. **b.** endotoxin. **c.** complement. **d.** interferon.	**14.** Why do you think that the intestines have one of the body's most well-developed sets of lymphoid tissues?	**15.** A less diverse gut microbiome has been linked to a reduction in healthy immune responses. Why might that be?
Impact of Microorganisms	**16.** The normal microbiota is part of the _______ line of defense. **a.** first **b.** second **c.** third **d.** None of these is correct.	**17.** People who smoke have higher rates of respiratory infections. Explain why this might be, invoking aspects of the first line of defense.	**18.** Recent studies suggest that even healthy blood contains a microbiome, most of which exists inside the white blood cells. Why are microorganisms and/or their DNA found there rather than in the red blood cells, and what might their role be?
Scientific Thinking	**19.** Figure 15.3 demonstrates that lymphatic fluid flows **a.** toward the heart. **b.** away from the heart. **c.** both toward and away from the heart. **d.** None of these is correct.	**20.** Suggest some reasons for so much redundancy and interactivity of immune responses.	**21.** There have been (controversial) attempts in the past to treat HIV+ people by infecting them with the malarial parasite. One of the reasons it was controversial is that it was ethically questionable for physicians to knowingly infect patients with a potentially serious organism (the malaria protozoan). Why did scientists think that a malaria infection might wipe out HIV in a patient's body?

Answers to the multiple-choice questions appear in Appendix A.

Visual Connections

This question uses visual images to connect content within and between chapters.

1. **From chapter 4, figure 4.16.**
   a. In both cell types shown, sketch where the membrane attack complex (MAC) would form.
   b. Speculate on whether gram-positive or gram-negative bacterial cells are more resistant to the formation of a membrane attack complex.

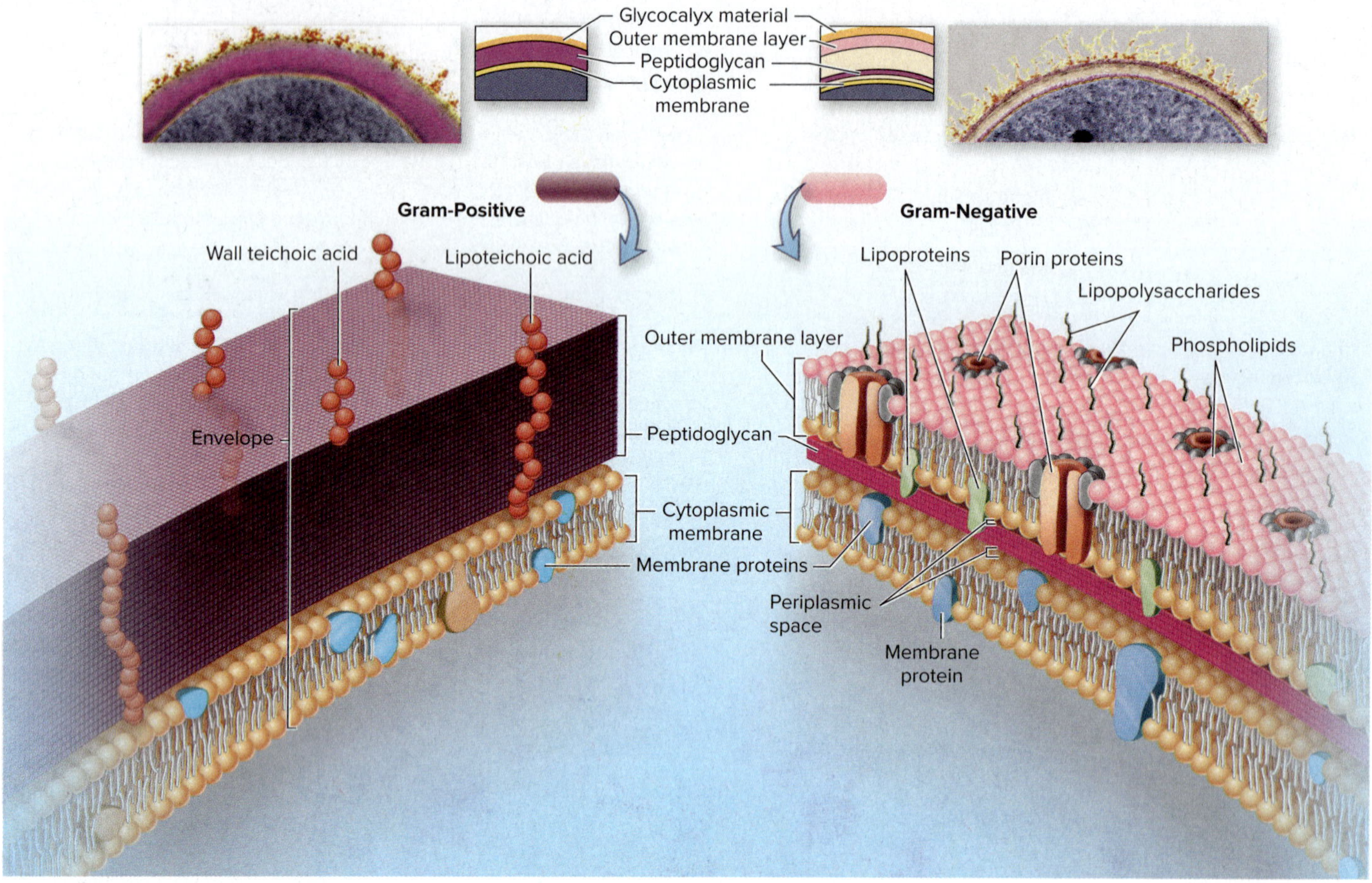
Glycocalyx material
Outer membrane layer
Peptidoglycan
Cytoplasmic membrane
Gram-Positive
Gram-Negative
Wall teichoic acid
Lipoteichoic acid
Envelope
Outer membrane layer
Peptidoglycan
Cytoplasmic membrane
Membrane proteins
Lipoproteins
Porin proteins
Lipopolysaccharides
Phospholipids
Periplasmic space
Membrane protein

Dr. Kari Lounatmaa/Science Source; Dennis Kunkel Microscopy, Inc./Science Source

High Impact Study

These terms and concepts are most critical for your understanding of this chapter—and may be the most difficult. Have you mastered them?

Concepts

- [ ] Three lines of defense
- [ ] Innate vs. adaptive
- [ ] Mononuclear phagocyte system
- [ ] Blood components
- [ ] Stages of inflammation
- [ ] Benefits of fever
- [ ] Complement activation

Terms

- [ ] Markers
- [ ] Lymphatic system
- [ ] Hematopoeisis
- [ ] Leukocytes
- [ ] Lymphocytes
- [ ] Monocytes
- [ ] Granulocytes
- [ ] Agranulocytes
- [ ] Phagocytosis
- [ ] PAMPs
- [ ] PRRs
- [ ] Cytokines
- [ ] Antimicrobial peptides
- [ ] Interferon

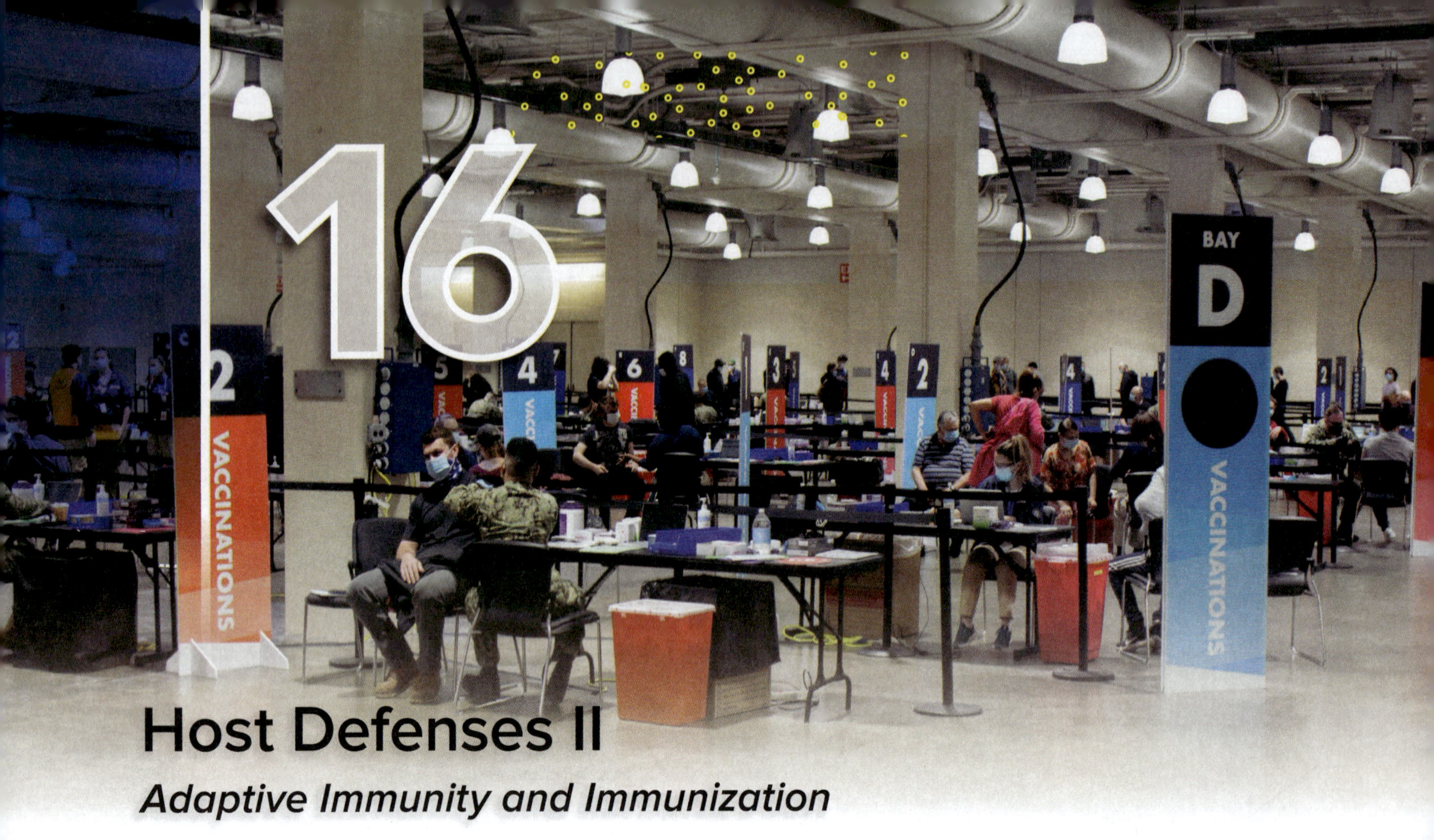
VACCINATIONS
BAY
D
VACCINATIONS

16 Host Defenses II

Adaptive Immunity and Immunization

Chuck Nacke/Alamy Stock Photo

MEDIA UNDER THE MICROSCOPE

Natural Immunity vs. Artificial Immunity for COVID-19

These case studies examine an article from the popular media to determine the extent to which it is factual and/or misleading. This case focuses on an October 2021 article from the Galveston County Daily News, *"Although Natural Immunity Exists, Health Experts Say Inoculation Is Safer."*

You will learn in this chapter that there are two ways to induce a long-lasting, specific immune response to a particular microorganism. One way, of course, is to experience and recover from the infection, which results in natural immunity. The other way, termed *artificial immunity,* is through vaccination. The article addresses a pertinent question: If you have already had COVID-19, do you need the vaccine?

The article presents a wealth of data about COVID diagnoses, vaccinations, hospitalizations, and deaths in Galveston County, Texas. They found that breakthrough cases (occurring in fully vaccinated individuals) occurred at a lower rate than reinfections (occurring in unvaccinated persons who have previously recovered from COVID). They do mention that people who are reinfected rather than experiencing breakthrough infections died at a lower rate.

The article also discusses the viewpoints of experts. The information they share includes the fact that the strength and longevity of natural immunity is difficult to predict. At times it is indeed stronger than artificial immunity, but the characteristics of some microbes allow them to only incompletely stimulate the hosts' immune response. That can mean that their immunity may not be highly effective or last for very long. At this point, there was not enough known about immunity from COVID-19 after initial infection. The bottom line is that the article recommends vaccination even for people who have recovered from the infection.

- What is the **intended message** of the article?
- What is your **critical reading** of the summary of the article presented above? Remember that in this context, "critical reading" does not necessarily mean *What criticism do you have?* but asks you to apply your knowledge to interpret whether the article is factual and whether the facts support the intended message.

- How would you **interpret** the news item for your nonmicrobiologist friends?
- What is your **overall grade** for the news item—taking into account its accuracy and the accuracy of its intended effect?

Media Under The Microscope Wrap-Up appears at the end of the chapter.

Outline and Learning Outcomes

16.1 Adaptive Immunity: The Third Line of Defense

1. Describe how the third line of defense is different from the other host defense mechanisms.
2. List the four stages of an adaptive immune response.
3. Discuss four major functions of immune system markers.
4. Define the role of the major histocompatibility complex (MHC), and list the three classes of MHC genes.
5. Compare and contrast the process of antigen recognition in T cells and B cells.

16.2 Step I: The Development of Lymphocyte Diversity

6. Summarize the maturation process of both B cells and T cells.
7. Draw a diagram illustrating how lymphocytes are capable of responding to nearly any antigen imaginable.
8. Outline the processes of clonal deletion and clonal selection.
9. Describe the structure of both a B-cell receptor and a T-cell receptor.

16.3 Step II: Presentation of Antigens

10. Compare the terms *antigen, immunogen,* and *epitope.*
11. List characteristics of antigens that optimize their immunogenicity.
12. Describe how the immune system responds to alloantigens and superantigens.
13. List the types of cells that can act as antigen-presenting cells.

16.4 Step III: Antigenic Challenge of T Cells and B Cells

14. Summarize the process of T-cell activation, and list major types of T cells produced in this process.
15. Diagram the steps of B-cell activation, and list the types of B cells produced in this process.

16.5 Step IV (1): The T-Cell Response

16. Describe the main functions of the major T-cell types and their subsets.
17. Explain the role of cytotoxic T cells in apoptosis, and list the potential targets of this process.

16.6 Step IV (2): The B-Cell Response

18. Diagram an antibody binding antigen, and list the possible end results of this process.
19. List the five types of antibodies and important characteristics of each.
20. Draw and label a graph illustrating the development of a secondary immune response.

16.7 Adaptive Immunity and Vaccination

21. List the four categories of acquired immunity, and provide examples of each.
22. Discuss the qualities of an effective vaccine.
23. List several types of vaccines, and discuss how they are utilized today.

16.1 Adaptive Immunity: The Third Line of Defense

In chapter 15, we described the capacity of the immune system to survey, recognize, and react to foreign cells and molecules. We also looked at the characteristics of innate host defenses, blood cells, phagocytosis, inflammation, and complement. In addition, we introduced the concepts of acquired immunity and specificity. In this chapter, we take a closer look at those topics.

When host barriers and innate defenses fail to control an infection, a person with a normally functioning immune system has a mechanism to resist the pathogen—the third, adaptive line of immunity. Immunity is the resistance developed after contracting diseases such as chickenpox or measles, and it provides long-term protection against future attacks. This sort of immunity is acquired only after an immunizing event such as an infection. The absolute need for acquired or adaptive immunity is impressively documented in children who have genetic defects in this system or in patients with AIDS who have lost it. Even with heroic measures to isolate the patient, combat infection, or restore lymphoid tissue, the victim is constantly vulnerable to life-threatening infections.

Acquired adaptive immunity is the product of a dual system that we have previously mentioned—the B and T lymphocytes. During fetal development, these lymphocytes undergo a selective process that specializes them for reacting mainly to one particular marker (on a microbe, for instance). During this time, **immunocompetence,** the ability of the body to react with countless foreign substances, develops. An infant is born with the theoretical potential to react to an immense array of different substances.

As we saw in chapter 15, antigens or markers are molecules that can be seen and identified by the immune system. They may or may not provoke an immune response after being "sensed" by the immune system. If they do provoke a response, they can be called **immunogens.** These molecules stimulate a response by T and B cells. They are usually protein or polysaccharide molecules on or inside all cells and viruses, including our own. (Environmental chemicals can also be immunogens. In fact, any exposed or released protein or polysaccharide is potentially an antigen or immunogen, even those on our own cells. For reasons we discuss later, our own antigens do not normally evoke a response from our own immune systems.)

We have already discussed pathogen-associated molecular patterns (PAMPs) that stimulate responses by phagocytic cells during an innate defense response. While PAMPs are molecules shared by many types of microbes that stimulate an innate response, immunogens are highly individual and stimulate adaptive immunity. The two types of molecules do share two characteristics: (1) They are "parts" of foreign cells (microbes or other foreign materials) and (2) they provoke a defensive reaction from the host. *Important note: As we move on in this chapter, it will simplify things if we use the term "antigen" for a specific part of a specific microbe that the adaptive immune response recognizes.*

The infographic on this page illustrates six characteristics that make adaptive immunity unique **(figure 16.1).** Two of the most important are **specificity** and **memory.** Unlike mechanisms such as anatomical barriers or phagocytosis, adaptive immunity is highly specific. For example, the antibodies produced during an infection against the chickenpox virus will function against that virus and not against the measles virus. The property of memory is the rapid mobilization of lymphocytes that have been programmed to "recall" their first engagement with the invader and rush to the attack once again.

The elegance and complexity of immune function are largely due to lymphocytes working closely together with phagocytes. To simplify the network of immunologic development and interaction, we present it here as a series of stages, with each stage covered in a separate section **(figure 16.2).** The principal stages are

I. Lymphocyte development and differentiation;
II. The presentation of markers (antigens);
III. Antigen challenge of B and T lymphocytes; and
IV. B-lymphocyte response (the production and activities of antibodies) and T-lymphocyte response (cell-mediated immunity).

This sequence is illustrated here and in figure 16.2. We will give an overview here and spend the rest of the chapter filling in the details.

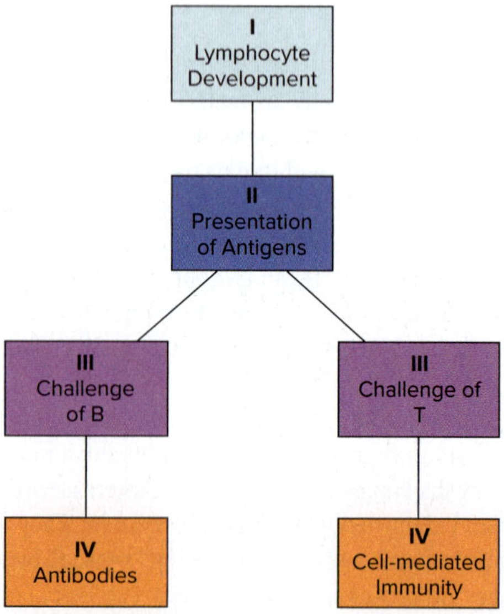
I
Lymphocyte Development
II
Presentation of Antigens
III
Challenge of B
III
Challenge of T
IV
Antibodies
IV
Cell-mediated Immunity

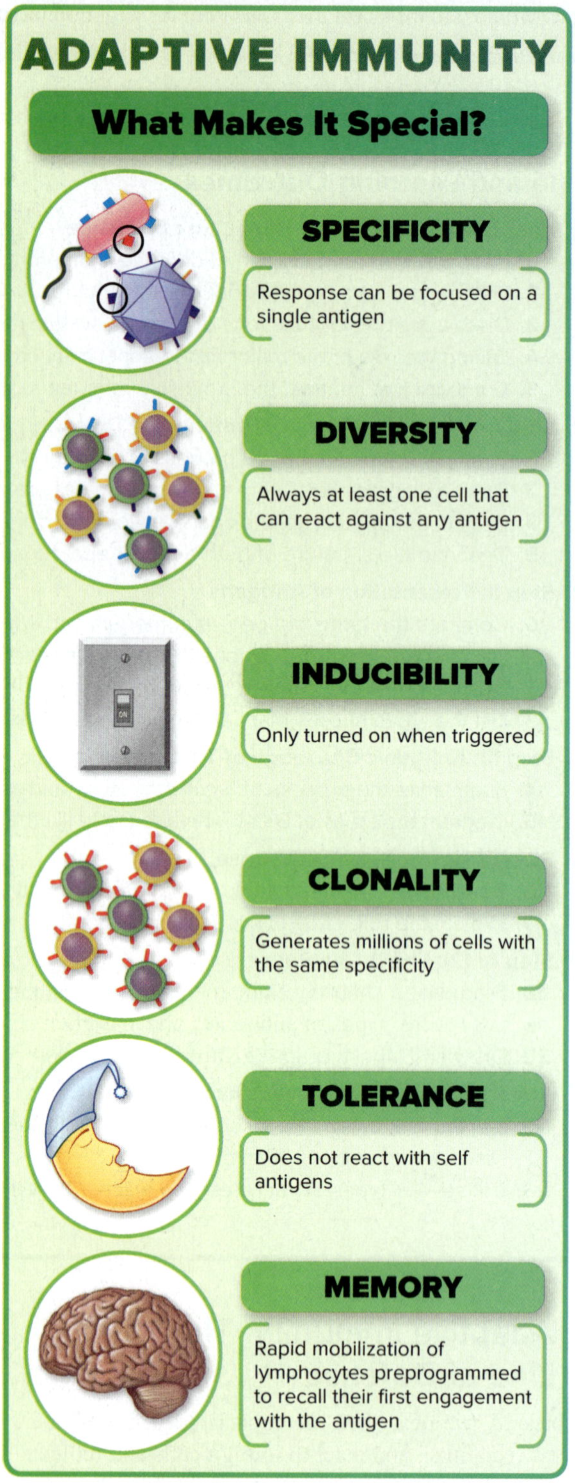
ADAPTIVE IMMUNITY
What Makes It Special?
SPECIFICITY
Response can be focused on a single antigen
DIVERSITY
Always at least one cell that can react against any antigen
INDUCIBILITY
Only turned on when triggered
CLONALITY
Generates millions of cells with the same specificity
TOLERANCE
Does not react with self antigens
MEMORY
Rapid mobilization of lymphocytes preprogrammed to recall their first engagement with the antigen

Figure 16.1 Special characteristics of adaptive immunity.

A Brief Overview of the Major Immune System Components

Lymphocyte Development

Lymphocytes are central to immune responsiveness. They undergo development that begins in the embryonic yolk sac and shifts to the liver and bone marrow. Although all lymphocytes arise from

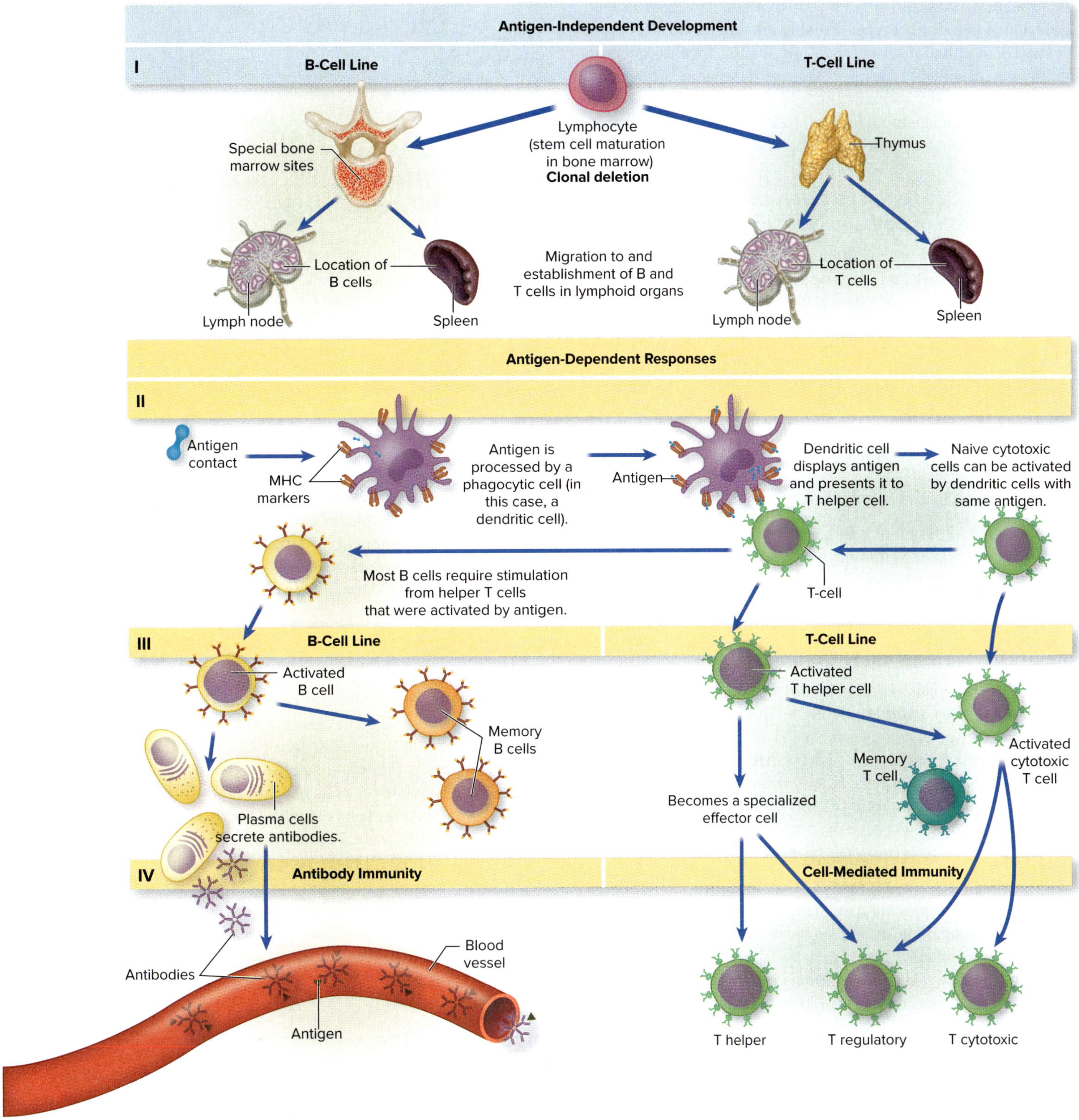
Antigen-Independent Development
I
B-Cell Line
T-Cell Line
Special bone marrow sites
Lymphocyte (stem cell maturation in bone marrow)
Clonal deletion
Thymus
Location of B cells
Lymph node
Spleen
Migration to and establishment of B and T cells in lymphoid organs
Location of T cells
Lymph node
Spleen
Antigen-Dependent Responses
II
Antigen contact
MHC markers
Antigen is processed by a phagocytic cell (in this case, a dendritic cell).
Antigen
Dendritic cell displays antigen and presents it to T helper cell.
Naive cytotoxic cells can be activated by dendritic cells with same antigen.
Most B cells require stimulation from helper T cells that were activated by antigen.
T-cell
III
B-Cell Line
T-Cell Line
Activated B cell
Memory B cells
Plasma cells secrete antibodies.
Activated T helper cell
Memory T cell
Activated cytotoxic T cell
Becomes a specialized effector cell
IV
Antibody Immunity
Cell-Mediated Immunity
Antibodies
Blood vessel
Antigen
T helper
T regulatory
T cytotoxic

Figure 16.2 Overview of the stages of lymphocyte development and function. **(I)** Development of B- and T-lymphocyte specificity and migration to lymphoid organs. **(II)** Dendritic cell displays antigen and presents it to naive T helper cell. **(III)** Lymphocyte activation, clonal expansion, and formation of memory B and T cells. **(IV)** End result of lymphocyte activation. *Left-hand side:* antibody release; *right-hand side:* cell-mediated immunity. Details of these processes are covered in each corresponding section heading.

the same basic stem cell type, at some point in development they diverge into two distinct types. Final maturation of B cells occurs in specialized bone marrow sites, and that of T cells occurs in the thymus. Both cell types then migrate to separate areas in the lymphoid organs (for example, nodes and spleen). B and T cells constantly recirculate through the circulatory system and lymphatics, migrating into and out of the lymphoid organs.

Markers on Cell Surfaces Involved in Recognition of Self and Nonself

Chapter 15 touched on the fundamental idea that cell markers determine specificity and identity. A given cell can express several different markers, each type playing a distinct and significant role in detection, recognition, and cell communication. Major functions of immune system markers are

1. attachment to nonself or foreign antigens;
2. binding to cell surface markers that indicate self, such as MHC molecules (discussed next);
3. receiving and transmitting chemical messages to coordinate the response; and
4. aiding in cellular development.

When a T or B cell recognizes a "self" marker, that clone of cells is generally deleted through a process called **clonal deletion.** Because of their importance in the immune response, we concentrate here on the major markers of lymphocytes and macrophages.

Major Histocompatibility Complex

One set of genes that codes for human cell markers is the **major histocompatibility complex (MHC).** This gene complex gives rise to a series of glycoproteins (called MHC molecules) found on all cells except red blood cells. The MHC may also be called the human leukocyte antigen (HLA) system. This marker complex plays a vital role in recognition of self by the immune system and in rejection of foreign tissue.

Three classes of MHC genes have been identified:

1. Class I genes code for markers that appear on all nucleated cells. They display unique characteristics of self and allow for the recognition of self molecules and the regulation of immune reactions. The system is rather complicated in its details, but in general, each human being inherits a particular combination of class I MHC (HLA) genes in a relatively predictable fashion. Although millions of different combinations and variations of these genes are possible among humans, the closer the blood relationship, the greater the probability for similarity in MHC profile.
2. Class II MHC genes also code for immune regulatory markers. These markers are found on macrophages, dendritic cells, and B cells and are involved in presenting antigens to T cells during cooperative immune reactions.
3. Class III MHC genes encode proteins involved with the complement system, among others. We will focus on classes I and II in this chapter. See **figure 16.3** for depictions of these two MHC classes.

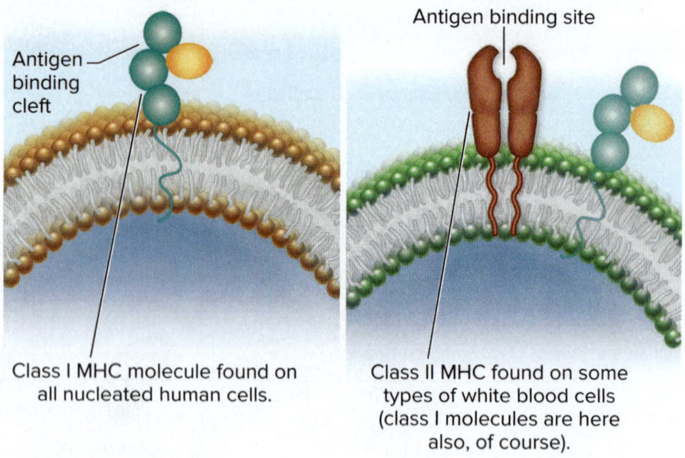
Antigen binding cleft
Antigen binding site
Class I MHC molecule found on all nucleated human cells.
Class II MHC found on some types of white blood cells (class I molecules are here also, of course).

Figure 16.3 Classes I and II molecules of the human major histocompatibility complex (MHC).

CD Molecules

Another set of markers that are important in immunity are the CD molecules. (CD stands for "cluster of differentiation.") CDs are molecules on the membranes of a variety of cells involved in the immune response. Close to 400 have been described. A few major ones will be discussed in the following sections.

Lymphocyte Receptors and Specificity to Antigen

The part lymphocytes play in immune surveillance and recognition emphasizes the essential role of their markers. These markers are even frequently called receptors, a name that emphasizes that their major role is to "accept" or "grasp" antigens in some form. B cells have receptors that pair up on the membrane with a CD molecule and bind antigens. T cells have receptors that bind antigens that have been processed and complexed with MHC molecules on the presenting cell surface. **Figure 16.4** illustrates the surfaces of B and T cells and their antigen receptors. Antigen molecules are very diverse; there are potentially billions of unique types. The many sources of antigens include microorganisms, cancer cells, and an endless array of chemical compounds in the environment. We will soon see how T and B cells recognize so many different antigens.

Entrance and Presentation of Antigens

When foreign cells, such as pathogens (carrying antigens), cross the first line of defense and enter the tissue, resident phagocytes migrate to the site. Tissue macrophages ingest the pathogens and induce an inflammatory response in the tissue if appropriate. Tissue dendritic cells ingest the antigen and migrate to the nearest lymphoid organ (often the draining lymph nodes). Here they process and present antigen to T lymphocytes. Pieces of the pathogens also drain into these lymph nodes. Along with dendritic cells, macrophages and B cells serve as antigen-presenting cells (APCs).

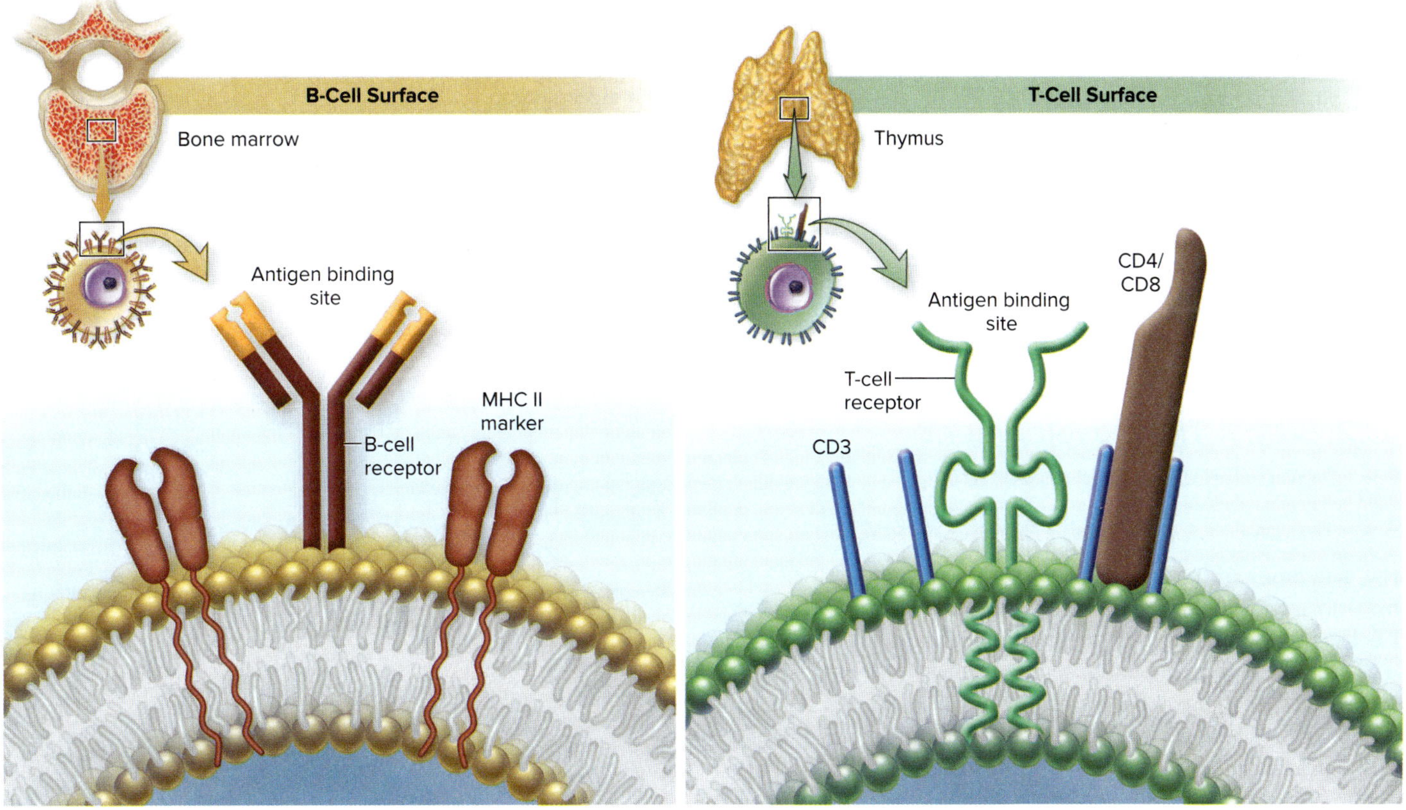
B-Cell Surface
Bone marrow
Antigen binding site
B-cell receptor
MHC II marker
T-Cell Surface
Thymus
Antigen binding site
CD4/ CD8
T-cell receptor
CD3

Figure 16.4 The surfaces of T cells and B cells.

Disease Connection

The palatine tonsils, located on either side of the back of the throat, help to guard against gastrointestinal and respiratory infections. This is possible because the tonsils contain B cells and different types of T cells, as well as antibodies that help to identify and attack harmful pathogens. This is the reason why doctors no longer routinely remove children's tonsils unless they chronically become infected.

Antigen Challenge and Clonal Selection

When challenged by antigen, both B cells and T cells further proliferate and differentiate. The multiplication of a particular lymphocyte creates a **clone,** or group of genetically identical cells, some of which are memory cells that will ensure future reactiveness against that antigen. Because the B-cell and T-cell responses differ significantly from this point on in the sequence, they are summarized separately.

How T Cells Respond to Antigen

T-cell types and responses are extremely varied. When activated (sensitized) by antigen, a T cell gives rise to a variety of different cell types with different roles in the immune response. They generally fall into three categories:

1. helper T cells that activate macrophages, assist B-cell processes, and help activate cytotoxic T cells;
2. regulatory T cells that control the T-cell response by secreting anti-inflammatory cytokines or preventing proliferation; and
3. cytotoxic T cells that lead to the destruction of infected host cells and other "foreign" cells.

One special class of T cells, called gamma-delta T cells, can be activated quickly by PAMPs, as seen in the innate response, or by specific antigens.

Although T cells secrete cytokines that help destroy pathogens and regulate immune responses, they do not produce antibodies.

How B Cells Respond to Antigen: Antibody Release

When activated by antigen, a B cell divides, giving rise to plasma cells, each with the same reactive profile. Plasma cells release antibodies into the tissue and blood. When these antibodies attach to the antigen for which they are specific, the antigen is marked for destruction or neutralization.

Just as there is one type of T cell that acts in a nonspecific way, there is one class of B cells, called the innate response activating B cells (IRA-B), that seems to have the job of alerting many components of the immune system to get active because a threat is on the way.

16.1 Learning Outcomes—Assess Your Progress

1. Describe how the third line of defense is different from the other host defense mechanisms.
2. List the four stages of an adaptive immune response.
3. Discuss four major functions of immune system markers.
4. Define the role of the major histocompatibility complex (MHC), and list the three classes of MHC genes.
5. Compare and contrast the process of antigen recognition in T cells and B cells.

16.2 Step I: The Development of Lymphocyte Diversity

Specific Events in T-Cell Development

The maturation of most T cells and the development of their specific receptors are directed by the thymus and its hormones. Other T cells reach full maturity in the gastrointestinal tract. In addition to the antigen-specific T-cell receptor, all mature T lymphocytes express CD3 markers. CD3 molecules surround the T-cell receptor and assist in binding. T cells also express either a CD4 or a CD8 coreceptor (see figure 16.4). So, T cells = CD3 + (CD4 or CD8). CD4 is an accessory receptor protein mostly found on T helper cells that helps the T cell receptor bind to MHC class II molecules. CD8 is mostly found on cytotoxic T cells, and it helps bind MHC class I molecules. T cells constantly circulate between the lymphatic and general circulatory systems, migrating to specific T-cell areas of the lymph nodes and spleen. It has been estimated that more than 10^9 T cells pass between the lymphatic and general circulations per day.

Specific Events in B-Cell Development

The site of B-cell maturation was first discovered in birds, which have an organ in the intestine called the bursa. In humans, B cells mature in the bone marrow. As a result of gene modification and selection, hundreds of millions of distinct B cells develop. These naive lymphocytes circulate through the blood, "homing" to specific sites in the lymph nodes, spleen, and other lymphoid tissue, where they take up residence. Here they will come in contact with antigens throughout life. B cells have immunoglobulins as surface receptors. **Table 16.1** highlights the differences between T cells and B cells.

Building Immunologic Diversity

Each naive lymphocyte bears an antigen receptor that recognizes a unique antigen. How is this possible? The mechanism, generally true for both B and T cells, can be summarized as follows: In the bone marrow, stem cells can become granulocytes, monocytes, or lymphocytes. The lymphocytes then become either T cells or B cells. Cells destined to become B cells stay in the bone marrow and T cells migrate to the thymus. Here they build their unique antigen receptors. Both B and T cells then migrate to secondary lymphoid tissues **(figure 16.5).**

Table 16.1 Contrasting Properties of B Cells and T Cells

	B Cells	T Cells
Site of Maturation	Bone marrow	Thymus
Specific Surface Markers	Immunoglobulin as receptor; Distinct CD molecules	T-cell receptor; Distinct CD molecules
Concentration in Blood	Low numbers	High numbers
Receptors for Antigen	B-cell receptor (immunoglobulin)	T-cell receptor
Location in Lymphoid Organs	Cortex (in follicles)	Paracortical sites (interior to the follicles)
Require Antigen Presented with MHC	No	Yes (gamma-delta T cells can be activated differently)
Result of Antigenic Stimulation	Plasma cells and memory cells	Several types of activated T cells and memory cells
General Functions	Production of antibodies to inactivate, neutralize, target antigens	Cells function in helping other immune cells, suppressing, killing abnormal cells; hypersensitivity; synthesize cytokines

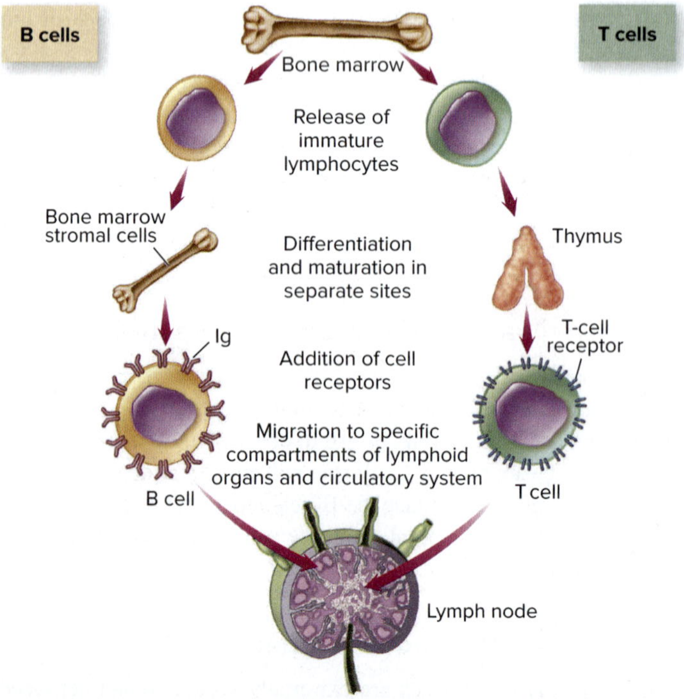
B cells
T cells
Bone marrow
Release of immature lymphocytes
Bone marrow stromal cells
Differentiation and maturation in separate sites
Thymus
Ig
T-cell receptor
Addition of cell receptors
Migration to specific compartments of lymphoid organs and circulatory system
B cell
T cell
Lymph node

Figure 16.5 Major stages in the development of B and T cells.

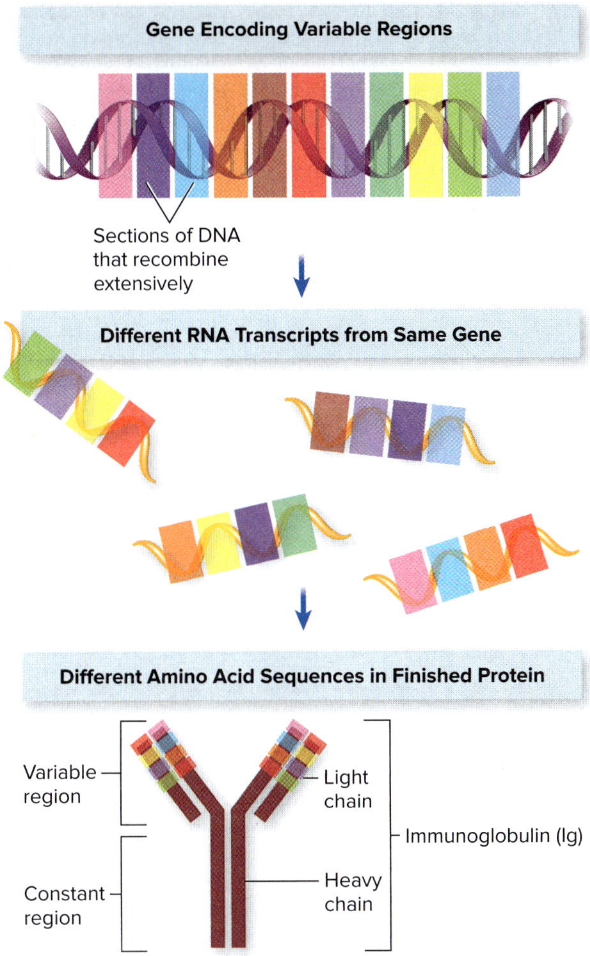
Gene Encoding Variable Regions
Sections of DNA that recombine extensively
Different RNA Transcripts from Same Gene
Different Amino Acid Sequences in Finished Protein
Variable region
Light chain
Immunoglobulin (Ig)
Constant region
Heavy chain

Figure 16.6 The mechanism behind antibody variability. The genes coding for the variable regions of antibody molecules have multiple different sections along their lengths. As a result of DNA recombination, very different RNA transcripts are created from the same original gene. When those transcripts are translated, the resulting protein will have extremely variable amino acid sequences—and, therefore, extremely variable shapes.

By the time T and B cells reach the lymphoid tissues, each one is already equipped to respond to a single unique antigen. This amazing diversity is generated by extensive DNA rearrangements of more than 500 gene segments that code for the antigen receptors on the T and B cells **(figure 16.6).** In time, every possible recombination occurs, leading to a huge assortment of lymphocytes. Estimates of the theoretical number of possible variations that may be created vary from 10^{14} to 10^{18} different specificities. Each genetically unique line of lymphocytes arising from these recombinations is termed a clone. Keep in mind that the rearranged genetic code is expressed as a protein receptor of unique configuration on the surface of the lymphocyte, something like a "sign post" announcing its specificity and reactivity for an antigen. This *proliferative* stage of lymphocyte development occurs prior to lymphocytes' contact with foreign antigens.

The Specific B-Cell Receptor: An Immunoglobulin Molecule

In the case of B lymphocytes, the genes that undergo the recombination described are those coding for **immunoglobulin** (im″-yoo-noh-glahb′-yoo-lin) **(Ig)** synthesis. Immunoglobulins are large glycoprotein molecules that serve as the antigen receptors of B cells and, if they are secreted, as antibodies. The basic immunoglobulin molecule is a composite of four polypeptide chains: a pair of identical heavy (H) chains and a pair of identical light (L) chains (see figure 16.6). One light chain is bonded to one heavy chain, and the two heavy chains are bonded to one another with disulfide bonds, creating a symmetrical, Y-shaped arrangement.

The ends of the forks formed by the light and heavy chains contain pockets called the **antigen binding sites.** These sites are highly variable in shape to fit a wide range of antigens. This extreme versatility is due to **variable regions (V)** in antigen binding sites, where amino acid composition is highly varied from one clone of B lymphocytes to another, a result of the genetic reassortment we discussed earlier. The remainder of the light chains and heavy chains consists of constant (C) regions whose amino acid content does not vary greatly from one antibody to another.

T-Cell Receptors

The T-cell receptor for antigen belongs to the same protein family as the B-cell receptor. It is similar to the B-cell receptor in being formed by DNA rearrangement events, having variable and constant regions, being inserted into the membrane, and having an antigen binding site formed from two parallel polypeptide chains (see figure 16.4). Unlike the B-cell receptor, the T-cell receptor is relatively small and is never secreted.

Clonal Deletion and Selection

The second stage of development—**clonal selection** and expansion—happens *after* exposure to the correct antigen such as a microbe. When this antigen enters the immune surveillance system, it encounters specific lymphocytes, ready to recognize it. Such contact stimulates that clone to undergo mitotic divisions and expands it into a larger population of lymphocytes, all bearing the same specificity **(figure 16.7).** This increases the capacity of the immune response to respond to that antigen. Two important facts about the phenomenon of clonal selection are that (1) lymphocyte specificity is preprogrammed, existing in the genetic makeup before an antigen has ever entered the tissues, and (2) each genetically distinct lymphocyte expresses only a single specificity and can react to only one type of antigen. Other important features of the lymphocyte response system are expanded in later sections.

Can you see that one potentially problematic outcome of random genetic assortment is the development of clones of lymphocytes able to react to *self?* This outcome could lead to severe damage when the immune system actually perceives self molecules as foreign and mounts a harmful response against the host's tissues. Under normal circumstances, any such clones are destroyed during fetal development through clonal deletion (step 1 in

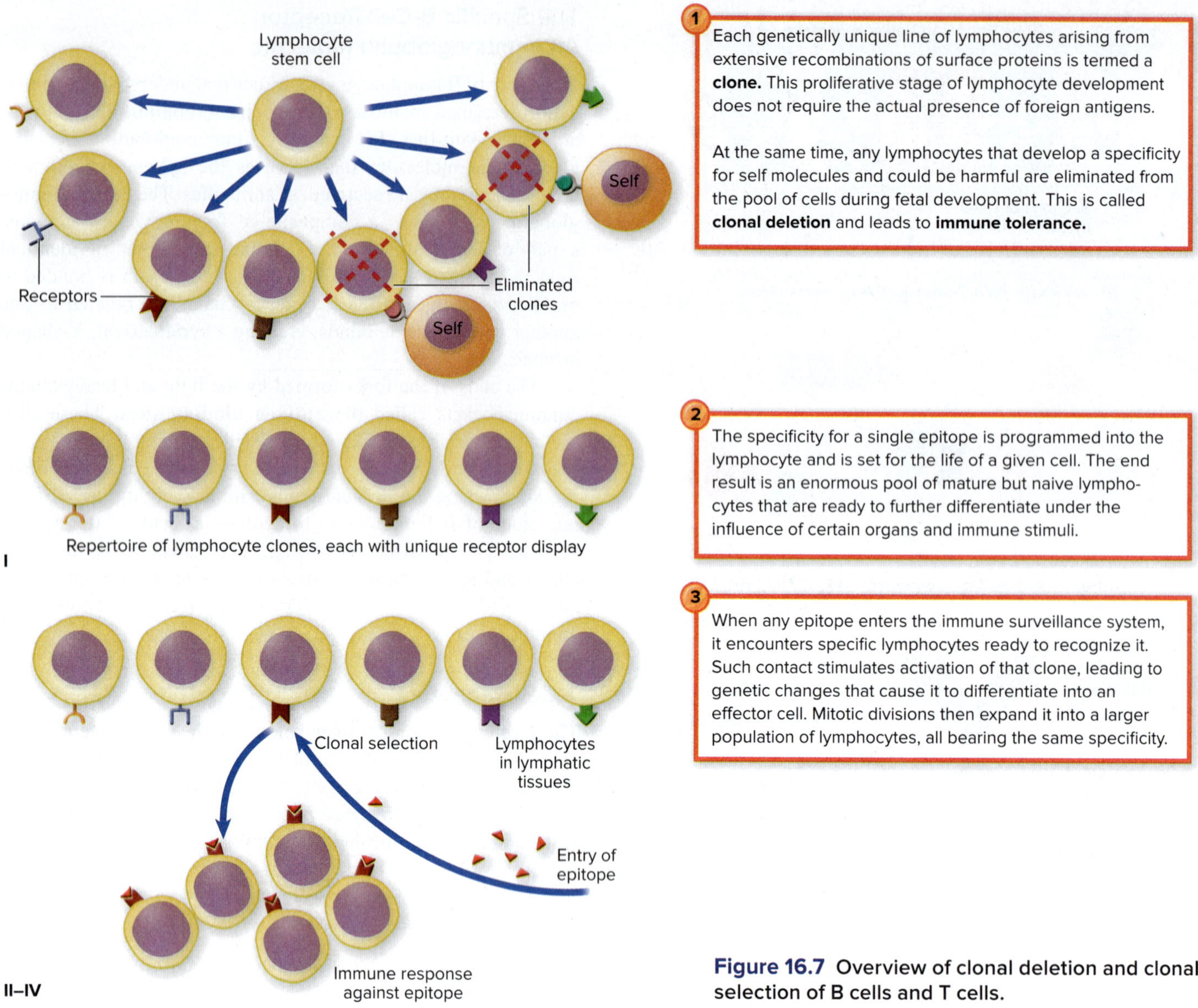
Lymphocyte stem cell
Self
Receptors
Eliminated clones
Self
Repertoire of lymphocyte clones, each with unique receptor display
I
Clonal selection
Lymphocytes in lymphatic tissues
Entry of epitope
Immune response against epitope
II–IV
1 Each genetically unique line of lymphocytes arising from extensive recombinations of surface proteins is termed a **clone.** This proliferative stage of lymphocyte development does not require the actual presence of foreign antigens.
At the same time, any lymphocytes that develop a specificity for self molecules and could be harmful are eliminated from the pool of cells during fetal development. This is called **clonal deletion** and leads to **immune tolerance.**
2 The specificity for a single epitope is programmed into the lymphocyte and is set for the life of a given cell. The end result is an enormous pool of mature but naive lymphocytes that are ready to further differentiate under the influence of certain organs and immune stimuli.
3 When any epitope enters the immune surveillance system, it encounters specific lymphocytes ready to recognize it. Such contact stimulates activation of that clone, leading to genetic changes that cause it to differentiate into an effector cell. Mitotic divisions then expand it into a larger population of lymphocytes, all bearing the same specificity.

Figure 16.7 Overview of clonal deletion and clonal selection of B cells and T cells.

figure 16.7). The removal of such potentially harmful clones is the basis of **immune tolerance,** or tolerance to self. Because humans are exposed to many new antigenic substances during their lifetimes, such as animal and plant cells that we consume as food, T cells and B cells in the periphery of the body have mechanisms for *not* reacting to innocuous antigens. Some diseases (autoimmune diseases) are thought to be caused by the loss of immune tolerance through the survival of certain "forbidden clones" or failure of these other systems.

16.2 Learning Outcomes—Assess Your Progress

6. Summarize the maturation process of both B cells and T cells.
7. Draw a diagram illustrating how lymphocytes are capable of responding to nearly any antigen imaginable.
8. Outline the processes of clonal deletion and clonal selection.
9. Describe the structure of both a B-cell receptor and a T-cell receptor.

16.3 Step II: Presentation of Antigens

Having reviewed the characteristics of lymphocytes, we can more deeply examine the properties of antigens, the substances that cause them to react. More properly, we will be examining immunogens because the definition of *immunogen* is an antigen that has been responded to by the immune system. For practical purposes, we often use the terms interchangeably. To be perceived as an antigen or immunogen, a substance must meet certain requirements in foreignness, shape, size, and accessibility.

Characteristics of Antigens

One important characteristic of an antigen is that it is perceived as foreign, meaning that it is not a normal constituent of the body. Whole microbes or their parts, cells, or substances that are found in other humans, animals, plants, and various molecules all possess this quality of foreignness. For this reason, they are potentially antigenic to the immune system of an individual **(figure 16.8).** Molecules of complex composition such as proteins and protein-containing compounds prove to be more immunogenic than repetitious polymers composed of a single type of unit. Most materials that serve as antigens fall into these chemical categories:

- Proteins and polypeptides (enzymes, cell surface structures, hormones, exotoxins),
- Lipoproteins (from cell membranes),
- Glycoproteins (blood cell markers),
- Nucleoproteins (DNA complexed to proteins but not pure DNA), and
- Polysaccharides (certain bacterial capsules) and lipopolysaccharides.

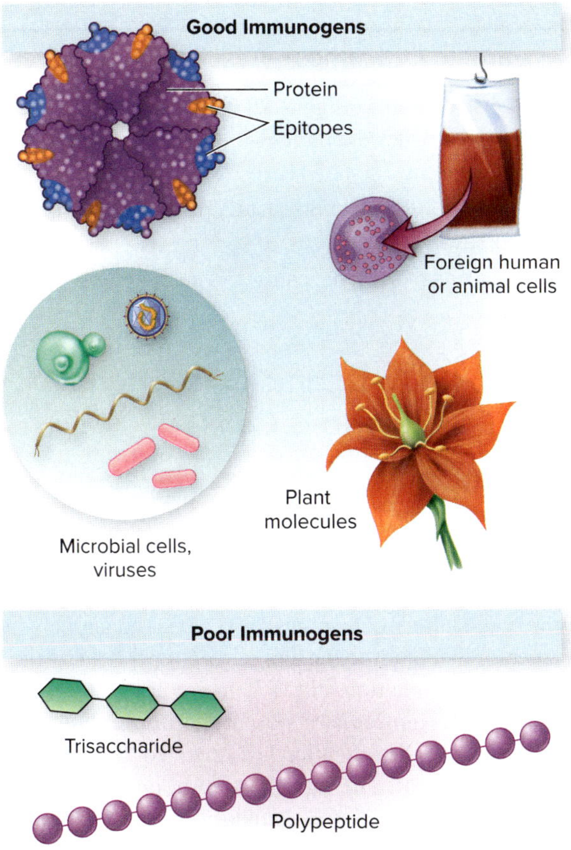
Good Immunogens
Protein
Epitopes
Foreign human or animal cells
Microbial cells, viruses
Plant molecules
Poor Immunogens
Trisaccharide
Polypeptide

Figure 16.8 A comparison of good immunogens and poor immunogens. *Top:* Good immunogens are large and complex. *Bottom:* Small molecules and linear molecules are less likely to be good immunogens.

"Good" Antigens

Antigens can have varying degrees of immunogenicity. Characteristics of good antigens (meaning they provoke a strong response) are (a) their chemical composition; (b) their context, meaning what types of cytokines are present; and (c) their size. We can generalize that large antigens are better than small antigens. However, large size alone is not sufficient for antigenicity. Glycogen, a polymer of glucose with a highly repetitious structure, has a molecular weight over 100,000 and is not normally antigenic, whereas insulin, a protein with a molecular weight of 6,000, can be antigenic.

A lymphocyte's capacity to discriminate differences in molecular shape is so fine that it recognizes and responds to only a portion of the antigen molecule. This molecular fragment, called the **epitope** (shown in figure 16.8), is the primary signal that the molecule is foreign.

A Note About Epitopes and Antigens

Although up to now we have been calling the immunogenic substance "the antigen," it is more precisely termed the *epitope*. You could say, for instance, "The antigenic portion of the protein on a microbe is the epitope." You will also note that, in practice, clinicians, and even other parts of this book, use the word *antigen* when the precise term is *epitope*. You will know, however, that the part of the molecule that is actually recognized by the immune system is the epitope. Each separate macromolecule on the surface of a cell may have several epitopes. This means that a single microbe can have thousands (or more) of epitopes. As a result, every epitope can be recognized by B- and T-cell receptors that were formed during genetic reassortment. The particular tertiary structure and shape of this determinant must conform like a key to the receptor "lock" of the lymphocyte, which then responds to it.

Small, foreign molecules that consist only of a determinant group and are too small by themselves to elicit an immune response are termed **haptens.** However, if such an incomplete antigen is linked to a larger carrier molecule, the combined molecule can develp immunogenicity **(figure 16.9).** The carrier group contributes to the size of the complex and enhances the proper spatial orientation of the determinative group, while the hapten serves as the epitope. Haptens include molecules such as drugs, metals, and ordinarily harmless household, industrial, and environmental chemicals. Many haptens inappropriately develop antigenicity in the body by combining with large carrier molecules such as serum proteins.

Because each human being is genetically and biochemically unique (except for identical twins), the proteins and other molecules of one person can be antigenic to another. **Alloantigens** are cell surface markers and molecules that occur in some members of the same species but not in others. Alloantigens are the basis for an individual's blood group and major histocompatibility profile, and they are responsible for incompatibilities that can occur in blood transfusion or organ grafting.

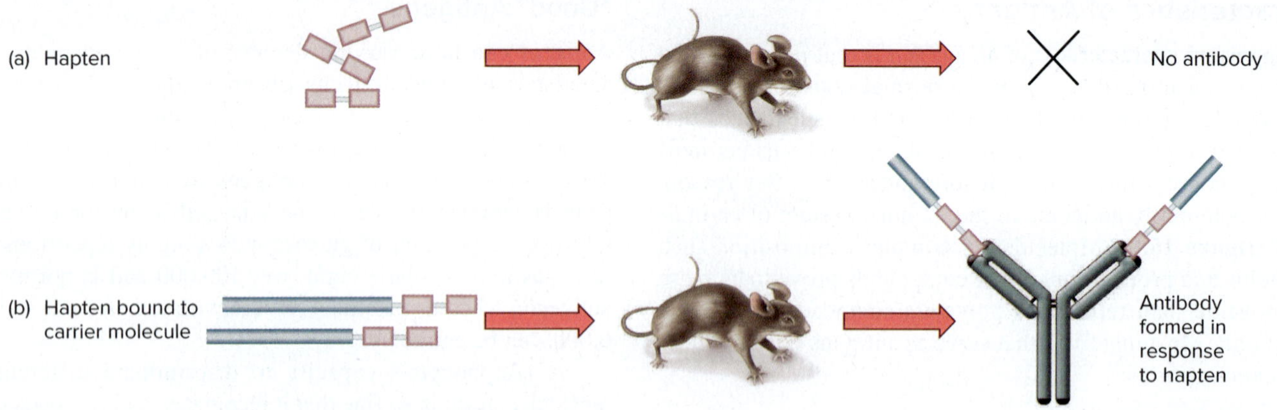

(a) Hapten
No antibody
(b) Hapten bound to carrier molecule
Antibody formed in response to hapten

Figure 16.9 The hapten-carrier phenomenon. **(a)** Haptens are too small to be discovered by an animal's immune system; no response. **(b)** A hapten bound to a large molecule will serve as an epitope and stimulate a response and an antibody that is specific for it.

Some bacterial chemicals, which belong to a group of immunogens called **superantigens,** are potent stimuli for T cells. Their presence in an infection activates T cells at a rate 100 times greater than ordinary antigens. The result can be an overwhelming release of cytokines and cell death. Such diseases as toxic shock syndrome and certain autoimmune diseases are associated with this class of antigens. The overwhelming release of cytokines is called a **cytokine storm,** and it plays a role in several very serious infectious diseases. Cytokine storms have been most recently cited in the symptoms seen in severe COVID-19 cases. They also play a role in influenza deaths and the pandemic form of the Black Death.

Antigens that evoke allergic reactions, called **allergens,** are discussed in detail in chapter 17.

Encounters with Antigens

The basis for most immune responses is the encounter between antigens and white blood cells. Microbes and other foreign substances enter most often through the respiratory or gastrointestinal mucosa and less frequently through other mucous membranes or the skin or across the placenta. Antigens introduced intravenously travel through the bloodstream and end up in the liver, spleen, bone marrow, kidney, and lung. If introduced by some other route, antigens are carried in lymphatic fluid and concentrated by the lymph nodes. The lymph nodes and spleen are important in concentrating the antigens and circulating them thoroughly through all areas populated by lymphocytes, so that they can come in contact with the proper clone.

The Role of Antigen Processing and Presentation

In most immune reactions, the antigen must be further acted upon and formally presented to lymphocytes by cells called antigen-presenting cells (APCs). Three different cells can serve as APCs: macrophages, B cells, and dendritic (den′-drih-tik) cells. Antigen-presenting cells grab the antigen-carrying microbe and ingest it. They degrade it and pass its antigens back out onto their membranes, complexed with either MHC-I or MHC-II markers. **Figure 16.10** illustrates how this process works. **Figure 16.11** illustrates the activity of one particular APC, a dendritic cell. After processing is complete, the antigen is inserted into a cleft on the MHC receptor, and the complex is moved to the surface of the APC so that it will be readily accessible to T lymphocytes during presentation.

16.3 Learning Outcomes—Assess Your Progress

10. Compare the terms *antigen, immunogen,* and *epitope.*
11. List characteristics of antigens that optimize their immunogenicity.
12. Describe how the immune system responds to alloantigens and superantigens.
13. List the types of cells that can act as antigen-presenting cells.

16.4 Step III: Antigenic Challenge of T Cells and B Cells

The Activation of T Cells and Their Differentiation into Subsets

Now that the antigen is presented on the surface of the APC, these cells are ready to activate T cells bearing CD4 markers. CD4 T cells are called the T helper class. They bear an antigen-specific T-cell receptor that binds to the antigen (epitope) held by the MHC molecule. At the same time, the T helper cell's CD4 marker also binds to the MHC molecule. (Look at the inset in the bottom of figure 16.10.) Once identification has occurred, the APC activates this T helper (T_H) cell. The T_H cell, in turn, produces a cytokine, **interleukin-2 (IL-2),** which stimulates T helper cells and cytotoxic T cells.

A stimulated T cell multiplies through successive mitotic divisions and produces a large population of genetically identical daughter cells in the process of clonal expansion **(figure 16.12).** Some cells that are activated stop short of becoming fully

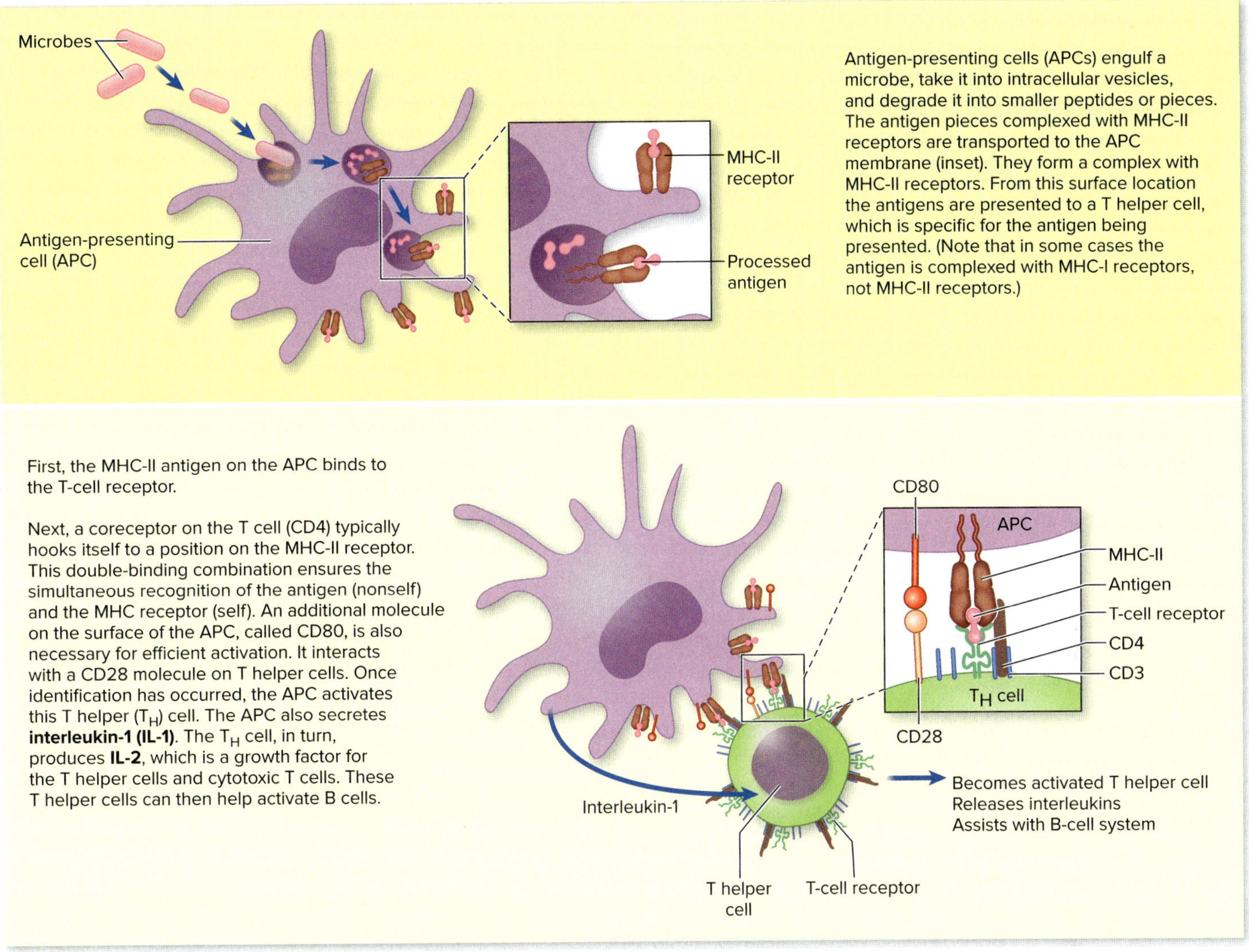
Microbes
Antigen-presenting cell (APC)
MHC-II receptor
Processed antigen
Antigen-presenting cells (APCs) engulf a microbe, take it into intracellular vesicles, and degrade it into smaller peptides or pieces. The antigen pieces complexed with MHC-II receptors are transported to the APC membrane (inset). They form a complex with MHC-II receptors. From this surface location the antigens are presented to a T helper cell, which is specific for the antigen being presented. (Note that in some cases the antigen is complexed with MHC-I receptors, not MHC-II receptors.)
First, the MHC-II antigen on the APC binds to the T-cell receptor.
Next, a coreceptor on the T cell (CD4) typically hooks itself to a position on the MHC-II receptor. This double-binding combination ensures the simultaneous recognition of the antigen (nonself) and the MHC receptor (self). An additional molecule on the surface of the APC, called CD80, is also necessary for efficient activation. It interacts with a CD28 molecule on T helper cells. Once identification has occurred, the APC activates this T helper (T_H) cell. The APC also secretes **interleukin-1 (IL-1)**. The T_H cell, in turn, produces **IL-2**, which is a growth factor for the T helper cells and cytotoxic T cells. These T helper cells can then help activate B cells.
CD80
APC
MHC-II
Antigen
T-cell receptor
CD4
CD3
T_H cell
CD28
Interleukin-1
Becomes activated T helper cell
Releases interleukins
Assists with B-cell system
T helper cell
T-cell receptor

Figure 16.10 **Interactions between antigen-presenting cells (APCs) and T helper (CD4) cells required for T-cell activation.** For T cells to recognize foreign antigens, they must have the antigen processed and presented by a professional APC such as a dendritic cell.

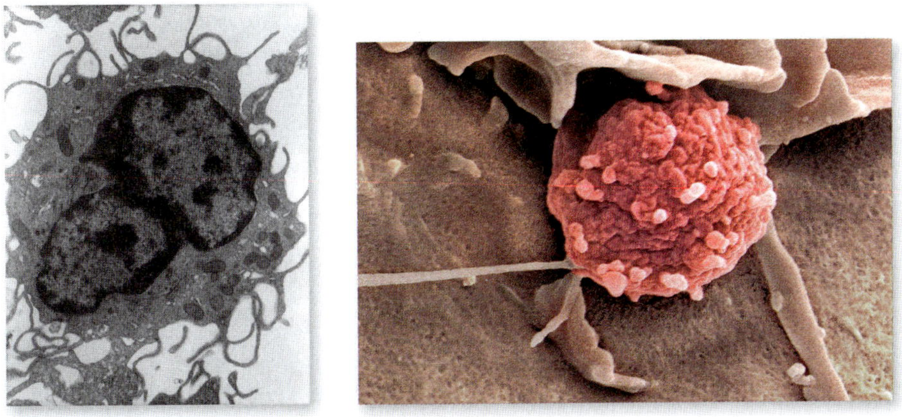

(a) (b)

Figure 16.11 **Dendritic cell.** **(a)** An electron micrograph of an entire dendritic cell. **(b)** This is a close-up view of a dendritic cell (brown) beginning to engulf a spore of the fungus *Aspergillus* (pink).

(a) David M. Phillips/Science Source; (b) Prof. Matthias Gunzer/Science Source

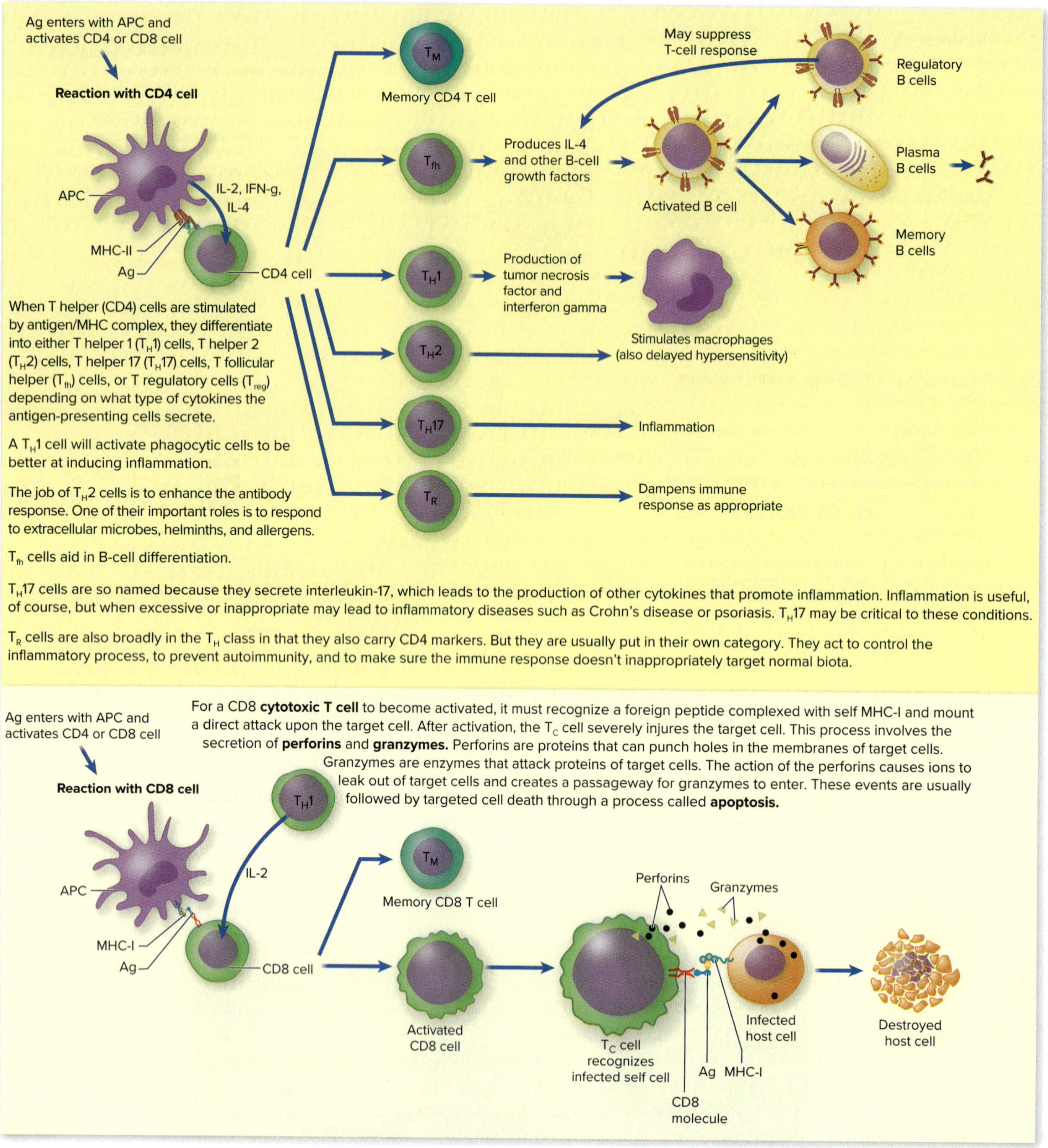
Ag enters with APC and activates CD4 or CD8 cell
Reaction with CD4 cell
APC
IL-2, IFN-g, IL-4
MHC-II
Ag
CD4 cell
T_M
Memory CD4 T cell
T_{fh}
Produces IL-4 and other B-cell growth factors
Activated B cell
May suppress T-cell response
Regulatory B cells
Plasma B cells
Memory B cells
T_H1
Production of tumor necrosis factor and interferon gamma
Stimulates macrophages (also delayed hypersensitivity)
T_H2
T_H17
Inflammation
T_R
Dampens immune response as appropriate
When T helper (CD4) cells are stimulated by antigen/MHC complex, they differentiate into either T helper 1 (T_H1) cells, T helper 2 (T_H2) cells, T helper 17 (T_H17) cells, T follicular helper (T_{fh}) cells, or T regulatory cells (T_{reg}) depending on what type of cytokines the antigen-presenting cells secrete.
A T_H1 cell will activate phagocytic cells to be better at inducing inflammation.
The job of T_H2 cells is to enhance the antibody response. One of their important roles is to respond to extracellular microbes, helminths, and allergens.
T_{fh} cells aid in B-cell differentiation.
T_H17 cells are so named because they secrete interleukin-17, which leads to the production of other cytokines that promote inflammation. Inflammation is useful, of course, but when excessive or inappropriate may lead to inflammatory diseases such as Crohn's disease or psoriasis. T_H17 may be critical to these conditions.
T_R cells are also broadly in the T_H class in that they also carry CD4 markers. But they are usually put in their own category. They act to control the inflammatory process, to prevent autoimmunity, and to make sure the immune response doesn't inappropriately target normal biota.
Ag enters with APC and activates CD4 or CD8 cell
Reaction with CD8 cell
For a CD8 **cytotoxic T cell** to become activated, it must recognize a foreign peptide complexed with self MHC-I and mount a direct attack upon the target cell. After activation, the T_C cell severely injures the target cell. This process involves the secretion of **perforins** and **granzymes.** Perforins are proteins that can punch holes in the membranes of target cells. Granzymes are enzymes that attack proteins of target cells. The action of the perforins causes ions to leak out of target cells and creates a passageway for granzymes to enter. These events are usually followed by targeted cell death through a process called **apoptosis.**
T_H1
IL-2
APC
MHC-I
Ag
CD8 cell
T_M
Memory CD8 T cell
Activated CD8 cell
Perforins
Granzymes
T_C cell recognizes infected self cell
CD8 molecule
Ag
MHC-I
Infected host cell
Destroyed host cell

Figure 16.12 Events in T-cell activation. Gamma-delta T cells are not pictured as the exact mechanism of their activation is not known.

differentiated, such as the memory cells. This cell population is activated but stays uninvolved in the current response. It will be ready to act in subsequent encounters with the same antigen, very quickly and very strongly.

A (non-memory) T helper cell activated in this way can then help activate B cells. The manner in which B and T cells subsequently become activated by the APC–T helper cell complex is addressed in later sections.

Not all antigens require T helper cell intervention to activate B cells. A few antigens can trigger a response from B lymphocytes without the cooperation of APCs or T helper cells. These are called T-cell-independent antigens and are usually simple molecules such as carbohydrates with many repeating and invariable determinant groups. Because so few antigens are of this type, most B-cell reactions require T helper cells. We call these T-cell-dependent antigens.

As you see at the bottom of figure 16.12, the activation of another type of T cell, which bears CD8—not CD4—markers, is very similar. CD8-bearing T cells are called T cytotoxic cells. They are activated by antigen on the surface of APCs that has been complexed with MHC-I (not MHC-II) molecules. They are then ready to do their job, which is very different than that of T helper cells, as we will see later.

In summary, mature T cells in lymphoid organs are primed to react with antigens that have been processed and presented to them by an antigen-presenting cell. They recognize an antigen only when it is presented in association with an MHC carrier. T cells with CD4 receptors recognize peptides presented on MHC-II, and T cells with CD8 receptors recognize peptides presented on MHC-I.

Activated T cells transform in preparation for mitotic divisions, and they differentiate into one of the subsets of effector cells and memory cells that can respond quickly upon subsequent contact **(table 16.2).**

The Activation of B Cells: Clonal Expansion and Antibody Production

The activation of B cells by most antigens (T-dependent antigens) involves a series of events **(figure 16.13):**

1. **Binding of antigen**
2. **Antigen processing and presentation**
3. **B cell/T_H cell cooperation and recognition**
4. **B-cell activation**
5. **Differentiation**
6. **Clonal expansion.** The primary action of plasma cells is to secrete copious amounts of antibodies with the same specificity as the original marker. Although an individual plasma cell can produce around 2,000 antibodies per second, production does not continue indefinitely. The plasma cells do not survive for long and deteriorate after they have released their antibodies.

As mentioned before, some antigens are able to stimulate a strong B-cell response without the involvement of T cells. These antigens are often very large polymers of repeating units. Examples include lipopolysaccharide from the cell wall of *Escherichia coli,* polysaccharide from the capsule of *Streptococcus pneumoniae,* and molecules from rabies and Epstein-Barr virus. They are capable of activating naive B cells simply by binding to their antigen receptors directly (left-hand side of figure 16.13).

Also, recently researchers identified another totally different kind of B cell—an early-warning B cell that detects bacterial infection and releases a chemical that initiates an immune response. It has been named the innate response activator B cell (IRA-B). It is a strange B cell because it participates at the front end of a response and activates the innate arm of the immune system.

Table 16.2 Characteristics of Subsets of T-Cell Types in the Classic T-Cell Response

Types	Co-Receptors on T Cell	Functions/Important Features
T helper cell 1 (T_H1)	CD4	Activates the cell-mediated immunity pathway; secretes tumor necrosis factor and interferon gamma; also responsible for delayed hypersensitivity (allergy occurring several hours or days after contact); secretes IL-2
T helper cell 2 (T_H2)	CD4	Can activate macrophages to expel helminths or protozoans, phagocytose extracellular antigens; contributes to type 1 (allergic) hypersensitivity; can encourage tumor development
T helper cell 17 (T_H17)	CD4	Promotes inflammation
T follicular helper cell	CD4, CD40L	Drives B-cell proliferation; aids B cells in antibody class switching
T regulatory cell (T_R)	CD25, CD4	Controls adaptive immune response; prevents autoimmunity; can contribute to cancer progression
T cytotoxic cell (T_C)	CD8	Destroys a target foreign cell by lysis; important in destruction of complex microbes, cancer cells, virus-infected cells; graft rejection; requires MHC-I for function
Gamma-delta T cells	—	React adaptively and innately; responsive to lipid antigens

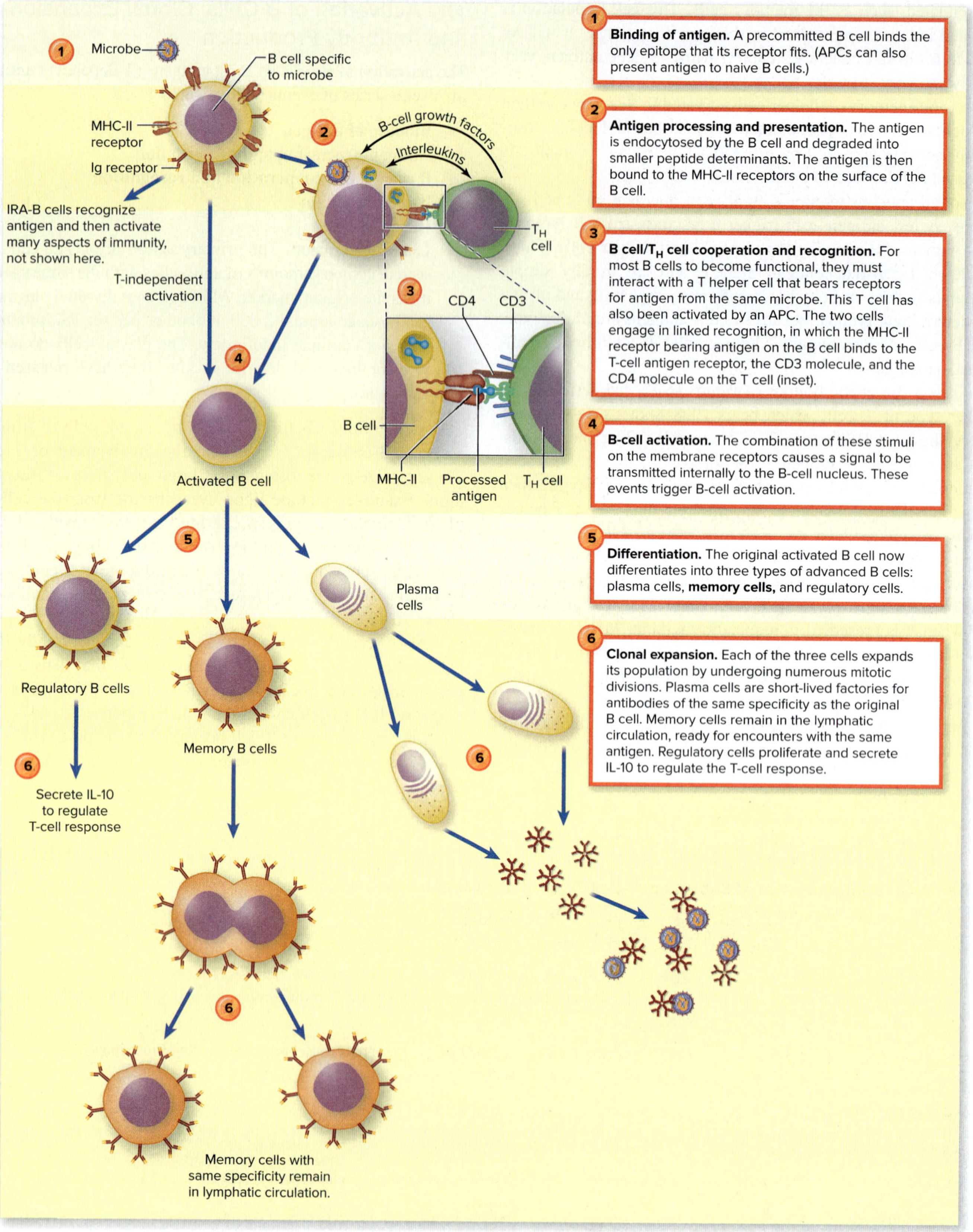
1
Microbe
B cell specific to microbe
MHC-II receptor
Ig receptor
IRA-B cells recognize antigen and then activate many aspects of immunity, not shown here.
T-independent activation
2
B-cell growth factors
Interleukins
T_H cell
3
CD4
CD3
B cell
MHC-II
Processed antigen
T_H cell
4
Activated B cell
5
Plasma cells
Regulatory B cells
Memory B cells
6
Secrete IL-10 to regulate T-cell response
6
6
Memory cells with same specificity remain in lymphatic circulation.
1 **Binding of antigen.** A precommitted B cell binds the only epitope that its receptor fits. (APCs can also present antigen to naive B cells.)
2 **Antigen processing and presentation.** The antigen is endocytosed by the B cell and degraded into smaller peptide determinants. The antigen is then bound to the MHC-II receptors on the surface of the B cell.
3 **B cell/T_H cell cooperation and recognition.** For most B cells to become functional, they must interact with a T helper cell that bears receptors for antigen from the same microbe. This T cell has also been activated by an APC. The two cells engage in linked recognition, in which the MHC-II receptor bearing antigen on the B cell binds to the T-cell antigen receptor, the CD3 molecule, and the CD4 molecule on the T cell (inset).
4 **B-cell activation.** The combination of these stimuli on the membrane receptors causes a signal to be transmitted internally to the B-cell nucleus. These events trigger B-cell activation.
5 **Differentiation.** The original activated B cell now differentiates into three types of advanced B cells: plasma cells, **memory cells,** and regulatory cells.
6 **Clonal expansion.** Each of the three cells expands its population by undergoing numerous mitotic divisions. Plasma cells are short-lived factories for antibodies of the same specificity as the original B cell. Memory cells remain in the lymphatic circulation, ready for encounters with the same antigen. Regulatory cells proliferate and secrete IL-10 to regulate the T-cell response.

Figure 16.13 Events in B-cell activation and antibody synthesis.

16.4 Learning Outcomes—Assess Your Progress

14. Summarize the process of T-cell activation, and list major types of T cells produced in this process.
15. Diagram the steps of B-cell activation, and list the types of B cells produced in this process.

16.5 Step IV (1): The T-Cell Response

The responses of T cells are **cell-mediated immunities,** which require the direct involvement of T lymphocytes throughout the course of the reaction. These reactions are among the most complex and diverse in the immune system and involve several subsets of T cells whose particular actions are dictated by the APCs that activate them. T cells require some type of MHC (self) recognition before they can be activated, and all produce cytokines with a spectrum of biological effects.

Rather than making antibodies to control foreign antigens, T cells stimulate other T cells, B cells, and phagocytes. This activity requires the cooperation of a variety of cell types (see table 16.2).

T Helper (T_H) Cells

T helper cells play a central role in regulating immune reactions to antigens, including those of B cells and other T cells. They are also involved in activating macrophages. They do this directly by receptor contact and indirectly by releasing cytokines, like interferon gamma (IFNγ). T helper cells secrete interleukin-2, which stimulates the primary growth and activation of many types of T cells, including cytotoxic T cells. Some T helper cells secrete interleukins-4, -5, and -6, which stimulate various activities of B cells. T helper cells are the most prevalent type of T cell in the blood and lymphoid organs, making up about 65% of this population. In HIV infection, the number of T helper cells is greatly reduced, which is a major contributor to the pathology of AIDS.

When T helper (CD4) cells are stimulated by an antigen/MHC complex, they differentiate into either T helper 1 (T_H1) cells, T helper 2 (T_H2) cells, or T helper 17 (T_H17) cells, depending on what type of cytokines the antigen-presenting cells secrete. A T helper 1 cell will activate phagocytic cells to be better at inducing inflammation, resulting in a delayed hypersensitivity reaction. The T helper 1 cell can also secrete interleukin-2, which activates T cytotoxic cells, as shown at the bottom of figure 16.12. If the APC secretes another set of cytokines, the T cell will differentiate into a T_H2 cell. These cells have the functions of (1) secreting substances that influence B-cell differentiation and (2) enhancing the antibody response. One of their important roles is to respond to extracellular microbes, helminths, and allergens. T helper 17 cells are so-named because they secrete interleukin-17, which leads to the production of other cytokines that promote inflammation. Inflammation is useful, of course, but when excessive or inappropriate, may lead to inflammatory diseases such as Crohn's disease or psoriasis; T_H17 cells may be critical to these conditions.

Regulatory T (T_R) Cells: Cells That Maintain the Happy Medium

Regulatory T cells are also broadly in the T_H class, in that they too carry CD4 markers. But they are not "helpers" in the sense that they encourage immune activity. They act to control the inflammatory process, to dampen autoimmunity, and to make sure the immune response does not inappropriately target normal biota. Regulatory B cells also regulate the degree of response from T cells. So B cells are involved in two ways in the T-cell response: (1) They can become activated to become plasma cells by cytokines from activated T cells and (2) regulatory B cells can secrete their own cytokines to dampen the T-cell response.

Cytotoxic T (T_C) Cells: Cells That Kill Other Cells

Cytotoxicity is the capacity of certain T cells to kill a target cell. It is a fascinating and powerful property that accounts for much of our immunity to foreign cells and cancer. Yet under some circumstances, it can lead to disease. For a CD8 **cytotoxic T cell** to become activated, it must recognize a foreign peptide complexed with self MHC-I presented to it and mount a direct attack upon the target cell. After activation, the T_C cell severely injures the target cell (see figure 16.12). This process involves the secretion of **perforins** and **granzymes.** Perforins are proteins that can punch holes in the membranes of target cells. Granzymes are enzymes that attack proteins of target cells. The action of the perforins causes ions to leak out of target cells and creates a passageway for granzymes to enter. These events are usually followed by targeted cell death through a process called **apoptosis.**

Target cells that T_C cells can destroy include the following:

- *Virally infected cells.* Cytotoxic T cells recognize these because of telltale virus peptides expressed on their surface. Cytotoxic defenses are an essential protection against viruses. They also recognize and target cells carrying intracellular bacteria.
- *Cancer cells.* T cells constantly survey the tissues and immediately attack any abnormal cells they encounter **(figure 16.14).** The importance of this function is clearly demonstrated in the susceptibility of T-cell-deficient people to cancer. **Insight 16.1** discusses how the microbiome influences our body's response to cancer.
- *Cells from other animals and humans.* Cytotoxic cell-mediated immunity is the most important factor in graft rejection. In this instance, the T_C cells attack the foreign tissues that have been implanted into a recipient's body.

Keep in mind that there are always T helper cells and T cytotoxic cells in any given immune response. When antigens activate both B cells and T cells, it is called **cross-presentation.**

Gamma-Delta T Cells

The subcategory of T cells called gamma-delta T cells is distinct from other T cells. They do have T-cell receptors that are rearranged to recognize a wide range of antigens, but they frequently respond to certain kinds of PAMPs on microorganisms, the

INSIGHT 16.1 MICROBIOME: Cancer and the Microbiome

We know that the gut microbiome is critical to the health of its human host. One of the ways it influences health is by having a profound effect on inflammation and immunity. In turn, inflammation and immunity—in all the complexity you have studied in these two chapters—profoundly affect the initiation and progression of tumor cells. The microbiome also impacts the effectiveness of cancer therapies and the susceptibility to toxic side effects.

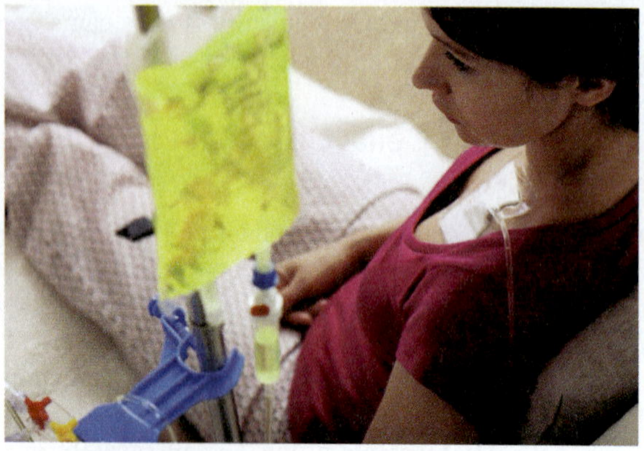

Image Point Fr/Shutterstock

One important factor is what is called the "crosstalk" among the gut microbiota, immune cells, and the mucosal surfaces. This refers to the close association and chemical signaling that occur among the three. When the microbiome is healthy, and this crosstalk is functioning well, it is in an optimal state to prevent the initiation of tumors. Disturbances to the microbiome, which can occur because of antibiotic treatment, lifestyle, diet, and disease, can lead to the loss of immune surveillance, allowing tumor growth to begin.

Further, it has become clear that an individual's reaction to cancer treatment is also profoundly affected by the microbiome. Because the microbiota prepare immune cells to release toxic oxygen species and also to provide an effective T-cell response, when the microbiome is dysfunctional, chemotherapy that works through those two mechanisms is less effective. A disturbed microbiome has even been shown to reduce the effectiveness—and increase the side effects—of radiation therapy. This field of research is only about 5 years old and has mostly been conducted in mouse models. But the prospects are good that we will eventually be able to improve the treatment of cancer, at least partially through the "engineering" of the microbiome.

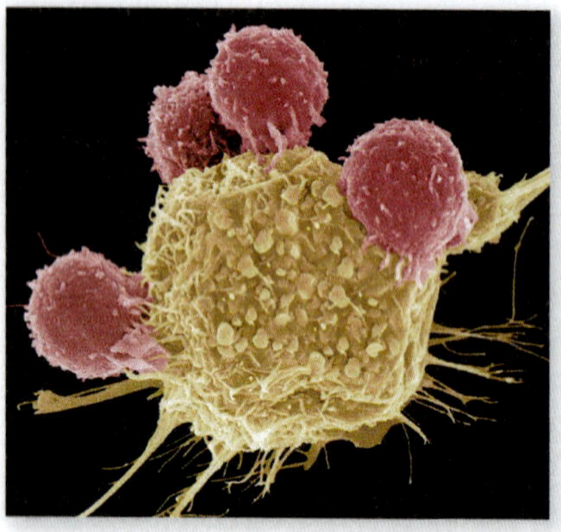

Figure 16.14 Cytotoxic T cells (pink cells) mount an attack on a tumor cell (large, yellow cell). These small killer cells perforate their cellular targets with holes that lead to lysis and death.
Steve Gschmeissner/Science Source

way WBCs do in the innate system. This allows gamma-delta T cells to respond more quickly. However, they still produce memory cells when they are activated. For these reasons, they bridge the innate and adaptive immune responses. They are particularly responsive to certain types of phospholipids and can recognize and react against tumor cells.

Additional Cells with Orders to Kill

Natural killer (NK) cells are a type of lymphocyte related to T cells that lack specificity for antigens. They circulate through the spleen, blood, and lungs and are probably the first killer cells to attack cancer cells and virus-infected cells. They destroy those cells by similar mechanisms as T_C cells. They are generally not considered part of adaptive cell-mediated immunity because they themselves do not possess antigen receptors.

Natural killer T cells (NKT cells) appear to be a hybrid type of cell because they share properties of both T cells and NK cells. They express both T-cell receptors and NK-cell markers and are stimulated by glycolipids on foreign cells. They exhibit the ability to rapidly produce cytokines as well as granzymes and perforins and, in turn, can trigger self-destruction in target cells.

As you can see, the T-cell system is very complex. In summary, T cells differentiate into many different types of cells (including memory cells), each of which contributes to the orchestrated immune response under the influence of a multitude of cytokines. The T-cell system is summarized in figure 16.12. Compare it with the B-cell system summary depicted in figure 16.13 for further study.

16.5 Learning Outcomes—Assess Your Progress

16. Describe the main functions of the major T-cell types and their subsets.
17. Explain the role of cytotoxic T cells in apoptosis, and list the potential targets of this process.

16.6 Step IV (2): The B-Cell Response

The Structure of Immunoglobulins

In section 16.4, you saw an overview of how B cells are activated. The end result of their activation is the secretion of highly specific

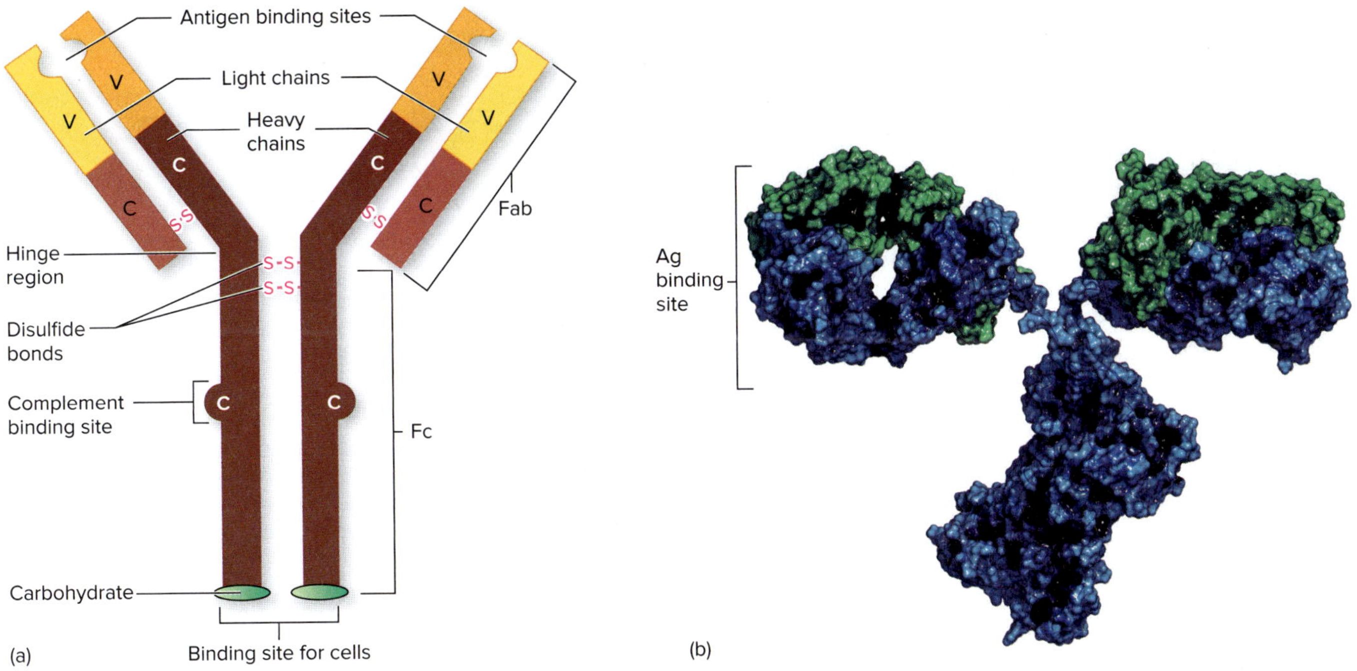
Antigen binding sites
Light chains
Heavy chains
V
V
V
V
C
C
C
C
S-S
S-S
Fab
Hinge region
S-S
S-S
Disulfide bonds
Complement binding site
C
C
Fc
Carbohydrate
Binding site for cells
(a)
Ag binding site
(b)

Figure 16.15 Working models of antibody structure. **(a)** Diagrammatic view of IgG depicts the principal functional areas (Fabs and Fc) of the molecule. **(b)** Realistic model of immunoglobulin shows the tertiary and quaternary structures achieved by additional intrachain and interchain bonds.

(b) Molekuul_be/Shutterstock

antibodies, also known as immunoglobulins. Earlier we saw that a basic immunoglobulin (Ig) molecule contains four polypeptide chains connected by disulfide bonds. We will view this structure once again using an IgG molecule as a model **(figure 16.15).** The two "arms" that bind antigen are called the antigen binding fragments (abbreviated "Fabs"), and the rest of the molecule is the crystallizable fragment (Fc). It is called this because it was the first part to be crystallized in a laboratory. The amino-terminal end of each Fab fragment (consisting of the variable regions of the heavy and light chains) folds into a groove that will accommodate one epitope. The Fc fragment (the "stem") serves as an anchor, involved in binding to various cells and molecules of the immune system itself.

Antibody–Antigen Interactions and the Function of the Fab

The site on the antibody where the epitope binds is composed of a hypervariable region whose amino acid content can be extremely varied. (Remember the gene rearrangements?) In an antibody, a minimal complementary fit is necessary for the antigen to be held effectively **(figure 16.16).** The specificity of antigen binding sites for antigens is similar to enzymes and substrates. Because the specificity of the two Fab sites is identical, an Ig molecule can bind two epitopes on the same cell or on two separate cells. If this happens, the two cells become linked.

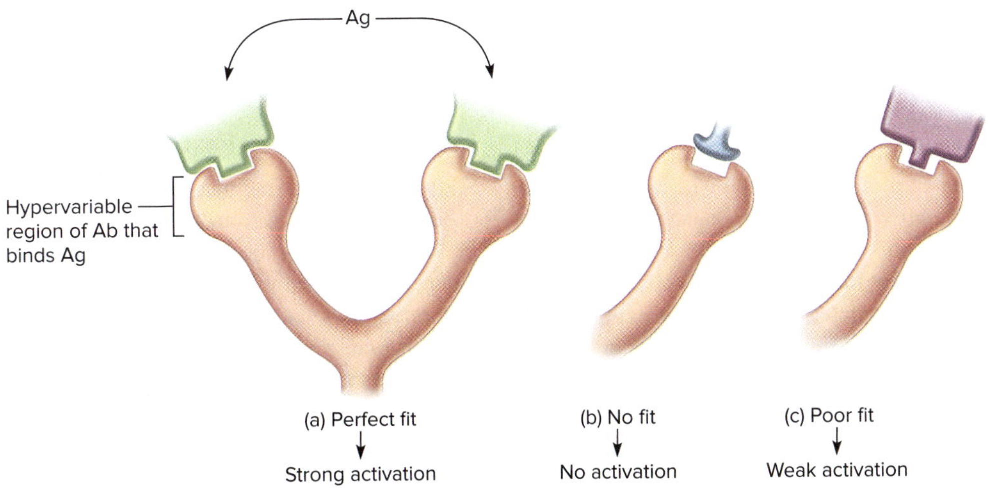
Ag
Hypervariable region of Ab that binds Ag
(a) Perfect fit
Strong activation
(b) No fit
No activation
(c) Poor fit
Weak activation

Figure 16.16 Antigen-antibody binding. The union of antibody (Ab) and antigen (Ag) is characterized by a certain degree of fit and is supported by a multitude of weak linkages, especially hydrogen bonds and electrostatic attraction. The better the fit—that is, antigen in **(a)** versus antigen in **(c)**—the stronger the stimulation of the lymphocyte during the activation stage.

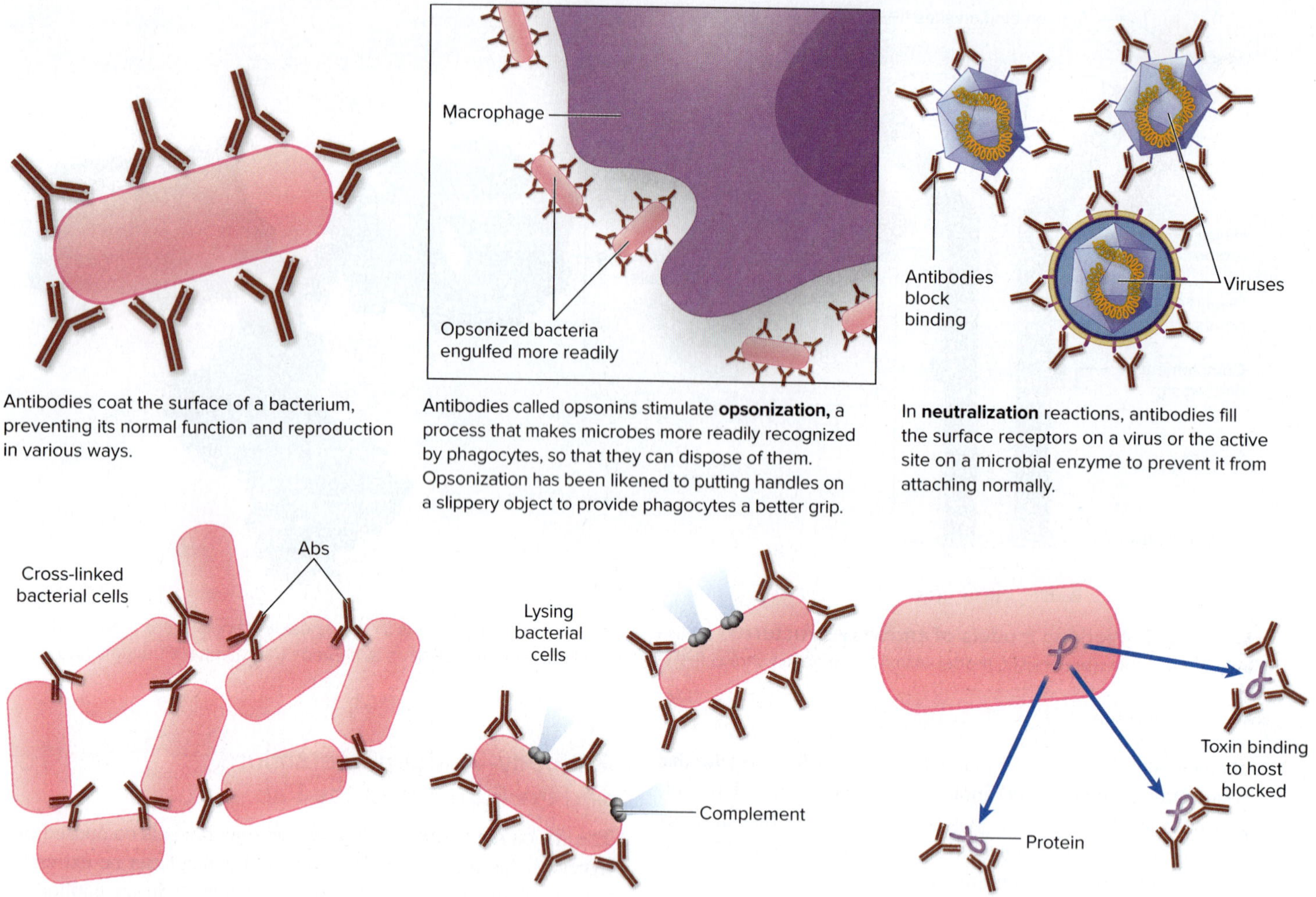
Macrophage
Opsonized bacteria engulfed more readily
Antibodies block binding
Viruses
Cross-linked bacterial cells
Abs
Lysing bacterial cells
Complement
Toxin binding to host blocked
Protein

Antibodies coat the surface of a bacterium, preventing its normal function and reproduction in various ways.

Antibodies called opsonins stimulate **opsonization,** a process that makes microbes more readily recognized by phagocytes, so that they can dispose of them. Opsonization has been likened to putting handles on a slippery object to provide phagocytes a better grip.

In **neutralization** reactions, antibodies fill the surface receptors on a virus or the active site on a microbial enzyme to prevent it from attaching normally.

The capacity for antibodies to aggregate, or **agglutinate,** antigens is the consequence of their cross-linking cells or particles into large clumps. Agglutination renders microbes immobile and enhances their phagocytosis. This is a principle behind certain immune tests discussed in chapter 18.

The interaction of an antibody with complement can result in the specific rupturing of cells and some viruses.

An **antitoxin** is a special type of antibody that neutralizes bacterial exotoxins.

Figure 16.17 Summary of antibody functions. Complement fixation, agglutination, and precipitation are covered further in chapter 18.

The principal activity of an antibody is to unite with, immobilize, call attention to, or neutralize the antigen for which it was formed **(figure 16.17).** Antibodies called *opsonins* stimulate **opsonization** (ahp″-son-uh-za′-shun), a process in which microorganisms or other particles are coated with antibodies so that they will be more readily recognized by phagocytes, which dispose of them. Opsonization has been likened to putting handles on a slippery object to provide phagocytes a better grip. The capacity for antibodies to aggregate, or **agglutinate,** antigens is the consequence of their cross-linking cells or particles into large clumps. Agglutination makes microbes immobile and enhances their phagocytosis. (This is also a principle behind certain immunologic tests we will discuss in chapter 18.) The interaction of an antibody with complement can result in the specific rupturing of cells and some viruses. In **neutralization** reactions, antibodies fill the surface receptors on a virus or the active site on a microbial enzyme to prevent it from attaching normally. An **antitoxin** is a special type of antibody that neutralizes bacterial exotoxins. Antibodies may also function to kill targets by inducing production of H_2O_2 and ozone.

Functions of the Fc Fragment

Although the Fab fragments bind antigen, the Fc fragment has a different binding function. In most classes of immunoglobulin, the Fc end can bind to the membranes of cells, such as macrophages, neutrophils, eosinophils, mast cells, basophils, and lymphocytes. The effect of an antibody's Fc fragment binding to a cell depends upon that cell's role. In the case of opsonization, the attachment of antibody to foreign cells and viruses exposes the epitopes to which they are bound to phagocytes. Certain antibodies have regions on the Fc portion for binding complement. In some immune reactions, the binding of Fc causes the release of cytokines. For example, the Fc end of the antibody of allergy (IgE) binds to basophils and mast cells, which causes the release of allergic mediators such as histamine. The size and amino acid composition of Fc also determine an antibody's permeability, its distribution in the body, and its class.

Table 16.3 Characteristics of the Immunoglobulin (Ig) Classes

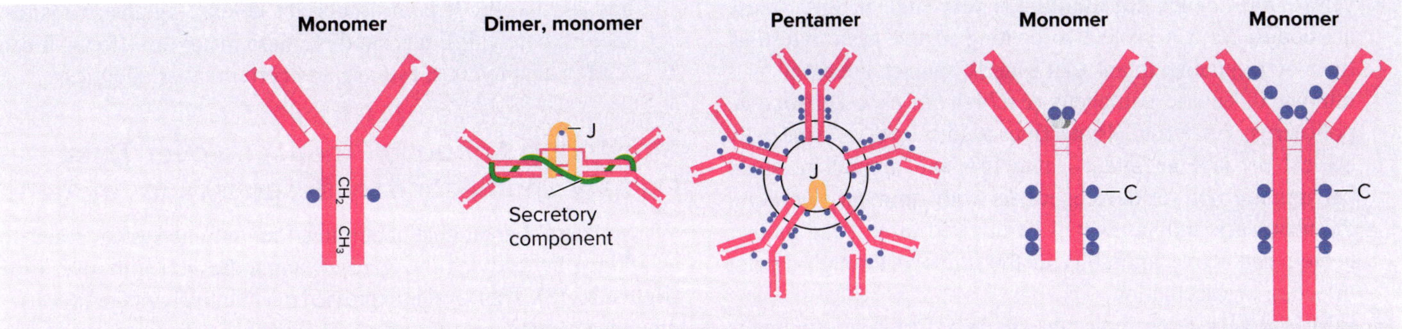

Monomer
Dimer, monomer
Pentamer
Monomer
Monomer
CH2
CH3
J
Secretory component
J
C
C

	IgG	IgA	IgM	IgD	IgE
Number of Antigen Binding Sites	2	4, 2	10	2	2
Molecular Weight	150,000	170,000–385,000	900,000	180,000	200,000
Percentage of Total Antibody in Serum	80%	13%	6%	1%	0.002%
Average Half-Life in Serum (Days)	23	6	5	3	2.5
Crosses Placenta?	Yes	No	No	No	No
Fixes Complement?	Yes	No	Yes	No	No
Fc Binds to	Phagocytes				Mast cells and basophils
Biological Function	Long-term immunity; memory antibodies; neutralizes toxins, opsonizes, fixes complement	Secretory antibody; on mucous membranes	Produced at first response to antigen; can serve as B-cell receptor	Receptor on B cells	Antibody of allergy; worm infections

Accessory Molecules on Immunoglobulins

All antibodies contain chemical groups in addition to the basic polypeptide structure. Varying amounts of carbohydrates are affixed to the constant regions in most instances **(table 16.3).** Two additional accessory molecules are the *J chain,* which joins the monomers[1] of IgA and IgM, and the *secretory component,* which helps move IgA across mucous membranes.

The Classes of Immunoglobulins

Immunoglobulins exist as structural and functional classes called *isotypes* (illustrated in table 16.3). The classes are differentiated with shorthand names (Ig, followed by a letter: IgG, IgA, IgM, IgD, IgE).

- The structure of IgG is the one we have been describing. It is a monomer of immunoglobulin, with the classic "Y" shape. It is produced by plasma cells in a primary response and by memory cells responding the second time to a given antigenic stimulus. It is by far the most prevalent antibody circulating throughout the tissue fluids and blood. It has numerous functions: It neutralizes toxins, opsonizes, and fixes (binds) complement. It is the only antibody capable of crossing the placenta.

1. *Monomer* means "one unit" or "one part." Accordingly, *dimer* means "two units," *pentamer* means "five units," and *polymer* means "many units."

Disease Connection

Immunotherapy harnesses the immune system to fight disease and is an emerging field in cancer treatment. CAR T-cell therapy is approved for treatment of certain cancers in adult and pediatric patients. (The acronym *CAR-T* refers to "chimeric antigen receptor.") The patient's T cells are collected and genetically engineered to recognize and attack cancer cell antigens. The engineered T cells are multiplied in the lab and administered back to the patient, where they seek out and destroy cancerous cells.

- The two forms of IgA are (1) a monomer that circulates in small amounts in the blood and (2) a dimer that is a significant component of the mucous and serous secretions of the salivary glands, intestine, nasal membrane, breast, lung, and genitourinary tract. The dimer, called secretory IgA, is formed by two monomers held together by a J chain. To facilitate the transport of IgA across membranes, a secretory piece is later added. IgA coats the surface of these membranes and is found in saliva, tears, colostrum, and mucus. It provides the most important adaptive local immunity to enteric, respiratory, and genitourinary pathogens. During lactation, the breast becomes a site for the proliferation of lymphocytes that produce IgA. The very earliest secretion of the breast—a thin, yellow milk called **colostrum**—is very high in IgA. These antibodies form a protective coating in the gastrointestinal tract of a nursing infant that guards against infection by a number of enteric pathogens (*Escherichia coli, Salmonella,* poliovirus, rotavirus). Protection at this level is especially critical because an infant's own IgA and natural intestinal barriers are not yet developed. As with immunity *in utero,* the necessary antibodies will be donated only if the mother herself has active immunity to the microbe through a prior infection or vaccination.

- IgM is a very large molecule composed of five monomers (making it a pentamer) attached by the Fc portions to a central J chain. With its 10 binding sites, this molecule has tremendous capacity for binding antigen. IgM is the first class synthesized following the host's first encounter with antigen. Its complement-fixing qualities make it an important antibody in many immune reactions. It circulates mainly in the blood and does not cross the placental barrier.
- IgD is a monomer found in minuscule amounts in the serum, and it does not fix complement, opsonize, or cross the placenta. Its main function is that it is the receptor for antigen on B cells, usually along with IgM. It seems to be the triggering molecule for B-cell activation.
- IgE is also an uncommon blood component unless one is allergic or has a parasitic worm infection. Its Fc region interacts with receptors on mast cells and basophils. Its biological role is to stimulate an inflammatory response through the release of potent physiological substances by the basophils and mast cells. It is an important defense against parasites. Unfortunately, IgE has another, more insidious effect—it can lead to anaphylaxis, asthma, and certain other allergies.

Monitoring Antibody Production over Time: Primary and Secondary Responses to Antigens

We can learn a great deal about how the immune system reacts to an antigen by studying the levels of antibodies in serum over time **(figure 16.18).** This level is expressed quantitatively as the **titer** (ty′-tur), or concentration of antibodies. Upon the first exposure to an antigen, the system undergoes a **primary response.** The earliest part of this response, the *latent period,* is marked by a lack of antibodies for that antigen, but much activity is occurring. During this time, the antigen is being concentrated in lymphoid tissue and is being processed by the correct clones of B lymphocytes. As plasma cells synthesize antibodies, the serum titer (concentration) increases to a certain plateau and then tapers off to a low level over a few weeks or months. It turns out that, early in the primary response, most of the antibodies are the IgM type, which is the first class to be secreted by plasma cells. Later, the class of the antibodies is switched to IgG or some other class (IgA or IgE). The specificity of the antibodies does not change, only the class (IgM vs. IgG or something else).

When the immune system is exposed again to the same immunogen within weeks, months, or even years, a **secondary response** occurs. The rate of antibody synthesis, the peak titer, and the length of antibody persistence are greatly increased over the

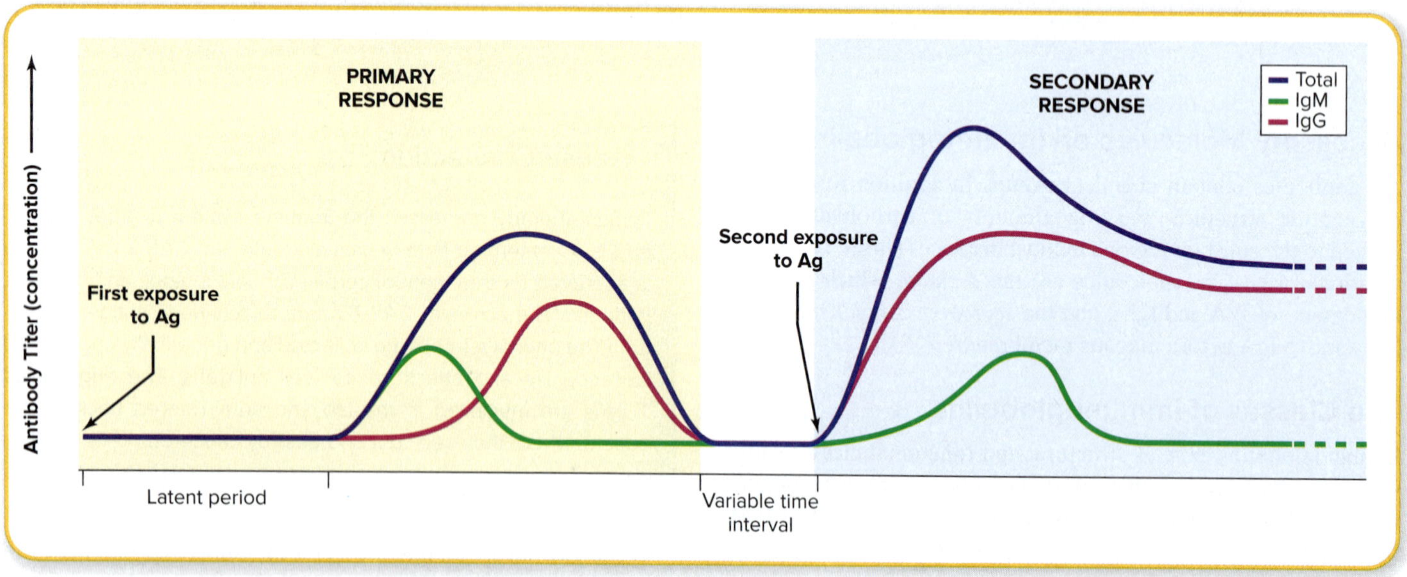
PRIMARY RESPONSE
SECONDARY RESPONSE
Total
IgM
IgG
Antibody Titer (concentration)
First exposure to Ag
Second exposure to Ag
Latent period
Variable time interval

Upon the first exposure to an antigen, the system undergoes a **primary response.** The earliest part of this response, the *latent period*, is marked by a lack of antibodies for that antigen, but much activity is occurring. During this time, the antigen is being concentrated in lymphoid tissue and is being processed by the correct clones of B lymphocytes. As plasma cells synthesize antibodies, the serum titer increases to a certain plateau and then tapers off to a low level over a few weeks or months. Early in the primary response, most of the antibodies are the IgM type, which is the first class to be secreted by plasma cells. Later, the class of the antibodies (but not their specificity) is switched to IgG or some other class (IgA or IgE).

After the initial response, there is no activity, but memory cells of the same specificity are located throughout the lymphatic system.

When the immune system is exposed again to the same immunogen within weeks, months, or even years, a **secondary response** occurs. The rate of antibody synthesis, the peak titer, and the length of antibody persistence are greatly increased over the primary response. The speed and intensity seen in this response are attributable to the memory B cells that were formed during the primary response. The secondary response is also called the **anamnestic response.** The advantage of this response is evident: It provides a quick and potent strike against subsequent exposures to infectious agents.

Figure 16.18 **Primary and secondary responses to antigens.**

primary response. The speed and intensity seen in this response are attributable to the memory B cells that were formed during the primary response. The secondary response is also called the **anamnestic response** (from the Greek word for "memory"). The advantage of this response is evident: It provides a quick and potent strike against subsequent exposures to infectious agents. This memory effect is the fundamental basis for vaccination, which we discuss later.

It is a well-accepted principle that memory B and T cells are only created from clones activated by a specific antigen. This provides a much quicker and more effective response on the second exposure, and all exposures afterward. But researchers are now investigating a phenomenon that has been suspected for some time and confirmed in rigorous studies. It seems that exposure to a particular antigen can result in memory cells to antigens that are *chemically related* to it, even if those antigens have not been seen by the host. This might explain the well-known phenomenon, seen most clearly in developing countries, that vaccines against one disease can provide some protection against others. In Africa, for example, vaccinating against measles also cuts deaths from pneumonia, sepsis, and diarrhea by one-third.

During the COVID-19 pandemic, scientists found that live attenuated vaccines (which we will learn about soon), such as those for measles, polio, and tuberculosis, can enhance a person's innate immunity against other microbes. In low-income countries that struggle to get the COVID-19 vaccines, re-vaccinating with one of these readily available live attenuated vaccines can be helpful.

16.6 Learning Outcomes—Assess Your Progress

18. Diagram an antibody binding antigen, and list the possible end results of this process.
19. List the five types of antibodies and important characteristics of each.
20. Draw and label a graph illustrating the development of a secondary immune response.

16.7 Adaptive Immunity and Vaccination

Adaptive immunity in humans and other mammals is categorized using two sets of criteria that, altogether, result in four specific descriptors of the immune state. Immunity can be either active or passive. Also, it can be either natural or artificial.

- **Active immunity** occurs when an individual receives an immune stimulus (antigen) that activates the B and T cells, causing the body to produce immune substances such as antibodies. Active immunity is marked by several characteristics: (1) It creates a memory that renders the person ready for quick action upon reexposure to that same antigen; (2) it requires several days to develop; and (3) it lasts for a relatively long time, sometimes for life. Active immunity can be stimulated by natural or artificial means.
- **Passive immunity** occurs when an individual receives immune substances (usually antibodies) that were produced actively in the body of another human or animal donor. The recipient is protected for a short time, even though he or she has not had prior exposure to the antigen. It is characterized by (1) lack of memory for the original antigen; (2) lack of production of new antibodies against that disease; (3) immediate protection; and (4) short-term effectiveness because antibodies have a limited period of function and, ultimately, the recipient's body disposes of them. Passive immunity can also be natural or artificial in origin.
- **Natural immunity** encompasses any immunity that is acquired during the normal biological experiences of an individual rather than through medical intervention.
- **Artificial immunity** is protection from infection obtained through medical procedures. This type of immunity is induced by immunization with vaccines or the administration of immune serum.

Table 16.4 illustrates the various possible combinations of acquired immunities. An outline summarizing the system of host defenses was presented in figure 15.1. You may want to use this resource to review major aspects of immunity.

Artificial Passive Immunization: Immunotherapy

The first attempts at passive immunization involved the transfusion of horse serum containing antitoxins to prevent tetanus and to treat patients exposed to diphtheria. Since then, antisera from animals have been replaced with products of human origin that function with various degrees of specificity. Intravenous immune globulin (IVIG), sometimes called *gamma globulin,* contains immunoglobulin extracted from the pooled blood of human donors. The method of processing IVIG concentrates the antibodies to increase potency and eliminates potential pathogens (such as the hepatitis B and HIV viruses). It is a treatment of choice in preventing measles and hepatitis A after a known exposure and in replacing antibodies in immunodeficient patients. Most forms of IVIG are injected intramuscularly to minimize adverse reactions, and the protection it provides lasts 2 to 3 months.

A preparation called specific immune globulin (SIG) is derived from a more defined group of donors. Companies that prepare SIG obtain serum from patients who are convalescing and in a hyperimmune state after such infections as pertussis, tetanus, chickenpox, and hepatitis B. These globulins are preferable to IVIG because they contain higher titers of specific antibodies obtained from a smaller pool of patients.

Disease Connection

The famous Iditarod sled dog race from Anchorage to Nome is run as a commemoration of a heroic trek made in 1925. At that time, 20 mushers and 100 dogs ran in relay fashion to deliver desperately needed IVIG to children in Nome, who were suffering from a diphtheria outbreak.

During the COVID-19 pandemic, a treatment was prepared using a laboratory copy of a "good" antibody to the virus found in a recovered patient. The method of preparation produced what is called a monoclonal antibody. It is a passive immunotherapy because it does not lead to immunity in the person receiving it. Although donated immunities only last a relatively short time, they act immediately and can protect patients for whom no other useful medication or vaccine exists.

Table 16.4 The Four Types of Acquired Immunity

Natural Immunity is acquired through the normal life experiences of a human and is not induced through medical means.

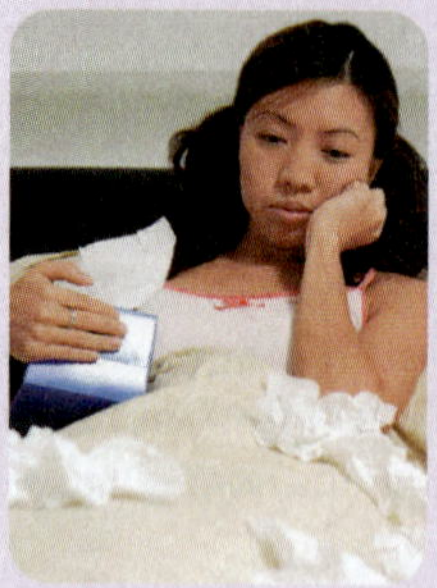

Floresco Productions/Corbis/Getty Images

Active

After recovering from infectious disease, a person will generally be actively resistant to reinfection for a period that varies according to the disease. In the case of childhood viral infections such as measles, mumps, and rubella, this natural active stimulus provides nearly lifelong immunity. Other diseases result in a less extended immunity of a few months to years (such as pneumococcal pneumonia and shigellosis), and reinfection is possible. Even a subclinical infection can stimulate natural active immunity. This probably accounts for the fact that some people are immune to an infectious agent without ever having been noticeably infected with or vaccinated for it.

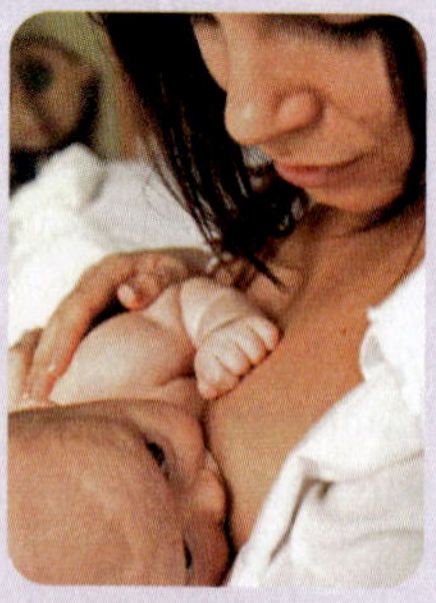

Ingram Publishing/Superstock

Passive

Natural, passively acquired immunity occurs only as a result of the prenatal and postnatal mother–child relationship. During fetal life, IgG antibodies circulating in the maternal bloodstream are small enough to pass or be actively transported across the placenta. This natural mechanism provides an infant with a mixture of many maternal antibodies that can protect it for the first few critical months outside the womb, while its own immune system is gradually developing active immunity. Depending on the microbe, passive protection lasts anywhere from a few months to a year.

Another source of natural passive immunity comes to the baby by way of the mother's milk. Although the human infant acquires 99% of natural passive immunity *in utero* and only about 1% through nursing, the milk-borne antibodies provide a special type of intestinal protection that is not available from transplacental antibodies.

Artificial Immunity is that produced purposefully through medical procedures.

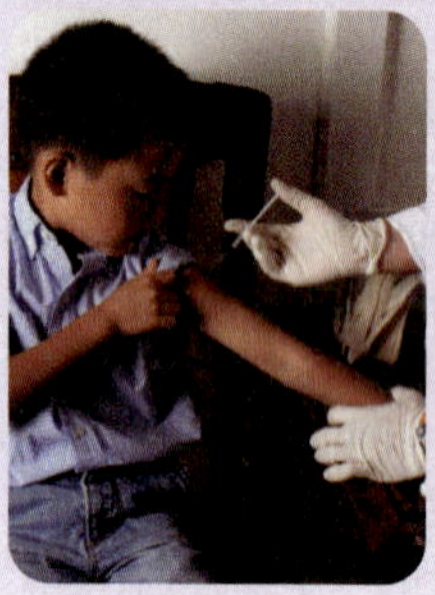

Jill Braaten/McGraw Hill

Active

Vaccination exposes a person to a specially prepared microbial (antigenic) stimulus, in a form that does not cause the disease. This then triggers the immune system to produce antibodies and memory lymphocytes to protect the person upon future exposure to that microbe. As with natural active immunity, the degree and length of protection vary.

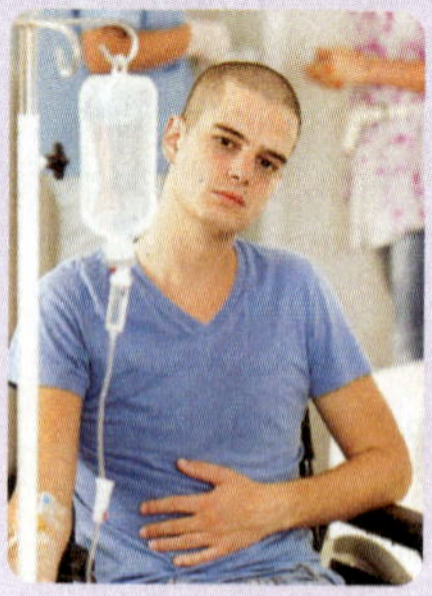

Miodrag Gajic/vgajic/E+/Getty Images

Passive

Passive immunotherapy involves a preparation that contains specific antibodies against a particular infectious agent. Pooled human serum from donor blood (gamma globulin) and immune serum globulins containing high quantities of antibodies are frequently used.

Artificial Active Immunity: Vaccination

Active immunity can be conferred artificially by **vaccination**—exposing a person to material that is antigenic but not pathogenic. The discovery of vaccination was one of the farthest reaching and most important developments in medical science. The basic principle behind vaccination is to stimulate a primary response that primes the immune system for future exposure to a virulent pathogen. If the actual pathogen later enters the body, the immune response—because it will be a secondary response—will be immediate, powerful, and sustained.

Principles of Vaccine Preparation

Factors that are considered in the design of a vaccine are antigen selection, effectiveness, ease in administration, safety, and cost. In natural immunity, an infectious agent stimulates a relatively long-term protective response. In artificial active immunity, the objective is to obtain this same response with a modified version of the microbe or its components. Qualities of an effective vaccine are listed in **table 16.5.** Vaccine preparations can be broadly categorized as either whole-organism or part-of-organism preparations. These categories also have subcategories:

Table 16.5 Checklist of Requirements for an Effective Vaccine

- It should have a low level of adverse side effects or toxicity and not cause serious harm.
- It should protect against infection with natural, wild forms of pathogen.
- It should stimulate both antibody (B-cell) response and cell-mediated (T-cell) response.
- It should have long-term, lasting effects (produce lasting memory).
- It should not require numerous doses or boosters.
- It should be inexpensive, have a relatively long shelf life, and be easy to administer.

1. Whole cells or viruses
   a. Live, attenuated cells or viruses.
   b. Killed cells or inactivated viruses
   c. Unrelated viruses (called viral vectors) that carry RNA or DNA for a surface component of a microbe.

2. Vaccines that deliver only an antigenic component of the microbe in question
   a. Microbial surface structures that are either isolated from cultures of cells or viruses, or structures that are produced using recombinant DNA technology.
   b. Vaccines containing DNA or mRNA that code for microbial surface components. These nucleic acids are enclosed in a lipid vesicle.
   c. Surface subunits conjugated with proteins (often from other microbes) to make them more immunogenic—called **conjugated vaccines.**

These categories are also shown in **table 16.6.**

In general, whole-organism vaccines may be more potent and long-lasting, for reasons you can appreciate now. First of all, if it is a whole killed cell or inactivated virus vaccine, it contains a wide range of surface markers that can activate the innate immune response as well as the adaptive immune response. On the other hand, killed or inactivated vaccines do not multiply inside the recipient. If it is a live, attenuated virus in the vaccine, it can continue to replicate in the vaccine recipient and, therefore, continue to stimulate strong immunity.

For a similar reason, viral vector vaccines are expected to be stronger stimulants of immunity. They wrap the nucleic acid in a real virus envelope, and therefore can stimulate innate immunity as well.

One special kind of vaccine is the toxoid vaccine. In this situation, it is an exotoxin of the microbe that is most dangerous in infection. To neutralize that toxin quickly, vaccines containing a nonharmful form of the toxin—termed a toxoid—are administered. It is a special case of an antigenic component vaccine (**2a**).

Booster Shots

The ideal situation, which is only sometimes achieved, is when a vaccine provides complete protection from a particular disease for a lifetime. But after a vaccine have been given to a large number of people, scientists can determine whether the vaccine needs to be "boosted"—which means another dose needs to be administered to shore up its long-term effectiveness. So we know that tetanus vaccines should be boosted every 10 years or so. The measles vaccine, given in infancy, calls for a booster at age 5 or 6. During the COVID-19 pandemic, we learned that the original vaccines should be followed by booster shots for maximum immunity. This raised an ethical question for wealthy countries, when there was a shortage of vaccines in less-wealthy nations. Should the precious vaccine doses be used on those who already were fully vaccinated, or made available to countries where a very low percentage of people had received the first vaccine regimen?

COVID-19 Vaccines

Usually textbooks try not to address current events, for obvious reasons. For example, we are writing this in 2022, but you may be reading it in 2024. But the COVID-19 pandemic has brought many infectious disease issues to light, including the introduction of a new kind of vaccine, an mRNA vaccine.

Three vaccines to be approved for use in the United States (and a fourth approved in Canada) are all vaccines that deliver a surface component to the recipient. Two of them are mRNA vaccines, meaning that mRNA that codes for the S (spike) protein of SARS-CoV-2 is enclosed in a lipid vesicle (not a cell or a virus). (This is **2b** on the list above.) The mRNA-in-a-vesicle enters human cells and uses our protein translation machinery to create many copies of the S protein, which is then presented on our cells and becomes visible to our immune systems. The immune system is trained to recognize that S protein, and from then on will react against the S protein on actual SARS-CoV-2 viruses it encounters. (See figure 6.17 in chapter 6.) The technology has been available for many years; it had just not yet been made into a licensed vaccine.

Other vaccines are viral vector vaccines. In the case of COVID-19, these vaccines consist of a non-disease-causing adenovirus that contains DNA that codes for the spike protein of SARS-CoV-2. After vaccination, the adenovirus invades human cells (harmlessly) and delivers the DNA to the nucleus of human cells. There the DNA is transcribed into mRNA, which is released into the cytoplasm and translated into the S protein. At that point, the S protein is transported to cell surfaces to alert the immune system much like the mRNA vaccines.

Vaccine Technologies in the Pipeline

Despite considerable successes, dozens of bacterial, viral, protozoan, and fungal diseases still remain without a functional vaccine. At the present time, no reliable vaccines are available for HIV/AIDS, various diarrheal diseases, respiratory diseases, and worm infections that affect over 200 million people per year worldwide. Worse than that, most existing vaccines are out of reach for much of the world's population.

New vaccine development is an active area of research. Some of the newer strategies are presented here. Almost all of the new strategies involve genetic engineering techniques. These capabilities have quickly revitalized the quest for improved vaccination.

One promising technology is the use of nanoparticles made of surface proteins of the virus of interest. If the proteins are manufactured inside eukaryotic cells, post-translational modifications can be successfully made so that the protein looks just like it would look in nature. In nanoparticle vaccines, the fully formed surface protein *is* the vaccine, and doesn't require the human cells it is injected into to do the translating or modification.

Another promising strategy is the use of gene editing techniques (CRISPR) to engineer our own B cells to express the antibodies we design.

Vaccination strategies are now under intense investigation for the prevention and treatment of noninfectious diseases. One of the best examples is vaccination for Alzheimer's disease. Researchers are examining whether administering a peptide that is found in the brain plaques of patients with Alzheimer's as a vaccine can lead to prevention of the disease. The principle is the same as with vaccines against microbial infection. If T and B memory cells that recognize these disease-inducing proteins can be created in a host via vaccination, once the proteins begin to form in the patient, the memory response eliminates them before they cause damage.

Table 16.6 Types of Vaccines

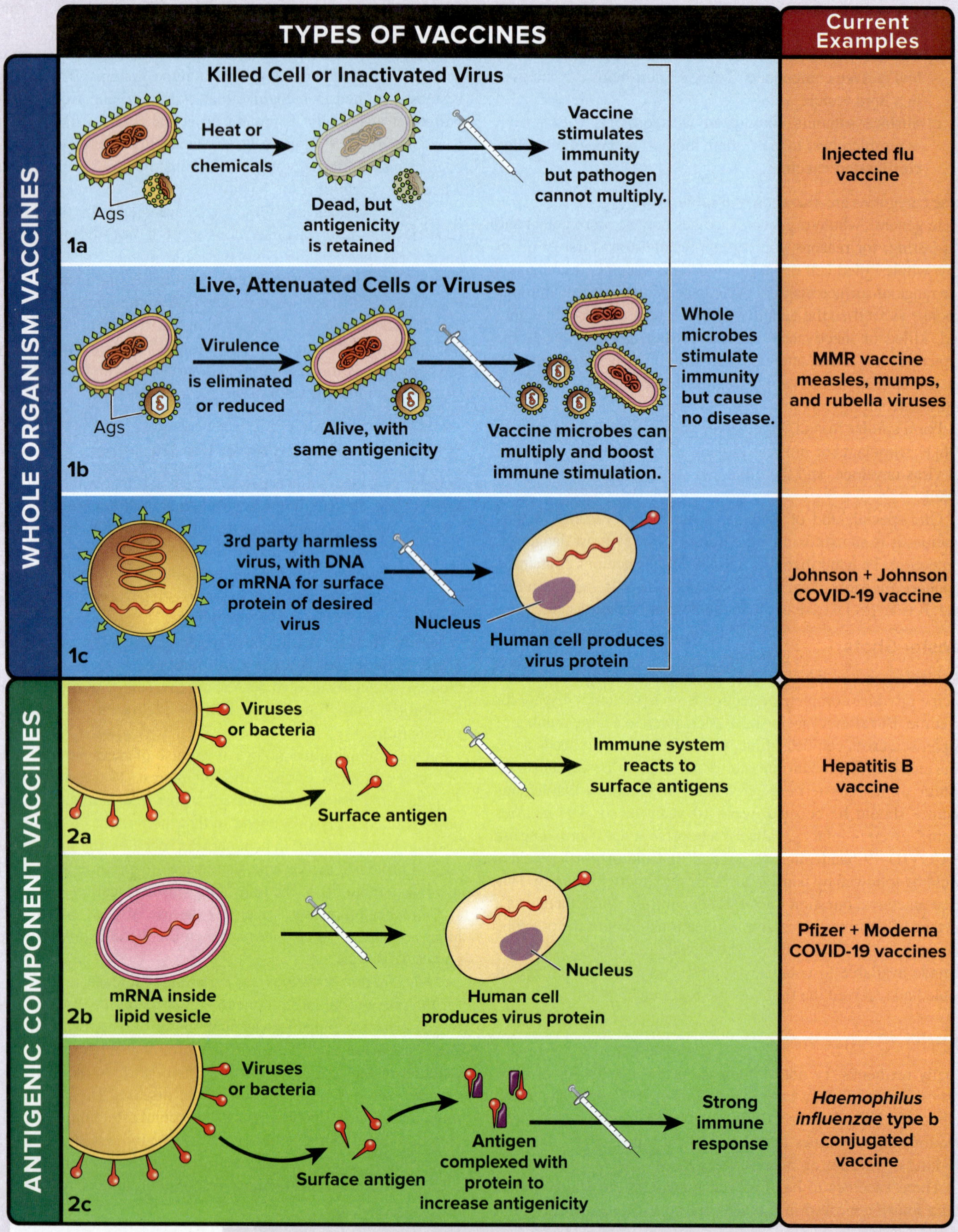
TYPES OF VACCINES
Current Examples
WHOLE ORGANISM VACCINES
Killed Cell or Inactivated Virus
Heat or chemicals
Ags
Dead, but antigenicity is retained
Vaccine stimulates immunity but pathogen cannot multiply.
1a
Injected flu vaccine
Live, Attenuated Cells or Viruses
Virulence is eliminated or reduced
Ags
Alive, with same antigenicity
Vaccine microbes can multiply and boost immune stimulation.
1b
MMR vaccine measles, mumps, and rubella viruses
3rd party harmless virus, with DNA or mRNA for surface protein of desired virus
Nucleus
Human cell produces virus protein
1c
Whole microbes stimulate immunity but cause no disease.
Johnson + Johnson COVID-19 vaccine
ANTIGENIC COMPONENT VACCINES
Viruses or bacteria
Surface antigen
Immune system reacts to surface antigens
2a
Hepatitis B vaccine
mRNA inside lipid vesicle
Nucleus
Human cell produces virus protein
2b
Pfizer + Moderna COVID-19 vaccines
Viruses or bacteria
Surface antigen
Antigen complexed with protein to increase antigenicity
Strong immune response
2c
Haemophilus influenzae type b conjugated vaccine

Route of Administration and Side Effects of Vaccines

Most vaccines are injected by subcutaneous, intramuscular, or intradermal routes. One (currently unused) form of the influenza vaccine comes in the form of a nasal spray. Oral (or nasal) vaccines are available for only a few diseases, but they have some distinct advantages. An oral or nasal dose of a vaccine can stimulate protection (IgA) on the mucous membrane of the portal of entry. Oral and nasal vaccines are also easier to give than are injections, are more readily accepted, and are well tolerated.

Some vaccines require the addition of a binding substance, or **adjuvant** (ad′-joo-vunt). An adjuvant is any compound that enhances immunogenicity and prolongs antigen retention at the injection site. The adjuvant precipitates the antigen and holds it in the tissues so that it will be released gradually. Its gradual release presumably facilitates contact with antigen-presenting cells and lymphocytes. This helps involve the innate immune system as well.

Disease Connection

During the COVID-19 pandemic, the health inequities in high-income vs. low-income countries were made abundantly clear. For example, as of October 2021, 46% of the world population had received at least one dose of a COVID-19 vaccine. Only 2.3% of people in low-income countries had received at least one dose. Of course, the way these numbers work is that the 46% number includes everyone in the world, and so it contains the rates for low-income countries as well as high-income countries, bringing the number down. But consider this: The top 5 countries on the list had from 80% to 94% of their citizens with at least one shot. The United States, which was number 18 on the ranked list of percent vaccinated, had 64% with at least one shot. This illustrates inequities globally, but the inequities within the United States are also quite large, based on socioeconomic and geographic factors.

All vaccines carry the risk of side effects. The vast majority of side effects are mild: soreness at the injection site, and possible short-term flu-like symptoms (since many of these symptoms are caused by the body's immune response, such as the release of interferons, rather than the microbe itself.) Serious side effects are very rare. The most common serious side effect is the risk of anaphylaxis. COVID-19 vaccines were seen to result in anaphylaxis in about five cases out of a million. This risk is the reason that you are asked to stay at the vaccination site for 15 minutes before leaving. On the other hand, by mid-2022, COVID-19 vaccination is thought to have saved approximately 20 million lives worldwide. Scientists also estimate that one-third of the over 1 million COVID-19 US deaths (as of mid-2022) could have been prevented by vaccination.

The World Health Organization named vaccine hesitancy as one of the top 10 threats to global health in 2019. The biggest cause of vaccine hesitancy seems to be vaccine misinformation. Scientists at Quinnipiac University recently conducted a study with 17,229 participants, with almost half of them identifying as health care students or health care workers. They found an alarming incidence of misconceptions, even among this population. For example, in the non–health care population, only 29% of respondents correctly agreed with this statement: *You cannot get the flu from the flu vaccine*. Among health care students or workers, only 49% agreed with that correct statement. In another example, 36% of all study participants agreed with or were unsure of this statement: *There is evidence that the mercury in the measles vaccine can cause autism*. In fact, the measles vaccine has never contained mercury and has been shown beyond a doubt that it does not cause autism.

After the first COVID-19 vaccines were approved, the amount of vaccine misinformation grew exponentially. Experience with now billions of administered doses has shown that it is one of the safest vaccines on the market.

Vaccinating: Who and When?

As you read earlier in the chapter, vaccination has traditionally been most common in childhood, for various reasons. Many of the diseases the vaccines protect against are most prominent and dangerous in childhood. The schedule for the most intense period of vaccinations, for children 0–15 months, is presented in **Table 16.7.** Some vaccines are mixtures of antigens from several pathogens, notably Pediarix (DTaP, IPV, and HB). With advanced understanding of disease control, it has become apparent to public health officials that vaccination of adults is often needed in order to boost an older immunization, protect against "adult" infections (such as pneumonia in elderly people), or provide special protection in people with certain medical conditions. Older child and adult vaccine schedules can be found at the CDC website.

16.7 Learning Outcomes—Assess Your Progress

21. List the four categories of acquired immunity, and provide examples of each.
22. Discuss the qualities of an effective vaccine.
23. List several types of vaccines, and discuss how they are utilized today.

Table 16.7 Recommended Immunization Schedules, Ages 0–15 Months, United States, 2020

Birth to 15 Months

Legend

- Range of recommended ages for all children
- Range of recommended ages for catch-up vaccination
- Range of recommended ages for certain high-risk groups

Vaccine	Birth	1 mo	2 mos	4 mos	6 mos	9 mos	12 mos	15 mos
Hepatitis B (HepB)	1st dose	←2nd dose→			←3rd dose→			
Rotavirus (RV) RV1 (2-dose series); RV5 (3-dose series)			1st dose	2nd dose				
Diphtheria, tetanus, & acellular pertussis (DTaP: <7 yrs)			1st dose	2nd dose	3rd dose			←4th dose→
Haemophilus influenzae type b (Hib)			1st dose	2nd dose			←3rd or 4th dose→	
Pneumococcal conjugate (PCV13)			1st dose	2nd dose	3rd dose		←4th dose→	
Inactivated poliovirus (IPV: <18 yrs)			1st dose	2nd dose	←3rd dose→			
Influenza (IIV4)					Annual vaccination 1 or 2 doses			
or Influenza (LAIV4)								
Measles, mumps, rubella (MMR)							←1st dose→	
Varicella (VAR)							←1st dose→	
Hepatitis A (HepA)							←2nd dose series→	
Tetanus, diphtheria, & acellular pertussis (Tdap: ≥7 yrs)								
Human papillomavirus (HPV)								
Meningococcal (MenACWY-D ≥9 mos, MenACWY-CRM ≥2 mos, MenACWY-TT ≥2years)								
Meningococcal B (MenB)								
Pneumococcal polysaccharide (PPSV23)								

Source: https://www.cdc.gov/vaccines/schedules/hcp/imz/child-adolescent.html

Media Under The Microscope Wrap-Up

Chuck Nacke/Alamy Stock Photo

The article from the *Galveston County Daily News* examines the question of whether recovering from COVID-19 provides enough immunity against being infected again, or whether a vaccine is still called for.

The **intended message** is to present evidence and expert opinion that vaccination is still preferred to natural immunity.

A **critical reading** of this article reveals that, while it ends up recommending vaccination, it does present contradictory data (higher death rates once infected among the vaccinated vs. the naturally immune). Presenting contradictory findings is a good indication of the integrity of the reporting. The article also quotes an expert and summarizes the collective findings of the public health community.

To **interpret** this article to friends, I would explain to them briefly about how adaptive immunity works and that both circumstances can lead to it (natural infection and "artificial" vaccination). Then the best approach would be to explain how sometimes one is better than the other, but it takes time to study the disease and the immune response to establish that. At this point in the pandemic, the expert opinion was that vaccination is the safer bet.

My **overall grade** for the article is an A. I didn't know what to expect with a local news agency's reporting, but it checked all the boxes for thorough and balanced reporting.

Source: Galveston County Daily News, "Although Natural Immunity Exists, Health Experts Say Inoculation Is Safer." Online article posted October 4, 2021. https://www.galvnews.com/news/free/article_eb41ff66-655d-5f7b-8921-2cf9ebb2117e.html

Study Smarter: Better Together

These activities are designed for you to use on your own with a study group—either a face-to-face group or a virtual one, consisting of 3–5 members. Studying together can be very helpful, but there are effective and ineffective ways to do it. For example, getting together without a clear structure is often not a good use of your time. Use your time efficiently by using one or more of the exercises below.

FACE-TO-FACE GROUPS

Use one or more of the activities below.

Peer Instruction: Assign numbers to your group members to use all semester long. Now look at these five concepts from this chapter. Each group member prepares a 5-minute lesson on the topic corresponding to their number. Don't worry if you have fewer than 5 members; just use however many you have! During your group study time, each member presents their lesson, and the group spends another 5–10 minutes discussing that lesson.

1. Clonal selection and expansion
2. Results of T-cell activation
3. Results of B-cell activation
4. Actions of antibodies
5. The memory response

Concept Maps: Each member of the group should use this list of terms from this chapter to generate their own concept map. This can be hand-drawn or created using software (see Appendix C for guidelines). During group study time, compare each other's concept maps and help each other make sure they are correct. Of course, there are many different "correct" maps. Examining each member's map will help you talk through the varied concepts and how they are related.

Concept Terms:

adaptive immunity	antibody secretion	inflammation
innate immunity	artificial immunity	natural immunity
vaccines	memory	activated T cells

(*continued*)

Table Topics: Each group member should identify a concept or topic from this week's class assignments with which they are having trouble and share it during group study time. The other group members can then help to clarify confusing issues or share how they figured it out. Aim for a maximum of 15 minutes per topic. If the topic remains unclear to the group, bring it up during class or use the instructor's office hours or e-mail to ask for help. Taking the time to struggle with a difficult concept first makes your questions much more specific and more likely to yield helpful answers.

VIRTUAL GROUPS

Not everyone has the time or opportunity to meet with group members outside of class time. You or your instructor can create a virtual group using e-mail or the course software.

Weekly Discussion Board: This forum can be used as a way for groups to discuss topics, via e-mail, or other learning management systems or online platforms, before they are covered in class. As each member of the group answers the current week's question, they should send their responses to every other member of their group. It's best to agree on a deadline based on how your class schedule works (Saturday for the next week's topics, for example). Then, after the topic is discussed in class, each member should send a response that all group members will see with a follow-up post on the same topic. If you cover more than one chapter in a week, someone can be designated to choose which chapter Discussion Board question you will use. Or simply decide up front that you will always use the first-chapter-of-the week's question, to keep the schedule simple.

Discussion Question

Theoretically, can humans develop adaptive immunity to a novel biological agent that arrives from outer space (say, on the surface of a space vehicle) that no human has ever encountered before? Explain your answer.

Chapter Summary

HOST DEFENSES PART II:
ADAPTIVE IMMUNITY AND IMMUNIZATION

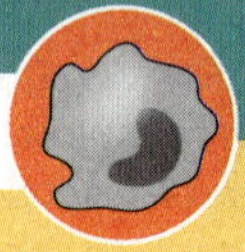

16.1 ADAPTIVE IMMUNITY: THE THIRD LINE OF DEFENSE

- **Adaptive immunity consists of four stages:**
  - **I. Lymphocytes differentiate into B cells and T cells.**
  - **II. Antigen-presenting cells detect invading pathogens and present these antigens to lymphocytes, which recognize the antigen and initiate the immune response.**
  - **III. Clones of lymphocytes respond to the antigen.**
  - **IV. Activated T cells participate directly in the response. Activated B cells release antibodies. Memory B and T cells are created for use in later exposures.**

16.2 STEP I: THE DEVELOPMENT OF LYMPHOCYTE DIVERSITY

- **During development, both B and T cells develop millions of genetically different clones.**
- **This level of receptor diversity is accomplished through reassortment of the DNA making up antigen receptor genes.**
- **Those that respond to self-antigens are mainly deleted (clonal deletion) during fetal development.**
- **Throughout life, an immune challenge that results in the binding of antigen to a particular clone is called clonal selection. That clone is exclusively amplified in the process of clonal expansion, which leads to an army of cells with that individual specificity.**

16.3 STEP II: PRESENTATION OF ANTIGENS

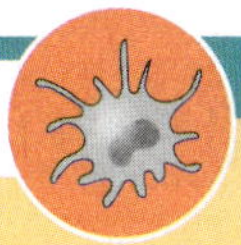

- Antigens are molecules that can be seen and identified by immune system cells. Sometimes they are called markers. In this context, if they do activate cells of the adaptive immune response, they are called immunogens as well.
- Antigen must be formally presented to lymphocytes by antigen-presenting cells (APCs) in most immune reactions.
- Antigen-presenting cells engulf and process foreign antigen and bind the epitope to MHC class II molecules on their surfaces for presentation to CD4 T-lymphocytes.

16.4 STEP III: ANTIGENIC CHALLENGE OF T CELLS AND B CELLS

- Physical contact between the APC, T cells, and B cells activates the lymphocytes to proceed with their responsive immune responses.
- Activated T cells differentiate into T memory cells, T helper cells, T regulatory cells, T cytotoxic cells, and NKT cells.
- B cells differentiate into memory B cells, regulatory B cells, and plasma cells. Plasma cells produce copious amounts of antibody.
- Both T cells and B cells, as well as antigen-presenting cells, produce cytokines of various types that either amplify or decrease the immune response.

16.5 STEP IV (1): THE T-CELL RESPONSE

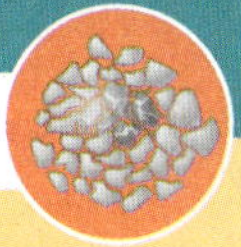

- T helper cells release cytokines that stimulate macrophages and B cells, among other functions.
- Regulatory T cells guard against excessive or inappropriate inflammation and immunity.
- Cytotoxic T cells induce apoptosis in target cells through the action of perforin and granzymes.

16.6 STEP IV (2): THE B-CELL RESPONSE

- B cells produce five classes of antibody: IgM, IgG, IgA, IgD, and IgE. IgM and IgG predominate in plasma. IgA predominates in body secretions. IgD is expressed on B cells as an antigen receptor. IgE binds to mast cells and basophils in tissues, promoting inflammation.
- Antibodies bind physically to the specific antigen that stimulates their production, thereby immobilizing the antigen and enabling it to be destroyed by other components of the immune system.

16.7 ADAPTIVE IMMUNITY AND VACCINATION

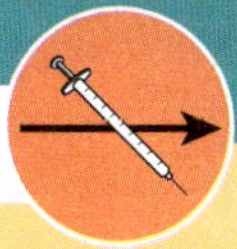

- Active immunity means that your body produces antibodies to a disease agent. If you contract the disease, you can develop natural active immunity. If you are vaccinated, your body will produce artificial active immunity.
- In passive immunity, you receive antibodies from another person or antibodies made in the laboratory. This can happen naturally as in breastfeeding or artificially through passive immunization.
- Vaccines are artificial active agents that provoke a protective immune response in the recipient, but they do not cause actual disease. Vaccination is the process of challenging the immune system with a specially selected antigen.
- Vaccines can be whole organisms that are either live attenuated or killed/inactivated. They can also consist of pieces of the organism that are either harvested from the microbe or created through recombinant DNA techniques. The antigenic pieces may be administered by themselves or packaged in a lipid vesicle or an unrelated viral vector.
- New vaccine types deployed during the COVID-19 pandemic were packaged mRNA coding for the spike protein of the virus in a lipid vesicle.

SmartGrid: From Knowledge to Critical Thinking

This *21 Question Grid* takes the topics from this chapter and arranges them with respect to the American Society for Microbiology's Undergraduate Curriculum guidelines—all six of the important "Concepts" as well as the important "Competency" of scientific literacy. Three questions are supplied, which cover chapter content referring to the Concept or Competency in increasing levels of Bloom's taxonomy for learning.

ASM Concept/ Competency	A. Bloom's Level 1, 2—Remember and Understand (Choose one)	B. Bloom's Level 3, 4—Apply and Analyze	C. Bloom's Level 5, 6—Evaluate and Create
Evolution	**1.** A single bacterium has ________ epitope(s). **a.** a specific **b.** multiple **c.** MHC **d.** clonal	**2.** Would you suspect that the innate or the adaptive arms of the immune response evolved first in mammals? Explain your answer.	**3.** Provide an explanation to refute the following statement: Humans cannot develop adaptive immunity to a novel biological agent created in a laboratory.
Cell Structure and Function	**4.** The primary B-cell receptor is **a.** IgD. **b.** IgA. **c.** IgE. **d.** IgG.	**5.** Name three antigen-presenting cells, and what other functions each of them has.	**6.** Major histocompatibility molecules are critical for adaptive immunity. But they are named after another major function they have. How does histocompatibility work?
Metabolic Pathways	**7.** In humans, B cells mature in the ____ and T cells mature in the ____. **a.** GALT; liver **b.** bursa; thymus **c.** bone marrow; thymus **d.** lymph nodes; spleen	**8.** Explain how the memory response is the cornerstone of vaccination.	**9.** Conduct research on clonal deletion and write a paragraph that explains it to a typical high school biology student.
Information Flow and Genetics	**10.** Which of the following cells is capable of specifically responding to a nearly infinite number of epitopes? **a.** B and T cells **b.** Plasma cells **c.** T cytotoxic cells **d.** All of these	**11.** Is antibody diversity generated at the DNA or RNA level? Explain.	**12.** In order for gene rearrangement of antigen receptors to be successful, the DNA sequence needs to remain in-frame after the rearrangement occurs. Explain what "in-frame" means and why it would be necessary in the generation of diversity in antigen receptors.
Microbial Systems	**13.** Some microbial products can activate B cells without the assistance of T cells. Which of the following can do this? **a.** Capsule of *S. pneumoniae* **b.** Lipopolysaccharide **c.** Some viral capsids **d.** All of these	**14.** Explain why it is useful that antibodies have two antigen-binding arms.	**15.** Using the details of T-cell activation, suggest a reason why T cells are so critical in recognizing and destroying virally infected host cells.
Impact of Microorganisms	**16.** A vaccine that contains parts of viruses is called **a.** acellular. **b.** recombinant. **c.** subunit. **d.** attenuated.	**17.** Why do you think adenoviruses are often chosen to be the viral vectors for vaccines?	**18.** Conduct research on tetanus and discuss how a toxoid is the best vaccine against it.

ASM Concept/ Competency	A. Bloom's Level 1, 2—Remember and Understand (Choose one)	B. Bloom's Level 3, 4—Apply and Analyze	C. Bloom's Level 5, 6—Evaluate and Create
Scientific Thinking	**19.** If you draw a blood sample from a patient to determine whether he or she has a herpes simplex infection, and the patient displays a large amount of IgG against the virus but low levels of IgM, what do you conclude? **a.** The patient is newly infected. **b.** The patient has had the infection for a while. **c.** The patient is not infected. **d.** It is impossible to draw conclusions.	**20.** Chronic lymphocytic leukemia leads to the production of cancerous B cells, and treatment often involves bone marrow transplantation. Based upon your knowledge of lymphocyte development, explain how this procedure can lead to therapeutic effects.	**21.** In 2021 the director of the World Health Organization issued an angry statement about developed countries giving booster shots to their citizens who were already fully vaccinated against COVID-19, when much of the world had not been vaccinated due to lack of vaccine. What are your thoughts on this?

Answers to the multiple-choice questions appear in Appendix A.

Visual Connections

This question uses visual images to connect content within and between chapters.

1. **From this chapter, figure 16.18.** In this figure describing primary and secondary responses to antigen, indicate where a vaccination may be most effective, and indicate where natural infection would play a role.

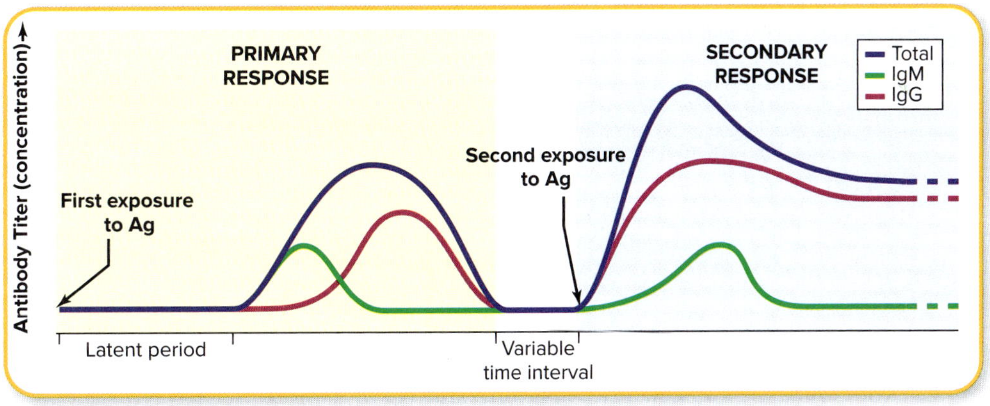
PRIMARY RESPONSE
SECONDARY RESPONSE
Total
IgM
IgG
Antibody Titer (concentration)
First exposure to Ag
Second exposure to Ag
Latent period
Variable time interval

Upon the first exposure to an antigen, the system undergoes a **primary response.** The earliest part of this response, the *latent period,* is marked by a lack of antibodies for that antigen, but much activity is occurring. During this time, the antigen is being concentrated in lymphoid tissue and is being processed by the correct clones of B lymphocytes. As plasma cells synthesize antibodies, the serum titer increases to a certain plateau and then tapers off to a low level over a few weeks or months. Early in the primary response, most of the antibodies are the IgM type, which is the first class to be secreted by plasma cells. Later, the class of the antibodies (but not their specificity) is switched to IgG or some other class (IgA or IgE).

After the initial response, there is no activity, but memory cells of the same specificity are located throughout the lymphatic system.

When the immune system is exposed again to the same immunogen within weeks, months, or even years, a **secondary response** occurs. The rate of antibody synthesis, the peak titer, and the length of antibody persistence are greatly increased over the primary response. The speed and intensity seen in this response are attributable to the memory B cells that were formed during the primary response. The secondary response is also called the **anamnestic response.** The advantage of this response is evident: It provides a quick and potent strike against subsequent exposures to infectious agents.

High Impact Study

These terms and concepts are most critical for your understanding of this chapter—and may be the most difficult. Have you mastered them?

Concepts

- [ ] Clonal selection and expansion
- [ ] Result of T-cell activation
- [ ] Result of B-cell activation
- [ ] Genetic rearrangement leading to antigen receptor diversity
- [ ] Antigen-presenting cells
- [ ] Five main types of T cells
- [ ] Actions of T cytotoxic cells
- [ ] Antigen recognition by B cells
- [ ] Antigen recognition by T cells
- [ ] Basic immunoglobulin structure
- [ ] Five types of immunoglobulins
- [ ] Six actions of antibodies
- [ ] The memory response
- [ ] Two main types of vaccine preparations

Terms

- [ ] Antigen
- [ ] Immunogen
- [ ] Major histocompatibility complex
- [ ] CD molecules
- [ ] Natural active immunity
- [ ] Natural passive immunity
- [ ] Artificial active immunity
- [ ] Artificial passive immunity

17

Disorders in Immunity

Cavan Images

MEDIA UNDER THE MICROSCOPE

Allergic to the Cold

These case studies examine an article from the popular media to determine the extent to which it is factual and/or misleading. This case focuses on the 2020 Insider *article "A Man with an Allergy to Cold Air Almost Died after Stepping out of a Hot Shower."*

In this chapter we will learn about what can go wrong when the immune system responds inappropriately. One of the most frequent issues is what is commonly known as allergy.

This article from the digital news outlet *Insider* tells the story of a man who experienced the very worst possible consequence of allergy. Anaphylactic shock occurs within minutes of an exposure to some substance or, in this case, atmospheric condition and leads to hives, nausea, difficulty breathing, and shock. If not treated quickly, it can result in death. Some substances are familiar allergens, such as peanuts, shellfish, or bee stings. But this article describes a man, originally from Micronesia (in the Philippine Sea), who had moved to Colorado.

The article tells us that his usual reaction was to break out in hives when he was exposed to cold air. But one day in 2020, he stepped out of a hot shower into the cold air of his bathroom and his family found him on the bathroom floor, covered in hives and struggling to breathe. This is typical anaphylaxis. He was treated with epinephrine and transported to the ER, where he was stabilized. Doctors diagnosed him with allergy to cold, sometimes known as cold urticaria. In this case, the name of the disorder was not quite correct as urticaria refers specifically to the development of hives. A more accurate name would be cold anaphylaxis in this case.

The article mentions that though it is a rare condition, it is important for emergency room personnel to recognize because these patients can also have severe reactions to the infusion of cold IV fluids.

- What is the **intended message** of the article?
- What is your **critical reading** of the summary of the article? Remember that in this context, "critical reading" does not necessarily mean *What criticism do you have?* but asks you to apply your knowledge to interpret whether the article is factual and whether the facts support the intended message.

- How would you **interpret** the news item for your nonmicrobiologist friends?
- What is your **overall grade** for the news item—taking into account its accuracy and the accuracy of its intended effect?

Media Under The Microscope Wrap-Up appears at the end of the chapter.

Outline and Learning Outcomes

17.1 The Immune Response: A Two-Sided Coin

1. Define *immunopathology,* and describe the two major categories of immune dysfunction.
2. Identify the four major categories of hypersensitivity, or overreaction to antigens.

17.2 Type I Allergic Reactions: Atopy and Anaphylaxis

3. Summarize genetic and environmental factors that influence allergy development.
4. Outline the steps of a type I allergic response, and discuss the effects on target organs and tissue.
5. Identify three conditions caused by IgE-mediated allergic reactions.
6. Describe the symptoms of anaphylaxis, and link them to physiological events.
7. Briefly describe two methods for diagnosing allergies.
8. List the three main ways to prevent or short-circuit type I allergic reactions.

17.3 Type II Hypersensitivities: Reactions That Lyse Cells

9. List the three immune components causing cell lysis in type II hypersensitivity reactions.
10. Explain the molecular basis for the ABO blood groups, and identify the blood type of a "universal donor" and the blood type of a "universal recipient."
11. Explain the role of Rh factor in hemolytic disease development and how the disease is prevented in newborns.

17.4 Type III Hypersensitivities: Immune Complex Reactions

12. Identify commonalities and differences between type II and type III hypersensitivities.
13. Describe how both antibodies and neutrophils contribute to acute post-streptococcal glomerulonephritis.

17.5 Type IV Hypersensitivities: Cell-Mediated (Delayed) Reactions

14. Identify one type IV delayed hypersensitivity reaction, and describe the role of T cells in the pathogenesis of this condition.
15. List four classes of grafts, and explain how host versus graft and graft versus host diseases develop.

17.6 An Inappropriate Response Against Self: Autoimmunity

16. Outline at least three different explanations for the origin of autoimmunity.
17. List three autoimmune diseases, and describe immunologic features most important to each.

17.7 Immunodeficiency Diseases: Hyposensitivities

18. Distinguish between primary and secondary immunodeficiencies, explaining how each develops.
19. Define *severe combined immunodeficiency,* and discuss current therapeutic approaches to this type of disease.

17.1 The Immune Response: A Two-Sided Coin

Humans possess a powerful and intricate system of defense, which by its very nature also carries the potential to cause injury and disease. Sometimes these reactions are excessive and uncontrolled. In most instances, misguided immune function is expressed in commonplace but miserable symptoms such as hay fever and dermatitis. Abnormal or undesirable immune functions are also involved in debilitating or life-threatening diseases such as asthma, anaphylaxis, diabetes, rheumatoid arthritis, and graft rejection.

Our previous discussions of the immune response have focused on its beneficial effects. In this chapter, we look at **immunopathology,** the study of disease states associated with overreactivity or underreactivity of the immune response **(figure 17.1).** Overreactivity takes the forms of allergy, hypersensitivity, and autoimmunity. In immunodeficiency or **hyposensitivity diseases,** immune function is incompletely developed, suppressed, or destroyed. Cancer falls into a special category because it can be both a cause and an effect of immune dysfunction.

Hypersensitivity: Four Types

The most widely accepted classification of the four types of allergy and hypersensitivity includes four major categories: type I ("common" allergy and anaphylaxis), type II (IgG- and IgM-mediated cell damage), type III (immune complex formation), and type IV (delayed hypersensitivity) **(table 17.1).** In general, types I, II, and III involve a B-cell/immunoglobulin response, and type IV involves a T-cell response (see figure 17.1). The antigens that elicit these reactions can be exogenous or endogenous. Exogenous means they originate from outside the body (microbes, pollen grains, and foreign cells and proteins). Endogenous means the antigens come from self tissue (autoimmunities).

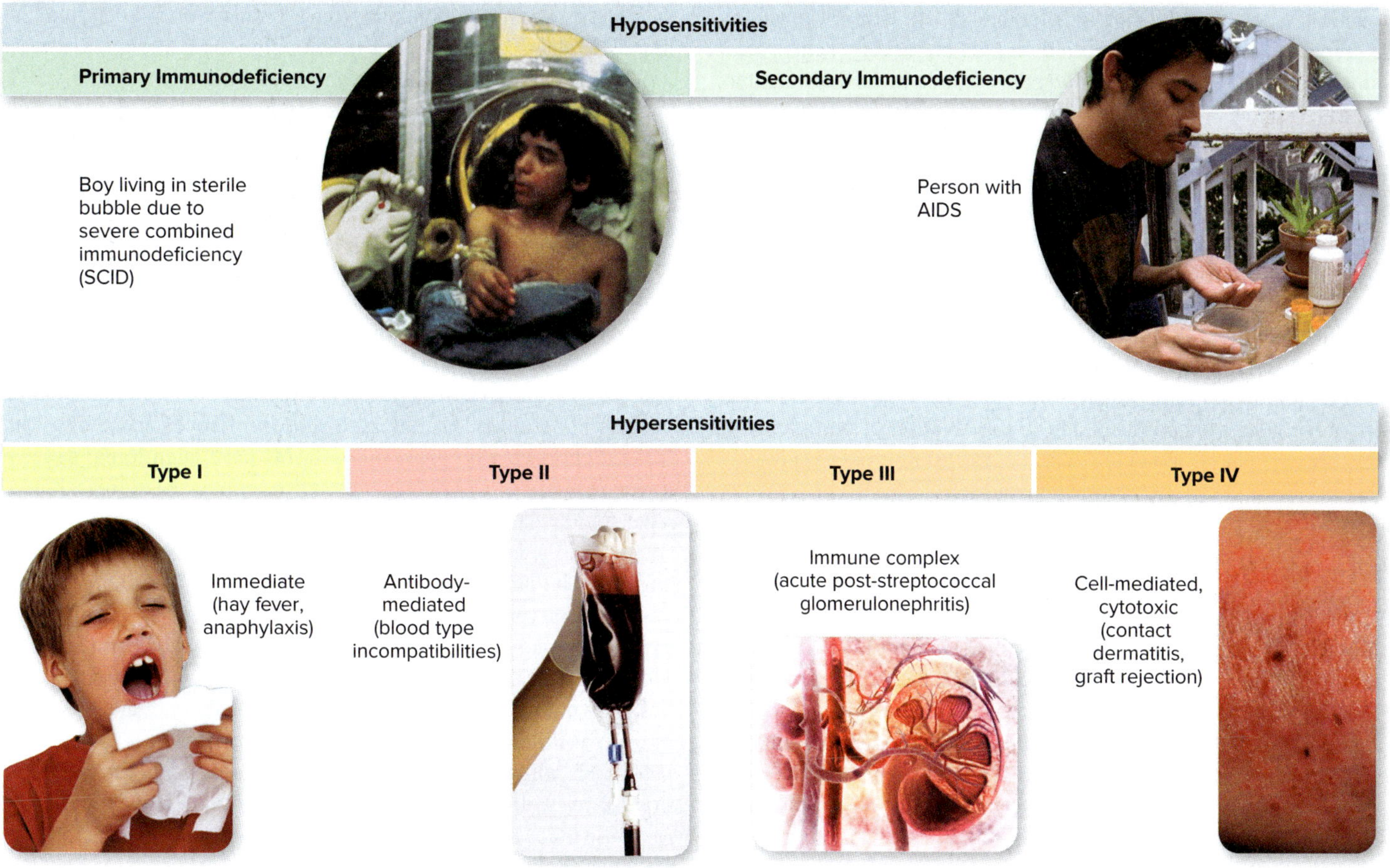
Hyposensitivities
Primary Immunodeficiency
Secondary Immunodeficiency
Boy living in sterile bubble due to severe combined immunodeficiency (SCID)
Person with AIDS
Hypersensitivities
Type I
Type II
Type III
Type IV
Immediate (hay fever, anaphylaxis)
Antibody-mediated (blood type incompatibilities)
Immune complex (acute post-streptococcal glomerulonephritis)
Cell-mediated, cytotoxic (contact dermatitis, graft rejection)

Figure 17.1 Overview of disorders of the immune system. Just as the system of T cells and B cells provides necessary protection against infection and disease, the same system can cause serious and debilitating conditions by overreacting or underreacting to antigens.

(a) Baylor College of Medicine, Public Affairs; (b) Christopher Kerrigan/McGraw Hill; (c) Pixtal/age fotostock; (d) Martin Barraud/age fotostock; (e) crystal light/Shutterstock; (f) blickwinkel/Alamy Stock Photo

Table 17.1 Hypersensitivity States

Type		Systems and Mechanisms Involved	Examples
I	Immediate hypersensitivity	IgE-mediated; involves mast cells, basophils, and allergic mediators	Anaphylaxis; allergies such as hay fever, asthma
II	Antibody-mediated	IgG, IgM antibodies plus complement act upon cells and cause cell lysis; includes some autoimmune diseases	Blood group incompatibility; pernicious anemia; myasthenia gravis
III	Immune complex–mediated	Antibody-mediated inflammation; circulating IgG complexes deposited in basement membranes of target organs; includes some autoimmune diseases	Systemic lupus erythematosus; rheumatoid arthritis; serum sickness; rheumatic fever
IV	T-cell-mediated	Delayed hypersensitivity and cytotoxic reactions in tissues; includes some autoimmune diseases	Infection reactions; contact dermatitis; graft rejection

17.1 Learning Outcomes—Assess Your Progress

1. Define *immunopathology,* and describe the two major categories of immune dysfunction.
2. Identify the four major categories of hypersensitivity, or overreaction to antigens.

17.2 Type I Allergic Reactions: Atopy and Anaphylaxis

The term **allergy** refers to an exaggerated immune response whose hallmark is inflammation. Although it is sometimes used interchangeably with hypersensitivity, most experts refer to immediate reactions such as hay fever as allergies and to delayed reactions as hypersensitivities. Allergic individuals are acutely sensitive to contact with antigens, called **allergens,** that do not affect nonallergic individuals. All type I allergies are immediate in onset and are associated with exposure to specific antigens. However, there are two levels of severity: **Atopy** is a chronic, local allergy such as hay fever, rashes or hives (called uticaria), or asthma. **Anaphylaxis** (an″-uh-fuh-lak′-sis) is a systemic—sometimes fatal—reaction that involves airway obstruction and circulatory collapse.

Although the general effects of hypersensitivity are detrimental, remember that it involves the very same types of immune reactions as those at work in normal protective immunities. These include humoral and cell-mediated actions, the inflammatory response, phagocytosis, and the activation of complement. This means that all humans have the potential to develop hypersensitivity under particular circumstances.

Who Is Affected, and How?

In the United States, nearly half of the population is affected by airborne allergens, such as dust, pollen, and mold. Particularly among children, the incidence of allergies has been increasing. The majority of type I allergies are relatively mild, but certain forms, such as asthma and anaphylaxis, may require hospitalization and can cause death. (**Insight 17.1** discusses asthma with respect to the microbiome.) In some individuals, atopic allergies last for a lifetime, others "outgrow" them, and still others only develop them later in life.

A predisposition to allergies seems to have a strong familial association. But what is inherited is a *generalized susceptibility,* not the allergy to a specific substance. For example, a parent who is allergic to ragweed pollen can have a child who is allergic to cat hair. The prospect of a child's developing atopic allergy is at least 25% if one parent exhibits symptoms, and it increases to nearly 50% if grandparents or siblings are also afflicted. The actual basis for atopy seems to be a genetic program that favors allergic antibody (IgE) production, increased reactivity of mast cells, and increased susceptibility of target tissue to allergic mediators.

The "hygiene hypothesis" provides one possible explanation for an environmental component to allergy development. This hypothesis suggests that the industrialized world has created a very hygienic environment, exemplified by antimicrobial products of all kinds and very well-insulated homes, and that this has been bad for our immune systems. It seems that our immune systems need to be "trained" by interaction with microbes as we develop. In fact, children who grow up on farms have been found to have lower incidences of several types of allergies. Also, researchers have found that the combination of being delivered by cesarean section and a maternal history of allergy elevates the risk that a child will be allergic to foods by a factor of eight. Scientists suggest that delivery by cesarean section keeps the baby from being exposed to vaginal and stool bacteria, which have been shown to jumpstart a healthy gut microbiome in the baby. Additional work has shown that babies need to be exposed to commensal bacteria in order for the IgA system to develop normally.

Might allergies reflect some beneficial evolutionary adaptation? Why would humans and other mammals evolve an allergic response that causes suffering, tissue damage, and even death? One possible explanation is that the components involved in an allergic response exist to defend against helminthic worms and other multicellular parasites in humans. It is only relatively recently in our evolutionary history that industrialized countries have seen

INSIGHT 17.1 MICROBIOME: Asthma and the Airway—and Gut—Microbiome

Asthma is considered a form of allergy, an inappropriate antibody response to epitopes that should not provoke a reaction. It affects more than 25 million people in the United States, 7 million of them children under the age of 18. Scientists are not sure what causes it. They have found factors associated with it, but these can be as varied as poor living conditions, exposure to roaches, and too *little* exposure to microbes.

Current research is showing that a disrupted microbiome in the airway is nearly always found in people who have asthma. This is surely not unexpected. But the gut microbiome also seems to be important. That gut microbiome is showing up everywhere! Scientists have been trying to tell us for years that the gut is almost like a second nervous system, and this asthma research bears that out.

Researchers looked at 319 babies in Canada and found a subgroup of 22 with a high risk of asthma. These babies were missing four types of bacteria that were present in healthy babies. The researchers concluded that the lack of these four bacterial groups put babies at a greatly increased risk for asthma. They have not yet teased out why the babies were missing those bacteria. They are investigating exposures such as cesarean birth, early antibiotic treatment, and breast-feeding.

Asthma prevalence in children by state (2019)

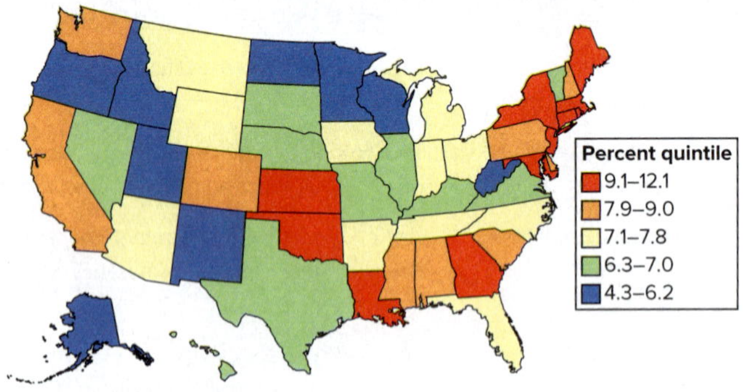
Percent quintile
9.1–12.1
7.9–9.0
7.1–7.8
6.3–7.0
4.3–6.2

CDC. Asthma Data Visualizations. Retrived: https://www.cdc.gov/asthma/data-visualizations/default.htm

Not only does the gut microbiome regulate and influence digestive functions, but it also is important in training the immune system to react appropriately throughout the body. Because the gut is the portal of entry for so many environmental microbes (through eating and drinking), it seems logical that it would be the "training center" for the body's defenses, saying: "Do not react to this. It is just a harmless bacterium coming in with this bite of carrot."

dramatically fewer infections with these parasites. One hypothesis is that the part of the immune system that fights helminthic worms is left idle in a population that has recently been "scrubbed" of these parasites, and that part starts acting inappropriately.

The Nature of Allergens and Their Portals of Entry

As with other antigens, allergens have certain immunogenic characteristics. Proteins are more allergenic than carbohydrates, fats, or nucleic acids. Some allergens are haptens, nonprotein substances with a molecular weight of less than 1,000, which can form complexes with carrier molecules in the body. Organic and inorganic chemicals found in industrial and household products, cosmetics, food, and drugs are commonly of this type. **Table 17.2** lists a number of common allergenic substances.

Allergens typically enter through epithelial portals in the respiratory tract, gastrointestinal tract, and skin **(figure 17.2).** The mucosal surfaces of the gut and respiratory system present a thin, moist surface that is fairly penetrable. The dry, tough keratin coating of skin is less permeable, but access still occurs through tiny breaks, glands, and hair follicles. It is worth noting that the organ where an allergy is expressed may or may not be the same as the portal of entry.

Allergens can be *inhalants, ingestants, injectants*, or *contactants.* Airborne environmental allergens such as pollen, house dust, dander (shed skin scales), or fungal spores are termed *inhalants* **(figure 17.2*a*).** Each geographic region harbors a particular combination of airborne substances that varies with the season and humidity. Pollen is given off seasonally by trees and other flowering plants, while mold spores are released throughout the year. Airborne animal hair and dander, feathers, and the saliva of dogs and cats are common sources of allergens. The component of house dust that appears to account for most dust allergies is not soil or other debris but the decomposed bodies and feces of tiny mites that commonly live in this dust.

Allergens that enter by mouth, called *ingestants,* often cause food allergies **(figure 17.2*b*).** *Injectant* allergies are triggered by drugs, vaccines, or hymenopteran (bee) venom **(figure 17.2*c*).** *Contactants* are allergens that enter through the skin **(figure 17.2*d*).** Many contact allergies are of the type IV (delayed) variety, discussed

Table 17.2 Common Allergens, Classified by Portal of Entry

Inhalants	Ingestants	Injectants	Contactants
Pollen Dust Mold spores Dander Animal hair Insect parts Formalin	Food (milk, peanuts, wheat, shellfish, soybeans, nuts, eggs, fruits) Food additives Drugs (aspirin, penicillin)	Bee, wasp venom Drugs Vaccines Serum Enzymes Hormones	Drugs Cosmetics Heavy metals Detergents Formalin Latex Glue Solvents Dyes

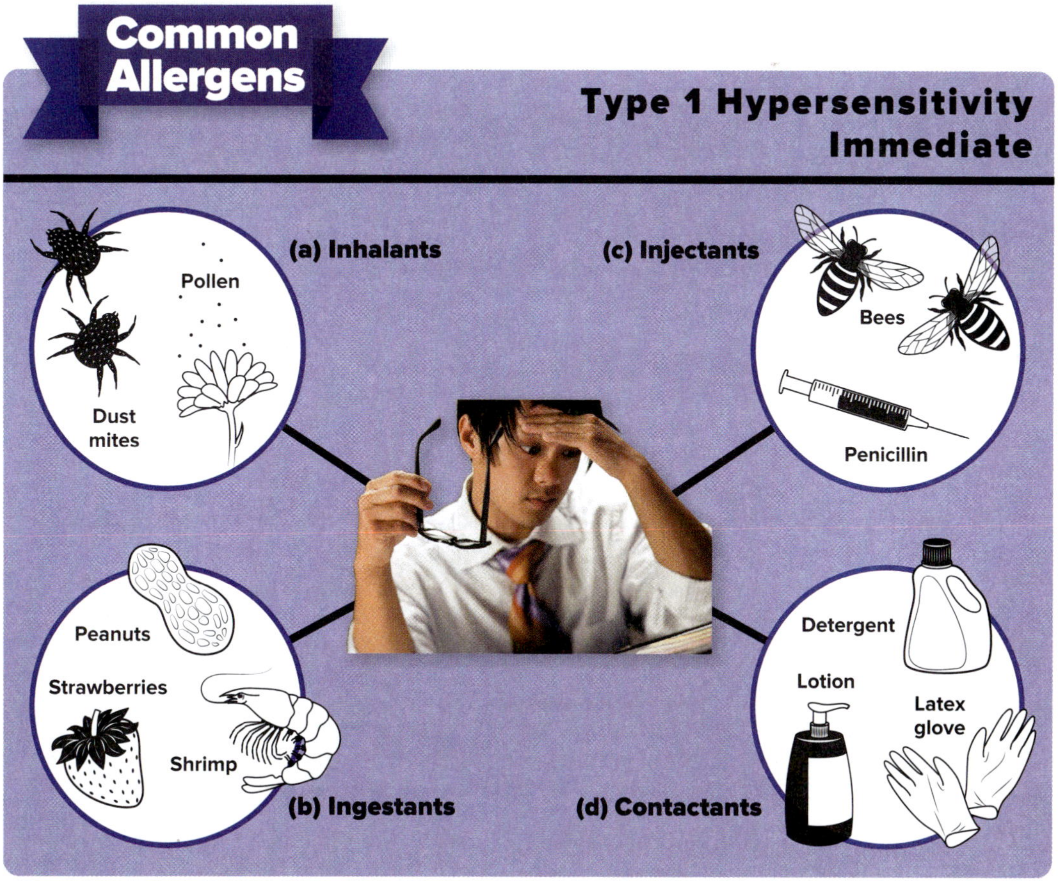
Common Allergens
Type 1 Hypersensitivity Immediate
(a) Inhalants
Pollen
Dust mites
(c) Injectants
Bees
Penicillin
Peanuts
Strawberries
Shrimp
(b) Ingestants
(d) Contactants
Detergent
Lotion
Latex glove

Figure 17.2 Common allergens, classified by portal of entry. **(a)** Common inhalants, or airborne environmental allergens, include pollen and insect parts. **(b)** Common ingestants, allergens that enter by mouth. **(c)** Common injectants, allergens that enter via the parenteral route. **(d)** Common contactants, allergens that enter through the skin.
Rubberball/Alamy Stock Photo

later in this chapter. It is also possible to be exposed to certain allergens—penicillin among them—during sexual intercourse due to the presence of allergens in the semen.

Mechanisms of Type I Allergy: Sensitization and Provocation

What causes some people to sneeze and wheeze every time they step out into the spring air, while others suffer no ill effects? In order to answer this question, we must examine what occurs in the tissues of the allergic individual that does not occur in the nonallergenic person. In general, type I allergies develop in stages **(figure 17.3).** This figure tells the whole story, beginning with the initial encounter with an allergen that sets up the conditions for the allergy to appear on subsequent encounters.

The Physiology of IgE-Mediated Allergies

During primary contact and sensitization, the allergen penetrates the portal of entry **(figure 17.3*a*).** When large particles such as pollen grains, hair, and spores encounter a mucous membrane, they release molecules of allergen that pass into the tissue fluids and lymphatics. The lymphatics then carry the allergen to the lymph nodes, where specific clones of B cells recognize it, are activated, and proliferate into plasma cells. These plasma cells produce immunoglobulin E (IgE), the antibody of allergy. IgE is different from other immunoglobulins in that it has an Fc region with great affinity for mast cells and basophils. The long-term binding of IgE to these cells in the tissues sets the scene for the reactions that occur upon repeated exposure to the same allergen **(figure 17.3*b*).**

The Role of Mast Cells and Basophils

Mast cells and basophils play an important role in allergy due to the following:

1. They are everywhere in tissues. Mast cells are located in the connective tissue of virtually all organs, but high concentrations exist in the lungs, skin, gastrointestinal tract,

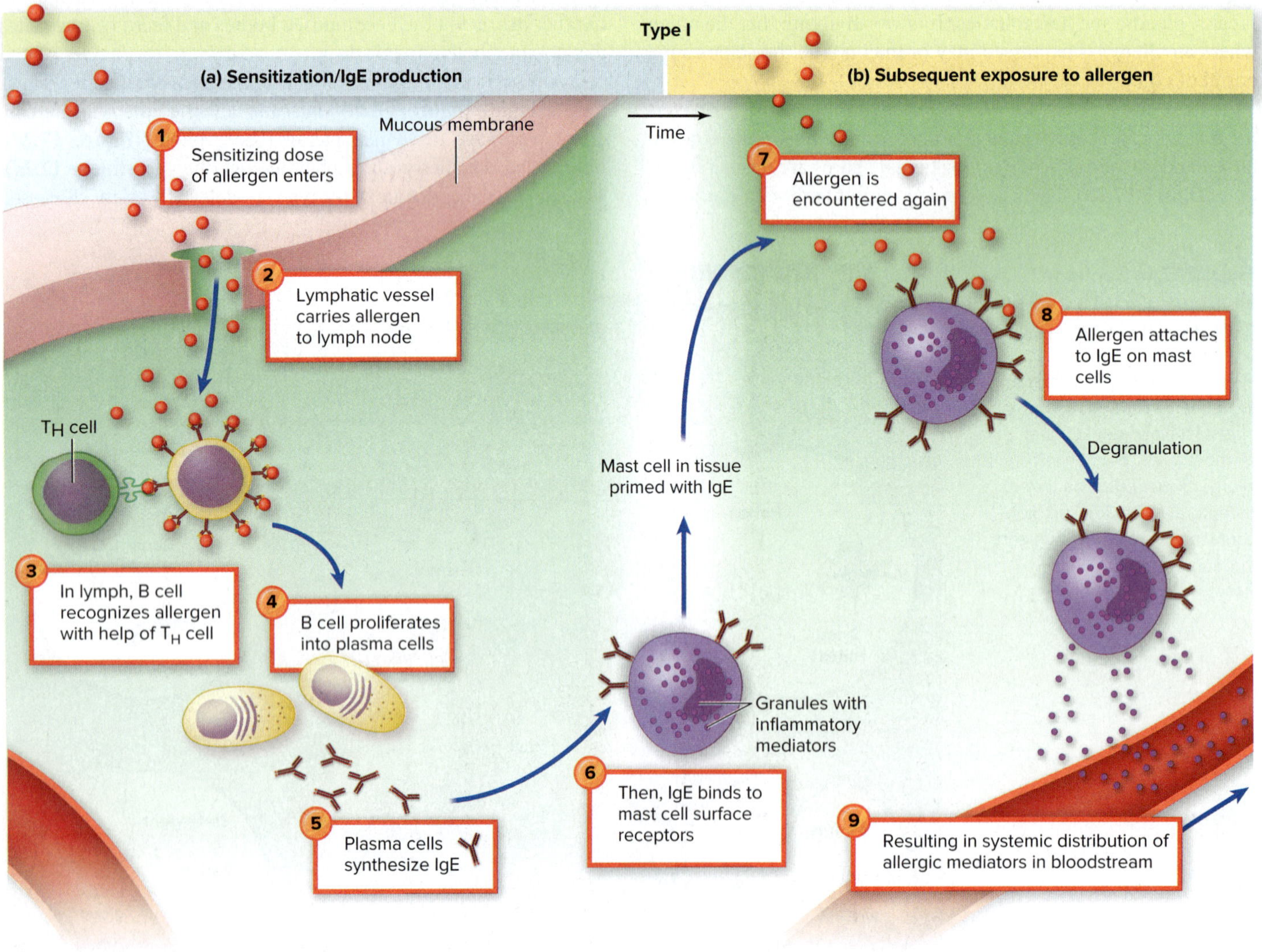
Type I
(a) Sensitization/IgE production
(b) Subsequent exposure to allergen
Time
Mucous membrane
1 Sensitizing dose of allergen enters
2 Lymphatic vessel carries allergen to lymph node
T_H cell
3 In lymph, B cell recognizes allergen with help of T_H cell
4 B cell proliferates into plasma cells
5 Plasma cells synthesize IgE
6 Then, IgE binds to mast cell surface receptors
Granules with inflammatory mediators
Mast cell in tissue primed with IgE
7 Allergen is encountered again
8 Allergen attaches to IgE on mast cells
Degranulation
9 Resulting in systemic distribution of allergic mediators in bloodstream

Figure 17.3 A schematic view of cellular reactions during the type I allergic response. **(a)** Sensitization (initial contact with sensitizing dose), 1–5. **(b)** Provocation (later contacts with provocative dose), 6–9.

and genitourinary tract. Basophils circulate in the blood but migrate easily into tissues.

2. They have a large capacity to bind IgE during sensitization (see figure 17.3) and *degranulate*. Each of these cells carries 30,000 to 100,000 receptors for IgE. When the IgE molecules on the surface encounter antigen and bind it, the release of inflammatory cytokines from cytoplasmic granules (secretory vesicles) is triggered.

We will now see what occurs when sensitized cells are challenged with the allergen a second time.

The Second Contact with an Allergen

After sensitization, the IgE-primed mast cells can remain in the tissues for years. Even after long periods without contact, a person can retain the capacity to react immediately upon reexposure. The next time allergen molecules contact these sensitized cells, the allergens bind across adjacent receptors and stimulate degranulation. As chemical mediators are released, they diffuse into the tissues and bloodstream. Cytokines give rise to numerous local and systemic reactions, many of which appear quite rapidly (see figure 17.3*b*). The symptoms of allergy are not caused by the direct action of allergens on tissues but by the physiological effects of mast cell chemicals on target organs.

In recent years researchers have also pinpointed a role for T-cell participation in allergic reactions to certain substances. This is not surprising because the orchestration of B-cell responses is usually influenced by T cells. So, as in most biological phenomena, it is more complicated than it appears on the surface. In this summary, we focused on the role of IgE released from B cells.

Cytokines, Target Organs, and Allergic Symptoms

Numerous substances involved in causing allergy have been identified. The principal chemical mediators produced by mast cells and basophils are histamine, serotonin, leukotriene, platelet-activating factor, prostaglandins, and bradykinin **(figure 17.4).**

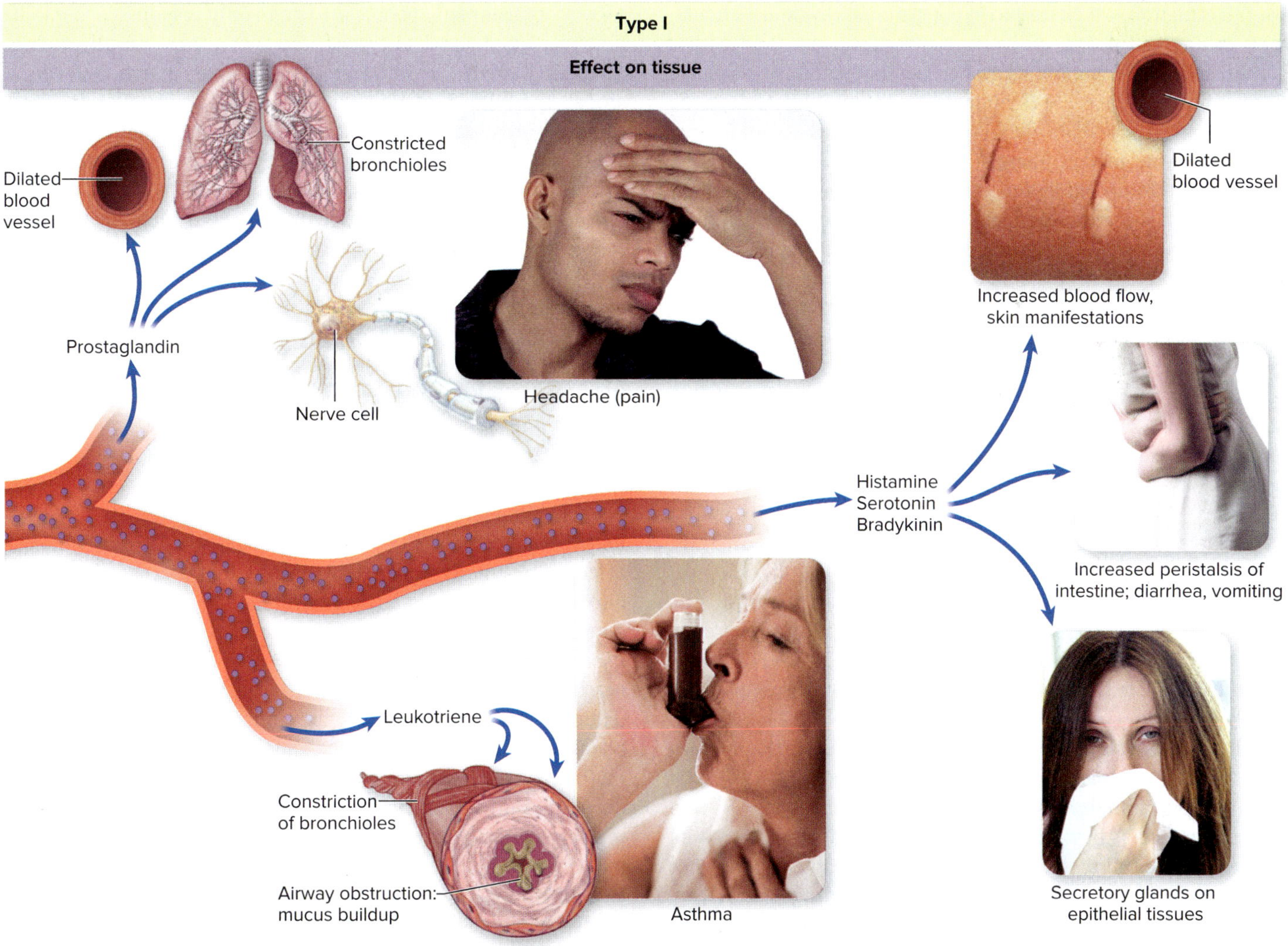
Type I
Effect on tissue
Dilated blood vessel
Constricted bronchioles
Prostaglandin
Nerve cell
Headache (pain)
Dilated blood vessel
Increased blood flow, skin manifestations
Histamine
Serotonin
Bradykinin
Increased peristalsis of intestine; diarrhea, vomiting
Leukotriene
Constriction of bronchioles
Airway obstruction: mucus buildup
Asthma
Secretory glands on epithelial tissues

Figure 17.4 The spectrum of reactions to inflammatory cytokines released by mast cells and the common symptoms they elicit in target tissues and organs. Note the extensive overlapping effects.

(a) Ingram Publishing/Superstock; (b) Ian Hooton/Science Photo Library/Science Source; (c) Southern Illinois University/Science Source; (d) Colin Anderson/Brand X Pictures/Getty Images; (e) Elenathewise/Panther Media GmbH/Alamy Stock Photo

These chemicals, acting alone or in combination, account for the tremendous scope of allergic symptoms. Targets of these chemicals include the skin, upper respiratory tract, gastrointestinal tract, and conjunctiva. The general responses of these organs include rashes, itching, redness, rhinitis, sneezing, diarrhea, and shedding of tears. Systemic targets include smooth muscle, mucous glands, and nervous tissue. Because smooth muscle is responsible for regulating the size of blood vessels and respiratory passageways, changes in its activity can profoundly alter blood flow, blood pressure, and respiration. Pain, anxiety, agitation, and lethargy are also attributable to the effects of these mediators on the nervous system.

Histamine is the most profuse and fastest-acting allergic mediator. It is a potent stimulator of smooth muscle, glands, and eosinophils. Histamine's actions on smooth muscle vary with location. It *constricts* the smooth muscle layers of the small bronchi and intestine, thereby causing labored breathing and increased intestinal motility. In contrast, histamine *relaxes* vascular smooth muscle and dilates arterioles and venules. It is responsible for hives, pruritus (itching), and headache. More severe reactions such as anaphylaxis can be accompanied by edema (swelling) and vascular dilation, which lead to hypotension, tachycardia, circulatory failure, and shock. Salivary, lacrimal, mucous, and gastric glands are also histamine targets. Histamine can also stimulate eosinophils to release inflammatory cytokines, escalating the symptoms.

Serotonin, another allergic mediator, acts similarly to histamine. Serotonin increases vascular permeability, capillary dilation, smooth muscle contraction, intestinal peristalsis, and respiratory rate; but it diminishes central nervous system activity.

Another class of compounds, the **leukotrienes** (loo′-koh-try″-eenz), contains a member that is known as the "slow-reacting substance of anaphylaxis" for its property of inducing gradual contraction of smooth muscle. This type of leukotriene is responsible for the prolonged bronchospasm, vascular permeability, and mucus secretion of the asthmatic individual. Other leukotrienes stimulate the activities of polymorphonuclear leukocytes, or granulocytes, which play a role in various immune functions.

Platelet-activating factor is a lipid released by basophils, neutrophils, monocytes, and macrophages. The physiological response to stimulation by this factor is similar to that of histamine, including increased vascular permeability, pulmonary smooth muscle contraction, pulmonary edema, hypotension, and a wheal-and-flare response (more commonly known as hives) in the skin.

Prostaglandins are a group of powerful inflammatory agents. Normally, these substances regulate smooth muscle contraction (they stimulate uterine contractions during delivery). In allergic reactions, they are responsible for vasodilation, increased vascular permeability, increased sensitivity to pain, and bronchoconstriction. Nonsteroidal anti-inflammatory drugs (NSAIDs), such as aspirin and ibuprofen, work by preventing the actions of prostaglandins. **Bradykinin** is part of a group of plasma and tissue peptides known as kinins that participate in blood clotting and chemotaxis. In allergic reactions, it causes prolonged smooth muscle contraction of the bronchioles, dilation of peripheral arterioles, increased capillary permeability, and increased mucus secretion.

IgE- and Mast-Cell-Mediated Allergic Conditions

The mechanisms just described are common to hay fever, allergic asthma, food allergy, drug allergy, eczema, and anaphylaxis. In this section, we cover the main characteristics of these conditions, followed by methods of detection and treatment/prevention.

Atopic Diseases

Hay fever is a generic term for **allergic rhinitis,** a seasonal reaction to inhaled plant pollen or molds, or a chronic, year-round reaction to a wide spectrum of airborne allergens or inhalants (see table 17.2). The targets are typically respiratory membranes, and the symptoms include nasal congestion; sneezing; coughing; profuse mucus secretion; itchy, red, and teary eyes; and mild bronchoconstriction.

Asthma is a respiratory disease characterized by episodes of impaired breathing due to severe bronchoconstriction. The airways of asthmatic people are exquisitely responsive to minute amounts of inhalant allergens, food, or other stimuli, such as infectious agents. The symptoms of asthma range from occasional, annoying bouts of difficult breathing to fatal suffocation. Labored breathing, shortness of breath, wheezing, cough, and ventilatory **rales** are present to a degree. The respiratory tract of a person with asthma is chronically inflamed and severely overreactive to allergy chemicals, especially leukotrienes and serotonin from pulmonary mast cells. Upon activation of the allergic response, natural killer T (NKT) cells are recruited and activated, adding to the cytokine storm brewing in the lungs. Other pathologic components are thick mucous plugs in the air sacs and lung damage that can result in long-term respiratory illness. An imbalance in the nervous control of the respiratory smooth muscles is involved in asthma, and the episodes can be influenced by the psychological state of the person, which suggests that there is a neurological connection.

The number of asthma sufferers in the United States is estimated at more than 25 million, with nearly 10% of all children affected by this disorder. For reasons that are not completely understood, asthma is on the increase, and deaths from it have doubled since 1982, even though effective agents to control it are more available now than they have ever been before. It has been suggested that more highly insulated buildings, mandated by energy efficiency regulations, have created indoor air conditions that harbor higher concentrations of contaminants, including insect remains and ozone. Decreasing air quality due to pollution and rising ambient temperatures may play an influential role as well. The increasingly common wildfires in the western United States pose a particularly dangerous problem as well. In the United States it has become apparent that certain groups of children have a higher rate of asthma, and therefore higher death rates from it, than others.

Atopic dermatitis is an intensely itchy inflammatory condition of the skin, sometimes also called **eczema.** Sensitization occurs through ingestion, inhalation, and occasionally skin contact with allergens. It usually begins in infancy with reddened, vesicular, weeping, encrusted skin lesions **(figure 17.5*a*).** It can progress in childhood and adulthood to a dry, scaly, thickened skin condition **(figure 17.5*b*).** Lesions can occur on the face, scalp, neck, limbs, and trunk. The itchy, painful lesions cause considerable discomfort, and they are often predisposed to secondary bacterial infections.

Food Allergy

The most common food allergens come from peanuts, fish, cow's milk, wheat, eggs, shellfish, and soybeans. Although the portal of entry is intestinal, food allergies can also affect the skin and respiratory tract. Gastrointestinal symptoms include vomiting, diarrhea, and abdominal pain. Other manifestations of food allergies include eczema, hives, rhinitis, asthma, and occasionally anaphylaxis. Classic food hypersensitivity involves IgE and degranulation of mast cells, but not all reactions involve this mechanism. (Do not confuse food allergy with food intolerance. Many people are lactose intolerant, for example, due to a deficiency in the enzyme that degrades the milk sugar.) Food (egg) allergies must be considered when vaccinating individuals, due to the presence of egg protein in some vaccine preparations.

Drug Allergy

Modern drug development has been responsible for many medical advances. Unfortunately, drugs are foreign compounds capable of stimulating allergic reactions in some people. In fact, drug allergies are one of the most common side effects of treatment (present in 5% to 10% of hospitalized patients). Depending on the allergen, route of entry, and individual sensitivities, virtually any tissue of the body can be affected, and reactions range from a mild rash **(figure 17.5*c*)** to fatal anaphylaxis. Compounds implicated most often are antibiotics (penicillin is number one in prevalence), synthetic antimicrobials (sulfa drugs), aspirin, opiates, and the contrast dye used in X rays. The actual allergen is not the intact drug itself but a hapten produced when the liver processes the drug. Some people become sensitized to penicillin because of the presence of small amounts of the drug in meat, milk, and other foods and from exposure to *Penicillium* mold in the environment.

Anaphylaxis: An Overpowering Systemic Reaction

The term *anaphylaxis,* or anaphylactic shock, was first used to denote a reaction of animals injected with a foreign protein. Although the animals showed no response during the first contact, upon reinoculation with the same protein at a later time, they exhibited acute symptoms—itching, sneezing, difficult breathing, prostration, and convulsions—and many died in a few minutes. Systemic anaphylaxis is characterized by sudden respiratory and circulatory disruption that can be fatal within a few minutes. In humans, the allergen and route of entry are variable, though bee stings and injections of antibiotics or serum are the allergens most often resulting in anaphylaxis. Bee venom is a complex material containing several allergens and enzymes that can create a sensitivity lasting for decades after exposure.

The underlying physiological events in anaphylaxis parallel those of atopy, but the concentration of chemical mediators and the strength of the response are greatly amplified. The immune system of a sensitized person exposed to a provocative dose of an allergen or allergens responds with a sudden, massive release of chemicals into the tissues and blood, which act rapidly on the target organs. Anaphylactic persons have been known to die within 15 minutes from complete airway blockage.

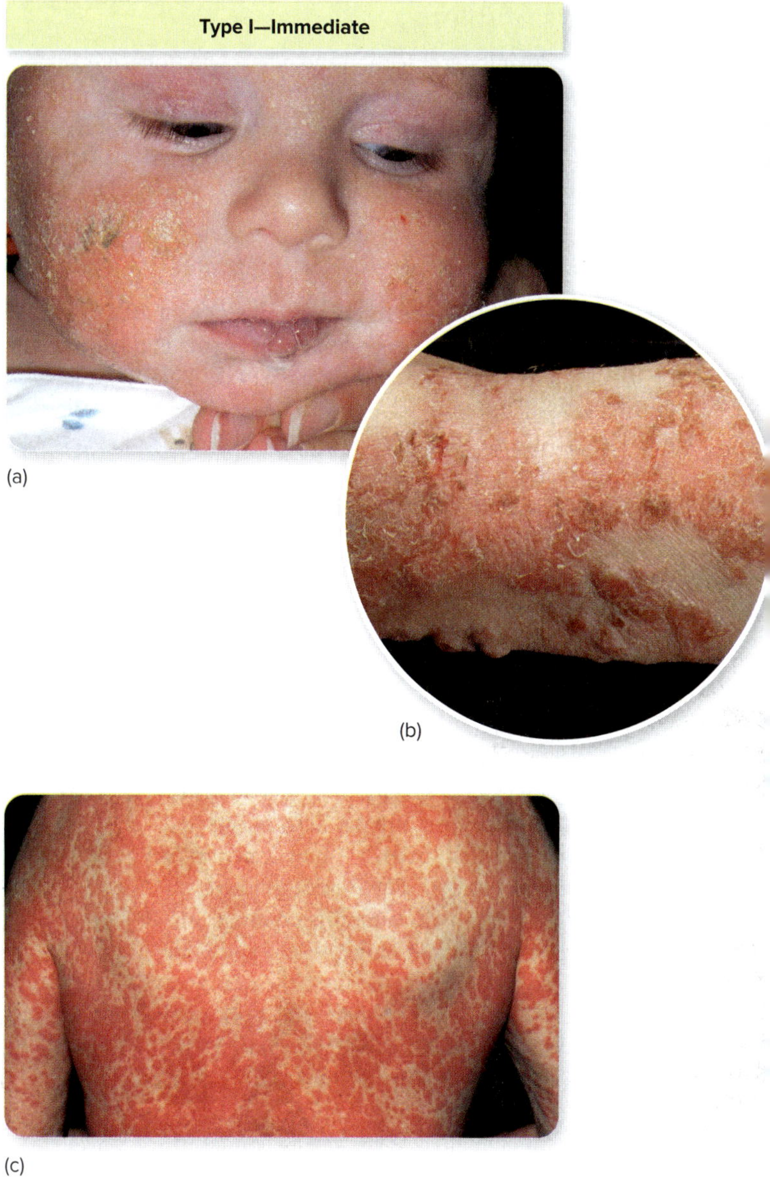
Type I—Immediate
(a)
(b)
(c)

Figure 17.5 Skin manifestations in atopic and drug allergies. **(a)** Vesicular, weepy, encrusted lesions are typical in afflicted infants. **(b)** In adulthood, lesions are more likely to be dry, scaly, and thickened. **(c)** A typical rash that develops in an allergic reaction to an antibiotic.

(a, c) Dr. P. Marazzi/Science Source; (b) Biophoto Associates/Science Source

Diagnosis of Allergy

Because allergy mimics infection and other conditions, it is important to determine if a person is actually allergic and to identify the specific allergen or allergens involved. Allergy diagnosis involves several different kinds of tests. Allergy testing can be done on blood samples or directly on the skin.

Blood Testing

The most widely used blood test is a test that measures the levels of IgE to specific allergens. Another test that can distinguish whether a patient has experienced an allergic attack measures elevated blood levels of tryptase, an enzyme released by mast cells that increases during an allergic response. Several types of specific *in vitro* tests can determine the allergic potential of a patient's blood sample. A differential blood cell count can indicate the levels of basophils and eosinophils, indicating allergy. The leukocyte histamine-release test (LHRT) measures the amount of histamine released from the patient's basophils when exposed to a specific allergen.

Skin Testing

A tried-and-true *in vivo* method to detect precise atopic or anaphylactic sensitivities is skin testing. With this technique, a patient's skin is injected, scratched, or pricked with a small amount of a pure allergen extract. There are hundreds of these allergen extracts containing common airborne allergens (plant and mold pollen) and more unusual allergens (mule dander, theater dust, bird feathers). The allergist maps the skin on the inner aspect of the forearms or back and injects the allergens intradermally according to this predetermined pattern **(figure 17.6*a*)**. Approximately 15 minutes after antigenic challenge, each site is appraised for a wheal response indicative of histamine release. The diameter of the wheal is measured and rated on a scale of 0 (no reaction) to 4 (greater than 15 mm) **(figure 17.6*b*)**.

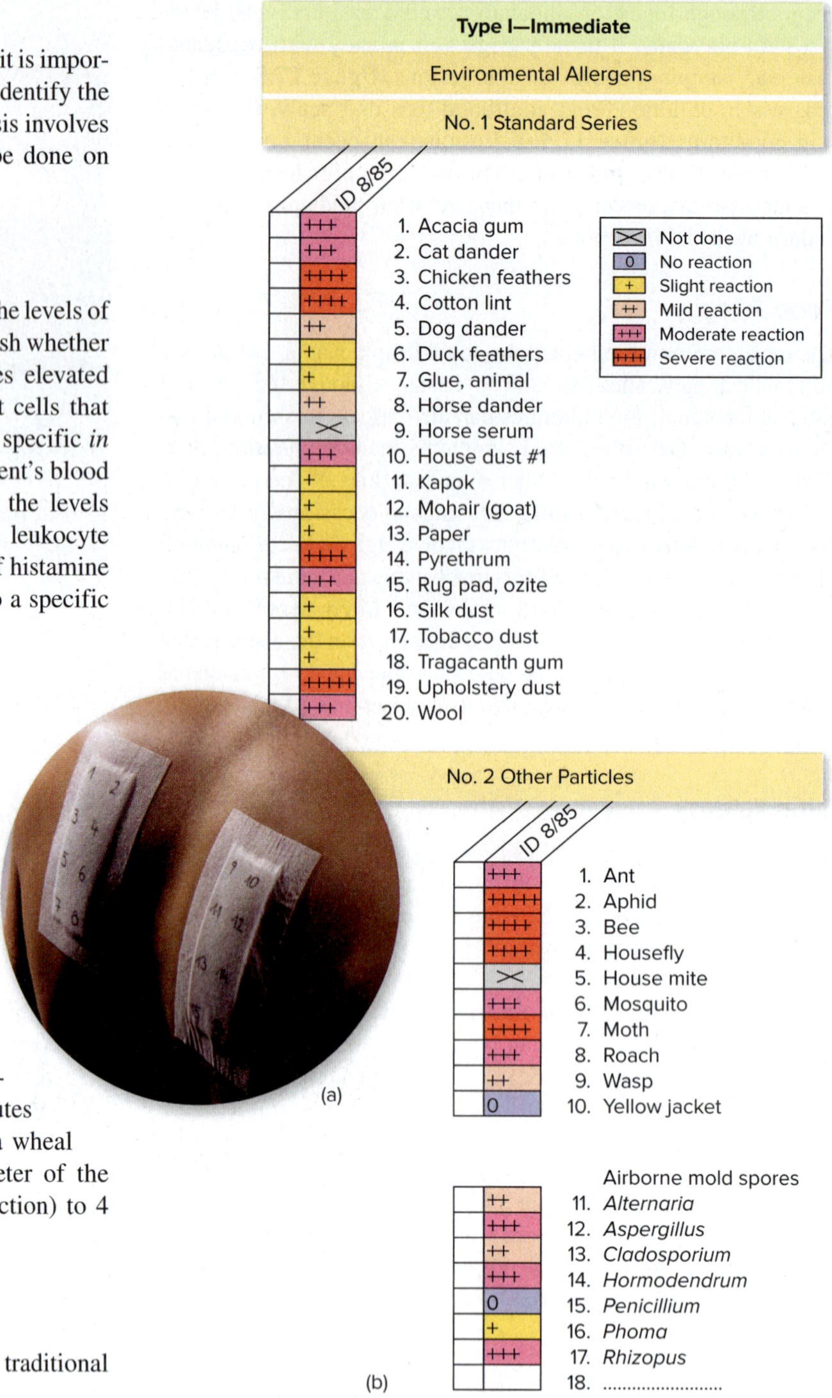
Type I—Immediate
Environmental Allergens
No. 1 Standard Series
ID 8/85
1. Acacia gum
2. Cat dander
3. Chicken feathers
4. Cotton lint
5. Dog dander
6. Duck feathers
7. Glue, animal
8. Horse dander
9. Horse serum
10. House dust #1
11. Kapok
12. Mohair (goat)
13. Paper
14. Pyrethrum
15. Rug pad, ozite
16. Silk dust
17. Tobacco dust
18. Tragacanth gum
19. Upholstery dust
20. Wool
Not done
No reaction
Slight reaction
Mild reaction
Moderate reaction
Severe reaction
No. 2 Other Particles
ID 8/85
1. Ant
2. Aphid
3. Bee
4. Housefly
5. House mite
6. Mosquito
7. Moth
8. Roach
9. Wasp
10. Yellow jacket
Airborne mold spores
11. Alternaria
12. Aspergillus
13. Cladosporium
14. Hormodendrum
15. Penicillium
16. Phoma
17. Rhizopus
18. ....
(a)
(b)

Figure 17.6 A method for conducting an allergy skin test. The forearm (or back) is mapped and then injected with a selection of allergen extracts. The allergist must be very aware of potential anaphylaxis attacks triggered by these injections. **(a)** Tests are frequently performed in a grid on the back. **(b)** An actual skin test record for some common environmental allergens [not related to **(a)**].

(a) Klaus Tiedge/Image Source

Treatment of Allergy

Once an allergy has appeared in a patient, these are the traditional methods of treating it:

1. Using drugs that block the action of lymphocytes, mast cells, or chemical mediators;
2. Avoiding the allergen;
3. Desensitization: controlled exposure to the antigen through ingestion, sublingual absorption, or injection to reset the allergic reaction.

Taking Drugs to Block Allergy

The aim of antiallergy medication is to block the progress of the allergic response somewhere along the route between IgE production and the appearance of symptoms. Oral anti-inflammatory drugs such as corticosteroids inhibit the activity of lymphocytes and thereby reduce the production of IgE, but they also have significant side effects and should not be taken for prolonged periods. Some drugs block the degranulation of mast cells and reduce the levels of inflammatory cytokines (Cromolyn). Asthma and rhinitis sufferers can find relief from a monoclonal antibody that binds to IgE in such a way that it is unable to dock with mast cells. Therefore, no degranulation takes place (omalizumab [Xolair®]).

Antihistamines are widely used medications for preventing symptoms of atopic allergy. These are the active ingredients in most over-the-counter allergy-control drugs. Antihistamines interfere with histamine activity by binding to histamine receptors on target organs. Other drugs that relieve inflammatory symptoms are aspirin and acetaminophen, which reduce pain by interfering with prostaglandin, and theophylline, a bronchodilator that reverses spasms in the respiratory smooth muscles. Persons who suffer from anaphylactic attacks are urged to carry at all times injectable or aerosolized epinephrine (adrenaline) and an identification tag indicating their sensitivity. Epinephrine reverses constriction of the airways and slows the release of allergic mediators. Although epinephrine works quickly and well, it has a very short half-life. It is very common to require more than one dose in anaphylactic reactions. Injectable epinephrine buys the individual time to get to a hospital for continuing treatment.

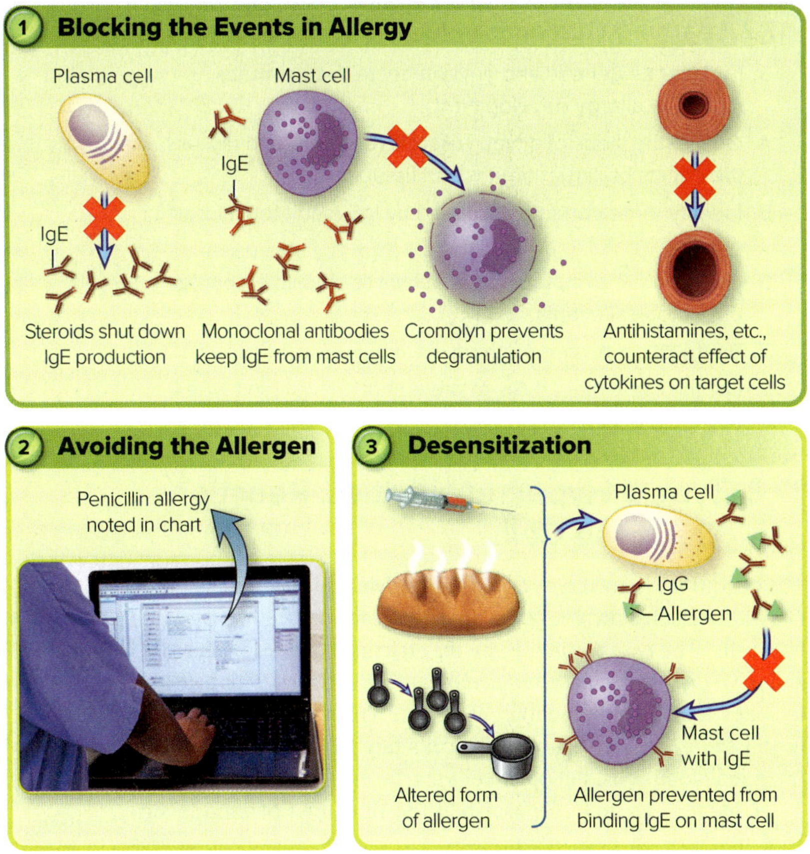
1 Blocking the Events in Allergy
Plasma cell
Mast cell
IgE
IgE
Steroids shut down IgE production
Monoclonal antibodies keep IgE from mast cells
Cromolyn prevents degranulation
Antihistamines, etc., counteract effect of cytokines on target cells
2 Avoiding the Allergen
Penicillin allergy noted in chart
3 Desensitization
Plasma cell
IgG
Allergen
Mast cell with IgE
Altered form of allergen
Allergen prevented from binding IgE on mast cell

Figure 17.7 Strategies for blocking allergy attacks.

Photo: McGraw Hill

Avoiding the Allergen

It is obvious that if you experience a negative health effect when exposed to a substance, then the first goal should be to avoid it. It is the most basic of all preventive measures and is practiced daily by millions of people.

Desensitization

In practice, avoiding an allergen can be very difficult, as you may have noticed if you're ever on a plane and the flight crew announces that there will be no peanuts available because someone in the cabin has a peanut allergy. But the field of allergy medicine is changing rapidly. Desensitization, in the form of "allergy shots," has been used for decades. It involves injecting specific amounts of the allergen under the skin.

The allergen preparations contain pure, preserved suspensions of plant antigens, venoms, dust mites, dander, and molds (but, so far, desensitization to foods has not proved very effective). The immunologic basis of this treatment is open to differences in interpretation. One hypothesis suggests that injected allergens stimulate the formation of allergen-specific IgG—**blocking antibodies**—that can remove an allergen from the system before it can bind to IgE **(figure 17.7).** It is also possible that an allergen delivered in this fashion combines with the IgE itself and takes it from circulation.

Researchers are now looking at desensitization treatments that are delivered sublingually (under the tongue) and orally in order to present the allergen through a mucosal surface, hoping to trigger IgA and IgG responses that would have a blocking effect on the allergic response.

Evidence also suggests that if children are allergic to fresh milk, having them consume it in small amounts in the form of baked goods (in which milk is an ingredient) can help desensitize them. As you know, heating proteins denatures them, changing their physical form. It is possible that the changed presentation of epitopes increases the IgG response—which blocks the antigen binding to the IgE on mast cells, and this blocks degranulation.

Peanut allergies are one of the scariest for parents, as children can develop anaphylactic symptoms quickly from this allergy. Recent research has shown that "oral immunotherapy," a treatment in which children were given increasing amounts of peanut protein daily, can effectively cure the peanut allergy in some children. Again, this suggests that altering the way the antigen is presented (in this case, the dose; in the previous example, the physical form) can correct an inappropriate response.

At the beginning of the chapter we inferred that the development of allergies might be the result of the sudden (in evolutionary time) disappearance of worm and protozoan pathogens in the developed world. The parts of immunity, such as IgE, that participate in allergy are the components that naturally react to helminths and larger microbes. So for several years now scientists (and some amateur allergy sufferers!) have experimented with using deliberate infections with *Trichuris suis,* a whipworm whose natural host is pigs. The worms can establish a brief colonization in humans without causing symptoms, during which time the immune system responds, and, in some cases, seems to reset itself so that it stops responding to the inappropriate antigen—the allergen.

Figure 17.7 supplies a summary of ways to interfere with allergy symptoms.

17.2 Learning Outcomes—Assess Your Progress

3. Summarize genetic and environmental factors that influence allergy development.
4. Outline the steps of a type I allergic response, and discuss the effects on target organs and tissue.
5. Identify three conditions caused by IgE-mediated allergic reactions.
6. Describe the symptoms of anaphylaxis, and link them to physiological events.
7. Briefly describe two methods for diagnosing allergies.
8. List the three main ways to prevent or short-circuit type I allergic reactions.

17.3 Type II Hypersensitivities: Reactions That Lyse Cells

The conditions termed *type II hypersensitivities* are a complex group of syndromes that involve complement-assisted destruction (lysis) of host cells. The lysis is directed by the attachment of antibodies (IgG and IgM) directed against those cells' surface antigens. This category includes transfusion reactions and some types of autoimmunities (discussed in a later section). The cells targeted for destruction are often red blood cells, but other cells can be involved.

Chapters 15 and 16 described the functions of unique surface markers on cell membranes. Ordinarily, these molecules play essential roles in transport, recognition, and development, but they become critically important when the tissues of one person are placed into the body of another person. **Alloantigens** are markers on an organism's cells that differ among individuals of that same species. Blood transfusions and organ donations introduce alloantigens on donor cells that are recognized by the lymphocytes of the recipient. These reactions are not really immune dysfunctions the way that allergy and autoimmunity are. The immune system is, in fact, working normally, but it is not equipped to distinguish between the desirable foreign cells of a transplanted tissue and the undesirable ones of a microbe.

The Basis of Human ABO Antigens and Blood Types

Like all cells, red blood cells have various markers on them. Some of these are alloantigens that are important to consider when exposing a person to someone else's red blood cells. While there are many antigens to consider, one set of antigens is usually the most influential on whether a recipient will react to them or not. This set of antigens places red blood cells into four groups and is called the ABO system.

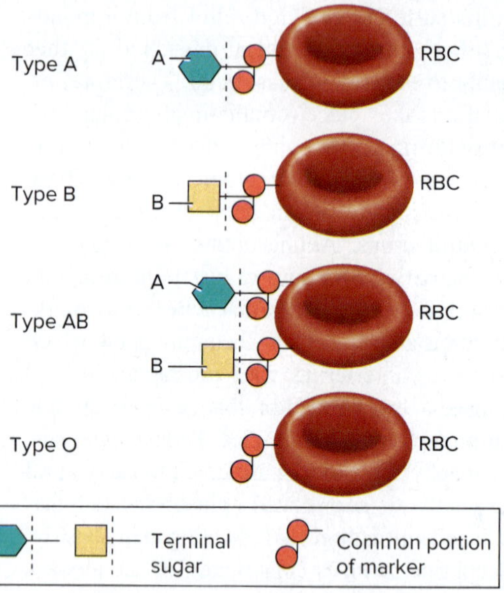
Type A
A
RBC
Type B
B
RBC
Type AB
A
B
RBC
Type O
RBC
Terminal sugar
Common portion of marker

Figure 17.8 The genetic/molecular basis for the A and B antigens (receptors) on red blood cells. In general, persons with blood types A, B, and AB inherit a gene for the enzyme that adds a certain terminal sugar to the basic RBC receptor. Type O persons do not have such an enzyme and lack the terminal sugar.

Like the MHC antigens on white blood cells, the ABO antigen markers on red blood cells are genetically determined and composed of glycoproteins. These ABO antigens are inherited as two (one from each parent) of three alternative **alleles:** A, B, or O. As **table 17.3** indicates, this mode of inheritance gives rise to four blood types (phenotypes), depending on the particular combination of genes. A person with an *AA* or *AO* genotype has type A blood; genotype *BB* or *BO* gives type B; genotype *AB* produces type AB; and genotype *OO* produces type O. Some important points about the blood types follow:

1. They are named for the dominant antigen(s).
2. The RBCs of type O persons have antigens but not A and B antigens.
3. Tissues other than RBCs carry A and B antigens.

A diagram of the AB antigens and blood types is shown in **figure 17.8.** Each of the A and B genes codes for an enzyme that adds a terminal carbohydrate to RBC surface molecules

Table 17.3 Characteristics of ABO Blood Groups

Genotype	Blood Type	Antigen Present on Erythrocyte Membranes	Antibody in Plasma	Incidence of Type in United States		
				Among Whites (%)	Among Asians (%)	Among Those of African and Caribbean Descent (%)
AA, AO	A	A	Anti-B	41	28	27
BB, BO	B	B	Anti-A	10	27	20
AB	AB	A and B	Neither anti-A nor anti-B	4	5	7
OO	O	Neither A nor B	Anti-A and anti-B	45	40	46

during maturation. RBCs of type A contain an enzyme that adds *N*-acetylgalactosamine to the molecule; RBCs of type B have an enzyme that adds D-galactose; RBCs of type AB contain both enzymes that add both carbohydrates; and RBCs of type O lack the genes and enzymes to add a terminal molecule.

Disease Connection

Since the 1980s, we have known that the bacterium *Helicobacter pylori* is a major cause of gastritis and stomach ulcers. Since that time, researchers have also discovered that people with the O blood type seem to have a higher susceptibility to the condition. This is probably due to the fact that the bacterium can bind to the glycoproteins on the stomach cell that resemble those found on O-type blood cells.

Antibodies Against A and B Antigens

Although an individual does not normally produce antibodies in response to his or her own RBC antigens, the serum can contain antibodies that react with blood of another antigenic type even though contact with this other blood type has *never* occurred. This is different than the immune response to most antigens, you will note. These preformed antibodies account for the immediate and intense quality of transfusion reactions. As a rule, type A blood contains antibodies (anti-B) that react against the B antigens on types B and AB red blood cells. Type B blood contains antibodies (anti-A) that react with A antigen on types A and AB red blood cells. Type O blood contains antibodies against both A and B antigens. Type AB blood does not contain antibodies against either A or B antigens (see table 17.3). What is the source of these anti-A and anti-B antibodies? It appears that they develop in early infancy because of exposure to certain antigens that are widely distributed in nature. These antigens are surface molecules on bacteria and plant cells that mimic the structure of A and B antigens. Exposure to these sources stimulates the production of corresponding antibodies.

Clinical Concerns in Transfusions

The presence of ABO antigens and A and B antibodies can present problems in giving blood transfusions. First, the individual blood types of donor and recipient must be determined. This is done with a standard technique, in which drops of donor and recipient blood are separately mixed with antisera that contain antibodies against the A and B antigens and are then observed for the evidence of agglutination **(figure 17.9).**

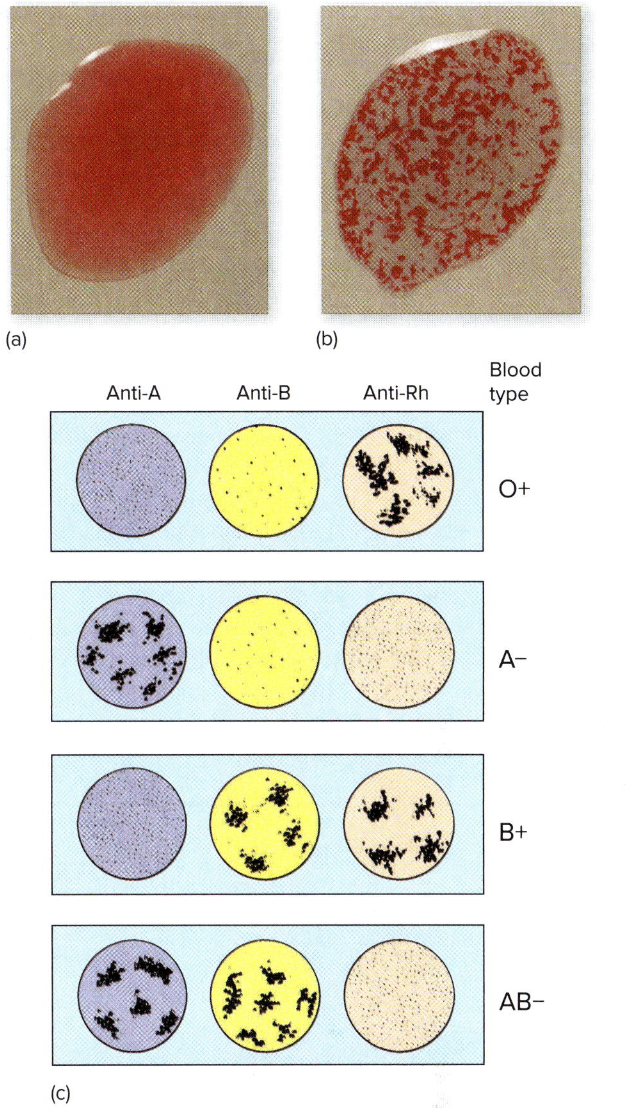
(a)
(b)
Anti-A
Anti-B
Anti-Rh
Blood type
O+
A–
B+
AB–
(c)

Figure 17.9 Interpretation of blood typing. In this test, a drop of blood is mixed with a specially prepared antiserum known to contain antibodies against the A, B, or Rh antigens. **(a)** If that particular antigen is not present, the red blood cells in that droplet do not agglutinate but form an even suspension. **(b)** If that antigen is present, agglutination occurs and the RBCs form visible clumps. **(c)** Several patterns and their interpretations. *Anti-A, anti-B,* and *anti-Rh* are shorthand for the antisera applied to the drops. (In general, O+ is the most common blood type, and AB– is the rarest.)

(a, b) Lisa Burgess/McGraw Hill

Knowing the blood types involved makes it possible to determine which transfusions are safe to perform. The general rule of compatibility is that the RBC antigens of the donor must not be agglutinated by antibodies in the recipient's blood. The ideal practice is to transfuse blood that is a perfect match (A to A, B to B). But even in this event, blood samples must be cross-matched before the transfusion because other blood group incompatibilities—involving other antigens—can exist. This test involves mixing the blood of the donor with the serum of the recipient to check for agglutination.

Under certain circumstances (emergencies, the battlefield), the concept of universal transfusions can be used. To appreciate how this works, remember that an individual generally does not produce antibodies to the RBC antigens they carry themselves. Type O blood lacks A and B antigens and will not be agglutinated by other blood types, so it could theoretically be used in any transfusion. Hence, a person with this blood type is called a *universal donor.* Because type AB blood lacks agglutinating antibodies, an individual with this blood could conceivably receive any type of blood. For that reason, type AB persons are called *universal recipients.* Although both types of transfusions involve some antigen-antibody incompatibilities, these are of less concern because of the dilution of the donor's blood in the body of the recipient.

Transfusion of the wrong blood type causes differing degrees of adverse reaction. The most severe reaction is massive hemolysis when the donated red blood cells react with recipient antibody and trigger the complement cascade. The resulting destruction of red cells leads to systemic shock and kidney failure brought on by the blockage of glomeruli (blood-filtering apparatuses) by cell debris. Death can follow. Other reactions caused by RBC destruction are fever, anemia, and jaundice. A transfusion reaction is managed by immediately halting the transfusion, administering drugs to remove hemoglobin from the blood, and beginning another transfusion with red blood cells of the correct type.

The Rh Factor and Its Clinical Importance

Another RBC antigen of major clinical concern is the **Rh factor** (or D antigen). This factor was first discovered in experiments exploring the genetic relationships among animals. Rabbits inoculated with the RBCs of rhesus monkeys produced an antibody that also reacted with human RBCs. Further tests showed that this monkey antigen (termed *Rh* for "rhesus") was present in about 85% of humans and absent in the other 15%. (This is an average of different ethnic groups; it varies widely.) The details of Rh inheritance are more complicated than those of ABO, but in simplest terms, a person's Rh type results from a combination of two possible alleles—a dominant one that codes for the factor and a recessive one that does not. A person inheriting at least one Rh gene will be Rh-positive (Rh+); only those persons inheriting two recessive genes are Rh-negative (Rh–). The "+" or "–" appearing after a blood type (such as O+) reflects the Rh status of the person (see figure 17.9*c*). Unlike the case with ABO antigens, the only way one can develop antibodies against this antigen is through coming into contact with blood containing the factor. Although the Rh factor should be matched for a transfusion to avoid this situation, it is acceptable to transfuse Rh– blood if the Rh type is not known.

Hemolytic Disease of the Newborn and Rh Incompatibility

The potential for problems is set up when a mother is Rh– and her unborn child is Rh+. It is possible for fetal RBCs to leak into the mother's circulation during late pregnancy and childbirth. The mother's immune system detects the foreign Rh factors on the fetal RBCs and is sensitized to them by producing antibodies and memory B cells. The first Rh+ child is usually not affected because the process begins so late in pregnancy that the child is born before maternal sensitization is completed. However, the mother's immune system has been strongly primed for a second contact with the Rh factor in a subsequent pregnancy **(figure 17.10)**.

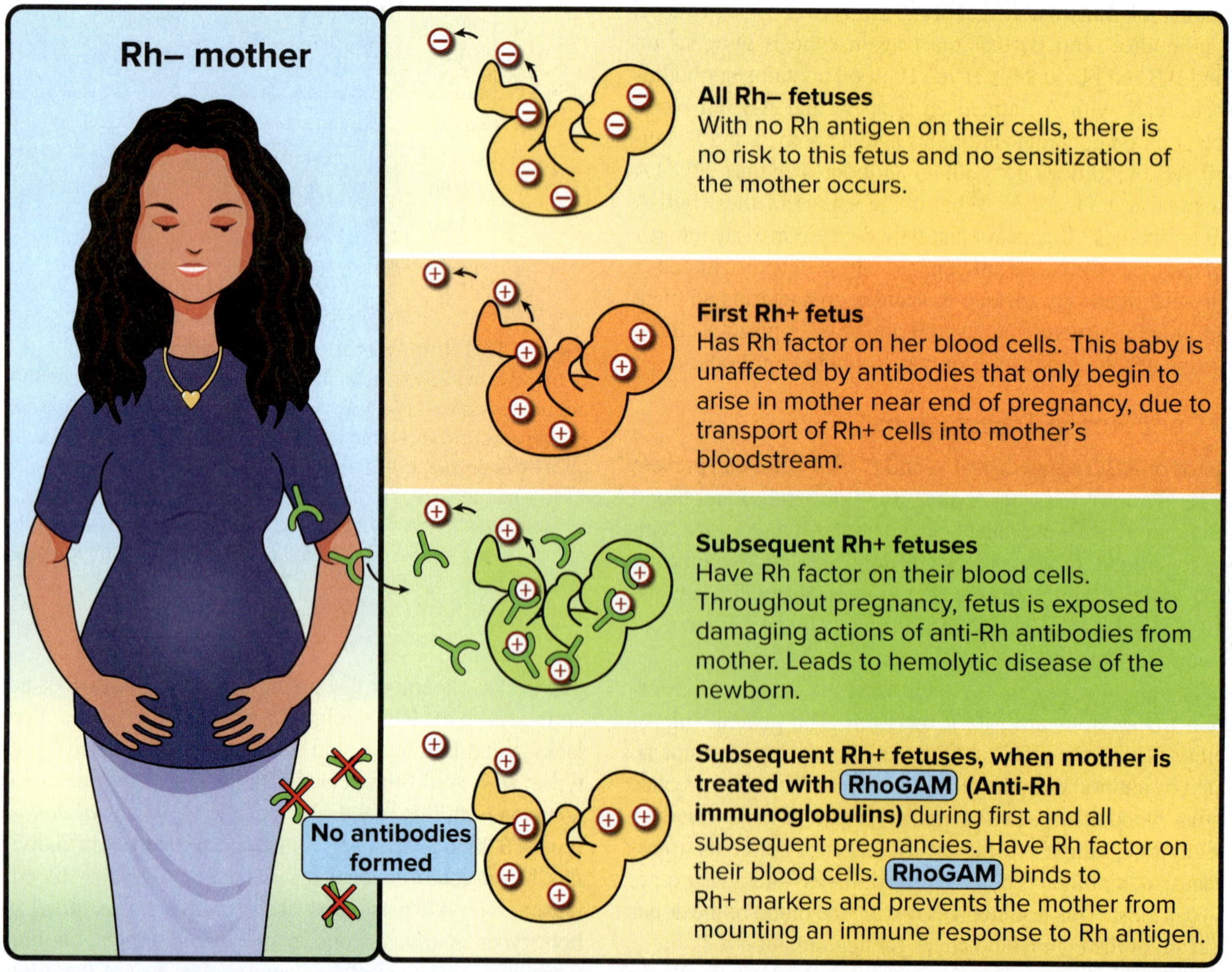
Rh– mother
All Rh– fetuses
With no Rh antigen on their cells, there is no risk to this fetus and no sensitization of the mother occurs.
First Rh+ fetus
Has Rh factor on her blood cells. This baby is unaffected by antibodies that only begin to arise in mother near end of pregnancy, due to transport of Rh+ cells into mother's bloodstream.
Subsequent Rh+ fetuses
Have Rh factor on their blood cells. Throughout pregnancy, fetus is exposed to damaging actions of anti-Rh antibodies from mother. Leads to hemolytic disease of the newborn.
No antibodies formed
Subsequent Rh+ fetuses, when mother is treated with RhoGAM (Anti-Rh immunoglobulins) during first and all subsequent pregnancies. Have Rh factor on their blood cells. RhoGAM binds to Rh+ markers and prevents the mother from mounting an immune response to Rh antigen.

Figure 17.10 Development and control of Rh incompatibility.

In the next pregnancy with an Rh+ fetus, fetal blood cells escape into the maternal circulation late in pregnancy and elicit a memory response. Maternal anti-Rh antibodies then cross the placenta into the fetal circulation, where they bind to fetal RBCs and cause complement-mediated lysis. The outcome is a potentially fatal **hemolytic disease of the newborn (HDN),** characterized by severe anemia and jaundice. It is also called *erythroblastosis fetalis* (eh-rith″-roh-blas-toh′-sis fee-tal′-is), reflecting the release of immature nucleated RBCs (erythroblasts) into the blood to compensate for destroyed RBCs. Maternal-fetal incompatibilities are also possible in the ABO blood group, but adverse reactions occur less frequently than with Rh sensitization because the antibodies to these blood group antigens are IgM rather than IgG and are unable to cross the placenta in large numbers.

Preventing Hemolytic Disease of the Newborn Once sensitization of the Rh– mother to Rh factor has occurred, all other Rh+ fetuses will be at risk for hemolytic disease of the newborn. Prevention requires a careful family history of an Rh– pregnant woman. It can predict the likelihood that she is already sensitized or is carrying an Rh+ fetus. It must take into account other children she has had, their Rh types, and the Rh status of the father. If the father is also Rh–, the child will be Rh– and free of risk, but if the father is Rh+, the probability that the child will be Rh+ is 50% or 100%, depending on the exact genetic makeup of the father. If there is any possibility that the fetus is Rh+, the mother must be passively immunized with antiserum containing antibodies against the Rh factor (*Rh_0 [D] immune globulin,* or RhoGAM).[1] This antiserum reacts with any fetal RBCs that have escaped into the maternal circulation, thereby preventing the sensitization of the mother's immune system to Rh factor (figure 17.10). Anti-Rh antibody must be given with each pregnancy that involves an Rh+ fetus. Before RhoGAM's development in the 1960s, 10,000 babies died from this condition every year.

Other RBC Antigens

Although the ABO and Rh systems are of greatest medical significance, about 20 other red blood cell antigen groups have been discovered. Examples are the MN, Ss, Kell, and P blood groups, some of which are unique membrane proteins, while others are simply carbohydrate antigens, as seen in the ABO system. Because of incompatibilities that these blood groups present, transfused blood is screened to prevent possible cross-reactions. The study of these blood antigens (as well as ABO and Rh) has given rise to other applications. For example, they can be useful in forensic medicine (crime detection), studying ethnic ancestry, and tracing prehistoric migrations in anthropology.

Later in this chapter, you will read about special cases in which type II hypersensitivity is directed against self.

1. RhoGAM: Immunoglobulin fraction of human anti-Rh serum, prepared from pooled human sera.

17.3 Learning Outcomes—Assess Your Progress

9. List the three immune components causing cell lysis in type II hypersensitivity reactions.
10. Explain the molecular basis for the ABO blood groups, and identify the blood type of a "universal donor" and the blood type of a "universal recipient."
11. Explain the role of Rh factor in hemolytic disease development and how the disease is prevented in newborns.

17.4 Type III Hypersensitivities: Immune Complex Reactions

Type III hypersensitivity involves the reaction of antigen with antibody and the deposition of the resulting complexes in basement membranes of epithelial tissue. It is similar to type II because it involves the production of IgG and IgM antibodies after repeated exposure to antigens and the activation of complement. Type III differs from type II because its antigens are not attached to the surface of a cell. The interaction of these antigens with antibodies produces free-floating complexes that can be deposited in the tissues, causing an **immune complex reaction** or disease. This category includes such as diseases such as acute post-streptococcal glomerulonephritis.

Mechanisms of Immune Complex Disease

After initial exposure to a large amount of antigen, the immune system produces great quantities of antibodies that circulate in the fluid compartments. When this antigen enters the system a second time, it reacts with the antibodies to form antigen–antibody complexes **(figure 17.11).** These complexes summon various inflammatory components such as complement and neutrophils, which would ordinarily eliminate Ag–Ab complexes as part of the normal immune response. In an immune complex disease, however, there is an excess of antigen and small amounts of antibody. This causes cross-linking of multiple antigens with the scarce antibodies, and the complexes can then deposit in the **basement membranes**[2] of epithelial tissues. In response to these events, neutrophils release lysosomal granules that digest tissues and cause a destructive inflammatory condition. This situation leads to the symptoms of type III hypersensitivities.

Types of Immune Complex Disease

Type III hypersensitivities

1. depend on IgG, IgM, or IgA rather than IgE;
2. require large doses of antigen (not a tiny dose as in anaphylaxis); and
3. show a delay in symptoms (a few hours to days).

A classic Type III–mediated condition is acute post-streptococcal glomerulonephritis (APSGN). We will learn about different

2. Basement membranes are the bottom layers of epithelia that normally filter out circulating antigen–antibody complexes.

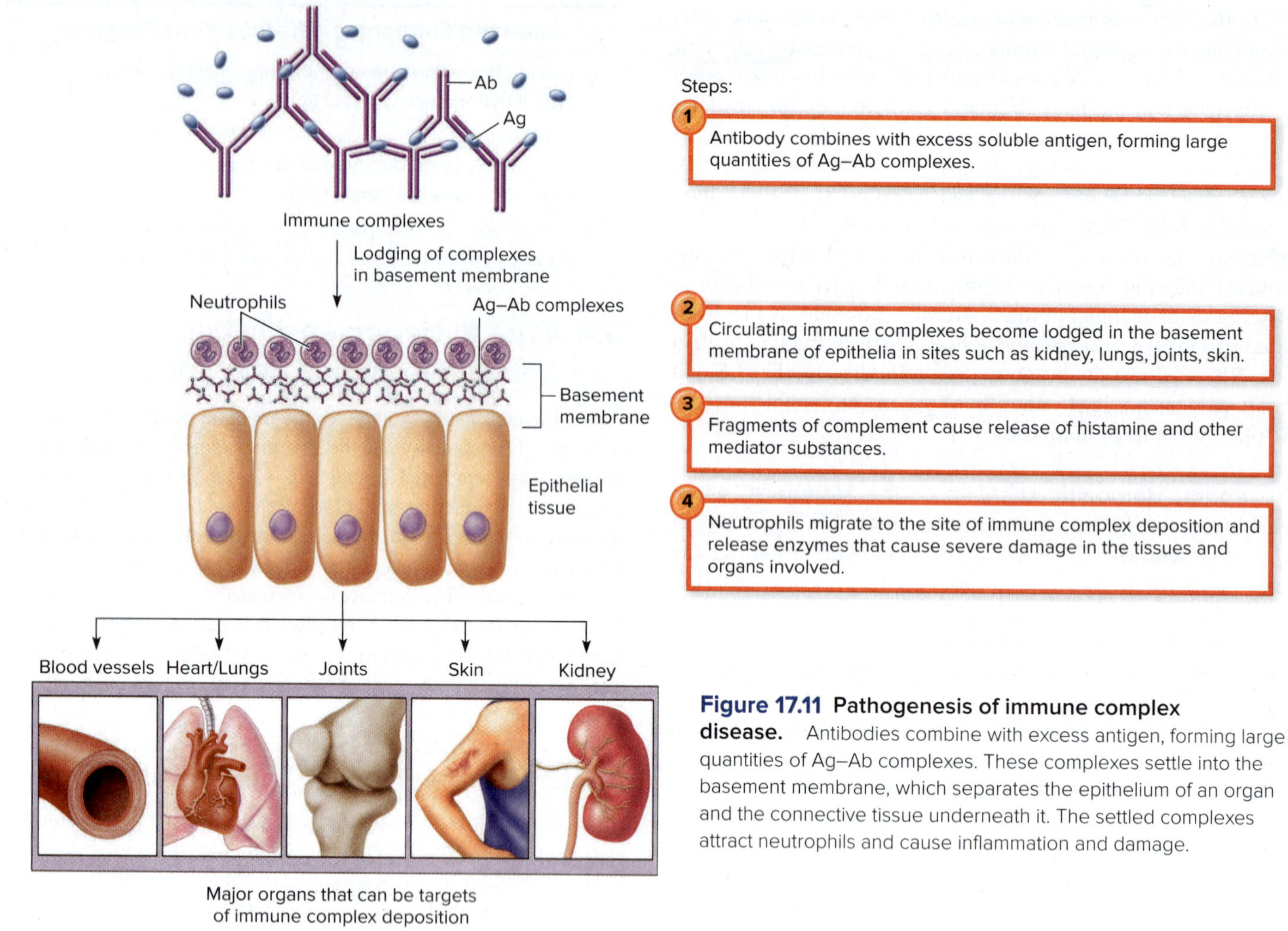
Ab
Ag
Immune complexes
Lodging of complexes in basement membrane
Neutrophils
Ag–Ab complexes
Basement membrane
Epithelial tissue
Blood vessels
Heart/Lungs
Joints
Skin
Kidney
Major organs that can be targets of immune complex deposition
Steps:
1 Antibody combines with excess soluble antigen, forming large quantities of Ag–Ab complexes.
2 Circulating immune complexes become lodged in the basement membrane of epithelia in sites such as kidney, lungs, joints, skin.
3 Fragments of complement cause release of histamine and other mediator substances.
4 Neutrophils migrate to the site of immune complex deposition and release enzymes that cause severe damage in the tissues and organs involved.

Figure 17.11 Pathogenesis of immune complex disease. Antibodies combine with excess antigen, forming large quantities of Ag–Ab complexes. These complexes settle into the basement membrane, which separates the epithelium of an organ and the connective tissue underneath it. The settled complexes attract neutrophils and cause inflammation and damage.

ways that primary infection with *Streptoccus pyogenes*, also known as Group A Strep, can cause a different set of symptoms some time after the first set of symptoms is resolved. One of these is APSGN.

APSGN occurs when particular types of Group A Strep cause disease such as a sore throat or, more commonly, when it causes a skin infection (covered in chapter 19). Complexes of antibody with streptococcal antigens are thought to settle onto the basement membranes on the glomeruli in the kidney. Activation of complement leads to the infiltration of neutrophils, resulting in damage to the kidney.

17.4 Learning Outcomes—Assess Your Progress

12. Identify commonalities and differences between type II and type III hypersensitivities.
13. Describe how both antibodies and neutrophils contribute to acute post-streptococcal glomerulonephritis.

17.5 Type IV Hypersensitivities: Cell-Mediated (Delayed) Reactions

Type IV diseases result when T cells respond to antigens displayed on self tissues or on transplanted foreign cells. Type IV immune dysfunction has traditionally been known as delayed hypersensitivity because the symptoms arise 2 to 3 days following the second contact with an antigen. Examples of type IV hypersensitivity include delayed reactions to infectious agents, contact dermatitis, and graft rejection. The basic mechanism is that sensitized T cells activate macrophages to cause an inflammatory response. While it is delayed in onset, it can last for days to weeks.

Delayed Hypersensitivity to Microbes

A classic example of delayed-type hypersensitivity occurs when a person sensitized by previous tuberculosis exposure is injected intradermally (very shallowly) with an extract (tuberculin) of the bacterium *Mycobacterium tuberculosis*. The so-called tuberculin reaction is an acute skin inflammation at the injection site appearing within 24 to 48 hours. Type IV hypersensitivity arises from time-consuming cellular events involving a specific class of T cells (T_H1) that receive the processed allergens from dendritic cells. Activated T_H cells release cytokines that recruit various inflammatory cells such as macrophages. The buildup of fluid and cells at the site gives rise to a red papule **(figure 17.12)**. Delayed hypersensitivity reactions can play a role in natural chronic infection (tertiary syphilis, for example), leading to extensive damage to organs through granuloma formation.

0 14.06
mm/in
0

Figure 17.12 Positive tuberculin test. Intradermal injection of tuberculin extract in a person sensitized to tuberculosis yields a slightly raised, red bump greater than 10 mm in diameter.

Contact Dermatitis

The most common delayed allergic reaction, contact dermatitis, is caused by exposure to resins in poison ivy or poison oak, to simple haptens in household and personal articles (jewelry, cosmetics, elasticized undergarments), or to certain drugs. Like immediate atopic dermatitis, the reaction to these allergens requires a sensitizing dose followed by a provocative dose. The allergen first penetrates the outer skin layers, is processed by Langerhans cells (skin dendritic cells), and is presented to T cells. When subsequent exposures attract lymphocytes and macrophages to this area, those cells give off enzymes and inflammatory cytokines that severely damage the epidermis in the immediate vicinity **(figure 17.13*a*).** This response accounts for the intensely itchy papules and blisters that are the early symptoms **(figure 17.13*b*).** As healing occurs, the epidermis is replaced by a thick, keratinized layer. Depending on the dose and the sensitivity of the individual, the time from initial contact to healing can be a week to 10 days.

T Cells and Their Role in Organ Transplantation

Transplantation or grafting of organs and tissues is a common medical procedure. Although it is life-saving, this technique is plagued by the natural tendency of lymphocytes to seek out foreign antigens and mount a campaign to destroy them. The bulk of the damage that occurs in graft rejections can be attributed to cytotoxic T-cell action. This section covers the mechanisms involved in graft rejection, tests for transplant compatibility, reactions against grafts, prevention of graft rejection, and types of grafts.

The Genetic and Biochemical Basis for Graft Rejection

In chapter 16, we learned that the genes and markers in major histocompatibility (MHC or HLA) classes I and II are extremely important in recognizing self and in regulating the immune response. These molecules also set the events of graft rejection in motion. Although the cells of different persons display different variants of these cell surface molecules, the markers will be identical in different cells of the same person. Similarity is seen among related siblings and parents, but the more distant the relationship, the less likely that the MHC genes and markers will be similar. When donor tissue (a graft) displays surface molecules of a different MHC class, the T cells of the recipient (the host) will recognize its foreignness and react against it.

Two Ways This Can Go Wrong

Host Rejection of Graft When the cytotoxic T cells of a host encounter foreign class I MHC markers on the surface of grafted cells, they release interleukin-2 as part of a general immune

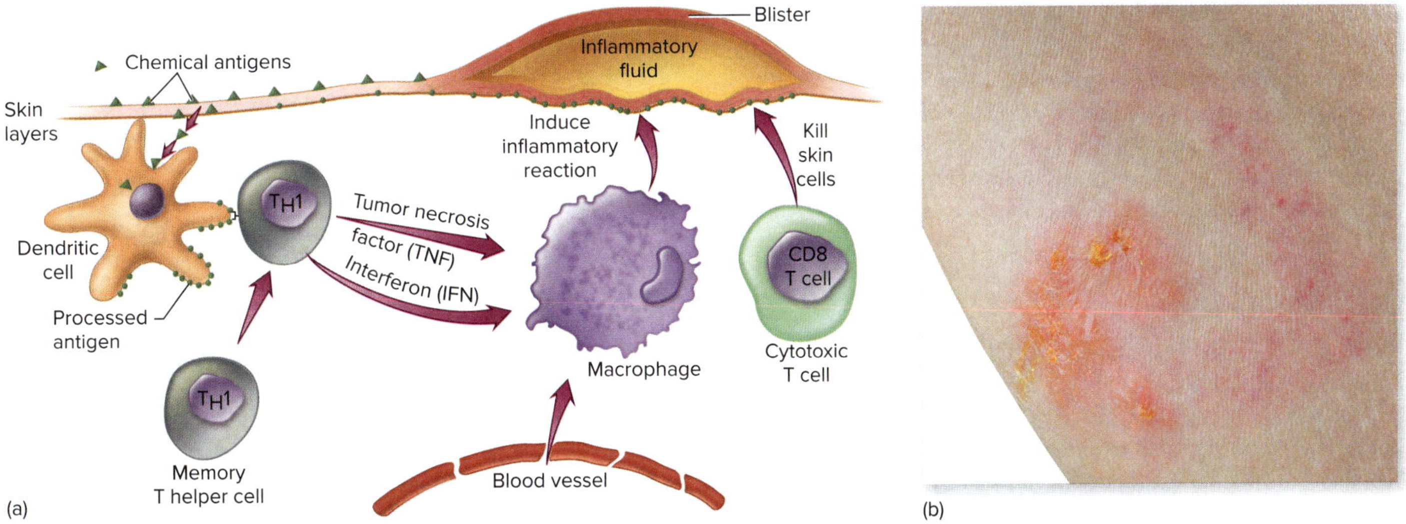
Blister
Chemical antigens
Inflammatory fluid
Skin layers
Induce inflammatory reaction
Kill skin cells
TH1
Tumor necrosis factor (TNF)
Interferon (IFN)
Dendritic cell
CD8 T cell
Processed antigen
Macrophage
Cytotoxic T cell
TH1
Memory T helper cell
Blood vessel
(a)
(b)

Figure 17.13 Contact dermatitis. **(a)** The process of contact dermatitis. For reference, look at figure 17.12. Here you see that the antigen is an environmental chemical on the skin, and that inflammatory events contribute to fluid buildup in skin blisters. **(b)** Contact dermatitis from poison ivy, showing various stages of involvement: blister, scales, and redness.

(b) carroteater/Shutterstock

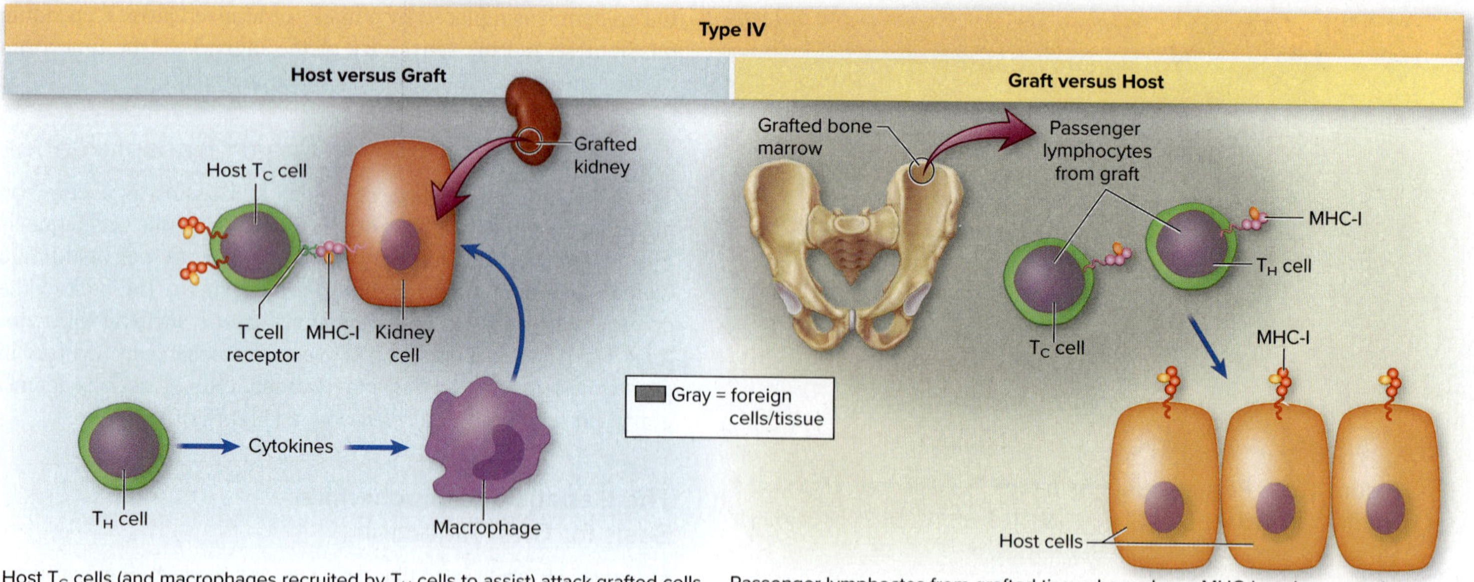
Type IV
Host versus Graft
Graft versus Host
Host T_C cell
Grafted kidney
T cell receptor
MHC-I
Kidney cell
T_H cell
Cytokines
Macrophage
Grafted bone marrow
Passenger lymphocytes from graft
MHC-I
T_H cell
T_C cell
MHC-I
Gray = foreign cells/tissue
Host cells

Host T_C cells (and macrophages recruited by T_H cells to assist) attack grafted cells with foreign MHC-I markers.

Passenger lymphoctes from grafted tissue have donor MHC-I markers; attack recipient cells with different MHC-I specificity.

Figure 17.14 **Development of type IV incompatible tissue graft reactions.**

mobilization **(figure 17.14).** Antigen-specific helper and cytotoxic T cells bind to the grafted tissue and secrete lymphokines that begin the rejection process within 2 weeks of transplantation. Late in this process, antibodies formed against the graft tissue contribute to immune damage, resulting in the destruction of the vascular supply and death of the graft.

Graft Rejection of Host In certain severe immunodeficiencies, the host cannot or does not reject a graft. But this failure may not protect the host from serious damage because graft incompatibility is a two-way phenomenon. Some grafted tissues (especially bone marrow) contain white blood cells from the donor, called passenger lymphocytes (as shown in figure 17.14). This makes it quite possible for the graft to reject the host, causing **graft versus host disease (GVHD).** Because any host tissue bearing MHC markers foreign to the graft can be attacked, the effects of GVHD are widely systemic and toxic. A bumpy, peeling skin rash is the most common symptom. Other organs affected are the liver, intestine, muscles, and mucous membranes. Lesions caused by this in the oral cavity can even become cancerous. GVHD typically occurs within 100 to 300 days of the graft. Overall such reactions are declining due to better screening and more sophisticated means of selecting tissues.

Classes of Grafts

Grafts are generally classified according to the genetic relationship between the donor and the recipient. Tissue transplanted from one site on an individual's body to another site on his or her body is known as an **autograft.** Typical examples are skin replacement in burn repair and the use of a vein to fashion a coronary artery bypass. In an **isograft,** tissue from an identical twin is used. Because isografts do not contain foreign antigens, they are not rejected. **Allografts,** the most common type of grafts, are exchanges between genetically different individuals belonging to the same species (two humans). A close genetic correlation is sought for most allograft transplants (see next section). A **xenograft** is a tissue exchange between individuals of different species, such as a pig heart valve grafted onto a human heart.

Types of Transplants

Transplantation involving every major organ, including parts of the brain and even faces, has been performed but most often involves skin, liver, heart, kidney, coronary artery, cornea, and bone marrow. The sources of organs and tissues are live donors, the recently deceased, and fetal tissues.

In the past decade, advancements in transplantation science have expanded the possibilities for treatment and survival. Fetal tissues have been used in the treatment of diabetes and Parkinson disease, while parents have successfully donated portions of their organs to help save their children suffering from the effects of cystic fibrosis or liver disease. Recent advances in stem cell technology have made it possible to isolate stem cells more efficiently from blood donors, and the use of umbilical cord blood cells has furthered progress in this area of science.

Bone marrow transplantation is a rapidly growing medical procedure for patients with immune deficiencies, aplastic anemia, leukemia and other cancers, and radiation damage. Before closely matched bone marrow can be infused, the patient is pretreated with chemotherapy and/or whole-body irradiation, a procedure designed to destroy the person's own blood stem cells to prevent rejection of the new marrow cells. While the donor is sedated, a bone marrow sample is obtained by inserting a needle into an accessible marrow cavity. The most favorable sites are the crest and spine of the ilium (major bone of the pelvis). In a few weeks, the depleted marrow will naturally replace itself. Implanting the harvested bone marrow is rather convenient because it is not necessary to place it directly into the marrow cavities of the recipient. Instead, it is dripped intravenously into the circulation, and the new marrow cells automatically settle in the appropriate bone marrow regions. Within 2 weeks to a month after infusion, the grafted cells are established in the host. Because donor lymphoid cells can still cause GVHD, antirejection drugs may be necessary. Interestingly, after bone marrow transplantation, a recipient's blood type may change to the blood type of the donor.

17.5 Learning Outcomes—Assess Your Progress

14. Identify one type IV delayed hypersensitivity reaction, and describe the role of T cells in the pathogenesis of this condition.
15. List four classes of grafts, and explain how host versus graft and graft versus host diseases develop.

17.6 An Inappropriate Response Against Self: Autoimmunity

Most of the immune diseases we have covered so far are caused by foreign antigens. In the case of autoimmunity, an individual develops hypersensitivity to him- or herself. This pathologic process accounts for **autoimmune diseases,** in which **autoantibodies,** T cells, and in some cases both mount an abnormal attack against self antigens. There are many different types of autoimmune disease. In general, they are either *systemic,* involving several major organs, or *organ-specific,* involving only one organ or tissue. There are more than 80 recognized autoimmune diseases affecting up to 23 million Americans. Many of these diseases result in catastrophic consequences. For example, people with rheumatoid arthritis have a 60% increased risk of death from cardiovascular disease. Some major diseases, their targets, and their basic pathology are presented in **table 17.4.** (For a reminder of hypersensitivity types, refer to table 17.1.)

Possible Causes of Autoimmune Disease

The root cause of autoimmune disease is still a frustrating mystery for the most part. We present some factors that influence—or may influence—the development of autoimmunity. We do know that susceptibility is determined by genetics and influenced by gender.

Women account for nearly 75% of all cases of diagnosed autoimmune disease, but the biological basis for this fact largely remains a mystery. A number of autoimmunities have been linked to genes on the X chromosome, and one hypothesis centers on the role of X-chromosome inactivation in the development of these diseases. One recent study showed that the immune systems of women respond more robustly to antigens. The researchers speculate that this has evolved as a strategy to protect developing fetuses. But this could also make women more likely to have autoimmune hyperreactivities.

The Origins of Autoimmune Disease

Genetics

Susceptibility can be influenced strongly by genetics and gender. Cases cluster in families, and even members without symptoms tend to develop the autoantibodies for that disease. Studies show particular genes in the class I and II major histocompatibility complex coincide with certain autoimmune diseases. For example, autoimmune joint diseases such as rheumatoid arthritis and ankylosing spondylitis are more common in persons with the B-27 HLA type. With the expansion of genomic technology and the screening of whole genomes, many novel genes have recently been found to play a role in the pathway to autoimmunity. Some research suggests that autoimmune symptoms are a side effect of the body's immune response to cancer that is developing in the body (and often does not come to fruition, though the autoimmune disease remains).

A moderate, regulated amount of autoimmunity is probably required to dispose of old cells and cellular debris. Disease apparently arises when this regulatory or recognition apparatus goes awry. Sometimes the processes go awry due to genetic irregularities or inherent errors in the host's physiological processes. In a large subset of cases, however, microbes are behind the malfunctioning.

Molecular Mimicry

This is a process in which microbial antigens bearing molecular markers similar to those on human cells induce the formation of antibodies that can cross-react with normal tissues. This is one possible explanation for the pathology of rheumatic fever. In this condition, it seems that a previous streptococcal infection "tricks" the immune response to react against healthy human tissues, such as cardiac and joint tissues. Similarly, T cells primed to react with streptococcal surface proteins also appear to react with keratin cells in the skin, causing them to proliferate. For this reason, patients with psoriasis often report flare-ups after a strep throat infection.

Table 17.4 Selected Autoimmune Diseases

Disease	Target	Type of Hypersensitivity	Characteristics
Systemic lupus erythematosus (SLE)	Systemic	III	Inflammation of many organs; antibodies against red and white blood cells, platelets, clotting factors, nucleus DNA.
Rheumatoid arthritis and ankylosing spondylitis	Systemic	II, III, and IV	Vasculitis; frequent target is joint lining; antibodies against other antibodies (rheumatoid factor); T-cell cytokine damage.
Graves' disease	Thyroid	II	Antibodies against thyroid-stimulating hormone receptors.
Myasthenia gravis	Muscle	II	Antibodies against the acetylcholine receptors on the nerve-muscle junction alter function.
Type I diabetes	Pancreas	IV	T cells attack insulin-producing cells.
Multiple sclerosis	Myelin	II and IV	T cells and antibodies sensitized to myelin sheath destroy neurons.

Infection

Autoimmune disorders such as type 1 diabetes and multiple sclerosis are possibly triggered by *viral infection*. Viruses can alter human cell receptors, thereby causing immune cells to attack the tissues bearing viral receptors.

The Gut Microbiome

By now you know the importance of the gut microbiome. We have also seen that the composition of the gut microbiome has changed over the last 50 years, due to the use of antibiotics, less exposure to the outdoors, and an altered, more artificial, diet. Many scientists connect the rise in autoimmune diseases over the same time period to this phenomenon. Studies have raised the idea that a healthy microbiome is fundamentally important to training our immune system what to react against and what to tolerate. Drastic changes in the microbiome would be expected to disrupt this training process.

Many researchers predict that altering the microbiome, or using chemicals from a healthy microbiome, will be a major treatment strategy for autoimmune diseases within 5 to 10 years.

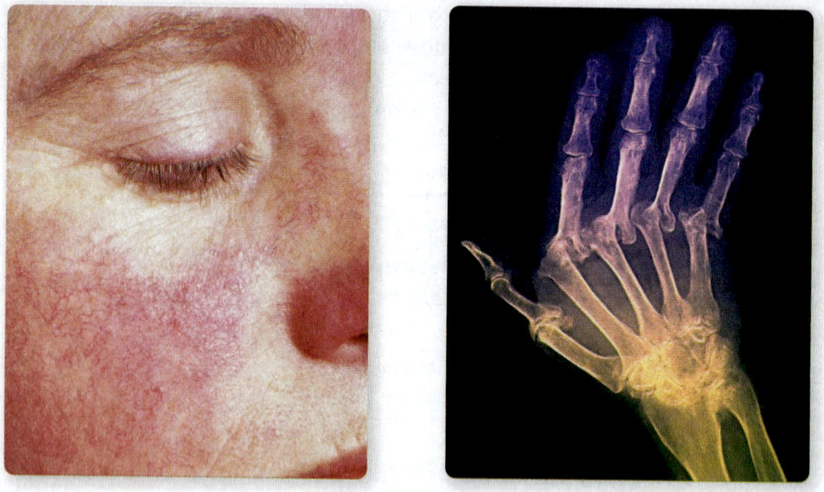

(a) (b)

Figure 17.15 Common autoimmune diseases. **(a)** Systemic lupus erythematosus. One symptom is a prominent rash across the bridge of the nose and on the cheeks. These papules and blotches can also occur on the chest and limbs. **(b)** Rheumatoid arthritis commonly targets the synovial membrane of joints. Over time, chronic inflammation causes thickening of this membrane, erosion of the articular cartilage, and fusion of the joint. These effects severely limit motion and can eventually swell and distort the joints.

(a) BSIP/Science Source; (b) Science Photo Library/Alamy Stock Photo

Examples of Autoimmune Disease

Systemic Autoimmunities

One of the most severe chronic autoimmune diseases is systemic lupus erythematosus (SLE), or lupus. This name originated from the characteristic butterfly-shaped rash that drapes across the nose and cheeks **(figure 17.15*a*)**, as ancient physicians thought the rash resembled a wolf bite on the face (*lupus* is Latin for "wolf"). Although the manifestations of the disease vary considerably, all SLE patients produce autoantibodies against a great variety of organs and tissues or intracellular materials, such as the nucleoprotein of the nucleus and mitochondria.

In SLE, autoantibody–autoantigen complexes appear to be deposited in the basement membranes of various organs. Kidney failure, blood abnormalities, lung inflammation, myocarditis, and skin lesions are the predominant symptoms. One form of chronic lupus (called discoid) is influenced by exposure to the sun and afflicts the skin. The etiology of lupus is still a puzzle, and it is not exactly clear how such a generalized loss of self-tolerance arises. Viral infection and the loss of normal suppression of immune response are suspected. Another possible cause is the inefficient clearing of normal cellular debris. Genomics research has also led to the identification of several genetic variations linked to SLE susceptibility and will most likely lead to the development of new targeted therapies for this disease. The diagnosis of SLE can usually be made with blood tests. Antibodies against the nucleus and various tissues are common, and a positive test for the lupus factor (an antinuclear factor) is also very indicative of the disease.

Rheumatoid arthritis, another systemic autoimmune disease, leads to progressive, debilitating damage to the joints. In some patients, the lungs, eyes, skin, and nervous system are also involved. In the joint form of the disease, autoantibodies form immune complexes that bind to the synovial membrane of the joints, activate phagocytes, and stimulate release of cytokines. Chronic inflammation leads to scar tissue and joint destruction. The joints in the hands and feet are affected first, followed by the knee and hip joints **(figure 17.15*b*)**. These cytokines (such as tumor necrosis factor [TNF]) can then trigger additional type IV delayed hypersensitivity responses. Treatment has traditionally involved the targeting of TNF or TNF-mediated pathways, but new drugs that target other immune system components are appearing frequently.

Autoimmunities of the Endocrine Glands

The underlying cause of **Graves' disease** is the attachment of autoantibodies to receptors on the thyroxin-secreting follicle cells of the thyroid gland. The abnormal stimulation of these cells causes the overproduction of this hormone and the symptoms of hyperthyroidism, which affect nearly every body system.

Type I diabetes is another condition that may be a result of autoimmunity. Insulin, secreted by the beta cells in the pancreas, regulates and is essential to the utilization of glucose by cells. Molecular mimicry has been implicated in the sensitization of cytotoxic T cells in type I diabetes. The inappropriate immune response then leads to the lysis of important pancreatic cells.

Neuromuscular Autoimmunities

An important example of a neuromuscular autoimmunity is multiple sclerosis (MS), a paralyzing neuromuscular disease associated with lesions in the insulating myelin sheath that surrounds neurons of the central nervous system. T-cell-induced and autoantibody-induced damage severely compromises the capacity of neurons to send impulses, resulting in muscular weakness and tremors, difficulties in speech and vision, and some degree of paralysis. Most patients with MS first experience symptoms as young adults, and they tend to experience remissions (periods of relief) alternating with recurrences of disease throughout their lives. Data suggest a possible association between infection with human herpesvirus 6 or some other virus and the onset of disease.

There are some early indications that people who recover from COVID-19 may experience worsening of preexisting autoimmune conditions, possibly due to excessive activation of inflammatory substances, including cytokines.

17.6 Learning Outcomes—Assess Your Progress

16. Outline at least three different explanations for the origin of autoimmunity.
17. List three autoimmune diseases, and describe immunologic features most important to each.

17.7 Immunodeficiency Diseases: Hyposensitivities

Occasionally, errors occur in the development of the immune system and a person is born with or develops weakened immune responses called immunodeficiencies. The predominant consequences of immunodeficiencies are recurrent, overwhelming infections, often with opportunistic microbes. Immunodeficiencies fall into two general categories: *primary diseases,* present at birth (congenital) and usually stemming from genetic errors, and *secondary diseases,* acquired after birth and caused by natural or artificial agents.

Primary Immunodeficiency Diseases

Primary deficiencies affect both adaptive immunities such as antibody production and less specific ones such as phagocytosis. Consult **figure 17.16** to survey the places in the normal

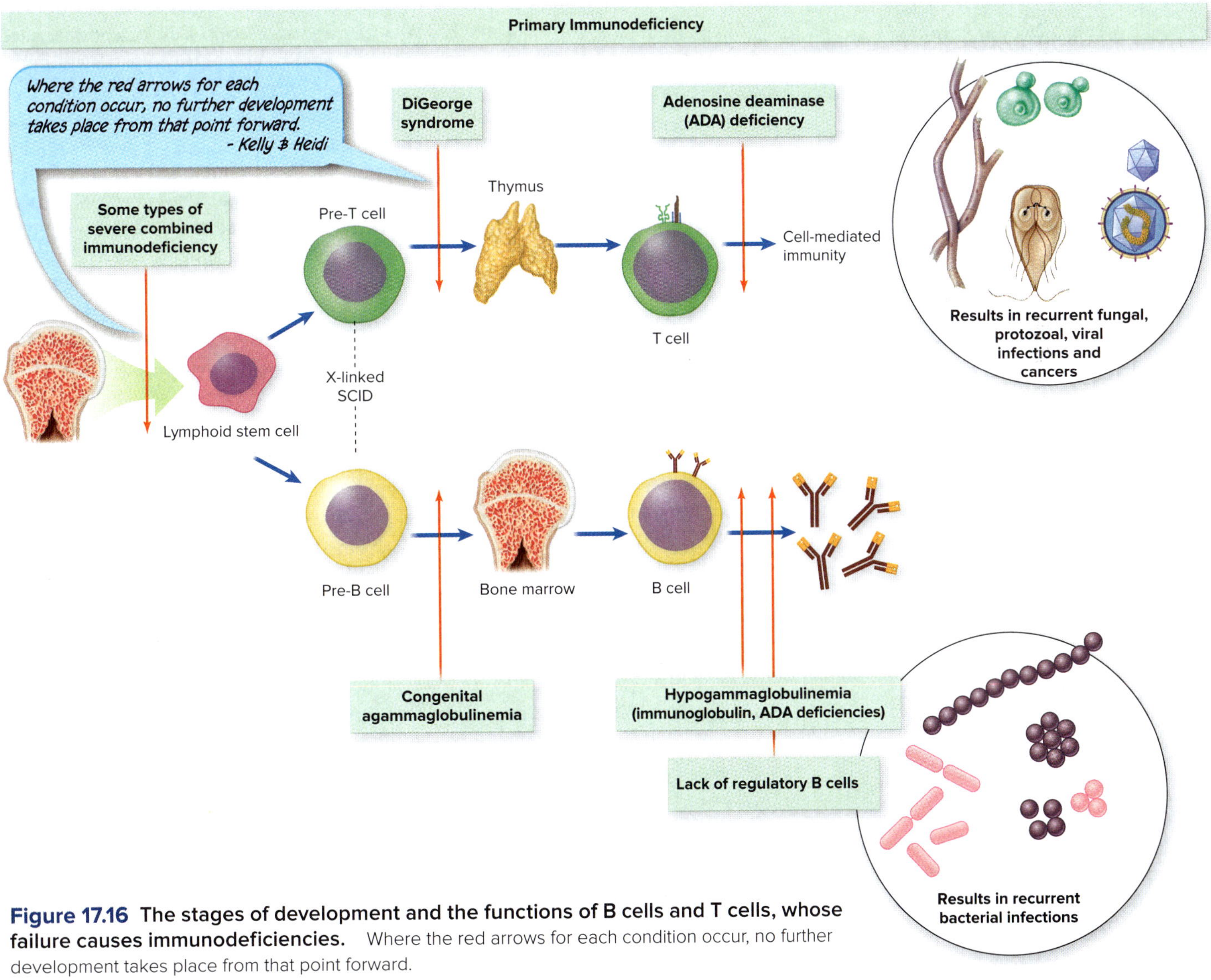

Primary Immunodeficiency
Where the red arrows for each condition occur, no further development takes place from that point forward. - Kelly & Heidi
DiGeorge syndrome
Adenosine deaminase (ADA) deficiency
Some types of severe combined immunodeficiency
Pre-T cell
Thymus
T cell
Cell-mediated immunity
X-linked SCID
Lymphoid stem cell
Results in recurrent fungal, protozoal, viral infections and cancers
Pre-B cell
Bone marrow
B cell
Congenital agammaglobulinemia
Hypogammaglobulinemia (immunoglobulin, ADA deficiencies)
Lack of regulatory B cells
Results in recurrent bacterial infections

Figure 17.16 The stages of development and the functions of B cells and T cells, whose failure causes immunodeficiencies. Where the red arrows for each condition occur, no further development takes place from that point forward.

sequential development of lymphocytes where defects can occur and the possible consequences. In many cases, the deficiency is due to an inherited abnormality, though the exact nature of the abnormality is not known for a number of diseases. In some deficiencies, the lymphocyte in question is completely absent or is present at very low levels; in others, lymphocytes are present but do not function normally. Because the development of B cells and T cells diverges at some point, an individual can lack one or both cell lines. Some deficiencies affect the function of other cells in the immune system as well.

Figure 17.17 Facial characteristics of a child with DiGeorge syndrome. Typical characteristics include low-set, deformed earlobes; wide-set, slanted eyes; a small, bowlike mouth; and the absence of a philtrum (the vertical furrow between the nose and upper lip).
Chris So/ZUMA Press/Newscom

Clinical Deficiencies in B-Cell Development or Expression

Genetic deficiencies in B cells usually appear as an abnormality in immunoglobulin expression. In some instances, only certain immunoglobulin classes are absent; in others, the levels of all types of immunoglobulins (Ig) are reduced. A significant number of B-cell deficiencies are X-linked (also called sex-linked) recessive traits, meaning that the gene occurs on the X chromosome and the disease appears primarily in male children.

The term **agammaglobulinemia** literally means "the absence of gamma globulin," the fraction of serum that contains immunoglobulins. Because it is very rare for Ig to be completely absent, some physicians prefer the term **hypogammaglobulinemia.** T-cell function in these patients is usually normal. The symptoms of recurrent, serious bacterial infections usually appear about 6 months after birth. The bacteria most often implicated are pyogenic cocci, *Pseudomonas,* and *Haemophilus influenzae.* The most common infection sites are the lungs, sinuses, meninges, and blood. Many Ig-deficient patients can have recurrent infections with viruses and protozoa, as well. Patients often manifest a wasting syndrome and have a reduced life span, but modern therapy has improved their prognosis. The current treatment for this condition is passive immunotherapy with immune serum globulin.

The lack of a particular class of immunoglobulin is a relatively common condition, though its underlying genetic mechanisms are not clear. IgA deficiency is the most prevalent form, in which patients have normal quantities of B cells and other immunoglobulins but are unable to synthesize IgA. Consequently, they lack protection against local microbial invasion of the mucous membranes and suffer frequent respiratory and gastrointestinal infections. There is no existing treatment for IgA deficiency because conventional preparations of immune serum globulin are high in IgG, not IgA.

Clinical Deficiencies in T-Cell Development or Expression

Due to the critical role of T cells in immune defenses, a genetic defect in T-cell development results in a broad spectrum of diseases, including severe opportunistic infections, wasting, and cancer. In fact, a defective T-cell line is usually more devastating than a defective B-cell line because T helper cells are required to assist in most adaptive immune reactions. The deficiency can occur anywhere along the developmental spectrum, from the thymus to mature, circulating T cells.

Abnormal Development of the Thymus The most severe of the T-cell deficiencies involve the congenital absence or immaturity of the thymus. Thymic aplasia, or **DiGeorge syndrome,** comes about due to errors during embryogenesis or deletions in chromosome 22. It is characterized by a lack of cell-mediated immunity, making children highly susceptible to persistent infections by fungi, protozoa, and viruses. Vaccinations using live, attenuated microbes pose a danger, and common childhood infections such as chickenpox, measles, and mumps can be overwhelming and fatal in these children. Other symptoms of thymic failure are reduced growth, wasting of the body, unusual facial characteristics **(figure 17.17),** and an increased incidence of lymphatic cancer. These children can have reduced antibody levels, and they are unable to reject transplants. The major therapy for them is a transplant of thymus tissue.

Severe Combined Immunodeficiencies: Dysfunction in B and T Cells

Severe combined immunodeficiencies (SCIDs) are the most serious and potentially lethal forms of immunodeficiency disease because they involve dysfunction in both lymphocyte systems. Some SCIDs are due to the complete absence of the lymphocyte stem cell in the marrow; others are attributable to the dysfunction of B cells and T cells later in development. Infants with SCIDs usually manifest the T-cell deficiencies within days after birth by developing candidiasis, sepsis, pneumonia, or systemic viral infections.

In the two most common forms, Swiss-type agammaglobulinemia and thymic alymphoplasia, genetic defects in the development of the lymphoid cell line result in extremely low numbers of all lymphocyte types and poorly developed humoral and cellular immunity. A rare form of SCID, called **adenosine deaminase (ADA) deficiency,** is caused by an autosomal recessive defect in the metabolism of adenosine. In this case, lymphocytes develop but a metabolic product builds up abnormally and selectively destroys them. A small number of SCID cases are due to a developmental defect in receptors for B and T cells, and other cases are due to X-linked deficiencies in interleukin receptors.

Because of their profound lack of adaptive immunities, children with SCID require the most rigorous kinds of aseptic techniques to protect them from opportunistic infections. Aside from life in a sterile plastic bubble, the only serious option for their longtime survival is total replacement or correction of dysfunctional lymphoid cells. Bone marrow transplants are usually performed, with varying degrees of success.

Secondary Immunodeficiency Diseases

Secondary acquired deficiencies in B cells and T cells are caused by one of four general agents:

1. infection,
2. noninfectious metabolic disease,
3. chemotherapy, or
4. radiation.

AIDS produces one well-recognized infection-induced immunodeficiency. This syndrome is caused when several types of immune cells, including T helper cells, monocytes, macrophages, and antigen-presenting cells, are infected by the human immunodeficiency virus (HIV). It is generally thought that the depletion of T helper cells and functional impairment of immune responses ultimately account for the cancers and opportunistic protozoan, fungal, and viral infections associated with this disease. Other infections that can deplete immunities are measles, leprosy, and malaria.

Cancers that target the bone marrow or lymphoid organs can be responsible for extreme malfunction of both humoral and cellular immunity. In leukemia, a massive number of cancer cells compete for space and literally displace the normal cells of the bone marrow and blood. Plasma cell tumors produce large amounts of nonfunctional antibodies, and thymus tumors cause severe T-cell deficiencies.

An ironic outcome of lifesaving medical procedures is the possible suppression of a patient's immune system. Drugs that prevent graft rejection or decrease the symptoms of rheumatoid arthritis can likewise suppress beneficial immune responses, while radiation and anticancer drugs can be extremely damaging to the bone marrow and other body cells.

17.7 Learning Outcomes—Assess Your Progress

18. Distinguish between primary and secondary immunodeficiencies, explaining how each develops.
19. Define *severe combined immunodeficiency,* and discuss current therapeutic approaches to this type of disease.

Media Under The Microscope Wrap-Up

The news article "A Man with an Allergy to Cold Air Almost Died after Stepping out of a Hot Shower" describes a research study published in the scientific literature. The report originally was published in the *Journal of Emergency Medicine*.

The **intended message** of the *Insider* article was to describe this very unusual incident. Along the way, it provided some insight into the nature of allergy and anaphylaxis. It also warns emergency room physicians to be aware that cold allergy can lead to negative consequences if patients are given IVs in which the fluid is cold.

A **critical reading** of the website article leads me to conclude that the author(s) did a good job with conveying the facts. It did not include comments from other scientists, which always increases the trust I have for such articles. I found it to be important, even though the title made me think it might be a little sensationalistic. Turns out it was not, and cold allergy can be a dangerous thing.

Cavan Images

I would explain to my **nonmicrobiologist friends** how allergy works, and that apparently cold activates something that the immune response responds to overwhelmingly in some people.

My **overall grade** for this article is a solid B. Good information that provided enough information to connect to basic science with which I am familiar.

Source: Insider, "A Man with an Allergy to Cold Air Almost Died after Stepping out of a Hot Shower," online article posted November 4, 2020.

Study Smarter: Better Together

These activities are designed for you to use on your own with a study group—either a face-to-face group or a virtual one, consisting of 3–5 members. Studying together can be very helpful, but there are effective and ineffective ways to do it. For example, getting together without a clear structure is often not a good use of your time. Use your time efficiently by using one or more of the exercises below.

FACE-TO-FACE GROUPS

Use one or more of the activities below.

Peer Instruction: Assign numbers to your group members to use all semester long. Now look at these five concepts from this chapter. Each group member prepares a 5-minute lesson on the topic corresponding to their number. Don't worry if you have fewer than 5 members; just use however many you have! During your group study time, each member presents his or her lesson, and the group spends another 5–10 minutes discussing that lesson.

1. Brief overview of four types of hypersensitivity
2. Type I hypersensitivity
3. Type II hypersensitivity
4. Type III hypersensitivity
5. Type IV hypersensitivity

Concept Maps: Each member of the group should use this list of terms from this chapter to generate their own concept map. This can be hand-drawn or created using software (see Appendix C for guidelines). During group study time, compare each other's concept maps and help each other make sure they are correct. Of course, there are many different "correct" maps. Examining each member's map will help you talk through the varied concepts and how they are related.

Concept Terms:

lysed cells	immune complexes	cell-bound antibody	allergens
damage by T cells	soluble antigens	degranulation	processed antigen

Table Topics: Each group member should identify a concept or topic from this week's class assignments with which they are having trouble and share it during group study time. The other group members can then help to clarify confusing issues or share how they figured it out. Aim for a maximum of 15 minutes per topic. If the topic remains unclear to the group, bring it up during class or use the instructor's office hours or e-mail to ask for help. Taking the time to struggle with a difficult concept first makes your questions much more specific and more likely to yield helpful answers.

VIRTUAL GROUPS

Not everyone has the time or opportunity to meet with group members outside of class time. You or your instructor can create a virtual group using e-mail or the course software.

Weekly Discussion Board: This forum can be used as a way for groups to discuss topics, via e-mail, or other learning management systems or online platforms, before they are covered in class. As each member of the group answers the current week's question, they should send their responses to every other member of their group. It's best to agree on a deadline based on how your class schedule works (Saturday for the next week's topics, for example). Then, after the topic is discussed in class, each member should send a response that all group members will see. If you cover more than one chapter in a week, someone can be designated to choose which chapter Discussion Board question you will use. Or simply decide up front that you will always use the first-chapter-of-the week's question, to keep the schedule simple.

Discussion Question

What feature of antibodies makes them particularly suited for forming complexes with antigen, as in type III hypersensitivities?

Chapter Summary

DISORDERS IN IMMUNITY

17.1 THE IMMUNE RESPONSE: A TWO-SIDED COIN

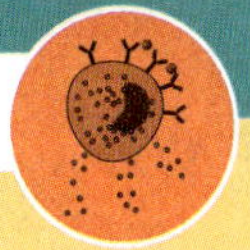

- Immunopathology is the study of diseases associated with excesses and deficiencies of the immune response. Such diseases include allergies, autoimmunity, graft rejections, transfusion reactions, and immunodeficiency disease.
- There are four categories of hypersensitivity reactions: type I (allergy and anaphylaxis), type II (IgG and IgM tissue destruction), type III (immune complex reactions), and type IV (delayed hypersensitivity reactions).

17.2 TYPE I ALLERGIC REACTIONS: ATOPY AND ANAPHYLAXIS

- Antigens that trigger hypersensitivity reactions are allergens. They can be either exogenous (originate outside the host) or endogenous (caused by the host's own tissue).
- Type I hypersensitivity reactions result from excessive IgE production in response to an exogenous antigen.
- Atopy is a chronic, local allergy, whereas anaphylaxis is a systemic, potentially fatal allergic response. Both result from excessive IgE production in response to exogenous antigens.
- Type I hypersensitivities are set up by a sensitizing dose of allergen and expressed when a second provocative dose triggers the allergic response.
- The primary participants in type I hypersensitivities are IgE, basophils, mast cells, and agents of the inflammatory response.
- Allergies are diagnosed by a variety of *in vitro* and *in vivo* tests that assay specific cells, IgE, and local reactions.
- Allergies are treated by medications that interrupt the allergic response at certain points. Allergic reactions can also be prevented by avoiding the allergen, or by injection of allergens in controlled amounts.

17.3 TYPE II HYPERSENSITIVITIES: REACTIONS THAT LYSE CELLS

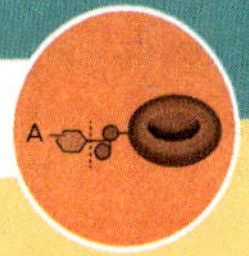
A

- Type II hypersensitivity reactions are complement-assisted reactions that occur when preformed antibodies (IgG or IgM) react with foreign cell–bound antigens, leading to membrane attack complex formation and lysis.
- ABO blood groups are based upon antigens present on red blood cells. These markers are genetically determined and composed of glycoproteins.
- The most common type II reactions occur when transfused blood is mismatched to the recipient's ABO type.
- Type II hypersensitivity disease can develop when Rh− mothers are sensitized to Rh+ RBCs of their unborn babies and the mother's anti-Rh antibodies cross the placenta in subsequent Rh+ pregnancies, causing hemolysis of the newborn's RBCs. This is called hemolytic disease of the newborn, or erythroblastosis fetalis.

17.4 TYPE III HYPERSENSITIVITIES: IMMUNE COMPLEX REACTIONS

- Type III hypersensitivity reactions occur when large quantities of antigen react with host antibody to form immune complexes that settle in tissue cell membranes, causing chronic destructive inflammation. The reactions appear hours or days after the antigen challenge.
- A common form of type III hypersensitivity is acute post-streptococcal glomerulonephritis, or APSGN.

17.5 TYPE IV HYPERSENSITIVITIES: CELL-MEDIATED (DELAYED) REACTIONS

- Type IV, delayed hypersensitivity reactions, like the tuberculin reaction and transplant reactions (host rejection and GVHD), occur when cytotoxic T cells attack either self tissue or transplanted foreign cells. Target cells must display both a foreign MHC and a nonself receptor site.

17.6 AN INAPPROPRIATE RESPONSE AGAINST SELF: AUTOIMMUNITY

- Autoimmune reactions occur when antibodies or host T cells mount an abnormal attack against self antigens.
- Susceptibility to autoimmune disease appears to be influenced by gender and by genes in the MHC complex.
- The composition of the microbiome may have an important influence on the occurrence of autoimmunity.
- Examples of autoimmune diseases include systemic lupus erythematosus, rheumatoid arthritis, and multiple sclerosis.

17.7 IMMUNODEFICIENCY DISEASES: HYPOSENSITIVITIES

- Immunodeficiency diseases occur when the immune response is reduced or absent.
- Primary immune diseases are genetically inherited deficiencies of B cells, T cells, the thymus, or combinations of these. SCIDs are the most severe forms of these diseases due to the loss of both humoral and cell-mediated immunity.
- Secondary immune diseases are caused by infection, organic disease, chemotherapy, or radiation. A well-known infection-induced immunodeficiency is AIDS.

SmartGrid: From Knowledge to Critical Thinking

This *21 Question Grid* takes the topics from this chapter and arranges them with respect to the American Society for Microbiology's Undergraduate Curriculum guidelines—all six of the important "Concepts" as well as the important "Competency" of scientific literacy. Three questions are supplied, which cover chapter content referring to the Concept or Competency in increasing levels of Bloom's taxonomy for learning.

ASM Concept/ Competency	A. Bloom's Level 1, 2—Remember and Understand (Choose one)	B. Bloom's Level 3, 4—Apply and Analyze	C. Bloom's Level 5, 6—Evaluate and Create
Evolution	**1.** Allergy and atopy might have evolved in human populations that have had a "sudden" decrease in **a.** vaccinations. **b.** exposure to vitamin D. **c.** exposure to helminth infections. **d.** intake of processed sugar.	**2.** Can you think of a reason why humans have evolved to mount an immune response to their own fetuses if they have a particular set of foreign antigens (as in the case of erythroblastosis fetalis)?	**3.** Do you suppose that if we continue to transplant allografts into human patients for thousands of years that the immune system will evolve to not react to those foreign antigens? Provide support for your argument.
Cell Structure and Function	**4.** Molecular mimicry is when ___________ on host cells resemble _________ on pathogens, causing the host to mount an immune response to host tissues. (Same word for both blanks.) **a.** antibodies **b.** markers **c.** antigens **d.** two of the above	**5.** What feature of antibodies makes them particularly suited for forming complexes with antigen, as in type III hypersensitivities?	**6.** An autoimmune disease called antiphospholipid syndrome manifests as thrombosis, or the formation of blood clots. It is characterized by the presence of antibodies directed against phospholipids. Connect the sign (elevated antibodies) to the symptom (thrombosis).

ASM Concept/ Competency	A. Bloom's Level 1, 2—Remember and Understand (Choose one)	B. Bloom's Level 3, 4—Apply and Analyze	C. Bloom's Level 5, 6—Evaluate and Create
Metabolic Pathways	**7.** Which substance is most likely to be allergenic? **a.** protein **b.** fat **c.** carbohydrate **d.** nucleic acid	**8.** One of the strategies for preventing type I hypersensitivity is to induce plasma cells to secrete IgG instead of IgE. Why would that block allergy symptoms?	**9.** EpiPens® are devices used to instantly inject a person with epinephrine in the event of an anaphylactic exposure. What does epinephrine do to stop anaphylaxis? Predict what would happen if too much epinephrine was administered.
Information Flow and Genetics	**10.** What percent of the human population is positive for the Rh antigen? **a.** 50% **b.** 65% **c.** 85% **d.** 99%	**11.** Steroids administered to stop an allergic response act in several ways. One of them is to shut down the synthesis of IgE by plasma cells. Identify some intracellular host macromolecules that may be bound (obstructed) by steroid molecules to result in this outcome.	**12.** A very small number of people have a condition called vibratory urticaria. In this condition, physical vibrations (bumpy bus rides, jackhammers, running) can cause skin rashes. Scientists have found a genetic mutation in a single gene for a protein that stabilizes the surface of immune cells. You take it from here: What would a mutation in that protein possibly lead to?
Microbial Systems	**13.** The hygiene hypothesis suggests that **a.** there are still too many microorganisms in the environment. **b.** we may need more contact with antimicrobials as our systems mature **c.** there are not enough microbes on farms. **d.** we may need more contact with microbes as our systems mature.	**14.** Probiotic supplements are sometimes promoted to alleviate the symptoms of hypersensitivity diseases such as rheumatoid arthritis and eczema. Trace the path to biological plausibility for that claim, if any.	**15.** Recent research shows that patients with asthma have a much higher load of bacteria in their lower airways than do healthy subjects. How could that contribute to what we see in the pathology of asthma?
Impact of Microorganisms	**16.** Which statement is true of autoimmunity? **a.** It involves misshapen antibodies. **b.** It refers to "automatic immunity." **c.** It often manifests as types II, III, and IV hypersensitivities. **d.** It has an acute course and then usually resolves itself.	**17.** It is often said that the normal microbiota of a human "trains" the immune system to react to foreign antigens and not react to self antigens. Use what you know about clonal deletion and clonal selection to suggest how this might happen.	**18.** Investigate the link between *Streptococcus pyogenes* infection and some types of obsessive-compulsive disorder. Identify similarities with some concepts in this chapter.
Scientific Thinking	**19.** Which disease would be most similar to AIDS in its pathology? **a.** X-linked agammaglobulinemia **b.** SCID **c.** ADA deficiency **d.** myasthenia gravis	**20.** Do you think people with B-cell deficiencies can be successfully vaccinated against various microbes? Why or why not?	**21.** The Insight box in this chapter tells us that the gut microbiomes of children prone to asthma differ in composition from those of healthy children. Does this prove that the type of microbiome in the gut causes asthma? Elaborate.

Answers to the multiple-choice questions appear in Appendix A.

Visual Connections

This question uses visual images to connect content within and between chapters.

1. **Figure 17.9*c*.** Use the third row below to draw and explain the expected agglutination pattern for a universal donor blood type. Two examples of blood types are provided here.

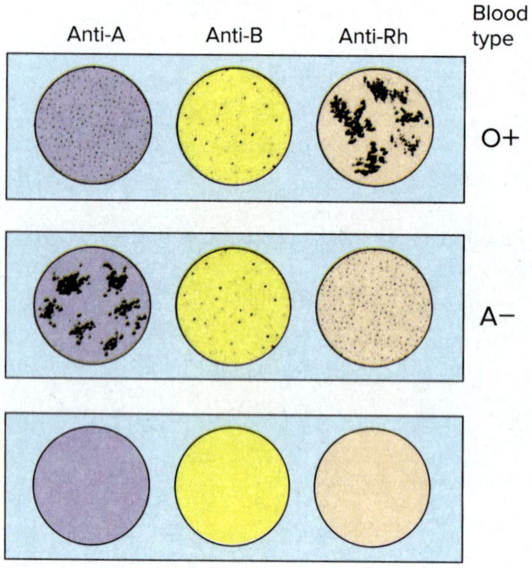
Anti-A
Anti-B
Anti-Rh
Blood type
O+
A−

High Impact Study

These terms and concepts are most critical for your understanding of this chapter—and may be the most difficult. Have you mastered them?

Concepts

- ☐ Hyposensitivity vs. hypersensitivity
- ☐ Four types of hypersensitivity
- ☐ Major components of immune response involved in each of four types
- ☐ Methods of blocking/treating allergy
- ☐ Blood typing
- ☐ Types of transplants
- ☐ Graft vs. host disease and host rejection of transplant
- ☐ Possible causes of autoimmune diseases
- ☐ Outcomes of B-cell or T-cell deficiency
- ☐ Severe combined immunodeficiency disease

Terms

- ☐ Atopy
- ☐ Allergy
- ☐ Anaphylaxis
- ☐ IgG blocking
- ☐ Rh factor
- ☐ Basement membrane

18 Diagnosing Infections

kwanchai.c/Shutterstock

Media Under The Microscope

Using the Microbiome to "Diagnose" Obesity?

This opening case examines an article from the popular media to determine the extent to which it is factual and/or misleading. This case focuses on a recent Daily Mail *article, "Swabbing a Child's Mouth for Bacteria Could Predict How Likely They Are to Become Obese."*

With over one-third of American children classified as overweight or obese, a lot of current research is aimed at understanding the causes or correlating factors of childhood obesity. This article reports on several well-established predictors of obesity, including speed of weight gain after birth, early diet and exercise habits, and the lack of diversity in the gut microbiome. The author also discusses a newly discovered potential link between the diversity of the mouth microbiome and the risk of obesity.

Researchers at Pennsylvania State University conducted a study of 226 two-year-olds by swabbing their mouths and identifying the composition of the microbiome. Children who gained weight rapidly after birth had a less diverse oral microbial population than other children with normal weight gain. The researchers suggest that with further research, an oral swab test just might become an easy screening test for obesity risk.

- What is the **intended message** of the article?
- What is your **critical reading** of the summary of the article? Remember that in this context, "critical reading" does not necessarily mean *What criticism do you have?* but asks you to apply your knowledge to interpret whether the article is factual and whether the facts support the intended message.
- How would you **interpret** the news item for your nonmicrobiologist friends?
- What is your **overall grade** for the news item—taking into account its accuracy and the accuracy of its intended effect?

Media Under the Microscope Wrap-Up appears at the end of the chapter.

Outline and Learning Outcomes

18.1 Preparation for the Survey of Microbial Diseases

1. List the three major categories of microbial identification techniques.
2. Provide a one-sentence description for each of these three categories.

18.2 First Steps: Specimen Collection

3. Identify factors that may affect the identification of an infectious agent from a patient sample.
4. Discuss the effect the bedside health professional, such as a nurse, can have on identification of disease agents.

18.3 Phenotypic Methods

5. List at least three different tests that fall in the phenotypic identification category.
6. Explain the main principle behind biochemical testing, and identify an example of such tests.
7. Discuss two major drawbacks of phenotypic testing methods that require culturing the pathogen.

18.4 Immunologic Methods

8. Define the term *serology,* and explain the immunologic principle behind serological tests.
9. Identify two immunological diagnostic techniques that rely on a secondary antibody, and explain how they work.

18.5 Genotypic Methods

10. Explain why nucleic acid amplification techniques are useful for infectious disease diagnosis.
11. Name two examples of techniques that employ hybridization.
12. Explain how whole-genome sequencing can be used for diagnosis.

18.6 Additional Diagnostic Technologies

13. Describe the benefits of "lab on a chip" technologies for global public health.
14. Identify an advantage presented by mass spectrometry and also by imaging techniques as diagnostic tools.

18.1 Preparation for the Survey of Microbial Diseases

For many professionals, the most pressing topic in microbiology is *how to identify unknown bacteria* in patient specimens or in samples from nature. As seen in **table 18.1,** methods microbiologists use to identify bacteria fall into three main categories: *phenotypic,* which includes a consideration of morphology (microscopic and macroscopic) as well as bacterial physiology or biochemistry; *immunologic,* which entails serological analysis; and *genotypic* (or genetic) techniques. Increasingly, genetic means of identification are being used as a sole resource for identifying bacteria. At the same time, some of the most useful techniques are simple, and "old school."

There has also been an increase in what is called point-of-care diagnosis. These are tests that can be performed either by the patients themselves or at the bedside, meaning that samples do not have to be "sent out" and patient care can continue without delay. During the COVID-19 pandemic, dozens of diagnostic tests were quickly developed. They included genotypic tests, such as polymerase chain reaction (PCR), and immunological tests measuring both antigen and antibody. Some of them return results within an hour and are considered point-of-care tests. Others are approved for at-home use.

It should be emphasized that identifying the antimicrobial sensitivities of microbes causing disease is also of utmost importance. Also, sometimes knowing the antimicrobial susceptibility profile of an isolate can help identify it.

Phenotypic Methods

You will remember that phenotype refers to the traits that an organism is expressing in the present. So, put simply, phenotypic methods of identifying microbes involve examining their appearance and their behavior. In this context, "behavior" is defined as what types of enzymatic activities it can carry out, what kind of physical conditions it thrives in, what antibiotics it is susceptible to, and the chemical composition of its walls and/or membranes.

Immunologic Methods

As you learned in chapter 16, one immune response to antigens is the production of antibodies, which are designed to bind tightly to specific antigens. The nature of the antibody response is exploited for diagnostic purposes when a patient sample is tested for the presence of specific antibodies to a suspected pathogen (antigen). Alternatively, microbial antigens in the patient's tissues can be tested with antibodies "off the shelf." These immunological methods can be easier than trying to isolate the microbe itself.

Genotypic Methods

Examining the genetic material itself (DNA and/or RNA) has revolutionized the identification and classification of bacteria. There are many advantages of genotypic methods over phenotypic methods, when they are available. The primary advantage is that culturing of the microorganisms is not always necessary. In recent decades, scientists have come to realize that there are many more

Table 18.1 Methods of Microbial Identification

Category	Description	Example
Phenotypic	Observation of microbe's microscopic and macroscopic morphology, physiology, antimicrobial susceptibility, and biochemical properties	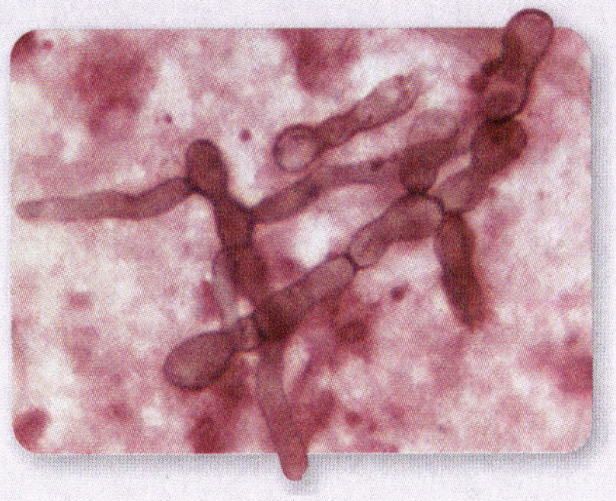
Immunologic	Analysis of microbe using antibodies, or of patients' antibodies using prepackaged antigens	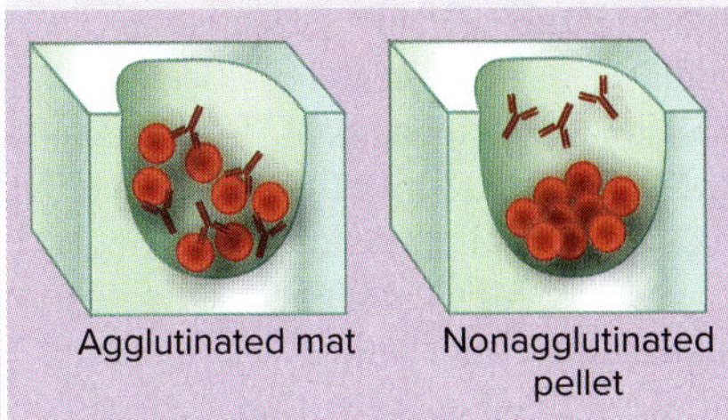 Agglutinated mat Nonagglutinated pellet
Genotypic	Analysis of microbe's DNA or RNA	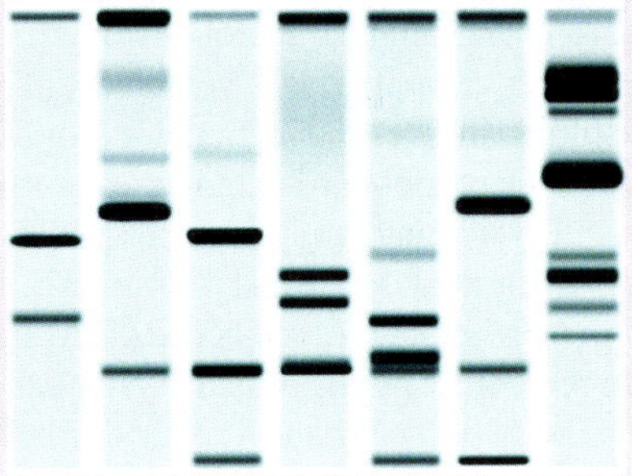

Arthur F. DiSalvo, M.D./CDC (phenotypic); Steven Puetzer/Photographer's Choice/Getty Images (genotypic)

microorganisms that we can't grow in the lab compared with those that we can **(Insight 18.1).**

Special Case: Bloodstream Infections

When microbes are actively multplying in the bloodstream, it is always an emergency. It is a major goal of research to find a way to find the causative agent and the antimicrobial susceptibilities of these infections immediately, rather than waiting for cultures to be grown. Traditionally, when bloodstream infections (also known as sepsis) are suspected, blood cultures are obtained and grown in the lab and then tested for antimicrobial susceptibilities. Since every hour a patient waits for these results can mean an extra 10% mortality rate, doctors typically prescribe a broad-spectrum antibiotic immediately after blood is drawn for the cultures. Good practice dictates that the antibiotic is switched to the correct (narrower-spectrum) drug after the lab results come in. In the era of the antibiotic resistance crisis, this is not ideal. We will highlight various methods to diagnose bloodstream infections, and to shorten the process, as we progress through this chapter.

18.1 Learning Outcomes—Assess Your Progress

1. List the three major categories of microbial identification techniques.
2. Provide a one-sentence description for each of these three categories.

18.2 First Steps: Specimen Collection

Regardless of the method of diagnosis, specimen collection is the common point that guides the health care decisions of every member of a clinical team. The success of identification and treatment depends on how specimens are collected, handled, stored, and cultured. Specimens can be taken by a clinical laboratory scientist or medical technologist, nurse, physician, or even the patient. However, it is vital that general aseptic procedures be used, including sterile sample containers and other tools to prevent contamination from the environment or the patient. **Figure 18.1** illustrates the most common sampling sites and procedures.

INSIGHT 18.1 MICROBIOME: The Human Microbiome Project and Diagnosis of Infection

For decades, scientists have realized that culturing microbes for identification is not only time-consuming and inefficient but also incomplete. Many microbes are fastidious, requiring highly specialized growth conditions, and many others are considered "viable but nonculturable" (VBNC). Some estimate that 99.9% of sequences found in the human body fall under the heading of VBNC, and little is understood about the role that these elusive microbes play in the human microbiome and in disease.

Using techniques developed by J. Craig Venter in mapping the genome of microbes in the oceans, scientists have been able to map the genomes of VBNC microorganisms in the Human Microbiome Project (HMP). Scientists at the Venter Institute developed multiple displacement amplification (MDA) technology. MDA is able to copy fragments of DNA from a single cell until they reach the equivalent of the billions required for analysis. MDA is able to capture 90% of the genes from a single cell.

Using MDA, scientists are able to obtain information about susceptibility to antibiotics, signaling proteins used, and even how a VBNC microbe lives and moves. All of this information is vital to the HMP. As you read in chapter 13, in the HMP, researchers sampled microbes from the mouth, nose, skin, intestine, and genitals of 242 volunteers (129 male and 113 female), resulting in over 5,000 samples from body sites. Using MDA, they collected about 3.5 terabases* of DNA sequences that have been entered into a gigantic comparative analysis system called the Integrated Microbial Genomes and Metagenomes for the Human Microbiome Project (IMG/M HMP).

Scientists point out that although there is only about a 0.01% difference between human genomes from one person to another, the genomes of our microbiome differ by as much as 50%. Researchers are beginning to use microbiome genome characteristics to predict who is at risk for type 2 diabetes, for example, and for leanness or obesity. While we should apply caution to "overinterpreting the microbiome," as some scientists say, there will clearly be important diagnostic tools coming from our knowledge of its genetics.

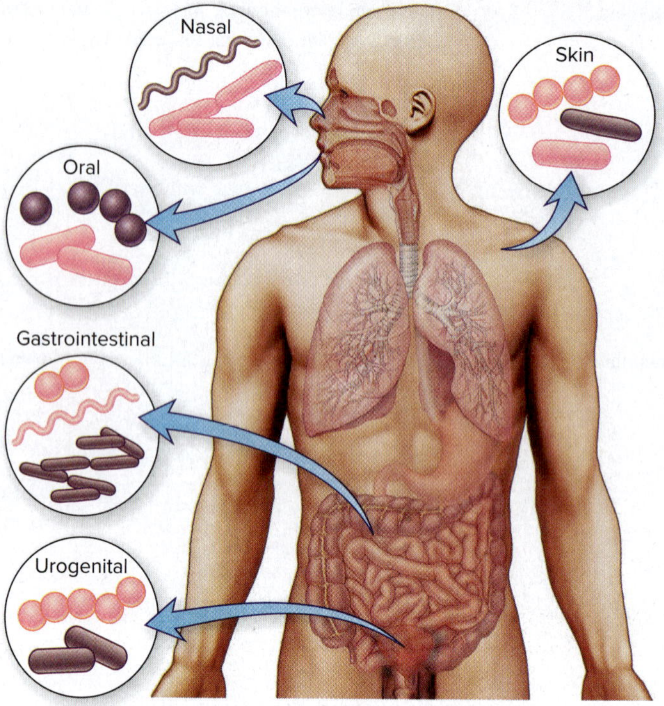
Nasal
Skin
Oral
Gastrointestinal
Urogenital

*A terabase is 10^{12} base pairs.

In sites that normally contain resident microbiota, care should be taken to sample only the infected site and not the surrounding areas. Saliva is an especially undesirable contaminant because it contains millions of bacteria per milliliter, most of which are normal biota. Sputum, the mucus secretion that coats the lower respiratory surfaces, especially the lungs, is collected by coughing or taken by a thin tube called a catheter to avoid contamination with saliva. In addition, throat and nasopharyngeal swabs should not touch the tongue, cheeks, or saliva. On occasion saliva samples are needed for dental diagnosis and are obtained by having the patient spit or drool into a container.

Urine is taken aseptically from the bladder with a catheter designed for that site. Another method, called a "clean catch," is taken by washing the external urethra and collecting the urine midstream. The latter method inevitably incorporates a few normal biota into the sample, but these can usually be differentiated from pathogens in an actual infection. Sometimes diagnostic techniques require first-voided "dirty catch" urine. The mucous lining of the urethra, vagina, or cervix can be sampled with a swab or applicator stick.

Depending on the nature of a skin lesion, skin can be swabbed or scraped with a scalpel to expose deeper layers. Wounds are sampled either by swabbing or by using a punch biopsy tool. Fluids such as blood, cerebrospinal fluid, and tissue fluids must be taken by sterile needle aspiration. Antisepsis of the puncture site is extremely important in these cases. Additional sources of specimens are the eye, ear canal, synovial fluid, nasal cavity (all by swab), and diseased tissue that has been surgically removed (biopsied).

After proper collection, the specimen is promptly transported to a lab and stored appropriately (usually refrigerated) if it must be held for a time. Nonsterile samples in particular, such as urine, feces, and sputum, are especially prone to deterioration at room temperature. Special swab and transport systems are designed to collect the specimen and maintain it in stable condition for several hours. These devices may provide nonnutritive maintenance media (so microbes survive but do not grow), a buffering system, and an anaerobic environment to prevent destruction of oxygen-sensitive bacteria.

Knowing how to properly collect, transport, and store specimens is an important aspect of many health professionals' jobs. In addition, labeling and identifying specimens, as well as providing an accurate patient history, are crucial to obtaining timely and accurate results.

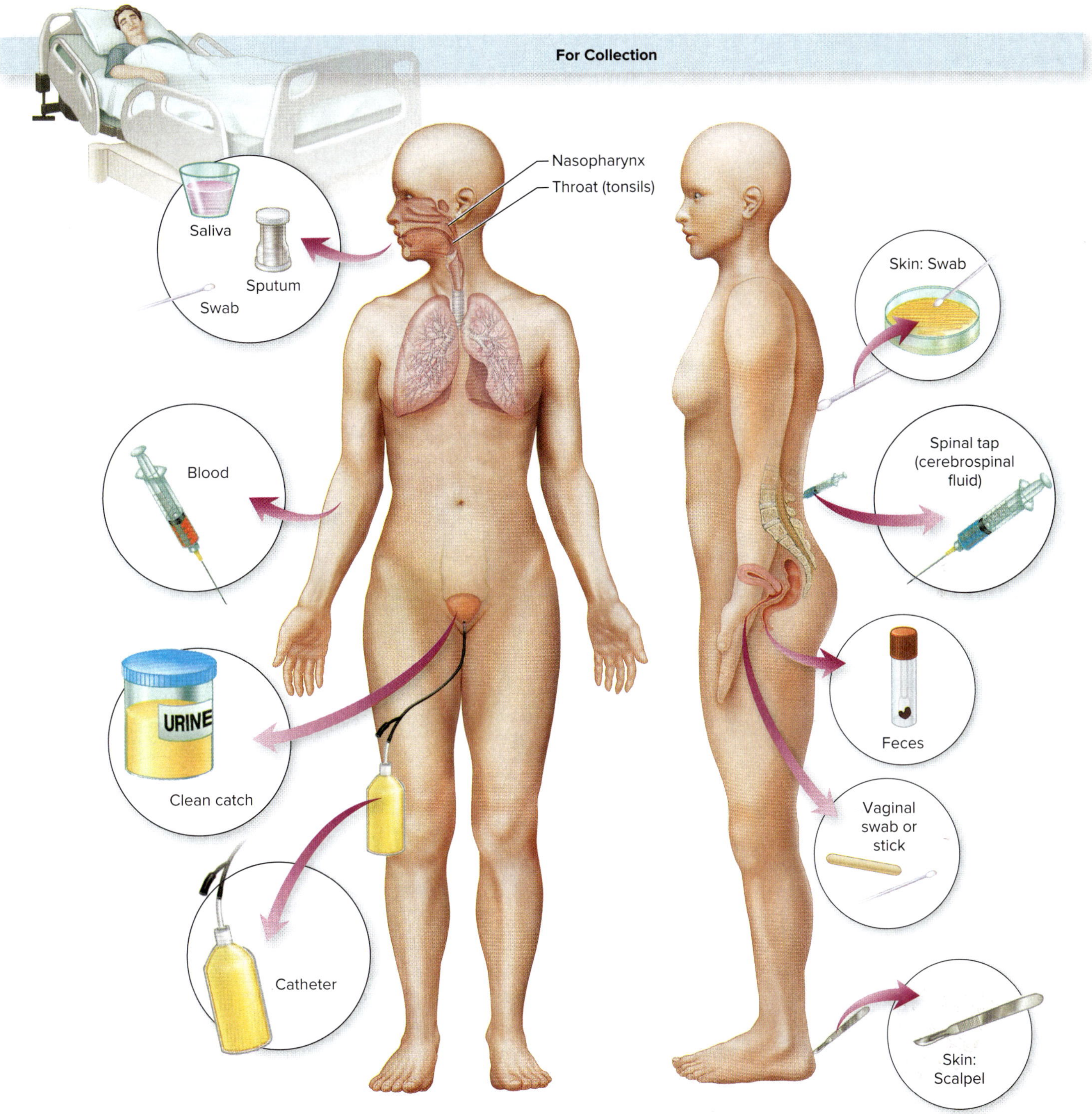
For Collection
Nasopharynx
Throat (tonsils)
Saliva
Sputum
Swab
Skin: Swab
Blood
Spinal tap (cerebrospinal fluid)
URINE
Clean catch
Feces
Vaginal swab or stick
Catheter
Skin: Scalpel

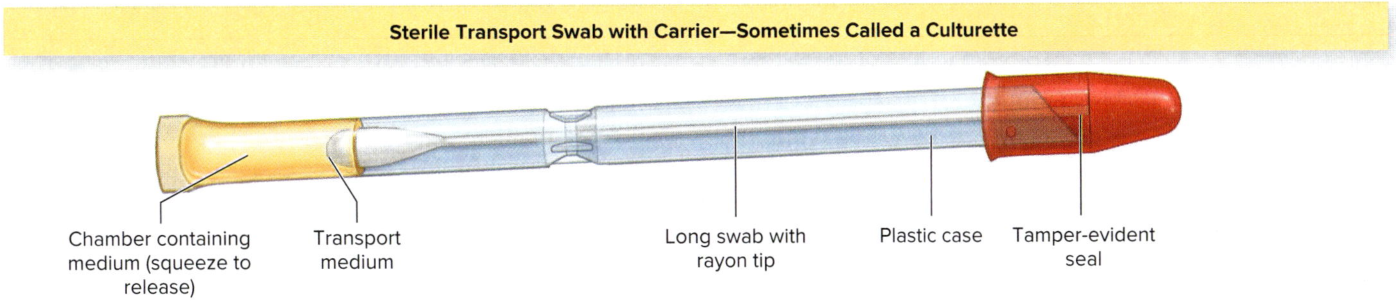
Sterile Transport Swab with Carrier—Sometimes Called a Culturette
Chamber containing medium (squeeze to release)
Transport medium
Long swab with rayon tip
Plastic case
Tamper-evident seal

Figure 18.1 Sampling sites and methods of collection for clinical laboratories.

Figure 18.2 Example of a clinical form used to report data on a patient's specimens.

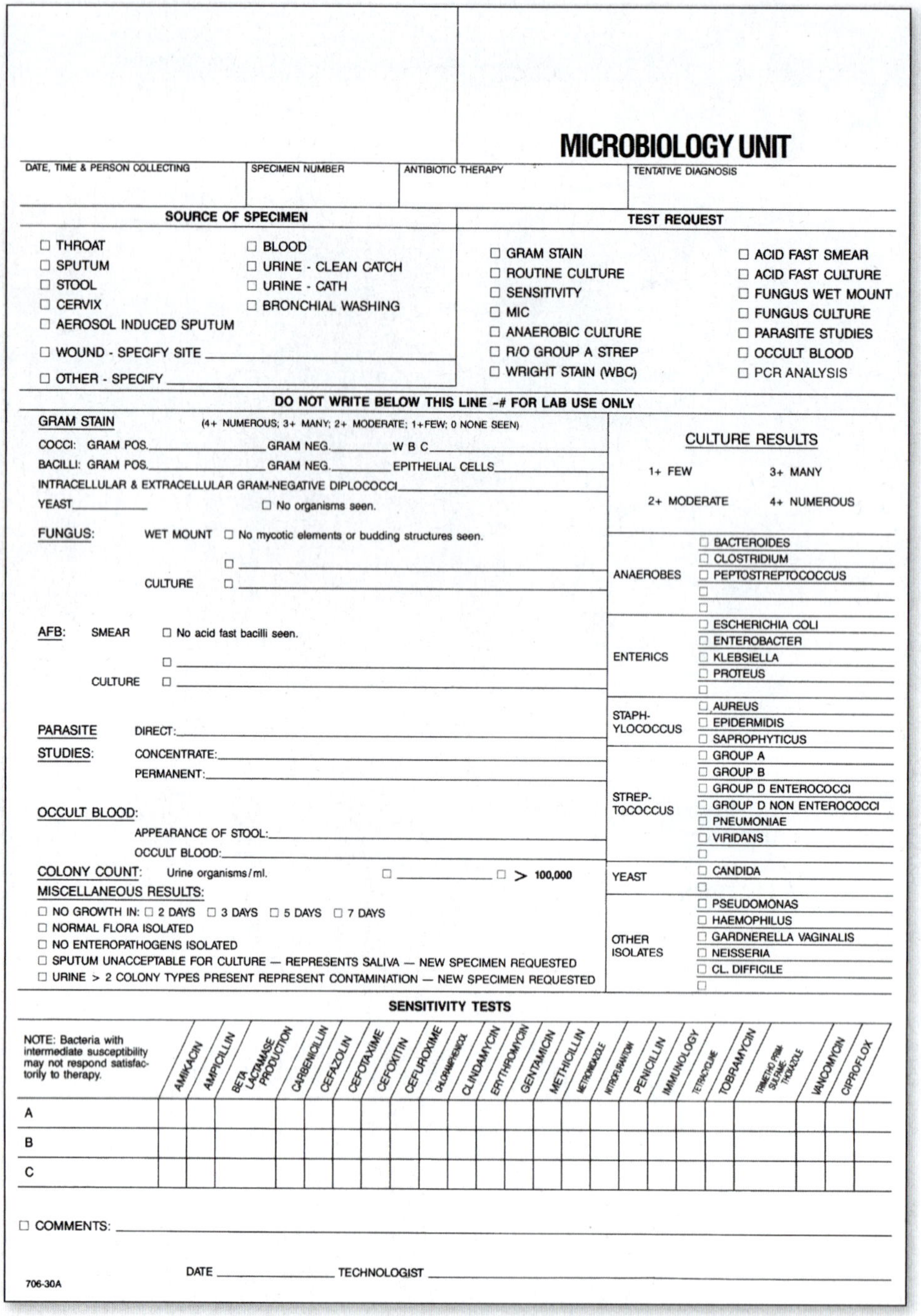
MICROBIOLOGY UNIT

DATE, TIME & PERSON COLLECTING | SPECIMEN NUMBER | ANTIBIOTIC THERAPY | TENTATIVE DIAGNOSIS

SOURCE OF SPECIMEN

☐ THROAT ☐ BLOOD
☐ SPUTUM ☐ URINE - CLEAN CATCH
☐ STOOL ☐ URINE - CATH
☐ CERVIX ☐ BRONCHIAL WASHING
☐ AEROSOL INDUCED SPUTUM
☐ WOUND - SPECIFY SITE
☐ OTHER - SPECIFY

TEST REQUEST

☐ GRAM STAIN ☐ ACID FAST SMEAR
☐ ROUTINE CULTURE ☐ ACID FAST CULTURE
☐ SENSITIVITY ☐ FUNGUS WET MOUNT
☐ MIC ☐ FUNGUS CULTURE
☐ ANAEROBIC CULTURE ☐ PARASITE STUDIES
☐ R/O GROUP A STREP ☐ OCCULT BLOOD
☐ WRIGHT STAIN (WBC) ☐ PCR ANALYSIS

DO NOT WRITE BELOW THIS LINE -# FOR LAB USE ONLY

GRAM STAIN (4+ NUMEROUS; 3+ MANY; 2+ MODERATE; 1+FEW; 0 NONE SEEN)
COCCI: GRAM POS. GRAM NEG. W B C
BACILLI: GRAM POS. GRAM NEG. EPITHELIAL CELLS
INTRACELLULAR & EXTRACELLULAR GRAM-NEGATIVE DIPLOCOCCI
YEAST ☐ No organisms seen.

FUNGUS: WET MOUNT ☐ No mycotic elements or budding structures seen.
☐
CULTURE ☐

AFB: SMEAR ☐ No acid fast bacilli seen.
☐
CULTURE ☐

PARASITE STUDIES: DIRECT: CONCENTRATE: PERMANENT:

OCCULT BLOOD: APPEARANCE OF STOOL: OCCULT BLOOD:

COLONY COUNT: Urine organisms/ml. ☐ ☐ > 100,000

MISCELLANEOUS RESULTS:
☐ NO GROWTH IN: ☐ 2 DAYS ☐ 3 DAYS ☐ 5 DAYS ☐ 7 DAYS
☐ NORMAL FLORA ISOLATED
☐ NO ENTEROPATHOGENS ISOLATED
☐ SPUTUM UNACCEPTABLE FOR CULTURE — REPRESENTS SALIVA — NEW SPECIMEN REQUESTED
☐ URINE > 2 COLONY TYPES PRESENT REPRESENT CONTAMINATION — NEW SPECIMEN REQUESTED

CULTURE RESULTS
1+ FEW 3+ MANY
2+ MODERATE 4+ NUMEROUS

ANAEROBES: ☐ BACTEROIDES ☐ CLOSTRIDIUM ☐ PEPTOSTREPTOCOCCUS ☐ ☐
ENTERICS: ☐ ESCHERICHIA COLI ☐ ENTEROBACTER ☐ KLEBSIELLA ☐ PROTEUS ☐
STAPHYLOCOCCUS: ☐ AUREUS ☐ EPIDERMIDIS ☐ SAPROPHYTICUS
STREPTOCOCCUS: ☐ GROUP A ☐ GROUP B ☐ GROUP D ENTEROCOCCI ☐ GROUP D NON ENTEROCOCCI ☐ PNEUMONIAE ☐ VIRIDANS ☐
YEAST: ☐ CANDIDA ☐
OTHER ISOLATES: ☐ PSEUDOMONAS ☐ HAEMOPHILUS ☐ GARDNERELLA VAGINALIS ☐ NEISSERIA ☐ CL. DIFFICILE ☐

SENSITIVITY TESTS

NOTE: Bacteria with intermediate susceptibility may not respond satisfactorily to therapy.

	AMIKACIN	AMPICILLIN	BETA LACTAMASE PRODUCTION	CARBENICILLIN	CEFAZOLIN	CEFOTAXIME	CEFOXITIN	CEFUROXIME	CHLORAMPHENICOL	CLINDAMYCIN	ERYTHROMYCIN	GENTAMICIN	METHICILLIN	METRONIDAZOLE	NITROFURANTOIN	PENICILLIN	IMMUNOLOGY	TETRACYCLINE	TOBRAMYCIN	TRIMETHOPRIM SULFAMETHOXAZOLE	VANCOMYCIN	CIPROFLOX
A																						
B																						
C																						

☐ COMMENTS:

DATE TECHNOLOGIST

706-30A

Overview of Laboratory Techniques

Results of specimen analysis are entered in a summary patient chart **(figure 18.2)** that can be used in assessment and treatment regimens. The form pictured here is an "old school" paper form. We're showing it because it gives a more comprehensive view of the various tests that can be requested than the more commonly used electronic record system, which shows different categories on separate screens. You may notice that it compiles information on tests with which you are already familiar. As a health care provider, your understanding of and ability to communicate lab results are critical steps in the successful treatment of patients.

18.2 Learning Outcomes—Assess Your Progress

3. Identify factors that may affect the identification of an infectious agent from a patient sample.
4. Discuss the effect the bedside health professional, such as a nurse, can have on identification of disease agents.

18.3 Phenotypic Methods

Immediate Direct Examination of Specimen

Using a microscope to examine a fresh or stained specimen is one of the most rapid methods of determining presumptive—and sometimes confirmatory—microbial characteristics. The Gram stain and the acid-fast stain (see figure 3.21) are most often used for bacterial identification. The acid-fast stain used to identify *Mycobacterium* species is often abbreviated AFB—for acid-fast bacillus stain. Another stain, the calcofluor-potassium hydroxide stain (also known as the KOH prep), is used to identify fungi in hair, nail, skin, and bodily fluids.

Methods Requiring Growth

Selective and Differential Growth

Such a wide variety of media exist for microbial isolation that a certain amount of preselection must occur, based on the nature of the specimen. In cases in which the suspected pathogen is present in small numbers or is easily overgrown by normal biota, the specimen can be initially enriched with specialized media. Nonsterile specimens containing a diversity of bacterial species (such as urine and feces) are cultured on selective media to encourage the growth of only the suspected pathogen. For example, a large percentage of urinary tract infections are known to be caused by *Escherichia coli,* so selective media that will be sure to allow the growth of this common pathogen are chosen. Specimens are often inoculated into differential media to identify definitive characteristics, such as reactions in blood (blood agar) and fermentation patterns (mannitol salt and MacConkey agar).

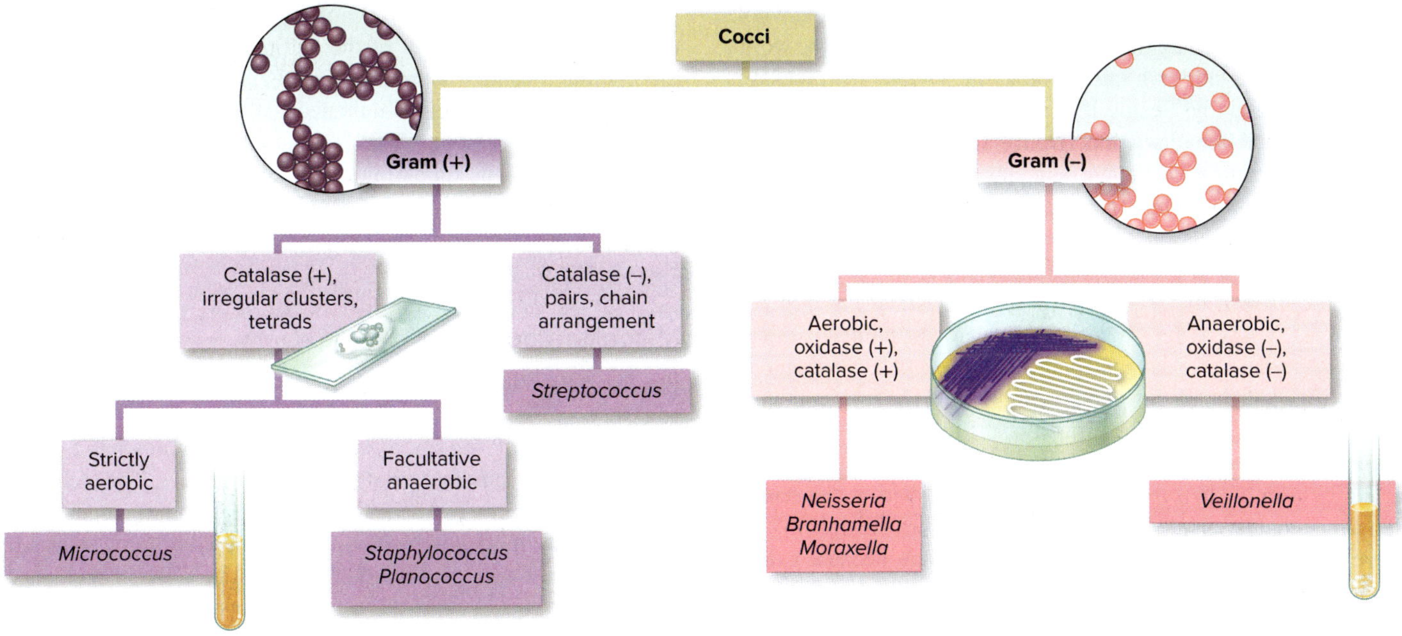
Cocci
Gram (+)
Gram (–)
Catalase (+), irregular clusters, tetrads
Catalase (–), pairs, chain arrangement
Streptococcus
Strictly aerobic
Facultative anaerobic
Micrococcus
Staphylococcus Planococcus
Aerobic, oxidase (+), catalase (+)
Anaerobic, oxidase (–), catalase (–)
Neisseria Branhamella Moraxella
Veillonella

Figure 18.3 **Flowchart (dichotomous key) to separate primary genera of gram-positive and gram-negative bacteria.** Identification scheme for cocci commonly involved in human diseases.

A straightforward way to phenotypically identify specimens is to combine the results of tests such as Gram staining, growth on different media, and simple enzymatic tests. Usually a Gram stain is the first test. In most cases, the specimen will be either gram-negative or gram-positive. This can form the starting point for a **dichotomous key,** a graphic method that is essentially a flowchart leading to the identification of specimens **(figure 18.3).** The name "dichotomous" refers to the fact that at each branch of the flowchart, there are two (di-) possible outcomes.

Biochemical Testing

The physiological reactions of bacteria to nutrients and other substrates reveal the types of enzyme systems present in a particular species. Knowing which enzymes an isolate has can often lead to its identity. Many of these tests are based on enzyme-mediated metabolic reactions that are visualized by a color change.

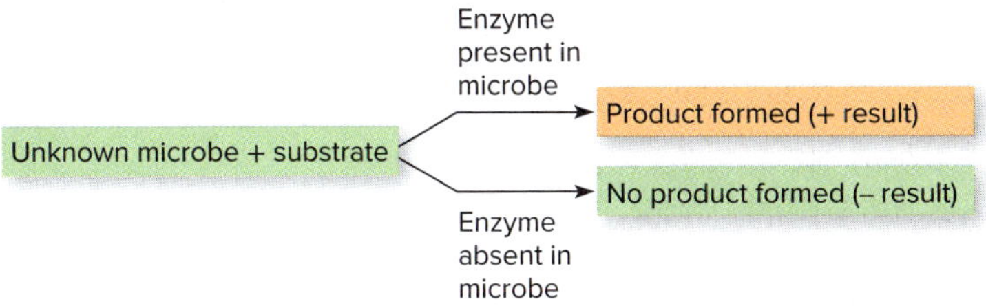
Enzyme present in microbe
Product formed (+ result)
Unknown microbe + substrate
No product formed (– result)
Enzyme absent in microbe

The microbe is cultured in a medium with a special substrate and then tested for a particular end product. Microbial expression of the enzyme is made visible by a colored dye. No coloration means it lacks the enzyme for utilizing the substrate in that particular way (see the illustration to the right). The biochemical testing process has been fully automated by diagnostic companies. Even the growth and incubation step can be performed by machines, and they can identify more than 2,500 different bacteria. With these machines, the sample is inoculated in some micro format,

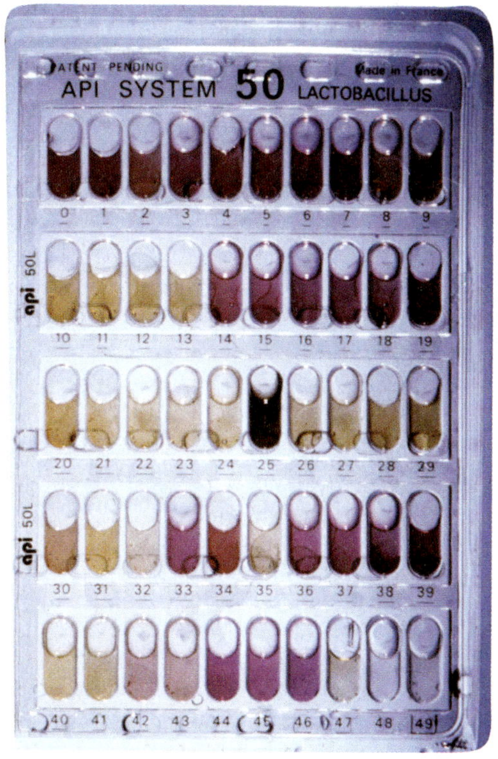
API SYSTEM 50 LACTOBACILLUS
api 50L

Figure 18.4 **Rapid tests—a biochemical system for microbial identification.** Each of these 50 wells acts as a tiny test tube. They are prefilled with liquid medium. In addition, a different enzymatic substrate, which will change color if it is acted upon, is present in each well. A single unknown bacterium is inoculated into each of the wells. If the bacterium has the enzyme that can metabolize the substrate in the well, the medium in the well will change color. The pattern of color changes is an easy way to detect which enzymes this bacterium has, and its identity can be deduced from that. This particular set of wells was inoculated with a bacterium that turned out to be *Clostridium ramosum*. (CDC)
CDC/ Dr. Gilda Jones

such as on a small card, and inserted in a machine, and the whole process is performed by the machine. However, all over the world, and probably in your own microbiology lab course, the process is performed manually. Seeing it done manually makes it easier to visualize how it actually works **(figure 18.4).** Additionally, direct biochemical testing of patient samples now can be performed without incubation, which produces results within hours instead of days.

Another type of phenotypic test is the MGIT system used to detect the growth of the slow-growing *Mycobacterium tuberculosis.* (MGIT stands for "mycobacterial growth indicator tube.") This system monitors oxygen levels in a tube that has been inoculated with a patient specimen (sputum, blood). The tube contains a medium that encourages the growth of the tuberculosis bacterium (which is excruciatingly slow-growing). There is a silicon chip at the bottom of the tube that is impregnated with a fluorescent substance that is sensitive to oxygen levels. When first inoculated, there will be a lot of free oxygen in the medium because even if the bacteria are present, there will not be many of them. But if the bacteria grow, they begin using the oxygen, and the decreased oxygen levels allow the silicon compound to fluoresce. In automated machines, when a threshold level of fluorescence occurs, visual and audible signals are given so a positive culture can be identified in the shortest possible time.

Antimicrobial Susceptibility Testing

As important as identifying the microbial pathogen is identifying what antimicrobial drugs it can be treated with. With the crisis-level rise in antimicrobial resistance, it is important not to waste valuable treatment time by simply guessing what drugs to use to treat an infection. Most of the automated phenotypic systems incorporate a panel of commonly used antimicrobials for the particular infection site, and simultaneously test susceptibility while identifying the pathogen. This is commonly accomplished using an adaptation of the tube-dilution method covered in chapter 12. Also, antimicrobial susceptibility tests can themselves provide the identity of some species, such as *Streptococcus, Clostridium,* and *Pseudomonas*.

Miscellaneous Tests

Phage typing relies on bacteriophages, viruses that attack bacteria in a very species-specific and strain-specific way. Such selection is useful in identifying some bacteria, primarily *Salmonella,* and is often used for tracing bacterial strains in epidemics. The technique of phage typing involves inoculating a lawn of bacterial cells onto agar, mapping it off into blocks, and applying a different phage to each sectioned area of growth **(figure 18.5).** Cleared areas corresponding to lysed cells indicate sensitivity to that phage, and a bacterial identification may be determined from this pattern.

Susceptible animals are needed to cultivate bacteria such as *Mycobacterium leprae* and significant quantities of *Treponema pallidum,* whereas avian embryos and cell cultures are used to grow host cell–dependent rickettsias, chlamydias, and viruses.

Determining Clinical Significance of Cultures

It is important to rapidly determine if an isolate from a specimen is clinically important or if it is merely a contaminant or normal

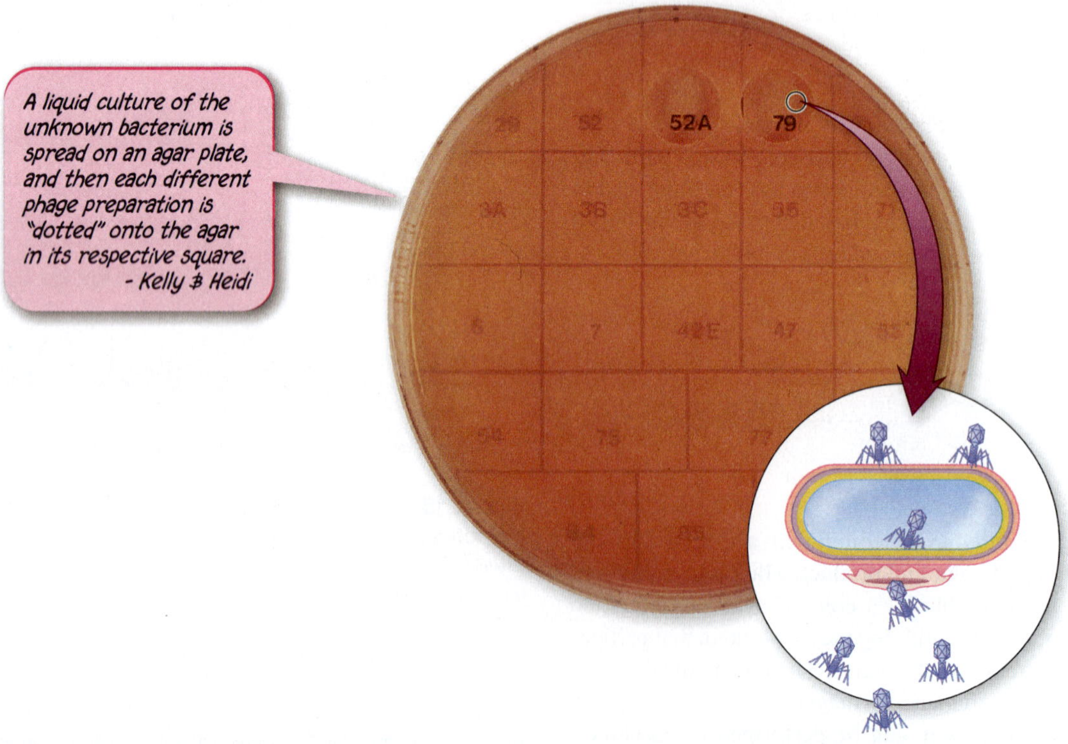
A liquid culture of the unknown bacterium is spread on an agar plate, and then each different phage preparation is "dotted" onto the agar in its respective square.
- Kelly & Heidi
52A
79

Figure 18.5 Phage typing of an unknown bacterium. A cleared area within a square of the bacterial lawn forms due to phage-induced lysis of cells, indicating sensitivity of the bacterium to the corresponding phage.
The University of Leeds

biota. Although answering these questions may prove difficult, one can first focus on the number of microbes in a specimen. For example, a few colonies of *Escherichia coli* in a urine sample can simply indicate normal biota, whereas several hundred can mean active infection. In contrast, the presence of a single colony of a true pathogen, such as *Mycobacterium tuberculosis* in a sputum culture or an opportunist in sterile sites such as cerebrospinal fluid or blood, is highly suggestive of its role in disease. Furthermore, the repeated isolation of a relatively pure culture of any microorganism can mean it is an agent of disease.

As reliable as phenotypic methods can be, they have two major drawbacks: (1) When the microbe has to be cultured, it takes a minimum of 18 to 24 hours and often longer; and (2) we are learning that many infectious conditions may be caused by nonculturable organisms, leaving open the possibility that the organism that we do culture is simply a bystander. **Table 18.2** summarizes phenotypic methods of diagnosis.

18.3 Learning Outcomes—Assess Your Progress

5. List at least three different tests that fall in the phenotypic identification category.
6. Explain the main principle behind biochemical testing, and identify an example of such tests.
7. Discuss two major drawbacks of phenotypic testing methods that require culturing the pathogen.

Table 18.2

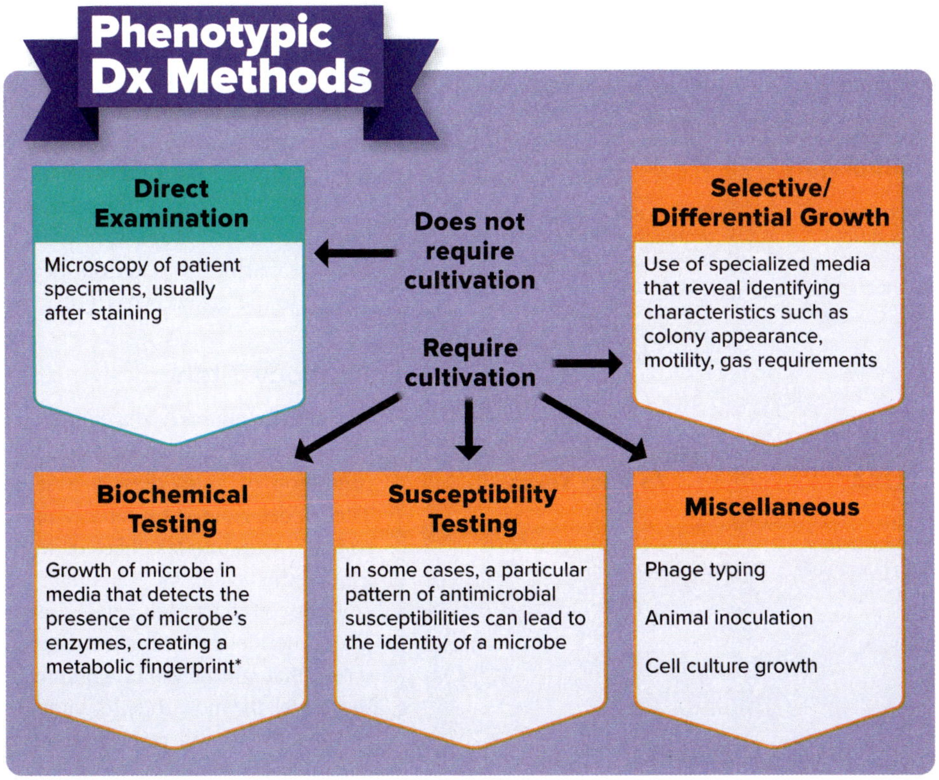
Phenotypic Dx Methods
Direct Examination
Microscopy of patient specimens, usually after staining
Does not require cultivation
Require cultivation
Selective/ Differential Growth
Use of specialized media that reveal identifying characteristics such as colony appearance, motility, gas requirements
Biochemical Testing
Growth of microbe in media that detects the presence of microbe's enzymes, creating a metabolic fingerprint*
Susceptibility Testing
In some cases, a particular pattern of antimicrobial susceptibilities can lead to the identity of a microbe
Miscellaneous
Phage typing
Animal inoculation
Cell culture growth

*Newer methods may not require cultivation.

18.4 Immunologic Methods

The antibodies formed during an immune reaction are important in combating infection, but they also form the basis of a powerful diagnostic strategy. Characteristics of antibodies (such as their quantity or specificity) can reveal the history of a patient's contact with microorganisms or other antigens. This is the underlying basis of serological testing. **Serology** is the branch of immunology that traditionally deals with diagnostic testing of serum.

Serological testing is based on the principle that antibodies have extreme specificity for antigens, so when a particular antigen is exposed to its specific antibody, it will fit like a hand in a glove. The ability to visualize this interaction macroscopically or microscopically provides a powerful tool for detecting, identifying, and quantifying antibodies—or, on the other hand, antigens. One can detect or identify an unknown antibody using a known antigen, or one can use an antibody of known specificity to help detect or identify an unknown antigen **(figure 18.6).** Even though the strategy is called "serology," we now test all kinds of body samples, not just sera. These include urine, cerebrospinal fluid, whole tissues, and saliva. These and other immune tests help to determine the immunologic status of patients, confirm a suspected diagnosis, or screen individuals for disease. The tests are summarized in **table 18.3.**

Different Kinds of Antigen–Antibody Interactions

The molecular basis of immunologic testing is the binding of an antibody (Ab) to a specific site (epitope) on an antigen (Ag). The binding reaction is not visible to the naked eye nor in light microscopy. But if the antigens and antibody are suspended in a solution, under the right circumstances, the presence of binding can be seen as clumps in the solution.

Agglutination and Precipitation Reactions

The essential differences between agglutination and precipitation, as seen in table 18.3, are in the size, solubility, and location of the antigen. In agglutination, the antigens are whole cells or organisms such as red blood cells, bacteria, or viruses displaying surface antigens **(figure 18.7).** In precipitation, the antigen you are looking for is a soluble molecule. In both reactions, when antigen and antibody concentrations are optimal, one antigen is interlinked by several antibodies to form insoluble aggregates that settle out in solution. Agglutination is easily seen because it forms visible clumps of cells, such as in the *Weil-Felix reaction* used in diagnosing rickettsial infections. Agglutination is also used to determine blood compatibility.

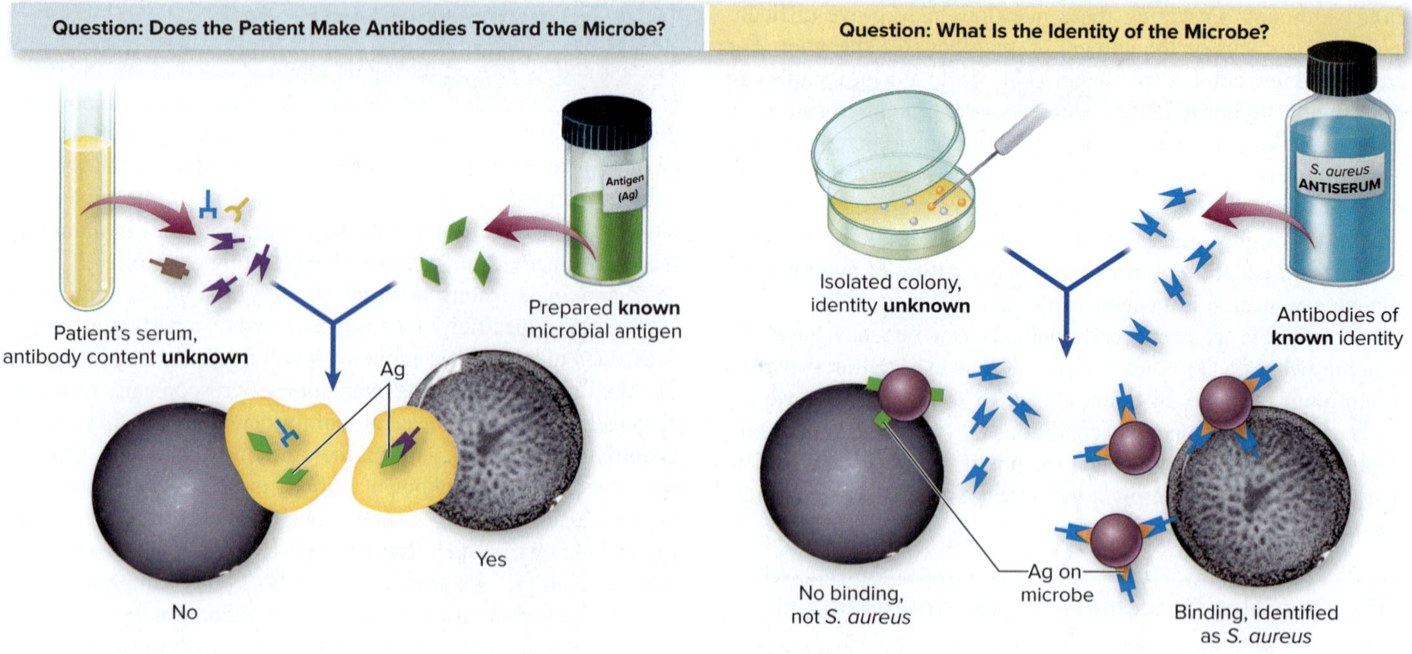
Question: Does the Patient Make Antibodies Toward the Microbe?
Question: What Is the Identity of the Microbe?
Antigen (Ag)
Patient's serum, antibody content unknown
Prepared known microbial antigen
Ag
No
Yes
Isolated colony, identity unknown
S. aureus ANTISERUM
Antibodies of known identity
No binding, not S. aureus
Ag on microbe
Binding, identified as S. aureus

Figure 18.6 **Basic principles of serological testing using antibodies and antigens.**
Lisa Burgess/McGraw Hill

These antigen-antibody binding reactions have been modified in many ways to make the reactions visible to the naked eye. One commonly used method is termed **immunochromatography,** often called the "lateral flow test." It can be found in drugstore pregnancy tests and in most of the at-home COVID-19 tests. It usually consists of a plastic cartridge that contains some kind of porous material or polymer that directs fluid to flow in a particular direction. The fluid (the patient sample) will encounter antibodies along its route of flow. If the sample contains the correct antigen, it will bind the antibodies and continue on to the next "station" in the cartridge. The next stage contains a third molecule that is impregnated on the paper in a stripe pattern. The third molecule binds the complexes, and eventually this causes the stripe to change color. Then a microscopic antigen–antibody reaction is made visible to the naked eye **(figure 18.8).**

Table 18.3

Immunologic Dx Methods

Agglutination/ Precipitation	Immuno-chromatography	Western Blot
Agglutination Antibody-mediated clumping of whole cells **Precipitation** Smaller complexes of antibody–antigen	Most common form is a lateral flow system, supplied in prepackaged cartridges that produce a colored stripe.	Electrophoresis separates proteins (either antigens or antibodies) and then labeled antibodies or antigens are used for detection.

Fluorescent Antibodies	ELISA	*In vivo* Tests
Direct (DFA) Unknown specimen is exposed to known fluorescent Ab. **Indirect (IFA)** Patient's antibody (its Fc portion) is probed with fluorescent Ab.	Sandwiching technique conducted in microwells using Ag, Ab, and a secondary Ab to produce a color change.	Antigen introduced into a patient to elicit a reaction, as in TB skin test.

Antibody Titers

An antigen–antibody reaction in tubes of any sort is read as a titer, or the concentration of antibodies in a sample. Titer is determined by serially diluting patient serum into test tubes or wells of a microtiter plate, all containing equal amounts of bacterial cells (antigen). Titer is defined as the highest dilution of serum that still produces agglutination. In general, the more a serum sample can be diluted and still react with antigen, the greater the concentration of antibodies and thus its titer. Antibody titers are often used to diagnose autoimmune

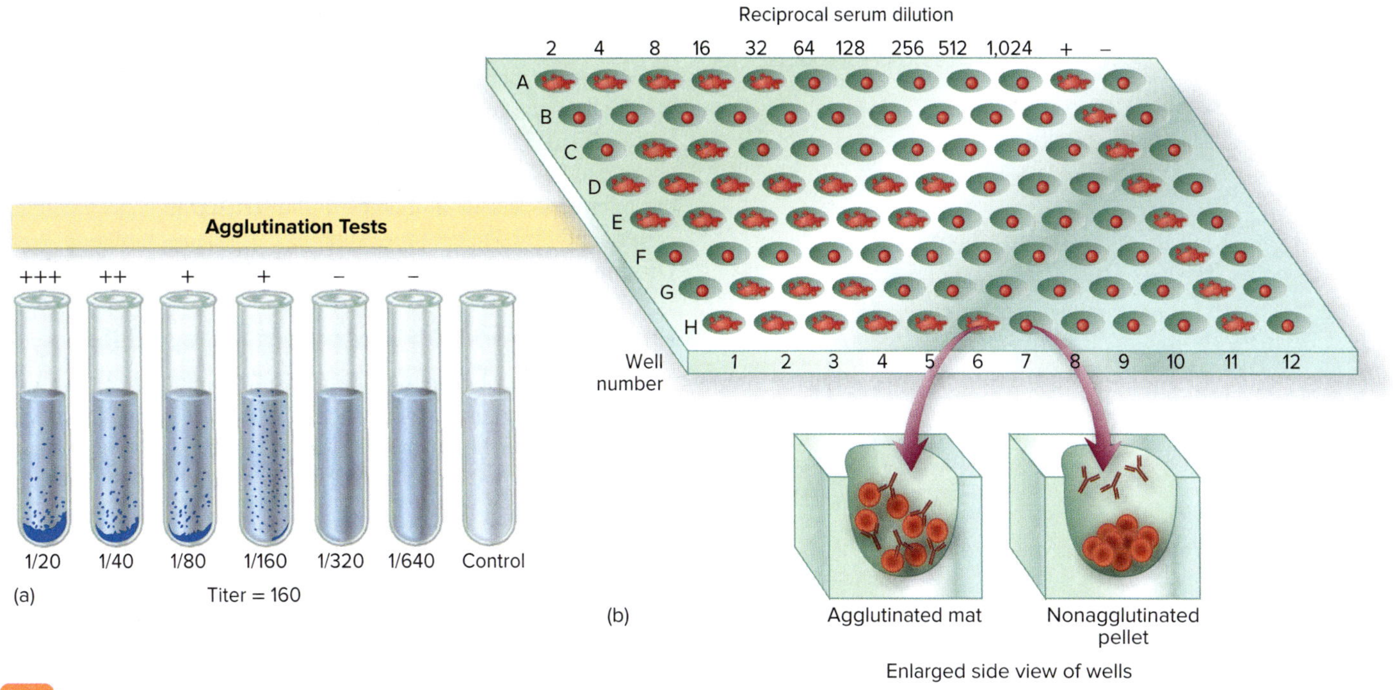
Agglutination Tests
+++
++
+
+
–
–
1/20
1/40
1/80
1/160
1/320
1/640
Control
(a)
Titer = 160
Reciprocal serum dilution
2 4 8 16 32 64 128 256 512 1,024 + –
A
B
C
D
E
F
G
H
Well number
1 2 3 4 5 6 7 8 9 10 11 12
(b)
Agglutinated mat
Nonagglutinated pellet
Enlarged side view of wells

Figure 18.7 Agglutination tests. **(a)** Tube agglutination test for determining antibody titer. The same number of cells (antigen) are added to each tube, and the patient's serum is diluted in series. The titer in this example is 160 because there is no agglutination in the next tube in the dilution series (1/320). **(b)** A microtiter plate illustrating hemagglutination. Serial dilutions of the antibody are placed in the wells of all rows in columns 1 through 10. Positive controls (column 11) and negative controls (column 12) are included for comparison. Red blood cells that display the antigen are added to each well. If sufficient antibody is present to agglutinate the cells, they sink as a diffuse mat to the bottom of the well. If insufficient antibody is present, they form a tight pellet at the bottom.

disorders such as rheumatoid arthritis and lupus, and also to determine past exposure to certain diseases such as rubella.

Serotyping

Serotyping is an antigen-antibody technique for identifying, classifying, and subgrouping certain bacteria into categories called serotypes. This method employs antisera against cell antigens such as the capsule, flagellum, and cell wall. Serotyping is widely used in identifying *Salmonella* species and strains, and is the basis for differentiating the numerous pneumococcal and streptococcal serotypes.

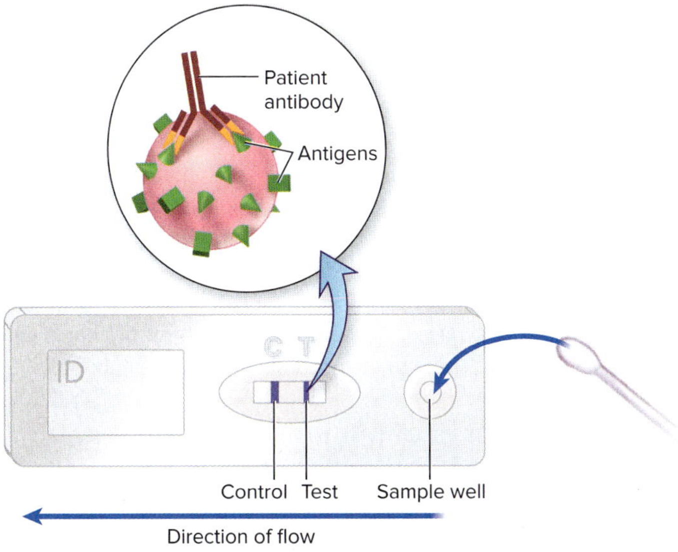
Patient antibody
Antigens
ID
C T
Control
Test
Sample well
Direction of flow

Figure 18.8 An immunochromatographic (lateral flow) test relying on antigen–antibody binding. This is a positive test for *Neisseria gonorrhoeae*.

The Western Blot Procedure

The **Western blot test** involves the separating of proteins by electrophoresis and then using the antibody–antigen binding principle to detect the proteins **(figure 18.9).** For example, a sample of proteins (antigens) obtained from microbes is separated via electrical charge within a gel. The proteins embedded in the gel are then transferred and immobilized to a special filter. The filter is next incubated with patient fluid, such as serum, to see if antibodies are present that are compatible with the test antigen. If there are antibodies, they will become visible when a secondary antibody, labeled with fluorescence and designed to bind the Fc portion of any antibodies bound to the filter, is added. After incubation, sites of specific antigen–antibody binding will appear as a pattern of bands that can be compared with known positive and negative controls. It is a highly specific and sensitive way to identify or verify the presence of microbial-specific antigens or antibodies in a patient sample. Western blotting is often the second (verification) test for preliminary antibody-positive HIV screening tests.

Immunofluorescence Testing

The fundamental tool in immunofluorescence testing is a fluorescent antibody—a monoclonal antibody labeled by a fluorescent dye. Fluorescent antibodies (FAbs) can be used for diagnosis in

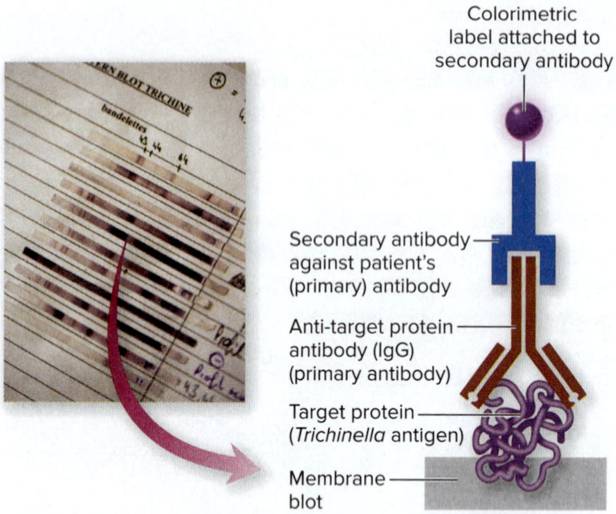
Colorimetric label attached to secondary antibody
Secondary antibody against patient's (primary) antibody
Anti-target protein antibody (IgG) (primary antibody)
Target protein (*Trichinella* antigen)
Membrane blot

Figure 18.9 The Western blot procedure. Major antigens from the nematode *Trichinella* were separated via electrophoresis and transferred to a filter. The filter was incubated with the sera of 10 separate patients (in the 10 rows). Sera contain primary antibodies. Secondary antibodies (blue in the drawing) and a colorimetric label were added to visualize the patient's bound antibody (maroon in the drawing). Complexes of *Trichinella* antigen and patient antibody are visible as bands.
Phanie/Science Source

two ways. In *direct testing* (DFA), an unknown test specimen or antigen is fixed to a slide and exposed to a FAb solution of known composition. If antibody–antigen complexes form, they will remain bound to the sample and will be visualized by fluorescence microscopy, indicating a positive result **(figure 18.10).** These tests are valuable for identifying and locating microbial antigens on cell surfaces or in tissues and in identifying the causative agents of syphilis, gonorrhea, rabies, and meningitis, among others.

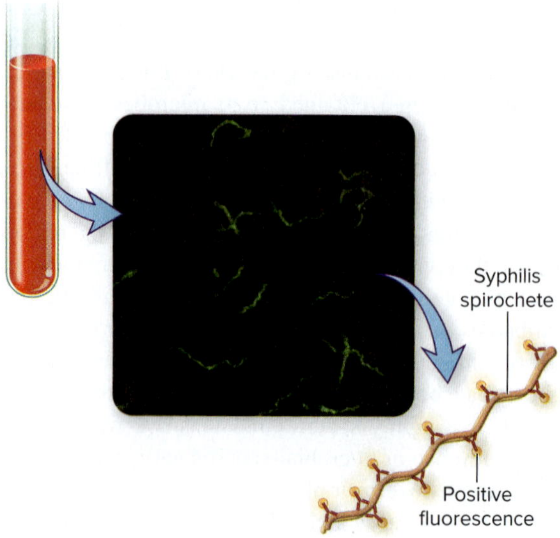
Syphilis spirochete
Positive fluorescence

Figure 18.10 Direct fluorescence antigen test. Photomicrograph of a direct fluorescence test for *Treponema pallidum*, the syphilis spirochete.
Russell/CDC

In contrast, FAbs used in *indirect testing* (IFA) recognize the Fc region of antibodies in patient sera (remember that antibodies are antigenic themselves!). Known antigen (i.e., bacterial cells) is added to the test serum, and binding of the fluorescent antibody is visualized through fluorescence microscopy. Fluorescing aggregates or cells indicate that the FAbs have complexed with the microbe-specific antibodies in the test serum. This technique is frequently used to diagnose syphilis and various viral infections. DFA and IFA can be used in a point-of-care setting.

Enzyme-Linked Immunosorbent Assay (ELISA)

The **ELISA** test, also known as enzyme immunoassay (EIA), uses an enzyme-linked indicator antibody to visualize antigen–antibody reactions. This technique relies on a solid support such as a plastic microtiter plate that can *adsorb* (bind to its surface) the reactants **(figure 18.11).**

The *indirect ELISA* test detects microbe-specific antibodies in patient sera. A known antigen is adsorbed to the surface of a well and mixed with a patient sample. If an antibody–antigen complex forms because antibodies are present in the patient, an added indicator antibody will bind, and subsequent development will produce a color change, indicating a positive result. This is the common test used for antibody screening for HIV, various rickettsial species, hepatitis A and C, and *Helicobacter.* Because false positives can occur, a verification test (Western blot) may be necessary.

There are multiple ways to perform an ELISA. One of them is known as the sandwich test. In this setup, a known antibody is adsorbed to the bottom of a well and incubated with an unknown antigen. If an antibody–antigen complex forms, it will attract the indicator antibody, and color will develop in these wells, indicating a positive result. New detection systems utilize computer chips that sense minute changes in electrical current that occur when antibody–antigen complexes are formed.

Complement Fixation Test

Many viral and fungal diseases are diagnosed using the principle that complement can lyse red blood cells. **Figure 18.12** illustrates how the test works. It uses the same kind of trays of miniwells seen in ELISA tests.

In Vivo Testing

In practice, *in vivo* tests employ principles similar to serological tests, except in this case an antigen is introduced into a patient to elicit some sort of visible reaction. The **tuberculin reaction,** where a small amount of purified protein derivative (PPD) from *Mycobacterium tuberculosis* is injected into the skin, is a classic example. The appearance of a red, raised, thickened lesion in 48 to 72 hours can indicate previous exposure to tuberculosis (shown in figure 17.12).

General Features of Immune Testing

The most effective serological tests have a high degree of two characteristics: specificity and sensitivity. *Specificity* is the property of a test to focus on only a certain antibody or antigen and not to react with unrelated or distantly related ones. Put another way, a test with high specificity will have a low false-positive rate. *Sensitivity* refers

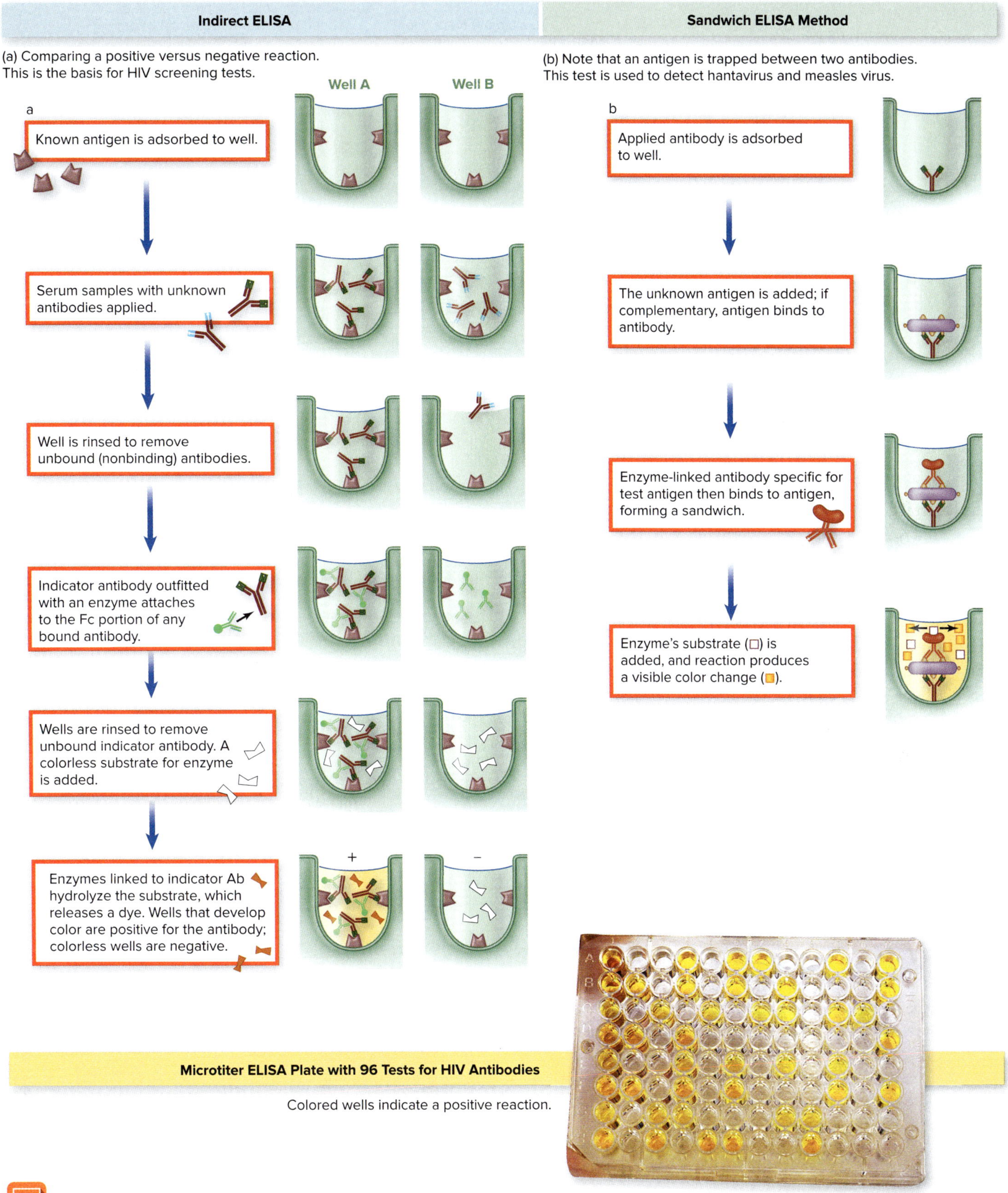
Indirect ELISA
(a) Comparing a positive versus negative reaction. This is the basis for HIV screening tests.
Well A
Well B
a
Known antigen is adsorbed to well.
Serum samples with unknown antibodies applied.
Well is rinsed to remove unbound (nonbinding) antibodies.
Indicator antibody outfitted with an enzyme attaches to the Fc portion of any bound antibody.
Wells are rinsed to remove unbound indicator antibody. A colorless substrate for enzyme is added.
Enzymes linked to indicator Ab hydrolyze the substrate, which releases a dye. Wells that develop color are positive for the antibody; colorless wells are negative.
+
–
Sandwich ELISA Method
(b) Note that an antigen is trapped between two antibodies. This test is used to detect hantavirus and measles virus.
b
Applied antibody is adsorbed to well.
The unknown antigen is added; if complementary, antigen binds to antibody.
Enzyme-linked antibody specific for test antigen then binds to antigen, forming a sandwich.
Enzyme's substrate (□) is added, and reaction produces a visible color change (■).
Microtiter ELISA Plate with 96 Tests for HIV Antibodies
Colored wells indicate a positive reaction.
A
B

Figure 18.11 Methods of ELISA testing. **(a)** Indirect ELISAs detect patient antibody. Here, both positive and negative tests are illustrated. **(b)** Sandwich ELISAs detect patient antigen (microbes). Here, only the positive reaction is illustrated.

Hank Morgan/Science Source

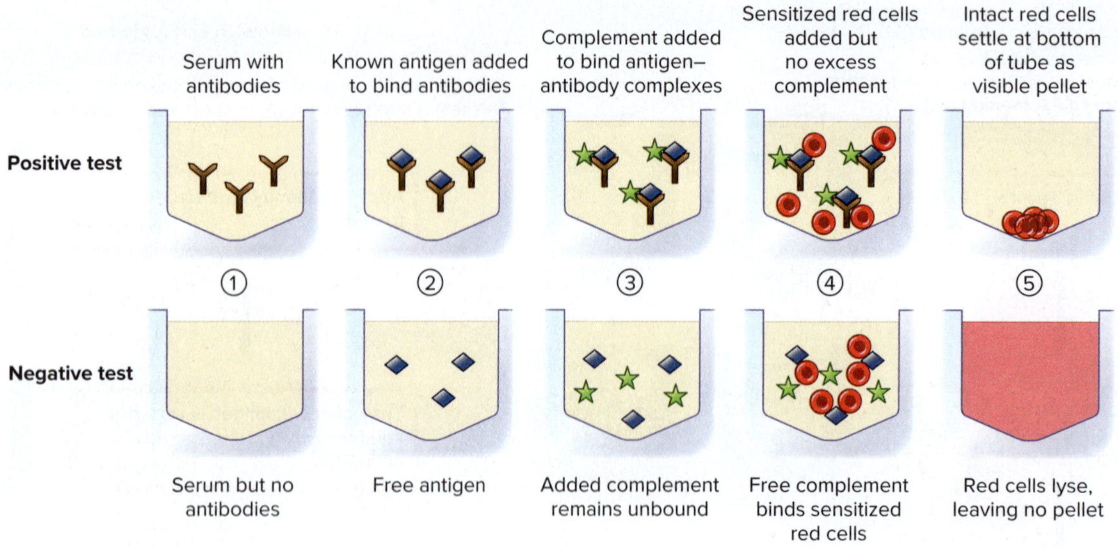
Serum with antibodies
Known antigen added to bind antibodies
Complement added to bind antigen–antibody complexes
Sensitized red cells added but no excess complement
Intact red cells settle at bottom of tube as visible pellet
Positive test
①
②
③
④
⑤
Negative test
Serum but no antibodies
Free antigen
Added complement remains unbound
Free complement binds sensitized red cells
Red cells lyse, leaving no pellet

Figure 18.12 Complement fixation test. **(1)** Patient serum is added to a microwell. **(2)** The antigen in question is added to the well. **(3)** Complement is added. **(4)** Red blood cells are added. **(5)** Result is read: If the serum contained the CORRECT antibodies, the antigen–antibody complex "soaks up" the complement, meaning it will not lyse the red blood cells.

to the detection of even minute quantities of antibodies or antigens in a specimen. A test with high sensitivity will have a low false-negative rate. Usually, a given test will be "better" at one or the other, and the trade-off must be considered when choosing a test. For example, in an epidemic of a deadly disease, such as Ebola, you might reason that you want to have a very high sensitivity to catch every case. People who test positive can then be screened using a different technique to determine if they actually have the disease. In other situations (for example, for that secondary screening test), you might want a higher specificity so that you don't have very many false positives. Other diagnostic methodologies besides serological are also characterized by a particular sensitivity and specificity.

18.4 Learning Outcomes—Assess Your Progress

8. Define the term *serology,* and explain the immunologic principle behind serological tests.
9. Identify two immunological diagnostic techniques that rely on a secondary antibody, and explain how they work.

18.5 Genotypic Methods

The sequence of nitrogenous bases within DNA or RNA is unique to every microorganism. Because of this and the many recent advances in genomic technology, nucleic acid–based tests have become a mainstay of microbial identification. Genotypic methods are summarized in **table 18.4** and discussed below.

Polymerase Chain Reaction (PCR): The Backbone of New Diagnostics

Many nucleic acid tests use the polymerase chain reaction (PCR). PCR results in the production of numerous identical copies of DNA or RNA molecules within hours (see figure 8.3). This method can amplify even minute quantities of nucleic acids present in a sample, which greatly improves the sensitivity of these tests. PCR amplification can be performed on genetic material from almost any bacteria, viruses, protozoa, and fungi. There are many variations on this method, and they are generally notated as NAATs, nucleic acid amplification tests. Often the tests are organized into

Table 18.4

Genotypic Dx Methods

Nucleic Acid Amplification Tests	Hybridization
Using primers to amplify specific DNA sequences Real-time PCR Multiplex PCR TMA Other NAAT	Fluorescent *in situ* hybridization (FISH)

Whole-Genome Sequencing
High-throughput methods have made whole-genome sequencing widely accessible

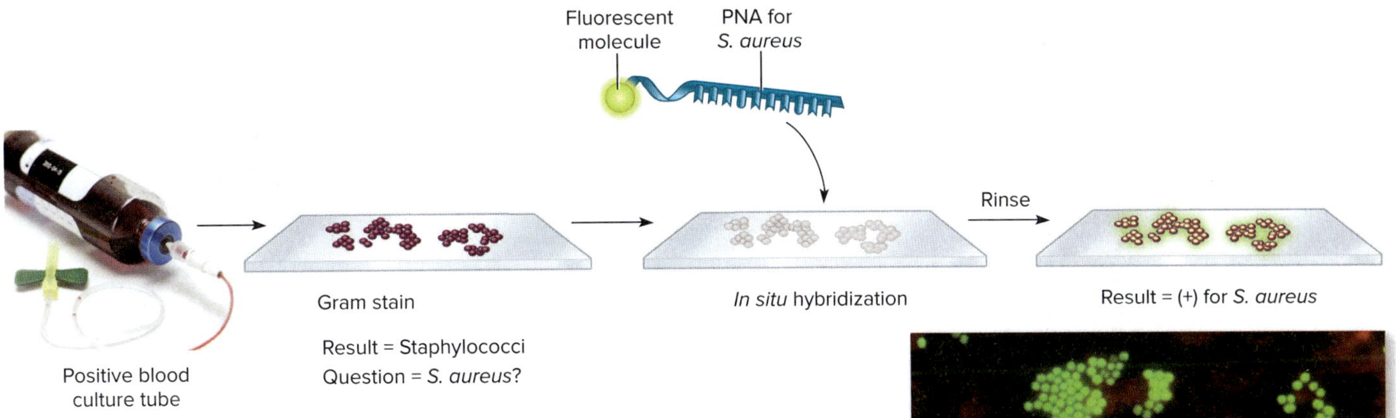
Fluorescent molecule
PNA for S. aureus
Rinse
Positive blood culture tube
Gram stain
Result = Staphylococci
Question = S. aureus?
In situ hybridization
Result = (+) for S. aureus

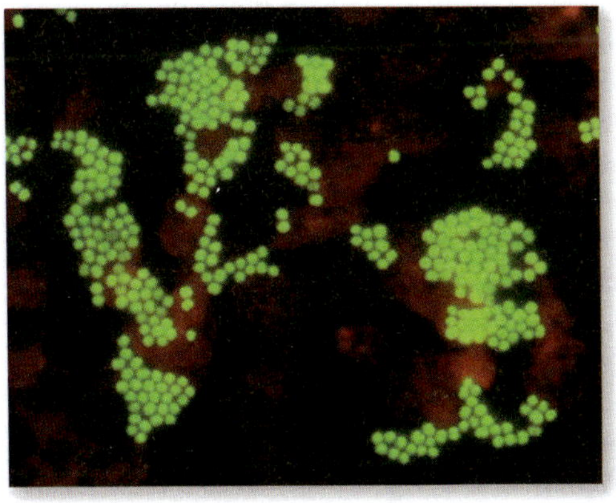

Figure 18.13 Peptide nucleic acid (PNA)–FISH testing for *S. aureus*. This type of testing can identify blood-borne pathogens more quickly than other methods. Photo to the right shows results.
MedstockPhotos/Shutterstock; Source: OpGen, Inc/AdvanDx, Inc..

syndromic panels, meaning several organisms in the differential diagnosis will be tested simultaneously. Many of these tests can be used as point-of-care tests. Some of the most important ones being used in diagnosis of infections are described next.

Real-Time PCR

This technique is also known as "qPCR" because there is a quantitative aspect to it. It uses fluorescent labeling during the amplification procedure, and the level of fluorescence is measured in real time as the reaction is running. It is fully automated and is faster than traditional PCR because analysis of the DNA after the reaction is finished is not necessary. Another advance is that qPCR assays often assess the antimicrobial susceptibilities at the same time they are identifying the organism. Note: Real-time PCR is often confused with reverse-transcriptase PCR (you can see why). Reverse-transcriptase PCR is rightly referred to as RT-PCR and involves the creation of DNA out of RNA ("reverse" transcription). An RT-PCR test has been a prominent test used for COVID-19 testing.

Multiplex PCR

This is a type of diagnostic PCR that contains primers for multiple organisms instead of just a single primer. As with biochemical panels, a multiplex PCR will contain primers for multiple organisms in the differential diagnosis for the patient's symptoms. Most often, multiplex PCR is also real-time PCR.

Panbacterial qPCR

The prefix *pan-* means "universal," and this real-time PCR technique uses a primer (16s rRNA) that is common to all bacteria. Every bacterium in a sample will be amplified. It is particularly important when assessing a wound or infection that is polymicrobial. This technique has the advantage that it can detect bacteria even after antibiotic treatment.

RT-LAMP Test

An amplification method called loop-mediated isothermal amplification is gaining traction. When combined with reverse transcription, it allows for more rapid detection of pathogens.

Hybridization: Probing for Identity

Hybridization is a technique that makes it possible to identify a microbe by analyzing segments of its genetic material. This requires small fragments of single-stranded DNA or RNA called **probes** that are known to be complementary to the specific sequences of nucleic acid isolated from a particular microbe. Base-pairing of the known probe to the nucleic acid can be observed, providing evidence of the microbe's identity. These tests have become more convenient and portable over the years. Probes are typically fluorescently labeled or attached to an enzyme that triggers a colorimetric change when hybridization occurs. This is based on the property of dyes such as fluorescein and rhodamine, which emit visible light in response to ultraviolet. This property of fluorescence has found numerous applications in diagnostic testing.

Fluorescent *in situ* hybridization (FISH) techniques involve the application of fluorescently labeled probes to intact cells within a patient specimen or an environmental sample **(figure 18.13).** Microscopic analysis is used to locate "glowing" cells and conclude the identity of a specific microbe. FISH is often used to confirm a diagnosis or to identify the microbial components within a biofilm. Due to the convenience and accuracy of new PCR methods, FISH methods are decreasing in use for diagnosis of infections but continue to be used in cancer diagnoses, especially in personalized medicine, where a patient's DNA is examined for particular characteristics that can make a certain drug a better choice for them.

Whole-Genome Sequencing

The cost of whole-genome sequencing, in terms of time and money, is becoming so low that this technique is now commonplace in clinical and epidemiological laboratories around the world. It is particularly useful for rapid analysis of outbreaks and drug-resistant organisms. More importantly, a single genome can be scanned and analyzed multiple times in a process called "deep sequencing," which minimizes errors. Some scientists suspect that these types of sequencing will become so cheap and so routine that we will soon "just sequence everything" from a patient sample to find the one or more microbes causing symptoms.

PulseNet is a CDC program that uses genotypic methods to assist in the investigation of possible infectious disease outbreaks. Until recently, it relied on the use of gel electrophoresis to compare patient samples and determine if the organisms from separate patients—and sometimes those found in a suspect food—are identical organisms. But in the spirit of "just sequencing everything," they have now switched to whole-genome sequencing as their standard methodology.

Special Case: Bloodstream Infections

Multiplex PCR (a NAAT technique) can be used to identify blood culture isolates, often within a single day.

18.5. Learning Outcomes—Assess Your Progress

10. Explain why nucleic acid amplification techniques are useful for infectious disease diagnosis.
11. Name two examples of techniques that employ hybridization.
12. Explain how whole-genome sequencing can be used for diagnosis.

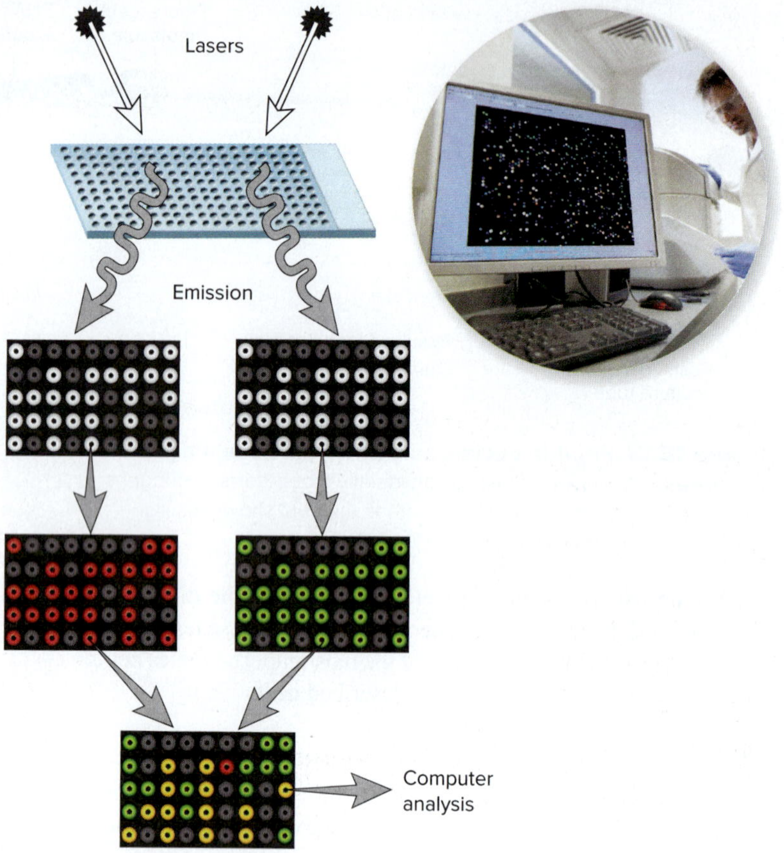
Lasers
Emission
Computer analysis

Figure 18.14 Microarrays. **(a)** Illustration of a microarray. **(b)** Digital view.
Monty Rakusen/Cultura RF/Getty Images

18.6 Additional Diagnostic Technologies

In addition to the traditional phenotypic and immunologic methods, you have seen in this chapter that highly automated processes for determining the genetic makeup of unknowns are being adopted at a rapid pace. In recent years, breakthroughs in other scientific disciplines have also been brought into infectious disease diagnosis. We'll look at a few of them: microarrays, lab on a chip, CRISPR, mass spectrometry, and imaging technologies.

Microarrays

Microarrays designed for infectious disease diagnosis are "chips" (absorbent plates) that contain gene sequences, mRNA, or antigens from potentially thousands of different possible infectious agents, selected based on the syndrome being investigated (such as respiratory infection or meningitis symptoms). The arrays are selected based on a very large differential diagnosis. In other words, what possible microbes could cause disease in this syndrome? Arrays can be made to contain bacterial, viral, and fungal identification in a single test. In this scenario, patient samples (sputum, cerebrospinal fluid) or the nucleic acids isolated from them are incubated with labeled gene sequences on the microarray. Matching sequences hybridize to the chip, and the label (in most cases, it is fluorescence) is detected by a computer program, which provides the identity of the isolate or isolates **(figure 18.14).**

Lab on a Chip

Lab on a chip devices take advantage of microfluidics, the movement of very small amounts of fluids through engineered troughs on horizontal slides or cartridges **(figure 18.15).** Chips are designed to diagnose a variety of conditions, including infectious diseases. The chips can simultaneously test for antimicrobial susceptibilities and the identity of the microbe, and because such tiny

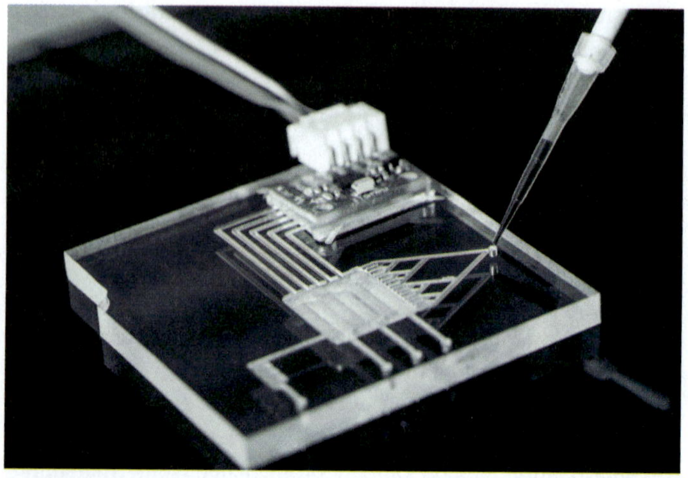

Figure 18.15 Lab on a Chip.
Science Photo Library/Alamy Stock Photo

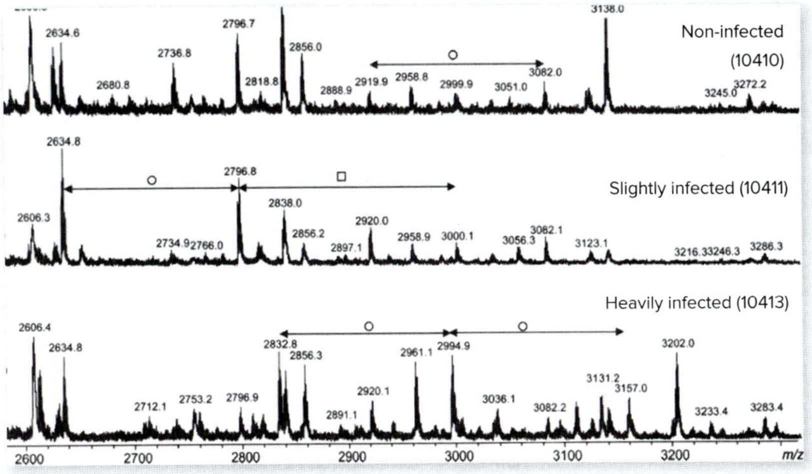
Non-infected (10410)
Slightly infected (10411)
Heavily infected (10413)
2600
2700
2800
2900
3000
3100
3200
m/z

Figure 18.16 Mass spectrometry of patient samples.

amounts are required, they can do so in hours rather than days. Lab on a chip tests are often available for point-of care use.

One new twist on the lab on a chip approach employs CRISPR technology—using both the LAMP methods and rtPCR to perform the CRISPR reaction on a chip. While it is useful in clinical laboratories in developed countries, the biggest impact of labs on chips is predicted to be in developing countries, where diagnosis is often not possible due to lack of supplies, expertise, or even the refrigeration required to store an array of diagnostic reagents.

Mass Spectrometry

Mass spectrometry has been utilized for years to determine the structure and composition of various chemical compounds and biological molecules. It is now being used, alone and in combination with other technologies such as PCR, to provide rapid and highly accurate microbial identification within just minutes. This technique, which is also often called MALDI-TOF, can be used to analyze a protein fingerprint from pure culture isolates or directly from samples isolated from patient specimens. It works by adding the patient sample to a metal plate and then striking it with a laser. This causes the sample to become ionized. The ions from the sample are guided into a machine that separates them and identifies them according to their mass-to-charge ratio. This technology has been applied to the identification of bacteria, viruses, and fungi so far and is becoming commonplace in many clinical and research laboratories due to its ability to produce rapid, precise, and cost-effective results compared to conventional phenotypic, genotypic, and immunologic methods. One distinct advantage of this technique is that clinical laboratories can construct databases of local strains of microorganisms by running profiles of known microbes. This provides for an infinitely customizable and expandable database.

Mass spectrometry technologies can also be used to simultaneously detect antibiotic susceptibilities—obviously, an important advantage **(figure 18.16).**

Imaging

For centuries, imaging, in the form of X rays, has been used to assist in the diagnosis of tuberculosis. Now, in the 21st century, a variety of new imaging techniques have been deployed to help localize infections. They offer a pair of eyes to inaccessible places and can save the patient an invasive biopsy. Infections associated with hip implants, for example, may be difficult to access through blood samples. The bacteria may be growing in biofilms on the implanted materials, or they may be growing in an abscess deep in the hip joint. Magnetic resonance imaging, computerized tomography (CT) scans, and positron emission tomography (PET) scans have been increasingly employed to find areas of localized infection in deep tissue, which can later be biopsied to aspirate samples for culture. In the event no infection is found on the image, the patient has been spared an invasive procedure **(figure 18.17).**

Special Case: Bloodstream Infections

MALDI-TOF testing is seen as a very useful methodology to replace traditional culture and antimicrobial susceptibility testing of blood cultures, reducing the time required to a few hours.

18.6 Learning Outcomes—Assess Your Progress

13. Describe the benefits of "lab on a chip" technologies for global public health.
14. Identify an advantage presented by mass spectrometry and also by imaging techniques as diagnostic tools.

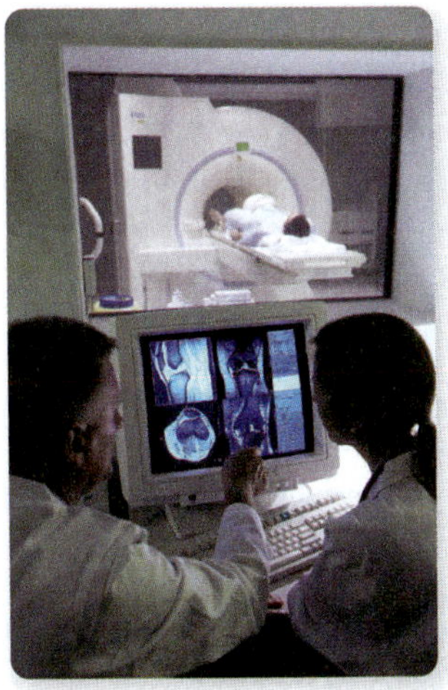

Figure 18.17 A patient undergoing MRI (magnetic resonance imaging) testing.

Javier Larrea/Pixtal/age fotostock

MEDIA UNDER THE MICROSCOPE WRAP-UP

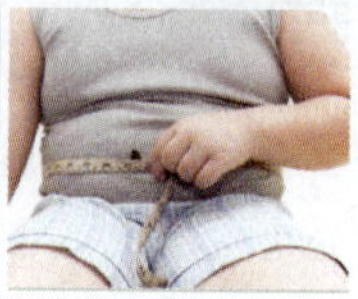
kwanchai.c/Shutterstock

The **intended message** of this article is to report a new research finding that suggests a lack of diversity in the oral microbiome of young toddlers may help to predict their risk for obesity.

When reading reports on microbiome research, it is not uncommon for the **critical reader** to be left with more questions than answers. Scientists are just beginning to scratch the surface when it comes to the microbiome and its impact on human health and disease. This article is no exception. While the article does highlight other factors that have been linked to obesity risk, the author doesn't clearly explain this new study and its impact for readers. As a reader, I might wonder, did these kids have a low-diversity microbiome since birth? Is it the lack of microbial diversity that led to the tendency for these children to gain weight rapidly after birth? Or did the microbial diversity diminish as a consequence of diet and other factors that are more strongly associated with obesity? How did the researchers control for many outside factors that could play a role in obesity or changes in microbial diversity? As a parent, I may wonder what I should do with this information to help prevent obesity in my children. The list of questions goes on.

The good news is that the article is about a specific research study at a reputable institution and quotes the lead researcher throughout the article. My next step would be to locate and read through the originally published research to see if I could find any answers to my questions before I shared the news with my **nonmicrobiologist friends.**

My **overall grade** for this article is a C. There is just too much left unsaid that could lead to misconceptions, and at this point, the reported information is of little value to the average reader. However, I do appreciate the author's use of terms such as "could" and "may" when discussing the potential of using oral swabs to predict obesity. At least she didn't use exaggerated language to add to the possibility for confusion.

Source: DailyMail.com, "Swabbing a Child's Mouth for Bacteria Could Predict How Likely They Are to Become Obese," online article posted 9/20/2018.

Study Smarter: Better Together

These activities are designed for you to use on your own with a study group—either a face-to-face group or a virtual one, consisting of 3–5 members. Studying together can be very helpful, but there are effective and ineffective ways to do it. For example, getting together without a clear structure is often not a good use of your time. Use your time efficiently by using one or more of the exercises below.

FACE-TO-FACE GROUPS

Use one or more of the activities below.

Peer Instruction: Assign numbers to your group members to use all semester long. Now look at these five concepts from this chapter. Each group member prepares a 5-minute lesson on the topic corresponding to their number. Do not worry if you have fewer than 5 members; just use however many you have! During your group study time, each member presents their lesson, and the group spends another 5–10 minutes discussing that lesson.

1. Examples of phenotypic diagnosis methods
2. Examples of immunologic diagnosis methods
3. Examples of genotypic diagnosis methods
4. Walk-through of an ELISA test
5. Explanation of *titer*

Concept Maps: Each member of the group should use this list of terms from this chapter to generate their own concept map. This can be hand-drawn or created using software (see Appendix C for guidelines). During group study time, compare each other's concept maps and help each other make sure they are correct. Of course, there are many different "correct" maps. Examining each member's map will help you talk through the varied concepts and how they are related.

Concept Terms:

sample collection	antimicrobial susceptibility testing	phenotypic methods
immunologic methods	genotypic methods	PCR
ELISA	microarrays	Gram staining

Table Topics: Each group member should identify a concept or topic from this week's class assignments with which they are having trouble and share it during group study time. The other group members can then help to clarify confusing issues or share how they figured it out. Aim for a maximum of 15 minutes per topic. If the topic remains unclear to the group, bring it up during class or use the instructor's office hours or e-mail to ask for help. Taking the time to struggle with a difficult concept first makes your questions much more specific and more likely to yield helpful answers.

VIRTUAL GROUPS

Not everyone has the time or opportunity to meet with group members outside of class time. You or your instructor can create a virtual group using e-mail or the course software.

Weekly Discussion Board: This forum can be used as a way for groups to discuss topics, via e-mail, or other learning management systems or online platforms, before they are covered in class. As each member of the group answers the current week's question, they should send their responses to every other member of their group. It's best to agree on a deadline based on how your class schedule works (Saturday for the next week's topics, for example). Then, after the topic is discussed in class, each member should send a response that all group members will see with a follow-up post on the same topic. If you cover more than one chapter in a week, someone can be designated to choose which chapter Discussion Board question you will use. Or simply decide up front that you will always use the first-chapter-of-the week's question, to keep the schedule simple.

Discussion Question

Explain how biochemical testing for the purpose of diagnosing an infectious disease works.

Chapter Summary

DIAGNOSING INFECTIONS

18.1 PREPARATION FOR THE SURVEY OF MICROBIAL DISEASES

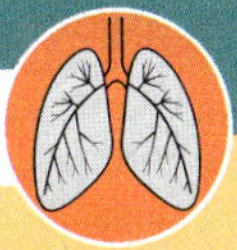

- **Microbiologists use three categories of techniques to diagnose infections: phenotypic, immunologic, and genotypic. Imaging and other specialty techniques are also used.**
- **Microbes causing bloodstream infections are particularly important to identify quickly and accurately.**

18.2 FIRST STEPS: SPECIMEN COLLECTION

- **The first step in diagnosis is collecting a specimen. If it is not performed correctly, the diagnosis will not be accurate no matter how good the test is.**

18.3 PHENOTYPIC METHODS

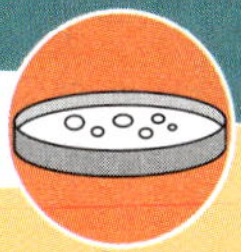

- **The main phenotypic methods include the direct examination of specimens, observation of the growth of specimen cultures on special media, and biochemical testing of pure cultures.**
- **Traditionally, most phenotypic methods required the culturing of the microbes. There are new technologies that can either shorten growth time or eliminate it altogether.**

18.4 IMMUNOLOGIC METHODS

- Serological testing can be performed on a variety of body fluids or tissues and is based on the principle that antibodies have extreme specificity for antigens.
- Agglutination reactions occur between antibodies and antigens bound to cells, resulting in visible clumping. It is the basis of determining titer, or antibody concentration, in patient sera.
- Immunochromatographic methods use cartridges and lateral flow technology to make antibody-antigen interactions easily visible.
- The ELISA test is widely used to detect antigens (sandwich method) or antibodies (indirect method) in patient samples.
- *In vivo* serological testing, such as the tuberculin reaction, involves injection of antigen to elicit a visible immune response in the host.
- Complement fixation uses red blood cells as indicators of the presence of antibody–antigen complexes.

18.5 GENOTYPIC METHODS

- The use of genotypic methods in microbial identification has grown exponentially.
- Nucleic acid amplification tests are based on the polymerase chain reaction (PCR).
- Nucleic acid hybridization exploits the ability of nucleic acids to bind (hybridize) to complementary base pairs.
- Increasingly, high-throughput whole genomes are sequenced using the more convenient and cheaper equipment that is now available.

18.6 ADDITIONAL DIAGNOSTIC TECHNOLOGIES

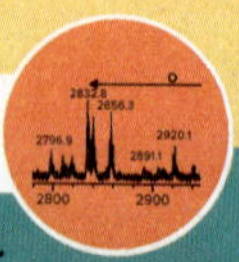

- Microarrays and labs on a chip miniaturize testing into a credit-card-sized "chip" using fewer reagents and requiring less microbial mass.
- Mass spectrometry detects microbes via their protein fingerprints.
- Imaging techniques such as magnetic resonance imaging (MRI) and computerized tomography (CT) scans are increasingly used to visualize deep-tissue infections.

SmartGrid: From Knowledge to Critical Thinking

This *21 Question Grid* takes the topics from this chapter and arranges them with respect to the American Society for Microbiology's Undergraduate Curriculum guidelines—all six of the important "Concepts" as well as the important "Competency" of scientific literacy. Three questions are supplied, which cover chapter content referring to the Concept or Competency in increasing levels of Bloom's taxonomy for learning.

ASM Concept/ Competency	A. Bloom's Level 1, 2—Remember and Understand (Choose one)	B. Bloom's Level 3, 4—Apply and Analyze	C. Bloom's Level 5, 6—Evaluate and Create
Evolution	**1.** Point-of-care testing **a.** is currently being used. **b.** is only used in research. **c.** will be licensed soon. **d.** is the most common currently used technique.	**2.** Explain why it is possible to identify some bacteria simply by knowing what antibiotic resistances they possess.	**3.** Serotyping identifies distinct members of the same species that are stably different from one another. Speculate about why these distinct serotypes are not different species instead.

ASM Concept/ Competency	A. Bloom's Level 1, 2—Remember and Understand (Choose one)	B. Bloom's Level 3, 4—Apply and Analyze	C. Bloom's Level 5, 6—Evaluate and Create
Cell Structure and Function	**4.** Mass spectrometry identifies microbes via **a.** fluorescent antibodies. **b.** cell surface carbohydrates. **c.** protein fingerprints. **d.** DNA profiling.	**5.** Name some bacterial structures that might be useful to target when using fluorescent antibodies designed for diagnosis, and justify your choices.	**6.** You perform a lumbar puncture on a patient with meningitis symptoms and see that the spinal fluid is cloudy. However, panbacterial PCR comes up with nothing. What is a likely explanation?
Metabolic Pathways	**7.** Which category of diagnosis is represented by studying a microbe's utilization of nutrients? **a.** phenotypic **b.** genotypic **c.** immunologic **d.** none of these	**8.** Write a paragraph that explains the mycobacterial growth indicator tube test to a middle school science class.	**9.** You inoculated a biochemical test kit with a patient isolate in a rural clinic you work in, and left it in the incubator overnight. When you came back in, you saw that there had been a power outage (with no backup power generator) and the temperature in the incubator was 25°C instead of 37°C. You examine the test strip and you see that the wells display growth. Do you trust the results? Why or why not?
Information Flow and Genetics	**10.** Which of the following diagnostic techniques is most likely to be affected by changes in growth conditions of the specimen? **a.** phenotypic **b.** immunologic **c.** genotypic	**11.** You perform a disc diffusion test to determine antibiotic susceptibilities of a clinical isolate. There is a clear zone of several millimeters around the penicillin disc, but there are two to three distinct colonies in the middle of the clear halo. What is the probable explanation for those colonies?	**12.** Why might culture conditions affect the results of a mass spectrometry method of diagnosis?
Microbial Systems	**13.** Which of the following techniques is most likely to reveal that an infection is in biofilm form? **a.** ELISA **b.** whole-genome sequencing **c.** lab on a chip **d.** imaging	**14.** Why is it more important to use selective media when diagnosing a GI tract infection than when diagnosing a bloodstream infection?	**15.** What type of diagnostic method do you think would be most suited to determining if a bacterium causing a human infection was lysogenized by a bacteriophage that contains a virulence gene?
Impact of Microorganisms	**16.** rtPCR stands for **a.** reverse transcriptase PCR. **b.** retested PCR. **c.** real-time PCR. **d.** run-time PCR.	**17.** Some HIV tests look for patient antibody to the virus rather than the virus itself. Identify some advantages to this approach.	**18.** Nonhealing wounds on the surface of the body are often extremely difficult to manage, in part because the microbial cause of the lack of healing is often extremely difficult to identify. Create a list of reasons this might be the case.
Scientific Thinking	**19.** A test that results in a very large number of false positives probably has an unacceptable level of **a.** sensitivity. **b.** specificity.	**20.** When PCR is performed "by hand" (not with a prepackaged test kit), the lab environment has to be scrupulously clean. Explain why that might be.	**21.** What kind of a control would be important to run when performing a metabolic diagnostic test such as seen in figure 18.4?

Answers to the multiple-choice questions appear in Appendix A.

Visual Connections

These questions use visual images or previous content to make connections to this chapter's concepts.

1. **From chapter 3, figure 3.5*b*.** What biochemical characteristic does this figure illustrate? How could this characteristic be used to begin the identification of these organisms? Explain your answer.

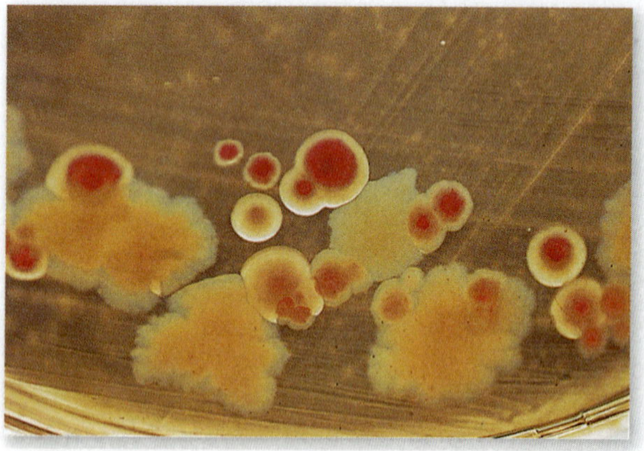

Kathleen Talaro

2. **From this chapter, figure 18.11.** Can you explain why one single bottle of secondary antibody can serve as the indicator antibody for any ELISA reaction?

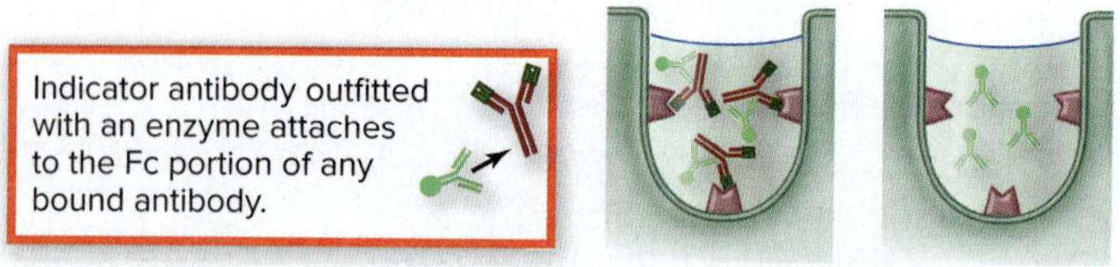
Indicator antibody outfitted with an enzyme attaches to the Fc portion of any bound antibody.

High Impact Study

These terms and concepts are most critical for your understanding of this chapter—and may be the most difficult. Have you mastered them?

Concepts

- ☐ Three categories of infectious diagnosis methods
- ☐ Sensitivity vs. specificity
- ☐ Nucleic acid amplification tests
- ☐ Lab on a chip
- ☐ Mass spectrometry and its use in diagnosis

Terms

- ☐ Dichotomous key
- ☐ Serology
- ☐ Agglutination
- ☐ Precipitation
- ☐ Western blot
- ☐ ELISA
- ☐ Titer
- ☐ Microarrays

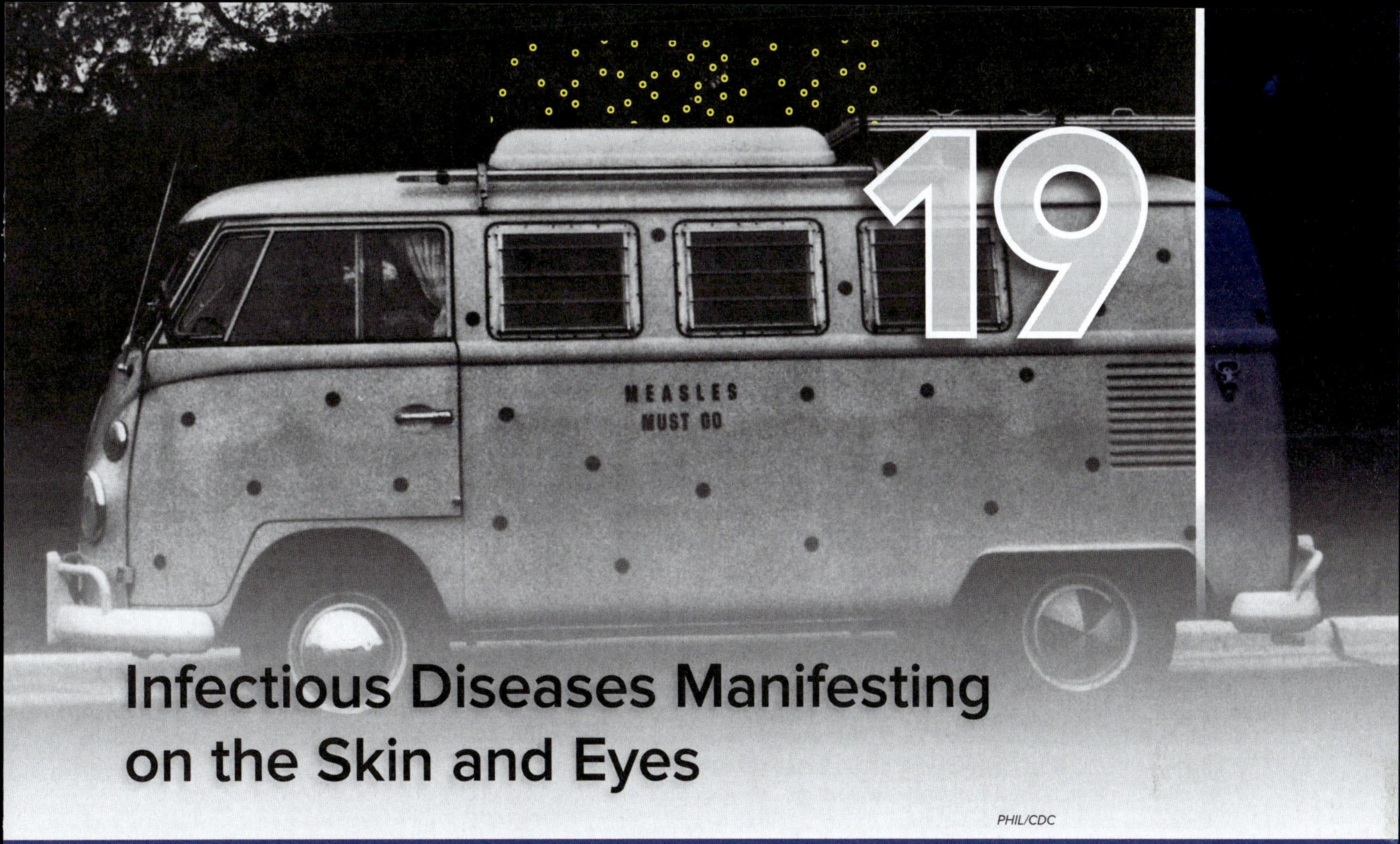
MEASLES
MUST GO

19 Infectious Diseases Manifesting on the Skin and Eyes

PHIL/CDC

MEDIA UNDER THE MICROSCOPE

Vaccinating Kids, 1960s and 2020s

This case study examines an article from a scientific journal to determine the extent to which it is factual and/or misleading. This case focuses on the 2021 New England Journal of Medicine *article "Vaccinating Children Against COVID-19—The Lessons of Measles."*

This is a "Perspective" article published in the *New England Journal of Medicine*. This generally means it is an opinion piece and not a research article. So our challenge here is to examine the facts and numbers that are cited in the piece to see if they support the authors' opinion, or perspective.

The authors make the case that children should be vaccinated against COVID-19 and use the measles vaccination campaign from the 1960s as a model for communicating. (Note: Always pay attention to the publication date of articles. This one was published in February 2021, which is after the adult version of the COVID-19 vaccines were in use and before the children's version was available.) First, they explain how important they believe it is for children to be vaccinated against COVID-19, for three main reasons:

1. to prevent the (infrequent) serious occurrence of disease in children;
2. to prevent the serious complications of multisystem inflammatory syndrome in children (MIS-C). and cardiac inflammation; and
3. to protect adults with whom they are in contact.

Next the authors recall the successful measles vaccination campaign in the 20th century. They note that before a vaccine was available, almost every child fell ill with measles. Four hundred to five hundred died every year in the United States, and there were 1,000 annual cases of encephalitis. Forty-eight thousand children were hospitalized annually. The authors compare measles to COVID-19, in that *most* children did not die of it, yet it had potentially serious consequences. Parents embraced the vaccine, much like they had the polio vaccine a decade earlier. Of course, children who had a pediatrician were more likely to be vaccinated, and like many other diseases, the burden of disease fell disproportionately on Black and Hispanic children, until laws were instituted that required children to have the measles shot to go to school.

The article ends by emphasizing that vaccinating children against COVID-19 will be the only way to end the pandemic, and that messaging that is sensitive to parents' concerns is the best way to get that done.

- What is the **intended message** of the article?
- What is your **critical reading** of the summary of the article? Remember that in this context, "critical reading" does not necessarily mean *What criticism do you have?* but asks you to apply your knowledge to interpret whether the article is factual and whether the facts support the intended message.
- How would you **interpret** the news item for your nonmicrobiologist friends?
- What is your **overall grade** for the news item—taking into account its accuracy and the accuracy of its intended effect?

Media Under The Microscope Wrap-Up appears at the end of the chapter.

Outline and Learning Outcomes

19.1 The Skin, Its Defenses, and Normal Biota

1. Describe the important anatomical features of the skin.
2. List the natural defenses present in the skin.
3. List the types of normal biota presently known to occupy the skin.

19.2 Infectious Diseases Manifesting on the Skin

4. List the possible causative agents for each of the infectious skin conditions: MRSA, impetigo, cellulitis, staphylococcal scalded skin syndrome, gas gangrene, vesicular or pustular rash diseases, maculopapular rash diseases, wartlike eruptions, large pustular skin lesions, and cutaneous and superficial mycosis.
5. Identify which of these conditions are transmitted to the respiratory tract through droplet contact.
6. List the skin conditions for which vaccination is recommended.
7. Summarize methods used to distinguish infections caused by *Staphylococcus aureus* and *Streptococcus pyogenes,* and discuss the spectrum of skin and tissue diseases caused by each.
8. Explain what MRSA stands for and its epidemiology.
9. Discuss the relative dangers of rubella and rubeola viruses in different populations.

19.3 The Surface of the Eye, Its Defenses, and Normal Biota

10. Describe the important anatomical features of the eye.
11. List the natural defenses present in the eye.
12. List the types of normal biota presently known to occupy the eye.

19.4 Infectious Diseases Manifesting in the Eye

13. List the possible causative agents for each of the infectious eye diseases: conjunctivitis, trachoma, keratitis, and river blindness.
14. Discuss why there are distinct differential diagnoses for neonatal and non-neonatal conjunctivitis.

A Note About the Chapter Organization

In a clinical setting, patients present themselves to health care practitioners with a set of symptoms, and the health care team makes an "anatomical" diagnosis—such as a *generalized vesicular rash.* The anatomical diagnosis allows practitioners to narrow down the list of possible causes to microorganisms that are known to be capable of creating such a condition (the differential diagnosis). Then the proper tests can be performed to arrive at an etiologic diagnosis (determining the exact microbial cause). The order of events is

1. anatomical diagnosis,
2. differential diagnosis, and
3. etiologic diagnosis.

In this book, we organize diseases according to anatomical diagnosis (which appears as a boxed heading). Then the agents in the differential diagnosis are each addressed. When we finish addressing each agent that could cause the condition, we sum them up in a Disease Table, whether there is only 1 possible cause or whether there are 9 or 10. Each organism is followed by a B, F, P, V, or H, standing for bacterium, fungus, protozoan, virus, or helminth. If it is a bacterium, its Gram status is also included (G+ or G−).

In the Disease Tables, you will also find a row featuring recommended treatment. Here we will identify the microbes that are on the CDC's "Threat" list for their antibiotic resistance (presented in a table in chapter 12).

19.1 The Skin and Its Defenses

The skin makes contact directly with the environment—not only with solid objects but also with water and other fluids—and with the atmosphere. Also, many infectious diseases include skin eruptions or lesions as part of the course of illness and often as a major symptom, even if they are not primarily infectious diseases of the skin.

The eye surface, like the skin, is also exposed constantly to the environment. For this reason, we include diseases of both organ systems in this chapter. The organs in this chapter form the boundary between the organism and the environment. The skin, together with the hair, nails, and sweat and oil glands, forms the **integument.** The skin has a total surface area of 1.5 to 2 square meters. Its thickness varies from 1.5 millimeters at places such as the eyelids to 4 millimeters on the soles of the feet. Several distinct layers can be found in this thickness, and we summarize them here. Follow **figure 19.1** as you read.

The outermost portion of the skin is the epidermis, which is further subdivided into four or five distinct layers. On top is a thick layer of epithelial cells called the stratum corneum, about 25 cells thick. The cells in this layer are dead and have migrated from the deeper layers during the normal course of cell division. They are packed with a protein called **keratin,** which the cells have been producing ever since they arose from the deepest level of the epidermis. Because this process is continuous, the entire epidermis is replaced every 25 to 45 days. Keratin gives the cells their ability to withstand damage and abrasion. The surface of the skin is called **keratinized** for this reason. The spaces between individual cells of the stratum corneum are packed with a special kind of lipid that has super water-repellent properties. Below the stratum corneum are three or four more layers of epithelial cells. The lowest layer, the stratum basale, or basal layer, is attached to the underlying dermis and is the source for all of the cells that make up the epidermis.

The dermis, underneath the epidermis, is composed of connective tissue instead of epithelium. This means that it is a rich matrix of fibroblast cells and fibers such as collagen, and it contains macrophages and mast cells. The dermis also harbors a dense network of nerves, blood vessels, and lymphatic vessels. Damage to the epidermis generally does not result in bleeding, whereas damage deep enough to penetrate the dermis results in broken blood vessels. Blister formation, the result of friction trauma or burns, is the result of a separation between the dermis and epidermis.

The "roots" of hairs, called follicles, are in the dermis. **Sebaceous** (oil) **glands** and scent glands are associated with the hair follicle. Separate sweat glands are also found in this tissue. All of these glands have openings on the surface of the skin, so they pass through the epidermis as well.

It could be said that the skin is its own defense—in other words, the very nature of its keratinized surface prevents most microorganisms from penetrating into sensitive deeper tissues. Millions of cells

Figure 19.1 A cross section of skin and the levels at which some disease processes take place. In the photomicrograph on the right, you can see the hair (dark purple) and a sebaceous gland (at its base).
Steve Gschmeissner/Science Source

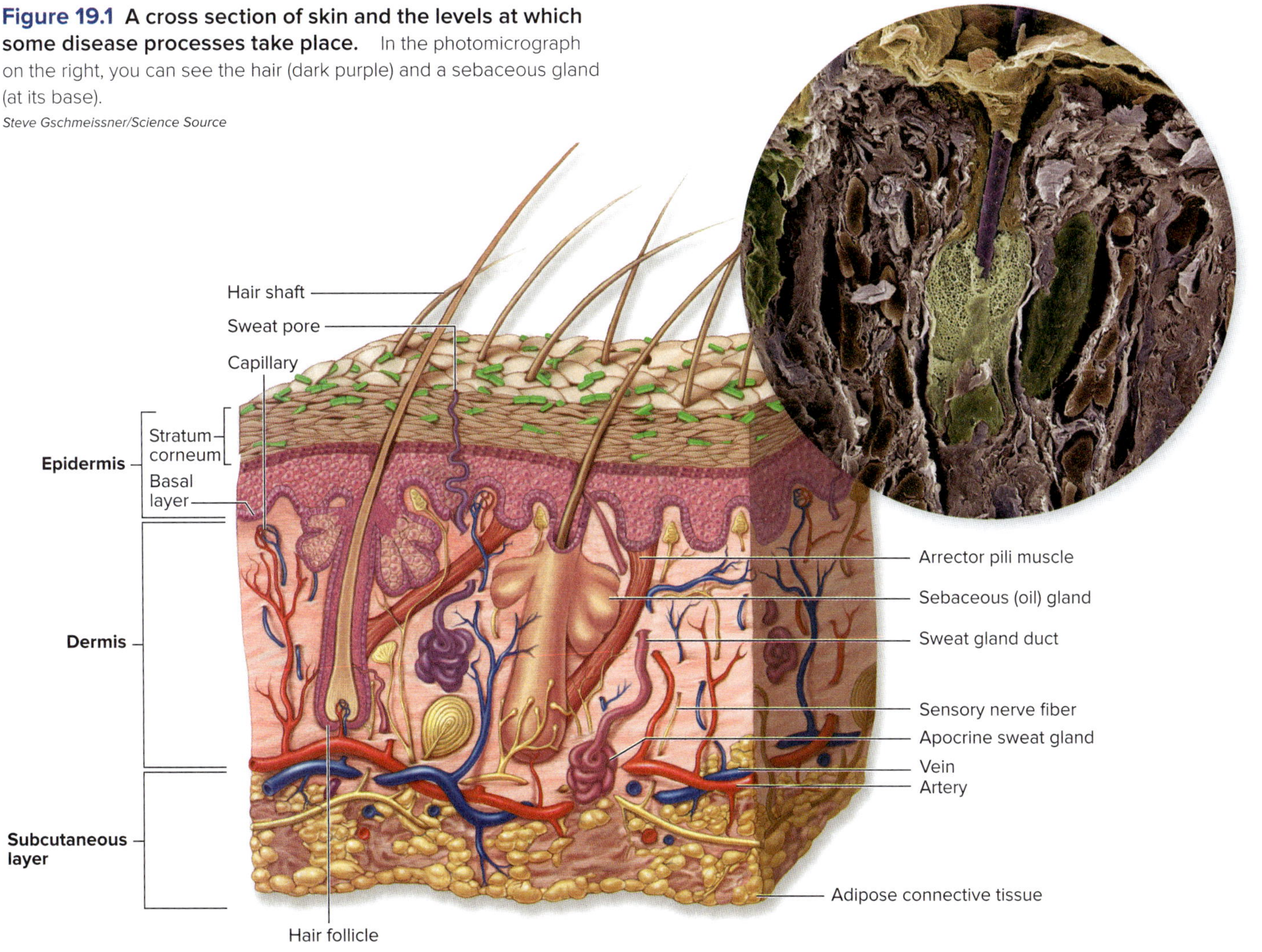

Hair shaft
Sweat pore
Capillary
Epidermis
Stratum corneum
Basal layer
Dermis
Subcutaneous layer
Hair follicle
Arrector pili muscle
Sebaceous (oil) gland
Sweat gland duct
Sensory nerve fiber
Apocrine sweat gland
Vein
Artery
Adipose connective tissue

from the stratum corneum slough off every day, and attached microorganisms slough off with them. The skin is also brimming with antimicrobial substances. Perhaps the most effective skin defense against infection is small molecules called **antimicrobial peptides** in epithelial cells. These are positively charged chemicals that act by disrupting (negatively charged) membranes of bacteria. There are many different types of these peptides, and they seem to be chiefly responsible for keeping the microbial count on skin relatively low.

The sebaceous glands' secretion, called **sebum,** has a low pH, which makes the skin inhospitable to most microorganisms. Sebum is oily due to its high concentration of lipids. The lipids can serve as nutrients for normal microbiota, but when lipids are broken down, the fatty acids in them lead to toxic by-products that inhibit other microorganisms. This mechanism helps control the growth of potentially pathogenic bacteria. Sweat is also inhibitory to microorganisms because of both its low pH and its high salt concentration. **Lysozyme** is an enzyme found in sweat (and tears and saliva) that specifically breaks down peptidoglycan, a unique component of bacterial cell walls.

Normal Biota of the Skin

Overall, the skin is inhospitable to the growth of many microorganisms. The vast dry, salty surfaces, many of which are coated in toxic antimicrobial lipids, were once thought to be only sparsely covered in biota. There are more dense microbial populations in moist areas and skin folds, such as the underarm and groin areas, and in the protected environment of the hair follicles and glandular ducts.

Results from recent microbiome studies have delivered some surprises, however. Some general observations about the skin microbiota follow:

- Hundreds of species of microbes are distributed over many different areas of the body.
- Although five major taxa are represented in the microbiota, the predominance of these groups varied in the different regions of the body sampled.
- There are large differences among people with respect to the types of microbes found on various skin sites.
- An individual's own skin microbiota seems to be relatively stable over time.

Staphylococcus epidermidis and *Propionibacterium acnes* have long been thought to be the most numerous normal biota on skin due to their tolerance of high salt conditions. Also, approximately 4% of the population carry *Staphylococcus aureus*—a potential pathogen—on their skin. Continuing studies look to establish differences among healthy microbiomes and those of patients affected by skin disorders and how a person's living environment may dictate the composition and distribution of microbes on the skin.

Many microbes have been identified living under the skin. The vast majority of microbes resided within the uppermost layers of the epidermis, and nearly 25% of all bacteria were localized to hair follicles. It will be important to determine the role of these microbes in dermatologic disease.

Insight 19.1 contains a story of how the mother's (and caretakers') skin microbiota become the intestinal biota of babies born by cesarean section—and why that might not be the best for the baby.

Defenses and Normal Biota of the Skin

	Defenses	Normal Biota
Skin	Keratinized surface, sloughing, low pH, high salt, lysozyme	Bacteria such as *Staphylococcus epidermidis, Propionibacterium, Corynebacterium, Lactobacillus, Bacteroides, Prevotella, Haemophilus*; yeasts such as *Malassezia, Candida*

INSIGHT 19.1 MICROBIOME: C-Section Babies vs. Vaginally Delivered Babies

Think about your circle of female family and friends: How many of them have delivered babies via C-section? In some regions of the United States, the rates of C-section approach 50%. In some other countries (such as Brazil), it can be as high as 90%. But microbiome science suggests that it might not be the best start for babies—at least with respect to their gut microbiome. And we know how important the gut microbiome is. We have talked about how it can influence many aspects of our health, even outside our guts.

Scientists studied the gut microbiome of babies who were delivered vaginally versus that of those delivered via C-section. The vaginally born babies were colonized with vaginal and fecal bacteria from the mother. That might sound bad, but the bacteria help babies digest milk and are generally protective to gut health. On the other hand, the guts of babies delivered through C-section are generally colonized by the skin bacteria of their mother and of their caregiver. Skin bacteria are not meant to become the microbiome in the gut. Several studies have shown that C-section babies are more prone to infections, and even more prone to allergies and obesity later in life.

Some scientists are researching a simple technique to change the initial colonization of C-section babies. An hour before the C-section surgery, a gauze packet is placed into the mother's vagina. Immediately after the baby is delivered, the gauze is wiped over the baby's mouth and skin in an attempt to preempt the establishment of skin microbes as the baby's microbiome. Other scientists caution that this is a dangerous practice and should not be recommended. Some recent research suggests that within a very short period of time, there is little difference in the gut microbiomes in babies treated in this way and those not treated. Like much research on the human microbiome, more work needs to be done before definitive conclusions can be drawn.

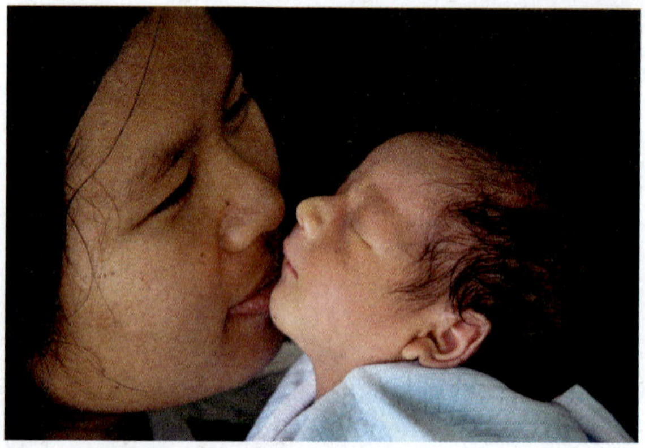
Kevin Landwer-Johan/E+/Getty Images

19.1 Learning Outcomes—Assess Your Progress

1. Describe the important anatomical features of the skin.
2. List the natural defenses present in the skin.
3. List the types of normal biota presently known to occupy the skin.

19.2 Infectious Diseases Manifesting on the Skin

Skin and soft tissue infections are the second most common infections encountered in primary care settings (behind respiratory infections).

MRSA Skin and Soft Tissue Infections

Methicillin-resistant *Staphylococcus aureus* (MRSA) is a common cause of skin lesions in nonhospitalized people. (The hospitalized population is more likely to acquire systemic bloodstream infections from MRSA, addressed in chapter 21.) Even though the name mentions "methicillin," these strains are usually resistant to multiple antibiotics.

Staphylococcus aureus is a gram-positive coccus that grows in clusters, like a bunch of grapes. It is nonmotile **(figure 19.2*a*).** Much of its destructiveness is due to its array of superantigens. It can be highly virulent, but it also appears as "normal" biota on the skin of a small but significant portion of the population. Strains that are methicillin-resistant are also found on healthy people.

This species is considered the sturdiest of all non-endospore-forming pathogens, with well-developed capacities to withstand high salt (7.5% to 10%), extremes in pH, and high temperatures (up to 60°C for 60 minutes). *S. aureus* also remains viable after months of air drying and resists the effect of many disinfectants and antibiotics.

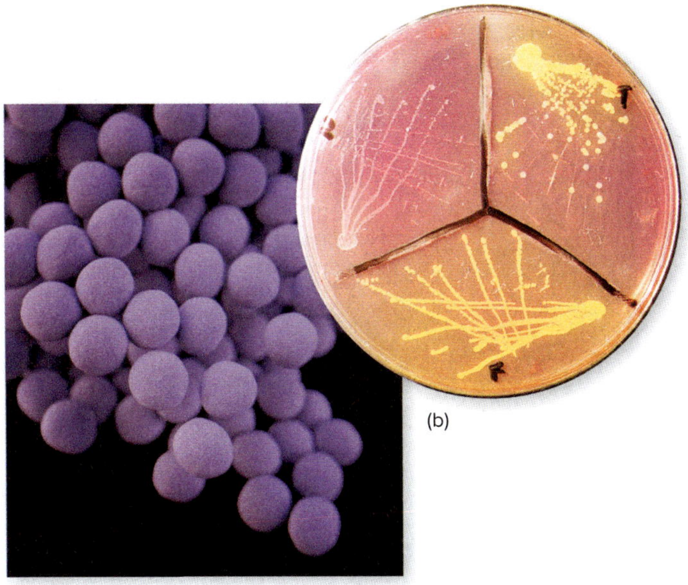

(a) (b)

Figure 19.2 *Staphylococcus aureus.* **(a)** Scanning electron micrograph of *S. aureus.* **(b)** Mannitol salt agar is both differential and selective for the growth of *S. aureus.* Unlike many organisms, *S. aureus* can withstand the relatively high levels of salt in this medium. In addition, it can ferment the mannitol found in the medium, leading to a yellow color change over time.

(a) CDC/Melissa Brower; (b) Kathy Park Talaro

Signs and Symptoms

MRSA infections of the skin tend to be raised, red, tender, localized lesions, often featuring pus and feeling hot to the touch **(figure 19.3).** They occur easily in breaks in the skin caused by injury, shaving, or even just abrasion. They may localize around a hair follicle. Fever is a common feature.

Transmission and Epidemiology

MRSA is a common contaminant of all kinds of surfaces you touch daily, especially if the surfaces are not routinely sanitized. Gym equipment, airplane tray tables, electronic devices, razors, and other fomites are all sources of indirect contact infection. Persons with active MRSA skin infections should keep them covered in order to avoid transmission to others.

Pathogenesis and Virulence Factors

Pathogenic *S. aureus* strains typically produce **coagulase**, an enzyme that coagulates plasma. Its presence is considered the most important characteristic for diagnosis.

Other enzymes expressed by *S. aureus* include hyaluronidase, which digests the intercellular "glue" (hyaluronic acid) that binds connective tissues together; staphylokinase, which digests blood clots; a nuclease that digests DNA (DNase); and lipases that help the bacteria colonize oily skin surfaces.

Culture and/or Diagnosis

Polymerase chain reaction (PCR) is routinely used to diagnose MRSA. Alternatively, primary isolation of *S. aureus* is achieved by inoculation on blood agar **(figure 19.4).** For heavily contaminated specimens, selective media such as mannitol salt agar are used (see **figure 19.2*b***). The production of catalase, an enzyme that breaks down hydrogen peroxide accumulated during oxidative metabolism, can be used to differentiate the staphylococci, which produce it, from the streptococci, which do not.

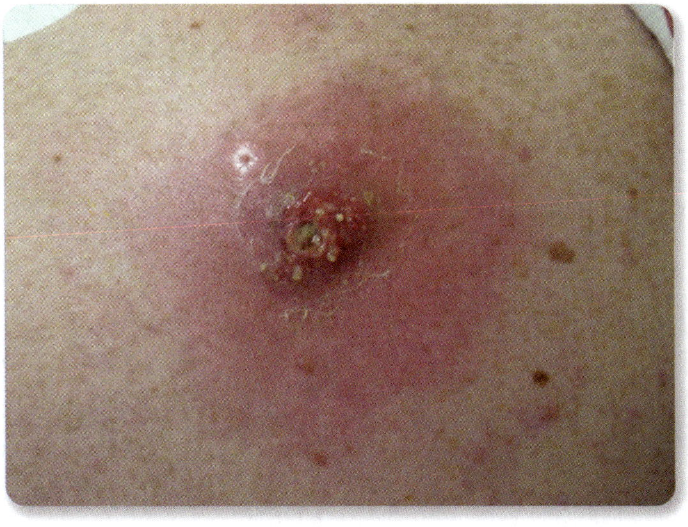

Figure 19.3 A typical MRSA lesion.

CDC/Gregory Moran, M.D.

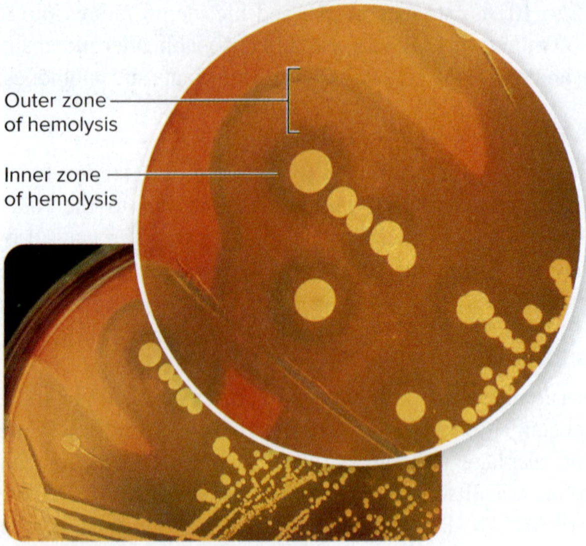
Outer zone
of hemolysis
Inner zone
of hemolysis

Figure 19.4 ***Staphylococcus aureus.*** Blood agar plate growing *S. aureus.* Some strains show two zones of hemolysis, caused by two different hemolysins. The inner zone is clear, whereas the outer zone is fuzzy and appears only if the plate has been refrigerated after growth.
Kathy Park Talaro

Based on the discussed virulence factors, one key technique for separating *S. aureus* from species of *Staphylococcus* is the coagulase test **(figure 19.5).** By definition, any isolate that coagulates plasma is *S. aureus;* all others are coagulase-negative.

▶ Prevention and Treatment

Prevention is only possible with good hygiene. Treatment of these infections often starts with incision of the lesion and drainage of the pus. Antimicrobial treatment should include more than one antibiotic. Current recommendations in the United States are for the use of vancomycin. These recommendations, of course, will change based on antibiotic-resistance patterns.

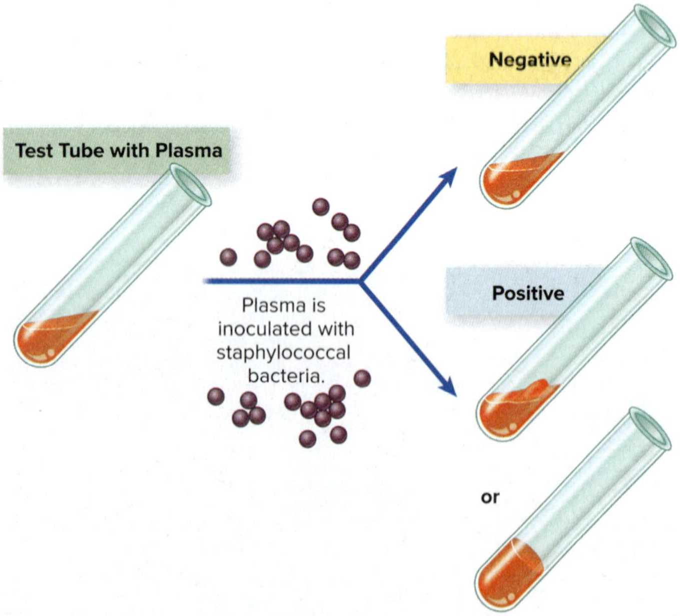
Test Tube with Plasma
Plasma is
inoculated with
staphylococcal
bacteria.
Negative
Positive
or

Figure 19.5 The coagulase test. Staphylococcal coagulase is an enzyme that reacts with factors in plasma to initiate clot formation. In the coagulase test, a tube of plasma is inoculated with the bacterium. If it remains liquid, the test is negative. If the plasma develops a lump or becomes completely clotted, the test is positive.

A Note About Statistics in the Disease Tables

Each condition we study is summarized in a Disease Table. The last row of each table contains information about the epidemiology of the disease. The type of epidemiological information that is most relevant to a particular disease can vary. For example, it is vital to know the numbers of new cases **(incidence)** of some diseases. This is the case for measles in the United States (see Disease Table 19.7), as we track the resurgence of a disease once controlled by vaccination. For other conditions, such as fifth disease, also discussed in Disease Table 19.7, it is more useful to know how many people have been affected by the time they reach a certain age. **Prevalence,** or the current number of people affected by the condition, is another common measure. And for some diseases, the most informative statistic is how deadly they are **(mortality rate).** The tables contain the most useful information about each condition.

Disease Table 19.1	MRSA Skin and Soft Tissue Infections
Causative Organism(s)	Methicillin-resistant *Staphylococcus aureus* B G+
Most Common Modes of Transmission	Direct contact, indirect contact
Virulence Factors	Coagulase, other enzymes, superantigens
Culture/Diagnosis	PCR, culture and Gram stain, coagulase and catalase tests, multitest systems
Prevention	Hygiene practices
Treatment	Vancomycin; in **Serious Threat** category in CDC Antibiotic Resistance Report
Epidemiology	Community-associated MRSA infections most common in children and young to middle-aged adults; incidence increasing in communities (decreasing in hospitals)

Impetigo

Impetigo is a superficial bacterial infection that causes the skin to flake or peel off **(figure 19.6).** It is not a serious disease but is highly contagious, and children are the primary victims. Impetigo can be caused by either *Staphylococcus aureus* or *Streptococcus pyogenes,* and a mixture of the two probably causes most cases. As you may know, these two bacteria cause a wide variety of skin conditions. These are summarized in the Note in the orange box. It has been suggested that *S. pyogenes* causes the initial infection of this disease, but in some cases, *S. aureus* later takes over and becomes the predominant bacterium cultured from lesions. Because *S. aureus* produces a bacteriocin (toxin) that can destroy *S. pyogenes,* it is possible that *S. pyogenes* is often missed in culture-based diagnosis of impetigo. Impetigo can also occur around local trauma to the skin.

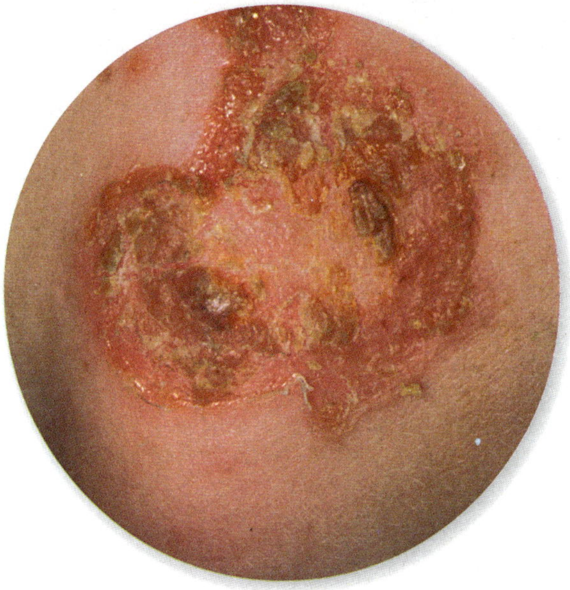

Figure 19.6 Impetigo lesions on a shoulder.
Hercules Robinson/Alamy Stock Photo

Signs and Symptoms

The "lesion" of impetigo looks variously like peeling skin, crusty and flaky scabs, or honey-colored crusts. Lesions are most often found around the mouth, face, and extremities, though they can occur anywhere on the skin. It is very superficial and it itches.

Impetigo Caused by *Staphylococcus aureus*

Pathogenesis and Virulence Factors

The most important virulence factors relevant to *S. aureus* impetigo are exotoxins called exfoliative toxins A and B, which are coded for by a phage that infects some *S. aureus* strains. At least one of the toxins attacks a protein that is very important for epithelial cell-to-cell binding in the outermost layer of the skin. Breaking up this protein leads to the characteristic blistering seen in the condition. The breakdown of skin architecture also facilitates the spread of the bacterium. All pathogenic *S. aureus* strains typically produce coagulase. It is thought that this enzyme causes fibrin to be deposited around the bacteria, concentrating the exotoxins in an area of local damage.

Impetigo Caused by *Streptococcus pyogenes*

S. pyogenes is a gram-positive coccus in Lancefield group A and is beta-hemolytic on blood agar. In addition to impetigo, it causes streptococcal pharyngitis (strep throat), scarlet fever, pneumonia, puerperal fever, necrotizing fasciitis, serious bloodstream infections, and poststreptococcal conditions such as rheumatic fever.

If the precise etiologic agent must be identified, there are well-established methods for identifying group A streptococci. **Figure 19.7*a*** illustrates a rapid direct test, while **figure 19.7*b*** depicts the beta-hemolysis the colonies display on blood agar, as well as their sensitivity to bacitracin.

Pathogenesis and Virulence Factors

The symptoms of *S. pyogenes* impetigo are indistinguishable from those caused by *S. aureus*. Like *S. aureus,* this bacterium possesses a huge arsenal of enzymes and toxins. As mentioned earlier, it anchors itself to surfaces (including skin) using a variety of adhesive elements on its surface (LTA, M protein and other proteins, and a hyaluronic acid capsule). M protein also protects it from phagocytosis. Like *S. aureus,* it possesses hyaluronidase.

Rarely, impetigo caused by *S. pyogenes* can be followed by acute poststreptococcal glomerulonephritis (see chapter 22).

Transmission and Epidemiology of Impetigo

Impetigo, whether it is caused by *S. pyogenes, S. aureus,* or both, is highly contagious and transmitted through direct contact but also via fomites and mechanical vectors. It affects mostly preschool children, but individuals of all ages can acquire the disease. The peak incidence is in the summer and fall.

Culture and/or Diagnosis

Doctors usually diagnose impetigo by visual inspection and typically treat it with antibiotics that target both probable causative agents. However, when an infection requires identification (for instance, if the first treatment fails), well-established methods exist to establish *S. aureus* as the etiologic agent, as described for MRSA infections. However, it is not possible to use colonial or microscopic characteristics to differentiate among gram-positive cocci, including *Staphylococcus epidermidis,* so additional biochemical testing is required. The production of catalase, an enzyme that breaks down hydrogen peroxide accumulated during oxidative metabolism, can be used to differentiate the staphylococci, which produce it, from the streptococci, which do not. The ability of *Staphylococcus* to grow anaerobically and to ferment sugars separates it from *Micrococcus,* a nonpathogenic genus that is a common specimen contaminant.

A Note on Skin: Staph and Strep

The skin is the largest organ of the body and has a unique landscape with folds, ridges, regions that are warm and moist, and regions that are cool and dry. It is also host to hundreds of different types of bacteria, some of which cannot be cultured in the laboratory and have only been discovered through 16S (ssu) rRNA sampling. Most bacteria on the skin are either harmless or beneficial and prevent pathogenic microbes from establishing themselves. However, there are a few pathogens in the mix, specifically *Staphylococcus* and *Streptococcus,* and these are the cause of the majority of bacterial skin infections. Some skin infections range from relatively mild skin eruptions such as folliculitis to deadly necrotizing fasciitis, caused by "flesh-eating bacteria." What makes the difference in these types of infections is the exact microbe causing them and the toxins it produces.

Skin Infections Caused by *Staphylococcus aureus*

Folliculitis, carbuncles, and furuncles (boils) are all infections of the hair follicle and are mainly caused by *Staphylococcus aureus,* although other microbes may be involved. **Folliculitis** is a relatively superficial infection of the hair follicle characterized by a rash or a pimple. **Furuncles,** also known as boils, are skin infections involving a hair follicle and the surrounding skin tissue. Furuncles may begin as a pea-sized lesion and then spread, oozing pus that may require drainage. **Carbuncles** involve a group of hair follicles and have similar symptoms as furuncles but are clustered and form a lump deep in the skin.

Skin Infections Caused by *Streptococcus pyogenes*

Streptococcus pyogenes is the main cause of cellulitis, impetigo, and erysipelas. Cellulitis and impetigo are covered in this chapter. In **erysipelas,** the bacteria enter the skin through a small cut or break in the skin and cause blisters or swollen lesions accompanied by fever, shaking, and chills. Twenty percent of erysipelas cases are seen on the face, with 80% affecting the legs. Similar to skin infections caused by *S. aureus,* erysipelas can result in more serious infections such as bacteremia and septic shock if it is not treated correctly or promptly.

Necrotizing Fasciitis

Several skin infections can be caused by either *Streptococcus pyogenes* or *Staphylococcus aureus.* Two of them are cellulitis and impetigo. There is another, much more serious one. Necrotizing fasciitis can be caused by numerous bacteria such as *Clostridium perfringens* and *Vibrio vulnificus,* but the main causes are *S. aureus* and *S. pyogenes.* Symptoms begin in a similar fashion as simple folliculitis or erysipelas—a minor infection at a hair follicle or break in the skin. However, the similarity ends there. Infection proceeds rapidly as toxins produced by the bacteria break down tissues, the tissues die, and the bacteria spread through the bloodstream. Additionally, these bacteria can produce **superantigens,** which produce powerful stimulation of T cells and a massive release of cytokines that can lead to shock and death. Necrotizing fasciitis is fatal in about 25% of those infected without treatment. It is treated with powerful broad-spectrum antibiotics and surgery to remove dead and damaged tissue and bone. Amputation may be necessary in severe cases.

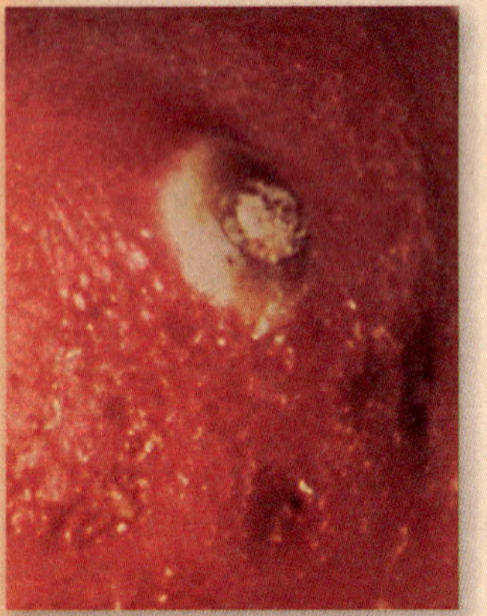

A furuncle, or boil, caused by *Staphylococcus aureus.*

Centers for Disease Control and Prevention/ Joe Miller.

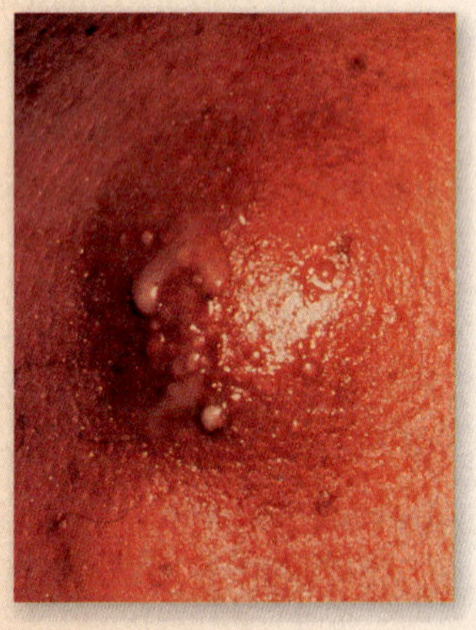

A carbuncle caused by *Staphylococcus aureus.*

John Watney/Science Source

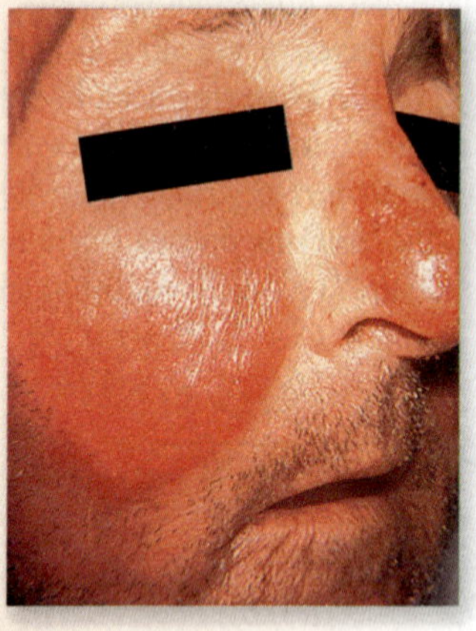

Facial erysipelas caused by *Streptococcus pyogenes.*

Dr Thomas F Sellers, Emroy University, Atlanta, GA./CDC

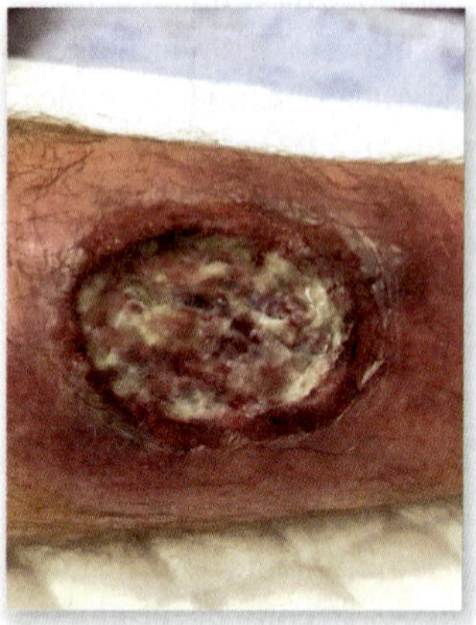

Necrotizing fasciitis, or "flesh-eating disease."

TisforThan/Shutterstock

▶ Prevention

The only prevention for impetigo is good hygiene.

▶ Treatment

Impetigo is often treated with a drug that will target either bacterium, *S. pyogenes* or *S. aureus,* eliminating the need to determine the exact etiologic agent. The drug of choice is topical mupirocin (brand name Bactroban), a protein synthesis inhibitor, although resistance to this drug is increasing in *S. aureus*. Retapamulin is also used topically. Often, cases caused by *S. aureus* require oral antibiotics as well. Dicloxacillin or cephalexin is used for sensitive strains. Clindamycin is the first alternative for methicillin-resistant

Disease Table 19.2 Impetigo		
Causative Organism(s)	*Staphylococcus aureus* B G+	*Streptococcus pyogenes* B G+
Most Common Modes of Transmission	Direct contact, indirect contact	Direct contact, indirect contact
Virulence Factors	Exfoliative toxin A, coagulase, other enzymes	Streptokinase, plasminogen-binding ability, hyaluronidase, M protein
Culture/Diagnosis	Routinely based on clinical signs; when necessary, culture and Gram stain, coagulase and catalase tests, multitest systems, PCR	Routinely based on clinical signs; when necessary, culture and Gram stain, coagulase and catalase tests, multitest systems, PCR
Prevention	Hygiene practices	Hygiene practices
Treatment	Topical mupirocin or retapamulin, oral cephalexin, dicloxacillin; MRSA is in **Serious Threat** category in CDC Antibiotic Resistance Report	Topical mupirocin or retapamulin; erythromycin-resistant *Streptococcus pyogenes* is in **Concerning Threat** category in CDC Antibiotic Resistance Report
Distinguishing Features	Clinically indistinguishable from streptococcal	Clinically indistinguishable from staphylococcal
Epidemiological Features	Prevalence approximately 1% of children in North America	

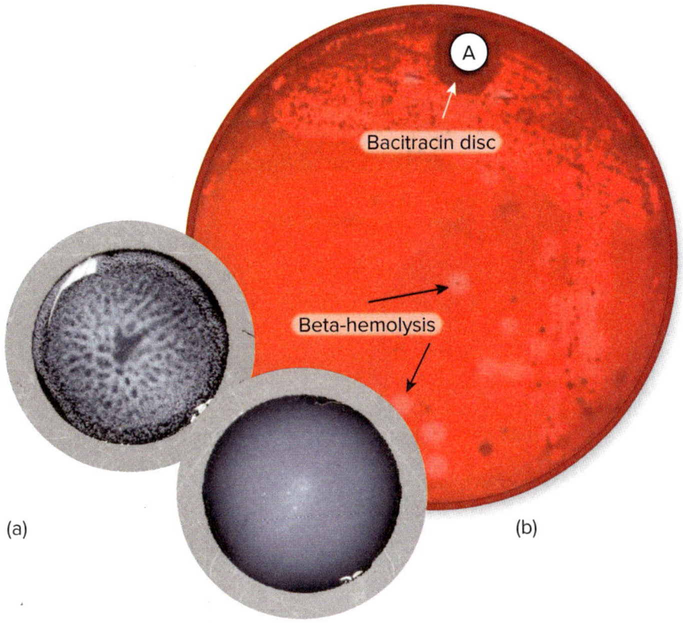
A
Bacitracin disc
Beta-hemolysis
(a)
(b)

Figure 19.7 Identification tests for *Streptococcus pyogenes*. **(a)** A rapid, direct test kit for diagnosis of group A infections. With this method, a patient's throat swab is introduced into a system composed of latex beads and monoclonal antibodies. (*Left*) In a positive reaction, the C carbohydrate on group A streptococci produces visible clumps. (*Right*) A smooth, milky reaction is negative. **(b)** Bacitracin disc test. With very few exceptions, only *Streptococcus pyogenes* is sensitive to a minute concentration (0.02 µg) of bacitracin. Any zone of inhibition around the A disc is interpreted as a presumptive indication of this species.

(a) Lisa Burgess/McGraw Hill; (b) Neal R. Chamberlin, PhD/McGraw Hill

S. aureus (MRSA), but these strains are often resistant to multiple drugs **(Disease Table 19.2).**

Cellulitis

Cellulitis is a condition caused by a fast-spreading infection in the dermis and in the subcutaneous tissues below. It causes pain, tenderness, swelling, and warmth. Fever and swelling of the lymph nodes draining the area may also occur. Frequently, red lines leading away from the area are visible (a phenomenon called *lymphangitis*); this symptom is the result of microbes and inflammatory products being carried by the lymphatic system. Although septicemia can develop, most cases of the disease are uncomplicated and patients have a good prognosis.

Cellulitis generally follows the introduction of bacteria or fungi into the dermis, either through trauma or by subtle means (with no obvious break in the skin). Cellulitis is very common on the lower leg, and it is thought the bacteria can enter through breaks in the skin between the toes caused by fungal infection (athlete's foot). Symptoms take several days to develop. The most common causes of the condition in healthy people are *Streptococcus pyogenes* and occasionally *Staphylococcus aureus,* although almost any bacterium and some fungi can cause this condition in an immunocompromised patient. In infants, group B streptococci are a frequent cause of this infection.

People who are immunocompromised or who have cardiac insufficiency are at higher risk for this condition compared to healthy individuals. They also risk complications, such as spread to the bloodstream, rapid spreading through adjacent tissues, and, especially in children, meningitis. Occasionally, cellulitis is a complication of varicella (chickenpox) infections.

Disease Table 19.3 Cellulitis

Causative Organism(s)	MRSA (B) G+	*Streptococcus pyogenes* (B) G+	Other bacteria or fungi
Most Common Modes of Transmission	Parenteral implantation	Parenteral implantation	Parenteral implantation
Virulence Factors	Exfoliative toxin A, coagulase, other enzymes	Streptokinase, plasminogen-binding ability, hyaluronidase, M protein	–
Culture/Diagnosis	Based on clinical signs	Based on clinical signs	Based on clinical signs
Prevention	–	–	–
Treatment	Vancomycin; surgery sometimes necessary; MRSA is in **Serious Threat** category in CDC Antibiotic Resistance Report	Oral or IV antibiotic (penicillin); surgery sometimes necessary; erythromycin-resistant *Streptococcus pyogenes* is in **Concerning Threat** category in CDC Antibiotic Resistance Report	Aggressive treatment with oral or IV antibiotic; surgery sometimes necessary
Distinguishing Features	–	–	More common in immunocompromised
Epidemiological Features	Incidence highest among males 45–64		

Mild cellulitis responds well to oral antibiotics chosen to be effective against both *S. aureus* (if it is *S. aureus,* it is nearly always MRSA) and *S. pyogenes.* More involved infections and infections in the immunocompromised require intravenous antibiotics. If there are extensive areas of tissue damage, surgical debridement (duh-breed′-munt) may be warranted **(Disease Table 19.3).**

Staphylococcal Scalded Skin Syndrome (SSSS)

This syndrome is another **dermolytic** condition caused by *Staphylococcus aureus.* Although children and adults can be affected, SSSS develops predominantly in newborns and babies. Newborns are susceptible when sharing a nursery with another newborn who is colonized with *S. aureus.* Transmission may occur when caregivers carry the bacterium from one baby to another. Adults in the nursery can also directly transfer *S. aureus* because approximately 30% of adults are asymptomatic carriers. Carriers can harbor the bacterium in the nasopharynx, axilla, perineum, and even the vagina. (Fortunately, only about 5% of *S. aureus* strains are lysogenized by the type of phage that codes for the toxins responsible for the pathogenesis of this disease.)

SSSS can be thought of as a systemic form of impetigo. Like impetigo, it is an exotoxin-mediated disease. The phage-encoded exfoliative toxins A and B are responsible for the damage. Unlike impetigo, the toxins enter the bloodstream from the site of initial infection (the throat, the eye, or sometimes an impetigo infection) and then travel throughout the body, interacting with the skin at many different sites. The A and B toxins cause **bullous** lesions, which often appear first around the umbilical cord (in neonates) or in the diaper or axilla area. The lesions begin as red areas, take on the appearance of wrinkled tissue paper, and then form very large blisters. Fever may precede the skin manifestations. Eventually, the top layers of epidermis peel off completely. The split occurs in the epidermal tissue layers just above the basal layer (see figure 19.1). Widespread **desquamation** of the skin follows, leading to the burned appearance referred to in the name of this condition **(figure 19.8).**

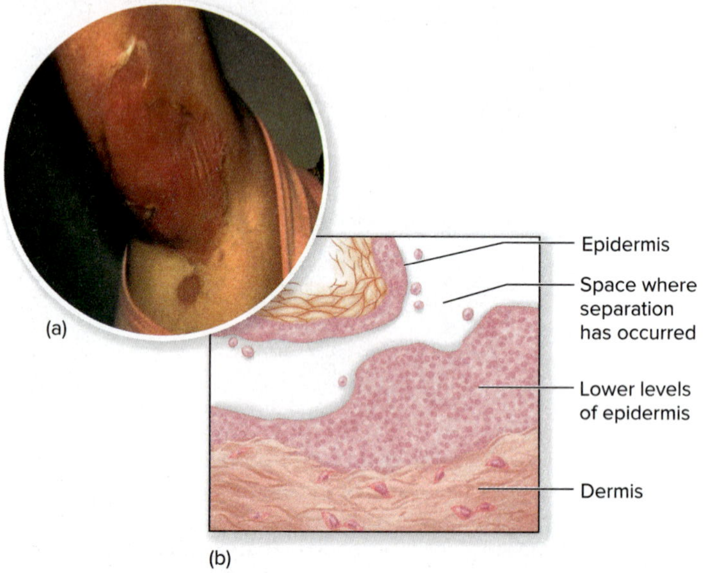
(a)
Epidermis
Space where separation has occurred
Lower levels of epidermis
Dermis
(b)

Figure 19.8 Staphylococcal scalded skin syndrome (SSSS) in an adult. **(a)** Exfoliative toxin produced in local infections causes blistering and peeling away of the outer layer of skin. **(b)** The point of epidermal shedding, or desquamation, is in the epidermis. The lesions will usually heal well because the level of separation is so superficial.

(a) DermPics/Science Source

At this point, the protective keratinized layer of the skin is gone, and the patient is vulnerable to secondary infections, cellulitis, and bacteremia. In the absence of these complications, young patients nearly always recover if treated promptly. Adult patients have a higher mortality rate—as high as 50%. Once a presumptive diagnosis of SSSS is made, immediate antibiotic therapy should be instituted.

It is important, however, to differentiate this disease from a similar skin condition called *toxic epidermal necrolysis* (*TEN*), which is caused by a reaction to antibiotics, barbiturates, or other drugs. TEN also has a significant mortality rate. The treatments for the two diseases are very different, so it is important to distinguish between them before instituting therapy. In TEN, the split in skin tissue occurs *between* the dermis and the epidermis, not within the epidermis as is the case with SSSS. Histological examination of tissue from a lesion is usually a better way to diagnose the disease than reliance on culture. Because SSSS is caused by the dissemination of exotoxin, *S. aureus* may not be found in lesions.

Disease Table 19.4	Scalded Skin Syndrome
Causative Organism(s)	*Staphylococcus aureus* B G+
Most Common Modes of Transmission	Direct contact, droplet contact
Virulence Factors	Exfoliative toxins A and B
Culture/Diagnosis	Histological sections; culture performed but false negatives common because toxins alone are sufficient for disease
Prevention	Eliminate carriers in contact with neonates
Treatment	Immediate systemic antibiotics (current recommendation is cephalexin); MRSA is in **Serious Threat** category in CDC Antibiotic Resistance Report
Distinguishing Features	Split in skin occurs *within* epidermis
Epidemiological Features	Mortality 1%–5% in children, 50%–60% in adults

Gas Gangrene

Clostridium perfringens, a gram-positive, endospore-forming bacterium, causes a serious condition called **gas gangrene,** or clostridial **myonecrosis** (my″-oh-neh-kro′-sis). The endospores of this species can be found in soil, on human skin, and in the human intestine and vagina. This bacterium is anaerobic and requires anaerobic conditions to manufacture and release the exotoxins that cause the damage in this disease. *Streptococcus pyogenes* and *Staphylococcus aureus* can also cause this condtion but less frequently.

Signs and Symptoms

Two forms of gas gangrene have been identified. In *anaerobic cellulitis,* the bacteria spread within damaged necrotic muscle tissue, producing toxins and gas as the infection proceeds. However, the infection remains localized and does not spread into healthy tissue. The pathology of *true myonecrosis* is more destructive. Toxins produced in large muscles, such as the thigh, shoulder, and buttocks, diffuse into nearby healthy tissue and cause local necrosis at these sites. This damaged tissue then serves as a focus for continued bacterial growth, toxin formation, and gas production. The disease can quickly progress through an entire limb or body area, destroying tissues as it goes **(figure 19.9).** Initial symptoms of pain, edema, and a bloody exudate in the lesion are followed by fever, tachycardia, and blackened necrotic tissue filled with bubbles of gas. Gangrenous infections of the uterus caused by septic abortions and clostridial septicemia are particularly serious complications that can arise. If treatment is not begun early, the disease is invariably fatal.

Pathogenesis and Virulence Factors

Because clostridia are not highly invasive, infection usually requires damaged or dead tissue, which supplies growth factors, and an anaerobic environment. The low-oxygen environment results from an interrupted blood supply and the presence of additional aerobic bacteria, which deplete oxygen. These conditions stimulate endospore germination, rapid vegetative growth

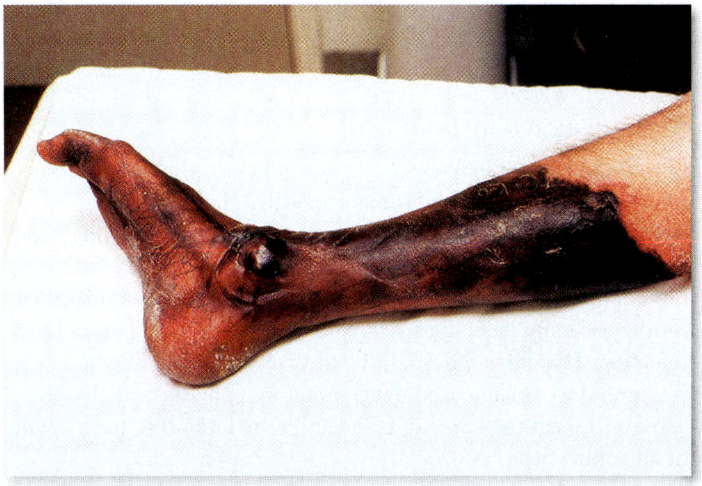

Figure 19.9 The clinical appearance of myonecrosis in a compound fracture of the leg.

John Watney/Science Source

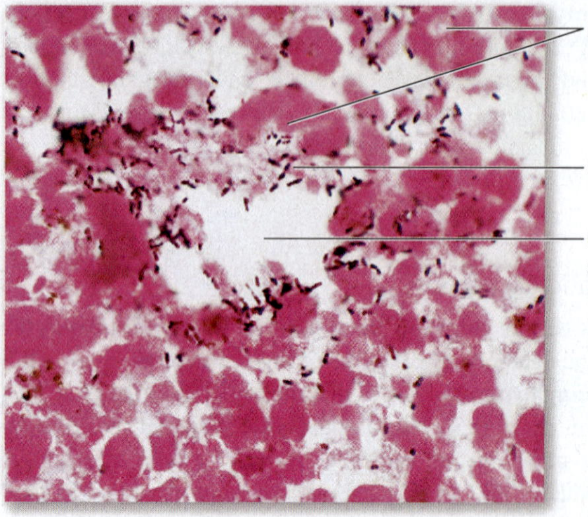
Muscle fibers
Clostridium
Gas-filled spaces

Figure 19.10 Growth of *Clostridium perfringens* (plump rods), causing gas formation and separation of the fibers.
A microscopic view of clostridial myonecrosis, showing a histological (tissue) section of gangrenous skeletal muscle.
Biophoto Associates/Science Source

Disease Table 19.5	Gas Gangrene
Causative Organism(s)	*Clostridium perfringens* B G+; other bacteria
Most Common Modes of Transmission	Vehicle (soil), endogenous transfer from skin, GI tract, and so on
Virulence Factors	Alpha toxin, other exotoxins, enzymes, gas formation
Culture/Diagnosis	Gram stain, CT scans, X ray, clinical picture
Prevention	Clean wounds, debride dead tissue
Treatment	Surgical removal, clindamycin + penicillin, oxygen therapy
Epidemiological Features	U.S. incidence 900–1,000 annually; mortality 25% but approaches 100% when treatment is delayed

in the dead tissue, and release of exotoxins. *C. perfringens* produces several active exotoxins. The most potent one, *alpha toxin,* causes red blood cell rupture, edema, and tissue destruction **(figure 19.10).** Virulence factors that add to the tissue destruction are collagenase, hyaluronidase, and DNase. The gas formed in tissues, resulting from fermentation of muscle carbohydrates, can also destroy muscle structure. Histology or MRI can visualize these disruptions.

Transmission and Epidemiology

The conditions that may predispose a person to gangrene are surgical incisions, compound fractures, diabetic ulcers, septic abortions, puncture and gunshot wounds, and crushing injuries contaminated by endospores from the environment.

Prevention and Treatment

One of the most effective ways to prevent clostridial wound infections is immediate and rigorous cleansing and surgical repair of deep wounds, decubitus ulcers (bedsores), compound fractures, and infected incisions. Debridement of diseased tissue eliminates the conditions that promote the spread of gangrenous infection. Surgery is supplemented by large doses of antibiotics to control infection. Hyperbaric oxygen therapy, in which the affected part is exposed to an increased oxygen mix in a pressurized chamber, can also lessen the severity of infection by inhibiting the growth of anaerobic bacteria.

Extensive myonecrosis of a limb may call for amputation. Because there are so many different antigenic subtypes in this bacterial group, active immunization is not possible.

Vesicular or Pustular Rash Diseases

Because this book groups diseases by their clinical presentation, it is important to have a shared vocabulary for the appearance of skin lesions. Consult **table 19.1** for common descriptions of skin rashes. There are three diseases that present as rashes on the body in which the individual lesions contain fluid. The lesions are often called *pox,* and two of the diseases are chickenpox and smallpox. The third disease is called hand, foot, and mouth disease (HFMD). All three are viral diseases. As this book goes to press in mid-2022, another pox disease is re-emerging. Monkeypox is similar to smallpox, but generally milder. Transmission routes and its epidemiology are still being worked out.

Chickenpox

Most people think of chickenpox as a mild disease, and in most people it is. However, in immunocompromised people, older adolescents, and adults, it can be life-threatening. Before the introduction of the vaccine in 1995, and now again in the midst of misunderstandings about vaccines, it is not unheard of for some families to hold "chickenpox parties." When one child in a group of acquaintances had chickenpox, other children would be brought together to play with them so that all the children could contract the infection at once so that it would run its course in a timely manner. Parents wanted to ensure that their

Table 19.1 Terms Used in Describing Skin Conditions and Infections

Descriptive Name	Appearance	Examples
Bulla	Large (wide) vesicle	Blister, gas blisters in gangrene
Cyst	Raised, encapsulated lesion, usually solid or semisolid when palpated	Severe acne
Macule	Flat, well-demarcated lesion characterized mainly by color change	Freckle, tinea versicolor (fungus infection)
Maculopapular rash	Flat to slightly raised colored bump	Measles, rubella, fifth disease, roseola
Papule	Small, elevated, solid bump	Warts, cutaneous leishmaniasis
Petechiae	Small purpura	Meningococcal bloodstream infection
Plaque	Elevated, flat-topped lesion larger than 1 cm (a wider papule)	Psoriasis
Purpura	Reddish-purple discoloration due to blood in small areas of tissue; does not blanch when pressed	Meningococcal bloodstream infection
Pustule	Small, elevated lesion filled with purulent fluid (pus)	Acne, smallpox, mucocutaneous leishmaniasis, cutaneous anthrax
Scale	Flaky portions of skin separated from deeper skin layers	Ringworm of body and scalp, athlete's foot
Vesicle	Elevated lesion with clear fluid	Chickenpox

children got the disease while they were young because they knew that getting the disease at an older age could lead to more serious disease. However, purposefully infecting a child with a potentially damaging pathogen is a bad idea when a safe, effective vaccine exists.

Signs and Symptoms

After an incubation period of 10 to 20 days, the first symptoms to appear are fever and an abundant rash that begins on the scalp, face, and trunk and radiates in sparse crops to the extremities. Skin lesions progress quickly from macules and papules to itchy vesicles filled with a clear fluid. In several days, they crust over and drop off, usually healing completely but sometimes leaving a tiny pit or scar. Lesions number from a few to hundreds and are more abundant in adolescents and adults than in young children. **Figure 19.11** contains images of chickenpox lesions. The lesion distribution is *centripetal,* meaning that there are more in the center of the body and fewer on the extremities, in contrast to the opposite distribution seen with smallpox. The illness usually lasts 4 to 7 days; new lesions stop appearing after about 5 days. Patients are considered contagious until all of the lesions have crusted over.

Most cases resolve without event within 2 to 3 weeks of onset. Some patients may experience secondary infections of the lesions caused by group A streptococci or staphylococci, and these require antibiotic therapy. Immunocompromised patients, as well as some adults and adolescents, may experience pneumonia as a result of chickenpox. The immunocompromised may also experience infection of the heart, liver, and kidney, resulting in a 20% mortality rate for this population.

Women who become infected with chickenpox during the early months of pregnancy are at risk for fetal infection. This virus can be teratogenic, and affected babies may be born with serious birth defects such as cataracts and missing limbs.

Chickenpox
Smallpox

Figure 19.11 Images of chickenpox and smallpox.

(chickenpox, top left): Picture Partners/Alamy Stock Photo; (chickenpox, bottom left): Centers for Disease Control and Prevention; (chickenpox, top right): David White/Alamy Stock Photo; (smallpox, left): Dr. Robinson/CDC; (smallpox, top right): Everett Collection Historical/Alamy Stock Photo; (smallpox, bottom right): Dr. John Noble, Jr./CDC

Shingles

After recuperation from chickenpox, the virus enters the sensory endings of nerves in dermatomes, which are regions of the skin supplied by the cutaneous branches of nerves, especially the thoracic nerves in the torso **(figure 19.12*a*)** and the trigeminal nerve in the head. From there, the virus becomes latent within the ganglia. Months or years later, the virus may reemerge from these cells, resulting in densely packed lesions on the portion of the skin served by that nerve. This reemergence is called **shingles** (also known as herpes zoster, or zoster). Shingles presents with a characteristic distribution of the lesions, usually along a single dermatome, which means it affects one side of the body and stops abruptly at the midline **(figure 19.12*b*)**.

Shingles develops abruptly after the virus is reactivated by a stimulus such as psychological stress, X-ray treatments, immunosuppressive and other drug therapy, surgery, or a developing malignancy. The virus is believed to migrate down the infected ganglia to the skin, where multiplication resumes and produces crops of tender, persistent vesicles. Inflammation of the ganglia and the associated pathways of nerves can cause pain and tenderness that can last for several months, a condition known as post-herpetic neuralgia. Involvement of cranial nerves can lead to eye inflammation and ocular and facial paralysis.

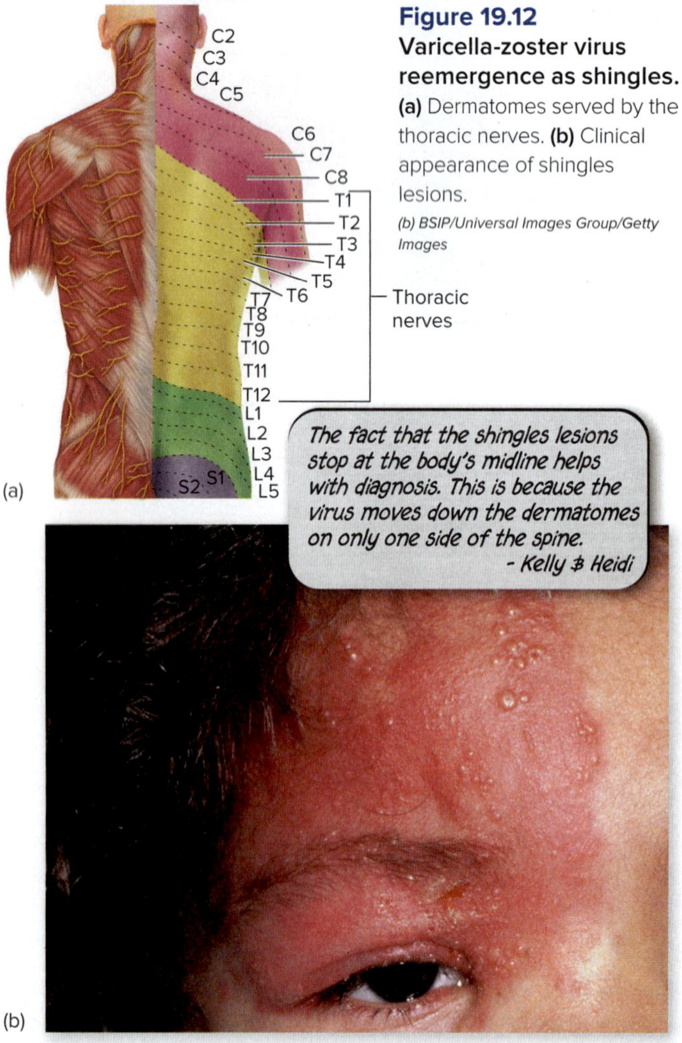
C2
C3
C4
C5
C6
C7
C8
T1
T2
T3
T4
T5
T6
T7
T8
T9
T10
T11
T12
L1
L2
L3
L4
L5
S1
S2
Thoracic nerves
(a)
The fact that the shingles lesions stop at the body's midline helps with diagnosis. This is because the virus moves down the dermatomes on only one side of the spine.
- Kelly $ Heidi
(b)

Figure 19.12 Varicella-zoster virus reemergence as shingles. **(a)** Dermatomes served by the thoracic nerves. **(b)** Clinical appearance of shingles lesions.

(b) BSIP/Universal Images Group/Getty Images

Causative Agent

Human herpesvirus 3 (HHV-3), also called **varicella** (var″-ih-sel′-ah) virus, causes chickenpox, as well as the condition called herpes zoster or shingles. The virus is sometimes referred to as the varicella-zoster virus (VZV). Like other herpesviruses, it is an enveloped DNA virus.

Pathogenesis and Virulence Factors

HHV-3 enters the respiratory tract, attaches to respiratory mucosa, and then enters the bloodstream. This viremia disseminates the virus to the skin, where the virus causes adjacent cells to fuse and eventually lyse, resulting in the characteristic lesions of this disease. The virus enters sensory nerves at this site, traveling to the ganglia.

The ability of HHV-3 to remain latent in ganglia is an important virulence factor because resting in this site protects it from attack by the immune system and provides a reservoir of virus for the reactivation condition of shingles.

Transmission and Epidemiology

Humans are the only natural hosts for HHV-3. The virus is harbored in the respiratory tract but is communicable via both respiratory droplets and the fluid of active skin lesions. People can acquire a chickenpox infection by being exposed to the fluid of shingles lesions. (It is not possible to "get" shingles from someone with shingles. If you are not immune to HHV-3, you can acquire HHV-3, which will manifest as chickenpox or, occasionally, as an asymptomatic infection. Once you have the virus, whether you experience shingles or not is dependent on your own host factors.)

Infected persons are most infectious a day or two prior to the development of the rash. Only in rare instances will a person acquire chickenpox more than once. Chickenpox is so contagious that if you are exposed to the virus and you do not have established immunity, you almost always become infected. Some people experience subclinical cases of the disease, meaning that lesions never appear. They will still develop lifelong immunity and will likely harbor the virus in their ganglia, making them subject to shingles in the future. When people think they have never had chickenpox yet they do not seem to get it when exposed to infected persons, it is likely that they have had a subclinical case at some time in their lives.

Epidemics of the disease used to occur in winter and early spring. The introduction of the varicella vaccine reduced the occurrence of the disease, so now cases develop sporadically.

Prevention

A live attenuated vaccine was licensed in 1995. It consists of a weakened form of the Oka strain of the HHV-3 virus, which was isolated from a Japanese boy named Oka. It is recommended that infants receive a first dose of the vaccine between the ages of 12 and 15 months and an additional dose between the ages of 4 and 6 years. Young adults should receive boosters. ProQuad is a multipathogen vaccine and can be used to immunize individuals against varicella in addition to measles, mumps, and rubella.

The preferred vaccine against shingles (for people over 50) is Shingrix. Two shots provide 90% protection against shingles. A large study has recently shown that vaccinated children have a much lower risk of developing shingles than do unvaccinated kids who contract chickenpox naturally.

▶ Treatment

Uncomplicated varicella is self-limiting and requires no therapy aside from alleviation of discomfort. Oral acyclovir or related antivirals should be administered within 24 hours of onset of the rash to people considered to be at risk for serious complications. The acyclovir may diminish viral load and prevent complications.

Smallpox

Largely through the World Health Organization's comprehensive global efforts, naturally occurring smallpox is now a disease of the past. However, after the terrorist attacks in the United States on September 11, 2001, and the anthrax bioterrorism event that followed shortly thereafter, the U.S. government began taking the threat of smallpox bioterrorism seriously. For a few years, American military personnel were vaccinated aganst it, but that was discontinued again for most military in 2016.

▶ Signs and Symptoms

Infection begins with fever and malaise. Later, a rash begins in the pharynx, spreads to the face, and progresses to the extremities. Initially, the rash is *macular,* evolving in turn to *papular, vesicular,* and *pustular* in appearance before eventually crusting over, leaving behind nonpigmented sites pitted with scar tissue. There are two principal forms of smallpox: variola minor and variola major. Variola major is a highly virulent form that causes toxemia, shock, and intravascular coagulation. People who have survived any form of smallpox nearly always develop lifelong immunity.

It is vitally important for health care workers to be able to recognize the early signs of smallpox. The diagnosis of even a single suspected case must be treated as a health and law enforcement emergency. A major distinguishing feature of this disease is that the pustules are indented in the middle **(Disease Table 19.6).** Also, patients report that the lesions feel as if they contain a BB pellet.

A Note About Bioterror Agents

The Centers for Disease Control and Prevention maintain a list of the most dangerous infectious agents that would be most logically used for a bioterror attack. There are three categories: A, B, and C. **Category A** agents (the most dangerous) are those that are (1) easily disseminated or transmissible person-to-person; (2) result in high mortality rates; (3) could incite panic and social disruption; and (4) require special or unique actions for preparedness. **Category A** agents will be pointed out in the Disease Tables in the next six chapters.

A patient with variola minor has a rash that is less dense and generally experiences weaker symptoms than someone affected by variola major.

▶ Causative Agent

The causative agent of smallpox, the variola virus, is an orthopoxvirus, an enveloped DNA virus. Variola is shaped like a brick and is 200 nanometers in diameter. Other members of this group are the monkeypox virus and the vaccinia virus from which smallpox vaccine is made. Variola is a hardy virus, surviving outside the host longer than most viruses.

▶ Pathogenesis and Virulence Factors

The infection begins by implantation of the virus in the nasopharynx. The virus invades the mucosa and multiplies in the regional lymph nodes, leading to viremia. Variola multiplies within white blood cells and then travels to the small blood vessels in the dermis. The lesions occur at the dermal level, which is the reason that scars remain after the lesions are healed.

▶ Transmission and Epidemiology

Before the eradication of smallpox, almost everyone contracted the disease over the course of his or her lifetime, either surviving with lifelong immunity or dying. In Victorian England, some people did not name their children until they had survived a bout of smallpox.

In the early 1970s, smallpox was endemic in 31 countries. Every year, 10 to 15 million people contracted the disease, and approximately 2 million people died from it. After 11 years of intensive effort by the world health community, the last natural case occurred in Somalia in 1977.

▶ Prevention

Earlier in this book, you read about Edward Jenner and his development of vaccinia virus to inoculate against smallpox. To this day, the vaccination for smallpox is based on the vaccinia virus. The United States stopped using the vaccine in 1972 after a massive effort to eradicate the virus worldwide. In 1980, the WHO declared that the war against smallpox was "won" and recommended that all laboratories destroy their stocks of the virus.

Even though officially smallpox does not exist anymore except in labs in Russia and the United States, there are fears that other entities may have stockpiled the virus for use in biological warfare or terror attacks. That is why the United States maintains a stockpile of smallpox vaccine to use in the case of a smallpox attack. A new smallpox vaccine was licensed in 2019, and goes by the name Jynneos. When monkeypox outbreaks started occurring in 2022, the CDC released some of that vaccine for use in high risk individuals, because it provides protection against monkeypox as well.

▶ Treatment

The first antiviral for pox viruses was approved by the FDA in 2018. It is called tecovirimat, and so far it has only been used on people acquring smallpox from laboratory accidents.

Hand, Foot, and Mouth Disease (HFMD)

This disease is most common in babies and children under the age of 5. It starts with a fever, sore throat, and malaise. The first spots erupt inside the mouth **(figure 19.13).** These are painful. Soon, red or blisterlike spots appear on the palms of the hands and soles of the feet and often the genitals, buttocks, knees and elbows (see Disease Table 19.6).

HFMD is caused by a number of different viruses in the *Enterovirus* group—most frequently, the Coxsackie virus. It is transmitted via secretions (saliva, sputum, blister fluid, feces) through direct contact. The sick person will feel worst before the lesions appear and is most contagious during the first week of illness, though many patients continue to shed contagious virus for days or weeks. The disease usually runs an uncomplicated course. There is no specific treatment and no vaccine for it. **(Disease Table 19.6).**

Disease Table 19.6 Vesicular/Pustular Rash Diseases

Disease	Chickenpox	Smallpox	Hand, Foot, and Mouth Disease
Causative Organism(s)	Human herpesvirus 3 (varicella-zoster virus) Ⓥ	Variola virus Ⓥ	Enteroviruses, usually Coxsackie Ⓥ
Most Common Modes of Transmission	Droplet contact, inhalation of aerosolized lesion fluid	Droplet contact, indirect contact	Direct and droplet contact
Virulence Factors	Ability to fuse cells, ability to remain latent in ganglia	Ability to dampen, avoid immune response	–
Culture/Diagnosis	Based largely on clinical appearance; PCR is available	Based largely on clinical appearance; if suspected, refer to CDC	Usually based on clinical presentation and history
Prevention	Live attenuated vaccine; there is also vaccine to prevent reactivation of latent virus (shingles)	Jynneos vaccine (approved 2019); older vaccine ACAM2000.	Hand hygiene
Treatment	None in uncomplicated cases; acyclovir for high risk	Cidofovir; vaccine within 7 days of exposure	None
Distinguishing Features	No fever prodrome; lesions are superficial; in centripetal distribution (more in center of body)	Fever precedes rash; lesions are deep and in centrifugal distribution (more on extremities)	Fever prodrome; lesions in mouth first
Epidemiological Features	Chickenpox: vaccine decreased hospital visits by 88%, ambulatory visits by 59%; shingles: 1 in 3 American adults will have it at least once	Last natural case worldwide was in 1977 Category A Bioterrorism Agent	Sporadic in most of world; possibly on the rise in United States
Appearance of Lesions	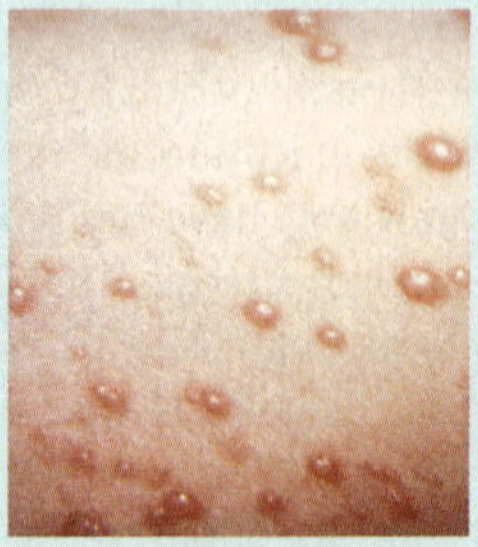 Centers for Disease Control	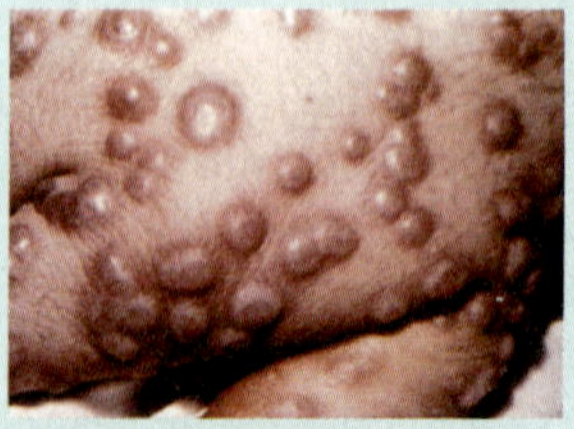 Dr. Charles Farmer, Jr./CDC	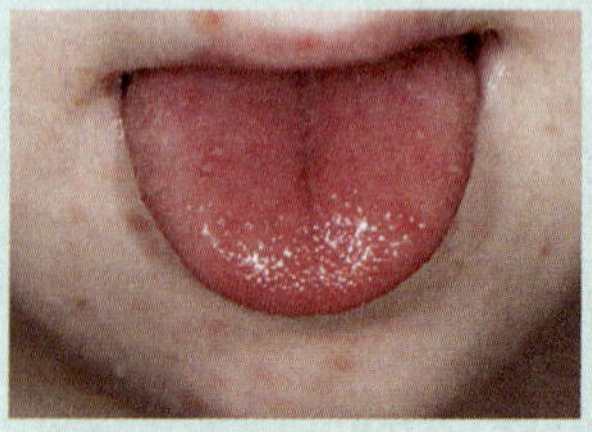 Dr. P. Marazzi/Science Source

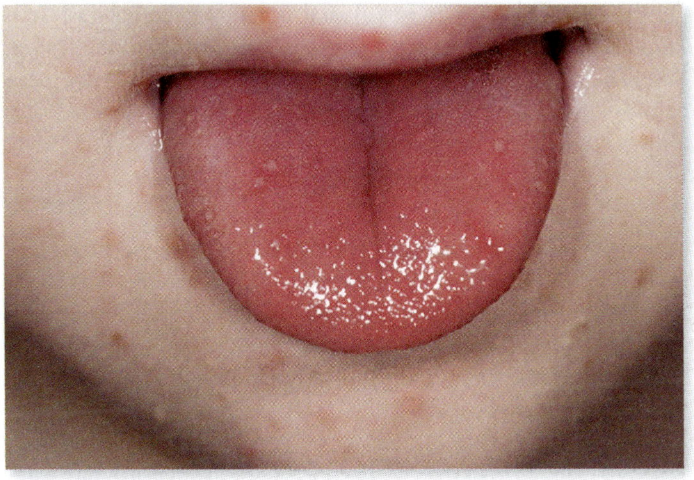

Figure 19.13 Hand, foot, and mouth disease sores in the mouth of a young child.
Dr. P. Marazzi/Science Source

Maculopapular Rash Diseases

The infectious conditions described in this section are those with their major manifestations on the skin. (Meningococcal meningitis, for instance, can result in a rash on the skin, but its major manifestations are in the central nervous system, so it is discussed in chapter 20.) In this section, we examine measles, rubella, fifth disease, and roseola. They all cause skin eruptions described as maculopapular.

Measles

In the United States, a sharp increase of measles cases had been occurring, prior to the onset of the COVID-19 pandemic. Many parents are opting not to have their children vaccinated, due to unfounded fears about the link between the vaccine and autism. We would do well to remember that before the vaccine was introduced, measles killed 6 million people each year worldwide.

Measles is also known as **rubeola.** Be very careful not to confuse it with the next maculopapular rash disease, rubella.

Signs and Symptoms

The initial symptoms of measles are sore throat, dry cough, headache, conjunctivitis, lymphadenitis, and fever. Shortly after infection, and before full-on symptoms, the prodromal finding of *Koplik's spots*—white spots on the inner cheeks of the mouth—appears as a prelude to the characteristic red, maculopapular **exanthem** (eg-zan′-thum) across the body. This begins on the head and then progresses to the trunk and extremities until most of the body is covered **(figure 19.14).** The rash gradually coalesces into red patches that fade to brown.

In a small number of cases, children develop laryngitis, bronchopneumonia, and bacterial secondary infections such as ear and sinus infections. Children afflicted with leukemia or thymic deficiency are especially predisposed to pneumonia because of their lack of a natural T-cell defense.

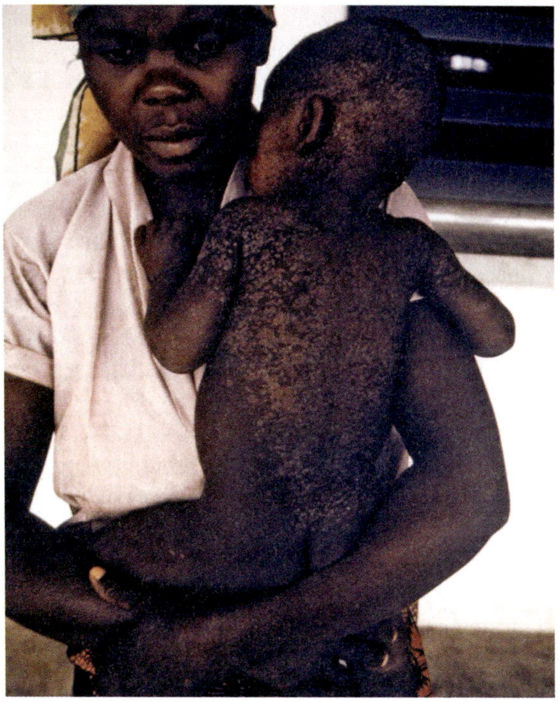

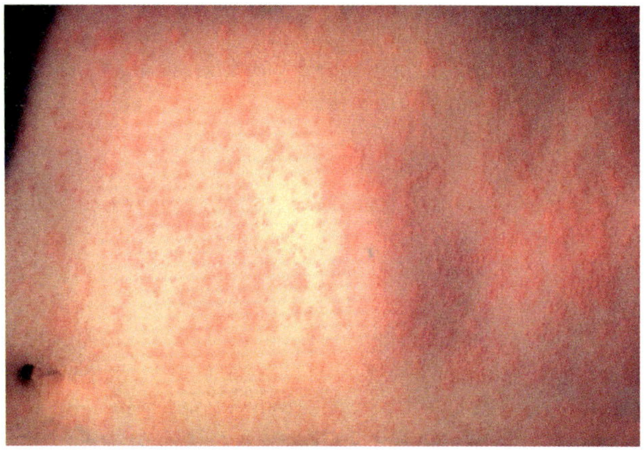

Figure 19.14 The rash of measles on dark skin and light skin.
(Top) *Dr. Lyle Conrad/CDC;* (Bottom) *Centers for Disease Control and Prevention*

A large number of measles patients experience secondary bacterial infections with *Haemophilus influenzae, Streptococcus pneumoniae,* or other streptococci or staphylococci. These can manifest as ear infections or pneumonia or upper respiratory tract complications. In about 1% of the cases, encephalitis can occur, with symptoms from disorientation to coma. Permanent brain damage or epilepsy can result. In 2019 researchers found the reason for the frequent secondary infections in measles patients. The measles virus infects and destroys both B and T memory cells, meaning that for some period of time, it essentially erases immune memory and leaves patients vulnerable once again to pathogens to which they had achieved immunity. Children who experienced measles were found to have higher rates of infection than their counterparts for up to five years after their infections.

The most serious complication is **subacute sclerosing panencephalitis (SSPE),** a progressive neurological degeneration of the cerebral cortex, white matter, and brain stem. It usually occurs years after the initial measles infection and is 100% fatal. Its incidence is approximately one case in every 600 measles infections, and it afflicts primarily male children and adolescents. The pathogenesis of SSPE appears to involve a defective virus, one that has lost its ability to form a capsid and be released from an infected cell. Instead, it spreads, unchecked, through the brain by cell fusion, gradually destroying neurons and accessory cells and breaking down myelin.

▶ Causative Agent

The measles virus is a member of the *Morbillivirus* genus. It is a single-stranded, enveloped RNA virus in the Paramyxovirus family.

▶ Pathogenesis and Virulence Factors

The virus implants in the respiratory mucosa and infects the tracheal and bronchial cells. From there it travels to the lymphatic system, where it multiplies and then enters the bloodstream. Viremia carries the virus to the skin and to various organs.

The measles virus induces the cell membranes of adjacent host cells to fuse into large **syncytia** (sin-sish′-uh), giant cells with many nuclei. These cells no longer perform their proper function. The host may be left vulnerable for many weeks after infection. This immune response disruption is one of the reasons that secondary bacterial infections are so common.

▶ Transmission and Epidemiology

Measles is one of the most contagious infectious diseases, transmitted principally by respiratory droplets. Epidemic spread is favored by crowding, low levels of herd immunity, malnutrition, and inadequate medical care. There is no reservoir other than humans, and a person is contagious during the periods of incubation, prodrome phase, and the skin rash phase but usually not during convalescence. Only relatively large, dense populations of susceptible individuals can sustain the continuous chain necessary for transmission.

Measles was considered eliminated in the United States in 2000, but it has made a comeback, due largely to the reluctance of parents to vaccinate their children due to discredited reports of possible harm the vaccine can cause.

▶ Culture and/or Diagnosis

The disease can be diagnosed on clinical presentation alone. If further identification is required, an ELISA test is available that tests for patient IgM to measles antigen, indicating a current infection. For best results, blood should be drawn on the third day of onset or later because before that time titers of IgM may not be high enough to be detected by the test. Also, the method of comparing acute and convalescent sera may be used to confirm a measles infection after the fact.

▶ Prevention

The MMR vaccine (for measles, mumps, and rubella) contains live attenuated measles virus, which confers protection for about 20 years. Because the disease is so contagious, 95% coverage with the vaccine in any given population is required to achieve herd immunity. Vaccination rates had dipped below that in the United States before the pandemic, due to vaccine hesitance. The pandemic caused rates to fall worldwide, as people were not accessing their normal health care services.

▶ Treatment

Treatment relies on reducing fever, suppressing cough, and replacing lost fluid. Complications require additional remedies to relieve neurological and respiratory symptoms and to sustain nutrient, electrolyte, and fluid levels. Therapy may include antibiotics for bacterial complications and doses of immune globulin. The antiviral ribavirin can be used.

Rubella

This disease is also known as German measles. Rubella is derived from the Latin for "little red," and that is a good way to remember it because it causes a relatively minor rash disease with few complications. Sometimes it is called the 3-day measles. The only exception to this mild course of events is when a fetus is exposed to the virus while in its mother's womb (*in utero*). Serious damage can occur, and for that reason women of childbearing years must be sure to have been vaccinated well before they plan to conceive.

▶ Signs and Symptoms

The two clinical forms of rubella are referred to as postnatal infection, which develops in children or adults, and **congenital** (prenatal) infection of the fetus, which can result in various types of birth defects.

Postnatal Rubella During an incubation period of 2 to 3 weeks, the rubella virus multiplies in the respiratory epithelium, infiltrates local lymphoid tissue, and enters the bloodstream. Early symptoms include malaise, mild fever, sore throat, and lymphadenopathy. The rash of pink macules and papules first appears on the face and progresses down the trunk and toward the extremities, advancing and resolving in about 3 days. The rash is milder-looking than the measles rash **(Disease Table 19.7).**

Congenital Rubella The mother is able to transmit the virus even if she is asymptomatic. The types of fetal injury vary according to the time of infection. Infection in the first trimester is most likely to induce miscarriage or multiple permanent defects in the newborn. The most common of these is deafness and may be the only defect seen in some babies. Other babies may experience cardiac abnormalities, ocular lesions, rashes, and mental and physical retardation in varying combinations **(figure 19.15).** Less drastic sequelae that usually resolve in time are anemia, hepatitis, pneumonia, carditis, and bone infection.

▶ Causative Agent

The rubella virus is a *Rubivirus,* in the family Togavirus. It is a nonsegmented, single-stranded RNA virus with a loose lipid envelope. There is only one known serotype of the virus, and humans are the only natural host. Its envelope contains two different viral proteins.

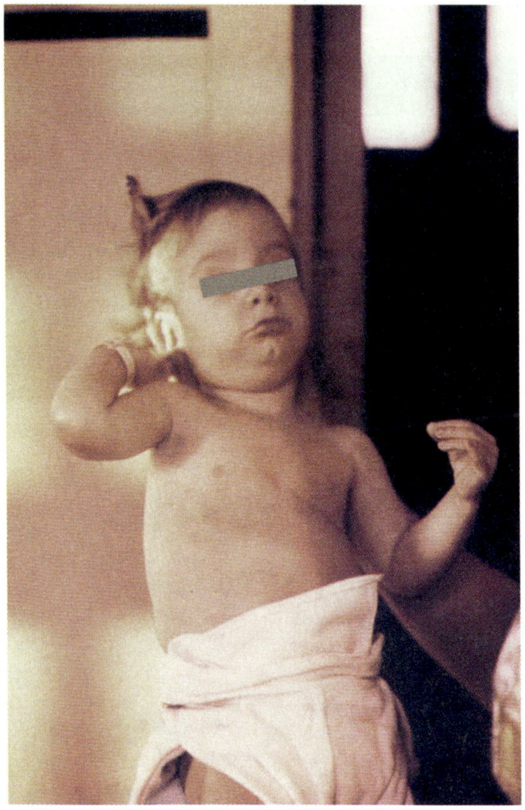

Figure 19.15 An infant born with congenital rubella can display a papular pink or purple rash. Congenital rubella can also cause cataracts, deafness, heart disease, and developmental defects.
CDC/Dr. Andre J. Lebrum

▶ Pathogenesis and Virulence Factors

The course of disease in postnatal rubella is mostly unremarkable, but when exposed to a fetus, the virus creates havoc. It has the ability to stop mitosis, which is an important process in a rapidly developing embryo and fetus. It also induces apoptosis of normal tissue cells. This inappropriate cell death can do irreversible harm to organs it affects. Last, the virus damages vascular endothelium, leading to poor development of many organs.

▶ Transmission and Epidemiology

Rubella is an endemic disease and is still common in many parts of the world. Infection is initiated through contact with respiratory secretions and occasionally urine. The virus is shed during the prodromal phase and up to a week after the rash appears. Congenitally infected infants are contagious for a much longer period of time.

▶ Culture and/or Diagnosis

Because it mimics other diseases, rubella should not be diagnosed on clinical grounds alone. IgM antibody to rubella virus can be detected early using an ELISA technique or a latex-agglutination card. Other conditions and infections can lead to false positives, however, and the IgM test should be augmented by an acute and convalescent measurement of IgG antibody. It is important to know whether the infection is in fact rubella, especially in women, because if so, they will be immune to reinfection in later pregnancies.

▶ Prevention

The attenuated rubella virus vaccine is usually given to children in the combined form (MMR or MMRV vaccination) at 12 to 15 months and a booster at 4 or 6 years of age.

▶ Treatment

Postnatal rubella is generally benign and requires only symptomatic treatment. No specific treatment is available for the congenital manifestations.

Fifth Disease

This disease, more precisely called *erythema infectiosum,* is so named because about 120 years ago, it was the fifth of the diseases recognized by doctors to cause rashes in children. The first four were scarlet fever, measles, rubella, and another rash called "fourth disease," which was thought to be a distinct illness but was later found to be misdiagnosed rubella or scarlet fever. The name "fifth disease" has stuck for this viral condition. It is a very mild disease that often results in a characteristic "slapped-cheek" appearance because of a confluent reddish rash that begins on the face. Within 2 days, the rash spreads on the body but is most prominent on the arms, legs, and trunk. The rash is maculopapular in appearance, and the blotches tend to run together rather than to appear as distinct bumps. The illness is rather mild, featuring low-grade fever and malaise and lasting 5 to 10 days. The virus may persist for days to weeks, and during that time the rash tends to recur under stress or with exposure to sunlight. As with almost any infectious agent, it can cause more serious disease in people with underlying immune disease.

The causative agent is parvovirus B19. You may have heard of "parvo" as a disease of dogs, but parvovirus B19 does not cause disease in dogs, nor does dog "parvo" cause disease in humans. Fifth disease is usually diagnosed by the clinical presentation, but sometimes it is helpful to rule out rubella by testing for IgM against rubella. Specific serological tests for fifth disease are available if they are considered necessary.

This infection is very contagious. It is transmitted through respiratory droplets or even direct contact. It can be transmitted through the placenta, with a range of possible effects from no symptoms to stillbirth. There is no vaccine and no treatment for this usually mild disease.

Roseola

This disease is common in young children and babies and is sometimes known as "sixth disease." It can result in a maculopapular rash, but a high percentage (up to 70%) of cases proceed without development of the rash stage. Children exhibit a high fever (up to 41°C, or 105°F) that comes on quickly and lasts for up to 3 days. Seizures may occur during this period, but other than that, patients remain alert and do not act terribly ill. On the fourth day, the fever disappears, and it is at this point that a rash can appear, first on the chest and trunk and less prominently on the face and limbs. By the time the rash appears, the disease is almost over.

Roseola is caused by a human herpesvirus called HHV-6. Like all herpesviruses, it can remain latent in its host indefinitely after the disease has cleared. Very occasionally, the virus

Disease Table 19.7 Maculopapular Rash Diseases

Disease	Measles (Rubeola)	Rubella	Fifth Disease
Causative Organism(s)	Measles virus V	Rubella virus V	Parvovirus B19 V
Most Common Modes of Transmission	Droplet contact	Droplet contact	Droplet contact, direct contact
Virulence Factors	Syncytium formation, ability to suppress cell-mediated immunity	Inhibition of mitosis, induction of apoptosis, and damage to vascular endothelium	–
Culture/Diagnosis	Clinical diagnosis; ELISA or PCR	Acute IgM, acute/ convalescent IgG	Usually diagnosed clinically
Prevention	Live attenuated vaccine (MMR or MMRV)	Live attenuated vaccine (MMR or MMRV)	–
Treatment	No antivirals; vitamin A, antibiotics for secondary bacterial infections	–	–
Distinguishing Features of the Rashes	Starts on head, spreads to whole body, lasts over a week	Milder red rash, lasts approximately 3 days	"Slapped-face" rash first, spreads to limbs and trunk, tends to be confluent rather than distinct bumps
Epidemiological Features	Incidence increasing in North America; in developing countries 100,000 deaths per year	10–12 cases in US (usually imported); worldwide: 100,000 infants/yr born with congenital rubella syndrome	60% of population seropositive by age 20
Appearance of Lesions	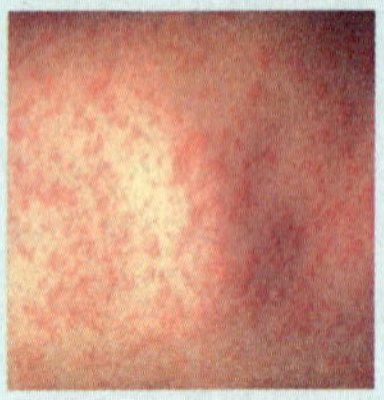 Centers for Disease Control and Prevention	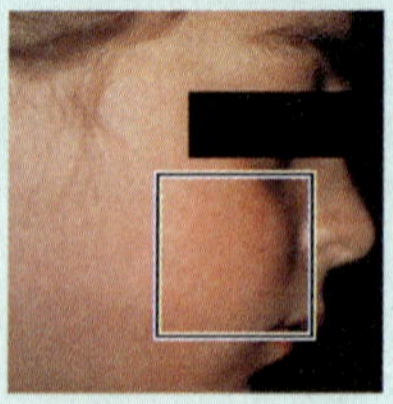 Centers for Disease Control and Prevention	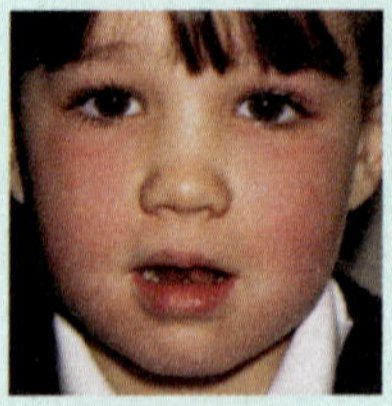 Dr. P. Marazzi/Science Source

reactivates in childhood or adulthood, leading to mononucleosis-like or hepatitis-like symptoms. Immunocompetent hosts generally do not experience reactivation. It is thought that 100% of the U.S. population becomes infected with this virus by adulthood, though many cases are subclinical or asymptomatic. (**Disease Table 19.7**).

Wartlike Eruptions

All types of warts are caused by viruses. Most common warts you have seen on yourself and others are probably caused by one of more than 130 human papillomaviruses, or HPVs. HPVs are also the cause of genital warts, described in chapter 24. Another virus causes a condition called **molluscum contagiosum,** which causes bumps that may look like warts.

Warts

Warts, also known as **papillomas,** can develop in nearly all individuals. Children seem to get them more frequently than adults, and there is speculation that people gradually build up immunity to the various HPVs they encounter over time, as is the case with the viruses that cause the common cold.

The warts caused by HPV are benign, squamous epithelial growths. Some HPVs can infect mucous membranes and others

Roseola	Other Conditions to Consider
Human herpesvirus 6 B	Other conditions resulting in a rash that may look similar to these maculopapular conditions should be considered:
Unknown	Scarlet fever (covered in chapter 22)
Ability to remain latent	Secondary syphilis (chapter 24)
Usually diagnosed clinically	Rocky Mountain spotted fever (chapter 21)
–	
–	
High fever precedes rash stage; rash not always present	
>90% of population seropositive; 90% of disease cases occur before age of 2	
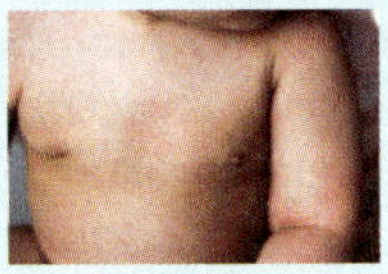 NeoStudio1/Shutterstock	

invade skin. The appearance and seriousness of the infection vary somewhat from one anatomical region to another. Painless, elevated, rough growths on the fingers and occasionally on other body parts are called common, or seed, warts **(Disease Table 19.8).** These growths commonly occur in children and young adults. Just as certain types of HPVs are associated with particular outcomes in the genital area, common warts are most often caused by HPV-2, -4, -27, and -29. **Plantar warts** are often caused by HPV-1. They are deep, painful papillomas on the soles of the feet. Flat warts (HPV types 3, 10, 28, and 49) are smooth, skin-colored lesions that develop on the face, trunk, elbows, and knees.

The warts contain variable amounts of virus. Transmission occurs through direct contact, and often warts are transmitted from one part of the body to another by autoinoculation. Because the viruses are fairly stable in the environment, they can also be transmitted indirectly from towels, shower stalls, or pedicure equipment, where they persist inside the protective covering of sloughed-off keratinized skin cells. The incubation period can be from 1 to 8 months. Almost all nongenital warts are harmless.

The warts caused by papillomaviruses are usually distinctive enough to permit reliable clinical diagnosis without much difficulty. However, a biopsy and histological examination can help clarify ambiguous cases. Warts disappear on their own 60% to 70% of the time, usually over the course of 2 to 3 years. Physicians do approve of home remedies for resolving warts, including nonprescription salicylic acid preparations. Physicians have other techniques for removing warts, including a number of drugs and/or cryosurgery. No treatment guarantees that the viruses are eliminated because the virus can integrate into the DNA of the host. For that reason, warts can grow back.

Molluscum Contagiosum

This disease is distributed throughout the world, with highest incidence occurring in certain regions of the Pacific Islands, although its incidence in North America has been increasing since the 1980s. Skin lesions take the form of smooth, waxy nodules on the face, trunk, and limbs. The firm nodules may be indented in the middle (see Disease Table 19.8), and they contain a milky fluid containing epidermal cells filled with virus particles in intracytoplasmic inclusion bodies. This condition is common in children, where it most often causes nodules on the face, arms, legs, and trunk. In adults, it appears mostly in the genital areas. In immunocompromised patients, the lesions can be more disfiguring and more widespread on the body. The disease is particularly common in AIDS patients and often presents as facial lesions.

The molluscum contagiosum virus is a poxvirus, containing double-stranded DNA and possessing an envelope. It is spread via direct contact and through fomites. Adults who acquire this infection usually acquire it through sexual contact. Autoinoculation can spread the virus from existing lesions to new places on the body, resulting in new nodules.

The condition may be diagnosed on clinical appearance alone, or a skin biopsy may be performed and histological analysis can be undertaken. A clinician can perform a more simple "squash procedure," in which fluid from the lesion is extracted onto a microscope slide, squashed by another microscope slide, stained, and examined for the presence of the characteristic inclusion bodies in the epithelial cells. PCR can also be used to detect the virus

Disease Table 19.8 Warts and Wartlike Eruptions		
Causative Organism(s)	Human papillomaviruses V	Molluscum contagiosum viruses V
Most Common Modes of Transmission	Direct contact, autoinoculation, indirect contact	Direct contact, including sexual contact, autoinoculation
Virulence Factors	–	–
Culture/Diagnosis	Clinical diagnosis, also histology, microscopy, PCR	Clinical diagnosis, also histology, microscopy, PCR
Prevention	Avoid contact	Avoid contact
Treatment	Home treatments, cryosurgery (virus not eliminated)	Usually none, although mechanical removal can be performed (virus not eliminated)
Epidemiological Features	United States: 6 million new cases per yr, prevalence 13%; worldwide HPV prevalence 12%	Worldwide incidence 2%–3%, with greater distribution in tropical areas and in overcrowded communities where there is poor hygiene
Appearance of Lesions	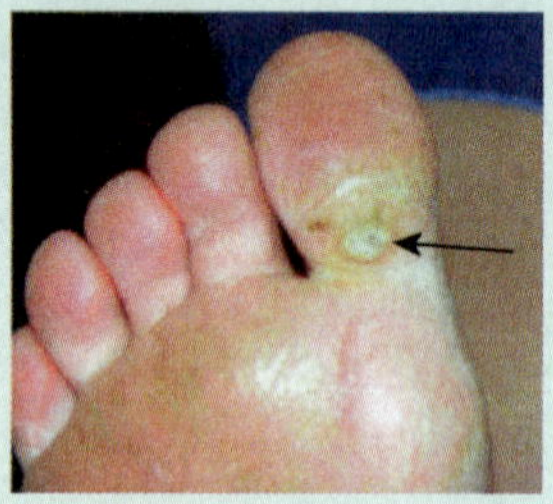 McGraw Hill	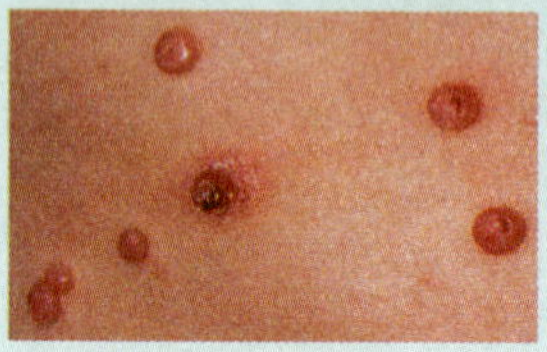 Dr. P. Marazzi/Science Source

in skin lesions. In most cases, no treatment is indicated, although a physician may remove the lesions or treat them with a topical chemical. Treatment of lesions, however, does not ensure elimination of the virus **(Disease Table 19.8).**

Large Pustular Skin Lesions

Leishmaniasis

Two infections that result in large lesions (greater than a few millimeters across) deserve mention in this chapter on skin infections. The first is leishmaniasis, a zoonosis transmitted among various mammalian hosts by female sand flies. This infection can express itself in several different forms depending on which species of the protozoan *Leishmania* is involved. Cutaneous leishmaniasis is a localized infection of the capillaries of the skin caused by *L. tropica,* found in Mediterranean, African, and Southeast Asian regions. A form of mucocutaneous leishmaniasis called espundia is caused by *L. braziliensis,* endemic to parts of Central and South America. It affects both skin and mucous membranes. Another form of this infection is systemic leishmaniasis.

Leishmania is transmitted to the mammalian host by the sand fly when it ingests the host's blood. The disease is endemic to equatorial regions that provide favorable conditions for the sand fly. Numerous wild and domesticated animals, especially dogs, serve as reservoirs for the protozoan. Although humans are usually accidental hosts, the flies freely feed on them. At particular risk are travelers or immigrants who have never had contact with the protozoan and therefore lack specific immunity.

In cutaneous leishmaniasis, a small, red papule occurs at the site of the bite and spreads laterally into a large ulcer **(Disease Table 19.9).** The edges of the ulcer are raised and the base is moist. It can be filled with a serous/purulent exudate or covered with a crust. Satellite lesions may occur. Mucocutaneous leishmaniasis usually begins with a skin lesion on the head or face and then progresses to single or multiple lesions, usually in the mouth and nose. Lesions can be quite extensive, eventually involving and disfiguring the hard palate, the nasal septum, and the lips.

Cutaneous Anthrax

This form of anthrax is the most common and least dangerous version of infection with *Bacillus anthracis*. (The spectrum of anthrax disease is discussed fully in chapter 21.) It is caused by endospores entering the skin through small cuts or abrasions.

Disease Table 19.9 Large Pustular Skin Lesions		
Disease	Leishmaniasis	Cutaneous Anthrax
Causative Organism(s)	*Leishmania* spp. (P)	*Bacillus anthracis* (B) G+
Most Common Modes of Transmission	Biological vector	Direct contact with endospores
Virulence Factors	Multiplication within macrophages	Endospore formation; capsule, lethal factor, edema factor (see chapter 21)
Culture/Diagnosis	Culture of protozoa, microscopic visualization	Culture on blood agar; serology, PCR performed by CDC
Prevention	Avoid sand fly	Avoid contact; vaccine available but not widely used
Treatment	Paromycin/methylbenzethonium chloride ointment or amphotericin B	Ciprofloxacin, plus two additional antibiotics
Distinguishing Features	Mucocutaneous and systemic forms	Can be fatal
Epidemiological Features	Untreated visceral leishmaniasis mortality rate 90%, 10% for cutaneous leishmaniasis	Untreated cutaneous anthrax mortality rate 20%, treated mortality rate less than 1% Category A Bioterrorism Agent
Appearance of Lesions	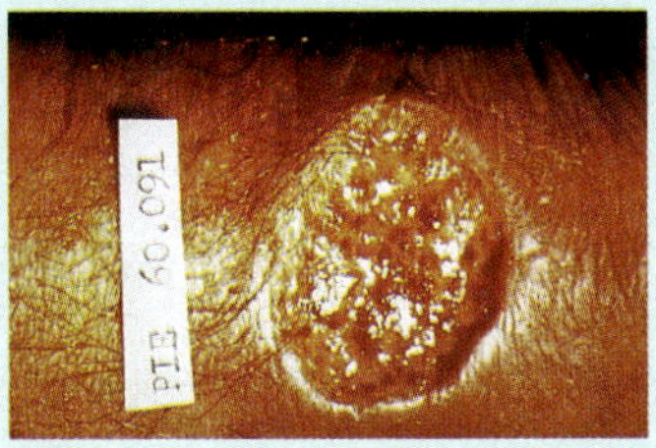 PIB 60.091 Centers for Disease Control and Prevention	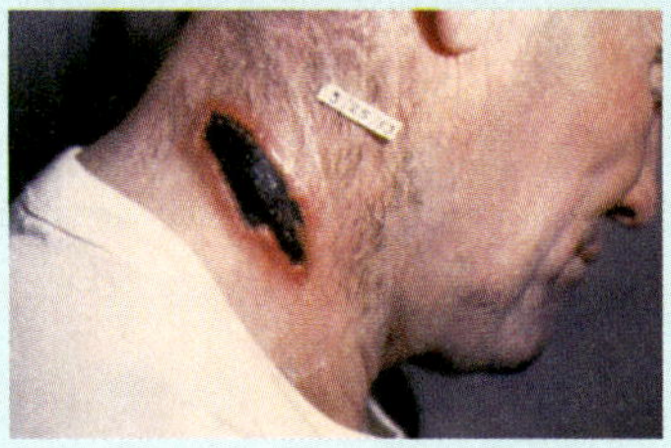 Centers for Disease Control and Prevention

Germination and growth of the pathogen in the skin are marked by the production of a papule that becomes increasingly necrotic and later ruptures to form a painless, black **eschar** (ess′-kar) (see Disease Table 19.9). In the fall of 2001, 11 cases of cutaneous anthrax occurred in the United States as a result of bioterrorism (along with 11 cases of inhalational anthrax). Mail workers and others contracted the infection when endospores were sent through the mail. The infection can be naturally transmitted by contact with hides of infected animals (especially goats).

Left untreated, even the cutaneous form of anthrax is fatal a small percentage of the time. A vaccine exists but is recommended only for high-risk persons and the military. Upon suspicion of cutaneous anthrax, ciprofloxacin, plus two other antibiotics, should be used initially. If the isolate is found to be sensitive to penicillin, patients can be switched to that drug. These patients should also receive the anthrax vaccine. **(Disease Table 19.9).**

Ringworm (Cutaneous Mycoses)

A group of fungi collectively termed **dermatophytes** causes a constellation of integument conditions. These mycoses are strictly confined to the nonliving epidermal tissues (stratum corneum) and their derivatives (hair and nails). All these conditions have different names that begin with the word **tinea** (tin′-ee-ah), which derives from the erroneous belief that they were caused by worms. (*Tinea* is the latin word for worm.) That misconception is also the reason these diseases are often called *ringworm*—ringworm of the scalp (tinea capitis), beard (tinea barbae), body (tinea corporis), groin (tinea cruris), foot (tinea pedis), and hand (tinea manuum). (Do not confuse these "tinea" terms with genus and species names. It is simply an old practice for naming conditions.) Most of these conditions are caused by one of three different dermatophytes. The signs and symptoms of ringworm infections are summarized in **Table 19.2.**

Causative Agents

There are about 39 species in the genera *Trichophyton, Microsporum,* and *Epidermophyton* that can cause the tinea conditions. The causative agent of a given type of ringworm varies from one geographic location to another and is not restricted to a particular genus and species. These fungi are so closely related and morphologically similar that they can be difficult to differentiate. Various species exhibit unique macroconidia, microconidia, and unusual types of hyphae. In general, *Trichophyton* produces thin-walled, smooth macroconidia and numerous microconidia **(figure 19.16*a*)**; *Microsporum* produces thick-walled, rough

Table 19.2 Signs and Symptoms of Cutaneous Mycoses

Ringworm of the Scalp (Tinea Capitis)	This mycosis results from the fungal invasion of the scalp and the hair of the head, eyebrows, and eyelashes.	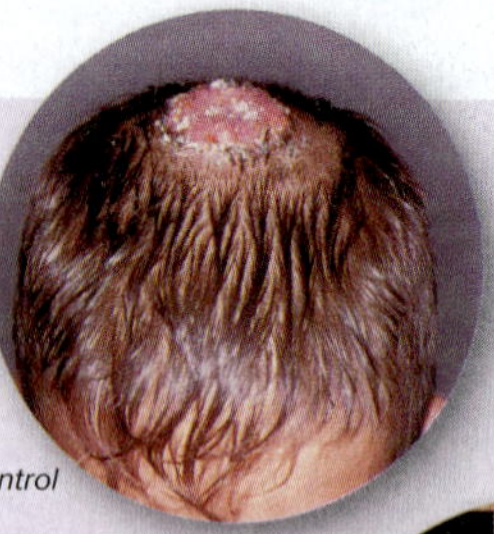 *Centers for Disease Control and Prevention*
Ringworm of the Beard (Tinea Barbae)	This tinea, also called *barber's itch*, affects the chin and beard of adult males. Although once a common aftereffect of unhygienic barbering, it is now contracted mainly from animals.	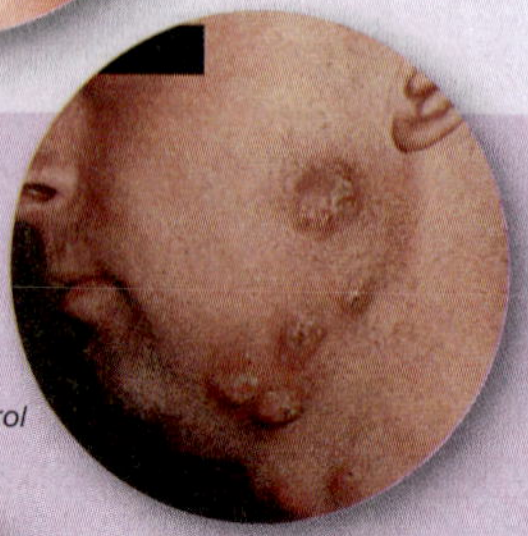 *Centers for Disease Control and Prevention*
Ringworm of the Body (Tinea Corporis)	This extremely prevalent infection of humans can appear nearly anywhere on the body's glabrous (smooth and bare) skin.	 *Centers for Disease Control and Prevention*
Ringworm of the Groin (Tinea Cruris)	Sometimes known as *jock itch,* crural ringworm occurs mainly in males on the groin, perianal skin, scrotum, and occasionally the penis. The fungus thrives under conditions of moisture and humidity created by sweating.	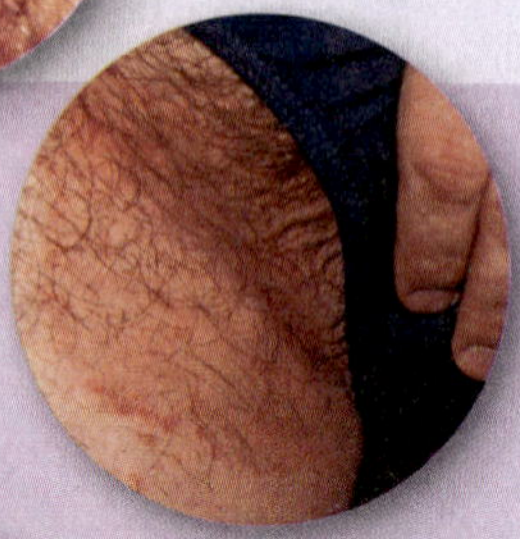 *Dr. Harout Tanielian/Science Source*
Ringworm of the Foot (Tinea Pedis)	Tinea pedis has more colorful names as well, including athlete's foot and jungle rot. Infections begin with blisters between the toes that burst, crust over, and can spread to the rest of the foot and nails.	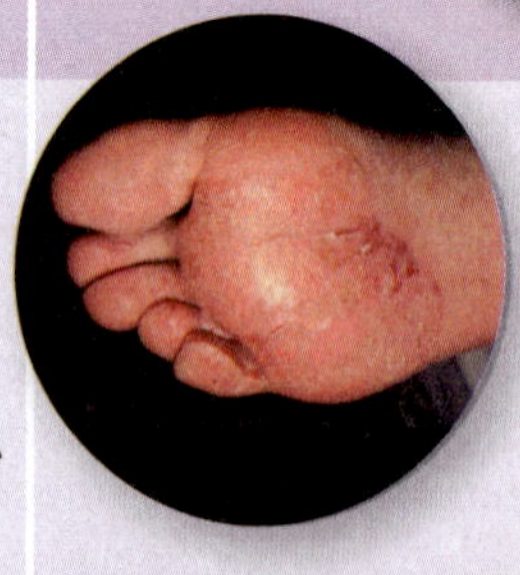 *Dr. P. Marazzi/Science Source*
Ringworm of the Nail (Tinea Unguium)	Fingernails and toenails, being masses of keratin, are often sites for persistent fungus colonization. The first symptoms are usually superficial white patches in the nail bed. A more invasive form causes thickening, distortion, and darkening of the nail.	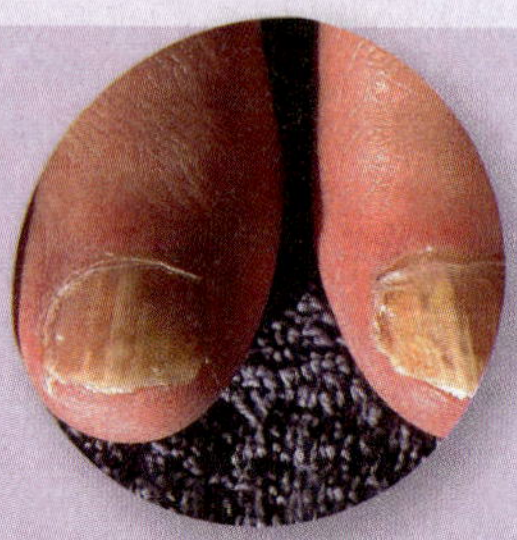 *Dr. Edwin P Ewing, Jr./Centers for Disease Control and Prevention*

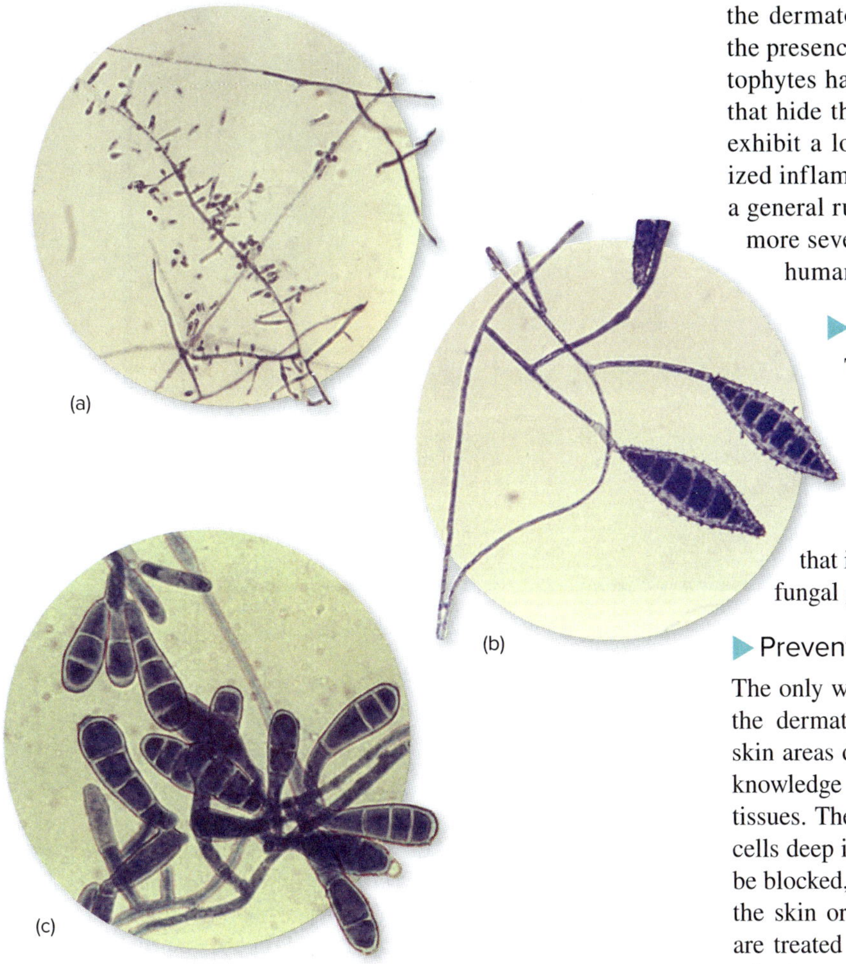

(a) (b) (c)

Figure 19.16 Examples of dermatophyte spores.
(a) Regular, numerous microconidia of *Trichophyton.* **(b)** Macroconidia of *Microsporum canis,* a cause of ringworm in cats, dogs, and humans. **(c)** Smooth-surfaced macroconidia in clusters, characteristic of *Epidermophyton.*
(all): Dr. Lucille K. George/CDC

macroconidia and sparser microconidia **(figure 19.16*b*)**; and *Epidermophyton* has ovoid, smooth, clustered macroconidia and no microconidia **(figure 19.16*c*)**.

The presenting symptoms of a cutaneous mycosis occasionally are so dramatic and suggestive of these genera that no further testing is necessary. In most cases, however, direct microscopic examination and culturing are required. Diagnosis of tinea of the scalp caused by some species is aided by use of a long-wave ultraviolet lamp that causes infected hairs to fluoresce. Samples of hair, skin scrapings, and nail debris treated with heated potassium hydroxide (KOH) show a thin, branching fungal mycelium if infection is present.

Pathogenesis and Virulence Factors

The dermatophytes have the ability to invade and digest keratin, which is naturally abundant in the cells of the stratum corneum. The fungi do not invade deeper epidermal layers. Important factors that promote infection are the hardiness of the dermatophyte spores (they can last for years on fomites), the presence of abraded skin, and intimate contact. The dermatophytes have recently been found to produce a set of proteins that hide them from the host immune system. Most infections exhibit a long incubation period (months), followed by localized inflammation and allergic reactions to fungal proteins. As a general rule, infections acquired from animals and soil cause more severe reactions than do infections acquired from other humans.

Transmission and Epidemiology

Transmission of the fungi that cause these diseases is direct and indirect contact with other humans or with infected animals. Some of these fungi can be acquired from the soil. Interestingly, people in cultures that walk barefoot most of the time have very low levels of athlete's foot, suggesting that it is the confined atmosphere of a shoe that encourages fungal growth.

Prevention and Treatment

The only way to prevent these infections is to avoid contact with the dermatophytes, which is impractical. Keeping susceptible skin areas dry is helpful. Treatment of ringworm is based on the knowledge that the dermatophyte is feeding on dead epidermal tissues. These regions undergo constant replacement from living cells deep in the epidermis, so if multiplication of the fungus can be blocked, the fungus will eventually be sloughed off along with the skin or nail. Unfortunately, this takes time. Most infections are treated with topical antifungal agents. Ointments containing tolnaftate, miconazole, itraconazole, terbinafine, or thiabendazole are applied regularly for several weeks. Some drugs work by speeding up loss of the outer skin layer. Often, tinea capitis is treated with oral terbinafine.

Superficial Mycosis

Superficial mycosis involves the outer epidermal surface and is ordinarily an innocuous infection with cosmetic rather than inflammatory effects. It is often called tinea versicolor. Tinea versicolor is caused by the yeast genus *Malassezia,* a genus that has at least 10 species living on human skin. The yeast feeds on the high oil content of the skin glands. Even though these yeasts are very common normal biota (carried by nearly 100% of humans tested), in some people their growth elicits mild, chronic scaling and interferes with production of pigment by melanocytes. The trunk, face, and limbs may take on a mottled appearance **(figure 19.17)**. The disease is most pronounced in young people who are frequently exposed to the sun because the area affected does not tan well. Other superficial skin conditions in which *Malassezia* is implicated are folliculitis, psoriasis, and seborrheic dermatitis (dandruff). It is also occasionally associated with systemic infections and catheter-associated sepsis in compromised patients. A recent study has found *Malassezia* species in the pancreas, and the possibility that it may be associated with pancreatic cancer.

Disease Table 19.10 Cutaneous and Superficial Mycoses

Disease	Cutaneous Infections	Superficial Infections (Tinea Versicolor)
Causative Organism(s)	*Trichophyton, Microsporum, Epidermophyton* F	*Malassezia* species F
Most Common Modes of Transmission	Direct and indirect contact, vehicle (soil)	Endogenous "normal biota"
Virulence Factors	Ability to degrade keratin, invoke hypersensitivity, avoidance of immune response	–
Culture/Diagnosis	Microscopic examination, KOH staining, culture	Usually clinical, KOH can be used
Prevention	Avoid contact	None
Treatment	Topical tolnaftate, itraconazole, terbinafine, miconazole, thiabendazole, oral terbinafine	Topical antifungals
Epidemiological Features	Among schoolchildren, 0%–19% prevalence, in humid climates up to 30%	Highest incidence among adolescents

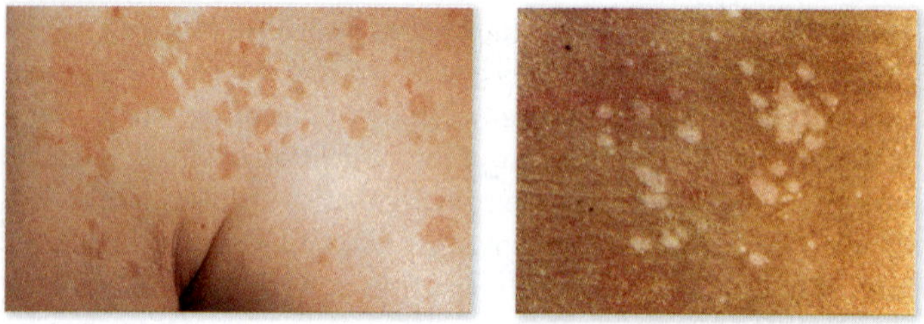

Figure 19.17 Tinea versicolor. Mottled, discolored skin pigmentation is characteristic of superficial skin infection by *Malassezia furfur.* Photo on the left is of light-colored skin. On the right you can see that if skin is darker, the infection can show up as lighter spots.

(left) Biophoto Associates/Science Source; (right) 4FR/iStock/Getty Images

19.2 Learning Outcomes—Assess Your Progress

4. List the possible causative agents for each of the infectious skin conditions: MRSA, impetigo, cellulitis, staphylococcal scalded skin syndrome, gas gangrene, vesicular or pustular rash diseases, maculopapular rash diseases, wartlike eruptions, large pustular skin lesions, and cutaneous and superficial mycosis.
5. Identify which of these conditions are transmitted to the respiratory tract through droplet contact.
6. List the skin conditions for which vaccination is recommended.
7. Summarize methods used to distinguish infections caused by *Staphylococcus aureus* and *Streptococcus pyogenes,* and discuss the spectrum of skin and tissue diseases caused by each.
8. Explain what MRSA stands for and its epidemiology.
9. Discuss the relative dangers of rubella and rubeola viruses in different populations.

19.3 The Surface of the Eye, Its Defenses, and Normal Biota

The eye is a complex organ with many different tissue types, but for the purposes of this chapter we consider only its exposed surfaces, the *conjunctiva* and the *cornea* **(figure 19.18).** The **conjunctiva** is a very thin, membranelike tissue that covers the eye (except for the cornea) and lines the eyelids. It secretes an oil- and mucus-containing fluid that lubricates and protects the eye surface. The **cornea** is the dome-shaped, central portion of the eye lying over the iris (the colored part of the eye). It has

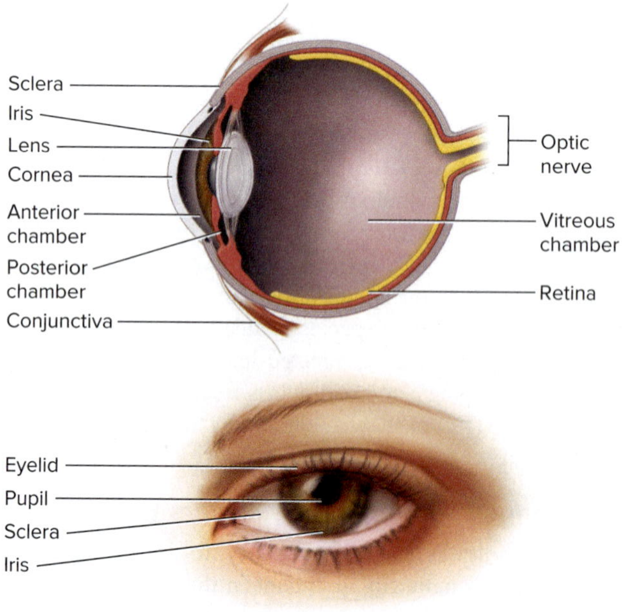
Sclera
Iris
Lens
Cornea
Anterior chamber
Posterior chamber
Conjunctiva
Optic nerve
Vitreous chamber
Retina
Eyelid
Pupil
Sclera
Iris

Figure 19.18 The anatomy of the eye.

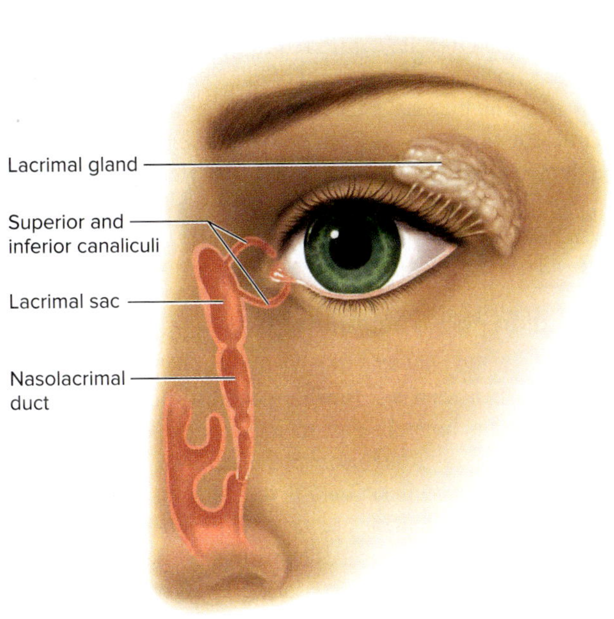
Lacrimal gland
Superior and
inferior canaliculi
Lacrimal sac
Nasolacrimal
duct

Figure 19.19 The lacrimal apparatus of the eye.

five to six layers of epithelial cells that can regenerate quickly if they are superficially damaged. The cornea has been called "the windshield of the eye."

The eye's best defense is the film of tears, which consists of an aqueous fluid, oil, and mucus. The tears are formed in the lacrimal gland at the outer and upper corner of each eye **(figure 19.19),** and they drain into the lacrimal duct at the inner corner. The aqueous portion of tears contains sugars, lysozyme, and lactoferrin. These last two substances have antimicrobial properties. The mucous layer contains proteins and sugars and plays a protective role. And, of course, the flow of the tear film prevents the attachment of microorganisms to the eye surface.

Because the eye's primary function is vision, anything that hinders vision would be counterproductive. For that reason, inflammation does not occur in the eye as readily as it does elsewhere in the body. Flooding the eye with fluid containing a large number of light-diffracting objects such as lymphocytes and phagocytes in response to every irritant would mean almost constantly blurred vision. So even though the eyes are relatively vulnerable to infection (not being covered by keratinized epithelium), the evolution of the vertebrate eye has, of necessity, favored reduced innate immunity. This characteristic is sometimes known as **immune privilege.**

The specific immune response, involving B and T cells, is also somewhat restricted in the eye. The anterior chamber (see figure 19.18) is largely cut off from the blood supply. Lymphocytes that do gain access to this area are generally less active than lymphocytes elsewhere in the body.

Normal Biota of the Eye

The eye was previously thought to be only sparsely populated by microbiota. 16s rRNA analysis of the healthy eye microbiome has revealed a more robust population, showing a lot of diversity in the bacteria found. In many cases, *Corynebacterium* is the dominant genus. To a large extent, the eye microbiome resembles that of the skin.

Defenses and Normal Biota of the Eyes

	Defenses	Normal Biota
Eyes	Mucus in conjunctiva and in tears; lysozyme and lactoferrin in tears	*Corynebacterium, Staphylococcus epidermidis, Micrococcus,* and *Streptococcus* species

19.3 Learning Outcomes—Assess Your Progress

10. Describe the important anatomical features of the eye.
11. List the natural defenses present in the eye.
12. List the types of normal biota presently known to occupy the eye.

19.4 Infectious Diseases Manifesting in the Eye

In this section, we cover the infectious agents that cause diseases of the surface structures of the eye—namely, the cornea and conjunctiva.

Conjunctivitis

Infection of the conjunctiva is relatively common. It can be caused by specific microorganisms that have a preference for eye tissues, by contaminants introduced by the presence of a contact lens or an eye injury, or by accidental inoculation of the eye by a traumatic event.

▶ Signs and Symptoms

Just as there are many different causes of conjunctivitis, there are many different clinical presentations. Inflammation of this tissue almost always causes a discharge. Most bacterial infections produce a milky discharge, whereas viral infections tend to produce a clear exudate. It is typical for a patient to wake up in the morning with an eye "glued" shut by secretions that have accumulated and solidified through the night. Some conjunctivitis cases are caused by an allergic response, and these often produce copious amounts of clear fluid as well. The pain generally is mild, although often patients report a gritty sensation in their eye(s). Redness and eyelid swelling are common, and in some cases patients report photophobia (sensitivity to light). The informal name for common conjunctivitis is pinkeye.

▶ Causative Agents and Their Transmission

Cases of neonatal eye infection caused by *Neisseria gonorrhoeae* or *Chlamydia trachomatis* are usually transmitted vertically from a genital tract infection in the mother (discussed in chapter 24). Either one of these eye infections can lead to serious eye damage if not treated promptly **(figure 19.20).** Note that herpes simplex can

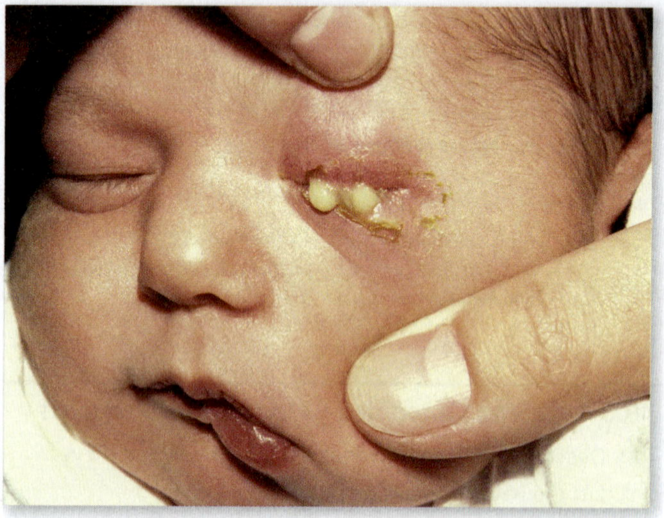

Figure 19.20 Neonatal conjunctivitis.
Medical-on-Line/Alamy Stock Photo

also cause neonatal conjunctivitis, but it is usually accompanied by generalized herpes infection (covered in chapter 24).

Bacterial conjunctivitis in other age groups is most commonly caused by *Staphylococcus aureus* or *Streptococcus pneumoniae,* although *Haemophilus influenzae* and *Moraxella* species are also frequent causes. *N. gonorrhoeae* and *C. trachomatis* can also cause conjunctivitis in adults. These infections may result from autoinoculation from a genital infection or from sexual activity, although *N. gonorrhoeae* can be part of the normal biota in the respiratory tract. A wide variety of bacteria, fungi, and protozoa can contaminate contact lenses and lens cases and then be transferred to the eye, resulting in disease that may be very serious. This means of infection is considered vehicle transmission, with the lens or the solution being the vehicle.

Viral conjunctivitis is commonly caused by adenoviruses, although other viruses may be responsible. (Herpesvirus infection of the eye is discussed later in this chapter.) Both bacterial and viral conjunctivitis are transmissible by direct and even indirect contact and are usually highly contagious.

Prevention and Treatment

Good hygiene is the only way to prevent conjunctivitis in adults and children other than neonates. Newborn children in the United States are administered antimicrobials in their eyes after delivery to prevent neonatal conjunctivitis from either *N. gonorrhoeae* or *C. trachomatis.* Treatment of those infections, if they are suspected, is started before lab results are available and usually is accomplished with oral erythromycin. If *N. gonorrhoeae* is confirmed, oral therapy is usually switched to ceftriaxone. If antibacterial therapy is prescribed for other conjunctivitis cases, it should cover all possible bacterial pathogens. Gatifloxacin or levofloxacin eye drops are a common choice. Erythromycin or gentamicin are also often used. Because conjunctivitis is usually diagnosed based on clinical signs, a physician may prescribe prophylactic antibiotics even if a viral cause is suspected. If symptoms do not begin improving within 48 hours, more extensive diagnosis may be performed. **Disease Table 19.11** lists the most common causes of conjunctivitis; keep in mind that other microorganisms can also cause conjunctival infections.

Disease Table 19.11 Conjunctivitis

Disease	Neonatal Conjunctivitis	Bacterial Conjunctivitis	Viral Conjunctivitis
Causative Organism(s)	*Chlamydia trachomatis* or *Neisseria gonorrhoeae* B G–	*Streptococcus pneumoniae, Staphylococcus aureus, Haemophilus influenzae, Moraxella,* and also *Neisseria gonorrhoeae, Chlamydia trachomatis* B G+, G–	Adenoviruses and others V
Most Common Modes of Transmission	Vertical	Direct, indirect contact	Direct, indirect contact
Virulence Factors	–	–	–
Culture/Diagnosis	Gram stain and culture	Clinical diagnosis	Clinical diagnosis
Prevention	Screen mothers, apply antibiotic to newborn eyes	Hygiene	Hygiene
Treatment	Oral antibiotics	Gatifloxacin or levofloxacin ophthalmic solution; several of these bacteria appear in the CDC Antibiotic Resistance Report	None, although antibiotics often given because type of infection not distinguished
Distinguishing Features	In babies <28 days old	Mucopurulent discharge	Serous (clear) discharge
Epidemiological Features	Less than 0.5% in developed world; higher incidence in developing world	More common in children	More common in adults

Trachoma

Ocular trachoma is a chronic *Chlamydia trachomatis* infection of the epithelial cells of the eye. It is an ancient disease and a major cause of blindness in certain parts of the world. Although a few cases occur annually in the United States, several million cases occur endemically in parts of Africa, Asia, the Middle East, Latin America, and the Pacific Islands. Transmission is favored by contaminated fingers, fomites, fleas, and a hot, dry climate. It is caused by a different *C. trachomatis* strain than the one that can cause simple conjunctivitis. Ongoing infection or many recurrent infections with this strain eventually lead to chronic inflammatory damage and scarring. It is also believed that other bacterial pathogens, such as *Streptococcus* and *Staphylococcus,* contribute to the scarring process once trachoma has begun.

The first signs of infection are a mild conjunctival discharge and slight inflammation of the conjunctiva. These symptoms are followed by marked infiltration of lymphocytes and macrophages into the infected area. As these cells build up, they impart a pebbled (rough) appearance to the inner upper eyelid **(figure 19.21).** In time, a vascular pseudomembrane of exudates and inflammatory leukocytes forms over the cornea, a condition called *pannus,* which lasts a few weeks. Chronic and secondary infections can lead to corneal damage and impaired vision. Early treatment of this disease with azithromycin is highly effective and prevents all of the complications. It is a tragedy that in this day of sophisticated preventive medicine, millions of children worldwide will develop blindness for lack of a single dose of azithromycin **(Disease Table 19.12).**

Disease Table 19.12	Trachoma
Causative Organism(s)	*C. trachomatis* serovars A–C B G–
Most Common Modes of Transmission	Indirect contact, mechanical vector
Virulence Factors	Intracellular growth
Culture/Diagnosis	Detection of inclusion bodies in stained preparations
Prevention	Hygiene, vector control, prompt treatment of initial infection
Treatment	Azithromycin or doxycycline
Epidemiological Features	Highest prevalence among children between ages 3 and 5, prevalence as high as 60% in endemic areas

Keratitis

Keratitis is a more serious eye infection than conjunctivitis. Invasion of deeper eye tissues occurs and can lead to complete corneal destruction. Any microorganism can cause this condition, especially after trauma to the eye. In developed countries, herpes simplex virus is the most common cause. It can cause keratitis in the absence of predisposing trauma. In developing countries, bacterial and fungal causes are more common.

The usual cause of herpetic keratitis is a "misdirected" reactivation of (oral) herpes simplex virus type 1 (HSV-1). The virus, upon reactivation, travels into the ophthalmic rather than the mandibular branch of the trigeminal nerve **(figure 19.22).** Infections with HSV-2 can also occur as a result of virus exposure during sexual activity, via transfer of the virus from the genital to eye area or through autoinoculation from a recurrent HSV-2 oral infection. Preliminary symptoms are a gritty feeling in the eye, conjunctivitis, sharp pain, and sensitivity to light. Some patients develop characteristic branched or opaque corneal lesions as well. In 25% to 50% of cases, this keratitis is recurrent and chronic and can interfere with vision. Blindness due to herpes is the leading infectious cause of blindness in the United States. The viral condition is treated with topical trifluridine, sometimes supplemented with oral acyclovir.

In the last few years, another form of keratitis has been increasing in incidence. An amoeba called *Acanthamoeba* has been causing serious keratitis cases, especially in people who wear contact lenses. This free-living amoeba is everywhere—it lives in tap water, freshwater lakes, and the like. The infections are usually associated with less-than-rigorous contact lens hygiene or previous trauma to the eye **(Disease Table 19.13).**

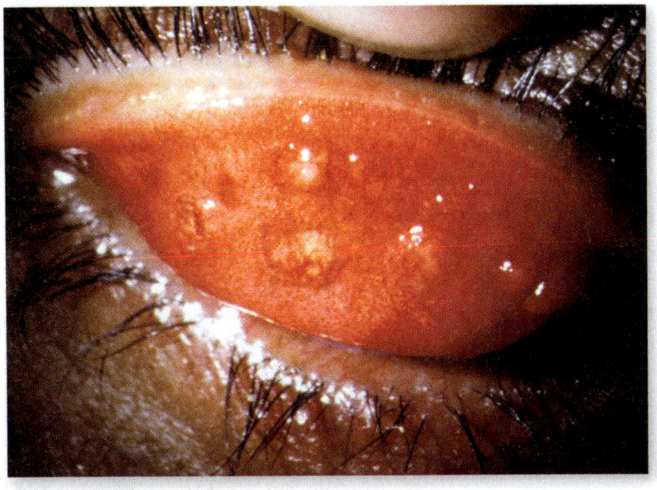

Figure 19.21 Ocular trachoma caused by *C. trachomatis.*
Western Ophthalmic Hospital/Science Source

Disease Table 19.13 Keratitis		
Causative Organism(s)	Herpes simplex virus V	Miscellaneous microorganisms, including *Acanthamoeba*
Most Common Modes of Transmission	Reactivation of latent virus, although primary infections can occur in the eye	Often traumatic introduction (parenteral)
Virulence Factors	Latency	Various
Culture/Diagnosis	Usually clinical diagnosis; viral culture or PCR if needed	Various
Prevention	–	–
Treatment	Topical trifluridine +/– oral acyclovir	Specific antimicrobials
Epidemiological Features	One-third of worldwide population infected; in United States, annual incidence of 500,000	

River Blindness

River blindness is a chronic parasitic (helminthic) infection. It is endemic in dozens of countries in Latin America, Africa, Asia, and the Middle East. At any given time, approximately 37 million people are infected with the worm called *Onchocerca volvulus* (ong″-koh-ser′-kah volv′-yoo-lus). This organism is a filarial (threadlike) helminthic worm transmitted by small, biting vectors called black flies. These voracious flies often attack in large numbers, and it is not uncommon in endemic areas to be bitten several hundred times a day. The disease gets its name from the habitat where these flies are most often found, rural settlements along rivers bordered with overhanging vegetation.

The *Onchocerca* larvae are deposited into a bite wound and develop into adults in the immediate subcutaneous tissues, where disfiguring nodules form within 1 to 2 years after initial contact. Microfilariae (immature worm forms) given off by the adult female migrate via the bloodstream to many locations but especially to the eyes. While the worms are in the blood, they can be transmitted to other feeding black flies.

Some cases of onchocerciasis result in a severe, itchy rash that can last for years. The worms eventually invade the entire eye, producing much inflammation and permanent damage to the retina and optic nerve. In fact, half a million people are blind due to this infection worldwide. In 1999, researchers first discovered large colonies of bacteria called *Wolbachia* living *inside* the *Onchocerca* worms. There is convincing evidence that the damage

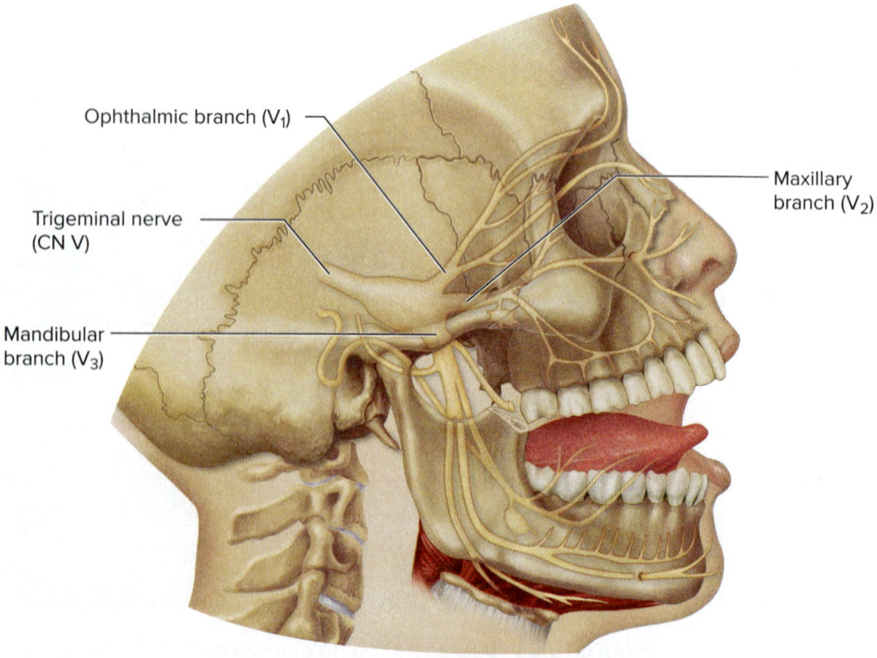
Ophthalmic branch (V_1)
Trigeminal nerve (CN V)
Mandibular branch (V_3)
Maxillary branch (V_2)

Figure 19.22 The trigeminal nerve. There are three branches of the nerve: the ophthalmic, maxillary, and mandibular.

caused to human tissues is induced by the bacteria rather than by the worms. Of course, the worms serve as the delivery system to the human, as it does not appear that the bacteria can infect humans on their own. These bacteria enjoy a mutualistic relationship with their hosts. They are essential for normal *Onchocerca* development.

In regions of high prevalence, it is not unusual for an ophthalmologist to see microfilariae wiggling in the anterior eye chamber during a routine eye checkup. Microfilariae die in several months, but adults can exist for up to 15 years in skin nodules.

River blindness has been a serious problem in many areas of Africa. In some villages, nearly half of the residents are affected by the disease. A campaign to eradicate onchocerciasis is under way, supported by the Carter Center, an organization run by former U.S. president Jimmy Carter. The approach is to treat people with *ivermectin,* a potent antifilarial drug, and to use insecticides to control the black flies. The drug company that manufactures ivermectin has promised to provide the drug for free as long as the need for it exists. Combined with work by the World Health Organization and the African Programme for Onchocerciasis Control (APOC), there is hope that this disease will be eradicated from these populations in the future **(Disease Table 19.14).**

19.4 Learning Outcomes—Assess Your Progress

13. List the possible causative agents for each of the infectious eye diseases: conjunctivitis, trachoma, keratitis, and river blindness.
14. Discuss why there are distinct differential diagnoses for neonatal and non-neonatal conjunctivitis.

Disease Table 19.14	**River Blindness**
Causative Organism(s)	*Wolbachia* B G+ plus *Onchocerca volvulus* H
Most Common Modes of Transmission	Biological vector
Virulence Factors	Induction of inflammatory response
Culture/Diagnosis	"Skin snips": small piece of skin in NaCl solution examined under microscope and microfilariae counted
Prevention	Avoiding black fly
Treatment	Ivermectin in United States
Distinguishing Features	Worms often visible in eye
Epidemiological Features	18–40 million afflicted worldwide; 99% of cases in 30 African countries; also in Central and South America

Media Under the Microscope Wrap-Up

PHIL/CDC

The article "Vaccinating Children Against COVID-19—The Lessons of Measles" is an opinion piece published in a scientific journal.

The **intended message** of this article is to prod its readers, who are mostly physicians, to use a little history as well as kindness to persuade parents to vaccinate their kids against COVID-19. The article highlights the campaign to vaccinate against measles in the 1960s.

A **critical reading** of this opinion article involves checking the facts used to build the case. Even when it is an opinion that is offered, if it is not bolstered by substantiated facts, it should be distrusted. Another critical skill to reading media articles is to clearly differentiate the authors' opinions and facts. The facts in this piece are easily verifiable, and so the question for the reader is whether you agree with the (well-substantiated) opinion or not.

To interpret this to my **nonmicrobiologist friends,** I would first be clear that it is an opinion written by doctors in a journal read by other doctors. Share the measles vaccine history with them and explain the similarities with COVID-19 to explain the authors' approach.

My **overall grade** for this article is A. I truly appreciate the use of history to gently inform current parents of the dangers that can be averted with vaccines. The facts are solid.

Source: *New England Journal of Medicine* "Vaccinating Children Against Covid-19—The Lessons of Measles," online article posted January 20, 2021..

Study Smarter: Better Together

These activities are designed for you to use on your own with a study group—either a face-to-face group or a virtual one, consisting of 3–5 members. Studying together can be very helpful, but there are effective and ineffective ways to do it. For example, getting together without a clear structure is often not a good use of your time. Use your time efficiently by using one or more of the exercises below.

FACE-TO-FACE GROUPS

Use one or more of the activities below.

Peer Instruction: Assign numbers to your group members to use all semester long. Now look at these five concepts from this chapter. Each group member prepares a 5-minute lesson on the topic corresponding to their number. Don't worry if you have fewer than 5 members; just use however many you have! During your group study time, each member presents their lesson, and the group spends another 5–10 minutes discussing that lesson.

1. The sequence of diagnosis
2. Defenses of the skin
3. MRSA in the community vs. MRSA in the hospital
4. Organisms in this chapter for which there are vaccines available
5. The public health significance of smallpox and anthrax

Concept Maps: Each member of the group should use this list of terms from this chapter to generate their own concept map. This can be hand-drawn or created using software (see Appendix C for guidelines). During group study time, compare each other's concept maps and help each other make sure they are correct. Of course, there are many different "correct" maps. Examining each member's map will help you talk through the varied concepts and how they are related.

Concept Terms:

Streptococcus pyogenes	*Clostridium perfringens*	*Staphylococcus aureus*	
exfoliative toxins A and B	SSSS	gas gangrene	
alpha toxin	scarlet fever	erythrogenic toxin	sandpaper-like rash
blistering of epidermis	myonecrosis		

Table Topics: Each group member should identify a concept or topic from this week's class assignments with which they are having trouble and share it during group study time. The other group members can then help to clarify confusing issues or share how they figured it out. Aim for a maximum of 15 minutes per topic. If the topic remains unclear to the group, bring it up during class or use the instructor's office hours or e-mail to ask for help. Taking the time to struggle with a difficult concept first makes your questions much more specific and more likely to yield helpful answers.

VIRTUAL GROUPS

Not everyone has the time or opportunity to meet with group members outside of class time. You or your instructor can create a virtual group using e-mail or the course software.

Weekly Discussion Board: This forum can be used as a way for groups to discuss topics, via e-mail or other learning management systems or online platforms, before they are covered in class. As each member of the group answers the current week's question, they should send their responses to every other member of their group. It's best to agree on a deadline based on how your class schedule works (Saturday for the next week's topics, for example). Then, after the topic is discussed in class, each member should send a response that all group members will see with a follow-up post on the same topic. If you cover more than one chapter in a week, someone can be designated to choose which chapter Discussion Board question you will use. Or simply decide up front that you will always use the first-chapter-of-the-week's question, to keep the schedule simple.

Discussion Question

Describe thoroughly at least four defenses the skin has against infection.

Deadliness and Communicability of Selected Diseases of the Skin and Eyes

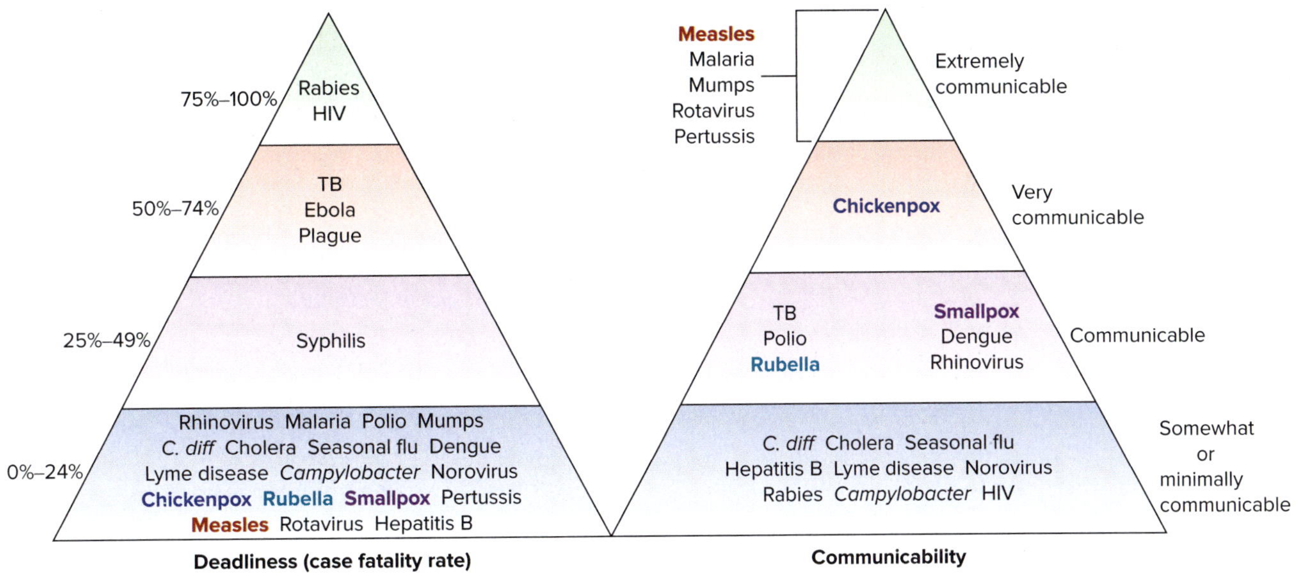
75%–100%
Rabies
HIV
50%–74%
TB
Ebola
Plague
25%–49%
Syphilis
0%–24%
Rhinovirus Malaria Polio Mumps
C. diff Cholera Seasonal flu Dengue
Lyme disease Campylobacter Norovirus
Chickenpox Rubella Smallpox Pertussis
Measles Rotavirus Hepatitis B
Deadliness (case fatality rate)
Measles
Malaria
Mumps
Rotavirus
Pertussis
Extremely communicable
Chickenpox
Very communicable
TB
Polio
Rubella
Smallpox
Dengue
Rhinovirus
Communicable
C. diff Cholera Seasonal flu
Hepatitis B Lyme disease Norovirus
Rabies Campylobacter HIV
Somewhat or minimally communicable
Communicability

▶ Summing Up

Taxonomic Organization Microorganisms Causing Diseases of the Skin and Eyes

Microorganism	Disease	Disease Table
Gram-positive bacteria		
Staphylococcus aureus	MRSA impetigo, cellulitis, scalded skin syndrome, folliculitis, abscesses (furuncles and carbuncles), necrotizing fasciitis, bacterial conjunctivitis	MRSA skin and soft tissue infections, 19.1; Impetigo, 19.2; Cellulitis, 19.3; Scalded skin syndrome, 19.4
Streptococcus pyogenes	Impetigo, cellulitis, erysipelas, necrotizing fasciitis, scarlet fever	Impetigo, 19.2; Cellulitis, 19.3; Maculopapular rash diseases, 19.7
Clostridium perfringens	Gas gangrene	Gas gangrene, 19.5
Bacillus anthracis	Cutaneous anthrax	Large pustular skin lesions, 19.9
Streptococcus pneumoniae	Conjunctivitis	Conjunctivitis, 19.11
Gram-negative bacteria		
Chlamydia trachomatis	Neonatal conjunctivitis, bacterial conjunctivitis	Conjunctivitis, 19.11
Neisseria gonorrhoeae	Neonatal conjunctivitis, bacterial conjunctivitis, trachoma	Conjunctivitis, 19.11; Trachoma, 19.12
Wolbachia (in combination with *Onchocerca*)	River blindness	River blindness, 19.14
Haemophilus influenzae and *Moraxella*	Conjunctivitis	Conjunctivitis, 19.11
DNA viruses		
Human herpesvirus 3 (varicella) virus	Chickenpox	Vesicular or pustular rash diseases, 19.6
Variola virus	Smallpox	Vesicular or pustular rash diseases, 19.6
Parvovirus B19	Fifth disease	Maculopapular rash diseases, 19.7
Human herpesvirus 6	Roseola	Maculopapular rash diseases, 19.7
Human papillomavirus	Warts	Warts and wartlike eruptions, 19.8
Molluscum contagiosum virus	Molluscum contagiosum	Warts and wartlike eruptions, 19.8
Herpes simplex virus	Keratitis	Keratitis, 19.13
Adenovirus	Conjunctivitis	Conjunctivitis, 19.11

(continued)

RNA viruses		
Enteroviruses (Coxsackie)	Hand, foot, and mouth disease	Vesicular or pustular rash diseases, 19.6
Measles virus	Measles	Maculopapular rash diseases, 19.7
Rubella virus	Rubella	Maculopapular rash diseases, 19.7
Fungi		
Trichophyton	Ringworm	Cutaneous and superficial mycoses, 19.10
Microsporum	Ringworm	Cutaneous and superficial mycoses, 19.10
Epidermophyton	Ringworm	Cutaneous and superficial mycoses, 19.10
Malassezia species	Superficial mycoses	Cutaneous and superficial mycoses, 19.10
Protozoa		
Leishmania spp.	Leishmaniasis	Large pustular skin lesions, 19.9
Acanthamoeba	Keratitis	Keratitis, 19.13
Helminths		
Onchocerca volvulus (in combination with *Wolbachia*)	River blindness	River blindness, 19.14

Chapter Summary

INFECTIOUS DISEASES MANIFESTING ON THE SKIN AND EYES

19.1 THE SKIN, ITS DEFENSES, AND NORMAL BIOTA

- The skin's outer layer is the epidermis, consisting of cells that contain keratin, which "waterproofs" the skin and protects it from microbial invasion.
- Other defenses include antimicrobial peptides, low pH sebum, high salt in sweat and lysozyme in sweat and saliva.
- Microbiome studies show that there is wide variation of normal biota among different people.

19.2 INFECTIOUS DISEASES MANIFESTING ON THE SKIN

- **MRSA skin and soft tissue infection**—Caused by methicillin-resistant *Staphylococcus aureus.*
- **Impetigo**—A highly contagious superficial bacterial infection that causes skin to peel or flake off. Causative organisms can be *Staphylococcus aureus* or *Streptococcus pyogenes* or both.
- **Cellulitis**— A fast-growing infection of the dermis and subcutaneous tissue below. Most commonly caused by the introduction of *S. aureus* or *S. pyogenes* into the dermis level.
- **Staphylococcal scalded skin syndrome (SSSS)**—Caused by *S. aureus.* Affects mostly newborns and babies and is similar to a systemic form of impetigo.
- **Gas gangrene**—Also called clostridial myonecrosis, it can manifest in two forms: anaerobic cellulitis or true myonecrosis. Most common cause is *Clostridium perfringens*, an endospore-forming anaerobe.

- **Vesicular or pustular rash diseases—Chickenpox:** Lesions with clear fluid, can lead to shingles later in life; caused by human herpesvirus 3. **Smallpox:** Natural infections eradicated; always a bioterror threat; caused by variola virus. **Hand, foot, and mouth disease:** Minor vesicles, caused by enteroviruses, most often Coxsackie.
- **Maculopapular rash diseases—Measles:** Also called rubeola; characteristic red maculopapular rash. Serious disease, leads to temporary erasure of some immune memory. Can lead, years later, to a fatal disease called subacute sclerosis panencenphalitis (SSPE). The MMR and MMRV vaccines contain attenuated viruses. The vaccines were falsely associated with autism, but that link has been definitively disproven. **Rubella:** Can appear in two forms: post-natal (usually mild) and congenital (extremely serious). **Fifth disease:** Also called *erythema infectiosum.* Mild but highly infectious. Characteristic "slapped-cheek" appearance. Caused by parvovirus B19. **Roseola:** May result in maculopapular rash; caused by human herpesvirus 6.
- **Wartlike eruptions**—Most common warts are caused by human papilloma virus or a pox virus known as molluscum contagiosum.
- **Large pustular skin lesions—Leishmaniasis:** A zoonotic protozoan infection transmitted by the female sand fly when it ingests its host's blood. Can be localized or systemic. **Cutaneous anthrax:** Most common and least dangerous version of infection with *Bacillus anthracis.*
- **Ringworm (cutaneous mycoses)**—A group of fungi, collectively termed dermatophytes, infect nonliving epidermal tissues, hair, and nails. Genera are *Trichophyton, Microsporum,* and *Epidermophyton.*
- **Superficial mycosis:**—"Tinea versicolor" caused by infection of outer epidermis. Caused by *Malassezia.*

19.3 THE SURFACE OF THE EYE, ITS DEFENSES, AND NORMAL BIOTA

- The flushing act of the tears, which contain lysozyme and lactoferrin, is a major protective feature of the eye.
- The eye has similar microbiota to skin, but in lower numbers.

19.4 INFECTIOUS DISEASES MANIFESTING IN THE EYE

- **Conjunctivitis**—In neonates is usually transmitted by maternal infection with *Neisseria gonorrhoeae* or *Chlamydia trachomatis*. In other age groups most commonly caused by the bacteria *Staphyloccus aureus, Streptococcus pneumoniae, Haemophilus influenzae, or Moraxella* species. Viral agents are often adenoviruses.
- **Trachoma**—A chronic infection with *Chlamydia trachomatis.* Major cause of blindness in some parts of the world. This infection and conjunctivitis are caused by different strains of *C. trachomatis.*
- **Keratitis**—Herpes simplex viruses (HSV-1 and HSV-2) and *Acanthamoeba* cause two different forms of the disease.
- **River blindness**—Chronic infection in Latin America, Africa, Asia, and the Middle East. Caused by a symbiotic pair, the bacterium *Wolbachia* living inside the helminth *Onchocerca.*

INFECTIOUS DISEASES AFFECTING
The Skin and Eyes

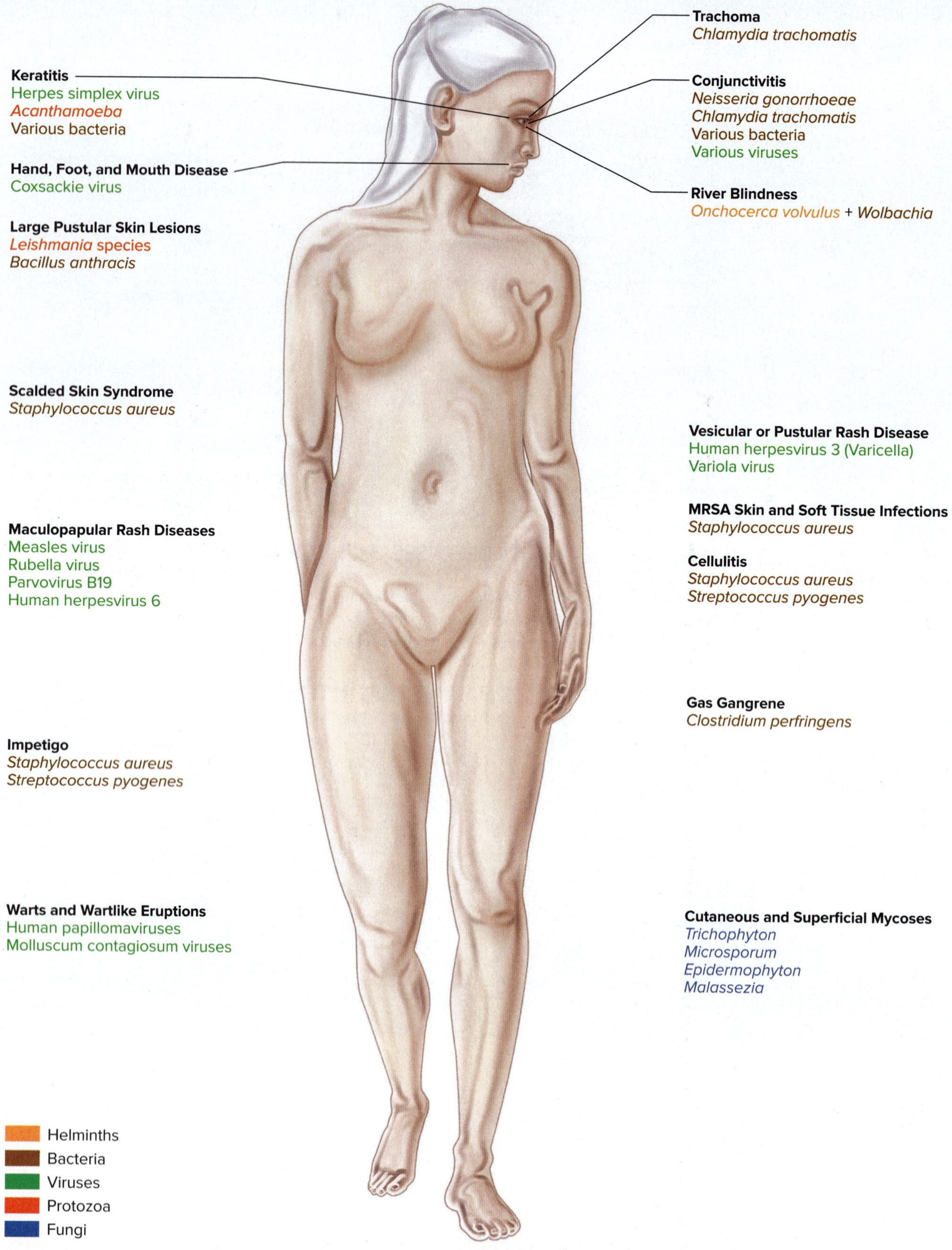

Keratitis
Herpes simplex virus
Acanthamoeba
Various bacteria
Hand, Foot, and Mouth Disease
Coxsackie virus
Large Pustular Skin Lesions
Leishmania species
Bacillus anthracis
Scalded Skin Syndrome
Staphylococcus aureus
Maculopapular Rash Diseases
Measles virus
Rubella virus
Parvovirus B19
Human herpesvirus 6
Impetigo
Staphylococcus aureus
Streptococcus pyogenes
Warts and Wartlike Eruptions
Human papillomaviruses
Molluscum contagiosum viruses
Trachoma
Chlamydia trachomatis
Conjunctivitis
Neisseria gonorrhoeae
Chlamydia trachomatis
Various bacteria
Various viruses
River Blindness
Onchocerca volvulus + Wolbachia
Vesicular or Pustular Rash Disease
Human herpesvirus 3 (Varicella)
Variola virus
MRSA Skin and Soft Tissue Infections
Staphylococcus aureus
Cellulitis
Staphylococcus aureus
Streptococcus pyogenes
Gas Gangrene
Clostridium perfringens
Cutaneous and Superficial Mycoses
Trichophyton
Microsporum
Epidermophyton
Malassezia
Helminths
Bacteria
Viruses
Protozoa
Fungi

System Summary Figure 19.23

SmartGrid: From Knowledge to Critical Thinking

This *21 Question Grid* takes the topics from this chapter and arranges them with respect to the American Society for Microbiology's Undergraduate Curriculum guidelines—all six of the important "Concepts" as well as the important "Competency" of scientific literacy. Three questions are supplied, which cover chapter content referring to the Concept or Competency in increasing levels of Bloom's taxonomy for learning.

ASM Concept/ Competency	A. Bloom's Level 1, 2—Remember and Understand (Choose one)	B. Bloom's Level 3, 4—Apply and Analyze	C. Bloom's Level 5, 6—Evaluate and Create
Evolution	**1.** Which of the following infectious agents has evolved to maintain a persistent state in its host? **a.** variola virus **b.** herpes virus **c.** vaccinia virus **d.** *Staphylococcus aureus*	**2.** Can you think of a plausible argument for why hospital cases of MRSA are declining and community cases of MRSA are increasing?	**3.** Smallpox has ravaged human populations for thousands of years. When it first came to the Americas in the 1500s, it had a high case-fatality rate. By the end of the 1800s, it manifested as a milder disease. Speculate on why that might be.
Cell Structure and Function	**4.** What is an antimicrobial enzyme found in sweat, tears, and saliva that can break down bacterial cell walls? **a.** lysozyme **b.** ß-lactamase **c.** catalase **d.** hyaluronidase	**5.** What cell structure of *Bacillus anthracis* makes it capable of surviving in powdered form, and naturally in soil, for indefinite periods of time?	**6.** Both *Staphylococcus aureus* and *Streptococcus pyogenes* are "pyogenic." What does that mean, and what attributes of the two bacteria are likely to contribute to that phenomenon?
Metabolic Pathways	**7.** Which of the following organisms produces an enzyme that breaks down hydrogen peroxide? **a.** *Streptococcus pyogenes* **b.** *Staphylococcus aureus* **c.** MRSA **d.** two of these	**8.** *Streptococcus pneumoniae,* which is a common inhabitant of healthy upper respiratory tracts, produces hydrogen peroxide to fight off microbial competitors. How is it possible then that *Staphylococcus aureus* is frequently a co-colonizer?	**9.** Some antifungal drugs inhibit an enzyme called squalene epoxidase. What does this enzyme do, and why is it an effective target for selective toxicity to the fungus and not the host?
Information Flow and Genetics	**10.** Which of these techniques has detected the larger number of normal microbiota on skin surfaces? **a.** culturing **b.** 16s rRNA sequencing **c.** antibody probing **d.** gel electrophoresis	**11.** The first person to die of measles in the United States in 12 years was a woman in Washington State in July 2015. What kind of laboratory tests would you perform to determine whether the virus was the same that circulated at Disney in 2014?	**12.** Herpesviruses cause more than one of the diseases in this chapter. Considering some features all of these conditions have in common, hypothesize what the viral DNA does after initial invasion of a host cell.
Microbial Systems	**13.** Which of the following conditions is most likely to be a polymicrobial infection? **a.** measles **b.** rubella **c.** leishmaniasis **d.** impetigo	**14.** Describe how a virus infection of human cells can lead to syncytia.	**15.** Research the bacterium *Micavibrio aeruginosavorus,* which is often found as normal biota of the eye. Formulate a theory on how it keeps pathogens from harming the eye.

(continued)

ASM Concept/ Competency	A. Bloom's Level 1, 2—Remember and Understand (Choose one)	B. Bloom's Level 3, 4—Apply and Analyze	C. Bloom's Level 5, 6—Evaluate and Create
Impact of Microorganisms	**16.** *Staphylococcus aureus* is part of the differential diagnosis of which of the following diseases? **a.** impetigo **b.** maculopapular rash **c.** both of these **d.** neither a nor b	**17.** Compare and contrast the effects of the rubeola virus and the rubella virus on humans.	**18.** In a single (unvaccinated) family one winter, the 12-year-old and 4-year-old came down with chickenpox. The mother and father had experienced the disease as children and were unaffected this time. The 3-month-old baby did not become ill, either. However, 4 months later the baby exhibited a shingles outbreak. Using immunological phenomena, explain what was likely going on with the baby.
Scientific Thinking	**19.** Which steps of the diagnostic process are in order? **a.** Differential diagnosis, anatomic diagnosis, etiologic diagnosis **b.** Anatomic diagnosis, etiologic diagnosis, differential diagnosis **c.** Anatomic diagnosis, differential diagnosis, etiologic diagnosis **d.** none of these	**20.** Explain why detecting a single case of smallpox would be considered a public health—and national security—emergency.	**21.** Construct a testable hypothesis for examining whether receiving the chickenpox vaccine results in fewer cases of shingles in the population.

Answers to the multiple-choice questions appear in Appendix A.

Visual Connections

This question uses visual images to connect content within and between chapters.

1. **From chapter 13, figure 13.7*a*.** Name at least two diseases from this chapter that utilize the illustrated pathogenic mechanism.

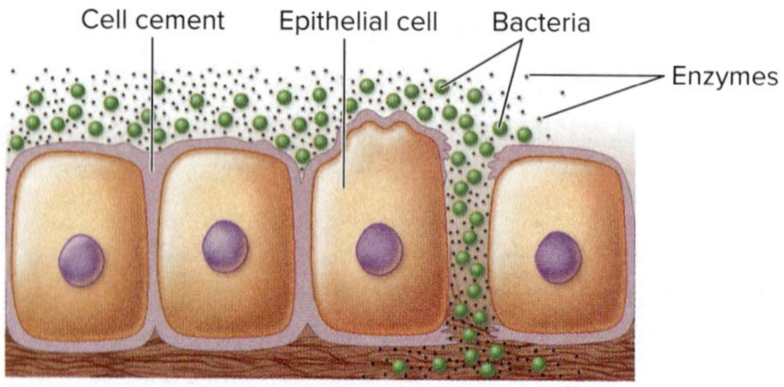
Cell cement
Epithelial cell
Bacteria
Enzymes

Microbes secrete enzymes and toxins.

High Impact Study

These terms and concepts are most critical for your understanding of this chapter—and may be the most difficult. Have you mastered them? In these disease chapters, the terms and concepts help you identify what is important in a different way than the comprehensive details found in the Disease Tables. Your instructor will help you understand what is important for your class.

Concepts

- [ ] The sequence of diagnosis
- [ ] Defenses of the skin
- [ ] Normal microbiota of the skin
- [ ] MRSA in the community vs. MRSA in the hospital
- [ ] Terms for skin lesions
- [ ] Spectrum of *Staphyloccus* and *Streptococcus* diseases
- [ ] Method of transmission for most vesicular or pustular maculopapular diseases
- [ ] The high contagiousness of measles/its relation to herd immunity
- [ ] The effects of the rubella virus on fetuses
- [ ] Variola vs. vaccinia
- [ ] The public health significance of smallpox and anthrax
- [ ] Three distinct etiologies of conjunctivitis
- [ ] Organisms in this chapter for which there are vaccines available
- [ ] Organisms in this chapter that display significant antibiotic resistance

Terms

- [ ] Superantigen
- [ ] Myonecrosis
- [ ] Congenital
- [ ] Teratogenic
- [ ] Dermatophyte
- [ ] Tineas
- [ ] Immune privilege

Design Element: (College students): Caia Image/Image Source

20

Infectious Diseases Manifesting in the Nervous System

Foodcollection/Martin Skultety/Image Source

MEDIA UNDER THE MICROSCOPE

Honey for Cancer?

These case studies examine an article from the popular media to determine the extent to which it is factual and/or misleading. This case focuses on the 2020 Green Living *article "Bee-cause I Said So!"*

This article is written by a gentleman who is touting the cancer-curing benefits of raw, unfiltered honey. He tells us that the most recent studies show that honey has anticancer properties in cell cultures and animal models. The studies, he reports, show that raw honey induces apoptosis, disrupts mitochondrial membranes, and stops the cell cycle, all of which will battle cancer.

The author tells us that honey is a proven immune booster, has antimicrobial properties, and is a natural anti-inflammatory agent. He further cites a study that found a negative correlation between rates of cancer in developing countries and their rates of honey consumption.

The author does not provide any references for the studies he cites and does not mention any possible hazards of ingesting raw honey, except for suggesting that you not buy commercially processed honey because it may be contaminated with questionable ingredients from China.

- What is the **intended message** of the article?
- What is your **critical reading** of the summary of the article? Remember that in this context, "critical reading" does not necessarily mean *What criticism do you have?* but asks you to apply your knowledge to interpret whether the article is factual and whether the facts support the intended message.
- How would you **interpret** the news item for your nonmicrobiologist friends?
- What is your **overall grade** for the news item—taking into account its accuracy and the accuracy of its intended effect?

Media Under The Microscope Wrap-Up appears at the end of the chapter.

Outline and Learning Outcomes

20.1 The Nervous System, Its Defenses, and Normal Biota

1. Describe the important anatomical features of the nervous system.
2. List the natural defenses present in the nervous system.
3. Discuss the current state of knowledge regarding the normal biota of the nervous system.

20.2 Infectious Diseases Manifesting in the Nervous System

4. List the possible causative agents for meningitis and neonatal/infant meningitis.
5. Identify which of the agents causing meningitis is the most common and which is the most deadly.
6. Discuss important features of meningoencephalitis, encephalitis, and subacute encephalitis.
7. Identify which encephalitis-causing viruses you should be aware of in your geographic area.
8. List the possible causative agents for each of the following conditions: Zika virus disease, rabies, poliomyelitis, tetanus, botulism, and African sleeping sickness.
9. Identify the conditions for which vaccination is available.
10. Explain the difference between the oral polio vaccine and the inactivated polio vaccine and the advantages and disadvantages of each.

20.1 The Nervous System, Its Defenses, and Normal Biota

The nervous system can be thought of as having two component parts: the central nervous system (CNS), consisting of the brain and spinal cord, and the peripheral nervous system (PNS), which contains the nerves that emanate from the brain and spinal cord to sense organs and to the periphery of the body **(figure 20.1).** The nervous system performs three important functions—sensory, integrative, and motor. The sensory function is fulfilled by sensory receptors at the ends of peripheral nerves. They generate nerve impulses that are transmitted to the central nervous system. There, the impulses are translated, or integrated, into sensation or thought, which in turn drives the motor function. The motor function necessarily involves structures outside of the nervous system, such as muscles and glands.

The brain and the spinal cord are dense structures made up of cells called **neurons.** They are both surrounded by bone. The brain is situated inside the skull, and the spinal cord lies within the spinal column **(figure 20.2),** which is composed of a stack of interconnected bones called vertebrae. The soft tissue of the brain and spinal cord is encased within a tough casing of three membranes called the **meninges.** The layers of membranes, from superficial to deep, are the dura mater, the arachnoid mater, and the pia mater. Between the arachnoid mater and pia mater is the subarachnoid space (that is, the space under the arachnoid mater). The subarachnoid space is filled with a clear, serumlike fluid called cerebrospinal fluid (CSF). The CSF provides nutrition to the CNS while acting as a liquid cushion for the sensitive brain and spinal cord. The meninges are a common site of infection, and microorganisms can often be found in the CSF when meningeal infection **(meningitis)** occurs.

The PNS consists of nerves and ganglia (see figure 20.1). A ganglion is a swelling in the nerve where the cell bodies of the neurons congregate. Nerves are bundles of neuronal axons that receive and transmit nerve signals. The axons and dendrites of adjacent

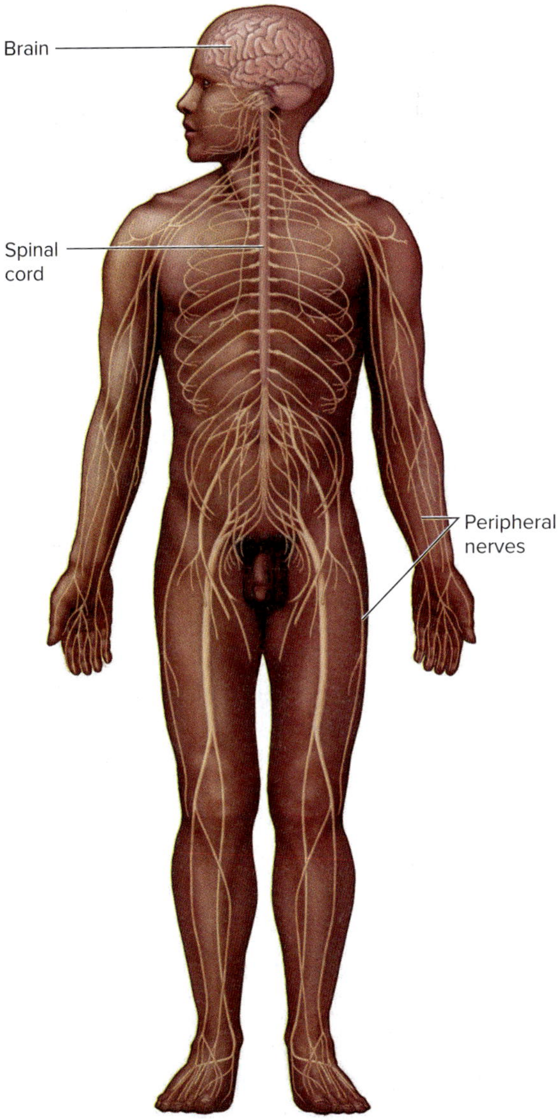
Brain
Spinal cord
Peripheral nerves

Figure 20.1 Nervous system. The central nervous system and the peripheral nerves.

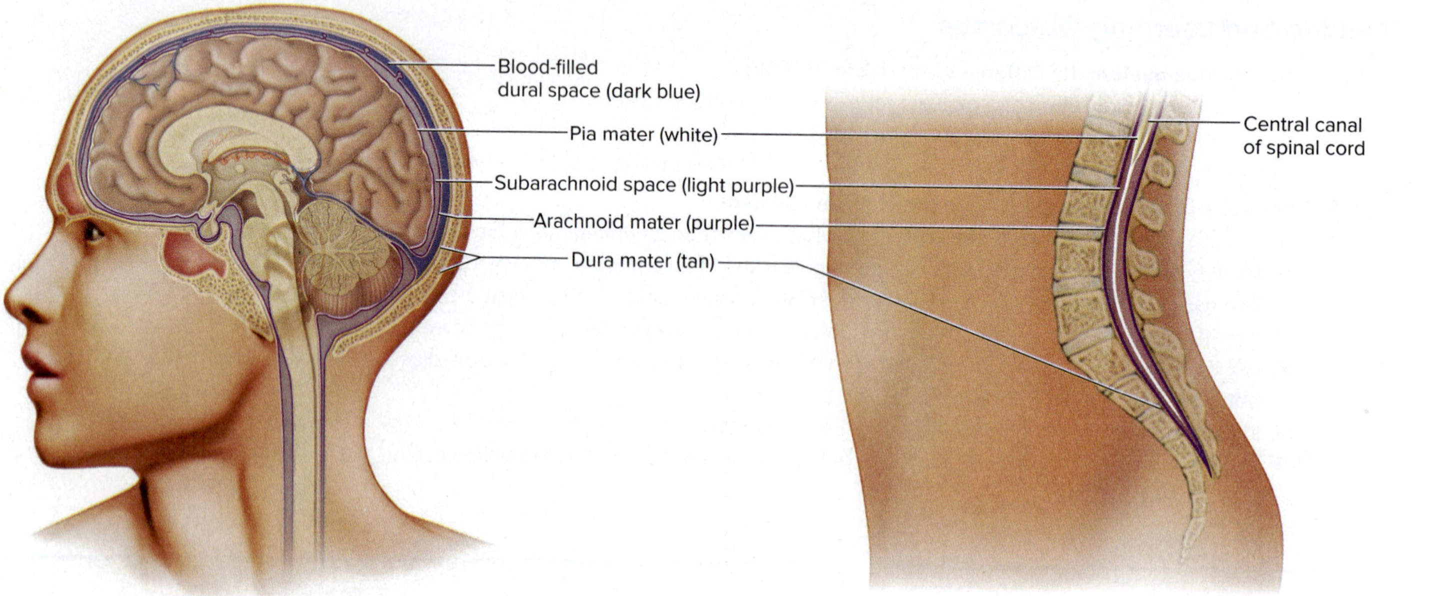
Blood-filled
dural space (dark blue)
Pia mater (white)
Subarachnoid space (light purple)
Arachnoid mater (purple)
Dura mater (tan)
Central canal
of spinal cord

Figure 20.2 Anatomy of the brain and spinal cord.

neurons communicate with each other over a very small space, called a synapse. Chemicals called neurotransmitters are released from one cell and act on the next cell in the synapse.

The defenses of the nervous system are mainly structural. The bony casings of the brain and spinal cord protect them from traumatic injury. The surrounding CSF also serves as a cushion against impact. The entire nervous system is served by the vascular system, but the interface between the blood vessels serving the brain and the brain itself is different from that of other areas of the body and provides a third structural protection. The cells that make up the walls of the blood vessels allow very few molecules to pass through. In other parts of the body, there is freer passage of ions, sugars, and other metabolites through the walls of blood vessels. The restricted permeability of blood vessels in the brain is called the **blood-brain barrier,** and it prohibits most microorganisms from passing into the central nervous system. The drawback of this feature is that drugs and antibiotics can be difficult to introduce into the CNS when needed.

The CNS is considered an "immunologically privileged" site. These sites are able to mount only a partial, or at least a different, immune response when exposed to immunologic challenge. The functions of the CNS are so vital for the life of an organism that even temporary damage that could result from "normal" immune responses would be very detrimental. The uterus and parts of the eye are also immunologically privileged sites. Cells in the CNS express lower levels of MHC antigens. Complement proteins are also in much lower quantities in the CNS. Researchers have established that MHC markers and complement proteins play a role in the development, regulation, and repair of neurons and nervous tissues through their signaling mechanisms. Other specialized cells in the central nervous system perform defensive functions. Microglial cells, part of the brain, exhibit phagocytic activity, which is beneficial in terms of both immunity and brain development. Other macrophages also exist in the CNS, although the activity of both of these types of cells is thought to be less than that of phagocytic cells elsewhere in the body.

It is still believed that the CNS and PNS both lack normal biota of any kind and that finding microorganisms of any type in these tissues represents a deviation from the healthy state. Viruses such as herpes simplex live in a dormant state in the nervous system between episodes of acute disease, but they are not considered normal biota. Emerging microbiome results are revealing a potential link between the gut microbiome and the nervous system. Gut microbiota may actually induce central nervous system autoimmunity and appear to cause changes in brain chemistry and behavior. This phenomenon is known as the gut-brain axis.

20.1 Learning Outcome—Assess Your Progress

1. Describe the important anatomical features of the nervous system.
2. List the natural defenses present in the nervous system.
3. Discuss the current state of knowledge regarding the normal biota of the nervous system.

Nervous System Defenses and Normal Biota

	Defenses	Normal Biota
Nervous System	Bony structures, blood-brain barrier, microglial cells, and macrophages	None

20.2 Infectious Diseases Manifesting in the Nervous System

Before we describe some well-understood diseases of the nervous system, it is worth mentioning that many people with SARS-CoV-2 infections, during the pandemic beginning in 2020, suffered various symptoms that are considered neurological, or part of the nervous system. These include altered smell and taste, headache, confusion, and movement problems. Time will tell how the virus triggered these symptoms and whether they might be chronic in some patients.

Another neurological condition is puzzling scientists. **See Insight 20.1** for information on acute flaccid myelitis (AFM) and how the COVID-19 pandemic has apparently thrown a wrench in its progression.

Meningitis

Meningitis, an inflammation of the meninges, is an excellent example of an anatomical syndrome. Many different microorganisms can cause an infection of the meninges, and they produce a similar constellation of symptoms. Noninfectious causes of meningitis exist as well, but they are much less common than the infections listed here.

The more serious forms of acute meningitis are caused by bacteria, but it is thought that their entrance to the CNS is often facilitated by coinfection or previous infection with respiratory viruses. Meningitis in neonates is most often caused by different microorganisms. For that reason, it is described separately in the following section.

Whenever meningitis is suspected, a lumbar puncture (spinal tap) is performed to obtain CSF, which is then examined by Gram stain and/or culture. Most physicians will begin treatment with a broad-spectrum antibiotic immediately and shift treatment if necessary after a diagnosis has been confirmed.

Signs and Symptoms

No matter the cause, meningitis results in these typical symptoms: photophobia (sensitivity to light), headache, painful or stiff neck, fever, and usually an increased number of white blood cells in the CSF. Many patients have described the headache associated with this disease as the "worst headache I have ever had." Specific microorganisms may cause additional, and sometimes characteristic, symptoms, which are described in the individual sections that follow.

INSIGHT 20.1 MICROBIOME: Mysterious Polio-Like Disease

The CDC is tracking a rare, but disturbing, condition that has been occurring mainly in children. They call it AFM, which stands for acute flaccid myelitis. It attacks the nervous system and causes extreme weakness in arms and legs. It is reminiscent of polio, though no poliovirus has been found in victims, most of whom have been children. Some of the children recover completely, and some have lingering paralysis. There has been an increase in people with this condition since 2014, with 670 confirmed cases since then (see the figure). Do you see how the incidence seems to peak in the fall of every *other* year? That is an interesting pattern. The year 2020's expected peak did not occur. Possibly, the measures taken to limit COVID-19 also limited transmission of this virus. So far, the most likely culprit is a type of enterovirus, called enterovirus D68. While previously scientists wondered if it might have been normal biota gone awry, the decrease in expected cases during the COVID-19 pandemic suggests that it is a transmitted disease since transmission of most childhood microbes was curtailed during that time.

Number of confirmed U.S. AFM cases reported to CDC by month of onset, August 2014–August 2021^*†

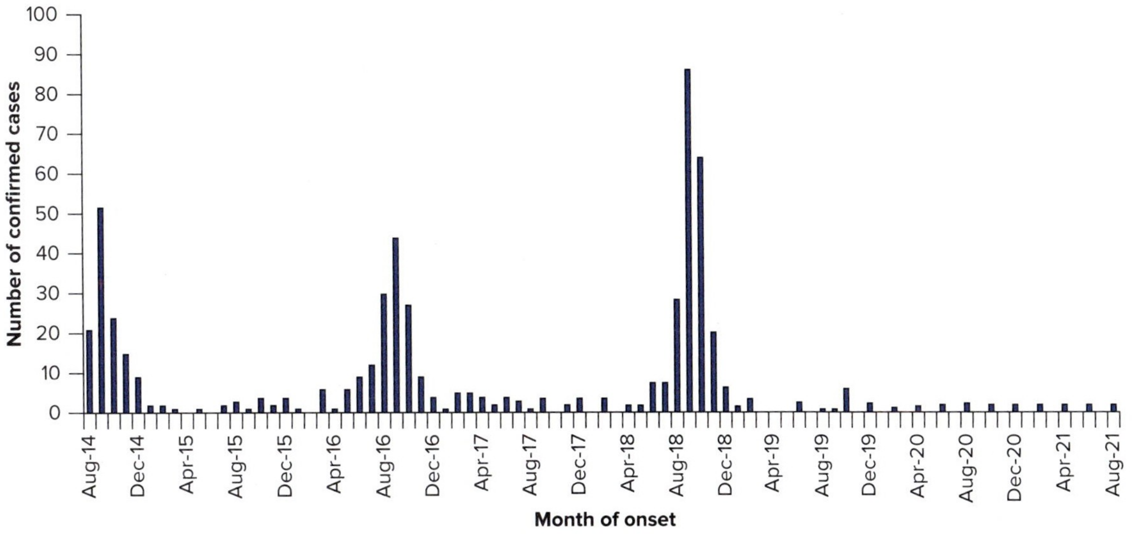
Number of confirmed cases
100
90
80
70
60
50
40
30
20
10
0
Aug-14
Dec-14
Apr-15
Aug-15
Dec-15
Apr-16
Aug-16
Dec-16
Apr-17
Aug-17
Dec-17
Apr-18
Aug-18
Dec-18
Apr-19
Aug-19
Dec-19
Apr-20
Aug-20
Dec-20
Apr-21
Aug-21
Month of onset

^Confirmed AFM cases that CDC has been made aware of as of August 2021. The case counts are subject to change.
*The data shown from August 2014 to July 2015 are based on the AFM investigation case definition: onset of acute limb weakness on or after August 1, 2014, and a magnetic resonance image (MRI) showing a spinal cord lesion largely restricted to gray matter in a patient age ≤21 years.
†The data shown from August 2015 to present are based on the AFM case definition adopted by the Council of State and Territorial Epidemiologists (CSTE): acute onset of focal limb weakness and an MRI showing spinal cord lesion largely restricted to gray matter and spanning one or more spinal segments, regardless of age.

CDC, Centers for Disease Control and Prevention

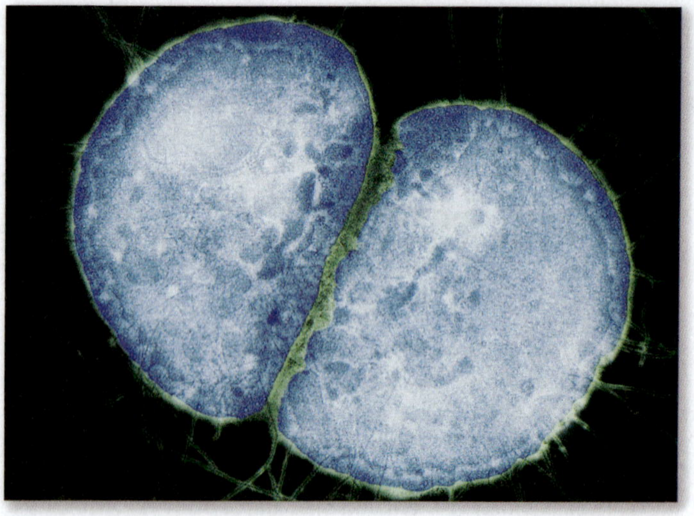

Figure 20.3 **Transmission electron micrograph of *Neisseria* (52,000×).**
Kwangshin Kim/Science Source

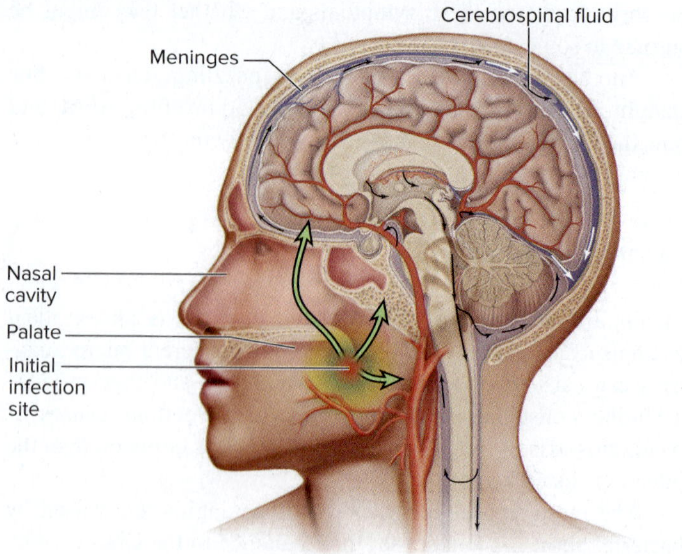
Cerebrospinal fluid
Meninges
Nasal cavity
Palate
Initial infection site

Figure 20.4 **Dissemination of the meningococcus from a nasopharyngeal infection.** Bacteria spread to the roof of the nasal cavity, which borders a highly vascular area at the base of the brain. From this location, they can enter the blood and escape into the cerebrospinal fluid, leading to infection of the meninges.

Like many other infectious diseases, meningitis can manifest as acute or chronic disease. Some microorganisms are more likely to cause acute meningitis, and others are more likely to cause chronic disease.

In a healthy person, it is very difficult for microorganisms to gain access to the nervous system. Those that are successful usually have specific virulence factors.

Neisseria meningitidis

Neisseria meningitidis appears as gram-negative diplococci lined up side by side in microscopic analysis **(figure 20.3)** and is commonly known as the meningococcus. It is often associated with epidemic forms of meningitis. This organism causes the most serious form of acute meningitis, and it is responsible for about 14% of all meningitis cases. Although 12 different strains of capsular antigens exist, serotypes A, B, C, Y, and W135 are responsible for most cases of the disease in the United States. In Africa other serotypes are prominent, and two serotypes (A and W) are associated with the Hajj, an annual pilgrimage made by Muslims to the city of Mecca in Saudi Arabia.

▶Pathogenesis and Virulence Factors

The portal of entry for this pathogen is the upper respiratory tract. The bacterium passes into surrounding blood vessels **(figure 20.4),** rapidly penetrating the meninges and producing symptoms of meningitis. Meningitis is marked by fever, sore throat, headache, stiff neck, convulsions, and vomiting. Severe stomach pain often is the first sign. The most serious complications of meningococcal infection are due to meningococcemia, which can accompany meningitis but can also occur on its own. The pathogen releases endotoxin within the bloodstream, which acts as a potent white blood cell stimulator. Damage to blood vessels caused by cytokines released by the white blood cells leads to vascular collapse, hemorrhage, and crops of red or purple lesions called **petechiae** (pee-tee′-kee-ay) on the trunk and appendages **(figure 20.5).**

Meningococcemia can become an overwhelming disease with a high mortality rate. The disease has a sudden onset, marked by fever higher than 40°C (104°F), chills, delirium, severe widespread ecchymoses (ek″-ih-moh′-seez) (areas of bleeding under the skin larger than petechiae), shock, and coma. Generalized intravascular clotting, cardiac failure, damage to the adrenal glands, and death can occur within a few hours. Recent evidence suggests a genetic role in this form of the disease, as a significant number of patients contain changes in the genes that encode toll-like receptors. These variations reduce the host's ability to initiate an early defensive response to the bacterium. The pathogen has a natural ability to avoid destruction through its production of IgA protease and the presence of a capsule.

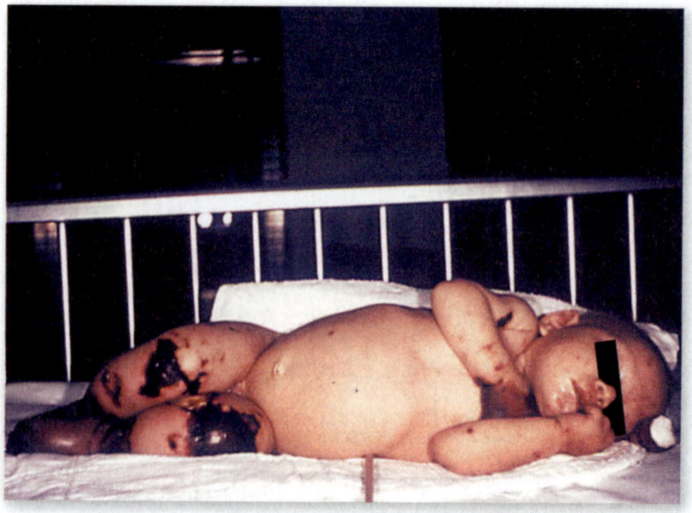

Figure 20.5 **Vascular damage (black areas) associated with meningococcal meningitis.**
Mr Gust/CDC

Transmission and Epidemiology

Because meningococci do not survive long in the environment, these bacteria are usually acquired through close contact with secretions or droplets. Upon reaching their portal of entry in the nasopharynx, the meningococci use attachment pili to adhere to mucosal membranes. In many people, this can result in simple asymptomatic colonization. In the more vulnerable individual, however, the meningococci are engulfed by epithelial cells of the mucosa and penetrate into the nearby blood vessels. Damage to the epithelium causes pharyngitis as the pathogen continues on its way to the meninges.

Meningococcal meningitis has a sporadic or epidemic incidence in late winter or early spring. The continuing reservoir of infection is humans who harbor the pathogen in the nasopharynx. The scene is set for transmission when carriers live in close quarters with nonimmune individuals, as may be expected in families, day care facilities, college dormitories, and military barracks. The carriage state, which can last from a few days to several months, exists in roughly 10% of the adult population. This rate can exceed 50% in institutional settings or endemic regions, and it appears to be influenced by the duration of time living in close quarters. The highest carriage of meningococci is seen in young adults (15 to 24 years old), with decreased rates occurring in young children (less than 4 years old) and individuals over the age of 50.

Every year, in what is called "the meningitis belt" in sub-Saharan Africa, a meningococcal epidemic sweeps through, coinciding with the dry season, which runs from approximately December to May.

Culture and/or Diagnosis

Suspicion of bacterial meningitis constitutes a medical emergency, and differential diagnosis must be done with great haste and accuracy. It is most important to confirm (or rule out) meningococcal meningitis because it can be rapidly fatal. Treatment is usually begun with this bacterium in mind until it can be ruled out. Cerebrospinal fluid, blood, or nasopharyngeal samples are stained and observed directly for the characteristic gram-negative diplococci. Cultivation may be necessary to differentiate the bacterium from other species. Specific rapid tests are also available for detecting the capsular polysaccharide or the cells directly from specimens without culturing.

Immediately after collection, specimens are streaked on Modified Thayer-Martin (MTM) medium or chocolate agar and incubated in a high CO_2 atmosphere. Presumptive identification of the genus is obtained by a Gram stain and oxidase testing on isolated colonies **(figure 20.6).** Further testing may be necessary to differentiate *N. meningitidis* and *N. gonorrhoeae* from one another, from other oxidase-positive species, and from normal biota of the oropharynx that may be mistaken for these pathogens. If no samples were obtained prior to antibiotic treatment, a PCR test is the best bet for identifying the pathogen. Susceptibility testing is also warranted to ensure that proper treatment protocols are used.

Prevention and Treatment

In the United States, immunization begins at the age of 11 with the conjugated MCV4 vaccine (Menveo or Menactra) that is effective against groups A, C, Y, and W135 but not B. A booster dose is needed to ensure protection through adolescence. This vaccine can also be used in young children who are at high risk for infection. At about the time a booster is needed (16 years of age), the CDC recommends additionally the first dose of the new vaccine effective against serotype B, the serotype that has caused several outbreaks on college campuses in recent years.

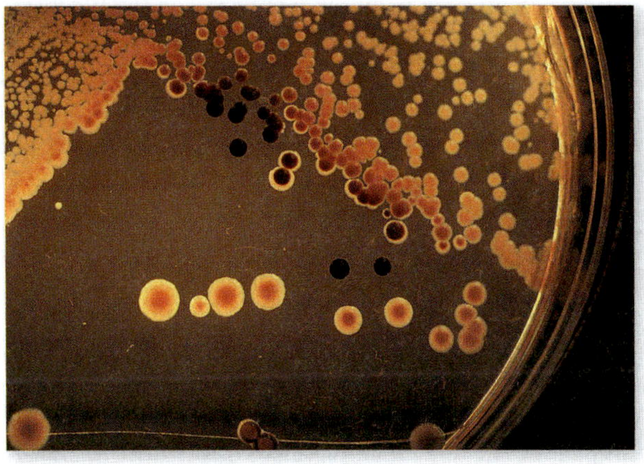

Figure 20.6 The oxidase test. A drop of oxidase reagent is placed on a suspected *Neisseria* or *Branhamella* colony. If the colony reacts with the chemical to produce a purple to black color, it is oxidase-positive; those that remain white to tan are oxidase-negative. Because several species of gram-negative rods are also oxidase-positive, this test is presumptive for these two genera only if a Gram stain has verified the presence of gram-negative cocci.
Kathy Park Talaro

Because even treated meningococcemial disease has a mortality rate of up to 15%, it is vital that antibiotic therapy begin as soon as possible with one or more drugs. Ceftriaxone is the first-line antibiotic for this condition; penicillin, aztreonam, or chloramphenicol may also be used. Patients may also require treatment for shock and intravascular clotting in addition to antibiotic therapy. When family members, medical personnel, or children in day care or school have come in close contact with infected people, preventive therapy with ciprofloxacin, rifampin, or ceftriaxone may be warranted.

Streptococcus pneumoniae

You will see in chapter 22 that *Streptococcus pneumoniae* causes the majority of bacterial pneumonias. (It is also referred to as the **pneumococcus.**) Pneumococcal meningitis is also caused by this bacterium. In fact, this pathogen is the most frequent cause of community-acquired meningitis and often causes a severe form of the disease. It does not cause the petechiae associated with meningococcal meningitis, and that difference is useful diagnostically. Pneumococcal meningitis is most likely to occur in patients with underlying susceptibility, such as patients with sickle-cell disease and those with absent or defective spleen function. Up to 25% of pneumococcal meningitis patients will also develop pneumococcal pneumonia. Pneumococcal infections occur worldwide and today are most prevalent in developing countries. But the bacterium exhibits the potential to be highly pathogenic. It can penetrate the respiratory mucosa; gain access to the bloodstream; and then, under certain conditions, enter the meninges.

The bacterium is a small, gram-positive, flattened coccus that appears in end-to-end pairs. Its appearance is distinctive in a Gram stain of cerebrospinal fluid. Testing of nasopharynx specimens is not useful because it is normal biota in many individuals. Like the meningococcus, this bacterium has a polysaccharide capsule that protects it against phagocytosis. Over 90 serotypes with varying capsular antigenicity have been identified so far. *S. pneumoniae* produces an alpha-hemolysin (observable on blood agar) and hydrogen peroxide, both of which have been shown to induce damage in the CNS, such as inducing brain cell apoptosis.

In pneumococcal meningitis, initial treatment with vancomycin + ceftriaxone is recommended. If the isolate comes back as penicillin-sensitive (the cerebrospinal fluid must be cultured before beginning antibiotic treatment), then treatment can be switched.

Two vaccines are available in the United States for protection against *S. pneumoniae* infection. We have the 13-valent conjugated vaccine (Prevnar 13), which protects children against 13 serotypes of the bacterium. A 23-valent polysaccharide vaccine (Pneumovax) is available for vaccination of adults aged 65 and older as well as at-risk patients. In many cases, older adults are also offered the 13-valent vaccine.

Haemophilus influenzae

Haemophilus influenzae is a gram-negative coccobacillus that causes one of the most severe forms of meningitis in humans. Humans are the only known reservoir, and the portal of entry for this bacterium is the nasopharynx. Asymptomatic carriage rates vary worldwide but have been greatly reduced in the United States due to successful vaccination. Disease caused by *H. influenzae* is often called "Hib" because it is due primarily to infection with the B serotype, though recently serotype A ("Hia") has been seen in the United States and Canada, causing aggressive disease. Routine vaccination with one of two subunit vaccines (both contain capsular polysaccharide conjugated to a protein) is recommended for all children, beginning at age 2 months, with the recommendation of a follow-up booster dose. Combination vaccines containing the Hib conjugate vaccine are also available for use in the current U.S. vaccine schedule. Invasive Hib disease has been virtually eliminated in the United States, a clear victory for successful vaccination programs. Physicians recognize, however, that this situation can rapidly change if vaccine coverage falls and herd immunity is compromised. Cases that do occur in the United States are now mostly caused by non–serotype B strains. Globally, it is still common and is an important cause of the disease in children under the age of 5.

Listeria monocytogenes

Listeria monocytogenes is a gram-positive bacterium **(figure 20.7)** that ranges in morphology from coccobacilli to long filaments in palisade formation. Cells do not produce capsules or endospores and have from one to four flagella. *Listeria* is not fastidious and is resistant to cold, heat, salt, pH extremes, and bile. It grows inside host cells and can move directly from an infected host cell to an adjacent healthy cell.

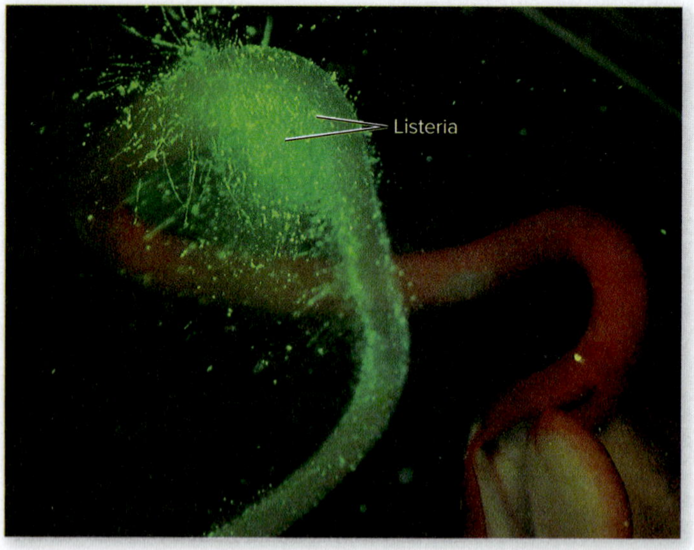
Listeria

Figure 20.7 *Listeria monocytogenes* on a radish shoot. The shoot was stained with a fluorescent dye that illustrates the root hairs (thinner threads) and the bacterium.

Photo by Lisa Gorski, USDA-ARS

Listeriosis in healthy adults is often a mild or subclinical infection with nonspecific symptoms of fever, diarrhea, and sore throat. However, listeriosis in elderly or immunocompromised patients, fetuses, and neonates usually affects the brain and meninges and results in septicemia. Some strains can target the heart. The death rate is around 20%. Pregnant women are especially susceptible to infection, which can be transmitted to the infant prenatally when the microbe crosses the placenta or postnatally through the birth canal. Intrauterine infections are systemic and usually result in premature abortion and fetal death.

L. monocytogenes has been isolated all over the world from water, soil, plant materials, and the intestines of healthy mammals (including humans), birds, fish, and invertebrates. Apparently, the primary reservoirs are soil and water. Animals, plants, and food are secondary sources of infection. Most cases of listeriosis are associated with ingesting contaminated dairy products, poultry, and meat. At almost any given time, the CDC website reports a current outbreak of foodborne disease with this pathogen.

In the summer of 2022, a multistate *Listeria* outbreak linked to one particular brand of ice cream occurred. At least one person died, and at least one pregnancy was lost as a result of this bacterial contamination.

Diagnosing listeriosis is hampered by the difficulty in isolating the microbe. The chances of isolation, however, can be improved by using a procedure called *cold enrichment,* in which the specimen is held at 4°C and periodically plated onto media, but this procedure can take 4 weeks. Rapid diagnostic kits using ELISA, immunofluorescence, and nucleic acid sequencing technologies are also available for direct testing of dairy products and cultures. Antibiotic therapy should be started as soon as listeriosis is suspected. Ampicillin and gentamicin are the first choices. Prevention can be improved by adequate pasteurization temperatures and by proper washing, refrigeration, and cooking of foods that are suspected of being contaminated with animal manure or sewage.

Cryptococcus neoformans

The fungus *Cryptococcus neoformans* causes a more chronic form of meningitis with a more gradual onset of symptoms, although in immunocompromised people the onset may be fast and the course of the disease more acute. It is sometimes classified as a meningoencephalitis (inflammation of both brain and meninges). Headache is the most common symptom, but nausea and neck stiffness are also very common. This fungus is a widespread resident of human habitats. It has a spherical to ovoid shape, with small, constricted buds and a large capsule that is important in its pathogenesis **(figure 20.8).**

▶ Transmission and Epidemiology

The primary ecological niche of *C. neoformans* is the bird population. It is prevalent in urban areas where pigeons congregate, and it proliferates in the high-nitrogen environment of droppings that accumulate on pigeon roosts. Masses of dried yeast cells are readily scattered into the air and dust. Its role as an opportunist is supported by evidence that healthy humans have strong resistance to it and that clinically obvious infection occurs primarily in debilitated patients. Most cryptococcal infections cause symptoms in the respiratory and central nervous systems.

This meningitis has a high fatality rate. Conditions that predispose individuals to infection are steroid treatment, diabetes, AIDS, and cancer. It is not considered communicable among humans.

The primary portal of entry for *C. neoformans* is the respiratory tract, but most lung infections are subclinical and rapidly resolved.

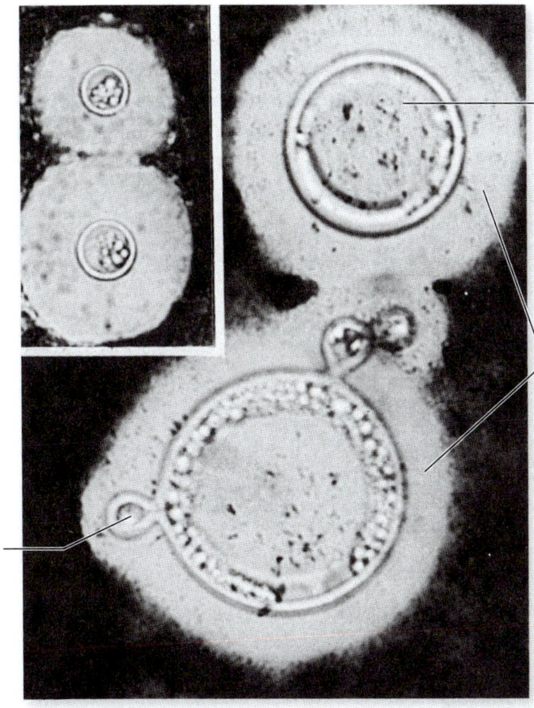
Cell body
Capsules
Bud

Figure 20.8 ***Cryptococcus neoformans*** **from infected spinal fluid stained negatively with India ink.** Halos around the large, spherical yeast cells are thick capsules. Also note the buds forming on one cell. Encapsulation is a useful diagnostic sign for cryptococcosis, although the capsule is fragile and may not show up in some preparations (150×).

Gordon Love, M.D. VA, North CA Healthcare System, Martinez, CA

▶ Pathogenesis and Virulence Factors

Cryptococcus shows an extreme affinity for the meninges and brain. The tumorlike masses formed in these locations can cause headache, mental changes, coma, paralysis, eye disturbances, and seizures. In some cases, the infection disseminates into the skin, bones, and viscera.

▶ Culture and/or Diagnosis

Cryptococcosis can be diagnosed via negative staining of specimens to detect encapsulated budding yeast cells that do not occur as pseudohyphae. Rapid tests, such as the cryptococcal antigen test, are also useful for disease diagnosis. Confirmatory results include a negative nitrate assimilation, pigmentation on birdseed agar, fluorescent antibody testing, and nucleic acid analysis.

▶ Prevention and Treatment

Systemic cryptococcosis requires immediate treatment with amphotericin B, flucytosine, and fluconazole over a period of weeks or months. There is no prevention.

Coccidioides

This fungus causes a condition that is often called "Valley Fever" in the American Southwest. The morphology of *Coccidioides* is very distinctive. At 25°C, it forms a moist white to brown colony with abundant, branching, septate hyphae. These hyphae fragment into thick-walled, blocklike **arthroconidia** (arthrospores) at maturity **(figure 20.9*a*).** On special media incubated at 37°C to 40°C, an arthrospore germinates into the parasitic phase, a small, spherical cell called a spherule **(figure 20.9*b*)** that can be found in infected tissues as well. This structure swells into a giant sporangium that cleaves internally to form numerous endospores that look like bacterial endospores but are not highly resistant to the environment. Remember, even though this organism produces endospores, it is not a bacterium.

▶ Pathogenesis and Virulence Factors

This is a true systemic fungal infection of high virulence, as opposed to an opportunistic infection. It usually begins with pulmonary infection but can disseminate quickly throughout the body. Coccidioidomycosis of the meninges is the most serious manifestation. All persons inhaling the arthrospores probably develop some degree of infection, but certain groups have a genetic susceptibility that gives rise to more serious disease. After the arthrospores are inhaled, they develop into spherules in the lungs. These spherules release scores of endospores

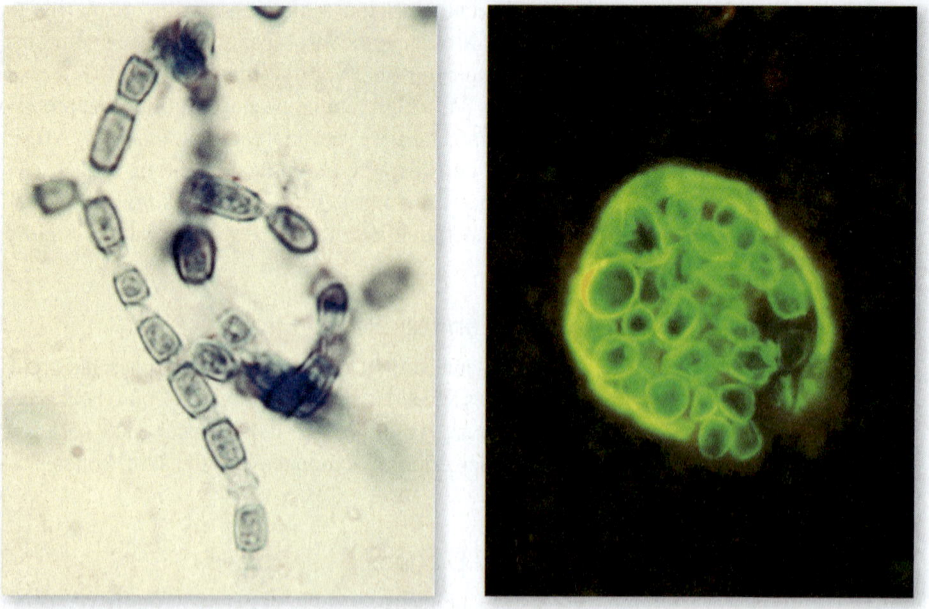

(a) Arthrospores (b) Spherules containing endospores

Figure 20.9 Two phases of *Coccidioides* infection. **(a)** Arthrospores are present in the environment and are inhaled. **(b)** In the lungs, the brain, or other tissues, arthrospores develop into spherules filled with endospores. Endospores are released and induce damage.
(a) Dr. Hardin/CDC; (b) CDC/Science Source

into the lungs. At this point, either the patient experiences mild respiratory symptoms, which resolve themselves, or the endospores lead to the development of disseminated disease. Disseminated disease can include meningitis, osteomyelitis, and skin granulomas.

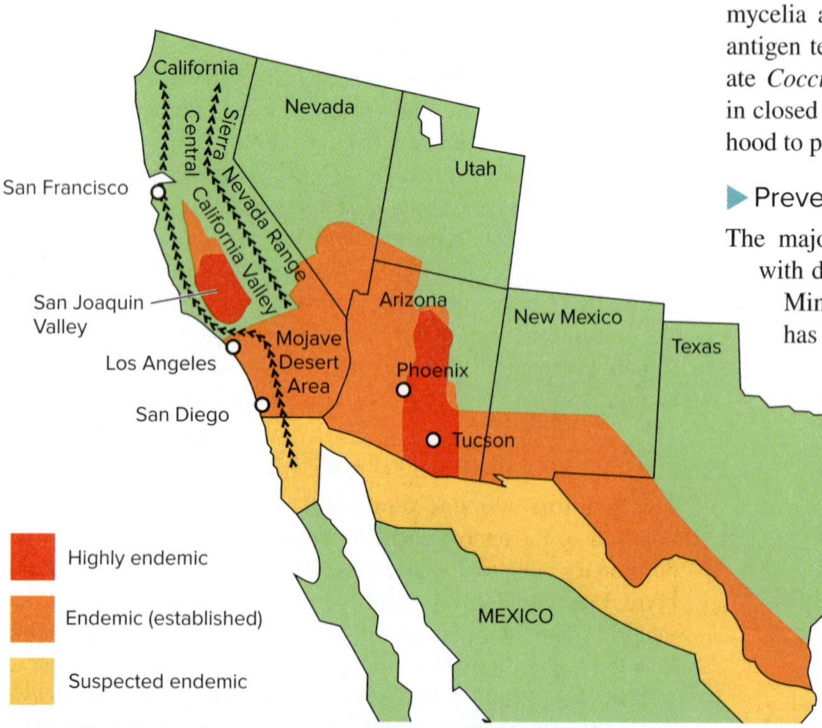
California
Nevada
Utah
Sierra Nevada Range
Central California Valley
San Francisco
San Joaquin Valley
Los Angeles
San Diego
Mojave Desert Area
Arizona
Phoenix
Tucson
New Mexico
Texas
MEXICO
Highly endemic
Endemic (established)
Suspected endemic

Figure 20.10 Areas in the United States endemic for *Coccidioides*.

▶ Transmission and Epidemiology

Coccidioides occurs endemically in various natural reservoirs and casually in areas where it has been carried by wind and animals. Conditions favoring its settlement include high carbon and salt content and a semiarid, relatively hot climate. The fungus has been isolated from soils, plants, and a large number of vertebrates. The natural history of the fungus follows a cyclic pattern—a period of dormancy in winter and spring, followed by growth in summer and fall. Growth and spread are greatly increased by cycles of drought and heavy rains.

There are two species that cause this disease, found in different locations. *C. immitis* causes disease in California. *C. posadasii* is found in northern Mexico, Central and South America, and the American Southwest, especially Arizona. Sixty percent of all U.S. infections occur in Arizona **(figure 20.10).** In the past 2 or 3 years, cases have popped up in Utah and Washington state, and warming temperatures seem to be the cause. Outbreaks are usually associated with farming activity, archaeological digs, construction, and mining.

▶ Culture and/or Diagnosis

Diagnosis of coccidioidomycosis is straightforward when the highly distinctive spherules are found in sputum, spinal fluid, and biopsies. This finding is further supported by isolation of typical mycelia and arthrospores on Sabouraud's agar. Newer specific antigen tests have been effective tools to identify and differentiate *Coccidioides* from other fungi. All cultures must be grown in closed tubes or bottles and opened in a biological containment hood to prevent laboratory infections.

▶ Prevention and Treatment

The majority of patients do not require treatment. In people with disseminated disease, however, oral fluconazole is used. Minimizing contact with the fungus in its natural habitat has been of some value. For example, oiling dirt roads and planting vegetation help reduce spore aerosols, and using dust masks while excavating soil prevents workers from inhaling as many spores.

Viruses

It is estimated that four out of five meningitis cases are caused by one of a wide variety of viruses. Because no bacteria or fungi are found in the CSF in viral meningitis, the condition is often called *aseptic meningitis.* Aseptic (nonbacterial, nonfungal, nonprotozoan) meningitis may also have noninfectious causes.

Note About Differential Diagnosis

Meningitis is a disease that illustrates how important it is for a clinician to think epidemiologically. When encountering a set of symptoms such as those of meningitis, a health care worker should be able to access a differential diagnosis in his/her mind. If the patient lives in Arizona, you definitely want to include *Coccidioides* in your differential diagnosis. Knowing the patient's vaccination history can help rule out some causative agents. And finally, if you look at our table of possible causative agents, you see that 80% of meningitis cases are caused by unidentified viruses and are mild. Here's where you see that simply knowing a list of possible causative agents is not enough. A little more information will tell you how likely each of them is. **Each Disease Table in this book serves to summarize the differential diagnosis.**

There's a saying among physicians: "When you hear hoofbeats, think horses, not zebras." This refers to the fact that you should consider the simplest possibility, which in the case of meningitis would be a virus, resulting in uncomplicated disease. But if the patient is very sick upon presentation or has signs of petechiae on his or her skin, you err on the side of safety and treat for *Neisseria* until it is ruled out.

By far the majority of cases of viral meningitis occur in children, and 90% are caused by enteroviruses. But many other viruses also gain access to the central nervous system on occasion. An initial infection with herpes simplex type 2 is sometimes known to cause meningitis. Other herpesviruses, such as HHV-6, HHV-7, and HHV-3 (the chickenpox virus), can infect the meninges as well. Research shows that herpesviruses can take over neuronal cells and use them as highways to travel throughout the central nervous system. Arboviruses, arenaviruses, and adenoviruses have also been identified as causative agents of meningitis. It is also recognized that HIV infection may manifest as meningitis even when the virus is controlled in the rest of the body.

Viral meningitis is generally milder than bacterial or fungal meningitis, and it is usually resolved within 2 weeks. The mortality rate is less than 1%. Diagnosis begins with the failure to find bacteria, fungi, or protozoa in CSF and can be confirmed, depending on the virus, by viral culture or specific antigen tests. In most cases, no treatment is indicated. Acyclovir can be used when the causative agent is a herpesvirus. **Disease Table 20.1** summarizes the agents causing meningitis.

Neonatal and Infant Meningitis

Meningitis in neonates is usually a result of infection transmitted by the mother, either *in utero* or (more frequently) during passage through the birth canal. (Infections caused by *Cronobacter* are the exception, as discussed soon.) As more premature babies survive, the rates of neonatal meningitis increase because the condition is more likely in patients with immature immune systems. Although morbidity has increased, mortality rates have significantly declined. In the United States, the two most common causes are *Streptococcus agalactiae* (known as Group B *Streptococcus*) and *Escherichia coli*. *Listeria monocytogenes* is also found frequently in neonates. It has already been covered here but is included in **Disease Table 20.2** as a reminder that it can cause neonatal cases as well. In the developing world, neonatal meningitis is more commonly caused by other organisms.

Streptococcus agalactiae

This species of *Streptococcus* belongs to group B of the streptococci. It colonizes 10% to 30% of female genital tracts and is the most frequent cause of neonatal meningitis. The treatment for neonatal disease is intravenous penicillin G, sometimes supplemented with an aminoglycoside.

Escherichia coli

The K1 strain of *Escherichia coli* is the second most common cause of neonatal meningitis. Most babies who suffer from this infection are premature, and their prognosis is poor. Twenty percent of them die, even with aggressive antibiotic treatment, and those who survive often have brain damage.

The bacterium is usually transmitted from the mother's birth canal. It causes no disease in the mothers but can infect the vulnerable tissues of a neonate. It seems to have a preference for the tissues of the central nervous system. Ceftazidime or cefepime +/– gentamicin is usually administered intravenously (Disease Table 20.2).

Cronobacter sakazakii

Cronobacter, formerly known as *Enterobacter sakazakii,* is a gram-negative bacillus. Found mainly in the environment, it can survive very dry conditions. It has been implicated in outbreaks of neonatal and infant meningitis transmitted through contaminated powdered infant formula. Although cases of *Cronobacter* meningitis are rare, mortality rates can reach 40%. The FDA and the CDC advise hospitals to use ready-to-feed or concentrated liquid formulas and that home caregivers make fresh formula for each feeding and discard any leftover formula. In 2022, the United States experienced a severe shortage of baby formula, because a major manufacturer was temporarily shut down due to violations of the strict *Cronobacter* inspection guidelines. **(Disease Table 20.2).**

Meningoencephalitis

Up to this point, we have described microorganisms causing meningitis (inflammation of the meninges). Next we discuss microorganisms that cause **encephalitis,** inflammation of the brain. Because the brain and the spinal cord (and the meninges) are so closely connected, infections of one of these structures may also involve the other.

Two microorganisms cause a distinct disease called *meningoencephalitis,* and they are both amoebas. *Naegleria fowleri* and *Acanthamoeba* are protozoans considered to be accidental parasites that invade the body only under unusual circumstances.

Disease Table 20.1 Meningitis

Causative Organism(s)	*Neisseria meningitidis* B G−	*Streptococcus pneumoniae* B G+	*Haemophilus influenzae* B G−
Most Common Modes of Transmission	Droplet contact	Droplet contact	Droplet contact
Virulence Factors	Capsule, endotoxin, IgA protease	Capsule, induction of apoptosis, hemolysin and hydrogen peroxide production	Capsule
Culture/Diagnosis	Gram stain/culture of CSF, blood, rapid antigenic tests, oxidase test, PCR	Gram stain/culture of CSF	Culture on chocolate agar
Prevention	Conjugated vaccine; ciprofloxacin, rifampin, or ceftriaxone used to protect contacts	Two vaccines: Prevnar (children and adults) and Pneumovax (adults)	Hib vaccine, ciprofloxacin, rifampin, or ceftriaxone
Treatment	Ceftriaxone, penicillin	Vancomycin + ceftriaxone or cefotaxime; resistant *S. pneumoniae* is categorized by the CDC as a **Serious Threat**	Ceftriaxone
Distinctive Features	Petechiae, meningococcemia, rapid decline	Serious, acute, most common meningitis in adults	Serious, acute, less common since vaccine became available
Epidemiological Features	United States: 14% of bacterial meningitis; increasing in MSM (men having sex with men); in meningitis belt: 1,000 cases per 100,000 annually	Most common cause of bacterial meningitis in United States (58% of cases)	Hia now becoming common in North America; before Hib vaccine, 300,000–400,000 deaths worldwide per year from b serotype

Disease Table 20.2 Neonatal and Infant Meningitis

Causative Organism(s)	*Streptococcus agalactiae* B G+	*Escherichia coli,* strain K1 B G−	*Listeria monocytogenes* B G+	*Cronobacter sakazakii* B G−
Most Common Modes of Transmission	Vertical (during birth)	Vertical (during birth)	Vertical	Vehicle (baby formula)
Virulence Factors	Capsule	–	Intracellular growth	Ability to survive dry conditions
Culture/Diagnosis	Culture mother's genital tract on blood agar; CSF culture of neonate	CSF Gram stain/ culture	Cold enrichment, rapid methods	Chromogenic differential agar or rapid detection kits
Prevention	Culture and treatment of mother	–	Cooking food, avoiding unpasteurized dairy products	Safe preparation and use of, or avoidance of, powdered formula
Treatment	Penicillin G plus aminoglycosides	Ceftazidime or cefepime +/– gentamicin	Ampicillin, trimethoprim-sulfamethoxazole	Begin with broad-spectrum drugs until susceptibilities determined
Distinctive Features	Most common; positive culture of mother confirms diagnosis	Suspected if infant is premature		–
Epidemiological Features	Before intrapartum antibiotics introduced in 1996: 1.8 cases per 1,000 live births; after intrapartum antibiotics: 0.32 case per 1,000 live births	Estimated at 0.2–5 per 1,000 live births; 20% of pregnant women colonized	Mortality as high as 33% in symptomatic cases	Rare (a handful of documented cases in United States annually) but deadly

Listeria monocytogenes B G+	*Cryptococcus neoformans* F	*Coccidioides* F	Viruses V
Vehicle (food)	Vehicle (air, dust)	Vehicle (air, dust, soil)	Droplet contact
Intracellular growth	Capsule, melanin production	Granuloma (spherule) formation	Lytic infection of host cells
Cold enrichment, rapid methods	Negative staining, biochemical tests, DNA probes, cryptococcal antigen test	Identification of spherules, cultivation on Sabouraud's agar	Initially, absence of bacteria/fungi/protozoa, followed by viral culture or antigen tests
Cooking food, avoiding unpasteurized dairy products	–	Avoiding airborne endospores	–
Ampicillin + gentamicin	Amphotericin B and flucytosine, followed by fluconazole	Fluconazole or ampthotericin B	Usually none (unless specific virus identified and specific antiviral exists)
Asymptomatic in healthy adults; meningitis in neonates, elderly, and immunocompromised	Acute or chronic, most common in AIDS patients	Almost exclusively in endemic regions	Generally milder than bacterial or fungal
Mortality as high as 33% in symptomatic cases	In United States, mainly a concern for HIV+ patients. 90% drop in incidence in the 1990s due to better management of AIDS; worldwide: 1 million new cases per year	Incidence in endemic areas: 200–300 annually	In the United States, 4 of 5 meningitis cases caused by viruses; 26,000–42,000 hospitalizations/year

Naegleria fowleri

The trophozoite of *Naegleria* is a small, flask-shaped amoeba that moves by means of a single, broad pseudopod. It can form a rounded, thick-walled cyst that is resistant to temperature extremes and mild chlorination.

Most cases of *Naegleria* infection reported worldwide occur in people who have been swimming in warm, natural bodies of freshwater. Infection can begin when amoebas are forced into human nasal passages as a result of swimming, diving, or other aquatic activities. The amoeba infects the olfactory epithelium and utilizes the olfactory nerve to travel to the brain. Infection then spreads as the pathogen enters the fluid-filled subarachnoid space, making diagnosis from CSF possible. The result is primary amoebic meningoencephalitis (PAM), a rapid, massive destruction of brain and spinal tissue that causes hemorrhage and coma and nearly always ends in death within a week of onset.

Although cases of disease are extremely rare, *Naegleria* meningoencephalitis advances so rapidly that treatment usually proves futile. Between 2011 and 2021, 29 people were infected by swimming, one person was infected by contaminated water used on a slip-n-slide, and three were infected using neti pots with nonsterile water. Note that in all cases, forcing water into the nose was the mode of transmission. Studies have indicated that early therapy with amphotericin B, sulfadiazine, or tetracycline in some combination can be of some benefit. Because of the wide distribution of the amoeba in nature and its hardiness, no general means of control exists. Public swimming pools and baths must be adequately chlorinated and checked periodically for the amoeba.

Acanthamoeba

The protozoan *Acanthamoeba* is characterized by a large, amoeboid trophozoite with spiny pseudopods and a double-walled cyst. It differs from *Naegleria* in its portal of entry; it invades broken skin, the conjunctiva, and occasionally the lungs and urogenital epithelia. Although it causes a meningoencephalitis somewhat similar to that of *Naegleria,* the course of infection is lengthier. The disease is called granulomatous amoebic meningoencephalitis (GAM) and has only a 2% to 3% survival rate. We discussed ocular infections caused by this pathogen in chapter 19. **(Disease Table 20.3).**

Acute Encephalitis

Encephalitis can present as acute or **subacute.** It is always a serious condition, as the tissues of the brain are extremely sensitive to damage by inflammatory processes. Acute encephalitis is almost always caused by viral infection. One category of viral encephalitis is caused by viruses borne by insects (arboviruses), including West Nile virus. Alternatively, other viruses, such as Nipah virus and members of the herpes family, are causative agents. Bacteria such as those covered under the meningitis section can also cause encephalitis, but the symptoms are usually more pronounced in the meninges than in the brain.

The signs and symptoms of encephalitis vary, but they may include behavior changes or confusion because of inflammation. Decreased consciousness and seizures frequently occur. Symptoms of meningitis are often also present. Few of these agents have specific treatments, but because swift initiation of acyclovir

Disease Table 20.3 Meningoencephalitis

	Primary Amoebic Meningoencephalitis (P)	Granulomatous Amoebic Meningoencephalitis (P)
Causative Organism(s)	*Naegleria fowleri*	*Acanthamoeba*
Most Common Modes of Transmission	Vehicle (exposure while swimming in water)	Direct contact
Virulence Factors	Invasiveness	Invasiveness
Culture/Diagnosis	Examination of CSF; brain imaging, biopsy	Examination of CSF; brain imaging, biopsy
Prevention	Limit warm freshwater or untreated tap water entering nasal passages	–
Treatment	Pentamidine, sulfadiazine	Surgical excision of granulomas; pentamidine
Epidemiological Features	United States: 0–7 cases a year; 97% case-fatality rate; spreading to northern states as climate warms	Predominantly occurs in immunocompromised patients

therapy can save the life of a patient suffering from herpesvirus encephalitis, most physicians will begin empiric therapy with acyclovir in all seriously ill neonates and most other patients showing evidence of encephalitis. That treatment will, in any case, do no harm in patients who are infected with other agents.

Arthropod-Borne Viruses (Arboviruses)

Most arthropods that serve as infectious disease vectors feed on the blood of hosts. Infections show a peak incidence when the arthropod is actively feeding and reproducing, usually from late spring through early fall in temperate zones. Warm-blooded vertebrates also maintain the virus during the cold and dry seasons. Humans can serve as dead-end, accidental hosts, as in equine encephalitis, or they can be a maintenance reservoir, as in yellow fever.

Arboviral diseases have a great impact on humans. Although exact statistics are unavailable, it is believed that millions of people acquire infections each year and thousands of them die. One common outcome of arboviral infection is an acute fever, often accompanied by rash. Viruses that primarily cause these symptoms are covered in chapter 21.

The arboviruses discussed in this chapter can cause encephalitis, and we consider them as a group because the symptoms and management are similar. The transmission and epidemiology of individual viruses are different, however, and are discussed for each virus in **figure 20.11.**

▶ Pathogenesis and Virulence Factors

Arboviral encephalitis begins with an arthropod bite, the release of the virus into tissues, and its replication in nearby lymphatic tissues. Prolonged viremia establishes the virus in the brain, where inflammation can cause swelling and damage to the brain, nerves, and meninges. Symptoms are extremely variable and can include coma, convulsions, paralysis, tremor, loss of coordination, memory deficits, changes in speech and personality, and heart disorders. In some cases, survivors experience some degree of permanent brain damage. Young children and the elderly are most sensitive to injury by arboviral encephalitis. Having said that, most people who are infected will actually show no symptoms.

▶ Culture and/or Diagnosis

Except during epidemics, detecting arboviral infections can be difficult. The patient's history of travel to endemic areas or contact with vectors, along with serum analysis, will help with diagnosis. Rapid serological and nucleic acid amplification tests are available for some of the viruses.

▶ Treatment

No satisfactory treatment exists for any of the arboviral encephalitides (plural of *encephalitis*). As mentioned earlier, empiric acyclovir treatment may be started in case the infection is actually caused by either herpes simplex virus or varicella zoster. Treatment of the other infections relies entirely on support measures to control fever, convulsions, dehydration, shock, and edema.

Most of the control safeguards for arbovirus disease are aimed at the arthropod vectors. Mosquito abatement by eliminating breeding sites and by broadcast-spreading insecticides has been highly effective in restricted urban settings. Birds play a role as reservoirs of the virus, but direct transmission between birds and humans does not occur.

Herpes Simplex Virus

Herpes simplex type I and II viruses can cause encephalitis in newborns born to HSV-positive mothers. In this case, the virus is disseminated and the prognosis is poor. Older children and young adults (ages 5 to 30), as well as older adults (over 50 years old), are also susceptible to herpes simplex encephalitis, caused most commonly by HSV-I. In these cases, the HSV encephalitis usually represents a reactivation of dormant HSV from the trigeminal ganglion.

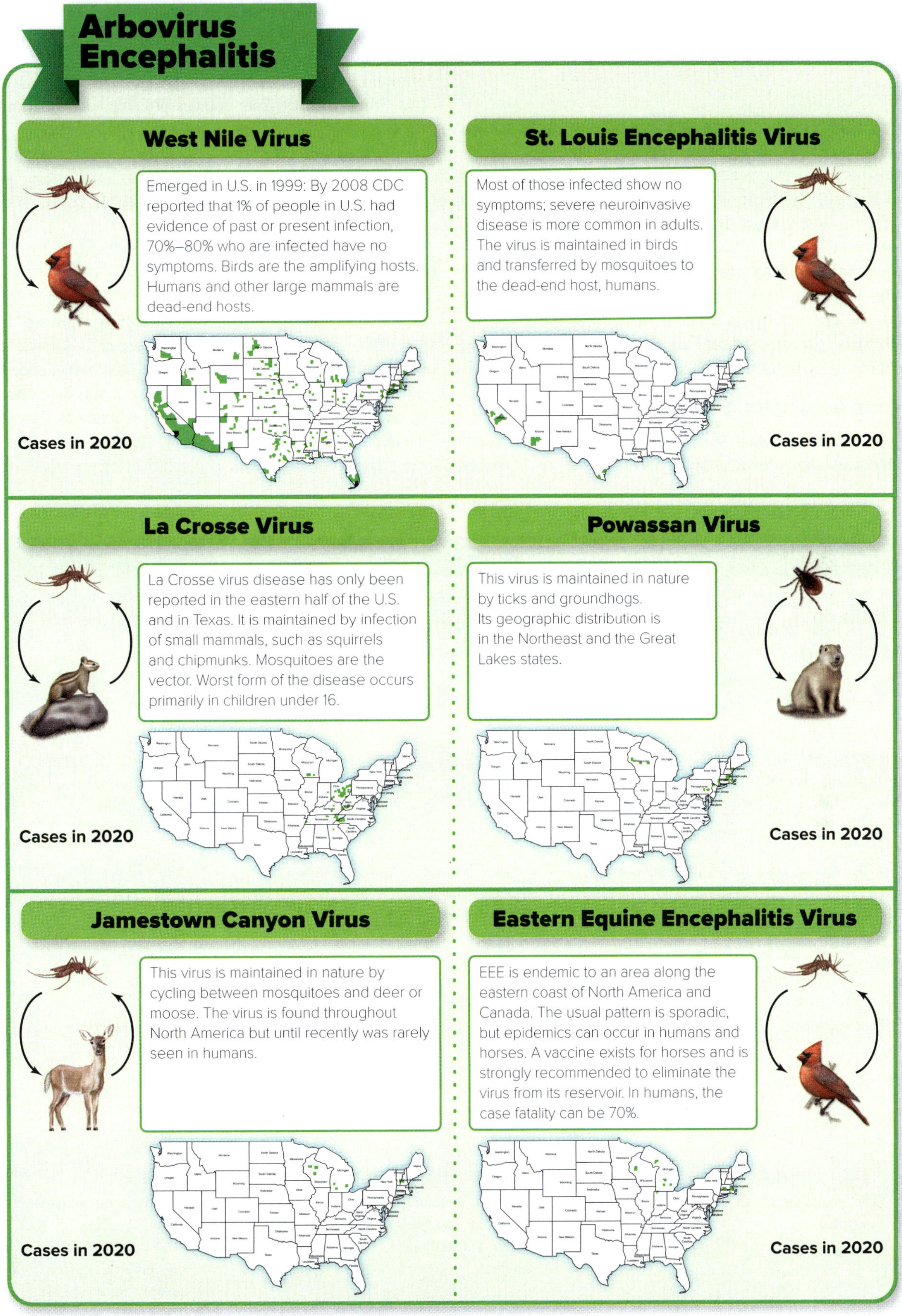
Arbovirus Encephalitis
West Nile Virus
Emerged in U.S. in 1999: By 2008 CDC reported that 1% of people in U.S. had evidence of past or present infection, 70%–80% who are infected have no symptoms. Birds are the amplifying hosts. Humans and other large mammals are dead-end hosts.
Cases in 2020
St. Louis Encephalitis Virus
Most of those infected show no symptoms; severe neuroinvasive disease is more common in adults. The virus is maintained in birds and transferred by mosquitoes to the dead-end host, humans.
Cases in 2020
La Crosse Virus
La Crosse virus disease has only been reported in the eastern half of the U.S. and in Texas. It is maintained by infection of small mammals, such as squirrels and chipmunks. Mosquitoes are the vector. Worst form of the disease occurs primarily in children under 16.
Cases in 2020
Powassan Virus
This virus is maintained in nature by ticks and groundhogs. Its geographic distribution is in the Northeast and the Great Lakes states.
Cases in 2020
Jamestown Canyon Virus
This virus is maintained in nature by cycling between mosquitoes and deer or moose. The virus is found throughout North America but until recently was rarely seen in humans.
Cases in 2020
Eastern Equine Encephalitis Virus
EEE is endemic to an area along the eastern coast of North America and Canada. The usual pattern is sporadic, but epidemics can occur in humans and horses. A vaccine exists for horses and is strongly recommended to eliminate the virus from its reservoir. In humans, the case fatality can be 70%.
Cases in 2020

NOTE: Alaska and Puerto Rico had no cases of any of these diseases in 2020.

Figure 20.11 The incidence and geographic distribution of arbovirus encephalitis.

CDC, Centers for Disease Control and Prevention

It should be noted the varicella-zoster virus can also reactivate from the dormant state, and it is responsible for rare cases of encephalitis.

JC Virus

The **JC virus (JCV)** gets its name from the initials of the patient in whom it was first diagnosed as the cause of illness. Seroprevalence of this polyomavirus nears 80% in many parts of the United States and Europe, though most infections are asymptomatic. In patients with immune dysfunction, especially in those with AIDS, this pathogen can cause a condition called **progressive multifocal leukoencephalopathy** (loo″-koh-en-sef″-uh-lop′-uh-thee) **(PML).** This uncommon but generally fatal infection is a result of JC virus attack of accessory brain cells. The infection demyelinizes certain parts of the cerebrum. This virus should be considered when encephalitis symptoms are observed in patients with AIDS.

Other Virus-Associated Encephalitides

Infection with measles and other childhood rash diseases can result 1 to 2 weeks later in an inappropriate immune response with consequences in the CNS. The condition is called postinfectious encephalitis (PIE), and it is thought to be a result of immune system action and not of direct viral invasion of neural tissue. Very rarely, PIE can occur after immunization with vaccines comprised of live attenuated virus. Note that PIE is distinct from another possible sequela of measles virus infection called SSPE (discussed later in this chapter).

Rabies and Zika should also be considered in the differential diagnosis. But their pathogenesis is so unique they get their own section later in this chapter **(Disease Table 20.4).**

Subacute Encephalitis

When encephalitis symptoms take longer to show up and when the symptoms are less striking, the condition is known as **subacute encephalitis.** The most common cause of subacute encephalitis is the protozoan *Toxoplasma*. A second form of subacute encephalitis can be caused by persistent measles virus as many as 7 to 15 years after the initial infection. Third, a class of infectious agents known as prions can cause a condition called spongiform encephalopathy. Finally, the

Disease Table 20.4 Acute Encephalitis

Causative Organism(s)	Arboviruses (West Nile virus, La Crosse virus, Jamestown Canyon virus, St. Louis encephalitis virus, Powassan virus, Eastern equine encephalitis virus) Ⓥ	Herpes simplex 1 or 2 Ⓥ	JC virus Ⓥ	Postinfection Encephalitis
Most Common Modes of Transmission	Vector (arthropod bites)	Vertical or reactivation of latent infection	Ubiquitous	Sequelae of measles, other viral infections, and, occasionally, vaccination
Virulence Factors	Attachment, fusion, invasion capabilities	–	–	–
Culture/ Diagnosis	History, rapid serological tests, nucleic acid amplification tests	Clinical presentation, PCR, Ab tests, growth of virus in cell culture	PCR of cerebrospinal fluid	History of viral infection or vaccination
Prevention	Insect control	Maternal screening for HSV	None	–
Treatment	None	Acyclovir	No drugs proven effective; mefloquine and others have been used	Steroids, anti-inflammatory agents
Distinctive Features	History of exposure to insect important; EEV—**Category B Bioterrorism Agent**	In infants, disseminated disease present; rare between 30 and 50 years	In severely immunocompromised, especially AIDS	History of virus/ vaccine exposure critical
Epidemiological Features	See incidence rates in figure 20.11	HSV-1 most common cause of encephalitis; 2 cases per million per year	Affects 5% of adults with untreated AIDS	Rare in United States due to vaccination; more common in developing countries, more common in boys than girls

differential diagnosis of subacute encephalitis should consider a variety of infections with primary symptoms elsewhere in the body.

Toxoplasma gondii

Toxoplasma gondii is a flagellated parasite with such extensive distribution that some experts estimate it affects the majority of the world's population at some time in their lives. Infection in the fetus and in immunodeficient people is severe and often fatal. Although infection in otherwise healthy people is generally unnoticed, recent data tell us it can have profound effects on their brain and the responses it controls. It seems that people with a history of *Toxoplasma* infection are more likely to display thrill-seeking behaviors and other significant changes in their brains.

T. gondii is a very successful parasite with so little host specificity that it can attack at least 200 species of birds and mammals. However, its primary reservoir and hosts are members of the feline family, both domestic and wild.

Signs and Symptoms

As just mentioned, most cases of toxoplasmosis are asymptomatic or marked by mild symptoms such as sore throat, lymph node enlargement, and low-grade fever. In patients whose immunity is suppressed by infection, cancer, or drugs, the outlook may be grim. The infection causes a more chronic or subacute form of encephalitis than do most viruses, often producing extensive brain lesions and fatal disruptions of the heart and lungs. A pregnant woman with toxoplasmosis has a 33% chance of transmitting the infection to her fetus. Congenital infection occurring in the first or second trimester is associated with stillbirth and severe abnormalities such as liver and spleen enlargement, liver failure, hydrocephalus, convulsions, and damage to the retina that can result in blindness.

Pathogenesis and Virulence Factors

Toxoplasma is an obligate intracellular parasite, making its ability to invade host cells an important factor for virulence.

Transmission and Epidemiology

To follow the transmission of toxoplasmosis, we will look at the general stages of the *Toxoplasma* life cycle in the cat **(figure 20.12*a*)**. The parasite undergoes a sexual phase in the intestine and is then released in feces, where it becomes an infective *oocyst* that survives in moist soil for several months. Ingested oocysts release an invasive asexual tissue phase called a *tachyzoite* that infects many different tissues and often causes disease in the cat. These forms eventually enter an asexual cyst state in tissues, called a *pseudocyst.*

Scientists have discovered that the protozoan crowds into a part of the rat brain that usually directs the rat to avoid the smell of cat urine (a natural defense against a domestic rat's major predator). When *Toxoplasma* infects rat brains, the rats lose their fear of cats. Infected rats are then easily eaten by cats, ensuring the continuing *Toxoplasma* life cycle. All other neurological functions in the rat are left intact.

Other vertebrates become a part of this transmission cycle **(figure 20.12*b*)**. Herbivorous animals such as cattle and sheep ingest oocysts that persist in the soil of grazing areas and then develop pseudocysts in their muscles and other organs. Carnivores such as canines are infected by eating pseudocysts in the tissues of carrier animals.

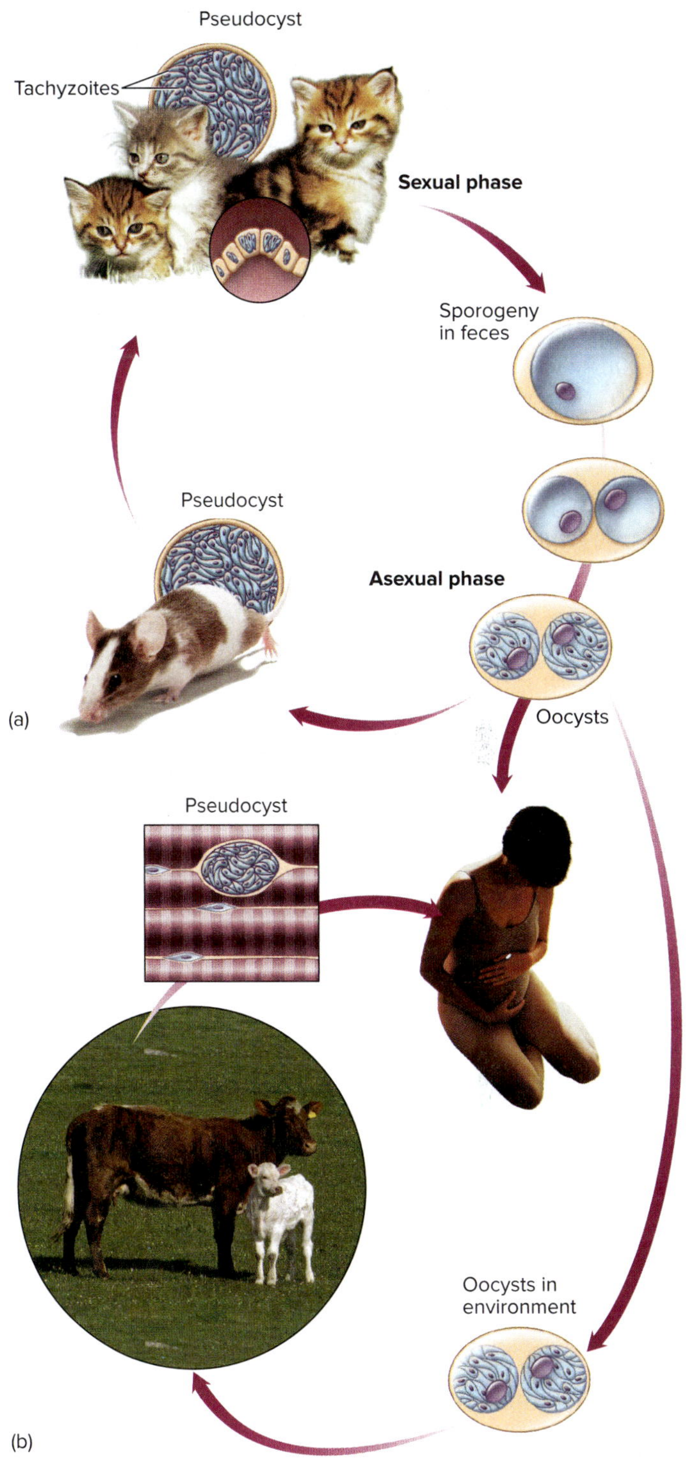
Pseudocyst
Tachyzoites
Sexual phase
Sporogeny in feces
Pseudocyst
Asexual phase
(a)
Oocysts
Pseudocyst
Oocysts in environment
(b)

Figure 20.12 The life cycle and morphological forms of *Toxoplasma gondii*. **(a)** The cycle in cats and their prey. **(b)** The cycle in other animal hosts. The zoonosis has a large animal reservoir (domestic and wild) that becomes infected through contact with oocysts in the soil. Humans can be infected through contact with cats or ingestion of pseudocysts in animal flesh. Infection in pregnant women is a serious complication because of the potential damage to the fetus.

(cat, mouse) Image Source; (cow) Keith Szafranski/iStock/Getty Images; (pregnant woman) Francisco Cruz/Purestock/SuperStock

Humans appear to be constantly exposed to the pathogen. The rate of prior infections, as detected through serological tests, can be as high as 90% in some populations. Many cases are caused by ingesting pseudocysts in contaminated meats. A common source is raw or undercooked meat. Also, the grooming habits of cats spread fecal oocysts on their body surfaces, and handling of them presents an opportunity to ingest oocysts. Infection can also occur when oocysts are inhaled in air or dust contaminated with cat droppings and when tachyzoites cross the placenta to the fetus.

▶ Culture and/or Diagnosis

This infection can be differentiated from viral encephalitides by means of serological tests that detect antitoxoplasma antibodies, especially those for IgM, which appears early in infection. Disease can also be diagnosed by culture or histological analysis for the presence of cysts or tachyzoites.

▶ Prevention and Treatment

The most effective drugs are pyrimethamine, leucovorin, and sulfadiazine—alone or in combination. Because these drugs do not destroy the cyst stage, they must be given for long periods to prevent recurrent infection.

In view of the fact that the oocysts are so widespread and resistant, hygiene is of paramount importance in controlling toxoplasmosis. Adequate cooking or freezing meats below −20°C destroys both oocysts and tissue cysts. Oocysts can also be avoided by washing the hands after handling cats or soil possibly contaminated with cat feces, especially sandboxes and litter boxes.

Measles Virus: Subacute Sclerosing Panencephalitis

Subacute sclerosing panencephalitis (SSPE) is sometimes called a "slow virus infection." The symptoms appear years after an initial measles episode (up to 1 in 6 are affected) and are different from those of immune-mediated postinfectious encephalitis, described earlier. SSPE seems to be caused by direct viral invasion of neural tissue. It is not clear what factors lead to persistence of the virus in some people. SSPE's important features are listed in **Disease Table 20.5.**

Prions

As you read in chapter 7, **prions** are *proteinaceous infectious particles* containing, apparently, no genetic material. They are known to cause diseases called **transmissible spongiform encephalopathies (TSEs),** neurodegenerative diseases with long incubation periods but rapid progressions once they begin. The human TSEs include **Creutzfeldt-Jakob disease (CJD),** Gerstmann-Sträussler-Scheinker disease, kuru, and fatal familial insomnia. TSEs are also found in animals and include a disease called scrapie in sheep and goats, transmissible mink encephalopathy, chronic wasting disease in deer and elk, and bovine spongiform encephalopathy (BSE).

▶ Signs and Symptoms of CJD

Symptoms of all forms of CJD include altered behavior, dementia, memory loss, impaired senses, delirium, and premature senility. Uncontrollable muscle contractions continue until death, which usually occurs within 1 year of diagnosis.

▶ Causative Agent of CJD

It is thought that 10% to 15% of CJD cases are due to an inherited mutation within a single gene. These are termed familial or hereditary CJD. In the 1980s, a protein called PrP^{C} was discovered that spontaneously transforms into a nonfunctional form in CJD. This altered protein (PrP^{SC}), which was termed a *prion,* triggers damage in the brain and other areas of the central nervous system. The PrP^{SC} actually becomes catalytic and able to spontaneously convert other normal human PrP^{C} proteins into the abnormal form. This becomes a self-propagating chain reaction that creates a massive accumulation of PrP^{SC}, leading to plaques, spongiform damage (that is, holes in the brain), and severe loss of brain function.

Further studies showed that prions could cause disease when transferred to a new host, confirming for the first time that a prion could function as an infectious transmissible agent. This led to the recognition of cases called iatrogenic CJD, in which

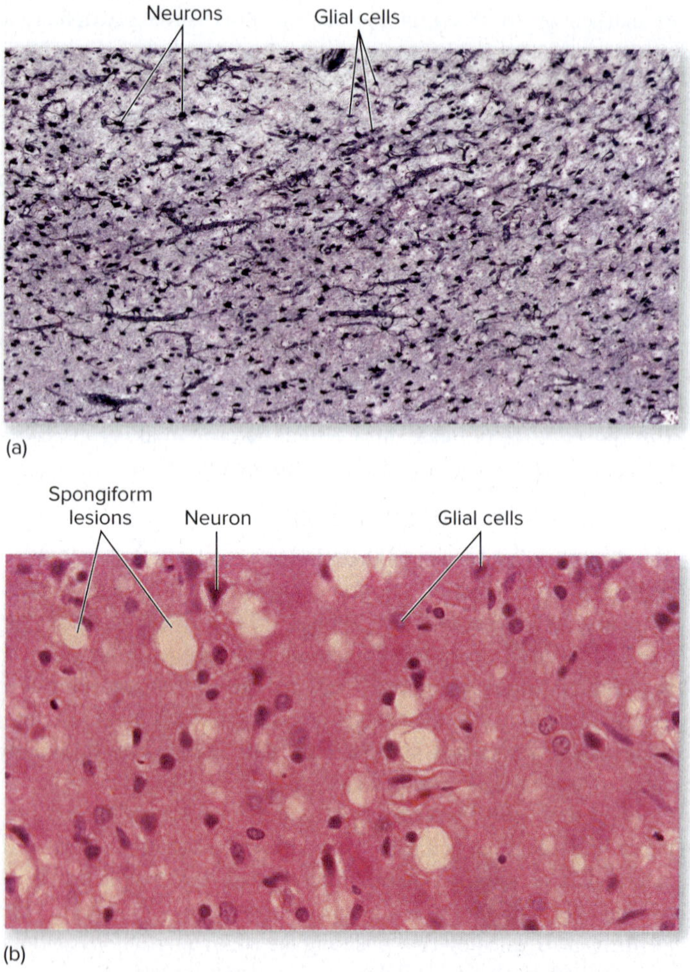
Neurons
Glial cells
(a)
Spongiform lesions
Neuron
Glial cells
(b)

Figure 20.13 The microscopic effects of spongiform encephalopathy. **(a)** Normal cerebral cortex section, showing neurons and glial cells. **(b)** Sectioned cortex in CJD patient shows numerous round holes, producing a "spongy" appearance. This destroys brain architecture and causes massive loss of neurons and glial cells.

(a) MicroScape/Science Source (b) Michael Abbey/Science Source

a patient acquired the disease through contaminated equipment during a medical procedure. In the late 1990s, it became apparent that humans were contracting a variant form of CJD (vCJD) after ingesting meat from cattle that had been afflicted by a related disease called bovine spongiform encephalopathy. Presumably, meat products had been contaminated with fluid or tissues infected with the prion. Cases of this disease were concentrated in Great Britain, where many cows were found to have BSE.

Pathogenesis and Virulence Factors

Autopsies of the brain of all CJD patients reveal spongiform lesions as well as tangled protein fibers (neurofibrillary tangles) and enlarged astroglial cells **(figure 20.13).** These changes affect the gray matter of the CNS and seem to be caused by the massive accumulation of altered PrP, which may be toxic to neurons. The altered PrPs apparently stimulate no host immune response. Prions are also incredibly hardy "pathogens." They are highly resistant to chemicals, radiation, and heat and can even withstand prolonged autoclaving.

Transmission and Epidemiology

Hereditary CJD and sporadic CJD are most common in elderly people. However, the median age at death of patients with vCJD is 28 years.

Aside from genetic transmission, prions can be spread through direct or indirect contact with infected brain tissue or cerebrospinal fluid. Ingestion of contaminated tissue has been documented to cause disease, and it is suggested that aerosols may represent another mode of transmission.

Health care professionals should be aware of the possibility of CJD in patients, especially when surgical procedures are performed, since iatrogenic cases have been reported due to transmission of CJD via contaminated surgical instruments. Due to the heat and chemical resistance of prions, normal disinfection and sterilization procedures are usually not sufficient to eliminate them from instruments and surfaces. The latest CDC guidelines for handling of CJD patients in a health care environment should be consulted. CJD has also been transmitted through corneal grafts and administration of contaminated human growth hormone.

Disease Table 20.5 Subacute Encephalitis

Causative Organism(s)	*Toxoplasma gondii* (P)	Subacute sclerosing panencephalitis (V)	Prions	Other conditions to consider
Most Common Modes of Transmission	Vehicle (meat) or fecal-oral	Persistence of measles virus	CJD = direct/parenteral contact with infected tissue, or inherited vCJD = vehicle (meat, parenteral)	Other conditions may display subacute encephalitis symptoms:
Virulence Factors	Intracellular growth	Cell fusion, evasion of immune system	Avoidance of host immune response	Rickettsial diseases (Rocky Mountain spotted fever) (chapter 21)
Culture/Diagnosis	Serological detection of IgM, culture, histology	EEGs, MRI, serology (Ab versus measles virus)	Biopsy, image of brain	Lyme disease (chapter 21)
Prevention	Personal hygiene, food hygiene	None	Avoiding infected meat or instruments; no prevention for inherited form	*Bartonella* or *Anaplasma* disease (chapter 21)
Treatment	Pyrimethamine and/or leucovorin and/or sulfadiazine	None	None	Tapeworm disease—*Taenia solium* (chapter 23)
Distinctive Features	Subacute, slower development of disease	History of measles	Long incubation period; fast progression once it begins	Syphilis—*Treponema pallidum* (chapter 24)
Epidemiological Features	15%–29% of U.S. population is seropositive; internationally, seroprevalence is up to 90%; disease occurs in 3%–15% of AIDS patients; considered a **neglected parasitic infection (NPI)** (see chapter 23)	Occurs in up to 1 in 6 people who have recovered from measles	CJD: 1 case per year per million worldwide; seen in older adults vCJD: 98% of cases originated in United Kingdom	

Culture and/or Diagnosis

It is very difficult to diagnose CJD. Definitive diagnosis requires examination of biopsied brain or nervous tissue, and this procedure is usually considered too risky because of both the trauma induced in the patient and the undesirability of contaminating surgical instruments and operating rooms. Electroencephalograms and magnetic resonance imaging can provide important clues. New tests are being developed to identify prions in cerebrospinal fluid samples, making diagnosis possible before the patient's death.

Prevention and Treatment

Prevention of this disease relies on avoiding infected tissues. Avoiding vCJD entails not ingesting tainted meats. No known treatment currently exists for any form of CJD. Patients inevitably die. There is active research in treatments for prion diseases. As you saw in the opening case in chapter 7, a new treatment might be available soon. Medical intervention focuses on easing symptoms and making the patient as comfortable as possible **(Disease Table 20.5).**

Zika Virus Disease

Starting in 2015 and continuing into 2016, an epidemic of babies born with abnormally small heads—a condition called microcephaly—became obvious in Brazil. The cause was quickly determined to be the Zika virus. Zika has been known to scientists since the mid-1900s, but the manifestation of hundreds of brain-damaged babies was new. Zika was first discovered in Uganda in Africa, where it seems to have circulated among human populations for some time. In the first half of the 20th century, the virus migrated to western Africa and then to Southeast Asia. In 2015 it was found to be circulating in South America, the first evidence of the virus in the Americas. In February 2016, the World Health Organization declared Zika an international health emergency. That emergency was ended in November of the same year. There have been no cases of the disease in the United States since 2019.

Signs and Symptoms

When adults are infected with Zika, they can experience a range of symptoms, from none at all to a skin rash, conjunctivitis, and muscle and joint pain. The virus also seems to trigger Guillain-Barré syndrome (GBS) in some adults. GBS is a neurological condition that can occur after infections with certain bacteria and viruses, and sometimes after exposure to vaccines. The syndrome is characterized by the immune system attacking peripheral nerves.

Far more serious consequences affect babies who acquire Zika during gestation. In these cases, it can cause *congenital Zika virus syndrome*. The most obvious manifestation is a characteristically small head, although it is not present in every case. Additional symptoms of the syndrome include vision problems, involuntary movements, seizures, and irritability. Symptoms of brainstem dysfunction such as swallowing problems are common as well.

Causative Agent

Zika virus is an RNA virus in the flavivirus family. It is closely related to the viruses causing dengue fever, West Nile fever, and yellow fever.

Disease Table 20.6	Zika Virus Infection
Causative Organism(s)	Zika virus (V)
Most Common Modes of Transmission	Vertical, vector-borne, sexual contact; likely through blood transfusions (not yet confirmed)
Virulence Factors	Protein that reduces innate immune response
Culture/Diagnosis	PCR testing
Prevention	Avoiding mosquitoes; no vaccine yet
Treatment	Supportive
Epidemiological Features	Originated in Africa but spread throughout world in 2016; small outbreaks in other parts of the world since

Rabies

Rabies is a slow, progressive zoonotic disease characterized by fatal encephalitis. It is so distinctive in its pathogenesis and its symptoms that we discuss it separately from the other encephalitides. It is distributed nearly worldwide, except for a few countries that have remained rabies-free by practicing rigorous animal control.

Signs and Symptoms

The average incubation period of rabies is 1 to 2 months or more, depending on the wound site, its severity, and the inoculation dose. The incubation period is shorter in facial, scalp, or neck wounds because of closer proximity to the brain. The prodromal phase begins with fever, nausea, vomiting, headache, fatigue, and other nonspecific symptoms.

In the form of rabies termed "furious," the first acute signs of neurological involvement are periods of agitation, disorientation, seizures, and twitching. Spasms in the neck and pharyngeal muscles lead to severe pain upon swallowing, leading to a symptom known as **hydrophobia** (fear of water). Throughout this phase, the patient is fully coherent and alert. With the "dumb" form of rabies, a patient is not hyperactive but is paralyzed, disoriented, and stuporous. Ultimately, both forms progress to the coma phase, resulting in death from cardiac or respiratory arrest. Until recently, humans were never known to survive rabies. But a handful of patients have recovered in recent years after receiving intensive, long-term treatment.

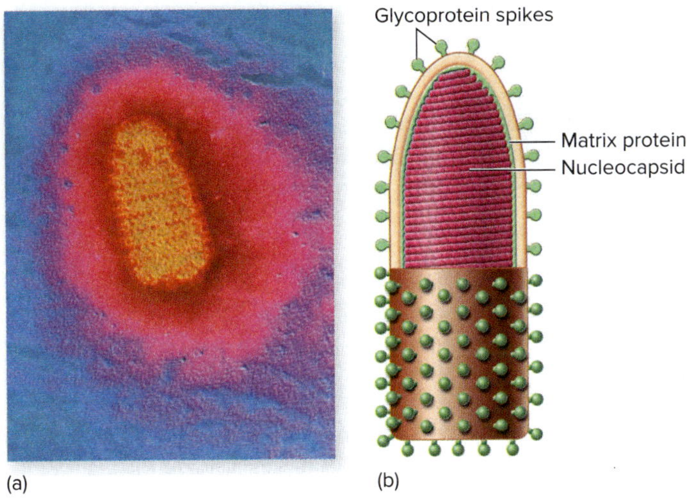
Glycoprotein spikes
Matrix protein
Nucleocapsid
(a)
(b)

Figure 20.14 The structure of the rabies virus. **(a)** Color-enhanced virion shows internal serrations, which represent the tightly coiled nucleocapsid. **(b)** A schematic model of the virus, showing its major features.

(a) CNRI/Science Source

Causative Agent

The rabies virus is in the family *Rhabdoviridae,* genus *Lyssavirus.* This virus has a distinctive, bulletlike appearance, round on one end and flat on the other. Additional features are a helical nucleocapsid and spikes that protrude through the envelope **(figure 20.14).** The family contains about 60 different viruses, but only the rabies *Lyssavirus* infects humans.

Pathogenesis and Virulence Factors

Infection with rabies virus typically begins when an infected animal's saliva enters a puncture site. The virus occasionally is inhaled or inoculated through the membranes of the eye. The rabies virus remains up to a week at the trauma site, where it multiplies. The virus then gradually enters nerve endings and advances toward the ganglia, spinal cord, and brain. Viral multiplication throughout the brain is eventually followed by migration to such diverse sites as the eye, heart, skin, and oral cavity. The infection cycle is completed when the virus replicates in salivary glands and is shed into the saliva. Clinical rabies proceeds through several distinct stages that almost inevitably end in death, unless vaccination is performed before symptoms begin.

Recent research has suggested that some people who sustained bites from vampire bats and lived do, indeed, have antibodies to the rabies virus in their blood, which may act in the same way that desensitization to allergens works: Low-level exposure to the virus may save them when they encounter a large dose of the virus. This may change the paradigm that *any* exposure to the virus is automatically deadly.

Transmission and Epidemiology

The primary reservoirs of the virus are wild mammals such as canines, skunks, raccoons, badgers, cats, and bats that can spread the infection to domestic dogs and cats. Both wild and domestic mammals can spread the disease to humans through bites, scratches, and inhalation of droplets. The annual worldwide total for human rabies is estimated to be from 35,000 to 50,000 cases, but only a tiny number of these occur in the United States. Most U.S. cases occur in wild animals (about 6,000 to 7,000 cases per year), while dog rabies is nearly nonexistent **(figure 20.15).** Globally, more than 90% of human rabies cases are still acquired from dogs. In the United States, 70% of human cases are acquired from bats.

In 2004, the first cases of rabies in recipients of donated organs occurred. The lungs, kidneys, and liver of a man were

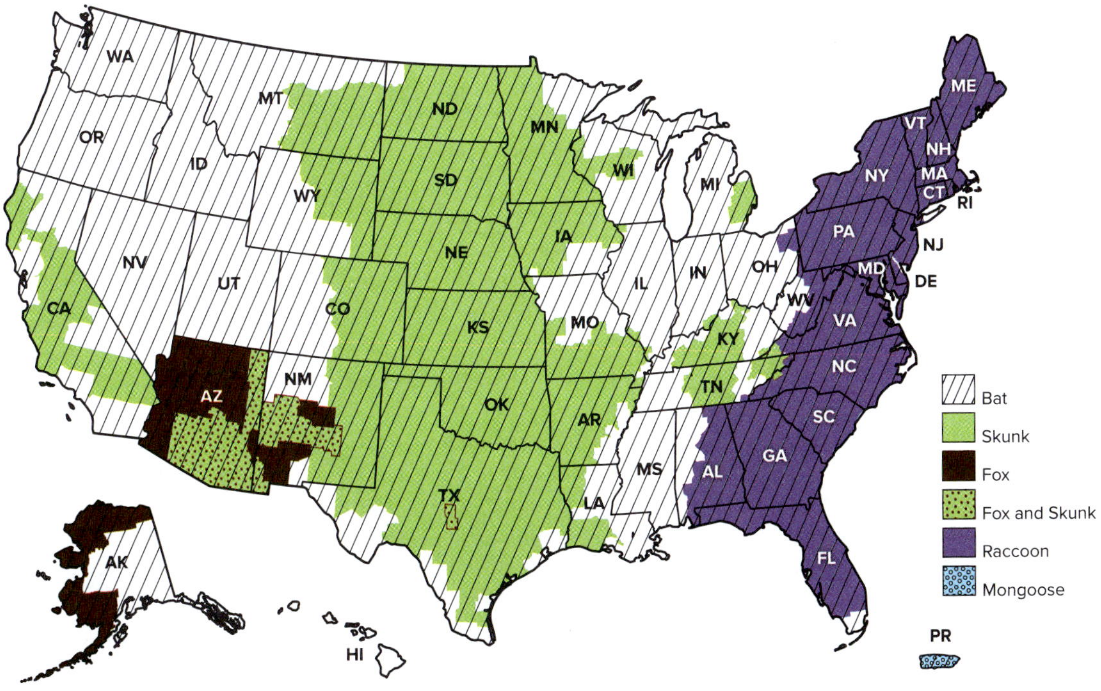
WA
OR
CA
NV
ID
MT
WY
UT
CO
AZ
NM
ND
SD
NE
KS
OK
TX
MN
IA
MO
AR
LA
WI
IL
MS
MI
IN
KY
TN
AL
OH
WV
VA
NC
SC
GA
FL
PA
NY
MD
DE
NJ
CT
RI
MA
NH
VT
ME
AK
HI
PR
Bat
Skunk
Fox
Fox and Skunk
Raccoon
Mongoose

Figure 20.15 Most common carriers of rabies in the United States. Note that the black hatched area covers the whole map. Bats are by far the most common source of human rabies infections in the United States.

CDC, Centers for Disease Control and Prevention

donated to four patients; three of them died of rabies (the fourth died of surgical complications). The virus has also been transmitted through cornea transplants.

Culture and/or Diagnosis

When symptoms appear after an attack by a rabid animal, the disease is readily diagnosed. But the diagnosis can be obscured when contact with an infected animal is not clearly defined or when symptoms are absent or delayed. Anxiety, agitation, and depression that are part of the disease are easily misunderstood. Muscle spasms resemble tetanus; and encephalitis with convulsions and paralysis mimics a number of other viral infections. Often the disease is diagnosed only at autopsy.

Diagnosis before death requires multiple tests. The gold standard is direct fluorescent antibody testing.

Prevention and Treatment

A bite from a wild or stray animal demands assessment of the animal, meticulous care of the wound, and a specific treatment regimen. A wild mammal—especially a skunk, raccoon, fox, or coyote—that bites without provocation is presumed to be rabid, and therapy is immediately begun. If the animal is captured, brain samples and other tissues are examined for verification of rabies. Healthy domestic animals are observed closely for signs of disease and sometimes quarantined. It is imperative that treatment begin after exposure and before symptoms develop. Only a few cases very recently have survived when treatment was delayed until after symptoms appeared. Rabies can be transmitted to humans in the absence of a bite. Aerosols of bat saliva are capable of transmitting the virus. For that reason, people who have found a bat in their house may be encouraged to undergo the postexposure prophylaxis regimen.

Rabies is one of the few infectious diseases for which a combination of passive and active postexposure immunization is indicated (and successful). Initially, the wound is infused with human rabies immune globulin (HRIG) to impede the spread of the virus, and globulin is injected intramuscularly to provide immediate systemic protection. A full course of vaccination is started simultaneously. The routine postexposure vaccination entails intramuscular or intradermal injection on the 1st, 3rd, 7th, and 14th days. Sometimes putting a patient in whom disease has already manifested in a drug-induced coma and on ventilator support can save their life. High-risk groups such as veterinarians, animal handlers, laboratory personnel, and travelers should receive three doses to protect against possible exposure.

Control measures such as vaccination of domestic animals, elimination of strays, and strict quarantine practices have helped reduce the virus reservoir. However, rabid dogs remain a dangerous source of infection in developing countries around the world where over 55,000 people still die of this preventable disease each year. In recent years, the United States and other countries have utilized a live oral vaccine made with a vaccinia virus that carries the gene for the rabies virus surface antigen for mass immunization of wild animals. The vaccine has been incorporated into bait (sometimes peanut butter sandwiches!) placed in the habitats of wild reservoir species such as skunks and raccoons.

Disease Table 20.7	Rabies
Causative Organism(s)	Rabies virus V
Most Common Modes of Transmission	Parenteral (bite trauma), droplet contact
Virulence Factors	Envelope glycoprotein
Culture/Diagnosis	Direct fluorescent antibody test (DFA)
Prevention	Inactivated vaccine
Treatment	Postexposure passive and active immunization; induced coma and ventilator support if symptoms have begun
Epidemiological Features	United States: 1–5 cases per year; worldwide: 35,000–55,000 cases annually

Poliomyelitis

Poliomyelitis (poh″-lee-oh-my″-eh-ly′tis) (polio) is an acute enteroviral infection of the spinal cord that can cause neuromuscular paralysis. Because it often affects small children, in the past it was called infantile paralysis. No civilization or culture has escaped the devastation of polio.

Signs and Symptoms

Most infections are contained as short-term, mild viremia. Some persons develop mild, nonspecific symptoms of fever, headache, nausea, sore throat, and myalgia. If the viremia persists, viruses can be carried to the central nervous system through its blood supply. The virus then spreads along specific pathways in the spinal cord and brain. Being **neurotropic,** the virus infiltrates the motor neurons of the anterior horn of the spinal cord, although it can also attack spinal ganglia, cranial nerves, and motor nuclei. Nonparalytic disease involves the invasion but not the destruction of nervous tissue. It gives rise to muscle pain and spasm, meningeal inflammation, and vague hypersensitivity.

In paralytic disease, invasion of motor neurons causes various degrees of flaccid paralysis over a period of a few hours to several days. Depending on the level of damage to motor neurons, paralysis of the muscles of the legs, abdomen, back, intercostals, diaphragm, pectoral girdle, and bladder can result. In rare cases of **bulbar poliomyelitis,** the brain stem, medulla, or even cranial nerves are affected. This situation leads to loss of control of cardiorespiratory regulatory centers, requiring mechanical respirators. In time, the unused muscles begin to atrophy, growth is slowed, and severe deformities of the trunk and limbs can develop. Common sites of deformities are the spine, shoulder, hips, knees, and feet.

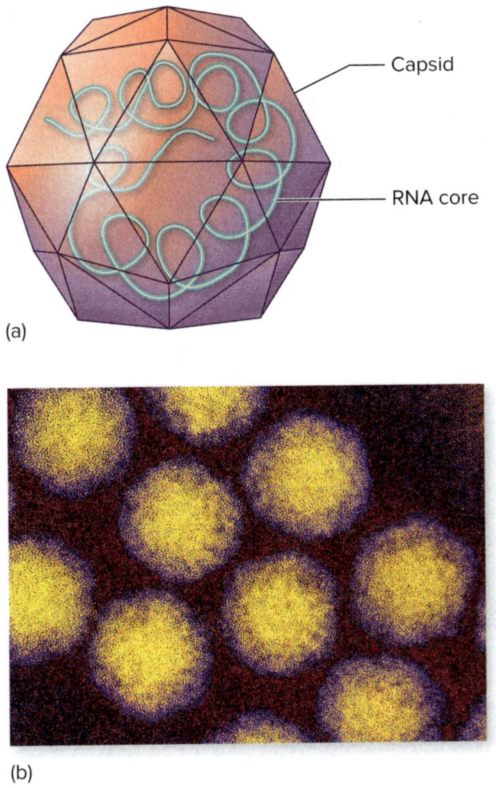
Capsid
RNA core
(a)
(b)

Figure 20.16 Typical structure of a picornavirus. **(a)** A poliovirus, a type of picornavirus that is one of the simplest and smallest viruses (30 nm). It consists of an icosahedral capsid shell around a molecule of RNA. **(b)** A crystalline mass of stacked poliovirus particles in an infected host cell (400,000×).

(b) NIBSC/Science Source

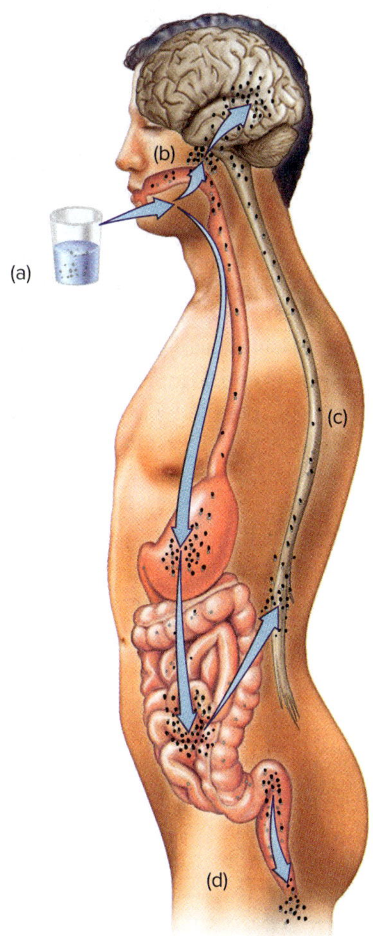
(a)
(b)
(c)
(d)

Figure 20.17 The stages of infection and pathogenesis of poliomyelitis. **(a)** First, the virus is ingested and carried to the throat and intestinal mucosa. **(b)** The virus then multiplies in the tonsils. Small numbers of viruses escape to the regional lymph nodes and blood. **(c)** The viruses are further amplified and cross into certain nerve cells of the spinal column and central nervous system. **(d)** Last, the intestine actively sheds viruses.

Because motor function but not sensation is compromised, the crippled limbs are often very painful.

In recent times, a condition called post-polio syndrome (PPS) has been diagnosed in long-term survivors of childhood infection. PPS manifests as a progressive muscle deterioration that develops in about 25% to 50% of patients several decades after their original polio attack.

Causative Agent

The poliovirus is in the family *Picornaviridae,* genus *Enterovirus*—named for its small size and its RNA core **(figure 20.16).** It is nonenveloped and nonsegmented. There are three strains, types I, II, and III. The naked capsid of the virus gives it chemical stability and resistance to acid, bile, and detergents. By this means, the virus survives the gastric environment and other harsh conditions, which contributes to its ease of transmission.

Pathogenesis and Virulence Factors

After being ingested, polioviruses adsorb to receptors of mucosal cells in the oropharynx and intestine **(figure 20.17).** There, they multiply in the mucosal epithelia and lymphoid tissue. Multiplication results in large numbers of viruses being shed into the throat and feces, and some of them leak into the blood.

Transmission and Epidemiology

Humans are the only known reservoir, and the virus is passed within the population through food, water, hands, objects contaminated with feces, and mechanical vectors. Although the 20th century saw a very large rise in paralytic polio cases, it was also the century during which effective vaccines were developed. The infection was eliminated from the Western Hemisphere in the late 20th century **(figure 20.18).** Because the virus cannot exist outside of a human, it is possible to totally eradicate the disease if

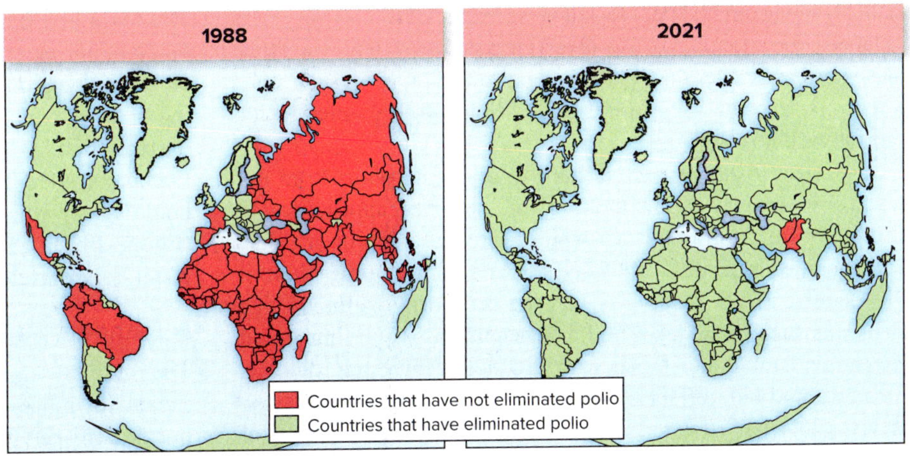
1988
2021
Countries that have not eliminated polio
Countries that have eliminated polio

Figure 20.18 **Progress in the elimination of polio.**

we eliminate (i.e., vaccinate) all susceptible hosts. Today an international partnership called the Global Polio Eradication Initiative (GPEI) is battling to eliminate polio. GPEI partners are the World Health Organization, Rotary International, the U.S. Centers for Disease Control and Prevention, the United Nations Children's Fund, and the Bill & Melinda Gates Foundation. It uses 20 million volunteers in 200 different countries and has vaccinated over 2.5 billion children since 1988.

In 2021 there were only two countries still experiencing transmission of wild poliovirus: Afghanistan and Pakistan. Polio workers are frequently targeted by violent attacks in these countries because false narratives have spread about the vaccine being used to sterilize Muslim children. Seventy polio workers have been killed in Pakistan alone since 2012.

Culture and/or Diagnosis

Poliovirus can usually be isolated by inoculating cell cultures with stool or throat washings in the early part of the disease. Viruses are sometimes then subjected to DNA fingerprinting or whole-genome sequencing to determine if they are wild strains or vaccine strains.

Prevention and Treatment

Treatment of polio rests largely on alleviating pain and suffering. During the acute phase, muscle spasm, headache, and associated discomfort can be alleviated by pain-relieving drugs. Respiratory failure may require artificial ventilation maintenance. Prompt physical therapy to diminish crippling deformities and to retrain muscles is recommended after the acute febrile phase subsides.

The mainstay of polio prevention is vaccination as early in life as possible, usually in four doses starting at about 2 months of age. The two forms of vaccine currently in use are inactivated poliovirus vaccine (IPV), developed by Jonas Salk in 1954, and oral poliovirus vaccine (OPV), developed by Albert Sabin in the 1960s. Both are prepared from animal cell cultures and are trivalent (combinations of the three serotypes of the poliovirus). Both vaccines are effective, but one may be favored over the other under certain circumstances.

For many years, the oral vaccine was used in the United States because it is easily administered by mouth, but it is not free of medical complications. It contains an attenuated virus that can multiply in vaccinated people and be spread to others. In very rare instances, the attenuated virus reverts to a neurovirulent strain that causes disease rather than protects against it. For this reason, IPV is the only vaccine used in the United States (see chapter 16 for current vaccine schedule).

Because it has proved so hard to eradicate polio from the world, researchers began looking for improved approaches to eliminating it from countries where it keeps cropping up. The oral vaccine continues to be used in parts of the world where it still occurs. The reasoning is that the oral form of the vaccine virus was shed in the feces and, because it was still living, could "passively" inoculate those who did not receive the vaccine. An unwanted side effect of this is the occasional transmission of virus that has reverted to its virulent form in some parts of central Africa and Madagascar.

Disease Table 20.8	Poliomyelitis
Causative Organism(s)	Poliovirus V
Most Common Modes of Transmission	Fecal-oral, vehicle
Virulence Factors	Attachment mechanisms
Culture/Diagnosis	Viral culture, serology
Prevention	Live attenuated (developing world); inactivated vaccine (developed world)
Treatment	None, palliative, supportive
Epidemiological Features	Polio has been 99% eradicated, except in Afghanistan and Pakistan, as of 2021

Tetanus

Tetanus is a neuromuscular disease whose alternate name, lockjaw, refers to an early effect of the disease on the jaw muscle. The etiologic agent, *Clostridium tetani,* is a common resident of soil and the gastrointestinal tracts of animals. It is a gram-positive, endospore-forming rod. The endospores it produces often swell the vegetative cell **(figure 20.19)** but are only produced under anaerobic conditions.

Signs and Symptoms

C. tetani releases a powerful exotoxin that is a neurotoxin, **tetanospasmin,** that binds to target sites on peripheral motor neurons, on the spinal cord and brain, and in the sympathetic nervous system. The toxin acts by blocking the inhibition of muscle contraction. Without inhibition of contraction, the muscles contract uncontrollably, resulting in spastic paralysis. The first symptoms are clenching of the jaw, followed in succession by extreme arching of the back, flexion of the arms, and extension of the legs **(figure 20.20).** Lockjaw confers the bizarre appearance of *risus sardonicus* (sardonic grin), which looks eerily as though the person is smiling **(figure 20.21*c*).** Death most often occurs due to paralysis of the respiratory muscles and respiratory arrest.

Pathogenesis and Virulence Factors

The mere presence of tetanus endospores in a wound is not sufficient to initiate infection because the bacterium is unable to invade damaged tissues readily. It is also a strict anaerobe, and the endospores cannot become established unless tissues at the site of

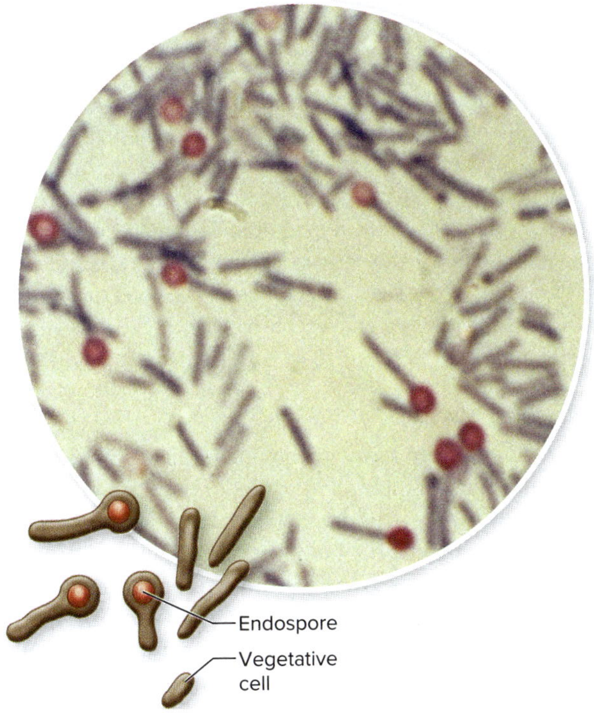
Endospore
Vegetative cell

Figure 20.19 ***Clostridium tetani.*** Its typical tennis racket morphology is created by terminal endospores that swell the end of the cell (1000×).
Biophoto Associates/Science Source

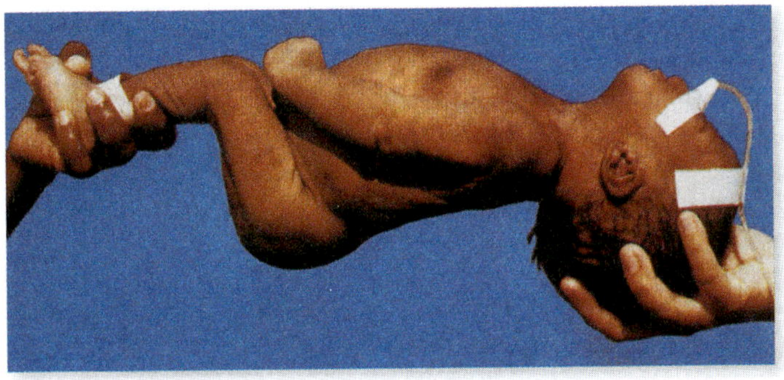

Figure 20.20 Neonatal tetanus. Baby with neonatal tetanus, showing spastic paralysis of the paravertebral muscles, which locks the back into a rigid, arched position. Also note the abnormal flexion of the arms and legs.
CDC

the wound are necrotic and poorly supplied with blood, conditions that favor germination.

As the vegetative cells grow, various metabolic products are released into the infection site, including the tetanospasmin toxin. The toxin spreads to nearby motor nerve endings in the injured tissue, binds to them, and travels via axons to the ventral horns of the spinal cord (**figure 20.21*b***). The toxin blocks the release of neurotransmitter. Like all bacterial exotoxins, only a small amount is required to initiate the symptoms. The incubation period varies from 4 to 10 days, and shorter incubation periods signify a more serious condition.

The muscle contractions are intermittent and extremely painful, and they may be forceful enough to break bones, especially the

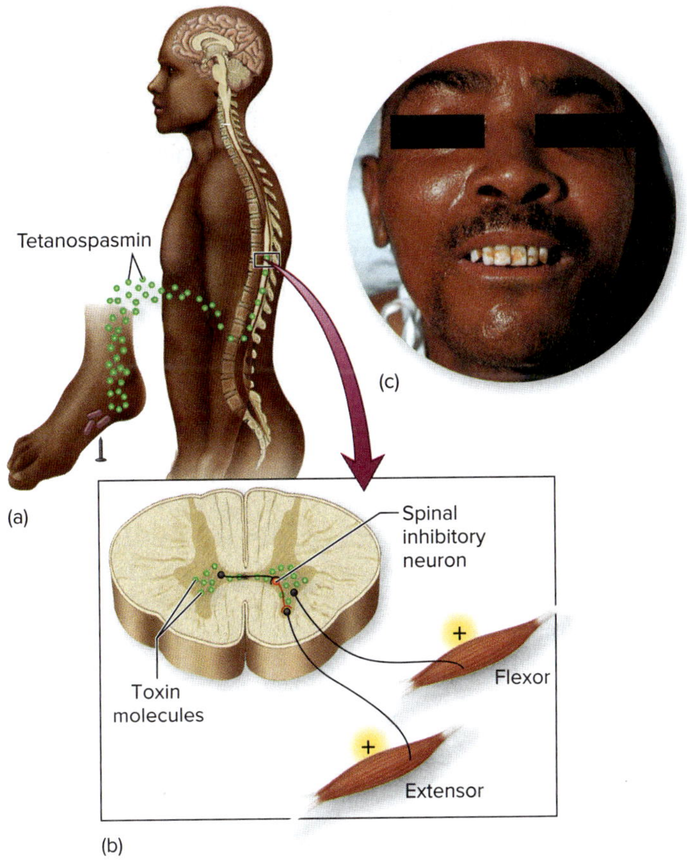
Tetanospasmin
(a)
Spinal inhibitory neuron
Toxin molecules
+
Flexor
+
Extensor
(b)
(c)

Figure 20.21 The events in tetanus. **(a)** After traumatic injury, bacteria infecting the local tissues secrete tetanospasmin, which is absorbed by the peripheral axons and is carried to the target neurons in the spinal column. **(b)** In the spinal cord, the toxin attaches to the junctions of regulatory neurons that inhibit inappropriate contraction. Released from inhibition, the muscles, even opposing members of a muscle group, receive constant stimuli and contract uncontrollably. **(c)** Muscles contract spasmodically, without regard to regulatory mechanisms or conscious control. Note the clenched jaw typical of *risus sardonicus*.
(c) C Farmer/AFIP/Centers for Disease Control and Prevention

vertebrae. The fatality rate, ranging from 10% to 70%, is highest in cases involving delayed medical attention, a short incubation time, or head wounds. Full recovery requires a few weeks, and no permanent damage to the muscles usually remains.

Transmission and Epidemiology

Endospores usually enter the body through accidental puncture wounds, burns, umbilical stumps, frostbite, and crushed body parts. The incidence of tetanus is low in North America. Most cases occur among geriatric patients, people who are intravenous drug abusers, and people who are unvaccinated. In developing countries, however, new mothers and neonates are at high risk for developing disease. Neonatal tetanus still kills approximately 25,000 newborns each year, but that number has decreased 88% since the year 2000. A majority of infections in these countries are a direct result of unhygienic practices during childbirth, including the use of dung, ashes, or mud to stop bleeding or for religious purposes during this process.

Prevention and Treatment

Tetanus treatment is aimed at deterring the degree of toxemia and infection and maintaining patient homeostasis. A patient with a clinical appearance suggestive of tetanus should immediately receive antitoxin therapy with human tetanus immune globulin (TIG). Penicillin G is also administered. Although the antitoxin inactivates circulating toxin, it will not counteract the effect of toxin already bound to neurons. Other methods include thoroughly cleansing and/or removing the affected tissue, controlling infection with penicillin or tetracycline, and administering muscle relaxants. The patient may require the assistance of a respirator, and a **tracheotomy**[1] is sometimes performed to prevent respiratory complications such as aspiration pneumonia or lung collapse. There are combination vaccines that provide protection against tetanus and additional infectious diseases (diphtheria, pertussis).

Immunized children are considered to be protected for 10 years. At that point, and every 10 years thereafter, they should receive a dose of Td, tetanus-diphtheria vaccine. Additional protection against neonatal tetanus may be achieved by vaccinating pregnant women, whose antibodies will be passed to the fetus. Toxoid should also be given to injured persons who have never been immunized, who have not completed the series, or whose last booster was received more than 10 years previously. The vaccine can be given simultaneously with passive TIG immunization to achieve immediate as well as long-term protection.

Disease Table 20.9 Tetanus

Causative Organism(s)	*Clostridium tetani* B G+
Most Common Modes of Transmission	Parenteral, direct contact
Virulence Factors	Tetanospasm exotoxin
Culture/Diagnosis	Symptomatic
Prevention	Tetanus toxoid immunization
Treatment	Combination of passive antitoxin and tetanus toxoid active immunization, metronidazole, muscle relaxants; sedation
Epidemiological Features	Worldwide, +/– 25,000 newborn deaths annually in developing countries

1. The surgical formation of an air passage by perforation of the trachea.

Botulism

Botulism is an **intoxication** (that is, caused by an exotoxin) associated with eating poorly preserved foods, although it can also occur as a true infection with the bacterium. Until recent times, it was relatively common and frequently fatal, but modern techniques of food preservation and medical treatment have reduced its incidence and its fatality rate. However, botulism is a common cause of death in livestock that have grazed on contaminated food and in aquatic birds that have eaten decayed vegetation.

Signs and Symptoms

There are three major forms of botulism, distinguished by their means of transmission and the population they affect. These are *foodborne botulism* (in children and adults), *infant botulism,* and *wound botulism.* Foodborne botulism in children and adults is an intoxication resulting from the ingestion of preformed toxin. The other two types of botulism are infections with the organism that are followed by the entrance of an exotoxin called **botulinum toxin** into the bloodstream (that is, toxemia). The symptoms are largely the same in all three forms, however. From the circulatory system, the toxin travels to its principal site of action, the neuromuscular junctions of skeletal muscles **(figure 20.22).** The effect of botulinum is to prevent the release of acetylcholine, the neurotransmitter that initiates the signal for muscle contraction. This results in what is called flaccid paralysis. In babies, it is sometimes called "floppy baby syndrome" **(figure 20.23).** The usual time before onset of symptoms is 12 to 72 hours, depending on the size of the dose. Neuromuscular symptoms first affect the muscles of the head and include double vision, difficulty in swallowing, and dizziness, but there is no sensory or mental lapse. Later symptoms are descending muscular paralysis and respiratory compromise. In the past, death resulted from respiratory arrest, but mechanical respirators have reduced the fatality rate to about 10%.

Surprisingly, doctors and scientists have been able to utilize the deadly effects of the botulinum toxin for effective medical treatments. In 1989, Botox was first approved to treat strabismus (cross-eyes) and uncontrollable blinking, two conditions resulting from the inappropriate contracting of muscles around the eye. Success in this first arena led to Botox treatment for a variety of neurological disorders that cause painful contraction of neck and shoulder muscles, as well muscle spasms caused by cerebral palsy. A much wider use of Botox occurred in so-called off-label uses, as doctors found that injecting facial muscles with the toxin inhibited contraction of these muscles and consequent wrinkling of the overlying skin. The "lunch-hour facelift" instantly became the most popular cosmetic procedure, even before winning FDA approval in 2002. The most common problem arising from Botox treatment is excessive paralysis of facial muscles resulting from poorly targeted injections. Drooping eyelids, facial paralysis, slurred speech, and drooling can also result from improperly placed injections.

In a surprise twist, patients undergoing Botox treatment for wrinkles reported fewer headaches, especially migraines. Clinical

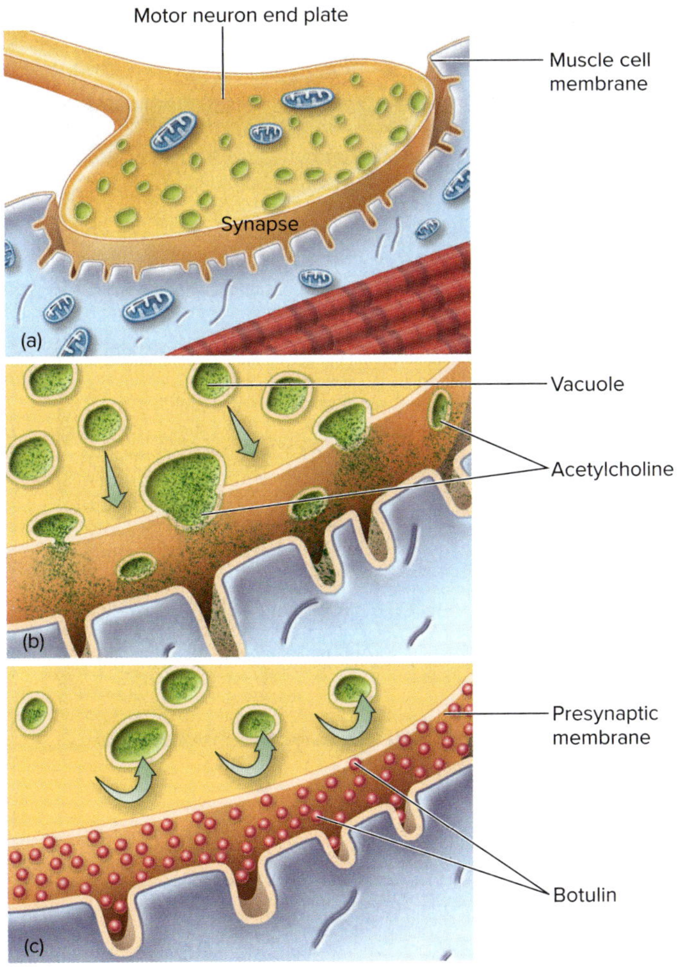
Motor neuron end plate
Muscle cell membrane
Synapse
(a)
Vacuole
Acetylcholine
(b)
Presynaptic membrane
Botulin
(c)

Figure 20.22 The physiological effects of botulism toxin (botulinum). **(a)** The relationship between the motor neuron and the muscle at the neuromuscular junction. **(b)** In the normal state, acetylcholine released at the synapse crosses to the muscle and creates an impulse that stimulates muscle contraction. **(c)** In botulism, the toxin enters the motor end plate and attaches to the presynaptic membrane, where it blocks release of the chemical. This prevents impulse transmission and keeps the muscle from contracting. This causes flaccid paralysis.

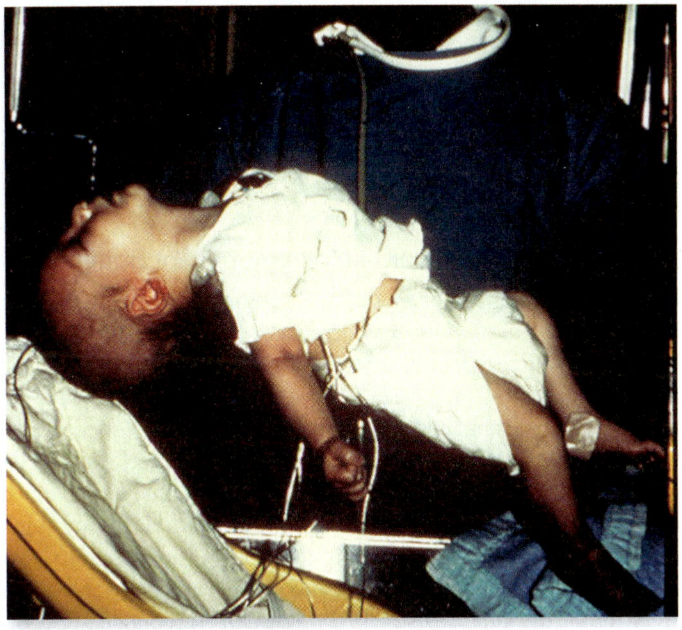

Figure 20.23 Infant botulism. This child is 6 weeks old and displays the flaccid paralysis characteristic of botulism. Compare this picture with that in figure 20.20.
CDC

trials have shown this result to be widespread and reproducible, but the exact mechanism by which Botox works to prevent headaches is unknown. New therapeutic uses for the toxin are being discovered. Botox treatment (in developed countries) is currently being used for several medical conditions that have nothing to do with vanity. People suffering from migraine, excessive sweating, urinary incontinence, spasmocity associated with multiple sclerosis, and even tennis elbow are experiencing relief from Botox injections.

Causative Agent

Clostridium botulinum, like *Clostridium tetani,* is an endospore-forming anaerobe that does its damage through the release of an exotoxin. *C. botulinum* commonly inhabits soil and water and occasionally the intestinal tract of animals. It is distributed worldwide but occurs most often in the Northern Hemisphere. The species has eight distinctly different types (designated A, B, C_a, C_b, D, E, F, and G) that vary in distribution among animals, regions of the world, and types of exotoxin. Human disease is usually associated with types A, B, E, and F. And animal disease is associated with types A, B, C, D, and E.

Both *C. tetani* and *C. botulinum* produce neurotoxins. Tetanospasmin, the toxin made by *C. tetani,* results in spastic paralysis (uncontrolled muscle contraction). In contrast, botulinum, the *C. botulinum* neurotoxin, results in flaccid paralysis.

Pathogenesis and Virulence Factors

As just described, the symptoms are caused entirely by the exotoxin botulinum. Its action is very potent in an affected individual, as the td50 (or median toxic dose) for botulinum toxin is only 0.03 μg per kilogram of body weight.

Transmission and Epidemiology of Foodborne Botulism in Children and Adults

In the United States, the disease is often associated with low-acid vegetables (green beans, corn), fruits, and occasionally meats, fish, and dairy products. Many botulism outbreaks occur in home-processed foods, including canned vegetables, smoked meats, and cheese spreads.

Several factors in food processing can lead to botulism. Endospores are present on the vegetables or meat at the time of gathering and are difficult to remove completely. When contaminated food is put in jars and steamed in a pressure cooker that does not reach reliable pressure and temperature, some endospores survive. At the same time, the pressure is sufficient to evacuate the air and create anaerobic conditions. Storage of the jars at room temperature favors endospore germination and vegetative growth, and one of the products of the cell's metabolism is botulinum, the most potent microbial toxin known.

Bacterial growth may not be evident in the appearance of the jar or can or in the food's taste or texture, and only minute amounts of toxin may be present. Botulism is never transmitted from person to person.

▶ Transmission and Epidemiology of Infant Botulism

Infant botulism is currently the most common type of botulism in the United States, with over 150 cases reported annually. The exact food source is not always known, although raw honey has been implicated in some cases. Apparently, the immature state of the neonatal intestine and microbial biota allows the endospores to gain a foothold, germinate, and give off neurotoxin. As in adults, babies exhibit flaccid paralysis, usually manifested as a weak sucking response, generalized loss of tone (the "floppy baby syndrome"), and respiratory complications. Although adults can also ingest botulinum endospores in contaminated vegetables and other foods, the adult intestinal tract normally inhibits this sort of infection.

▶ Transmission and Epidemiology of Wound Botulism

Perhaps three or four cases of wound botulism occur each year in the United States. In this form of the disease, endospores enter a wound or puncture, much as in tetanus, but the symptoms are similar to those of foodborne botulism. Occasionally, this form of botulism is reported in people who are intravenous drug users as a result of needle puncture. The toxin is considered a possible bioterror agent on the CDC's list.

▶ Culture and/or Diagnosis

Diagnostic standards are slightly different for the three different presentations of botulism. In foodborne botulism, some laboratories attempt to identify the toxin in the offending food. Alternatively, if multiple patients present with the same symptoms after ingesting the same food, a presumptive diagnosis can be made. The cultivation of *C. botulinum* in feces is considered confirmation of the diagnosis because the healthy carrier rate is very low.

In infant botulism, finding the toxin or the organism in the feces confirms the diagnosis. In wound botulism, the toxin should be demonstrated in the serum, or the organism should be grown from the wound. Because minute amounts of the toxin are highly dangerous, laboratory testing should only be performed by experienced personnel. A suspected case of botulism should trigger a phone call to the state health department or the CDC before proceeding with diagnosis or treatment.

▶ Prevention and Treatment

The CDC maintain a supply of type A, B, and E trivalent antitoxin, which, when administered early, can prevent the worst outcomes of the disease. Wound botulism is also treated with penicillin G. Patients are also managed with respiratory and cardiac support systems. In all cases, hospitalization is required and recovery takes weeks. There is an overall 5% mortality rate.

Disease Table 20.10	Botulism
Causative Organism(s)	*Clostridium botulinum* B G+
Most Common Modes of Transmission	Vehicle (foodborne toxin, airborne organism); direct contact (wound); parenteral (injection)
Virulence Factors	Botulinum exotoxin
Culture/Diagnosis	Culture of organism; demonstration of toxin
Prevention	Food hygiene; toxoid immunization available for laboratory professionals
Treatment	Antitoxin, penicillin G for wound botulism, supportive care
Epidemiological Features	United States: 75% of botulism is infant botulism; over 150 cases annually Category A Bioterrorism Agent

African Sleeping Sickness

This condition is caused by *Trypanosoma brucei,* a member of the protozoan group known as hemoflagellates because of their propensity to live in the blood and tissues of the human host. The disease, also called trypanosomiasis, has greatly affected the living conditions of Africans since ancient times. Millions of individuals residing in 36 countries are at risk for disease. Although the total number of new cases has dropped significantly over the past 10 years, many more cases are thought to go unreported in these areas.

▶ Signs and Symptoms

Trypanosomiasis affects the lymphatics and areas surrounding blood vessels. Usually, a long asymptomatic period precedes onset of symptoms. Symptoms include intermittent fever, enlarged spleen, swollen lymph nodes, and joint pain. There are two variants of the disease, caused by two different subspecies of the protozoan. In both forms, the central nervous system is affected, the initial signs being personality and behavioral changes that progress to extreme fatigue and sleep disturbances. The disease is commonly called *sleeping sickness* but, in fact, uncontrollable sleepiness occurs primarily in the day and is followed by sleeplessness at night. Signs of advancing neurological deterioration are muscular tremors, shuffling gait, slurred speech, seizures, and local paralysis. Death results from coma, secondary infections, or heart damage.

▶ Causative Agent

Trypanosoma brucei is a flagellated protozoan, an obligate parasite that is spread by a blood-sucking insect called the tsetse fly, which serves as its intermediate host. It shares a complicated life cycle with other hemoflagellates. *T. brucei gambiense* is found in west and central Africa, is associated with chronic disease, and accounts for over 95% of total reported cases. *T. brucei rhodesiense* is found in eastern and southern Africa. It is much less common (<5% of all cases) and causes acute infection that leads to rapid disease development **(figure 20.24*a*).** Note that in chapter 5, we first described the trypanosome life cycle using the example of *T. cruzi,* the agent that causes Chagas disease.

▶ Transmission and Epidemiology

The cycle begins when a tsetse fly becomes infected after feeding on an infected reservoir host, such as a wild animal (antelope, pig, lion, hyena), domestic animal (cow, goat), or human **(figure 20.24*b*).** In the fly's gut, the trypanosome multiplies, migrates to the salivary glands, and develops into the infectious stage. When the fly bites a new host, it releases the large, fully formed stage of the parasite into the wound. At this site, the trypanosome multiplies and produces a sore called the *primary chancre.* From there, the pathogen moves into the lymphatics and the blood (as shown in the figure). The trypanosome can also cross the placenta and damage a developing fetus.

African sleeping sickness occurs only in sub-Saharan Africa. Tsetse flies exist elsewhere, and it is not known why they do not support *Trypanosoma* in other regions. Cases seen in the United States are only those that were acquired by travelers or emigrants from Africa.

▶ Pathogenesis and Virulence Factors

The protozoan manages to flourish in the blood even though it stimulates a strong immune response. The immune response is counteracted by an unusual adaptation of the trypanosome. As soon as the host begins manufacturing IgM antibodies to the trypanosome, surviving organisms change the structure of their surface glycoprotein antigens. This change in specificity (sometimes referred to as an antigenic shift) renders the existing IgM ineffective so that the parasite eludes control and multiplies in the blood. The host responds by producing IgM of a new specificity, but the protozoan changes its antigens again. The host eventually becomes exhausted and overwhelmed by repeated efforts to catch up with this trypanosome masquerade.

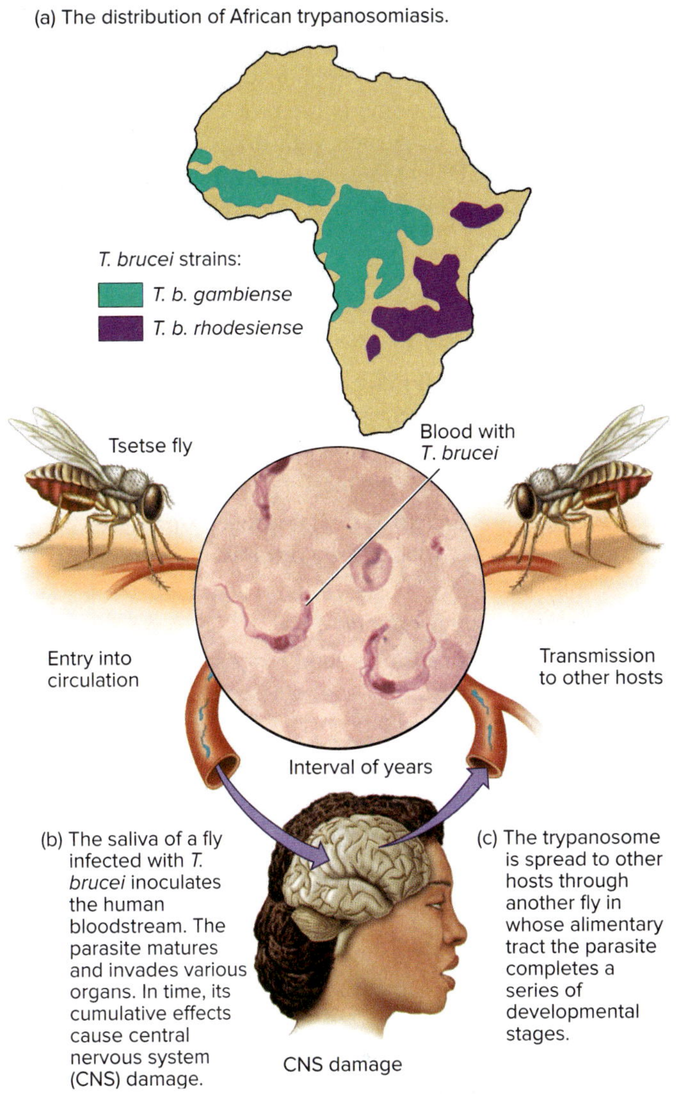
(a) The distribution of African trypanosomiasis.
T. brucei strains:
T. b. gambiense
T. b. rhodesiense
Tsetse fly
Blood with T. brucei
Entry into circulation
Transmission to other hosts
Interval of years
(b) The saliva of a fly infected with T. brucei inoculates the human bloodstream. The parasite matures and invades various organs. In time, its cumulative effects cause central nervous system (CNS) damage.
(c) The trypanosome is spread to other hosts through another fly in whose alimentary tract the parasite completes a series of developmental stages.
CNS damage

Figure 20.24 The generalized cycle between humans and the tsetse fly vector.
CDC/Dr. Mae Melvin

▶ Culture and/or Diagnosis

Trypanosomes are readily visible in blood smears, as well as in spinal fluid or lymph nodes. Serological tests are available for diagnosis of *T. brucei gambiense* infection in endemic areas of Africa.

▶ Prevention and Treatment

Control of trypanosomiasis in western Africa, where humans are the main reservoir hosts, involves eliminating tsetse flies by applying insecticides, trapping flies, or destroying the shelter and breeding sites. In eastern regions, where cattle herds and large wildlife populations are reservoir hosts, control is less practical because large mammals are the hosts and flies are less concentrated in specific sites. The antigenic shifting exhibited by the trypanosome makes the development of a vaccine very difficult.

Drug treatment is most successful if administered prior to nervous system involvement. Two drugs are available for the early stages of the disease. Suramin works against *T. brucei rhodesiense,* and pentamidine is used for *T. brucei gambiense.* Brain infection must be treated with drugs that can cross the blood-brain barrier. One of these is a highly toxic, arsenic-based drug called melarsoprol; another is called eflornithine (**Disease Table 20.11**).

20.2 Learning Outcomes—Assess Your Progress

4. List the possible causative agents for meningitis and neonatal/infant meningitis.
5. Identify which of the agents causing meningitis is the most common and which is the most deadly.
6. Discuss important features of meningoencephalitis, encephalitis, and subacute encephalitis.
7. Identify which encephalitis-causing viruses you should be aware of in your geographic area.
8. List the possible causative agents for each of the following conditions: Zika virus disease, rabies, poliomyelitis, tetanus, botulism, and African sleeping sickness.
9. Identify the conditions for which vaccination is available.
10. Explain the difference between the oral polio vaccine and the inactivated polio vaccine and the advantages and disadvantages of each.

Disease Table 20.11 African Sleeping Sickness

Causative Organism(s)	*Trypanosoma brucei* subspecies *gambiense* or *rhodesiense* (P)
Most Common Modes of Transmission	Vector, vertical
Virulence Factors	Immune evasion by antigen shifting
Culture/Diagnosis	Microscopic examination of blood, CSF
Prevention	Vector control
Treatment	Suramin or pentamidine (early), eflornithine or melarsoprol (late)
Epidemiological Features	*T. brucei gambiense:* 7,000–10,000 cases reported annually; actual occurrence estimated at 600,000; *T. brucei rhodesiense:* estimated 30,000 cases occur annually

Media Under The Microscope Wrap-Up

This article from the online outlet *Green Living* is authored by someone who believes that raw honey cured his cancer.

The **intended message** of the article is to tout the anticancer properties of raw, unfiltered honey. The author presents a lot of assertions about the health properties of the honey and claims that it cured his cancer.

My **critical reading** of the article forced *me* to examine *my* biases. There were so many red flags (no references, a sample size of one, disparaging products from China without citing evidence, etc.) that I had to be extra careful to not dismiss it out of hand, and to truly look at it with an open mind.

I would interpret this for my **nonmicrobiologists** in the following way: This author believes that raw honey is a potent anticancer agent, but he does not mention or recognize that raw honey can also contain dangerous toxins as well as disease-causing microorganisms. I would point out that this is someone's personal story. Even though he talks about research, he does not provide references, and therefore his sample size (n = 1) is very small.

Foodcollection/Martin Skultety/Image Source

My **overall grade** for the article is an F. Despite my efforts to read with an open mind, the article makes some broad claims about raw honey and its anticancer properties, with no references and no warnings about the dangers of raw honey, especially to infants.

Source: GreenLiving.com, "Bee-cause I Said So!," online article posted December 2020.

Study Smarter: Better Together

These activities are designed for you to use on your own with a study group—either a face-to-face group or a virtual one, consisting of 3–5 members. Studying together can be very helpful, but there are effective and ineffective ways to do it. For example, getting together without a clear structure is often not a good use of your time. Use your time efficiently by using one or more of the exercises below.

FACE-TO-FACE GROUPS

Use one or more of the activities below.

Peer Instruction: Assign numbers to your group members to use all semester long. Now look at these five concepts from this chapter. Each group member prepares a 5-minute lesson on the topic corresponding to their number. Don't worry if you have fewer than 5 members; just use however many you have! During your group study time, each member presents their lesson, and the group spends another 5–10 minutes discussing that lesson.

1. The anatomy and defenses of the nervous system
2. All of the possible causes of meningitis, including what type of organism each one is
3. Organisms in this chapter for which there are vaccines available
4. Explanation of what a prion is
5. Description of the blood-brain barrier

Concept Maps: Each member of the group should use this list of terms from this chapter to generate their own concept map. This can be hand-drawn or created using software (see Appendix C for guidelines). During group study time, compare each other's concept maps and help each other make sure they are correct. Of course, there are many different "correct" maps. Examining each member's map will help you talk through the varied concepts and how they are related.

Concept Terms:

bacteria	viruses	fungi	meningitis	vaccination	droplets
vehicles	colonization	transmission			

Table Topics: Each group member should identify a concept or topic from this week's class assignments with which they are having trouble and share it during group study time. The other group members can then help to clarify confusing issues or share how they figured it out. Aim for a maximum of 15 minutes per topic. If the topic remains unclear to the group, bring it up during class or use the instructor's office hours or e-mail to ask for help. Taking the time to struggle with a difficult concept first makes your questions much more specific and more likely to yield helpful answers.

VIRTUAL GROUPS

Not everyone has the time or opportunity to meet with group members outside of class time. You or your instructor can create a virtual group using e-mail or the course software.

Weekly Discussion Board: This forum can be used as a way for groups to discuss topics, via e-mail, or other learning management systems or online platforms, before they are covered in class. As each member of the group answers the current week's question, they should send their responses to every other member of their group. It's best to agree on a deadline based on how your class schedule works (Saturday for the next week's topics, for example). Then, after the topic is discussed in class, each member should send a response that all group members will see with a follow-up post on the same topic. If you cover more than one chapter in a week, someone can be designated to choose which chapter Discussion Board question you will use. Or simply decide up front that you will always use the first-chapter-of-the-week's question, to keep the schedule simple.

Discussion Question
Why is it only necessary to use a form of the exotoxin in the vaccines for botulism and tetanus?

Deadliness and Communicability of Selected Diseases Affecting the Nervous System

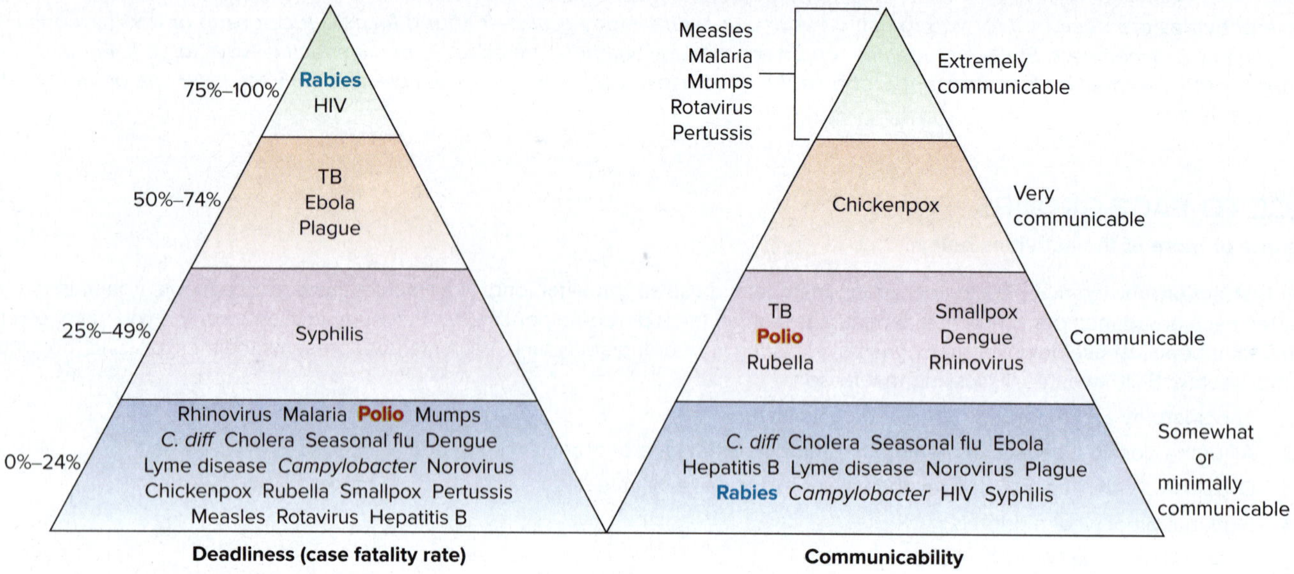
75%–100%
Rabies
HIV
50%–74%
TB
Ebola
Plague
25%–49%
Syphilis
0%–24%
Rhinovirus Malaria Polio Mumps
C. diff Cholera Seasonal flu Dengue
Lyme disease Campylobacter Norovirus
Chickenpox Rubella Smallpox Pertussis
Measles Rotavirus Hepatitis B
Deadliness (case fatality rate)
Measles
Malaria
Mumps
Rotavirus
Pertussis
Extremely communicable
Chickenpox
Very communicable
TB
Polio
Rubella
Smallpox
Dengue
Rhinovirus
Communicable
C. diff Cholera Seasonal flu Ebola
Hepatitis B Lyme disease Norovirus Plague
Rabies Campylobacter HIV Syphilis
Somewhat or minimally communicable
Communicability

▶ Summing Up

Taxonomic Organization Microorganisms Causing Disease in the Nervous System

Microorganism	Disease	Disease Table
Gram-positive endospore-forming bacteria		
Clostridium tetani	Tetanus	Tetanus, 20.9
Clostridium botulinum	Botulism	Botulism, 20.10
Gram-positive bacteria		
Streptococcus pneumoniae	Meningitis	Meningitis, 20.1
Listeria monocytogenes	Meningitis, neonatal meningitis	Meningitis, 20.1; Neonatal and infant meningitis, 20.2
Streptococcus agalactiae	Neonatal meningitis	Neonatal and infant meningitis, 20.2
Gram-negative bacteria		
Neisseria meningitidis	Meningococcal meningitis	Meningitis, 20.1
Haemophilus influenzae	Meningitis	Meningitis, 20.1
Escherichia coli, strain K1	Neonatal meningitis	Neonatal and infant meningitis, 20.2
Cronobacter sakazakii	Neonatal and infant meningitis	Neonatal and infant meningitis, 20.2
DNA viruses		
Herpes simplex virus 1 and 2	Encephalitis	Acute encephalitis, 20.4
JC virus	Progressive multifocal leukoencephalopathy	Acute encephalitis, 20.4
RNA viruses		
Arboviruses: West Nile virus, La Crosse virus, Jamestown Canyon virus, St. Louis encephalitis virus, Powassan virus, Eastern equine encephalitis virus	Encephalitis	Acute encephalitis, 20.4
Measles virus	Subacute sclerosing panencephalitis	Subacute encephalitis, 20.5
Zika virus	Zika virus disease	Zika virus infection, 20.6
Rabies virus	Rabies	Rabies, 20.7
Poliovirus	Poliomyelitis	Poliomyelitis, 20.8

Microorganism	Disease	Disease Table
Fungi		
Cryptococcus neoformans	Meningitis	Meningitis, 20.1
Coccidioides species	Meningitis	Meningitis, 20.1
Prions		
Creutzfeldt-Jakob prion	Creutzfeldt-Jakob disease	Subacute encephalitis, 20.5
Protozoa		
Naegleria fowleri	Meningoencephalitis	Meningoencephalitis, 20.3
Acanthamoeba	Meningoencephalitis	Meningoencephalitis, 20.3
Toxoplasma gondii	Subacute encephalitis	Subacute encephalitis, 20.5
Trypanosoma brucei subspecies *gambiense* and *rhodesiense*	African sleeping sickness	African sleeping sickness, 20.11

Chapter Summary

INFECTIOUS DISEASES MANIFESTING IN THE NERVOUS SYSTEM

20.1 THE NERVOUS SYSTEM, ITS DEFENSES, AND NORMAL BIOTA

- The nervous system has two parts: the central nervous system (the brain and spinal cord) and the peripheral nervous system (nerves and ganglia).
- The soft tissue of the brain and spinal cord is encased within the tough covering of three membranes called meninges. The subarachnoid space is filled with a clear, serumlike fluid called cerebrospinal fluid (CSF). The blood-brain barrier limits the passage of substances from the bloodstream to the brain and spinal cord.

20.2 INFECTIOUS DISEASES MANIFESTING IN THE NERVOUS SYSTEM

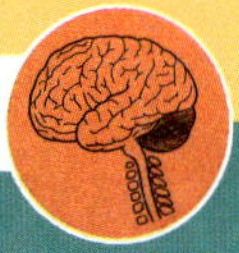

- There is technically no normal biota in the nervous system. However, normal biota of the gut may indirectly influence the function of the nervous system
- **Meningitis**—Commonly the most serious forms are caused by bacteria, and sometimes fungi; the most frequent infections are caused by viruses. Agents in the differential diagnosis: ***Neisseria meningitidis:*** Gram-negative diplococcus; causes most acutely serious form. ***Streptococcus pneumoniae:*** Gram-positive diplococcus, most frequent cause of community-acquired bacterial meningitis. ***Haemophilus influenzae:*** Cases have declined sharply due to vaccination. ***Listeria monocytogenes:*** Most cases are associated with ingesting contaminated dairy products, poultry, and meat. ***Cryptococcus neoformans:*** Fungus causing more chronic form of meningitis. ***Coccidioides:*** True systemic fungal infection. Begins in lungs but can disseminate quickly throughout the body; highest incidence occurs in southwestern United States, Mexico, and Central and South America. ***Viruses:*** Very common, especially in children.
- **Neonatal and infant meningitis**—Usually transmitted vertically. Primary causes in United States are *Streptococcus agalactiae, Escherichia coli, and Listeria monocytogenes. Cronobacter* is a rare but deadly cause that is associated with infant formula.
- **Meningoencephalitis**—Caused mainly by two amoebas, *Naeglaria fowleri* and *Acanthamoeba.*
- **Acute encephalitis**—Often caused by viruses carried by arthropods (arboviruses); other viruses can also cause it. **Arboviruses** are West Nile virus, La Crosse virus, Jamestown Canyon virus, St. Louis encephalitis virus, Powassan virus, and Eastern equine

encephalitis virus. **Herpes simplex viruses type 1 and 2** cause encephalitis in newborns born to HSV-positive mothers as well as in older children and adults. **JC virus** can cause progressive multifocal leukoencephalopathy.

- **Subacute encephalitis**—Symptoms take longer to manifest than the acute form. ***Toxoplasma gondii:*** Protozoan. Relatively asymptomatic in the healthy, severe in the immunodeficient and fetuses. **Measles virus:** Can produce subacute sclerosis panencephalitis (SSPE) years after initial measles infection. **Prions:** Proteinaceous infectious particles containing no genetic material. Cause neurodegenerative diseases with long incubation periods but rapid progressions to death once they begin.
- **Rabies**—Slow, progressive zoonotic disease leading to death in the vast majority of cases. Caused by rabies virus.
- **Poliomyelitis**—Acute enterovirus infection of spinal cord. Two vaccines are used: Inactivated polio vaccine (IPV) is used in developed countries; attenuated live oral vaccine (OPV) is used in the developing world where endemic spread is still possible.
- **Tetanus**—Neuromuscular disease, caused by *Clostridium tetani* neurotoxin tetanospasmin, which blocks inhibition of muscle contraction.
- **Botulism**—Caused by exotoxin of *C. botulinum*. Three types: foodborne botulism, infant botulism, and wound botulism.
- **African sleeping sickness**—Caused by two subspecies of the protozoan *Trypanosoma brucei*.

INFECTIOUS DISEASES AFFECTING
The Nervous System

Encephalitis
Arboviruses
Herpes simplex virus 1 or 2
JC virus

Subacute Encephalitis
Toxoplasma gondii
Measles virus
Prions

Rabies
Rabies virus

Zika virus disease
Zika virus

Tetanus
Clostridium tetani

African Sleeping Sickness
Trypanosoma brucei

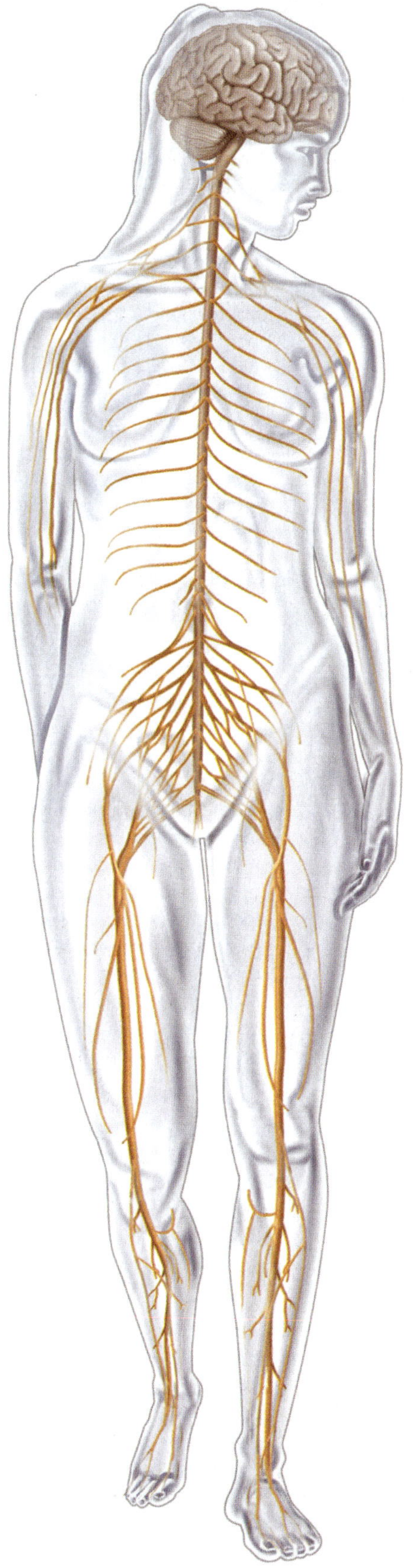

Creutzfeldt-Jakob Disease
Prion

Meningoencephalitis
Naegleria fowleri
Acanthamoeba

Meningitis
Neisseria meningitidis
Streptococcus pneumoniae
Haemophilus influenzae
Listeria monocytogenes
Cryptococcus neoformans
Coccidioides
Various viruses

Neonatal and Infant Meningitis
Streptococcus agalactiae
Escherichia coli
Listeria monocytogenes
Cronobacter sakazakii

Polio
Poliovirus

Botulism
Clostridium botulinum

Bacteria
Viruses
Protozoa
Fungi
Prions

System Summary Figure 20.25

SmartGrid: From Knowledge to Critical Thinking

This *21 Question Grid* takes the topics from this chapter and arranges them with respect to the American Society for Microbiology's Undergraduate Curriculum guidelines—all six of the important "Concepts" as well as the important "Competency" of scientific literacy. Three questions are supplied, which cover chapter content referring to the Concept or Competency in increasing levels of Bloom's taxonomy for learning.

ASM Concept/ Competency	A. Bloom's Level 1, 2—Remember and Understand (Choose one)	B. Bloom's Level 3, 4—Apply and Analyze	C. Bloom's Level 5, 6—Evaluate and Create
Evolution	**1.** Which pathogen has evolved to make its rodent host less avoidant of cats? **a.** *Cryptococcus neoformans* **b.** *Neisseria meningitidis* **c.** rabies virus **d.** *Toxoplasma gondii*	**2.** *Neisseria meningitidis* occurs as noninvasive normal biota in the nares (noses) of approximately 35% of the population. Those that have developed the ability to be invasive, entering the nervous system, are said to be at an "evolutionary dead end." Speculate on why the invasive strains are characterized with that label.	**3.** Speculate on why the poliovirus has changed over evolutionary time to cause severe disease less often in humans, while a disease such as rabies has not become less virulent for humans over time.
Cell Structure and Function	**4.** What cellular structure do several of the organisms that cause meningitis share? **a.** capsule **b.** pilus **c.** fimbria **d.** endospore	**5.** Why is it necessary only to include the exotoxin (in toxoid form) in vaccines for tetanus and botulism?	**6.** Justify the use of only a Gram stain in diagnosing some types of meningitis.
Metabolic Pathways	**7.** Which of the following organisms is anaerobic? **a.** poliovirus **b.** *Cryptococcus* **c.** *Clostridium* **d.** *Coccidioides*	**8.** When cultivating specimens for *N. meningitidis,* nonselective chocolate agar is often used. But when culturing for another *Neisseria* species, *N. gonorrhoeae,* a different medium, Modified Thayer-Martin (MTM) medium, which contains multiple antibiotics, is preferred. Speculate on why this is.	**9.** What do you expect to happen to the geographical distribution of *Coccidioides* cases over the next 10 years? Justify your answer.
Information Flow and Genetics	**10.** Which disease is caused by an infectious agent that carries no nucleic acid? **a.** CJD **b.** rabies **c.** polio **d.** meningitis	**11.** Why would PCR be the best method for identifying microbes in the cerebrospinal fluid if antibiotic treatment was begun before the fluid was sampled?	**12.** Scientists have analyzed the "transcriptome" (the RNA transcripts) of mouse brains that have been infected with *Toxoplasma.* In what ways might that be informative?
Microbial Systems	**13.** The normal gut microbiota in adults, but not infants, inhibit the growth of which pathogen? **a.** *Neisseria meningitidis* **b.** *Clostridium botulinum* **c.** *Clostridium tetani* **d.** *Naegleria fowleri*	**14.** Botulinum toxin has been proposed as a treatment for the symptoms of tetanus. Discuss why this might be useful.	**15.** Recent research shows that when scientists infect *Aedes aegypti* with a bacterium called *Wolbachia,* they become less likely to spread Zika and other viruses to humans. Speculate on what *Wolbachia* may be doing in the mosquito.

ASM Concept/ Competency	A. Bloom's Level 1, 2—Remember and Understand (Choose one)	B. Bloom's Level 3, 4—Apply and Analyze	C. Bloom's Level 5, 6—Evaluate and Create
Impact of Microorganisms	**16.** Subacute encephalitis can be caused by **a.** *Toxoplasma gondii.* **b.** *Streptococcus agalactiae.* **c.** *Naegleria fowleri.* **d.** *Haemophilus influenzae.*	**17.** "Accidental herd immunity" conferred by the oral polio vaccine in the developing world has aided the elimination of polio from many countries. Explain what accidental herd immunity is and how it works.	**18.** Conduct research about the multistage meningitis outbreak linked to steroid injections in 2012. Speculate as to why this "nonpathogenic" fungal species caused such severe disease and death.
Scientific Thinking	**19.** Mosquito eradication could change the epidemiology of **a.** polio. **b.** Zika. **c.** West Nile. **d.** two of these.	**20.** The paralytic form of poliomyelitis began to occur in epidemics in North America and Europe in the early part of the 1900s. Many scientists believe this increase in incidence and in disease manifestations was due to significant improvements in sanitation that occurred around that time. Use what you know about how the virus is transmitted to justify this seemingly puzzling outcome. (Hint: Maternal antibodies might be involved.)	**21.** Using laboratory mice as your subjects, design an experiment that could show that prolonged antibiotic exposure disrupts their gut microbiome and causes altered brain development.

Answers to the multiple-choice questions appear in Appendix A.

Visual Connections

This question uses visual images to connect content within and between chapters.

1. **From chapter 3, figure 3.21.** Two microbes that cause disease in this chapter are described as diplococci, *Streptococcus pneumoniae and Neisseria meningitidis.* Which of the following staining procedures would be most useful for telling them apart?

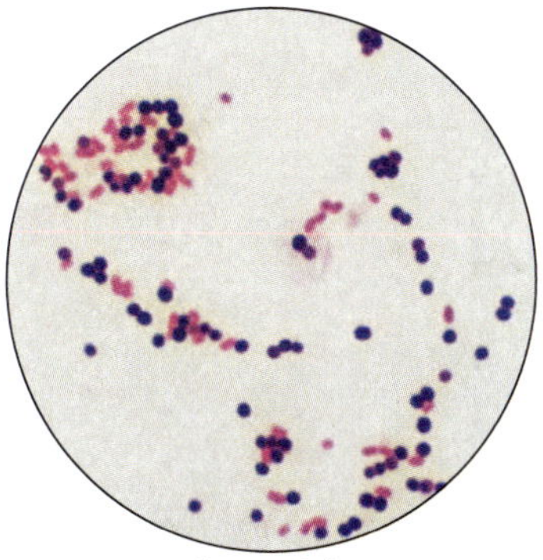

Gram stain
Lisa Burgess/McGraw Hill

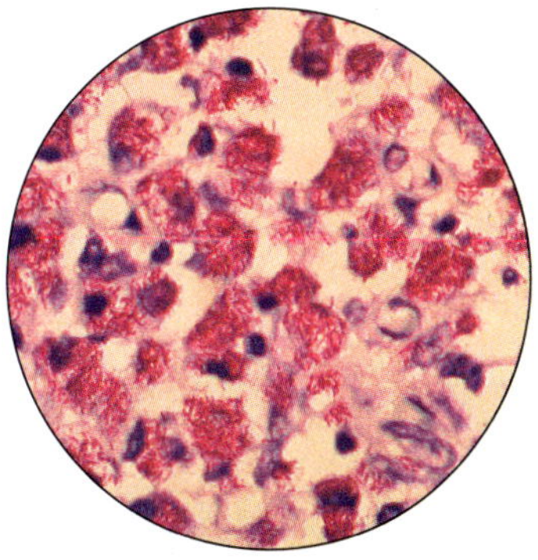

Acid-fast stain
Dr. Edwin P. Ewing, Jr./Centers for Disease Control and Prevention

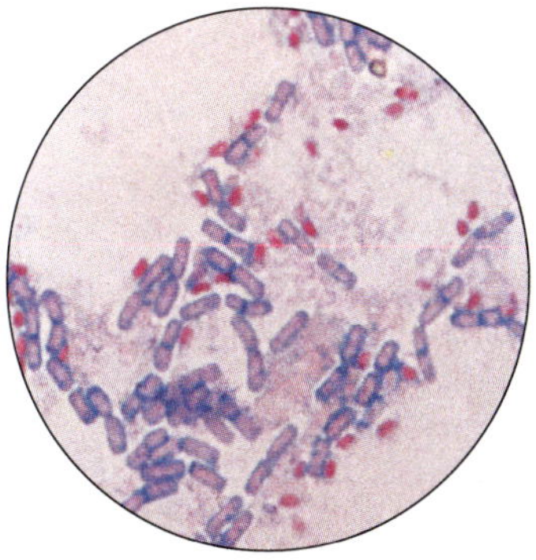

Endospore stain
Manfred Kage/Science Source

High Impact Study

These terms and concepts are most critical for your understanding of this chapter—and may be the most difficult. Have you mastered them? In these disease chapters, the terms and concepts help you identify what is important in a different way than the comprehensive details in the Disease Tables. Your instructor will help you understand what is important for your class.

Concepts

- ☐ Defenses of nervous system
- ☐ Normal microbiota of nervous system
- ☐ Four bacterial causes of meningitis
- ☐ Other causes of meningitis
- ☐ Foodborne cause of meningitis
- ☐ Meningitis vaccines
- ☐ Gram-negative diplococci vs. gram-positive diplococci
- ☐ Difference between CJD and vCJD
- ☐ Global polio eradication
- ☐ Three types of botulism
- ☐ Differences and similarities between tetanus and botulism
- ☐ Organisms in this chapter for which there are vaccines available
- ☐ Organisms in this chapter that display significant antibiotic resistance

Terms

- ☐ Meninges
- ☐ Cerebrospinal fluid
- ☐ Blood-brain barrier
- ☐ Arbovirus
- ☐ Dead-end host
- ☐ Prion
- ☐ Progressive multifocal leukoencephalopathy
- ☐ Postinfection encephalitis
- ☐ Subacute sclerosing parencephalitis

Design Element: (College students): Caia Image/Image Source

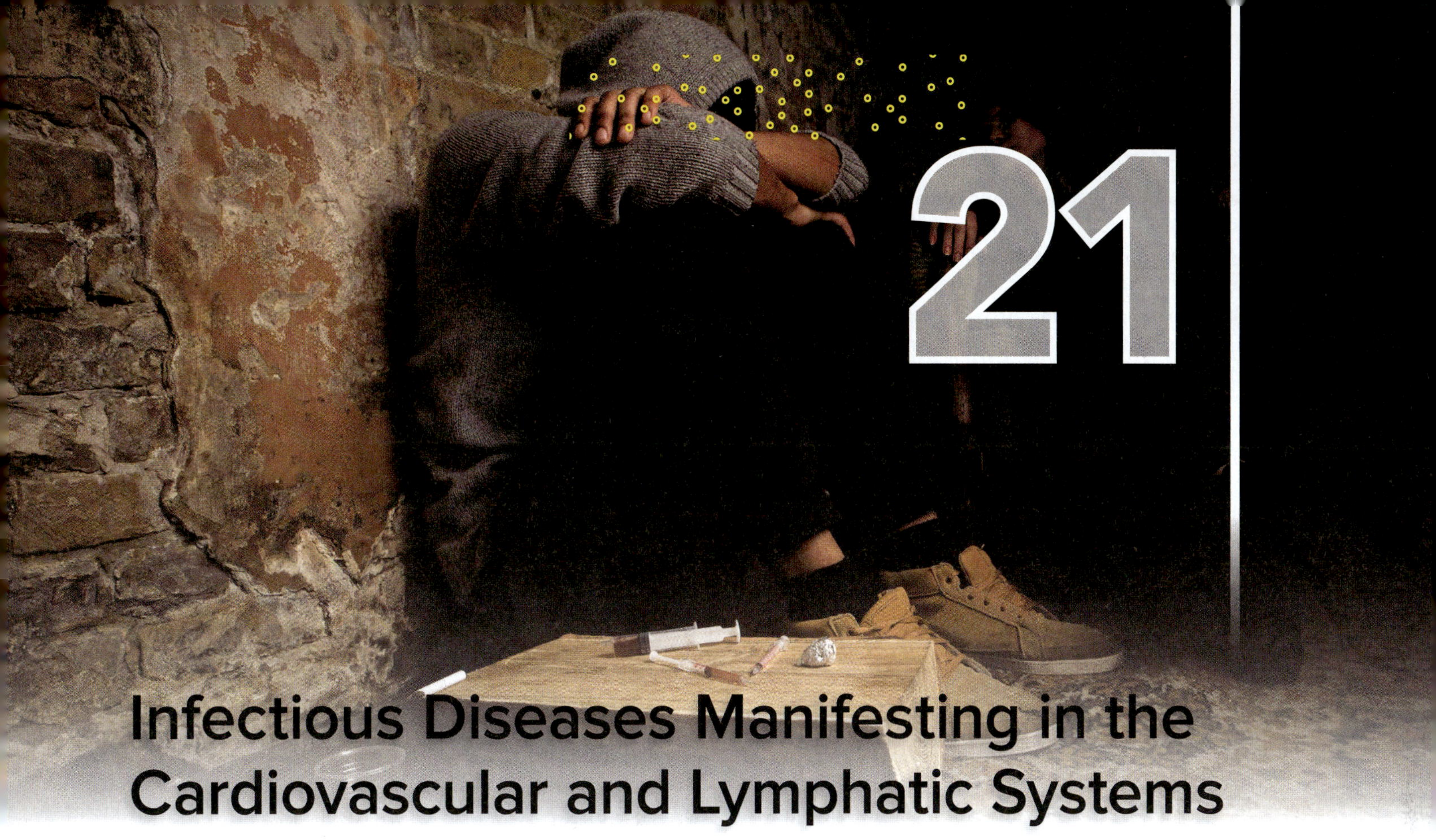

21

Infectious Diseases Manifesting in the Cardiovascular and Lymphatic Systems

Lipik/Shutterstock

MEDIA UNDER THE MICROSCOPE

The Opioid Epidemic Fuels Heart Valve Infections

This opening case examines an article from the popular media to determine the extent to which it is factual and/or misleading. This case focuses on a story from a news site called Newswise: *"Infective Endocarditis Increases Tenfold in North Carolina."*

In this chapter you will read about two kinds of infective endocarditis. Subacute and acute endocarditis are infections that target heart valves in particular. Acute endocarditis occurs when the bloodstream contains an overwhelming number of bacteria and/or fungi. That most often occurs when the microbes are introduced parenterally. This article reports that in one state, North Carolina, the number of hospitalizations and surgeries has increased by a factor of 10 during the last decade, due almost entirely to illicit injecting drug use. The most significant increase has occurred in the last four years. Before the opioid epidemic, fewer than 10 valve replacement surgeries were performed annually in the state. Now, according to the article, 109 surgeries are performed every year. This kind of increase is occurring all over the country.

One of the problems accompanying this increase in hospitalizations and surgeries is the increased cost to hospitals and to programs like Medicaid. The author of the article says that 73% of patients with drug-associated endocarditis are uninsured or on Medicaid. And it is a very expensive illness. The median cost of a hospitalization and valve surgery is $251,000.

- What is the **intended message** of the article?
- What is your **critical reading** of the summary of the article? Remember that in this context, "critical reading" does not necessarily mean *What criticism do you have?* but asks you to apply your knowledge to interpret whether the article is factual and whether the facts support the intended message.
- How would you **interpret** the news item for your nonmicrobiologist friends?
- What is your **overall grade** for the news item—taking into account its accuracy and the accuracy of its intended effect?

Media Under The Microscope Wrap-Up appears at the end of the chapter.

Outline and Learning Outcomes

21.1 The Cardiovascular and Lymphatic Systems, Their Defenses, and Normal Biota

1. Describe the important anatomical features of the cardiovascular system.
2. List the natural defenses present in the cardiovascular and lymphatic systems.
3. Discuss the current state of knowledge of the normal biota of the cardiovascular and lymphatic systems.

21.2 Infectious Diseases Manifesting in the Cardiovascular and Lymphatic Systems

4. List the possible causative agents for each of the following infectious cardiovascular conditions: COVID-19, endocarditis, plague, tularemia, Lyme disease, infectious mononucleosis, anthrax, Chagas disease, and malaria.
5. Explain why COVID-19 is placed in the cardiovascular chapter.
6. Discuss what series of events may lead to sepsis and how it should be prevented and treated.
7. Discuss the difference between hemorrhagic and nonhemorrhagic fever diseases.
8. List the possible causative agents and modes of transmission for hemorrhagic fever diseases.
9. List the possible causative agents and modes of transmission for nonhemorrhagic fever diseases.
10. Identify the five or six most relevant facts about malaria.
11. Describe important events in the course of an HIV infection in the absence of treatment.
12. Discuss the epidemiology of HIV infection in the United States and in the developing world.

21.1 The Cardiovascular and Lymphatic Systems, Their Defenses, and Normal Biota

The Cardiovascular System

The cardiovascular system is the pipeline of the body. It is composed of the blood vessels, which carry blood to and from all regions of the body, and the heart, which pumps the blood. This system moves the blood in a closed circuit, and it is therefore known as the *circulatory system*. The cardiovascular system provides tissues with oxygen and nutrients and carries away carbon dioxide and waste products, delivering them to the appropriate organs for removal. A closely related but largely separate system, the **lymphatic system** is a major source of immune cells and fluids, and it serves as a one-way passage, returning fluid from the tissues to the cardiovascular system.

The heart is a fist-size, muscular organ that pumps blood through the body. It is divided into two halves, each of which is divided into an upper and a lower chamber **(figure 21.1).** The upper chambers are called atria (singular, *atrium*), and the lower are ventricles. The entire organ is encased in a fibrous covering, the pericardium, which is occasionally a site of infection. The actual wall of the heart has three layers: from outer to inner, they are the epicardium, the myocardium, and the endocardium. The endocardium also covers the valves of the heart, and it is a relatively common target of microbial infection.

The right atrium receives blood coming from the body. This blood, which is low in oxygen and high in carbon dioxide, then passes through to the right ventricle. From there, it is pumped through the pulmonary arteries to the lungs, where it becomes oxygenated and reenters the heart through the left atrium. Finally, the blood moves into the left ventricle

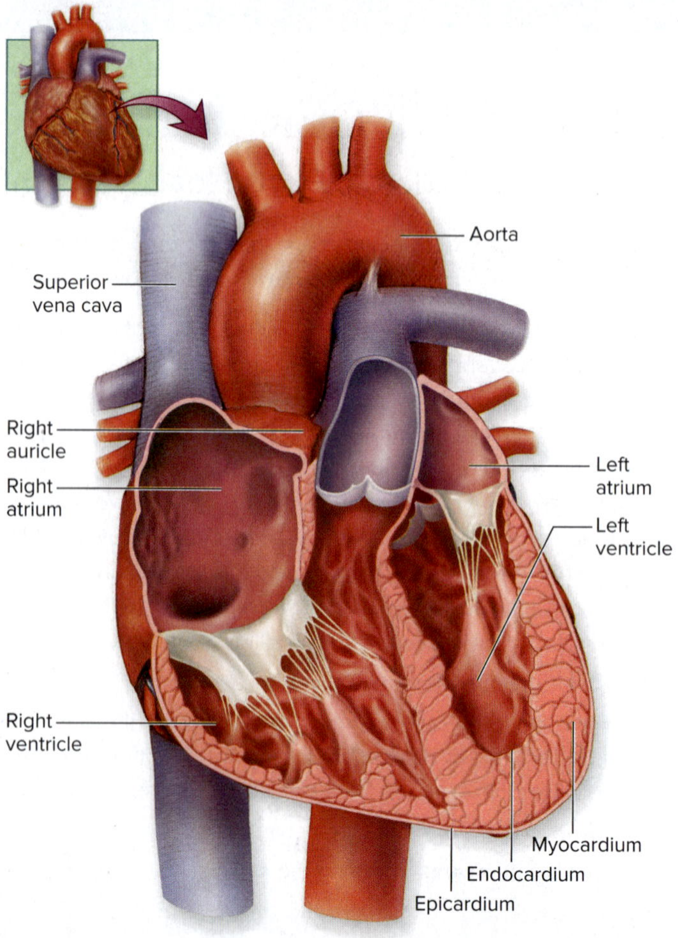
Aorta
Superior vena cava
Right auricle
Right atrium
Left atrium
Left ventricle
Right ventricle
Myocardium
Endocardium
Epicardium

Figure 21.1 The heart.

and is pumped into the aorta and the rest of the body. Valves control the movement of blood into and out of the chambers of the heart.

The blood vessels consist of *arteries, veins,* and *capillaries.* Systemic arteries carry oxygenated blood away from the heart under relatively high pressure. They branch into smaller vessels called arterioles. Veins actually begin as smaller venules in the periphery of the body and coalesce into veins. The smallest blood vessels, the capillaries, connect arterioles to venules. Both arteries and veins have walls made of three layers of tissue. The innermost layer is composed of a smooth epithelium called endothelium. Its smooth surface encourages the smooth flow of cells and platelets through the system. The next layer is composed of connective tissue and muscle fibers. The outside layer is a thin layer of connective tissue. Capillaries, the smallest vessels, have walls made of only one layer of endothelium. **Figure 21.2** illustrates the complete cardiovascular system.

The Lymphatic System

Chapter 15 provided a detailed description of the lymphatic system. You may want to review figure 15.3 before continuing. In short, the lymphatic system consists mainly of the lymphatic vessels, which roughly parallel the blood vessels; lymph nodes, which cluster at body sites such as the groin, neck, armpit, and intestines; and the spleen. This system collects fluid that has left the blood vessels and entered tissues, filters it of impurities and infectious agents, and returns it to the blood.

Defenses of the Cardiovascular and Lymphatic Systems

The cardiovascular system is highly protected from microbial infection. Microbes that successfully invade the system, however, gain access to every part of the body, and any system may be affected. For this reason, bloodstream infections are called **systemic infections.**

Multiple defenses against infection reside in the bloodstream. The blood is full of leukocytes, with approximately 5,000 to 10,000 white blood cells per microliter of blood. The various types of white blood cells include the lymphocytes, responsible for adaptive immunity, and the phagocytes, which are critical to innate as well as adaptive immune responses. Only special microbes can survive in the blood with so many defensive elements. That said, a handful of infectious agents have nonetheless evolved exquisite mechanisms for avoiding blood-borne defenses.

Medical conditions involving the blood often have the suffix *-emia.* For instance, viruses that cause meningitis can travel to the nervous system via the bloodstream. Their presence in the blood is called **viremia.** When fungi are in the blood, the condition is termed **fungemia.** Bacterial presence in the blood is called **bacteremia,** a general term denoting only their *presence.* Bacteria frequently are introduced into the bloodstream during the

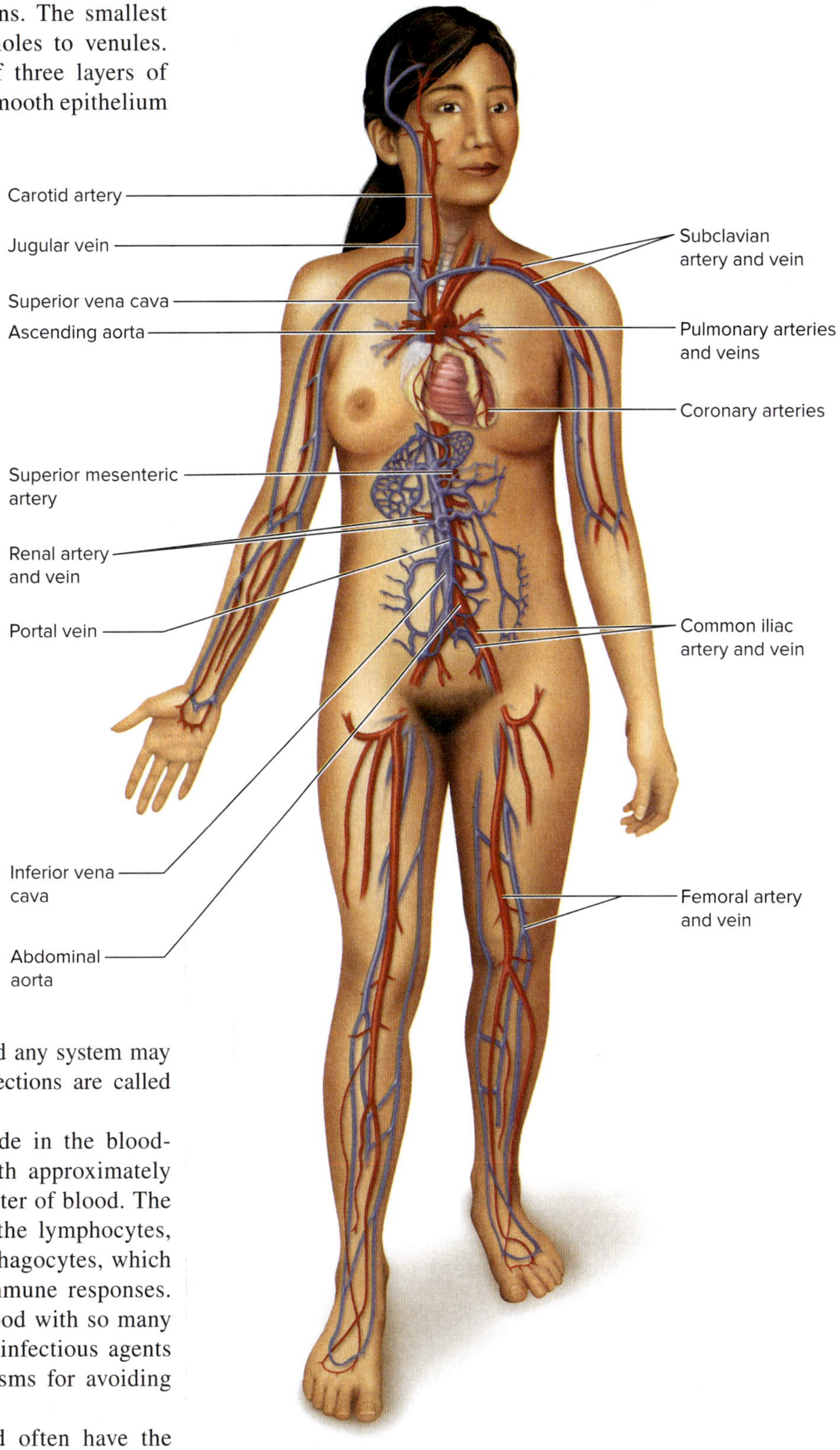
Carotid artery
Jugular vein
Superior vena cava
Ascending aorta
Subclavian artery and vein
Pulmonary arteries and veins
Coronary arteries
Superior mesenteric artery
Renal artery and vein
Portal vein
Common iliac artery and vein
Inferior vena cava
Abdominal aorta
Femoral artery and vein

Figure 21.2 The anatomy of the cardiovascular system.

course of daily living. Brushing your teeth or tearing a hangnail can introduce bacteria from the mouth or skin into the bloodstream. This situation is usually temporary, but there is mounting evidence of the development of more long-term effects as oral microbes have been localized in arterial plaques of heart disease patients. When bacteria flourish and grow in the bloodstream, the condition is termed **septicemia,** or sepsis. (There is a technical difference between bacteremia, the presence of bacteria in the blood, and septicemia, which refers to bacteria actively multiplying there.) Sepsis can very quickly lead to cascading immune responses, resulting in decreased systemic blood pressure, which can lead to **septic shock,** a life-threatening condition.

Normal Biota of the Cardiovascular and Lymphatic Systems

Like the nervous system, the cardiovascular and lymphatic systems are "closed" systems with no normal access to the external environment. But even in the absence of disease, microorganisms may be transiently present in either system as just described. The lymphatic system filters microbes and their products out of tissues. So, of course, microbes are found there transiently. It is less clear if the lymphatic system is home to a microbiome of its own. Recent findings from microbiome studies suggest that the bloodstream is not completely sterile, even during periods of apparent health. Bacterial DNA is mainly associated with the buffy layer, which contains the white blood cells, although there are trace amounts in red blood cells and in the plasma. It is tempting to speculate that these low-level microbial "infections" may contribute to diseases for which no infectious cause has been found.

Since we are talking about the cardiovascular system, it is worth noting that a loss of diversity in the gut microbiome has been associated with hypertension (high blood pressure).

Cardiovascular and Lymphatic System Defenses and Normal Biota

	Defenses	Normal Biota
Cardiovascular System	Blood-borne components of innate and adaptive immunity—including phagocytosis, adaptive immunity	Sparse; mostly in white blood cells
Lymphatic System	Numerous immune defenses reside here.	Unknown

21.1 Learning Outcomes—Assess Your Progress

1. Describe the important anatomical features of the cardiovascular system.
2. List the natural defenses present in the cardiovascular and lymphatic systems.
3. Discuss the current state of knowledge of the normal biota of the cardiovascular and lymphatic systems.

21.2 Infectious Diseases Manifesting in the Cardiovascular and Lymphatic Systems

Categorizing cardiovascular and lymphatic infections according to clinical presentation is somewhat difficult because most of these conditions are systemic, with effects on multiple organ systems. We begin with COVID-19, whose most acute symptoms are usually in the respiratory system, but whose effects show it to be a systemic infection of the cardiovascular system.

COVID-19

COVID-19 is a pandemic that began in December 2019 caused by a coronavirus named SARS-CoV-2. Right there in the name (SARS = severe acute respiratory syndrome), we see that its most acute and noticeable symptoms are in the respiratory system. But it became clear quickly that the virus accesses the cardiovascular system and, therefore, can spread throughout the body. With the intense attention on this virus and its effects, it has helped scientists and clinicians understand that quite a few viral illnesses are actually spread throughout the body by the cardiovascular system and can cause varying degrees of damage for some time after the initial respiratory symptoms. For that reason, we place COVID-19 in the cardiovascular chapter.

The second part of the virus name, CoV-2, stands for "coronavirus #2." The disease that the virus causes is called COVID-19, which stands for coronavirus disease 2019. It is the second coronavirus named SARS to cause a dangerous outbreak. The first one occurred in 2003, starting in China and spreading to several other countries. Rapid and effective measures were taken, and it was brought under control within several months.

To make things just a bit more confusing, a third coronavirus caused a dangerous disease starting in 2012. It was first discovered in Saudi Arabia and spread to several other Middle Eastern countries. It was given the name MERS-CoV, for Middle East Respiratory Syndrome. It has a very high case fatality rate but a low transmission rate. Of course, these things were not known at the beginning of an outbreak, so it was quite a cause for concern. A few cases have been seen in the United States, but they were seen in health care workers who had worked in Saudi Arabia, and they did not transmit it to contacts in the U.S.

Four other coronaviruses have infected humans for decades, and probably longer. They are a common cause of minor upper respiratory infections such as colds. The first dangerous coronavirus infection was that SARS outbreak in 2003.

SARS-CoV-2 is thought to have been circulating in bats for a very long time, and then experienced one or more mutations to make it capable of infecting, multiplying, and transmitting among humans after humans contacted it (via an intermediate animal) in a wildlife market in Wuhan, China, in 2019 **(figure 21.3).** The first COVID-19 case seems to have been identified in Wuhan in December 2019, and the virus probably began circulating in the United States in January 2020 and in Canada in March of that year.

Figure 21.3 **A wildlife market in Southeast Asia like the one thought to have triggered the leap of SARS-CoV-2 from animals to humans.**
Michele Burgess/Alamy Stock Photo

Signs and Symptoms

Of course, acute respiratory symptoms are prominent in this disease and can lead to rapid death in severe cases. The infection can set off a cytokine storm (see chapter 16), leading to what experts call hyperinflammation, both in the lungs and in other organs, including the digestive tract. A common symptom of acute COVID-19 is vomiting and diarrhea. The systemic effects of the virus can lead to skin rashes, muscle pain, heart damage, joint pain, and blood clots.

The virus is also considered neurotropic (infects the cells of the nervous system). A common symptom is the loss of taste and smell. This seems to be caused by viral disruption of the cells supporting olfactory neurons. In mild cases, the loss of smell and taste is the only symptom. In other cases, the virus can travel to the hippocampus in the brain, leading to headaches and even long-term cognitive decline and dementia. Researchers believe that brain irregularities caused by the infection are at least partially responsible for dysregulation in the lungs and in the digestive tract.

Surprisingly, up to one-third of infections can be asymptomatic. Many others are very mild. But those who have few or no symptoms are still susceptible to the long-term effects of COVID, which include many of the brain, heart, and digestive system symptoms described. This is often termed long-COVID and is still not well understood. Studies in 2022 showed that over 75 percent of long-COVID patients were not hospitalized for their initial illness, indicating that mild infections can still lead to the long term effects. Research hospitals around the country have set up long-haul COVID clinics, with practitioners from many specialties coming together to try to understand and treat these conditions.

Another condition that has been associated with COVID-19 infection is called MIS, multisystem inflammatory syndrome. There are a children's version and an adult's version (abbreviated MIS-C and MIS-A). It is characterized by fever and severe inflammation in the heart, lungs, kidneys, brain, skin, eyes, and/or gastrointestinal organs. It is more severe than the chronic long-COVID condition. It can be serious, requiring hospitalization, and occasionally can lead to death. New evidence also suggests that babies born to women infected with SARS-Co-V-2 during certain parts of pregnancy have a higher rate of neurological damage.

Causative Agents

The causative agent, SARS-CoV-2, is one of the seven members of the *Coronaviridae* family that infects humans. Hundreds more infect other animals. Since this virus started causing COVID-19, it has continued to accumulate mutations in its genome, which have given rise to variants. These have received distinguishing names. The first variant was termed Alpha, and it was more contagious than the original virus. The Delta variant arose in India and was first documented in the United States in March 2021. It was almost twice as transmissible as the original, and it was more deadly. Remember: the dynamics of both transmissibility and deadliness determine how a disease affects a population.

A variant called Omicron followed Delta. Omicron variant BA2 dominated in North America for the first half of 2022. Omicron variants BA4 and BA5 arrived in the summer of 2022. The variants will continue to come, and each will have its own transmissibility and virulence characteristics.

The viruses are enveloped, positive-sense RNA viruses (class IV in table 7.5 in chapter 7). They use spike proteins (S proteins) to attach to particular macromolecules on host cell surfaces.

Pathogenesis and Virulence Factors

The S protein of the virus attaches to a host cell protein called ACE-2 and uses it to initiate its entry into the cell. When it binds the protein, it also blocks its functions, one of which is to dampen the inflammatory response. So, the brakes on inflammation are lifted, leading to the hyperinflammation discussed earlier. ACE-2 proteins are found on the surface of cells in the lungs, brain, heart, blood vessels, kidneys, liver, placenta, and gastrointestinal tract. So if the virus leaves the upper respiratory tract, it will potentially damage those tissues, resulting in the reported symptoms.

Some research suggests that infection with COVID-19 "breaks" the immune system's tolerance to self, producing large quantities of auto-antibodies that contribute to acute damage and may lead to autoimmune conditions post-infection.

Transmission and Epidemiology

The virus is transmitted via respiratory droplets and is also airborne; thus, it can infect by direct and indirect contact. Early in the pandemic it was mistakenly thought that it was not necessarily airborne, and that misperception plagues us still. Face shields and plexiglass partitions in restaurants are less effective than we think. However, plexiglass barriers between people making a transaction, such as a patient checking in at a receptionist's desk, are useful for stopping droplets. Some of the variants are more transmissible than others, and that is due to two main factors: how well viruses can attach to the epithelial cells in the upper respiratory tract of a new host and how quickly they replicate, creating a larger or smaller load of virus to be expelled from the infected host.

As the virus and its variants have washed over the planet in waves, we have witnessed how a highly communicable microbe reaches pandemic level. Public health officials acted with unprecedented speed to try to find measures that would stop the spread. Because it was a brand new strain of coronavirus, and we started out knowing very little about it, mistakes were made and the advice changed over time. This led some poeple to distrust what was being said by scientists, but that is a misunderstanding of how science works. Scientists are constantly revising their ideas as more information arises. It is the very nature of the field.

Before the first vaccines were made available in December 2020, the only preventive measures we had were the use of masks, social distancing, and hand washing. The societal upheaval that this virus—and the divisive responses to mitigation efforts—caused worldwide had not been experienced since the flu pandemic of 1918. Seemingly overnight in March 2020, schools transitioned to online (at-home) learning, workers were sent home to work, and refrigerated trucks were deployed as temporary morgues in cities like New York when there was no more room for the deceased in the regular facilities. In some countries, including the United States, there was a resistance movement among people who did not believe the virus was real and/or dangerous. Intense disagreements broke out over the use of masks. The vaccine was also refused by a significant percentage of the population.

We have now found that this zoonotic virus can cycle back into animal populations, undergo multiple mutations, and reinfect humans. The virus has been found in deer and mink populations in large numbers.

Culture and/or Diagnosis

Creating a test for COVID-19 infection was a very high priority once the virus emerged. The genome was sequenced within days and enabled a nucleic acid amplification test to be developed quickly. In the end, dozens of diagnostic tests were quickly developed. They included genotypic tests, such as RT-PCR, and immunological tests measuring both antigen and antibody.

Prevention

Physical measures to prevent transmission include anything that prevents a virus leaving someone's respiratory tract and reaching yours. Simple as that. Social distancing, which means keeping 6 feet between people, can eliminate the droplet transmission. Because the virus is also airborne, wearing a mask over the mouth and nose is important. Masks of varying types provide varying levels of protection. The very best masks are called respirators. N95 and KN95 come in second. The common blue surgical masks and cloth masks are not highly effective but if you wear one over the other their effectiveness is improved. Any mask is better than none.

Vaccines are key. Scientists and pharmaceutical companies worked with unprecedented speed to produce them. At around the same time at the end of 2020, three vaccines were given emergency use authorization approval by the Food and Drug Administration (FDA) in the United States. (Emergency use authorization is discussed in chapter 16.) Other vaccines were approved in other countries as well. Two of the vaccines, produced by Pfizer BioNTech and Moderna, consist of mRNA coding for the spike protein of SARS-CoV-2. The mRNA is encased in a lipid envelope that penetrates host cells, delivers the mRNA, and causes the cells to produce and display the spike protein, educating the immune system so that it will respond forcefully to natural infection.

A third vaccine, made by Johnson and Johnson, is called a viral vector vaccine. It consists of a modified adenovirus that is harmless to humans. The virus contains a snippet of genetic material coding for the SARS-CoV-2 spike protein, so the adenovirus produces the spike protein and, as with the other vaccines, this educates the vaccine recipient's immune system to recognize the actual SARS-CoV-2 if it invades.

All three of the vaccines were found to have a very good safety profile. They all showed very high effectiveness. As new variants continued to arise, vaccinations became less effective at preventing cases, but still were highly protective against the worst outcomes of the infection. Because of the constant race to create vaccines that are effective against new variants, and because the later variants seem to be able to infect people infected with previous variants and the vaccinated, masks and social distancing remain a vital strategy.

Treatment

In the early days of the epidemic, a variety of ineffective and even dangerous treatments such as ivermectin, an animal de-wormer, were encouraged by the media, but they were never approved and were shown to be ineffective.

The drug Paxlovid, approved in late 2021 for patients with high risk for COVID complications, proved effective in controlling the infection if started soon after infection. After widespread use, though, it was found that a certain percentage of people experienced rebounding symptoms when the drug was stopped.

Disease Table 21.1	COVID-19
Causative Organism(s)	SARS-CoV-2 V
Most Common Modes of Transmission	Droplet, airborne
Virulence Factors	Attachment to ACE-2; induction of autoimmunity
Culture/Diagnosis	RT-PCR, Ab and Ag tests
Prevention	Vaccine, mitigation efforts
Treatment	Antivirals such as paxlovid™
Epidemiological Features	Spreading constantly worldwide

Infective Endocarditis

Endocarditis is an inflammation of the endocardium, or inner lining of the heart. Infectious endocarditis refers to an infection of the valves of the heart, such as the mitral or aortic valves **(figure 21.4).** Two variations of infectious endocarditis have been described: acute and subacute. Each has distinct groups of possible causative agents, most of which are bacterial. Fungi have been

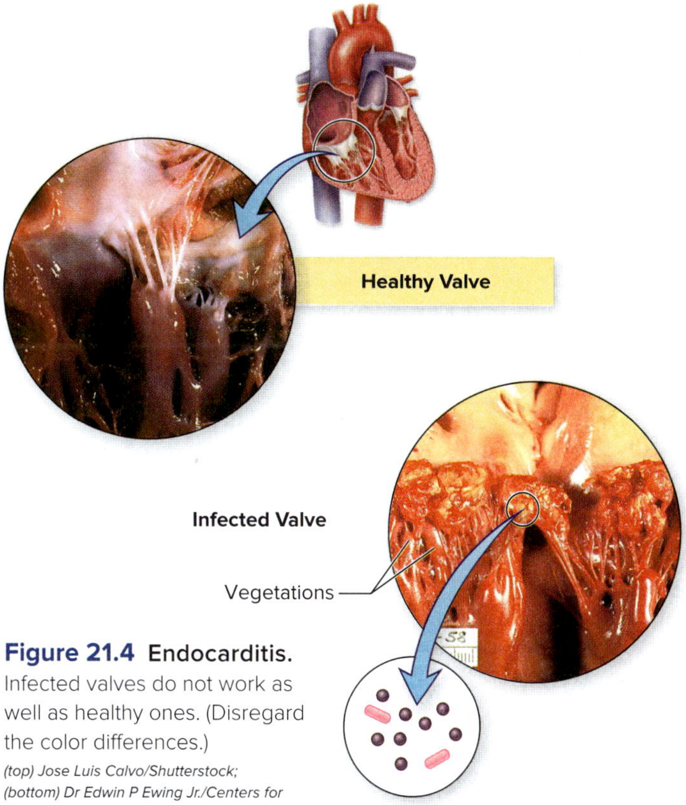
Healthy Valve
Infected Valve
Vegetations

Figure 21.4 **Endocarditis.** Infected valves do not work as well as healthy ones. (Disregard the color differences.)
(top) Jose Luis Calvo/Shutterstock; (bottom) Dr Edwin P Ewing Jr./Centers for Disease Control and Prevention

documented to cause rare cases of endocarditis, while the role of viruses in this disease is still under investigation.

In acute forms of the disease, an overwhelming number of bacteria are introduced into the body, through severe injury or injecting drug use. In this situation, bacteria can colonize healthy valves. In subacute forms, a trickle of bacteria in the bloodstream, introduced through daily activities such as toothbrushing or minor injuries, encounter damaged or scarred valves and find that a hospitable place to attach and accumulate. Accumulations of bacteria on the valves (vegetations) hamper their function and can lead directly to cardiac malfunction and death. Alternatively, pieces of the bacterial vegetation can break off and create emboli (blockages) in vital organs. These bacterial biofilms can also provide a constant source of blood-borne bacteria, with the accompanying systemic inflammatory response and shock. Bacteria that are attached to surfaces bathed by blood (such as heart valves) quickly become covered with a mesh of fibrin and platelets that protects them from the immune components in the blood.

Patients with prosthetic valves can acquire acute endocarditis if bacteria are introduced during the surgical procedure to install them, and such infections result in high rates of morbidity and mortality. Alternatively, the prosthetic valves can serve as infection sites for the subacute form of endocarditis long after the surgical procedure.

Signs and Symptoms

The signs and symptoms are similar for both types of endocarditis, except that in the subacute condition they develop more slowly and are less pronounced than with the acute disease. Symptoms include fever, anemia, abnormal heartbeat, and sometimes symptoms similar to myocardial infarction (heart attack). Shortness of breath is a common symptom. Chills may also develop. Abdominal or side pain is sometimes reported.

The patient may look very ill and may have petechiae (small, red-to-purple spots) over the upper half of the body and under the fingernails. Red, painless skin spots on the palms and the soles (Janeway lesions) and small, painful nodes on the pads of fingers and toes (Osler's nodes) may also be apparent on examination. In subacute cases, an enlarged spleen may have developed over time. Longstanding cases can lead to clubbed fingers and toes due to lack of oxygen in the blood.

Culture and/or Diagnosis

The diagnostic procedures for the two forms of endocarditis are essentially the same. One of the most important diagnostic tools is a high index of suspicion. A history of risk factors or behaviors, such as abnormal valves, intravenous drug use, recent surgery, or bloodstream infections, should lead one to consider endocarditis when the symptoms just described are observed. Blood cultures, if positive, are the gold standard for diagnosis, but negative blood cultures do not rule out endocarditis. Echocardiography is an important adjunct to blood cultures as well.

In acute endocarditis, the symptoms may be magnified. The patient may also display central nervous system symptoms suggestive of meningitis, such as stiff neck or headache.

Causative Agents

The acute form of endocarditis is most often caused by *Staphylococcus aureus.* Other agents that cause it are *Streptococcus pyogenes, Streptococcus pneumoniae, Enterococcus,* and *Pseudomonas aeruginosa* as well as a host of other bacteria.

Most commonly, subacute endocarditis is caused by bacteria of low pathogenicity, often coming from the oral cavity. Alpha-hemolytic streptococci, such as *Streptococcus sanguinis, S. oralis,* and *S. mutans,* are most often responsible, although normal biota from the skin and other bacteria can also colonize abnormal valves and lead to this condition. People who have suffered rheumatic fever and the accompanying damage to heart valves are particularly susceptible to this condition.

Transmission and Epidemiology

The most common route of transmission for acute endocarditis is parenteral—that is, via direct entry into the body. Intravenous or subcutaneous drug users are a significant risk group for the condition. As highlighted in the opening Media Under The Microscope story, the heroin epidemic currently sweeping the United States has led to a large increase in the incidence of acute endocarditis, usually caused by *Staphylococcus aureus.* In many hospitals, on any given day, the majority of an infectious disease physician's list of consults is for endocarditis secondary to injecting drug use. These patients often also suffer from infections of the vertebrae and epidural abscesses, caused by bacteria from the bloodstream forming abscesses along the spine. Traumatic injuries and surgical procedures can also introduce the large number of bacteria required for the acute form of endocarditis.

As mentioned above, subacute forms of the infection almost always require damaged or misshapen valves, in which case normal transient bloodstream bacteremias can cause endocarditits.

Disease Table 21.2 Infective Endocarditis

Disease	Acute Endocarditis B	Subacute Endocarditis B
Causative Organism(s)	*Staphylococcus aureus, Streptococcus pyogenes, S. pneumoniae, Enterococcus, Pseudomonas aeruginosa,* others	Alpha-hemolytic streptococci, others
Most Common Modes of Transmission	Parenteral	Endogenous transfer of normal biota to bloodstream
Culture/Diagnosis	Blood culture	Blood culture
Prevention	Aseptic surgery, injections	Prophylactic antibiotics before invasive procedures
Treatment	Vancomycin; surgery	Broad-spectrum antibiotics; surgery may be necessary
Distinctive Features	Acute onset, high fatality rate	Slower onset
Epidemiological Features	Greatly increased incidence due to heroin epidemic	–

Prevention and Treatment

Prevention is based on avoiding the introduction of bacteria into the bloodstream during surgical procedures or injections. Vancomycin is the antibiotic to start with until bacteria and their susceptibilities are identified. High, continuous blood levels of antibiotics are required to resolve the infection because the bacteria tend to exist in biofilm vegetations. In addition to the decreased access of antibiotics to bacteria deep in the biofilm, these bacteria often express a phenotype of lower susceptibility to antibiotics. Surgical debridement of the valves, accompanied by antibiotic therapy, may be required.

Prophylactic antibiotics are prescribed for patients before surgery, in order to prevent subacute endocarditis. The same practice was used for decades in advance of routine dental procedures but is no longer used unless there is a high risk for endocarditis. **(Disease Table 21.2).**

Sepsis

Many different bacteria (and a few fungi) can cause this condition. Because organisms are actively multiplying in the bloodstream, sepsis is also called septicemia. Patients suffering from these infections are sometimes described as "septic." The condition is also called BSI, or bloodstream infection.

Signs and Symptoms

Fever is a prominent feature of sepsis. The patient appears very ill and may have an altered mental state, shaking chills, and gastrointestinal symptoms. Often an increased breathing rate is exhibited, accompanied by respiratory alkalosis (increased tissue pH due to breathing disorder). Low blood pressure is a hallmark of this condition and is caused by the inflammatory response to infectious agents in the bloodstream, which leads to a loss of fluid from the vasculature. This condition is the most dangerous feature of the disease, often culminating in death.

Causative Agents

Bacteria cause the vast majority of septicemias and are evenly divided between gram-positive and gram-negative organisms. Perhaps 10% are caused by fungal infections. Polymicrobial bloodstream infections involving more than one microorganism are increasingly being identified.

An important emerging cause of sepsis is the yeast *Candida auris*. This microbe was just discovered in 2009 but has quickly become a great concern to public health specialists. Most isolates are resistant to several classes of antifungal drugs, and some of them have become resistant to all of them.

Pathogenesis and Virulence Factors

Gram-negative bacteria multiplying in the blood release large amounts of endotoxin into the bloodstream, stimulating a massive inflammatory response mediated by a host of cytokines. This response invariably leads to a drastic drop in blood pressure, a condition called **endotoxic shock.** Gram-positive bacteria can instigate a similar cascade of events when fragments of their cell walls are released into the blood.

Transmission and Epidemiology

In some cases, sepsis can be traced to parenteral introduction of microorganisms via intravenous lines or surgical procedures. Other infections may arise from serious urinary tract infections or from renal, prostatic, pancreatic, gallbladder, skin, or bone abscesses. Patients with underlying spleen malfunction may be predisposed to multiplication of microbes in the bloodstream. Meningeal infections, bone infections (osteomyelitis), or pneumonia can all occasionally lead to sepsis. It should be noted that hospitalization for sepsis has more than doubled in recent years.

More alarming is the fact that there is a 20% to 50% mortality rate associated with septic infections, reflecting the high risks of this disease even when treatment is available. Also, one in three patients who die in a hospital dies from sepsis.

Culture and/or Diagnosis

Because the infection is in the bloodstream, a blood culture is the obvious route to diagnosis. As we saw in the diagnostics chapter, faster methods of identification and susceptibility testing are being developed. These include nucleic acid amplification tests, MALDI-TOF, and others. BSIs are at the intersection of two serious problems: an infection that can quickly kill and the desire not to use broad-spectrum antibiotics. Traditionally the broad-spectrum drugs are prescribed immediately, until the microbe's identity and susceptibility can be determined. The newer techniques can enable us to skip this step.

Prevention and Treatment

Empiric therapy, which is begun immediately after blood cultures are taken, often begins with a broad-spectrum antibiotic. Once the organism is identified and its antibiotic susceptibility is known, treatment can be adjusted accordingly. Statistics show that successful patient outcomes depend on rapid diagnosis and treatment. Increasingly, deep sequencing techniques are being used for faster and more appropriate treatment.

Disease Table 21.3	Sepsis
Causative Organism(s)	Bacteria or fungi
Most Common Modes of Transmission	Parenteral, endogenous transfer
Virulence Factors	Cell wall or membrane components
Culture/Diagnosis	Blood culture, deep sequencing
Prevention	–
Treatment	Broad-spectrum antibiotic until identification and susceptibilities tested. *C. auris* is in **Urgent Threat** category in CDC Antibiotic Resistance Report
Epidemiological Features	In United States: 1.7 million cases and 270,000 deaths per year

Plague

Although pandemics of plague have probably occurred since ancient times, the first one that was reliably chronicled killed an estimated 100 million people in the sixth century AD. The last great pandemic occurred in the late 1800s and was transmitted around the world, primarily by rat-infested ships. The disease was brought to the United States through the port of San Francisco around 1906. Infected rats eventually mingled with native populations of rodents and gradually spread the disease throughout the West and Southwest, where it is endemic today.

Signs and Symptoms

Three possible kinds of infection occur with the bacterium causing plague. **Pneumonic plague** is a respiratory disease. In **bubonic plague,** the bacterium, which is injected by the bite of a flea, enters the lymph and is filtered by a local lymph node. Infection causes inflammation and necrosis of the node, resulting in a swollen lesion called a **bubo,** usually in the groin or axilla **(figure 21.5*a*).** The incubation period lasts 2 to 8 days, ending abruptly with the onset of fever, chills, headache, nausea, weakness, and tenderness of the bubo.

These cases often progress to massive bacterial growth in the blood termed **septicemic plague.** The presence of the bacteria in the blood results in disseminated intravascular coagulation, subcutaneous hemorrhage, and purpura (red or purple discolorations of the skin) that may degenerate into necrosis and gangrene. Mortality rates, once the disease has progressed to this point, are 30% to 50% with treatment and 100% without treatment. Because of the visible darkening of the skin, the plague is often called the "Black Death."

Causative Agent

The cause of this dreadful disease is a tiny, gram-negative rod, *Yersinia pestis,* a member of the family *Enterobacteriaceae. Y. pestis* displays unusual bipolar staining that makes it look like a safety pin **(figure 21.5*b*).**

Pathogenesis and Virulence Factors

The number of bacteria required to initiate a plague infection is small—perhaps only 3 to 50 cells for bubonic or septicemic cases. *Y. pestis* carries genes that help it to cause disease in mice and to survive in the flea vector. These genes include a gene for capsule formation and a gene for plasminogen activation (similar to the streptokinase expressed by *S. pyogenes*). Plasminogen activation assists with *Yersinia* dissemination and avoidance of the immune system.

Transmission and Epidemiology

The principal agents in the transmission of the plague bacterium are fleas. These tiny, bloodsucking insects have a special relationship with the bacterium. After a flea ingests a blood meal from an infected animal, the bacteria multiply in its gut. In fleas that effectively transmit the bacterium, the esophagus becomes blocked due to coagulation factors produced by the pathogen. Being unable to feed properly, the ravenous flea jumps from animal to animal in a futile attempt to get nourishment. During this process, regurgitated infectious material is inoculated into the bite wound.

The plague bacterium exists naturally in many animal hosts. Plague still exists endemically in large areas of Africa, South America, the Mideast, Asia, and the former Soviet Union. A large

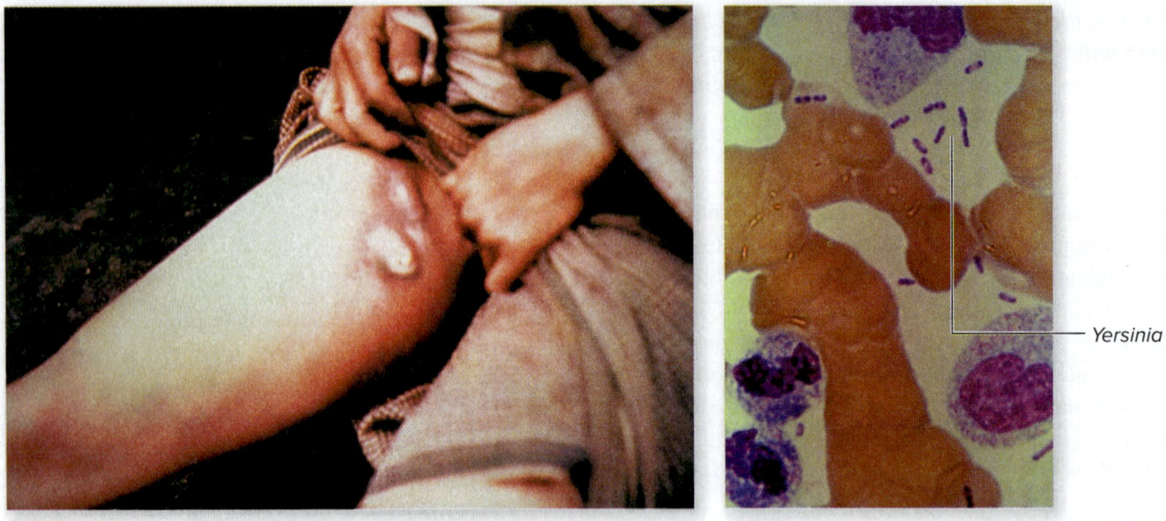
Yersinia

(a) (b)

Figure 21.5 Bubonic plague. **(a)** A classic inguinal bubo of bubonic plague. This hard nodule is very painful and can rupture onto the surface. **(b)** *Yersinia pestis.* These bacteria are said to have a "safety pin" appearance.

(a, b) Centers for Disease Control and Prevention

outbreak involving over 2000 cases and 200 deaths occurred in 2017 in Madagascar. In the United States, sporadic cases (usually fewer than 10 per year) occur as a result of contact with wild and domestic animals **(figure 21.6).** In 2012, a young girl in Colorado was diagnosed with plague after attempting to bury a dead squirrel while camping. Her sweatshirt was on the ground during the procedure and at some point she tied it around her waist, where later several insect bites were found. Persons with the pneumonic form of the disease can spread *Y. pestis* through respiratory droplets.

The Animal Reservoirs The plague bacterium occurs in 200 different species of mammals. The primary long-term *endemic reservoirs* are various rodents, such as mice and voles, that harbor the organism but do not develop the disease. These hosts spread the disease to other mammals, called *amplifying hosts,* that become infected with the bacterium and experience massive die-offs during epidemics.

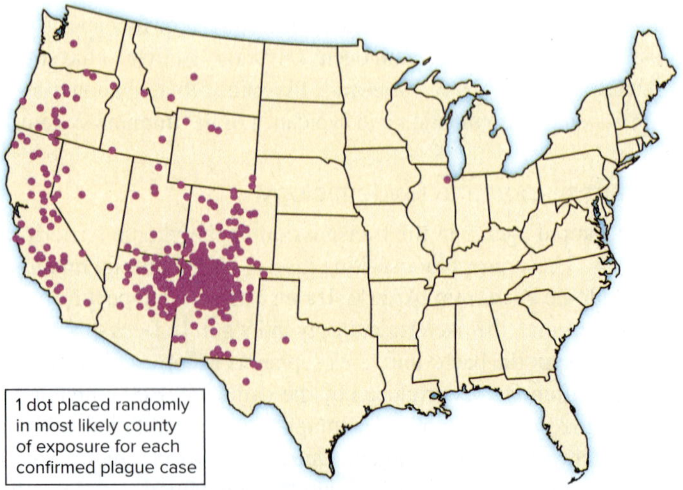
1 dot placed randomly
in most likely county
of exposure for each
confirmed plague case

Figure 21.6 Reported cases of human plague—United States, 1970–2017 (latest data). The single case in Illinois was lab-associated.

Culture and/or Diagnosis

Today, rapid genomic and immunochromatographic tests have been developed to quickly diagnose infection. These are critical tools in light of the potential use of *Y. pestis* as a biological weapon.

Prevention and Treatment

Plague is one of a handful of internationally quarantinable diseases (other examples are cholera and yellow fever). Currently, there

Disease Table 21.4 Plague

Causative Organism(s)	*Yersinia pestis* B G–
Most Common Modes of Transmission	Biological vector (flea), also droplet contact (pneumonic) and direct contact with body fluids
Virulence Factors	Capsule, plasminogen activator
Culture/Diagnosis	Rapid genomic methods
Prevention	Flea and/or animal control; vaccine available for high-risk individuals
Treatment	Streptomycin or ciprofloxacin
Epidemiological Features	United States: endemic in all western and southwestern states; internationally, 95% of human cases occur in Africa, including Madagascar **Category A Bioterrorism Agent**

is no vaccine available in the United States. This would be an important step in protecting the human population from potential bioterrorism attacks involving plague. Streptomycin or ciprofloxacin can be used in the treatment of disease in most cases, though the looming threat of drug resistance in *Y. pestis* could make these drugs ineffective in the future **(Disease Table 21.4).**

Tularemia

Signs and Symptoms

Tularemia is a zoonotic disease that is endemic throughout the Northern Hemisphere. After an incubation period ranging from a few days to 3 weeks, acute symptoms of headache, backache, fever, chills, malaise, and weakness appear. Further clinical manifestations are tied to the portal of entry. They might be ulcerative skin lesions **(figure 21.7),** swollen lymph glands, conjunctival inflammation, sore throat, intestinal disruption, or pulmonary involvement. The death rate in the most serious forms of disease is 30%, but proper treatment reduces mortality to almost zero.

Causative Agent

The causative agent of tularemia is a facultative intracellular gram-negative bacterium called *Francisella tularensis.* It has several characteristics in common with *Yersinia pestis,* and the two species were previously included in a single genus called *Pasteurella.* It is a zoonotic disease of assorted mammals endemic to the Northern Hemisphere. Because it has been associated with outbreaks of disease in wild rabbits, it is sometimes called "rabbit fever." It is currently listed as a pathogen of concern on the list of Category A bioterrorism agents.

Transmission and Epidemiology

Tularemia is abundantly distributed through numerous animal reservoirs and vectors in northern Europe, Asia, and North America but not in the tropics. This disease is noteworthy for its complex epidemiology and spectrum of symptoms. Although rabbits and rodents (muskrats and ground squirrels) are the chief reservoirs, other wild animals (skunks, beavers, foxes, opossums) and some domestic animals are implicated as well. The chief route of transmission in the past had been through the activity of skinning rabbits, but with the decline of rabbit hunting, transmission via tick bites is more common. Ticks are the most frequent arthropod vector, followed by biting flies, mites, and mosquitoes. Due to this shift in transmission, disease cases now occur more often in summer months rather than the winter, as was previously seen.

Tularemia is strikingly varied in its portals of entry and disease manifestations. Although bites by a vector are the most common source of infection, in many cases infection results when the skin or eye is inoculated through contact with infected animals, animal products, contaminated water, and dust. Pulmonary forms of the infection can result from aerosolized soils or animal fluids and from spread of the bacterium in the bloodstream. The disease is not spread from human to human. With disease developing after exposure to just 10 to 50 organisms, *F. tularensis* is often considered one of the most infectious of all bacterial pathogens. The name "lawnmower" tularemia refers to tularemia acquired when people have accidentally run over dead rabbits while lawn mowing, presumably from inhaling aerosolized bacteria. In recent years, two people in Alaska acquired tularemia after pulling infected rabbits from their dogs' mouths.

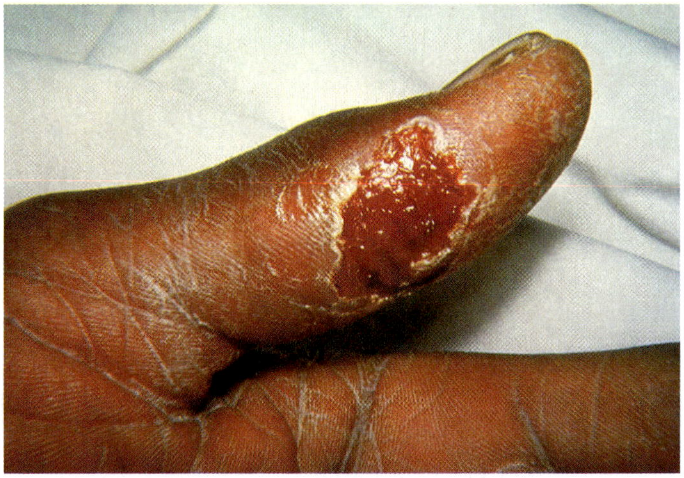

Figure 21.7 A skin ulcer caused by *F. tularemia.*
Dr Thomas F Sellers, Emroy University, Atlanta GA/ CDC

Prevention and Treatment

Treatment typically involves the use of gentamicin or streptomycin. Because the intracellular persistence of *F. tularensis* can lead to relapses, antimicrobial therapy must not be discontinued prematurely.

Disease Table 21.5	Tularemia
Causative Organism(s)	*Francisella tularensis* B G–
Most Common Modes of Transmission	Biological vector (tick); also direct contact with body fluids from infected animal; airborne
Virulence Factors	Intracellular growth
Culture/Diagnosis	Culture dangerous to lab workers and not reliable; serology most often used; fine needle aspirations of lymph nodes sometimes used
Prevention	–
Treatment	Gentamicin or streptomycin
Epidemiological Features	United States: several hundred cases per year; internationally, 500,000 cases per year **Category A Bioterrorism Agent**

Lyme Disease

In the 1970s, a mysterious cluster of arthritis cases appeared in the town of Old Lyme, Connecticut. The process of its discovery began in the home of Polly Murray, who, along with her family, was beset for years by recurrent bouts of stiff neck, swollen joints, malaise, and fatigue that seemed vaguely to follow a rash from tick bites. When Mrs. Murray's son was diagnosed as having juvenile rheumatoid arthritis, she became skeptical. Conducting her own literature research, she began to discover inconsistencies. Rheumatoid arthritis was described as a rare, noninfectious disease, yet over an 8-year period, she found that 30 of her neighbors had experienced similar illnesses. Ultimately, this cluster of cases and others were reported to state health authorities. Eventually, Lyme disease was shown to be caused by *Borrelia burgdorferi.* It is now recognized that Lyme disease has been around for centuries. Lyme disease and other tickborne diseases are being watched carefully because climate change has led to these insect vectors living in places in which they previously had not.

Signs and Symptoms

Lyme disease is slow-acting, but it often evolves into a progressive syndrome that mimics neuromuscular and rheumatoid conditions. An early symptom in some cases is a rash at the site of a tick bite. The lesion, called *erythema migrans,* can look like a bull's-eye, with a raised, erythematous (reddish) ring that gradually spreads outward and a pale central region **(figure 21.8).** Until recently, this lesion was thought to be the most common presentation of Lyme disease. But the lesion only has this appearance in about 10% of cases. It can also be flat and scaly with no clear areas, or it can be pustular. It can mimic the appearance of ringworm. Clinicians now realize that many Lyme disease cases have been missed because of the unpredictable nature of the initial lesion. Other early symptoms are fever, headache, stiff neck, and dizziness. If not treated or if treated too late, the disease can advance to the second stage, during which cardiac and neurological symptoms, such as facial palsy, can develop. After several weeks or months, a crippling polyarthritis can attack joints. Some people experience chronic neurological complications that are severely disabling.

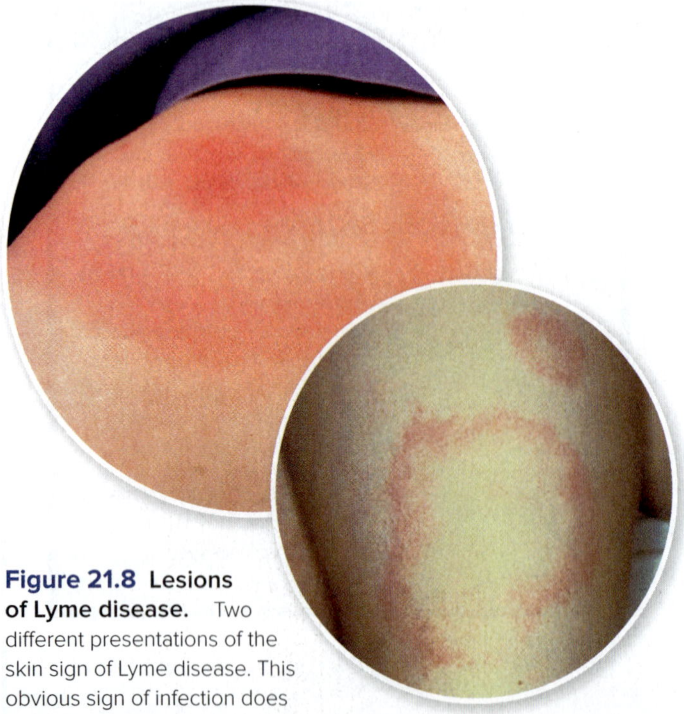

Figure 21.8 Lesions of Lyme disease. Two different presentations of the skin sign of Lyme disease. This obvious sign of infection does not always occur.
(arm, leg) James Gathany/CDC

Causative Agent

Although *Borrelia burgdorferi* is a spirochete bacterium, it is morphologically distinct from other pathogenic spirochetes. *B. burgdorferi* is comparatively larger, ranging from 0.2 to 0.5 micrometer in width and from 10 to 20 micrometers in length, and it exhibits 3 to 10 irregularly spaced and loose coils **(figure 21.9).** The nutritional requirements of *Borrelia* are complex, and culturing the bacterium in artificial media is difficult at best. Closely related *Borrelia burgdorferi*-like strains can also cause the disease.

Pathogenesis and Virulence Factors

The bacterium is very good at evading the immune system. It changes its surface antigens while it is in the tick and again after it has been transmitted to a mammalian host. It provokes a strong humoral and cellular immune response, but this response is mainly ineffective, perhaps because of the bacterium's ability to switch its antigens. In fact, it is possible that the immune response contributes to the pathology of the infection.

B. burgdorferi also has multiple proteins for attachment to host cells. These are considered virulence factors as well.

Transmission and Epidemiology

B. burgdorferi is transmitted primarily by hard ticks of the genus *Ixodes.* In the northeastern part of the United States, *Ixodes*

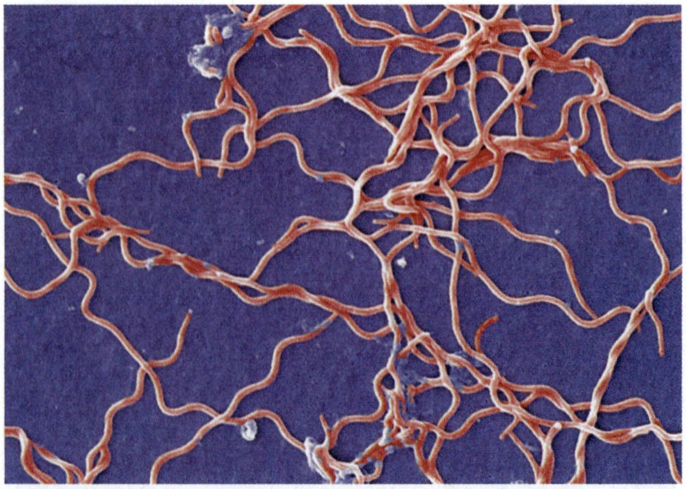

Figure 21.9 ***Borrelia burgdorferi.*** This spirochete has 3–10 loose, irregular coils.
Janice Haney Carr/CDC

scapularis (the black-legged deer tick) passes through a complex 2-year cycle that involves two principal hosts **(figure 21.10).** As a larva or nymph, it feeds on the white-footed mouse, birds, or raccoons, where it picks up the infectious agent. The nymph is relatively nonspecific and will try to feed on nearly any type of vertebrate—for that reason, it is the form most likely to bite humans. The adult tick reproductive phase of the cycle is completed on deer. In California, the transmission cycle involves *Ixodes pacificus,* another black-legged tick, and the dusky-footed woodrat as a reservoir.

The greatest concentrations of Lyme disease are found in areas having high deer populations. Most of the cases have occurred in New York, Pennsylvania, Connecticut, New Jersey, Rhode Island, and Maryland, but the numbers in the Midwest and West are growing. Highest-risk groups include hikers, backpackers, and people living in newly developed communities near woodlands and forests. A study in 2021 showed that ticks carrying Lyme disease are plentiful on California beaches, which is a surprising new finding.

Culture and/or Diagnosis

Culture of the organism is not useful. Diagnosis in the early stages, while the rash is present, is usually accomplished based on symptoms and a history of possible exposure to ticks because the organism is not easily detectable at this stage. Acute and convalescent sera may be helpful.

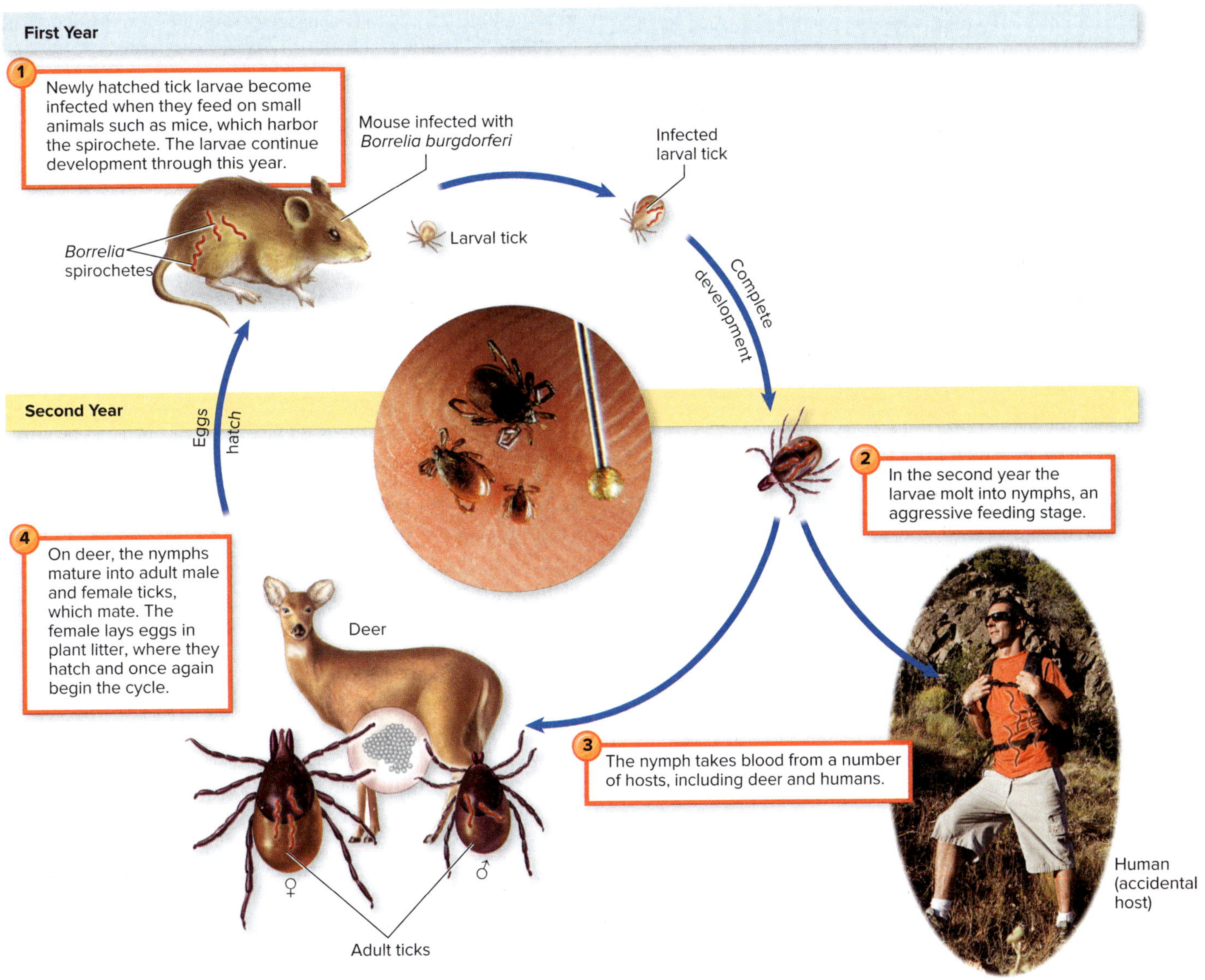
First Year
1
Newly hatched tick larvae become infected when they feed on small animals such as mice, which harbor the spirochete. The larvae continue development through this year.
Mouse infected with *Borrelia burgdorferi*
Infected larval tick
Borrelia spirochetes
Larval tick
Complete development
Second Year
Eggs hatch
2
In the second year the larvae molt into nymphs, an aggressive feeding stage.
4
On deer, the nymphs mature into adult male and female ticks, which mate. The female lays eggs in plant litter, where they hatch and once again begin the cycle.
Deer
3
The nymph takes blood from a number of hosts, including deer and humans.
♀
♂
Adult ticks
Human (accidental host)

Figure 21.10 The cycle of Lyme disease in the northeastern United States.
The disease is tied intimately into the life cycle of a tick vector, which generally is completed over a 2-year period. The exact hosts and species of tick vary from region to region but still display this basic pattern. The photograph in the middle gives an idea of the actual size of the nymph and adult black-legged deer ticks displayed on a human finger. Many people may not realize how small and difficult to detect the feeding nymph can be.

(ticks) Scott Camazine/Science Source; (hiker) Javier Perini CM/Image Source

It is important to consider coinfection with *Anaplasmosis* or *Babesia* because these organisms are transmitted by the same kind of tick that transmits Lyme disease, and the tick may be coinfected. Lingering cases of Lyme disease may be due to failure to consider these microbes.

Prevention and Treatment

There is currently no vaccine for humans. Anyone involved in outdoor activities should wear protective clothing, boots, leggings, and insect repellant containing DEET.[1] Individuals exposed to heavy infestation should routinely inspect their bodies for ticks and remove ticks gently without crushing, preferably with forceps or fingers protected with gloves, because it is possible to become infected by tick feces or body fluids.

Early, prolonged (3 to 4 weeks) treatment with doxycycline and amoxicillin is effective, and other antibiotics such as ceftriaxone and penicillin are used in late Lyme disease therapy. Roughly 10% to 20% of treated patients, however, go on to develop chronic Lyme disease, sometimes known as posttreatment Lyme disease syndrome (PTLDS). Previously PTLDS was often treated with prolonged antibiotic therapy. The National Institute of Allergy and Infectious Disease, after rigorous study, no longer supports that treatment. Evidence suggests that some form of the spirochete—even if just its peptidoglycan wall—can elicit antibodies. Either the spirochete or the antibodies may be responsible for the lingering symptoms.

Disease Table 21.6	Lyme Disease
Causative Organism(s)	*Borrelia burgdorferi* and closely related species B G–
Most Common Modes of Transmission	Biological vector (tick)
Virulence Factors	Antigenic shifting, adhesins
Culture/Diagnosis	Acute and convalescent sera testing
Prevention	Tick avoidance
Treatment	Doxycycline and/or amoxicillin (3–4 weeks), also cephalosporins and penicillin
Epidemiological Features	In United States, 25,000–30,000 cases/yr; endemic in North America, Europe, and Asia

1. *N,N*-Diethyl-*m*-toluamide—the active ingredient in OFF! and Cutter brand insect repellants.

Infectious Mononucleosis

This lymphatic system disease, which is often simply called "mono" or the "kissing disease," is most commonly caused by the **Epstein-Barr virus (EBV),** a member of the family *Herpesviridae.*

Signs and Symptoms

The symptoms of mononucleosis are sore throat, high fever, and cervical lymphadenopathy, which develop after a long incubation period (30 to 50 days). Many patients also have a gray-white exudate in the throat, a skin rash, and enlarged spleen and liver. A notable sign of mononucleosis is sudden leukocytosis, consisting initially of infected B cells and later T cells. Fatigue is a hallmark of the disease. Patients remain fatigued for a period of weeks. During that time, they are advised to not engage in strenuous activity due to the possibility of injuring their enlarged spleen (or liver).

Eventually, a strong, cell-mediated immune response is decisive in controlling the infection and preventing complications. But after recovery, people usually remain chronically infected with EBV.

Causative Agent

Most cases are caused by the Epstein-Barr virus. It shares morphological and antigenic features with other herpesviruses. It contains a circular form of DNA that is readily spliced into the host cell DNA. Other viruses, including cytomegalovirus, can sometimes cause this condition.

Pathogenesis and Virulence Factors

The latency of the Epstein-Barr virus and its ability to splice its DNA into host cell DNA make it an extremely versatile virus that can avoid the host's immune response.

Transmission and Epidemiology

More than 90% of the world's population has been infected with EBV. In general, the virus causes no noticeable symptoms, but the time of life when the virus is first encountered seems to matter. In the case of EBV, infection during the teen years results in disease about 30% to 77% of the time, whereas infection before or after this period may be entirely asymptomatic.

Direct oral contact and contamination with saliva are the principal modes of transmission, although transfer through blood transfusions, sexual contact, and organ transplants is possible. True outbreaks of this disease rarely occur.

Culture and/or Diagnosis

A test called the "Monospot test" detects *heterophile antibodies*—which are antibodies that are not directed against EBV but are seen when a person has an EBV infection. This test is not reliable in children younger than age 4, in which case a specific EBV antigen/antibody test is conducted.

Prevention and Treatment

The usual treatments for infectious mononucleosis are directed at symptomatic relief of fever and sore throat. Hospitalization is rarely needed. Occasionally, rupture of the spleen necessitates immediate surgery to remove it.

Disease Table 21.7	Infectious Mononucleosis
Causative Organism(s)	Epstein-Barr virus (EBV) V
Most Common Modes of Transmission	Direct, indirect contact; parenteral
Virulence Factors	Latency, ability to incorporate into host DNA
Culture/Diagnosis	Differential blood count, Monospot test for heterophile antibody, specific ELISA
Prevention	–
Treatment	Supportive
Distinctive Features	Lifelong persistence
Epidemiological Features	United States: 500 cases per 100,000 per year

Anthrax

Anthrax is discussed in other chapters as well as in this one. We discuss anthrax in this chapter because it multiplies in large numbers in the blood and because septicemic anthrax is a possible outcome of all forms of anthrax.

For centuries, anthrax has been known as a zoonotic disease of herbivorous livestock (sheep, cattle, and goats). It has an important place in the history of medical microbiology because Robert Koch used anthrax as a model for developing his postulates in 1877 and, later, Louis Pasteur used the disease to prove the usefulness of vaccination.

Signs and Symptoms

An anthrax infection can exhibit its primary symptoms in various locations of the body: on the skin (cutaneous anthrax), in the lungs (pulmonary anthrax), in the gastrointestinal tract (acquired through ingestion of contaminated foods), and in the central nervous system (anthrax meningitis). The cutaneous and pulmonary forms of the disease are the most common. In all of these forms, the anthrax bacterium gains access to the bloodstream, and death, if it occurs, is usually a result of an overwhelming septicemia. Pulmonary anthrax—and the accompanying pulmonary edema and hemorrhagic lung symptoms—can sometimes be the primary cause of death, although it is difficult to separate the effects of sepsis from the effects of pulmonary infection.

In addition to symptoms specific to the site of infection, septicemic anthrax results in headache, fever, and malaise. Bleeding in the intestine and from mucous membranes and orifices may occur in late stages of septicemia.

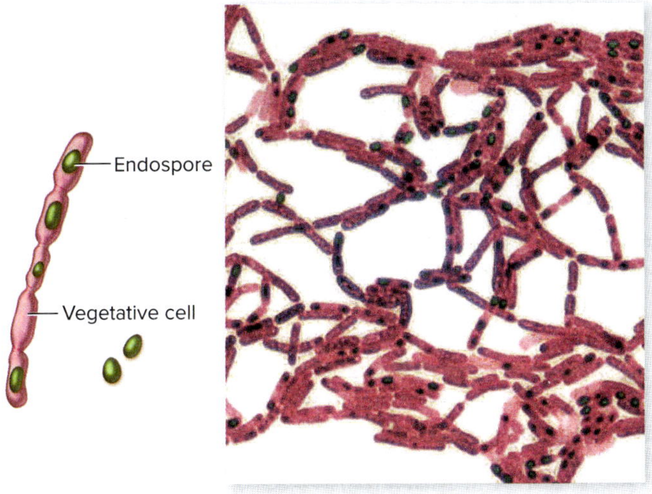
Endospore
Vegetative cell

Figure 21.11 ***Bacillus anthracis.*** Note the centrally placed endospores and streptobacillus arrangement (600×).

Larry Stauffer/Oregon State Public/CDC

Causative Agent

Bacillus anthracis is a gram-positive, endospore-forming rod. It is composed of block-shaped, angular rods 3 to 5 micrometers long and 1 to 1.2 micrometers wide. Central endospores develop under all growth conditions except in the living body of the host **(figure 21.11).** The genus *Bacillus* is aerobic and catalase-positive, and none of the species are fastidious. *Bacillus* as a genus is noted for its versatility in degrading complex macromolecules, and it is a common source of antibiotics. Because the primary habitat of many species, including *B. anthracis,* is the soil, endospores are continuously dispersed by means of dust into water and onto the bodies of plants and animals.

Pathogenesis and Virulence Factors

The main virulence factor of *B. anthracis* is what is referred to as a "tripartite" toxin—an exotoxin complex composed of three separate proteins. One of the proteins is called *edema factor,* which increases cellular cyclic AMP levels, leading to disruption of water balance and ultimately edema. Another part of the toxin is *protective antigen,* so named because it is a good target for vaccination, not because it protects the bacterium or the host during actual infection. It helps the edema factor get to its target site. The third exotoxin is called *lethal factor.* It combines with edema factor to form lethal toxin, which appears to target alveolar epithelium, triggering massive inflammation and initiation of shock—especially in cases of pulmonary anthrax.

Transmission and Epidemiology

The anthrax bacillus is a facultative bacterium that undergoes its cycle of vegetative growth and sporulation in the soil. Animals become infected while grazing on grass contaminated with endospores. When the pathogen is returned to the soil in animal excrement or carcasses, it can sporulate and become a long-term reservoir of infection for the animal population. The majority of natural anthrax cases are reported in livestock from Africa, Asia, and the Middle East. Most recent (natural) cases in the United States have occurred in textile workers handling imported animal hair, hide, or products made from them. Because of effective control procedures, the number of cases in the United States is extremely low (fewer than 10 per year).

In 2001, starting one week after the 9/11 attacks in the United States, envelopes containing freeze-dried endospores of *B. anthracis* were sent through the mail to two U.S. senators and several media outlets. It was an act of bioterrorism. During that attack, 22 people acquired anthrax and 5 people died.

Culture and/or Diagnosis

Diagnosis requires a high index of suspicion. This means that anthrax must be present as a possibility in the clinician's mind or it is likely not to be diagnosed because it is such a rare disease in the developed world and because, in all of its manifestations, it can mimic other infections that are not so rare. First-level (presumptive) diagnosis begins with culturing the bacterium on blood agar and performing a Gram stain. Further tests can be performed to provide evidence of the presence of *B. anthracis* as opposed to other *Bacillus* species. These tests include motility (*B. anthracis* is nonmotile) and a lack of hemolysis on blood agar. Ultimately, samples should be handled by the Centers for Disease Control and Prevention, which will perform confirmatory tests, usually involving direct fluorescent antibody testing and phage lysis tests.

Prevention and Treatment

A vaccine containing live endospores and a toxoid prepared from a special strain of *B. anthracis* are used to protect livestock in areas of the world where anthrax is endemic. Humans should be vaccinated with the purified toxoid—AVA (BioThrax), currently the only licensed human vaccine—if they have occupational contact with aerosols of *B. anthracis*. The vaccine can also be administered to persons believed to be exposed to it, in conjunction with antibiotic treatment. At any rate, treatment of human cases of disease or exposures to the agent will be conducted in consultation with the CDC.

Carcasses of animals that have died from anthrax must be burned or chemically decontaminated before burial to prevent establishing the microbe in the soil. Imported items containing animal hides, hair, and bone should be gas sterilized **(Disease Table 21.8).**

Disease Table 21.8	Anthrax
Causative Organism(s)	*Bacillus anthracis* B G+
Most Common Modes of Transmission	Vehicle (air, soil), indirect contact (animal hides), vehicle (food)
Virulence Factors	Triple exotoxin
Culture/Diagnosis	Culture, direct fluorescent antibody tests
Prevention	Vaccine for high-risk population; used in conjuction with antibiotics post-exposure
Treatment	In consultation with the CDC
Epidemiological Features	Internationally, 2,000–20,000 cases annually, most cutaneous Category A Bioterrorism Agent

Hemorrhagic Fever Diseases

A number of agents that infect the blood and lymphatics cause extreme fevers, some of which are accompanied by internal hemorrhaging. The presence of the virus in the bloodstream causes capillary fragility and disrupts the blood-clotting system, which leads to various degrees of pathology, including death. All of these viruses are RNA enveloped viruses, the distribution of which is restricted to their natural host's distribution.

The fast-spreading Ebola epidemics of 2014 and 2018 demonstrate that diseases that may get minimal amounts of space in a textbook can suddenly become worldwide threats. While sporadic infections are still occurring, the Ebola epidemic has been contained through the often heroic efforts of medical professionals in Africa and many from around the world to help fight the disease.

The diseases yellow fever, chikungunya, and dengue fever are all notable for the fact that they are spread by the *Aedes* genus of mosquito. All three diseases are common in South America

Disease Table 21.9	Hemorrhagic Fevers
Disease	Yellow Fever
Causative Organism(s)	Yellow fever virus V
Most Common Modes of Transmission	Biological vector (*Aedes* mosquito)
Virulence Factors	Disruption of clotting factors
Culture/Diagnosis	ELISA, PCR
Prevention	Live attenuated vaccine available
Treatment	Supportive
Distinctive Features	Accompanied by jaundice
Epidemiological Features	United States: only sporadic cases in travelers; internationally, 200,000 cases annually, 30,000 deaths; 90% of cases in Africa

and Africa. Dengue fever has exploded into the Americas in recent years. One of the diseases in chapter 20 is Zika virus disease. Zika is also transmitted by the *Aedes* mosquito. When we discussed that disease, we mentioned that Zika virus spread rapidly through the Americas, and continues to do so. A warming environment is affecting the distribution of the mosquito greatly. **Figure 21.12** depicts the predicted change in population of *Aedes aegypti,* the most common species that carries the infectious agents. It shows that in the early 2000s, there was some increase in the population compared to 1950, but also quite a lot of areas with decrease (the blue shading). Using a model projecting moderate control of greenhouse gas emissions, the year 2050 will show greatly enhanced habitat area (red shading) of the mosquito carrier. Changing temperatures inevitably lead to new disease patterns.

Currently in several places in the world, trials are being conducted in which male *A. aegypti* mosquitoes infected with the bacterium *Wolbachia* are being released in areas with heavy dengue burden. When these males mate with females, the resultant eggs are not delivered. Then the mosquito population is greatly reduced or eliminated. This is viewed as a much better approach than killing mosquitoes with insecticide, a technique we have relied on for decades. Mosquitoes, like bacteria, acquire resistance to chemicals that are initially lethal to them. Note that chikungunya and dengue fever have now spread to the Americas.

These diseases are summarized in **Disease Table 21.9.**

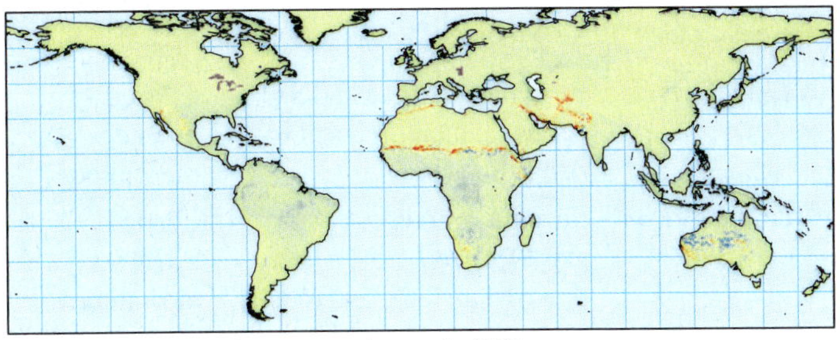

(a) Difference in 2000s compared to early 1950s.

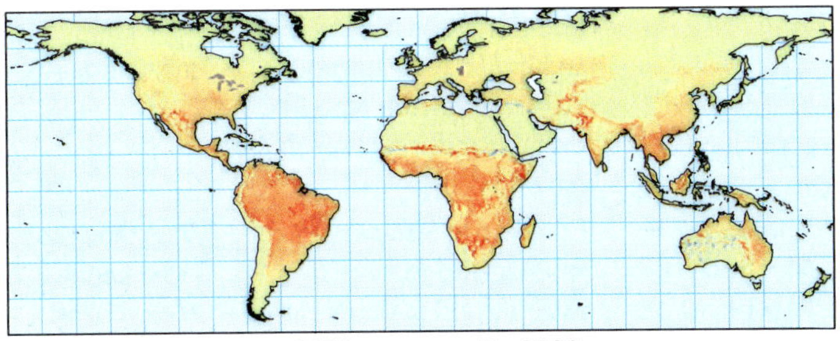

(b) Projected difference in 2050s compared to 2000s.

Figure 21.12 Scientists' prediction of the difference in the population of *Aedes aegypti.* Decreases are indicated by blue color; increases indicated by red. The projected scenario for 2050 relies on a relatively optimistic, moderate level of increase in climate temperature that will require significant decreases in greenhouse gas emissions.

Takuya Iwamura, Adriana Guzman-Holst, and Kris A. Murray. "Accelerating invasion potential of disease vector Aedes aegypti under climate change" Nature Communications. Volume 11(2130), 2020. Used with permission of the authors.

Dengue Fever	Chikungunya	Ebola and/or Marburg	Lassa Fever
Dengue fever virus V	Chikungunya virus V	Ebola virus, Marburg virus V	Lassa fever virus V
Biological vector (*Aedes* mosquito)	Biological vector (*Aedes* mosquito)	Direct contact, body fluids	Droplet contact (aerosolized rodent excretions), direct contact with infected fluids
Disruption of clotting factors	Disruption of clotting factors	Disruption of clotting factors	Disruption of clotting factors
Rise in IgM titers	PCR	PCR, viral culture (conducted at CDC)	ELISA
New vaccine approved in 2019 for use in children aged 9–16 with previous Dengue infection living in endemic areas in U.S.	—	New Ebola vaccine suitable for epidemic situations tested successfully in 2016	Avoiding rats, safe food storage
Supportive	Supportive	New drugs developed for ongoing outbreaks in Africa	Ribavirin
"Breakbone fever"—so named due to severe pain in some forms	Arthritic symptoms	Massive hemorrhage; rash sometimes present	Chest pain, deafness as long-term sequelae
United States: most cases in Puerto Rico, the U.S. Virgin Islands, Samoa, and Guam; internationally, 50–300 million people infected every year and tens of thousands of deaths occur, mostly among children.	First local transmission in United States in 2014; has exploded in the Americas since its arrival in 2013 with an estimated 1.7 million suspected cases	United States: only imported infections; internationally, sporadic outbreaks in Africa; major Ebola outbreak 2014–2016 **Category A Bioterrorism Agent**	United States: no reported cases; internationally, estimated 100,000–300,000 cases annually in West Africa **Category A Bioterrorism Agent**

Here is a quick run-down on the hemorhagic fever diseases.

Yellow Fever:	Endemic in Africa and South America; more frequent in rainy climates	Carried by *Aedes* mosquitoes
Dengue Fever:	Endemic in southeast Asia and Africa; epidemics have occurred in South and Central America and the Caribbean	Carried by *Aedes* mosquitoes
Chikungunya:	Endemic in Africa; arrived in Central America in 2013 and in U.S. and Europe in 2014; went from zero to over 1.7 million cases in the Americas in 3 years	Carried by *Aedes* mosquitoes
Ebola and Marburg Fevers:	Endemic to Africa; capillary fragility is extreme and patients can bleed from their orifices and mucous membranes	Bats thought to be natural reservoir of Ebola
Lassa Fever:	Endemic to West Africa; asymptomatic in 80% of cases; in others, severe symptoms develop	Reservoir of virus is multimammate rat

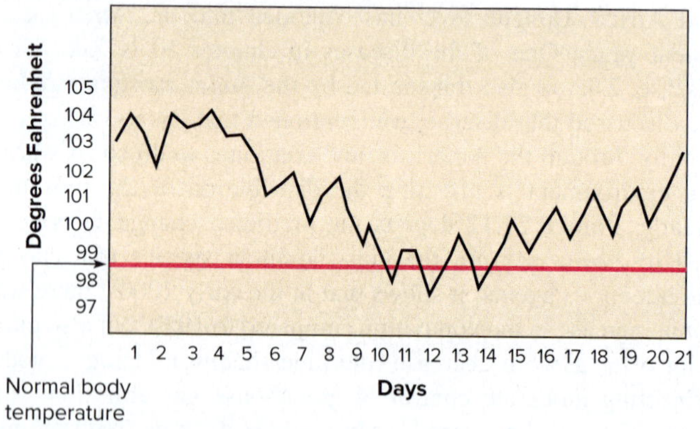
Degrees Fahrenheit
105
104
103
102
101
100
99
98
97
1 2 3 4 5 6 7 8 9 10 11 12 13 14 15 16 17 18 19 20 21
Days
Normal body temperature

Figure 21.13 The temperature cycle in classic brucellosis. Body temperature undulates between day and night and between elevated, normal, and subnormal.

Source: Alice Lorraine Smith, Principles of Microbiology, 10th ed., 1985. Times Mirror/Mosby College Pub.

Nonhemorrhagic Fever Diseases

In this section, we examine some infectious diseases that result in a syndrome characterized by high fever but without the capillary fragility that leads to hemorrhagic symptoms. All but one of the diseases in this section are caused by bacteria.

Brucellosis

This disease goes by several different names (besides brucellosis): Malta fever, undulant fever, and Bang's disease.[2]

Signs and Symptoms

The *Brucella* species responsible for this disease live in phagocytic cells. These cells carry the bacteria into the bloodstream, creating focal lesions in the liver, spleen, bone marrow, and kidney. The most prominent manifestation of human brucellosis is a fluctuating pattern of fever, which is the origin of the common name *undulant fever* **(figure 21.13).** It is accompanied by chills, profuse sweating, headache, muscle pain and weakness, and weight loss. Fatalities are not common, although the syndrome can last for a few weeks to a year, even with treatment.

Causative Agent

The bacterial genus *Brucella* contains tiny, aerobic gram-negative coccobacilli. At least six species are known to cause disease in humans: *B. abortus* (cattle), *B. melitensis* (goats, sheep), *B. canis* (canines), *B. suis* (pigs), and at least two species living in marine mammals. In humans, infection with *B. melitensis* is most common. Even though a principal manifestation of the disease in animals is an infection of the placenta and fetus, human placentas do not become infected. The CDC list *Brucella* species as possible bioterror agents, though they are not designated as being "of highest concern."

Pathogenesis and Virulence Factors

Brucella enters through damaged skin or via mucous membranes of the digestive tract, conjunctiva, and respiratory tract. From there, it is taken up by phagocytic cells. Because it is able to avoid destruction in the phagocytes, the bacterium is transported easily through the bloodstream and to various organs, such as the liver, kidney, breast tissue, or joints.

Transmission and Epidemiology

Brucellosis is one of the most common zoonotic diseases, as more than 500,000 human cases are reported worldwide each year in areas of Europe, Africa, India, and Latin America. It is associated predominantly with occupational contact in slaughterhouses, livestock handling, and the veterinary profession. Infection takes place through contact with blood, urine, and placentas and through consumption of raw milk and cheese. Human-to-human transmission is rare, but brucellosis is considered the most common laboratory-acquired infection. In 2020, over 3,000 people were infected in China after a factory accident in a facility that produced *Brucella* vaccine for livestock.

Brucellosis is also a common disease of wild herds of bison and elk. Cattle that share grazing land with these wild herds often suffer severe outbreaks of the placental infections (Bang's disease). Along with the toll on human health, the worldwide economic impact from animal loss due to disease is immense.

Culture and/or Diagnosis

The patient's history can be very helpful in diagnosis, as are serological tests of the patient's blood and blood culture of the pathogen.

2. After B. L. Bang, a Danish physician.

Prevention and Treatment

Prevention is achieved by testing and elimination of infected animals, quarantine of imported animals, and pasteurization of milk. Several types of animal vaccines are available. The status of this pathogen as a potential germ warfare agent makes a reliable human vaccine even more urgent. A reemergence of brucellosis has occurred recently in countries that have been free of disease for decades.

A combination of doxycycline and gentamicin or streptomycin taken for 3 to 6 weeks is usually effective in controlling infection.

Q Fever

The name of this disease arose from the frustration created by not being able to identify its cause. The *Q* stands for "query." Its cause, a bacterium called *Coxiella burnetii,* was finally identified in the mid-1900s. The clinical manifestations of acute Q fever are abrupt onset of fever, chills, head and muscle ache, and, occasionally, a rash. The disease is sometimes complicated by pneumonitis (30% of cases), hepatitis, and endocarditis. About a quarter of the cases are chronic rather than acute and result in vascular damage and endocarditis-like symptoms.

C. burnetii is a very small, pleomorphic, gram-negative bacterium. It is an intracellular parasite, and it produces an unusual type of endospore-like structure. *C. burnetii* is harbored by a wide assortment of vertebrates and arthropods, especially ticks, which play an essential role in transmission between wild and domestic animals. Ticks do not transmit the disease to humans, however. Humans acquire infection largely by means of environmental contamination and airborne spread. Birth products such as placentas of infected domestic animals contain large numbers of bacteria. Other sources of infectious material include urine, feces, milk, and airborne particles from infected animals. The primary portals of entry are the lungs, skin, conjunctiva, and gastrointestinal tract.

C. burnetii has been isolated from most regions of the world. The mid- and western United States have the highest numbers of cases in this country, although most cases probably go undetected. People at highest risk are farm workers, meat cutters, veterinarians, laboratory technicians, and consumers of raw milk products.

Mild or subclinical cases resolve spontaneously, and more severe cases respond to long-term therapy with tetracycline or trimethoprim/sulfamethoxazole. A vaccine is available in many parts of the world but not in the United States. Q fever is of potential concern as a bioterror agent because it is very resistant to heat and drying, it can be inhaled, and even a single bacterium is enough to cause disease. It is an organism that the U.S. military worked with during the period when potential biowarfare agents were being developed in this country (the 1950s and 1960s).

Cat-Scratch Disease

This disease is one of a group of diseases caused by different species of the small, gram-negative rod *Bartonella. Bartonella* species are considered to be emerging zoonotic pathogens. They are fastidious but not obligate intracellular parasites, so they will grow on blood agar. In addition to cat-scratch disease and trench fever, discussed next, a new species of *Bartonella* (*B. rochalimae*) that causes high fever and life-threatening anemia has recently been identified.

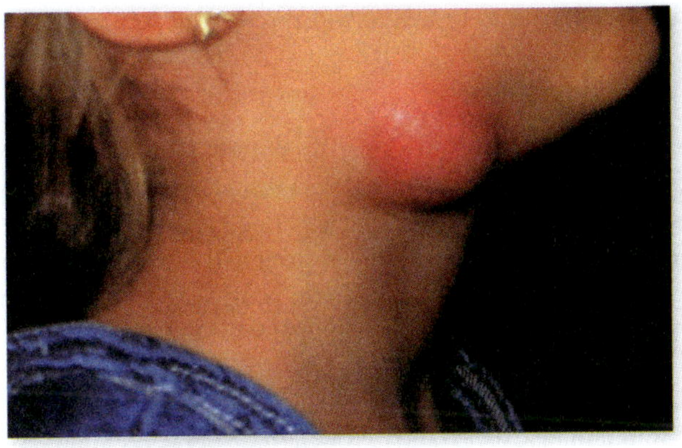

Figure 21.14 Cat-scratch disease. A primary nodule appears at the site of the scratch in about 21 days. In time, large quantities of pus collect and the regional lymph nodes swell.
Neal R. Chamberlin/McGraw Hill

Bartonella henselae is the agent of cat-scratch disease (CSD), an infection connected with being clawed or bitten by a cat. The pathogen is present in over 40% of cats, especially kittens. There are approximately 25,000 cases per year in the United States, 80% of them in children 2 to 14 years old. The symptoms start after 1 to 2 weeks, with a cluster of small papules at the site of inoculation. In a few weeks, the lymph nodes along the lymphatic drainage swell and can become pus-filled **(figure 21.14).** Only about one-third of patients experience high fever. Most infections remain localized and resolve in a few weeks, but drugs such as azithromycin, erythromycin, and rifampin can be effective therapies. The disease can be prevented by thorough antiseptic cleansing of a cat bite or scratch.

Trench Fever

This disease has a long history. Trench fever was once a common condition of soldiers in battle. The causative agent, *Bartonella quintana,* is carried by lice. The feces of the lice contain the bacterium, and transmission usually occurs when the feces enter the bite wound. Most cases occur in endemic regions of Europe, Africa, and Asia, although the disease is reemerging in poverty-stricken areas of large cities in the developed world. This version of the disease is called "urban trench fever," and recent studies documented that 33% of homeless individuals in San Francisco carried body lice harboring the bacterium. Highly variable symptoms can include a 5- to 6-day fever (the species epithet, *quintana,* refers to a 5-day fever). Symptoms also include leg pains, especially in the tibial region (the disease is sometimes called "shinbone fever"); headache; chills; and muscle aches. A macular rash can also occur. (See Table 19.1 for definitions of skin lesions.) Endocarditis can develop, especially in the urban version of the disease.

Trench fever may be treated with doxycycline +/– erythromycin.

Ehrlichiosis

Ehrlichia is a small, intracellular bacterium with a strict parasitic existence and association with ticks (*Ixodes* species). The species of tick varies with the geographic location in the United States and Europe. Its distribution is in the eastern half of the United States. The signs and symptoms include an acute febrile state resulting in headache, muscle pain, and rigors. Most patients recover rapidly with no lasting effects, but about 5% of older, chronically ill patients can die.

Rapid diagnosis is done through PCR tests and indirect fluorescent antibody tests. It can be critical to differentiate or detect coinfection with the Lyme disease agent *Borrelia,* which is carried by the same tick. Doxycycline will clear up most infections within 7 to 10 days.

Anaplasmosis

Anaplasma is another small, intracellular bacterium. It shares lifestyle characteristics with *Ehrlichia* and causes nearly identical clinical manifestations. But the two bacteria have different geographic distributions. *Anaplasma* occurs mainly in the Northeast and along the Great Lakes, while *Ehrlichia* is distributed throughout the eastern half of the United States. Also, they are carried by two different species of ticks. Treatment of anaplasmosis is also with doxycycline.

Babesiosis

Babesia is a protozoan organism that infects red blood cells. It produces symptoms similar to those of *Ehrlichia* and *Anaplasma.* It is also carried by ticks and is often diagnosed via a blood smear. The protozoan is visible inside red blood cells.

Because it is a protozoan, treatment is different than for *Ehrlichia* and *Anaplasma.* Combined therapy of either atovaquone (an antiprotozoal) + azithromycin, or clindamycin + quinine (another antiprotozoal) is recommended.

Spotted Fever Rickettsioses (SFR)

This is a category of diseases of which Rocky Mountain Spotted Fever (RMSF) is the most well-known. It is caused by the bacterium *Rickettsia rickettsii.* In recent years it has become clear that other *Rickettsia* species cause similar diseases. The various *Rickettsia* species are not distinguishable with lab tests, so the spectrum of diseases has been renamed spotted fever rickettsiosis (SFR). Interestingly, while RMSF is named for the region in which it was first detected in the United States—the Rocky Mountains of Montana and Idaho—the disease does not occur frequently in the western United States. The majority of cases are concentrated in the Southeast and eastern seaboard regions. It also occurs in Canada and Central and South America. Infections occur most frequently in the spring and summer, when the tick vectors are most active. The rate of infections with the SFRs in the United States has increased dramatically—10-fold—since 2000.

Rickettsia species are transmitted by hard ticks such as the wood tick (*Dermacentor andersoni*), the American dog tick (*D. variabilis,* among others), and the Lone Star tick (*Amblyomma americanum*). The dog tick is probably most responsible for transmission to humans because it is the major vector in the southeastern United States.

After 2 to 4 days of incubation, the first symptoms are sustained fever, chills, headache, and muscular pain. A distinctive, spotted rash usually comes on within 2 to 4 days after the prodrome **(figure 21.15)** and develops first on the wrists, forearms, and ankles before spreading to other areas of the body. Early lesions are slightly mottled, like measles, but later ones are macular, maculopapular, and even petechial. In the most severe untreated cases, the enlarged lesions merge and can become necrotic, predisposing to gangrene of the toes or fingertips.

Although the spots are the most obvious symptom of the disease, the most serious manifestations are cardiovascular disruption, including hypotension, thrombosis, and hemorrhage. Conditions of restlessness, delirium, convulsions, tremor, and coma are

Disease Table 21.10 Nonhemorrhagic Fever Diseases

Disease	Brucellosis
Causative Organism(s)	*Brucella melitensis, B. abortus,* or *B. suis* B G–
Most Common Modes of Transmission	Direct contact, airborne, parenteral (needlesticks)
Virulence Factors	Intracellular growth; avoidance of destruction by phagocytes
Culture/Diagnosis	Gram stain of biopsy material; PCR
Prevention	Animal control, pasteurization of milk
Treatment	Doxycycline plus gentamicin or streptomycin
Distinctive Features	Undulating fever, muscle aches
Epidemiological Features	United States: fewer than 100 cases per year; internationally, 500,000 cases per year Category B Bioterrorism Agent

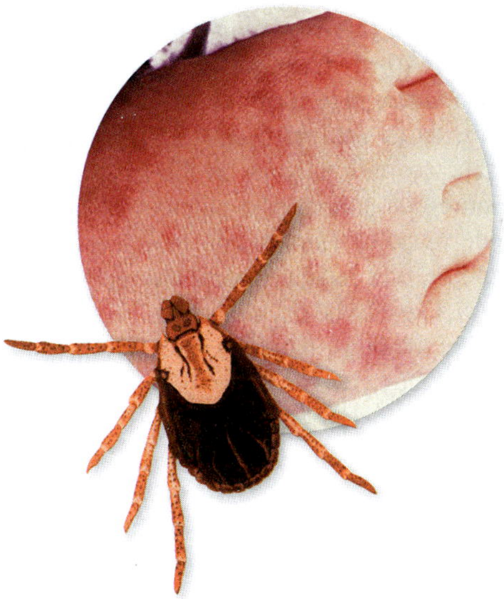

Figure 21.15 The rash in RMSF. This case occurred in a child several days after the onset of fever. Also pictured is an example of the hard ticks that transmit the SFR infections.

(hand) Centers for Disease Control and Prevention; (tick) Andrew J. Brooks/CDC

signs of the often overwhelming effects on the central nervous system. Fatalities occur in an average of 20% of untreated cases and 5% to 10% of treated cases.

Suspected cases of an SFR require immediate treatment even before laboratory confirmation. A recent aid to early diagnosis is a method for staining *Rickettsia* directly in a tissue biopsy using fluorescent antibodies. Isolating *Rickettsia* from the patient's blood or tissues is desirable, but it is expensive and requires specially qualified lab personnel and lab facilities. Specimens taken from the rash lesions are suitable for PCR assay, which is very specific and sensitive and can circumvent the need for culture.

The drug of choice for suspected and known cases is doxycycline administered for 1 week. Preventive measures parallel those for Lyme disease: wearing protective clothing, using insect sprays, and fastidiously removing ticks **(Disease Table 21.10).**

Q Fever	Cat-Scratch Disease	Trench Fever	Ehrlichiosis	Anaplasmosis	Babesiosis	Spotted Fever Rickettsiosis
Coxiella burnetii B G–	*Bartonella henselae* B G–	*Bartonella quintana* B G–	*Ehrlichia* species B G–	*Anaplasma* species B G–	*Babesia* species P	*Rickettsia* species B G–
Airborne, direct contact, food-borne	Parenteral (cat scratch or bite)	Biological vector (lice)	Biological vector (tick)	Biological vector (tick)	Biological vector (tick)	Biological vector (tick)
Endospore-like structure	Endotoxin	Endotoxin	–	–	–	Induces apoptosis in cells lining blood vessels
Serological tests for antibody; PCR	Biopsy of lymph nodes plus Gram staining; ELISA (performed by CDC)	ELISA (performed by CDC)	PCR, indirect antibody test	PCR, indirect antibody test	Blood smear	Fluorescent antibody, PCR
Vaccine for high-risk population	Clean wound sites	Avoid lice	Avoid ticks	Avoid ticks	Avoid ticks	Avoid ticks
Tetracycline or TMP/SMZ	Azithromycin or rifampin	Azithromycin +/– doxycycline	Doxycycline	Doxycycline	Combination therapy with antibacterial + antiprotozoal	Doxycycline
Airborne route of transmission, variable disease presentation	History of cat bite or scratch; fever not always present	Endocarditis common, 5-day fever	Southeast, south central United States	Upper Midwest and northeastern United States	Northeastern and upper midwestern United States	Rocky Mountain spotted fever is most severe of the rickettsioses
One-third of cases occur in four states: Colorado, California, Texas, and Illinois **Category B Bioterrorism Agent**	United States: estimated incidence is 9.3 cases per 100,000; internationally, seroprevalence from 0.6%–37% depending on cat population	Most infections asymptomatic; found on every continent except Antarctica	Great increase in incidence since mid-1990s	Great increase in incidence since mid-1990s	–	Only in Americas; tenfold increase since 2000.

Chagas Disease

Chagas disease is sometimes called "the American trypanosomiasis." The causative agent is *Trypanosoma cruzi*. A different trypanosome, *T. brucei,* causes sleeping sickness on the African continent.

▶ Signs and Symptoms

Once the trypanosomes are transmitted by a group of insects called the triatomines **(figure 21.16),** they multiply in muscle and blood cells. From time to time, the blood cells rupture and large numbers of trypanosomes are released into the bloodstream. The disease manifestations are divided into acute and chronic phases. Soon after infection, the acute phase begins; symptoms are relatively nondescript and range from mild to severe fever, nausea, and fatigue. A swelling called a "chagoma" at the site of the bug bite may be present. If the bug bite is close to the eyes, a distinct condition called Romana's sign, swelling of the eyelids, may appear. The acute phase lasts for weeks or months, after which the condition becomes chronic and virtually asymptomatic for a period of years or indefinitely. Eventually, the trypanosomes are found in numerous sites around the body, which in later years may lead to inflammation and disruption of function in organs such as the heart, the brain, and the intestinal tract.

▶ Causative Agent

T. cruzi is a flagellated protozoan.

▶ Pathogenesis and Virulence Factors

T. cruzi is equipped with antioxidant enzymes that act to neutralize the lysosomal attack of cells they infect. This allows it to live inside host cells without being killed by them. In addition, it can cloak itself in host proteins, disguising itself from the immune system. It can also induce autoimmunity, so that the same immune cells trained to recognize it begin to react with host tissues, causing the symptoms of late-stage Chagas disease.

▶ Transmission and Epidemiology

This disease has been called "the new AIDS of the Americas" as it has a long incubation time and is difficult to cure. Estimates put the prevalence of this disease at 8 million people, 230,000 of whom live in the United States. Argentina, Brazil, and Mexico are the top three countries in terms of numbers of cases.

Triatomine bugs are often called "kissing bugs" because of their tendency to bite humans on the face. The bugs become infected by biting an animal or a human that has *T. cruzi* in its blood. A wide range of animals can be infected, including raccoons, armadillos, and rodents. Domesticated animals, like dogs and guinea pigs, can also be infected. The disease multiplies in the intestinal tract of the bugs. After the insect finishes a blood meal, it defecates. Its feces contain the trypanosome, which can gain access to the bloodstream via the bite puncture. This process is often facilitated by the subject's scratching of the bite site.

The trypanosome can also be transmitted vertically because it crosses the placenta and via blood transfusion with infected blood. Recently, the United States began screening all donated blood for this disease.

▶ Culture and/or Diagnosis

During the acute phase of the disease, there are large numbers of trypanosomes in the blood, so a peripheral blood smear will detect the organism **(figure 21.17).** In later stages of the disease, it can be diagnosed with serological methods.

▶ Prevention

There is no vaccine for Chagas disease. In endemic areas, pesticides and improved building materials in houses are used to minimize the presence of the bug.

▶ Treatment

Treatment is most successful if begun during the acute phase. However, it is often not accomplished because the acute phase of the disease is not necessarily suggestive of Chagas. Drugs for treatment are only available through the CDC. During the chronic phase of the disease, symptomatic treatment of cardiac and other problems may also be indicated.

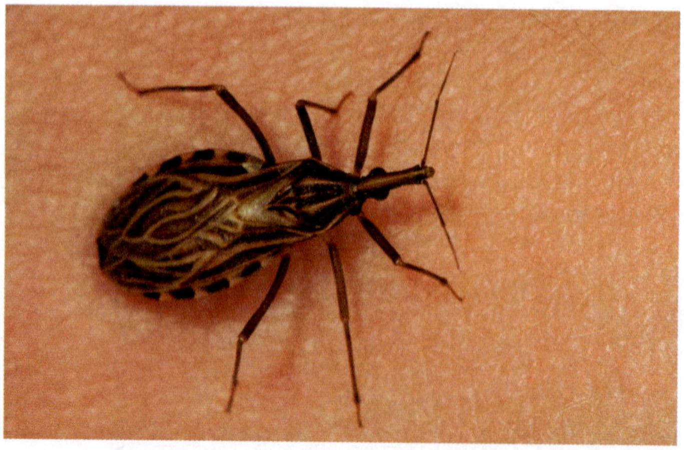

Figure 21.16 The triatomine bug, the carrier of *T. cruzi.*
Ray Wilson/Alamy Stock Photo

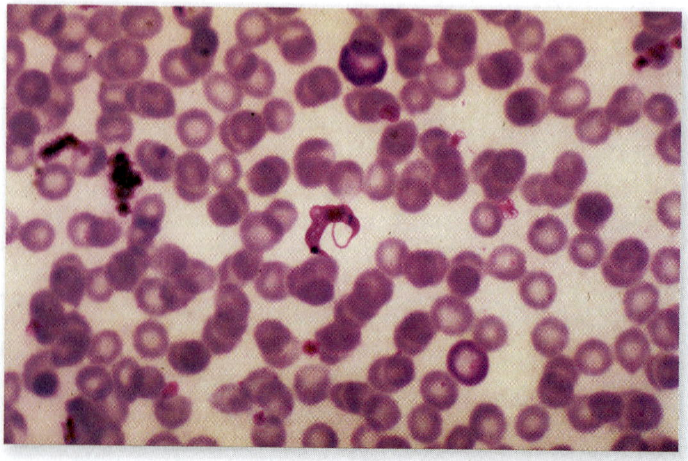

Figure 21.17 *T. cruzi* seen in a blood smear. The round purplish objects are red blood cells.
Dr. Mae Melvin/CDC

Disease Table 21.11	Chagas Disease
Causative Organism	*Trypanosoma cruzi* (P)
Most Common Modes of Transmission	Biological vector (triatomine bug), vertical
Virulence Factors	Antioxidant enzymes, co-opting host antigens; induces autoimmunity
Culture/Diagnosis	Blood smear in acute phase; serological methods in later stages
Prevention	Insect control
Treatment	Consult CDC
Epidemiological Features	Endemic in Central and South America; 230,000 cases present in United States; considered a **neglected parasitic infection (NPI)**

Malaria

Throughout history, including prehistoric times, malaria has been one of humankind's greatest afflictions, in the same rank as bubonic plague, influenza, and tuberculosis. Even now, as the dominant protozoan disease, it threatens 40% of the world's population every year. The origin of the name is from the Italian words *mal,* "bad," and *aria*, "air." The superstitions of the Middle Ages alleged that evil spirits or mists and vapors arising from swamps caused malaria because many victims came down with the disease after this sort of exposure. We now know that swamps were involved—but as a habitat for the mosquito vector.

Signs and Symptoms

After a 10- to 16-day incubation period, the first symptoms are malaise, fatigue, vague aches, and nausea with or without diarrhea, followed by bouts of chills, fever, and sweating. These symptoms occur at 48- or 72-hour intervals, as a result of the synchronous rupturing of red blood cells. The interval, length, and regularity of symptoms reflect the type of malaria (described next). Patients with falciparum malaria, the most virulent type, often manifest persistent fever, cough, and weakness for weeks without relief. Complications of malaria are hemolytic anemia from lysed blood cells and organ enlargement and rupture due to cellular debris that accumulates in the spleen, liver, and kidneys. One of the most serious complications of falciparum malaria is termed *cerebral malaria*. In this condition, small blood vessels in the brain become obstructed due to the increased ability of red blood cells (RBCs) to adhere to vessel walls (a condition called *cytoadherence* induced by the infecting protozoan). The resulting decrease in oxygen in brain tissue can result in coma and death. In general, malaria has the highest death rate in the acute phase, especially in children. Certain kinds of malaria (those caused by *Plasmodium vivax* and *P. ovale*) are subject to relapses because some infected liver cells harbor dormant protozoans for up to 5 years.

Causative Agent

Plasmodium species are protozoans in the sporozoan group. They are **apicomplexans,** which live in animal hosts and lack locomotor organelles in the mature state (chapter 5 describes protozoan classification). The genus *Plasmodium* contains five species causing disease in humans: *P. malariae, P. vivax, P. falciparum, P. ovale*, and *P. knowlesi.* Humans are the primary vertebrate hosts for most of the species.

Development of the malarial parasite is divided into two distinct phases: the asexual phase, carried out in the human **(figure 21.18),** and the sexual phase, carried out in the mosquito. **Figure 21.19** depicts the entire infection cycle.

Pathogenesis and Virulence Factors

The invasion of the merozoites into RBCs leads to the release of fever-inducing chemicals into the bloodstream. *P. vivax* and *P. ovale* have a propensity to persist in the liver. Without sufficient treatment, they can reemerge over the course of several years to cause recurrent bouts of malarial symptoms.

The fact that the protozoan has several different life stages within a host helps it escape immune responses mounted against any single life stage. This evasion is only strengthened by the

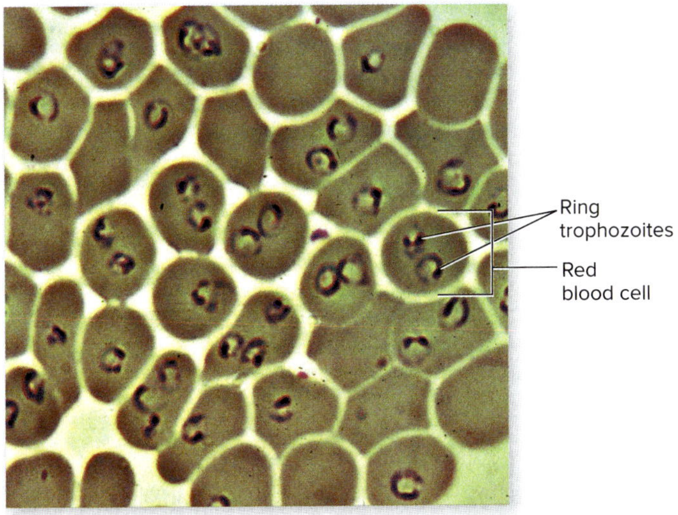
Ring
trophozoites
Red
blood cell

Figure 21.18 The ring trophozoite stage in a *Plasmodium falciparum* infection. A smear of peripheral blood shows ring forms in red blood cells. Some RBCs have multiple trophozoites.
Stephen B. Aley, PhD

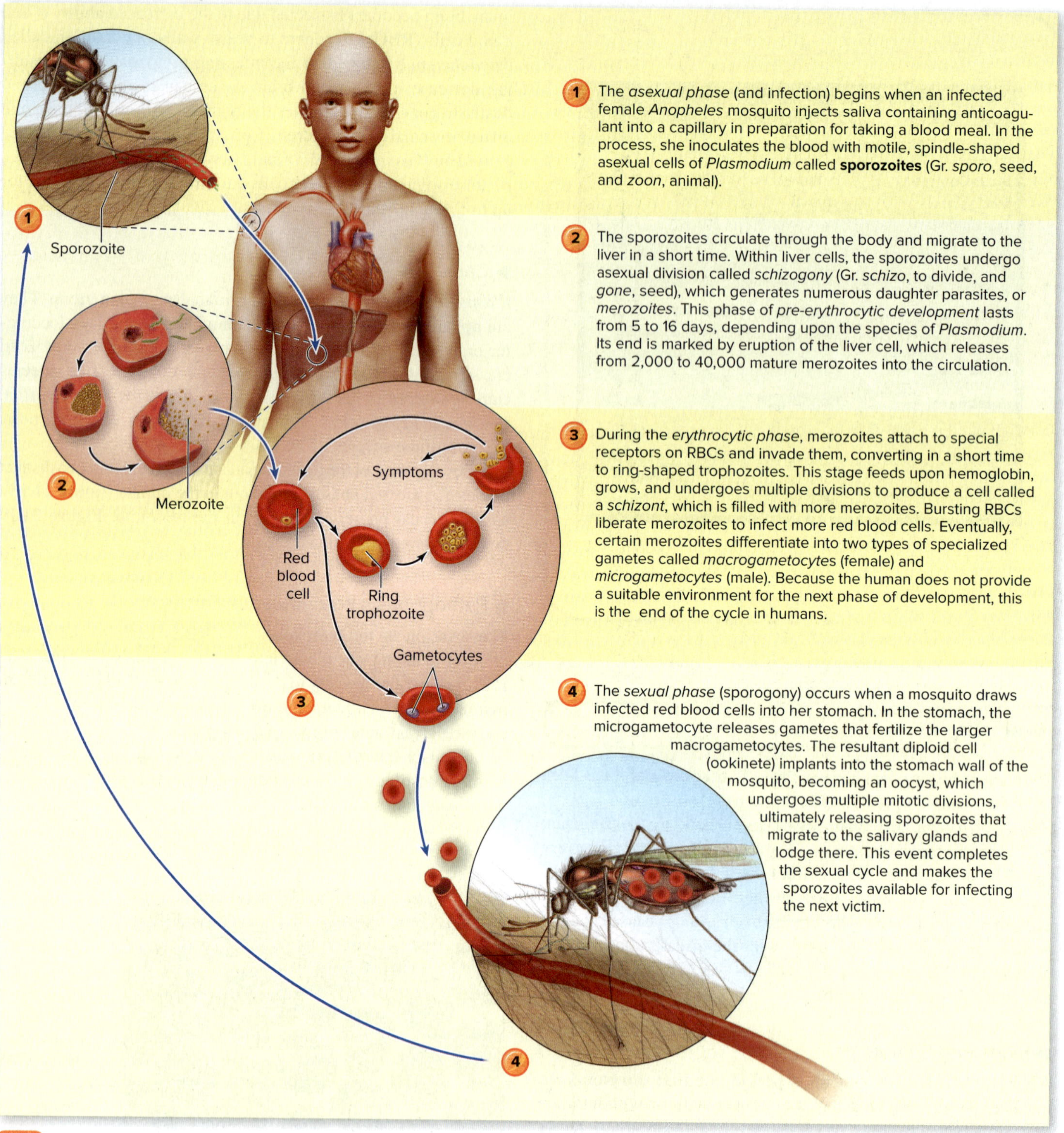
1 The asexual phase (and infection) begins when an infected female Anopheles mosquito injects saliva containing anticoagulant into a capillary in preparation for taking a blood meal. In the process, she inoculates the blood with motile, spindle-shaped asexual cells of Plasmodium called sporozoites (Gr. sporo, seed, and zoon, animal).
2 The sporozoites circulate through the body and migrate to the liver in a short time. Within liver cells, the sporozoites undergo asexual division called schizogony (Gr. schizo, to divide, and gone, seed), which generates numerous daughter parasites, or merozoites. This phase of pre-erythrocytic development lasts from 5 to 16 days, depending upon the species of Plasmodium. Its end is marked by eruption of the liver cell, which releases from 2,000 to 40,000 mature merozoites into the circulation.
3 During the erythrocytic phase, merozoites attach to special receptors on RBCs and invade them, converting in a short time to ring-shaped trophozoites. This stage feeds upon hemoglobin, grows, and undergoes multiple divisions to produce a cell called a schizont, which is filled with more merozoites. Bursting RBCs liberate merozoites to infect more red blood cells. Eventually, certain merozoites differentiate into two types of specialized gametes called macrogametocytes (female) and microgametocytes (male). Because the human does not provide a suitable environment for the next phase of development, this is the end of the cycle in humans.
4 The sexual phase (sporogony) occurs when a mosquito draws infected red blood cells into her stomach. In the stomach, the microgametocyte releases gametes that fertilize the larger macrogametocytes. The resultant diploid cell (ookinete) implants into the stomach wall of the mosquito, becoming an oocyst, which undergoes multiple mitotic divisions, ultimately releasing sporozoites that migrate to the salivary glands and lodge there. This event completes the sexual cycle and makes the sporozoites available for infecting the next victim.
1
Sporozoite
2
Merozoite
Symptoms
Red blood cell
Ring trophozoite
Gametocytes
3
4

Figure 21.19 The life and transmission cycle of *Plasmodium,* the cause of malaria.

parasite's ability to undergo antigenic variation, in which the pattern of gene transcription varies among organisms within a single population. This leads to changes in surface antigens, which constantly changes the appearance of the microbe to the human immune system—a true master of disguise!

Transmission and Epidemiology

All forms of malaria are spread primarily by the female *Anopheles* mosquito and occasionally by shared hypodermic needles and blood transfusions. One group of researchers has found that the composition of your skin microbiome makes you more or less

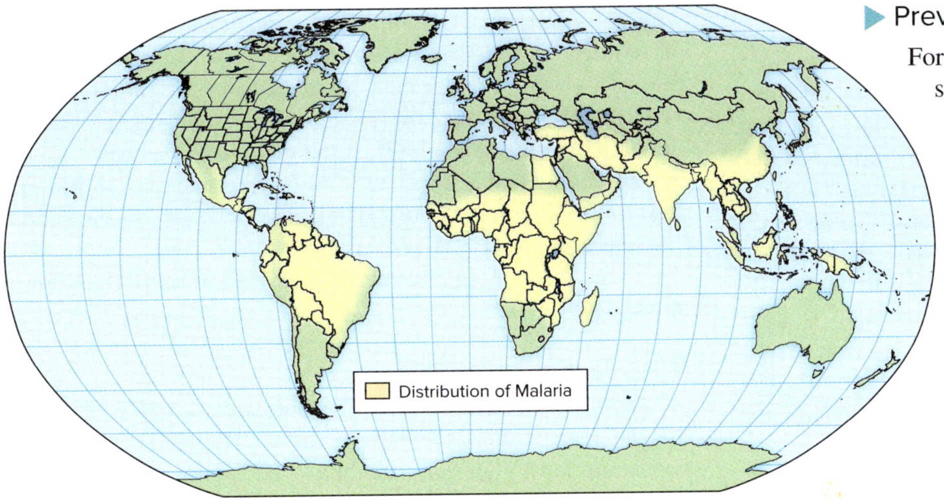
Distribution of Malaria

Figure 21.20 The malaria belt. Yellow zones outline the major regions that harbor malaria. The malaria belt corresponds to a band around the equator.

attractive to the mosquitoes carrying malaria. Although malaria was once distributed throughout most of the world, the control of mosquitoes in temperate areas has successfully restricted it mostly to a belt extending around the equator **(figure 21.20).** Despite this achievement, approximately 230 million new cases are still reported each year, about 90% of them in Africa. The most frequent victims are children and young adults, of whom approximately 400,000 die annually. A particular form of the malarial protozoan causes damage to the placenta in pregnant women, leading to excess mortality among fetuses and newborns. The total case rate in the United States is about 1,000 to 2,000 new cases a year, most of which occur in immigrants or travelers who have visited endemic areas.

Culture and/or Diagnosis

Malaria can be diagnosed definitively by the discovery of a typical stage of *Plasmodium* in stained blood smears (see figure 21.18). Newer Ag-specific tests have been developed, but the smears are still considered the gold standard.

A Note About the Global Fund to Fight AIDS, Tuberculosis, and Malaria

In 2002, a unique partnership, known as the Global Fund, was formed among governments, the private sector, civil society, and people affected by AIDS, tuberculosis, or malaria. Its aim is to raise money and distribute it to programs that work in countries where the epidemics are the worst, and to save lives. The Bill and Melinda Gates Foundation is a major contributor to the Global Fund. And it has been very successful. Now, as of 2022, there are one-third fewer deaths due to those "Big 3" diseases in countries where the Global Fund is active. The Global Fund reports that 44 million lives have been saved since 2002, at the rate of about 2 million a year. This chapter covers malaria and AIDS, and tuberculosis will be addressed in chapter 22.

Prevention

For decades, malaria prevention was attempted strictly through long-term mosquito abatement and human chemoprophylaxis. A malaria vaccine for children was finally approved by the WHO in 2021. By mid-2022 at least 1 million children in Africa had been immunized. It is a huge step forward, but will not replace traditional abatement methods.

Abatement includes elimination of standing water that could serve as a breeding site and spraying of insecticides to reduce populations of adult mosquitoes, especially in and near human dwellings. Scientists have also tried introducing sterile male mosquitoes into endemic areas. Another promising area of research is to use the CRISPR/Cas 9 system to introduce mutations into the mosquito offspring that disallow infection of the insect with the protozoan. (Western travelers to endemic areas are often prescribed antimalarials before and during their trips as a prophylactic measure. People with a recent history of malaria must be excluded from blood donations. The WHO and other international organizations focus on efforts to distribute bed nets and to teach people how to dip the nets twice a year into an insecticide **(figure 21.21).** The use of bed nets has been estimated to reduce childhood mortality from malaria by 20%.

Figure 21.21 An African family sits under a treated mosquito net acquired through the UNICEF mosquito nets program.
Anthony Asael/Art in All of Us/Corbis News/Getty Images

Treatment

The malarial protozoan has developed resistance to nearly every drug used for its treatment. A treatment called Artemisinin Combination Therapy (ACT) is recommended as a first-line treatment. Artemisin is a plant-derived compound from the wormwood tree, discovered in 1972 by a Chinese scientist. She was later awarded the Nobel Prize for this life-saving finding **(Disease Table 21.12).** Clinicians in the United States should be aware of where the infection might have been acquired and then consult the WHO's online map depicting antimalarial resistance.

A growing problem contributing to the development of antimalarial resistance, particularly in Southeast Asia, seems to be the sale of counterfeit antimalarial drugs, many of which contain only a fraction of the appropriate dosage.

Disease Table 21.12 Malaria	
Causative Organism(s)	*Plasmodium falciparum, P. vivax, P. ovale, P. malariae, P. knowlesi* (P)
Most Common Modes of Transmission	Biological vector (mosquito), vertical
Virulence Factors	Multiple life stages; multiple antigenic types, ability to scavenge glucose, GPI toxin, cytoadherence
Culture/Diagnosis	Blood smear; serological methods
Prevention	Mosquito control; use of bed nets; for children in endemic areas now beginning use of RTS,S vaccine; prophylactic antiprotozoal agents
Treatment	Artemisinin, combination therapy; consult WHO
Epidemiological Features	United States: cases are generally in travelers or immigrants; internationally, 300 million cases in "malaria belt"; half million deaths per year; more deadly in children

HIV Infection and AIDS

The sudden emergence of AIDS in the early 1980s focused an enormous amount of public attention, research studies, and financial resources on the virus and its disease. Physicians in Los Angeles, San Francisco, and New York City saw the first cases of AIDS. They observed clusters of young male patients with one or more of a complex of symptoms: severe pneumonia caused by *Pneumocystis jirovecii* (ordinarily a harmless fungus), a rare vascular cancer called Kaposi's sarcoma, sudden weight loss, swollen lymph nodes, and general loss of immune function. Eventually, virologists at the Pasteur Institute in France isolated a novel retrovirus, later named the **human immunodeficiency virus (HIV).** This cluster of symptoms was therefore clearly a communicable infectious disease, and the medical community termed it **acquired immunodeficiency syndrome (AIDS).**

Signs and Symptoms

A spectrum of clinical disease is associated with HIV infection. To understand the progression, follow **figure 21.22** closely. Symptoms in HIV infection are directly tied to two things: the level of virus in the blood and the level of T cells in the blood. This figure shows two different lines that correspond to virus and T-cell levels in the blood, in addition to a line depicting the amount of circulating antibody against the virus. Note that the figure depicts the course of HIV infection in the absence of medical intervention or chemotherapy.

Initial symptoms may be fatigue, diarrhea, weight loss, and neurological changes, but most patients first notice this infection because of one or more opportunistic infections or other diseases. These are detailed in **Table 21.1.** Other disease-related symptoms appear to accompany severe immune deregulation, hormone imbalances, and metabolic disturbances. Pronounced wasting of body mass is a consequence of weight loss, diarrhea, and poor nutrient absorption. Both a rash and generalized lymphadenopathy in several chains of lymph nodes are the presenting symptoms in many AIDS patients.

Causative Agent

HIV is a retrovirus in the genus *Lentivirus*. Many retroviruses have the potential to cause cancer and produce dire, often fatal diseases and are capable of altering the host's DNA in profound ways. They are named "retroviruses" because they reverse the usual order of transcription. They contain an unusual enzyme called **reverse transcriptase (RT)** that catalyzes the replication of double-stranded DNA from single-stranded RNA. The association of retroviruses with their hosts can be so intimate that viral genes are permanently integrated into the host genome.

There are two major types of HIV—namely HIV-1, which is the dominant form in most of the world, and HIV-2. Genetic sequencing of HIV-1 shows that it is most related to simian immunodeficiency viruses in chimpanzees, while HIV-2 evolved from related viruses in sooty mangabeys, a type of monkey found in Africa. Both highlight the evidence that HIV in humans was derived from a zoonotic primate virus. HIV and other retroviruses display structural features typical of enveloped RNA viruses. The outermost component is a lipid envelope with transmembrane glycoprotein spikes and knobs that mediate viral adsorption to the host cell. HIV can only infect host cells that present the required receptors, which is a combination receptor consisting of the CD4 marker plus a coreceptor called CCR-5. The virus uses these receptors to gain entrance to several types of leukocytes and tissue cells.

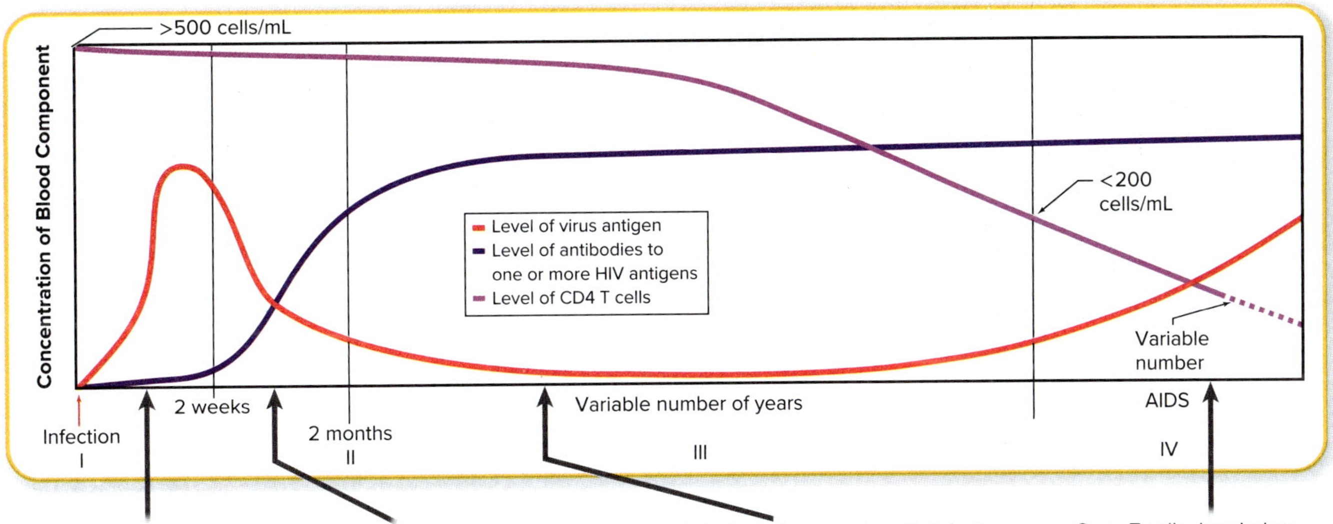
>500 cells/mL
Concentration of Blood Component
Level of virus antigen
Level of antibodies to one or more HIV antigens
Level of CD4 T cells
<200 cells/mL
Variable number
Infection
I
2 weeks
2 months
II
Variable number of years
III
AIDS
IV

Initial infection is often attended by vague, mononucleosis-like symptoms that soon disappear. This phase corresponds to the initial high levels of virus (the green line above). Antibodies are not yet abundant.

In the second phase, virus numbers in blood drop dramatically and antibody begins to appear. CD4 T cells begin to decrease in number.

A long period of mostly asymptomatic infection ensues. During this time, which can last from 2 to 15 years, swollen lymph nodes may be the prominent symptom. During the mid- to late-asymptomatic period, the number of T cells in the blood is steadily decreasing. Once the T-cell level reaches a (low) threshold, the symptoms of AIDS ensue.

Once T cells drop below 200 cells/mL, AIDS results. Note that even though antibody levels remain high, virus levels in the blood begin to rise.

Figure 21.22 The progression of HIV in the absence of treatment.

Table 21.1 Common complications of AIDS

Opportunistic Infections	Diseases of the Nervous System	Cancers	Wasting Syndrome
Skin, mucous membranes, and eyes Cytomegalovirus retinitis (loss of vision) Herpes simplex chronic ulcers **Nervous system** HIV encephalopathy Progressive multifocal leukoencephalopathy Cerebral toxoplasmosis Cryptococcus meningitis **Cardiovascular and lymphatic systems** Coccidioidomycosis Cytomegalovirus Histoplasmosis *Salmonella* septicemia **Respiratory system** Candidiasis of the trachea, bronchi, and lungs *P. jirovicecii* pneumonia *Mycobacterium tuberculosis* *Mycobacterium avium* complex Cryptococcosis Pneumonia, recurrent **Gastrointestinal system** Candidiasis of the esophagus and GI tract Intestinal isosporiasis Chronic cryptosporidiosis CMV colitis **Genitourinary tract and reproductive tract** Herpes simplex chronic ulcers Vaginal candidiasis Genital warts	AIDS dementia complex Peripheral neuropathy Lymphoma, primarily in the brain Toxoplasmosis of the brain Progressive, multifocal leukoencephalopathy	Kaposi's sarcoma Burkitt's lymphoma Immunoblastic lymphoma Invasive cervical cancer carcinoma Lymphoma Non-Hodgkin's lymphoma	Inadequate nutrition Metabolic disturbances Involuntary loss of more than 10% of body weight 30 days of weakness, diarrhea, or fever

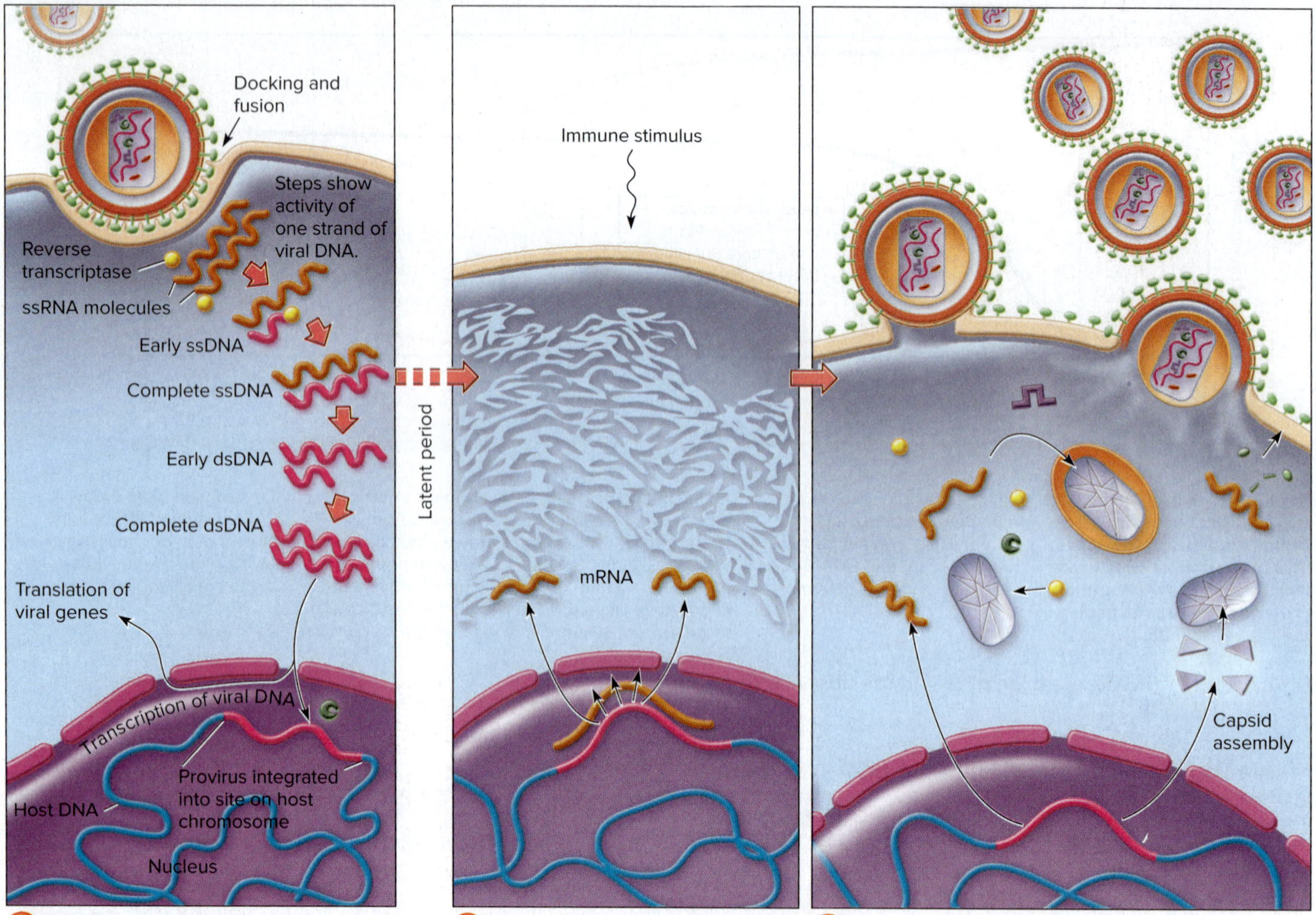
Docking and fusion
Steps show activity of one strand of viral DNA.
Reverse transcriptase
ssRNA molecules
Early ssDNA
Complete ssDNA
Early dsDNA
Complete dsDNA
Translation of viral genes
Transcription of viral DNA
Provirus integrated into site on host chromosome
Host DNA
Nucleus
Latent period
Immune stimulus
mRNA
Capsid assembly

1 The virus is adsorbed and endocytosed, and the twin RNAs are uncoated. Reverse transcriptase catalyzes the synthesis of a single complementary strand of DNA (ssDNA). This single strand serves as a template for synthesis of a double strand (ds) of DNA. In latency, dsDNA is inserted into the host chromosome as a provirus.

2 After a latent period, various immune activators stimulate the infected cell, causing reactivation of the provirus genes and production of viral mRNA.

3 HIV mRNA is translated by the cell's synthetic machinery into virus components (capsid, reverse transcriptase, spikes), and the viruses are assembled. Budding of mature viruses lyses the infected cell.

Figure 21.23 The general multiplication cycle of HIV.

Pathogenesis and Virulence Factors

As summarized in **figure 21.23,** HIV enters a mucous membrane or the skin and travels to dendritic cells, a type of phagocyte living beneath the epithelium. In the dendritic cell, the virus grows and is shed from the cell without killing it. The virus is amplified by macrophages in the skin, lymphoid organs, bone marrow, and blood. One of the great ironies of HIV is that it infects and destroys many of the very cells needed to combat it, including the helper (T4 or CD4) class of lymphocytes, monocytes, macrophages, and even B lymphocytes. The virus is adapted to docking onto its host cell's surface receptors. It then induces viral fusion with the cell membrane and creates syncytia.

Once the virus is inside the cell, its reverse transcriptase converts its RNA into DNA. Although initially it can produce a lytic infection, in many cells, it enters a latent period in the nucleus of the host cell and integrates its DNA into host DNA (as shown in figure 21.23). This latency accounts for the lengthy course of the disease. Despite being described as a "latent" stage, new viruses are constantly being produced and new T cells are constantly being manufactured, in an ongoing race that ultimately the host cells lose (in the absence of treatment).

The primary effects of HIV infection—those directly due to viral action—are harm to T cells and the central nervous system. The death of T cells and other white blood cells results in extreme **leukopenia** and loss of essential CD4 memory clones and stem cells. The viruses also cause formation of giant T cells and other syncytia, which allow the spread of viruses directly from cell to cell, followed by mass destruction of the syncytia. The destruction of CD4 lymphocytes paves the way for invasion by opportunistic agents and malignant cells. The central nervous system is affected when infected macrophages cross the blood-brain barrier and release viruses, which then invade nervous tissue. Studies have indicated that some of the viral envelope proteins can have a direct toxic effect on the brain's glial cells and other cells.

The secondary effects of HIV infection are the opportunistic infections and malignancies that occur as the immune system becomes progressively crippled by viral attack **(figure 21.24).**

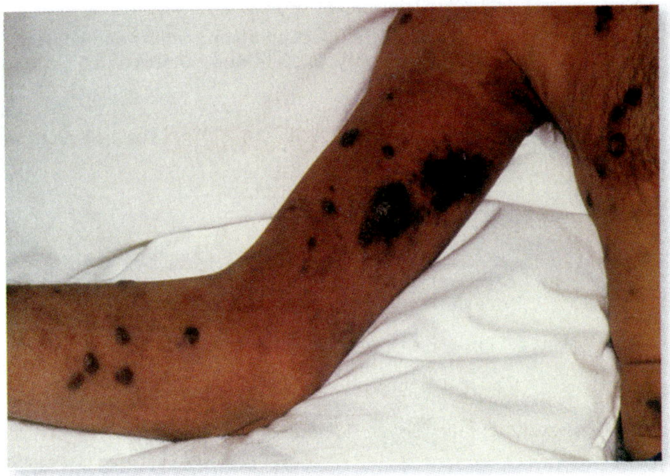

Figure 21.24 Kaposi's sarcoma in an AIDS patient.
SPL/Science Source

Transmission

In general, HIV is spread only by direct and rather specific routes **(Table 21.2).** Because the blood of HIV-infected individuals harbors high levels of free virus in both very early and very late stages of infection and high levels of infected leukocytes throughout infection, any form of intimate contact involving transfer of blood (trauma, needle sharing) can be a potential source of infection. Semen and vaginal secretions also harbor free virus and infected white blood cells. For that reason, they are significant factors in sexual transmission. Because breast milk contains significant numbers of leukocytes, neonates who escaped infection prior to and during birth can still become infected through nursing.

Epidemiology

HIV has probably infected humans since the 1920s. Scientists propose that a simian immunodeficiency virus (in monkeys) infected humans during the butchering of bush meat. The adaptation of the virus to humans was slow, initially. Then, in the 1980s, the disease, caused by the (now-adapted) *human* immunodeficiency virus, became more virulent and more communicable suddenly due to a number of factors. It was initially recognized in the United States when a high number of gay men were stricken with a previously unknown combination of symptoms. Since the beginning of the HIV epidemic in the early 1980s, more than 33 million have died of HIV-worldwide.

In the United States, only about 15% of those infected are unaware of it. The highest rate of transmission is among men having sex with men. But heterosexual contact is the second most

Table 21.2 HIV Transmission

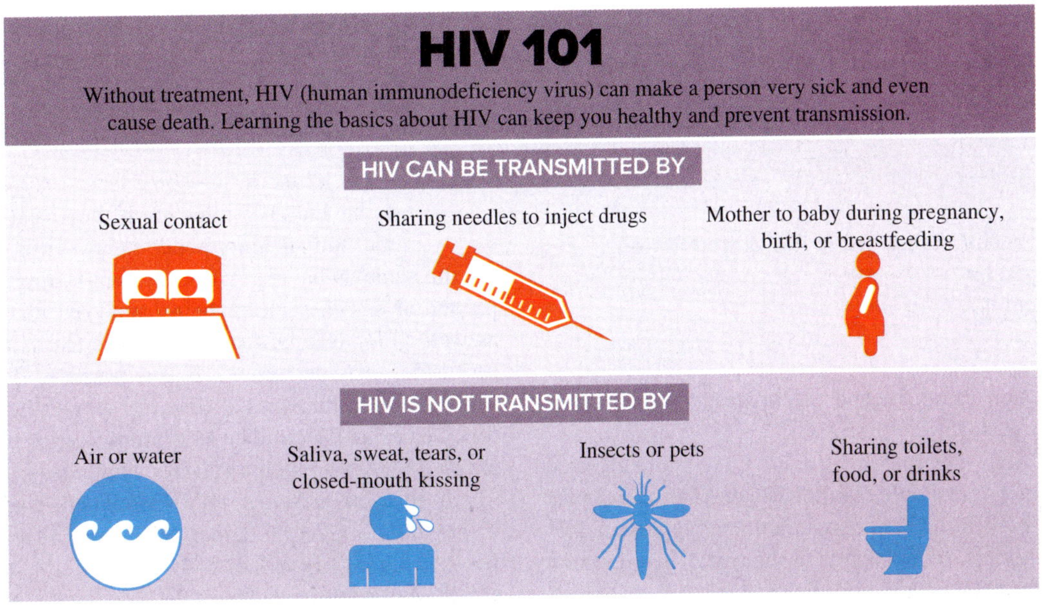
HIV 101
Without treatment, HIV (human immunodeficiency virus) can make a person very sick and even cause death. Learning the basics about HIV can keep you healthy and prevent transmission.
HIV CAN BE TRANSMITTED BY
Sexual contact
Sharing needles to inject drugs
Mother to baby during pregnancy, birth, or breastfeeding
HIV IS NOT TRANSMITTED BY
Air or water
Saliva, sweat, tears, or closed-mouth kissing
Insects or pets
Sharing toilets, food, or drinks

Source: CDC, Centers for Disease Control and Prevention

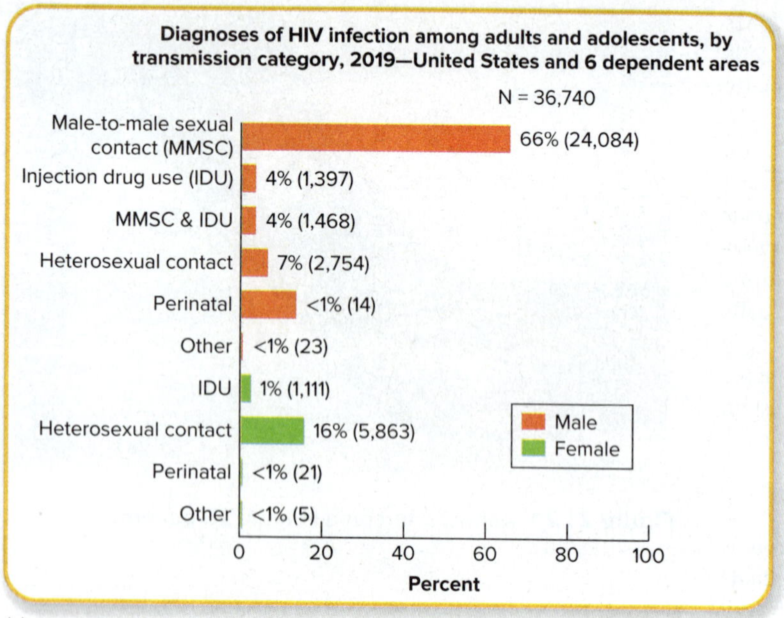
Diagnoses of HIV infection among adults and adolescents, by transmission category, 2019—United States and 6 dependent areas
N = 36,740
Male-to-male sexual contact (MMSC)
66% (24,084)
Injection drug use (IDU)
4% (1,397)
MMSC & IDU
4% (1,468)
Heterosexual contact
7% (2,754)
Perinatal
<1% (14)
Other
<1% (23)
IDU
1% (1,111)
Heterosexual contact
16% (5,863)
Perinatal
<1% (21)
Other
<1% (5)
Male
Female
0
20
40
60
80
100
Percent

(a)

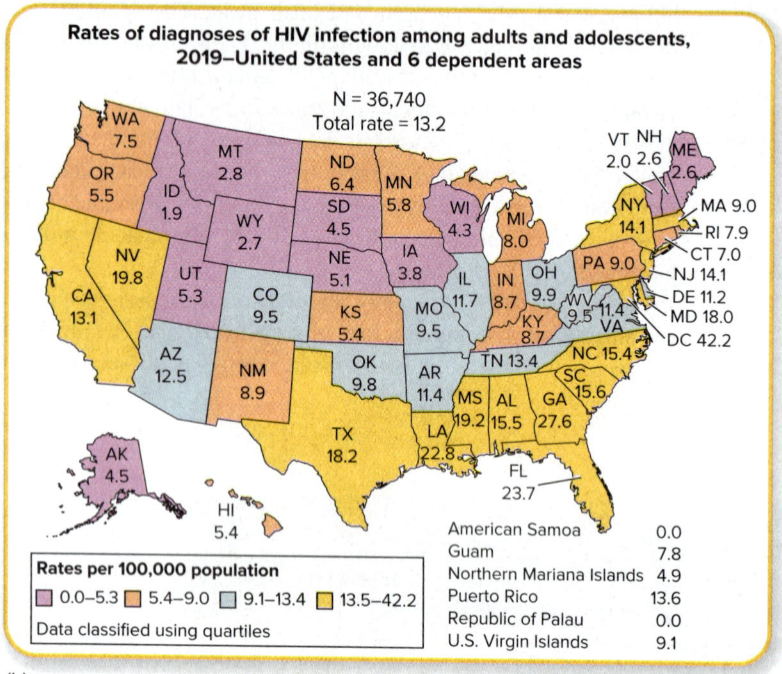
Rates of diagnoses of HIV infection among adults and adolescents, 2019–United States and 6 dependent areas
N = 36,740
Total rate = 13.2
WA 7.5
OR 5.5
ID 1.9
MT 2.8
WY 2.7
ND 6.4
SD 4.5
MN 5.8
WI 4.3
MI 8.0
VT 2.0
NH 2.6
ME 2.6
MA 9.0
RI 7.9
CT 7.0
NJ 14.1
DE 11.2
MD 18.0
DC 42.2
NY 14.1
PA 9.0
OH 9.9
WV 9.5
VA 11.4
NC 15.4
SC 15.6
GA 27.6
FL 23.7
AL 15.5
MS 19.2
LA 22.8
TN 13.4
KY 8.7
IN 8.7
IL 11.7
IA 3.8
MO 9.5
AR 11.4
NE 5.1
KS 5.4
OK 9.8
TX 18.2
NM 8.9
CO 9.5
UT 5.3
AZ 12.5
NV 19.8
CA 13.1
AK 4.5
HI 5.4
American Samoa 0.0
Guam 7.8
Northern Mariana Islands 4.9
Puerto Rico 13.6
Republic of Palau 0.0
U.S. Virgin Islands 9.1
Rates per 100,000 population
0.0–5.3
5.4–9.0
9.1–13.4
13.5–42.2
Data classified using quartiles

(b)

Figure 21.25 New HIV diagnoses in the United States.

(a, b) CDC, Centers for Disease Control and Prevention

common transmission category, and the biggest reason women acquire HIV **(figure 21.25).**

In the United States, people of color are much more affected by HIV than others. For example, while African Americans make up approximately 13% of the U.S. population, they represent 42% of annual new diagnoses. In most parts of the world, heterosexual intercourse is the primary mode of transmission.

Blood products (serum, coagulation factors) were once implicated in HIV transmission. Thousands of hemophiliacs died from the disease in the 1980s and 1990s. It is now standard practice to heat-treat any therapeutic blood products to destroy all viruses.

Not everyone who becomes infected or is antibody-positive develops AIDS. About 0.2% of people who are antibody-positive remain free of disease, indicating that functioning immunity to the virus can develop. Any person who remains healthy despite HIV infection is termed a *nonprogressor*. These people are the object of intense scientific study. Some have been found to lack the cytokine receptors that HIV requires for entry. Others are infected by a strain of virus weakened by genetic mutation. Since the beginning of the epidemic in 1981, two people have appeared to clear the virus on their own, the latest in 2021. This gives researchers hope that they can identify factors that shut down the virus and mimic them therapeutically.

Treatment of HIV-infected mothers with a simple anti-HIV drug has dramatically decreased the rate of maternal-to-infant transmission of HIV during pregnancy. Current treatment regimens result in a transmission rate of approximately 1%. In 2015, Cuba became the first country to completely eliminate mother-to-child transmission of the virus, through comprehensive testing and treatment of mothers. Several other countries, including Armenia, Belarus, and Malaysia, have eliminated mother-to-child transmission. The United States has greatly reduced the transmission but has not eliminated it.

A health care worker involved in an accident in which gross inoculation with contaminated blood occurs (as in the case of a needlestick) has a less than 1 in 1,000 chance of becoming infected. We should emphasize that transmission of HIV will not occur through casual contact or routine patient care procedures and that universal precautions for infection control were designed to give full protection for both worker and patient.

Culture and/or Diagnosis

First, let us define some terms. A person is diagnosed as having HIV infection if they have tested positive for the human immunodeficiency virus. This diagnosis is not the same as having AIDS. Current testing guidelines call for plasma or serum samples being analyzed for antibodies to HIV as well as for HIV p24 antigen. If the test is negative, there are no further tests performed. If it is positive, further tests are run to determine if the virus is HIV-1 or HIV-2. Equivocal results are followed by an HIV nucleic acid amplification test (NAAT). There are also FDA-approved, over-the-counter testing kits.

In the United States, HIV^+ people are diagnosed with AIDS if they have a CD4 (T helper cell) count of fewer than 200 cells per microliter of blood, or if they have one of the AIDS-defining conditions (ADIs) listed in table 21.1.

▶ Prevention

Avoidance of sexual contact with infected persons is a cornerstone of HIV prevention. Although avoiding intravenous drugs is an obvious deterrent, many drug addicts do not, or cannot, choose this option. In such cases, risk can be decreased by not sharing syringes or needles or by cleaning needles with bleach and then rinsing before another use.

Pre-exposure prophylaxis, called PrEP, is recommended for people who are at risk for becoming infected—for example, if their partner is positive and they are negative. Potentially it can reduce the risk of transmission to zero. Studies have shown it can lower the risk of infection through IV drug use by almost 75% and the risk of infection through sex by 99%. It consists of a two-drug combination marketed as Truvada.

Coming up with a successful HIV vaccine has been a difficult task because of the characteristics of the virus. Among them, HIV becomes latent in cells; its cell surface antigens mutate rapidly; and although it does elicit immune responses, it is apparently not completely controlled by them. In view of the great need for a vaccine, however, none of those facts has stopped the medical community from moving ahead. Currently, there are about 11 clinical trials of various vaccine candidates, one of which uses the mRNA technology used in some COVID-19 vaccines.

▶ Treatment

Clear-cut guidelines exist for treating people who test HIV-positive. These guidelines are updated regularly. Treatment always includes multiple drugs to help prevent acquisition of resistance. In addition to antiviral chemotherapy, AIDS patients can receive a wide array of drugs to prevent or treat a variety of opportunistic infections and other complications such as wasting disease. These treatment regimens vary according to each patient's profile and needs.

Figure 21.26 illustrates the different targets for antiretroviral drugs **(Disease Table 21.13).**

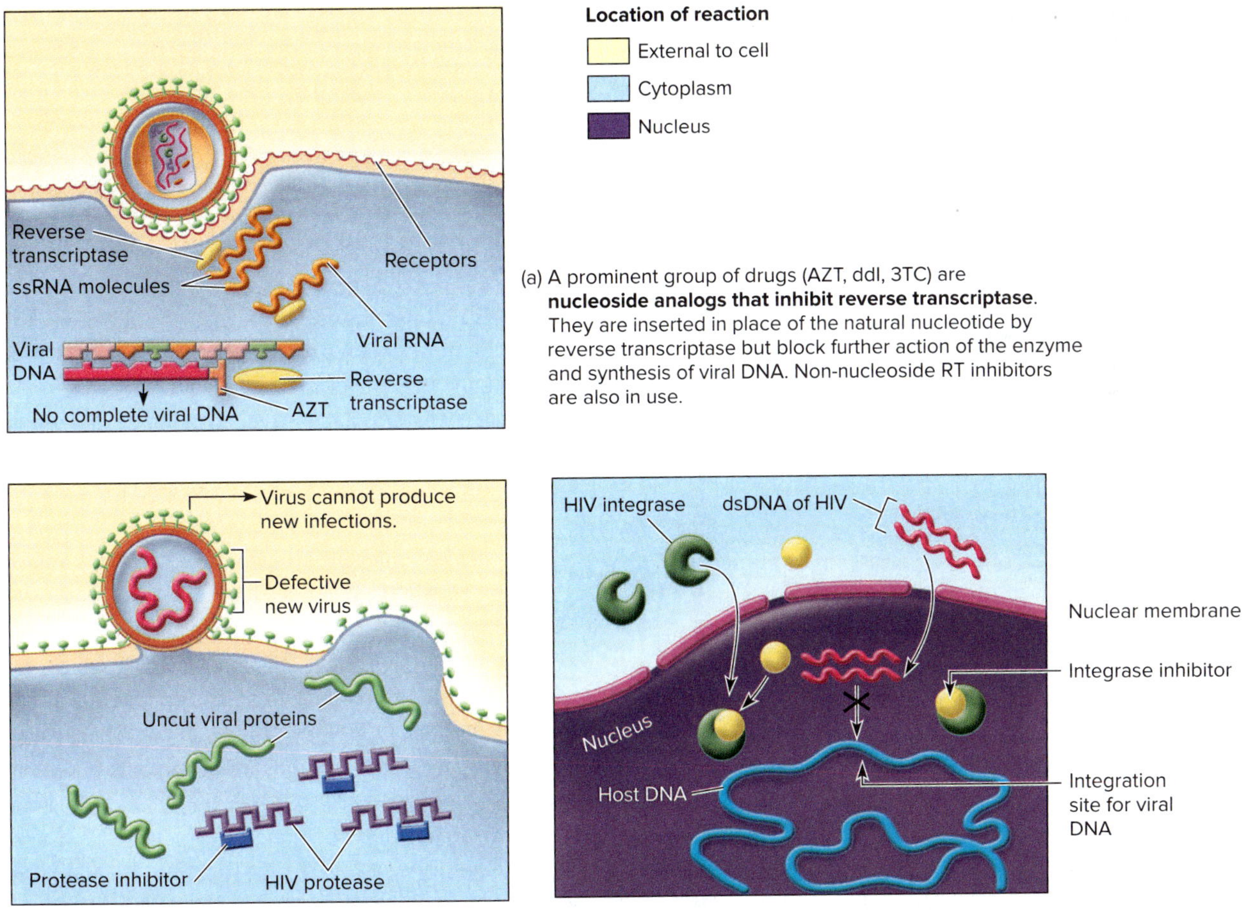
Location of reaction
External to cell
Cytoplasm
Nucleus
Reverse transcriptase
ssRNA molecules
Receptors
Viral RNA
Viral DNA
Reverse transcriptase
No complete viral DNA
AZT
Virus cannot produce new infections.
Defective new virus
Uncut viral proteins
Protease inhibitor
HIV protease
HIV integrase
dsDNA of HIV
Nuclear membrane
Integrase inhibitor
Nucleus
Host DNA
Integration site for viral DNA

(a) A prominent group of drugs (AZT, ddI, 3TC) are **nucleoside analogs that inhibit reverse transcriptase.** They are inserted in place of the natural nucleotide by reverse transcriptase but block further action of the enzyme and synthesis of viral DNA. Non-nucleoside RT inhibitors are also in use.

(b) **Protease inhibitors** plug into the active sites on HIV protease. This enzyme is necessary to cut elongated HIV protein strands and produce functioning smaller protein units. Because the enzyme is blocked, the proteins remain uncut, and abnormal defective viruses are formed.

(c) **Integrase inhibitors** are a class of experimental drugs that attach to the enzyme required to splice the dsDNA from HIV into the host genome. This will prevent formation of the provirus and block future virus multiplication in that cell.

Figure 21.26 Mechanisms of action of anti-HIV drugs.

INSIGHT 21.1 MICROBIOME: Gut Microbiome and High Blood Pressure

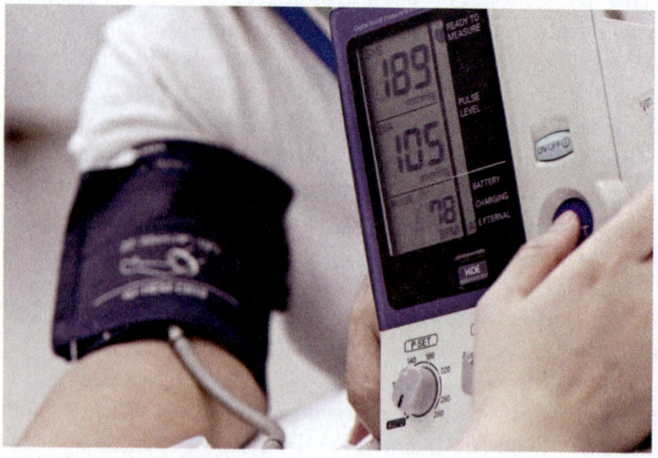
189
105
78

anucha maneechote/Shutterstock

In this chapter we are examining obvious infectious causes of disease in the cardiovascular system. One of the most prevalent causes of disease in that system is not directly caused by infectious agents, however. Hypertension, commonly called high blood pressure, affects about 1 in 3 adults in the United States, and can lead to heart attacks and strokes.

A recent study published in the journal *Microbiome* suggests a link between the gut microbiome and hypertension. Researchers looked at the microbiomes of 41 healthy people, 56 people with prehypertension, and 99 individuals with actual hypertension. They used metagenomic methods to assess the gut microbiomes. They found that in the prehypertensive and hypertensive individuals, the gut microbiome was much less diverse than in healthy subjects. Keep in mind that this is correlation, not causation, particularly because the two parameters (degree of hypertension and gut microbiome) were assessed simultaneously. In that situation, it is not possible to say that one caused the other, only that they are correlated. The fact that the prehypertensive subjects shared the low-diversity microbiome may point to the possibility that the change in microbiome preceded the hypertension.

The authors tried to address this inability to show causation by using a form of Koch's postulates. They inoculated germ-free mice, who had low to normal blood pressure, with the fecal material from hypertensive humans and found that the mice did indeed go on to experience high blood pressure. Still, this research, like most microbiome research, is preliminary—and highly intriguing.

Sources: 2021. *Circulation Research.* The Gut Microbiome in Hypertension: Recent Advances and Future Perspectives. https://doi.org/10.1161/CIRCRESAHA.121.318065Circulation Research. 2021;128:934–950: 2017. *Microbiome.* https://doi.org/10.1186/s40168-016-0222-x

Disease Table 21.13 HIV Infection and AIDS

Causative Organism(s)	Human immunodeficiency virus 1 or 2
Most Common Modes of Transmission	Direct contact (sexual), parenteral (blood-borne), vertical (perinatal and via breast milk)
Virulence Factors	Attachment, syncytia formation, reverse transcriptase, high mutation rate
Culture/Diagnosis	Immunoassay to detect antibodies as well as HIV antigen
Prevention	Avoidance of contact with infected sex partner, contaminated blood, breast milk; pre-exposure prophylaxis (PreP) for high-risk individuals
Treatment	Anti-retroviral regimen begun as early as possible
Epidemiological Features	United States: HIV infection = 1.2 million; internationally, HIV infection = 38 million

21.2 Learning Outcomes—Assess Your Progress

4. List the possible causative agents for each of the following infectious cardiovascular conditions: COVID-19, endocarditis, plague, tularemia, Lyme disease, infectious mononucleosis, anthrax, Chagas disease, and malaria.
5. Explain why COVID-19 is placed in the cardiovascular systems chapter.
6. Discuss what series of events may lead to sepsis and how it should be prevented and treated.
7. Discuss the difference between hemorrhagic and nonhemorrhagic fever diseases.
8. List the possible causative agents and modes of transmission for hemorrhagic fever diseases.
9. List the possible causative agents and modes of transmission for nonhemorrhagic fever diseases.
10. Identify the five or six most relevant facts about malaria.
11. Describe important events in the course of an HIV infection in the absence of treatment.
12. Discuss the epidemiology of HIV infection in the United States, and in the developing world.

Media Under The Microscope Wrap-Up

Lipik/Shutterstock

Because I was not familiar with the website *Newswise*, I clicked on the "About" tab and found that it considers itself a news site for journalists, publishing stories on many different topics, medicine being only one of them. Many of the stories come from universities. This one came from the University of North Carolina School of Medicine. That is a good sign that the information is credible.

The **intended message** is simply to report the 10-fold increase in hospitalizations and surgeries due to acute infective endocarditis.

My **critical reading** of the article is that it is sound. We know that injecting drug users very often reuse needles or use dirty needles. The substances they inject are not sterile, either. And they inject them directly into the cardiovascular system. Such an overwhelming dose of microbes is likely to colonize valvular tissue of the heart. I have a quibble with the article's statement that "109 surgeries are performed every year." It seems unlikely that 109 surgeries are performed every year, and more likely that 109 surgeries were performed *last* year. A scientist certainly wouldn't make this mistake in reporting her research, but the article was no doubt written by someone in the university's public affairs office.

I would **interpret** the article to my friends (after I read this chapter) by explaining that heart valves are particularly prone to be attachment sites for microorganisms, and then explain the facts I included in the paragraph above.

My **overall grade** for the article is an A–.

Source: 2018. *Newswise*. Article ID: 701783. Released October 8, 2018.

Study Smarter: Better Together

These activities are designed for you to use on your own with a study group—either a face-to-face group or a virtual one, consisting of 3–5 members. Studying together can be very helpful, but there are effective and ineffective ways to do it. For example, getting together without a clear structure is often not a good use of your time. Use your time efficiently by using one or more of the exercises below.

FACE-TO-FACE GROUPS

Use one or more of the activities below.

Peer Instruction: Assign numbers to your group members to use all semester long. Now look at these five concepts from this chapter. Each group member prepares a 5-minute lesson on the topic corresponding to their number. Don't worry if you have fewer than 5 members; just use however many you have! During your group study time, each member presents their lesson, and the group spends another 5–10 minutes discussing that lesson.

1. Cardiovascular disease(s) transmitted by fleas (with summary of the disease[s])
2. Cardiovascular disease(s) transmitted by ticks (with summary of the disease[s])
3. Defenses of the cardiovascular system
4. Cardiovascular disease(s) transmitted by mosquitoes (with summary of the disease[s])
5. The timeline of the COVID-19 pandemic, starting in December 2019

Concept Maps: Each member of the group should use this list of terms from this chapter to generate their own concept map. This can be hand-drawn or created using software (see Appendix C for guidelines). During group study time, compare each other's concept maps and help each other make sure they are correct. Of course, there are many different "correct" maps. Examining each member's map will help you talk through the varied concepts and how they are related.

Concept Terms:

HIV	AIDS	CD4 lymphocytes	CD8 lymphocytes
adaptive immunity	opportunistic infections	B lymphocytes	PrEP
latency	reverse transcriptase		

Table Topics: Each group member should identify a concept or topic from this week's class assignments with which they are having trouble and share it during group study time. The other group members can then help to clarify confusing issues or share how they figured it out. Aim for a maximum of 15 minutes per topic. If the topic remains unclear to the group, bring it up during class or use the instructor's office hours or e-mail to ask for help. Taking the time to struggle with a difficult concept first makes your questions much more specific and more likely to yield helpful answers.

(continued)

VIRTUAL GROUPS

Not everyone has the time or opportunity to meet with group members outside of class time. You or your instructor can create a virtual group using e-mail or the course software.

Weekly Discussion Board: This forum can be used as a way for groups to discuss topics, via e-mail, or other learning management systems or online platforms, before they are covered in class. As each member of the group answers the current week's question, they should send their responses to every other member of their group. It's best to agree on a deadline based on how your class schedule works (Saturday for the next week's topics, for example). Then, after the topic is discussed in class, each member should send a response that all group members will see with a follow-up post on the same topic. If you cover more than one chapter in a week, someone can be designated to choose which chapter Discussion Board question you will use. Or simply decide up front that you will always use the first-chapter-of-the-week's question, to keep the schedule simple.

Discussion Question

Compare and contrast the cardiovascular system and the lymphatic system.

Deadliness and Communicability of Selected Diseases of the Cardiovascular and Lymphatic Systems

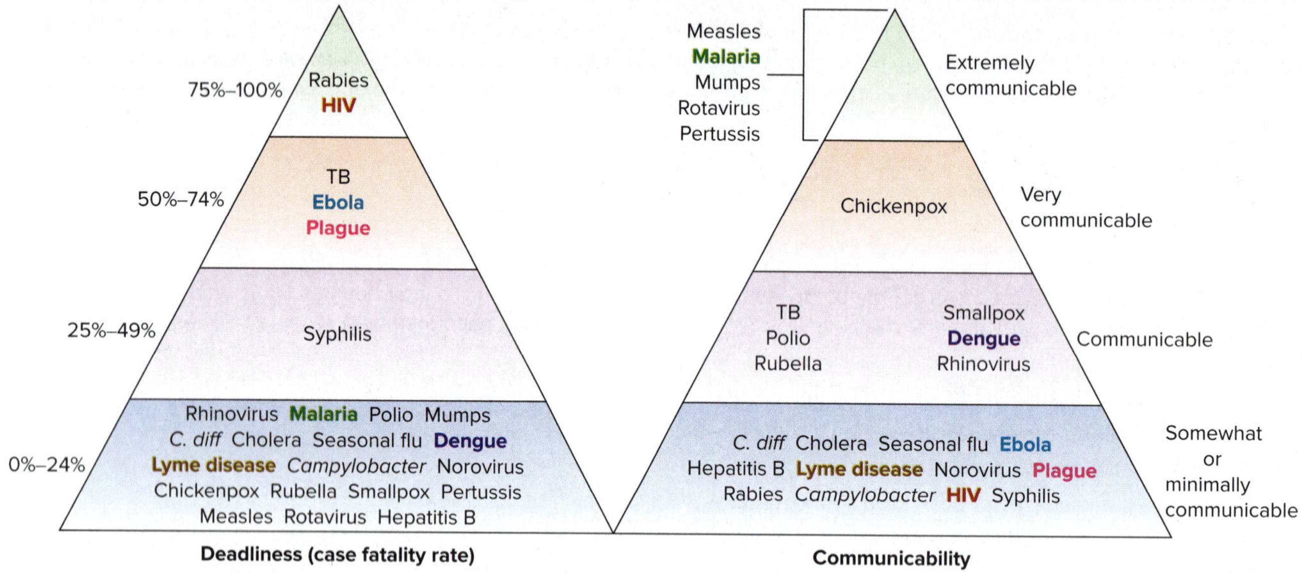
75%–100%
Rabies
HIV
50%–74%
TB
Ebola
Plague
25%–49%
Syphilis
0%–24%
Rhinovirus Malaria Polio Mumps
C. diff Cholera Seasonal flu Dengue
Lyme disease Campylobacter Norovirus
Chickenpox Rubella Smallpox Pertussis
Measles Rotavirus Hepatitis B
Deadliness (case fatality rate)
Measles
Malaria
Mumps
Rotavirus
Pertussis
Extremely communicable
Chickenpox
Very communicable
TB
Polio
Rubella
Smallpox
Dengue
Rhinovirus
Communicable
C. diff Cholera Seasonal flu Ebola
Hepatitis B Lyme disease Norovirus Plague
Rabies Campylobacter HIV Syphilis
Somewhat or minimally communicable
Communicability

▶ Summing Up

Taxonomic Organization Microorganisms Causing Diseases in the Cardiovascular and Lymphatic System

Microorganism	Disease	Disease Table
Gram-positive endospore-forming bacteria		
Bacillus anthracis	Anthrax	Anthrax, 21.8
Gram-positive bacteria		
Staphylococcus aureus	Acute endocarditis	Endocarditis, 21.2
Streptococcus pyogenes	Acute endocarditis	Endocarditis, 21.2
Streptococcus pneumoniae	Acute endocarditis	Endocarditis, 21.2
Enterococcus	Acute endocarditis	Endocarditis, 21.2

Gram-negative bacteria		
Pseudomonas aeruginosa	Acute endocarditis	Endocarditis, 21.2
Yersinia pestis	Plague	Plague, 21.4
Francisella tularensis	Tularemia	Tularemia, 21.5
Borrelia burgdorferi	Lyme disease	Lyme disease, 21.6
Brucella abortus, B. suis	Brucellosis	Nonhemorrhagic fever diseases, 21.10
Coxiella burnetii	Q fever	Nonhemorrhagic fever diseases, 21.10
Bartonella henselae	Cat-scratch disease	Nonhemorrhagic fever diseases, 21.10
Bartonella quintana	Trench fever	Nonhemorrhagic fever diseases, 21.10
Ehrlichia species	Ehrlichiosis	Nonhemorrhagic fever diseases, 21.10
Anaplasma species	Anaplasmosis	Nonhemorrhagic fever diseases, 21.10
Rickettsia species	Spotted fever rickettsiosis	Nonhemorrhagic fever diseases, 21.10
DNA viruses		
Epstein-Barr virus	Infectious mononucleosis	Infectious mononucleosis, 21.7
RNA viruses		
SARS-CoV-2	COVID-19	COVID-19, 21.1
Yellow fever virus	Yellow fever	Hemorrhagic fevers, 21.9
Dengue fever viruses	Dengue fever	Hemorrhagic fevers, 21.9
Chikungunya virus	Hemorrhagic fever	Hemorrhagic fevers, 21.9
Ebola and Marburg viruses	Ebola and Marburg hemorrhagic fevers	Hemorrhagic fevers, 21.9
Lassa fever virus	Lassa fever	Hemorrhagic fevers, 21.9
Retroviruses		
Human immunodeficiency virus 1 and 2	HIV infection and AIDS	HIV infection and AIDS, 21.13
Protozoa		
Babesia species	Babesiosis	Nonhemorrhagic fever diseases, 21.10
Trypanosoma cruzi	Chagas disease	Chagas disease, 21.11
Plasmodium falciparum, P. vivax, P. ovale, P. malariae	Malaria	Malaria, 21.12

Chapter Summary

INFECTIOUS DISEASES MANIFESTING IN THE CARDIOVASCULAR AND LYMPHATIC SYSTEMS

21.1 THE CARDIOVASCULAR AND LYMPHATIC SYSTEMS, THEIR DEFENSES, AND NORMAL BIOTA

- The cardiovascular system is composed of the blood vessels and the heart. The lymphatic system is a one-way system returning fluid from the tissues to the cardiovascular system. The systems are well-protected from microbial infection, as they are not open body systems and they contain many components of the host's immune system.
- Research suggests that even in the healthy state, the cardiovascular and lymphatic systems contain low levels of microbes, particularly in the leukocytes.

21.2 INFECTIOUS DISEASES MANIFESTING IN THE CARDIOVASCULAR AND LYMPHATIC SYSTEMS

- **COVID-19**—A global pandemic caused by the coronavirus SARS-CoV-2, which was first found in humans in December of 2019.
- **Endocarditis**—Usually manifests as an infection of the heart valves. Can be acute, which is usually caused by direct parental implantation of bacteria into the bloodstream, or subacute, which is usually preceded by damage to the heart valves and can involve the introduction of even small numbers of normal biota into the bloodstream.
- **Sepsis**—Organisms actively multiplying in the blood. Caused by bacteria or fungi.
- **Plague**—Comes in three possible forms: bubonic, septicemic, and pneumonic. Caused by *Yersinia pestis*.
- **Tularemia**—Sometimes called rabbit fever, caused by *Francisella tularensis*.
- **Lyme disease**—Caused by *Borrelia burgdorferi*, a spirochete transmitted primarily by *Ixodes* ticks.
- **Infectious mononucleosis**—Caused by the Epstein-Barr virus (EBV). Cell-mediated immunity can control the infection, but people usually remain chronically infected.
- **Anthrax**—Exhibits primary symptoms in various locations: skin, lungs, gastrointestinal tract, central nervous system. Caused by *Bacillus anthracis*, an organism that lends itself to use in bioterror attacks due to its deadliness and its ability to form endospores.
- **Hemorrhagic fever diseases**—Extreme fevers, often accompanied by internal hemorrhaging. **Yellow fever:** Endemic in rainy climates in Africa and South America; virus carried by *Aedes aegyptii* mosquitoes. **Dengue fever:** Virus also carried by *Aedes* mosquitoes. Hemorrhagic forms can be lethal. **Chikungunya:** Mosquito-borne virus; found throughout Americas. **Ebola** and **Marburg** viruses are filoviruses endemic to central Africa. Lead to extreme capillary fragility. **Lassa fever:** Caused by arenavirus whose reservoir is a rodent in west Africa.
- **Nonhemorrhagic fever diseases**—Characterized by high fever without the capillary fragility in hemorrhagic diseases. **Brucellosis:** Transmitted through contact with livestock products. **Q fever:** Caused by *Coxiella burnetii*; most common in Colorado, California, Texas, and Illinois. **Cat-scratch disease:** Caused by *Bartonella henselae*; history of cat bite or scratch. **Trench fever:** *Bartonella quintana*, transmitted by lice. **Ehrlichiosis:** Transmitted by ticks, great increase in incidence in southern U.S. **Anaplasmosis:** *Anaplasma* bacterium transmitted by ticks; upper Midwest and northeastern U.S. **Babesiosis:** *Babesia* protozoan carried by ticks; roughly same distribution as *Anaplasma*. **Spotted fever rickettsiosis:** Another tick-borne group of diseases; causes rashes and internal symptoms.
- **Chagas disease**—*Trypanosoma cruzi* transmitted by insects; disease has acute and chronic phases.
- **Malaria**—Threatens 40% of world's population every year. Caused by four different species of *Plasmodium* protozoan. Vaccine for children now authorized for use in endemic areas.
- **HIV infection and AIDS**—Caused by HIV, a retrovirus. Destruction of T4 lymphocytes paves way for invasion by other (opportunistic) microbes and malignant cells.

INFECTIOUS DISEASES AFFECTING
The Cardiovascular and Lymphatic Systems

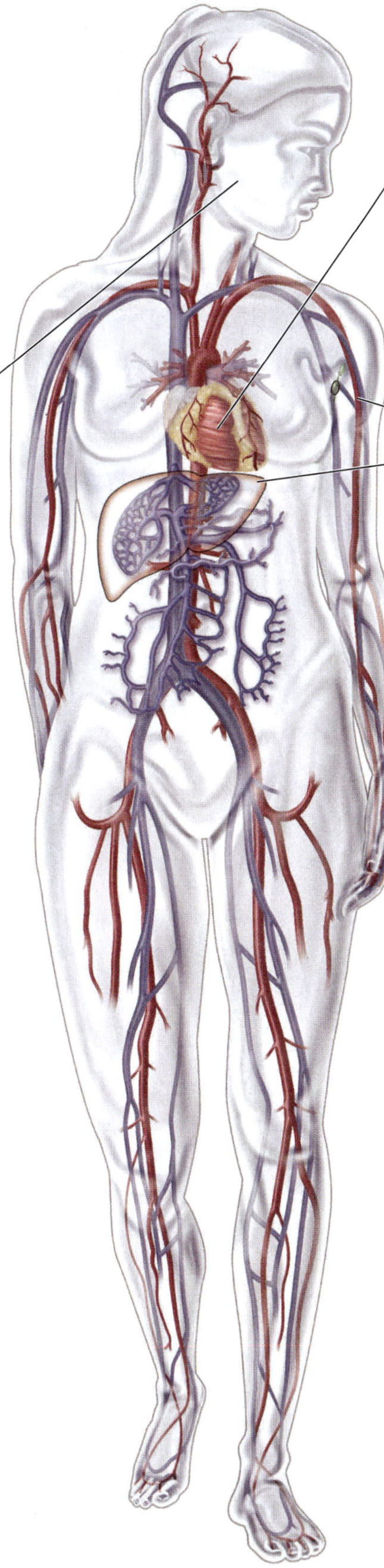

COVID-19
SARS-CoV-2
Nonhemorrhagic Fever Diseases
Brucella abortus
Brucella suis
Coxiella burnetii
Bartonella henselae
Bartonella quintana
Ehrlichia species
Anaplasma species
Babesia species
Infectious Mononucleosis
Epstein-Barr virus
Other viruses
Tularemia
Francisella tularensis
Lyme Disease
Borrelia burgdorferi
Hemorrhagic Fever Diseases
Yellow fever virus
Dengue fever virus
Ebola virus
Marburg virus
Lassa fever virus
Chikungunya virus
Endocarditis
Various bacteria
Plague
Yersinia pestis
Sepsis
Various bacteria
Various fungi
Malaria
Plasmodium species
Anthrax
Bacillus anthracis
HIV Infection and AIDS
Human immunodeficiency virus 1 or 2
Chagas Disease
Trypanosoma cruzi
Helminths
Bacteria
Viruses
Protozoa
Fungi

System Summary Figure 21.27

SmartGrid: From Knowledge to Critical Thinking

This *21 Question Grid* takes the topics from this chapter and arranges them with respect to the American Society for Microbiology's Undergraduate Curriculum guidelines—all six of the important "Concepts" as well as the important "Competency" of scientific literacy. Three questions are supplied, which cover chapter content referring to the Concept or Competency in increasing levels of Bloom's taxonomy for learning.

ASM Concept/ Competency	A. Bloom's Level 1, 2—Remember and Understand (Choose one)	B. Bloom's Level 3, 4—Apply and Analyze	C. Bloom's Level 5, 6—Evaluate and Create
Evolution	**1.** Which of the following microbes have evolved a lifestyle suited to the inside of host cells? **a.** *Bacillus anthracis* **b.** *Coxiella burnetii* **c.** MRSA **d.** two of these	**2.** Why do you think that malarial infection is more often fatal in children than in adults in areas where it is endemic?	**3.** In chapter 20 you were asked to speculate on why a disease such as polio has become less virulent over time, while rabies has not. Malaria has also not become less virulent over time. There is evidence that HIV is becoming less virulent in humans. Is there a pattern you can discern in microbes that do not become less virulent over time?
Cell Structure and Function	**4.** Which of the following is a gram-positive bacterium? **a.** *Staphylococcus aureus* **b.** *Borrelia burgdorferi* **c.** *Coxiella burnetii* **d.** *Trypanosome cruzi*	**5.** What characteristic(s) of *Bacillus anthracis* make it a good candidate for bioterror?	**6.** Construct an immunological argument about why Lyme disease might continue to cause symptoms after the bacterium has been cleared.
Metabolic Pathways	**7.** Which of the following diseases is characterized by the formation of a biofilm? **a.** plague **b.** HIV **c.** endocarditis **d.** Chagas disease	**8.** Name some ways in which you think intracellular bacteria such as *Coxiella* and *Brucella* might resemble viruses.	**9.** Argue for the bloodstream being an advantageous environment for microbial pathogens. Now argue the opposite.
Information Flow and Genetics	**10.** Which of the following diseases is caused by a retrovirus? **a.** Lassa fever **b.** Ebola **c.** anthrax **d.** HIV	**11.** Discuss whether human genetics plays a role in HIV infection, providing at least one example to illustrate your position.	**12.** Remdesivir, an antiviral developed for hepatitis C infection, was initially used to try to treat COVID-19 patients. What biological reason can you find for this hope that it would work against SARS-CoV-2?
Microbial Systems	**13.** The bite of a tick can cause **a.** ehrlichiosis. **b.** Lyme disease. **c.** anaplasmosis. **d.** all of these.	**14.** Discuss the merits and problems inherent in eliminating all mosquitoes in an effort to battle diseases such as yellow fever, dengue fever, and chikungunya.	**15.** This text frequently discusses the increase in infectious diseases in humans caused by the warming climate. Can you describe a hypothetical microbial pathogen that would cause less disease in a warming climate?

ASM Concept/ Competency	A. Bloom's Level 1, 2—Remember and Understand (Choose one)	B. Bloom's Level 3, 4—Apply and Analyze	C. Bloom's Level 5, 6—Evaluate and Create
Impact of Microorganisms	**16.** Normal biota found in the oral cavity are most likely to cause **a.** acute endocarditis. **b.** subacute endocarditis. **c.** malaria. **d.** tularemia.	**17.** In the Middle Ages, during a massive plague epidemic, one of the control measures instituted was the quarantine of infected people. Why was this not successful?	**18.** For decades public health officials have been trying to eliminate malaria. There are some ecologists and anthropologists who suggest that eliminating malaria will lead to overpopulation and famine in "the malaria belt." Defend one of these viewpoints.
Scientific Thinking	**19.** Lyme disease is most likely to occur in **a.** North Dakota. **b.** Connecticut. **c.** Oklahoma. **d.** Arkansas.	**20.** New variants of SARS-CoV-2 are characterized by changes in the amino acids in their spike proteins. Why is there so much focus on the spike protein?	**21.** When female mosquitoes mate with males infected with the bacterium *Wolbachia,* they no longer produce viable eggs. This is being tested as a control strategy for dengue fever and other mosquito-borne diseases. What questions would be important to answer before making this a widespread practice?

Answers to the multiple-choice questions appear in Appendix A.

Visual Connections

This question uses visual images to connect content within and between chapters.

1. a. **From chapter 15, figure 15.16.** Imagine that the WBCs shown in this illustration are unable to control the microorganisms. Could the change that has occurred in the vessel wall help the organism spread to other locations? If so, how?

   b. If the organisms are able to survive phagocytosis, how could that impact the progress of this disease? Explain your answer.

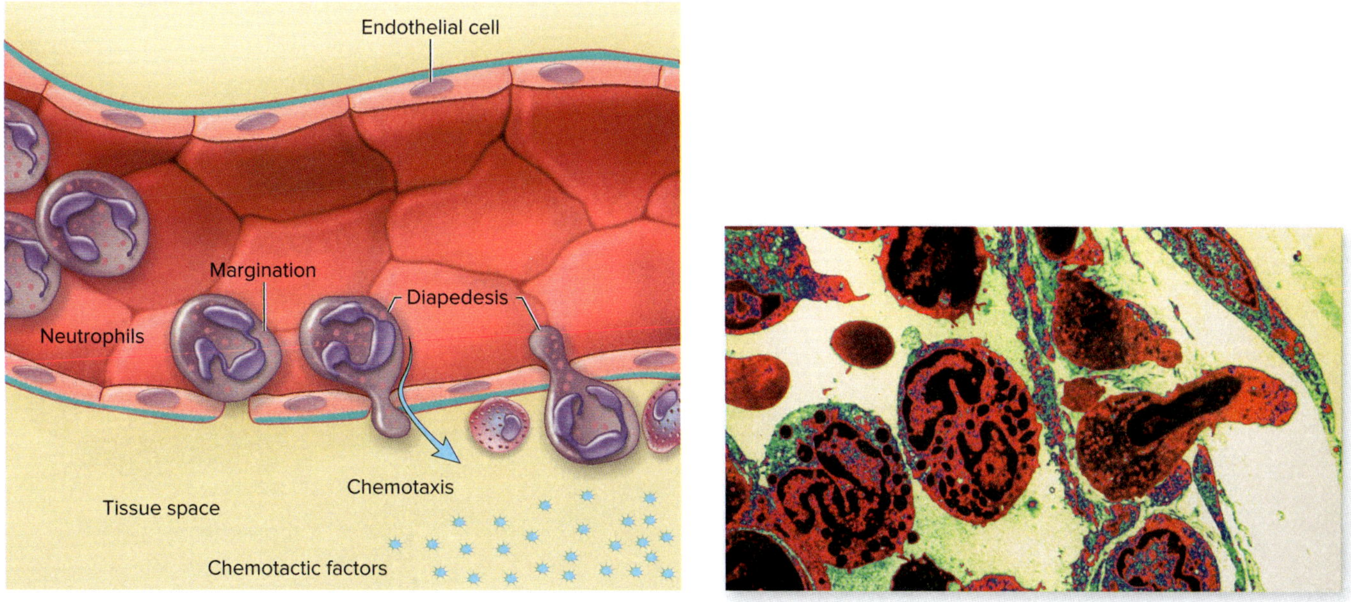
Endothelial cell
Margination
Diapedesis
Neutrophils
Chemotaxis
Tissue space
Chemotactic factors

(a) (b)

Joubert/Science Source

High Impact Study

These terms and concepts are most critical for your understanding of this chapter—and may be the most difficult. Have you mastered them? In these disease chapters, the terms and concepts help you identify what is important in a different way than the comprehensive details found in the Disease Tables. Your instructor will help you understand what is important for your class.

Concepts

- [ ] Defenses of cardiovascular system
- [ ] Normal microbiota of cardiovascular system
- [ ] Differing epidemiology of subacute and acute endocarditis
- [ ] Three forms of plague
- [ ] Diseases transmitted by fleas
- [ ] Diseases transmitted by ticks
- [ ] Diseases transmitted by mosquitoes
- [ ] Diseases in Category A for bioterrorism
- [ ] The COVID-19 pandemic
- [ ] Organisms in this chapter for which there are vaccines available
- [ ] Organisms in this chapter that display significant antibiotic resistance

Terms

- [ ] -emias
- [ ] Erythema migrans
- [ ] Reverse transcriptase

Design Element: (College students): Caia Image/Image Source

LightField Studios/Shutterstock

22

Infectious Diseases Manifesting in the Respiratory System

MEDIA UNDER THE MICROSCOPE

Man Flu

This opening case examines an article from the popular media to determine the extent to which it is factual and/or misleading. This case focuses on this recent article from AARP, "Yes, 'Man Flu' Is Real."

Some people believe that men are bigger wimps when it comes to colds and flu than women are. This has come to be referred to as "Man Flu." This article reported on two different research papers that seem to suggest that, in fact, men do suffer more from respiratory viruses.

The first research paper discussed by the AARP article was conducted in mice. Scientists injected mice with a bacterium that evokes a similar immune response as do viruses. The male mice suffered more signs and symptoms than the female mice. Their body temperatures dropped more than those of females, and they had more signs of generalized inflammation. As the article states, "They even looked more miserable—droopy-eyed, shivering, and huddled together." The male mice took 48 hours to recover from their infections, while on average the females got better in 24 hours. The author corresponded with the study author and included quotes from her in the article.

The second research paper was a type of meta-analysis, which is a study of dozens or hundreds of previous research papers to try to capture an overarching picture of the state of the science around a particular question. The meta-analysis looked at studies of men and women, and concluded that the evidence supports an increased severity of symptoms in males. There is biological plausibility to this conclusion, as testosterone tends to decrease the immune response, and estrogen tends to enhance it, according to our article. The author of the article could not resist poking a bit of fun at men by saying, "The one thing the researchers couldn't detect, though, was whether those male mice whined more about their symptoms."

- What is the **intended message** of the article?
- What is your **critical reading** of the summary of the article? Remember that in this context, "critical reading" does not necessarily mean *What criticism do you have?* but asks you to apply your knowledge to interpret whether the article is factual and whether the facts support the intended message.

Media Under The Microscope Wrap-Up appears at the end of the chapter.

- How would you **interpret** the news item for your nonmicrobiologist friends?
- What is your **overall grade** for the news item—taking into account its accuracy and the accuracy of its intended effect?

Outline and Learning Outcomes

22.1 The Respiratory Tract, Its Defenses, and Normal Biota

1. Draw or describe the anatomical features of the respiratory tract.
2. List the natural defenses present in the respiratory tract.
3. List the main genera of normal biota presently known to occupy the respiratory tract.

22.2 Infectious Diseases Manifesting in the Upper Respiratory Tract

4. List the possible causative agents, modes of transmission, virulence factors, diagnostic techniques, and prevention and treatment for each of the diseases of the upper respiratory tract: the common cold, sinusitis, otitis media, and pharyngitis.
5. Identify which disease is often caused by a mixture of microorganisms.
6. Identify two bacteria that can cause dangerous pharyngitis cases.

22.3 Infectious Diseases Manifesting in Both the Upper and Lower Respiratory Tracts

7. List the possible causative agents for each of the infectious conditions affecting both the upper and lower respiratory tracts: whooping cough, RSV disease, and influenza.
8. Describe the symptoms appearing in each stage of whooping cough.
9. Discuss reasons for the increase of pertussis cases over the past three decades.
10. Identify the age groups at most risk for serious disease from RSV.
11. Plan a response for a situation in which a patient declines influenza vaccination because they believe that the warnings about pandemics are always exaggerated.
12. Compare and contrast *antigenic drift* and *antigenic shift* in influenza viruses.

22.4 Infectious Diseases Manifesting in the Lower Respiratory Tract

13. List the possible causative agents for each of the diseases affecting the lower respiratory tract: tuberculosis, community-acquired pneumonia, healthcare-associated pneumonia, and hantavirus pulmonary syndrome.
14. Discuss the problems associated with MDR-TB and XDR-TB.
15. Demonstrate an in-depth understanding of the current epidemiology of tuberculosis infection.
16. Explain why so many diverse microorganisms can cause the condition of pneumonia.
17. Identify three important causes of serious community-acquired pneumonia.
18. List the distinguishing characteristics of healthcare-associated pneumonia compared to community-acquired pneumonia.
19. Outline the chain of transmission of the causative agent of hantavirus pulmonary syndrome.

22.1 The Respiratory Tract, Its Defenses, and Normal Biota

The respiratory tract is the most common place for infectious agents to gain access to the body. Obviously, we breathe 24 hours a day, and anything in the air we breathe passes at least temporarily into this organ system.

The structure of the system is illustrated in **figure 22.1*a*.** Most clinicians divide the system into two parts, the *upper* and *lower respiratory tracts*. The upper respiratory tract includes the mouth, the nose, the nasal cavity and sinuses above it, the throat (pharynx), and the epiglottis and larynx. The lower respiratory tract begins with the trachea, which feeds into the bronchi and bronchioles in the lungs. At the end of the bronchioles are small, balloonlike structures called alveoli, which inflate and deflate with inhalation and exhalation. These are the sites of oxygen exchange in the lungs.

Several anatomical features of the respiratory system protect it from infection. As described in chapter 15, nasal hair traps particles. Cilia **(figure 22.1*b*)** on the epithelium of the trachea and bronchi (the ciliary escalator) propel particles upward and out of the respiratory tract. Mucus on the surface of the mucous membranes lining the respiratory tract is a natural trap for invading microorganisms. Once the microorganisms are trapped, involuntary responses such as coughing, sneezing, and swallowing can move them out of sensitive areas. These are first-line defenses.

The second and third lines of defense also help protect the respiratory tract. Complement action in the lungs helps to protect against invading pathogens, and increased levels of cytokines and antimicrobial peptides further reduce the ability of microbes to cause disease. Macrophages inhabit the alveoli of the lungs and the clusters of lymphoid tissue (tonsils) in the throat. Secretory IgA that targets specific pathogens can be found in the mucus secretions as well.

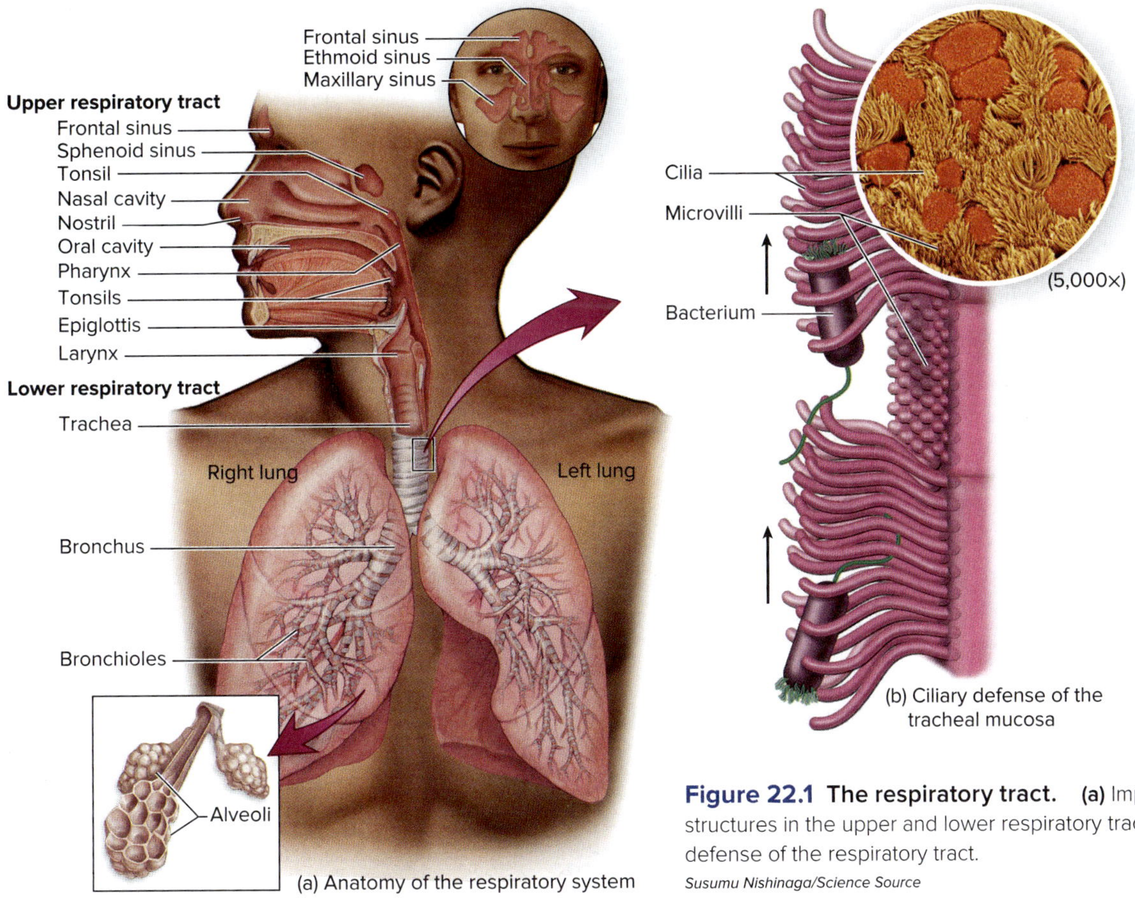
Frontal sinus
Ethmoid sinus
Maxillary sinus
Upper respiratory tract
Frontal sinus
Sphenoid sinus
Tonsil
Nasal cavity
Nostril
Oral cavity
Pharynx
Tonsils
Epiglottis
Larynx
Lower respiratory tract
Trachea
Right lung
Left lung
Bronchus
Bronchioles
Alveoli
(a) Anatomy of the respiratory system
Cilia
Microvilli
Bacterium
(5,000×)
(b) Ciliary defense of the tracheal mucosa

Figure 22.1 The respiratory tract. **(a)** Important structures in the upper and lower respiratory tracts. **(b)** Ciliary defense of the respiratory tract.
Susumu Nishinaga/Science Source

Because of its constant contact with the external environment, the respiratory system harbors a large number of commensal microorganisms. It was previously thought that such microbes were only localized to the upper respiratory tract. Recent metagenomic analysis of sputum samples, however, has shown that healthy lungs are a virtual tapestry of microorganisms **(Insight 22.1).** A significant portion of the normal biota belongs to nine major bacterial genera: *Prevotella, Sphingomonas, Pseudomonas, Acinetobacter, Fusobacterium, Megasphaera, Veillonella, Staphylococcus,* and *Streptococcus.* Yeasts, especially *Candida albicans,* also colonize the mucosal surfaces of the mouth in the upper respiratory tract.

Note that some bacteria considered "normal biota" in the respiratory tract can cause serious disease, especially in immunocompromised individuals. These include *Streptococcus pyogenes, Haemophilus influenzae, Streptococcus pneumoniae, Neisseria meningitidis,* and *Staphylococcus aureus.* In addition, researchers have discovered that the overall composition of the lung microbiome is altered in patients suffering from lung disorders such as chronic obstructive pulmonary disease (COPD), asthma, and cystic fibrosis. Further studies may reveal new treatments that will involve reestablishing the normal biota composition within the lungs of these patients.

In the respiratory system, as in some other organ systems, the normal biota performs the important function of microbial antagonism. This reduces the chances of pathogens establishing themselves in the same area by competing with them for resources and space. To illustrate this point, *Lactobacillus sakei,* a known member of the sinus microbiome, can suppress the pathogenic potential of another normal biota organism, *Corynebacterium tuberculostearicum,* reducing the incidence of sinus infection.

22.1 Learning Outcomes—Assess Your Progress

1. Draw or describe the anatomical features of the respiratory tract.
2. List the natural defenses present in the respiratory tract.
3. List the main genera of normal biota presently known to occupy the respiratory tract.

Respiratory System Defenses and Normal Biota

	Defenses	Normal Biota
Respiratory System	Nasal hair, ciliary escalator, mucus, involuntary responses such as coughing and sneezing, secretory IgA, alveolar macrophages, cytokines and complement	Large number of genera. Most abundant: *Streptococcus, Prevotella, Sphingomonas, Pseudomonas, Acinetobacter, Fusobacterium, Megasphaera, Veillonella,* and *Staphylococcus* Note: *Streptococcus pyogenes, Streptococcus pneumoniae, Haemophilus influenzae, Neisseria meningitidis,* and *Staphylococcus aureus* often present as "normal" biota.

INSIGHT 22.1 MICROBIOME: The Lungs Are Not Sterile

For decades, textbooks have been telling us that the lungs are basically sterile, with the exception of the odd microbe that falls down there and gets gobbled up by the ever-vigilant alveolar macrophages. Well, let us look at two facts:

- The lungs are less than a half meter away from our "microbe intake region"—the mouth and nose.
- Bacteria and other microbes are incredibly adaptive. They live in hot springs; they live in permafrost.

How could we expect them *not* to be living in an environment that offers moisture, protection from the elements, and a wide variety of temperatures (based on how close you are to the oral cavity)? To be perfectly clear, there are two ways that bacteria can get into the lungs, and three ways they can get out. The balance of these factors determines how many bacteria are actually there.

Getting into the Lungs

1. Inhalation. Air contains between 10,000 and 10 million bacteria per milliliter, or "cc." You are inhaling a lot of bacteria.
2. You are constantly "micro-aspirating" upper respiratory contents. In plain English, you are breathing deeply of the microbiome that lives in your nose and throat.

Getting out of the Lungs

1. Mucociliary clearance of microbes.
2. Coughing. You would be surprised how often you do this, even when you are not sick.
3. Innate and adaptive defenses.

In the end, it appears that the lung microbiome is sparse (when compared to the gut microbiome, for example), but it is most definitely there. And recent research suggests that the lungs of people who are ill with acute respiratory distress syndrome (ARDS) have gut microbiota in their lungs. Because the treatment of ARDS currently focuses exclusively on inflammation and tissue damage, some doctors predict the treatment of ARDS will soon include antibiotics.

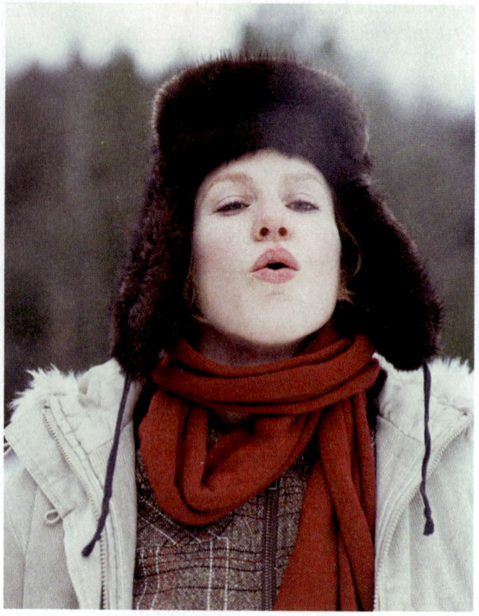

Dynamic Graphics Group/Creatas/Alamy stock Photo

Source: Dickson, Robert P., and Gary B. Huffnagle. "The Lung Microbiome: New Principles for Respiratory Bacteriology in Health and Disease." PLOS, online article, 7/9/2015.

22.2 Infectious Diseases Manifesting in the Upper Respiratory Tract

The Common Cold

In the course of a year, people in the United States suffer from about 67 million colds caused by viruses. The common cold is often called *rhinitis,* from the Latin word *rhin,* meaning "nose," and the suffix *-itis,* meaning "inflammation." Many people have several episodes a year, and economists estimate that this fairly innocuous infection costs the United States $40 billion a year in trips to the doctors, medications, and 22 million missed days of work.

Signs and Symptoms

Everyone is familiar with the symptoms of a cold: sneezing, scratchy throat, and runny nose (rhinorrhea), which usually begin 2 or 3 days after infection. An uncomplicated cold generally is not accompanied by fever, although children can experience low fevers (less than 102°F). Note that people with asthma and other underlying respiratory conditions (chronic obstructive pulmonary disease, or COPD) often suffer more severe symptoms triggered by the common cold.

A Note About COVID-19

In this book we included the detailed discussion of COVID-19 in the cardiovascular/lymphatic chapter, chapter 21. Although the most obvious acute symptoms often center in the respiratory tract, we presented evidence there that the infection takes advantage of the cardiovascular system to access many organs in the body—thus meeting the definition of a systemic, cardiovascular condition.

Causative Agents

The common cold is caused by one of over 200 different kinds of viruses. The particular virus in an infection is almost never identified, and the symptoms and handling of the infection are the same no matter which of the viruses is responsible.

The most common type of virus leading to the common cold is the group called rhinoviruses, of which there are 99 serotypes. Coronaviruses and adenoviruses are also major causes. Most viruses causing the common cold never lead to any serious consequences, but some of them can be serious for some patients. The respiratory syncytial virus (RSV) causes colds in most people, but in some, especially infants, infection with this virus can lead to more serious respiratory tract symptoms (discussed later in the chapter). In this section, we consider all cold-causing viruses together as a group because they are treated similarly. Rhinoviruses are in the genus *Enterovirus*. Since 2014 there has been another enterovirus, called enterovirus-D68 (EV-D68), that has caused cold-like symptoms in children and that seems to sometimes progress into polio-like symptoms. Severe colds should include EV-D68 in the differential diagnosis.

Viral infection of the upper respiratory tract can predispose a patient to secondary infections by other microorganisms, such as bacteria. Secondary infections may explain why some people report that their colds improved when they were given antibiotics. A virus originally caused the cold, but a bacterial infection might have followed.

Pathogenesis and Virulence Factors

Viruses that induce the common cold do not have many virulence mechanisms. They must penetrate the mucus that coats the respiratory tract and then find firm attachment points. Once they are attached, they use host cells to produce more copies of themselves. The symptoms we experience as the common cold are mainly the result of our body fighting back against the viral invaders. Virus-infected cells in the upper respiratory tract release chemicals that attract certain types of white blood cells to the site, and these cells release cytokines and other inflammatory mediators.

Transmission and Epidemiology

Cold viruses are transmitted by droplet contact, but indirect transmission may be more common, such as when a healthy person touches a fomite and then touches one of his or her own vulnerable surfaces, such as the mouth, nose, or an eye. In some cases, the viruses can remain airborne in droplet nuclei and aerosols and can be transmitted via the respiratory route.

The epidemiology of the common cold is fairly simple: Practically everybody gets colds—and fairly frequently. Children have more frequent infections than adults, probably because nearly every virus they encounter is a new one and they have no secondary immunity to it. People can acquire some degree of immunity to a cold virus that they have encountered before, but because there are more than 200 viruses, this immunity does not provide much overall protection.

Prevention

There is no vaccine for the common cold. A traditional vaccine would need to contain antigens from about 200 viruses to provide complete protection. But for now, the best prevention is to stop the transmission between hosts. The best way to prevent transmission is frequent hand washing, followed closely by stopping droplets from traveling away from the mouth and nose by covering them when sneezing or coughing. It is better to do this by covering the face with the crook of the arm rather than the hand because you are less likely to touch other surfaces with the inside of your elbow than with your hand.

Treatment

No antimicrobial agents cure the common cold. A wide variety of over-the-counter agents, such as antihistamines and decongestants, improve symptoms by blocking inflammatory mediators and their action. The use of these agents may also cut down on transmission to new hosts because fewer virus-loaded secretions are produced. Researchers are looking at a protein in human cells that cold viruses need for their replication and considering an agent to block that protein as a treatment for the common cold.

Disease Table 22.1	The Common Cold
Causative Organism(s)	Approximately 200 viruses (rhinoviruses, adenoviruses, and coronaviruses) V
Most Common Modes of Transmission	Indirect contact, droplet contact
Virulence Factors	Attachment proteins; most symptoms induced by host response
Culture/Diagnosis	Not necessary
Prevention	Hygiene practices
Treatment	For symptoms only
Epidemiological Features	Highest incidence among preschool and elementary schoolchildren, with average of three to eight colds per year; adults and adolescents: two to four colds per year

Sinusitis

Commonly called a *sinus infection,* this inflammatory condition of any of the four pairs of sinuses in the skull (see figure 22.1) can be caused by allergy (most common) or infections. The infectious agents that may be responsible for the condition include a variety of viruses or bacteria and, less commonly, fungi. Infections of the sinuses often follow a bout with the common cold. The inflammatory symptoms of a cold produce a large amount of fluid and mucus, and when trapped in the sinuses, these secretions provide an excellent growth medium for bacteria or fungi. This is why it is common for patients suffering from the common cold to then develop sinusitis caused by bacteria or fungi.

Causative Agents

Viruses Viral infection is probably the most common cause of mild sinusitis. The viruses involved are the same as with the common cold.

Bacteria Any number of bacteria that are normal biota in the upper respiratory tract may cause sinus infections. Many cases are caused by *Streptococcus pneumoniae, Streptococcus pyogenes, Staphylococcus aureus, Corynebacterium tuberculostearicum,* and *Haemophilus influenzae.* The causative organism is usually not identified, but treatment is begun empirically, based on the symptoms.

Broad-spectrum antibiotics may be prescribed when the physician feels that the sinusitis is bacterial in origin. In fact, one out of five antibiotic prescriptions in the United States is for sinusitis. However, recent studies have shown that only about 2% of sinusitis cases are caused by bacteria, and even those cases are not significantly improved with antibiotic treatment. Most uncomplicated cases may be best treated by having the patient "wait it out" while his or her own immune system clears the infection.

Fungi Fungal sinusitis is rare, but it is often recognized when antibacterial drugs fail to alleviate symptoms. Simple fungal infections may normally be found in the maxillary sinuses and are noninvasive in nature. These colonies are generally not treated with antifungal agents but instead are simply mechanically removed by a physician. *Aspergillus fumigatus* is a common fungus involved in this type of infection, but *Bipolaris* species are an emerging cause of fungal sinusitis today. The growth of fungi in this type of sinusitis may be encouraged by trauma to the area.

More serious invasive fungal infections of the sinuses may be found in severely immunocompromised patients. Fungi such as *Aspergillus* and *Mucor* species may invade the bony structures in the sinuses and even travel to the brain or eye. These infections are treated aggressively with a combination of surgical removal of the fungus and intravenous antifungal therapy **(Disease Table 22.2).**

Acute Otitis Media (Ear Infection)

This condition is another common sequela of the common cold. Viral infections of the upper respiratory tract lead to inflammation of the eustachian tubes and the buildup of fluid in the middle ear, which can lead to bacterial multiplication in those fluids. Bacteria can migrate along the eustachian tube from the upper respiratory tract **(figure 22.2).** When bacteria encounter mucus and fluid buildup in the middle ear, they multiply rapidly. Their presence increases the inflammatory response, leading to pus production and continued fluid secretion. This fluid is referred to as *effusion.*

Another condition, known as chronic otitis media, occurs when fluid remains in the middle ear for indefinite periods of time. This form of otitis media is usually caused by a mixed biofilm of bacteria that are attached to the membrane of the inner ear.

Signs and Symptoms

Otitis media often causes a sensation of fullness or pain in the ear and loss of hearing. Younger children may exhibit irritability,

Disease Table 22.2 Sinusitis

Causative Organism(s)	Viruses (V)	Various bacteria, often mixed infection (B)	Various fungi (F)
Most Common Modes of Transmission	Direct contact, indirect contact	Endogenous (opportunism)	Introduction by trauma or opportunistic overgrowth
Virulence Factors	–	–	–
Culture/Diagnosis	Culture not usually performed; diagnosis based on clinical presentation	Culture not usually performed; diagnosis based on clinical presentation, occasionally X rays or other imaging technique used	Same
Prevention	Hygiene	–	–
Treatment	None	Recommendation is for no antibiotics unless it remains unresolved for some weeks	Physical removal of fungus; in severe cases, antifungals used
Distinctive Features	Viral and bacterial much more common than fungal	Viral and bacterial much more common than fungal	Suspect in immunocompromised patients
Epidemiological Features	Commonly follows the common cold	United States: affects 1 of 7 adults; between 12 and 30 million diagnoses per year	Fungal sinusitis varies with geography; in United States, more common in SE and SW; internationally: more common in India, North Africa, Middle East

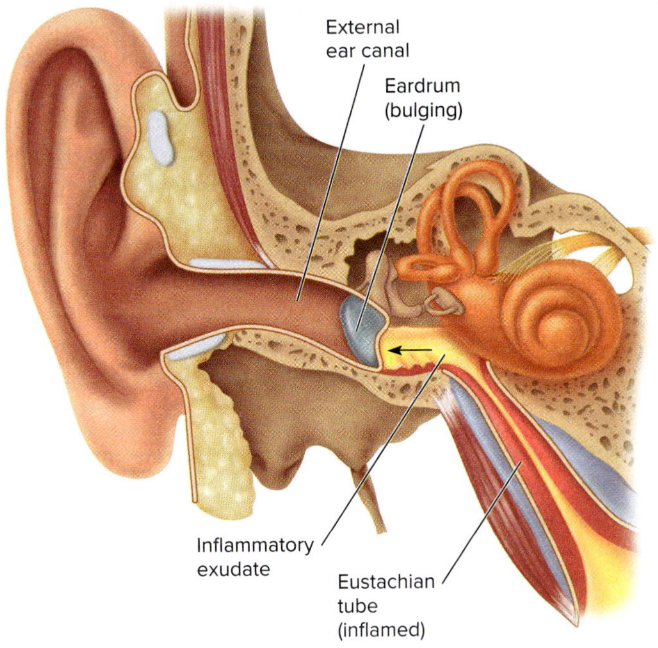
External
ear canal
Eardrum
(bulging)
Inflammatory
exudate
Eustachian
tube
(inflamed)

Figure 22.2 An infected middle ear.

fussiness, and difficulty in sleeping, eating, or hearing. Severe or untreated infections can lead to rupture of the eardrum because of pressure of pus buildup, or to internal breakthrough of these infected fluids, which can lead to more serious conditions such as mastoiditis, meningitis, or intracranial abscess.

Causative Agents

Many different viruses and bacteria, as well as a fungus called *Candida auris,* can cause acute otitis media. The most common cause is *Streptococcus pneumoniae*. *Streptococcus pneumoniae* appears as pairs of elongated, gram-positive cocci joined end to end. It is often called by the familiar name *pneumococcus,* and diseases caused by it are termed *pneumococcal.*

An emerging, and worrying, cause of ear infections is the fungus *Candida auris*. This microbe was mentioned in chapter 21 as a cause for sepsis. The fungus causes ear infections and wound and bloodstream infections. It is worrisome because it has quickly become resistant to most antifungals used to treat it.

Transmission and Epidemiology

Otitis media is a sequela of upper respiratory tract infection and is not communicable, although the upper respiratory infection preceding it is. Children are particularly susceptible, and boys have a slightly higher incidence than do girls.

Prevention

A vaccine against *S. pneumoniae* has been a part of the recommended childhood vaccination schedule since 2000. The vaccine (PCV13) is a 13 valent conjugated vaccine. It contains polysaccharide capsular material from 13 different strains of the bacterium complexed with a chemical that makes it more antigenic. There is an older vaccine—called Pneumovax—for the same bacterium.

Treatment

The current recommendation is to use a broad-spectrum antibiotic in babies less than 6 months old. In older children with uncomplicated acute otitis media and a fever below 104°F, "watchful waiting" for 72 hours to allow the body to clear the infection is best practice. When antibiotics are used, antibiotic resistance must be considered. Children who experience frequent recurrences of ear infections sometimes have small tubes placed through the tympanic membranes into their middle ears to provide a means of keeping fluid out of the site when inflammation occurs.

Disease Table 22.3 Otitis Media

Causative Organism(s)	*Streptococcus pneumoniae* B G+	*Candida auris* F
Most Common Modes of Transmission	Endogenous (may follow upper respiratory tract infection by *S. pneumoniae* or other microorganisms)	Not known
Virulence Factors	Capsule, hemolysin	Biofilm formation
Culture/Diagnosis	Usually relies on clinical symptoms and failure to resolve within 72 hours	MALDI-TOF or PCR; CDC will identify if requested
Prevention	Pneumococcal conjugate vaccine (PCV13)	None
Treatment	Wait for resolution; if needed, amoxicillin (high rates of resistance) or amoxicillin + clavulanate or cefuroxime; antibiotic usage recommended for babies less than 6 months old	Consult with CDC; in **Urgent** Category in CDC Antibiotic Resistance Report
Distinctive Features	–	–
Epidemiological Features	30% of cases in United States	First appeared in 2009; increasing in United States and elsewhere

Pharyngitis

Signs and Symptoms

The name says it all—this is an inflammation of the throat, which the host experiences as pain and swelling. The severity of pain can range from moderate to severe, depending on the causative agent. Viral sore throats are generally mild and sometimes lead to hoarseness. Sore throats caused by bacteria are generally more painful than those caused by viruses, and they are more likely to be accompanied by fever, headache, and nausea.

Clinical signs of a sore throat are reddened mucosa, swollen tonsils, and sometimes white packets of inflammatory products visible on the walls of the throat, especially in streptococcal disease **(figure 22.3).** The mucous membranes may be swollen, affecting speech and swallowing. Often, pharyngitis results in foul-smelling breath. The incubation period for most sore throats is generally 2 to 5 days.

Causative Agents

The same viruses causing the common cold most commonly cause a sore throat. It can also accompany other diseases, such as infectious mononucleosis. Pharyngitis may simply be the result of mechanical irritation from prolonged shouting or from drainage of an infected sinus cavity. The most serious cases of pharyngitis are caused by the bacteria *Streptococcus pyogenes* and *Fusobacterium necrophorum.*

Streptococcus pyogenes

S. pyogenes is a gram-positive coccus that grows in chains. It does not form endospores, is nonmotile, and forms capsules and slime layers. *S. pyogenes* is a facultative anaerobe that ferments a variety of sugars. It does not produce catalase, but it does have a peroxidase system for inactivating hydrogen peroxide, which allows its survival in the presence of oxygen.

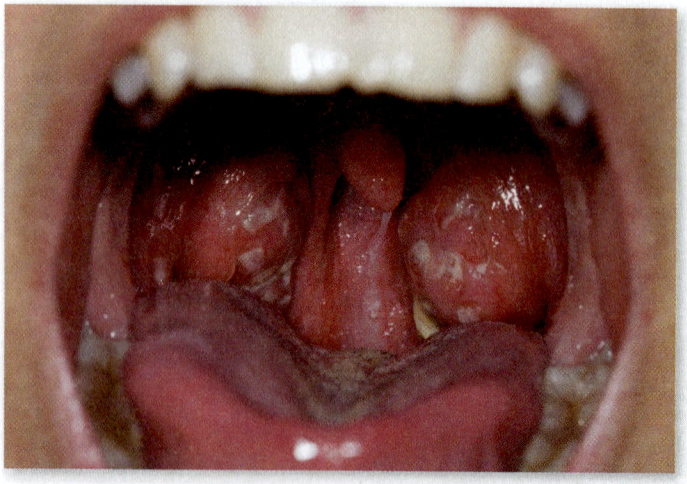

Figure 22.3 The appearance of the throat in pharyngitis and tonsillitis. The pharynx and tonsils become bright red and produce pus. Whitish pus nodules may also appear on the tonsils.
Stefan Sollfors/Alamy Stock Photo

Pathogenesis

Untreated streptococcal throat infections occasionally can result in serious complications, either right away or days to years after the throat symptoms subside. These complications include scarlet fever, rheumatic fever, and glomerulonephritis. More rarely, invasive and deadly conditions—such as necrotizing fasciitis, which is described in a Note in chapter 19—can result from infection by *S. pyogenes.* The apparent mechanism behind some of these complications is autoimmunity. The bacterium has antigens on its surface that resemble heart, joint, and brain proteins. This initially allows the bacterium to evade immune detection. Eventually, though, the immune system learns to respond to these antigens and then also may attack the human analogs.

Scarlet Fever Scarlet fever is the result of infection with an *S. pyogenes* strain that is itself infected with a bacteriophage. This lysogenic virus confers on the streptococcus the ability to produce an exotoxin called erythrogenic toxin. Scarlet fever is characterized by a sandpaper-like rash, most often on the neck, chest, elbows, and inner surfaces of the thighs. High fever accompanies the rash. It most often affects school-age children and was a source of great suffering in the United States in the early part of the 20th century. In epidemic form, the disease can have a fatality rate of up to 95%. Most cases seen today are mild. They are easily recognizable and amenable to antibiotic therapy. Because of the fear elicited by the name "scarlet fever," the disease is often called "scarlatina" in North America.

Rheumatic Fever Rheumatic fever is thought to be due to an immunologic cross-reaction between the streptococcal M protein and heart muscle. It tends to occur approximately 3 weeks after pharyngitis has subsided. It can result in permanent damage to heart valves. Other symptoms include arthritis in multiple joints and the appearance of nodules over bony surfaces just under the skin.

Glomerulonephritis Glomerulonephritis is thought to be the result of streptococcal proteins participating in the formation of antigen–antibody complexes, which then are deposited in the basement membrane of the glomeruli of the kidney. It is characterized by **nephritis** (appearing as swelling in the hands and feet and low urine output), blood in the urine, increased blood pressure, and occasionally heart failure. It can result in permanent kidney damage. The incidence of poststreptococcal glomerulonephritis has been declining in the United States, but it is still common in Africa, the Caribbean, and South America.

Virulence Factors

As already noted, surface antigens of *S. pyogenes* mimic host proteins, leading to collateral damage by the immune system. Its virulence is also enhanced by the substantial array of surface antigens, toxins, and enzymes it can generate.

Streptococci display numerous surface antigens **(figure 22.4).** Specialized polysaccharides on the surface of the cell wall help to protect the bacterium from being dissolved by the lysozyme of the host. Lipoteichoic acid (LTA) contributes to the adherence of *S. pyogenes* to epithelial cells in the pharynx. A spiky surface projection called *M protein* contributes to

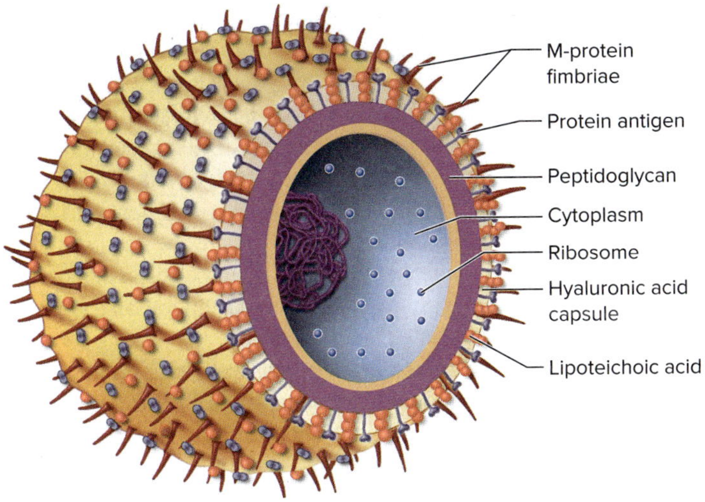
M-protein fimbriae
Protein antigen
Peptidoglycan
Cytoplasm
Ribosome
Hyaluronic acid capsule
Lipoteichoic acid

Figure 22.4 Cutaway view of group A streptococcus.

virulence by resisting phagocytosis and possibly by contributing to adherence. A capsule made of *hyaluronic acid* (HA) is formed by most *S. pyogenes* strains. It probably contributes to the bacterium's adhesiveness.

Extracellular Toxins Group A streptococci owe some of their virulence to the effects of hemolysins called streptolysins. The two types are streptolysin O (SLO) and streptolysin S (SLS).[1] Both types cause beta-hemolysis of sheep blood agar (see "Culture and / or Diagnosis"). Both hemolysins rapidly injure many cells and tissues, including leukocytes and liver and heart muscle.

A key toxin in the development of scarlet fever is **erythrogenic** (eh-rith″-roh-jen′-ik) **toxin.** This toxin is responsible for the bright red rash typical of this disease **(figure 22.5*a*),** and it induces fever by acting upon the temperature regulatory center in the brain.

1. In *SLO, O* stands for *oxygen* because the substance is inactivated by oxygen. In *SLS, S* stands for *serum* because the substance has an affinity for serum proteins. SLS is oxygen-stable.

Only lysogenic strains of *S. pyogenes* that contain genes from a temperate bacteriophage can synthesize this toxin. (For a review of the concept of lysogeny, see chapter 7.)

Some of the streptococcal toxins (erythrogenic toxin and streptolysin O) contribute to increased tissue injury by acting as *superantigens.* These toxins elicit excessively strong reactions from monocytes and T lymphocytes. When activated, these cells proliferate and produce *tumor necrosis factor* (*TNF*), which leads to a cascade of immune responses, resulting in vascular injury. This is the likely mechanism for the severe pathology of toxic shock syndrome and necrotizing fasciitis.

Transmission and Epidemiology

Physicians estimate that 10% to 20% of sore throats may be caused by *S. pyogenes,* adding up to several million cases each year. Most transmission of *S. pyogenes* is via respiratory droplets or direct contact with mucus secretions. This bacterium is carried as "normal" biota by 15% of the population, but transmission from this reservoir is less likely than from a person who is experiencing active disease from the infection because of the higher number of bacteria present in the disease condition. It is less common but possible to transmit this infection via fomites. Humans are the only significant reservoir of *S. pyogenes.*

More than 80 serotypes of *S. pyogenes* exist, meaning that people can experience multiple infections throughout their lives because immunity is serotype-specific. Even so, only a minority of encounters with the bacterium result in disease.

Culture and/or Diagnosis

The failure to recognize group A streptococcal infections can have serious effects. Rapid cultivation and diagnostic techniques to ensure proper treatment and prevention measures are essential. Several different rapid diagnostic test kits are used in clinics and doctors' offices to detect group A streptococci from pharyngeal swab samples. These tests are based on antibodies that react with the outer carbohydrates of group A streptococci **(figure 22.5*b*).** Newer tests using NAAT techniques are more accurate and do not require backup cultures.

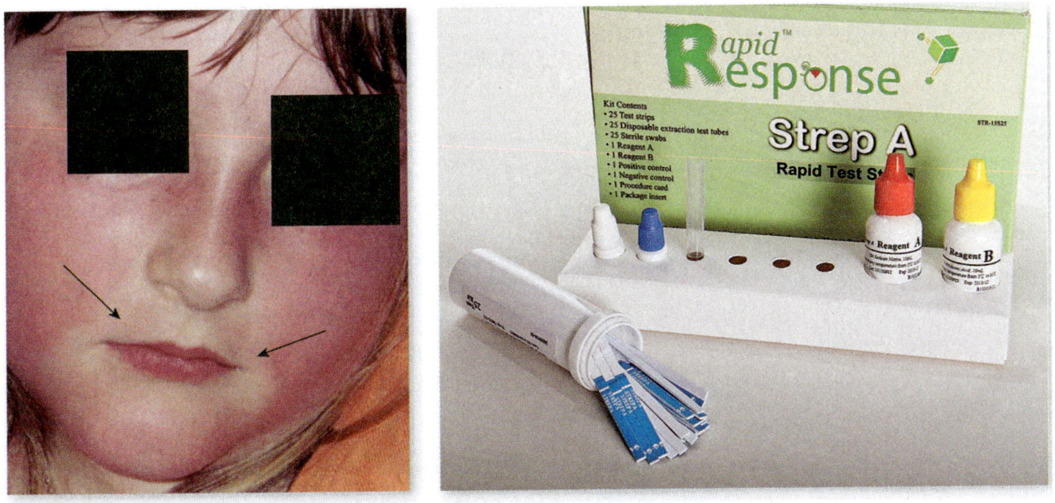
Rapid Response
Strep A
Reagent A
Reagent B

(a) (b)

Figure 22.5 Streptococcal infections. **(a)** The bright red rash characteristic of scarlet fever. The area around the mouth remains unaffected. **(b)** A rapid immunologic test for diagnosis of group A infections. With this method, a patient's throat swab is introduced into an immunochromatographic (lateral flow) system.

(a) Neal R. Chamberlin, PhD/McGraw Hill; (b) Science Source

Cultures are still widely used. *S. pyogenes* displays a beta-hemolytic pattern due to its streptolysins (and hemolysins) (see figure 19.7). Group A streptococci are by far the most common beta-hemolytic isolates in human diseases, but lately an increased number of infections by group B streptococci (also beta-hemolytic), as well as the existence of beta-hemolytic enterococci, have made it important to use differentiation tests. A positive bacitracin disc test provides additional evidence for group A organisms.

▶ Prevention

No vaccine exists for group A streptococci, although many researchers are working on the problem. A vaccine against this bacterium would also be a vaccine against rheumatic fever, and, for that reason, it is in great demand. In the meantime, infection can be prevented by good hand washing, especially after coughing and sneezing and before preparing foods or eating.

▶ Treatment

The antibiotic of choice for *S. pyogenes* is penicillin. In patients with penicillin allergies, a first-generation cephalosporin, such as cephalexin, is prescribed.

Fusobacterium necrophorum

In recent years, another potentially dangerous bacterium has been recognized to cause pharyngitis. *Fusobacterium necrophorum* is a gram-negative bacterium that can cause up to 15% of acute pharyngitis cases in the older adolescent (>15 years old) group. It can cause a peritonsillar abscess, which can lead to a life-threatening invasion of the bloodstream, and sepsis. This is called Lemierre's syndrome.

Until recently, sore throats were routinely treated with penicillin with *Streptococcus pyogenes* in mind. Fortunately, *Fusobacterium* is also sensitive to penicillin. Unfortunately, it is not sensitive to second-line antibiotics used for *S. pyogenes.* And now that physicians routinely culture for *S. pyogenes* before prescribing antibiotics, *Fusobacterium* infections can be missed because negative culture results are mistaken to indicate viral pharyngitis. Awareness is rising, however, and physicians are starting to consider that severe sore throats may be caused by this other bacterium. Notably, lingering pain after an episode of pharyngitis seen in late adolescents or young adults should raise suspicions about *Fusobacterium* **(Disease Table 22.4).**

22.2 Learning Outcomes—Assess Your Progress

4. List the possible causative agents, modes of transmission, virulence factors, diagnostic techniques, and prevention and treatment for each of the diseases of the upper respiratory tract: the common cold, sinusitis, otitis media, and pharyngitis.
5. Identify which disease is often caused by a mixture of microorganisms.
6. Identify two bacteria that can cause dangerous pharyngitis cases.

Disease Table 22.4 Pharyngitis

Causative Organism(s)	*Streptococcus pyogenes* B G+	*Fusobacterium necrophorum* B G–	Viruses V
Most Common Modes of Transmission	Droplet or direct contact	Usually endogenous	All forms of contact
Virulence Factors	LTA, M protein, hyaluronic acid capsule, SLS and SLO, superantigens, induction of autoimmunity	Invasiveness, endotoxin	–
Culture/Diagnosis	Beta-hemolytic on blood agar, sensitive to bacitracin, rapid antigen tests	Culture anaerobically; CT scan for abscess(es)	Goal is to rule out *S. pyogenes* (and *F. necrophorum*); further diagnosis usually not performed
Prevention	Hygiene practices	?	Hygiene practices
Treatment	Penicillin, cephalexin in penicillin-allergic	Penicillin	Symptom relief only
Distinctive Features	Generally more severe than viral pharyngitis	Can lead to Lemierre's syndrome	Hoarseness frequently accompanies viral pharyngitis
Epidemiological Features	United States: 10% to 20% of all cases of pharyngitis	Causes up to 15% of acute pharyngitis in teens/young adults	Ubiquitous; responsible for 40% to 60% of all pharyngitis

22.3 Infectious Diseases Manifesting in Both the Upper and Lower Respiratory Tracts

A number of infectious agents affect both the upper and lower respiratory tract regions. We discuss the more common diseases in this section; specifically, they are whooping cough, respiratory syncytial virus (RSV), and influenza.

Whooping Cough

Whooping cough is also known as *pertussis* (the suffix *-tussis* is Latin for "cough"). A vaccine for this potentially serious infection has been available since 1926. The disease is still troubling to the public health community because some parents have recently become concerned about the safety of the vaccine. It is vitally important for health care professionals to convey accurate information about this disease and the safety of the vaccine.

Signs and Symptoms

The disease has two distinct symptom phases called the catarrhal and paroxysmal stages, which are followed by a long recovery (or convalescent) phase, during which a patient is particularly susceptible to other respiratory infections. After an incubation period of from 3 to 21 days, the **catarrhal** stage begins when bacteria present in the respiratory tract cause what appear to be cold symptoms, most notably a runny nose. This stage lasts 1 to 2 weeks. The disease worsens in the second **(paroxysmal)** stage, which is characterized by severe and uncontrollable coughing (a *paroxysm* is a convulsive attack). The common name for the disease comes from the whooping sound a patient makes when trying to grab a breath between uncontrollable bouts of coughing. The violent coughing spasms can result in burst blood vessels in the eyes or even vomiting. In the worst cases, seizures result from small hemorrhages in the brain.

As in any disease, the **convalescent period** is the time when numbers of bacteria are decreasing and no longer cause ongoing symptoms. But the active stages of the disease damage the cilia on respiratory tract epithelial cells, and complete recovery of these surfaces requires weeks or even months. During this time, other microorganisms can more easily colonize and cause secondary infection.

Causative Agent

Bordetella pertussis is a very small, gram-negative rod. Sometimes it looks like a coccobacillus. It is strictly aerobic and fastidious, having specific nutritional requirements for successful culture.

Pathogenesis and Virulence Factors

The progress of this disease can be clearly traced to the virulence mechanisms of the bacterium. It is absolutely essential for the bacterium to attach firmly to the epithelial cells of the mouth and throat, and it does so using specific adhesive molecular structures on its surface. One of these structures is called *filamentous hemagglutinin (FHA)*. It is a fibrous structure that surrounds the bacterium like a capsule and is secreted in soluble form. In that form, it can act as a bridge between the bacterium and the epithelial cell.

Once the bacteria are attached in large numbers, production of mucus increases and localized inflammation ensues, resulting in the early stages of the disease. Then the real damage begins: The bacteria release multiple exotoxins that damage ciliated respiratory epithelial cells and cripple other components of the host defense, including phagocytic cells.

The two most important exotoxins are *pertussis toxin* and *tracheal cytotoxin.* Pertussis toxin triggers excessive amounts of cyclic AMP to accumulate in affected cells. This results in copious production of mucus and a variety of other effects in the respiratory tract and the immune system.

Tracheal cytotoxin results in more direct destruction of ciliated cells. The cells become incapable of clearing mucus and secretions, leading to the extraordinary coughing required to get relief. Another important contributor to the pathology of the disease is *B. pertussis* endotoxin. As always with endotoxins, its release leads to the production of a host of cytokines that have direct and indirect effects on physiological processes and on the host response.

Transmission and Epidemiology

B. pertussis is transmitted via respiratory droplets. It is highly contagious during both the catarrhal and paroxysmal stages. The disease manifestations are most serious in infants. Twenty-five percent of infections occur in older children and adults, who generally have milder symptoms—but who can unwittingly pass it to infants.

Pertussis outbreaks continue to occur in the United States and elsewhere. Although high vaccination coverage has kept the incidence of pertussis low in the United States, there have been peaks of incidence of the disease in recent years **(figure 22.6).** We have learned over time that the vaccine does not provide lifelong protection. Immunity begins to wane a few years after the childhood series of vaccinations is completed. That has resulted in adults becoming infected with the bacterium, often with no or mild symptoms. They then pass it on to infants who are not yet fully immunized, and the infants experience serious illness. Second, there has been a rise in the percentage of unvaccinated individuals. In many cases, the states with the highest increase in pertussis cases are also the states with the lowest rates of vaccination. During the COVID-19 pandemic, preliminary data suggest that the incidence of pertussis decreased due to mitigation measures put in place in 2020 and 2021.

Culture and/or Diagnosis

This disease is often diagnosed based solely on its symptoms because they are so distinctive. When culture confirmation is desired, nasopharyngeal swabs can be inoculated on specific media—Bordet-Gengou (B-G) medium, charcoal agar, or potato-glycerol agar. A PCR assay is available.

Prevention and Treatment

The current vaccine for pertussis is an acellular formulation of important *B. pertussis* antigens. It is generally given in the form of the DTaP vaccine. A single booster with Tdap after the age of 11

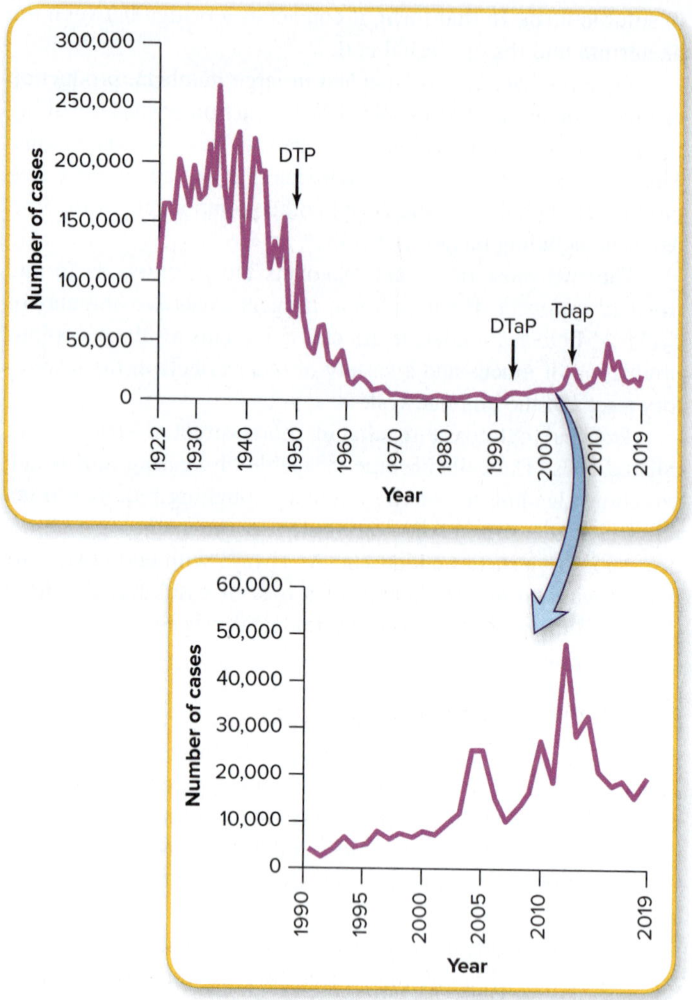
300,000
250,000
200,000
150,000
100,000
50,000
0
Number of cases
DTP
DTaP
Tdap
1922
1930
1940
1950
1960
1970
1980
1990
2000
2010
2019
Year
60,000
50,000
40,000
30,000
20,000
10,000
0
Number of cases
1990
1995
2000
2005
2010
2019
Year

Figure 22.6 Reported pertussis incidence over time. The upper figure shows when the DTP vaccine was introduced, eventually reducing the incidence altogether for almost 30 years. The lower figure shows the recent increase in greater detail. During COVID-19, mitigation measures seem to have reduced the incidence of pertussis for the years 2020 and 2021, but those data are not finalized yet.

Disease Table 22.5	Pertussis (Whooping Cough)
Causative Organism(s)	*Bordetella pertussis* B G–
Most Common Modes of Transmission	Droplet contact
Virulence Factors	FHA (adhesion), pertussis toxin and tracheal cytotoxin, endotoxin
Culture/Diagnosis	PCR or growth on B-G, charcoal, or potato-glycerol agar; diagnosis can be made on symptoms
Prevention	Acellular vaccine (DTaP), azithromycin for contacts
Treatment	Azithromycin; drug-resistant *B. pertussis* is a **Concerning** threat on the CDC's list of antibiotic resistance threats
Epidemiological Features	United States: 19,000 cases in 2017, 14,000 in 2018; internationally: hundreds of millions of cases annually

is especially important to maintain immunity against this disease. A second prevention strategy is the administration of antibiotics to contacts of people who have been diagnosed with the disease to prevent disease in those who may have been infected.

To treat an active case of the disease, azithromycin is the drug of choice **(Disease Table 22.5).**

Respiratory Syncytial Virus Disease

As its name indicates, respiratory syncytial virus (RSV) infects the respiratory tract and produces giant, multinucleated cells (syncytia). It is a member of the paramyxovirus family and contains single-stranded, negative-sense RNA. It is an enveloped virus.

Outbreaks of droplet-spread RSV disease occur regularly throughout the world, with peak incidence in the winter and early spring. Children 6 months of age or younger, as well as premature babies, are especially susceptible to serious disease caused by this virus. RSV is the most prevalent cause of respiratory infection in the newborn age group, and nearly all children have experienced it by age 2. Infections peak in the fall and winter. In 2020, with all of the mitigation procedures put in place for COVID-19, RSV infections were very low. But in spring 2021, many people stopped wearing masks and social distancing. Until the Delta variant of COVID-19 emerged in the United States in early summer, precautions were relaxed. It is probable that this is the reason that hospitals saw an unusual wave of RSV infections in the summer of 2021.

The first symptoms are fever that lasts for approximately 3 days, rhinitis, pharyngitis, and otitis. More serious infections progress to the bronchial tree and lung parenchyma, giving rise to symptoms that include acute bouts of coughing, wheezing, difficulty in breathing (called **dyspnea**), and abnormal breathing sounds (called rales). This condition is often called "croup" and bronchiolitis. Be aware that both of these terms are clinical descriptions of diseases caused by a variety of viruses (in addition to RSV) and sometimes bacteria.

The virus is highly contagious and is transmitted through droplet contact but also through fomite contamination. The best diagnostic procedures are those that demonstrate the viral antigen directly from specimens (direct and indirect fluorescent staining, ELISA, and DNA probes).

There is no RSV vaccine available yet, but a humanized monoclonal antibody preparation called palivizumab is recommended for high-risk babies. Ribavirin, an antiviral drug, can be administered together with passive antibody (IVIG) as an inhaled aerosol to very sick children as a treatment.

Disease Table 22.6 RSV Disease

Causative Organism(s)	Respiratory syncytial virus (RSV) V
Most Common Modes of Transmission	Droplet and indirect contact
Virulence Factors	Syncytia formation
Culture/Diagnosis	RT-PCR
Prevention	Passive antibody (humanized monoclonal) in high-risk children
Treatment	Ribavirin plus passive antibody in severe cases
Epidemiological Features	United States: general population, less than 1% mortality rates, 3% to 5% mortality in premature infants or those with congenital heart defects; internationally: 7 times higher fatality rate in children in developing countries

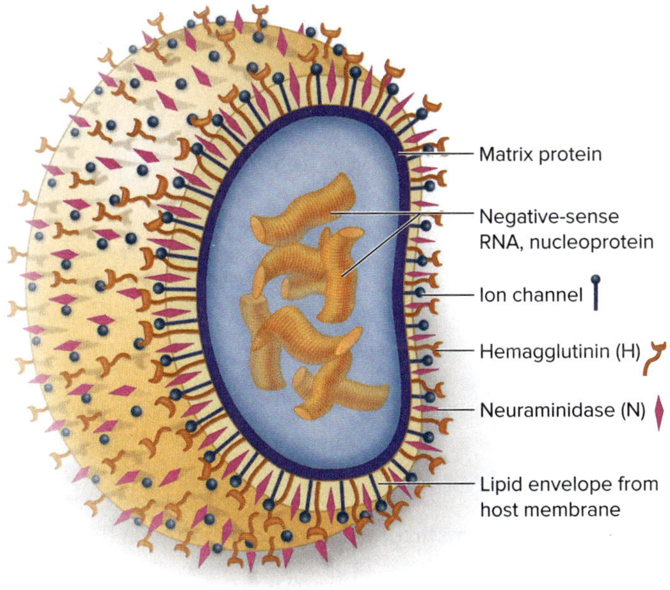
Matrix protein
Negative-sense RNA, nucleoprotein
Ion channel
Hemagglutinin (H)
Neuraminidase (N)
Lipid envelope from host membrane

Figure 22.7 **Schematic drawing of influenza virus.**

Influenza

The "flu" is a very important disease to study for several reasons. First of all, the familiar annual "flu seasons" have the potential of turning deadly for very many people very quickly. Second, many conditions are erroneously termed the "flu," while in fact only diseases caused by influenza viruses are actually the flu. Third, the way that influenza viruses behave provides an excellent illustration of the way other viruses can, and do, change to cause more serious diseases than they did previously.

Influenza viruses that circulate every year are called "seasonal" flus. Often, these are the only influenza viruses that circulate each year. Occasionally, another influenza strain appears, one that is new and may cause worldwide pandemics. In some years, such as in 2009, both types of influenza infection are problematic. They are characterized by different symptoms, affect different age groups, and require separate vaccine protocols.

Signs and Symptoms

Influenza begins in the upper respiratory tract but in serious cases may also affect the lower respiratory tract. There is a 1- to 4-day incubation period, after which symptoms begin very quickly. These include headache, chills, dry cough, body aches, fever, stuffy nose, and sore throat. Even the sum of all these symptoms cannot describe how a person actually feels: lousy. The flu is known to "knock you off your feet." Extreme fatigue can last for a few days or even a few weeks. An infection with influenza can leave patients vulnerable to secondary infections, often bacterial. Influenza infection often leads to a pneumonia that can cause rapid death, even in young, healthy adults. Patients with emphysema or cardiopulmonary disease, along with very young, elderly, or pregnant patients, are more susceptible to serious complications.

Causative Agent

All cases of influenza are caused by one of three influenza viruses: A, B, or C. They belong to the family *Orthomyxoviridae*. They are spherical particles with an average diameter of 80 to 120 nanometers. Each virion is covered with a lipoprotein envelope that is studded with glycoprotein spikes acquired during viral maturation **(figure 22.7).** Also note that the envelope contains proteins that form a channel for ion transport into the virus. The two glycoproteins that make up the spikes of the envelope and contribute to virulence are called hemagglutinin (H) and neuraminidase (N). The name hemagglutinin is derived from this glycoprotein's agglutinating action on red blood cells, which is the basis for viral assays used to identify the viruses. Hemagglutinin contributes to infectivity by binding to host cell receptors of the respiratory mucosa, a process that enables viral penetration. Neuraminidase breaks down the protective mucus coating of the respiratory tract, assists in viral budding and release, and participates in host cell fusion.

The ssRNA genome of the influenza virus is known for its extreme variability. It is subject to constant genetic changes that alter the structure of its envelope glycoproteins. Research has shown that genetic changes are very frequent in the area of the glycoproteins recognized by the host immune response but very rare in the areas of the glycoproteins used for attachment to the host cell **(figure 22.8).** In this way, the virus can continue to attach to host cells while managing to decrease the effectiveness of the host response to its presence. This constant mutation of the glycoproteins is called **antigenic drift**—the antigens gradually change their amino acid composition, resulting in decreased ability of host memory cells to recognize them.

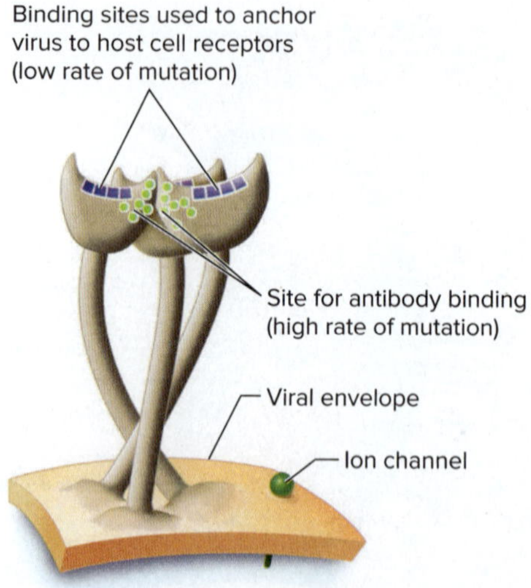
Binding sites used to anchor virus to host cell receptors (low rate of mutation)
Site for antibody binding (high rate of mutation)
Viral envelope
Ion channel

Figure 22.8 Schematic drawing of hemagglutinin (H) of influenza virus. Blue boxes depict site used to attach virus to host cells; green circles depict sites for anti-influenza antibody binding. Researchers are working on a vaccine that will send antibodies to the long stem of the hemagglutin, which experiences far fewer mutations.

An even more serious phenomenon is known as **antigenic shift.** The genome of the virus consists of eight separate RNA strands, except for influenza C, which has seven. Antigenic shift is the swapping out of one of those strands with a gene or strand from a different influenza virus. Some explanation is in order. First, we know that certain influenza viruses infect both humans and swine (pigs). Other influenza viruses infect birds (ducks) and swine. All of these viruses have genes coding for the same important influenza proteins (including H and N)—but the actual sequence of the genes is different in the different types of viruses. Second, when the two viruses just described infect a single swine host, with both virus types infecting the same host cell, the viral packaging step can accidentally produce a human influenza virus that contains seven human influenza virus RNA strands plus a single duck influenza virus RNA strand **(figure 22.9).** When that virus infects a human, no immunologic recognition of the protein that came from the duck virus occurs. Experts have traced the flu pandemics of 1918, 1957, 1968, 1977, and 2009 to strains of a virus that came from pigs (swine flu). In fact, sequencing of the 2009 strain showed it to contain genes from swine, avian (bird), and human viruses. Influenza A viruses are named according to the different types of H and N spikes they display on their surfaces. Influenza B viruses are not divided into subtypes because they are thought to undergo only antigenic drift and not antigenic shift. Influenza C viruses are thought to cause only minor respiratory disease and are probably not involved in epidemics.

Figure 22.10 illustrates the emergence of pandemic strains of influenza over time.

Pathogenesis and Virulence Factors

The influenza virus binds primarily to ciliated cells of the respiratory mucosa. Infection causes the rapid shedding of these cells along with a load of viruses. Stripping the respiratory epithelium to the basal layer eliminates protective ciliary clearance. Combine that with what is often called a "cytokine storm" caused by the viral stimulus and the lungs experience severe inflammation and irritation.

As just noted, the glycoproteins and their structure are important virulence determinants because of their ability to change. But they also mediate the adhesion of the virus to host cells. One feature of the 2009 H1N1 virus is that it bound to cells lower in the respiratory tract—and at a much higher rate, leading to massive damage, and often death, in the worst-affected patients. There was a total of around 14,000 deaths worldwide in the 2009 pandemic, either from primary influenza pneumonia

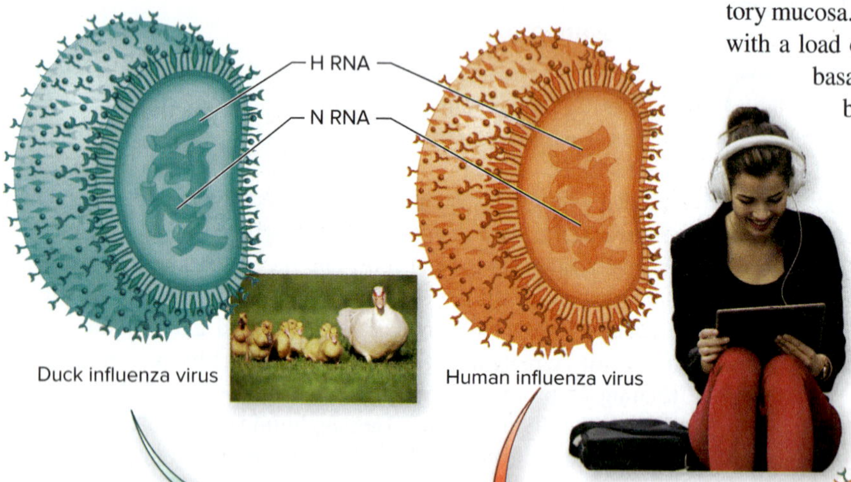
H RNA
N RNA
Duck influenza virus
Human influenza virus

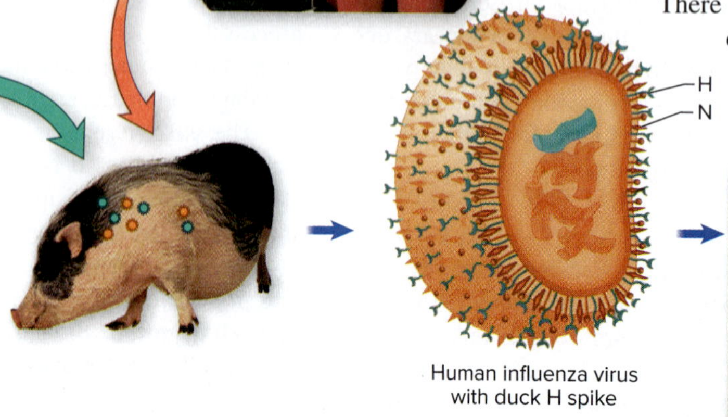
H
N
Human influenza virus with duck H spike

Figure 22.9 Antigenic shift event. Where ducks, swine, and humans live close together, the swine can serve as a melting pot for creating "hybrid" influenza viruses that are not recognized by the human immune system.

(duck and ducklings) Image Source; (healthy young woman) Westend61/Getty Images; (pot-bellied pig) G. K. & Vikki Hart/Getty Images; (ill young woman) Image Source/Getty Images

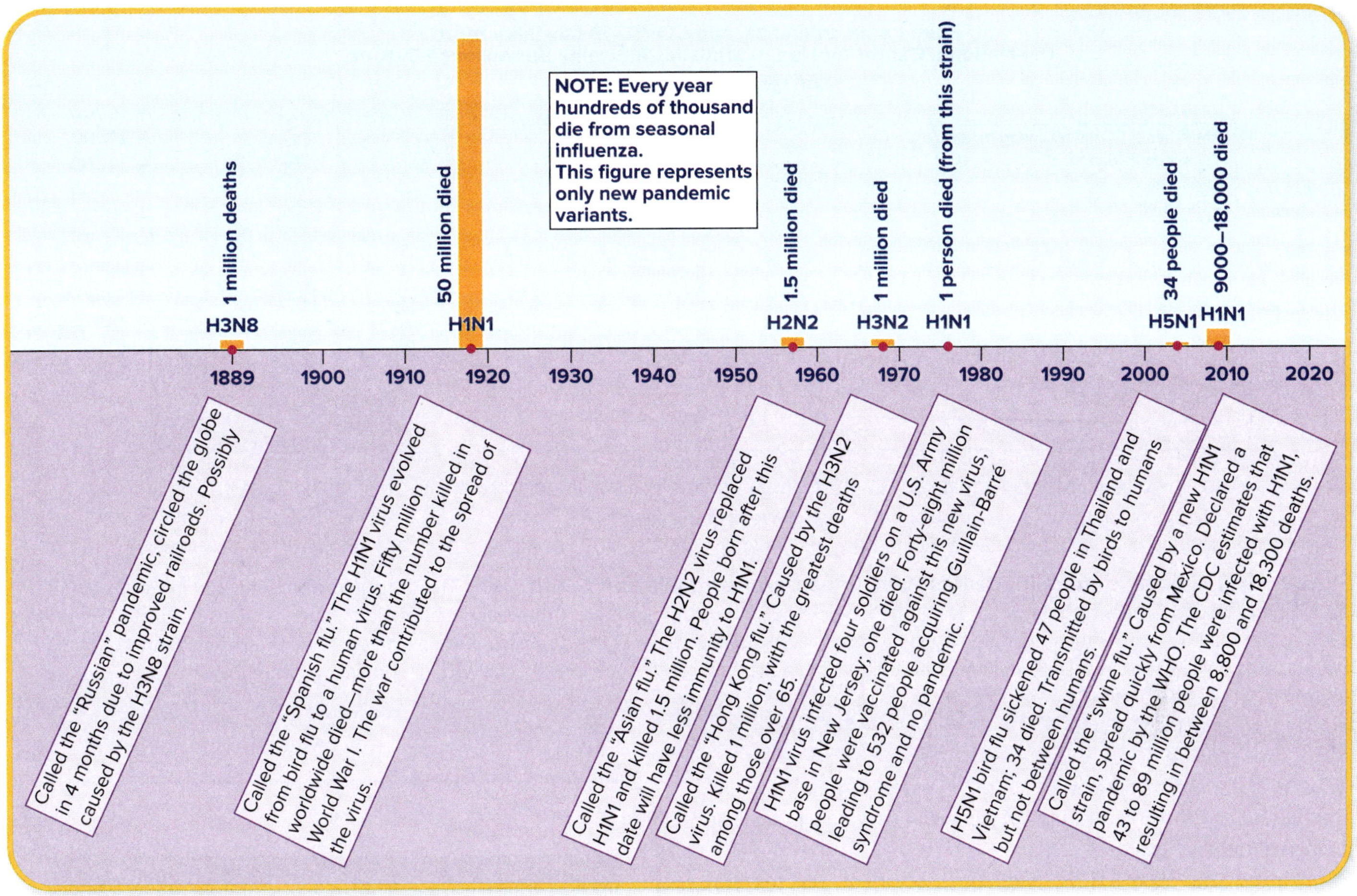
NOTE: Every year hundreds of thousand die from seasonal influenza. This figure represents only new pandemic variants.
1 million deaths
H3N8
50 million died
H1N1
1.5 million died
H2N2
1 million died
H3N2
1 person died (from this strain)
H1N1
34 people died
H5N1
9000–18,000 died
H1N1
1889
1900
1910
1920
1930
1940
1950
1960
1970
1980
1990
2000
2010
2020
Called the "Russian" pandemic, circled the globe in 4 months due to improved railroads. Possibly caused by the H3N8 strain.
Called the "Spanish flu." The H1N1 virus evolved from bird flu to a human virus. Fifty million worldwide died—more than the number killed in World War I. The war contributed to the spread of the virus.
Called the "Asian flu." The H2N2 virus replaced H1N1 and killed 1.5 million. People born after this date will have less immunity to H1N1.
Called the "Hong Kong flu." Caused by the H3N2 virus. Killed 1 million, with the greatest deaths among those over 65.
H1N1 virus infected four soldiers on a U.S. Army base in New Jersey; one died. Forty-eight million people were vaccinated against this new virus, leading to 532 people acquiring Guillain-Barré syndrome and no pandemic.
H5N1 bird flu sickened 47 people in Thailand and Vietnam; 34 died. Transmitted by birds to humans but not between humans.
Called the "swine flu." Caused by a new H1N1 strain, spread quickly from Mexico. Declared a pandemic by the WHO. The CDC estimates that 43 to 89 million people were infected with H1N1, resulting in between 8,800 and 18,300 deaths.

Figure 22.10 The emergence of pandemic strains and their death tolls—global.

or secondary bacterial infection that resulted from opportunistic microbes infecting already compromised lungs. It is nearly impossible to predict whether a new pandemic virus will cause 50 million deaths, as in 1918, or 14,000, as in 2009.

Of course, the latest pandemic was caused by another ssRNA virus, a coronavirus. This highlights that even though scientists know what type of microbe is most likely to cause a sudden worldwide pandemic (ssRNA viruses have a particularly high mutation rate), they can't predict which specific respiratory virus will emerge and when.

Transmission and Epidemiology

Inhalation of virus-laden aerosols and droplets constitutes the major route of influenza infection, although fomites can play a secondary role. Transmission is greatly facilitated by crowding and poor ventilation in classrooms, barracks, nursing homes, dormitories, and military installations in the late fall and winter. The drier air of winter facilitates the spread of the virus, as the moist particles expelled by sneezes and coughs become dry very quickly, helping the virus remain airborne for longer periods of time. In addition, the dry, cold air makes respiratory tract mucous membranes more brittle, with microscopic cracks that facilitate invasion by viruses. Influenza is highly contagious and affects people of all ages. Annually, from 17,000 to 52,000 deaths occur in the United States from seasonal influenza and its complications, mainly among the very young and the very old.

Figure 22.11 shows a hybrid graph. For years the National Center for Health Statistics has kept track of the percentage of deaths in the United States that is caused by what they call "P and I"—pneumonia and influenza. Because there was a massive increase in pneumonia deaths due to another virus—SARS-CoV-2, they included those deaths in their graph and called it "PIC"—pneumonia, influenza, and COVID-19 deaths. During the pandemic, those refusing the COVID-19 vaccine often said, "It's just another flu." This graph drives home that it is not just another flu, but much more deadly than the seasonal flu we see every year.

Culture and/or Diagnosis

A wide variety of culture-based and non-culture-based methods may be used to diagnose the infection. RT-PCR is by far the preferred method for diagnosis. Culture can be useful for identifying the particular subtype of influenza involved.

Prevention

Preventing influenza infections and epidemics is one of the top priorities for public health officials. The flu vaccine is the best way to do that. The CDC recommends that all persons above the age of 6 months get vaccinated every year. The particular forms the vaccine takes may change from year to year, and the recommendations also change. The influenza page at the CDC

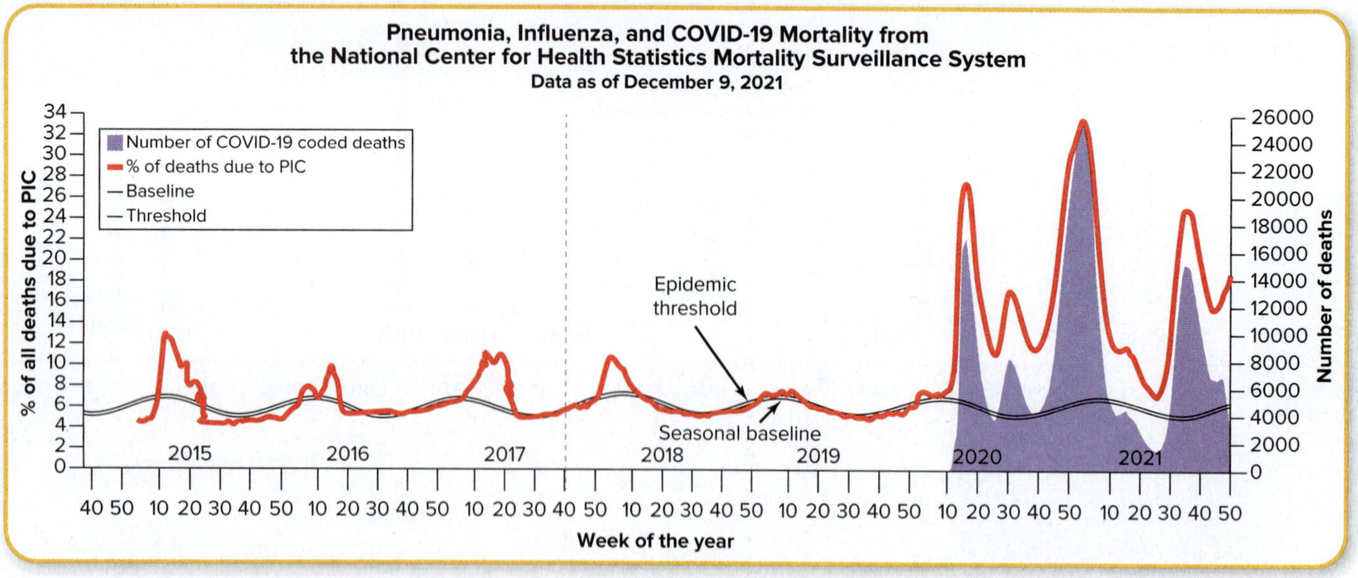
Pneumonia, Influenza, and COVID-19 Mortality from
the National Center for Health Statistics Mortality Surveillance System
Data as of December 9, 2021
Number of COVID-19 coded deaths
% of deaths due to PIC
Baseline
Threshold
% of all deaths due to PIC
Number of deaths
Epidemic threshold
Seasonal baseline
2015
2016
2017
2018
2019
2020
2021
Week of the year

Figure 22.11 Epidemic curve for deaths from influenza/pneumonia and COVID-19. The horizontal axis is marked in weeks of the year. Clearly, influenza peaks in midwinter. The bottom black line in the graph indicates the seasonal baseline number of cases calculated from years of data. The top black line is the threshold beyond which influenza or the cause of the deaths is declared an epidemic. You can see that in most winters, epidemic levels of seasonal influenza were seen, but none as deadly as the arrival of COVID in the early winter of 2020.
Adapted from CDC, Centers for Disease Control and Prevention

website is the most trustworthy source of information. There are flu vaccines with three strains of the virus (trivalent) and also quadrivalent vaccines.

During the 2009 H1N1 pandemic, seasonal vaccine had already been distributed when officials realized a different pandemic strain was causing disease. A new vaccine containing the pandemic strain was quickly prepared. The existence of two vaccines confused the public and made flu vaccination more confusing.

The vaccine cannot cause the flu, despite what many people believe. If you experience upper respiratory tract symptoms shortly after receiving a flu vaccine, it is likely to be a cold virus that is taking advantage of your body's immune response being busy responding to the antigens in the vaccine. That is a small price to pay for protection against a potentially fatal disease.

We may be on the brink of a revolutionary change to how we vaccinate against influenza. Several "universal" flu vaccines are in clinical trials. These vaccines target the parts of the hemagglutinin molecule that do not mutate, and that are shared among nearly all flu viruses known globally (see figure 22.8). When one or more of these comes to market, it should be possible to vaccinate once (or with a limited schedule of boosters) instead of vaccinating every year.

Treatment

Antiviral treatment for influenza must be taken early in the infection, preferably by the second day. This requirement is an inherent difficulty because most people do not realize until later that they may have the flu. The most commonly used drug is Tamiflu (oseltamivir).

Disease Table 22.7	Influenza
Causative Organism(s)	Influenza A, B, and C viruses V
Most Common Modes of Transmission	Droplet contact, direct contact, or indirect contact
Virulence Factors	Glycoprotein spikes, antigenic drift and shift.
Culture/Diagnosis	Gold standard is RT-PCR tests
Prevention	A variety of vaccines are available and should be received annually
Treatment	Oseltamivir (Tamiflu), baloxavir (Xofluza)
Epidemiological Features	For seasonal flu, deaths vary from year to year; United States: range from 17,000–52,000; internationally: range from 250,000–500,000

22.3 Learning Outcomes—Assess Your Progress

7. List the possible causative agents for each of the infectious conditions affecting both the upper and lower respiratory tracts: whooping cough, RSV disease, and influenza.

8. Describe the symptoms appearing in each stage of whooping cough.
9. Discuss reasons for the increase of pertussis cases over the past three decades.
10. Identify the age groups at most risk for serious disease from RSV.
11. Plan a response for a situation in which a patient declines influenza vaccination because they believe that the warnings about pandemics are always exaggerated.
12. Compare and contrast *antigenic drift* and *antigenic shift* in influenza viruses.

22.4 Infectious Diseases Manifesting in the Lower Respiratory Tract

In this section, we consider microbial diseases that primarily affect the lower respiratory tract—namely, the bronchi, bronchioles, and lungs—with minimal involvement of the upper respiratory tract. Our discussion focuses on tuberculosis and pneumonia.

Tuberculosis

Mummies from the Stone Age, ancient Egypt, and Peru provide evidence that tuberculosis (TB) is an ancient human disease. After the discovery of streptomycin in 1943, the rates of tuberculosis in the developed world declined rapidly, although that did not happen in the developing world. Tuberculosis reemerged in the United States and other developed countries in the mid-1980s, fueled by the HIV epidemic and the bacterium's resistance to multiple antibiotics. That resurgence was eventually quelled again in the developed world. However, in Southeast Asia, the Western Pacific region, and Africa, a multidrug-resistant TB epidemic is raging. The cause of tuberculosis is primarily the bacterial species *Mycobacterium tuberculosis,* informally called the tubercle bacillus. In this discussion, we will first address the general aspects of the infection, then turn to its most troubling form today, multidrug-resistant TB (MDR-TB).

Signs and Symptoms

A clear-cut distinction can be made between infection with the TB bacterium and the disease it causes. In general, humans are rather easily infected with the bacterium but are resistant to the disease. Estimates project that only about 5% to 10% of infected people actually develop a clinical case of tuberculosis. The majority of TB cases are contained in the lungs, even though disseminated TB bacteria can give rise to tuberculosis in any organ of the body. Clinical tuberculosis is divided into primary tuberculosis, secondary (reactivation or reinfection) tuberculosis, and disseminated tuberculosis.

Primary Tuberculosis The minimum infectious dose for lung infection in primary tuberculosis is around 10 bacterial cells. Alveolar macrophages phagocytose these cells, but the bacteria are not killed and continue to multiply inside the macrophages. This period of hidden infection is asymptomatic or accompanied by mild fever. Some bacteria escape from the lungs into the blood and lymphatics. After 3 to 4 weeks, the immune system mounts a complex, cell-mediated assault against the bacteria. The large

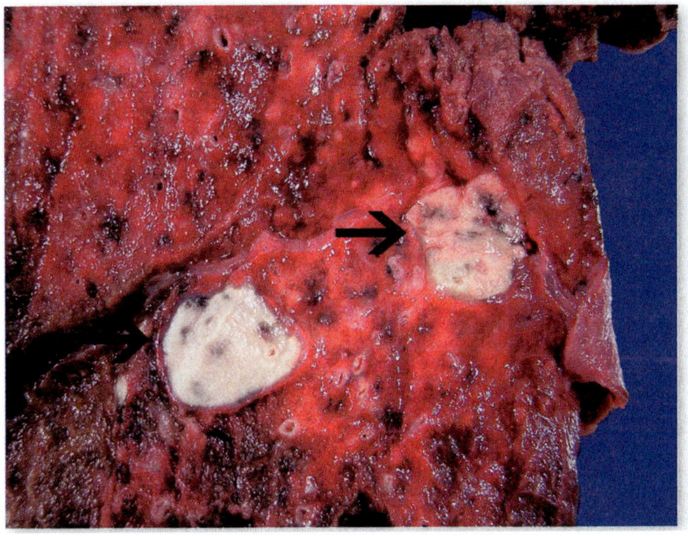

Figure 22.12 Tubercle formation. Two areas of caseous necrosis in a lung.
McGraw Hill

influx of mononuclear cells into the lungs plays a part in the formation of specific infection sites called tubercles. Tubercles are granulomas that consist of a central core containing TB bacteria in enlarged macrophages and an outer wall made of fibroblasts, lymphocytes, and macrophages. The center of tubercles contains a soft-white material known as caseous necrosis **(figure 22.12).** Although this response further checks the spread of infection and helps prevent the disease, it also carries a potential for damage. Frequently, as neutrophils come on the scene and release their enzymes, the centers of tubercles break down into necrotic caseous (kay′-see-us) lesions that gradually heal by calcification—normal lung tissue is replaced by calcium deposits. The response of T cells to *M. tuberculosis* proteins also causes a cell-mediated immune response evident in the skin test called the tuberculin reaction, a valuable diagnostic and epidemiological tool **(figure 22.13).**

TB infection outside of the lungs is more common in immunosuppressed patients and young children. Organs most commonly involved in **extrapulmonary TB** are the regional lymph nodes, intestines, kidneys, long bones, genital tract, brain, and meninges. Because of the debilitation of the patient and the high load of TB bacteria, these complications are usually grave. Renal tuberculosis results in necrosis and scarring of the kidney and the pelvis,

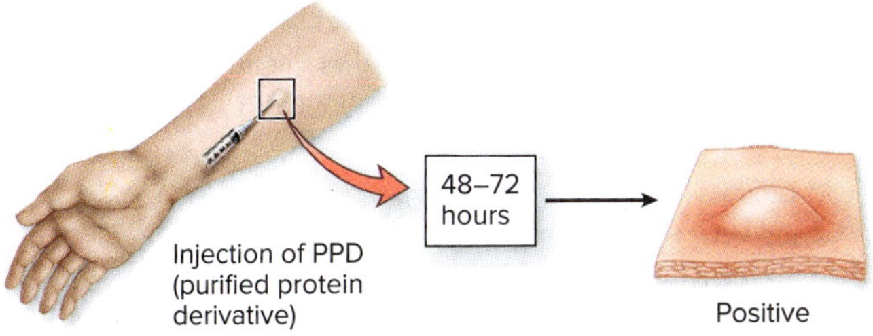
48–72
hours
Injection of PPD
(purified protein
derivative)
Positive

Figure 22.13 Skin testing for tuberculosis. The Mantoux test. Tuberculin is injected into the dermis. A small bleb from the injected fluid develops but will be absorbed in a short time. After 48 to 72 hours, the skin reaction is rated by the degree (or size) of the raised area. The surrounding red area is not counted in the measurement.

ureters, and bladder. This damage is accompanied by painful urination, fever, and the presence of blood and the TB bacterium in urine. Genital tuberculosis in males damages the prostate gland, epididymis, seminal vesicles, and testes. In females, the fallopian tubes, ovaries, and uterus can be affected. Tuberculosis of the bones and joints is a common complication. The spine is a frequent site of infection, although the hip, knee, wrist, and elbow can also be involved. Advanced infiltration of the vertebral column produces degenerative changes that collapse the vertebrae, resulting in abnormal curvature of the thoracic region (humpback) or of the lumbar region (swayback). Neurological damage stemming from compression on nerves can cause extensive paralysis and sensory loss.

Tubercular meningitis is the result of an active brain lesion seeding bacteria into the meninges. Over a period of several weeks, the infection of the cranial compartments can create mental deterioration, permanent developmental disability, blindness, and deafness. Untreated tubercular meningitis is invariably fatal, and even treated cases can have a 30% to 50% mortality rate.

Secondary (Reactivation) Tuberculosis Although the majority of adequately treated TB patients recover more or less completely from the primary episode of infection, live bacteria can remain dormant (latent) and become reactivated weeks, months, or years later, especially in people with weakened immunity. In chronic tuberculosis, tubercles filled with masses of bacteria expand, cause cavities in the lungs, and drain into the bronchial tubes and upper respiratory tract. The patient gradually experiences more severe symptoms, including violent coughing, greenish or bloody sputum, low-grade fever, anorexia, weight loss, extreme fatigue, night sweats, and chest pain. It is the gradual wasting of the body that accounts for an older name for tuberculosis—*consumption.* Untreated secondary disease has nearly a 60% mortality rate.

▶ Causative Agents

Mycobacterium tuberculosis is the cause of tuberculosis in most patients. It is a long and thin acid-fast rod. It is a strict aerobe and is not referred to as gram-positive or gram-negative because its acid-fast nature is much more relevant in a clinical setting. It grows very slowly, with a generation time of 15 to 20 hours. A period of up to 6 weeks is required for colonies to appear in culture. (*Note:* The prefix *Myco-* may make you think of fungi, but this is a bacterium. The prefix in the name came from the mistaken impression that colonies growing on agar **(figure 22.14)** resembled fungal colonies. And be sure to differentiate this bacterium from *Mycoplasma*—they are unrelated.)

Robert Koch identified that *M. tuberculosis* often forms serpentine cords while growing, and he called the unknown substance causing this style of growth "cord factor." Cord factor appears to be associated with virulent strains, and it is a lipid component of the mycobacterial cell wall. All mycobacterial species have walls that have a very high content of complex lipids, including mycolic acid and waxes. This chemical characteristic makes them relatively impermeable to stains and difficult to decolorize (acid-fast) once they are stained. The lipid wall of the bacterium also influences its virulence and makes it resistant to drying and disinfectants.

In recent decades, tuberculosis-like conditions caused by *Mycobacterium avium,* and related mycobacterial species (sometimes referred to as the *M. avium* complex, or MAC), have been found in AIDS patients and other immunocompromised people. In this section, we consider only *M. tuberculosis.*

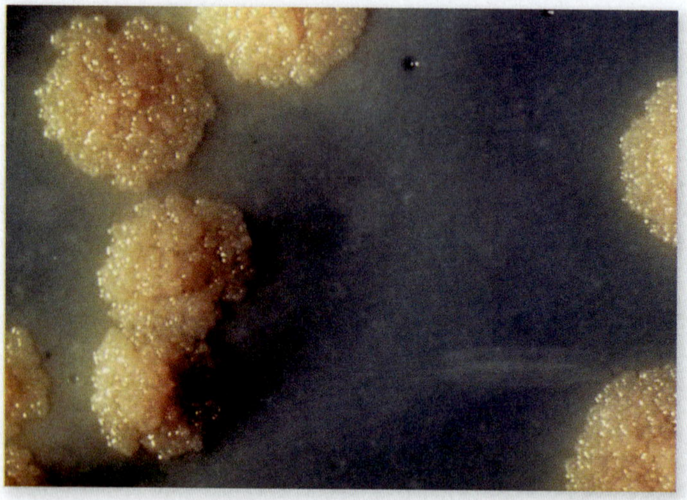

Figure 22.14 Cultural appearance of *Mycobacterium tuberculosis.* Colonies with a typical granular, waxy pattern of growth.

CDC/Dr. George Kubica

Before routine pasteurization of milk, humans acquired bovine TB, caused by a species called *Mycobacterium bovis,* from the milk they drank. It was recently thought to be very rare. But recently scientists published a paper showing that acquiring tuberculosis (caused by *M. bovis*) from animals is much more common than previously thought. Estimates of how many cases of tuberculosis come from animals vary widely, from 9% in one study, to 45% in another.

▶ Pathogenesis and Virulence Factors

The course of the infection—and all of its possible variations—was previously described under "Signs and Symptoms." Important characteristics of the bacterium that contribute to its virulence are its waxy surface (contributing to both its survival in the environment and its survival within macrophages) and its ability to stimulate a strong cell-mediated immune response that contributes to the pathology of the disease. The tubercle bacilli are able to survive attack by the macrophages that tried to engulf and kill them, allowing the pathogen to avoid immune recognition. This activity is due to a special enzyme that blocks the proper formation of phagosomes, acidic cellular compartments used to degrade debris.

▶ Transmission and Epidemiology

The agent of tuberculosis is transmitted almost exclusively by fine droplets of respiratory mucus suspended in the air. The TB bacterium is highly resistant and can survive for 8 months in fine aerosol particles. Although larger particles become trapped in mucus and are expelled, tinier ones can be inhaled into the bronchioles and alveoli. This effect is especially pronounced among people sharing small, closed rooms with limited access to sunlight and fresh air.

The epidemiological patterns of *M. tuberculosis* infection vary with the living conditions in a community or an area of the world. Factors that significantly affect people's susceptibility to tuberculosis are inadequate nutrition, debilitation of the immune system, poor access to medical care, lung damage, and their own genetics. Put simply, TB is an infection of poverty. People in developing countries are often infected as infants and harbor the microbe for many years until the disease is reactivated in young adulthood. In 2020, 1.5 million people worldwide died from TB. About 1.7 billion people, an amazing 23% of the world's population, are believed to have latent TB infection. People who are HIV+ are at particular risk for TB infection.

▶ Culture and/or Diagnosis

Skin Testing Because infection with the TB bacillus can lead to delayed hypersensitivity to tuberculoproteins, testing for hypersensitivity has been an important way to screen populations for latent tuberculosis infection. Although newer methods are available, the most widely used test is still the tuberculin skin test, called the **Mantoux test.** It involves local injection of purified protein derivative (PPD), a standardized solution taken from culture fluids of *M. tuberculosis.* The injection is done intradermally into the forearm to produce an immediate, small bleb. After 48 hours, the site is observed for a red wheal called an **induration,** which is measured and interpreted as positive or negative according to size (see figure 22.13). The disadvantage to skin testing is that a second visit to the health care provider is required.

Subgroups with severely compromised immune systems, such as those with AIDS, advanced age, or chronic disease, may be unable to mount a reaction even though they are infected. Skin testing is not a reliable diagnostic indicator in these populations. False positives may result when a person born in another country is tested if that person received a common TB vaccine called BCG (discussed in the "Prevention" section). This vaccine is not used in the United States.

IGRA In the IGRA test, a patient's blood is drawn and incubated in test kits that detect the presence of T cells that react with *M. tuberculosis* antigens. If they have been so sensitized, they will release interferon-gamma (IFN-γ) after binding the antigens. High levels of IFN-γ trigger a positive response. The advantage of these tests is that no return visit is required.

PCR Test Recently, a PCR method has become available. Known as *Xpert MTB/RIF,* it simultaneously detects *M. tuberculosis* and determines its rifampin sensitivity within 100 minutes.

Sputum Smears and Chest X Rays Smears stained with Ziehl-Neelsen stain display the acid-fast bacteria as bright red against a blue background **(figure 22.15).** Chest X rays can reveal abnormal opaque patches **(figure 22.16),** the appearance and location of which can help with diagnosis.

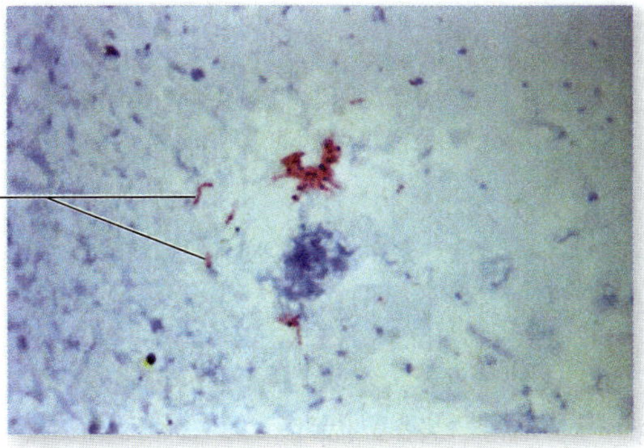
M. tuberculosis

Figure 22.15 Ziehl-Neelsen staining of *Mycobacterium tuberculosis* in sputum.
Centers for Disease Control and Prevention

▶ Prevention

Preventing TB in the United States is accomplished by limiting exposure to infectious airborne particles. Extensive precautions, such as isolation in negative-pressure rooms, are used in health care settings when a person with active TB is identified. Vaccine is not used in the United States, although an attenuated vaccine, called BCG, is used in many countries. *BCG* stands for "Bacille Calmette-Guerin," named for two French scientists who created the vaccine in the early 1900s. It is a live strain of a bovine tuberculosis bacterium that has been made avirulent by long passage through artificial media.

Prevention in the context of tuberculosis may also refer to preventing a person with latent TB from experiencing reactivation. This strategy is more accurately referred to as "treatment of latent infection" and is considered in the next section.

▶ Treatment

Treatment of latent TB infection is effective in preventing full-blown disease in persons who have positive skin tests and who are at risk for reactivated TB. Treatment of latent TB is usually with isoniazid for a long period of time or with isoniazid plus one other antibiotic in order to shorten the course of treatment.

Treatment of active TB infection occurs in two phases. In the first phase, four drugs—rifampin, isoniazid, ethambutol, and pyrazinamide—are used for 2 months. The second phase uses only two drugs that susceptibility testing have shown to be effective and lasts either 4 or 7 months, decided on a case-by-case basis.

Multidrug-Resistant and Extensively Drug-Resistant *Mycobacterium tuberculosis* (MDR-TB and XDR-TB)

One of the biggest problems with TB therapy is noncompliance on the part of the patient. It is very difficult, even under the best of circumstances, to keep to a regimen of multiple antibiotics daily

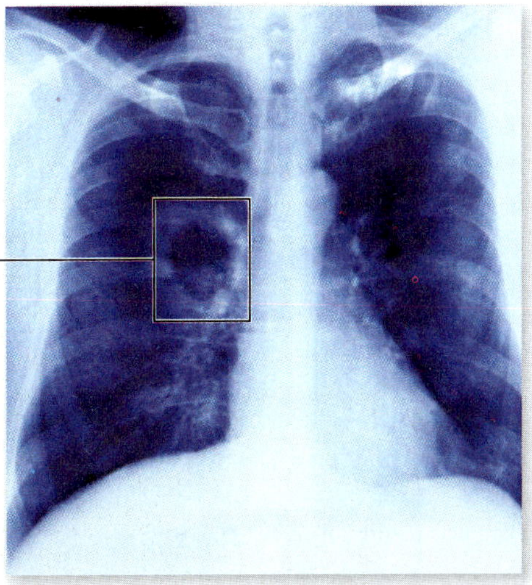
Area of infection

Figure 22.16 After primary infection, lungs reinfected by *Mycobacterium tuberculosis.*
BSIP/Newscom

for months—and most TB patients are not living under the best of circumstances. Failure to adhere to the antibiotic regimen leads to antibiotic resistance in the slow-growing microorganism. Many *M. tuberculosis* isolates are now found to be **MDR-TB,** or multidrug-resistant TB. The threat to public health is so great when patients do not adhere to treatment regimens that the United States and other countries have occasionally incarcerated people—and isolated them—for not following their treatment schedules.

▶ Multidrug-Resistant Tuberculosis (MDR-TB)

When *Mycobacterium tuberculosis* is defined as being resistant to at least isoniazid and rifampin, it is called multidrug-resistant tuberculosis (MDR-TB). It requires 18 to 24 months of treatment with four to six drugs. In some parts of the world, the rate of MDR-TB among previously treated TB patients is 50% to 60%. Among those being treated for TB for the first time, the MDR-TB rate is closer to 4%.

People with MDR-TB are generally sicker and have higher mortality rates than those infected with non-MDR-TB. This is true even when they are being treated, as the multiple-drug combination has severe side effects. The current treatment recommendations are the use of three drugs: pretomanid, bedaquiline, and linezolid.

▶ Extensively Drug-Resistant Tuberculosis (XDR-TB)

MDR-TB strains exhibiting resistance to two additional drugs are designated as XDR-TB. These strains have been reported in 84 countries. Worldwide, 9% of the MDR-TB cases also qualify as XDR-TB. Patients with XDR-TB have few treatment options, and their mortality rate is estimated to be about 70% within months of diagnosis. The epidemiology is difficult to document, but India and China have the highest burden of XDR-TB. A few cases are seen in the United States every year (**Disease Table 22.8**.)

Pneumonia

Pneumonia is a classic example of an *anatomical diagnosis.* It is defined as an inflammatory condition of the lung in which fluid fills the alveoli. The set of symptoms that we call pneumonia can be caused by a wide variety of microorganisms. In a sense, the microorganisms need only to have appropriate characteristics to allow them to circumvent the host's defenses and to penetrate and survive in the lower respiratory tract. In particular, the microorganisms must avoid being phagocytosed by alveolar macrophages, or at least avoid being killed once inside the macrophage. Bacteria and a wide variety of viruses can cause pneumonias. Viral pneumonias are usually—but not always—milder than those caused by bacteria. At the same time, some bacterial pneumonias are very serious, but others are not. In addition, fungi such as *Histoplasma* can cause pneumonia. Every year in the United States, there are 2 to 3 million cases of pneumonia, resulting in more than 50,000 deaths. It is much more common in the winter. Globally, more than 1.5 million children younger than 5 years of age die from pneumonia every year. In short, pneumonia kills more children than any other infectious disease in the world today.

Physicians distinguish between two forms of pneumonia, each characterized by different modes of transmission and pathogenic agents. Community-acquired pneumonias (CAP) are those

Disease Table 22.8 Tuberculosis

Causative Organism(s)	*Mycobacterium tuberculosis* B gram-variable	MDR-TB and XDR-TB
Most Common Modes of Transmission	Vehicle (airborne)	
Virulence Factors	Lipids in wall, ability to stimulate strong cell-mediated immunity (CMI)	
Culture/Diagnosis	Culture, PCR test (Xpert®), IGRA, complemented by skin test and chest X ray	
Prevention	Avoiding airborne *M. tuberculosis;* BCG vaccine in other countries	
Treatment	Isoniazid, rifampin, and pyrazinamide + ethambutol or streptomycin for varying lengths of time (always lengthy)	Multiple-drug regimen, which may include pretomanid, bedaquiline, and linezolid in **Serious Threat** category in CDC Antibiotic Resistance Report
Distinctive Features	Remains airborne for long periods; extremely slow-growing, which has implications for diagnosis and treatment	Much higher fatality rate over shorter duration
Epidemiological Features	United States: approx. 10,000 cases/year, 16% of cases Whites, 84% ethnic minorities; internationally: 1.3 million deaths/yr	United States: a few cases per year; worldwide: approximately 500,000 new infections with MDR-TB in 2020

experienced by persons in the general population. Healthcare-associated pneumonias (HCAP) develop in individuals receiving treatment at health care facilities, including hospitals. All pneumonias have similar symptoms, which we describe next, followed by separate sections for each cause of pneumonia.

Signs and Symptoms

Pneumonias of all types usually begin with upper respiratory tract symptoms, including congestion. Headache is common. Fever is often present, and the onset of lung symptoms follows. These symptoms are chest pain, fever, cough, and the production of discolored sputum. Because of the pain and difficulty of breathing, the patient appears pale and presents an overall sickly appearance. The severity and speed of onset of the symptoms vary according to the etiologic agent.

Community-Acquired Pneumonia

Causative Agents

Pneumonias that occur in the community have a very distinct differential diagnosis than do those that occur in a health care setting such as a hospital or rehabilitation facility. Recently the CDC released a large study of community-acquired pneumonia among 2,300 patients hospitalized for the condition, documenting the most common causative agents. In 62% of the cases, no pathogen was able to be detected. The most commonly isolated causative agents were the rhinovirus (9% of cases), influenza (6%), and *Streptococcus pneumoniae* (5%). In less serious cases not requiring hospitalization, *Haemophilus influenzae, Moraxella catarrhalis, Chlamydophila pneumoniae,* and *Mycoplasma pneumoniae* can be causative agents. These agents can cause what is commonly called "walking pneumonia," meaning a pneumonia that still allows you to carry on with everyday activities.

In this section, we will consider the bacterium *Streptococcus pneumoniae* and a cause of walking pneumonia, *Mycoplasma pneumoniae*. In addition, we will cover some less-common but important possible causes of pneumonia, *Legionella* species, and two fungi, *Histoplasma capsulatum* and *Pneumocystis jirovecii*. (Be aware that *Bacillus anthracis,* covered in chapters 19 and 21, can also cause a pneumonic form of anthrax.)

During the COVID-19 pandemic, any pneumonia should trigger a COVID-19 test. Even though COVID-19 is situated in the cardiovascular chapter, its acute symptoms include severe pneumonia, usually in both lungs. It is yet to be seen whether the infection will settle into an endemic infection that causes colds or pneumonia or nothing at all.

Streptococcus pneumoniae

This bacterium, which is often simply called the pneumococcus, is a small, gram-positive, flattened coccus that often appears in pairs, lined up end to end **(figure 22.17*a*).** It is alpha-hemolytic on blood agar **(figure 22.17*b*).** *S. pneumoniae* is part of the normal biota in the upper respiratory tract of up to 50% of healthy people. Infection can occur when the bacterium is inhaled into deep areas of the lung or by transfer of the bacterium between two people via respiratory

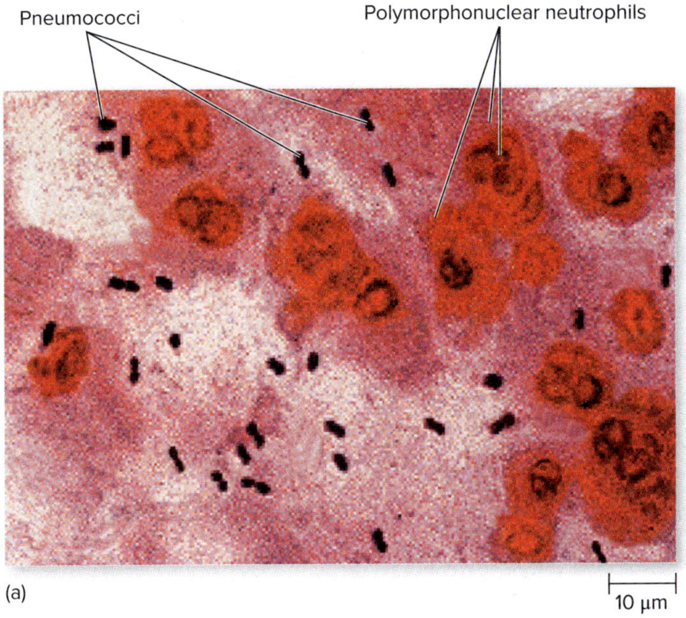

Pneumococci
Polymorphonuclear neutrophils
(a)
10 µm

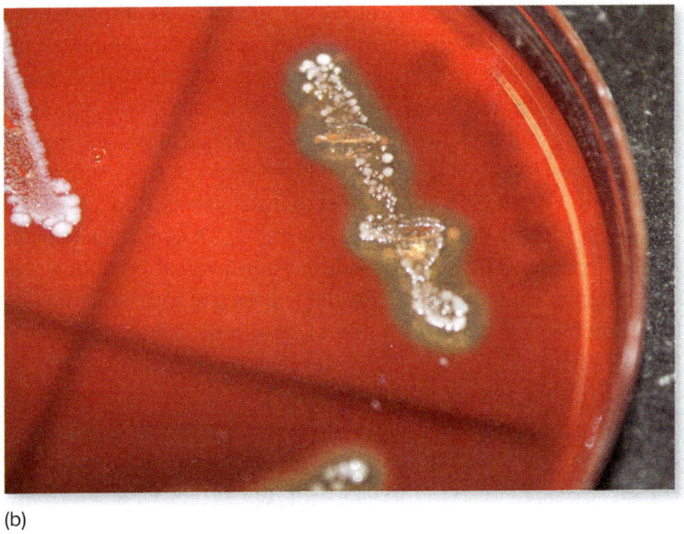

(b)

Figure 22.17 ***Streptococcus pneumoniae.*** **(a)** Gram stain of sputum. **(b)** Alpha-hemolysis of *S. pneumoniae* on blood agar. The "halo" is greenish.

(a) Neal R. Chamberlin, PhD/McGraw Hill; (b) Lisa Burgess/McGraw Hill

droplets. *S. pneumoniae* is very delicate and does not survive long out of its habitat. Factors that favor the ability of the pneumococcus to cause disease are old age, the season (rate of infection is highest in the winter), underlying viral respiratory disease, diabetes, and chronic abuse of alcohol or narcotics. Healthy people commonly inhale this and other microorganisms into the respiratory tract without serious consequences because of the host defenses present there.

This pneumonia is likely to occur when mucus containing a load of bacterial cells passes into the bronchi and alveoli. The pneumococci multiply and induce an overwhelming inflammatory response. The polysaccharide capsule of the bacterium prevents efficient phagocytosis, apparently by blocking the attachment of complement, with the result that the fluids of inflammation are continuously released into the lungs. As the infection and inflammation spread rapidly through the lung, the patient can actually "drown" in his or her own secretions. If this mixture of exudates,

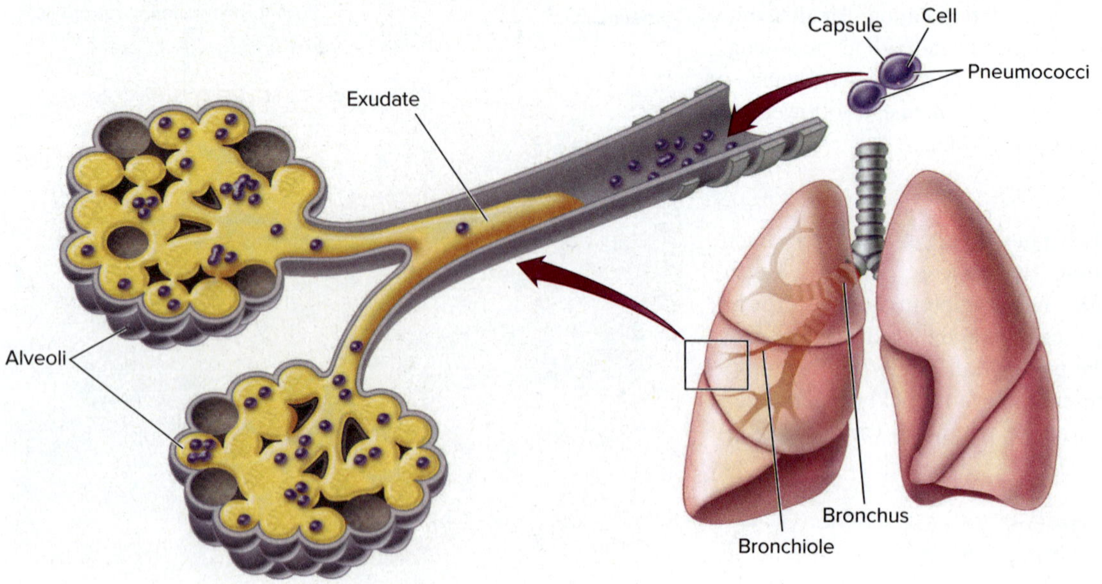
Capsule
Cell
Pneumococci
Exudate
Alveoli
Bronchus
Bronchiole

Figure 22.18 The course of bacterial pneumonia. As the pneumococcus traces a pathway down the respiratory tree, it provokes intense inflammation and exudate formation. The blocking of the bronchioles and alveoli by consolidation of inflammatory cells and products is evident.

cells, and bacteria solidifies in the air spaces, a condition known as *consolidation* occurs **(figure 22.18).** Systemic complications of pneumonia are pleuritis and endocarditis, but pneumococcal bacteremia and meningitis are the greatest danger to the patient.

Many, if not most, deaths stemming from influenza are caused by pneumonia, either from the original virus or from secondary infection. Vaccination with either the 13-valent PCV13 or the 23-valent PPSV23 vaccine is recommended for children and older adults. People aged 6 to 64 with certain medical conditions, or who smoke, should also be vaccinated. Active disease is treated with antibiotics, but the choice of antibiotic is often difficult. Many isolates of *S. pneumoniae* are resistant to penicillin and its derivatives, as well as to the macrolides, tetracyclines, and fluoroquinolones. For that reason, broad-spectrum cephalosporins are now typically prescribed for drug therapy, with or without vancomycin. Treatment also varies based on whether the patient receives outpatient or inpatient care. This bacterium is clearly capable of rapid development of resistance, and effective treatment requires that the practitioner be familiar with local resistance trends.

Mycoplasma pneumoniae

Pneumonias caused by a handful of bacteria, most often *Mycoplasma,* are often called atypical pneumonia—atypical in the sense that the symptoms do not resemble those of pneumococcal or other severe pneumonia. They are more like a bad case of the common cold, except that there is mucus accumulation in the lungs.

Mycoplasmas, as you learned in chapter 4, are among the smallest known self-replicating microorganisms. They naturally lack a cell wall and are therefore irregularly shaped. They may resemble cocci, filaments, doughnuts, clubs, or helices. They are free-living but fastidious, requiring complex medium to grow in the lab. (This genus should not be confused with *Mycobacterium.*)

Mycoplasma pneumonia is transmitted by aerosol droplets among people confined in close living quarters, especially families, students, and the military. Lack of acute illness in most patients has given rise to the name "walking pneumonia." A cyclical incidence of *Mycoplasma* pneumonia seems to occur every 3 to 6 years in the United States.

Diagnosis of *Mycoplasma* may begin with ruling out other bacteria or viral agents. Serological or PCR tests confirm the diagnosis. These bacteria do not stain with Gram stain and are not visible in direct smears of sputum. More advanced genotypic testing is performed today to monitor outbreaks and the development of macrolide resistance in this pathogen.

Legionella pneumophila

Legionella is a weakly gram-negative bacterium that displays a range of shapes, from coccus to filaments. Several species or subtypes have been characterized, but *L. pneumophila* (lung-loving) is the one most frequently isolated from infections.

Although the organisms were originally described in the late 1940s, they were not clearly associated with human disease until 1976. The incident that brought them to the attention of medical microbiologists was a sudden and mysterious epidemic of pneumonia that afflicted 200 American Legion members attending a convention in Philadelphia and killed 29 of them. After 6 months of painstaking analysis, epidemiologists isolated the pathogen and traced its source to contaminated air-conditioning vents in the hotel hosting the Legionnaires' convention.

Legionella is widely distributed in aqueous habitats as diverse as tap water, cooling towers, spas, ponds, and other freshwaters. It is resistant to chlorine. It is released during aerosol formation and can be carried for long distances. Cases have been traced to supermarket vegetable sprayers, hotel fountains, and even the fallout from the Mount St. Helens volcano eruption in 1980. Cases of this disease increased by a factor of 5 in the United States between 2000 and 2021, possibly due to the aging population, because it more often causes disease in the elderly. One example will serve to illustrate how these outbreaks occur. In the summer of 2015, a cluster of Legionnaires' cases occurred in the Bronx in New York City. Cooling towers were suspected early on, and the health commissioner ordered all New York City cooling towers to be disinfected. In all, 133 people became ill and 16 died. Eventually, one cooling tower was identified as the source of the outbreak. **Figure 22.19** is a map of the area, showing the offending cooling tower and the location of the infected persons.

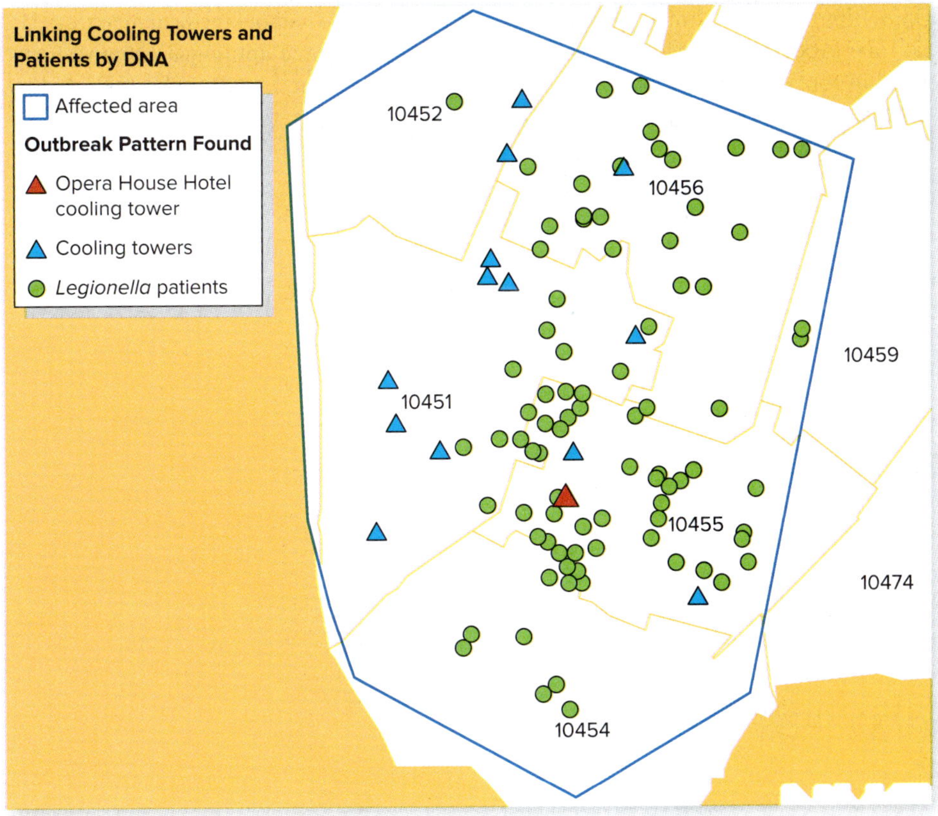
Linking Cooling Towers and Patients by DNA
Affected area
Outbreak Pattern Found
Opera House Hotel cooling tower
Cooling towers
Legionella patients
10452
10456
10459
10451
10455
10474
10454

Figure 22.19 An epidemiological map of *Legionella* patients and suspected cooling towers in the Bronx, New York, during the summer 2015 outbreak. The source tower turned out to be the Opera House Hotel cooling tower (red triangle).

Source: City of New York City.

Although this bacterium can cause another disease called Pontiac fever, pneumonia is the more serious disease, with a fatality rate of 3% to 30%. *Legionella* pneumonia is thought of as an opportunistic disease, usually affecting elderly people and rarely being seen in children and healthy adults. It is difficult to diagnose, even with specific antibody tests. It is not transmitted person to person.

Histoplasma capsulatum

Pulmonary infections with this dimorphic fungus have probably afflicted humans since ancient times, but it was not described until 1905 by Dr. Samuel Darling. Through the years, it has been known by various names: Darling's disease, Ohio Valley fever, and spelunker's disease. Certain aspects of its current distribution and epidemiology suggest that it has been an important disease for as long as humans have practiced agriculture.

Pathogenesis and Virulence Factors

Histoplasmosis presents a formidable array of manifestations. It can be benign or severe, acute or chronic; and it can cause pulmonary, systemic, or cutaneous lesions. Inhaling a small dose of microconidia into the deep recesses of the lung establishes a primary pulmonary infection that is usually asymptomatic. Its primary location of growth is in the cytoplasm of phagocytes such as macrophages. It flourishes within these cells and is carried to other sites. Some people experience mild symptoms such as aches, pains, and coughing. A few develop more severe symptoms, including fever, night sweats, and weight loss.

The most serious systemic forms of histoplasmosis occur in patients with defective cell-mediated immunity such as AIDS patients. In these cases, the infection can lead to lesions in the brain, intestines, heart, liver, spleen, bone marrow, and skin. Persistent colonization of patients with emphysema and bronchitis causes *chronic pulmonary histoplasmosis,* a complication that has signs and symptoms similar to those of tuberculosis.

Transmission and Epidemiology

The organism is endemically distributed on all continents except Australia. Its highest rates of incidence occur in the eastern and central regions of the United States, especially in the Ohio Valley. This fungus appears to grow most abundantly in moist soils high in nitrogen content, especially those supplemented by bird and bat droppings **(figure 22.20).** The organism is most often transmitted from soils or the environment to humans. Human-to-human transmission has not been documented.

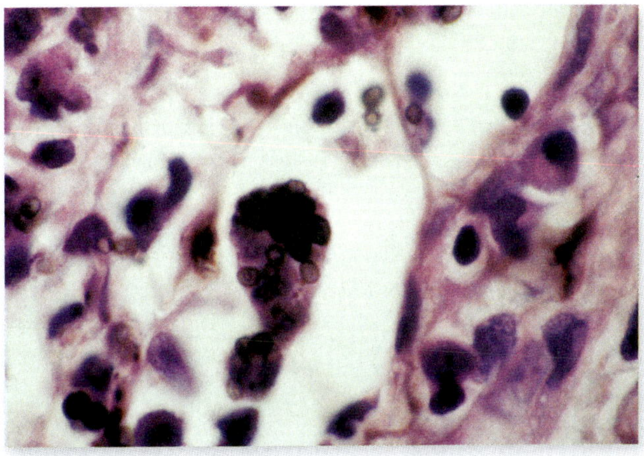

Figure 22.20 *Histoplasma* in lung tissue. The dark purple blobs are fungi inside phagocytic cells.

Dr. Cecil H Fox/Science Source/Getty Images

In high-prevalence areas such as southern Ohio, Illinois, Missouri, Kentucky, Tennessee, Michigan, Georgia, and Arkansas, 80% to 90% of the population shows signs of prior infection. Histoplasmosis incidence in the United States is estimated at about 250,000 cases per year, with several thousand of them requiring hospitalization and a small number resulting in death.

▶ Culture and/or Diagnosis

Discovering *Histoplasma* in clinical specimens is diagnostic. Usually, it appears as spherical, "fish-eye" yeasts intracellularly in macrophages and occasionally as free yeasts in samples of sputum and cerebrospinal fluid. A urine antigen test is also available.

▶ Prevention and Treatment

Avoiding the fungus is the only way to prevent this infection, and in many parts of the country this is impossible. Luckily, undetected or mild cases of histoplasmosis resolve without medical management. More severe disease calls for systemic antifungal chemotherapy. Treatment is with amphotericin B for 1 to 2 weeks, followed by several weeks of itraconazole. Surgery to remove affected masses in the lungs or other organs is sometimes also useful.

Pneumocystis jirovecii

This is a fungus that is probably normal biota in healthy people, but in severely immunocompromised persons, it can cause a serious pneumonia. It is commonly called PCP, which stands for *Pneumocystis carinii* pneumonia, because the species used to be classified as *carinii*.

▶ Symptoms, Pathogenesis, and Virulence Factors

In people with intact immune defenses, *P. jirovecii* is usually held in check by lung phagocytes and lymphocytes, but in those with

Disease Table 22.9 Community-Acquired Pneumonia

Causative Organism(s)	Rhinoviruses V	*Streptococcus pneumoniae* B G+	*Mycoplasma pneumoniae* B	*Legionella species* B G−	*Histoplasma capsulatum* F	*Pneumocystis jirovecii* F
Most Common Modes of Transmission	Droplet contact or endogenous transfer	Droplet contact or endogenous transfer	Droplet contact	Vehicle (water droplets)	Vehicle—inhalation of fungal spores in contaminated soil	Vehicle—inhalation of fungal spores
Virulence Factors	–	Capsule	Adhesins	–	Survival in phagocytes	–
Culture/ Diagnosis	Failure to find bacteria or fungi	Gram stain often diagnostic, alpha-hemolytic on blood agar	Rule out other etiologic agents; serology; PCR	Urine antigen test; culture requires selective charcoal yeast extract agar	Rapid antigen tests, microscopy	Microscopy
Prevention	Hygiene	PCV-13 or PPSV23 vaccine	No vaccine, no permanent immunity	–	Avoid soil contaminated with bird and bat droppings	Antibiotics given to AIDS patients to prevent this
Treatment	None	Doxycycline, ceftriaxone, with or without vancomycin; much resistance	Erythromycin	Fluoroquinolone, azithromycin, clarithromycin	Itraconazole	Trimethoprim-sulfamethoxazole
Distinctive Features	Usually mild	Patient usually severely ill	Usually mild; "walking pneumonia"	Mild pneumonias in healthy people; can be severe in elderly or immunocompromised	Many infections asymptomatic	Vast majority occur in patients with AIDS
Epidemiological Features	9% of CAP cases	5% of CAP cases; Drug-resistant strains in **CDC Serious Threat category**	–	United States: on the rise; 6,000 to 8,000 cases annually	In United States, 250,000 infected per year; 5% to 10% have symptoms	Almost exclusively in severely immunocompromised patients

deficient immune systems, it multiplies intracellularly and extracellularly. The massive numbers of fungi adhere tenaciously to the lung pneumocytes and cause an inflammatory condition. The lung epithelial cells slough off, and a foamy exudate builds up. Symptoms are nonspecific and include cough, fever, shallow respiration, and cyanosis (sy″-uh-noh′-sis).

▶ Transmission and Epidemiology

There is some debate about how *Pneumocystis* is acquired. Inhalation of spores is probably common. Healthy people may even harbor it as normal biota in their lungs. Contact with the agent is so widespread that in some populations, a majority of people show serological evidence of infection by the age of 3 or 4. Until the AIDS epidemic, symptomatic infections by this organism were very rare, occurring only among elderly people, premature infants, and patients that were severely debilitated or malnourished.

▶ Culture and/or Diagnosis

Definitive diagnosis requires visualizing cysts in a sputum specimen. There is a PCR test, but it is considered too sensitive, resulting in too many false positives.

▶ Prevention and Treatment

Traditional antifungal drugs are ineffective against *Pneumocystis* pneumonia because the chemical makeup of the organism's cell wall differs from that of most fungi. The primary treatment is trimethoprim-sulfamethoxazole. This combination should be administered even if disease appears mild or is only suspected. It is sometimes given to patients with low T-cell counts to prevent the disease. The airways of patients in the active stage of infection often must be suctioned to reduce the symptoms **(Disease Table 22.9).**

Healthcare-Associated Pneumonia

▶ Causative Agents

About 1% of hospitalized or institutionalized people experience the complication of pneumonia. It is most commonly associated with mechanical ventilation, via an endotracheal or tracheostomy tube. This is sometimes labeled "ventilator-associated pneumonia," or VAP. The mortality rate is quite high—between 30% and 50%. The most frequent causes of all forms of HCAP today are MRSA strains of *Staphylococcus aureus,* as well as nonresistant strains. MRSA is on the CDC's list as a **Serious Threat.** After MRSA, gram-negative bacteria are most common. These include *Klebsiella pneumoniae, Enterobacter, E. coli, Pseudomonas aeruginosa,* and *Acinetobacter.* Some strains are highly antibiotic resistant. When members of the family *Enterobacteriaceae* such as *Klebsiella* and *E. coli* are resistant to a last-line antibiotic (carbapenem), they are designated CRE and are in the **Urgent Threat** category from the CDC. Likewise, *Acinetobacter* is likely to be multidrug resistant. In those cases, it is in the **Urgent Threat** category. Further complicating matters, many cases of HCAP appear to be polymicrobial in origin—meaning that there are multiple microorganisms multiplying in the alveolar spaces.

▶ Prevention and Treatment

Because microorganisms aspirated from the upper respiratory tract cause many healthcare-associated pneumonias, measures that discourage the transfer of microbes into the lungs are very useful for preventing the condition. Elevating patients' heads to a 30- to 45-degree angle helps reduce aspiration of secretions. Good preoperative education of patients about the importance of deep breathing and frequent coughing can reduce postoperative infection rates. Proper care of mechanical ventilation and respiratory therapy equipment is essential as well.

Studies have shown that delaying antibiotic treatment of suspected healthcare-associated pneumonia leads to a greater likelihood of death. Even in this era of conservative antibiotic use, empiric therapy should be started as soon as healthcare-associated pneumonia is suspected, using multiple antibiotics that cover both gram-negative and gram-positive organisms.

Disease Table 22.10	Healthcare-Associated Pneumonia
Causative Organism(s)	Gram-negative and gram-positive bacteria from upper respiratory tract or stomach; environmental contamination of ventilator
Most Common Modes of Transmission	Endogenous (aspiration)
Virulence Factors	–
Culture/Diagnosis	Culture of lung fluids
Prevention	Elevating patient's head, preoperative education, care of respiratory equipment
Treatment	Varies by etiology
Epidemiological Features	United States: 300,000 cases per year; occurs in 0.5% to 1.0% of admitted patients; mortality rate in United States and internationally is 20% to 50%

Hantavirus Pulmonary Syndrome

In 1993, hantavirus suddenly burst into the American consciousness. A cluster of unusual cases of severe lung edema among healthy, young adults arose in the Four Corners area of New Mexico. Most of the patients died within a few days. They were later found to have been infected with hantavirus, an agent that had previously only been known to cause severe kidney disease and hemorrhagic fevers in other parts of the world. The new condition was named hantavirus pulmonary syndrome (HPS). Since 1993, the disease has occurred sporadically, but it has a mortality rate of at least 33%. It is considered an emerging disease.

▶ Symptoms, Pathogenesis, and Virulence Factors

Common features of the prodromal phase of this infection include fever, chills, myalgias (muscle aches), headache, nausea, vomiting, and diarrhea or a combination of these symptoms. A cough is common but is not a prominent early feature. Initial symptoms resemble those of other common viral infections. Soon, a severe pulmonary edema occurs and causes acute respiratory distress (acute respiratory distress syndrome [ARDS] has many microbial and nonmicrobial causes; this is only one of them).

The acute lung symptoms appear to be due to the presence of large amounts of hantavirus antigen, which becomes disseminated throughout the bloodstream (including the capillaries surrounding the alveoli of the lung). Massive amounts of fluid leave the blood vessels and flood the alveolar spaces in response to the inflammatory stimulus, causing severe breathing difficulties and a drop in blood pressure. The propensity to cause a massive inflammatory response could be considered a virulence factor for this organism.

▶ Transmission and Epidemiology

Hantavirus is transmitted via airborne dust contaminated with the urine, feces, or saliva of infected rodents. Deer mice and other rodents can harbor one of the multiple strains of hantavirus identified throughout the world today, exhibiting few apparent symptoms. Small outbreaks of the disease are usually correlated with increases in the local rodent population. In 2012, there was a small outbreak of HPS among campers at Yosemite National Park. Nine campers who stayed in special tent cabins in the park contracted the virus, and three adult males died of the infection. The tent cabins were insulated and had double walls, into which deer mice had burrowed and made their nests.

Epidemiologists suspect that rodents have been infected with this pathogen for centuries. It has no doubt been the cause of sporadic cases of unexplained acute respiratory distress and disease in humans for decades, but the incidence seems to be increasing, especially in areas of the United States west of the Mississippi River **(figure 22.21)**.

▶ Treatment and Prevention

The diagnosis is established by detection of IgM to hantavirus in the patient's blood or by using PCR techniques to find hantavirus genetic material in clinical specimens. Treatment consists mainly of supportive care. Mechanical ventilation is often required.

There is no specific treatment other than supportive care. **Disease Table 22.11.**

22.4 Learning Outcomes—Assess Your Progress

13. List the possible causative agents for each of the diseases affecting the lower respiratory tract: tuberculosis, community-acquired pneumonia, healthcare-associated pneumonia, and hantavirus pulmonary syndrome.
14. Discuss the problems associated with MDR-TB and XDR-TB.
15. Demonstrate an in-depth understanding of the current epidemiology of tuberculosis infection.
16. Explain why so many diverse microorganisms can cause the condition of pneumonia.

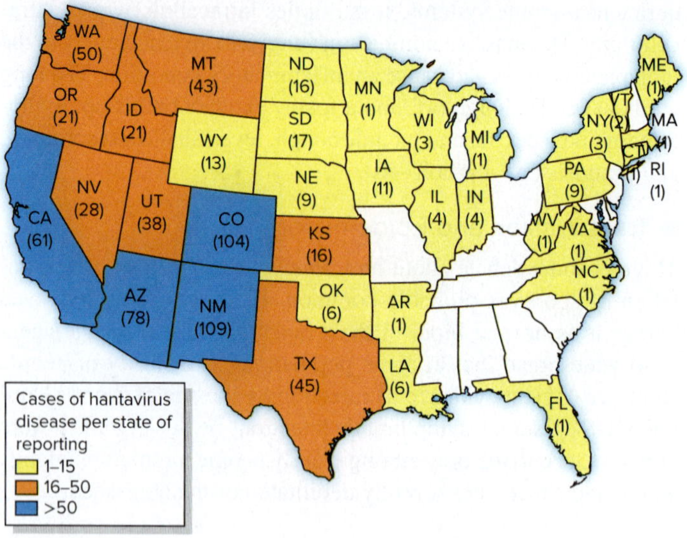
WA (50)
OR (21)
ID (21)
MT (43)
ND (16)
MN (1)
SD (17)
WY (13)
WI (3)
MI (1)
ME (1)
NY (2)
MA (1)
RI (1)
PA (9)
IA (11)
NE (9)
IL (4)
IN (4)
NV (28)
UT (38)
CA (61)
CO (104)
KS (16)
WV (1)
VA (1)
NC (1)
OK (6)
AR (1)
AZ (78)
NM (109)
TX (45)
LA (6)
FL (1)
Cases of hantavirus disease per state of reporting
1–15
16–50
>50

Figure 22.21 Distribution of hantavirus pulmonary syndrome (HPS) cases in the United States. Data for the states in white had no reported data.

17. Identify three important causes of serious community-acquired pneumonia.
18. List the distinguishing characteristics of healthcare-associated pneumonia compared to community-acquired pneumonia.
19. Outline the chain of transmission of the causative agent of hantavirus pulmonary syndrome.

Disease Table 22.11	Hantavirus Pulmonary Syndrome
Causative Organism(s)	Hantavirus V
Most Common Modes of Transmission	Vehicle—airborne virus emitted from rodents
Virulence Factors	Ability to induce inflammatory response
Culture/Diagnosis	Serology (IgM), PCR identification of antigen in tissue
Prevention	Avoid mouse habitats and droppings
Treatment	Supportive
Epidemiological Features	United States: 10 to 25 cases per year; similar rates internationally. Previously thought to be universally lethal, but now known fatality rate is 25% to 50%. Category C Bioterrorism Agent

MEDIA UNDER THE MICROSCOPE WRAP-UP

The **intended message** of the AARP article about the "man flu" was that it may well be true that men suffer more with respiratory infections than do women.

My **critical reading** of the article is that it does a decent job describing the two studies that led to this conclusion, the mouse study and the meta-analysis. It also laid the groundwork by stating that previous studies have already shown that the immune systems of men and women respond differently to pathogens.

LightField Studios/Shutterstock

I would **interpret** this article by saying that perhaps characterizing a man's infection as "man flu" is unnecessarily harsh. Then I would briefly describe the two studies.

My **overall grade** for the AARP article is an A. It provides a simple overview of the studies and includes commentary from the author of one of the original studies.

Source: Candy Sagon, "Yes, 'Man Flu' Is Real," AARP.org, December 12, 2017.

Study Smarter: Better Together

These activities are designed for you to use on your own with a study group—either a face-to-face group or a virtual one, consisting of 3–5 members. Studying together can be very helpful, but there are effective and ineffective ways to do it. For example, getting together without a clear structure is often not a good use of your time. Use your time efficiently by using one or more of the exercises below.

FACE-TO-FACE GROUPS

Use one or more of the activities below.

Peer Instruction: Assign numbers to your group members to use all semester long. Now look at these five concepts from this chapter. Each group member prepares a 5-minute lesson on the topic corresponding to their number. Don't worry if you have fewer than 5 members; just use however many you have! During your group study time, each member presents their lesson, and the group spends another 5–10 minutes discussing that lesson.

1. The spectrum of post-infection consequences (sequelae) from *Streptococcus pyogenes*
2. Defenses of the respiratory system
3. Community-acquired pneumonia vs. healthcare-associated pneumonia
4. Explanation of the "H" and "N" naming system of influenza viruses
5. Explanation of both antigenic drift and antigenic shift in influenza viruses

Concept Maps: Each member of the group should use this list of terms from this chapter to generate their own concept map. This can be hand-drawn or created using software (see Appendix C for guidelines). During group study time, compare each other's concept maps and help each other make sure they are correct. Of course, there are many different "correct" maps. Examining each member's map will help you talk through the varied concepts and how they are related.

Concept Terms:

Bordetella pertussis	pertussis toxin	FHA	cilia
mucus	tracheal cytotoxin	coughing	multiplication
endotoxin	secondary infection		

Table Topics: Each group member should identify a concept or topic from this week's class assignments with which they are having trouble and share it during group study time. The other group members can then help to clarify confusing issues or share how they figured it out. Aim for a maximum of 15 minutes per topic. If the topic remains unclear to the group, bring it up during class or use the instructor's office hours or e-mail to ask for help. Taking the time to struggle with a difficult concept first makes your questions much more specific and more likely to yield helpful answers.

VIRTUAL GROUPS

Not everyone has the time or opportunity to meet with group members outside of class time. You or your instructor can create a virtual group using e-mail or the course software.

Weekly Discussion Board: This forum can be used as a way for groups to discuss topics, via e-mail, or other learning management systems or online platforms, before they are covered in class. As each member of the group answers the current week's question, they should send their responses to every other member of their group. It's best to agree on a deadline based on how your class schedule

(continued)

works (Saturday for the next week's topics, for example). Then, after the topic is discussed in class, each member should send a response that all group members will see with a follow-up post on the same topic. If you cover more than one chapter in a week, someone can be designated to choose which chapter Discussion Board question you will use. Or simply decide up front that you will always use the first-chapter-of-the-week's question, to keep the schedule simple.

Discussion Question
Provide an explanation for the need for annual vaccination against influenza.

Deadliness and Communicability of Selected Diseases of the Respiratory System

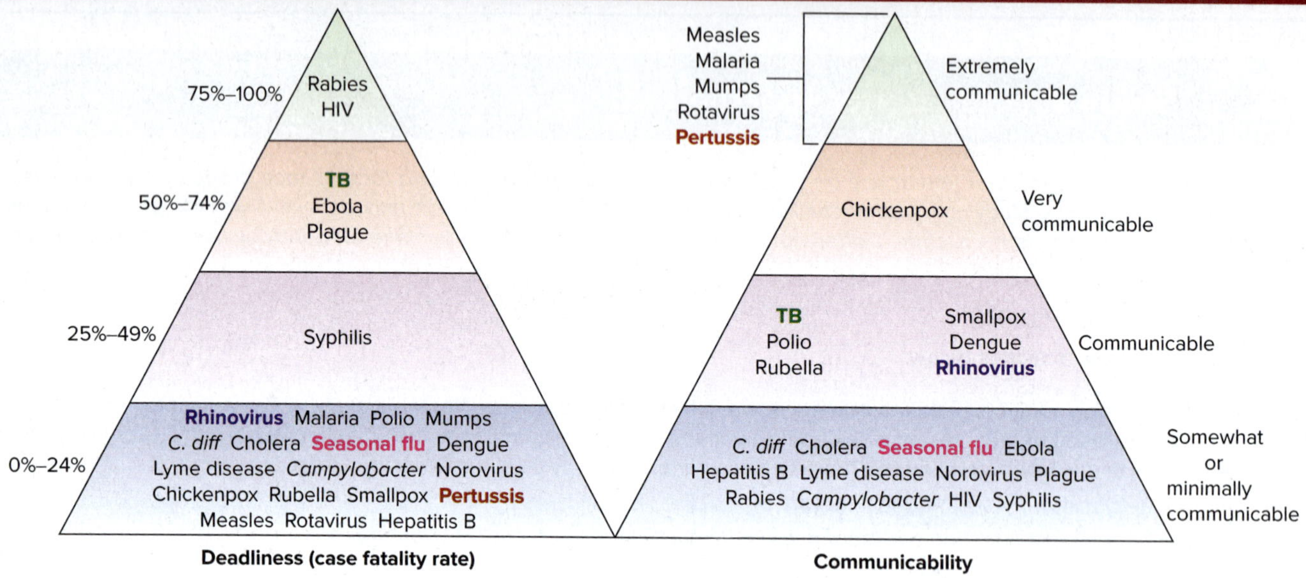
75%–100%
Rabies
HIV
50%–74%
TB
Ebola
Plague
25%–49%
Syphilis
0%–24%
Rhinovirus Malaria Polio Mumps
C. diff Cholera Seasonal flu Dengue
Lyme disease Campylobacter Norovirus
Chickenpox Rubella Smallpox Pertussis
Measles Rotavirus Hepatitis B
Deadliness (case fatality rate)
Measles
Malaria
Mumps
Rotavirus
Pertussis
Extremely communicable
Chickenpox
Very communicable
TB
Polio
Rubella
Smallpox
Dengue
Rhinovirus
Communicable
C. diff Cholera Seasonal flu Ebola
Hepatitis B Lyme disease Norovirus Plague
Rabies Campylobacter HIV Syphilis
Somewhat or minimally communicable
Communicability

Summing Up

Taxonomic Organization Microorganisms Causing Disease in the Respiratory Tract

Microorganism	Disease	Disease Table
Gram-positive bacteria		
Streptococcus pneumoniae	Otitis media, pneumonia	Otitis media, 22.3; Community-acquired pneumonia, 22.9
Streptococcus pyogenes	Pharyngitis	Pharyngitis, 22.4
Gram-negative bacteria		
Fusobacterium necrophorum	Pharyngitis	Pharyngitis, 22.4
Bordetella pertussis	Whooping cough	Pertussis (whooping cough), 22.5
Legionella spp.	Pneumonia	Community-acquired pneumonia, 22.9
Other bacteria		
*Mycobacterium tuberculosis**	Tuberculosis	Tuberculosis, 22.8
Mycoplasma pneumoniae	Pneumonia	Community-acquired pneumonia, 22.9
RNA viruses		
Respiratory syncytial virus	RSV disease	RSV disease, 22.6
Influenza virus A, B, and C	Influenza	Influenza, 22.7
Hantavirus	Hantavirus pulmonary syndrome	Hantavirus pulmonary syndrome, 22.11
Rhinoviruses	Pneumonia	Community-acquired pneumonia, 22.9
Fungi		
C. auris	Otitis media	Otitis media, 22.3
Histoplasma capsulatum	Histoplasmosis	Community-acquired pneumonia, 22.9
Pneumocystis jirovecii	*Pneumocystis* pneumonia	Community-acquired pneumonia, 22.9

*There is some debate about the Gram status of the genus *Mycobacterium*; it is generally not considered gram-positive or gram-negative.

Chapter Summary

INFECTIOUS DISEASES MANIFESTING IN THE RESPIRATORY SYSTEM

22.1 THE RESPIRATORY TRACT, ITS DEFENSES, AND NORMAL BIOTA

- The upper respiratory tract includes the mouth, nose, nasal cavity and sinus, throat (pharynx), epiglottis, and larynx. The lower respiratory tract begins with the trachea, which feeds into the bronchi, bronchioles, and alveoli in the lungs.
- The ciliary escalator and mucus are physical features of the respiratory tract's defenses. Involuntary responses such as coughing, sneezing, and swallowing also protect it. The alveoli and the tonsils contain macrophages, and IgA is present on mucous membranes.
- Normal biota is abundant and includes organisms from several bacterial genera: *Prevotella, Sphingomonas, Pseudomonas, Acinetobacter, Fusobacterium, Staphylocccus,* and *Streptococcus.* The fungus *Candida* is also a normal inhabitant.

22.2 INFECTIOUS DISEASES MANIFESTING IN THE UPPER RESPIRATORY TRACT

- **The common cold**—Caused by over 200 different viruses, symptoms from body's response.
- **Sinusitis**—Inflammatory condition most commonly caused by allergy or by a variety of bacteria, viruses, or fungi.
- **Acute otitis media**—Most common cause is *Streptococcus pneumoniae*, though many infections are polymicrobial.
- **Pharyngitis**—Usually the same viruses causing the common cold cause a sore throat. However, *Streptococcus pyogenes* and *Fusobacterium necrophorum* are potentially serious causes. Untreated streptococcal throat infections can result in scarlet fever, rheumatic fever, glomerulonephritis, and necrotizing fasciitis. *Fusobacterium* infections can be deadly.

22.3 INFECTIOUS DISEASES MANIFESTING IN BOTH THE UPPER AND LOWER RESPIRATORY TRACTS

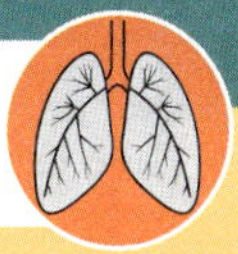

- **Whooping cough**—Causative agent, *Bordetella pertussis,* releases exotoxins—pertussis toxin and tracheal cytotoxin—that damage ciliated respiratory epithelial cells and cripple other components of host defense.
- **Respiratory syncytial virus disease**—Infects the respiratory tract and produces giant, multinucleate cells (syncytia). RSV is the most prevalent cause of respiratory disease in newborns.
- **Influenza**—Caused by one of three influenza viruses, A, B, or C. The ssRNA genome is subject to constant genetic changes that alter the structure of its envelope glycoprotein. Antigenic drift and antigenic shift change the body's ability to recognize and protect against the virus.

22.4 INFECTIOUS DISEASES MANIFESTING IN THE LOWER RESPIRATORY TRACT

- **Tuberculosis**—Primary cause is *Mycobacterium tuberculosis.* Vaccine not used in the United States, although an attenuated vaccine, BCG, is used in many countries. MDR-TB and XDR-TB: Multidrug-resistant and extensively drug-resistant tuberculosis are widespread in the world and have worse outcomes than antibiotic-sensitive TB.
- **Pneumonia**—Inflammatory condition of the lung in which fluid fills the alveoli. Caused by a wide variety of microorganisms. The differential diagnoses for community-acquired and healthcare-associated pneumonia are distinct.
- **Hantavirus pulmonary syndrome**—Causes a form of severe acute respiratory distress. Spread in mouse excretions.

INFECTIOUS DISEASES AFFECTING

The Respiratory Tract

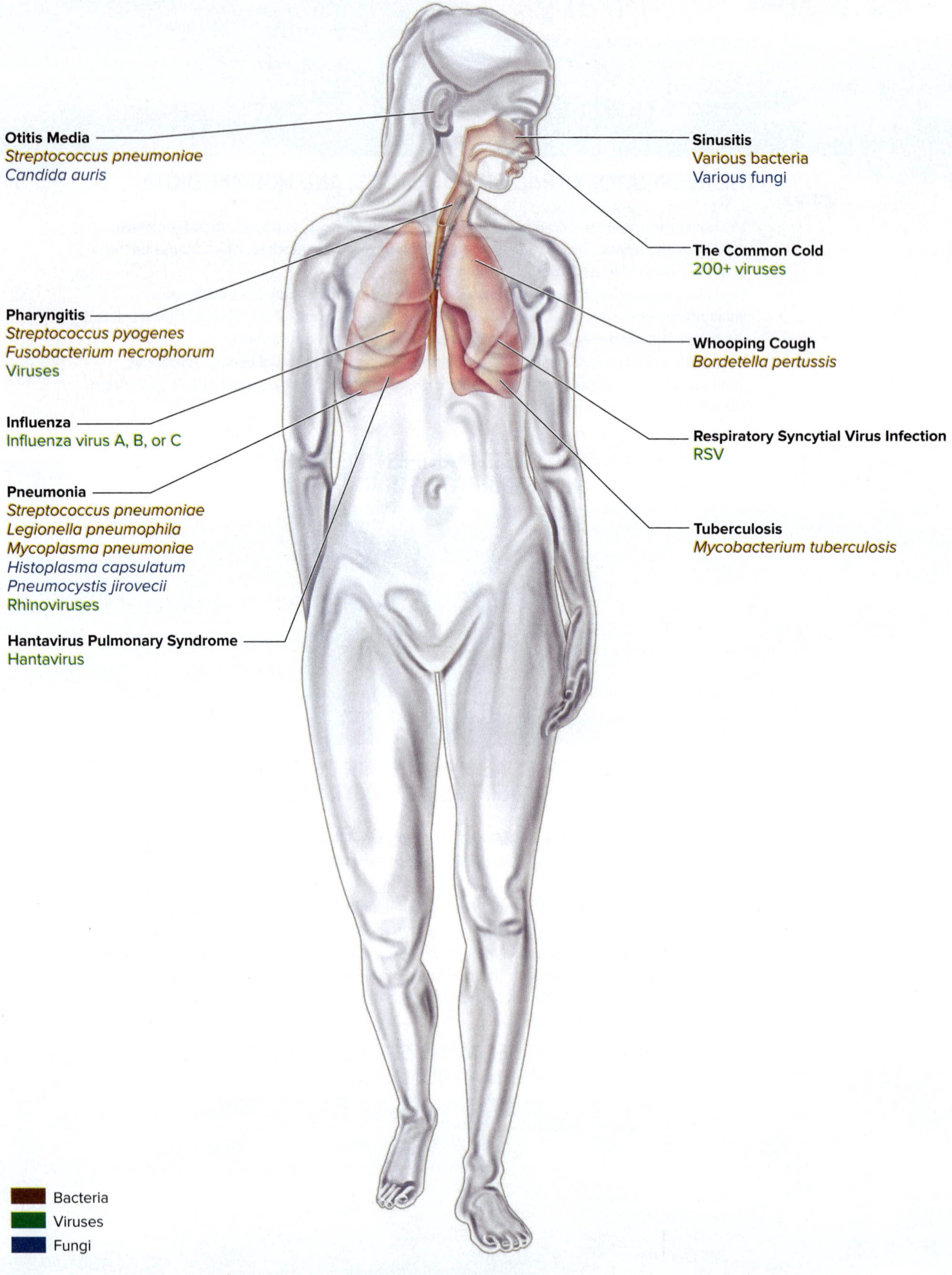

Otitis Media
Streptococcus pneumoniae
Candida auris
Sinusitis
Various bacteria
Various fungi
The Common Cold
200+ viruses
Pharyngitis
Streptococcus pyogenes
Fusobacterium necrophorum
Viruses
Whooping Cough
Bordetella pertussis
Influenza
Influenza virus A, B, or C
Respiratory Syncytial Virus Infection
RSV
Pneumonia
Streptococcus pneumoniae
Legionella pneumophila
Mycoplasma pneumoniae
Histoplasma capsulatum
Pneumocystis jirovecii
Rhinoviruses
Tuberculosis
Mycobacterium tuberculosis
Hantavirus Pulmonary Syndrome
Hantavirus
Bacteria
Viruses
Fungi

System Summary Figure 22.22

SmartGrid: From Knowledge to Critical Thinking

This *21 Question Grid* takes the topics from this chapter and arranges them with respect to the American Society for Microbiology's Undergraduate Curriculum guidelines—all six of the important "Concepts" as well as the important "Competency" of scientific literacy. Three questions are supplied, which cover chapter content referring to the Concept or Competency in increasing levels of Bloom's taxonomy for learning.

ASM Concept/ Competency	A. Bloom's Level 1, 2—Remember and Understand (Choose one)	B. Bloom's Level 3, 4—Apply and Analyze	C. Bloom's Level 5, 6—Evaluate and Create
Evolution	**1.** Which of the following organisms is most problematic with respect to its resistance to antibiotics? **a.** *Streptococcus pyogenes* **b.** *Haemophilus influenzae* **c.** *Mycobacterium tuberculosis* **d.** *Bordetella pertussis*	**2.** Explain why an antigenic shift event in the influenza virus can lead to a catastrophic pandemic.	**3.** It has recently been established that the lower respiratory tract—the lungs—has a microbiome. Use what you know about human evolution to explain why this is not surprising, and how it is probably highly advantageous to health.
Cell Structure and Function	**4.** Which of these microbes has an unusual, waxy wall structure that contributes to its virulence? **a.** *Streptococcus pyogenes* **b.** *Haemophilus influenzae* **c.** *Mycobacterium tuberculosis* **d.** *Bordetella pertussis*	**5.** What cell structure of *Streptococcus pneumoniae* makes it such a successful and versatile pathogen, in your estimation? Defend your answer.	**6.** Scientists have worked for years to create a vaccine for the common cold. One effort attempts to create a vaccine that is targeted to the host epithelial cell instead of viral antigens. What is the logic behind this approach? What antigens would you include in such a vaccine?
Metabolic Pathways	**7.** The beta-hemolysis of blood agar observed with *Streptococcus pyogenes* is due to the presence of **a.** streptolysin. **b.** M protein. **c.** hyaluronic acid. **d.** a capsule.	**8.** Would you expect the microbiome in the lung to be aerobes, strict anaerobes, or facultative anaerobes? Explain your answer.	**9.** In some diseases and outbreaks of respiratory diseases caused by certain strains of hantavirus and influenza, healthy young adults suffer the most pathology and mortality. This is an unusual pattern; it is more common for younger and older vulnerable patients to suffer the worst effects of infectious diseases. Speculate on what is happening when the most robust suffer the most pathology.
Information Flow and Genetics	**10.** Which of the following pathogens is never treated with antibiotics? **a.** *Streptococcus pneumoniae* **b.** *Mycobacterium tuberculosis* **c.** *Bordetella pertussis* **d.** *Respiratory syncytial virus*	**11.** Explain why we are advised to get a flu vaccine every year. Then explain why there may be years when we are advised to get two flu vaccines.	**12.** The vaccine for pertussis is an acellular vaccine. We have learned that current circulating strains of *B. pertussis* may not "match" the vaccine as well as previous strains did. Suggest some molecular events that would have led to this phenomenon.
Microbial Systems	**13.** Which of the following infections often has/have a polymicrobial cause? **a.** otitis media **b.** healthcare-associated pneumonia **c.** sinusitis **d.** all of these	**14.** Outline how a progression from influenza infection to pneumonia to death might occur, focusing on the microbes responsible and the symptoms that ensue.	**15.** Explain why it is unlikely that we will ever see a vaccine for sinusitis.

(*continued*)

ASM Concept/ Competency	A. Bloom's Level 1, 2—Remember and Understand (Choose one)	B. Bloom's Level 3, 4—Apply and Analyze	C. Bloom's Level 5, 6—Evaluate and Create
Impact of Microorganisms	**16.** The vast majority of pneumonias caused by this organism occur in AIDS patients. **a.** hantavirus **b.** *Histoplasma capsulatum* **c.** *Pneumocystis jirovecii* **d.** *Mycoplasma pneumoniae*	**17.** A distraught mother recently posted on Facebook that although she had not been sick since the birth of her child 3 months ago, she was responsible for her baby being in the ICU with whooping cough. Explain this (accurate) assessment.	**18.** Conducting your own research, describe the total impact on human population from the two diseases malaria and tuberculosis.
Scientific Thinking	**19.** "Watchful waiting" is often the recommended response to which infection? **a.** otitis media **b.** tuberculosis **c.** hantavirus pulmonary syndrome **d.** healthcare-associated pneumonia	**20.** Why might it be inadvisable to use tuberculin skin testing to determine the TB status of a person who is HIV+?	**21.** Explain everything we can learn from figure 22.11, the PIC surveillance graph. Be sure to explain the meaning of the black lines as well as the red.

Answers to the multiple-choice questions appear in Appendix A.

Visual Connections

This question uses visual images to connect content within and between chapters.

1. **From this chapter, figure 22.10.** Find the latest data for worldwide deaths due to COVID-19 and draw a bar to the right of this graph to represent that.

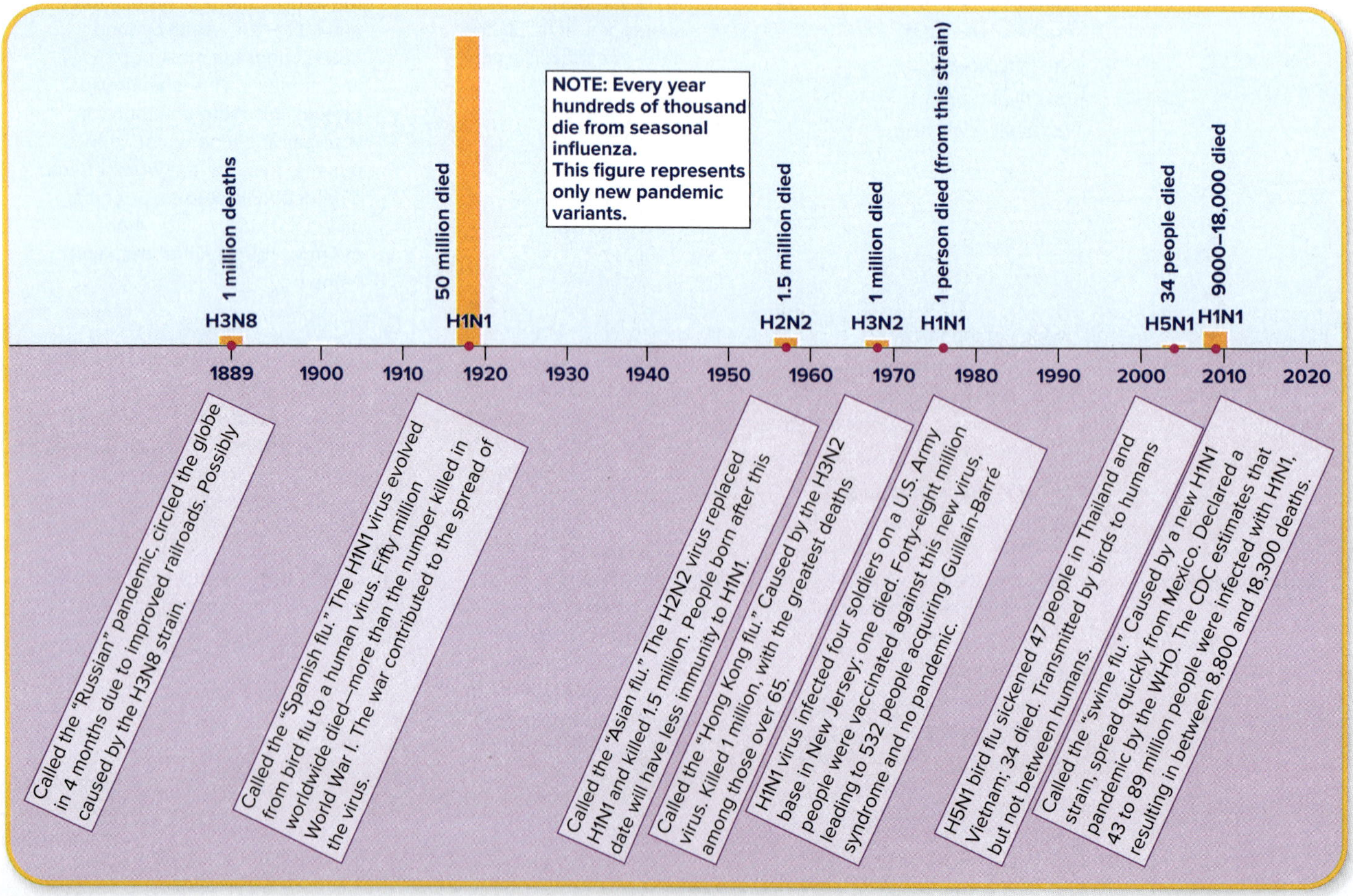

NOTE: Every year hundreds of thousand die from seasonal influenza. This figure represents only new pandemic variants.
1 million deaths
H3N8
50 million died
H1N1
1.5 million died
H2N2
1 million died
H3N2
1 person died (from this strain)
H1N1
34 people died
H5N1
9000–18,000 died
H1N1
1889
1900
1910
1920
1930
1940
1950
1960
1970
1980
1990
2000
2010
2020
Called the "Russian" pandemic, circled the globe in 4 months due to improved railroads. Possibly caused by the H3N8 strain.
Called the "Spanish flu." The H1N1 virus evolved from bird flu to a human virus. Fifty million worldwide died—more than the number killed in World War I. The war contributed to the spread of the virus.
Called the "Asian flu." The H2N2 virus replaced H1N1 and killed 1.5 million. People born after this date will have less immunity to H1N1.
Called the "Hong Kong flu." Caused by the H3N2 virus. Killed 1 million, with the greatest deaths among those over 65.
H1N1 virus infected four soldiers on a U.S. Army base in New Jersey; one died. Forty-eight million people were vaccinated against this new virus, leading to 532 people acquiring Guillain-Barré syndrome and no pandemic.
H5N1 bird flu sickened 47 people in Thailand and Vietnam; 34 died. Transmitted by birds to humans but not between humans.
Called the "swine flu." Caused by a new H1N1 strain, spread quickly from Mexico. Declared a pandemic by the WHO. The CDC estimates that 43 to 89 million people were infected with H1N1, resulting in between 8,800 and 18,300 deaths.

High Impact Study

These terms and concepts are most critical for your understanding of this chapter—and may be the most difficult. Have you mastered them? In these disease chapters, the terms and concepts help you identify what is important in a different way than the comprehensive details in the Disease Tables. Your instructor will help you understand what is important for your class.

Concepts

- ☐ Defenses of respiratory system
- ☐ Normal microbiota of respiratory system
- ☐ How 200 viruses can cause the common cold
- ☐ The spectrum of sequelae from *Streptococcus pyogenes* infection
- ☐ Three stages of pertussis
- ☐ The "HxNx" naming system of influenza viruses
- ☐ Seasonal flu vs. pandemic flu
- ☐ Community-acquired vs. healthcare-associated pneumonia
- ☐ Methods of transmission of the various community-acquired and healthcare-associated pneumonias
- ☐ Organisms in this chapter for which there are vaccines available
- ☐ Organisms in this chapter that display significant antibiotic resistance

Terms

- ☐ Streptolysins
- ☐ Antigenic drift
- ☐ Antigenic shift
- ☐ MDR-TB
- ☐ XDR-TB

Design Element: (College students): Caia Image/Image Source

23

Infectious Diseases Manifesting in the Gastrointestinal Tract

Gerhard Zwerger-Schoner/imageBROKER/Alamy Stock Photo

MEDIA UNDER THE MICROSCOPE

Ancient Stomach Ulcer?

This opening case examines an article from the popular media to determine the extent to which it is factual and/or misleading. This case focuses on a recent Washington Post *article, "New Study on Otzi the Iceman Reveals Humanity's Intimate Affair with One Microbe."*

The Iceman is a mummified, well-preserved 5,300-year-old man who was found in the Alps in 1991. (The photo above is of the place where he was found.) For the past 33 years, scientists have been studying his clothing, the way he died, and his tattoos to get a taste of what life was like in the Copper Age. This *Washington Post* story reports on a new detail: the genetic details of one particular bacterium found in his stomach—*Helicobacter pylori.* You may think, "Wow! Five thousand years sounds like a long time for a bacterium that is still colonizing humans to have been around." In fact, *H. pylori* has been associated with humans for at least 100,000 years, according to the article. And at least half of humans alive today are infected with it.

The article's main point is that the unique genetic makeup of the *H. pylori* gives us important clues about who the Iceman interacted with, where he had been, and so on. *Helicobacter* strains from different people mingle together when people have direct contact and even when they are just engaged in activities together. It is transmitted fecal-orally, so, particularly in Iceman's time, there was a lot of *Helicobacter* exchange. The strains commingle in the gut and exchange genetic elements. With enough work, scientists can unravel the lineages of the strains. Because bacteria mutate so often and exchange DNA so well, they can provide more finely tuned information than can studying human genes. The *Helicobacter* DNA in Otzi's stomach was not consistent with European strains, even though he had died in Europe, but rather contained DNA from Asian and African strains. This was a clear indication that the Iceman, and probably other hominids of the time, intermingled and traveled extensively.

The scientists are careful to mention that this was a study size of 1. But they called it "a nice, solid, data point."

- What is the **intended message** of the article?
- What is your **critical reading** of the summary of the article provided above? Remember that in this context, "critical reading" does not necessarily mean *What criticism do you have?* but asks you to apply your knowledge to interpret whether the article is factual and whether the facts support the intended message.
- How would you **interpret** the news item for your nonmicrobiologist friends?
- What is your **overall grade** for the news item—taking into account its accuracy and the accuracy of its intended effect?

Media Under The Microscope Wrap-Up appears at the end of the chapter.

Outline and Learning Outcomes

23.1 The Gastrointestinal Tract, Its Defenses, and Normal Biota

1. Draw or describe the anatomical features of the gastrointestinal tract.
2. List the natural defenses present in the gastrointestinal tract.
3. List the types of normal biota presently known to occupy the various regions of the gastrointestinal tract.
4. Summarize the known functions of the gastrointestinal microbiota and the role they may play in disease development.

23.2 Infectious Diseases Manifesting in the Gastrointestinal Tract (Nonhelminthic)

5. List the possible causative agents for the following infectious gastrointestinal conditions: dental caries, periodontal diseases, mumps, and gastric ulcers.
6. Name nine bacterial and three nonbacterial causes of acute diarrhea, and identify the most common cause of foodborne illness in the United States.
7. Name one distinct feature for each of the acute diarrhea pathogens.
8. Differentiate between food poisoning and foodborne infection.
9. Identify three causative agents for chronic diarrhea.
10. Differentiate among the main types of hepatitis and discuss causative agents, modes of transmission, diagnostic techniques, prevention, and treatment of each.

23.3 Helminthic Diseases Manifesting in the Gastrointestinal Tract

11. Describe some distinguishing characteristics and commonalities seen in helminthic infections.
12. List four helminths that cause primarily intestinal symptoms, and identify which life cycle each follows and one unique fact about each helminth.
13. List three helminths that cause intestinal symptoms that may be accompanied by migratory symptoms, identifying which life cycle each follows and one unique fact about each helminth.
14. Identify the most dangerous outcome of *Taenia solium* infection.
15. List the modes of transmission for each of the helminthic infections resulting in liver and intestinal symptoms. These are infections caused by *Opisthorchis sinensis, Clonorchis sinensis,* and *Fasciola hepatica.*
16. Describe the type of disease caused by *Trichinella* species.
17. Diagram the life cycle of *Schistosoma mansoni* and *S. japonicum,* and describe the importance of these organisms in world health.

23.1 The Gastrointestinal Tract, Its Defenses, and Normal Biota

The gastrointestinal (GI) tract can be thought of as a long tube, extending from mouth to anus. It is a very sophisticated delivery system for nutrients, composed of *eight* main sections and *four* accessory organs. The eight sections are the mouth, pharynx, esophagus, stomach, small intestine, large intestine, rectum, and anus. Along the way, the salivary glands, liver, gallbladder, and pancreas add digestive fluids and enzymes to assist in digesting and processing the food we take in **(figure 23.1).** The GI tract is often called the *digestive tract* or the *enteric tract.*

The GI tract has a very heavy load of microorganisms, and it encounters millions of new ones every day. Because of this, defenses against infection are extremely important. All intestinal surfaces are coated with a layer of mucus, which provides mechanical protection. Secretory IgA can also be found on most intestinal surfaces. The muscular walls of the GI tract keep food (and microorganisms) moving through the system through the action of peristalsis. Various fluids in the GI tract have antimicrobial properties. Saliva contains the antimicrobial proteins lysozyme and lactoferrin. The stomach fluid is antimicrobial by virtue of its extremely low pH. Bile is also antimicrobial.

The entire system is outfitted with cells of the immune system, collectively called gut-associated lymphoid tissue (GALT). The tonsils and adenoids in the oral cavity and pharynx, small areas of lymphoid tissue in the esophagus, Peyer's patches in the small intestine, and the appendix are all packets of lymphoid tissue

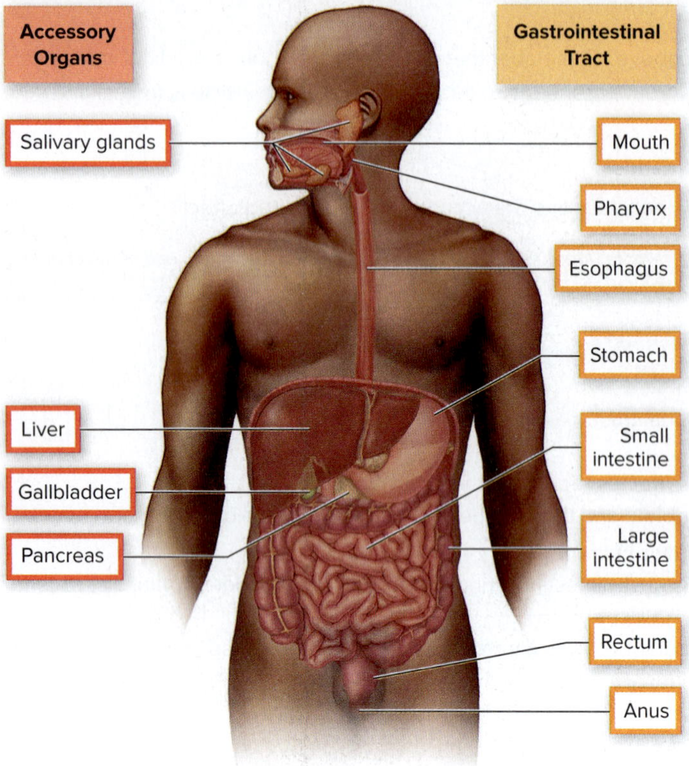
Accessory Organs
Gastrointestinal Tract
Salivary glands
Mouth
Pharynx
Esophagus
Stomach
Liver
Small intestine
Gallbladder
Pancreas
Large intestine
Rectum
Anus

Figure 23.1 Major organs of the digestive system.

consisting of T and B cells as well as cells of innate immunity. One of their jobs is to produce IgA, but they perform a variety of other immune functions.

The GI tract is home to a very large variety of normal biota. Every portion of it has a distinct microbial population. Antonie van Leeuwenhoek was one of the first to observe and describe microbes in the mouth, by observing tooth scrapings using his newly developed lenses. Since then, the oral cavity has been found to harbor a vast number and variety of microorganisms, which is not surprising, as it is in constant contact with the external environment. Breast-fed babies obtain at least part of their oral microbiome from breast milk, which contains abundant normal biota from the mother. Other babies pick up their oral biota from contact with their caregivers. The predominant bacterial types appear to be *Prevotella, Treponema, Streptococcus, Actinomyces, Neisseria, Veillonella,* and *Lactobacillus* species. Numerous species of normal biota bacteria live on the teeth in large accumulations called dental plaque, which is a type of biofilm.

Although species from the domain Bacteria clearly dominate the oral microbiome, methane-producing archaea have also been identified and may play a role in certain diseases. Research has also identified 85 fungal genera in the oral cavity, with *Candida albicans* being the most common member. A few protozoa (*Trichomonas tenax, Entamoeba gingivalis*) also call the mouth home. The extent of the oral virome (human viruses and bacteriophages) is still being uncovered.

Both the esophagus and stomach were thought to be sterile portions of the GI tract, due to the presence of physical and chemical barriers as well as immune defenses. Recent genetic analysis of these regions, however, has revealed the presence of nearly 200 different species of microorganisms. The most common types belong to the Firmicutes (*Streptococcus, Staphylococcus, Clostridium,* and *Bacillus* species). Though some of these are likely to be "just passing through," many of the microbes are true colonizers.

The large intestine has always been known to be a haven for billions of microorganisms (10^{11} per gram of contents), including the bacteria *Bacteroides, Fusobacterium, Bifidobacterium, Clostridium, Streptococcus, Peptostreptococcus, Lactobacillus, Escherichia,* and *Enterobacter;* the fungus *Candida;* and several protozoa. Researchers have also found archaeal species there. Recent studies have identified distinct overall gut microbiota profiles, what they call "enterotypes," in humans. This new evidence may pave the way for the development of personalized digestive medicine for purposes of prevention, diagnosis, and patient treatment.

Currently, the accessory organs of the GI tract (salivary glands, gallbladder, liver, and pancreas) are considered to be free of a natural microbiome but can be exposed to microbes when normal barriers in the gut are disrupted by a condition broadly called dysbiosis. This simply refers to an unhealthy mix of gut microbes in the intestinal tract. It can result in leakage of bacteria or their metabolic products into internal organs. Research has shown that gut dysbiosis can play a role in various diseases in the accessory organs.

The normal gut biota provide a protective function and can "teach" our immune system to react appropriately to microbial antigens. They also perform other jobs as well, such as aiding in digestion or providing nutrients that we cannot produce ourselves. *E. coli,* for instance, synthesizes vitamin K. Its mere presence in the large intestine seems to be important for the proper formation of epithelial cell structure. One thing is becoming clear: A diverse gut microbiome is associated with health. When the gut microbiome loses its diversity, deviations from gastrointestinal—and systemic—health can occur. Disruptions may come from antibiotic treatment, illness, pregnancy, or dietary changes. This can result in such diverse conditions as obesity, diabetes, or other seemingly noninfectious disorders. You recall from the chapter about the nervous system that it is now clear that the gut microbiome influences the development of the nervous system, including portions of the brain, and can have effects on behavior and neurological wellness. Within the gut itself, there are some obvious conditions that seem to be influenced by the gut microbiome, such as inflammatory bowel disease and Crohn's disease. **Insight 23.1** tells the story of how difficult it is to tease out causation as opposed to association with respect to an altered gut microbiome.

23.1 Learning Outcomes—Assess Your Progress

1. Draw or describe the anatomical features of the gastrointestinal tract.
2. List the natural defenses present in the gastrointestinal tract.
3. List the types of normal biota presently known to occupy the various regions of the gastrointestinal tract.
4. Summarize the known functions of the gastrointestinal microbiota and the role they may play in disease development.

INSIGHT 23.1 MICROBIOME: Crohn's Disease and the Gut Microbiome

Throughout this book, you have read about the importance of the gut microbiome and its possible influence on mood, body weight, autoimmune diseases, and many other things. Let us examine for a moment the role the gut microbiome might play in a mostly gut-focused disease: Crohn's disease.

Physicians consider Crohn's to be idiopathic, meaning they are not sure what causes it. Genetic analysis of Crohn's sufferers' DNA compared to that of healthy subjects shows some difference in some Crohn's patients, but not all. It would not be unreasonable to expect the microbiome to have some influence on gut health.

The important question is, if there *is* a difference in microbiota between Crohn's patients and healthy people, is that what causes the Crohn's? Or does Crohn's pathology lead to a different microbiome? Which came first?

Many studies have been conducted in an effort to characterize the microbiome in these two circumstances (Crohn's and health). So far, what has been found is that there are fewer different types of microbes in Crohn's patients' than in healthy guts. This decrease in diversity is seen in many different health conditions, but in only a few of them has it been found to *precede* the condition. So the question remains. And all we can say is that there is an *association*—a term that does not imply which causes which.

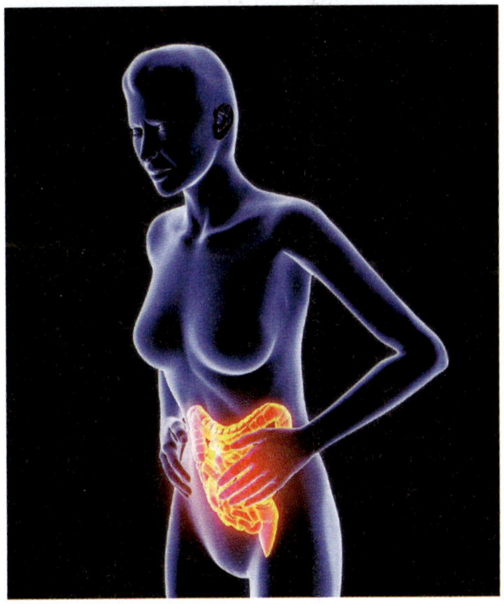

Roger Harris/Science Photo Library RF/Science Source

Defenses and Normal Biota of the Gastrointestinal Tract

	Defenses	Normal Biota
Oral Cavity	Saliva, sIgA, lysozyme, tonsils, adenoids	*Prevotella, Treponema, Streptococcus, Actinomyces, Neisseria, Veillonella, Lactobacillus*
Rest of GI Tract	GALT, lymphoid tissue, Peyer's patches, appendix, sIgA, rich normal biota	Esophagus, stomach: *Streptococcus, Staphylococcus, Clostridium, Bacillus* Large intestine: *Bacteroides, Fusobacterium, Bifidobacterium, Clostridium, Streptococcus, Peptostreptococcus, Lactobacillus, Escherichia,* and *Enterobacter; Candida;* and protozoa

23.2 Infectious Diseases Manifesting in the Gastrointestinal Tract (Nonhelminthic)

Tooth and Gum Infections

It is difficult to pinpoint exactly when the "normal biota biofilm" described for the oral environment becomes a "pathogenic biofilm." If left undisturbed, the biofilm structure eventually contains anaerobic bacteria that can damage the soft tissues and bones (referred to as the periodontium) surrounding the teeth. Also, the introduction of carbohydrates to the oral cavity can result in breakdown of hard tooth structure (the dentition) due to the production of acid by certain bacteria in the biofilm. These two circumstances are discussed separately in the following sections.

Dental Caries (Tooth Decay)

Dental caries, or tooth decay, is the most common infectious disease of human beings. The process of decay involves the dissolution of solid tooth surface due to the metabolic action of bacteria. (**Figure 23.2** depicts the structure of a tooth.) The symptoms are often not noticeable but range from minor disruption in the outer (enamel) surface of the tooth to complete destruction of the enamel and then destruction of deeper layers **(figure 23.3).** Deeper lesions can result in infection to the soft tissue inside the tooth, called the pulp, which contains blood vessels and nerves. These deeper infections lead to pain, referred to as a "toothache."

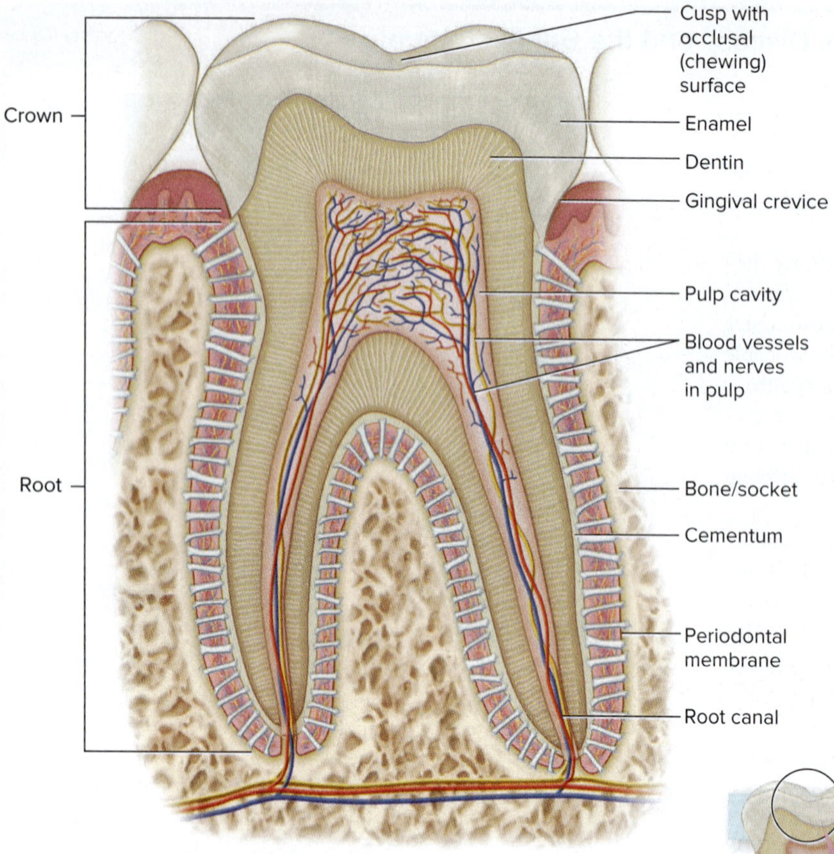
Cusp with occlusal (chewing) surface
Crown
Enamel
Dentin
Gingival crevice
Pulp cavity
Blood vessels and nerves in pulp
Root
Bone/socket
Cementum
Periodontal membrane
Root canal

Figure 23.2 The anatomy of a tooth.

Causative Agents

Two representatives of oral alpha-hemolytic streptococci, *Streptococcus mutans* and *Streptococcus sobrinus,* seem to be the main causes of dental caries, although a mixed species consortium, consisting of other *Streptococcus* species and some lactobacilli, is probably the best route to caries. A specific condition called *early childhood caries* may also be caused by a newly identified species, *Scardovia wiggsiae*. Note that in the absence of dietary carbohydrates, bacteria do not cause decay.

Pathogenesis and Virulence Factors

In the presence of sucrose and, to a lesser extent, other carbohydrates, *S. mutans* and other streptococci produce sticky polymers of glucose called fructans and glucans. These adhesives help bind them to the smooth enamel surfaces and contribute to the sticky bulk of the plaque biofilm **(figure 23.4).** If mature plaque is not removed from sites that readily trap food, it can result in a carious lesion. This is due to the action of the streptococci and other bacteria that produce acid as they ferment the carbohydrates. If the acid is immediately flushed from the plaque and diluted in the mouth, it has little effect. However, in the denser regions of plaque, the acid can accumulate in direct contact with the enamel surface and lower the pH to below 5, which is acidic enough to begin to dissolve (decalcify) the calcium phosphate of the enamel in that spot. This initial lesion can remain localized in the enamel and can be repaired with various inert materials (fillings). Once the deterioration has reached the level of the dentin, tooth destruction speeds up and the tooth can be rapidly destroyed. Exposure of the pulp leads to severe tenderness and toothache, and the chance of saving the tooth is diminished.

Teeth become vulnerable to caries as soon as they appear in the mouth at around 6 months of age. Early childhood caries, defined as caries in a child between birth and 6 years of age, can extensively damage a child's primary teeth and affect the proper eruption of the permanent teeth. The practice of putting a baby down to nap with a bottle of fruit juice or formula can lead to rampant dental caries in the vulnerable primary teeth. This condition is called *nursing bottle caries*.

Transmission and Epidemiology

The bacteria that cause dental caries are transmitted to babies and children by their close contacts,

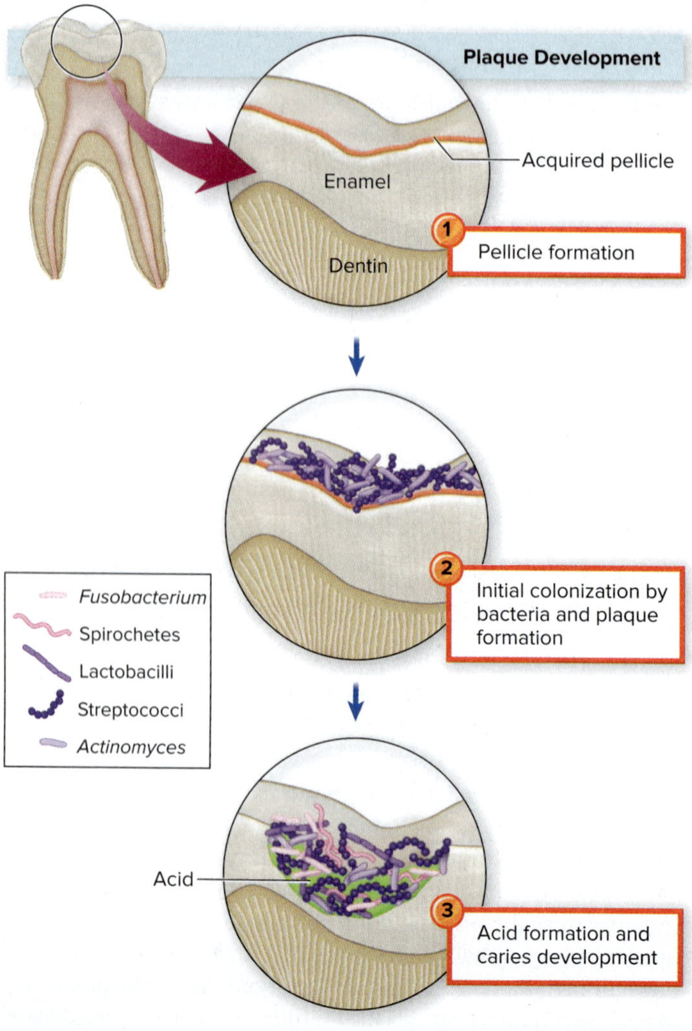
Plaque Development
Acquired pellicle
Enamel
Dentin
1 Pellicle formation
2 Initial colonization by bacteria and plaque formation
Fusobacterium
Spirochetes
Lactobacilli
Streptococci
Actinomyces
Acid
3 Acid formation and caries development

Figure 23.3 Stages in plaque development and cariogenesis.

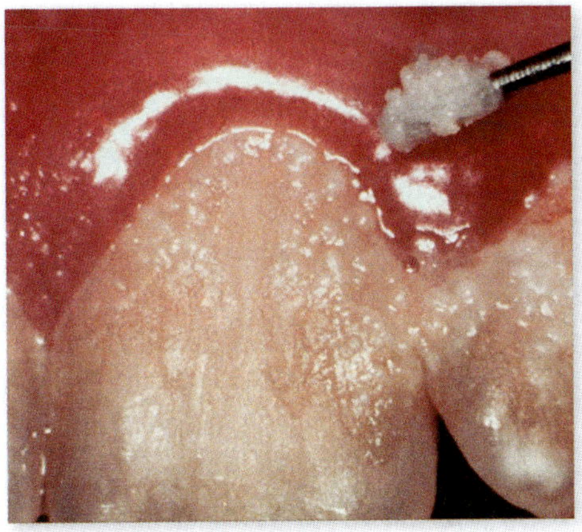

(a)

(b)

Figure 23.4 The macroscopic and microscopic appearance of plaque. **(a)** Heavy plaque accumulations at the junction of the tooth and gingiva. **(b)** Scanning electron micrograph of the plaque biofilm with long, filamentous forms and plumper coccobacilli (blue).

(a) BSIP SA/Alamy Stock Photo; (b) Steve Gschmeissner/SPL/Getty Images

especially the mother or closest caregiver. There is evidence for transfer of oral bacteria between children in day care centers as well. Although it was previously believed that humans do not acquire *S. mutans* or *S. sobrinus* until the eruption of teeth in the mouth, it now seems likely that both of these species may survive in the infant's oral cavity prior to appearance of the first teeth.

Dental caries has a worldwide distribution. Its incidence varies according to many factors, including amount of carbohydrate consumption, hygiene practices, and host genetic factors. Susceptibility to caries generally decreases with age, possibly due to the fact that grooves and fissures—common sites of dental caries—tend to become more shallow as teeth are worn down.

▶ Culture and/or Diagnosis

Dental professionals diagnose caries based on the tooth condition. Culture of the lesion is not routinely performed.

▶ Prevention and Treatment

The best way to prevent dental caries is through dietary restriction of sucrose and other refined carbohydrates. Regular brushing and flossing to remove plaque are also important. Most municipal communities in the United States add trace amounts of fluoride to their drinking water because fluoride, when incorporated into the tooth structure, can increase tooth (as well as bone) hardness. Fluoride can also encourage the remineralization of teeth that have begun the demineralization process. These and other proposed actions of fluoride could make teeth less susceptible to decay. Fluoride is also added to toothpastes and mouth rinses and can be applied in gel form. Many European countries do not fluoridate their water due to concerns over additives in drinking water, and the same controversy exists in some segments of the American population.

Treatment of a carious lesion involves removal of the affected part of the tooth (or the whole tooth, in the case of advanced caries), followed by restoration of the tooth structure with an artificial material.

Disease Table 23.1	Dental Caries
Causative Organism(s)	A polymicrobial mixture of acid-producing bacteria B
Most Common Modes of Transmission	Direct contact
Virulence Factors	Adhesion, acid production
Culture/Diagnosis	–
Prevention	Oral hygiene, fluoride supplementation
Treatment	Removal of diseased tooth material
Epidemiological Features	Globally, 60% to 90% prevalence in school-age children

Periodontal Disease

Periodontal disease is so common that 97% to 100% of the population have some manifestation of it by age 45. Most kinds are due to bacterial colonization and varying degrees of inflammation that occur in response to gingival damage.

Periodontitis

▶ Signs and Symptoms

The initial stage of periodontal disease is **gingivitis,** the signs of which are swelling, loss of normal gingival contour, patches of redness, and increased bleeding of the gingiva. There is a natural space or pocket between the tooth and the gingiva that is part of the

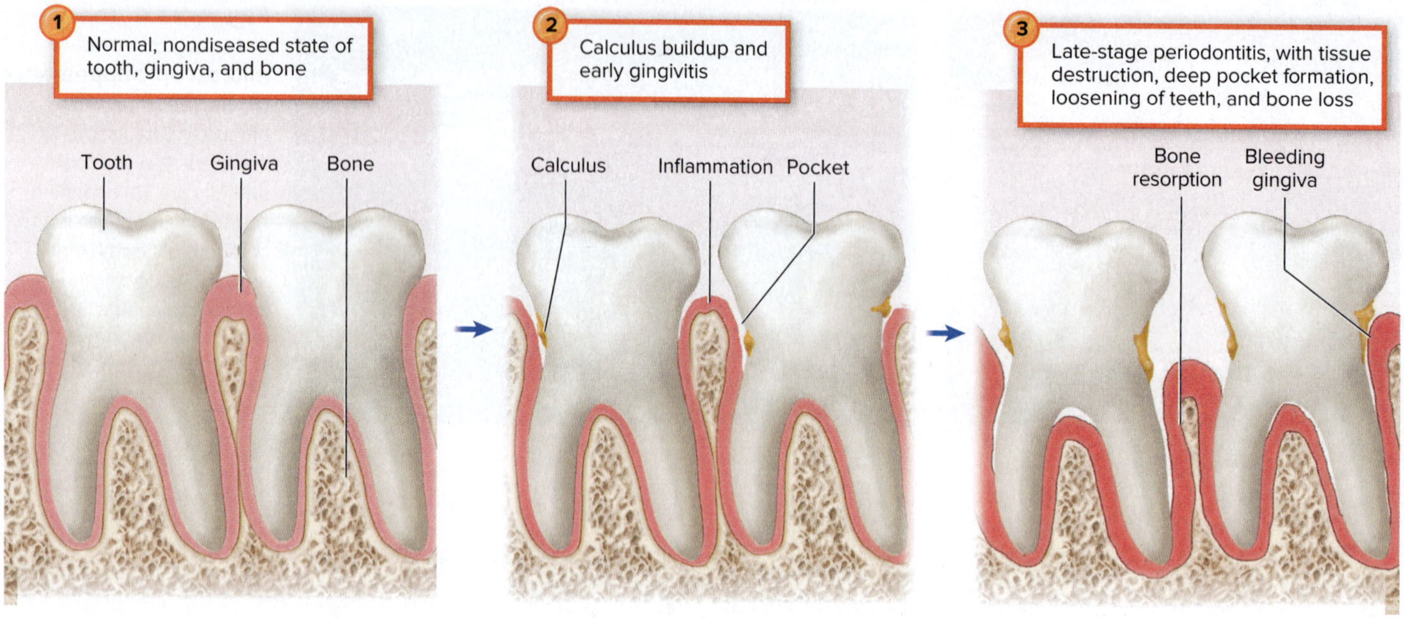
1
Normal, nondiseased state of tooth, gingiva, and bone
Tooth
Gingiva
Bone
2
Calculus buildup and early gingivitis
Calculus
Inflammation
Pocket
3
Late-stage periodontitis, with tissue destruction, deep pocket formation, loosening of teeth, and bone loss
Bone resorption
Bleeding gingiva

Figure 23.5 Stages in soft tissue infection, gingivitis, and periodontitis.

normal anatomy. In gingivitis, the depth of the pocket increases, from a healthy depth of 1–3 mm, to 5 mm or more. If this condition persists, a more serious disease called periodontitis results. This is the natural extension of the disease into the periodontal membrane and cementum. The deeper involvement increases the size of the pockets and can cause bone resorption severe enough to loosen the tooth in its socket. If the condition is allowed to progress, the tooth can be lost **(figure 23.5).**

Causative Agent

Dental scientists stop short of stating that particular bacteria cause periodontal disease because not all of the criteria for establishing causation have been satisfied. In fact, dental diseases (in particular, periodontal disease) provide an excellent model of disease caused by communities of microorganisms rather than single organisms. When the polymicrobial biofilms consist of the right combination of bacteria, such as the anaerobes *Tannerella forsythia* (formerly *Bacteroides forsythus*), *Aggregatibacter actinomycetemcomitans, Porphyromonas gingivalis,* and perhaps *Fusobacterium* and spirochete species, the periodontal destruction process begins. An individual's risk for dental caries or periodontitis is directly related to the composition of their personal oral microbiome. The presence of *Methanobrevibacter oralis* in the gingival crevice seems to be an important contributor to periodontal disease, which marks the first association between archaeal species and the development of human disease. Other factors are also important in the development of periodontal disease, such as behavioral and genetic influences as well as tooth position. The most common predisposing condition occurs when the plaque becomes mineralized (calcified) with calcium and phosphate crystals. This process produces a hard, porous substance called **calculus** (tartar in older usage) above and below the gingival margin (edge) that can induce varying degrees of periodontal damage **(figure 23.6).**

Pathogenesis and Virulence Factors

Calculus and plaque accumulating in the gingival sulcus cause abrasions in the delicate gingival membrane, and the chronic trauma causes a pronounced inflammatory reaction. The damaged tissues become a portal of entry for a variety of bacterial residents. The bacteria have an arsenal of enzymes, such as proteases, that destroy soft oral tissues. In response to the mixed infection, the damaged area becomes infiltrated by neutrophils and macrophages and, later, by lymphocytes, which cause additional inflammation and tissue damage. There is a great deal of evidence that people with high numbers of the bacteria associated with periodontitis also have thicker carotid arteries and increased rates of cardiovascular disease, further supporting the systemic effects of oral inflammation.

Transmission and Epidemiology

As with caries, the resident oral bacteria, acquired from close oral contact, are responsible for periodontal disease. Dentists refer to

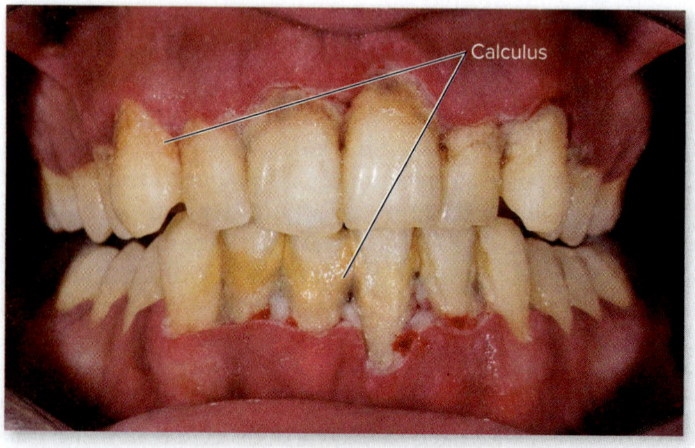
Calculus

Figure 23.6 Advanced periodontal disease.
Daniel Zgombic/E+/Getty Images

a wide range of risk factors associated with periodontal disease, especially deficient oral hygiene. But because it is so common in the population, it is evident that most of us could use some improvement in our oral hygiene.

Culture and/or Diagnosis

Like caries, periodontitis is generally diagnosed by the appearance of the oral tissues.

Prevention and Treatment

Regular brushing and flossing to remove plaque automatically reduce both caries and calculus production. Once calculus has formed on teeth, it cannot be removed by brushing but can be dislodged only by mechanical procedures (scaling) in the dental office. Because much of the pathology results from inflammation, some scientists are testing the use of new anti-inflammatory peptides to control disease progression. As already noted, the identification of high-risk microbiome profiles may also lead to the development of early preventive measures in patients.

Most periodontal disease is treated by removal of calculus and plaque and maintenance of good oral hygiene. Often, surgery to reduce the depth of periodontal pockets is required. Antibiotic therapy, either systemic or applied in periodontal packings, may also be utilized. Steroid use may also benefit the patient by reducing inflammation.

Oral microbes causing dental caries and periodontitis can enter the bloodstream at the site of periodontal infection, spreading to distant sites within the body. Recently, a wide variety of oral bacteria were detected via PCR in plaques in coronary arteries. Live cells of *Porphyromonas gingivalis* and *Aggregatibacter actinomycetemcomitans,* two periodontal pathogens, have also been localized to atherosclerotic tissue. A newly discovered oral bacterium, *Streptococcus tigurinus,* can escape into the bloodstream, increasing the risk for endocarditis and even meningitis. Evidence seems to be increasing that allowing our oral health to slide can have more widespread consequences than once thought.

Necrotizing Ulcerative Gingivitis and Periodontitis

The most destructive periodontal diseases are necrotizing ulcerative gingivitis (NUG) and necrotizing ulcerative periodontitis (NUP). The two diseases were formerly lumped under one name, acute necrotizing ulcerative gingivitis (ANUG). It was commonly referred to as "trench mouth," reflecting the poor dental health of soldiers in the battlefield trenches of World War I. These diseases are synergistic infections involving *Treponema vincentii, Prevotella intermedia,* and *Fusobacterium* species. These pathogens together produce several invasive factors that cause rapid advancement into the periodontal tissues. The condition is associated with severe pain, bleeding, pseudomembrane formation, and necrosis. Scientists believe that NUP may be an extension of NUG, but the conditions can be distinguished by the advanced bone destruction that results from NUP. Both diseases seem to result from poor oral hygiene, altered host defenses, or prior gum disease rather than being communicable. The diseases are common in AIDS patients and other immunocompromised populations. Diabetes and cigarette smoking can predispose people to these conditions. NUG and NUP usually respond well to targeted antibiotics after removal of damaged periodontal tissue **(Disease Table 23.2).**

Mumps

The word *mumps* is Old English for "lump" or "bump." The symptoms of this viral disease are so distinctive that Hippocrates clearly characterized it in the fifth century BC as a self-limited, mildly

Disease Table 23.2 Periodontal Diseases

Disease	Periodontitis	Necrotizing Ulcerative Gingivitis and Periodontitis
Causative Organism(s)	Polymicrobial community including some or all of *Tannerella forsythia, Aggregatibacter actinomycetemcomitans, Porphyromonas gingivalis,* others B	Polymicrobial community (*Treponema vincentii, Prevotella intermedia, Fusobacterium* species) B
Most Common Modes of Transmission	–	–
Virulence Factors	Induction of inflammation, enzymatic destruction of tissues	Inflammation, invasiveness
Culture/Diagnosis	–	–
Prevention	Oral hygiene	Oral hygiene
Treatment	Removal of plaque and calculus, gum reconstruction, possibly anti-inflammatory treatments	Debridement of damaged tissue, possibly antibiotics
Epidemiological Features	United States: smokers = 11%, nonsmokers = 2%; internationally: 10% to 15% of adults	–

Figure 23.7 Mumps. The external appearance of swollen parotid glands in mumps (parotitis).
CDC

epidemic illness associated with painful swelling at the angle of the jaw. In recent years, the incidence of mumps has spiked in the United States.

Signs and Symptoms

After an average incubation period of 2 to 3 weeks, symptoms of fever, nasal discharge, muscle pain, and malaise develop. These may be followed by inflammation of the salivary glands (especially the parotids), producing the classic gopherlike swelling of the cheeks on one or both sides **(figure 23.7).** Swelling of the gland is called parotitis, and it can cause considerable discomfort. Viral multiplication in salivary glands is followed by invasion of other organs, especially the testes, ovaries, thyroid gland, pancreas, meninges, heart, and kidney. Despite the invasion of multiple organs, the prognosis of most infections is complete, uncomplicated recovery with permanent immunity.

Complications in Mumps In 20% to 30% of infected young adult males, mumps infection localizes in the epididymis and testis, usually on one side only. The resultant syndrome of orchitis and epididymitis may be rather painful, but no permanent damage usually occurs. The popular belief that mumps readily causes sterilization of adult males is still held, despite medical evidence to the contrary. Perhaps this notion has been reinforced by the tenderness that continues long after infection and by the partial atrophy of one testis that occurs in about half the cases. Permanent sterility due to mumps is very rare.

In mumps pancreatitis, the virus replicates in beta cells and pancreatic epithelial cells. Mumps meningitis, characterized by fever, headache, and stiff neck, appears 2 to 10 days after the onset of parotitis, lasts for 3 to 5 days, and then dissipates, leaving few or no adverse side effects. Another rare event is infection of the inner ear that can lead to deafness.

Causative Agent

Mumps is caused by an enveloped, single-stranded RNA virus (mumps virus) from the genus *Paramyxovirus,* which is part of the family *Paramyxoviridae*. Other members of this family that infect humans are *Morbillivirus* (measles virus) and the respiratory syncytial virus.

Pathogenesis and Virulence Factors

A virus-infected cell is modified by the insertion of proteins called HN spikes into its cell membrane. The HN spikes immediately bind an uninfected neighboring cell, and in the presence of another type of spike called F spikes, the two cells permanently fuse. A chain reaction of multiple cell fusions then produces a *syncytium* (sin-sish′-yum) with cytoplasmic inclusion bodies, which is a diagnostically useful cytopathic effect **(figure 23.8).** The ability to induce the formation of syncytia is characteristic of the family *Paramyxoviridae*.

Transmission and Epidemiology

Humans are the exclusive natural hosts for the mumps virus. It is communicated primarily through salivary and respiratory secretions. Transmission occurs readily among populations living in close proximity, at home or in dormitories, and the virus has a greater chance of spreading the longer one is in contact with an infected individual. Infection occurs worldwide, with increases in the late winter and early spring in temperate climates.

High rates of infection arise among crowded populations or in communities with poor herd immunity. Most cases occur in children under the age of 15, and as many as 40% are subclinical. Public health officials keep a wary eye on mumps because the vaccine against it is the MMR vaccine, and measles outbreaks have become increasingly common in recent years.

Culture and/or Diagnosis

Diagnosis is usually based on ELISA tests for IgM or PCR on cheek swab. Negative results can be overruled if clinical signs are suggestive.

Prevention and Treatment

The general pathology of mumps is mild enough that symptomatic treatment to relieve fever, dehydration, and pain is usually adequate. The new vaccine recommendations call for a dose of MMR at 12 to 15 months and a second dose at 4 to 6 years. Health care workers and college students who have not already had both doses are advised to do so. Even though the vaccine provides 80% to 90% protection against disease, this still leaves a susceptible population that may become infected with mumps virus even if adequately vaccinated. Researchers have determined that this resurgence of mumps infections is not due to the evolution of novel mutant strains of the virus, but is, in fact, due to a reduced secondary immune response even after vaccination.

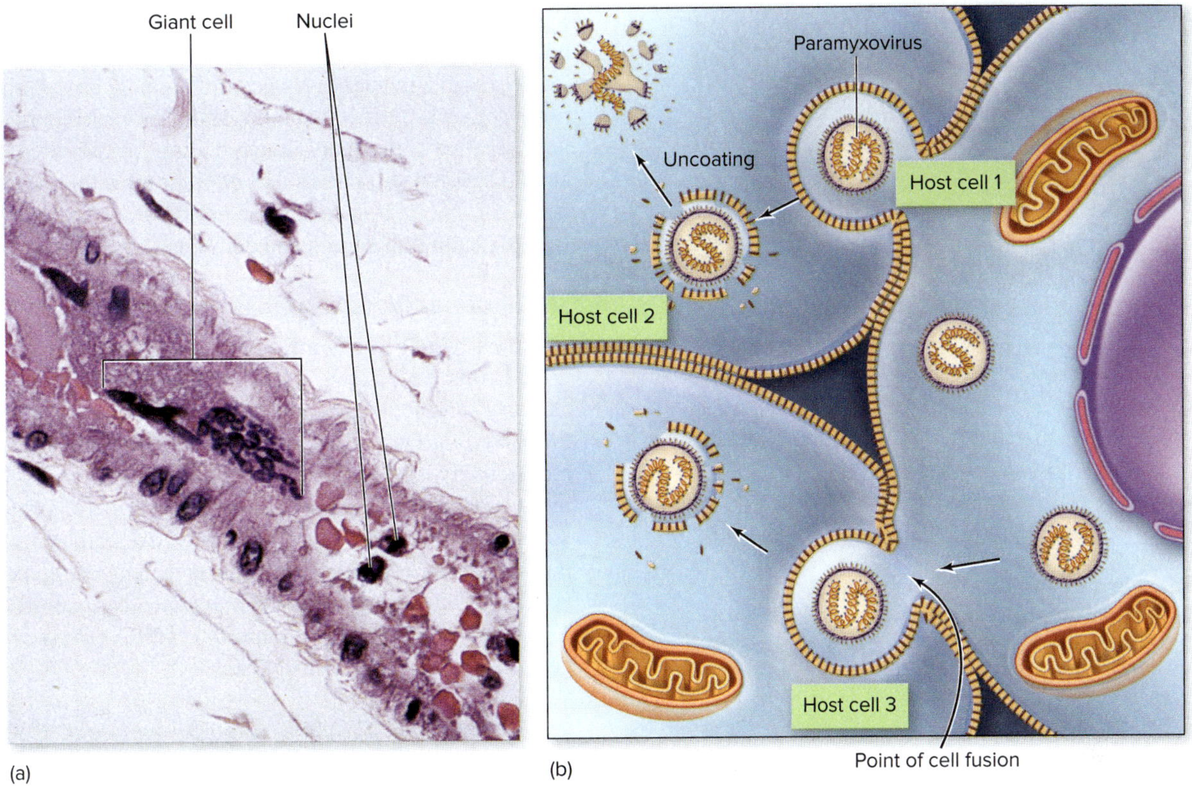
Giant cell
Nuclei
(a)
Paramyxovirus
Uncoating
Host cell 1
Host cell 2
Host cell 3
Point of cell fusion
(b)

Figure 23.8 The effects of paramyxoviruses. **(a)** When they infect a host cell, paramyxoviruses induce the cell membranes of adjacent cells to fuse into large multinucleate giant cells, or syncytia. This particular paramyxovirus is the Nipah virus. **(b)** This fusion allows direct passage of viruses from an infected cell to uninfected cells by communicating membranes. Through this means, the virus evades antibodies.

(a) Brian W.J. Mahy, BSc, MA, PhD, ScD, DSc/Centers for Disease Control

Disease Table 23.3	Mumps
Causative Organism(s)	Mumps virus (genus *Paramyxovirus*) V
Most Common Modes of Transmission	Droplet contact
Virulence Factors	Spike-induced syncytium formation
Culture/Diagnosis	ELISA for Ab; PCR
Prevention	MMR live attenuated vaccine
Treatment	Supportive
Epidemiological Features	United States: fluctuates between a few hundred cases a year and a few thousand; internationally: epidemic peaks every 2 to 5 years

Gastritis and Gastric Ulcers

The curved cells of *Helicobacter* were first detected by J. Robin Warren in 1979 in stomach biopsies from ulcer patients. He and an assistant, Barry J. Marshall, isolated the microbe in culture and even served as guinea pigs themselves by swallowing a large quantity of the liquid culture to prove that it would cause gastric ulcers. Warren and Marshall won the Nobel Prize in Medicine in 2005 for their discovery.

▶ Signs and Symptoms

Gastritis is experienced as sharp or burning pain emanating from the abdomen. Gastric or peptic ulcers are actual lesions in either the mucosa of the stomach (gastric ulcers) or the uppermost portion of the small intestine (duodenal ulcers) **(Figure 23.9).** Severe ulcers can be accompanied by bloody stools, vomiting, or both. The symptoms are often worse at night, after eating, or under conditions of psychological stress.

The fifth most common cancer in the world is stomach cancer (although it has been declining in the United States), and ample evidence suggests that long-term infection with *H. pylori* is a major contributing factor.

▶ Causative Agent

Helicobacter pylori is a curved, gram-negative rod, closely related to *Campylobacter*, which we study later in this chapter.

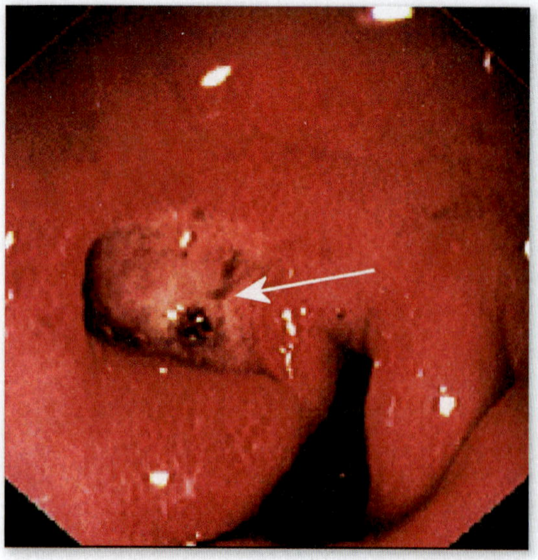

Figure 23.9 View of a duodenal ulcer caused by *Helicobacter*, seen on endoscopy.
Neal R. Chamberlin, PhD/McGraw Hill

Pathogenesis and Virulence Factors

Once the bacterium passes into the gastrointestinal tract, it bores through the outermost mucous layer that lines the stomach epithelial tissue. Then it attaches to specific binding sites on the cells and entrenches itself. One receptor specific for *Helicobacter* is the same molecule on human cells that confers the O blood type. This finding accounts for the higher rate of ulcers in people with this blood type. Another protective adaptation of the bacterium is the formation of urease, an enzyme that converts urea into ammonium and bicarbonate, both alkaline compounds that can neutralize stomach acid. As the immune system recognizes and attacks the pathogen, infiltrating white blood cells damage the epithelium to some degree, leading to chronic active gastritis. In some people, these lesions lead to deeper erosions and ulcers that can lay the groundwork for cancer to develop.

Before the bacterium was discovered, spicy foods, high-sugar diets (which increase acid levels in the stomach), and psychological stress were considered to be the cause of gastritis and ulcers. Now it appears that these factors merely aggravate the underlying infection.

Transmission and Epidemiology

Helicobacter occurs in the stomachs of 25% of healthy middle-age adults and in more than 60% of adults over 60 years of age. *H. pylori* is probably transmitted from person to person by the oral-oral or fecal-oral route. It seems to be acquired early in life mainly through what is called "familial transfer"—the microbe is acquired from family members, especially from infected mothers to their children. As you read in the case file at the beginning of the chapter, the bacterium has been in human stomachs for at least 100,000 years. The evidence shows that the percentage of people with *Helicobacter* in their stomachs is decreasing with each generation. Scientists suspect that this is due to our more sanitary environment and the use of antibiotics. While you might think this is a good thing in light of the fact that *Helicobacter* can cause ulcers and stomach cancer, there is evidence to suggest that its absence leads to higher incidences of acid reflux and even asthma. It should not surprise us by now that a single microbe can have both positive and negative consequences, depending on the circumstances. It should also not surprise us that obliterating a long-standing member of our microbiome in a short period of (evolutionary) time can lead to health imbalances and consequences.

Culture and/or Diagnosis

The urea breath test is a noninvasive method that is sometimes used. In this test, patients ingest urea that has a radioactive tag on its carbon molecule. If *Helicobacter* is present in a patient's stomach, the bacterium's urease breaks down the urea and the patient exhales radioactively labeled carbon dioxide. In the absence of urease, the intact urea molecule passes through the digestive system. Patients whose breath is positive for the radioactive carbon are considered positive for *Helicobacter*. This test is usually recommended when verifying that the organism was eradicated by treatment.

A stool test is also available. The HpSA (*H. pylori* stool antigen) test is an ELISA format test.

Prevention and Treatment

There is no vaccine for *Helicobacter* and no obvious way to avoid colonization. As discussed earlier, in many circumstances, colonization may be beneficial. For symptomatic infection, many over-the-counter remedies offer symptom relief by acting to neutralize stomach acid. The best treatment is clarithromycin, an acid suppressor, and amoxicillin.

Disease Table 23.4	Gastritis and Gastric Ulcers
Causative Organism(s)	*Helicobacter pylori* B G–
Most Common Modes of Transmission	Oral-oral or oral-fecal route likely
Virulence Factors	Adhesins, urease
Culture/Diagnosis	Direct antigen test on stool; urea breath test
Prevention	None
Treatment	Clarithromycin + acid suppression
Epidemiological Features	United States: infection (not disease) rates at 35% of adults; internationally: infection rates at 50%

Acute Diarrhea (With or Without Vomiting)

Diarrhea—usually defined as three or more loose stools in a 24-hour period—needs little explanation. In recent years, on average, citizens of the United States experienced 1.2 to 1.9 cases of diarrhea per person per year. Among children, that number is twice as high. In tropical countries, children may experience more than 10 episodes of diarrhea a year. In fact, more than 700,000 children a year—mostly in developing countries—die from a diarrheal disease. In developing countries, the high mortality rate is not the only issue. Children who survive dozens of bouts with diarrhea during their developmental years are likely to have permanent physical and cognitive effects. The effect on the overall well-being of these children is hard to estimate, but it is very significant.

In the United States, up to a third of all acute diarrhea is transmitted by contaminated food. New food safety measures are being implemented all the time, including the development of more rapid testing methods. The use of ultraviolet light and food irradiation, in addition to more novel methods such as the application of bacteriophage, is also helping to create a safer food supply in the United States. Even with all of these measures in place, it is still necessary for the consumer to be aware of and to practice good food-handling techniques. Accurate numbers are hard to come by, but the CDC estimates that 48 million people are sickened each year by a foodborne illness, and 3,000 of them die. Dozens of microbes are capable of causing foodborne illness. The most common culprits in the United States are norovirus, *Salmonella, Clostridium perfringens, Campylobacter,* and *Staphylococcus aureus* **(figure 23.10).**

Although most diarrhea episodes are self-limiting and therefore do not require treatment, others (such as shiga-toxin-producing *E. coli*—known as STEC) can have devastating effects. In most diarrheal illnesses, antimicrobial treatment is contraindicated (inadvisable), but some, such as shigellosis, call for quick treatment with antibiotics. For public health reasons, it is important to know which agents are causing diarrhea in the community, but in many cases identification of the agent is not performed.

In this section, we describe acute diarrhea having infectious agents as the cause. In the sections following this one, we discuss acute diarrhea and vomiting caused by toxins, commonly known as food poisoning, and chronic diarrhea and its causes.

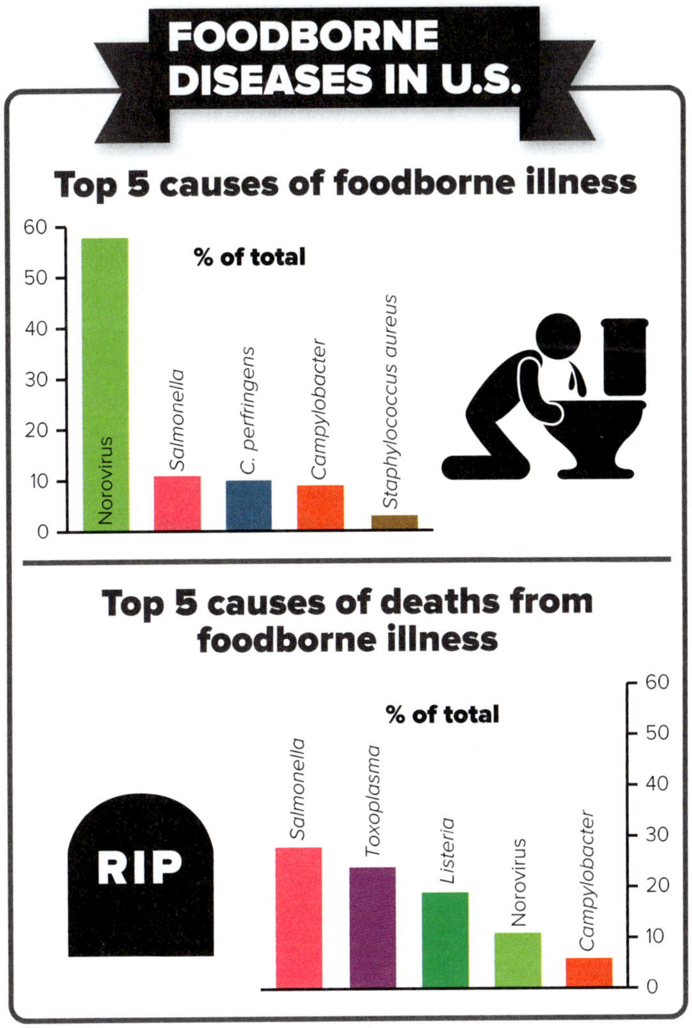
FOODBORNE DISEASES IN U.S.
Top 5 causes of foodborne illness
% of total
Norovirus
Salmonella
C. perfringens
Campylobacter
Staphylococcus aureus
0
10
20
30
40
50
60
Top 5 causes of deaths from foodborne illness
% of total
RIP
Salmonella
Toxoplasma
Listeria
Norovirus
Campylobacter

Figure 23.10 The most common (top) and the most deadly (bottom) causes of foodborne illness in the United States. The *Salmonella* species referenced here are non-typhoidal strains.
CDC

Salmonella

Salmonella is a very large genus of bacteria, but only one species is of interest to us: *S. enterica* is divided into many serotypes, based on variation in the major surface antigens. Many gram-negative enteric bacteria are named and designated according to the following antigens: H, the flagellar antigen; K, the capsular antigen; and O, the cell wall antigen. Not all enteric bacteria carry the H and K antigens, but all have O, the polysaccharide portion of the lipopolysaccharide implicated in endotoxic shock. Most species of gram-negative enterics exhibit a variety of subspecies, variants, or serotypes caused by slight variations in the chemical structure of the HKO antigens. Some bacteria in this chapter (for example, *E. coli* O157:H7) are named according to their surface antigens.

Salmonellae are motile. They ferment glucose with acid and sometimes gas; and most of them produce hydrogen sulfide (H_2S) but not urease. They grow readily on most laboratory media and can survive outside the host in inhospitable environments such as freshwater and freezing temperatures. These pathogens are resistant to chemicals such as bile and dyes, which are the bases for isolation on selective media.

Signs and Symptoms

The genus *Salmonella* causes a variety of illnesses in the GI tract and beyond. Roughly 1 million cases of illness are reported each year in the United States, with about 400 deaths attributed to *Salmonella* infection. Until the mid-1900s, its most severe manifestation was typhoid fever. Since that time, a milder disease, usually called salmonellosis, has been much more common. Sometimes the condition is also called enteric fever or gastroenteritis. Whereas typhoid fever is caused by *Salmonella enterica* serotype Typhi, gastroenteritises in the United States are generally caused by the serotypes known as Typhimurium, Enteritidis, Heidelberg, Newport, and Javiana. *Salmonella* bacteria are normal intestinal biota in cattle, poultry, rodents, and reptiles, and each (including domesticated pets) has been documented as a source of infection in humans.

Salmonellosis can be relatively severe, with an elevated body temperature and septicemia as more prominent features than GI tract disturbance. But it can also be fairly mild, with gastroenteritis—vomiting, diarrhea, and mucosal irritation—as its major feature. Blood can appear in the stool. In otherwise healthy adults, symptoms spontaneously subside after 2 to 5 days.

Pathogenesis and Virulence Factors

The ability of *Salmonella* to cause disease seems to be highly dependent on its ability to adhere effectively to the gut mucosa. Recent research has uncovered an "island" of genes in *Salmonella* that seems to confer enhanced attachment capabilities. Other pathogenicity islands encoding proteins allowing for immune evasion have also been identified. It is also believed that endotoxin is an important virulence factor for *Salmonella*.

Transmission and Epidemiology

An important factor to consider in all diarrheal pathogens is how many organisms must be ingested to cause disease (their ID_{50}). It varies widely. *Salmonella* has a high ID_{50}, meaning many organisms have to be ingested in order for disease to result. Animal products such as meat and milk can be readily contaminated with *Salmonella* during slaughter, collection, and processing. Inherent risks are involved in eating poorly cooked chicken or unpasteurized fresh or dried milk, ice cream, and cheese.

Most cases are traceable to a common food source such as milk or eggs. Some cases may be due to poor sanitation. In one outbreak, about 60 people became infected after visiting the Komodo dragon exhibit at the Denver Zoo. They had picked up the infection by handling the rails and fence of the dragon's cage.

Prevention and Treatment

The only prevention for salmonellosis is avoiding contact with the bacterium. A vaccine is used in poultry. A vaccine against typhoid fever is available for travelers to endemic areas.

Uncomplicated cases of salmonellosis are treated with fluid and electrolyte replacement. If the patient has underlying immunocompromise or if the disease is severe, antibiotics are recommended. However, multidrug-resistant *Salmonella* strains have evolved, due in large part to the prophylactic use of antibiotics in animal herds. If treatment is indicated due to the severity of the disease, ciprofloxacin is recommended.

Shigella

The *Shigella* bacteria are gram-negative, straight rods, nonmotile, and non-endospore-forming. They do not produce urease or hydrogen sulfide, traits that help in their identification. They are primarily human parasites, though they can infect primates. All produce a similar disease, including bloody diarrhea, that can vary in intensity. These bacteria resemble some types of pathogenic *E. coli* very closely. Stool culture is still the gold standard for identification in the case of *Shigella* infections.

Although *Shigella dysenteriae* causes the most severe form of dysentery, it is uncommon in the United States and occurs primarily in the Eastern Hemisphere.

Signs and Symptoms

The symptoms of shigellosis include frequent, watery stools; fever; and often intense abdominal pain. Nausea and vomiting are common. Stools often contain obvious blood and even more often are found to have occult (not visible to the naked eye) blood. Diarrhea containing blood is also called **dysentery.** Mucus from the GI tract will also be present in the stools.

Pathogenesis and Virulence Factors

Shigellosis is different from many GI tract infections in that *Shigella* invades the villus cells of the large intestine rather than the small intestine. In addition, it is not as invasive as *Salmonella* and does not perforate the intestine or invade the blood. It enters the intestinal mucosa by means of lymphoid cells in Peyer's patches. Once in the mucosa, *Shigella* instigates an inflammatory response that causes extensive tissue destruction. The release of endotoxin causes fever. **Enterotoxin,** an exotoxin that affects the enteric (or GI) tract, damages the mucosa and villi. Local areas of erosion give rise to bleeding and heavy secretion of mucus **(figure 23.11).** *Shigella dysenteriae* (and perhaps some of the other species) produces a heat-labile exotoxin called **shiga toxin,** which seems to be responsible for the more serious damage to the intestine as well as any systemic effects, including injury to nerve cells. You will encounter shiga toxin again when we discuss shiga-toxin-producing *E. coli*.

Transmission and Epidemiology

In addition to the usual oral route, shigellosis is also acquired through direct person-to-person contact, largely because of the small infectious dose required (from 10 to 200 bacteria). The disease is mostly associated with lax sanitation, malnutrition, and crowding. It is spread epidemically in day care centers, prisons, mental institutions, nursing homes, and military camps. As in other enteric infections, *Shigella* can establish a chronic carrier condition in some people that lasts several months.

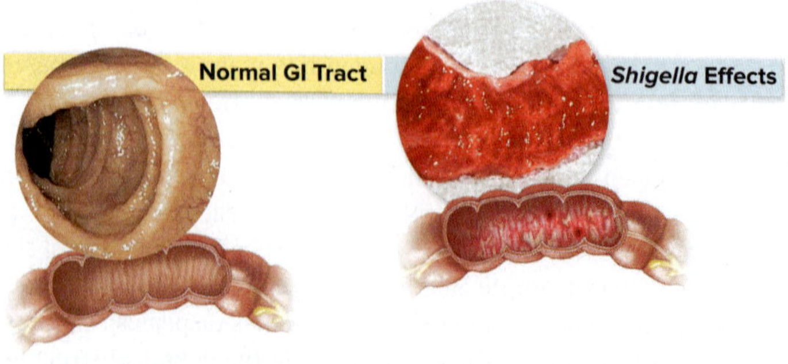
Normal GI Tract
Shigella Effects

Figure 23.11 The appearance of the large intestinal mucosa in *Shigella* dysentery. Note the patches of blood and mucus, the erosion of the lining, and the absence of perforation.
(left) Gastrolab/Science Source; (right) CDC

▶ Prevention and Treatment

The only prevention of this and most other diarrheal diseases is practicing good hygiene and avoiding contact with infected persons. Most physicians recommend prompt treatment of shigellosis with amoxicillin or ciprofloxacin.

Shiga-Toxin-Producing *E. coli* (STEC)

In January 1993, a new *E. coli* strain burst into the public's consciousness when four children died after eating undercooked hamburgers at a fast-food restaurant in Washington State. The cause of their illness was determined to be *E. coli* O157:H7. Since then, this strain and related *E. coli* strains have caused numerous and ongoing contamination of foods as diverse as romaine lettuce, hazelnuts, and flour.

Dozens of different strains of *E. coli* exist, most of which cause no disease at all. A handful of them cause various degrees of intestinal symptoms, as described in this and the following section. Some of them cause urinary tract infections (see chapter 24). *E. coli* O157:H7 and its close relatives are the most virulent of them all. This collection of organisms, of which this *E. coli* strain is the most famous representative, is generally referred to as **shiga-toxin-producing *E. coli* (STEC).**

▶ Signs and Symptoms

STEC is the agent of a spectrum of conditions, ranging from mild gastroenteritis with fever to bloody diarrhea. About 10% of patients develop **hemolytic uremic syndrome (HUS),** a severe hemolytic anemia that can cause kidney damage and failure. Neurological symptoms such as blindness, seizure, and stroke (and long-term debilitation) are also possible.

▶ Pathogenesis and Virulence Factors

These *E. coli* owe much of their virulence to shiga toxins (so named because they are identical to the shiga exotoxin secreted by virulent *Shigella* species). The shiga toxin genes are present on bacteriophage in *E. coli* but are on the chromosome of *Shigella dysenteriae,* suggesting that STEC strains of *E. coli* acquired the virulence factor through phage-mediated transfer. As described earlier for *Shigella,* the shiga toxin interrupts protein synthesis in its target cells. It seems to be responsible especially for the systemic effects of this infection.

Another important virulence determinant for STEC is the ability to efface (rub out or destroy) enterocytes, which are gut epithelial cells. The net effect is a lesion in the gut (effacement), usually in the large intestine. The microvilli are lost from the gut epithelium, and the lesions produce bloody diarrhea.

▶ Transmission and Epidemiology

Contaminated ground beef used to be a primary source of STEC. The USDA began testing ground beef for the pathogen in 2012, and since then, other foodstuffs have more commonly been a source. The bacteria inhabit the GI tracts of cattle, and any plant or product that is contaminated with even trace amounts of cow manure can be a source.

Products that are eaten raw, such as lettuce, vegetables, and apples used in unpasteurized cider, are particularly problematic.

▶ Culture and/or Diagnosis

Infection with this type of *E. coli* should be confirmed with stool culture or with PCR.

▶ Prevention and Treatment

The best prevention for this disease is to avoid raw or even rare hamburger and to wash raw vegetables well. The shiga toxin is heat-labile, and the *E. coli* is killed by heat. If you are thinking "I used to be able to eat rare hamburgers," you are correct, but things have changed. The emergence of this pathogen in the early 1990s, probably resulting from a regular *E. coli* picking up the shiga toxin from *Shigella,* has changed the rules for proper food handling.

No human vaccine exists for *E. coli* O157:H7 or other STEC strains. Some countries vaccinate cattle against *E. coli* O157:H7 as a means to protect human populations.

Antibiotics are contraindicated for this infection. Even with severe disease manifestations, antibiotics may increase the pathology by releasing more toxin, leading to HUS. Supportive therapy, including plasma transfusion to dilute toxin in the blood, is the best option.

Other *E. coli*

At least five other categories of *E. coli* can cause diarrheal diseases. In clinical practice, most physicians are interested in differentiating shiga-toxin-producing *E. coli* (STEC) from all the others. Each of the five other categories is considered separately and briefly here. In **Disease Table 23.5,** the non-shiga-toxin-producing *E. coli* are grouped together in one column.

- Enterotoxigenic *E. coli* (ETEC). The presentation varies depending on which type of *E. coli* is causing the disease. Traveler's diarrhea, characterized by watery diarrhea, low-grade fever, nausea, and vomiting, is usually caused by enterotoxigenic *E. coli* (ETEC). These strains also cause a great deal of illness in infants in developing countries.

  Most infections with ETEC are self-limiting, however miserable they make you feel. They are treated only with fluid replacement, often due to the high rate of drug resistance. In infants, ETEC can be life-threatening, and fluid replacement is vital to survival.
- Enteroinvasive *E. coli* (EIEC). These strains cause bacillary dysentery, which is often mistaken for *Shigella* dysentery. The bacteria invade gut mucosa and cause widespread destruction. Blood and mucus will be found in the stool. Significant fever is often present. EIEC does not produce the heat-labile or heat-stable exotoxins just described and does not have a shiga toxin, despite the clinical similarity to *Shigella* disease.

  Disease caused by this bacterium is more common in developing countries. It is transmitted primarily through contaminated food and water. Treatment is supportive (including rehydration).
- Enteropathogenic *E. coli* (EPEC). These strains result in a profuse, watery diarrhea. Fever and vomiting are also common. The EPEC bacteria are very similar to the

shiga-toxin-producing *E. coli* (STEC) described earlier—they produce effacement of gut surfaces. The important difference between EPEC and STEC is that EPEC does not produce a shiga toxin and, therefore, does not produce the systemic symptoms characteristic of those bacteria.

- Enteroaggregative *E. coli* (EAEC). These bacteria are most notable for their ability to cause chronic diarrhea in young children and in severely immunocompromised people. EAEC is considered in the section on chronic diarrhea.
- Diffusely adherent *E. coli* (DAEC). These bacteria are identified based on virulence factors used to attach to host cells. They are typically associated with the development of urinary tract infections in addition to acute diarrhea in the developing world.

Campylobacter

Although you may never have heard of *Campylobacter,* it is considered to be the most common bacterial cause of diarrhea in the United States. It probably causes more diarrhea than *Salmonella* and *Shigella* combined, with 1.3 million cases of diarrhea in the United States credited to it per year.

The symptoms of campylobacteriosis are frequent, watery stools, as well as fever, vomiting, headaches, and severe abdominal pain. The symptoms may last longer than most acute diarrheal episodes, sometimes extending beyond 2 weeks. They may subside and then recur over a period of weeks.

Campylobacter jejuni is the most common cause, although there are other pathogenic *Campylobacter* species. *Campylobacter* are slender, curved or spiral, gram-negative bacteria propelled by polar flagella at one or both poles, often appearing in S-shaped or gull-winged pairs **(figure 23.12).** These bacteria tend to be microaerophilic inhabitants of the intestinal tract, genitourinary tract, and oral cavity of humans and animals. A close relative, *Helicobacter pylori,* is the causative agent of most stomach ulcers (described earlier). Transmission of this pathogen takes place via the ingestion of contaminated beverages and food, especially water, milk, meat, and chicken. In late 2019 and early 2020, an outbreak occured in at least 30 states in which people were infected by contact with the stool of healthy-looking puppies they had purchased or worked with as pet store employees.

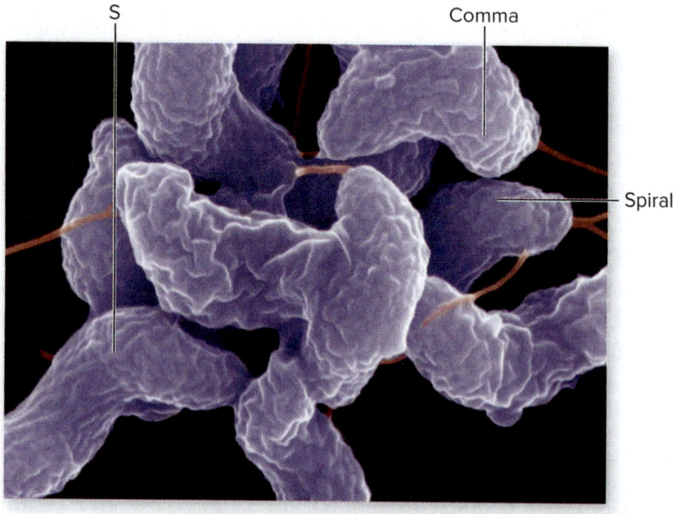
S
Comma
Spiral

Figure 23.12 Scanning electron micrograph of *Campylobacter*.
Photo by De Wood, digital colorization by Stephen Ausmus, USDA-ARS

Once ingested, *C. jejuni* cells reach the mucosa at the last segment of the small intestine (ileum) near its junction with the colon. They adhere, burrow through the mucus, and multiply. Symptoms commence after an incubation period of 1 to 7 days. The mechanisms of pathology appear to involve a heat-labile enterotoxin that stimulates a secretory diarrhea like that of cholera. In a small number of cases, infection with this bacterium can lead to a serious neuromuscular paralysis called Guillain-Barré syndrome.

Guillain-Barré syndrome (GBS) (pronounced gee″-luhn-buh-ray′) is the leading cause of acute paralysis in the United States since the eradication of polio. The good news is that many patients recover completely from this paralysis. The condition is still mysterious in many ways, but it seems to be an autoimmune reaction that can be brought on by infection with viruses and bacteria, by vaccination in rare cases, and even by surgery. The single most common precipitating event for the onset of GBS is *Campylobacter* infection. Twenty to forty percent of GBS cases are preceded by infection with *Campylobacter.* The reasons for this are not clear. (Note that even though 20% to 40% of GBS cases are preceded by *Campylobacter* infection, only about 1 in 1,000 cases of *Campylobacter* infection results in GBS.)

Resolution of infection with *Campylobacter* occurs in most instances with simple, nonspecific rehydration and electrolyte-balance therapy. In more severely affected patients, it may be necessary to administer azithromycin. Antibiotic resistance is growing in these bacteria, in large part due to the use of fluoroquinolones in the treatment of poultry destined for human consumption. Prevention depends on rigid sanitary control of water and milk supplies and care in food preparation. Vaccine development for use in poultry is ongoing.

Clostridioides difficile

Clostridioides difficile is a gram-positive, endospore-forming rod found as normal biota in the intestine. It was once considered relatively harmless but now is known to cause a condition called pseudomembranous colitis, also known as antibiotic-associated colitis. In many cases, this infection is precipitated by therapy with broad-spectrum antibiotics. It is a major cause of diarrhea in hospitals, although community-acquired infections have been on the rise in the last few years. Also, new studies suggest that the use of gastric acid inhibitors for the treatment of heartburn can predispose patients to this infection. Although *C. difficile* is relatively noninvasive, it is able to superinfect the large intestine when drugs have disrupted the normal biota. It produces two enterotoxins, toxins A and B, that cause areas of necrosis in the wall of the intestine. The predominant symptom is diarrhea. More severe cases exhibit abdominal cramps, fever, and leukocytosis. The colon is inflamed and gradually sloughs off loose, membranelike patches called pseudomembranes consisting of fibrin and cells **(figure 23.13).** If the condition is not stopped, perforation of the cecum and death can result.

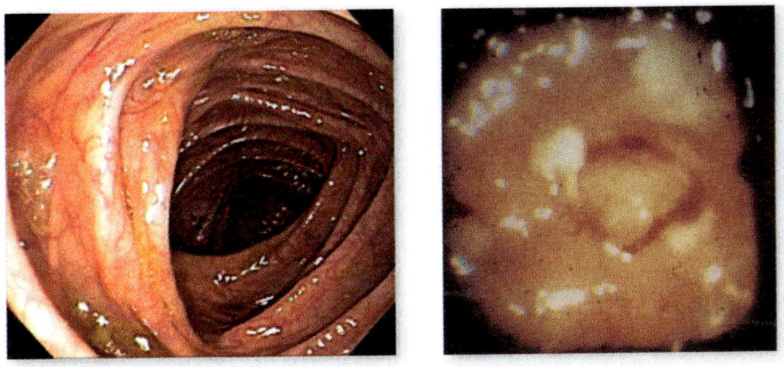
(a)
(b)

Figure 23.13 The colon in *C. diff* infection. **(a)** Normal colon. **(b)** A mild form of colitis with diffuse, inflammatory patches.
(a) David Musher/Science Source; (b) Fred Pittman

The epidemiology of *C. diff* is changing. Previously, most cases were seen in hospitalized patients on IV antibiotics. That situation does lead to *C. diff*—for example, the colonization rate of hospitalized patients is 20% to 40% compared to just 3% of the community population. But increasingly community-acquired transmission is occurring. If a patient is receiving clindamycin, ceftriaxone, or a fluoroquinolone for a different infection and displays *C. diff* symptoms, the first step is to withdraw the offending antibiotic. *C. diff* infections should be treated with fidaxomicin or vancomycin. It can be very difficult to eradicate, and stubborn infections can significantly degrade a patient's quality of life. Fecal implants, or stool transplantation from healthy donors, is the treatment with the most success in these patients.

A key point to remember is that *C. diff* releases endospores, which contaminate the environment. Hospitalized patients must be put in isolation conditions, and constant attention to disinfection and infection control is required.

Vibrio cholerae

Cholera has been a devastating disease for centuries. It is not an exaggeration to say that the disease has shaped a good deal of human history in Asia and Latin America, where it has been endemic. These days, we have come to expect outbreaks of cholera to occur after natural disasters, war, or large refugee movements, especially in underdeveloped parts of the world.

Vibrios are curved rods with a single polar flagellum. They belong to the family *Vibrionaceae*. A freshly isolated specimen of *Vibrio cholerae* contains quick, darting cells that resemble a comma **(figure 23.14).** *Vibrio* shares many cultural and physiological characteristics with members of the *Enterobacteriaceae,* a closely related family. Vibrios are fermentative and grow on ordinary or selective media containing bile at 37°C. They possess unique O and H antigens and membrane receptor antigens that provide some basis for classifying members of the family. There are two major biotypes, called classic and *El Tor.*

Signs and Symptoms

After an incubation period of a few hours to a few days, symptoms begin abruptly with vomiting, followed by copious watery feces called secretory diarrhea. The intestinal contents are lost very quickly, leaving only secreted fluids. This voided fluid contains flecks of mucus—hence, the description "rice-water stool." Fluid losses of nearly 1 liter per hour have been reported in severe cases, and an untreated patient can lose up to 50% of body weight during the course of this disease. The diarrhea causes loss of blood volume, acidosis from bicarbonate loss, and potassium depletion, which manifest in muscle cramps, severe thirst, flaccid skin, sunken eyes, and—in young children—coma and convulsions. Secondary cardiovascular consequences can include hypotension, tachycardia, cyanosis, and collapse from shock within 18 to 24 hours. If cholera is left untreated, death can occur in less than 48 hours. The mortality rate is between 55% and 70%.

Pathogenesis and Virulence Factors

After being ingested with food or water, *V. cholerae* travels through the stomach to the small intestine. At the junction of the duodenum and jejunum, the vibrios penetrate the mucous barrier using their flagella, adhere to the microvilli of the epithelial cells, and multiply there. The bacteria never enter the host cells or invade the mucosa. The virulence of *V. cholerae* lies mainly in the action of an enterotoxin called cholera toxin (CT), which disrupts the normal physiology of intestinal cells. When this toxin binds to specific intestinal receptors, a secondary signaling system is activated. Under the influence of this system, the cells shed large

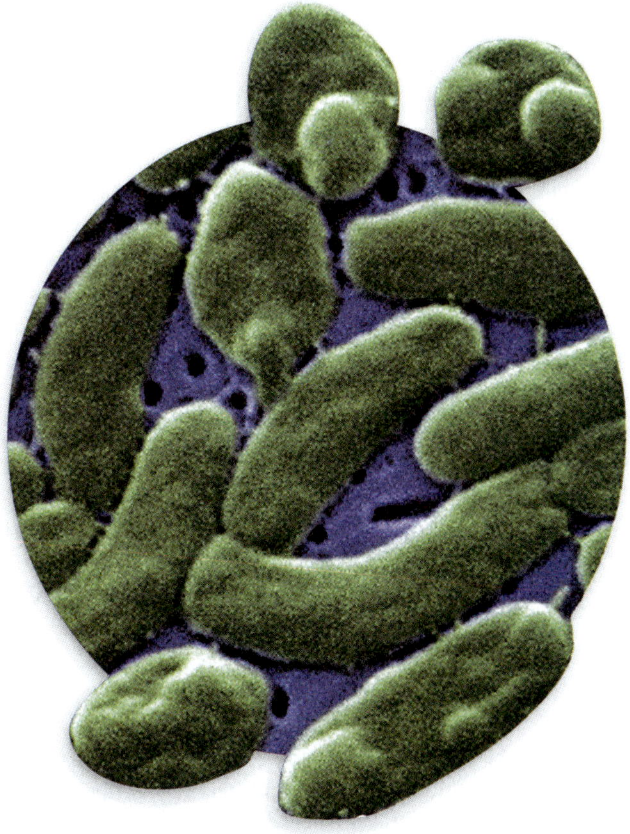

Figure 23.14 *Vibrio cholerae.* Note the characteristic curved shape of this bacterium.
Janice Carr/CDC

amounts of electrolytes into the intestine, an event accompanied by profuse water loss. It was recently discovered that *V. cholerae* uses quorum sensing to regulate the precise expression of its virulence factors, making proteins used in this process potential targets for drug therapy.

▶ Transmission and Epidemiology

The pattern of cholera transmission and the onset of epidemics are greatly influenced by the season of the year and the climate. Cold, acidic, dry environments inhibit the migration and survival of *Vibrio,* whereas warm, monsoon, alkaline, and saline conditions favor them. The bacteria survive in water sources for long periods of time. Recent outbreaks in several parts of the world have been traced to giant cargo ships that pick up ballast water in one port and empty it in another elsewhere in the world. One of the world's preeminent microbiologists, Rita Colwell **(figure 23.15*a*),** has been studying *Vibrio cholerae* and safe drinking water for over 50 years. She is particularly interested in cholera in third-world countries, such as Bangladesh, where there are seasonal epidemics twice a year. She discovered that the majority of *V. cholerae* are attached to a type of plankton called copepods in the water. It would be impossible to filter the *V. cholerae* out of water with only everyday supplies, but because they are mainly attached to the larger copepods, she tested whether the material used to make saris could serve as a filter. It turns out that if a sari is folded over and the raw water is poured over it, 99% of the cholera could be removed, making the water much safer to drink **(figure 23.15*b*).** This was a huge innovation in providing safe drinking water and decreased the incidence of cholera by 50%.

In nonendemic areas such as the United States, the microbe is spread by water and food contaminated by asymptomatic carriers, but it is relatively uncommon. Sporadic outbreaks occur along the Gulf of Mexico, and *V. cholerae* is sometimes isolated from shellfish in that region. Because climate change is leading to warmer waters and increasing copepod populations, scientists are concerned about the risk of more frequent cholera epidemics in the future.

▶ Culture and/or Diagnosis

V. cholerae can be readily isolated and identified in the laboratory from stool samples. Direct dark-field microscopic observation reveals characteristic curved cells with brisk, darting motility as confirmatory evidence. Immobilization or fluorescent staining of feces with group-specific antisera is supportive as well. Real-time PCR methods are now being employed to rapidly test food and water supplies for the presence of *V. cholerae* and related pathogenic species (*V. vulnificus* and *V. parahaemolyticus*).

▶ Prevention and Treatment

Effective prevention is contingent on proper sewage treatment and water purification. Detecting and treating carriers with mild or asymptomatic cholera are serious goals, but they are difficult to accomplish because of inadequate medical provisions in those countries where cholera is endemic. Vaccines are available for travelers and people living in endemic regions. There is a vaccine available in the United States, but its routine use is not recommended.

The key to cholera therapy is prompt replacement of water and electrolytes because their loss accounts for the severe morbidity and mortality. This therapy can be accomplished by various rehydration techniques that replace the lost fluid and electrolytes. One of these, oral rehydration therapy (ORT), has been instrumental in saving lives since 1978.

Cases in which the patient is unconscious or has complications from severe dehydration require intravenous replenishment as well. Oral antibiotics such as doxycycline are given as an adjunct to rehydration. They also diminish the period of vibrio excretion.

(a)

(b)

Figure 23.15 Rita Colwell (a) and her sari solution (b).

(a) Jim Lo Scalzo/EPA/Shutterstock; (b) Dr. Anwar Huq/University of Maryland

Non-Cholera *Vibrio* Species

In the United States, it is more common to experience infection with a non-cholera species of *Vibrio* than a *V. cholerae* infection. These infections are called vibrioses, as opposed to cholera. Two species are most prominent, *V. vulnificus* and *V. parahaemolyticus.* Infection can be from exposure to seawater but more often is associated with eating contaminated shellfish. It is a relatively rare infection but on the increase. Scientists suspect that the increase is due to three factors: (1) increased consumption of raw oysters; (2) increased awareness, meaning that more people are diagnosed; and (3) a warming climate, causing a wider habitat for these bacteria in bodies of water. In people who are immunocompromised, the infections can be fatal, especially in the case of *V. vulnificus.* These infections, along with those by *V. cholerae,* are nationally notifiable diseases.

Cryptosporidium

Cryptosporidium is an intestinal protozoan of the apicomplexan type (see chapter 5) that infects a variety of mammals, birds, and reptiles. The organism's life cycle includes a hardy intestinal oocyst as well as a tissue phase. Humans accidentally ingest the oocysts with water or food that has been contaminated by feces from infected animals. The oocyst "excysts" once it reaches the intestines and releases sporozoites that attach to the epithelium of the small intestine **(figure 23.16).**

The organism penetrates the intestinal cells and lives intracellularly within them. It undergoes asexual and sexual reproduction in these cells and produces more oocysts, which are released into the gut lumen, excreted from the host, and after a short time become infective again. The oocysts are highly infectious and extremely resistant to treatment with chlorine and other disinfectants. The prominent symptoms mimic other types of gastroenteritis, with headache, sweating, vomiting, severe abdominal cramps, and diarrhea. The agent can be detected in fecal samples or in biopsies using ELISA or acid-fast staining. Stool cultures should be performed to rule out other (bacterial) causes of infection.

Half of the outbreaks of diarrhea associated with swimming pools are caused by *Cryptosporidium.* Because chlorination is not entirely successful in eradicating the cysts, most treatment plants and recreational water parks utilize a combination of ultraviolet light treatment and filtration, but even this method is not foolproof.

Treatment is not usually required for otherwise healthy patients. Antidiarrheal agents (antimotility drugs) may be used. Although no curative antimicrobial agent exists for *Cryptosporidium,* physicians will often try a drug called nitazoxanide.

Rotavirus

Rotavirus is a member of the *Reovirus* group, which consists of an unusual double-stranded RNA genome with both an inner and an outer capsid. Globally, rotavirus is the primary viral cause of morbidity and mortality resulting from diarrhea, accounting for nearly 50% of all cases. It is estimated that there are 2 to 3 million cases of rotavirus infection in the United States every year, leading to 70,000 hospitalizations. Peak occurrences of this infection are seasonal; in the Southwest, the peak is often in the late fall; in the Northeast, the peak comes in the spring.

Diagnosis of rotavirus infections is not always performed, as it is treated symptomatically. Nevertheless, studies are often conducted so that public health officials can maintain surveillance of how prevalent the infection is. Stool samples from infected persons contain large amounts of virus, which is readily visible using electron microscopy **(figure 23.17).** The virus gets its name from its physical appearance, which is said to resemble a "spoked wheel." A rapid antigen test for stool specimens is commonly used in clinical settings, and an ELISA test is available.

The virus is transmitted by the fecal-oral route, including through contaminated food, water, and fomites. For this reason, disease is most prevalent in areas of the world with poor sanitation.

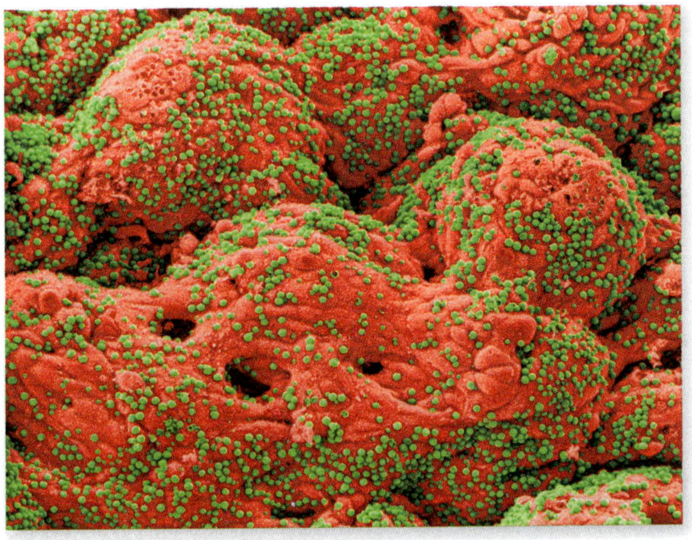

Figure 23.16 Scanning electron micrograph of *Cryptosporidium* (green) attached to the intestinal epithelium.
Moredun Animal Health Ltd/Science Source

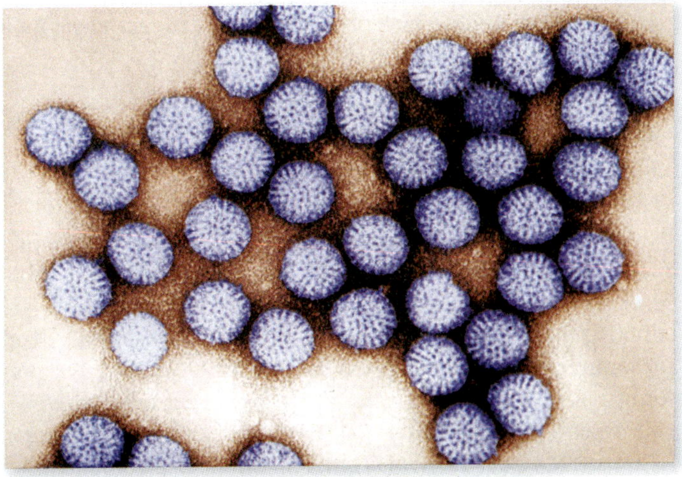

Figure 23.17 Rotavirus visible in a sample of feces from a child with gastroenteritis. Note the unique "spoked-wheel" morphology of the virus.
Dr. Erskine Palmer & Byron Skinner/CDC

In the United States, rotavirus infection is common, but its course is generally mild.

The effects of infection vary with the age, nutritional state, general health, and living conditions of the patient. Babies from 6 to 24 months of age lacking maternal antibodies have the greatest risk for fatal disease. These children present symptoms of watery diarrhea, fever, vomiting, dehydration, and shock. The intestinal mucosa can be damaged in a way that chronically compromises nutrition, and long-term or repeated infections can retard growth. Newborns seem to be protected by maternal antibodies. Adults can also acquire this infection, but it is generally mild and self-limiting.

Children are treated with oral replacement fluid and electrolytes. Hospital admissions due to rotavirus infection have decreased by nearly 90% since vaccines were introduced in the 2000s. There are two vaccines, both live attenuated: Rota Teq and Rotarix.

Norovirus

A bewildering array of viruses can cause gastroenteritis, including adenoviruses, rotavirus, astroviruses, and noroviruses (sometimes known as Norwalk viruses). Norovirus is the most common of these and, as we saw in figure 23.10, the most common cause of foodborne illness in the United States.

Transmission is fecal-oral or via contamination of food and water. Viruses generally cause a profuse, watery diarrhea of 3 to 5 days' duration. Severe vomiting is a feature of the disease, especially in the early phases. Mild fever is often seen. Scientists consider this an exquisitely tuned pathogen because it has a very low infectious dose (1 to 20 viruses), causes the host to expel enormous amounts of the virus during illness, and survives for days on countertops and in the air. In one study, scientists found 21 different types of norovirus on a single hospital countertop.

One particular setting that is frequently affected by norovirus outbreaks is the cruise ship industry. This has led to the CDC's development of the Vessel Sanitation Program, aimed at protecting passengers and populations living in ports of call.

Treatment of these infections always focuses on rehydration **(Disease Table 23.5).**

Acute Diarrhea with Vomiting Caused by Exotoxins (Food Poisoning)

If a patient presents with severe nausea and frequent vomiting accompanied by diarrhea and reports that companions with whom he or she shared a recent meal (within the last 1 to 6 hours) are suffering the same fate, food poisoning should be suspected. **Food poisoning** refers to symptoms in the gut that are caused by a preformed exotoxin. In many cases, the toxin comes from *Staphylococcus aureus.* In others, the source of the toxin is *Bacillus cereus* or *Clostridium perfringens.* The toxin occasionally comes from nonmicrobial sources such as fish, shellfish, or mushrooms. In any case, if the symptoms are violent and the incubation period is very short, *intoxication* (the effects of a toxin) rather than *infection* should be considered.

Staphylococcus aureus Exotoxin

This illness is associated with eating foods such as custards, sauces, cream pastries, processed meats, chicken salad, or ham that have been contaminated by handling and then left unrefrigerated for a few hours. Because of the high salt tolerance of *S. aureus,* even foods containing salt as a preservative are not exempt. The toxins produced by the multiplying bacteria do not noticeably alter the food's taste or smell. The exotoxin (which is an enterotoxin, meaning that it acts on the enteric, or gastrointestinal, system) is heat-stable. Inactivation requires 100°C for at least 30 minutes. Thus, heating the food after toxin production may not prevent disease.

The illness produced by exotoxin-producing strains is caused by the toxin itself and does not require *S. aureus* to be present or alive in the contaminated food. The ingested toxin acts upon the gastrointestinal epithelium and stimulates nerves, with acute symptoms of cramping, nausea, vomiting, and diarrhea. Recovery is also rapid, usually within 24 hours. The disease is not transmissible person to person. Often, a single source will contaminate several people, leading to a small point-source outbreak.

As you learned earlier, many diarrheal diseases have symptoms caused by bacterial exotoxins. In most cases, the bacteria take up temporary residence in the gut and then start producing exotoxin, so the incubation period is longer than the 1 to 6 hours seen with the onset of *S. aureus* food poisoning. Proper food handling, preparation, and storage are required to prevent this form of food poisoning. This condition is almost always self-limiting, and antibiotics are usually not warranted.

Bacillus cereus Exotoxin

Bacillus cereus is a sporulating gram-positive bacterium that is naturally present in soil. As a result, it is a common resident on vegetables and other products in close contact with soil. It produces two exotoxins, one of which causes a diarrheal-type disease and the other of which causes an **emetic** (ee-met′-ik), or vomiting, disease. The type of disease that takes place is influenced by the type of food that is contaminated by the bacterium. The emetic form is most frequently linked to fried rice or pasta, especially when it has been cooked and kept warm for long periods of time. These conditions are apparently ideal for the expression of the low-molecular-weight, heat-stable exotoxin having an emetic effect. The diarrheal form of the disease is usually associated with cooked meats or vegetables that are held at a warm temperature for long periods of time. These conditions apparently favor the production of the high-molecular-weight, heat-labile exotoxin. The symptom in these cases is a watery, profuse diarrhea that lasts only for about 24 hours.

Diagnosis of the emetic form of the disease is accomplished by finding the bacterium in the implicated food source. Microscopic examination of stool samples is used to diagnose the diarrheal form of the disease. Of course, in everyday practice, neither diagnosis nor treatment is performed because of the short duration of the disease.

For many years, *B. cereus* has been regarded as a relatively harmless bacterium in light of its ability to cause limited disease. However, the rise of *B. cereus* as a formidable pathogen in immunocompromised individuals has recently made microbiologists take a closer look at its pathogenic capabilities.

Clostridium perfringens Exotoxin

Another sporulating gram-positive bacterium that causes intestinal symptoms is *Clostridium perfringens.* You first read about this bacterium as the causative agent of gas gangrene in chapter 19. Endospores from *C. perfringens* can also contaminate many kinds of foods. Those most frequently implicated in disease are animal flesh (meat, fish) and vegetables such as beans that have not been cooked thoroughly enough to destroy endospores. When these foods are cooled, endospores germinate and the germinated cells multiply, especially if the food is left unrefrigerated. If the food is eaten without adequate reheating, live *C. perfringens* cells enter the small intestine and release exotoxin. The toxin, acting upon epithelial cells, initiates acute abdominal pain, diarrhea, and nausea in 8 to 16 hours. Recovery is rapid, and deaths are rare. A recent outbreak at a psychiatric facility, however, led to a nearly 6% fatality rate, leading clinicians to recognize the increased risk of *Clostridium* food poisoning in patients receiving psychiatric medication. This is due to the fact that these drugs can slow down the functioning of the GI tract, enhancing the pathogen's ability to cause disease.

C. perfringens also causes an enterocolitis infection similar to that caused by *C. difficile.* This infectious type of diarrhea is acquired from contaminated food, or it may be transmissible by inanimate objects **(Disease Table 23.6).**

Chronic Diarrhea

Chronic diarrhea is defined as lasting longer than 14 days. It can have infectious causes or can reflect noninfectious conditions. Most of us are familiar with diseases such as irritable bowel syndrome and ulcerative colitis. When the presence of an infectious agent is ruled out by a negative stool culture or other tests, these conditions are suspected.

Next we examine a few of the microbes that can be responsible for chronic diarrhea in otherwise healthy people. Keep in mind that practically any disease of the intestinal tract has a sexual mode of transmission in addition to the ones that are commonly stated. For example, any kind of oral-anal sexual contact efficiently transfers pathogens to the "oral" partner.

Enteroaggregative *E. coli* (EAEC)

In the section on acute diarrhea, you read about the various categories of *E. coli* that can cause disease in the gut. One type, the enteroaggregative *E. coli* (EAEC), is particularly associated with chronic disease, especially in children. This bacterium secretes neither the heat-stable nor the heat-labile exotoxins previously described for enterotoxigenic *E. coli* (ETEC). It is distinguished by its ability to adhere to human cells in aggregates rather than as single cells **(figure 23.18).** Its presence appears to stimulate secretion of large amounts of mucus in the gut, which may be part of its role in causing chronic diarrhea. The bacterium also seems capable of exerting toxic effects on the gut epithelium, although the mechanisms are not well understood.

Transmission of the bacterium is through contaminated food and water. It is difficult to diagnose in a clinical lab because EAEC is not easy to distinguish from other *E. coli,* including normal biota. Genotypic methods such as PCR are needed for accurate identification during outbreaks. The designation EAEC is not actually a serotype but is functionally defined as an *E. coli* that adheres in an aggregative pattern.

This bacterium seems to be associated with chronic diarrhea in people who are malnourished. It is not exactly clear whether the malnutrition predisposes patients to this infection or whether this infection contributes to malnutrition. Probably both possibilities are operating in patients, who are usually children in developing countries. More recently, the bacterium has been associated with acute diarrhea in industrialized countries, perhaps providing a clue to this question. It may be that in well-nourished hosts, the bacterium produces acute, self-limiting disease.

Cyclospora

Cyclospora cayetanensis is an emerging protozoan pathogen. Since the first occurrence in 1979, hundreds of outbreaks of cyclosporiasis have been reported in the United States and Canada. Its mode of transmission is fecal-oral—though not through the ingestion of cysts themselves. This differentiates this pathogen from its relative, *Cryptosporidium.* Infection occurs only after the oocyst sporulates, a process that begins the formation of the infectious form of the pathogen. In most cases, sporulated oocysts are ingested through the consumption of fresh produce and water presumably contaminated with feces. This disease occurs worldwide and, although primarily of human origin, is not spread directly from person to person. Outbreaks have been traced to imported raspberries, salad made with fresh greens, and drinking water. Since 2017, its incidence has increased dramatically for unknown reasons in the United States.

The organism is 8 to 10 micrometers in diameter **(figure 23.19)** and can usually be seen with a bright-field or phase contrast microscope. Diagnosis can be complicated by the lack of recognizable oocysts in the feces. Techniques that improve identification of the parasite are examination of fresh preparations under a fluorescent microscope and an acid-fast stain of a processed stool specimen. Autofluorescence of the oocysts can be visualized when specimens are exposed to ultraviolet (UV) light. A PCR-based test can also be used to identify *Cyclospora* and differentiate it from other parasites. This form of analysis is more sensitive and can detect protozoan genetic material even in the absence of actual cysts.

After an incubation period of about 1 week, symptoms of watery diarrhea, stomach cramps, bloating, fever, and muscle aches appear. Patients with prolonged diarrheal illness experience anorexia and weight loss.

Most cases of infection have been effectively controlled with trimethoprim-sulfamethoxazole lasting 1 week. Traditional antiprotozoan drugs are not effective. Some cases of disease may be prevented by cooking or freezing food to kill the oocysts.

Giardia duodenalis

Giardia duodenalis (also known as *Giardia intestinalis and Giardia lamblia*) is a pathogenic flagellated protozoan first observed by Antonie van Leeuwenhoek in his own feces.

Disease Table 23.5 Acute Diarrhea (With or Without Vomiting)

	Bacterial Causes			
Causative Organism(s)	*Salmonella* B G−	*Shigella* B G−	Shiga-toxin-producing *E. coli* (O157:H7 and others) B G−	Other *E. coli* (non-shiga-toxin-producing) B G−
Most Common Modes of Transmission	Vehicle (food, beverage), fecal-oral	Fecal-oral, direct contact	Vehicle (food, beverage), fecal-oral	Vehicle, fecal-oral
Virulence Factors	Adhesins, endotoxin	Endotoxin, enterotoxin, shiga toxins in some strains	Shiga toxins; proteins for attachment, secretion, effacement	Various: proteins for attachment, secretion, effacement; heat-labile and/or heat-stable exotoxins; invasiveness
Culture/Diagnosis	Stool culture, not usually necessary	Stool culture; antigen testing for shiga toxin	Stool culture, antigen testing for shiga toxin	Stool culture not usually necessary in absence of blood, fever
Prevention	Food hygiene and personal hygiene	Food hygiene and personal hygiene	Avoid live *E. coli* (cook meat and clean vegetables); stay aware of food recalls	Food and personal hygiene
Treatment	Rehydration; no antibiotic for uncomplicated disease; in complicated disease ciprofloxacin; resistant *Salmonella* is considered a **Serious Threat** by the CDC	Azithromycin or ciprofloxacin; drug-resistant *Shigella* is in the CDC's **Serious Threat** category	Antibiotics contraindicated, supportive measures	Rehydration, antimotility agent
Fever Present	Usually	Often	Often	Sometimes
Blood in Stool	Sometimes	Often	Usually	Sometimes
Distinctive Features	Often associated with chickens, reptiles	Very low ID_{50}	Hemolytic uremic syndrome (HUS)	ETEC, EIEC, EPEC, DAEC, EAEC
Epidemiological Features	United States: +/− 1.2 million cases/yr; 20% of all cases require hospitalization; death rate of 0.6% Category B Bioterrorism Agent	United States: estimated 450,000 cases per year; internationally: 165 million cases per year Category B Bioterrorism Agent	Internationally: causes HUS in 10% of patients; 25% of HUS patients suffer neurological complications, 50% have chronic renal sequelae Category B Bioterrorism Agent	–

For 200 years, it was considered a harmless or weak intestinal pathogen. Only since the 1950s has its prominence as a cause of diarrhea been recognized. In fact, it is the most common flagellate isolated in clinical specimens. Observed straight on, the trophozoite has a unique symmetrical heart shape with organelles positioned in such a way that it resembles a face **(figure 23.20).** Four pairs of flagella emerge from the ventral surface, which is concave and acts as a suction cup for attachment to a substrate. *Giardia* cysts are small and compact, and they contain four nuclei.

				Nonbacterial Causes		
Campylobacter B G−	*Clostridioides difficile* B G+	*Vibrio cholerae* B G−	Non-cholera *Vibrio* species B G−	*Cryptosporidium* P	Rotavirus V	Norovirus V
Vehicle (food, water), fecal-oral	Endogenous (normal biota)	Vehicle (water and some foods), fecal-oral	Vehicle (food or natural bodies of water)	Vehicle (water, food), fecal-oral	Fecal-oral, vehicle, fomite	Indirect, vehicle (food), direct contact
Adhesins, exotoxin, induction of autoimmunity	Enterotoxins A and B	Cholera toxin (CT)	–	Intracellular growth	–	Limited immunity to reinfection
Stool culture not usually necessary; dark-field microscopy	Stool culture, PCR, ELISA demonstration of toxins in stool	Clinical diagnosis, microscopic techniques, serological detection of antitoxin	Culture of stool or blood	Acid-fast staining, ruling out bacteria	Rapid antigen test	Rapid antigen test
Food and personal hygiene	–	Water and food hygiene	Avoiding raw shellfish	Water treatment, proper food handling	Oral live-virus vaccines	Hygiene
Rehydration; azithromycin in severe cases (antibiotic resistance rising; resistant *Campylobacter* is in CDC's **Serious Threat** category)	Metronidazole in mild cases; vancomycin for severe; fecal transplants; resistant strains are in the CDC's **Urgent Threat** category	Rehydration and possibly doxycycline	Doxycycline	None or nitazoxanide	Rehydration	Rehydration
Usually	Sometimes	No	Yes	Often	Often	Sometimes
No	Not usually; mucus prominent	No	No	Not usually	No	No
Guillain-Barré syndrome	Associated with disruption of normal biota	Rice-water stools	Sepsis can follow	Resistant to chlorine disinfection	Severe in infants	Resistant to disinfection
United States: 1.3 million cases per year; internationally: 400 million cases per year	United States: 500,000 cases per year	Global estimate: 21,000 to 143,000 deaths annually Category B Bioterrorism Agent	Cause 90% of seafood-related deaths in United States	United States: estimated 748,000 cases per year; 30% seropositive Category B Bioterrorism Agent	United States: 2 to 3 million cases per year; internationally: 125 million cases of infantile diarrhea annually	United States: second most common cause of foodborne illness hospitalization

▶ Signs and Symptoms

Typical symptoms include diarrhea of long duration, abdominal pain, and flatulence. Stools have a greasy, distinctly foul-smelling quality. Fever is usually not present.

▶ Pathogenesis and Virulence Factors

Ingested *Giardia* cysts enter the duodenum, germinate, and travel to the jejunum to feed and multiply. Some trophozoites remain on the surface, while others invade the deeper crypts to varying

Disease Table 23.6	Acute Diarrhea with Vomiting Caused by Exotoxins (Food Poisoning)		
Causative Organism(s)	*Staphylococcus aureus* exotoxin	*Bacillus cereus* exotoxin	*Clostridium perfringens* exotoxin
Most Common Modes of Transmission	Vehicle (food)	Vehicle (food)	Vehicle (food)
Virulence Factors	Heat-stable exotoxin	Heat-stable toxin, heat-labile toxin	Heat-labile toxin
Culture/Diagnosis	Usually based on epidemiological evidence	Microscopic analysis of food or stool	Detection of toxin in stool
Prevention	Proper food handling	Proper food handling	Proper food handling
Treatment	Supportive	Supportive	Supportive
Fever Present	Not usually	Not usually	Not usually
Blood in Stool	No	No	No
Distinctive Features	Suspect in foods with high salt or sugar content	Two forms: emetic and diarrheal	Acute abdominal pain
Epidemiological Features	United States: estimated 240,000 cases per year **Category B Bioterrorism Agent**	United States: estimated 63,000 cases per year	United States: estimated 966,000 cases per year **Category B Bioterrorism Agent**

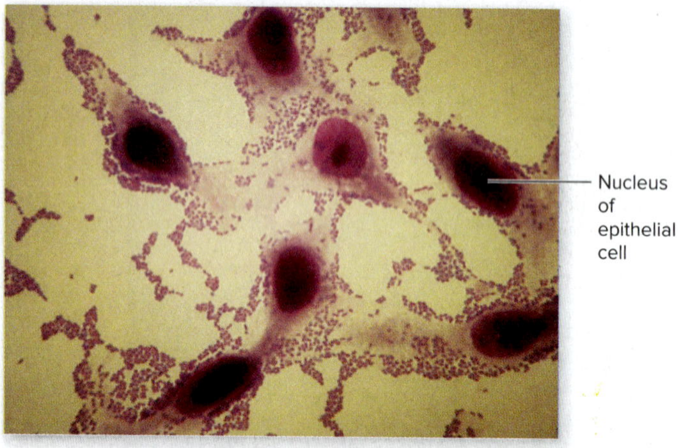
Nucleus of epithelial cell

Figure 23.18 Enteroaggregative *E. coli* adhering to epithelial cells.
Iruka Okeke

degrees. Superficial invasion by trophozoites causes damage to the epithelial cells, edema, and infiltration by white blood cells, but these effects are reversible. The presence of the protozoan leads to malabsorption (especially of fat) in the digestive tract and can cause significant weight loss.

Transmission and Epidemiology of Giardiasis

Giardiasis has a complex epidemiological pattern. The protozoan has been isolated from the intestines of beavers, cattle, coyotes, cats, and human carriers. Although both trophozoites and cysts escape in the stool, the cysts play a greater role in transmission. Unlike other pathogenic flagellates, *Giardia* cysts can survive for 2 months in the environment. Cysts are usually ingested with water and food or swallowed after close contact with infected people or contaminated objects. Infection can occur with a dose of only 10 to 100 cysts.

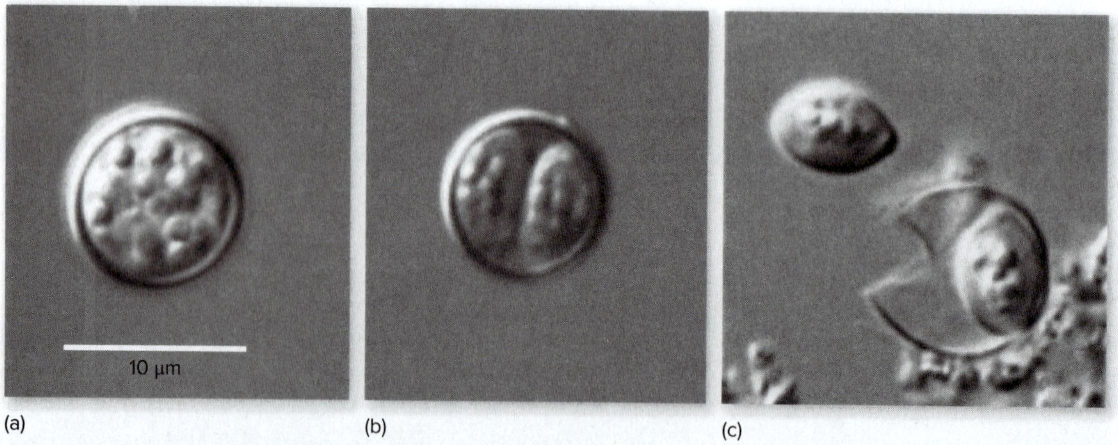
10 µm
(a)
(b)
(c)

Figure 23.19 ***Cyclospora cayetanensis.*** **(a)** The immature oocyst shed by infected people in their feces; **(b)** the oocyst is beginning to sporulate; two sporocysts are visible; **(c)** the oocyst has ruptured, releasing one of the sporocysts.
(a–c) CDC/DPDM

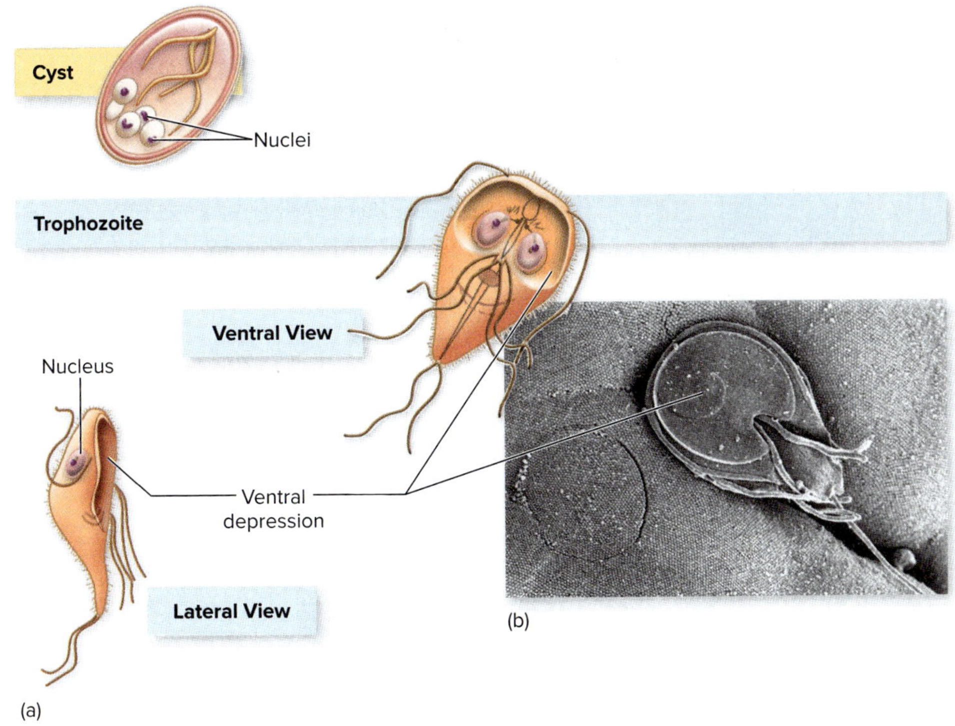
Cyst
Nuclei
Trophozoite
Ventral View
Nucleus
Ventral depression
Lateral View
(a)
(b)

Figure 23.20 ***Giardia lamblia*** **trophozoite.** **(a)** Schematic drawing. **(b)** Scanning electron micrograph of intestinal surface, revealing (on the left) the lesion left behind by an adhesive disk of a *Giardia* that has detached. The trophozoite on the right is lying on its "back" and is revealing its adhesive disk.
Dr. Stan Erlandsen/CDC

Giardia outbreaks have been traced to water from fresh mountain streams as well as chlorinated municipal water supplies in several states. Infections are not uncommon in hikers and campers who used what they thought was clean water from ponds, lakes, and streams in remote mountain areas. Because wild mammals such as muskrats and beavers are intestinal carriers, they could account for cases associated with drinking water from these sources.

Cases of fecal-oral transmission have been documented in day care centers. Food contaminated by infected persons has also transmitted the disease.

▶Culture and/or Diagnosis

Diagnosis of giardiasis can be difficult because the organism is shed in feces only intermittently. Sometimes ELISA tests are used to screen fecal samples for *Giardia* antigens, and PCR tests are available, although they are mainly used for detection of the protozoan in environmental samples.

▶Prevention and Treatment

There is a vaccine against *Giardia* that can be given to animals, including dogs. No human vaccine is available. Avoiding drinking from freshwater sources is the major preventive measure that can be taken. The agent is killed by boiling, ozone, and iodine; however, the amount of chlorine used in municipal water supplies does not destroy the cysts. Because cysts can be present in treated municipal water supplies, water agencies have had to rethink their policies on water maintenance and testing.

Treatment is with tinidazole or nitazoxanide.

Entamoeba

Amoebas are widely distributed in aqueous habitats and are frequent parasites of animals, but only a small number of them have the necessary virulence to invade tissues and cause serious pathology. One of the most significant pathogenic amoebas is *Entamoeba histolytica* (en″-tah-mee′-bah his″-toh-lit′-ih-kuh). The relatively simple life cycle of this parasite alternates between a large trophozoite that is motile by means of pseudopods and a smaller, compact, nonmotile cyst **(figure 23.21*a–d*).** The trophozoite lacks most of the organelles of other eukaryotes, and it has a single large nucleus that contains a prominent nucleolus called a *karyosome.* Amoebas from fresh specimens are often packed with food vacuoles containing host cells and bacteria. The mature cyst is encased in a thin yet tough wall and contains four nuclei as well as distinctive cigar-shaped bodies called *chromatoidal bodies,* which are actually dense clusters of ribosomes.

▶Signs and Symptoms

As hinted to by its species name, tissue damage is one of the formidable characteristics of untreated *E. histolytica* infection. Clinical amoebiasis exists in intestinal and extraintestinal forms. The initial targets of intestinal amoebiasis are the cecum, appendix, colon, and rectum. The amoeba secretes enzymes that dissolve tissues, and it actively penetrates deeper layers of the mucosa, leaving erosive ulcerations. This phase is marked by dysentery (bloody, mucus-filled stools), abdominal pain, fever, diarrhea, and weight loss. The most life-threatening manifestations of intestinal infection are hemorrhage, perforation, appendicitis, and tumorlike growths called amebomas. Lesions in the mucosa of the colon have a characteristic flasklike shape.

Extraintestinal infection occurs when amoebas invade the viscera of the peritoneal cavity. The most common site of invasion is the liver. Here, abscesses containing necrotic tissue and trophozoites develop and cause amoebic hepatitis. Another, rarer complication is pulmonary amoebiasis. Other infrequent targets of infection are the spleen, adrenals, kidney, skin, and brain. Severe forms of the disease result in about a 10% fatality rate.

▶Pathogenesis and Virulence Factors

Amoebiasis begins when viable cysts are swallowed and arrive in the small intestine, where the alkaline pH and digestive juices of this environment stimulate excystment. Each cyst releases four trophozoites, which are swept into the cecum and large intestine. There, the trophozoites attach by fine pseudopods, multiply,

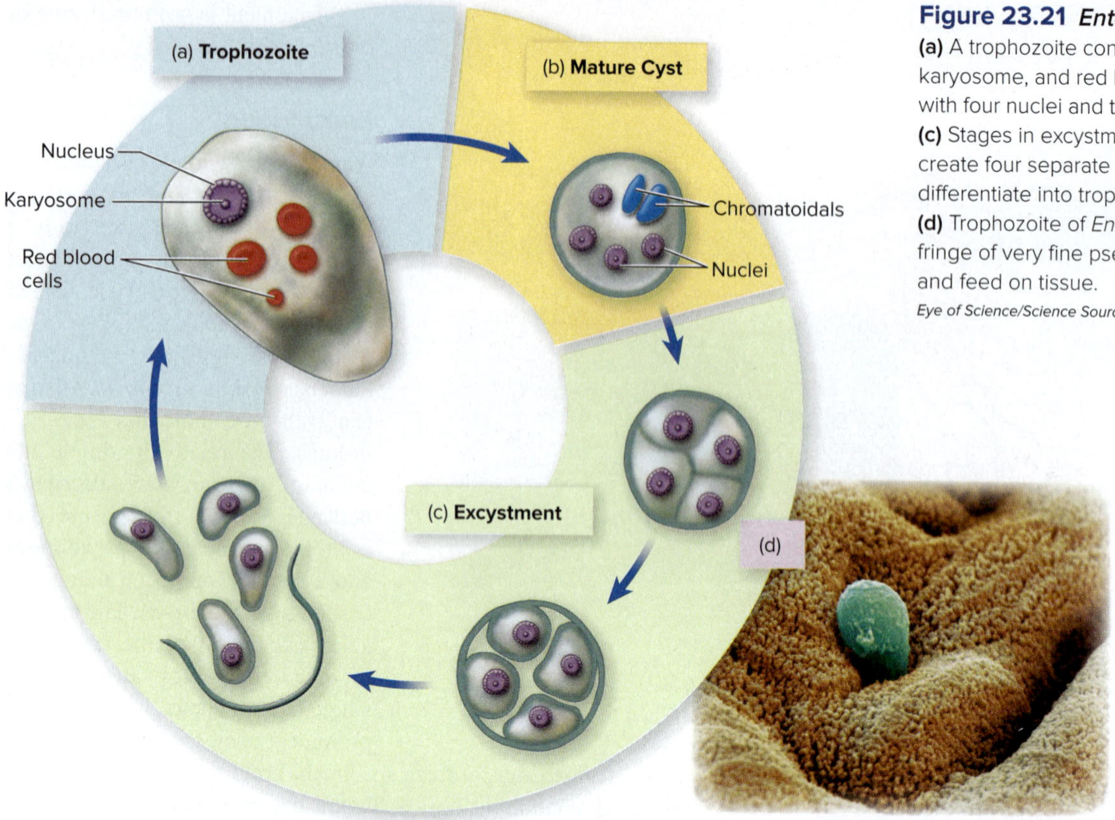
(a) Trophozoite
(b) Mature Cyst
Nucleus
Karyosome
Red blood cells
Chromatoidals
Nuclei
(c) Excystment
(d)

Figure 23.21 ***Entamoeba histolytica.*** **(a)** A trophozoite containing a single nucleus, a karyosome, and red blood cells. **(b)** A mature cyst with four nuclei and two blocky chromatoidals. **(c)** Stages in excystment. Divisions in the cyst create four separate cells, or metacysts, that differentiate into trophozoites and are released. **(d)** Trophozoite of *Entamoeba histolytica.* Note the fringe of very fine pseudopods it uses to invade and feed on tissue.
Eye of Science/Science Source

actively move about, and feed. In about 90% of patients, infection is asymptomatic or very mild, and the trophozoites do not invade beyond the most superficial layer. The severity of the infection can vary with the strain of the parasite, inoculum size, diet, and host resistance.

The secretion of lytic enzymes by the amoeba seems to induce apoptosis of host cells. This means that the host is contributing to the process by destroying its own tissues on cue from the protozoan. The invasiveness of the amoeba is also a clear contributor to its pathogenicity.

▶ Transmission and Epidemiology

Entamoeba is harbored by chronic carriers whose intestines favor the encystment stage of the life cycle. Cyst formation cannot occur in active dysentery because the feces are so rapidly flushed from the body, but after recuperation, cysts are continuously shed in feces.

Humans are the primary hosts of *E. histolytica.* Infection is usually acquired by ingesting food or drink contaminated with cysts released by an asymptomatic carrier. The amoeba is thought to be carried in the intestines of one-tenth of the world's population, and it kills up to 100,000 people a year. Its geographic distribution is partly due to local sewage disposal and fertilization practices. Occurrence is highest in tropical regions (Africa, Asia, and Latin America), where "night soil" (human excrement) or untreated sewage is used to fertilize crops and the sanitation of water and food can be substandard. Although the prevalence of the disease is lower in the United States, as many as 10 million people may harbor the agent.

▶ Culture and/or Diagnosis

Diagnosis of this protozoal infection relies on a combination of tests, including microscopic examination of stool for the characteristic cysts or trophozoites, ELISA tests of stool for *E. histolytica* antigens, and serological testing for the presence of antibodies to the pathogen. PCR testing is the current gold standard. It is important to differentiate *E. histolytica* from the similar *Entamoeba coli* and *Entamoeba dispar,* which occur as normal biota.

▶ Prevention and Treatment

No vaccine yet exists for *E. histolytica.* Prevention of the disease therefore relies on purification of water. Because regular chlorination of water supplies does not kill cysts, more rigorous methods such as boiling or iodine are required.

Effective treatment usually involves the use of drugs such as metronidazole (Flagyl) or paramomycin. Dehydroemetine is used to control symptoms, but it will not cure the disease. Other drugs are given to relieve diarrhea and cramps, while lost fluid and electrolytes are replaced by oral or intravenous therapy. Infection with *E. histolytica* provokes antibody formation against several antigens, but permanent immunity is unlikely and reinfection can occur.

Disease Table 23.7 Chronic Diarrhea

Causative Organism(s)	Enteroaggregative *E. coli* (EAEC) B G–	*Cyclospora cayetanensis* P	*Giardia duodenalis* P	*Entamoeba histolytica* P
Most Common Modes of Transmission	Vehicle (food, water), fecal-oral	Fecal-oral, vehicle	Vehicle, fecal-oral, direct and indirect contact	Vehicle, fecal-oral
Virulence Factors	?	Invasiveness	Attachment to intestines alters mucosa	Lytic enzymes, induction of apoptosis, invasiveness
Culture/Diagnosis	Difficult to distinguish from other *E. coli*	Stool examination, PCR	Stool examination, ELISA	PCR, stool examination, ELISA, serology
Prevention	?	Washing, cooking food, personal hygiene	Water hygiene, personal hygiene	Water hygiene, personal hygiene
Treatment	Rehydration, or ciprofloxacin	TMP-SMZ	Tinidazole, nitazoxanide	Metronidazole or paramomycin
Fever Present	No	Usually	Not usually	Yes
Blood in Stool	Sometimes, mucus also	No	No, mucus present (greasy and foul smelling)	Yes
Distinctive Features	Chronic in the malnourished	–	Frequently occurs in backpackers, campers	–
Epidemiological Features	Developing countries: 87% of chronic diarrhea in children >2 years old	United States: estimated 16,000 cases per year; internationally: endemic in 27 countries, mostly tropical	United States: estimated 1.2 million cases per year; internationally: prevalence rates from 2% to 5% in industrialized world	Internationally: 40,000 to 100,000 deaths annually

Hepatitis

When certain viruses infect the liver, they cause **hepatitis,** an inflammatory disease marked by necrosis of hepatocytes and a response by mononuclear white blood cells that swells and disrupts the liver architecture. This pathologic change interferes with the liver's excretion of bile pigments such as bilirubin into the intestine. When bilirubin, a greenish-yellow pigment, accumulates in the blood and tissues, it causes **jaundice,** a yellow tinge in the skin and eyes. The condition can be caused by a variety of viruses, including cytomegalovirus and Epstein-Barr virus. The others are all called "hepatitis viruses," but only because they all can cause this inflammatory condition in the liver, not because they belong to a "hepatitis" family of viruses. They are quite different from one another. In this section we will cover five hepatitis-causing viruses that are well understood, named hepatitis A through hepatitis E.

Note that noninfectious conditions can also cause inflammation and disease in the liver, including some autoimmune conditions, drugs, and alcohol abuse.

Hepatitis A Virus

Hepatitis A virus (HAV) is a nonenveloped, single-stranded RNA enterovirus. It belongs to the family *Picornaviridae.* In general, HAV disease is milder and shorter-term than the other forms.

Signs and Symptoms

Most infections by this virus are either subclinical or accompanied by vague, flulike symptoms. In more overt cases, the presenting symptoms may include jaundice and swollen liver. Darkened urine is often seen in this and other hepatitises. Jaundice is present in only about 10% of the cases. Hepatitis A occasionally occurs as a fulminant disease (meaning it is sudden and very acute) and causes liver damage, but this manifestation is relatively rare. The virus is not oncogenic (cancer causing), and complete, uncomplicated recovery can be expected.

Pathogenesis and Virulence Factors

Most of the pathogenic effects are thought to be the result of host response to the presence of virus in the liver.

Transmission and Epidemiology

There is an important distinction between this virus and hepatitis B and C viruses: Hepatitis A virus is spread through the fecal-oral route (and is sometimes known as infectious hepatitis). In general, the disease is associated with deficient personal hygiene and lack of public health measures. In countries with inadequate sewage control, most outbreaks are associated with feces-contaminated water and food. In the United States before 2016, 1,000 to 2,000 cases occurred annually. Most of these were a result of close institutional contact, unhygienic food handling, consumption of shellfish, sexual transmission, or travel to other countries. In 2016 the epidemiology of the infection changed and the rates increased drastically. The majority of new cases are transmitted through injecting drug use.

Hepatitis A can be spread by blood or blood products, but this is the exception rather than the rule. In developing countries, children are the most common victims because exposure to the virus tends to occur early in life, whereas in North America and Europe, more cases appear in adults.

Culture and/or Diagnosis

Diagnosis of the disease is aided by detection of anti-HAV IgM antibodies produced early in the infection and by tests to identify HA antigen or virus directly in stool samples.

Prevention and Treatment

Prevention of hepatitis A is based primarily on immunization. Two inactivated viral vaccines (Havrix and VAQTA) are used in the United States today. Short-term protection can be conferred by passive immune globulin. This treatment is useful for people who have come in contact with HAV-infected individuals or who have eaten at a restaurant that was the source of a recent outbreak. It has also recently been discovered that administering Havrix after exposure can prevent symptoms. A combined hepatitis A/hepatitis B vaccine, called Twinrix, is recommended for people who may be at risk for both diseases, such as people with chronic liver dysfunction, intravenous drug users, and men who have sex with men. Travelers to areas with high rates of both diseases should obtain vaccine coverage as well. Development of active natural immunity toward hepatitis A virus leads to lifelong protection from reinfection.

No specific medication is available for hepatitis A once the symptoms begin. Drinking lots of fluids and avoiding liver irritants such as aspirin or alcohol will speed recovery. Patients who receive immune globulin early in the disease usually experience milder symptoms than patients who do not receive it.

Hepatitis B Virus

Hepatitis B virus (HBV) is an enveloped DNA virus in the family *Hepadnaviridae*. Intact viruses are often called Dane particles. The genome is partly double-stranded and partly single-stranded **(figure 23.22).**

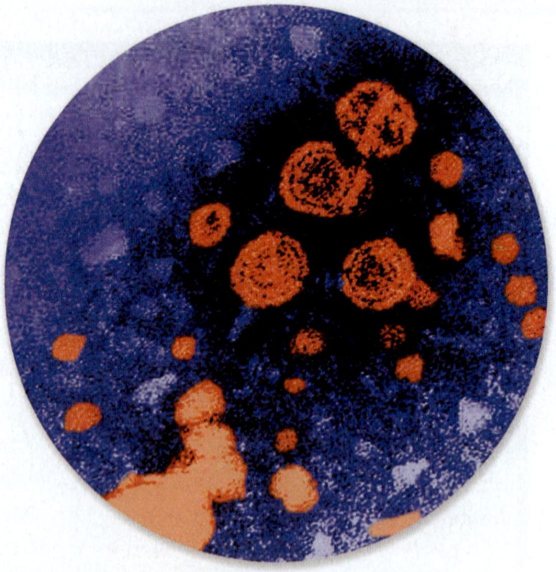

Figure 23.22 Hepatitis B virions.
Dr. Erskine Palmer/CDC

Signs and Symptoms

In addition to the direct damage to liver cells, the spectrum of hepatitis disease may include fever, chills, malaise, anorexia, abdominal discomfort, diarrhea, and nausea. Rashes may appear and arthritis may occur. Hepatitis B infection can be very serious, even life-threatening. A small number of patients develop glomerulonephritis and arterial inflammation. Complete liver regeneration and restored function occur in most patients. A small number of patients develop chronic liver disease in the form of necrosis or cirrhosis (permanent liver scarring and loss of tissue). In some cases, chronic HBV infection can lead to liver cancer.

Patients who become infected as children have significantly higher risks of long-term infection and disease. In fact, 90% of neonates infected at birth develop chronic infection, as do 30% of children infected between the ages of 1 and 5, but only 6% of persons infected after the age of 5. This finding is one of the major justifications for the routine vaccination of children. Also, infection becomes chronic more often in men than in women. The mortality rate is 15% to 25% for people with chronic infection.

Some patients infected with hepatitis B are coinfected with a particle called the delta agent, sometimes also called a **hepatitis D** virus. This agent is a defective RNA virus that cannot produce infection unless a cell is also infected with HBV. Hepatitis D virus invades host cells by "borrowing" the outer receptors of HBV. When HBV infection is accompanied by the delta agent, the disease becomes more severe and is more likely to progress to permanent liver damage.

Pathogenesis and Virulence Factors

The hepatitis B virus enters the body through a break in the skin or mucous membrane or by injection into the bloodstream. Eventually, it reaches the liver cells (hepatocytes), where it multiplies and releases viruses into the blood during an incubation period of 4 to 24 weeks (7 weeks average). Surprisingly, the majority of people infected exhibit few overt symptoms and eventually develop an immunity to HBV, but some people experience the symptoms described earlier. The precise mechanisms of virulence are not clear. The ability of HBV to remain latent in some patients contributes to its pathogenesis. Chronic infection without overt symptoms sometimes leads to a condition called necroinflammation, in which protracted inflammation caused by the presence of the virus leads to liver disease.

▶ Transmission and Epidemiology

An important factor in the transmission pattern of hepatitis B virus is that it multiplies exclusively in the liver, which continuously seeds the blood with viruses. Electron microscopic studies have revealed up to 10^7 virions per milliliter of infected blood. Even a minute amount of blood (a *millionth* of a milliliter) can transmit infection. The abundance of circulating virions is so high and the minimal dose so low that such simple practices as sharing a toothbrush or a razor can transmit the infection. HBV has also been detected in semen and vaginal secretions, and it can be transmitted by these fluids. Growing concerns about virus spread through donated organs and tissue are prompting increased testing prior to surgery. Spread of the virus by means of close contact in families or institutions is also well documented. Vertical transmission is possible, and it predisposes the child to development of the carrier state and increased risk of liver cancer. The disease is sometimes known as *serum hepatitis.*

This virus is one of the major infectious concerns for health care workers. Needlesticks can easily transmit the virus. For that reason, most workers are required to have the full series of HBV vaccinations. Unlike HIV, HBV remains infective for days in dried blood, for months when stored in serum at room temperature, and for decades if frozen. Although it is not inactivated after 4 hours of exposure to 60°C, boiling for the same period can destroy it. Disinfectants containing chlorine, iodine, and glutaraldehyde show potent anti–hepatitis B activity.

Tattooing and ear or body piercing can expose a person to infection if the instruments are not properly sterilized. The only reliable method for destroying HBV on reusable instruments is autoclaving.

▶ Culture and/or Diagnosis

Serological tests can detect either virus antigen or antibodies. ELISA testing permits detection of the important surface antigen of HBV very early in infection. These tests are essential for screening blood destined for transfusions, semen in sperm banks, and organs intended for transplant. Antibody tests are most valuable in patients who are negative for the antigen.

▶ Prevention and Treatment

The primary prevention for HBV infection is vaccination. The most widely used vaccines are recombinant, containing the pure surface antigen cloned in yeast cells. Vaccines are given in three doses over 18 months, with occasional boosters. Vaccination is a must for medical and dental workers and students, patients receiving multiple transfusions, immunodeficient persons, and cancer patients. The vaccine is also now strongly recommended for all newborns as part of a routine immunization schedule. As just mentioned, a combined vaccine for HAV/HBV may be appropriate. A vaccine called Pediarix contains protection against hepatitis B, diphtheria, tetanus, pertussis, and polio. It is recommended for children between 6 weeks and 7 years.

Passive immunization with hepatitis B immune globulin (HBIG) gives significant immediate protection to people who have been exposed to the virus through needle puncture, broken blood containers, or skin and mucosal contact with blood. Another group for whom passive immunization is highly recommended is neonates born to infected mothers.

Mild cases of hepatitis B are managed by symptomatic treatment and supportive care. Chronic infection can be controlled with recombinant human interferon, tenofovir, or entecavir. Each of these can help to slow virus multiplication and prevent liver damage in many but not all patients. None of the drugs are considered curative. Different drug regimens are called for when a patient is coinfected with HBV and HIV.

Hepatitis C Virus

Hepatitis C is sometimes referred to as the "silent epidemic" because 2.4 million Americans are infected with the virus, but it takes many years to cause noticeable symptoms. Liver failure from hepatitis C is one of the most common reasons for liver transplants in this country. Hepatitis C is an RNA virus in the *Flaviviridae* family. It is closely related to viruses causing West Nile fever and yellow fever. It used to be known as "non-A non-B" virus. It is usually diagnosed with a blood test for antibodies to the virus.

▶ Signs and Symptoms

People have widely varying experiences with this infection. It shares many characteristics of hepatitis B disease, but it is much more likely to become chronic. Of those infected, 75% to 85% will remain infected indefinitely. (In contrast, only about 6% of persons who acquire hepatitis B after the age of 5 will be chronically infected.) With HCV infection, it is possible to have severe symptoms without permanent liver damage, but it is more common to have chronic liver disease even if there are no overt symptoms. Cancer may also result from chronic HCV infection. Worldwide, HBV infection is the most common cause of liver cancer, but in the United States it is more likely to be caused by HCV. Current CDC recommendations are for adults between 18 and 79 years old to be screened for hepatitis C.

▶ Pathogenesis and Virulence Factors

The virus is so adept at establishing chronic infections that researchers are studying the ways that it evades immunologic

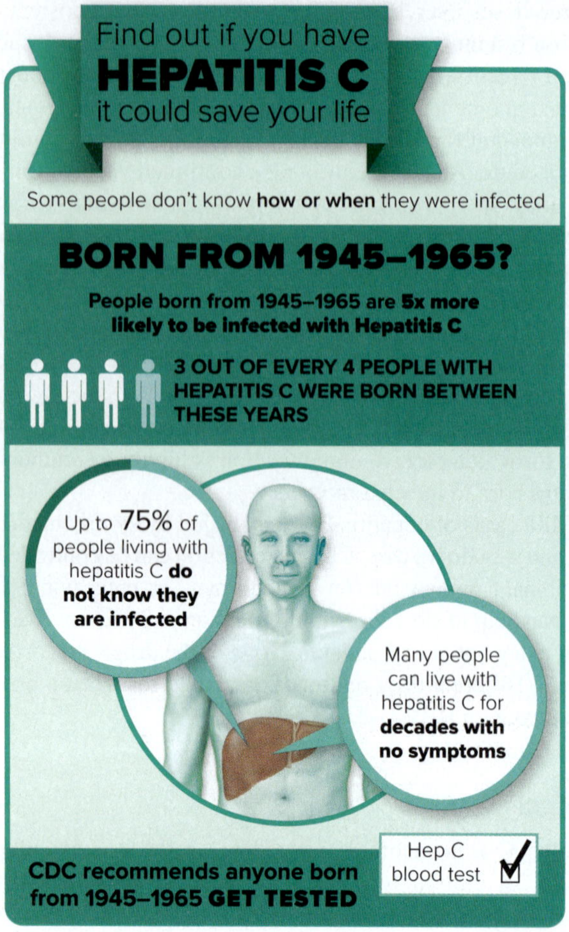
Find out if you have
HEPATITIS C
it could save your life
Some people don't know how or when they were infected
BORN FROM 1945–1965?
People born from 1945–1965 are 5x more likely to be infected with Hepatitis C
3 OUT OF EVERY 4 PEOPLE WITH HEPATITIS C WERE BORN BETWEEN THESE YEARS
Up to 75% of people living with hepatitis C do not know they are infected
Many people can live with hepatitis C for decades with no symptoms
CDC recommends anyone born from 1945–1965 GET TESTED
Hep C blood test

Figure 23.23 Testing recommendations for hepatitis C.
Source: CDC

detection and destruction. The virus's core protein seems to play a role in the suppression of cell-mediated immunity as well as in the production of various cytokines. Scientists recently identified that HCV enters liver cells using the same receptors utilized for cholesterol entry. This is a potential target for therapeutic drug development.

Transmission and Epidemiology

This virus is acquired in ways similar to HBV. It is more commonly transmitted through blood contact (both "sanctioned," such as in blood transfusions, and "unsanctioned," such as needle sharing by injecting drug users) than through transfer of other body fluids. Vertical transmission is also possible. Anyone with a history of exposure to blood products or organs before 1992 (when effective screening became available) is at higher risk for this infection, as is anyone with a history of injecting drug use.

Because HCV was not recognized sooner, a relatively large percentage of the population is infected **(figure 23.23).** Eighty percent of the roughly 3.5 million affected in this country are suspected to have no symptoms. It has a very high prevalence in parts of South America, Central Africa, and China.

Prevention and Treatment

There is currently no vaccine for hepatitis C. In 2013, a two-drug regimen became available that has excellent results. Sofosbuvir is a nucleotide analog that "fools" the RNA polymerase of the virus. Simeprevir acts as a protease inhibitor.

Disease Table 23.8 Hepatitis

Causative Organism(s)	Hepatitis A or E virus V	Hepatitis B virus V	Hepatitis C virus V
Most Common Modes of Transmission	Fecal-oral, vehicle	Parenteral (blood contact), direct contact (especially sexual), vertical	Parenteral (blood contact), vertical
Virulence Factors	–	Latency	Core protein suppresses immune function?
Culture/Diagnosis	IgM serology	ELISA	Serology, also PCR
Prevention	Hepatitis A vaccine or combined HAV/HBV vaccine	HBV recombinant vaccine	–
Treatment	HAV: hepatitis A vaccine or immune globulin; HEV: immune globulin	Interferon, tenofovir, or entecavir	Sofosbuvir + simeprevir
Incubation Period	2 to 4 weeks	1 to 6 months	2 to 8 weeks
Epidemiological Features	Hepatitis A, United States: 20,000 cases annually and 40% of adults show evidence of prior infection; internationally: 1.4 million cases per year; hepatitis E, internationally: 20 million infections per year; 60% in East and Southeast Asia	United States: 19,000 new cases per year; 800,000 to 2.2 million have chronic infection; internationally: 240 million	United States: estimated 30,000 new diagnoses per year; internationally: 150 million chronically infected

A Note About Hepatitis E Virus

Another RNA virus, called hepatitis E, causes a type of hepatitis very similar to that caused by hepatitis A. Like hepatitis A virus, it is a single-stranded, nonenveloped RNA virus. The disease it causes is usually self-limiting and often goes undiagnosed. The jaundice that is typical of other forms of hepatitis is not seen in hepatitis E infection. In rare cases, hepatitis E disease can lead to acute liver failure. Pregnant women in their third trimester are at highest risk for this severe form of disease, and the fatality rate is nearly 20%. Overall, hepatitis E infection is more common in developing countries. A majority of the cases reported in the United States occur in people who have traveled to these endemic regions. Hepatitis E virus is transmitted by the fecal-oral route, mainly through contaminated water and food. The infection does not seem to be transmitted from person to person, though blood transfusions have been documented to transmit the pathogen to uninfected patients. There is currently no vaccine.

23.2 Learning Outcomes—Assess Your Progress

5. List the possible causative agents for the following infectious gastrointestinal conditions: dental caries, periodontal diseases, mumps, and gastric ulcers.
6. Name nine bacterial and three nonbacterial causes of acute diarrhea, and identify the most common cause of foodborne illness in the United States.
7. Name one distinct feature for each of the acute diarrhea pathogens.
8. Differentiate between food poisoning and foodborne infection.
9. Identify three causative agents for chronic diarrhea.
10. Differentiate among the main types of hepatitis and discuss causative agents, modes of transmission, diagnostic techniques, prevention, and treatment of each.

23.3 Helminthic Diseases Manifesting in the Gastrointestinal Tract

Helminths that parasitize humans are amazingly diverse, ranging from barely visible roundworms (0.3 mm) to huge tapeworms (25 m long). In the introduction to these organisms in chapter 5, we grouped them into three categories—nematodes (roundworms), trematodes (flukes), and cestodes (tapeworms)—and discussed basic characteristics of each group. You may want to review those sections before continuing. In this section, we examine the intestinal diseases caused by helminths. Although they can cause symptoms that may be mistaken for some of the diseases discussed elsewhere in this chapter, helminthic diseases are usually accompanied by an additional set of symptoms that arise from the host response to helminths. Helminthic infection usually provokes an increase in granular leukocytes called eosinophils, which have a specialized capacity to destroy multicellular parasites. This increase, termed **eosinophilia,** is a hallmark of helminthic infection and is detectable in blood counts. If the following symptoms occur coupled with eosinophilia, helminthic infection should be suspected. Many of these infections are considered "neglected tropical infections"—infections that cast a large burden of disease in the poorest countries of the world yet receive the least recognition and research funding today (see note in chapter 5). Due to the efforts of dedicated tropical disease medicine specialists and organizations, such as both the Carter and the Gates foundations, some of these helminthic diseases are on the decline.

Helminthic infections may be acquired through the fecal-oral route or through penetration of the skin, but most of them spend part of their lives in the intestinal tract. **Figure 23.24** depicts the four different types of life cycles of the helminths. While the worms are in the intestines, they can produce a range of intestinal symptoms. Some of them also produce symptoms outside of the intestine. They are considered in separate categories.

General Clinical Considerations

Up to this point, the diseases in this book have been arranged in a particular way, based on how the disease appears in terms of signs and symptoms (how the patient appears upon presentation to the health care provider). But this section on helminthic diseases will take a different approach. We talk about diagnosis, pathogenesis and prevention, and treatment of the helminths as a group in the next subsections. Each type of infection is then described in the sections that follow.

Pathogenesis and Virulence Factors in General

Helminths have numerous adaptations that allow them to survive in their hosts. They have specialized mouthparts for attaching to tissues and for feeding, enzymes with which they liquefy and penetrate tissues, and a cuticle or other covering to protect them from host defenses. In addition, their organ systems are usually reduced to the essentials: getting food and processing it, moving, and reproducing. The damage they cause in the host is very often the result of the host's response to the presence of the invader.

Many helminths have more than one host during their lifetimes. If this is the case, the host in which the adult worm reproduces sexually is called the **definitive host** (usually a vertebrate).

Sometimes the actual definitive host is not the host usually used by the parasite but an accidental bystander. Humans often become the accidental definitive hosts for helminths whose normal definitive host is a cow, pig, or fish. Larval stages of helminths are found in intermediate hosts. Humans can serve as intermediate hosts, too. Helminths may require no intermediate host at all or may need one or more intermediate hosts for their entire life cycle.

Diagnosis in General

Diagnosis of almost all helminthic infections follows a similar series of steps. A differential blood count showing eosinophilia and serological tests indicating sensitivity to helminthic antigens all provide indirect evidence of worm infection. A history of travel

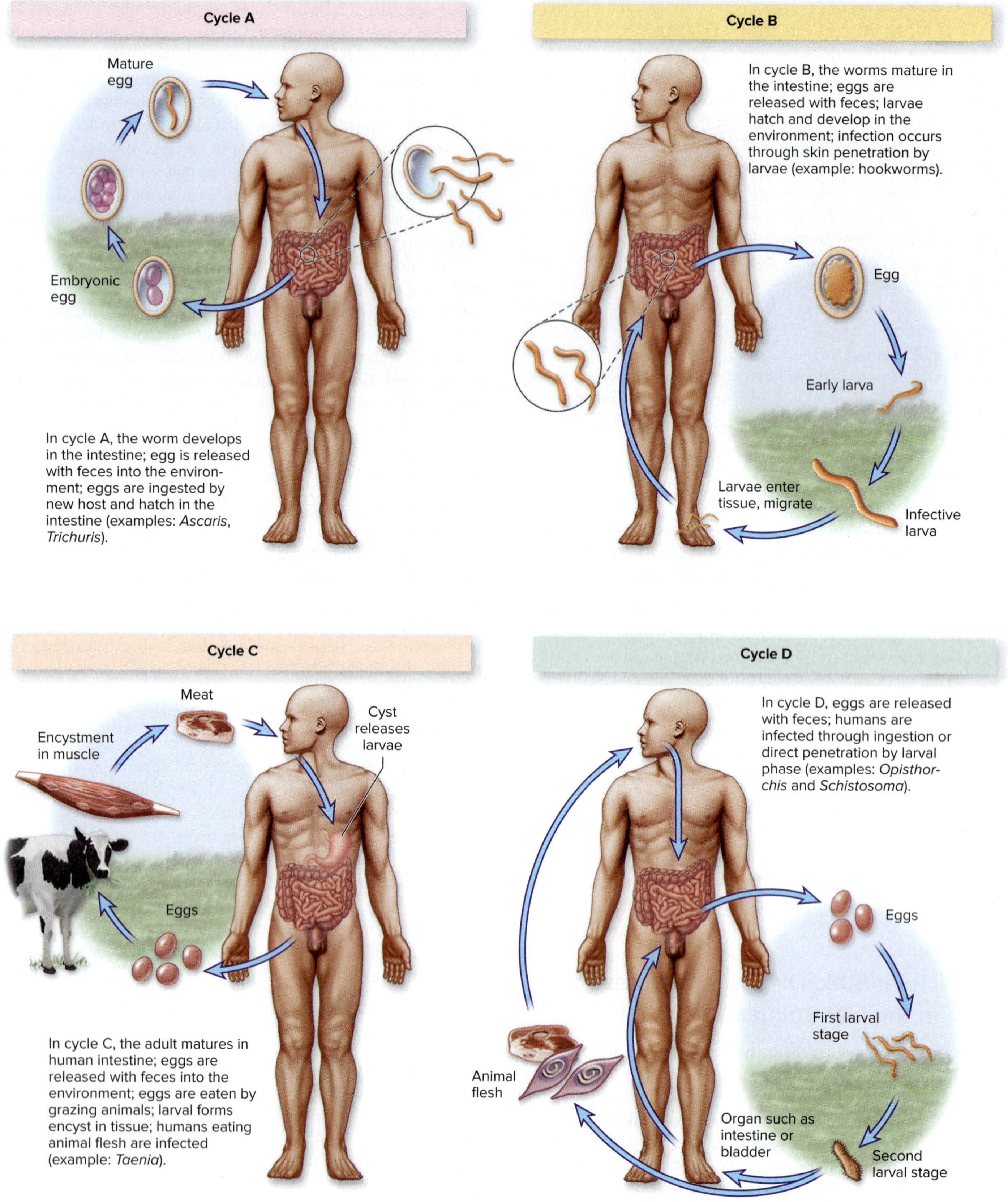
Cycle A
Mature egg
Embryonic egg
In cycle A, the worm develops in the intestine; egg is released with feces into the environ-ment; eggs are ingested by new host and hatch in the intestine (examples: *Ascaris*, *Trichuris*).
Cycle B
In cycle B, the worms mature in the intestine; eggs are released with feces; larvae hatch and develop in the environment; infection occurs through skin penetration by larvae (example: hookworms).
Egg
Early larva
Larvae enter tissue, migrate
Infective larva
Cycle C
Meat
Encystment in muscle
Cyst releases larvae
Eggs
In cycle C, the adult matures in human intestine; eggs are released with feces into the environment; eggs are eaten by grazing animals; larval forms encyst in tissue; humans eating animal flesh are infected (example: *Taenia*).
Cycle D
In cycle D, eggs are released with feces; humans are infected through ingestion or direct penetration by larval phase (examples: *Opisthor-chis* and *Schistosoma*).
Eggs
First larval stage
Animal flesh
Organ such as intestine or bladder
Second larval stage

Figure 23.24 Four basic helminth life and transmission cycles.

to the tropics or immigration from those regions is also helpful, even if it occurred years ago, because some flukes and nematodes persist for decades. The most definitive evidence, however, is the discovery of eggs, larvae, or adult worms in stools or in tissues. The worms are sufficiently distinct in morphology that positive identification can be based on any stage, including eggs. Stool is commonly examined in a microscopic procedure called "an O & P," an ova and parasite smear. That said, not all of these diseases result in eggs or larval stages that can easily be found in stool.

Prevention and Treatment in General

No vaccines are available to prevent any of the helminthic infections described here. Treatment twice a year with one of the antihelminthic drugs has been shown to keep people in endemic regions healthy. In areas where worms are transmitted by fecally contaminated soil and water, disease rates are significantly reduced through proper sewage disposal, using sanitary latrines, avoiding human feces as fertilizer, and disinfection of the water supply. In cases in which the larvae invade through the skin, people should avoid direct contact with infested water and soil. Foodborne disease can be avoided by thoroughly washing and cooking vegetables and meats. Also, because adult worms, larvae, and eggs are sensitive to cold, freezing foods is a highly satisfactory preventive measure. These methods work best if humans are the sole host of the parasite. If they are not, control of reservoirs or vector populations may be necessary.

Some helminths have developed resistance to the drugs used to treat them. In some cases, surgery may be necessary to remove worms or larvae, although this procedure can be difficult if the parasite load is high or is not confined to one area.

Helminth Disease: Intestinal Distress as the Primary Symptom

Both tapeworms and roundworms can infect the intestinal tract in such a way as to cause primary symptoms there. The pinworm *Enterobius vermicularis* and the whipworm *Trichuris trichiura* are two of these. In addition to these nematodes, two tapeworm genera can be responsible: *Hymenolepis* (species *nana* and *dimunata*) and *Diphyllobothrium latum*.

Enterobius vermicularis

This nematode is often called the pinworm, or threadworm. It is the most common worm disease of children in temperate zones. Some estimates put the prevalence of this infection in the United States at 5% to 15%, amounting to up to 40 million cases a year. The transmission of this roundworm is of the cycle A type.

Freshly deposited eggs have a sticky coating that causes them to lodge beneath the fingernails and to adhere to fomites. Upon drying, the eggs become airborne and settle in house dust. Eggs are ingested from contaminated food or drink and from self-inoculation from one's own fingers. Eggs hatch in the small intestine and release larvae that migrate to the large intestine. There the larvae mature into adult worms and mate.

The hallmark symptom of this condition is pronounced anal itching when the mature female emerges from the anus and lays eggs. Although infection is not fatal and most cases are asymptomatic, the afflicted child can suffer from disrupted sleep and sometimes nausea, abdominal discomfort, and diarrhea. A simple rapid test can be performed by pressing a piece of transparent adhesive tape against the anal skin and then applying it to a slide for microscopic examination. The eggs are not excreted with the stool, so a standard O & P is not useful. When one member of the family is diagnosed, the entire family should be tested and/or treated because it is likely that multiple members are infected.

Trichuris trichiura

The common name for this nematode—whipworm—refers to its likeness to a miniature buggy whip. Its life cycle and transmission are of the cycle A type (see figure 23.24). Humans are the sole host. Trichuriasis has its highest incidence in areas of the tropics and subtropics that have poor sanitation. Embryonic eggs deposited in the soil are not immediately infective and continue development for 3 to 6 weeks in this habitat. Ingested eggs hatch in the small intestine, where the larvae attach, penetrate the outer wall, and go through several molts. The mature adults move to the large intestine and gain a hold with their long, thin heads, while the thicker tail dangles free in the intestinal lumen. Following sexual maturation and fertilization, the females eventually lay 3,000 to 5,000 eggs daily into the bowel. The entire cycle requires about 90 days, and untreated infection can last up to 2 years.

Symptoms of this infection may include localized hemorrhage of the bowel caused by worms burrowing and piercing intestinal mucosa. This can also provide a portal of entry for secondary bacterial infection. Heavier infections can cause dysentery, loss of muscle tone, and rectal prolapse, which can prove fatal in children.

Diphyllobothrium latum

This tapeworm has an intermediate host in fish. It is common in the Great Lakes, Alaska, and Canada. Humans are its definitive host. It develops in the intestine and can cause long-term symptoms. It can be transmitted in raw food such as sushi and sashimi made from salmon. (Reputable sushi restaurants employ authentic sushi chefs who are trained to carefully examine fish for larvae and other signs of infection.) It must also be recognized that transmission of this pathogen can occur in the United States through the consumption of undercooked or lightly smoked trout, perch, or pike—all very common fish to the American freshwater sport fisherman.

As is the case with most tapeworms, symptoms are minor and usually vague and include possible abdominal discomfort or nausea. The tapeworm seems to have the ability to absorb and use vitamin B_{12}, making it unavailable to its human host. Anemia is therefore sometimes reported with this infection. You should be aware that certain people of Scandinavian descent have a genetic

Disease Table 23.9 Intestinal Distress

Causative Organism(s)	*Enterobius vermicularis* (pinworm) H	*Trichuris trichiura* (whipworm) H	*Diphyllobothrium latum* (fish tapeworm) H	*Hymenolepis nana* and *H. diminuta* H
Most Common Modes of Transmission	Cycle A: vehicle (food, water), fomites, self-inoculation	Cycle A: vehicle (soil), fecal-oral	Cycle C: vehicle (seafood)	Cycle C: vehicle (ingesting insects), fecal-oral
Virulence Factors	–	Burrowing and invasiveness	Vitamin B_{12} usage	–
Culture/Diagnosis	Adhesive tape + microscopy	Blood count, serology, egg or worm detection	Blood count, serology, egg or worm detection	Blood count, serology, egg or worm detection
Prevention	Hygiene	Hygiene, sanitation	Cook meat	Hygienic environment
Treatment	Mebendazole, piperazine	Mebendazole	Praziquantel	Praziquantel
Distinctive Features	Common in United States	Humans sole host	Largest human tapeworm; can be up to 30 feet long; anemia	Most common tapeworm infection
Epidemiological Features	United States: up to 40 million cases per year	United States: prevalence approx. 0.1%; internationally: prevalence as high as 80% in Southeast Asia, Africa, the Caribbean, and Central and South America	Estimated 20 million infections worldwide	United States: prevalence approximately 0.4%; internationally: the single most prevalent tapeworm infection

predisposition for not adsorbing B_{12}. In these patients, *Diphyllobothrium latum* infection can be quite dangerous.

Hymenolepis species

Hymenolepis species are small tapeworms and are the most common human tapeworm infections in the world. They follow cycle C. There are two species: *Hymenolepis nana,* known as the dwarf tapeworm because it is only 15 to 40 mm in length, and *H. diminuta,* the rat tapeworm, which is usually 20 to 60 cm in length as an adult **(Disease Table 23.9).**

Helminth Disease: Intestinal Distress Accompanied by Migratory Symptoms

A diverse group of helminths enter the body as larvae or eggs, mature to the worm stage in the intestine, and then migrate into the circulatory and lymphatic systems, after which they travel to the heart and lungs, migrate up the respiratory tree to the throat, and are swallowed. This journey returns the mature worms to the intestinal tract, where they then take up residence. All of these conditions, in addition to causing symptoms in the digestive tract, may induce inflammatory reactions along their migratory routes, resulting in eosinophilia and, during their lung stage, pneumonia. Only *Ascaris lumbricoides* will be covered in any depth. Other causes of this type of infection appear in **Disease Table 23.10.**

Ascaris lumbricoides

Ascaris lumbricoides is a large intestinal roundworm (up to 300 mm—a foot or more—long) that probably accounts for the greatest number of worm infections worldwide (estimated at 1 billion cases). *Ascaris* spends its larval and adult stages in humans and releases embryonic eggs in feces, which are then spread to other humans through food, drink, or contaminated objects placed in the mouth. The eggs thrive in warm, moist soils and resist cold and chemical disinfectants, but they are sensitive to sunlight, high temperatures, and drying. After ingested eggs hatch in the human intestine, the larvae embark upon an odyssey in the tissues. First, they penetrate the intestinal wall and enter the lymphatic and circulatory systems. They are swept into the heart and eventually arrive at the capillaries of the lungs. From this point, the larvae migrate up the respiratory tree to the glottis. Worms entering the throat are swallowed and returned to the small intestine, where they reach adulthood and reproduce, producing up to 200,000 fertilized eggs a day.

Even as adults, male and female worms are not attached to the intestine and retain some of their exploratory ways. They are known to invade the biliary channels of the liver and gallbladder, and on occasion the worms emerge from the nose and mouth. Severe inflammatory reactions mark the migratory route, and allergic reactions such as bronchospasm, asthma, or skin rash can occur. Heavy worm loads can slow down the physical and mental development of children **(figure 23.25).** One factor that contributes to intestinal worm infections is self-reinoculation due to poor personal hygiene **(Disease Table 23.10).**

Disease Table 23.10 Intestinal Distress Plus Migratory Symptoms			
Causative Organism(s)	*Toxocara* species (H)	*Ascaris lumbricoides* (intestinal roundworm) (H)	*Necator americanus* and *Ancylostoma duodenale* (hookworms) (H)
Most Common Modes of Transmission	Cycle A: dog or cat feces	Cycle A: vehicle (soil/fecal-oral), fomites, self-inoculation	Cycle B: vehicle (soil), fomite
Virulence Factors	–	Induction of hypersensitivity, adult worm migration, abdominal obstruction	Induction of hypersensitivity, adult worm migration, abdominal obstruction
Culture/ Diagnosis	Blood count, serology, egg or worm detection	Blood count, serology, egg or worm detection	Blood count, serology, egg or worm detection
Prevention	Hygiene	Hygiene	Sanitation
Treatment	Albendazole	Albendazole	Albendazole
Distinctive Features	Can cause migration symptoms or blindness	Most cases mild, unnoticed	Penetrates skin, serious intestinal symptoms
Epidemiological Features	Nearly 100% of newborn puppies in United States infected; 14% of people in United States have been infected; considered a **neglected parasitic infection (NPI)**	Internationally: up to 25% prevalence, 80,000 to 100,000 deaths per year	United States: widespread in Southeast until early 1900s; internationally: 800 million infected

Figure 23.25 **A mass of *Ascaris lumbricoides* worms.** These worms were passed by a child in Kenya in 2007.
James Gatheny/Centers for Disease Control

Cysticercosis

Taenia solium

This helminth is a tapeworm. The adult worms are usually around 5 meters long and have a scolex with hooklets and suckers to attach to the intestine **(figure 23.26).** Disease caused by *T. solium* (the pig tapeworm) is distributed worldwide but is mainly concentrated in areas where humans live in close proximity with pigs or eat undercooked pork. In pigs, the eggs hatch in the small intestine and the released larvae migrate throughout the organs. Ultimately, they encyst in the muscles, becoming *cysticerci,* young tapeworms that are the infective stage for humans. When humans ingest a live cysticercus in pork, the coat is digested and the organism is flushed into the intestine, where it firmly attaches by the scolex and develops into an adult tapeworm. This transmission and life cycle are shown in cycle C in figure 23.24. The pork tapeworm is not the same as the more commonly known pork helminthic infection, *trichinosis.* It is discussed in a later section.

Infection with *T. solium* can take another form when humans ingest the tapeworm eggs rather than cysticerci. Although humans are not the usual intermediate hosts, the eggs can still hatch in the intestine, releasing tapeworm larvae that migrate to all tissues. They form bladderlike sacs throughout the body that can cause serious damage. This leads to a condition called **cysticercosis,** one of the five neglected parasitic infections (NPIs) in the United States. It is estimated that tens of thousands of Latinos living in the United States are affected by cysticercosis, but it is not often recognized because American physicians may not know to look for it. A particularly nasty form of this condition is neurocysticercosis, in which the larvae encyst in the brain **(figure 23.27).** It is estimated to be responsible for 10% of seizures requiring emergency room visits in some U.S. cities **(Disease Table 23.11).**

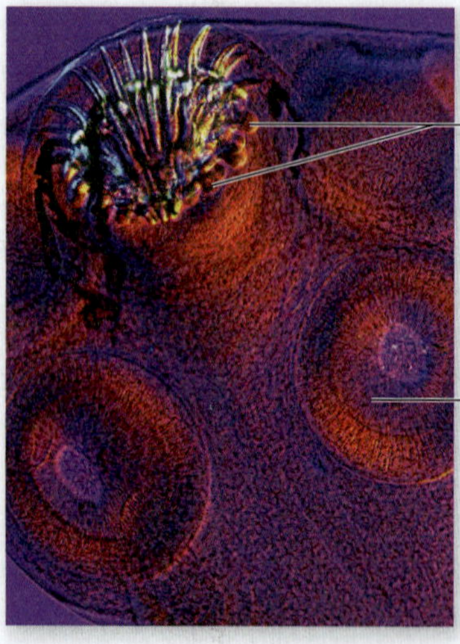
Hooklets
Sucker

(a) Tapeworm scolex showing sucker and hooklets.

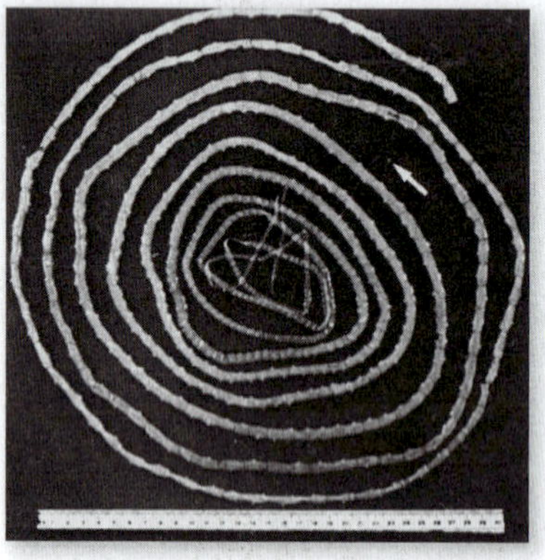

(b) Adult *Taenia saginata*. The arrow points to the scolex; the remainder of the tape, called the strobila, has a total length of 5 meters.

Figure 23.26 Tapeworm characteristics.

(a) Michael J. Klein, M.D./Cultura Creative/Alamy Stock Photo; (b) Dr. Dickson D. Despommier

Disease Table 23.11	Cysticercosis
Causative Organism(s)	*Taenia solium* (pork tapeworm) H
Most Common Modes of Transmission	Cycle C: vehicle (pork), fecal-oral
Virulence Factors	–
Culture/Diagnosis	Blood count, serology, egg or worm detection
Prevention	Cook meat, avoid pig feces
Treatment	Praziquantel
Distinctive Features	Ingesting larvae embedded in pork leads to intestinal tapeworms; ingesting eggs (fecal-oral route) causes cysticercosis, larval cysts embedded in tissue of new host
Epidemiological Features	United States: considered a **neglected parasitic infection,** common cause of seizures; internationally: very common in Latin America and Asia

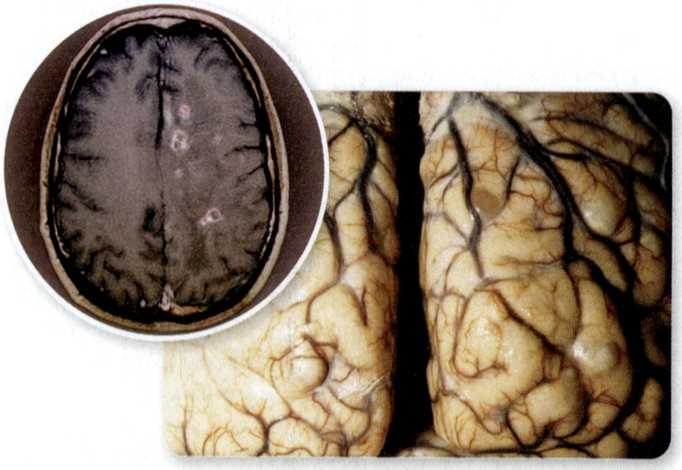

Figure 23.27 Cysticerci (white spots) in the brain caused by *Taenia solium*.

PR Bouree/age fotostock; (Inset) Science Photo Library/Getty Images

Helminth Disease: Liver and Intestinal Damage

One group of worms that lands in the intestines has a particular affinity for the liver. Three of these worms are trematodes (flatworms), and they are categorized as liver flukes.

Opisthorchis sinensis and *Clonorchis sinensis*

Opisthorchis sinensis and *Clonorchis sinensis* are two worms known as liver flukes. They complete their sexual development in mammals such as humans, cats, dogs, and swine. Their intermediate development occurs in snail and fish hosts. Humans ingest metacercariae in inadequately cooked or raw freshwater fish (see cycle D in figure 23.24). Larvae hatch and crawl into the bile duct, where they mature and shed eggs into the intestinal tract. Feces containing eggs are passed into standing water that harbors the intermediate snail host. The cycle is complete when infected snails release cercariae that invade fish living in the same water.

Symptoms of *Opisthorchis* and *Clonorchis* infection are slow to develop but include thickening of the lining of the bile duct and possible granuloma formation in areas of the liver if eggs enter the stroma of the liver. If the infection is heavy, the bile duct can be blocked.

Fasciola hepatica

This liver fluke **(figure 23.28)** is a common parasite in sheep, cattle, goats, and other mammals and is occasionally transmitted to humans. Periodic outbreaks in temperate regions of Europe and South America are associated with eating wild watercress. The life cycle is very complex, involving the mammal as the definitive host, the release of eggs in the feces, the hatching of eggs in the water into *miracidia,* invasion of freshwater snails, development and release of cercariae, encystment of metacercariae on a water plant, and ingestion of the cyst by a mammalian host eating the plant. The cysts release young flukes into the intestine that wander to the liver, lodge in the gallbladder, and develop into adults. Humans develop symptoms of vomiting, diarrhea, hepatomegaly, and bile obstruction if they are chronically infected by a large number of flukes **(Disease Table 23.12).**

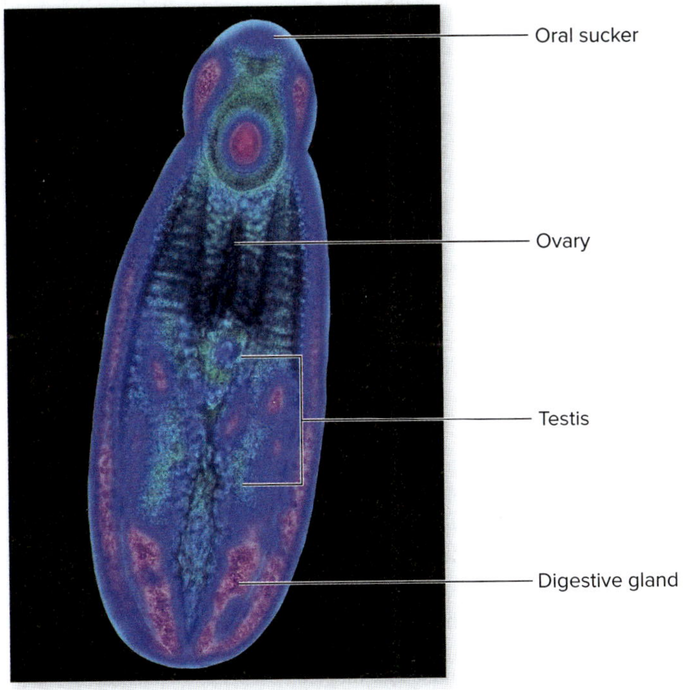
Oral sucker
Ovary
Testis
Digestive gland

Figure 23.28 An immature fluke. Magnification 15×.
Science Photo Library/Alamy Stock Photo

Helminth Disease: Muscle and Neurological Symptoms

Trichinosis is an infection transmitted by eating pork (and sometimes other wildlife) that have the cysts of *Trichinella* species embedded in the meat. The life cycle of this nematode is spent entirely within the body of a mammalian host such as a pig, bear, cat, dog, or rat. In nature, the parasite is maintained in an encapsulated (encysted) larval form **(figure 23.29)** in the muscles of these animal reservoirs and is transmitted when other animals prey upon them. The disease cannot be transmitted from one human to another (except in the case of cannibalism).

The cyst envelope is digested in the stomach and small intestine, which liberates the larvae. After burrowing into the intestinal mucosa, the larvae reach adulthood and mate. The larvae that result

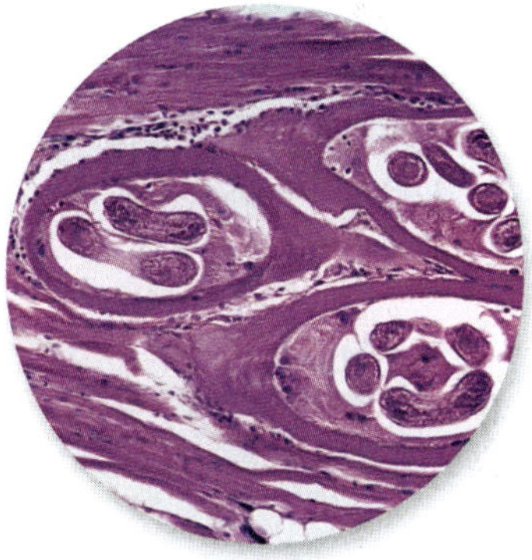

Figure 23.29 *Trichinella* cysts embedded in pork muscle.
Ed Reschke/Photolibrary/Getty Images

Disease Table 23.12	Liver and Intestinal Disease	
Causative Organism(s)	*Opisthorchis sinensis, Clonorchis sinensis* H	*Fasciola hepatica* H
Most Common Modes of Transmission	Cycle D: vehicle (fish or crustaceans)	Cycle D: vehicle (water and water plants)
Virulence Factors	–	–
Culture/Diagnosis	Blood count, serology, egg or worm detection	Blood count, serology, egg or worm detection
Prevention	Cook food, sanitation of water	Sanitation of water
Treatment	Praziquantel	Triclabendazole
Distinctive Features	Live in liver	Live in liver and gallbladder
Epidemiological Features	United States: most cases imported; internationally: 56 million infected	

from this union penetrate the intestine and enter the lymphatic channels and blood. All tissues are at risk for invasion, but final development occurs when the coiled larvae are encysted in the skeletal muscle. At maturity, the cyst is about 1 mm long and can be observed by careful inspection of meat. Although larvae can deteriorate over time, they have also been known to survive for years.

Symptoms may be unnoticeable, or they could be life-threatening, depending on how many larvae were ingested in the tainted meat. The first symptoms, when present, mimic influenza or other viral fevers, with diarrhea, nausea, abdominal pains, fever, and sweating. The second phase, brought on by the mass migration of larvae and their entrance into muscle, produces puffiness around the eyes, intense muscle and joint pain, shortness of breath, and pronounced eosinophilia. The most serious life-threatening manifestations are heart and brain involvement. Although the symptoms eventually subside, a cure is not available once the larvae have encysted in muscles.

The most effective preventive measures for trichinosis are to adequately store and cook pork and wild meats. In the United States, wild meats such as bear are the major source of infection.

Disease Table 23.13 Muscle and Neurological Symptoms

Causative Organism(s)	*Trichinella* species (H)
Most Common Modes of Transmission	Vehicle (food)
Virulence Factors	–
Culture/Diagnosis	Serology combined with clinical picture; muscle biopsy
Prevention	Cook meat
Treatment	Mebendazole, steroids
Distinctive Features	Brain and heart involvement can be fatal
Epidemiological Features	United States: 20 cases per year; internationally: 10,000 cases per year

Schistosomiasis Liver Disease

When liver swelling or malfunction is accompanied by eosinophilia, **schistosomiasis** should be suspected. The disease is caused by the blood flukes *Schistosoma mansoni* and *S. japonicum*, species that are morphologically and geographically distinct but share similar life cycles, transmission methods, and general disease manifestations. Schistosomiasis is one of the few infectious agents that can invade intact skin.

Signs and Symptoms

The first symptoms of infection are itchiness in the area where the worm enters the body, followed by fever, chills, diarrhea, and cough. The most severe consequences, associated with chronic infection, are hepatomegaly, liver disease, and splenomegaly. Other serious conditions caused by a different schistosome occur in the urinary tract—bladder obstruction and blood in the urine. This condition is discussed in chapter 24 (genitourinary tract diseases). Occasionally, eggs from the worms are carried into the central nervous system and heart and create a severe granulomatous response. Adult flukes can live for many years and, by eluding the immune defenses, cause a chronic affliction.

Causative Agent

Schistosomes are trematodes, or flukes (see chapter 5), but they are more cylindrical than flat **(figure 23.30).** They are often called blood flukes. Flukes have digestive, excretory, neuromuscular, and reproductive systems, but they lack circulatory and respiratory systems. Humans are the definitive hosts for the blood fluke, and snails are the intermediate host.

Pathogenesis and Virulence Factors

This parasite is extremely clever. Once inside the host, it coats its outer surface with proteins from the host's bloodstream, basically "cloaking" itself from the host defense system. This coat reduces its surface antigenicity and allows it to remain in the host indefinitely.

Transmission and Epidemiology

The life cycle of the schistosome is of the D type and is very complex (as shown in figure 23.30). The cycle begins when infected humans release eggs into irrigated fields or ponds, either by deliberate fertilization with excreta or by defecating or urinating directly into the water. The egg hatches in the water and gives off an actively swimming ciliated larva called a **miracidium,** which instinctively swims to a snail and burrows into a vulnerable site, shedding its ciliated covering in the process. In the body of the snail, the miracidium multiplies into a larger, fork-tailed swimming larva called a **cercaria.** Cercariae are given off by the thousands into the water by infected snails.

Upon contact with a human wading or bathing in water, cercariae attach themselves to the skin by ventral suckers and penetrate into hair follicles. They pass into small blood and lymphatic vessels and are carried to the liver. Here, the schistosomes achieve sexual maturity, and the male and female worms remain permanently entwined to facilitate mating (see figure 23.30). In time, the pair migrates to and lodges in small blood vessels at specific sites. *Schistosoma mansoni* and *S. japonicum* end up in the mesenteric venules of the small intestine. While attached to these intravascular sites, the worms feed upon blood, and the female lays eggs that are eventually voided in feces or urine.

The disease is endemic to 74 countries located in Africa, South America, the Middle East, and Asia. *S. mansoni* is found throughout these regions but not in Asia. *S. japonicum* has a much smaller geographic distribution than *S. mansoni,* being found

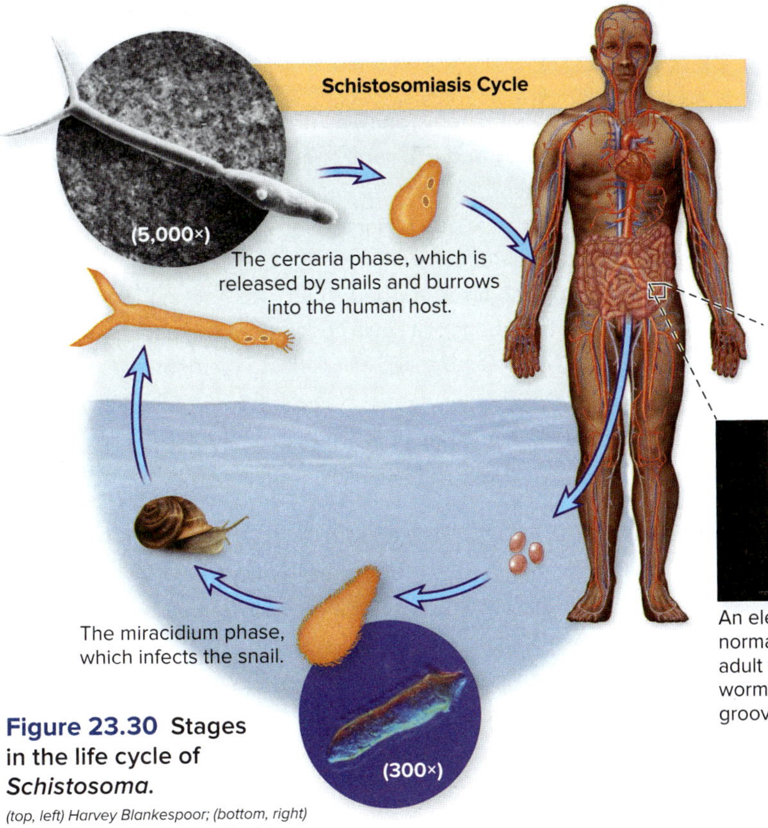
Schistosomiasis Cycle
(5,000×)
The cercaria phase, which is released by snails and burrows into the human host.
The miracidium phase, which infects the snail.
(300×)

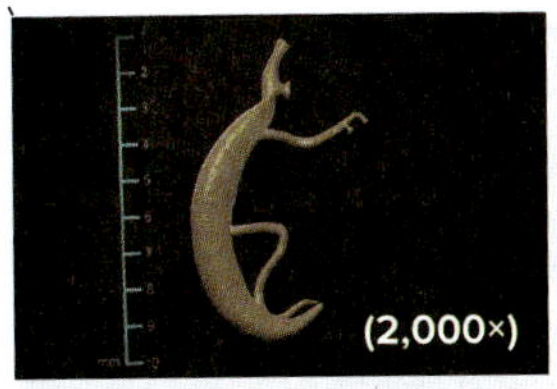
(2,000×)

An electron micrograph of normal mating position of adult worms. The larger male worm holds the female in a groove on his ventral surface.

Figure 23.30 Stages in the life cycle of *Schistosoma.*

(top, left) Harvey Blankespoor; (bottom, right) MedicalRF.com; (bottom, centre) Sinclair Stammers/Science Source

only in Asia. Schistosomiasis (including the urinary tract form) is the second most prominent parasitic disease after malaria, probably affecting 200 million people at any one time worldwide. Recent increases in its occurrence in Africa have been attributed to new dams on the Nile River, which have provided additional habitats for snail hosts.

Culture and/or Diagnosis

Diagnosis depends on identifying the eggs in urine or feces. The clinical pictures of hepatomegaly, splenomegaly, or both also contribute to the diagnosis.

Prevention and Treatment

The cycle of infection cannot be broken as long as people are exposed to untreated sewage in their environment. It is quite common for people to be cured and then to be reinfected because their village has no sewage treatment. A vaccine would provide widespread control of the disease, but so far none is licensed. More than one vaccine is in development, however.

Praziquantel is the drug treatment of choice. It works by crippling the worms, making them more antigenic and thereby allowing the host immune response to eliminate them. Clinicians use an "egg-hatching test" to determine whether an infection is current and whether treatment is actually killing the eggs. Urine or feces containing eggs are placed in room-temperature water, and if miracidia emerge, the infection is still "active" (**Disease Table 23.14**).

Disease Table 23.14 Schistosomiasis Liver Disease

Causative Organism(s)	*Schistosoma mansoni, S. japonicum* H
Most Common Modes of Transmission	Cycle D: vehicle (contaminated water)
Virulence Factors	Antigenic "cloaking"
Culture/Diagnosis	Identification of eggs in feces, scarring of intestines detected by endoscopy
Prevention	Avoiding contaminated vehicles
Treatment	Praziquantel
Distinctive Features	Penetrates skin, lodges in blood vessels of intestine, damages liver
Epidemiological Features	Internationally: 230 million new infections per year by these and the urinary schistosome

23.3 Learning Outcomes—Assess Your Progress

11. Describe some distinguishing characteristics and commonalities seen in helminthic infections.
12. List four helminths that cause primarily intestinal symptoms, and identify which life cycle each follows and one unique fact about each helminth.
13. List three helminths that cause intestinal symptoms that may be accompanied by migratory symptoms, identifying which life cycle each follows and one unique fact about each helminth.
14. Identify the most dangerous outcome of *Taenia solium* infection.
15. List the modes of transmission for each of the helminthic infections resulting in liver and intestinal symptoms. These are infections caused by *Opisthorchis sinensis, Clonorchis sinensis,* and *Fasciola hepatica.*
16. Describe the type of disease caused by *Trichinella* species.
17. Diagram the life cycle of *Schistosoma mansoni* and *S. japonicum,* and describe the importance of these organisms in world health.

MEDIA UNDER THE MICROSCOPE WRAP-UP

Gerhard Zwerger-Schoner/image BROKER/Alamy Stock Photo

The **intended message** of the Iceman article is that genetic patterns in gut bacteria can give us clues about the movements of the humans they inhabited. The article tells the story well, explaining how *Helicobacter* species exchanged genetic information that could be used as puzzle pieces to put together their lineages. My **critical reading** is that it is well done and informative enough for people with no background to understand it.

I would **interpret** it by first explaining that *Helicobacter* is a common member of human stomach biota and has been for at least 100,000 years. And because it would be a common contaminant in the environment (from fecal matter), it is easily and constantly passed from person to person, where its quick genetic changes make it a good marker for human movements and interactions.

My **grade?** An A, a good, solid job. And any article that includes the limitations of its own conclusions (where it noted that the Iceman was a sample size of 1) is doing the right thing.

Source: *Washington Post*, "New Study on Otzi the Iceman Reveals Humanity's Intimate Affair with One Microbe," online article posted 1/8/2016.

Study Smarter: Better Together

These activities are designed for you to use on your own with a study group—either a face-to-face group or a virtual one, consisting of 3–5 members. Studying together can be very helpful, but there are effective and ineffective ways to do it. For example, getting together without a clear structure is often not a good use of your time. Use your time efficiently by using one or more of the exercises below.

FACE-TO-FACE GROUPS

Use one or more of the activities below.

Peer Instruction: Assign numbers to your group members to use all semester long. Now look at these five concepts from this chapter. Each group member prepares a 5-minute lesson on the topic corresponding to their number. Don't worry if you have fewer than 5 members; just use however many you have! During your group study time, each member presents their lesson, and the group spends another 5–10 minutes discussing that lesson.

1. Compare and contrast hepatitis caused by the hepatitis A, hepatitis B, and hepatitis C viruses
2. Important features of *C. difficile* disease
3. Anatomy and defenses of the gastrointestinal system
4. The life cycle of *Enterobius vermicularis*
5. Organisms in this chapter for which there are vaccines available

Concept Maps: Each member of the group should use this list of terms from this chapter to generate their own concept map. This can be hand-drawn or created using software (see Appendix C for guidelines). During group study time, compare each other's concept maps and help each other make sure they are correct. Of course, there are many different "correct" maps. Examining each member's map will help you talk through the varied concepts and how they are related.

Concept Terms:

exotoxins	*E. coli*	shiga toxin	STEC	EAEC	bacteriophage
protein synthesis	*Shigella*	EIEC			

Table Topics: Each group member should identify a concept or topic from this week's class assignments with which they are having trouble and share it during group study time. The other group members can then help to clarify confusing issues or share how they figured it out. Aim for a maximum of 15 minutes per topic. If the topic remains unclear to the group, bring it up during class or use the instructor's office hours or e-mail to ask for help. Taking the time to struggle with a difficult concept first makes your questions much more specific and more likely to yield helpful answers.

VIRTUAL GROUPS

Not everyone has the time or opportunity to meet with group members outside of class time. You or your instructor can create a virtual group using e-mail or the course software.

(continued)

Weekly Discussion Board: This forum can be used as a way for groups to discuss topics, via e-mail, or other learning management systems or online platforms, before they are covered in class. As each member of the group answers the current week's question, they should send their responses to every other member of their group. It's best to agree on a deadline based on how your class schedule works (Saturday for the next week's topics, for example). Then, after the topic is discussed in class, each member should send a response that all group members will see with a follow-up post on the same topic. If you cover more than one chapter in a week, someone can be designated to choose which chapter Discussion Board question you will use. Or simply decide up front that you will always use the first-chapter-of-the-week's question, to keep the schedule simple.

Discussion Question
What is the most common treatment for acute diarrhea and why?

Deadliness and Communicability of Selected Diseases of the Gastrointestinal Tract

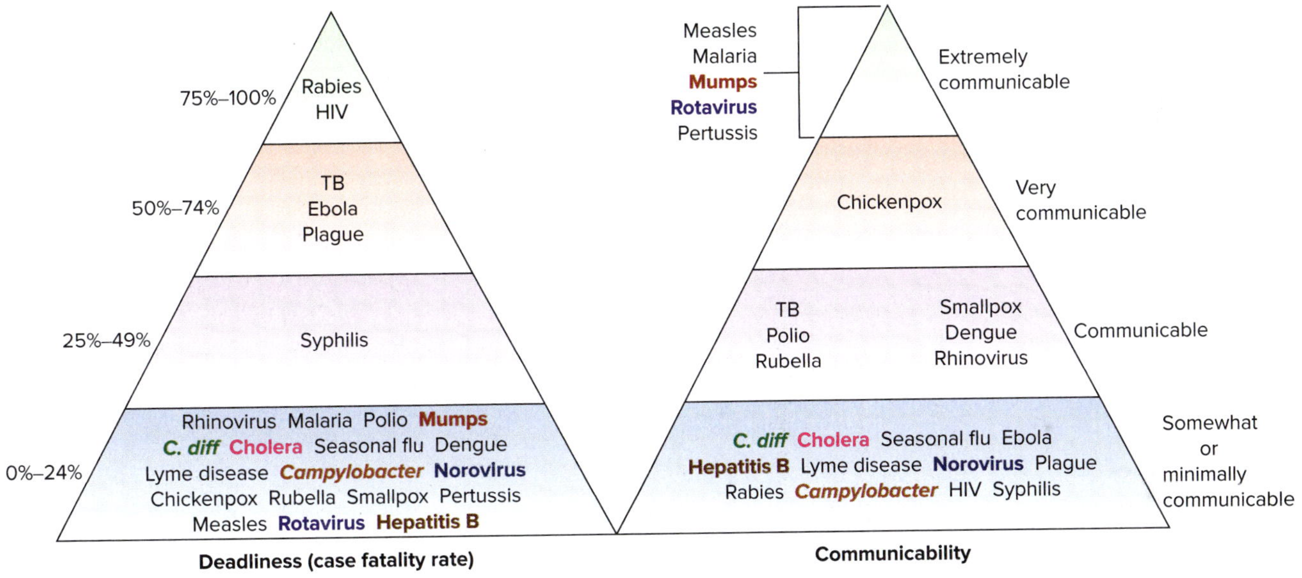
75%–100%
Rabies
HIV
50%–74%
TB
Ebola
Plague
25%–49%
Syphilis
0%–24%
Rhinovirus Malaria Polio Mumps
C. diff Cholera Seasonal flu Dengue
Lyme disease Campylobacter Norovirus
Chickenpox Rubella Smallpox Pertussis
Measles Rotavirus Hepatitis B
Deadliness (case fatality rate)
Measles
Malaria
Mumps
Rotavirus
Pertussis
Extremely communicable
Chickenpox
Very communicable
TB
Polio
Rubella
Smallpox
Dengue
Rhinovirus
Communicable
C. diff Cholera Seasonal flu Ebola
Hepatitis B Lyme disease Norovirus Plague
Rabies Campylobacter HIV Syphilis
Somewhat or minimally communicable
Communicability

Summing Up

Taxonomic Organization Microorganisms Causing Disease in the GI Tract

Microorganism	Disease	Disease Table
Gram-positive, endospore-forming bacteria		
Clostridioides difficile	Antibiotic-associated diarrhea	Acute diarrhea (with or without vomiting), 23.5
Bacillus cereus	Food poisoning	Acute diarrhea with vomiting caused by exotoxins (food poisoning), 23.6
Clostridium perfringens	Food poisoning	Acute diarrhea with vomiting caused by exotoxins (food poisoning), 23.6
Gram-positive bacteria		
Streptococcus mutans	Dental caries	Dental caries, 23.1
Streptococcus sobrinus	Dental caries	Dental caries, 23.1
Staphylococcus aureus	Food poisoning	Acute diarrhea with vomiting caused by exotoxins (food poisoning), 23.6

Summing Up

Taxonomic Organization Microorganisms Causing Disease in the GI Tract

Microorganism	Disease	Disease Table
Gram-negative bacteria		
Tannerella forsythia, Aggregatibacter actinomycetemcomitans, Porphyromonas gingivalis, Treponema vincentii, Prevotella intermedia, Fusobacterium	Periodontal disease	Periodontal diseases, 23.2
Helicobacter pylori	Gastritis/gastric ulcers	Gastritis and gastric ulcers, 23.4
Salmonella	Acute diarrhea	Acute diarrhea (with or without vomiting), 23.5
Shigella	Acute diarrhea and dysentery	Acute diarrhea (with or without vomiting), 23.5
Escherichia coli STEC	Acute diarrhea plus hemolytic syndrome	Acute diarrhea (with or without vomiting), 23.5
Other *E. coli*	Acute or chronic diarrhea	Acute diarrhea (with or without vomiting), 23.5; Chronic diarrhea, 23.7
Campylobacter jejuni	Acute diarrhea	Acute diarrhea (with or without vomiting), 23.5
Vibrio cholerae	Cholera	Acute diarrhea (with or without vomiting), 23.5
Non-cholera *Vibrio* species	Vibrioses	Acute diarrhea (with or without vomiting), 23.5
DNA viruses		
Hepatitis B virus	"Serum" hepatitis	Hepatitis, 23.8
RNA viruses		
Mumps virus	Mumps	Mumps, 23.3
Rotavirus	Acute diarrhea	Acute diarrhea (with or without vomiting), 23.5
Norovirus	Acute diarrhea	Acute diarrhea (with or without vomiting), 23.5
Hepatitis A virus	"Infectious" hepatitis	Hepatitis, 23.8
Hepatitis E virus	"Infectious" hepatitis	Hepatitis, 23.8
Hepatitis C virus	"Serum" hepatitis	Hepatitis, 23.8
Protozoa		
Cryptosporidium	Acute diarrhea	Acute diarrhea (with or without vomiting), 23.5
Cyclospora cayetanensis	Chronic diarrhea	Chronic diarrhea, 23.7
Giardia duodenalis	Chronic diarrhea	Chronic diarrhea, 23.7
Entamoeba histolytica	Chronic diarrhea	Chronic diarrhea, 23.7
Helminths—nematodes		
Enterobius vermicularis	Intestinal distress	Intestinal distress, 23.9
Trichuris trichiura	Intestinal distress	Intestinal distress, 23.9
Toxocara species	Intestinal distress plus migratory symptoms	Intestinal distress plus migratory symptoms, 23.10
Ascaris lumbricoides	Intestinal distress plus migratory symptoms	Intestinal distress plus migratory symptoms, 23.10
Necator americanus	Intestinal distress plus migratory symptoms	Intestinal distress plus migratory symptoms, 23.10
Ancylostoma duodenale	Intestinal distress plus migratory symptoms	Intestinal distress plus migratory symptoms, 23.10
Trichinella species	Muscle and neurological symptoms	Muscle and neurological symptoms, 23.13
Helminths—cestodes		
Diphyllobothrium latum	Intestinal distress	Intestinal distress, 23.9
Hymenolepis nana, H. diminuta	Intestinal distress	Intestinal distress, 23.9
Taenia solium	Cysticercosis	Cysticercosis, 23.11
Opisthorchis sinensis	Liver and intestinal disease	Liver and intestinal disease, 23.12
Clonorchis sinensis	Liver and intestinal disease	Liver and intestinal disease, 23.12
Helminths—trematodes		
Fasciola hepatica	Liver and intestinal disease	Liver and intestinal disease, 23.12
Schistosoma mansoni, S. japonicum	Schistosomiasis	Schistosomiasis liver disease, 23.14

Chapter Summary

INFECTIOUS DISEASES MANIFESTING IN THE GASTROINTESTINAL TRACT

23.1 THE GASTROINTESTINAL TRACT, ITS DEFENSES, AND NORMAL BIOTA

- The gastrointestinal (GI) tract is composed of *eight* main sections—the mouth, pharynx, esophagus, stomach, small intestine, large intestine, rectum, and anus—and *four* accessory organs—the salivary glands, liver, gallbladder, and pancreas.
- The GI tract has a very heavy load of microorganisms, and it encounters millions of new ones every day. There are significant mechanical, chemical, and antimicrobial defenses to combat microbial invasion.
- Bacteria are abundant in all of the eight main sections of the gastrointestinal tract. Even the highly acidic stomach is colonized.
- The bacteria in the intestines—the gut microbiome—is absolutely critical in regulating all kinds of body functions.

23.2 INFECTIOUS DISEASES MANIFESTING IN THE GASTROINTESTINAL TRACT (NONHELMINTHIC)

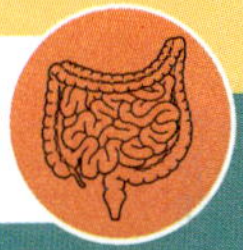

- **Tooth and gum infections—Dental caries:** Caused by alpha-hemolytic streptococci. **Periodonal diseases:** Diseases causing destruction of tooth supporting structures, generally caused by a polymicrobial infection of anaerobic bacteria.
- **Mumps**—Swelling of salivary gland; caused by RNA virus from the genus *Paramyxovirus*.
- **Gastritis and gastric ulcers**—*Helicobacter pylori*, a curved gram-negative rod, is responsible.
- **Acute diarrhea (with or without vomiting)**—In the United States, a third of all acute diarrhea is transmitted by contaminated food. **Salmonella:** Frequent contaminant of animal and dairy products. **Shigella:** Symptoms of frequent, watery, bloody stools; dysentery. Produces a heat-labile exotoxin called shiga toxin. **Shiga-toxin producing *E. coli* (STEC):** Agents of severe diarrhea that can be accompanied by hemolytic uremic syndrome, a life-threatening condition. **Other *E. coli:*** At least five other strains of *E. coli* cause various types of disease: enterotoxigenic, enteroinvasive, enteropathogenic, enteroaggregative, and diffusely adherent *E. coli*. ***Campylobacter:*** Infrequently, infection can lead to a serious neuromuscular condition called Guillain-Barré syndrome. ***Clostridioides difficile:*** Causes pseudomembranous colitis. ***Vibrio cholera:*** Secretory diarrhea and severe fluid loss. Can be quickly deadly. **Other *Vibrio* species:** Foodborne illnesses caused by other *Vibrio* species are increasing. ***Cryptosporidium:*** Waterborne protozoan that infects mammals, birds, and reptiles. **Rotavirus:** Most dangerous to babies age 6–24 months. **Norovirus:** Most common foodborne illness in United States; also transmitted via fecal-oral route.
- **Acute diarrhea with vomiting caused by exotoxins (food poisoning)**—Refers to symptoms in the gut caused by a preformed toxin. ***Staphylococcus aureus* exotoxin:** Heat-stable enterotoxin requires 100° C for 30 minutes for inactivation. Acute symptoms of cramping, nausea, vomiting, and diarrhea. ***Bacillus cereus* exotoxin:** Common resident on vegetables and in soil. Produces two exotoxins; one causes diarrhea and the other emesis. ***Clostridium perfringens* exotxoin:** Acute abdominal pain, diarrhea.
- **Chronic diarrhea**—Lasts longer and generally less severe than acute diarrhea. **Enteroaggregative *E. coli* (EAEC).** Associated with chronic disease, especially in children. Transmitted through contaminated food and water. ***Cyclospora:*** A protozoan transmitted via the fecal-oral route; associated with fresh produce and water. ***Giardia:*** Protozoan that causes long-term diarrhea, abdominal pain, and flatulence. Freshwater is a common vehicle of infection. ***Entamoeba histolytica:*** A freshwater protozoan leading to abdominal pain, fever, diarrhea, and weight loss.
- **Hepatitis**—Can be caused by a variety of viruses (as well as other exposures). **Hepatitis A:** Nonenveloped single-stranded RNA enterovirus of low virulence. Spread via fecal-oral route. Inactivated vaccine available. **Hepatitis B:** Enveloped DNA virus. Can be very serious; some patients develop chronic liver disease and cancer. HBV is transmitted by blood and other body fluids. **Hepatitis C:** RNA virus in *Flavidiridae* family shares characteristics with hepatitis B disease but is more likely to become chronic.

23.3 HELMINTHIC DISEASES MANIFESTING IN THE GASTROINTESTINAL TRACT

- **Helminths are low on virulence factors and high on the ability to persist in the host.**
- **Helminth disease: intestinal distress as the primary symptom**—Both tapeworms and roundworms can be causes. ***Enterobius vermicularis:*** Pinworm. ***Trichuris trichiura:*** Worms bury into intestinal mucosa. ***Diphyllobothrium latum:*** Intermediate host is fish; transmitted in fresh uncooked fish such as sushi. ***Hymenolopis* species:** Most prevalent tapeworm infection.
- **Helminth disease: intestinal distress accompanied by migratory symptoms—*Toxocara* species:** Infect 100% of newborn puppies; frequently infect humans. ***Ascaris lumbricoides:*** Intestinal roundworm that releases eggs through feces. ***Necator amercanus* and *Ancylostoma duodenale:*** Both called "hookworm." Infect by penetrating skin. **Cysticercosis:** Larval tapeworm cysts embedded in brain and other tissues; can cause seizures. *Taenia solium* is responsible.
- **Helminth disease: liver and intestinal damage**—One group of worms with a particular affinity for the liver is called liver flukes. ***Opisthorchis sinensis* and *Clonorchis sinensis:*** Humans infected by eating inadequately cooked freshwater fish and crustaceans. ***Fasciola hepatica:*** Common parasite in sheep, cattle, goats, etc.
- **Helminth disease: muscle and neurological symptoms—*Trichinella:*** Transmitted by eating undercooked pork containing cysts of protozoan.
- **Liver disease—Schistosomiasis:** Caused by blood flukes.

SmartGrid: From Knowledge to Critical Thinking

This *21 Question Grid* takes the topics from this chapter and arranges them with respect to the American Society for Microbiology's Undergraduate Curriculum guidelines—all six of the important "Concepts" as well as the important "Competency" of scientific literacy. Three questions are supplied, which cover chapter content referring to the Concept or Competency in increasing levels of Bloom's taxonomy for learning.

ASM Concept/ Competency	A. Bloom's Level 1, 2—Remember and Understand (Choose one)	B. Bloom's Level 3, 4—Apply and Analyze	C. Bloom's Level 5, 6—Evaluate and Create
Evolution	**1.** Which of the following causes of acute diarrhea is of the most concern to the CDC due to its antibiotic resistance characteristics? **a.** *Salmonella* **b.** *Clostridioides difficile* **c.** *Cryptosporidium* **d.** *Shigella*	**2.** Healthy teeth are important for good nutrition and overall human health. Yet despite the fact that humans and their oral microbiome have coevolved over thousands of years, there was a vast increase in the incidence of human dental caries with the advent of Western civilization. What factor(s) are likely responsible for that?	**3.** Considering that *Helicobacter* has probably been colonizing human stomachs for thousands of years, would you suspect that its primary role is as a pathogen or as normal biota? Support your answer.
Cell Structure and Function	**4.** This endospore-forming bacterium contaminates meat and vegetables and, in other situations, causes gas gangrene. **a.** *Bacillus cereus* **b.** *Clostridium perfringens* **c.** *Shigella* **d.** *Staphylococcus aureus*	**5.** Why is heating food contaminated with *Staphylococcus aureus* no guarantee that potential food poisoning will be prevented?	**6.** Several pathogens in this chapter have the ability to form endospores or cysts. Create an argument, using what you know about transmission mechanisms, for why these organisms are particularly suited for GI tract disease.

ASM Concept/ Competency	A. Bloom's Level 1, 2—Remember and Understand (Choose one)	B. Bloom's Level 3, 4—Apply and Analyze	C. Bloom's Level 5, 6—Evaluate and Create
Metabolic Pathways	**7.** Which of the following bacteria produces an enzyme that breaks down urea? **a.** *Helicobacter pylori* **b.** *Salmonella* **c.** *Shigella* **d.** *Shiga-toxin-producing E. coli*	**8.** You probably think of the oral cavity as an aerobic environment. Considering the anatomy of the teeth and gums, where would you expect the anaerobes that colonize the mouth to survive?	**9.** *Vibrio cholerae* can survive in a wide range of aquatic conditions, from brackish water to seawater, and in habitats that have scarce nutrients. It readily forms biofilms on submerged surfaces, including other organisms such as phytoplankton. How would living in a biofilm protect *Vibrio* in a low-nutrient environment?
Information Flow and Genetics	**10.** Which of these organisms has an unusual double-stranded RNA genome? **a.** *norovirus* **b.** *Schistosoma* **c.** *hepatitis C* **d.** *rotavirus*	**11.** Why is it thought that the shiga toxin in shiga-toxin-producing *E. coli* originated in *Shigella*, and not the other way around? Also, explain how this characteristic was transferred between the two species.	**12.** Conduct research on how the hepatitis B vaccine is manufactured, and write a paragraph responding to a patient who is unhappy about receiving a recombinant vaccine.
Microbial Systems	**13.** The normal biota of the GI tract is most diverse in the **a.** *pharynx.* **b.** *stomach.* **c.** *small intestine.* **d.** *large intestine.*	**14.** The normal biota of the GI tract seems to include a lot of disease-causing organisms. How can that be? That is, if they are "normal," why are they also potential pathogens?	**15.** The *Salmonella* strain causing typhoid fever is transmitted human-to-human. The *Salmonella* strains causing gastroenteritis generally are transmitted from animals. Sketch a scenario for why gastroenteritis is now the more common outcome of *Salmonella* infection and not typhoid fever.
Impact of Microorganisms	**16.** Which of these microorganisms is associated with Guillain-Barré syndrome? **a.** *E. coli* **b.** *Salmonella* **c.** *Campylobacter* **d.** *Shigella*	**17.** Describe the populations that are most at risk for hepatitis C, and why.	**18.** Use what you know about inflammation to explain how periodontal organisms might be responsible for diseases elsewhere in the body, such as the heart.
Scientific Thinking	**19.** You are examining a patient who is a recent immigrant from Southeast Asia. She is complaining of diarrhea and vomiting. Which of these would you add to your differential diagnosis in light of her history? **a.** *Vibrio cholerae* **b.** *Salmonella* **c.** *Shigella* **d.** *Campylobacter*	**20.** In Disease Table 23.5, in the *Cryptosporidium* column, it is stated that 30% of the population is seropositive. What does this mean, and also what does it not mean?	**21.** Consider the function of the liver. Then speculate on why so many different types of viruses can cause pathology there.

Answers to the multiple-choice questions appear in Appendix A.

INFECTIOUS DISEASES AFFECTING
The Gastrointestinal Tract

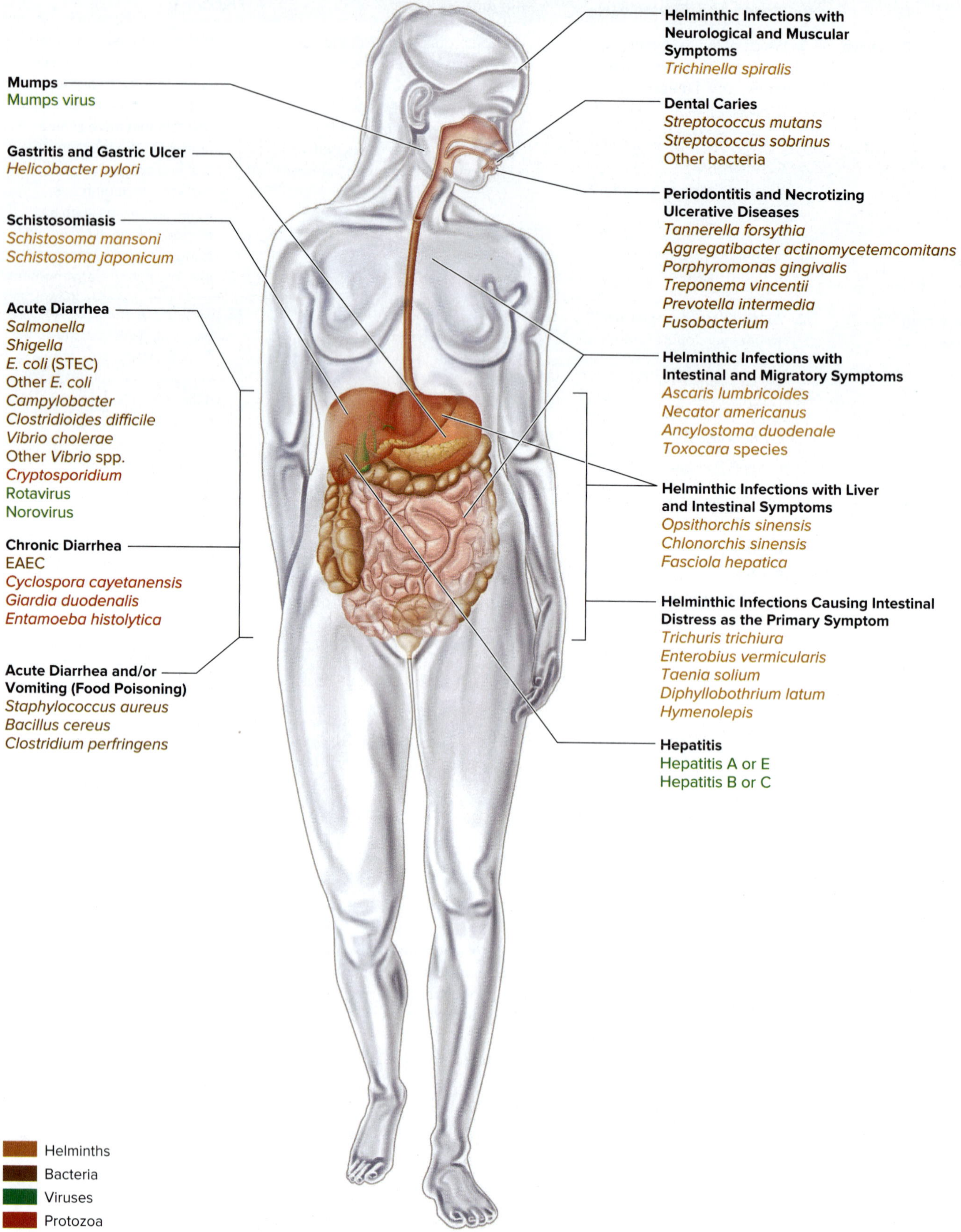

Mumps
Mumps virus
Gastritis and Gastric Ulcer
Helicobacter pylori
Schistosomiasis
Schistosoma mansoni
Schistosoma japonicum
Acute Diarrhea
Salmonella
Shigella
E. coli (STEC)
Other *E. coli*
Campylobacter
Clostridioides difficile
Vibrio cholerae
Other *Vibrio* spp.
Cryptosporidium
Rotavirus
Norovirus
Chronic Diarrhea
EAEC
Cyclospora cayetanensis
Giardia duodenalis
Entamoeba histolytica
Acute Diarrhea and/or Vomiting (Food Poisoning)
Staphylococcus aureus
Bacillus cereus
Clostridium perfringens
Helminthic Infections with Neurological and Muscular Symptoms
Trichinella spiralis
Dental Caries
Streptococcus mutans
Streptococcus sobrinus
Other bacteria
Periodontitis and Necrotizing Ulcerative Diseases
Tannerella forsythia
Aggregatibacter actinomycetemcomitans
Porphyromonas gingivalis
Treponema vincentii
Prevotella intermedia
Fusobacterium
Helminthic Infections with Intestinal and Migratory Symptoms
Ascaris lumbricoides
Necator americanus
Ancylostoma duodenale
Toxocara species
Helminthic Infections with Liver and Intestinal Symptoms
Opsithorchis sinensis
Chlonorchis sinensis
Fasciola hepatica
Helminthic Infections Causing Intestinal Distress as the Primary Symptom
Trichuris trichiura
Enterobius vermicularis
Taenia solium
Diphyllobothrium latum
Hymenolepis
Hepatitis
Hepatitis A or E
Hepatitis B or C
Helminths
Bacteria
Viruses
Protozoa

System Summary Figure 23.31

Visual Connections

This question uses visual images to connect content within and between chapters.

1. **From chapter 6, figure 6.21.** This process resulted in the recipient bacterium being able to cause a more serious form of disease than previously. Which diarrhea pathogen from this chapter are we referring to?

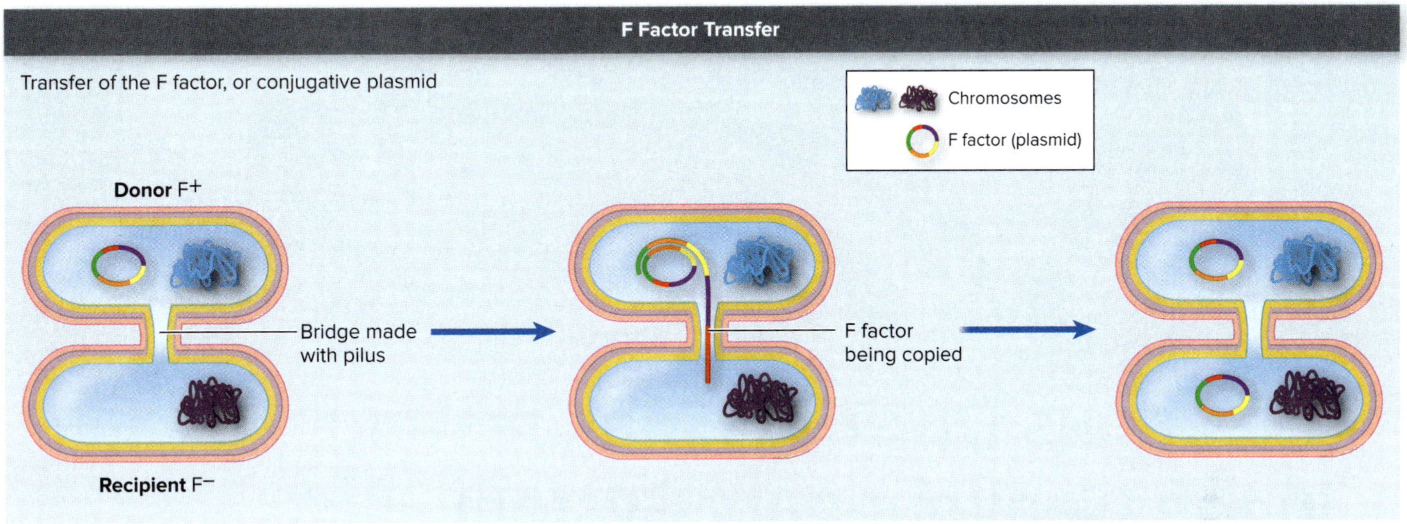
F Factor Transfer
Transfer of the F factor, or conjugative plasmid
Chromosomes
F factor (plasmid)
Donor F+
Bridge made with pilus
F factor being copied
Recipient F−

High Impact Study

These terms and concepts are most critical for your understanding of this chapter—and may be the most difficult. Have you mastered them? In these disease chapters, the terms and concepts help you identify what is important in a different way than the comprehensive details found in the Disease Tables. Your instructor will help you understand what is important for your class.

Concepts

- ☐ Defenses of gastrointestinal system
- ☐ Normal microbiota of gastrointestinal system
- ☐ Types of oral disease
- ☐ Positive and negative aspects of *Helicobacter* colonization
- ☐ Neglected parasitic infections (NPIs)
- ☐ Four cycles of transmission for helminth infections
- ☐ Organisms in this chapter for which there are vaccines available
- ☐ Organisms in this chapter that display significant antibiotic resistance

Terms

- ☐ Enteric tract
- ☐ GALT
- ☐ ANUG
- ☐ STEC
- ☐ Guillain-Barré syndrome
- ☐ Oral rehydration therapy

Design Element: (College students): Caia Image/Image Source

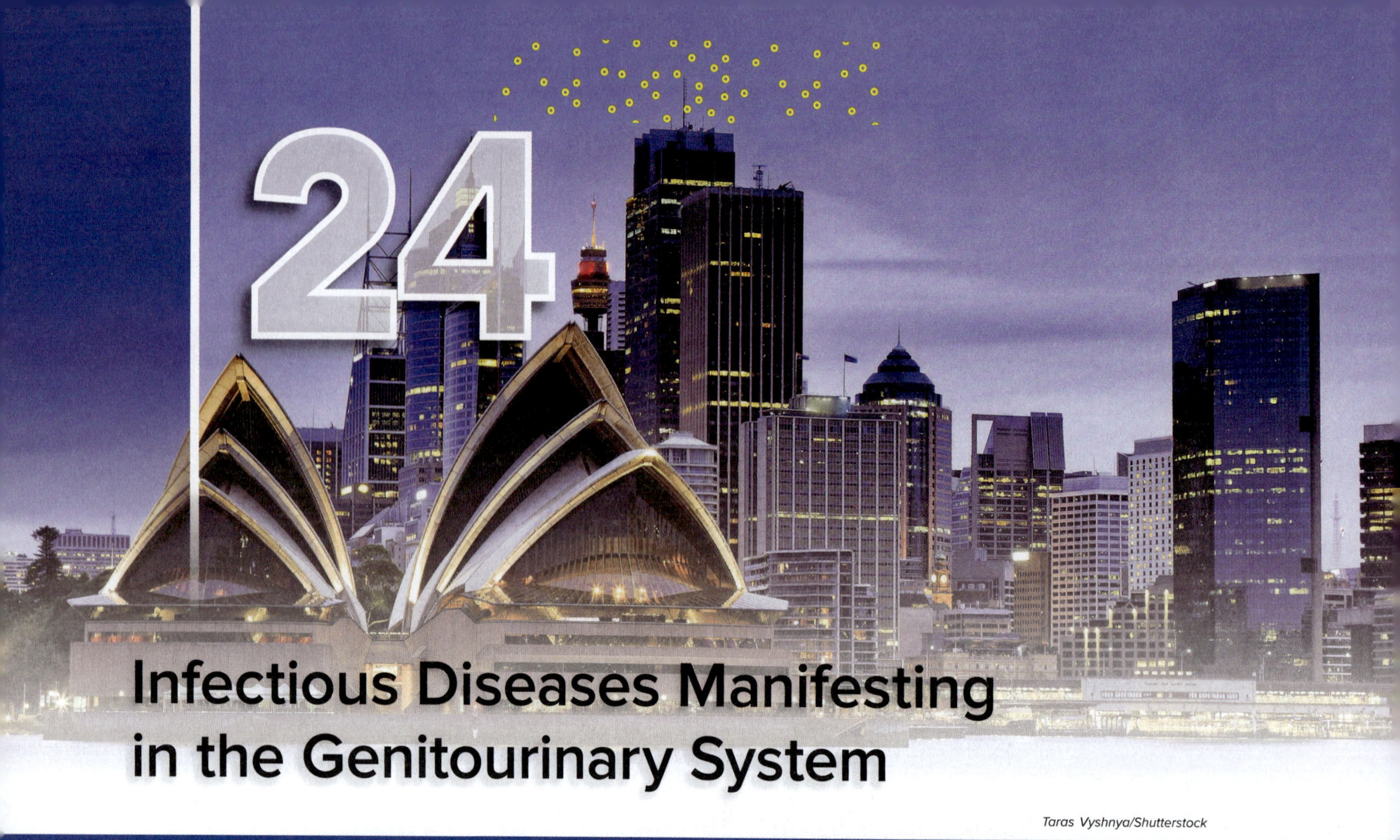

24 Infectious Diseases Manifesting in the Genitourinary System

Taras Vyshnya/Shutterstock

Media Under The Microscope

Cervical Cancer on Track for Elimination in Australia

This opening case examines an article from the popular media to determine the extent to which it is factual and/or misleading. This case focuses on the recent ScienceAlert *article "Australia's on Track to Eliminate Cervical Cancer, the First Country to Ever Do So."*

Cancer is a worldwide scourge. Cervical cancer is the fourth most common cause of cancer in women worldwide. This article points out that it is particularly deadly because it produces few symptoms and therefore goes undetected until it is too late. The central point of this article is that Australia is on track to fully eliminate the cancer in its population. Researchers predict that by 2028 the rate of cervical cancer in Australia will fall to 4 in 100,000 cases, which the researchers say qualifies as "elimination." The article's authors point out that this is a rather arbitrary designation, as no country has ever eliminated a cancer before, so the World Health Organization has not ruled on what "elimination" might mean.

The authors explain that the country accomplished this by offering free vaccinations for the human papilloma virus, which causes 99% of cervical cancers, starting in 2007. In contrast, the vaccine series can be very expensive in the United States.

- What is the **intended message** of the article?
- What is your **critical reading** of the summary of the article provided above? Remember that in this context, "critical reading" does not necessarily mean *What criticism do you have?* but asks you to apply your knowledge to interpret whether the article is factual and whether the facts support the intended message.
- How would you **interpret** the news item for your nonmicrobiologist friends?
- What is your **overall grade** for the news item—taking into account its accuracy and the accuracy of its intended effect?

Media Under The Microscope Wrap-Up appears at the end of the chapter.

Outline and Learning Outcomes

24.1 The Genitourinary Tract, Its Defenses, and Normal Biota

1. Draw or describe the anatomical features of the genitourinary tracts of both genders.
2. List the natural defenses present in the genitourinary tracts.
3. List the types of normal biota presently known to occupy the genitourinary tracts of both genders.
4. Summarize how the microbiome of the female reproductive tract changes over time.

24.2 Infectious Diseases Manifesting in the Urinary Tract

5. List the possible causative agents for each type of urinary tract infection: cystitis/pyelonephritis, leptospirosis, and schistosomiasis.
6. Discuss the epidemiology of the three types of urinary tract infection.

24.3 Infectious Diseases Manifesting in the Reproductive Tract

7. List the possible causative agents for each of the following infectious reproductive tract conditions: vaginitis, vaginosis, prostatitis, genital discharge diseases, genital ulcer diseases, and wart diseases.
8. Identify which of the preceding conditions can cause disease through vertical transmission.
9. Distinguish between vaginitis and vaginosis.
10. Summarize important aspects of prostatitis.
11. Discuss pelvic inflammatory disease, and identify which organisms are most likely to cause it.
12. Provide some detail about HPV vaccination.
13. Identify the most important risk group for group B *Streptococcus* infection, and discuss why these infections are so dangerous in this population.

24.1 The Genitourinary Tract, Its Defenses, and Normal Biota

As suggested by the name, the structures considered in this chapter are really two distinct organ systems. The *urinary tract* has the job of removing substances from the blood, regulating certain body processes, and forming urine and transporting it out of the body. The *genital system* has reproduction as its major function. It is also called the *reproductive system.*

The urinary tract includes the kidneys, ureters, bladder, and urethra **(figure 24.1).** The kidneys remove metabolic wastes from the blood, acting as a sophisticated filtration system. Ureters are tubular organs extending from each kidney to the bladder. The bladder is a collapsible organ that stores urine and empties it into the urethra, which is the conduit of urine to the exterior of the body. In males, the urethra is also the terminal organ of the reproductive tract, but in females the urethra is separate from the vagina, which is the outermost organ of the reproductive tract.

One obvious defensive mechanism is the flushing action of the urine flowing out of the system. The flow of urine also encourages the **desquamation** (shedding) of the epithelial cells lining the urinary tract. For example, each time a person urinates, he or she loses hundreds of thousands of epithelial cells. Any microorganisms attached to them are also shed, of course. Probably the most common microbial threat to the urinary tract is the group of microorganisms that constitute the normal biota in the gastrointestinal tract because the two organ systems are in close proximity. But the cells of the epithelial lining of the urinary tract have different chemicals on their surfaces than do those lining the GI tract. For that reason, most bacteria that are adapted to adhere to the chemical structures in the GI tract cannot gain a foothold in the urinary tract.

Urine, in addition to being acidic, also contains two antibacterial proteins, lysozyme and lactoferrin. You may recall that lysozyme is an enzyme that breaks down peptidoglycan. Lactoferrin is an iron-binding protein that inhibits bacterial growth. Finally, secretory IgA specific for previously encountered microorganisms can be found in the urine.

The male reproductive system produces, maintains, and transports sperm cells and is the source of male sex hormones. It consists of the *testes,* which produce sperm cells and hormones, and the *epididymides,* which are coiled tubes leading out of the testes. Each epididymis terminates in a *vas deferens,* which

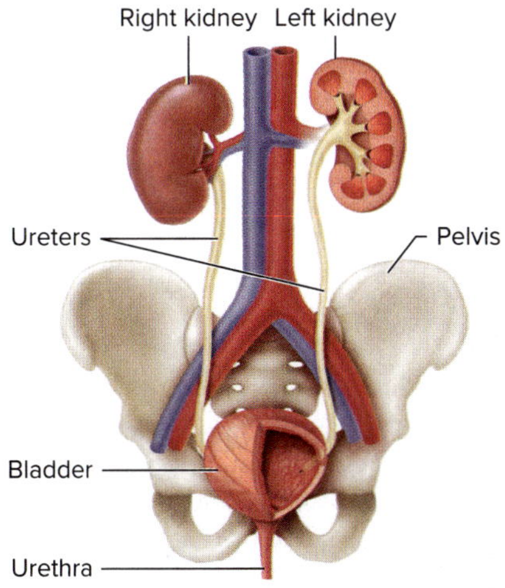
Right kidney
Left kidney
Ureters
Pelvis
Bladder
Urethra

Figure 24.1 The urinary system.

combines with the seminal vesicle and terminates in the ejaculatory duct **(figure 24.2).** The contents of the ejaculatory duct empty into the urethra during ejaculation. The *prostate gland* is a walnut-shaped structure at the base of the urethra. It also contributes to the released fluid (semen). The external organs are the scrotum, containing the testes, and the *penis,* a cylindrical organ that houses the urethra. As for its innate defenses, the male reproductive system also benefits from the flushing action of the urine, which helps move microorganisms out of the system.

The female reproductive system consists of the *uterus, fallopian tubes* (also called uterine tubes), *ovaries,* and *vagina* **(figure 24.3).** During childbearing years, an egg is released from one of the ovaries approximately every 28 days. It enters the fallopian tubes, where fertilization by sperm may take place if sperm are present. The fertilized egg moves through the fallopian tubes to the uterus, where it is implanted in the uterine lining. If fertilization does not occur, the lining of the uterus degenerates and sloughs off. This is the process of menstruation. The terminal portion of the female reproductive tract is the vagina, which is a tube about 9 cm long. The vagina is the exit tube for fluids from the uterus, the channel for childbirth, and the receptive chamber for the penis during sexual intercourse. One very important tissue of the female reproductive tract is the *cervix,* which is the lower one-third of the uterus and the part that connects to the vagina. The opening of the uterus is part of the cervix. The cervix is a common site of infection in the female reproductive tract.

The natural defenses of the female reproductive tract vary over the lifetime of the woman. The vagina is lined with mucous membranes so it has the protective covering of secreted mucus. During childhood and after menopause, this mucus is the major innate defense of this system. Secretory IgA antibodies specific for any previously encountered infections would be present on these surfaces. During a woman's reproductive years, a major portion of the defense is provided by changes in the pH of the vagina brought about by the release of estrogen. This hormone stimulates the vaginal mucosa to secrete glycogen, which certain bacteria can ferment into acid, lowering the pH of the vagina to about 4.5. Before puberty, a girl produces little estrogen and little glycogen and has a vaginal pH of about 7. The change in pH beginning in adolescence results in a vastly different normal biota in the vagina, described later. The healthy biota of women in their childbearing years is thought to prevent the establishment and invasion of microbes that might have the potential to harm a developing fetus.

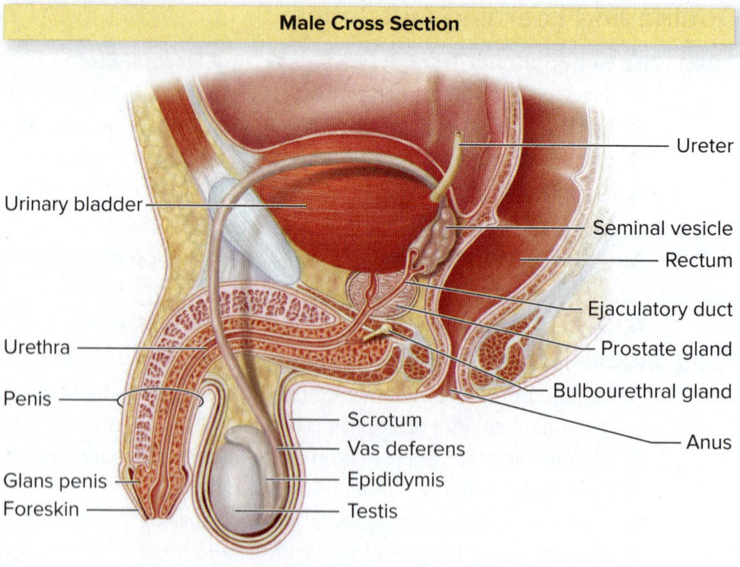
Male Cross Section
Ureter
Urinary bladder
Seminal vesicle
Rectum
Ejaculatory duct
Urethra
Prostate gland
Bulbourethral gland
Penis
Scrotum
Anus
Vas deferens
Glans penis
Epididymis
Foreskin
Testis

Figure 24.2 The male reproductive system.

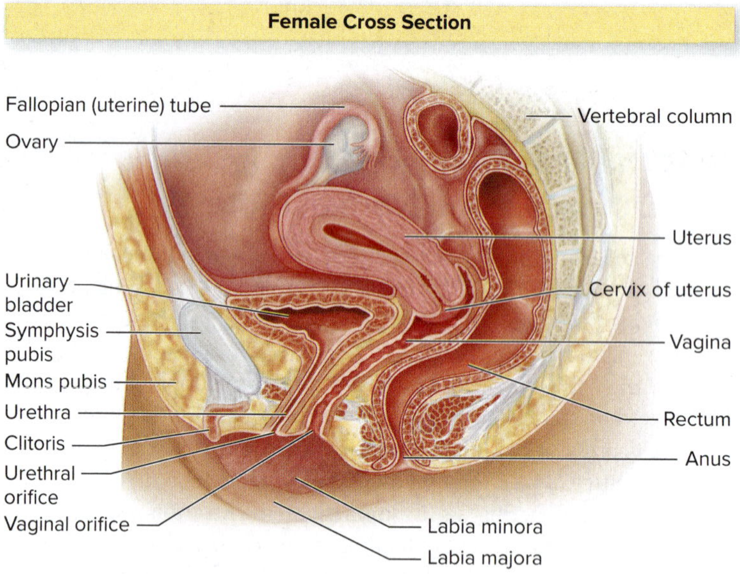
Female Cross Section
Fallopian (uterine) tube
Vertebral column
Ovary
Uterus
Urinary bladder
Cervix of uterus
Symphysis pubis
Vagina
Mons pubis
Urethra
Rectum
Clitoris
Anus
Urethral orifice
Vaginal orifice
Labia minora
Labia majora

Figure 24.3 The female reproductive system.

Normal Biota of the Genitourinary Tract

In both genders, the outer region of the urethra harbors some normal biota. The kidney, ureters, bladder, and upper urethra were previously thought to be sterile. However, recent data suggest that some of these areas may actually contain microbiota that are simply unculturable using currently available methods. Genomic analysis of aseptically obtained urine samples from women showed the presence of a variety of microorganisms. These included known residents of the urethra (nonhemolytic streptococci, staphylococci, corynebacteria, and some lactobacilli) and additionally *Prevotella, Veillonella,* and *Gardnerella* species. However, the exact microbial composition varied among men and among women, indicating that other variables play a role in establishing the normal biota within the urinary tract. Because the urethra in women is so short (about 3.5 cm long) and is in such close proximity to the anus, it can act as a pipeline for bacteria from the GI tract to the bladder, resulting in urinary tract infections. In men, removal of the penile foreskin triggers a change in the composition of the known normal biota on the outer surface of the penis and perhaps in the urethra as well.

Genitourinary Tract Defenses and Normal Biota

	Defenses	Normal Biota
Urinary Tract (Both Genders)	Flushing action of urine; specific attachment sites not recognized by most nonnormal biota; shedding of urinary tract epithelial cells, secretory IgA, lysozyme, and lactoferrin in urine	Nonhemolytic *Streptococcus, Staphylococcus, Corynebacterium, Lactobacillus, Prevotella, Veillonella, Gardnerella*
Female Genital Tract (Childhood and Postmenopausal)	Mucus secretions, secretory IgA	Same as for urinary tract
Female Genital Tract (Childbearing Years)	Acidic pH, mucus secretions, secretory IgA	Variable, but often *Lactobacillus* predominates; also *Prevotella, Sneathia, Streptococcus,* and *Candida albicans*
Male Genital Tract	Same as for urinary tract	Urethra: same as for urinary tract; outer surface of penis: *Pseudomonas* and *Staphylococcus;* sulcus of uncircumcised penis: anaerobic gram-negatives

Normal Biota of the Male Genital Tract

Recent studies have revealed that the normal biota of the male genital tract (that is, in the urethra) is composed of many of the same residents colonizing the external portions of the penis. In addition, *Lactobacillus* and *Streptococcus* species can be found in the urethra of most healthy men. What is interesting is that the microbiota of the urethra in males apparently shifts once sexual activity begins, and microbes associated with sexually transmitted infections (STIs) can begin to take up residence in the genital tract. In fact, men engaging in vaginal, anal, or even oral intercourse can often harbor bacteria that produce bacterial vaginosis in females.

Normal Biota of the Female Genital Tract

Like other body systems we have studied, the most internal portions—in this case, the uterus and above—were long thought to be sterile. Next-generation sequencing has suggested otherwise. We do not know for sure how much and what kind of microbes colonize the upper female reproductive tract, but there are almost certainly either occasional "trespassers" or possibly more permanent residents. We do know that the vaginal canal is colonized by a diverse array of microorganisms. Before puberty and after menopause, the pH of the vagina is close to neutral, and the vagina harbors a biota that is similar to that found in the urethra. After the onset of puberty, estrogen production leads to glycogen release in the vagina, resulting in an acidic pH. The physical and chemical barriers of the vagina select for the growth of normal biota such as *Lactobacillus* species, which thrive in the acidic environment. But these microbes also contribute to the low pH environment, converting sugars to acid. Their predominance in the vagina, combined with the acidic environment, discourages the growth of many microorganisms and actually plays a major role in developing the overall composition of the vaginal biota. Even though the *Lactobacilli* dominate the normal biota of the vagina in most women, studies show that this is not the case in all women **(Insight 24.1).** Others show higher percentages of anaerobic bacteria such as *Prevotella, Sneathia,* or *Streptococcus* species. Scientists have shown that the microbial makeup can actually shift dramatically during the menstrual cycle and during pregnancy, and changes can even occur over just a few days. All of this information reflects the fact that there is no "core" or common vaginal biota composition during childbearing years, and this microbiome is not always stable. Future studies will provide a better understanding of how these microorganisms maintain a healthy, disease-free vaginal canal over time.

The estrogen-glycogen effect continues throughout the childbearing years until menopause. Genomic techniques have led to new findings about the normal biota in postmenopausal women. In contrast to women in their childbearing years, the normal biota composition in postmenopausal women appears to be stable over time. Although *Lactobacillus* and *Gardnerella* species are still common, there is a drop in other characteristic microbial species seen in premenopausal women. It has also been noted that the number of *Lactobacilli* decreases as vaginal dryness increases, which opens a new door to investigate shifts in microbial composition in women suffering from this common symptom of menopause. Note that the very common fungus *Candida albicans* is also present at low levels in the healthy female reproductive tract.

24.1 Learning Outcomes—Assess Your Progress

1. Draw or describe the anatomical features of the genitourinary tracts of both genders.
2. List the natural defenses present in the genitourinary tracts.
3. List the types of normal biota presently known to occupy the genitourinary tracts of both genders.
4. Summarize how the microbiome of the female reproductive tract changes over time.

INSIGHT 24.1 MICROBIOME: Save the World with the Vaginal Microbiome

A few years ago, a scientist named Gregor Reid made a provocative statement to a meeting of the American Society for Microbiology. He said, "To not place a huge focus on the human vaginal microbiome is like putting human survival at risk."

It is estimated that one-third of American women have abnormal vaginal microbiomes, and this increases with douching, multiple sex partners, and other factors like smoking and obesity. A major result of the disruption is referred to as bacterial vaginosis (BV). Aerobic vaginitis (AV) and vulvovaginal candidiasis (VVC) can also result. It can present as fishy malodor, but in many women there are no signs or symptoms at all. If that's the case, why is it still problematic? The answer is it can increase the risk of sexually transmitted infection, cervical cancer, and endometriosis and it can decrease fertility.

Arthimedes/Shutterstock

In pregnant women, this condition can cause the fertilized egg to grow outside the uterus, and also increase the risk of preterm delivery. This is why Gregor Reid refers to the vaginal microbiome as "the cradle of civilization" because dysbiosis puts reproduction in jeopardy. Although this is a worst-case scenario, it places a focus on the beneficial microbes, particularly lactobacilli and bifidobacteria that help retain vaginal health and pass on benefits to the newborns delivered vaginally.

For physicians and researchers, this knowledge is important for several reasons. Diagnosis of these conditions, especially BV, is not straightforward based on Gram stain and microscopy. DNA sequencing or using chromatography to identify certain metabolites (gamma hydroxybutyrate and other compounds) are much better but are not always available in everyday practice. Antimicrobial agents used to treat these conditions were never designed for them specifically. Thus, they do not always work, but they always cause loss of the beneficial bacteria. With this in mind, probiotic lactobacilli have been proposed to aid in treatment and recovery. However, only those strains/products with a scientific rationale, tested appropriately and shown to confer health benefits, with findings published in peer-reviewed journals, should be considered.

Thanks to Gregor Reid, Distinguished Professor Emeritus, Western University, and Scientist, Lawson Health Research Institute, for contributions.

24.2 Infectious Diseases Manifesting in the Urinary Tract

We consider three types of diseases in this section. **Urinary tract infections (UTIs)** result from invasion of the urinary system by bacteria or other microorganisms. Leptospirosis, by contrast, is a spirochete-caused disease transmitted by contact of broken skin or mucous membranes with contaminated animal urine. Lastly, we discuss a helminth disease, urinary schistosomiasis, that is very common in many developing countries.

Urinary Tract Infections (UTIs)

Even though the flushing action of urine helps to keep infections to a minimum in the urinary tract, urine itself is a good growth medium for many microorganisms. When urine flow is reduced, or bacteria are accidentally introduced into the bladder, an infection of that organ (known as *cystitis*) can occur. Occasionally, the infection can also affect the kidneys, in which case it is called *pyelonephritis*. If an infection is limited to the urethra, it is called *urethritis*.

▶ Signs and Symptoms

Cystitis is a disease of sudden onset. Symptoms include pain, frequent urges to urinate even when the bladder is empty, and burning pain accompanying urination (called *dysuria*). The urine can be cloudy due to the presence of bacteria and white blood cells. It may have an orange tinge from the presence of red blood cells (*hematuria*). Low-grade fever and nausea are frequently present. If back pain is present and fever is high, it is an indication that the kidneys may also be involved (pyelonephritis). Pyelonephritis is a serious infection that can result in permanent damage to the kidneys if improperly or inadequately treated. If only the bladder is involved, the condition is sometimes called acute uncomplicated UTI.

Recurrent UTIs are very common in elderly women. When they experience this infection, it is often accompanied by confusion and delirium. Those complications can quickly lead to other problems, such as falls, and also the premature termination of their rights to make their own healthcare decisions due to the incorrect assumption that they have dementia.

▶ Causative Agents

As we saw in the discussion of pneumonia, it is important to distinguish between UTIs that are acquired in health care facilities and those acquired outside of the health care setting. When they occur in health care facilities, they are almost always a result of catheterization and are therefore called *catheter-associated UTIs (CA-UTIs)*. Be careful: The abbreviation "CA" in other infections often refers to "community-acquired"—just the opposite of what is meant here! We will spell out "community" in referring to non-healthcare-associated UTIs.

In 95% of UTIs, the cause is bacteria that are normal biota in the gastrointestinal tract. *Escherichia coli* is by far the most common of these, accounting for approximately 80% of community-acquired urinary tract infections. *Staphylococcus saprophyticus* and *Enterococcus* are also common culprits. These last two are only referenced following the discussion of *E. coli.*

The *E. coli* species that cause UTIs are ones that exist as normal biota in the gastrointestinal tract. They are not the ones that cause diarrhea and other digestive tract diseases.

Transmission and Epidemiology

Community-acquired UTIs are nearly always "transmitted" *not* from one person to another but from one organ system to another, namely from the GI tract to the urinary system. They are much more common in women than in men because of the nearness of the female urethral opening to the anus (see figure 24.3). Many women experience what have been referred to as "recurrent urinary tract infections," although it is now known that some *E. coli* can invade the deeper tissue of the urinary tract and therefore avoid being destroyed by antibiotics. They can emerge later to cause symptoms again. It is not clear how many "recurrent" infections are actually infections that reactivate in this way.

Catheter-associated UTIs are also most commonly caused by *E. coli, S. saprophyticus,* and *Enterococcus*. *Klebsiella* species are another common cause. The National Healthcare Safety Network is now recommending minimizing the use of urinary catheters as much as possible to limit the incidence of these infections.

Treatment

A drug called nitrofurantoin (Macrobid) is most often used for UTIs of various etiologies. Often, another nonantibiotic drug called phenazopyridine (Pyridium) is administered simultaneously. This drug relieves the very uncomfortable symptoms of burning and urgency. Pyridium is an azo dye and causes the urine to turn a dark orange to red color. It may also color contact lenses being worn by people using this drug. A new strain of *E. coli* (ST131) has arisen that is highly virulent and, more troubling, resistant to multiple antibiotics. Medical professionals are ringing alarm bells about this strain, saying that if it acquires resistance to one more class of antibiotics, it will become virtually untreatable. Recently some physicians are avoiding the use of antibiotics by prescribing urine acidifying agents which can thwart bacterial growth **(Disease Table 24.1).**

Leptospirosis

This infection is a zoonosis associated with wild animals and domesticated animals. It can affect the kidneys, liver, brain, and eyes. It is considered in this section because it can have major effects on the kidneys and its presence in animal urinary tracts causes it to be shed into the environment through animal urine.

Signs and Symptoms

Leptospirosis has two phases. During the early—leptospiremic—phase, the pathogen appears in the blood and cerebrospinal fluid. Symptoms are sudden high fever, chills, headache, muscle aches, conjunctivitis, and vomiting. During the second—immune—phase, the blood infection is cleared by natural defenses. This period is marked by milder fever; headache due to leptospiral meningitis; and *Weil's syndrome,* a cluster of symptoms characterized by kidney invasion, hepatic disease, jaundice, anemia, and neurological disturbances. Long-term disability and even death can result from damage to the kidneys and liver, but they occur primarily with the most virulent strains and in elderly persons.

Causative Agent

Leptospires are typical spirochete bacteria marked by tight, regular, individual coils with a bend or hook at one or both ends **(figure 24.4).** *Leptospira interrogans* (lep″-toh-spy′-rah in-terr′-oh-ganz) is the species that causes leptospirosis in humans and animals. There are nearly 200 different serotypes of this species distributed among various animal groups, which accounts for extreme variations in the disease manifestations in humans.

Disease Table 24.1 Urinary Tract Infections

Causative Organism(s)	*Escherichia coli* B G–	*Staphylococcus saprophyticus* B G+	*Enterococcus* B G+
Most Common Modes of Transmission	Opportunism: transfer from GI tract (community-acquired) or environment or GI tract (via catheter)		
Virulence Factors	Adhesins, motility	–	–
Culture/Diagnosis	Usually culture-based; antimicrobial susceptibilities always checked		
Prevention	Hygiene practices; in case of CA-UTIs, limit catheter usage		
Treatment	Usually nitrofurantoin	Usually nitrofurantoin	Based on susceptibility testing; vancomycin-resistant *Enterococcus* is in **Serious Threat** category in CDC Antibiotic Resistance Report
Epidemiological Features	Causes 90% of community UTIs and 50% to 70% of CA-UTIs	Causes small percentage of community UTIs and even lower percentage of CA-UTIs	Frequent cause of CA-UTIs

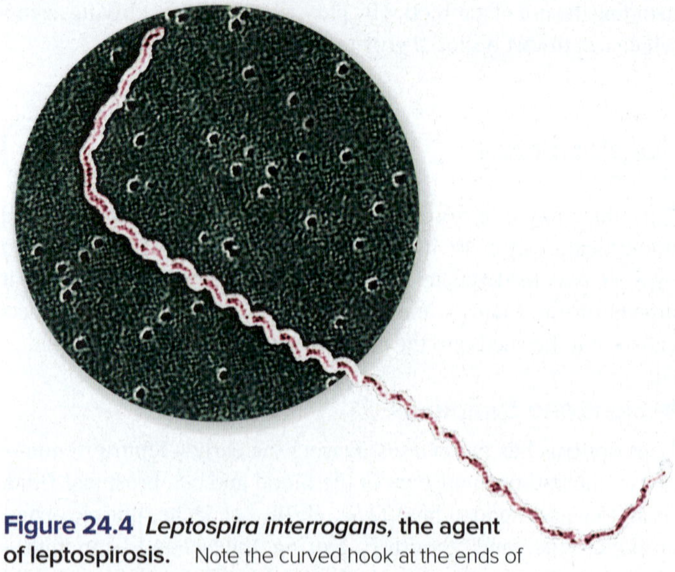

Figure 24.4 ***Leptospira interrogans*, the agent of leptospirosis.** Note the curved hook at the ends of the spirochete. The green disk is a piece of filter paper, and the holes are the pores of the filter.
Janice Haney Carr/NCID/HIP/CDC

▶ Pathogenesis and Virulence Factors

This organism has genes that code for virulence factors such as adhesins and invasion proteins. These factors allow the pathogen to rapidly penetrate host cells and enter into the bloodstream and enable the bacterium to cause cell death in kidney tissue. Because it appears that the bacterium evolved from its close relatives, which are free-living and cause no disease, finding out how the bacterium acquired these genes will be useful in understanding its pathogenesis.

▶ Transmission and Epidemiology

Leptospirosis is a zoonosis, affecting wild animals such as rodents, skunks, raccoons, and foxes and some domesticated animals—particularly horses, dogs, cattle, and pigs. It is found throughout the world, although it is more common in the tropics. It is an occupational hazard of people who work with animals or in the outdoors. Leptospires that are shed in the urine of an infected animal can survive for several months in neutral or alkaline soil or water. Infection occurs almost entirely through contact of skin abrasions or mucous membranes with animal urine or some environmental source containing urine. In 1998, dozens of athletes competing in the swimming phase of a triathlon in Illinois contracted leptospirosis from the water. In late 2009, the Philippines experienced a major outbreak after a series of typhoons flooded the country. Today, leptospirosis is becoming an increasingly significant disease in urban slums around the world. The disease does not appear to be easily transmissible from person to person.

▶ Prevention

A preventive vaccine for humans is being investigated, though a protective vaccine for animals is used to reduce the spread of the pathogen. For now, the best prevention is to wear protective footwear and clothing and to avoid swimming and wading in natural water sources that are frequented by livestock. Anyone participating in aquatic recreational activities should be aware of this infection, especially in more tropical regions of the world.

▶ Treatment

Early treatment with doxycycline rapidly reduces symptoms and shortens the course of disease, but delayed therapy is less effective. In severe disease, penicillin G or ceftriaxone should be used. Other spirochete diseases, such as syphilis (described later), also exhibit this pattern of reduced antibiotic susceptibility over time.

Disease Table 24.2	Leptospirosis
Causative Organism(s)	*Leptospira interrogans* B G–
Most Common Modes of Transmission	Vehicle: contaminated soil or water
Virulence Factors	Adhesins, invasion proteins
Culture/Diagnosis	Slide agglutination test of patient's blood for antibodies; in U.S., CDC will culture specimens
Prevention	Avoiding contaminated vehicles
Treatment	Doxycycline, penicillin G, or ceftriaxone
Epidemiological Features	United States: 100 to 200 cases per year, half in Hawaii; internationally: 80% of people in tropical areas are seropositive

Urinary Schistosomiasis

In chapter 23, we talked about schistosomiasis because one of its two distinct disease manifestations occurs in the liver and spleen, both parts of the digestive system. One particular species of this trematode (helminth) lodges in the blood vessels of the bladder. This may or may not result in symptoms. Blood in the urine and, eventually, bladder obstruction can occur.

▶ Signs and Symptoms

As with the other forms of schistosomiasis, the first symptoms of infestation are itchiness in the area where the helminth enters the body, followed by fever, chills, diarrhea, and cough. Urinary tract symptoms occur at a later date. Remember that adult flukes can live for many years and, by eluding the immune defenses, cause chronic infection.

▶ Causative Agent

The urinary manifestations occur if a host is infected with a particular species of schistosome, *Schistosoma haematobium*. It is found throughout Africa, the Caribbean, and the Middle East. (*S. mansoni* and *S. japonicum* are the species responsible for liver manifestations.) *Schistosomes* are trematodes, or flukes (illustrated in figure 23.30). Humans are the definitive hosts for schistosomes, and snails are the intermediate hosts.

▶ Pathogenesis and Virulence Factors

Like the other species, *S. haematobium* is able to invade intact skin and attach to vascular endothelium. It engages in the same antigenic cloaking behavior as the other two species. The disease manifestations occur when the eggs in the bladder induce a massive granulomatous response that leads to leakage in the blood vessels and blood in the urine. Significant portions of the bladder eventually can be filled with granulomatous tissue and scar tissue. Function of the bladder is decreased or halted altogether. Chronic infection with *S. haematobium* can also lead to bladder cancer.

▶ Transmission and Epidemiology

The life cycle of the schistosome is described completely in chapter 23. After the helminths pass into small blood and lymphatic vessels, they are carried to the liver. Eventually, *S. haematobium* enters the venous plexus of the bladder. While attached to these intravascular sites, the helminths feed upon blood, and the female lays eggs that are eventually voided in urine. The appropriate snail vector does not exist in the United States, so cases found here are virtually all imported.

▶ Culture and/or Diagnosis

Diagnosis depends on identifying the eggs in urine. Newly developed genotypic tests may prove to be more sensitive in the detection of disease.

▶ Prevention and Treatment

The cycle of infection cannot be broken as long as people are exposed to untreated sewage in their environment. It is quite common for people to be cured and then to be reinfected because their village has no sewage treatment. A vaccine would provide widespread control of the disease, but so far none is licensed. More than one vaccine is in development, however.

Praziquantel is the drug treatment of choice and is quite effective at eliminating the helminths, though drug resistance is developing in many areas of the world.

Disease Table 24.3	Urinary Schistosomiasis
Causative Organism(s)	*Schistosoma haematobium* (H)
Most Common Modes of Transmission	Vehicle: contaminated water
Virulence Factors	Antigenic "cloaking," induction of granulomatous response
Culture/Diagnosis	Identification of eggs in urine, PCR methods
Prevention	Avoiding contaminated vehicles
Treatment	Praziquantel
Epidemiological Features	Endemic in Africa, Middle East, India, and Turkey; in sub-Saharan Africa: 120 million infected

24.2 Learning Outcomes—Assess Your Progress

5. List the possible causative agents for each type of urinary tract infection: cystitis/pyelonephritis, leptospirosis, and schistosomiasis.
6. Discuss the epidemiology of the three types of urinary tract infection.

24.3 Infectious Diseases Manifesting in the Reproductive Tract

We saw earlier that reproductive tract diseases in men almost always involve the urinary tract as well, and this is sometimes but not always the case with women. Not all reproductive tract diseases are sexually transmitted, though many are.

We begin this section with a discussion of infections that are symptomatic primarily in women: *vaginitis* and *vaginosis.* Men may also harbor similar infections with or without symptoms. We next consider three broad categories of sexually transmitted infections (STIs): *discharge diseases,* in which increased fluid is released in male and female reproductive tracts; *ulcer diseases,* in which microbes cause distinct open lesions; and the *wart diseases.* The discharge diseases are responsible for large numbers of infertility cases. Herpes (ulcer disease) and human papillomavirus (HPV) (wart disease) infections are incurable and therefore simply increase in their prevalence over time. The section concludes with a neonatal disease caused by group B *Streptococcus* colonization.

Vaginitis

▶ Signs and Symptoms

Vaginitis, an inflammation of the vagina, is a condition characterized by some degree of vaginal itching, depending on the etiologic agent. Symptoms may also include burning, and sometimes a discharge, which may take different forms as well.

▶ Causative Agents

The most common cause of vaginitis is *Candida albicans.* The vaginal condition caused by this fungus is known as a *yeast infection.* Most women experience this condition one or multiple times during their lives. Other bacteria—and even protozoa, such as *Trichomonas*—can also cause vaginal infections.

Candida albicans

Candida albicans is a dimorphic fungus that is normal biota in from 50% to 100% of humans, living in low numbers on many mucosal surfaces such as the mouth, gastrointestinal tract, vagina, and so on. The vaginal condition it causes is often called vulvovaginal candidiasis. The yeast is easily detectable on a wet prep or a Gram stain of material obtained during a pelvic exam **(figure 24.5).** The presence of pseudohyphae in the smear is a clear indication that the yeast is growing rapidly and causing a yeast infection.

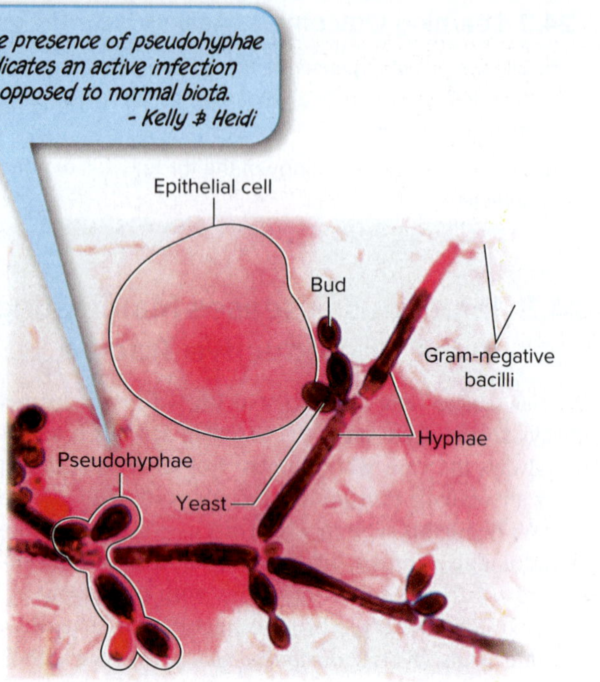
The presence of pseudohyphae indicates an active infection as opposed to normal biota.
- Kelly & Heidi
Epithelial cell
Bud
Gram-negative bacilli
Hyphae
Pseudohyphae
Yeast

Figure 24.5 **Gram stain of *Candida albicans* in a vaginal smear.**
Danny L. Wiedbrauk

▶ Pathogenesis and Virulence Factors

The fungus grows in thick, whitish colonies on the walls of the vagina. The colony debris contributes to a white vaginal discharge. In otherwise healthy people, the fungus is not invasive and limits itself to this surface infection. However, *Candida* infections of the bloodstream do occur and have high mortality rates. They do not normally stem from vaginal infections with the fungus, however, and are seen most frequently in hospitalized patients. People with AIDS are also at risk of developing systemic *Candida* infections.

▶ Transmission and Epidemiology

Vaginal infections with this organism are nearly always opportunistic. Disruptions of the normal bacterial biota or even minor damage to the mucosal epithelium in the vagina can lead to overgrowth by this fungus. Disruptions may be mechanical, such as wearing very tight pants, or they may be chemical, as when broad-spectrum antibiotics taken for some other purpose temporarily diminish the vaginal bacterial population. Diabetics and pregnant women are also predisposed to vaginal yeast overgrowths. Some women are prone to this condition during menstruation.

It is possible to transmit this microbe through sexual contact, especially if a woman is experiencing active infection. The recipient's immune system may well subdue the potential pathogen so that it acts as normal biota in them. But the yeast may be passed back to the original partner during further sexual contact after treatment. Because of this, it is recommended that a patient's sexual partner also be treated to short-circuit the possibility of retransmission. The important thing to remember is that *Candida* is an opportunistic fungus. However, a small percentage of women with no underlying immune disease experience chronic or recurrent vaginal infection with *Candida* for reasons that are not clear.

▶ Prevention and Treatment

No vaccine is available for *C. albicans*. Topical and oral azole drugs are used to treat vaginal candidiasis, and they are now available over the counter. If infections recur frequently or fail to resolve, it is important to see a physician for evaluation. While *Candida albicans*, the causative agent here, is the most frequent *Candida* species to infect humans, it is not as likely to be drug-resistant as other *Candida* species.

Trichomonas vaginalis

Trichomonads are small, pear-shaped protozoa with four anterior flagella and an undulating membrane **(figure 24.6).** *Trichomonas vaginalis* seems to cause asymptomatic infections in approximately 50% of females and also males, despite its species name. Trichomonads are considered asymptomatic infectious agents rather than normal biota because of evidence that some people experience long-term negative effects. Even though *Trichomonas* is a protozoan, it has no cyst form and does not survive long outside of the host.

▶ Pathogenesis and Virulence Factors

Many cases are asymptomatic, and men seldom have symptoms. Women often have vaginitis symptoms, which can include a white to green, frothy discharge. Chronic infection can make a person more susceptible to other infections, including HIV. Also, women who become infected during pregnancy are predisposed to premature labor and low-birthweight infants. Chronic infection may also lead to infertility. Recent research has suggested a link between *Trichomonas* and prostate cancer. Scientists have found that the protozoan can activate a set of proteins whose cascade of effects can increase prostate cancer risk.

▶ Transmission and Epidemiology

Because *Trichomonas* is common biota in so many people, it is easily transmitted through sexual contact. It has been called the most common nonviral sexually transmitted infection, and it is estimated that 10 million Americans have the infection. It does not appear to undergo opportunistic shifts within its host (that is, becoming symptomatic under certain conditions). Instead, the protozoan causes symptoms when transmitted to a noncarrier. Some

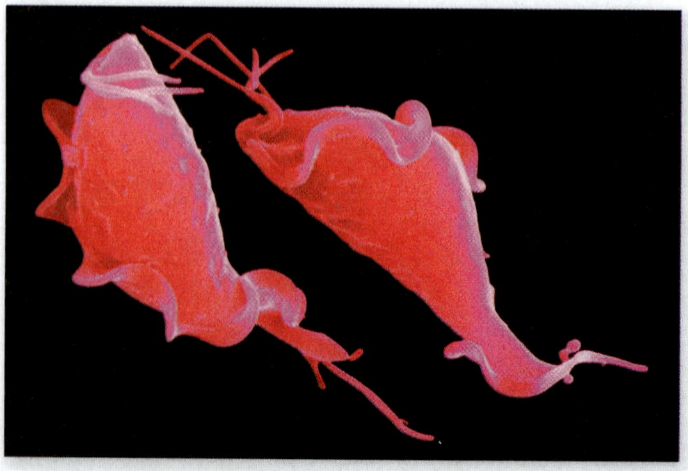

Figure 24.6 ***Trichomonas vaginalis.***
David M. Phillips/Science Source

Disease Table 24.4	Vaginitis	
Causative Organism(s)	*Candida albicans* F	*Trichomonas vaginalis* P
Most Common Modes of Transmission	Opportunism	Direct contact (STI)
Virulence Factors	–	–
Culture/Diagnosis	Wet prep or Gram stain	Protozoa seen on Pap smear or Gram stain; culture is the gold standard
Prevention	–	Barrier use during intercourse
Treatment	Topical or oral azole drugs, some over-the-counter drugs	Metronidazole, tinidazole
Distinctive Features	White, curdlike discharge	Discharge may be greenish
Epidemiological Features	United States: causes 20% of all vaginitis cases; 75% of women reported to have had at least one infection in their lifetimes	7 to 8 million women infected per year

recent data suggest that the protozoan can be transmitted through communal bathing, in public facilities, and from mother to child, but these types of transmission are rare in most populations.

Prevention and Treatment

There is no vaccine for *Trichomonas*. The antiprotozoal drug metronidazole is the drug of choice, although some isolates are resistant to it **(Disease Table 24.4).**

Vaginosis

There is a particularly common—and misunderstood—condition in women in their childbearing years. This condition is usually called vaginosis rather than vaginitis because it does not appear to induce inflammation in the vagina. It is also known as BV, or bacterial vaginosis. Despite the absence of an inflammatory response, a vaginal discharge is associated with the condition. It is often characterized by a fishy odor, and itching is common. But it is also true that many women have this condition with no noticeable symptoms.

Vaginosis is most likely a result of a reduction in the number of "good bacteria" (lactobacilli) in the vagina. The growth of additional microbes plays a role in the development of vaginosis, and new research shows that the diversity of microbiota in cases of BV is much higher than in the healthy vagina. It appears that this condition should be considered the result of a mixed infection. Many of these bacteria are normally found in low numbers in a healthy vagina, including *Gardnerella vaginalis,* a facultatively anaerobic bacterium, as well as *Atopobium,* which is an aerobe, and *Mobiluncus* species, which are anaerobic. The imbalance of the normal biota can leave the vagina open for infection by other opportunistic pathogens, including a newly recognized organism associated with preterm labor (*Leptotrichia amnionii*). There does not appear to be a common set of microbes associated with all cases of BV, and the often-mentioned fishy odor comes from the metabolic by-products produced by many of these anaerobic bacteria.

Pathogenesis and Virulence Factors

The mechanism of damage in this disease is not well understood, but some of the outcomes are. Besides the symptoms just mentioned, vaginosis can lead to complications such as pelvic inflammatory disease (PID), to be discussed later in the chapter; infertility; and, more rarely, ectopic pregnancies. Babies born to some mothers with vaginosis have low birthweights.

Transmission and Epidemiology

This mixed infection is not considered to be sexually transmitted, although women who have never had sex rarely develop the condition. It is very common in sexually active women. We do not know exactly what causes the off-kilter balance of biota in the vagina. The low pH typical of the vagina is usually higher in vaginosis, but it is not clear whether this causes or is caused by the change in bacterial biota. (See Insight 24.1.)

Culture and/or Diagnosis

The condition can be diagnosed by a variety of methods. Sometimes a simple stain of vaginal secretions is used to examine sloughed vaginal epithelial cells. In vaginosis, some cells will appear to be nearly covered with adherent bacteria. (In normal times, vaginal epithelial cells are sparsely covered with bacteria.) These cells are called clue cells and are a helpful diagnostic indicator **(figure 24.7).** They can also be found on Pap smears. Due to the complex nature of the infection, genomic analysis of vaginal swabs is often necessary to diagnose disease.

Prevention and Treatment

No known prevention exists. Asymptomatic cases are generally not treated. Women who find the condition uncomfortable or who are

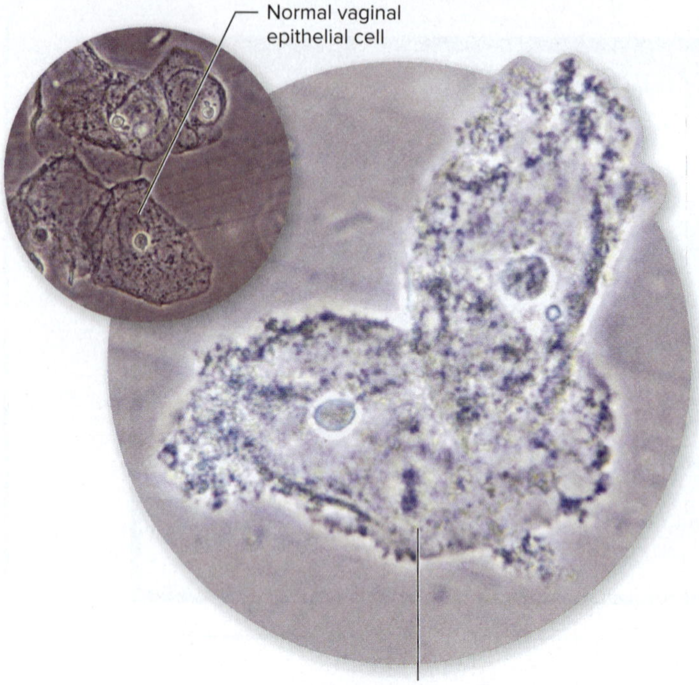
Normal vaginal epithelial cell
Vaginal epithelial cell with numerous bacteria (clue cell)

Figure 24.7 Clue cell in bacterial vaginosis. These epithelial cells came from a pelvic exam. The cells in the large circle have an abundance of bacteria attached to them.
M. Rein/CDC

planning on becoming pregnant should be treated. Women who use intrauterine devices (IUDs) for contraception should also be treated because IUDs can provide a passageway for the bacteria to gain access to the upper reproductive tract. The usual treatment is oral or topical metronidazole or clindamycin **(Disease Table 24.5).**

Prostatitis

Prostatitis is an inflammation of the prostate gland (see figure 24.2). It can be acute or chronic. Acute prostatitis is virtually always caused by bacterial infection. The bacteria are usually normal biota from the intestinal tract or may have caused a previous urinary tract infection. Chronic prostatitis is also often caused by bacteria. Researchers have found that chronic prostatitis, often unresponsive to antibiotic treatment, can be caused by mixed biofilms of bacteria in the prostate.

Symptoms may include pain in the groin and lower back, frequent urge to urinate, difficulty in urinating, blood in the urine, and painful ejaculation. Acute prostatitis is accompanied by fever, chills, and flulike symptoms. Patients appear to be quite ill with the acute form of the disease.

Treatment involves ciprofloxacin or levofloxacin. Also, muscle relaxers or drugs called alpha blockers, which relax the neck of the bladder, may be prescribed. Prostatitis is distinct from prostate cancer, although some of the symptoms may be similar.

Disease Table 24.5	Vaginosis
Causative Organism(s)	Mixed infection
Most Common Modes of Transmission	Opportunism or STI
Virulence Factors	–
Culture/Diagnosis	Visual exam of vagina, or clue cells seen in Pap smear or other smear
Prevention	–
Treatment	Metronidazole or clindamycin
Distinctive Features	Discharge may have fishy smell
Epidemiological Features	United States: estimated 7.4 million new cases per year; internationally: prevalence rates vary by country from 20% to 51%

Disease Table 24.6	Prostatitis
Causative Organism(s)	GI tract biota
Most Common Modes of Transmission	Endogenous transfer from GI tract; otherwise unknown
Virulence Factors	Various
Culture/Diagnosis	Digital rectal exam to examine prostate; culture of urine or semen
Prevention	None
Treatment	Antibiotics, muscle relaxers, alpha blockers
Distinctive Features	Pain in genital area and/or back, difficulty urinating
Epidemiological Features	United States: 50% of men experience during lifetime

A Note About HIV and Hepatitis B and C

This chapter is about diseases whose *major* (presenting) symptoms occur in the genitourinary tract. But some sexually transmitted infections do not have their major symptoms in this system. HIV and hepatitis B and C can all be transmitted in several ways, one of them being through sexual contact. HIV is considered in chapter 21 because its major symptoms occur in the cardiovascular and lymphatic systems. Because the major disease manifestations of hepatitis B and C occur in the gastrointestinal tract, these diseases are discussed in chapter 23. Anyone diagnosed with any sexually transmitted infection should also be tested for HIV.

Discharge Diseases with Major Manifestation in the Genitourinary Tract

Discharge diseases are those in which the infectious agent causes an increase in fluid discharge in the male and female reproductive tracts. Examples are trichomoniasis, gonorrhea, and *Chlamydia* infection. The causative agents are transferred to new hosts when the fluids in which they live contact the mucosal surfaces of the receiving partner. Trichomoniasis was described in the preceding section because its main disease manifestation is considered to be vaginitis. In this section, we cover the other two major discharge diseases: gonorrhea and *Chlamydia* infection.

Gonorrhea

Gonorrhea has been known as a sexually transmitted disease since ancient times. For a fairly long period in history, gonorrhea was confused with syphilis. Later, microbiologists went on to cultivate *Neisseria gonorrhoeae,* also known as the **gonococcus,** and proved conclusively that it alone was the etiologic agent of gonorrhea.

▶ Signs and Symptoms

In the male, infection of the urethra elicits urethritis, painful urination and a yellowish discharge, although a relatively large number of cases are asymptomatic. In most cases, infection is limited to the distal urogenital tract, but it can spread from the urethra to the prostate gland and epididymis (see figure 24.2). Scar tissue formed in the spermatic ducts during healing of an invasive infection can render a man infertile. This outcome is becoming increasingly rare with improved diagnosis and treatment regimens.

In the female, it is likely that both the urinary and genital tracts will be infected during sexual intercourse. A mucopurulent (containing mucus and pus) or bloody vaginal discharge occurs in about half of the cases, along with painful urination if the urethra is affected. Major complications occur when the infection ascends from the vagina and cervix to higher reproductive structures such as the uterus and fallopian tubes **(figure 24.8).** One disease resulting from this progression is **salpingitis** (sal″-pin-jy′-tis). This inflammation of the fallopian tubes may be isolated, or it may also include inflammation of other parts of the upper reproductive tract, called pelvic inflammatory disease (PID). It is not unusual for the microbe that initiates PID to become involved in mixed infections with

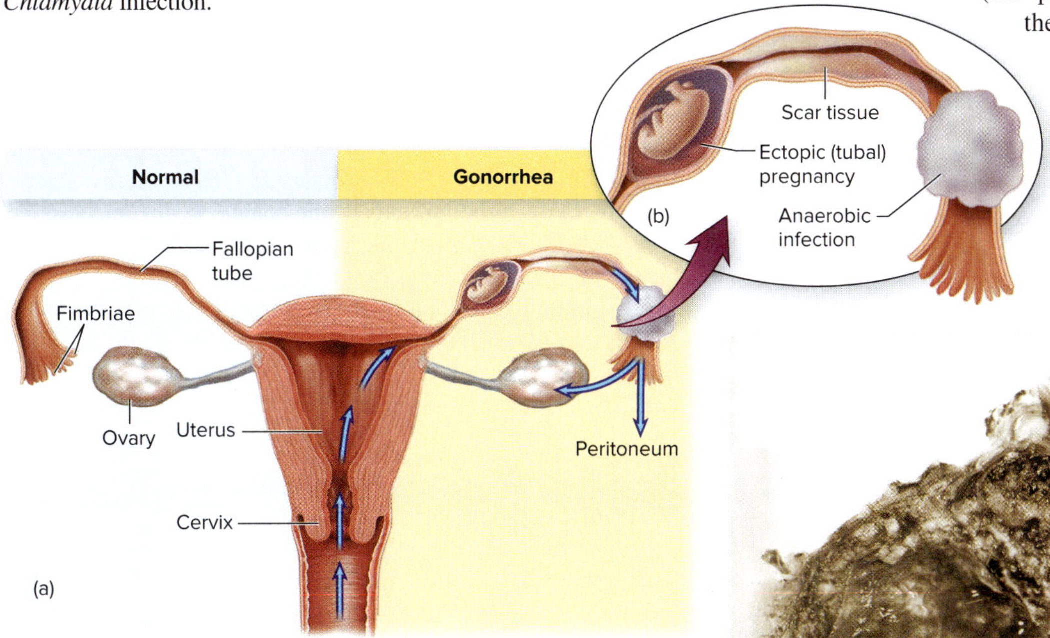

Normal
Gonorrhea
Fallopian tube
Fimbriae
Ovary
Uterus
Cervix
Peritoneum
(a)
Scar tissue
Ectopic (tubal) pregnancy
Anaerobic infection
(b)

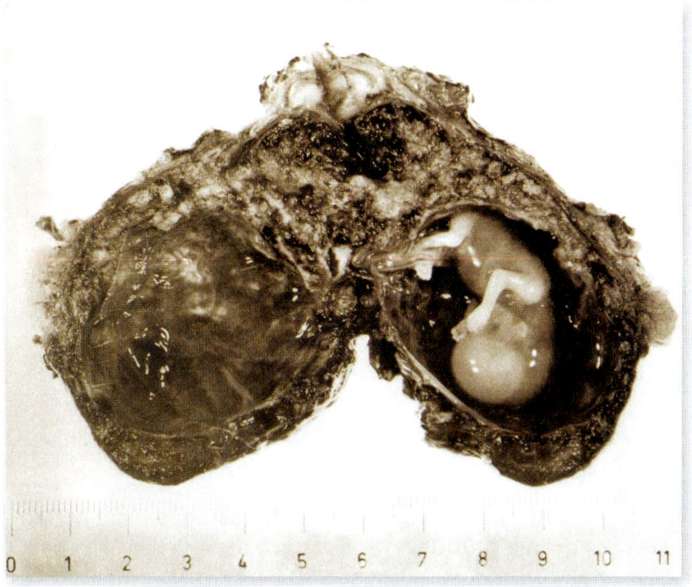

0 1 2 3 4 5 6 7 8 9 10 11
(c)

Figure 24.8 Invasive gonorrhea in women. **(a)** Normal state. **(b)** In ascending gonorrhea, the gonococcus is carried from the cervical opening up through the uterus and into the fallopian tubes. Pelvic inflammatory disease (PID) is a serious complication that can lead to scarring in the fallopian tubes, ectopic pregnancies, and mixed anaerobic infections. **(c)** Pregnancies can also rupture the fallopian tubes and establish in the abdominal cavity, as seen in this tissue removed from a pregnant woman. Ectopic pregnancies in the fallopian tubes or the abdominal cavity are always fatal to the fetus, and sometimes to the mother.

(c) Robert S. Craig/CDC

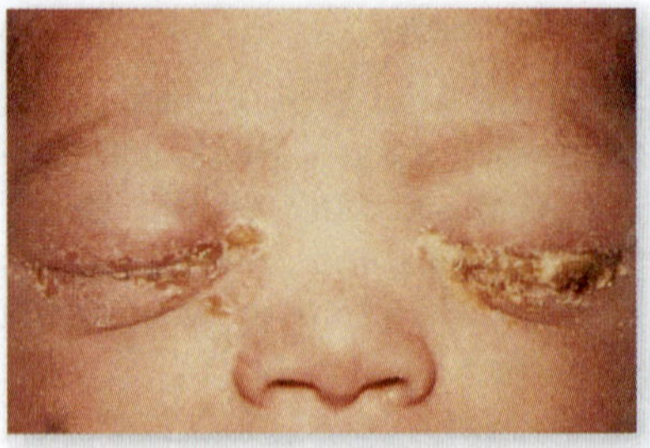

Figure 24.9 Gonococcal ophthalmia neonatorum in a week-old infant. The infection is marked by intense inflammation and edema; if allowed to progress, it causes damage that can lead to blindness. Fortunately, this infection is completely preventable and treatable.

J. Pledger/Centers for Disease Control and Prevention

anaerobic bacteria. The buildup of scar tissue from PID can block the fallopian tubes, causing sterility or ectopic pregnancies.

Serious consequences of gonorrhea can occur outside of the reproductive tract. In a small number of cases, the gonococcus enters the bloodstream and is disseminated to the joints and skin. Involvement of the wrist and ankle can lead to chronic arthritis and a painful, sporadic, papular rash on the limbs. Rare complications of gonococcal bacteremia are meningitis and endocarditis.

Children born to gonococcus carriers are also in danger of being infected as they pass through the birth canal. Because of the potential harm to the fetus, physicians usually screen pregnant mothers for its presence. Gonococcal eye infections are very serious and often result in keratitis, ophthalmia neonatorum, and even blindness **(figure 24.9).** A universal precaution to prevent such complications is the use of antibiotic eyedrops or ointments (usually erythromycin) for newborn babies. The pathogen may also infect the pharynx and respiratory tract of neonates. Finding gonorrhea in children other than neonates is strong evidence of sexual abuse by infected adults, and it calls for child welfare consultation along with thorough bacteriologic analysis.

Causative Agent

N. gonorrhoeae is a pyogenic, gram-negative diplococcus. It appears as pairs of kidney bean–shaped bacteria, with their flat sides touching **(figure 24.10).**

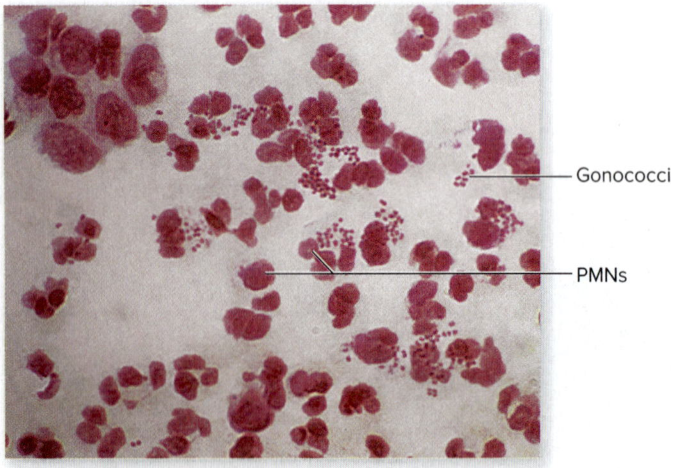
Gonococci
PMNs

Figure 24.10 Gram stain of urethral pus from a male patient with gonorrhea (1,000×). Note the intracellular (phagocytosed) gram-negative diplococci (arranged side-to-side) in polymorphonuclear leukocytes (neutrophils).

Dr. Norman Jacobs/CDC

Pathogenesis and Virulence Factors

Successful attachment is key to the organism's ability to cause disease. Gonococci use specific chemicals on the tips of fimbriae to anchor themselves to mucosal epithelial cells. They only attach to nonciliated cells of the urethra and the cervix, for example. Once the bacterium attaches, it invades the cells and multiplies within the basement membrane.

The fimbriae may also play a role in slowing down effective immunity. The fimbrial proteins are controlled by genes that can be turned on or off, depending on the bacterium's situation. This phenotypic change is called phase variation. In addition, the genes can rearrange themselves to put together fimbriae of different configurations. This antigenic variation confuses the body's immune system. Antibodies that previously recognized fimbrial proteins may not recognize them once they are rearranged.

The gonococcus also possesses an enzyme called IgA protease, which can cleave IgA molecules stationed for protection on mucosal surfaces. In addition, the bacterium pinches off pieces of its outer membrane. These blebs, containing endotoxin, probably play a role in pathogenesis because they can stimulate portions of the innate defense response, resulting in localized damage.

Transmission and Epidemiology

N. gonorrhoeae does not survive more than 1 or 2 hours on fomites and is most infectious when transferred to a suitable mucous membrane. Except for neonatal infections, the gonococcus spreads through some form of sexual contact. The pathogen requires an appropriate portal of entry that is genital or extragenital (rectum, eye, or throat).

Gonorrhea is a strictly human infection that occurs worldwide and ranks among the most common sexually transmitted infections. Although more than 200,000 cases are reported in the United States each year, it is estimated that the actual incidence is much higher—in the millions if one counts asymptomatic infections. Most infections—of both gonorrhea and chlamydia—occur between the ages of 15 and 24.

It is important to consider the reservoir of asymptomatic males and females when discussing the transmission of the infection. Because approximately 10% of infected males and 50% of infected females experience no symptoms, it is often spread unknowingly.

A Note About STI Statistics

It is difficult to compare the incidence of different STIs to one another, for several reasons. The first is that many infections are "silent." That means that infected people do not access the health care system and do not get counted. Of course, we know that many silent infections are actually causing damage that will not be noticed for years; when it is, the original causative organism is almost never sought out. The second reason is that only some STIs are officially reportable to health authorities. *Chlamydia* infection and gonorrhea are, for example, but herpes and HPV are not (see table 14.1). In each section, we will try to present accurate estimates of the prevalence and/or incidence of the diseases as we know them. Finally, the best data come from the CDC, but the verified data usually run 2 years behind.

▶ Culture and/or Diagnosis

The best method for diagnosis is a PCR test of secretions. A Gram stain of male secretions usually yields visible gonococci inside polymorphonuclear cells, but this procedure is not considered sensitive enough to rule out infection if no bacteria are found. Gonorrhea is a reportable disease.

▶ Prevention

Currently, no vaccine is available for gonorrhea, although finding one is a priority for government health agencies. This has become even more important because gonorrhea infection greatly enhances one's risk of HIV infection, and there has been a dramatic increase in antibiotic-resistant gonorrhea infections worldwide. Using condoms is an effective way to avoid transmission of this and other discharge diseases.

▶ Treatment

The CDC runs a program called the Gonococcal Isolate Surveillance Project (GISP) to monitor the occurrence of antibiotic resistance in *N. gonorrhoeae*. Every month in 28 local STI clinics around the country, *N. gonorrhoeae* isolates from the first 25 males diagnosed with the infection are sent to regional testing labs, their antibiotic sensitivities are determined, and the data are provided to the GISP program at the CDC. The CDC considers the antibiotic resistance of this bacterium to be highly dangerous. The CDC has recently changed its overall recommendation for the treatment of gonorrhea worldwide. The CDC now advises the use of ceftriaxone + azithromycin. This is a major change in patient treatment, but it is hoped that making this switch will slow down the spread of the highly resistant strain.

Because those infected with *N. gonorrhoeae* are frequently co-infected with *Chlamydia,* treatment recommendations include treating for that bacterium as well, unless its presence has been ruled out.

For a long time, genital discharge diseases were divided into those caused by *N. gonorrhoeae* and those that were not. These latter were commonly called NGUs, or non-gonococcal urethritis. *Chlamydia trachomatis* infections rose in prevalence over the last few decades, and the "NGU" term lost usage because it was usually *Chlamydia* if it was not gonorrhea. The term *NGU* is coming back into usage, however, as new infectious agents come to light. One of these is *Mycoplasma genitalium.* It causes an increasing number of urethritis/discharge diseases. Another emerging agent for urethritis is another species of *Neisseria, N. meningitidis,* which we studied as an agent of meningitis in chapter 20. In the next section, though, we will focus on *Chlamydia.*

Chlamydia Disease

Genital *Chlamydia* infection is the most common reportable infectious disease in the United States. Annually, more than 2 million cases are reported, but the actual infection rate may be five to seven times that number. The overall prevalence among sexually active young women ages 14 to 19 years is 6.8%, according to the CDC. It is at least two to three times more common than gonorrhea and is also the most commonly reported STI, even though the vast majority of cases are asymptomatic. When we consider the serious consequences that may follow *Chlamydia* infection, those facts are very disturbing.

▶ Signs and Symptoms

In males who experience *Chlamydia* symptoms, the bacterium causes an inflammation of the urethra. The symptoms mimic gonorrhea—namely, discharge and painful urination. Untreated infections may lead to epididymitis. Females who experience symptoms have cervicitis, a discharge, and often salpingitis. Pelvic inflammatory disease is a frequent sequela of female *Chlamydia* infection. A woman is even more likely to experience PID as a result of a *Chlamydia* infection than as a result of gonorrhea. (The electron micrograph in **figure 24.11** depicts *Chlamydia* bacteria

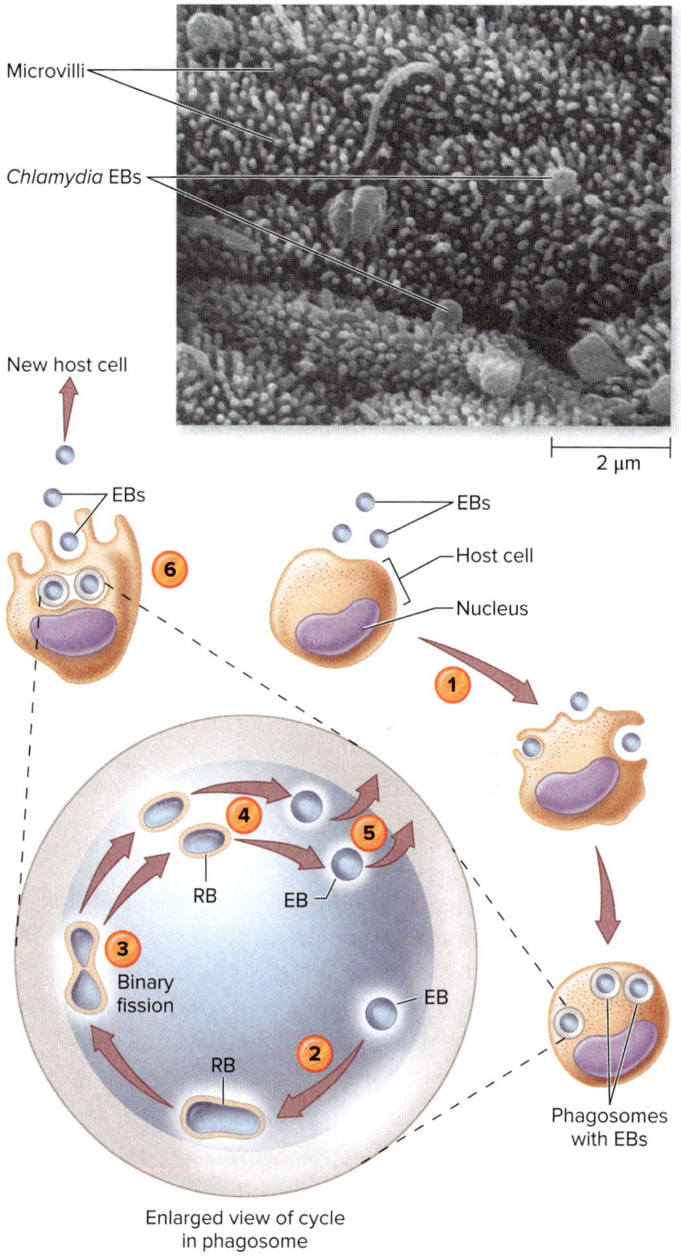
Microvilli
Chlamydia EBs
New host cell
2 µm
EBs
EBs
Host cell
Nucleus
6
1
4
5
RB
EB
3
Binary fission
EB
2
RB
Phagosomes with EBs
Enlarged view of cycle in phagosome

Figure 24.11 The life cycle of *Chlamydia*. The infectious stage, or elementary body (EB), is taken into phagocytic vesicles by the host cell. (1) In the phagosome, each elementary body develops into a reticulate body (RB). (2) Reticulate bodies multiply by regular binary fission. (3) and (4) Mature RBs become reorganized into EBs. (5) Completed EBs are released from the host cell. (6) EBs inside the phagosome of a phagocyte. The top photo is a micrograph of *C. trachomatis* adhering to a fallopian tube.

Morris D. Cooper, Ph.D., Professor of Medical Microbiology, Southern Illinois University School of Medicine, Springfield, IL

adhering inside a fallopian tube.) Up to 75% of *Chlamydia* infections are asymptomatic, which puts women at risk for developing PID because they do not seek treatment for initial infections. The PID itself may be acute and painful, or it may be relatively asymptomatic, allowing damage to the upper reproductive tract to continue unchecked.

Certain strains of *C. trachomatis* can invade the lymphoid tissues, resulting in another condition called lymphogranuloma venereum. This condition is accompanied by headache, fever, and muscle aches. The lymph nodes near the lesion begin to fill with granuloma cells and become enlarged and tender. These "nodes" can cause long-term lymphatic obstruction that leads to chronic, deforming edema of the genitalia or anus. The disease is endemic to South America, Africa, and Asia but occasionally occurs in other parts of the world. Its incidence in the United States is about 500 cases per year, with the number of cases increasing in men who have sex with men.

Babies born to mothers with *Chlamydia* infections can develop eye infections and pneumonia if they become infected during passage through the birth canal. Infant conjunctivitis caused by contact with maternal *Chlamydia* infection is the most prevalent form of conjunctivitis in the United States. Antibiotic drops or ointment applied to newborns' eyes is used to eliminate both *Chlamydia* and *N. gonorrhoeae*.

Causative Agent

C. trachomatis is a very small, gram-negative bacterium. It lives inside host cells as an obligate intracellular parasite. All *Chlamydia* species alternate between two distinct stages: (1) a small, metabolically inactive infectious form called the elementary body, which is released by the infected host cell, and (2) a larger, noninfectious, actively dividing form called the reticulate body, which grows within the host cell vacuoles (see figure 24.11). Elementary bodies are tiny, dense spheres shielded by a rigid, impervious envelope that ensures survival outside the eukaryotic host cell. Studies of reticulate bodies indicate that they are "energy parasites," entirely lacking enzyme systems for synthesizing ATP, although they do possess ribosomes and mechanisms for synthesizing proteins, DNA, and RNA. Reticulate bodies ultimately become elementary bodies during their life cycle.

Pathogenesis and Virulence Factors

Chlamydia's ability to grow intracellularly contributes to its virulence because it escapes certain aspects of the host's immune response. Also, the bacterium has a unique cell wall that apparently prevents the phagosome from fusing with the lysosome inside phagocytes. The presence of the bacteria inside cells causes the release of cytokines that provoke intense inflammation. This defensive response leads to most of the actual tissue damage in *Chlamydia* infection. Of course, the last step of inflammation is repair, which often results in scarring. This can have disastrous effects on a narrow tube like the fallopian tube.

Transmission and Epidemiology

The reservoir of pathogenic strains of *C. trachomatis* is the human body. The microbe shows an astoundingly broad distribution within the population. More alarming is the fact that *Chlamydia* infections have risen steadily over the past few years to reporting levels that have never been seen with any other CDC-notifiable disease. Adolescent women are more likely than older women to harbor the bacterium because it prefers to infect cells that are particularly prevalent on the adolescent cervix. This, along with increased screening rates in women, may in part explain why disease incidence is nearly three times higher in females than in males. It is transmitted through sexual contact as well as vertically. Fifty percent of babies born to infected mothers will acquire conjunctivitis (more common) or pneumonia (less common). **Figure 24.12** shows the distribution of *Chlamydia* among men and women in different age groups.

Culture and/or Diagnosis

Infection with this microorganism is usually detected initially using a rapid technique such as PCR or ELISA. NAAT tests are now the gold standard. There is a high rate of coinfection with gonorrhea in many patients testing positive for *Chlamydia* infection.

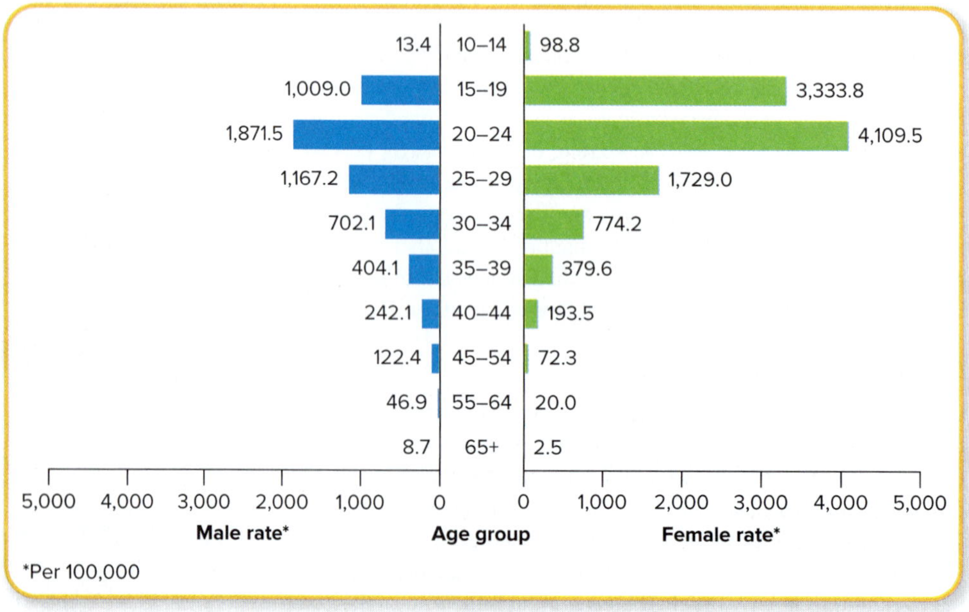
13.4
10–14
98.8
1,009.0
15–19
3,333.8
1,871.5
20–24
4,109.5
1,167.2
25–29
1,729.0
702.1
30–34
774.2
404.1
35–39
379.6
242.1
40–44
193.5
122.4
45–54
72.3
46.9
55–64
20.0
8.7
65+
2.5
5,000 4,000 3,000 2,000 1,000 0
0 1,000 2,000 3,000 4,000 5,000
Male rate*
Age group
Female rate*
*Per 100,000

Figure 24.12 Rates of *Chlamydia* infection, 2019. This is called a demographic chart. It is frequently used in epidemiologic and demographic studies. The age groups are depicted in the central vertical bar. Males are represented on the left and females on the right. The bars represent new cases in 2019 in each age group.

Centers for Disease Control and Prevention

▶ Prevention

As yet, no vaccine exists for *Chlamydia*. Researchers have developed several types of experimental vaccines, including a DNA vaccine, but none has been approved for use to date. Avoiding contact with infected tissues and secretions through abstinence or barrier protection (condoms) is the only means of prevention.

▶ Treatment

Treatment for this infection relies on being aware of it, so part of the guidelines issued by the CDC is a recommendation for annual screening of young women for presence of the bacterium. It is recommended that older women with some risk factor (new sexual partner, for instance) also be screened. If infection is found, treatment is usually with doxycycline or azithromycin. Coinfection with gonorrhea should be assumed and treated similarly. Note that according to public health officials, many patients become reinfected soon after treatment. The recommendation is that patients be rechecked for *Chlamydia* infection 3 to 4 months after treatment. Treatment of all sexual partners of the patient is also recommended to prevent reinfection.

Repeated infections with *Chlamydia* increase the likelihood of PID and other serious sequelae. *Pelvic inflammatory disease (PID)* is a generalized term for infection of the upper reproductive structures of women most often caused by *Chlamydia trachomatis* or *Neisseria gonorrhoeae*. According to CDC statistics, 80% to 90% of all infections caused by *C. trachomatis* and 50% of infections caused by *N. gonorrhoeae* are asymptomatic, which can then progress to PID. Most often, the uterus, fallopian tubes, and ovaries are involved. Because there is no normal biota in these organs, inflammation resulting from infection with these organisms can lead to scar tissue and pelvic adhesions, which can cause pelvic pain, discharge, fever, nausea, diarrhea, painful urination, and pain during intercourse. There is great variation in the symptoms, and often women are misdiagnosed or show no symptoms at all. PID can be treated with broad-spectrum antibiotics, but undiagnosed, subclinical, or recurrent infection can result in scarring in the uterus and fallopian tubes, which can lead to ectopic pregnancy and infertility.

It is estimated that 62% of high-school students have had sexual intercourse. A CDC report shows that nearly half of the 25 million new STIs each year are among young people ages 15 to 24 years. These statistics indicate that potentially millions of young teenage girls are at risk for *Chlamydia* and gonorrhea, which can lead to PID and infertility later in life. A study conducted by Johns Hopkins University showed that young teenage girls are unlikely to seek treatment for the early symptoms of PID on their own with their family doctor or at an outpatient clinic. It is likely that they are afraid to tell their parents about their sexual activity or ask for help in seeking medical treatment for an STI. The study showed that more often, teenaged girls and young women are hospitalized with symptoms of PID, indicating that they wait until symptoms are so severe that either the parents notice or they are in so much

Disease Table 24.7 Genital Discharge Diseases (in Addition to Vaginitis/Vaginosis)

	Gonorrhea	*Chlamydia*
Causative Organism(s)	*Neisseria gonorrhoeae* B G–	*Chlamydia trachomatis* B G–
Most Common Modes of Transmission	Direct contact (STI), also vertical	Direct contact (STI), vertical
Virulence Factors	Fimbrial adhesins, antigenic variation, IgA protease, membrane blebs/endotoxin	Intracellular growth resulting in avoiding immune system and cytokine release, unusual cell wall preventing phagolysosome fusion
Culture/Diagnosis	Gram stain in males, rapid tests (PCR, ELISA) for females, culture on Thayer-Martin agar	PCR or ELISA, can be followed by cell culture
Prevention	Avoid contact; condom use	Avoid contact; condom use
Treatment	Coinfection by gonorrhea and *Chlamydia* should be assumed; treat with ceftriaxone + azithromycin; antibiotic-resistant strains on **Urgent Threat** list from CDC	Coinfection by *Chlamydia* and gonorrhea should be assumed; treat with doxycycline or azithromycin
Distinctive Features	Rare complications include arthritis, meningitis, endocarditis	More commonly asymptomatic than gonorrhea
Effects on Fetus	Eye infections, blindness	Eye infections, pneumonia
Epidemiological Features	United States: increased 56% since 2015; internationally: 26 million cases	United States: 92% increase since 2009; internationally: eye infection (trachoma) has 90% prevalence rate in developing world

pain that they are forced to ask for help. Unfortunately, costs for emergency room and hospital visits are 6 to 12 times higher than for an outpatient visit, and the delay may result in greater damage to reproductive organs **(Disease Table 24.7).**

Genital Ulcer Diseases

Three common infectious conditions can result in lesions on a person's genitals: syphilis, chancroid, and genital herpes. In this section, we consider each of these.

Syphilis

The disease was first recognized at the close of the 15th century in Europe, a period coinciding with the return of Columbus from the West Indies. From this, some medical scholars have concluded that syphilis was introduced to Europe (the Old World) from the New World. However, recent analysis of data points to the fact that the predecessor of the spirochete that causes the disease actually traveled in reverse—from the Old World to the New World. This predecessor, which was a *non*–sexually transmitted pathogen, evolved in the Old World—through a combination of the immunologically naive population of Europe, the European wars, and sexual promiscuity—and set the stage for the worldwide transmission of syphilis that continues to this day.

A disturbing chapter of syphilis history in the United States is worth noting. Beginning in 1932, the U.S. government conducted a study called the Tuskegee Study of Untreated Syphilis in the Negro Male, which eventually involved indigent African American men living in the South. Three hundred ninety-nine infected men were recruited into the study (along with 201 uninfected men), which sought to document the natural progression of the disease. These men were never told that they had syphilis and were never treated for it, even after penicillin was shown to be an effective cure. The study ended in 1972 after it became public. Much later, in 1997, President Bill Clinton issued a public apology on behalf of the U.S. government, and the government has paid millions of dollars in compensation to the victims and their heirs.

Signs and Symptoms

Untreated syphilis is marked by distinct clinical stages designated as *primary, secondary,* and *tertiary syphilis*. The disease also has latent periods of varying duration during which it is quiescent. The spirochete appears in the lesions and blood during the primary and secondary stages and is transmissible at these times. During the early latency period between secondary and tertiary syphilis, it is also transmissible. Syphilis is largely nontransmissible during the "late latent" and tertiary stages. Symptoms of each of these stages and congenital syphilis are briefly described here.

Primary Syphilis The earliest indication of syphilis infection is the appearance of a hard **chancre** (shang′-ker) at the site of entry of the pathogen **(figure 24.13*a*).** A chancre appears after an incubation period that varies from 9 days to 3 months. The chancre begins as a small, red, hard bump that enlarges and breaks down, leaving a shallow crater with firm margins. The base of the chancre beneath the encrusted surface swarms with spirochetes. Most chancres appear on the internal and external genitalia, but about 20% occur on the lips, oral cavity, nipples, or fingers, or around the anus. Because these ulcers tend to be painless, they may escape notice, especially when they are on internal surfaces. Lymph nodes draining the affected region become enlarged and firm, but systemic symptoms are absent at this point. The chancre heals spontaneously without scarring in 3 to 6 weeks, but the healing is deceptive because the spirochete has escaped into the circulation and is entering a period of tremendous activity.

Secondary Syphilis About 3 weeks to 6 months after the chancre heals, the secondary stage appears. By then, many systems of the body have been invaded, and the signs and symptoms are more profuse and intense. Initial symptoms are fever, headache, and sore throat, followed by lymphadenopathy and a peculiar red or brown rash that breaks out on all skin surfaces, including the palms of the hands and the soles of the feet **(figure 24.13*b*).** A person's hair often falls out. Like the chancre, the lesions contain viable spirochetes and disappear spontaneously in a few weeks. The major complications of this stage, occurring in the bones, hair follicles, joints, liver, eyes, and brain, can linger for months and years.

Latency and Tertiary Syphilis After resolution of secondary syphilis, about 30% of infections enter a highly variable latent period that can last for 20 years or longer. During latency, although antibodies to the bacterium are readily detected, the

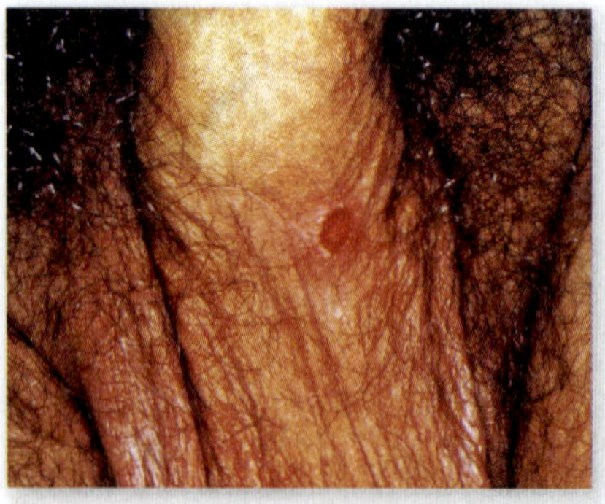
(a)

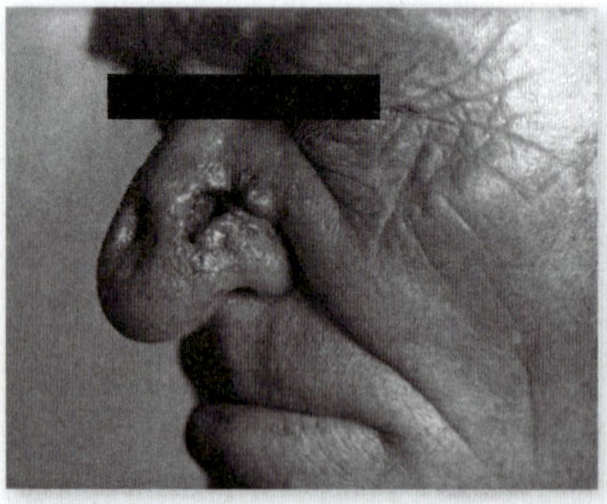
(c)

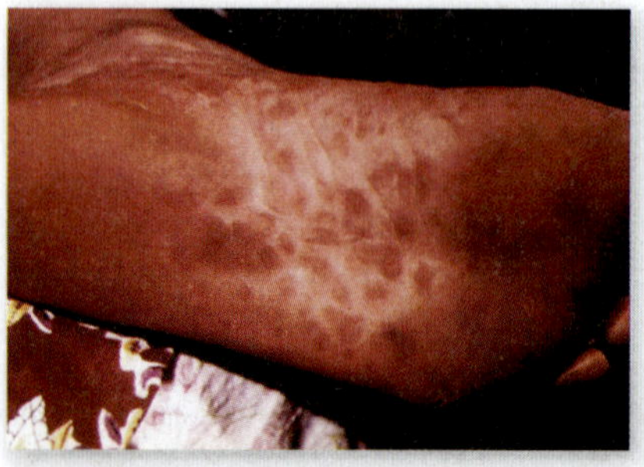
(b)

Figure 24.13 Primary, secondary, and tertiary syphilis. **(a)** The lesion (chancre) of primary syphilis. **(b)** Secondary syphilis produces a generalized rash that also appears on the palms of the hands and soles of the feet. **(c)** Tertiary syphilis patient with gummas on his nose.

(a) Dr. N.J. Fiumara, Dr. Gavin Hart/CDC; (b) CDC; (c) Susan Lindsley/CDC

bacterium itself is not. The final stage of the disease, tertiary syphilis, is relatively rare today because of widespread use of antibiotics. But it is so damaging that it is important to recognize. By the time a patient reaches this phase, numerous pathologic complications occur in susceptible tissues and organs. Cardiovascular syphilis results from damage to the small arteries in the aortic wall. As the fibers in the wall weaken, the aorta is subject to distension and fatal rupture. The same pathologic process can damage the aortic valves, resulting in insufficiency and heart failure.

In one form of tertiary syphilis, painful, swollen syphilitic tumors called **gummas** (goo-mahz′) develop in tissues such as the liver, skin, bone, and cartilage **(figure 24.13*c*).** Gummas are usually benign and only occasionally lead to death, but they can impair function. Neurosyphilis can involve any part of the nervous system, but it shows particular affinity for the blood vessels in the brain, cranial nerves, and dorsal roots of the spinal cord. The diverse results include severe headaches, convulsions, atrophy of the optic nerve, blindness, dementia, and a sign called the Argyll-Robertson pupil—a condition caused by adhesions along the inner edge of the iris that fix the pupil's position into a small, irregular circle.

Congenital Syphilis The syphilis bacterium can pass from a pregnant woman's circulation into the placenta and can be carried throughout the fetal tissues. An infection leading to congenital syphilis can occur in any of the three trimesters, but it is most common in the second and third. The pathogen inhibits fetal growth and disrupts critical periods of development with varied consequences, ranging from mild effects to the extremes of spontaneous miscarriage or stillbirth. Early congenital syphilis encompasses the period from birth to 2 years of age and is usually first detected 3 to 8 weeks after birth. Infants often demonstrate such signs as profuse nasal discharge **(figure 24.14),** skin eruptions, bone deformation, and nervous system abnormalities. The late form gives rise to an unusual assortment of problems in the bones, eyes, inner ear, and joints and causes the malformation of teeth. The number of congenital syphilis cases is closely tied to the incidence in adults. In 2020, 2100 cases of congenital syphilis were reported, up from 500 in 2015.

Causative Agent

Treponema pallidum, a spirochete, is a thin, regularly coiled cell with a gram-negative cell wall. It is a strict parasite with complex growth requirements that require it to be cultivated in living host cells. Most spirochete bacteria are nonpathogenic. *Treponema* and *Leptospira*, described earlier, are among the pathogens of this group.

Syphilis is a complicated disease to diagnose. Not only do the stages each mimic other diseases, but their appearance can also be so separated in time as to seem unrelated. The chancre and secondary lesions must be differentiated from bacterial, fungal, and parasitic infections; tumors; and even allergic reactions. Overlapping symptoms of sexually transmitted infections that the patient is concurrently experiencing, such as gonorrhea or *Chlamydia,* can further complicate diagnosis. The disease can be diagnosed using two different strategies: either by detecting the bacterium in patient lesions or by looking for antibodies in the patient's blood.

Pathogenesis and Virulence Factors

Brought into direct contact with mucous membranes or abraded skin, *T. pallidum* binds avidly by its hooked tip to the epithelium **(figure 24.15).** At the binding site, the spirochete multiplies and penetrates the capillaries nearby. Within a short time, it moves into the circulation, and the body is transformed into a large receptacle for incubating the pathogen. Virtually any tissue is a potential target.

The specific factor that accounts for the virulence of the syphilis spirochete appears to be outer membrane lipoproteins.

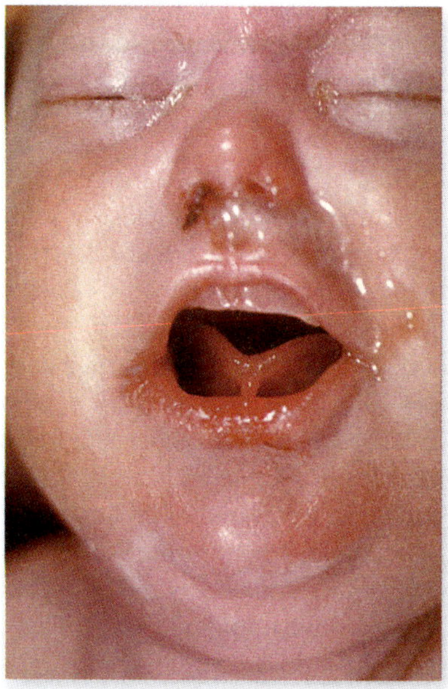

Figure 24.14 **Congenital syphilis.** An early sign is snuffles, a profuse nasal discharge that obstructs breathing. The nasal discharge is loaded with *T. pallidum.* Other congenital defects can include malformations of the nose, forehead, and hard palate.
Dr. Norman Cole/CDC

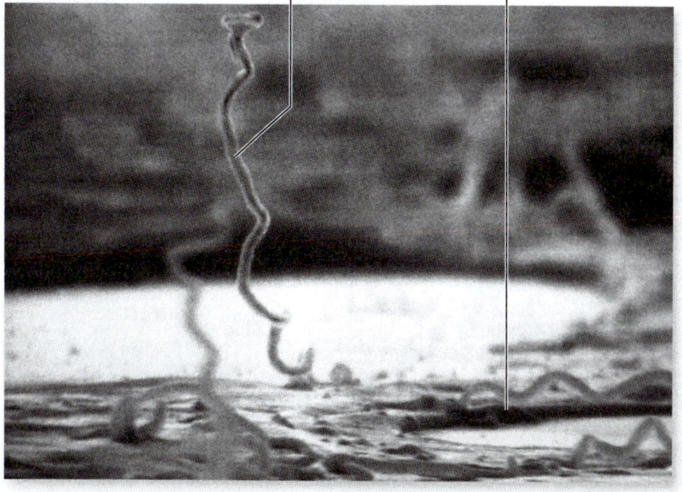

Tip of spirochete
Host cell

Figure 24.15 **Electron micrograph of the syphilis spirochete attached to cells.**
Dr. David Cox/CDC

These molecules stimulate a strong inflammatory response, which is helpful in clearing the organism but can produce damage as well. *T. pallidum* produces no toxins and does not appear to kill cells directly. Studies have shown that, although phagocytes seem to act against it and several types of antitreponemal antibodies are formed, immune responses are unable to contain it. The primary lesion occurs when the spirochetes invade the spaces around arteries and stimulate an inflammatory response. Organs are damaged when granulomas form at these sites and block circulation.

Transmission and Epidemiology

Humans are the sole natural hosts and source of *T. pallidum*. The bacterium is extremely fastidious and sensitive and cannot survive for long outside the host, being rapidly destroyed by heat, drying, disinfectants, soap, high oxygen tension, and pH changes. It survives a few minutes to hours when protected by body secretions and about 36 hours in stored blood. Research with human subjects has demonstrated that the risk of infection from an infected sexual partner is 12% to 30% per encounter. The bacterium can also be transmitted to the fetus in utero.

For centuries, syphilis was a common and devastating disease in the United States, so much so that major medical centers had "Departments of Syphilology." Its effect on social life was enormous. This effect diminished quickly when antibiotics were discovered. If you examine **figure 24.16,** you will see that syphilis has been increasing for years. One of the reasons syphilis rates in the United States have increased is the ongoing drug use epidemic. Recently the CDC reported a significant link between drug use and syphilis among women and heterosexual men. As mentioned previously, persons with syphilis often suffer concurrent infections with other STIs, including HIV.

Culture and/or Diagnosis

Syphilis can be detected in patients most rapidly by using dark-field microscopy of a suspected lesion. A wet mount is then observed for the characteristic size, shape, and motility of *T. pallidum*. Another microscopic test for discerning the spirochete directly in samples is direct immunofluorescence staining with monoclonal antibodies.

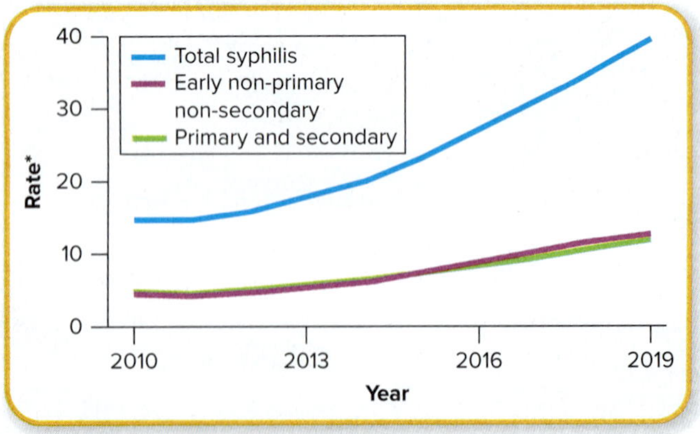

40
30
20
10
0
Rate*
Total syphilis
Early non-primary non-secondary
Primary and secondary
2010
2013
2016
2019
Year

Figure 24.16 Incidence of syphilis cases. Early non-primary non-secondary refers to patients in the latent stage. *per 100,000

Centers for Disease Control and Prevention

Very commonly, blood tests are used for this diagnosis. These tests are based on detection of antibody formed in response to *T. pallidum* infection. The best test is one that specifically reacts with treponemal antigens. Additional specific tests are available when considered necessary. One of these is the indirect immunofluorescent method called the FTA-ABS (fluorescent treponemal antibody absorbance) test. The test serum is first allowed to react with treponemal cells and then reacted with antihuman globulin antibody labeled with fluorescent dyes. If antibodies to the treponeme are present, the fluorescently labeled antibody will bind to the human antibody bound to the treponemal cells. The result is highly visible with a fluorescence microscope. A PCR test is available for syphilis, but its accuracy is dependent on the type of tissue being tested.

Prevention

The core of an effective prevention program depends on detection and treatment of the sexual contacts of syphilitic patients. Public health departments and physicians are charged with the task of questioning patients and tracing their contacts. All individuals identified as being at risk, even if they show no signs of infection, are given immediate prophylactic penicillin in a single long-acting dose.

The barrier effect of a condom provides excellent protection during the primary phase. Protective immunity apparently does arise in humans, allowing the prospect of an effective immunization program in the future, although no vaccine exists currently.

Treatment

Throughout most of history, the treatment for syphilis was a dose of mercury or even a "mercurial rub" applied to external lesions. In 1906, Paul Ehrlich discovered that a derivative of arsenic called salvarsan could be very effective. The fact that toxic compounds, like mercury and arsenic, were used to treat syphilis gives some indication of how dreaded the disease was and to what lengths people would go to rid themselves of it.

Once penicillin became available, it replaced all other treatments, and penicillin G retains its status as a wonder drug in the treatment of acute-stage syphilis. Alternative drugs (tetracycline and doxycycline) are less effective, and they are indicated only if penicillin allergy has been documented. It is important that patients be monitored for successful clearance of the spirochete.

Chancroid

This ulcerative disease usually begins as a soft papule, or bump, at the point of contact. It develops into a "soft chancre" (in contrast to the hard syphilis chancre), which is very painful in men but may be unnoticed in women **(Disease Table 24.8).** Inguinal lymph nodes can become very swollen and tender.

Chancroid is caused by a **pleomorphic** gram-negative rod called *Haemophilus ducreyi.* Recent research indicates that a hemolysin (exotoxin) is important in the pathogenesis of chancroid disease. It is very common in the tropics and subtropics and is becoming more common in the United States. Chancroid is transmitted exclusively through direct contact and is considered a sexually transmitted infection. This disease is associated with sex workers and poor hygiene. Uncircumcised men seem to be more commonly infected than those who have been circumcised. People may carry this bacterium asymptomatically.

No vaccine exists. Prevention of chancroid is the same as for other sexually transmitted infections: Avoid contact with infected tissues, either by abstaining from sexual contact or by properly using barrier protection.

Antibiotics such as azithromycin and ceftriaxone are effective, but patients should be reexamined after a course of treatment to ensure that the bacterium has been eliminated.

Genital Herpes

Virtually everyone becomes infected with a herpesvirus at some time because this large family of viruses can infect a wide range of host tissues. Genital herpes is caused by herpes simplex viruses (HSVs). Two types of HSV have been identified, HSV-1 and HSV-2. Other members of the *Herpesviridae* family are herpes zoster (causing chickenpox and shingles), cytomegalovirus (associated with congenital disease and with HIV-associated disease), Epstein-Barr virus (causing infection of the lymphoid tissue, as in infectious mononucleosis), and more recently identified viruses (herpesvirus-6, -7, and -8).

Genital herpes is much more common than most people think.

Signs and Symptoms

Genital herpes infection has multiple presentations. After initial infection, a person may notice no symptoms. Alternatively, herpes can cause the appearance of single or multiple vesicles on the genitalia, perineum, thigh, and buttocks. The vesicles are small and are filled with a clear fluid (see Disease Table 24.8). They are intensely painful to the touch. The appearance of lesions the first time one gets them can be accompanied by malaise, anorexia, fever, and bilateral swelling and tenderness in the groin. Occasionally, central nervous system symptoms such as meningitis or encephalitis can develop. Thus, initial infection can be either completely asymptomatic or serious enough to require hospitalization.

After recovery from initial infection, a person may have recurrent episodes of lesions. They are generally less severe than the original symptoms, although the whole gamut of possible severity is seen as well. Some people never have recurrent lesions. Others have nearly constant outbreaks, with little recovery time between them. On average, the number of recurrences is four or five a year. Their frequency tends to decrease over the course of years.

In most cases, patients remain asymptomatic or experience recurrent "surface" infections indefinitely. Very rarely, complications can occur. Every year, one or two persons per million with chronic herpes infections develop encephalitis. The virus disseminates along nerve pathways to the brain (although it can also infect the spinal cord). The effects on the central nervous system begin with headache and stiff neck and can progress to mental disturbances and coma. The fatality rate in untreated encephalitis cases is 70%, although treatment with acyclovir is effective. Patients with underlying immunodeficiency are more prone to severe, disseminated herpes infection than are immunocompetent patients. Of greatest concern are patients receiving organ grafts, patients with cancer who are on immunosuppressive therapy, those with congenital immunodeficiencies, and patients with AIDS. Recent data suggest that people with HSV-1 are more prone to Alzheimer's disease, particularly if they carry a particular variant of a particular gene. This is quite sobering when you think that approximately 80% of elderly people are HSV-1-positive, and up to 30% of them carry the gene variant. However, there is hope: Anti-herpes drugs may make a difference in Alzheimer's in these people.

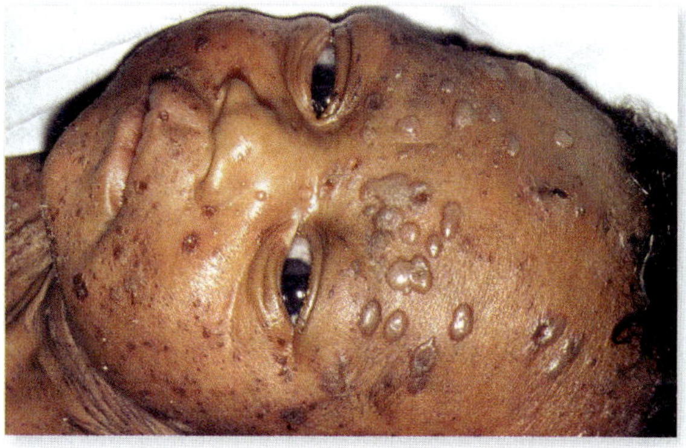

Figure 24.17 Neonatal herpes simplex. Babies can be born with these lesions or develop them 1 to 2 weeks after birth.
Dr. M.A. Ansary/Science Source

Herpes of the Newborn Although HSV infections in healthy adults are annoying and unpleasant, only rarely are they life-threatening. However, in the neonate and the fetus **(figure 24.17),** HSV infections are very destructive and can be fatal. Most cases occur when infants are contaminated by the mother's reproductive tract immediately before or during birth, but they have also been traced to hand transmission from the mother's lesions to the baby. Because HSV-2 is more often associated with genital infections, it is more frequently involved. However, HSV-1 infection has similar complications. In infants whose disease is confined to the mouth, skin, or eyes, the mortality rate is 30%, but disease affecting the central nervous system has a 50% to 80% mortality rate.

Because of the danger of herpes to fetuses and newborns and because of the increase in the number of cases of genital herpes, it is now standard procedure to screen pregnant women for the herpesvirus early in their prenatal care. (Do not forget that most women who are infected do not even know it.) Pregnant women with a history of recurrent infections should be constantly monitored for any signs of viral shedding, especially in the last 4 weeks of pregnancy. If no evidence of recurrence is seen, vaginal birth is indicated, but any evidence of an outbreak at the time of delivery necessitates a cesarean section.

Causative Agent

Both HSV-1 and HSV-2 can cause genital herpes if the virus contacts the genital epithelium, although HSV-1 is thought of as a virus that infects the oral mucosa, resulting in "cold sores" or "fever blisters" **(figure 24.18),** and HSV-2 is thought of as the genital virus. In reality, either virus can infect either region, depending on the type of contact that transmits the infectious agent.

HSV-1 and HSV-2 are DNA viruses with icosahedral capsids and envelopes containing glycoprotein spikes. Like other enveloped viruses, herpesviruses are prone to deactivation by organic solvents or detergents and are unstable outside the host's body.

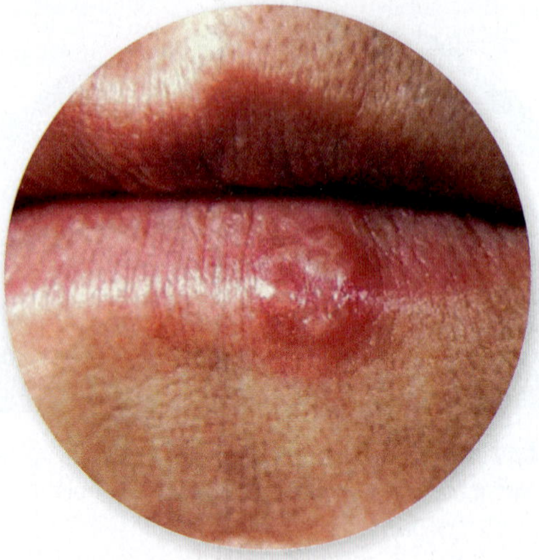

Figure 24.18 Oral herpes infection. Tender, itchy papules erupt around the mouth and progress to vesicles that burst, drain, and scab over. These sores and fluid are highly infectious and should not be touched.
Centers for Disease Control and Prevention

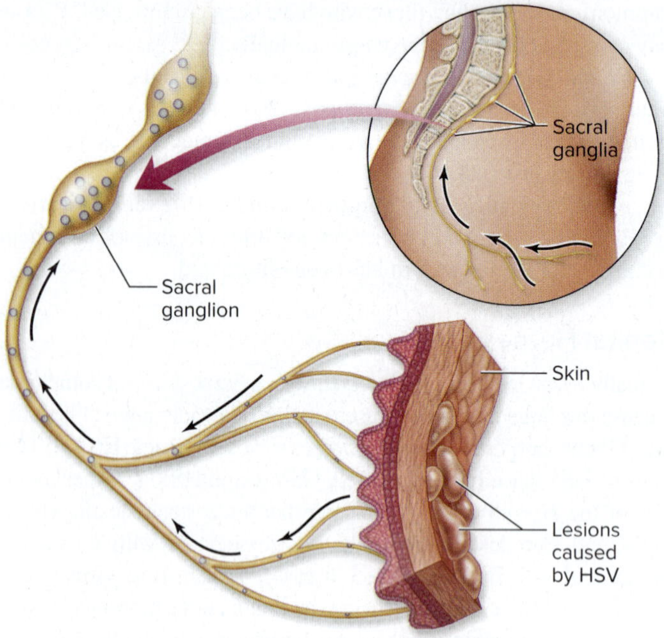

Sacral ganglia
Sacral ganglion
Skin
Lesions caused by HSV

Figure 24.19 Latent HSV in lumbosacral ganglion. The ganglion is the nerve root near the base of the spine. When the virus is reactivated, it travels down the neuron to the body's surface.

Pathogenesis and Virulence Factors

Herpesviruses have a tendency to become latent. During latency, some type of signal causes most of the HSV genome not to be transcribed. This allows the virus to be maintained within cells of the nervous system between episodes. Recent research has found that microRNAs are in part responsible for the latency of HSV-1. It is further suggested that in some peripheral cells, viral replication takes place at a constant, slow rate, resulting in constant, low-level shedding of the virus without lesion production.

HSV-2 (or HSV-1, if it has infected the genital region) usually becomes latent in the ganglion of the lumbosacral spinal nerve trunk **(figure 24.19).** Reactivation of the virus can be triggered by a variety of stimuli, including stress, UV radiation (sunlight), injury, menstruation, or another microbial infection. At that point, the virus begins manufacturing large numbers of entire virions, which cause new lesions on the surface of the body served by the neuron, usually in the same site as previous lesions.

HSV-1 (or HSV-2, if it is in the oral region) behaves in a similar way, but it becomes latent in the trigeminal nerve, which has extensive innervations in the oral region.

Transmission and Epidemiology

Herpes simplex infection occurs globally in all seasons and among all age groups. Because these viruses are relatively sensitive to the environment, transmission is primarily through direct exposure to secretions containing the virus. People with active lesions are the most significant source of infection, but studies indicate that genital herpes can be transmitted even when no lesions are present.

Fifty to eighty percent of adults in the United States are seropositive for HSV-1; 20% to 40% are positive for HSV-2. Seropositivity does not indicate whether the infection is oral or genital. It is estimated that about 25% of American adults have genital herpes. As many as 50% to 90% of people who are infected do not even know it, either because they have rare symptoms that they fail to recognize or because they have no symptoms at all.

Culture and/or Diagnosis

These two viruses—HSV-1 and HSV-2—are sometimes diagnosed based on the characteristic lesions alone. PCR tests are available to test for these viruses directly from lesions. Alternatively, antibody to either of the viruses can be detected from blood samples.

Herpes-infected mucosal cells display notable characteristics in a Pap smear **(figure 24.20).** Laboratory culture and specific tests

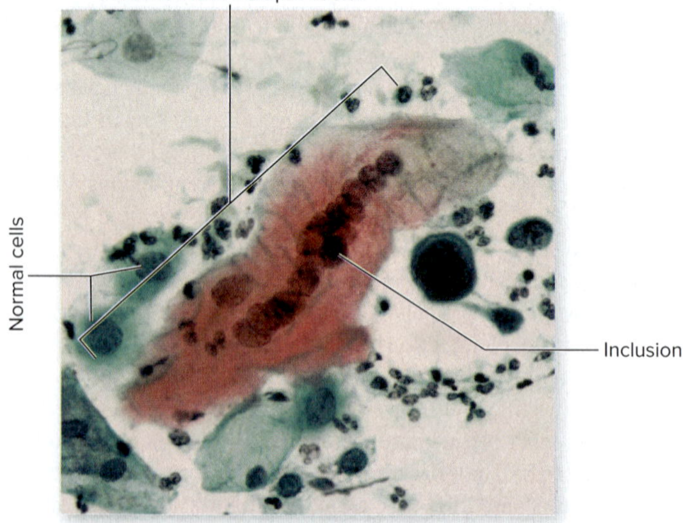

Giant cell with multiple nuclei
Normal cells
Inclusion

Figure 24.20 The appearance of herpesvirus infection in a Pap smear. A Pap smear of a cervical scraping shows enlarged (multinucleate giant) cells and intranuclear inclusions typical of herpes simplex, type 2. This appearance is not specific for the simplex form of herpesvirus, but most other herpesviruses do not infect the reproductive mucosa. This figure also highlights the fact that Pap smears, while intended primarily to detect cervical cancer, can also provide information about other infections.
Dr. Walker/Science Source

are essential for diagnosing severe or complicated herpes infections. They are also used when screening pregnant women for the presence of virus on the vaginal mucosa. A specimen of tissue or fluid is inoculated into a primary cell culture line and then observed for cytopathic effects that are characteristic for specific viruses.

Prevention

No vaccine is currently licensed for HSV, but more than one is being tested in clinical trials, meaning that vaccines may become available soon. In the meantime, avoiding contact with infected body surfaces is the only way to avoid HSV. Condoms provide good protection when they actually cover the site where the lesion is, but lesions can occur outside of the area covered by a condom. In general, people experiencing active lesions should avoid sex. Because the virus can be shed when no lesions are present, barrier protection should be practiced at all times by persons infected with HSV.

Mothers with oral cold sores should be careful in handling their newborns; they should never kiss their infants on the mouth. Some of the drugs used to "treat" genital herpes really function to prevent recurrences of lesions. In this way, they serve as protection for potential partners of people with herpes. It is important to remember that herpes infection is a lifetime infection.

Treatment

Several agents are available for treatment. These agents often result in reduced viral shedding and a decrease in the frequency of lesion occurrence. They are not curative. Acyclovir and its derivatives (famciclovir or valacyclovir) are very effective. Topical formulations can be applied directly to lesions, and pills are available. Sometimes medicines are prescribed on an ongoing basis to decrease the frequency of recurrences, and sometimes they are prescribed to be taken at the beginning of a recurrence to shorten it.

Disease Table 24.8 Genital Ulcer Diseases

	Syphilis	Chancroid	Herpes
Causative Organism(s)	*Treponema pallidum* B G–	*Haemophilus ducreyi* B G–	Herpes simplex 1 and 2 V
Most Common Modes of Transmission	Direct contact and vertical	Direct contact (vertical transmission not documented)	Direct contact, vertical
Virulence Factors	Lipoproteins	Hemolysin (exotoxin)	Latency
Culture/Diagnosis	Direct tests (immunofluorescence, dark-field microscopy), blood tests for treponemal and nontreponemal antibodies, PCR	Rule out other ulcer diseases	Clinical presentation, PCR, Ab tests, growth of virus in cell culture
Prevention	Antibiotic treatment of all possible contacts, avoiding contact	Avoiding contact	Avoiding contact, antivirals can reduce recurrences
Treatment	Penicillin G	Ceftriaxone or azithromycin	Acyclovir and derivatives
Distinctive Features	Three stages of disease plus latent period, possibly fatal	No systemic effects	Ranges from asymptomatic to frequent recurrences
Effects on Fetus	Congenital syphilis	None	Blindness, disseminated herpes infection
Appearance of Lesions	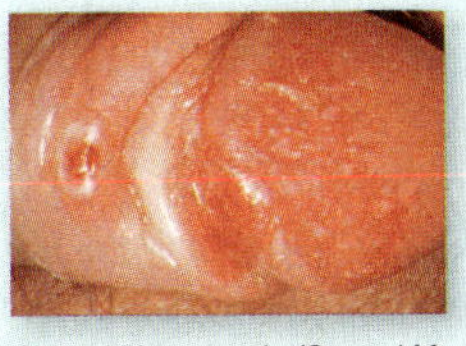 Clinical Photography/Central Manchester University Hospitals NHS Foundation Trust, UK/Science Source	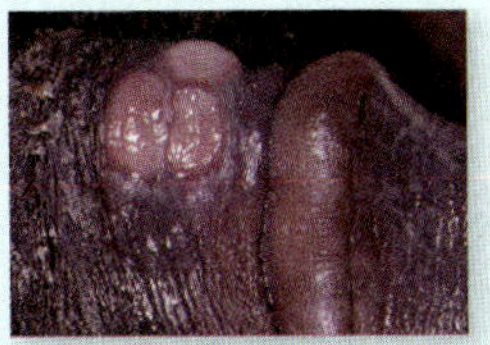 Dr. M.A. Ansary/Science Source	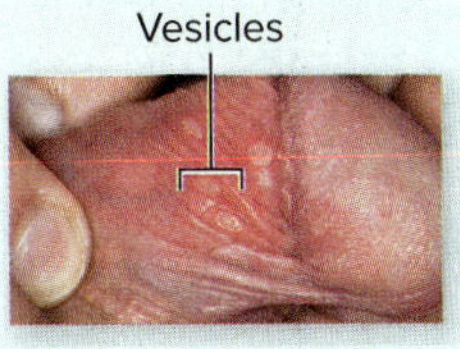 Vesicles Dr. P. Marazzi/Science Source
Epidemiological Features	United States: estimated 120,000 new cases per year; internationally: estimated 12 million new infections per year	United States: no more than a handful per year; internationally: estimated 7 million cases annually	United States: 25% prevalence in adults; internationally: estimated 536 million infected in 15–49 age group

Wart Diseases

In this section, we describe two viral STIs that cause wart-like growths. The more serious disease is caused by the *human papillomavirus* (*HPV*); the other condition, called *molluscum contagiosum,* apparently has no serious effects outside of the growths themselves.

Human Papillomaviruses

These viruses are the causative agents of genital warts. But an individual can be infected with these viruses without having any warts, while still risking serious consequences.

▶ Signs and Symptoms

Symptoms, if present, may manifest as warts—outgrowths of tissue on the genitals **(Disease Table 24.9).** In females, these growths can occur on the vulva and in and around the vagina. In males, the warts can occur in or on the penis and the scrotum. In both sexes, the warts can appear in or on the anus and even on the skin around the groin, such as the area between the thigh and the pelvis. The warts themselves range from tiny, flat, inconspicuous bumps to extensively branching, cauliflower-like masses called **condyloma acuminata.** The warts are unsightly and can be obstructive, but they do not generally lead to more serious symptoms.

Other types of HPV can lead to more subtle symptoms. Certain types of the virus infect cells on the female cervix. This infection may be "silent," or it may lead to abnormal cell changes in the cervix. Some of these cell changes can eventually result in cervical cancer. Ninety-nine percent of cervical cancers are caused by HPV infection. Approximately 4,000 women die each year in the United States from cervical cancer. In recent years, a link between oral sexual activity and an increased risk of throat cancer, presumably due to HPV, was established. An alarming trend has also been documented in men—HPV-16, a strain associated with cervical cancer, was found in over 70% of throat cancer specimens.

Disease Table 24.9 Wart Diseases

	HPV	Molluscum Contagiosum
Causative Organism(s)	Human papillomaviruses V	Poxvirus, sometimes called the molluscum contagiosum virus (MCV) V
Most Common Modes of Transmission	Direct contact (STI), also autoinoculation, indirect contact	Direct contact (STI), also indirect and autoinoculation
Virulence Factors	Oncogenes (in the case of malignant types of HPV)	–
Culture/Diagnosis	PCR tests for certain HPV types, clinical diagnosis; Pap smear	Clinical diagnosis, also histology, PCR
Prevention	Vaccine available; avoid direct contact; prevent cancer by screening cervix	Avoid direct contact
Treatment	Warts or precancerous tissue can be removed; virus not treatable	Warts can be removed; virus not treatable
Distinguishing Features	Infection may or may not result in warts; infection may result in malignancy	Wartlike growths are only known consequence of infection
Effects on Fetus	May cause laryngeal warts	–
Appearance of Lesions	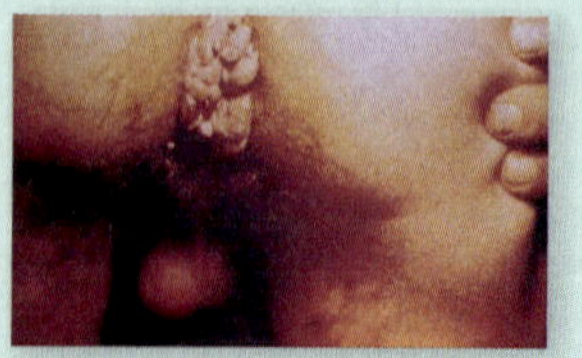 *Dr. Wiesner/Centers for Disease Control and Prevention*	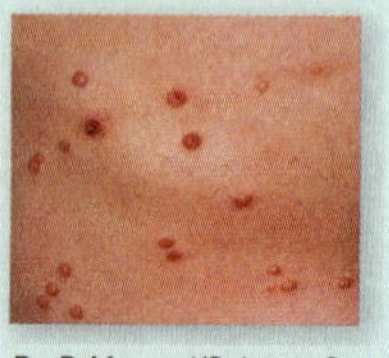 *Dr. P. Marazzi/Science Source*
Epidemiological Features	United States: estimated 6 million new infections per year, 12,000 new cases of HPV-associated cervical cancer	United States: affects 2% to 10% of children annually

Males can also get genitourinary tract cancer from infection with these viruses. The sites most often affected are the penis and the anus.

Causative Agent

The human papillomaviruses are a group of nonenveloped DNA viruses belonging to the *Papovaviridae* family. There are more than 100 different types of HPV. Some types are specific for the mucous membranes; others invade the skin. Some of these viruses are the cause of plantar warts, which often occur on the soles of the feet. Other HPVs cause the common or "seed" warts and flat warts.

Among the HPVs that infect the genital tract, some are more likely to cause the appearance of warts. Ninety percent of genital warts are associated with HPV-6 and HPV-11 infection. Others that have a preference for growing on the cervix can lead to cancerous changes. Two types in particular, HPV-16 and HPV-18, are most closely associated with development of cervical cancer. Other types put you at higher risk for vulvar or penile cancer.

Pathogenesis and Virulence Factors

The major virulence factors for cancer-causing HPVs are **oncogenes,** which code for proteins that interfere with normal host cell function, resulting in uncontrolled growth.

Transmission and Epidemiology

Young women have the highest rate of HPV infections. Twenty-five percent to 46% of women under the age of 25 become infected with genital HPV. It is estimated that 14% of female college students become infected with this condition each year. It is safe to assume that any unprotected sex carries a good chance of encountering either HSV or HPV.

The mode of transmission is direct contact. Autoinoculation is also possible—meaning that the virus can be spread to other parts of the body by touching warts. Indirect transmission occurs but is more common for nongenital warts caused by HPV.

Culture and/or Diagnosis

PCR-based screening tests can be used to test samples from a pelvic exam for the presence of dangerous HPV types. These tests are now recommended for women over the age of 30. A Greek-born physician named George Papanicolaou developed what is now known as the "Pap smear" in the early part of the 20th century, in which changes in vaginal smears are evaluated for precancerous changes. The Pap smear is still the single best screening procedure available for cervical cancer and has caused a 74% decrease in the incidence of cervical cancer since 1955. The procedure is simple and relatively painless. During a pelvic exam, a sample of cells is taken from the cervix using a wooden spatula or small cervical brush. Then the sample is "smeared" onto a glass microscope slide and preserved with a fixative. In a newer method, the fluid is saved, and later it is automatically applied in a thin layer to a microscope slide called a "thin prep." Whether the slide is produced with a "smear" or a "thin prep," it is then viewed microscopically by a technician or by a computer so that abnormal cells can be detected. Nearly all cervical cancer can be prevented if women get Pap smears on the recommended schedule.

Prevention

When discussing HPV prevention, we should consider two possibilities. One of these is infection with the viruses, which is prevented the same way other sexually transmitted infections are prevented—by avoiding direct, unprotected contact—but also by being vaccinated with the 9-valent Gardasil vaccine. This vaccine protects against the two most common strains causing cervical cancer and the two most common causes of warts, as well as five additional strains, which cause the remaining cases of cervical cancer. Health officials emphasize that vaccinated women should still get Pap smears to screen for cervical cancer because other strains of the HPV can also cause the condition.

In 2019, the CDC extended the vaccine recommendations to include men up to the age of 26 because HPV-caused cancers of the male oral cavity, anus, and penis are common. At the same time the CDC suggested that all people up to the age of 45 discuss with their doctor whether to receive the vaccine if they had not previously. They emphasize, however, that the best time to get the vaccine for the greatest protection is at the ages of 11 to 12.

A recent study from the United Kingdom showed that the HPV vaccine decreased the rates of cervical cancer by 87%. And as seen in the chapter opening case, Australia, which has provided free vaccine since 2007, is on track to eliminate it altogether.

Treatment

Many infections are eventually cleared by the immune system, but this is very unpredictable and may take up to 2 years. Current studies show that even when tests for viral DNA are negative, the HPV may still reside latently in an infected female—for up to 20 years or more.

Treatment of cancerous cell changes is an important part of HPV therapy, and it can only be instituted if the changes are detected through Pap smears. Again, the *results* of the infection are treated (cancerous cells removed), but the viral infection is not amenable to treatment and relies upon the activity of the host immune system.

Molluscum Contagiosum

An unclassified virus in the family *Poxviridae* can cause a condition called molluscum contagiosum. This disease can take the form of skin lesions, and it can be transmitted sexually. The wartlike growths that result from this infection can be found on the mucous membranes or the skin of the genital area (see Disease Table 24.9). Few problems are associated with these growths beyond the warts themselves. In severely immunocompromised people, the disease can be more serious, resulting in extensive growth of warts.

The virus causing these growths can also be transmitted through fomites such as clothing or towels and through autoinoculation **(Disease Table 24.9).**

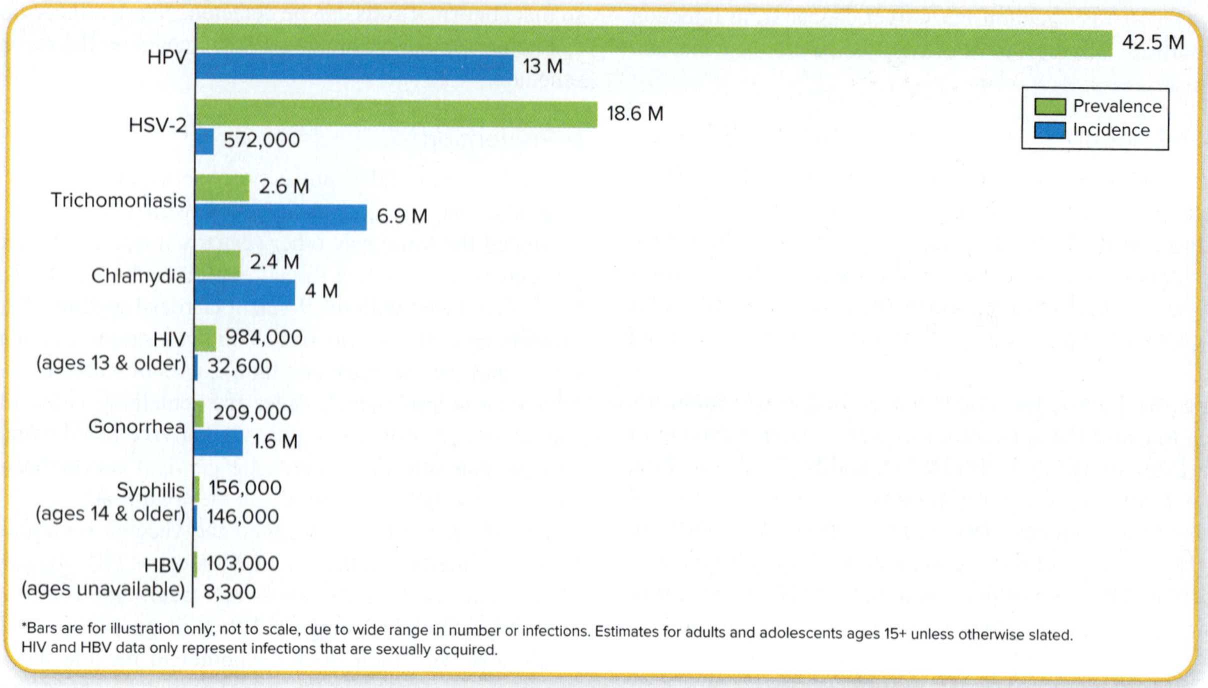
HPV
42.5 M
13 M
HSV-2
18.6 M
572,000
Trichomoniasis
2.6 M
6.9 M
Chlamydia
2.4 M
4 M
HIV (ages 13 & older)
984,000
32,600
Gonorrhea
209,000
1.6 M
Syphilis (ages 14 & older)
156,000
146,000
HBV (ages unavailable)
103,000
8,300
Prevalence
Incidence
*Bars are for illustration only; not to scale, due to wide range in number or infections. Estimates for adults and adolescents ages 15+ unless otherwise slated.
HIV and HBV data only represent infections that are sexually acquired.

Figure 24.21 Sexually transmitted infections in the United States.
https://www.cdc.gov/std/stats17/national-2017.pdf

Group B *Streptococcus* "Colonization"—Neonatal Disease

Ten to forty percent of women in the United States are colonized, asymptomatically, by a beta-hemolytic *Streptococcus* in Lancefield group B (GBS). Nonpregnant women experience no ill effects from this colonization. But when these women become pregnant and give birth, about half of their infants become colonized by the bacterium during passage through the birth canal or by ascension of the bacteria through ruptured membranes. For that reason this colonization is considered a reproductive tract disease.

A small percentage of infected infants experience life-threatening bloodstream infections, meningitis, or pneumonia. If they recover from these acute conditions, they may have permanent disabilities such as developmental disabilities, hearing loss, or impaired vision. In some cases, the mothers also experience disease, such as amniotic infection or subsequent stillbirths. Although GBS infections have declined in the United States, they remain a major threat to infant morbidity and mortality worldwide.

The CDC recommends that all pregnant women be screened for group B *Streptococcus* colonization at 35 to 37 weeks of pregnancy. Because colonization has been associated with preterm birth, recommendations for earlier testing are sometimes warranted. Women positive for the bacterium should be treated with penicillin or ampicillin unless the bacterium is found to be resistant to these—and unless allergy to penicillin is present, in which case erythromycin may be used. Development of a protective vaccine is under way, but a major hurdle is the ability to design one vaccine that will protect all populations against the many serotypes of this bacterium **(Disease Table 24.10)**.

Figure 24.21 illustrates the incidence and prevalence of various STIs in the United States.

Disease Table 24.10	Group B *Streptococcus* Colonization
Causative Organism(s)	Group B *Streptococcus* B G+
Most Common Modes of Transmission	Vertical
Virulence Factors	–
Culture/Diagnosis	Culture of mother's genital tract
Prevention/Treatment	Treat mother with penicillin/ampicillin
Epidemiological Features	United States: vaginal carriage rates 15% to 45%, neonatal sepsis due to this occurs in 1.8 to 3.2 per 1,000 live births; internationally: vaginal carriage rates 12% to 27%

24.3 Learning Outcomes—Assess Your Progress

7. List the possible causative agents for each of the following infectious reproductive tract conditions: vaginitis, vaginosis, prostatitis, genital discharge diseases, genital ulcer diseases, and wart diseases.
8. Identify which of the preceding conditions can cause disease through vertical transmission.
9. Distinguish between vaginitis and vaginosis.
10. Summarize important aspects of prostatitis.
11. Discuss pelvic inflammatory disease, and identify which organisms are most likely to cause it.
12. Provide some detail about HPV vaccination.
13. Identify the most important risk group for group B *Streptococcus* infection, and discuss why these infections are so dangerous in this population.

Media Under The Microscope Wrap-Up

Taras Vyshnya/ Shutterstock

The **intended message** of the *Science Alert* article is to report that Australia is on track to eliminate a very common cancer. The subtitle of the article also is important: "The First Country to Ever Do So." This is useful to draw your eye to the article, as it *is* indeed a landmark accomplishment.

My **critical reading** is that the information is reliable as it cites the World Health Organization and quotes one of the study's authors. It also makes it clear what led to the predicted elimination and that "elimination" may well still mean that 4 in 100,000 women will acquire the cancer. So it avoids sensationalism.

I would **interpret** the article to my friends in 3 steps: I would explain that (1) 99% of cervical cancers are caused by HPV, (2) there is a vaccine against HPV, and (3) near universal immunization can virtually eliminate the cancer. They might wonder what the rate of the cancer is in the United States, so I looked that up on the CDC website: 7.8 cases per 100,000 women.

The **grade** I would give the article is a solid A.

Source: Signe Dean and Fiona MacDonald, "Australia's on Track to Eliminate Cervical Cancer, the First Country to Ever Do So," www.sciencealert.com, October 4, 2018.

Study Smarter: Better Together

These activities are designed for you to use on your own with a study group—either a face-to-face group or a virtual one, consisting of 3–5 members. Studying together can be very helpful, but there are effective and ineffective ways to do it. For example, getting together without a clear structure is often not a good use of your time. Use your time efficiently by using one or more of the exercises below.

FACE-TO-FACE GROUPS

Use one or more of the activities below.

Peer Instruction: Assign numbers to your group members to use all semester long. Now look at these five concepts from this chapter. Each group member prepares a 5-minute lesson on the topic corresponding to their number. Don't worry if you have fewer than 5 members; just use however many you have! During your group study time, each member presents their lesson, and the group spends another 5–10 minutes discussing that lesson.

1. Infections in this chapter having implications for the fetus
2. The role of pH in the vagina
3. Community-acquired vs. catheter-associated UTIs
4. Infections in this chapter that are not curable
5. Pelvic inflammatory disease

Concept Maps: Each member of the group should use this list of terms from this chapter to generate their own concept map. This can be hand-drawn or created using software (see Appendix C for guidelines). During group study time, compare each other's concept maps and help each other make sure they are correct. Of course, there are many different "correct" maps. Examining each member's map will help you talk through the varied concepts and how they are related.

Concept Terms:

genital warts	bacteria	ulcers	cancer	discharge
molluscum contagiosum	warts	syphilis	incurable	virus
chancroid	herpes	curable		

Table Topics: Each group member should identify a concept or topic from this week's class assignments with which they are having trouble and share it during group study time. The other group members can then help to clarify confusing issues or share how they figured it out. Aim for a maximum of 15 minutes per topic. If the topic remains unclear to the group, bring it up during class or use the instructor's office hours or e-mail to ask for help. Taking the time to struggle with a difficult concept first makes your questions much more specific and more likely to yield helpful answers.

VIRTUAL GROUPS

Not everyone has the time or opportunity to meet with group members outside of class time. You or your instructor can create a virtual group using e-mail or the course software.

Weekly Discussion Board: This forum can be used as a way for groups to discuss topics, via e-mail, or other learning management systems or online platforms, before they are covered in class. As each member of the group answers the current week's question, they should send their responses to every other member of their group. It's best to agree on a deadline based on how your class schedule works (Saturday for the next week's topics, for example). Then, after the topic is discussed in class, each member should send a response that group members will see with a follow-up post on the same topic. If you cover more than one chapter in a week, someone can be designated to choose which chapter Discussion Board question you will use. Or simply decide up front that you will always use the first-chapter-of-the-week's question, to keep the schedule simple.

Discussion Question

Discuss the disease in this chapter that can result in cancer, including its transmission and its prevention.

Deadliness and Communicability of Selected Diseases of the Genitourinary Tract

HIV is included here even though it is covered in chapter 21

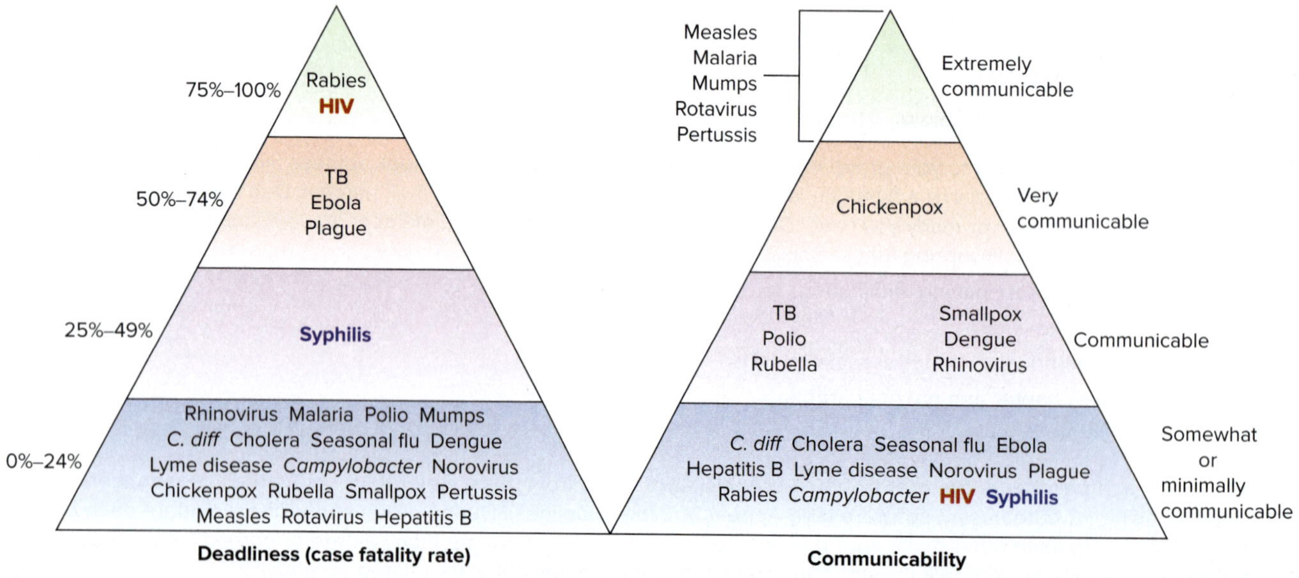
75%–100%
Rabies
HIV
50%–74%
TB
Ebola
Plague
25%–49%
Syphilis
0%–24%
Rhinovirus Malaria Polio Mumps
C. diff Cholera Seasonal flu Dengue
Lyme disease Campylobacter Norovirus
Chickenpox Rubella Smallpox Pertussis
Measles Rotavirus Hepatitis B
Deadliness (case fatality rate)
Measles
Malaria
Mumps
Rotavirus
Pertussis
Extremely communicable
Chickenpox
Very communicable
TB
Polio
Rubella
Smallpox
Dengue
Rhinovirus
Communicable
C. diff Cholera Seasonal flu Ebola
Hepatitis B Lyme disease Norovirus Plague
Rabies Campylobacter HIV Syphilis
Somewhat or minimally communicable
Communicability

Summing Up

Taxonomic Organization Microorganisms Causing Disease in the Genitourinary Tract

Microorganism	Disease	Disease Table
Gram-positive bacteria		
Staphylococcus saprophyticus	Urinary tract infection	Urinary tract infections, 24.1
Group B *Streptococcus*	Neonatal disease	Group B *Streptococcus* colonization, 24.10
Gram-negative bacteria		
Escherichia coli	Urinary tract infection	Urinary tract infections, 24.1
Enterococcus	Urinary tract infection	Urinary tract infections, 24.1
Leptospira interrogans (spirochete)	Leptospirosis	Leptospirosis, 24.2
Neisseria gonorrhoeae	Gonorrhea	Genital discharge diseases (in addition to vaginitis/vaginosis), 24.7
Chlamydia trachomatis	*Chlamydia*	Genital discharge diseases (in addition to vaginitis/vaginosis), 24.7
Treponema pallidum (spirochete)	Syphilis	Genital ulcer diseases, 24.8
Haemophilus ducreyi	Chancroid	Genital ulcer diseases, 24.8
DNA viruses		
Herpes simplex viruses 1 and 2	Genital herpes	Genital ulcer diseases, 24.8
Human papillomaviruses	Genital warts, cervical carcinoma	Wart diseases, 24.9
Poxviruses	Molluscum contagiosum	Wart diseases, 24.9
Fungi		
Candida albicans	Vaginitis	Vaginitis, 24.4
Protozoa		
Trichomonas vaginalis	Trichomoniasis (vaginitis)	Vaginitis, 24.4
Helminth—trematode		
Schistosoma haematobium	Urinary schistosomiasis	Urinary schistosomiasis, 24.3

Chapter Summary

INFECTIOUS DISEASES MANIFESTING IN THE GENITOURINARY SYSTEM

24.1 THE GENITOURINARY TRACT, ITS DEFENSES, AND NORMAL BIOTA

- **The reproductive tract in males and females is composed of structures and substances that allow for sexual intercourse and the creation of a new fetus; protected by normal mucosal defenses and specialized features (such as low pH).**
- **The urinary system allows the excretion of fluid and wastes from the body. It has mechanical and chemical defense mechanisms.**
- **Normal biota in the male reproductive and urinary systems resemble skin biota. The same is generally true for the female urinary system. The normal biota in the female reproductive tract changes over the course of her lifetime.**

24.2 INFECTIOUS DISEASES MANIFESTING IN THE URINARY TRACT

- **Urinary tract infections (UTIs)**—Can occur in the bladder, kidneys, and/or urethra. Most common causes are *Escherichia coli, Staphylococcus saprophyticus,* and *Enterococcus.* Community-acquired UTIs are most often transmitted from GI tract to urinary system. UTIs are the most common cause of healthcare-associated infections.
- **Leptospirosis**—Zoonosis associated with wild animal urine and affects kidneys, liver, brain, and eyes. The causative agent is spirochete *Leptospira interrogans.*
- **Urinary schistosomiasis**—This form of schistosomiasis is caused by *S. haematobium.* Bladder is damaged by trematode eggs and the granulomatous response they induce.

24.3 INFECTIOUS DISEASES MANIFESTING IN THE REPRODUCTIVE TRACT

- **Vaginitis**—Can be either opportunistic (from GI tract) or sexually transmitted. Caused by *Candida albicans* (commonly called a yeast infection) or *Trichomonas vaginalis.*
- **Vaginosis**—Vaginosis has a discharge but no inflammation in the vagina. It is characterized by the presence of diverse species of bacteria instead of the healthy vagina, which is dominated by *Lactobacillus* species.
- **Prostatitis**—Inflammation of the prostate; can be acute or chronic. It has not been established that all cases have a microbial cause, but most do.
- **Discharge diseases**—These are diseases in which there is an increase in fluid discharge in the male and female reproductive tracts. Both major causes can lead to pelvic inflammatory disease and dangerous outcomes for fetuses. **Gonorrhea:** More common in females, although male symptoms are more obvious when present. ***Chlamydia:*** Most common reportable infectious disease in the United States.
- **Genital ulcer diseases—Syphilis:** Caused by the spirochete *Treponema palladium.* It has 3 distinct clinical stages: primary, secondary, and tertiary syphilis, with a latent period between secondary and tertiary. Can lead to congenital syphilis. Incidence increasing. **Chancroid:** Caused by *Haemophillus ducreyi.* Not common, but incidence increasing. **Genital herpes:** Caused by herpes simplex viruses 1 and 2. Symptoms range from none to life-threatening. Virus remains in body indefinitely, and in many, it causes recurrent outbreaks of lesions.
- **Wart diseases—Human papilloma viruses:** Certain types infect cells on the female cervix that eventually result in malignancies of the cervix. Males can also get cancer from these viral types. Genital warts can be removed, but the virus will remain. Treatment of cancerous cell changes—detected through Pap smears in females—is an important part of HPV therapy. Vaccine for the cancer-causing types of HPV is available. **Molluscum contagiosum:** Can cause harmless warts that are transmitted sexually or through fomite contact.
- **Group B *Streptococcus* colonization/neonatal disease**—Asymptomatic colonization of women by a beta-hemolytic *Streptococcus* in Lancefield group B is very common. It can cause preterm delivery and infections in newborns.

INFECTIOUS DISEASES AFFECTING
The Female Genitourinary Tract

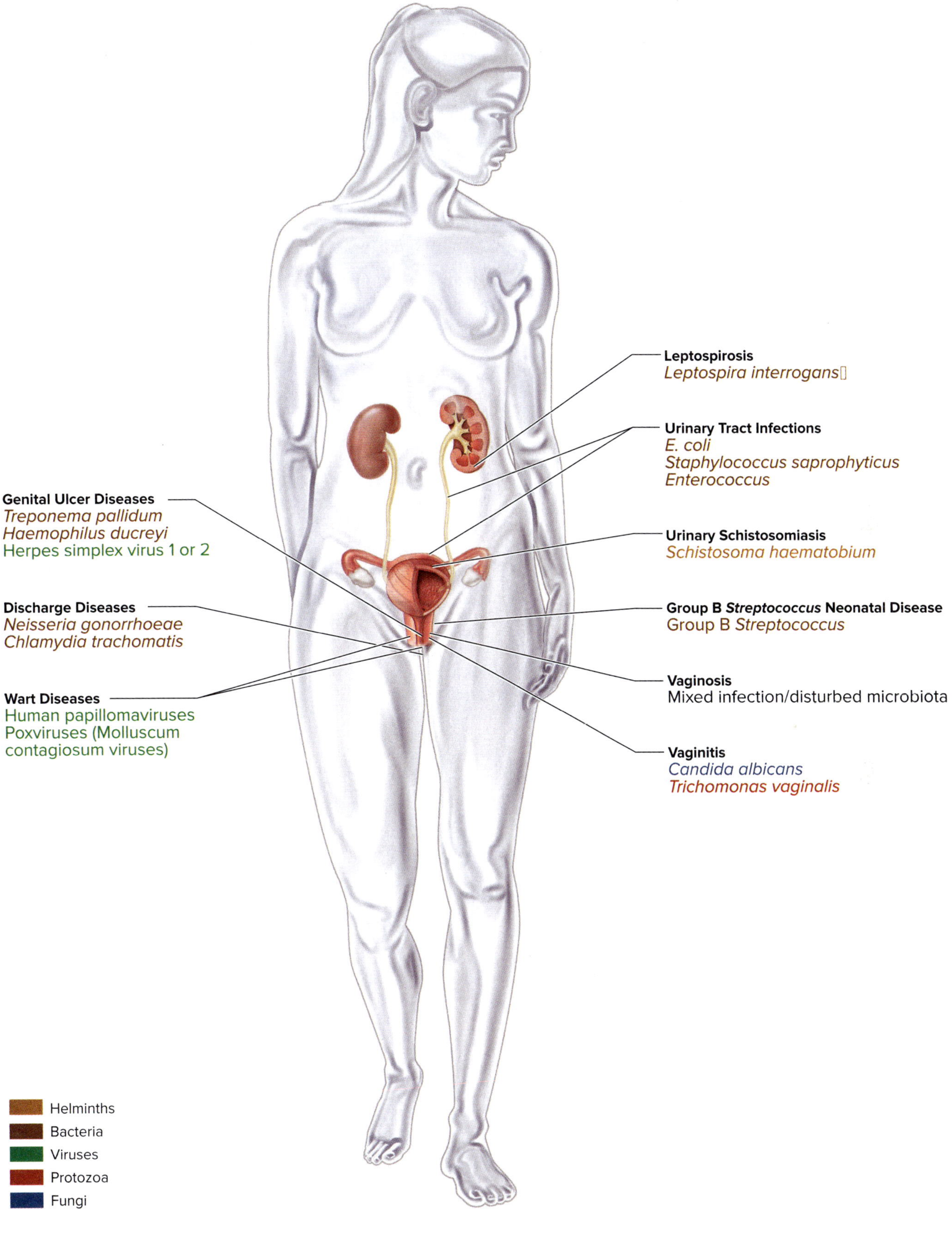
Leptospirosis
Leptospira interrogans
Urinary Tract Infections
E. coli
Staphylococcus saprophyticus
Enterococcus
Genital Ulcer Diseases
Treponema pallidum
Haemophilus ducreyi
Herpes simplex virus 1 or 2
Urinary Schistosomiasis
Schistosoma haematobium
Discharge Diseases
Neisseria gonorrhoeae
Chlamydia trachomatis
Group B Streptococcus Neonatal Disease
Group B Streptococcus
Vaginosis
Mixed infection/disturbed microbiota
Wart Diseases
Human papillomaviruses
Poxviruses (Molluscum contagiosum viruses)
Vaginitis
Candida albicans
Trichomonas vaginalis
Helminths
Bacteria
Viruses
Protozoa
Fungi

System Summary Figure 24.22

INFECTIOUS DISEASES AFFECTING
The Male Genitourinary Tract

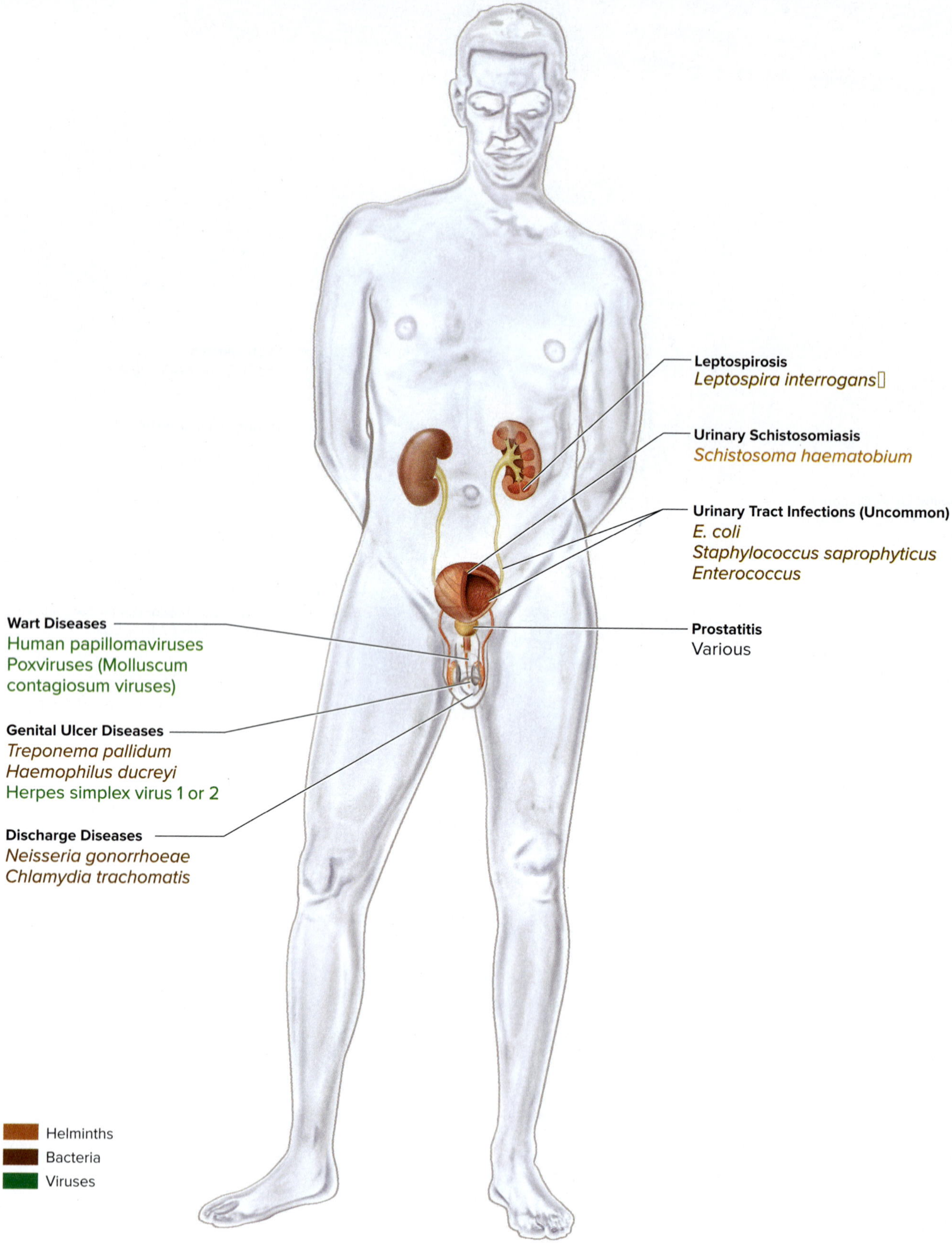
Leptospirosis
Leptospira interrogans
Urinary Schistosomiasis
Schistosoma haematobium
Urinary Tract Infections (Uncommon)
E. coli
Staphylococcus saprophyticus
Enterococcus
Prostatitis
Various
Wart Diseases
Human papillomaviruses
Poxviruses (Molluscum contagiosum viruses)
Genital Ulcer Diseases
Treponema pallidum
Haemophilus ducreyi
Herpes simplex virus 1 or 2
Discharge Diseases
Neisseria gonorrhoeae
Chlamydia trachomatis
Helminths
Bacteria
Viruses

System Summary Figure 24.23

SmartGrid: From Knowledge to Critical Thinking

This *21 Question Grid* takes the topics from this chapter and arranges them with respect to the American Society for Microbiology's Undergraduate Curriculum guidelines—all six of the important "Concepts" as well as the important "Competency" of scientific literacy. Three questions are supplied, which cover chapter content referring to the Concept or Competency in increasing levels of Bloom's taxonomy for learning.

ASM Concept/ Competency	A. Bloom's Level 1, 2—Remember and Understand (Choose one)	B. Bloom's Level 3, 4—Apply and Analyze	C. Bloom's Level 5, 6—Evaluate and Create
Evolution	**1.** Which of the following is on the CDC's "Urgent Threat" list because of its potential resistance to antibiotics? **a.** *Staphylococcus saprophyticus* **b.** *Trichomonas vaginalis* **c.** Herpesvirus **d.** *Enterobacteriaceae*	**2.** Some of the *E. coli* strains that are normal biota in the GI tract have evolved adhesive structures that can attach to the urinary tract and cause disease there. What sequence of molecular events (in the bacteria) do you suppose led to this?	**3.** Many STIs are asymptomatic, or remain latent for long periods of time. What evolutionary advantage do you think that gives them?
Cell Structure and Function	**4.** Which of the following microbes is described as "dimorphic"? **a.** *Trichomonas vaginalis* **b.** *Gardnerella* **c.** *Candida albicans* **d.** *E. coli*	**5.** Three microorganisms in this chapter can be identified via a Gram stain. Name them and explain why this is possible for each of them.	**6.** The drug Flagyl (metronidazole) has the unusual characteristic of being effective against some bacteria as well as some protozoa. What cellular structure(s) would it be most likely to target?
Metabolic Pathways	**7.** Which of the following bacteria is known to produce substances that lower the local pH? **a.** *Gardnerella* **b.** *E. coli* **c.** *Lactobacillus* **d.** *Leptospira*	**8.** What advantage does the particular life cycle of *Chlamydia* give to the pathogen?	**9.** Infection with *Gardnerella* seems not to induce inflammation in the vagina. Speculate on characteristics of the infection or the organism that could result in lack of inflammation.
Information Flow and Genetics	**10.** Which of the following bacteria behaves most similarly to viruses? **a.** *E. coli* **b.** *Chlamydia* **c.** *Neisseria* **d.** *Haemophilus ducreyi*	**11.** Using your knowledge of viral oncogenesis, outline the steps that the human papillomavirus might take from infection to cervical cancer.	**12.** The bacterium *Treponema palladium* is known to express an unusually low density of proteins in its outer membrane. Speculate on why that might contribute to its ability for long-term infection of the host.
Microbial Systems	**13.** Which part of the urinary tract has the most abundant and diverse microbiome? **a.** lower urethra **b.** upper urethra **c.** bladder **d.** kidneys	**14.** Why do many of the infections discussed in this chapter make the host more vulnerable to HIV infection? Provide detail in your answer.	**15.** What are some of the stimuli that can trigger reactivation of a latent herpesvirus infection? Speculate on why.

ASM Concept/ Competency	A. Bloom's Level 1, 2—Remember and Understand (Choose one)	B. Bloom's Level 3, 4—Apply and Analyze	C. Bloom's Level 5, 6—Evaluate and Create
Impact of Microorganisms	**16.** Bacterial vaginosis is commonly associated with which organism? **a.** *Candida albicans* **b.** *Gardnerella* **c.** *Trichomonas* **d.** any of these	**17.** A 38-year-old male living in Hawaii tried to rescue some of the family's belongings as the basement flooded with water from a nearby flooded stream. Within a few days, he developed flulike symptoms, which cleared in a few days. Several weeks later he developed a painful headache and jaundice. The physician requested a urinalysis. What disease do you suspect and why?	**18.** Use science to explain this statement: Forensic microbiology can help identify sexual abuse of minors.
Scientific Thinking	**19.** Which of the following STIs is curable with antimicrobials? **a.** HPV **b.** herpes **c.** *Chlamydia* **d.** two of these	**20.** Use figure 24.12 to determine at which age group *Chlamydia* infection rates for males are greater than for females.	**21.** In figure 24.21 explain why the prevalence of some diseases is so much greater than their incidence, and is the opposite for other diseases.

Answers to the multiple-choice questions appear in Appendix A.

Visual Connections

This question uses visual images to connect content within and between chapters.

1. **From chapters 21 and 24, figures 21.15 and 24.13*b*.** Compare these two rashes. What kind of information would help you determine the diagnosis in both cases?

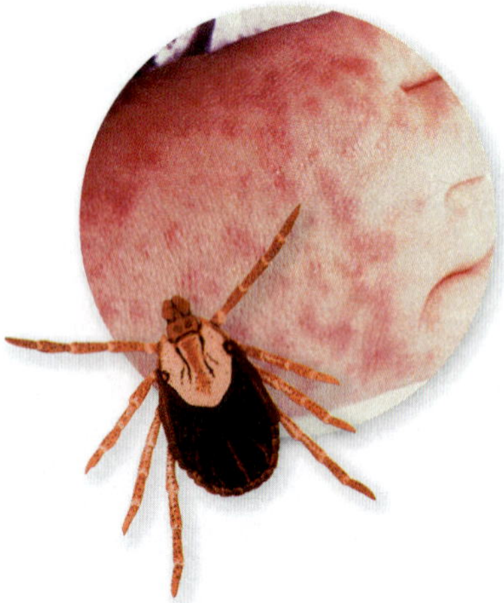

(hand) Centers for Disease Control and Prevention; Andrew J. Brooks/CDC

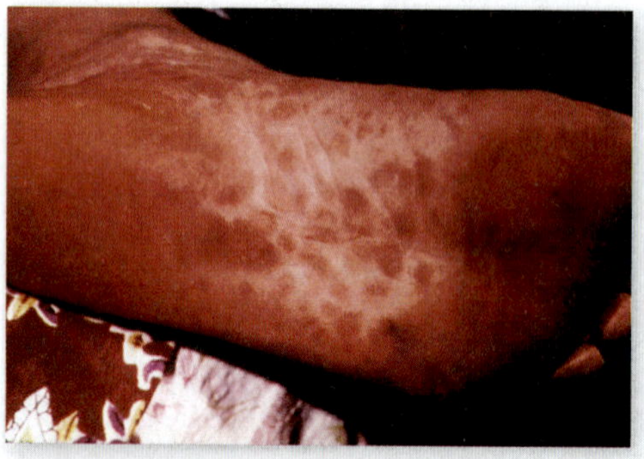

(b)
CDC

High Impact Study

These terms and concepts are most critical for your understanding of this chapter—and may be the most difficult. Have you mastered them? In these disease chapters, the terms and concepts help you identify what is important in a different way than the comprehensive details found in the Disease Tables. Your instructor will help you understand what is important for your class.

Concepts

- ☐ Defenses of genitourinary system
- ☐ Normal microbiota of genitourinary system
- ☐ Community-acquired vs. catheter-associated UTIs
- ☐ The role of pH of the vagina
- ☐ Differences between vaginitis and vaginosis
- ☐ Infections in this chapter having implications for the fetus
- ☐ Infections in this chapter that are not curable
- ☐ Organisms in this chapter for which there are vaccines available
- ☐ Organisms in this chapter that display significant antibiotic resistance

Terms

- ☐ Pelvic inflammatory disease
- ☐ Cystitis
- ☐ Pyelonephritis
- ☐ Clue cell
- ☐ Condyloma acuminata
- ☐ Oncogene

25 Microbes in Our Environment and the Concept of One Health

Getty Images/Moment Open

Media Under The Microscope

Climate Change and Infections

These case studies examine an article from the popular media to determine the extent to which it is factual and/or misleading. This case focuses on the 2021 Bloomberg article "Climate Change Threatens to Spread Viruses Through an Unprepared World."

This article summarizes research from a study published in the British medical journal *Lancet* that warns that climate change is creating the conditions for greatly increased transmission of infectious diseases, some old and some new, to humans. Its secondary message is that health care systems are not prepared.

The *Bloomberg* article, published in the midst of the COVID-19 pandemic in October 2021, mentions that the world has done a very poor job handling that pandemic, leaving health care workers and systems stretched to their limits in rich countries and with very little vaccine in underresourced countries. This does not spark optimism for the probable resurgence of diseases such as malaria, dengue fever, and cholera, which is likely as the climate warms. Readers in the United States may dismiss these diseases as not relevant to them, but both malaria and cholera were rampant in the United States before proper public health measures were taken, and dengue fever cases have been steadily increasing in the United States, especially in the South.

The article states that the *Lancet* report traces 44 indicators of health impacts that are "directly linked to climate change and highlights worsening social inequalities." It tells us that malaria infections are increasing in traditionally cooler areas and that coastal waters around Europe and the United States are producing more gastroenteritis and sepsis cases caused by bacteria.

The article concludes that the mishandling of the COVID-19 pandemic, similarly to the handling of the climate crisis, can lead to dire consequences for the future of infectious diseases in a warming world.

- What is the **intended message** of the article?
- What is your **critical reading** of the summary of the article? Remember that in this context, "critical reading" does not necessarily mean *What criticism do you have?* but asks you to apply your knowledge to interpret whether the article is factual and whether the facts support the intended message.

- How would you **interpret** the news item for your nonmicrobiologist friends?
- What is your **overall grade** for the news item—taking into account its accuracy and the accuracy of its intended effect?

Media Under The Microscope Wrap-Up appears at the end of the chapter.

Outline and Learning Outcomes

25.1 Microbial Ecology and Element Cycling

1. Define *microbial ecology.*
2. Summarize why our view of the abundance of microbes on earth has changed in recent years.
3. Diagram a carbon cycle.
4. Explain the role of methanogens within the carbon cycle.
5. List the four reactions involved in the nitrogen cycle.
6. Describe the process of nitrogen fixation, and explain how microbes play a role in this biochemical reaction.

25.2 Microbes on Land and in Water

7. Outline the basic process used to perform metagenomic analysis of the environment.
8. Discuss how metagenomic sampling has changed our view of deep subsurface and oceanic microbiology.
9. Define *eutrophication,* and explain how microbes are responsible for its impact on aquatic life.

25.3 Applied Microbiology and Biotechnology

10. Compose a sentence about the history of applied microbiology.
11. Define *biotechnology*, and explain how its uses have changed in modern times.
12. Outline the steps in water purification.
13. Differentiate water purification from sewage treatment.
14. List five important pathogens of drinking water.
15. Provide examples of indicator bacteria, and describe their role in the survey of water quality.
16. Summarize methods for identifying and quantifying microbial contaminants in water supplies.
17. Summarize the microbial process that leads to leavening in bread.
18. Write the equation showing how yeasts convert sugar into alcoholic beverages.

25.4 The Concept of "One Health"

19. Name the three components that interact for "one health."
20. Discuss three ways that changes in global climate can impact the reach and impact of infectious diseases.

25.1 Microbial Ecology and Element Cycling

The study of microbes in their natural habitats is known as **microbial ecology.** In chapter 9, we first touched upon the widespread distribution of microorganisms and their adaptations to most habitats of the world, from extreme environments to more welcoming locations. Although we have known for a long time that geologic features on the earth, including coal and limestone, are formed in small or large part by microbes, it is only recently that we have come to understand the sheer mass of microbial life present on our planet. Remember that the vast majority of microbial life has not yet been cultured. With the development of genomic techniques that do not rely on cultivating bacteria, we have discovered abundant microbial life all over—and within and around—our planet **(figure 25.1).** We are learning about the planet-shaping effects of bacteria deep in the earth's core and deep within glaciers. The sheer abundance of viral life in the oceans has been a huge surprise, as well.

In this chapter we will look briefly at the world of microbes in the environment, how we study them, and how we exploit and control them for the sake of human convenience and health. We will also introduce the concept of One Health, an important scientific approach to understanding human, environmental, and animal health as critically interconnected.

Carbon and Nitrogen Cycling

Because of the finite supply of life's building blocks, the long-term sustenance of the biosphere results from continuous **recycling** of elements and nutrients. Essential elements such as carbon, nitrogen, sulfur, phosphorus, oxygen, and iron are cycled through biological, geologic, and chemical mechanisms called **biogeochemical cycles.** Although these cycles vary in certain specific characteristics, they share several general qualities, as summarized in the following list:

- All elements ultimately originate from a nonliving, long-term reservoir in the atmosphere, the lithosphere, or the hydrosphere. They cycle in pure form (N_2) or as compounds (PO_4). Their cycling is carried out through redox reactions.
- Elements cycle between the nonliving environment and the living environment.

Figure 25.1 A deep sea hydrothermal vent in New Zealand. This is called a black smoker. The vent effluent is rich in sulfides, and it feeds chemolithotrophic bacteria in the vicinity. The temperatures around these vents can approach 750°F.

Image courtesy of Submarine Ring of Fire 2006 Exploration, NOAA Vents Program

- Recycling maintains a necessary balance of nutrients in the biosphere so that they do not build up or become unavailable.
- Cycles are complex systems that rely on the interplay of many types of organisms. Often, the waste products of one organism become a source of energy or building material for another.
- All organisms participate in recycling, but only certain categories of microorganisms have the metabolic pathways for converting inorganic compounds from one nutritional form to another.

For billions of years, microbes have played prominent roles in the formation and maintenance of the earth's crust, the development of rocks and minerals, and the formation of fossil fuels. This revolution in understanding the biological involvement in geologic processes has given rise to a field called *geomicrobiology.* A logical extension of this discipline is **astromicrobiology,** also known as *exobiology,* which is the study of life on planets and bodies other than earth.

In the next sections, we examine how two elemental chemicals, carbon and nitrogen, cycle in and out of the living and nonliving parts of our planet.

The Carbon Cycle

Because carbon is the fundamental atom in all biomolecules and accounts for at least one-half of the dry weight of biomass, the **carbon cycle** is more intimately associated with the energy transfers in the biosphere than are other elements. Carbon exists predominantly in the mineral state and as an organic reservoir in the bodies of organisms. A much smaller amount of carbon also exists in the gaseous state as carbon dioxide (CO_2), carbon monoxide (CO), and methane (CH_4). In general, carbon is recycled through ecosystems via carbon fixation, respiration, or fermentation of organic molecules; limestone decomposition; and methane production. A convenient starting point from which to trace the movement of carbon is with carbon dioxide, which occupies a central position in the cycle and represents a large, common pool that diffuses into all parts of the ecosystem **(figure 25.2).** As a general rule, the cycles of oxygen and hydrogen are closely allied to the carbon cycle.

The principal users of the atmospheric carbon dioxide pool are photosynthetic autotrophs (photoautotrophs) such as plants, algae, and bacteria. An estimated 165 billion tons of organic material per year are produced by terrestrial and aquatic photosynthesis, with phytoplankton contributing to nearly half of the earth's overall photosynthetic output. Although we do not yet know exactly how many autotrophs exist in the earth's crust, a small amount of CO_2 is used by bacteria (chemolithoautotrophs) that derive their energy from bonds in inorganic chemicals. A review of the general equation for photosynthesis in figure 10.24 reveals that phototrophs use energy from the sun to fix CO_2 into organic compounds, such as glucose, that can be used in synthesis. Photosynthesis is also the primary means by which the atmospheric supply of O_2 is regenerated.

While photosynthesis removes CO_2 from the atmosphere and converts solar energy into stored chemical energy, respiration and

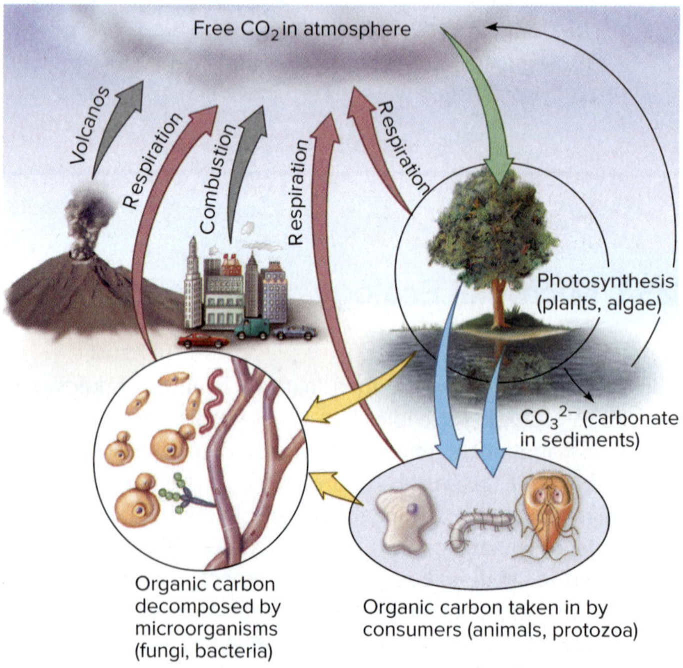

Free CO_2 in atmosphere
Volcanos
Respiration
Combustion
Respiration
Respiration
Photosynthesis (plants, algae)
CO_3^{2-} (carbonate in sediments)
Organic carbon decomposed by microorganisms (fungi, bacteria)
Organic carbon taken in by consumers (animals, protozoa)

Figure 25.2 The carbon cycle. This cycle traces carbon from the CO_2 pool in the atmosphere to the photosynthesizers (green arrow), where it is fixed into protoplasm. Organic carbon compounds are taken in by consumers (blue arrows) and decomposers (yellow arrows) that produce CO_2 through respiration and return it to the atmosphere (pink arrows). Combustion of fossil fuels and volcanic eruptions also add to the CO_2 pool. Some of the CO_2 is carried into inorganic sediments by organisms that synthesize carbonate (CO_3) skeletons. In time, natural processes acting on exposed carbonate skeletons can liberate CO_2.

fermentation release CO_2 and convert stored chemical energy into kinetic energy and work. As you may recall from the discussion of aerobic respiration in chapter 10, in the presence of O_2, organic compounds such as glucose are degraded completely to CO_2 with the release of energy and the formation of H_2O. Carbon dioxide is also released by anaerobic respiration and by certain types of fermentation reactions.

A small but important phase of the carbon cycle involves certain limestone deposits composed primarily of calcium carbonate ($CaCO_3$). Limestone is produced when marine organisms such as mollusks, corals, protozoa, and algae form hardened shells by combining carbon dioxide and calcium ions from the surrounding water. When these organisms die, the durable skeletal components accumulate in marine deposits. As these immense deposits are gradually exposed by geologic upheavals or receding ocean levels, various decomposing agents liberate CO_2 and return it to the CO_2 pool of the water and atmosphere.

The complementary actions of photosynthesis and respiration, along with other natural CO_2-releasing processes such as limestone erosion and volcanic activity, have maintained a relatively stable atmospheric pool of carbon dioxide. Recently we have found that this balance is being disturbed as humans burn *fossil fuels* and other organic carbon sources. Fossil fuels, including coal, oil, and natural gas, were formed through millions of years of natural biological and geologic activities. Humans are so dependent upon this energy source that within the past 60 years, the proportion of CO_2 in the atmosphere has steadily increased from 312 to 418 parts per million (ppm) **(figure 25.3).**

Compared with carbon dioxide, methane gas (CH_4) plays a secondary part in the carbon cycle, though it can be a significant product in anaerobic ecosystems dominated by **methanogens** (methane producers). In general, when methanogens reduce CO_2 by means of various oxidizable substrates, they give off CH_4. The practical applications of methanogens are covered in this chapter in a section on sewage treatment.

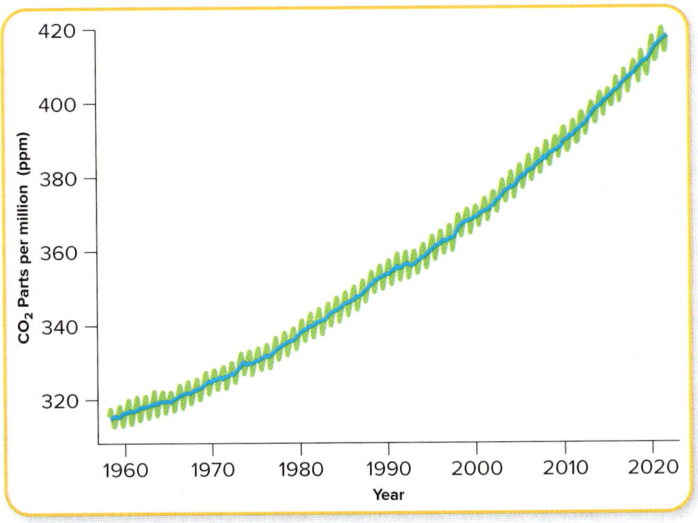

420
400
380
360
340
320
CO2 Parts per million (ppm)
1960
1970
1980
1990
2000
2010
2020
Year

Figure 25.3 Global monthly mean CO_2 amounts since 1960. CO_2 as measured at Mauna Loa Observatory in Hawaii, the site conducting the longest-standing CO_2 sampling.

Source: Scripps Oceanography Institute and the National Oceanic and Atmospheric Administration.

Today methane is recognized as a more potent greenhouse gas than CO_2, as it can trap nearly 20 times more heat in the atmosphere. Much of this comes from gas—intestinal gas, that is. Methane released from the gastrointestinal tracts of ruminant animals such as cattle, goats, and sheep contributes to an estimated 20% of global methane production. Methanogens in the GI tract of humans as well as tiny termites also add to this value each year.

The Nitrogen Cycle

Nitrogen (N_2) gas is the most abundant component of the atmosphere, accounting for nearly 79% of air volume. As we will see, this extensive reservoir in the air is largely unavailable to most organisms. Only about 0.03% of the earth's nitrogen is combined (or fixed) in some other form such as nitrates (NO_3^-), nitrites (NO_2^-), ammonium ion (NH_4^+), and organic nitrogen compounds (proteins, nucleic acids).

The **nitrogen cycle** is more intricate than other cycles because it involves such a diversity of specialized microbes to maintain the flow of the cycle. In many ways, it is actually more of a nitrogen "web" because of the array of reactions that occur. Higher plants can utilize NO_3^- and NH_4^+; animals must receive nitrogen in organic form from plants or other animals. Microorganisms can use all forms of nitrogen: NO_2^-, NO_3^-, NH_4^+, N_2, and organic nitrogen. The cycle includes four basic types of reactions: nitrogen fixation, ammonification, nitrification, and denitrification.

Root Nodules: Natural Fertilizer Factories

A significant symbiotic association occurs between **rhizobia** (ry-zoh′-bee-uh) (bacteria in genera such as *Rhizobium, Bradyrhizobium,* and *Azorhizobium*) and **legumes,** (plants, such as soybeans, peas, alfalfa, and clover, that characteristically produce seeds in pods). The infection of legume roots by these gram-negative, motile, rod-shaped bacteria causes the formation of nitrogen-fixing organs called **root nodules (figure 25.4).** Nodulation begins when rhizobia colonize specific sites on root hairs. From there, the bacteria invade deeper root cells and induce the cells to form tumorlike masses. The bacterium's enzyme system supplies a constant source of reduced nitrogen to the plant, and the plant furnishes nutrients and energy for the activities of the bacterium. The legume uses the NH_4^+ to aminate (add an amino group to) various carbohydrate intermediates and thereby synthesize amino acids and other nitrogenous compounds that are used in plant and animal biosynthesis.

Plant-bacteria associations have great practical importance in agriculture because an available and usable source of nitrogen is often a limiting factor in the growth of crops. The self-fertilizing nature of legumes makes them valuable food plants in areas with poor soils and in countries with limited resources. It has been shown that crop health and yields can be improved by inoculating legume seeds with pure cultures of rhizobia because the soil is often deficient in the proper strain of bacteria for forming nodules.

Ammonification, Nitrification, and Denitrification

In another part of the nitrogen cycle, nitrogen-containing organic matter is decomposed by various bacteria (*Clostridium, Proteus*)

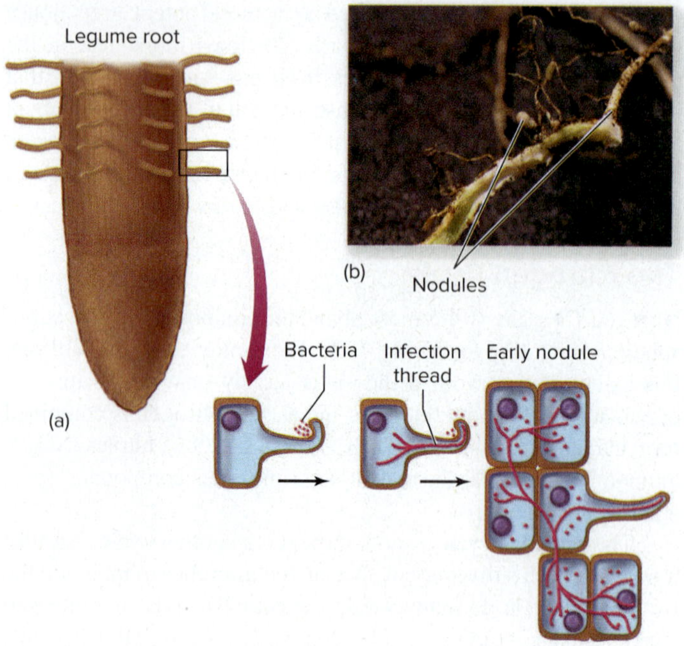
Legume root
(a)
(b)
Nodules
Bacteria
Infection thread
Early nodule

Figure 25.4 Nitrogen fixation through symbiosis. **(a)** Events leading to formation of root nodules. Cells of the bacterium *Rhizobium* attach to a legume root hair and cause it to curl. Invasion of the legume root proper by *Rhizobium* initiates the formation of an infection thread that spreads into numerous adjacent cells. The presence of bacteria in cells causes nodule formation. **(b)** Mature nodules that have developed in a sweet clover plant.

(b) Lisa Burgess/McGraw Hill

that live in the soil and water. Organic debris consists of large amounts of protein and nucleic acids from dead organisms and nitrogenous animal wastes such as urea and uric acid. The decomposition of these substances splits off amino groups and produces NH_4^+. For that reason this process is known as **ammonification.** The ammonium released can be reused by certain plants or converted to other nitrogen compounds, as discussed next.

The oxidation of NH_4^+ to NO_2^- and NO_3^- is a process called **nitrification.** It is an essential conversion process for generating the most oxidized form of nitrogen (nitrate, NO_3^-). This reaction occurs in two phases and involves two different kinds of lithotrophic bacteria in soil and water. In the first phase, certain gram-negative genera such as *Nitrosomonas, Nitrosospira,* and *Nitrosococcus* oxidize NH_3 to NO_2^- as a means of generating energy. Nitrite is rapidly acted upon by a second group of nitrifiers, including *Nitrobacter, Nitrosospira,* and *Nitrococcus,* which perform the *final* oxidation of NO_2^- to NO_3^-. Nitrates can be assimilated through several routes by a variety of organisms (plants, fungi, and bacteria). Nitrate and nitrite are also important in anaerobic respiration, where they serve as terminal electron acceptors; some bacteria use them as a source of oxygen as well.

The nitrogen cycle is complete when nitrogen compounds are returned to the reservoir in the air by a reaction series that converts NO_3^- through intermediate steps to atmospheric nitrogen. The first step, which involves the reduction of nitrate to nitrite, is so common that hundreds of bacterial species can do it. Several genera such as *Bacillus, Pseudomonas, Spirillum,* and *Thiobacillus* can carry out this **denitrification** process to completion as follows:

$$NO_3^- \rightarrow NO_2^- \rightarrow NO \rightarrow N_2O \rightarrow N_2 \text{ (gas)}$$

When this process is not carried through to completion, the greenhouse gas nitrous oxide (N_2O) is added to the atmosphere.

25.1 Learning Outcomes—Assess Your Progress

1. Define *microbial ecology*.
2. Summarize why our view of the abundance of microbes on earth has changed in recent years.
3. Diagram a carbon cycle.
4. Explain the role of methanogens within the carbon cycle.
5. List the four reactions involved in the nitrogen cycle.
6. Describe the process of nitrogen fixation, and explain how microbes play a role in this biochemical reaction.

25.2 Microbes on Land and in Water

As you have read several times already in this book, until fairly recently, our knowing which microbes inhabited a place, whether it was the human gut or your backyard pond, relied on culturing them. Just as there is a Human Microbiome Project, scientists have been busy using these techniques to identify the microbes living in the environment, which includes land, water, and air. The field of environmental genomics has revealed many surprising things, such as bacteria living in glaciers and deep under the seafloor—places once thought to be uninhabitable by any life forms.

Since 2010, a project called the Earth Microbiome Project (EMP) was launched as a massively collaborative effort to analyze microbial communities across the globe. The effort is characterizing microbes using a standardized set of methods and makes the data publicly available to all researchers. So far there are more than 2,000 research papers published using the EMP's methodology. Like the human microbiome studies before it, it promises to revolutionize our view of organismal life on this planet.

Environmental Sampling in the Genomic Era

The methods for identifying bacteria and genes in the environment are evolving rapidly. When the genes of all microbes in a habitat are sampled, it is called metagenomics. The process employed is usually high-throughput sequencing, as described in chapter 8. The process always begins with an environmental sample, such as a large volume of seawater or soil. Techniques are available to extract the DNA from such samples. Then, deep sequencing is carried out as summarized in **figure 25.5.**

Soil Microbiology

At the microscopic level, soil is a dynamic ecosystem that supports complex interactions between numerous geologic, chemical, and biological factors. This rich region teems with

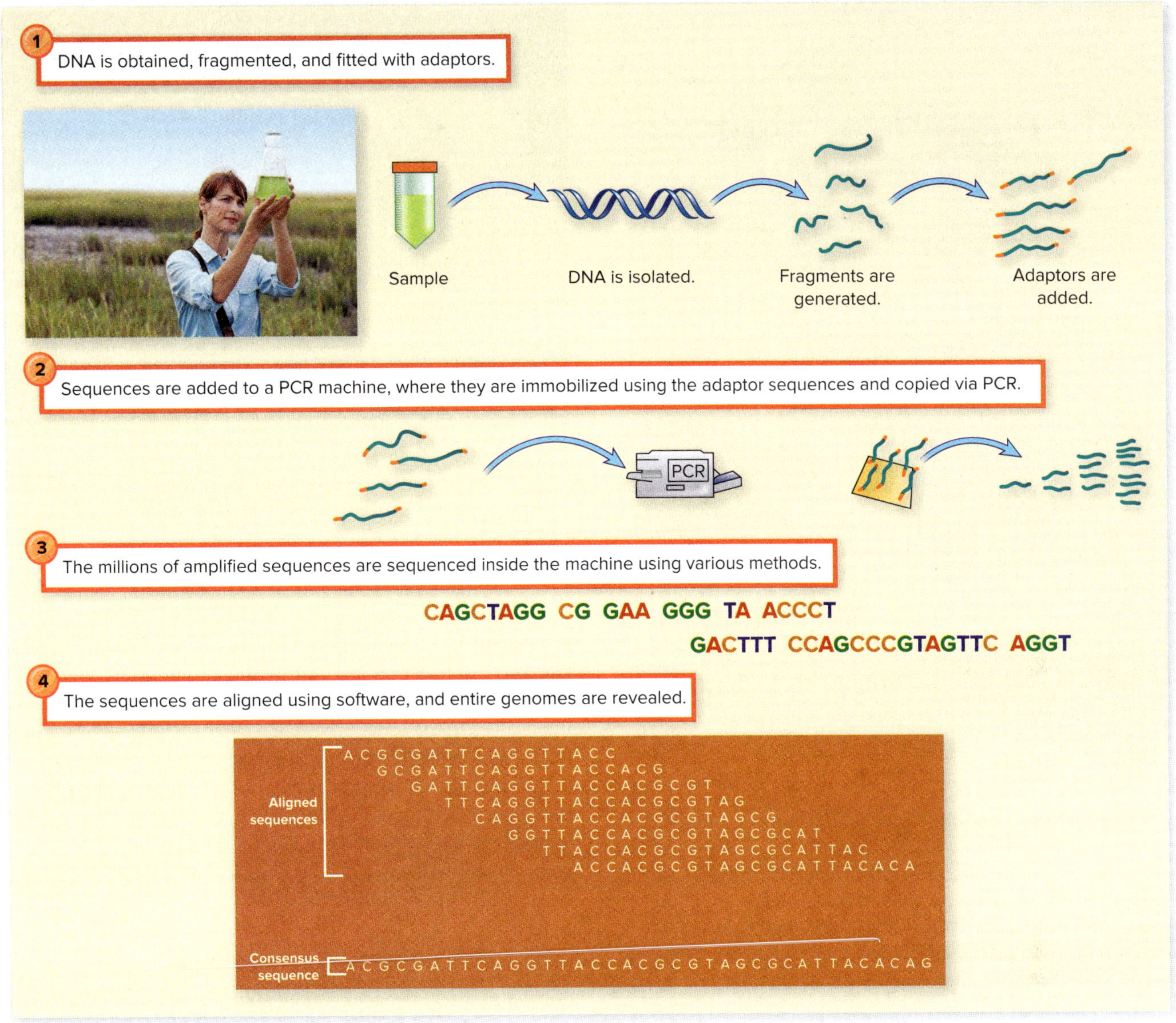
1
DNA is obtained, fragmented, and fitted with adaptors.
Sample
DNA is isolated.
Fragments are generated.
Adaptors are added.
2
Sequences are added to a PCR machine, where they are immobilized using the adaptor sequences and copied via PCR.
PCR
3
The millions of amplified sequences are sequenced inside the machine using various methods.
CAGCTAGG CG GAA GGG TA ACCCT
GACTTT CCAGCCCGTAGTTC AGGT
4
The sequences are aligned using software, and entire genomes are revealed.
Aligned sequences
ACGCGATTCAGGTTACC
GCGATTCAGGTTACCACG
GATTCAGGTTACCACGCGT
TTCAGGTTACCACGCGTAG
CAGGTTACCACGCGTAGCG
GGTTACCACGCGTAGCGCAT
TTACCACGCGTAGCGCATTAC
ACCACGCGTAGCGCATTACACA
Consensus sequence
ACGCGATTCAGGTTACCACGCGTAGCGCATTACACAG

Figure 25.5 Construction and screening of genomic libraries directly from the environment.

(Photo) Ariel Skelley/Blend Images LLC

microbes, serves a dynamic role in biogeochemical cycles, and is an important repository for organic debris and dead terrestrial organisms. Both antibiotic-producing and antibiotic-degrading bacteria are present in these soils. The latter actually metabolize the compounds and use them for energy, providing new insight into how the environment shapes the development of antibiotic resistance.

Rock decomposition releases various-size particles, ranging from rocks, pebbles, and sand grains to microscopic morsels that lie in a loose aggregate **(figure 25.6).** The porous structure of soil creates various-size pockets or spaces that provide numerous microhabitats. Some spaces trap moisture and form a liquid phase in which mineral ions and other nutrients are dissolved. Other spaces trap air that will provide gases to soil microbes, plants, and animals. Water-saturated soils contain less oxygen, and dry soils have more. Gas tensions in soil can also vary vertically. In general, the concentration of O_2 decreases and that of CO_2 increases with the depth of soil. Aerobic and facultative organisms tend to occupy looser, drier soils, whereas anaerobes are adapted to waterlogged, poorly aerated soils.

Disease Connection

The vast majority of microbes that are found in soil are not human pathogens. The biota in soil thrive in cool, dry conditions—quite a different environment than the human body. However, some human pathogens can survive in soil. Bacteria that are capable of forming endospores survive in the endospore form for years in soil. These include the bacteria causing tetanus (*Clostridium tetani*) and anthrax (*Bacillus anthracis*).

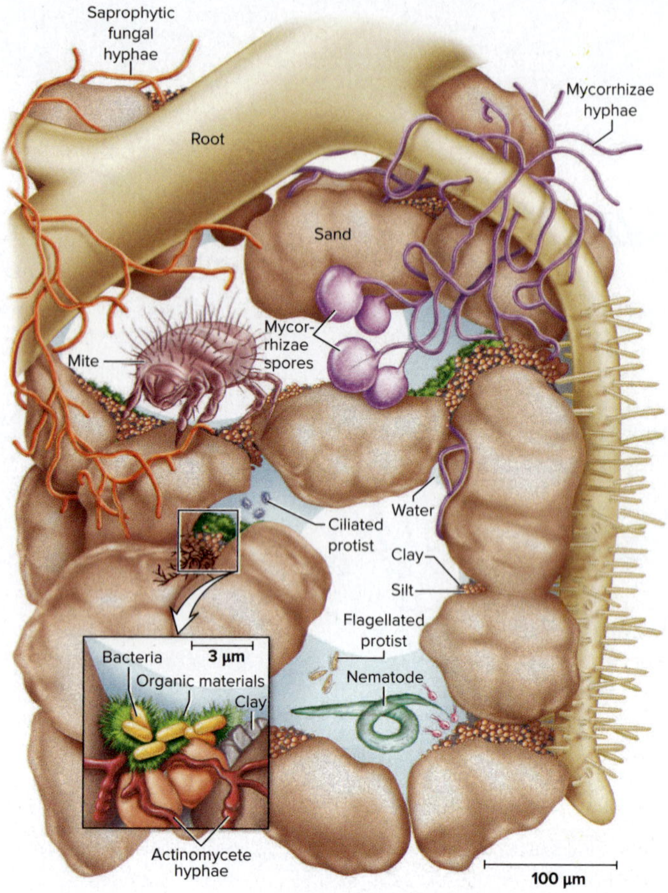
Saprophytic fungal hyphae
Root
Mycorrhizae hyphae
Sand
Mycor-rhizae spores
Mite
Water
Ciliated protist
Clay
Silt
Flagellated protist
Nematode
Bacteria
3 µm
Organic materials
Clay
Actinomycete hyphae
100 µm

Figure 25.6 The soil habitat. A typical soil habitat contains a mixture of clay, silt, and sand along with soil organic matter. Roots and animals (such as nematodes and mites), as well as protozoa and bacteria, consume oxygen, which rapidly diffuses into the soil pores where the microbes live. Note that two types of fungi are present: mycorrhizal fungi, which derive their organic carbon from plant roots, and saprophytic fungi, which help degrade organic material.

Living Activities in Soil

The rich culture medium of the soil supports a fantastic array of microorganisms (bacteria, fungi, algae, protozoa, and viruses). A gram of moist loam soil with high humus content can have a microbe count as high as 10 billion. Some of the most distinctive biological interactions occur in the **rhizosphere,** the zone of soil surrounding the roots of plants, which contains associated bacteria, fungi, and protozoa (see figure 25.6). Plants interact with soil microbes in a synergistic fashion. Studies have shown that a rich microbial community grows in a biofilm around the root hairs and other exposed surfaces. Their presence stimulates the plant to exude growth factors such as carbon dioxide, sugars, amino acids, and vitamins. These nutrients are released into fluid spaces, where they can be readily captured by microbes. Bacteria and fungi likewise contribute to plant survival by releasing hormonelike growth factors and protective substances. They are also important in converting minerals into forms usable by plants. We saw numerous examples in the nitrogen, sulfur, and phosphorus cycles.

We previously observed that plants can form close symbiotic associations with microbes to fix nitrogen. Other mutualistic partnerships between plant roots and microbes are **mycorrhizae** (my″-koh-ry′-zee). These associations occur when various species of basidiomycetes, ascomycetes, or zygomycetes attach themselves to the roots of vascular plants **(figure 25.7).** The plant feeds the fungus through photosynthesis, and the fungus sustains the relationship in several ways. By extending its mycelium into the rhizosphere, it helps anchor the plant and increases the surface area for capturing water from dry soils and minerals from poor soils. Plants with mycorrhizae can inhabit severe habitats more successfully than plants without them.

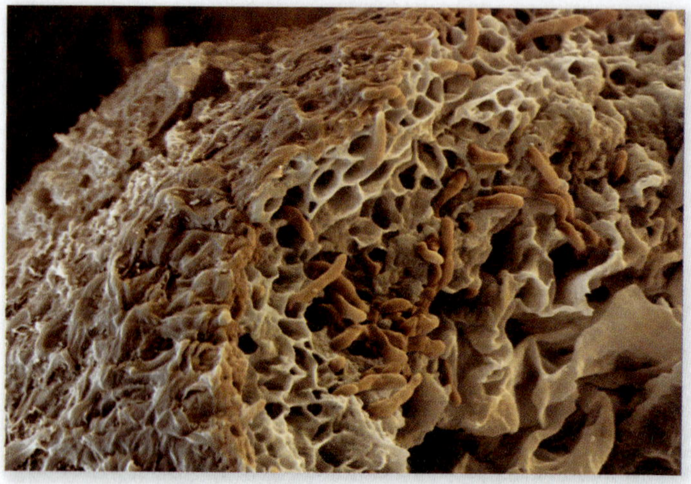

Figure 25.7 Mycorrhizae. These symbiotic associations between fungi and plant roots favor the absorption of water and minerals from the soil. The fungus (darker brown) surrounds the outside of the root and penetrates inside it.
Eye of Science/Science Source

Deep Subsurface Microbiology

For the past few decades, scientists have been sampling the deep subsurface—2 miles and more below the surface. From the very beginning of these studies, the results have been astounding. With the advent of genomic sampling, the discoveries are piling up. The most fascinating information has been found in deep ocean sediments, where the microbial diversity is immense and extends to depths thought unimaginable. For example, bacteria have been found 30 meters beneath the seafloor in clay that was deposited there 86 million years ago. The clay contains infinitesimally small levels of nutrients, yet bacteria and archaea were found. These extreme environments around the globe, characterized by high pressures, cold temperatures, and the complete absence of light, may provide new insights into biogeochemical processes and more. Scientists search to determine how bacteria survive in very hostile environments and along the way are discovering new metabolic capabilities that may turn out to provide clues to the very origin of life.

Aquatic Microbiology

Water occupies nearly three-fourths of the earth's surface. **Figure 25.8** shows how water is distributed on the earth. Surface water collects in extensive subterranean pockets produced by the

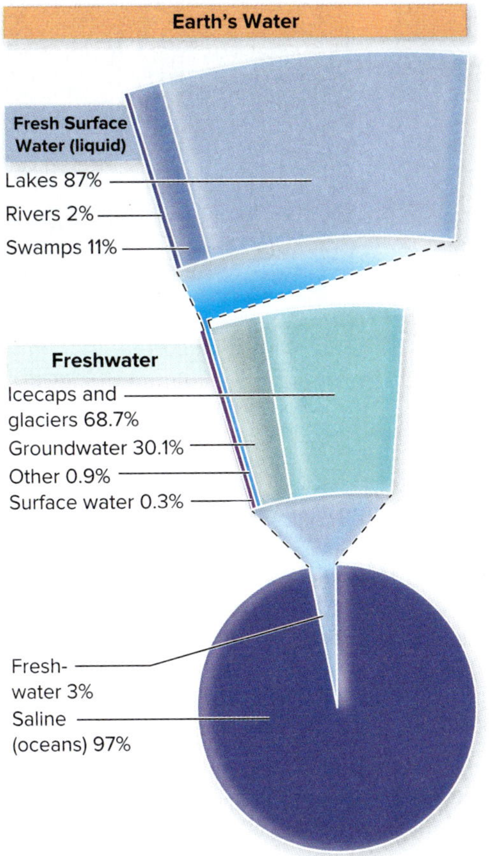
Earth's Water
Fresh Surface Water (liquid)
Lakes 87%
Rivers 2%
Swamps 11%
Freshwater
Icecaps and glaciers 68.7%
Groundwater 30.1%
Other 0.9%
Surface water 0.3%
Fresh-water 3%
Saline (oceans) 97%

Figure 25.8 The distribution of water on earth.

underlying layers of rock, gravel, and sand. This process forms a deep groundwater source called an **aquifer.** The water in aquifers circulates very slowly and is an important replenishing source for surface water. It can resurface through springs, geysers, and hot vents. It is also tapped as the primary supply for one-fourth of all water used by humans.

Marine Environments

The ocean exhibits extreme variations in salinity, depth, temperature, hydrostatic pressure, and mixing. Even so, it supports a great abundance of bacteria and viruses, the extent of which has only been appreciated in very recent years. In 2004, J. Craig Venter, one of the leaders of the Human Genome Project, set sail on a 100-ft yacht to determine the DNA profile of an entire ecosystem—the earth's oceans. His group sailed around the world for more than 2 years, collecting ocean samples along the way. They eventually discovered 6 million new genes and thousands of new proteins—essentially doubling the number of known proteins. Proteins, of course, are responsible for nearly all of the activities of cells.

Another scientific sailing expedition is a French project called the Tara Expedition. Labs from around the world participate. It uses a schooner to travel around the oceans, collecting samples and analyzing them with state-of-the-art techniques **(figure 25.9).** The expedition has produced countless new findings and a wealth of important information about its three major topic areas: oceans and humankind, oceans and biodiversity, and oceans and climate.

We know that high salinity is an effective physical measure to control microbial growth. The Dead Sea is a hypersaline body of water whose salinity is nearly 35%, which led microbiologists to believe that it would be devoid of all microbial life. It was very shocking, then, to find that it was teeming with microbes thriving in the very water that should prevent their survival. Most of the organisms were identified as salt-tolerant archaea. Oceans also contain an estimated 10 million viruses per milliliter. Most of these viruses are bacteriophages and therefore pose no danger to humans, but as parasites of bacteria, they appear to be a natural control mechanism for these populations. Plus, their lysis of bacteria plays an important role in the turnover of nutrients in the ocean. From an evolutionary standpoint, however, an interesting discovery has been made: The oceans represent the most extensive gene pool known to humankind, a virtual toolbox of nature-engineered genes. Because bacteriophages can swap any number of these microbial genes, their ability to acquire new traits that increase their survival plays a major role in their biology and their evolution. In one example, scientists found that cyanophages contain genes that can alter the light-harvesting abilities of their cyanobacterial hosts. This means that the bacteriophages can use the genes they acquire from microbes in their environment to boost the photosynthetic rate of their hosts. But why? It is all about survival and, more specifically, energy; experiments show that this boost in photosynthesis provides the energy the phages require to complete their life cycle. It will take many years to fully uncover the intricacies of the oceanic virome, but these studies are sure to reveal even more exciting information along the way.

Some microbial inhabitants of the ocean produce the periodic emergence of *red tides* around ocean coastlines **(figure 25.10).** Scientists call these "harmful algal blooms" (HABs). Environmental factors cause an increase in the number of these algae, leading to an increase in food for organisms farther up the food chain but also resulting in toxin production that can harm fish, shellfish, and even humans who may ingest the seafood or swim in the water. These algae produce a potent muscle toxin that can be concentrated by shellfish through filtration feeding. When humans eat clams, mussels, or oysters that contain the toxin, they

Figure 25.9 The Tara Expedition schooner.
Thierry Creux/ZUMA Press/Newscom

(a) (b) (c)

Figure 25.10 Red tides. **(a)** Single-celled red algae called dinoflagellates (*Gymnodinium* shown here) bloom in high-nutrient, warm seawater and give it a noticeable red or brown color. **(b)** An aerial view of California coastline in the midst of a red tide. **(c)** Fish washed ashore during a red tide bloom in Florida.

(a) David M. Phillips/Science Source; (b) Carleton Ray/Science Source; (c) J. G. Domke/Alamy stock Photo

can develop paralytic shellfish poisoning. The coasts of Florida are sites of frequent red tides and brown tides. The majority of medical complaints associated with these events are respiratory irritation due to the aerosolization of toxins.

There is an emerging realization that the amount of plastic floating in the ocean is staggering, and that it affects the ecology dramatically. This has been termed the **plastisphere.** It is estimated that 88% of the ocean's surface is coated with microplastic particles. This establishes a reef of sorts for the microorganisms in the ocean, it can obstruct sunlight from reaching into the ocean, and it poses a health risk for invertebrates and fish. Researchers are looking into the implications for the delicate balances in the ocean ecosystem, as well as for ways to remove the plastic.

Freshwater Communities

The freshwater environment is a site of tremendous microbiological activity. Microbial distribution is associated with sunlight, temperature, oxygen levels, and nutrient availability. The uppermost portion is the most productive self-sustaining region because it contains large amounts of **plankton,** a floating microbial community that drifts with wave action and currents. A major member of this assemblage is the phytoplankton, containing a variety of photosynthetic algae and cyanobacteria. The phytoplankton provides nutrition for **zooplankton,** composed of microscopic consumers such as protozoa and invertebrates that filter, feed, prey, or scavenge. The plankton supports numerous other trophic levels such as larger invertebrates and fish. With their high nutrient content, the deeper regions also support an extensive variety and concentration of organisms, including aquatic plants, aerobic bacteria, and anaerobic bacteria actively involved in recycling organic detritus.

Because oxygen is not very soluble in water and is rapidly used up by the plankton, its concentration forms a gradient from highest in the epilimnion to lowest at the bottom. In general, the amount of oxygen that can be dissolved is dependent on temperature. Warmer strata on the surface tend to carry lower levels of this gas. But of all the characteristics of water, the greatest range occurs in nutrient levels. Nutrient-deficient aquatic ecosystems are called **oligotrophic** (ahl″-ih-goh-trof′-ik). Species that can make a living on such starvation rations are *Hyphomicrobium* and *Caulobacter.* These bacteria have special stalks that capture even minuscule amounts of hydrocarbons present in oligotrophic habitats. The addition of excess quantities of nutrients to aquatic ecosystems, called **eutrophication,** often wreaks havoc on the communities involved. The sudden influx of abundant nutrients, along with warm temperatures, encourages a heavy surface growth of cyanobacteria and algae similar to red tides in oceans **(figure 25.11).** This heavy mat of biomass effectively shuts off the oxygen supply to the lake below. The oxygen content below the surface is further depleted by aerobic heterotrophs that actively decompose the organic matter. The lack of oxygen greatly disturbs the ecological balance of the community. It causes massive die-offs of strict aerobes (fish, invertebrates), and only anaerobic or facultative microbes will survive. This effect can be triggered by the addition of industrial wastes, detergents in household wastewater, or runoff from manure and fertilizer-rich fields and can be remediated over long periods of time.

Figure 25.11 Heavy surface growth of algae and cyanobacteria in a eutrophic pond.

Pat Watson/McGraw Hill

25.2 Learning Outcomes—Assess Your Progress

7. Outline the basic process used to perform metagenomic analysis of the environment.
8. Discuss how metagenomic sampling has changed our view of deep subsurface and oceanic microbiology.
9. Define *eutrophication,* and explain how microbes are responsible for its impact on aquatic life.

25.3 Applied Microbiology and Biotechnology

This section emphasizes the ways that humans have learned to use the diverse capabilities of microbes to improve our lives and environment. It includes the artificial applications of microbes in waste remediation; water treatment; and the manufacture of food, medical, biochemical, drug, and agricultural products. Microbes have evolved by responding to functional pressures, such as when nutrients are limited or unevenly available, or when other organisms are competing for the nutrients. Applied and industrial microbiologists have learned from microbes' own survival mechanisms and have devised ways to manipulate them for use by people.

The practical applications of microorganisms in manufacturing products or carrying out a particular decomposition process belong to the large and diverse area of **biotechnology.** Biotechnology has an ancient history, dating back nearly 6,000 years to those first observant humans who discovered that grape juice left to sit resulted in wine or that bread dough properly infused with a starter culture of yeast or bacteria would rise. Modern biotechnology includes the use of recombinant DNA technology methods to boost or augment the naturally occurring abilities of microbes. The field of biotechnology is providing hundreds of applications in industry, medicine, agriculture, food sciences, and environmental protection. Most biotechnological systems involve the actions of bacteria, yeasts, molds, and algae that have been selected or altered to synthesize a certain food, drug, organic acid, alcohol, or vitamin. Many such food and industrial end products are obtained through **fermentation**—a general term used here to refer to the mass, controlled culture of microbes to produce desired organic compounds. Biotechnology also includes the use of microbes in sewage control, pollution control, metal mining, and bioremediation **(Insight 25.1).**

Microorganisms in Water and Wastewater Treatment

Most drinking water comes from rivers, aquifers, and springs. Only in remote, undeveloped, or high mountain areas can water be used in its natural form. In most cities, it must be treated before it is supplied to consumers. Water supplies such as deep wells that are relatively clean and free of contaminants require less treatment than those from surface sources laden with wastes. The stepwise process in water purification as carried out by most cities is shown in **figure 25.12.** The figure outlines what happens to water between its natural source and the point at which it flows through your faucet at home. It involves filtration and chemical disinfection processes that make the water safe to drink.

In many parts of the world, the same water that serves as a source of drinking water is also used as a dump for solid and liquid wastes **(figure 25.13).** Continued pressure on the earth's finite water resources may require reclaiming and recycling contaminated water such as sewage. Sewage is the used wastewater draining out of homes and industries that contains a wide variety

INSIGHT 25.1 MICROBIOME: Bioremediation

Oil spills are relatively common occurrences in the Gulf of Mexico, but on April 20, 2010, the largest oil spill in U.S. waters occurred. The *Deepwater Horizon* oil rig off the coast of Louisiana exploded and sank, and 11 workers were killed. The rig had successfully dug the deepest oil well in history, with a depth of 35,050 ft. It took several months to cap the oil spill, and it spewed into the ocean from April to June of that year. Approximately 4.9 million barrels, or 779 million liters, were released. Within 2 to 6 days, microbes already present in the Gulf of Mexico biodegraded about half the dispersed oil released from the BP *Deepwater Horizon.* Even in the cold, deep waters of the Gulf, microbes consumed over 90% of the oil within a month of its release. Molecular analyses showed that diverse microbes, particularly *Oceanospirillum* and *Colwellia,* were responsible for the rapid biodegradation of this oil.

Bioremediation quickly grew to include the use of already present or added microbes to break down a wide variety of undesirable compounds in natural environments. Perhaps the most intriguing use is to remediate, or clean up, antibiotics present in farmland and water sources. The antibiotics are there because of excretion by livestock (and humans, to a lesser extent) as well as waste released by pharmaceutical companies. So research is focused on seeding those environments with microbes that can degrade the antibiotics.

Oil burning in the Gulf of Mexico after the Deepwater Horizon explosion.

Chief Petty Officer John Kepsimelis/US Coast Guard

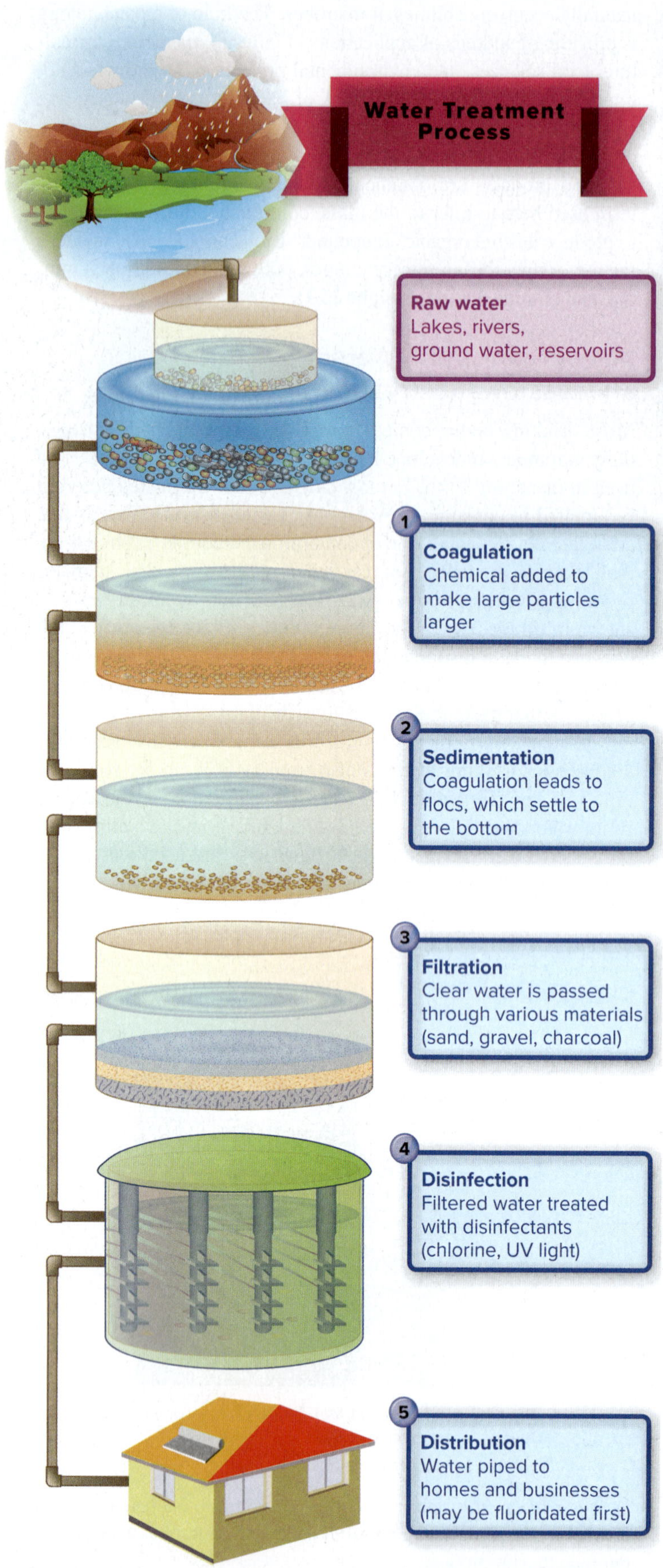

Water Treatment Process
Raw water
Lakes, rivers, ground water, reservoirs
1
Coagulation
Chemical added to make large particles larger
2
Sedimentation
Coagulation leads to flocs, which settle to the bottom
3
Filtration
Clear water is passed through various materials (sand, gravel, charcoal)
4
Disinfection
Filtered water treated with disinfectants (chlorine, UV light)
5
Distribution
Water piped to homes and businesses (may be fluoridated first)

Figure 25.12 A common pathway for water purification.

Figure 25.13 Water: In many places in the world, there is a single source (such as a river or lake) that is used for washing clothes, for bathing, and as a source for drinking water.
Paula Bronstein/Getty Images

of chemicals, debris, and microorganisms. Sewage contains large amounts of solid wastes, dissolved organic matter, and toxic chemicals that pose a health risk. To remove all potential health hazards, treatment typically requires three phases: The *primary stage* separates out large matter; the *secondary stage* reduces remaining matter and can remove some toxic substances; and the *tertiary stage* completes the purification of the water. Microbial activity is an integral part of the overall process. The systems for sewage treatment are massive engineering marvels.

In the primary phase of treatment, bulkier, floating materials such as paper, plastic waste, and bottles are skimmed off. The remaining smaller, suspended particulates are allowed to settle. Sedimentation in settling tanks usually takes 2 to 10 hours and leaves a mixture rich in organic matter. This aqueous portion is carried into a secondary phase of active microbial decomposition, or biodegradation. In this phase, a diverse community of natural bioremediators (bacteria, algae, and protozoa) aerobically decomposes the remaining particles of wood, paper, fabrics, petroleum, and organic molecules inside a large digester tank. This forms a suspension of material called *sludge* that tends to settle out and slow the process. To hasten aerobic decomposition of the sludge, most processing plants have systems to *activate* it by injecting air, mechanically stirring it, and recirculating it. A large amount of organic matter is mineralized into sulfates, nitrates, phosphates, carbon dioxide, and water. Certain volatile gases such as hydrogen sulfide, ammonia, nitrogen, and methane may also be released. Water from this process is siphoned off and carried to the tertiary phase, which involves further filtering and chlorinating prior to discharge. Such reclaimed sewage water is usually used to water golf courses and parks for use in aquaculture, or it is gradually released into large bodies of water. The safety of this practice has recently come into question, due to the persistence of antibiotics in the released water and the risk of spreading antibiotic-resistant bacteria that have managed to survive the process.

In some cases, the solid waste that remains after aerobic decomposition is harvested and reused. Its rich content of nitrogen, potassium, and phosphorus makes it a useful fertilizer. It is estimated that 50% of the 8 million tons of sludge (also called "biosolids") made in the United States annually is recycled and applied to land. This has been viewed as a "green" alternative to burying or burning the sludge. However, scientists are now raising concerns that hundreds of thousands of pounds of potent antimicrobial substances are also being spread on the ground because these chemicals accumulate in the sludge and are not degraded by the typical process of wastewater treatment.

The COVID-19 pandemic has revived the practice of wastewater surveillance for pathogenic microbes that were used in the past to assess the presence of polio in communities. Because COVID-19, like poliovirus, is excreted in feces, the presence of COVID-19 in a particular neighborhood or community can be pinpointed. Because it does not rely on people becoming ill and accessing the health care system, it can be a leading indicator of the presence of the virus because the virus is shed in feces a few days before any symptoms appear. Of course, this type of surveillance does not distinguish between asymptomatic and symptomatic cases. It is a low-cost way to monitor the appearance and disappearance of the virus in a country or community. The sewage is tested for the RNA of SARS-CoV-2.

Microbiology of Drinking Water Supplies

In a large segment of the world's population, the lack of sanitary water is responsible for billions of cases of diarrheal illness that kill millions of children each year. In the United States, hundreds of thousands of people develop waterborne illness every year.

Good health is dependent on a clean, potable (drinkable) water supply. This means the water must be free of pathogens; dissolved toxins; and disagreeable turbidity, odor, color, and taste. As we shall see, water of high quality does not come easily, and microbes are a part of the problem *and* part of the solution.

Disease Connection

After a devastating earthquake in Haiti in 2010, a deadly and long-lasting cholera epidemic began, apparently taken to the struggling nation by some United Nations peacekeepers. The outbreak lasted 9 years, killing more than 10,000 people and making 820,000 ill. Because *V. cholerae* thrives in fresh and brackish water, a single infected person expelling stool into a broken water system can spark an epidemic of these proportions.

Through ordinary exposure to air, soil, and effluents, surface waters usually acquire harmless, saprobic microorganisms. But along its course, water can also pick up pathogenic contaminants. Among the most prominent waterborne pathogens of recent times are the protozoa *Giardia* and *Cryptosporidium;* the bacteria *Campylobacter, Salmonella, Shigella, Vibrio,* and *Mycobacterium;* and hepatitis A and Norwalk viruses. Some of these agents (especially encysted protozoa) can survive in natural waters for long periods without a human host, whereas others are present only transiently and are rapidly lost. The microbial content of drinking water must be continuously monitored to ensure that the water is free of infectious agents.

Attempting to survey water for specific pathogens can be very difficult and time-consuming, so most assays of water purity are more focused on detecting fecal contamination. High fecal levels can mean the water contains pathogens and is consequently unsafe to drink. So, wells, reservoirs, and other water sources can be analyzed for the presence of various **indicator bacteria.** These species are intestinal residents of birds and mammals, and they are readily identified using routine lab procedures.

In the late 1800s, it was suggested that a good way to determine if water or its products had been exposed to feces was to test for *E. coli.* Although most *E. coli* strains are not pathogenic, they almost always come from a mammal's intestinal tract, so their presence in a sample is a clear indicator of fecal contamination. Because at the time it was too difficult to differentiate *E. coli* from the closely related species of *Citrobacter, Klebsiella,* and *Enterobacter,* laboratories instead simply reported whether a sample contained one of these isolates. (All of these bacteria ferment lactose and are phenotypically similar.) The terminology adopted was *coliform-positive* or *coliform-negative* (*coliform* means "*E. coli*–like"). Coliforms, then, are gram-negative, lactose-fermenting, gas-producing bacteria such as those previously mentioned.

The use of this coliform assay has been the standard procedure since 1914, and it is still in widespread use. Pick up a newspaper in the summer, and you will likely find a report about a swimming pool or a river with a high coliform count. Coliform counts are also used to monitor food production and to trace the causes of foodborne outbreaks. Recently, microbiologists have noted serious problems with the use of coliforms to indicate fecal contamination. The main issue is that the three other bacterial species already mentioned, among others, are commonly found growing in fecal-free environments such as freshwater and plants that eventually become food. In other words, if you are not looking specifically for *E. coli,* you cannot be sure you are looking for feces.

Microbiologists are now advocating that *E. coli* alone be used as an indicator of fecal contamination. Newer identification techniques make this as simple as, if not simpler than, the standard coliform tests. Very commonly what happens is routine testing is used to detect coliforms, and if coliforms are present, additional testing to identify *E. coli* are performed.

Water Quality Assays A rapid method for testing the total bacterial levels in water is the standard plate count. In this technique, a small sample of water is spread over the surface of a solid medium. The numbers of colonies that develop provide an estimate of the total viable population without differentiating coliforms from other microbial species. This information is particularly helpful in evaluating the effectiveness of various water purification stages. Another general indicator of water quality is the level of dissolved oxygen it contains. It is established that water containing high levels of organic matter and bacteria will have a significantly lower concentration of oxygen due to the metabolism of aerobic microorganisms.

Counting Coliforms The membrane filter method is a widely used rapid method that can be used in the field or lab to process and test larger quantities of water (100 to 200 mL). This method is more suitable for dilute fluids, such as drinking water, that are relatively free of particulate matter. It is less suitable for water containing heavy microbial growth or debris. This technique is related to the method described in chapter 11 for sterilizing fluids by filtering out microbial contaminants, except that in this system, the filter containing the trapped microbes is the desired end product. The steps in membrane filtration are diagrammed in **figure 25.14.** After filtration, the membrane filter is placed in a Petri dish containing selective medium. After incubation, both nonfecal and fecal coliform colonies can be counted and often presumptively identified by their distinctive characteristics on these media.

Another, more time-consuming but useful technique is the **most probable number (MPN)** procedure, which detects coliforms by a series of *presumptive, confirmatory,* and *completed* tests. The presumptive test involves three subsets of fermentation tubes, each containing a different amount of lactose or lauryl tryptose broth. The three subsets are inoculated with various-size water samples. After 24 hours of incubation, the tubes are evaluated for gas production. A positive test for gas formation is presumptive evidence of coliforms; negative for gas means no coliforms. The number of positive tubes in each subset is tallied, and this set of numbers is applied to a statistical table to estimate the most likely or probable concentration of coliforms.

When a test is negative for coliforms, the water is considered generally fit for human consumption, but even slight coliform levels are allowable under some circumstances. For example, municipal waters can have a maximum of 4 coliforms per 100 mL; private wells can have an even higher count. There is no acceptable level for fecal coliforms, enterococci, viruses, or pathogenic protozoa in drinking water. Waters that will not be consumed but are used for fishing or swimming are permitted to have counts of 70 to 200 coliforms per 100 mL. If the coliform level of recreational water reaches 1,000 coliforms per 100 mL, health departments usually bar its usage.

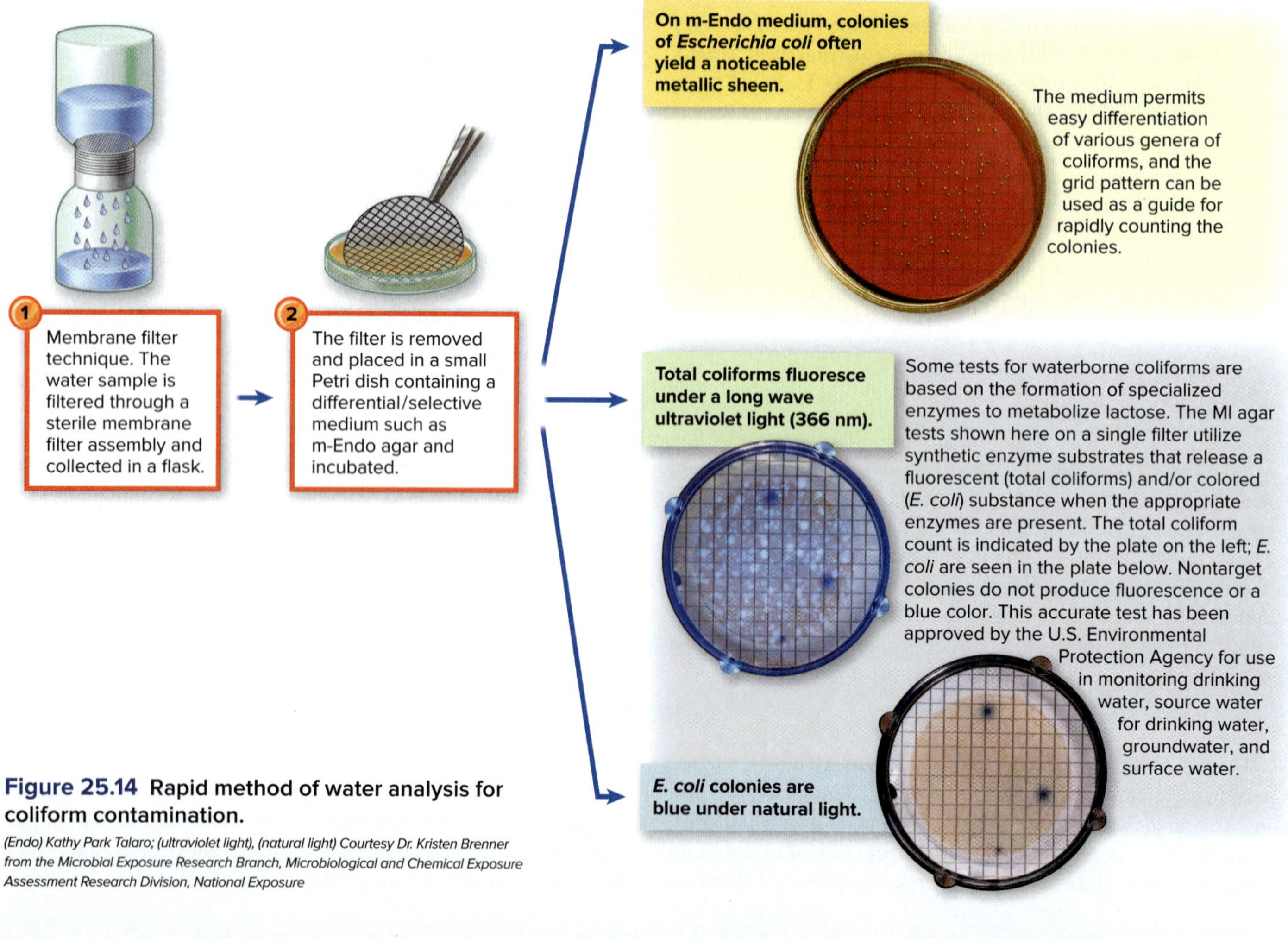
1
Membrane filter technique. The water sample is filtered through a sterile membrane filter assembly and collected in a flask.
2
The filter is removed and placed in a small Petri dish containing a differential/selective medium such as m-Endo agar and incubated.
On m-Endo medium, colonies of *Escherichia coli* often yield a noticeable metallic sheen.
The medium permits easy differentiation of various genera of coliforms, and the grid pattern can be used as a guide for rapidly counting the colonies.
Total coliforms fluoresce under a long wave ultraviolet light (366 nm).
Some tests for waterborne coliforms are based on the formation of specialized enzymes to metabolize lactose. The MI agar tests shown here on a single filter utilize synthetic enzyme substrates that release a fluorescent (total coliforms) and/or colored (*E. coli*) substance when the appropriate enzymes are present. The total coliform count is indicated by the plate on the left; *E. coli* are seen in the plate below. Nontarget colonies do not produce fluorescence or a blue color. This accurate test has been approved by the U.S. Environmental Protection Agency for use in monitoring drinking water, source water for drinking water, groundwater, and surface water.
E. coli colonies are blue under natural light.

Figure 25.14 Rapid method of water analysis for coliform contamination.

(Endo) Kathy Park Talaro; (ultraviolet light), (natural light) Courtesy Dr. Kristen Brenner from the Microbial Exposure Research Branch, Microbiological and Chemical Exposure Assessment Research Division, National Exposure

Microorganisms Making Food and Drink

All human food—from vegetables to caviar to cheese—comes from some other organism, and rarely is it obtained in a sterile, uncontaminated state. Somewhere along the route of growth, procurement, processing, or preparation, food becomes contaminated with microbes from the soil, the bodies of plants and animals, water, air, food handlers, or utensils. The final effects depend on the types and numbers of microbes and whether the food is cooked or preserved. In some cases, specific microbes can even be added to food to obtain a desired effect.

As long as food contains no harmful substances or organisms, its suitability for consumption is largely a matter of taste. But what tastes like rich flavor to some may seem like decay to others. The test of whether certain foods are edible is guided by culture, experience, and preference. The flavors, colors, textures, and aromas of many cultural delicacies are supplied by bacteria and fungi. Kombucha, pickled cabbage, Norwegian fermented fish, and Limburger cheese are notable examples. If you examine the foods of most cultures, you will find some foods that derive their delicious and unique flavor from microbes.

Microbial Fermentations in Food Products from Plants

In contrast to methods that destroy or keep out unwanted microbes, many culinary procedures deliberately add microorganisms and encourage them to grow. Common substances such as bread, cheese, beer, wine, yogurt, and pickles are the result of **food fermentations.** These reactions actively encourage biochemical activities that impart a particular taste, smell, or appearance to food. The microbe or microbes can occur naturally on the food substrate, as in sauerkraut, or they can be added as pure or mixed samples of known bacteria, molds, or yeasts called **starter cultures.** Many food fermentations are synergistic, with a series of microbes acting together to convert a starting substrate to the desired end product. Because large-scale production of fermented milk, cheese, bread, alcoholic brews, and vinegar depends upon inoculation with starter cultures, considerable effort is spent selecting, maintaining, and preparing these cultures and getting rid of contaminants that can spoil the fermentation. In this section we will examine two very common consumer products, bread and beer, that are created through the action of microbes.

Bread Microorganisms accomplish three functions in bread making:

1. leavening the flour-based dough,
2. providing flavor and odor, and
3. conditioning the dough to make it workable.

Leavening is achieved primarily through the release of gas to produce a porous and spongy product. Without leavening, bread dough remains dense, flat, and hard. Although various microbes and leavening agents can be used, the most common ones are various strains of the baker's yeast *Saccharomyces cerevisiae.* Other gas-forming microbes such as coliform bacteria, certain *Clostridium* species, heterofermentative lactic acid bacteria, and wild yeasts can be used, depending on the type of bread desired.

Yeast metabolism requires a source of fermentable sugar such as maltose or glucose. Because the yeast respires aerobically in bread dough, the chief products of maltose fermentation are carbon dioxide and water, rather than alcohol (the main product in beer and wine). Other contributions to bread texture come from kneading, which incorporates air into the dough, and from microbial enzymes, which break down flour proteins (gluten) and give the dough elasticity.

Besides carbon dioxide production, bread fermentation generates other volatile organic acids and alcohols that impart delicate flavors and aromas. These are especially well-developed in handmade bread, which is leavened more slowly than commercial bread. Yeasts and bacteria can also impart unique flavors, depending upon the culture mixture and baking techniques used. The pungent flavor of rye bread, for example, comes in part from starter cultures of lactic acid bacteria such as *Lactobacillus plantarum, L. brevis, L. bulgaricus, Leuconostoc mesenteroides,* and *Streptococcus thermophilus.* Sourdough bread gets its unique tang from *Lactobacillus sanfranciscensis.*

Beer The production of alcoholic beverages takes advantage of another useful property of yeasts. By fermenting carbohydrates in fruits or grains anaerobically, they produce ethyl alcohol, as shown by this equation:

$$C_6H_{12}O_6 \rightarrow 2C_2H_5OH + 2CO_2$$
$$\text{Yeast} + \text{Sugar} = \text{Ethanol} + \text{Carbon dioxide}$$

Depending on the starting materials, the type of yeast used, and the processing method, alcoholic beverages vary in alcohol content and flavor. The principal types of fermented beverages are beers, wines, and spirit liquors.

The earliest evidence of beer brewing appears in ancient tablets by the Sumerians and Babylonians around 6000 BC. The starting ingredients for both ancient and present-day versions of beer, ale, stout, porter, and other variations are water, malt (barley grain), hops, and special strains of yeasts. The steps in brewing include malting, mashing, adding hops, fermenting, aging, and finishing.

For brewer's yeast to convert the carbohydrates in grain into ethyl alcohol, the barley must first be sprouted and softened to make its complex nutrients available to yeasts. This process, called **malting,** releases amylases that convert starch to dextrins and maltose, and proteases that digest proteins. Other sugar and starch supplements added in some forms of beer are corn, rice, wheat, soybeans, potatoes, and sorghum. After the sprouts have been separated, the remaining malt grain is dried and stored in preparation for mashing.

The malt grain is soaked in warm water and ground up to prepare a **mash.** Sugar and starch supplements are then introduced to the mash mixture, which is heated to a temperature of about 65°C to 70°C. During this step, the starch is hydrolyzed by amylase and simple sugars are released. Heating this mixture to 75°C stops the activity of the enzymes. Solid particles are next removed by settling and filtering (the result is what can then be added to an anaerobic digester). **Wort,** the clear fluid that comes off, is rich in dissolved carbohydrates. It is boiled for about 2.5 hours with **hops,** the dried scales of the female flower of *Humulus lupulus,* to extract the bitter acids and resins that give aroma and flavor to the finished product. Boiling also caramelizes the sugar and imparts a golden or brown color, destroys any bacterial contaminants that can destroy flavor, and concentrates the mixture. The filtered and cooled supernatant is then ready for the addition of yeasts and fermentation.

Fermentation begins when wort is inoculated with a species of *Saccharomyces* that has been specially developed for beer making. Top yeasts such as *Saccharomyces cerevisiae* function at the surface and are used to produce the higher alcohol content of *ales.* Bottom yeasts such as *S. uvarum* (*carlsbergensis*) function deep in the fermentation vat and are used to make other beers. In both cases, the initial inoculum of yeast starter is aerated briefly to promote rapid growth and increase the load of yeast cells. Shortly thereafter, an insulating blanket of foam and carbon dioxide develops on the surface of the vat and promotes anaerobic conditions **(figure 25.15).** During 8 to 14 days of fermentation, the wort sugar is converted chiefly to ethanol and carbon dioxide. The diversity of flavors in the finished product is partly due to the release of small amounts of glycerol, acetic acid, and esters. Fermentation is self-limited, and it essentially ceases when a concentration of 3% to 6% ethyl alcohol is reached.

Freshly fermented, or "green," beer is **lagered,** meaning it is held for several weeks to months in vats near 0°C. During this maturation period, yeast, proteins, resin, and other materials settle, leaving behind a clear, yellow fluid. Lager beer is subjected to a final filtration step to remove any residual yeasts that could spoil it. Finally, it is carbonated with carbon dioxide collected during fermentation and packaged in kegs, bottles, or cans.

For the past few decades, a craft brewing community has been growing. Small cooperatives or even individuals are creating their own special flavors in beers by creatively combining various malts—which generally are responsible for beer's sweetness—with hops to balance the sweetness with bitterness.

Figure 25.15 Anaerobic conditions in homemade beer production. A layer of carbon dioxide foam keeps oxygen out.
Ryan Hatch

25.3 Learning Outcomes—Assess Your Progress

10. Compose a sentence about the history of applied microbiology.
11. Define *biotechnology,* and explain how its uses have changed in modern times.
12. Outline the steps in water purification.
13. Differentiate water purification from sewage treatment.
14. List five important pathogens of drinking water.
15. Provide examples of indicator bacteria, and describe their role in the survey of water quality.
16. Summarize methods for identifying and quantifying microbial contaminants in water supplies.
17. Summarize the microbial process that leads to leavening in bread.
18. Write the equation showing how yeasts convert sugar into alcoholic beverages.

25.4 The Concept of "One Health"

It is obvious by this point in the book that human health, animal health, and environmental health are closely interrelated—and that the health of all life on earth is connected. Scientists call the concept *one health* and emphasize three interacting parts of the biosphere: the environment, humans, and other animals. The reasoning here is that microorganisms circulate among human hosts, animal hosts, and environmental reservoirs. Changes in the environment can lead to transmission of pathogens to animals and humans that previously were not exposed to them.

It might be easiest to visualize one health as three overlapping spheres **(figure 25.16).** A change in any one of the spheres impacts the others, and it happens continuously. The mixing of microbes in different animal hosts and under different environmental conditions can lead to the evolution of new and potentially dangerous pathogens. Human activities, in particular, can promote the emergence of infectious diseases—for example, through ecological disturbances and the movement of animals. This occurs frequently for some microbes. Look at the example of influenza viruses, in which the mixing of different strains in birds, swine, and humans results in the evolution of new recombinant strains with the potential to spread globally in any given year (refer to chapter 22 for the details of how this works).

A classic case of this occurred in the 1990s. A frightening outbreak of acute respiratory distress disease introduced us to the hantavirus. It had suddenly started causing human disease and deaths because of an increase in its reservoir, the deer mouse. The deer mouse population in a region of the southwestern United States experienced explosive growth in 1993, and experts blame an El Niño–associated heavy rainfall that year, ending several years of drought. The rain simultaneously decreased the numbers of deer mouse predators, such as snakes, owls, and coyotes, and increased the growth of the piñon nut, a favorite food of the mice. The increased population of mice led to increased contact of humans

Figure 25.16 "One Health."

(rooster) Siede Preis/Getty Images; (tree) OGphoto/Getty Images

with the feces and urine of the mice, the inhalation of which leads to human respiratory infection.

How Infectious Diseases Move into New Populations, and New Species

Both behavioral changes, on the part of animals or humans, and environmental changes can cause a known infectious disease to adopt a new reservoir or a new set of hosts. These influences can also cause an animal infection to "jump hosts" and infect humans. This is called a **spillover event**. These changes can also cause a microbe to increase or decrease its virulence, or to cause an entirely new disease manifestation in an already-known host. The emergence of hantavirus in the Southwest in 1993 appears to follow this last pattern.

Hantavirus

Previously, hantaviruses had been known to cause a condition known as hemorrhagic fever with renal syndrome (HFRS), a relatively uncommon disease with a 15% mortality rate. The virus name comes from the Hantaan River in South Korea. It is endemic in Asia and Europe but not in the United States. But then hantavirus made an explosive debut in the United States with the outbreak in 1993. Disease here was very different, and much more deadly. Now dubbed New World hantavirus infections, these cases caused hantavirus pulmonary syndrome (HPS), a severe pneumonia with a 50–70% mortality rate. Both forms of the disease are transmitted via mouse urine and excrement. Scientists say that the sudden growth in the mouse population, influenced by the piñon nut abundance caused by abnormally heavy rainfall in the dry Southwest, created easily aerosolized mouse products that overwhelmed the respiratory systems of even young and healthy humans who encountered them.

HIV

HIV infections provide an example of a virus that was probably at some level endemic, with low pathogenicity, from the time of its introduction to humans from primates. There are two main groups of HIV, HIV-1 and HIV-2. HIV-1 is responsible for the vast majority of infections and so we will focus on how it crossed into humans and then became a deadly pandemic.

Scientists believe that HIV existed for thousands of years as SIV (simian immunodeficiency virus) in chimpanzees in the rainforests of Cameroon, in Africa. The best estimate for when it first infected a human—becoming human immunodeficiency virus, or HIV—was in the 1920s. It is most likely that the first human was infected with the virus while butchering primates as bush meat and coming in contact with their blood. HIV is a highly mutable virus, and a variant in this mix was able to colonize humans. Over the next few decades, it is likely that it spread infrequently to other humans, during which time it had relatively low virulence. Cultural and government shifts in Central Africa led to the rise of cities, most notably Leopoldville, the capital of the Belgian Congo (the city is now Kinshasa and the country is the Democratic Republic of the Congo). Central Africa became a major trading center of the continent, bringing people from rural areas to live and work there, and establishing major routes of transportation (mainly rail) and a robust sex trade **(figure 25.17).** Health services were improving, and large-scale vaccination campaigns against smallpox and sleeping sickness took place, often using the same needles for multiple patients. The increased sexual mixing and use of used needles probably caused the virus to become more virulent as it was transmitted. This means that virus variants that were more deadly were no longer weeded out of the population, whereas in the past, slower rates of sexual exchange and blood contamination selected for the milder forms of the virus.

The changeability of virus variants based on environmental or human factors is known as the **evolution of virulence.** In the very late 1970s and early 1980s, the virus was brought to the United States and introduced into the male homosexual population, where frequent partner exchange was common. The virulence of the virus continued to increase, and when the first cases were discovered in the United States they were exclusively in the gay male community. In fact, in the beginning the disease was referred to as GRID—Gay Related Immunodeficiency Disease. But it quickly became clear that heterosexual sexual encounters, as well as exposure to blood and blood products, could also transmit the virus. In this case, the spillover event was the human acquiring infection from a primate carrying a variant capable of multiplying in humans. The push to a pandemic came about from the high contact circumstances favoring high virulence variants.

Mosquito-Borne Diseases

As we discussed in chapters 20 and 21, viruses transmitted by arthropods change their epidemiology as the habitat of the arthropods changes. In particular, mosquito-borne viruses such as dengue, Zika, yellow fever, and West Nile, carried by the *Aedes*

MALTA
TUNISIA
GREECE
CYPRUS
LEBANON
ISRAEL
SYRIA
IRAQ
IRAN
JORDAN
KUWAIT
QATAR
UNITED ARAB EMIRATES
MOROCCO
ALGERIA
LIBYA
EGYPT
SAUDI ARABIA
WESTERN SAHARA
MAURITANIA
MALI
NIGER
CHAD
SUDAN
ERITREA
YEMEN
SENEGAL
THE GAMBIA
GUINEA-BISSAU
GUINEA
BURKINA FASO
BENIN
TOGO
NIGERIA
DJUOUTI
ETHIOPIA
SIERRA LEONE
COTE D`IVOIRE
GHANA
LIBERIA
CENTRAL AFRICAN REPUBLIC
CAMEROON
SOMALIA
UGANDA
EQUATORIAL GUINEA
KENYA
GABON
CONGO
DEMOCRATIC REPUBLIC OF THE CONGO
Kinshasa
TANZANIA
COMOROS
MAYOTTE (FRANCE)
ANGOLA
MOZAMBIQUE
ZAMBIA
MALAWI
ZIMBABWE
NAMIBIA
BOTSWANA
MADAGASCAR
SOUTH AFRICA

Figure 25.17 Present-Day Africa. Note the central location of Kinshasa (formerly Leopoldville), which is the largest city in Africa, and a transportation and commerce hub. The inset is a photo of the city.

(map) max776/123RF; (inset photo) David Keith Jones/Images of Africa Photobank/Alamy Stock Photo

genus of mosquito, are expanding their reach because of warming climates that allow the mosquitoes to live in more northern locales **(figure 25.18).** In this case, changes in the environment allow a virus to find new hosts. As an example, **figure 25.19** depicts the increase in the average number of days that mosquitoes are active and able to transmit disease in San Francisco starting in 1970 until recent years.

SARS-CoV-2

Now let's look at the emergence of SARS-CoV-2 in late 2019 in Wuhan, China. It is believed to be a bat virus that then jumped species into another animal, and then into humans. As it happens, bats are particularly rich reservoirs of viruses. While each individual bat species is estimated to be host to an average of

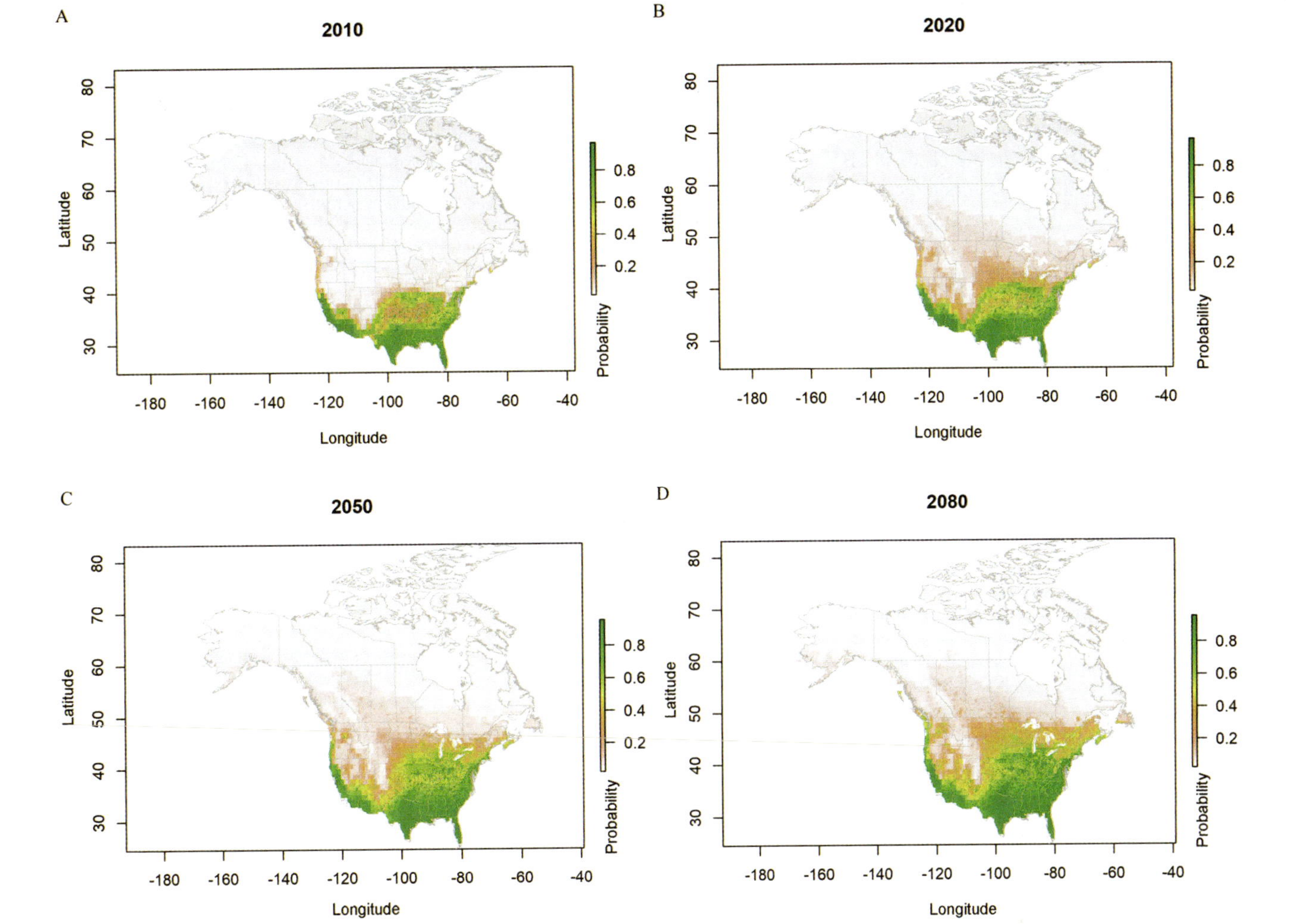
A
2010
B
2020
C
2050
D
2080
Latitude
Longitude
Probability
0.8
0.6
0.4
0.2
80
70
60
50
40
30
-180
-160
-140
-120
-100
-80
-60
-40

Figure 25.18 Probability of presence of *Aedes aegypti* based on a climate model that accounts for moderate efforts to stem climate-changing emissions. (There are also models using "minimal" and "maximum" efforts.) The shading, from gray to green, represents the probability of *A. aegypti* establishing a niche in a particular area (see vertical bar to right of graphs).

Khan ei. al. 2020

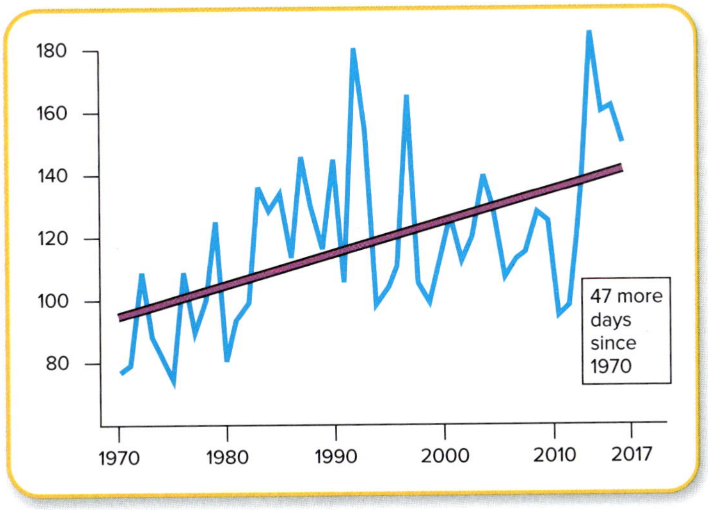
180
160
140
120
100
80
1970
1980
1990
2000
2010
2017
47 more days since 1970

Figure 25.19 Number of days per year in which mosquitoes are active and able to transmit disease. Data are for the city of San Francisco.

Climate Central, US Faces a Rise in Mosquito "Disease Danger Days." https://www.climatecentral.org/news/us-faces-a-rise-in-mosquito-danger-days. 21903.

2 different viruses, their colony lifestyle means that they frequently exchange viruses with bats of other species. Scientists consider them rich sources of new viruses—particularly coronaviruses—that will jump species to infect other mammals. Bats are known to transmit rabies, Ebola, and the first SARS virus to humans. Some researchers suggest that changes in climate in Yunnan province, in southwestern China, as well as the adjacent countries Laos and Myanmar, led to vast increases in the number of bats, and a corresponding increase in the coronaviruses they carry. They document large-scale changes in the type of vegetation in the area. Where tropical shrub lands were previously dominant, increases in temperature, sunlight, and carbon dioxide fostered the growth of tropical savannah and deciduous woodland. These habitats are very favorable for bats. A study showed that 40 new bat species have moved into Yunnan province in recent decades, harboring approximately 100 novel bat-borne coronaviruses. In fact, one research group said that the region "form(s) a global hotspot of climate change–driven increase in bat richness." Bats and other animals are regularly trucked to the wildlife markets of large cities nearby, including Wuhan, where the first cases of COVID-19

were found. For those reasons, it is not surprising that another novel coronavirus jumped species into an animal (of still unknown identity) and then made the leap into humans.

Keep in mind that while we have the benefit of hindsight and years of research to have examined the circumstances of spillover and evolution of microbes as in the cases of HIV, hantavirus, and the first SARS event in the 1990s, conclusions about SARS-CoV-2 are still preliminary.

Other Effects of Climate Change

Another, perhaps unexpected, consequence of climate change is the increasing frequency and intensity of wildfires in the American West **(figure 25.20).** The fires lead to change in the forest floor and soil community structure. This includes a long-term change in the microbiome of the soil, accompanied by changes in pH and nutrient availability that will likely affect the ecosystems for years to come. From a one health perspective, disruptions in such an important habitat are predicted to have downstream effects on vegetation, humans, and other animals. One immediate consequence of the wildfires in 2021 was a worsening of COVID symptoms in some infected people living near areas where the smoke was dense.

The tangling of two major crises, climate change and the COVID-19 pandemic, was made evident in September 2021, when an editorial was simultaneously published in over 200 leading medical journals. It was an emergency call to action for world leaders to limit climate change, restore biodiversity, and protect health. The authors argued that we cannot ignore climate change while we are dealing with the more immediate challenge of a global pandemic because they are closely intertwined.

We summarize emerging and reemerging infectious agents in **table 25.1.** It also contains a list, created by the World Health Organization, of the emerging diseases that constitute the greatest threats of causing epidemics in the future and for which there are currently no effective countermeasures. **Figure 25.21** summarizes the threats to health that go along with climate change.

Figure 25.20 Wildfire. The Environmental Protection Agency reports that climate change has led to an increase in wildfire season length, wildfire frequency, and burned area.
Elmer Frederick Fischer/Corbis

CLIMATE CHANGE & HEALTH
WEATHER EXTREMES
Heavy rains, floods, hurricanes, wildfires, etc., endanger health as well as destroy property and livelihoods. In 2021, deadly wildfires destroyed 7.13 million acres. Historically, developing countries are hardest hit by major weather events.
RECORD HIGH AND LOW TEMPERATURES
Short-term fluctuations in temperature can affect health—causing heat stress or extreme cold—and lead to increased death rates from heart and respiratory disease. Record high temperatures lead to wildfires.
ALLERGEN LEVELS
Pollen and other airborne allergen levels are higher in extreme heat. These can trigger asthma, which affects hundreds of millions of people worldwide. Ongoing temperature increases are expected to increase this burden.
RISING SEA LEVELS
Rising sea levels increase the risk of coastal flooding and could cause population displacement, which puts people at increased risk for epidemic disease. More than half of the world's population now lives within 60 kilometers (about 37 miles) of shorelines.
MORE VARIABLE RAINFALL PATTERNS
Fresh water can become more scarce. Globally, water scarcity affects 4 of 10 people. Lack of water can compromise hygiene and health and lead to diarrhea, and death. Alternatively, excess rain can lead to deadly flooding.
WATER- AND MOSQUITO-BORNE DISEASES
Climate-sensitive diseases are among the largest global killers. Diarrhea, malaria, and malnutrition alone cause millions of deaths annually and are expected to be affected by climate changes.

Figure 25.21
Source: World Health Organization

25.4 Learning Outcomes—Assess Your Progress

19. Name the three components that interact for "one health."
20. Discuss three ways that changes in global climate can impact the reach and impact of infectious diseases.

Table 25.1 Emerging Infectious Diseases

Pathogens Newly Recognized in Recent Decades	Reemerging Pathogens	WHO List of Most Concerning Emerging Pathogens
Acanthamebiasis	Chikungunya	Chikungunya
Australian bat lyssavirus	*Clostridioides difficile*	Crimean Congo hemorrhagic fever
Babesia, atypical	Crimean Congo hemorrhagic fever	Ebola virus disease
Bartonella henselae	Enterovirus 71	Lassa fever
Ebola	Mumps virus	Marburg
Ehrlichiosis	*Staphylococcus aureus*	MERS
Encephalitozoon cuniculi	*Streptococcus,* group A	Nipah
Encephalitozoon hellem	Zika virus	Rift Valley fever
Enterocytozoon bieneusi	*Candida auris*	SARS CoV-2
Hendra virus		Zika virus
Human herpesvirus-6	Monkeypox	*Candida auris*
Human herpesvirus-8		
Lassa fever		
Lyme disease		
Marburg		
MERS		
Nipah		
Parvovirus B19		
Rift Valley fever		
SARS viruses		

Source: World Health Organization.

Media Under The Microscope Wrap-Up

Marmolejos/Shutterstock

The **intended message** of the *Bloomberg* article is to summarize the warnings in a report published in the medical journal *Lancet.* The article ties together two global crises—the COVID-19 pandemic and climate change—making the point that we need to more actively address climate change or additional infectious disease threats will arise.

My **critical reading** of the article is that it is factual and reiterates points about both COVID-19 and climate change that have been discussed extensively, although not often together. The news source, *Bloomberg*, is trustworthy and the medical journal cited, *Lancet,* is one of the most-respected. Of course, summaries of even the most solid research can get it wrong, so a critical eye is always warranted. I do have a quibble with the *Bloomberg* article. After its headline "Climate Change Threatens to Spread Viruses Through and Unprepared World," it discusses well-known facts about COVID-19, and some in-depth information about climate change, but does not say much more about the 44 health indicators it refers to from *Lancet*, so I'm not sure it would be convincing to someone who was already aware of (a) COVID-19 and (b) climate change.

I would **interpret** the article to my nonmicrobiologist friends by pointing out that warming temperatures lead to the expansion of the habitat of microbes and their vectors that prefer warmer climates. The Bloomberg article doesn't do a good job actually connecting the dots between climate change and the COVID-19 pandemic. But we do know that deforestation, and environmental disruption tied to that led to changes in bat habitat, eventually resulted in a bat virus causing a global pandemic.

My **overall grade** for the Bloomberg article is a B–. It left me eager to read the *Lancet* report, but there was not even a reference for it.

Source: *Bloomberg News* "Climate Change Threatens to Spread Viruses Through an Unprepared World," online article posted October 20, 2021.

Study Smarter: Better Together

These activities are designed for you to use on your own with a study group—either a face-to-face group or a virtual one, consisting of 3–5 members. Studying together can be very helpful, but there are effective and ineffective ways to do it. For example, getting together without a clear structure is often not a good use of your time. Use your time efficiently by using one or more of the exercises below.

FACE-TO-FACE GROUPS

Use one or more of the activities below.

Peer Instruction: Assign numbers to your group members to use all semester long. Now look at these five concepts from this chapter. Each group member prepares a 5-minute lesson on the topic corresponding to their number. Do not worry if you have fewer than 5 members; just use however many you have! During your group study time, each member presents their lesson, and the group spends another 5–10 minutes discussing that lesson.

1. Three stages of sewage treatment
2. Carbon cycle
3. Environmental metagenomics
4. One Health
5. Spillover events

Concept Maps: Each member of the group should use this list of terms from this chapter to generate their own concept map. This can be hand-drawn or created using software (see Appendix C for guidelines). During group study time, compare each other's concept maps and help each other make sure they are correct. Of course, there are many different "correct" maps. Examining each member's map will help you talk through the varied concepts and how they are related.

Concept Terms:

climate	vegetation	virulence	human behavior
bats	wildfires	habitat	emerging disease

Table Topics: Each group member should identify a concept or topic from this week's class assignments with which they are having trouble and share it during group study time. The other group members can then help to clarify confusing issues or share how they figured it out. Aim for a maximum of 15 minutes per topic. If the topic remains unclear to the group, bring it up during class or use the instructor's office hours or e-mail to ask for help. Taking the time to struggle with a difficult concept first makes your questions much more specific and more likely to yield helpful answers.

VIRTUAL GROUPS

Not everyone has the time or opportunity to meet with group members outside of class time. You or your instructor can create a virtual group using e-mail or the course software.

Weekly Discussion Board: This forum can be used as a way for groups to discuss topics, via e-mail, or other learning management systems or online platforms, before they are covered in class. As each member of the group answers the current week's question, they should send their responses to every other member of their group. It's best to agree on a deadline based on how your class schedule works (Saturday for the next week's topics, for example). Then, after the topic is discussed in class, each member should send a response that all group members will see with a follow-up post on the same topic. If you cover more than one chapter in a week, someone can be designated to choose which chapter Discussion Board question you will use. Or simply decide up front that you will always use the first-chapter-of-the-week's question, to keep the schedule simple.

Discussion Question
Does refrigeration of food prevent the growth of all bacteria? Explain.

Chapter Summary

MICROBES IN OUR ENVIRONMENT AND THE CONCEPT OF ONE HEALTH

25.1 MICROBIAL ECOLOGY AND ELEMENT CYCLING

- Microbial ecology is the study of microbes in their natural habitats, as they naturally occur.
- The earth's totality of life forms is maintained by the recycling of elements through biogeochemical cycles.
- The carbon cycle and the nitrogen cycle are both atmospheric cycles, two important examples of how elements are cycled through redox reactions in various parts of the biosphere.
- The carbon cycle maintains a steady level of carbon dioxide in the atmosphere, but the burning of fossil fuels is increasing atmospheric CO_2 levels markedly.
- The nitrogen cycle is complex and involves multiple microorganisms to make nitrogen available to plants and animals. One notable relationship in the nitrogen cycle is that between rhizobia bacteria and the roots of legumes.

25.2 MICROBES ON LAND AND IN WATER

- The earth's land, water, and air are colonized by more microbes than we ever imagined. We have discovered the magnitude of their numbers through metagenomics, the sampling of the environment for DNA sequences.
- The deep subsurface, below land and sea, is colonized by a rich array of microbes that have a wide variety of metabolic capabilities.
- The ocean is populated by millions of microorganisms per milliliter. Bacteriophages are abundant and play an important role in marine ecosystems.
- Eutrophication of freshwater and marine systems is caused by the addition of excess nutrients. It causes major disruptions in the ecology of these systems.

25.3 APPLIED MICROBIOLOGY AND BIOTECHNOLOGY

- The use of microorganisms for practical purposes to benefit humans is called biotechnology.
- Wastewater or sewage is treated in three stages to remove organic material, microorganisms, and chemical pollutants.
- Significant waterborne pathogens include protozoa, bacteria, and viruses.
- Water quality assays screen for coliforms as indicator organisms or may assess the most probable number of microorganisms. As these results may be misleading, more emphasis is being placed on identifying *E. coli* itself to signal fecal contamination.
- Food fermentation processes utilize bacteria or yeasts to produce desired components such as alcohols and organic acids in foods and beverages. Beer and bread are examples of such processes.

25.4 THE CONCEPT OF "ONE HEALTH"

- "One health" is a concept that captures the interrelated nature of microbes, the environment, and animals (including humans).
- Human activity that alters one of the three components of one health can trigger unintended consequences in infectious disease epidemiology.
- When microbes that infect animals jump species to infect and then become transmissible between humans, it is called a spillover event.
- Weather events and a warming climate can cause microbes to cause different manifestations in their hosts, as was the case when the New World hantavirus emerged in the 1990s.
- The spillover event with HIV probably occurred in the first half of the 1900s. Its eventual adaptation to spread quickly and cause deadly symptoms illustrates the evolution of virulence.
- Diseases carried by mosquitoes are expected to infect their hosts in more and more northern latitudes due to a warming climate.
- The COVID-19 pandemic and the warming climate constitute an alarming combination of threats to human health.

SmartGrid: From Knowledge to Critical Thinking

This *21 Question Grid* takes the topics from this chapter and arranges them with respect to the American Society for Microbiology's Undergraduate Curriculum guidelines—all six of the important "Concepts" as well as the important "Competency" of scientific literacy. Three questions are supplied, which cover chapter content referring to the Concept or Competency in increasing levels of Bloom's taxonomy for learning.

ASM Concept/ Competency	A. Bloom's Level 1, 2—Remember and Understand (Choose one)	B. Bloom's Level 3, 4—Apply and Analyze	C. Bloom's Level 5, 6—Evaluate and Create
Evolution	**1.** Which of the following is unlikely to be a waterborne pathogen? **a.** *Giardia lamblia* **b.** *Salmonella* **c.** *Vibrio* **d.** *Staphylococcus*	**2.** In Malaysia, a form of malaria that was previously only seen in macaque monkeys is being seen in humans because deforestation activities bring the monkeys and humans closer together. Diagram the chain of events that leads to human infection, including plant life, mosquitoes, monkeys, and humans.	**3.** Develop an argument about how a sexually transmitted microbe might benefit from evolving toward lower virulence while a mosquito-borne virus would not.
Cell Structure and Function	**4.** What cellular component is analyzed during environmental metagenomic studies? **a.** plant cell walls **b.** proteins **c.** mitochondria **d.** DNA	**5.** Why do you think microbes are so adept at breaking down sewage and wastewater?	**6.** Look at chapter 9. Use table 9.2 to explain why microbes might be good sources of protein.
Metabolic Pathways	**7.** Contaminated water has **a.** lower oxygen content than pure water. **b.** higher oxygen content than pure water. **c.** the same oxygen content as pure water. **d.** no demand for oxygen.	**8.** Why do microbes exist that are capable of cleaning up oil spills in the environment?	**9.** What features of fermentation make it particularly suited to producing organic compounds?

ASM Concept/ Competency	A. Bloom's Level 1, 2—Remember and Understand (Choose one)	B. Bloom's Level 3, 4—Apply and Analyze	C. Bloom's Level 5, 6—Evaluate and Create
Information Flow and Genetics	**10.** In the ocean, bacteriophages have been found that "borrow" genes from their bacterial hosts to increase the amount of **a.** viruses **b.** photosynthesis **c.** sunlight **d.** plankton	**11.** Until the 1993 outbreak, hantaviruses were only known to cause hemorrhagic fevers accompanied by kidney damage. Genetic changes may have made it capable of causing lung disease. What aspect of its genetic makeup would suggest that it mutates easily?	**12.** With what you have learned about increasingly affordable high-throughput DNA sequencing for diagnosing infectious diseases, speculate on the pros and cons of using that technology to replace fecal coliform screening.
Microbial Systems	**13.** The rhizosphere consists of **a.** bacteria **b.** fungi **c.** protozoa **d.** all of the above.	**14.** Explain how a warming climate is likely to impact the epidemiology of mosquito-borne diseases.	**15.** Conduct research to find out where the microorganisms used in wastewater treatment come from.
Impact of Microorganisms	**16.** The action of already-present microbes in an environment to clean up pollution and chemical spills is called **a.** bioremediation. **b.** bioprocessing. **c.** biostimulation. **d.** biofilming.	**17.** Create an infographic or visual representation that describes the similarities and differences between *E. coli* and other coliforms.	**18.** Speculate on how people thousands of years ago discovered that microbes could produce alcoholic beverages.
Scientific Thinking	**19.** What element has been removed in the blue line in figure 25.3, in contrast to the green line? **a.** fake data **b.** seasonal variation **c.** carbon **d.** oxygen	**20.** Defend or refute this statement with information from this chapter: The environment would be a safer place for human health if all microbes could be eliminated from it.	**21.** Research the Coalition for Epidemic Preparedness Innovations, and write a paragraph describing the experiences that form the group's motivation for creating vaccines for emerging diseases.

Answers to the multiple-choice questions appear in Appendix A.

Visual Connections

These questions use visual images to connect content within and between chapters.

1. **From chapter 3, figure 3.5*b*.** If this MacConkey agar plate was inoculated with well water, would you report that coliforms were present in the water?

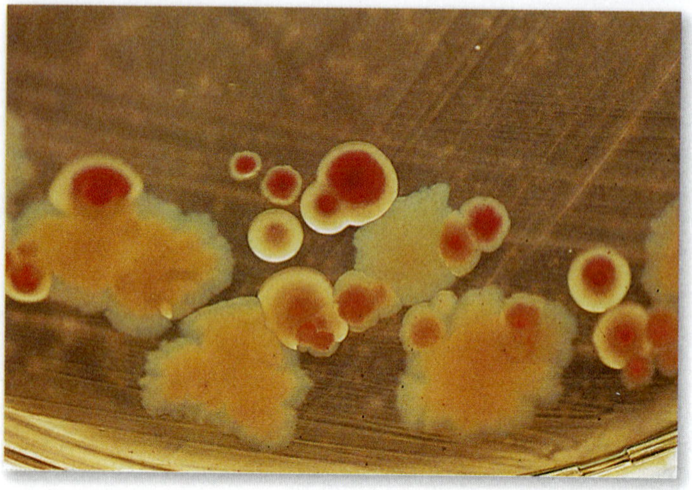

Kathy Park Talaro

2. **From chapter 10.** During which portion of metabolism shown here does fermentation occur?

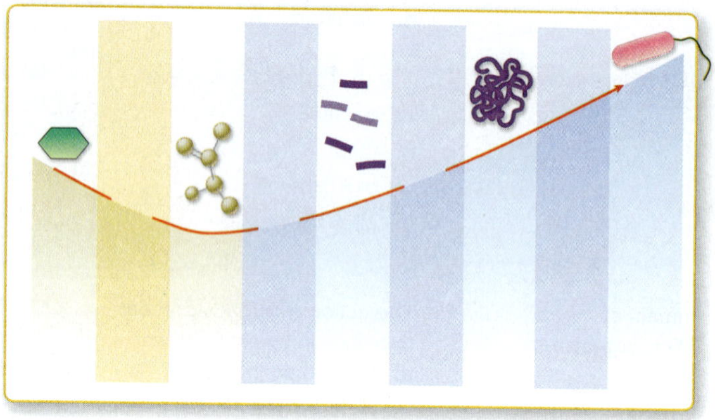

High Impact Study

These terms and concepts are most critical for your understanding of this chapter—and may be the most difficult. Have you mastered them?

Concepts

- ☐ Three stages of sewage treatment
- ☐ Role of fermentation in food and beverage creation
- ☐ Carbon cycle
- ☐ Nitrogen cycle
- ☐ Environmental metagenomics
- ☐ Red tides
- ☐ One health
- ☐ Spillover event
- ☐ Evolution of virulence

Terms

- ☐ Applied microbiology
- ☐ Biotechnology
- ☐ Bioremediation
- ☐ Coliforms
- ☐ Ecology
- ☐ Plastisphere
- ☐ Nitrification
- ☐ Rhizosphere
- ☐ Eutrophication

Design Element: (College students): Caia Image/Image Source

Viaframe/Corbis/Getty Images

APPENDIX A

Answers to Multiple-Choice Questions in SmartGrid

Chapter 1
1. a
4. c
7. d
10. d
13. d
16. c
19. c

Chapter 2
1. c
4. b
7. a
10. a
13. b
16. b
19. c

Chapter 3
1. d
4. a
7. d
10. a
13. b
16. b
19. d

Chapter 4
1. d
4. b
7. c
10. b
13. d
16. d
19. b

Chapter 5
1. a
4. b
7. c
10. c
13. a
16. c
19. a

Chapter 6
1. c
4. d
7. b
10. b
13. a
16. c
19. d

Chapter 7
1. d
4. a
7. a
10. c
13. a
16. a
19. c

Chapter 8
1. a
4. d
7. b
10. d
13. d
16. d
19. c

Chapter 9
1. d
4. a
7. a
10. c
13. b
16. a
19. b

Chapter 10
1. b
4. a
7. b
10. d
13. d
16. c
19. d

Chapter 11
1. a
4. c
7. c
10. c
13. d
16. a
19. d

Chapter 12
1. d
4. b
7. b
10. b
13. d
16. b
19. b

Chapter 13
1. d
4. d
7. d
10. b
13. d
16. c
19. b

Chapter 14
1. d
4. b
7. d
10. b
13. a
16. c
19. c

Chapter 15
1. b
4. d
7. d
10. b
13. b
16. a
19. a

Chapter 16
1. b
4. a
7. c
10. c
13. d
16. c
19. b

Chapter 17
1. c
4. d
7. a
10. c
13. d
16. c
19. b

Chapter 18
1. a
4. c
7. a
10. a
13. d
16. a
19. a

Chapter 19
1. b
4. a
7. b
10. b
13. d
16. a
19. c

Chapter 20
1. d
4. a
7. c
10. a
13. b
16. a
19. d

Chapter 21
1. b
4. a
7. c
10. d
13. d
16. b
19. b

Chapter 22
1. c
4. c
7. a
10. d
13. d
16. c
19. a

Chapter 23
1. b
4. b
7. a
10. d
13. d
16. c
19. a

Chapter 24
1. d
4. c
7. c
10. b
13. a
16. b
19. c

Chapter 25
1. b
4. d
7. a
10. b
13. d
16. a
19. b

APPENDIX B

Viaframe/Corbis/Getty Images

Exponents

Dealing with concepts such as microbial growth often requires working with numbers in the billions, trillions, and even greater. Exponents are a mathematical shorthand for expressing such numbers. The exponent of a number indicates how many times (designated by a superscript) that number is multiplied by itself. These exponents are also called common *logarithms,* or logs. The following chart, based on multiples of 10, summarizes this system.

Exponential Notation for Base 10

Number	Quantity	Exponential Notation*	Number Arrived at By:	One Followed By:
1	One	10^0	Numbers raised to zero power are equal to one	No zeros
10	Ten	10^1**	10×1	One zero
100	Hundred	10^2	10×10	Two zeros
1,000	Thousand	10^3	$10 \times 10 \times 10$	Three zeros
10,000	Ten thousand	10^4	$10 \times 10 \times 10 \times 10$	Four zeros
100,000	Hundred thousand	10^5	$10 \times 10 \times 10 \times 10 \times 10$	Five zeros
1,000,000	Million	10^6	10 times itself 6 times	Six zeros
1,000,000,000	Billion	10^9	10 times itself 9 times	Nine zeros
1,000,000,000,000	Trillion	10^{12}	10 times itself 12 times	Twelve zeros
1,000,000,000,000,000	Quadrillion	10^{15}	10 times itself 15 times	Fifteen zeros
1,000,000,000,000,000,000	Quintillion	10^{18}	10 times itself 18 times	Eighteen zeros

Other large numbers are sextillion (10^{21}), septillion (10^{24}), and octillion (10^{27}).

*The proper way to say the numbers in this column is 10 raised to the *n*th power, where *n* is the exponent. The numbers in this column can also be represented as 1×10^n, but for brevity, the $1 \times$ can be omitted.

**The exponent 1 is usually omitted.

Converting Numbers to Exponent Form

As the chart shows, using exponents to express numbers can be very economical. When simple multiples of 10 are used, the exponent is always equal to the number of zeros that follow the 1, but this rule will not work with numbers that are more varied. Other large whole numbers can be converted to exponent form by the following operation: First, move the decimal (if there isn't one, we assume it to be at the end of the number) to the left until it sits just behind the first number in the series (example: 3568. = 3.568). Then count the number of spaces (digits) the decimal has moved; that number will be the exponent. (The decimal has moved from 8. to 3., or 3 spaces.) In the final notation, the converted number is multiplied by 10 with its appropriate exponent: 3568 is now 3.568×10^3.

Rounding Off Numbers

The notation in the previous example has not actually been shortened, but it can be reduced further by rounding off the decimal fraction to the nearest thousandth (three digits), hundredth (two digits), or tenth (one digit). To round off a number, drop its last digit and either increase the one next to it or leave it as it is. If the number dropped is 5, 6, 7, 8, or 9, the subsequent digit is increased by one (rounded up); if it is 0, 1, 2, 3, or 4, the subsequent digit remains as is. Using the example of 3.528, removing the 8 rounds off the 2 to a 3 and produces 3.53 (two digits). If further rounding is desired, the same rule of thumb applies, and the number becomes 3.5 (one digit). Other examples of exponential conversions follow.

Number	Is the Same As	Rounded Off, Placed in Exponent Form
16,825.	$1.6825 \times 10 \times 10 \times 10 \times 10$	1.7×10^4
957,654.	$9.57654 \times 10 \times 10 \times 10 \times 10 \times 10$	9.58×10^5
2,855,000.	$2.855000 \times 10 \times 10 \times 10 \times 10 \times 10 \times 10$	2.86×10^6

Negative Exponents

The numbers we have been using so far are greater than 1 and are represented by positive exponents. But the correct notation for numbers less than 1 involves negative exponents (10 raised to a negative power, or 10^{-n}). A negative exponent says that the number has been divided by a certain power of 10 (10, 100, 1,000). This usage is handy when working with concepts such as pH that are based on very small numbers otherwise needing to be represented by large decimal fractions—for example, 0.003528. Converting this and other such numbers to exponential notation is basically similar to converting positive numbers, except that you work from left to right and the exponent is negative. Using the example of 0.003528, first convert the number to a whole integer followed by a decimal fraction and keep track of the number of spaces the decimal point moves (example: 0.003528 = 3.528). The decimal has moved three spaces from its original position, so the finished product is 3.528×10^{-3}. Other examples follow.

Number	Is the Same As	Rounded Off, Expressed with Exponents
0.0005923	$\frac{5.923}{10 \times 10 \times 10 \times 10}$	5.92×10^{-4}
0.00007295	$\frac{7.295}{10 \times 10 \times 10 \times 10 \times 10}$	7.3×10^{-5}

Viaframe/Corbis/Getty Images

APPENDIX C

An Introduction to Concept Mapping

Concept maps are visual tools for presenting and organizing what you have learned. They can take the place of an outline, though for most people they contain much more meaning and can illustrate connections and interconnections in ways that ordinary outlines cannot. They are also very flexible. If you are creating a concept map, there are nearly an infinite number of ways that it can be put together and still be "correct." Concept maps are also a way to incorporate and exploit your own creative impulses so that you are not stuck inside a rigid framework but can express your understanding of concepts and their connections in ways that make sense to you.

This is an example of a relatively large concept map:

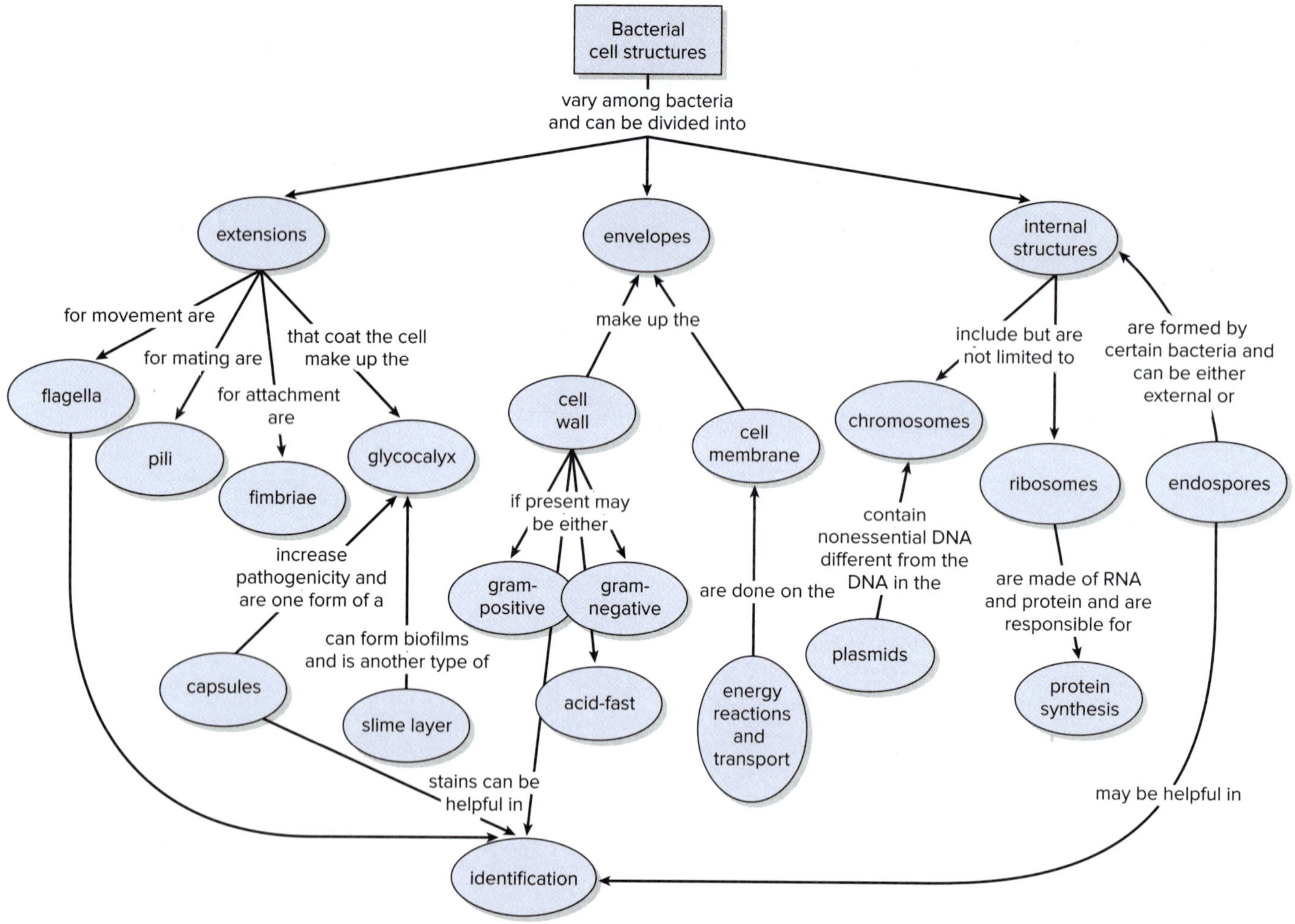
Bacterial cell structures
vary among bacteria and can be divided into
extensions
envelopes
internal structures
for movement are
for mating are
that coat the cell make up the
for attachment are
flagella
pili
fimbriae
glycocalyx
increase pathogenicity and are one form of a
capsules
can form biofilms and is another type of
slime layer
make up the
cell wall
cell membrane
if present may be either
gram-positive
gram-negative
acid-fast
are done on the
energy reactions and transport
stains can be helpful in
identification
include but are not limited to
chromosomes
contain nonessential DNA different from the DNA in the
plasmids
ribosomes
are made of RNA and protein and are responsible for
protein synthesis
are formed by certain bacteria and can be either external or
endospores
may be helpful in

There are a lot of ways to work with concept maps, such as using them as an introductory overview of material or using them as an evaluation tool. There are even software programs that enable concept mappers to create elaborate maps, complete with sound bites and photos. Some of these will even convert an outline into a concept map for you.

All concept maps are made of two basic components:

1. Boxes or circles, each containing a single *concept,* which is most often a noun. The boxes are arranged on the page in vertical, horizontal, or diagonal rows or arrangements. They may also be arranged in a more free-form manner. You can also draw an object (like a bacterium or a body part) instead of a box and put the concept word inside of that!
2. Connecting lines that join each concept box to at least one other box. Each connecting line has a word or a phrase associated with it—a linking word. These words/phrases are almost never nouns—but are verbs (like "requires") or adjectives or adverbs (like "underneath").

In the end, a picture is created that maps what you know about a subject. It illustrates which concepts are bigger and which are details. It illustrates that multiple concepts may be connected. Experts say that concept maps can lead us to conclude that all concepts in a subject can be connected in some way. This is true! And nowhere is it truer than in biology. The trick is to get used to finding the right connecting word to show how two concepts are, indeed, related. When you succeed, you will know the material in a deeper way than is possible by simply answering a single question or even a series of questions.

Many students report that their first experiences with concept mapping can be frustrating. But when they have invested some time in their first few concept maps, many of them find they can never "go back" to organizing information in linear ways. Maps can make the time fly when you are studying. And creating concept maps with a partner or a group is also a great way to review material in a meaningful way. Give concept maps a try. Let your creative side show!

Viaframe/Corbis/Getty Images

Glossary

A

abiogenesis The belief in spontaneous generation as a source of life.

abscess An inflamed, fibrous lesion enclosing a core of pus.

acid-fast stain A solution containing carbol fuchsin, which, when bound to lipids in the envelopes of *Mycobacterium* species, cannot be removed with an acid wash.

acidic A solution with a pH value below 7 on the pH scale.

acidic fermentation An anaerobic degradation of pyruvic acid that results in organic acid production.

acidophilic An organism that prefers environments of high acidity.

acquired immunodeficiency syndrome See *AIDS*.

actin Protein component of long filaments of protein arranged under the cell membrane of bacteria; contributes to cell shape and division.

actin cytoskeleton Part of cell cytoskeleton, fibers inside the cytoplasmic membrane.

actin filaments Long, thin, protein strands found throughout a eukaryotic cell—but mainly concentrated just inside the cell membrane.

activation energy The amount of energy required to overcome initial resistance to an enzymatic reaction.

active immunity Immunity acquired through direct stimulation of the immune system by antigen.

active site The specific region on an apoenzyme that binds substrate. The site for reaction catalysis.

active transport Nutrient transport method that requires carrier proteins in the membranes of the living cells and the expenditure of energy.

acute infection A condition that appears relatively quickly after exposure and is of short duration.

adenine (A) One of the nitrogen bases found in DNA and RNA, with a purine form.

adenosine deaminase (ADA) deficiency An immunodeficiency disorder and one type of SCIDs that is caused by an inborn error in the metabolism of adenine. The accumulation of adenine destroys both B and T lymphocytes.

adenosine triphosphate (ATP) A nucleotide that is the primary source of energy to cells.

adhesion The process by which microbes gain a more stable foothold at the portal of entry; often involves a specific interaction between the molecules on the microbial surface and the receptors on the host cell.

adjuvant In immunology, a chemical vehicle that enhances antigenicity, presumably by prolonging antigen retention at the injection site.

adsorption A process of adhering one molecule onto the surface of another molecule.

aerobe A microorganism that lives and grows in the presence of free gaseous oxygen (O_2).

aerobic respiration Respiration in which the final electron acceptor in the electron transport chain is oxygen (O_2).

aerosols Suspensions of fine dust or moisture particles in the air that contain live pathogens.

aerotolerant The state of not utilizing oxygen but not being harmed by it.

agammaglobulinemia Also called *hypogammaglobulinemia*. The absence of or severely reduced levels of antibodies in serum.

agar A polysaccharide found in seaweed and commonly used to prepare solid culture media.

agglutination The aggregation by antibodies of suspended cells or similar-size particles (agglutinogens) into clumps that settle.

agranulocyte One form of leukocyte (white blood cell) having globular, nonlobed nuclei and lacking prominent cytoplasmic granules.

AIDS Acquired immunodeficiency syndrome. The complex of signs and symptoms characteristic of the late phase of human immunodeficiency virus (HIV) infection.

akaryote A microbe that has no nucleus.

alcoholic fermentation An anaerobic degradation of pyruvic acid that results in alcohol production.

algae Photosynthetic, plantlike organisms that generally lack the complex structure of plants; they may be single-celled or multicellular and inhabit diverse habitats such as marine and freshwater environments, glaciers, and hot springs.

allele A gene that occupies the same location as other alternative (allelic) genes on paired chromosomes.

allergen A substance that provokes an allergic response.

allergic rhinitis Respiratory symptoms caused by immune sensitivity to environmental antigens

allergy The altered, usually exaggerated, immune response to an allergen. Also called *hypersensitivity*.

alloantigen An antigen that is present in some but not all members of the same species.

allograft Relatively compatible tissue exchange between nonidentical members of the same species. Also called *homograft*.

allosteric Pertaining to the altered activity of an enzyme due to the binding of a molecule to a region other than the enzyme's active site.

Ames test A method for detecting mutagenic and potentially carcinogenic agents based upon the genetic alteration of nutritionally defective bacteria.

amino acids The building blocks of protein. Amino acids exist in 20 naturally occurring forms that impart different characteristics to the various proteins they compose.

aminoglycoside A complex group of drugs derived from soil actinomycetes that impairs ribosome function and has antibiotic potential. Example: streptomycin.

ammonification Phase of the nitrogen cycle in which ammonia is released from decomposing organic material.

amphibolism Pertaining to the metabolic pathways that serve multiple functions in the breakdown, synthesis, and conversion of metabolites.

amphipathic Relating to a compound that has contrasting characteristics, such as hydrophilic-hydrophobic or acid-base.

amphitrichous Having a single flagellum or a tuft of flagella at opposite poles of a microbial cell.

anabolism The energy-consuming process of incorporating nutrients into protoplasm through biosynthesis.

anaerobe A microorganism that grows best, or exclusively, in the absence of oxygen.

anaerobic respiration Respiration in which the final electron acceptor in the electron transport chain is an inorganic molecule containing sulfate, nitrate, nitrite, carbonate, and so on.

anamnestic response In immunology, an augmented response or memory related to a prior stimulation of the immune system by antigen. It boosts the levels of immune substances.

anaphylaxis The unusual or exaggerated allergic reaction to antigen that leads to severe respiratory and cardiac complications.

anion A negatively charged ion.

annotating In the context of genome sequencing, it is the process of assigning biological function to genetic sequence.

anoxygenic Non-oxygen-producing.

antagonism Relationship in which microorganisms compete for survival in a common environment by taking actions that inhibit or destroy another organism.

antibody A large protein molecule evoked in response to an antigen that interacts specifically with that antigen.

anticodon The trinucleotide sequence of transfer RNA that is complementary to the trinucleotide sequence of messenger RNA (the codon).

antigen (Ag) Any cell, particle, or chemical that induces a specific immune response by B cells or T cells and can stimulate resistance to an infection or a toxin. See *immunogen.*

antigen binding site Specific region at the ends of the antibody molecule that recognizes specific antigens. These sites have numerous shapes to fit a wide variety of antigens.

antigenic drift Minor antigenic changes in the influenza A virus due to mutations in the spikes' genes.

antigenic shift Major changes in the influenza A virus due to recombination of viral strains from two different host species.

antihistamine A drug that counters the action of histamine and is useful in allergy treatment.

antimicrobial peptides (AMPs) Short protein molecules found in epithelial cells; have the ability to kill bacteria.

antiparallel A description of the two strands of DNA, which are parallel to each other, but the orientation of the deoxyribose and phosphate groups run in the opposite directions, with the 5′ carbon at the top of the leading strand and the 3′ carbon at the top of the lagging strand.

antiseptic A growth-inhibiting agent used on tissues to prevent infection.

antitoxin Globulin fraction of serum that neutralizes a specific toxin. Also refers to the specific antitoxin antibody itself.

apicomplexans A group of protozoans that lack locomotion in the mature state.

apoenzyme The protein part of an enzyme, as opposed to the nonprotein or inorganic cofactors.

apoptosis The genetically programmed death of cells that is both a natural process of development and the body's means of destroying abnormal or infected cells.

appendages Accessory structures that sprout from the surface of bacteria. They can be divided into two major groups: those that provide motility and those that enable adhesion.

aqueous Referring to solutions in which water is used as the solvent.

aquifer A subterranean water-bearing stratum of permeable rock, sand, or gravel.

archaea Prokaryotic single-celled organisms of primitive origin that have unusual anatomy, physiology, and genetics and live in harsh habitats; when capitalized **(Archaea),** the term refers to one of the three domains of living organisms as proposed by Woese.

arthroconidia Reproductive body of *Coccidioides immitis;* also *arthrospore.*

artificial immunity Immunity that is induced as a medical intervention, either by exposing an individual to an antigen or administering immune substances to the individual.

aseptic technique Methods of handling microbial cultures, patient specimens, and other sources of microbes in a way that prevents infection of the handler and others who may be exposed.

assay medium Microbiological medium used to test the effects of specific treatments to bacteria, such as antibiotic or disinfectant treatment.

assembly (viral) The step in viral multiplication in which capsids and genetic material are packaged into virions.

asthma A type of chronic local allergy in which the airways become constricted and produce excess mucus in reaction to allergens, exercise, stress, or cold temperatures.

astromicrobiology A branch of microbiology that studies the potential for and the possible role of microorganisms in space and on other planets.

asymptomatic An infection that produces no noticeable symptoms even though the microbe is active in the host tissue.

asymptomatic carrier A person with an inapparent infection who shows no symptoms of being infected yet is able to pass the disease agent on to others.

atom The smallest particle of an element to retain all the properties of that element.

atopy Allergic reaction classified as type I, with a strong familial relationship; caused by allergens such as pollen, insect venom, food, and dander; involves IgE antibody; includes symptoms of hay fever, asthma, and skin rash.

ATP synthase A unique enzyme located in the mitochondrial cristae and chloroplast grana that harnesses the flux of hydrogen ions to the synthesis of ATP.

AUG (start codon) The codon that signals the point at which translation of a messenger RNA molecule is to begin.

autoantibody An "anti-self" antibody having an affinity for tissue antigens of the subject in which it is formed.

autoclave A sterilization chamber that allows the use of steam under pressure to sterilize materials. The most common temperature/pressure combination for an autoclave is 121°C and 15 psi.

autograft Tissue or organ surgically transplanted to another site on the same subject.

autoimmune disease The pathologic condition arising from the production of antibodies against autoantigens. Example: rheumatoid arthritis. Also called *autoimmunity.*

autotroph A microorganism that requires only inorganic nutrients and whose sole source of carbon is carbon dioxide.

axenic A sterile state such as a pure culture. An axenic animal is born and raised in a germ-free environment. See *gnotobiotic.*

axial filament A type of flagellum (called an *endoflagellum*) that lies in the periplasmic space of spirochetes and is responsible for locomotion. Also called *periplasmic flagellum.*

B

bacitracin Antibiotic that targets the bacterial cell wall; component of over-the-counter topical antimicrobial ointments.

back-mutation A mutation that counteracts an earlier mutation, resulting in the restoration of the original DNA sequence.

bacteremia The presence of viable bacteria in circulating blood.

Bacteria When capitalized can refer to one of the three domains of living organisms proposed by Woese, containing all nonarchaea prokaryotes.

bacteria (singular, *bacterium*) Category of prokaryotes with peptidoglycan in their cell walls and circular chromosome(s). This group of small cells is widely distributed in the earth's habitats.

bacterial chromosome A circular body in bacteria that contains the primary genetic material. Also called *nucleoid.*

bactericide An agent that kills bacteria.

bacteristatic Any process or agent that inhibits bacterial growth.

bacterium A tiny unicellular prokaryotic organism that usually reproduces by binary fission and usually has a peptidoglycan cell wall, has various shapes, and can be found in virtually any environment.

barophile A microorganism that thrives under high (usually hydrostatic) pressure.

basement membrane A thin layer (1–6 μm) of protein and polysaccharide found at the base of epithelial tissues.

basic A solution with a pH value above 7 on the pH scale.

beta oxidation The degradation of long-chain fatty acids. Two-carbon fragments are formed as a result of enzymatic attack directed against the second or beta carbon of the hydrocarbon chain. Aided by coenzyme A, the fragments enter the Krebs cycle and are processed for ATP synthesis.

β-lactam A four-member ring cyclic amide that characterizes a certain class of antibiotics, such as penicillin.

β-lactamase An enzyme secreted by certain bacteria that cleaves the beta-lactam ring of penicillin and cephalosporin and thus provides for resistance against the antibiotic. See *penicillinase.*

binary fission The formation of two new cells of approximately equal size as the result of parent cell division.

binomial system Scientific method of assigning names to organisms that employs two names to identify every organism—genus name plus species name.

biochemistry The study of organic compounds produced by (or components of) living things. The four main categories of biochemicals are carbohydrates, lipids, proteins, and nucleic acids.

biofilm A complex association that arises from a mixture of microorganisms growing together on the surface of a habitat.

biogeochemical cycle A process by which matter is converted from organic to inorganic form and returned to various nonliving reservoirs on earth (air, rocks, and water), where it becomes available for reuse by living things. Elements such as carbon, nitrogen, and phosphorus are constantly cycled in this manner.

bioinformatics The use of computer software to determine the function of genes through analysis of the DNA and protein sequences.

biological vector An animal that not only transports an infectious agent but plays a role in the life cycle of the pathogen, serving as a site in which it can multiply or complete its life cycle. It is usually an alternate host to the pathogen.

bioremediation Decomposition of harmful chemicals by microbes or consortia of microbes.

biotechnology The use of microbes or their products in the commercial or industrial realm.

blocking antibody The IgG class of immunoglobulins that competes with IgE antibody for allergens, thus blocking the degranulation of basophils and mast cells.

blood cells Cellular components of the blood consisting of red blood cells, primarily responsible for the transport of oxygen and carbon dioxide, and white blood cells, primarily responsible for host defense and immune reactions.

blood-brain barrier Decreased permeability of the walls of blood vessels in the brain, restricting access to that compartment.

botulinum toxin An exotoxin produced by *Clostridium botulinum* that causes flaccid muscle paralysis.

bradykinin An active polypeptide that is a potent vasodilator released from IgE-coated mast cells during anaphylaxis.

broad-spectrum Denotes drugs that have an effect on a wide variety of microorganisms.

bubo The swelling of one or more lymph nodes due to inflammation.

bubonic plague The form of plague in which bacterial growth is primarily restricted to the lymph and is characterized by the appearance of a swollen lymph node referred to as a *bubo.*

budding See *exocytosis.*

bulbar poliomyelitis Complication of polio infection in which the brain stem, medulla, or cranial nerves are affected. Leads to loss of respiratory control and paralysis of the trunk and limbs.

bullous Consisting of fluid-filled blisters.

C

calculus Dental deposit formed when plaque becomes mineralized with calcium and phosphate crystals. Also called *tartar.*

Calvin cycle A series of reactions in the second phase of photosynthesis that generates glucose.

capsid The protein covering of a virus's nucleic acid core. Capsids exhibit symmetry due to the regular arrangement of subunits called *capsomeres.* See *icosahedron.*

capsomere A subunit of the virus capsid shaped as a triangle or disc.

capsule In bacteria, the loose, gel-like covering or slime made chiefly of polysaccharides. This layer is protective and can be associated with virulence.

capsule staining Any staining method that highlights the outermost polysaccharide and/or protein structure on a bacterial, fungal, or protozoal cell.

carbohydrate A compound containing primarily carbon, hydrogen, and oxygen in a 1:2:1 ratio.

carbohydrate fermentation medium A growth medium that contains sugars that are converted to acids through fermentation. Usually contains a pH indicator to detect acid protection.

carbon cycle That pathway taken by carbon from its abiotic source to its use by producers to form organic compounds (biotic), followed by the breakdown of biotic compounds and their release to a nonliving reservoir in the environment (mostly carbon dioxide in the atmosphere).

carbon fixation Reactions in photosynthesis that incorporate inorganic carbon dioxide into organic compounds such as sugars. This occurs during the Calvin cycle and uses energy generated by the light reactions. This process is the source of all production on earth.

carbuncle A deep staphylococcal abscess joining several neighboring hair follicles.

carotenoid Yellow, orange, or red photosynthetic pigments.

carrier A person who harbors infections and inconspicuously spreads them to others. Also, a chemical agent that can accept an atom, chemical radical, or subatomic particle from one compound and pass it on to another.

catabolism The chemical breakdown of complex compounds into simpler units to be used in cell metabolism.

catalyst A substance that alters the rate of a reaction without being consumed or permanently changed by it. In cells, enzymes are catalysts.

catalytic site The niche in an enzyme where the substrate is converted to the product (also *active site*).

catarrhal A term referring to the secretion of mucus or fluids; term for the first stage of pertussis.

cation A positively charged ion.

cell An individual membrane-bound living entity; the smallest unit capable of an independent existence.

cell membrane A thin sheet of lipid and protein that surrounds the cytoplasm and controls the flow of materials into and out of the cell pool.

cell wall In bacteria, a rigid structure made of peptidoglycan that lies just outside the cytoplasmic membrane; eukaryotes also have a cell wall, but it may be composed of a variety of materials.

cell-mediated immunity The type of immune responses brought about by T cells, such as cytotoxic and helper effects.

cellulose A long, fibrous polymer composed of β-glucose; one of the most common substances on earth.

cercaria The free-swimming larva of the schistosome trematode that emerges from the snail host and can penetrate human skin, causing schistosomiasis.

cestode The common name for tapeworms that parasitize humans and domestic animals.

chancre The primary sore of syphilis that forms at the site of penetration by *Treponema pallidum.* It begins as a hard, dull red, painless papule that erodes from the center.

chemical bond A link formed between molecules when two or more atoms share, donate, or accept electrons.

chemical mediators Small molecules that are released during inflammation and specific immune reactions that allow communication between the cells of the immune system and facilitate surveillance, recognition, and attack.

chemiosmosis The generation of a concentration gradient of hydrogen ions (called the *proton motive force*) by the pumping of hydrogen ions to the outer side of the membrane during electron transport.

chemoautotroph An organism that relies upon inorganic chemicals for its energy and carbon dioxide for its carbon. Also called a *chemolithotroph.*

chemoheterotroph Microorganisms that derive their nutritional needs from organic compounds.

chemolithoautotrophs Bacteria that rely on inorganic minerals to supply their nutritional needs. Sometimes referred to as *chemoautotrophs.*

chemostat A growth chamber with an outflow that is equal to the continuous inflow of nutrient media. This steady-state growth device is used to study such events as cell division, mutation rates, and enzyme regulation.

chemotaxis The tendency of organisms to move in response to a chemical gradient (toward an attractant or to avoid adverse stimuli).

chemotroph Organism that oxidizes compounds to feed on nutrients.

chitin A polysaccharide similar to cellulose in chemical structure. This polymer makes up the horny substance of the exoskeletons of arthropods and certain fungi.

chlorophyll A group of mostly green pigments that are used by photosynthetic eukaryotic organisms and cyanobacteria to trap light energy to use in making chemical bonds.

chloroplast An organelle containing chlorophyll that is found in photosynthetic eukaryotes.

cholesterol Best-known member of a group of lipids called *steroids.* Cholesterol is commonly found in cell membranes and animal hormones.

chromatin The genetic material of the nucleus. Chromatin is made up of nucleic acid and stains readily with certain dyes.

chromosome The tightly coiled bodies in cells that are the primary sites of genes.

chronic carrier An individual who has recovered from an initial infection but continues to harbor and shed infectious agents for a long period of time.

chronic infection A condition that appears slowly, can last a long time, and can have muted symptoms.

cilium (plural, *cilia*) Eukaryotic structure similar to a flagellum that propels a protozoan through the environment.

class In the levels of classification, the division of organisms that follows phylum.

clonal deletion The process of destroying lymphocytes that are responsive toward "self" antigens.

clonal selection The recognition by a single clone of a B or T cell of a foreign antigen.

clone A colony of cells (or group of organisms) derived from a single cell (or single organism) by asexual reproduction. All units share identical characteristics. *Clone* is also used as a verb to refer to the process of producing a genetically identical population of cells or genes.

cloning host An organism such as a bacterium or a yeast that receives and replicates a foreign piece of DNA inserted during a genetic engineering experiment.

coagulase A plasma-clotting enzyme secreted by *Staphylococcus aureus.* It contributes to virulence and is involved in forming a fibrin wall that surrounds staphylococcal lesions.

coccus A spherical-shaped bacterial cell.

codon A specific sequence of three nucleotides in mRNA (or the sense strand of DNA) that constitutes the genetic code for a particular amino acid.

coenzyme A complex organic molecule, several of which are derived from vitamins (e.g., nicotinamide, riboflavin). A coenzyme operates in conjunction with an enzyme. Coenzymes serve as transient carriers of specific atoms or functional groups during metabolic reactions.

coevolution A biological process whereby a change in the genetic composition in one organism leads to a change in the genetics of another organism.

cofactor An enzyme accessory. It can be organic, such as coenzymes, or inorganic, such as Fe^{2+}, Mn^{2+}, or Zn^{2+} ions.

cold sterilization The use of nonheating methods such as radiation or filtration to sterilize materials.

collectins Host pattern recognition receptors that circulate throughout the body and bind to pathogen-associated molecular patterns, or PAMPs, and mark them for destruction.

colonize The ability of microbes to become resident on a particular host (such as the human body).

colony A macroscopic cluster of cells appearing on a solid medium, each arising from the multiplication of a single cell.

colostrum A thin secretion from the breast that precedes the production of milk.

commensalism An unequal relationship in which one species derives benefit without harming the other.

common-source epidemic An outbreak of disease in which all affected individuals were exposed to a single source of the pathogen, even if they were exposed at different times.

communicable Capable of being transmitted from one individual to another.

competent Referring to bacterial cells that are capable of absorbing free DNA in their environment either naturally or through induction by exposure to chemicals or electrical currents.

competition Situation in which two molecules of similar shape can bind to the same binding site on a carrier protein.

competitive inhibition Control process that relies on the ability of metabolic analogs to control microbial growth by successfully competing with a necessary enzyme to halt the growth of bacterial cells.

complement In immunology, serum protein components that act in a definite sequence when set in motion either by an antigen-antibody complex or by factors of the alternative (properdin) pathway.

complementary DNA (cDNA) DNA created by using reverse transcriptase to synthesize DNA from RNA templates.

compounds Molecules that are a combination of two or more different elements.

concentration The expression of the amount of a solute dissolved in a certain amount of solvent. It may be defined by weight, volume, or percentage.

condyloma acuminata Extensive, branched masses of genital warts caused by infection with human papillomavirus.

congenital rubella Transmission of the rubella virus to a fetus *in utero.* Injury to the fetus is generally much more serious than it is to the mother.

conidia Asexual fungal spores shed as free units from the tips of fertile hyphae.

conidiospore A type of asexual spore in fungi; not enclosed in a sac.

conjugation In bacteria, the contact between donor and recipient cells associated with the transfer of genetic material such as plasmids. Can involve special (sex) pili. Also a form of sexual recombination in ciliated protozoans.

conjunctiva The thin fluid-secreting tissue that covers the eye and lines the eyelid.

constitutive enzyme An enzyme present in bacterial cells in constant amounts, regardless of the presence of substrate. Enzymes of the central catabolic pathways are typical examples.

contagious Communicable; transmissible by direct contact with infected people and their fresh secretions or excretions.

contaminant An impurity; any undesirable material or organism.

contaminated culture A medium that once held a pure (single or mixed) culture but now contains unwanted microorganisms.

contigs Contiguous sets of overlapping nucleotide sequences determined by sequencing fragments of a genome in a genomic library.

continuation The phase in the course of an infection in which *either* the organism lingers for months, years, or indefinitely after the patient is completely well or the organism is gone but symptoms continue.

convalescent carriers People who have recovered from an infectious disease but are still carrying the infectious agent and may be capable of spreading it.

convalescent period The phase after the period of invasion in which the patient's immune system responds to the infection, signs and symptoms gradually decline, and the patient's health gradually returns.

corepressor A molecule that combines with inactive repressor to form active repressor, which attaches to the operator gene site and inhibits the activity of structural genes subordinate to the operator.

cornea The transparent, dome-shaped tissue covering the iris, pupil, and anterior chamber of the eye composed of five to six layers of quickly regenerating epithelial cells.

cortex The outermost layer of a lymph node.

covalent bond A chemical bond formed by the sharing of electrons between two atoms.

Creutzfeldt-Jakob disease (CJD) A spongiform encephalopathy caused by infection with a prion. The disease is marked by dementia, impaired senses, and uncontrollable muscle contractions.

CRISPR Clustered regularly interspaced short palindromic repeats in DNA that are being used by scientists in genetic engineering applications.

crista The infolded inner membrane of a mitochondrion that is the site of the respiratory chain and oxidative phosphorylation.

cross-presentation The activation of both T cytotoxic and T helper cells in cell-mediated immunity.

culture The visible accumulation of microorganisms in or on a nutrient medium. Also, the propagation of microorganisms with various media.

curved One of the basic shapes of bacteria.

cyst The resistant, dormant but infectious form of protozoans. Can be important in spread of infectious agents such as *Entamoeba histolytica* and *Giardia lamblia.*

cysteine A sulfide-containing amino acid that usually produces covalent disulfide bonds in an amino acid sequence, contributing to the tertiary structure of the protein.

cysticercosis A condition in which larvae of the tapeworm *Taenia solium* become embedded in human tissues.

cytochrome A group of heme protein compounds whose chief role is in electron and/or hydrogen transport occurring in the last phase of aerobic respiration.

cytokine A chemical substance produced by white blood cells and tissue cells that regulates development, inflammation, and immunity.

cytokine storm An overwhelming release of immune-modulating chemicals known as cytokines.

cytopathic effects (CPEs) The degenerative changes in cells associated with virus infection.

cytoplasm Dense fluid encased by the cell membrane; the site of many of the cell's biochemical and synthetic activities.

cytoplasmic membrane Lipid bilayer that encloses the cytoplasm of bacterial cells.

cytosine (C) One of the nitrogen bases found in DNA and RNA, with a pyrimidine form.

cytotoxicity The ability to kill cells; in immunology, certain T cells are called *cytotoxic T cells* because they kill other cells.

D

deamination The removal of an amino group from an amino acid.

death phase End of the cell growth due to lack of nutrition, depletion of environment, and accumulation of wastes. Population of cells begins to die.

decomposition The breakdown of dead matter and wastes into simple compounds that can be directed back into the natural cycle of living things.

definitive host The organism in which a parasite develops into its adult or sexually mature stage. Also called the *final host.*

degerm To physically remove surface oils, debris, and soil from skin to reduce the microbial load.

dehydration synthesis During the formation of a carbohydrate bond, the step in which one carbon molecule gives up its OH group and the other loses the H from its OH group, thereby producing a water molecule. This process is common to all polymerization reactions.

denaturation The loss of normal characteristics resulting from some molecular alteration. Usually in reference to the action of heat or chemicals on proteins whose function depends upon an unaltered tertiary structure.

denitrification The end of the nitrogen cycle when nitrogen compounds are returned to the reservoir in the air.

deoxyribonucleic acid (DNA) The nucleic acid often referred to as the "double helix." DNA carries the master plan for an organism's heredity.

deoxyribose A 5-carbon sugar that is an important component of DNA.

dermatophytes A group of fungi that cause infections of the skin and other integument components. They survive by metabolizing keratin.

dermolytic Capable of damaging the skin.

desensitization See *hyposensitization.*

desiccation To dry thoroughly. To preserve by drying.

desquamate To shed the cuticle in scales; to peel off the outer layer of a surface.

diapedesis The migration of intact blood cells between endothelial cells of a blood vessel such as a venule.

dichotomous keys Flowcharts that offer two choices or pathways at each level.

differential medium A single substrate that discriminates between groups of microorganisms on the basis of differences in their appearance due to different chemical reactions.

differential stain A technique that utilizes two dyes to distinguish between different microbial groups or cell parts by color reaction.

differentiation The maturation process that cells go through that results in their ultimate set of characteristics.

diffusion The dispersal of molecules, ions, or microscopic particles propelled down a concentration gradient by spontaneous random motion to achieve a uniform distribution.

DiGeorge syndrome A birth defect usually caused by a missing or incomplete thymus that results in abnormally low or absent T cells and other developmental abnormalities.

dimorphic In mycology, the tendency of some pathogens to alter their growth form from mold to yeast in response to rising temperature.

dipicolinic acid An organic acid found in the walls of endospores; contributes to their extreme resistance to chemicals, drying, and heat.

direct (total) cell count 1. Counting total numbers of individual cells being viewed with magnification. 2. Counting isolated colonies of organisms growing on a plate of media as a way to determine population size.

disaccharide A sugar containing two monosaccharides. Example: sucrose (fructose + glucose).

disease Any deviation from health, as when the effects of microbial infection damage or disrupt tissues and organs.

division In the levels of classification, an alternate term for phylum.

DNA See *deoxyribonucleic acid.*

DNA polymerase Enzyme responsible for the replication of DNA. Several versions of the enzyme exist, each completing a unique portion of the replication process.

DNA profile A pattern of restriction enzyme fragments that is unique for an individual organism.

domain In the levels of classification, the broadest general category to which an organism is assigned. Members of a domain share only one or a few general characteristics.

doubling time Time required for a complete fission cycle—from parent cell to two new daughter cells. Also called *generation time.*

droplet nuclei The dried residue of fine droplets produced by mucus and saliva sprayed while sneezing and coughing. Droplet nuclei are less than 5 μm in diameter (large enough to bear a single bacterium and small enough to remain airborne for a long time) and can be carried by air currents. Droplet nuclei are drawn deep into the air passages.

drug resistance An adaptive response in which microorganisms begin to tolerate an amount of drug that would ordinarily be inhibitory.

dysentery Diarrheal illness in which stools contain blood and/or mucus.

dyspnea Difficulty in breathing.

E

ecosystem A collection of organisms together with its surrounding physical and chemical factors.

ectoplasm The outer, more viscous region of the cytoplasm of a phagocytic cell such as an amoeba. It contains microtubules but not granules or organelles.

eczema An acute or chronic allergy of the skin associated with itching and burning sensations. Typically, red, edematous, vesicular lesions erupt, leaving the skin scaly and sometimes hyperpigmented.

edema The accumulation of excess fluid in cells, tissues, or serous cavities. Also called *swelling.*

electrolyte Any compound that ionizes in solution and conducts current in an electrical field.

electron A negatively charged subatomic particle that is distributed around the nucleus in an atom.

electronegativity The tendency of an atom to attract a bonding pair of electrons.

element A substance comprising only one kind of atom that cannot be degraded into two or more substances without losing its chemical characteristics.

ELISA Abbreviation for **e**nzyme-**l**inked **i**mmuno**s**orbent **a**ssay, a very sensitive serological test used to detect antibodies in diseases such as AIDS.

emetic Inducing vomit.

empirical A discovery made through experience, not planned experimentation.

encephalitis An inflammation of the brain, usually caused by infection.

endemic disease A native disease that prevails continuously in a geographic region.

endergonic reaction A chemical reaction that occurs with the absorption and storage of surrounding energy. Antonym: *exergonic.*

endocytosis The process whereby solid and liquid materials are taken into the cell through membrane invagination and engulfment into a vesicle.

endoenzyme An intracellular enzyme, as opposed to enzymes that are secreted.

endogenous Originating or produced within an organism or one of its parts.

endoplasm The granular inner region of the cytoplasm of a eukaryotic cell that contains the nucleus, mitochondria, and vacuoles.

endoplasmic reticulum (ER) An intracellular network of flattened sacs or tubules with or without ribosomes on their surfaces.

endospore A small, dormant, resistant derivative of a bacterial cell that germinates under favorable growth conditions into a vegetative cell. The bacterial genera *Bacillus* and *Clostridium* are typical sporeformers.

endosymbiosis Relationship in which a microorganism resides within a host cell and provides a benefit to the host cell.

endotoxic shock A massive drop in blood pressure caused by the release of endotoxin from gram-negative bacteria multiplying in the bloodstream.

endotoxin A bacterial toxin that is not ordinarily released (as is exotoxin). Endotoxin is composed of a phospholipid-polysaccharide complex that is an integral part of gram-negative bacterial cell walls. Endotoxin can cause severe shock and fever.

enriched medium A nutrient medium supplemented with blood, serum, or some growth factor to promote the multiplication of fastidious microorganisms.

enterotoxin A bacterial toxin that specifically targets intestinal mucous membrane cells. Enterotoxigenic strains of *Escherichia coli* and *Staphylococcus aureus* are typical sources.

enumeration medium Microbiological medium that does not encourage growth and allows for the counting of microbes in food, water, or environmental samples.

enzyme A protein biocatalyst that facilitates metabolic reactions.

enzyme induction One of the controls on enzyme synthesis. This occurs when enzymes appear only when suitable substrates are present.

enzyme repression The inhibition of enzyme synthesis by the end product of a catabolic pathway.

eosinophil A leukocyte whose cytoplasmic granules readily stain with red eosin dye.

eosinophilia Marked increase in the number of eosinophils in circulating blood.

epidemic A sudden and simultaneous outbreak or increase in the number of cases of disease in a community.

epigenetic Referring to changes in the way DNA is transcribed, not actual changes in the DNA sequence.

epimutation The change in phenotype brought about by an epigenetic change.

epitope The precise molecular group of an antigen that defines its specificity and triggers the immune response.

Epstein-Barr virus (EBV) Herpesvirus linked to infectious mononucleosis, Burkitt's lymphoma, and nasopharyngeal carcinoma.

erysipelas An acute, sharply defined inflammatory disease specifically caused by hemolytic *Streptococcus*. The eruption is limited to the skin but can be complicated by serious systemic symptoms.

erythrogenic toxin An exotoxin produced by lysogenized group A strains of β-hemolytic streptococci that is responsible for the severe fever and rash of scarlet fever in the nonimmune individual. Also called a *pyrogenic toxin.*

eschar A dark, sloughing scab that is the lesion of anthrax and certain rickettsioses.

essential nutrient Any ingredient such as a certain amino acid, fatty acid, vitamin, or mineral that cannot be formed by an organism and must be supplied in the diet. A growth factor.

ethylene oxide A potent, highly water-soluble gas invaluable for gaseous sterilization of heat-sensitive objects such as plastics, surgical and diagnostic appliances, and spices.

etiologic agent The microbial cause of disease; the pathogen.

Eukarya One of the three domains (sometimes called *superkingdoms*) of living organisms, as proposed by Woese; contains all eukaryotic organisms.

eukaryote A member of the domain Eukarya whose cells have a well-defined nucleus and membrane-bound organelles; includes plants, animals, fungi, protozoa, and algae.

eutrophication The process whereby dissolved nutrients resulting from natural seasonal enrichment or industrial pollution of water cause overgrowth of algae and cyanobacteria to the detriment of fish and other large aquatic inhabitants.

evolution Scientific principle that states that living things change gradually through hundreds of millions of years, and these changes are expressed in structural and functional adaptations in each organism. Evolution presumes that those traits that favor survival are preserved and passed on to following generations, and those traits that do not favor survival are lost.

evolution of virulence The ways in which a microbe alters its genotype and phenotype in response to pressures from the environment, including plants, weather, humans, and other animals.

exanthem An eruption or rash of the skin.

exergonic A chemical reaction associated with the release of energy to the surroundings. Antonym: *endergonic.*

exocytosis The process that releases enveloped viruses from the membrane of the host's cytoplasm.

exoenzyme An extracellular enzyme chiefly for hydrolysis of nutrient macromolecules that are otherwise impervious to the cell membrane. It functions in saprobic decomposition of organic debris and can be a factor in invasiveness of pathogens.

exogenous Originating outside the body.

exon A stretch of eukaryotic DNA coding for a corresponding portion of mRNA that is translated into peptides. Intervening stretches of DNA that are not expressed are called *introns.* During transcription, exons are separated from introns and are spliced together into a continuous mRNA transcript.

exotoxin A toxin (usually protein) that is secreted and acts upon a specific cellular target. Examples: botulin, tetanospasmin, diphtheria toxin, and erythrogenic toxin.

exponential Pertaining to the use of exponents, numbers that are typically written as a superscript to indicate how many times a factor is to be multiplied. Exponents are used in scientific notation to render large, cumbersome numbers into small workable quantities.

exponential growth phase The period of maximum growth rate in a growth curve. Cell population increases logarithmically.

extrapulmonary tuberculosis A condition in which tuberculosis bacilli have spread to organs other than the lungs.

exudate Fluid that escapes cells into the extracellular spaces during the inflammatory response.

F

facultative Pertaining to the capacity of microbes to adapt or adjust to variations; not obligate. Example: the presence of oxygen is not obligatory for a facultative anaerobe to grow. See *obligate.*

family In the levels of classification, a midlevel division of organisms that groups more closely related organisms than previous levels. An order is divided into families.

fastidious Requiring special nutritional or environmental conditions for growth. Said of bacteria.

fermentation The extraction of energy through anaerobic degradation of substrates into simpler, reduced metabolites. In large industrial processes, *fermentation* can mean any use of microbial metabolism to manufacture organic chemicals or other products.

fertility (F) factor Donor plasmid that allows synthesis of a pilus in bacterial conjugation. Presence of the factor is indicated by F^+, and lack of the factor is indicated by F^-.

filament A helical structure composed of proteins that is part of bacterial flagella.

fimbria A short, numerous-surface appendage on some bacteria that provides adhesion but not locomotion.

Firmicutes Taxonomic category of bacteria that have gram-positive cell envelopes.

flagellar staining A staining method that highlights the flagellum of a bacterium.

flagellum A structure that is used to propel the organism through a fluid environment.

fluorescence The property possessed by certain minerals and dyes to emit visible light when excited by ultraviolet radiation. A fluorescent dye combined with specific antibody provides a sensitive test for the presence of antigen.

fluorescent *in situ* hybridization (FISH) A technique in which a fluorescently labeled DNA or RNA probe is used to locate a specific sequence of DNA in an organism without removing it from its natural environment.

focal infection Occurs when an infectious agent breaks loose from a localized infection and is carried by the circulation to other tissues.

folliculitis An inflammatory reaction involving the formation of papules or pustules in clusters of hair follicles.

fomite Virtually any inanimate object an infected individual has contact with that can serve as a vehicle for the spread of disease.

food fermentations Addition to and growth of known cultures of microorganisms in foods to produce desirable flavors, smells, or textures. Includes cheeses, breads, alcoholic beverages, and pickles.

food poisoning Symptoms in the intestines (which may include vomiting) induced by preformed exotoxin from bacteria.

formalin A 37% aqueous solution of formaldehyde gas; a potent chemical fixative and microbicide.

formyl methionine In bacteria, the first amino acid on the N-terminus of a growing polypeptide.

frameshift mutation An insertion or deletion mutation that changes the codon reading frame from the point of the mutation to the final codon. Almost always leads to a nonfunctional protein.

free energy Energy in a chemical system that can be used to do work.

fructose One of the carbohydrates commonly referred to as sugars. Fructose is commonly fruit sugars.

functional group In chemistry, a particular molecular combination that reacts in predictable ways and confers particular properties on a compound. Examples: —COOH, —OH, —CHO.

fungemia The condition of fungi multiplying in the bloodstream.

fungi (singular, *fungus*) Macroscopic and microscopic heterotrophic eukaryotic organisms that can be uni- or multicellular.

furuncle A boil; a localized pyogenic infection arising from a hair follicle.

G

GALT See *gut-associated lymphoid tissue*.

gas gangrene Disease caused by a clostridial infection of soft tissue or wound. The name refers to the gas produced by the bacteria growing in the tissue. Unless treated early, it is fatal. Also called *myonecrosis*.

gastritis Pain and/or nausea, usually experienced after eating; result of inflammation of the lining of the stomach.

gel electrophoresis A laboratory technique for separating DNA fragments according to length by employing electricity to force the DNA through a gel-like matrix typically made of agarose. Smaller DNA fragments move more quickly through the gel, thereby moving farther than larger fragments during the same period of time.

gene A site on a chromosome that provides information for a certain cell function. A specific segment of DNA that contains the necessary code to make a protein or RNA molecule.

gene drive A process by which the rate of mutations is greatly increased, using CRISPR technology.

gene editing Making changes in an organism's genome, using a process called CRISPR.

gene probe Short strands of single-stranded nucleic acid that hybridize specifically with complementary stretches of nucleotides on test samples and thereby serve as a tagging and identification device.

general-purpose media Material used to promote the growth of a broad array of microbes.

generation time Time required for a complete fission cycle—from parent cell to two new daughter cells. Also called *doubling time*.

genetic engineering A field involving deliberate alterations (recombinations) of the genomes of microbes, plants, and animals through special technological processes.

genctics The science of heredity.

genome The complete set of chromosomes and genes in an organism.

genomic libraries Collections of DNA fragments representing the entire genome of an organism inserted into plasmids and stored in vectors such as bacteria or yeast.

genomics The systematic study of an organism's genes and their functions.

genotype The genetic makeup of an organism. The genotype is ultimately responsible for an organism's phenotype, or expressed characteristics.

genus In the levels of classification, the second most specific level. A family is divided into several genera.

germ free See *axenic*.

germ theory of disease A theory first originating in the 1800s proposed that microorganisms can be the cause of diseases. The concept is actually so well established in the present time that it is considered a fact.

germicide An agent lethal to non-endospore-forming pathogens.

gingivitis Inflammation of the gum tissue in contact with the roots of the teeth.

gluconeogenesis The formation of glucose (or glycogen) from noncarbohydrate sources such as protein or fat. Also called *glyconeogenesis*.

glucose One of the carbohydrates commonly referred to as sugars. Glucose is characterized by its 6-carbon structure.

glycan A type of carbohydrate or polysaccharide that is combined with another organic molecule such as a lipid or protein; examples include peptidoglycan and glycocalyx.

glycerol A 3-carbon alcohol, with three OH groups that serve as binding sites.

glycocalyx A filamentous network of carbohydrate-rich molecules that coats cells.

glycogen A glucose polymer stored by cells.

glycolysis The energy-yielding breakdown (fermentation) of glucose to pyruvic or lactic acid. It is often called *anaerobic glycolysis* because no molecular oxygen is consumed in the degradation.

glycopeptide A chemical form of antibiotic, combining a carbohydrate and a protein, that inhibits the synthesis of peptidoglycan.

glycosidic bond A bond that joins monosaccharides to form disaccharides and polymers.

gnotobiotic Referring to experiments performed on germ-free animals.

Golgi apparatus An organelle of eukaryotes that participates in packaging and secretion of molecules.

gonococcus Common name for *Neisseria gonorrhoeae*, the agent of gonorrhea.

Gracilicutes Taxonomic category of bacteria that have gram-negative envelopes.

graft versus host disease (GVHD) A condition associated with a bone marrow transplant in which T cells in the transplanted tissue mount an immune response against the recipient's (host) normal tissues.

Gram stain A differential stain for bacteria useful in identification and taxonomy. Gram-positive organisms appear purple from crystal violet mordant retention, whereas gram-negative organisms appear red after loss of crystal violet and absorbance of the safranin counterstain.

gram-negative A category of bacterial cells that describes bacteria with an outer membrane, a cytoplasmic membrane, and a thin cell wall.

gram-positive A category of bacterial cells that describes bacteria with a thick cell wall and no outer membrane.

grana Discrete stacks of chlorophyll-containing thylakoids within chloroplasts.

granulocyte A mature leukocyte that contains noticeable granules in a Wright stain. Examples: neutrophils, eosinophils, and basophils.

granuloma A solid mass or nodule of inflammatory tissue containing modified macrophages and lymphocytes. Usually a chronic pathologic process of diseases such as tuberculosis or syphilis.

granzymes Enzymes secreted by cytotoxic T cells that damage proteins of target cells.

Graves' disease A malfunction of the thyroid gland in which autoantibodies directed at thyroid cells stimulate an overproduction of thyroid hormone (hyperthyroidism).

group translocation A form of active transport in which the substance being transported is altered during transfer across a plasma membrane.

growth curve A graphical representation of the change in population size over time. This graph has four periods known as lag phase, exponential or log phase, stationary phase, and death phase.

growth factor An organic compound such as a vitamin or amino acid that must be provided in the diet to facilitate growth. An essential nutrient.

guanine (G) One of the nitrogen bases found in DNA and RNA in the purine form.

Guillain-Barré syndrome A neurological complication of infection or vaccination.

gumma A nodular, infectious granuloma characteristic of tertiary syphilis.

gut-associated lymphoid tissue (GALT) A collection of lymphoid tissue in the gastrointestinal tract that includes the appendix, the lacteals, and Peyer's patches.

H

halogens A group of related chemicals with antimicrobial applications. The halogens most often used in disinfectants and antiseptics are chlorine and iodine.

halophile A microbe whose growth is either stimulated by salt or requires a high concentration of salt for growth.

hapten An incomplete or partial antigen. Although it constitutes the determinative group and can bind antigen, hapten cannot stimulate a full immune response without being carried by a larger protein molecule.

hay fever A form of atopic allergy marked by seasonal acute inflammation of the conjunctiva and mucous membranes of the respiratory passages. Symptoms are irritative itching and rhinitis.

healthcare-associated infection (HAI) Formerly referred to as "nosocomial infection," any infection acquired as a direct result of a patient's presence in a hospital or health care setting.

helical Having a spiral or coiled shape. Said of certain virus capsids and bacteria.
helminth A term that designates all parasitic worms.
hematopoiesis The process by which the various types of blood cells are formed, such as in the bone marrow.
hemolysin Any biological agent that is capable of destroying red blood cells and causing the release of hemoglobin. Many bacterial pathogens produce exotoxins that act as hemolysins.
hemolytic disease of the newborn (HDN) Incompatible Rh factor between mother and fetus causes maternal antibodies to attack the fetus and trigger complement-mediated lysis in the fetus.
hemolytic uremic syndrome (HUS) Severe hemolytic anemia leading to kidney damage or failure; can accompany *E. coli* O157:H7 intestinal infection.
hemolyze When red blood cells burst and release hemoglobin pigment.
hepatitis Inflammation and necrosis of the liver, often the result of viral infection.
hepatitis A virus (HAV) Enterovirus spread by contaminated food responsible for short-term (infectious) hepatitis.
hepatitis B virus (HBV) DNA virus that is the causative agent of serum hepatitis.
hepatitis D The delta agent; a defective RNA virus that cannot reproduce on its own unless a cell is also infected with the hepatitis B virus.
herd immunity The status of collective acquired immunity in a population that reduces the likelihood that nonimmune individuals will contract and spread infection. One aim of vaccination is to induce herd immunity.
heredity Genetic inheritance.
hermaphroditic Containing the sex organs for both male and female in one individual.
heterotroph An organism that relies upon organic compounds for its carbon and energy needs.
hexose A 6-carbon sugar such as glucose and fructose.
histamine A cytokine released when mast cells and basophils release their granules. An important mediator of allergy, its effects include smooth muscle contraction, increased vascular permeability, and increased mucus secretion.
histiocyte Another term for *macrophage.*
histone Proteins associated with eukaryotic DNA. These simple proteins serve as winding spools to compact and condense the chromosomes.
holobiont The human host and all of its resident microbiota.
holoenzyme An enzyme complete with its apoenzyme and cofactors.
hops The ripe, dried fruits of the hop vine (*Humulus lupulus*) that are added to beer wort for flavoring.
horizontal gene transfer Transmission of genetic material from one cell to another through nonreproductive mechanisms, such as from one organism to another living in the same habitat.
host range The limitation imposed by the characteristics of the host cell on the type of virus that can successfully invade it.
hot-air oven A method for achieving disinfection or sterilization.
human immunodeficiency virus (HIV) A retrovirus that causes acquired immunodeficiency syndrome (AIDS).
Human Microbiome Project (HMP) A project of the National Institutes of Health to identify microbial inhabitants of the human body and their role in health and disease; uses metagenomic techniques instead of culturing.
hybridization A process that matches complementary strands of nucleic acid (DNA-DNA, RNA-DNA, RNA-RNA). Used for locating specific sites or types of nucleic acids.
hydration The addition of water as in the coating of ions with water molecules as ions enter into aqueous solution.
hydrogen bond A weak chemical bond formed by the attraction of forces between molecules or atoms—in this case, hydrogen and either oxygen or nitrogen. In this type of bond, electrons are not shared, lost, or gained.
hydrolysis A process in which water is used to break bonds in molecules. Usually occurs in conjunction with an enzyme.
hydrophilic The property of attracting water. Molecules that attract water to their surface are called *hydrophilic.*
hydrophobia, literally "fear of water" A symptom of rabies disease in which the patient avoids swallowing due to the pain it causes.
hydrophobic The property of repelling water. Molecules that repel water are called *hydrophobic.*
hyperthermophile An organism whose optimal growth temperature is above 80°C (176°F), with a temperature range from 60°C to 113°C (140°F to 235°F).
hypertonic Having a greater osmotic pressure than a reference solution.
hyphae The tubular threads that make up filamentous fungi (molds). This web of branched and intertwining fibers is called a *mycelium.*
hypogammaglobulinemia An inborn disease in which the gamma globulin (antibody) fraction of serum is greatly reduced. The condition is associated with a high susceptibility to pyogenic infections.
hyposensitivity diseases Diseases in which there is a diminished or lack of immune reaction to pathogens due to incomplete immune system development, immune suppression, or destruction of the immune system.
hyposensitization A therapeutic exposure to known allergens designed to build tolerance and eventually prevent allergic reaction.
hypothesis A tentative explanation of what has been observed or measured.
hypotonic Having a lower osmotic pressure than a reference solution.

I

icosahedron A regular geometric figure having 20 surfaces that meet to form 12 corners. Some virions have capsids that resemble icosahedral crystals.
immune complex reaction Type III hypersensitivity of the immune system. It is characterized by the reaction of soluble antigen with antibody, and the deposition of the resulting complexes in basement membranes of epithelial tissue.
immune privilege The restriction or reduction of immune response in certain areas of the body that reduces the potential damage to tissues that a normal inflammatory response could cause.
immune tolerance Tolerance to self; the inability of one's immune system to react to self proteins or antigens.
immunity An acquired resistance to an infectious agent due to prior contact with that agent.
immunochromatography A technique used in diagnosis (sometimes called lateral flow test) in which antigens and antibodies flow together and provide visible reactions if they bind to each other.
immunocompetence The ability of the body to recognize and react with multiple foreign substances.
immunocompromised The state of having an underlying condition that diminishes the capability to respond to microbial invasion.
immunogen Any substance that induces a state of sensitivity or resistance after processing by the immune system of the body.
immunoglobulin (Ig) The chemical class of proteins to which antibodies belong.
immunopathology The study of disease states associated with overreactivity or underreactivity of the immune response.
in utero Literally means "in the uterus"; pertains to events or developments occurring before birth.
in vitro Literally means "in glass," signifying a process or reaction occurring in an artificial environment, as in a test tube or culture medium.
in vivo Literally means "in a living being," signifying a process or reaction occurring in a living thing.
inapparent Referring to an infection in which infectious agents have entered the body and some signs of infection are present but no disease symptoms are manifest; also described as *subclinical* or *asymptomatic.*
incidence In epidemiology, the number of new cases of a disease occurring during a period.
incineration The high-temperature combustion of materials leaving only ash and gases.
inclusion A relatively inert body in the cytoplasm such as storage granules, glycogen, fat, or some other aggregated metabolic product.
inclusion body One of a variety of different storage compartments in bacterial cells.
incubate To isolate a sample culture in a temperature-controlled environment to encourage growth.
incubating carriers Persons with an infection that is in the incubation phase but are able to transmit the infection.

incubation period The period from the initial contact with an infectious agent to the appearance of the first symptoms.

index case The first case of a disease identified in an outbreak or epidemic.

indicator bacteria In water analysis, any easily cultured bacteria that may be found in the intestine and can be used as an index of fecal contamination. The category includes coliforms and enterococci. Discovery of these bacteria in a sample means that pathogens may also be present.

induced mutation Any alteration in DNA that occurs as a consequence of exposure to chemical or physical mutagens.

inducer A molecule in an inducible operon responsible for initiating transcription of the operon by removing the repressor from the operator section of the DNA, allowing transcription to proceed.

induction The process whereby a bacteriophage in the prophage state is activated and begins replication and enters the lytic cycle.

induration Area of hardened, reddened tissue associated with the tuberculin test.

infection The entry, establishment, and multiplication of pathogenic organisms within a host.

infectious disease The state of damage or toxicity in the body caused by an infectious agent.

inflammation A natural, nonspecific response to tissue injury that protects the host from further damage. It stimulates immune reactivity and blocks the spread of an infectious agent.

inoculating loop A tool used in the microbiology laboratory sometimes comprised of a platinum or nichrome wire loop attached to a heat-proof handle.

inoculation The implantation of microorganisms into or upon culture media.

inorganic chemicals Molecules that lack the basic framework of the elements of carbon and hydrogen.

insertion elements The smallest transposable elements, consisting only of tandem repeats that are capable of inserting themselves into DNA but do not carry any genes.

integrase inhibitors A class of drugs that attach to the enzyme required to splice the dsDNA from HIV into the host genome.

integument The outer surfaces of the body: skin, hair, nails, sweat glands, and oil glands.

interferon (IFN) Natural human chemical that inhibits viral replication; used therapeutically to combat viral infections and cancer.

interleukins A class of chemicals released from host cells that have potent effects on immunity.

intermediate filament Proteinaceous fibers in eukaryotic cells that help provide support to the cells and their organelles.

interstitium Extracellular spaces and the lymph fluid they contain.

intoxication Poisoning that results from the introduction of a toxin into body tissues through ingestion or injection.

intron The segments on split genes of eukaryotes that do not code for polypeptide. They can have regulatory functions. See *exon.*

iodophor A combination of iodine and an organic carrier that is a moderate-level disinfectant and antiseptic.

ion An unattached, charged particle.

ionic bond A chemical bond in which electrons are transferred and not shared between atoms.

ionization The aqueous dissociation of an electrolyte into ions.

ionizing radiation Radiant energy consisting of short-wave electromagnetic rays (X ray) or high-speed electrons that cause dislodgment of electrons on target molecules and create ions.

irradiation The application of radiant energy for diagnosis, therapy, disinfection, or sterilization.

isograft Transplanted tissue from one monozygotic twin to the other; transplants between highly inbred animals that are genetically identical.

isoniazid (INH) Older drug that targets the bacterial cell wall; used against *M. tuberculosis.*

isotonic Two solutions having the same osmotic pressure such that, when separated by a semipermeable membrane, there is no net movement of solvent in either direction.

isotope A version of an element that is virtually identical in all chemical properties to another version except that their atoms have slightly different atomic masses.

J

jaundice The yellowish pigmentation of skin, mucous membranes, sclera, deeper tissues, and excretions due to abnormal deposition of bile pigments. Jaundice is associated with liver infection, as with hepatitis B virus and leptospirosis.

JC virus (JCV) Causes a form of encephalitis (progressive multifocal leukoencephalopathy), especially in AIDS patients.

K

keratin Protein produced by outermost skin cells that provide protection from trauma and moisture.

kingdom In the levels of classification, the second division from more general to more specific. Each domain is divided into kingdoms.

Koch's postulates A procedure to establish the specific cause of disease. In all cases of infection, (1) the agent must be found; (2) inoculations of a pure culture must reproduce the same disease in animals; (3) the agent must again be present in the experimental animal; and (4) a pure culture must again be obtained.

L

labile In chemistry, molecules or compounds that are chemically unstable in the presence of environmental changes.

lactose One of the carbohydrates commonly referred to as sugars. Lactose is commonly found in milk.

lactose (*lac*) operon Control system that manages the regulation of lactose metabolism. It is composed of three DNA segments, including a regulator, a control locus, and a structural locus.

lag phase The early phase of population growth during which no signs of growth occur.

lager The maturation process of beer, which is allowed to take place in large vats at a reduced temperature.

lagging strand The newly forming 5′ DNA strand that is discontinuously replicated in segments (Okazaki fragments).

latency The state of being inactive. Example: a latent virus or latent infection.

leading strand The newly forming 3′ DNA strand that is replicated in a continuous fashion without segments.

leaven To lighten food material by entrapping gas generated within it. Example: the rising of bread from the CO_2 produced by yeast or baking powder.

legumes Plants that produce seeds in pods. Examples include soybeans and peas.

lesion A wound, injury, or some other pathologic change in tissues.

leukocidin A heat-labile substance formed by some pyogenic cocci that impairs and sometimes lyses leukocytes.

leukocytes White blood cells. The primary infection-fighting blood cells.

leukocytosis An abnormally large number of leukocytes in the blood, which can be indicative of acute infection.

leukopenia A lower-than-normal leukocyte count in the blood that can be indicative of blood infection or disease.

leukotriene An unsaturated fatty acid derivative of arachidonic acid. Leukotriene functions in chemotactic activity, smooth muscle contractility, mucus secretion, and capillary permeability.

library In biotechnology, a collection of DNA fragments obtained by exposing the DNA to restriction enzymes, separating the fragments through gel electrophoresis, and inserting the fragments into plasmids.

ligase An enzyme required to seal the sticky ends of DNA pieces after splicing.

light-dependent reactions The series of reactions in photosynthesis that are driven by the light energy (photons) absorbed by chlorophyll. They involve splitting of water into hydrogens and oxygen, transport of electrons by NADP, and ATP synthesis.

light-independent reactions The series of reactions in photosynthesis that can proceed with or without light. It is a cyclic system that uses ATP from the light reactions to incorporate or fix carbon dioxide into organic compounds, leading to the production of glucose and other carbohydrates (also called the *Calvin cycle*).

lipase A fat-splitting enzyme. Example: triacylglycerol lipase separates the fatty acid chains from the glycerol backbone of triglycerides.

lipid A term used to describe a variety of substances that are not soluble in polar solvents such as water but will dissolve in nonpolar solvents such as benzene and chloroform. Lipids include triglycerides, phospholipids, steroids, and waxes.

lipopeptide A chemical containing both lipids and protein. Some of these have antibiotic effects.

lipopolysaccharide A molecular complex of lipid and carbohydrate found in the bacterial cell wall. The lipopolysaccharide (LPS) of gram-negative bacteria is an endotoxin with generalized pathologic effects such as fever.

lipoteichoic acid Anionic polymers containing glycerol that are anchored in the cytoplasmic membranes of gram-positive bacteria.

liquid media Growth-supporting substance in fluid form.

localized infection Occurs when a microbe enters a specific tissue, infects it, and remains confined there.

log phase Maximum rate of cell division during which growth is geometric in its rate of increase. Also called *exponential growth phase.*

lophotrichous Describing bacteria having a tuft of flagella at one or both poles.

lymphadenitis Inflammation of one or more lymph nodes. Also called *lymphadenopathy.*

lymphatic system A system of vessels and organs that serve as sites for the development of immune cells and immune reactions. It includes the spleen, thymus, lymph nodes, and GALT.

lymphocyte The second most common form of white blood cells.

lyophilization A method for preserving microorganisms (and other substances) by freezing and then drying them directly from the frozen state.

lysogenic conversion A bacterium acquires a new genetic trait due to the presence of genetic material from an infecting phage.

lysogeny The indefinite persistence of bacteriophage DNA in a host without bringing about the production of virions.

lysosome A cytoplasmic organelle containing lysozyme and other hydrolytic enzymes.

lysozyme An enzyme found in sweat, tears, and saliva that breaks down bacterial peptidoglycan.

M

macromolecules Large, molecular compounds assembled from smaller subunits, most notably biochemicals.

macronutrient A chemical substance required in large quantities (phosphate, for example).

macroorganism Organism visible to the naked eye.

major histocompatibility complex (MHC) A set of genes in mammals that produces molecules on surfaces of cells that differentiate among different individuals in the species.

MALT See *mucosa-associated lymphoid tissue.*

malting The step in beer brewing in which the grain is allowed to sprout, releasing important enzymes.

maltose One of the carbohydrates referred to as sugars. A fermentable sugar formed from starch.

Mantoux test An intradermal screening test for tuberculin hypersensitivity. A red, firm patch of skin at the injection site greater than 10 mm in diameter after 48 hours is a positive result that indicates current or prior exposure to the TB bacillus.

marker Any trait or factor of a cell, virus, or molecule that makes it distinct and recognizable. Example: a genetic marker.

mash In making beer, the malt grain is steeped in warm water, ground up, and fortified with carbohydrates to form mash.

matrix The dense ground substance between the cristae of a mitochondrion that serves as a site for metabolic reactions.

matter All tangible materials that occupy space and have mass.

maximum temperature The highest temperature at which an organism will grow.

MDR-TB Multidrug-resistant tuberculosis.

mechanical vector An animal that transports an infectious agent but is not infected by it, such as houseflies whose feet become contaminated with feces.

medium (plural, *media*) A nutrient used to grow organisms outside of their natural habitats.

medullary sinus Inner portion of lymph nodes containing B cells and macrophages.

meiosis The type of cell division necessary for producing gametes in diploid organisms. Two nuclear divisions in rapid succession produce four gametocytes, each containing a haploid number of chromosomes.

memory (immunologic memory) The capacity of the immune system to recognize and act against an antigen upon second and subsequent encounters.

Mendosicutes Taxonomic category of bacteria that have unusual cell walls; archaea.

meninges The tough tri-layer membrane covering the brain and spinal cord. Consists of the dura mater, arachnoid mater, and pia mater.

meningitis An inflammation of the membranes (meninges) that surround and protect the brain. It is often caused by bacteria such as *Neisseria meningitidis* (the meningococcus) and *Haemophilus influenzae.*

mesophile Microorganisms that grow at intermediate temperatures.

messenger RNA (mRNA) A single-stranded transcript that is a copy of the DNA template that corresponds to a gene.

metabolism A general term for the totality of chemical and physical processes occurring in a cell.

metabolomics The study of the complete complement of small chemicals present in a cell at any given time.

metachromatic granules A type of inclusion in storage compartments of some bacteria that stain a contrasting color when treated with colored dyes.

metagenomics The study of all the genomes in a particular ecological niche, as opposed to individual genomes from single species.

methanogens Methane producers.

methanotrophs Certain species of bacteria that derive carbon and energy through the oxidation of methane.

MHC See *major histocompatibility complex.*

MIC See *minimum inhibitory concentration.* Abbreviation for **m**inimum **i**nhibitory **c**oncentration. The lowest concentration of antibiotic needed to inhibit bacterial growth in a test system.

micro RNA Short sequences of RNA that are capable of binding to mRNA, ultimately repressing the production of a particular protein; found in all eukaryotes, viruses, and many bacteria.

microaerophile An aerobic bacterium that requires oxygen at a concentration less than that in the atmosphere.

microarray A glass, silicon, or nylon chip that contains sequences from tens of thousands of different genes in cDNA form that fluoresce when complementary DNA binds to them, indicating what mRNA molecules are present in a cell under varying conditions.

microbe See *microorganism.*

microbial antagonism Relationship in which microorganisms compete for survival in a common environment by taking actions that inhibit or destroy another organism.

microbial ecology The study of microbes in their natural habitats.

microbicides Chemicals that kill microorganisms.

microbiology A specialized area of biology that deals with living things ordinarily too small to be seen without magnification, including bacteria, archaea, fungi, protozoa, and viruses.

micronutrient A chemical substance required in small quantities (trace metals, for example).

microorganism A living thing ordinarily too small to be seen without magnification; an organism of microscopic size.

microscopic Invisible to the naked eye.

microscopy Science that studies structure, magnification, lenses, and techniques related to use of a microscope.

microtubules Long hollow tubes in eukaryotic cells; maintain the shape of the cell and transport substances from one part of the cell to another; involved in separating chromosomes in mitosis.

minimum inhibitory concentration (MIC) The smallest concentration of drug needed to visibly control microbial growth.

minimum temperature The lowest temperature at which an organism will grow.

miracidium The ciliated first-stage larva of a trematode. This form is infective for a corresponding intermediate host snail.

missense mutation A mutation in which a change in the DNA sequence results in a different amino acid being incorporated into a protein, with varying results.

mitochondrion A double-membrane organelle of eukaryotes that is the main site for aerobic respiration.

mitosis Somatic cell division that preserves the somatic chromosome number.

mixed acid fermentation An anaerobic degradation of pyruvic acid that results in more than one organic acid being produced (e.g., acetic acid, lactic acid, succinic acid).
mixed culture A container growing two or more different, known species of microbes.
mixed infection Occurs when several different pathogens interact simultaneously to produce an infection. Also called a *synergistic infection.*
molecule A distinct chemical substance that results from the combination of two or more atoms.
molluscum contagiosum Poxvirus-caused disease that manifests itself by the appearance of small lesions on the face, trunk, and limbs. Can be associated with sexual transmission.
monocyte A large mononuclear leukocyte normally found in the lymph nodes, spleen, bone marrow, and loose connective tissue. This type of cell makes up 3% to 7% of circulating leukocytes.
monomer A simple molecule that can be linked by chemical bonds to form larger molecules.
mononuclear phagocyte system (MPS) A collection of monocytes and macrophages scattered throughout the extracellular spaces that function to engulf and degrade foreign molecules.
monosaccharide A simple sugar such as glucose that is a basic building block for more complex carbohydrates.
monotrichous Describing a microorganism that bears a single flagellum.
morbidity A diseased condition.
morbidity rate The number of persons afflicted with an illness under question or with illness in general, expressed as a numerator, with the denominator being some unit of population (as in *x*/100,000).
mortality rate The number of persons who have died as the result of a particular cause or due to all causes, expressed as a numerator, with the denominator being some unit of population (as in *x*/100,000).
most probable number (MPN) Test used to detect the concentration of contaminants in water and other fluids.
motility Self-propulsion.
mucosa-associated lymphoid tissue (MALT) Small patches of lymphoid tissue situated in and on mucosal surfaces, containing T cells, B cells, phagoctyes, and other immune cells.
mutagen Any agent that induces genetic mutation. Examples: certain chemical substances, ultraviolet light, radioactivity.
mutant strain A subspecies of microorganism that has undergone a mutation, causing expression of a trait that differs from other members of that species.
mutation A permanent inheritable alteration in the DNA sequence or content of a cell.
mutualism Organisms living in an obligatory but mutually beneficial relationship.
mycelium The filamentous mass that makes up a mold. Composed of hyphae.
mycolic acid A thick, waxy, long-chain fatty acid found in the cell wall of *Mycobacterium* and *Nocardia* that confers resistance to chemicals and dyes.
mycorrhizae Various species of fungi adapted in an intimate, mutualistic relationship to plant roots.
mycosis Any disease caused by a fungus.
myonecrosis Death of muscle tissue.

N

nanowire/nanotube Long membrane-enclosed appendages that can exchange materials between bacterial cells.
narrow-spectrum Denotes drugs that are selective and limited in their effects. For example, they inhibit either gram-negative or gram-positive bacteria but not both.
natural immunity Any immunity that arises naturally in an organism via previous experience with the antigen.
negative stain A staining technique that renders the background opaque or colored and leaves the object unstained so that it is outlined as a colorless area.
negative-sense RNA Single-stranded viral RNA that is complementary to positive-sense RNA and must be converted into positive-sense RNA before it can be translated.
nematode A common name for helminths called *roundworms.*
nephritis Inflammation of the kidney.
neurons Cells that make up the tissues of the brain and spinal cord that receive and transmit signals to and from the peripheral nervous system and central nervous system.
neurotropic Having an affinity for the nervous system. Most likely to affect the spinal cord.
neutralization The process of combining an acid and a base until they reach a balanced proportion, with a pH value close to 7.
neutron An electrically neutral particle in the nuclei of all atoms except hydrogen.
neutrophil A mature granulocyte present in peripheral circulation, exhibiting a multilobular nucleus and numerous cytoplasmic granules that retain a neutral stain. The neutrophil is an active phagocytic cell in bacterial infection.
nitrification Phase of the nitrogen cycle in which ammonium is oxidized.
nitrogen base A ringed compound of which pyrimidines and purines are types.
nitrogen cycle The pathway followed by the element nitrogen as it circulates from inorganic sources in the nonliving environment to living things and back to the nonliving environment. The longtime reservoir is nitrogen gas in the atmosphere.
nitrogenous base A nitrogen-containing molecule found in DNA and RNA that provides the basis for the genetic code. Adenine, guanine, and cytosine are found in both DNA and RNA, while thymine is found exclusively in DNA and uracil is found exclusively in RNA.
nomenclature A set system for scientifically naming organisms, enzymes, anatomical structures, and so on.
noncommunicable An infectious disease that does not arrive through transmission of an infectious agent from host to host.
noncompetitive inhibition Form of enzyme inhibition that involves binding of a regulatory molecule to a site other than the active site.
nonionizing radiation Method of microbial control, best exemplified by ultraviolet light, that causes the formation of abnormal bonds within the DNA of microbes, increasing the rate of mutation. The primary limitation of nonionizing radiation is its inability to penetrate beyond the surface of an object.
nonpolar A term used to describe an electrically neutral molecule formed by covalent bonds between atoms that have the same or similar electronegativity.
nonself Molecules recognized by the immune system as containing foreign markers, indicating a need for immune response.
nonsense mutation A mutation that changes an amino-acid-producing codon into a stop codon, leading to premature termination of a protein.
nucleocapsid In viruses, the close physical combination of the nucleic acid with its protective covering.
nucleoid The basophilic nuclear region or nuclear body that contains the bacterial chromosome.
nucleolus A granular mass containing RNA that is contained within the nucleus of a eukaryotic cell.
nucleoside analog A chemical or drug that resembles a particular nucleoside and can be used to stop replication of viruses.
nucleotide The basic structural unit of DNA and RNA; each nucleotide consists of a phosphate, a sugar (ribose in RNA, deoxyribose in DNA), and a nitrogenous base such as adenine, guanine, cytosine, thymine (DNA only), or uracil (RNA only).
nutrient Any chemical substance that must be provided to a cell for normal metabolism and growth. Macronutrients are required in large amounts and micronutrients in small amounts.
nutrition The acquisition of chemical substances by a cell or organism for use as an energy source or as building blocks of cellular structures.

O

obligate Without alternative; restricted to a particular characteristic. Example: an obligate parasite survives and grows only in a host; an obligate aerobe must have oxygen to grow; an obligate anaerobe is destroyed by oxygen.
Okazaki fragment In replication of DNA, a segment formed on the lagging strand in which biosynthesis is conducted in a discontinuous manner dictated by the $5' \rightarrow 3'$ DNA polymerase orientation.
oligodynamic action A chemical having antimicrobial activity in minuscule amounts. Example: certain heavy metals are effective in a few parts per billion.
oligotrophic Nutrient-deficient ecosystem.
oncogene A naturally occurring type of gene that when activated can transform a normal cell into a cancer cell.
operator In an operon sequence, the DNA segment where transcription of structural genes is initiated.
operon A genetic operational unit that regulates metabolism by controlling mRNA production. In sequence, the unit consists of a regulatory gene, inducer or repressor control sites, and structural genes.

opportunistic In infection, ordinarily nonpathogenic or weakly pathogenic microbes that cause disease primarily in an immunologically compromised host.

opsonization The process of stimulating phagocytosis by affixing molecules (opsonins such as antibodies and complement) to the surfaces of foreign cells or particles.

optimum temperature The temperature at which a species shows the most rapid growth rate.

orbitals The pathways of electrons as they rotate around the nucleus of an atom.

order In the levels of classification, the division of organisms that follows class. Increasing similarity may be noticed among organisms assigned to the same order.

organelle A small component of eukaryotic cells that is bounded by a membrane and specialized in function.

organic chemicals Molecules that contain the basic framework of the elements carbon and hydrogen.

osmophile A microorganism that thrives in a medium having high osmotic pressure.

osmosis The diffusion of water across a selectively permeable membrane in the direction of lower water concentration.

outer membrane An additional membrane possessed by gram-negative bacteria; a lipid bilayer containing specialized proteins and polysaccharides. It lies outside of the cell wall.

oxazolidinones A class of antibiotic, synthesized in the lab, that acts on the 50S subunit of the ribosome to inhibit protein synthesis.

oxidative phosphorylation The synthesis of ATP using energy given off during the electron transport phase of respiration.

oxidized The state of a reactant when it has lost one or more electrons.

oxygenic Any reaction that gives off oxygen; usually in reference to the result of photosynthesis in eukaryotes and cyanobacteria.

P

palindrome A word, verse, number, or sentence that reads the same forward or backward. Palindromes of nitrogen bases in DNA have genetic significance as transposable elements, as regulatory protein targets, and in DNA splicing.

PAMPs See *pathogen-associated molecular patterns.*

pandemic A disease afflicting an increased proportion of the population over a wide geographic area (often worldwide).

pangenome The entire set of genes in a species, including ones not present in all members of the species.

papilloma Benign, squamous epithelial growth commonly referred to as a *wart.*

paracortical area Portion of a lymph node that contains T lymphocytes.

parasitic The relationship between a parasite and its host in which the parasite lives on or within the host and damages the host in some way; characteristic of an organism considered to be a parasite.

parasitism A relationship between two organisms in which the host is harmed in some way, while the colonizer benefits.

parenteral Administering a substance into a body compartment other than through the gastrointestinal tract, such as via intravenous, subcutaneous, intramuscular, or intramedullary injection.

paroxysmal Events characterized by sharp spasms or convulsions; sudden onset of a symptom such as fever and chills.

passive carrier Person who mechanically transfers a pathogen without ever being infected by it; for example, a health care worker who doesn't wash their hands adequately between patients.

passive immunity Specific resistance that is acquired indirectly by donation of preformed immune substances (antibodies) produced in the body of another individual.

pasteurization Heat treatment of perishable fluids such as milk, fruit juices, or wine to destroy heat-sensitive vegetative cells, followed by rapid chilling to inhibit growth of survivors and germination of spores. It prevents infection and spoilage.

pathogen Any agent (usually a virus, bacterium, fungus, protozoan, or helminth) that causes disease.

pathogen-associated molecular patterns (PAMPs) Molecules on the surfaces of many types of microbes that are not present on host cells that mark the microbes as foreign.

pathogenicity islands Areas of the genome containing multiple genes that contribute to a new trait for the organism that increases its ability to cause disease.

pattern recognition receptors (PRRs) Molecules on the surface of host defense cells that recognize pathogen-associated molecular patterns (PAMPs) on microbes.

pellicle A membranous cover; a thin skin, film, or scum on a liquid surface; a thin film of salivary glycoproteins that forms over newly cleaned tooth enamel when exposed to saliva.

penetration (viral) The step in viral multiplication in which virus enters the host cell.

penicillinase An enzyme that hydrolyzes penicillin; found in penicillin-resistant strains of bacteria.

pentose A monosaccharide with five carbon atoms per molecule. Examples: arabinose, ribose, xylose.

peptide Molecule composed of short chains of amino acids, such as a dipeptide (two amino acids), a tripeptide (three), and a tetrapeptide (four).

peptide bond The covalent union between two amino acids that forms between the amine group of one and the carboxyl group of the other. The basic bond of proteins.

peptidoglycan A network of polysaccharide chains cross-linked by short peptides that forms the rigid part of bacterial cell walls. Gram-negative bacteria have a smaller amount of this rigid structure than do gram-positive bacteria.

perforin Proteins released by cytotoxic T cells that produce pores in target cells.

periodontal Of or pertaining to the gums and supporting structures of the teeth.

peritrichous In bacterial morphology, having flagella distributed over the entire cell.

persisters Bacteria that grow more slowly than others so that they are less affected by antibiotics and can reestablish infection when the antibiotic is removed.

petechiae Minute hemorrhagic spots in the skin that range from pinpoint- to pinhead-size.

Peyer's patches Oblong lymphoid aggregates of the gut located chiefly in the wall of the terminal and small intestine. Along with the tonsils and appendix, Peyer's patches make up the gut-associated lymphoid tissue that responds to local invasion by infectious agents.

pH The symbol for the negative logarithm of the H ion concentration; p (power) or $[H^+]_{10}$. A system for rating acidity and alkalinity.

phage A bacteriophage; a virus that specifically parasitizes bacteria.

phagocyte A class of white blood cells capable of engulfing other cells and particles.

phagocytosis A type of endocytosis in which the cell membrane actively engulfs large particles or cells into vesicles.

phagolysosome A body formed in a phagocyte, consisting of a union between a vesicle containing the ingested particle (the phagosome) and a vacuole of hydrolytic enzymes (the lysosome).

phagosome A vacuole formed within a phagocytic cell when it extends its pseudopods to enclose a cell or particle.

phase variation The process of bacteria turning on or off a group of genes that changes its phenotype in a heritable manner.

phenotype The observable characteristics of an organism produced by the interaction between its genetic potential (genotype) and the environment.

phosphate An acidic salt containing phosphorus and oxygen that is an essential inorganic component of DNA, RNA, and ATP.

phospholipid A class of lipids that compose a major structural component of cell membranes.

phosphorylation Process in which inorganic phosphate is added to a compound.

photoautotroph An organism that utilizes light for its energy and carbon dioxide chiefly for its carbon needs.

photolysis The splitting of water into hydrogen and oxygen during photosynthesis.

photon A subatomic particle released by electromagnetic sources such as radiant energy (sunlight). Photons are the ultimate source of energy for photosynthesis.

photophosphorylation The process of electron transport during photosynthesis that results in the synthesis of ATP from ADP.

photosynthesis A process occurring in plants, algae, and some bacteria that traps the sun's energy and converts it to ATP in the cell. This energy is used to fix CO_2 into organic compounds.

phototaxis The movement of organisms in response to light.

phototrophs Microbes that use photosynthesis to feed.

phycobilin Red or blue-green pigments that absorb light during photosynthesis.

phylum In the levels of classification, the third level of classification from general to more specific. Each kingdom is divided into numerous phyla. Sometimes referred to as a *division.*

pili (singular, *pilus*) Long, tubular structures made of pilin protein produced by gram-negative bacteria and used for conjugation.

pilin A class of protein that makes up bacterial pili.

pinocytosis The engulfment, or endocytosis, of liquids by extensions of the cell membrane.

plankton Minute animals (zooplankton) or plants (phytoplankton) that float and drift in the limnetic zone of bodies of water.

planktonic Term referring to microbes that are free-floating in a liquid medium.

plantar warts Deep, painful warts on the soles of the feet as a result of infection by human papillomavirus.

plaque In virus propagation methods, the clear zone of lysed cells in tissue culture or chick embryo membrane that corresponds to the area containing viruses. In dental application, the filamentous mass of microbes that adheres tenaciously to the tooth and predisposes to caries, calculus, or inflammation.

plasma The carrier fluid element of blood.

plasmids Extrachromosomal genetic units characterized by several features. A plasmid is a double-stranded DNA that is smaller than and replicates independently of the cell chromosome; it bears genes that are not essential for cell growth; it can bear genes that code for adaptive traits; and it is transmissible to other bacteria.

pleomorphism Normal variability of cell shapes in a single species.

pneumococcus Common name for *Streptococcus pneumoniae,* the major cause of bacterial pneumonia.

pneumonia An inflammation of the lung leading to accumulation of fluid and respiratory compromise.

pneumonic plague The acute, frequently fatal form of pneumonia caused by *Yersinia pestis.*

point mutation A change that involves the loss, substitution, or addition of one or a few nucleotides.

point-source epidemic An outbreak of disease in which all affected individuals were exposed to a single source of the pathogen at a single point in time.

polar Term to describe a molecule with an asymmetrical distribution of charges. Such a molecule has a negative pole and a positive pole.

poliomyelitis An acute enteroviral infection of the spinal cord that can cause neuromuscular paralysis.

polymer A macromolecule made up of a chain of repeating units. Examples: starch, protein, DNA.

polymerase An enzyme that produces polymers through catalyzing bond formation between building blocks (polymerization).

polymerase chain reaction (PCR) A technique that amplifies segments of DNA for testing. Using denaturation, primers, and heat-resistant DNA polymerase, the number can be increased several-million-fold.

polymicrobial Involving multiple distinct microorganisms.

polymyxin A mixture of antibiotic polypeptides from *Bacillus polymyxa* that are particularly effective against gram-negative bacteria.

polypeptide A relatively large chain of amino acids linked by peptide bonds.

polyribosomal complex An assembly line for mass production of proteins composed of a chain of ribosomes involved in mRNA transcription.

polysaccharide A carbohydrate that can be hydrolyzed into a number of monosaccharides. Examples: cellulose, starch, glycogen.

porin Transmembrane protein of the outer membrane of gram-negative cells that permits transport of small molecules into the periplasmic space but bars the penetration of larger molecules.

portal of entry Route of entry for an infectious agent; typically a cutaneous or membranous route.

portal of exit Route through which a pathogen departs from the host organism.

positive stain A method for coloring microbial specimens that involves a chemical that sticks to the specimen to give it color.

positive-sense RNA Single-stranded viral RNA that has the same sequence as host cell mRNA and can be directly translated into viral proteins.

posttranslational Referring to modifications to the protein structure that occur after protein synthesis is complete, including removal of formyl methionine, further folding of the protein, addition of functional groups, or addition of the protein to a quaternary structure.

prebiotics Nutrients used to stimulate the growth of favorable biota in the intestine.

prevalence The total number of cases of a disease in a certain area and time period.

primary infection An initial infection in a previously healthy individual that is later complicated by an additional (secondary) infection.

primary lymphatic organs Sites in the body in which T cells and B cells are produced and mature.

primary response The first response of the immune system when exposed to an antigen.

primary structure Initial protein organization described by type, number, and order of amino acids in the chain. The primary structure varies extensively from protein to protein.

primers Synthetic oligonucleotides of known sequence that serve as landmarks to indicate where DNA amplification will begin.

prion A concocted word to denote "proteinaceous infectious agent"; a cytopathic protein associated with the slow-virus spongiform encephalopathies of humans and animals.

probes Small fragments of single-stranded DNA (RNA) that are known to be complementary to the specific sequence of DNA being studied.

probiotics Preparations of live microbes used as a preventive or therapeutic measure to displace or compete with potential pathogens.

prodromal stage A short period of mild symptoms occurring at the end of the period of incubation. It indicates the onset of disease.

product(s) In a chemical reaction, the substance(s) that is(are) left after a reaction is completed.

progressive multifocal leukoencephalopathy (PML) An uncommon, fatal complication of infection with JC virus (polyomavirus).

promoter Part of an operon sequence. The DNA segment that is recognized by RNA polymerase as the starting site for transcription.

propagated epidemic An outbreak of disease in which the causative agent is passed from affected persons to new persons over the course of time.

prophage A lysogenized bacteriophage; a phage that is latently incorporated into the host chromosome instead of undergoing viral replication and lysis.

prostaglandin A hormonelike substance that regulates many body functions. Prostaglandin comes from a family of organic acids containing 5-carbon rings that are essential to the human diet.

protease Enzymes that act on proteins, breaking them down into component parts.

protease inhibitors Drugs that act to prevent the assembly of functioning viral particles.

protein Predominant organic molecule in cells, formed by long chains of amino acids.

proteomics The study of an organism's complement of proteins (its *proteome*) and functions mediated by the proteins.

proton An elementary particle that carries a positive charge. It is identical to the nucleus of the hydrogen atom.

protozoa A group of single-celled, eukaryotic organisms.

provirus The genome of a virus when it is integrated into a host cell's DNA.

PRRs See *pattern recognition receptors.*

pseudohypha A chain of easily separated, spherical to sausage-shaped yeast cells partitioned by constrictions rather than by septa.

pseudopods Protozoan appendage responsible for motility. Also called "false feet."

psychrophile A microorganism that thrives at low temperature (0°C–20°C), with a temperature optimum of 0°C–15°C.

pure culture A container growing a single species of microbe whose identity is known.

purine A nitrogen base that is an important encoding component of DNA and RNA. The two most common purines are adenine and guanine.

pus The viscous, opaque, usually yellowish matter formed by an inflammatory infection. It consists of serum exudate, tissue debris, leukocytes, and microorganisms.

pyogenic Pertains to pus formers, especially the pyogenic cocci: pneumococci, streptococci, staphylococci, and neisseriae.

pyrimidine Nitrogen bases that help form the genetic code on DNA and RNA. Uracil, thymine, and cytosine are the most important pyrimidines.

pyrimidine dimer The union of two adjacent pyrimidines on the same DNA strand, brought about by exposure to ultraviolet light. It is a form of mutation.

pyrogen A substance that causes a rise in body temperature. It can come from pyrogenic microorganisms or from polymorphonuclear leukocytes (endogenous pyrogens).

Q

quaternary structure Most complex protein structure characterized by the formation of large, multiunit proteins by more than one of the polypeptides. This structure is typical of antibodies and some enzymes that act in cell synthesis.

quats A word that pertains to a family of surfactants called *quaternary ammonium compounds.* These detergents are only weakly microbicidal and are used as sanitizers and preservatives.

quinine A substance derived from cinchona trees that was used as an antimalarial treatment; has been replaced by synthetic derivatives.

quinolone Broad-spectrum antibiotic that works on gram-positive and gram-negative bacteria.

quorum sensing The ability of bacteria to regulate their gene expression in response to sensing bacterial density.

R

rabies The only rhabdovirus that infects humans. Zoonotic disease characterized by fatal meningoencephalitis.

radiation Electromagnetic waves or rays, such as those of light given off from an energy source.

rales Sounds in the lung, ranging from clicking to rattling; indicate respiratory illness.

reactants Molecules entering or starting a chemical reaction.

real image An image formed at the focal plane of a convex lens. In the compound light microscope, it is the image created by the objective lens.

recombinant An organism that contains genes that originated in another organism, whether through deliberate laboratory manipulation or natural processes.

recombinant DNA technology A technology, also known as genetic engineering, that deliberately modifies the genetic structure of an organism to create novel products, microbes, animals, plants, and viruses.

recombination A type of genetic transfer in which DNA from one organism is donated to another.

recycling A process that converts unusable organic matter from dead organisms back into their essential inorganic elements and returns them to their nonliving reservoirs to make them available again for living organisms. This is a common term that means the same as mineralization and decomposition.

redox Denoting an oxidation-reduction reaction.

reduced The state of a reactant when it has gained one or more electrons.

reducing medium A growth medium that absorbs oxygen and allows anaerobic bacteria to grow.

redundancy The property of the genetic code that allows an amino acid to be specified by several different codons.

refraction In optics, the bending of light as it passes from one medium to another with a different index of refraction.

refractive index The measurement of the degree of light that is bent, or refracted, as it passes between two substances such as air, water, or glass.

regulated enzymes Enzymes whose extent of transcription or translation is influenced by changes in the environment.

regulator DNA segment that codes for a protein capable of repressing an operon.

release The final step in the multiplication cycle of viruses in which the assembled viral particle exits the host cell and moves on to infect another cell.

replication fork The Y-shaped point on a replicating DNA molecule where the DNA polymerase is synthesizing new strands of DNA.

reportable disease Those diseases that must be reported to health authorities by law.

repressor The protein product of a repressor gene that combines with the operator and arrests the transcription and translation of structural genes.

reservoir In disease communication, the natural host or habitat of a pathogen.

resistance (R) factor Plasmids, typically shared among bacteria by conjugation, that provide resistance to the effects of antibiotics.

resolving power The capacity of a microscope lens system to accurately distinguish between two separate entities that lie close to each other. Also called *resolution.*

restriction endonuclease An enzyme present naturally in cells that cleaves specific locations on DNA. It is an important means of inactivating viral genomes, and it is also used to splice genes in genetic engineering.

restriction factors Host cell molecules that inhibit viruses in some way.

restriction fragment length polymorphisms (RFLPs) Variations in the lengths of DNA fragments produced when a specific restriction endonuclease acts on different DNA sequences.

restriction fragments Short pieces of DNA produced when DNA is exposed to restriction endonucleases.

retrotransposon A transmissible element capable of translating itself from DNA to RNA to make many copies of itself and then translating itself back into DNA in order to insert itself into a new location on the chromosome.

reverse transcriptase (RT) The enzyme possessed by retroviruses that carries out the reversion of RNA to DNA—a form of reverse transcription.

Rh factor An isoantigen that can trigger hemolytic disease in newborns due to incompatibility between maternal and infant blood factors.

rhizobia Bacteria that live in plant roots and supply supplemental nitrogen that boosts plant growth.

rhizosphere The zone of soil, complete with microbial inhabitants, in the immediate vicinity of plant roots.

ribonucleic acid (RNA) The nucleic acid responsible for carrying out the hereditary program transmitted by an organism's DNA.

ribose A 5-carbon monosaccharide found in RNA.

ribosomal RNA (rRNA) A single-stranded transcript that is a copy of part of the DNA template.

ribosome A bilobed macromolecular complex of ribonucleoprotein that coordinates the codons of mRNA with tRNA anticodons and, in so doing, constitutes the peptide assembly site.

ribozyme A part of an RNA-containing enzyme in eukaryotes that removes intervening sequences of RNA called *introns* and splices together the true coding sequences (exons) to form a mature messenger RNA.

RNA polymerase Enzyme process that translates the code of DNA to RNA.

rod One of the basic shapes of bacteria.

root nodules Small growths on the roots of legume plants that arise from a symbiotic association between the plant tissues and bacteria (rhizobia). This association allows fixation of nitrogen gas from the air into a usable nitrogen source for the plant.

rough endoplasmic reticulum (RER) Microscopic series of tunnels that originates in the outer membrane of the nuclear envelope and is used in transport and storage. Large numbers of ribosomes, partly attached to the membrane, give the rough appearance.

rubeola (red measles) Acute disease caused by infection with morbillivirus.

run Bacteria moving toward a stimulus in a straight line.

S

S layer Single layer of thousands of copies of a single type of protein linked together on the surface of a bacterial cell that is produced when the cell is in a hostile environment.

saccharide Scientific term for *sugar*. Refers to a simple carbohydrate with a sweet taste.

salpingitis Inflammation of the fallopian tubes.

SALT See *skin-associated lymphoid tissue*.

sanitize To clean inanimate objects using soap and degerming agents so that they are safe and free of high levels of microorganisms.

saprobe A microbe that decomposes organic remains from dead organisms. Also known as a *saprophyte* or *saprotroph*.

sarcina A cubical packet of 8, 16, or more cells; the cellular arrangement of the genus *Sarcina* in the family *Micrococcaceae*.

saturation The complete occupation of the active site of a carrier protein or enzyme by the substrate.

schistosomiasis Infection by blood fluke, often as a result of contact with contaminated water in rivers and streams. Symptoms appear in liver, spleen, or urinary system depending on species of *Schistosoma*. Infection may be chronic.

scientific method Principles and procedures for the systematic pursuit of knowledge, involving the recognition and formulation of a problem, the collection of data through observation and experimentation, and the formulation and testing of a hypothesis.

screening Testing designed to detect pathological changes before they cause obvious symptoms.

sebaceous glands The sebum- (oily, fatty) secreting glands of the skin.

sebum Low pH, oil-based secretion of the sebaceous glands.

secondary infection An infection that compounds a preexisting one.

secondary lymphatic organs Locations in the body where immune cells reside or exert their actions.

secondary response The rapid rise in antibody titer following a repeat exposure to an antigen that has been recognized from a previous exposure. This response is brought about by memory cells produced as a result of the primary exposure.

secondary structure Protein structure that occurs when the functional groups on the outer surface of the molecule interact by forming hydrogen bonds. These bonds cause the amino acid chain either to twist, forming a helix, or to pleat into an accordion pattern called a *β-pleated sheet.*

selective media Nutrient media designed to favor the growth of certain microbes and to inhibit undesirable competitors.

selectively permeable Describes a property of cell membranes in which certain substances are able to pass through the membrane, while other substances cannot pass through unaided and require special carrier proteins in order to enter or exit the cell.

selectively toxic Property of an antimicrobial agent to be highly toxic against its target microbe, while being far less toxic to other cells, particularly those of the host organism.

self Natural markers of the body that are recognized by the immune system.

semiconservative replication In DNA replication, the synthesis of paired daughter strands, each retaining a parent strand template.

semisolid media Nutrient media with a firmness midway between that of a broth (a liquid medium) and an ordinary solid medium; motility media.

semisynthetic Drugs that, after being naturally produced by bacteria, fungi, or other living sources, are chemically modified in the laboratory. Compare to *synthetic.*

sensitivity (of a test) The probability that a screening test will identify all of the positive cases.

sepsis The state of putrefaction; the presence of pathogenic organisms or their toxins in tissue or blood.

septa (singular, *septum*) A partition or cellular cross wall, as in certain fungal hyphae.

septic shock Blood infection resulting in a pathologic state of low blood pressure accompanied by a reduced amount of blood circulating to vital organs. Endotoxins of all gram-negative bacteria can cause shock, but most clinical cases are due to gram-negative enteric rods.

septicemia Systemic infection associated with microorganisms multiplying in circulating blood.

septicemic plague A form of infection with *Yersinia pestis* occurring mainly in the bloodstream and leading to high mortality rates.

sequela A morbid complication that follows a disease.

sequence map A map that shows the exact order of DNA bases in an organism determined by whole-genome shotgun sequencing.

serious threat category A designation from the Centers for Disease Control and Prevention indicating that a particular microbe exhibits a degree of antimicrobial resistance that is *serious.*

serology The branch of immunology that deals with *in vitro* diagnostic testing of serum.

serotonin A vasoconstrictor that inhibits gastric secretion and stimulates smooth muscle.

serotyping The subdivision of a species or subspecies into an immunologic type, based upon antigenic characteristics.

serous Referring to serum, the clear fluid that escapes cells during the inflammatory response.

severe acute respiratory syndrome (SARS) A severe respiratory disease caused by infection with a newly described coronavirus.

severe combined immunodeficiencies (SCIDs) A collection of syndromes occurring in newborns caused by a genetic defect that knocks out both B- and T-cell types of immunity. There are several versions of this disease, termed *SCIDs* for short.

sexually transmitted infection (STI), also sexually transmitted disease (STD) Infection resulting from pathogens that enter the body via sexual intercourse or intimate, direct contact.

shiga toxin Heat-labile exotoxin released by some *Shigella* species and by *E. coli* O157:H7; responsible for worst symptoms of these infections.

shiga-toxin-producing *E. coli* (STEC) A strain of *E. coli* that produces the shiga toxin.

shingles Lesions produced by reactivated human herpesvirus 3 (chickenpox) infection; also known as herpes zoster.

sign Any abnormality uncovered upon physical diagnosis that indicates the presence of disease. A *sign* is an objective assessment of disease, as opposed to a *symptom,* which is the subjective assessment perceived by the patient.

silent mutation A mutation that, because of the degeneracy of the genetic code, results in a nucleotide change in both the DNA and mRNA but not the resultant amino acid and, thus, not the protein.

simple stain Type of positive staining technique that uses a single dye to add color to cells so that they are easier to see. This technique tends to color all cells the same color.

single nucleotide polymorphism (SNP) An alteration in a gene sequence in which a single nucleotide base is altered.

skin-associated lymphoid tissue (SALT) Discrete bundles of lymphatic tissue located just under the surface of the skin.

slime layer A diffuse, unorganized layer of polysaccharides and/or proteins on the outside of some bacteria.

smooth endoplasmic reticulum (SER) A microscopic series of tunnels lacking ribosomes that functions in the nutrient processing function of a cell.

solute A substance that is uniformly dispersed in a dissolving medium or solvent.

solution A mixture of one or more substances (*solutes*) that cannot be separated by filtration or ordinary settling.

solvent A dissolving medium.

source The person or item from which an infection is directly acquired. See *reservoir.*

species In the levels of classification, the most specific level of organization.

specificity In immunity, the concept that some parts of the immune system only react with antigens that originally activated them.

specificity (of a test) The probability that a screening test will not call any negative cases positive.

spillover event The point at which an animal virus becomes able to infect humans and be transmitted human to human.

spliceosome A molecule composed of RNA and protein that removes introns from eukaryotic mRNA before it is translated by forming a loop in the intron, cutting it from the mRNA, and joining exons together.

spontaneous generation Early belief that living things arose from vital forces present in nonliving, or decomposing, matter.

spontaneous mutation A mutation in DNA caused by random mistakes in replication and not known to be influenced by any mutagenic agent. These mutations give rise to an organism's natural, or background, rate of mutation.

sporadic Description of a disease that exhibits new cases at irregular intervals in unpredictable geographic locales.

sporangiospore A form of asexual spore in fungi; enclosed in a sac.

sporangium A fungal cell in which asexual spores are formed by multiple cell cleavage.

spore A differentiated, specialized cell form that can be used for dissemination, for survival in times of adverse conditions, and/or for reproduction. Spores are usually unicellular and may develop into gametes or vegetative organisms.

sporozoite One of many minute elongated bodies generated by multiple division of the oocyst. It is the infectious form of the malarial parasite that is harbored in the salivary gland of the mosquito and inoculated into the victim during feeding.

sporulation The process of spore formation.

starch A polysaccharide composed of glucose monomers

start codon The nucleotide triplet AUG that codes for the first amino acid in protein sequences.

starter culture The sizable inoculation of pure bacterial, mold, or yeast sample for bulk processing, as in the preparation of fermented foods, beverages, and pharmaceuticals.

stationary growth phase Survival mode in which cells either stop growing or grow very slowly.

stem cells Pluripotent, undifferentiated cells.

sterile Completely free of all life forms, including spores and viruses.

sterilization Any process that completely removes or destroys all viable microorganisms, including viruses, from an object or habitat. Material so treated is *sterile*.

stop codon One of the codons UAA, UAG, and UGA, which have no corresponding tRNA and thus signal the end of transcription; also known as a nonsense codon.

strain In microbiology, a set of descendants cloned from a common ancestor that retain the original characteristics. Any deviation from the original is a different strain.

streptogramin A class of antibiotic, derived from *Streptomyces,* that targets the 50S subunit of the ribosome.

stroma The matrix of the chloroplast that is the site of the dark reactions.

subacute Indicates an intermediate status between acute and chronic disease.

subacute encephalitis A brain infection and inflammation that is not acute but long-lasting.

subacute sclerosing panencephalitis (SSPE) A complication of measles infection in which progressive neurological degeneration of the cerebral cortex invariably leads to coma and death.

subclinical A period of inapparent manifestations that occurs before symptoms and signs of disease appear.

subculture To make a second-generation culture from a well-established colony of organisms.

subspecies Bacteria of the same species that have differing characteristics; also known as a bacterial *type*.

substrate The specific molecule upon which an enzyme acts.

sucrose One of the carbohydrates commonly referred to as sugars. Common table or cane sugar.

superantigens Bacterial toxins that are potent stimuli for T cells and can be a factor in diseases such as toxic shock.

superficial mycosis A fungal infection located in hair, nails, and the epidermis of the skin.

superinfection An infection occurring during antimicrobial therapy that is caused by an overgrowth of drug-resistant microorganisms.

surfactant A surface-active agent that forms a water-soluble interface. Examples: detergents, wetting agents, dispersing agents, and surface tension depressants.

surveillance Monitoring health or behaviors on an ongoing basis, to be able to quickly recognize deviations from normal rates of occurrence.

symbiosis An intimate association between individuals from two species; used as a synonym for *mutualism*.

symptom The subjective evidence of infection and disease as perceived by the patient.

syncytium A multinucleated protoplasmic mass formed by the consolidation of individual cells.

syndrome The collection of signs and symptoms that, taken together, paint a portrait of the disease.

synergism The coordinated or correlated action by two or more drugs or microbes that results in a heightened response or greater activity.

synthesis (viral) The step in viral multiplication in which viral genetic material and proteins are made through replication and transcription/translation.

synthetic Referring to a chemotherapeutic agent manufactured entirely through chemical processes in the laboratory that mimics the actions of antibiotics. Compare to *semisynthetic*.

systemic Occurring throughout the body; said of infections that invade many compartments and organs via the circulation.

systemic infection An infection spread by the blood to multiple sites in the body at some distance from the site of initial infection.

T

tartar See *calculus.*

taxa Taxonomic categories.

taxonomy The formal system for organizing, classifying, and naming living things.

teichoic acid Anionic polymers containing glycerol that appear in the walls of gram-positive bacteria.

telomeres Areas of repeating DNA sequences at the ends of a linear chromosome that protect the chromosome from being deteriorated during rounds of DNA replication.

temperate phage A bacteriophage that enters into a less virulent state by becoming incorporated into the host genome as a prophage instead of in the vegetative or lytic form that eventually destroys the cell.

template strand The strand in a double-stranded DNA molecule that is used as a model to synthesize a complementary strand of DNA or RNA during replication or transcription.

Tenericutes Taxonomic category of bacteria that lack cell walls.

tertiary structure Protein structure that results from additional bonds forming between functional groups in a secondary structure, creating a three-dimensional mass.

tetanospasmin The neurotoxin of *Clostridium tetani,* the agent of tetanus. Its chief action is directed upon the inhibitory synapses of the anterior horn motor neurons.

tetracyclines A group of broad-spectrum antibiotics with a complex 4-ring structure.

tetrads Groups of four.

theory A collection of statements, propositions, or concepts that explain or account for a natural event.

theory of evolution The evidence cited to explain how evolution occurs.

therapeutic index The ratio of the toxic dose to the effective therapeutic dose that is used to assess the safety and reliability of the drug.

therapeutic window The range of blood level of the drug in which it is producing the desired effect without toxicity.

thermal death point The lowest temperature that achieves sterilization in a given quantity of broth culture upon a 10-minute exposure. Examples: 55°C for *Escherichia coli,* 60°C for *Mycobacterium tuberculosis,* and 120°C for spores.

thermal death time The least time required to kill all cells of a culture at a specified temperature.

thermophile A microorganism that thrives at a temperature of 50°C or higher.

thylakoid Vesicles of a chloroplast formed by elaborate folding of the inner membrane to form "discs." Solar energy trapped in the thylakoids is used in photosynthesis.

thymine (T) One of the nitrogen bases found in DNA but not in RNA. Thymine is in a pyrimidine form.

thymus Butterfly-shaped organ near the tip of the sternum that is the site of T-cell maturation.

tincture A medicinal substance dissolved in an alcoholic solvent.

tinea Ringworm; a fungal infection of the hair, skin, or nails.

titer In immunochemistry, a measure of antibody level in a patient, determined by agglutination methods.

toll-like receptors (TLRs) A category of pattern recognition receptors that binds to pathogen-associated molecular patterns on microbes.

tonsils A ring of lymphoid tissue in the pharynx that acts as a repository for lymphocytes.

TORCH Common infections of the fetus and neonate. Stands for toxoplasmosis, other diseases, rubella, cytomegalovirus, and herpes simplex.

toxemia Condition in which a toxin (microbial or otherwise) is spread throughout the bloodstream.

toxigenicity The tendency for a pathogen to produce toxins. It is an important factor in bacterial virulence.

toxin A specific chemical product of microbes, plants, and some animals that is poisonous to other organisms.

toxinosis Disease whose adverse effects are primarily due to the production and release of toxins.

toxoid A toxin that has been rendered nontoxic but is still capable of eliciting the formation of protective antitoxin antibodies; used in vaccines.

trace elements Micronutrients (zinc, nickel, and manganese) that occur in small amounts and are involved in enzyme function and maintenance of protein structure.

tracheotomy A procedure in which a cut is made in the front of the neck and into the trachea in order to open the airway.

transcript A newly transcribed RNA molecule.

transcription mRNA synthesis; the process by which a strand of RNA is produced against a DNA template.
transduction The transfer of genetic material from one bacterium to another by means of a bacteriophage vector.
transfer RNA (tRNA) A transcript of DNA that specializes in converting RNA language into protein language.
transformation In microbial genetics, the transfer of genetic material contained in "naked" DNA fragments from a donor cell to a competent recipient cell.
transgenic Referring to genetically modified organisms that contain foreign genes inserted into their genome through recombinant DNA technologies.
translation Protein synthesis; the process of decoding the messenger RNA code into a polypeptide.
translocation The movement of the ribosome from one codon to the next during translation of the mRNA sequence after the peptide bond has been formed between amino acids.
transmissible spongiform encephalopathies (TSEs) Diseases caused by proteinaceous infectious particles (also known as *prions*).
transport The passage of nutrients across a membrane.
transport medium Microbiological medium that is used to transport specimens.
transposon A DNA segment with an insertion sequence at each end, enabling it to migrate to another plasmid, to the bacterial chromosome, or to a bacteriophage.
trematode A category of helminth; also known as flatworm or fluke.
trichinosis Infection by the *Trichinella spiralis* parasite, usually caused by eating the meat of an infected animal. Early symptoms include fever, diarrhea, nausea, and abdominal pain that progress to intense muscle and joint pain and shortness of breath. In the final stages, heart and brain function are at risk, and death is possible.
triglyceride A type of lipid composed of a glycerol molecule bound to three fatty acids.
trophozoite A vegetative protozoan (feeding form) as opposed to a resting (cyst) form.
tuberculin reaction A diagnostic test in which PPD, or purified protein derivative (of *M. tuberculosis*), is injected superficially under the skin and the area of reaction measured; also called the *Mantoux test.*
tumble The way that bacteria move when they are not attracted to a substance.
turbid Cloudy appearance of nutrient solution in a test tube due to growth of microbe population.
type Bacteria of the same species that have differing characteristics; also known as a bacterial *subspecies.*

U

ubiquitous Present everywhere at the same time.
ultraviolet (UV) radiation Radiation with an effective wavelength from 240 to 260 nm. UV radiation induces mutations readily but has very poor penetrating power.
uncoating The process of removal of the viral coat and release of the viral genome by its newly invaded host cell.
uracil (U) One of the nitrogen bases in RNA but not in DNA. Uracil is in a pyrimidine form.
urinary tract infection (UTI) Invasion and infection of the urethra and bladder by bacterial residents, most often *E. coli.*

V

vaccination Exposing a person to the antigenic components of a microbe without its pathogenic effects for the purpose of inducing a future protective response.
vacuoles In the cell, membrane-bounded sacs containing fluids or solid particles to be digested, excreted, or stored.
valence The combining power of an atom based upon the number of electrons it can either take on or give up.
van der Waals forces Weak attractive interactions between molecules of low polarity.
vancomycin Antibiotic that targets the bacterial cell wall; used often in antibiotic-resistant infections.
variable region The antigen-binding fragment of an immunoglobulin molecule, consisting of a combination of heavy and light chains whose molecular conformation is specific for the antigen.
varicella Informal name for virus responsible for chickenpox as well as shingles; also known as *human herpesvirus 3* (*HHV-3*).
vector An animal that transmits infectious agents from one host to another, usually a biting or piercing arthropod like the tick, mosquito, or fly. Infectious agents can be conveyed mechanically by simple contact or biologically whereby the parasite develops in the vector. A genetic element such as a plasmid or a bacteriophage used to introduce genetic material into a cloning host during recombinant DNA experiments.
vegetative In describing microbial developmental stages, a metabolically active feeding and dividing form, as opposed to a dormant, seemingly inert, nondividing form. Examples: a bacterial cell versus its spore; a protozoan trophozoite versus its cyst.
vehicle An inanimate material (solid object, liquid, or air) that serves as a transmission agent for pathogens.
viremia The presence of viruses in the bloodstream.
virion An elementary virus particle in its complete morphological and thus infectious form. A virion consists of the nucleic acid core surrounded by a capsid, which can be enclosed in an envelope.
viroid An infectious agent that, unlike a virion, lacks a capsid and consists of a closed circular RNA molecule. Although known viroids are all plant pathogens, it is conceivable that animal versions exist.
virome The genome sequences of the entire complement of viruses that live on or in an organism.
virtual image In optics, an image formed by diverging light rays; in the compound light microscope, the second, magnified visual impression formed by the ocular from the real image formed by the objective.
virulence In infection, the relative capacity of a pathogen to invade and harm host cells.
virulence factors A microbe's structures or capabilities that allow it to establish itself in a host and cause damage.
virus Microscopic, acellular agent composed of nucleic acid surrounded by a protein coat.
virus particle A more specific name for a virus when it is outside of its host cells.
vitamins A component of coenzymes critical to nutrition and the metabolic function of coenzyme complexes.

W

Western blot test A procedure for separating and identifying antigen or antibody mixtures by two-dimensional electrophoresis in polyacrylamide gel, followed by immune labeling.
whole blood A liquid connective tissue consisting of blood cells suspended in plasma.
wild type The natural, nonmutated form of a genetic trait.
wobble A characteristic of amino acid codons in which the third base of a codon can be altered without changing the code for the amino acid.
wort The clear fluid derived from soaked mash that is fermented for beer.

X

xenograft The transfer of a tissue or an organ from an animal of one species to a recipient of another species.

Z

zoonosis An infectious disease indigenous to animals that humans can acquire through direct or indirect contact with infected animals.
zooplankton The collection of nonphotosynthetic microorganisms (protozoa, tiny animals) that float in the upper regions of aquatic habitat and together with phytoplankton comprise the plankton.

Viaframe/Corbis/Getty Images

Index

Note: In this index page numbers set in **boldface** refer to boxed material, page numbers followed by an f refer to figures, page numbers set in *italics* refer to definitions or introductory discussions, page numbers followed by a t refer to tables.